The Geologic

volume 2

The Geologic Time Scale 2012

Volume 2

Editors

Felix M. Gradstein

James G. Ogg

Mark D. Schmitz

Gabi M. Ogg

AMSTERDAM • BOSTON • HEIDELBERG • LONDON • NEW YORK • OXFORD
PARIS • SAN DIEGO • SAN FRANCISO • SINGAPORE • SYDNEY • TOKYO

The Boulevard, Langford Lane, Kidlington, Oxford OX5 1GB, UK
Radarweg 29, PO Box 211, 1000 AE Amsterdam, The Netherlands
225 Wyman Street, Waltham, MA 02451, USA

First edition 2012

British Library Cataloguing in Publication Data
A catalogue record for this book is available from the British Library

Library of Congress Cataloging-in-Publication Data
A catalog record for this book is availabe from the Library of Congress

Volume 1 ISBN: 978-0-44-459390-0,
This Volume (2) ISBN: 978-0-44-459434-1,
Set ISBN: 978-0-44-459425-9

Printed and bound in China

12 13 14 15 16 10 9 8 7 6 5 4 3 2

Quote

'To place all the scattered pages of Earth history in their proper chronological order is by no means an easy task'

(Arthur Holmes)

Dedication

We dedicate this edition of the Geologic Time Scale book to the many participants in the CHRONOS Project, Geological Time Scale Next (GTSNext), the Earth Time Projects and the International Commission of Stratigraphy (ICS).

These programs and others have championed the compilation, standardization, enhancement, numerical age-dating, and international public distribution of our progressive unraveling of Earth's fascinating history.

Contents

Volume 2

ORGANIZATIONS

CGMW	Commission for the Geological Map of the World
DNAG	Decade of North American Geology
DSDP	Deep Sea Drilling Project
GSC	Geological Survey of Canada
ICS	International Commission of Stratigraphy
IODP	International Ocean Drilling Project
IGC	International Geological Congress
IGCP	International Geological Correlation Project
INQUA	International Quaternary Association
IUGS	International Union of Geological Sciences
IUPAC	International Union of Pure and Applied Chemistry
ODP	Ocean Drilling Project
SNS	Subcommision (of ICS) on Neogene Stratigraphy
ISPS	Subcommission (of ICS) on Paleogene Stratigraphy
SQS	Subcommission (of ICS) on Quaternary Stratigraphy
STS	Subcommission (of ICS) on Triassic Stratigraphy
SOS	Subcommission (of ICS) on Ordovician Stratigraphy
SCS	Subcommision (of ICS) on Cambrian Stratigraphy
UNESCO	United Nations Education, Scientific, and Cultural Organization
USGS	United States Geological Survey

TIME SCALE PUBLICATIONS (SEE REFERENCES FOR DETAILS)

NDS82	*Numerical Dating in Stratigraphy* (Odin et al., 1982)
GTS82	*A Geologic Time Scale* (Harland *et al.*, 1982)
DNAG83	*Geologic Time Scale, Decade of North American Geology* (Palmer, 1983)
KG85	Kent and Gradstein (1985)
EX88	Exxon 1988 (Haq *et al.*, 1988)
GTS89	*A Geologic Time Scale 1989* (Harland *et al.*, 1990)
OB93	Obradovich (1993)
JGR94	*Journal of Geophysical Research* 1994 (Gradstein *et al.*, 1994)
SEPM95	Society for Sedimentary Geology 1995 (Gradstein *et al.*, 1995)
GO96	Gradstein and Ogg (1996)
GTS2004 (GTS04)	Gradstein, Ogg and Smith (2004)
GTS2008 (GTS08)	Ogg, Ogg and Gradstein (2008)
GTS2012	Gradstein, Ogg, Schmitz and Ogg (2012)

GEOSCIENTIFIC CONCEPTS

CA-TIMS	Chemical abrasion - thermal ionization mass spectrometry (in U-Pb dating)
FAD	First appearance datum
FCT	Fish Canyon Tuff sanidine monitor standard (in Ar—Ar dating)
FOD	First occurrence datum
FCT (FCs)	Fish Canyon Tuff sanidine monitor standard (in Ar—Ar dating)
GPTS	Geomagnetic polarity time scale
GSSP	Global Stratotype Section and Point
GSSA	Global Standard Stratigraphic Age (in Precambrian)
HO	Highest occurrence level
HR—SIMS	High-resolution secondary ion mass spectrometry (in U—Pb dating)
ID-TIMS	Isotope dilution thermal ionization mass spectrometry (in U—Pb dating)
LAD	Last appearance datum
LA-ICPMS	Laser ablation-inductively coupled plasma mass spectrometry (in U-Pb dating)
LO	Lowest occurrence level
LOD	Last occurrence datum
LA2004	Laskar 2004 numerical solution of orbital periodicities
LA2010	Laskar 2010 numerical solution of orbital periodicities (Laskar et al., 2011)
MMhb-1	McClure Mountain hornblende monitor standard (in Ar—Ar dating)
SL13	Sri Lanka 13 monitor zircon standard (in HR—SIMS dating)
TCs	Taylor Creek Rhyolite sanidine monitor standard (in Ar—Ar dating)

SYMBOLS

ka	10^3 years ago (kilo annum)
kyr	10^3 years duration
Ma	10^6 years ago (mega annum)
myr	10^6 years duration
Ga	10^9 years ago (giga annum)
gyr	10^9 years duration
SI	Le Système Internationale d'Unités
a	annus (year)
s	second

S. Peng, L.E. Babcock and R.A. Cooper

Chapter 19

The Cambrian Period

Abstract: Appearance of metazoans with mineralized skeletons, "explosion" in biotic diversity and disparity, infaunalization of the substrate, occurrence of metazoan Konservat Fossil-Lagerstätten, establishment of most invertebrate phyla, strong faunal provincialism, dominance of trilobites, globally warm climate (greenhouse conditions), opening of the Iapetus Ocean, progressive equatorial drift and separation of Laurentia, Baltica, Siberia, and Avalonia from Gondwana all characterize the Cambrian Period.

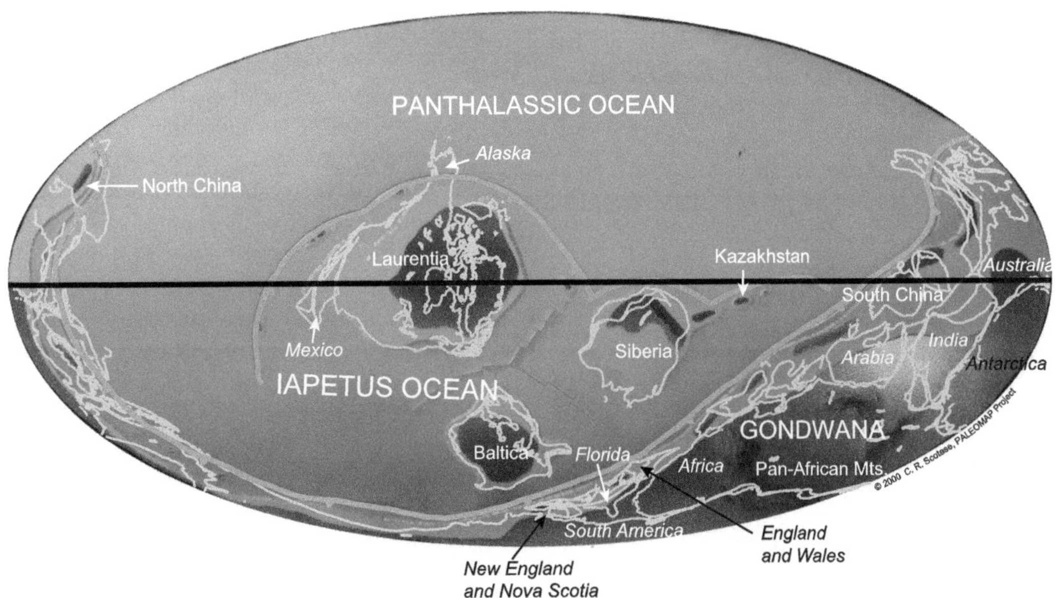

Chapter Outline

The Geologic Time Scale 2012. DOI: 10.1016/B978-0-444-59425-9.00019-6

19.1. HISTORY AND SUBDIVISIONS

The name "Cambrian" is derived from *Cambria*, the classical name for Wales. *Cambria* is latinized from the Welsh name *Cymru*, which refers to the Welsh people. The term "Cambrian" was first used by Adam Sedgwick (in Sedgwick and Murchison, 1835) for the "Cambrian successions" in North Wales and Cumberland (northwestern United Kingdom). Sedgwick, (portrayed in Figure 19.1) divided strata of the area into three groups; Lower, Middle, and Upper Cambrian. Some strata originally defined as Cambrian were included by Murchison (1839) in the lower part of the Silurian. This led to a long drawn-out conflict on the boundary position between the two systems. To end the dispute, Lapworth (1879) excluded the disputed strata from both systems and proposed the interval as a new system, the Ordovician (Stubblefield, 1956; Cowie *et al.*, 1972; Bassett, 1985). In 1960, the term Cambrian was officially accepted at the 21st International Geological Congress (IGC) in Copenhagen, Denmark, as the lowest system of the Paleozoic. As used today, the name Cambrian applies only to part of the "Lower Cambrian" as defined by Sedgwick in 1852. The modern usage excludes the Tremadoc and Arenig slates from the system; the lower boundary has been shifted downward considerably (Peng *et al.*, 2006; Babcock *et al.*, 2011: Figure 1). Figure 19.2 gives a chart showing the chronostratigraphic subdivisions of the Cambrian System adopted by the Cambrian Subcommission of the International Commission on Stratigraphy (ISCS), global standard stages, series, and GSSPs ratified by the ICS and IUGS through March 2012. Other provisional subdivisions are identified with numbers. Potential GSSP levels are also indicated (revised from Babcock *et al.*, 2005).

FIGURE 19.1 **Portrait of Adam Sedgwick (1785–1873) at 82 years of age by Lowes (Cato) Dickensen, 1867.** *Courtesy of Sedgwick Museum, UK.*

The Cambrian marks an important phase in the history of life on Earth. The system is characterized by the appearance of numerous animals (metazoans) bearing mineralized skeletons, and by a rapid diversification of animals commonly referred to as the "Cambrian explosion". Nearly all animal phyla known from the fossil record appeared during the Cambrian Period. The biostratigraphically most useful fossil group is the trilobites, especially agnostoid trilobites, which show a remarkable evolutionary diversification, particularly in the upper half of the Cambrian. Inarticulate brachiopods, archaeocyaths, conodonts, acritarchs, and diverse skeletal remains referred to as "small shelly fossils" also provide good biostratigraphic control in appropriate facies. Trace fossils have been used to zone the lowermost part of the system and to identify its base. The principal regional biostratigraphic zonal schemes of the Cambrian are shown in Figure 19.3. Also shown on this three-part figure are the carbon isotope curve and its (named) excursions, the principal bioevents and geomagnetic polarity reversal trends.

Among non-biostratigraphic correlation criteria, excursions of stable isotopes, particularly carbon ($\delta^{13}C$), play an increasingly important role in recognizing global or regional stratigraphic tie points and boundary positions. Many widely recognizable carbon isotopic excursions in the Cambrian seem to correlate with important biotic events (Peng *et al.*, 2004a; Babcock *et al.*, 2005; Zhu *et al.*, 2006). For example, the base of the Paibian Stage, which coincides with the first appearance of the cosmopolitan agnostoid trilobite *Glyptagnostus reticulatus*, is closely associated with the onset of the SPICE excursion (Steptoean PositIve Carbon isotopic Excursion; Saltzman *et al.*, 2000; Peng *et al.*, 2004a; Figure 19.3).

For many years there was no international agreement on standard chronostratigraphic (or geochronologic) subdivisions of the Cambrian, nor was there international agreement on positions of series and stage boundaries within the system. Following Sedgwick's (1852) practice, the Cambrian has traditionally been divided into lower, middle, and upper parts (corresponding to series/epochs) in most parts of the world. Unfortunately, due to the absence of an internationally accepted standard, the boundary of each series (epoch) was placed at a chronostratigraphic level that varied from region to region. The placement of the base of the Middle Cambrian was especially subject to widely differing interpretations (Geyer, 1990, 1998, 2005; Geyer *et al.*, 2000). Numerous

System	Series	Stage	Boundary Horizons (GSSPs) or Provisional Stratigraphic Tie Points
Ordo-vician	Lower	Tremadocian	FAD of *Iapetognathus fluctivagus* (GSSP)
Cambrian	Furon-gian	*Stage 10*	FAD of *Lotagnostus americanus*
		Jiangshanian	FAD of *Agnostotes orlentalis* (GSSP)
		Paibian	FAD of *Glyptagnostus reticulatus* (GSSP)
	Series 3	Guzhangian	FAD of *Lejopyge laevigata* (GSSP)
		Drumian	FAD of *Ptychagnostus atavus* (GSSP)
		Stage 5	FAD of *Oryctocephalus indicus / Ovatoryctocara granulata*
	Series 2	*Stage 4*	?FAD of *Olenellus, Redlichia, Judomia,* or *Bergeroniellus*
		Stage 3	?FAD of trilobites
	Terre-neuvian	*Stage 2*	?FAD of *Watsonella crosbyi* or *Aldanella attleborensis*
		Fortunian	FAD of *Trichophycus pedum* (GSSP)
Ediacaran			

FIGURE 19.2 **Chart showing chronostratigraphic subdivisions of the Cambrian System adopted by the International Subcommission on Cambrian Stratigraphy (ISCS).** Global standard stages, series, and GSSPs ratified by the ICS and IUGS through March 2012 are indicated in black and red lettering. Other provisional subdivisions are identified with numbers. Potential GSSP levels are also indicated. *(revised from Babcock et al., 2005).*

different nomenclatures for regional stages were established over the years, and there was little uniformity from region to region. Sometimes terminology even varied within individual regions or according to stratigraphic practice (Babcock *et al.*, 2011).

Through the years, differing stratigraphic philosophies have been used for definition of series and stages. Most older definitions of series and stages were based on the unit-stratotype concept (see Salvador, 1994), in which a unit is defined and characterized with reference to a type section. The lower and upper boundaries of a unit are normally specified by reference to a type section. Some of the more recently established definitions, however, have been based on the boundary-stratotype concept, in which a point in strata is used to define the base of a series or stage, and in which the upper limit of each chronostratigraphic unit is automatically defined by the base of the overlying chronostratigraphic unit. The recently introduced global chronostratigraphic units all have definitions based on the boundary-stratotype concept.

The first steps toward achieving internationally acceptable subdivisions of the Cambrian, and toward definition of those subdivisions, were taken at the 1960 IGC. At this conference, the idea of the subcommission on Cambrian

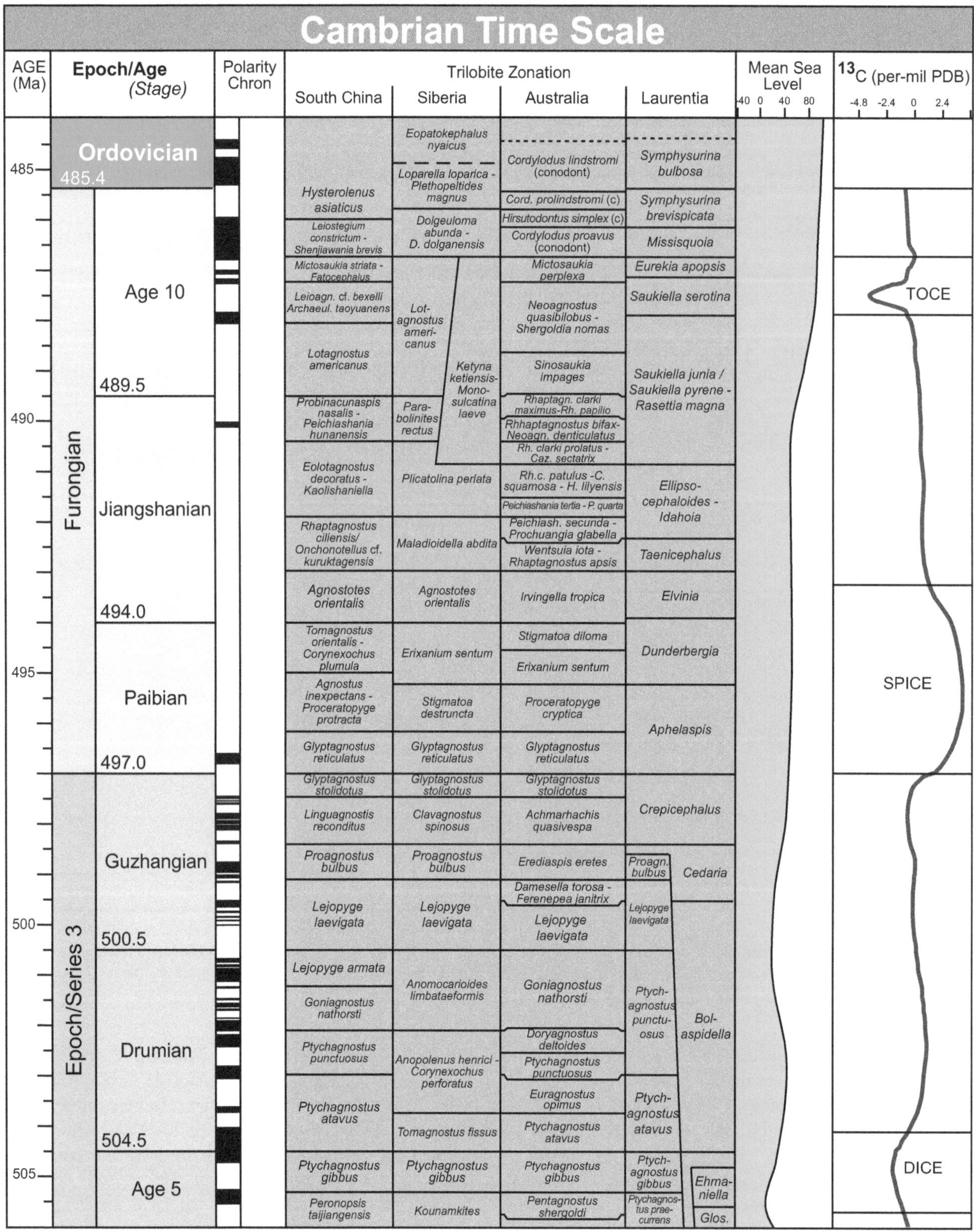

FIGURE 19.3 Principal regional biostratigraphic zonal schemes of the Cambrian. Also shown are the carbon isotope curve and its (named) excursions, principal bioevents and the trend of geomagnetic polarity reversals. Black in Polarity Chron column = normal polarity; white = reversed polarity.

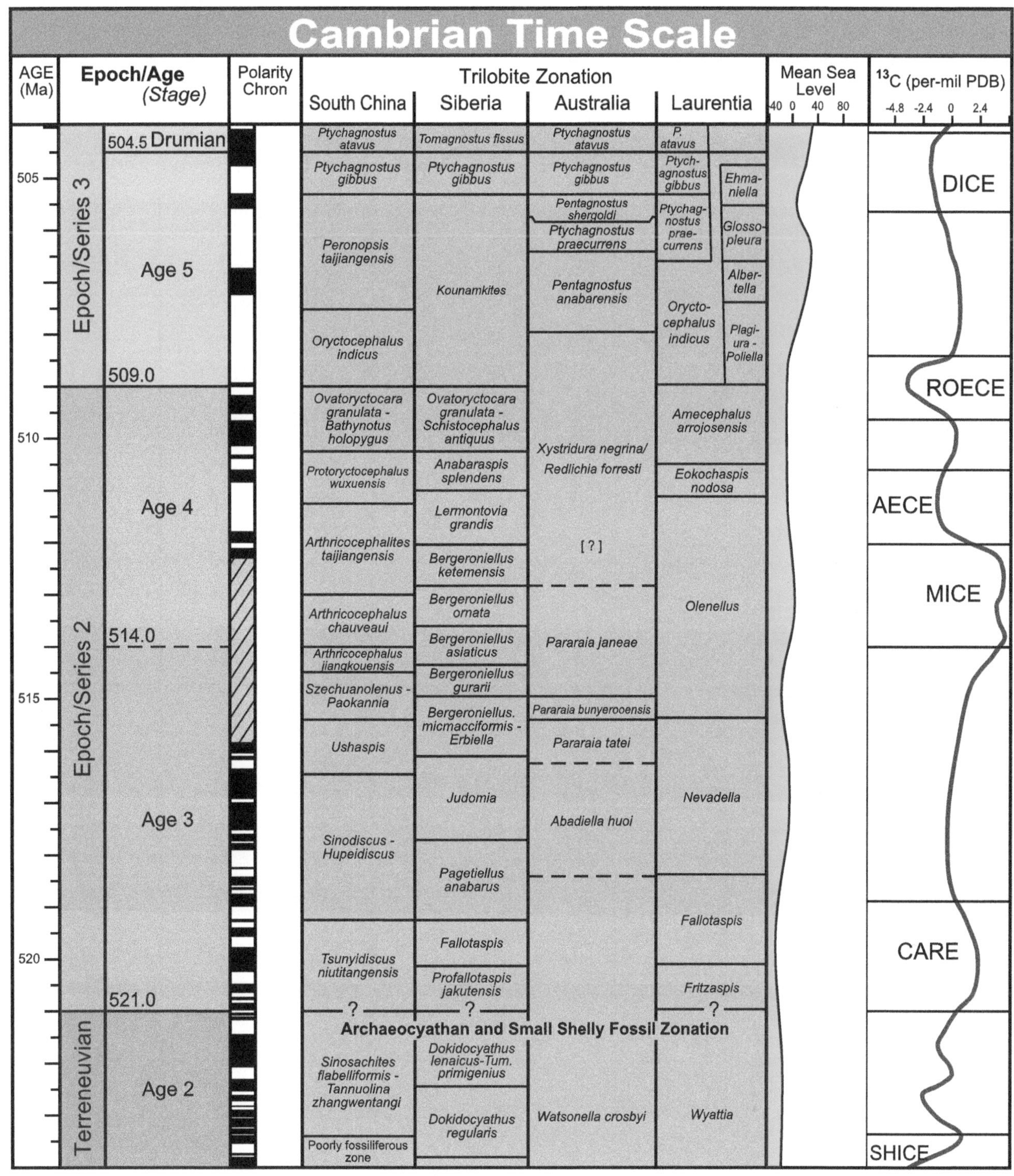

FIGURE 19.3 (*Continued*).

stratigraphy was born. Since its foundation in 1964, the International Subcommission on Cambrian Stratigraphy (ISCS) has worked to develop an internationally applicable, standard chronostratigraphic scale for the Cambrian System. From 1964 on, extensive studies of Cambrian stratigraphy have been carried out throughout the world to resolve correlation problems in various facies and biogeographic realms, to identify the stratigraphic horizons having best correlation potential on intracontinental and global scales, and to establish global boundary stratotype sections and points (GSSPs) of formal chronostratigraphic units. In 1968, the ISCS decided that resolving the problem

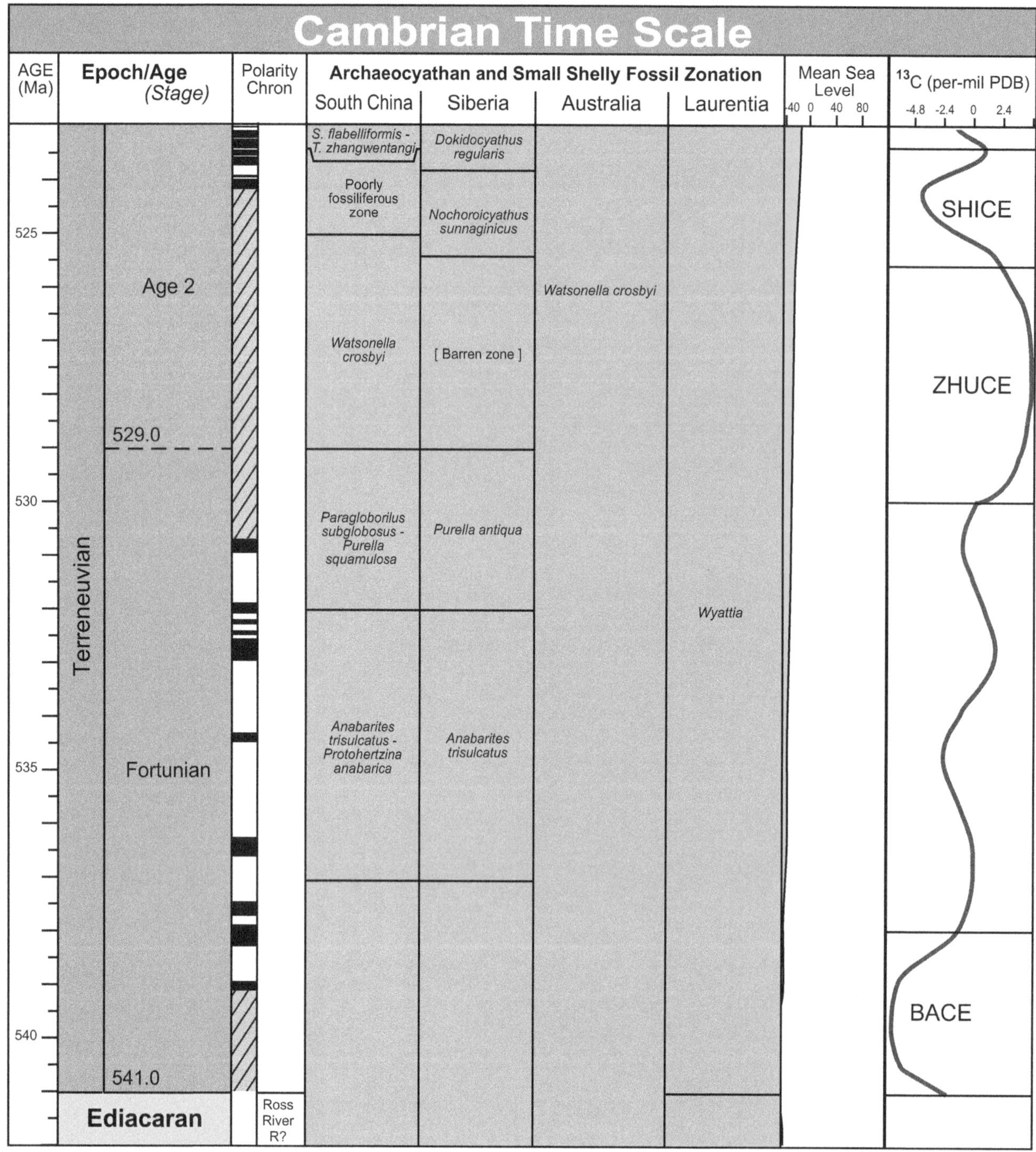

FIGURE 19.3 (*Continued*).

of the "base of the Cambrian System" should be one of its first tasks towards a precise definition of the system. In 1972, the first working group of the Cambrian Subcommission (called the Working Group on the Precambrian–Cambrian Boundary) was established. After two decades of study, a GSSP for the base of Cambrian, the first GSSP for the system, was erected in the Fortune Head section, eastern Newfoundland, Canada, in 1992 (Narbonne *et al.*, 1987; Landing, 1991, 1994; Brasier *et al.*, 1994). The boundary position is identified by a significant change in trace fossil associations (see Figure 19.4(c)) This criterion supplanted the other historic criteria for marking the boundary level, namely the appearance of trilobites (Walcott, 1890a; Wheeler, 1947), and the appearance of pre-trilobitic skeletal faunas (Rozanov, 1967; Cowie, 1978). It also extended the base of the system to a level

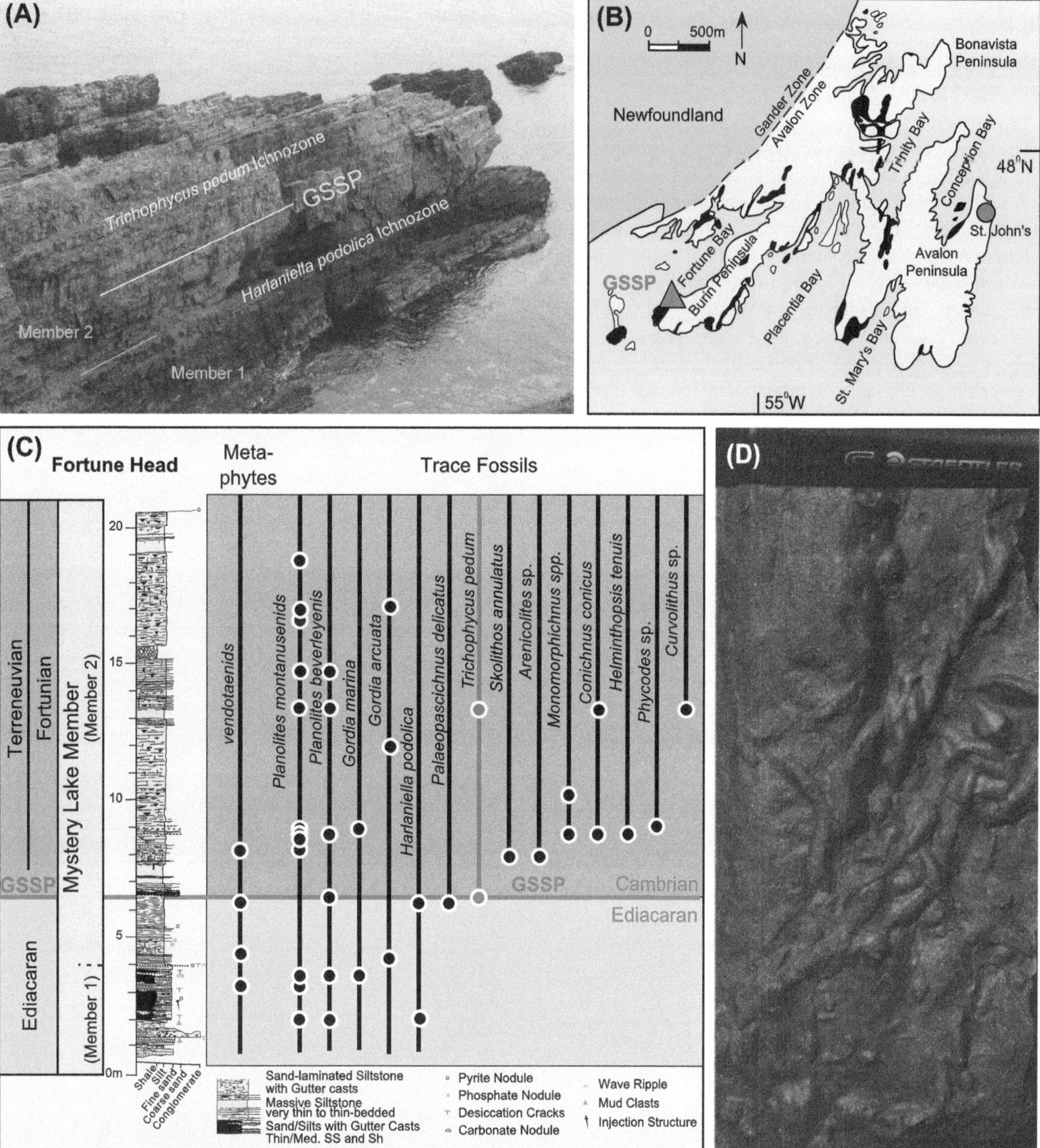

FIGURE 19.4 (A) and (B) GSSP at the base of the Cambrian System, Terreneuvian Series, and Fortunian Stage; Fortune Head section, Newfoundland, Canada; (C) zonation of trace fossils and body fossils associated with the basal Cambrian GSSP as recognized at the time of boundary ratification (1992); (D) the trace fossil *Trichophycus pedum* (Seilacher, 1955), whose first appearance datum (FAD) at the time of boundary ratification coincides with the base of the Fortunian Stage.

well below both trilobites and the earliest small shelly fossils (SSFs). With formal definition of the base of the Cambrian in the section at Fortune Head, Newfoundland, Canada, the system "added" a thick pre-trilobite interval bearing trace fossils of Phanerozoic aspect and small shelly fossils. Prior to ratification of the Cambrian base in 1992, this "added" section was correlated with the upper part of the Proterozoic.

The Cambrian–Ordovician boundary was defined in 1997 with ratification of the Ordovician base (Cooper and Nowlan, 1999; Cooper *et al.*, 2001). This was followed by an acceleration in the pace of the work of the ISCS toward subdividing the Cambrian System. A detailed, region-by-region correlation chart of the Cambrian was published by Geyer *et al.* (2000). On this chart, 14 stratigraphic levels were recognized as having strong correlation potential. Voting members of the ISCS identified six of those levels as being recognizable on global or intercontinental scales (see GSSP levels on Figure 19.2), and therefore potentially useful for defining global stages (Geyer and Shergold, 2000). This led to the establishment of a number of working groups charged with further, detailed study of these levels. Following further investigation, four intra-Cambrian GSSPs have so far been erected in northwestern Hunan and western Zhejiang, China, and the Great Basin, USA (Peng *et al.*, 2004a, 2009a; Babcock *et al.*, 2007). Other GSSPs will be decided upon in the near future.

Addition of a thick pre-trilobitic interval to the traditional Lower Cambrian opened discussions on the possibility of dividing the system (period) into four series (epochs) (see Figure 19.2). Important to the discussion were two significant facts:

1) the suggestion from geochronologic dating that the expanded Early Cambrian Epoch represents a duration of time that is longer than the traditional Middle and Late Cambrian combined; and
2) longstanding recognition that the traditional Early Cambrian bears a clear and important bioevent, the appearance of trilobites. This event has long been viewed as useful for defining an epoch boundary (Figure 19.2).

Four-fold regional subdivisions of the Cambrian were proposed for Laurentia by Palmer (1998) and South China by Peng (2000a,b). In 2004, it was the unanimous opinion of participants in a Cambrian Subcommision workshop held in South Korea that four series should be established for the global chronostratigraphic scale of the Cambrian System. Subsequently, the subcommission approved a subdivision of the system with one sub-trilobitic series and three trilobite-dominated series (Peng and Babcock, 2005a; Babcock *et al.*, 2005; Figures 19.2 and 19.3).

In the current conceptual model of the Cambrian, the system is further subdivided into 10 global stages (Figure 19.2). Because of strong faunal provincialism in the earlier part of the Cambrian Period, the lower half of the Cambrian System (i.e. the lower two series) bears only a few levels that have potential for global or intercontinental correlation. For this reason, these two series may each be subdivided into two stages. More diverse faunas in the upper half of the Cambrian System enable subdivision of each series into three stages. As illustrated in Figure 19.2, the levels used for defining stages in the upper half of the Cambrian are based on the first appearance datum (FAD) horizons of key species of agnostoid trilobites. Apart from the base of the system, no agreement has yet been reached on the criteria to be used for defining stages of the lower half of the Cambrian, but attention is now focusing on several horizons deemed to have intercontinental correlation potential.

In accordance with ICS standards, units of the new Cambrian chronostratigraphic time scale are to be defined by Global boundary Stratotype Sections and Points (GSSPs). To avoid any possible confusion with regional series and stage concepts applied previously, the Cambrian Subcommission decided to introduce a set of new globally applicable names for all series (epochs) and stages (ages) as new GSSPs are established. The new terms are based on geographic features, preferably ones associated with the GSSP-bearing sections, and all are defined according to the boundary-stratotype concept. By the end of 2011, five Cambrian stages (Fortunian, Drumian, Guzhangian, Paibian, and Jiangshanian) and two series (Terreneuvian and Furongian) had received formal names (Figure 19.2; Babcock *et al.*, 2007; Peng *et al.*, 2004a, 2009a; Landing *et al.*, 2007). Other series and stages are as yet undefined, and have received provisional numerical designations.

19.1.1. Terreneuvian Series

The GSSP for the base of the Terreneuvian Series is located in the lower Mystery Lake Member of the Chapel Island Formation in the Fortune Head section of the Burin Peninsula, eastern Newfoundland, Canada (Figure 19.4(b); Landing, 1994, 1996, 2004). The Terreneuvian Series is a fully sub-trilobite-bearing succession, characterized in its lower part by complex, substrate-penetrating (i.e. Phanerozoic-type) trace fossils, and in its higher part by diverse, biomineralized (calcareous and phosphatic) or secondarily phosphatized small shelly fossils (Landing *et al.*, 1989).

The name "Terreneuvian" evokes "Terre Neuffve", which, prior to a spelling reform, was the formal name for the 17th century French colony in Newfoundland that essentially corresponds to the Burin Peninsula. The base of the series is conterminous with the base of the Phanerozoic Eonothem, the Paleozoic Erathem, the Cambrian System, and the Fortunian Stage. The GSSP for the Cambrian System was ratified by the ICS and the IUGS at the 29th International Geological Congress at Kyoto, Japan, in 1992 (Brasier *et al.*, 1994; Landing, 1994). The Cambrian's lowest series and stage were not formally named at that time. Afterward, they were provisionally termed Series 1 and Stage 1 of the Cambrian System, respectively. In 2006, the Cambrian Subcommission voted to apply the name "Terreneuvian" to the lowermost series and "Fortunian" to the lowermost stage of the series. The new terms were approved by the ICS and ratified by the IUGS in 2007 (Landing *et al.*, 2007).

The current concept is that the Terreneuvian Series comprises two stages, the Fortunian Stage and an overlying stage provisionally called Cambrian Stage 2. The Terreneuvian Series is overlain by a series provisionally termed "Cambrian Series 2". The base of Series 2 has not been defined yet, but is expected to be close to the horizon marking the first appearance of trilobites in Gondwana. That horizon and its GSSP, once they are ratified, will automatically define the top of the Terreneuvian Series.

19.1.1.1. Fortunian Stage and the Base of the Cambrian System

The Fortunian Stage was named for the Fortune Head section, which contains the GSSP, on the southern Burin Peninsula, eastern Newfoundland, Canada. The Fortunian is the lowest stage of the Terreneuvian Series and the Cambrian System. The GSSP is a point that lies 2.4 m above the base of what was earlier referred to as "Member 2" of the Chapel Island Formation (Narbonne *et al.*, 1987). This horizon is exposed in coastal cliffs low in the 440-m-thick Fortune Head section, and just above the transition to a storm-influenced facies (Narbonne *et al.*, 1987; Landing, 1994; Brasier *et al.*, 1994) (Figures 19.4(a),(c)). The units earlier termed members 1 and 2 of the Chapel Island Formation (Narbonne *et al.*, 1987) constitute the lower part of what is now called the Mystery Lake Member of the Chapel Island Formation (Landing, 1996). The GSSP coincides with the first appearance datum (FAD) of the ichnofossil *Phycodes pedum* (now referred to by various authors as *Trichophycus pedum*, *Treptichnus pedum*, or *Manykodes pedum*; Figure 19.4(d)), as recognized at the time of boundary ratification. The FAD of *T. pedum* defines the base of the *T. pedum* Ichnozone, an assemblage zone based on trace fossils. It reflects the appearance of complex sediment-disturbing behavior by multiple epifaunal and infaunal animals. Since the time of boundary ratification, it has been shown that the lowest occurrence of *T. pedum* is slightly below the GSSP level (Gehling *et al.*, 2001). Strata below the GSSP in the lower part of the Mystery Lake Member include uppermost Proterozoic (Ediacaran) layers assigned to the *Harlaniella podolica* Ichnozone. The highest observed occurrence of *H. podolica* is 0.2 m below the GSSP, and the lowest observed occurrence of a low-diversity SSF assemblage is about 400 m higher in the succession (Narbonne *et al.*, 1987; Landing *et al.*, 1989). The lowest trilobite-bearing strata (assigned to the *Callavia broeggeri* Zone) lie approximately 1400 m above the GSSP (Landing, 1996). However, a major regional unconformity (a type-1 sequence boundary) separates the Terreneuvian Series from the overlying trilobite-bearing strata (provisional Series 2) across the Avalon paleocontinent, including areas both in eastern North America (such as eastern Newfoundland), and in the United Kingdom (Landing, 1996, 2004).

19.1.1.2. Stage 2 (Undefined)

The base of the second stage of the Cambrian has not yet been defined. Likewise, a criterion for defining the boundary has not yet been determined. Few biostratigraphically defined levels within the Terreneuvian Series have good potential for intercontinental or even global correlation. Ultimately, the FAD of a small shelly fossil (SSF) or an archaeocyath may be used to define the base (Babcock *et al.*, 2005). SSFs of the Terreneuvian Series have been studied extensively during the last two decades, especially in Siberia, South China, Avalonia, and Laurentia. Although most seem to be highly endemic, a few species have potential in long distance correlation (Li *et al.*, 2007). The micromollusk (rostroconch) *Watsonella crosbyi*, for example, has wide distribution in the sub-trilobitic strata of Siberia, South China, Australia, Mongolia, Kyrgyzstan, Avalonian North America (eastern Newfoundland and Massachusetts), and France (Demidenko, 2001). It ranges through a narrow stratigraphic interval and occurs in both carbonate and siliciclastic successions. The mollusk *Aldanella attleborensis*, which usually co-occurs with *W. crosbyi*, is also widely distributed. Specimens are known from the sub-trilobite strata of Siberia, South China, Tarim, Iran, and possibly Spain. Archaeocyaths have excellent application in biostratigraphic zonation of the lower part of the traditional Lower Cambrian of Siberia, but high faunal provincialism and endemism limit the potential of this group for interregional correlation (Zhuravlev, 1995; Kruse and Shi, 2000).

19.1.2. Series 2 (Undefined)

Series 2 is the first trilobite-dominated series. It has fewer levels having intercontinental correlation potential than the two younger series because of strong provincialism and endemism among trilobite faunas. The series is likely to be divided into two stages. The base of the series is expected to be placed at a horizon close to the first appearance of trilobites.

During Epoch 2 of the Cambrian Period the world oceans experienced an explosive diversification of metazoans, but the faunas, particularly the polymerid trilobites and archaeocyaths, were highly provincial (Kobayashi, 1971, 1972; Debrenne, 1992; Palmer and Repina, 1993; Kruse and Shi, 2000; Álvaro *et al.*, in press). Olenellid and redlichiid trilobites, which characterize two separate faunal provinces, diversified during Epoch 2. The olenellids, eodiscids, and most redlichiids became extinct close to the end of the epoch, a significant bioevent recognized as the oldest mass extinction of trilobites.

19.1.2.1. Stage 3 (Undefined)

The base of Cambrian Stage 3, which is also the base of Cambrian Series 2, has not been defined. This boundary is expected to be marked by a significant and widely recognizable bioevent that will divide the lower half of the Cambrian as equally as possible. The FAD of trilobites, with its historic aspects, is a possibility for the boundary position, although the FAD of an SSF could eventually be selected as the primary marker of the boundary. In any case, definition of the base of provisional Stage 3 will divide the lower half of the Cambrian subequally into a sub-trilobitic series and a trilobite-dominated series. Such a distinction of two series has been introduced in Laurentia as the Begadean and Waucobian series (Palmer, 1998), in South China as the Diandongian and Qiandongian series (Peng, 2000a,b, 2003), in western Gondwana (Morocco and Iberia) as the Cordubian and Atlasian series (Geyer and Landing, 2004), and in Avalonia as the Placentian and Branchian series (Landing, 1992).

Precise correlation of a horizon marked solely by the FAD of trilobites will be hard to achieve, because trilobites appear in strata at slightly different times in different regions. Differences in the point of appearance among different regions are due to diachroneity, and to the appearance of the earliest trilobites immediately above unconformities in many regions. If the first appearance of trilobites were accepted as a marker for the lower boundary of provisional Stage 3 (and Series 2), the most likely regions for defining the GSSP are Siberia, western Gondwana (Morocco and Iberia), and western Laurentia. The earliest occurrences of trilobites as recognized currently on other paleocontinents appear to be younger in age. The earliest trilobites include *Profallataspis jakutensis* in Siberia, *Hupetina antiqua* in Morocco, and *Fritzaspis* in Laurentia. In Siberia, *P. jakutensis* occurs 10 m above the base of the Atdabanian Stage (Astashkin *et al.*, 1990, 1991; Shergold *et al.*, 1991; Varlamov *et al.*, 2008a); in Morocco, the FAD of *H. antiqua* coincides with the base of the Issendelenian Stage (Geyer, 1990, 1995); and in western Laurentia *Fritzaspis*, *Profallotaspis*?, *Amplifallotaspis*, and *Repinaella* all occur below the lower boundary of the Montezuman Stage, and the Waucobian Series (Palmer, 1998, Hollingsworth, 2007, 2011).

To sidestep issues associated with use of a trilobite as the correlation tool coinciding with the base of Stage 3, it is possible that an SSF such as *Pelagiella subangulata* or *Microdictyon effusum* might be selected. The micromollusk *P. subangulata* has a narrow stratigraphic range and is widely distributed through Siberia, South China, Australia, Antarctica, Iran, India, continental Europe (Germany and Sardinia), the United Kingdom, Canada (Nova Scotia), and Kazakhstan. Sclerites of the armored lobopod *M. effusum* are known from Siberia, South China, Australia, the United Kingdom, Baltica, Kazakhstan, and Laurentia. At least eight species of *Microdictyon* have been named, and some may be synonyms, so a taxonomic overhaul of the genus is required.

19.1.2.2. Stage 4 (Undefined)

The base of provisional Stage 4 has not been formally defined. One level, the base of the interval bearing the eodiscoid trilobites *Hebediscus, Calodiscus, Serrodiscus*, and *Triangulaspis*, referred to as the HCST band (Geyer, 2005), was proposed as a potential stage marker in the upper part of the traditional Lower Cambrian (Geyer and Shergold, 2000), but it has not received broad support from members of the Cambrian Subcommission. Instead, the subcommission has favored placing the base of Stage 4 at a level coinciding with the FAD of a single trilobite species. Possibilities include a species of *Olenellus* (s.l.), *Redlichia* (s.l.), *Judomia,* or *Bergeroniellus* (Babcock *et al.*, 2011). Such a position would be at a level roughly corresponding to the base of the Dyeran Stage of Laurentia (Palmer, 1998), the base of the Duyunian of South China (Peng, 2000a,b, 2003), and the base of the Botoman Stage of Siberia (Repina *et al.*, 1964; Khomentovsky and Repina, 1965).

Stage 4 spans almost the entire range of *Olenellus* (s.l.) and *Redlichia* (s.l.), the representative forms of the "olenellid" and "redlichiid" faunal realms. Both olenellid and redlichiid trilobites became extinct near the end of the stage.

19.1.3. Series 3 (Undefined)

The conterminous base of provisional (undefined) Series 3 and Stage 5 has also been referred to as the "Lower–Middle Cambrian boundary". The traditional usage stems from separate works by Brøgger (1878, 1882, 1886) and Walcott (1889a,b; 1890a,b), who developed concepts of the "Early Cambrian" and "Middle Cambrian" that were of different duration and that were applied in different faunal provinces (Robison *et al.*, 1977; Fletcher, 2003; Geyer, 2005). Up to the present, there has been considerable variance in the horizon interpreted as the "Lower–Middle Cambrian boundary" (Geyer and Shergold, 2000; Shergold and Cooper, 2004).

The base of Series 3 has not been defined but will possibly be placed at a level close to one of the traditional, regional usages of the "Lower–Middle Cambrian boundary", either the boundary concept as used in Laurentia and eastern Gondwana, or that as used in Baltica. The top of the series was defined automatically as the lower boundary of the Furongian Series. That horizon coincides with the FAD of the agnostoid trilobite *Glyptagnostus reticulatus*, a level that lies two full biozones (the *Linguagnostus reconditus* and *Glyptagnostus stolidotus* zones) above the traditional "Middle–Upper Cambrian boundary" (Peng *et al.*, 2004a). Therefore Cambrian Series 3 does not equal the traditional Middle Cambrian any more but is greatly expanded in concept.

The series is divided into three stages, of which only the middle and the upper stages have been formally defined. Cambrian Epoch 3 was a time in which trilobite faunas recovered from the mass extinction near the end of Epoch 2 (Álvaro *et al.*, in press). Trilobite diversity increased dramatically, so there is a major change in faunal composition on Cambrian continents. The series is characterized by the predominance of widespread ptychagnostid and diplagnostid agnostoids, and by endemic paradoxidid and ptychopariid polymerids. In addition, oryctocephalid and bathynotid polymerids predominate in the lower part of the series, whereas anomocarid and damesellid polymerids are more common in the upper part of the series. Ceratopygid polymerids emerged in the uppermost stage of the series.

19.1.3.1. Stage 5 (Undefined)

The base of Cambrian Stage 5 is undefined at present. Recent investigations indicate that oryctocephalid trilobites evolved quickly through the time interval of the traditional "Lower–Middle Cambrian boundary". On account of this, they will have an important role in correlating the base of the stage. In 2006, during its meeting in South Australia, the Cambrian Subcommission decided to investigate two levels, both marked by first appearances of oryctocephalid trilobite species, which could be used to define the conterminous base of Cambrian Series 3 and Stage 5. The levels are at the FAD of *Oryctocephalus indicus* and the FAD of *Ovatoryctocara granulata*, which is slightly lower in stratigraphic position (Figures 19.2, 19.3 and 19.5). Both levels appear to have good potential for interprovincial and intercontinental correlation (Sundberg and McCollum, 1997; McCollum and Sundberg, 1999; Geyer and Shergold, 2000; Fletcher, 2003, 2007; Geyer, 2005; Geyer and Peel, 2011; Naimark *et al.*, 2011), and the associated trilobites make it possible to recognize the levels widely in most Cambrian regions (Geyer, 2005). A final decision on the primary boundary criterion and a GSSP will hinge largely on investigations of sections in the Great Basin (USA), South China, and Siberia. If an *O. indicus*-based GSSP is erected for the base of Stage 5 and Series 3, it would correspond closely to the traditional "Lower–Middle Cambrian boundary" of Laurentia and eastern Gondwana, as it would coincide with the base of the Delamaran Stage (Lincolnian Series) of Laurentia (Palmer, 1998) and the base of the Taijiangian Stage (Wulingian Series) of South China (Peng, 2000a,b; Peng and Babcock, 2001). If an *O. granulata*-based GSSP is erected, it would correspond closely to the "Lower–Middle Cambrian boundary" as recognized in northern Siberia (Korovnikov, 2005).

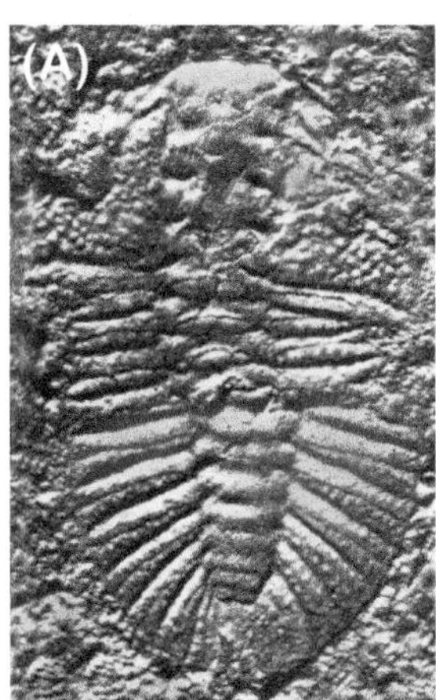

FIGURE 19.5 **Key polymerid trilobite guide fossils associated with the conterminous base of provisional Series 3 and Stage 5.** (A) *Ovatoryctocare granulata* (B) *Oryctocephalus indicus*.

Regardless of the choice of primary correlation criterion, the base of Stage 5 will correspond to a sharp, high-magnitude, negative shift in carbon isotopic ($\delta^{13}C$) values ($\geq 4‰$). This, the ROECE excursion (Redlichiid–Olenellid Extinction Carbon isotope Excursion), is one of the largest negative carbon isotope excursions known from the Cambrian (Zhu *et al.*, 2006; Figure 19.3 and Section 19.2.7). The excursion begins just below the traditional "Lower–Middle Cambrian boundary" as recognized in Laurentia, and suggests the onset of major paleoceanographic and paleoclimatic changes associated with the extinction of olenellid trilobites in Laurentia and redlichiid trilobites in Gondwana (Montañez *et al.*, 2000).

19.1.3.2. Drumian Stage

The Drumian Stage is the middle stage of Series 3. The name, which was ratified in 2006, is derived from the Drum Mountains in northern Millard County, western Utah, USA (Figure 19.6(c)). The base of the stage is defined by a GSSP coinciding with the FAD of a cosmopolitan agnostoid trilobite *Ptychagnostus atavus* (Figures 19.6(b), (d)). The GSSP is located 62 m above the base of the Wheeler Formation in the Stratotype Ridge section of the Drum Mountains (Babcock *et al.*, 2004, 2007; Figures 19.6(a),(d)). *Ptychagnostus atavus* has been identified from almost all major Cambrian regions of the world (e.g., Robison *et al.*, 1977; Rowell *et al.*, 1982; Robison, 1999; Peng and Robison, 2000; Geyer and Shergold, 2000; Shergold and Geyer, 2003; Babcock *et al.*, 2004, 2007), and, even long before definition of the boundary level, had been in use as a zonal guide fossil in deposits of Baltica, Gondwana, Kazakhstania, and Laurentia (e.g., Westergård, 1946; Robison, 1976, 1984; Öpik, 1979; Ergaliev, 1980; Geyer and Shergold, 2000; Peng and Robison, 2000). In addition to the FAD of *P. atavus*, two other related, cosmopolitan agnostoid species help constrain the base of the Drumian Stage. The first appearance of *P. atavus* always

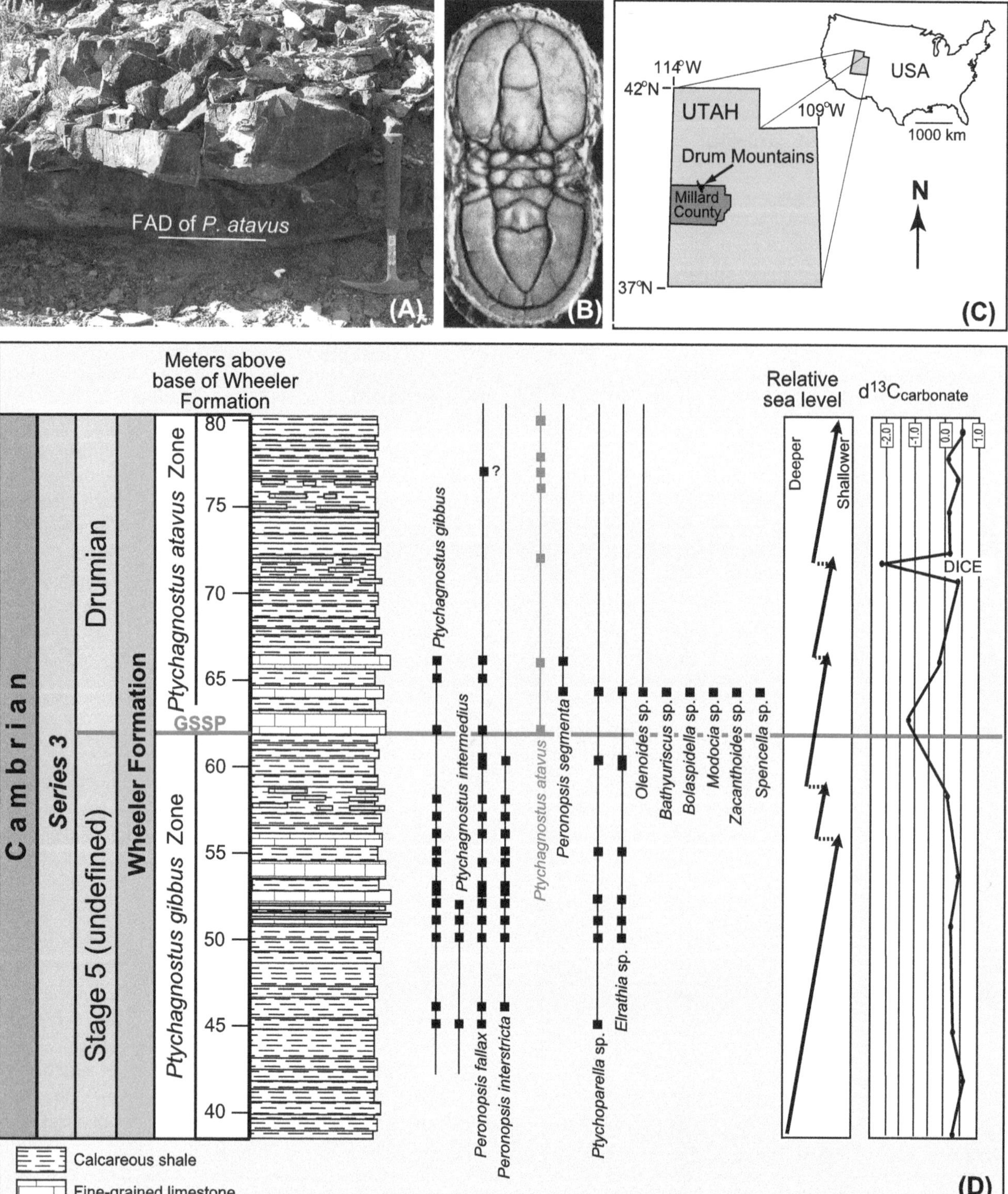

FIGURE 19.6 **(A) and (C) Drumian Stage GSSP in the Stratotype Ridge section, Drum Mountains, Utah, USA.** Stratigraphic distribution of trilobites close to the base of the Drumian Stage in the Stratotype Ridge section, Drum Mountains, Utah, USA; (B) *Ptychagnostus atavus* (Tullberg, 1880), an agnostoid trilobite whose first appearance datum (FAD) coincides with the base of the Drumian Stage; (D) Stratigraphic distribution of trilobites close to the base of the Drumian Stage in the Stratotype Ridge section, Utah, USA (redrawn after Babcock *et al.*, 2007).

succeeds the first appearance of *Ptychagnostus gibbus*, and it precedes the first appearance of *Ptychagnostus punctuosus*. Both *P. gibbus* and *P. punctuosus* have been widely used as zonal guide fossils (Geyer and Shergold, 2000).

The base of the Drumian Stage can be recognized on a global scale not only by the FAD of *Ptychagnostus atavus*, but also by significant changes in polymerid trilobite faunas, conodont faunas, and by a carbon isotope excursion. The position coincides closely with the bases of two biozones based on polymerid trilobites: the *Bolaspidella* Zone of Laurentia (Robison, 1976; Palmer, 1998, 1999; Babcock *et al.*, 2004, 2007) and the *Dorypyge richthofeni* Zone of South China (Peng *et al.*, 2004b). It also corresponds closely to a turnover in conodonts near the base of *Gapparodus bisulcatus-Westergaardodina brevidens* Assemblage Zone (Dong and Bergström, 2001a,b). The base of the stage corresponds to the onset of a medium-scale negative carbon excursion (Babcock *et al.*, 2007; Brasier and Sukhov, 1998), referred to as the DICE excursion (DrumIan Carbon isotope Excursion (Zhu *et al.*, 2006; Figures 19.3 and 19.6; see also Section 19.2.7).

19.1.3.3. Guzhangian Stage

The Guzhangian Stage is the third and last stage of Series 3. The name is derived from Guzhang County in the Wuling Mountains of northwestern Hunan Province, China. The base of the Guzhangian Stage is defined by a GSSP that coincides with the FAD of the cosmopolitan agnostoid trilobite *Lejopyge laevigata* (Figures 19.7C, D). The GSSP, which was ratified in 2007, is 121.3 m above the base of the Huaqiao Formation in the Luoyixi section, exposed along a roadcut on the south bank of the Youshui River, 4 km northwest of the town of Luoyixi (Peng *et al.*, 2009a; Figure 19.7(b)).

Lejopyge laevigata, which is the primary correlation tool for the base of the Guzhangian Stage, has been recognized in all Cambrian paleocontinents (e.g., Westergård, 1946; Cowie *et al.*, 1972; Robison, 1976, 1984; Öpik, 1979; Geyer and Shergold, 2000; Peng and Robison, 2000; Axheimer *et al.*, 2006; Peng *et al.*, 2006). The species has been used as a zonal guide fossil in Baltica, Gondwana, Kazakhstania, Siberia, Laurentia, and eastern Avalonia. In addition to the FAD of *L. laevigata*, two other congeneric species, *Lejopyge armata* and *L. calva*, can be used to help constrain the position of the Guzhangian base. *Lejopyge laevigata* consistently appears slightly above the first appearance of either *Lejopyge armata* (in Australia, South China, North China, Kazakhstan, Siberia, and Sweden) or *L. calva* (in Laurentia).

The Guzhangian GSSP can be tightly constrained by stratigraphic criteria other than just the ranges of *Lejopyge* species. The Guzhangian base is above the last occurrence datum (LAD) of the agnostoid *Goniagnostus nathorsti*, and below the FAD of the agnostoid *Proagnostus bulbus*. The first appearance of *L. laevigata* closely corresponds to the first appearance of damesellid trilobites (e.g., *Palaeodotes*, *Parablackwelderia*, *Blackwelderia*) in South China and Kazakhstan (Ergaliev, 1980, Ergaliev and Ergaliev, 2004; Peng *et al.*, 2004b, 2006). It also approximately coincides with a faunal change associated with the base of the *Paradoxides forchhammeri* Zone in western Avalonia (Geyer and Shergold, 2000).

The range of *L. laevigata* has long been used regionally as a major stratigraphic marker. The FAD of *L. laevigata* defines the base of the Ayusokian Stage in Kazakhstan, the base of the Boomerangian Stage in Australia (Öpik, 1967; Geyer and Shergold, 2000), and the base of the *Aldanaspis* Zone in Siberia (Egorova *et al.*, 1982). By using the first appearance of *L. laevigata*, rather than its local abundance, the base of the Scandinavian *L. laevigata* Zone has been moved downward to the base of the *Solenopleura? brachymetopa* Zone (Axheimer *et al.*, 2006). This change makes the base of the Guzhangian Stage identical with the base of the revised *L. laevigata* Zone in Scandinavia (a zone that now embraces the traditional *Solenopleura? brachymetopa* Zone; Ahlberg *et al.*, 2009).

The Guzhangian Stage embraces the interval of the traditional "Middle–Upper Cambrian boundary". In Sweden, the traditional "Upper Cambrian" was marked by the base of the *Agnostus pisiformis* Zone (Westergård, 1946). Peng and Robison (2000) stated that the "Middle–Upper Cambrian boundary" in Sweden is commonly drawn at a position marking the local abundance of *A. pisiformis* rather than its first appearance. Recently, *Linguagnostus reconditus* was reported from the *A. pisiformis* Zone in Sweden, and it is associated with abundant *A. pisiformis* (Ahlberg and Ahlgren, 1996; Ahlberg, 2003). This suggests that the traditional "Middle–Upper Cambrian boundary" of Sweden is very close to, if not identical with, the base of the *Linguagnostus reconditus* Zone. The *L. reconditus* Zone is the third of four agnostoid zones recognized within Guzhangian Stage strata of South China.

19.1.4. Furongian Series

The Furongian Series is the uppermost series of the Cambrian System. Its name is derived from Furong, which means lotus, in reference to Hunan Province, the Lotus State of China. The GSSP for the conterminous base of the Furongian Series and its lowermost stage, the Paibian Stage, is located in the Wuling Mountains of northwestern Hunan, China. The GSSP coincides with the FAD of *Glyptagnostus reticulatus*, a level that is stratigraphically much higher than the traditional "Middle–Upper Cambrian boundary" as recognized earlier in Sweden. Definition of a GSSP for the base of the Forungian Series resulted in an "Upper Cambrian" that was restricted in length compared to that of traditional usage because the *Agnostus pisiformis* Zone of Sweden, which corresponds to two biozones as used in South China, was excluded (Ahlberg,

Base of the Guzhangian Stage of the Cambrian System in the Louyixi Section in the Wuling Mountains, NW Hunan Province, China

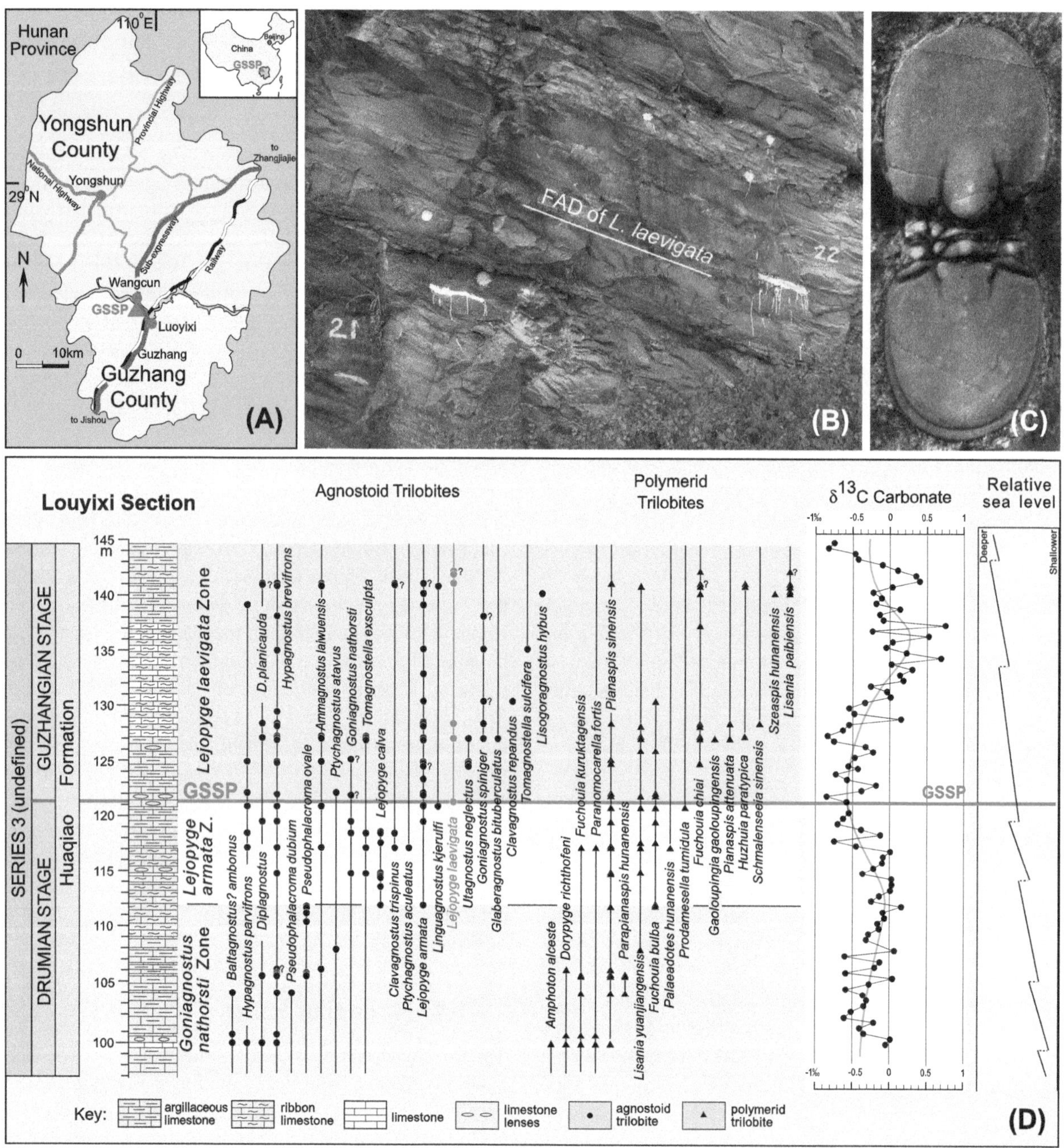

FIGURE 19.7 (A) and (B) Guzhangian Stage GSSP in the Luoyixi section, Hunan, China; (C) *Lejopyge laevigata* (Linnarsson, 1869), an agnostoid trilobite whose first appearance datum coincides with the base of the Guzhangian Stage; (D) Stratigraphic distribution of trilobites close to the base of the Guzhangian Stage in the Luoyixi section, Hunan, China (redrawn after Peng *et al*., 2009a).

2003; Axmeimer *et al*., 2006; see also preceding discussion of the Guzhangian Stage).

The upper boundary of the Furongian Series is automatically defined by the base of the Tremadocian Series (Lower Ordovician). The basal Tremadocian GSSP coincides with the FAD of the conodont *Iapetognathus fluctivagus*. Currently, the majority opinion within the Cambrian Subcommission is that the Furongian should be subdivided into three subequal stages. The lowermost stage (Paibian) and the middle stage (Jiangshanian) have already been ratified. The base of the upper stage is expected to coincide with, or be close to, the FAD of an agnostoid, *Lotagnostus americanus* (Babcock *et al*., 2005).

The Furongian Epoch marks a time of great faunal turnover among polymerid trilobites. Stepwise extinction of polymerids in Gondwana and Laurentia at the end of Epoch 3 was followed by recovery during the early part of the Furongian. In eastern Gondwana, damesellid trilobites were replaced by leiostegiid (e.g., *Chuangia* and *Prochuangia*) and ceratopygid trilobites. In Laurentia, the Furongian Series corresponds to the Pterocephaliid Biomere, and strata are zoned according to the record of evolution in pterocephaliid, elviniid, and saukiid polymerids. In Baltica, evolutionary changes in olenid trilobites allows for fine regional zonation of the series. Agnostid and pseudagnostid agnostoids are important for regional and global subdivision of the series.

19.1.4.1. Paibian Stage

The base of the Paibian Stage, and the base of the Furongian Series, is defined by a GSSP coinciding with the first appearance of the cosmopolitan agnostoid trilobite *Glyptagnostus reticulatus* (Figure 19.8(b),(d)). The stage is named for Paibi, a village in Huayuan County, about 35 km west of Jishou, northwestern Hunan, China. The GSSP, which was ratified in 2003, is 369 m above the base of the Huaqiao Formation along a south-facing hill in the Paibi section (Figure 19.8(a),(d)).

The FAD of *Glyptagnostus reticulatus* is one of the most widely recognizable stratigraphic horizons in the Cambrian. Even before its selection as the criterion for marking the base of the Paibian Stage and Furongian Series, *G. reticulatus* was used as a zonal guide fossil in Siberia, Kazakhstan, South China, Australia, and Laurentia. The interval containing the FAD of *G. reticulatus* marks a time of significant faunal change, and that change has been formalized in various regional stratigraphic schemes. The position corresponds to the base of the Pterocephaliid Biomere or the base of the Steptoean Stage in Laurentia, the base of the Idamean Stage in Australia, the base of the Sakian Stage in Kazakhstan, the base of the Maduan Horizon in Siberia, and the base of the Furongian Series in Scandinavia (Terfelt *et al.*, 2008). The level is also near the base of a large positive shift in $\delta^{13}C$ values, referred to as the Steptoean PositIve Carbon isotope Excursion (SPICE excursion) (Saltzman *et al.*, 2000; Figure 19.3; see also Section 19.2.7).

19.1.4.2. Jiangshanian Stage

The base of the Jiangshanian Stage is defined by a GSSP coinciding with the first appearance of the cosmopolitan agnostoid trilobite *Agnostotes orientalis* (Figure 19.9(c),(d)). The stage is named for Jiangshan County, Zhejiang Province, China. The GSSP, which was ratified in 2011, is 108.12 m above the base of the Huayansi Formation in the Duibian B section (Figure 19.9(a),(d)). The stratotype section is exposed in natural outcrops situated at the base of Dadoushan Hill, west of Duibian Village (Peng *et al.*, 2009a).

The holotype of *A. orientalis* (Kobayashi, 1935) is poorly preserved, and because the species was poorly characterized originally, a number of junior synonyms have been proposed. In some regions, the species is better known by these synonyms, such as *Pseudoglyptagnostus clavatus*, *Agnostotes (Pseudoglyptagnostus) clavatus*, *A. clavata*, and *Glyptagnostotes elegans*. Currently the species is recognized from South China (Hunan and Zhejiang), South Korea, Siberia (Kharaulakh Ridge, northeast Siberian Platform, Chopko River of the Norilsk Region, northwest Siberian Platform), southern Kazakhstan (Malyi Karatau), and Laurentia (Mackenzie Mountains and southeastern British Columbia).

In many regions *A. orientalis* and the polymerid trilobite *Irvingella* co-occur (Lazarenko, 1966; Öpik, 1967; Ergaliev, 1980; Peng, 1992; Pratt, 1992; Chatterton and Ludvigsen, 1998; Choi, 2004; Hong *et al.*, 2003; Varlamov *et al.*, 2005; Peng *et al.*, 2009a; Ergaliev and Ergaliev, 2008; Varlamov and Rozova, 2009), and both trilobites can be used to constrain the base of the Jiangshanian Stage. Together, the two trilobites have been used as zonal guide fossils in South China, northeastern Siberia, northwestern Siberia, and South Korea. *Irvingella*, however, has a wider paleogeographic distribution (Geyer and Shergold, 2000), and it allows close correlation into Australia, Baltica, Avalonia, eastern and western Laurentia, Argentina, and probably Antarctica. In South China and Canada, *A. orientalis* and *Irvingella angustilimbata* make their first appearances at the same stratigraphic level (Peng, 1992; Pratt, 1992; Peng *et al.*, 2009b). This is true even in the Duibian B section, which contains the GSSP. The horizon corresponds to the base of the Taoyuanian Stage as used previously in South China (Peng and Babcock, 2001, 2008; Shergold and Cooper, 2004) and the base of the *Proceratopyge rectispinata* Zone in the Mackenzie Mountains. The base of the *P. rectispinata* Zone lies somewhat below the base of the Sunwaptan Stage of Laurentia, but above the base of the *Elvinia* Zone, the uppermost zone of the stage. The FAD of *A. orientalis* corresponds closely to the base of the Iverian Stage in Australia, the base of the *Parabolina spinulosa* Zone of Sweden and England, the base of the *Pseudagnostus vastulus–Irvingella tropica* Zone of Kazakhstan (Ergaliev and Ergaliev, 2008), and a level that is somewhat below the base of the "Tukalandyan Stage" of Rozova (1963) or within the Chekurovian Stage of Lazarenko and Nikiforiv (1972) in Siberia. The FAD of *A. orientalis* lies in a position corresponding to the upper part of the SPICE excursion, a large positive shift in $\delta^{13}C$ values (Figure 19.3; see also Section 19.2.7; Peng *et al.*, 2009b).

19.1.4.3. Stage 10 (Undefined)

The base of Cambrian Stage 10, the uppermost stage of the Furongian Series and the Cambrian System, is undefined. The Cambrian Subcommission favors marking the base at or close

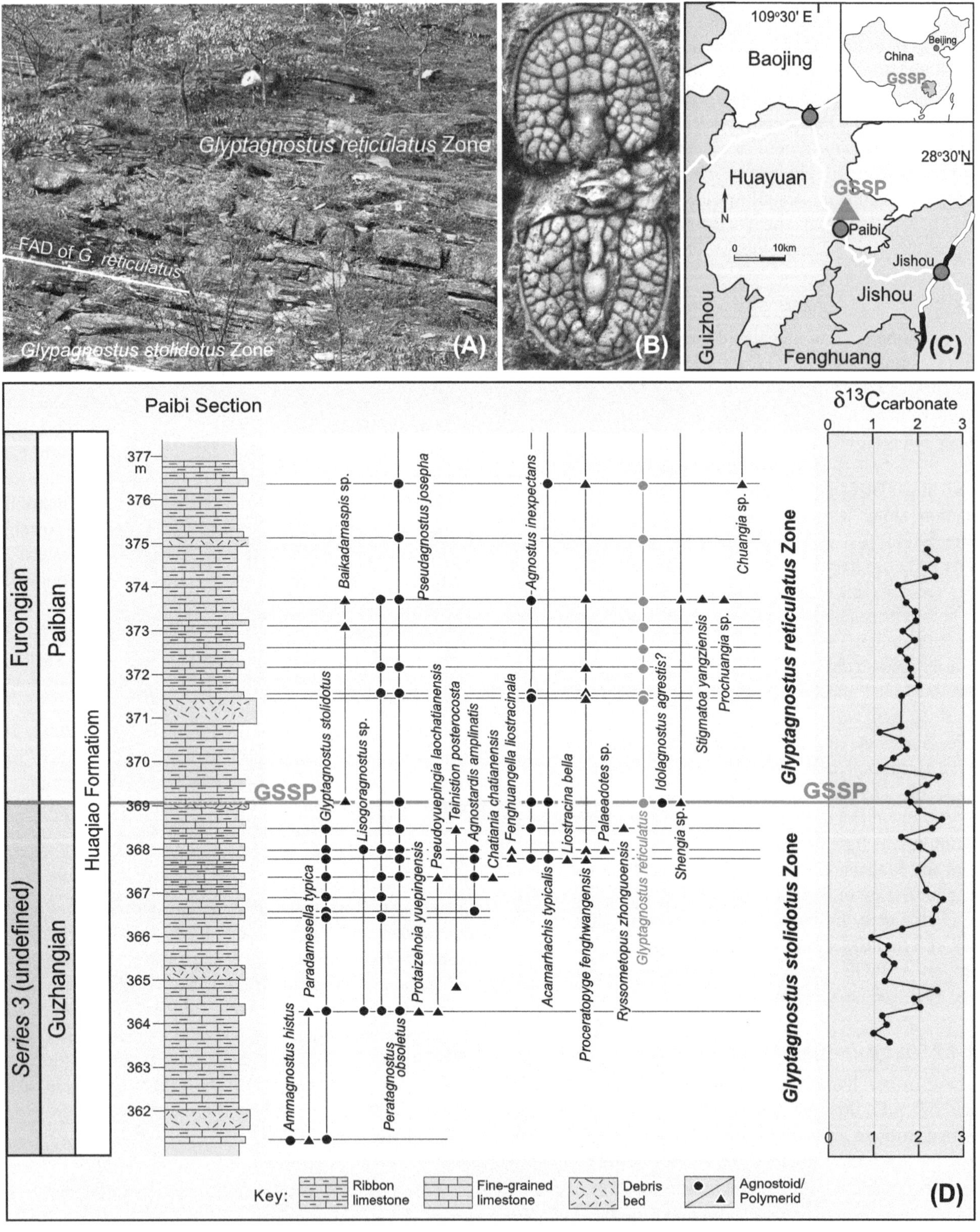

FIGURE 19.8 (A) and (C) GSSP of the Furongian Series and Paibian Stage in the Paibi section, Hunan, China; (B) *Glyptagnostus reticulatus* (Angelin, 1851), an agnostoid trilobite whose first appearance datum coincides with the base of the Furongian Series and Paibian Stage; (D) Stratigraphic distribution of trilobites close to the base of the Furongian Series and Paibian Stage in the Paibi section, Hunan, China (redrawn after Peng *et al.*, 2004a).

Base of the Jiangshanian Stage of the Cambrian System in the Duibian B Section, W Zhejiang Province, China

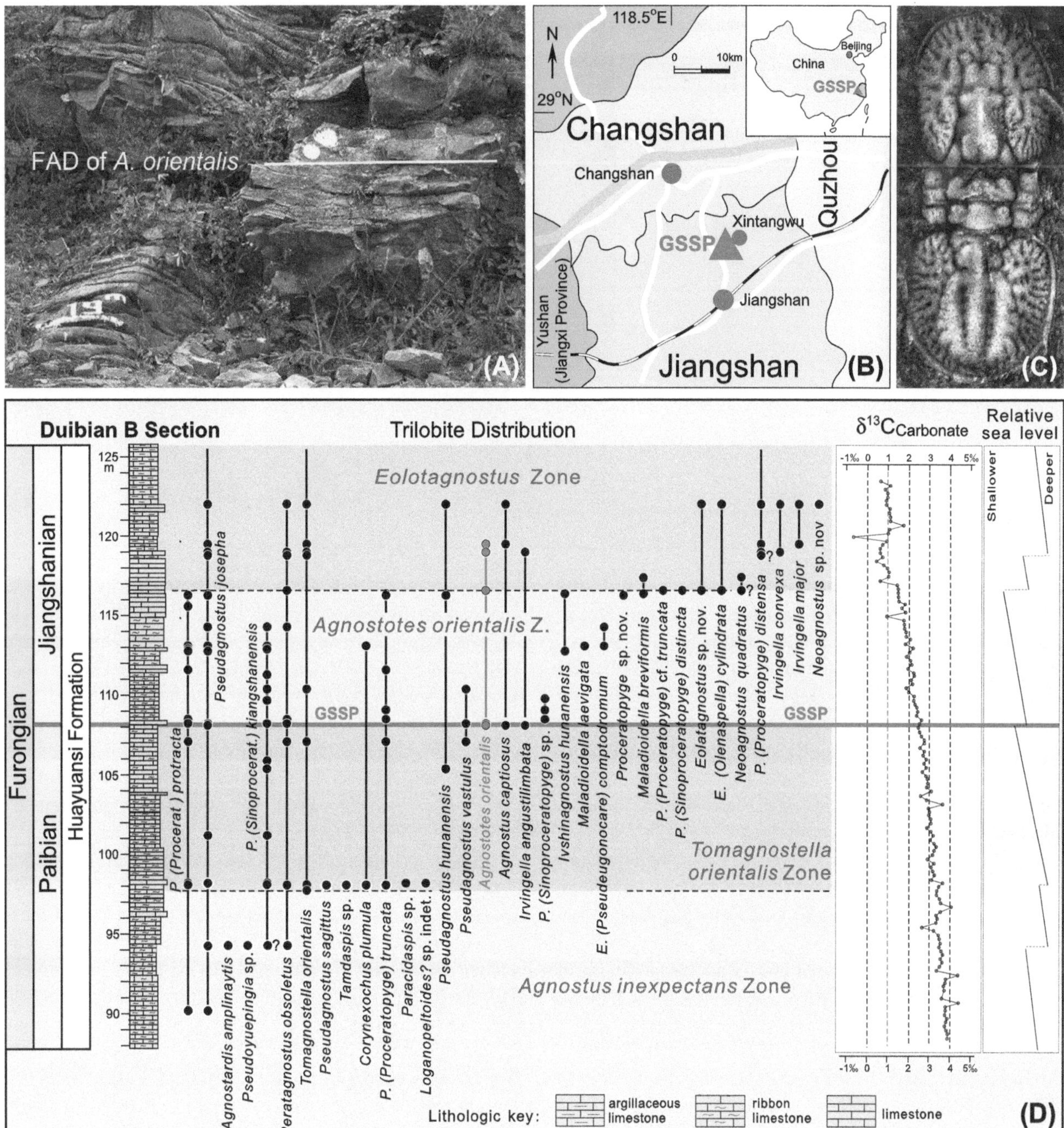

FIGURE 19.9 (A) and (B) GSSP of the Jiangshanian Stage in the Duibian B section, Zhejiang, China; (C) *Agnostotes orientalis* (Kobayashi, 1935), an agnostoid trilobite whose first appearance datum coincides with the base of the Jiangshanian Stage; (D) Stratigraphic distribution of trilobites close to the base of the Jiangshanian Stage in the Duibian B section, Zhejiang, China (redrawn after Peng *et al.*, 2009b).

to the FAD of the cosmopolitan agnostoid trilobite *Lotagnostus americanus* (Figures 19.2 and 19.10). The *L. americanus* level seems to be widely recognizable, as the species has been recognized (commonly with names of junior synonyms; Peng and Babcock, 2005b) from open-shelf lithofacies of all major Cambrian paleocontinents. The species, as interpreted by most recent workers (Peng and Babcock, 2005b; Terfelt *et al.*, 2008; Lazarenko *et al.*, 2011), is known

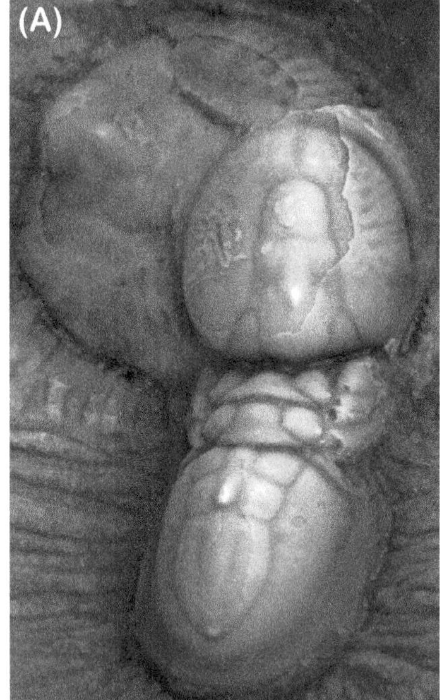

FIGURE 19.10 ***Lotagnostus americanus*** **(Billings, 1860), an agnostoid trilobite key to recognizing the base of provisional Stage 10.** *(A) a specimen showing weak scrobiculation; (B) a specimen showing well-developed scrobiculation.*

from South China (Hunan, Zhejiang, and Anhui), Northwest China (Kuruktagh and northern Tianshan), Siberia, Kazakhstan, Uzbekistan, eastern Avalonia (England and Wales), western Avalonia (eastern Newfoundland), Baltica, Australia (Tasmania), New Zealand, and Laurentia (Canada and Nevada). In China and Kazakhstan, *L. americanus* is used as a zonal guide fossil (Xiang and Zhang, 1985; Lu and Lin, 1989; Peng, 1992; Ergaliev, 1992). The co-occurrence of *L. americanus* with *Hedinaspis* in certain regions enables correlations to be extended even further. If a GSSP for Stage 10 is established at the FAD of *L. americanus*, the boundary level will correspond to the bases of the *L. americanus* Zone (agnostoids) and *Ctenopyge spectabilis* Zone (polymerids) in Sweden (Terfelt *et al.*, 2008, 2011), the base of the Aksayan Stage in Kazakstan, the base of the Niuchehean Stage in South China, and possibly the bases of both the Payntonian Stage in Australia and the Ketyan Horizon in Siberia.

On account of the importance of *Lotagnostus* for intercontinental correlation, species concepts and the paleogeographic distribution of included species are in the process of re-evalution. Peng and Babcock (2005b) considered *L. americanus* to show moderate intraspecific variation, and to have a widespread distribution. Rushton (2009) recognized two subspecies of *L. americanus*: *L. americanus americanus* and *L. americanus trisectus*. Lazarenko *et al.* (2011) recognized *Lotagnostus obscurus* as separate from *L. americanus*. Westrop *et al.* (2011) adopted the view that agnostoids should be considered to have very little intraspecific variation, and in doing so, rejected earlier interpretations of intraspecific variability in *L. americanus*. As a consequence of this view, which is at odds with most recent interpretations of species concepts in agnostoids (e.g., Pratt, 1992; Robison, 1984, 1994; Ahlberg and Ahlgren, 1996; Peng and Robison, 2000; Ahlberg *et al.*, 2004; Lazarenko *et al.*, 2011), including statistical studies based on large populations (Rowell *et al.*, 1982), Westrop *et al.* (2011) recommended recognition of several described forms including *L. trisectus* as species distinct from *L. americanus*. In the view of Westrop *et al.* (2011), *L. americanus* should be restricted to specimens from a single locality in Quebec, Canada.

Two conodont species, *Eoconodontus notchpeakensis* and *Cordylodus andresi*, have been suggested as possible markers for the base of the uppermost Cambrian stage. One of these is the euconodont *E. notchpeakensis*, which is recognizable in western Utah, USA (Miller *et al.*, 2006, 2011; Landing *et al.*, 2010, 2011) and elsewhere. Transport of the elements of this and other euconodont species allows for the identification of *E. notchpeakensis* through peritidal, platform, and slope deposits of Laurentia, Gondwana, Baltica, and Kazakhstan (Dubinina, 2009; Landing *et al.*, 2010, 2011). This species has its first observed appearance in Laurentia at the base of the *E. notchpeakensis* Subzone of the *Eoconodontus* Zone (conodonts), equivalent to the middle of the *Saukia* Zone (*Saukiella junia* Subzone, polymerid trilobites). The FAD coincides with the onset of the TOCE (Top of Cambrian carbon isotope Excursion), a $\delta^{13}C$ excursion alternatively referred to as the HERB Event (Ripperdan, 2002; Miller *et al.*, 2006; but see Landing *et al.*, 2011, who recognized two separate excursions/events). *Eoconodontus notchpeakensis* has a long stratigraphic range, extending into the

Iapetognathus fluctivagus Zone of the Lower Ordovician. The onset of the TOCE excursion (Figure 19.3; see also Section 19.2.7) is at present regarded as occurring about halfway through provisional Stage 10, and the return to positive $\delta^{13}C$ values occurs a little below the Ordovician base. A stage whose base would be identified by the FAD of *E. notchpeakensis* would be about half the stratigraphic thickness of Stage 10 as currently envisioned. The FAD of the euconodont *Cordylodus andresi* at the base of the *Cordylodus proavus* Zone in Utah, USA, marks another horizon that could serve as the base of the uppermost Cambrian stage (Miller *et al.*, 2006). This position is recognizable intercontinentally, and occurs above the first appearance of *E. notchpeakensis.* A stage whose base would be identified by the FAD of *C. andresi* would represent a thickness less than half that currently envisioned for Stage 10. If either *E. notchpeakensis* or *C. andresi* were selected as the marker for the base of provisional Stage 10, this stage would represent a much shorter span of time than any other Cambrian stage/age. If the *E. notchpeakensis* level were selected, the terminal Cambrian stage would represent a time duration of about 2 myr, and if the *C. andresi* level were selected, the stage would represent even less time.

19.1.5. Regional Cambrian Stage Suites

Regional stages and series have been erected for many parts of the Cambrian world. Examples of intensively studied areas are South China, Australia, Siberia, and Laurentia (Figure 19.11).

19.1.5.1. Cambrian Stages of South China

A chronostratigraphy for South China with boundary-stratotype-based stages and series has been developed recently (Peng *et al.*, 1998, 1999, 2000a,b; Peng, 2000a,b, 2003, 2008; Peng and Babcock, 2001; Figure 19.11). Apart from the lowermost two stages, the stages are based on sections in the Jiangnan Slope Belt, where the Cambrian succession yields rich agnostoids having significance for global or intercontinental correlation. The lower boundary of the Cambrian System in South China has been difficult to identify using *Trichophycus pedum* (Zhu et al., 2001; Steiner *et al.*, 2007). Instead, the BACE (BAse of Cambrian isotope Excursion) $\delta^{13}C$ excursion is a more reliable stratigraphic marker. South China is home to the GSSPs of three global stages, the Guzhangian, Paibian, and Jiangshanian. Once these global stages were established, it became advantageous to replace regional stage names having essentially the same concept and content.

Jinningian: The lowest stage of the Cambrian System in South China was proposed by Peng (2000a,b). It refers to the strata in the Yuhucun Formation below the "Marker B" (or "China B Point") in the Meishucun section, a GSSP candidate for the boundary, near Meishucun town, Jinning County, eastern Yunnan Province, (Luo *et al.*, 1984, 1990, 1991, 1992, 1994; Cowie *et al.* 1989; Brasier *et al.*, 1994). "Marker B" is one of the key levels proposed in the late 1980s as a possibility for the global Precambrian–Cambrian boundary. The Jinningian Stage embraces the oldest small shelly fossil zone, the *Anabarites trisulcatus–Protohertzina anabarica* Zone, which is characterized by the predominance of simple, low-diversity hyolithids. The base of the SSF zone was originally defined at "Marker A", the observed lowest occurrence of SSFs in the Meishucun section. In the Meishucun section an unconformity lies slightly below "Marker A", and a relatively thick interval, corresponding to the Daibu Member of northeastern Yunnan (Zhu *et al.*, 2001), is missing. The Daibu Member is regarded as part of the Cambrian System (Li *et al.*, 2001). *Trichophycus pedum* was reported as occurring within the *A. trisulcatus–P. anabarica* Zone in the Meishucun section, much higher than "Marker A", but this is apparently not its lowest occurrence. As originally defined, the base of the Jinningian Stage is the base of the Cambrian System. The horizon equivalent to the basal Cambrian GSSP, however, remains unknown in South China (Zhu *et al.*, 2001).

Meishucunian: The name of the stage was originally proposed as a lithologic unit, the Meishucun Formation (Jiang *et al.*, 1964), which comprises sub-trilobite sequences of "phosphate-bearing beds" in the basal part of the Meishucun section. Qian (1997) regarded this formation as the first stage of the Cambrian System of China, but failed to define its base at that time. Subsequently, Luo *et al.* (1994) revised the stage concept, drawing the base of the stage at the base of the *Paragloborites–Siphogonuchites* Zone (i.e. "Marker B"), which coincides with a major change in SSF fauna marked by the abrupt appearances of phosphatized micromollusks and problematica. It is also the base of Bed 7 in the section. "Marker B" was selected as a possibility for the base of the Cambrian System by the Precambrian–Cambrian boundary Working Group of the Cambrian Subcommission (Xing *et al.*, 1991; Cowie, 1985; Luo *et al.*, 1994). The point is positioned in the Xiaowaitoushan Formation in the Meishucun section, and is currently identified by the FAD of the hyolithid *Paragloborilus subglobosus*. The Meishucunian Stage covers an interval occupied by three biozones with abundant and diverse small shelly fossils and an interzone that is poorly fossiliferous (Luo *et al.*, 1994; Peng and Babcock, 2001; Steiner *et al.*, 2007).

Nangaoan: As originally defined (Peng, 2000a), the base of the stage is placed at the FAD of trilobites, as this level represents an important event in biotic development. This criterion has also been provisionally adopted by the Cambrian Subcommission to define the base of provisional global Stage 3 (Peng and Babcock, 2005a, 2008; Babcock *et al.*, 2005; Babcock and Peng, 2007). In practice, the base of the Nangaoan Stage is drawn at the lowest observed occurrence of *Tsunyidiscus niutitangensis* in the middle part of Bed 5 of the

TABLE 19.1 GSSPs of the Cambrian Stages, with Location and Primary Correlation Criteria (Status Jan. 2012)

Stage	GSSP Location	Latitude, Longitude	Boundary Level	Correlation Events	Reference
Stage 10				*Trilobite FAD of* Lotagnostus americanus. *An internal substage division might be FAD of* Cordylodus andresi *conodont*	
Jiangshanian	Duibian B Section, Jiangshan County, Zhejing Province, China	28°48.958′N 118°36.896′E	108.12m above base of the Huayansi Formation	Trilobite, FAD *of Agnostotes orientalis*	Science in China, Ser. D, Earth Science 52/4, 2009
Paibian	Wuling Mountains, Huayuan County, NW Hunan Province, China	28°23.37′N 109°31.54′E	At 396 m above the base of the Huaqiao Formation	Trilobite FAD *Glyptagnostus reticulatus*	Lethaia 37, 2004
Guzhangian	Louyixi, Guzhang County, NW Hunan Province, S. China	28°43.20′ N; 109°57.88′ E	121.3 m above the base of the Huaqiao Formation	Trilobite FAD *Lejopyge laevigata*	Episodes 32/1, 2009
Drumian	Drum Mountains, Millard County, Utah, USA	39°30.705′N 112°59.489′W	At the base of a dark-gray thinly laminated calcisiltite layer, 62 m above the base of the Wheeler Formation	Trilobite FAD *Ptychagnostus atavus*	Episodes 30/2, 2007
Stage 5	*candidate sections are Wuliu-Zengjiayan (east Guizhou, China) and Split Mountain (Nevada, USA)*			*Trilobite, potentially FAD of* Oryctocephalus indicus/Ovatorycara granulata	
Stage 4				*Trilobite, FAD of* Olenellus *or* Redlichia	
Stage 3				*Trilobites – their FAD*	
Stage 2				*Small Shelly Fossil, or Archaeocyath species*	
Fortunian *(base Cambrian)*	Fortune Head, Burin Peninsula, E Newfoundland, Canada	47°4′34.47″N 55°49′51.71″W*	2.4 m above the base of Member 2 in the Chapel Island Formation	Trace fossil FAD *Trichophycus pedum*	Episodes 17/1&2, 1994; Episodes, 30/3, 2007

**according to Google Earth*

Xiaosai section near Xiaosai, Yuqing County, eastern Guizhou Province (Zhang *et al.*, 1979). The bed belongs to the Niutitang Formation, which is composed of black shale. Detailed work is expected to define the base of the stage more precisely.

The Nangaoan Stage is the oldest trilobite-bearing stage in South China, with its lower part characterized by the occurrences of diverse eodiscoids (*Tsunyidiscus, Neocobboldia, Sinodiscus,* and *Hupeidiscus*) associated with protolenids (*Paraichangia*) in eastern Guizhou and the occurrence of the Chengjiang Biota in eastern Yunnan. The redlichiid *Parabadiella* has been reported from South China (Zhang, 1987). It is one of the earliest trilobites, but the redlichiids as a whole are regarded as having evolved from fallotaspidoids (Jell, 2003). As defined, the base of the stage corresponds closely to base of the traditional Chiungchussuan Stage of Yunnan.

Duyunian: Peng (2000a) defined the base of the Duyunian Stage by the FAD of *Arthricocephalus duyunensis*, a junior

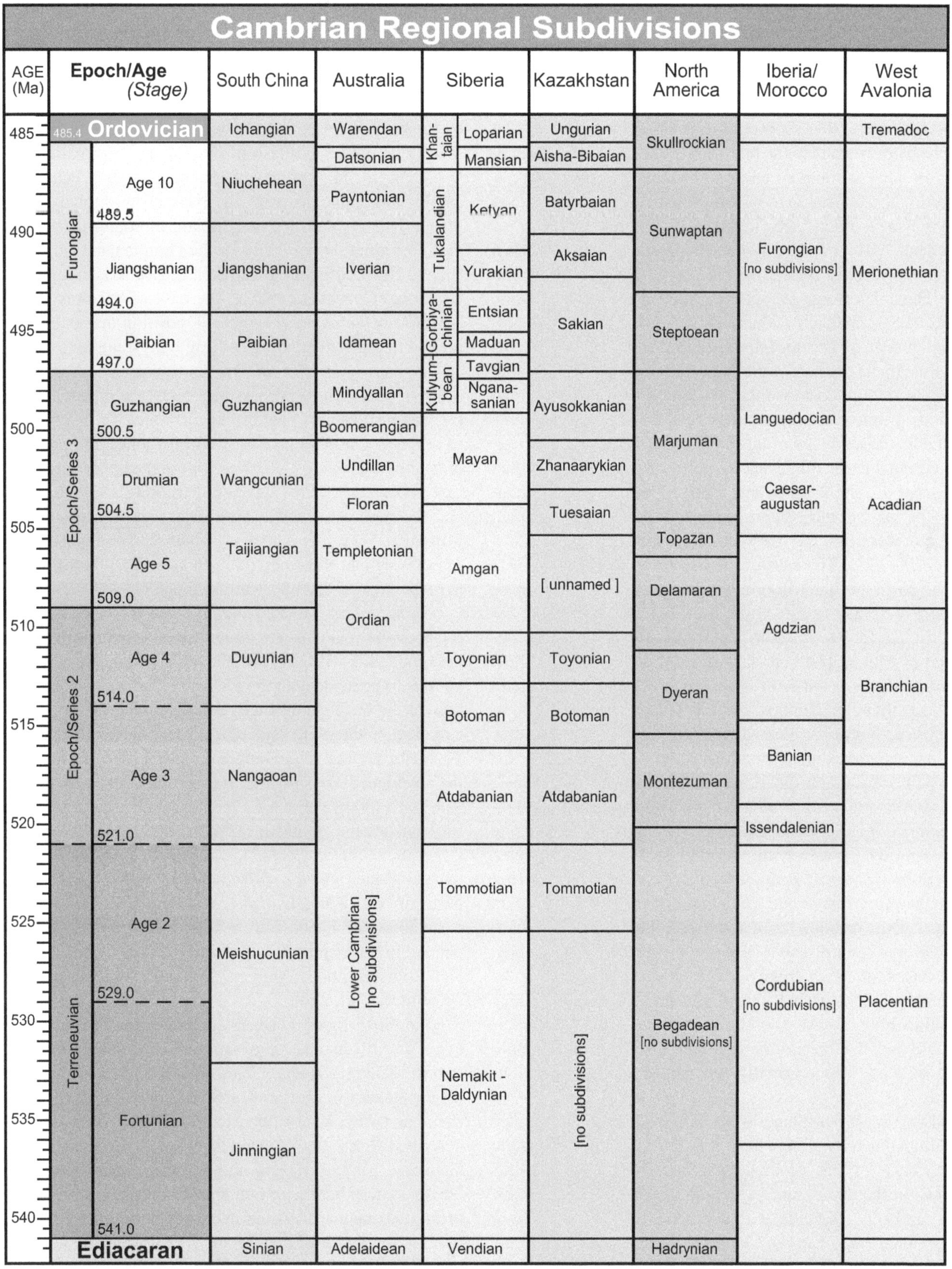

Cambrian Regional Subdivisions

AGE (Ma)	Epoch	Age (Stage)	South China	Australia	Siberia	Kazakhstan	North America	Iberia/ Morocco	West Avalonia
485	485.4 Ordovician		Ichangian	Warendan	Khantaian: Loparian	Ungurian	Skullrockian		Tremadoc
	Furongian	Age 10 (489.5)	Niuchehean	Datsonian; Payntonian	Khantaian: Mansian; Tukalandian: Ketyan	Aisha-Bibaian; Batyrbaian	Skullrockian; Sunwaptan	Furongian [no subdivisions]	Merionethian
490	Furongian	Jiangshanian (494.0)	Jiangshanian	Iverian	Tukalandian: Yurakian	Aksaian	Sunwaptan	Furongian [no subdivisions]	Merionethian
495	Furongian	Paibian (497.0)	Paibian	Idamean	Gorbiyachinian: Entsian; Maduan	Sakian	Steptoean	Furongian [no subdivisions]	Merionethian
	Epoch/Series 3	Guzhangian (500.5)	Guzhangian	Mindyallan; Boomerangian	Kulyumbean: Tavgian; Nganasanian	Ayusokkanian	Marjuman	Languedocian	Merionethian; Acadian
500	Epoch/Series 3	Drumian (504.5)	Wangcunian	Undillan; Floran	Mayan	Zhanaarykian; Tuesaian	Marjuman	Caesar-augustan	Acadian
505	Epoch/Series 3	Age 5 (509.0)	Taijiangian	Templetonian	Amgan	[unnamed]	Topazan; Delamaran	Agdzian	Acadian
510	Epoch/Series 2	Age 4 (514.0)	Duyunian	Ordian; Lower Cambrian [no subdivisions]	Toyonian	Toyonian	Dyeran	Agdzian	Branchian
515	Epoch/Series 2	Age 3 (521.0)	Nangaoan	Lower Cambrian [no subdivisions]	Botoman; Atdabanian	Botoman; Atdabanian	Montezuman	Banian; Issendalenian	Branchian; Placentian
520–525	Terreneuvian	Age 2 (529.0)	Meishucunian	Lower Cambrian [no subdivisions]	Tommotian; Nemakit - Daldynian	Tommotian; [no subdivisions]	Begadean [no subdivisions]	Cordubian [no subdivisions]	Placentian
530–540	Terreneuvian	Fortunian (541.0)	Jinningian	Lower Cambrian [no subdivisions]	Nemakit - Daldynian	[no subdivisions]	Begadean [no subdivisions]	Cordubian [no subdivisions]	Placentian
	Ediacaran		Sinian	Adelaidean	Vendian		Hadrynian		

FIGURE 19.11 **Principal regional stage schemes of the Cambrian, and the four-fold division of the system into series adopted by the International Commission on Stratigraphy (ICS).** See text for discussion of caveats in applying the numerical scale to stage boundaries.

synonym of *A. chauveaui* as pointed out by some authors (Lane *et al.*, 1988, Blaker and Peel, 1997; McNamara *et al.*, 2003). The lower boundary of the stage lies in the Niutitang Formation, about 25 m above the base of Bed 10 in the Jiumenchong section near Nangao, Danzhai, eastern Guizhou (Zhang *et al.*, 1979). Subsequent investigations suggest that *A. chauveaui* occurs in the boundary stratotype section earlier than in other sections of eastern Guizhou (Zhou Zhiyi, pers. comm.). *A. chauveaui* is widely distributed in South China and is also recorded from Greenland (Lane *et al.*, 1988; Geyer and Peel, 2011). In the Jiumenchong section, its observed lowest occurrence is near the base of the Balang Formation. The Duyunian Stage occupies an interval with four trilobite zones (Peng, 2000a,b), which are characterized by the early development of oryctocephalids and the flourishing of redlichiids. In the top part of the stage, the cosmopolitan trilobite *Bathynotus*, the cosmopolitan trilobite *Ovatoryctocara*, and various primitive ptychopariids occur. They comprise much of the first assemblage of the Kaili Biota. There is a medium-scale faunal extinction event at the end of the Duyunian Age (Yuan *et al.*, 2002; Zhen and Zhou, 2008), and only about 25% of the trilobite genera present in the upper Duyunian range upward into the overlying Taijiangian Stage.

Taijiangian: The base of the stage is defined by the FAD of the cosmopolitan trilobite *Oryctocephalus indicus*, which occurs in the lower part of the Kaili Formation, close to the traditional Lower–Middle Cambrian boundary as recognized in China. The boundary interval is exposed in a hillside section between Wuliu and Zengjiayan, near Balangcun Village, Jianghe County, eastern Guizhou. The FAD of *O. indicus* lies at the base of Bed 10, which is 52.8 m above the base of the Kaili Formation (Zhao *et al.*, 2001). Originally the Taijiangian Stage was represented by three zones, the *Oryctocephalus indicus* Zone (lowermost), the *Ptychagnostus gibbus* Zone, and the *Ptychagnostus atavus* Zone (uppermost). However, the concept of the stage is revised herein by restricting it to the two lower zones and referring the *Ptychagnostus atavus* Zone to the overlying Wangcunian Stage. This revision brings the regional stages of South China into conformance with global chronostratigraphy. The base of the *P. atavus* Zone coincides with the base of the Drumian Stage (global usage) and the Wangcunian Stage (South China usage). Yao *et al.* (2009) recognized an additional zone, the *Peronopsis taijiangensis* Zone, in the upper part of what was originally recognized as the *O. indicus* Zone.

The Taijiangian Stage is characterized by the abundance and high diversity of oryctocephids in the lower zone, and the occurrence of the cosmopolitan agnostoid *Ptychagnostus gibbus* in the upper zone. The base of the Taijiangian Stage will probably coincide with, or be close to, the base of provisional Stage 5 of the Cambrian System if the base of the global stage is finally selected at a position close to the FAD of *O. indicus.*

Wangcunian: The base of the Wangcunian Stage, defined originally at the FAD of the cosmopolitan agnostoid trilobite *Ptychagnostus punctuosus* (Peng *et al.*, 1998) in the Huaqiao Formation near Wangcun, Yongshun County, northwestern Hunan Province, was shifted downward to the level marked by the FAD of *Ptychagnostus atavus* (Peng, 2009). This level lies 1.2 m above the base of the Huaqiao Formation in the Wangcun section (Peng and Robison, 2000; Peng *et al.*, 2004b) and coincides with the base of the global Drumian Stage. This revision has been made to conform to the new global correlation standards for the Drumian and Guzhangian stages. The succeeding Youshuian Stage has been replaced by the global Guzhangian Stage, the base of which is drawn at the FAD of *Lejopyge laevigata*, a position that is below the base of the Youshuian Stage as originally defined (at the FAD of *Linguagnostus reconditus*). As revised, the Wangcunian Stage in South China is characterized by the diversification of ptychagnostid and hypagnostid agnostoids, the diversification of corynexochid and proasaphiscid trilobites, the occurrence of lisaniids (*Lisania, Qiandonaspis*) in the upper part, and the first appearance of the agnostoid trilobite *Lejopyge armata* and the damesellid *Palaeodotes* near the top of the stage. More than 90 trilobite taxa, of which 25 are agnostoid trilobites (Peng and Robison, 2000; Peng *et al.*, 2004b), occur in the type area of the Wangcunian Stage.

Guzhangian: The Guzhangian Stage is a global stage that replaces the regional Youshuian Stage (as revised, with the base moved downward from the FAD of *Linguagnostus reconditus* to coincide with the FAD of *Lejopyge laevigata*). The base of the stage is defined by a GSSP in the Luoyixi section, a roadcut on the south bank of the Youshui River, opposite the Wangcun section. The Wangcun section is the stratotype for the abandoned Youshuian Stage. As recognized by Peng and Robison (2000), the stage is represented in its type area by four successive agnostoid zones: the *Lejopyge laevigata* Zone, the *Proagnostus bulbus* Zone, the *Linguagnostus reconditus* Zone, and the *Glyptagnostus stolidotus* Zone. The base of the *Linguagnostus reconditus* Zone is closely correlative with the base of the traditional Upper Cambrian as defined in Sweden by Westergård (1922, 1947; see also Peng and Robison, 2000; Ahlberg, 2003; Ahlberg *et al.*, 2004; Axheimer *et al.*, 2006).

The Guzhangian Stage is characterized by a high abundance and diversification of trilobites, especially ones in the families Clavagnostidae, Damesellidae, and Lisaniidae, and by the genera *Lejopyge, Linguagnostus, Proagnostus, Torifera*, and *Fenghuangella.* More than 150 taxa have been described from the Guzhangian Stage, of which 45 are agnostoid trilobites (Peng and Robison, 2000; Peng *et al.*, 2004b). There is a major faunal extinction event near the end of the Guzhangian Age, an event resulting in the extinction of more than 90 % of taxa. In the type area, all but one of the dameseliid trilobites are confined to the stage; only one species ranges upward. The extinction event is also recognized as the faunal crisis at the beginning of the Idamean Stage of Australia (Öpik, 1966; Shergold and Cooper, 2004), and the top of the Marjumiid Biomere of Laurentia (Palmer, 1979, 1984, 1998; Ludvigsen and Westrop, 1985; Saltzman *et al.*, 2000).

Paibian: The Paibian Stage is also a global stage, applied now as a regional stage in South China because of its origin. It replaces the abandoned Waergangian Stage as used previously for South China. The GSSP that defines the base of the stage coincides with the FAD of the cosmopolitan agnostoid *Glyptagnostus reticulatus* in the upper part of the Huaqiao Formation of the Paibi section, near Paibi, Huayuan, northwestern Hunan. A remarkable faunal turnover occurs at the beginning of the stage with the first co-appearances of a number of taxa – i.e. the leiostegiid *Chuangia*, the pagodiid *Prochuangia*, the eulomiid *Stigmatoa*, the olenid *Olenus,* the lisaniid *Shengia quadrata*, the agnostoid *Glyptagnostus reticulatus*, and diverse species of *Pseudagnostus* and *Proceratopyge*, of which only a few species range upward from the underlying Guzhangian Stage (Peng, 1992; Peng *et al.*, 2004b). The Paibian trilobite fauna marks a recovery period following the end-Guzhangian extinction. Only about 30 taxa occur in three successive zones.

Jiangshanian: The Jiangshanian Stage is another global chronostratigraphic unit applied as a regional stage for South China. It replaces the revised Taoyuanian Stage (Peng, 2008), which is restricted to the lower part of the original Taoyuanian Stage, corresponding to the interval below the level marked by the FAD of *Lotagnostus americanus*. The upper part of the original Taoyuanian Stage was proposed as a new stage, the Niuchehean Stage (see Niuchehean Stage below).

The GSSP that defines the base of the Jiangshanian Stage coincides with the FAD of the cosmopolitan agnostoid *Agnostotes orientalis* in the upper part of the Huayansi Formation of the Duibian B section, at Duibian, western Zhejiang. In the boundary stratotype section, and also the Wa'ergang section, northwestern Hunan, the FAD of *A. orientalis* coincides with that of *Irvingella angustilimbata* (Peng, 1992; Peng *et al.*, 2009b). As a cosmopolitan polymerid trilobite, *I. angustilimbata* can also be used to constrain the base of the Jiangshanian Stage. The stage is characterized by the diversification of the agnostioid subfamily Pseudagnostinae, with successive separation of *Pseudagnostus, Rhaptagnostus*, and *Neoagnostus*; the diversification of the superfamily Leiostegioidae; and the first occurrences of Dikelocephaloidea, Saukiidea, Hapalopleuridae, Shumardiidae, and Macropygiinae. In northwestern Hunan the Jiangshanian Stage contains 70 trilobite taxa in four biozones.

Niuchehean: The Niuchehean Stage was proposed by Peng (2008) for the upper part of the abandoned Taoyuanian Stage as originally defined. It is the uppermost stage of the Cambrian System in South China. The base of the stage is defined at the base of the *Lotagnostus americanus* Zone in the Wa'ergang section, Taoyuan, northwestern Hunan. The top of the stage in the stratotype section is the level marked by the first appearance of the conodont *Iapetognathus fluctivagus* within the conodont *Cordylodus lindstromi* Zone (Dong *et al.*, 2004). This level also lies within the trilobite *Hysterolenus* Zone, and coincides with the base of the global Tremadocian Stage of the Ordovician System. The stage is named for Niuchehe, a township that governs Wa'ergang Village. The base of the uppermost global stage of the Cambrian System will probably be defined at or close to the FAD of *L. americanus* (Peng and Babcock, 2005a; Babcock *et al.*, 2005). The stage embraces four and a half assemblage zones, collectively containing more than 80 trilobite taxa in its type area (Peng, 1984, 1990, 1992). It is characterized by the abundance and diversification of ceratopygiids (*Charchaqia, Diceratopyge, Hedinaspsis, Hunanopyge, Macropyge, Promacropyge, Yuepingia, Hysterolenus, Onychopyge)*; the diversification of eulomimids (*Archaeuloma, Proteuloma, Euloma, Karataspis, Ketyna)*; the appearance of remopleuriids (*Fatocephalus, Ivshinanspis*), nileids (*Trodssonia, Shenjiawania*), pilekids (*Parapilekia*) and harpidiids (*Eotrinucleus*); the separation of true asaphids from ceratopygids; and the occurrence of leiostegiids and saukiids. Agnostoids of the stage are dominated by members of the family Agnostidae. Apart from trilobites, the Niuchehean Stage in the type area is also characterized by the occurrences of euconodonts, including *Cordylodus proavus*, and by primitive nautiloids (Peng, 1984).

19.1.5.2. Australian Cambrian Stages

Australian stages were summarized by Shergold (1995), Young and Laurie (1996), and Kruse *et al.* (2009), on which the following outline is based (Figure 19.11). The stages are described as "biochronological" units (see also Chapter 3, Section 3.4.3) and are defined in terms of their contained fauna (Shergold, 1995). Boundary stratotypes therefore have not been designated. Apart from the Ordian, the stages discussed below were all erected in the Georgina Basin of western Queensland.

Pre-Ordian: Stages have not yet been designated for most of the traditional Lower Cambrian of Australia. Archaeocyaths, small shelly fossils, and trilobite correlations indicate that the Atdabanian to Toyonian Stages of the Siberian Platform and the Altay–Sayan Foldbelt of Russia can be recognized through southern and central Australia. Ichnofaunas in central and southern Australia are thought to possibly correlate with the Tommotian and Nemakit-Daldynian (Walter *et* al., 1989; Bengtson *et al.*, 1990; Shergold, 1996).

Ordian: The Ordian Stage was originally proposed by Öpik (1968) as a time and time-rock division of the Cambrian characterized by the occurrence of the *Redlichia chinensis* faunal assemblage. The Templetonian Stage, "a liberal interpretation of Whitehouse's (1936) Templetonian series" (Öpik, 1968), was originally conceived by Öpik as containing the *Xystridura templetonensis* assemblage of western Queensland, followed by faunas of the *Ptychagnostus gibbus* (under the name of *Triplagnostus gibbus*) Zone. In practice, it is difficult to distinguish the *Redlichia* and *Xystridura* faunas because four species of *Xystridura*, similar eodiscoid and ptychoparioid trilobites, some bradoriid ostracodes, and chancelloriids occur in rocks of both Ordian and early Templetonian ages. Accordingly Shergold (1995) regarded the Ordian–early Templetonian as a single stadial unit, but recently Laurie (2004, 2006) redefined the base of Templetonian Stage based on recognition of three pre-*P. gibbus* agnostoid zones in southern Georgina Basin drillholes, and thereby defined the top of the Ordian Stage. The Ordian Stage had long been regarded as the earliest traditional Middle Cambrian Stage in Australia even though it apparently correlates with the Longwangmiaoan Stage of China (Shergold, 1997; Chang, 1998; Geyer *et al.*, 2000, 2003; Peng, 2003) and with the Toyonian Stage of the Siberian Platform (Zhuravlev, 1995), both of which are traditionally regarded as terminal "Lower" Cambrian. Based on recent biostratigraphic information, Kruse *et al.* (2009) considered the Ordian Stage to be at least partly (and possibly entirely) equivalent to Stage 4 of Series 2 (or the uppermost part of the traditional Lower Cambrian).

Templetonian: Laurie (2004, 2006) redefined the base of the Templetonian Stage, fixing its base at the base of the agnostoid trilobite *Pentagnostus anabarensis* Zone. That biozone is succeeded by two additional agnostoid zones, the *Ptychagnostus praecurrens* (under the name *Pentagnostus praecurrens*) Zone and the *Pentagnostus shergoldi* Zone. These three agnostoid zones equal the interval recognized as early Templetonian Stage by Shergold (1995), who divided Templetonian Stage into lower and upper portions, and annexed each portion to adjacent stages to produce the Ordian–Lower Templetonian Stage and the Upper Floran Stage respectively. As revised (Laurie, 2004, 2006), the Templetonian Stage includes four agnostoid zones: the *Pentagnostus anabarensis*, *Ptychagnostus praecurrens*, *Pentagnostus shergoldi*, and *Ptychagnostus gibbus* zones. The Templetonian is an important stage because it contains cosmopolitan agnostoid trilobites (e.g., *Ptychagnostus praecurrens* and *Ptychagnostus gibbus),* and oryctocephalid trilobites (Shergold, 1969), both of which are significant for international correlation.

Floran: As originally defined (Öpik, 1979), the Floran Stage contained the agnostoid trilobite zones of *Ptychagnostus atavus* (under the name of *Acidusus atavus*) and *Euagnostus opimus*. This concept was revised by Shergold (1995) to include subjacent strata of the late Templetonian zone of *Ptychagnostus gibbus*, argued on the grounds of faunal continuity (overlap in the ranges of *P. atavus* and *P. gibbus* in the Georgina Basin, western Queensland) and sequence stratigraphy (Southgate and Shergold, 1991). By redefining the Templetonian Stage, the base of the Floran has been restored to its original proposed level (Laurie, 2004, 2006; Kruse *et al.*, 2009). The Floran Stage embraces only two agnostoid zones, the *Ptychagnostus atavus* Zone and the overlying *Euagnostus opimus* Zone, and its base coincides with the base of the global Drumian Stage, the base of the revised Wangcunian Stage of South China, and the base of the Marjuman Stage of North America.

Undillan: The Undillan Stage, defined by Öpik (1979), is based on the fauna of two agnostoid zones, the *Ptychagnostus punctuosus* Zone and the succeeding *Goniagnostus nathorsti* Zone. A third zone, based on *Doryagnostus deltoides*, containing 15 agnostoid species including *P. punctuosus* and *G. nathorsti*, was recognized by Öpik (1979) in the Undilla region of the Georgina Basin. The agnostoid fauna of the Undillan Stage has a cosmopolitan distribution. Agnostoids apart, the trilobites include ptychoparioids, anomocarids, mapaniids, damesellids, conocoryphids, corynexochids, nepeiids, and dolichometopids, all of widespread distribution.

Boomerangian: The Boomerangian Stage (Öpik, 1979) is essentially the *Lejopyge laevigata* Zone divided into three. A *Ptychagnostus cassis* Zone at the base is overlain by zones defined by the polymerid trilobites *Proampyx agra* and *Holteria arepo*. Boomerangian agnostoids are accompanied by a range of polymerid trilobites including species of *Centropleura*, dolichometopids, olenids, mapaniids, corynexochids, and damesellids. A "Zone of Passage", characterized by the occurrence of *Damesella torosa* and *Ascionepea janitrix*, was interposed by Öpik (1966, 1967) between the Boomerangian (uppermost Middle Cambrian) and Mindyallan (considered at that time to mark the beginning of the Upper Cambrian) stages. Subsequently, Daily and Jago (1975) restricted this zone to the Boomerangian, and placed the Middle–Upper Cambrian boundary within the early Mindyallan Stage.

Mindyallan: Originally, Öpik (1963) defined the Mindyallan Stage to include a *Glyptagnostus stolidotus* Zone (above) and a "pre-*stolidotus*" Zone (below). Subsequently Öpik (1966, 1967) revised the stage, placing the *G. stolidotus* Zone in the upper Mindyallan and dividing the underlying strata into an initial Mindyallan *Erediaspis eretes* Zone and an overlying *Acmarhachis quasivespa* (under the name of *Cyclagnostus quasivespa*) Zone. The *E. eretes* Zone contains 45 trilobites, including 18 agnostoid genera. The polymerid trilobites belong to a wide variety of families: anomocarids, asaphiscids, catillicephalids, damesellids, leiostegiids, lonchocephalids, menomoniids, nepeiids, norwoodiids, rhyssometopids, and tricrepicephalids are represented. The *A. quasivespa* Zone has 18 species of trilobites confined to it, but many other species range upward from lower zones. Daily and Jago (1975) proposed subdividing the *A. quasivespa* Zone into two assemblages based on the occurrence of *Leiopyge cos* and *Blackwelderia sabulosa*. Because *L. cos* appears to be synonymous with *L. armata*, a late Middle Cambrian taxon, they drew the Middle–Upper Cambrian boundary between these two assemblages. Only eight species range from Öpik's (1966, 1967) *A. quasivespa* Zone into the overlying *G. stolidotus* Zone. The *G. stolidotus* Zone contains 75 species; some (asaphiscids, auritamids, catillicephalids, norwoodiids, and raymondinids) have Laurentian biogeographic affinities, and some (damesellids and liostracinids) have Chinese biogeographic affinities.

Idamean: The Idamean Stage, as introduced by Öpik (1963), encompassed five successive assemblage zones: the *Glyptagnostus reticulatus* with *Olenus ogilviei* Zone, the *Glypagnostus reticulatus* with *Proceratopyge nectans* Zone, the *Corynexochus plumula* Zone, the *Erixanium sentum* Zone, and the *Irvingella tropica* with *Agnostotes inconstans* Zone. This biostratigraphic scheme was criticized by Henderson (1976, 1977), who proposed an alternative zonation in which the two zones with *Glyptagnostus* were united into a single *G. reticulatus* Zone, the *Corynexochus* Zone was renamed the *Proceratopyge cryptica* Zone, and the *Erixanium sentum* Zone was subdivided into a zone of *E. sentum* followed by a zone of *Stigmatoa diloma*. The name *Irvingella tropica–Agnostotes inconstans* Zone was changed to *Irvingella tropica* Zone. Henderson's scheme was adopted by Shergold (1982). The *Irvingella tropica* Zone is now regarded as the lowermost zone of the succeeding Iverian Stage (see Shergold, 1982, for justification; Shergold, 1993).

There was a major faunal crisis at the beginning of the Idamean. Few Mindyallan trilobite genera and no Mindyallan species survived the extinction (Öpik, 1966). There was also a major reorganization of trilobite families, as outer-shelf communities dominated by agnostoids, olenids, pterocephaliids, leiostegiids, eulomids, and ceratopygids abruptly replaced those of the shallow shelf Mindyallan biota. Shergold (1982) recorded a total of 69 Idamean taxa, which permit a highly resolved biochronology enabling precise international correlations.

Iverian: The Iverian Stage (Shergold, 1993) was proposed for the concept of a post-Idamean–pre-Payntonian interval in the eastern Georgina Basin, western Queensland, the only region where a probable complete sequence has so far been described (Shergold, 1972, 1975, 1980, 1982, 1993). Paleontologically the Iverian Stage is clearly distinguished. On the basis of trilobites, it is characterized by:

- The occurrence of the cosmopolitan genus *Irvingella* in Australia;
- The diversification of the agnostoid subfamily Pseudagnostinae during which *Pseudagnostus*, *Rhaptagnostus*, and *Neoagnostus* separate and become biostratigraphically important;
- Diversification of the Leiostegioidea, especially the families Kaolishaniidae and Pagodiidae;
- The first occurrence of the Dikelocephaloidea, Remopleuridoidea, and Shumardiidae; and
- The separation of the true asaphids from ceratopygids.

As a result, nine trilobite assemblage zones have been recognized based on successive species of *Irvingella*, *Peichiashania*, *Hapsidocare*, and *Lophosaukia* (Shergold, 1993). Subsequently the *Hapsidocare lilyensis* and the *Rhaptagnostus clarki patulus–Caznaia squamosa* assemblage zones were united into a single *R. c. patulus–C. squamosa–H. lilyensis* Assemblage Zone (Shergold and Cooper, 2004; Kruse *et al.*, 2009). More than 160 trilobite taxa occur in the type area of the Iverian Stage.

Payntonian: As defined by Jones *et al.* (1971), the Payntonian Stage is recognized on the basis of its trilobite assemblages (Shergold, 1975), its base lying at the point in its type section (Black Mountain, western Queensland), where the co-mingled Laurentian–Asian assemblages of the Iverian are replaced by others of only Asian biogeographic affinity. These are dominated by tsinaniid, leiostegioidean, saukiid, ptychaspidid, dikelocephaloidean, and remopleuridoidean trilobites. A tripartite zonal scheme is applicable following biostratigraphic revisions suggested by Nicoll and Shergold (1992), Shergold and Nicoll (1992), and Shergold (1993). In ascending order, these zones are based on *Sinosaukia impages*, *Neoagnostus quasibilobus* with *Shergoldia nomas*, and *Mictosaukia perplexa*. These zones are fully calibrated by a comprehensive conodont biostratigraphy (Nicoll, 1990, 1991; Shergold and Nicoll, 1992). The Payntonian Stage contains a total of 30 trilobite taxa.

Datsonian: The concept of the Datsonian Stage remains as defined by Jones *et al.* (1971), with its base located at the FAD of the conodont *Cordylodus proavus*. Only rare trilobites, *Onychopyge* and leiostegiids, occur, and these are insufficient for the establishment of a trilobite biostratigraphy. Accordingly, the Datsonian Stage is defined solely on the basis of conodonts and embraces three successive zones: the *C. proavus* Zone, the *Hirsutodontus simplex* Zone, and the *C. prolindstromi* Zone.

Warendan: The Warendan (corrected from Warendian by Kruse *et al.*, 2009) Stage was originally defined at the base of the conodont *Cordylodus prion–Scolopodus* Assemblage Zone (Jones *et al.*, 1971). Revision of the cordylodids by Nicoll (1990, 1991) resulted in introduction of a single zone of *C. lindstromi* to replace the assemblage zone as the lowermost zone of the Warendan Stage because *C. prion* was recognized as a part of the septimembrate apparatus of the eponymous species (Shergold and Nicoll, 1992).

The Warendan Stage of Australia is inferred to be Cambrian in the basal part, and Ordovician through most of its extent. In the Ordovician stratotype at Green Point, Newfoundland, Canada, the FAD of the conodont *Iapetognathus fluctivagus*, which is the guide event for the base of the Ordovician, lies within the interval containing the co-ranging species *C. lindstromi* and *C. prion* (Cooper *et al.*, 2001). The Ordovician base thus correlates to a level above the base of the *C. lindstromi* Zone in the Green Point section. At Wa'ergang, South China, *C. prion* also first occurs within the *C. lindstromi* Zone (Dong *et al.*, 2004). Although *I. fluctivagus* has not been identified from Warendan Stage strata of Australia, correlations based on co-occurring taxa from the sections in Canada and China suggest that the lowermost part of the stage is correlative to the uppermost Cambrian.

19.1.5.3. Siberian Cambrian Stages

The first attempts to develop a Cambrian chronostratigraphy in the former USSR were based on sections of the Siberian Platform (Pokrovskaya, 1954; Suvorova, 1954). A scale with four Lower Cambrian stages and two Middle Cambrian stages was accepted as the national standard of the Soviet Union at the All-Union Stratigraphic Meeting in 1982. In addition to the Lower–Middle Cambrian stages, the Upper Cambrian standard scale for the Soviet Union adopted the stages established by Ergaliev (1980, 1981) in the Malyi Karatau Range, Kazakhstan, which then was a republic of the USSR. Presently there is no officially accepted standard for the traditional Upper Cambrian of Russia (including the Siberian Platform). Upper Cambrian stages used here for Siberia are only a regional standard, recognized as stages ("super-horizons") in a section measured on the Kulymbe River in the Koyuy–Igarka Region, northwestern Siberia (Rozova, 1963, 1964, 1968, 1970), each of which includes two regional "horizons". The lower boundary of the Cambrian in Siberia is commonly drawn between the Tommotian Stage and the underlying Nemakit-Daldynian Stage. In Russia, the Nemakit-Daldynian Stage usually has been regarded as Precambrian (Vendian) (Khomentovsky, 1974, 1976, 1984; Khomentovsky and Karlova, 1993, 2005; Rozanov *et al.*, 2008; Varlamov *et al.*, 2008b), although on biostratigraphic and carbon isotopic evidence it has been widely regarded internationally as the lowermost Cambrian. All traditional Siberian stages (Figure 19.11) are unit-stratotype-based, with the stage boundaries defined at the base of either a biozone or a lithologic unit.

Nemakit-Daldynian: The Nemakit-Daldyn Stage was named for the Nemakit-Daldyn River, Kotuy River Basin, northwestern Siberia, Russia. The stage name was introduced directly from a lithologic unit, the Nemakit-Daldyn "horizon". As originally proposed by Savitsky (1962), the horizon comprises a carbonate succession cropping out on the upper reaches of the Nemakit-Daldyn River, and both its lower and upper boundaries are defined at disconformities. The stage was proposed by Khomentovsky (1976, 1984), who recognized the Nemakit-Daldynian Stage as the uppermost of three successive stages ("horizons") that he proposed for the Vendian System (Yudomian Series) of the Yudoma-Anabar facies region. The Nemakit-Daldynian Stage is characterized by the appearance of the first skeletal fossils belonging to the *Anabarites trisulcatus* Zone, the oldest small shelly fossil (SSF) assemblage, and by fauna of the succeeding *Purella antiqua* Zone (Khomentovsky and Karlova, 1993). No lower boundary can be reliably defined at the type section of the Manukai (Nemakit-Daldyn) Formation of the eastern Pre-Anabar Area, which was formally regarded as the stratotype of the stage (V.V. Khomentovsky, pers. comm., 2007). Practically, the lower boundary of the stage is recognized by the change of strata bearing Ediacaran fossils to the strata bearing SSFs of the *A. trisulcatus* Zone; the upper boundary is placed at the boundary between the *P. antiqua* Zone and the overlying *Nochoroicyathus sunnaginicus* Zone.

Tommotian: The Tommotian Stage was named for the town of Tommot on the Aldan River, Russia. As originally defined (Rozanov, 1966; Rozanov and Missarzhevsky, 1966; Rozanov *et al.*, 1969), the Tommotian was the lowermost stage of the Cambrian, represented by a stage stratotype with 85-m-thick successions in the middle reaches of the Aldan River, cropping out from Dvortsy to Ulakhan-Sulugur Creek. The stage is characterized by the abrupt appearance of SSFs including hyoliths, gastropods, inarticulate brachiopods, and problematica, and by the occurrence of primitive archaeocyath sponges having simple systems of porous walls and septae. Trilobites have not been found in the Tommotian Stage. The pre-trilobite stage embraces three successive zones based on archaeocyath assemblages. In ascending order, they are the *Nochoroicyathus sunnaginicus* Zone, the *Dokidocyathus regularis* Zone, and the *Dokidocyathus lenaicus* Zone (previously *Dokidocyathus lenaicus–Tumuliolynthus primigenius* Zone). The base of the stage is drawn at the base of the *N. sunnaginicus* Zone.

Atdabanian: The Atdabanian Stage, named for Atdaban Village on the Lena River, Russia, was named by Zhuravleva *et al.* (1969) for strata cropping out between the mouth of the Negyurchyune River and the mouth of Achchagyy-Kyyry-Taas Creek (Zhuravlev and Repina, 1990). The lower half of the stage is not exposed in the stratotype area, and the lower boundary of the stage is defined elsewhere (at the base of Bed 4 in the Zhurinsky Mys section, along the Lena River; Varlamov *et al.*, 2008a). *Profallotaspis*, the oldest trilobite known from Siberia, first appears 2.6 m above the boundary in Bed 4. The stage is characterized by the first appearance and the early development of trilobites, dominated by fallotaspidids, in the lower half, and increasing diversity of trilobites in the upper half. The stage embraces four trilobite zones; in ascending order they are the *Profallotaspis jakutensis* Zone, the *Fallotaspis* Zone, the *Pagetiellus anabarus* Zone, and the *Judomia–Uktaspis (Prouktaspis)* Zone. The stage is also characterized by a sharp increase in archaeocyaths that bear compound skeletal elements; this is regarded as the second evolutionary stage in this fossil group. Four archaeocyath-based zones have been established for the stage (Rozanov and Sokolov, 1984). Mollusk and SSF diversity is rather low in the lower part of the stage, and the upper part of the stage is characterized by SSFs having intercontinental distributions.

Botoman: The Botoman Stage (alternatively referred to in literature as the Botomian Stage) was named by Repina *et al.* (1964) for the Botoma River, a tributary in the middle reaches of the Lena River, Russia. The stage stratotype section lies on the left bank of the Botoma River (Rozanov and Sokolov, 1984), and the lower boundary of the stage is at the base of unit III, within the Perekhod Formation, in the Ulakhan-Kyyry-Taas section located 1.5 km downstream from the Ulakhan-Kyyry-Taas Creek mouth (Rozanov and Sokolov, 1984). The stage base corresponds to the conterminous base of a trilobite biozone (the *Bergeroniellus micmacciformis–Erbiella* Zone) and an archaeocyathan biozone (the *Porocyathus squamosus–Botomocyathus zelenovi* Zone). The stage is characterized by a high diversity of trilobites, archaeothyaths, brachiopods, rare mollusks, and various hyoliths. Trilobites first appearing above the base of the stage include *Neocobboldia, Protolenus, Bergeroniellus, Bergeroniaspis, Micmaccopsis, Erbiella,* and *Judomiella.* The trilobites are numerous and diverse, especially in the lower half, in which protolenids predominate. In contrast, trilobite diversity in the upper half of the stage is greatly reduced. The archaeocyaths, which are abundant and diverse, but restricted to only the basal part of the stage in the stratotype section, are characterized mainly by forms having complex walls.

Toyonian: The Toyonian Stage was introduced as the national standard stage of the uppermost Lower Cambrian in the former USSR (Spizharsky *et al.*, 1983). The name is derived from Ulakhan-Toyon Island on the Lena River, Russia. The stage stratotype is in a carbonate succession that crops out in the middle reaches of the Lena River between Tit-Ary and Elanka villages. Previously this interval of strata was named the Lenan Stage by Repina *et al.* (1964) and the Elankan Stage by Rozanov (1973). The lower boundary of the stage is drawn at the base of the Keteme Formation, which is also the base of the *Bergeroniellus ketemensis* Zone, in the stratotype section. However, the principal guide fossil, *B. ketemensis*, and other trilobites first appear 6 m above the lower boundary (Zhuravlev and Repina, 1990). The stage embraces three trilobite zones: the *Bergeroniellus ketemensis* Zone, the *Lermontovia grandis* Zone, and the *Anabaraspis splendens* Zone. Trilobites of the Toyonian Stage include edelsteinaspidids, dorypygids (*Kootenia, Kooteniella*), and dinesidians (*Erdia*), and these forms predominate in the Anabar-Sinsk facies region. In the Yudoma-Olenek facies, menneraspidid, lermontoviine, and paramicmaccine trilobites are more common. In general, archaeocyaths occur throughout the stage, but in the stratotype section they are confined to the *L. grandis* Zone. The Toyonian Age is regarded as the fourth (and last) evolutionary stage of archaeocyaths. The archaeocyaths became extinct before the end of the Toyonian. Hyolithids are represented by forms having shells with polygonal cross-sections.

Amgan: The Amgan Stage was named for the Amga River, a tributary of the Aldan River, Russia. Its stratotype is in the middle reaches of the Amga River (Chernysheva, 1961). The base of the Amgan Stage is placed at the base of the *Schistocephalus* Zone, 27 m above the base of the Elanka Formation on the Lena River, near Elanka Village. The *Schistocephalus* Zone corresponds to the *Oryctocara* Zone of the Yudoma-Olenek facies region (Egorova *et al.*, 1976). In the Yudona-Olenek facies region, the zone is succeeded by the *Kounamkites* Zone, the *Ptychagnostus gibbus* Zone, and the *Tomagnostus fissus–Paradoxides sacheri* Zone. More than 100 trilobite taxa of Amgan age have been documented from the stratotype area (Chernysheva, 1961; Egorova *et al.*, 1976). The stage is characterized by a turnover of faunas, and includes the first appearances of paradoxidids (*Pardoxides, Schistocephalus*), oryctocephalids (*Oryctocara, Ovatoryctocara, Oryctocephalus, Tonkinella*), ptychopariids (*Kounamkites, Ptychoparia*), and agnostoids after the extinction of the "Lower Cambrian" ellipsocephalids (*Bergeroniellus, Lermotovia, Paramicmacca, Protolenus, Protolenellus*), and redlichiids (*Redlichia, Redlichina*).

Mayan: The Mayan Stage was named for the Maya River, a tributary of the Aldan River, Russia, by Egorova *et al.* (1982). The stage stratotype comprises a number of outcrops on the Yudoma and Maya rivers. The lower boundary of the stage was previously drawn at the base of the *Anopolenus henrici* Zone (Chernysheva, 1967), but Egorova *et al.* (1982) advocated shifting the boundary downward to the base of the underlying *Tomagnostus fissus–Paradoxides sacheri* Zone. The upper boundary of the stage was also defined, but imprecisely, by the disappearance of *Lejopyge*, *Goniagnostus*, etc., and by the abundant appearance of "Late Cambrian" trilobites such as *Homagnostus fecundus*, *Buttsia pinga*, and *Toxitis* in the Kharaulakh area. In northwestern Siberia, the top of the stage is characterized by the disappearance of *Maiaspis, Aldanaspis, Buitella*, etc., and by the appearance of *Pauciella prima, Nganasanella*, and *Homagnostus paraobesus* (Egorova *et al.*, 1982). More than 230 trilobite species were documented by Egorova *et al.* (1982), including various agnostoids, most of which are widely distributed.

Kulyumbean: The Kulyumbean Stage is named for the middle reaches of the Kulyumbe River (Rozova, 1963), on the northwestern Siberian platform, Russia. The stage is subdivided into two "horizons", the Nganasanian Horizon (below) and the Tavgian Horizon (above). The lower boundary of the stage is drawn at the base of a 10-m-thick limestone breccia that occurs at the bottom of the Nganasanian Horizon. The trilobite fauna in the Kulyumbean Stage is characterized by a high diversity of polymerid trilobites comprising acrocephalitids, eoacidaspidids, lonchocephalids, crepicephalids, pterocephalids, catillicephalids, etc., all of which are basically endemic forms. A few agnostoids, such as the widely distributed *Nahannagnostus nganasanicus,* are also present in the stage. With wide distributions, the agnostoids suggest a correlation to the upper Marjuman Stage in the Mackenzie Mountains, Canada (Pratt, 1992), and to the upper Guzhangian Stage of South China and Northwest China (Zhou *et al.*, 1996; Peng and Robison, 2000).

Gorbiyachinian: The Gorbiyachinian Stage was named for the Gorbiyachin River, a tributary of the Kulyumbe River (Rozova, 1963, 1968), Russia. The stage stratotype directly overlies the Kulyumbean Stage stratotype, both of which are in the middle reaches of the Kulyumbe River. The lower boundary of the Gorbiyachinian Stage is identified by the abundance of the plethopeltid trilobite *Koldinia mino*. As originally defined by Rozova (1963), the Gorbiyachinian Stage embraces two "horizons", the Maduan Horizon, and the overlying Entsian Horizon. The upper boundary of the stage is drawn at a level where the illaenurid trilobite *Yurakia yurakiensis* occurs in abundance. The stage is characterized by a turnover of trilobite faunas, with only a single species ranging upward from the underlying Kulyumbean Stage into the stage in the stratotype section (Rozova, 1968). The trilobite fauna comprises various endemic polymerids, and characteristic forms are *Acidaspidina, Maduiya, Kulyumbopeltis, Taenicephalus,* and *Parakoldinia*. The equivalent of the stage in the Chopko River section (Varlamov *et al.*, 2005) is characterized by the occurrence of the agnostoid *Agnostotes orientalis* and the polymerid *Irvingella*, both of which have cosmopolitan distributions.

Tukalandian: The Tukalandian Stage derives its name from the Tukalandy River, a tributary of the Khantai River (Rozova, 1963, 1968), Russia. The base of the Tukalandian Stage is defined by the local abundance of *Yurakia yurakiensis* in the stratotype section, where Tukalandian strata lie in succession over strata of the Gorbiyachinian Stage. The upper boundary of the stage is defined by the occurrence of the eulomid trilobite *Dolgeuloma abunda*. The Tukalandian is characterized by the presence of various endemic polymerid trilobites, primarily eulomids (*Kujandaspis* = *Ketyna*), aphelaspidids (*Amorphella*), advanced eoacidaspidids (*Eoacidaspis*), lonchocephalids (*Graciella, Monosulcatina, Nordia,*), and illaenurids (*Polyariella, Yurakia*), and by a turnover of trilobite faunas with no species ranging upward from the underlying Gorbiyachinian Stage (Rozova, 1968). The stage is subdivided into two horizons, the Yurakian Horizon below and the Ketyan Horizon above.

Khantaian: The Khantaian Stage is named for the Khantai River, a tributary of the Yenisey River, in northwestern Siberia, Russia. Its lower boundary with the Tukalandian Stage is defined by the appearance of the trilobite *Dolgeuloma abunda* (Rozova, 1963, 1968).

The Khantaian Stage of Russia is inferred to be uppermost Cambrian in the lower part and Lower Ordovician (Tremadocian Stage) in the upper part. Originally (Rozova, 1963) the Khantaian Stage was not subdivided. Later, Rozova (1968) subdivided the stage into two "horizons", the Mansian Horizon and the Loparian Horizon, and assigned both to the Upper Cambrian. Present in the upper part of the Loparian Horizon is the graptolite *Dictyonema flabelliforme* (a cosmopolitan species that elsewhere first occurs in the Tremadocian Stage). This inference of an Ordovician age for that part of the Loparian Horizon is supported by the presence of the polymerid trilobite *Plethopeltides*, which has an Ordovician aspect. The trilobites that characterize the stage include *Pseudoacrocephalites, Kaninia, Dolgeuloma,* and *Mansiella*, all of which are endemic and of low diversity.

19.1.5.4. Laurentian Cambrian Stages

The use of Cambrian stage and series nomenclature in Laurentia was recently reviewed by Babcock *et al.* (2011). Two sets of regional names have been used, but neither extends through the entire system.

The development of a stadial nomenclature for Laurentia has until recently (Palmer, 1998) been complicated by the concept of the biomere ("segment of life") introduced by Palmer (1965a) and subsequently discussed by Stitt (1975), Palmer (1979, 1984), Taylor (1997, 2006) and others. As originally defined, a biomere is a regional biostratigraphic unit bounded by an abrupt extinction event on the shallow cratonic shelf. When this happens to trilobite faunas, an evolving shelf fauna is replaced by a new, low-diversity fauna dominated by simple ptychoparioid trilobites invading from the outer-shelf or shelf break. The new fauna then evolves until another extinction event occurs. Six such cycles were suggested by Palmer (1981), but the lower two are as yet undefined. In ascending order they are the "Olenellid", "Corynexochid", Marjumiid, Pterocephaliid, Ptychaspid, and Symphysurinid biomeres.

Ludvigsen and Westrop (1985) considered biomeres to be stages because they were based on an aggregate of trilobite zones and subzones. They named three stages in the "Upper" Cambrian, Marjuman, Steptoean, and Sunwaptan, using reference sections in western North America showing the biomere pattern. These stage names were intended to replace the obsolete terms Dresbachian, Franconian, and Trempealeauan, which were based on trilobite assemblages from formations in the Upper Mississippi Valley area that had been in long-term use for the Upper or Middle–Upper Cambrian (e.g., Lochman-Balk and Wilson, 1958). These terms continue to be used in certain circumstances, however, particularly in subsurface studies. Palmer (1998) considered biomeres to be retained as units subtly different from stages, and extended Ludvigsen and Westrop's (1985) proposed sequence of stages for the Laurentian Cambrian based on trilobites (Figure 19.11, column 6). The portion of the Cambrian named by Palmer (1998) as the Begadean Series, and regarded as pre-trilobitic, still lacks defined stages. However, Hollingsworth (2007, 2011) subsequently reported a small assemblage of polymerid trilobites from the upper part of this interval.

Montezuman: The Montezuman Stage (Palmer, 1998) was named for the Montezuma Range, Nevada, USA. Its base is defined by the appearance of characteristic fallotaspidid trilobites. As in Morocco and Siberia, the fallotaspidids are followed by nevadiids and holmiids. At least three families of Olenellina, which are different from olenellines of the succeeding stage (see generic range charts in Palmer and Repina, 1993), are present in the Montezuman Stage. The Montezuman Stage also contains the oldest Laurentian archaeocyaths.

Dyeran: The Dyeran Stage (Palmer, 1998) was named for the town of Dyer, Nevada, USA, and covers the biostratigraphic interval that for many years was assigned to the *Olenellus* Zone. The *Olenellus* Zone of historic usage has subsequently been regarded as multizonal (Palmer and Repina, 1993; Palmer, 1998; Webster, 2011). The base of the stage coincides with a major change in the olenelloid fauna following the nevadiid-bearing late Montezuman. A similar change was documented by Fritz (1992) from British Columbia, Canada. Olenellid trilobites are characteristic of the Dyeran Stage.

Delamaran: The stratotype of the Delamaran Stage (Palmer, 1998) is the Oak Spring Summit section, Delamar Mountains, Nevada, USA. The stage embraces the "Corynexochid" Biomere and the *Plagiura—Poliella*, *Albertella*, and *Glossopleura* Zones in restricted-shelf environments (Palmer and Halley, 1979; Eddy and McCollum, 1998). It is characterized by ptychoparioid, corynexochid, zacanthoidid, dolichometopid, and oryctocephalid trilobites.

Topazan: The Topazan Stage was erected (Sundberg, 2005) by restriction of the Marjuman Stage. Recognition of the Topazan Stage resulted from restoration of the base of the overlying Marjuman Stage to its original proposed level (Ludvigsen and Westrop, 1985; *contra* Palmer, 1998), the FAD of the cosmopolitan agnostoid *Ptychagnostus atavus* (see Marjuman Stage, below). The Topazan Stage is defined as the interval between the top of the Delamaran Stage and the base of the Marjuman Stage (as restricted; Sundberg, 2005). The stage was named for the Topaz Internment Camp (active during World War II), located some 25 km to the southwest of the stratotype. The base of the stage is defined within a shale sequence, 2.6 m above the base of the upper shale member of the Chisholm Formation at section Do2 of Sundberg (1990, 1994) in the Drum Mountains, Utah, USA. The basal 10 cm of the stage contains the FAD of the polymerid trilobite *Proehmaniella basilica*.

The Topazan Stage embraces only a single polymerid zone, the *Ehmaniella* Zone, which is subdivided into four subzones: the *Proehmaniella* Subzone, the *Elrathiella* Subzone, the *Ehmaniella* Subzone, and the *Altioccullus* Subzone. In outer-shelf facies, the agnostoid *Ptychagnostus praecurrens* and *P. gibbus* zones characterize the Topazan Stage. Almost 70 trilobite taxa of Topazan age have been documented from Nevada and Utah (Sundberg, 1994, 2005).

Marjuman: The Marjuman Stage (Ludvigsen and Westrop, 1985; emended by Palmer, 1998; restricted by Sundberg, 2005) takes its name from Marjum Pass in the House Range, Utah, USA, and was intended to replace the Marjumiid Biomere. Ludvigsen and Westrop (1985) originally defined the base of the Marjuman Stage at the base of the *Ptychagnostus atavus* Zone (which closely corresponds to the base of the *Bolaspidella* Zone based on polymerid trilobites) but this was not a major extinction event according to Palmer (1998). Ludvigsen and Westrop (1985) equated the Marjuman Stage with the Marjumiid Biomere (Palmer, 1981), but the biomere event occurred earlier on the inner-shelf with a major change from trilobites of the *Glossopleura* Zone to those of the *Ehmaniella* Zone (*Proehmaniella* Subzone; Sundberg, 1994). In open-shelf environments, this event corresponds to the base of the *Bathyuriscus—Elrathina* Zone. Palmer (1998) moved the base of the Marjuman Stage downward to coincide with the lowest occurrence of *Proehmaniella basilica*, which marks the base of the Marjumiid Biomere as he (Palmer, 1981) envisioned it. The revised stage embraced three polymerid trilobite zones; in ascending order they are the *Ehmaniella* Zone (with the *P. basilica* Subzone at the base), the *Bolaspidella* Zone, and the *Crepicephalus* Zone (Palmer, 1999).

Sundberg (2005) restored the original concept of Ludvigsen and Westrop's (1985) Marjuman, and proposed a new Topazan Stage (see Topazan Stage, above) for the interval between the top of the underlying Delamaran Stage and the base of the Marjuman Stage as defined at the base of the *Ptychagnostus atavus* Zone. Subsequently the base of the *P. atavus* Zone, which coincides with the FAD of the eponymous species in the Drum Mountains of northern Millard County, Utah, was designated as the primary stratigraphic marker coinciding with the GSSP for the Drumian Stage of global chronostratigraphy (Babcock *et al.*, 2007). The Marjum Pass, Utah, section, for which the Marjuman Stage was named, shows considerable structural complications, and the true first appearance of *P. atavus* there is unknown.

The Marjuman is characterized in open-shelf environments by cedariid trilobites, four zones of which were documented by Pratt (1992). Cedariid and crepicephalid trilobites characterize inner-shelf facies of the Marjuman Stage. The Marjuman Stage, as conceptualized by Ludvigsen and Westrop (1985) and Sundberg (2005), embraces agnostoid zones from the *Ptychagnostus atavus* Zone to the *Glyptagnostus stolidotus* Zone, and corresponds to the Drumian through Guzhangian stages of global usage.

Steptoean: The Steptoean Stage (Ludvigsen and Westrop, 1985) was named for Steptoe Valley, in the Duck Creek Range, near McGill, eastern Nevada, USA, and was intended to replace the Pterocephaliid Biomere. The base of the stage is defined at the base of the *Aphelaspis* Zone, which also corresponds to the base of the Pterocephaliid Biomere (Palmer, 1965b). The *Aphelaspis* Zone contains *Glyptagnostus reticulatus*, which allows precise correlation globally. Above the *Aphelaspis* Zone, the Steptoean Stage embraces the *Dicanthopyge*, *Prehousia*, *Dunderbergia*, and lower *Elvinia* zones in restricted-shelf environments. The

Glyptagnostus reticulatus, *Olenaspella regularis*, and *O. evansi* zones characterize the Steptoean Stage in outer-shelf facies. The *Parabolinoides calvilimbata* and *Proceratopyge rectispinata* faunas, documented by Pratt (1992), are typical of open-shelf environments.

Sunwaptan: The Sunwaptan Stage was named (Ludvigsen and Westrop, 1985) for Sunwapta Creek, Wilcox Peak, Jasper National Park, in southern Alberta, Canada, and was intended to replace the Ptychaspid Biomere (see Longacre, 1970; Stitt, 1975). The base of the Sunwaptan Stage is at the base of the *Irvingella major* Subzone of the *Elvinia* Zone, which Chatterton and Ludvigsen (1998) argued should be regarded as a separate zone. This is succeeded by the *Taenicephalus* Zone, the *Stigmacephalus oweni* fauna, and the *Ellipsocephaloides* Zone in the lower Sunwaptan, and the *Illaenurus* Zone, and most of the *Saukia* Zone in the upper Sunwaptan. More than 130 trilobite taxa of Sunwaptan age have been documented from Alberta by Westrop (1986), and from the District of Mackenzie, Northwest Territories, Canada, by Westrop (1995). Characteristic are dikelocephalid, ptychaspidid, parabolinoidid, saukiid, ellipsocephaloid, illaenurid, and elviniid trilobites.

Skullrockian: The Skullrockian Stage (Ross *et al.*, 1997) was named from Skull Rock Pass in the House Range, Utah, USA. It was originally conceived of as the lowermost stage of the Ibexian Series, which at the time was considered to be Lower Ordovician. The base of the Skullrockian is defined by conodonts at the base of the *Hirsutodontus hirsutus* Subzone of the *Cordylodus proavus* Zone. On the polymerid trilobite zonal scale this level corresponds to the base of the *Eurekia apopsis* Zone (Ross *et al.*, 1997; Miller *et al.*, 2006). The *E. apopsis* Zone has a limited trilobite fauna, as does the overlying *Missisquoia* Zone, and the primary group used for high-resolution correlation is conodonts.

The Skullrockian Stage of Laurentia is Cambrian in the lower part, and Lower Ordovician through most of its stratigraphic extent. Following definition of the Ordovician GSSP at the FAD of the euconodont *Iapetognathus fluctivagus*, a position that is partway up through the Skullrockian Stage, the lower part of the stage (through the *Cordylodus lindstromi* Zone of conodont zonation and the *Symphysurina brevispicata* Subzone of the *Symphysurina* Zone of polymerid trilobite zonation) was automatically reassigned to the Cambrian. Most of the stage remained assigned to the Ordovician, however (Miller *et al.*, 2003, 2006). Difficulties in achieving precise correlation between the horizon containing *I. fluctivagus* from the Ordovician stratotype at Green Point, Newfoundland, Canada, and western Utah, USA, where the Skullrockian was defined, were discussed by Miller *et al.* (2003). As a result, redefinition of the Skullrockian Stage has not taken place, nor has a replacement stage whose base corresponds to the base of the Ordovician, been proposed.

19.2. CAMBRIAN STRATIGRAPHY

19.2.1. Faunal Provinces

The Cambrian Period is noteworthy from a biologic standpoint because it marks the appearance of most multicellular phyla that have populated the Earth. Faunal provincialism tended to be strong, and biostratigraphic zonal schemes based on benthic and nektobenthic taxa generally cannot be applied beyond their provincial boundaries.

Álvaro *et al.* (in press) summarized the history of studies on Cambrian trilobite biogeography, and provided a comprehensive review based on an updated database of Cambrian genera. Most authors have recognized biogeographic differentiation into two main provinces during Cambrian Epoch 2 (Kobayashi, 1972; Palmer, 1973; Cowie, 1971; Lu et al., 1974; Lu, 1981; Chang, 1989; Palmer and Repina, 1993; Álvaro *et al.*, in press). One faunal province, the Redlichiid Province of Gondwana, is characterized by endemic redlichiids, pandemic ellipsocephaloids, and eodiscids. The other faunal province, the Olenellid Province, comprising much of Baltica, Laurentia, and Siberia, is characterized by endemic olenellids, pandemic ellipsocephaloids, and eodiscids. An overlap in the geographic ranges of taxa characteristic of both major provinces in some peri-Gondwanan margins led Pillola (1991) to erect the intermediate Bigotinid Province.

For trilobites of Epoch 3 and the Furongian Epoch, Palmer (1973) and others (e.g., Sdzuy, 1972; Jell, 1974; Chang, 1989) have recognized more complicated biogeographic schemes. Terms such as Pacific (or North American) and Atlantic (or Acado–Baltic) have often been used to distinguish biogeographic units. Chang (1989) characterized the Pacific Province using an assemblage of centropleurid, xystridurid, and olenid trilobites, and characterized the Atlantic Province using an assemblage of paradoxidid and olenid trilobites. The Acado–Baltic Province (*sensu* Sdzuy, 1972), is characterized by the persistent presence of a paradoxidid–solenopleurid–conocoryphid assemblage, and was widespread through Avalonia, the Mediterranean and central-European areas, and Baltica (Álvaro and Vizcaïno, 2003). Babcock (1994a,b) showed that this trilobite assemblage was widely distributed in cool marine waters of various latitudes, including in deep water surrounding tropical Laurentia. Quantitative analysis of a large data set of Cambrian genera led Jell (1974) to recognize three trilobite provinces: 1. Columban in North and South America; 2. Viking in Europe, maritime North America, and northwestern Africa; and 3. Tollchuticook in Asia, Australia, and Antarctica. Palmer (1973), Robison (1976), and Pegel (2000), among others, have recognized differentiation between trilobite faunas of outer-shelf and inner-shelf areas of low-latitude continents such as Laurentia and Siberia.

Biogeographic studies on Cambrian trilobites have played an integral role in the recognition of tectonostratigraphic terranes. In general, the juxtaposition of trilobites representing two distinct faunal units in neighboring strata has been used to help infer the boundary of an accreted terrane (e.g., Secor *et al.*, 1983; Samson *et al.*, 1990). However, Babcock (1994b) advised caution in such interpretations, as warm water shelf faunas can occur in close association with cooler water faunas of adjacent deep water in tropical regions. Stratigraphic and structural/tectonic evidence must be used to supplement biogeographic information to arrive at a conclusion as to a terrane's provenance. Today, a complex mosaic of tectonostratigraphic terranes is recognized, particularly for areas such as Europe, Asia, and the margins of North America. Álvaro *et al.* (in press) analyzed the biogeographic affinities of trilobites among all Cambrian continents and numerous terranes. These results are in general agreement with more classical interpretations of biogeographic provinces, but provide considerable additional information about biogeographic links between regions.

Archaeocyaths also showed provincialism during the Cambrian Period. Debrenne (1992) identified three archaeocyathan faunal provinces that existed in the early half of the Cambrian:

1. An Afro-European Province, which possibly extends to China, characterized by Anthomorphidae;
2. An Australo-Antarctica Province characterized by Flindersicyathidae, Metacyathidae, and Syringocnemidae; and
3. A Siberian Province characterized by genera belonging to all of these families.

Kruse and Shi (2000), who analyzed the distributions of archaeocyaths statistically, recognized five provinces:

1. Siberia–Mongolia;
2. Europe–Morocco;
3. Central Asia–East Asia;
4. Australia–Antarctica; and
5. North America–Koryakia.

19.2.2. Trilobite Zones

The most widely used fossil group for biostratigraphic zonation of the Cambrian are the trilobites, the best known group of Paleozoic arthropods. Beginning in provisional Series 2, they enable fine stratigraphic subdivision and good correlation reliability. In general, polymerid and agnostoid trilobites have different biogeographic distributions and correlation value. Polymerid species and genera tend to be endemic to individual regions or paleocontinents and are thus of greatest use in correlating deposits of the continental shelf and platform (Robison, 1976, Babcock, 1994a; Peng *et al.*, 2004b; Babcock *et al.*, 2007, 2011). Agnostoid species tend to be much more widespread, and many are cosmopolitan. They are of great value in correlating open-shelf to shelf-margin deposits intercontinentally (Westergård, 1946; Robison, 1976, 1984, 1994; Öpik, 1979; Peng and Robison, 2000; Ahlberg, 2003; Ahlberg *et al.*, 2004; Peng *et al.*, 2004a; Babcock *et al.*, 2007, 2011).

In the latter half of the Cambrian, trilobite diversification and evolutionary turnover was extreme. For this reason, fine zonations of polymerids and agnostoids have been established on the major paleocontinents (Figure 19.3). Pelagic agnostoids enable global correlation of Series 3 and Furongian strata (Westergård, 1946; Öpik, 1979; Peng and Robison, 2000; Peng and Babcock, 2005b; Babcock *et al.*, 2011). Thirteen agnostoid zones have been defined, with the zonal bases being placed at the first appearances of eponymous species. In ascending order, these are the *Ptychagnostus gibbus*, *Ptychagnostus atavus*, *Ptychagnostus punctuosus*, *Goniagnostus nathorsti*, *Lejopyge armata*, *Lejopyge laevigata*, *Proagnostus bulbus*, *Linguagnostus reconditus*, *Glyptagnostus stolidotus*, *Glyptagnostus reticulatus*, *Agnostotes orientalis*, and *Lotagnostus americanus* zones (Robison, 1984; Peng and Robison, 2000; Peng and Babcock, 2005b). Some zones, such as the *G. nathorsti* and *L. armata* zones, are not recognized on all paleocontinents (Robison and Babcock, 2011; Babcock *et al.*, 2011).

China, Russia, North America, Scandinavia, and Australia have the most complete Cambrian trilobite zonal successions. Those of South China, Siberia, North America, and Australia are shown in Figure 19.3. Historically, differing biostratigraphic philosophies have been applied in different regions, and these have resulted in differing concepts of trilobite zones. In North America, for example, the concept of a zone was commonly based on the range of a characteristic species or genus (an interval-zone; Robison, 1994). In Australia, China, Scandinavia, and Russia, species zones or assemblage zones have been most commonly applied. In Scandinavia, the pre-Furongian agnostoid zones were, until recently (Terfelt *et al.*, 2008; Ahlberg *et al.*, 2009), based on local abundance of eponymous species (Westergård, 1946; Peng and Robison, 2000; Axheimer *et al.*, 2006). Increasingly in recent years, zones based on the first appearances of characteristic species have been replacing the older, regional concepts of zones. Particularly where widespread species, such as agnostoids, are the characteristic species of zones, this practice has led to precise correlation regionally and intercontinentally.

19.2.3. Archaeocyathan Zones

More than 300 genera of regular Archaeocyatha and Radiocyatha are known from carbonate platforms in the lower half of the Cambrian (Kruse and Shi, 2000). Archaeocyaths have been used extensively for biostratigraphy in certain regions. The most detailed archaeocyathan biostratigraphy has been developed in Siberia, where the Tommotian Stage embraces three successive assemblage zones, the Atdabanian four, the Botoman three, and the Toyonian three (Debrenne and Rozanov, 1983; Zhuravlev, 1995). Archaeocyathan zones have also been established in South Australia (five), Laurentia (nine), Spain (eleven), Morocco (four) and South China (four) (Zhuravlev, 1995; Yang *et al.*, 2005). Problems associated with correlation of these areas on the basis of archaeocyaths are primarily due to regional endemism. For example, Kruse and Shi (2000) noted that of the 240 archaeocyathan species occurring in Australia and Antarctica, only 26 are shared between the two continents and only genera with wide stratigraphic distributions are common to Australia and Siberia (Zhuravlev and Gravestock, 1994).

19.2.4. Small Shelly Fossil Zones

The primarily and secondarily phosphatic skeletonized microfossils, termed small shelly fossils (SSFs), occur in differing levels of abundance in the lower half of the Cambrian. In South China, Siberia, and Australia, SSFs are usually used in regional biostratigraphy. A detailed regional SSF biostratigraphy with three assemblage zones embracing eight subzones has been developed for the Diandongian Series of South China with the Jinningian Stage embracing five subzones and the Meishucunian Stage (s.s.) three (Luo *et al.*, 1984). However, Qian *et al.* (1999) subsequently recognized only four assemblage zones for the series but added four assemblage zones for the overlying Qiandongian Series. In Siberia, SSF biostratigraphy with two zones has only been developed for the Nemakit-Daldynian Stage, although a diverse SSF assemblage occurs in the basal part of the succeeding Tommotian Stage (Khomentovsky and Karlova, 1993). In Australia, three informal SSF zones have been established in the Arrowie and Stansbury basins (Demidenko *et al.*, 2001; Jago *et al.*, 2006). An SSF biostratigraphy or succession has also been developed for the Terreneuvian Series of England, Poland, Iran, southern France, and Mongolia (Brasier, 1984, Keber, 1988; Orłowski, 1992, Hamdi *et al.*, 1989; Khomentovsky and Ginsher, 1996). Because of apparent regional endemism, small shelly fossil correlation is more or less limited and problematic. It is not entirely certain at present how much of the apparent endemism of SSFs is related to the development of separate taxonomic nomenclature in separate regions of Cambrian exposure.

19.2.5. Conodont Zones

Conodont elements, including slender, simple cones referred to as protoconodonts, range through Cambrian strata beginning about the middle of the Terreneuvian Series (Bengtson, 1976). In both Siberia and South China, the protoconodont *Protohertzina* has been used as an eponymous genus of the *Anabarites–Protohertzina* Zone that occurs in the basal part of a regional stage (Nemakit-Daldynian in Siberia; Jinningian in South China). Conodonts begin to diversify in the middle of Epoch 3, and in the Furongian they are sufficiently common and differentiated to be used biostratigraphically.

Conodonts of the Furongian Series, including protoconodonts, paraconodonts, and euconodonts, have been most intensively studied in the Great Basin (e.g., Miller, 1980, 1988; Landing *et al.*, 2011; Miller *et al.*, 2011), western Queensland, Australia (Black Mountain; Druce and Jones, 1971; Nicoll and Shergold, 1992; Shergold and Nicoll, 1992), and South China (Dong and Bergström, 2001a,b; Dong *et al.*, 2004). In Utah, 11 subzones have been defined through the interval of the upper Sunwaptan Stage through the lower Skullrockian Stage (*Saukia junia* Subzone of the *Saukia* Zone through the *Symphysurina bulbos*a Subzone of the *Symphysurina* Zone) (Miller, 1980; Miller *et al.*, 2006). The subzones are named for *Proconodontus posterocostatus, Proconodontus muelleri, Eoconodontus notchpeakensis, Cambrooistodus minutus, Hirsutodontus hirsutus, Fryxellodontus inornatus, Clavohamulus elongatus, Hirsutodontus simplex, Clavohamulus hintzei* and *Cordylodus lindstromi* Zone (which has lower and upper subzones). In Australia, a little more than seven conodont assemblages have been recognized through provisional stages 9 and 10. These assemblage zones are, in ascending order, based on *Teridontus nakamurai, Hispidodontus resimus, Hispidodontus appressus, Hispidodontus discretus, Cordylodus proavus, Hirsutodontus simplex*, *Cordylodus prolindstromi*, and *Cordylodus lindstromi* (basal part only) (Shergold and Nicoll, 1992; Kruse *et al.*, 2009). Dong and Bergström (2001a,b) and Dong *et al.* (2004) developed a comprehensive conodont biostratigraphy with a little more than 11 zones ranging through the interval of the Drumian Stage through the Furongian Series in Hunan, China. In ascending order, the zones are the *Gapparodus bisulcatus–Westergaardodina brevidens* Zone, *Shandongodus priscus–Hunanognathus tricuspidatus* Zone, *Westergaardodina quadrata* Zone, *Westergaardodina matsushitai–W. grandidens* Zone, *Westergaardodina lui–W. ani* Zone, *Westergaardodina* cf. *calix–Prooneotodus rotundatus* Zone, *Proconeotodus tenuiserratus* Zone, *Proconeotodus* Zone, *Eoconodontus* Zone, *Cordylodus proavus* Zone, *Cordylodus intermedius* Zone, and *C. lindstromi* Zone (lower part).

19.2.6. Magnetostratigraphy

Two types of magnetostratigraphic information have been used for correlation of Cambrian strata, magnetic polarity studies and magnetic susceptibility studies. Most work has involved development of a magnetic polarity time scale (Figure 19.3). To the present, such a time scale remains incomplete, for reasons summarized by Trench (1996). Detailed studies however, have been undertaken through parts of all four series, with the most intense research being concentrated on the Cambrian–Ordovician boundary interval. So far, magnetic susceptibility work has been applied only to strata within the Drumian Stage and to strata near the base of the stage.

Kirschvink and Rozanov (1984), Kirschvink *et al.* (1991), and Varlamov *et al.*, (2008b) provided a detailed magnetostratigraphic polarity scale for the uppermost Terreneuvian Series and lower part of Series 2 derived from studies along the Lena River of Siberia. In the Tommotian and Atdabanian stages, as used regionally on the Siberian Platform, Kirschvink and Rozanov (1984) and Kirschvink *et al.* (1991) found many polarity reversals. The Tommotian correlates approximately to the upper part of Stage 2, and the Atdabanian correlates approximately to the lower to middle part of Stage 3. Pavlov and Gallet (2001) challenged the interpretation of Kirschvink and Rozanov's (1984) results because the paleomagnetic pole they obtained differs significantly from other pole positions obtained from the Siberian Platform, and because a predominant reversed polarity is most often observed for this time interval (Khramov and Rodionov, 1980; Pisarevsky *et al.*, 1997). Nevertheless, the information from Siberia is essentially in agreement with results obtained from Morocco and South China near the equivalents of the Tommotian–Atdabanian boundary (Kirschvink *et al.*, 1991, 1997), where several magnetic reversals were discovered. Magnetostratigraphic information, calibrated with chemostratigraphic results, can be used in a general way to correlate among these three areas (Kirschvink *et al.*, 1991).

Rudimentary magnetostratigraphic polarity results are available for the Botoman and Toyonian stages of Siberian usage (middle to upper part of Series 2), derived from studies along the Lena River of Siberia (Kirschvink and Rozanov, 1984; Kirschvink *et al.*, 1991; Varlamov *et al.*, 2008b). These results show two long episodes each of normal and reversed polarity. The Botoman correlates approximately to upper Stage 3–lowermost Stage 4, and the Toyonian correlates approximately to lower-middle Stage 4. In contrast, magnetostratigraphic data from the Yuanshan Member of the Chiungchussu Formation (also known as Maotianshan Shale) from the Chengjiang area, eastern Yunnan, China, reveal a relatively high frequency of magnetic pole reversals (Yin, 2002). In the Yuanshan Member, which is the unit containing the Chengjiang Biota, at least 29 magnetic polarity intervals are recognized. This member correlates to the lowermost part of Stage 3 (lower Nangaoan Stage of South China regional usage).

Magnetic polarity studies have been conducted on several paleocontinents in the Cambrian–Ordovician boundary interval, and studies from the middle of Series 3 through the lower Furongian have been reported from Siberia. Early, and rather limited, investigations in strata adjacent to the Ordovician base were made by Kirschvink (1978a,b) and Klootwijk (1980) in central and South Australia. More detailed information comes from studies across the Cambrian–Ordovician boundary interval at Black Mountain in western Queensland, Australia (Ripperdan and Kirschvink, 1992; Ripperdan *et al.*, 1992), at Batyrbai, southern Kazakhstan (Apollonov *et al.*, 1992), at Dayangcha and Tangshan, North China (Ripperdan *et al.*, 1993; Yang *et al.*, 2002), and along the Kulyumbe River, northwestern Siberian Platform (Pavlov and Gallet, 2001, 2005; Kouchinsky *et al.*, 2008). Geomagnetic results for the Drumian through Paibian stages were reported from the Kulyumbe River section of Siberia by Pavlov and Gallet (1998, 2001, 2005), Kischvink and Raub (2003), Pavlov *et al.* (2008) and Kouchinsky *et al.* (2008). Combining results, a composite magnetic polarity time scale is now available from the Drumian Stage upward into the Tremadocian Stage of the Ordovician System (Kouchinsky *et al.*, 2008).

According to Kouchinsky *et al.* (2008), the "Middle Cambrian" (presumably equivalent to Series 3) has up to 100 geomagnetic reversals, although data were presented only for the Drumian and Guzhangian stages. A total of 100 geomagnetic intervals correspond to a reversal frequency of 10 per million years, an extremely high rate. In the "Upper Cambrian" (Furongian Series), only 10 to 11 magnetic intervals were recognized, and this corresponds to a reversal rate of about 1 per million years. This is an order of magnitude lower rate than for Series 3. The Furongian is dominated by intervals of reversed polarity, most of them relatively long, and mostly short intervals of normal polarity. Except for a short interval of normal polarity in the lower part of the stage, the Paibian shows a long interval of continuous reversed polarity. The longest interval of normal polarity in the Furongian Series is close to the top of the series. The initial Ordovician is dominated by periods of normal polarity with a couple intervals of reversed polarity (Ripperdan *et al.*, 1993; Yang *et al.*, 2002; Pavlov and Gallet, 2005). Only two to three geomagnetic intervals were recognized from the Tremadacian Stage (Ordovician) by Kouchinsky *et al.* (2008).

Studies of magnetic susceptibility have recently been applied to limited intervals of the Drumian Stage, and to strata adjacent to the base of the stage. Magnetic susceptibility shows excellent potential for high-resolution correlation, and adds to the magnetostratigraphic information obtained through magnetic polarity studies. A detailed profile across the interval containing the Drumian Stage GSSP in the Drum Mountains, Utah, USA, shows a positive deflection in magnetic susceptibility that correlates precisely to the beginning of the DICE $\delta^{13}C$ excursion (Babcock *et al.*, 2009). Halgedahl *et al.* (2009) showed that higher in the Drumian Stage, peaks in magnetic susceptibility can be matched among sections within the same general area of Utah.

19.2.7. Chemostratigraphy

A significant and stratigraphically important body of chemostratigraphic information now exists. Particularly important are stable isotopes of carbon ($\delta^{13}C$) and strontium ($^{87}Sr/^{86}Sr$) (Figure 19.12).

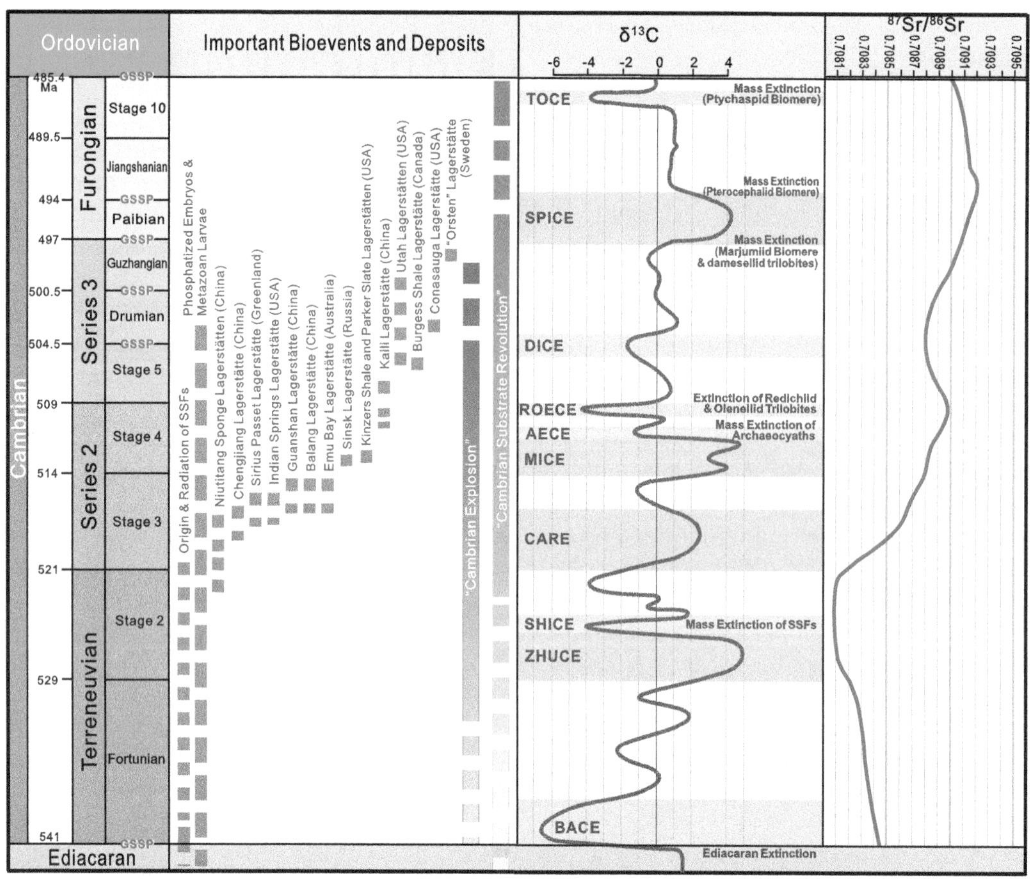

FIGURE 19.12 **Carbon isotope ($\delta^{13}C$) and strontium isotope ($^{87}Sr/^{86}Sr$) chemostratigraphy of the Cambrian System and comparison to biotic events.** The carbon isotope chemostratigraphy and comparison to biotic events is modified from Zhu *et al.* (2006) and Miller *et al.* (2006). The strontium curve is a composite derived from curves for the upper Terreneuvian and lower Series 2 (Derry *et al.*, 1994); upper Series 2 through Series 3 (Montañez *et al.*, 2000); and much of the Furongian (Saltzman *et al.*, 1995; Kouchinsky *et al.*, 2008) except the uppermost part of Stage 10 (Ebneth *et al.*, 2001).

Zhu *et al.* (2006) synthesized previous studies of $\delta^{13}C$ isotopic values in Cambrian deposits (e.g., Derry *et al.*, 1994; Zhang *et al.*, 1997; Brasier, 1993; Brasier and Sukhov, 1998; Saltzman *et al.*, 1998; Montañez *et al.*, 2000; Corsetti and Hagadorn, 2001; Buggisch *et al.*, 2003; Peng *et al.*, 2004a, 2006; Zhu *et al.*, 2004; Babcock *et al.*, 2005; Guo *et al.*, 2005; Maloof *et al.*, 2005; Kouchinsky *et al.*, 2005) and added new information from carbonates of South China to develop a generalized curve encompassing the entire Cambrian System. They recognized 10 distinct isotopic excursions, many of which coincide with important biotic events such as evolutionary radiations and extinctions, or with times of taphonomic windows (Figure 19.12). Three positive $\delta^{13}C$ excursions recorded in the Terreneuvian Series correspond to times when faunas of small shelly fossils (SSFs) radiated on the Yangtze Platform. Only the last of these three excursions has been named (the ZHUCE excursion, or ZHUjiaqing Carbon isotope Excursion; Zhu *et al.*, 2006). The CARE excursion (Cambrian Arthropod Radiation isotope Excursion; Zhu *et al.*, 2006) is a positive shift in $\delta^{13}C$ values associated with the appearance of a wide variety of arthropod fossils, particularly in major Konservat-lagerstätten. The extinction of acritarchs and other organisms near the end of the Ediacaran Period corresponds to a strong negative shift in $\delta^{13}C$ values recorded in carbonate sediments (BACE excursion, or Basal Cambrian Carbon isotope Excursion; Zhu *et al.*, 2006). Mass extinctions of SSFs during Cambrian Stage 2 (SHICE excursion, or SHIyantou Carbon isotope Excursion; Zhu *et al.*, 2006), extinction of archaeocyaths in Stage 4 (AECE excursion, or Archaeocyathid Extinction Carbon isotope Excursion; Zhu *et al.*, 2006), and extinction of redlichiid and olenellid trilobites at the end of Stage 4 (ROECE excursion, or Redlichiid–Olenellid Extinction Carbon isotope Excursion; Zhu *et al.*, 2006) are all associated with strong shifts toward negative $\delta^{13}C$ values. The SPICE excursion (Steptoean PositIve Carbon isotope Excursion; Saltzman *et al.*, 1998) corresponds to the Pterocephaliid Biomere of Laurentia (Saltzman *et al.*, 1998; Peng *et al.*, 2004a; Zhu *et al.*, 2006). Onset of this large positive excursion marks the extinction of marjumiid trilobites (at the top of the Marjumiid Biomere) in Laurentia. A similar biotic turnover is recognized at an equivalent horizon in eastern Gondwana (Peng *et al.*, 2004a). Subsequent extinction of pterocephaliid trilobites in Laurentia (at the top of the Pterocephaliid Biomere) is reflected in a return of $\delta^{13}C$ values to near zero.

Each Cambrian series is bracketed by a pair of distinct carbon isotopic excursions. The base of the Cambrian corresponds to the onset of the BACE excursion, which reaches a peak value exceeding -6 ‰. The base of Series 2 is associated with the onset of the CARE excursion, which reaches a peak value of about $+2.5$ ‰. The base of Series 3 is associated with the ROECE excursion, which reaches a peak value exceeding -4 ‰. The base of the Furongian Series is associated with the onset of the SPICE excursion, which reaches a peak value exceeding $+4$ ‰. The peak of the TOCE excursion (Top of Cambrian isotope Excursion; Zhu *et al.*, 2006) is just below the base of the Ordovician, and its peak value is about -3.5 ‰. The TOCE excursion has also previously been referred to as the HERB Event (Ripperdan, 1992).

The $\delta^{13}C$ curve has emerged as an increasingly powerful tool for intercontinental and intracontinental correlation of Cambrian strata, especially in regions where the primary biologic marker for a key horizon is absent. In the upper half of the Cambrian, all GSSPs defined to date are in outer-shelf to slope environments, and correlation into shallow epeiric seas of the continental interiors has in the past been hindered by strong differentiation of trilobite faunas collected from these ecologically different environments. In outer-shelf and slope environments, agnostoids are the primary guide fossils, and in inner-shelf environments, endemic polymerids are the primary guide fossils. Recognizable carbon isotopic excursions such as the SPICE (Saltzman *et al.*, 1998) and DICE (DrumIan Carbon isotope Excursion; Howley *et al.*, 2006; Zhu *et al.*, 2006; Babcock *et al.*, 2007; Howley and Jiang, 2010) excursions have been used to overcome the problem of extending intercontinental correlations based on outer-shelf- and slope-dwelling agnostoids into shallow shelf seas. At the base of the Cambrian, because of difficulty recognizing the horizon marked by the FAD of *Trichophycus pedum* outside of the Avalonian paleocontinent, constraining the base of the system intercontinentally is more commonly performed by means of the BACE excursion (e.g., Corsetti and Hagadorn, 2001; Zhu *et al.*, 2001, 2006; Amthor *et al.*, 2003; Babcock *et al.*, 2011).

Studies suggest a relationship between eustatic sea-level history and the carbon isotopic curve for the Cambrian. Fine-scale eustatic information is not yet available for the entire system, but well-resolved interpretations (e.g., Miller *et al.*, 2003; Peng *et al.*, 2004a; Babcock *et al.*, 2005; Howley *et al.*, 2006; Howley and Jiang, 2010) suggest a close correspondence in the timing of sea-level change and changes in $\delta^{13}C$ values. The base of the Drumian Stage, which is in the lower phase of a eustatic rise, is associated with the end of the DICE excursion (Howley *et al.*, 2006; Babcock *et al.*, 2007; Howley and Jiang, 2010). The base of the Paibian Stage is also in the lower part of a eustatic rise, and it is associated with the onset of the SPICE excursion (Peng *et al.*, 2004a).

Data on the temporal variation of strontium isotopes ($^{87}Sr/^{86}Sr$) are now available through most of the Cambrian, although the scale of resolution is variable. The most detailed information exists for the lowermost Terreneuvian Series, and from about the base of Series 3 through the Ordovician. Reconstruction of a high-resolution seawater $^{87}Sr/^{86}Sr$ curve for the entire Cambrian has been hindered, principally because of the perceived lack of suitable, well-preserved materials for analysis. Until recently, vague intercontinental age constraints, particularly in the lower half of the system, contributed to the difficulty of generalizing results to a global scale.

For much of the Cambrian, the fossils preferred for $^{87}Sr/^{86}Sr$ analyses are rare. Suitably robust low-Mg calcite brachiopod shells and apatite conodont elements are not particularly common except in some of the uppermost Cambrian strata. Phosphatic small shelly fossils and brachiopods, which have potential for $^{87}Sr/^{86}Sr$ studies, are essentially untested, as are low-Mg calcite trilobite exoskeletons. All of these remain as potentially promising alternatives for Cambrian studies.

To date, bulk micrite or early marine cements generally have been the preferred materials for constraining Cambrian seawater $^{87}Sr/^{86}Sr$ ratios. The heterogeneous nature of bulk carbonate necessitates, however, that diagenetic alteration needs to be assessed geochemically on a sample-by-sample basis. Samples having least-altered compositions are extrapolated from diagenetic trends. Few studies report enough data from any one stratigraphic horizon to enable a thorough diagenetic analysis, which means that inferred secular trends may in fact be the result of post-depositional effects rather than the result of the isotopic evolution of seawater. This needs to be borne in mind when considering the Sr isotope record shown in Figure 19.12.

Several $^{87}Sr/^{86}Sr$ studies span the Proterozoic–Cambrian boundary interval. The most comprehensive work is that of Brasier *et al.* (1996). Results of that study and others (Derry *et al.*, 1994; Kaufman *et al.*, 1996; Nicholas, 1996; Valledares *et al.*, 2006; Jiang *et al.*, 2007; Sawaki *et al.*, 2008) constrain latest Ediacaran and earliest Cambrian $^{87}Sr/^{86}Sr$ to about $0.708\,45 \pm 0.0005$. Least-altered samples from Mongolia and Siberia (Brasier *et al.*, 1996; Kaufman *et al.*, 1996) reveal a decreasing trend to a low of $0.708\,05 \pm 0.0005$ by the end of the Terreneuvian, before rising through Series 2.

Three studies provide data from close to the base of Series 3. Values from least-altered samples reported by Montañez *et al.* (2000) from the Great Basin (USA) and Wotte *et al.* (2007) from France and Spain are mutually consistent, whereas high Mg/Ca ratios indicate that the slightly lower values reported by Derry *et al.* (1994) from Siberia, Russia, developed at the time of dolomitization. The values of 0.708 91–0.708 98 on least-altered samples are not significantly different from those reported by Kouchinsky *et al.* (2008) for samples from higher in Series 3 (Drumian and Guzhangian stages).

Published data show an increasing trend of $^{87}Sr/^{86}Sr$ values from the middle to upper part of Series 3 through most of the Furongian Series. The trend is abruptly reversed with a decrease in $^{87}Sr/^{86}Sr$ values near the top of the series. Values near the base of the Guzhangian Stage are 0.708 93 ± 0.0002, and they increase until close to the top of the Paibian Stage (Furongian Series). Values on least-altered samples in the SPICE interval (Paibian Stage) reach 0.709 10 ± 0.0001 (Montañez *et al.*, 2000; Kouchinsky *et al.*, 2008). Isotopic data reported from western North America (Saltzman *et al.*, 1995) for the *Elvinia–Taenicephalus* biozone boundary (lower part of the Jiangshanian Stage) are internally consistent and imply that seawater $^{87}Sr/^{86}Sr$ rose to its highest ever value (0.709 25) in the Jiangshanian Stage, before falling sharply to 0.709 14 near the top of the Jiangshanian Stage, and to 0.709 10–0.709 11 in Stage 10. A study by Ebneth *et al.* (2001) on samples from conodont elements confirms that this decrease continued to 0.709 00 near the base of the Ordovician.

19.2.8. Sequence Stratigraphy

Mei *et al.* (2007) provided a summary of second- and third-order eustatic changes in South China during the Cambrian (Figure 19.13). They recognized two second-order sequences that correspond to firstly the Terreneuvian Series plus Series 2; and secondly Series 3 plus the Furongian Series. Within the Terreneuvian–Series 2 sequence, five third-order sequences were recognized, a large one terminating in the upper part of Stage 2, another large one extending to near the top of Stage 3, and three cycles of short duration through upper Stage 3 and Stage 4. Seven third-order cycles of short to moderately long duration occupy the Series 3–Furongian sequence.

More detailed sea-level histories have been provided by, among others, Babcock *et al.* (2005, 2007), Howley *et al.* (2006), Miller *et al.* (2003, 2006), Jago *et al.* (2006), Peng *et al.* (2009a,b), and Howley and Jiang (2010). Babcock *et al.* (2005, 2007) and Peng *et al.* (2009a,b) showed that the first appearances of some agnostoid guide fossils in the upper half of the Cambrian closely follow small-scale eustatic rises of sea level. Agnostoid biozones therefore begin in the lower parts of transgressive systems tracts.

19.2.9. Cambrian Evolutionary Events

The Cambrian records two important, and evidently linked, evolutionary events; the "Cambrian explosion" (Cloud, 1968) and the "Cambrian substrate revolution" (e.g., Bottjer *et al.*,

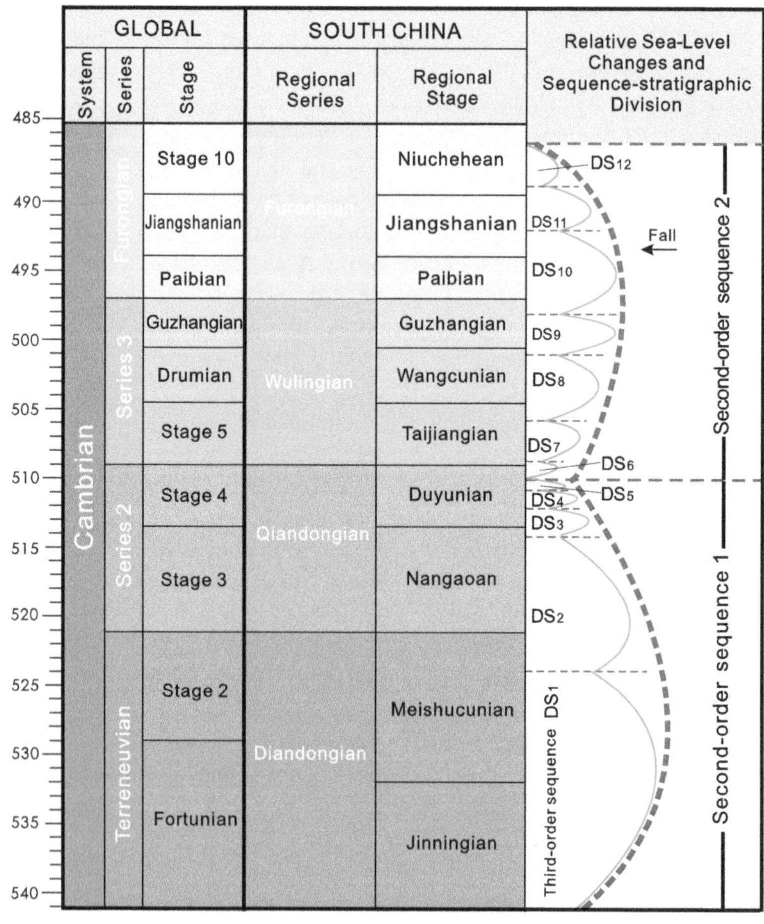

FIGURE 19.13 **Cambrian sequence stratigraphy, showing second- and third-order cycles recognized from Guizhou, South China** (redrawn after Mei *et al.*, 2007).

2000). These biotic transformations are inseparably linked as components of a larger-scale, sweeping reorganization of marine ecosystems referred to as the Early Paleozoic Marine Revolution (Babcock, 2003).

The Cambrian explosion refers to the first great evolutionary radiation during the Phanerozoic (e.g., Cloud, 1968; Runnegar, 1982; Bengtson, 1994; Briggs *et al.*, 1994; Chen *et al.*, 1996; Fortey *et al.*, 1996; Conway Morris, 1998; Hou *et al.*, 1999). The radiation, which largely consisted of metazoans, essentially began in the Terreneuvian Series with the introduction of the first biota of small shelly fossils (SSFs), although it is perhaps more appropriate to view the Cambrian explosion as a starting point for reorganization of biotic systems that had a longer Proterozoic history (e.g., Conway Morris, 1998; Babcock, 2005). All skeletonized metazoan phyla have a fossil record dating to the Furongian Epoch or earlier (Landing *et al.*, 2010), and most invertebrate phyla were established by the Drumian Age. Biomineralized skeletons had evolved in a limited number of animals, *Cloudina* (Hua *et al.*, 2005), *Namacalathus* (Hoffman and Mountjoy, 2001; Amthor *et al.*, 2003), and possibly the inferred hexactinellid sponge *Palaeophragmodictya* (Gehling and Rigby, 1996), during the Ediacaran Period but most clades that evolved biomineralized skeletons did so during the Cambrian Period (Terreneuvian Epoch or Epoch 2). Through the Terreneuvian, Series 2, and Series 3, there was a spectacular burst in diversity (number of species and genera) and in disparity (number of distinct body plans). Brasier (1979) gave an extensive review of the fossil record during the early half of the Cambrian.

Fossil groups involved in the Cambrian explosion include prokaryotes; eukaryotic protoctists, acritarchs, and chitinozoans; larger algae and vascular plants; Parazoa (Porifera, Chancelloriida, Radiocyatha, Archaeocyatha, Stromatoporoidea); Radiata; Bilateria (Priapulida, Sipunculida, Mollusca, Annelida, Arthropoda including Lobopoda and Tardigrada, Pogonophora, Brachiopoda, Ectoprocta, Phoronida, Mitrosagophora and Tommotiida, mobergellids, Echinodermata, Hemichordata, Chaetognatha, Conodontophorida, and Chordata). Brasier (1979) also commented on phyletic changes, skeletal changes, niche changes, size changes, and environment-related changes. In attempting to explain the "Cambrian explosion", Brasier (1982, 1995a) attempted to link Cambrian "bioevents" to sea-level fluctuation and oxygen depletion, to nutrient enrichment (Brasier, 1992a,b) and, finally, to eutrophy and oligotrophy (Brasier, 1995b,c).

The Cambrian explosion was not a single evolutionary burst. Increasingly detailed resolution of the stratigraphic record, made possible in large measure by application of carbon isotope chemostratigraphy, shows a series of radiations, often punctuated by extinctions (Zhu *et al.*, 2006; Figure 19.12). At least two major waves of radiation of small shelly fossils, a radiation of archaeocyaths, a radiation of non-trilobite arthropods, and several waves of radiation of trilobites, characterize the Cambrian fossil record.

A lengthy review (Zhuravlev and Riding, 2001) discussed the ecology of the Cambrian radiation in the context of life environments, community patterns and dynamics, and "ecologic radiation" of major fossil groups, with important chapters on paleomagnetically and tectonically based maps of global facies distribution, and supercontinental amalgamation as a trigger for the "explosion", climate change, and biotic diversity and structure. Babcock (2003) provided further discussion of the ecologic context of the early Phanerozoic body fossil and trace fossil record, attributing much of the increasing preservability of fossils to factors linked with escalation in predator–prey systems.

Insight into Cambrian diversity patterns is provided by "Konservat Fossil-lagerstätten"; deposits containing exquisitely well-preserved fossils, particularly of non-biomineralized ("soft") body parts. They are known globally from approximately 40 localities (Conway Morris, 1985; Babcock *et al.*, 2001) if the "Orsten"-type preservation style of Sweden, China, and elsewhere (e.g., Maas *et al.*, 2006) is included. The richest and most spectacular lagerstätten are in the Buen Formation of Greenland (e.g., Conway Morris *et al.*, 1987; Conway Morris and Peel, 1990; Budd, 1997; Budd and Peel, 1998; Babcock and Peel, 2007), the Yuanshan Member of the Chiungchussu Formation or Maotianshan Shale (Chengjiang Biota) of Yunnan, China (e.g., Chen *et al.*, 1996; Hou *et al.* 1999; Luo *et al.*, 1999; Babcock *et al.*, 2001), the Burgess Shale of British Columbia (e.g., Conway Morris, 1977, 1985; Whittington, 1977; Briggs *et al.*, 1994), and the Kaili Formation of Guizhou, China (e.g., Zhao *et al.*, 1999, 2002, 2005, 2011). Collectively, deposits of the Great Basin (e.g., Gunther and Gunther 1981; Robison, 1991; Briggs *et al.*, 2005, 2008; Robison and Babcock, 2011; Stein *et al.*, 2011) have also produced spectacular material, but in lower numbers than these other localities.

The Cambrian substrate revolution (e.g., Seilacher and Pfluger, 1994; Bottjer *et al.*, 2000) has been used to describe the changes in marine substrates through Cambrian time, changes that were evidently coupled to biotic radiation. Marine substrates of the Ediacaran Period were largely stabilized by microbial mat communities (Gehling, 1999). The relatively few known Ediacaran traces fossils were essentially surface traces. More penetrative traces appear in the Cambrian, especially in Series 3. In the upper Furongian, traces penetrating up to several centimeters are not unusual, and some sedimentary layers are well bioturbated (Droser and Bottjer, 1989). Thus, the Cambrian substrate revolution involves a transition from mat-dominated substrates to more fluidized, bioturbated substrates. Correlated with this change is a decline in helicoplacoid echinoderms and other organisms having a "mat sticking" life habit, or other life habit that was dependent on microbial mats (Bottjer *et al.*, 2000). Among the available bioturbators of the Cambrian were a variety of priapulid and other worms (Conway Morris, 1977). Babcock (2003) showed that polymerid trilobites increasingly burrowed into Cambrian substrates, sometimes in search of prey, which themselves were burrowers.

19.3. CAMBRIAN TIME SCALE

19.3.1. Age of the Ediacaran–Cambrian Boundary

The base of the Cambrian Period has received considerable attention from geochronologists, in part because of the major biotic changes that occurred during the late Ediacaran–Cambrian interval. For this reason, calibration of the lower boundary of the Cambrian, and thus of the Paleozoic Era and Phanerozoic Eon, is relatively well constrained. Further refinement in calibration of the boundary probably depends more on advances in biostratigraphy than in geochronology, particularly on the discovery of new stratigraphic sections and fossil occurrences that help in the definition and correlation of the boundary. Dated samples used for calibration of the Cambrian Period are listed in Appendix 2 of this volume. The age and two sigma error range given in the original cited references are adjusted here, following the guidelines and procedures outlined in Chapter 6. In the following account the adjusted ages are used.

A U-Pb date of 540.61 ± 0.88 (originally given as 543.3 ± 1) Ma on volcanic ashes in the upper Spitskopf Member of the Schwarzrand Subgroup in Namibia is assigned to the latest Ediacaran (Grotzinger *et al.*, 1995) and provides a maximum age constraint on the base of the Cambrian. The Spitskopf Member is overlain, with erosional contact, by the Nomtsas Formation, U-Pb dated at 538.18 ± 1.24 Ma, the basal beds of which contain the trace fossil *Trichophycus pedum*. Interestingly, some elements of the globally distributed Ediacaran biota are found stratigraphically just above the dated ash bed in the Spitskopf Member, indicating that, at least locally, Ediacaran-grade organisms range into the lowermost Cambrian (Grotzinger *et al.*, 1995). Similar faunal relationships have been found in South Australia (Jensen *et al.*, 1998). Additional reports of putative Ediacaran-grade organisms in Cambrian strata of other regions (e.g., Conway Morris, 1993; Jensen *et al.*, 1998; Hagadorn *et al.*, 2000; Shu *et al.*, 2006; Babcock and Ciampaglio, 2007) suggest that some but not all members of the biota became extinct prior to the end of the Ediacaran Period.

On the data from Namibia the adjusted age of the base of the Cambrian is close to 540 Ma (Brasier *et al.*, 1994; Grotzinger *et al.*, 1995), an age consistent with other zircon dates, stratigraphically less constrained, from Siberia (Bowring *et al.*, 1993), and from the upper Ediacaran (Grotzinger *et al.*, 1995; Tucker and McKerrow, 1995).

In the Ara group of Oman, chemostratigraphic and paleontologic data on subsurface samples are interpreted to indicate the simultaneous occurrence of an extinction of Neoproterozoic biomineralized, skeletonized fossils (*Namacalathus* and *Cloudina*) and a large-magnitude negative excursion in carbon isotopes (Amthor *et al.*, 2003; Bowring *et al.*, 2003, 2007; Figure 19.14), the BACE excursion (Zhu *et al.*, 2006), which is widely equated with the boundary (Grotzinger *et al.*, 1995; Bartley *et al.*, 1998; Corsetti and Hagadorn, 2001; Kimura and Watanabe, 2001; Zhu *et al.*, 2006). Following Bowring *et al.* (2007) the ash bed at the peak of the isotope excursion (sample BB-5, 1 m above the base of the A4C carbonate unit) gave an adjusted age of 541.00 ± 0.63 Ma (2-sigma). Nine meters below the top of the next lowest carbonate unit (A3C, sample MKZ-11B) an ash bed has an adjusted age of 542.37 ± 0.63 Ma. From 3 m above the base of the same unit (A3C), a further ash bed has yielded an adjusted age of 542.9 ± 0.63 Ma (sample Minha-1A). If we accept the peak of the negative carbon isotope excursion as coinciding with the Ediacaran/Cambrian boundary, then the Oman sequence suggests an age of at least 541 Ma for the base of the Cambrian.

The Namibian sequence, which has the best paleontological constraints, suggests an age close to 540 Ma for base of the Cambrian, whereas the Oman sequence, which probably has better radiometric control, suggests 541 Ma. The two ages overlap in their uncertainty brackets, indicating good control on the age of this important stratigraphic boundary. Bowring and Schmitz (2003) regarded the Oman data as providing the best constraints on the age of the boundary. Pending further information, we regard 541.0 ± 0.63 Ma to be the maximum age for the Ediacaran/Cambrian boundary and for the base of the Paleozoic and Phanerozoic. The difference between the age of the top of the Cambrian Period (i.e. of the beginning of the Ordovician) at 485.4 Ma, and the bottom, 541.0 Ma, gives 55.6 myr for the duration of the period.

19.3.2. Age of Internal Boundaries

Ages of internal Cambrian boundaries are generally poorly constrained, especially in its lower part, and caution should be exercised when using the numerical scale in Figures 19.3, 19.11, 19.12, 19.13 and 19.15. Within the Furongian, which is finely zoned by trilobite biostratigraphy, stages are here proportioned approximately according to the number of trilobite zones they contain. In Australia, Siberia, and South China, three to seven zones are recognized. This method is also used for stages of Series 3, where agnostoid trilobites provide reliable zonation and inter-regional correlation. In South China the series is divided into 11 biozones, with a single polymerid zone at the base and 10 agnostoid zones above. The method, however, assumes a more or less constant rate of evolutionary turnover and a uniformity in

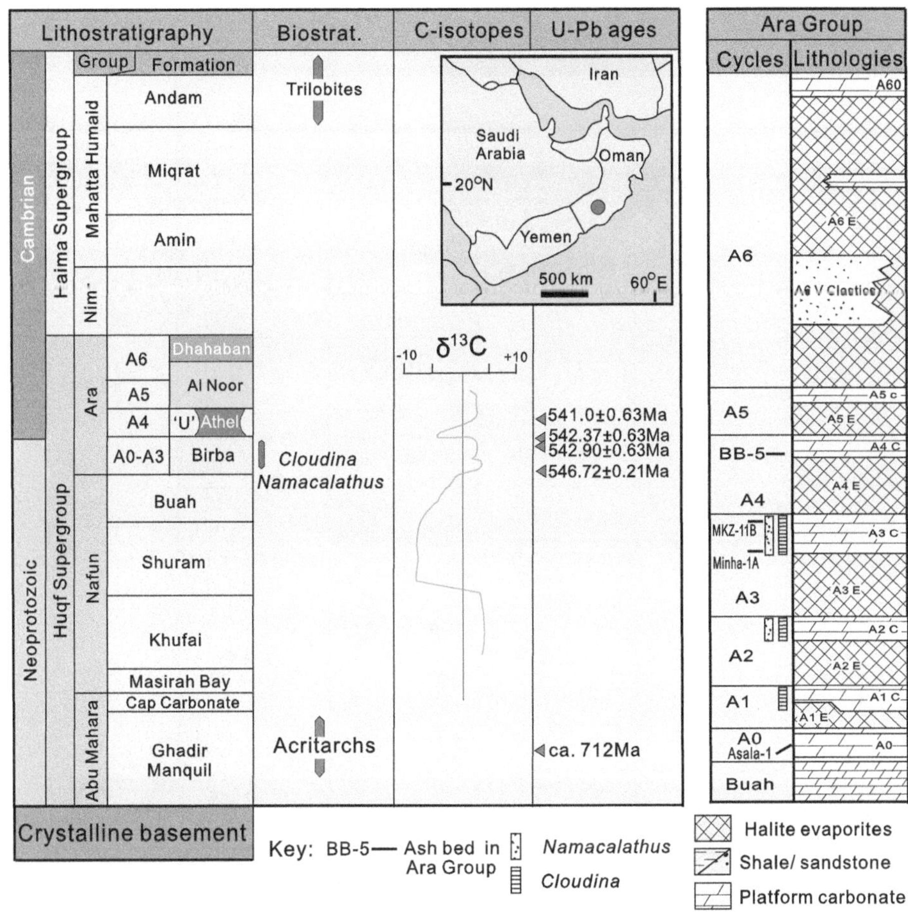

FIGURE 19.14 **Stratigraphy of the Huqf Supergroup, Oman.** Inset map shows location (red circle) of the subsurface basin in the Sultanate of Oman where U-Pb dates were obtained close to the Ediacaran–Cambrian boundary (redrawn after Amthor *et al.*, 2003, with permission of the authors). The numerical age dates were revised by Bowring *et al.* (2007). Chemostratigraphic and paleontologic data indicate the near-simultaneous occurrence of an extinction of Ediacaran fossils (*Namacalathus* and *Cloudinia*) and the BACE $\delta^{13}C$ excursion, a large-magnitude, short-lived negative excursion in carbon isotopes that is widely equated with the boundary.

paleontologic practice in zonal designation, which are not only unproven, but are unlikely to be true.

Fossil diversity and abundance diminish passing downwards through the lower part of Cambrian Series 3, Cambrian Series 2, and the Terreneuvian Series and, as a result, the biostratigraphic framework becomes increasingly vague. In the lower part of the Terreneuvian Series, resolution of the time scale is limited as much by lack of biostratigraphically useful fossils as by lack of radio-isotopic data. Our estimates of stage durations become correspondingly intuitive and the estimated ages of stage boundaries in the Terreneuvian Series shown in Figures 19.3, 19.11 and 19.15 should be regarded as highly approximate.

In the uppermost Furongian, a volcanic sandstone in North Wales gives a maximum age for the *Ctenopyge bisulcata* Subzone, the lowermost subzone of the *Peltura scarabaeoides* Zone, of 488.71 ± 2.78 Ma (Davidek *et al.*, 1998; Landing *et al.*, 2000). In Sweden, all Furongian polymerid subzones, as recognized traditionally, have been elevated to zones that correspond to a set of newly proposed agnostoid zones, and all traditional "superzones", including the *P. scarabaeoides* Zone, have been abandoned (Terfelt *et al.*, 2008, 2011). The elevated *Ctenopyge bisulcata* Zone overlies the agnostoid *Lotagnostus americanus* Zone, which corresponds to three elevated polymerid zones. The base of the *Lotagnostus americanus* Zone is defined by the FAD of the eponymous species and coincides with the base of Cambrian Stage 10. This level lies a single agnostoid zone, or three polymerid zones (elevated from subzones), below the *C. bisulcata* Zone. If we assume the average duration of an agnostoid zone is about 1 myr (Peng and Robison, 2000), or slightly less, the base of Stage 10 is likely to be about 489.5 Ma.

The Taylor Formation in Antarctica (Encarnación *et al.*, 1999) has yielded zircons with a recalculated weighted mean age of 502.1 ± 3.50 Ma on ashes interbedded with trilobite-bearing limestones assigned to the Undillan Stage. Based on the number of agnostoid zones, we can estimate the age of the base of the Furongian Series. This boundary lies six agnostoid zones above the base of the Undillan Stage of Australia, and six zones above the base of the Wangcunian Stage in South China. The base of the Furongian Series, and Paibian Stage, is therefore likely to be about 496 or 497 Ma; we take 497 Ma to be the best estimated age. The base of the Jiangshanian Stage is estimated to be about 494 Ma, as it lies three agnostoid zones above that boundary. The base of the traditional Upper

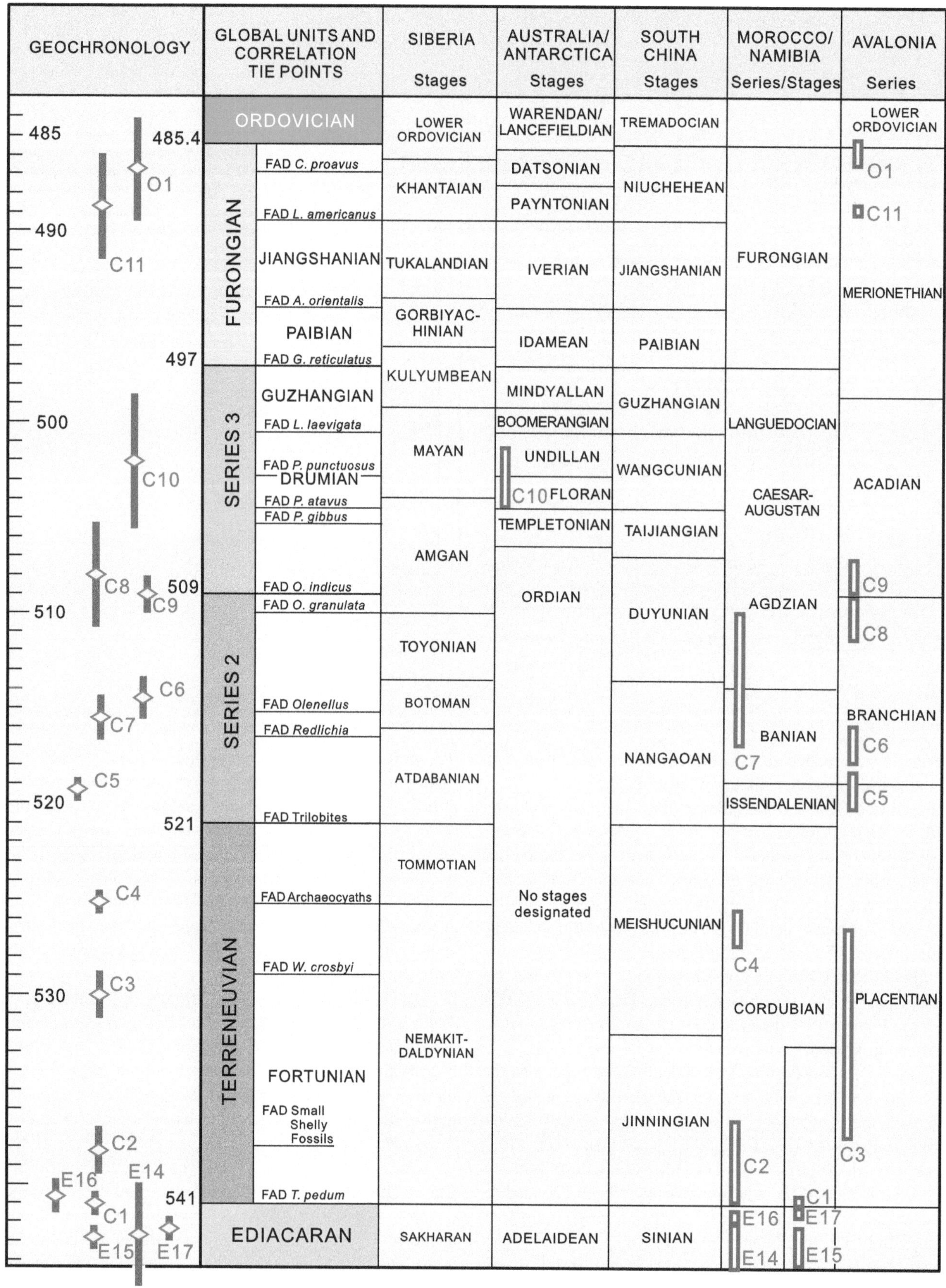

FIGURE 19.15 Distribution of the numerical age dates utilized for Cambrian geochronology based on $^{206}Pb/^{238}U$ and $^{207}Pb/^{206}Pb$ radiometric analyses (see text of Section 19.3 and Appendix 2 for stratigraphic and/or method details).

Cambrian as commonly applied in most regions is at a position below the base of the Furongian Series. It is equivalent to the base of the *Agnostus pisiformis* Zone, which is close to the base of the *Linguagnostus reconditus* Zone of South China (Peng and Robison, 2000; Ahlberg, 2003; Ahlberg *et al.*, 2004) or to the base of its equivalent *Acmarhachis quasivespa* Zone of Australia. The base of the traditional Upper Cambrian is about 499 Ma.

Ash beds associated with "upper Lower Cambrian" protolenid trilobites (*O. granulata*–"*P*". *howleyi* Zones) in southern New Brunswick, Canada (Landing *et al.*, 1998), have yielded zircons that give a recalculated weighted mean age of 508.05 ± 2.75 Ma. The beds were correlated with the Toyonian Stage of Siberia but are better correlated with the *Ovatoryctocara granulata* Zone in the basal part of the Amgan Stage. An ash bed in the Upper Comley Sandstone of Shropshire, United Kingdom, has given a weighted mean $^{206}Pb/^{238}U$ age of 509.02 ± 0.79 Ma on four (of six) single-grain fractions (Harvey *et al.*, 2011). Beds immediately overlying this position yield trilobites including *Paradoxides harlani*, which indicate that the *P. harlani* Zone of Newfoundland, Canada, is correlative with the *Oryctocephalus indicus* Zone of South China and Laurentia (Geyer, 2005; Fletcher, 2006) and the base of the traditional "Middle Cambrian" (St. David's Series) in Shropshire. The ash bed thus helps to constrain the age of the base of Series 3. Although it conflicts with the age of the stratigraphically lower New Brunswick ash bed (508.05 ± 2.75 Ma) reported by Landing *et al.* (1998), the conflict is easily accommodated within the error ranges for the two dates (Figure 19.15). Taken together, the two dates give a reasonably well-corroborated age for the base of Series 3 close to 509 Ma.

Also from Shropshire, an ash bed in the Lower Comley Sandstone has given a weighted mean $^{206}Pb/^{238}U$ age of 514.5 ± 0.81 (Harvey *et al.*, 2011). The ash bed lies within the *Callavia* trilobite Zone, which was correlated by the authors with Stage 3 (Series 2) of the Cambrian. It is based on only two (of seven) concordant single-grain zircon fractions. However, it is generally consistent with the Moroccan ash bed ages of Landing *et al.* (1998) and corroborates our scale. In Pembrokeshire in South Wales, an ash bed at Cwm Bach in the Caerfai Bay Shales has given an age of 519.38 ± 0.28 Ma, based on five (of six) concordant single-grain zircon fractions, consistent with an age of 519 ± 1 Ma for the same formation by Landing *et al.* (1998). The dated ash bed is correlated with a horizon within the Lower Comley Sandstone Formation on the presence of the bradoriid arthropod *Indiana lentiformis* (Siveter and Williams, 1995; Harvey *et al.*, 2011). Although biostratigraphic control is not good, the radio-isotopic age is well corroborated and is used here.

Ash beds from Morocco are taken as representing the "middle Botomian to Toyonian" (Landing *et al.*, 1998). Five single-grain zircon analyses give an adjusted weighted mean age of 515.56 ± 1.16 Ma. The age of the base of Series 2 is poorly constrained and is here taken to be about 521 Ma.

The upper Adoudou Formation ash bed in the Anti-Atlas Mountains of Morocco gives a good U-Pb weighted mean zircon age of 525.23 ± 0.61 Ma (Maloof *et al.*, 2005). On the basis of the correlation of a $+7‰$ $\delta^{13}C$ excursion at the dated level with a global excursion at the Tommotian/Nemakit-Daldynian boundary, an age of 525 Ma can be applied to this boundary. Ash beds in New Brunswick, Canada, with an age of 530.02 ± 1.20 Ma, from the "Placentian Series", the upper part of the trace fossil *Rusophycus avalonensis* Zone, are regarded as equivalent to the Nemakit-Daldynian Stage of Siberia (Isachsen *et al.*, 1994), providing a rather loose numerical calibration for the Nemakit-Daldynian.

To summarize, the duration of the Cambrian is regarded as 55.6 myr, ranging from 541.0 to 485.4 Ma (Figures 19.3, 19.11, 19.12, 19.13, 19.15). The base of the Furongian Series is near 497 Ma, the base of Cambrian Series 3 is near 509 Ma, and the base of Series 2 is approximately 521 Ma. The Furongian lasted for approximately 10 myr, Series 3 lasted for approximately 12 myr, Series 2 lasted for approximately 12 myr, and the Terreneuvian Series lasted for approximately 20 myr. Although only a crude approximation, we estimate uncertainty of stage boundary ages to be approximately 2.0 Ma, except for the near-direct or direct date at the base and top of the Cambrian and the Paibian (see Figure 1.7 of Chapter 1). More U-Pb (CA-TIMS) intra-Cambrian radiometric dates are urgently required.

REFERENCES

Ahlberg, P., 2003. Trilobites and international tie points in the Upper Cambrian of Scandinavia. Geologica Acta 1, 127–134.

Ahlberg, P., Ahlgren, J., 1996. Agnostids from the Upper Cambrian of Västergötland, Sweden. GFF 118, 129–140.

Ahlberg, P., Axheimer, N., Eriksson, M., Terfelt, F., 2004. Agnostoids and intercontinental tie points in the middle and upper Cambrian of Scandinavia. GFF 126, 108.

Ahlberg, P., Axheimer, N., Babcock, L.E., Eriksson, M.E., Schmitz, B., Terfelt, F., 2009. Cambrian high-resolution biostratigraphy and carbon isotope chemostratigraphy in Scania, Sweden: First record of the SPICE and DICE excursions in Scandinavia. Lethaia 42, 2–16.

Álvaro, J.J., Ahlberg, P., Babcock, L.E., Bordonaro, O.L., Choi, D.K., Cooper, R.A., Ergaliev, G.Kh., Gapp, I.W., Pour, M.G., Hughes, N.C., Jago, J.B., Korovnikov, I., Laurie, J.R., Lieberman, B.S., Paterson, J.R., Pegel, T.V., Popov, L.E., Rushton, A.W.A., Sukhov, S.S., Tortello, M.F., Zhou, Z.Y., and Żylińska, A., in press, Global Cambrian trilobite palaeobiogeography assessed using parsimony analysis of endemicity. *In* Harper, D.A.T., and Servais, T. (Eds), Early Palaeozoic Palaeobiogeography and Palaeogeography. London, Geological Society of London.

Álvaro, J.J., Vizcaïno, D., 2003. The conocoryphid biofacies, a benthic assemblage of normal-eyed and blind trilobites. Special Papers in Palaeontology 70, 127–140.

Amthor, J.E., Grotzinger, J.P., Schröder, S., Bowring, S.A., Ramezani, J., Martin, M.W., Matter, A., 2003. Extinction of *Cloudina* and *Namacalathus* at the Precambrian boundary in Oman. Geology 31, 431–434.

Apollonov, M.K., Bekteleuov, A.K., Kirschvink, J.L., Ripperdan, R.L., Tursunov, B.N., Tynbaev, K.L., 1992. Paleomagnetic scale of Upper Cambrian and Lower Ordovician in Batyrbai section (Lesser Karatau, southern Kazakhstan). Izvestiya Akademii Nauk Kazakhskoy SSR, Seriya Geologicheskaya 4 (326), 51–57 [In Russian].

Astashkin, V.A., Varlamov, A.I., Esakova, N.V., Zhuravlev, A.Yu., Repina, L.N., Rozanov, A.Yu., Fedorov, A.V., Shabanov, Yu.Ya., 1990. Guide on Aldan and Lena Rivers, Siberian Platform. Third International Symposium on Cambrian System. Institute of Geology and Geophysics, Siberian Branch, Academy of Sciences of USSR, Novosibirsk, p. 115. [In Russian].

Astashkin, V.A., Pegel, T., Shabanov, Yu.Ya., Sukhov, S.S., Sundukov, V.M., Repina, L.N., Rozanov, A.Yu., Zhuravlev, A.Yu., 1991. The Cambrian System of the Siberian Platform. IUGS Publication 27, p.133.

Axheimer, N., Eriksson, M.E., Ahlberg, P., Bengtsson, A., 2006. The middle Cambrian cosmopolitan key species *Lejopyge laevigata* and its biozone: New data from Sweden. Geological Magazine 143, 447–455.

Babcock, L.E., 1994a. Systematics and phylogenetics of polymeroid trilobites from the Henson Gletscher and Kap Stanton formations (Middle Cambrian), North Greenland. Grønlands Geologiske Undersøgelse Bulletin 169, pp. 79–12.

Babcock, L.E., 1994b. Biogeography and biofacies patterns of Middle Cambrian polymeroid trilobites from North Greenland: Palaeogeographic and palaeo-oceanographic implications. Grønlands Geologiske Undersøgelse Bulletin 169, 129–147.

Babcock, L.E., 2003. Trilobites in Paleozoic predator-prey systems, and their role in reorganization of early Paleozoic ecosystems. In: Kelley, P.A., Kowalewski, M., Hansen, T.A. (Eds.), Predator-Prey Interactions in the Fossil Record. Kluwer Academic/Plenum Publishers, New York, pp. 55–92.

Babcock, L.E., 2005. Interpretation of biological and environmental changes across the Neoproterozoic-Cambrian boundary: Developing a refined understanding of the radiation and preservational record of early multicellular organisms. Palaeogeography, Palaeoclimatology, Palaeoecology 220, 1–5.

Babcock, L.E., Ciampaglio, C.N., 2007. Frondose fossil from the Conasauga Formation (Cambrian: Drumian Stage) of Georgia, USA. Memoirs of the Association of Australasian Palaeontologists 34, 555–562.

Babcock, L.E., Peel, J.S., 2007. Palaeobiology, taphonomy and stratigraphic significance of the trilobite *Buenellus* from the Sirius Passet Biota, Cambrian of North Greenland. Memoirs of the Association of Australasian Palaeontologists 34, 401–418.

Babcock, L.E., Peng, S.C., 2007. Cambrian chronostratigraphy: Current state and future plans. Palaeogeography, Palaeoclimatology, Palaeoecology 254, 62–66.

Babcock, L.E., Zhang, W.T., Leslie, S.A., 2001. The Chengjiang Biota: Record of the Early Cambrian diversification of life and clues to exceptional preservation. GSA Today 11 (2), 4–9.

Babcock, L.E., Rees, M.N., Robison, R.A., Langenburg, E.S., Peng, S.C., 2004. Potential Global Standard Stratotype-section and Point (GSSP) for a Cambrian stage boundary defined by the first appearance of the trilobite *Ptychagnostus atavus*, Drum Mountains, Utah, USA. Geobios 37, 149–158.

Babcock, L.E., Peng, S.C., Geyer, G., Shergold, J.H., 2005. Changing perspectives on Cambrian chronostratigraphy and progress toward subdivision of the Cambrian System. Geoscience Journal 9, 101–106.

Babcock, L.E., Robison, R.A., Rees, M.N., Peng, S.C., Saltzman, M.R., 2007. The Global boundary Stratotype Section and Point (GSSP) of the Drumian Stage (Cambrian) in the Drum Mountains, Utah, USA. Episodes 30, 85–95.

Babcock, L.E., Peng, S.C., Ellwood, B.B., 2009. Progress toward completion of a Cambrian chronostratigraphic scale. In: Ergaliev, G.Kh., Nikitina, O.I., Zhemchuzhnikov, V.G., Popov, L.E., Basset, M.G. (Eds.), Stratigraphy, Fossils and Progress of International Stratigraphic Scale of Cambrian System. Materials of the 14th International Field Conference of Excursion of the Cambrian Stage Subdivision Working Group. Gylym, Almaty, pp. 7–8.

Babcock, L.E., Robison, R.A., Peng, S.C., 2011. Cambrian stage and series nomenclature of Laurentia and the developing global chronostratigraphic scale. Museum of Northern Arizona Bulletin 67, 12–26.

Bartley, J.K., Pope, M., Knoll, A.H., Semikhatov, M.A., Petrov, P.Y., 1998. A Vendian-Cambrian boundary succession from the northwestern margin of the Siberian Platform: Stratigraphy, paleontology, and correlation. Geological Magazine 135, 473–494.

Bassett, M., 1985. Towards a "common language" in stratigraphy. Episodes 8, 87–92.

Bengtson, S., 1976. The structure of some Middle Cambrian conodonts, and the early evolution of conodont structure and function. Lethaia 9, 185–206.

Bengtson, S., 1994. The advent of animal skeletons. In: Bengtson, S. (Ed.), Early Life on Earth. Columbia University Press, New York, pp. 412–425.

Bengtson, S., Conway Morris, S., Cooper, B.J., Jell, P.A., Runnegar, B.N., 1990. Early Cambrian Fossils from South Australia. Memoirs of the Association of Australasian Palaeontologists 9, 1–364.

Blaker, M.R., Peel, J.S., 1997. Lower Cambrian trilobites from North Greenland. Meddelelser om Grønland Geoscience 35, 1–145.

Bottjer, D.J., Hagadorn, J.W., Dornbos, S.Q., 2000. The Cambrian substrate revolution. GSA Today 10 (9), 1–7.

Bowring, S.A., Schmitz, M.D., 2003. High-precision U-Pb Zircon geochronology and the stratigraphic record. Reviews in Mineralogy and Geochemistry 53, 305–326.

Bowring, S.A., Grotzinger, J.P., Isachsen, C.E., Knoll, A.H., Pelechaty, S.M., Kolosov, P., 1993. Calibrating rates of Early Cambrian evolution. Science 261, 1293–1298.

Bowring, S.A., Grotzinger, J.P., Condon, D.J., Ramezani, J., Newall, M.J., Allen, P.A., 2007. Geochronologic constraints on the chronostratigraphic framework of the Neoproterozoic Huqf Supergroup, Sultanate of Oman. American Journal of Science 307, 1097–1145.

Brasier, M.D., 1979. The Cambrian radiation event. In: House, M.R. (Ed.), The Origin of Major Invertebrate Groups. Systematics Association Special Volume, 12, pp. 103–159.

Brasier, M.D., 1982. Sea-level changes, facies changes and the late Precambrian-early Cambrian evolutionary explosion. Precambrian Research 17, 105–123.

Brasier, M.D., 1984. Microfossils and small shelly fossils from the Lower Cambrian Hyolithes Limestone at Nuneaton, English Midlands. Geological Magazine 121, 229–253.

Brasier, M.D., 1992a. Paleoceanography and changes in the biological cycling of phosphorus across the Precambrian-Cambrian boundary. In: Lipps, J.H., Signor, P.W. (Eds.), Origin and Early Evolution of the Metazoa. Plenum Press, New York, pp. 483–523.

Brasier, M.D., 1992b. Nutrient-enriched waters and the early skeletal fossil record. Journal of the Geological Society 149, 621–629.

Brasier, M.D., 1993. Towards a carbon isotope stratigraphy of the Cambrian System: Potential of the Great Basin succession. In: Hailwood, E.A., Kidd, R.B. (Eds.), High Resolution Stratigraphy. Geological Society Special Publication 70, pp. 341–350.

Brasier, M.D., 1995a. Fossil indicators of nutrient levels. 1: Eutrophication and climate change. In: Bosence, D.W.J., Allison, P.A. (Eds.), Marine Palaeoenvironmental Analysis from Fossils. Geological Society Special Publication 83, pp. 113–132.

Brasier, M.D., 1995b. Fossil indicators of nutrient levels. 2: Evolution and extinction in relation to oligotrophy. In: Bosence, D.W.J., Allison, P.A. (Eds.), Marine Palaeoenvironmental Analysis from Fossils. Geological Society Special Publication 83, pp. 133–150.

Brasier, M.D., 1995c. The basal Cambrian transition and Cambrian bio-events (from terminal Proterozoic extinctions to Cambrian biomeres). In: Walliser, O.H. (Ed.), Global Events and Event Stratigraphy in the Phanerozoic. Springer, Berlin, pp. 113–118.

Brasier, M.D., Sukhov, S.S., 1998. The falling amplitude of carbon isotopic oscillations through the Lower to Middle Cambrian: Northern Siberia data. Canadian Journal of Earth Sciences 35, 353–373.

Brasier, M.D., Cowie, J., Taylor, M., 1994. Decision on the Precambrian-Cambrian boundary stratotype. Episodes 17, 3–8.

Brasier, M.D., Shields, G., Kuleshov, V.N., Zhegallo, E.A., 1996. Integrated chemo- and biostratigraphic calibration of early animal evolution of southwest Mongolia. Geological Magazine 133, 445–485.

Briggs, D.E.G., Erwin, D.H., Collier, F.J., 1994. The Fossils of the Burgess Shale. Smithsonian Institution Press, Washington, p. 238.

Briggs, D.E.G., Lieberman, B.S., Halgedahl, S.L., Jarrard, R.D., 2005. A new metazoan from the Middle Cambrian of Utah and the nature of the Vetulicolia. Palaeontology 48, 681–686.

Briggs, D.E.G., Lieberman, B.S., Hendricks, J.R., Halgedahl, S.L., Jarrard, R.D., 2008. Middle Cambrian arthropods from Utah. Journal of Paleontology 82, 238–254.

Brøgger, W.C., 1878. Om Paradoxidesskifrene ved Krekling. Nyt Magazin Naturvidenskap 24, 18–88.

Brøgger, W.C., 1882. Die silurischen Etagen 2 und 3 im Kristianiagebiet und auf Eker, ihre Gliederung, Fossilien, Schichtenstörungen und Kontact-metamorfosen. Kristiania Universitäts-Programm 32, 376.

Brøgger, W.C., 1886. Om alderen af Olenelluszone i Nordamerika. Geologiska Föreningens i Stockholm Förhandlingar 8, 182–213.

Budd, G.E., 1997. Stem group arthropods from the Lower Cambrian Sirius Passet fauna of North Greenland. In: Fortey, R.A., Thomas, R.H. (Eds.), Arthropod Relationships. Systematics Association Special Volume, 55, pp. 125–138.

Budd, G.E., Peel, J.S., 1998. A new xenusiid lobopod from the Early Cambrian Sirius Passet fauna of North Greenland. Palaeontology 41, 1201–1213.

Buggisch, W., Keller, M., Lehnert, O., 2003. Carbon isotope record of Late Cambrian to Early Ordovician carbonates of Argentine Pre-cordillera. Palaeogeography, Palaeoclimatology, Palaeoecology 195, 357–373.

Chang, W.T., 1989. World Cambrian biogeography. Developments in Geo-science, Chinese Academy of Sciences. Science Press, Beijing. pp. 209–220.

Chang, W.T., 1998. Cambrian biogeography of the Perigondwana Faunal Realm. Revista Española de Paleontología, Extra no. Homenaje al Prof. Gonzalo Vidal, pp. 35–49.

Chatterton, B.D.E., Ludvigsen, R., 1998. Upper Steptoean (Upper Cambrian) trilobites from the McKay Group of southeastern British Columbia, Canada. Paleontological Society Memoir 49, 43.

Chen, J.Y., Zhou, G.Q., Zhu, M.Y., Yeh, K.Y., 1996. The Chengjiang Biota, a Unique Window of the Cambrian Explosion. National Museum of Natural Sciences, Taichun, p. 222. [In Chinese].

Chernysheva, N.E., 1961. Stratigraphy of the Cambrian of the Aldan Anti-cline and the palaeontological basis for separation of the Amgan Stage. Trudy Vsesoyuznogo Nauchno-Issledobatelskogo Geologicheskogo Instituta (VSEGEI), new series 49, 1–347.

Choi, D.K., 2004. Stop 12, Machari Formation at the Gonggiri section. In: Choi, D.K., Chough, S.K., Fitches, W.R., Kwon, Y.K., Lee, S.B., Kang, I., Woo, J., Shinn, Y.J., Lee, H.S., Sohn, J.W. (Eds.), Korea, 2004, Cambrian in the Land of Morning Calm. Seoul National University, Seoul, pp. 51–56.

Chernysheva, N.E., 1967. Stratigraphical section of Mayan Stage of Middle Cambrian. Bulletin of Science and Technology information, ONTI VIEMS, Series Geology, Ore Deposits, Useful Mineral, and Regional Geology 7, 38–42 [In Russian].

Cloud, P.E., 1968. Pre-metazoan evolution and the origins of the Metazoa. In: Drake, E.T. (Ed.), Evolution and Environment. Yale University Press, New Haven, pp. 1–72.

Conway Morris, S., 1977. Fossil priapulid worms. Special Papers in Palae-ontology 20, 159.

Conway Morris, S., 1985. Cambrian Lagerstätten: Their distribution and significance. Philosophical Transactions of the Royal Society Series B 311, 49–65.

Conway Morris, S., 1993. Ediacaran fossils in Cambrian Burgess Shale-type faunas of North America. Palaeontology 36, 539–635.

Conway Morris, S., 1998. The Crucible of Creation. The Burgess Shale and the Rise of Animals. Oxford University Press, Oxford, p. 276.

Conway Morris, S., Peel, J.S., 1990. Articulated halkieriids from the Lower Cambrian of North Greenland. Nature 345, 802–805.

Conway Morris, S., Peel, J.S., Higgins, A.K., Soper, N.J., Davis, N.C., 1987. A Burgess Shale-like fauna from the Lower Cambrian of Greenland. Nature 326, 181–183.

Cooper, R.A., Nowlan, G.S., 1999. Proposed global stratotype section and point for base of the Ordovician System. Acta Universitatis Carolinae Geologica 43, 61–64.

Cooper, R.A., Nowlan, G.S., Williams, H.S., 2001. Global Stratotype Section and Point for base of the Ordovician System. Episodes 24, 19–28.

Corsetti, F.A., Hagadorn, J.W., 2001. The Precambrian-Cambrian transition in the southern Great Basin, USA. The Sedimentary Record 1 (1), 4–8.

Cowie, J.W., 1971. Lower Cambrian faunal provinces. In: Middlemiss, F.A., Rawson, P.F., Newall, G. (Eds.), Faunal Provinces in Space and Time. Geological Journal Special Issue, 4, pp. 31–46.

Cowie, J.W., 1978. I.U.G.S./I.G.C.P. Project 29, Precambrian-Cambrian Boundary Working Group in Cambridge, 1978. Geological Magazine 115, 151–152.

Cowie, J.W., 1985. Continuing work on the Precambrian-Cambrian boundary. Episodes 8 (2), 93–97.

Cowie, J.W., Rushton, A.W.A., Stubblefield, C.J., 1972. A correlation of Cambrian rocks in the British Isles. Geological Society Special Reports 2, 1–42.

Cowie, J.W., Ziegler, W., Remane, J., 1989. Stratigraphic Commission accelerates progress, 1984–1989. Episodes 12 (2), 79–83.

Daily, B., Jago, J.B., 1975. The trilobite *Leiopyge* Hawle and Corda and the Middle-Upper Cambrian boundary. Palaeontology 18, 527–550.

Davidek, K., Landing, E., Bowring, S.A., Westrop, S.R., Rushton, A.W.A., Fortey, R.A., Adrain, J.M., 1998. New uppermost Cambrian U-Pb date from Avalonian Wales and age of the Cambrian-Ordovician boundary. Geological Magazine 135, 305–309.

Debrenne, F., 1992. Diversification of Archaeocyatha. In: Lipps, J.H., P.W. Signor, P.W. (Eds.), Origin and Early Evolution of the Metazoa. Plenum Press, New York, pp. 425–443.

Debrenne, F., Rozanov, A.Yu., 1983. Paleogeographic and stratigraphic distribution of regular Archaeocyatha (Lower Cambrian fossils). Geobios 16 (6), 727–736.

Demidenko, Yu.E., 2001. Small shelly fossils. In: Alexander, E.M., Jago, J.B., Rozanov, A.Yu., Zhuravlev, A.Yu. (Eds.), The Cambrian Biostratigraphy of the Stansbury Basin, South Australia. Transactions of the Palaeontological Institute, Russian Academy of Sciences, 282, pp. 85–117.

Demidenko, Yu.E., Jago, J.B., Lin, T.R., Parkhaev, P., Rozanov, A.Yu., Ushatinskaya, G.T., Zang, W.L., Zhuravlev, A.Yu., 2001. Conclusions. In: Alexander, E.M., Jago, J.B., Rozanov, A.Yu., Zhuravlev, A.Yu. (Eds.), The Cambrian Biostratigraphy of the Stansbury Basin, South Australia. Transactions of the Palaeontological Institute, Russian Academy of Sciences, 282, pp. 66–73.

Derry, L.A., Brasier, M.D., Corfield, R.M., Rozanov, A.Yu., Zhuravlev, A.Yu., 1994. Sr and C isotopes in Lower Cambrian carbonates from the Siberian craton: A paleoenvironmental record during the "Cambrian explosion". Earth and Planetary Science Letters 128, 671–681.

Dong, X.P., Bergström, S.M., 2001a. Middle and Upper Cambrian protoconodonts and paraconodonts from Hunan, South China. Palaeontology 44, 949–985.

Dong, X.P., Bergström, S.M., 2001b. Stratigraphic significance of Middle and Upper Cambrian protoconodonts and paraconodonts from Hunan, South China. Palaeoworld 13, 307–309.

Dong, X.P., Repetski, J.E., Bergström, S.M., 2004. Conodont biostratigraphy of the Middle Cambrian through lowermost Ordovician in Hunan, South China. Acta Geologica Sinica 78, 1185–1260.

Droser, M.L., Bottjer, D.J., 1988. Trends in depth and extent of bioturbation in Cambrian carbonate marine environments, western United States. Geology 16, 233–236.

Druce, E.C., Jones, P.-J., 1971. Cambro-Ordovician conodonts from the Burke River Structural Belt, Queensland. Bureau of Mineral Resources of Australia Bulletin 110, 158.

Dubinina, S.V., 2009. Proposed new terminal stage of the Cambrian System in Kazakhstan. In: Ergaliev, G.Kh., Nikitina, O.I., Zhemchuzhnikov, V.G., Popov, L.E., Basset, M.G. (Eds.), Stratigraphy, Fossils and Progress of International Stratigraphic Scale of Cambrian System. Materials of the 14th International Field Conference of Excursion of the Cambrian Stage Subdivision Working Group. Gylym, Almaty, pp. 18–19.

Ebneth, S., Shields, G.A., Veizer, J., Miller, J.F., Shergold, J.H., 2001. High resolution strontium isotope stratigraphy across the Cambrian-Ordovician transition. Geochimica et Cosmochimica Acta 65, 2273–2292.

Eddy, J.D., McCollum, L.B., 1998. Early Middle Cambrian *Albertella* Biozone Trilobites of the Pioche Shale, Southeastern Nevada. Journal of Paleontology 72, 864–887.

Egorova, L.I., Shabanov, Yu.Ya., Rozanov, A.Yu., Savitsky, V.E., Chernysheva, N.E., Shishkin, B.B., 1976. Elansk and Kuonamsk facies stratotypes in the lower part of the Middle Cambrian of Siberia. Sibirskogo Nauchno-issledovatel'skogo Instituta Geologii, Geofiziki i Mineral'nogo Syr'ya, Seriya Paleontologiya i Stratigrafiya 211, 167 [In Russian].

Egorova, L.I., Shabanov, Yu.Ya., Pegel, T.V., Savitsky, V.E., Sukhov, S.S., Chernysheva, N.E., 1982. Type section of the Mayan Stage (Middle Cambrian of the southeastern Siberian Platform. Academy of Sciences of the USSR, Ministry of Geology of the USSR, Interdepartmental Stratigraphic Committee of the USSR, Transactions 8, 146 [In Russian].

Encarnación, J., Rowell, A.J., Grunow, A.M., 1999. A U-Pb age for the Cambrian Taylor Formation, Antarctica: Implications for the Cambrian time scale. Journal of Geology 107, 497–504.

Ergaliev, G.Kh, 1980. Middle and Upper Cambrian trilobites from Malyi Karatau. Akademiya Nauk Kazakhskoi SSR, Alma-Ata, p. 211. [In Russian].

Ergaliev, G.Kh, 1981. Upper Cambrian biostratigraphy of the Kyrshabakty section, Malyi Karatau, southern Kazakhstan. In: Taylor, M.E. (Ed.), Short Papers for the Second International Symposium on the Cambrian System. U.S. Geological Survey Open-file Report, 81–743, pp. 82–88.

Ergaliev, G.Kh, 1992. On palaeogeography of Karatau (southern Kazakhstan) in Middle Cambrian to Early Ordovician. Proceedings of Academy of Sciences, Republic of Kazakhstan, Series Geology 6, 51–56.

Ergaliev, G.Kh., Ergaliev, F.G., 2004. Stages and zones of the Middle and Upper portions of Cambrian of Malyi Karatau for the Project of International Stratigraphic Chart. In: Daukeev, S.Zh (Ed.), Geology of Kazakhstan. Kazakhstan Geological Society, Almaty, pp. 36–52 [In Russian].

Ergaliev, G.Kh., Ergaliev, F.G., 2008. Middle and Upper Cambrian Agnostida of the Aksai National Geological Reserve, South Kazakhstan (Kyrshabakty River, Malyi Karatau Range), part 1. Almaty. Gylym Press, p. 376.

Fletcher, T.P., 2003. *Ovatoryctocara granulata*: The key to a global Cambrian stage boundary and the correlation of the olenellid, redlichiid and paradoxidid realms. Special Papers in Palaeontology 70, 73–102.

Fletcher, T.P., 2006. Bedrock Geology of the Cape St. Mary's Peninsula, Southwest Avalon Peninsula, Newfoundland (includes parts of NTS map sheets 1M/1, 1N/4, 1L/6 and 1K/13). Government of Newfoundland and Laborador, Geological Survey, Department of Natural Resources, St. John's, Report 06-02, 117.

Fletcher, T.P., 2007. The base of Cambrian Series 3: The global significance of key oryctocephalid trilobite sanges in the Kaili Formation of South China. Memoir of the Association of Australasian Palaeontologists 33, 29–33.

Fortey, R.A., Briggs, D.E.G., Wills, M.A., 1996. The Cambrian evolutionary 'explosion': Decoupling cladogenesis from morphological disparity. Biological Journal of the Linnaean Society 57, 13–33.

Fritz, W.H., 1992. Walcott's Lower Cambrian olenellid trilobite collection 61K, Mount Robson area, Canadian Rocky Mountains. Geological Survey of Canada Bulletin 432, 1–65.

Gehling, J.G., 1999. Microbial mats in Proterozoic siliciclastics: Ediacaran death masks. Palaios 14, 40–57.

Gehling, J.G., Rigby, L.K., 1996. Long expected sponges from the Neoproterozoic Ediacara fauna of South Australia. Journal of Paleontology 70, 185–195.

Gehling, J.G., Jensen, S., Droser, M.L., Myrow, P.M., Narbonne, G.M., 2001. Burrowing below the basal Cambrian GSSP, Fortune Head, Newfoundland. Geological Magazine 138, 213–218.

Geyer, G., 1990. Correlation along the Lower/Middle Cambrian boundary– a puzzling story with an elusory end? In: Repina, L.N., Zhuravlev, A.Yu. (Eds.), Third International Symposium on the Cambrian System. Institute of Geology and Geophysics, Novosibirsk, pp. 100–102.

Geyer, G., 1995. The Fish River Subgroup in Namibia: Stratigraphy, depositional environments and the Proterozoic-Cambrian boundary problem revisited. Geological Magazine 142, 465–498.

Geyer, G., 1998. Intercontinental, trilobite-based correlation of the Moroccan early Middle Cambrian. Canadian Journal of Earth Sciences 35, 374–401.

Geyer, G., 2005. The base of a revised Middle Cambrian: Are suitable concepts for a series boundary in reach? Geosciences Journal 9 (2), 81–99.

Geyer, G., Landing, E., 2004. A unified Lower-Middle Cambrian chronostratigraphy for West Gondwana. Acta Geologica Polonica 54, 179–218.

Geyer, G., Shergold, J., 2000. The quest for internationally recognized divisions of Cambrian time. Episodes 23, 188–195.

Geyer, G., Peel, J.S., 2011. The Henson Gletscher Formation, North Greenland, and its bearing on the global Cambrian Series 2-Series 3 boundary. Bulletin of Geosciences 86, 465–534.

Geyer, G., Peng, S.C., Shergold, J., 2000. Correlation chart for major Cambrian areas. In: Geyer, G., Shergold, J. (Eds.), The Quest for Internationally Recognized Divisions of Cambrian Time. Episodes, 23, pp. 190–191.

Geyer, G., Peng, S.C., Shergold, J., 2003. Correlation chart for major Cambrian areas (revised version, 2002). In: Shergold, J., Geyer, G. (Eds.), The Subcommission on Cambrian Stratigraphy: The Status Quo. Geologica ActaI, 1, pp. 6–7.

Grotzinger, J.P., Bowering, S.A., Saylor, B.Z., Kaufman, A.J., 1995. Biostratigraphic and geochronologic constraints on early animal evolution. Science 270, 598–604.

Gunther, L.F., Gunther, V.G., 1981. Some Middle Cambrian fossils of Utah. Brigham Young University Geology Studies 28, 1–81.

Guo, Q.J., Strauss, H., Liu, C.Q., Zhao, Y.L., Pi, D.H., Fu, P.Q., Zhu, L.J., Yang, R.D., 2005. Carbon and oxygen isotopic composition of Lower to Middle Cambrian sediments at Taijiang, Guizhou Province, China. Geological Magazine 142, 723–733.

Hagadorn, J.W., Fedo, C.M., Waggoner, B.M., 2000. Early Cambrian Ediacaran-type fossils from California. Journal of Paleontology 74, 731–740.

Halgedahl, S.L., Jarrard, R.D., Brett, C.E., Allison, P.A., 2009. Geophysical and geological signatures of relative sea level change in the upper Wheeler Formation, Drum Mountains, west-central Utah: A perspective into exceptional preservation of fossils. Palaeogeography, Palaeoclimatology, Palaeoecology 277, 34–56.

Hamdi, B., Brasier, M.D., Jiang, Z., 1989. Earliest skeletal fossils from Precambrian-Cambrian boundary strata, Elburz Mountains, Iran. Geological Magazine 126, 283–289.

Harvey, T.H.P., Williams, M., Condon, D.J., Wilby, P.R., Siveter, D.J., Rushton, A.W.A., Leng, M.J., Gabbott, S.E., 2011. A refined chronology for the Cambrian succession of southern Britain. Journal of the Geological Society 168, 705–716.

Henderson, R.A., 1976. Upper Cambrian (Idamean) trilobites from western Queensland, Australia. Palaeontology 19, 325–364.

Henderson, R.A., 1977. Stratigraphy of the Georgina Limestone and a revised zonation for the early Upper Cambrian Idamean Stage. Journal of the Geological Society of Australia 23, 423–433.

Hoffman, H.J., Mountjoy, E.W., 2001. *Namacalathus-Cloudina* assemblage in Neoproterozoic Miette Group (Byng Formation), British Columbia: Canada's oldest shelly fossils. Geology 29, 1091–1094.

Hollingsworth, J.S., 2007. Fallotaspidoid trilobite assemblage (Lower Cambrian) from the Esmeralda Basin (western Nevada, U.S.A.): The oldest trilobites from Laurentia. Memoirs of the Association of Australasian Palaentologists 33, 123–140.

Hollingsworth, J.S., 2011. Lithostratigraphy and biostratigraphy of Cambrian Stage 3 in western Nevada and eastern California. Museum of Northern Arizona Bulletin 67, 26–42.

Hong, P., Lee, J.G., Choi, D.K., 2003. Late Cambrian trilobite *Irvingella* from the Machari Formation, Korea: Evolution and correlation. Special Papers in Palaeontology 70, 175–196.

Hou, X.G., Bergström, J., Wang, H.F., Feng, X.H., Chen, A.L., 1999. The Chengjiang Fauna: Exceptionally Well-Preserved Animals from 530 Million Years Ago. Yunnan Science and Technology Press, Kunming. p. 170. [In Chinese].

Howley, R.A., Jiang, G.Q., 2010. The Cambrian Drumian carbon isotope excursion (DICE) in the Great Basin, western United States. Palaeogeography, Palaeoclimatology, Palaeoecology 296, 138–150.

Howley, R.A., Rees, M.N., Jiang, G.Q., 2006. Significance of Middle Cambrian mixed carbonate-siliciclastic units for global correlation: Southern Nevada, USA. Palaeoworld 15, 360–366.

Hua, H., Chen, Z., Yuan, X.L., Zhang, L.Y., Xiao, S.H., 2005. Skeletogenesis and asexual reproduction in the earliest biomineralizing animal *Cloudina*. Geology 33, 277–280.

Isachsen, C.E., Bowring, S.A., Landing, E., Samson, S.D., 1994. New constraint on the division of Cambrian time. Geology 22, 496–498.

Jago, J.B., Zang, W., Sun, X., Brock, G.A., Paterson, J.R., Skovsted, C.B., 2006. A review of the Cambrian biostratigraphy of South Australia. Palaeoworld 15, 406–423.

Jell, P.A., 1974. Faunal provinces and possible planetary reconstruction of the Middle Cambrian. Journal of Geology 82, 319–350.

Jell, P.A., 2003. Phylogeny of Early Cambrian trilobites. Special Papers in Palaeontology 70, 45–57.

Jensen, S., Gehling, J.G., Droser, M.L., 1998. Ediacara-type fossils in Cambrian sediments. Nature 393, 567–569.

Jiang, G.Q., Kaufman, A.J., Christie-Blick, N., Zhang, S.H., Wu, H.C., 2007. Carbon isotope variability across the Ediacaran Yangtze platform in South China: Implications for a large surface-to-deep ocean $\delta^{13}C$ gradient. Earth and Planetary Science Letters 261, 303–320.

Jiang, N.R., Wang, Z.Z., Chen, Y.G., 1964. Cambrian stratigraphy of eastern Yunnan. Acta Geologica Sinica 44 (2), 137–155 [In Chinese].

Jones, P.J., Shergold, J.H., Druce, E.C., 1971. Late Cambrian and Early Ordovician Stages in western Queensland. Journal of the Geological Society of Australia 18, 1–32.

Kaufman, A.J., Knoll, A.H., Semikhatov, M.A., Grotzinger, J.P., Jacobsen, S.B., Adams, W., 1996. Integrated chronostratigraphy of Proterozoic-Cambrian boundary beds in the western Anabar region, northern Siberia. Geological Magazine 133, 509–533.

Keber, M., 1988. Mikrofossilien aus unterkambrischen Gesteinen der Montagne Noire, Frankreich. Palaeontographica A 202, 127–203.

Khomentovsky, V.V., 1974. Principles for the recognition of the Vendian as a Paleozoic system. In: Yanshin, A.L. (Ed.), Essays on Stratigraphy. Nauka, Moscow, pp. 33–70 [In Russian].

Khomentovsky, V.V., 1976. Vendian. Transactions of the Institute of Geology and Geophysics, Siberian Branch of Russian Academy of Sciences 234, 1–271 [In Russian].

Khomentovsky, V.V., 1984. Vendian. In: Yanshin, A.L. (Ed.), Phanerozoic of Siberia, vol. 1. Nauka, Novosibirsk, pp. 5–35 [In Russian].

Khomentovsky, V.V., Repina, L.N., 1965. The Lower Cambrian Stratotype Sections in Siberia. Nauka, Moscow, p. 200. [In Russian].

Khomentovsky, V.V., Karlova, G.A., 1993. Biostratigraphy of the Vendian-Cambrian beds and the Lower Cambrian boundary in Siberia. Geological Magazine 130, 29–45.

Khomentovsky, V.V., Gibsher, A.S., 1996. The Neoproterozoic-Lower Cambrian in northern Govi-Altai, western Mongolia: Regional setting, lithostratigraphy and biostratigraphy. Geological Magazine 133, 371–390.

Khomentovsky, V.V., Karlova, G.A., 2005. The Tommotian Stage base as the Cambrian lower boundary in Siberia. Stratigraphy and Geological Correlation 13, 21–34.

Khramov, A., Rodionov, V., 1980. The geomagnetic field during the Palaeozoic time. Journal of Geomagnetism and Geoelectricity 32 (Suppl. III), 99–115.

Kimura, H., Watanabe, Y., 2001. Oceanic anoxia at the Precambrian-Cambrian boundary. Geology 29, 995–998.

Kirschvink, J.L., 1978a. The Precambrian-Cambrian boundary problem: Magnetostratigraphy of the Amadeus Basin, central Australia. Geological Magazine 115, 139–150.

Kirschvink, J.L., 1978b. The Precambrian-Cambrian boundary problem: Palaeomagnetic directions from the Amadeus Basin, central Australia. Earth and Planetary Science Letters 40, 91–100.

Kirschvink, J.L., Raub, T.D., 2003. A methane fuse for the Cambrian explosion: Carbon cycles and true polar wander. Geoscience 335 (1), 65–78.

Kirschvink, J.L., Rozanov, A.Yu., 1984. Magnetostratigraphy of Lower Cambrian strata from the Siberian Platform: A palaeomagnetic pole and a preliminary polarity time-scale. Geological Magazine 121, 189–203.

Kirschvink, J.L., Margaritz, M., Ripperdan, R.L., Zhuravlev, A.Yu., Rozanov, A.Yu., 1991. The Precambrian-Cambrian boundary: Magnetostratigraphy and carbon isotopes resolve correlation problems between Siberia, Morocco and South China. GSA Today 1 (4), 61–91.

Kirschvink, J.L., Ripperdan, R.L., Evans, D., 1997. Evidence for a large scale reorganisation of Early Cambrian continental masses by inertial interchange true polar wander. Science 227, 541–545.

Klootwijk, C.T., 1980. Early Palaeozoic magnetism in Australia. Tectonophysics 64, 249–332.

Kobayashi, T., 1935. The Cambro-Ordovician formations and faunas of South Chosen, Palaeontology, Part 3, Cambrian faunas of South Chosen with a special study on the Cambrian trilobite genera and families. Journal of the Faculty of Science, University of Tokyo, Section II 4, 49–344.

Kobayashi, T., 1971. The Cambro-Ordovician faunal provinces and the interprovincial correlation discussed with special reference to the trilobites in eastern Asia. Journal of the Faculty of Science, University of Tokyo, Section II 18 (1), 129–299.

Kobayashi, T., 1972. Three faunal provinces in the Early Cambrian Period. Proceedings of Japan Academy 48, 242–247.

Korovnikov, I.V., 2005. Lower and Middle Cambrian boundary in open shelf facies of the Siberian Platform. Acta Micropalaeontologica Sinica 22 (Supplement), 80–84.

Kouchinsky, A., Bengtson, S., Pavlov, V., Runnergar, B., Val'kov, A., Young, E., 2005. Pre-Tommotian age of the lower Pestrotsvet Formation in the Selinde section on the Siberian Platform: Carbon isotope evidence. Geological Magazine 142, 319–325.

Kouchinsky, A., Bengtson, S., Gallet, Y., Korovnikov, I., Pavlov, V., Runnegar, B., Shields, G., Veizer, J., Young, E., Ziegler, K., 2008. The SPICE carbon isotope excursion in Siberia: A combined study of the upper Middle Cambrian-lowermost Ordovician Kulyumbe River section, northwestern Siberian Platform. Geological Magazine 145, 609–622.

Kruse, P., Shi, G.R., 2000. Archaeocyaths and radiocyaths. In: Brock, G., Engelbresten, M.J., Jago, J.B., Kruse, P.D., Laurie, J.R., Shergold, J.H., Shi, G.R., Sorauf, J.E. (Eds.), Palaeo-biogeographic Affinities of Australian Cambrian Faunas. Memoirs of the Association of Australasian Palaeontologists, 23, pp. 13–20.

Kruse, P., Jago, J.B., Laurie, J.R., 2009. Recent developments in Australian Cambrian biostratigraphy. Journal of Stratigraphy 33, 35–47.

Landing, E., 1991. A unified uppermost Precambrian through Lower Cambrian lithostratigraphy for Avalonian North America. Geological Society of America Abstracts with Programs 23, 56.

Landing, E., 1992. Lower Cambrian of southeastern Newfoundland: Epeirogeny and Lazarus faunas, lithofacies-biofacies linkages, and the myth of a global chronostratigraphy. In: Lipps, J.H., Signor, P.W. (Eds.), Origin and Early Evolution of Metazoa. Plenum Press, New York, pp. 283–309.

Landing, E., 1994. Precambrian-Cambrian boundary global stratotype ratified and a new perspective of Cambrian time. Geology 22, 179–182.

Landing, E., 1996. Avalon: Insular continent by the latest Precambrian. In: Nance, R.D., Thomlpson, M.D. (Eds.), Avalon and Related Peri-Gondwana Terranes of the Circum-North Atlantic. Geological Society of America, Special Paper, 304, pp. 29–63.

Landing, E., 2004. Precambrian-Cambrian boundary interval deposition and the marginal platform of the Avalon microcontinent. Journal of Geodynamics 37, 411–435.

Landing, E., Myrow, P., Benus, A.P., Narbonne, G.M., 1989. The Placentian Series: Appearance of the oldest skeletalized faunas in southeastern Newfoundland. Journal of Paleontology 63, 739–769.

Landing, E., Bowring, S.A., Davidek, K., Westrop, S.R., Geyer, G., Heldmaier, W., 1998. Duration of the Early Cambrian: U-Pb ages of volcanic ashes from Avalon and Gondwana. Canadian Journal of Earth Sciences 35, 329–338.

Landing, E., Bowring, S.A., Davidek, K.L., Rushton, A.W.A., Fortey, R.A., Wimbledon, W.A.P., 2000. Cambrian-Ordovician boundary age and duration of the lowest Ordovician Tremadoc Series based on U-Pb zircon dates from Avalonian Wales. Geological Magazine 137, 485–494.

Landing, E., Peng, S.C., Babcock, L.E., Geyer, G., Moczydlowska-Vidal, M., 2007. Global standard names for the Lowermost Cambrian Series and Stage. Episodes 30, 287–299.

Landing, E., English, A.M., Keppie, J.D., 2010. Cambrian origin of all skeletalized metazoan phyla–discovery of Earth's oldest bryozoans (Upper Cambrian, southern Mexico). Geology 38, 547–550.

Landing, E., Westrop, S.R., Miller, J.F., 2010. Globally practical base for the uppermost Cambrian (Stage 10) – FAD of the conodont *Eoconodontus notchpeakensis* and the Housian Stage. In: Fatka, P., Budil, P. (Eds.), Prague 2010, The 15th Field Conference of the Cambrian Stage Subdivision Working Group, International Subcommission on Cambrian Stratigraphy. Czech Geological Survey, Prague, p. 18.

Landing, E., Westrop, S.R., Adrain, J.M., 2011. The Lawsonian Stage – the *Eoconodontus notchpeakensis* (Miller, 1969) FAD and HERB carbon isotope excursion define a globally correlatable terminal Cambrian stage. Bulletin of Geosciences 86, 621–640.

Lane, P.D., Blaker, M.R., Zhang, W.T., 1988. Redescription of the Early Cambrian trilobite *Arthricocephalus chauveaui* Bergeron, 1899. Acta Palaeontologica Sinica 27, 553–560 [In Chinese and English].

Lapworth, C., 1879. On the tripartite classification of the Lower Paleozoic rocks. Geological Magazine 6, 1–15. new series.

Laurie, J.R., 2004. Early Middle Cambrian trilobite faunas from NTGS Elkedra 3 corehole, southern Georgina Basin. Northern Territory. Memoirs of the Association of Australasian Palaeontologists 30, 221–260.

Laurie, J.R., 2006. Early Middle Cambrian trilobites from Pacific Oil 8L Gas Baldwin 1 well, southern Georgina Basin, Northern Territory. Memoirs of the Association of Australasian Palaeontologists 32, 127–204.

Lazarenko, N.P., 1966. Biostratigraphy and some new trilobites of the Upper Cambrian Olenek Rise and Kharaulakh Mountains. Uchenie Zapisky Paleontologiya i Biostratigrafiya (NIIGA) 11, 33–78.

Lazarenko, N.P., Nikiforiv, N.I., 1972. Middle and Upper Cambrian of the northern Siberian Platform and adjacent folded areas. Stratigraphy, Paleogeography, and Mineral Deposits of the North Arctic. NIIGA Press, Leningrad, pp. 4–9.

Lazarenko, N.P., Gogin, I.Y., Pegel, T.V., Sukhov, S.S., Abaimova, G.P., 2011. The Khos-Nelege section of the Ogon'or Formation: A potential candidate for the GSSP of Stage 10, Cambrian System. Bulletin of Geosciences 86, 555–568.

Li, G.X., Zhang, J.M., Zhu, M.Y., 2001. Litho- and biostratigraphy of the Lower Cambrian Meishucunian Stage in the Xiaotan section, Eastern Yunnan. Acta Palaeontologica Sinica 40 (Suppl.), 40–53.

Li, G.X., Steiner, M., Zhu, M.Y., Zhu, X.J., Erdtmann, B.-D., 2007. Early Cambrian fossil record of metazoans in South China: Generic diversity and radiation patterns. Palaeogeography, Palaeoclimatology, Palaeoecology 254, 226–246.

Lochman-Balk, C., Wilson, J.L., 1958. Cambrian biostratigraphy in North America. Journal of Paleontology 32, 312–350.

Longacre, S.A., 1970. Trilobites of the Upper Cambrian Ptychaspid Biomere, Wilberns Formation, central Texas. Journal of Paleontology 44 (Suppl. 2), 1–70.

Lu, Y.H., 1981. Provincialism, dispersal, development, and phylogeny of trilobites. Geological Society of America, Special Paper 187, 143–150.

Lu, Y.H., Lin, H.L., 1989. The Cambrian trilobites of western Zhejiang. Palaeontologia Sinica 25, 1–287.

Lu, Y.H., Zhu, Z.L., Qian, Y.Y., Lin, H.L., Zhou, Z.Y., Yuan, K.X., 1974. The bio-environmental control hypothesis and its application to the Cambrian biostratigraphy and palaeozoo-geography. Memoir of Nanjing Institute of Geology and Palaeontology, Academia Sinica 5, 27–110.

Ludvigsen, R., Westrop, S.R., 1985. Three new Upper Cambrian stages for North America. Geology 13, 139–143.

Luo, H.L., Jiang, Z.W., Wu, X.C., Song, X.L., Ou, Y.L., Xin, Y.S., Liu, G.Z., Zhang, S.S., Tao, Y.H., 1984. Sinian-Cambrian Boundary Stratotype Section at Meishucun, Jinning, Yunnan, China. Yunnan People's Publishing House, Kunming, p. 154. [In Chinese].

Luo, H.L., Jiang, Z.W., Wu, X.C., Song, X.L., Ou, Y.L., 1990. The global biostratigraphic correlation of the Meishucunian Stage and the Precambrian-Cambrian boundary. Science in China, Series B 1990, 313–318 [In Chinese].

Luo, H.L., Wu, X.C., Ou, Y.L., 1991. Facies changes and transverse correlation of the Sinian-Cambrian boundary strata in eastern Yunnan. Sedimentary Facies and Palaeogeography 1991 (4), 27–35 [In Chinese].

Luo, H.L., Jiang, Z.W., Wu, X.C., Ou, Y.L., Song, X.L., Xue, X.F., 1992. A further research on the Precambrian-Cambrian boundary at Meishucun section of Jinning, Yunnan, China. Acta Geologica Sinica 5, 197–207.

Luo, H.L., Jiang, Z.W., Tang, L.D., 1994. Stratotype sections for Lower Cambrian stages in China. Yunnan Science and Technology Press, Kunming, p. 183. [In Chinese].

Luo, H.L., Hu, S.X., Chen, L.Z., Zhang, S.S., Tao, Y.H., 1999. Early Cambrian Chengjiang Fauna from Kunming Region, China. Yunnan Science and Technology Press, Kunming, p. 129. [In Chinese].

Maas, A., Braun, A., Dong, X.P., Donoghued, P.C.J., Müller, K.J., Olempska, E., Repetski, J.E., Siveter, D.J., Stein, M., Waloszek, D., 2006. The 'Orsten'— More than a Cambrian Konservat-lagerstätte yielding exceptional preservation. Palaeoworld 15, 266–282.

Maloof, A.C., Schrag, D.P., Crowley, J.L., Bowring, S.A., 2005. An expanded record of Early Cambrian carbon cycling from the Anti-Atlas Margin, Morocco. Canadian Journal of Earth Sciences 42, 2195–2216.

McCollum, L.B., Sundberg, F.A., 1999. Stop 9. Biostratigraphy of the traditional Lower-Middle Cambrian boundary interval in the outer shelf Emigrant Formation, Split Mountain East section, Esmeralda County, Nevada. In: Palmer, A.R. (Ed.), Laurentia 99, V Field Conference of the Cambrian Stage Subdivision Working Group. Institute for Cambrian Studies, Boulder, pp. 29–34.

McNamara, K.J., Yu, F., Zhou, Z.Y., 2003. Ontogeny and heterochrony in the oryctocephalid trilobite *Arthricocephalus* from the Early Cambrian of China. Special Papers in Palaeontology 70, 103–126.

Mei, M.X., Ma, Y.S., Zhang, H., Meng, X.Q., Chen, Y.H., 2007. Sequence-stratigraphic framework for the Cambrian of the Upper Yangtze region: Ponder on the sequence stratigraphic background of the Cambrian biological diversity events. Journal of Stratigraphy 31, 68–78 [In Chinese].

Miller, J.F., 1980. Taxonomic revisions of some Upper Cambrian and Lower Ordovician conodonts with comments on their evolution. University of Kansas Paleontological Contributions, Paper 99, 1–43.

Miller, J.F., 1988. Conodonts as biostratigraphic tools for redefinition and correlation of the Cambrian-Ordovician boundary. Geological Magazine 125, 349–362.

Miller, J.F., Evans, K.E., Loch, J.D., Ethington, R.L., Stitt, J.H., Holmer, L.E., Popov, L.E., 2003. Stratigraphy of the Sauk III interval (Cambrian-Ordovician) in the Ibex area, western Millard County, Utah. Brigham Young University Geology Studies 47, 23–118.

Miller, J.F., Ethington, R.L., Evans, K.R., Holmer, L.E., Loch, J.D., Popov, L.E., Repetski, J.E., Ripperdan, R.L., Taylor, J.F., 2006. Proposed stratotype for the base of the highest Cambrian stage at the first appearance datum of *Cordylodus andresi*, Lawson Cove section, Utah, USA. Palaeoworld 15, 384–405.

Miller, J.F., Evans, K.R., Freeman, R.L., Ripperdan, R.L., Taylor, J.F., 2011. Proposed stratotype for the base of the Lawsonian Stage (Cambrian Stage 10) at the first appearance datum of *Eoconodontus notchpeakensis* (Miller) in the House Range, Utah, USA. Bulletin of Geosciences 86, 595–620.

Montañez, I.P., Osleger, D.A., Banner, J.L., 2000. Evolution of the Sr and C isotope composition of Cambrian oceans. GSA Today 10 (5), 1–7.

Murchison, R.I., 1839. The Silurian System, Founded on Geological Researches in the Counties of Salop, Hereford, Radnor, Montgomery, Caermarthen, Brecon, Pembroke, Monmouth, Gloucester, Worcester, and Stafford: With Descriptions of the Coal-fields and Overlying Formations. John Murray, London, vol. 1: pp. 1–576; vol. 2: pp. 577–768.

Naimark, E., Shabanov, Y., Korovnikov, I., 2011. Cambrian trilobite *Ovatoryctocara* Tchernysheva, 1962 from Siberia. Bulletin of Geosciences 86, 405–422.

Narbonne, G.M., Myrow, P., Landing, E., Anderson, M.M., 1987. A candidate stratotype for the Precambrian-Cambrian boundary, Fortune Head,

Burin Peninsula, southeastern Newfoundland. Canadian Journal of Earth Sciences 24, 277–293.

Nicholas, C.J., 1996. The Sr isotopic evolution of the oceans during the "Cambrian explosion". Journal of the Geological Society 153, 243–254.

Nicoll, R.S., 1990. The genus *Cordylodus* and a latest Cambrian-earliest Ordovician conodont biostratigraphy. Bureau of Mineral Resources of Australia. Journal of Australian Geology and Geophysics 11, 529–558.

Nicoll, R.S., 1991. Differentiation of Late Cambrian-Early Ordovician species of *Cordylodus* (Conodonta) with biapical basal cavities. Bureau of Mineral Resources of Australia. Journal of Australian Geology and Geophysics 12, 223–244.

Nicoll, R.S., Shergold, J.H., 1992. Revised Late Cambrian (pre-Payntonian-Datsonian) conodont stratigraphy at Black Mountain, Georgina Basin, western Queensland, Australia. Bureau of Mineral Resources of Australia. Journal of Australian Geology and Geophysics 12, 93–118.

Öpik, A.A., 1963. Early Upper Cambrian fossils from Queensland. Bureau of Mineral Resources of Australia Bulletin 64, 1–133.

Öpik, A.A., 1966. The Early Upper Cambrian crisis and its correlation. Journal and Proceedings of the Royal Society of New South Wales 100, 9–14.

Öpik, A.A., 1967. The Mindyallan fauna of north-western Queensland. Bureau of Mineral Resources of Australia Bulletin 74 (1), p. 404 and (2), p. 167.

Öpik, A.A., 1968. The Ordian Stage of the Cambrian and its Australian Metadoxididae. Bureau of Mineral Resources of Australia Bulletin 92, 133–170.

Öpik, A.A., 1979. Middle Cambrian agnostids: Systematics and biostratigraphy. Bureau of Mineral Resources of Australia Bulletin 172 (1), p. 188 and (2), p. 67.

Orłowski, S., 1992. Cambrian stratigraphy and stage subdivision in the Holy Cross Mountains, Poland. Geological Magazine 129, 471–474.

Palmer, A.R., 1965a. Biomere – a new kind of biostratigraphic unit. Journal of Paleontology 39, 149–153.

Palmer, A.R., 1965b. Trilobites of the Late Cambrian Pterocephaliid Biomere in the Great Basin. United States. U.S. Geological Survey Professional Paper 493, 105.

Palmer, A.R., 1973. Cambrian trilobites. In: Hallam, A. (Ed.), Atlas of Palaeobiogeography. Elsevier, New York, pp. 3–18.

Palmer, A.R., 1979. Biomere boundaries re-examined. Alcheringa 3, 33–41.

Palmer, A.R., 1981. Subdivision of the Sauk Sequence. In: Taylor, M.E. (Ed.), Short Papers for the Second International Symposium on the Cambrian System. U.S. Geological Survey Open-file Report, 81–743, pp. 160–162.

Palmer, A.R., 1984. The biomere problem: Evolution of an idea. Journal of Paleontology 58, 599–611.

Palmer, A.R., 1998. A proposed nomenclature for stages and series for the Cambrian of Laurentia. Canadian Journal of Earth Sciences 35, 323–328.

Palmer, A.R., 1999. Introduction. In: Palmer, A.R. (Ed.), Laurentia 99, V Field Conference of the Cambrian Stage Subdivision Working Group. International Subcommission on Cambrian Stratigraphy, Institute for Cambrian Studies, Boulder (Colorado), pp. 1–4.

Palmer, A.R., Halley, R.B., 1979. Physical stratigraphy and trilobite biostratigraphy of the Carrara Formation (Lower and Middle Cambrian) in the southern Great Basin. U.S. Geological Survey, Professional Paper 1047, 131.

Palmer, A.R., Repina, L.N., 1993. Through a glass darkly: Taxonomy, phylogeny, and biostratigraphy of the Olenellina. University of Kansas Paleontological Contributions, new series 3, 1–35.

Pavlov, V., Gallet, Y., 1998. Upper Cambrian to Middle Ordovician magnetostratigraphy from the Kulumbe River section (northwestern Siberia). Physics of the Earth and Planetary Interiors 108, 49–59.

Pavlov, V., Gallet, Y., 2001. Middle Cambrian high magnetic reversal frequency (Kulumbe River section, northwestern Siberia) and reversal behaviour during the Early Palaeozoic. Earth and Planetary Science Letters 185, 173–183.

Pavlov, V., Gallet, Y., 2005. Third superchron during the early Paleozoic. Episodes 28, 78–84.

Pavlov, V., Bachtadse, V., Mikhailov, V., 2008. New Middle Cambrian and Middle Ordovician palaeomagnetic data from Siberia: Llandelian magnetostratigraphy and relative rotation between the Aldan and Anabar-Angara blocks. Earth and Planetary Science Letters 276, 229–242.

Pegel, T.V., 2000. Evolution of trilobite biofacies in Cambrian basins of the Siberian Platform. Journal of Paleontology 74, 1000–1019.

Peng, S.C., 1984. Cambrian-Ordovician boundary in the Cili-Taoyuan border area, northwestern Hunan with descriptions of relative trilobites. In: Nanjing Institute of Geology and Palaeontology, Academia Sinica (Ed.), Stratigraphy and Palaeontology of Systemic Boundaries in China, Cambrian-Ordovician Boundary (1). Anhui Science and Technology Publishing House, Hefei, pp. 285–405.

Peng, S.C., 1990. Tremadoc stratigraphy and trilobite faunas of northwestern Hunan. Beringeria 2, 1–171.

Peng, S.C., 1992. Upper Cambrian biostratigraphy and trilobite faunas of Cili-Taoyuan area, northwestern Hunan, China. Memoirs of the Association of Australasian Palaentologists 13, 1–119.

Peng, S.C., 2000a. Cambrian of slope facies (of China). In: Nanjing Institute of Geology and Palaeontology (Ed.), Stratigraphical Studies in China (1979–1999). Anhui Science and Technology Publishing House, Hefei, pp. 23–38 [In Chinese].

Peng, S.C., 2000b. A new chronostratigraphic subdivision of Cambrian for China. In: Aceñolaza, G.F., Peralta, S. (Eds.), Cambrian from Southern Edge. INSUGEO, 6, pp. 119–122.

Peng, S.C., 2003. Chronostratigraphic subdivision of the Cambrian of China. Geologica Acta 1 (1), 135–144.

Peng, S.C., 2008. Revision on Cambrian chronostratigraphy of South China and related remarks. Journal of Stratigraphy 32 (3), 239–245 [In Chinese].

Peng, S.C., 2009. The newly developed Cambrian biostratigraphic succession and chronostratigraphic scheme for South China. Chinese Science Bulletin 54, 2691–2698.

Peng, S.C., Babcock, L.E., 2001. Cambrian of the Hunan-Guizhou Region, South China. In: Peng, S.C., Babcock, L.E., Zhu, M.Y. (Eds.), Cambrian System of South China. Palaeoworld, 13, pp. 3–51.

Peng, S.C., Babcock, L.E., 2005a. Towards a new global subdivision of the Cambrian System. Journal of Stratigraphy 29, 171–177, p. 204. [In Chinese].

Peng, S.C., Babcock, L.E., 2005b. Two Cambrian agnostoid trilobites, *Agnostotes orientalis* (Kobayashi, 1935) and *Lotagnostus americanus* (Billings, 1860): Key species for defining global stages of the Cambrian System. Geoscience Journal 9, 107–115.

Peng, S.C., Babcock, L.E., 2008. Cambrian Period. In: Ogg, J.G., Ogg, G.J., Gradstein, F.M. (Eds.), The Concise Geologic Time Scale. Cambridge University Press, Cambridge, pp. 37–46.

Peng, S.C., Robison, R.A., 2000. Agnostoid biostratigraphy across the Middle-Upper Cambrian Boundary in China. Paleontological Society Memoir 53, 1–104.

Peng, S.C., Zhou, Z.Y., Lin, T.R., 1998. Late Middle-late Upper Cambrian chronostratigraphy of China. In: Alhberg, P., Eriksson, M., Olsson, I. (Eds.), IV Field Conference of the Cambrian Stage Subdivision Working Group, International Subcommission on Cambrian Stratigraphy, Abstracts. Lund University, Lund, p. 20.

Peng, S.C., Zhou, Z.Y., Lin, T.R., 1999. A proposal of the Cambrian chronostratigraphic scale in China. Geosciences 31 (2), 242 [In Chinese].

Peng, S.C., Zhou, Z.Y., Lin, T.R., Yuan, J.L., 2000a. Research on Cambrian chronostratigraphy: Present and tendency. Journal of Stratigraphy 24, 8–17 [In Chinese].

Peng, S.C., Yuan, J.L., Zhao, Y.L., 2000b. Taijiangian Stage: A new chronostratigraphic unit for the traditional Lower Middle Cambrian in South China. Journal of Stratigraphy 24, 56–57 [In Chinese].

Peng, S.C., Babcock, L.E., Robison, R.A., Lin, H.L., Ress, M.N., Saltzman, M.R., 2004a. Global Standard Stratotype-section and Point (GSSP) of the Furongian Series and Paibian Stage (Cambrian). Lethaia 37, 365–379.

Peng, S.C., Babcock, L.E., Lin, H.L., 2004b. Polymerid Trilobites from the Cambrian of Northwestern Hunan, China. Volume 1. Corynexochida, Lichida, and Asaphida. Science Press, Beijing, p. 333.

Peng, S.C., Babcock, L.E., Geyer, G., Maczydlowska, M., 2006. Nomenclature of Cambrian epochs and series based on GSSPs –comments on an alternative proposal by Rowland and Hicks. Episodes 29, 130–132.

Peng, S.C., Babcock, L.E., Zuo, J.X., Lin, H.L., Zhu, X.J., Yang, X.F., Robison, R.A., Qi, Y.P., Bagnoli, G., Chen, Y., 2009a. The Global boundary Stratotype Section and Point of the Guzhangian Stage (Cambrian) in the Wuling Mountains, northwestern Hunan, China. Episodes 32, 41–55.

Peng, S.C., Babcock, L.E., Zuo, J.X., Lin, H.L., Zhu, X.J., Yang, X.F., Qi, Y.P., Bagnoli, G., Wang, L.W., 2009b. Proposed GSSP for the base of Cambrian Stage 9, coinciding with the first appearance of *Agnostotes orientalis*, at Duibian, Zhejiang, China. Science in China, Series D, Earth Science 52 (4), 1–18.

Pillola, G.L., 1991. Trilobites du Cambrien inférieur du SW de la Sardaigne, Italie. Palaeontographica Italica 78, 1–174.

Pisarevsky, S., Gurevich, E., Khramov, A., 1997. Paleomagnetism of the Lower Cambrian sediments from the Olenek River section (northern Siberia): Paleopoles and the problem of the magnetic polarity in the Early Cambrian. Geophysics Journal International 130, 746–756.

Pokrovskaya, N.V., 1954. Stratigraphy of the Cambrian sediments in the south of the Siberian Platform. In: Shatsky, N.S. (Ed.), Issue of Geology in Asia, vol. 1. Academy of Sciences Press of USSR, Moscow, pp. 444–465.

Pratt, B.R., 1992. Trilobites of the Marjuman and Steptoean stages (Upper Cambrian), Rabbitkettle Formation, southern Mackenzie Mountains, northwest Canada. Palaeontographica Canadiana 9, 1–179.

Qian, Y., 1997. Hyolitha and some problematica from the Lower Cambrian Meishucun Stage in Central and SW China. Acata Palaeontologica Sinica 16, 255–278 [In Chinese].

Qian, Y., Chen, M.E., He, T.G., Zhu, M.Y., Yin, G.Z., Fen, W.M., Xu, J.T., Jiang, Z.W., Liu, D.Y., Li, G.X., Ding, L.F., Mao, Y.Q., Xiao, B., 1999. Taxonomy and Biostratigraphy of Small Shelly Fossils in China. Science Press, Beijing, p. 247. [In Chinese].

Repina, L.N., Khomentovsky, V.V., Zhuravleva, I.T., Rozanov, A.Yu., 1964. Lower Cambrian biostratigraphy of the Saya-Altay folded zone. Nauka, Moscow, p. 378. [In Russian].

Ripperdan, R.L., 2002. The HERB Event: End of Cambrian carbon cycle paradigm? Geological Society of America Abstracts with Programs 34 (6), 413.

Ripperdan, R.L., Kirschvink, J.L., 1992. Paleomagnetic results from the Cambrian-Ordovician boundary section at Black Mountain, Georgina Basin, western Queensland, Australia. In: Webby, B.D., Laurie, J.R. (Eds.), Global Perspectives on Ordovician Geology. A.A. Balkema, Rotterdam, pp. 381–394.

Ripperdan, R.L., Magaritz, M., Nicoll, R.S., Shergold, J.H., 1992. Simultaneous changes in carbon isotopes, sea level, and conodont biozones within the Cambrian-Ordovician boundary interval at Black Mountain, Australia. Geology 20, 1039–1042.

Ripperdan, R.L., Magaritz, M., Kirschvink, J.L., 1993. Magnetic polarity and carbon isotope evidence for non-depositional events within the Cambrian-Ordovician boundary section near Dayangcha, Jilin Province, China. Geological Magazine 130, 443–452.

Robison, R.A., 1976. Middle Cambrian trilobite biostratigraphy of the Great Basin. Brigham Young University Geology Studies 23, 93–109.

Robison, R.A., 1984. Cambrian Agnostida of North America and Greenland, part 1, Ptychagnostidae. University of Kansas Paleontological Contributions, Paper 109, 1–59.

Robison, R.A., 1991. Middle Cambrian biotic diversity: Examples from four Utah Lagerstätten. In: Simonetta, A.M., Conway Morris, S. (Eds.), The Early Evolution of Metazoa and the Significance of Problematic Taxa. Cambridge University Press, Cambridge, U.K, pp. 77–98.

Robison, R.A., 1994. Agnostoid trilobites from the Henson Gletscher and Kap Stanton formations (Middle Cambrian), North Greenland. Grønlands Geologiske Undersøgelse Bulletin 169, 25–77.

Robison, R.A., 1999. Base of *Ptychagnostus atavus* Zone, candidate stratotype for base of unnamed international series. In: Palmer, A.R. (Ed.), Laurentia 99, V Field Conference of the Cambrian Stage Subdivision Working Group, International Subcommission on Cambrian Stratigraphy. Institute for Cambrian Studies, Boulder (Colorado), pp. 15–17.

Robison, R.A., and Babcock, L.E., 2011. Systematics, paleobiology, and taphonomy of some exceptionally preserved trilobites from Cambrian Lagerstätten of Utah. Paleontological Contributions, 5, 1–47.

Robison, R.A., Rosova, A.V., Rowell, A.J., Fletcher, T.P., 1977. Cambrian boundaries and divisions. Lethaia 10, 257–262.

Ross Jr., R.J., Hintze, L.F., Ethington, R.L., Miller, J.F., Taylor, M.E., Repetski, J.E., Sprinkle, J., Guensburg, T.E., 1997. The Ibexian, lowermost series in the North American Ordovician. In: Taylor, M.E. (Ed.), Early Paleozoic Biochronology of the Great Basin, Western United States. U.S. Geological Survey Professional Paper, 1579, p. 50.

Rowell, A.J., Robison, R.A., Strickland, D.K., 1982. Aspects of Cambrian agnostoid phylogeny and chronocorrelation. Journal of Paleontology 56, 161–182.

Rozanov, A.Yu., 1966. The problem of the lower boundary of the Cambrian. In: Keller, B.M. (Ed.), Results of Science, General Geology, Stratigraphy. VINTTI Press, Moscow, pp. 92–111 [In Russian].

Rozanov, A.Yu., 1967. The Cambrian lower boundary problem. Geological Magazine 104, 415–434.

Rozanov, A.Yu., 1973. Regularities of the morphological evolution of archaeocyaths and the issues of stage subdivision of Lower Cambrian. Transactions of the Geological Institute, Academy of Sciences SSSR 241, 164 [In Russian].

Rozanov, A.Yu., Missarzhevsky, V.V., 1966. Biostratigraphy and fauna of the Cambrian lower horizons. Transactions of Geological Institute, Academy of Sciences of USSR 148, 1–127 [In Russian].

Rozanov, A.Yu., Sokolov, B.S. (Eds.), 1984. Lower Cambrian stage subdivision, Stratigraphy. Nauka, Moscow, p. 184 [in Russian].

Rozanov, A.Yu., Missarzhevsky, V.V., Volkova, N.A., Voronova, L.G., Krylov, L.N., Keller, B.M., Korolyuk, I.K., Lendzion, K., Michniak, R., Pykhova, I.G., Sidorov, A.D., 1969. The Tommotian Stage and the problem of the lower boundary of the Cambrian. Transactions of Geological Institute, Academy of Sciences of USSR 206, 380 [In Russian; English translation: Raaben, M.E. (ed), 1981. New Delhi, Amerind Publishing.].

Rozanov, A.Yu., Khomentovsky, V.V., Shabanov, Yu.Ya., Karlova, G.A., Varlamov, A.I., Parkhaev, P., Korovnikov, I.V., Skorlotova, N.A., 2008. To the problem of stage subdivision of the Lower Cambrian. Stratigraphy and Geological Correlation 16 (1), 3–21.

Rozova, A.V., 1963. Biostratigraphic chart of the zonation of the Upper Cambrian and the terminal Middle Cambrian of the northwestern Siberian Platform and New Upper Cambrian trilobites from the Kulyumbe River. Geology and Geophysics 1963 (9), 3–19 [In Russian].

Rozova, A.V., 1964. Biostratigraphy and Description of Trilobites from the Middle and Upper Cambrian of the Northwestern Siberian Platform. Nauka, Moscow, p. 148 [In Russian].

Rozova, A.V., 1968. Cambrian and Lower Ordovician of the Northwestern Siberian Platform. Nauka, Moscow, p. 196 [in Russian].

Rozova, A.V., 1970. On the biostratigraphic charts of the Upper Cambrian and Lower Ordovician of the northwestern Siberian Platform. Geology and Geophysics 1970 (5), 26–31 [in Russian].

Runnegar, B., 1982. The Cambrian explosion: Animals or fossils? Journal of the Geological Society of Australia 29, 395–411.

Rushton, A.W.A., 2009. Revision of the Furongian agnostoid *Lotagnostus trisectus* (Salter). Memoirs of the Association of Australasian Palaeontologists 37, 273–279.

Saltzman, M.R., Davidson, J.P., Holden, P., Runnegar, B., Lohmann, K.C., 1995. Sea-level-driven changes in ocean chemistry at an upper Cambrian extinction horizon. Geology 23, 893–896.

Saltzman, M.R., Runnegar, B., Lohmann, K.C., 1998. Carbon isotope stratigraphy of Upper Cambrian (Steptoean Stage) sequences of the eastern Great Basin: Record of a global oceanographic event. Geological Society of America Bulletin 110, 285–297.

Saltzman, M.R., Ripperdan, R.L., Brasier, M.D., Lohmann, K.C., Robison, R.A., Chang, W.T., Peng, S.C., Ergaliev, G.Kh., Runnegar, B., 2000. Global carbon isotope excursion (SPICE) during the Late Cambrian: Relation to trilobite extinctions, organic-matter burial and sea level. Palaeogeography, Palaeoclimatology, Palaeoecology 162, 211–223.

Salvador, A. (Ed.), 1994. International Stratigraphic Guide, second ed. International Union of Geological Sciences and the Geological Society of America, Trondheim, p. 214.

Samson, S., Palmer, A.R., Robison, R.A., Secor Jr., D.T., 1990. Biogeographical significance of Cambrian trilobites from the Carolina slate belt. Geological Society of America Bulletin 102, 1459–1470.

Savitsky, V.E., 1962. Relationships between Upper Precambrian and Cambrian in the Anabar Shield. Proceedings of the Conference on Upper Precambrian Stratigraphy in Siberia and Far East. Institute of Geology and Geophysics, Novosibirsk, pp. 53–44.

Sawaki, Y., Ohno, T., Fukushi, Y., Komiya, T., Ishikawa, T., Hirata, T., Maruyama, S., 2008. Sr isotope excursion across the Precambrian-Cambrian boundary in the Three Gorges area, South China. Gondwana Research 14, 134–147.

Sdzuy, K., 1972. Das Kambrium der Acadobaltischen Faunenprovinz – Gegenwärtiger Kenntnisstand und Probleme. Zentralblatt für Geologie und Paläontologie (II) 1972, 1–91.

Secor Jr., D.T., Samson, S.L., Snoke, A.W., Palmer, A.R., 1983. Confirmation of the Carolina slate belt as an exotic terrane. Science 221, 649–650.

Sedgwick, A., 1852. On the classification and nomenclature of the Lower Palaeozoic rocks of England and Wales. Quarterly Journal of the Geological Society 8, 136–168.

Sedgwick, A., Murchison, R.I., 1835. On the Silurian and Cambrian Systems, exhibiting the order in which the older sedimentary strata succeed each other in England and Wales. The London and Edinburgh Philosophical Magazine and Journal of Science 7, 483–535.

Seilacher, A., Pfluger, E., 1994. From biomats to benthic agriculture: A biohistoric revolution. In: Krumbein, W.S., Paterson, D.M., Stal, L.J. (Eds.), Biostabilization of Sediments. Bibliotheks und Informationssystem der Universität Oldenberg, Oldenberg, pp. 97–105.

Shergold, J.H., 1969. Oryctocephalidae (Trilobita: Middle Cambrian) of Australia. Bureau of Mineral Resources of Australia Bulletin 104, 1–66.

Shergold, J.H., 1972. Late Upper Cambrian trilobites from the Gola Beds, western Queensland. Bureau of Mineral Resources of Australia Bulletin 112, 1–126.

Shergold, J.H., 1975. Late Cambrian and Early Ordovician trilobites from the Burke River structural belt, western Queensland, Australia. Bureau of Mineral Resources of Australia Bulletin 153 (1), p. 251.

Shergold, J.H., 1980. Late Cambrian trilobites from the Chatsworth Limestone, western Queensland. Bureau of Mineral Resources of Australia Bulletin 186, 1–111.

Shergold, J.H., 1982. Idamean (Late Cambrian) trilobites, Burke River structural belt, western Queensland. Bureau of Mineral Resources of Australia Bulletin 187, 1–69.

Shergold, J.H., 1993. The Iverian, a proposed Late Cambrian stage, and its subdivision in the Burke River structural belt, western Queensland. Bureau of Mineral Resources of Australia. Journal of Australian Geology and Geophysics 13, 345–358.

Shergold, J.H., 1995. Timescales 1. Cambrian. Australian Phanerozoic timescales, biostratigraphic charts and explanatory notes. Second series. Australian Geological Survey Organisation (Record 1995) 30, 1–32.

Shergold, J.H., 1996. Cambrian (Chart 1). In: Young, G.C., Laurie, J.R. (Eds.), An Australian Phanerozoic Timescale. Oxford University Press, Melbourne, pp. 63–76.

Shergold, J.H., 1997. Explanatory notes for the Cambrian correlation chart. In: Whittington, H.B., Chatterton, B.D.E., Speyer, S.E., Fortey, R.A., Owens, R.M., Chang, W.T., Dean, W.T., Jell, P.A., Laurie, J.R., Palmer, A.R., Repina, L.N., Rushton, A.W.A., Shergold, J.H., Clarkson, E.N.K., Wilmot, N.V., Kelly, S.R.A. (Eds.), Treatise on Invertebrate Paleontology, Part O Arthropoda 1, Trilobita (Revised), (Volume 1) Introduction, Order Agnostida, Order Redlichiida. Geological Society of America and University of Kansas, Boulder and Lawrence, pp. 303–311.

Shergold, J.H., Nicoll, R.S., 1992. Revised Cambrian-Ordovician boundary biostratigraphy, Black Mountain, western Queensland. In: Webby, B.D., Laurie, J.R. (Eds.), Global Perspectives on Ordovician Geology. Balkema, Rotterdam, pp. 81–92.

Shergold, J.H., Rozanov, A.Yu., Palmer, A.R., 1991. The Cambrian System on the Siberian Platform. International Union of Geological Sciences Publications, Trondheim, p. 133.

Shergold, J.H., Geyer, G., 2003. The Subcommission on Cambrian Stratigraphy: The status quo. Geologica Acta 1, 5–9.

Shergold, J.H., Cooper, R.A., 2004. The Cambrian Period. In: Gradstein, F.M., Ogg, J.G., Smith, A.G. (Eds.), A Geologic Time Scale 2004. Cambridge University Press, Cambridge, pp. 147–164.

Shu, D.G., Conway Morris, S., Han, J., Li, Y., Zhang, X.L., Hua, H., Zhang, Z.F., Liu, J.N., Guo, J.F., Yao, Y., Yasui, K., 2006. Lower Cambrian vendobionts from China and early diploblast evolution. Science 312, 731–734.

Siveter, D.J., Williams, M., 1995. An Early Cambrian assignment for the Caerfai Group of South Wales. Journal of the Geological Society 152, 221–224.

Southgate, P.N., Shergold, J.H., 1991. Application of sequence stratigraphic concepts to Middle Cambrian phosphogenesis, Georgina Basin, Australia. Bureau of Mineral Resources of Australia. Journal of Australian Geology and Geophysics 12, 119–144.

Spizharsky, T.N., Ergaliev, G.Kh., Zhuravleva, I.T., Repina, L.N., Rozanov, A.Yu., 1983. Stage scale of the Cambrian System. Soviet Geology 1983 (8), 57–72.

Stein, M., Church, S.B., Robison, R.A., 2011. A new Cambrian arthropod, *Emeraldella brutoni*, from Utah. Paleontological Contributions 3, 1–9.

Steiner, M., Li, G.X., Qian, Yi., Zhu, M.Y., Erdtmann, B.-D., 2007. Neoproterozoic to early Cambrian small shelly fossil assemblages and a revised biostratigraphic correlation of the Yangtze Platform (China). Palaeogeography, Palaeoclimatology, Palaeoecology 254, 67–99.

Stitt, J.H., 1975. Adaptive radiation, trilobite paleoecology and extinction, Ptychaspidid Biomere, Late Cambrian of Oklahoma. Fossils and Strata 4, 381–390.

Stubblefield, G.J., 1956. Cambrian palaeogeography in Britian. In: Rodgers, J. (Ed.), El Sistema Cámbrico, su Paleogeografiá y el Problema de su Base. 20th International Geological Congress, Mexico City, 1, pp. 1–43.

Sundberg, F.A., 1990. Morphological Diversification of Ptychopariid Trilobites in the Marjumiid Biomere (Middle to Upper Cambrian). Ph.D. dissertation. Virginia Polytechnic Institute and State University, Blacksburg, p. 674.

Sundberg, F.A., 1994. Corynexochida and Ptychopariida (Trilobita, Arthropoda) of the *Ehmaniella* Biozone (Middle Cambrian), Utah and Nevada. Natural History Museum of Los Angeles County, Contributions in Science 446, 1–137.

Sundberg, F.A., 2005. The Topazan Stage, a new Laurentian stage (Lincolnian Series – "Middle" Cambrian). Journal of Paleontology 79, 63–71.

Sundberg, F.A., McCollum, L.B., 1997. Oryctocephalids (Corynexochida: Trilobita) of the Lower-Middle Cambrian boundary interval from California and Nevada. Journal of Paleontology 71, 1065–1090.

Suvorova, N.P., 1954. On the Lenian Stage of the Lower Cambrian in Yakutia. In: N.S. Shatsky, N.S. (Ed.), Problem of the Geology of Asia, vol. 1. Academy of Sciences Press of USSR, Moscow, pp. 466–483 [In Russian].

Taylor, J.F., 1997. Upper Cambrian biomeres and stages, two distinctly different and equally vital stratigraphic units. 2nd International Trilobite Conference, August 1997, St. Catharines, Ontario, Abstracts with Program, 47.

Taylor, J.F., 2006. History and status of the biomere concept. Memoirs of the Association of Australasian Palaeontologists 32, 247–265.

Terfelt, F., Eriksson, M.E., Ahlberg, P., Babcock, L.E., 2008. Furongian (Cambrian) biostratigraphy of Scandinavia – a revision. Norwegian Journal of Geology 88, 73–87.

Terfelt, F., Ahlberg, P., Eriksson, M.E., 2011. Complete record of Furongian polymerid trilobites and agnostoids of Scandinavia – a biostratigraphical scheme. Lethaia 44, 8–14.

Trench, A., 1996. Magnetostratigraphy: Cambrian to Silurian. In: Young, G.C., Laurie, J.R. (Eds.), An Australian Phanerozoic Timescale. Oxford University Press, Melbourne, pp. 23–29.

Tucker, R.D., McKerrow, W.S., 1995. Early Paleozoic chronology: A review in light of new U-Pb zircon ages from Newfoundland and Britain. Canadian Journal of Earth Sciences 32, 368–379.

Valledares, M.I., Ugidos, J.M., Barba, P., Fallick, A.E., Ellam, R.M., 2006. Oxygen, carbon and strontium isotope records of Ediacaran carbonates in central Iberia (Spain). Precambrian Research 147, 354–365.

Varlamov, A.I., Rozova, A.V., 2009. New Upper Cambrian (Evenyiskie) Regional Stage of Siberia. In: Efimov, A.C. (Ed.), New Data on Lower Palaeozoic Stratigraphy and Palaeontology of Siberia. SNIIGGiMS, Novosibirsk, pp. 3–61.

Varlamov, A.I., Pak, K.L., Rosova, A.V., 2005. The Upper Cambrian of the Chopko River Section, Norilsk Region, Northwestern Siberian Platform: Stratigraphy and Trilobites. Nauka, Novosibirsk, p. 84 [In Russian; English edition: Paleontological Journal, 40 (Suppl.1), 1–56].

Varlamov, A.I., Rozova, A.V., Khamentovsky, Yu.Ya., *et al.*, 2008a. The Zhurinsky Mys section. In: Rozanov, A.Yu., Varlamov, A.I. (Eds.), The Cambrian System of the Siberian Platform. Part 1: The Aldan-Lena Region. XIII International Field Conference of the Cambrian Stage Subdivision Working Group. Paleontological Institute, Russian Academy of Sciences, Moscow and Novosibirsk, pp. 71–92.

Varlamov, A.I., Rozova, A.V., Khamentovsky, Yu.Ya., *et al.*, 2008b. Introduction. In: Rozanov, A.Yu., Varlamov, A.I. (Eds.), The Cambrian System of the Siberian Platform. Part 1: The Aldan-Lena Region. XIII International Field Conference of the Cambrian Stage Subdivision Working Group. Paleontological Institute, Russian Academy of Sciences, Moscow and Novosibirsk, pp. 6–11.

Walcott, C.D., 1889a. Stratigraphic position of the *Olenellus* fauna in North America and Europe. American Journal of Science 37, 374–392.

Walcott, C.D., 1889b. Stratigraphic position of the *Olenellus* fauna in North America and Europe (continued). American Journal of Science 38, 29–42.

Walcott, C.D., 1890a. The fauna of the Lower Cambrian or *Olenellus* Zone. In: Tenth Annual Report of the Director, 1888–1889. Part 1. United States Geological Survey, pp. 509–774.

Walcott, C.D., 1890b. Descriptive notes on new genera and species from the Lower Cambrian or *Olenellus* Zone of North America. Proceedings of the U.S. National Museum 12, 33–46.

Walter, M.R., Elphinstone, R., Heys, G.R., 1989. Proterozoic and Early Cambrian trace fossils from the Amadeus and Georgina Basins, central Australia. Alcheringa 13, 209–256.

Webster, M., 2011. Upper Dyeran trilobite biostratigraphy in the southern Great Basin, U.S.A. Museum of Northern Arizona Bulletin 67, 121–154.

Westergård, A.H., 1922. Sveriges Olenidskiffer. Sveriges Geologiska Undersökning Ca18, 1–205.

Westergård, A.H., 1946. Agnostidea of the Middle Cambrian of Sweden. Sveriges Geologiska Undersökning C477, 1–141.

Westergård, A.H., 1947. Supplementary notes on the Upper Cambrian trilobites of Sweden. Sveriges Geologiska Undersökning C489, 1–34.

Westrop, S.R., 1986. Trilobites of the Upper Cambrian Sunwaptan Stage, southern Canadian Rocky Mountains, Alberta. Palaeontographica Canadiana 3, 1–179.

Westrop, S.R., 1995. Sunwaptan and Ibexian (Upper Cambrian-Lower Ordovician) trilobites of the Rabbitkettle Formation, Mountain River region, northern Mackenzie Mountains, northwest Canada. Palaeontographica Canadiana 12, 1–75.

Westrop, S.R., Adrain, J.M., Landing, E., 2011. The Cambrian (Sunwaptan, Furongian) agnostoid arthropod *Lotagnostus* Whitehouse, 1936, in Laurentian and Avalonian North America: Systematics and biostratigraphic significance. Bulletin of Geosciences 86, 569–594.

Wheeler, H.E., 1947. Base of the Cambrian System. Journal of Geology 55, 153–159.

Whitehouse, F.W., 1936. The Cambrian faunas of northeastern Australia. Parts 1 and 2. Memoirs of the Queensland Museum 11, 59–112.

Whittington, H.B., 1977. The Middle Cambrian trilobite *Naraoia*, Burgess Shale, British Columbia. Philosophical Transactions of the Royal Society, Series B 280, 409–443.

Wotte, T., Álvaro, J.J., Shields, G.A., Brown, B., Brasier, M.D., Veizer, J., 2007. C-, O- and Sr-isotope stratigraphy across the Lower-Middle Cambrian transition of the Cantabrian Zone (Spain) and the Montagne Noire (France), West Gondwana. Palaeogeography, Palaeoclimatology, Palaeoecology 256, 47–70.

Xiang, L.W., Zhang, T.R., 1985. Stratigraphy and trilobite faunas of the Cambrian in the western part of northern Tianshan, Xinjiang. Ministry of Geology and Mineral Resources, Geological Memoirs, series 2 4, 243.

Xing, Y.S., Luo, H.L., Jiang, Z.W., Zhang, S.S., 1991. A candidate Global Stratotype Section and Point for the Precambrian-Cambrian boundary at Meishucun, Yunnan, China. Journal of China University of Geosciences 2 (1), 47–57.

Yang, A.H., Zhu, M.Y., Debrenne, F., Yuan, K.X., Vannier, J., Zhang, J.M., Li, G.X., 2005. Early Cambrian archaeocyathan zonation of the Yangtze Platform and biostratigraphic implications. Acta Macropalaeontologica Sinica 22 (Suppl.), 205–210.

Yao, L., Peng, J., Fu, X.P., Zhao, Y.L., 2009. Ontogenesis of *Tuzoia bispinosa* (Arthropoda) from the Middle Cambrian Kaili biota, Guizhou, China. Acta Palaeontologica Sinica 48, 56–64 [In Chinese].

Yang, Z.Y., Otofuji, Y., Sun, Z.M., Huang, B.C., 2002. Magnetostratigraphic constraints on the Gondwanan origin of North China: Cambrian/Ordovician boundary results. Geophysical Journal International 151, 1–10.

Yin, J.Y., 2002. Research of Palaeomagenetism. In: Chen, L.X., Luo, H.L., Hu, S.X., Yin, J.Y., Jiang, Z.W., Wu, Z.L., LI, F., Chen, A.L. (Eds.), Early Cambrian Chengjiang Fauna in eastern Yunnan, China. Science and Technology Press, Yunnan, pp. 98–113 [In Chinese].

Young, G.C., Laurie, J.R., 1996. An Australian Phanerozoic Timescale. Oxford University Press, Melbourne, p. 279.

Yuan, J.L., Zhao, Y.L., Li, Y., Huang, Y.Z., 2002. Trilobite Fauna of the Kaili Formation (Uppermost Lower Cambrian-Lower Middle Cambrian) from Southeastern Guizhou, South China. Shanghai Science and Technology Press, Shanghai, p. 422 [In Chinese].

Zhang, J.M., Li, G.X., Zhou, C.M., Zhu, M.Y., Yu, Z.Y., 1997. Carbon isotope profiles and their correlation across the Neoproterozoic-Cambrian boundary interval on the Yangtze Platform, China. Bulletin of National Museum of Natural Sciences 10, 107–116.

Zhang, W.T., 1987. World's oldest Cambrian trilobites from eastern Yunnan. In: Nanjing Institute of Geology and Palaeontology (Ed.), Stratigraphy and Palaeontology of Systemic Boundaries in China. Precambrian-Cambrian Boundary (1). Nanjing University Publishing House, Nanjing, pp. 1–16.

Zhang, W.T., Yuan, K.X., Zhou, Z.Y., Qian, Y., Wang, Z.Z., 1979. Cambrian of southwestern China. In: Nanjing Institute of Geology and Palaeontology (Ed.), Carbonate Biostratigraphy of Southwest China. Science Press, Beijing, pp. 39–107 [In Chinese].

Zhao, Y.L., Yuan, J.L., Zhu, M.Y., Yang, R.D., Guo, Q.J., Qian, Y., Huangg, Y.Z., Pan, Y., 1999. A progress report on research on the early Middle Cambrian Kaili biota, Guizhou, PRC. Acta Palaeontologica Sinica 38 (Suppl.), 1–14 [In Chinese].

Zhao, Y.L., Yang, R.D., Yuan, J.L., Zhu, M.Y., Guo, Q.J., Yang, X.L., Tai, T.S., 2001. Cambrian stratigraphy at Balang, Guizhou Province, China: Candidate section for a global unnamed series and stratotype section for the Taijiangian Stage. Palaeoworld 13, 189–208.

Zhao, Y.L., Yuan, J.L., Zhu, M.Y., Yang, R.D., Guo, Q.J., Peng, J., Yang, X.L., 2002. Progress and significance in research on the early Middle Cambrian Kaili biota, Guizhou Province, China. Progress in Natural Science 12, 649–654.

Zhao, Y.L., Yuan, J.L., Zhu, M.Y., Babcock, L.E., Peng, J., Wang, Y., Yang, X.L., Guo, Q.J., Yang, R.D., Tai, T.S., 2005. Balang section, Guizhou, China: Stratotype section for the Taijiangian Stage and candidate for GSSP of an unnamed Cambrian series. In: Peng, S.C., Babcock, L.E., Zhu, M.Y. (Eds.), Cambrian System of China and Korea, Guide to Field Excursions. University of Science and Technology of China Press, Hefei, pp. 62–83.

Zhao, Y.L., Zhu, M.Y., Babcock, L.E., Peng, J., 2011. The kaili Biota. Marine Organisms from 508 Million Years Ago. Guizhou Publishing Group, Guyang, China, p. 251.

Zhen, Y.Y., Zhou, Z.Y., 2008. History of trilobite biodiversity: A Chinese perspective. In: Zhou, Z.Y., Zhen, Y.Y. (Eds.), Trilobite Record of China. Science Press, Beijing, pp. 301–330.

Zhou, Z.Q., Cao, X.D., Hu, Y.X., Zhao, J.T., 1996. Early Palaeozoic stratigraphy and sedimentary-tectonic evolution in eastern Qilian Mountains, China. Northwest Geosciences 17 (1), 1–58 [In Chinese].

Zhu, M.Y., Li, G.X., Zhang, J.M., Steiner, M., Qian, Y., Jiang, Z.W., 2001. Early Cambrian stratigraphy of East Yunnan, southwestern China: A synthesis. Acta Palaeontologica Sinica 40 (Suppl.), 4–39.

Zhu, M.Y., Zhang, J.M., Li, G.X., Yang, A.H., 2004. Evolution of C isotopes in the Cambrian of China: Implications for Cambrian subdivision and trilobite mass extinctions. Geobios 37, 287–310.

Zhu, M.Y., Babcock, L.E., Peng, S.C., 2006. Advances in Cambrian stratigraphy and paleontology: Integrating correlation techniques, paleobiology, taphonomy and paleoenvironmental reconstruction. Palaeoworld 15, 217–222.

Zhuravlev, A.Yu., 1995. Preliminary suggestions on the global Early Cambrian zonation. In: Landing, E., Geyer, G. (Eds.), Morocco '95: The Lower-Middle Cambrian Standard of Western Gondwana. Beringeria, Special Issue 2, pp. 147–160.

Zhuravlev, A.Yu., Repina, L.N., 1990. Guidebook for Excursion on the Aldan and Lena Rivers, Siberian Platform. Institute of Geology and Geophysics, Siberian Branch of Academy of Sciences of USSR, Novosibirsk, p. 115.

Zhuravlev, A.Yu., Gravestock, D.I., 1994. Archaeocyaths from Yorke Peninsula, South Australia and archaeocyathan Early Cambrian zonation. Alcheringa 18, 1–54.

Zhuravlev, A.Yu., Riding, R., 2001. The Ecology of the Cambrian Radiation. Perspectives in Paleobiology and Earth History 7, p. 525.

Zhuravleva, I.T., Korshunov, V.I., Rozanov, A.Yu., 1969. Atdabanian Stage and its archaeocyathan basis in the stratotype section. In: Zhuravleva, I.T. (Ed.), Lower Cambrian Biostratigraphy and Palaeontology of Siberia and Far East. Nauka, Moscow, pp. 5–59.

R.A. Cooper and P.M. Sadler
Contributors: O. Hammer and F.M. Gradstein

Chapter 20

The Ordovician Period

Abstract: Rapid and sustained biotic diversification reached its highest levels in the Paleozoic. A prolonged "hot-house" climate through Early Ordovician, cooling through Middle Ordovician and changing to "ice-house" conditions in Late Ordovician, global glaciation, oceanic turnover and mass extinction at end of period, strong fluctuations in eustatic sea level, appearance and diversification of pandemic planktonic graptolites and conodonts important for correlation, moderate to strong benthic faunal provincialism, reorganization and rapid migration of tectonic plates surrounding the Iapetus Ocean and migration of the South Pole from North Africa to central Africa all characterize the Ordovician Period. All seven Ordovician stages have formalized GSSPs.

458 Ma Ordovician

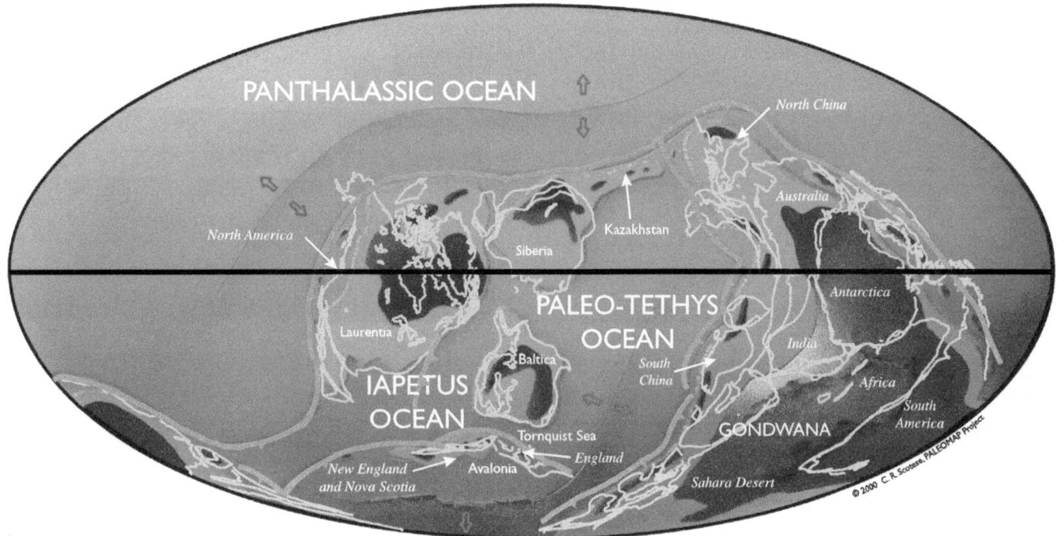

Chapter Outline

The Geologic Time Scale 2012. DOI: 10.1016/B978-0-444-59425-9.00020-2

20.1. HISTORY AND SUBDIVISIONS

Named after the Ordovices, a northern Welsh tribe, the Ordovician was proposed as a new system by Lapworth in 1879. It was a compromise solution to the controversy over strata in North Wales that had been included by Adam Sedgwick in his Cambrian System, but which were also included by Murchison as constituting the lower part of his Silurian System. Although it was initially slow to be accepted in Britain, where it was instead generally called Lower Silurian well into the twentieth century, the Ordovician was soon recognized and used elsewhere, such as in the Baltic region and Australia. The name Ordovician was officially adopted at the 1960 International Geological Congress in Copenhagen.

Black graptolite-bearing shales are widely developed in Ordovician sedimentary successions around the world. Lapworth (1879–1880) described the stratigraphic distribution of British graptolites at the same time that he proposed the Ordovician System, and graptolites have played a major role in the recognition and correlation of Ordovician rocks since that time. Lapworth demonstrated as long ago as 1878, in southern Scotland, the fine biostratigraphic precision that can be achieved with this group. In the last several decades, conodonts have proved to be of comparable global biostratigraphic value in the carbonate facies. In the shelly facies, developed mainly on the continental shelf and platform, trilobites and brachiopods are used extensively for local and regional zonation, and coral–stromatoporoid communities enable biostratigraphic subdivision in the Upper Ordovician. Many other biotic groups became established and diversified during the Ordovician and enable biostratigraphic subdivision in local regions. Chitinozoan and acritarch zonations are being developed and both groups are useful for providing long-range correlation with good precision.

Subdivision of the Ordovician into Upper and Lower, or Upper, Middle and Lower parts has been very inconsistent (Jaanusson, 1960; Webby, 1998). The International Subcommission on Ordovician Stratigraphy (Ordovician subcommission) voted to recognize a three-fold subdivision of the system (Webby, 1995), now accepted by the International Commission on Stratigraphy (ICS) as the series, Lower, Middle and Upper Ordovician.

Because of marked faunal provincialism and facies differentiation throughout most of the Ordovician, no existing regional suite of stages or series has been found to be satisfactory in its entirety for global application. The Ordovician subcommission therefore undertook to identify the best fossil-based datums, wherever they are found, for global correlation, and to use these for definition of global chronostratigraphic (and chronologic) units (Webby, 1995, 1998). In this respect, it deviated from the course followed by the Silurian and Devonian subcommissions, both of which recommended the adoption of pre-existing (regional) stage or series schemes for global use.

During the early 1990s, the Ordovician subcommission established a number of working groups to investigate and recommend levels within the period that were suitable for international correlation, and therefore for defining international stages (Webby, 1995, 1998). Seven general chronostratigraphic levels were certified as primary correlation levels for the seven international stages. They are based on the first appearance of key graptolite or conodont species. All boundaries have been formally voted on and are defined by a global stratigraphic section and point (GSSP). From the bottom upwards they are Tremadocian, Floian, Dapingian, Darriwilian, Sandbian, Katian, and Hirnantian (Figure 20.1(a) and (b)).

20.1.1. Stages of the Lower Ordovician

20.1.1.1. Cambrian–Ordovician Boundary and the Tremadocian Stage

The base of the Tremadocian Stage, the Lower Ordovician Series, and the Cambrian–Ordovician boundary is defined by a GSSP in the Green Point section of western Newfoundland (Figure 20.2), in Bed 23 of the measured section (lower Broom Point Member, Green Point Formation) (Cooper and Nowlan, 1999; Cooper *et al.*, 2001). This level coincides with the appearance of the conodont *Iapetognathus fluctivagus* (base of the *I. fluctivagus* Zone) at Green Point, and is just 4.8 m above the appearance of planktonic graptolites, which therefore can be taken as a proxy for the boundary in shale sections (Figure 20.2). This definition enables both graptolites and conodonts to be used in correlation of the boundary, and resolved a controversy extending back for at least 90 years (Henningsmoen, 1972).

The boundary also coincides with the appearance of the cosmopolitan trilobite *Jujuyaspis borealis* and lies within the *Symphysurina bulbosa* species complex (Loch *et al.*, 2009), both of which are useful for correlation in carbonate successions. It lies at the peak of the largest positive excursion in the $\delta^{13}C$ curve through the boundary interval (Ripperdan and

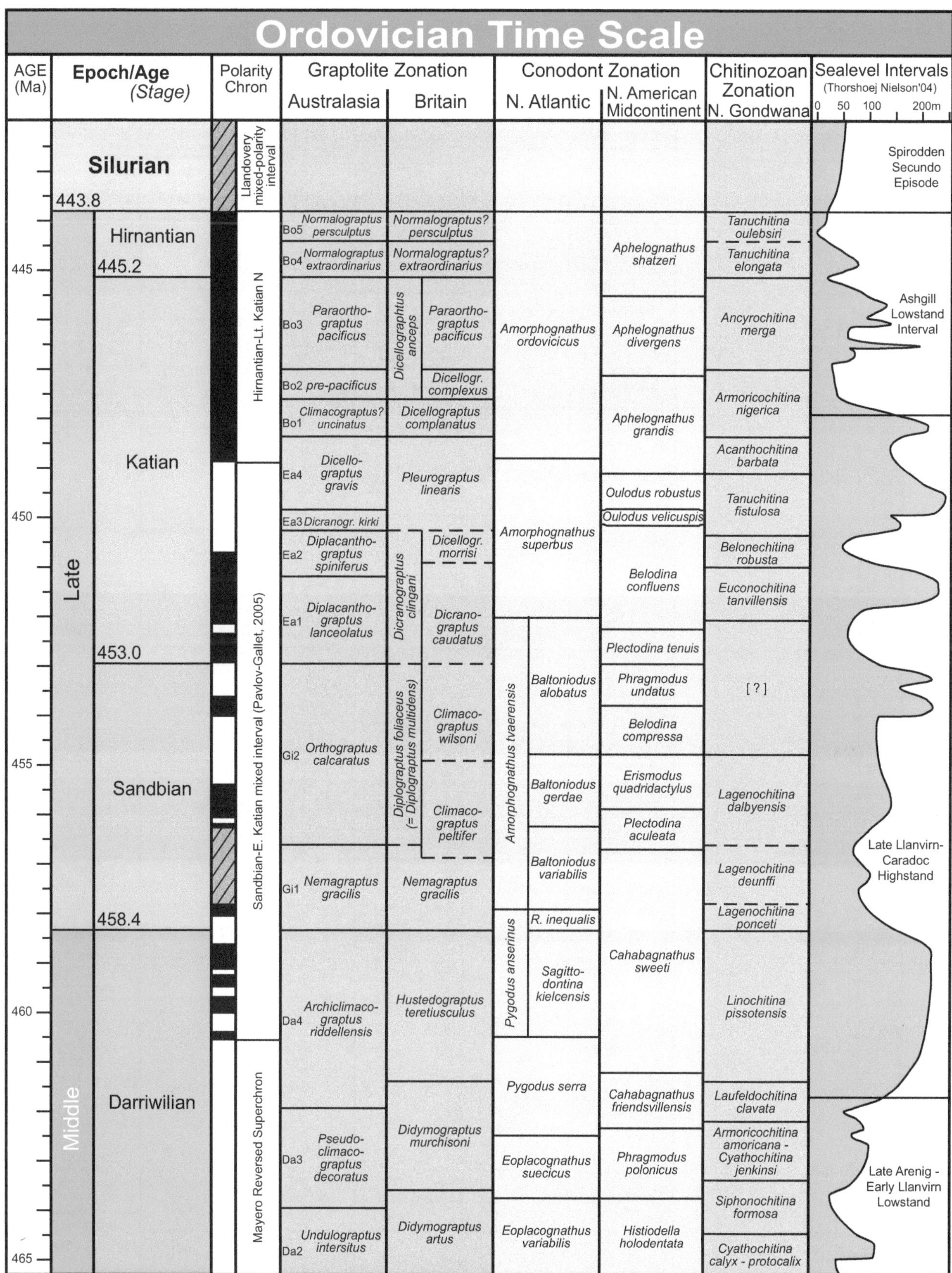

FIGURE 20.1 **Ordovician time scale, geomagnetic polarity scheme, graptolite, conodont and chitinozoan zonal schemes, and main transgressive-regressive trends**. Zonal suites and correlations are after Löfgren (1993), Webby *et al.* (2004b), Bergström and Wang (1995), Goldman *et al.* (2007), Nõlvak *et al.* (2006) and Paris (pers. comm., 2010). The magnetic polarity scale is from Pavlov and Gallet (2005), and the eustatic sea level curve from Nielsen (2004). Dashed lines at zonal boundaries indicate significant uncertainty in placement or correlation of the boundaries relative to the composite scale.

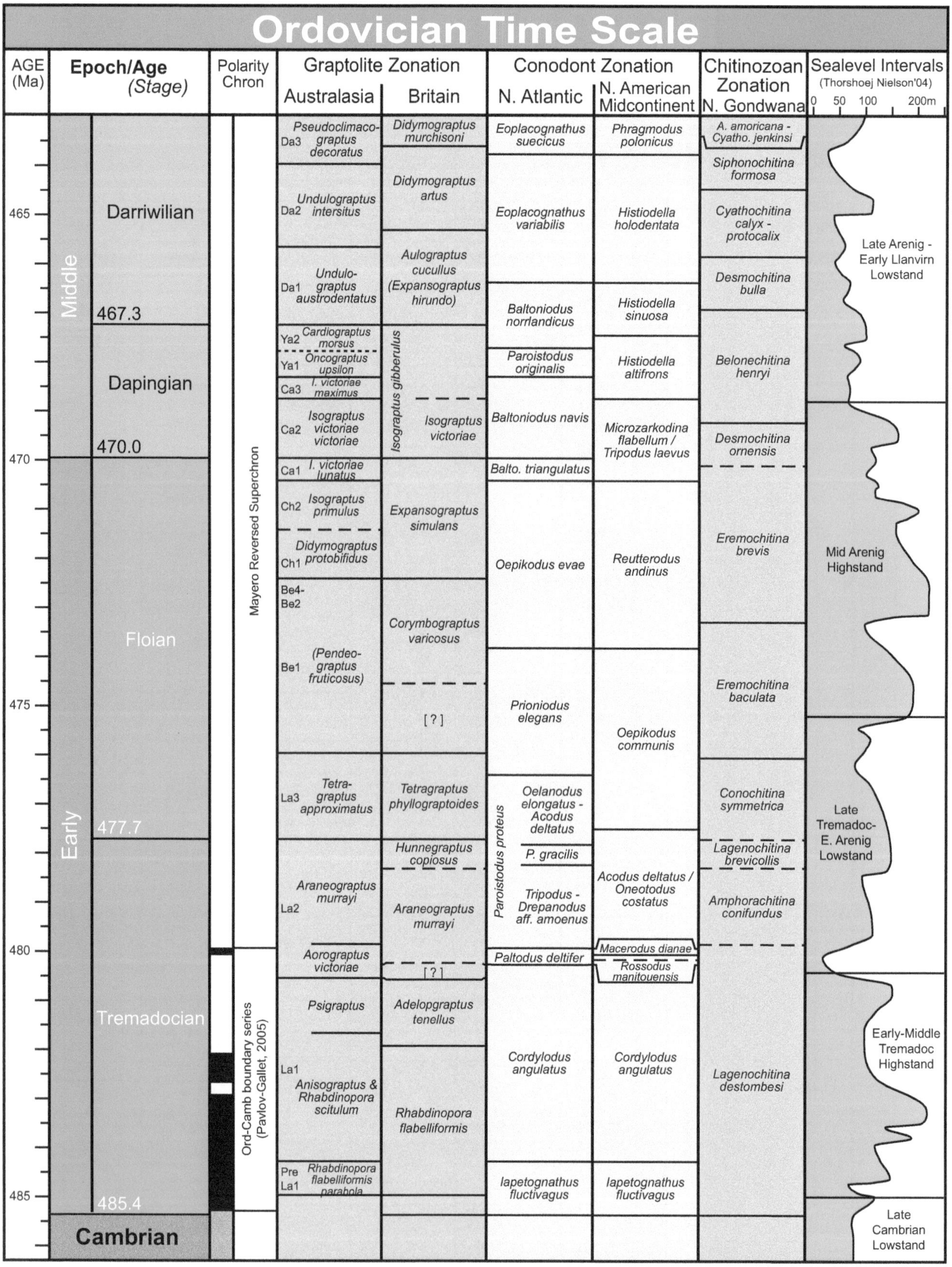

Ordovician Time Scale

AGE (Ma): 465, 470, 475, 480, 485

Epoch/Age (Stage):
- Middle
  - Darriwilian
  - 467.3
  - Dapingian
  - 470.0
- Early
  - Floian
  - 477.7
  - Tremadocian
  - 485.4
- Cambrian

Polarity Chron:
- Mayero Reversed Superchron
- Ord-Camb boundary series (Pavlov-Gallet, 2005)

Graptolite Zonation – Australasia:
- Da3 *Pseudoclimacograptus decoratus*
- Da2 *Undulograptus intersitus*
- Da1 *Undulograptus austrodentatus*
- Ya2 *Cardiograptus morsus*
- Ya1 *Oncograptus upsilon*
- Ca3 *I. victoriae maximus*
- Ca2 *Isograptus victoriae victoriae*
- Ca1 *I. victoriae lunatus*
- Ch2 *Isograptus primulus*
- Ch1 *Didymograptus protobifidus*
- Be4-Be2
- Be1 *(Pendeograptus fruticosus)*
- La3 *Tetragraptus approximatus*
- La2 *Araneograptus murrayi*
- *Aorograptus victoriae*
- *Psigraptus*
- La1 *Anisograptus & Rhabdinopora scitulum*
- Pre La1 *Rhabdinopora flabelliformis parabola*

Graptolite Zonation – Britain:
- *Didymograptus murchisoni*
- *Didymograptus artus*
- *Aulograptus cucullus (Expansograptus hirundo)*
- *Isograptus gibberulus*
- *Isograptus victoriae*
- *Expansograptus simulans*
- *Corymbograptus varicosus*
- [?]
- *Tetragraptus phyllograptoides*
- *Hunnegraptus copiosus*
- *Araneograptus murrayi*
- [?]
- *Adelograptus tenellus*
- *Rhabdinopora flabelliformis*

Conodont Zonation – N. Atlantic:
- *Eoplacognathus suecicus*
- *Eoplacognathus variabilis*
- *Baltoniodus norrlandicus*
- *Paroistodus originalis*
- *Baltoniodus navis*
- *Balto. triangulatus*
- *Oepikodus evae*
- *Prioniodus elegans*
- *Paroistodus proteus*
  - *Oelanodus elongatus - Acodus deltatus*
  - *P. gracilis*
  - *Tripodus - Drepanodus aff. amoenus*
- *Paltodus deltifer*
- *Cordylodus angulatus*
- *Iapetognathus fluctivagus*

Conodont Zonation – N. American Midcontinent:
- *Phragmodus polonicus*
- *Histiodella holodentata*
- *Histiodella sinuosa*
- *Histiodella altifrons*
- *Microzarkodina flabellum / Tripodus laevus*
- *Reutterodus andinus*
- *Oepikodus communis*
- *Acodus deltatus / Oneotodus costatus*
- *Macerodus dianae*
- *Rossodus manitouensis*
- *Cordylodus angulatus*
- *Iapetognathus fluctivagus*

Chitinozoan Zonation – N. Gondwana:
- *A. americana - Cyatho. jenkinsi*
- *Siphonochitina formosa*
- *Cyathochitina calyx - protocalix*
- *Desmochitina bulla*
- *Belonechitina henryi*
- *Desmochitina ornensis*
- *Eremochitina brevis*
- *Eremochitina baculata*
- *Conochitina symmetrica*
- *Lagenochitina brevicollis*
- *Amphorachitina conifundus*
- *Lagenochitina destombesi*

Sealevel Intervals (Thorshoej Nielson'04): 0, 50, 100, 200m
- Late Arenig - Early Llanvirn Lowstand
- Mid Arenig Highstand
- Late Tremadoc-E. Arenig Lowstand
- Early-Middle Tremadoc Highstand
- Late Cambrian Lowstand

FIGURE 20.1 (*Continued*).

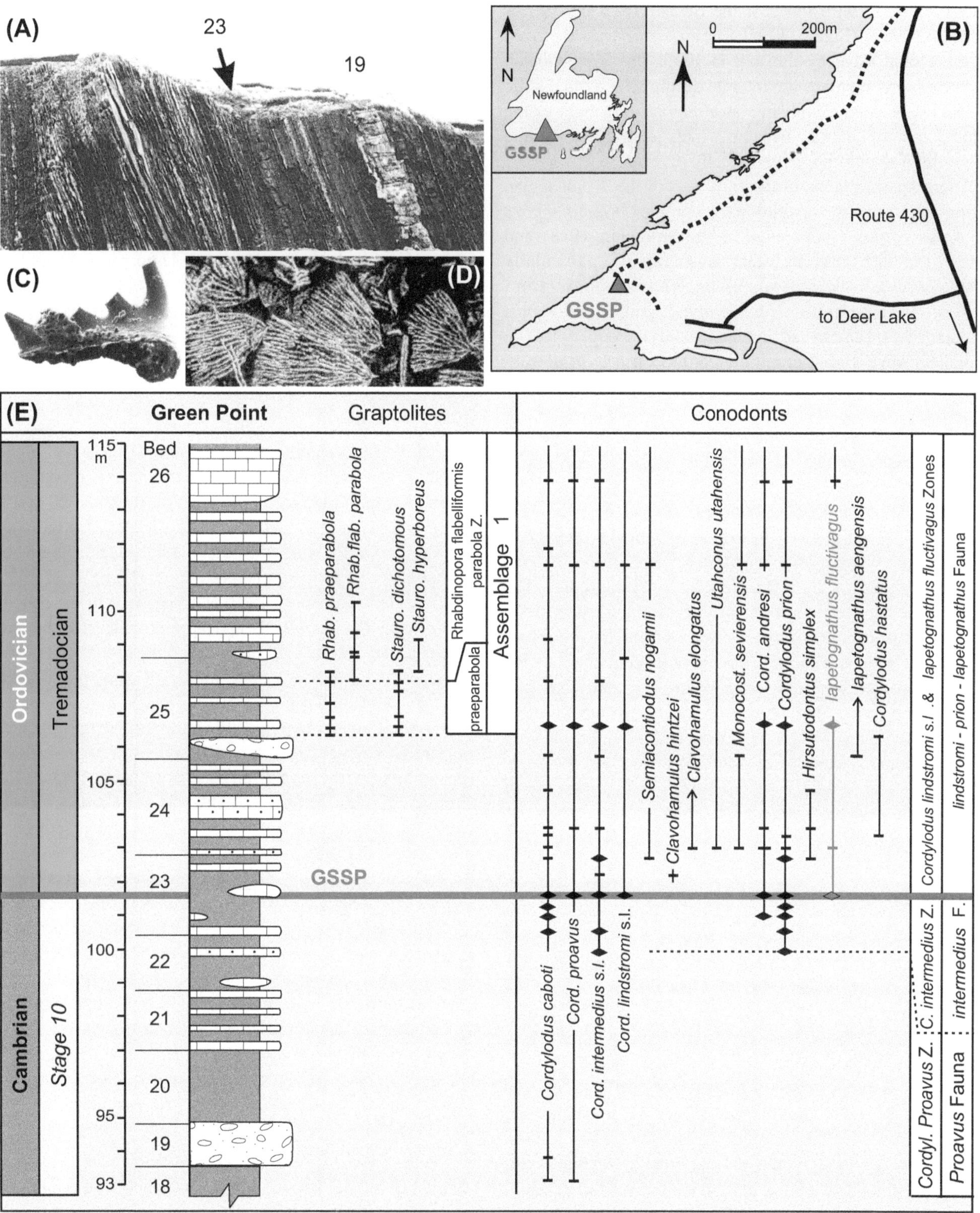

FIGURE 20.2 **Stratigraphic ranges of graptolites and conodonts at Green Point, the global stratotype section for base of the Tremadocian Stage and of the Ordovician System, western Newfoundland, Canada.** The boundary is placed at the first appearance of the conodont *Iapetognathus fluctivagus* (Nicoll *et al.*, 1992) (subfigure (c), specimen is 0.5 mm long), 4.8 m below the first appearance of planktonic graptolites, and the zonal graptolite taxa *Rhabdinopora praeparabola and Rhabdinopora flabelliformis parabola*, (bottom specimen is 17 mm long; from Cooper *et al.*, 2001). The photograph under (A) shows the cliff exposure and extensive shore platform, with the GSSP level indicated in Bed 23. The prominent massive limestone conglomerate bed at right (Bed 19) is also shown. Note that the sequence youngs from right to left and is overturned (photograph by S.H. Williams).

Miller, 1995; Cooper *et al.*, 2001) and during the global marine transgression that followed the Acerocare regressive event.

The upper and lower boundaries of the stage almost exactly coincide with those of the British Tremadoc Series (Rushton, 1982) (Figure 20.1(b)). This name was approved for the global stage by the ICS in 1999. The Tremadocian Stage thus encompasses the interval during which planktonic graptolites became established as a major component of the oceanic macroplankton, became widely distributed around the world, and became taxonomically diverse.

The graptolite fauna of the early part of the Tremadocian is dominated by the evolutionary complex *Rhabdinopora* and *Anisograptus* (Cooper *et al.*, 1998; Erdtmann, 1988), and that of the later part by other anisograptids, particularly *Paradelograptus, Paratemnograptus, Kiaerograptus, Aorograptus, Araneograptus, Hunnegraptus*, and *Clonograptus* (Maletz, 1999; Maletz and Egenhoff, 2001). Cooper (1999a) recognized nine global graptolite chronozones – in upward sequence, the zones of *Rhabdinopora praeparabola, R. flabelliformis parabola, Anisograptus matanenesis, R. f. anglica, Adelograptus, Paradelograptus antiquus, Kiaerograptus, Araneograptus murrayi* and *Hunnegraptus copiosus* (see also Maletz, 1999).

The early Tremadocian contains two widespread conodont zones (Figure 20.1): the zones of *Iapetognathus fluctivagus* and *Cordylodus angulatus,* equivalent to the graptolite zones of *R. praeparabola* to *P. antiquus.* The middle and late Tremadocian is finely subdivided into six conodont subzones (Löfgren, 1993), within the zones of *Paltodus deltifer* and *Paroistodus proteus,* equivalent to the three graptolite zones, *Kiaerograptus, A. murrayi,* and *H. hunnebergensis.* The two sets of zones have been closely integrated by means of the conodont- and graptolite-bearing shale–carbonate sections of southwest Sweden (Löfgren, 1993, 1996), providing a global correlation framework of high precision for the Tremadocian Stage.

20.1.1.2. Floian Stage

The base of the Floian, the second stage of the Ordovician, is defined by a GSSP at the first appearance of the graptolite *Tetragraptus approximatus* in the section exposed in the Diabasbrottet quarry on the northwestern slope of Mount Hunneberg, in the province of Västergötland, southern Sweden (Figure 20.3) (Maletz *et al.*, 1996; Bergström *et al.*, 2004). This biostratigraphic datum can be widely recognized throughout low- to middle-paleolatitude regions and has proved a distinctive and reliable marker. It is also the level adopted for the base of the revised British Arenig Series (Fortey *et al.*, 1995) and thus is employed in a high-paleolatitude region. The name is derived from the village of Flo, located about 5 km southeast of the GSSP. The GSSP was ratified by the ICS in 2002 and the stage name in 2005.

The stage base coincides with the base of the graptolite zone of *T. approximatus*, of global distribution. It lies just above the base of the conodont subzone of *Oelandodus elongatus–Acodus deltatus*, the highest subzone of the *P. proteus* Zone in the Baltic (mid-paleolatitude) sequence (Bergström *et al.*, 2004). It lies at, or very close to, the base of the *deltatus–costatus* Zone of the midcontinent (low-paleolatitude) realm. The boundary lies within the *Megistaspis (Paramegistaspis) planilimbata* trilobite zone.

During the Floian Stage there is a spectacular increase in the diversity and abundance of graptolites, driven by expansion of the Dichograptidae and Sigmagraptidae, and anisograptid graptolites become rare. Graptolite species diversity reaches its highest level in the Lower to Middle Ordovician. There is also a peak in the faunal turnover rate at this time. The Floian is a relatively finely zoned part of the Ordovician in low-paleolatitude regions such as Australasia and Laurentia.

20.1.2. Stages of the Middle Ordovician

20.1.2.1. Dapingian Stage

The base of the Dapingian Stage, and of the Middle Ordovician Series, is defined by the first appearance of the conodont, *Baltoniodus? triangularis* (Wang *et al.*, 2005). The GSSP is at the base of SHod16, 10.57 m. above the base of the Dawan formation, in the Huanghuachang section near Yichang, southern China (Figure 20.4) (Wang *et al.*, 2005). The level is 0.2 m. below the first appearance of the conodont *Microzarkodina flabellum*, and approximates the boundary between the lower and upper graptolite subzones of the *Azygograptus suecicus* Zone. This level is nearly coincident with the base of the *Belonechitina henryi* chitinozoan Biozone. Well known phylogenetic conodont lineages, including *Baltoniodus, Trapezognathus, Periodon* and *Microzarkodina* span the boundary, enabling good correlation (Figure 20.4) (Bergström and Löfgren, 2009). The boundary level is close to the first appearance of the global correlation graptolite, *Isograptus victoriae victoriae*. It lies within the *Ampullula–Barakella felix* acritarch assemblage zone. The Dapingian Stage was ratified by the ICS in 2007.

The Dapingian spans the evolutionary development of *Isograptus* and its derivatives (particularly the *Isograptus victoriae* and *Parisograptus caduceus* groups), providing fine zonal subdivision and correlation. Two Australasian stages and five zones are represented – in upward sequence, the zones of *Isograptus victoriae victoriae, I. v. maximus, I. v. maximodivergens* (Castlemainian stage), and *Oncograptus upsilon* and *Cardiograptus morsus* (Yapeenian stage).

20.1.2.2. Darriwilian Stage

The GSSP for the base of the Darriwilian Stage lies at the first appearance of the graptolite *Undulograptus austrodentatus* in the Huangnitang section, near Changshan, Zhejiang Province, southeast China (Figure 20.5) (Chen and Bergström, 1995;

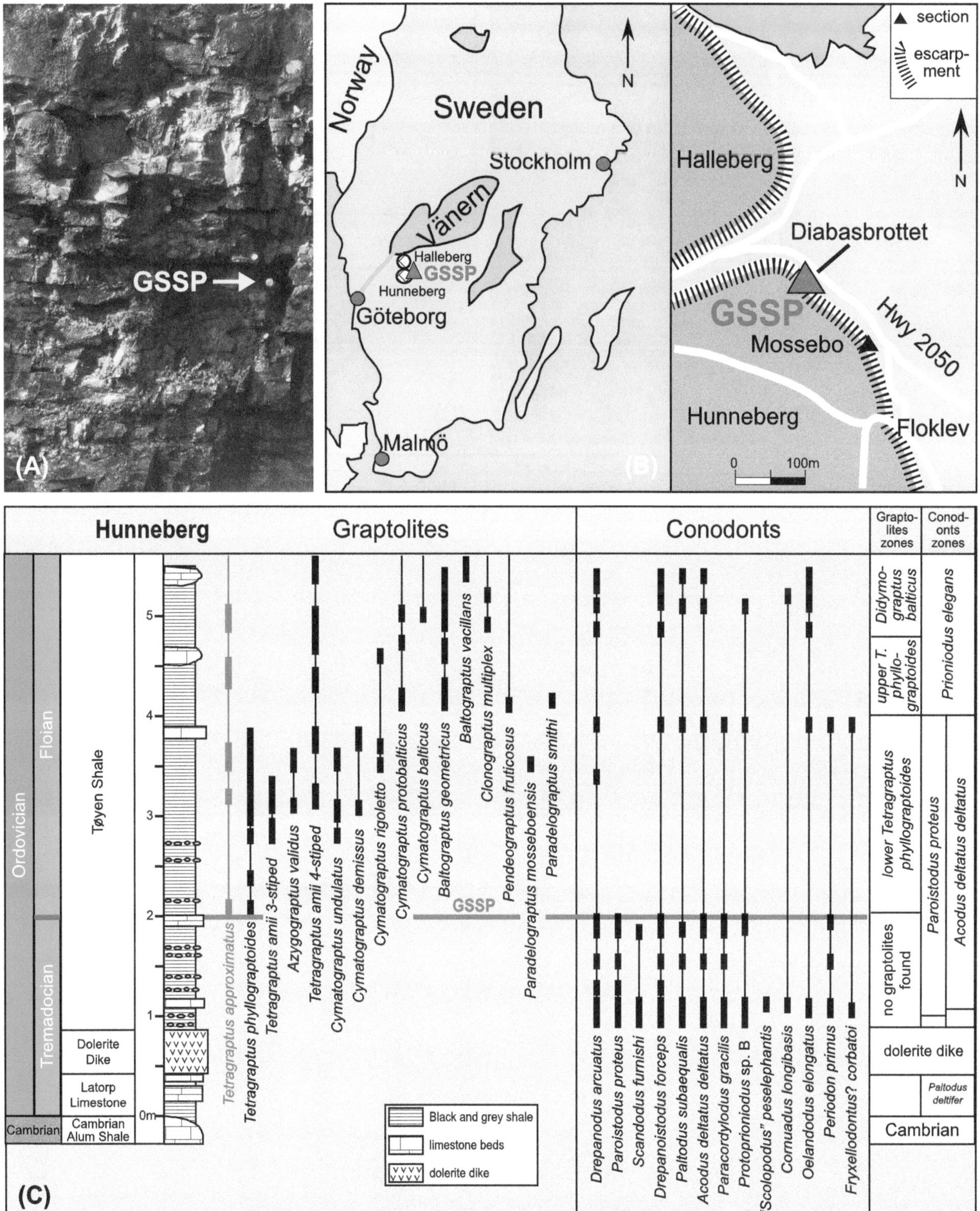

FIGURE 20.3 **Stratigraphic ranges of graptolites and conodonts in the Diabasbrottet section, Mt Hunneberg, Sweden, the global stratotype section for the base of the Floian Stage.** The boundary point is placed at the first appearance of *Tetragraptus approximatus* (arrow) in this section. *After Bergström et al. (2004: Figure7).*

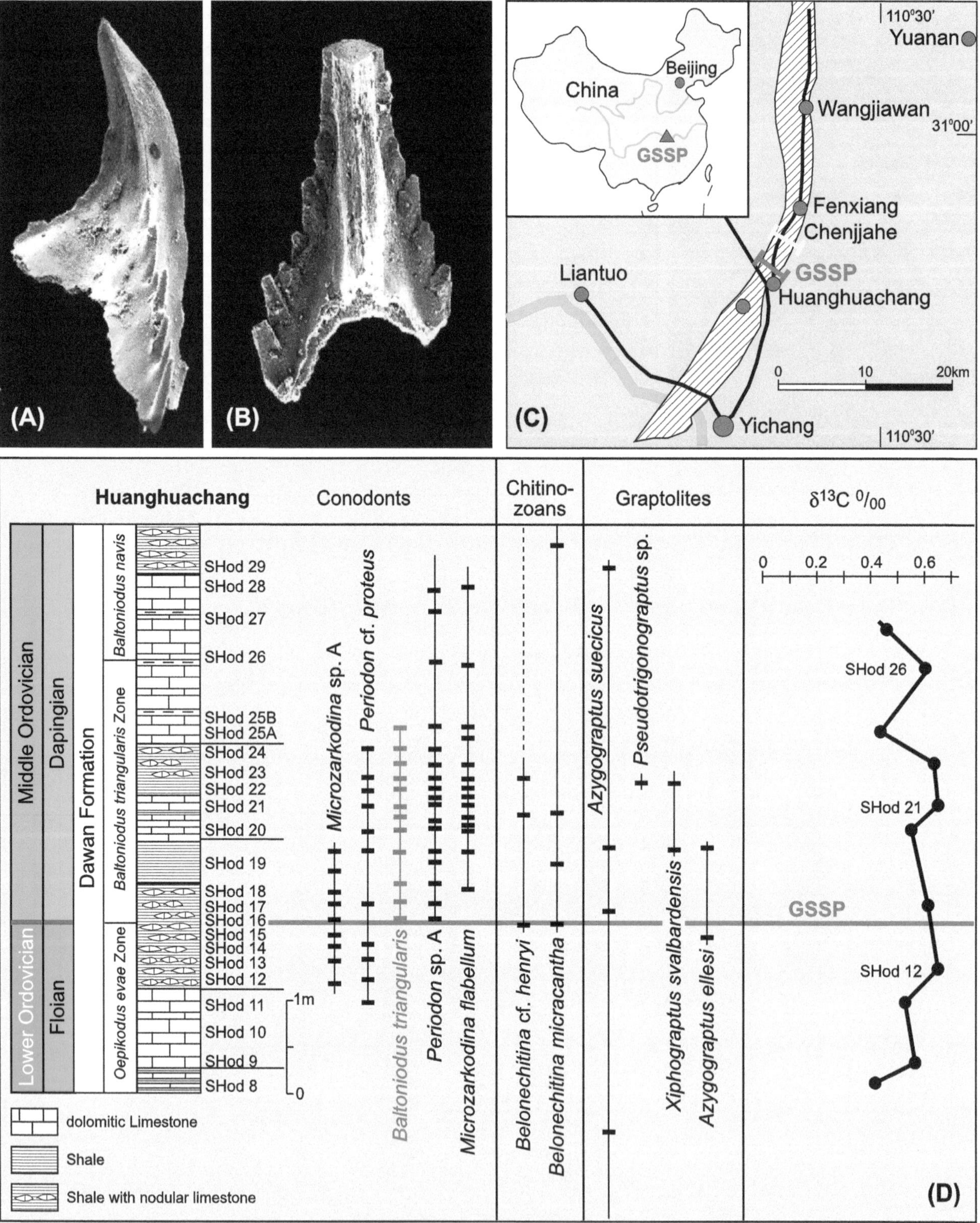

FIGURE 20.4 **Stratigraphic succession across the boundary interval for the formal base of the Dapingian Stage, Huanghuachang GSSP section, Hubei, China, showing stratigraphic ranges of conodonts, chitinozoans and graptolites, and the $\delta^{13}C_{carb}$ isotope curve.** *Baltoniodus triangularis*, the conodont used for locating the GSSP for the Dapingian Stage, Huanghuachang section, Yichang, southern China. (A): Pa element, lateral view, specimen is 0.23 mm in height; (B): Sa element, posterior view, specimen is 0.16 mm in height. *From Wang et al. (2005: Figure 4).*

Base of the Darriwilian Stage of the Ordovician System at Huangnitang, Changshan County, Zhejiang Province, southeast China

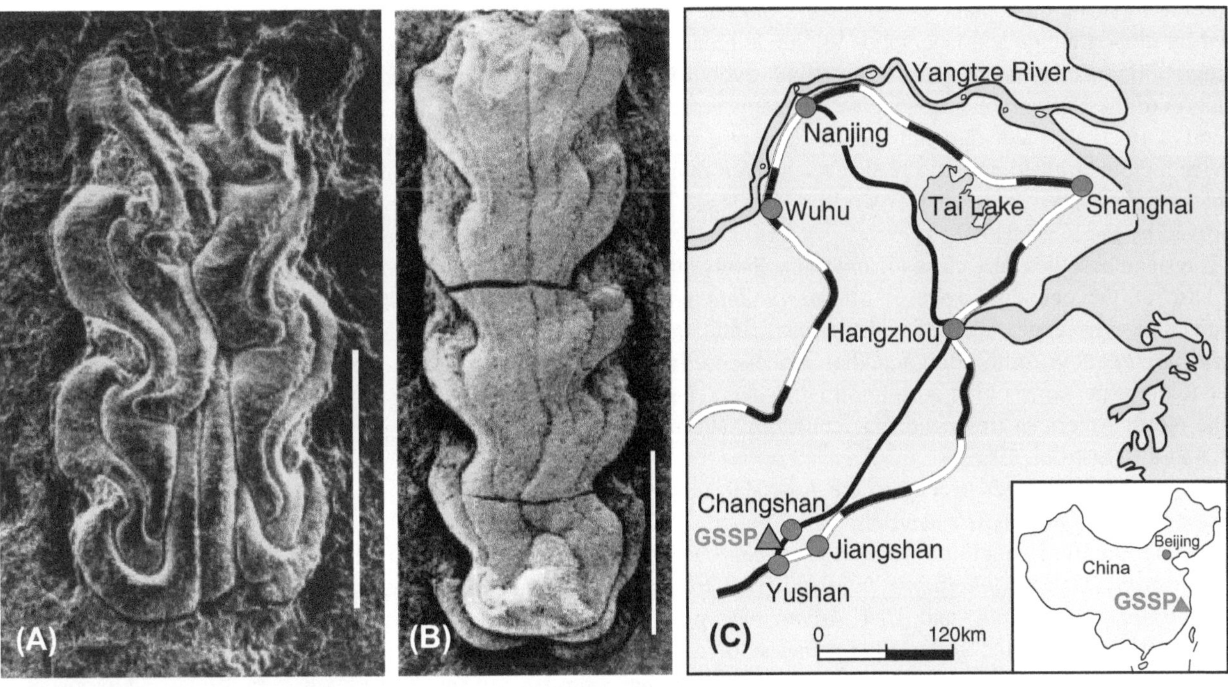

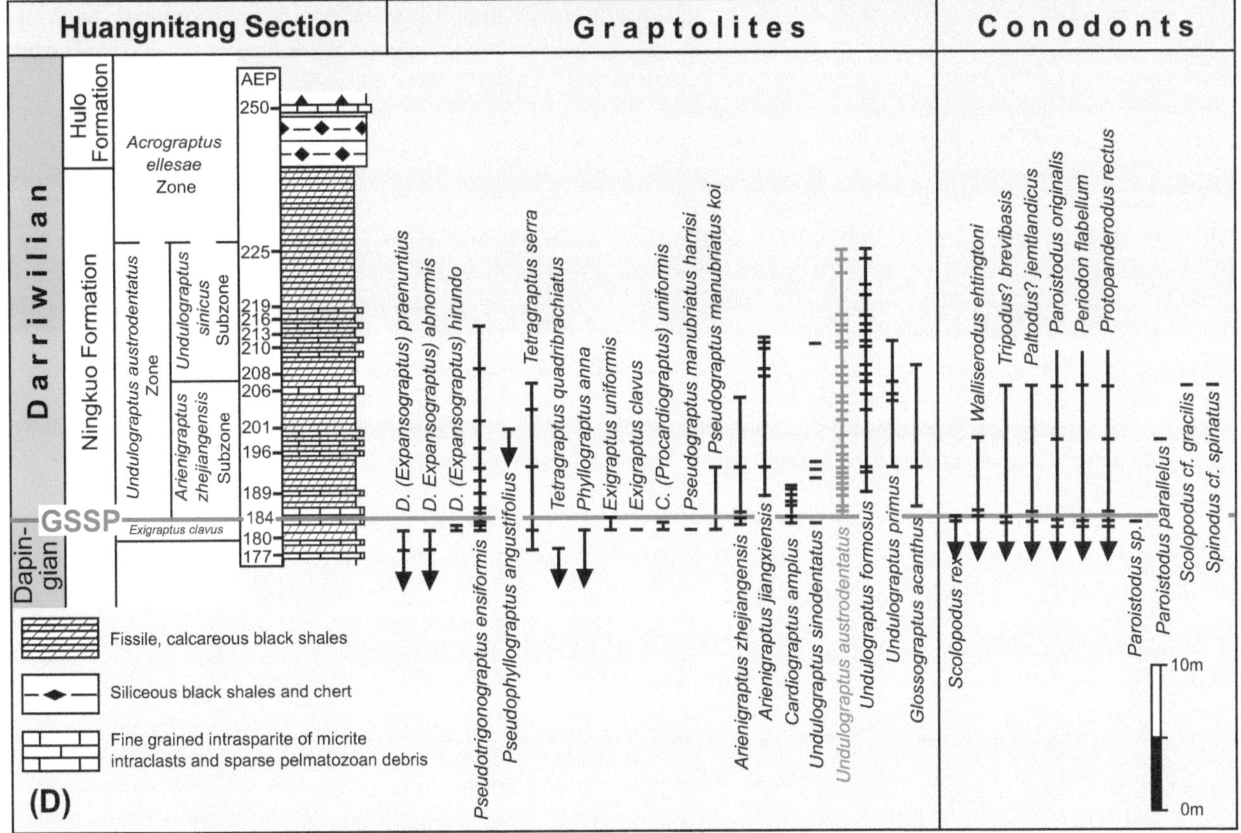

FIGURE 20.5 **Stratigraphic succession across the boundary interval for the formal base of the Darriwilian Stage, Huangnitang GSSP, Changshan, China**, showing stratigraphic ranges of graptolites (left) and conodonts (right), for the *U. austrodentatus* and adjacent zones (from Mitchell *et al.*, 1997: Figure 5). Pictures (A) and (B) are scanning electron micrographs of three-dimensionally preserved, pyritized graptolites from the Ningkuo Formation, all internal molds. A: *Undulograptus sinicus*. B: *Undulograptus austrodentatus*, the first appearance of which coincides with the GSSP. Scale bars 1 mm. *Figures supplied by Zhang Yuandong.*

Mitchell *et al.*, 1997). This level lies just above the first appearance of *Arienigraptus zhejiangensis* and marks the onset of a major change in taxonomic composition of the graptolite fauna in most parts of the world, from one dominated by dichograptids and isograptids to one dominated by diplograptids and glossograptids. The rapid evolutionary radiation of the Diplograptacea along with the appearance of several distinctive pseudisograptid and glossograptid species provide a global transition that is readily recognized in many graptolite sequences around the world. The stage was approved by the ICS in 1997.

The stage base lies very close to, and immediately above, the base of the conodont zone of *Microzarkodina parva*, which marks the appearance of both *M. parva* and *Baltoniodus norrlandicus*, in the Baltoscandian succession. It lies close to the appearance of the zone fossil *Histiodella sinuosa* in the North American (midcontinent) conodont succession (Chen and Bergström, 1995).

The stage corresponds exactly to the Australasian stage after which it is named (VandenBerg and Cooper, 1992), comprising the four graptolite zones Da1 *Undulograptus austrodentatus,* Da2 *Undulograptus intersitus,* Da3 *Pseudoclimacograptus decoratus,* and Da4 *Archiclimacograptus riddellensis*. It marks the progressive increase in relative abundance of diplograptid graptolites. In the middle Darriwilian Da3 Zone, a narrow diversity peak marks a sharp and short-lived expansion of a diverse dichograptacean and glossograptacean assemblage, including didymograptids sigmagraptids, sinograptids, isograptids, and glossograptids (Cooper *et al.*, 2004; Sadler *et al.*, 2011). It is followed by an extinction event, at the end of the Da3 Zone, one of the most severe within the Ordovician prior to the Hirnantian. In the upper Darriwilian Stage, graptolite diversity was greatly reduced.

In terms of the Baltoscandian conodont zones, the Darriwilian Stage embraces the (middle and upper) *norrlandicus* Zone, the *variabilis, suecicus,* and *serra* Zones, and the lower *anserinus* Zone. In terms of the midcontinent conodont zones, the Darriwilian Stage ranges from the *sinuosa* Zone, through *holodentata, polystrophos,* and *friendsvillensis* Zones to the lower *sweeti* Zone.

20.1.3. Stages of the Upper Ordovician

20.1.3.1. Sandbian Stage

The base of the Sandbian Stage, and of the Upper Ordovician Series, is defined at the first appearance of the globally distributed zonal graptolite, *Nemagraptus gracilis*. The GSSP is in outcrop E14b, located on the south bank of Sularp Brook at the locality known as Fågelsång in the Province of Scania, Sweden, 1.4 m below the Fågelsång Phosphorite marker bed in the Dicellograptus Shale (Figure 20.6) (Bergström *et al.*, 2000). This level coincides with the base of the *N. gracilis* graptolite zone (Finney and Bergström, 1986). *Nemagraptus gracilis* is also used to define the base of the British Caradoc Series (Fortey *et al.*, 1995), the Australasian Gisbornian Stage (VandenBerg and Cooper, 1992) and the Chinese Hanjiang Series. The base lies near the middle of the conodont zone of *Pygodus anserinus* (at approximately the base of the *Amorphognathus inaequalis* Subzone), which has global correlation value. The level also corresponds to the middle part of the *Laufeldochitina stentor* chitinozoan zone. The stage name is derived from the community of South Sandby, where the GSSP lies. The GSSP was approved by the ICS in 2002 and the name in 2005 (Nõlvak *et al.*, 2006)

Following an abrupt decline in the later part of the Darriwilian, graptolites expanded in diversity through the Sandbian, a trend driven by proliferation of the Dicranograptidae. Dichograptids are rare and diplograptids abundant.

20.1.3.2. Katian Stage

The base of the Katian Stage is taken at the first appearance of the global correlation graptolite, *Diplacanthograptus caudatus* (Figure 20.7). The GSSP is at a point 4.0 m above the base of the Bigfork Chert in a quarried exposure along Black Knob Ridge, about 5 km northeast of Atoka in southeastern Oklahoma (Goldman *et al.*, 2007). The level is very close to the first appearances of several other widespread graptolite species: *D. lanceolatus, Corynoides americanus, Orthograptus pageanus, O. quadrimucronatus* and *Neurograptus margaritatus* (Figure 20.7). The base of the stage lies high in the North American conodont zone of *Amorphognathus tvaerensis*, just below the base of the *Plectodina tenuis* Zone, and in the middle-upper part of the chitinozoan *Spinachitina cervicornis* Zone. Other important marker horizons are the Millbrig and Kinnekulle K-bentonite complexes in Eastern North America and Scandinavia respectively, which lie just below the GSSP level, and the beginning of the Gutenberg (GICE) positive δ^{13}C excursion (Figure 20.7).

The name Katian is derived from Katy Lake (now drained) near the southern end of Black Knob Ridge (Bergström *et al.*, 2006a). The GSSP and stage name were ratified by the ICS in 2005. The upper part of the Katian, equivalent to the Australasian zones Bo1-Bo3 (*C. uncinatus–P. pacificus*), contains a rich graptolite fauna including dicellograptids, climacograptids (*Appendispinograptus* and *Euclimacograptus*) and orthograptids (the 'DCO' fauna), and species diversity reached a peak (Chen *et al.*, 2005). It was followed, at the end of the Katian, by the most severe depletion in diversity of the entire Ordovician, during which the DCO fauna, along with a wide range of other fossil groups (see below), was completely extinguished (Melchin and Mitchell, 1991; Chen *et al.*, 2005; Finney *et al.*, 2007).

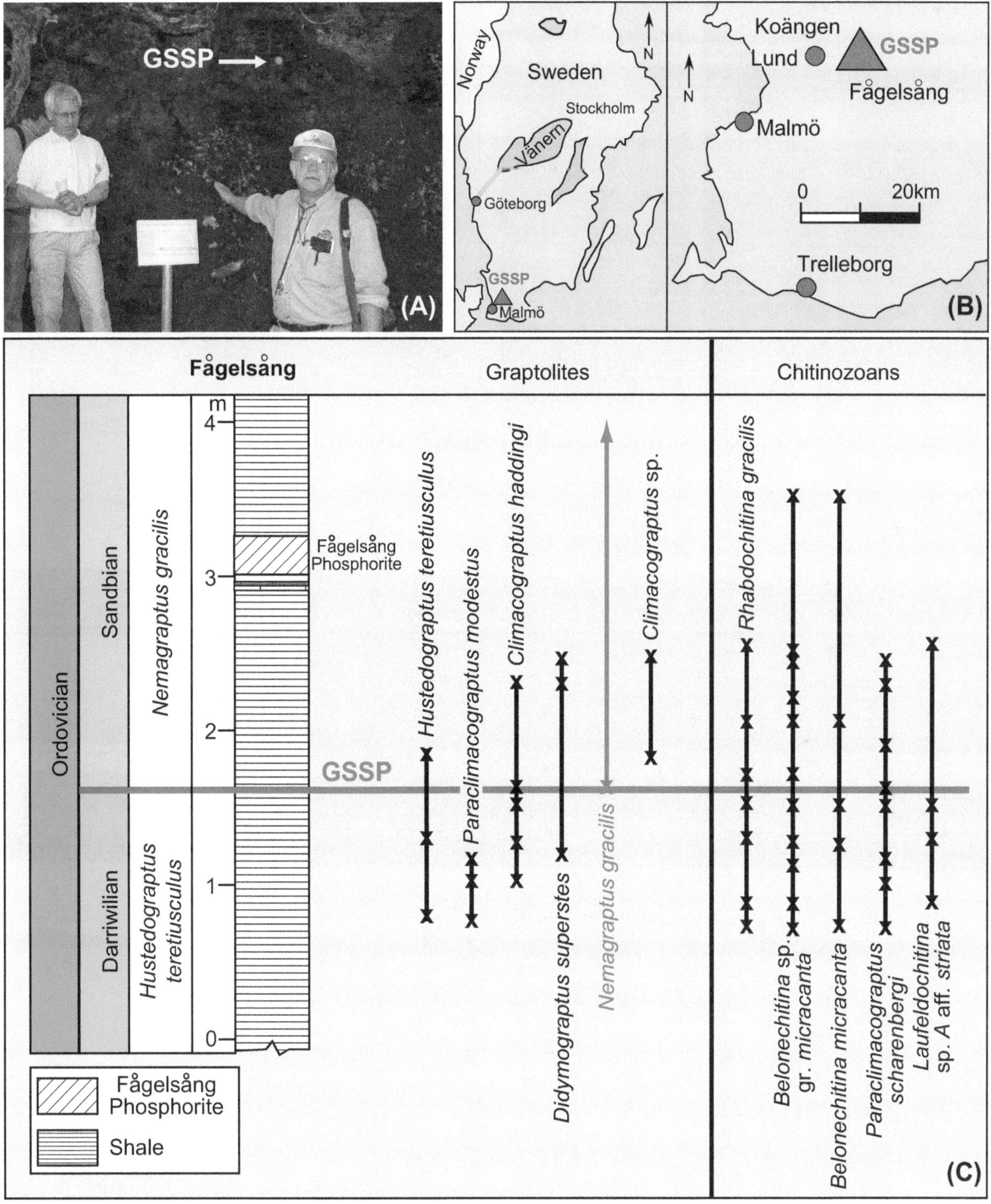

FIGURE 20.6 **Stratigraphic ranges of graptolites and chitinozoans across the base of the *Nemagraptus gracilis* Biozone at the E14b outcrop, Fågelsång, Sweden, global stratotype section for the Sandbian Stage**. Position of samples are marked with x's along the range bars. *From Bergström et al. (2000: Figure 5).*

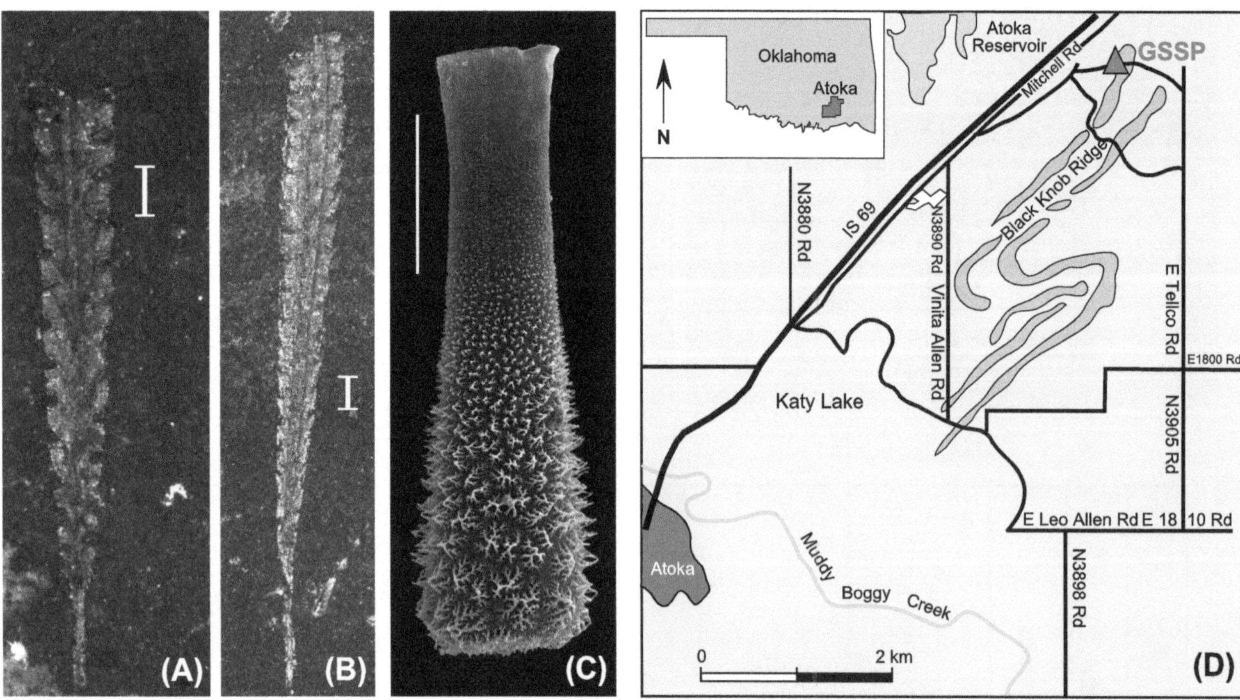

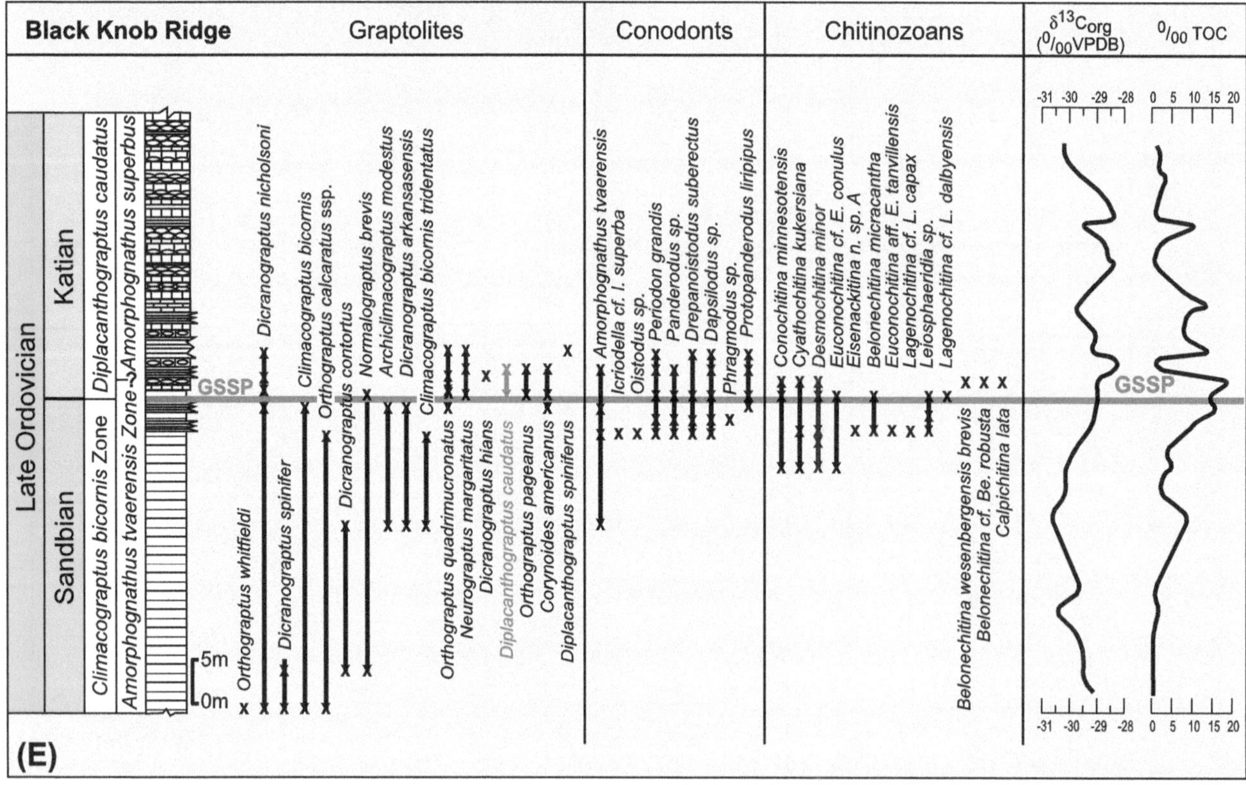

FIGURE 20.7 **Stratigraphic column (in subfigure E) with a range chart of graptolites, conodonts, and chitinozoans for the Black Knob Ridge section, Oklahoma, USA, global stratotype section for the Katian Stage** (from Goldman et al., 2007: Figure 5). The carbon isotope stratigraphy of the Sandbian–Katian transition shows the sharply defined GICE positive excursion considered to have inter-regional, possibly global, correlation value (see also Figure 20.10). *Diplacanthograptus caudatus* from the Big Fork Chert at Black Knob Ridge is shown in subfigures A and B; the first appearance of this species in this section coincides with the GSSP for the base of the Katian Stage (photograph by D. Goldman), scale bar is 1 mm. Subfigure C is the chitinozoan index *Belonechitina robusta* (Eisenack); the specimen is from the lowermost Viola Springs Formation, Sect. D near Fittstown, Oklahoma, USA (figure supplied by J. Nõlvak). Length of scale bar is 0.1 mm. Recent active quarrying near the GSSP will require an assessment of possible damage to the section and whether an alternative section needs to be considered. D. Goldman pers. comm. February 2012.

20.1.3.3. Hirnantian Stage

The uppermost stage of the Ordovician is the Hirnantian. The stage closely corresponds to the classical Hirnantian, the uppermost subdivision of the Ashgill regional series of Britain, and is named after the Hirnant Limestone at Bala in Wales. The lower stage boundary is placed at the first appearance of the graptolite, *Normalograptus extraordinarius,* and the GSSP is at Wangjiawan, 42 km north of Yichang City, western Hubei, China (Figure 20.8) (Chen *et al.*, 2006). The Hirnantian Stage coincides with major climatic oscillations and eustatic events, and one or more strong positive carbon isotope excursions (Figure 20.8), events associated with the end Ordovician glaciation (Brenchley *et al.*, 2003; Brenchley, 2004; Melchin and Holmden, 2006). Graptolites, along with many other fossil groups (see below), were drastically reduced in diversity to only a few lineages and were almost completely wiped out, in the graptolite zones of *Normalograptus extraordinarius* and *N. persculptus*. A distinctive shelly fossil association, the *Hirnantia*–*Dalmanitina* fauna (commonly referred to as the '*Hirnantia* fauna') (Figure 20.8), is widely distributed around the world in strata of latest Ordovician (generally Hirnantian) age (Rong and Harper, 1988). The Hirnantian stage spans the chitinozoan biozones of *Spinachitina taugourdeaui* and *Chonochitina scabra* (Paris *et al.*, 2004).

The top of the stage is defined by the base of the overlying Silurian System at the base of the *Akidograptus ascensus* Zone, marked by the first appearance of the graptolite *Akidograptus ascensus* in the Dob's Linn section of southern Scotland (Melchin and Williams, 2000).

20.2. PREVIOUS STANDARD DIVISIONS

The British series – Tremadoc, Arenig, Llanvirn, Llandeilo, Caradoc, and Ashgill – established in North Wales and England, have been the most widely used chronostratigraphic units around the world prior to the present internationally defined stages, and it is likely that they will continue to be useful, particularly in high-paleolatitude regions. The classical British series divisions that came into use at the turn of the century were applied to the British graptolite zonal succession by Elles in 1925. It is in the sense of Elles (1925) that the series have been most widely applied and correlated around the world (Skevington, 1963; Fortey *et al.*, 1995), but there has always been a problem of how to relate the graptolite-based divisions to the classical divisions which were tied to "type areas" containing shelly faunas. This problem has been considerably alleviated by redefining the series using graptolite bioevents (Fortey *et al.*, 1995, 2000).

The series have been fully reviewed in several papers (Whittington *et al.*, 1984; Fortey *et al.*, 1991, 1995, 2000). The following summary accepts the recommendations in these works. Fortey *et al.* (1995) recommend boundary stratotype sections and levels and, for all but the Ashgill, boundaries based on a graptolite datum. They expanded the Llanvirn to include the Llandeilo of Elles, relegating the Llandeilo Series to stage status and reducing the number of series in the Ordovician to five. This classification is outlined here.

20.2.1. Tremadoc

Sedgwick (1847) introduced the stratigraphical term Tremadoc Group for trilobite-, graptolite-, and mollusk-bearing strata in Wales. Salter (1866) excluded the beds with *Rhabdinopora sociale*, referring them to the underlying Lingula flags, and Marr (1905), who introduced the term Tremadoc Series, followed Salter's classification. Most authors (Fearnsides, 1905, 1910; Whittington *et al.*, 1984; Fortey *et al.*, 1995), however, have included the *Rhabdinopora*-bearing beds in the Tremadoc, the "type area" for which is in North Wales around the town of Tremadog (Fortey *et al.*, 1995). Reviews are given in Whittington *et al.* (1984).

A base for the British Tremadoc Series has been proposed by Fortey *et al.* (1991) to be at Bryn-llyn-fawr in North Wales, at the horizon of appearance of *Rhabdinopora flabelliforme, sensu lato,* about 2 m above the base of the Dol-cyn-afon Member of the Cwmhesgen Formation. This level was adopted by Fortey *et al.* (1995). The level equates to that taken as a proxy for base of the Ordovician System in shale sections (Cooper *et al.*, 2001). The Tremadoc Series thus closely corresponds in scope with the international Tremadocian Stage.

20.2.2. Arenig

Following Sedgwick's (1852) reference to rocks around the mountain Arenig Fawr as characterizing a broad lithological division of the Lower Paleozoic overlying the Tremadoc, the Arenig Series became established as the second division in the Ordovician of Britain (Fearnsides, 1905). The base is an unconformity in the type area, and there has been much debate about its international correlation (Whittington *et al.*, 1984). Following Fortey *et al.* (1995) the base is taken at the base of the *Tetragraptus approximatus* Zone, thereby bringing definition of the boundary into accord with general international usage (Skevington, 1963). The best section in Britain that spans this level is Trusmador in the Lake District, but the zone base cannot yet be located with any precision. The base, as redefined by Fortey *et al.* (1995), also accords with the base of the international Floian Stage.

20.2.3. Llanvirn

The Llanvirn Group was erected by Hicks (1881) for rocks previously referred to as "Upper Arenig" and "Lower Llandeilo" in sections near Llanvirn Farm, southwest Wales.

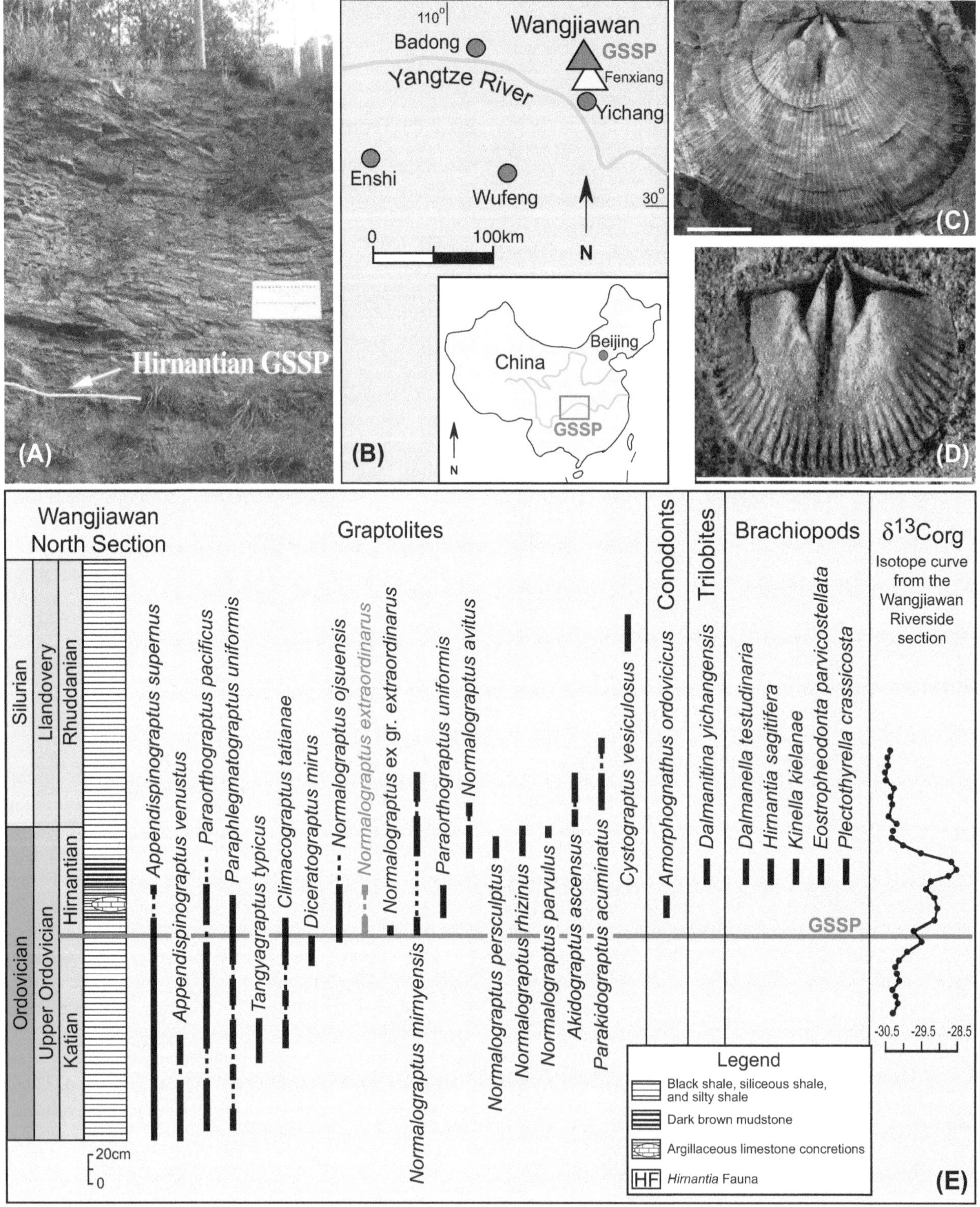

FIGURE 20.8 **Stratigraphic column in subfigure E through the Ordovician–Silurian transition and the formal base of the Hirnantian Stage, with ranges of graptolites and shelly faunas, Wangjiawan North section, Yichang, China (see subfigure B).** From Chen *et al.* (2006: Figure 6). The organic carbon isotopic curve of isolated kerogen with the GICE positive excursion at the Wangjiawan Riverside section, Yichang, China is also shown in subfigure E (from Chen *et al.*, 2006: Figure 11). The GSSP level in the Wangjiawan outcrop section (green arrow) is in subfigure A. The *Hirnantia* fauna brachiopods from the Kuanyinchiao Bed (upper *Normalograptus extraordinarius* Biozone to lower *Normalograptus persculptus* Biozone), Wangjiawan, western Hubei, central China, are part of a distinctive *Hirnantia* shelly fauna that is known worldwide. C: *Hirnantia sagittifera* (M'Coy), dorsal internal mold. D: *Kinnella kielanae* (Temple), dorsal internal mold. Scale bar is 5 mm for both figures (supplied and identified by Rong Jiayu).

Its base has generally been taken at the base of the *Didymograptus artus* Zone (Elles, 1925) and Fortey *et al.* (1995) indicate an appropriate section for boundary stratotype in the Llanfallteg Formation in South Wales. This boundary cannot be precisely located in low paleolatitudes and has not been favored as an international correlation datum by the Ordovician subcommission. Fortey *et al.* (1995) incorporated as a stage within the Llanvirn the entire Llandeilo Series, as it has generally been used internationally. The "historical" Llandeilo lies largely within the redefined Caradoc (see below).

20.2.4. Caradoc

The Caradoc Sandstone (Murchison, 1839), exposed adjacent to the hill Caer Caradoc, near Church Stretton, south Shropshire, is the basis of the name Caradoc Series. The type section became accepted as the Onny River section but, unfortunately, in this region the base of the series is an unconformity (Whittington *et al.*, 1984). Consequently, there has been uncertainty about age and correlation of the base. Internationally, the base of the Caradoc Series has generally been taken at the base of the graptolite zone of *Nemagraptus gracilis*, following Elles (1925) and the British Geological Survey. However, this level has been found to lie well down in the "historical" Llandeilo Series, causing much confusion. Fortey *et al.* (1995) take the base to be at the base of the *gracilis* Zone thus removing the bulk of the historical Llandeilo Series to the Caradoc. The base, thus defined, accords with the base of the Sandbian Stage.

20.2.5. Ashgill

The "Ashgillian Series" was erected by Marr (1905) who subsequently (1913) designated the Cautley district of northwest Yorkshire as the "type area". The base of the Ashgill Series was taken by Fortey *et al.* (1991) to be at Foggy Gill in the Cautley district, at a level marking the appearance of a number of shelly fossils, including the trilobites *Brongniartella bulbosa* and *Gravicalymene jugifera* and the brachiopods *Onniella* cf. *argentea* and *Chonetoidea* aff. *radiatula*. The absence of diagnostic graptolites and conodonts in the Foggy Gill section has led to continuing debate about correlation of this level with the graptolite and conodont successions, even within Britain (Williams and Bruton, 1983; Whittington *et al.*, 1984). Fortey *et al.* (1995) correlate the base with a level within the *Pleurograptus linearis* Zone, and close to, but just above, the base of the conodont zone of *Amorphognathus ordovicicus*. On the basis of chitinozoan correlations, Vandenbroucke *et al.* (2005) suggest a much lower level, in the middle Katian. Here we provisionally follow Webby *et al.* (2004a) and align its base with the base of the Bolindian Stage and the *A. ordovicicus* zone.

20.2.6. Australasian Stages

Graptolite-based stages were established in Victoria, Australia, in the late nineteenth and early twentieth centuries (Hall, 1895; Harris and Keble, 1932; Harris and Thomas, 1938). A suite of nine stages has been used for the Ordovician of Australia and New Zealand for over 70 years (Harris and Thomas, 1938; Thomas, 1960; VandenBerg and Cooper, 1992). In upward sequence they are the Lancefieldian, Bendigonian, Chewtonian, Castlemainian, Yapeenian, Darri wilian, Gisbornian, Eastonian, and Bolindian Stages. They have proved to be widely applicable in graptolite successions around the world, particularly those representing low-paleolatitude regions (30° N to 30° S of the paleoequator), such as North America, Cordilleran South America, Greenland, and Spitsbergen. As originally defined and used, the nine stages were, in effect, groupings of graptolite zones (see below) and their lower boundaries were taken at zone boundaries (Figure 20.1).

Only one stage, the Lancefieldian, has a lower boundary formally defined by a boundary stratotype (Cooper and Stewart, 1979). In a move toward establishing the remaining stages as chronostratigraphic units, bioevents and reference sections for defining their lower boundaries have been given by VandenBerg and Cooper (1992). One stage, the Darriwilian, has since been adopted for international use, with a lower boundary stratotype (GSSP) established in China (see above).

The Australian stages have been used by Cooper and Sadler (2004), Sadler *et al.* (2009) and herein (Figure 20.1) for calibration of the Ordovician time scale because they are readily correlated with other regional stage successions and with the international stage GSSPs.

20.2.7. Ordovician Stage Slices

In an attempt to define correlation units finer than stages for the analysis of biodiversification, Webby *et al.* (2004b) proposed a set of what they called "time slices". Six primary divisions (labeled 1–6, essentially equivalent to the six Ordovician stages) and 19 secondary divisions (labeled 1a–d, 2a–c, 3a–b, 4a–c, 5a–d, and 6a–c) were listed. As Bergström *et al.* (2009) have pointed out, the designation "time slice" implies that these are chronostratigraphic units; however, their original definitions were rather broad and their status is not entirely clear. They have nevertheless proved to be useful for infra stage-level correlation and have become increasingly common in the literature (e.g. Trotter *et al.* 2008; Shields *et al.* 2003; Nõlvak *et al.* 2006). Bergström *et al.* (2009) revised and redefined the units, calling them "stage slices", and recognizing 20 in all (Figures 20.9 and 20.10). They (Bergström *et al.*, 2009, p. 102) stated that, in terms of stratigraphic scope, "a stage slice falls between a stage and a faunal zone, that is, it corresponds to a substage or superzone". On the same page, stage slices are said to be "informal,

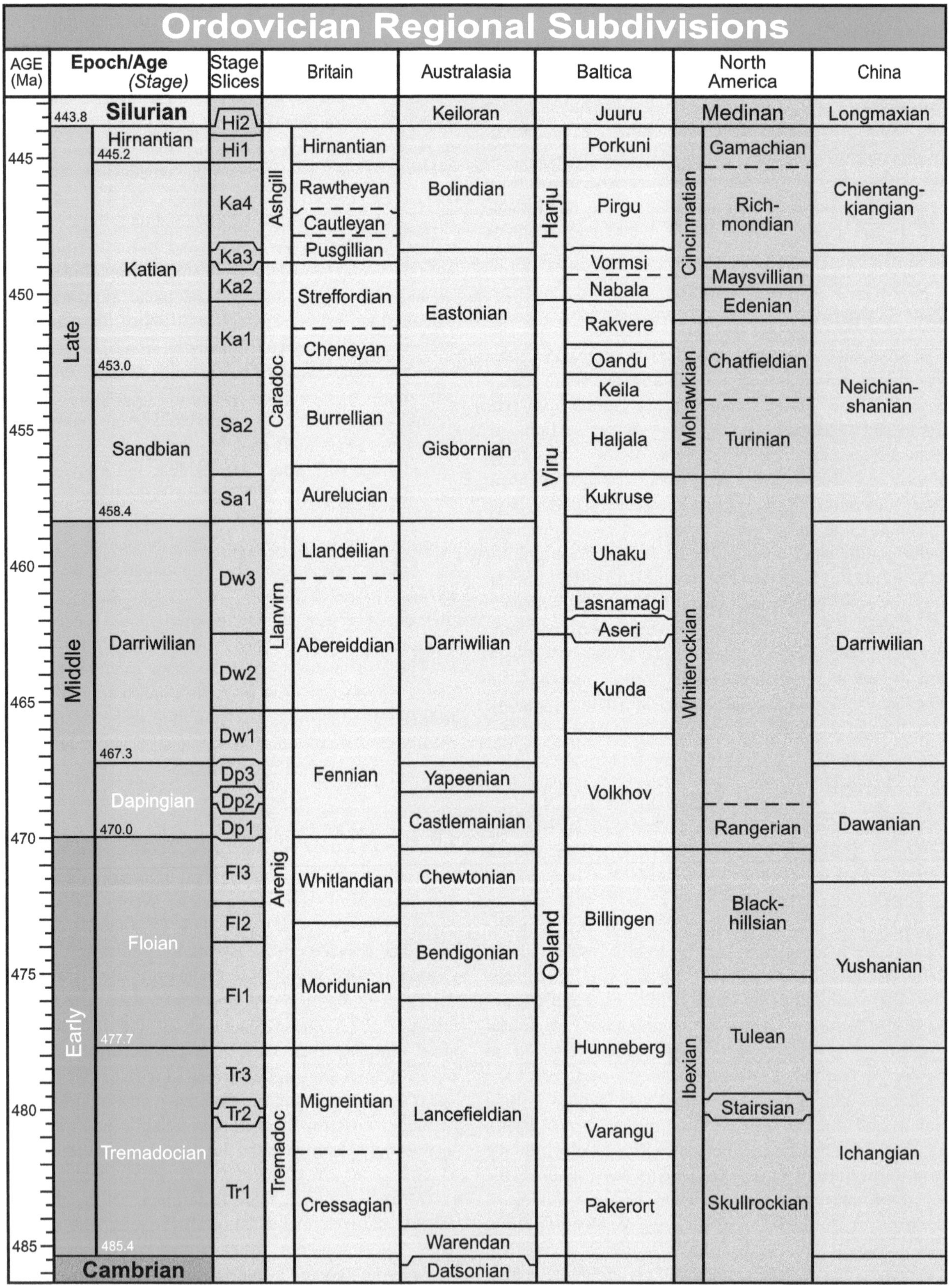

Ordovician Regional Subdivisions

AGE (Ma): 445, 450, 455, 460, 465, 470, 475, 480, 485

Epoch/Age (Stage):
- Silurian (443.8)
- Late: Hirnantian (445.2), Katian (453.0), Sandbian (458.4)
- Middle: Darriwilian (467.3), Dapingian (470.0)
- Early: Floian (477.7), Tremadocian (485.4)
- Cambrian

Stage Slices: Hi2, Hi1, Ka4, Ka3, Ka2, Ka1, Sa2, Sa1, Dw3, Dw2, Dw1, Dp3, Dp2, Dp1, Fl3, Fl2, Fl1, Tr3, Tr2, Tr1

Britain:
- Ashgill: Hirnantian, Rawtheyan, Cautleyan, Pusgillian
- Caradoc: Streffordian, Cheneyan, Burrellian, Aurelucian
- Llanvirn: Llandeilian, Abereiddian
- Arenig: Fennian, Whitlandian, Moridunian
- Tremadoc: Migneintian, Cressagian

Australasia: Keiloran, Bolindian, Eastonian, Gisbornian, Darriwilian, Yapeenian, Castlemainian, Chewtonian, Bendigonian, Lancefieldian, Warendan, Datsonian

Baltica:
- Juuru
- Harju: Porkuni, Pirgu, Vormsi
- Viru: Nabala, Rakvere, Oandu, Keila, Haljala, Kukruse, Uhaku, Lasnamagi, Aseri
- Oeland: Kunda, Volkhov, Billingen, Hunneberg, Varangu, Pakerort

North America:
- Medinan
- Cincinnatian: Gamachian, Rich-mondian, Maysvillian, Edenian
- Mohawkian: Chatfieldian, Turinian
- Whiterockian: Rangerian
- Ibexian: Black-hillsian, Tulean, Stairsian, Skullrockian

China: Longmaxian, Chientang-kiangian, Neichian-shanian, Darriwilian, Dawanian, Yushanian, Ichangian

FIGURE 20.9 **International Ordovician stages and selected regional suites of Ordovician stages and series**. The calibration, taken from Figures 20.11, 12, and 13, applies to the Australasian stage boundaries. Other regions are calibrated by correlation with the Australasian stages. *Stage slices are from Bergström et al. (2009).*

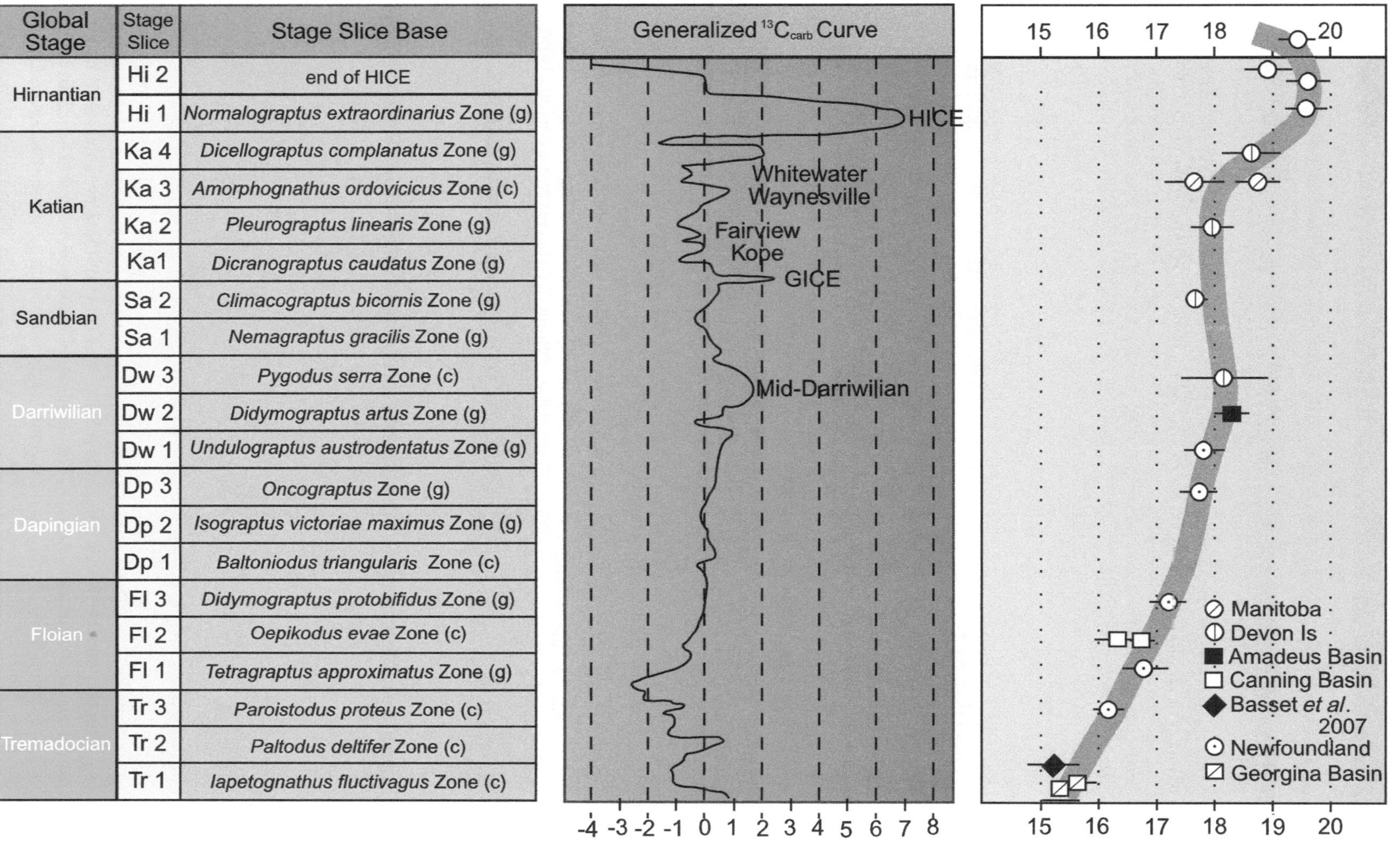

FIGURE 20.10 Left panel, Ordovician stage slices, their lower boundary definitions, and their relationship to the stages (from Bergström *et al.*, 2009: Figure 2). Centre panel, generalized, composite $^{13}C_{carb}$ curve compiled by Bergström *et al.* (2009: Figure 2), based on the following sources: Floian and Tremadocian (Buggisch *et al.*, 2003), Sandbian, Darriwilian and Katian (Kaljo *et al.*, 2007), Hirnantian (Finney *et al.*, 1999). Right panel, oxygen isotope compositions of bioapatite of conodonts from Canada and Australia, 2-sigma error bars shown (Trotter *et al.*, 2008: Figure 2). Note: stage slices are given equal duration, vertical axis not to scale. For details see text.

but defined chronostratigraphic units". Although it remains debatable as to whether they are essentially chronostratigraphic or biostratigraphic in nature, the clearer definitions in Bergström *et al.* render the units less ambiguous and more useful. We regard the units as informal and follow the definitions of Bergström *et al.* (2009), all but one of which define a lower boundary based on a geographically widespread graptolite or conodont bioevent. The exception is the Hirnantian Hi2 stage slice, the lower boundary of which is at the top of the global $\delta^{13}C$ excursion (HICE). In terms of definition, the stage slices are directly equivalent to the Australasian graptolite zones which have been used widely for inter-regional correlation of graptolite sequences. The 20 stage slices, and the bioevents on which they are based, are listed in Figure 20.10.

20.3. ORDOVICIAN STRATIGRAPHY

20.3.1. Biostratigraphy

The two most reliable and cosmopolitan fossil groups for correlation in the Ordovician are graptolites and conodonts. Graptolites are the most abundant in shale sections, particularly those of the outer continental shelf, slope and ocean, whereas conodonts are most abundant in carbonate sections of the shelf and platform. Together they provide a biostratigraphic correlation framework that can be applied with high precision across a wide range of facies and latitudinal zones (Bergström, 1986; Cooper, 1999b). Other fossil groups that are useful for regional and inter-regional correlation include trilobites, brachiopods, and, in Upper Ordovician carbonate facies, corals and stromatoporoids (Webby *et al.*, 2000, 2004a). Chitinozoans and acritarchs are increasingly being used for global correlation (Achab, 1989; Paris, 1996; Webby *et al.*, 2004b; Grahn and Nõlvak, 2007). Chitinozoans first appear at the base of the Ordovician and diversify through the period, reaching a total of 35 genera and a peak in raw diversity in the late Darriwilian (Paris *et al.*, 2004).

20.3.1.1. Graptolite Zones

Graptolites (Phylum Hemichordata) were a major component of the Ordovician macroplankton. They lived at various depths in the ocean waters (Chen, 1990; Cooper *et al.*, 1991; Chen *et al.*, 2001), were particularly abundant along continental margins in upwelling zones (Fortey and Cocks, 1986; Finney and Berry, 1997), and are found in a wide range of sedimentary facies. Many graptolite species dispersed rapidly, are widespread globally, and are of relatively short stratigraphic duration, lasting for 2 myr or less. These attributes combine to make them extremely valuable fossils for zoning and correlating Ordovician and Silurian strata (Skevington, 1963; Cooper and Lindholm, 1990; Webby *et al.*, 2004a). Together with conodonts, they are the primary fossil group for global correlation of Ordovician successions.

One of the most detailed and best-established regional zonal schemes, spanning almost the entire Ordovician, is that of Australasia (Figure 20.1; Harris and Thomas, 1938; Thomas, 1960; VandenBerg and Cooper, 1992), widely applicable in low-paleolatitude regions ("Pacific Province", 30° N–30° S). Thirty zones, two of which are divided into subzones, are recognized giving an average duration of 1.5 myr each. Zones are based on the stratigraphic ranges of species, most zone boundaries being tied to first appearance events. The zone boundaries have been calibrated by Sadler *et al.* (2009) by the CONOP (constrained optimization) method using a data set of 430 stratigraphic sections from around the world. This method seeks the earliest occurrence of each taxon in any section and builds a composite that can differ in detail from the regional successions of first and last appearance events. The most representative zonal scheme for middle to high paleolatitudes ("Atlantic Province") is that of southern Britain (Zalasiewiscz *et al.*, 2009) where some 16 zones span the Tremadoc to middle Caradoc, averaging 2.3 myr each. Many species were cosmopolitan and correlation between high- and low-paleolatitude regions is well controlled throughout most of the period. Other important graptolite zonal schemes for correlation are those of Cordilleran North America, eastern North America, Scandinavia, and South China (Figure 20.1).

20.3.1.2. Conodont Zones

Conodonts are tooth-like structures of primitive chordates, small eel-like animals that were predators, living in shoals in Paleozoic seas (Aldridge and Briggs, 1989). They were most abundant on continental shelves and were readily preserved in shelf carbonates. Conodonts are composed of calcium phosphate and can be extracted from the carbonate rock by acid digestion. The conodont animal roamed widely, and some species are found in a wide range of sedimentary environments and geographical regions, making them valuable fossils for long-range correlation. Conodonts range from the early Cambrian to the Triassic and are used as zone fossils in all these periods.

In the Ordovician, conodont fauna, like graptolite fauna, are distributed in two major biogeographic provinces (Sweet and Bergström, 1976, 1984). A warm-water province ranged about 30° N and S of the Equator, and a cold-water province extended poleward from 30–40° latitude. The warm-water province, typified by the North American midcontinent region (Sweet and Bergström, 1976), contains a diverse and rich fauna that can be finely zoned – 26 zones are listed in Figure 20.1 and some of these are subdivided into subzones. The cold-water province is best known from the North Atlantic region in which 17 zones and several subzones are recognized. The two zonal successions provide good resolution through the Lower and Middle Ordovician. In the Upper

Ordovician, there are 11 warm-water zones and only three cold-water zones.

20.3.2. Evolutionary Events – The Ordovician Radiation and Mass Extinction

The Ordovician Period encompasses one of the greatest evolutionary radiations recorded in the history of life (Webby *et al.*, 2004a). This is generally known as the Great Ordovician Biodiversification Event or GOBE. Marine biodiversity increased three-fold between the end of the Cambrian and the middle Sandbian, to reach a level (about 1600 genera) that was not significantly exceeded during the remainder of the Paleozoic and the early Mesozoic (Sepkoski, 1995) or possibly, up to the Paleocene (Alroy, 2008). The cause of this radiation is not clear but several factors favoring biodiversification were evident during the Ordovician, including the widespread development of shallow continental seas, marked geographic provincialism, strong magmatic and tectonic activity and a relatively low-diversity starting base in the late Cambrian. Two spectacular bursts in diversity took place, one in the late Arenig and the other in the late Llanvirn to early Caradoc, mainly through expansion in what Sepkoski has termed the Paleozoic Evolutionary Fauna. A bottom dwelling community dominated by suspension-feeders was installed and many new megaguilds introduced (Bottjer *et al.*, 2001). Bioturbation increased and biotic tiering above and below the sediment–water interface expanded the range of niche space. Brachiopods, trilobites, corals, echinoderms, bryozoans, gastropods, bivalves, nautiloids, graptolites, and conodonts all show major generic increases through the Ordovician (Sepkoski, 1995; Webby *et al.*, 2004a). They replace the trilobite-dominated communities of the Cambrian Evolutionary Fauna. Successive community replacements in an onshore-to-offshore direction saw the nature of the marine fauna change almost completely, from the Cambrian to the Silurian (Sepkoski and Sheehan, 1983), and a pattern established that was to last for the following 200 myr of the Paleozoic.

Barnes *et al.* (1995) recognized five major evolutionary events through the Ordovician, based on the faunal histories of conodonts, trilobites and graptolites. The events are marked by marked depletion of diversity for each group, followed by a radiation into a more diversified fauna. These all lie at stage or series boundaries:

(1) The base of the Tremadocian Stage (and base of the Ordovician),
(2) The base of the Floian Stage,
(3) The base of the Darriwilian Stage,
(4) The base of the Sandbian Stage, and
(5) The base of the Hirnantian Stage.

The basal-Ordovician event marks the origination of euconodonts and planktonic graptolites, the abundant appearance of new platform trilobites and early radiation of inarticulate brachiopods and nautiloid cephalopods. Significant faunal turnover in all three groups, as well as other groups such as brachiopods and nautiloid cephalopods, mark the following three evolutionary events. The last event marks the second greatest mass depletion in diversity in the Paleozoic. Each evolutionary event is accompanied by a significant eustatic event.

The late Katian to Hirnantian depletion in diversity was driven by a mass extinction (Bambach *et al.*, 2004). Estimates of losses range from 22–26% of all marine animal families, 49–61% of genera and about 85% of species, making it one of the largest in the Phanerozoic (Brenchley, 1989; Jablonski, 1991; Sepkoski, 1995; Sheehan, 2001). Trilobites, brachiopods, graptolites, echinoderms, conodonts, coral, and chitinozoans were drastically reduced in generic diversity, and the event caused a major faunal turnover. The extinction event coincides with the climatic and sea-level fluctuations associated with the latest Ordovician glaciation, and was probably brought about, at least in part, by a combination of factors such as reduced shelf- and platform-habitable space, cold and fluctuating temperature, and perturbation of the ocean stratification and circulation systems (Brenchley, 1989).

20.3.3. Magnetostratigraphy

The Ordovician is dominated by a ~20-myr reversed-polarity "Moyero" superchron that spans the upper Tremadocian through Darriwilian stages (Pavlov and Gallet, 2005). Magnetostratigraphic studies, mainly in sedimentary sequences from Siberia, have documented only two significant normal-polarity intervals in the lower Tremadocian, but potentially a higher frequency of reversals in the uppermost Darriwilian through Katian (summarized in Pavlov and Gallet, 2005). The majority of the Ordovician magnetic polarity scale awaits detailed calibration to biostratigraphy and verification in multiple sections.

20.3.4. Carbon Isotope Stratigraphy

A continuous $\delta^{13}C$ curve for the entire Ordovician is not yet available from any single continent. A synthetic curve, compiled from a number of regional curves, has recently been published by Bergström *et al.* (2009) and is reproduced in Figure 20.10. These authors calibrate the curve to the Ordovician stage slices which they redefine (see above), and which are readily correlated to many regional stage schemes. The curve is built from the following regional curves:

The Hirnantian curve from Monitor Range, central Nevada (Finney *et al.*, 1999; Berry *et al.*, 2002);
The Katian curve from the eastern Midcontinent of North America (Bergström *et al.*, 2007);

The Sandbian–Darriwilian curve from Estonia (Kaljo *et al.*, 2007); and
The Dapingian–Tremadocian curve from Argentina (Buggisch *et al.*, 2003).

Although it is composite and generalized, Bergström *et al.* (2009) believe the synthetic curve is representative of the changes in $\delta^{13}C$ values through the period.

The carbon isotope stratigraphy of the Katian and Hirnantian stages has been most intensively studied. In these stages the curve shows several sharply defined positive excursions, several of which (including the Hirnantian HICE and Katian GICE) have been shown to be of high inter-regional, possibly global, correlation value (Ainsaar *et al.*, 2004, 2010; Barta *et al.*, 2007; Kaljo *et al.*, 2004, 2007; Ludvigsen *et al.*, 2004; Young *et al.*, 2005; Melchin and Holmden, 2006; Bergström *et al.*, 2006b, 2007, 2010; Schmitz and Bergström, 2007). A prominent positive excursion in the Darriwilian is widely recognized in Baltoscandia (Kaljo *et al.*, 2004, 2007; Ainsaar *et al.*, 2010) but may be missing from Cordilleran North America (Saltzman and Young, 2005). In the Tremadocian Stage and latest Cambrian, a number of short-lived excursions have been recognized, with apparently good correlation value, in Precordilleran Argentina, Australia, South China and Utah (Ripperdan *et al.*, 1992; Ripperdan and Miller, 1995; Buggisch *et al.*, 2003).

20.3.5. Eustatic and Climatic Events

Eustatic curves for the Ordovician have been proposed by several authors (e.g., Woodcock, 1990; Thorshøj Nielsen, 1992, 2004; Nicoll *et al.*, 1992; Ross and Ross, 1995; Kanygin *et al.*, 2010). Although consensus on a global eustatic pattern has yet to be achieved (Munnecke *et al.*, 2010), several significant, abrupt eustatic changes are widely recognized (Figure 20.1), and include the *Acerocare*, Black Mountain and *Ceratopyge* Regressive Events in the Tremadocian, the major transgression in the Floian, the regression in the Dapingian to early Darriwilian, the transgression in the Sandbian, and the regression(s) in the Hirnantian. This last event probably had multiple phases and was associated with glaciation in high paleolatitudes (Brenchley, 1984, 1989; Brenchley *et al.*, 1994, Ghienne, 2003, Melchin and Holmden, 2006) and a mass extinction event (Sepkoski, 1995). A distinctive cold-water shelly fauna known as the "*Hirnantia* fauna" can be recognized globally in the Hirnantian Stage.

Using oxygen isotopic composition of apatite in conodonts determined by the SHRIMP II ion microprobe, Trotter *et al.* (2008) present evidence that tropical seawater temperatures changed through the Ordovician in four main phases:

i) sustained cooling during the Early Ordovician (~15.3 to 18.3‰ $\delta^{18}O$ V-SMOW) reflecting a shift from warm greenhouse conditions (~42°C) to modern equatorial temperatures over about 20 myr;
ii) climate stability for about 20 myr from Darriwilian to middle Katian;
iii) a rapid temperature drop during the late Katian and Hirnantian marking the Hirnantian glaciation;
iv) a rapid return to equatorial temperatures by the early Silurian.

When changes in the isotopic composition of seawater are allowed for (Finnegan *et al.*, 2011), somewhat higher Late Ordovician sea surface temperatures result. Using $\delta^{18}O_{carb}$ determined in brachiopods, bryozoans, crinoids, corals and trilobites, these authors found tropical sea surface temperatures stayed within the range 32–37°C from the early Katian to the late Aeronian. $\delta^{18}O_{water}$ values indicate the buildup of pre-Hirnantian ice sheets at least intermittently during the Katian. This initial glaciation of Gondwana in the Katian occurred with little or no cooling of the tropical oceans, and tropical sea surface temperatures of 34–37°C exceeded the present day range. During the Hirnantian glacial maximum they cooled but remained in excess of 30°C, indicating a steep latitudinal temperature gradient. In the Aeronian, temperatures were rapidly restored to Katian levels.

Biological, sedimentological, geochemical and isotopic evidence suggests a long interval of predominantly greenhouse conditions from the Early Ordovician to early Katian, with N-limiting oceans and widespread anoxia. It was during this interval that global biodiversity dramatically expanded. The global ocean–atmosphere system changed during the Katian to more volatile and predominantly icehouse conditions, with P-limiting and more ventilated oceans, which persisted through the late Katian, Hirnantian and Silurian (Saltzman and Young 2005; Saltzman 2005; Trotter *et al.*, 2008; Ainsaar *et al.*, 2010; Munnecke *et al.*, 2010; Finnegan *et al.*, 2011). The Hirnantian represents a short, intense glacial episode within this protracted interval of icehouse conditions.

20.3.6. Sr Isotope Stratigraphy

The $^{87}Sr/^{86}Sr$ curve for the interval is based on the data of Denison *et al.* (1998), Shields *et al.* (2003) and Young *et al.* (2009). Marine $^{87}Sr/^{86}Sr$ was around 0.7090 at the beginning of the Ordovician and decreased slowly through the Tremadocian, Floian, and Dapingian. Around the time represented by the *Pygodus serra* Zone (North Atlantic conodont zonation) of the Darriwilian, values of $^{87}Sr/^{86}Sr$ began a more rapid decrease that continued through the Sandbian and early Katian, before beginning a long-term increase in $^{87}Sr/^{86}Sr$ through the Silurian (Chapters 7 and 21). The rate of rapid decrease in $^{87}Sr/^{86}Sr$ is difficult to assess, owing to uncertainty over whether, in the ash-rich sediments of the North American Sandbian and Katian, alteration has increased or decreased the $^{87}Sr/^{86}Sr$ of bulk micrite (Denison *et al.*, 1999; Young *et al.*, 2009) and macrofossils (Shields *et al.*, 2003). The rate of change is between 0.000 045 and 0.000 080 per myr, making it the highest, or amongst the highest, in the Phanerozoic. According to Shields *et al.* (2003) and Young *et al.* (2009) the decrease arose through rapid weathering of volcanics of mantle affinity *viz.* island-arc volcanic rocks in Kazakhstan, and the island-arc volcanic rocks associated with the Taconic Orogeny of eastern Laurentia.

20.4. ORDOVICIAN TIME SCALE

20.4.1. Radiogenic Isotope Dates

Although many radiogenic isotope dates have been obtained from Ordovician rocks (e.g., Gale, 1985; Kunk *et al.*, 1985; McKerrow *et al.*, 1985; Odin, 1985; Tucker and McKerrow, 1995; Harland *et al.*, 1990; Compston and Williams, 1992), only few reach the standard of precision, analytical quality, and biostratigraphic constraint desirable for time scale calibration. An initiative by R.J. Ross and his colleagues in North America and Britain in the late 1980s systematically sampled and dated British volcanic ash beds which can be biostratigraphically constrained by, or reliably correlated with, the British graptolite zones. Zircon separations from the samples were initially dated by the fission track method (Ross *et al.*, 1982), but the resulting dates had unacceptably long uncertainty intervals for refining the time scale. However, zircons from many of the samples were subsequently dated by the U-Pb method and have produced the high-precision dates used in the present calibration. In addition, volcanogenic zircons have been recovered by other workers from ash beds in North America and Baltica, and other localities in Britain.

Tucker and McKerrow (1995) reviewed the available dates and rejected all but those based on zircon crystals and determined using the U-Pb isotope system, except for a selected number of dates by the $^{40}Ar/^{39}Ar$ and Sm-Nd methods, using other minerals. They produced a calibration for the Early Paleozoic that became widely accepted. We follow Tucker and McKerrow in accepting only the high-resolution, biostratigraphically constrained dates. Eleven of the dated Ordovician samples that were used by Tucker and McKerrow are used here. They range in age from middle Darriwilian to earliest Silurian. The stratigraphic and radiogenic isotope ages have been revised as follows and are shown in Appendix 2 of this book, which summarizes the isotopic data, dating method, and biostratigraphic constraints.

Dates previously reported under various uranium and lead isotopic decay systems ($^{206}Pb/^{238}U$, $^{207}Pb/^{235}U$, and $^{207}Pb/^{206}Pb$), have, where possible, here been mapped onto the $^{206}Pb/^{238}U$ decay system (see Chapter 6). Similarly, error propagation of analytical, tracer/standard and decay constant errors has been treated consistently and rigorously for each published age. All dates have been reappraised for analytical and biostratigraphic reliability. The few $^{40}Ar/^{39}Ar$ ages have also been recalibrated to the Kuiper *et al.* (2008) FCs standard age of 28.201 $\pm$ 0.046 with associated systematic error propagation. The ages for many of the dated samples in Appendix 2 therefore differ from those given in the original publications, and several of the dates used for time scale calibration by Cooper and Sadler (2004) and Sadler *et al.* (2009) are here revised or rejected. Also, the biostratigraphic age assignments of some samples are revised following a reassessment of the associated fauna or of its correlation.

For reasons discussed by Villeneuve (2004), zircon dates determined with the high-resolution secondary ion mass spectrometry (HR-SIMS) method in the Canberra Geochronological Laboratory (Compston and Williams, 1992; Compston, 2000a,b), using standard SL13, are not employed here in Ordovician time scale construction.

Two dates accepted by Tucker and McKerrow were rejected in GTS2004; their dated item 16, a Sm-Nd date on the Borrowdale Volcanics in the Lake District, has a large error range and poor biostratigraphic constraint, and their dated item 17, from Llanwrtyd Wells in Wales, has poor biostratigraphic constraint. Two further dates used by Tucker and McKerrow and in GTS2004 are rejected here. An $^{40}Ar/^{39}Ar$ date of 454.1 $\pm$ 2.0 Ma (item 20a) on biotite from the Millbrig K-bentonite (Kunk *et al.*, 1985) and an $^{40}Ar/^{39}Ar$ date of 455 $\pm$ 3 Ma (item 18) on biotite and sanidine from K-bentonite at Kinnekulle, Sweden (Kunk and Sutter, 1984) are analytically inferior to newer age determinations of these bentonites.

Since 1995, three $^{40}Ar/^{39}Ar$ age determinations on biotites have been published that are acceptable for calibration of the Ordovician time scale. The K-bentonite in the Dalby Limestone at Mossen Quarry at Kinnekulle, Sweden, has been dated by Min *et al.* (2001). Five concordant $^{40}Ar/^{39}Ar$ plateau spectra give a weighted mean age of 457.4 $\pm$ 2.9 Ma when recalibrated to FCs 28.201 $\pm$ 0.046 (Kuiper *et al.*, 2008), in good agreement with the $^{206}Pb/^{238}U$ zircon age of this bentonite reported by Tucker and McKerrow (1995), here recalculated as 456.9 $\pm$ 2.10 Ma. The same authors (Min *et al.*, 2001) also dated the Dieke K-bentonite at Lexington Quarry, Kentucky. Nine concordant $^{40}Ar/^{39}Ar$ plateau spectra gave a weighted mean age of 452.4 $\pm$ 2.7 Ma when recalibrated to FCs 28.201 $\pm$ 0.046 Ma (Kuiper *et al.*, 2008). This age is within error of the $^{206}Pb/^{238}U$ zircon age determination on the same unit reported in Tucker (1992) and here recalculated as 454.5 $\pm$ 1.10 Ma. The stratigraphically higher Millbrig K-bentonite has given a (recalculated) weighted mean age of 450.6 $\pm$ 2.60 Ma based on three concordant $^{40}Ar/^{39}Ar$ plateau spectra (Min *et al.*, 2001). This age is within error of the $^{206}Pb/^{238}U$ zircon age reported for the Millbrig K-bentonite (Huff *et al.*, 1992) here recalculated as 452.92 $\pm$ 1.35 Ma. The Kinnekulle, Diecke and Millbrig K-bentonites are of similar biostratigraphic age (*Phragmodus undatus* conodont zone, late *Diplograptus multidens* graptolite zone) and the spread of their radiogenic isotope ages (450.6–457.4 Ma) is large. However, the dated Millbrig samples come from different localities (Kentucky and Missouri) and this may also apply to the Diecke samples. In view of the profusion of K-bentonite horizons of about this age in North America and Baltica (Kolata *et al.*, 1996; Sell and Samson, 2011), some dates may not apply to their identified source bentonite, which may, in part, account for the spread of ages.

In addition, five zircon age determinations are here added to the data set. Zircons in two K-bentonite beds in the lower member of the Los Azules Formation in the Precordillera of Argentina have been dated by the isotope dilution method (Huff *et al.*, 1997). One of these (ARG-1) has yielded three multigrain zircon fractions (14 grains in total) which give a weighted mean $^{206}Pb/^{238}U$ age of 465.46 ± 3.53 Ma. A rich graptolite fauna of early Darriwilian (Da2) Zone age is interbedded with the bentonites (Mitchell *et al.*, 1998) making it one of the biostratigraphically best-constrained dated samples in the Ordovician.

A volcanic arc-related rhyolite body in the Tyrone Volcanic Group of Northern Ireland has been dated by the isotope dilution method (Cooper *et al.*, 2008). Zircon crystals from very fine-grained rhyolite at Formil Hill were chemically abraded. Three (out of four) single grain fractions gave a weighted mean $^{206}Pb/^{238}U$ age of 473 ± 0.94 Ma. The rhyolite lies stratigraphically below graphitic pelite and black chert which, at Slieve Gallion nearby, has yielded the global correlation graptolite, *Isograptus victoriae lunatus*. This horizon lies in the *I.v. lunatus* Zone, just below the base of the Dapingian Stage.

A K-bentonite in the Chelsey Drive Group on Cape Breton Island, Nova Scotia has given a weighted mean $^{207}Pb/^{206}Pb$ age of 481.13 ± 2.76 Ma, based on six multigrain zircon samples (Landing *et al.*, 1997). The olenid trilobite *Peltocare rotundifrons* is found below and above the dated bentonite bed, and the conodont *Scandodus* is found 3.5 m below. These taxa suggest an age in the late Tremadocian. *Peltocare rotundifrons* is found elsewhere in the Chelsey Drive Group associated with the graptolites *Hunnegraptus* sp. cf. *copiosus* and *Adelograptus* of *quasimodo* type, supporting the late Tremadocian assignment which is accepted here. Although the biostratigraphic constraint on this bentonite bed is not tight, the bentonite is important in lying within what is otherwise a long, undated interval in the Early Ordovician.

Two closely spaced pyritic, tuffaceous sandstone beds in the Bryn-Llyn-Fawr road section, in the Dolgellau Formation of North Wales, have given a weighted mean $^{207}Pb/^{206}Pb$ age of 486.78 ± 2.57 Ma based on 14 single zircon grains (Landing *et al.*, 2000). Although the zircon crystals provide only a maximum age for the volcanic sandstone, their age is inferred to be close to that of the sandstone. The sandstone overlies strata with *Acerocare* Zone trilobites (late Cambrian) and is immediately overlain by strata with planktonic graptolites, including *Rhabdinopora flabelliformis parabola*, a zonal indicator in the lower Tremadocian. The dated zircon-bearing bed lies very close to the Cambrian–Ordovician boundary.

Also from the Dolgellau Formation, at Ogof-Ddu near Criccieth in North Wales, a crystal-rich volcanic sandstone has yielded nine multigrain zircon fractions (Davidek *et al.*, 1998). The weighted mean $^{207}Pb/^{206}Pb$ age is 488.71 ± 2.78 Ma for the zircons, which themselves are inferred to be of similar age to that of the enclosing volcanic sandstone. The sandstone lies 17 m below the appearance level of the early Ordovician graptolite *Rhabdinopora flabelliformis sociale* and is interbedded with mudstone and siltstone with calcareous nodules that contain a rich trilobite fauna representing the *Peltura scarabaeoides* Zone (late Cambrian). Thus, this date provides a maximum age for the volcanic sandstone that underlies the Cambrian–Ordovician boundary, and is consistent with the preceding zircon date on Dolgellau volcanic sandstone.

Altogether, 26 radiogenic isotopically dated samples are regarded as of sufficient analytical quality and biostratigraphic constraint to be used for calibration of the Ordovician and Silurian time scales (Appendix 2). They range from latest Cambrian to earliest Devonian in age. The 16 intra-Ordovician dates are unevenly distributed, and the Early Ordovician, in particular, is sparsely populated. All Ordovician dated samples, except for two using the $^{40}Ar/^{39}Ar$ method, are based on zircons that are analyzed by the ID TIMS method.

20.4.2. Building the Ordovician and Silurian Composite Standard

Quantitative methods for interpolating period boundaries between nearby radiogenic isotope dates have been used by Harland *et al.* (1990) and, for interpolating Mesozoic stage boundaries, by Gradstein *et al.* (1994), both of whom also estimate the associated error. But where radiogenic isotope dates are sparse, as in much of the Early Paleozoic, these methods cannot be applied and some way of estimating stage durations is necessary. The compromise procedures have received little rigorous attention in the past. Generally, some arbitrary assumption is made, explicitly or implicitly.

In some previous time scales, stages are simply assumed to be of more-or-less uniform duration or scaled according to the number of biozones that they contain (Boucot, 1975; McKerrow *et al.*, 1985; Harland *et al.*, 1990). Another approach assumes constancy in sedimentation rate, so that stage duration can be estimated from its mean stratigraphic thickness (Churkin *et al.*, 1977). This assumption requires judicious selection of the reference section. Graphic correlation has been employed to utilize the stratigraphic thickness of one reference section in which the fossil ranges have been adjusted according to collections made elsewhere (Sweet, 1984, 1988, 1995; Kleffner, 1989; Fordham, 1992). Cooper (1999b) compared graphic correlations from different regions as a test for steadiness of depositional rate. Generally, some unspecified combination of these methods and assumptions, plus a measure of intuition, is employed (Tucker and McKerrow, 1995).

As discussed in Chapters 1 and 3 of this book, a number of quantitative, or semi-quantitative, methods of biostratigraphic correlation can be adapted to derive *relative* time scales – that is, scales that estimate the relative proportions of stages without recourse to radiogenic isotope dates. These are, of course, uncalibrated in millions of years. The *con*strained *op*timization (CONOP) correlation technique, developed by Kemple *et al.* (1995) and Sadler (2001), uses evolutionary programming techniques to find a composite range chart with optimal fit to all the field observations. Thus, it resembles a multidimensional graphic correlation in the sense that it considers all the local stratigraphic sections. It differs, however, in treating all sections simultaneously and avoids the need to select an initial "standard reference section". A closer analogy exists between CONOP and algorithms that search for the most parsimonious cladogram. For time scale work, a well-studied, pandemic fossil group with good biostratigraphic utility is required. Dated tuff beds that are associated with these fossils may be included in the optimization process. For building Ordovician and Silurian time scales, the good documentation and high resolution of graptolite biostratigraphy and the scattered radiogenic isotope dated volcaniclastic beds render the CONOP method suitable. The principles and methodology of applying constrained optimization to time scale development are summarized by Sadler and Cooper (2003) and described in detail by Sadler *et al.* (2009; see also Chapter 3 in this book), who use the graptolite successions of the Ordovician and Silurian to develop and demonstrate the method.

Five hundred and twelve measured stratigraphic sections in graptolitic shales from around the world, containing more than 2000 species, have been compiled in a database that spans the latest Cambrian–Early Devonian. Ten composite sequences were developed at various stages as the database increased in size from 177 sections to 430 sections. They provide a test of the stability of the time scale as additional data are added. With the exception of the very basal portion, every biostratigraphic event level in the Ordovician composite is spanned by at least 10, and at most 80, measured sections, averaging 57. Because planktonic graptolites do not range below the base of the Ordovician, eight trilobite and seven conodont species which range down into the Late Cambrian and which are present in graptolite-bearing sections in the boundary interval, have been included in the database. These help compensate for the lack of graptolites.

In the global composite the order and spacing of events provide a proxy for a biochronology, once calibrated with the age dates. The order and spacing are determined separately, because the spacing requires simplifying assumptions that need not compromise determination of the optimal order (a crucial difference from graphic correlation). First, the optimal order of events is established by minimizing misfit in sequence between the composite and all of the individual sections. Misfit is determined primarily by the net number of event horizons through which the observed range ends must be extended in order to make all the observed range charts fit the same composite sequence. The composite, at this stage, is only an ordinal sequence of events, and is based purely on the sequences of events observed in the measured sections.

Next, the spaces between events in the optimal composite sequence are scaled by the following procedure. The observed fossil ranges in the individual sections are extended as necessary to match the composite sequence, and missing taxa are added. The total thickness of each section is then rescaled according to the number of events that it spans in the composite sequence (assumes that net biologic change is a more reasonable guide to relative duration, *in the long term*, than raw stratigraphic thickness). Finally, the spacing of every pair of adjacent events in the composite is determined from the *average* of the *rescaled* spacings of events in the sections (this assumes that relative thickness is a reasonable guide to relative duration *in the short term*).

The scaling of the composite is therefore derived from all of the sections, and it is the ratio of the thicknesses between events that is used, not the absolute thickness. The influence of aberrant sections, incomplete preservation, and non-uniform depositional rates is thus minimized. Where evolutionary change is rapid, many range-end events fall at the same horizon in measured sections; these "zero spacings" are included in the averaging process and prevent high diversity from being misinterpreted in terms of long time intervals. The procedure is more vulnerable to intervals of extraordinarily low diversity. The graptolite clade survived long after the close of the Ordovician Period, but had very low diversity at the start of the Ordovician. We should anticipate that the composite spacings are least reliable near the base of the system.

Events in the scaled composite are spaced along a scale of arbitrary "composite units". The scale spans the entire graptolite clade, and includes the Silurian and Ordovician periods. This scaled composite is itself a proxy time scale. Graptolite zonal boundaries and stage boundaries are located in the scaled composite to produce a relative time scale for the Ordovician. The zones thus recognized are chronozones rather than biozones in the sense that they are tied to global datums rather than local or regional bioevents.

20.4.2.1. Calibration of Stage Boundaries by Composite Standard Optimization

The Floian Tyrone Volcanics, the three Wenlock bentonites (Hörsne, Djupvik and Wren's Nest) and the Devonian Kalkberg dated bentonites, which became available, or whose biostratigraphic age has been revised, since the Ordovician–Silurian composite for GTS2012 was built, have been located in the composite by the graptolite taxa associated with them, either directly in outcrop or by correlation. The stratigraphic range in the composite is taken as the stratigraphic uncertainty and the mid-point in this range is the level. All other radiogenic isotope dated ash beds used for calibration were included in the optimizing and compositing processes and have been built into the

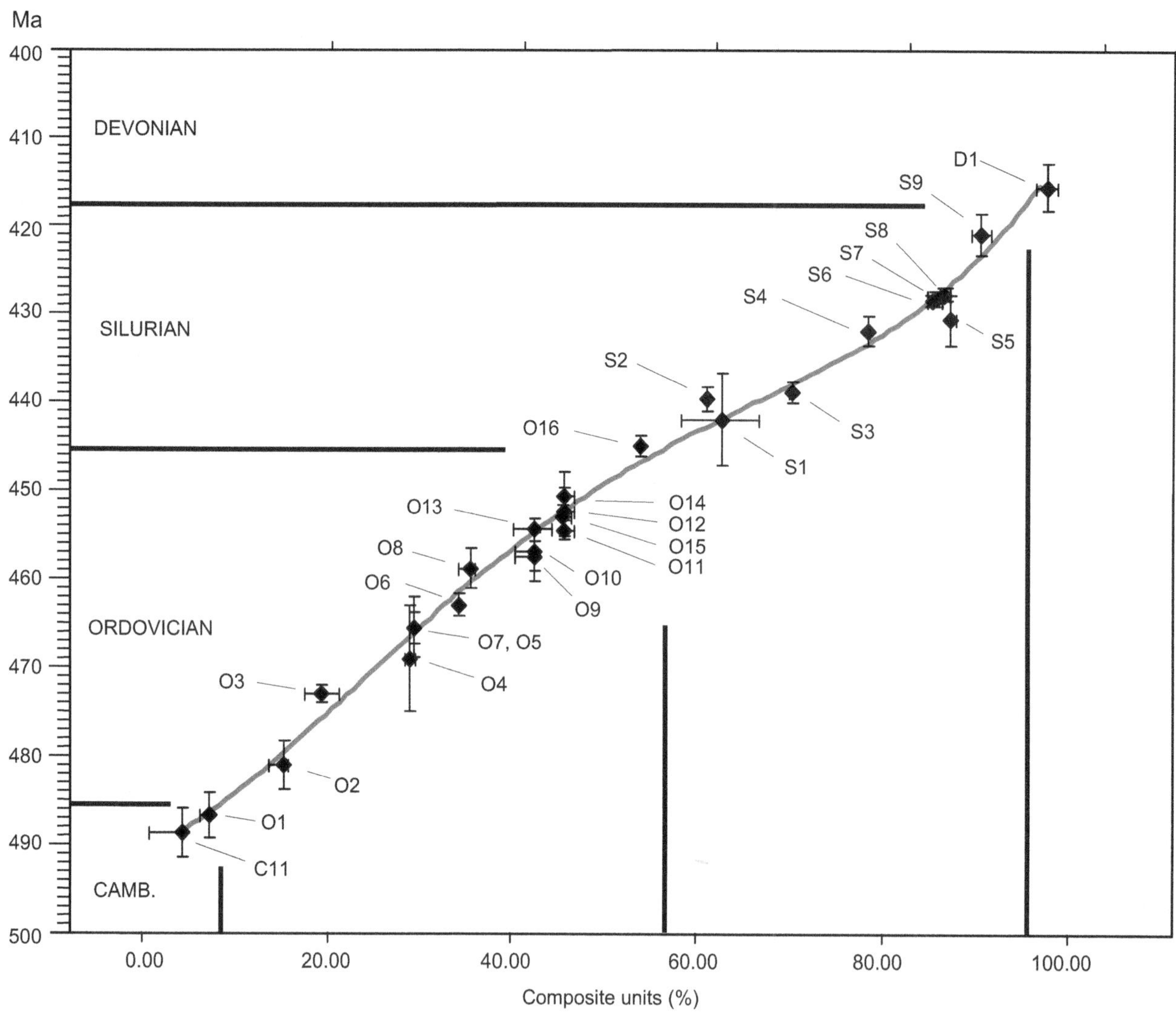

FIGURE 20.11 Calibration of the Ordovician–Silurian graptolite composite standard of Table 20.1 using regression. The y-axis is the chronometric scale in Ma, the x-axis is the composite section scaled in 100 composite units. The radio-isotopically dated ash-fall events are located in the composite section along the x-axis by the compositing procedure, using their associated fossils, and along the y-axis by their age as listed in Appendix 2 of this volume. The dated sample numbers correspond to those listed in Appendix 2. The fitted regression line is a fourth-order polynomial ($y = -0.000\,002\,793\,213\,6x^4 + 0.000\,508\,636\,382\,2x^3 - 0.026\,838\,259\,116\,5x^2 - 0.405\,750\,195\,566\,6x + 490.895\,920\,211\,070\,0$; $R^2 = 0.9924$). Vertical error bars represent analytical error (2 standard deviations); horizontal error bars depict stratigraphic uncertainty. For discussion of the Silurian radio-isotopic dates see Chapter 21.

composite sequence, along with their stratigraphic uncertainty ranges.

A bivariate plot of the ages of dated samples against their position in the composite (Figure 20.11) shows the distribution of dated samples to be generally rectilinear (R^2 = 0.9805). Although there is no *a priori* reason that this best fit should approach linearity, the latter generally is taken to indicate good accordance between the compositing and scaling procedures on the one hand and the radiogenic isotope dating on the other. The chronometric error for radiogenic isotopically dated samples shown is the 2-sigma error associated with the dates, as re-evaluated by Schmitz and Villeneuve (Chapter 6 of this book) and given in Appendix 2. The stratigraphic error shown is as explained above. An improvement in fit is achieved by a polynomial curve (R^2 = 0.9924), which suggests some non-linearity in the scaled composite. The polynomial equation (Figure 20.11) enables the precise level of all taxon range-end events in the composite, including the zone and stage boundary-defining events, to be converted to chronometric ages.

The compositing and scaling process therefore has most influence on the relative duration of shorter time intervals – stages and zones – and the dated events have most influence on the durations of longer time intervals – series and periods – and on the overall fit of the regression line and the resulting calibration. The calculated durations and boundary ages using the polynomial equation for the Ordovician stages and zones, are given in Table 20.1. Note that rounding procedures can

TABLE 20.1 Ordovician Australasian graptolite zones in the CONOP composite and ages of their lower boundaries.

Stage		Graptolite Zone	Level in Composite (%)	Polynomial Age (Ma)	GTS 2012 (Spline) Age (Ma)	2-Sigma	Duration myr
Base Silurian		*Akidograptus ascensus*	56.927	445.33	443.83	1.546	
Hirnantian	Bo5	*Normalograptus persculptus*	56.065	445.84	444.43	1.490	0.6
Hirnantian	Bo4	*Normalogr. extraordinarius*	55.060	446.44	445.16	1.378	0.73
Katian	Bo3	*Paraorthograptus pacificus*	52.692	447.92	447.02	1.138	1.86
Katian	Bo2	pre-*pacificus* zone	51.968	448.38	447.62	1.077	0.6
Katian	Bo1	*Climacograptus? uncinatus*	51.079	448.96	448.38	0.964	0.76
Katian	Ea4	*Dicellograptus gravis*	49.381	450.10	449.86	0.770	1.48
Katian	Ea3	*Dicranograptus kirki*	48.899	450.44	450.28	0.735	0.42
Katian	Ea2	*Diplacanthogr. spiniferus*	47.850	451.17	451.21	0.719	0.93
Katian	Ea1	*Diplacanthogr. lanceolatus*	45.855	452.61	452.97	0.679	1.76
Sandbian	Gi2	*Orthograptus calcaratus*	41.525	455.93	456.63	0.857	3.66
Sandbian	Gi1	*Nemagraptus gracilis*	39.389	457.68	458.36	0.893	1.73
Darriwilian	Da4	*Archiclimacogr. riddellensis*	34.879	461.56	461.95	0.765	3.59
Darriwilian	Da3	*Pseudoclimacogr. decoratus*	32.320	463.87	463.97	0.796	2.02
Darriwilian	Da2	*Undulograptus intersitus*	30.200	465.83	465.66	0.867	1.69
Darriwilian	Da1	*Undulograptus austrodentatus*	28.232	467.69	467.25	1.107	1.59
Dapingian	Ya1-2	*Oncograptus upsilon*	26.935	468.92	468.31	1.224	1.06
Dapingian	Ca4	*I. victoriae maximodivergens*	26.577	469.26	468.61	1.260	0.3
Dapingian	Ca3	*Isograptus victoriae maximus*	26.402	469.43	468.75	1.270	0.14
Dapingian	Ca2	*Isograptus victoriae victoriae*	24.959	470.82	469.96	1.418	1.21
Floian	Ca1	*Isograptus victoriae lunatus*	24.423	471.33	470.42	1.429	0.46
Floian	Ch1-2	*Didymograptus protobifidus*	22.118	473.55	472.41	1.454	1.99
Floian	Be1-4	*Pendeograptus fruticosus*	18.197	477.30	475.97	1.352	3.56
Floian	La3	*Tetragraptus approximatus*	16.342	479.03	477.72	1.388	1.75
Tremadocian	La2b	*Araneograptus murrayi*	14.101	481.07	479.86	1.561	2.14
Tremadocian	La2a	*Aorograptus victoriae*	13.369	481.72	480.55	1.638	0.69
Tremadocian	La1b	*Psigraptus*	12.202	482.74	481.67	1.653	1.12
Tremadocian	La1b	*R.scitulum & Anisograptus*	9.479	485.01	484.28	1.821	2.61
Tremadocian	pre-La	*R. flabelliformis parabola*	8.753	485.59	484.97	1.857	0.69
Tremadocian	–	*Iapetognathus fluctivagus*	8.334	485.91	485.37	1.862	0.4

The age derived from the polynomial regression is given together with the spline age used here. From this calibration, the ages of other zonal suites, and of the Ordovician stage boundaries, are derived by correlation.

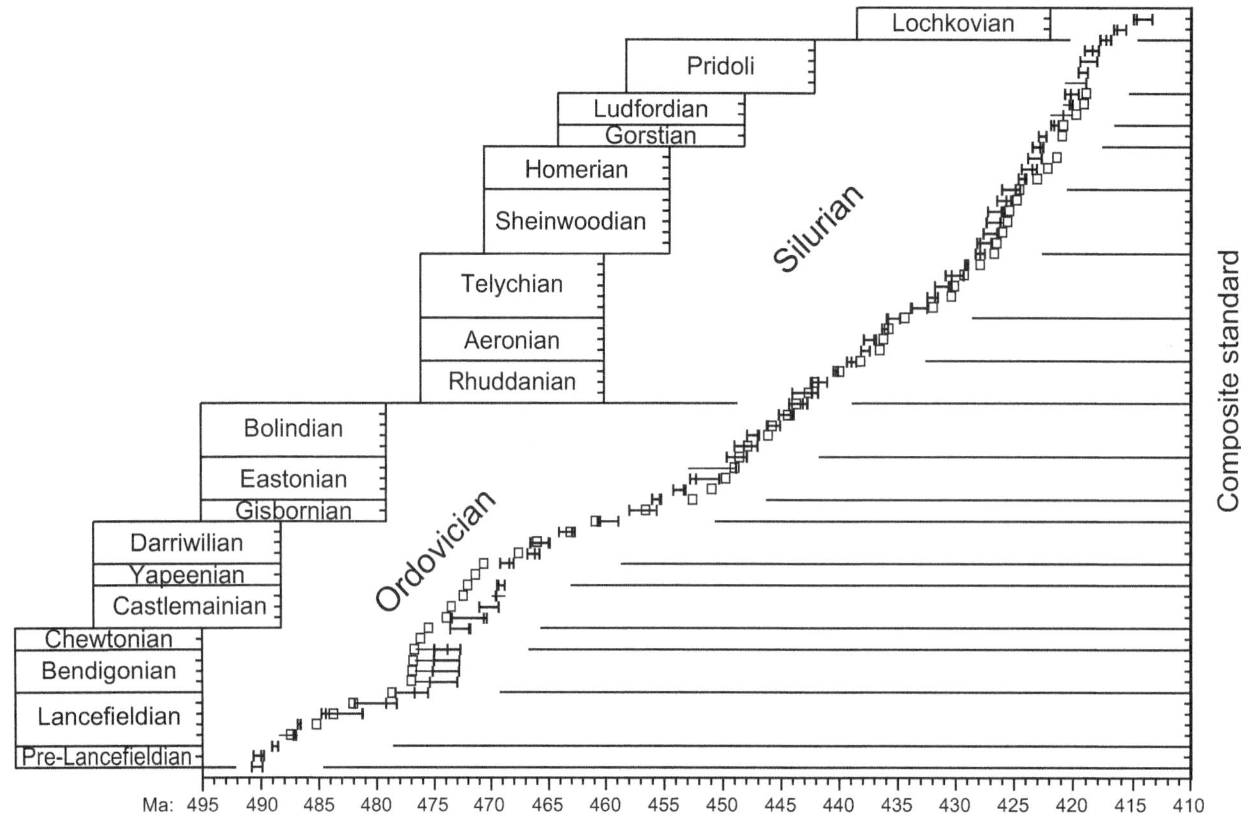

FIGURE 20.12 **Estimation of between-composite uncertainty – range of age of bases of graptolite zones from four composite sequences** (from Sadler *et al.* 2009, Fig. 12). Horizontal lines with vertical bars represent composites 198, 256 and 430; boxes represent composite 214. Zones are given equal spacing along the Y-axis. The four composites represent different data sets and optimization rules (see Sadler *et al.* 2009 for numerical values and other details). This gives the most conservative estimate of uncertainty (between-composite uncertainty). See text for further explanation.

result in slight discrepancy between stage durations and ages of boundaries.

20.4.2.2. Uncertainty in the Composite Standard

Four sources of uncertainty in the Ordovician and Silurian biochronology, chronostratigraphy and time scale itself were recognized by Sadler *et al.* (2009). First, radio-isotopic uncertainty (usually expressed as 2σ) has little effect on our scale because dated events are used as an ordered ensemble and stage boundary ages are derived by regression. Second, optimization uncertainty ("best-fit" intervals of Sadler and Cooper, 2003), which derives from variation in successive composites based on the same data set, is minor; events are relatively stable in repeat re-runs of the composite.

Third, boundary-placement uncertainty derives from the range of positions in the composite within which a stage or zone boundary could equally well be placed. Boundary placement is subjective and depends on matching a local succession, to which the zone applies, with the composite, which is global. The age regression enables conversion of this uncertainty to a time interval in myr. Although boundary-placement uncertainty is not cumulative, the relative duration of zones within stages is quite uncertain for some zones, particularly in the Lower Ordovician (see also Sadler *et al.*, 2009).

Fourthly, the most conservative estimate of uncertainty (between-composite uncertainty) derives from variation in age among 10 successive composites, built as the database expanded (from 177 sections and 570 taxa to 430 sections and 1928 taxa) and as alternative optimizing rules were tried. Regressions were fitted in four of these, and were all nearly linear (R^2 greater than 0.95: Sadler *et al.*, 2009: Figure 5). The between-composite uncertainties on ages of zone bases in four composites are shown in Figure 20.12. These uncertainties are very conservative estimates of stability.

20.4.3. Age of Stage Boundaries

In the final analysis for the age of the stage boundaries and stage duration in the Ordovician and Silurian, the chronometric and stratigraphic ages of the radiogenic isotope dates were subjected to a cubic spline fit as described in Chapter 14, based on the approach used in GTS2004. Points were weighted according to the procedure given in that chapter, taking both radiometric and stratigraphic errors into account. A smoothing factor of 1.45 was calculated by

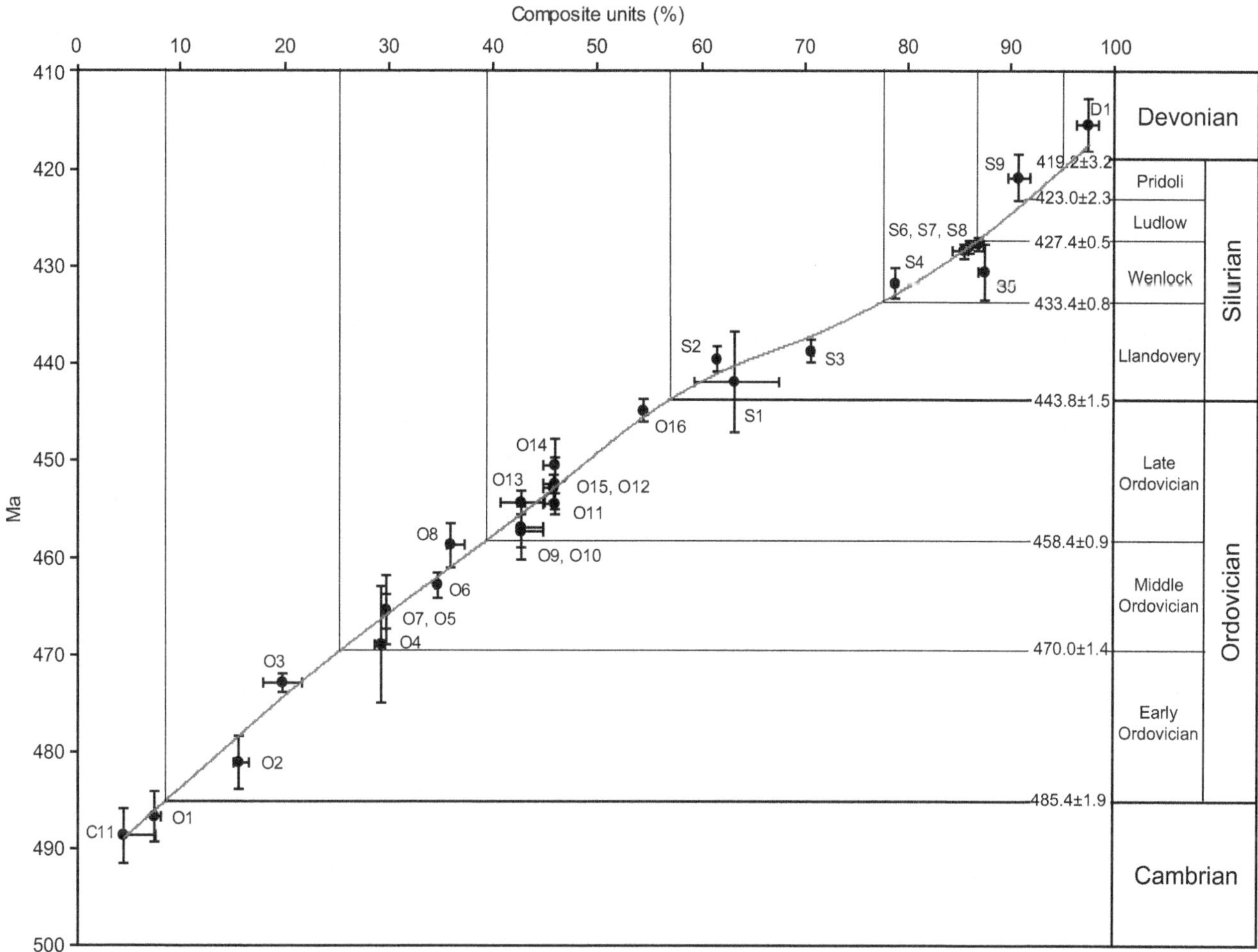

FIGURE 20.13 **Geochronology of the Ordovician and Silurian series with boundary ages and error bars** using a smoothing spline fitted through the 26 radiometrically dated ash beds used for calibrating the Ordovician and Silurian scales (Appendix 2) plotted on the CONOP composite (X-axis) against their analytical age (Y-axis). Compared to GTS2004 the base Ordovician is younger (485.4 ± 1.9 versus 488.3 ± 1.7 Ma) and the base Devonian older (419.2 ± 3.2 versus 416.0 ± 2.8 Ma). Base Silurian stays virtually identical in age. Using the regression of Figure 20.11 on the data, most of the stage boundary ages are (almost) within the error bars of the spline fit. Error bars are larger near the base Ordovician and top Silurian because of the lack of high-resolution constraining points in the Cambrian and Devonian.

cross-validation. Two of the 26 points did not pass the chi-squared goodness-of-fit test at a significance level of $p<0.01$. These points were O4 ($\chi^2=7.01$; $p=0.0081$) and S5 ($\chi^2=6.66$; $p=0.0099$). No post-hoc adjustments were made to the weighting of these points. The resulting spline (Figure 20.13) is close to linear through the Ordovician. A lessening of slope in the Llandovery has the effect of shortening the duration of this series relative to its duration in the composite.

Interpolated stage boundary ages are given in Tables 20.1 and 20.2. The ages computed from the polynomial fit described earlier in this chapter are within the error bars of the ages computed from the spline. The only exception is for the base Aeronian, where the polynomial interpolation age is 1.3 myr older than the spline age, which has a 2-sigma error bar of 1.2 myr.

The Tremadocian lasted from 485.4 to 477.7 Ma, for a duration of 7.7 ± 0.7 myr. The Floian lasted 7.8 ± 1.2 myr, from 477.7 to 470.0 Ma (duration calculated from 2-decimals ages and then rounded up). The Dapingian Stage ranged from 470.0 to 467.3 Ma, for a duration of 2.7 ± 0.3 myr. The Darriwilian lasted for 8.9 ± 0.9 myr, from 467.3 to 458.4 Ma, and is thus the longest stage in the Ordovician and Silurian. The Sandbian Stage started at 458.4 and ended at 453.0 Ma, lasting 5.4 ± 0.4 myr. The Katian ran from 453.0 to 445.2 Ma for 7.8 ± 1.3 myr and the Hirnantian lasted only 1.3 ± 0.2 myr, from 445.2 Ma to base Silurian at 443.8 Ma.

The Early Ordovician had a duration of 15.4 ± 1.3 myr, the Middle Ordovician 11.6 ± 1.0 myr and the Late Ordovician 14.5 ± 1.5 myr.

The scale differs from that of GTS2004 and Sadler *et al.* (2009), mainly because of the revision of the chronometric

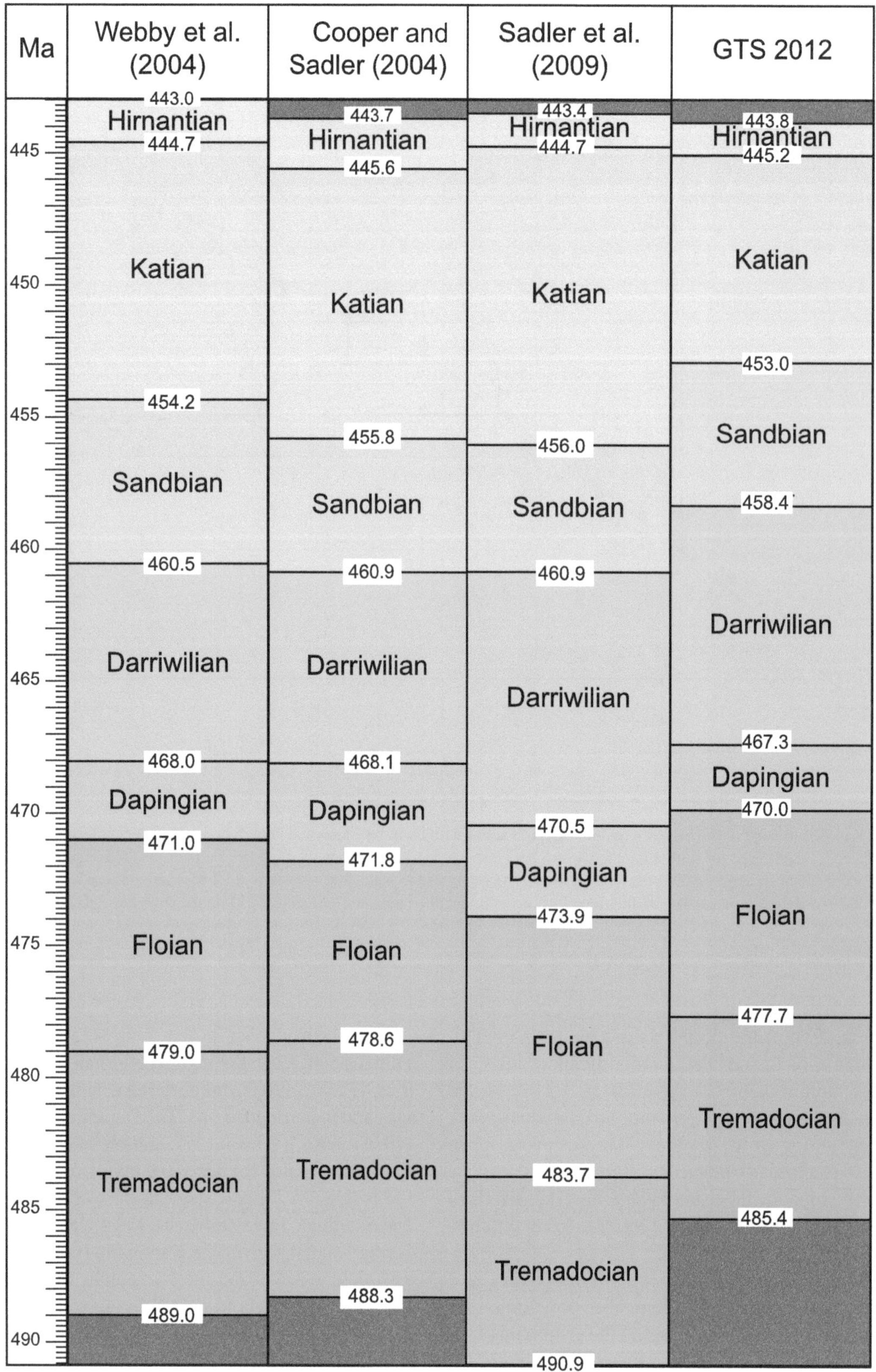

FIGURE 20.14 **Comparison of GTS2012 with previous time scales**; Cooper and Sadler (2004) is GTS2004 (see text for discussion).

TABLE 20.2 Geochronology of the Ordovician–Silurian Stages for GTS2012, with a Comparison to GTS2004

Period	Epoch	Stage	Base Ma GTS2012	Est. ± myr (2σ)	Epoch Duration (myr)	Est. ± myr (2σ)	Stage Duration (myr)	Est. ± myr (2σ)	Base Ma GTS2004
Devonian	Early	Lochkovian	419.2	3.2					416
Silurian	Pridoli		423.0	2.3	3.8	1.0	3.8	1.0	
	Ludlow	Ludfordian	425.6	0.9			2.6	1.1	421.3
		Gorstian	427.4	0.5	4.4	1.6	1.8	0.5	422.9
	Wenlock	Homerian	430.5	0.7			3.1	0.6	426.2
		Sheinwoodian	433.4	0.8	6.0	0.7	2.9	0.3	428.2
	Llandovery	Telychian	438.5	1.1			5.1	1.0	436
		Aeronian	440.8	1.2	10.4	1.4	2.3	1.3	439
		Rhuddanian	443.8	1.5			3.0	0.5	443.7
Ordovician	Late	Hirnantian	445.2	1.4			1.4	0.2	445.6
		Katian	453.0	0.7	14.6	1.5	7.8	1.3	455.8
		Sandbian	458.4	0.9			5.4	0.4	460.9
	Middle	Darriwilian	467.3	1.1			8.9	0.9	468.1
		Dapingian	470.0	1.4	11.6	1.0	2.7	0.3	471.8
	Early	Floian	477.7	1.4			7.8	1.2	478.6
		Tremadocian	485.4	1.9	15.4	1.3	7.7	0.7	488.3

TABLE 20.3 GSSPs of the Ordovician stages, with location and primary correlation criteria (status Jan. 2012)

Stage	GSSP Location	Latitude, Longitude	Boundary Level	Correlation Events	Reference
Hirnantian	Wangjiawan North section, N of Yichang city, Western Hubei Province, China	30°59′2.68"N 111°25′10.76"E	0.39 m below the base of the Kuanyinchiao Bed	Graptolite FAD *Normalograptus extraordinarius*	Episodes **29**/3, 2006
Katian	Black Knob Ridge Section, Atoka, Oklahoma (USA)	34°25.829′N 96°4.473′W	4.0 m above the base of the Bigfork Chert	Graptolite FAD *Diplacanthograptus caudatus*	Episodes **30**/4, 2007
Sandbian	Sularp Brook, Fågelsång, Sweden	55°42′49.3"N 13°19′31.8"E*	1.4 m below a phosphorite marker bed in the E14b outcrop	Graptolite FAD *Nemagraptus gracilis*	Episodes **23**/2, 2000
Darriwilian	Huangnitang section, Changshan, Zhejiang Province, SE China	28°51′14"N 118°29′23"E*	base of Bed AEP 184	Graptolite FAD *Undulograptus austrodentatus*	Episodes **20**/3, 1997
Dapingian	Huanghuachang section, NE of Yichang city, Hubei Province, S, China	30°51′ 37.8"N 110°22′ 26.5"E	10.57 m above base of the Dawan Formation	Conodont FAD of *Baltoniodus triangularis*	*Proposal in Episodes* **28**/2, 2005
Floian	Diabasbrottet, Hunneberg, Sweden	58°21′32.2"N 12°30′08.6"E	in the lower Tøyen Shale, 2.1 m above the top of the Cambrian	Graptolite FAD *Tetragraptus approximatus*	Episodes **27**/4, 2004
Tremadocian *(base Ordovician)*	Green Point Section, western Newfoundland	49°40′58.5"N 57°57′55.09"W*	at the 101.8 m level, within Bed 23, in the measured section	Conodont FAD *Iapetognathus fluctivagus*	Episodes **24**/1, 2001

*according to Google Earth

and stratigraphic age of many of the dated samples, discussed above, the addition of eight new age dates to the data set, and the rejection of four age dates used in previous scales. For the present scale, a much larger biostratigraphic data set has been utilized (512 sections and 2000 taxa, versus 256 sections and 1400 taxa) giving higher stratigraphic resolution (Figure 20.14).

The error bars on boundary ages are large near the base Ordovician and top Silurian because of the lack of constraining points in the Cambrian and Devonian. This is unavoidable at this stage because creating a long spline from base Cambrian through Devonian, or even younger, is not warranted. Cambrian and Devonian stage scaling is not well constrained, and there are fewer stratigraphically precise radiogenic isotope ages. Hence, the endpoints using the spline of the internal Ordovician–Silurian scaling are respectively the top Cambrian and base Devonian constraining boundary ages.

Most of the GTS2004 ages are outside the spline error bars, such that the Ordovician ages have here generally been moved significantly up (younger) and the Silurian significantly down (older) relative to GTS2004. The Ordovician–Silurian boundary stays almost in position at 443.8 ± 1.5 Ma (443.7 ± 1.5 in GTS2004). The Cambrian–Ordovician boundary is here estimated at 485.4 ± 1.9 Ma, considerably younger than in GTS2004 (488.3 ± 1.7 Ma). This new estimate is well within the error limit of age date O1 in Appendix 2 (see also Figure 19.15 in the previous chapter on the Cambrian Period) close to the Cambrian–Ordovician boundary. The base Devonian is here 419.2 ± 3.2 Ma, compared with 416.0 ± 2.8 Ma in GTS2004. The consequence of this is that both the Ordovician and Silurian have been considerably shortened. The new duration of the Ordovician is 41.5 ± 2.2 myr (was 44.6 myr in GTS2004), while the duration of the Silurian is 24.6 ±1.6 myr (was 27.7 myr in GTS2004).

ACKNOWLEDGMENTS

For assistance with figures, GSSP charts, fossil and outcrop photos, we thank the following; G. Nowlan, S.M. Bergström, J. Nõlvak, Rong Jiayu, Chen Xu, D. Goldman, C.M. Mitchell, Zang Yuandong, and Zhou Zhiyi. D.A.T. Harper kindly commented on the text. For access to radiogenic isotope data in advance of their publication, we thank D. Condon and U. Söderholm.

REFERENCES

Achab, A., 1989. Ordovician chitinozoan zonation of Quebec and western Newfoundland. Journal of Paleontology 63, 14–24.

Ainsaar, L., Meidla, T., Martma, T., 2004. The middle Caradoc facies and faunal turnover in the late Ordovician Baltoscandian paleobasin. Palaeogeography, Palaeoclimatology, Palaeoecology 210, 119–133.

Ainsaar, L., Kaljo, D., Martma, T., Meidla, T., Mannik, J., Nolvak, J., Tinn, O., 2010. Middle and Upper Ordovician carbon isotope chemostratigraphy in Baltoscandia: A correlation standard and clues to environmental history. Palaeogeography, Palaeoclimatology, Palaeoecology 294, 189–201.

Aldridge, R.J., Briggs, D.E.G., 1989. A soft body of evidence. Natural History 5/89, 6–11.

Alroy, J., 2008. Phanerozoic trends in the global diversity of marine invertebrates. Science 321 97–11.

Bambach, R.K., Knoll, A.H., Wang, S.C., 2004. Origination, extinction, and mass depletions of marine diversity. Paleobiology 30, 522–542.

Barnes, C.R., Fortey, R.A., Williams, H., 1995. The pattern of global bioevents during the Ordovician Period. In: Walliser, O.H. (Ed.), Global Events and Event Stratigraphy in the Phanerozoic. Springer-Verlag, Berlin, pp. 139–172.

Barta, N.C., Bergström, S.M., Saltzman, M.R., Schmitz, B., 2007. First record of the Ordovician Guttenberg $\delta^{13}C$ excursion (GICE) in New York State and Ontario: Local and regional chronostratigraphic implications. Northeastern Geology and Environmental Sciences 29, 276–298.

Bergström, S.M., 1986. Biostratigraphic integration of Ordovician graptolite and conodont zones – a regional review. In: Rickards, R.B., Hughes, C.P. (Eds.), Paleobiology and biostratigraphy of graptolites. Blackwell Scientific, Oxford, pp. 61–78.

Bergström, S.M., Wang, Z., 1995. Global correlation of Castlemainian to Darriwilian conodont faunas and their relation to the graptolite zone succession. In: Chen, X., Bergström, S.M. (Eds.), The Base of the *Austrodentatus* Zone as a Level for Global Subdivision of the Ordovician System. Nanjing University Press, Nanjing, pp. 92–98.

Bergström, S.M., Löfgren, A., 2009. The base of the Dapingian Stage (Ordovician) in Baltoscandia: Conodonts, graptolites and unconformities. Earth and Environmental Science Transactions of the Royal Society of Edinburgh 99, 189–212.

Bergström, S.M., Finney, S.C., Xu, C., Palsson, C., Zhi-Gao, W., Grahn, Y., 2000. A proposed global boundary stratotype for the base of the Upper Series of the Ordovician System: The Fågelsång section, Scania, southern Sweden. Episodes 23 (2), 102–109.

Bergström, S.M., Löfgren, A., Maletz, J., 2004. The GSSP of the second (upper) stage of the Lower Ordovician Series: Diabasbrottet at Hunneberg, Province of Vaastergötland, southwest Sweden. Episodes 27 (4), 265–272.

Bergström, S.M., Finney, S.M., Chen, X., Goldman, D., Leslie, S.A., 2006a. Three new Ordovician global stage names. Lethaia 39, 287–288.

Bergström, S.M., Saltzman, M.R., Schmitz, B., 2006b. First record of the Hirnantian (Upper Ordovician) $\delta^{13}C$ excursion in the North American Midcontinent and its regional implications. Geological Magazine 143, 657–678.

Bergström, S.M., Young, S.A., Schmitz, B., Saltzman, M.R., 2007. Upper Ordovician (Katian) $\delta^{13}C$ chemostratigraphy: A trans-Atlantic comparison. Acta Palaeontologica Sinica 46 (Suppl.), 37–39.

Bergström, S.M., Xu, C., Guteirrez-Marco, J.-C., Dronov, A., 2009. The new chronostratigraphic classification of the Ordovician System and its relations to major regional series and stages and to $\delta^{13}C$ chemostratigraphy. Lethaia 42, 97–107.

Bergström, S.M., Schmitz, B., Saltzman, M.R., Huff, W.D., 2010. The Upper Ordovician Guttenberg excursion (GICE) in North America and Baltoscandia: Occurrence, chronostratigraphic significance, and paleoenvironmental relationships. In: Finney, S.C., Berry, W.B.N. (Eds.), The Ordovician Earth System. Geological Society of America Special Paper, 466, pp. 37–67.

Berry, W.B.N., Ripperdan, R.L., Finney, S.M., 2002. Late Ordovician extinction: A laurentian view. Geological Society of America Special Paper 356, 463–471.

Bottjer, D.J., Droser, M.L., Sheehan, P.M., McGhee, G.R., 2001. The ecological architecture of major events in the Phanerozoic history of marine invertebrate life. In: Allmon, W.D., Bottjer, D.J. (Eds.), Evolutionary Paleoecology: The Ecological Context of Macroevolutionary Change. Columbia University Press, New York, pp. 35–61.

Boucot, A.J., 1975. Evolution and Extinction Rate Controls. Elsevier, Amsterdam, pp. 427.

Brenchley, P.J., 1984. Late Ordovician extinctions and their relationship to the Gondwana glaciation. In: Brenchley, P.J. (Ed.), Fossils and Climate. Wiley, Chichester, pp. 291–315.

Brenchley, P.J., 1989. The Late Ordovician extinction. In: Donovan, S.K. (Ed.), Mass Extinctions: Processes and Evidence. Belhaven Press, London, pp. 104–132.

Brenchley, P.J., 2004. End Ordovician glaciation. In: Webby, B.D., Droser, M.L., Paris, F., Percival, I.G. (Eds.), The Great Ordovician Biodiversification Event. Columbia University Press, New York, pp. 81–83.

Brenchley, P.J., Marshall, J.D., Carden, G.A.F., Robertson, D.B.R., Longe, D.G.F., Meidla, T., Hintes, L., Anderson, T.F., 1994. Bathymetric and isotopic evidence for a short-lived Late Ordovician glaciation in a greenhouse period. Geology 22, 295–298.

Brenchley, P.J., Carden, G.A., Hints, L., Kaljo, D., Marshall, J.D., Martma, T., Meidla, T., Nõlvak, J., 2003. High-resolution stable isotope stratigraphy of Upper Ordovician sequences: Constraints on the timing of bioevents and environmental changes associated with mass extinction and glaciation. Bulletin of the Geological Society of America 115 (1), 89–104.

Buggisch, W., Keller, M., Lehnert, O., 2003. Carbon isotope record of Late Cambrian to Early Ordovician carbonates of the Argentine Precordillera. Palaeogeography, Palaeoclimatology, Palaeoecology 195, 357–373.

Chen, X., 1990. Graptolite depth zonation. Acta Palaeontologica Sinica 29 (5), 507–526.

Chen, X., Bergström, S.M., 1995. The Base of the *Austrodentatus* Zone as a Level for Global Subdivision of the Ordovician System. Nanjing University Press, Nanjing, p. 117.

Chen, X., Zhang, Y., Mitchell, C.E., 2001. Early Darriwilian graptolites from central and western China. Alcheringa 25, 191–210.

Chen, X., Melchin, M.J., Sheets, H.D., Mitchell, C.E., Fan., J.-X., 2005. Patterns and processes of latest Ordovician graptolite extinction and recovery based on data from south China. Journal of Paleontology 79, 842–861.

Chen, X., Rong, J., Fan, J., Zhan, R., Mitchell, C.E., Harper, D.A.T., Melchin, M.J., Peng, P., Finney, S.F., Wang, X., 2006. The global boundary stratotype section and point (GSSP) for the base of the Hirnantian Stage (the uppermost of the Ordovician System). Episodes 29, 183–196.

Churkin, M., Carter, C., Johnson, B.R., 1977. Subdivision of Ordovician and Silurian time scale using accumulation rates of graptolitic shale. Geology 5, 452–456.

Compston, W., 2000a. Interpretations of SHRIMP and isotope dilution zircon ages for the geological time-scale: I. The early Ordovician and late Cambrian. Mineralological Magazine 64, 43–57.

Compston, W., 2000b. Interpretations of SHRIMP and isotope dilution zircon ages for the Palaeozoic time-scale: II. Silurian to Devonian. Mineralological Magazine 64, 1127–1146.

Compston, W., Williams, I.S., 1992. Ion probe ages for the British Ordovician and Silurian stratotypes. In: Webby, B.D., Laurie, J. (Eds.), Global Perspectives on Ordovician Geology. Balkema, Rotterdam, pp. 59–67.

Cooper, M.R., Crowley, Q.G., Rushton, A.W.A., 2008. New age constraints for the Ordovician Tyrone Volcanic Group, Northern Ireland. Journal of the Geological Society 165, 333–339.

Cooper, R.A., Fortey, R.A., Lindholm, K., 1991. Latitudinal and depth zonation of early Ordovician graptolites. Lethaia 24, 199–218.

Cooper, R.A., 1999a. Ecostratigraphy, zonation and global correlation of earliest planktic graptolites. Lethaia 32, 1–16.

Cooper, R.A., 1999b. The Ordovician time scale – calibration of graptolite and conodont zones. Acta Universitatis Carolinae – Geologica 43 (1/2), 1–4.

Cooper, R.A., Lindholm, K., 1990. A precise worldwide correlation of Early Ordovician graptolite sequences. Geological Magazine 127, 497–525.

Cooper, R.A., Stewart, I.R., 1979. The Tremadoc graptolite sequence of Lancefield, Victoria. Palaeontology 22, 767–797.

Cooper, R.A., Nowlan, G.S., 1999. Proposed global stratotype section and point for base of the Ordovician System. Acta Universitatis Carolinae – Geologica 43 (1/2), 61–64.

Cooper, R.A., Sadler, P.M., 2004. Ordovician. In: Gradstein, F., Ogg, J.G., Smith, A.G. (Eds.), A Geologic Time Scale. Cambridge University Press, Cambridge, pp. 165–187.

Cooper, R.A., Maletz, J., Wang, H., Erdtmann, B.-D., 1998. Taxonomy and evolution of the earliest Ordovician graptolites. Norsk Geologisk Tiddskrift 78, 3–32.

Cooper, R.A., Nowlan, G.S., Williams, H.S., 2001. Global stratotype section and point for base of the Ordovician System. Episodes 24, 19–28.

Cooper, R.A., Maletz, J., Taylor, L., Zalasiewicz, J.A., 2004. Estimates of Ordovician mean standing diversity in low, middle and high paleolatitudes. In: Webby, B.D., Paris, F., Droser, M.L., Percival, I.G. (Eds.), The Great Ordovician Biodiversification Event. Columbia University Press, New York, pp. 281–293.

Davidek, K., Landing, E., Westrop, S.R., Rushton, A.W.A., Fortey, R.A., Adrain, J.M., 1998. New uppermost Cambrian U-Pb date from Avalonian Wales and the age of the Cambrian-Ordovician boundary. Geological Magazine 133, 305–309.

Denison, R.E., Koepnick, R.B., Burke, W.H., Hetherington, E.A., 1998. Construction of the Cambrian and Ordovician seawater $^{87}Sr/^{86}Sr$ curve. Chemical Geology 152, 325–340.

Elles, G.L., 1925. The characteristic assemblages of the graptolite zones of the British Isles. Geological Magazine 62, 337–347.

Erdtmann, B.-D., 1988. The earliest Ordovician nematophorous graptolites: Taxonomy and correlation. Geological Magazine 125, 327–348.

Fearnsides, W.G., 1905. On the geology of Arenig Fawr and Moel Llyfnant. Quarterly Journal of the Geological Society of London 61, 608–637.

Fearnsides, W.G., 1910. The Tremadoc Slates of south-east Carnarvonshire. Quarterly Journal of the Geological Society of London 66, 142–188.

Finnegan, S., Bergmann, K., Eiler, J.M., Jones, D.S., Fike, D.A., Eisenman, I., Hughes, N.C., Tripati, A.K., Woodward, W.F., 2011. The magnitude and duration of Late Ordovician-Early Silurian glaciation. Science 331, 2011.

Finney, S.C., Bergström, S.M., 1986. Biostratigraphy of the Ordovician *Nemagraptus gracilis* Zone. In: Hughes, C.P., Rickards, R.B. (Eds.), Palaeoecology and Biostratigraphy of Graptolites. Geological Society Special Publication 20, pp. 47–59.

Finney, S.C., Berry, W.B.N., 1997. New perspectives on graptolite distributions and their use as indicators of platform margin dynamics. Geology 25, 919–922.

Finney, S.C., Berry, W.B.N., Cooper, J.D., Ripperdan, R.L., Sweet, W.C., Jacobson, S.R., Soufiane, A., Achab, A., Noble, P.J., 1999. Late Ordovician mass extinction: A new perspective from stratigraphic sections in central Nevada. Geology 27, 215–218.

Finney, S.C., Berry, W.B.N., Cooper, J.D., 2007. The influence of denitrifying seawater on graptolite extinction and diversification during the Hirnantian (latest Ordovician) mass extinction event. Lethaia 40, 281–291.

Fordham, B.G., 1992. Chronometric calibration of mid-Ordovician to Tournaisian conodont zones: A compilation from recent graphic correlation and isotope studies. Geological Magazine 129, 709–721.

Fortey, R.A., Cocks, L.R.M., 1986. Marginal faunal belts and their structural implications, with examples from the Lower Palaeozoic. Journal of the Geological Society 143, 151–160.

Fortey, R.A., Bassett, M.G., Harper, D.A.T., Hughes, R.A., Ingham, J.K., Molyneux, A.W., Owen, A.W., Owens, R.M., Rushton, A.W.A., Sheldon, P.R., 1991. Progress and problems in the selection of stratotypes for the bases of series in the Ordovician System of the historical type area in the U.K. In: Barnes, C.R., Williams, S.H. (Eds.), Advances in Ordovician Geology. Geological Survey of Canada Paper, 90–9, pp. 5–25.

Fortey, R.A., Harper, D.A.T., Ingham, J.K., Owen, A.W., Rushton, A.W.A., 1995. A revision of Ordovician series and stages from the historical type area. Geological Magazine 132, 15–30.

Fortey, R.A., Harper, D.A.T., Ingham, J.K., Owen, A.W., Parkes, M.A., Rushton, A.W.A., Woodcock, N.H., 2000. A revised correlation of the Ordovician rocks of the British Isles. The Geological Society, Special Report 24, 83.

Gale, N.H., 1985. Numerical calibration of the Paleozoic time-scale: Ordovician, Silurian and Devonian periods. In: Snelling, N.J. (Ed.), The Chronology of the Geological Record. Memoir of the Geological Society of London, 10, pp. 81–88.

Ghienne, J.F., 2003. Late Ordovician sedimentary environments, glacial cycles, and post-glacial transgression in the Taoudeni basin, West Africa. Palaeogeography, Palaeoclimatology, Palaeoecology 189, 117–145.

Goldman, D., Leslie, S.A., Nõlvak, J., Young, S., Bergström, S.M., Huff, W.D., 2007. The global stratotype section and point (GSSP) for the base of the Katian Stage of the Upper Ordovician Series at Black Knob Ridge, southeastern Oklahoma, USA. Episodes 30, 258–270.

Gradstein, F.M., Agterberg, F.P., Ogg, J.G., Hardenbol, J., van Veen, P., Thierry, J., Huang, Z., 1994. A Mesozoic time scale. Journal of Geophysical Research 99 (B12), 24,051–24,074.

Grahn, Y., Nõlvak, J., 2007. Ordovician Chitinozoa and biostratigraphy from Skåne and Bornholm, southernmost Scandinavia – an overview and update. Bulletin of Geosciences 82, 11–26.

Hall, T.S., 1895. The geology of Castlemaine, with a subdivision of part of the Lower Silurian rocks of Victoria, and a list of minerals. Proceedings of the Royal Society of Victoria 7, 55–88.

Harland, W.B., Armstrong, R.L., Cox, A.V., Craig, L.E., Smith, A.G., Smith, G., 1990. A Geologic Time Scale 1989. Cambridge University Press, Cambridge, p. 263.

Harris, W.J., Keble, R.A., 1932. Victorian graptolite zones, with correlations and description of species. Proceedings of the Royal Society of Victoria 44, 25–48.

Harris, W.J., Thomas, D.E., 1938. A revised classification and correlation of the Ordovician graptolite beds of Victoria. Mining and Geological Journal 1, 62–72.

Henningsmoen, G., 1972. The Cambro-Ordovician boundary. Lethaia 6, 423–439.

Hicks, H., 1881. The classification of the Eozoic and Lower Paleozoic rocks of the British Isles. Popular Science Review 5, 289–308.

Huff, W.D., Bergström, S.M., Kolata, D.R., 1992. Gigantic Ordovician volcanic ash fall in North America and Europe: Biological, tectonomagmatic, and event-stratigraphic significance. Geology 20, 875–878.

Huff, W.D., Davis, D., Bergström, S.M., Krekeler, M.P.S., Kolata, D.R., Cingolani, C., 1997. A biostratigraphically well-constrained K-bentonite U-Pb zircon age of the lowermost Darriwilian Stage (Middle Ordovician) from the Argentine Precordillera. Episodes 20, 29–33.

Jaanusson, V., 1960. On series of the Ordovician System. Reports of the 21st International Geological Congress. Copenhagen 7, 70–81.

Jablonski, D., 1991. Extinctions: A paleontological perspective. Science 253, 754–757.

Kaljo, D.L., Hints, L., Martma, T., Nõlvak, J., Oraspõld, A., 2004. Late Ordovician carbon isotope trend in Estonia, its significance in stratigraphy and environmental analysis. Palaeogeography, Palaeoclimatology, Palaeoecology 210 (2–4), 165–185.

Kaljo, D.L., Martma, T., Saadre, T., 2007. Post-Hunnebergian Ordovician carbon isotope trend in Baltoscandia, its environmental implications and some similarities with that of Nevada. Palaeogeography, Palaeoclimatology, Palaeoecology 245, 138–145.

Kanygin, A., Dronov, A., Timokhin, A., Gonta, T., 2010. Depositional sequences and palaeoceanographic change in the Ordovician of the Siberian craton. Palaeogeography, Palaeoclimatology, Palaeoecology 296, 285–296.

Kemple, W.G., Sadler, P.M., Strauss, D.J., 1995. Extending graphic correlation to many dimensions: Stratigraphic correlation as constrained optimisation. In: Mann, K.O., Lane, H.R., Scholle, P.A. (Eds.), Graphic Correlation. SEPM Special Publication, 53, pp. 65–82.

Kleffner, M.A., 1989. A conodont-based Silurian chronostratigraphy. Geological Society of America Bulletin 101, 904–912.

Kolata, D.R., Huff, W.D., Bergström, S.M., 1996. Ordovician K-bentonites of eastern North America. Geological Society of America Special Paper 313, 1–84.

Kuiper, K.F., Deino, A., Hilgen, F.J., Krijgsman, W., Renne, P.R., Wijbrans, J.R., 2008. Synchronising rock clocks of Earth history. Science 320 (5875), 500–504.

Kunk, M.J., Sutter, J.F., 1984. $^{40}Ar/^{39}Ar$ age spectrum dating of biotites from Middle Ordovician bentonites, eastern North America. In: Brutton, D.L. (Ed.), Aspects of the Ordovician System. Paleontological Contributions from the University of Oslo, 295, pp. 11–22.

Kunk, M.J., Sutter, J., Obradovitch, J.D., Lanphere, M.A., 1985. Age of biostratigraphic horizons within the Ordovician and Silurian Systems. In: Snelling, N.J. (Ed.), The Chronology of the Geological Record. Memoir of the Geological Society of London, 10, pp. 89–92.

Landing, E., Bowring, S.A., Fortey, R.A., Davidek, K., 1997. U-Pb zircon date from Avalonian Cape Breton Island and geochronological calibration of the Early Ordovician. Canadian Journal of Earth Sciences 34, 724–730.

Landing, E., Bowring, S.A., Davidek, K.L., Rushton, A.W.A., Fortey, R.A., Wimbledon, W.A.P., 2000. Cambrian-Ordovician boundary age and duration of the lowest Ordovician Tremadoc Series based on U-Pb zircon dates from Avalonian Wales. Geological Magazine 137, 485–494.

Lapworth, C., 1879. On the tripartite classification of the Lower Paleozoic rocks. Geological Magazine 6, 1–15.

Lapworth, C., 1879–80. On the geological distribution of the Rhabdophora. Annals and Magazine of Natural History, series 5 (3): 245–257. 449–455; (4): 333–341, 423–431; (5): 45–62, 273–285, 359–369; (6): 16–29, 185–207 [Volumes 3–4, 1879, volumes 5–6, 1880].

Loch, J.D., Taylor, J.F., Miller, J.F., Repetski, J., 2009. Finding the Cambrian-Ordovician boundary in Laurentian platform carbonates: Refined species concepts in the trilobite genus, Sympysurina. 2009 North American Paleontological Convention, abstract, Cincinnati, 224.

Löfgren, A., 1993. Conodonts from the lower Ordovician at Hunneberg, south-central Sweden. Geological Magazine 130, 215–232.

Löfgren, A., 1996. Lower Ordovician conodonts, reworking, and biostratigraphy of the Orreholmen quarry, Vastergötland, south-central Sweden. Lethaia 118, 169–183.

Ludvigson, G.A., Witzke, B.J., Schneider, C.L., Smith, E.A., Emerson, N.R., Carpenter, S.J., Gonzalez, L.A., 2004. Late Ordovician (Turinian-Chatfieldian) carbon isotope excursions and their stratigraphic and paleoceanic significance. Paleogeography, Paleoclimatology. Paleoecology 210, 187–214.

Maletz, J., 1999. Late Tremadoc graptolites and the base of the *Tetragraptus approximatus* Zone. Acta Universitatis Carolinae – Geologica 43, 25–28.

Maletz, J., Egenhoff, S.O., 2001. Late Tremadoc to Early Arenig graptolite faunas of southerm Bolivia and their implications for a worldwide biozonation. Lethaia 34, 47–62.

Maletz, J., Löfgren, A., Bergström, S.M., 1996. The base of the *Tetragraptus approximatus* Zone at Mt. Hunneberg, S.W. Sweden: A proposed Global Stratotype for the Base of the Second Series of the Ordovician System. Newsletters in Stratigraphy 34, 129–159.

Marr, J.E., 1905. The classification of the sedimentary rocks. Quarterly Journal of the Geological Society of London 61 lxi–lxxxvi.

Marr, J.E., 1913. The Lower Paleozoic rocks of the Caughtley District. Quarterly Journal of the Geological Society of London 69, 1–18.

McKerrow, W.S., Lambert, R.St.J., Cocks, L.R.M., 1985. The Ordovician, Silurian and Devonian Periods. In: Snelling, N.J. (Ed.), The Chronology of the Geological Record. Memoir of the Geological Society of London, 10, pp. 73–80.

Melchin, J.M., Mitchell, C.E., 1991. Late Ordovician extinction in the Graptoloidea. In: Barnes, C.R., Williams, S.H. (Eds.), Advances in Ordovician Geology. Geological Survey of Canada, Paper, 90–9, pp. 143–156.

Melchin, M.J., Williams, S.H., 2000. A restudy of the akidograptine graptolites from Dob's Linn and a proposed redefined zonation of the Silurian Stratotype. In: Cockle, P., Wilson, G.A., Brock, G.A., Engerbretsen, M.J., Simpson, A. (Eds.), Palaeontology Down-Under 2000. Geological Society of Australia, pp. 61–63. Abstracts.

Melchin, M.J., Holmden, C., 2006. Carbon isotope chemostratigraphy of the Llandovery in Arctic Canada: Implications for global correlation and sea-level change. GFF 128, 173–180.

Min, K., Renne, P.R., Huff, W.D., 2001. ^{40}Ar/^{39}Ar dating of Ordovician K-bentonites in Laurentia and Baltoscandia. Earth and Planetary Science Letters 185, 121–134.

Mitchell, C.E., Chen, X., Bergström, S.M., Zhang, Y., Wang, Z., Webby, B.D., Finney, S.C., 1997. Definition of a global boundary stratotype for the Darriwilian Stage of the Ordovician System. Episodes 20, 158–166.

Mitchell, C.E., Brussa, E.D., Astini, R.A., 1998. A diverse Da2 fauna preserved within an altered volcanic ash fall, Eastern Precordillera, Argentina: Implications for graptolite paleoecology. In: Gutiérrez-Marco, J.C., Rábano, I. (Eds.), Proceedings of the Sixth International Graptolite Conference of the Graptolite Working Group (IPA) and the 1998 Field Meeting of the International Subcommission on Silurian Stratigraphy (ICS-IUGS), pp. 222–223. Madrid.

Munnecke, A., Calner, M., Harper, D.A.T., Servais, T., 2010. Ordovician and Silurian sea-water chemistry, sea level, and climate: A synopsis. Palaeogeography, Palaeoclimatology, Palaeoecology 296, 389–413.

Murchison, R.I., 1839. The Silurian System Founded on Geological Researches in the Counties of Salop, Hereford, Radnor, Montgomery, Caermarthen, Brecon, Pembroke, Monmouth, Gloucester, Worcester, and Stafford: With Descriptions of the Coal-fields, and Overlying Formations. John Murray, London. 2 vols.

Nielsen, A.T., 2004. Ordovician sea level changes: A Baltoscandian perspective. In: Webby, B.D., Paris, F., Droser, M.L., Percival, I.G. (Eds.), The Great Ordovician Biodiversity Event. Columbia University Press, New York, pp. 84–93.

Nicoll, R.S., Thorshøj Nielsen, A., Laurie, J.R., Shergold, J.H., 1992. Preliminary correlation of latest Cambrian to Early Ordovician sea level events in Australia and Scandinavia. In: Webby, B.D., Laurie, J.R. (Eds.), Global Perspectives on Ordovician Geology: Proceedings of the 6th International Symposium on the Ordovician System, University of Sydney, Australia, 15–19 July 1991. Balkema Press, Rotterdam, pp. 381–394.

Nõlvak, J., Hints, O., Männik, P., 2006. Ordovician timescale in Estonia: Recent developments. Proceedings of the Estonian Academy of Sciences, Geology 55, 98–108.

Odin, G.S., 1985. Remarks on the numerical scale of Ordovician to Devonian times. In: Snelling, N.J. (Ed.), The chronology of the geological record. Memoir of the Geological Society of London, 10, pp. 93–97.

Paris, F., 1996. Chitinozoan biostratigraphy and paleoecology. In: Jansonius, J., McGregor, D.C. (Eds.), Palynology: Principles and Applications. American Association of Stratigraphic Palynologists Foundation, 2, pp. 531–552.

Paris, F., Achab, A., Asselin, E., Chen, X.-H., Grahn, Y., Nõlvak, J., Obut, O., Samuelsson, J., Sennikov, N., Vecoli, M., Verniers, J., Wang, X.-F., Winchester-Seeto, T., 2004. Chitinozoans. In: Webby, B.D., Paris, F., Droser, M.L., Percival, I.G. (Eds.), The Great Ordovician Biodiversification Event. Columbia University Press, New York, pp. 281–293.

Pavlov, V., Gallet, Y., 2005. A third superchron during the Early Paleozoic. Episodes 28, 78–84.

Ripperdan, R.L., Miller, J.F., 1995. Carbon isotope ratios from the Cambrian-Ordovician boundary section Lawson Cove. Ibex area, Utah. In: Cooper, J.D., Droser, M.L., Finney, S.F. (Eds.), Ordovician Odyssey: Short Papers for the Seventh International Symposium on the Ordovician System. The Pacific Section for the Society for Sedimentary Geology (SEPM), Fullerton, pp. 129–132.

Ripperdan, R.L., Magaritz, M., Nicoll, R.S., Shergold, J.H., 1992. Simultaneous changes in carbon isotopes, sea level, and conodont biozones within the Cambrian-Ordovician boundary interval at Black mountain, Australia. Geology 20, 1039–1042.

Rong, J.-Y., Harper, D.A.T., 1988. A global synthesis of latest Ordovician Hirnantian brachiopod faunas. Transactions of the Royal Society of Edinburgh: Earth Sciences 79, 383–402.

Ross, C.A., Ross, J.P.R., 1995. North American depositional sequences and correlations. In: Cooper, J.D., Droser, M.L., Finney, S.F. (Eds.),

Ordovician Odyssey: Short Papers for the Seventh International Symposium on the Ordovician System. The Pacific Section for the Society for Sedimentary Geology (SEPM), Fullerton, pp. 309–314.

Ross, R.J., Naeser, C.W., Izett, G.A., Obradovich, J.D., Bassett, M.J., Hughes, C.P., Cocks, L.R.M., Dean, W.T., Ingham, J.K., Jenkins, C.J., Rickards, R.B., Sheldon, P.R., Toghill, P., Whittington, H.B., Zalasiewicz, J., 1982. Fission track dating of British Ordovician and Silurian stratotypes. Geological Magazine 119, 135–153.

Rushton, A.W.A., 1982. The biostratigraphy and correlation of the Merioneth-Tremadoc Series boundary in North Wales. In: Bassett, M.G., Dean, W.T. (Eds.), The Cambrian-Ordovician Boundary: Sections, Fossil Distributions, and Correlations. National Museum of Wales Geological Series, 3, pp. 41–59.

Sadler, P.M., 2001. Constrained optimisation – approaches to the paleobiologic correlation and seriation problems: A users' guide and reference manual to the CONOP program family. Unpublished. P.M. Sadler, Riverside, pp. 165.

Sadler, P.M., Cooper, R.A., 2003. Best-fit intervals and consensus sequences; Comparison of the resolving power of traditional biostratigraphy and computer-assisted correlation. In: Harries, P. (Ed.), High Resolution Approaches in Stratigraphic Paleontology. Kluwer Academic, Dordrecht, pp. 49–94.

Sadler, P.M., Cooper, R.A., Melchin, M.R., 2009. High resolution, early Paleozoic (Ordovician-Silurian) time scales. Geological Society of America Bulletin 121, 887–906.

Sadler, P.M., Cooper, R.A., Melchin, M.J., 2011. Sequencing the graptolite clade: Building a global diversity curve from local range-charts, regional composites and global time-lines. Proceedings of the Yorkshire Geological Society 58 (4), 329–343.

Salter, J.W., 1866. On the fossils of North Wales. In: Ramsay, A.C. (Ed.), The Geology of North Wales. Memoirs of the Geological Survey of Great Britain, 3, pp. 240–381.

Saltzman, M.R., 2005. Phosphorus, nitrogen, and the redox evolution of the Paleozoic oceans. Geology 33, 573–576.

Saltzman, M.R., Young, S.A., 2005. Long-lived glaciation in the Late Ordovician? Isotopic and sequence-stratigraphic evidence from western Laurentia. Geology 33, 109–112.

Schmitz, B., Bergström, S.M., 2007. Chemostratigraphy in the Swedish Upper Ordovician: Regional significance of the Hirnantian $\delta^{13}C$ excursion (HICE) in the Boda Limestone of the Siljan region. GFF 129, 133–140.

Sedgwick, A., 1847. On the classification of the fossiliferous slates of North Wales, Cumberland, Westmoreland and Lancashire. Quarterly Journal of the Geological Society of London 3, 133–164.

Sedgwick, A., 1852. On the classification and nomenclature of the Lower Paleozoic rocks of England and Wales. Quarterly Journal of the Geological Society of London 8, 136–168.

Sell, B.K., Samson, S.D., 2011. Apatite phenocryst compositions demonstrate a miscorrelation between the Millbrig and Kinnekulle K-bentonites of North America and Scandinavia. Geology 39, 303–306.

Sepkoski, J.J., 1995. The Ordovician radiations: Diversification and extinction shown by global genus-level taxonomic data. In: Cooper, J.D., Droser, M.L., Finney, S.F. (Eds.), Ordovician Odyssey: Short Papers for the Seventh International Symposium on the Ordovician System. The Pacific Section for the Society for Sedimentary Geology (SEPM), Fullerton, pp. 393–396.

Sepkoski, J.J., Sheehan, P., 1983. Diversification, faunal change, and community replacement during the Ordovician radiations. In: Tevesz, M.J.S., McCall, P.L. (Eds.), Biotic Interactions in Recent and Fossil Benthic Communities. Plenum Press, New York, pp. 673–717.

Sheehan, P.M., 2001. The Late Ordovician mass extinction. Annual review of Earth and Planetary Sciences 29, 331–364.

Shields, G.A., Carden, G.A.F., Vezier, J., Meidla, T., Jia-yu, R., 2003. Sr, C, and O isotope geochemistry of Ordovician brachiopods: A major isotopic event around the Middle-Late Ordovician transition. Geochimica et Cosmochimica Acta 67 2205–2035.

Skevington, D., 1963. A correlation of Ordovician graptolite-bearing sequences. GFF 85, 298–319.

Sweet, W.C., 1984. Graphic correlation of upper Middle and Upper Ordovician rocks, North American Midcontinental Province, USA. In: Bruton, D.L. (Ed.), Aspects of the Ordovician System. Paleontological Contributions of the University of Oslo, 295, pp. 23–35.

Sweet, W.C., 1988. Mohawkian and Cincinnatian chronostratigraphy. Bulletin of the New York State Museum 462, 84–90.

Sweet, W.C., 1995. A conodont-based composite standard for the North American Ordovician: Progress report. In: Cooper, J.D., Droser, M.L., Finney, S.F. (Eds.), Ordovician Odyssey: Short Papers for the Seventh International Symposium on the Ordovician System. The Pacific Section for the Society for Sedimentary Geology (SEPM), Fullerton, pp. 15–20.

Sweet, W.C., Bergström, S.M., 1976. Conodont biostratigraphy of the Middle and Upper Ordovician of the United Staes Midcontinent. In: Bassett, M.G. (Ed.), The Ordovician System: Proceedings of a Palaeontological Association Symposium, Birmingham. University of Wales Press and National Museum of Wales, Cardiff, pp. 121–151. September 1974.

Sweet, W.C., Bergström, S.M., 1984. Conodont provinces and biofacies of the Late Ordovician. Geological Society of America Special Paper 196, 69–87.

Thomas, D.E., 1960. The zonal distribution of Australian graptolites. Journal and Proceedings of the Royal Society of New South Wales 94, 1–58.

Thorshøj Nielsen, A.T., 1992. Ecostratigraphy and the recognition of Arenigian (Early Ordovician) sea-level changes. In: Webby, B.D., Laurie, J.R. (Eds.), Global Perspectives on Ordovician Geology. Balkema Press, Rotterdam, pp. 355–366.

Thorshoej Nielsen, A.T., 2004. Ordovician sea level changes: A Balto-scandian perspective. In: Webby, B.D., Paris, F., Droser, M.L., Percival, I.G. (Eds.), The Great Ordovician Biodiversification Event. Columbia University Press, New York, pp. 84–93.

Trotter, J.A., Williams, I.S., Barnes, C.R., Lecuyer, C., Nicoll, R.S., 2008. Did cooling oceans trigger Ordovician biodiversification? Evidence from conodont thermometry. Science 321, 550–554.

Tucker, R.D., 1992. U-Pb dating of Plinian-eruption ashfalls by the isotope dilution method: A reliable and precise tool for time-scale calibration and biostratigraphic correlation. Geological Society of America, Abstracts with Programs 24 (7), A198.

Tucker, R.D., McKerrow, W.S., 1995. Early Paleozoic chronology: A review in light of new U-Pb zircon ages from Newfoundland and Britain. Canadian Journal of Earth Sciences 32, 368–379.

VandenBerg, A.H.M., Cooper, R.A., 1992. The Ordovician graptolite sequence of Australasia. Alcheringa 16, 33–65.

Villeneuve, M., 2004. Radiogenic isotope geochronology. In: Gradstein, F.M., Ogg, J.G., Smith, A.G. (Eds.), A Geologic Time Scale. Cambridge University Press, Cambridge, pp. 87–95.

Wang, X., Stouge, S., Erdtmann, B.-D., Chen, X., Li, Z., Wang, C., Zeng, Q., Zhou, Z., Chen, H., 2005. A proposed GSSP for the base of the Middle Ordovician Series: The Huanghuachang section, Yichang, China. Episodes 28 (2), 105–117.

Webby, B.D., 1995. Towards an Ordovician timescale. In: Cooper, J.D., Droser, M.L., Finney, S.F. (Eds.), Ordovician Odyssey: Short Papers for the Seventh International Symposium on the Ordovician System. The Pacific Section for the Society for Sedimentary Geology (SEPM), Fullerton, pp. 5–9.

Webby, B.D., 1998. Steps towards a global standard for Ordovician stratigraphy. Newsletters in Stratigraphy 36, 1–33.

Webby, B.D., Young, G.C., Talent, J.A., Laurie, J.R., 2000. Palaeobiogeography of Australasian faunas and floras. Association of Australasian Palaeontologists, Memoir 23, 63–126.

Webby, B.D., Paris, F., Droser, M.L., Percival, I.G., 2004a. The Great Ordovician Biodiversification Event. Columbia University Press, New York, p. 496.

Webby, B.D., Cooper, R.A., Bergström, S.M., Paris, F., 2004b. Stratigraphic framework and time slices. In: Webby, B.D., Paris, F., Droser, M.L., Percival, I.G. (Eds.), The Great Ordovician Biodiversification Event. Columbia University Press, New York, pp. 41–47.

Whittington, H.B., Dean, W.T., Fortey, R.A., Rickards, R.B., Rushton, A.W.A., Wright, A.D., 1984. Definition of the Tremadoc series and the series of the Ordovician System in Britain. Geological Magazine 121, 17–33.

Williams, S.H., Bruton, D.L., 1983. The Caradoc-Ashgill boundary in the central Oslo Region and associated graptolite faunas. Norsk Geologisk Tidsskrift 63, 147–191.

Woodcock, N.H., 1990. Sequence stratigraphy of the Palaeozoic Welsh Basin. Journal of the Geological Society 147, 537–547.

Young, S.A., Saltzman, M.R., Bergström, S.M., 2005. Upper Ordovician (Mohawkian) carbon isotope ($\delta^{13}C$) stratigraphy in eastern and central North America: Regional expression of a perturbation of the global carbon cycle. Palaeogeography, Palaeoclimatology, Palaeoecology 222, 53–76.

Young, S.A., Saltzman, M.R., Foland, K.A., Linder, J.S., Kump, L., 2009. A major drop in seawater $^{87}Sr/^{86}Sr$ during the middle Ordovician (Darriwilian): Links to volcanism and climate? Geology 37, 951–954.

Zalasiewicz, J.A., Taylor, L., Rushton, A.W.A., Loydell, D.K., Rickards, R.B., Williams, M., 2009. Graptolites in British stratigraphy. Geological Magazine 146, 785–850.

Chapter 21

M.J. Melchin, P.M. Sadler and B.D. Cramer
With contributions by R.A. Cooper, F.M. Gradstein and O. Hammer

The Silurian Period

Abstract: The features of the Silurian Period include a rapid recovery in biodiversity after the end-Ordovician extinction event, a highly dynamic climate accentuated by multiple short-lived events, strong eustatic sea level fluctuations and oceanic turnover, associated with extinction of moderate scale. Land colonization occurred, alongside general convergence of continental plates and low levels of faunal provincialism, closure of Iapetus Ocean, narrowing of Rheic Ocean and migration of the South Pole over South American and South African Gondwana.

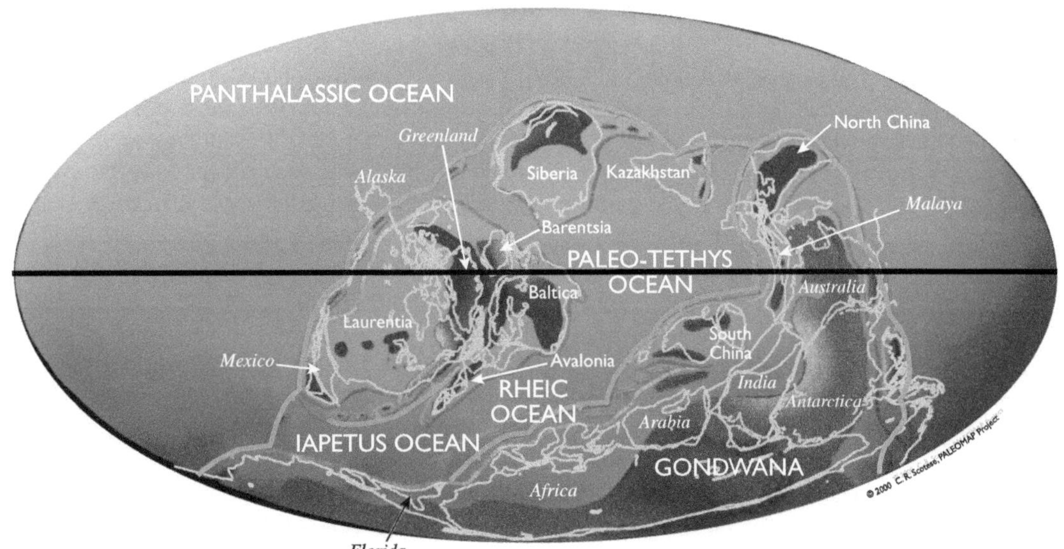

Chapter Outline

The Geologic Time Scale 2012. DOI: 10.1016/B978-0-444-59425-9.00021-4

21.1. HISTORY AND SUBDIVISIONS

The Silurian System was erected by Murchison (1839) and named after the Silures, a Welsh borderland tribe. As originally conceived, the Silurian embraced rocks that were claimed as Cambrian by Sedgwick, leading to a protracted debate. The disputed rocks were separated out as the new Ordovician System by Lapworth in 1879, but the debate lingered and Lower Silurian was used in Britain in parallel with Ordovician for many years, while Gotlandian was used in parallel with upper Silurian in some regions. Eventually, the name Silurian was officially adopted in its restricted (upper Silurian) sense by the IGC in Copenhagen in 1960 (Sorgenfrei, 1964). The rather complex nomenclatural history of definition and subdivision of the Silurian System has been reviewed by Whittard (1961), Cocks *et al.* (1971) and Holland (1989). Melchin *et al.* (2004) provided a recent general summary of the Silurian System.

As with the Ordovician, black shales are widely developed in Silurian sedimentary successions around the world and graptolites have proved to be valuable fossils for correlation. However, compared to the Ordovician it was a time of relatively low faunal provincialism. Over 30 successive graptolite zones are recognized widely around the world, providing a subdivision and correlation framework of extraordinary precision. Based on the present scale and generalized graptolite zonation there are 36 globally recognizable zones, within a span of slightly less than 25 myr. Although variable in duration, the zones represent an average of approximately 0.68 myr. Conodonts have proved to be of considerable global biostratigraphic value in shallow-water carbonate facies and the conodont biostratigraphic scale is rapidly becoming increasingly well resolved. A rich and diverse fauna is present in the shelly facies where trilobites and brachiopods are used extensively for zonation, and coral–stromatoporoid communities enable local biostratigraphic subdivision and correlation. Chitinozoan and acritarch zonations have been developed for several regions and are proving to be useful for correlation in many circumstances. Vertebrate microfossil, sporomorph and radiolarian zonations are also being developed for the Silurian.

Very recently, variations and excursions in the record of stable isotopes, particularly carbon, have proved to be an important tool for correlation, as well as for understanding the record of paleoenvironmental changes through Silurian time.

21.2. SILURIAN SERIES AND STAGES

The Silurian comprises four series, the Llandovery, Wenlock, Ludlow and Pridoli in upward sequence. All series and their constituent stages (Figure 21.1) have designated lower boundary stratotype sections and points (Bassett, 1985; Cocks, 1985). All of the GSSPs were biostratigraphically defined with reference to standard graptolite zones, although the index taxa for these zones (Figure 21.2) are not found at some of the GSSP localities.

21.2.1. Llandovery Series

Named from the type area in Dyfed, southern Wales, the Llandovery Series comprises three stages, approved by the ICS (Bassett, 1985); the Rhuddanian, Aeronian and Telychian stages. The Aeronian Stage is approximately, but not exactly, equivalent to the previously employed Idwian and Fronian stages.

21.2.1.1. Rhuddanian Stage

Although the stage is named for the Cefn-Rhuddan Farm in the Llandovery area, its lower boundary stratotype section and point are at Dob's Linn in the southern uplands of Scotland, defined at a point 1.6 m above the base of the Birkhill Shale in the Linn Branch Trench section (Figure 21.3). This point was previously regarded as coincident with the local base of the *Parakidograptus acuminatus* Zone (Cocks, 1985). Recent resampling and systematic revisions have shown, however, that *Parakidograptus acuminatus* has its first occurrence datum 1.6 m above this level and that the succession can be readily subdivided, both at this section and globally, into a lower *Akidograptus ascensus* Zone and upper *Parakidograptus acuminatus* Zone (Melchin and Williams, 2000). Melchin and Williams (2000) therefore proposed that the base *Akidograptus ascensus* Zone, marked by the first occurrences of *A. ascensus* (Figure 21.2(a)) and *Parakidograptus praematurus* (the latter was identified by Williams (1983) as *P. acuminatus sensu lato*), be regarded as the biostratigraphic horizon that is coincident with the base of the Silurian System. This proposal was recently ratified by the International Union of Geological Sciences (Rong *et al.*, 2008). Thus redefined, the Rhuddanian Stage spans the *Akidograptus ascensus* to *Coronograptus cyphus* zones.

21.2.1.2. Aeronian Stage

The Aeronian Stage is named for the Cwm-coed-Aeron Farm in the Llandovery area. The stratotype section and point are located in the Trefawr Formation, in the Trefawr track section 500 m north of the Cwm-coed-Aeron Farm, between Locality 71 and 72 of Cocks *et al.* (1984). The stratotype point is just below the level of occurrence of *Monograptus austerus sequens* which indicates the *Demirastrites triangulatus* Zone (Bassett, 1985; Cocks, 1989) (Figure 21.4). However, *M. austerus sequens* has previously been reported from only one other locality, also in Wales (Sudbury, 1958; Hutt, 1974), where its level of first occurrence is within, but significantly higher than, the base of the *D. triangulatus* Zone. In addition, at the stratotype section, the *D. triangulatus* Zone is confidently represented by only a single fossil horizon (Davies *et al.*, 2011). Thus, although the stratotype point can be shown

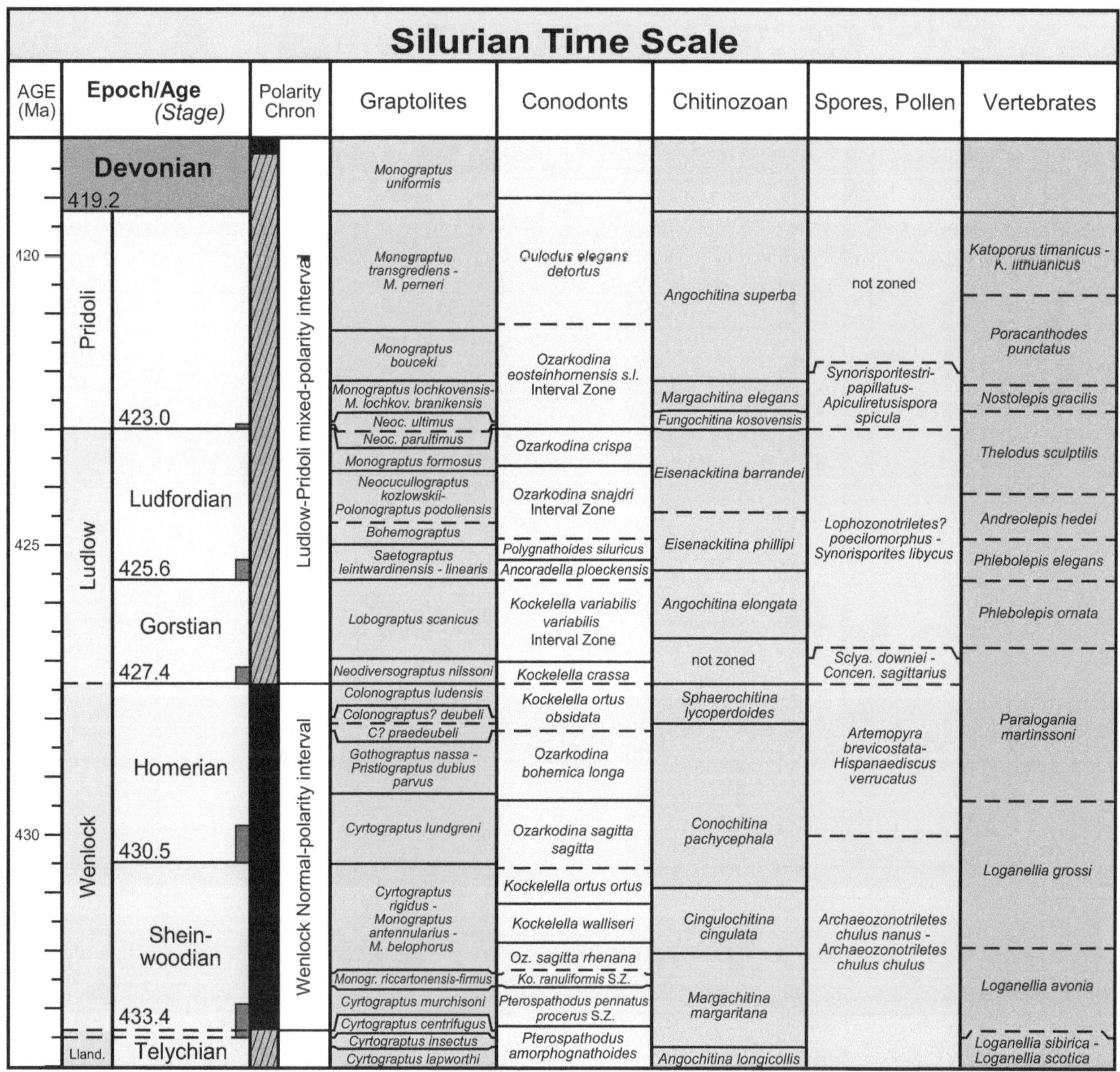

FIGURE 21.1 **Silurian epochs and ages time scale, geomagnetic polarity scheme, graptolite, conodont, chitinozoan, sporomorph and vertebrate zonal schemes**. Graptolite zonation based on Cramer *et al.* (2011a); conodont zonation based on Cramer *et al.* (2011b); Chitinozoan zonation slightly modified from Verniers *et al.* (1995); spore zonation from Subcommission on Silurian Stratigraphy (1995), modified after Burgess and Richardson (1995); and vertebrate zonation from Märss *et al.* (1995). Lower dashed line in epoch and age columns at the base of Wenlock indicates the level originally intended in the GSSP definition (see text for explanation) and the second dashed line immediately above indicates the GSSP level as suggested by Loydell (2008a). Gray boxes at stage boundaries indicate the interval of uncertainty in correlation between stratotype points and the graptolite zonation (see text for explanation). Dashed lines at zonal boundaries indicate significant uncertainties in placement or correlation of the zonal boundaries relative to the composite scale. Note that the name of the conodont zone that spans the Silurian–Devonian boundary differs somewhat between this figure and that shown in Figure 22.10 as a result of differences in taxonomic usage between the Silurian and Devonian conodont literature.

to occur between levels representing the *C. cyphus* and *D. triangulatus* zones, it cannot be shown to correlate precisely with the boundary between those zones, and it probably represents a level within the *D. triangulatus* Zone.

The Aeronian Stage is generally regarded as extending through the *Stimulograptus sedgwickii* Zone, although in parts of Wales this zone can be subdivided into a lower *Stimulograptus sedgwickii* Zone and upper *Stimulograptus halli* Zone (Loydell, 1991). In this case, the latter is regarded as the uppermost graptolite zone of the Aeronian Stage (although see discussion of the Telychian Stage, below).

21.2.1.3. Telychian Stage

The Telychian Stage is named for the Pen-lan-Telych Farm. The stratotype section and point are located in an abandoned quarry that forms part of the Cefn-cerig Road section, at

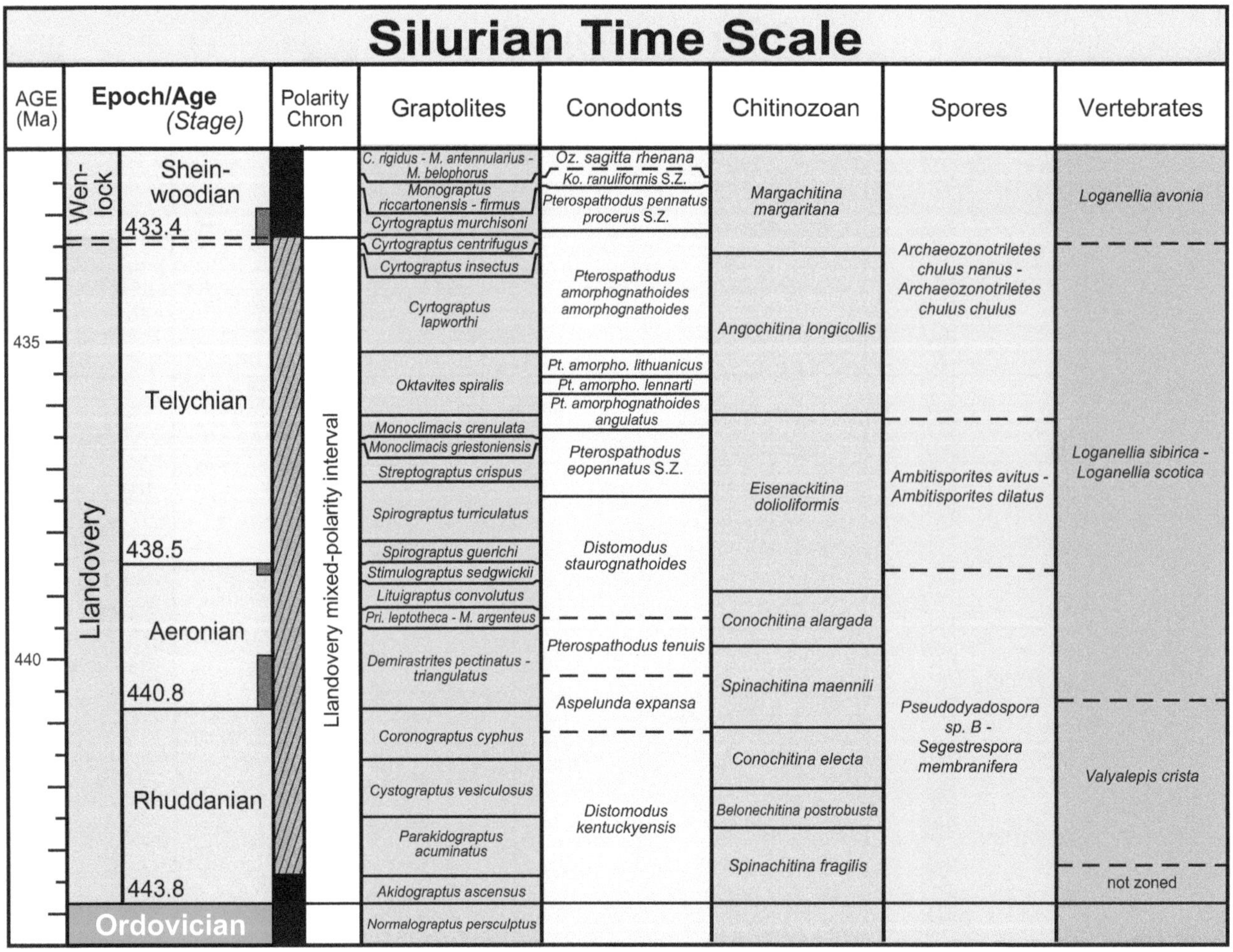

FIGURE 21.1 (Continued).

locality 162 of Cocks *et al.* (1984) and Cocks (1989), approximately 31 m below the top of the Wormwood Formation. Biostratigraphically, it is marked at a level above the highest occurrence of the brachiopod *Eocoelia intermedia* and below the first appearance of *Eocoelia curtisi* (Bassett 1985) (Figure 21.5). This was regarded as corresponding with the base of the *Spirograptus turriculatus* Zone and this was considered to be supported by the occurrence of *Paradiversograptus runcinatus* in the beds above the stratotype level, although not at the stratotype section.

Loydell *et al.* (1993) revised the species of the genus *Spirograptus* and found that the stratigraphically lower specimens previously assigned to *S. turriculatus* and *S. turriculatus minor* actually belong to a distinct, new species, *S. guerichi* (Figure 21.2(c)). Accordingly, those authors found that the strata that had previously been assigned to the lower part of the *S. turriculatus* Zone (the *S. turriculatus minor* Zone of some authors and the *Rastrites linnaei* Zone of others) could be regarded as belonging to a globally correlatable *S. guerichi* Zone. This zone is now regarded as the lowest graptolite zone of the Telychian. However, it should be noted that *Paradiversograptus runcinatus*, the only identifiable graptolite that has been found in the lowest Telychian beds in the vicinity of the stratotype, is known to have its first occurrence in upper Aeronian strata elsewhere in Wales (Loydell, 1991), although it reaches its acme in the lower *S. guerichi* Zone. In addition, Doyle *et al.* (1991) showed that elsewhere in Britain, the last occurrence of *Eocoelia intermedia* occurs within the upper part of the *Stimulograptus sedgwickii* Zone rather than at the base of the *S. guerichi* Zone. Thus, there remains some uncertainty regarding the precise correlation of the stratotype point with the graptolite zonation, but it appears to occur within the upper part of the *S. sedgwickii* Zone.

Recent work has documented the occurrence of *S. guerichi* in strata above the level of the GSSP, at approximately the same level of the previous record of *P. runcinatus* (Davies *et al.*, 2010). However, the same study and also that of Davies *et al.* (2011) showed that the strata in between the last occurrence of *Eocoelia intermedia* and below the first appearance of *Eocoelia curtisi* at the GSSP locality are

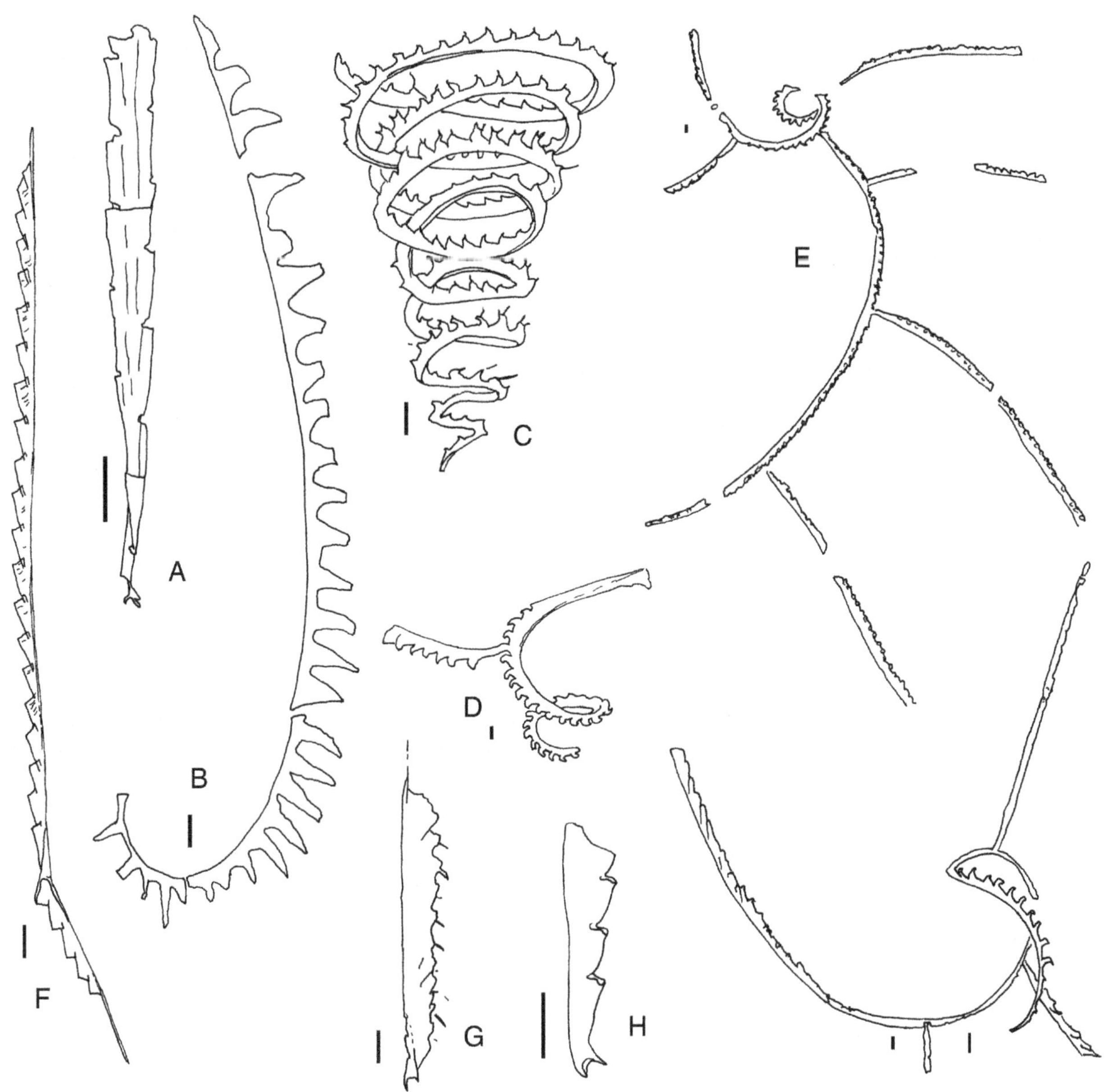

FIGURE 21.2 **Illustrations of graptolite zonal index species for zones indicative of stage boundaries (see text for explanation).** A) *Akidograptus ascensus* Davies, holotype SM A10021, camera lucida drawing by Michael Melchin. B) *Demirastrites triangulatus triangulatus* (Harkness), lectotype BGS GSM6941, redrawn from Zalasiewicz (2008). C) *Spirograptus guerichi* (Loydell, Štorch and Melchin), holotype CGS PŠ359/1, redrawn from Loydell *et al.* (1993, text-Figure 6D). D) *Cyrtograptus centrifugus* Bouček, CGS PŠ572, redrawn from Štorch (1984, Figure 7H). E) *Cyrtograptus murchisoni* Carruthers, counterpart of holotype BGS GSM10718, redrawn from Zalasiewicz and Williams (2008). F) *Neodiversograptus nilssoni* (Barrande), neotype TCD 9735D, redrawn from Palmer (1971, Figure 8A). G) *Saetograptus leintwardinensis leintwardinensis* (Hopkinson), lectotype BU1527, redrawn from Zalasiewicz (2000). H) *Neocolonograptus parultimus* (Jaeger), holotype PMHUg607.1, redrawn from Jaeger (in Kříz *et al.*, 1986, Figure 31). I) *Cyrtograptus lundgreni* Tullberg, lectotype LU O546T, redrawn from Williams and Zalasiewicz (2004, text-Figure 8A). Abbreviations for type numbers: SM – Sedgwick Museum, Cambridge University; BGS – British Geological Survey; CGS – Czech Geological Survey; TDC – Trinity College Dublin; BU – Birmingham University; PMHU – Paläontologisches Museum, Museum für Naturkunde, Humboldt-Universität; LU – Lund University. All scale bars are 1 mm.

structurally disrupted, and mudrocks within that interval and in higher strata contain graptolites and chitinozoans indicative of a mid-Sheinwoodian age intermixed with sediments containing typically Telychian brachiopods in a sedimentary mélange. This evidence shows that the Telychian stratotype locality does not record a continuous stratigraphic succession through the upper Aeronian and lower Telychian.

Standard graptolite zonations have previously shown the Telychian extending upward through the *Monoclimacis crenulata* Zone, overlain by the *Cyrtograptus centrifugus*

Base of the Rhuddanian Stage of the Silurian System in Dob's Linn, near Moffat in the Southern Uplands of Scotland, U.K.

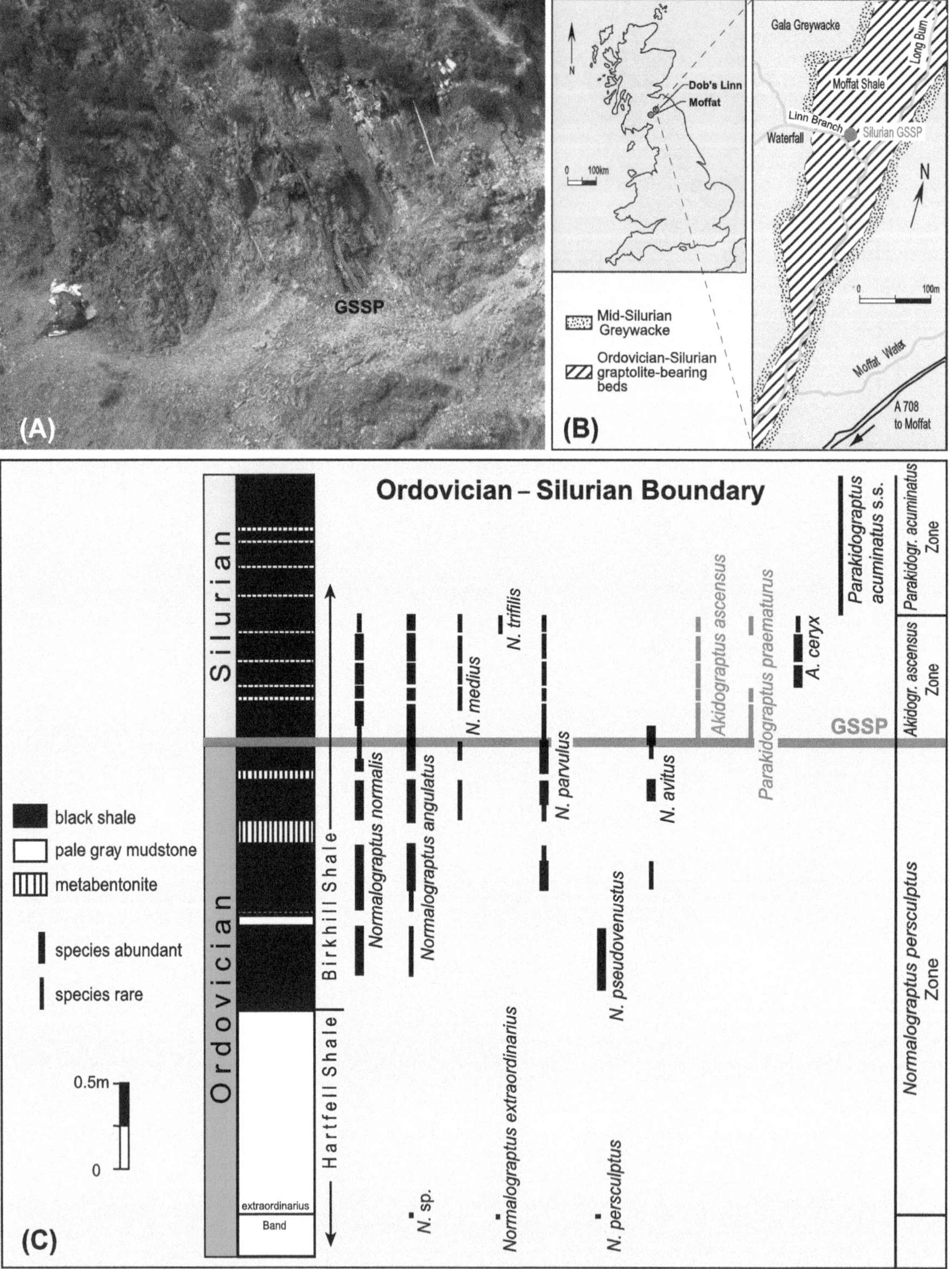

FIGURE 21.3 A) Photograph of the GSSP for the base of the Silurian System at Dob's Linn, Scotland. Yellow stick is 1 m in length. Strata are overturned. Photo by Michael Melchin. B) Map of the GSSP for the base of the Silurian System at Dob's Linn, Scotland. C) Stratigraphy and ranges of selected graptolite taxa at the GSSP for the base of the Silurian System at Dob's Linn, Scotland. Ranges after Williams (1983), Melchin *et al.* (2003) and Rong *et al.* (2008). Taxon names updated after Chen *et al.* (2005).

Base of the Aeronian Stage of the Silurian System in the Trefawr Track Section, Wales, Great Britain

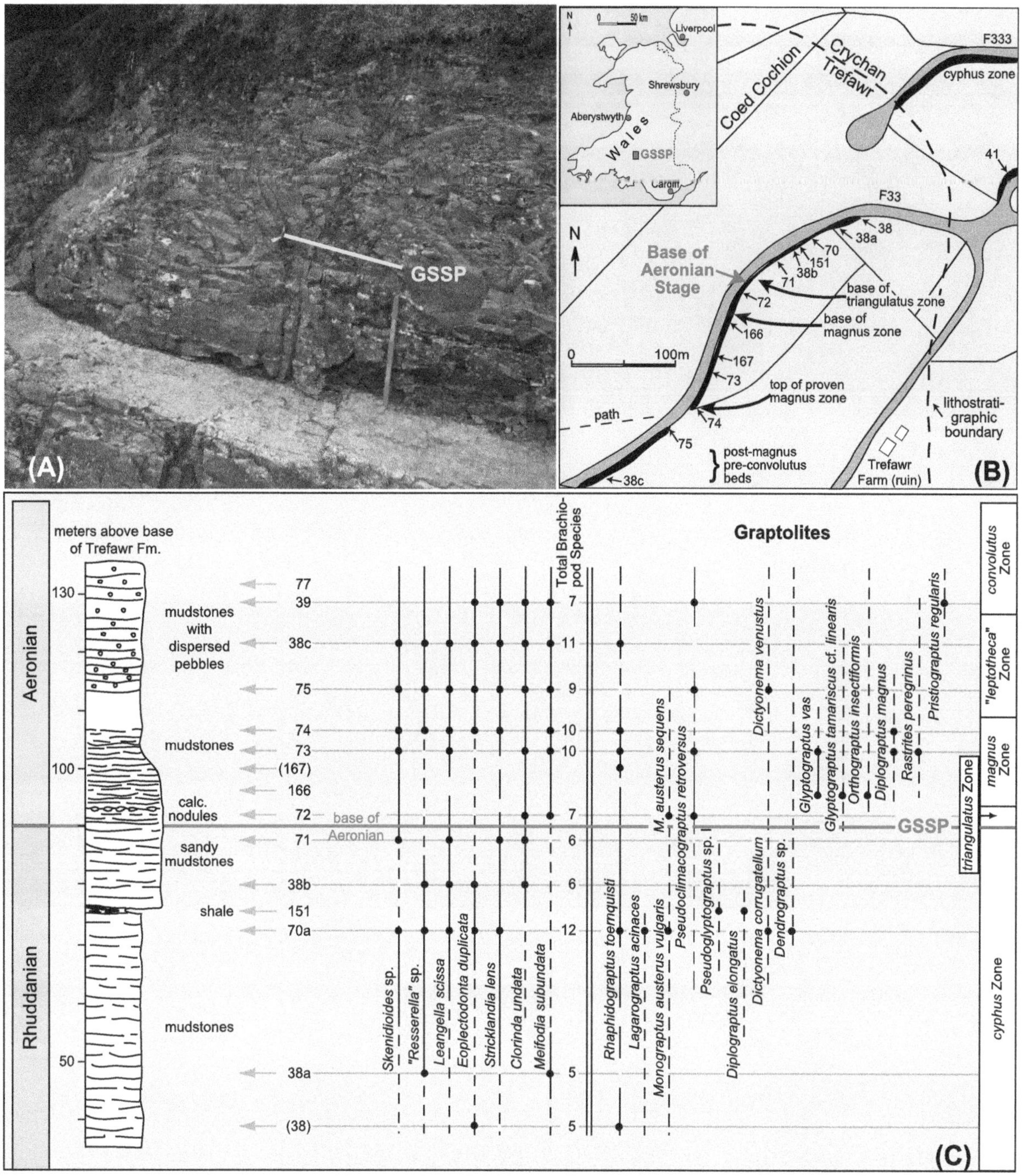

FIGURE 21.4 A) Photograph of the GSSP for the base of the Aeronian Stage at Trefawr track cutting, Crychan Forest, Wales. Photo by Michael Melchin. B) Map of the GSSP for the base of the Aeronian Stage. Exposures are shown in dense black. The locality numbers correspond to the numbers in the profile. C) Stratigraphy and ranges of key taxa at the GSSP for the base of the Aeronian Stage (from Cocks *et al*., 1984).

Zone. However, recent work, especially in Bohemia (Štorch, 1994) and Wales (Loydell and Cave, 1996) has shown that the "standard" British zonation was incomplete and that between the *M. crenulata* and *C. centrifugus* zones there occur strata readily assignable to the *Oktavites spiralis*, *Stomatograptus grandis*–*Cyrtograptus lapworthi* and *Cyrtograptus insectus* zones. Therefore, the Telychian is now regarded as extending from the *S. guerichi* Zone through the *C. insectus* Zone (although see discussion of the Sheinwoodian Stage, below).

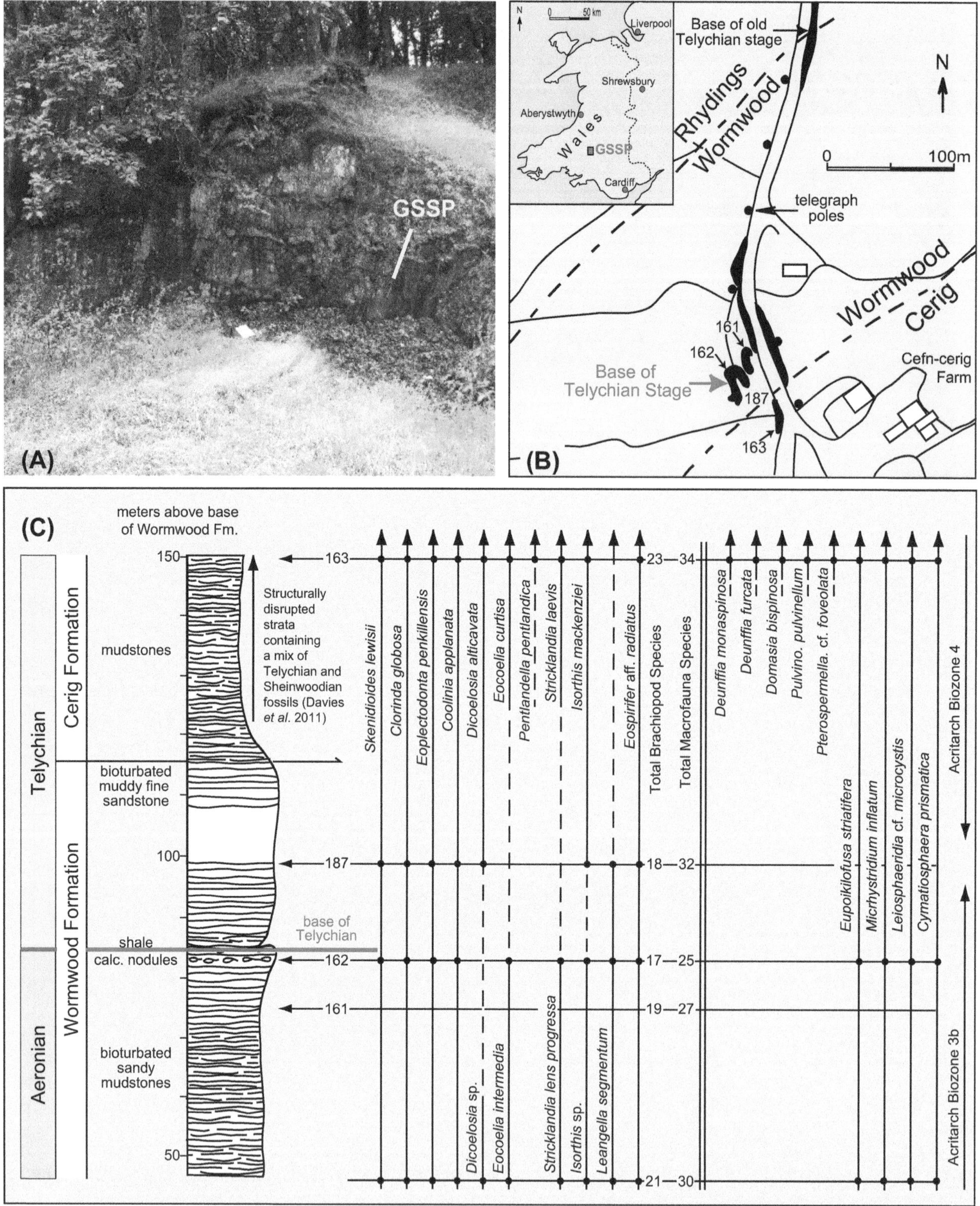

FIGURE 21.5 A) Photograph of the GSSP for the base of the Telychian Stage at Cefn-cerig Quarry, near Llandovery, Wales. Geologic hammer indicates the scale. Photo by Michael Melchin. Strata are not overturned. B) Map of the GSSP for the base of the Telychian Stage, locality numbers refer to numbers on the profile. C) Stratigraphy and ranges of key taxa at the GSSP for the base of the Telychian Stage (from Cocks, 1989). Note that level thought to represent Wormwood-Cerig formational contact has been recently reinterpreted as a regional synsedimentary slide surface (Davies *et al.*, 2010, 2011) and the overlying strata as a sedimentary mélange containing a mix of sediments yielding both Telychian and Sheinwoodian fossils.

21.2.2. Wenlock Series

The Wenlock Series is named for the type area, Wenlock Edge, in the Welsh borderlands of England. It has been divided into two stages, the lower Sheinwoodian Stage and the upper Homerian Stage.

21.2.2.1. Sheinwoodian Stage

The type locality for the Sheinwoodian Stage occurs in Hughley Brook, 200 m southeast of Leasowes Farm and 500 m northeast of Hughley Church. The stratotype point is the base of the Buildwas Formation as described by Bassett *et al.* (1975) and Bassett (1989). The stratotype point occurs within the *Pterospathodus amorphognathoides amorphognathoides* conodont zonal group (sensu Jeppsson, 1997a), between the base of acritarch zone 5 and the last occurrence of *P. am. amorphognathoides* (Mabillard and Aldridge, 1985) (Figure 21.6). This level was considered to be approximately correlative with the base of the *Cyrtograptus centrifugus* graptolite zone, although no graptolites are known to occur in the boundary interval at the stratotype section.

The occurrence of *Monoclimacis* aff. *vomerina* and *Pristiograptus watneyae* higher in the Buildwas Formation, together with species indicative of the *M. greistoniensis* and *M. crenulata* zones in the underlying Purple Shales Formation were regarded as indicating that the stratotype point was near the *centrifugus–crenulata* zonal boundary. However, as noted above, it has since been demonstrated that three graptolite zones can be identified between these zones.

Loydell (2008a, 2011a) reviewed the data pertaining to the biostratigraphic position of the GSSP for the base of the Wenlock. He noted that the conodont data suggested that the GSSP was within a few centimeters of the base of the Upper *Pseudooneotodus bicornis* Zone (Jeppsson, 1997a), or Datum 2 of the Ireviken Event (Jeppsson, 1998). He also noted that data from Estonia and Latvia (Loydell *et al.*, 1998, 2003) indicated that this conodont datum was correlative with a level near the base of the *Cyrtograptus murchisoni* Zone, the graptolite zone above the *C. centrifugus* Zone. Chitinozoan data from the GSSP section (Mullins and Aldridge, 2004) also strongly supported the conclusion that the GSSP was correlative with a level near the base of the *C. murchisoni* Zone. However, more recent conodont data from the east Baltic region suggest that the base of the upper *Ps. bicornis* Zone (Ireviken Datum 2) occurs at a level high within the *C. murchisoni* Zone, rather than near its base (Mannik, 2007a), and this correlation is supported by a recent global graphic correlation study by Kleffner and Barrick (2010). Most recently, in a global study of the biostratigraphy and carbon isotope chemostratigraphy of this interval, Cramer *et al.* (2010a) provided evidence that both Ireviken Datum 2 and the base of the *C. murchisoni* Zone typically occur within less than 200 kyr of the position of the base of the Wenlock Series at its GSSP and, for purposes of correlation, either could be regarded as a good proxy for that level. However, given that the evidence clearly suggests that the GSSP for the base of the Wenlock does not coincide with the graptolite biostratigraphic level originally intended (base of *C. centrifugus* Zone), the International Subcommission on Silurian Stratigraphy is considering whether the biostratigraphic correlation should be revised, or if a new GSSP should be sought that coincides with the base of the *C. centrifugus* Zone. As a result, Figures 20.1 and 20.11 show two tentative levels for the correlation of this boundary, base of *C. centrifugus* Zone as originally intended, and a level near the base of *C. murchisoni* Zone and base of the Upper *Ps. bicornis* conodont zone, as indicated by the current GSSP.

21.2.2.2. Homerian Stage

The stratotype locality for the base of the Homerian Stage is in the north bank of a small stream that flows into a tributary of Sheinton Brook in Whitwell Coppice, which is 500 m north of the hamlet of Homer. The stratotype point is within the Apedale Member of the Coalbrookdale Formation, at the point of first appearance of a graptolite fauna containing *Cyrtograptus lundgreni* (Figure 21.7). Underlying strata contain graptolites of the *C. ellesae* Zone (Bassett *et al.*, 1975; Bassett, 1989). Although the faunas of the *C. ellesae* and *C. lundgreni* zones seem to be stratigraphically and taxonomically distinct in this and some other regions, recent work in Wales has shown a succession where the first occurrence of *C. lundgreni* is below that of *C. ellesae* (Zalasiewicz *et al.*, 1998; Williams and Zalasiewicz, 2004). In addition, Loydell (2011b) suggested that *C. ellesae* and *C. lundgreni* may be the same species. Whether or not this is true, the currently available data suggest that the ranges of the zonal index taxa may be incomplete at the stratotype section and that the stratotype point for the Homerian is likely to be within the *C. lundgreni* Zone rather than at its base.

21.2.3. Ludlow Series

The Ludlow Series is named for the type area near the town of Ludlow, in Shropshire, England. It has been divided into two stages, the lower Gorstian Stage and the upper Ludfordian Stage.

21.2.3.1. Gorstian Stage

The stratotype locality for the base of the Gorstian Stage is in a disused quarry, the Pitch Coppice Quarry, 4.5 km west-south-west of the town of Ludlow. The stratotype point is the base of the Lower Elton Formation where it overlies the Much Wenlock Limestone Formation (Holland *et al.*, 1963; Lawson and White, 1989). Graptolites questionably assigned to the zonal index species *Neodiversograptus nilssoni* (Figure 21.2(f)) and *Saetograptus varians* collected immediately above the base of the Lower Elton Formation indicate

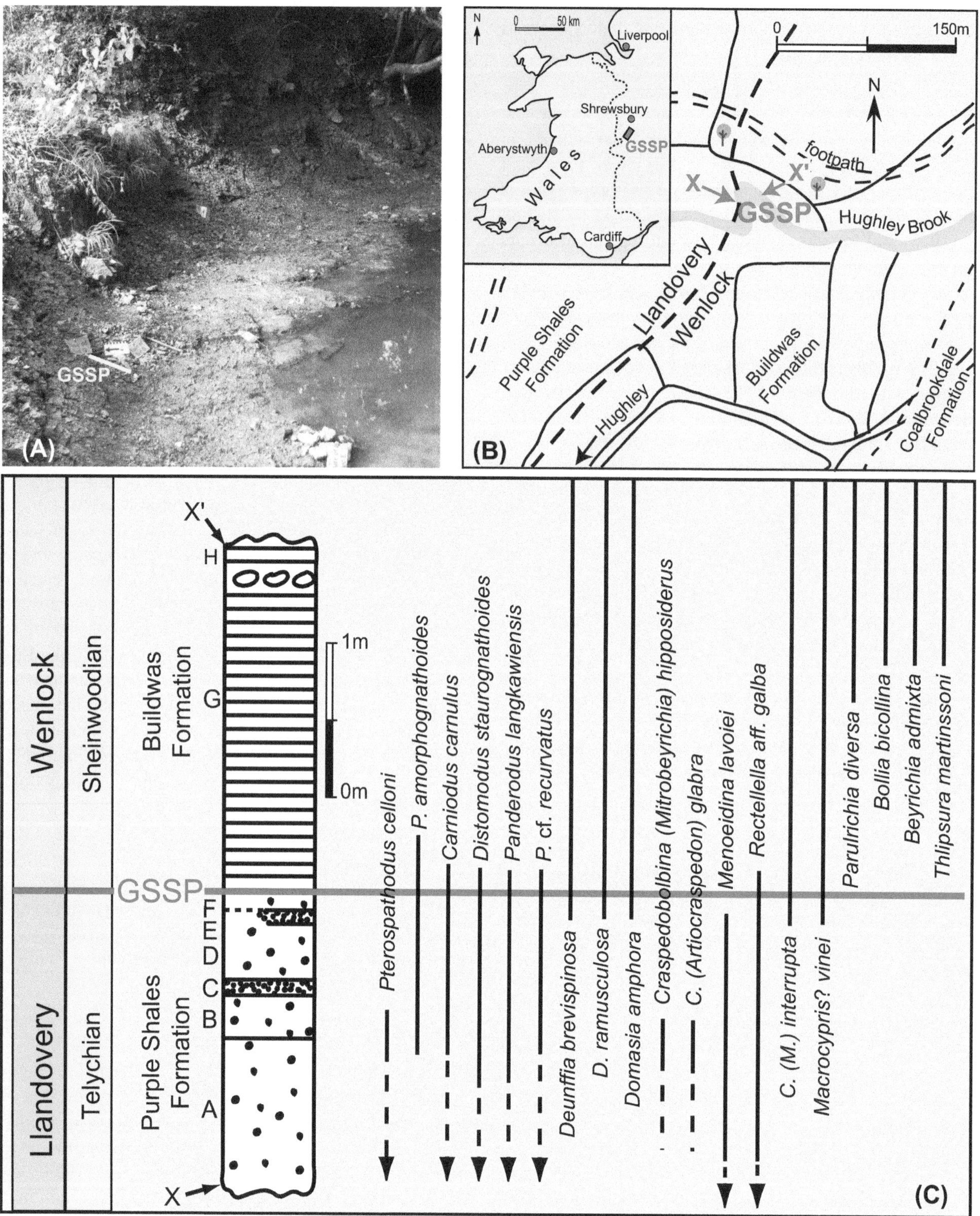

FIGURE 21.6 A) Photograph of the GSSP for the base of the Wenlock Series and Sheinwoodian Stage at Hughley Brook, Shropshire. White scale card is 15 cm in length. Photo by Michael Melchin. (B) Map of the GSSP for the base of the Sheinwoodian Stage. C) Stratigraphy and ranges of key taxa at the GSSP for the base of the Wenlock Series (from Bassett, 1989).

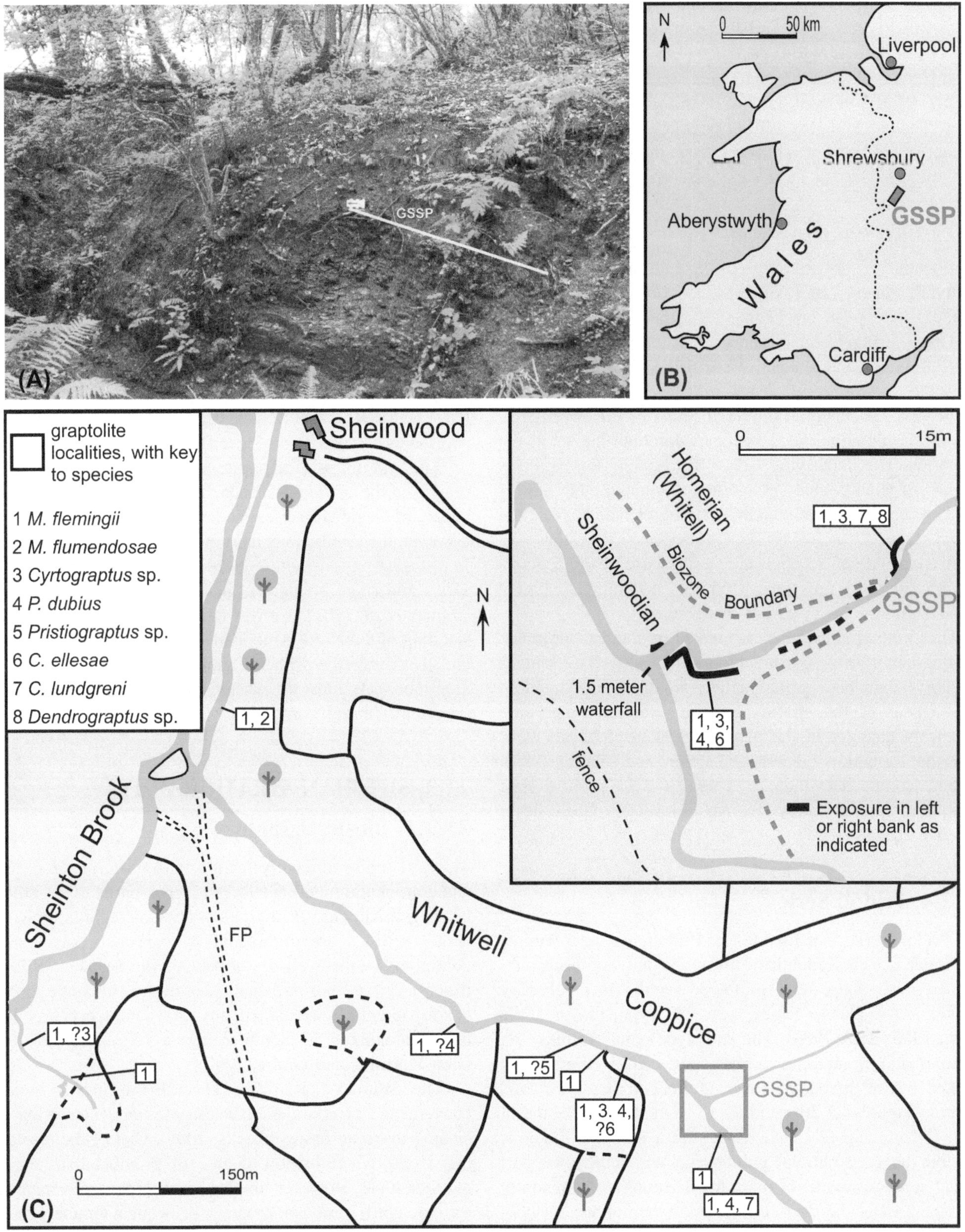

FIGURE 21.7 A) Photograph of the GSSP for the base of the Homerian Stage at Whitwell Coppice, Shropshire. White scale card is 15 cm in length. Photo by Michael Melchin. B), C) Maps showing location of the GSSP for the base of the Homerian Stage, including positions of key biostratigraphic levels (from Bassett, 1989).

that this unit occurs within the *N. nilssoni* Zone (Figure 21.8). However, the absence of graptolites from other parts of the Homerian–Gorstian interval in the type area makes it impossible to precisely correlate the stratotype point with the base of that zone. Lawson and White (1989) noted that neither the shelly fossils nor the conodonts are useful in providing a more refined biostratigraphic definition of the stratotype point. Thomas and Ray (2011) provided isotope data indicating that the GSSP level occurs above the upper peak of the Homerian (Mulde) positive carbon isotope excursion.

21.2.3.2. Ludfordian Stage

The locality of the stratotype of the Ludfordian Stage is at Sunnyhill Quarry, approximately 2.5 km southwest of the town of Ludlow. The level coincides with the contact between the Upper Bringewood Formation and the Lower Leintwardine Formation (Lawson and White, 1989) (Figure 21.9). The graptolite *Saetograptus leintwardinensis leintwardinensis* (Figure 21.2G) occurs in the basal beds of the Lower Leintwardine Formation, and it becomes common higher in the formation. The underlying Upper Bringewood Formation is devoid of identifiable graptolites, although the Lower Bringewood Formation contains graptolites indicative of the *P. tumescens/S. incipiens* Zone. Therefore, the stratotype point is considered to approximate the base of the *S. leintwardinensis* Zone (Lawson and White, 1989), although it may occur within the lower part of that zone.

The stratotype level is also marked by the disappearance of a number of distinctive brachiopod taxa, as well as changes in relative abundances among others. No distinctive conodont taxa appear at the stratotype point. However, there are significant changes in the palynological assemblages at or near the formational contact (Lawson and White, 1989). Carbon isotope data indicate a weak (~1‰) positive shift beginning approximately 50 cm below the GSSP level (Cherns, 2011).

21.2.4. Pridoli Series

The Pridoli Series is named for the Přídolí area, near Prague, Bohemia, Czech Republic. This series has not been subdivided into stages. The stratotype section is the Požáry section in the Daleje Valley, near Řeporyjie, Prague (Kříz *et al.*, 1986; Kříz, 1989). The stratotype point is in bed 96, approximately 2 m above the base of the Požáry Formation, marked by the first appearance of *Neocolonograptus parultimus* (Figures 21.2(h) and 21.10). Graptolites are absent from the immediately underlying strata, however, so it is possible that the stratotype point occurs within the lower part of that zone. A number of other fossil groups are common in the type area of the Pridoli besides the graptolites, but only chitinozoans show potential for more detailed biostratigraphic correlation in this region. The base of the *Fungochitina kosovensis* chitinozoan zone occurs approximately 20 cm above the stratotype point.

21.2.5. Other Important Stage Classifications

Although a number of regions of the world have had a regional series and stage classification, the majority of these have fallen out of usage since the global standard series and stage stratotype sections and points were defined. The generally high degree of faunal cosmopolitanism has greatly facilitated the global usage of the standard time scale. However, there remain significant intervals in which correlation between the graptolite biostratigraphic scale and that of conodonts and carbonate shelf facies are imprecise. As a result, in the East Baltic Region, which has a long history of detailed faunal and stratigraphic study in mainly carbonate strata, workers continue to refer to a scale of regional stages (East Baltic Regional Stages, Figure 21.11) based on faunal and facies changes in that area (Bassett *et al.*, 1989).

The interior of Laurentia, including the midcontinent regions of the USA and Canada, has also seen some continued usage of regional series and stages. Unlike the Baltic, however, the terminology has not been consistently employed in different regions within North America, and many of the series and stage boundaries have not been clearly defined. Cramer *et al.* (2011a) reviewed the current state of understanding of the North American series and stage terminology and its correlation with the global standard, and identified a number of intervals where there remains considerable uncertainty in this correlation.

21.3. SILURIAN STRATIGRAPHY

21.3.1. Biostratigraphy

Much of the sedimentary record of the Silurian in basinal, and continental margin settings is represented by graptolite-bearing mudrocks. In the paleo-tropical regions, epicontinental settings are dominated by carbonate successions, commonly with well-developed reefs. In later Silurian time, large evaporite basins developed in some epeiric basins. Glacial deposits, mainly of Llandovery age, have been recognized in South America (e.g., Caputo 1998; Díaz-Martinez and Grahn, 2007).

The Silurian Period was generally a time of relative convergence of continental landmasses and narrowing and closing of ocean basins (Cocks, 2000). One of the results of this is the generally low degree of provincialism seen in marine faunas. However, another result of this convergence is tectonic uplift in several orogenic belts and a number of areas that were dominated by marine sedimentation through much of Ordovician and Early Silurian time become sites of

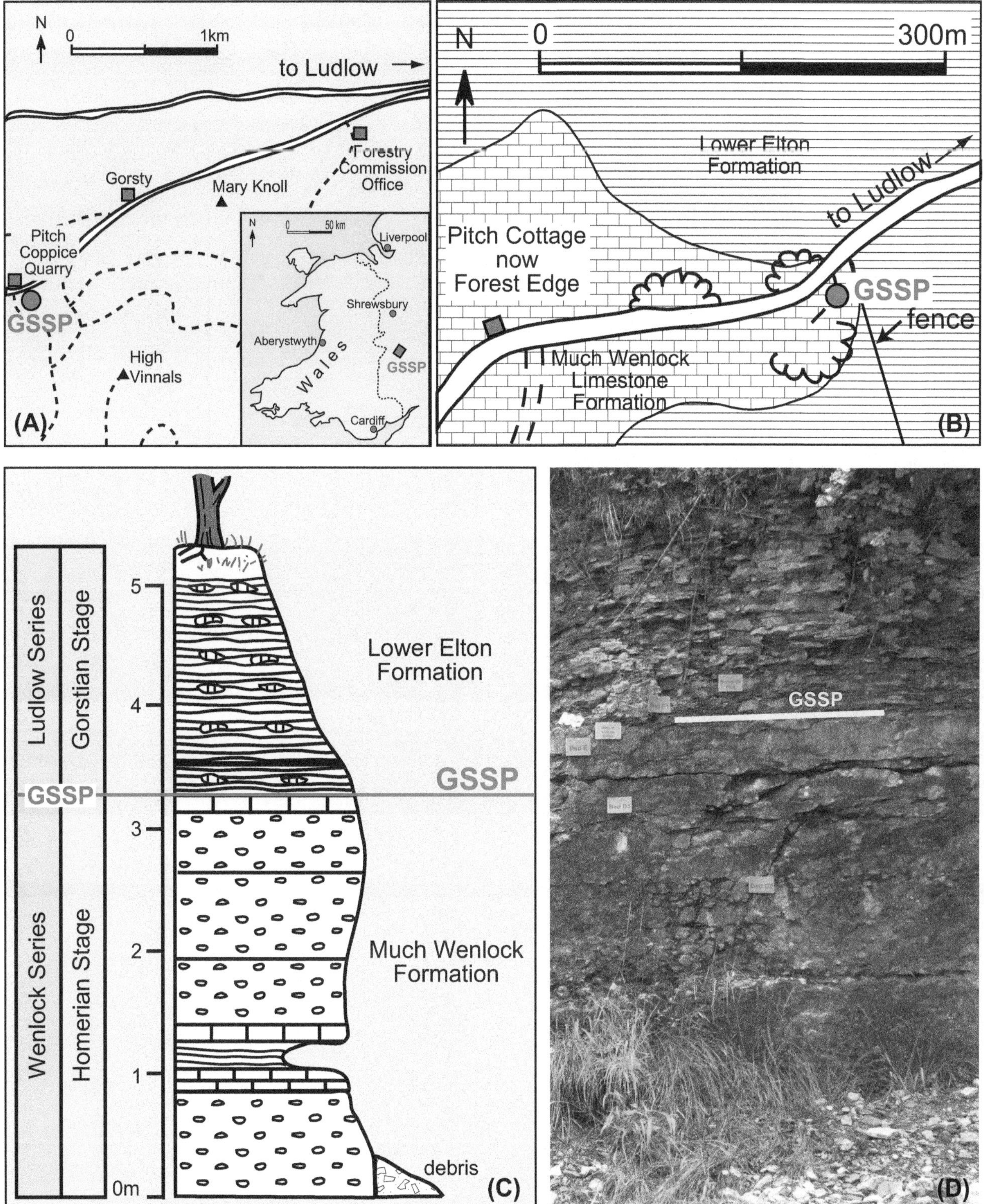

FIGURE 21.8 A), B), Map of the GSSP for the base of the Ludlow Series and Gorstian Stage at Pitch Coppice near Ludlow, Shropshire, Great Britain. C) Lithologic log of GSSP section (from Lawson and White, 1989). D) Photograph of the GSSP for the base of the Gorstian Stage. White scale card is 15 cm in length. Photo by Michael Melchin.

Base of the Ludfordian Stage of the Silurian System at Sunnyhill Quarry, United Kingdom.

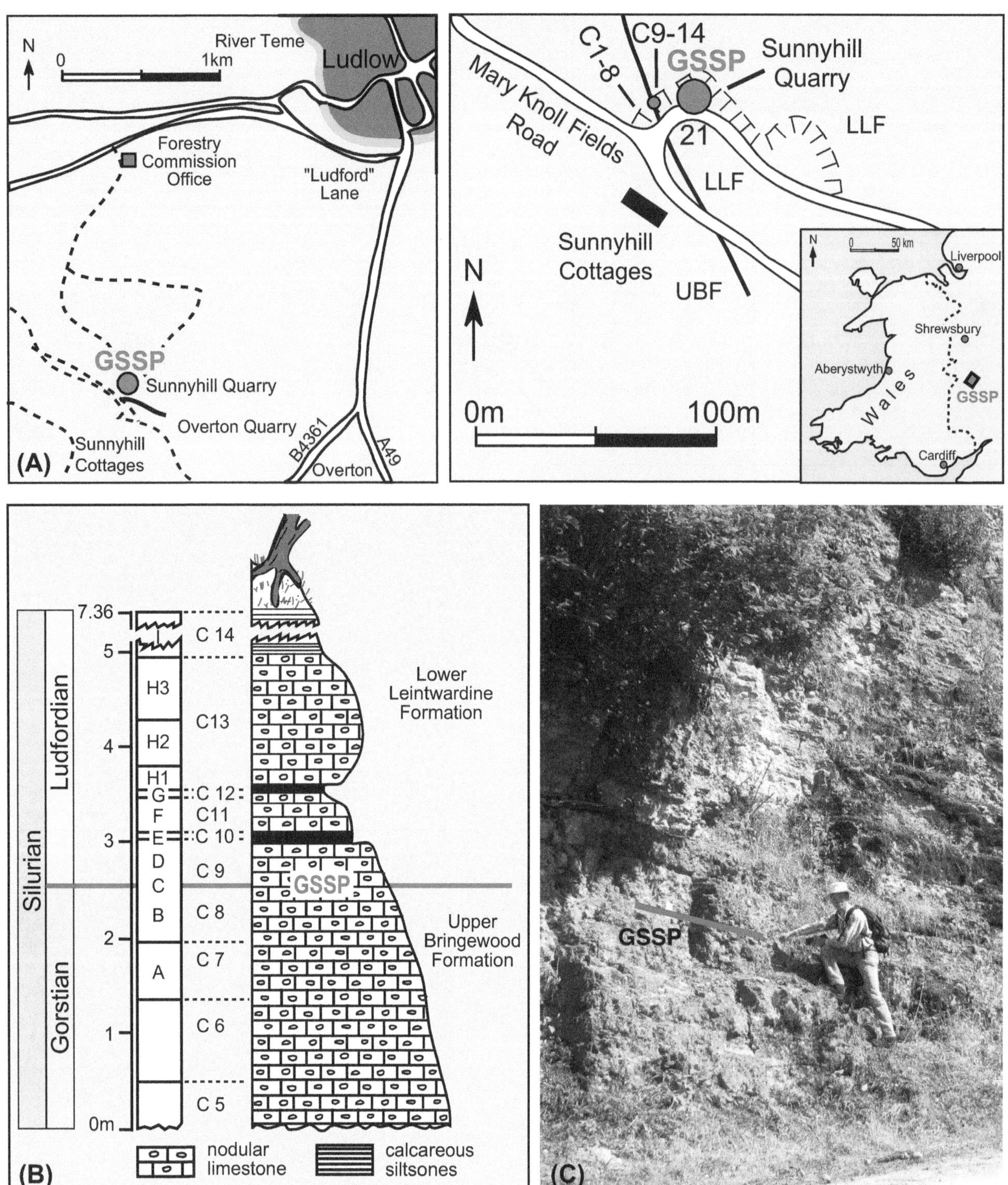

FIGURE 21.9 A), B), Map of the GSSP for the base of the Ludfordian Stage at Sunnyhill Quarry near Ludlow, Shropshire, Great Britain. C) Lithologic log of GSSP section (from Lawson and White, 1989). D) Photograph of the GSSP for the base of the Ludfordian Stage. *Photo by Michael Melchin.*

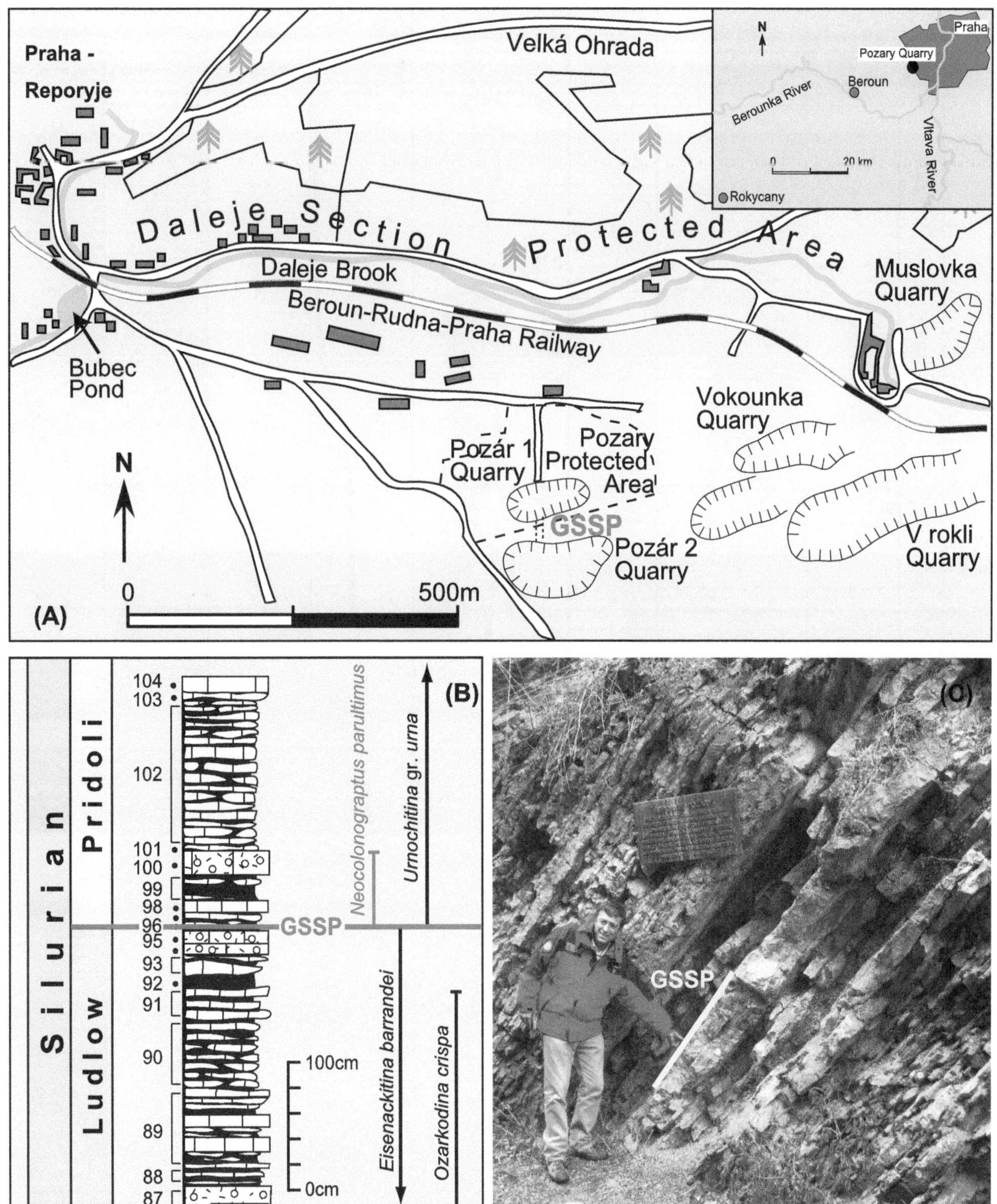

FIGURE 21.10 A) Map of the GSSP for the base of the Pridoli Series in the Daleje Valley, Prague, Czech Republic. B) Lithologic log of the GSSP section showing numbered sample levels and ranges of some key taxa (from Kříz, 1989). C) Photograph of the GSSP for the base of the Pridoli Series. *Photo by Michael Melchin.*

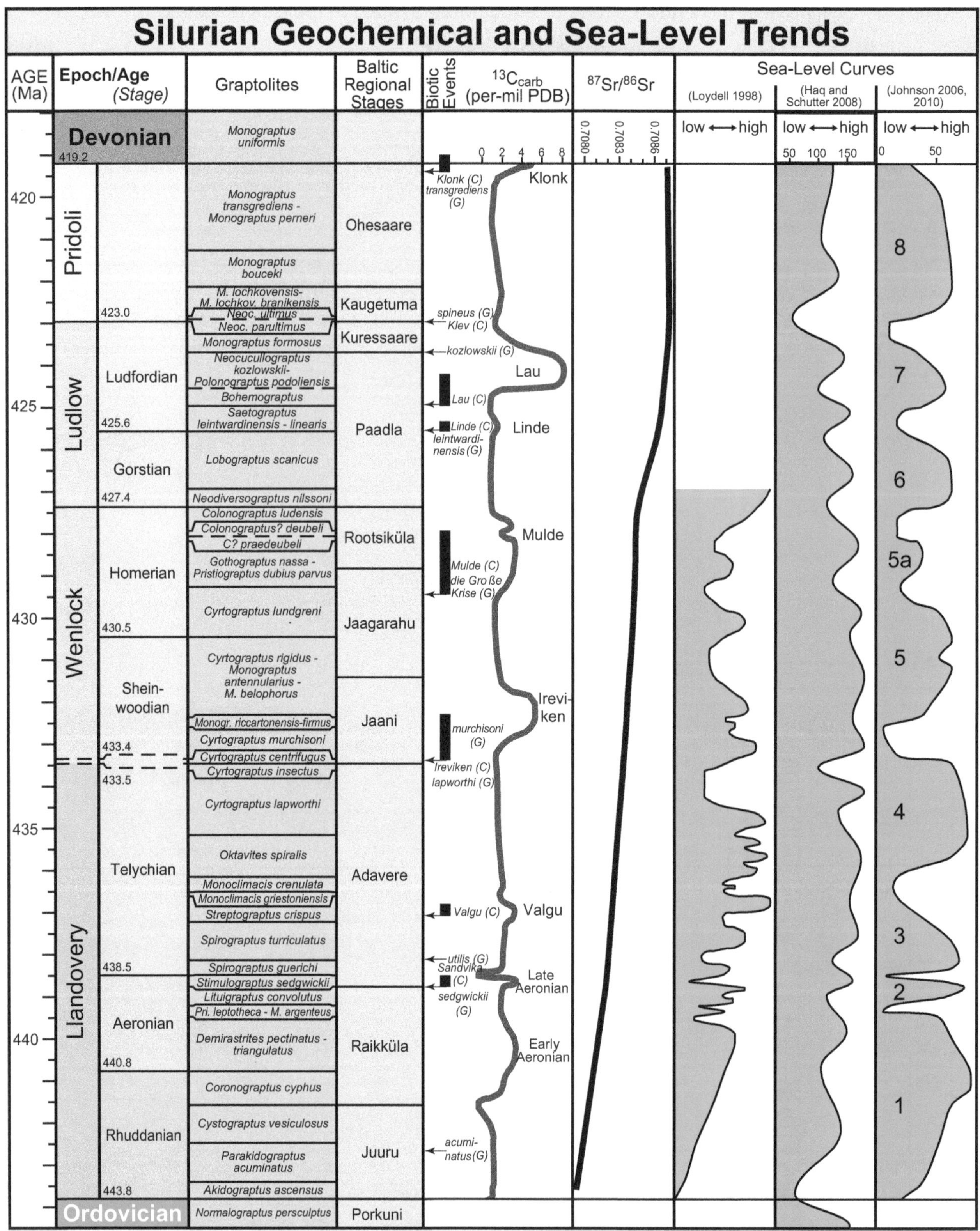

FIGURE 21.11 Silurian time scale with graptolite biozonation, Baltic regional stages, graptolite and conodont bioevents (Jaeger, 1991; Jeppsson, 1998; Melchin *et al.*, 1998, Jeppsson *et al.*, 2006); stable isotope chemostratigraphy for carbon (Cramer *et al.*, 2011a) and strontium (Cramer *et al.*, 2011c), and relative eustasy scales, with sources as indicated. Sea-level curves of Loydell (1998) and Johnson (2006) were scaled to the new time scale using levels of graptolite zonal boundaries. Sea-level curve of Haq and Schutter (2008) was rescaled using radiometric ages of stage boundaries, and ages of lowstand events were proportionately rescaled within each stage.

continental sedimentation or non-deposition by the end of the Silurian.

As in the Ordovician, the two fossil groups that have been most widely used in Silurian biostratigraphic correlation are the graptolites and conodonts. Reference to the GSSPs for the series and stages is mainly by relation to the graptolite zones, although conodonts, chitinozoans, and carbon isotopes are increasingly being used for correlation with GSSPs. A number of shelly fossil groups, especially brachiopods, have proved to be useful for regional correlation, as have acritarchs, microvertebrate remains, and, in some circumstances, land plant spores.

21.3.1.1. Graptolite zones

Graptolites (Phylum Hemichordata) were a component of the Silurian macroplankton. As in the Ordovician, they lived at various depths in the ocean waters (e.g., Underwood, 1993; Cooper and Sadler, 2010), were particularly abundant in upwelling zones along continental margins (Finney and Berry, 1997) and are found in a wide range of sedimentary facies. Most graptolite species dispersed rapidly, are geographically widespread, and are of relatively short stratigraphic duration (0.5–4 myr). These attributes combine to make them extremely valuable fossils for the zonation and correlation of strata. Together with conodonts, they are the primary fossil group for global correlation of Silurian sequences and the biostratigraphic levels that are used to correlate the Silurian GSSPs are all based on graptolite zones (Figure 21.1).

The most established zonal scheme as the "standard" for Silurian graptolite biostratigraphy (Harland *et al.*, 1990) has traditionally been based on the British zonation (e.g., Rickards, 1976), which was recently updated and refined by Zalasiewicz *et al.* (2009). The exception is the Pridoli zonation, which is based on the succession in Bohemia (e.g., Kříz *et al.*, 1986). However, recent efforts have been undertaken to establish a globally recognizable, standard zonation, based on widely recognizable episodes of faunal change rather than the succession of any particular region.

A first step toward this was the publication of the "generalized graptolite zonal sequence" (Subcommission on Silurian Stratigraphy, 1995; Koren' *et al.*, 1996), which was assembled for the purpose of a coordinated study of global paleogeography, but was also used for a study of patterns of global diversity and survivorship in Silurian graptolites (Melchin *et al.*, 1998). In the course of the latter study it was found that a number of the generalized zones recognized by Koren' *et al.* (1996) could be readily subdivided and still recognized in several different paleogeographic regions of the world. Further refinements of the Generalized Graptolite Zonation were presented in Melchin *et al.* (2004), Sadler *et al.* (2009) and Cramer *et al.* (2011a,b). The graptolite zonation presented here represents a slight refinement of the one published by Cramer *et al.* (2011a). The criterion used in this zonation, as was the case for the original Generalized Graptolite Zonation (Koren' *et al.*, 1996), was that zonal base should be recognizable in at least three different paleogeographic regions of the world.

Figure 21.1 shows the remarkable precision in correlation obtainable using Silurian graptolites. The 36 graptolite zonal divisions of the Silurian span 24.6 myr and thus average approximately 683 000 years each in duration. At the regional level, where more zones are recognized in some intervals, the precision is even better.

21.3.1.2. Conodont Zones

Conodonts are tooth-like structures in the feeding apparatus of small lamprey-like primitive chordates (Aldridge and Briggs, 1989). They are most readily preserved in, and extracted from, carbonate facies; however, the free-swimming conodont animal roamed widely and some species are found in a broad variety of lithologies and depositional environments, making them an important tool for global stratigraphic correlation. The fluorapatite (calcium phosphate) conodont elements are typically extracted via acid digestion using either acetic or (preferably) formic acid. In particularly dolomitic Silurian carbonate rocks, the use of buffered formic acid has greatly improved the yield (number of elements/kg of rock) during digestion.

Exceptionally detailed conodont biozonations have been constructed for portions of the Silurian (Wenlock and Ludlow: Jeppsson, 1997a; Jeppsson *et al.*, 2006; Telychian: Männik, 1998, 2007b; Ludfordian to Pridoli: Corradini and Serpagli, 1999; Corriga and Corradini, 2009), although other portions of the Silurian remain much less well refined (e.g., Rhuddanian and Aeronian). At the finest scale of resolution (<500 kyr), it is becoming clear that the first appearances of some zonally important conodont species are not always synchronous globally (Cramer *et al.*, 2010b). Although similar difficulties can be found among Silurian graptolites, these diachroneities must be addressed when working at extremely high resolution (i.e. approaching the Milankovitch band). The conodont zonation of the Baltic Basin remains the most highly refined of any Silurian paleocontinent, and the recognition of these zones is in its infancy globally. In the comparatively few studies published from outside of Baltica that have been conducted at sufficient resolution and using suitable digestion techniques, the vast majority of the Baltic zones appear to be recognizable. The composite biozonation used here (Figure 21.1) is from Cramer *et al.* (2011a), which was constructed specifically to include the entire Silurian and to be more globally applicable than previously published zonations.

21.3.1.3. Chitinozoan Zones

Chitinozoa are organic-walled microfossils of unknown biological affinities, although many accept the hypothesis that they were the planktonic egg capsules of some metazoan. They occur in a variety of marine facies and many species were geographically widespread and relatively short-lived. Verniers *et al.* (1995) proposed a global biozonation for Silurian chitinozoa, based on correlation of well-known successions in Laurentia, Avalonia, Baltica and Gondwana (Figure 21.1). Global biozonal levels are defined by well-established taxa whose first appearances are regarded as synchronous in two or more distinct paleogeographic regions. Many of these biozonal levels have been defined or recognized in direct reference to global stratotype sections and points. A new, refined, global chitinozoan biozonation is currently in preparation.

21.3.1.4. Other Zonal Groups

The Subcommission on Silurian Stratigraphy (1995), which produced generalized zonations for graptolites, conodonts and chitinozoans, also provided Silurian zonations for spores and vertebrates (Figure 21.1). The former, which has been slightly revised by Burgess and Richardson (1995), is particularly important in that it provides the possibility for biostratigraphic correlation in terrestrial strata and between the terrestrial and marine realm. The vertebrate zonation, based mainly on disarticulated remains (ichthyoliths), was refined by Märss *et al.* (1995).

21.3.2. Physical Stratigraphy

21.3.2.1. Magnetostratigraphy

Our understanding of the magnetostratigraphic scale for the Silurian System is still in a very preliminary state and is based on incomplete data from only a few localities. Based on the presently available data it appears that much of the Silurian is characterized by a mixed polarity, with a predominantly normal phase through much of the Wenlock (Figure 21.1; Trench *et al.*, 1993).

21.3.2.2. Chemostratigraphy

Chemostratigraphic analyses are now available at fairly high levels of stratigraphic resolution for all or most of the Silurian System for the $\delta^{13}C$ and Sr isotopic systems (Figure 21.11). Curves for the $\delta^{18}O$ system are less complete and/or provide lower resolution than those of $\delta^{13}C$ and Sr (Munnecke *et al.*, 2010) (see further discussion below).

Carbon Isotope Stratigraphy

The carbon isotopic ratio of Silurian seawater, measured either in carbonate ($\delta^{13}C_{carb}$) or in organic matter ($\delta^{13}C_{org}$), was highly variable through time and indicates that the global carbon cycle and global climate system were less stable during the Silurian than in almost any other period during the Phanerozoic. As is the case with most of the Paleozoic, the $\delta^{13}C_{carb}$ record is better known than the $\delta^{13}C_{org}$ record, and throughout the rest of this chapter we will be referring to the $\delta^{13}C_{carb}$ record unless otherwise indicated. Significant positive excursions to the carbon isotopic ratio of Silurian seawater occurred during the early and late Aeronian, early Telychian, early to middle Sheinwoodian, middle to late Homerian, middle Ludfordian, and during the latest Pridoli – crossing the base of the Devonian (see Cramer *et al.*, 2011a and references therein; Figure 21.11). The Sheinwoodian (Samtleben *et al.*, 1996), Ludfordian (Samtleben *et al.*, 1996) and Silurian–Devonian boundary (Saltzman, 2002) positive $\delta^{13}C_{carb}$ excursions have all been shown to have exceeded +5.0‰ (VPDB). Although the late Aeronian excursion has been well recorded in deep-water strata (Melchin and Holmden, 2006), it has rarely been documented from carbonate facies. Wenzel (1997) recorded values >4.5‰ during this excursion. Similarly, the Homerian excursion tends to be of lower amplitude than those that occurred during the Sheinwoodian or Ludfordian in any given basin; however, values greater than +4.0‰ have been recorded from Homerian strata and this relationship is not always the case in every basin (e.g., Podolia : Kaljo *et al.*, 2007). The Ludfordian excursion is exceptional in its amplitude (typically >+8.0‰) in that it is likely the largest post-Cambrian excursion of the entire Phanerozoic. By the middle Ludfordian, it had been more than 100 million years since the carbon isotopic ratio of the global ocean had been at such an elevated level.

Carbon isotope chemostratigraphy can be used to investigate changes in the global carbon cycle and the ocean–atmosphere–biosphere system, and, in addition, can be used as a purely stratigraphic tool. It is in the latter capacity that the Silurian $\delta^{13}C_{carb}$ record has become among the most highly resolved of any period of the Paleozoic. Because $\delta^{13}C$ chemostratigraphy can be recovered from strata that contain conodonts or graptolites (i.e. from carbonate or clastic settings), carbon isotope chemostratigraphy has become an invaluable tool for comparing and refining the correlation between biostratigraphically useful groups, such as graptolite and conodont biostratigraphy, which are infrequently found together in the stratigraphic record (e.g., Kaljo and Martma, 2006; Cramer *et al.*, 2010a).

Whereas the utility of carbon isotope chemostratigraphy is broadly recognized throughout the Silurian community, the relationship between the carbon isotopic record and Silurian biotic/global climate events remains a highly contentious issue (e.g., Cramer and Saltzman, 2007; Loydell, 2007, 2008b; Cramer and Munnecke, 2008; Munnecke *et al.*, 2010). As is the case with the majority of the positive carbon isotope excursions during the Paleozoic, paleobiodiversity events occurred immediately prior to and/or during the onset of positive excursions, whereas the majority of any given excursion coincides with the post-extinction/recovery phase of the biotic event. Quantitative marine paleoproductivity data only exist for the Ludfordian (Lau) positive excursion and indicate a crash in marine productivity during the positive excursion (Sticanne *et al.*, 2006). Reconstruction of Silurian sea levels remains problematic (Munnecke *et al.*, 2010). However, it is clear that the positive excursions persisted through several sea-level cycles that were likely on the order of eccentricity (400 kyr), given that the excursions themselves were typically on the order of 1–1.5 myr in duration. The $\delta^{13}C_{org}$ record typically mirrors that of $\delta^{13}C_{carb}$ during the Silurian, although the magnitudes of the excursions in $\delta^{13}C_{org}$ are typically muted compared to the same isotopic event in the $\delta^{13}C_{carb}$ record.

Oxygen Isotope Stratigraphy

A number of published oxygen isotope ($\delta^{18}O$) curves for the Silurian show very similar trends (Wenzel and Joachimski, 1996; Samtleben *et al.*, 1996; Azmy *et al.*, 1998; Heath *et al.*, 1998; Lehnert *et al.*, 2010; Munnecke *et al.*, 2010). There is a slight general trend toward reduction in $\delta^{18}O$ values through Silurian time, and superimposed on this trend are a number of significant positive excursions of approximately 1–3.5‰ magnitude, that roughly match in timing with positive excursions in the $\delta^{13}C_{carb}$ record (e.g., Žigaitė *et al.*, 2008; Lehnert *et al.*, 2010; Munnecke *et al.*, 2010). Bickert *et al.* (1997) demonstrated that the oxygen isotope variations within the Silurian are too large to be interpreted as being solely the result of either temperature, ice-volume or salinity alone. Similarly, the Llandovery with its unequivocal record of continental glaciations should correspond with some of the largest variations in $\delta^{18}O$; however, this is not the case (see Munnecke *et al.*, 2010). There is considerable controversy surrounding the reliability of different media (e.g., brachiopod carbonate, whole-rock carbonate, conodont phosphate, vertebrate phosphate) of analysis for the $\delta^{18}O$ record, as well as the significance of long-term, secular trends in terms of estimation of seawater temperatures (e.g., Came *et al.*, 2007; Trotter *et al.*, 2008; Pucéat *et al.*, 2010). Considerable further research is required, particularly independent methods to verify changes in the parameters that can control $\delta^{18}O$ (salinity, temperature, ice-volume, etc.); for example, by applying clumped-isotope paleothermometry for temperature, fluid inclusion analysis for salinity, etc.

Strontium Isotope Stratigraphy

The strontium isotopic ratio ($^{87}Sr/^{86}Sr$) of Silurian seawater shows a generally increasing trend from roughly 0.707 95 during the early Rhuddanian to roughly 0.708 70 during the latest Pridoli (Cramer *et al.*, 2011c; Figure 21.11). This Silurian trend is a continuation from the latest Hirnantian, following the Katian–early Hirnantian low recorded from the Ordovician. Previous composites have indicated that $^{87}Sr/^{86}Sr$ values continue to increase throughout the Silurian (e.g., Ruppel *et al.*, 1996; Azmy *et al.*, 1999; Veizer *et al.*, 1999), although recent recalibrations of available data indicate that the inflection from rising to more stable values may have taken place within the Pridoli. It is clear, however, that the descending limb of the Late Ordovician–Silurian–Early Devonian $^{87}Sr/^{86}Sr$ record (Veizer *et al.*, 1999) did not begin in earnest until the Devonian. Uncertainty in the duration of stages within the Silurian time scale prohibits meaningful discussion of the rates of change in the Silurian $^{87}Sr/^{86}Sr$ record; however, the Ludlow portion of the $^{87}Sr/^{86}Sr$ curve appears to be an interval of interest (Cramer *et al.*, 2011c). The successive reduction in the length of the Ludlow since the time scale of Harland *et al.* (1990) (see Melchin *et al.*, 2004; Ogg *et al.*, 2008; Sadler *et al.*, 2009) forces the Silurian $^{87}Sr/^{86}Sr$ curve to have a significant inflection point in the rate of change in $^{87}Sr/^{86}Sr$ at or near the base of the Ludlow Series to more rapidly increasing values throughout the Ludlow. Further refinement to the radiometric age calibration of the Silurian time scale is required to fully interpret the Silurian $^{87}Sr/^{86}Sr$ record.

The Silurian trend toward higher $^{87}Sr/^{86}Sr$ values throughout the period is most readily explained by the paleogeographic/paleotectonic history of the Silurian. The paleocontinents of Baltica, Avalonia and Laurentia were colliding throughout the Silurian period as the three paleocontinents amalgamated during this time. Laurentia alone had a collisional margin many thousands of kilometers long, including its northern, eastern, and to a lesser extent, southern margins. All else being equal however, a reduction in the rate of sea-floor spreading and/or increased weatherability and weathering of old sialic strata likely contributed to the overall increasing trend (Cramer *et al.*, 2011c). Improved Silurian $^{87}Sr/^{86}Sr$ data, particularly in conjunction with detailed chronostratigraphy of graptolites, conodonts or carbon isotopes is sorely needed. $^{87}Sr/^{86}Sr$ chemostratigraphy can provide a useful tool for chronostratigraphic correlation at the stage level throughout the Silurian, and is particularly useful in settings where biostratigraphy and chemostratigraphy are either unavailable or unlikely to be produced.

21.3.2.3. Eustasy

A considerable body of literature exists on the effort to construct a eustatic curve for the Silurian Period. There have been several different approaches to the problem of estimating eustatic changes. Johnson (1996) and Munnecke *et al.* (2010) summarized a series of earlier papers in which the sea-level histories of individual regions have been reconstructed mainly based on the use of benthic assemblages and sedimentary structures. Sea-level curves for each region are then correlated and compared to identify global signals. Johnson *et al.* (1998) added to this the study of submergence of paleotopographic features as measures of absolute sea-level change. Ross and Ross (1996) also employed biofacies data, but also incorporated some lithofacies information, although their curve is based almost entirely on sections in Laurentia. Johnson (2006) provided an update of the Silurian eustatic record, based on this approach (Figure 21.11), and included a discussion of the relationship between sea-level change and the patterns of extinction and oceanic events.

In contrast, Loydell (1998; see Figure 21.11) defined episodes of sea-level rise and fall based primarily on identification of the oxidation state of the strata under investigation and the graptolite fauna contained therein. For example, graptolitic shales, in which, he argued, transgressive intervals are recognized as being generally condensed and organic-rich with diverse graptolite faunas, whereas regressive intervals are thicker and less organic-rich with more depauperate graptolite assemblages. This method has some advantages in that the graptolite biostratigraphy provides a more precise and globally correlatable temporal control. In addition, it is also based mainly on deeper-water successions, which are less susceptible to truncation by subaerial exposure and erosion. On the other hand, this method provides no means of deriving quantitative estimates of magnitudes of sea-level change, and the oxidation state of the sea floor need not necessarily be related to sea level.

A recent paper that interprets global eustatic patterns during the Silurian is by Haq and Schutter (2008), and is based upon sequence stratigraphic interpretation of records from cratonic basin successions. Intriguingly, Haq and Schutter's record proposes a higher number of short-term global sea-level fluctuations than any of the previous studies (Figure 21.11). Johnson (2010) provided a detailed discussion and comparison of the methodologies and results of the Johnson (2006) and Haq and Schutter (2008) sea-level studies.

Reconstruction of Silurian sea levels is complicated especially by two issues: 1) the interpretation of the stratigraphic record, and 2) the chronostratigraphic correlation of the studied sections. It is often the case that the inability to sufficiently correlate stratigraphic sections, either with each other or with the Silurian time scale, produces diametrically opposing curves (e.g., Munnecke *et al.*, 2010, Figure 2).

21.3.2.4. Climatic Events

There is considerable controversy surrounding the various means used to estimate climatic conditions during the Silurian (e.g., Landing and Johnson, 1998; Munnecke *et al.*, 2003; Cramer and Saltzman, 2007; Calner, 2008; Loydell, 2008b; Cramer and Munnecke, 2008; Munnecke *et al.*, 2010). Nevertheless, it is now clear that the history of Silurian climate was extremely dynamic, as recorded in the high frequency and high-magnitude fluctuations in C and O isotopic values (Munnecke *et al.*, 2010; Cramer *et al.*, 2011a), as well as depositional patterns indicating pronounced changes in ocean state (e.g., Jeppsson, 1990; Brunton *et al.*, 1998; Page *et al.*, 2007).

Following the major glacial event of the Late Ordovician (e.g., Brenchley *et al.*, 1994) there were several episodes of glacial advance in the Llandovery (e.g., Caputo, 1998; Díaz-Martinez and Grahn, 2007) and the Hirnantian glacial interval is perhaps better thought of as the Late Ordovician to Early Silurian glacial interval (e.g., the "Early Palaeozoic Icehouse" interval of Page *et al.*, 2007). As noted above, the existence of significant excursions in the record of $\delta^{18}O$ and $\delta^{13}C$ that persist through several sea-level cycles and that, in many ways, resemble the glacially related events of the Late Ordovician (Munnecke *et al.*, 2003; Žigaitė *et al.*, 2008), indicate that it is possible that there were later Silurian glacial episodes that have yet to be documented in the physical stratigraphic record. Regardless, the frequency and magnitude of several of the Silurian isotopic shifts clearly indicate that this time interval was a period of rapidly evolving climatic and environmental conditions. In addition, many of these events are closely associated with intervals of biotic change, most clearly seen in the record of graptolites and conodonts (Figure 21.11).

21.3.2.5. Volcanism and K-Bentonite Stratigraphy

Volcanic ash beds or K-bentonites have been widely reported through the Silurian, particularly from Europe and eastern North America (e.g., Bergström *et al.*, 1998a). Geochemical studies of these bentonites suggest that those distributed over Laurentia, Avalonia, and Baltica can be attributed to at least three different volcanic centers within the closing Iapetus and Rheic ocean basins (Bergström *et al.*, 1997; Huff *et al.*, 2000). However, some of these individual K-bentonite units have been shown by geochemical fingerprinting to be geographically very widespread and serve as excellent marker beds for high-resolution, regional correlations (e.g., Bergström *et al.*, 1998b; Batchelor *et al.*, 2003; Ray, 2007; Kiipli *et al.*, 2008a,b; Ray *et al.*, 2011). Bentonites also provide the principal source of crystals for radiogenic isotope dates within stratigraphic successions for calibration of the Silurian time scale (see below).

21.4. SILURIAN TIME SCALE

As with the Ordovician, there have been many published radiogenic isotope dates for calibrating the Silurian time scale (Gale, 1985; Kunk *et al.*, 1985; McKerrow *et al.*, 1985; Odin, 1985; Snelling, 1985; Harland *et al.*, 1990; Tucker *et al.*, 1990; Compston and Williams, 1992; Tucker and McKerrow, 1995; Compston, 2000a,b; Bergström *et al.*, 2008). Unfortunately, few of these dates are of sufficient analytical reliability, precision and biostratigraphic constraint to meet modern standards for time scale calibration. Following the practice of Paleozoic time scale geochronologists (e.g., Tucker and McKerrow, 1995; Tucker *et al.*, 1998; Gradstein *et al.*, 2004 and numerous publications referred to therein), we use mainly those dates that are based on volcanogenic minerals preserved in ash beds that are interbedded with age-diagnostic fossiliferous strata and dated using either U-Pb or Ar-Ar methods.

21.4.1. Radiogenic Isotope Age Dates

Under the new standardized treatment (see Chapter 6), dates previously reported under various uranium and lead isotopic decay systems ($^{206}Pb/^{238}U$, $^{207}Pb/^{235}U$, and $^{207}Pb/^{206}Pb$) have, where possible, here been mapped onto the $^{206}Pb/^{238}U$ decay system. Similarly, error propagation of analytical, tracer/standard and decay constant errors has been treated consistently and rigorously for each published age. All dates used in our calibration have been reappraised since Gradstein *et al.* (2004) for analytical and biostratigraphic reliability. The few $^{40}Ar/^{39}Ar$ ages have also been recalibrated to the Kuiper *et al.* (2008) FCs standard age of 28.201 ± 0.046 (2σ) Ma with associated systematic error propagation. The radiogenic isotope ages for many of the dated samples in Appendix 2, therefore, differ from those given in the original publications. All of the dates used for time scale calibrations by Melchin *et al.* (2004) are here revised or rejected. For reasons discussed by Villeneuve (2004), zircon age determinations determined with the high-resolution secondary ion mass spectrometry (HR-SIMS) method in the Canberra Geochronological Laboratory (Compston and Williams, 1992; Compston, 2000a,b), using standard SL13, are not employed in Silurian time scale construction. Details of the dated samples used here are given in Appendix 2. The following comments refer mainly to the reliability of the biostratigraphic constraints on the dated samples.

Of the six Silurian radiogenic isotope dates used for calibration by Tucker and McKerrow (1995), three are accepted here (their dated items 22, 23, 26). Date No. 22 is from the Birkhill Shales, Scotland ($^{206}Pb/^{238}U$ age of 439.57 ± 1.13 Ma: Toghill, 1968; Ross *et al.*, 1982; Tucker *et al.*, 1990). Although the exact level of the bentonite in the section is uncertain (Cooper and Sadler, 2004; Bergström *et al.*, 2008), we accept the assignment of Ross *et al.* (1982) of this bentonite to the *Coronograptus cyphus* Zone. Date No. 24 is from the Telychian in the Buttington brick pit at Welshpool, Wales, and previously assigned to the *Monoclimacis crenulata Zone* ($^{206}Pb/^{238}U$ age of 429.20 ± 2.67 Ma) by Tucker and McKerrow (1995). This "zone" at Buttington is now known to span at least four graptolite zones, the zones of *M. crenulata* (*sensu stricto*), *Oktavites spiralis*, *Cyrtograptus lapworthi* and *Cyrtograptus insectus* (Loydell and Cave 1993, 1996). It is also uncertain which of several bentonite beds in the quarry was sampled, and how broadly Tucker and McKerrow interpreted the concept of the *M. crenulata* zone. There is therefore a wide range of stratigraphic uncertainty about the dated bentonite and the "safest" range to use is the zonal range *Spirograptus turriculatus* to middle *Cyrtograptus lapworthi* Zone, which encompasses the Tarannon Shales Formation from which the bentonite was collected (D. Loydell, pers. comm.). However, because of its biostratigraphic uncertainty, we do not use this date for calibration.

The dated item No. 26 of Tucker and McKerrow, from the upper Whitcliffe Formation, Ludlow ($^{206}Pb/^{238}U$ age of 420.88 ± 2.39 Ma: Tucker and McKerrow, 1995; Tucker *et al.*, 1998), was rejected by Melchin *et al.* (2004) and by Sadler *et al.* (2009). Although analytically acceptable, it is not tied to a graptolite or conodont horizon and therefore to a level in the CONOP composite. It is included here, however, with a wide stratigraphic uncertainty. From regional correlations, the age is taken as post-*Saetograptus leintwardinensis* Zone to pre-*Neocolonograptus ultimus–N. parultimus* Zone in the Ludfordian.

Two samples dated by the Ar-Ar method are included here. An Ar-Ar date on a sedimentary breccia in the Descon Formation of Esquibel Island, southeastern Alaska (Tucker and McKerrow item No. 23: Churkin and Carter, 1970; Ross *et al.*, 1982; Kunk *et al.*, 1985) is rather loosely constrained by biostratigraphy. The breccia is "autobrecciated", contains juvenile volcanic material including crystals of hornblende, pyroxene and plagioclase, but no age-diagnostic fossils. An Ar-Ar total gas age on hornblende gave 436.2 ± 5 Ma, here revised to 441.9 ± 5.2 Ma. The bed lies 4 m above graptolites of the *Coronograptus cyphus* Zone, and this zone was taken as the age of the breccia by Tucker and McKerrow (1995). Bergström *et al.* (2008) point out that the breccia could be as young as strata equivalent to *S. sedgwickii* Zone, which is reported from a younger part of the Descon Formation. Here we take a broad stratigraphic age uncertainty interval for the breccia, extending from the base of the *Coronograptus cyphus* Zone to the top of the *S. sedgwickii* Zone, and the level as the midpoint in this range. An Ar-Ar date on biotite in a bentonite bed in the Middle Elton Formation, Shropshire, UK gave 423.7 ± 1.7 Ma (Kunk *et al.*, 1985), here revised to 430.6 ± 2.9 Ma. The stratigraphic correlation of the bentonite given by Ross *et al.* (1982), *Neodiversograptus nilssoni* to *Lobograptus scanicus* zones, Gorstian Stage, is here taken as the stratigraphic uncertainty interval for the sample and the level as the midpoint in this range. The Laidlaw Volcanics of Canberra, Australia, dated by the K-Ar and Rb-Sr methods (420.7 ± 2.2 Ma: Wyborn *et al.*, 1982) and accepted by Melchin *et al.* (2004), fails the analytical quality screening protocols of the present database and is here rejected.

Five high-quality zircon dates have become available since publication of the Silurian time scale of Melchin *et al.* (2004). The first is from the Osmundsberget K-bentonite of Dalarna, Sweden (Bergström *et al.*, 2008). Four out of five zircons yielded a weighted mean $^{206}Pb/^{238}U$ age of 438.74 ± 1.11 Ma; one zircon was affected by inheritance. The graptolites *Spirograptus turriculatus* and *Streptograptus johnsoni* are found in the section about 0.5 m below the K-bentonite, and *S. turriculatus* is also present 0.5 m above it (Loydell and Maletz, 2002). We take the overlap in range of *Spirograptus turriculatus* and *Streptograptus johnsoni* in the CONOP composite as the stratigraphic uncertainty range and the midpoint in this range as the level. A second K-bentonite from Dalarna, in the Kallholn Shale, gave a $^{206}Pb/^{238}U$ age of 437.0 ± 0.5; however, uncertainty about its biostratigraphic age and correlation led Bergström *et al.* (2008) to reject it for time scale calibration, and we have not included it here.

The other four of the five new high-quality zircon $^{206}Pb/^{238}U$ dates since Melchin *et al.* (2004), come from Cramer *et al.* (submitted). Preliminary dating of the Ireviken Bentonite from Gotland, Sweden, was reported by Jeppsson *et al.* (2005), but the complete analytical details remained unavailable until recently. Cramer *et al.* (submitted), using the CA-ID-TIMS method of Mattinson (2005) and Condon *et al.* (2007), show that 11 out of 15 zircons yielded a weighted mean $^{206}Pb/^{238}U$ age of 431.80 ± 0.71 Ma for the Ireviken Bentonite from the Lusklint 1 outcrop (for locality guide see Eriksson and Calner, 2005). At this well-studied section (e.g., Jeppsson, 1997a,b), the Ireviken Bentonite occurs 2.85 m above the Lusklint Bentonite (the local reference level), which is roughly 8.28 m above sea level. The Ireviken Bentonite occurs within the Lower Visby Formation and is within the Upper *Pseudooneotodus bicornis* conodont zone between Datum 2 and Datum 3 of the Ireviken Event (Jeppsson, 1997a,b). Based on our present knowledge of the base Wenlock GSSP, Datum 2 of the Ireviken Event is likely within a few centimeters of the "golden spike" at Hughley Brook (see above; Mabillard and Aldridge, 1985; Jeppsson, 1997a; Cramer *et al.*, 2010a), and the Ireviken Bentonite is 14 cm above Datum 2 in the Lusklint 1 outcrop. Based upon biostratigraphic correlation of the conodont zonations with the graptolite zonations (e.g., Cramer *et al.*, 2010a), as well as the correlation of bentonites throughout the Baltic where they co-occur with graptolites (e.g., Aizpute core: Kiipli *et al.*, 2008b), the Ireviken Bentonite occurs within the *Cyrtograptus murchisoni* graptolite zone, and is below the onset of the Sheinwoodian (Ireviken) positive $\delta^{13}C$ excursion (Munnecke *et al.*, 2003).

The second new date from the Swedish island of Gotland comes from the outcrop Hörsne 3 (Laufeld, 1974; Kiipli *et al.*, 2008b). The name "Grötlingbo Bentonite" was first given to a 38 cm-thick bentonite recognized in the Grötlingbo drill core (Jeppsson and Calner, 2003), where it occurs within the *Gothograptus nassa–Pristiograptus dubius parvus* graptolite zone (Calner *et al.*, 2006). The only identified exposure of this bentonite on Gotland is the Hörsne 3 locality where it is 30 cm thick (Kiipli *et al.*, 2008b). At Hörsne 3, the Grötlingbo Bentonite is within strata equivalent to the Mulde Brickclay Member of the Halla Formation and within the *Ozarkodina bohemica longa* conodont zone (Jeppsson *et al.*, 2006; Kiipli *et al.*, 2008b). A bentonite from the Hörsne 3 locality was previously dated using K-Ar at 425.8 ± 6.0 Ma by Odin *et al.* (1986: sample B126). The Grötlingbo Bentonite has been correlated more broadly around the East Baltic by Kiipli *et al.* (2008a,b), who demonstrated that the identification and correlation based on the mineralogy of magmatic phenocrysts is consistent with its biostratigraphic position throughout the Baltic Basin. This sample (G05-335LJ of Kiipli *et al.*, 2008b) comes from the upper part of the 30 cm-thick exposure and five out of nine zircons yielded a weighted mean $^{206}Pb/^{238}U$ age of 428.47 ± 0.72 Ma (Cramer *et al.*, submitted). The Grötlingbo Bentonite occurs within the rising limb of the first peak of the Homerian "Mulde" positive $\delta^{13}C$ excursion (compare Calner *et al.*, 2006; Kaljo *et al.*, 2007; and Kiipli *et al.*, 2008b).

The third and final new date from Gotland comes from the Djupvik 1 locality (Jeppsson, 1982; Eriksson and Calner, 2005), and was previously studied by Odin *et al.* (1986: sample B108) and Batchelor and Jeppsson (1999). Odin *et al.* (1986) determined a K-Ar age of 427.4 ± 6.0 Ma from this locality. The lower of two bentonites separated by 16 cm was sampled (G03-340LJ of Kiipli *et al.*, 2008b) by Cramer *et al.* (submitted) and six out of nine zircons yielded a weighted mean $^{206}Pb/^{238}U$ age of 428.06 ± 0.68 Ma. This bentonite occurs within the Djupvik Member of the Halla Formation, within the *Kockelella ortus absidata* conodont zone (Kiipli *et al.*, 2008b) and likely correlates to a position near the boundary between the *Colonograptus praedeubeli* and *C. deubeli* graptolite zones. The Djupvik Member, and the bentonites contained therein, roughly marks the position of the low point between the two peaks of the Homerian "Mulde" positive $\delta^{13}C$ excursion (Calner *et al.*, 2006).

The final new date included by Cramer *et al.* (submitted) comes from Wren's Nest Hill, Dudley, England. This bentonite occurs in Lion's Mouth Cavern within the uppermost part of the Upper Quarried Limestone Member of the Much Wenlock Limestone Formation, near the base of PS12 (Parasequence 12) of Ray *et al.* (2010). By correlation to the base Ludlow GSSP, this bentonite (Wren's Nest Hill 15: Ray *et al.*, 2011) likely corresponds with a level only a few centimeters below the golden spike at Pitch Coppice Quarry (see above). There is little direct biostratigraphic information available from this outcrop; however, through the detailed correlation of Ray *et al.* (2010), and the position of the bentonite with respect to the base of the Lower Elton Formation, this bentonite can be correlated to a position very near the top of the *Colonograptus ludensis* graptolite zone and the *Kockelella ortus absidata* conodont zone, just above the final conclusion of the Homerian "Mulde" positive $\delta^{13}C$ excursion (Ray *et al.*, 2010). Cramer *et al.* (submitted) report that six out of eleven zircons yielded a weighted mean $^{206}Pb/^{238}U$ age of 427.77 ± 0.68 Ma.

Two further dated samples – a bentonite from the Late Ordovician and one from the Early Devonian – help constrain the age of the lower and upper boundaries of the Silurian respectively. A tuff bed in the Late Ordovician (Ashgill) Hartfell Shales of Scotland (dated item No. 21 of Tucker and McKerrow, 1995), gave a U-Pb age (Tucker *et al.*, 1990) here revised to 448.88 ± 1.17 Ma. Similarly, Tucker *et al.* (1998) report a weighted mean $^{207}Pb/^{206}Pb$ age here recalculated using the U decay constant ratio of Schoene *et al.* (2006) as 415.48 ± 2.71 Ma based on nine multigrain zircon fractions in a K-bentonite from the Kalkberg Formation (Helderberg Group), Cherry Valley, New York, that was used to help constrain the age of the composite in the Early Devonian (Melchin *et al.*, 2004; Sadler *et al.*, 2009), and for calibration of the Devonian time scale (Kaufmann, 2006). The Kalkberg Formation, and most of the lower Helderberg Group, contains the zonal conodont *Icriodus woschmidti*, indicating a level that can be correlated with the *M. uniformis* graptolite zone. However, work in progress (M. Kleffner and J. Ebert, pers. comm.) indicates that this zonal assignment needs revision. The "Kalkberg" bentonite lies most probably not in the Kalkberg Formation but in the overlying New Scotland Formation, and the bentonite is probably at least one conodont zone above the *woschmidti* and *postwoschmidti* levels. Further, Carls *et al.* (2007) have revised the correlation of the *woschmidti* Zone and suggest that it lies above, not at, the base of the Devonian. Its exact equivalent in the graptolite sequence is uncertain but the zone base lies in the early part of the early Lochkovian in their view. Taken together, these revisions suggest that the "Kalkberg" bentonite lies somewhere in the middle or late Lochkovian. In this study we therefore use the graptolite proxy taxa *M. praehercynicus* + *M. hercynicus nevadensis* for biostratigraphic position of the bentonite; their combined range in the composite gives the stratigraphic error and the midpoint of this combined range gives the stratigraphic age.

Thus, we have listed a total of eleven radiogenic isotope dates, nine of which lie within the Silurian Period (Appendix 2; see also Figures 20.11 and 20.13). The five most recently dated samples (Osmundsberg, Ireviken, Hörsne, Djupvik and Wren's Nest) stand out as having both reliable biostratigraphic constraint and conforming with the modern CA-ID-TIMS analytical technique. Of the remaining six, several are far from ideal where analytical procedure is concerned although, as mentioned above, all analyses have passed the screening protocols adopted in this work for quality control on analytical procedure. Most of the remaining six also have biostratigraphic problems and Bergström *et al.* (2008) questioned the suitability of several of them for time scale calibration. Because of the dearth of reliable dates for calibration of the Silurian scale, we have retained as many as possible. The middle part of the Silurian Period is now well constrained and the main areas of concern are the earliest and latest parts of the period. The numerical ages of both the Silurian–Devonian boundary and the Ordovician–Silurian boundary are likely to change when high-quality dating in these intervals becomes available.

21.4.2. Methods of Estimating Relative Duration of Zones and Stages

Because there are so few high-quality radiogenic isotope dates available for the calibration of the Silurian, the quantitative methods of interpolating stage boundaries used in other parts of the time scale by Harland *et al.* (1990) and Gradstein *et al.* (1994) cannot be applied. As with the Ordovician, various proxy time scales have been devised by previous workers, in which the stage durations are estimated by some other method. Almost all of them rely on assumptions about either uniformity of depositional rate or stage or zone duration.

The chronogram (minimized misfit) method of boundary calibration developed by Harland *et al.* (1990) is a means of estimating the age of stage boundaries from radiogenic isotope dates. Twenty-four radiogenic isotope dates in Silurian rocks were used, spanning a wide range of dating methods and quality. Almost none of these dates are accepted here as of sufficient analytical reliability and biostratigraphic control. It is not surprising, therefore, that Harland *et al.* (1990) found that most chronograms for Silurian stage boundaries were:

'either poor or meaningless and [stage durations] must be estimated using chron interpolation.'

For chrons, they used graptolite zones. Thus, the number of graptolite zones present was a primary guide to the duration of each stage. This rule, however, was not consistently applied. The method assumes uniformity in graptolite zone duration and in paleontological practice. Similar assumptions about zone duration in Silurian time scale construction were employed, in whole or in part, by Gale (1985), McKerrow *et al.* (1985), and Tucker and McKerrow (1995).

In an alternative approach, the thicknesses of fossiliferous stratigraphic sections through Silurian zones and stages are used to estimate their duration. Kleffner (1989) developed a composite section through the Silurian, using the graphic correlation method of Shaw (1964). Forty-two conodont- and graptolite-bearing carbonate and carbonate-shale sections, mostly in North America but also from Europe, were used, and the Cellon section in Austria was selected as the standard reference section. In this exercise, the stratigraphic ranges of species in the Cellon section are extended to match the maximum ranges observed among the correlated sections, and species not present in the Cellon section are similarly composited into it based on their maximum ranges in other sections. The Cellon section thus becomes a composite, scaled section of all compared sections. The conodont and graptolite zone boundaries recognized in the composite section are regarded as defining chronozones because they are based on global rather than local stratigraphic ranges. The composite section was then regarded by Kleffner (1989) as a proxy time scale. It is graduated in "composite time units", which are derived from the original stratigraphic thickness units of the Cellon section. Zone and stage boundaries are located within it based on the composite stratigraphic ranges of the defining species. Fordham (1998) illustrated how dependent the graphic correlation time scale is on choice of the initial standard reference section, the Cellon section in Austria. Some method of minimizing, or normalizing for, thickness bias is therefore needed, and composite sections derived by graphic correlation, without a test for steadiness of depositional rate may be unreliable.

21.4.3. Relative Duration of Stages from Constrained Optimization

As in the Ordovician, Silurian deep-water shales have the prerequisite for a high-resolution time scale, i.e. rich successions of graptolite faunas, inter-bedded ash layers with dateable zircon crystals, and minimally interrupted accumulation. The constrained optimization method of Kemple *et al.* (1995) and Sadler (1999, 2004), using the CONOP family of programs, can be used to combine graptolite ranges in many measured sections into a scaled composite sequence. The method is fully described in Sadler *et al.* (2009), and is briefly outlined in Chapters 3 and 20 of this book. The Silurian and Ordovician time scales have been built from a composite derived from 512 graptolitic, primarily deep-water stratigraphic successions from around the world from latest Cambrian to Early Devonian in age, containing over 2000 species. Approximately half of the sections range wholly or partly in the Silurian.

The method used for developing a relative (proxy) time scale for the Silurian is the same as that used for the Ordovician, outlined in more detail in Chapter 20 of this volume. All sections have been treated simultaneously, avoiding the bias introduced by choice of the initial standard reference section as in graphic correlation. Otherwise, our automated compositing procedure closely resembles Edwards' (1979) "no-space" variant of graphic correlation. The result is an optimized, scaled, global composite sequence of events. The order and spacing of events in the scaled composite serve as a proxy time scale (Table 21.1). Stage boundaries are located in the composite, as they are tied to graptolite first-appearance events, although, as noted in the discussions of the GSSPs (above), there is significant uncertainty in the precision with which some of the GSSPs can be tied to the relevant graptolite events.

21.4.4. Calibration of Stage Boundaries by Composite Standard Optimization

The eleven ash beds dated by radiogenic isotopes discussed above, in combination with the CONOP scaled composite, enable calculation of the duration of Silurian global graptolite zones and the time scale. The level within the CONOP composite of each dated ash bed is determined from the graptolite fossils associated with it, either directly in outcrop or by correlation. When the composite levels for the dated ash beds are plotted against the geochronometric scale (Figure 20.13), the overall linearity of the CONOP composite can be tested. Both the polynomial and spline fitted curves pass close to all five of the precise and accurate Silurian dates ($R^2 = 0.9924$, Figure 20.11). The slightly sinuous curve suggests that the composite deviates only slightly from linear.

Although there is no *a priori* reason that this particular best fit of x and y axis data should approach linearity, the latter generally is taken to indicate good accordance between the compositing and scaling procedures on the one hand and the radiogenic isotope dating on the other. The chronometric error for radiogenic isotopically dated samples shown is the 2-sigma error associated with the dates as re-evaluated by Mark Schmitz (herein) and given in Appendix 2. The stratigraphic error shown is as explained in Chapter 20.4.2.2. The fitted curves enable the precise age of all events in the composite to be converted to chronometric ages. Global graptolite zones vary in duration from 0.10 myr or less to 2.69 myr (Table 21.1).

21.4.5. Age of Stage Boundaries

In the final analysis for the age of the stage boundaries and stage duration in the Silurian (and Ordovician), the chronometric and stratigraphic ages of the radiogenic isotope dates were subjected to a cubic spline fit as described in Chapter 14, based on the standard approach used in GTS2004. Points were weighted according to the procedure given in that chapter, taking both radiogenic isotope and stratigraphic errors into account. A smoothing factor of 1.45 was calculated by cross-validation. Two of the 27 points did not pass the chi-squared goodness-of-fit test at a significance level of $p<0.01$. These points were O4 ($\chi^2=7.01$; $p=0.0081$) and S5 ($\chi^2=6.66$; $p=0.0099$). No post-hoc adjustments were made to the weighting of these points. The resulting spline (Figure 20.13 in Chapter 20 on the Ordovician Period) is close to linear through the Ordovician, but a lessening of slope in the Llandovery has the effect of shortening the duration of this series relative to its duration in the composite.

Interpolated Silurian stage boundary ages are given in Table 21.2. The ages computed from the polynomial fit described earlier in this chapter are within the error bars of the ages computed from the spline. The only exception is for the base Aeronian, where the polynomial interpolation age is 1.3 myr older than the spline age, which has a 2-sigma error bar of 1.2 myr.

The base of the Silurian is 443.8 ± 1.5 Ma, and its top is 419.2 ± 3.2 Ma; the duration of the Silurian is thus 24.6 ± 1.6 myr. The durations of the four series (and age of their bases) are as follows: Llandovery 10.5 ± 1.4 myr (base at 443.8 ± 1.5 Ma), Wenlock 6.0 ± 0.7 myr (base at 433.4 ± 0.8 Ma), Ludlow 4.4 ±1.6 myr (base at 427.4 ± 0.5 Ma) and Pridoli 3.8 ± 1.0 myr (base at 423.0 ± 2.3 Ma). The Rhuddanian Stage lasted from 443.8 to 440.8 Ma, for a duration of 3.0 ± 0.5 myr. Aeronian lasted 2.3 ±1.3 myr, from 440.8 to 438.5 Ma.

TABLE 21.1 **Silurian graptolite zone and stage levels in the CONOP composite and computed ages of their lower boundaries**

Stage	Zone	Level in Composite (%)	Polynomial Age (Ma)	GTS 2012 (Spline) Age (Ma)	2-Sigma	Duration (myr)
Lochkovian	*Monograptus yukonensis*	98.805	413.28	416.25	3.8622	
Lochkovian	*M. hercynicus hercynicus*	97.146	415.74	417.86	3.3827	1.61
Base Devonian	*Monograptus uniformis*	95.764	417.64	419.2	3.1939	1.34
Pridoli	*Monograptus perneri–M. transgrediens*	93.634	420.33	421.26	2.7449	2.06
Pridoli	*Monograptus bouceki*	92.720	421.40	422.13	2.5765	0.87
Pridoli	*Neocolonogr. lochkovensis –N. lochkovensis branikensis*	91.970	422.24	422.83	2.3214	0.69
Pridoli	*Neocolonogr. parultimus–N. ultimus*	91.835	422.39	422.96	2.2755	0.13
Ludfordian	*Monograptus formosus*	91.054	423.24	423.69	1.9388	0.73
Ludfordian	*B. cornutus–P. podoliensis–Neocolonogr. kozlowskii*	89.668	424.65	424.96	1.1531	1.27
Ludfordian	*S. leintwardinensis–S. linearis*	88.986	425.32	425.57	0.8673	0.61
Gorstian	*Lobograptus scanicus*	87.409	426.78	426.93	0.5663	1.36
Gorstian	*Neodiversograptus nilssoni*	86.886	427.24	427.36	0.5306	0.43
Homerian	*Colonograptus ludensis*	86.196	427.83	427.92	0.4949	0.56
Homerian	*Colonograptus? praedeubeli–C? deubeli*	85.854	428.12	428.18	0.4796	0.26
Homerian	*Pristiograptus dubius parvus–Gothogr. nassa*	84.428	429.27	429.26	0.5714	1.08
Homerian	*Cyrtograptus lundgreni*	82.749	430.54	430.45	0.7347	1.19
Sheinwoodian	*Cyrtogr. rigidus–Monogr. antennularius–M. belophorus*	79.828	432.56	432.36	0.8929	1.59
Sheinwoodian	*Monograptus firmus–M. riccartonensis*	79.460	432.80	432.59	0.8827	0.23
Sheinwoodian	*Cyrtograptus murchisoni*	78.194	433.60	433.35	0.8418	0.76
Telychian	*Cyrtograptus centrifugus*	78.015	433.71	433.46	0.8469	0.11
Telychian	*Cyrtograptus insectus*	77.715	433.90	433.63	0.8418	0.17
Telychian	*Cyrtograptus lapworthi*	75.021	435.49	435.15	0.9439	1.52
Telychian	*Oktavites spiralis interval*	73.099	436.57	436.15	1.1020	1.00
Telychian	*Monoclimacis crenulata*	72.369	436.97	436.51	1.1633	0.36
Telychian	*Monoclimacis griestoniensis*	72.156	437.09	436.61	1.1786	0.10
Telychian	*Monograptus crispus*	70.871	437.78	437.2	1.2500	0.59
Telychian	*Spirograptus turriculatus*	68.668	438.95	438.13	1.1939	0.93
Telychian	*Spirograptus guerichi*	67.777	439.42	438.49	1.0867	0.36
Aeronian	*Stimulograptus sedgwickii*	67.075	439.78	438.76	0.9949	0.27
Aeronian	*Lituigraptus convolutus*	65.940	440.38	439.21	0.9031	0.45

(Continued)

TABLE 21.1 **Silurian graptolite zone and stage levels in the CONOP composite and computed ages of their lower boundaries—cont'd**

Stage	Zone	Level in Composite (%)	Polynomial Age (Ma)	GTS 2012 (Spline) Age (Ma)	2-Sigma	Duration (myr)
Aeronian	*Monogr. argenteus–Pri. leptotheca*	65.299	440.72	439.47	0.8724	0.26
Aeronian	*Demi. pectinatus–D. triangulatus*	62.282	442.33	440.77	1.2296	1.30
Rhuddanian	*Coronograptus cyphus*	60.677	443.20	441.57	1.5051	0.80
Rhuddanian	*Cystograptus vesiculosus*	59.077	444.09	442.47	1.5510	0.90
Rhuddanian	*Parakidograptus acuminatus*	57.576	444.95	443.4	1.5663	0.93
Base Silurian	*Akidograptus ascensus*	56.927	445.33	443.83	1.5459	0.43

Note that recent work has shown that the combined *Colonograptus? praedeubeli–C? deubeli, Bohemograptus cornutus–Polonograptus podoliensis–Neocucullograptus kozlowskii,* and *Neocolonograptus parultimus–N. ultimus* can each be readily subdivided in two for the purposes of global correlation (e.g., Cramer *et al.*, 2011a,b) but the boundaries between the split zones have not been precisely located within the CONOP composite and are shown at approximate levels within the correlation charts (Figs 21.1, 21.11).

TABLE 21.2 **Geochronology of the Ordovician–Silurian Stages for GTS2012, with a Comparison to GTS2004**

Period	Epoch	Stage	Base Ma GTS2012	Est. ± myr (2σ)	Epoch Duration myr	Est. ± myr (2σ)	Stage Duration myr	Est. ± myr (2σ)	Base Ma GTS2004
Devonian	Early	Lochkovian	419.2	3.2					416
Silurian	Pridoli		423.0	2.3	3.8	1.0	3.8	1.0	
	Ludlow	Ludfordian	425.6	0.9			2.6	1.1	421.3
		Gorstian	427.4	0.5	4.4	1.6	1.8	0.5	422.9
	Wenlock	Homerian	430.5	0.7			3.1	0.6	426.2
		Sheinwoodian	433.4	0.8	6.0	0.7	2.9	0.3	428.2
	Llandovery	Telychian	438.5	1.1			5.1	1.0	436
		Aeronian	440.8	1.2	10.5	1.4	2.3	1.3	439
		Rhuddanian	443.8	1.5			3.1	0.5	443.7
Ordovician	Late	Hirnantian	445.2	1.4			1.3	0.2	445.6
		Katian	453.0	0.7	14.5	1.5	7.8	1.3	455.8
		Sandbian	458.4	0.9			5.4	0.4	460.9
	Middle	Darriwilian	467.3	1.1			8.9	0.9	468.1
		Dapingian	470.0	1.4	11.6	1.0	2.7	0.3	471.8
	Early	Floian	477.7	1.4			7.8	1.2	478.6
		Tremadocian	485.4	1.9	15.4	1.3	7.7	0.7	488.3

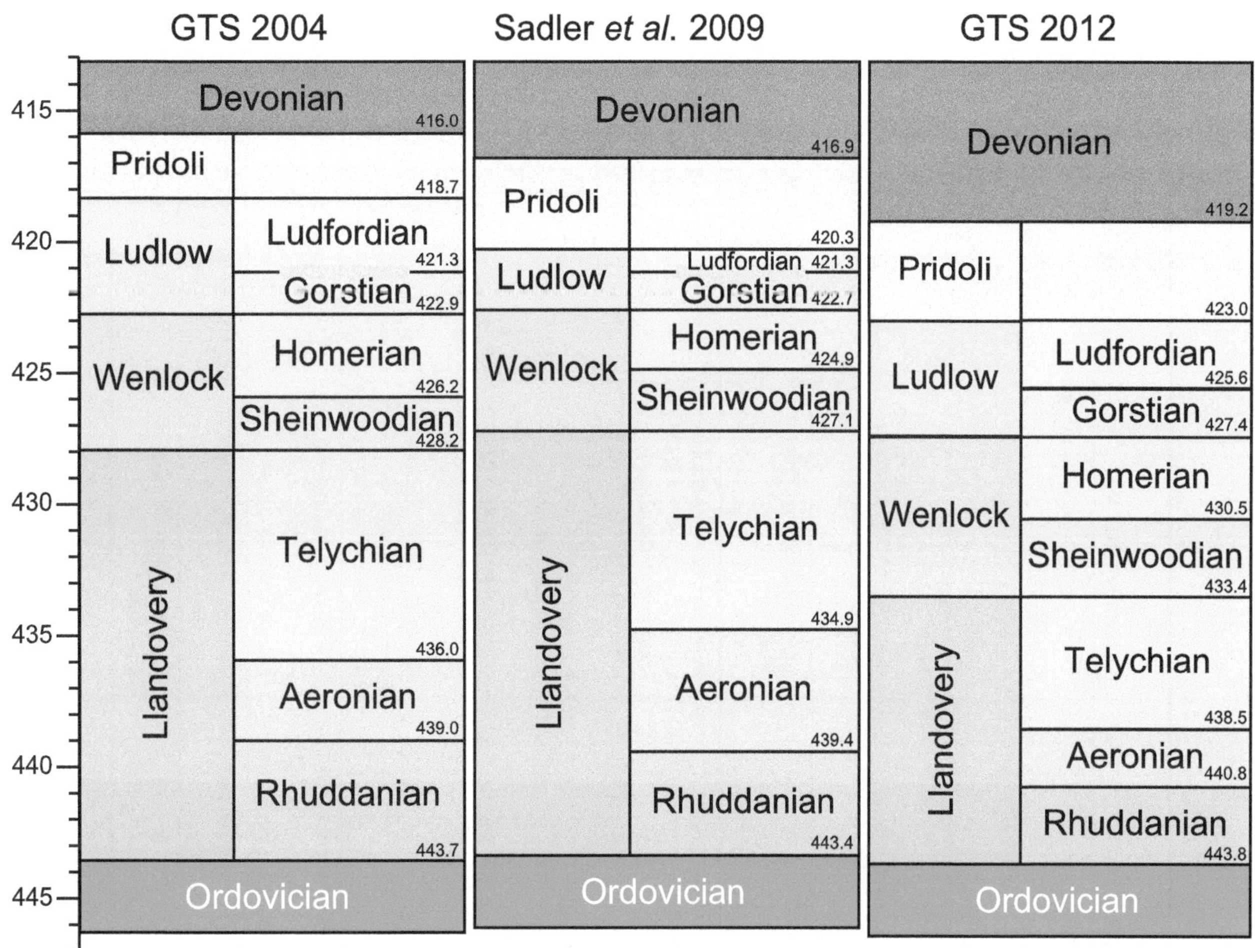

FIGURE 21.12 **Comparison of the CONOP time scales of Melchin *et al.* (2004, GTS2004), Sadler *et al.* (2009), and present (GTS2012), showing series and stages.** Note that in the Melchin *et al.* (2004) and Sadler *et al.* (2009) papers, the base of the Wenlock was drawn at a level correlative with the base of the *C. centrifugus* Zone, whereas in the GTS2012 column it is drawn at the base of the overlying *C. murchisoni* Zone (see text for explanation).

The Telychian Stage ranged from 438.5 to 433.4 Ma for a duration of 5.1 ± 1.0 myr, being thus the longest stage in the Silurian Period. The Sheinwoodian lasted for 2.9 ± 0.3 myr, from 433.4 to 430.5 Ma. The Homerian Stage started at 430.5 and ended at 427.4 Ma, lasting 3.1 ± 0.6 myr. The Gorstian ran from 427.4 to 425.6 Ma for 1.8 ± 0.5 myr and is the shortest stage in the Silurian. The Ludfordian lasted 2.6 ± 1.1 myr, from 425.6 to 423.0 Ma and the Pridoli spans 3.8 ±1.0 myr to the base of the Devonian from 423.0 to 419.2 Ma.

The error bars on boundary ages are large near the top of the Silurian; this is due to a lack of well-constrained data points in the Pridoli and lower part of the Devonian. More and stratigraphically better constrained age dates will reduce uncertainty.

GSSPs of the Silurian Stages, with Location and Primary Correlation Criteria (Status Jan. 2012)

Stage	GSSP Location	Latitude, Longitude	Boundary Level	Correlation Events	Reference
Pridoli (Series)	Požáry Section, Řeporyjie, Prague, Czech Republic	50°01′39.82"N 14°19′29.56"E*	Within Bed 96	Graptolite FAD *Neocolonograptus parultimus*	Episodes 8/2, 1985; Geol. Series, Nat. Mus. Wales 9, 1989
Ludfordian	Sunnyhill Quarry, near Ludlow, UK	52°21′33"N 2°46′38"W*	Coincident with the base of the Leintwardine Formation	Imprecise. May be near base of *Saetograptus leintwardinensis* graptolite zone	Lethaia 14; Episodes 5/3, 1982; Geol. Series, Nat. Mus. Wales 9, 1989
Gorstian	near Ludlow, UK	52°21′33"N 2°46′38"W*	Coincident with the base of the Lower Elton Formation	Graptolite FAD *Saetograptus varians*	Lethaia 14; Episodes 5/3, 1982; Geol. Series, Nat. Mus. Wales 9, 1989
Homerian	Sheinton Brook, Homer, UK	52°36′56"N 2°33′53"W*	Within upper part of the Apedale Member of the Coalbrookdale Formation	Graptolite FAD *Cyrtograptus lundgreni*	Lethaia 14; Episodes 5/3, 1982; Geol. Series, Nat. Mus. Wales 9, 1989
Sheinwoodian	Hughley Brook, UK	52°34′52"N 2°38"20"W*	Base of the Buildwas Formation	Imprecise. Between the base of acritarch biozone 5 and LAD of conodont *Pterospathodus amorphognathoides*. The current GSSP does not coincide with the base of the *Cyrtograptus centrifugus* Biozone, as was supposed when the GSSP was defined. Restudy recommends a slightly higher and correlatable level on conodonts –the Ireviken datum 2, which coincides approximately with the base of the *murchisoni* graptolite biozone	Lethaia 14; Episodes 5/3, 1982; Geol. Series, Nat. Mus. Wales 9, 1989
Telychian	Cefn-cerig Road Section, Wales, UK	51.97°N 3.79°W**	Approximately 31 m below the top of the Wormwood Formation	Just above LAD of brachiopod *Eocoelia intermedia* and below FAD of *Eocoelia curtisi*	Episodes 8/2, 1985; Geol. Series, Nat. Mus. Wales 9, 1989
Aeronian	Trefawr Track Section, Wales, UK	52.03°N 3.70°W**	Within Trefawar Formation	Graptolite FAD *Monograptus austerus sequens*	Geol. Series, Nat. Mus. Wales 9, 1989
Rhuddanian *(base Silurian)*	Dob's Linn, Scotland	55.44°N 3.27°W**	1.6m above the base of the Birkhill Shale Formation	Graptolite FAD *Akidograptus ascensus*	Episodes 8/2, 1985; Episodes 31/3, 2008

**according to Google Earth*
***derived from map*

ACKNOWLEDGMENTS

The research leading to this chapter was made possible by the following sources of research funding: a Natural Sciences and Engineering Research Council of Canada Discovery Grant to MJM; National Science Foundation Grant #EAR-0948277 to BDC. Valuable discussions with many collaborators have led to the refined correlations and zonations presented in this chart. We particularly thank J. Verniers for insights pertaining to the chitinozoan zonation as well as D. Condon and U. Söderlund for early access to new Silurian age dates.

REFERENCES

Aldridge, R.J., Briggs, D.E.G., 1989. A soft body of evidence: A puzzle that has confused paleontologists for more than a century has finally been solved. Natural History 89, 6–9.

Azmy, K., Veizer, J., Bassett, M.G., Copper, P., 1998. Oxygen and carbon isotopic composition of Silurian brachiopods: Implications for coeval seawater and glaciations. Geological Society of America Bulletin 110, 1499–1512.

Azmy, K., Veizer, J., Wenzel, B., Bassett, M.G., Copper, P., 1999. Silurian strontium isotope stratigraphy. Geological Society of America Bulletin 111, 475–483.

Bassett, M.G., 1985. Towards a "common language" in stratigraphy. Episodes 8, 87–92.

Bassett, M.G., 1989. The Wenlock Series in the type area. In: Holland, C.H., Bassett, M.G. (Eds.), A Global Standard for the Silurian System. National Museum of Wales, Geological Series, 9, pp. 51–73.

Bassett, M.G., Cocks, L.R.M., Holland, C.H., Rickards, R.B., Warren, P.T., 1975. The type Wenlock Series. Report of the Institute of Geological Sciences 75/13, 1–19.

Bassett, M.G., Kaljo, D., Teller, L., 1989. The Baltic region. In: Holland, C.H., Bassett, M.G. (Eds.), A Global Standard for the Silurian System. National Museum of Wales, Geological Series, 9, pp. 158–170.

Batchelor, R.A., Jeppsson, L., 1999. Wenlock metabentonites from Gotland, Sweden: Geochemistry, sources and potential as chemostratigraphic markers. Geological Magazine 136, 661–669.

Batchelor, R.A., Harper, D.A.T., Anderson, T.B., 2003. Geochemistry and potential correlation of Silurian (Telychian) metabentonites from Ireland and SW Scotland. Geological Journal 38, 161–174.

Bergström, S.M., Huff, W.D., Kolata, D.R., Melchin, M.J., 1997. Occurrence and significance of Silurian K-bentonite beds at Arisaig, Nova Scotia, eastern Canada. Canadian Journal of Earth Sciences 34, 1630–1643.

Bergström, S.M., Huff, W.D., Kolata, D.R., 1998a. Silurian K-bentonites in North America and Europe. In: Gutiérrez-Marco, J.C., Rábano, I. (Eds.), Proceedings of the Sixth International Graptolite Conference on the GWG (IPA) and the 1998 Field Meeting of the International Subcommission on Silurian Stratigraphy (ICS-IUGS). Instituto Technológico Geominero de España, Temas Geologico-Mineros, 23, pp. 54–56.

Bergström, S.M., Huff, W.D., Kolata, D.R., 1998b. The Lower Silurian Osmundsberg K-bentonite. Part I: Stratigraphic position, distribution, and paleogeographic significance. Geological Magazine 135, 1–13.

Bergström, S.M., Toprak, F.O., Huff, W.D., Mundil, R., 2008. Implications of a new, biostratigraphically well-controlled, radio-isotopic age for the lower Telychian Stage of the Llandovery Series (Lower Silurian, Sweden). Episodes 31, 309–314.

Bickert, T., Pätzold, J., Samtleben, C., Munnecke, A., 1997. Paleoenvironmental changes in the Silurian indicated by stable isotopes in brachiopod shells from Gotland, Sweden. Geochimica et Cosmochimica Acta 61, 2717–2730.

Brenchley, P.J., Marshall, J.D., Carden, G.A.F., Robertson, D.B.R., Long, D.G.F., Meidla, T., Hints, L., Anderson, T.F., 1994. Bathymetric and isotopic evidence for a short-lived Late Ordovician glaciation in a greenhouse period. Geology 22, 295–298.

Brunton, F.R., Smith, L., Dixon, O.A., Copper, P., Nestor, H., Kershaw, S., 1998. Silurian reef episodes, changing seascapes, and paleobiogeography. In: Landing, E., Johnson, M.E. (Eds.), Silurian Cycles: Linkages of Dynamic Stratigraphy with Atmospheric, Oceanic, and Tectonic Changes. New York State Museum Bulletin, 491, pp. 265–282.

Burgess, N.D., Richardson, J.B., 1995. Late Wenlock to early Přídolí cryptospores and miospores from south and southwest Wales, Great Britain. Palaeontographica Abt. B 236, 1–44.

Calner, M., 2008. Silurian global events – at the tipping point of climate change. In: Elewa, A.M.T. (Ed.), Mass Extinctions. Springer-Verlag, Heidelberg, pp. 21–58.

Calner, M., Kozlowska, A., Masiak, M., Schmitz, B., 2006. A shoreline to deep basin correlation chart for the middle Silurian coupled extinction-stable isotopic event. GFF 128 (2), 79–84.

Came, R.E., Eiler, J.M., Veizer, J., Azmy, K., Brand, U., Weidman, C.R., 2007. Coupling of surface temperatures and atmospheric CO_2 concentrations during the Palaeozoic era. Nature 449, 198–201.

Caputo, M.V., 1998. Ordovician-Silurian glaciations and global sea-level changes. In: Landing, E., Johnson, M.E. (Eds.), Silurian Cycles: Linkages of Dynamic Stratigraphy with Atmospheric, Oceanic and Tectonic Changes. New York State Museum Bulletin, 491, pp. 15–25.

Carls, P., Slavik, L., Valenzuela-Rios, J.I., 2007. Revisions of conodont biostratigraphy across the Silurian-Devonian boundary. Bulletin of Geosciences 82 (2), 145–164.

Chen, X., Fan, J.-X., Melchin, M.J., Mitchell, C.E., 2005. Hirnantian (latest Ordovician) graptolites from the Upper Yangtze region, China. Palaeontology 48, 235–280.

Cherns, L., 2011. The GSSP for the base of the Ludfordian Stage, Sunnyhill Quarry. In: Ray, D.C. (Ed.), Siluria Revisited, a Field Guide, International Subcommission on Silurian Stratigraphy Field Meeting 2011, pp. 75–81.

Churkin, M., Carter, C., 1970. Early Silurian graptolites from southeastern Alaska and their correlation with graptolitic sequences in North America and the Arctic. United States Geological Survey Professional Papers 653, 51.

Cocks, L.R.M., 1985. The Ordovician-Silurian boundary. Episodes 8, 98–100.

Cocks, L.R.M., 1989. The Llandovery Series in the Llandovery area. In: Holland, C.H., Bassett, M.G. (Eds.), A Global Standard for the Silurian System. National Museum of Wales, Geological Series, 9, pp. 36–50.

Cocks, L.R.M., 2000. Ordovician and Silurian global geography. Journal of the Geological Society 158, 197–210.

Cocks, L.R.M., Holland, C.H., Rickards, R.B., Strachan, I., 1971. A correlation of the Silurian rocks on the British Isles. Journal of the Geological Society 127, 103–136.

Cocks, L.R.M., Woodcock, N.H., Rickards, R.B., Temple, J.T., Lane, P.D., 1984. The Llandovery Series of the type area. Bulletin of the British Museum of Natural History, Geology 38, 131–182.

Compston, W., 2000a. Interpretations of SHRIMP and isotope dilution zircon ages for the geological time-scale: 1. The early Ordovician and late Cambrian. Mineralogical magazine 64 (1), 43–57.

Compston, W., 2000b. Interpretations of SHRIMP and isotope dilution zircon ages for the Palaeozoic time-scale: II. Silurian to Devonian. Mineralological magazine 64 (6), 1127–1146.

Compston, W., Williams, I.S., 1992. Ion probe ages for the British Ordovician and Silurian stratotypes. In: Webby, B.D., Laurie, J. (Eds.), Global Perspectives on Ordovican Geology. Balkema Press, Rotterdam, pp. 59–67.

Condon, D., Schoene, B., Bowring, S., Parrish, R., McLean, N., Noble, S., Crowley, Q., 2007. EARTHTIME: Isotopic tracers and optimized solutions for high-precision U-Pb ID-TIMS geochronology. American Geophysical Union, Fall Meeting 2007 abstract #V41E–06.

Cooper, R.A., Sadler, P.M., 2004. Ordovician. In: Gradstein, F.M., Ogg, J.G., Smith, A.G. (Eds.), A Geologic Time Scale 2004. Cambridge University Press, Cambridge, pp. 165–187.

Cooper, R.A., Sadler, P.M., 2010. Facies preference predicts extinction risk in Ordovician graptolites. Paleobiology 36 (2), 167–187.

Corriga, M.G., Corradini, C., 2009. Upper Silurian and Lower Devonian conodonts from the Monte Cocco II section (Carnic Alps, Italy). Bulletin of Geosciences 84, 155–168.

Corradini, C., Serpagli, E., 1999. A Silurian conodont zonation from late Llandovery to end Přídolí in Sardinia. In: Serpagli, E. (Ed.), Studies on Conodonts, Proceedings of the Seventh European Conodont Symposium. Bollettino della Societê Paleontologica Italiana, 37, pp. 255–283.

Cramer, B.D., Saltzman, M.R., 2007. Fluctuations in epeiric sea carbonate production during Silurian positive carbon isotope excursions: A review of proposed paleoceanographic models. Palaeogeography, Palaeoclimatology, Palaeoecology 245, 37–45.

Cramer, B.D., Munnecke, A., 2008. Early Silurian positive $\delta^{13}C$ excursions and their relationship to glaciations, sea-level changes and extinction events: Discussion. Geological Journal 43, 517–519.

Cramer, B.D., Loydell, D.K., Samtleben, C., Munnecke, A., Kaljo, D., Männik, P., Martma, T., Jeppsson, L., Kleffner, M.A., Barrick, J.E., Johnson, C.A., Emsbo, P., Joachimski, M.M., Bickert, T., Saltzman, M.R., 2010a. Testing the limits of Paleozoic chronostratigraphic correlation via high-resolution (<500,000 yrs) integrated conodont, graptolite, and carbon isotope ($\delta^{13}C_{carb}$) biochemostratigraphy across the Llandovery-Wenlock (Silurian) boundary: Is a unified Phanerozoic timescale achievable? Geological Society of America Bulletin 122, 1700–1716.

Cramer, B.D., Kleffner, M.A., Brett, C.E., McLaughlin, P.I., Jeppsson, L., Munnecke, A., Samtleben, C., 2010b. Paleobiogeography, high-resolution stratigraphy, and the future of Paleozoic biostratigraphy: Fine-scale diachroneity of the Wenlock (Silurian) conodont *Kockelella walliseri*. Palaeogeography, Palaeoclimatology, Palaeoecology 294, 232–241.

Cramer, B.D., Brett, C.E., Melchin, M.A., Männik, P., Kleffner, M.A., McLaughlin, P.I., Loydell, D.K., Munnecke, A., Jeppsson, L., Corradini, C., Brunton, F.R., Saltzman, M.R., 2011a. Revised chronostratigraphic correlation of the Silurian System of North America with global and regional chronostratigraphic units and $\delta^{13}C_{carb}$ chemostratigraphy. Lethaia 44, 185–202.

Cramer, B.D., Davies, J.R., Ray, D.C., Thomas, A.T., Cherns, L., 2011b. Siluria Revisited: An introduction. In: Ray, D.C. (Ed.), Siluria Revisited, a Field Guide, International Subcommission on Silurian Stratigraphy Field Meeting 2011, pp. 6–27.

Cramer, B.D., Munnecke, A., Schofield, D.I., Hasse, K.M., Haase-Schramm, A., 2011c. A revised $^{87}Sr/^{86}Sr$ curve for the Silurian: Implications for global ocean chemistry and the Silurian timescale. Journal of Geology 119, 335–349.

Cramer, B.D., Condon, D.J., Söderlund, U., Marshall, C., Worton, G.J., Thomas, A.T., Calner, M., Ray, D.C., Perrier, V., Boomer, I., Patchett, P.J., and Jeppsson, L., submitted for review, U-Pb (zircon) age constraints on the timing and duration of Wenlock (Silurian) paleocommunity collapse and recovery during the 'Big Crisis'. Geological Society of America Bulletin.

Davies, J.R., Waters, R.A., Zalasiewicz, J.A., Molyneux, S.G., Vandenbroucke, T.R.A., Williams, M., 2010. A revised sedimentary and biostratigraphical architecture for the type Llandovery and Garth areas, Central Wales: A field guide. British Geological Survey Open Report OR/10/037.

Davies, J.R., Molyneux, S.G., Vandenbroucke, T.R.A., Verniers, J., Waters, R.A., Williams, M., Zalasiewicz, J.A., 2011. Pre-conference field trip to the type Llandovery area. In: Ray, D.C. (Ed.), Siluria Revisited, a Field Guide, International Subcommission on Silurian Stratigraphy Field Meeting 2011, pp. 29–72.

Díaz-Martínez, E., Grahn, Y., 2007. Early Silurian glaciation along the western margin of Gondwana (Peru, Bolivia, and northern Argentina): Palaeogeographic and geodynamic setting. Palaeogeography, Palaeoclimatology, Palaeoecology 245, 62–81.

Doyle, E.N., Hoey, A.N., Harper, D.A.T., 1991. The rhynchonellide brachiopod *Eocoelia* from the Upper Llandovery of Ireland and Scotland. Palaeontology 34, 439–454.

Edwards, L.E., 1979. Range charts and no-space graphs. Computers & Geosciences 4, 247–255.

Eriksson, M.E., Calner, M., 2005. The Dynamic Silurian Earth. Subcommission on Silurian Stratigraphy Field Meeting 2005. Field Guide and Abstracts. Geological Survey of Sweden, Rapporter och Meddelanden 121, 99.

Finney, S.C., Berry, W.B.N., 1997. New perspectives on graptolite distributions and their use as indicators of platform margin dynamics. Geology 25, 919–922.

Fordham, B.G., 1998. Silurian time: How much of it was Pridoli? In: Gutiérrez-Marco, J.C., Rábano, I. (Eds.), Proceedings of the Sixth International Graptolite Conference on the GWG (IPA) and the 1998 Field Meeting of the International Subcommission on Silurian Stratigraphy (ICS-IUGS). Instituto Technológico Geominero de España, Temas Geologico-Mineros, 23, pp. 80–84.

Gale, N.H., 1985. Numerical calibration of the Paleozoic time-scale: Ordovician, Silurian and Devonian periods. In: Snelling, N.J. (Ed.), 1985, The Chronology of the Geological Record. Geological Society Memoir, 10, pp. 81–88.

Gradstein, F.M., Agterberg, F.P., Ogg, J.G., Hardenbol, J., van Veen, P., Thierry, J., Huang, Z., 1994. A Mesozoic time scale. Journal of Geophysical Research 99 (B12), 24051–24074.

Gradstein, F.M., Ogg, J.G., Smith, A.G. (Eds.), A Geologic Time Scale 2004. Cambridge University Press, Cambridge, p. 589.

Haq, B.U., Schutter, S.R., 2008. A chronology of Paleozoic sea-level changes. Science 322, 64–68.

Harland, W.B., Armstrong, R.L., Cox, A.V., Craig, L.E., Smtih, A.G., Smith, D.G., 1990. A Geologic Time Scale. Cambridge University Press, Cambridge, p. 263.

Heath, R.J., Brenchley, P.J., Marshall, J.D., 1998. Early Silurian carbon and oxygen stable-isotope stratigraphy of Estonia: Implications for climate change. In: Landing, E., Johnson, M.E. (Eds.), Silurian Cycles: Linkages of Dynamic Stratigraphy with Atmospheric, Oceanic and Tectonic changes. New York State Museum Bulletin, 491, pp. 313–327.

Holland, C.H., 1989. History of classification of the Silurian System. In: Holland, C.H., Bassett, M.G. (Eds.), A Global Standard for the Silurian System. National Museum of Wales Geological Series, 9, pp. 7–26.

Holland, C.H., Lawson, J.D., Walmsley, V.G., 1963. The Silurian rocks of the Ludlow District, Shropshire. Bulletin of the British Museum of Natural History, Geology 8, 93–171.

Huff, W.D., Bergström, S.M., Kolata, D.R., 2000. Silurian K-bentonites of the Dnestr Basin, Podolia, Ukraine. Journal of the Geological Society 157, 493–504.

Hutt, J.E., 1974. A new group of Llandovery biform monograptids. In: Rickards, R.B., Jackson, D.E., Hughes, C.P. (Eds.), Graptolite Studies in Honour of O.M.B. Bulman. Special Papers in Paleontology, 13, pp. 189–203.

Jaeger, H., 1991. Neue Standard-Graptolithenzonenfolge nach der 'Grossen Krise' an der Wenlock/Ludlow Grenze (Silurs). Neues Jarhbuch für Geologie und Palaontologie Abhandlungen 182 (3), 303–354.

Jeppsson, L., 1982. Third European Conodont Symposium (Ecos III) guide to excursion. Publications from the Institutes of Mineralogy, Palaeontology and Quaternary Geology, University of Lund 239, 1–32.

Jeppsson, L., 1990. An oceanic model for lithological and faunal changes tested on the Silurian record. Journal of the Geological Society 147, 663–674.

Jeppsson, L., 1997a. A new latest Telychian, Sheinwoodian and early Homerian (early Silurian) standard conodont zonation. Transactions of the Royal Society of Edinburgh, Earth Sciences 88, 91–114.

Jeppsson, L., 1997b. Recognition of a probable secundo-primo event in the early Silurian. Lethaia 29, 311–315.

Jeppsson, L., 1998. Silurian oceanic events: Summary of general characteristics. In: Landing, E., Johnson, M.E. (Eds.), Silurian Cycles: Linkages of Dynamic Stratigraphy with Atmospheric, Oceanic and Tectonic changes. New York State Museum Bulletin, 491, pp. 239–257.

Jeppsson, L., Calner, M., 2003. The Silurian Mulde event and a scenario for secundo—secundo events. Transactions of the Royal Society of Edinburgh 93, 135–154.

Jeppsson, L., Eriksson, M.E., Calner, M., 2005. Locality descriptions. In: Eriksson, M.E., Calner, M. (Eds.), The Dynamic Silurian Earth. Subcommission on Silurian Stratigraphy Field Meeting 2005. Field Guide and Abstracts. Geological Survey of Sweden, Rapporter och Meddelanden 121, pp. 22–56.

Jeppsson, L., Eriksson, M.E., Calner, M., 2006. A latest Llandovery to latest Ludlow high-resolution biostratigraphy based on the Silurian of Gotland – a summary. GFF 128, 109–114.

Johnson, M.E., 1996. Stable cratonic sequences and a standard for Silurian eustasy. In: Witzke, B.J., Ludvigson, G.A., Day, J. (Eds.), Paleozoic Sequence Stratigraphy: Views from the North American Craton. Geological Society of America Special Paper, 306, pp. 203–211.

Johnson, M.E., 2006. Relationship of Silurian sea-level fluctuations to oceanic episodes and events. GFF 128, 123–129.

Johnson, M.E., 2010. Tracking Silurian eustasy: Alignment of empirical evidence or pursuit of deductive reasoning? Palaeogeography, Palaeoclimatology, Palaeoecology 296, 276–284.

Johnson, M.E., Rong, J., Kershaw, S., 1998. Calibrating Silurian eustacy against the erosion and burial of coastal paleotopography. In: Landing, E., Johnson, M.E. (Eds.), Silurian Cycles: Linkages of Dynamic Stratigraphy with Atmospheric, Oceanic and Tectonic changes. New York State Museum Bulletin, 491, pp. 3–13.

Kaljo, D., Martma, T., 2006. Application of carbon isotope stratigraphy to dating the Baltic Silurian rocks. GFF 128, 123–129.

Kaljo, D., Grytsenko, V., Martma, T., Mõtus, M.-A., 2007. Three global carbon isotope shifts in the Silurian of Podolia (Ukraine): Stratigraphical implications. Estonian Journal of Earth Sciences 56, 205–220.

Kaufmann, B., 2006. Calibrating the Devonian time scale: A synthesis of U-Pb ID-TIMS ages and conodont stratigraphy. Earth and Planetary Science Letters 76, 175–190.

Kemple, W.G., Sadler, P.M., Strauss, D.J., 1995. Extending graphic correlation to many dimensions: Stratigraphic correlation as constrained optimisation. In: Mann, K.O., Lane, H.R., Scholle, P.A. (Eds.), Graphic Correlation, SEPM Special Publication 53, pp. 65–82.

Kiipli, T., Orlova, K., Kiipli, E., Kallaste, T., 2008a. Use of immobile trace elements for the correlation of Telychian bentonites on Saaremaa Island, Estonia, and mapping of volcanic ash clouds. Estonian Journal of Earth Sciences 57, 39–52.

Kiipli, T., Radzevičius, S., Kallaste, T., Motuza, V., Jeppsson, L., Wickström, L.M., 2008b. Wenlock bentonites in Lithuania and correlation with bentonites from sections in Estonia, Sweden and Norway. GFF 130, 203–210.

Kleffner, M.A., 1989. A conodont-based Silurian chronostratigraphy. Geological Society of America Bulletin 101, 904–912.

Kleffner, M.A., Barrick, J.E., 2010. Telychian-early Sheinwoodian (early Silurian) conodont-, graptolite-, chitinozoan- and event-based chronostratigraphy developed using the graphic correlation method. Memoirs of the Association of Australasian Palaeontologists 39, 191–210.

Koren', T.N., Lenz, A.C., Loydell, D.K., Melchin, M.J., Štorch, P., Teller, L., 1996. Generalized graptolite zonal sequence defining Silurian time intervals for global paleogeographic studies. Lethaia 29, 59–60.

Kříz, J., 1989. The Přídolí Series in the Prague Basin (Barrandian area, Bohemia). In: Holland, C.H., Bassett, M.G. (Eds.), A Global Standard for the Silurian System. National Museum of Wales, Geological Series, 9, pp. 90–100.

Kříz, J., Jaeger, H., Paris, F., Schönlaub, H.-P., 1986. Přídolí – the Fourth Series of the Silurian. Jahrbuch der Geologischen Bundesanstalt, Wein 129, 291–360.

Kuiper, K.F., Deino, A., Hilgen, F.J., Krijgsman, W., Renne, P.R., Wijbrans, J.R., 2008. Synchronizing Rock Clocks of Earth History. Science 320, 500–504.

Kunk, M.J., Sutter, J., Obradovitch, J.D., Lanphere, M.A., 1985. Age of biostatigraphic horizons within the Ordovician and Silurian Systems. In: Snelling, N.J. (Ed.), The Chronology of the Geological Record. Geological Society Memoir, 10, pp. 89–92.

Lapworth, C., 1879. On the tripartite classification of Lower Palaeozoic rocks. Geological Magazine 6, 1–15.

Landing, E., Johnson, M., 1998. Silurian cycles: Linkages of dynamic stratigraphy with atmospheric, oceanic, and tectonic changes. New York State Museum Bulletin 491, 327.

Laufeld, S., 1974. Silurian Chitinozoa from Gotland. Fossils and Strata 5, 130.

Lawson, J.D., White, D.E., 1989. The Ludlow Series in the type area. In: Holland, C.H., Bassett, M.G. (Eds.), A Global Standard for the Silurian System. National Museum of Wales, Geological Series, 9, pp. 73–90.

Lehnert, O., Mannik, P., Joachimski, M.M., Calner, M., Fryda, J., 2010. Palaeoclimate perturbations before the Sheinwoodian glaciation: A trigger for extinctions during the 'Ireviken Event'. Palaeogeography, Palaeoclimatology, Palaeoecology 296, 320–331.

Loydell, D.K., 1991. The biostratigraphy and formational relationships of the upper Aeronian and lower Telychian (Llandovery, Silurian) formations of western mid-Wales. Geological Journal 26, 209–244.

Loydell, D.K., 1998. Early Silurian sea-level changes. Geological Magazine 135, 447–471.

Loydell, D.K., 2007. Early Silurian positive $d^{13}C$ excursions and their relationship to glaciations, sea-level changes and extinction events. Geological Journal 42, 531–546.

Loydell, D.K., 2008a. The base of the Wenlock Series. Silurian Times 15, 10–12.

Loydell, D.K., 2008b. Reply to 'Early Silurian positive $\delta^{13}C$ excursions and their relationship to glaciations, sea-level changes and extinction events: discussion' by Bradley D. Cramer and Axel Munnecke. Geological Journal 43, 511–515.

Loydell, D.K., 2011a. The GSSP for the base of the Wenlock Series, Hughley Brook. In: Ray, D.C. (Ed.), Siluria Revisited, a Field Guide, International Subcommission on Silurian Stratigraphy Field Meeting 2011, pp. 91–99.

Loydell, D.K., 2011b. The GSSP for the base of the Homerian Stage, Whitwell Coppice. In: Ray, D.C. (Ed.), Siluria Revisited, a Field Guide, International Subcommission on Silurian Stratigraphy Field Meeting 2011, pp. 85–90.

Loydell, D.K., Cave, R., 1993. The Telychian (Upper Llandovery) stratigraphy of Buttington Brick Pit, Wales. Newsletters on Stratigraphy 29, 91–103.

Loydell, D.K., Cave, R., 1996. The Llandovery-Wenlock boundary and related stratigraphy in eastern mid-Wales with special reference to the Banwy River section. Newsletters on Stratigraphy 34, 39–64.

Loydell, D.K., Maletz, J., 2002. Isolated *'Monograptus' gemmatus* from the Silurian of Osmundsberget, Sweden. GFF 124, 193–196.

Loydell, D.K., Štorch, P., Melchin, M.J., 1993. Taxonomy, evolution and biostratigraphical importance of the Llandovery graptolite *Spirograptus*. Palaeontology 36, 909–926.

Loydell, D.K., Kaljo, D., Männik, P., 1998. Integrated biostratigraphy of the lower Silurian of the Ohesaare core, Saaremaa, Estonia. Geological Magazine 135, 769–783.

Loydell, D.K., Männik, P., Nestor, V., 2003. Integrated biostratigraphy of the lower Silurian of the Aizpute-41 core, Latvia. Geological Magazine 140, 205–229.

Mabillard, J.E., Aldridge, R.J., 1985. Microfossil distribution across the base of the Wenlock Series in the type area. Palaeontology 28, 89–100.

Männik, P., 1998. Evolution and taxonomy of the Silurian conodont *Pterospathodus*. Palaeontology 41, 1001–1050.

Männik, P., 2007a. Some comments on Telychian-early Sheinwoodian conodont faunas, events and stratigraphy. Acta Palaeontologica Sinica 46 (Suppl.), 305–310.

Männik, P., 2007b. An updated Telychian (Late Llandovery, Silurian) conodont zonation based on Baltic faunas. Lethaia 40, 45–60.

Märss, T., Fredholm, D., Talimaa, V., Turner, S., Jeppsson, L., Nowlan, G.S., 1995. Silurian vertebrate biozonal scheme. In: Lelievre, H., Wenz, S., Blieck, A., Cloutier, R. (Eds.), Premiers vertebrés et vertebrés inférieurs. Géobios, 19, pp. 369–372.

Mattinson, J.M., 2005. Zircon U-Pb chemical abrasion ("CA-TIMS") method: Combined annealing and multi-step partial dissolution analysis for improved precision and accuracy of zircon ages. Chemical Geology 220, 47–66.

McKerrow, W.S., Lambert, R.St.J., Cocks, L.R.M., 1985. The Ordovician, Silurian and Devonian Periods. In: Snelling, N.J. (Ed.), The Chronology of the Geological Record. Geological Society Memoir, 10, pp. 73–80.

Melchin, M.J., Williams, S.H., 2000. A restudy of the akidograptine graptolites from Dob's Linn and a proposed redefined zonation of the Silurian Stratotype. In: Cockle, P., Wilson, G.A., Brock, G.A., Engerbretsen, M.J., Simpson, A. (Eds.), Palaeontology Down-under 2000, 11th–15th July 2000, Orange, NSW, Geological Society of Australia, Abstracts, pp. 61–63.

Melchin, M.J., Holmden, C., 2006. Carbon isotope chemostratigraphy of the Llandovery in Arctic Canada: Implications for global correlation and sea-level change. GFF 128, 173–180.

Melchin, M.J., Koren', T.N., Štorch, P., 1998. Global diversity and survivorship patterns of Silurian graptoloids. In: Landing, E., Johnson, M.E. (Eds.), Silurian Cycles: Linkages of Dynamic Stratigraphy with Atmospheric, Oceanic and Tectonic changes. New York State Museum Bulletin, 491, pp. 165–182.

Melchin, M.J., Holmden, C., Williams, S.H., 2003. Correlation of graptolite biozones, chitinozoan biozones, and carbon isotope curves through the Hirnantian. In: Albanesi, G.L., Beresi, M.S., Peralta, S.H. (Eds.), Ordovician from the Andes. INSUGEO, Serie Correlación Geológica, 17, pp. 101–104.

Melchin, M.J., Cooper, R.A., Sadler, P.M., 2004. The Silurian Period. In: Gradstein, F.M., Ogg, J.G., Smith, A.G. (Eds.), A Geologic Time Scale 2004. Cambridge University Press, Cambridge, pp. 188–201.

Mullins, G.L., Aldridge, R.J., 2004. Chitinozoan biostratigraphy of the basal Wenlock Series (Silurian) global stratotype section and point. Palaeontology 47, 745–773.

Munnecke, A., Samtleben, C., Bickert, T., 2003. The Ireviken Event in the lower Silurian of Gotland, Sweden – relation to similar Palaeozoic and Proterozoic events. Palaeogeography, Palaeoclimatology, Palaeoecology 195, 99–124.

Munnecke, A., Calner, M., Harper, D.A.T., Servais, T., 2010. Ordovician and Silurian sea-water chemistry, sea level, and climate: A synopsis. Palaeogeography Palaeoclimatology Palaeoecology 296 (3–4), 389–413.

Murchison, R.I., 1839. The Silurian System, Founded on Geological Researches in the counties of Salop, Hereford, Radnor, Montgomery, Caermarthen, Brecon, Pembroke, Monmouth, Gloucester, Worcester and Stafford: With Descriptions of the Coal-fields and Overlying Formations. John Murray, London, p. 768.

Odin, G.S., 1985. Remarks on the numerical scale of Ordovician to Devonian times. In: Snelling, N.J. (Ed.), 1985, The Chronology of the Geological Record. Geological Society Memoir, 10, pp. 93–97.

Odin, G.S., Hunziker, J.C., Jeppsson, L., Spjeldnaes, N., 1986. Ages radiometriques K-Ar de biotites pyroclatiques sedimentées dans le Wenlock de Gotland (Suède). In: Odin, G.S. (Ed.), Calibration of the Phanerozoic Time Scale. Chemical Geology, 59, pp. 117–125.

Ogg, J.G., Ogg, G.J., Gradstein, F.M. (Eds.), 2008. The Concise Geologic Time Scale. Cambridge University Press, Cambridge, p. 177.

Page, A.A., Zalasiewicz, J.A., Williams, M., Popov, L.E., 2007. Were transgressive black shales a negative feedback modulating glacioeustasy during the Early Palaeozoic Icehouse? In: Williams, M., Haywood, A.M., Gregory, F.J., Schmidt, D.N. (Eds.), Deep Time Perspectives on Climate Change: Marrying the Signal from Computer Models and Biological Proxies. The Geological Society for the Micropalaeontological Society, London, pp. 123–156.

Palmer, D., 1971. The Ludlow graptolites *Neodiversograptus nilssoni* and *Cucullograptus (Lobograptus) progenitor*. Lethaia 4, 357–394.

Pogson, D.J., 2009. The Siluro-Devonian time scale: A critical review and interim revision. Geological Survey of NSW, Quarterly Notes 130, 1–11.

Pucéat, E., Joachimski, M.M., Bouilloux, A., Monna, A., Bonin, A., Motreuil, S., Morinière, P., Hénard, S., Mourin, J., Dera, G., Quesne, D., 2010. Revised phosphate-water fractionation equation reassessing paleotemperatures derived from biogenic apatite. Earth and Planetary Science Letters 298, 135–142.

Ray, D.C., 2007. The correlation of Lower Wenlock Series (Silurian) bentonites from the Lower Hill Farm and Eastnor Park boreholes, Midland Platform, England. Proceedings of the Geologists' Association 118, 175–185.

Ray, D.C., Brett, C.E., Thomas, A.T., Collings, A.V.J., 2010. Late Wenlock sequence stratigraphy in central England. Geological Magazine 147, 123–144.

Ray, D.C., Collings, A.V.J., Worton, G.J., Jones, G., 2011. Upper Wenlock bentonites from Wren's Nest Hill, Dudley: Comparisons with prominent bentonites along Wenlock Edge, Shropshire, England. Geological Magazine 148, 670–681.

Rickards, R.B., 1976. The sequence of Silurian graptolite zones in the British Isles. Geological Journal 11, 153–188.

Rong, J., Melchin, M.J., Williams, S.H., Koren, T.N., Verniers, J., 2008. Report of the restudy of the defined global stratotype of the base of the Silurian System. Episodes 31, 315–318.

Ross, C.A., Ross, J.P.R., 1996. Silurian sea-level fluctuations. In: Witzke, B.J., Ludvigson, G.A., Day, J. (Eds.), Paleozoic Sequence Stratigraphy: Views from the North American Craton. Geological Society of America Special Paper, 306, pp. 187–192.

Ross, R.J., Naeser, C.W., Izett, G.A., Obradovich, J.D., Bassett, M.J., Hughes, C.P., Cocks, L.R.M., Dean, W.T., Ingham, J.K., Jenkins, C.J., Rickards, R.B., Sheldon, P.R., Toghill, P., Whittington, H.B., Zalasiewicz, J., 1982. Fission track dating of British Ordovician and Silurian stratotypes. Geological Magazine 119 (2), 135–153.

Ruppel, S.C., James, E.W., Barrick, J.E., Nowlan, G.S., Uyeno, T.T., 1998. High-resolution $^{87}Sr/^{86}Sr$ record: Evidence for eustatic control of seawater chemistry? In: Landing, E., Johnson, M.E. (Eds.), Silurian Cycles: Linkages of Dynamic Stratigraphy with Atmospheric, Oceanic and Tectonic changes. New York State Museum Bulletin, 491, pp. 285–295.

Sadler, P.M., 1999. The influence of hiatuses on sediment accumulation rates. In: Bruns, P., Hass, H.C. (Eds.), On the Determination of Sediment Accumulation Rates. GeoResearch Forum, 5, pp. 15–40.

Sadler, P.M., 2004. Quantitative biostratigraphy – achieving finer resolution in global correlation. Annual Review of Earth and Planetary Science 32, 187–213.

Sadler, P.M., Cooper, R.A., Melchin, M., 2009. High-resolution, early Paleozoic (Ordovician-Silurian) time scales. Geological Society of America Bulletin 121, 887–906.

Saltzman, M.R., 2002. Carbon isotope $\delta^{13}C$ stratigraphy across the Silurian-Devonian transition in North America: Evidence for a perturbation of the global carbon cycle. Palaeogeography, Palaeoclimatology, Palaeoecology 187, 83–100.

Samtleben, C., Munnecke, A., Bickert, T., Pätzold, J., 1996. The Silurian of Gotland (Sweden): Facies interpretation based on stable isotopes in brachiopod shells. Geologische Rundeschau 85, 278–292.

Schoene, B., Crowley, J.L., Condon, D.J., Schmitz, M.D., Bowring, S.A., 2006. Reassessing the uranium decay constants for geochronology using ID-TIMS U-Pb data. Geochimica et Cosmochimica Acta 70 (2), 426–445.

Shaw, A.B., 1964. Time in Stratigraphy. McGraw-Hill, New York, p. 365.

Snelling, N.J. (Ed.), 1985. The Chronology of the Geological Record. Geological Society Memoir, 10, p. 343.

Sorgenfrei, T. (Ed.), 1964. Report of the 21st International Geological Congress. Copenhagen, Norden, 1960, Part 28: General proceedings, p. 277.

Štorch, P., 1994. Graptolite biostratigraphy of the Lower Silurian (Llandovery and Wenlock) of Bohemia. Geological Journal 29, 137–165.

Stricanne, L., Munnecke, A., Pross, J., 2006. Assessing mechanisms of environmental change: Palynological signals across the Late Ludlow (Silurian) positive isotope excursion ($\delta^{13}C$, $\delta^{18}O$) on Gotland, Sweden. Palaeogeography, Palaeoclimatology Palaeoecology 230, 1–31.

Subcommission on Silurian Stratigraphy, 1995. Left hand column for correlation charts. Silurian Times 2, 7–8.

Sudbury, M., 1958. Triangulate monograptids from the *Monograptus gregarius* Zone (Lower Llandovery) in the Rheidol Gorge (Cardiganshire). Philosophical Transactions of the Royal Society, Series B 241, 485–554.

Thomas, A.T., Ray, D.C., 2011. Pitch Coppice: GSSP for the base of the Ludlow Series and Gorstian Stage, Whitwell Coppice. In: Ray, D.C. (Ed.), Siluria Revisited, a Field Guide, International Subcommission on Silurian Stratigraphy Field Meeting 2011, pp. 80–84.

Toghill, P., 1968. The graptolite assemblages and zones of the Birkhill Shales (Lower Silurian) at Dobb's Linn. Palaeontology 11 (5), 654–668.

Trotter, J.A., 2008. Did cooling oceans trigger Ordovician biodiversification? Evidence from conodont thermometry. Science 321, 550–554.

Tucker, R.D., McKerrow, W.S., 1995. Early Paleozoic chronology: A review in light of new U-Pb zircon ages from Newfoundland and Britain. Canadian Journal of Earth Sciences 32 (4), 368–379.

Tucker, R.D., Krogh, T.E., Ross, R.J., Williams, S.H., 1990. Time-scale calibration by high-precision U-Pb zircon dating of interstratified volcanic ashes in the Ordovician and lower Silurian stratotypes of Britain. Earth and Planetary Sciences Letters 1000, 51–58.

Tucker, R.D., Bradlety, D.C., Ver Straeten, C.A., Harris, A.G., Ebert, J.R., McCutcheon, S.R., 1998. New U-Pb zircon ages and the duration and division of Devonian time. Earth and Planetary Science Letters 158, 175–186.

Trench, A., McKerrow, W.S., Torsvik, T.H., Li, X., McCracken, S.R., 1993. The polarity of the Silurian magnetic field: Indications from global data compilation. Journal of the Geological Society 150, 823–831.

Underwood, C.J., 1993. The position of graptolites within Lower Palaeozoic planktic ecosystems. Lethaia 26, 189–202.

Veizer, J., Ala, D., Azmy, K., Bruckschen, P., Buhl, D., Bruhn, F., Carden, G.A.F., Diener, A., Ebneth, S., Godderis, Y., Jasper, T., Korte, C., Pawellek, F., Podlaha, O.G., Strauss, H., 1999. $^{87}Sr/^{86}Sr$, $\delta^{13}O$ evolution of Phanerozoic seawater. Chemical Geology 161, 59–88.

Verniers, J., Nestor, V., Paris, F., Dufka, P., Sutherland, S., van Grootel, G., 1995. A global Chininozoa biozonation for the Silurian. Geological Magazine 132, 651–666.

Villeneuve, M., 2004. Radiogenic isotope geochronology. In: Gradstein, F.M., Ogg, J.G., Smith, A.G. (Eds.), A Geologic Time Scale 2004. Cambridge University Press, Cambridge, pp. 87–95.

Wenzel, B., 1997. Isotopenstratigraphische Untersuchungen an silurischen Abfolgen und deren paläozeanographische Interpretation. Erlanger geologische Abhandlungen 129, 1–117.

Wenzel, B., Joachimski, M.M., 1996. Carbon and oxygen isotopic composition of Silurian brachiopods (Gotland/Sweden): Palaeoceanographic implications. Palaeogeography, Palaeoclimatology, Palaeoecology 122, 143–166.

Whittard, W.F., 1961. Lexique Stratigraphique International, vol. 1. Europe, Fasc. 3a. Pays de Galles, Ecosse, V. Silurien. Centre National de la Recherche Scientifique, Paris, p. 273.

Williams, M., Zalasiewicz, J.A., 2004. The Wenlock *Cyrtograptus* species of the Builth Wells district, central Wales. Palaeontology 47, 223–263.

Williams, S.H., 1983. The Ordovician-Silurian boundary graptolite fauna of Dob's Linn, southern Scotland. Palaeontology 26, 605–639.

Wyborn, D., Owen, N., Compston, W., McDougall, I., 1982. The Laidlaw Volcanics: A late Silurian point on the geological time scale. Earth and Planetary Science Letters 59, 90–100.

Zalasiewicz, J.A., 2000. *Saetograptus (Saetograptus) leintwardinensis leintwardinensis* (Hopkinson MS, Lapworth, 1880). In: Zalasiewicz, J.A., Rushton, A.W.A., Hutt, J.E., Howe, M.P.A. (Eds.), Atlas of Graptolite Type Specimens. Palaeontographical Society, London, Folio 1.90.

Zalasiewicz, J.A., 2008. *Monograptus triangulatus triangulatus* (Harkness, 1851). In: Zalasiewicz, J.A., Rushton, A.W.A., Hutt, J.E., Howe, M.P.A. (Eds.), Atlas of Graptolite Type Specimens. Palaeontographical Society, London, Folio 2.92.

Zalasiewicz, J.A., Williams, M., 2008. *Cyrtograptus murchisoni* Carruthers, 1867. In: Zalasiewicz, J.A., Rushton, A.W.A., Hutt, J.E., Howe, M.P.A. (Eds.), Atlas of Graptolite Type Specimens. Palaeontographical Society, London, Folio 2.60.

Zalasiewicz, J., Williams, M., Verniers, J., Jachowicz, M., 1998. A revision of the graptolite biozonation and calibration with the chitinozoa and acritarch biozonations for the Wenlock succession of the Builth Wells district, Wales, U. K. In: Gutiérrez-Marco, J.C., Rábano, I. (Eds.), Proceedings of the Sixth International Graptolite Conference of the GWG (IPA) and the 1998 Field Meeting of the International Subcommission on Silurian Stratigraphy (ICS-IUGS). Instituto Tecnológico Geominero de España, Temas Geológico-Mineros, 23, p. 141.

Zalasiewicz, J.A., Taylor, L., Rushton, A.W.A., Loydell, D.K., Rickards, R.B., Williams, M., 2009. Graptolites in British stratigraphy. Geological Magazine 146, 785–850.

Žigaitė, Ž, Joachimski, M.M., Lehnert, O., Brazauskas, A., 2008. $\delta^{18}O$ composition of conodont apatite indicates climatic cooling during the Middle Pridoli. Palaeogeography, Palaeoclimatology, Palaeoecology 294, 242–247.

R.T. Becker, F.M. Gradstein and O. Hammer

Chapter 22

The Devonian Period

Abstract: All seven Devonian stages have been defined by GSSPs, but revisions of the base of the Emsian and of the Devonian–Carboniferous boundary are currently underway. Most of the Devonian Period was a time of exceptionally high sea-level stand and inferred widespread equable climates, but glaciations occurred immediately before its end in the south polar areas of Gondwana (South America, Central and South Africa). There is even evidence for contemporaneous mountain glaciers in tropical latitudes (in the Appalachians of eastern North America). The cold-water Malvinocaffric Province of southern Gondwana existed throughout the Early Devonian but disappeared stepwise in the Middle Devonian. Most present-day continental areas and shelves were grouped in one hemisphere, creating a giant "Proto-Pacific" or Panthalassia Ocean, whose margins are poorly preserved in allochthonous terrains. Following the tectonic events of the Caledonian orogeny of Laurasia, many "Old Red Sandstone" terrestrial deposits formed. After the closure of the narrow Rheic Ocean early in the Devonian, Eovariscan tectonic movements occurred in the Upper Devonian in the western Proto-Tethys of Europe and North Africa. Other active fold belts existed in western North America (early Antler orogeny), Polar Canada, in the Appalachians, in the Urals, along the southern margin of Siberia, in NW China, and in eastern Australia. The Devonian is the time of greatest carbonate production, with a peak of reef growth, and of the greatest diversity of marine fauna in the Paleozoic. Vascular plants became established near the end of the Middle Devonian and early in the Middle Devonian tetrapods appeared, which spread and diversified in the Upper Devonian.

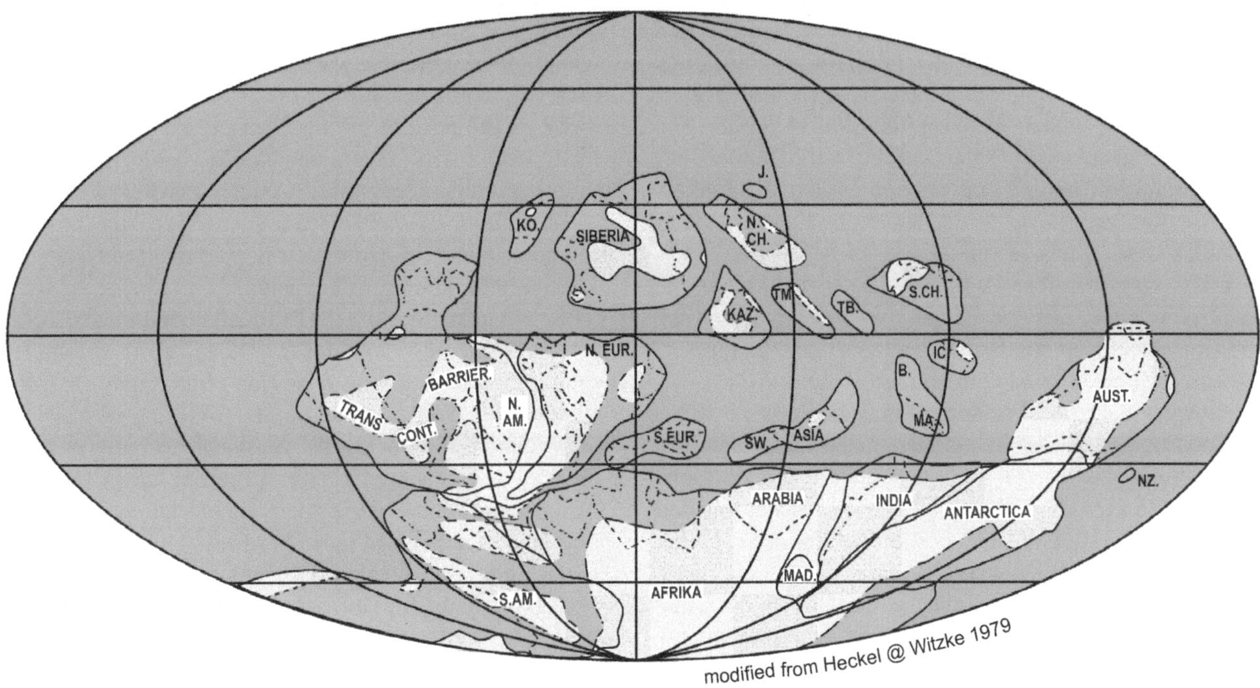

Chapter Outline

The Geologic Time Scale 2012. DOI: 10.1016/B978-0-444-59425-9.00022-6

22.1. HISTORY AND SUBDIVISIONS

Until recently, there was considerable confusion over the use of the term Devonian because the boundaries were used differently in various parts of the world. A broad, international review of the Devonian Period is given in *Devonian of the World* (McMillan *et al.*, 1988) and in House (1991).

The Devonian System was established by Sedgwick and Murchison (1839) when it was recognized through the then unpublished work of Lonsdale (1840) that marine rocks in southwest England were the equivalent of terrestrial Old Red Sandstone deposits in Wales, the north of England, and Scotland; an early recognition of facies change. Murchison's definition of the boundary between the Silurian System and the Old Red Sandstone in Wales and the Welsh Borders has some ambiguities, but general opinion is that the Ludlow Bone Bed was very close to the intention. However, other boundaries were used over the next century (White, 1950). The result was that there was no clear definition of the boundary in what may be called the type area, and no consistent practice among British geologists. All this was of little help for international correlations.

Following the detailed work on British graptolites by Elles and Wood (1901–1918), it was recognized that graptolites were last present in the late Ludlow below the Ludlow Bone Bed. Therefore, the extinction of graptolites was considered to be the major guide to the position of the base of the Devonian elsewhere in the world. It was not until 1960 that it became clear from evidence outside the British Isles that graptolites continued long after the time equivalent of the Ludlow Bone Bed. This evidence came from the work of R. Thorsteinson (Geological Survey of Canada) on Ellesmere Island, and the association of monograptids with basal Emsian (late Early Devonian) rocks in Europe. Thus the base of the Devonian, as defined by the extinction of graptolites, especially in continental Europe and Asia, belongs to a level well above the Ludlow Bone Bed. This definition raised the Silurian–Devonian boundary to near the *Psammosteus* Limestone (Richardson *et al.*, 1981).

As to the Devonian–Carboniferous boundary, it could be argued that Sedgwick and Murchison (1840) placed the top of the Devonian at a fairly unambiguous boundary in north Devon, but faunal and floral studies were not then precise enough for accurate correlation. As a result, stratigraphic levels were taken that were subsequently demonstrated to be inaccurate, or other names were used for strata where there was some uncertainty of assignment. In the latter category were names such as Kinderhookian in North America, and Etroungt and Strunian in continental Europe.

In 1960, a committee was established by the International Union of Geological Sciences (IUGS) to make recommendations on the position of the Silurian–Devonian boundary, which led to its recommendations being accepted at the International Geological Congress (IGC) in Montréal in 1972. Martinsson (1977) has summarized the procedure and conclusions. A separate working group considered the Devonian–Carboniferous boundary.

Recommendations of GSSPs for all boundaries of system, series, and stage divisions for the Devonian were completed by the Subcommission on Devonian Stratigraphy (SDS) and ratified by IUGS in 1996, and in the following year accounts of all decisions were published. Summary accounts have been published subsequently in two special volumes (Bultynck, 2000a,b). Since some of the stages are much longer than those in other systems, and since there are important times of second-order global extinction and sedimentary perturbations, giving natural breaks within stages (e.g., House, 1985; Walliser, 1996), the SDS is currently working on formal substage definitions.

The Devonian is divided into the Lower, Middle, and Upper Series. The Lower Devonian is divided into the Lochkovian, Pragian, and Emsian stages, the Middle into the Eifelian and Givetian stages, and the Upper into the Frasnian and Famennian stages. Recently it has become clear that the basal Emsian GSSP of Uzbekistan lies far below the base of the classical Emsian of German usage, and that it correlates with a level in the lower half of the classical Pragian of Bohemia (Carls *et al.*, 2008). Kaiser (2009) provided new conodont data from the Devonian/Carboniferous boundary stratotype section, and showed that the current boundary level lies higher than originally thought, and that it cannot be correlated with precision into any other section. Consequently, both GSSPs are subject to ongoing revision.

The current standard international chronostratigraphic divisions for the Devonian are given in Figure 22.1. There are also many terms widely used as local and regional stages, e.g., for the Lower Devonian, the Gedinnian and Siegenian. For the Emsian, a subdivision into formal Lower and Upper substages has been decided by the SDS but the precise level is still under discussion. The goal is to find an appropriate level for international correlation near the Zlichovian/Dalejan boundary of Bohemian terminology. In the past, it was the miscorrelation of the Dalejan deepening event with the Eifelian deepening that led to many complications in the definition of the Lower–Middle Devonian boundary. The term Couvinian (Figure 22.1) is now

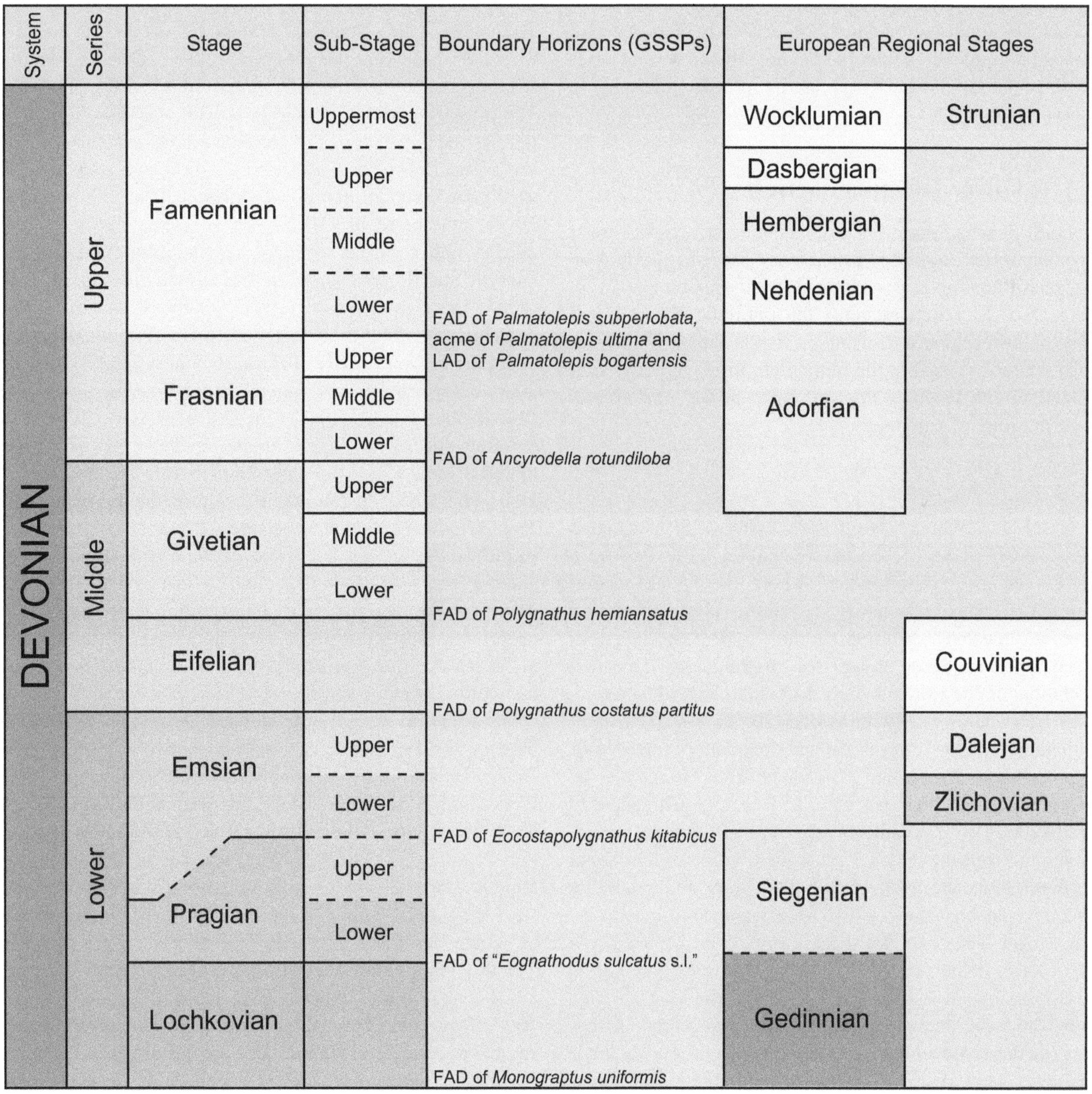

System	Series	Stage	Sub-Stage	Boundary Horizons (GSSPs)	European Regional Stages	
DEVONIAN	Upper	Famennian	Uppermost		Wocklumian	Strunian
			Upper		Dasbergian	
			Middle		Hembergian	
			Lower		Nehdenian	
				FAD of *Palmatolepis subperlobata*, acme of *Palmatolepis ultima* and LAD of *Palmatolepis bogartensis*		
		Frasnian	Upper		Adorfian	
			Middle			
			Lower			
				FAD of *Ancyrodella rotundiloba*		
	Middle	Givetian	Upper			
			Middle			
			Lower			
				FAD of *Polygnathus hemiansatus*		
		Eifelian				Couvinian
				FAD of *Polygnathus costatus partitus*		
	Lower	Emsian	Upper			Dalejan
			Lower			Zlichovian
				FAD of *Eocostapolygnathus kitabicus*		
		Pragian	Upper		Siegenian	
			Lower			
				FAD of "*Eognathodus sulcatus* s.l."		
		Lochkovian			Gedinnian	
				FAD of *Monograptus uniformis*		

FIGURE 22.1 **Current standard international chronostratigraphic divisions for the Devonian, giving the stage definitions according to recent taxonomic and biostratigraphic updates.** Regional European stages on the right (for details see text).

a regional term of the Ardennes only. In 2006, the SDS decided to subdivide the Givetian into Lower, Middle, and Upper substages, with the latter positioned close to the old base of the Upper Devonian of German tradition, which was defined by the ammonoid succession (base of *Pharciceras* faunas). The German Upper Devonian Stufen, Adorfian, Nehdenian, Hembergian, Dasbergian, and Wocklumian (Figure 22.1), are now regional terms. The Frasnian was decided by the SDS in 2006 to be formally subdivided into Lower, Middle and Upper substages and precise levels have been selected (Becker, 2007a) but are not yet ratified by the ICS. The rather long Famennian will include four substages named as Lower, Middle, Upper and Uppermost Famennian. The latter will roughly equal the outdated Strunian of Belgian regional chronostratigraphy. It is important to note that in several cases (sub)stage names may have been used somewhat differently in the past.

22.1.1. Lower Devonian Series

The GSSP for the Silurian–Devonian boundary, basal-Lower Devonian Series, and basal-Lochkovian Stage is at Klonk in the Czech Republic as documented in Martinsson (1977), in which D. J. McLaren recounts the scientific and political problems in reaching a decision on its placement. Important faunal characters used in the initial definition were the entry of the graptolite *Monograptus uniformis* and the *Warburgella rugulosa* group of trilobites.

22.1.1.1. Lochkovian

The basal-Devonian and basal-Lochkovian GSSP is situated southwest of Prague, in the Czech Republic, in the Paleozoic area known as the Barrandian, where the Klonk section (Figure 22.2) near Suchomasty is a natural 34 m cliff section embracing the latest Silurian (Pridoli) and the early Lochkovian. The sequence comprises rhythmically deposited allochthonous limestones with autochthonous intervening shales. The GSSP is within Bed 20, a 7–10 cm unit "immediately below the sudden and abundant occurrence of *M. uniformis uniformis* and *M. uniformis angustidens* in the upper part of that bed" (Jaeger, 1977, p. 21). Chlupáč and Hladil (2000) reviewed the stratigraphy of the type section and summarized the detailed subsequent work on the faunal and floral sequence at the GSSP. Conodont and graptolite fauna have enabled this stage to be recognized in most parts of the world. The conodont scale is now revised, since Carls *et al.* (2007) demonstrated that the base of the Devonian is best approximated by the entry of *Caudicriodus hesperius*, not by the traditional *woschmidti* Zone. True *Caud. woschmidti* (in the sense of its type) enter higher in the Lochkovian, even later than the previously supposed descendent *Caud. postwoschmidti.* Richardson and McGregor (1986) and Richardson *et al.* (2000) provided spore evidence for correlation of the Lochkovian areas and facies not covered by conodonts and graptolites. Trilobite lineages (*Acastella* and *Warburgella*) provide precise correlation markers in neritic facies. Saltzman (2002), Buggisch and Mann (2004), Kleffner *et al.* (2009), and Zhao *et al.* (2011) show that the Silurian–Devonian boundary can be approximated by a significant positive carbon isotope excursion. It can be separated from a topmost Silurian isotopic peak and extinction interval (*transgrediens* Event) in the Barrandian (Manda and Frýda, 2010).

22.1.1.2. Pragian

The base of this stage is defined by the GSSP at Velka Chuchle near Prague (Figure 22.3), Czech Republic (Chlupáč and Oliver, 1989; reviewed by Chlupáč, 2000). The primary correlation marker for this boundary used to be the first occurrence of the conodont *Eognathodus sulcatus,* but recent taxonomic restrictions now place early Pragian specimens in the genus *Gondwania* (Bardashev *et al.*, 2002), and more specifically in *G. juliae* (Yolkin *et al.*, 2011). Other forms which are important include the dacryoconarid *Nowakia (Turkestanella) acuaria.* Murphy (2005) and Slavík *et al.* (2007) suggested that the oldest eognathodids should be assigned to *Eo. irregularis* (which also falls in *Gondwania* of Bardashev *et al.*, 2002) and they commence in the Barrandian slightly below the GSSP level (Slavík and Hladil, 2004). *Eo. sulcatus* is no longer the defining species for the Pragian; in the sense of its type material the species does not enter before the middle Pragian. The GSSP is sandwiched within the range of early *Eognathodus/Gondwania* between the first occurrence of *Caudicriodus steinachensis* morphotype beta slightly below and the entry of *Now. (T.) acuaria* just above. Both taxa have a wide geographical distribution. Chitinozoans are useful for recognizing the boundary interval and provide a link to the spore zonation. Ranges of other invertebrate taxa around the GSSP are given by Chlupáč (2000). As the spore zonation is currently defined (Steemans, 1989), the base of the Pragian falls within a spore zone such that it cannot be accurately defined using palynology but, nevertheless, fairly accurate placing has been achieved in several areas elsewhere (Richardson *et al.*, 2000). Slavík *et al.* (2007) and Carls *et al.* (2008) emphasize that the problematic Emsian GSSP places up to two thirds of the classical Pragian of Bohemia into the Emsian. Unintentionally, the stage has lost more than half of its original duration. The Pragian conodont zonation is still in a state of flux after all three traditional zones have been discarded (Slavík, 2004a; Murphy, 2005; Slavík *et al.*, 2007). Currently there is no precise correlation between Europe and North America. Yolkin *et al.* (2011) proposed a succession with a new subzone for Uzbekistan, which has the potential to become a standard.

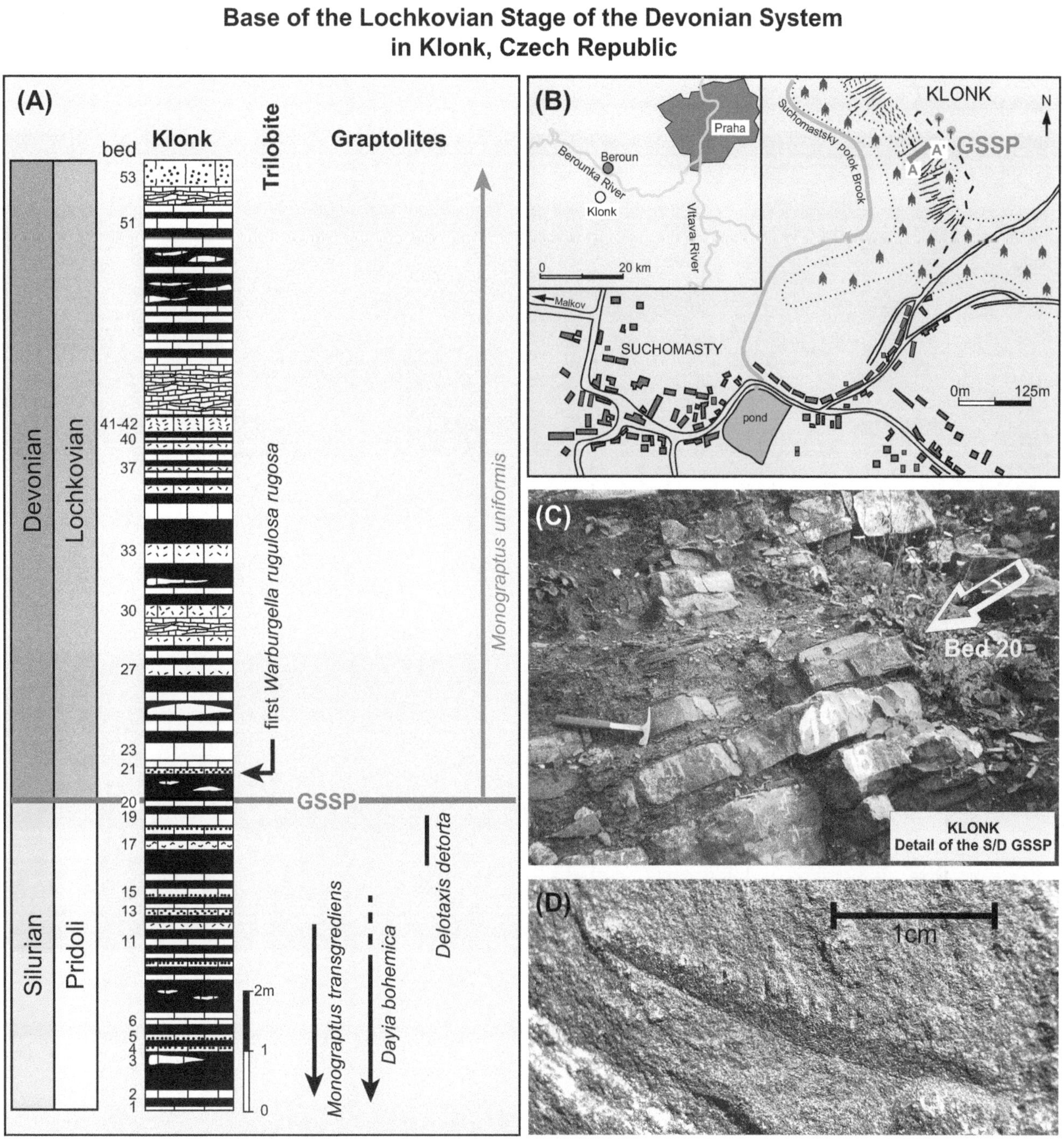

Base of the Lochkovian Stage of the Devonian System in Klonk, Czech Republic

FIGURE 22.2 **The basal-Devonian (basal-Lochkovian) GSSP at Klonk near Suchomasty, Barrandian, Czech Republic.** A) Lithological log and ranges of important biostratigraphical markers in relation to the GSSP, B) Geographic position of the GSSP section, C) Section detail with the GSSP in the upper part of Bed 20 (photo by L. Slavik), D) Photo of the defining index fossil, *Monograptus uniformis uniformis (photo by L. Slavik).*

22.1.1.3. Emsian

The GSSP for the basal Emsian is in the Zinzi'ban Gorge (Figure 22.4) of the Kitab National Park in Uzbekistan (Yolkin *et al.*, 1998). The key conodont used for the stage definition is *Polygnathus kitabicus,* which falls in the genus *Eocostapolygnathus* of Bardashev *et al.* (2002) (see comments in Becker and Aboussalam, 2001). However, the "*kitabicus* boundary" lies much lower than the base of the Emsian in the classical region of Germany (Carls and Valenzuela-Ríos, 2007; Jansen, 2008; Carls *et al.*, 2008) and it correlates with a position in the lower half of the Praha Limestone of Bohemia (Slavík *et al.*, 2007). Consequently, and after completion of the required 10 year moratorium, the SDS decided in 2008 to revise the Emsian base. The entry of *Eoc. kitabicus* and the current GSSP shall

Base of the Pragian Stage of the Devonian System in the Velká Chuchle Quarry in the southwest part of Prague, Czech Republic

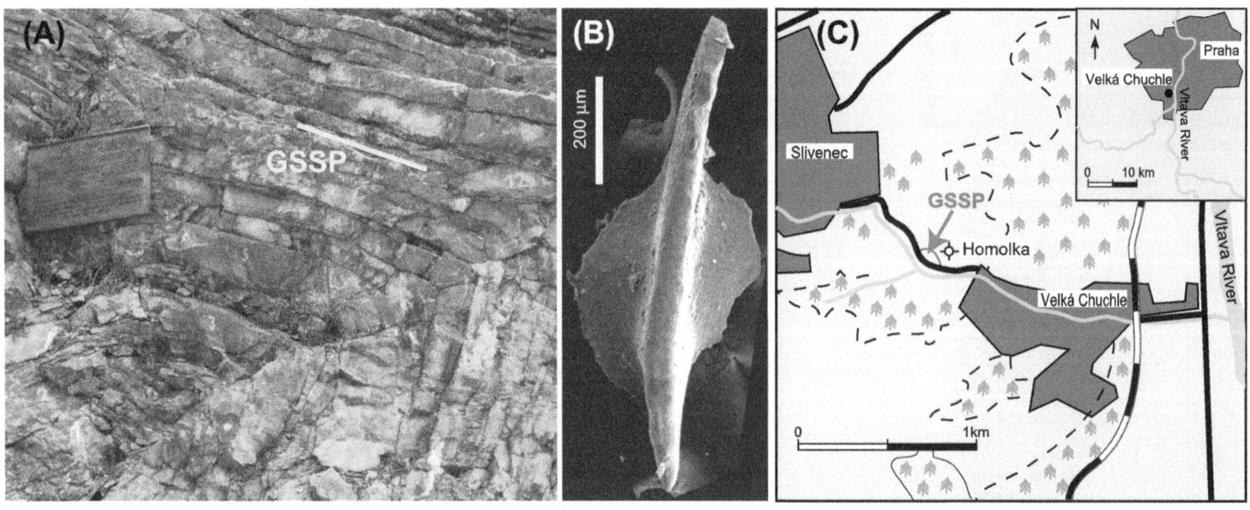

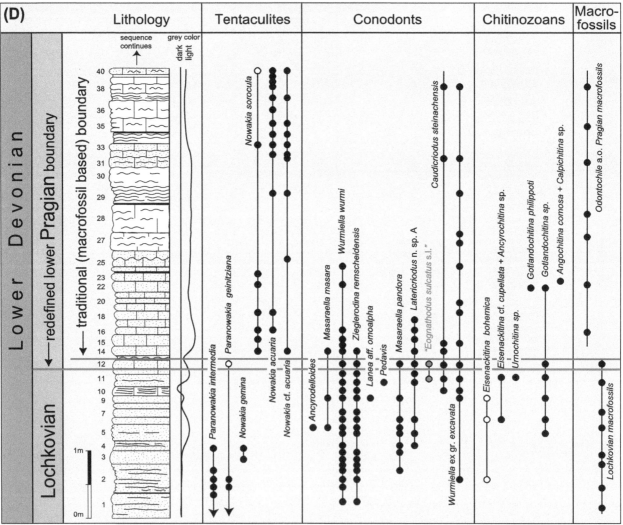

FIGURE 22.3 **The basal-Pragian GSSP at Velka Chuchle, S of Prague, Barrandian, Czech Republic.** A) Section detail around the GSSP level at the base of Bed 12, with the plate explaining the stratotype (photo by L. Slavik), B) Photo of an "*Eognathodus sulcatus* s.l." (or specimen of *Gondwania*) from the GSSP Bed (Slavik and Hladil, 2004, pl. 1, Figure 9), C) Locality map, D) Lithological log with ranges of important index fossils in relation to the GSSP *(based on data from Chlupáč and Hladil, 2000, and Slavík and Hladil, 2004; with updated taxonomy)*

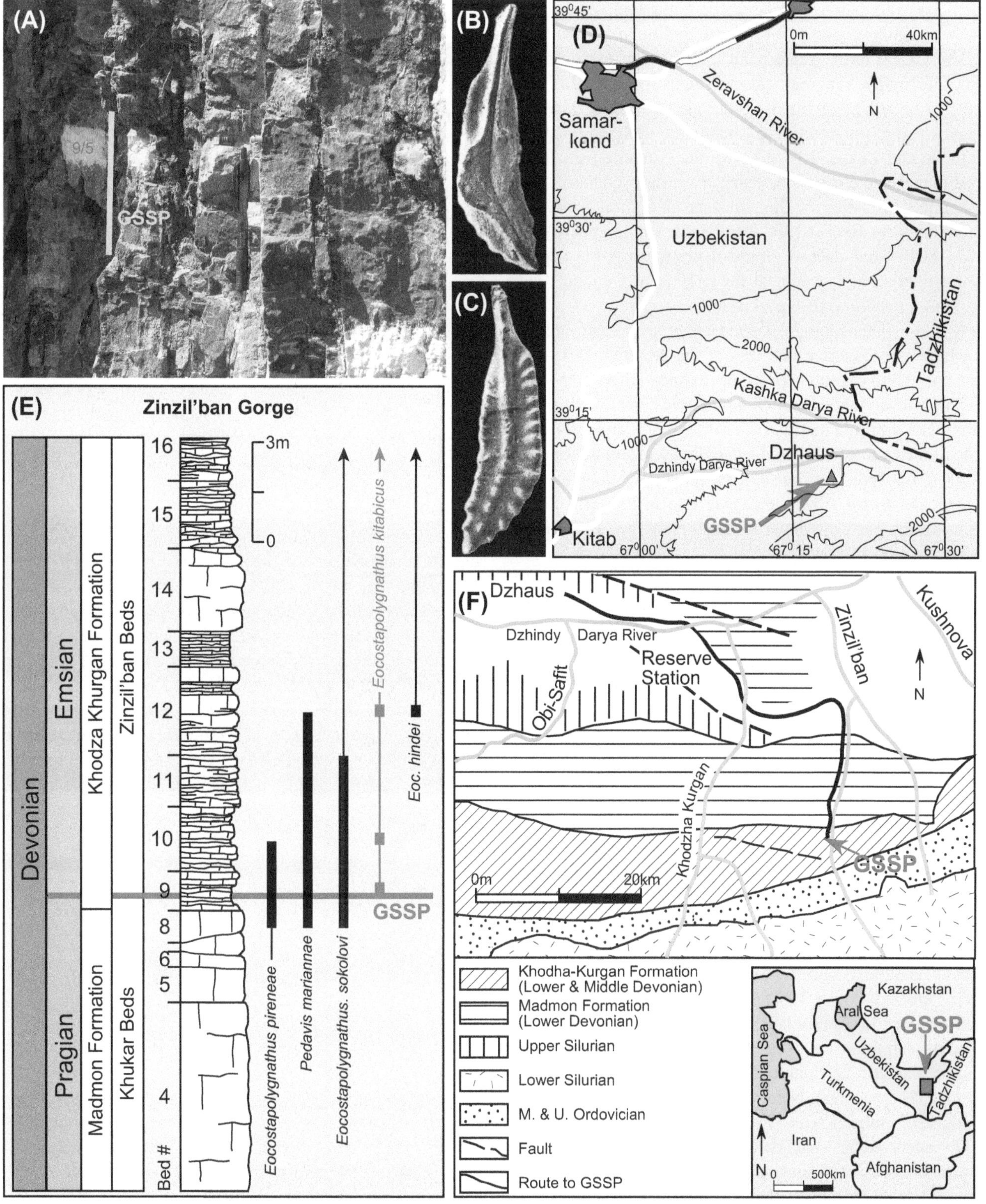

FIGURE 22.4 **The current basal-Emsian GSSP in the Zinzil'ban Gorge, Kitab State Reserve, Uzbekistan.** A) Section detail with the GSSP position at the base of Bed 9-5, B–C) Lower and upper views of the defining taxon, *Eocostapolygnathus kitabicus*, from the GSSP Bed (Yolkin *et al.*, 1994, pl. 1, Figures 1–2), D) Position of the Kitab region and GSSP area in eastern Uzbekistan, E) Lithological log with ranges of marker conodonts in relation to the GSSP (based on data in Yolkin *et al.*, 2008), F) Geographic position of the GSSP in the Zinzil'ban Gorge.

define a future Upper Pragian substage whilst the Emsian base shall be raised to a level near the entry of *P. excavatus*, with a new GSSP in the Zinzi'ban Gorge (Becker, 2009). Based on brachiopod–conodont correlations in Celtiberia (Spain), this new conodont level will be close to the traditional Emsian base.

The Czech terms Zlichovian and Dalejan are currently used informally for regional substages (Figure 22.1). A formal Lower/Upper Emsian substage boundary shall be placed close to the traditional Zlichovian–Dalejan boundary (Chlupáč and Lukes, 1999), above the extinction of *Anetoceras* faunas and close to the entry of the dacryoconarid *Now. cancellata*. Problems of this substage definition have been summarized by Becker (2007b). Recent ammonoid (Becker *et al.*, 2010) and dacryoconarid data (Kim, 2011) suggest a short overlap of last typical lower Emsian goniatites with first *Now. cancellata* in Uzbekistan.

A major paleoecological changeover in pelagic areas is shown by the reduction and loss of the graptolites, with the uniserial monograptids becoming extinct within the basal Emsian. Closely above, coiled ammonoids enter and become a dominant group in marine facies until their extinction at the close of the Cretaceous. The Daleje Shale, within the Emsian, marks a transgressive pulse that is widely recognized in Europe and North Africa and gave rise to the term Daleje Event. In the past, this level in many areas was erroneously assigned to the Eifelian.

22.1.2. Middle Devonian Series

22.1.2.1. Eifelian

The base of the Middle Devonian Series and of the Eifelian Stage is drawn at a GSSP at Wetteldorf Richtschnitt (Figure 22.5) in the Prüm Syncline of the Eifel District of Germany (Ziegler and Werner, 1982; Ziegler, 2000). The recommendation of the SDS was ratified at meetings of the IUGS held with the ICS in Moscow in 1984. A primary correlation marker for this boundary is the junction of the *Polygnathus costatus patulus* and *P. costatus partitus* conodont zones (Weddige, 1977, 1982), which lies somewhat below the anoxic pulse of the global Choteč Event that occurred near the end of the *partitus* Zone (e.g., Berkyová, 2009). The GSSP is in a trench, and the locality is protected by a building (the Happel Hut) erected in 1990 by the Senckenbergische Naturforschende Gesellschaft. As this boundary is in the Eifel Hills, its definition involved a minor change from historical usage of the term Eifelian in the area, but the upper boundary, discussed below, had been defined in several ways. The multiphase, global Kačak Event (e.g., House, 1996; Schöne, 1997) occurred at the end of the Eifelian. Based on Appalachian Foreland successions, DeSantis and Brett (2011) proposed two additional smaller bioevents associated with eustatic changes between the Choteč and Kačák intervals, the Bakoven and Stony Hollow events.

22.1.2.2. Givetian

The base of the Givetian Stage is drawn at a GSSP in southern Morocco at Jebel Mech Irdane (Figure 22.6) in the Tafilalt area of the eastern Anti Atlas, 12 km southwest of Rissani (Walliser *et al.*, 1996; Walliser, 2000; Ellwood *et al.*, 2003, 2011a; Schmitz *et al.*, 2006, Walliser and Bultynck, 2011). The primary correlation marker for this boundary is the base of the *Polygnathus hemiansatus* conodont zone, corresponding to the upper part of the former *P. ensensis* Zone. The oldest maenioceratid ammonoids (*Bensaidites koeneni*) commence one bed below the boundary, and, based on evidence from the Eifel region, the spore *Geminospora lemurata* slightly above it (Streel *et al.*, 2000a).

The SDS's recommendation was ratified at an IUGS meeting in London in 1994. The original Givet Limestone, and Assisse de Givet, in the Ardennes, gave the name to the stage, which had been defined in several different ways, based either on local neritic characters or on inferred correlation with fauna in pelagic areas. The lower boundary of the present stage is not too far below the base of the Givet Limestone and very close to the entry of the classical Givetian brachiopod *Stringocephalus* (Bultynck and Hollevoet, 1999). In Morocco, the boundary falls in the top part of the polyphase Kačák Event interval.

Following a proposal by Bultynck and Gouwy (2002), the Givetian is formally subdivided into three substages. The SDS decided in 2006 to accept the proposal (e.g., Bultynck, 2005) to place the base of the Middle Givetian at the base of the *P. rhenanus–varcus* conodont zone. Following the proposal by Aboussalam and Becker (2002), the Upper Givetian includes most of the interval with pharciceratid ammonoid faunas that traditionally were placed in the Upper Devonian; it is defined by the base of the (Lower) *Schmidtognathus hermanni* conodont zone (Becker, 2007a) and coincides with a global eustatic rise. The SDS Givetian substage definitions have not yet been ratified by the ICS. The global Taghanic Event or crisis interval lies at the top of the Middle Givetian (e.g., Aboussalam, 2003; Aboussalam and Becker, 2011; Marshall *et al.* 2011). Two international small-scale bioevents, the Lower and Upper *pumilio* events (Lotmann, 1990), fall in the lower and middle part of the Middle Givetian.

22.1.3. Upper Devonian Series

22.1.3.1. Frasnian

There have been considerable historic differences in the definitions of the base of the Upper Devonian, which hampered international correlations. These disputes are now rendered irrelevant by the selection of a GSSP for the base of the Upper Devonian Series and of the Frasnian Stage at Col du Puech de la Suque in the southern Montagne Noire, France (Feist and Klapper, 1985; Klapper *et al.*, 1987; House *et al.*, 2000a). The section is overturned and the GSSP lies at the base of Bed 42a' of the succession (Figure 22.7). Guides to

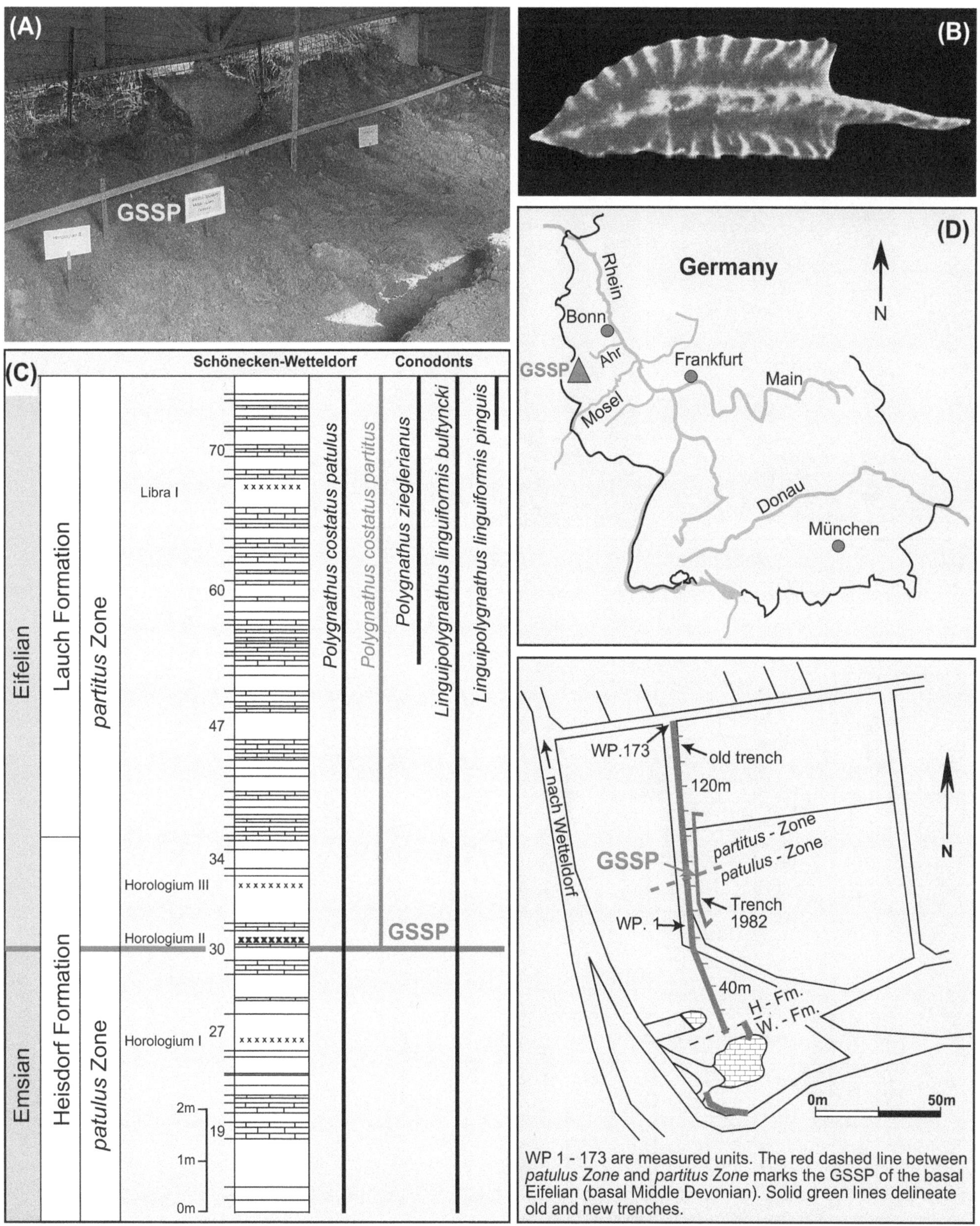

FIGURE 22.5 **The basal-Eifelian (basal-Middle Devonian) GSSP in the trench at Wetteldorf, Eifel Mountains, Germany.** A) Section details within the Happel Hut, with the marked GSSP position (photo by K. Weddige), B) Specimen of the index conodont, *Polygnathus costatus partitus*, from the GSSP bed (photo by K. Weddige), C) Lithological log showing the range of marker conodonts in relation to the GSSP, D) Geographic position of the trench close to Wetteldorf.

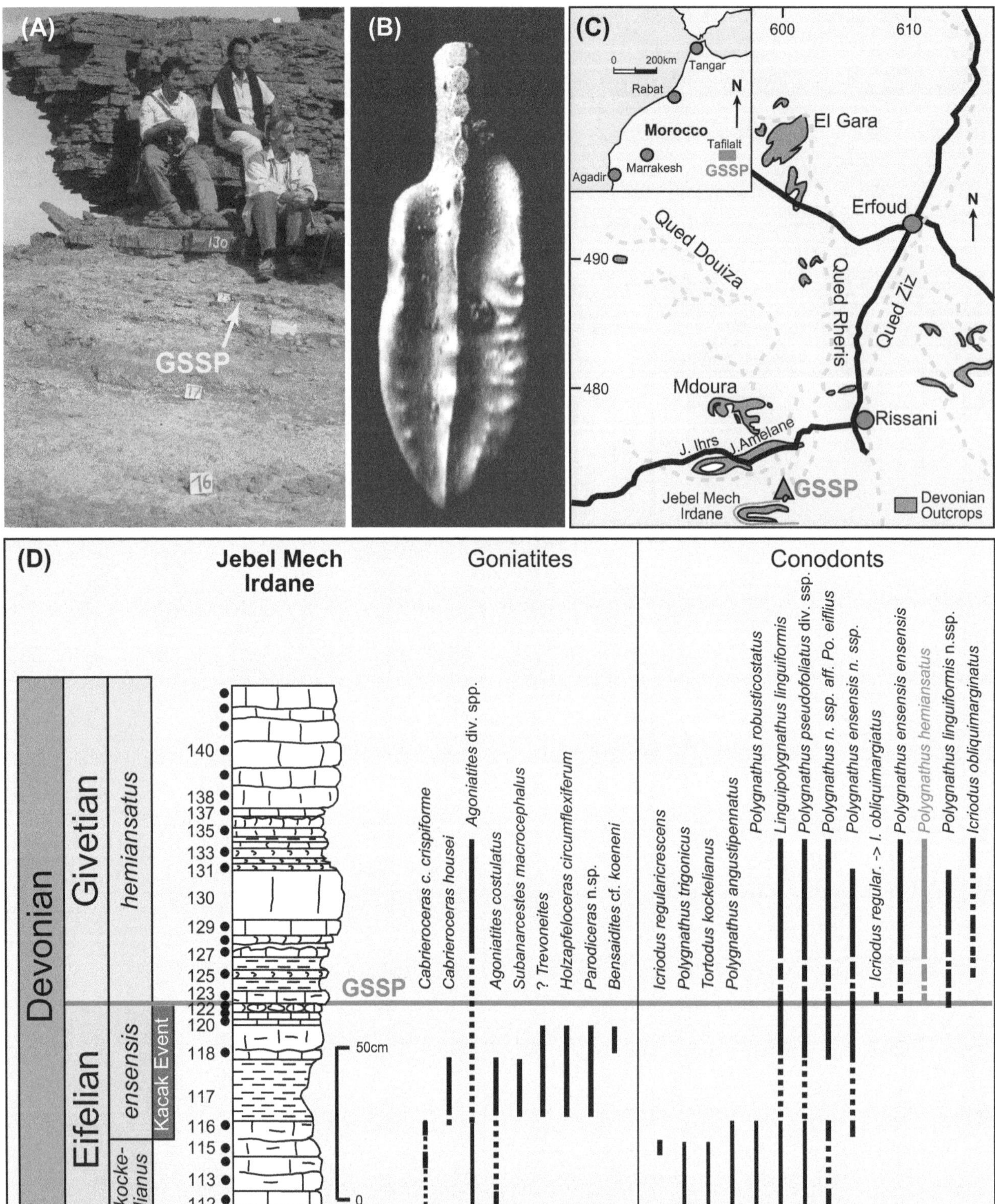

FIGURE 22.6 **The basal-Givetian GSSP in the western part of Jebel Mech Irdane, western Tafilalt, Anti-Atlas, SE Morocco.** A) Section details showing the position of the GSSP. B) Specimen of the index conodont, *Polygnathus hemiansatus*, from the GSSP Bed (Walliser, 2000, Figure 3C3), C) Geographic position of Jebel Mech Irdane SW of Rissani, Tafilalt, D) Lithological log showing the range of marker goniatites and conodonts around the GSSP.

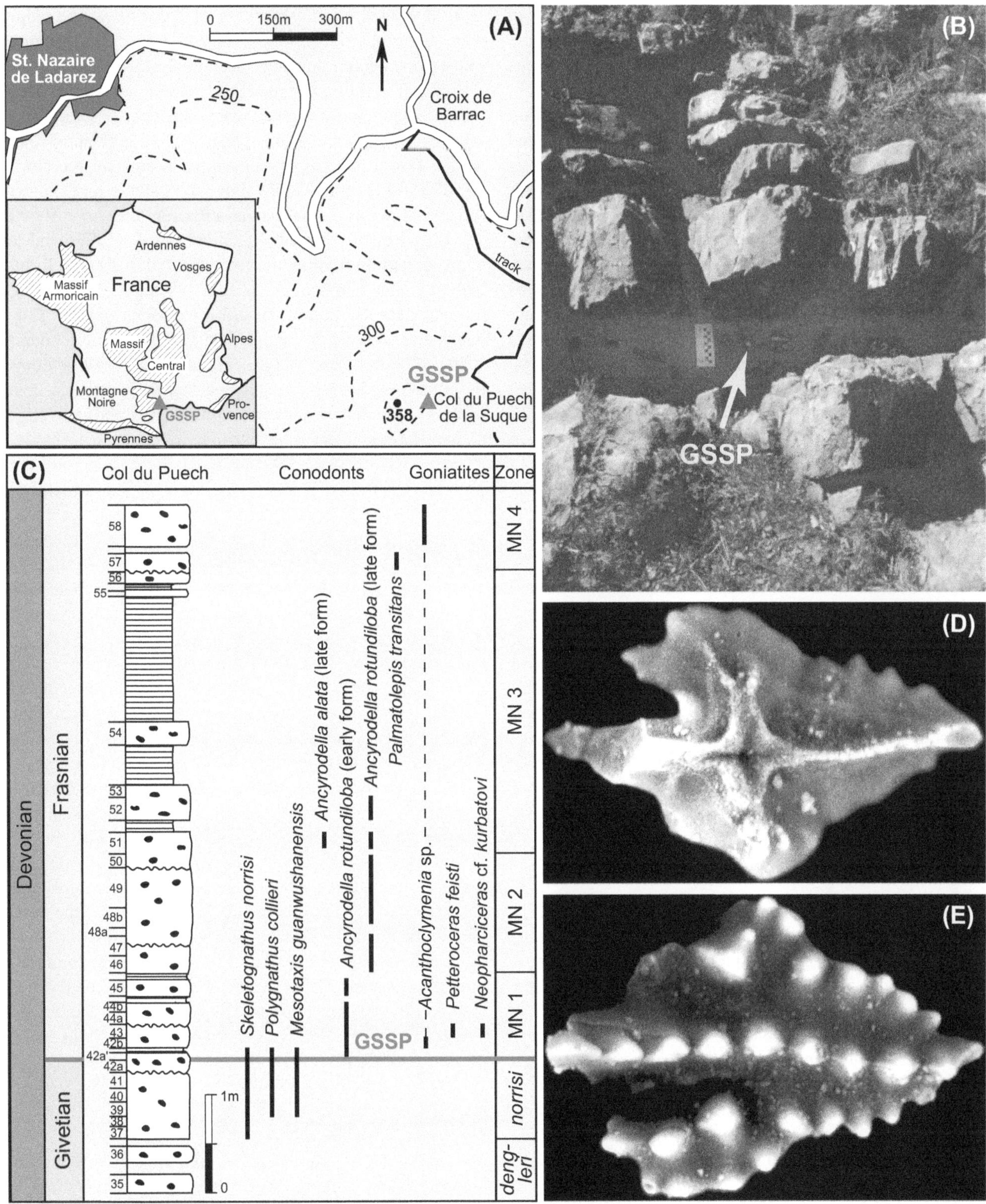

FIGURE 22.7 **The basal Frasnian (basal-Upper Devonian) GSSP at Col du Puech de la Suque, Montagne Noire, southern France.** (A) Geographic position of the GSSP in the Mont Peyroux Nappe SE of St. Nazaire de Ladarez, (B) Section details with the GSSP in a recessive part of the succession (photo by R. Feist), (C) Lithological log with the ranges of marker conodonts and goniatites around the GSSP (with updated taxonomy). (D–E) Representative adult specimen of the index conodont, *Ancyrodella rotundiloba* (early morphotype = *rot. pristina*) from the GSSP (photos by G. Klapper).

the boundary are the first representatives of the conodont genus *Ancyrodella* that are variably named as an early morphotype of *Ad. rotundiloba* (Klapper, 1985), as *Ad. pristina* (e.g., Sandberg *et al.*, 1989), or as *Ad. rotundiloba pristina* (e.g., Aboussalam and Becker, 2007). The taxonomic dispute has been summarized by Klapper (2000a). The GSSP forms the base of the Montagne Noire zone MN 1 (Klapper, 1989), which equals the *rotundiloba pristina* Zone and falls within the lower part of the *Mesotaxis falsiovalis* Zone of Ziegler and Sandberg (1990). The zonal index species of the latter is an invalid synonym of the Chinese *M. guanwushanensis* (Aboussalam and Becker, 2007), which questions its use. In the type area for the naming of the stage (Frasnes, Belgium), the GSSP level almost coincides with the base of the Frasnes Group. The entry of the goniatite genus *Neopharciceras* occurs immediately above the GSSP in Bed 43 (Korn, in House *et al.*, 2000a), and this provides a useful correlation with Asian successions. There are no sufficient brachiopod or palynomorph data that enable an easy correlation into neritic and terrestrial clastic deposits. Some polygnathid conodonts have potential for the correlation into more shallow-water carbonates.

The SDS decided to formally subdivide the Frasnian into three substages (Becker, 2007a). Following a proposal by Becker and House (1999), the base of the Middle Frasnian is placed at the base of the MN 5 or *Palmatolepis punctata* conodont zone, which coincides with the transgressive and hypoxic Middlesex Event, isotopic signals (e.g., Balinski *et al.* 2006), and the first entry of the goniatite *Sandbergeroceras* (Becker *et al.*, 1993). Based on a proposal by Ziegler and Sandberg (1997), the base of the Upper Frasnian coincides with the eustatic *semichatovae* Transgression that allowed the wide spread of the name-giving palmatolepid conodont low in MN Zone 11. At the same time a significant goniatite radiation occurred. Streel and Loboziak (2000) summarize the potential of miospores to follow these levels into the nearshore and terrestrial realm. The ICS has not yet ratified the SDS decisions on Frasnian substages.

22.1.3.2. Famennian

In the past, the base of the Famennian Stage has been placed at different levels, reflecting, in part, problems of correlation between faunal groups. The new definition for the base of the Famennian proposed by the SDS was ratified by the IUGS in 1993. The GSSP is very close to the base of the Famennian as formerly used in the Famenne area in Belgium (Bultynck and Martin, 1995). It lies above the Upper Coumiac Quarry (Figure 22.8), N of Cessenon, Montagne Noire, France (Becker *et al.*, 1989; Klapper *et al.*, 1993; House *et al.*, 2000b). The boundary falls at the junction of the *Palmatolepis linguiformis* (or MN 13c, Girard *et al.*, 2005) and Lower *Pa. triangularis* conodont zones (Klapper, 2000a; zonal revision by Klapper *et al.*, 2004, and Klapper, 2000a). The GSSP does not coincide with the entry (FAD) of *Pa. triangularis*, but with the extinction of Frasnian palmatolepids (especially of *Pa. bogartensis*), ancyrodellids, and ancyrognathids, and the FAD of *Pa. subperlobata* in beds with a sudden flood occurrence of *Pa. ultima*. The Gephuroceratidae and Beloceratidae become extinct at the boundary; most tornoceratid lineages become Lazarus Taxa (Becker, 1993a). The Lower Famennian marker genus *Cheiloceras* enters significantly higher, which suggests a long delay of post-event recovery. At the GSSP section, both anoxic pulses of the Lower and Upper Kellwasser Events are present. The GSSP level is immediately above the latter, a level known to be one of major extinctions in other regions.

Based on a proposal by Streel *et al.* (1998) and after a prolonged discussion, the SDS decided in 2003 to formally subdivide the Famennian into four substages. The position of the substage boundaries is still a matter of debate. The old German classification with Upper Devonian II (Nehdenian), III/IV (Hembergian), V (Dasbergian), and VI (Wocklumian) is not adopted internationally but has been used in pelagic settings of widely distant basins in several continents (e.g., in Australia and in Russian basins). It has been proposed to place the Uppermost Famennian at the base of the Upper *Palmatolepis gracilis expansa* conodont zone (Streel, 2005), which allows an approximation with the widely used but not formally recognized Strunian stage of regional Ardenne chronostratigraphy (Streel *et al.*, 2006). *P. gracilis gonioclymeniae, Bispathodus ultimus*, and *Pseudopolygnathus trigonicus* are the main index species that allow an easy identification of the proposed substage level. Streel (2009) provided an update of Famennian miospore—conodont correlation, and Becker *et al.* (2002) and Becker and House (2009) provided an update of Famennian ammonoid zones and their conodont correlation. The two pulses of the global Condroz Events (Becker, 1993a) ended a long phase of global high sea level ("Nehden-Event") at the end of the Lower Famennian, followed by a short, significant transgression in the Lower *marginifera* Zone. The Enkeberg Event (House, 1985) is an interval of small-scale overturn in pelagic faunas. Three global hypoxic and transgressive events interrupted the overall shallowing trend of the Middle/Upper Famennian, the Lower and Upper *Annulata* Events high in the Upper *P. rugosa trachytera* conodont zone and the biphasic Dasberg Crisis around the boundary of the Lower and Middle *expansa* zones (Becker, 1993a; Becker *et al.*, 2004a; Hartenfels *et al.*, 2009; Hartenfels and Becker, 2009; Racka *et al.* 2010).

22.1.3.3. Base of the Carboniferous

Following the recommendations of a special working group on the boundary, the IUGS, in 1989, accepted a GSSP for the Devonian—Carboniferous boundary at La Serre (Figure 22.9), near Cabrières, Montagne Noire, southern France (Paproth *et al.*, 1991; Feist *et al.*, 2000). The current GSSP guide is the

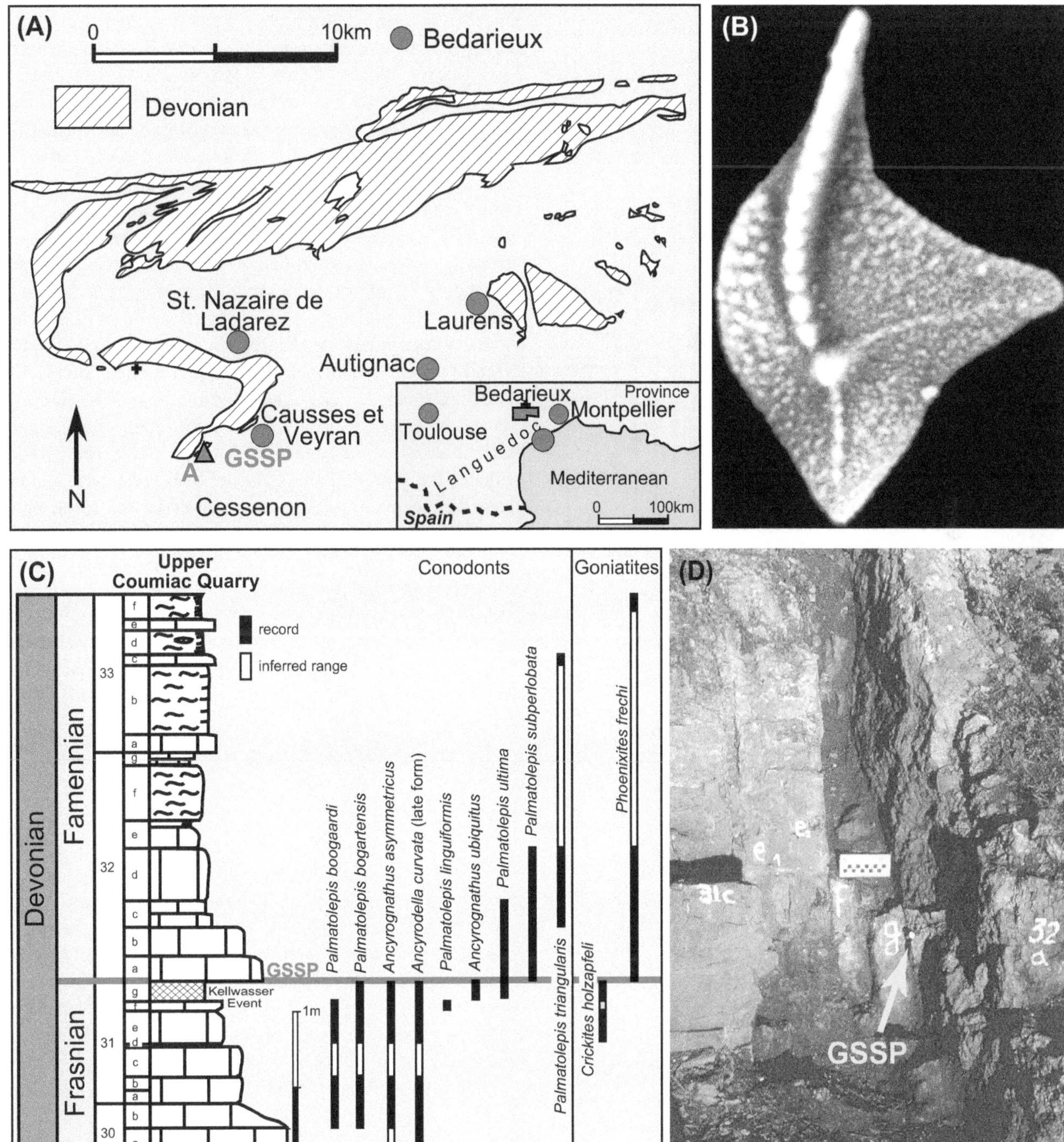

FIGURE 22.8 **The basal-Famennian GSSP in the Upper Quarry at Coumiac, Montagne Noire, southern France.** A) Geographic position of the GSSP N of Cessenon, B) Typical specimen of *Palmatolepis ultima* from the GSSP Bed (photo by G. Klapper) just above the Upper Kellwasser level, C) Lithological log with the ranges of marker conodonts and goniatites around the GSSP (based on combined records of Klapper *et al.* 1993, Schülke, 1995, House *et al.*, 2000b, and Girard *et al.*, 2005), D) Outcrop of the GSSP section at Upper Coumiac Quarry.

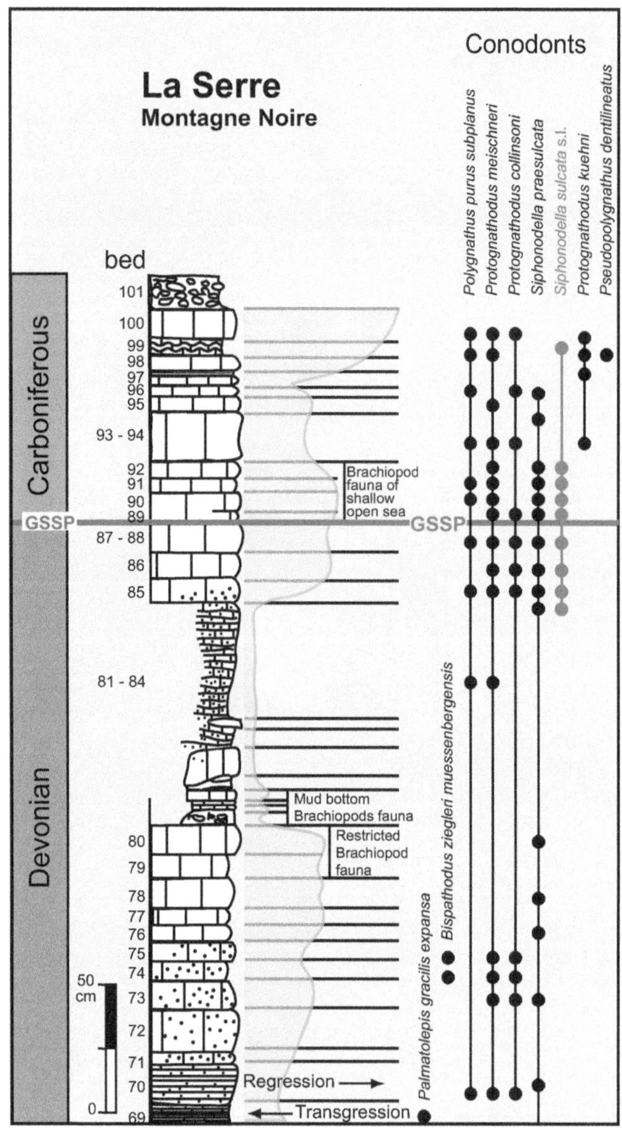

FIGURE 22.9 **Log of the current Devonian–Carboniferous boundary at La Serre, Trench E, stratotype section in the Montagne Noire, southern France.** More section details are in Figure 23.2 of Chapter 23.

entry of the conodont *Siphonodella sulcata* in a supposed gradual transition from *S. praesulcata* (Flajs and Feist, 1988), but the taxonomy of the group is problematic (Kaiser and Corradini, 2011), and the type of the index species is poorly known, much younger, and lost. Kaiser (2009) confirmed that *S. sulcata* morphotypes identical to those that were used to define the GSSP occur much lower than previously known at La Serre (Ziegler and Sandberg, 1996), and the assumed *S. praesulcata–sulcata* lineage is not preserved in mostly reworked faunas from oolites. The current GSSP level lies high in the *sulcata* Zone and cannot be correlated with precision into any other section but most likely with a level above the entry of the classical Carboniferous goniatite *Gattendorfia* elsewhere. Hence, a GSSP revision is inevitable.

The Devonian–Carboniferous transition is well-defined palynologically by the last appearance of *Retispora lepidophyta*, which disappears at the LN/VI zone boundary just beneath the first occurrence of *Siphonodella sulcata* in the Hasselbachtal auxiliary stratotype section (Becker and Paproth, 1993; Becker, 1996) in Germany (Higgs *et al.*, 1993). A second auxiliary stratotype was designated in more neritic facies at Nanbiancun, NW of Guilin, Guangxi Province, South China (Yu, 1988) but there is also a dispute concerning its conodont stratigraphy and correlation with the La Serre GSSP (e.g., Ji, 1989; Gong *et al.*, 1991). A first-order global mass extinction, especially of goniatites, clymeniids, trilobites, corals, and stromatoporoids, occurred below the boundary and coincided with the initial anoxic pulse (main Hangenberg Event) of the prolonged Hangenberg Crisis, and with the onset of a positive carbon isotope excursion (Kaiser *et al.*, 2006, 2011). The short-termed but significant main glacial phase (e.g., Streel *et al.*, 2000a; Caputo *et al.*, 2008) and major sea-level fall (Hangenberg Regression) followed the mass extinction and black shale event. Wicander *et al.* (2011), however, provided evidence for several glacial episodes in Bolivia, with an initial phase that seems to predate the mass extinction level. Spores from the event interval (LN miospore zone) are partly abnormal (Filipiak and Racki 2010). Subsequently, the post-glacial topmost Famennian transgression and peak of isotopic excursion (Cramer *et al.*, 2008) in the *Protognathodus kockeli* Zone (formerly Upper *praesulcata* Zone) saw the initial radiation of Carboniferous-type ostracods, conodonts, ammonoids, trilobites, brachiopods, and corals (e.g., Korn *et al.*, 1994).

22.2. DEVONIAN STRATIGRAPHY

22.2.1. Biostratigraphy

There are refined zonations in pelagic facies using especially ammonoids, conodonts, entomozoid ostracods, dacryoconarids, and for the Lower Devonian, monograptid graptolites. The acritarch zonation is less detailed. For neritic facies, brachiopods, trilobites, and ostracods are regionally important, but faunas tend to have many endemic characteristics. Chitinozoans occur both in the pelagic and in the neritic facies. In the terrestrial facies, miospores, macroplants, and fish are stratigraphically useful. Many problems still remain on the correlation of the refined pelagic zonations with those of neritic and terrestrial facies (e.g., Becker and Kirchgasser, 2007). In near-shore facies, miospores provide an important correlation tool. A listing of zones and their general correlation is given in Figures 22.10 and 22.11, and is elaborated in Weddige (1996) and Bultynck (2000b).

22.2.1.1. Conodont Zonations

Fundamental studies by Bischoff and Ziegler (1957) and Ziegler (1962) led to a very refined conodont zonation of considerable value in Devonian correlation. The terminology

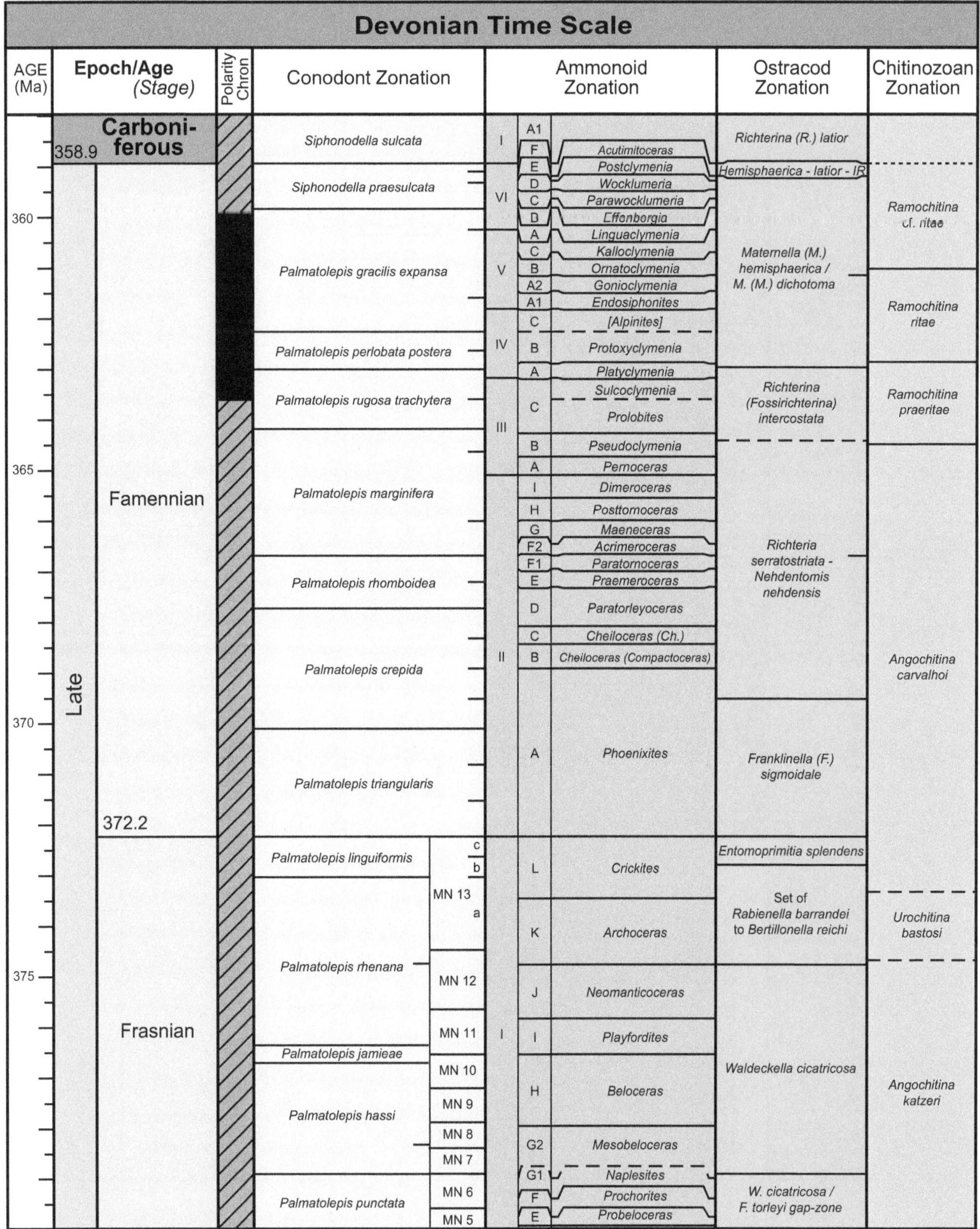

FIGURE 22.10 **Devonian chronostratigraphy and major marine biostratigraphy**. Biostratigraphic scales include conodonts, ammonoid zonation with selected guide taxa, graptolites (in the Lochkovian/Pragian), entomozoid ostracods, dacryoconarids and chitinozoans. Details are given in the text, and in House (2002) and Bultynck (2000a,b).

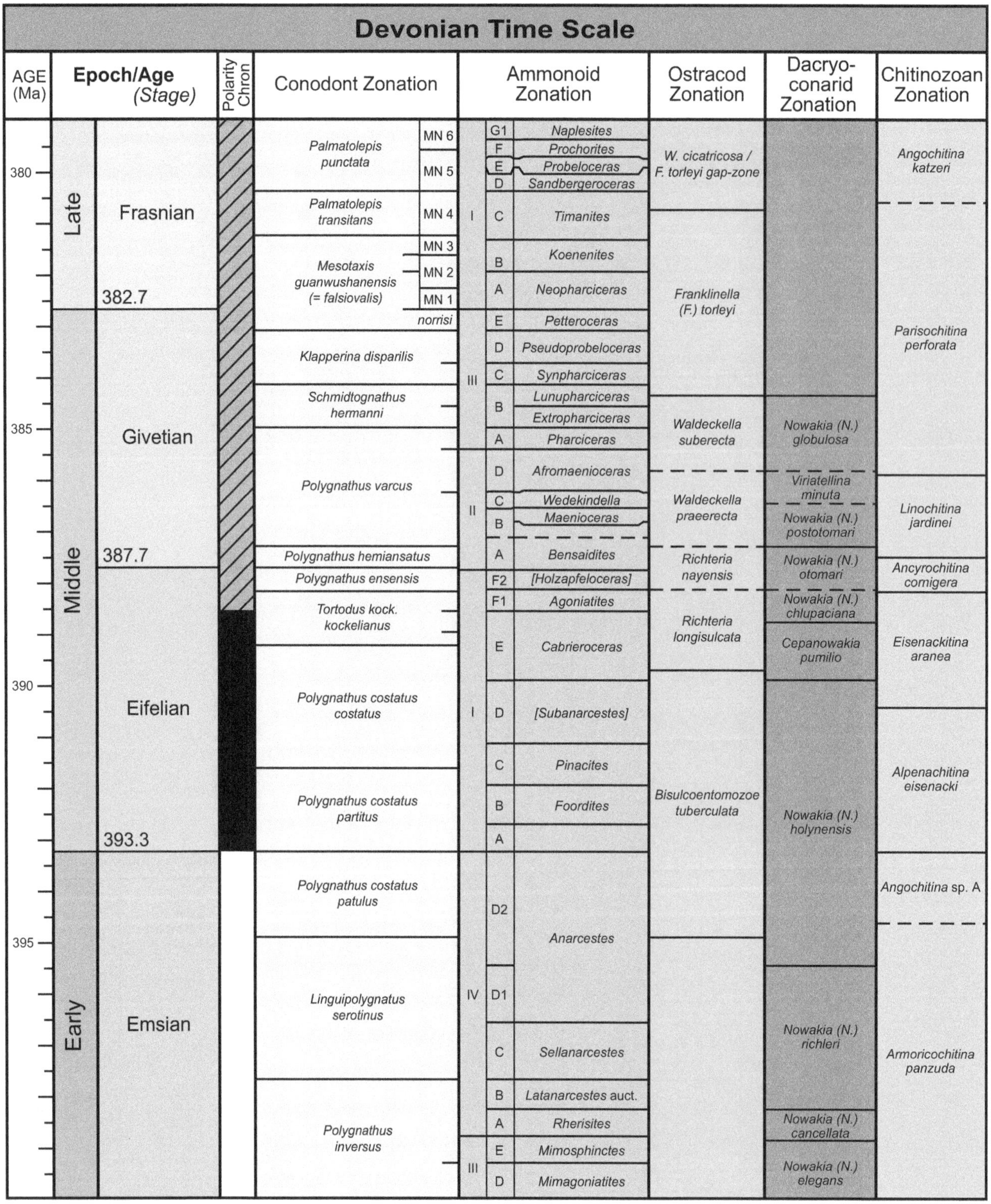

Devonian Time Scale

AGE (Ma)	Epoch/Age *(Stage)*	Polarity Chron	Conodont Zonation	Ammonoid Zonation	Ostracod Zonation	Dacryo-conarid Zonation	Chitinozoan Zonation

AGE (Ma): 380, 385, 390, 395

Epoch/Age (Stage):
- Late: Frasnian (base 382.7)
- Middle: Givetian (base 387.7); Eifelian (base 393.3)
- Early: Emsian

Conodont Zonation (top to bottom):
- *Palmatolepis punctata* — MN 6, MN 5
- *Palmatolepis transitans* — MN 4
- *Mesotaxis guanwushanensis* (= *falsiovalis*) — MN 3, MN 2, MN 1
- *norrisi*
- *Klapperina disparilis*
- *Schmidtognathus hermanni*
- *Polygnathus varcus*
- *Polygnathus hemiansatus*
- *Polygnathus ensensis*
- *Tortodus kock. kockelianus*
- *Polygnathus costatus costatus*
- *Polygnathus costatus partitus*
- *Polygnathus costatus patulus*
- *Linguipolygnatus serotinus*
- *Polygnathus inversus*

Ammonoid Zonation (top to bottom):
- I: G1 *Naplesites*; F *Prochorites*; E *Probeloceras*; D *Sandbergeroceras*; C *Timanites*; B *Koenenites*; A *Neopharciceras*
- III: E *Petteroceras*; D *Pseudoprobeloceras*; C *Synpharciceras*; B *Lunupharciceras*, *Extropharciceras*; A *Pharciceras*
- II: D *Afromaenioceras*; C *Wedekindella*; B *Maenioceras*; A *Bensaidites*
- I: F2 *[Holzapfeloceras]*; F1 *Agoniatites*; E *Cabrieroceras*; D *[Subanarcestes]*; C *Pinacites*; B *Foordites*; A
- IV: D2, D1 *Anarcestes*; C *Sellanarcestes*; B *Latanarcestes* auct.; A *Rherisites*
- III: E *Mimosphinctes*; D *Mimagoniatites*

Ostracod Zonation
- *W. cicatricosa / F. torleyi* gap-zone
- *Franklinella (F.) torleyi*
- *Waldeckella suberecta*
- *Waldeckella praeerecta*
- *Richteria nayensis*
- *Richteria longisulcata*
- *Bisulcoentomozoe tuberculata*

Dacryoconarid Zonation (top to bottom):
- *Nowakia (N.) globulosa*
- *Viriatellina minuta*
- *Nowakia (N.) postotomari*
- *Nowakia (N.) otomari*
- *Nowakia (N.) chlupaciana*
- *Cepanowakia pumilio*
- *Nowakia (N.) holynensis*
- *Nowakia (N.) richleri*
- *Nowakia (N.) cancellata*
- *Nowakia (N.) elegans*

Chitinozoan Zonation (top to bottom):
- *Angochitina katzeri*
- *Parisochitina perforata*
- *Linochitina jardinei*
- *Ancyrochitina cornigera*
- *Eisenackitina aranea*
- *Alpenachitina eisenacki*
- *Angochitina* sp. A
- *Armoricochitina panzuda*

FIGURE 22.10 *(Continued)*.

Devonian Time Scale

AGE (Ma)	Epoch/Age (Stage)	Polarity Chron	Conodont Zonation	Ammonoid Zonation	Graptolite Zonation	Dacryoconarid Zonation	Chitinozoan Zonation
400	Early – Emsian		Polygnathus inversus	III – E Mimosphinctes; D Mimagoniatites		Nowakia (N.) elegans	Armoricochitina panzuda
			Eocostapolygnathus nothoperbonus	C Anetoceras		Nowakia (N.) barrandei	
			Eocostapolygnathus gronbergi	B Metabactrites		Nowakia (Dmitriella) praecursor	
405			Eocostapolygnathus excavatus	A Bactrites		Nowakia (N.) zlichovensis	
			Eocostapolygnathus kitabicus		Neomonograptus pacificus	Guerichina strangulata	Bursach. bursa
	407.6 – Pragian		Eocostapolygnathus pireneae	II	Neomonograptus yukonensis	Nowakia (Turkestanella) acuaria acuaria	Bulbochitina bulbosa
410			Gondwania kindlei		Neomonograptus thomasi		Angochitina caeciliae - Ang. comosa
	410.8 – Lochkovian		Gondwania irregularis		Neomonograptus falcarius	Styliacus bedbouceki	
			Pedavis gilberti		Neomonograptus hercynicus	Paranowakia intermedia	Urochitina simplex
			Masaraella pandora morpho. Beta			Homoctenowakia bohemica	
			Ancyrodelloides trigonicus			Homoctenowakia senex	
			Lanea transitans		Monograptus praehercynicus		Fungochitina lata
			Lanea eleanorae				
415			Lanea omoalpha	I			
			Caudicriodus postwoschmidti		Monograptus uniformis		Eisenackitina bohemica
			Caudicriodus hesperius				
	419.2 – Silurian		Delotaxis detorta		Monograptus transgrediens, M. bouceki-perneri; Monograptus lochkovensis		

FIGURE 22.10 (*Continued*).

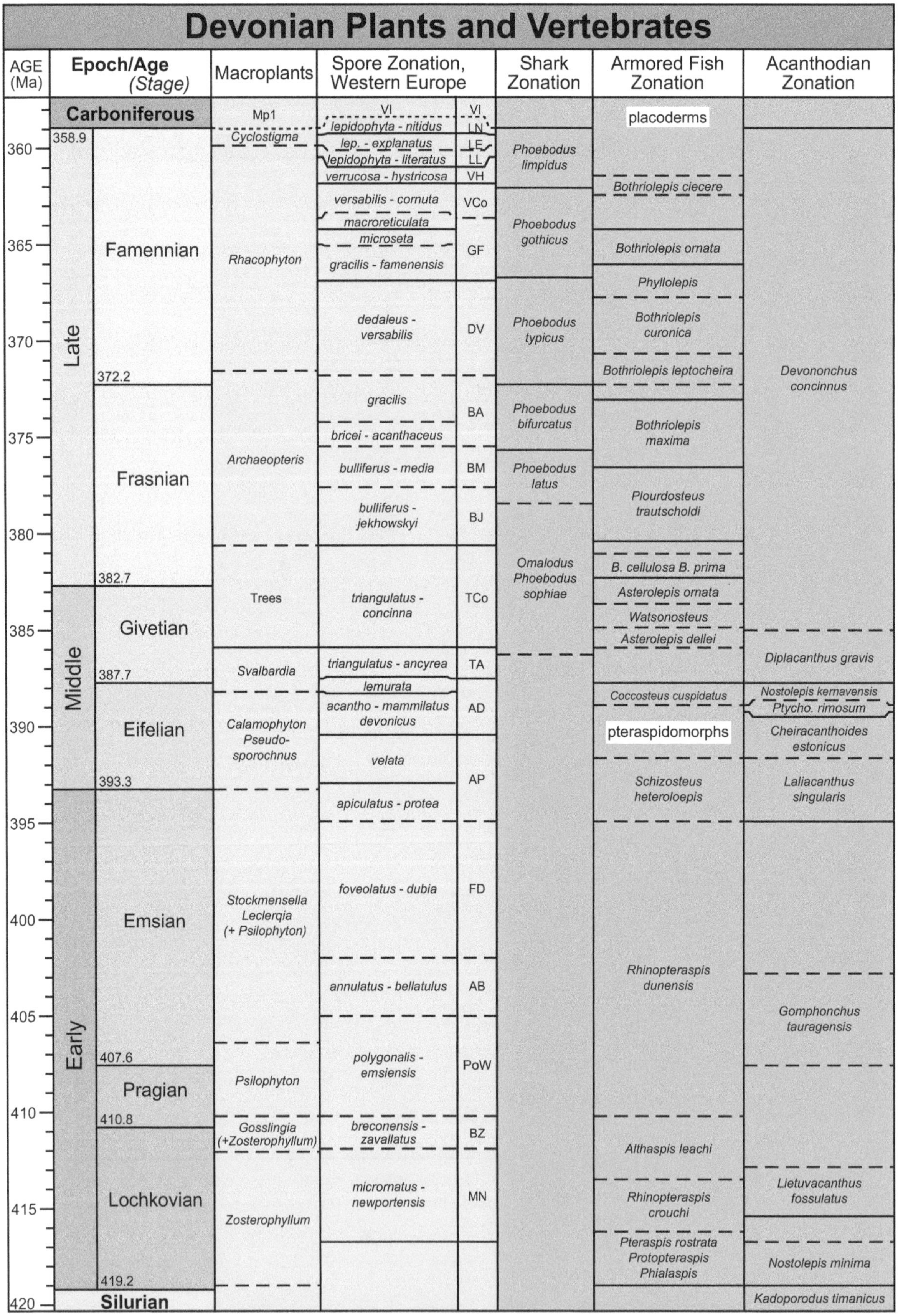

Devonian Plants and Vertebrates

AGE (Ma) | Epoch/Age (Stage) | Macroplants | Spore Zonation, Western Europe | Shark Zonation | Armored Fish Zonation | Acanthodian Zonation

AGE (Ma): 360, 365, 370, 375, 380, 385, 390, 395, 400, 405, 410, 415, 420

Epoch/Age (Stage)
- Carboniferous
- Late: Famennian (358.9), Frasnian (372.2); 382.7
- Middle: Givetian (382.7), Eifelian (387.7); 393.3
- Early: Emsian (393.3), Pragian (407.6), Lochkovian (410.8); 419.2
- Silurian

Macroplants
- Mp1
- Cyclostigma
- Rhacophyton
- Archaeopteris
- Trees
- Svalbardia
- Calamophyton Pseudo-sporochnus
- Stockmensella Leclerqia (+ Psilophyton)
- Psilophyton
- Gosslingia (+Zosterophyllum)
- Zosterophyllum

Spore Zonation, Western Europe
- VI — VI
- lepidophyta - nitidus — LN
- lep. - explanatus — LE
- lepidophyta - literatus — LL
- verrucosa - hystricosa — VH
- versabilis - cornuta — VCo
- macroreticulata
- microseta; gracilis - famenensis — GF
- dedaleus - versabilis — DV
- gracilis; bricei - acanthaceus — BA
- bulliferus - media — BM
- bulliferus - jekhowskyi — BJ
- triangulatus - concinna — TCo
- triangulatus - ancyrea — TA
- lemurata; acantho - mammilatus devonicus — AD
- velata; apiculatus - protea — AP
- foveolatus - dubia — FD
- annulatus - bellatulus — AB
- polygonalis - emsiensis — PoW
- breconensis - zavallatus — BZ
- micrornatus - newportensis — MN

Shark Zonation
- Phoebodus limpidus
- Phoebodus gothicus
- Phoebodus typicus
- Phoebodus bifurcatus
- Phoebodus latus
- Omalodus Phoebodus sophiae

Armored Fish Zonation
- placoderms
- Bothriolepis ciecere
- Bothriolepis ornata
- Phyllolepis
- Bothriolepis curonica
- Bothriolepis leptocheira
- Bothriolepis maxima
- Plourdosteus trautscholdi
- B. cellulosa B. prima
- Asterolepis ornata
- Watsonosteus
- Asterolepis dellei
- Coccosteus cuspidatus
- pteraspidomorphs
- Schizosteus heteroloepis
- Rhinopteraspis dunensis
- Althaspis leachi
- Rhinopteraspis crouchi
- Pteraspis rostrata Protopteraspis Phialaspis

Acanthodian Zonation
- Devononchus concinnus
- Diplacanthus gravis
- Nostolepis kernavensis
- Ptycho. rimosum
- Cheiracanthoides estonicus
- Laliacanthus singularis
- Gomphonchus tauragensis
- Lietuvacanthus fossulatus
- Nostolepis minima
- Kadoporodus timanicus

FIGURE 22.11 **Devonian terrestrial facies zonations for macroplants and spores, and the zonations for sharks, armored fish and Acanthodians (spiny fish).** For details see text.

for the Upper Devonian has been revised by Sandberg and Ziegler (1973) and Ziegler and Sandberg (1984, 1990). Important revisions for the Lower Devonian were published by Valenzuela-Ríos and Murphy (1997), Murphy and Valenzuela-Ríos (1999), Slavík (2004a,b), Slavík and Hladil (2004), Murphy (2005), Slavík *et al.* (2007), and Yolkin *et al.* (2011). The Givetian zonation was revised and further developed by Bultynck (1985, 1987), Aboussalam (2003), Aboussalam and Becker (2007), and Narkiewicz and Bultynck (2010). A Frasnian zonation, based originally on the Montagne Noire (southern France) succession, (Klapper, 1989, 2000b), is more detailed then the so-called "standard zonation", and can be applied globally. Its correlation with the zonation of Ziegler and Sandberg (1990) is given in Klapper and Becker (1999). Important additional data for the Frasnian can be found in Klapper *et al.* (1996), Klapper (2007b), and Ovnatanova and Kononova (2008); the latter monograph enables the correlation into more shallow-water facies dominated by polygnathids. Schülke (1999), Klapper *et al.* (2004), Girard *et al.* (2005), and Klapper (2007a) improved the precise conodont biostratigraphy around the Frasnian–Famennian boundary. Famennian zonation problems are emphasized by Corradini (2008), Hartenfels *et al.* (2009), Hartenfels and Becker (2009), and Hartenfels (2011), with emphasis on the classical *Scaphignathus velifer* and *Polygnathus styriacus* zones. Changes near the Devonian–Carboniferous boundary are proposed by Kaiser *et al.* (2009). The updated zonation scheme is shown in Figure 22.11. Apparent zone subdivisions (e.g., Lower *expansa* Zone) are full zones, not subzones. Principal problems of "standard zonations" were recently discussed by Bultynck (2007). Significant progress in graphic correlation (e.g., Klapper *et al.*, 1995; Klapper 1997; Belka *et al.* 1997; Gouwy and Bultynck 2000, 2003; Gouwy *et al.* 2007) allows the establishment of the complete ranges and reliability of many conodont taxa. Bultynck (2003) summarized the potential of icriodids for alternative zonations, especially in neritic settings.

22.2.1.2. Ammonoid Zonations

Coiled ammonoids appeared low in the Emsian of its classical sense (e.g., Klug, 2001), above a level with first abundant bactritids (Klug *et al.*, 2008). The Upper Devonian zonation compiled by Wedekind in 1917 established that goniatites and clymeniids provide an important correlation tool in marine, and especially pelagic, facies. Much further precision has been added: for the Lower Devonian by Chlupáč and Turek (1983), Becker and House (1994), Ruan (1996), Klug (2001, 2002), De Baets *et al.* (2010), Becker *et al.* (2010), Becker and Aboussalam (2011), and Ebbighausen *et al.* (2011); for the Middle Devonian by House (1978), House *et al.* (1985), Becker and House (1994), Klug (2002), Becker *et al.* (2004b), and Becker (2005); for the Upper Devonian (Frasnian) by Matern (1931), House and Ziegler (1977), Becker *et al.* (1993, 2000), Becker and House (1994), House and Kirchgasser (1993, 2008), and others; and for the Famennian by the work of Lange (1929), Schindewolf (1937), Korn and Luppold (1987), Clausen *et al.* (1989), Becker (1992, 1993a,b), Korn *et al.* (1994, 2000), Korn (1999, 2004), Becker *et al.* (2002), Nikolaeva (2007), Hartenfels and Becker (2009), and others. A review of zones was published by Becker and House (2000), who extended the zonal scheme of Wedekind, using Latin numbering into the Middle and Lower Devonian. The zonation, recognizing updates in Becker and House (2009), is shown on Figure 22.10.

22.2.1.3. Ostracod Zonation

The development of pelagic ostracods in the Devonian has contributed a very detailed biostratigraphic tool (Figure 22.10). Major initial studies were conducted by Rabien (1954) and Blumenstengel (1965). Benthic ostracods are excellent indicators of ecotypes, but a global biozonation has not been established, although the evolution of several groups (e.g., polyzygiids) provides time markers that can be widely used. The biostratigraphic role of ostracods in the Devonian has been reviewed by Groos-Uffenorde *et al.* (2000).

22.2.1.4. Dacryoconarid Zonation

The use of planktonic tentaculitids for stratigraphy has resulted in the recognition of their importance for biostratigraphy. Major contributions result from the work of Alberti (e.g., 1982, 1998, 2000). In the Lower Devonian, taxonomic advance is required to solve correlation problems and to improve the base for the revision and subdivision of the Pragian and Emsian. Kim (2011) provided new data for the Emsian stratotype region. For the Middle Devonian and Frasnian, ranges of stratigraphically important forms are summarized in Liashenko (1967) and Sauerland (1983).

22.2.1.5. Radiolarian Zonation

Radiolaria may be abundant in Devonian deeper-water, outer shelf to oceanic siliceous sediments but their zonation is still in a stage of development (e.g., Holdsworth and Jones, 1980; Schwartzapfel and Holdsworth, 1996; Aitchison *et al.*, 1999; Wang *et al.*, 2000) with some focus in the Upper Devonian. Especially useful for correlation are recently described joint radiolarian–conodont assemblages (e.g., Zhang *et al.*, 2008; Obut and Shcherbanenko, 2008; Izokh *et al.*, 2008), for example at the Frasnian–Famennian boundary.

22.2.1.6. Palynomorph Zonations

Several basic spore zonations have been proposed. A scheme by Streel *et al.* (1987) used lettered abbreviations, while that of Richardson and McGregor (1986) used a more standard form (Figure 22.11). For the Lower Devonian the work of Steemans (1989) is fundamental. An analysis of spore records in relation to the conodont zonation of eastern Europe is given by Avkhimovitch *et al.* (1993). The miospore zonation in relation to GSSP boundaries has been treated in detail by Mark-Kurik *et al.* (1999) and Streel *et al.* (2000b); the Frasnian–Famennian boundary is reviewed by Streel *et al.* (2000a) and Obukhovskaya *et al.* (2000). Matyja and Turnau (2008) provided a Middle Devonian miospore–conodont correlation, which was extended by Turnau and Narkiewicz (2011). Significant advances in the study of miospores from South America (e.g., Melo and Loboziak, 2003) allow a floral correlation with the Euramerican zonation (Loboziak and Melo, 2002). The fast plant radiation in the Upper Devonian and around the Devonian–Carboniferous boundary provides an especially detailed zonation (revised in Maziane *et al.*, 1999), which has been correlated with the conodont succession (Streel, 2009).

Acritarchs are organic-walled microfossils of unknown and probably varied biological affinities, but their stratigraphical use, especially in the Middle and Upper Devonian has increased. At present, the current zonation falls behind that of some other groups in resolving power. Reviews of zonations are given by Molyneux *et al.* (1996) and Le Hérissé *et al.* (2000).

Progress towards a global Devonian chitinozoan zonation was last summarized by Paris *et al.* (2000). Based on detailed studies in South America (e.g., Grahn, 2003; Grahn and Melo, 2004), it has been significantly improved in subsequent years and, at least in western Gondwana, can be correlated with the miospore zonation (Grahn, 2005; Grahn *et al.*, 2010).

22.2.1.7. Plant Megafossil Zonation

Vascular plants, which began well before the Devonian, rise in dominance during the period and the oldest forests occur in the Upper Givetian. A broad division of the Devonian into seven zones was suggested by Banks (1980) and this has been further refined by Edwards *et al.* (2000), whose zonation, extending Banks' numbers, is shown in Figure 22.11.

22.2.1.8. Vertebrate Zonations

The Devonian is marked by the evolution of tetrapods; vertebrates with limbs and digits, which are only of very broad stratigraphical use due to the rarity of specimens and relative isolation of most known genera. Tetrapod trackways have recently been discovered in the Eifelian of Poland (Niedzwiedzki *et al.*, 2010). The earliest body fossils are known from the Frasnian of Latvia, Scotland, and China (Clack, 2002; latest global review in Blieck *et al.*, 2007, 2010). During the Devonian these vertebrates are in many ways fish-like and aquatic, and fully terrestrial tetrapods are not known before the earliest Carboniferous (Tournaisian).

In terrestrial facies, fish are helpful in age determination (Figure 22.11). Several zonations have developed using various fish groups and teeth and other micro-vertebrate remains. Major zonations include those for thelodonts and heterostracans and for placoderm and acanthodian fish, and are of importance for the difficult problem of marine to non-marine correlation. A summary of achievements in Paleozoic vertebrate biostratigraphy was published by Blieck and Turner (2000). In open shelf environments, shark teeth assemblages give a widely applicable zonation (e.g., Ginter and Ivanov, 2000; Ginter *et al.*, 2002), which, however, is less detailed than that of the co-occurring conodonts or ammonoids.

22.2.2. Physical Stratigraphy

22.2.2.1. Extinction and Hypoxic Event Stratigraphy

Several time-specific facies have been recognized in the Devonian and often coincide with extinction events. Many of these are associated with contemporaneous dark limestone or shale episodes in widely separated basins of different continents and are interpreted as global hypoxic/anoxic events. The events are generally named after type sections or lithostratigraphic units showing characteristic facies development (Figure 22.12). General reviews of these episodes include those by Walliser (1996: with a different event terminology based on index taxa) and House (1985, 2002). Manda and Frýda (2010) explored extinctions around the Silurian/Devonian boundary. Emsian events were investigated by García-Alcalde (1997), Klug *et al.* (2008), and Becker and Aboussalam (2011); the lower Eifelian Choteč Event by Koptíková *et al.* (2008), Koptíková (2011) and Elrick *et al.* (2009); the upper Eifelian Kačák Events by Budil (1995), House (1996), Schöne (1997), Marshall *et al.* (2007), DeSantis *et al.* (2007), Ellwood *et al.* (2011a), and Troth *et al.* (2011); the Givetian *pumilio* Events by Lottmann (1990); the Taghanic Crisis by Aboussalam (2003), Baird and Brett (2008), and Aboussalam and Becker (2011); the Frasnes Events by Becker and Aboussalam (2004) and Aboussalam and Becker (2007); events around the Lower/Middle Frasnian boundary in thematic volumes by Balinski *et al.* (2006) and Racki *et al.* (2008) as well as by Sliwinski *et al.* (2011); the Upper Frasnian Kellwasser Crisis by, amongst many others, Buggisch (1972, 1991), Sandberg *et al.* (1988), Schindler (1990, 1993), Joachimski and Buggisch (1993), Becker and House (1994), Racki (1999, 2005), Murphy *et al.* (2000), Morrow (2000), Joachimski *et al.* (2001), Ma and Bai (2002), Racki and House (2002), Balinski *et al.* (2002), Becker (2006), Hartkopf-Fröder *et al.* (2007), and Riquier *et al.* (2007); the *Annulata* Events by Becker (1992), Becker *et al.* (2004a), Korn (2004), Hartenfels *et al.* (2009), and Racka *et al.* (2010); the Dasberg Crisis by Hartenfels and Becker (2009), Marynowski *et al.* (2010), and Myrow *et al.* (2011); and the Hangenberg Events by Bless *et al.* (1993), Becker (1996), Caplan and Bustin (1999), Streel *et al.* (2000b),

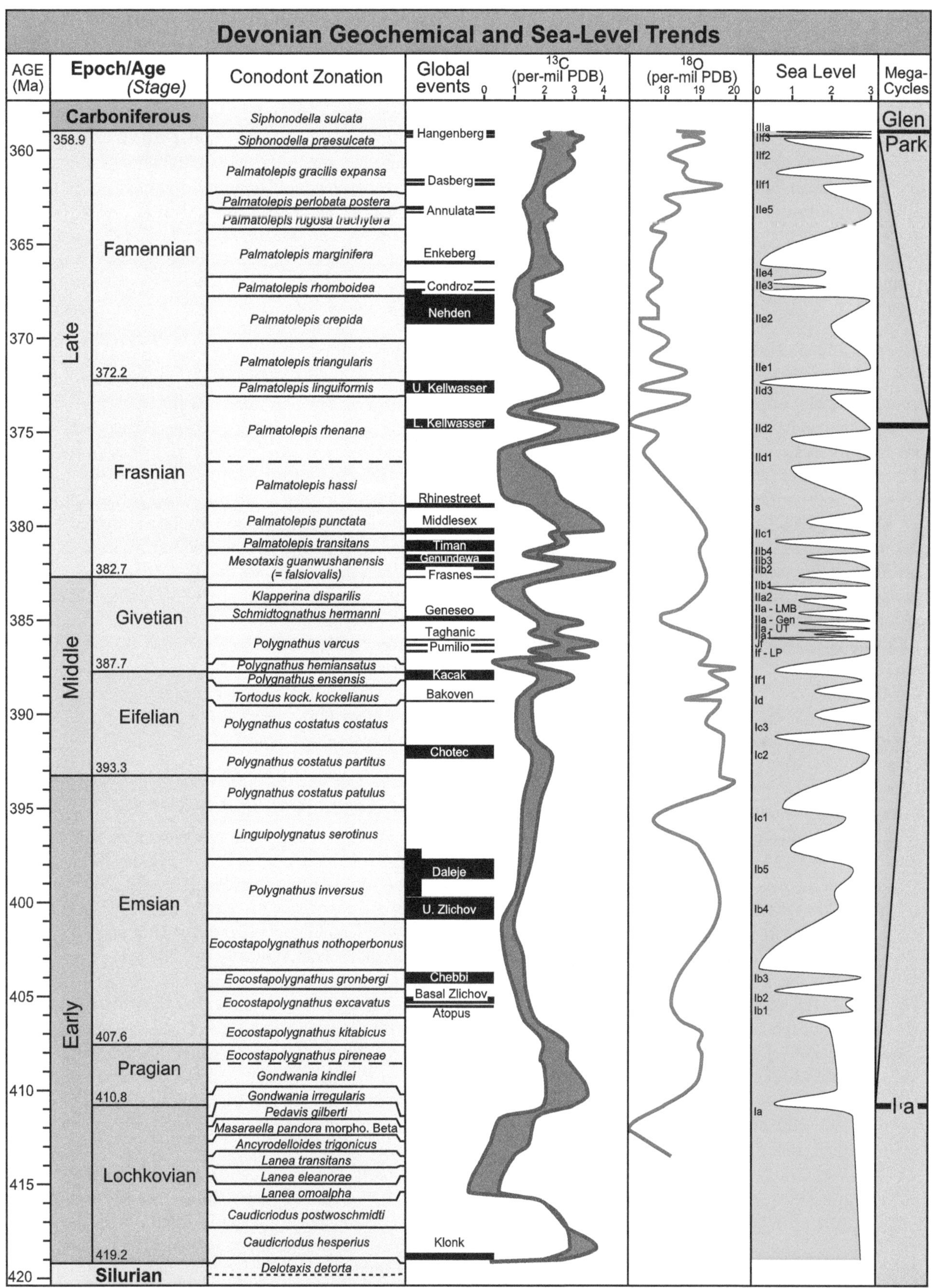

FIGURE 22.12 Generalized isotope (carbon, oxygen, and strontium) stratigraphy in the Devonian Period based on Denison *et al.* (1997), Aboussalam (2003), Buggisch and Mann (2004), Buggisch and Joachimski (2006), Joachimski *et al.* (2004), and Kaiser *et al.* (2006, 2008). Also shown are the standard conodont zonation (on the left), and the relative global sea-level trends and cycles (on the right).

Marynowski and Filipiak (2007), Brezinski *et al.* (2010), Filipiak and Racki (2010), and Kaiser *et al.* (2006, 2008, 2009).

All these include pulses of rapid sea-level changes and, using sequence stratigraphy, can be used for correlation across facies barriers. Although much emphasized in numerous publications, the Upper Kellwasser Event (Frasnian—Famennian boundary) may not be the greatest extinction event at invertebrate family level in the Devonian. The Taghanic, Hangenberg, and, perhaps, Kačák Events are almost similar in their geographic extent, and the Hangenberg mass extinction was at least as severe as the Upper Kellwasser extinction.

22.2.2.2. Cyclostratigraphy

Numerous publications deal with Devonian cyclic sediments but few case studies have tried to use these to improve and calibrate the time scale. Cycles in thick carbonate platforms (e.g., Chen and Tucker, 2003) or mostly clastic successions (Clark *et al.*, 2009) often lack sufficient biostratigraphic control. Cyclic pelagic carbonates and carbonate—shale alternations offer the best correlation and tuning potential due to their calm, relative complete deposition and conodont and/or ammonoid content. Milankovitch orbital cyclicity has been applied for determining times scales in the Devonian for the Lower Devonian (Chlupáč, 2000), Givetian (6.5 myr duration; see House, 1995), Givetian to Frasnian (Bai, 1995), and Famennian (Bai *et al.*, 1995). In neither case has detailed harmonic analysis been applied to sections, but the presumption has been made that dominant cycles may be precession or eccentricity cycles. The inadequacies of the radiometric scale make checking this difficult. One result is to demonstrate that conodont zones are not of equal duration.

A preliminary study of the cyclicity in the *Hollardops* Limestone of southern Morocco (Brett *et al.*, 2012) produces the same number of cycles (a little less than 160) in two outcrops at c. 80 km distance. If these are based on precession, then the upper part of the *laticostatus* conodont zone (the basal part of the upper Emsian or the *Icriodus fusiformis* Zone) alone could have lasted c. 3 myr. This is in agreement with considerable thicknesses in some regions. Currently the number of cycles of the supposedly very long *serotinus* and *patulus* zones has not yet been counted with precision in SW Morocco. Chlupáč (2000) gave an estimate of cycles within the Daleje Shale of Bohemia, a correlative of the *Hollardops* Limestone, as 100—130; not too far off. The number of cycles in the subsequent limestone, covering the *serotinus and patulus* zones, was estimated as 150—250. If the cyclic signal remains the same, then one can assume another c. 4 myr for the higher part of the upper Emsian, which gives a total of 7 myr.

Concerning the duration of conodont zones in the Famennian, there are data from a pelagic section in the northern Renish Massif (Oese: Hartenfels, 2011 and unpublished data), where the microfacies of cycles have been studied, and where zonal boundaries were fixed with great precision. This data set is better than Kaufmann's (2006) estimates based on the thickness of zones in a single section, i.e. the Lali Section (Ji and Ziegler, 1993), where the sedimentology is virtually unstudied. Tentative estimates of the number of cycles and durations for the Upper/Uppermost Famennian in the Oise Road Section, northern Rhenish Massif, are as follows: the Hangenberg Event Interval may represent c. 300 kyr (a guess, based on three rough glacial cycles), *praesulcata* Zone — 44 cycles (base Lower *praesulcata* Zone to base Hangenberg Event), Upper *expansa* Zone — 29 cycles, Middle *expansa* Zone — 84 cycles (certainly much longer than the other zones), Lower *expansa* Zone — 23 cycles, Upper *postera/styriacus* Zone — 21 cycles, Lower *postera/styriacus* Zone — 19 cycles, Upper *trachytera* Zone — c. 20 cycles, Lower *trachytera* Zone — c. 43 cycles, Uppermost *marginifera* Zone — >30 cycles (zonal base still open, cycles less pronounced, higher sedimentation rate), within the upper part of the *praesulcata* Zone the *Wocklumeria sphaeroides* ammonoid zone has 16 cycles. Kaufmann (2006) gives c. 5 myr for the interval from the base of the Lower *trachytera* Zone to the D/C boundary. The c. 280 cycles of the same interval at Oese, therefore, can translate well into precession.

Above relative subdivisions of stages using a combination of observed sedimentary cycles and regional sedimentary thickness comparison of parts of stages is input in the scaling shown in Table 22.2; these data complement the relative zonal scaling shown in this table.

22.2.2.3. Chemostratigraphy

Many publications have used inorganic geochemistry data to fingerprint, interpret and correlate Devonian black shale units (e.g., Robl and Barron, 1988; Rimmer *et al.*, 2004) and anoxic event levels, especially the Kellwasser Beds (e.g., McGhee *et al.*, 1986; Goodfellow *et al.*, 1988; Bratton *et al.*, 1999; Ma and Bai, 2002; Tribovillard *et al.*, 2004; Pujol *et al.*, 2006; Riquier *et al.*, 2005, 2007; Ma *et al.*, 2009) or the Kačák (Ellwood *et al.*, 2011a), "*punctata*" (= Middlesex, Sliwinski *et al.*, 2011) and *Annulata* events (Racka *et al.*, 2010). Paleoproductivity proxies, redox-sensitive and trace elements can contribute to detailed event stratigraphy, whereas rare earth element (REE) studies (e.g., Grandjean-Lécuyer *et al.*, 1993; Girard and Albarède, 1996; Girard and Lecuyer, 2002) are less conclusive.

Supposed impact signatures near the Eifelian—Givetian (Ellwood *et al.*, 2003; contra: Racki and Koeberl, 2004) and Frasnian—Famennian boundaries that might be used for correlation, including iridium anomalies (e.g., Playford *et al.*, 1984; Wang *et al.*, 1991), could not be substantiated in subsequent studies (McGhee *et al.*, 1986; Claeys *et al.*, 1996;

Girard *et al.*, 1997; Schmitz *et al.*, 2006), or are local features and do not coincide with extinction levels (Wang, 1992; Wang *et al.*, 1994; Marini *et al.*, 1997). The well-dated Frasnian Flynn Creek (e.g., Schieber and Over, 2005) and Alamo impacts (e.g., Warme and Sandberg, 1996; Morrow *et al.*, 2009) were regional events in the Frasnian of North America but they occurred close to the global Timan and Middlesex events.

Measurements of total organic carbon (TOC) contents of carbonates and shales aid the recognition of Devonian anoxic events (e.g., Joachimski *et al.*, 2001; Kaiser *et al.*, 2006; Hartkopf-Fröder *et al.*, 2007; Marynowski and Filipiak, 2007) and are useful in event stratigraphy. Based on the mostly high thermal maturity of Devonian sediments, organic geochemistry analyses, including biomarkers and sterane/hopane ratios, are geographically restricted but document episodic changes in primary producers (e.g., Joachimski *et al.*, 2001) that can be used for correlation. For example, biomarkers of green sulfur bacteria give evidence for anoxia reaching the open ocean photic zone in Upper Kellwasser time (Hartkopf-Fröder *et al.*, 2007), in the subsequent lower Famennian (Marynowski *et al.*, 2011), and during the *Annulata* (Racka *et al.*, 2010) and Dasberg events (Marynowski *et al.*, 2010).

The stratigraphic use of stable isotope data has increased significantly through numerous studies. Initially (Talent *et al.*, 1993; Hladiková *et al.*, 1997), carbon isotopes were analyzed in order to identify changes in the global carbon reservoirs in association with global events. Complete carbon isotope curves for the Devonian of Europe (Figure 22.12) based on whole rock data or brachiopods are available through the work of Buggisch and Mann (2004), Buggisch and Joachimski (2006), and Van Geldern *et al.* (2006). Positive isotopic $\delta^{13}C$ spikes can be recognized, partly in individual biomarkers (Joachimski *et al.*, 2002). They characterize the Silurian–Devonian boundary or Klonk Event (Saltzman, 2002; Buggisch and Mann, 2004; Kleffner *et al.*, 2009; Manda and Frýda, 2010; Zhao *et al.*, 2011), the lower Pragian (Buggisch and Mann, 2004), the Basal Zlichov (Buggisch and Mann, 2004), Kačák (Hladiková *et al.*, 1997), *pumilio* (Buggisch and Joachimski, 2006), Taghanic (Aboussalam, 2003), Frasnes (Buggisch and Joachimski, 2006), Middlesex (Yans *et al.*, 2007, Sliwinski et al., 2011), Kellwasser (e.g., Joachimski and Buggisch, 1993; Joachimski *et al.*, 2001, 2002; Stephens and Sumner, 2003; Chen *et al.*, 2005), ? Dasberg (Myrow *et al.*, 2011), and Hangenberg events (Schönlaub *et al.*, 1992; Brand *et al.*, 2004; Kaiser, 2007; Kaiser *et al.*, 2006, 2008; Cramer *et al.*, 2008). Diagenetic overprinting and locally variable primary producers do not allow us to use the magnitude of isotopic excursions for correlation (e.g., Da Silva and Boulvain, 2008) and may cause misleading local negative spikes (e.g., Zheng *et al.*, 1993).

Widespread diagenetic alteration of carbonates prevents the use of whole rock oxygen isotope data (e.g., Halas *et al.*, 1992; Hladiková *et al.*, 1997; Chen *et al.*, 2005) for correlation and paleotemperature reconstructions (Brand, 2004) in the Devonian. At least in the Middle and Upper Devonian, oxygen data from conodont phosphate (Elrick *et al.*, 2009; Joachimski *et al.*, 2009) are superior to measurements of supposed pristine brachiopod calcite (Joachimski *et al.*, 2004; Van Geldern *et al.*, 2006). There are oxygen isotopic peaks that suggest sudden paleoclimate changes associated with the Kellwasser (Joachimski and Buggisch, 2002) and Hangenberg events (Kaiser *et al.*, 2006). The available data set is still too incomplete to use oxygen isotope data for the construction of the Devonian time scale.

The $^{87}Sr/^{86}Sr$ of marine Sr through the interval (Diener *et al.*, 1996; Denison *et al.*, 1997; Van Geldern *et al.*, 2006) shows a broad trough, centered on the basal Eifelian to Middle Givetian that reaches 0.7078 at its minima between an early Devonian high of 0.7087 and an early Carboniferous high of 0.7083. This range is large enough to offer much potential for using Sr isotope stratigraphy in the interval. However, reliable data (Brand, 2004) are best obtained from pristine brachiopod calcite, not from conodont phosphate (Kürschner *et al.*, 1993; Ebneth *et al.*, 1997; Veizer *et al.*, 1997) or whole rock measurements (e.g., Carpenter *et al.*, 1991; Chen *et al.*, 2005).

Based on fluctuations in oceanic circulation, productivity, and bacterial sulfate reduction rates, sulfur isotope data have proved to be useful event stratigraphic markers at the Frasnian–Famennian boundary (e.g., Geldsetzer *et al.*, 1987; Halas *et al.*, 1992; Wang *et al.*, 1996; Joachimski *et al.*, 2001). Simon *et al.* (2007) used a still limited data set to reconstruct a curve for the sulfur isotope composition of Devonian marine sulfate, which shows a significant depression centered in the middle Emsian and with a broad high in the Lower/Middle Famennian.

Only few studies have so far utilized neodymium isotopic changes for Devonian stratigraphy. Water mass movements with sea-level fluctuations are recorded in values measured from conodont phosphate. Dopieralska *et al.* (2006) showed that the basal Upper Frasnian *semichatovae* Transgression can be recognized by the sudden incursion of oceanic water with more radiometric values on the northern Gondwana shelf, whilst the eustatic pre-Hangenberg shallowing is recorded in the Sudetes by an unusual fall of $^{\varepsilon}Nd(t)$ values.

22.2.2.4. Eustatic Changes / Sequence Stratigraphy

In a broad sense, sea-level changes in the Devonian have been commented upon at least since the collative work of French in the nineteenth century. A more systematic comparison between New York (USA) and Europe was published as an initial attempt to create a Devonian eustatic curve by House (1983). This was improved using detailed conodont evidence for the Laurussian area, leading to the sea-level curve and depophase terminology of Johnson *et al.* (1985), which soon became a widely quoted standard that is still recycled in recent publications. However, the "Johnson *et al.* curve" is not based on a sequence stratigraphic approach and concentrates on the correlation of times of fastest and maximum transgression.

Many publications have summarized regional relative sea-level curves, with some focus on short-termed eustatic fluctuations reflected by sudden hypoxic events. Prominent and regular sea-level oscillations in thick shallow-water carbonate platforms and siliciclastics, unfortunately, mostly lack precise age control. Regional reviews correlated against the standardized conodont time scale were bundled in a thematic volume by House and Ziegler (1997). The clear separation of sequence boundaries and transgressive maxima and the biostratigraphically tuned correlation of regional curves will soon result in a revised curve with third- and fourth-order global sequences. Age revisions and refinements have been proposed for the Lower Devonian by García-Alcalde (1997) and Ver Straeten (2001), for the Givetian by Witzke *et al.* (1988), Klapper (1992), Day *et al.* (1996), Aboussalam (2003), Baird and Brett (2003), Brett *et al.* (2011), and Aboussalam and Becker (2011), for the Frasnian by Klapper (1992), Day *et al.* (1996), Racki *et al.* (2004), and Whalen and Day (2008), and for the Famennian by Becker *et al.* (1993), Bless *et al.* (1993), Becker (1993b), and Hartenfels and Becker (2009), and are recognized in Figure 22.12.

22.2.2.5. Magnetic Susceptibility (MS)

The usage of magnetic susceptibility signatures for the Devonian has been advanced much through the work of B.B. Ellwood and co-authors (e.g., Crick *et al.*, 1997; Ellwood *et al.*, 2000, 2001). Normalized against sample mass and in the absence of strong diagenetic overprint, cyclic changes in the magnetic susceptibility reflect changes in the detrital input of dia-, para- and ferromagnetic minerals. This is especially useful in carbonate facies with good biostratigraphical control (e.g., Whalen and Day, 2008). However, the interpretation of involved sea-level and climate changes and the correct correlation of MS values require microfacies analysis, the identification of possibly fluctuating mineral contents that carry the magnetic signal, and an understanding of detrital depositional factors (e.g., Hladil *et al.*, 2006). The importance of MS work has been demonstrated, especially in relation to GSSPs and the events associated with them. The Silurian−Devonian boundary interval has been analyzed by Crick *et al.* (2001), and the Lower Devonian generally by Ellwood *et al.* (2001) and Koptíková *et al.* (2010). Work on the Choteč Event (Koptíková, 2011) and the Eifelian−Givetian boundary and associated Kačák Event (Crick *et al.*, 1997; Hladil *et al.*, 2002) has established correlations between North Africa southern France, and the Czech Republic, and led to the proposal to establish a "magnetostratotype" (Crick *et al.*, 2000) and formal magnetozones (chrons and subchrons; Ellwood *et al.*, 2006). Ellwood *et al.* (2011a,b) extended this approach to use cycles of MS values and the MS zones to propose a floating time scale for the whole Givetian, with a duration of c. 200 kyr for the Kačák Event. Gersl and Hladil (2004) and Nawrocki *et al.* (2008) studied the Lower/Middle Frasnian boundary interval in order to trace Middlesex Event equivalents in European carbonate platforms. The Frasnian−Famennian boundary and Kellwasser Events have been analyzed by Crick *et al.* (2000) and Riquier *et al.* (2007). Studies on shallow-water carbonate platforms (Da Silva and Boulvain, 2002, 2006; Hladil *et al.*, 2006) provide detailed cyclic signals that, in the absence of fine biostratigraphic control, are still difficult to correlate with the conodont scale. Within IGCP 580 on "Application of magnetic susceptibility as a paleoclimate proxy on Paleozoic sedimentary rocks and characterization of the magnetic signal" many Devonian successions are investigated and many new results were published in 2010 in a special volume (13/4) of *Geologica Belgica.*

22.2.2.6. Gamma Ray Spectrometry

Correlations based on outcrop γ-ray spectrometry have been introduced in the Devonian stratigraphic literature only rather recently. They provide cyclic signals that are correlated with MS data and interpreted in terms of sea-level changes, sometimes also with color analysis (spectral reflectance, Koptíková *et al.*, 2010). Several studies have contributed to event and cyclostratigraphy in the Middle Devonian to Frasnian (Gersl and Hladil, 2004; Hladil *et al.*, 2006; Bábek *et al.*, 2007; Koptíková, 2011). High-resolution data across the Frasnian−Famennian boundary are less distinctive (Bond *et al.*, 2004) but identify the *Annulata* Event in Poland.

22.3. DEVONIAN TIME SCALE

22.3.1. Previous Scales

Devonian geological time scales have been composed by Tucker *et al.* (1998), Compston (2000), Williams *et al.* (2000), and more recently by GTS2004 (House and Gradstein, 2004) and Kaufmann (2006). Noticeable in reviews of the Devonian time scale are the limited number of good quality and chronostratigraphically fixed radiometric age dates within the period. As a result, various assumptions have been made to arrive at estimates of duration and boundary ages of the stages, such as equal duration of stages or of zones or subzones, particularly for conodonts, even though the available cyclostratigraphic data suggest that these assumptions have no logical base. Despite these stratigraphic misgivings, one does not have to look far back in literature to realize that the relative number of subzones per stage was often utilized for their relative scaling (see for example chapters on the Mesozoic periods in GTS2004). The finer the subdivisions on this stratigraphic ruler, and the more successive age dates, the better the scaling. In those cases,

TABLE 22.1 Comparison of Age and Duration of the Devonian Period in Selected Time Scales Published

	Base (Ma)	Top (Ma)	Duration (myr)
Harland *et al.* (1990)	408.5	362.5	46
Young and Laurie (1996)	410	354	56
Tucker *et al.* (1998)	418	362	56
Compston (2000b) (1)	411.6	359.6	52
Compston (2000b) (2)	418	362	56
Williams *et al.* (2000a)	418	362	56
Gradstein *et al.* (2004)	416 ± 2.8	359.2 ± 2.5	56.8
Kaufmann (2006)	418.1 ± 3.0	360.7 ± 2.7	57.4
This study	419.2 ± 3.2	358.9 ± 0.4	60.3

there simply was no alternative scaling method available for geochronology.

Older time scales that assumed constant sedimentation rates, or relied on the subjective views of specialists on probable durations, have now generally been discarded. A comparison of the age of the base and/or top of the Devonian Period with the corresponding duration, from selected time scales published during the last two decades is given in Table 22.1.

Although a substantial number of key dates were published after Harland *et al.* (1990) and Young and Laurie (1996) presented their limited evidence and interpolations, the Devonian time scale is still not well established. Recently, new Devonian age dates were published for the Pragian (Parry *et al.*, 2011), Emsian (Kaufmann *et al.*, 2005), Givetian/Frasnian (Compston, 2004), Frasnian–Famennian boundary (Kaufmann *et al.*, 2004; Turgeon *et al.*, 2007), and Devonian–Carboniferous boundary (Trapp *et al.*, 2004; Selby and Creaser, 2005). A problem is the discrepancy between U-Pb zircon dates for key Devonian levels using TIMS and HR-SIMS (SHRIMP) dates as discussed by Tucker *et al.* (1998), Compston (2000), Williams *et al.* (2000), and House and Gradstein (2004). SHRIMP ages on average are 1.3% younger ages (~5 myr) for Devonian stages than TIMS ages. Using such information, GTS2004 "normalized" SHRIMP ages that use the SL13 monitor standard, but not those that use the QGNG monitor standard. In addition to the problems posed by accurate laboratory procedures, by standards, and by constants in radiometric age dating, there is the matter of critical analysis of the biostratigraphic assignments of the sample levels (e.g. Streel, 2000). For example, the Australian Phanerozoic time scale (Young and Laurie, 1996) was analyzed by Klapper (2000b; response by Young, 2000), who criticized biostratigraphic assumptions in relation to the series boundary data points. In the case of the Los Frailes Stockwork, Rio Tinto, an age just above the Devonian–Carboniferous boundary was plotted, although there is no exact biostratigraphic placement. Similarly, Williams *et al.* (2000) show a U-Pb age for the Horses Glen Lower Tuffs of 387.5 ± 0.2 Ma. The unusually low analytical error strongly contrasts with a stratigraphic range spanning late Frasnian and possibly well into Famennian time; thus, the date does not contribute to the Devonian time scale. The age published for central Victorian volcanics by Compston (2004) has no biostratigraphic control since the age of the associated Taggerty fish fauna is absolutely unclear in the long Givetian–Frasnian interval.

For the construction of the Devonian geologic time scale in GTS2004, radiometric ages, mainly generated with the U/Pb ID-TIMS method, were plotted against the relative (sub) zone-stage scale ("House scale"). Approximate radiometric and stratigraphic errors were also taken into account. The relative scale "juggled" the high-resolution ammonoid and conodont (sub) zonations, such that shortest segments were fairly close to equal duration. The latter is not unlike the abandoned chron concept of GTS1989 by Harland *et al.* (1990: see Chapter 1). At the same time a fairly linear two-way fit was achieved. In the final analysis, the age obtained from the Devonian data of 416.7 ± 2.9 Ma for the top of the Silurian was abandoned in favor of 416.0 ± 1.4 Ma, which resulted from the better-constrained data for the Ordovician and Silurian periods.

The best-fit line for the two-way plot (Figure 14.5 in GTS2004) of radiometric ages against the zones and stages was calculated by a cubic spline fitting method that combines stratigraphic uncertainty estimates with the 2-sigma error bars of the radiometric data. This is achieved with Ripley's MLFR procedure (Chapter 8 in GTS2004); a smoothing factor of about 1.4 was calculated with cross-validation. The chi-square test of residuals indicated that the 0.5-myr error bar on the 390.0 Ma monazite date (No. 6 in Table 14.2 of GTS2004) was too narrow, and an "average" error bar was used instead. This, together with the large error bars on some other Emsian dates, reduces its duration. A straight-line fit, as produced by Tucker *et al.* (1998), would substantially increase the duration of the Emsian Stage.

To get some idea of the sensitivity of the scaling to these simplifying assumptions, the Devonian time scale was also calculated in GTS2004 using the relative scale of Tucker *et al.* (1998), where the stages are proportioned to an empirical scheme of graphic correlation and/or biostratigraphic intuition. The *pesavis* conodont zone is now placed in the early Pragian. There is no significant difference between the ages in the "House" and the Tucker scales, but this might have been different had there been more detailed age dates to match the much higher resolution in the zonal schemes. For GTS2004, the House scale was preferred because it reduces the duration of the Emsian and Lochkovian stages in line with biostratigraphic reasoning and intuition. The scale also increases the duration of the Pragian Stage, which agrees with the analysis of cyclicity in the limestones in the classical Devonian sections of the Barrandian (Czech Republic) that suggest that the Pragian is not much shorter than the underlying Lochkovian (Chlupáč, 2000). This, however, does not consider the significant shortening (by more than 50%) of the Pragian by the too low Emsian GSSP (in relation to the traditional Pragian boundaries of the Prague Basin).

Kaufmann (2006) made an ambitious attempt to improve the Devonian time scale, assisted by 13 biostratigraphic U/Pb ID-TIMS zircon and monazite ages. New are two Emsian age dates, and one basal Tournaisian date (Trapp *et al.*, 2004; Kaufmann *et al.*, 2005). No $^{40}Ar/^{39}Ar$ or HR-SIMS dates were used, and the radiometric age dates were confined solely to the Devonian and the earliest Carboniferous. No use was made of Silurian and Carboniferous scaling in GTS2004, which hampers the placement of this new scale in the standard Phanerozoic geologic time scale, and substantially weakens ages and precision of top and base Devonian. For scaling the Devonian stages along the y-axis of a two way graph, Kaufmann (2006) employed a mixture of simple graphic correlation, thickness of single sedimentary sections, and "biostratigraphic intuition" on duration of zones, shuffled such that a linear fit channel was arrived at (see Figure 22.13). In some cases, like Emsian and Upper/Uppermost Famennian zones and stage, relative durations are quite uncertain (B. Kaufmann, pers. comm., 2006), since there is no support for zonal durations from graphic correlation of comparable sedimentary sections, tuned sedimentary cycles, or subzonal duration constraints. Although there is no *a priori* reason that this particular best fit of x and y axis data should approach linearity, the latter often is taken to indicate good agreement between the compositing and scaling procedures on the one hand and the radiometric dating on the other. But one can reasonably question why a linear fit should result.

As commented upon by F.P. Agterberg (pers. comm., 2006), adding 2 myr to the 2-sigma error of individual ages by B. Kaufmann is subjective, but interestingly it is comparable to results achieved along different lines in GTS2004. In that study, Devonian stage boundary uncertainties were obtained by multiplying age by 0.000 685, after an elaborate analysis considering various sources of uncertainty. The mean Kaufmann uncertainty is 3.2 myr and mean GTS2004 uncertainty is 2.7 myr, where rounding-off would give the same results.

The major difference between the Devonian scales in GTS2004 and in Kaufman (2006) is that the base of the Eifelian is nearly 6 myr younger in the latter, and consequently the Emsian Stage is much longer than in GTS2004. Contributing to this long duration are the 392.2 ± 1.5 Ma age used for the Wetteldorf "Hercules I" zircon tips, plus the very long "*serotinus*" Zone, which together bring the Tioga ash dates (items D6 and D7 in Appendix 2 of this volume, and "Tioga age dates" in Figure 22.13) on the linear channel fit. Hydrogen fluoride leaching of the zircons without annealing may have changed the U-Pb content (F. Corfu, pers. comm. 2007), and the real "Hercules I" age may be between 400 and 390 Ma, with an uncertainty that covers 10 myr or less. Figure 22.13 shows the Devonian time scaling employed by Kaufmann (2006).

22.3.2. Radiometric Data

The detailed and high-resolution conodont–ammonoid zonation for the Devonian, with over 60 zones (Figures 22.10 and 22.11), is in stark contrast to the less than eighteen radiometric age dates employed in Devonian time scale building, some of which have a ± 6 myr uncertainty. The duration of the conodont and ammonoid zones are estimated to vary between less than 200 kyr to more than 2.5 myr (Table 22.2). The best radiometric coverage is in the upper part of the Devonian and in the lowermost Carboniferous. Spacing and quality of the radiometric dates is such that the middle Emsian and part of the Frasnian and lower Famennian are not covered. The Givetian Stage does not contain any dates.

Appendix 2 of this book lists 16 Devonian through lowermost Carboniferous U/Pb TIMS, and 2 Lower Devonian HR-SIMS (SHRIMP) radiometric age dates that were judged suitable for time scale construction (see items D1-18 and cb1 and 2). Unfortunately, there is a large spread in analytical and stratigraphic uncertainty. Three Re-Os dates are listed that have useful biostratigraphic constraints, but were not incorporated in the spline fit of Figure 22.14. The U/Pb age dates of Red Deer Creek Tuff, the Nordegg Tuff, the Caldera Complex rhyolite and the Zeigler Pit monazites have analytical uncertainties below 1.0 myr, and appear to have good stratigraphic control.

In order to provide constraints on the age of the base of the Devonian, we included the extrapolated age of 419.2 ± 3.2 Ma for the Silurian–Devonian boundary (see Chapter 21). To increase resolution across the Devonian–Carboniferous boundary two earliest Carboniferous age dates were initially considered for time scale construction. Trapp *et al.* (2004) used U-Pb isotope dilution-thermal ionization mass spectrometry (ID-TIMS) for zircons from two successive basal Tournaisian bentonites of the Hasselbachtal Auxiliary Stratotype to reinterpolate the boundary at 360.7 ± 0.7 Ma. However, our fitting suggests these ages to be slightly old. This is independently confirmed by three high-resolution CA-TIMS age dates closely below and above the Devonian–Carboniferous boundary in the Rhenish Mountains (Items D18, cb1 and cb2 in Appendix 2 of this book). The interpolated age of the Devonian–Carboniferous boundary using the three new dates is 358.9 ± 0.4 Ma. This new high-precision age estimate confirms that the GTS2004 age of 359.2 Ma (with a small uncertainty tentatively set at 1 myr) for the Devonian–Carboniferous boundary was reasonable.

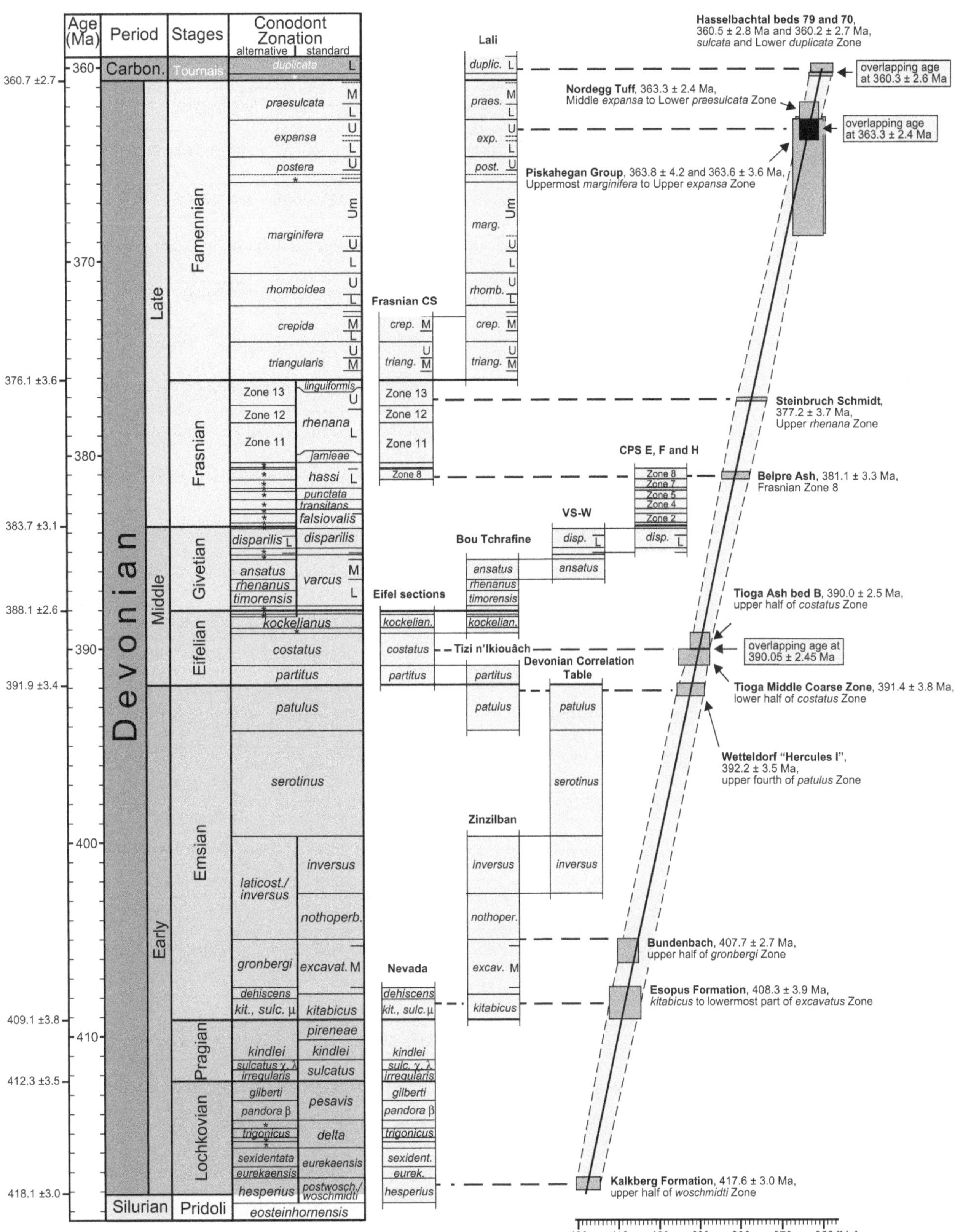

FIGURE 22.13 **Devonian geologic time scale constructed by Kaufmann (2006).** For details see Section 22.3.1 of this chapter.

TABLE 22.2 Scaling and durations of the Devonian conodont zonal subdivisions ("Becker scale") as input for the spline fit of zones + stages versus radiometric ages in Figure 22.14

Stage	Conodont Zone	Zonal Position (Ma)	GTS2012 (Spline) age (Ma)	2-Sigma	Duration myr	Curve	
Tournaisian	*sulcata*	359.20	358.94	0.393		Spline	
Famennian	*kockeli*	359.79	359.10	0.408	0.16	Spline	
Famennian	*praesulcata*	361.54	359.84	0.571	0.74	Spline	
Famennian	*expansa*	365.10	362.27	0.974	2.43	Spline	
Famennian	*styriacus*	366.15	363.01	1.096	0.74	Spline	
Famennian	*trachytera*	367.80	364.19	1.439	1.18	Spline	
Famennian	*marginifera*	371.06	366.69	1.806	2.50	Spline	
Famennian	*rhomboidea*	372.24	367.73	1.837	1.04	Spline	
Famennian	*crepida*	374.61	370.11	1.796	2.38	Spline	
Famennian	*triangularis*	376.39	372.24	1.628	2.13	Spline	
Frasnian	*linguiformis*	376.99	373.04	1.510	0.80	Spline	
Frasnian	*bogartensis*	377.99	374.48	1.469	1.44	Spline	
Frasnian	*winchelli*	378.76	375.65	1.500	1.17	Spline	
Frasnian	*feisti*	379.36	376.54	1.628	0.89	Spline	
Frasnian	*plana*	379.95	377.40	1.750	0.86	Spline	
Frasnian	*proversa–housei*	380.72	378.43	1.939	1.03	Spline	
Frasnian	*nonaginta–primus*	381.73	379.58	2.138	1.15	Spline	
Frasnian	*punctata*	382.56	380.38	2.189	0.80	Spline	
Frasnian	*transitans*	382.97	381.25	0.995	0.87	Straight-line	These error bars
Frasnian	*rotundiloba*	384.99	382.69	0.883	1.44	Straight-line	
Givetian	*norrisi*	385.58	383.10	0.847	0.41	Straight-line	not directly
Givetian	*disparilis*	387.06	384.15	0.765	1.05	Straight-line	
Givetian	*hermanni*	388.25	384.99	0.709	0.84	Straight-line	
Givetian	*semialternans*	388.84	385.41	0.704	0.42	Straight-line	comparable
Givetian	*ansatus*	390.03	386.25	0.679	0.84	Straight-line	
Givetian	*rhenanus–varcus*	390.98	386.92	0.684	0.67	Straight-line	with
Givetian	*timorensis*	391.51	387.30	0.694	0.38	Straight-line	
Givetian	*hemiansatus*	392.10	387.72	0.694	0.42	Straight-line	spline errors
Eifelian	*ensensis*	392.75	388.18	0.791	0.46	Straight-line	
Eifelian	*eiflius*	393.29	388.56	0.821	0.38	Straight-line	
Eifelian	*kockelianus*	393.88	388.97	0.847	0.41	Straight-line	
Eifelian	*australis*	394.47	389.23	1.000	0.26	Spline	
Eifelian	*costatus*	395.95	391.62	0.929	2.39	Spline	
Eifelian	*partitus*	396.84	393.25	1.245	1.63	Spline	
Emsian	*patulus*	397.73	394.91	1.648	1.66	Spline	
Emsian	*serotinus*	399.21	397.68	2.133	2.77	Spline	

TABLE 22.2 Scaling and durations of the Devonian conodont zonal subdivisions ("Becker scale") as input for the spline fit of zones + stages versus radiometric ages in Figure 22.14—cont'd

Stage	Conodont Zone	Zonal Position (Ma)	GTS2012 (Spline) age (Ma)	2-Sigma	Duration myr	Curve
Emsian	*laticostatus*	400.11	399.30	2.357	1.62	Spline
Emsian	*inversus*	400.99	400.86	2.526	1.56	Spline
Emsian	*nothoperbonus*	402.18	402.80	2.571	1.94	Spline
Emsian	*gronbergi*	403.37	404.55	2.587	1.75	Spline
Emsian	*excavatus*	404.55	406.11	2.622	1.56	Spline
Emsian	*kitabicus*	405.74	407.57	2.582	1.46	Spline
Pragian	*profunda*	408.11	410.19	2.755	2.62	Spline
Pragian	base Pragian (within *irregularis* zone)	408.70	410.78	2.776	0.59	Spline
Lochkovian	*irregularis*	409.29	411.35	2.776	0.57	Spline
Lochkovian	*gilberti*	409.89	411.90	2.730	0.55	Spline
Lochkovian	*pandora*	410.77	412.70	2.684	0.80	Spline
Lochkovian	*trigonicus*	411.67	413.48	2.592	0.78	Spline
Lochkovian	*transitans*	412.26	414.01	2.587	0.53	Spline
Lochkovian	*eleanorae*	412.85	414.54	2.505	0.53	Spline
Lochkovian	*omoalpha*	413.74	415.38	2.357	0.84	Spline
Lochkovian	*postwoschmidti*	415.52	417.27	2.224	1.89	Spline
Lochkovian	*hesperius*	417.00	418.98	2.847	1.71	Spline

22.3.3. Age of Stage Boundaries

For the Devonian and lowermost Carboniferous segments of GTS2012, we used the radiometric dates discussed in the previous section, including their lambda errors (Appendix 2). The three Re-Os dates D11, D12 and D17 were not included. Their inclusion produced undesirable curvature in the spline through the Lower Famennian, which has no age dates to corroborate such curvature. Stratigraphic positions with their rectangular error bars were assigned for all age dates according to the chronostratigraphic scale in Table 22.2. The zonal position of the conodont zonation in the Devonian stages is based on our initial scaling estimate of stages and zones (see also discussion of cyclostratigraphy in Section 22.2.2). The age of the base Devonian, with error bars, was specified according to the boundary age and error of the Ordovician–Silurian spline (Chapter 21).

In a first run of the smoothing spline procedure outlined in Chapter 14, the cross-validation suggested an optimal smoothing factor of 1.85. With such a high smoothing factor, six of the data points did not pass the χ^2 test (D4, χ^2=8.12, p=0.004; D5, χ^2=21.7, p=0.000; D6, χ^2=5.45, p=0.020; D8, χ^2=7.85, p=0.005; D9, χ^2=4.86, p=0.032; and D15, χ^2=4.43, p=0.035). For each of these points, the combined stratigraphic and radiometric error s_{xy} was therefore increased to enforce a p value of 0.5.

The adjusted errors were used for a second computation of the spline. In this new run, the optimal smoothing factor was 0.45. The resulting final spline is seen in Figure 22.14.

The Givetian does not contain any radiometric dates, and the shape of the spline is relatively unconstrained in this interval as long as it is monotone. To estimate the upper and lower boundaries of the Givetian, we therefore made a pragmatic exception to the standard procedure, reverting to straight-line interpolation between the closest points along the scale (D7 and D9). As for the spline, error bars for this straight line were estimated by a Monte Carlo procedure, randomly moving D7 and D9 within their stratigraphic and radiometric errors. This method gives a duration for the Givetian of 5.0 myr, closer to the 6.5 myr of House (1995), based on assumed precessional cycles, than the 4.3 myr resulting from the spline. Interestingly, Ellwood *et al.* (2011b) recently suggested 5.6 myr, based on the analysis of cyclic magnetosusceptibility data.

The interpolated Devonian stage boundaries and their bootstrapped confidence intervals are given in Table 22.3. The

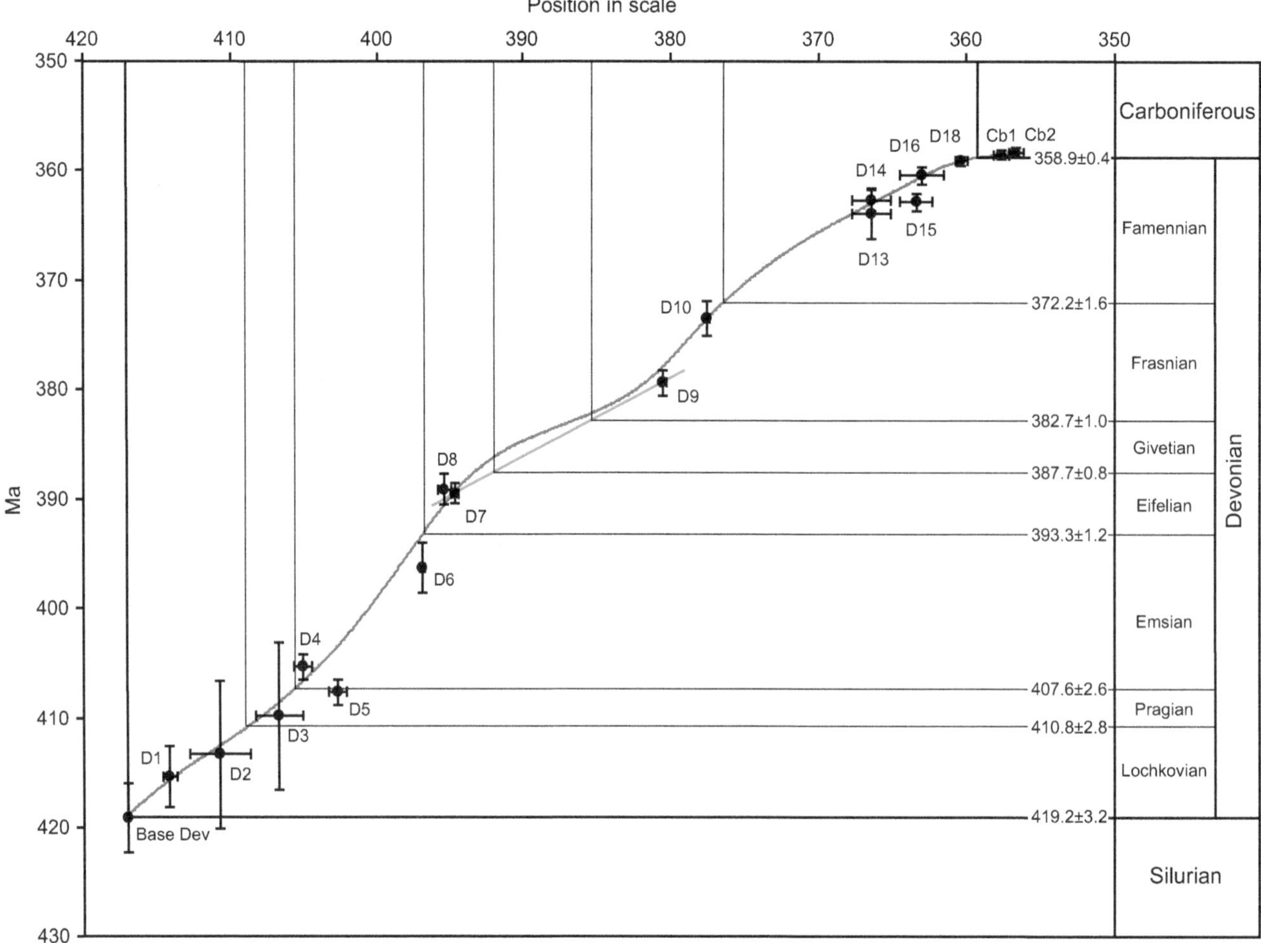

FIGURE 22.14 Construction of the Devonian time scale using a cubic spline fit on 17 selected radiometric ages for Devonian and lowermost Carboniferous scaled chronostratigraphically listed in Appendix 2 of this book. Radiometric error bars are 2-sigma, stratigraphic (scaling) error bars represent total uncertainty. The blue line shows the exception made for the Givetian, which is not properly constrained by the spline. Spline construction details are in Section 22.3 of this chapter.

TABLE 22.3 Interpolated Ages of Devonian Epoch and Stage Boundaries, with Bootstrapped Error Bars (One-Sided)

Period	Epoch	Stage	Base Ma GTS2012	Est. ± myr (2σ)	Epoch Duration (myr)	Est. ± myr (2σ)	Stage Duration (myr)	Est. ± myr (2σ)	GTS2004
Carboniferous	Early Miss.	Tournaisian	358.9	0.4					359.2
Devonian	Late	Famennian	372.2	1.6			13.3	0.8	374.5
		Frasnian	382.7	1.0	23.8	2.0	10.5	2.5	385.3
	Middle	Givetian	387.7	0.8			5.0	*not calculated*	391.8
		Eifelian	393.3	1.2	10.6	*not calculated*	5.6	1.9	397.5
	Early	Emsian	407.6	2.6			14.3	2.6	407.0
		Pragian	410.8	2.8	25.9	2.0	3.2	1.2	411.2
		Lochkovian	419.2	3.2			8.4	3.5	416.7

GSSPs of the Devonian Stages, with Location and Primary Correlation Criteria (Status Jan. 2012)

Stage	GSSP Location	Latitude, Longitude	Boundary Level	Correlation Events	Reference
Famennian	Coumiac Quarry, near Cessenon, Montagne Noire, France	43°27′40.6"N 3°02′25"E*	Base of Bed 32a	Conodont FAD *Palmatolepis subperlobata* and flood occurrence of *Palmatolepis ultima*	Episodes 16/4, 1993
Frasnian	Col du Puech de la Suque, Montage Noire, France	43°30′11.4"N 3°05′12.6"E*	Base of Bed 42a′ at Col du Puech de la Suque section E	Conodont FAD *Ancyrodella rotundiloba*	Episodes 10/2, 1987
Givetian	Jebel Mech Irdane, Morocco	31°14′14.7"N 4°21′14.8"W*	Base of Bed 123	Conodont FAD *Polygnathus hemiansatus*	Episodes 18/3, 1995
Eifelian	Wetteldorf, Eifel Hills, Germany	50°08′58.6"N 6°28′17.6"E*	21.25m above the base of the exposed section, base of unit WP30	Conodont FAD *Polygnathus costatus partitus*	Episodes 8/2, 1985
Emsian	Zinzil'ban Gorge, Uzbekistan	39°12′N 67°18′20"E	Base of Bed 9/5 in the Zinzil'ban Gorge in the Kitab State Geological Reserve	Conodont FAD *Eocostapolygnathus kitabicus*	Episodes 20/4, 1997
Pragian	Velká Chuchle, Prague, Czech Republic	50°00′53"N 14°22′21.5"E*	Base of Bed 12 in Velká Chuchle Quarry	Conodont FAD *Eognathodus sulcatus sulcatus* or, better, of *Caudicriodus steinachensis* Morph beta	Episodes 12/2, 1989
Lochkovian *(base Devonian)*	Klonk, near Prague, Czech Republic	48.855°N 13.792°E**	Within Bed 20	Graptolite FAD *Monograptus uniformis*	IUGS Series A, 5, 1977

**according to Google Earth*
***derived from map*

base of the Devonian is at 419.2 ± 3.2 Ma, and its top is 358.9 ± 0.4 Ma. The durations of the three epochs and seven stages are as follows:

Early Devonian lasted 25.9 ± 2.0 myr, with the Lochkovian Stage lasting 8.4 ± 3.5 myr (base at 419.2 ± 3.2 Ma), the Pragian Stage 3.2 ± 1.2 myr (base at 410.8 ± 2.8 Ma), and the Emsian Stage 14.3 ± 2.6 myr, starting at 407.6 ± 2.6 Ma. The Pragian is the shortest stage in the Devonian at 3.2 myr, and the (unrevised) Emsian is the longest stage in the Devonian, followed by the Famennian (13.3 myr).

The Middle Devonian has a duration of 10.6 myr, with the Eifelian lasting 5.6 ± 1.9 myr (starting at 393.3 ± 1.2 Ma), and the Givetian 5.0 myr (starting at 387.7 ± 0.8 Ma). The latter stage was estimated to be 4.5 myr in duration by Kaufmann (2006). Because of the relatively unconstrained cubic spline in this interval, no confidence limits are calculated for the Middle Devonian Epoch and the Givetian Stage.

The Late Devonian started at 382.7 ± 1.0 Ma and ended at 358.9 ± 0.4 Ma, lasting 23.8 ± 2.0 myr, almost as long as the Early Devonian. The Frasnian Stage lasted 10.5 ± 2.5 myr, and the Famennian Stage 13.3 ± 0.8 myr; the latter starts at 372.2 ± 1.6 Ma.

The error bars decrease in magnitude through the Devonian, reflecting more accurate and more precise (U/Pb) dates in the younger part of the period. But, looking at the spline fit of Figure 22.14, one cannot avoid the consideration that the Tioga monazite dates D7 and D8, of the *P. costatus* conodont zone of the middle Eifelian, would fit better if they were from a chronostratigraphic level that is one or two conodont zones higher. This deserves investigation, with the GTS2004 results suggesting this even more strongly (see Figure 14.5 of the chapter on Devonian in Gradstein *et al.*, 2004).

The GTS2004 ages for the Frasnian and Famennian lower stage boundaries are outside the GTS2012 spline error bars for these levels. This is significant, since the accuracy and precision of the dates in this interval are considered to be good. We consider that this part of the Devonian scale is maturing, although more infra-stage dates are desirable, not only focusing on the Hangenberg Event but also on the zones and biochronology leading up to that. The Givetian needs age dates to corroborate its 5 myr duration. The new and total duration of the Devonian is 60.3 myr (compared to 57.5 myr in GTS2004).

Note: After completion of this text, Parry *et al.* (2011) published a high-resolution U-Pb zircon age of 411.5 ± 1.3 Ma for an andesite lava within the Rhynie Outlier of NE Scotland. Using spores the chronostratigraphy of this level is in the interval from? lowermost Pragian through Lower Emsian. Accordingly, the Lochkovian–Pragian boundary should predate 411.5 Ma, and the Pragian–Emsian boundary should be younger or could approximate this age. Uncertainty about the Pragian–Emsian boundary definition should be resolved prior to refinement of the boundary age.

REFERENCES

Aboussalam, Z.S., 2003. Das "Taghanic-Event" im höheren Mittel-Devon von West-Europa und Marokko. Münstersche Forschungen zur Geologie und Paläontologie 97, 1–332.

Aboussalam, Z.S., Becker, R.T., 2002. The base of the *hermanni* Zone as the base of an Upper Givetian substage. Subcommission on Devonian Stratigraphy Newsletter 19, 25–34.

Aboussalam, Z.S., Becker, R.T., 2007. New upper Givetian to basal Frasnian conodont faunas from the Tafilalt (Anti-Atlas, Southern Morocco). Geological Quarterly 51 (4), 345–374.

Aboussalam, Z.S., Becker, R.T., 2011. The global Taghanic Biocrisis (Givetian) in the eastern Anti-Atlas, Morocco. Palaeogeography, Palaeoclimatology, Palaeoecology 304, 136–164.

Aitchison, J.C., Davis, A.M., Stratford, J.M.C., Spiller, F.C.P., 1999. Lower and Middle Devonian radiolarian biozonation of the Gamilaroi terrane, New England Orogen, eastern Australia. Micropaleontology 45 (2), 138–162.

Alberti, G.K.B., 1982. Nowakiidae (Dacryoconarida) aus dem Hunsrück-schiefer von Bundenbach (Rheinisches Schiefergebirge). Senckenbergiana lethaea 63 (5/6), 451–463.

Alberti, G.K.B., 1998. Planktonische Tentakuliten des Devon. III. Dacryoconarida Fisher 1962 aus dem Unter-Devon und oberen Mitteldevon. Palaeontographica, Abteilung A 250 (1–3), 1–46.

Alberti, G.K.B., 2000. Planktonische Tentakuliten des Devon. IV. Dacryoconarida Fisher 1962 aus Unter-Devon. Palaeontographica, Abteilung A 254, 1–23.

Avkhimovitch, V.I., Tchibrikova, E.V., Obukhovskaya, T.G., Nazarenko, A.M., Umnova, V.T., Raskatova, L.G., Mantsurova, V.N., Loboziak, S., Streel, M., 1993. Middle and Upper Devonian miospore zonation of eastern Europe. Bulletin des Centres Recherches Exploration-Production Elf-Aquitaine 17, 79–147.

Bábek, O., Prikryl, T., Hladil, J., 2007. Progressive drowning of carbonate platform in the Moravo-Silesian Basin (Czech Republic) before the Frasnian/Famennian event: Facies, compositional variation and gamma-ray spectrometry. Facies 53, 293–316.

Bai, S.L., 1995. Milankovitch cyclicity and time scale of the Middle and Upper Devonian. International Geology Review 37, 1109–1114.

Bai, S.L., Bai, Z.Q., Ma, X.P., Wang, D.R., Sun, Y.L., 1995. Devonian Events and Biostratigraphy of South China. Peking University Press, Beijing, p. 303.

Baird, G.C., Brett, C.E., 2003. Shelf and off-shelf deposits of the Tully Formation in New York and Pennsylvania: Faunal incursions, eustasy and tectonics. Courier Forschungsinstitut Senckenberg 242, 141–156.

Baird, G.C., Brett, C.E., 2008. Late Givetian Taghanic bioevents in New York State: New discoveries and questions. Bulletin of Geosciences 83 (4), 357–370.

Balinski, A., Olempska, E., Racki, G., 2002. Biotic responses to the Late Devonian global events. Acta Palaeontologica Polonica 47 (2), 186–404.

Balinski, A., Olempska, E., Racki, G., 2006. Biotic aspects of the Early-Middle Frasnian eventful transition. Acta Palaeontologica Polonica 51 (4), 606–832.

Banks, H.P., 1980. Floral assemblages in the Siluro - Devonian. In: Dilcher, D.L., Taylor, T.N. (Eds.), Biostratigraphy of Fossil Plants. Dowden, Hutchinson & Ross Inc, Stroudsville, pp. 1–24.

Bardashev, I.A., Weddige, K., Ziegler, W., 2002. The phylomorphogenesis of some Early Devonian Platform Conodonts. Senckenbergiana lethaea 82 (2), 375–451.

Becker, R.T., 1992. Zur Kenntnis von Hemberg-Stufe und *Annulata*-Event im Nordsauerland (Oberdevon, Rheinisches Schiefergebirge, GK 4611 Hohenlimburg). Berliner geowissenschaftliche Abhandlungen E3, 3–41.

Becker, R.T., 1993a. Stratigraphische Gliederung und Ammonoideen-Fauna im Nehdenium (Oberdevon II) von Europe und Nord-Afrika. Courier Forschungsinstitut Senckenberg 155, 1–405.

Becker, R.T., 1993b. Anoxia, eustatic changes, and Upper Devonian to lowermost Carboniferous global ammonoid diversity. Systematics Association Special Volume 47, 115–163.

Becker, R.T., 1996. New faunal records and holostratigraphic correlation of the Hasselbachtal D/C-Boundary Auxiliary Stratotype. Annales de la Société Géologique de Belgique 117 (1), 19–45.

Becker, R.T., 2005. Ammonoids and substage subdivisions in the Givetian open shelf facies. In: Yolkin, E.A., Izokh, N.G., Obut, O.T., Kipriyanova, T.P. (Eds.), Devonian Terrestrial and Marine Environments: From Continent to Shelf. Contributions from the IGCP 499 Project / SDS joint field meeting, Novosibirsk, Russia, July 25 – August 9, 2005, pp. 29–31.

Becker, R.T., 2006. Myths and facts concerning the Frasnian/Famennian boundary mass extinction. In: Yang, Q., Wang, Y., Weldon, E.A. (Eds.), Ancient Life and Modern Approaches. Abstracts of the Second International Palaeontological Congress, June 17–21, 2006, Beijing, China. University of Science and Technology of China Press, Beijing, p. 352.

Becker, R.T., 2007a. Results of the voting on Givetian and Frasnian substages. Subcommission on Devonian Stratigraphy Newsletter 22, 2.

Becker, R.T., 2007b. Emsian substages and the Daleje Event – a consideration of conodont, dacryoconarid, ammonoid and sealevel data. Subcommission on Devonian Stratigraphy Newsletter 22, 29–32.

Becker, R.T., 2009. Minutes of the SDS Business Meeting, Kitab State Geological Reserve, Uzbekistan. Subcommission on Devonian Stratigraphy Newsletter 24, 12–15.

Becker, R.T., Paproth, E., 1993. Auxiliary stratotype section for the Global Stratotype Section and Point (GSSP) for the Devonian-Carboniferous boundary: Hasselbachtal. Annales de la Société Géologique de Belgique 115 (2), 703–706.

Becker, R.T., House, M.R., 1994. International Devonian goniatite zonation, Emsian to Givetian, with new records from Morocco. Courier Forschungsinstitut Senckenberg 169, 79–135.

Becker, R.T., House, M.R., 1999. Proposals for an international substage subdivision of the Frasnian. Subcommission on Devonian Stratigraphy Newsletter 15, 17–22.

Becker, R.T., House, M.R., 2000. Devonian ammonoid zones and their correlation with established series and stage boundaries. Courier Forschungsinstitut Senckenberg 220, 113–151.

Becker, R.T., Aboussalam, Z.S., 2004. The Frasnes Event – a phased 2nd order global crisis and extinction period around the Middle-Upper Devonian boundary. In: El Hassani, A. (Ed.), Devonian Neritic-Pelagic Correlation and Events in the Dra Valley (Western Anti-Atlas, Morocco). International Meeting on Stratigraphy, Rabat, March 1–10, 2004, Abstracts, pp. 8–10.

Becker, R.T., Kirchgasser, W.T., 2007. Devonian Events and Correlations. The Geological Society Special Publication 278, 1–280.

Becker, R.T., House, M.R., 2009. Devonian ammonoid biostratigraphy of the Canning Basin. Geological Survey of Western Australia, Bulletin 145, 415–439.

Becker, R.T., Aboussalam, Z.S., 2011. Emsian chronostratigraphy – preliminary new data and a review of the Tafilalt (SE Morocco). Subcommission on Devonian Stratigraphy Newsletter 26, 33–43.

Becker, R.T., Feist, R., Flajs, G., House, M.R., Klapper, G., 1989. Frasnian-Famennian extinction events in the Devonian at Coumiac, southern France. Comptes Rendus de l'Academie des Sciences Paris, Series II 309, 259–266.

Becker, R.T., House, M.R., Kirchgasser, W.T., 1993. Devonian goniatite biostratigraphy and timing of facies movements in the Frasnian of the Canning Basin, Western Australia. Geological Society Special Publication 70, 293–321.

Becker, R.T., House, M.R., Menner, V.V., Ovnatanova, N.S., 2000. Revision of ammonoid biostratigraphy in the Frasnian (Upper Devonian) of the Southern Timan (Northeast Russian Platform). Acta Geologica Polonica 50 (1), 67–97.

Becker, R.T., House, M.R., Bockwinkel, J., Ebbighausen, V., Aboussalam, Z.S., 2002. Famennian ammonoid zones of the eastern Anti-Atlas (southern Morocco). Münstersche Forschungen zur Geologie und Paläontologie 93, 159–205.

Becker, R.T., Ashouri, A.R., Yazdi, M., 2004a. The Upper Devonian *Annulata* Event in the Shotori Range (eastern Iran). Neues Jahrbuch für Geologie und Paläontologie, Abhandlungen 231 (1), 119–143.

Becker, R.T., Jansen, U., Plodowski, G., Schindler, E., Aboussalam, Z.S., Weddige, K., 2004b. Devonian litho- and biostratigraphy of the Dra Valley area – an overview. Documents de l'Institut Scientifique 19, 3–18.

Becker, R.T., Aboussalam, Z.S., Bockwinkel, J., Ebbighausen, V., El Hassani, A., Nübel, H., 2004b. The Givetian and Frasnian at Oued Mzerreb (Tata region, eastern Dra Valley). Documents de l'Institut Scientifique 19, 29–43.

Becker, R.T., De Baets, K., Nikolaeva, S., 2010. New ammonoid records from the lower Emsian of the Kitab Reserve (Uzbekistan) – preliminary results. Subcommission on Devonian Stratigraphy Newsletter 25, 20–28.

Belka, Z., Kaufmann, B., Bultynck, P., 1997. Conodont-based quantitative biostratigraphy for the Eifelian of the eastern Anti-Atlas, Morocco. Geological Society of America, Bulletin 109 (6), 643–651.

Berkyová, S., 2009. Lower-Middle Devonian (upper Emsian-Eifelian, *serotinus-kockelianus* zones) conodont faunas from the Prague Basin, Czech Republic. Bulletin of Geosciences 84 (4), 667–686.

Bischoff, G., Ziegler, W., 1957. Die Conodontenchronologie des Mitteldevons und des tiefsten Oberdevons. Abhandlungen des Hessischen Geologischen Landesamtes für Bodenforschung 22, 1–136.

Bless, M.J.M., Becker, R.T., Higgs, K., Paproth, E., Streel, M., 1993. Eustatic cycles around the Devonian-Carboniferous Boundary and the sedimentary and fossil record in Sauerland (Federal Republic of Germany). Annales de la Société Géologique de Belgique 115 (2), 689–702.

Blieck, A., Turner, S., 2000. Palaeozoic Vertebrate Biochronology and Global Marine/Non-Marine Correlation – Final Report of IGCP 328 (1991–1996). Courier Forschungsinstitut Senckenberg 223, 1–575.

Blieck, A., Clement, G., Blom, H., Lelievre, H., Luksevics, E., Streel, M., Thorez, J., Young, G.C., 2007. The biostratigraphical and palaeogeographical framework of the earliest diversification of tetrapods (Late Devonian). In: Becker, R.T., Kirchgasser, W.T. (Eds.), Devonian Events and Correlations. The Geological Society Special Publication, 278, pp. 219–235.

Blieck, A., Clément, G., Streel, M., 2010. The biostratigraphical distribution of earliest tetrapods (Late Devonian) – a revised version with comments on biodiversification. In: Vecolli, M., Clément, G., Meyer-Berthaud, B. (Eds.), The Terrestrialization Process: Modelling Complex Interactions at the Biosphere-Geosphere Interface. The Geological Society Special Publication, 339, pp. 129–138.

Blumenstengel, H., 1965. Zur Taxionomie und Biostratigraphie verkieselter Ostracoden aus dem Thüringer Oberdevon. Freiberger Forschungshefte C183, 1–127. Acta Geologica Polonica 53(2): 93–99.

Bond, D., Wignall, P.B., Racki, G., 2004. Extent and duration of marine anoxia during the Frasnian-Famennian (Late Devonian) mass extinction in Poland, Germany, Austria and France. Geological Magazine 141 (2), 173–193.

Brand, U., 2004. Carbon, oxygen and strontium isotopes in Paleozoic carbonate components: An evaluation of original seawater-chemistry proxies. Chemical Geology 204, 23–24.

Brand, U., Legrand-Blain, M., Streel, M., 2004. Biochemostratigraphy of the Devonian-Carboniferous boundary global stratotype section and point, Griotte Formation, La Serre, Montagne Noire, France. Palaeogeography, Palaeoclimatology, Palaeoecology 195, 99–124.

Bratton, J.F., Berry, W.B.N., Morrow, J.R., 1999. Anoxia pre-dates Frasnian-Famennian boundary mass extinction horizon in the Great Basin, USA. Palaeogeography, Palaeoclimatology, Palaeoecology 154, 275–292.

Brett, C.E., Baird, G.C., Bartholomew, A., DeSantis, M., Ver Straeten, C., 2011. Sequence stratigraphy and revised sea level curve for the Middle Devonian of Eastern North America. Palaeogeography, Palaeoclimatology, Palaeoecology 304, 21–53.

Brett, C.E., Zambito, J.J., Schindler, E., and Becker, R.T., 2012. Diagenetically-enhanced trilobite obrution deposits in concretionary limestones: The paradox of "rhythmic events beds". Palaeogeography, Palaeoclimatology, Palaeoecology in press doi : 10.1016/j.palaeo.2011.12.004.

Brezinski, D.K., Cecil, C.B., Skema, V.W., 2010. Late Devonian glacigenic and associated facies from the central Appalachian Basin, eastern United States. Geological Society of America Bulletin 122 (1/2), 265–281.

Budil, P., 1995. Demonstrations of the Kacák event (Middle Devonian, uppermost Eifelian) at some Barrandian localities. Vestnik Ceského geologeckého ústavu 70, 1–24.

Buggisch, W., Mann, W., 2004. Carbon isotope stratigraphy of Lochkovian to Eifelian limestones from the Devonian of central and southern Europe. International Journal of Earth Sciences 93, 521–541.

Buggisch, W., Joachimski, M.M., 2006. Carbon isotope stratigraphy of the Devonian of Central and Southern Europe. Palaeogeography, Palaeoclimatology, Palaeoecology 240, 68–88.

Bultynck, P., 1985. Lower Devonian (Emsian)-Middle Devonian (Eifelian and lowermost Givetian) conodont successions from the Ma'der and the Tafilalt, southern Morocco. Courier Forschungsinstitut Senckenberg 75, 261–286.

Bultynck, P., 1987. Pelagic and neritic conodont successions from the Givetian of pre-Sahara Morocco and the Ardennes. Bulletin de l'Institut Royal des Sciences Naturelles de Belgique, Sciences de la Terre 57, 149–181.

Bultynck, P., 2000a. Fossil groups important for boundary definition. Courier Forschungsinstitut Senckenberg 220, 1–205.

Bultynck, P., 2000b. Recognition of Devonian series and stage boundaries in geological areas. Courier Forschungsinstitut Senckenberg 225, 1–347.

Bultynck, P., 2003. Devonian Icriodontidae: Biostratigraphy, classification and remarks on palaeoecology and dispersal. Revista Española de Paleontología 35 (3), 295–314.

Bultynck, P., 2005. The Givetian Working Group. Subcommission on Devonian Stratigraphy Newsletter 21, 20–22.

Bultynck, P., 2007. Limitations on the application of the Devonian standard conodont zonation. Geological Quarterly 51 (4), 339–344.

Bultynck, P., Martin, F., 1995. Assessment of an old stratotype: The Frasnian/Famennian boundary at Senzeilles, Southern Belgium. Bulletin de l'Institut Royal des Sciences Naturelles de Belgique, Sciences de la Terre 65, 5–34.

Bultynck, P., Hollevoet, C., 1999. The Eifelian-Givetian boundary and Struve's Middle Devonian Great Cap in the Couvin area (Ardennes, southern Belgium). Senckenbergiana lethaea 79 (1), 3–11.

Bultynck, P., Gouwy, S., 2002. Towards a standardization of global Givetian substages. In: Yushkin, N.P., Tsyganko, V.S., Männik, P. (Eds.), Geology of the Devonian System: Proceedings of the International Symposium, July 9–12 2002, Syktyvkar, Komi Republic, pp. 142–144.

Caplan, M.L., Bustin, R.M., 1999. Devonian-Carboniferous Hangenberg mass extinction event, widespread organic-rich mudrock and anoxia: Causes and consequences. Palaeogeography, Palaeoclimatology, Palaeoecology 148 (4), 187–207.

Caputo, M.V., Melo, J.H.G., Streel, M., Isbell, J.L., 2008. Late Devonian and Early Carboniferous glacial records of South America. Geological Society of America, Special Paper 441, 161–173.

Carls, P., Valenzuela-Ríos, J.I., 2007. From the Emsian GSSP to the early late-Emsian correlations with historical boundaries. Subcommission on Devonian Stratigraphy Newsletter 22, 24–28.

Carls, P., Slavík, L., Valenzuela-Ríos, J.I., 2007. Revisions of conodont biostratigraphy across the Silurian-Devonian boundary. Bulletin of Geosciences 82 (2), 145–164.

Carls, P., Slavík, L., Valenzuela-Ríos, J.I., 2008. Comments on the GSSP for the basal Emsian stage boundary: The need for its redefinition. Bulletin of Geosciences 83 (4), 383–390.

Carpenter, S.J., Lohmann, K.C., Holden, P., Walter, L.M., Huston, T.J., Haliday, A.N., 1991. $\delta^{18}O$ values, $^{87}Sr/^{86}Sr$ and Sr/Mg ratios of Late Devonian abiotic calcite: Implications for the composition of ancient seawater. Geochimica et Cosmochimica Acta 55, 1991–2010.

Chen, D., Tucker, M.E., 2003. The Frasnian-Famennian mass extinction: Insights from high-resolution sequence stratigraphy and cyclostratigraphy in South China. Palaeogeography, Palaeoclimatology, Palaeoecology 193, 87–111.

Chen, D., Qing, H., Li, R., 2005. The Late Devonian Frasnian-Famennian (F/F) biotic crisis: Insights from $\delta^{13}C_{carb}$, $\delta^{13}C_{org}$ and $^{87}Sr/^{86}Sr$ isotopic systematics. Earth and Planetary Science Letters 235, 151–166.

Chlupáč, I., 2000. Cyclicity and duration of Lower Devonian stages: Observations from the Barrandian area, Czech Republic. Neues Jahrbuch für Geologie und Paläotologie, Abhandlungen 215 (1), 97–124.

Chlupáč, I., Turek, V., 1983. Devonian goniatites from the Barrandian area, Czechoslovakia. Rozpravy Ùstrédního ústavu geologického 46, 1–159.

Chlupác, I., Oliver Jr., W.A., 1989. Decision on the Lochkovian - Pragian boundary stratotype (Lower Devonian). Episodes 12, 109–113.

Chlupáč, I., Lukes, P., 1999. Pragian/Zlichovian and Zlichovian/Dalejan boundary sections in the Lower Devonian of the Barrandian area, Czech Republic. Newsletters on Stratigraphy 37 (1/2), 75–100.

Chlupáč, I., Hladil, J., 2000. The global stratotype section and point of the Silurian-Devonian boundary. Courier Forschungsinstitut Senckenberg 225, 1–7.

Clack, J.A., 2002. Gaining Ground: The Origin and Evolution of Tetrapods. Indiana University Press, Bloomington, p. 369.

Claeys, P., Kyte, F.T., Herbosch, A., Casier, J.-G., 1996. Geochemistry of the Frasnian-Famennian boundary in Belgium: Mass extinction, anoxic oceans and microtektite layer, but not much iridium. Geological Scoiety of America, Special Paper 307, 491–504.

Clark, S., Day, J., Ellwood, B., Harry, R., Tomkin, J., 2009. Astronomical tuning of integrated Upper Famennian-Early Carboniferous faunal, carbon isotope and high resolution magnetic susceptibility records: Western Illinois Basin. Subcommission on Devonian Sratigraphy Newsletter 24, 27–35.

Clausen, C.-D., Lorn, D., Luppold, F.W., Stoppel, D., 1989. Untersuchungen zur Devon/Karbon-Grenze auf dem Müssenberg (Nördliches Rheinisches Schiefergebirge). Bulletin de la Société Belge de Géologie 98 (3/4), 353–369.

Compston, W., 2000. Interpretations of SHRIMP and isotope dilution zircon ages for the Palaeozoic time-scale: II. Silurian to Devonian. Mineralogical Magazine 64 (6), 1127–1146.

Compston, W., 2004. SIMS U-Pb zircon ages for the Upper Devonian Snobs Creek and Cerberean Volcanics from Victoria, with age uncertainty based on UO_2/UO v. UO/U precision. Journal of the Geological Society 161, 223–228.

Corradini, C., 2008. Revision of Famennian-Tournaisian (Late Devonian – Early Carboniferous) conodont biostratigraphy of Sardinia, Italy. Revue de Micropaléontologie 51, 123–132.

Cramer, B.D., Saltzman, M.R., Day, J., Witzke, B.J., 2008. Record of the Late Devonian Hangenberg global positive carbon isotope excursion in epeiric sea setting: Carbonate production, organic carbon burial, and paleoceanography during the Late Famennian. In: Pratt, B., Holmden, C. (Eds.), Epeiric Seas. Geological Association of Canada, Paper, 48, pp. 103–118.

Crick, R.E., Ellwood, B.B., Feist, R., El Hassani, A., Feist, R., Hladil, J., 1997. MagnetoSusceptibility Event and Cyclostratigraphy (MSEC) of the Eifelian-Givetian GSSP and associated boundary sequences in north Africa and Europe. Episodes 20, 167–174.

Crick, R.E., Ellwood, B.B., Feist, R., El Hassani, A., Feist, R., 2000. Proposed magnetostratigraphy susceptibility stratotype for the Eifelian-Givetian GSSP (Anti-Atlas Morocco). Episodes 23, 93–101.

Crick, R.E., Ellwood, B.B., El Hassani, A., Hladil, J., Hrouda, F., Chlupáč, I., 2001. Magnetostratigraphy susceptibility of the Pridoli-Lochkovian (Silurian-Devonian) GSSP (Klonk, Czech Republic) and a coeval sequence in Anti-Atlas Morocco. Palaeogeography, Palaeoclimatology, Palaeoecology 167, 73–100.

Da Silva, A.-C., Boulvain, F., 2002. Sedimentology, magnetic susceptibility and isotopes of a Middle Frasnian carbonate platform: Talifer Section, Belgium. Facies 46, 89–102.

Da Silva, A.-C., Boulvain, F., 2006. Upper Devonian carbonate platform correlations and sea level variations recorded in magnetic susceptibility. Palaeogeography, Palaeoclimatology, Palaeoecology 240, 373–388.

Da Silva, A.-C., Boulvain, F., 2008. Carbon isotope lateral variability in a Middle Frasnian carbonate platform (Belgium): Significance of facies, diagenesis and sea-level history. Palaeogeography, Palaeoclimatology, Palaeoecology 269, 189–204.

Day, J., Uyeno, T., Norris, W., Witzke, B.J., Bunker, B.J., 1996. Middle-Upper Devonian relative sea-level histories of central and western North American interior basin. Geological Society of America, Special Paper 306, 259–275.

De Baets, K., Klug, C., Plusquellec, Y., 2010. Zlíchovian faunas with early ammonoids from Morocco and their use for the correlation of the eastern Anti-Atlas and the western Dra Valley. Bulletin of Geosciences 85 (2), 317–352.

Denison, R.E., Koepnick, R.B., Burke, W.H., Hetherington, E.A., Fletcher, A., 1997. Construction of the Silurian and Devonian seawater $^{87}Sr/^{86}Sr$ curve. Chemical Geology 140, 109–121.

DeSantis, M.K., Brett, C.E., Ver Straeten, C.A., 2007. Persistent depositional sequences and bioevents in the Eifelian (early Middle Devonian) of eastern Laurentia: North American evidence of the Kacák Events? In: Becker, R.T., Kirchgasser, W.T. (Eds.), Devonian Events and Correlations. Geological Society Special Publication 278, pp. 83–104.

DeSantis, M.K., Brett, C.E., 2011. Late Eifelian (Middle Devonian) biocrises: Timing and signature of the pre-Kačak Bakoven and Stony Hollow Events in eastern North America. Palaeogeography, Palaeoclimatology, Palaeoecology 304, 113–135.

Diener, A., Ebneth, S., Veizer, J., Buhl, D., 1996. Strontium isotope stratigraphy of the Middle Devonian: Brachiopods and conodonts. Geochimica et Cosmochimica Acta 60, 639–652.

Dopieralska, J., Belka, Z., Haack, U., 2006. Geochemical decoupling of water masses in the Variscan oceanic system during Late Devonian times. Palaeogeography, Palaeoclimatology, Palaeoecology 240, 108–119.

Ebbighausen, V., Becker, R.T., Bockwinkel, J., 2011. Emsian and Eifelian ammonoids from Oufrane, eastern Dra Valley (Anti-Atlas, Morocco) – Taxonomy, stratigraphy and correlation. Neues Jahrbuch für Geologie und Paläontologie 259 (3), 313–379.

Ebneth, S., Diener, A., Buhl, D., Veizer, J., 1997. Strontium isotope systematics of conodonts: Middle Devonian, Eifel Mountains, Germany. Palaeogeography, Palaeoclimatology, Palaeoecology 132, 79–96.

Edwards, D., Fairon-Demaret, M., Berry, C.M., 2000. Plant megafossils in Devonian stratigraphy: A progress report. Courier Forschungsinstitut Senckenberg 220, 25–37.

Elles, G.L., Wood, E.M.R., 1901–1918. A monograph of British graptolites. Monograph of the Palaeontographical Society, 11 parts, p. 539.

Ellwood, B.B., Crick, R.E., El Hassani, A., Benoist, S., Young, R., 2000. Magnetosusceptibility event and cyclostratigraphy method applied to marine rocks: Detrital input versus carbonate productivity. Geology 28, 1135–1138.

Ellwood, B.B., Crick, R.E., Garcia-Alcalde Fernandez, J.L., Soto, F.M., Truyols-Massoni, M., El Hassani, A., Kovas, E.J., 2001. Global correlation using magnetic susceptibility data from Lower Devonian rocks. Geology 29, 583–586.

Ellwood, B.B., Benoist, S.L., El Hassani, A., Wheeler, C., Crick, R.E., 2003. Impact ejecta layer from the mid-Devonian: Possible connection to global mass extinctions. Science 300, 1734–1737.

Ellwood, B.B., García-Alcalde, J.L., El Hassani, A., Hladil, J., Soto, F.M., Truyóls-Massoni, M., Weddige, K., Koptikova, L., 2006. Stratigraphy of the Middle Devonian boundary: Formal definition of the susceptibility magnetostratigraphy in Germany with comparisons to sections in the Czech Republic, Morocco and Spain. Tectonophysics 418, 31–49.

Ellwood, B.B., Algeo, T.J., El Hassani, A., Tomkin, J.H., Rowe, H.D., 2011a. Defining the timing and duration of the Kačak Interval within the Eifelian/Givetian boundary GSSP, Mech Irdane, using geochemical and magnetic susceptibility patterns. Palaeogeography, Palaeoclimatology, Palaeoecology 304, 74–84.

Ellwood, B.B., Tomkin, J.H., El Hassani, A., Bultynck, P., Brett, C.E., Schindler, E., Feist, R., Bartholomew, A.J., 2011b. A climate-driven model and development of a floating point time scale for the entire Middle Devonian Givetian Stage: A test using magnetostratigraphy susceptibility as a climate proxy. Palaeogeography, Palaeoclimatology, Palaeoecology 304, 85–95.

Elrick, M., Berkyová, S., Klapper, G., Sharp, Z., Joachimski, M., Frýda, J., 2009. Stratigraphic and oxygen isotope evidence for My-scale glaciation driving eustasy in the Early-Middle Devonian greenhouse world. Palaeogeography, Palaeoclimatology, Palaeoecology 276, 170–181.

Feist, R., Klapper, G., 1985. Stratigraphy and conodonts in pelagic sequences across the Middle–Upper Devonian boundary, Montagne Noire, France. Palaeontographica A188, 1–8.

Feist, R., Flajs, G., Girard, C., 2000. The stratotype section of the Devonian-Carboniferous boundary. Courier Forschungsinstitut Senckenberg 225, 77–82.

Filipiak, P., Racki, G., 2010. Proliferation of abnormal palynoflora during the end-Devonian biotic crisis. Geological Quarterly 54 (1), 1–14.

Flajs, G., Feist, R., 1988. Index conodonts, trilobites and environment of the Devonian-Carboniferous Boundary beds at La Serre (Montagne Noire, France). Courier Forschungsinstitut Senckenberg 100, 53–107.

García-Alcalde, J.L., 1997. North Gondwanan Emsian events. Episodes 20 (4), 241–246.

Geldsetzer, H.H.J., Goodfellow, W.D., McLaren, D.J., Orchard, M.J., 1987. Sulfur-isotope anomaly associated with the Frasnian-Famennian extinction, Medicine Lake, Alberta, Canada. Geology 15, 393–396.

Gersl, M., Hladil, J., 2004. Gamma-ray and magnetic susceptibility correlation across a Frasnian carbonate platform and the search for "*punctata* equivalents in stromatoporoid-coral limestone facies of Moravia. Geological Quarterly 48 (3), 283–292.

Ginter, M., Ivanov, A., 2000. Stratigraphic distribution of chondrichthyans in the Devonian on the East European Platform margin. Courier Forschungsinstitut Senckenberg 223, 325–339.

Ginter, M., Hairapetian, V., Klug, C., 2002. Famennian chondrichthyans from the shelves of North Gondwana. Acta Geologica Polonica 52 (2), 169–215.

Girard, C., Albarède, F., 1996. Trace elements in conodont phosphates from the Frasnian/Famennian boundary. Palaeogeography, Palaeoclimatology, Palaeoecology 126, 195–209.

Girard, C., Lecuyer, C., 2002. Variations in Ce anomalies of conodonts through the Frasnian/Famennian boundary of Poland (Kowala – Holy Cross Mountains: Implications for the redox state of seawater and biodiversity. Palaeogeography, Palaeoclimatology, Palaeoecology 181, 209–311.

Girard, C., Robin, E., Rocchia, R., Froget, L., Feist, R., 1997. Search for impact remains at the Frasnian-Famennian boundary in the stratotype area, southern France. Palaeogeography, Palaeoclimatology, Palaeoecology 132, 391–397.

Girard, C., Klapper, G., Feist, R., 2005. Subdivision of the terminal Frasnian *linguiformis* conodont Zone, revision of the correlative interval of Montagne Noire Zone 13, and discussion of stratigraphically significant associated trilobites. In: Over, D.J., Morrow, J.R., Wignall, P.B. (Eds.), Understanding Late Devonian and Permian-Triassic Biotic and Climatic Events: Towards an Integrated Approach. Developments in Palaeontology & Stratigraphy, 20, pp. 181–198.

Gong, X., Huang, H., Zhang, M., Huang, Q., 1991. The Stratigraphic Classification and Correlation of Carbonate Rocks of Upper Devonian and Lower Carboniferous in Guilin Karst Region. Guangxi Science and Technology Publishing House, Guilin, p. 106.

Goodfellow, W.D., Geldsetzer, H.H.J., McLaren, D.J., Orchard, M.J., Klapper, G., 1988. The Frasnian-Famennian extinction: Current results and possible causes. Canadian Society of Petroleum Geologists, Memoir 14 (III), 9–21.

Gouwy, S., Bultynck, P., 2000. Graphic correlation of Frasnian sections (Upper Devonian) in the Ardennes, Belgium. Bulletin de l'Institut Royal des Sciences Naturelles de Belgique, Sciences de la Terre 70, 25–52.

Gouwy, S., Bultynck, P., 2003. Conodont based graphic correlation of the Middle Devonian formations of the Ardennes (Belgium): Implications for stratigraphy and construction of a regional composite. Revista Española de Micropaleontologia 35 (3), 315–344.

Gouwy, S., Haydukiewicz, J., Bultynck, P., 2007. Conodont-based graphic correlation of upper Givetian-Frasnian sections of the Eastern Anti-Atlas (Morocco). Geological Quarterly 51 (4), 375–392.

Grahn, Y., 2003. Silurian and Devonian chitinozoan assemblages from the Chaco-Paraña Basin, northeastern Argentina and Central Uruguay. Revista Española de Micropaleontología 35, 1–8.

Grahn, Y., 2005. Devonian chitinozoan biozones of western Gondwana. Acta Geologica Polonica 55, 211–227.

Grahn, Y., Melo, J.H.G., 2004. Integrated Middle Devonian chitinozoan and miospore zonation of the Amazonas Basin, northern Brazil. Revue de Micropaléontologie 47, 71–85.

Grahn, Y., Mauller, P.M., Pereira, E., Loboziak, S., 2010. Palynostratigraphy of the Chapada Group and its significance in the stratigraphy of the Paraná Basin, south Brazil. Journal of South American Earth Sciences 29, 354–370.

Grandjean-Lécuyer, P., Feist, R., Albarède, F., 1993. Rare earth elements in old biogenic apatites. Geochimica et Cosmochimica Acta 57, 2507–2514.

Groos-Uffenorde, H., Lethiers, F., Blumenstengel, H., 2000. Ostracodes and Devonian stage boundaries. Courier Forschungsinstitut Senckenberg 220, 99–111.

Halas, S., Balínski, A., Gruszczynski, M., Hoffman, A., Malkowski, K., Narkiewicz, M., 1992. Stable isotope record at the Frasnian/Famennian boundary in southern Poland. Neues Jahrbuch für Geologie und Paläontologie, Monatshefte 1992 (3), 129–138.

Harland, W.B., Armstrong, R.L., Cox, A.V., Craig, L.A., Smith, A.G., Smith, D.G., 1990. A Geologic Time Scale 1989. Cambridge University Press, Cambridge, p. 263.

Hartenfels, S., 2011. Die globalen *Annulata*-Events und die Dasberg-Krise (Famennium, Oberdevon) in Europa und Nord-Afrika – hochauflösende Conodonten-Stratigraphie, Karbonat –Mikrofazies, Paläoökologie und Paläodiversität. Münstersche Forschungen zur Geologie und Paläontologie 105, 17–527.

Hartenfels, S., Becker, R.T., 2009. Timing of the global Dasberg Crisis – implications for Famennian eustasy and chronostratigraphy. Palaeontographica Americana 63, 69–95.

Hartenfels, S., Becker, R.T., Tragelehn, H., 2009. Marker conodonts around the global *Annulata* Events and the definition of an Upper Famennian substage. Subcommission on Devonian Stratigraphy Newsletter 24, 40–48.

Hartkopf-Fröder, C., Kloppisch, M., Mann, U., Neumann-Mahlkau, P., Schaefer, R.G., Wilkes, H., 2007. The end-Frasnian mass extinction in the Eifel Mountains, Germany: New insights from organic matter composition and preservation. In: Becker, R.T., Kirchgasser, W.T. (Eds.), Devonian Events and Correlations. Geological Society Special Publication, 278, pp. 173–196.

Higgs, K.T., Streel, M., Korn, D., Paproth, E., 1993. Palynological data from the Devonian-Carboniferous boundary beds in the new Stockum trench II and the Hasselbach borehole, Northern Rhenish Massif, Germany. Annales de la Société Géologique de Belgique 115, 551–557.

Hladíková, J., Hladil, J., Kríbek, B., 1997. Carbon and oxygen isotope record across Pridoli to Givetian stage boundaries in the Barrandian basin (Czech Republic). Palaeogeography, Palaeoclimatology, Palaeoecology 132, 225–241.

Hladil, J., Pruner, P., Venhodová, D., Hladíkova, T., Man, O., 2002. Toward an exact age of Middle Devonian Celechovice corals – past problems in biostratigraphy and present solutions complemented by new magnetosusceptibility measurements. Coral Research Bulletin 7, 65–71.

Hladil, J., Gersl, M., Strnad, L., Frana, J., Langrova, A., Spisiak, J., 2006. Stratigraphic variation of complex impurities in platform limestones and possible significance of atmospheric dust: A study with emphasis on gamma-ray spectrometry and magnetic susceptibility outcrop logging (Eifelian-Frasnian, Moravia, Czech Republic). International Journal of Earth Sciences 95, 703–723.

Holdsworth, B.K., Jones, D.L., 1980. Preliminary radiolarian zonation for Late Devonian through Permian time. Geology 8, 281–285.

House, M.R., 1978. Devonian ammonoids from the Appalachians and their bearing on international zonation and correlation. Special Papers in Palaeontology 21, 1–70.

House, M.R., 1983. Devonian eustatic events. Proceedings of the Ussher Society 5, 396–405.

House, M.R., 1985. Correlation of mid-Palaeozoic ammonoid evolutionary events with global sedimentary perturbations. Nature 313, 17–22.

House, M.R., 1991. Devonian Period. Encyclopaedia Britannica 19, 804–814.

House, M.R., 1995. Devonian precessional and other signatures for establishing a Givetian timescale. Geological Society Special Publication 85, 37–49.

House, M.R., 1996. The Middle Devonian Kačak Event. Proceedings of the Ussher Society 9, 79–84.

House, M.R., 2002. Strength, timing, setting and cause of mid-Palaeozoic extinctions. Palaeogeography, Palaeoclimatology, Palaeoecology 181, 5–25.

House, M.R., Ziegler, W., 1977. The goniatite and conodont sequences in the early Upper Devonian at Adorf, Germany. Geologica et Palaeontologica 11, 69–108.

House, M.R., Kirchgasser, W.T., 1993. Devonian goniatite biostratigraphy and timing of facies movements in the Frasnian of eastern North America. In: Hailwood, E.A., Kidd, R.B. (Eds.), High Resolution Stratigraphy. Geological Society Special Publication, 70, 267–292.

House, M.R., Ziegler, W., 1997. On sea-level fluctuations in the Devonian. Courier Forschungsinstitut Senckenberg 199, 1–146.

House, M.R., Gradstein, F.M., 2004. The Devonian Period. In: Gradstein, F.M., Ogg, J.G., Smith, A.G. (Eds.), A Geologic Time Scale 2004. Cambridge University Press, Cambridge, pp. 202–221.

House, M.R., Kirchgasser, W.T., 2008. Late Devonian Goniatites (Cephalopoda, Ammonoidea) from New York State. Bulletins of American Paleontology 374, 1–288.

House, M.R., Kirchgasser, W.T., Price, J.D., Wade, G., 1985. Goniatites from Frasnian (Upper Devonian) and adjacent strata of the Montagne Noire. Hercynica 1, 1–21.

House, M.R., Feist, R., Korn, D., 2000a. The Middle/Upper Devonian boundary at Puech de la Suque, Southern France. Courier Forschungsinstitut Senckenberg 225, 49–58.

House, M.R., Becker, R.T., Feist, R., Flajs, G., Girard, C., Klapper, G., 2000b. The Frasnian/Famennian boundary GSSP at Coumiac, southern France. Courier Forschungsinstitut Senckenberg 225, 59–75.

Izokh, N.G., Obut, O.T., Morrow, J., Sandberg, C.A., 2008. New findings of Upper Frasnian conodonts and radiolarians. In: Königshof, P., Linnemann, U. (Eds.), From Gondwana and Laurussia to Pangea: Dynamics of Oceans and Supercontinents, 20th International Senckenberg Conference and 2nd Geinitz Conference, Final Meeting of IGCP 497 and IGCP 499, Abstracts and Programme, pp. 186–188.

Jaeger, H., 1977. Graptolites. In: Martinsson, A. (Ed.), The Silurian-Devonian Boundary. International Union of Geological Sciences, Series A, 5, pp. 337–345.

Jansen, U., 2008. Biostratigraphy and correlation of the traditional Emsian stage. In: Kim, A.I., Salimova, F.A., Meshchankina, N.A. (Eds.), Global Alignments of Lower Devonian Carbonate and Clastic Sequences: Contributions from the IGCP 499 Project/SDS Joint Field Meeting, Kitab State Geological Reserve, Uzbekistan, August 25 – September 3, 2008, pp. 42–45.

Ji, Q., 1989. A comparison between the Dapoushang section and the other candidate sections for the Devonian-Carboniferous boundary stratotype. In: Ji, Q. (Ed.), The Dapoushang Section, an Excellent Section for the Devonian-Carboniferous Boundary Stratotype in China. Science Press, Beijing, pp. 66–79.

Ji, Q., Ziegler, W., 1993. Lali section: An excellent reference section for Upper Devonian in south China. Courier Forschungsinstitut Senckenberg 157, 183.

Joachimski, M.M., 1997. Comparisons of organic and inorganic carbon isotope patterns across the Frasnian-Famennian boundary. Palaeogeography, Palaeoclimatology, Palaeoecology 132, 133–145.

Joachimski, M.M., Buggisch, W., 1993. Anoxic events in the late Frasnian – Causes of the Frasnian-Famennian faunal crisis? Geology 21, 657–678.

Joachimski, M.M., Buggisch, W., 2002. Conodont apatite $\delta^{18}O$ signatures indicate climatic cooling as a trigger of the Late Devonian mass extinction. Geology 30 (8), 711–714.

Joachimski, M.M., Ostertag-Henning, C., Pancost, R.D., Strauss, H., Freeman, K.H., Littke, R., Sinninghe Damsté, J.S., Racki, G., 2001. Water column anoxia, enhanced productivity and concomitant changes in $\delta^{13}C$ and $\delta^{34}S$ across the Frasnian-Famennian boundary (Kowala – Holy Cross Mountains / Poland). Chemical Geology 175, 109–131.

Joachimski, M.M., Pancrost, R.D., Freeman, K.A., Ostertag-Henning, C., Buggisch, W., 2002. Carbon isotope geochemistry of the Frasnian - Famennian transition. Palaeogeography, Palaeoclimatology Palaeoecology 181, 91–109.

Joachimski, M.M., van Geldern, R., Breisig, S., Buggisch, W., Day, J., 2004. Oxygen isotope evolution of biogenic calcite and apatite during the Middle and Late Devonian. International Journal of Earth Sciences 93, 542–553.

Joachimski, M.M., Breisig, S., Buggisch, W., Talent, J.A., Mawson, R., Gereke, M., Morrow, J.R., Day, J., Weddige, K., 2009. Devonian climate and reef evolution: Insights from oxygen isotopes in apatite. Earth and Planetary Science Letters 284, 599–609.

Johnson, J.G., Klapper, G., Sandberg, C.A., 1985. Devonian eustatic fluctuations in Euramerica. Geological Society of America Bulletin 96, 567–587.

Kaiser, S.I., 2007. Conodontenstratigraphie und Geochemie ($\delta^{13}C_{carb}$, $\delta^{13}C_{org}$ $\delta^{18}O_{phosph}$) aus dem Devon/Karbon-Grenzbereich der Karnischen Alpen. Jahrbuch der Geologischen Bundesanstalt 147 (1/2), 301–314.

Kaiser, S.I., 2009. The Devonian/Carboniferous boundary stratotype section (La Serre, France) revisited. Newsletters on Stratigraphy 43 (2), 195–205.

Kaiser, S.I., Corradini, C., 2011. The early siphonodellids (Conodonta, Late Devonian-Early Carboniferous): Overview and taxonomic state. Neues Jahrbuch für Geologie und Paläontologie, Abhandlungen 261 (1), 19–35.

Kaiser, S.I., Steuber, T., Becker, R.T., Joachimski, M.M., 2006. Geochemical evidence for major environmental change at the Devonian-Carboniferous boundary in the Carnic Alps and the Rhenish Massif. Palaeogeography, Palaeoclimatology, Palaeoecology 240 (1–2), 146–160.

Kaiser, S.I., Steuber, T., Becker, R.T., 2008. Environmental change during the Late Famennian and Early Tournaisian (Late Devonian-Early Carboniferous): Implications from stable isotopes and conodont biofacies in southern Europe. Geological Journal 43, 241–260.

Kaiser, S.I., Becker, R.T., Spaletta, C., 2009. High-resolution conodont stratigraphy, biofacies and extinctions around the Hangenberg Event in pelagic successions from Austria, Italy, and France. Palaeontographica Americana 63, 97–139.

Kaiser, S.I., Becker, R.T., Steuber, T., Aboussalam, Z.S., 2011. Climate-controlled mass extinctions, facies, and sea-level changes around the Devonian-Carboniferous boundary in the eastern Anti-Atlas (SE Morocco). Palaeogeography, Palaeoclimatology, Palaeoecology 310, 340–364. doi:10.1016/j.palaeo.2011.07.026.

Kaufmann, B., 2006. Calibrating the Devonian Time Scale: A synthesis of U-Pb ID-TIMS ages and conodont stratigraphy. Earth-Science Reviews 76, 175–190.

Kaufmann, B., Trapp, E., Mezger, K., 2004. The numerical age of the Upper Frasnian (Upper Devonian) Kellwasser Horizons: A new U-Pb Zircon date from Steinbruch Schmidt (Kellerwald, Germany). Journal of Geology 112, 495–501.

Kaufmann, B., Trapp, E., Mezger, K., Wedige, K., 2005. Two new Emsian (Early Devonian) U-Pb zircon ages from volcanic rocks of the Rhenish Massif (Germany): Implications for the Devonian time scale. Journal of the Geological Society 162, 363–371.

Kim, A.I., 2011. Devonian tentaculites from the Kitab State Geological Reserve (Zeravshan-Gissar mountainous area, Uzbekistan). News on Paleontology and Stratigraphy, Supplement to Geologiya I Geofizika 52 (15), 65–81.

Klapper, G., 1985. Sequence in conodont genus *Ancyrodella* in Lower *asymmetricus* Zone (earliest Frasnian, Upper Devonian) of the Montagne Noire, France. Palaeontographica A188, 19–34.

Klapper, G., 1989. The Montagne Noire Frasnian (Upper Devonian) conodont succession. In: McMillan, N.J., Embry, A.F., Glass, D.J. (Eds.), Devonian of the World. Canadian Society of Petroleum Geology, Memoir, 14, pp. 449–468.

Klapper, G., 1992. North American midcontinent Devonian T-R cycles. Oklahoma Geological Survey Bulletin 145, 127–135.

Klapper, G., 1997. Graphic correlation of Frasnian (Upper Devonian) sequences in Montagne Noire, France, and western Canada. Geological Society of America, Special Paper 321, 113–129.

Klapper, G., 2000a. Species of Spathiognathodontidae and Polygnathidae (Conodonta) in the recognition of Upper Devonian stage boundaries. Courier Forschungsinstitut Senckenberg 220, 153–159.

Klapper, G., 2000b. A flawed Devonian timescale. The Australian Geologist 114, 12.

Klapper, G., 2007a. Conodont taxonomy and the recognition of the Frasnian/Famennian (Upper Devonian) Stage Boundary. Stratigraphy 4 (1), 67–76.

Klapper, G., 2007b. Frasnian (Upper Devonian) conodont succession at Horse Spring and correlative sections, Canning Basin, Western Australia. Journal of Paleontology 81 (3), 513–537.

Klapper, G., Becker, R.T., 1999. Comparison of Frasnian (Upper Devonian) conodont zonations. Bolletino della Società Paleontologica Italiana 37, 339–347.

Klapper, G., Feist, R., House, M.R., 1987. Decision on the Boundary Stratotype for the Middle/Upper Devonian Series Boundary. Episodes 10, 97–101.

Klapper, G., Feist, R., Becker, R.T., House, M.R., 1993. Definition of the Frasnian / Famennian Stage boundary. Episodes 16, 433–441.

Klapper, G., Kirchgasser, W.T., Baesemann, J.F., 1995. Graphic correlation of a Frasnian (Upper Devonian) composite standard. SEPM Special Publication 53, 177–184.

Klapper, G., Kuz'min, A., Ovnatanova, N.S., 1996. Upper Devonian conodonts from the Timan-Pechora region, Russia, and correlation with a Frasnian composite standard. Journal of Paleontology 70 (1), 131–152.

Klapper, G., Uyeno, T.T., Armstrong, D.K., Telford, P.G., 2004. Conodonts of the Williams Island and Long Rapids Formations (Upper Devonian, Frasnian-Famennian) of the Onakawana B Drillhole, Moose River Basin, Northern Ontario, with a revision of Lower Famennian species. Journal of Paleontology 78 (2), 371–387.

Kleffner, M.A., Barrick, J.E., Ebert, J.R., Matteson, D.K., Karlsson, H.R., 2009. Conodont biostratigraphy, $\delta^{13}C$ chemostratigraphy, and recognition of Silurian/Devonian boundary in the Cherry Valley, New York region of the Appalachian Basin. In: Over, D.J. (Ed.), Conodont Studies Commemorating the 150th Anniversary of the First Conodont Paper (Pander, 1856) and the 40th Anniversary of the Pander Society. Palaeontographica Americana, 62, pp. 57–73.

Klug, C., 2001. Early Emsian ammonoids from the eastern Anti-Atlas (Morocco) and their succession. Paläontologische Zeitschrift 74, 479–515.

Klug, C., 2002. Quantitative stratigraphy and taxonomy of late Emsian and Eifelian ammonoids of the eastern Anti-Atlas (Morocco). Courier Forschungsinstitut Senckenberg 238, 1–109.

Klug, C., Kröger, B., Rücklin, M., Korn, D., Schemm-Gregory, M., De Baets, K., Mapes, R.H., 2008. Ecological change during the early Emsian (Devonian) in the Tafilalt (Morocco), the origin of the Ammonoidea, and the first African pyrgocystid edrioasterioids, machaerids, and phyllocarids. Palaeontographica, Abteilung A 283, 83–176.

Koptíková, L., 2011. Precise position of the Basal Choteč event and evolution of sedimentary environments near the Lower-Middle Devonian boundary: The magnetic susceptibility, gamma-ray spectrometric, lithological, and geochemical record of the Prague Synform. Palaeogeography, Palaeoclimatology, Palaeoecology 304, 96–112.

Koptíková, L., Berkyova, S., Hladil, J., Slavik, L., Schnabl, P., Frana, J., Bohmova, V., 2008. Long-distance correlation of basal Choteč Event sections using magnetic susceptibility (Barrandian vs. Nevada) and lateral and vertical variations in fine-grained non-carbonate mineral phases. In: Kim, A.I., Salimova, F.A., Meshchankina, N.A. (Eds.), Global Alignments of Lower Devonian Carbonate and Clastic Sequences: Contributions from the IGCP 499 Project/SDS Joint Field Meeting, Kitab State Geological Reserve, Uzbekistan, August 25 – September 3, 2008, pp. 60–62.

Koptíková, L., Bábek, O., Hladil, J., Kalvoda, J., Slavik, L., 2010. Stratigraphic significance and resolution of spectral reflectance logs in Lower Devonian carbonates of the Barrandian area, Czech Republic: A correlation with magnetic susceptibility and gamma ray logs. Sedimentary Geology 225, 83–98.

Korn, D., 1999. Famennian ammonoid stratigraphy of the Ma'der and Tafilalt (Eastern Anti-Atlas, Morocco). Abhandlungen der Geologischen Bundesanstalt 54, 147–179.

Korn, D., 2004. The mid-Famennian ammonoid succession in the Rhenish Mountains: The "*annulata* Event" reconsidered. Geological Quarterly 48 (3), 245–252.

Korn, D., Luppold, F.W., 1987. Nach Clymenien und Conodonten gegliederte Profile des oberen Famennium im Rheinischen Schiefergebirge. Courier Forschungsinstitut Senckenberg 92, 199–223.

Korn, D., Clausen, C.-D., Luppold, F.W., 1994. Die Devon/Karbon-Grenze im Rheinischen Schiefergebirge. Geologie und Paläontologie in Westfalen 29, 1–221.

Korn, D., Klug, C., Reisdorf, A., 2000. Middle Famennian ammonoid stratigraphy in the Amessoui Syncline (Late Devonian; eastern Anti-Atlas, Morocco). Travaux de l'Institut Scientifique, Rabat, Série Géologie & Géographie Physique 20, 69–77.

Kürschner, W., Becker, R.T., Buhl, D., Veizer, J., 1993. Strontium isotopes in conodonts: Devonian/Carboniferous transition, the northern Rhenish Slate Mountains, Germany. Annales de la Société Géologique de Belgique 115 (2), 595–622.

Lange, W., 1929. Zur Kenntnis des Oberdevons am Enkeberg und bei Balve (Sauerland). Abhandlungen der Preußischen Geologischen Landesanstalt, Neue Folge 119, 1–132.

Le Hérissé, A., Servais, T., Wicander, R., 2000. Devonian acritarchs and related forms. Courier Forschungsinstitut Senckenberg 220, 195–204.

Liashenko, G.P., 1967. Coniconchia (Tentaculitida, Nowakiida, Styliolinida) and their importance in Devonian biostratigraphy. In: Oswald, D.H. (Ed.), International Symposium on the Devonian System, Calgary 1967, Vol. II, pp. 897–903.

Loboziak, S., Melo, J.H.G., 2002. Devonian miospore successions of Western Gondwana: Update and correlation with Southern Euramerican miospore zones. Review of Palaeobotany and Palynology 121, 133–148.

Lonsdale, W., 1840. Notes on the age of the Limestones of South Devonshire. Transactions of the Geological Society of London, Series 2, 5, 721–738.

Lottmann, J., 1990. Die *pumilio*-Events (Mittel-Devon). Göttinger Arbeiten zur Geologie und Paläontologie 44, 1–98.

Ma, X.-P., Bai, S.-L., 2002. Biological, depositional, microspherule, and geochemical records of the Frasnian/Famennian boundary beds, South China. Palaeogeography, Palaeoclimatology, Palaeoecology 181, 325–346.

Ma, X.-P., Wang, C.-Y., Racki, G., Racka, M., 2008. Facies and geochemistry across the Early-Middle Frasnian transition (Late Devonian) on South China carbonate shelf: Comparison with the Polish reference successions. Palaeogeography, Palaeoclimatology, Palaeoecology 269, 130–151.

Manda, S., Frýda, J., 2010. Silurian-Devonian boundary events and their influence on cephalopod evolution: Evolutionary significance of cephalopod egg size during mass extinction. Bulletin of Geosciences 85 (3), 513–540.

Marini, F., Casier, J.-G., Claude, J.-M., Théry, J.-M., 1997. Cosmic magnetic sphaerules in the Famennian of the Bad Windsheim borehole (Germany): Preliminary study and implications. Sphaerula 1 (1), 4–19.

Mark-Kurik, E., Blieck, A., Lobaziak, S., Candilier, A.-M., 1999. Miospore assemblage from the Lode member (Gauja Formation) in Estonia and the Middle-Upper Devonian boundary problem. Proceedings of the Estonian Academy of Sciences, Geology 48, 86–98.

Marshall, J.E.A., Astin, T.R., Brown, J.F., Mark-Kurik, E., Lazauskiene, J., 2007. Recognizing the Kacák Event in the Devonian terrestrial environment and its implications for understanding land – sea interactions. In: Becker, R.T., Kirchgasser, W.T. (Eds.), Devonian Events and Correlations. Geological Society Special Publication 278, pp. 133–155.

Marshall, J.E.A., Brown, J.F., Astin, T.R., 2011. Recognising the Taghanic Crisis in the Devonian terrestrial environment and its implications for understanding land-sea interactions. Palaeogeography, Palaeoclimatology, Palaeoecology 304, 165–183.

Martinsson, A. (Ed.), 1977. The Silurian-Devonian boundary: final report of the Committee of the Siluro-Devonian Boundary within IUGS Commission on Stratigraphy and a state of the art report for Project Ecostratigraphy. International Union of Geological Sciences, Series A, 5, p. 347.

Marynowski, L., Filipiak, P., 2007. Water column euxinia and wildfire evidence during deposition of the Upper Famennian Hangenberg event horizon from the Holy Cross Mountains (central Poland). Geological Magazine 144 (3), 569–595.

Marynowski, L., Filipiak, P., Zaton, M., 2010. Geochemical and palynological study of the Upper Famennian Dasberg event horizon from the Holy Cross Mountains (central Poland). Geological Magazine 147 (4), 527–550.

Marynowski, L., Rakocinski, M., Borcuch, E., Kremer, B., Schubert, B.A., Jahren, A.H., 2011. Molecular and petrographic indicators of redox conditions and bacterial communities after the F/F mass extinction (Kowala, Holy Cross Mountains, Poland). Palaeogeography, Palaeoclimatology, Palaeoecology 306, 1–14.

Matern, H., 1931. Das Oberdevon der Dill-Mulde. Abhandlungen der Preußischen Geologischen Landesanstalt, Neue Folge 134, 1–139.

Matyja, H., Turnau, E., 2008. Integrated analyses of miospores and conodonts: A tool for correlation of shallow-water, mixed siliciclastic-carbonate successions of Lower? and Middle Devonian (Pommeranian Basin, NW Poland). In: Kim, A.I., Salimova, F.A., Meshchankina, N.A. (Eds.), Global Alignments of Lower Devonian Carbonate and Clastic Sequences: Contributions from the IGCP 499 Project/SDS Joint Field Meeting, Kitab State Geological Reserve, Uzbekistan, August 25 – September 3, 2008, pp.67–71.

Maziane, N., Higgs, K.T., Streel, M., 1999. Revision of the late Famennian miospore zonation scheme in eastern Belgium. Journal of Micropalaeontology 18, 17–25.

McGhee Jr., G.R., Orth, C.J., Quintana, L.R., Gilmore, J.S., Olsen, E.J., 1996. Late Devonian "Kellwasser Event" mass-extinction horizon in Germany: No geochemical evidence for a large-body impact. Geology 14, 776–779.

McMillan, N.J., Embry, A.F., Glass, D.J., 1988. Devonian Geology of the World. 3 vol. Canadian Society of Petroleum Geologists, Calgary.

Melo, J.H.G., Loboziak, S., 2003. Devonian-Early Carboniferous miospore biostratigraphy of the Amazon Basin, Northern Brazil. Review of Palaeobotany and Palynology 124, 131–202.

Molyneux, S.G., Le Hérissé, A., Wicander, R., 1996. Paleozoic phytoplankton. In: Jansonius, J., McGregor, D.C. (Eds.), Palynology: Principles and Applications. American Association of Stratigraphic Palynologists, Dallas, pp. 493–529.

Morrow, J.R., 2000. Shelf-to-basin lithofacies and conodont paleoecology across Frasnian-Famennian (F-F, mid-Late Devonian) boundary, Central Great Basin (Western U.S.A.) Courier Forschungsinstitut Senckenberg 219, 1–57.

Morrow, J.R., Sandberg, C.A., Malkowski, K., Joachimski, M.M., 2009. Carbon isotope chemostratigraphy and precise dating of middle Frasnian (lower Upper Devonian) Alamo Breccia, Nevada, USA. Palaeogeography, Palaeoclimatology, Palaeoecology 282, 105–118.

Murphy, E.M., Sageman, B.B., Hollander, D.J., 2000. Eutrophication by decoupling of the marine biogeochemical cycles of C, N, and P: A mechanism for the Late Devonian mass extinction. Geology 28 (5), 427–430.

Murphy, M.A., 2000. Conodonts first occurrences in Nevada. In: Weddige, K. (Ed.), Devonian Correlation Table. Senckenbergiana Lethaea, 80, p. 695.

Murphy, M.A., 2005. Pragian conodont zonal classification in Nevada, Western North America. Revista Española de Paleontología 20 (2), 177–206.

Murphy, M.A., Valenzuela-Ríos, J.I., 1999. *Lanea* new genus, lineage of Early Devonian conodonts. Bolletino della Società Paleontologica Italiana 37 (2/3), 321–334.

Myrow, P.M., Strauss, J.V., Creveling, J.R., Sicard, K.R., Ripperdan, R., Sandberg, C.A., Hartenfels, S., 2011. A carbon isotopic and sedimentological record of the latest Devonian (Famennian) from the Western U.S. and Germany. Palaeogeography, Palaeoclimatology, Palaeoecology 306, 147–159.

Narkiewicz, K., Bultynck, P., 2010. The Upper Givetian (Middle Devonian) *subterminus* conodont zone in North America, Europe and North Africa. Journal of Paleontology 84 (4), 588–625.

Nawrocki, J., Polechonska, O., Werner, T., 2008. Magnetic susceptibility and selected geochemical-mineralogical data as proxies for Early to Middle Frasnian (Late Devonian) carbonate depositional settings in the Holy Cross Mountains, southern Poland. Palaeogeography, Palaeoclimatology, Palaeoecology 269, 176–188.

Niedzwiedzki, G., Szrek, P., Narkiewicz, M., Ahlberg, P.E., 2010. Tetrapod trackways from the early Middle Devonian period of Poland. Nature 463, 43–48.

Nikolaeva, S., 2007. New data on the clymeniid faunas of the Urals and Kazakhstan. In: Landman, N.H., Davis, R.A., Mapes, R.H. (Eds.), Cephalopods – Present and Past: New Insights and Fresh Perspectives. Springer, New York, pp. 317–343.

Obukhovskaya, T.G., Avkhimovitch, V.I., Streel, M., Loboziak, S., 2000. Miospores from the Frasnian-Famennian boundary deposits in Eastern Europe (the Pripyat Depression, Belarus and the Timan–Pechora Province, Russia) and comparisons with Western Europe (Northern France). Review of Palaeobotany and Palynology 112, 229–246.

Obut, O.T., Shcherbanenko, T.A., 2008. Late Devonian radiolarians from the Rudnyi Altai (SW Siberia). Bulletin of Geosciences 83 (4), 371–382.

Ovnatanova, N.S., Kononova, L.I., 2008. Frasnian conodonts from the Eastern Russian Platform. Paleontological Journal 42 (10), 997–1166.

Paproth, E., Feist, R., Flajs, G., 1991. Decision on the Devonian-Carboniferous boundary stratotype. Episodes 14 (4), 331–336.

Paris, F., Winchester-Seeto, T., Boumendjel, K., Grahn, Y., 2000. Towards a global biozonation of Devonian chitinozoans. Courier Forschungsinstitut Senckenberg 220, 39–55.

Parry, S.F., Noble, S.R., Crowley, Q.G., Wellman, C.H., 2011. A high-precision U-Pb age constraint on the Rhynie Chert Konservat-Lagerstätte: Time scale and other implications. Journal of the Geological Society 168, 863–872.

Playford, P.E., McLaren, D.J., Orth, C.J., Gilmore, J.S., Goodfellow, W.D., 1984. Iridium anomaly in the Upper Devonian of the Canning Basin, Western Australia. Science 226, 437–439.

Pujol, F., Berner, Z., Stüben, D., 2006. Palaeoenvironmetal changes at the Frasnian/Famennian boundary in key European sections: Chemostratigraphic constraints. Palaeogeography, Palaeoclimatology, Palaeoecology 240, 120–145.

Rabien, A., 1954. Zur Taxionomie und Chronologie der Oberdevonischen Ostracoden. Abhandlungen des Hessischen Landesamtes für Bodenforschung 9, 1–268.

Racka, M., Marynowski, L., Filipiak, P., Sobstel, M., Pisarzowska, A., Bond, D.P.G., 2010. Anoxic *Annulata* Events in the Late Famennian of the Holy Cross Mountains (Southern Poland): Geochemical and palaeontological record. Palaeogeography, Palaeoclimatology, Palaeoecology 297 (3/4), 549–575.

Racki, G., 1999. The Frasnian-Famennian biotic crisis: How many (if any) bolide impacts? Geologische Rundschau 87, 617–632.

Racki, G., 2005. Toward understanding Late Devonian global events: Few answers, many questions. In: Over, D.J., Morrow, J.R., Wignall, P.B. (Eds.), Understanding Late Devonian and Permian-Triassic Biotic and Climatic Events: Towards an Integrated Approach. Developments in Palaeontology and Stratigraphy, 20, pp. 5–36.

Racki, G., House, M.R., 2002. The Frasnian/Famennian boundary extinction event. Palaeogeography, Palaeoclimatology, Palaeoecology 181, 374.

Racki, G., Koeberl, C., 2004. Comment on impact ejecta layer from the mid-Devonian: Possible connection to global mass extinctions. Science 303, 471.

Racki, G., Piechota, A., Bond, D., Wignall, P.B., 2004. Geochemical and ecological aspects of lower Frasnian pyrite-ammonoid level at Kostomloty (Holy Cross Mountains, Poland). Geological Quarterly 48 (3), 267–282.

Racki, G., Joachimski, M.M., Morrow, J.R., 2008. A major perturbation of the global carbon budget in the Early-Middle Frasnian transition (Late Devonian). Palaeogeography, Palaeoclimatology, Palaeoecology 269, 127–204.

Richardson, J.B., Rasul, S.M., Al-Ameri, T., 1981. Acritarchs, miospores and correlation of the Ludlovian-Downtonian and Silurian-Devonian boundaries. Reviews of Palaeobotany and Palynology 34, 209–224.

Richardson, J.B., McGregor, D.C., 1986. Silurian and Devonian spore zones of the Old Red Sandstone Continent and adjacent areas. Geological Survey of Canada Bulletin 364, 1–79.

Richardson, J.B., Rodriguex, R.M., Sutherland, J.E., 2000. Palynology and recognition of the Silurian/Devonian boundary in some British terrestrial sediments by correlation with other European marine sequences - a progress report. Courier Forschungsinstitut Senckenberg 220, 1–7.

Rimmer, S.M., Thompson, J.A., Goodnight, S.A., Robl, T.L., 2004. Multiple controls on the preservation of organic matter in Devonian-Mississippian marine black shales: Geochemical and petrographic evidence. Palaeogeography, Palaeoclimatology, Palaeoecology 215, 125–154.

Riquier, L., Tribovillöard, N., Averbuch, O., Joachimski, M.M., Racki, G., Devleeschouwer, X., El Albani, A., Riboulleau, A., 2005. Productivity and bottom water redox conditions at the Frasnian-Famennian boundary on both sides of the Eovariscan Belt: Constraints from trace-element geochemistry. In: Over, D.J., Morrow, J.R., Wignall, P.B. (Eds.), Understanding Late Devonian and Permian-Triassic Biotic and Climatic Events: Towards an Integrated Approach. Developments in Palaeontology and Stratigraphy, 20, pp. 199–224.

Riquier, L., Averbuch, O., Tribovillard, N., El Albani, A., Lazreq, N., Chakiri, S., 2007. Environmental changes at the Frasnian-Famennian boundary in Central Morocco (Northern Gondwana): Integrated rock-magnetic and geochemical studies. In: Becker, R.T., Kirchgasser, W.T. (Eds.), Devonian Events and Correlations. Geological Society Special Publication 278, pp. 197–217.

Robl, T.L., Barron, L.S., 1988. The geochemistry of Devonian black shales in central Kentucky and its relationship to inter-basinal correlation and depositional environments. Canadian Society of Petroleum Geologists, Memoir 14 (II), 377–392.

Ruan, Y., 1996. Zonation and distribution of the early Devonian primitive ammonoids in South China. In: Wang, H., Wang, X. (Eds.), Centennial Memorial Volume of Professor Sun Yunzhu (Y.C. Sun): Palaeontology and Stratigraphy. China University of Geosciences Press, Wuhan, pp. 104–112.

Saltzman, M.R., 2002. Carbon isotope (δ^{13}C) stratigraphy across the Silurian-Devonian transition in North America: Evidence for a perturbation of the global carbon cycle. Palaeogeography, Palaeoclimatology, Palaeoecology 187, 83–100.

Sandberg, C.A., Ziegler, W., 1973. Refinement of standard Upper Devonian conodont zonation based on sections in Nevada and West Germany. Geologica et Palaeontologica 7, 97–122.

Sandberg, C.A., Ziegler, W., Dreesen, R., Butler, J.L., 1988. Late Frasnian mass extinction: Conodont event stratigraphy, global changes, and possible causes. Courier Forschungsinstitut Senckenberg 102, 263–307.

Sandberg, C.A., Ziegler, W., Bultynck, P., 1989. New standard conodont zones and Early *Ancyrodella* Phylogeny across Middle-Upper Devonian boundary. Courier Forschungsinstitut Senckenberg 110, 195–230.

Sauerland, U., 1983. Dacryoconariden und Homocteniden der Givet- und Adorf-Stufe aus dem Rheinischen Schiefergebirge (Tentaculitoidea, Devon). Göttinger Arbeiten zur Geologie und Paläontologie 25, 1–86.

Schieber, J., Over, D.J., 2005. Sedimentary infill of the Late Devonian Flynn Creek crater: A hard target marine impact. In: Over, D.J., Morrow, J.R., Wignall, P.B. (Eds.), Understanding Late Devonian and Permian–Triassic Biotic and Climatic Events: Towards an Integrated Approach. Developments in Palaeontology and Stratigraphy, 20, pp. 51–69.

Schindewolf, O.H., 1937. Zur Stratigraphie und Paläontologie der Wocklemer Schichten (Oberdevon). Abhandlungen der Preussischen geologischen Landesanstalt, Neue Folge 178, 1–132.

Schindler, E., 1990. Die Kellwasser-Krise (hohe Frasne-Stufe, Ober-Devon). Göttinger Arbeiten zur Geologie und Paläontologie 46, 1–115.

Schindler, E., 1993. Event-stratigraphic markers within the Kellwasser crisis near the Frasnian/Famennian boundary (Upper Devonian) in Germany. Palaeogeography, Palaeoclimatology, Palaeoecology 104, 115–125.

Schmitz, B., Ellwood, B.B., Peucker-Ehrenbrink, B., El Hassani, A., Bultynck, P., 2006. Platinum group elements and $^{187}Os/^{188}Os$ in a purported impact ejecta layer near the Eifelian-Givetian stage boundary, Middle Devonian. Earth and Planetary Science Letters 249, 162–172.

Schöne, B.R., 1997. Der *otomari*-Event und seine Auswirkungen auf die Fazies des Rhenoherzynischen Schelfs (Devon, Rheinisches Schiefergebirge). Göttinger Arbeiten zur Geologie und Paläontologie 70, 1–140.

Schönlaub, H.P., Attrep, M., Boeckelmann, K., Dreesen, R., Feist, R., Fenninger, A., Hahn, G., Klein, P., Korn, D., Kratz, R., Magaritz, M., Orth, C.J., Schramm, J.-M., 1992. The Devonian/Carboniferous boundary in the Carnic Alps – a multidisciplinary approach. Jahresberbericht der Geologischen Bundesanstalt Wien 135, 57–98.

Schülke, I., 1995. Evolutive Prozesse bei *Palmatolepis* in der frühen Famenne-Stufe (Conodonta, Ober-Devon). Göttinger Arbeiten zur Geologie und Paläontologie 67, 1–108.

Schülke, I., 1999. Conodont multielement reconstructions from the early Famennian (Late Devonian) of the Montagne Noire. Geologica et Palaeontologica SB 3, 1–124.

Schwartzapfel, J.A., Holdsworth, B.K., 1996. Upper Devonian and Mississippian radiolarian zonation and biostratigraphy of the Woodford, Sycamore, Caney and Goddard formations, Oklahoma. Cushman Foundation for Foraminiferal Research, Special Publication 33, 1–275.

Sedgwick, A., Murchison, R.I., 1839. Stratification of the older stratified deposits of Devonshire and Cornwall. Philosophical Magazine, Series 3 (14), 241–260.

Sedgwick, A., Murchison, R.I., 1840. On the physical structure of Devonshire, and on the subdivisions and geological relations of its older stratified deposits. Transactions of the Geological Society of London, Series 2, 5, 633–704.

Selby, D., Creaser, R.A., 2005. Direct radiometric dating of the Devonian-Mississippian time-scale boundary using the Re-Os black shale geochronometer. Geology 33 (7), 545–548.

Simon, L., Godderis, Y., Buggisch, W., Strauss, H., Joachimski, M.M., 2007. Modeling the carbon and sulfur isotope compositions of marine sediments: Climate evolution during the Devonian. Chemical Geology 246, 19–38.

Slavík, L., 2004a. A new conodont zonation of the Pragian Stage (Lower Devonian) in the stratotype area (Barrandian, central Bohemia). Newsletters on Stratigraphy 40 (1/2), 39–71.

Slavík, L., 2004b. The Pragian-Emsian conodont succession of the Barrandian area: Search of an alternative to the GSSP polygnathid-based correlation concept. Geobios 37, 454–470.

Slavík, L., Hladil, J., 2004. Lochkovian/Pragian GSSP revisited: Evidence about conodont taxa and their stratigraphic distribution. Newsletters on Stratigraphy 40 (3), 137–153.

Slavík, L., Valenzuela-Ríos, J.I., Hladil, J., Carls, P., 2007. Early Pragian conodont-based correlations between the Barrandian area and the Spanish Central Pyrenees. Geological Journal 42, 499–512.

Sliwinski, M.G., Whalen, M.T., Newberry, R.J., Payne, J.H., Day, J.E., 2011. Stable isotope $\delta^{13}C_{carb\ and\ org}$, $\delta^{15}N_{org}$) and trace element anomalies during the Late Devonian '*punctata* Event' in the Western Canada Sedimentary Basin. Palaeogeography, Palaeoclimatology, Palaeoecology 307, 245–271.

Steemans, P., 1989. Etude palynostratigraphique du Devonien Inferieur dans l'ouest de l'Europe. Mémoire, Explication des Cartes Géologiques et Mineralogiques de Belgique 27, 1–453.

Stephens, N.P., Sumner, D.Y., 2003. Late Devonian carbon isotope stratigraphy and sea level fluctuations, Canning Basin, Western Australia. Palaeogeography, Palaeoclimatology, Palaeoecology 191, 203–219.

Streel, M., 2000. The late Famennian and early Frasnian datings given by Tucker and others (1998) are biostratigraphically poorly constrained. Subcommission on Devonian Stratigraphy Newsletter 17, 59.

Streel, M., 2005. Subdivision of the Famennian stage into four substages and correlation with the neritic and continental miospore zonation. Subcommission on Devonian Stratigraphy Newsletter 21, 14–17.

Streel, M., 2009. Upper Devonian miospore and conodont zone correlation in western Europe. In: Königshof, P. (Ed.), Devonian Change: Case Studies in Palaeogeography and Palaeoecology. Geological Society Special Publication 314, pp. 163–176.

Streel, M., Loboziak, S., 2000. Correlation of the proposed conodont based Upper Devonian substage boundary levels into the neritic and terrestrial miospore zonation. Subcommission on Devonian Stratigraphy Newsletter 17, 12–14.

Streel, M., Higgs, K., Loboziak, S., Riegel, W., Steemans, P., 1987. Spore stratigraphy and correlation with faunas and floras in the type marine Devonian of the Ardenne – Rhenish regions. Review of Palaeobotany and Palynology 50, 211–229.

Streel, M., Brice, D., Degardin, J.-M., Derycke, C., Dreesen, R., Groessens, E., Hance, L., Legrand-Blain, M., Lethiers, F., Loboziakl, S., Maziane, N., Milhau, B., Mistiaen, B., Poty, E., Rohart, J.-C., Sartenaer, P., Thorez, J., Vachard, D., Blieck, A., 1998. Proposal for a Strunian substage and a subdivision of the Famennian Stage into four substages. Subcommission on Devonian Stratigraphy Newsletter 15, 47–52.

Streel, M., Caputo, M.V., Loboziak, S., Melo, J.H.G., 2000a. Late Frasnian-Famennian climates based on palynomorph analyses and the question of the Late Devonian glaciations. Earth-Science Reviews 52 (1–3), 121–173.

Streel, M., Loboziak, S., Steemanns, P., Bultynck, P., 2000b. Devonian miospore stratigraphy and correlation with the global stratotype sections and points. Courier Forschungsinstitut Senckenberg 220, 9–23.

Streel, M., Brice, D., Mistiaen, B., 2006. Strunian. Geologica Belgica 9 (1/2), 105–109.

Talent, J.A., Mawson, R., Andrew, A.S., Hamilton, P.J., Whitford, D.J., 1993. Middle Palaeozoic extinction events: Faunal and isotopic data. Palaeogeography, Palaeoclimatology, Palaeoecology 104, 139–152.

Trapp, E., Kaufmann, B., Mezger, K., Korn, D., Weyer, D., 2004. Numerical calibration of the Devonian-Carboniferous boundary: Two new U-Pb isotope dilution-thermal ionization mass spectrometry single-zircon ages from Hasselbachtal (Sauerland, Germany). Geology 32 (10), 857–860.

Tribovillard, N., Averbusch, O., Devleeschouwer, X., Racki, G., Riboulleau, A., 2004. Deep-water anoxia over the Frasnian-Famennian boundary (La Serre, France): A tectonically induced oceanic anoxic event? Terra Nova 16, 288–295.

Troth, I., Marshall, J.E.A., Racey, A., Becker, R.T., 2011. Devonian sea-level change in Bolivia: A high palaeolatitude biostratigraphical calibration of the global sea-level curve. Palaeogeography, Palaeoclimatology, Palaeoecology 304, 3–20.

Tucker, R.D., Bradley, D.C., Ver Straeten, C.A., Harris, A.G., Ebert, J.R., McCutcheon, S.R., 1998. New U-Pb zircon ages and the duration and division of Devonian time. Earth and Planetary Science Letters 158 (3–4), 175–186.

Turgeon, S.C., Creaser, R.A., Algeo, T.J., 2007. Re-Os depositional ages and seawater Os estimates for the Frasnian-Famennian boundary: Implications for weathering rates, land plant evolution, and extinction mechanisms. Earth and Planetary Science Letters 261, 649–661.

Turnau, E., Narkiewicz, K., 2011. Biostratigraphical correlation of spore and conodont zonations within Givetian and ?Frasnian of the Lublin area (SE Poland). Review of Palaeobotany and Palynology 164, 30–38.

Valenzuela-Ríos, J.I., Murphy, M.A., 1997. A new zonation of middle Lochkovian (Lower Devonian) conodonts and evolution of *Flajsella* n. gen. (Conodonta). Geological Society of America, Special Paper 321, 131–144.

Van Geldern, R., Joachimski, M.M., Day, J., Jansen, U., Alvarez, F., Yolkin, E.A., Ma, X.-P., 2006. Carbon, oxygen and strontium isotope records of Devonian brachiopod shell calcite. Palaeogeography, Palaeoclimatology, Palaeoecology 240, 47–67.

Veizer, J., Buhl, D., Diener, A., Ebneth, S., Podlaha, O.G., Bruckschen, P., Jasper, T., Korte, C., Schaaf, M., Ala, D., Azmy, K., 1997. Strontium isotope stratigraphy: Potential resolution and event correlation. Palaeogeography, Palaeoclimatology, Palaeoecology 132, 65–77.

Ver Straeten, C.A., 2001. Emsian-Eifelian (Lower-Middle Devonian) of Virginia-West Virginia, and a basinwide synthesis and sequence stratigraphy. Geological Society of America, Abstracts with Programs 33 (2), A61.

Walliser, O.H., 1996. Global events in the Devonian and Carboniferous. In: Walliser, O.H. (Ed.), Global Events and Event Stratigraphy in the Phanerozoic. Springer–Verlag, Berlin, pp. 225–250.

Walliser, O.H., 2000. The Eifelian-Givetian Stage boundary. Courier Forschungsinstitut Senckenberg 225, 37–47.

Walliser, O.H., Bultynck, P., 2011. Extinctions, survival and innovations of conodont species during the Kacák Episode (Eifelian-Givetian) in south-eastern Morocco. Bulletin de l'Institut Royal des Sciences Naturelles de Belgique 81, 5–25.

Walliser, O.H., Bultynck, P., Weddige, K., Becker, R.T., House, M.R., 1996. Definition of the Eifelian-Givetian Stage boundary. Episodes 18, 107–115.

Wang, K., 1992. Glassy microspherules (microtectites) from an Upper Devonian limestone. Science 256, 1547–1550.

Wang, K., Orth, C.J., Attrep Jr., M., Chatterton, B.D.E., Hou, H., Geldsetzer, H.H., 1991. Geochemical evidence for a catastrophic biotic event at the Frasnian-Famennian boundary in south China. Geology 19, 776–779.

Wang, K., Geldsetzer, H.H.J., Chatterton, B.D.E., 1994. A Late Devonian extraterrestrial impact and extinction in eastern Gondwana: Geochemical, sedimentological, and faunal evidence. Geological Society of America, Special Paper 293, 111–120.

Wang, K., Geldsetzer, H.H.J., Goodfellow, W.D., Krouse, H.R., 1996. Carbon and sulfur isotope anomalies across the Frasnian-Famennian extinction boundary, Alberta, Canada. Geology 24 (2), 187–191.

Wang, Y.-J., Fang, Z.-J., Yang, Q., Zhou, Z.-C., Cheng, Y.-N., Duan, Y.-X., Xiao, Y.-W., 2000. Middle-Late Devonian strata of cherty facies and radiolarian faunas from West Yunnan. Acta Micropalaeontologica Sinica 17 (3), 235–254.

Warme, J.E., Sandberg, C.A., 1996. Alamo Megabreccia: Record of a Late Devonian impact in southern Nevada. Geology Today 6 (1), 1–7.

Weddige, K., 1977. Die Conodonten der Eifel-Stufe im Typusgebiet und in benachbarten Faziesgebieten. Senckenbergiana Lethaea 58, 271–419.

Weddige, K., 1982. The Wetteldorf Richtschnitt as boundary stratotype from the view point of conodont stratigraphy. Courier Forschungsinstitut Senckenberg 55, 26–37.

Weddige, K., 1996. Devon-Korrelationstabelle. Senckenbergiana Lethaea 76, 267–286.

Whalen, M.T., Day, J.E., 2008. Magnetic susceptibility, biostratigraphy, and sequence stratigraphy: Insights into Devonian carbonate platform development and basin infilling, western Alberta, Canada. SEPM Special Publication 89, 291–314.

White, E.I., 1950. The vertebrate faunas of the Lower Old Red Sandstone of the Welsh Borders. Bulletin of the British Museum of Natural History (Geology) 1, 51–67.

Wicander, R., Clayton, G., Marshall, J.E.A., Troth, I., Racey, A., 2011. Was the latest Devonian glaciation a multiple event? New palynological evidence from Bolivia. Palaeogeography, Palaeoclimatology, Palaeoecology 305, 84–92.

Williams, E.A., Friend, P.F., Williams, P.J., 2000. A review of Devonian time scales: Databases, construction and new data. In: Friend, P.F., Williams, B.P.J. (Eds.), New Perspectives on the Old Red Sandstone. Geological Society Special Publications 180, pp. 1–21.

Witzke, B.J., Bunker, B.J., Rogers, F.S., 1988. Eifelian through Lower Frasnian stratigraphy and deposition in the Iowa area, central mid-continent, U.S.A. In: McMillan, N.J., Embry, A.F., Glass, D.J. (Eds.), Devonian of the World, Vol. I. Canadian Society of Petroleum Geologists, Calgary, pp. 221–250.

Yans, J., Corfield, R., Racki, G., Préat, A., 2007. Evidence for perturbations of the carbon cycle in the Middle Frasnian *punctata* Zone (Late Devonian). Geological Magazine 144, 263–370.

Yolkin, E.A., Weddige, K., Izokh, N.G., Erina, M.V., 1994. New Emsian conodont zonation (Lower Devonian). Courier Forschungsinstitut Senckenberg 168, 139–157.

Yolkin, E.A., Kim, A.I., Weddige, K., Talent, J.A., House, M.R., 1998. Definition of the Pragian/Emsian Stage boundary. Episodes 20, 235–240.

Yolkin, E.A., Kim, A.I., Talent, J.A., 2008. Devonian Sequences of the Kitab Reserve Area. Field Excursion Guidebook, International Conference Global Alignments of Lower Devonian Carbonate and Clastic Sequences (SDS/IGCP 499 Project joint field meeting), Kitab State Geological Reserve, Uzbekistan, August 25 – September 3, 2008. Publishing House of SB RAS, Novosibirsk, p. 97.

Yolkin, E.A., Izokh, N.G., Weddige, K., Erina, V., Valenzuela-Rios, J.I., Apekina, L.S., 2011. Eognathodid and polygnathid lineages from the Kitab State Geological Reserve sections (Zeravshan-Gissar mountainous area, Uzbekistan) as the bases for improvements of the Pragian-Emsian standard conodont zonation. News on Paleontology and Stratigraphy, Supplement to Geologiya I Geofizika 52 (15), 37–47.

Young, G.C., 2000. Flawed timescale, or flawed logic. The Australian Geologist 115, 6.

Young, G.C., Laurie, J.R., 1996. An Australian Phanerozoic Timescale. Oxford University Press, Oxford, p. 279.

Yu, C.-M. (Ed.), 1988. Devonian-Carboniferous boundary in Nanbiancun, Guilin, China – Aspects and Records. Science Press, Beijing, p. 379.

Zhang, N., Xia, W., Dong, Y., Shang, H., 2008. Conodonts and radiolarians from pelagic cherts of the Frasnian-Famennian boundary interval at Bancheng, Guangxi, China: Global recognition of the upper Kellwasser event. Marine Micropaleontology 67, 180–190.

Zhao, W., Wang, N., Zhu, M., Mann, U., Herten, U., Lücke, A., 2011. Geochemical stratigraphy and microvertebrate assemblage sequences across the Silurian/Devonian transition in South China. Acta Geologica Sinica 85 (2), 340–353.

Zheng, Y., Hou, H.-F., Ye, L.-F., 1993. Carbon and oxygen isotope event markers near the Frasnian-Famennian boundary, Louxiu section, South China. Palaeogeography, Palaeoclimatology, Palaeoecology 104, 97–104.

Ziegler, W., 1962. Taxionomie und Phylogenie Oberdevonischer Conodonten un ihre stratigraphische Bedeutung. Abhandlungen des Hessischen Geologischen Landesamtes für Bodenforschung 38, 1–166.

Ziegler, W., 2000. The Lower Eifelian Boundary. Courier Forschungsinstitut Senckenberg 225, 27–36.

Ziegler, W., Werner, R., 1982. On Devonian stratigraphy and palaeontology of the Ardenno-Rhenish mountains and related Devonian matters. Courier Forschungsinstitut Senckenberg 55, 1–505.

Ziegler, W., Sandberg, C.A., 1984. *Palmatolepis*-based revision of upper part of standard Late Devonian conodont zonation. In: Clark, D.L. (Ed.), Conodont biofacies and provincialism. Geological Society of America, Special Paper, 196, pp. 179–194.

Ziegler, W., Sandberg, C.A., 1990. The Late Devonian standard zonation. Courier Forschungsinstitut Senckenberg 121, 115.

Ziegler, W., Sandberg, C.A., 1996. Reflexions on Frasnian and Famennian Stage boundary decisions as a guide to future deliberations. Newsletters on Stratigraphy 33 (3), 157–180.

Ziegler, W., Sandberg, C.A., 1997. Proposal of boundaries for a late Frasnian substage and for subdivision of the Famennian Stage into three substages. Subcommission on Devonian Stratigraphy Newsletter 14, 11–12.

V.I. Davydov, D. Korn, M.D. Schmitz,
With contributions from F.M. Gradstein and O. Hammer

Chapter 23

The Carboniferous Period

Abstract: Only the GSSPs for the Bashkirian (base of the Pennsylvanian), Visean and Tournaisian (base of the Mississippian) have been formalized, although the latter now has complications.

The supercontinent Pangea caused major changes in ocean circulation, biogeographic differentiation, high bio-provincialism, diversification of land plants and increased continental weathering rates and storage of organic carbon as coal, strong fluctuations of atmospheric carbon doxide, significant global cooling and warming and sharp sea-level fluctuations, cyclic marine sequences, appearance of reptiles (with amniotic egg reproduction) and occupation of new (dry-land) niches, extinction or decreasing role of early Paleozoic biota such as stromatoporids, tabulate corals, trilobites, ostracods, heavily armored marine fish, appearance or very rapid diversification of Foraminifera, ammonoids, fresh-water pelecypods, gastropods, sharks, ray-finned fishes, and wingless insects. The Late Carboniferous-Early Permian Kiaman Superchron is the longest known period of predominantly reversed polarity.

306 Ma Carboniferous

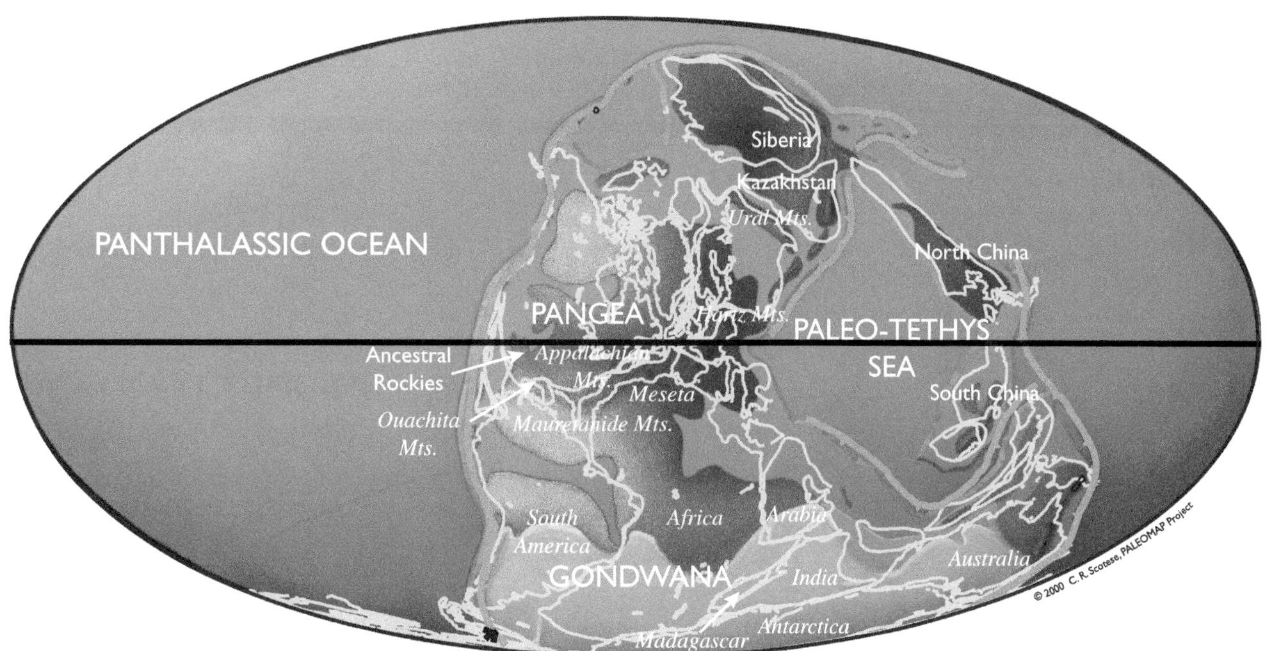

Chapter Outline

The Geologic Time Scale 2012. DOI: 10.1016/B978-0-444-59425-9.00023-8

23.1. HISTORY AND SUBDIVISIONS

Due to climatic variability and tectonic and volcanic activity, the Carboniferous was a time of incredible diversification and abundant terrestrial biota. It signifies Earth's first episode of widespread, massive coal formation and consequently oxygen and CO_2 variability. The commercial production of coal led to the early development of Carboniferous stratigraphic classifications in three major regions; Western Europe, Eastern Europe, and North America. Indeed, the name Carboniferous is derived from the Italian *Carbonarium* (charcoal producer) or Latin *carbo* (charcoal) and *ferrous* (i.e. bearing). The term Carboniferous was first used in an adjectival sense by Kirwan (1799) and as an informal term for the section heading to describe "Coal-measures or Carboniferous Strata" by Farey in 1811 (Ramsbottom, 1984). Later it became a common term for coal-producing sediments in Great Britain and Western Europe. Four stratigraphic units, in ascending order, were included:

1. The Old Red Sandstone, later assigned to the Devonian.
2. The Mountain, or Carboniferous Limestone, first listed by William Phillips in 1818.
3. The Millstone Grit, proposed by Whitehurst in 1778.
4. The Coal Measures, proposed by Farey in 1807.

Conybeare and Phillips (1822) constituted these units as the Carboniferous or Medial Order, and Phillips (1835) coined the term Carboniferous System (Ramsbottom, 1984).

Even though the Carboniferous was one of the first-established geological periods, it is one of the most complicated and confusing in terms of stratigraphic classification and correlation. The reasons for this confusion are:

1. The assembly of the supercontinent Pangea by the collision of Laurussia with Gondwana, the Variscan, and Hercynian orogenies, and the subsequent nearly complete separation of tropical and subtropical shelves by this supercontinent.
2. The Gondwana superglaciation and the consequent drastic climatic changes, sea-level fluctuations, and significant biogeographic differentiation.

The current subdivision of the Carboniferous Period as favored by the Subcommission on Carboniferous Stratigraphy of the ICS is shown in Figure 23.1, and selected regional stages are illustrated in the same figure. Some progress in stage boundary definition has been achieved since GTS2004 (Davydov *et al.*, 2004), but new complications have arisen. As of March 2010, the major division of the Carboniferous Period into the Mississippian and Pennsylvanian subperiods and the base of Visean Global Stage within the Mississippian have been officially ratified. At the same time the GSSP established for the Devonian–Carboniferous boundary (Paproth *et al.*, 1991) has a problem with its defining event, the FAD of the conodont *Siphonodella sulcata,* and thus with the defined boundary (Kaiser, 2009).

23.1.1. Evolution of Traditional European and Russian Subdivisions

In Western Europe, the Carboniferous System was historically divided into the Lower Series (marine Mountain Limestone or Dinantian) and the Upper Series (predominantly terrestrial Millstone Grit and Coal Measures). Two independent regional stages were developed in the Dinantian of Belgium and Great Britain (Figure 23.1).

Several parallel chronostratigraphic schemes have been developed in Europe. Traditionally, the Carboniferous was subdivided into an older (**Dinantian**) and a younger (**Silesian**) portion, separated by the first occurrence of the ammonoid species *Cravenoceras leion*. With the definition of the mid-Carboniferous boundary at the FAD of *Declinognathodus noduliferus*, which marks a much younger horizon, the old subdivision scheme became obsolete.

In Belgium, the subdivision of the Early Carboniferous can be traced back to Dumont (1832), who named the Calcaires de Tournai and Visé (named after two towns in western and eastern

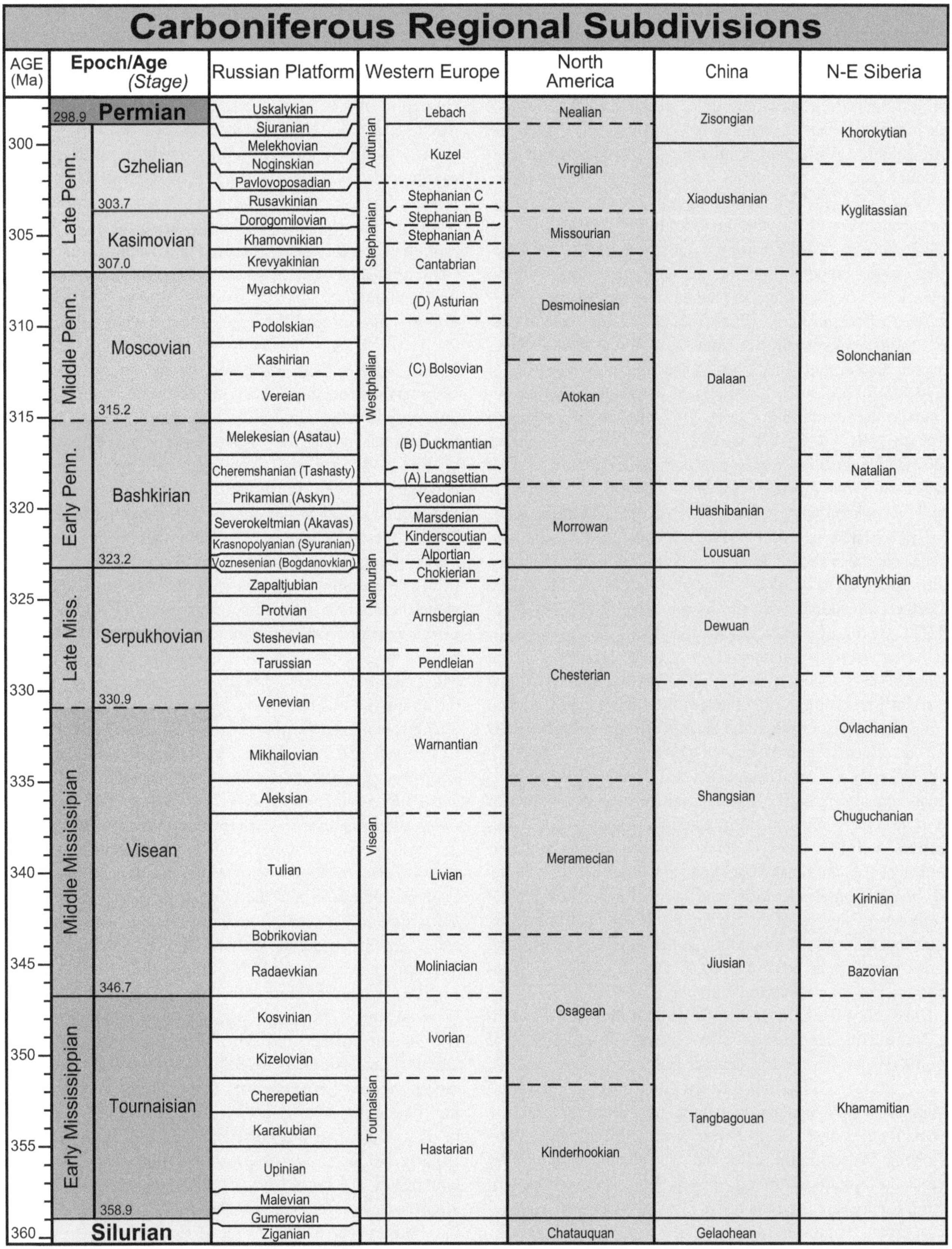

Carboniferous Regional Subdivisions

AGE (Ma): 300, 305, 310, 315, 320, 325, 330, 335, 340, 345, 350, 355, 360

Epoch/Age (Stage):
- Permian (298.9)
- Late Penn.: Gzhelian (303.7); Kasimovian (307.0)
- Middle Penn.: Moscovian (315.2)
- Early Penn.: Bashkirian (323.2)
- Late Miss.: Serpukhovian (330.9)
- Middle Mississipian: Visean (346.7)
- Early Mississippian: Tournaisian (358.9)
- Silurian

Russian Platform: Uskalykian; Sjuranian; Melekhovian; Noginskian; Pavlovoposadian; Rusavkinian; Dorogomilovian; Khamovnikian; Krevyakinian; Myachkovian; Podolskian; Kashirian; Vereian; Melekesian (Asatau); Cheremshanian (Tashasty); Prikamian (Askyn); Severokeltmian (Akavas); Krasnopolyanian (Syuranian); Voznesenian (Bogdanovkian); Zapaltjubian; Protvian; Steshevian; Tarussian; Venevian; Mikhailovian; Aleksian; Tulian; Bobrikovian; Radaevkian; Kosvinian; Kizelovian; Cherepetian; Karakubian; Upinian; Malevian; Gumerovian; Ziganian

Western Europe:
- Autunian: Lebach; Kuzel
- Stephanian: Stephanian C; Stephanian B; Stephanian A; Cantabrian
- Westphalian: (D) Asturian; (C) Bolsovian; (B) Duckmantian; (A) Langsettian
- Namurian: Yeadonian; Marsdenian; Kinderscoutian; Alportian; Chokierian; Arnsbergian; Pendleian
- Visean: Warnantian; Livian; Moliniacian
- Tournaisian: Ivorian; Hastarian

North America: Nealian; Virgilian; Missourian; Desmoinesian; Atokan; Morrowan; Chesterian; Meramecian; Osagean; Kinderhookian; Chatauquan

China: Zisongian; Xiaodushanian; Dalaan; Huashibanian; Lousuan; Dewuan; Shangsian; Jiusian; Tangbagouan; Gelaohean

N-E Siberia: Khorokytian; Kyglitassian; Solonchanian; Natalian; Khatynykhian; Ovlachanian; Chuguchanian; Kirinian; Bazovian; Khamamitian

FIGURE 23.1 **Correlation of regional Carboniferous stages and other stratigraphic subdivisions.**

Belgium, respectively), which later became the official chronostratigraphic stages. The base of the **Tournaisian** coincides with the base of the Carboniferous, and the base of the **Viséan** is defined by the FAD of *Eoparastaffella simplex*. Later, substages were introduced, in ascending order subdividing the Tournaisian into the Hastarian and Ivorian, and the Viséan into the Moliniacian, Livian, and Warnantian. The **Hastarian** was introduced on the base of an outcrop at the Meuse River, northeast of the Hastière-Lavaux church (Hance *et al.*, 2006a). The substage correlates with the foraminiferal zones MFZ1 to MFZ3 and part of MFZ4, rugose coral zones RC1 and RC2, *Siphonodella* conodont interval, and with the ammonoid *Gattendorfia*–*Eocanites* Zone as well as with part of the *Goniocyclus*–*Protocanites* Zone (Poty *et al.*, 2006). The **Ivorian** with the stratotype Yvoir on the right bank of the Meuse River between Namur and Dinant has a base defined by the FAD of *Polygnathus communis carina* (Hance *et al.*, 2006b). It corresponds to the foraminiferal zones MFZ4 to MFZ8, the rugose coral zones RC3 to RC4β1, the *Polygnathus communis carina* and *Scaliognathus anchoralis* conodont zones, and part of the ammonoid *Goniocyclus*–*Protocanites* Zone and *Fascipericyclus*–*Ammonellipsites* Zone. The base of the **Moliniacian** is coupled with the base of the Viséan; it is named after the stratotype near the village of Salet, on the left flank of the Molignée valley, about 4 km WSW of Yvoir (Devuyst *et al.*, 2006). The substage corresponds with the foraminiferal zones MFZ9 to MFZ11, the rugose coral zones RC4β2 to RC5, the lower part of the *Fascipericyclus*–*Ammonellipsites* Zone, and part of the *Bollandites*–*Bollandoceras* Zone. The **Livian**, after a cliff and quarry at Lives, on the right bank of the Meuse valley (Hance *et al.*, 2006c) has a base at the FAD of *Koskinotextularia and Pojarkovella nibelis*, at the base of the "Banc d'or de Bachant". It is obviously a short substage and correlates with the foraminiferal zones MFZ12, the rugose coral zones RC6, and part of the *Bollandites*–*Bollandoceras* ammonoid Zone. The **Warnantian**, with the stratotype Camp de César quarry at Thon-Samson near Namur, has a base defined by the base of the thick bed of dark bioclastic limestone overlying the beige algal limestones of the top of the Livian. It has a long stratigraphic range including the foraminiferal zones MFZ13-MFZ15, the rugose coral zones RC7 and RC8, and the *Entogonites* to *Lusitanoceras*–*Lyrogoniatites* ammonoid zones.

Classically, the Late Carboniferous in Belgium began with the **Namurian** stage, named after outcrops in the vicinity of the citadel at Namur in central Belgium (Purves, 1883; Dusar, 2006). This unit, which now spans the Mississippian–Pennsylvanian boundary, has been replaced by international stages (e.g., Serpukhovian, Bashkirian). The following **Westphalian** (after the Westphalian coal mining area) and **Stephanian** (after the city of Saint-Étienne in eastern central France) are used mainly in the coal mining areas.

In Great Britain, the subdivision of the Early Carboniferous was initiated in the late 19th Century, and the pioneering work of Vaughan (1905) provided the first biostratigraphic scheme based on the occurrence of corals and brachiopods. Based on this, later investigations – e.g., by Bisat (1924) – led to a finer subdivision, also using ammonoids. The first formal chronostratigraphic scheme was then introduced by George *et al.* (1976) and discussed by Ramsbottom (1981), Riley (1993), and Waters and Davies (2006).

The oldest of the Carboniferous substages is the **Courceyan** with the stratotype at the Old Head of Kinsale in Ireland. Its base coincides with the Devonian–Carboniferous boundary. It correlates with the Hastarian and older part of the Ivorian of the Belgium scheme. The **Chadian** is named after Chatburn near Clitheroe, Lancashire; its base is defined by the first lithological change below the entry of the genus *Eoparastaffella*. As Riley (1993) found when reinvestigating the type locality, the FOD of the proposed foraminifer *Eoparastaffella* does not coincide with the proposed stage boundary but occurred much higher in the type section. He thus proposed the subdivision into early and late Chadian for the Tournaisian and the Viséan portions, respectively.

The following four substages belong in the Viséan, beginning with the **Arundian**. Its stratotype is at Hobbyhorse Bay, Dyfed (South Wales) and its base is defined by the base of the Pen-y-Holt Limestone. The **Holkerian**, with a stratotype at Barker Scar, south Cumbria, has the same duration as the Livian in Belgium. It begins with the base of the Park Limestone Formation; coinciding with the FOD of the corals *Carcinophyllum vaughani* and *Lithostrotion minus* as well as the brachiopods *Davidsonia carbonaria*, *Composita ficoides*, and *Linoprotonia corrugatohemispherica*. The **Asbian**, with the stratotype at Little Asby Scar, Ravenstonedale, Cumbria, already belongs to the upper Viséan. The base of the Potts Beck Limestone and the FOD of the corals *Dibunophyllum bourtonense*, *Siphonodendron pauciradiale*, *S. juncerum*, and *Palaeosmila murchisoni,* as well as the brachiopods *Linoprotonia hemispherica* and *Daviesiella llangollensis* mark the base of the substage. Finally, the **Brigantian**, with the stratotype Janny Wood near Dent, Cumbria, begins with the Peghorn Limestone and the FOD of the corals *Diphyphyllum lateseptatum*, *Actinocyathus floriformis*, *Nemistium edmondsi*, and *Palastrea regia* as well as the brachiopods *Productus hispidus*, *P. productus*, and *Pugilis pugilis*.

The two following substages composed the Serpukhovian stage. The **Pendleian**, which is named after Pendle Hill in Lancashire, and which has its stratotype in Light Clough, Lancashire, is traditionally defined by the FAD of the ammonoid *Cravenoceras leion*. The first occurrence of the conodont species *Lochriea ziegleri*, which is supposed to be the index fossil for the base of the Serpukhovian, is slightly below *Cravenoceras leion*, meaning that the two boundaries are not identical. The **Arnsbergian** is named after the town of Arnsberg in the Rhenish Mountains, but has its stratotype at Slieve Anierin, Co. Leitrim (Ireland). Its base is defined by the FAD of the ammonoid *Cravenoceras cowlingense*.

Seven British regional substages compose the Bashkirian. These are the following units, together with their stratotypes and index ammonoid species defining their bases:

Chokierian (Stonehead Beck, Yorkshire; *Isohomoceras subglobosum*),
Alportian (Blake Brook, Derbyshire; *Hudsonoceras proteus*),
Kinderscoutian (Samlesbury Bottoms near Blackburn, Lancashire; *Hodsonites magistrorus*),
Marsdenian (Park Clough, Derbyshire; *Bilinguites gracilis*),
Yeadonian (Orchard Farm, Derbyshire; *Cancelloceras cancellatum*),
Langsettian (Langsett, Derbyshire; *Gastrioceras subcrenatum* marine band), and
Duckmantian (Duckmanton, Derbyshire; *Anthracoceratites vanderbeckei* marine band).

The **Bolsovian** (stratotype is Bolsover, Derbyshire), defined by the *Donetzoceraas aegirianum* marine band, already belongs in the Moscovian.

In Germany, chronostratigraphic units have been proposed for the Dinantian (i.e. Lower Carboniferous in old terminology), but these names, which derive from localities in the Rhenish Mountains, have been used only sporadically. These are **Balvian** (after the Devonian–Carboniferous boundary beds exposed near Balve), corresponding to the *Gattendorfia* Stufe, **Erdbachian** (after the Erdbach Limestone), corresponding to the "*Pericyclus* Stufe" in previous understanding, and **Aprathian** (after the trilobite locality in the Aprath Shales), corresponding to the "*Goniatites* Stufe" in previous understanding (Paproth *et al.*, 1963; Korn, 2006).

Between 1841 and 1845, Murchison and others (Murchison, 1845), in collaboration with Russian geologists and based on Russian sections, divided the Carboniferous into Lower, Middle and Upper stages, which were later elevated to series (Möller, 1878, 1880). In terms of recent chronostratigraphy, Möller's Upper Series include the Upper Pennsylvanian and most of the Cisuralian (Lower Permian) Series, excluding the Kungurian Stage, and therefore all three proposed series were approximately equal. Subsequently, the Cisuralian portion of the succession was excluded from Möller's Upper Series. The remaining Lower, Middle, and revised Upper series then constituted a three-fold subdivision of the Carboniferous that became the tradition in Russia and surrounding territories in Eastern Europe and Asia.

Munier-Chalmas and Lapparent (1893) named these three Carboniferous subdivisions the Dinantian (Mountain Limestone), Moscovian, and Uralian stages. The Westphalian (Millstone Grit and Coal Measures combined) and Stephanian Stages were also established by these geologists in terrestrial successions of Western Europe as equivalents to the Moscovian and Uralian in the marine facies. This was the first attempt to build a dual marine–terrestrial classification for the Carboniferous, which is still advocated by some geologists (Wagner and Winkler Prins, 1997), but is not accepted by the ICS. In the 19th century, the majority of Russian geologists accepted a two-fold subdivision of the Carboniferous. Nikitin (1890) proposed the Serpukhovian (type locality near Serpukhov City) and placed it at the highest position in the Lower Carboniferous. He proposed to divide the Upper Carboniferous into the Moscovian and Gzhelian stages with type localities around Moscow (Moscovian) and a series of exposures near the villages of Gzhel, Pavlovo-Posad, and Noginsk (Gzhelian), along the Klyazma River and in the Oksko-Tsna Swell (Nikitin, 1890, pp. 77–78). The Moscovian originally included all the present-day Kasimovian, and the Gzhelian included the rest of the Carboniferous, and also the Asselian Stage (Lower Permian) of the modern scale. The Kasimovian (initially known as the *Tiguliferina* Horizon) was separated from the Moscovian by Ivanov (1926) and was named by Dan'shin (1947). Teodorovich (1949) suggested this unit should be a chronostratigraphic stage and regarded the Gzhelian (in the modern sense) to be the uppermost stage of the Carboniferous. Ruzhencev (1945) proposed the Orenburgian Stage, equal to the "*Pseudofusulina*" Horizon of Rauser-Chernousova (1937), in the southern Urals, as the terminal Carboniferous stage. Because of a miscorrelation to the Russian Platform, the Orenburgian was thought to be the equivalent of the lower Asselian (Pnev *et al.*, 1975; Harland *et al.*, 1990) and was removed from the Russian stratigraphic scale. However, this miscorrelation was recognized and the Orenburgian succession was returned to the Carboniferous once again (Davydov and Popov, 1986), where it is generally merged with the Gzhelian Stage. Development of Carboniferous stratigraphy during the 19th and early 20th centuries in Western Europe was summarized in the Heerlen Congresses in 1927 and 1935 (Heerlen Classification), in which a two-fold division of the Carboniferous was formalized and the Lower Carboniferous was replaced by the Dinantian. The latter was divided into two stages; the Tournaisian (proposed by de Koninck in 1872) and the Visean (proposed by Dupont in 1883). The Upper Carboniferous (later named the Silesian) was also divided into two stages; the Namurian (name proposed by Purves in 1883 for the equivalent of the Millstone Grit), and the Westphalian, each with three divisions (A, B, and C; Ramsbottom, 1984).

Further biostratigraphic studies in marine and terrestrial successions in Western Europe, particularly of ammonoids and plants, resulted in significant refinement of the regional stratigraphic scale into 16 stages from the Tournaisian through Westphalian with separate British and Belgium scales (George *et al.*, 1976). Details on the history of the Carboniferous classification in Western Europe can be found in George *et al.* (1976), Ramsbottom (1984), and Wagner and Winkler Prins (1997).

23.1.2. Evolution of Traditional North American Subdivisions

In the USA, the Carboniferous was subdivided into the Mississippian, or lower Carboniferous, proposed by Alexander Winchell in 1870, and the Pennsylvanian, or upper Carboniferous, proposed by J. J. Stevenson in 1888, each of which was put forward as an independent system by Williams (1891). Predominantly marine rocks of early Carboniferous age were assigned to the Mississippian, with type localities in the upper Mississippi Valley. Kinderhookian, Osagean, Meramecian and Chesterian are the subdivisions most commonly used in the US Geological Survey and other geological communities in the USA (Lane and Brenckle, 2005). Originally established as unconformable bounding units, they later received comprehensive biostratigraphic characteristics and became regional chronostratigraphic units in North America (Lane and Brenckle, 2005).

Kinderhookian was first designated near the village of Kinderhook, Pike County, Illinois, as a sequence between the upper Devonian Louisiana Limestone and lower Osagean Burlington Limestone (Meek and Worthen, 1861). The proposed international definition of the Devonian–Carboniferous boundary at the FAD of *Siphonodella sulcata* conodont brings the base of the Kinderhookian up to the base of Horton Creek Formation (Collinson, 1961; Conkin and Conkin, 1974). The top of the Kinderhookian is an unconformity between the Chouteau Formation and the Meppen Limestone that corresponds with the base of **Osagean** Regional Stage in the North American Mississippian type area. The term was proposed for the sequence from the base of Meppen Limestone to the top of lower Warsaw Formation (Williams, 1891) in the area along the Osage River in St. Clair County, Missouri (Lane and Brenckle, 2005). The base of the Osagean chronostratigraphic unit corresponds to the *Gnathodus punctatus* conodont zone that is missing in the type area (Lane, 1974).

The **Meramecian** was originally defined as a Meramec Group at the Meramec Highlands Quarry along Meramec River, southwest of St. Louis, Missouri (Ulrich, 1904). The group includes the Warsaw Formation, Spergen Hill (Salem) and the St. Louis Limestones. In terms of biostratigraphy the base of the Meramecian in the type area is only characterized with local brachiopods and foraminifers, and appears within the *Gnathodus texanus* conodont zone (Lane and Brenckle, 2005).

Chesterian is named after the Chester Group that was established near the town of Chester in southern Illinois (Worthen, 1860, 1866). The original group has recently been adjusted, putting the Grove Church Shale in the upper part and Ste. Genevieve Formation near the bottom. The Meramecian–Chesterian boundary is quite prominent, and is associated with the boundary between the *Hindeodus scitulus – Apatognathodus scalenus* and *Gnathodus bilineatus* conodont zones, and with the appearance of the asteroarchaediscids Foraminifera. The Chesterian Stage in the type area is incomplete and approximately the upper third of this chronostratigraphic unit is missing (Lane and Brenckle, 2005).

Coal-producing beds in the state of Pennsylvania were termed Pennsylvanian and considered as the stratigraphic equivalents of the western European "Coal Measures". The Pennsylvanian included the Pottsville Conglomerate, Lower Productive Coal Measure, and Upper Barren Coal Measures. The type localities for the Pennsylvanian stages occur in marine cyclic sequences in the North American midcontinent basin in Arkansas (Morrowan), Oklahoma (Atokan), central Iowa (Desmoinesian), along the Missouri River in Iowa and Missouri (Missourian), and in east-central Kansas (Virgilian).

The **Morrowan** Stage was first recognized as a formation in northwestern Arkansas (Adams and Ulrich, 1904). It was regarded as a series (Moore, 1944) and later as a stage equal to a zone of primitive fusulinid *Millerella* (Moore and Thompson, 1949). The base of this stage in the type area, and in many other places in North America, appears to be discomformable with an exception at the mid-Carboniferous GSSP in Nevada and some basinal facies in Arkansas (Lane *et al.*, 1999). In the United States and South America this stage is recognized by a very low diversity and highly provincial foraminiferal assemblage. The equivalent of the Morrowan in western Canada and Canadian Arctic has more diverse fauna, but belongs to another Boreal (Uralian) province. Two parallel conodont zonations, a shallow-marine and a deeper marine one, have been developed within Morrowan sequences (Lane, 1977; Barrick *et al.*, 2004).

The **Atokan** Stage was originally designated as a formation exposed near the town of Atoka in southwestern Oklahoma (Taff and Adams, 1900). Like the Morrowan, it was regarded as an unconformably bounded series and then as a stage equal to the fusulinid genozones *Profusulinella* and *Fusulinella* (Moore, 1948; Moore and Thompson, 1949). It was later found that the base of the Atokan coincides with the FAD of *Eoschubertella* fusulinid (Groves, 1986), and this position is now commonly accepted. Thus, the Atokan Stage is divided into three fusulinid zones in order: *Eoschubertella*, *Profusulinella* and *Fusulinella* (Wilde, 1990). The taxonomic composition of these zones is highly provincial. The Atokan conodont zonation is poorly developed with two zones in shallow-water (*Neognathodus*) and deeper water settings (Barrick *et al.*, 2004).

The name of the **Desmoinesian** Stage derives from the Des Moines Formation that is exposed along the Des Moines River in Central Iowa (Keyes, 1893). It was later commonly regarded as a series and then a stage (Moore and Thompson, 1949). The Desmoinesian in the type area is a classic succession of cyclic shales, limestone, coals and sandstone in the northern Midcontinent that is recognized together with the Missourian and the Virgilian as a glacio-eustatic cyclothem (Wanless and Shepard, 1936; Heckel, 1986). The stage in the type locality is unconformable upon the Atokan (Lambert and Heckel, 1990) but in the Ardmore Basin, Oklahoma the contact between Atokan and Desmoinesian stages is transitional (Clopine, 1992). The Desmoinesian has also been defined as a zone of *Beedeina* (and/or *Fusulina*) and *Wedekindellina* characterized by a very provincial species (Moore and Thompson, 1949). The stage lower boundary is not precisely defined, and is questionable due to the transitional character of evolution between *Fusulinella* and *Beedeina* and between *Fusulinella* and *Wedekindellina* (Clopine, 1992). The exact position of the Atokan–Desmoinesian boundary in the conodont succession remains unresolved. The basal Desmoinesian possesses a conodont fauna in its type region that includes the first occurrences of *Neognathodus caudatus* Lambert, *Idiognathodus amplificus* Lambert, and *Idiognathodus obliquus* Kozitskaya and Kossenko. Two parallel shallow-water *Neognathodus* and deeper water *Idiognathodus/Swadelina* zonations were proposed for the Desmoinesian Stage (Barrick *et al.*, 2004).

The name **Missourian** was proposed at the same time and in the same formational status as the Desmoinesian (Keyes, 1893), and is part of the classic limestone dominated cyclothems in the Northern Midcontinent Shelf (Heckel, 1986). It was considered formally as a stage at the same time as the Desmoinesian (Moore and Thompson, 1949). The original base of the stage coincided with an unconformity at the base of the Pleasanton Group, and also with one of the major extinction events in the North American province, with the loss of most Desmoinesian invertebrates and a slow recovery that took a significant time (>1 myr). The unit was considered as the lower part of the fusulinid *Triticites* zone, although the genus appears above the lower third of the stage (Thompson *et al.*, 1956). Later, the base of the stage was redefined at the base of the provincial fusulinid *Eowaeringella* Zone (Stewart, 1968), as well as by the appearance of the ammonoid *Pennoceras* (Boardman *et al.*, 1990), and the conodont *Streptognathodus* (s.l.) (Barrick *et al.*, 2004). Recently the base of the Missourian was defined by the FAD of *Idiognathodus eccentricus* in the basinal shales that are equivalent to the Exline Limestone, which lies within the Pleasanton Group in the Northern Midcontinent shelf (Barrick *et al.*, 2004).

The **Virgilian** Stage, named after a town in east-central Kansas, was established as a series (Moore, 1932) and later as a terminal stage of the Pennsylvanian "System" (Moore and Thompson, 1949). Like the Missourian Stage, it was an unconformity-bounded unit that included sets of limestone-dominated cyclothems of the Douglas, Shawnee and Wabaunsee groups. The Admire and the lower part of Council Grove groups that now are part of the Carboniferous were originally included in the Permian. The biostratigraphic content of the stage was neither precise nor prominent. The stage was considered as the upper part of the fusulinid *Triticites* Genozone and ammonoid *Uddenites* Genozone (Moore and Thompson, 1949). Recently, the Virgilian was divided into five conodont zones (Barrick *et al.*, 2004) with its base fixed by the FAD of the conodont *Streptognathodus simulator* and the FAD of the ammonoid genus *Schumardites* (Boardman *et al.*, 2006). This boundary appears above the top of the Douglas Group, i.e. much higher than the traditional boundary at the base of this group (Moore and Thompson, 1949). Usage of the Pennsylvanian stages is only feasible for the USA and South America (N. American province). More details on the establishment and internal units of the Pennsylvanian stages can be found in Heckel (1999).

23.1.3. Status of the International Scale

The first international attempt to build a global Carboniferous scale integrating the various chronostratigraphic classifications was made during the Eighth International Congress on Carboniferous Stratigraphy and Geology in Moscow (Bouroz *et al.*, 1975). Two subsystems of the Carboniferous, the Mississippian and the Pennsylvanian, were subdivided into two series to successfully merge the system used in North America with the three-fold divisions employed in the former USSR (Wagner and Winkler Prins, 1994). Tournaisian, Visean, Serpukhovian, Bashkirian, Moscovian, Kasimovian, and Gzhelian were proposed as global Carboniferous stages (Figure 23.1). Bouroz *et al.*'s proposal significantly advanced the development of an international Carboniferous stratigraphy. Within a few years several working groups were founded (Brenckle and Manger, 1990), and two major Carboniferous boundaries – the base of the Carboniferous Period and the mid-Carboniferous (or Mississippian–Pennsylvanian) boundary – were accepted as GSSPs (Paproth *et al.*, 1991; Lane *et al.*, 1999). Recently, the base of the Visean Stage was ratified at Penchong, south China. The top of the Carboniferous is established by the GSSP for the base of the Permian (Davydov *et al.*, 1998).

23.1.4. Mississippian Subsystem (Lower Carboniferous)

After several glaciation events during the Late Devonian (Famennian), with corresponding extinction events in marine and terrestrial biota (e.g., the Kellwasser and Hangenberg Events), the beginning of the Carboniferous (late Tournaisian) was probably free of ice sheets (Buggisch *et al.*, 2008; Kaiser *et al.*, 2008; Joachimski *et al.*, 2009). However, except for late Visean (middle Chesterian) and late Gzhelian (late Virgilian), most of the rest of the Carboniferous was a time of extensive pulses of glaciation (Fielding *et al.*, 2008a; Birgenheier *et al.*, 2009) accompanied by high-frequency sea-level fluctuations and corresponding global transgressive–regressive sequences (Ramsbottom, 1973; Ramsbottom *et al.*, 1978; Veevers and Powell, 1987).

During the Mississippian, there was generally unobstructed marine communication between paleo-Tethyan and

Panthalassan shelves. Therefore, Mississippian marine faunas are generally worldwide in their distribution and latitudinal differences are more strongly developed than are longitudinal differences (Ross and Ross, 1988).

23.1.4.1. Devonian–Carboniferous Boundary and the Tournaisian Stage

The Devonian–Carboniferous boundary interval was a time of global regression with a major sequence boundary located slightly below it (Paproth *et al.*, 1991). The GSSP for the base of the Tournaisian Stage, Mississippian Subperiod and Carboniferous Period has been defined in trench E at La Serre section, Montagne Noire, southern France (Paproth *et al.*, 1991) (Figure 23.2). The section lies on the southern slope of the La Serre hill, 2.5 km southwest of the village of Cabrières, near the "classic" base of the *Gattendorfia* ammonoid Stufe.

The boundary is defined in Bed 89 at the FAD of the conodont *Siphonodella sulcata*, within the evolutionary lineage of *S. praesulcata–S. sulcata*. The transitional boundary beds in the section consist of platy and nodular cephalopod-bearing and oolitic limestone, and contain a mixture of pelagic and near-shore biota that show signs of minor transport before lithification (Flajs and Feist, 1988). The La Serre section has recently been restudied and *S. sulcata* was found 0.45 m below the proposed boundary. Also, the co-occurrence of *S. praesulcata* and *S. sulcata* above a facies break shows that the evolutionary lineage from *S. praesulcata* to *S. sulcata,* which was previously claimed for the definition of the D/C boundary, is obviously absent at La Serre (Kaiser, 2009). All these new data result in issues that affect not only the Carboniferous time scale but also the GSSP concept.

Several solutions have already been proposed to resolve these issues, including re-evaluating the defining species and phylogenies (Corradini and Kaiser, 2009; Kaiser, 2009; Tragelehn, 2009; Corradini *et al.*, 2010). The Devonian and Carboniferous subcommissions of the IGC have formed a task group to work on the Devonian–Carboniferous boundary problem (Richards and Task Group, 2009), but for the moment the Devonian/Carboniferous boundary remains at bed 89 in La Serre section.

Auxiliary stratotype sections (ASS) for the Devonian–Carboniferous boundary are located in the Hasselbachtal, Sauerland area, Germany (Becker and Paproth, 1993) and in Nanbiancun, southern China (Wang, 1993). The former is the more important because of the presence of several volcanic ashes within the Devonian–Carboniferous transition that were dated previously (Claoue-Long *et al.*, 1995). These Rhenish Mountain ashes have now been re-collected and re-dated with ID-TIMS (Davydov *et al.*, 2008a). The results are presented below (see Section 23.3).

The last appearance datum (LAD) of the miospore *Retispora lepidophyta* (*Retispora lepidophyta–Verrucosisporites nitidus* Biozone [LN]) almost coincides with the FAD of the conodont *Siphonodella sulcata*. In the ammonoid zonation, the base of the Tournaisian coincides with the first entry of *Gattendorfia* (Kullmann *et al.*, 1990). The major extinctions of ammonoids (particularly clymeniids) occurred during the Hangenberg Event of the latest Famennian extinction crisis (House, 1993; Korn, 1993). In the type area in the Franco–Belgian basin, the basal Tournaisian is characterized by foraminiferal MFZ1 unilocular zone (Poty *et al.*, 2006). The FAD of the foraminiferal species *Chernyshinella glomiformis* and *Tournayella beata* appear slightly above.

23.1.4.2. Visean

Historically, the Visean Stage, like the Tournaisian, was established in southern Belgium. The base of the Visean was first officially defined in 1967 by the International Congress on Carboniferous Stratigraphy and Geology at the lowest *marbre noir* (i.e. first black limestone) intercalation in the Leffe facies at the Bastion Section in the Dinant Basin (George *et al.*, 1976; Hance *et al.*, 1997). This boundary coincided with the first occurrence of the foraminiferal genus *Eoparastaffella*, and occurs less than 1 m below the FAD of the conodont *Gnathodus homopunctatus* (Devuyst *et al.*, 2003).

Because of the historical priority of achieving stratigraphic stability in international stratigraphic practice, the Tournaisian–Visean boundary working group decided to keep the base of the Visean as close as possible to the classical level (Hance *et al.*, 1997). A significant turnover in pelagic fossils (either conodonts or ammonoids) does not occur near the classical Tournaisian–Visean boundary, so it was proposed to define the boundary within the foraminiferal phylogenetic lineage of *Eoparastaffella* genus. The disappearance of deeper-water "Tournaisian" conodonts and the entry of shallow-water "Visean" Foraminifera in the classic Franco–Belgian basin appears to be an ecological event related to a significant sea-level drop in late Tournaisian–early Visean time. Hence, this faunal change may be of questionable chronostratigraphic significance (Hance *et al.*, 1997). Simple morphological parameters of *Eoparastaffella*, identifying two successive morphotypes (types 1 and 2, corresponding, respectively, to *E. rotunda typica* and *E. simplex typica*) can be used by non-foraminiferal specialists to recognize this evolution and identify the base of the Visean. This proposal has been accepted as the most appropriate definition of the classic base of the Visean. The definition is the first benthic marker within the entire Paleozoic. The GSSP section for the base of the Visean Stage is the base of bed 83 in Pengchong section, Guangxi, South China (Figure 23.3), that corresponds with the FAD of *Eoparastaffella simplex* (Devuyst *et al.*, 2003; Richards and Aretz, 2009). A final report on the ratification of this GSSP and publication in *Episodes* will be issued.

23.1.4.3. Serpukhovian

The Serpukhovian Stage was proposed by Nikitin (1890) as the terminal stage of the lower Carboniferous "Series" and

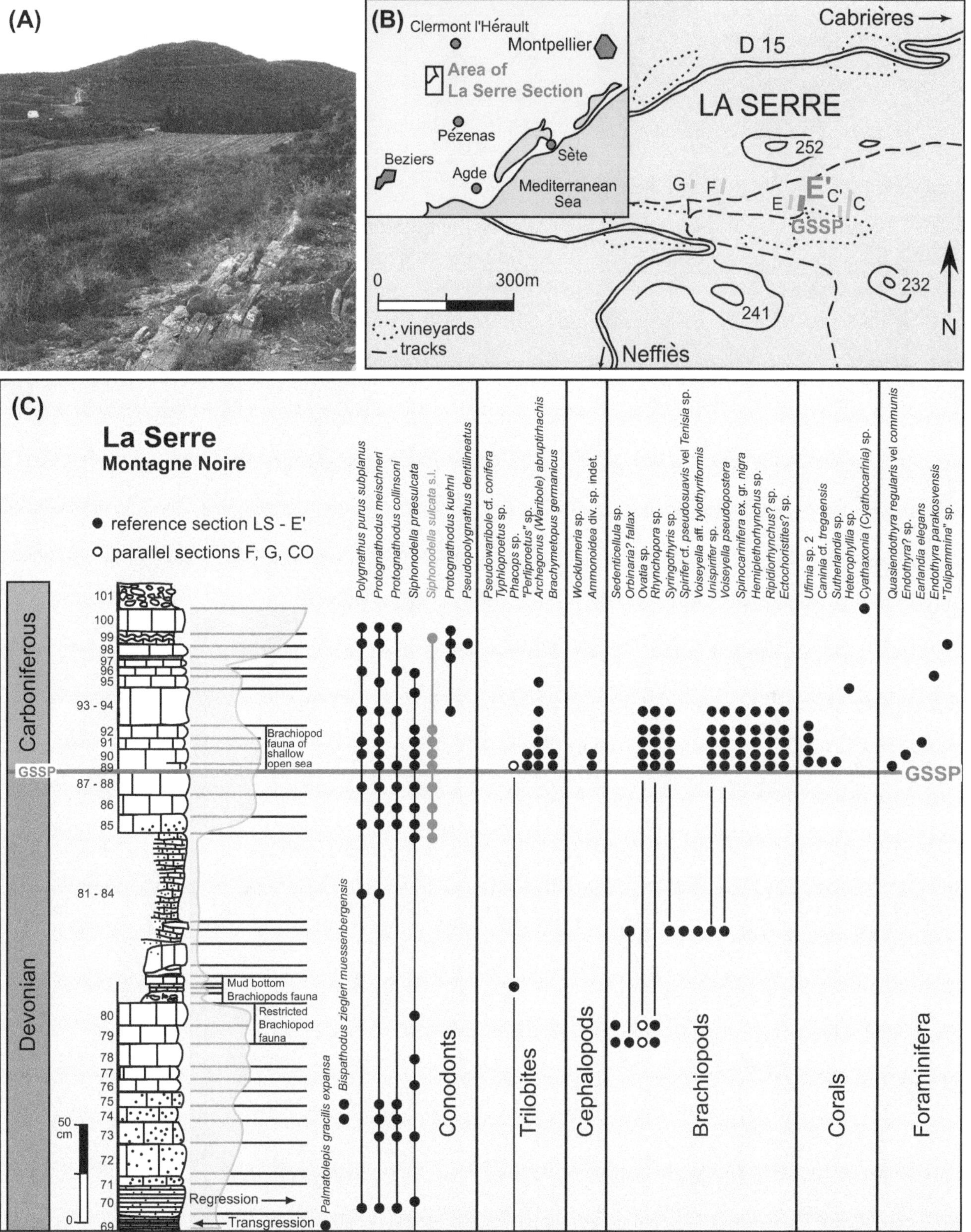

FIGURE 23.2 **The GSSP for the base of the Tournaisian Stage, Mississippian Subperiod and Carboniferous Period has been defined in trench E at La Serre section, Montagne Noire, southern France**. The section lies on the southern slope of the La Serre hill, 2.5 km southwest of the village of Cabrières, near the "classic" base of the *Gattendorfia* ammonoid Stufe. (The photo of GSSP is courtesy of Markus Aretz.)

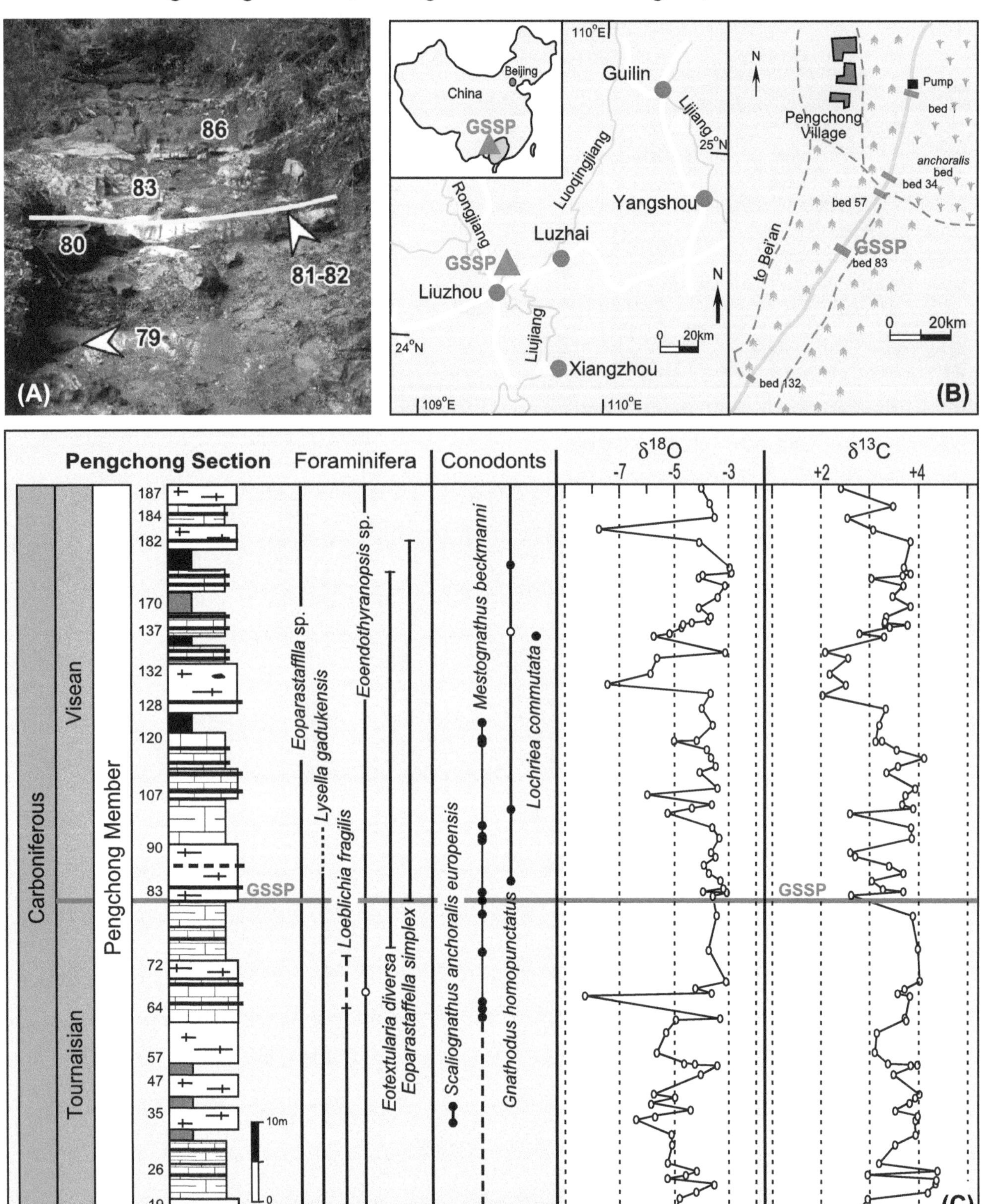

FIGURE 23.3 GSSP section for the base of the Visean Stage of the Carboniferous System in the Pengchong section, Guangxi Autonomous Region, South China.

was named after Serpukhov City, where the stage is exposed in a series of quarries and localities along the Oka River. The Serpukhovian is the shortest chronostratigraphic unit in the Mississippian. In the type area of the Moscow syncline, it is represented by an unconformably bounded sequence of open- to restricted-marine shallow carbonate and mixed carbonate–siliciclastic rocks.

The Serpukhovian in the Russian Platform and surrounding regions is informally divided into lower and upper substages and into five horizons (regional stages); Tarusian (Tarusky), Steshevian (Steshevsky), Protvinian (Protvinsky), Zapaltubinian (Zapaltubinsky), and Voskresenian (Voskresensky), the last two being almost completely absent in the type area (Makhlina *et al.*, 1993).

More complete successions of the upper Serpukhovian and its transition to the Bashkirian are present in the Donets Basin, in the northern portion of Timan-Pechora, and in the Urals as well in Central Asia, Cantabrian Mountains and China. In North America it comprises an upper Chesterian Stage (Figure 23.1).

The Visean–Serpukhovian transition coincides with a major Gondwanan glaciation (Mii *et al.*, 2001; Saltzman, 2003; Davydov *et al.*, 2010; Gulbranson *et al.*, 2010) and sharp climatic changes coupled with severe marine biotic endemism. In the type area in the Moscow Basin, the Visean–Serpukhovian transition deposits were restricted, and connections with open basins in the east (Urals), south (Precaspian), and north (Timan-Pechora) were short lived and/or limited. Benthic faunas in the Russian Platform are diachronous in their distribution, which is controlled mostly by ecological rather than by evolutionary factors (Makhlina *et al.*, 1993).

The lower boundary of the Serpukhovian still remains uncertain. In the type section in the Moscow Basin the traditional base of the stage coincided with the base of Tarusian Horizon (Shvetsov, 1932). The fauna around the Visean–Serpukhovian boundary is somewhat transitional and often diachronous laterally (Makhlina *et al.*, 1993). Although the Visean–Serpukhovian boundary is fixed at the base of the local *Pseudoendothyra globosa–Neoarchasediscus parvus* foraminiferal zone, this is an acme zone. The index taxa and other species range from the Venevian Horizon of the Visean into the Serpukhovian (Gibshman, 2003). The zone corresponds with the Cf7 Foraminifera Zone and/or the MFZ16 Foraminifera Zone in Franco–Belgian basins (Conil *et al.*, 1991; Poty *et al.*, 2006).

The Task Group of the Subcommission on the Carboniferous of the ICS, responsible for defining the GSSP close to the Visean–Serpukhovian (Namurian) boundary, proposed the conodont species *Lochriea ziegleri* as an index for the base of global Serpukhovian Stage (Nigmadganov and Nemirovskaya, 1992; Skompski *et al.*, 1995). The occurrence of this species was reported in the upper Visean in many sections throughout Europe (Brigantian Stage) and Asia (Skompski *et al.*, 1995; Belka and Lehmann, 1998; Kulagina *et al.*, 2006; Akhmetshina *et al.*, 2008; Kullmann *et al.*, 2008) and in the middle Venevian Horizon in the Moscow Basin (Makhlina *et al.*, 1993; Davydov *et al.*, 2004). Thus, the proposed boundary appears within the uppermost Visean. Recent data from the Donets Basin suggests the FAD of *Lochriea ziegleri* to be 1–1.5 myr older than the traditional base of the Serpukhovian in the Moscow Basin (Davydov *et al.*, 2010). Several sections now are under detailed study, and are being considered as potential candidates for the GSSP of the Serpukhovian Stage (Richards and Task Group, 2008) in the Urals (Nikolaeva *et al.*, 2009a,b), S. China (Yuping and Wang, 2005) and Spain (Nemyrovska, 2005; Sanz-Lopez *et al.*, 2007).

23.1.5. Pennsylvanian Subsystem (Upper Carboniferous)

23.1.5.1. Mississippian–Pennsylvanian Boundary and the Bashkirian Stage

The beginning of the Pennsylvanian coincides with a significant mid-Carboniferous glaciation and consequent sea-level drop forming a sequence boundary in many sections. Veevers and Powell (1987) proposed that this boundary marks the beginning of a major Gondwanan glaciation and climate cooling. However, several pulses of glaciation within the mid Carboniferous have now been recognized. One of them is in the early-middle Serpukhovian and one of the largest was in the early-middle Bashkirian (Saltzman, 2003; Fielding *et al.*, 2008a; Grossman *et al.*, 2008; Bishop *et al.*, 2009).

The Bashkirian Stage was established in the early 1930s (Semikhatova, 1934) in the mountains of Bashkiria (currently the Bashkortostan Republic) in the southern Urals, Russia. Several sections have been suggested as a stratotype for the Bashkirian, but since the mid 1970s, the Askyn section, one of the best exposed and studied sections in the type area, has been accepted as a hypostratotype for the stage (Semikhatova *et al.*, 1979; Sinitsyna and Sinitsyn, 1987; Nemirovskaya and Alekseev, 1994; Kulagina *et al.*, 2001; Kulagina and Pazukhin, 2002). The basal beds of the Bashkirian as it was defined, were originally characterized by the appearance of the Foraminifera species *Pseudostaffella antique,* i.e. the base of the Bashkirian within the chronostratigraphic succession was much higher than its modern position.

The GSSP for the beginning of the Pennsylvanian and of the Bashkirian Stage is defined within the transition from *Gnathodus girty simplex* to *Declinognathodus noduliferus s.l.*, and has been fixed in Arrow Canyon, in the Great Basin, Nevada, USA (Lane *et al.*, 1999) (Figure 23.4). The type section is located approximately 75 km northwest of Las Vegas, on the east side of a strike valley immediately north of Arrow Canyon gorge. The GSSP boundary horizon is located

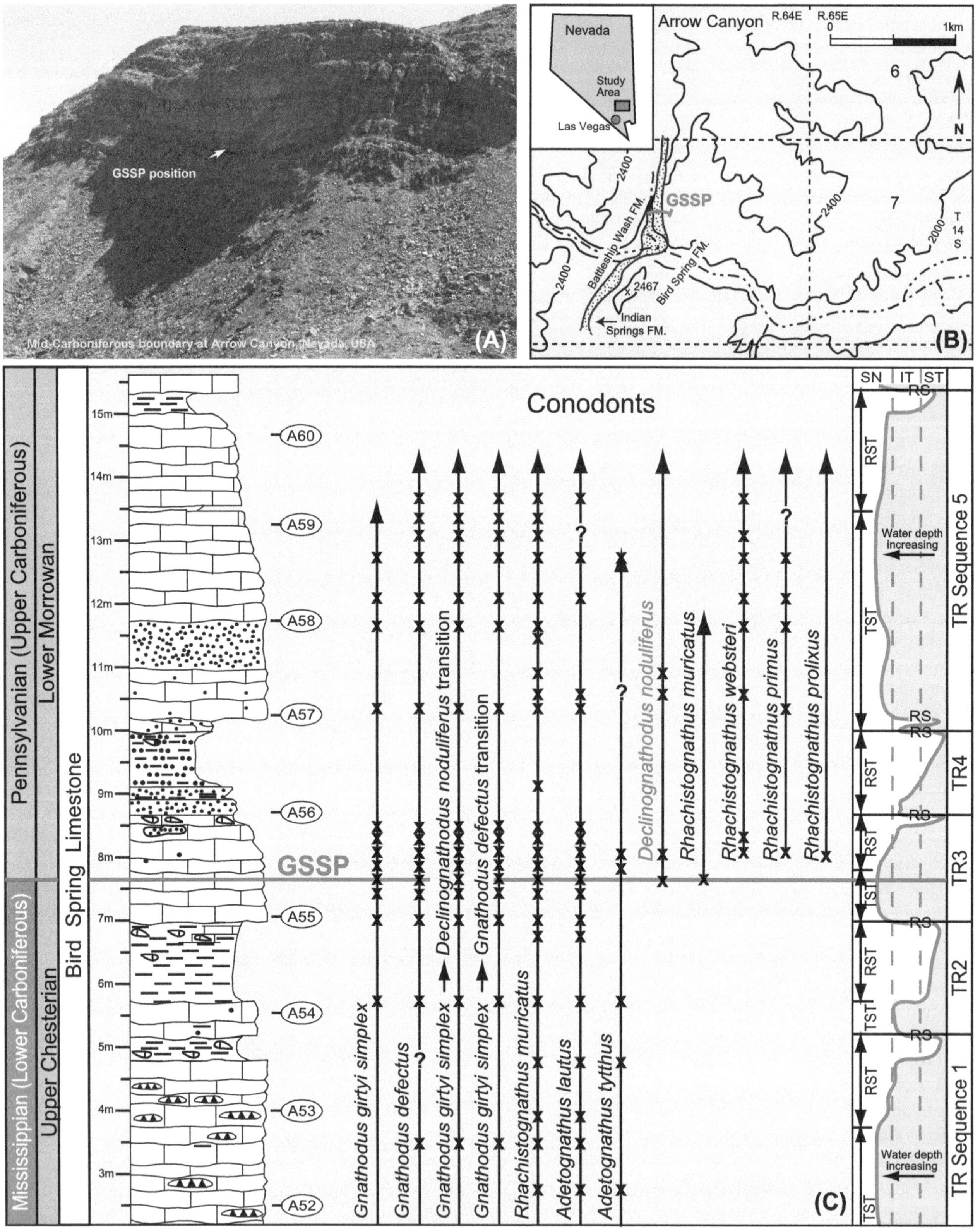

FIGURE 23.4 **The GSSP for the beginning of the Pennsylvanian and of the Bashkirian Stage in Arrow Canyon, in the Great Basin, Nevada, USA.** The type section is located approximately 75 km northwest of Las Vegas.

within the lower part of the Bird Spring Formation. During Carboniferous time, Arrow Canyon was situated near the paleoequator in a pericratonic tropical to subtropical seaway that extended from southern California through western Canada and into Alaska (Lane *et al.*, 1999). The succession which contains the mid-Carboniferous boundary interval comprises numerous high-order transgressive–regressive sequences driven by glacio-eustatic fluctuations caused by ongoing glaciation in Gondwana. Transgressive bioclastic and mixed bioclastic–siliciclastic limestones and fine sandstone sequences are separated by regressive mudstone, oolitic, and brecciated limestone intervals (Richards *et al.*, 2002; Barnett and Wright, 2008; Bishop *et al.*, 2009).

An alternative model of evolution of *Declinognathodus noduliferus s.l.*, in the eastern Hemisphere has been proposed within the chronocline *Gnathodus postbilineatus–Declinognathodus praenoduliferus–Declinognathodus noduliferus* (Nemyrovska, 1999). It suggests a polyphyletic origin for *Declinognathodus noduliferus,* and thus a potential diachroneity of the boundary between Eastern and Western Hemispheres cannot be excluded. Several subspecies are recognized within the *Declinognathodus noduliferus s.l.* The GSSP in Arrow Canyon is defined at the FAD of *D. noduliferus inaequalis*, whereas in Spain *D. noduliferus bernesgae* appears in the strata that are dated with ammonoids and conodonts as Serpukhovian (Sanz-Lopez *et al.*, 2006).

The mid-Carboniferous boundary GSSP is coincident with the base of the *noduliferus–primus* conodont zone of Baesemann and Lane (Baesemann and Lane, 1985). Beds below the mid-Carboniferous boundary contain an abundant archediscacean foraminiferal assemblage that is dominated by the eosigmoilines *Eosigmoilina robertsoni* and *Brenckleina rugosa*, and these species disappear slightly above the mid-Carboniferous boundary (samples 62 and 63 at Arrow Canyon). The foraminifer *Globivalvulina bulloides* is a useful marker to approximate the boundary, although in Japan, Arctic Alaska, the Pyrenees, and the Donets Basin it appears slightly below the boundary. The Foraminifera *Millerella pressa* and *M. marblensis* are also informal markers for the boundary, although primitive *Millerella* appears already in the late Mississippian (Brenckle *et al.*, 1997).

In the eastern hemisphere (Arctic, Pyrenees, Russian Platform, Urals, Donets Basin, Central Asia, Japan), the base of the Bashkirian is marked by the occurrence of species of the foraminiferal genus *Plectostaffella: Pl. varvariensis, P. jakhensis, P. posochovae,* and *P. bogdanovkensis*; however, the first primitive representatives of the genus appear slightly below the mid-Carboniferous boundary (Aizenverg *et al.*, 1983). The base of the Bashkirian in the Urals and surrounding regions coincides with the boundary between the *Monotaxinoides transitorius* and *Plectostaffella varvariensis* (or *Plectostaffella posokhovae*) foraminiferal zones (Kulagina *et al.*, 1992).

In the ammonoid zonation, the mid-Carboniferous boundary is identified at the base of the *Homoceras* Zone or *Isohomoceras subglobosum* Zone of Great Britain, Nevada, and Central Asia (Ramsbottom and Saunders, 1985; Kullmann and Nikolaeva, 2002). However, it is most likely that the first *Isohomoceras subglobosum* in Nevada and Central Asia occurs slightly earlier, in the latest Mississippian (Serpukhovian) (Nigmadganov and Nemirovskaya, 1992; Titus and Manger, 2001).

23.1.5.2. Moscovian

The classic sedimentary sequence of shallow-marine bioclastic limestone intercalated with colorful clays outcropping in and around the Moscow area gave rise to the Moscovian Stage (Nikitin, 1890). The quarry near the village of Myachkovo has been proposed as the type section of the stage. However, as mentioned earlier, the original Moscovian unit also included strata at locations around the villages of Khamovniki and Dorogomilovo, now within Moscow, and near the villages of Voskresensk and Yauza, east of Moscow. These sediments are now referred to the early-late Kasimovian Stage in terms of modern chronostratigraphy. Fortunately, the sequence at Myachkovo quarry belongs to the late Moscovian, keeping the Moscovian Stage valid.

Ivanov (1926) was the first to recognize the significant difference between fauna in the limestone of Khamovniki, Dorogomilovo, Voskresensk, and Yauza and fauna from Moscovian strata. He named these different sequences the "*Teguliferina*" Horizon (i.e. a regional stage in the Russian sense) after an attractive and common brachiopod. Dan'shin (1947) named this horizon Kasimovian, as these sediments are well exposed near the city of Kasimov in the Oksk-Tsna Swell.

Ivanova and Khvorova (1955) and Makhlina *et al.* (1979) provided detailed sedimentology and stratigraphy of Moscovian through Gzhelian Stages in the Moscow Basin. Four horizons with chronostratigraphic meaning in the Moscovian Stage were established by Ivanov (1926), in ascending order: Vereisky, Kashirsky, Podol'sky, and Myachkovsky. Later these units were proposed to be substages or stages (Ivanova and Khvorova, 1955), and widely used in this sense in equatorial and subequatorial belts of the eastern hemisphere. However, these units were not recognized elsewhere.

In the type area of the Moscow Basin, strata of the Moscovian Stage unconformably overlie limestones assigned to the Mississippian, with Bashkirian strata missing. The lowermost Moscovian (Vereian Horizon) is made up of three formations: Shatskaya, Aljutovskaya, and Ordynskaya (Ivanova and Khvorova, 1955). The lower formation is a siliciclastic and largely terrestrial sequence with no marine fossils recovered. Recently, the Aljutovskaya Formation has been extended downwards to include a siliciclastic unit commonly referred to as the Shatskaya Formation (Makhlina *et al.*, 2001). The latter has been excluded from the practice due to nomenclature problems. In the type section of the Aljutovskaya Formation, a characteristic fusulinid species *Aljutovella aljutovica* and the conodont species *Declinognathodus donetzianus* were found 3 m above its base (Makhlina *et al.*, 2001). *Aljutovella aljutovica* is an index species of the foraminiferal zone, commonly accepted as defining the base of the Moscovian in the former Soviet Union. The *Aljutovella aljutovica* fusulinid zone correlates widely through the entire northern and eastern margins of Pangea (Kulagina *et al.*, 2009). However, because of significant provincialism of fusulinid assemblages between eastern and western hemispheres, only conodonts can be used to define the base of the Moscovian Stage in the global chronostratigraphic scale. Nemirovska (1999) proposed the *Declinognathodus donetzianus* conodont species as indicative of the base of the Moscovian Stage. This species is widely distributed geographically (in the Moscow Basin, Urals, northwestern Europe, Spain, Canadian Arctic, Alaska, and Japan) and its evolutionary position is clearly recognized (Nemyrovska *et al.*, 1999; Goreva and Alekseev, 2007). Both conodont and fusulinid species occur in the basal beds of the Moscovian Stage in its type area, and hence can be used for defining the traditional Bashkirian–Moscovian boundary. However, in the sections where the Bashkirian–Moscovian transition is complete *Declinognathodus donetzianus* appears earlier than *Aljutovella aljutovica* (Ueno and Nemirovska, 2008; Davydov, 2009; Kulagina *et al.*, 2009). Thus, the defined base of the Moscovian Stage will be slightly older than the traditional boundary (Davydov, 2009).

Recently the FAD of *Diplognathodus ellesmerensis* has been proposed as another potential marker event to define the boundary (Groves and Task Group, 2007). This species possesses a distinct morphology and has global distribution. However the majority of conodont workers noted that it is infrequent elsewhere, is usually represented by young specimens, and its taxonomy and phylogeny is so poorly known that this species is not the best choice for the boundary definition (Groves and Task Group, 2008). Besides, the FAD of the species seems to be about 0.5 Ma younger than the traditional boundary (Davydov *et al.*, 2010).

It is most likely that the base of the Moscovian in North America approximates with the first appearance of the genus *Profusulinella* in the region (Moore and Thompson, 1949; Groves *et al.*, 1999).

In the ammonoid zonation, the Bashkirian–Moscovian boundary coincides with either the base of the *Winslowoceras–Diaboloceras* Zone (a zone based on genera, rather than species) on the Russian Platform and Urals (Ruzhenzev and Bogoslovskaya, 1978), or with the *Eowellerites* Zone in western Europe and North America (Ramsbottom and Saunders, 1985). It is most likely that the base of the *Eowellerites* Zone is slightly older than the base of the *Winslowoceras–Diaboloceras* Zone.

23.1.5.3. Kasimovian

The Kasimovian Stage was originally included in the Moscovian (Nikitin, 1890). It was the last stage in the Pennsylvanian Series to be established in the Moscow Basin (Teodorovich, 1949). Ivanov (1926) was the first to recognize the independence of the "*Teguliferina*" Horizon (Kasimovian Stage) from the Moscovian, and considered it to be an "Upper Carboniferous" unit (in the sense of the three-fold Russian classification of the Carboniferous) as opposed to the "Middle Carboniferous" Moscovian Stage. He placed the base of his "*Teguliferina*" Horizon at the limestone conglomerate in the base of the local "sharsha" lithostratigraphic unit.

Macrofossil assemblages of the *Teguliferina* Horizon in the type region, comprising mostly brachiopods and pelecypods, have only local correlative potential. A more widely recognizable definition of the Kasimovian Stage was proposed by Rauser-Chernousova and Reitlinger (1954), using the first occurrence of the fusulinid species *Obsoletes obsoletes* and *Protriticites pseudomontiparus*. Although these species were only found in the "sharsha" unit of the upper Suvorovskya Formation of the lowermost Kasimovian (i.e. a few meters above the original boundary proposed by Ivanov, 1926), the base of the Kasimovian has conventionally been placed at the conglomerate marking the base of the local "garnasha" lithostratigraphic unit of the lower upper Suvorovskya Formation, i.e. slightly below its original position.

The fusulinid definition proposed by Rauser-Chernousova and Reitlinger (1954) for the Kasimovian has been accepted for some years in the shallow-marine sections in Eurasia. After the evolution of *Obsoletes* and *Protriticites* became better understood (Kireeva, 1964; Davydov, 1990), it was suggested that the Moscovian–Kasimovian boundary should be placed between the latest Moscovian *Praeobsoletes burkemensis* and *Protriticites ovatus* fusulinid zone and the early Kasimovian *Obsoletes obsoletes* and *Protriticites pseudomontiparus* fusulinid zone (Davydov, 1990). In the Moscow Basin, the boundary was placed 1.0–1.5 m below the base of the "garnasha" lithostratigraphic unit, at the base of the local "lyska" lithostratigraphic unit within the upper Peskovskaya Formation, which traditionally has been included in the uppermost Moscovian (Davydov, 1997). This boundary coincides with a eustatic sea-level low and, consequently, with disconformities represented in many sections around the world (Davydov, 1997; Ehrenberg *et al.*, 1998; Davydov and Krainer, 1999; Leven and Davydov, 2001).

In the North American midcontinent this boundary occurs two major cycles below the previously proposed upper Pennsylvanian major sequence boundary at the base of the Hepler Shales of the Pleasanton Group (Ross and Ross, 1988; Heckel *et al.*, 1998).

Although the working group of the ICS Subcommission for the Carboniferous that was set up to establish the GSSP for the Moscovian–Kasimovian boundary has been active for nearly 20 years, and has gained significant insight into boundary correlations (Ueno and Task Group, 2009), no decision regarding the definition of this boundary has been forthcoming. The conodont species *Streptognathodus subexcelsus,* that characterized the basal beds of the Kasimovian Stage in the Moscow Basin, has been described recently (Goreva and Alekseev, 2006), and its correlation potential is not clear. *Idiognathodus saggitalis* (Kozitskaya *et al.*, 1978) and *Idiognathodus turbatus* (Rosscoe and Barrick, 2009) are proposed as potential index taxa for a GSSP of the Moscovian–Kasimovian boundary, but appear well above the existing boundary within the middle Kasimovian (Goreva *et al.*, 2009; Rosscoe and Barrick, 2009). If this boundary position is accepted, significant redefinition of both the Moscovian and Kasimovian stages would be required. According to new Donets Basin data, from where *Idiognathodus saggitalis* was originally described, the new boundary will be nearly 1 Ma younger than that in traditional use (Davydov *et al.*, 2010). Consequently, the duration of the Kasimovian will be more than 5 times shorter than the Moscovian (1.8 myr vs 9.5 myr).

An alternative definition of the GSSP within the evolutionary chronocline of primitive and advanced fusulinids belonging in the genus *Protriticites* has been proposed recently (Davydov, 2007; Davydov and Khodjanyazova, 2009). These fauna are widely distributed throughout the Tethyan and Boreal biogeographic provinces and have now also been found in the North American Great Basin within mid to upper Desmoinesian Strata (Wahlman *et al.*, 1997; Davydov *et al.*, 1999). Although using benthic markers to define a GSSP could be considered a disadvantage, it is successfully used for the base of the Visean (Devuyst *et al.*, 2003). The proposed fusulinid definition will keep the Moscovian–Kasimovian boundary at its classical level, and will give stability to the generally accepted Pennsylvanian scale.

In the ammonite zonation, the Moscovian–Kasimovian boundary is set at the base of the *Dunbarites–Parashumardites* Zone (Ruzhenzev, 1974). However, this zone is different from the *Parashumardites* Zone of Ramsbottom and Saunders (1985), which characterizes the base of the Missourian. The *Dunbarites–Parashumardites* Zone occurs at the base of the Wewoka Formation of the Desmoinesian (Ruzhenzev, 1974). The base of the *Dunbarites–Parashumardites* Zone most probably correlates with the *Wewokites* Subzone of the *Wellerites* Zone (Boardman *et al.*, 1994). A long-standing suggestion from ammonoid study is that the base of the Kasimovian Stage lies within the upper Desmoinesian (Ruzhenzev, 1974), and this is now supported by evidence from fusulinids (Davydov, 1997) and conodonts (Rosscoe and Barrick, 2009).

23.1.5.4. Gzhelian

The Gzhelian Stage, proposed in the late 19th century (Nikitin, 1890), only became widely used in the middle of the 20th century, in connection with the fusulinid zonation established in the Moscow Basin (Rauser-Chernousova, 1941) and elsewhere in the Boreal and Tethyan faunal provinces. The name of the stage comes from the small village of Gzhel', famous for porcelain teapots and earthenware, produced from the lower Gzhelian clay, which is still mined around the area. A series of localities proposed by Nikitin as types of the stage near the villages of Rusavkino, Amerevo, Pavlovo-Posad and Noginsk, became type sections for a regional division (i.e. horizons) of the stage (Rosovskaya, 1950; Makhlina *et al.*, 1979). The top of the Gzhelian has changed position through time and has stabilized only recently, when the GSSP for the base of the Permian was ratified (Davydov *et al.*, 1998).

The traditional lower boundary of the Gzhelian Stage is at the base of the Rusavkino Formation, exposed near the village of Gzhel'. It is an unconformity (Makhlina *et al.*, 1979) interpreted to be a second-order sequence boundary (Briand *et al.*, 1998). The first appearance of the fusulinid species *Rauserites rossicus* and *R. stuckenbergi* were traditionally considered as the operational definition of the boundary (Rauser-Chernousova, 1941; Rosovskaya, 1950). However, this definition, due to foraminiferal provincialism, was not applicable in Pantalassa and the eastern margin of Pangaea. Additionally, in the Moscow Basin these taxa occur slightly higher than the traditional base of the Gzhelian, in the upper Rusavkino Formation (Isakova and Ueno, 2007; Davydov, *et al.*, 2008b; Alekseev *et al.*, 2009). The conodont *Streptognathodus simulator* (*sensu stricto*) Ellison has recently been accepted as the best global marker by the ICS boundary Task Group (Heckel *et al.*, 2008). In the Moscow Basin, this species occurs in the upper Rusavkino Formation, 5–6 m above the traditional boundary and co-incident with the occurrence of traditional fusulinid markers (Davydov, *et al.*, 2008b; Goreva *et al.*, 2009). In the North American Mid-Continent this species occurs at the highstand within the Oread cyclothem (Barrick *et al.*, 2008). Two alternative phylogenetic chronoclines of the appearance of *St. simulator* have been proposed as the boundary definition (Chernykh, 2005; Barrick *et al.*, 2008), although there is an opinion that ancestral forms in both lineages are synonyms (Chernykh *et al.*, 2006).

Two sections have been proposed recently to fix the Kasimovian–Gzhelian boundary in the gobal scale. The first one is the Usolka section, where the conodont chronocline *St. praenuntius–St. simulator* has been established and well documented. This section also possesses several horizons with fusulinids and numerous volcanic ash beds that are particularly frequent within the boundary transition (Chernykh *et al.*, 2006; Davydov *et al.*, 2008b). High-precision U/Pb radiometric dating of zircons has been performed from 25 ashes in Usolka (Schmitz and Davydov, 2011) as well as Sr/Sr isotopes from conodonts (Schmitz *et al.*, 2009). The second candidate is a Nashui section located in S. China and still under current study (Ueno and Task Group, 2009). In the ammonoid succession, the lower boundary of the *Shumardites–Vidrioceras* ammonoid zone has been conventionally placed at the base of the Gzhelian Stage (Bogoslovskaya, 1984). Both indexes appear at the base of the newly defined global Gzhelian Stage in North America (Finis Shale Member of the Graham Formation in north-central Texas) and in Central Asia (Popov *et al.*, 1989; Popov, 1999; Boardman *et al.*, 2006). In the southern Urals, however, *Shumardites* ranges from the upper Gzhelian to the top of the Carboniferous (Popov *et al.*, 1985; Davydov *et al.*, 1994).

23.2. CARBONIFEROUS STRATIGRAPHY

23.2.1. Biostratigraphy

23.2.1.1. Biostratigraphic Zonations of Ammonoids, Foraminifers and Conodonts

Brachiopods were the first fossil group that was used to calibrate the Carboniferous stratigraphic sequence for the Mississippian in western Europe (Delépine, 1911) and for the Pennsylvanian in eastern Europe (Nikitin, 1890). Although brachiopods are still used within certain regions (e.g., Carter, 1990; Legrand-Blain, 1990; Poletaev and Lazarev, 1994), their correlation potential is considered to be essentially local.

Since the beginning of the 20th century, the ammonoid successions in the Mississippian and early Pennsylvanian of western Europe (Bisat, 1924, 1928; Schmidt, 1925; Ramsbottom and Saunders, 1985; Korn 1996, 2006) and in the entire Carboniferous in eastern Europe (Ruzhenzev, 1965; Ruzhenzev and Bogoslovskaya, 1978) have served as chronostratigraphic standards in inter-regional and global correlation (Figure 23.5).

Together with the ammonoid zonations, the benthic foraminiferal zonation established in the Mississippian (Lipina and Reitlinger, 1970; Mamet and Skipp, 1970; Conil *et al.*, 1977) and Pennsylvanian (Rauser-Chernousova, 1941, 1949; Rosovskaya, 1950; Solovieva, 1977; Ross and Ross, 1988) is the most practical inter-regional biostratigraphic standard (Figure 23.5). However, because of significant provincialism in both groups of fossils, particularly at the time of assembly of Pangea, and the beginning of the Carboniferous Gondwana glaciation, different standard zonations are used for eastern and western hemispheres (i.e. for Eurasia and surrounding areas in the east, versus America in the west).

Over the last three decades, the conodont succession, although being the least studied and also to some degree provincial, has become the most reliable tool for calibration and geochronological boundary definition within the Carboniferous (Figure 23.5). The zonations established in the Mississippian and Pennsylvanian in North America and Europe (Higgins, 1975; Sandberg *et al.*, 1978; Dunn, 1970; Lane *et al.*, 1971, 1980; Barskov and Alekseev, 1975; Barskov *et al.*, 1980) are used worldwide and are actively being refined (Skompski *et al.*, 1995; Nemyrovska, 1999; Chernykh, 2002; Lambert *et al.*, 2002; Chernykh, 2005; Alekseev and Goreva, 2007).

23.2.1.2. Other Marine Micro- and Macrofauna (Corals, Radiolarians, Ostracods)

The majority of the other biostratigraphic events are calibrated relative to ammonoid, conodont, and foraminiferal scales. Although a global coral zonation has been proposed (Sando, 1990), Mississippian corals have generally only regional biostratigraphic significance (Poty, 1985; Kossovaya, 1996; Bamber and Fedorowski, 1998), but good biogeographic and paleo-environmental utility.

Radiolaria, as a pelagic microfossil group, have high biostratigraphic potential (Nazarov and Ormiston, 1985; Holdsworth and Murchey, 1988; Gourmelon, 1987; Nazarov, 1988; Braun and Schmidt-Effing, 1993; Won, 1998; Chuvashov *et al.*, 1999). However, their zonation in the Carboniferous is poorly developed and has been utilized, generally, only at regional levels.

Ostracods are widely distributed in the Carboniferous and in several regions were utilized for detailed regional biostratigraphy (Chizhova, 1977; Gorak, 1977; Crasquin, 1985; Abushik *et al.*, 1990). However, the chronostratigraphic value of ostracods is limited by paleoecological, facies, and paleoclimatic factors.

Crinoids, because of the rare occurrence of complete specimens, are not commonly used in Carboniferous biostratigraphy. However, an alternative classification of crinoid columnals and stems and biostratigraphic zonation at the stadial level has been proposed by Stukalina (1988). Because of the ontogenetical approach of this classification, it can potentially be used for detailed biostratigraphy over large areas. Crinoid stems occur in a wide variety of facies and environmental conditions.

23.2.1.3. Plants and Palynology

Plants were one of the first fossil groups utilized for dating Carboniferous coal basins. The Carboniferous was a time of considerable evolutionary activity in the creation of large taxonomic clusters with provincial differentiations. Zonal floristic successions have been proposed in several regions along the equatorial climatic belt and were unified by Wagner (1984) into 16 floral zones for the entire Carboniferous. Floral successions in the higher latitudes, such as the Angara Province in the northern hemisphere (Meyen *et al.*, 1996) and the Gondwana Province in the southern hemisphere (Archangelsky *et al.*, 1995), are much less developed.

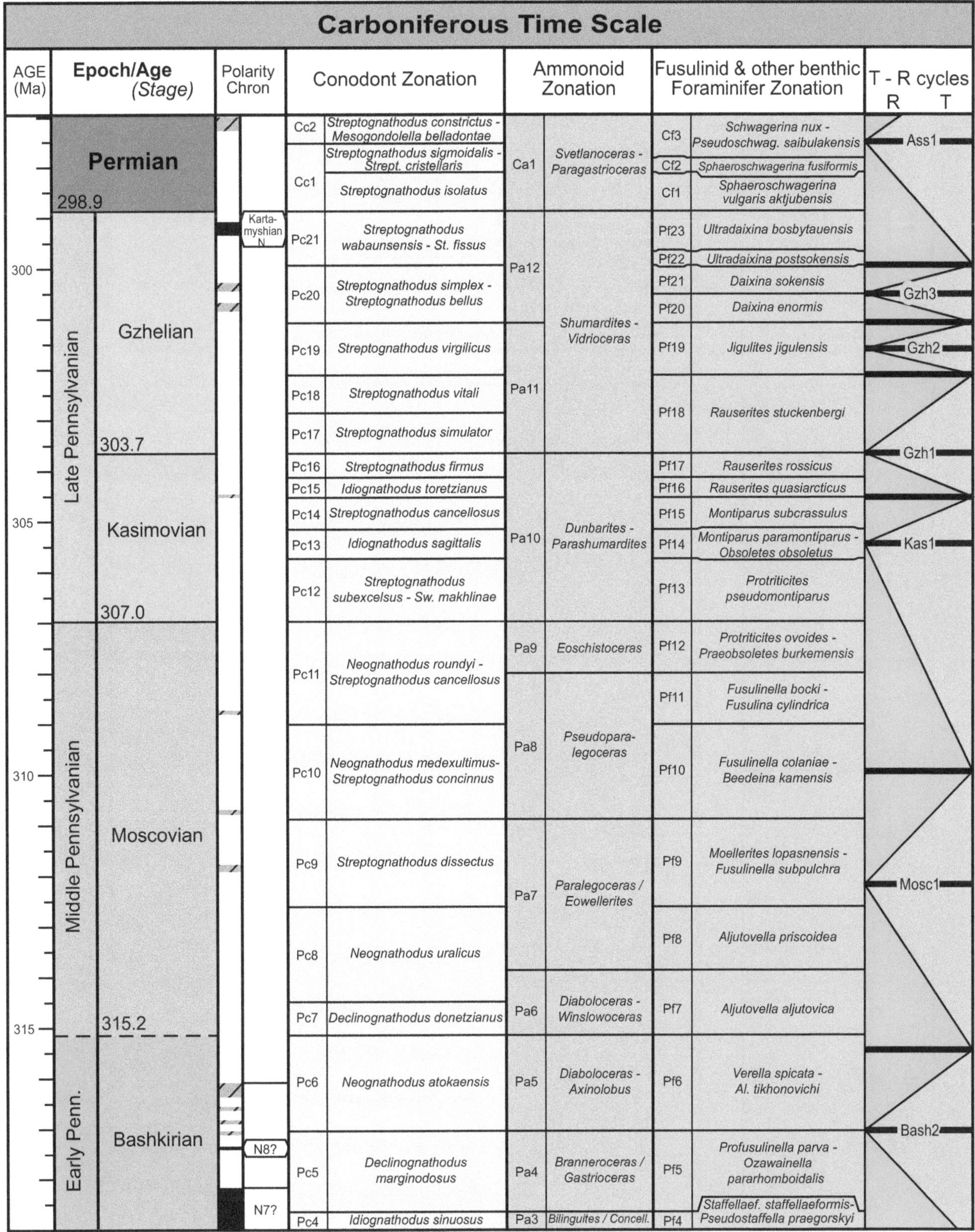

Carboniferous Time Scale
AGE (Ma)
Epoch/Age (Stage)
Polarity Chron
Conodont Zonation
Ammonoid Zonation
Fusulinid & other benthic Foraminifer Zonation
T - R cycles
R T
300
305
310
315
Permian
298.9
Late Pennsylvanian
Gzhelian
303.7
Kasimovian
307.0
Middle Pennsylvanian
Moscovian
315.2
Early Penn.
Bashkirian
Karta-myshian N
N8?
N7?
Cc2 Streptognathodus constrictus - Mesogondolella belladontae
Streptognathodus sigmoidalis - Strept. cristellaris
Cc1 Streptognathodus isolatus
Pc21 Streptognathodus wabaunsensis - St. fissus
Pc20 Streptognathodus simplex - Streptognathodus bellus
Pc19 Streptognathodus virgilicus
Pc18 Streptognathodus vitali
Pc17 Streptognathodus simulator
Pc16 Streptognathodus firmus
Pc15 Idiognathodus toretzianus
Pc14 Streptognathodus cancellosus
Pc13 Idiognathodus sagittalis
Pc12 Streptognathodus subexcelsus - Sw. makhlinae
Pc11 Neognathodus roundyi - Streptognathodus cancellosus
Pc10 Neognathodus medexultimus- Streptognathodus concinnus
Pc9 Streptognathodus dissectus
Pc8 Neognathodus uralicus
Pc7 Declinognathodus donetzianus
Pc6 Neognathodus atokaensis
Pc5 Declinognathodus marginodosus
Pc4 Idiognathodus sinuosus
Ca1 Svetlanoceras - Paragastrioceras
Pa12
Pa11
Shumardites - Vidrioceras
Pa10 Dunbarites - Parashumardites
Pa9 Eoschistoceras
Pa8 Pseudopara-legoceras
Pa7 Paralegoceras / Eowellerites
Pa6 Diaboloceras - Winslowoceras
Pa5 Diaboloceras - Axinolobus
Pa4 Branneroceras / Gastrioceras
Pa3 Bilinguites / Concell.
Cf3 Schwagerina nux - Pseudoschwag. saibulakensis
Cf2 Sphaeroschwagerina fusiformis
Cf1 Sphaeroschwagerina vulgaris aktjubensis
Pf23 Ultradaixina bosbytauensis
Pf22 Ultradaixina postsokensis
Pf21 Daixina sokensis
Pf20 Daixina enormis
Pf19 Jigulites jigulensis
Pf18 Rauserites stuckenbergi
Pf17 Rauserites rossicus
Pf16 Rauserites quasiarcticus
Pf15 Montiparus subcrassulus
Pf14 Montiparus paramontiparus - Obsoletes obsoletus
Pf13 Protriticites pseudomontiparus
Pf12 Protriticites ovoides - Praeobsoletes burkemensis
Pf11 Fusulinella bocki - Fusulina cylindrica
Pf10 Fusulinella colaniae - Beedeina kamensis
Pf9 Moellerites lopasnensis - Fusulinella subpulchra
Pf8 Aljutovella priscoidea
Pf7 Aljutovella aljutovica
Pf6 Verella spicata - Al. tikhonovichi
Pf5 Profusulinella parva - Ozawainella pararhomboidalis
Pf4 Staffellaef. staffellaeformis- Pseudostaffella praegorskyi
Ass1
Gzh3
Gzh2
Gzh1
Kas1
Mosc1
Bash2

FIGURE 23.5 **Subperiod and stage boundaries for the Carboniferous Period**, with zonations of conodonts, foraminifers and ammonoids. Also displayed are the geomagnetic polarity scale with Russian-based nomenclature of super- and hyperchrons (see text) and the principal eustatic trends. In blue the numbered composite standard units (see text).

Carboniferous Time Scale

AGE (Ma)	Epoch/Age (Stage)	Polarity Chron	Conodont Zonation		Ammonoid Zonation		Fusulinid & other benthic Foraminifer Zonation		T - R cycles (R T)
	Early Pennsylvanian – Bashkirian	N7?	Pc5	*Declinognathodus marginodosus*	Pa4	*Branneroceras / Gastrioceras*	Pf5	*Profusulinella parva - Ozawainella pararhomboidalis*	
			Pc4	*Idiognathodus sinuosus*	Pa3	*Bilinguites / Concelloceras*	Pf4	*Staffellaeformes staffellaeformis - Pseudostaffella praegorskyi*	
320			Pc3	*Neognathodus askynensis*			Pf3	*Pseudostaffella antiqua*	
			Pc2	*Idiognathoides sinuatus*	Pa2	*Baschkortoceras / Reticuloceras*	Pf2	*Semistaffella variabilis*	Bash1
	323.2	N5	Pc1	*Declinognathodus noduliferus*	Pa1	*Homoceras / Hudsonoceras*	Pf1	*Plectostaffella bogdanovkensis*	
	Late Mississippian – Serpukhovian	Inferred GAP; N2?	Mc17	*Gnathodus postbilineatus*	Ma9	*Delepinoceras / Fayettevillea*	Mf18	*Monotaxinoides transitoriusis - Eosigmolina explicata*	
325			Mc16	*Gnathodus bollandensis*			Mf17	*Eostaffellina protvae*	
							Mf16	*Eostaffella mirifica - Eost. decurta*	
		N1?	Mc15	*Lochriea cruciformis*	Ma8	*Cravenoceras / Uralopronorites*	Mf15	*Pseudoendothyra globosa - Neoarchaediscus parvus*	
330			Mc14	*Lochriea ziegleri*	Ma7	*Hypergoniatites / Ferganoceras*	Mf14	*Eostaffella tenebrosa - Endothyranopsis sphaericus*	
	330.9								
	Middle Mississippian – Visean		Mc13	*Lochriea nodosa*					
			Mc12	*Lochriea mononodosa*			Mf13	*Eostaffella ikensis - Bradyina rotula*	
335			Mc11	*Gnathodus bilineatus*	Ma6	*Beyrichoceras / Goniatites*	Mf12	*Eostaffella proikensis - Archaediscus gigas*	Vis2
			Mc10	*Gnathodus praebilineatus*			Mf11	*Endothyranopsis compressa - Archaediscus krestovnikovi*	

FIGURE 23.5 (*Continued*).

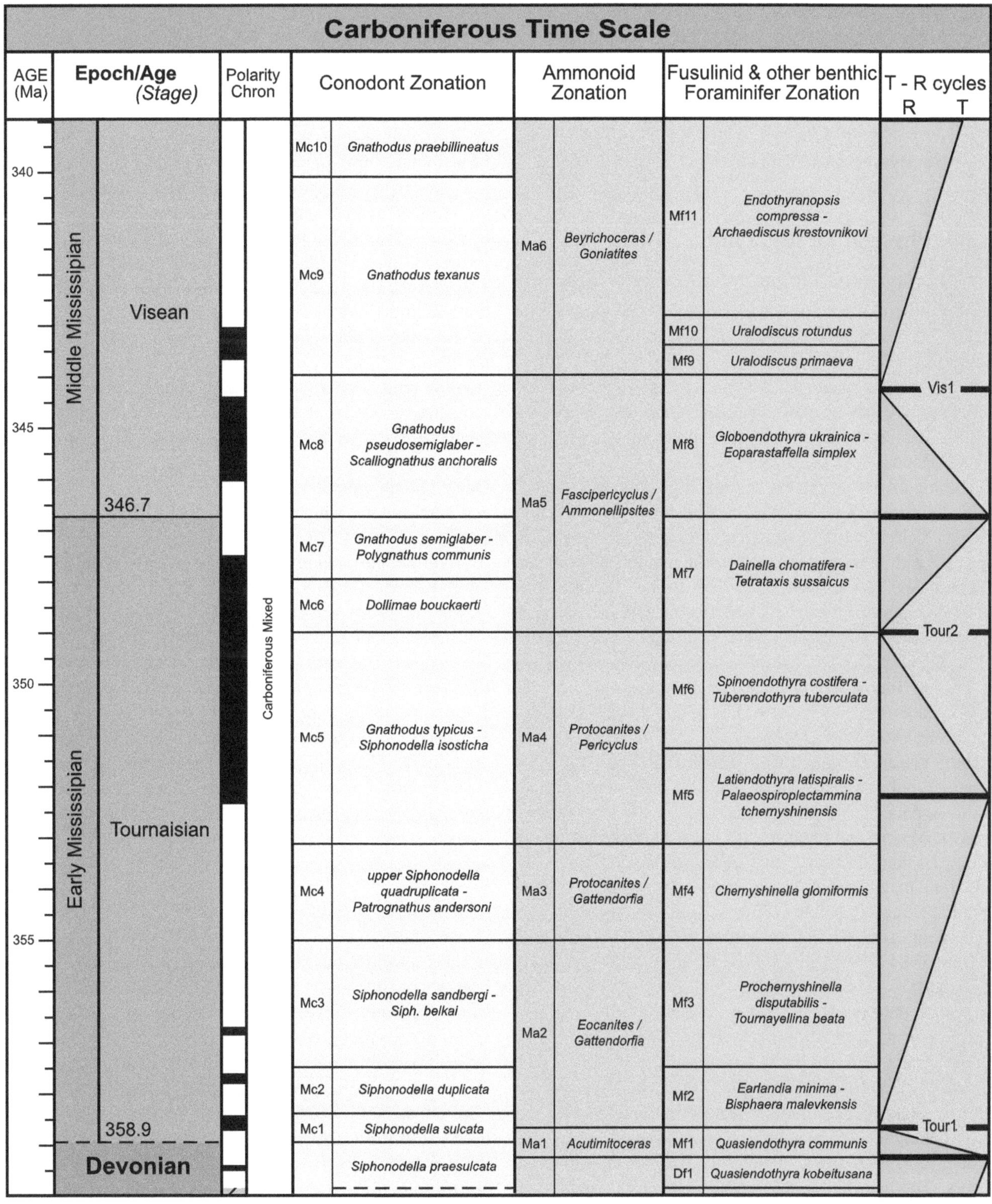

FIGURE 23.5 (*Continued*).

Palynological assemblages provide one of the most common dating methods for Carboniferous coal fields and hydrocarbon deposits. Although the palynological spore assemblages are different from region to region, they are successfully used in several basins (Inosova *et al.*, 1976; Clayton *et al.*, 1977; Byvsheva *et al.*, 1979; Owens, 1984; Higgs *et al.*, 1988; Byvsheva and Umnova, 1993; Peppers, 1996).

23.2.2. Physical Stratigraphy

23.2.2.1. Magnetostratigraphy

The Carboniferous polarity scale is not yet well known, but current views based on composite sections are summarized in Figure 23.5. The most significant general feature of the Carboniferous magnetic field is a long late Carboniferous interval of reverse polarity that continues into the early Permian, forming the Kiaman chron of predominantly reverse polarity (Hounslow *et al.*, 2004).

In terms of the magnetostratigraphic classification developed in the former Soviet Union, the Mississippian and Pennsylvanian belong to the Donetzian mixed normal–reversal megazone with two superzones: Tikhvinian and Debaltzevian (Khramov and Rodionov, 1981). The former includes the stratigraphic interval from the Upper Devonian (mid Frasnian) up to the base of the Serpukhovian. The Debaltzevian Superzone is equal to the Serpukhovian–Moscovian stratigraphic interval. The Tournaisian magneto-zonal portion of the scale is based on data from the far east of Russia and Belgian classical sections (Kolesov, 1984, 2001).

The Visean–Serpukhovian–Bashkirian part of the magneto-zonal scale is based on sections from northern and central Appalachian Pennsylvania, New Brunswick, Nova Scotia (DiVenere and Opdyke, 1991a,b), Donets Basin, and Moscow Basin (Khramov *et al.*, 1974; Khramov, 2000). Post-Bashkirian magnetostratigraphy is based on sections from Australia, Donets Basin, the southern Urals, North Caucasus, and Central Asia (Irving and Parry, 1963; Khramov and Davydov, 1984, 1993; Davydov and Khramov, 1991; Opdyke *et al.*, 2000).

The late Pennsylvanian belongs to the Permo–Carboniferous Reversed Hyperchron interval, or "Kiaman", as originally described in Australia by Irving and Parry (1963). The Kiaman Hyperchron is recognized as a long period of reversed polarity of Earth's magnetic field, spanning approximately 50–60 myr, and it has been considered as one of the most important Paleozoic magnetostratigraphic markers. The age of the base of the Kiaman has not been resolved in terms of the chronological scale. In Australia, the base of the Kiaman was originally defined immediately above the Paterson Volcanics beds, and dated as 309–310 Ma (Palmer *et al.*, 1985) as it was thought to be close to the base of Westphalian of the western European stratigraphic scale, or to the base of Atokan in terms of the North American scale. However, the age of the Paterson Volcanics has been reinterpreted as 328.5 ± 1.4 Ma (Claouè-Long *et al.*, 1995), which would place the base of the Kiaman in the type area within the upper Visean or lower Serpukhovian (Opdyke and Channell, 1996).

The youngest pre-Kiaman normal polarity zone age has been found within approximately Westphalian A strata (i.e. mid Atokan and uppermost Bashkirian) in Nova Scotia and in the Northern Appalachians (DiVenere and Opdyke, 1991a). Similarly, a normal polarity zone radiometrically constrained within 317 and 313 Ma has been found in eastern New South Wales, Australia (Opdyke *et al.*, 2000), essentially at the same level as in the Northern Appalachians.

Khramov (1987) found three normal polarity zones in the Donets Basin within the lower–middle Moscovian. Opdyke *et al.* (1993) re-studied the Bashkirian–Moscovian sequences in Donets Basin but did not confirm the proposed magneto-zones as Moscovian. However, Khramov (2000) suggested that the zones in the Bashkirian–Moscovian in Donets Basin are present in the siliciclastic rocks (siltstones and sandstones), which were not studied by Opdyke *et al.* (1993).

A single but pronounced normal polarity zone within the lower Kiaman Hyperzone, originally found in Donets Basin (Khramov, 1963), was reported in southwest USA, Germany, Caucasus, southern Urals, and Central Asia (Peterson and Nairn, 1971; Dachroth, 1976; Khramov and Davydov, 1984). This zone has been named the "*Kartamyshian*" (Davydov and Khramov, 1991). In the southern Urals and Central Asia, the *Kartamyshian* Zone has been precisely constrained within the uppermost Carboniferous fusulinid zone *Ultradaixina bosbytauensis–Schwagerina robusta* and therefore is an important magnetostratigraphic marker for separation of the Carboniferous and Permian periods.

23.2.2.2. Chemical Stratigraphy: Stable Isotopes of Sr, O and C

In the Carboniferous, chemostratigraphy has only recently become useful, and Chapters 7, 10 and 11 in this book summarize and illustrate current trends in stable isotopes ratios of strontium, oxygen and carbon (Figure 23.6). The degree of lithification in many Paleozoic rocks makes alteration of original chemical signals a significant problem (Grossman *et al.*, 1996). For $^{87}Sr/^{86}Sr$ stratigraphy, one of the most useful fossils are brachiopods with thick, large non-luminescent shells (Bruckschen *et al.*, 1999), but whole rock samples have also proven useful in establishing coarse trends (Denison *et al.*, 1994). Conodonts were initially thought to have little potential for chemostratigraphy (Martin and Macdougall, 1995; Ruppel *et al.*, 1996), but stratigraphic sections of the Southern Urals contain abundant and well-preserved conodont fauna for precise biostratigraphic correlation. Common volcanic ash beds

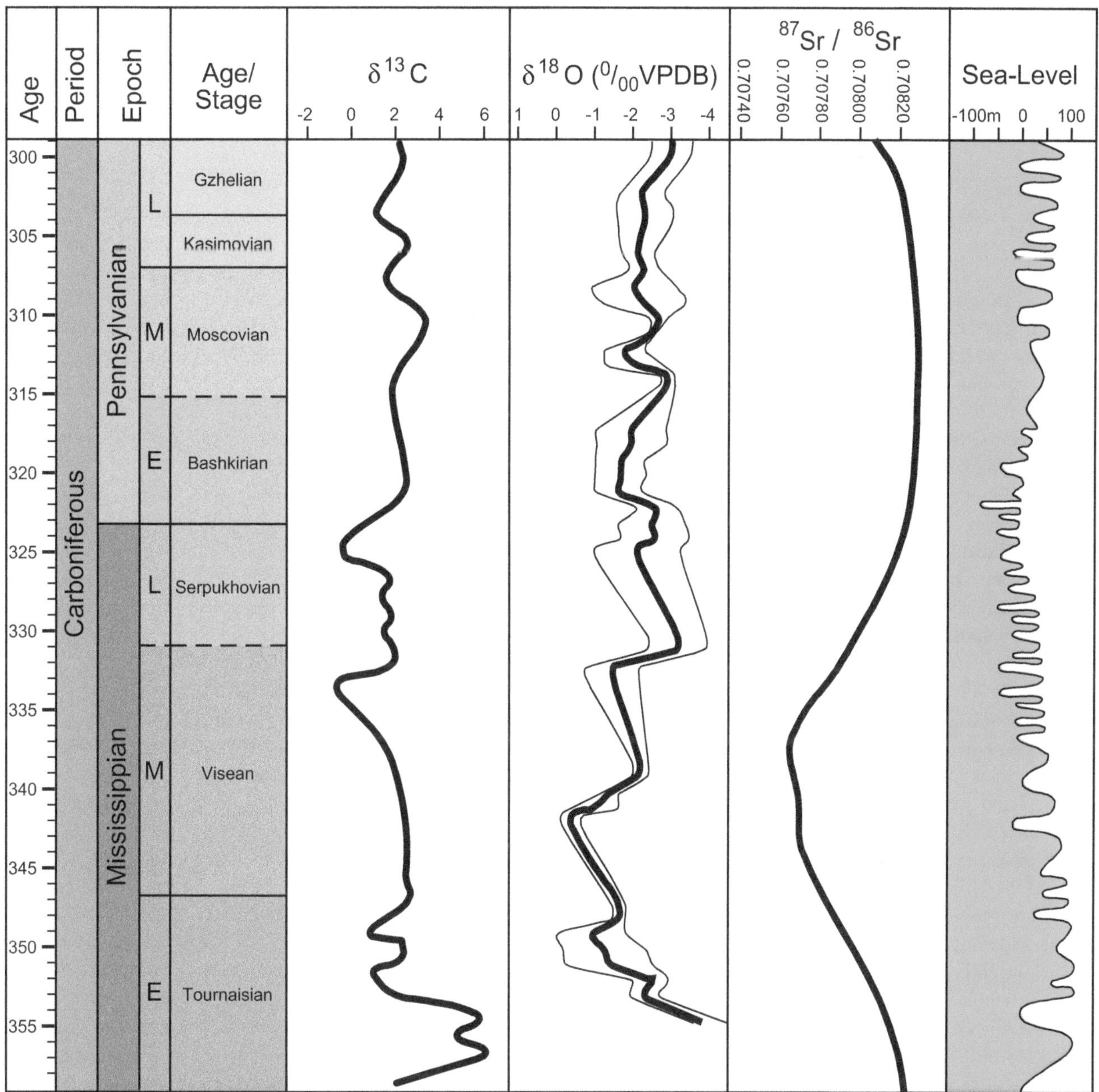

FIGURE 23.6 **Geochemical trends during the Carboniferous Period.** The generalized carbon-isotope curve was compiled by Saltzman and Thomas (Chapter 11, this book), the oxygen-isotope by Grossman (Chapter 10, this book) and schematic $^{87}Sr/^{86}Sr$ curve by McArthur *et al.* (Chapter 7, this book). *The sea-level curve is modified from Haq and Schutter (2008).*

intercalated in the conodont series and dated by U-Pb zircon geochronology (Schmitz and Davydov, 2011) were used to produce a seawater Sr curve (Schmitz *et al.*, 2009) that shows a significant reduction in data scatter compared to earlier curves (Denison *et al.*, 1994; Veizer *et al.*, 1999; Bruckschen *et al.*, 1999). The relatively flat late Moscovian through mid-Gzhelian seawater Sr curve is generally consistent with that of Bruckschen *et al.* (1999). The data from the Urals define a decreasing trend in $^{87}Sr/^{86}Sr$, beginning in the mid-Gzhelian through the mid-Sakmarian (Schmitz *et al.*, 2009). The range of $^{87}Sr/^{86}Sr$ in the Carboniferous is large (from 0.707 67 to 0.708 31; Figure 23.6). The poor quality of the $^{87}Sr/^{86}Sr$ curve for most of the period owes much to the difficulty of correlating sections used in curve construction.

Reconnaissance curves for $\delta^{13}C$ and $\delta^{18}O$ have been produced by Veizer *et al.* (1999). Refined curves of these isotopes precisely tied to the chronostratigraphic scale have been published (Mii *et al.*, 1999; Bruckschen *et al.*, 1999; Grossman *et al.*, 2002; Saltzman, 2003). The most recent studies of $\delta^{18}O$ from conodont biogenic apatite show three major positive shifts of +2‰ and +1.5‰ V-SMOW in the earliest and middle-late Tournaisian and in middle Serpukhovian, respectively, that are interpreted as reflecting climatic cooling and changes in ice volume (Buggisch *et al.*, 2008).

Three major, possibly global, peaks in $\delta^{13}C$ distribution were found in the latest Devonian and Carboniferous studies. Two sharp but low amplitude positive and negative excursions in the late *praesulcata* conodont zone have been reported recently in the Carnic Alps (Kaiser *et al.*, 2006) and North American Mid-Continent (Cramer *et al.*, 2008) that suggests the global nature of these changes. Two negative shifts in early *Siphonodella crenulata* and in the middle of *Gnathodus typicus* conodont zones were reported in southern Europe (Kaiser *et al.*, 2006; Buggisch *et al.*, 2008). A brief but sharp positive excursion of $\delta^{13}C$ up to +7‰ corresponds to the middle Tournaisian (upper Kinderhookian) *Siphonodella isosticha* conodont zone (lower Mc5 Zone) in southeast Idaho, southeast Nevada, northeastern Utah, northern Iowa, and western Europe (Bruckschen *et al.*, 1999; Mii *et al.*, 1999; Saltzman, 2003; Buggisch *et al.*, 2008).

A moderate but distinct rise in $\delta^{13}C$ was found recently in the middle Serpukhovian *Gnathodus bilineatus bollandensis* conodont zone (Buggisch *et al.*, 2008). A large $\delta^{13}C$ shift of approximately +6‰ has been documented in the early Pennsylvanian in the paleo-Tethyan regions (Popp *et al.*, 1986; Bruckschen *et al.* 1999; Mii *et al.*, 1999; Grossman *et al.*, 2002; Saltzman, 2003). In North America, this shift is late Chesterian and early Morrowan (Serpukhovian–Bashkirian) and is reduced by +1.5–2.5‰, which most probably reflects changes in ocean circulation patterns associated with the closing of the Rheic-Tethys equatorial seaway.

Based on the timing of the $\delta^{13}C$ divergence between North America and Europe, the isolation of Paleo-Tethys began in the late Serpukhovian. Similarly, the onset of the provincialism of benthic fauna (western Euramerica versus Paleo-Tethys) is coincident with this paleotectonic event (equatorial seaway closure). All these data support a scenario in which the closure of a subequatorial oceanic gateway during the assembly of Pangea altered the oceanic distribution of nutrients and led to the enhanced poleward transport of heat and moisture. This change marks the transition from a cool, moisture-starved Gondwana to the ice-house world of the Pennsylvanian and Early Permian (Saltzman, 2003).

Pennsylvanian $\delta^{13}C$ data are still scarce and not well constrained chronostratigraphically. It should be noted that $\delta^{13}C$ shifts preceded major coal burial. The middle Tournaisian positive $\delta^{13}C$ excursion is 5–6 myr earlier than late Tournaisian and early Visean coal burial in the Moscow Basin, East Siberia, and the Urals; the large $\delta^{13}C$ shift across the Mississippian–Pennsylvanian transition also appears approximately 5–6 myr earlier than the major coal burial in western Europe, Donets Basin, the Appalachian Basin, and northern China. If this is a general rule, then it might be expected that for the Permian coal-forming episodes, a $\delta^{13}C$ shift in the Artinskian–Kungurian would have preceded the burial of Kungurian–Roadian coal and a similar shift in Capitanian time would have preceded Wuchiapingian coal burial.

23.2.2.3. Sequence Stratigraphy and Cyclostratigraphy

Sequence stratigraphy is widely applicable to the Carboniferous cyclic sequences because the cycles are generally considered to be glacio-eustatic and, therefore, global in distribution (Ramsbottom, 1977; Caputo and Crowell, 1985; Veevers and Powell, 1987). Ross and Ross (1987, 1988); Haq and Schutter (2008) proposed global frameworks of transgressive–regressive depositional cycles in the marine shallow-water cratonic shelves of the Carboniferous–Permian tropical and subtropical regions (Figure 23.6).

The Late Paleozoic hosts some of the most dramatic manifestations of sedimentary cyclicity in response to global climate changes. The Pennsylvanian cyclothems of the Euramerican basins have been hypothesized for over seven decades as the sedimentary response to glacio-eustatic fluctuations driven by changes in the Earth's orbital parameters (Wanless and Shepard, 1936; Ramsbottom, 1977; Heckel, 1986, 1994). Various authors have imposed assumed periods on the fourth- and fifth-order sequences comprising cyclothemic successions, in order to estimate the durations of stage and period divisions of the time scale (Heckel, 1986; Algeo and Wilkinson, 1988; Boardman and Heckel, 1989; Connolly and Stanton, 1992; Dickinson *et al.*, 1994).

While astrochronology and its cyclostratigraphic application have matured as powerful means of calibrating the absolute age of younger Cenozoic sediments (Shackleton *et al.*, 1999; Lourens *et al.*, 2004; Westerhold *et al.*, 2008), these methods have seen little confident application to Paleozoic strata. This is in part due to limiting assumptions of tidal dissipation and chaotic diffusion in the inner solar system for the accurate theoretical calculation of precession and obliquity variations which prevent their extrapolation to deep time (Laskar, 1990, 1999; Berger *et al.*, 1992); however, eccentricity-driven orbital variations – particularly the long period, 404 ka cycle controlled by gravitational interaction between Earth, Jupiter, and Venus – are recognized as theoretically stable throughout the Phanerozoic (Berger *et al.*, 1992; Laskar, 1999; Hinnov, 2000, 2004) and may be observed in some well-preserved and continuous sedimentary strata (Herbert, 1999; Olsen and Kent, 1999).

Despite its potential, the application of quantitative cyclostratigraphy to create a robust chronostratigraphic framework for the Late Paleozoic has been historically stymied by the lack of independent radiometric constraints – and tests – for cycle periods. Recent studies of the biostratigraphic correlation of cyclothems across Euramerica from the Russian platform from the Donets Basin of the Ukraine to the Midcontinent USA have successfully demonstrated the relative age correlations and synchrony between cycles across basins (Heckel *et al.*, 2007). However, these correlations hinge upon the assumption of simultaneous first occurrences of taxa between basins – correlations which to date have been untestable within a radiometrically constrained global time scale. Neither are such correlations by themselves able to establish the absolute age and quantitative periods of cyclothems.

Recent studies on high-precision radiometric dating of ash beds within bio- and cyclostratigraphical-rich Carboniferous successions in eastern Europe for the first time have quantitatively demonstrated that the fourth-order and higher-frequency sequences exhibited by cyclothemic sedimentation represent transgressive–regressive eustatic changes, due to both long- and short-period orbital eccentricity forcing. In the Serpukhovian Ostrava Formation of the Upper Silesian Coal Basin of the Czech Republic, Gastaldo *et al.* (2009) calibrated the duration of high-frequency coal-bearing terrestrial-paralic cyclothems at 0.83 ± 0.24 myr, overlapping at the 95% confidence interval with the short period eccentricity cycle. For this calibration the authors used a pair of U-Pb isotope dilution thermal ionization mass spectrometry (ID-TIMS) zircon tuff ages. In paralic, cyclothemic Moscovian strata of the Donets Basin of the Ukraine, Davydov *et al.* (2010) demonstrated, on the basis of seven U-Pb ID-TIMS zircon ages, that the individual high-frequency cyclothems and bundles of cyclothems into fourth-order sequences are the eustatic response to orbital eccentricity forcing (c. 100 and 404 ka). In the case of the Donets Basin, the long period eccentricity cycle is a dominant signal, as has been proposed observationally for "major" cyclothems of the Midcontinent USA (Heckel, 1986; Veevers and Powell, 1987; Connolly and Stanton, 1992). These findings are elaborated on in Section 23.3.4.

23.2.2.4. Paleoclimate

The Late Paleozoic is commonly regarded as a time of global greenhouse–icehouse climate transition. The most recent data suggest multiple phases of glaciation and deglaciation, although the number of these phases, their extent and age constraints are debated (Isbell *et al.*, 2003; Fielding *et al.*, 2008b; Montanez *et al.*, 2007). Marine biotas are sensitive to local, regional or global environmental changes and the exceptionally well-studied benthic Foraminifera are among the best indices of paleoenvironments. The Cenozoic record of benthic Foraminifera diversity (Hallock *et al.*, 1991) shows a strong correlation with well-studied climatic changes for that time (Zachos *et al.*, 2001), and this model of the interaction between tropical benthic Foraminifera and climate may be applied to the Late Paleozoic.

Recent symbiont-bearing larger Foraminifera are restricted to the euphotic zone of tropical and warm-temperate seas. Their species distribution alters with changes in primary limiting factors; temperature, light, water movement, substrate and nutrients. Temperature determines geographic distribution and affects depth distribution of larger Foraminifera by the development of a shallow thermocline that truncates the distribution of shallower species and excludes species adapted to the deepest euphotic zone. Within these constraints, light is the most important primary factor, because larger Foraminifera are at least partly dependent upon photosynthesis by their algal endosymbionts for growth and calcification (Hohenegger, 2004). As pointed out by this author, temperature restricts larger Foraminifera to those geographical regions or water depths characterized by temperatures never falling below 14°C for several weeks. These are important observations relevant to our data, suggesting that larger Foraminifera in warm-water tropical conditions may occupy a wider variety of depth environments, whereas in cooler-water, upwelling, or higher latitude cooler climates, larger foraminifers will be limited to very shallow conditions which experience the appropriate range of temperature (>14–15°C). Shallow water assemblages of recent larger Foraminifera with optimal water temperatures (20–30°C) generally are much more diverse then those with temperatures greater than 30°C and/or less than 20°C (Murray, 2006; Beavington-Penney and Racey, 2004).

The data on diversity of larger benthic and planktonic Foraminifera with symbionts (Hallock *et al.*, 1991) when plotted against the data on Paleocene–Miocene greenhouse–icehouse transition (Zachos *et al.*, 2001) clearly suggest that diversity minima correspond to icehouse conditions, and diversity maxima correlate with greenhouse, warm climates. Our model for the climatically controlled diversity of foraminifers in Late Paleozoic borrows from Cenozoic observations.

Foraminifera were advanced single-cell organisms in the Late Paleozoic, and were distributed in tropical–subtropical belts (up to 40–45 degrees south/north latitude) within carbonate to mixed carbonate–siliciclastic shallow water settings. This paleogeographic distribution and their known sensitivity to paleoenvironment, coupled with their high-resolution spatial and temporal framework, provides the basis for the paleobiogeography and paleoclimate model presented here. Although most Paleozoic benthic Foraminifera (including all larger Foraminifera such as fusulinids) became extinct at the end of the Permian, their paleobiology and paleoecology is inferred from studies of Recent benthic taxa.

The GraphCor quantitative biostratigraphic method was used to create a composite section and adjust ranges of taxa (Davydov *et al.*, 2003). The data sets for the Mississippian and Pennsylvanian include 96 sections, including type sections in the Urals and Russian Platform with additional data from Central Asia and the Arctic. Taxonomy is internally consistent. The data set includes information from limited areas of northern Pangaea, but it most likely reflects global climatic change patterns. Plotting the diversity of taxa against time, using the linearly scaled composite section, shows clear trends that reflect changes in climate (Figure 23.7). The scale

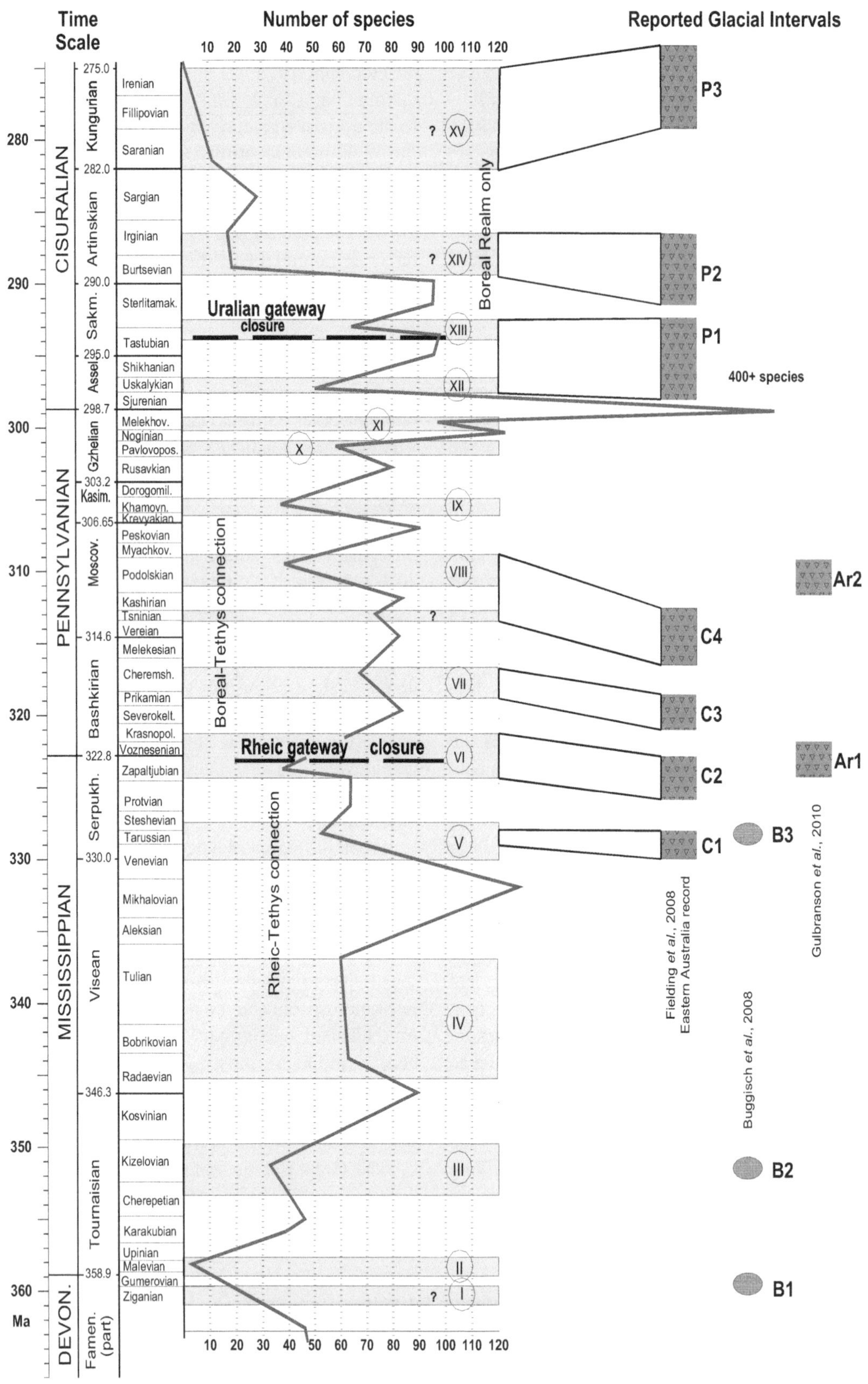
Time Scale
Number of species
Reported Glacial Intervals
CISURALIAN
PENNSYLVANIAN
MISSISSIPPIAN
DEVON.
Kungurian
Artinskian
Sakm.
Assel
Gzhelian
Kasim.
Moscov.
Bashkirian
Serpukh.
Visean
Tournaisian
Famen. (part)
Irenian
Fillipovian
Saranian
Sargian
Irginian
Burtsevian
Sterlitamak.
Tastubian
Shikhanian
Uskalykian
Sjurenian
Melekhov.
Noginian
Pavlovopos.
Rusavkian
Dorogomil.
Khamovn.
Krevyakian
Peskovian
Myachkov.
Podolskian
Kashirian
Tsninian
Vereian
Melekesian
Cheremsh.
Prikamian
Severokelt.
Krasnopol.
Voznesenian
Zapaltjubian
Protvian
Steshevian
Tarussian
Venevian
Mikhalovian
Aleksian
Tulian
Bobrikovian
Radaevian
Kosvinian
Kizelovian
Cherepetian
Karakubian
Upinian
Malevian
Gumerovian
Ziganian
Uralian gateway closure
Rheic gateway closure
Boreal-Tethys connection
Rheic-Tethys connection
Boreal Realm only
400+ species
P3
P2
P1
C4
C3
C2
C1
Ar2
Ar1
B3
B2
B1
Fielding et al., 2008 Eastern Australia record
Buggisch et al., 2008
Gulbranson et al., 2010
Ma

and dynamics of foraminiferal diversity change (lower during cool, higher during warmer periods) must be further tested with other climate proxies (e.g., stable isotopes, paleosoil proxies, etc.). Some of the diversity minima (e.g., Early Tournaisian) are not supported by current geochemical data (Kaiser *et al.*, 2006; Buggisch *et al.*, 2008) and might have been caused by several factors, including local/regional climate changes, oceanic shelf anoxia, sea-level changes etc. Other diversity minima (e.g., III cooling event, Late Tournaisian) match well with geochemical evidence of glaciations (Saltzman, 2003). In the Mississippian, the diversity of Foraminifera progressively rose and reached a maximum in the late Visean (over 120−160 species). At that time, Foraminifera displayed the lowest level of provincialism and nearly global distribution within the tropical−subtropical belts, and correspond to what has been suggested to be the warmest climate in Mississippian time (Reitlinger, 1975).

There is a significant negative shift in diversity in the early Serpukhovian, with one small positive shift in the middle Serpukhovian. The first negative shift approximately corresponds with the onset of glaciations inferred recently in Australian successions (Fielding *et al.*, 2008a). The glaciation events in Australia, however, are only imprecisely constrained with palynomorph, provincial brachiopod biostratigraphy, and SHRIMP ages, and hence, age uncertainty is high (>5−7 myr). The second negative shift at the Serpukhovian−Bashkirian boundary corresponds with a pronounced glaciation event that is widely recognized (e.g., Mii *et al.*, 2001; Saltzman, 2003; Fielding *et al.*, 2008b).

During the late Bashkirian through Asselian time, the frequency and amplitude of negative/positive diversity shifts increased (Figure 23.7). The minima in diversity are associated with cooling events constrained in Australia and Argentina (Fielding *et al.*, 2008a; Gulbranson *et al.*, 2010) and surrounding regions (Scheffler *et al.*, 2003) and with cooling events suggested from geochemical proxies (Montanez *et al.*, 2007) that match reasonably well with minimum Foraminifera diversity peaks .

The diversity in the Moscovian−Asselian shifts between a minimum of 40 species and a maximum of 90−120 species, with one large peak (over 400 species) near the Carboniferous−Permian transition. This large peak coincides with major evolutionary events and the rapid migration of tropical taxa into the high latitudes (*Sphaeroschwagerina*, *Likharevites*, etc.). The Carboniferous−Permian transition is associated with a short, but sharp and significant, global warming event. The proposed model of biota/climate interaction should be tested further with different approaches and methods.

23.3. CARBONIFEROUS TIME SCALE

23.3.1. Radiometric Data

The detailed and high-resolution conodont−foraminifer−ammonoid zonation for the Carboniferous, with over 35−40 zones (Figure 23.5) is now constrained with 36 radiometric dates employed in time scale building for this period (Appendix 2, items d18 and cb1-35). This is a significant improvement over GTS2004 with only three reliable ID-TIMS age dates. With the duration of the Carboniferous Period being 60 myr, the average resolution of a zone is equivalent to 1.5−1.0 myr, although individual zone duration varies widely. The Mississippian is still poorly characterized (with few, stratigraphically scattered, high-precision ages) whereas the many high-precision ages in the Pennsylvanian are distributed evenly throughout.

There is disagreement in estimates of the age of Carboniferous stage boundaries because of the local use of older $^{40}Ar/^{39}Ar$ and SHRIMP ages in western Europe and Australia (Menning *et al.*, 2006; see Table 23.1). The new ID-TIMS data from the Rhenish Mountains, Donets Basin and the Urals provide a solid and more reliable framework for the Carboniferous, and contribute to numerical age stability (Davydov *et al.*, 2010; Schmitz and Davydov, in press).

Analysis at Boise State University of bentonite samples from the Devonian−Carboniferous transition from the Rhenish Mountains (Aprike and Hasselbachtal sections) in Germany now suggests the age of the D/C boundary to be close to what was proposed in GTS2004 (Davydov *et al.*, 2004). Two of the bentonite horizons (ω2 and ω3) are preserved in the upper Wocklum Limestone, the upper of which was sampled and studied (age date d18 of 359.24 Ma in Appendix 2 of this volume). Three bentonite horizons (α1, α2, and α3) are found in the Hangenberg Limestone and the first two (α1 and α2) were sampled and analyzed (age dates cb1 of 358.71 Ma and cb2 of 358.43 Ma in Appendix 2). All three dates (d18, cb1 and cb2) have less than 0.5 myr external uncertainties. As discussed in more detail below, our revised age estimate for the Devonian−Carboniferous boundary and base Tournaisian Stage is 358.88 ± 0.4 Ma.

FIGURE 23.7 **Dynamics of warm-water benthic Foraminifera species diversity in northern Pangaea during Mississippian−Cisuralian.** *The maximal peaks in diversity associated with warming events, whereas minimal peaks represent cooling (glaciation) events labeled with Latin numbers (I−XV). Previously proposed glaciation/cooling events: I − Buggisch et al., 2008 (B1); II − Kalvoda, 2002; III − Saltzman, 2003; Kalvoda, 2002; Buggisch et al., 2008 (B2); IV − Kalvoda, 2002; V−VIII, XII−XV − Fielding et al., 2008a (Eastern Australia); Buggisch et al., 2008 (B3), Gulbranson et al., 2010 (Ar1); VI− Mii et al., 2001; Saltzman, 2003; VIII − Gulbranson et al., 2010 (Ar2) − the age is reinterpreted here as the glacial event appears below the age 310 and above 312 Ma; IX−X − Scheffler et al., 2003; XII−XV − Montanez et al., 2007. Regional stage nomenclature and numerical ages slightly differ from GTS2012.*

TABLE 23.1 **Comparison of Age Estimates and Durations of Carboniferous Subperiods from Selected Time Scales Published in Recent Years**

	Mississippian			Pennsylvanian	
	Base (*Ma*)	Top (*Ma*)	Duration (*myr*)	Top (*Ma*)	Duration (*myr*)
Harland *et al.* (1990)	362.5	322.8	39.7	290	32.8
Young and Laurie (1996)	354	314	40	298	16
Tucker *et al.* (1998)	362				
Compston (2000)	359.6				
Ross and Ross (1988)	360	320	40	286	34
Jones (1995)	356	317	39	300	17
Menning *et al.* (2000)	354	312	42	292	20
Menning *et al.* (2000)	354	320	34	292	28
Menning *et al.* (2006)	358	320	38	296	24
Heckel (2002)		320		290	30
GTS2004	359.2	318.1	41.1	299.0	19.1
GTS2012	358.9	323.2	35.7	298.9	24.3

Twelve new radiometric ages in the master GTS2012 radiometric data set in Appendix 2, from early Tournaisian through late Moscovian, were obtained by us in the Donets Basin. A new age constraint from the lower Mokrovolnovakhskaya [C_1^t (A)] Series goes some way towards remedying the dearth of radiometric (and particularly ID-TIMS U-Pb zircon) ages for the global Mississippian subperiod (Davydov *et al.*, 2004; Menning *et al.*, 2006). The age of 357.26 ± 0.08 Ma obtained from an ash near the base of the $C_1^t b_2$ biostratigraphic zone is consistent with an age for the Devonian–Carboniferous boundary of 358.88 Ma, and supports the proposition of a much shorter duration of the Hastarian Substage of the Tournaisian in western Europe (Davydov *et al.*, 2004; Haq and Schutter, 2008; Menning *et al.*, 2006). Two samples collected within and at the top of the $C_1^v c$ zone – which reliably correlates with the lower Moliniacian (Late Chadian) of the lowermost Visean in western Europe – provide a minimum age of 346.3 Ma for the base of the global Visean Stage; i.e. one million years older than proposed in the most recent time scale compilations (Davydov *et al.*, 2004; Menning *et al.*, 2006). An age of 342.01 ± 0.10 Ma from a bentonite in the lower Styl'skaya Formation extends the duration of the Tulian and consequently the Holkerian Substage up to 6 myr (Davydov *et al.*, 2010).

A dramatic change in the standard Carboniferous time scale is provided by the new age of 328.14 ± 0.11 Ma obtained from the c_{11} coal sample, in the $C_1^v g_2$ biozone. According to foraminifers (*Betpakodiscus cornuspiroides*) and ammonoids (*Eumorphoceras*), this coal correlates with the lower Steshevian Horizon of the Serpukhovian in the Russian Platform, and the Pendleian (E1 Zone) of western Europe. This age pushes the lower boundary of the Serpukhovian down to approximately 330 Ma – i.e. about 4 myr older than in previous global time scale compilations. Similar ages were recently obtained from the aforementioned tonstein in the Upper Silesian Basin (Gastaldo *et al.*, 2009) with dates of 328.84 ± 0.38 and 328.01 ± 0.36Ma; their host strata correlate with the lower and middle Pendleian Substage of western Europe.

Radiometric calibrations of the Pennsylvanian time scale have relied mainly upon a series of $^{40}Ar/^{39}Ar$ sanidine ages from the Donets Basin (Hess *et al.*, 1999), the Upper Silesian Basin and several central European (Sahr, Ruhr, Bohemian, Intra-Sudetic) basins (Hess and Lippolt, 1986; Burger *et al.*, 1997). A new radiometric date for the l_3 coal of the Donets Basin of 312.01 ± 0.08 Ma (Davydov *et al.*, 2010) may be directly compared to the result of Hess *et al.* (1999) for a correlative sample of the l_3 coal of 305.5 ± 1.5 Ma. Beyond the obvious contrast in precision, the accuracy of the significantly younger $^{40}Ar/^{39}Ar$ sanidine age is clearly called into question. Even taking into account systematic errors associated with decay constants and monitor standards (Min *et al.*, 2000; Renne *et al.*, 1998; Villeneuve *et al.*, 2000; Kuiper *et al.*, 2008), this sanidine age is anomalously young. Although this phenomenon was noted and interpreted as indicating systematic problems with biostratigraphic correlation (Hess *et al.*, 1999), it is apparent from new results from the Donets Basin that instead this age suffers from a systematic analytical or geological bias. The stratigraphic fidelity of $^{40}Ar/^{39}Ar$ sanidine ages from other European basins is similarly suspect, although the large errors on these ages make it generally difficult to assess the degree of bias (Davydov *et al.*, 2004). In summary, these imprecise Carboniferous sanidine ages appear to be plagued by one or a combination of systematic analytical errors and open system behavior, and are thus superseded by the new accurate and precision U-Pb ages from Donets Basin for Pennsylvanian time scale calibration.

Current studies in the Urals that have provided a new age date of 333.9 Ma from the Visean (late Aleksian) strata at Verkhnyaya Kardailovka and three new Bashkirian ages from the Kljuch section are the first robust radiometric constraints in this stage (Schmitz and Davydov, in press).

Although the Kljuch ages reside in the middle-upper Bashkirian, they strengthen the ability of the CONOP composite (see below) to indicate an age for base Bashkirian near 323 Ma. Similarly, the age of the top of the Bashkirian (base Moscovian) can be estimated from the Urals composite via the occurrence of the proposed index of the global Moscovian stage, *Declinognathodus donetzianus*, in the Basu section (Kulagina *et al.*, 2009). The interpolated radiometric age of this horizon in the Urals composite is in excellent agreement (±0.1 Ma) with our interpretation of the age of the first occurrence of *D. donetzianus* in the K_1 Limestone of the Donets Basin (Ueno and Nemirovska, 2008) at 314.6 Ma, based on a tonstein age of 314.40 ± 0.06 Ma in the overlying k_3 coal.

Most extensive dating of tonsteins has been made from the Moscovian Stage, as many shafts in Donets Basin are actively mining coal of this age. Six ages were obtained from coals k_3, k_7, l_1, l_3 (two samples from different shafts), m_3, and n_1. These samples (items cb14–19 in Appendix 2) and their associated cyclostratigraphic calibration of the Late Pennsylvanian dramatically change our understanding of the distribution of time in the Moscovian Stage (see below).

23.3.2. Carboniferous and Permian Composite Standard

In order to integrate and calibrate zonal successions of foraminifers, conodonts, and ammonoids relative to each other, and to construct the Carboniferous and Permian chronostratigraphic (relative) scale, quantitative zonation and correlation were employed. Modern methods of quantitative biostratigraphic analysis hold promise for a robust ordination of bioevents (and other unique events like dated ash beds) and estimation of the true ranges of taxa.

We have utilized the CONOP9 program (Sadler and Cooper, 2003; Sadler *et al.*, 2003) to develop a stratigraphic composite for the Donets Basin and Urals, Guadalupian Mountains and S. China. As described in Sadler *et al.* (2009), the compositing process goes in several steps:

1) Construction of an overall ordering of events (ordinal composite sequence) compiled from a database that involves both paired events (first and last occurrences of taxa, for which range extensions are allowed) and unpaired dated events (radiometrically dated ash beds, for which relative separation is fixed) in a variety of stratigraphic sections;
2) Scaling of the intervals between events in the composite sequence according to the relative position of events in all the measured sections (and constrained by the absolute positions of dated events), after the observed ranges have been extended and missing taxa inserted to match the best-fit composite sequence; and
3) Identifying key index taxa to locate the relative age of standard zone and stage boundaries within the scaled composite section.

In Donets Basin, the composite from late Devonian to Asselian was built with 2641 species of foraminifers, conodonts, ammonoids and algae (5282 paired bioevents) and 357 unpaired events (coals and limestone) including 9 ash beds from 46 wells and sections (Brazhnikova *et al.*, 1967; Aizenverg *et al.*, 1963; Alekseeva *et al.*, 1983; Aizenverg *et al.*, 1983; Vdovenko, 2001; Nemirovskaya, 1999; Fohrer *et al.*, 2006; Davydov, 1992, 2009; Davydov and Khodjaniyazova, 2009; Davydov *et al.*, 2010). The composite in the Urals is based on 1156 Visean through Kungurian species of foraminifers (including fusulinids), conodonts and ammonoids (2312 paired bioevents) in 27 stratigraphic sections, and include 29 ash beds coded as dated unpaired events (Schmitz and Davydov, in press). The Guadalupian composite is based on the biostratigraphic data from the Glass and Del Norte Mountains, and Apache Mountains in the type area, Texas (Wardlaw, 2000; Yang and Yancey, 2000; Nestell *et al.*, 2006; Wardlaw and Nestell, 2010). In addition, four sections (Tieqiao, Maoershan, Pingxiang and Dachongling) from S. China also were integrated in the composite (Shen *et al.*, 2007; Zhang *et al.*, 2008, 2010). The Guadalupian composite includes 286 species of conodonts and foraminifera and one volcanic ash from Monzanita bed in the Nipple Hill (Bowring *et al.*, 1998; Nicklen, 2003).

The Lopingian composite is built with 33 sections in S. China, Iran, Azerbaijan, Armenia, Pakistan, Italy, Austria, Turkey, Canadian Artic and Greenland and includes 316 species of conodonts, foraminifera, ammonoids and radiolaria. Eight ash beds from well biostratigraphically constrained sections (Meishan and Shangsi sections) dated with ID-TIMS (Mundil *et al.*, 2004) integrated in the Lopingian composite. Paleomagnetic events turned out to be frequently inconsistent with biostratigraphy for the Lopingian composite and were not involved in the compositing process.

Our methods follow those of Chapters 20 and 21 in this volume for the global Ordovician–Silurian graptolite clade, although in contrast we have worked with a multi-taxa (conodont, fusulinid, ammonoid) database within a fairly restricted geographic province. Because the analyzed data are derived from a series of closely connected basins, problems of provincialism and migration are mitigated. Secondly, we have used consistent taxonomies provided by well-recognized regional specialists (Brazhnikova and co-workers in Donets Basin; Kulagina and Pazukhin – Mississippian–lower Pennsylvanian foraminifers and conodonts; Chernykh and Davydov – middle Pennsylvanian–Cisuralian conodonts and foraminifers; Ruzhenzev, Bogoslovskaya and Nikolaeva – Mississippian–Cisuralian ammonoids).

Figure 23.8 shows the Carboniferous composite standard with its main zonal units. Table 23.2 lists the composite standard values of the zones (before rescaling with radiometric ages), and of the incorporated ash beds with their numbered radiometric age dates, with 95% confidence limits and external error estimates. Table 23.2 covers both Carboniferous and Permian and is further discussed in Chapter 24 dealing with the Permian Period.

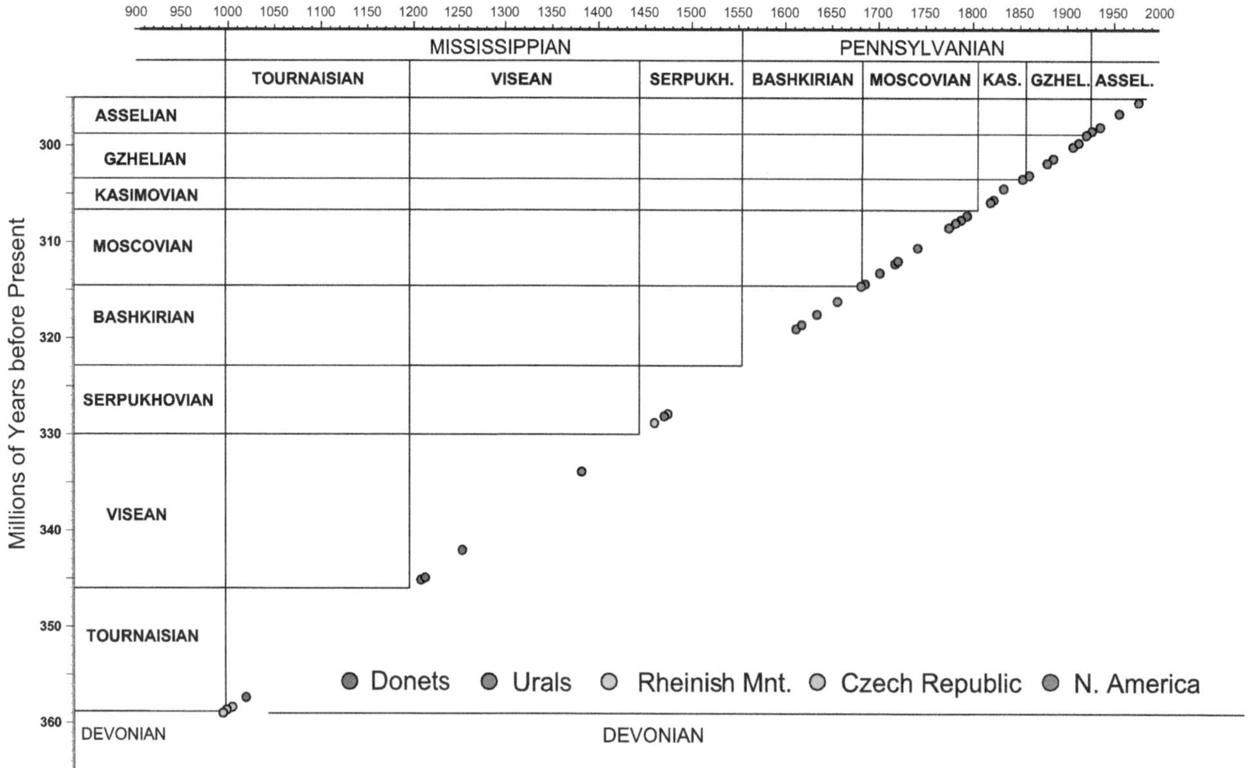

FIGURE 23.8 The Carboniferous composite standard may be rescaled using radiometric age estimates of its events (coloured dots on the best-fit line). The composite standard is rescaled (rubber-banded) to a linear scale such that the best-fit line is straight, allowing calibration of zones and stages discussed in Section 23.3.4. Regional stage nomenclature and numerical ages are retained as per composite standard and slightly differ from the finalized scale of GTS2012.

23.3.3. Calibration of Carboniferous Stage Boundaries by Composite Standard Optimization

The Carboniferous composite standard may be rescaled using age estimates of its events, as shown in Figure 23.8. As discussed in the previous section, the composite is based on the order and relative spacing of a large number of fossil events in many stratigraphic sections for the Urals, with all radiometric ages included; the composite is complemented by a Visean to Serpukhovian composite from the Donets Basin. The composite standard is scaled (rubber-banded) to a linear scale such that the best-fit line is straight in Figure 23.8, allowing calibration of zones and stages discussed below. Theoretically, a linear fit would only ensue, without rubber banding the composite, with *a priori* assumptions about either uniformity of depositional rate or stage or zone duration.

The composite is well constrained for late Serpukhovian through Bashkirian, and late Moscovian through early Artinskian time. By contrast, the middle Moscovian in the Urals remains relatively poorly studied, while the Mississippian is sparser in its compiled bioevent density compared with the Pennsylvanian and Cisuralian. Continued refinement of these intervals is likely as new and reliable biostratigraphic data become available.

From the Urals composite we extracted the necessary range extensions of key index taxa for the foraminiferal and conodont interval zone assemblages of the Visean through Artinskian. In fact, we found that most first appearance levels for the interval zone assemblages in our composite were not adjusted, or adjusted only slightly (<0.1 myr), compared with the first observed datums of the index taxa in our intensively studied, taxonomically rich sections with numerous dated ash beds (e.g., Usolka, DTR, DTQ). Notable exceptions were: the late Moscovian (Myachkovian) FAD of *Neognathodus roundyi*, which adjusted downward 0.6 Ma with respect to its first occurrence in the DTQ section; the early Gzhelian (Rusavkian) FAD of *St. vitali*, which shifted downward 0.4 myr with respect to its first occurrence in the Usolka section; and the early Sakmarian (Tastubian) FAD of *Sw. merrilli* which adjusted 0.6 myr relative to its first occurrence in the Usolka section, and to a level similar to the first occurrence of *M. uralensis*. In rare instances, for example the latest Bashkirian *Aljutovella tikhonovich–Verella spicata* Zone and middle Moscovian *Fusulinella priscoidea* Zone, the foraminiferal indexes were not present in the data sets; these zones were defined by their position within the neighboring zones and by the zonal assemblages.

The composite is scaled by our radiometric ages to directly establish the absolute ages of the FADs of index taxa

TABLE 23.2 Listing of the Composite Standard Values of the Zones and of the Incorporated Ash Beds with their Radiometric Age Dates, with 95% Confidence Limits and External Error Estimates

Stage	Conodont Zone Base	Age in Ma with 95% C.L. and λ Error	Composite Level
Triassic	*Hindeodus parvus*		
Triassic		Base Triassic at 252.2 ± 0.5	2656.73
Changhsingian		252.5 ± 0.42	2658.97
Changhsingian	*Clarkina zhejiangensis–C. meishanensis*		2650.18
Changhsingian	*Clarkina deflecta–yini*		2642.84
Changhsingian	*Clarkina changxingensis*		2634.36
Changhsingian		253.24 ± 0.46	2627.25
Changhsingian	*Clarkina subcarinata*		2624.21
Changhsingian		253.7 ± 0.46	2616.07
Changhsingian	*Clarkina wangi*		2610.63
Wuchiapingian	*Clarkina longicuspidata*		2602.74
Wuchiapingian	*Clarkina orientalis*		2589.19
Wuchiapingian	*Clarkina transcaucasica*		2568.28
Wuchiapingian		257.3 ± 0.51	2560.59
Wuchiapingian	*Clarkina guangyuanensis*		2557.66
Wuchiapingian	*Clarkina leveni*		2548.11
Wuchiapingian	*Clarkina asymmetrica*		2536.38
Wuchiapingian	*Clarkina dukouensis*		2527.15
Wuchiapingian	*Clarkina postbitteri postbitteri*		2515.97
Capitanian	*Jinogondolella granti*		2500.23
Capitanian	*Jinogondolella xuanhanensis*		2484.46
Capitanian	*Jinogondolella prexuahanensis*		2471.35
Capitanian	*Jinogondolella altudaensis*		2453.61
Capitanian	*Jinogondolella shannoni*		2446.09
Capitanian	*Jinogondolella postserrata*		2438.82
		265.35 ± 0.46	2435.07
Wordian	*Jinogondolella aserrata*		2393.78
Roadian	*Jinogondolella nankingensis*		2354.18
Kungurian	*Neostreptognathodus prayi*		2318.76
Kungurian	*Neostreptognathodus pnevi*		2277.23
Artinskian	*Neostreptognathodus pequopensis*		2247.87
Artinskian	*Sweetognathus clarki*		2222.49
Artinskian		288.21 ± 0.34	2518.44
Artinskian		288.36 ± 0.35	2513.92
Artinskian	*Sweetognathus whitei*		2141.53

(Continued)

TABLE 23.2 **Listing of the Composite Standard Values of the Zones and of the Incorporated Ash Beds with their Radiometric Age Dates, with 95% Confidence Limits and External Error Estimates—cont'd**

Stage	Conodont Zone Base	Age in Ma with 95% C.L. and λ Error	Composite Level
Sakmarian		290.5 ± 0.35	2138.88
Sakmarian		290.81 ± 0.36	2133.26
Sakmarian		291.50 ± 0.36	2104.31
Sakmarian	*Sweetognathus anceps*		2045.88
Sakmarian	*Sweetognathus binodosus*		2028.15
Sakmarian	*Sweetognathus merrilli*		1987.05
Asselian	*Streptognathodus postfusus*–Meso. pseudostriata–Meso. striata		1963.27
Asselian		296.69 ± 0.37	1956.62
Asselian	*Streptognathodus fusus–Mesogondolella simulata*		1951.14
Asselian	*Streptognathodus constrictus–Mesogondolella belladontae*		1940.86
Asselian		298.05 ± 0.56	1931.16
Asselian	Streptognathodus sigmoidalis–Strep. cristellaris		1929.93
Asselian		298.49 ± 0.37	1921.55
Asselian	*Streptognathodus isolatus*		1915.96
Gzhelian		299.22 ± 0.37	1912.24
Gzhelian	*Streptognathodus wabaunsensis*–St. fissus		1897.67
Gzhelian		300.22 ± 0.37	1889.68
Gzhelian	*Streptognathodus simplex–Streptognathodus bellus*		1878.96
Gzhelian		301.29 ± 0.36	1876.25
Gzhelian		301.82 ± 0.36	1873.47
Gzhelian	*Streptognathodus virgilicus*		1862.09
Gzhelian	*Streptognathodus vitali*		1850.17
Gzhelian		303.1 ± 0.36	1842.09
Gzhelian	*Streptognathodus simulator*		1838.22
Kasimovian		303.54 ± 0.39	1837.87
Kasimovian	*Streptognathodus praenuntius*		1833.48
Kasimovian		304.42 ± 0.36	1830.75
Kasimovian	*Streptognathodus firmus*		1831.54
Kasimovian	*Idiognathodus toretzianus*		1826.12
Kasimovian		304.83 ± 0.36	1823.94
Kasimovian	*Streptognathodus cancellosus*		1817.76
Kasimovian		305.52 ± 0.36	1815.63
Kasimovian	*Idiognathodus sagittalis*		1810.48
Kasimovian		305.95 ± 0.37	1808.17
Kasimovian		305.96 ± 0.36	1803.58
Kasimovian	*Streptognathodus subexcelsus–Swadelina makhlinae*		1796.11

TABLE 23.2 Listing of the Composite Standard Values of the Zones and of the Incorporated Ash Beds with their Radiometric Age Dates, with 95% Confidence Limits and External Error Estimates—cont'd

Stage	Conodont Zone Base	Age in Ma with 95% C.L. and λ Error	Composite Level
Moscovian		307.26 ± 0.36	1792.33
Moscovian		307.66 ± 0.37	1787.42
Moscovian		308.36 ± 0.30	1781.19
Moscovian		308.5 ± 0.36	1778.65
Moscovian	*Neognathodus roundyi–Streptognathodus cancellosus*		1775.64
Moscovian		310.55 ± 0.38	1764.26
Moscovian	*Neognathodus medexultimus–Streptognathodus concinnus*		1758.17
Moscovian		312.01 ± 0.37	1749.94
Moscovian		312.23 ± 0.37	1746.73
Moscovian		313.16 ± 0.37	1736.65
Moscovian	*Streptognathodus dissectus*		1741.31
Moscovian	*Neognathodus uralicus*		1722.03
Moscovian		314.4 ± 0.37	1719.12
Moscovian	*Declinognathodus donetzianus*		1715.12
Bashkirian	*Neognathodus atokaensis*		1694.87
		317.54 ± 0.38	1688.24
		318.63 ± 0.40	1681.38
Bashkirian	*Declinognathodus marginodosus*		1676.49
		319.09 ± 0.38	1670.76
Bashkirian	*Idiognathodus sinuosus*		1658.96
Bashkirian	*Neognathodus askynensis*		1646.68
Bashkirian	*Idiognathoides sinuatus*		1619.93
Bashkirian	*Declinognathodus noduliferus*		1607.79
Serpukhovian	*Gnathodus postbilineatus*		1563.82
Serpukhovian	*Gnathodus bollandensis*		1523.07
		328.14 ± 0.40	1501.14
Serpukhovian	*Lochriea cruciformis*		1472.26
Serpukhovian	*Lochriea ziegleri*		1445.34
		333.95 ± 0.39	1392.63
Visean	*Lochriea nodosa*		1425.61
Visean	*Lochriea mononodosa*		1380.45
Visean	*Gnathodus bilineatus*		1355.06
Visean	*Gnathodus praebilineatus*		1307.06
		342.01 ± 0.42	1281.96
Visean	*Gnathodus texanus*		1231.97

(Continued)

TABLE 23.2 Listing of the Composite Standard Values of the Zones and of the Incorporated Ash Beds with their Radiometric Age Dates, with 95% Confidence Limits and External Error Estimates—cont'd

Stage	Conodont Zone Base	Age in Ma with 95% C.L. and λ Error	Composite Level
		345.0 ± 0.41	1208.19
		345.17 ± 0.41	1206.37
Visean	*Gnathodus pseudosemiglaber–Scaliognathus anchoralis*		1187.23
Tournaisian	*Gnathodus semiglaber–Polygnathus communis*		1171.98
Tournaisian	*Dollimae bouckaerti*		1159.77
Tournaisian	*Gnathodus typicus–Siphonodella isosticha*		1113.32
Tournaisian	*upper Siphonodella quadruplicata–Patrognathus andersoni*		1090.48
		357.26 ± 0.42	1059.84
Tournaisian	*Siphonodella sandbergi–Siph. belkai*		1051.66
		358.43 ± 0.42	1037.63
		358.71 ± 0.42	1010.44
Tournaisian	*Siphonodella duplicata*		1027.91
Tournaisian	*Siphonodella sulcata*	Base Tournaisian at 358.88 ± 0.4	991.64
		359.25 ± 0.42	987.56
Famennian (Devonian)	*Siphonodella praesulcata*	359.44 ± 0.42	981.972

This table covers both Carboniferous and Permian. The table is built the same way as Tables 20.2 and 21.2 in the chapters dealing with the Ordovician and Silurian periods, the only difference being that for the Carboniferous–Permian a multi-fossil database was developed, not only based on one group (graptolites).

defining the regional conodont and fusulinid biozones for the Urals. Given the common usage of the Russian Platform regional substages for correlation throughout eastern Euramerica, we also illustrate their best scaling with respect to the Urals composite, and the global stages of the Pennsylvanian and Early Permian. Prior iterations of the Pennsylvanian–Cisuralian global time scale have suffered from a paucity of radiometric calibration points, except an ID-TIMS zircon-based age for the Permo–Carboniferous boundary from the Usolka section (Ramezani *et al.*, 2007). Significant changes in the durations of the constituent stages are thus unsurprising; we now examine these changes to the time scale resulting from integration of our Uralian results with those of the correlative Donets Basin (Davydov *et al.*, 2010).

As mentioned above, analysis of new bentonite samples from Devonian–Carboniferous transition sections provides a revised age estimate for the Devonian–Carboniferous boundary and base Tournaisian Stage of 358.88 ± 0.4 Ma. This age was arrived at from CONOP analysis integrating our three age dates within standard chronostratigraphy, biostratigraphic, sedimentologic and geochemical data compiled from sections in Rhenish Mountains, Carnic Alps, Italy, France, N. America Midcontinent, Urals and China (Ji *et al.*, 1989; Korn, 1992, 1993; Over, 1992; Schonlaub *et al.*, 1992; Becker and Paproth, 1993; Becker, 1996; Kaiser *et al.*, 2006, 2008; Kaiser, 2009). In total, 25 sections were involved in the quantitative CONOP analyses with 209 events, including FADs LADs of the taxa, ashes, and the base of the Hangenberg black shale. In addition, carbon peak cycles in Rhenish sections were run as unpaired events. The events that were involved, including the FAD of the index-species of the base of Carboniferous *S. sulcata*, were then calibrated in terms of radiometric ages, leading to the age of the current Devonian–Carboniferous boundary being 258.88 Ma (± 0.06 internal error, and ± 0.4 and/myr external error) (Davydov *et al.*, 2011). Since the current definition of the boundary requires revision (Kaiser, 2009), the age of the boundary may undergo adjustment.

Using six new ages for Moscovian tonstein (Appendix 2), the base of the Moscovian Stage shifts down to near 314.6 Ma (one fourth-order cycle below coal k_3 from which the oldest Moscovian age was obtained). The top of the

Moscovian (e.g., the base of the traditional Kasimovian in the N_3 limestone) is calibrated as 306.6 Ma, thus the duration of the Moscovian Stage increases to 8 myr as opposed to 6–7 myr in Davydov *et al.* (2004) and Menning *et al.* (2006). The duration of the Moscovian approximately equals the duration of the Kasimovian and Gzhelian together.

Both the traditional and alternative definitions for the base of the Kasimovian Stage can be assigned numerical ages based upon our tuned cyclostratigraphic model for the Late Pennsylvanian. As noted above, the traditional base of the Kasimovian (at the FADs of fusulinid *Protriticites pseudomontiparus* and the conodont *Streptognathodus subexcelsus*) in the N_3 limestone is calibrated as 306.6 Ma. On the other hand, the FAD of the conodont *Streptognathodus saggitalis* in the O_2 limestone is calibrated at 305.2 Ma. In the latter case, the durations of the Moscovian and Kasimovian stages change to 9.4 and 2.0 myr, respectively.

The base of the Gzhelian Stage, taken as the FAD of the conodont *Streptognathodus simulator,* is calibrated via the extension of our tuned cyclostratigraphic model at 303.4 Ma, thus constraining the durations of the traditional Kasimovian (3.2 myr) and Gzhelian stages (4.7 myr). This age of the base of the Gzhelian Stage in the Donets Basin is in good agreement with independent geochronological data for the same FAD in the Usolka section of the southern Urals (Schmitz and Davydov, in press).

Davydov *et al.* (2010) have proposed a high-resolution (± 0.1 myr) age model for the late Pennsylvanian based upon the direct dating and tuning of fourth-order sequences of the Donets Basin (Ukraine) to the 404 ka long-period eccentricity cycle. U-Pb zircon ages for seven tonsteins in the Moscovian strata of the basin provided multiple affirmative tests of this orbital control on the cyclothem depositional framework compiled by Izart *et al.* (1996). The Donets cyclostratigraphic model was then extended upward through the Kasimovian and Gzhelian succession to the limits of marine sedimentation at the base of the Permian in the Donets Basin, utilizing the sequence stratigraphic interpretations of Izart *et al.* (2006), with occasional modifications. Unfortunately, no dateable ash beds were recovered from the Kasimovian and Gzhelian of the Donets Basin with which to test this cyclostratigraphic extrapolation. The abundance of ash bed ages for the Kasimovian–Gzhelian in the Urals reported in this study, as well as the relative ease of biostratigraphic correlation between the two basins, provides an opportunity for testing and comparing the two independently derived, cyclicity (Donets) versus radiometrically (Urals) calibrated chronostratigraphic models for the Late Pennsylvanian.

Figures 23.9(a) and (b) illustrate these tests by comparing the absolute ages of the FADs of various conodont taxa, estimated for the Donets Basin via cyclostratigraphic calibration of its lithostratigraphic (and integrated biostratigraphic) framework, and for the Urals Basin via the radiometrically calibrated composite. In Figure 23.9(a), the six Moscovian U-Pb zircon ages calibrating the Donets Basin cyclostratigraphy are illustrated, as are their projections through the sequence stratigraphic interpretations of Izart *et al.* (1996, 2006) to the lithostratigraphic framework of the Donets Basin.

The base of the Gzhelian Stage, as defined by the first occurrence of *St. simulator* (Ellison) in the O_6 limestone (Goreva and Alekseev, 2007), was previously estimated at 303.2 Ma based on the Donets Basin cyclostratigraphic model (Davydov *et al.*, 2010). In the Urals composite, this taxon's first occurrence is assigned an age of 303.4 Ma as constrained by bracketing ash beds in the Usolka section; this slightly older age for the base of the global Gzhelian Stage is preferred, given its more direct relationship to dated ash beds.

There is a remarkable agreement between the two independent age models (Donets Basin and Urals) as evidence that our Pennsylvanian time scale is remarkably accurate, and resolved at a high level of precision of the calibrating ash bed ages. The radiometric age near 299 Ma of the Carboniferous roof is constrained at the FAD of the Permian base index conodont *Streptognathodus isolatus* with several closely spaced ashes in the Usolka section (Ramezani *et al.*, 2007; Schmitz and Davydov, in press).

23.3.4. Age of Stage Boundaries

In the final analysis for the age of the stage boundaries and the stage durations in the Carboniferous (and Permian) periods, the chronometric and stratigraphic ages of the radiogenic isotope dates were subjected to a cubic spline fit (Figure 23.10). The method is described in Chapter 14, largely based on the approach used in GTS2004. Composite standard values are those before linear rescaling (Table 23.2). Points were weighted according to the procedure given in Chapter 14, taking radiometric errors into account. A smoothing factor of 1.2 was calculated by cross-validation.

As mentioned above, analysis of new bentonite samples from Devonian–Carboniferous transition sections provides a revised age estimate for the Devonian–Carboniferous boundary and base Tournaisian Stage of 358.9 ± 0.4 Ma. This age estimate anchors the lower end of the cubic spline; the anchor for the upper end of the spline is the age of the Permian–Triassic boundary at 252.2 ± 0.5 (see also Chapters 24 and 25).

All of the 46 points passed the chi-squared goodness-of-fit test at a significance level of $p < 0.01$. The resulting spline (Figure 23.10) is fairly linear through the Carboniferous, apart from a slight increase in the duration of scaling zones (decreased turnover rate) in the Bashkirian.

Interpolated stage boundary ages with their error bars and durations are given in Table 23.3. Interestingly, the ages

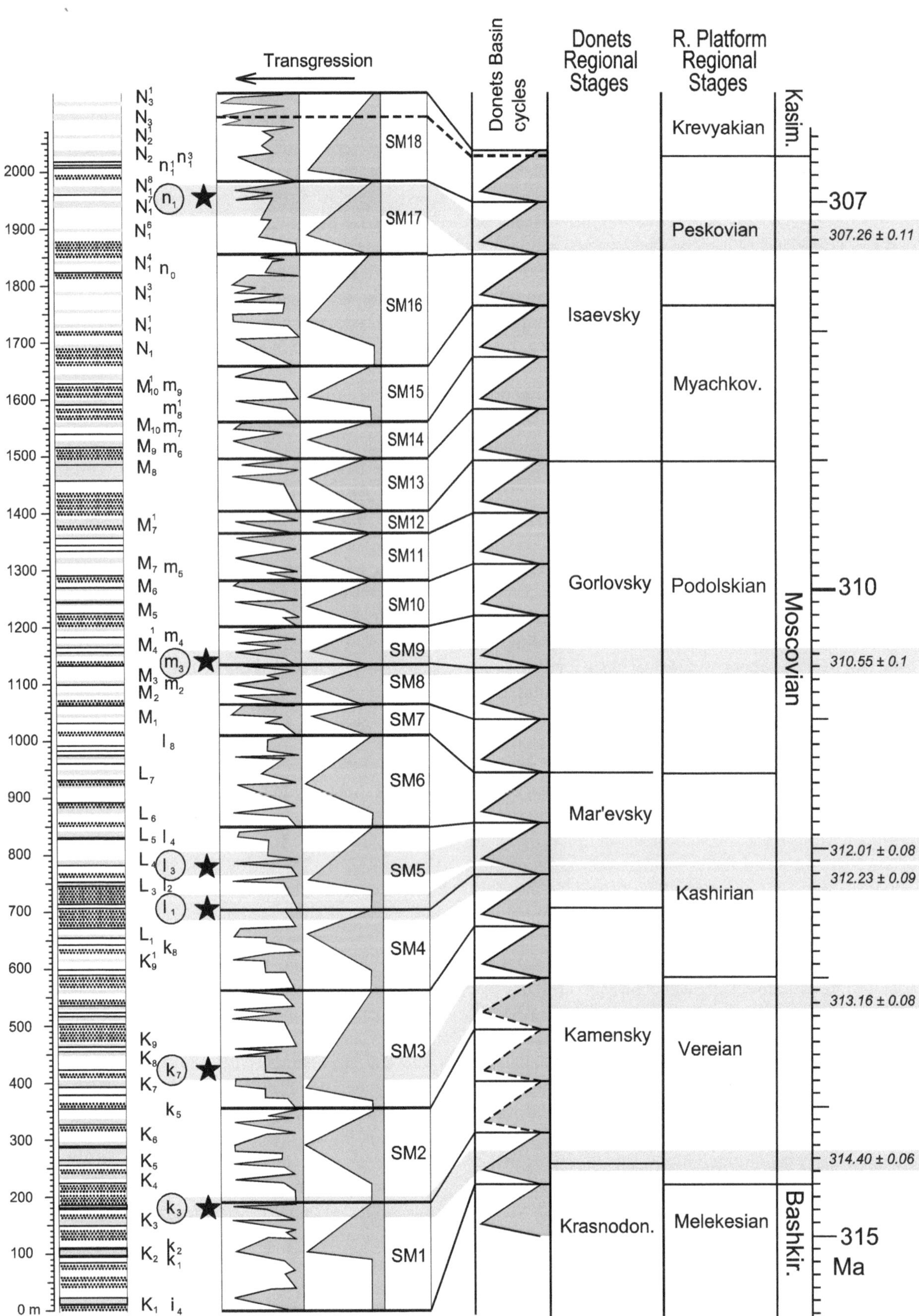

FIGURE 23.9A Lithostratigraphy and sequence stratigraphy through the Moscovian succession of the central exposed Donets Basin (modified from figure 3 of Izart *et al.*, (1996)), with position of six radiometric ages obtained in our study. Projection of stratal architecture onto a time linear scale constrained by ash bed ages reveals the consistent c. 400 ka tempo of the 4th order sequences of Izart and coworkers. Only a few high frequency cycles in the lowermost Moscovian must be reinterpreted as 4th order major transgressions to maintain consistency with the model. Tuning of these 4th order sequences to the long eccentricity cycle allows calibration of the biostratigraphic record at a resolution of c. 100 ka.

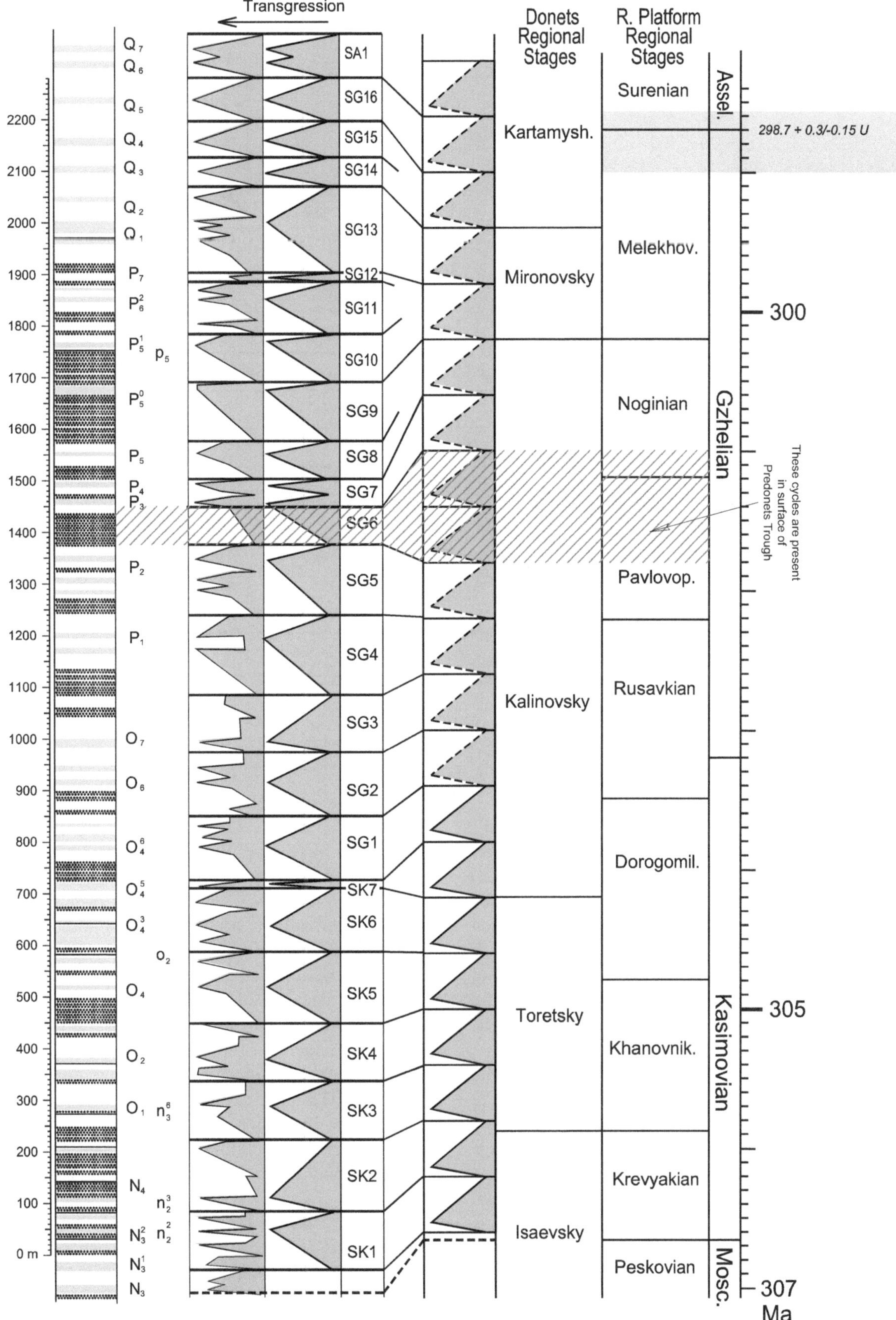

FIGURES 23.9B Lithostratigraphy and sequence stratigraphy through the Kasimovian–Gzhelian succession of the central exposed Donets Basin (modified from figure 12 of Izart *et al.*, (2006)). The long eccentricity cycle tuning of 4th order sequences derived for the Moscovian succession is extrapolated upward to the Carboniferous–Permian boundary constrained at 298.7 Ma (Ramezani *et al.*, 2007) in the Usolka parastratotype section of the Urals. High frequency and 4th order sequences of the Kasimovian and early Gzhelian are well developed, lending more confidence to the cyclostratigraphic calibration. Although cyclicity becomes more ambiguous in the increasingly continental upper Gzhelian succession, only modest reinterpretation of Izart *et al*'s (2006) 4th order sequences as higher frequency cycles is necessary to align the base of the Asselian in the Donets Basin with the radiometric constraint from the Urals.

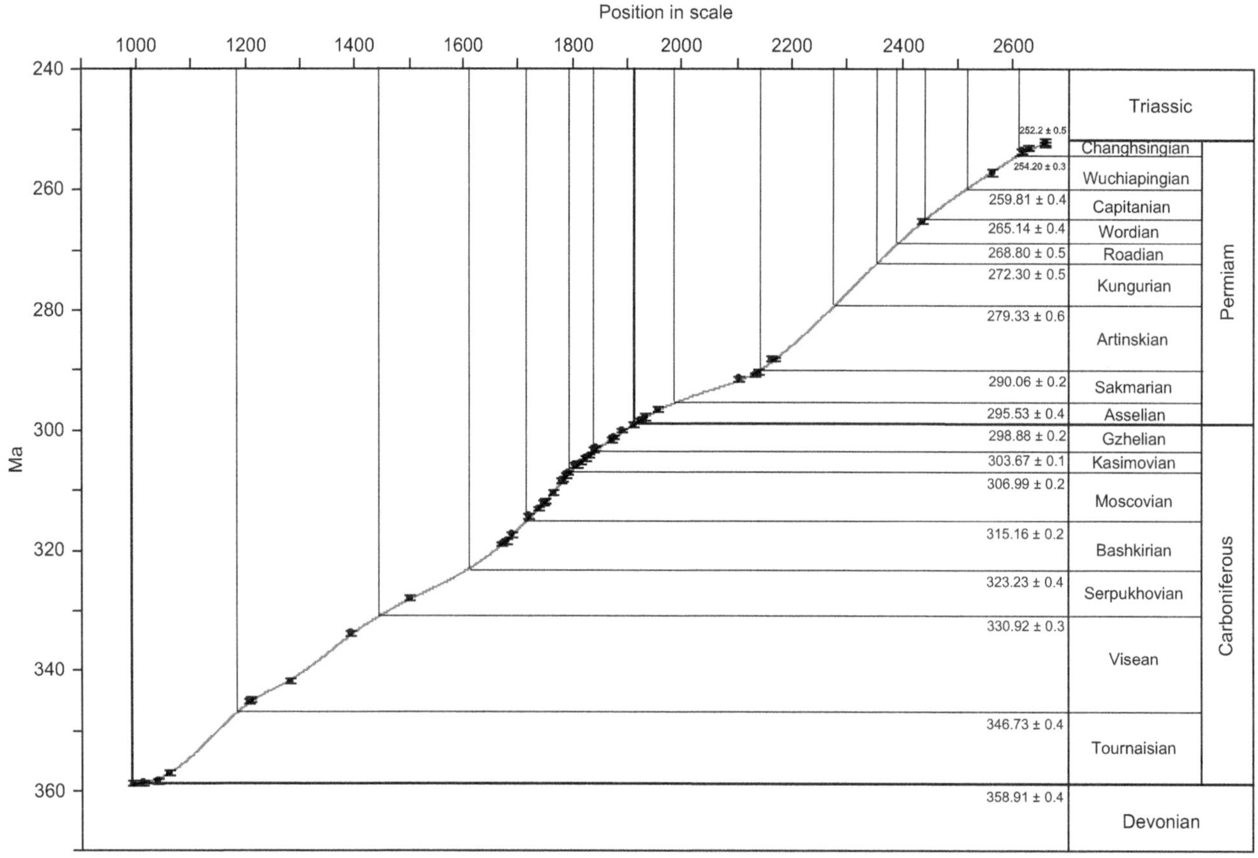

FIGURE 23.10 **Geochronology of the Carboniferous and Permian stages with error bars on the radiometric ages.** Use is made of a smoothing spline fit on the two way plot between the composite standard scaling of the stages (Table 23.2) and radiometric ages. There are 32 carefully selected and standardized radiometric age dates for the Carboniferous and 14 for the Permian (Table 23.2), including lower (D/C) and upper (P/T) boundary age assignments. For details see text.

computed through the linear regression fit described earlier in this chapter (Section 23.3.3, Figure 23.8 and Table 23.3) only differ within decimal values from the ages computed with the cubic spline. Although a majority of the regression ages are outside the error estimate for the cubic spline age of the stage boundaries used for the standard geochronology of the Carboniferous, this does not detract from the discussion on the relation between sequences and cycles in Section 23.3.3.

Tournaisian lasted from 358.9 ± 0.2 to 346.7 ± 0.4 Ma, for a duration of 12.2 ± 0.2 myr Visean lasted 15.8 ± 0.2 myr from 346.7 ± 0.4 to 330.9 ± 0.3 Ma, thus being the longest stage in the Phanerozoic. The Serpukhovian Stage ranged from 330.9 ± 0.3 to 323.2 ± 0.4 Ma, for a duration of 7.7 ± 0.2 myr Bashkirian lasted for 8.1 ± 0.2 myr from 323.2 ± 0.4 to 315.2 ± 0.2 Ma. The Moscovian Stage started at 315.2 ± 0.2 and ended at 307 ± 0.2 Ma, lasting 8.2 ± 0.1 myr Kasimovian ran from 307 ± 0.2 to 303.7 ± 0.1 Ma for only 3.3 ± 0.1 myr, thus being the shortest Carboniferous stage, and the Gzhelian lasted 4.8 ± 0.1 myr from 303.7 ± 0.1 Ma to base Permian at 298.9 ± 0.2 Ma (Table 23.4). Mississippian had a duration of 35.7 myr, and Pennsylvanian 23.3 myr. The Carboniferous Period lasted from 358.9 to 298.9 Ma, for a duration of 60 myr.

The geochronologic scale for the Carboniferous differs from that in GTS2004 (Table 23.3) because of the dramatic increase in the number, accuracy and precision of radiometric age dates discussed above, and changes in the definition of stages. For the present scale, a more detailed and a much larger biostratigraphic data set has been used than in GTS2004, giving higher stratigraphic resolution. Relative to GTS2004, the base Bashkirian is almost 5 myr older, and the base Moscovian almost 4 myr; base Serpukhovian is 3.1 myr older and base Visean 1.4 myr. For the other stage boundaries, age changes are in decimal values only.

The error bars on boundary ages are relatively small, mainly because radiometric ages have become precise. No estimate was provided on the chronostratigraphic uncertainty of radiometric ages, or the uncertainty in the composite standard itself; error estimates may slightly increase if this information was to be taken into account.

TABLE 23.3 Geochronology of the Carboniferous and Permian Periods

	GTS2012					Linear Regression	GTS2004
Stage Base	Scale	Ma	2-Sigma	Duration	2-Sigma	Ma	Ma
Changhsingian	2610.63	254.2	0.3	2.0	...	254.2	253.8
Wuchiapingian	2515.97	259.8	0.4	5.6	0.2	260.0	260.4
Capitanian	2438.82	265.1	0.4	5.3	0.2	265.0	265.8
Wordian	2393.78	268.8	0.5	3.7	0.1	270.0	268.0
Roadian	2354.18	272.3	0.5	3.5	0.1	275.0	270.6
Kungurian	2277.23	279.3	0.6	7.0	0.1	282.0	275.6
Artinskian	2141.53	290.1	0.2	10.7	0.2	290.0	284.4
Sakmarian	1987.05	295.5	0.4	5.5	0.1	295.0	294.6
Asselian	1915.96	298.9	0.2	3.4	0.1	298.7	299.0
Gzhelian	1838.22	303.7	0.1	4.8	0.1	303.4	303.9
Kasimovian	1796.11	307.0	0.2	3.3	0.1	306.7	306.5
Moscovian	1715.12	315.2	0.2	8.2	0.1	314.6	311.6
Bashkirian	1607.79	323.2	0.4	8.0	0.2	322.8	318.1
Serpukhovian	1445.34	330.9	0.3	7.7	0.2	330.0	326.4
Visean	1187.23	346.7	0.4	15.8	0.2	346.0	345.3
Tournaisian	991.64	358.9	0.4	12.2	0.2	358.9	359.2

The age of the stage boundaries, the duration of the stages and the 95% confidence limits are derived from cubic spline interpolation shown in Figure 23.10. Composite standard values of stage boundaries are also listed. Error estimates on stage boundary ages use the external errors on radiometric ages in Appendix 2, and the internal (laboratory) errors on stage durations (see Chapter 14). To the right are columns with the GTS2004 age of the stage boundaries, and stage boundary ages using linear regression on the composite standard of Figure 23.8.

TABLE 23.4 Geochronology of the Carboniferous Period

Period	Subperiod	Stage	Age of Base (Ma)	Est. ± myr (2-Sigma)	Duration (myr)	Est. ± myr (2-Sigma)
Permian			298.9	0.2		
Carboniferous						
	Pennsylvanian					
		Gzhelian	303.7	0.1	4.8	0.1
		Kasimovian	307.0	0.2	3.3	0.1
		Moscovian	315.2	0.2	8.2	0.1
		Bashkirian, base Pennsylvanian	323.2	0.4	8.0	0.2
	Mississippian					
		Serpukhovian	330.9	0.3	7.7	0.2
		Visean	346.7	0.4	15.8	0.2
		Tournaisian, base Mississippian, (base Carboniferous)	358.9	0.4	12.2	0.2
Devonian						

GSSPs of the Carboniferous Stages, with Location and Primary Correlation Criteria (Status Jan. 2012)

Stage	GSSP Location	Latitude, Longitude	Boundary Level	Correlation Events	Reference
Gzhelian	*Candidates are in southern Urals or Nashui (South China).*			*FAD of conodont* Idiognathodus simulator *(s.str.). Close to FAD of ammonoid* Shumardites.	
Kasimovian	*Candidates are in southern Urals, southwest USA and south-central China.*			*FAD of fusulinid Protriticites, which is near FAD of ammonoid Eothalossoceras. Alternative (higher) base is FAD of fusulinid Montiparus montiparus, which is near FAD of conodont Idiognathodus sagittalis. Age given here is the higher version; the lower one is about 1 myr older.*	
Moscovian	*Candidates are in southern Urals or Nashui (South China).*			*Either FAD of conodont* Idiognathoides postsulcatus *or* Declinognathodus donetzianus.	
Bashkirian	Arrow Canyon, Nevada	36°44′00″ N, 114°46′40″ W**	82.9m above the top of the Battleship Formation in the lower Bird Spring Formation	Conodont FAD *Declinognathodus noduliferus*	Episodes 22/4, 1999
Serpukhovian	*Candidates are Verkhnyaya Kardailovka (Urals) or Nashui (South China)*			*FAD of conodont* Lochriea ziegleri	
Visean	Pengchong, south China	24°26′N, 109°27′E	Base of bed 83 in the Pengchong Section	Foraminifer, lineage *Eoparastaffella simplex* morphotype 1/morphotype 2	
Tournaisian *(base Carboniferous)*	La Serre, France	43°33′19.9"N 3°21′26.3"E*	Base of Bed 89 in Trench E′ at La Serre, *(but FAD now known to be at base of Bed 85)*	Conodont FAD *Siphonodella sulcata* IMPRECISE (GSSP discovered in 2006 to have biostrat problems, and can not be correlated wih precision).	Episodes 14/4, 1991; Kölner Forum Geol. Paläont., 15, 2006

**according to Google Earth*
***derived from map*

ACKNOWLEDGMENTS

The authors gratefully acknowledge long-term support for Permo–Carboniferous chronostratigraphy through NSF grants EAR-0003343, EAR-0106796, EAR-0418703, EAR-0746107 and EAR-1004079 and instrumentation support via EAR-0451802. We thank our colleagues Boris Chuvashov, Valery Chernykh, Vladimir Pazukhin, Vladislav Poletaev and Victor Puchkov for their facilitation, advice, and cooperation during intensive study of the Uralian and Donets sections. Analytical innovations throughout the course of the study were inspired and facilitated by the EARTHTIME Initiative and community. Felix Gradstein and Øyvind Hammer provided extremely valuable input that improved this chapter. Phil Heckel is thanked for a detailed text review.

REFERENCES

Abushik, A.F., Chizhova, V.A., Guseva, E.A., Sidaravichene, N.V., 1990. Evolution and biostratigraphy of ostracods in Paleozoic. In: Abushik, A.F. (Ed.), Practical Guide in Microfauna of USSR. VSEGEI, Leningrad, pp. 1–35 [In Russian].

Adams, G.I., Ulrich, E.O., 1904. Zinc and lead deposits of Northern Arkansas. United States Geological Survey Professional Paper 24, 1–118.

Aisenverg, D.E., Brazhnikova, N.E., Novik, K.O., Rotay, A.P., Shulga, P.L., 1963. Carboniferous Stratigraphy of the Donets Basin, 37. Publishing House of Ukranian Academy of Sciences, Kiev, p. 183.

Aizenverg, D.E., Astakhova, T.V., Berchenko, O.I., Brazhnikova, N.E., Vdovenko, M.V., Dunaeva, N.N., Zernetskaya, N.V., Poletaev, V.I., Sergeeva, M.T., 1983. Upper Serpukhovian Stages of the Donets Basin: Paleontological characteristics. Naukova Dumka, Kiev, p. 164 [In Russian].

Akhmetshina, L., Nikolaeva, S., Konovalova, V., Korobkov, V., Zainakaeva, G., 2008. Visean-Serpukhovian boundary carbonates in western Kazakhstan: A case study of Lower Carboniferous sedimentation. 33rd International Geological Congress 2008, Oslo. Abstracts on CD-ROM.

Alekseev, A.S., Goreva, N.V., 2007. Conodont zonation for the type Kasimovian and Gzhelian stages in the Moscow Basin, Russia. In: Wong, T.E. (Ed.), Proceedings of the XVth International Congress on Carboniferous and Permian Stratigraphy, Utrecht, 10–16 August 2003, pp. 229–242.

Alekseev, A.S., Goreva, N.V., Isakova, T.N., Kossovaya, O.L., 2009. The stratotype of the Gzhelian Stage (Upper Carboniferous) in Moscow Basin, Russia. In: Puchkov, V.N. (Ed.), The Carboniferous Type Sections in Russia, Potential and Proposed Stratotypes. Proceedings of the International Conference Ufa - Sibai, 13–18 August, 2009. Institute of Geology, Bashkirian Academy of Science, Ufa, pp. 165–177.

Alekseeva, I.A., Glushenko, N.V., Ivanov, V.K., Inosova, K.I., Kalmykova, M.A., Kashik, D.S., Kozitskaya, R.I., Kruzina, A.K.H., Movshovich, E.V., Redichkin, N.A., Schvartsman, E.G., 1983. Key section of the Carboniferous-Permian boundary beds of the Southern part of Eastern-European Platform (Gzhelian and Asselian, borehole 4199, Skosyrskaya). Transactions of Stratigraphic Committee of the U.S.S.R. 12, 5–135 [In Russian].

Algeo, T.J., Wilkinson, B.H., 1988. Periodicity of mesoscale Phanerozoic sedimentary cycles and the role of Milankovitch orbital modulation. Journal of Geology 96, 313–322.

Archangelsky, S., Arrondo, O.G., Leguizamon, R.R., 1995. Floras Paleozoicas. Contribuciones a la palaeophytologia Argentina y revision y actualizacion de la obra paleobotanica de Kurtz en la Republica Argentina. Actas de la Academia Nacional de Ciencias de la Republica Argentina 11 (1–4), 85–125.

Baesemann, J.F., Lane, H.R., 1985. Taxonomy and evolution of the genus *Rhachistognathus* Dunn (Conodonta; Late Mississippian to early Middle Pennsylvanian). Courier Forschungsinstitut Senckenberg 74, 93–135.

Bamber, E.W., Fedorowski, J., 1998. Biostratigraphy and systematics of Upper Carboniferous cerioid rugose corals, Ellesmere Island, Arctic Canada. Geological Survey of Canada Bulletin 511, p. 127.

Barnett, A.J., Wright, V.P., 2008. A sedimentological and cyclostratigraphic evaluation of the completeness of the Mississippian-Pennsylvanian (Mid-Carboniferous) global stratotype section and point, Arrow Canyon, Nevada, USA. Journal of the Geological Society 165, 859–873.

Barrick, J.E., Lambert, L.L., Heckel, P.H., Boardman II, D.R., 2004. Pennsylvanian conodont zonation for Midcontinent North America. Revista Española de Micropaleontologia 36, 231–250.

Barrick, J.E., Heckel, P.H., Boardman II, D.R., 2008. Revision of the conodont *Idiognathodus simulator* (Ellison 1941), the marker species for the base of the Late Pennsylvanian global Gzhelian Stage. Micropaleontology 54, 125–137.

Barskov, I.S., Alekseev, A.S., 1975. Conodonts of the middle and upper Carboniferous of Moscow Basin. News of the Academy of Sciences of USSR, Geology Series 6, 84–99.

Barskov, I.S., Alekseev, A.S., Goreva, N.V., 1980. Conodonts and stratigraphic scale of the Carboniferous. News of the Academy of Sciences of USSR, Geology Series 3, 43–45.

Beavington-Penney, S.J., Racey, A., 2004. Ecology of extant nummulitids and other larger benthic foraminifera: applications in palaeoenvironmental analysis. Earth-Science Reviews 67, 219–265.

Becker, R.T., Paproth, E., 1993. Auxiliary stratotype sections for the Global Stratotype Section and Point (GSSP) for the Devonian-Carboniferous boundary. Hasselbachtal. Annales de la Société Géologique de Belgique 115, 703–706.

Becker, R.T., 1996. New faunal records and holostratigraphic correlation of the Hasselbachtal D/C-boundary auxiliary stratotype (Germany). Annales de la Société Géologique de Belgique 117, 19–45.

Belka, Z., Lehmann, J., 1998. Late Visean/early Namurian conodont succession from the Esla area of the Cantabrian Mountains, Spain. Acta Geologica Polonica 48, 31–41.

Berger, A., Loutre, M.F., Laskar, J., 1992. Stability of the astronomical frequencies over the Earth's history for paleoclimate studies. Science 255, 560–566.

Birgenheier, L.P., Fielding, C.R., Rygel, M.C., Frank, T.D., Roberts, J., 2009. Evidence for dynamic climate change on sub-10^6-year scales from the late Paleozoic glacial record, Tamworth Belt, New South Wales, Australia. Journal of Sedimentary Research 79, 56–82.

Bisat, W.S., 1924. The Carboniferous goniatites of the north of England and their zones. Proceedings of the Yorkshire Geological Society, new series 20 (1), 40–124.

Bisat, W.S., 1928. The Carboniferous goniatite zones of England and their continental equivalents. 1st International Congress on the Carboniferous Stratigraphy and Geology 1927, Herlen, pp. 117–133.

Bishop, J.W., Montañez, I.P., Gulbranson, E.L., Brenckle, P.L., 2009. The onset of mid-Carboniferous glacio-eustasy: Sedimentologic and diagenetic constraints, Arrow Canyon, NV. Palaeogeography, Palaeoclimatology, Palaeoecology 276, 217–243.

Boardman II, D.R., Heckel, P.H., 1989. Glacial-eustatic sea-level curve for early Late Pennsylvanian sequence in north-central Texas and biostratigraphic correlation with curve for Midcontinent North America. Geology 17, 802–805.

Boardman II, D.R., Heckel, P.H., Barrick, J.E., Nestell, M., Peppers, R.A., 1990. Middle-Upper Pennsylvanian chronostratigraphic boundary in the Midcontinent region of North America. In: Brenkle, P.L., Manger, W.L. (Eds.), Intercontinental Division and Correlation of the Carboniferous System. Courier Forschungsinstitut Senckenberg, 130, pp. 319–337.

Boardman II, D.R., Work, D.M., Mapes, R.H., Barrick, J.E., 1994. Biostratigraphy of Middle and Late Pennsylvanian (Desmoinesian-Virgilian) ammonoids. Kansas Geological Survey, Report 232, 121.

Boardman II, D.R., Heckel, P.H., Work, D.M., 2006. Conodont and ammonoid distribution across position of proposed Kasimovian-Gzhelian boundary in lower Virgilian strata in North American Midcontinent. Newsletter on Carboniferous Stratigraphy 24, 29–34.

Bogoslovskaya, M.F., 1984. The main stages of the ammonoid evolution in the Late Carboniferous. In: Menner, V.V., Grigorjeva, A.D. (Eds.), Upper Carboniferous of the USSR. Mezhvedomstrennyi stratigraficheskii komitet, Trudy, 11, pp. 88–91.

Bouroz, A., Wagner, R.H., Winkler, P.C., 1975. Report and proceedings of the IUGS Subcomission on Carboniferous Stratigraphy Meeting in Moscow, 8–12 September, 1975. Compte Rendu, 8ème Congrès International de Stratigraphie et de Géologie Carbonifère. In: General Problems of the Carboniferous Stratigraphy, Vol. 1. Nauka, Moscow. pp. 27–35.

Bowring, S.A., Erwin, D.H., Jin, Y.G., Martin, M.W., Davidek, K., Wang, W., 1998. U/Pb zircon geochronology and tempo of the end-Permian mass extinction. Science 280, 1039–1045.

Braun, A., Schmidt-Effing, R., 1993. Biozonation, diagenesis and evolution of radiolarians in the Lower Carboniferous of Germany. Marine Micropaleontology 21 (4), 369–383.

Brazhnikova, N.E., Vakarchuk, G.I., Vdovenko, M.V., Vinnichenko, L.V., Karpova, M.A., Kolomiyets, Y.I., Potiyevskaya, P.D., Rostovtseva, L.F., Shevchenko, G.D., 1967. Mikrofaunisticheskiye markiruyushchiye gorizonty kamennougol'nykh i permskikh otlozheniy Dneprovsko-Donetskoy vpadiny. Microfaunal marker horizons in the Carboniferous and Permian deposits of the Dnieper-Donets basin. Akad. Nauk Ukr. SSR, Inst. Geol. Nauk, Kiev.

Brenckle, P.L., Manger, W.L., 1990. Intercontinental correlation and division of the Carboniferous System: Contributions from the Carboniferous

Subcommission meeting Provo, Utah, September 1989. Courier Forschungsinstitut Senckenberg 130, 350.

Brenckle, P.L., Baesemann, J.F., Lane, H.R., West, R.R., Webster, G.D., Langenheim, R.L., Brand, U., Richards, B.C., 1997. Arrow Canyon, the mid-Carboniferous boundary stratotype. Cushman Foundation Foraminiferal Research Supplement to Special Publication 36, pp. 45–33.

Briand, C., Izart, A., Vaslet, D., Vachard, D., Makhlina, M., Goreva, N., Isakova, T., Kossovaya, O., Jaroshenko, A., 1998. Stratigraphy and sequence stratigraphy of the Moscovian, Kasimovian and Gzhelian in the Moscow Basin. Bulletin de la Société Géologique de France 169, 35–52.

Bruckschen, P., Oesmann, S., Veizer, J., 1999. Isotope stratigraphy of the European Carboniferous: Proxy signals for ocean chemistry, climate and tectonics. Chemical Geology 161, 127–163.

Buggisch, W., Joachimski, M.M., Sevastopulo, G., Morrow, J.R., 2008. Mississippian $\delta^{13}C_{carb}$ and conodont apatite $\delta^{18}O$ records: Their relation to the late Palaeozoic glaciation. Palaeogeography, Palaeoclimatology, Palaeoecology 268, 273–292.

Burger, K., Hess, J.C., Lippolt, H.J., 1997. Tephrochronologie mit Kaolin-Kohlentonsteinen: Mittel zur Korrelation paralischer und limnischer Ablagerungen des Oberkarbons. Geologische Jahrbucher A 147, 3–39.

Byvsheva, T.V., Owens, B., Teteriuk, V.K., 1979. Stratigraphical palynology of the Tournaisian to Stephanian deposits of the USSR and western Europe. Congrès International de Stratigraphie et de Géologie du Carbonifère, Compte Rendu 8, 209–215.

Byvsheva, T.V., Umnova, N.I., 1993. Zonal scale on spores. In: Makhlina, M.K., Vdovenko, M.V., Alekseev, A.S., Byvsheva, T.V., Donakova, L.M., Zhulitova, V.E., Kononova, L.I., Umnova, N.I., Shik, E.M. (Eds.), Lower Carboniferous in the Mosocow Syncline and Voronezh Anticline. Mosocow. Nauka, pp. 136–141.

Caputo, M.V., Crowell, J.C., 1985. Migration of glacial centers across Gondwana during the Paleozoic era. Geological Society of America Bulletin 96 (8), 1020–1036.

Carter, J.L., 1990. Subdivision of the Lower Carboniferous in North America by means of articulate brachiopod generic ranges. In: Brenkle, P.L., Manger, W.L. (Eds.), Intercontinental Correlation and Division of the Carboniferous System; Contributions from the Carboniferous Subcommission Meeting. Courier Forschungsinstitut Senckenberg, 130, pp. 145–155.

Chernykh, V.V., 2002. Zonal scale of the Gzhelian and Kasimovian Stages based on conodonts of genus Streptognathodus. In: Chuvashov, B.I., Amon, E.A. (Eds.), Stratigraphy and Paleogeography of the Carboniferous of Eurasia. Institute of Geology and Geochemistry, Uralian Branch of the Russian Academy of Sciences. Ekaterinburg, pp. 302–306 [In Russian].

Chernykh, V.V., 2005. Zonal method in biostratigraphy: Zonal conodont scale of the lower Permian in the Urals. Institute of Geology and Geochemistry, Uralian Branch of the Russian Academy of Sciences, Ekaterinburg, pp. 217 [In Russian].

Chernykh, V.V., Chuvashov, B.I., Davydov, V.I., Schmitz, M.D., Snyder, W.S., 2006. Usolka section (southern Urals, Russia): A potential candidate for GSSP to define the base of the Gzhelian Stage in the global chronostratigraphic scale. Geologija 49, 205–217.

Chizhova, V.A., 1977. Ostracods from Devonian-Carboniferous boundary beds of the Russian Platform. VNIGRI 49, 3–255 [In Russian].

Chuvashov, B.I., Amon, E.O., Karidrua, M., Prust, Z.N., 1999. Late Paleozoic radiolarians from the polyfacies formations of Uralian Foreland. Stratigrafiya, Geologicheskaya Korrelyatsiya 7 (1), 41–55 [In Russian and English].

Claouè-Long, J.C., Compston, W., Roberts, J., Fanning, C.M., 1995. Two Carboniferous ages: A comparison of SHRIMP zircon dating with conventional zircon ages and $^{40}Ar/^{39}Ar$ analysis. Society for Sedimentary Geology, Special Publication 54, 3–21.

Clayton, G., Coquel, R., Doubinger, J., Gueinn, K.J., Loboziak, S., Owens, B., Streel, M., 1977. Carboniferous miospores of western Europe. Report of Comission Internationale de Microflore du Paleozoic Working Group on Carboniferous Stratigraphic Palynology. Mededelingen vans Rijks Geologische Dienst 29, 1–11.

Clopine, W.W., 1992. Lower and Middle Pennsylvanian fusulinid biostratigraphy of southern New Mexico and westernmost Texas. New Mexico Bureau of Geology and Mineral Resources, Report 143, p. 68.

Collinson, C.W., 1961. The Kinderhookian Series in the Mississippi Valley. In Guidebook, 26th Regional Field Conference, Kansas Geological Society, northeastern Missouri and west-central Illinois. Missouri Geological Survey and Water Resources, Report of Investigations 27, 100–109.

Compston, W., 2000. Interpretation of SHRIMP and isotope dilution zircon ages for the Palaeozoic time-scale, II: Silurian to Devonian. Mineralogical Magazine 64 (6), 1127–1146.

Conil, R., 1977. The use of foraminifera for the biostratigraphy of the Dinantian in Moravia. In: Holub, V.M., Wagner, R.H. (Eds.), Symposium on Carboniferous Stratigraphy. Geological Survey, Prague, pp. 377–398.

Conil, R., Groessens, E., Laloux, M., Poty, E., Tourneur, F., 1991. Carboniferous guide Foraminifera, corals and conodonts in the Franco-Belgian and Campine Basins: Their potential for widespread correlation. Courier Forshungsinstitut Senckenberg 130, 15–30.

Conkin, J.E., Conkin, B.M., 1974. The paracontinuity and the determination of the Devonian-Mississippian boundary in the Lower Mississippian type area of North America. University of Louisville Studies in Paleontology and Stratigraphy 1, 1–36.

Connolly, W.M., Stanton Jr., R.J., 1992. Interbasinal cyclostratigraphic correlation of Milankovitch band transgressive-regressive cycles: Correlation of Desmoinesian-Missourian strata between southeastern Arizona and the midcontinent of North America. Geology 20, 999–1002.

Conybeare, W.D., Phillips, W., 1822. Outlines of the geology of England and Wales, with an introduction compendium of the general principles of that science, and comparative views of the structure of foreign countries. Part 1. William Phillips, London, p. 470.

Corradini, C., Kaiser, S.I., 2009. Morphotypes in the early *Siphonodella* lineage: Implications for the defintion of the Devonian/Carboniferous boundary. Permophiles (supplement) 53, 13.

Corradini, C., Kaiser, S.I., Perri, M.C., Spaletta, C., 2010. Conodont genus *Protognathus* as a possible tool for recognizing the Devonian/Carboniferous boundary. Third International Paleontological Congress (IPC3), Program with Abstracts S9, 131.

Cramer, B.D., Saltzman, M.R., Day, J., Witzke, B.J., 2008. Lithological expression of global positive carbon isotope excursions in epeiric sea settings: Carbonate production, organic carbon burial, and oceanography during the late Famennian. In: Pratt, B.R., Holmden, C. (Eds.), Dynamics of Epeiric Seas. Geological Association of Canada Special Paper, 48, pp. 1–16.

Crasquin, S., 1985. Zonation par les Ostracodes dans le Mississippien de l'Ouest canadien. Revue de Paléobiologie 4 (1), 45–52.

Dachroth, W., 1976. Gesteinsmagnetische Marken im Perm Mitteluropas. Geologisches Jahrbuch 10, 3–63.

Dan'shin, V.M., 1947. Geology and mineral resources of the region of the Moscow Basin. Transactions of the Moscow Society of Natural Studies, Moscow, p. 308 [In Russian].

Davydov, V.I., 1990. On the clarification of the origin and phylogeny of triticitids and the Mid-Upper Carboniferous boundary. Paleontological Journal 24, 13–25 [In Russian].

Davydov, V.I., 1992. Subdivision and correlation of Upper Carboniferous and Lower Permian deposits in Donets Basin according to fusulinid data. Soviet Geology 5, 53–61 [In Russian].

Davydov, V.I., 1997. Middle-Upper Carboniferous Boundary: The problem of definition and correlation. In: Podemski, M., Dybova-Jachowicz, S., Jureczka, J., Wagner, R. (Eds.), Proceeding of the XIII International Congress on the Carboniferous and Permian, vol. 1. Polish Geological Institute, Warszawa, pp. 113–122.

Davydov, V.I., 2007. Protriticites foraminiferal fauna and its utilization in the Moscovian-Kasomovian boundary definition. In: Wong, T.E. (Ed.), Proceedings of the XVth International Congress on Carboniferous and Permian Stratigraphy. Utrecht, 10–16 August 2003. Royal Netherlands Academy of Arts and Sciences, Amsterdam, pp. 456–466.

Davydov, V.I., 2009. Bashkirian-Moscovian transition in Donets Basin: The key for Tethyan-Boreal correlation. In: Puchkov, V.N. (Ed.), The Carboniferous Type Sections in Russia, Potential and Proposed Stratotypes. Proceedings of the International Conference Ufa - Sibai, 13–18 August, 2009. Institute of Geology, Bashkirian Academy of Science, Ufa, p. 188.

Davydov, V.I., Popov, A.V., 1986. Upper Carboniferous and Lower Permian sections of the southern Urals. In: Chuvashov, B.I., Leven, E.Y., Davydov, V.I. (Eds.), Carboniferous-Permian Boundary Beds of the Urals, Pre-Urals and Central Asia. Nauka Publishing House, Moscow, pp. 29–33 [In Russian].

Davydov, V.I., Khramov, A.N., 1991. Paleomagnetism of Upper Carboniferous and Lower Permian in the Karachatyr region (southern Ferhgana) and the problems of correlation of the Kiama hyperzone. In: Kramov, A.N. (Ed.), Paleomagnetism and Paleogeodynamics of the Territory of USSR. Transactions of VNIGRI, St. Petersburg, pp. 45–53 [In Russian].

Davydov, V., Krainer, K., 1999. Fusulinid assemblages and facies of the Bombaso Fm. and basal Meledis Fm. (Moscovian-Kasimovian) in the central Carnic Alps (Austria/Italy). Facies 40, 157–196.

Davydov, V.I., Khodjanyazova, R., 2009. Moscovian-Kasimovian transition in Donets Basin: Fusulinid taxonomy, biostratigraphy correlation and paleobiogeography. In: Puchkov, V.N. (Ed.), The Carboniferous Type Sections in Russia, Potential and Proposed Stratotypes. Proceedings of the International Conference Ufa - Sibai, 13–18 August, 2009. Institute of Geology, Bashkirian Academy of Science, Ufa, pp. 193–196.

Davydov, V.I., Barskov, I.S., Bogoslovskaya, M.F., Leven, E.Y., 1994. Carboniferous - Permian boundary in the stratotype sections of the southern Urals and its correlation. Stratigraphy and Geological Correlations 2, 32–45.

Davydov, V.I., Glenister, B.F., Spinosa, C., Ritter, S.M., Chernykh, V.V., Wardlaw, B.R., Snyder, W.S., 1998. Proposal of Aidaralash as GSSP for base of the Permian System. Episodes 21, 11–18.

Davydov, V.I., Schiappa, T.A., Snyder, W.S., 2003. Testing the sequence stratigraphy model: Response of fusulinacean fauna to sea level fluctuations (examples from Pennsylvanian and Cisuralian of Pre-Caspian-southern Urals Region). In: Olson, H.C., Leckie, R.M. (Eds.), Micropaleontologic Proxies for Sea-Level Changes and Stratigraphic Discontinuities. SEPM Special Publication 75, pp. 359–375.

Davydov, V.I., Snyder, W.W., Schiappa, T.A., 1999. Moscovian-Kasimovian transition in Nevada and problems of its intercontinental correlation. XIV International Congress on Carboniferous and Permian, Program with Abstracts, Calgary, p. 27.

Davydov, V.I., Wardlaw, B.R., Gradstein, F.M., 2004. The Carboniferous Period. In: Gradstein, F.M., Ogg, J.G., Smith, A.G. (Eds.), A Geologic Time Scale 2004. Cambridge University Press, Cambridge, pp. 222–248.

Davydov, V.I., Schmitz, M.D., Chernykh, V.V., Crowley, J., Henderson, C.M., Korn, D., 2008a. Carboniferous and Permian geologic timescale: State of the art, chronostratigraphic, biostratigraphic and radiometric calibration and integration. 33rd International Geological Congress 2008, Oslo. Abstracts on CD-ROM.

Davydov, V.I., Chernykh, V.V., Chuvashov, B.I.M.D.S., Snyder, W.S., 2008b. Faunal assemblage and correlation of Kasimovian-Gzhelian transition at Usolka section, southern Urals, Russia (a potential candidate for GSSP to define base of Gzhelian Stage). Stratigraphy 5, 113–135.

Davydov, V.I., Crowley, J.L., Schmitz, M.D., Poletaev, V.I., 2010. High-precision U-Pb zircon age calibration of the global Carboniferous time scale and Milankovitch-band cyclicity in the Donets Basin, eastern Ukraine. Geochemistry, Geophysics, Geosystems 11 Q0AA04. doi: 10.1029/2009GC002736.

Davydov, V.I., Schmitz, M., Korn, D., 2011. The Hangenberg Event was abrupt and short at the global scale: The quantitative intergration and intercalibration of biotic and geochronologic data within the Devonian-Carboniferous transition. Geological Society of America, Abstracts with Programs vol. 43 (5), 128.

Delepine, G., 1911. Recherches sur le Calcaire carbonifère de la Belgique. Travaux des Facultés Catholiques, University of Lille, Mémoire. 8, p. 419.

Denison, R.E., Koepnick, R.B., Burke, W.H., Hetherington, E.A., Fletcher, A., 1994. Construction of the Mississippian, Pennsylvanian and Permian seawater $^{87}Sr/^{86}Sr$ curve. Chemical Geology 112, 145–167.

Devuyst, F.-X., Hance, L., Hou, H., Wu, X., Tian, S., Coen, M., Sevastopulo, G., 2003. A proposed global stratotype section and point for the base of the Visean Stage (Carboniferous): The Pengchong Section, Guangzi, south China. Episodes 26, 105–115.

Devuyst, F.-X., Hance, L., Poty, E., 2006. Moliniacian. Geologica Belgica 9 (1–2), 123–131.

Dickinson, W.R., Soreghan, G.S., Giles, K.A., 1994. Glacio-eustatic origin of Permo-Carboniferous stratigraphic cycles: Evidence from the southern Cordilleran foreland region. In: Dennison, J.M., Ettensohn, F.R. (Eds.), Concepts in Sedimentology and Paleontology, vol. 4. Society for Sedimentary Geology, Tulsa, pp. 25–34.

DiVenere, V.J., Opdyke, N.D., 1991a. Magnetic polarity stratigraphy and Carboniferous paleopole position from the Joggins Section, Cumberland Basin, Nova Scotia. Journal of Geophysical Research 96, 4051–4064.

DiVenere, V.J., Opdyke, N.D., 1991b. Magnetic polarity stratigraphy in the uppermost Mississippian Mauch Chunk Formation, Pottsville, Pennsylvania Series/Source. Geology 19, 127–130.

Dumont, A., 1832. Mémoire sur la constitution géologique de la Province de Liège. Académie royale de Belgique, Bruxelles, p. 372.

Dunn, D.L., 1970. Conodont zonation near the Mississippian-Pennsylvanian boundary in western United States. Geological Society of America Bulletin 81 (10), 2959–2974.

Dusar, M., 2006. Namurian. In: Dejonghe, L. (Ed.), Current Status of Chronostratigraphic Units Named from Belgium and Adjacent Areas. Geologica Belgica 9 (1–2), 163–175.

Ehrenberg, S.N., Nielsen, E.B., Svana, T.A., Stemmerik, L., 1998. Depositional evolution of the Finnmark carbonate platform, Barents Sea: Results from wells 7128/6–1 and 7128/4–1. Norsk Geologisk Tidsskrift 78, 185–224.

Fielding, C.R., Frank, T.D., Birgenheier, L.P., Rygel, M.C., Jones, A.T., Roberts, J., 2008a. Stratigraphic imprint of the late Palaeozoic ice age in eastern Australia: A record of alternating glacial and nonglacial climate regime. Journal of the Geological Society 165, 129–140.

Fielding, C.R., Frank, T.D., Isbell, J.L., 2008b. The late Paleozoic ice age: A review of current understanding and synthesis of global climate patterns. Geological Society of America Special Paper 441, 343–354.

Flajs, G., Feist, R., 1988. Index conodonts, trilobites and environment of the Devonian/Carboniferous boundary beds La Serre (Montagne Noire, France). Courier Forschungsinstitut Senckenberg 100, 53–101.

Fohrer, B., Nemyrovska, T.I., Samankassou, E., Ueno, K., 2007. The Pennsylvanian (Moscovian) Izvarino Section, Donets Basin, Ukraine: a multidisciplinary study on microfacies, biostratigraphy (conodonts, foraminifers, and ostracodes), and paleoecology. Journal of Paleontology 81 (Suppl. 69), 1–85.

Gastaldo, R.A., Purkyňová, E., Šimůnek, Z., Schmitz, M.D., 2009. Ecological persistence in the Late Mississippian (Serpukhovian - Namurian A) megafloral record of the Upper Silesian Basin, Czech Republic. Palaios 24, 336–350.

George, T.N., Johnson, G.A.L., Mitchell, M., Prentice, J.E., Ramsbottom, W.H.C., Sevastopulo, G.D., Wilson, R.B., 1976. A correlation of Dinantian rocks in the British Isles. Geological Society of London, Special Report 7, 1–87.

Gibshman, N.B., 2003. Characteristics of foraminifers from the Serpukhovian stratotype, Zabor'ye Quarry; Moscow region. Stratigrafiya, Geologicheskaya Korrelyatsiya 11, 39–63 [In Russian].

Gorak, S.V., 1977. Carboniferous ostracods of the Great Donets Basin: Paleoecology, paleozoogeography and biostratigraphy. Naukova Dumka, Kiev, p. 148.

Goreva, N.V., Alekseev, A.S., 2006. New conodont species from the Kasimovian Stage (Upper Carboniferous) of Moscow and Moscow Basin. Paleontological Journal 40, 193–197.

Goreva, N.V., Alekseev, A.S., 2007. Correlation of upper Carboniferous (Pennsylvanian) deposits in the Moscow Syneclise and Donets Basin by conodonts. In: Gozhik, P.F. (Ed.), Paleontological Investigations in Ukraine: History, Current Status and Perspectives. Naukova Dumka, Kiev, pp. 110–114 [In Russian].

Goreva, N.V., Alekseev, A.S., Isakova, T.N., Kossovaya, O.L., 2009. Biostratigraphical analysis of the Moscovian-Kasimovian transition at the neostratotype of Kasimovian Stage (Afanasievo section, Moscow Basin, Russia). Palaeoworld 18, 102–113.

Gourmelon, F., 1987. Les Radiolaires tournaisiens des nodules phosphates de la Montagne Noire et des Pyrenees centrales; systematique; biostratigraphie, paleobiogeographie. Biostratigraphie du Paleozoique 6, 172.

Grossman, E.L., Mii, H.-S., Zhang, C., Yancey, T.E., 1996. Chemical variations in Pennsylvanian brachiopod shells - effect of diagenesis, taxonomy, microstructure, and paleoenvironments. Journal of Sedimentary Research 66, 1011–1022.

Grossman, E.L., Bruckschen, P., Mii, H.-S., Chuvashov, B.I., Yancey, T.E., Veizer, J., 2002. Carboniferous paleoclimate and global changes: Isotopic evidence from the Russian Platform. In: Chuvashov, B.I., Amon, E.O. (Eds.), Carboniferous Stratigraphy and Paleogeography in Eurasia. Uralian Branch of the Russian Academy of Sciences, Ekaterinburg, pp. 61–71.

Grossman, E.L., Yancey, T.E., Jones, T.E., Bruckschen, P., Chuvashov, B., Mazzullo, S.J., Mii, H.-S., 2008. Glaciation, aridification, and carbon sequestration in the Permo-Carboniferous: The isotopic record from low latitudes. Palaeogeography, Palaeoclimatology, Palaeoecology 268, 222–233.

Groves, J.R., 1986. Foraminiferal characterization of the Morrowan-Atokan (lower Middle Pennsylvanian) boundary. Geological Society of America Bulletin 97, 346–353.

Groves, J.R., Task Group, 2007. Report of the Task Group to establish a GSSP close to the existing Bashkirian–Moscovian boundary. Newsletter on Carboniferous Stratigraphy 25, 6–7.

Groves, J.R., Task Group, 2008. Report of the Task Group to establish a GSSP close to the existing Bashkirian-Moscovian boundary. Newsletter on Carboniferous Stratigraphy 26, 10–11.

Gulbranson, E.L., Montanez, I.P., Schmitz, M.D., Limarino, C.O., Isbell, J.L., Marenssi, S.A., Crowley, J.L., 2010. High-resolution U-Pb calibration of Carboniferous glacigenic deposits, Rio Blanco and Paganzo basins, northwest Argentina. Geological Society of America Bulletin 122, 1480–1498.

Hallock, P., Premoli Silva, I., Boersma, A., 1991. Similarities between planktonic and larger foraminiferal evolutionary trends through Paleogene paleoceanographic changes. Paleogeography. Paleoclimatology and Paleogeography 83, 49–64.

Hance, L., Brenckle, P.L., Coen, M., Hou, H., Liao, Z., Muchez, P., Paproth, E., Nicholas, T.P., Riley, J., Roberts, J., Wu, X., 1997. The search for a new Tournaisian-Visean boundary stratotype. Episodes 20, 176–180.

Hance, L., Poty, E., Devuyst, F.-X., 2006a. Tournaisian. Geologica Belgica 9, 47–53.

Hance, L., Poty, E., Devuyst, F.-X., 2006b. Ivorian. Geologica Belgica 9, 117–122.

Hance, L., Poty, E., Devuyst, F.-X., 2006c. Visean. Geologica Belgica 9, 55–62.

Haq, B.U., Schutter, S.R., 2008. A chronology of Paleozoic sea-level changes. Science 322 (5898), 64–68.

Harland, W.B., Armstrong, R.L., Cox, A.V., Craig, L.A., Smith, A.G., Smith, D.G., 1990. A Geologic Time Scale 1989. Cambridge University Press, Cambridge, p. 263.

Heckel, P.H., 1986. Sea-level curve for Pennsylvanian eustatic marine transgressive-regressive depositional cycles along Midcontinent outcrop belt, North America. Geology 14, 330–334.

Heckel, P.H., 1994. Evaluation of evidence for glacio-eustatic control over marine Pennsylvanian cyclothems in North America and consideration of possible tectonic events. Concepts in Sedimentology and Paleontology 4, 65–87.

Heckel, P.H., 1999. Overview of Pennsylvanian (Upper Carboniferous) stratigraphy in Midcontinent region of North America. In: Heckel, P.H. (Ed.), Middle and Upper Pennsylvanian (Upper Carboniferous) Cyclothem Succession in Midcontinent Basin. Kansas Geological Survey Open-file Report, 99–27, pp. 68–102.

Heckel, P.H., 2002. Overview of Pennsylvanian cyclothems in Midcontinent North America and brief summary of those elsewhere in the world. In: Hills, L.V., Charles, M., Bamber, E.W. (Eds.), Carboniferous and Permian of the World. XIV ICCP Proceedings, Memoir, 19, pp. 79–98.

Heckel, P.H., Alekseev, A.S., Nemirovskaya, T.I., 1998. Preliminary conodont correlations of late Middle to early Upper Pennsylvanian rocks between North America and eastern Europe. Newsletters on Carboniferous Stratigraphy 16, 8–12.

Heckel, P.H., Alekseev, A.S., Barrick, J.E., Boardman II, D.R., Goreva, N.V., Nemyrovska, T.I., Ueno, K., Villa, E., Work, D.M., 2007. Cyclothem ["digital"] correlation and biostratigraphy across the global Moscovian-Kasimovian-Gzhelian stage boundary interval (Middle-Upper Pennsylvanian) in North America and eastern Europe. Geology 35, 607–610.

Heckel, P.H., Alekseev, A.S., Barrick, J.E., Boardman II, D.R., Goreva, N.V., Isakova, T.N., Nemyrovska, T.I., Ueno, K., Villa, E., Work, D.M., 2008. Choice of conodont *Idiognathodus simulator* (sensu stricto) as the event marker for the base of the global Gzhelian Stage (Upper Pennsylvanian Series, Carboniferous System). Episodes 31, 319–325.

Herbert, T., 1999. Toward a composite orbital chronology for the Late Cretaceous and Early Palaeocene GPTS. Philosophical Transactions of the Royal Society, Series A 357, 1891–1905.

Hess, J.C., Lippolt, H.J., 1986. $^{40}Ar/^{39}Ar$ ages of tonstein and tuff sanidines: New calibration points for the improvement of the Upper Carboniferous time scale. Chemical Geology 59, 143–154.

Hess, J.C., Lippolt, H.J., Burger, K., 1999. High-precision $^{40}Ar/^{39}Ar$ spectrum dating on sanidine from the Donets Basin, Ukraine: Evidence for correlation problems in the the Upper Carboniferous. Journal of the Geological Society 156, 527–533.

Higgins, A.C., 1975. Conodont zonation of the late Visean-early Westphalian strata of the south and central Pennines of northern England. Geological Survey of Great Britain Bulletin 53, 1–90.

Higgs, K.T., McPhilemy, B., Keegan, J.B., Clayton, G., 1988. New data on palynological boundaries within the Irish Dinantian. Review of Palaeobotany and Palynology 56 (1–2), 61–68.

Hinnov, L., 2000. New perspectives on orbitally forced stratigraphy. Annual Review of Earth and Planetary Sciences 28, 419–475.

Hinnov, L.A., 2004. Earth's orbital parameters and cyclostratigraphy. In: Gradstein, F.M., Ogg, J.G., Smith, A.G. (Eds.), A Geologic Time Scale 2004. Cambridge University Press, Cambridge, pp. 55–62.

Hohenegger, J., 2004. Depth coenoclines and environmental considerations of western Pacific larger Foraminifera. Journal of Foraminiferal Research 34, 9–33.

Holdsworth, B.K., Murchey, B.L., 1988. Paleozoic radiolarian biostratigraphy of the northern Brooks Range, Alaska. Geology and exploration of the National Petroleum Reserve in Alaska, 1974 to 1982. US Geological Survey Professional Paper 1399, 777–797.

Hounslow, M.W., Davydov, V.I., Klootwijk, C.T., Turner, P., 2004. Magnetostratigraphy of the Carboniferous: A review and future prospects. Newsletter on Carboniferous Stratigraphy 22, 35–41.

House, M.R., 1993. Earliest Carboniferous goniatite recovery after the Hangenberg Event. Annales de la Société géologique de Belgique 115, 559–579.

Inosova, K.I., Kruzina, A.K., Shvartsman, E.G., 1976. Atlas of the Miospores and Pollen of the Upper Carboniferous and Lower Permian of the Donets Basin. Nedra, Moscow, pp. 154 [In Russian].

Irving, E., Parry, L.G., 1963. The magnetism of some Permian rocks from New South Wales. Geophysical Journal of the Royal Astronomical Society 7 (4), 395–411.

Isakova, T.N., Ueno, K., 2007. About the lectotype *Rauserites rossicus* Schellwien 1908 (Foraminifera) from the Gzhelian Stage of the Donets and Moscow Basins. Zbirnik Naukovikh Prats Institutu Geologichnikh Nauk NAN Ukraini: Biostratigraphy and Paleontology. Naukova Dumka, Kiev, pp. 105–109.

Isbell, J.L., Miller, M.F., Wolfe, K.L., Lenaker, P.A., 2003. Timing of late Paleozoic glaciation in Gondwana: Was glaciation responsible for the development of northern hemisphere cyclothems? In: Chan, M.A., Archer, A.A. (Eds.), Extreme Epositional Environments: Mega End Members in Geologic Time, pp. 5–24. Geological Society of America Special Paper 370.

Ivanov, A.P., 1926. Middle-Upper Carboniferous deposits of the Moscow province. Bulletin of the Moscow Society of Natural Studies, Geological. Series 4, 133–180 [In Russian].

Ivanova, E.A., Khvorova, I.V., 1955. Stratigraphy of the Middle and Upper Carboniferous of the western part of the Moscow syneclise. Transactions of the Paleontological Institute of the Academy of Sciences of the USSR 53, 3–279 [In Russian].

Izart, A., Briand, C., Vaslet, D., Vachard, D., Coquel, R., Maslo, A., 1996. Stratigraphy and sequence stratigraphy of the Moscovian in the Donets Basin. Tectonophysics 268, 189–209.

Izart, A., Sachsenhofer, R.F., Privalov, V.A., Elie, M., Panova, E.A., Antsiferov, V.A., Alsaab, D., Rainer, T., Sotirov, A., Zdravkov, A., Zhykalyak, M.V., 2006. Stratigraphic distribution of macerals and biomarkers in the Donets Basin: Implications for paleoecology, paleoclimatology and eustacy. International Journal of Coal Geology 66, 69–107.

Ji, Q., Wang, Z., Sheng, H., Hou, J., Feng, R., Wei, J., Wang, S., Wang, H., Xiang, L., Fu, G., 1989. The Dapoushang Section: An excellent Section for the Devonian-Carboniferous Boundary Stratotype in China. Science Press, Beijing, p. 184.

Joachimski, M.M., Breisig, S., Buggisch, W., Talent, J.A., Mawson, R., Gereke, M., Morrow, J.R., Day, J., Weddige, K., 2009. Devonian climate and reef evolution: Insights from oxygen isotopes in apatite. Earth and Planetary Science Letters 284, 599–609.

Jones, C.E., 1995. Timescale 5: Carboniferous. Australian Geological Survey Organisation Record 1995/34, 3–45.

Kalvoda, J., 2002. Late Devonian-Early Carboniferous Foraminiferal Fauna: Zonations, Evolutionary events, paleobiogeography and tectonic implications. Masaryk University, Brno. Czech Republic, 39.

Kaiser, S.I., 2009. The Devonian/Carboniferous boundary stratotype section (La Serre, France) revisited. Newsletters on Stratigraphy 43, 195–205.

Kaiser, S.I., Steuber, T., Becker, R.T., Joachimski, M.M., 2006. Geochemical evidence for major environmental change at the Devonian-Carboniferous boundary in the Carnic Alps and the Rhenish Massif. Palaeogeography, Palaeoclimatology, Palaeoecology 240, 146–160.

Kaiser, S.I., Becker, R.T., Steuber, T., 2008. Environmental change during the Late Famennian and Early Tournaisian (Late Devonian-Early Carboniferous): Implications from stable isotopes and conodont biofacies in southern Europe. Geological Journal 40, 241–260.

Keyes, C.R., 1893. Geological Formations of Iowa. Iowa Geological Survey Annual Report 1, 11–144.

Khramov, A.N., 1963. Paleomagnetism of Paleozoic. VNIGRI, St. Petersburg, p. 283 [In Russian].

Khramov, A.N., 1987. Paleomagnetology. Springer-Verlag, Berlin, p. 308.

Khramov, A.N., 2000. The General magnetostratigraphic scale of the Phanerozoic. In: Zhamoida, A.I. (Ed.), Supplements to the Stratigraphic Code of Russia. Transactions of VSEGEI, St. Petersburg, pp. 24–45 [In Russian].

Khramov, A.N., Rodionov, V.P., 1981. The geomagnetic field during Palaeozoic time. In: McElhinny, M.W., Khramov, A.N., Ozima, M., Valencio, D.A. (Eds.), Global Reconstruction and the Geomagnetic Field during the Palaeozoic. Advances in Earth and Planetary Science, 10, pp. 99–115.

Khramov, A.N., Davydov, V.I., 1984. Paleomagnetism of Upper Carboniferous and Lower Permian in the South of the U.S.S.R. and the problems of structure of the Kiama hyperzone. Transactions of VNIGRI, St. Petersburg. 55–73. [In Russian].

Khramov, A.N., Davydov, V.I., 1993. Results of Paleomagnetic investigations. Permian System, Guides to Geological Excursions in the Uralian Type Localities. Occasional Publications of the Earth Sciences and Resources Institute 10, 34–42.

Khramov, A.N., Goncharov, G.I., Komisssarova, R.A., Osipova, E.P., Pogarskaya, I.A., Rodionov, V., Stautsilais, I.P., Smirnov, L.S., Forsh, N.N., 1974. Paleozoic Paleomagnetism. Transactions of VNIGRI, St. Petersburg, p. 238 [In Russian].

Kireeva, G.D., 1964. Taxonomical analyses of the wall structure of some fusulinids within the Moscovian-Kasimovian transition. Problems of micropaleontology. Academy of Science of the USSR 8, 53–56 [In Russian].

Kirwan, R., 1799. Additional Observations on the Proportion of Real Acid in the Three Ancient Known Mineral Acids and on the Ingredients in Various Neutral Salts and other Compaunds. Georg Bonham, Dublin. p. 137.

Kolesov, E.V., 1984. Paleomagnetic stratigraphy of the Devonian-Carboniferous boundary beds in the Soviet North-East and in the Franco-Belgian Basin. Annales de la Société Géologique de Belgique 107, 135–136.

Kolesov, E.V., 2001. The Paleomagnetism of the Upper Paleozoic volcanogenic-sedimentary sequences of Prikolym uplift. In: Simakov, K.K. (Ed.), Paleomagnetic Studies of Geological Rocks on the North-Eastern of Russia. Far East Branch of the Russian Academy of Sciences, Magadan, p. 32–44 [In Russian].

Korn, D., 1992. Ammonoideen vom Devon/Karbon-Grenzprofil an der Grünen Schneid (Karnische Alpen, Österreich). Jahrbuch der Geologischen Bundesanstalt 135, 7–19.

Korn, D., 1993. The ammonoid faunal change near the Devonian-Carboniferous boundary. Annales de la Société Géologique de Belgique 115, 581–593.

Korn, D., 1996. Revision of the Late Visean goniatite stratigraphy. Annales de la Société Géologique de Belgique 117 (2), 205–212.

Korn, D., 2006. Ammonoideen. Schriftenreihe der Deutschen Gesellschaft für Geowissenschaften 41, 147–170.

Kossovaya, O.L., 1996. Correlation of uppermost Carboniferous and lowermost Permian rugose coral zones from the Urals to Western North America. Palaios 11 (1), 71–82.

Kozitskaya, R.I., Kosenko, Z.A., Lipnyagov, O.M., Nemirovskaya, T.I., 1978. Konodonty karbona Donetskogo Basseyna. Conodonts from the Carboniferous of the Donets Basin. Naukova Dumka, Kiev [In Russian] 1–130.

Kuiper, K.F., Deino, A., Hilgen, F.J., Krijgsman, W., Renne, P.R., Wijbrans, J.R., 2008. Synchronizing rock clocks of Earth history. Science 320 (5875), 500–504.

Kulagina, E.I., Pazukhin, V.N., 2002. The Bashkirian stage as a global stratigraphic scale member of the Carboniferous. Memoir, Canadian Society of Petroleum Geologists 19, 776–779.

Kulagina, E.I., Rumjantseva, Z.S., Pazukhin, B.N., Kochetova, N.N., 1992. The mid-Carboniferous boundary in Southern Urals and Middle Tian-Shan. Nauka, Moscow, p. 112 [In Russian].

Kulagina, E.I., Pazukhin, V.N., Kochetkova, N.M., Synytsina, Z.A., Kochetova, N.N., 2001. The Type and the Key Sections of Bashkirian Stage (Carboniferous) in the Southern Urals. Gilem, Ufa, p. 138 [In Russian].

Kulagina, E.I., Nikolaeva, S.V., Akhmetshina, L.Z., Zainakaeva, G.F., Konovalova, V.A., Korobkov, V.F., Kan, A.N., 2006. The Dombar Limestone in the South Urals and the Visean-Serpukhovian boundary. Newsletter on Carboniferous Stratigraphy 24, 9–11.

Kulagina, E.I., Pazukhin, V.N., Davydov, V.I., 2009. Pennsylvanian biostratigraphy of the Basu River section with emphasis on the Bashkirian-Moscovian transition. In: Puchkov, V.N. (Ed.), The Carboniferous Type Sections in Russia, Potential and Proposed Stratotypes. Proceedings of the International Conference Ufa - Sibai, 13–18 August, 2009. Institute of Geology, Bashkirian Academy of Science, Ufa, pp. 42–63.

Kullmann, J., Nikolaeva, S.V., 2002. Mid-Carboniferous boundary and the global lower Bashkirian ammonoid biostratigraphy. Memoir, Canadian Society of Petroleum Geologists 19, 780–795.

Kullmann, J., Korn, D., Weyer, D., 1990. Ammonoid zonation of the Lower Carboniferous subsystem. Courier Forschungsinstitut Senckenberg 130, 127–131.

Kullmann, J., Perret Mirouse, M.-F., Delvolve, J.-J., 2008. Goniatites et conodontes du Viséen/Serpukhovien dans les Pyrénées centrales et occidentales, France. Geobios 41, 635–656.

Lambert, L.L., Heckel, P.H., 1990. The Atokan/Desmoinesian boundary in North America: Preliminary considerations for selecting a boundary horizon. In: Brenkle, P.L., Manger, W.L. (Eds.), Intercontinental Division and Correlation of the Carboniferous System. Courier Forschungsinstitut Senckenberg, 130, pp. 307–318.

Lambert, L.L., Barrick, J.E., Heckel, P.H., 2002. Lower and Middle Pennsylvanian conodont zonation for Midcontinent North America. Geological Society of America Abstracts with Programs 34, 27.

Lane, H.R., 1974. Mississippian of southeastern New Mexico and West Texas: A wedge-on-wedge relation. American Association of Petroleum Geologists Bulletin 58, 269–282.

Lane, H.R., 1977. Morrowan (Early Pennsylvanian) conodonts of northwestern Arkansas and northeastern Oklahoma. In: Sutherland, P.K., Manger, W.L. (Eds.), Upper Chesterian-Morrowan stratigraphy and the Mississippian-Pennsylvanian boundary in northeastern Oklahoma and northwestern Arkansas. Oklahoma Geological Survey Guidebook, 18, pp. 177–180.

Lane, H.R., Brenckle, P.L., 2005. Type Mississippian subdivisions and biostratigraphic succession. Guidebook Series Illinois State Geological Survey, Report 34, 76–98.

Lane, H.R., Merrill, G.K., Straka II, J.J., Webster, G.D., 1971. North American Pennsylvanian conodont biostratigraphy. Geological Society of North America, Memoir 127, 395–414.

Lane, H.R., Sandberg, C.A., Ziegler, W., 1980. Taxonomy and phylogeny of some Lower Carboniferous conodonts and preliminary standard post-*Siphonodella* zonation. Geologica et Palaeontologica 14, 117–164.

Lane, H.R., Brenckle, P.L., Baesemann, J.F., Richards, B., 1999. The IUGS boundary in the middle of the Carboniferous; Arrow Canyon, Nevada, USA. Episodes 22, 272–283.

Laskar, J., 1990. The chaotic motion of the solar system: A numerical estimate of the chaotic zones. Icarus 88, 266–291.

Laskar, J., 1999. The limits of Earth orbital calculations for geological time-scale use. Philosophical Transactions of the Royal Society, Series A 357, 1735–1759.

Legrand-Blain, M., 1990. Brachiopods as potential boundary-defining organisms in the Lower Carboniferous of Western Europe: Recent data and productid distribution. In: Brenkle, P.L., Manger, W.L. (Eds.),

Intercontinental Correlation and Division of the Carboniferous System: Contributions from the Carboniferous Subcommission Meeting. Courier Forschungsinstitut Senckenberg, 130, pp. 157–171.

Leven, E.J., Davydov, V.I., 2001. Stratigraphy and fusulinids of the Kasimovian and lower Gzhelian (Upper Carboniferous) in the southwestern Darvaz (Pamir). Rivista Italiana di Paleontologia e Stratigrafia 107, 3–45.

Lipina, O.A., Reitlinger, E.A., 1970. Stratigraphie zonale et paléozoogéographie du carbonifére inférieur d'aprés les foraminiféres. Congrès International de Stratigraphie et de Géologie du Carbonifère, Compte Rendu 6, 1101–1112.

Lourens, L., Hilgen, F.J., Shackleton, N.J., Laskar, J., Wilson, D., 2004. The Neogene Period. In: Gradstein, F.M., Ogg, J.G., Smith, A.G. (Eds.), A Geologic Time Scale 2004. Cambridge University Press, Cambridge, pp. 409–440.

Makhlina, M.K., Kulikova, A.M., Nikitina, T.A., 1979. Stratigraphy, biostratigraphy and paleogeography of Upper Carboniferous of the Moscow Syneclise. In: Makhlina, M.K., Shik, C.M. (Eds.), Stratigraphy, Paleontology and Paleogeography of the Carboniferous of the Moscow Syneclise. Geological Foundation, Ministry of Geology of Russian Federation, Moscow, pp. 25–69 [In Russian].

Makhlina, M.K., Vdovenko, M.V., Alekseev, A.S., Byvsheva, T.V., Donakova, L.M., Zhulitova, V.E., Kononova, L.I., Umnova, N.I., Shik, E.M., 1993. Lower Carboniferous of the Moscow Syneclise and Voronezh Anteclise. Nauka, Moscow, p. 219 [In Russian].

Makhlina, M.K., Alekseyev, A.S., Goreva, N.V., Isakova, T.N., Drutskoy, S.N., 2001. Stratigraphy. In: Alekseev, A.S., Shik, S.M. (Eds.), Middle Carboniferous of the Moscow Syneclise, vol. 1. Rossiyskaya Akademiya Nauk, Paleontologicheskiy Institut, Moscow p. 244 [In Russian].

Mamet, B., Skipp, B., 1970. Lower Carboniferous calcareous foraminifera: Preliminary zonation and stratigraphic implications for the Mississippian of North America. Congrès International de Stratigraphie et de Géologie du Carbonifère, Compte Rendu 6, 1129–1146.

Martin, E.E., Macdougall, J.D., 1995. Sr and Nd isotopes at the Permian/Triassic boundary: A record of climate change. Chemical Geology 125, 73–99.

Meek, F.B., Worthen, A.H., 1861. Notes to the Paper of the paper of Messrs. Meek and Worthen on the Age of the Goniutite Limestone. American Journal of Science and Arts 32, 288.

Menning, M., Weyer, D., Drozdzewski, G., van Amerom, H.W.J., Wendt, I., 2000. A Carboniferous Time Scale 2000. Discussion and use of geological parameters as time indicators from Central and Western Europe. Geologische Jahresberichte A156, 3–44.

Menning, M., Alekseev, A., Chuvashov, B., Davydov, V., Devuyst, F., Forke, H., Grunt, T., Hance, L., Heckel, P., Izokh, N., et al., 2006. Global time scale and regional stratigraphic reference scales of Central and West Europe, East Europe, Tethys, South China, and North America as used in the Devonian-Carboniferous-Permian Correlation Chart 2003 (DCP 2003). Palaeogeography, Palaeoclimatology, Palaeoecology 240, 318–372.

Meyen, S.V., Afanasieva, G.A., Betekhtina, O.A., Durante, M.V., Ganelin, V.G., Gorelova, S.G., Graizer, M.I., Kotlyar, G.V., Maximova, S.V., Tschernjak, G.E., Yuzvitsky, A.Z., 1996. Angara and surrounding marine basins. In: Martinez Diaz, C., Wagner, R.H., Winkler Prins, C.F., Granados, L.F. (Eds.), The Carboniferous of the World. The former USSR, Mongolia, Middle Eastern Platform, Afghanistan, & Iran, vol. 3. Instituto Geológico y Minero de España, Madrid, pp. 180–237.

Mii, H.-S., Grossman, E.L., Yancey, T.E., 1999. Carboniferous isotope stratigraphies of North America: Implications for Carboniferous paleoceanography and Mississippian glaciation. Geological Society of America Bulletin 111, 960–973.

Mii, H.-S., Grossman, E.L., Yancey, T.E., Chuvashov, B., Egorov, A., 2001. Isotopic records of brachiopod shells from the Russian Platform: Evidence for the onset of Mid-Carboniferous glaciation. Chemical Geology 175, 133–147.

Min, K., Mundil, R., Renne, P.R., Ludwig, K.R., 2000. A test for systematic errors in $^{40}Ar/^{39}Ar$ geochronology through comparison with U/Pb analysis of a 1.1-Ga rhyolite. Geochimica et Cosmochimica Acta 64, 72–98.

Möeller, V., 1878. Spirally coiled foraminifers in the Carboniferous limestone of Russia. Materials on the Geology of Russia 8, 1–129.

Möeller, V., 1880. Ueber einige foraminiferenführende Gesteine Persiens. Jahrbuch Österreich Geologische Reichanstalt 30, 573–586.

Montanez, I.P., Tabor, N.J., Niemeier, D., DiMichele, W.A., Frank, T.D., Fielding, C.R., Isbell, J.L., Birgenheier, L.P., Rygel, M.C., 2007. CO_2 - forced climate and vegetation instability during late Paleozoic deglaciation. Science 315, 87–91.

Moore, R.C., 1932. Reclassification of the Pennsylvanian System in the northern midcontinent region. Guidebook. 6th Annual Field Conference, Kansas Geological Society, pp. 79–98.

Moore, R.C., 1944. Correlation of Pennsylvanian formations of North America. Geological Society of America Bulletin 55, 657–706.

Moore, R.C., 1948. Classification of Pennsylvanian rocks in Iowa, Kansas, Missouri, Nebraska, and northern Oklahoma. Bulletin of the American Association of Petroleum Geologists 32, 2011–2040.

Moore, R.C., Thompson, M.L., 1949. Main divisions of Pennsylvanian period and system. Bulletin of the American Association of Petroleum Geologists 33, 275–302.

Mundil, R., Ludwig, K.R., Metcalfe, I., Renne, P.R., 2004. Age and timing of the end Permian mass extinctions: U/Pb geochronology on closed-system zircons. Science 305, 1760–1763.

Munier-Chalmas, E., Lapparent, d.A., 1893. Note sur la nomenclature des terrains sédimentaires. Bulletin de la Societe Geologique de France 3, 479–480.

Murchison, R., 1845. Geology of Russia in Europe and the Ural Mountains. John Murray, London, p. 700.

Murray, J.W., 2006. Ecology and Applications of Benthic Foraminifera. Cambridge University Press, Cambridge.

Nazarov, B.B., 1988. Radiolaria of the Paleozoic. In: Zhamoida, A.I. (Ed.), Practical Guidebook in Microfauna of the USSR. Nedra, Leningrad p. 232 [In Russian].

Nazarov, B.B., Ormiston, A.R., 1985. Radiolaria from the late Paleozoic of the Southern Urals, USSR and West Texas, USA. Micropaleontology 31 (1), 1–54.

Nemyrovska, T.I., 1999. Bashkirian conodonts of the Donets Basin, Ukraine. Scripta Geologica 119, 115.

Nemyrovska, T.I., Perret-Mirouse, M.-F., Alekseev, A., 1999. On Moscovian (Late Carboniferous) conodonts of the Donets Basin, Ukraine. Neues Jahrbuch fuer Geologie und Palaeontologie. Abhandlungen 214, 169–194.

Nemyrovska, T.I., 2005. Late Visean/early Serpukhovian conodont succession from the Triollo section, Palencia (Cantabrian Mountains, Spain). Scripta Geologica 129, 13–89.

Nemirovska, T.I., Alekseev, A.S., 1994. The Bashkirian conodonts of the Askyn Section, Bashkirian mountains, Russia. Bulletin de la Société Belge de Géologie 103, 109–133.

Nestell, M.K., Nestell, G.P., Wardlaw, B.R., Sweatt, M.J., 2006. Integrated biostratigraphy of foraminifers, radiolarians and conodonts in shallow and deep water Middle Permian (Capitanian) deposits of the "Rader slide", Guadalupe Mountains, West Texas. Stratigraphy 3, 161–194.

Nicklen, B.L., 2003. Middle Guadalupian (Permian) Bentonite Beds, Manzanita Member, Cherry Canyon Formation, West Texas: Stratigraphic and Tectonomagmatic Applications. M.Sc Thesis, University of Cincinnati, Cincinnati, p. 74.

Nigmadganov, I.M., Nemirovskaya, T.I., 1992. Mid-Carboniferous boundary conodonts from the Gissar Ridge, South Tienshan, Middle Asia. Courier Forschungsintitut Senckenberg 154, 253–275.

Nikitin, S.N., 1890. Carboniferous deposits of Moscow Basin and artesian water around Moscow. Transactions of Geological Committee 5, 1–182 [In Russian].

Nikolaeva, S.V., Akhmetshina, L.Z., Konovalova, V.A., Korobkov, V.F., Zainakaeva, G.F., 2009a. The Carboniferous carbonates of the Dombar Hills (western Kazakhstan) and the problem of the Viséan–Serpukhovian boundary. Palaeoworld 18, 80–93.

Nikolaeva, S.V., Kulagina, E.I., Pazukhin, V.N., Kochetova, N.N., Konovalova, V.A., 2009b. Paleontology and microfacies of the Serpukhovian in the Verkhnyaya Kardailovka Section, South Urals, Russia: Potential candidate for the GSSP for the Visean-Serpukhovian boundary. Newsletters on Stratigraphy 43, 165–193.

Olsen, P., Kent, D., 1999. Long-period Milankovitch cycles from the Late Triassic and Early Jurassic of eastern North America and their implications for the calibration of the Early Mesozoic time-scale and the long-term behaviour of the planets. Philosophical Transactions of the Royal Society, Series A 357, 1761–1786.

Opdyke, N.D., Channell, J.E.T., 1996. Magnetic Stratigraphy. Academic Press, New York, p. 346.

Opdyke, N.D., Khramov, A.N., Gurevitch, E., Iosifidi, A.G., Makarov, I.A., 1993. A paleomagnetic study of the middle Carboniferous of the Donets Basin, Ukraine (abstract). EOS Transactions American Geophysical Union, Spring Meeting, San Francisco, p. 118.

Opdyke, N.D., Roberts, J., Claoue-Long, J., Irving, E., Jones, P.J., 2000. Base of Kiaman: Its definition and global significance. Geological Society of America Bulletin 112 (9), 1315–1341.

Over, D.J., 1992. Conodonts and the Devonian-Carboniferous boundary in the upper Woodford Shale, Arbuckle Mountains, south-central Oklahoma. Journal of Paleontology 66, 293–311.

Owens, B., 1984. Miospore zonation of the Carboniferous. In: Sutherland, P.K., Manger, W.L. (Eds.), Compte Rendu - Neuvieme Congres International de Stratigraphie et de Geologie du Carbonifere, vol. 2, pp. 90–102.

Palmer, J.A., Perry, S.P.G., Tarling, D.H., 1985. Carboniferous magnetostratigraphy. Journal of the Geological Society 142, 945–955.

Paproth, E., Teichmüller, R., Remy, W., 1963. Allemagne, Carbonifère. Lexique Stratigraphique International, Europe. Fascicule 5c1, 1–307.

Paproth, E., Feist, R., Flajs, G., 1991. Decision on the Devonian-Carboniferous boundary stratotype. Episodes 14, 331–336.

Peppers, R.A., 1996. Palynological correlation of major Pennsylvanian (Middle and Upper Carboniferous) chronostratigraphic boundaries in the Illinois and other coal basins. Geological Society of America Memoir 188, p. 111.

Peterson, D.N., Nairn, A.E.M., 1971. Palaeomagnetism of Permian redbeds from the South-western United States. Geophysical Journal of the Royal Astronomical Society 23 (2), 191–205.

Phillips, J., 1835. Illustrations of the geology of Yorkshire; or, a description of the strata and organic remains: accompanied by a geological map, sections, and plates of the fossil plants and animals. John Murray, London, p. 236.

Pnev, V.P., Polozova, A.N., Pavlov, A.M., Faddeyeva, I.Z., 1975. Stratotype section of the Orenburgian Stage, southern Urals (Nikol'skoe village). News of Academy of Sciences of USSR, Geology Series 6, 100–109 [In Russian].

Poletaev, V.I., Lazarev, S.S., 1994. General stratigraphic scale and brachiopod evolution in the Late Devonian and Carboniferous subequatorial belt. Bullletin de la Société Belge de Géologie 103 (1–2), 99–107.

Popov, A.V., 1999. Gzhelian ammonoids from Karachatyr, Central Asia. Voprosy Paleontologii (St.Petersburg) 11, 75–87 [In Russian].

Popov, A.V., Davydov, V.I., Donakova, L.M., Kossovaya, O.L., 1985. Gzhelian stratigraphy in Southern Urals. Sovetskaya Geologiya 3, 57–67 [In Russian].

Popov, A.V., Davydov, V.I., Kossovaya, O.L., 1989. Gzhelian stratigraphy in Central Asia. Sovetskaya Geologiya 3, 64–76 [In Russian].

Popp, B.N., Anderson, T.F., Sandberg, P.A., 1986. Brachiopods as indicators of original isotopic compositions in some Paleozoic limestones. Geological Society of America Bulletin 97 (10), 1262–1269.

Poty, E., 1985. A rugose coral biozonation for the Dinantian of Belgium as a basis for a coral biozonation of the Dinantian of Eurasia. VII Congrès International de Stratigraphie et de Géologie du Carbonifère 1983, Compte Rendu, 29–31.

Poty, E., Devuyst, F.-X., Hance, L., 2006. Upper Devonian and Mississippian foraminiferal and rugose coral zonations of Belgium and northern France: A tool for Eurasian correlations. Geological Magazine 143, 829–857.

Purves, J.C., 1883. Sur la delimitation et la constitution de l' étage houiller inferieur de la Belgique. Bulletin de l'Académie Royale de Belgique 3, 514–568.

Ramezani, J., Schmitz, M.D., Davydov, V.I., Bowring, S.A., Snyder, W.S., Northrup, C.J., 2007. High-precision U-Pb zircon age constraints on the Carboniferous–Permian boundary in the Southern Urals stratotype. Earth and Planetary Science Letters 256, 244–257.

Ramsbottom, W.H.C., 1973. Transgressions and regressions in the Dinantian: A new synthesis of British Dinantian stratigraphy. Proceedings of the Yorkshire Geological Society 39, 567–607.

Ramsbottom, W.H.C., 1977. Major cycles of transgression and regression (mesothems) in the Namurian. Proceedings of the Yorkshire Geological Society 41, 261–291.

Ramsbottom, W.H.C., 1981. Eustacy, sea level and local tectonism, with examples from the British Carboniferous. Proceedings of the Yorkshire Geological Society 43 (4), 473–482.

Ramsbottom, W.H.C., 1984. The founding of the Carboniferous System. In: Mackenzie, G. (Ed.), Comptu Rendu 9ème Congres International de Stratigraphie et de Géologie du Carbonifère. South Illinois University Press, Carbondale, pp. 109–112.

Ramsbottom, W.H.C., Saunders, W.B., 1985. Evolution and evolutionary biostratigraphy of Carboniferous ammonoids. Journal of Paleontology 59, 123–139.

Ramsbottom, W.H.C., Calver, M.A., Eagar, R.M.C., Hodson, F., Holliday, D.W., Stubblefield, C.J., Wilson, R.B., 1978. A correlation of Silesian rocks in the British Isles. Geological Society, Special Report 10, 81.

Rauser-Chernousova, D.M., 1937. About fusulinids and stratigraphy of Upper Carboniferous and Artinskian of the western slope of the Urals. Bulletin Moscow Society of Nature Studies. Geological series 15 (5), 478 [In Russian].

Rauser-Chernousova, D.M., 1941. New Upper Carboniferous stratigraphic data from the Oksko-Tsninskyi Dome. Reports of the Academy of Science of the USSR 30, 434–436 [In Russian].

Rauser-Chernousova, D.M., 1949. Stratigraphy of Upper Carboniferous and Artinskian deposits of Bashkirian Preurals. In: Rauser-Chernousova, D.M. (Ed.), Foraminifers of Upper Carboniferous and Artinskian Deposits of Bashkirian Pre-Urals. Transactions of the Geological Institute of the Academy of Sciences of USSR 105 (35), 3–21 [In Russian].

Rauser-Chernousova, D.M., Reitlinger, E.A., 1954. Biostratigraphic distribution of foraminifers in Middle Carboniferous depositis of southern limb of Moscow Syneclise. In: NalivkinD.V., MennerV.V. (Eds.), Regional Stratigraphy of the U.S.S.R. Stratigraphy of Middle Carboniferous deposits of Central and Eastern parts of Russian Platforms (based on foraminifers study), Vol. 2. Transactions of the Geological Institute of the Academy of Science of the USSR, Moscow, pp. 7–120 [In Russian].

Reitlinger, E.A., 1975. Paleozoogeografiya vizeyskikh i rannenamyurskikh basseynov po foraminiferam. Paleozoogeography of Visean and early Namurian basins based on foraminifera. Voporsy Mikropaleontologii 18, 3–20 [In Russian].

Renne, P.R., Swisher III, C.C., Deino, A.L., Karner, D.B., Owens, T.L., DePaolo, D.J., 1998. Intercalibration of standards, absolute ages and uncertainties in $^{40}Ar/^{39}Ar$ dating. Chemical Geology 145, 117–152.

Richards, B.C., Task Group, 2008. Report of the Task Group to establish a GSSP close to the existing Visean-Serpukhovian boundary. Newsletter on Carboniferous Stratigraphy 26, 8–9.

Richards, B.C., Aretz, M., 2009. Report of the task group to establish a GSSP for the Tournaisian-Visean boundary. Newsletter on Carboniferous Stratigraphy 27, 9–10.

Richards, B.C., Lane, H.R., Brenckle, P.L., 2002. The IUGS Mid-Carboniferous (Mississippian-Pennsylvanian) global boundary stratotype section and point at Arrow Canyon, Nevada, USA. Memoir, Canadian Society of Petroleum Geologists 19, 802–831.

Richards, B.C., Task Group, 2009. Report of the Joint Devonian-Carboniferous boundary GSSP reappraisal Task Group. Newsletter on Carboniferous Stratigraphy 27, 7–9.

Riley, N.J., 1993. Dinantian (Lower Carboniferous) biostratigraphy and chronostratigraphy in the British Isles. Journal of the Geological Society 150, 427–446.

Rosovskaya, S.E., 1950. *Triticites* genus, its evolution and stratigraphic significance. Transaction of the Paleontological Institute of Academy of Sciences of USSR 26, 1–79 [In Russian].

Ross, C.A., Ross, J.R.P., 1987. Biostratigraphic zonation of Late Paleozoic depositional sequences. Cushman Foundation for Foraminiferal Research Special Publication, 24, 151–168.

Ross, C.A., Ross, J.R.P., 1988. Late Paleozoic transgressive-regressive deposition. In: Wilgus, C.K., Hastings, B.S., Posamentier, H., Van Wagoner, J., Ross, C.A., Kendall, C.G. (Eds.), Sea level changes – an integrated approach. Society of Economic Paleontologists and Mineralogists Special Publication, 42, pp. 227–247.

Rosscoe, S., Barrick, J.E., 2009. Revision of *Idiognathodus* species from the Middle-Upper Pennsylvanian boundary interval in the Midcontinent basin, North America. In: Over, J.D. (Ed.), Conodont Studies Commemorating the 150th Anniversary of the First Conodont Paper (Pander, 1856) and the 40th Anniversary of the Pander Society. Palaeontographica Americana, 62, pp. 115–147.

Ruppel, S.C., James, E.W., Barrick, J.E., Nowlan, G., Uyeno, T.T., 1996. High-resolution $^{87}Sr/^{86}Sr$ chemostratigraphy of the Silurian: Implications for event correlation and strontium flux. Geology 24, 831–834.

Ruzhenzev, V.E., 1945. Suggestions for subdividing Upper Carboniferous into Stages. Reports of the Academy of Sciences of the USSR 46 (7), 314–317 [In Russian].

Ruzhenzev, V.E., 1965. The major ammonoid assemblages of the Carboniferous Period. Paleontological Journal 2, 3–17 [In Russian].

Ruzhenzev, V.E., 1974. Late Carboniferous ammonoids of the Russian Platform and Cisuralia. Paleontological Journal 8, 311–323 [In Russian].

Ruzhenzev, V.E., Bogoslovskaya, M.F., 1971. The Namurian Stage in the evolution of the Ammonoidea: Early Namurian Ammonoidea. Transactions of the Paleontological Institute of the Academy of Sciences of the USSR 133, 1–382 [In Russian].

Ruzhenzev, V.E., Bogoslovskaya, M.F., 1978. The Namurian Stage in the evolution of the Ammonoidea: Late Namurian Ammonoidea. Transactions of the Paleontological Institute of the Academy of Sciences of the USSR 167, 1–338 [In Russian].

Sadler, P.M., Cooper, R.A., 2003. Best-fit intervals and consensus sequences–a comparison of the resolving power of traditional biostratigraphy and computer assisted correlation. In: Harries, P.J. (Ed.), High Resolution Approaches in Stratigraphic Paleontology. Topics in Geobiology, 21, pp. 49–94.

Sadler, P.M., Kemple, W.G., Kooser, M.A., 2003. Contents of the compact disk–CONOP9 programs for solving the stratigraphic correlation and seriation problems as constrained optimization. In: Harries, P.J. (Ed.), High Resolution Approaches in Stratigraphic Paleontology. Topics in Geobiology, 21, pp. 461–465.

Sadler, P.M., Cooper, R.A., Melchin, M., 2009. High-resolution, early Paleozoic (Ordovician-Silurian) time scales. Geological Society of American Bulletin 121, 887–906.

Saltzman, M.R., 2003. Late Paleozoic ice age: Oceanic gateway or pCO_2? Geology 31, 151–154.

Sandberg, C.A., Ziegler, W., Leuteritz, K., Brill, S.M., 1978. Phylogeny, speciation, and zonation of *Siphonodella* (Conodonta, Upper Devonian and Lower Carboniferous). Newsletters on Stratigraphy 7 (2), 102–120.

Sando, W.J., 1990. Global Mississippian coral zonation. In: Brenkle, P.L., Manger, W.L. (Eds.), Intercontinental Correlation and Division of the Carboniferous System, Contributions from the Carboniferous Subcommission Meeting. Courier Forschungsinstitut Senckenberg, 130, pp. 179–187.

Sanz-Lopez, J., Blanco-Ferrera, S., Garsia-Lopez, S., Sanchez de Posada, L.C., 2006. The Mid-Carboniferous boundary in Northern Spain: Difficulties for correlation of the Global Stratotype Section and Point. Revista Italiana di Paleontologia e Stratigrafia 112, 3–22.

Sanz-Lopez, J., Blanco-Ferrera, S., Sanchez de Posada, L.C., Garcia-Lopez, S., 2007. Serpukhovian conodonts from northern Spain and their biostratigraphic application. Palaeontology 50, 883–904.

Scheffler, K., Hoernes, S., Schwark, L., 2003. Global changes during Carboniferous-Permian glaciation of Gondwana: Linking polar and equatorial climate evolution by geochemical proxies. Geology 31 (7), 605–608.

Schmidt, H., 1925. Die carbonischen Goniatiten Deutschlands. Jahrbuch der Preussischen Geologischen Landesanstalt 45, 489–609.

Schmitz, M.D., Davydov, V.I., 2012. Quantitative radiometric and biostratigraphic calibration of the Pennsylvanian–Early Permian (Cisuralian) time scale, and pan-Euramerican chronostratigraphic correlation. Geological Society of America Bulletin 124, 549–577.

Schmitz, M.D., Davydov, V.I., Snyder, W.S., 2009. Permo-Carboniferous conodonts and tuffs: High-precision marine Sr isotope geochronology. Permophiles, ICOS 2009 Abstracts 53, 48.

Schonlaub, H.P., Attrep, M., Boeckelmann, K., Dreesen, R., Feist, R., Fenninger, A., Hahn, G., Klein, P., Korn, D., Kratz, R., Magaritz, M., Orth, C.J., Schramm, J.M., 1992. The Devonian/Carboniferous boundary in the Carnic Alps (Austria) – a multidisciplinary approach. Jahrbuch Geologische Bundesanstalt 135, 57–98.

Semikhatova, S.V., 1934. The deposits of Moscovian in lower and middle Povolzhie region and the position of Moscovian Stage in the Carboniferous System. Problems of Soviet Geology 3, 73–90 [In Russian].

Semikhatova, S.V., Eynor, O.L., Kireyeva, G.D., Gubareva, V.S., Grozdilova, L.P., Degtyarev, D.D., Lebedeva, N.S., Sinitsyna, Z.A., 1979. Bashkirian Stage in the Urals; stratotype. In: Mackenzie, G. (Ed.), Compte Rendu, 9ème Congrès International de Stratigraphie et de Géologie du Carbonifère, Vol. 3. Southern Illinois University Press, Carbondale, pp. 112–118.

Shackleton, N., Crowhurst, S., Weedon, G., Laskar, J., 1999. Astronomical calibration of Oligocene-Miocene time. Philosophical Transactions of the Royal Society, Series A 357, 1907–1929.

Shen, S.-Z., Wang, Y., Henderson, C.M., Cao, C.-Q., Wang, W., 2007. Biostratigraphy and lithofacies of the Permian System in the Laibin-Heshan area of Guangxi, South China. Palaeoworld 16, 120–139.

Shvetsov, M.S., 1932. Obshchaya geologicheskava karta evropeiskoi chasti SSSR. L. 58, Severo-zapanaya chast' lista. Gosnauchtekhizdat, Moscow, p. 184 [In Russian].

Sinitsyna, Z.A., Sinitsyn, I.I., 1987. Biostratigraphy of Bashkirian Stage in the stratotype. Bashkirian Branch of Academy of Sciences of USSR, Ufa, p. 71 [In Russian].

Skompski, S., Alekseev, A., Meischner, D., Nemirovskaya, T., Perret, M.-F., Varker, W.J., 1995. Conodont distribution across the Viséan/Namurian boundary. Courier Forschungsinstitut Senckenberg 188, 177–209.

Solovieva, M.N., 1977. Zonal fusulinid stratigraphy of Middle Carboniferous of the USSR. Questions of Micropaleontology 28, 3–23 [In Russian].

Stewart, W.J., 1968. The stratigraphic and phylogenetic significance of the fusulinid genus *Eowaeringella*, with several new species. Cushman Foundation for Foraminiferal Research Special Publication, 10, p. 29.

Stukalina, G.A., 1988. Studies in Paleozoic crinoid columnals and stems. Palaeontographica Abteilung A: Palaeozoologie-Stratigraphie 204 (1–3), 1–66.

Taff, J.A., Adams, G.I., 1900. Geology of the eastern Choctaw coal field, Indian Territory (Oklahoma). United States Geological Survey Annual Report 21, 257–311.

Teodorovich, G., 1949. On subdividing of the Upper Carboniferous into stages. Reports of the Academy of Sciences of the U.S.S.R 67 (3), 537–540 [In Russian].

Thompson, M.L., Verville, G.J., Lokke, D.H., 1956. Fusulinids of the Desmoinesian-Missourian contact. Journal of Paleontology 30, 793–810.

Titus, A.L., Manger, W.L., 2001. Mid-Carboniferous Ammonoid biostratigraphy, Southern Nye County, Nevada: Implications of the first North American Homoceras. Journal of Paleontology 75, 1–31.

Tragelehn, H., 2009. Short notes of the origin of the conodont genus *Siphonodella* in the uppermost Famennian. Newsletter of the Subcommission on Devonian Stratigraphy 25, 41–43.

Tucker, R.D., Bradley, D.C., Ver Straeten, C.A., Harris, A.G., Ebert, J.R., McCutcheon, S.R., 1998. New U-Pb zircon ages and the duration and division of Devonian time. Earth and Planetary Science letters 1000, 51–58.

Ueno, K., Nemirovska, T.I., 2008. Bashkirian-Moscovian (Pennsylvanian, Upper Carboniferous) boundary in the Donets Basin, Ukraine. Journal of Geography 117, 919–932.

Ueno, K., Task Group, 2009. Report of the Task Group to establish the Moscovian-Kasimovian and Kasimovian-Gzhelian boundaries. Newsletter on Carboniferous Stratigraphy 27, 14–18.

Ulrich, E.O., 1904. Preliminary notes on classification and nomeclature of certain Paleozoic rock units in eastern Missoury. In: Buckley, E.R., Buehler, H.A. (Eds.), The Quarrying Industry in Missouri. Missouri Bureau of Geology and Mines, Reports, 2nd series, 2, pp. 109–111.

Vaughan, A., 1915. The palaeontological sequence in the Carboniferous Limestone of the Bristol area. Quarterly Journal of the Geological Society 61, 181–307.

Vdovenko, M.V., 2001. Atlas of Foraminifera from the upper Visean and lower Serpukhovian (Lower Carboniferous) of the Donets Basin (Ukraine). Abhandlungen und Berichte fuer Naturkunde 23, 93–178.

Veevers, J.J., Powell, M., 1987. Late Paleozoic glacial episodes in Gondwanaland reflected in transgressive-regressive depositional sequences in Euramerica. Geological Society of America Bulletin 98, 475–487.

Veizer, J., Ala, D., Azmy, K., Bruckschen, P., Buhl, D., Bruhn, F., Carden, G.A.F., Diener, A., Ebneth, S., Godderis, Y., Jasper, T., Korte, C., Pawellek, F., Podlaha, O.G., Strauss, H., 1999. $^{87}Sr/^{86}Sr$, $\delta^{13}C$ and $\delta^{18}O$ evolution of Phanerozoic seawater. In: Veizer, J. (Ed.), Earth System Evolution, Geochemical Perspective. Chemical Geology 161 (1–3), 59–88.

Villeneuve, M.E., Sandeman, H., Davis, W.J., 2000. Intercalibration of U-Pb and $^{40}Ar/^{39}Ar$ ages in the Phanerozoic. Geochimica et Cosmochimica Acta 64, 4017–4030.

Wagner, R.H., 1984. Megafloral zones of the Carboniferous. Congrès International de Stratigraphie et de Géologie du Carbonifère 9, 109–134.

Wagner, R.H., Winkler Prins, C.F., 1994. General overview of Carboniferous stratigraphy. Annales de la Société Géologique de Belgique 116, 163–174.

Wahlman, G.P., Verville, G.J., Sanderson, G.A., Ross, C.A., Ross, J.R.P., Brenckle, P.L., 1997. Biostratigraphic significance of the fusulinacean *Protriticites* in the Desmoinesian (Pennsylvanian) of the Rocky Mountains, Western U.S.A. Cushman Foundation for Foraminiferal Research Special Publication 36, 163–168.

Wang, C.Y., 1993. Auxiliary stratotype sections for the global stratotype section and point (GSSP) for the Devonian - Carboniferous boundary: Nanbiancun. Annales de la Société Géologique de Belgique 115, 707–708.

Wanless, H.R., Shepard, F.P., 1936. Sea level and climatic changes related to late Paleozoic cycles. Geological Society of America Bulletin 47, 1177–1206.

Wardlaw, B.R., 2000. Guadalupian conodont biostratigraphy of the Glass and Del Norte Mountains. In: Wardlaw, B.R., Grant, R.E., Rohr, D.M. (Eds.), The Guadalupian Symposium. Smithsonian Contribution to the Earth Sciences, 32, pp. 37–87.

Wardlaw, B.R., Nestell, M.K., 2010. Latest Middle Permian conodonts from the Apache Mountains, West Texas. Micropaleontology 56, 149–184.

Waters, C.N., Davies, S.J., 2006. Carboniferous extensional basins, advancing deltas and coal swamps. In: Brechley, P.J., Rawson, P.F. (Eds.), The Geology of England and Wales. The Geological Society, London, Second ed., pp. 173–223.

Westerhold, T., Rohl, U., Raffi, I., Fornaciari, E., Monechi, S., Reale, V., Bowles, J., Evans, H.F., 2008. Astronomical calibration of the Paleocene time. Palaeogeography, Palaeoclimatology, Palaeoecology 257, 377–403.

Wilde, G.L., 1990. Practical fusulinid zonation: the species concept; with Permian basin emphasis. West Texas Geological Society 29, 5–34.

Williams, H.S., 1891. Correlation papers: Devonian and Carboniferous. United States Geological Survey, Bulletin 80, 1–279.

Won, M.-Z., 1998. A Tournaisian (Lower Carboniferous) radiolarian zonation and radiolarians of the *A. pseudoparadoxa* zone from Oese (Rheinische Schiefergebirge), Germany. Journal of the Korean Earth Science Society 19 (2), 216–259.

Worthen, A.H., 1860. Remarks on the discovery of a terrestrial flora in the Mountain Limestone of Illinois. American Association of the Advancement of the Scientific Process 13, 312–313.

Worthen, A.H., 1866. Stratigraphical geology, Tertiary deposits and coal measure; Sub-Carboniferous. In: Worthen, A.H. (Ed.), Geological Survey of Illinois: Geology, Vol. 1, pp. 40–118.

Yang, Z., Yancey, T.E., 2000. Fusulinid biostratigraphy and paleontology of the Middle Permian (Guadalupian) strata of the Glass Mountains and Del Norte Mountains, West Texas. In: Wardlaw, B.R., Grant, R.E., Rohr, D.M. (Eds.), The Guadalupian Symposium. Smithsonian Contribution to the Earth Sciences, 32, pp. 185–260.

Young, G.C., Laurie, J.R., 1996. An Australian Phanerozoic Timescale. Oxford University Press, Melbourne, p. 279.

Yuping, Q., Wang, Z., 2005. Serpukhovian conodont sequence and the Visean-Serpukhovian boundary in south China. Rivista Italiana di Paleontologia e Stratigrafia 111, 3–10.

Zachos, J., Pagani, M., Sloan, L., Thomas, E., Billups, K., 2001. Trends, rhythms, and aberrations in global climate 65 Ma to present. Science 292, 686–693.

Zhang, L., Zhang, N., Xia, W., 2008. Conodont succession in the Guadalupian-Lopingian boundary interval (upper Permian) of the Maoershan section, Hubei Province, China. Micropaleontology 53, 433–446.

Zhang, N., Henderson, C.M., Xia, W., Wang, G., Shang, H., 2010. Conodonts and radiolarians through the Cisuralian-Guadalupian boundary from the Pingxiang and Dachongling sections, Guangxi region, south China. Alcheringa 34, 135–160.

C.M. Henderson, V.I. Davydov and B.R. Wardlaw
Contributors: F.M. Gradstein and O. Hammer

Chapter 24

The Permian Period

Abstract: Pangea moves north. Ice-house to greenhouse (humid to arid) climate transition; dramatic reduction of coal swamps and amphibian habitat; some spore-bearing plants extinct; major evaporites; changes in internal and external carbonate invertebrate skeletons; major diversification of fusulinacean foraminifers, ammonoids, bryozoans, and brachiopods, then major end-Permian extinction of fusulinacean foraminifers, trilobites, rugose and tabulate corals, blastoids, acanthodians, placoderms, and pelycosaurs; dramatic reduction of bryozoans, brachiopods, ammonoids, sharks, bony fish, crinoids, eurypterids, ostracodes, and echinoderms.

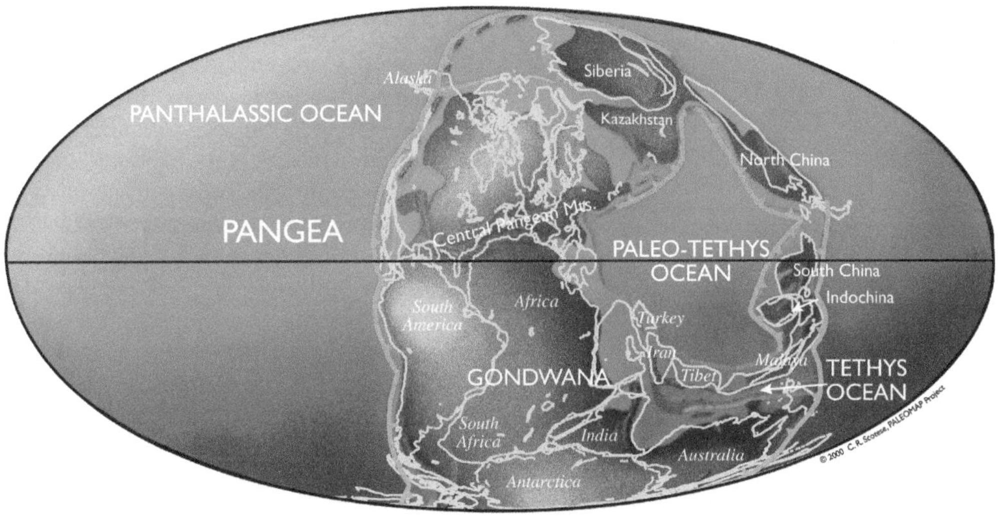

Chapter Outline

The Geologic Time Scale 2012. DOI: 10.1016/B978-0-444-59425-9.00024-X

24.1. HISTORY AND SUBDIVISIONS

In 1841, after a tour of Russia with French paleontologist Edouard de Verneuil, Roderick I. Murchison, in collaboration with Russian geologists, named the Permian System to take in the "vast series of beds of marl, schist, limestone, sandstone and conglomerate" that surmounted the Carboniferous System throughout a great arc stretching from the Volga eastwards to the Urals, and from the Sea of Archangel to the southern steppes of Orenburg. He proposed the name "Permian" based on the extensive region that composed the ancient kingdom of Permia; the city of Perm lies on the flanks of the Urals. Murchison (1872, p. 309) indicated that:

'the animals and plants of the Permian era, though chiefly of new species, are generically connected with those of the preceding or Carboniferous epoch, whilst they are almost wholly dissimilar to those of the next succeeding period, the Trias.'

In 1845 he included rocks now known as Kungurian–Tatarian in age and for a time the underlying strata (Artinskian, etc.) were known as Permo-Carboniferous, i.e. intermediate between Carboniferous and Permian (Dunbar, 1940).

As early as 1822 (e.g., Conybeare and Phillips, 1822), the Magnesian Limestone and New Red Sandstone of England were well known, as were the equivalent German Rotliegendes and Zechstein (a traditional miner's name) with its valuable Kupferschiefer. However, all these rocks lacked richly fossiliferous strata, were difficult to correlate, and unsuitable to justify the erection of a new system in Western Europe.

In North America, J. Marcou recognized Permian rocks between 1853 and 1867 in a large area from the Mississippi to the Rio Colorado, and noted two divisions analogous to those in western Europe. He accordingly suggested the name Dyassic as more suitable than Permian and proposed a combined Dyas and Trias as a major period. Murchison (1872, p. 442) summarized Marcou's reports as follows:

'the discovery of strata containing many true Permian species in the northwestern part of Texas and along the eastern edge of the Rocky Mountains teaches us that all of the seas of Paleozoic time, even the very last of them, had a very great extension and were inhabited by similar groups of animals over enormously wide areas.'

Murchison (1872, p. 309) also rejected the notion of Dyas saying:

'I resist the introduction of a new name founded upon a local binary division which, though good in Saxony, is inapplicable to many other countries, in several of which the separation is tripartite...and thus I prefer the simpler geographic name 'Permian', which, like 'Silurian', involves no theory.'

For further historical details on the history of the Permian Period see GTS1989 (Harland *et al.*, 1990, pp. 46–47).

The Permian divides itself naturally into three series (Figure 24.1). In the classic area of the southern Urals, the Upper Carboniferous and Lower Permian (or Cisuralian) are well represented by marine deposits and abundant biota. This marine dominance disappears in the Kungurian, and the Middle Permian (or Guadalupian) and Upper Permian (or Lopingian) are dominated by terrestrial–marginal-marine deposits. The Guadalupian deposits of West Texas are dominated by diversified and well-studied marine fossil assemblages, and the deposits are the subject of seminal studies in sequence stratigraphy. China, Iran, and the Trans-Caucasian region are exemplary for their Upper Permian deposits and biota.

Permian biostratigraphy has been greatly refined over the last three decades, especially through a detailed understanding of the distribution of conodonts in relation to ammonoids and fusulinaceans; these refined correlations form the basis for the following discussion.

24.1.1. The Cisuralian Series: Lower Permian

The base of the Permian was originally defined in the Ural Mountains of Russia to coincide with strata marking the initiation of evaporite deposition (Murchison, 1841), now recognized as within the Kungurian Stage. Since 1841, the base has been lowered repeatedly to include a succession of faunas with post-Carboniferous affinities. Karpinsky (1874) identified clastic successions that Murchison had included in the British Millstone Grit, as being younger, transitional between Carboniferous and Permian, and termed them the Artinskian Series. His subsequent classic study of the abundant ammonoid fauna (Karpinsky, 1889) led him to add the interval to the Permian. Further study, especially of ammonoids, led Ruzhenzev (1936) to recognize the Sakmarian as an independent lower subdivision of the Artinskian. In turn, he subdivided the Sakmarian, and referred the lower interval to the Asselian Stage (Ruzhenzev, 1954). The base of the Asselian and of the Permian Period was defined by the appearance of the ammonoid families Paragastrioceratidae, Metalegoceratidae, and Popanoceratidae, concurrent with the first inflated fusulinaceans referable to "*Schwagerina*" (i.e. *Sphaeroschwagerina*).

The base of the Asselian Stage, defined by reference to both ammonoids and fusulinaceans, received progressively greater recognition, and eventually official Russian status (Resolutions of the Interdepartmental Stratigraphic Committee of Russia and its Permanent Commissions, 1992) following Ruzhenzev's original proposal.

The Cisuralian was proposed by Waterhouse (1982) to comprise the Asselian, Sakmarian, and Artinskian stages. The Kungurian was included in the Cisuralian (Jin *et al.*, 1997), so that it corresponded to the Lower Permian as recognized in Russia (Likharew, 1966; Kotlyar and Stepanov, 1984) and corresponded better to the Rotliegendes of

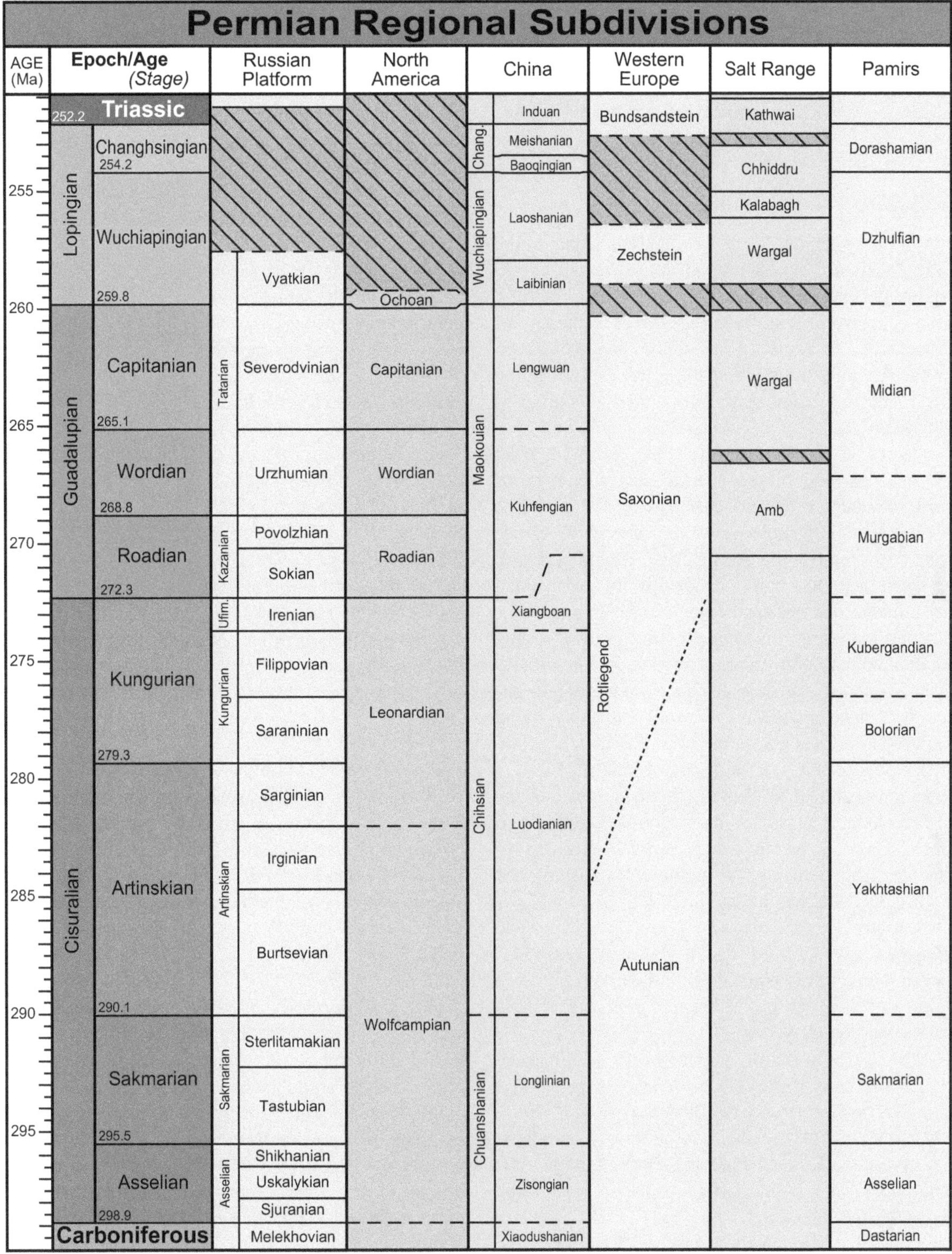

FIGURE 24.1 **Correlation chart of the Permian with international subdivisions and selected regional stage and substage nomenclature.** Vertical pattern indicates a widespread regional hiatus.

Harland *et al.* (1990). The Uralian Series, named by de Lapparent in 1900, interpreted by Gerasimov (1937) to include pre-Kungurian stages of the Lower Permian, and utilized by Jin *et al.* (1994), has been abandoned because the name caused too much confusion from a history of varied usage.

24.1.1.1. Asselian

The GSSP for the beginning of the Permian Period and for the base of the Asselian Stage is located at Aidaralash Creek, Atobe region, northern Kazakhstan (Davydov *et al.*, 1998). The section is approximately 50 km southeast of the city of Atobe. A stone and concrete marker with a plaque has been erected at the Aidaralash section, marking the exact location of the GSSP, and the boundary between the Carboniferous and Permian Periods.

The strata of Late Paleozoic age at Aidaralash Creek were deposited on a narrow, shallow-marine shelf that formed the western boundary of the orogenic zone to the east. The fluvial–deltaic conglomerate–sandstone successions grade upward into transgressive, marginal marine sequences (beach and upper shore face) that, in turn, grade upward into massive mudstone–siltstone and fine sandstone beds with ammonoids, conodonts, and radiolaria, interpreted as maximum flooding units. The maximum flooding zone is overlain by a regressive sequence (progressively, offshore to shoreface to delta front), which in turn is capped by an unconformity with an overlying conglomerate. The critical GSSP interval is completely within a maximum flooding unit, free of disconformities.

The position of the GSSP is at the first occurrence of the conodont *Streptognathodus isolatus*, which developed from an advanced morphotype in the *S. wabaunsensis* chronomorphocline. This is located 27 m above the base of Bed 19, Aidaralash Creek (Davydov *et al.*, 1998).

The first occurrences of *Streptognathodus invaginatus* and *S. nodulinearis*, also species within the "*wabaunsensis*" chronomorphocline, nearly coincide with the first occurrence of *S. isolatus* in many sections, and can be used as accessory indicators for the boundary.

The GSSP is 6.3 m below the traditional fusulinacean boundary, i.e. the base of the *Sphaeroschwagerina vulgaris aktjubensis–S. fusiformis* Zone (Davydov *et al.*, 1998; Figures 24.2 and 24.3). The latter can be widely correlated with Spitsbergen, the Russian Platform, Urals, Central Asia China, and Japan, and is of practical value in identifying proximity to the Asselian Stage base.

The traditional ammonoid boundary, 26.8 m above the GSSP, includes the termination of the *Prouddenites–Uddenites* lineage at the top of Bed 19, and the introduction of the Permian taxa *Svetlanoceras primore* and *Prostacheoceras principale* in Bed 20 (Davydov *et al.*, 1998; Figures 24.1 and 24.2). The evolution from *Artinskia irinae* to *A. kazakhstanica* may be a chronomorphocline that crosses the Carboniferous–Permian boundary. Difficulties with the ammonoid taxa are that they are relatively rare and many taxa are endemic.

Utilization of magnetostratigraphy to assist with recognition and correlation of the Carboniferous–Permian boundary is difficult because it is within the Kiaman Long Reversed-Polarity Chron (see the geological time scale chart insert of this book). However, Davydov *et al.* (1998) cite reports that show that most of the *Ultradaixina bosbytauensis –Schwagerina robusta* fusulinacean zone, just below the Carboniferous–Permian boundary in Aidaralash, is characterized by normal polarity. That same stratigraphic polarity relationship is also known elsewhere in the southern Urals, and the northern Caucasus and Donetz Basin, and possibly correlates with the normally polarized magnetic zone in the Manebach Formation of the Thuringian Forest (Menning, 1987).

The conodont succession observed at Aidaralash is displayed in several sections in the southern Urals, especially the basinal reference section at Usolka. It is also displayed in the Red Eagle cyclothem of the midcontinent of the USA (Boardman *et al.*, 1998, 2009), in the West Texas regional stratotype Wolfcamp Hills (Wardlaw and Davydov, 2000), and in south China (Wang, 2000), as well as in many other intervening localities, and, therefore, serves as an excellent boundary definition.

24.1.1.2. Sakmarian

Two alternative definitions for the base-Sakmarian have been considered. The first proposed boundary for the Sakmarian Stage is near the level originally proposed by Ruzhenzev (1950) in the same Kondurovsky section, Orenburg Province, Russia. A conodont succession exhibiting the evolutionary change within the chronomorphocline from *Sweetognathus expansus* to *S. merrilli* at 115 m above the base (uppermost Bed 11 of Chuvashov *et al.*, 1993a,b) is the proposed definition.

The boundary originally proposed by Ruzhenzev (1950) was at the base of Bed 11, at an unconformable formation break and based on the change in fusulinacean faunas, with *Sakmarella* (*Pseudofusulina*) *moelleri* occurring above the break. The actual introduction of the *S. moelleri* group occurs in Beds 6–12 (Wardlaw *et al.*, 1999), with traditional *S. moelleri* occurring in Bed 12, just a few meters above the first occurrence of *S. merrilli*. *Sweetognathus merrilli* is widespread and its FAD is well constrained throughout Kansas in the upper part of the Eiss Limestone of the Bader Limestone, but the presence of associated *Streptognathodus* spp. at the latter locality may indicate diachroneity within this "lineage".

Furthermore, the local first occurrences of *S. merrilli* are clearly diachronous (communication between Henderson and Kotlyar, 2009, *Permophiles*, **54**: 4), likely because in many sections this taxon is very rare and may be present as a migration

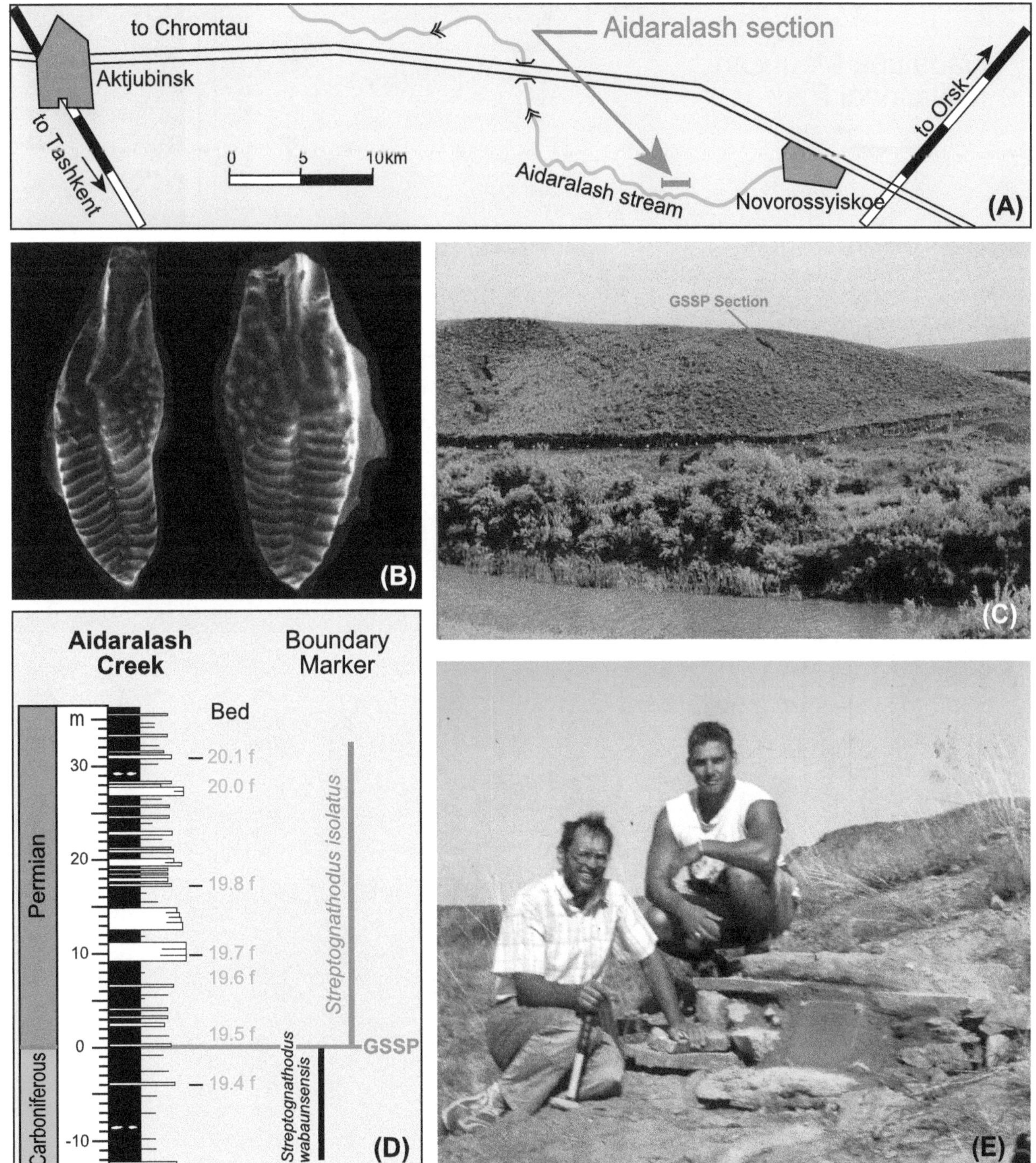

FIGURE 24.2 **Global boundary and stratotype section (GSSP) for the Carboniferous–Permian boundary at Aidalarash in Kazakhstan.** Semi-permanent plaque marks the boundary; the index conodont shows the holotype of *Streptognathodus isolatus* (left, 1.2mm long and right 0.8mm long).

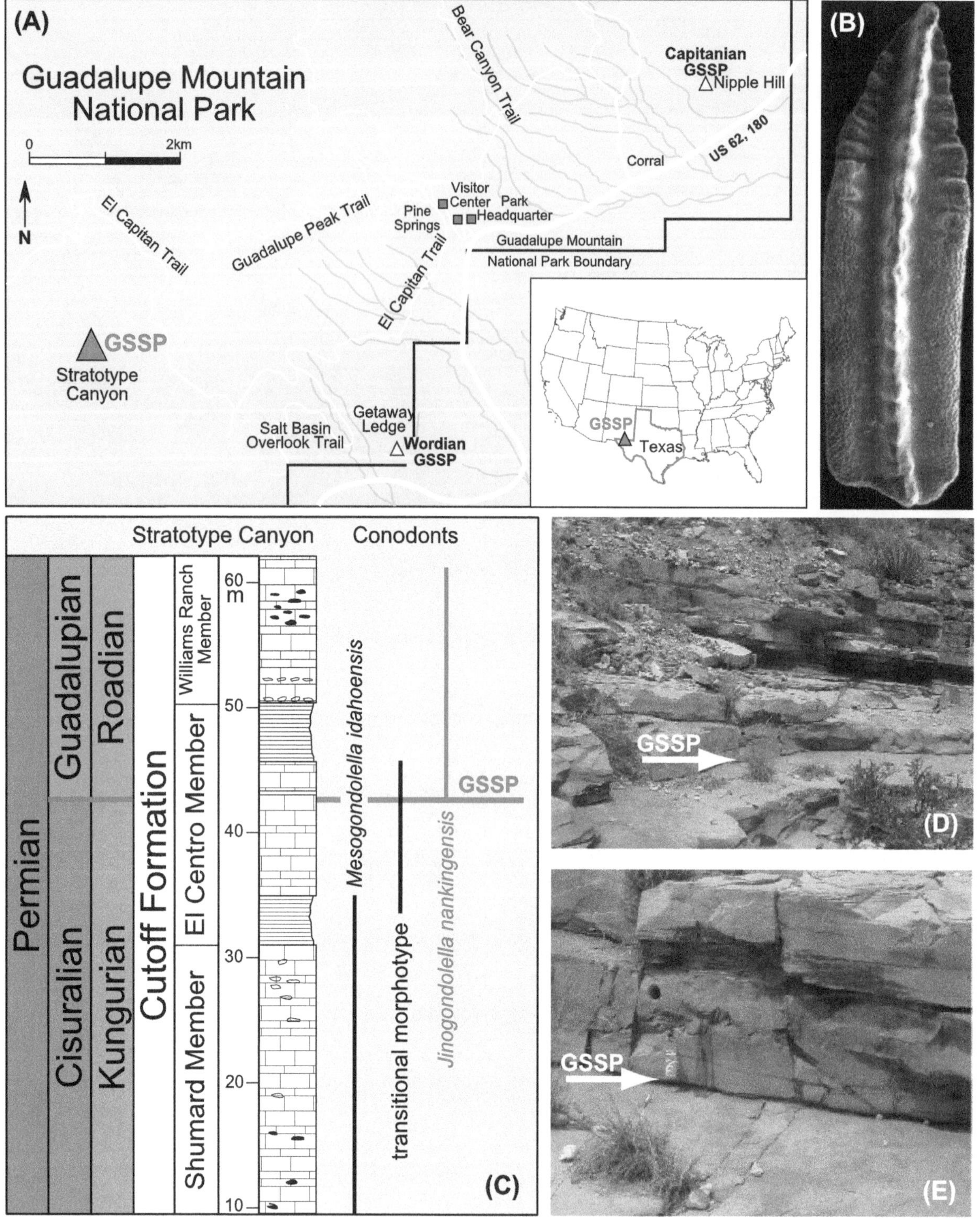

FIGURE 24.3 **Global boundary and stratotype section (GSSP) for the Roadian Stage of the Guadalupian Series at Stratotype Canyon, Texas, USA.** Holotype of *Jinogondolella nankingensis*; it is 1.0 mm long.

event; test samples at Kondurovsky failed to produce any specimens. As a result, a second definition at the FAD of *Mesogondolella uralensis* within the chronomorphocline of *M. pseudostriata* to *M. arcuata* to *M. uralensis* at 51.6 meters above base (mab) of the Usolka section is now preferred (Schmitz and Davydov, 2011). This taxon is present at Kondurovsky at 104.2 meters above base just below the FO of *S. merrilli*.

24.1.1.3. Artinskian

The Artinskian Stage was proposed by Karpinsky in 1874, for the sandstone of the Kashkabash Mountain on the right bank of the Ufa River, near the village of Arty. This is the stratotype, and Karpinsky (1891) studied abundant and diverse ammonoids in several exposures and small quarries along the Ufa River. The taxonomically diverse ammonoid assemblage from the Arty area was distinctly more advanced than the Sakmarian one in terms of cephalopod evolution and this stimulated Karpinsky (1874) to define two regional belts with ammonoids: the lower, at Sakmara River, and the upper, at Ufa River.

The Sakmarian–Artinskian boundary deposits are well represented in the Dal'ny Tulkus section, a counterpart of the Usolka section. The upper part of the Sakmarian Stage (Beds 28–31) at the Usolka River and Bed 18 at the Dal'ny Tulkus section are composed of dark-colored marl, argillite, and carbonate mudstone, or less commonly, detrital limestone with fusulinacean foraminifers, radiolaria, rare ammonoids, and bivalves. The upper part of the Sakmarian includes fusulinaceans characteristic of the Sterlitamakian Horizon including *Pseudofusulina longa*, *P. fortissima*, *P. plicatissima*, *Leeina urdalensis*, and *L. urdalensis abnormis*.

The best GSSP section is the Dal'ny Tulkus section, in Russia, at a point defined by the FAD of *Sweetognathus* "*whitei*" within the chronomorphocline *S. binodosus* to *S. anceps* to S. "*whitei*" at 2.7 m above bed 4 (Schmitz and Davydov, 2011). The succession of *S. binodosus* to *S.* "*whitei*" can also be recognized in the lower Great Bear Cape Formation, southwest Ellesmere Island (Henderson, 1988; Beauchamp and Henderson, 1994; Mei *et al.*, 2002) and possibly in the Schroyer-Florence limestones of the Chase Group Kansas (Boardman *et al.*, 2009), but at this latter locality species of *Streptognathodus* spp. suggest an older age. The defining taxon is in quotes to denote some taxonomic problems associated with comparisons to type specimens of *S. whitei* from Wyoming (Rhodes, 1963). Specimens of *S.* "*whitei*" in the Canadian Arctic and in the Urals are found above high frequency cyclothems. Henderson (2010) summarized progress on the definition and indicated additional correlation tools in strontium isotopes and radiometric ages.

24.1.1.4. Kungurian

The stratotype of the Kungurian Stage was not defined when the stage itself was established (Stuckenberg, 1890). Sometime later, the carbonate–sulfate section exposed along the Sylva River, upstream of the town of Kungur, was arbitrarily accepted for the stratotype. In line with the new position of the Kungurian lower boundary at the base of the Sarana Horizon (Chuvashov *et al.*, 1999), the original stratotype section in this area consists of:

1) The Saraninian Horizon, including reefal limestone of the Sylva Formation and its lateral equivalent the Shurtan Formation, composed of marls and clayey limestone;
2) The Filippovian Horizon; and
3) The Irenian Horizon.

A disadvantage of the section is the poor fossil content of the limey Kamai Formation underlying the Sarana Horizon; it contains only small benthic foraminifers, bryozoans, and brachiopods, with taxa unsuitable for age determination. However, another section of Artinskian–Kungurian boundary deposits, located near the Mechetlino settlement on the Yuryuzan' River, has a better fossil assemblage both below and above the boundary interval, and had been selected as a possible stratotype for the base-Kungurian at the FAD of the conodont *Neostreptognathodus pnevi* in bed 19 (Chuvashov *et al.*, 2002). Subsequent test collections, however, did not yield any conodonts, nor did sampled volcanic ash beds provide minerals useful for dating. Therefore a second section is currently under consideration for the base-Kungurian GSSP. The Rockland section occurs in the Pequop Mountains of northeastern Nevada (p. 5 of Permophiles 54; 2009) and demonstrates the chronomorphocline from *N. pequopensis* to *N. pnevi*; detailed sampling was conducted in July 2010 to determine the first occurrence.

The Cisuralian is divided into stages based on the FADs of specific species from three lineages of conodonts. The first, the base of the Permian, and for that matter the zones of the Asselian, is based on the widespread occurrence of *Streptognathodus* species, as is most of the upper Pennsylvanian. Specimens of the genus become progressively rare and less widespread after the beginning of the Sakmarian, and are extinct by the mid-Sakmarian. The Sakmarian will be defined within a *Mesogondolella* lineage, and the Artinskian and Kungurian stages will be based on a lineage of *Sweetognathus* and its derived descendant *Neostreptognathodus*.

24.1.2. The Guadalupian Series: Middle Permian

The Guadalupian was first proposed by Girty at the turn of the last century for the spectacular fossils found in the Guadalupe and Glass Mountains of West Texas. These faunas have been well documented, and represent an unprecedented display in an exhumed, well-preserved backreef, reef, and basin facies. The West Texas depositional basins represent a tropical North American faunal suite, well removed from the more typical tropical Tethyan fauna of Asia and Europe. The Middle Permian was a time of strong provincialism and presents some complexities for correlation. The formal establishment of the Guadalupian and its constituent stages is based on the evolution of a single genus of conodont, *Jinogondolella*. The genus has a limited distribution, though it is common in West Texas and South China.

24.1.2.1. Roadian

The GSSP for the base of the Roadian Stage, Guadalupian Series, Middle Permian, is in Stratotype Canyon, Guadalupe Mountains National Park, Texas, USA (Figure 24.3). The marker horizon is the first evolutionary appearance of the conodont *Jinogondolella nankingensis* from its ancestors *Mesogondolella idahoensis idahoensis* and *M. idahoensis lamberti* (Mei and Henderson, 2002), at 42.7 m above the base of the black, thin-bedded limestone of the Cutoff Formation, and 29 cm below a prominent shale band in the upper part of the El Centro Member (Glenister *et al.*, 1999). This member consists of skeletal carbonate mudstone with one shale bed, deposited in a basinal setting, proximal to the slope. In terms of magnetostratigraphy, the Cutoff Formation indicates reversed polarity, and may fall in the Kiaman reversed superchron. The GSSP of the Roadian Stage was ratified in 2001.

24.1.2.2. Wordian

The GSSP for the beginning of the Wordian Stage in the Guadalupian series is located in Guadalupe Pass, Texas; a short distance from Stratotype Canyon (Figure 24.4). The marker horizon for this stage is the first evolutionary appearance of *Jinogondolella aserrata* from its ancestor *J. nankingensis* at 7.6 m above the base of the Getaway Ledge outcrop section in Guadalupe Pass, Guadalupe Mountains National Park, Texas, USA. This level is just below the top of the Getaway Limestone Member of the Cherry Canyon Formation, a succession of skeletal carbonate mudstone in a base of slope depositional setting (Glenister *et al.*, 1999). Like the Roadian sediments in the type area, Wordian Stage limestone of Guadalupian National Park also displays reversed polarity. The GSSP of the Wordian Stage was ratified in early 2001.

24.1.2.3. Capitanian

Like the GSSPs for the Roadian and Wordian stages in the Middle Permian, the GSSP for the Capitanian Stage was also selected in the Guadalupe National Park (Figure 24.5). The marker horizon for the Capitanian Stage is the first evolutionary appearance of the conodont *Jinogondolella postserrata* within the lineage *J. nankingensis* to *J. aserrata* to *J. postserrata*. This level is at 4.5 m in the outcrop section at Nipple Hill, in the upper Pinery Limestone Member of the Bell Canyon Formation (Glenister *et al.*, 1999). The GSSP is in a monotonous succession of pelagic carbonate, representing a lower slope depositional setting. An ashbed at 37.2 m below the base of the *J. postserrata* entry (Bowring *et al.*, 1998) yields a date of 265.35 ± 0.46 Ma (Item P9 in Appendix 2 of this volume). Some samples in the Pinery Limestone and overlying Lamar Limestone of the Bell Canyon Formation display normal polarity, with the first normal polarity indicative of the approximate position of the Illawarra reversal occurring below the FAD of *J. postserrata* in the Upper Wordian. The GSSP of the Capitanian Stage was ratified in early 2001.

The abundant and well-preserved conodont faunas of West Texas show that the genus of *Jinogondolella* and its species evolved through short-lived transitional morphotypes, generally through a mosaic of paedomorphogenesis (retention of juvenile characters in later growth stages). The first species of the genus, *J. nankingensis*, is also the marker for the Guadalupian and its basal stage, the Roadian. The species is abundant in West Texas and South China, but occurs rarely in several other sites (i.e. Canadian Arctic, Pamirs); however, its distribution along the western coast of Pangea represents a geographic cline from the tropical Delaware Basin (West Texas) to the upwelling-influenced Phosphoria Basin (Idaho) to temperate Canadian Arctic and exhibits overlap with several genera, especially within the Phosphoria Basin where it is abundant, and provides excellent correlation globally (Henderson and Mei, 2007).

The Illawarra geomagnetic reversal is an important tie point for the Guadalupian and proximity to the base of the Capitanian. The Illawarra reversal is well known from the lower part of the Tatarian in the Volga region of Russia (Gialanella *et al.*, 1997). It has also been documented from Pakistan. Haag and Heller (1991) show that normal polarity starts at the base of the Wargal Formation in the Nammal Gorge, Salt Range, which is the base of the Illawarra reversal (F. Heller, pers. comm; 1998). Peterson and Nairn (1971) record a reversal in West Texas–New Mexico that has been interpreted with additional study by Menning (2000) to occur just below the upper Pinery Limestone Member of the Bell Canyon Formation.

The ammonoid genus *Waagenoceras* has long been associated with the Guadalupian, and, in particular, the Wordian, but it also occurs within the Upper Roadian according to GSSP definitions above. The distribution of the conodont *Jinogondolella*, which characterizes the Guadalupian, is common in the tropical zone of South China and West Texas and the upwelling area of the Phosphoria Basin. Its rarer appearances around the margin of the Tethyan tropical zone and temperate zones bordering the Tethys and the northern margin of Pangea link it to other faunal provinces, making the Guadalupian an appropriate standard for the Middle Permian.

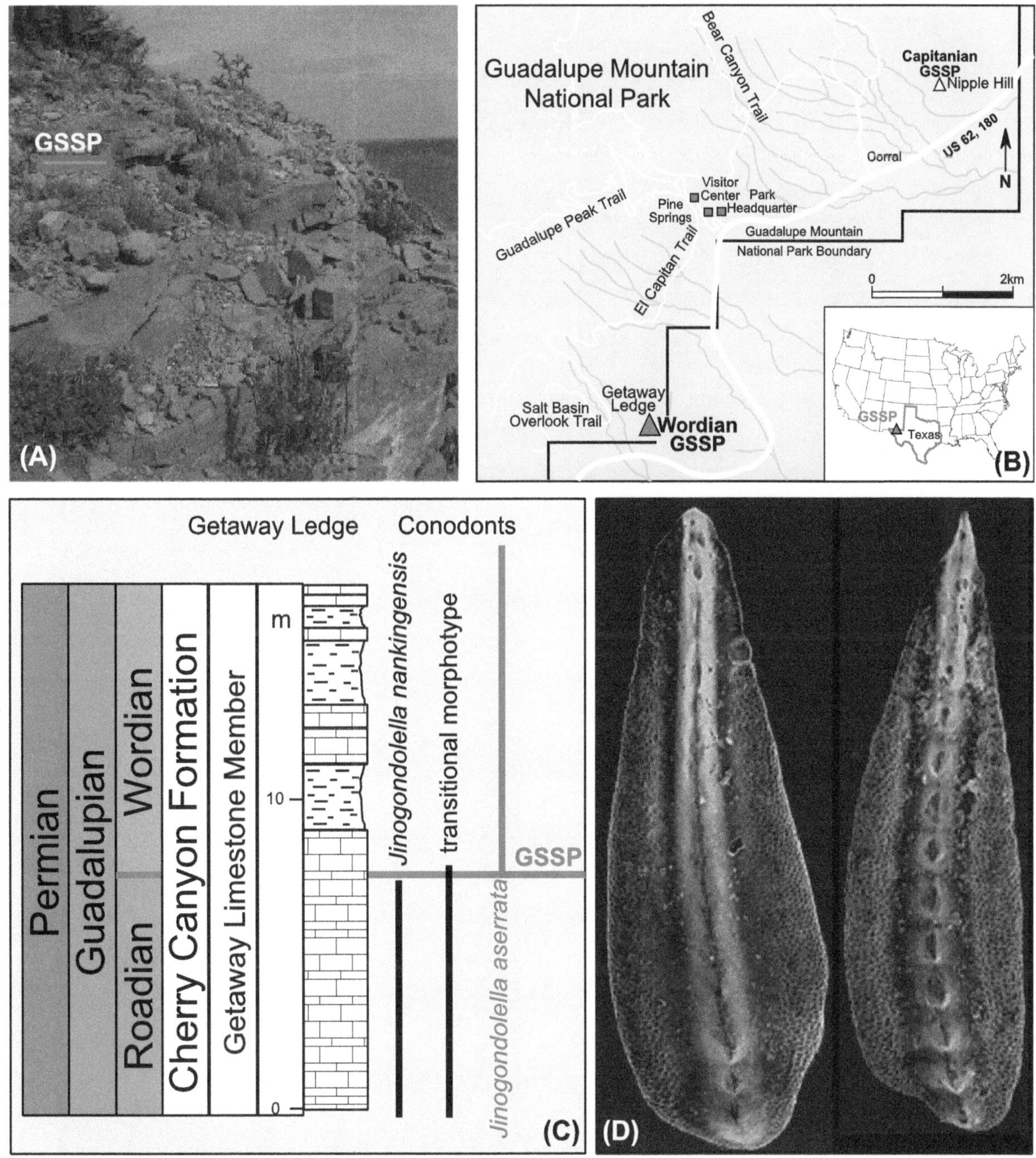

FIGURE 24.4 **Global boundary and stratotype section (GSSP) for the Wordian Stage of the Guadalupian Series at Getaway Ledge section, Texas, USA**. Holotype of *Jinogondolella aserrata*; it is 1.0 mm long on the left and 0.9 mm long on the right.

24.1.3. The Lopingian Series: Upper Permian

The Lopingian (Huang, 1932), Dzhulfian (Furnish, 1973), Transcaucasian, and Yichangian (Waterhouse, 1982) have been proposed for the uppermost Permian series. Of these, the Lopingian appears to be the first formally designated series name to be based on relatively complete marine sequences. The Lopingian Series comprises two stages: the Wuchiapingian and the Changhsingian. Shen *et al.* (2010) provided a high-resolution Lopingian time scale.

Base of the Capitanian Stage of the Guadalupian Series at Nipple Hill, Texas, USA.

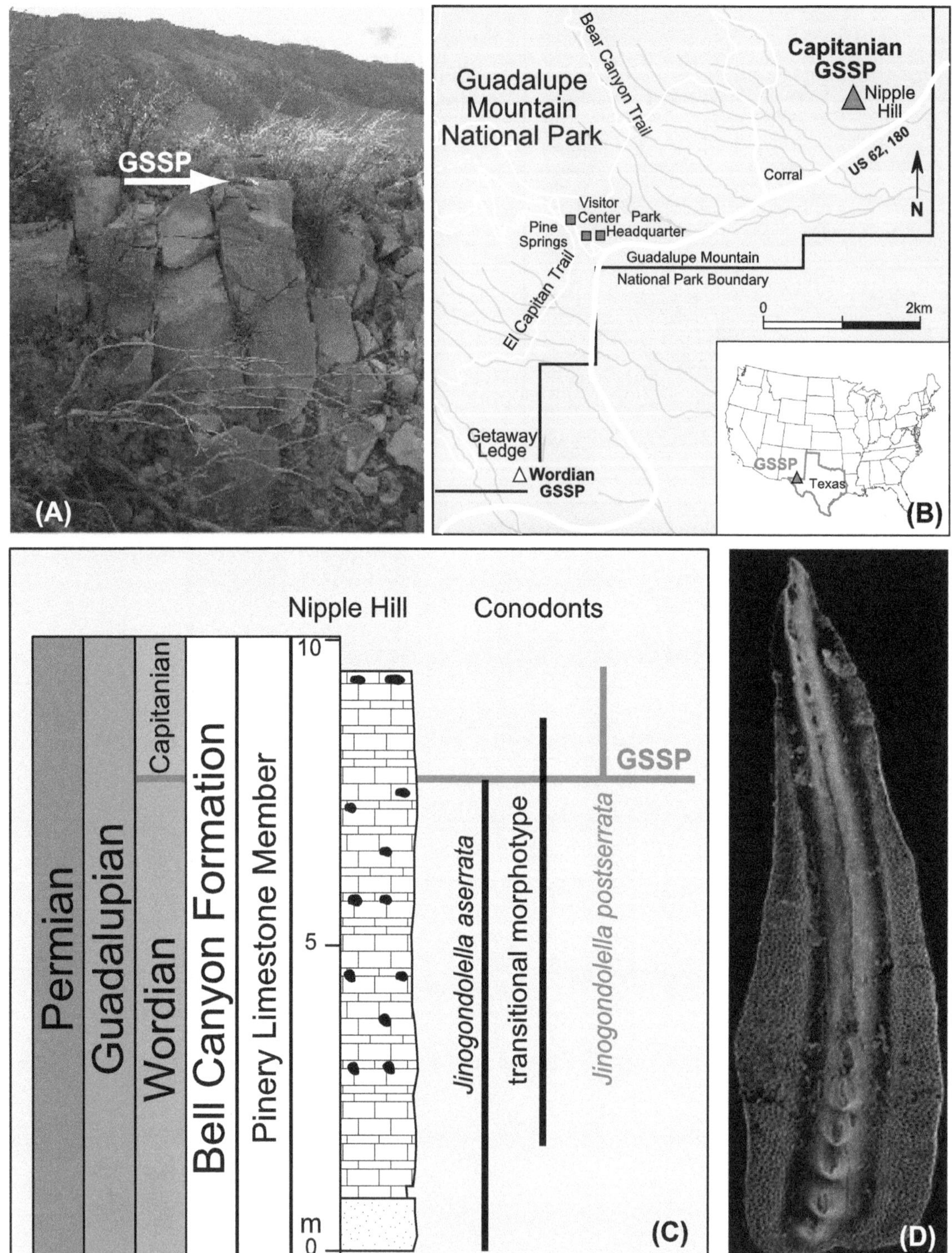

FIGURE 24.5 **Global boundary and stratotype section (GSSP) for the Capitanian Stage of the Guadalupian Series at Nipple Hill, Texas, USA.** Holotype of *Jinogondolella postserrata*; it is 1.1 mm long.

The Tatarian of the traditional Volga region of Russia is mostly a continental deposit and corresponds largely to the upper Guadalupian; it does not serve as a comprehensive subdivision of the Upper Permian.

The upper boundary of the Permian (i.e. the base of the Triassic) in the original type area, the Buntsandstein of Germany, and in the Urals is non-marine and unsuitable for worldwide correlation. The functional definition for the base of the Triassic was for a long time the base of the ammonoid *Otoceras* Zone of the Himalayas (Griesbach, 1880). The first appearance of the conodont *Hindeodus parvus* is more widespread than *Otoceras* and provides a precise basis for base-Triassic (Chapter 25).

24.1.3.1. Wuchiapingian

The boundary between the Guadalupian and Lopingian series and the base of the Wuchiapingian Stage was historically designated to coincide with a global regression, i.e. with the boundary surface between the Middle and the Upper Absaroka Megasequences. Extensive surveys of marine sections demonstrate that few sections can be considered to be continuous across the Guadalupian–Lopingian boundary. Sections with a complete succession of open-marine fauna are particularly rare. Guadalupian–Lopingian boundary successions were reported from Abadeh and Jolfa in Central Iran, from southwestern USA, and the Salt Range. The Laibin Syncline in Guangxi Province, China, is unique among these sections in that it contains a complete and inter-regionally correlatable succession of open-marine conodont zones and other diverse fossils. The GSSP for the Lopingian Series (Figure 24.6) coincides with the first occurrence of *Clarkina postbitteri postbitteri* within an evolutionary lineage from *C. postbitteri hongshuiensis* to *C. dukouensis* at the base of Bed 6k of the Penglaitan section. The Tieqiao (Rail-Bridge) section on the western slope of the syncline is proposed as a supplementary reference section (Jin *et al.*, 2001). The Wuchiapingian GSSP was ratified in 2004 (Jin *et al.*, 2006a).

24.1.3.2. Changhsingian

Initially, the GSSP for the Changhsingian Stage was formally recommended as the horizon between the *Clarkina orientalis* and the *Clarkina subcarinata* zones, which was located at the base of Bed 2, the base of the Changxing Limestone in Section D at Meishan, Changxing County, Zhejiang Province, China (Zhao *et al.*, 1981). Further research revealed that the base of the Changhsingian Stage should be defined within the *Clarkina longicuspidata*–*Clarkina wangi* lineage based on a better understanding of conodont taxonomy and evolution (Mei *et al.*, 2004). The GSSP for the Changhsingian (Figure 24.7) at the first occurrence of *C. wangi* within Bed 4 in Section D at Meishan, Changxing County, China was ratified in 2005 (Jin *et al.*, 2006b).

The basal part of the Changxing Limestone, and, therefore, the boundary interval of the Wuchiapingian–Changhsingian, is marked by the occurrence of advanced forms of *Palaeofusulina*, and the ammonoid families Tapashanitidae and Pseudotirolitidae, which still approximate the lower Changhsingian in the new definition.

The Dzhulfian and Dorashamian stages of Transcaucasia correspond respectively, to the Wuchiapingian and Changhsingian. However, the successions in the basal part of the Dzhulfian Stage and the top portion of the Dorashamian Stage are not as well developed in their type areas as corresponding intervals in the standard succession of South China (Iranian-Chinese Research Group, 1995).

24.2. REGIONAL CORRELATIONS

24.2.1. Russian Platform

Increased research on the Permian of Russia (e.g., Chuvashov and Nairn, 1993; Esaulova *et al.*, 1998; Chuvashov *et al.*, 2002) has led to significant changes from the traditional stratigraphic scheme for the Russian platform discussed earlier. The scheme was largely based on the distribution of ammonoids. For the Lower Permian, i.e. Cisuralian, the much more common fusulinids and conodonts now provide a refined zonation. The traditional "Upper" Permian units of Russia have much less common ammonoids, fusulinaceans, and conodonts than the Cisuralian, but detailed work has greatly improved biostratigraphic correlation (Figure 24.8).

The Kungurian is sparsely fossiliferous at best. To improve its correlation potential, the upper horizon of the traditional Artinskian, the Saranian, the last unit with a well-developed, fully marine fauna, was taken as a reliable horizon for world-wide correlation. Further, the conodont succession from *Neostreptognathodus pequopensis* to *Neostreptognathodus pnevi* was taken as a reliable evolutionary event to establish a base for the Kungurian (Chuvashov *et al.*, 2002). This boundary was discussed previously under the Kungurian of the Cisuralian, and corresponds to a *Neostreptognathodus* evolutionary event in the regional stratotype of the Leonardian, which indicates that the newly revised Kungurian and the Leonardian are nearly equivalent.

The Ufimian has been abandoned by the All Russian Stratigraphic Commission because it represents terrestrial and marginal-marine facies of the upper Kungurian (the Irenian, lower Ufimian) and the lower Kazanian (the Sokian, upper Ufimian). However, there is no known section that shows Sokian (lower Kazanian) lying on Irenian (upper Kungurian), so problems still exist in interpreting this boundary interval.

The Tatarian, basically a series of stacked soils in its type area, is difficult to correlate. It does contain the Illawarra

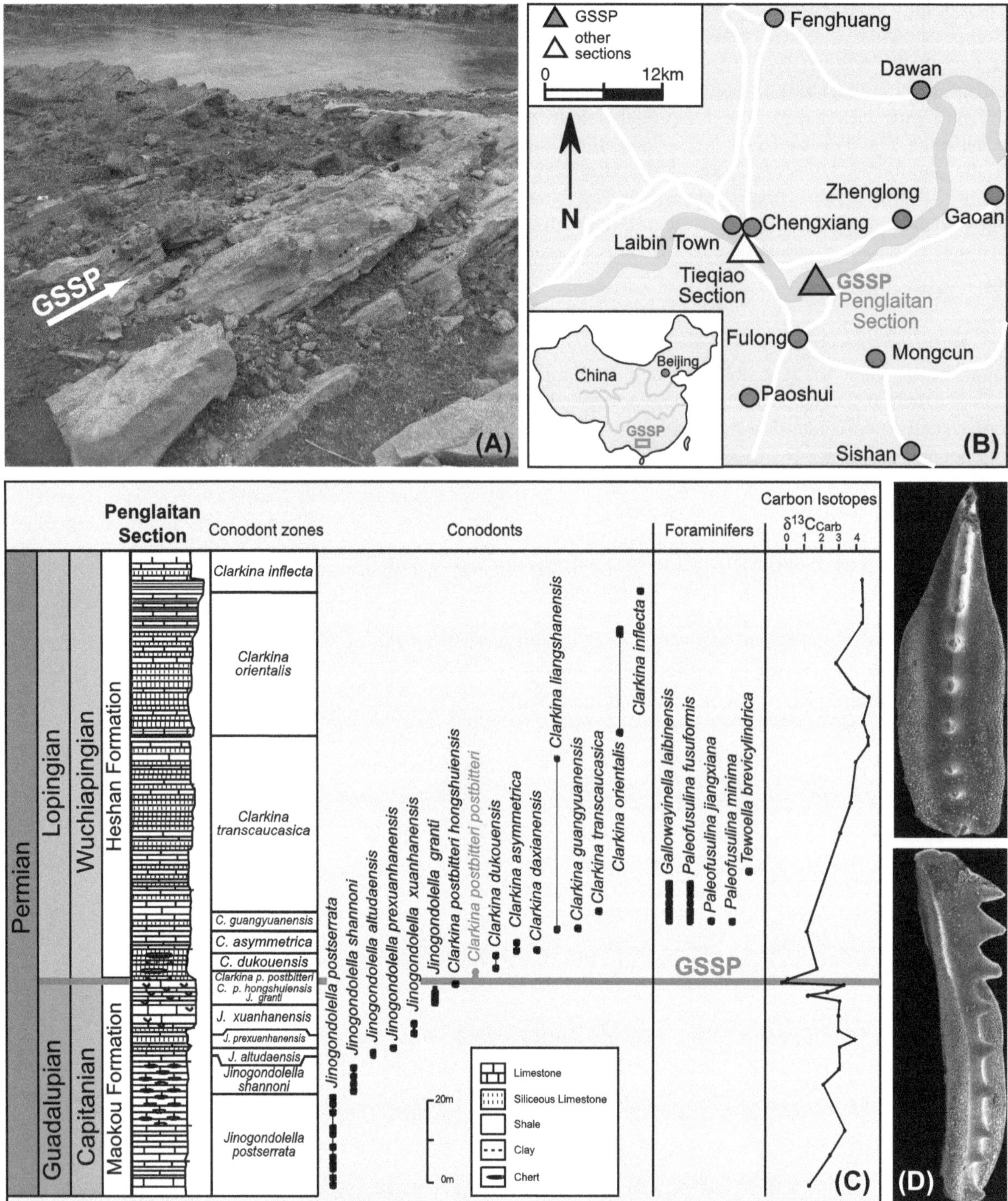

FIGURE 24.6 Global boundary and stratotype section (GSSP) for the Wuchiapingian Stage of the Permian System at Penglaitan section, southern China. Holotype of *Clarkina postbitteri postbitteri*; it is 1 mm long.

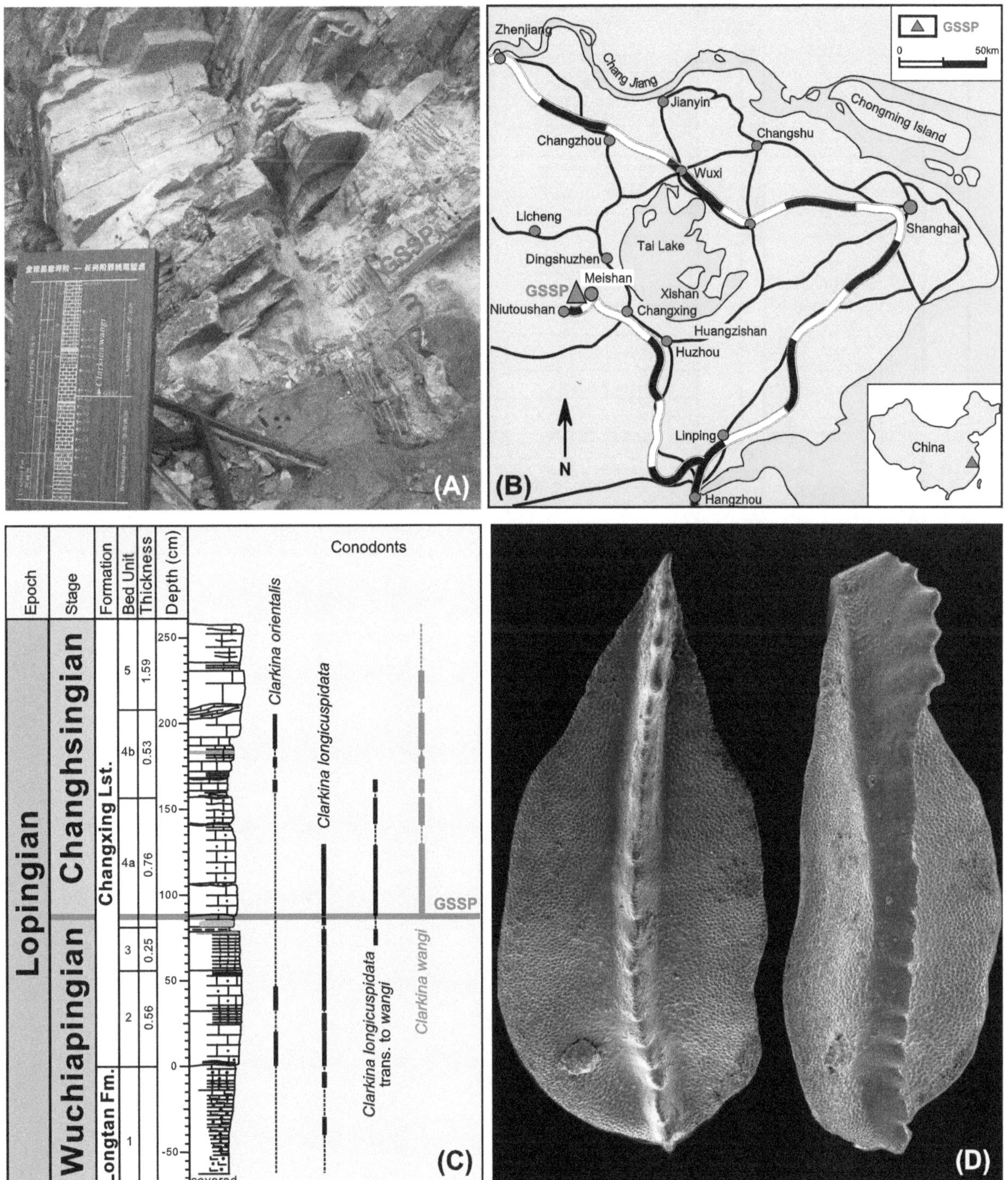

FIGURE 24.7 Global boundary and stratotype section (GSSP) for the Changhsingian Stage of the Permian System at Meishan section D, Zhejiang Province, China. Holotype of *Clarkina wangi*; it is approx. 1 mm long.

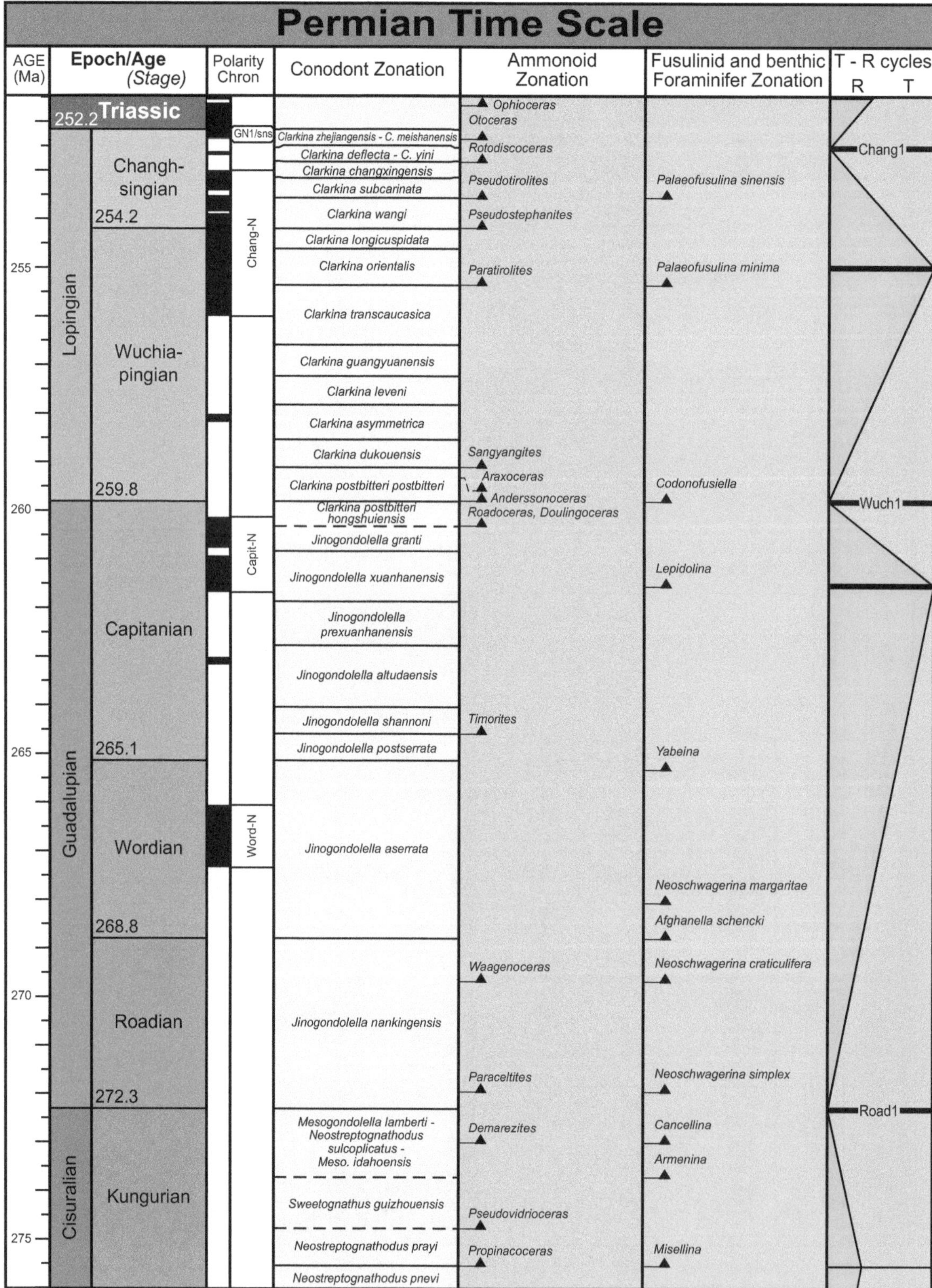

FIGURE 24.8 **Permian time scale and magnetic polarity, conodont, fusulinacean, ammonoid, and general transgressive–regressive sequences.** Magnetic polarity scale compiled from Irving (1963), Khramov, (1963), Peterson and Nairn (1971), Khramov and Davydov (1984), Steiner (1988), Haag and Heller (1991), Heller *et al.* (1995), Opdyke and Channell (1996), and Opdyke *et al.*, (2000).

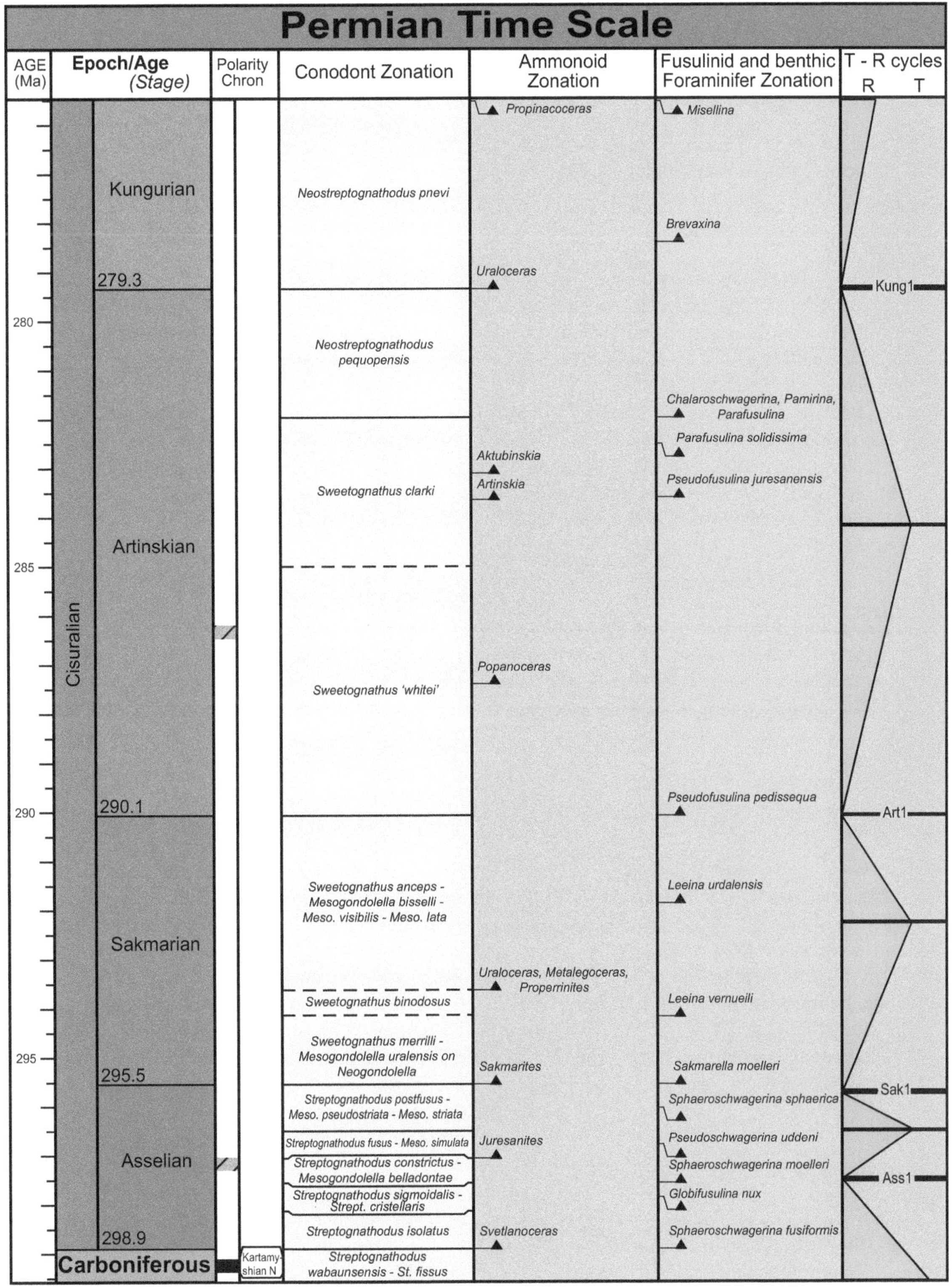

FIGURE 24.8 (*Continued*).

geomagnetic reversal in its lower part, which ties that part to the uppermost Wordian. The age of the youngest Tatarian is an open question. Sequence stratigraphy suggests that the upper Tatarian and Capitanian are roughly equivalent, but the Tatarian top may be a little younger. There is a sharp changeover in fossil taxa at the Tatarian Vetluzhian boundary suggesting a significant unconformity (Figure 24.1).

24.2.2. Germanic Basin

The Germanic Basin is only briefly dealt with here. Zechstein 1 (Figures 24.1 and 24.8) contains a fairly diverse marine microfauna that includes the conodonts *Mesogondolella britannica* and *Merrillina divergens*. This assemblage occurs above the Illawarra reversal, which is within the upper part of the Rotliegendes. Both tie Zechstein 1 to the early Wuchiapingian (Mei and Henderson, 2001). How much time is reflected in the remaining Zechstein units is unknown; the evaporites could represent very short depositional intervals between long hiatuses, or a very short interval similar to that of the Ochoan of west Texas.

24.2.3. Pamirs

The fusulinaceans and, to a lesser extent, the ammonoids are well known from the Pamirs (Figure 24.8). The fusulinacean zonation serves as the standard for the shallow-water Permian of the Tethyan. Conodonts indicate correlation potential, and require major investigation.

24.2.4. Salt Range

The Salt Range (Figure 24.1) does not serve as a regional standard, but has inter-bedded temperate and Tethyan fauna. Of major importance are the overlapping of ranges of *Merrillina praedivergens*, *Neoschwagerina margaritae*, and the Illawarra reversal within the lower part of the Wargal Limestone, below a significant unconformity within that formation, indicating that all are at least part upper Wordian and Capitanian. Above the unconformity occur common *Iranognathus* and a succession of *Clarkina* species.

24.2.5. South China

The regional stages developed for South China (Figure 24.1) are very much defined by their fossil constituents. From Sheng and Jin (1994) the following can be discerned:

Zisongian–is based on the fusulinid *Pseudoschwagerina uddeni–P. texana* Zone and the *Sphaeroschwagerina* Zone.

Longlinian – is based on the biostratigraphic sequence between the last *Pseudoschwagerina* and the first *Misellina* (not an easy concept to establish as a stage). The Zisongian and Longlinian are collectively referred to as the Chuanshanian Series and are approximately equal to the Asselian and Sakmarian (Figure 24.1).

Luodianian – is based on the fusulinid *Misellina* Zone.

Xiangboan – is based on the fusulinid *Cancellina* Zone. The Luodianian and Xiangboan are collectively referred to the Chihsian Series; the base of the Chihsian as discussed in many papers is diachronous, but nearly approximates the base-Artinskian (Figure 24.1).

Kuhfengian – is based on the first occurrence of the conodont *Jinogondolella nankingensis* (the same definition of the Roadian), but in many areas of South China the local first occurrence of *J. nankingensis* is almost certainly younger (upper Roadian) than the FAD in west Texas. The Kuhfengian also includes the range of *J. aserrata*.

Lengwuan – includes the ranges of the conodonts *Jinogondolella shannoni* and *J. xuanhanensis* and is based on the first occurrence of *J. postserrata* (the same definition of the Capitanian). The Kuhfengian and Lengwuan are collectively referred to the Maokovian (Figure 24.1) and are equivalent to the Guadalupian.

Wuchiapingian – is based on the first occurrence of the conodont species *Clarkina postbitteri postbitteri*, just above the origin of this genus.

Changhsingian – is based on the first occurrence of the conodont *Clarkina wangi* near the base of the Changxing Limestone.

Both the Wuchiapingian and Changhsingian definitions have been modified, voted, and ratified as GSSPs and are dealt with under the Lopingian (Upper Permian).

24.2.6. West Texas

Both the Wolfcampian and Ochoan (Figure 24.1), the former based on a sequence of delta front conglomerate, sandstone, and siltstone, and the latter based on basin-filling evaporites, pose problems for correlation. Conodonts from scattered units in the Wolfcampian suggest that the top bed of the Grey Limestone Member of the Gaptank Formation contains *Streptognathodus isolatus* Zone conodonts and, therefore, signifies the base of the Permian. The overlying Neal Ranch Formation has conodonts and fusulinaceans in scattered limestones, and contains *Streptognathodus isolatus* and *S. barskovi* (Wardlaw and Davydov, 2000). The upper part of the Neal Ranch Formation and the lower part of the overlying Lenox Hills Formation yield only sparse fauna. *Sweetognathus whitei* is present along with common fusulinaceans in the upper part of the Lenox Hills Formation. The Skinner Ranch and Cathedral Mountain formations (Leonardian) yield an abundance of fauna. In the base of the Skinner Ranch Formation, below the first occurrence of common *Neostreptognathodus "exsculptus"* is a sparse fauna of *N. pequopensis*, which correlates with the Upper Artinskian. Similarly, the Road Canyon and Word formations yield abundant and diverse fauna. Slope and basinal equivalents of the Vidrio Formation and Capitan Limestone also yield excellent microfossil assemblages.

Recent studies have documented conodonts and radiolarians in the Reef Trail Member of the Bell Canyon Formation that indicate proximity to the Guadalupian–Lopingian boundary (Lambert *et al.*, 2010; Maldonado and Noble, 2010). In the Tansill Formation (the shelf equivalent of the Lamar Limestone Member of the Bell Canyon Formation), there is a sparse fauna dominated by a species of *Sweetina* (Croft, 1978). Virtually the same species is found overlying the evaporites of the Ochoan stage, within the Rustler Formation (above the Castille and Salado formations), implying that the deposition of the basin-filling evaporites occurred in much less than one conodont zone.

24.3. PERMIAN STRATIGRAPHY

24.3.1. Biostratigraphy

24.3.1.1. Conodont Zonation

Since conodont biostratigraphy is critical to the standard Permian stratigraphy, it is briefly reviewed here. The species listed in Figure 24.8 are arranged according to first appearance; zone boundaries are not included to reflect that some taxa range through more than one zone.

The Lower Permian (Cisuralian Series) conodont zonation is derived from a variety of published and unpublished sources. The Asselian and Lower Sakmarian zonation, based on the succession of *Streptognathodus* species, is from Chernykh *et al.* (1997), Boardman *et al.* (1998), and Wardlaw *et al.* (1999) as well as two books by Valeri Chernykh published in Russian (Chernykh, 2005, 2006). This succession is well represented in Kansas and the southern Ural Mountains of Russia. The Sakmarian succession is based on a lineage of *Sweetognathus* and *Mesogondolella* species; the latter is best known in the Urals of Russia (Chernykh, 2005, 2006), but is also being found in other areas now.

The Artinskian zonation is based on species of *Sweetognathus* and *Neostreptognathodus* and reflects the major changeover in the forms dominating shelf fauna during this interval. It is largely based on material from the southern Urals (Chernykh, 2005, 2006).

The Kungurian zonation is based on the succession of *Neostreptognathodus* species modified from Wardlaw and Grant (1987) from West Texas, USA.

The Middle Permian (Guadalupian Series) conodont zonation is from Wardlaw and Lambert (1999) and Wardlaw (2000), except that the rapid succession of upper Guadalupian *Jinogondolella* (*altudaensis, prexuanhanensis,* and *xuanhanensis*) are all overlapped by *J. altudaensis* and considered as subzone indicators of that zone. This succession is well represented in West Texas and South China.

The Upper Permian (Lopingian Series) conodont zonation of successive *Clarkina* species from Mei *et al.* (1994, 1998), as modified by Wardlaw and Mei (1998), reflects the compromise (Henderson *et al.*, 2001) for definition of the base-Lopingian using redefined *C. postbitteri hongshuiensis* and *C. postbitteri postbitteri* (Henderson *et al.*, 2002). Also, even though the classic definition of the Changhsingian was the first appearance of *C. subcarinata*, a dramatic changeover in fauna occurs within Bed 4 in the Changxing Limestone from one dominated by *C. orientalis* and C. *longicuspidata* (Wuchiapingian) to one dominated by *C. wangi* (Mei *et al.*, 2001), which marks a boundary closer to the base of the formation and more acceptable to non-conodont workers. This succession is well represented in both South China and the Dzhulfa area of Iran and Transcaucasia.

24.3.1.2. Fusulinacean Zones

The species shown in Figure 24.8 are listed as close as possible to their point of origin. Zone boundaries are not defined because it is not possible in many cases to precisely correlate them with the conodont zonation. Significant turnovers occur in the Kungurian (from schwagerinids to neoschwagerinids) and in the Upper Capitanian (loss of large fusulinaceans). In the southern Urals, sequence boundaries coincide with the bases of several fusulinacean zones. Eustatic lowstands correspond to significant fusulinacean extinction events. The base of the Asselian (i.e. the base of the Permian), and the base of the Sakmarian coincide with lowstands, and some fusulinacean zones coincide with highstands. Therefore, fusulinacean speciation appears to be associated with both highstands and lowstands. Sea-level lowstands may have been very stressful for global fusulinacean assemblages, and may have been a catalyst for both speciation and extinction. Highstands also may have created environmental opportunities and appear to be more closely associated with fusulinacean speciation than extinction. Sequence boundaries located within fusulinacean zones perhaps reflect local tectonism or local climatic changes.

24.3.2. Physical Stratigraphy

24.3.2.1. The Illawarra Geomagnetic Polarity Reversal

As mentioned earlier, the Illawarra geomagnetic reversal (Figure 24.1) is an important tie point for the Guadalupian Series and proximity to the base of the Capitanian Stage. The reversal is near the top of the *J. aserrata* Zone in West Texas and near the top of the *M. praedivergens* Zone in the Salt Range.

In the Guadalupe Mountains, a tuff near the projected position of the Illawarra, within the top of the range of *J. aserrata*, just below the first occurrence of *J. postserrata*, the indicator for the Capitanian, yields a date of 265.35 ± 0.46 Ma (Item P9 in Appendix 2 of this volume), but this date may be too young due to Pb-loss in zircons unmitigated by physical abrasion methods, and is awaiting reanalysis.

24.3.2.2. Geochemistry

The seawater $^{87}Sr/^{86}Sr$ curve (Figure 24.9) indicates that a minimum value for all of Phanerozoic time occurred during the Capitanian. The driving mechanism of the Late Permian variations appears to have been climate change rather than tectonic; Pangea was assembled and fairly stable through this interval with little mid-ocean ridge activity that would normally produce a decrease in the isotopic values. The period of decreasing seawater $^{87}Sr/^{86}Sr$ for the Early–Middle Permian begins with a waning ice age, and continues in association with high continental aridity and low external runoff attributed to the huge Pangean landmass. The increase in the isotopic values during the Late Permian and Early Triassic may be related to increased continental weathering by climate changes associated with volcanism from E-Meishan and Siberian Traps volcanism (Algeo *et al.*, 2012). The Asselian is marked by high-frequency cyclothems associated with glacial eustasy, and this is consistent with higher values in oxygen isotopes indicating cooler temperature. A fluctuating trend around a mean $\delta^{18}O$ value of $-2‰$ begins in the mid-Sakmarian and continues through much of the remaining Permian indicating an overall warmer climate following glaciation. Though not apparent in the oxygen-18 curve, there is faunal evidence for significant cooling after the early Roadian in the Sverdrup Basin continuing for most of the rest of the Permian (Henderson, 2002) and cooling in the Capitanian in the Salt Range, indicating bipolarity in cooling during much of the Middle Permian. There is considerable evidence for a runaway greenhouse and significant increase in temperatures in the latest Permian and Early Triassic.

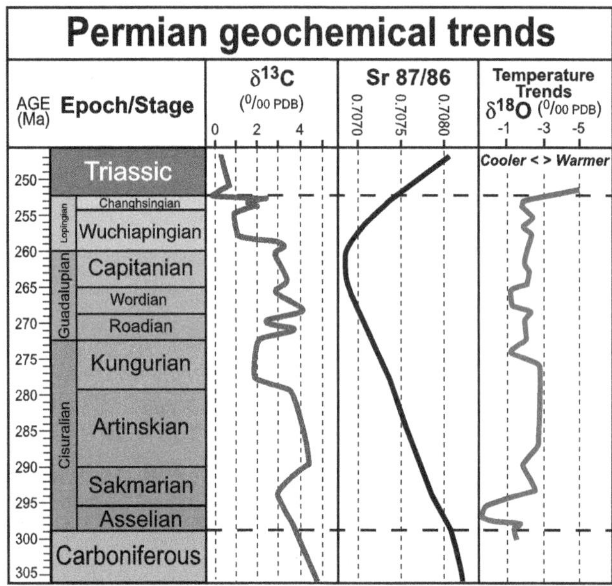

FIGURE 24.9 Geochemical trends during the Permian Period. The schematic carbon isotope curve is a composite of two segments. The first is from Buggisch *et al.* (2011) for the Upper Carboniferous to mid-Capitanian based on the Naqing slope to basinal carbonate succession in South China (Kungurian-Roadian boundary has been shifted down to about 560 meters based correlation with the Murgabian). The second is from Shen *et al.* (2010) based on South China sections of Late Capitanian to Early Triassic age. The schematic $^{87}Sr/^{86}Sr$ curve is also based on two sets of data including Schmitz *et al.* (2009) for the Upper Carboniferous through Artinskian with a projection into the Kungurian and Early Guadalupian and Korte *et al.* (2003) for the Late Permian to Early Triassic. The schematic oxygen isotope curve (inverted scale) is derived from Korte *et al.* (2008). Large-scale global shifts to higher oxygen-18 values in carbonates are generally interpreted as cooler seawater or glacial episodes, but there are many other contributing factors (e.g., Veizer *et al.*, 1999; Wallman, 2001).

Marine extinctions began in a step-wise function following the Guadalupian. This suggests significant temperature fluctuations and the beginning of mass extinction coincided with the increase in $^{87}Sr/^{86}Sr$. This further suggests an amelioration of the climate, leading to a decrease in continentality, an increase in precipitation, and an overall declining sea level in the Upper Permian, with more area exposed to erosional processes.

There are many excursions in the global carbon cycle that overall shows a negative trend, but include significant positive excursions in the Sakmarian, Roadian and early Changhsingian and significant negative excursions in the Kungurian, early Wuchiapingian and the most significant negative carbon excursion just below the Permian–Triassic boundary. The latter excursion is related to the final extinction event of the Permian indicating a sharp decrease in productivity and burial of organic carbon at the very end of the Permian. Although there is still some uncertainty associated with the causes leading to Earth's greatest extinction (Erwin, 1993, 1995), there is a general consensus emerging that it is associated with effects related to Siberian Traps volcanism.

24.4. PERMIAN TIME SCALE

24.4.1. Radiometric Age Dates

Although precise age constraints are in place for the base and top of the Permian, the Permian time scale is among the least internally constrained in the Phanerozoic. Most of the current precise radiometric dates are available for the lower and upper parts of the Permian, but few for the Middle Permian. Many stratigraphically precise and analytically accurate radiometric dates have been acquired for GTS2012, and new high-temperature annealing techniques and interlaboratory correlation are leading to very precise results. As a result, many age dates presented in GTS2004 are now considered inaccurate. Furthermore, the intra-Permian scale is still heavily dependent on relative zonal scaling and is approximate at best. Recent research in the Urals has produced numerous new dates within the Asselian to Early Artinskian interval that has changed stage boundaries significantly from GTS2004 (Schmitz and Davydov, 2011; see Chapter 23).

Below, we discuss current additions to the Permian radiometric database; all age dates are listed with internal (lab based) uncertainties at the 95% level. The complete radiometric data set for the Permian Period used for GTS2012 is in Appendix 2 of this volume.

Ramezani *et al.* (2003) collected volcanic ash beds within mid–ramp carbonate as well as offshore mixed carbonate–siliciclastic successions in three southern Urals sections: Usolka, Dal'ny Tulkus road cut, and Dal'ny Tulkus quarry. One ash layer in the Usolka section is 0.6 m above the Carboniferous–Permian boundary. It contains numerous zircons and the conodont *Streptognathodus isolatus* – the index species for the base of the Permian. Preliminary results from TIMS analyses of the U-Pb ratios in zircons from this ash, and others stratigraphically just above and below the Carboniferous–Permian boundary, suggest an age of 299 Ma for the boundary, to which an uncertainty of 1 myr was assigned. Ramezani *et al.* (2007) reported four U-Pb zircon ages at the Usolka section that bracket the Carboniferous–Permian boundary (299.22 ± 0.13 Ma and 298.49 ± 0.13 Ma) placing the boundary at 298.7 ± 0.15 Ma. Three ash beds from Usolka constrain the Asselian–Sakmarian boundary at 295.0 ± 0.18 Ma, and three ash beds at Dal'ny Tulkas constrain the Sakmarian–Artinskian boundary at 290.0 ± 0.26 Ma (see Appendix 2). These boundaries are respectively 0.4 myr and 5.6 myr older than reported in GTS2004. Continuing paleontologic and geochronologic studies in the southern Urals provide an even more robust data set for precise calibration of the Upper Carboniferous through the Cisuralian interval of the time scale (e.g., Schmitz and Davydov, 2011).

Bowring *et al.* (1998) discussed several important TIMS dates, of which a few are well constrained within the proposed stratotypes for the Middle and Late Permian. In particular, an age of 265.3 ± 0.2 Ma from just below the ratified GSSP for the base of the Capitanian in the lower unit of the Bell Canyon Formation below the Pinery Limestone Member nearly coincides with the estimated age of 265 Ma by Menning (1995) for the Illawarra reversal, although it is possible that Pb-loss might mean that this age is too young. Menning (in Glenister *et al.*, 1999) places the Illawarra reversal within this important section between the isotopically-dated horizon and the conodont-defined base of the Capitanian. The Illawarra is suggested to date near 266 Ma in Figure 24.1.

Bowring *et al.* (1998) reported several TIMS dates from the Meishan section in China, including the lower boundary of the Changhsingian Stage. A date of 253.4 ± 0.2 Ma was derived from Bed 7, above the first occurrence of *Clarkina wangi* (top of Bed 4) and immediately below the first occurrence of *Clarkina subcarinata* (*sensu strictu*) (Bed 8). Mundil *et al.* (2001, 2004) discussed several important TIMS dates, of which several are well constrained within the proposed stratotypes for the Late Permian. Controversy regarding estimates of the timing and tempo of the Permo–Triassic extinction at the time of compilation of GTS2004 has been largely resolved through the application of the chemical abrasion method to zircons from ash beds in the Meishan and Shangsi sections of south China. New and revised dates reported in Shen *et al.* (2010) from Meishan and Shangsi sections that bracket the *wangi* Zone and the base of the Changhsingian Stage indicate an age of 254.2 ± 0.07 Ma.

Computation of the age of the Permian–Triassic boundary in GTS2004 was based on two series of consecutive TIMS age dates of Bowring *et al.* (1998) at the GSSP sections at Meishan and a correlative locality near Heshan in China, The final estimate for the age of the boundary at that time was reported at 251.0 ± 0.2 Ma. New dates reported in Shen *et al.* (2010), together with some recently released, indicate that the PTB is now dated to 252.17 ± 0.06 Ma (Shen *et al.*, 2011).

24.4.2. Permian Composite Standard

In order to integrate and calibrate zonal successions of foraminifers, conodonts, and ammonoids, and to construct a Permian composite biostratigraphic scale, composite standard techniques were undertaken using CONOP-9. The composite standard technique was applied to a joint Carboniferous–Permian data set that is discussed in Section 23.3.3 of Chapter 23, and included here, in part, as Table 24.1 and Figure 24.10. Stability and resolution of the composite will increase with more TIMS-type radiometric dates; at least one or two per stage is a desirable goal.

With the base of the Permian near 299 Ma, at the base of the *S. isolatus* conodont zone, and the top of the Permian, at the base of the *parvus* conodont zone near 252 Ma, the Permian lasted about 47 myr. The Artinskian and Kungurian are the longest stages, and the Changhsingian is the shortest stage (Table 24.2). Permian conodont zones appear to range in duration from 0.35 myr in the Changhsingian to 3.5 myr in the Roadian and Wordian, with many around 1 myr.

24.4.3. Age of Stage Boundaries

In the final analysis for the age of the stage boundaries and stage durations in the Permian (and Carboniferous) periods, the chronometric and stratigraphic ages of the radiogenic isotope dates of Appendix 2 of this volume were subjected to a cubic spline fit. The method is described in Chapter 14,

TABLE 24.1 Listing of the Composite Standard Values of the Zones and of the Incorporated Ash Beds with their Radiometric Age Dates, with 95% Confidence Limits and External Error Estimates

Stage	Conodont Zone Base	Age in Ma with 95% C.L. and λ error	Composite Level
Triassic	*Hindeodus parvus*		
Triassic		Base Triassic at 252.2 ± 0.5	2656.73
Changhsingian		252.5 ± 0.42	2658.97
Changhsingian	*Clarkina zhejiangensis–C. meishanensis*		2650.18
Changhsingian	*Clarkina deflecta–yini*		2642.84
Changhsingian	*Clarkina changxingensis*		2634.36
Changhsingian		253.24 ± 0.46	2627.25
Changhsingian	*Clarkina subcarinata*		2624.21
Changhsingian		253.7 ± 0.46	2616.07
Changhsingian	*Clarkina wangi*		2610.63
Wuchiapingian	*Clarkina longicuspidata*		2602.74
Wuchiapingian	*Clarkina orientalis*		2589.19
Wuchiapingian	*Clarkina transcaucasica*		2568.28
Wuchiapingian		257.3 ± 0.51	2560.59
Wuchiapingian	*Clarkina guangyuanensis*		2557.66
Wuchiapingian	*Clarkina leveni*		2548.11
Wuchiapingian	*Clarkina asymmetrica*		2536.38
Wuchiapingian	*Clarkina dukouensis*		2527.15
Wuchiapingian	*Clarkina postbitteri postbitteri*		2515.97
Capitanian	*Jinogondolella granti*		2500.23
Capitanian	*Jinogondolella xuanhanensis*		2484.46
Capitanian	*Jinogondolella prexuanhanensis*		2471.35
Capitanian	*Jinogondolella altudaensis*		2453.61
Capitanian	*Jinogondolella shannoni*		2446.09
Capitanian	*Jinogondolella postserrata*		2438.82
		265.35 ± 0.46	2435.07
Woardian	*Jinogondolella aserrata*		2393.78
Roadian	*Jinogondolella nankingensis*		2354.18
Kungurian	*Neostreptognathodus prayi*		2318.76
Kungurian	*Neostreptognathodus pnevi*		2277.23
Artinskian	*Neostreptognathodus pequopensis*		2247.87
Artinskian	*Sweetognathus clarki*		2222.49
Artinskian		288.21 ± 0.34	2518.44

TABLE 24.1 Listing of the Composite Standard Values of the Zones and of the Incorporated Ash Beds with their Radiometric Age Dates, with 95% Confidence Limits and External Error Estimates—cont'd

Stage	Conodont Zone Base	Age in Ma with 95% C.L. and λ error	Composite Level
Artinskian		288.36 ± 0.35	2513.92
Artinskian	*Sweetognathus whitei*		2141.53
Sakmarian		290.5 ± 0.35	2138.88
Sakmarian		290.81 ± 0.36	2133.26
Sakmarian		291.50 ± 0.36	2104.31
Sakmarian	*Sweetognathus anceps*		2045.88
Sakmarian	*Sweetognathus binodosus*		2028.15
Sakmarian	*Sweetognathus merrilli*		1987.05
Asselian	*Streptognathodus postfusus –Meso. pseudostriata –Meso. striata*		1963.27
Asselian		296.69 ± 0.37	1956.62
Asselian	*Streptognathodus fusus–Mesogondolella simulata*		1951.14
Asselian	*Streptognathodus constrictus–Mesogondolella belladontae*		1940.86
Asselian		298.05 ± 0.56	1931.16
Asselian	*Streptognathodus sigmoidalis–Strep. cristellaris*		1929.93
Asselian		298.49 ± 0.37	1921.55
Asselian	*Streptognathodus isolatus*		1915.96
Gzhelian		299.22 ± 0.37	1912.24
Gzhelian	*Streptognathodus wabaunsensis–St. fissus*		1897.67
Gzhelian		300.22 ± 0.37	1889.68
Gzhelian	*Streptognathodus simplex–Streptognathodus bellus*		1878.96
Gzhelian		301.29 ± 0.36	1876.25
Gzhelian		301.82 ± 0.36	1873.47
Gzhelian	*Streptognathodus virgilicus*		1862.09
Gzhelian	*Streptognathodus vitali*		1850.17
Gzhelian		303.1 ± 0.36	1842.09
Gzhelian	*Streptognathodus simulator*		1838.22
Kasimovian		303.54 ± 0.39	1837.87
Kasimovian	*Streptognathodus praenuntius*		1833.48
Kasimovian		304.42 ± 0.36	1830.75

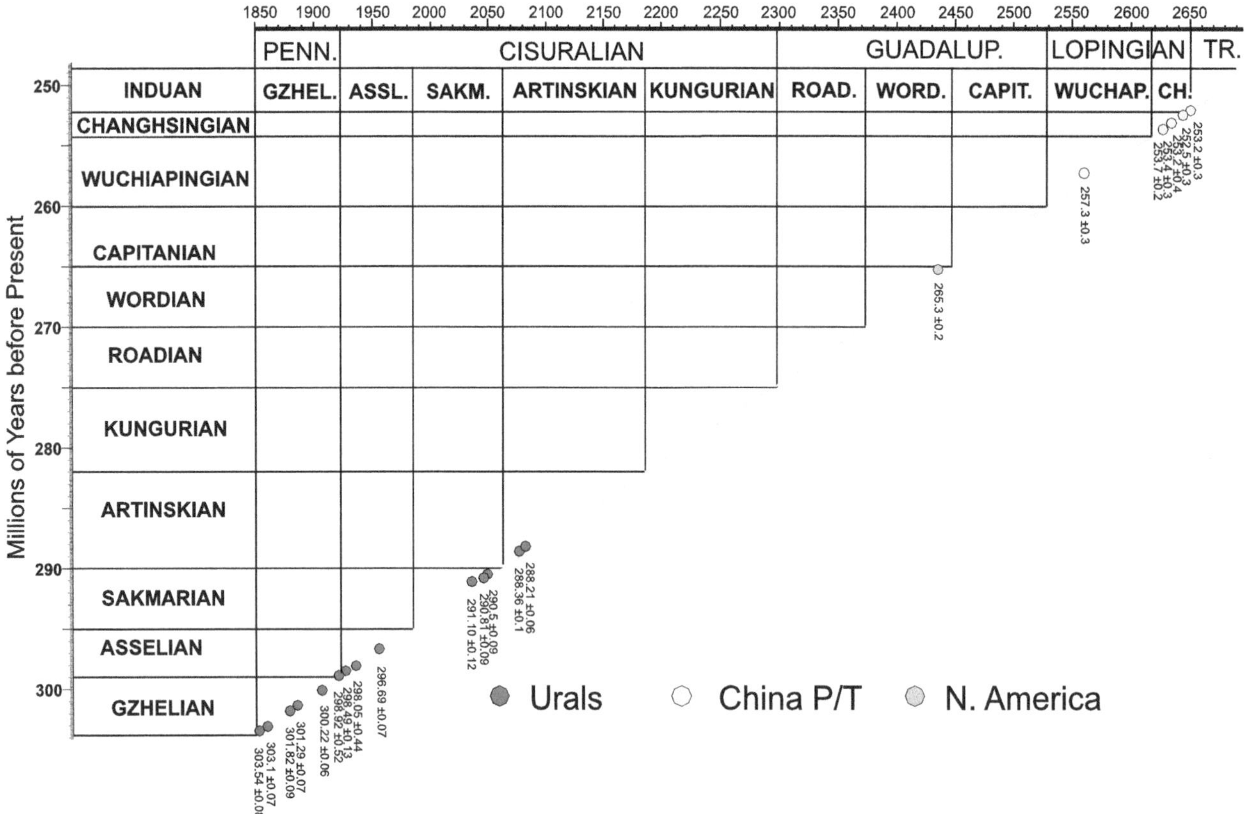

FIGURE 24.10 **Construction of the Permian time scale composite.** Radiometric age dates for the Permian in Appendix 2 are plotted against the relative zonal scale. For details see text.

based on the approach used in GTS2004. Composite standard values are those before linear rescaling (Table 24.1). Points were weighted according to the procedure given in Chapter 14, taking external radiometric errors into account for the stage boundary age estimates. A smoothing factor of 1.2 was calculated by cross-validation.

As mentioned above, analysis of new bentonite samples from Devonian–Carboniferous transition sections provides a revised age estimate for the Devonian–Carboniferous boundary and base of the Tournaisian Stage: 358.9 ± 0.4 Ma. This estimate anchors the lower end of the cubic spline; the anchor for the upper end of the spline is the age of the Permian–Triassic boundary at 252.2 ± 0.5 (see also Chapters 23 and 25).

All of the 46 points passed the chi-squared goodness-of-fit test at a significance level of $p<0.01$. The resulting spline (see Figure 23.10 in Chapter 23 on the Carboniferous Period) is fairly linear through the Carboniferous, apart from a slight increase in the duration of scaling zones (decreased turnover rate) in the Bashkirian.

Interpolated stage boundary ages with their error bars and durations are given in Tables 24.2 and 24.3. The lower boundary ages for the Kungurian and Roadian, when computed through the linear regression fit described earlier in this chapter (Section 24.4.1), differ substantially from the ages computed with the cubic spline, and less so for the Wordian. This is no surprise, given the absence of Middle Permian radiometric dates. The other regression ages are either well within or just outside the error estimate for the cubic spline age of the Permian stage boundaries. The discrepancies will decrease with a more detailed and more stable composite standard and more suitable radiometric ages.

The base-Permian and base-Asselian interpolate at 298.9 ± 0.2 Ma. Asselian lasted from 298.9 ± 0.2 to 295.5 ± 0.4 Ma, for a duration of 3.4 ± 0.1 myr. Sakmarian lasted 5.5 ± 0.1 myr, from 295.5 ± 0.4 to 290.1 ± 0.2 Ma (durations calculated from 2-decimal values). The Artinskian Stage spanned 290.1 ± 0.2 to 279.3 ± 0.6 Ma for a duration of 10.8 ± 0.2 myr, thus being the longest ranging in the Permian Period. Kungurian lasted for 7 ± 0.1 myr, from 279.3 ± 0.6 to 272.3 ± 0.5 Ma. The Roadian Stage started at 272.3 ± 0.5 Ma and ended at 268.8 ± 0.5 Ma, lasting 3.5 ± 0.1 myr. Wordian ran from 268.8 ± 0.5 to 265.1 ± 0.4 Ma for 3.7 ± 0.1 myr, and Capitanian from 265.1 ± 0.4 to 259.8 ± 0.4 Ma for 5.3 ± 0.2 myr. Wuchiapingian lasted 5.6 ± 0.8 myr, from 259.8 ± 0.4 Ma to 254.2 ± 0.3 Ma, and Changhsingian started at 254.2 ± 0.3 Ma and ended at the Permian–Triassic boundary at 252.2 ± 0.5 Ma. The Changhsingian Stage is the

TABLE 24.2 Geochronology of the Carboniferous and Permian Periods

	GTS2012					Linear Regression	GTS2004
Stage Base	Scale	Ma	2-Sigma	Duration	2-Sigma	Ma	Ma
Changhsingian	2610.63	254.2	0.3	2.0	...	254.2	253.8
Wuchiapingian	2515.97	259.8	0.4	5.6	0.2	260.0	260.4
Capitanian	2438.82	265.1	0.4	5.3	0.2	265.0	265.8
Wordian	2393.78	268.8	0.5	3.7	0.1	270.0	268.0
Roadian	2354.18	272.3	0.5	3.5	0.1	275.0	270.6
Kungurian	2277.23	279.3	0.6	7.0	0.1	282.0	275.6
Artinskian	2141.53	290.1	0.2	10.7	0.2	290.0	284.4
Sakmarian	1987.05	295.5	0.4	5.5	0.1	295.0	294.6
Asselian	1915.96	298.9	0.2	3.4	0.1	298.7	299.0
Gzhelian	1838.22	303.7	0.1	4.8	0.1	303.4	303.9
Kasimovian	1796.11	307.0	0.2	3.3	0.1	306.7	306.5
Moscovian	1715.12	315.2	0.2	8.2	0.1	314.6	311.6
Bashkirian	1607.79	323.2	0.4	8.0	0.2	322.8	318.1
Serpukhovian	1445.34	330.9	0.3	7.7	0.2	330.0	326.4
Visean	1187.23	346.7	0.4	15.8	0.2	346.0	345.3
Tournaisian	991.64	358.9	0.4	12.2	0.2	358.9	359.2

The age of the stage boundaries, the duration of the stages and the 95% confidence limits are derived from cubic spline interpolation shown in Figure 23.10. Composite standard values of stage boundaries are also listed. Error estimates on stage boundary ages use the external errors on radiometric ages in Appendix 2, and the internal (laboratory) errors on stage durations. All durations calculated from 2-decimal values. To the right are columns with the GTS2004 age of the stage boundaries and stage boundary ages using linear regression on the composite standard of Figure 24.10.

TABLE 24.3 Geochronology of the Permian Period

Period	*Epoch*	Stage	Age of Base (Ma)	Est. ± Ma (2-Sigma)	Duration (myr)	Est. ± myr (2-Sigma)
Triassic		Induan (top of Permian)	252.20	0.5		
Permian						
	Lopingian					
		Changhsingian	254.2	0.3	2.0	
		Wuchiapingian	259.8	0.4	5.6	0.2
	Guadalupian					
		Capitanian	265.1	0.4	5.3	0.2
		Wordian	268.8	0.5	3.7	0.1
		Roadian	272.3	0.5	3.5	0.1
	Cisuralian					
		Kungurian	279.3	0.6	7.0	0.1
		Artinskian	290.1	0.2	10.7	0.2
		Sakmarian	295.5	0.4	5.5	0.1
		Asselian (base of Permian)	298.9	0.2	3.4	0.1
Carboniferous						

Stage	GSSP Location	Latitude, Longitude	Boundary Level	Correlation Events	Reference
Changhsingian	Meishan, Zhejiang Province, China	31° 4′55″N 119°42′22.9″E	Base of Bed 4a-2, 88 cm above the base of the Changxing Limestone at the Meishan D Section	Conodont FAD *Clarkina wangi*	Episodes 29/3, 2006
Wuchiapingian	Penglaitan, Guangxi Province, South China	23°41″43″N 109°19″16″E	Base of Bed 6k in the Penglaitan Section	Conodont FAD *Clarkina postbitteri postbitteri*	Episodes 29/4, 2006
Capitanian	Nipple Hill, SE Guadalupe Mountains, Texas, USA	31°54′32.8″N 104°47′21.1″W	4.5m above the base of the outcrop section of the Pinery Limestone Member of the Bell Canyon Formation	Conodont FAD *Jinogondolella postserrata*	
Wordian	Guadalupe Pass, Texas, USA	31°51′56.9″N 104°49′58.1″W	7.6m above the base of the Getaway Ledge outcrop section of the Getaway Limestone Member of the Cherry Canyon Formation	Conodont FAD *Jinogondolella aserrata*	
Roadian	Stratotype Canyon, Texas, USA	31°52′36.1″N 104°52′36.5″W	42.7m above the base of the Cutoff Formation	Conodont FAD *Jinogondolella nankingensis*	
Kungurian	*Candidates are in southern Ural Mtns., Russia, or Pequop Mtns., Nevada*			*Near FAD of conodont* Neostreptognathus pnevi	
Artinskian	*Candidates are in southern Ural Mtns.*			*FAD of conodont* Sweetognathus whitei	
Sakmarian	*Candidates are Kondurovsky or Usolka section, Russia.*			*Near FAD of conodont* Sweetognathus merrilli *or* Mesogondolella uralensis	
Asselian *(base Permian)*	Aidaralash Creek, Kazakhstan	50°14′45″N 57°53′29"E*	27m above the base of Bed 19, Aidaralash Creek	Conodont FAD *Streptognathodus isolatus*	Episodes 21/1, 1998

**derived from map
**according to Google Earth*

shortest stage in the Permian with just 2 myr duration, and it is one of the shortest in the Phanerozoic. The Cisuralian Epoch or Early Permian stretches between 298.9 and 272.3 Ma, and has a duration of 26.6 myr. Guadalupian started at 272.3 Ma and its upper boundary has an age of 259.8 Ma, it being 12.5 myr long. The upper epoch of the Permian, the Lopingian, started at 259.8 Ma to the base Triassic at 252.2 Ma, thus lasting 7.6 myr. The total duration of the Permian Period is 46.7 myr, from 298.9 to 252.2 Ma.

The geochronologic scale for the Permian differs substantially from that in GTS2004 for the lower boundary ages of Artinskian, Kungurian and Roadian; and less so for the other six stages. Nevertheless, the latter changes are outside the GTS2004 error limits. Particularly for the Middle Permian there is an urgent need for stratigraphically meaningful radiometric information.

REFERENCES

Algeo, T.J., Henderson, C., Ellwood, B., Rowe, H., Elswick, E., Bates, S., Lyons, T., Hower, J.C., Smith, C., Maynard, B., Hays, L., Summons, R., Fulton, J., Freeman, K., 2012. Evidence for a diachronous Late Permian marine crisis from the Canadian Arctic region. Geological Society of America Bulletin. doi:10.1130/B30505.1.

Beauchamp, B., Henderson, C.M., 1994. The Lower Raanes, Great Bear Cape and Trappers Cove formations, Sverdrup Basin, Canadian Arctic: Stratigraphy and conodont zonation. Canadian Society of Petroleum Geologists Bulletin 42, 562–597.

Boardman II, D.R., Nestell, M.K., Wardlaw, B.R., 1998. Uppermost Carboniferous and lowermost Permian deposition and conodont biostratigraphy of Kansas, USA. In: Jin, Y., Wardlaw, B.R., Wang, Y. (Eds.), Permian Stratigraphy, Environments and Resources, Vol. 2. Stratigraphy and Environments. Palaeoworld, 9, pp. 19–32.

Boardman II, D.R., Wardlaw, B.R., Nestell, M.K., 2009. Stratigraphy and conodont biostratigraphy of the uppermost Carboniferous and Lower Permian from the North American midcontinent. Kansas Geological Survey Bulletin 255, 253.

Bowring, S.A., Erwin, D.H., Jin, Y.G., Martin, M.W., Davidek, K., Wang, W., 1998. U/Pb zircon geochronology and tempo of the end-Permian mass extinction. Science 280, 1039–1045.

Buggisch, W., Wang, X.D., Alekseev, A.S., Joachimski, M.M., 2011. Carboniferous-Permian carbon isotope stratigraphy of successions from China (Yangtze platform), USA (Kansas) and Russia (Moscow Basin and Urals). Palaeogeography, Palaeoclimatology, Palaeoecology 301, 18–38.

Chernykh, V.V., 2005. Zonal methods of biostratigraphy – Conodont zonal scheme for the Lower Permian of the Urals. Ekaterinburg: Russian Academy of Sciences, 217 [In Russian].

Chernykh, V.V., 2006. Lower Permian conodonts of the Urals. Ekaterinburg: Russian Academy of Sciences, 130 [In Russian].

Chernykh, V.V., Ritter, S.M., Wardlaw, B.R., 1997. *Streptognathodus isolatus* new species (Conodonta): Proposed index for the Carboniferous-Permian boundary. Journal of Paleontology 71, 162–164.

Chuvashov, B.I., Nairn, A.E.M., 1993. Permian System: Guides to geological excursions in the Uralian type localities. Occasional Publications of the Earth Sciences and Resources Institute 10, 303.

Chuvashov, B.I., Chernykh, V.V., Davydov, V.I., Pnev, V.P., 1993a. Kondurovsky Section, Permian System, Guides to Geological Excursion in the Uralian type localities. Occasional Publications of the Earth Sciences and Resources Institute 10, 4–31.

Chuvashov, B.I., Chernykh, V.V., Mizens, G.A., 1993b. Zonal divisions of the boundary deposits of Carboniferous and Permian in sections of different facies in the south Urals. Permophiles 22, 11–16.

Chuvashov, B.I., Amon, E.O., Karidrua, M., Prust, Z.N., 1999. Late Paleozoic radiolarians from the polyfacies formations of Uralian Foreland. Stratigrafiya, Geologicheskaya Korrelyatsiya 7 (1), 41–55 [In Russian and English].

Chuvashov, B.I., Chernykh, V.V., Bogoslovskaya, M.F., 2002. Biostratigraphic characteristic of stage stratotypes of the Permian System. Stratigraphy and Geological Correlation 10 (4), 317–333.

Conybeare, W.D., Phillips, W., 1822. Outlines of the Geology of England and Wales, with an introduction compendium of the general principles of that science, and comparative views of the structure of foreign countries. Part 1. William Phillips, London. 470.

Croft, J.S., 1978. Upper Permian conodonts and other microfossils from the Pinery and Lamar Limestones Members of the Bell Canyon Formation and from the Rustler Formation, West Texas. M.Sc. thesis. The Ohio State University, Columbus, p. 176.

Davydov, V.I., Glenister, B.F., Spinosa, C., Ritter, S.M., Chernykh, V.V., Wardlaw, B.R., Snyder, W.S., 1998. Proposal of Aidaralash as GSSP for base of the Permian System. Episodes 21, 11–18.

Dunbar, C.O., 1940. The type Permian: Its classification and correlation. Bulletin of the American Association of Petroleum Geologists 24 (2), 237–281.

Erwin, D.H., 1993. The Great Paleozoic Crisis: Life and Death in the Permian. Columbia University Press, New York, p. 327.

Erwin, D.H., 1995. The end-Permian mass extinction. In: Scholle, P.A., Peryt, T.M., Ulmer-Scholle, D.S. (Eds.), The Permian of Northern Pangea: Vol. I, Paleogeography, Paleoclimates, Stratigraphy. Springer-Verlag, Berlin, 20–34.

Esaulova, N.K., Lozovsky, V.R., Rozanov, A.Y. (Eds.), 1998. Stratotypes and References Sections of the Upper Permian in the Regions of the Volga and Kama Rivers. Ekotsentr, Kazan, p. 300.

Furnish, W.M., 1973. Permian Stage names. In: Logan, A., Hills, L.V. (Eds.), The Permian and Triassic systems and their mutual boundary. Canadian Society of Petroleum Geologists Memoir, 2, 522–549.

Gerasimov, L.P., 1937. The Uralian series of the Permian System. Scientific reports of Kazan University, Geology 97 (3–4) fasc. 8–9: 68.

Gialanella, P.R., Heller, F., Haag, M., Nurgaliev, D., Borisov, A., Burov, B., Jasonov, P., Khasanov, D., Ibragivom, S., Zharkov, I., 1997. Late Permian magnetostratigraphy on the eastern Russian Platform. In: Dekkers, M.J., Langereis, C.G., Van der Voo, R. (Eds.), Analysis of Paleomagnetic Data: A Tribute to Hans Zijderveld. Geologie en Mijnbouw, 76, 145–154.

Glenister, B.F., Wardlaw, B.R., Lambert, L.L., Spinosa, C., Bowring, S.A., Erwin, D.H., Menning, M., Wilde, G.L., 1999. Proposal of Guadalupian and component Roadian, Wordian and Capitanian Stages as International Standards for the Middle Permian Series. Permophiles 34, 3–11.

Griesbach, C.L., 1880. Paleontological notes on the Lower Trias on the Himalayas. Records of the Geological Survey of India 13 (2), 94–113.

Haag, M., Heller, F., 1991. Late Permian to Early Triassic magnetostratigraphy. Earth and Planetary Science Letters 107 (1), 42–54.

Harland, W.B., Armstrong, R.L., Cox, A.V., Craig, L.A., Smith, A.G., Smith, D.G., 1990. A Geologic Time Scale 1989. Cambridge University Press, Cambridge, p. 263.

Heller, F., Chen, H., Dobson, J., Haag, M., 1995. Permian-Triassic magnetostratigraphy: New results from South China. Physics of the Earth and Planetary Interiors 89 (3–4), 281–295.

Henderson, C.M., Mei, S.L., 2007. Geographical clines in Permian and lower Triassic gondolellids and its role in taxonomy. Palaeoworld 16, 190–201.

Henderson, C.M., 1988. Conodont paleontology and biostratigraphy of the Upper Carboniferous to Lower Permian Canyon Fiord, Belcher Channel, Nansen, unnamed, and Van Hauen formations, Canadian Arctic Archipelago. Unpublished Ph.D. thesis. University of Calgary, p. 287.

Henderson, C.M., 2002. Kungurian to Lopingian correlations along western Pangea. Geological Society of America Abstracts with Programs 34 (3) A–30.

Henderson, C.M., 2010. Update on base-Artinskian GSSP. Permophiles 55, 15–17.

Henderson, C.M., Wardlaw, B., Mei, S., 2001. New conodont definitions at the Guadalupian–Lopingian boundary. Permophiles 38, 35–36.

Henderson, C.M., Mei, S.L., Wardlaw, B.R., 2002. New conodont definitions at the Guadalupian-Lopingian boundary. In: Hills, L.V., Henderson, C.M., Bamber, E.W. (Eds.), Carboniferous and Permian of the World. Canadian Society of Petroleum Geologist Memoir, 19, pp. 725–735.

Huang, T.K., 1932. The Permian formations of Southern China. Memoirs of the Geological Survey of China, Series A 10, 1–40.

Iranian-Chinese Research Group, 1995. Field work on the Lopingian stratigraphy in Iran. Permophiles 27, 5–6.

Jin, Y.G., 1994. Two phases of the end-Permian mass extinction. Canadian Society of Petroleum Geologists. Memoir 17, 813–822.

Jin, Y.G., Glenister, B.F., Kotlyar, G.V., Sheng, J.-Z., 1994. An operational scheme of Permian Chronostratigraphy. In: Jin, Y., Utting, J., Wardlaw, B.R. (Eds.), Permian Stratigraphy, Environments and Resources. Palaeoworld, 4, 1–14.

Jin, Y.G., Wardlaw, B.R., Glinister, B.F., Kotlyar, G.V., 1997. Permian chronostratigraphic subdivisions. Episodes 20, 10–15.

Jin, Y.G., Henderson, C.M., Wardlaw, B.R., Glenister, B.F., Mei, S.L., Shen, S.H., Wang, X.D., 2001. Proposal for the Global Stratotype Section and Point (GSSP) for the Guadalupian-Lopingian boundary. Permophiles 39, 32–42.

Jin, Y.G., Shen, S.Z., Henderson, C.M., Wang, X.D., Wang, W., Wang, Y., Cao, C.Q., Shang, Q.H., 2006a. The Global Stratotype Section and Point (GSSP) for the boundary between the Capitanian and Wuchiapingian stage (Permian). Episodes 29, 253–262.

Jin, Y.G., Wang, Y., Henderson, C.M., Wardlaw, B.R., Shen, S.Z., Cao, C.Q., 2006b. The Global Boundary Stratotype Section and Point (GSSP) for the base of the Changhsingian Stage (Upper Permian). Episodes 29, 175–182.

Karpinsky, A.P., 1874. Geological investigation of the Orenburg area. Zapiski Imperatorskargo S. Peterburgskago Mineralogicheskoe Obshchestvo, series 2 (9), 212–310 [In Russian].

Karpinsky, A.P., 1889. Über die Ammoneen der Artinsk-Stufe und einige mit denselben verwandte carbonische Forman. Memoir of the Imperial Academy of Science St. Petersburg, series 7 37 (2), 104.

Karpinsky, A.P., 1891. On Artinskian ammonites and some similar Carboniferous forms. Zapiski Imperatorskargo S. Peterburgskago Mineralogicheskoe Obshchestvo, series 2 (27), 1–192 [In Russian].

Khramov, A.N., 1963. Paleomagnetism of Paleozoic. VNIGRI, St. Petersburg, p. 283 [In Russian].

Khramov, A.N., Davydov, V.I., 1984. Paleomagnetism of Upper Carboniferous and Lower Permian in the South of the U.S.S.R. and the problems of structure of the Kiama hyperzone. Transactions of VNIGRI, St. Petersburg, pp. 55–73. [In Russian].

Korte, C., Kozur, H.W., Bruckschen, P., Veizer, J., 2003. Strontium isotope evolution of Late Permian and Triassic seawater. Geochimica et Cosmochimica Acta 67 (1), 47–62.

Korte, C., Jones, P.J., Brand, U., Mertmann, D., Veizer, J., 2008. Oxygen isotope values from high-latitudes: Clues for Permian sea-surface temperature gradients and Late Paleozoic deglaciation. Palaeogeography, Palaeoclimatology, Palaeoecology 269, 1–16.

Kotlyar, G.V., Stepanov, D.L. (Eds.), 1984. Main Features of the Stratigraphy of the Permian System in USSR. Nedra, Leningrad, p. 233.

Lambert, L.L., Bell, G.L., Fronimos, J.A., Wardlaw, B.R., Yisa, M.O., 2010. Conodont biostratigraphy of a more complete Reef Trail Member section near the type section, latest Guadalupian Series type region. Micropaleontology 56 (1–2), 233–253.

Likharew, B.K. (Ed.), 1966. The Stratigraphy of USSR: The Permian System. Nedra, Moscow, p. 536.

Maldonado, A.L., Nobel, P.J., 2010. Radiolarians from the upper Guadalupian (Middle Permian) Reef Trail Member of the Bell Canyon Formation, West Texas and their biostratigraphic implications. Micropaleontology 56 (1–2), 69–115.

Mei, S.L., Henderson, C.M., 2001. Evolution of Permian conodont provincialism and its significance in global correlation and paleoclimate implication. Palaeogeography, Palaeoclimatology, Palaeoecology 170, 237–260.

Mei, S.L., Henderson, C.M., 2002. Conodont definition of the Kungurian (Cisuralian) and Roadian (Guadalupian) boundary. In: Hills, L.V., Henderson, C.M., Bamber, E.W. (Eds.), Carboniferous and Permian of the World. Canadian Society of Petroleum Geologists Memoir, 19, 529–551.

Mei, S.L., Jin, Y.G., Wardlaw, B.R., 1994. Succession of conodont zones from the Permian Kufeng Formation, Xuanhan, Sichuan and its implications in global correlation. Acta Palaeontologica Sinica 33 (1), 1–23.

Mei, S.L., Jin, Y.G., Wardlaw, B.R., 1998. Conodont succession of the Guadalupian-Lopingian boundary strata in Laibin of Guangxi. Chian and West Texas, USA. Palaeoworld 9, 53–76.

Mei, S.L., Henderson, C.M., Wardlaw, B., 2001. Progress on the definition for the base of the Changhsingian. Permophiles 38, 36–37.

Mei, S.L., Henderson, C.M., Wardlaw, B.R., 2002. Evolution and distribution of the conodonts *Sweetognathus* and *Iranognathus* and related genera during the Permian, and their implications for climate change. Palaeogeography, Palaeoclimatology, Palaeoecology 180, 57–91.

Mei, S.L., Henderson, C.M., Cao, C.Q., 2004. Conodont sample-population approach to defining the base of the Changhsingian Stage, Lopingian Series, Upper Permian. In: Beaudoin, A., Head, M. (Eds.), The Palynology and Micropalaeontology of Boundaries. Geological Society Special Publications, 230, 105–121.

Menning, M., 1987. Problems of stratigraphic correlation: Magnetostratigraphy. In: Lützner, H. (Ed.), Sedimentary and Volcanic Rotliegendes of the Saale Depression: Excursion Guidebook. Academy of Sciences of the GDR, Central Institute for Physics of the Earth, Potsdam, 92–96.

Menning, M., 1995. A numerical time scale for the Permian and Triassic Periods: An integrated time analysis. In: Scholle, P.A., Peryt, T.M., Ulmer-Scholle, D.S. (Eds.), The Permian of Northern Pangea. Vol. I, Paleogeography, Paleoclimates, Stratigraphy. Springer-Verlag, Berlin, 77–97.

Menning, M., 2000. Magnetostratigraphic results from the Middle Permian type section, Guadalupe Mountains, West Texas. Permophiles 37, 16.

Mundil, R., Metcalfe, I., Ludwig, K.R., Renne, P.R., Oberli, F., Nicoll, R.S., 2001. Timing of the Permian-Triassic biotic crisis: Implications from new zircon U/Pb age data (and their limitations). Earth and Planetary Science Letters 187, 131–145.

Mundil, R., Ludwig, K.R., Metcalfe, I., Renne, P.R., 2004. Age and timing of the Permian mass extinctions: U/Pb dating of closed-system zircons. Science 305, 1760–1763.

Murchison, R.I., 1841. First sketch of the principal results of a second geological survey of Russia. Philosophical Magazine, series 3 (19), 417–422.

Murchison, R.I., 1872. Siluria: A history of the oldest rocks in the British Isles and other countries, fifth ed. John Murray, London, p. 566.

Opdyke, N.D., Channell, J.E.T., 1996. Magnetic Stratigraphy. Academic Press, New York, p. 346.

Opdyke, N.D., Roberts, J., Claoue-Long, J., Irving, E., Jones, P.J., 2000. Base of Kiaman: Its definition and global significance. Geological Society of America Bulletin 112 (9), 1315–1341.

Peterson, D.N., Nairn, A.E.M., 1971. Palaeomagnetism of Permian redbeds from the South-western United States. Geophysical Journal of the Royal Astronomical Society 23 (2), 191–205.

Ramezani, J., Davydov, V.I., Northrup, C.J., Bowering, S.A., Chuvashov, B., Snyder, W.S., 2003. Volcanic ashes in the Upper Paleozoic of the Southern Urals: Opportunities for high-precision calibration of the Upper Carboniferous Cisuralian interval. Abstracts of the XVth International Congress on Carboniferous and Permian Stratigraphy, 431–432. Utrecht.

Ramezani, J., Schmitz, M.D., Davydov, V.I., Bowring, S.A., Snyder, W.S., Northrup, C.J., 2007. High-precision U-Pb zircon age constraints on the Carboniferous-Permian boundary in the Southern Urals stratotype. Earth and Planetary Science Letters 256, 244–257.

Resolutions of the Interdepartmental Stratigraphic Committee of Russia and its Permanent Commissions, 1992. Resolution on Carboniferous/Permian boundary. Interdepartmental Stratigraphic Committee of Russia, St. Petersburg, pp. 52–56 [In Russian].

Rhodes, F.H.T., 1963. Conodonts from the topmost Tensleep Sandstone of the eastern Big Horn Mountains, Wyoming. Journal of Paleontology 37 (2), 401–408.

Ruzhenzev, V.E., 1936. New data on the stratigraphy of the Carboniferous and Lower Permian deposits of the Orenburg and Aktyubinsk regions. Problems of Soviet Geology 6, 470–506 [In Russian].

Ruzhenzev, V.E., 1950. Type section and biostratigraphy of the Sakmarian Stage. Doklady Akademiya Nauk USSR, Novaya Seriya 71 (6), 1101–1104 [In Russian].

Ruzhenzev, V.E., 1954. Asselian Stage of the Permian System. Doklady Akademiya Nauk USSR, Seriya Geologicheskaya 99 (6), 1079–1082 [In Russian].

Schmitz, M.D., Davydov, V.I., 2011. Quantitative radiometric and biostratigraphic calibration of the Pennsylvanian–Early Permian (Cisuralian) time scale, and pan-Euramerican chronostratigraphic correlation. Geological Society of America Bulletin, B30385. doi: 10.1130/830385.1

Schmitz, M.D., Davydov, V.I., Snyder, W.S., 2009. Permo-Carboniferous Conodonts and Tuffs: High precision marine Sr isotope geochronology. Permophiles 53 (Suppl.1), 48.

Sheng, J.Z., Jin, Y.G., 1994. Correlation of Permian deposits of China. In: Jin, Y.G., Utting, J., Wardlaw, B.R. (Eds.), Permian Stratigraphy, Environments and Resources, vol. 1: Palaeontology and Stratigraphy. Palaeoworld, 4, pp. 14–113.

Shen, S.Z., Henderson, C.M., Bowring, S.A., Cao, C.Q., Wang, Y., Wang, W., Zhang, H., Zhang, Y.C., Mu, L., 2010. High Resolution Lopingian (Late Permian) timescale of South China. Geological Journal 45, 122–134.

Shen, S.Z., Crowley, J.L., Wang, Y., Bowring, S.A., Erwin, D.H., Sadler, P.M., Cao, C.Q., Rothman, D.H., Henderson, C.M., Ramezani, J., Zhang, H., Shen, Y., Wang, X.-D., Wang, W., Mu, L., Li, W.-Z., Tang, Y.G., Liu, X.L., Liu, L.J., Zeng, Y., Jiang, Y.F., Jin, Y.G., 2011. Calibrating the end-Permian mass extinction. Science 334, 1367–1372.

Steiner, M.B., 1988. Paleomagnetism of the Late Pennsylvanian and Permian, a test of the rotation of the Colorado Plateau. Journal of Geophysical Research 93 (3), 2201–2215.

Stuckenberg, A.A., 1890. Geological map of Russia: Sheet 138, Geological Studies in the Northwestern Part of Sheet 138. Trudy Geologicheskogo Komiteta 4, 1–115 [In Russian].

Veizer, J., Ala, D., Azmy, K., Bruckschen, P., Buhl, D., Bruhn, F., Carden, G.A.F., Diener, A., Ebneth, S., Godderis, Y., Jasper, T., Korte, C., Pawellek, F., Podlaha, O.G., Strauss, H., 1999. $^{87}Sr/^{86}Sr$, $\delta^{13}C$ and $\delta^{18}O$ evolution of Phanerozoic seawater. In: Veizer, J. (Ed.), Earth System Evolution, Geochemical Perspective. Chemical Geology 161 (1–3), 59–88.

Wallmann, K., 2001. The geological water cycle and the evolution of marine $\delta^{18}O$ values. Geochimica et Cosmochimica Acta 65, 2469–2485.

Wang, C.Y., 2000. The base of the Permian System in China defined by. *Streptognathodus isolatus*. Permophiles 36, 14–15.

Wardlaw, B.R., 2000. Guadalupian conodont biostratigraphy of the Glass and Del Norte Mountains. In: Wardlaw, B.R., Grant, R.E., Rohr, D.M. (Eds.), The Guadalupian Symposium. Smithsonian Contribution to the Earth Sciences, 32, pp. 37–87.

Wardlaw, B.R., Grant, R.E., 1987. Conodont biostratigraphy of the Cathedral Mountain and Road Canyon Formations, Glass Mountains, West Texas. In: Cromwell, D., Mazzullo, L. (Eds.), The Leonardian Facies in W. Texas and S.E. New Mexico and Guidebook to the Glass Mountains, West Texas Permian Basin Section. Special Publication of the Society of Economic Geologists and Paleontologists, 27, pp. 63–66.

Wardlaw, B.R., Mei, S.L., 1998. *Clarkina* (conodont) zonation for the Upper Permian of China. Permophiles 31, 3–4.

Wardlaw, B.R., Lambert, L.L., 1999. Evolution of *Jinogondolella* and the definition of Middle Permian stage boundaries. Permophiles 33, 12–13.

Wardlaw, B.R., Davydov, V.I., 2000. Preliminary placement of the International Lower Permian Working Standard to the Glass Mountains, Texas. Permophiles 36, 11–14.

Wardlaw, B.R., Leven, E.Y., Davydov, V.I., Schiappa, T.A., Snyder, W.S., 1999. The base of the Sakmarian Stage: Call for discussion (possible GSSP in the Kondurovsky Section, Southern Urals, Russia). Permophiles 34, 19–26.

Waterhouse, J.B., 1982. An early Djulfian (Permian) brachiopod faunule from Upper Shyok Valley, Karakorum Range, and the implications for dating of allied faunas from Iran and Pakistan. Contribution to Himalayas Geology 2, 188–233.

Zhou, Z., Glenister, B.F., Spinosa, C., 1996. Multi-episodal extinction and ecological differentiation of Permian ammonoids. Permophiles 29, 52–62.

Zhao, J.K., Sheng, J.Z., Yao, Z.Q., Liang, X.L., Chen, C.Z., Rui, L., Liao, Z.T., 1981. The Changhsingian and Permian-Triassic boundary of South China. Bulletin of the Nanjing Institute of Geology and Palaeontology, Academica Sinica 2, 1–112.

J.G. Ogg

Chapter 25

Triassic

Abstract: The Triassic is bound by two mass extinctions that coincide with vast outpourings of volcanic flood basalts. The Mesozoic begins with a gradual recovery of plant and animal life after the end-Permian mass extinction. Conodonts and ammonoids are the main correlation tools for marine deposits. The Pangea supercontinent has no known glacial episodes during the Triassic, but the modulation of its monsoonal climate by Milankovitch cycles left sedimentary signatures useful for high-resolution scaling. Dinosaurs begin to dominate the terrestrial ecosystems in latest Triassic. In contrast to the rapid evolution and pronounced environmental changes that characterize the Early Triassic through Carnian, the Norian-Rhaetian of the Late Triassic was an unusually long interval of stability in Earth history.

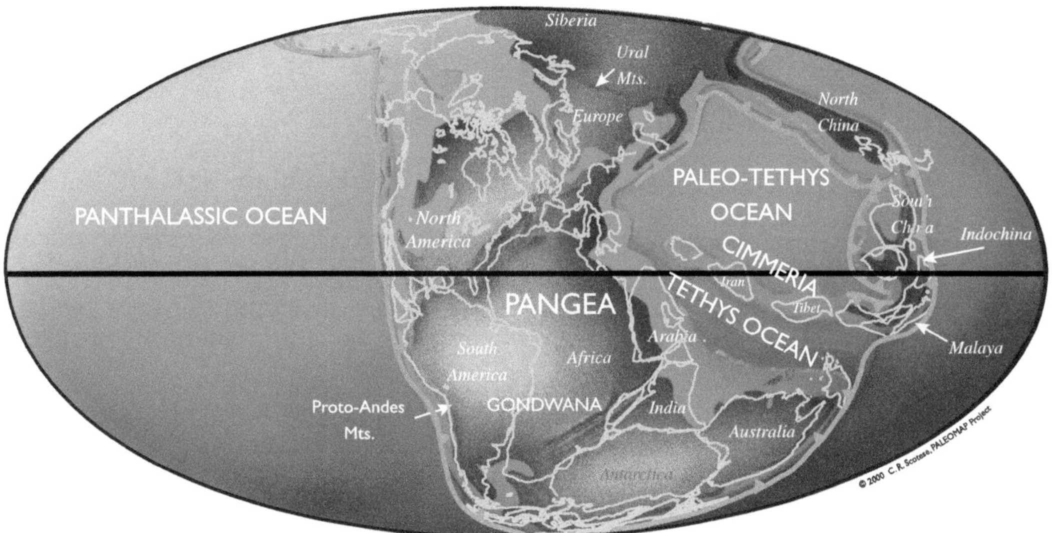

Chapter Outline

The Geologic Time Scale 2012. DOI: 10.1016/B978-0-444-59425-9.00025-1

Our understanding of the Triassic underwent a revolution after 2002. Collaborations among geochemists, paleomagnetists, paleontologists and other stratigraphers have concentrated on its exciting boundary intervals, but have also enabled compilation of a comparatively detailed bio-mag-geochem-cyclostratigraphy scale for much of the Triassic. Abundant high-precision radio-isotopic dates, coupled with enhanced methods for global correlation and detailed compilations of geochemical oscillations, have revealed a startling inequality in duration of Triassic subdivisions and pace of evolutionary and environmental change. The Norian of Late Triassic apparently spans nearly three times the duration of the entire Early Triassic (Induan and Olenekian). The early half of that brief Early Triassic has been revealed as a brief interval of pronounced environmental stress and extraordinarily rapid evolutionary turnover.

This knowledge and the inevitable new questions and debates have been partially summarized in review articles and special volumes. In particular, the book *The Triassic Timescale* (Lucas, 2010a) contains separate papers on every main stratigraphic topic (e.g., history and status of chronostratigraphy, biostratigraphy of different marine and terrestrial groups, magnetic polarity time scale, radio-isotopic age database, etc.).

25.1. HISTORY AND SUBDIVISIONS

The *Trias* of Friedrich August von Alberti (1834) united a trio of formations widespread in southern Germany – a lower Buntsandstein ("colored sandstone"), Muschelkalk ("clam limestone") and an upper Keuper (non-marine reddish beds). These continental and shallow-marine formations were difficult to correlate beyond Germany and therefore most of the traditional stages (Anisian, Ladinian, Carnian, Norian, Rhaetian) were named from ammonoid-rich successions of the Northern Calcareous Alps of Austria. However, the stratigraphy of these Austrian tectonic slices proved unsuitable for establishing formal boundary stratotypes, or even deducing the sequential order of the stages (Tozer, 1984). For example, the Norian was originally considered to underlie the Carnian stage, but after a convoluted scientific–political debate (reviewed in Tozer, 1984), the Norian was established as the younger stage. Over 50 different stage names have been proposed for subdividing the Triassic (tabulated in Tozer, 1984).

The Subcommission on Triassic Stratigraphy (International Commission on Stratigraphy) adopted seven standard Triassic stages in 1991 (Visscher, 1992); but the general lack of unambiguous historical precedents for placement of Triassic stage boundaries slowed the establishment of formal GSSPs (Figure 25.1). European stratigraphers commonly use substages with geographic names, whereas North American stratigraphers prefer a generic lower/middle/upper nomenclature (e.g., Fassanian substage versus Lower Ladinian).

Details of the history, definitions and correlation of Triassic subdivisions include the *Albertiana* newsletters and annual reports of the Subcommission on Triassic Stratigraphy (ICS), Tozer (1967, 1984) and Lucas (2010b).

25.1.1. The Permian–Triassic Boundary

25.1.1.1. End-Permian Ecological Catastrophes and Defining the Base of the Mesozoic

The Paleozoic terminated in a complex environmental catastrophe and mass extinction of life. This sharp

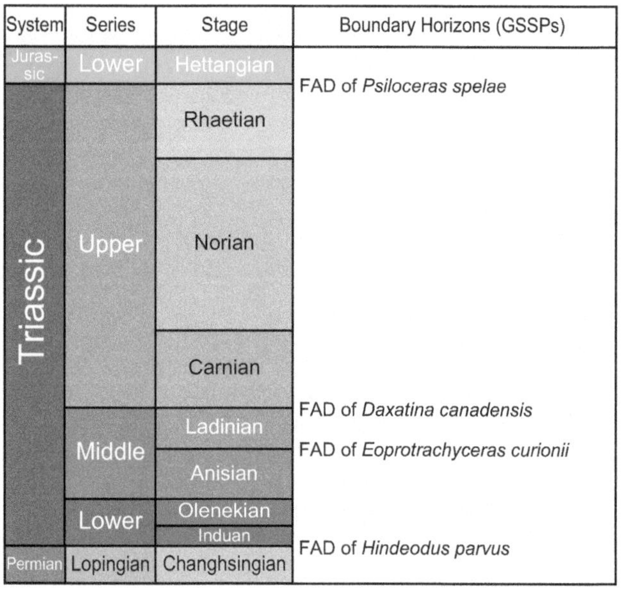

System	Series	Stage	Boundary Horizons (GSSPs)
Jurassic	Lower	Hettangian	
			FAD of *Psiloceras spelae*
Triassic	Upper	Rhaetian	
		Norian	
		Carnian	
			FAD of *Daxatina canadensis*
	Middle	Ladinian	
			FAD of *Eoprotrachyceras curionii*
		Anisian	
	Lower	Olenekian	
		Induan	
			FAD of *Hindeodus parvus*
Permian	Lopingian	Changhsingian	

FIGURE 25.1 **Summary of Triassic stage nomenclature and status.** Stages with ratified GSSPs are indicated by the first-appearance datum (FAD) of the main correlation marker, which are ammonoids except for the conodont *Hindeodus parvus* at the beginning of the Triassic. Numerical age model for this Triassic diagram is "option 2" (long Rhaetian; shorter Carnian).

evolutionary division led J. Phillips (1840, 1841) to introduce the Mesozoic ("*middle life*", with Triassic at the base) between the Paleozoic ("*old life*", ending with the Permian) and the Kainozoic ("*recent life*", after the Cretaceous). The latest Permian to earliest Triassic events include the progressive disappearance of up to 80% of marine genera, pronounced negative carbon isotope and strontium isotope anomalies, the massive flood basalts of the Siberian Traps, widespread anoxic oceanic conditions, a major sea-level regression and exposure of shelves followed by a major transgression, a "chert gap" and "coal gap", and replacement of reefal ecosystems with microbial-dominated carbonate precipitation (e.g., reviews by Holser and Magaritz, 1987; Erwin, 1993, 2006; Kozur, 1998; Hallam and Wignall, 1999; Erwin *et al.*, 2002; Wignall, 2007; Knoll *et al.*, 2007; Metcalfe and Isozaki, 2009a; Korte *et al.*, 2010; and many other compilations). The majority of ecosystems did not fully recover until near the end of the Early Triassic.

A common hypothesis is that the onset of the enormous Siberian continental flood basalts was the main contributor to the end-Permian wave of extinctions. Release of aerosols and/or carbon dioxide, coupled with their cooling/warming feedbacks on ocean circulation and stratification and on terrestrial systems, precipitated a progression of environmental and ecological stresses. A plethora of models have been published, and there is not yet a consensus on whether cooling ("plume winter") or super-greenhouse warming was the most important climate change producing the extinctions. Debate also continues regarding the role of sea-level change, whether the end-Permian mass extinction is two close-spaced events caused by different environmental factors, and the relative timing and magnitudes of marine and terrestrial turnovers (e.g., Yin *et al.*, 2007; Wignall, 2001, 2007; Korte *et al.*, 2010; Isozaki, 2009; Lucas, 2009; Metcalfe and Isozaki, 2009b; Preto *et al.*, 2010; etc.). Even though many of the ecological features resemble the aftermath of the asteroid impact that concluded the Mesozoic, the end-Paleozoic strata have not yielded unambiguous signatures of a bolide catastrophe. Radio-isotopic dating of the Siberian Trap succession and paleontology of inter-trap levels has not yet provided definitive correlations to Permian–Triassic boundary interval levels. Indeed, the published, high-resolution ages of the main Siberian Traps when compared to the dated interbedded tuffs in South China that bracket the main episodes of mass extinction would imply that the volcanism did not begin until after the mass extinctions (e.g., Burgess and Bowring, 2010). Additional, high-resolution analyses with unambiguous stratigraphic control are therefore desired (Mundil *et al.*, 2010).

The mass disappearance of the Paleozoic fauna and flora, coupled with the widespread occurrence of a major regression–transgression unconformity in most regions, led to a dilemma. It was easy to recognize the bleak final act of the Permian, but how should the beginning of the Mesozoic be defined? The base of the Buntsandstein in SW Germany defined the original Trias concept (von Alberti, 1834), but it is a diachronous boundary within continental beds, now assigned to the upper Permian. Similarly, the base of the Werfen Group (base of Tesero Oolite) in the Italian Alps is a diachronous facies boundary. Ammonoids are the common biostratigraphic tool throughout the Mesozoic, and the *Otoceras* ammonoid genus was long considered to be the first "Triassic" form. Therefore, Griesbach (1880) assigned the Triassic base to the base of the *Otoceras woodwardi* Zone in the Himalayan region, but this species is only known from the Peri-Gondwana paleomargin of eastern Tethys (e.g., Iran to Nepal). The first occurrence of *Otoceras* species in the Arctic realm (*Otoceras concavum* Zone) was used by Tozer (1967, 1986, 1994) as a Boreal marker of the base of the Triassic, but is now known to appear significantly prior to *Otoceras woodwardi* in the Tethyan realm (Krystyn and Orchard, 1996). The progressive evolution of the conodont *Hindeodus* genus through the Permian–Triassic boundary interval provided global correlation markers with no obvious facies dependence; however, conodont biostratigraphy requires special processing and identification experience. Non-biological correlation markers, such as carbon isotope excursions or magnetic polarity changes are conclusive when preserved (e.g., Newell, 1994), but can suffer from diagenetic overprints.

In 2000, the Triassic subcommission chose the first occurrence of the conodont *Hindeodus parvus* (= *Isarcicella parva* of some earlier conodont studies) within the evolutionary lineage *Hindeodus typicalis–H. latidentatus praeparvus–H. parvus–H. postparvus* as the primary correlation marker for the base of the Mesozoic and Triassic. This biostratigraphic event is the first cosmopolitan correlation level associated with the initial stages of recovery following the end-Permian mass extinctions and environmental changes. Global correlations indicate that this conodont species appears just after the carbon isotope ($\delta^{13}C$) minimum, although its lowest occurrence may be slightly earlier in some local successions (e.g., Payne *et al.*, 2009). This level is slightly lower than the base of the *Otoceras woodwardi* ammonoid zone of the Himalayas. The revised definition for the base-Triassic assigns the *Otoceras concavum* and lowermost portion of *Otoceras boreale* ammonoid zones of the Arctic (the lower part of the "Griesbachian" substage of Tozer, 1967) into the Permian (e.g., Baud, 2001; Baud and Beauchamp, 2001).

In continental settings, the correlated level to this conodont event is close to the disappearance of typical Permian *Dicynodon* tetrapods after an interval of co-occurrence with "Triassic" dicynodont *Lystrosaurus* (Kozur, 1998).

The choice of the first appearance of this conodont to serve as the primary marker for the beginning of the Triassic implies that former traditional concepts of the Permian–Triassic boundary, such as the disappearance of

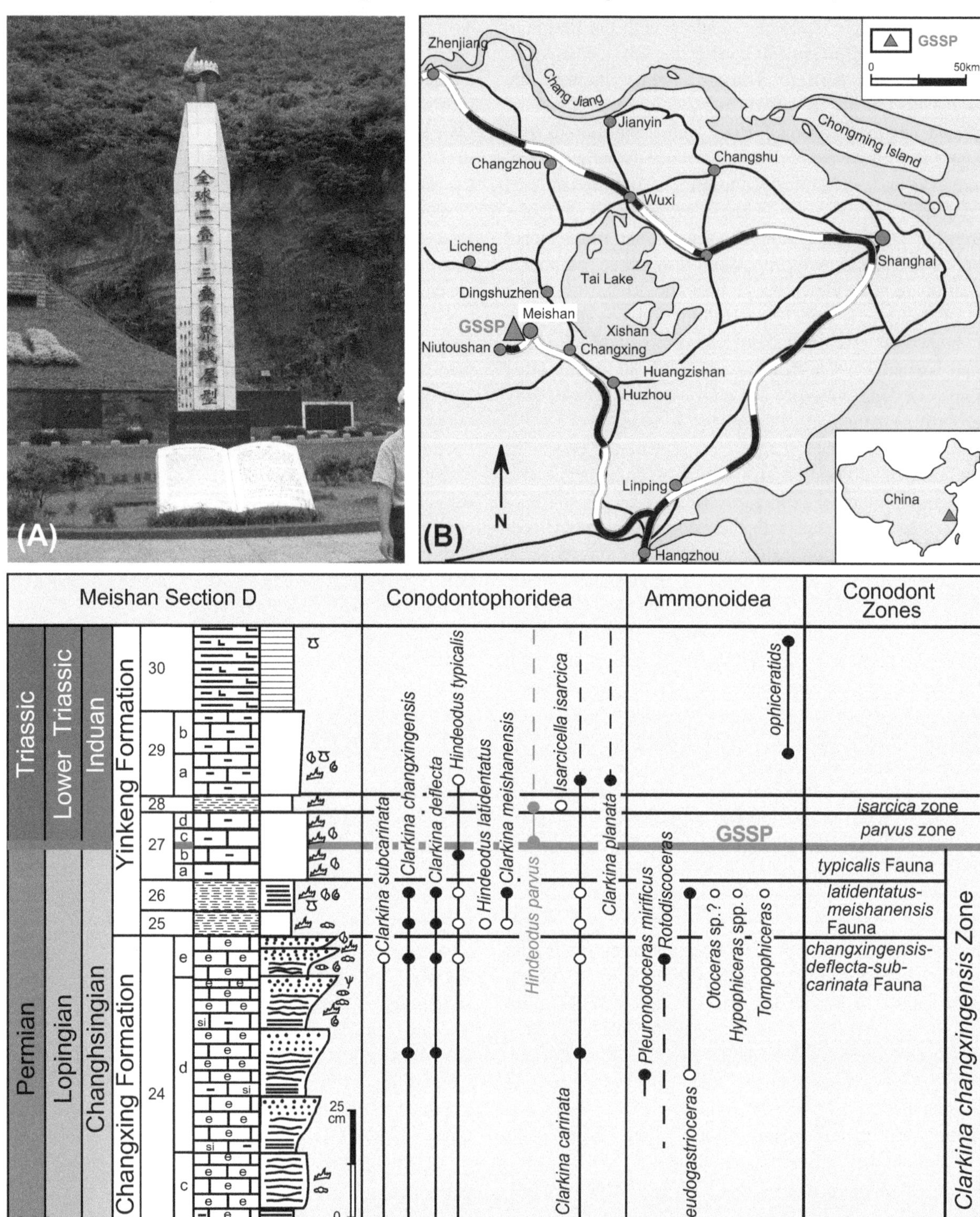

FIGURE 25.2 **GSSP for the base of Triassic (base of Mesozoic; base of Induan Stage) at Meishan, China.** The GSSP is preserved within a GeoPark, with the GSSP site located at the top of the stairs on the right side of the decorative wall. The GSSP level coincides with the lowest occurrence of conodont *Hindeodus parvus*, and is within a major negative excursion in carbon isotopes.

typical Permian marine fauna, rapid facies changes, extensive volcanism and onset of isotope anomalies are now assigned to the latest Permian.

25.1.1.2. Paleozoic–Mesozoic Boundary Stratotype (Base of Triassic)

The GSSP for the base of the Mesozoic Erathem, the Triassic System and the Induan Stage is at the base of Bed 27c at a section near Meishan, Zhejiang Province, southern China (Figure 25.2). This level coincides with the lowest occurrence of the conodont *Hindeodus parvus* (Yin *et al.*, 2001, 2005). This Meishan section also hosts the GSSP for the underlying Changhsingian Stage of the uppermost Permian, has been supplemented by complete coring of the Changhsingian–earliest Triassic interval (220 m; e.g., Cao *et al.*, 2009), and is now within a special GeoPark that includes a museum of Earth's history. Indeed, the formal park-like setting with sculptures and educational exhibitions is probably the most impressive GSSP site for the geologic record.

Approximately 18 cm below the GSSP is the former "boundary clay", bentonite (Bed 25), which has been sampled extensively as methods for Ar/Ar and U/Pb radio-isotopic dating evolved in methodology and precision (e.g., Renne *et al.*, 1995; Bowring, 1998; Metcalfe *et al.*, 1999; Erwin *et al.*, 2002; Mundil *et al.*, 2001, 2004, 2010; Shen *et al.*, 2010, 2011). This "boundary clay" is now placed approximately two brief conodont zones below the base of the Triassic. Another volcanic-ash clay, approximately 8 cm above the GSSP (Bed 28), enables a narrow bracketing of the age of the Permian–Triassic boundary. Therefore, this is one of the few boundary GSSPs that can be precisely dated by radio-isotope methods. As techniques and calibrations have evolved, the measured date for this boundary level has progressively shifted to older ages, from c. 248 Ma to c. 252 Ma.

Other important reference sections for the events across the Permian–Triassic boundary are in the Dolomites of Italy (e.g., Broglio Loriga and Cassinis, 1992; Wignall and Hallam, 1992), in the Canadian Arctic (e.g., Tozer, 1967), in the Salt Ranges of Pakistan (e.g., Baud *et al.*, 1996), in sections within Iran (e.g., Kozur, 2007), in marine and terrestrial strata in South China (e.g., Metcalfe *et al.*, 2009c; Glen *et al.*, 2009) and in terrestrial beds of the Karoo Basin.

25.1.2. Subdivisions of the Lower Triassic

A multitude of stage and substage nomenclatures have been applied to the Lower Triassic interval. The Triassic subcommission adopted the current subdivision into a lower Induan stage and an upper Olenekian stage in 1991. The Induan and Olenekian stages of Kiparisova and Popov (1956, revised in 1964) were named after exposures in the Indus river basin in the Hindustan region of Asia and in the lower reaches of the Olenek river basin of northeast Siberia, respectively.

A suite of four substages is widely used. In an imaginative procedural twist, these Griesbachian, Dienerian, Smithian and Spathian substages are named after exposures along associated small creeks on Ellesmere and Axel Heiberg islands in the Canadian Arctic, which in turn were named after the Triassic paleontologists – Carl L. Griesbach (1847–1907), Carl Diener (1862–1928), James Perrin Smith (1864–1931) and Leon Spath (1888–1957) – who played important roles in Lower Triassic biostratigraphy (Tozer, 1965). These substages were originally defined by grouping of ammonoid zones.

25.1.2.1. Induan

The Induan Stage is informally divided into two substages. The lower substage, Griesbachian, is named after Griesbach Creek on northwest Axel Heiberg Island. The definition of the Permian–Triassic boundary implies that the lower portion of the original Griesbachian of Tozer (1965, 1967) is now assigned to the uppermost Permian.

The Dienerian substage is named after Diener Creek of northwest Ellesmere Island. The Griesbachian/Dienerian boundary is marked by the appearance of Gyronitidae ammonoids. This substage boundary is recognized in Canada and in the Himalayas as the boundary between *Otoceras* and *Meekoceras* ammonoid-bearing beds of Diener (1912) and in the Salt Range of Pakistan at the base of the Lower Ceratite Limestone (Tozer, 1967).

25.1.2.2. Olenekian

History, Definition, and Boundary Stratotype Candidates

The Olenekian Stage was originally proposed from sections in Arctic Siberia, whereas the stratotype for the Induan stage was in the Hindustan region of Pakistan–India. Neither region has fossiliferous strata spanning their mutual boundary – the Induan in the Olenek River basin is marginal marine to lagoonal, and ammonoids in the transitional interval in the Hindustan region are rare or absent (Zakharov, 1994). The lower Olenekian is marked by the appearance of a diverse ammonoid assemblage of *Hedenstroemia*, *Meekoceras*, *Juvenites*, *Pseudoprospingites*, *Arctoceras*, *Flemingites* and *Euflemingites*. A sea-level regression caused a scarcity of age-diagnostic conodonts and bivalves during the latest Induan to earliest Olenekian, but the transition seems to be within the lower portion of the *Neospathodus pakistanensis* conodont zone (Zakharov, 1994; Paull, 1997; Orchard and Tozer, 1997). Proposed ammonoid-based biostratigraphic definitions of the stage boundary were the highest occurrence of the ammonoid *Gyronites subdharmus* and the lowest occurrence of the representatives of the *Meekoceras* or *Hedenstroemia* ammonoid genera (Zahkarov *et al.*, 2000, 2002). The base of its lower Smithian substage was originally defined as the base of a broad *Euflemingites romunderi* ammonoid zone (Tozer, 1965, 1967), then the biostratigraphy was revised to add a *Hedenstroemia hedenstroemi* ammonoid zone (e.g., Orchard and Tozer, 1997). Conodonts were undergoing a pronounced evolutionary evolution at the beginning of the Olenekian; and, even though some taxonomic details remain to be resolved, the widespread distribution and resolution of *Neospathodus* species provide the main method for inter-regional correlations (Orchard, 2010).

Therefore, the base-Olenekian task group selected the lowest occurrence of the conodont *Neospathodus waageni sensu lato* as the primary boundary marker. Correlation of ammonoid and conodont events among paleogeographic provinces indicates equivalence of the first occurrence of *N. waageni* with the lowermost part of the *Rohillites rohilla* ammonoid zone (Spiti region of Tethyan realm), slightly below the lowest occurrence of *Flemingites* and *Euflemingites* ammonoid genera (S. China, Tethyan realm), and lower part of *Lepiskites kolyhmensis* Zone (Siberia, Boreal realm; which is just above the regional *Hedenstroemia hedenstromia* Zone) (Zakharov *et al.*, 2009). This level is also just prior to the peak of the first Triassic positive excursion in δ^{13}C, slightly below the top of the second major Triassic reversed-polarity magnetozone (LT2r of Hounslow and Muttoni, 2010), and just above a widely recognizable sequence boundary (Subcommission on Triassic Stratigraphy, annual reports of 2008 and 2009; Krystyn *et al.*, 2007a).

Two leading candidate GSSPs for this transition are a roadside outcrop near Chaohu city in the Anhui Province of eastern China (Tong *et al.*, 2004; Sun *et al.*, 2007; Chinese Triassic Working Group, 2007) and a section near Mud (Muth) village in the Spiti valley of northwest India (Krystyn *et al.*, 2007a,b). A preliminary vote to select Mud as the GSSP was put on hold in 2008 when the desired conodont marker was identified in strata lower than the proposed GSSP level.

Smithian and Spathian Substages

A major environmental and evolutionary event occurs within the Olenekian. The boundary between its two substages, the Smithian and the succeeding Spathian, has been termed the "biggest crisis in Triassic conodont history" (Orchard, 2007a, Goudemand *et al.*, 2008). It is marked by a sudden reduction of ammonoid diversity back to initial Triassic conditions, and shift from latitudinal to cosmopolitan distributions (Brayard *et al.*, 2009a), and coincides with a major positive peak in carbon-13 (δ^{13}C) and climatic shift (e.g., Galfetti *et al.*, 2007a).

These two informal substages of the Olenekian Stage were named after Smith and Spath creeks on Ellesmere Island of the Canadian Arctic. The Spathian substage is characterized by *Tirolites*, *Columbites*, *Subcolumbites*, *Prohungarites* and *Keyserlingites* ammonoid genera. In these original stratotypes, the Smithian–Spathian boundary was placed at the base of the *Olenekites pilaticus* ammonoid zone, but there appears to be a missing biostratigraphic interval in the type region (Tozer, 1967; Orchard and Tozer, 1997). The ammonoids recovered in the early Spathian, with a dramatic evolutionary radiation accompanied by the development of a pronounced latitudinal gradient of diversity (Brayard *et al.*, 2009a). The exciting discovery of major geochemical, climatic and paleontological events at the global scale that coincide with this substage boundary has led to speculations of concurrent volcanic release of carbon dioxide, and/or enhanced storage and preservation of organic matter in the ocean (e.g., Payne and Kump, 2007; Galfetti *et al.*, 2007a,b,c).

25.1.3. Subdivisions of the Middle Triassic

25.1.3.1. Anisian

History, Definition, and Boundary Stratotype Candidates

The Anisian Stage was named after limestone formations near the Enns (= *Anisus*) River at Grossreifling, Austria (Waagen and Diener, 1895). The original Anisian stratotype lacks ammonoids in the lower portion, and its lower limit was later clarified in the Mediterranean region (Assereto, 1974). The appearance of a number of ammonoid genera, including *Aegeiceras*, *Japonites*, *Paracrochordiceras* and *Paradanubites*, may be used to define the base of the Anisian within different regions (e.g., Gaetani, 1993). However, other markers are suggested that may provide a more global correlation value. The lowest occurrence of the *Chiosella timorensis* conodont slightly precedes the ammonoid level and can be correlated to North American and Asian stratigraphy (Orchard and Tozer, 1997; Orchard, 2010). The boundary interval is close to a peak in carbon-13 (δ^{13}C) values. A shift from reversed-polarity- to normal-polarity-dominated magnetostratigraphy (base of normal-magnetozone MT1n of Hounslow and Muttoni, 2010) has been formally proposed as a primary global boundary marker that can be unambiguously correlated between Boreal and Tethyan faunal realms (Hounslow *et al.*, 2007).

The current candidate for the base-Anisian GSSP at Desli Caira Hill in north Dobrogea, Romania (Gradinaru *et al.*, 2007) is within a condensed Hallstatt limestone facies that may be missing a partial ammonoid zone in the boundary interval (Subcommission on Triassic Stratigraphy annual report, 2009). A potential alternative is a conodont–magnetic–isotope reference section at Guandao in the Nanpanjiang Basin (Guizhou Province, South China) where volcanic ashes within the boundary interval also produced radio-isotopic dates (c. 247 Ma; e.g., Lehrmann *et al.*, 2006). However, this section lacks a good ammonoid record and the chronostratigraphic reliability of the lowest occurrence of the conodont *Chiosella timorensis* marker is questioned (Ovcharova *et al.*, 2010). An additional complication arose when *C. timorensis* was reported from within the *Neopopanoceras haugi* ammonoid zone in Nevada, one zone below the base of the Anisian in the western United States ammonoid zonation (Goudemand *et al.*, submitted to

Geobios; M. Orchard, pers. comm. to J. Ogg, 2011). Therefore, even when the correlations among different stratigraphic methods are clarified, no single section spanning the boundary interval has yielded satisfactory records for the combined suite of proposed markers.

Anisian Substages

The Anisian Stage has three to four informal substages. Assereto (1974) proposed a stratotype for the Lower Anisian (also called "Aegean" or "Egean") in beds with *Paracrochordiceras* ammonoids at Mount Marathovouno on Chios Island (Aegean Sea, Greece). The Middle Anisian is sometimes subdivided into two substages: a lower "Bithynian", named by Assereto (1974) after the Kokaeli Peninsula (Bithynia) of Turkey, and an upper "Pelsonian", from the Latin name for the region around Lake Balaton in Hungary (Pia, 1930) spanning the *Balatonites balatonicus* ammonoid zone (Assereto, 1974). The Upper Anisian is also called "Illyrian" after the Latin term for Bosnia (Pia, 1930).

25.1.3.2. Ladinian

History, Definition, and Boundary Stratotype Candidates

The Ladinian Stage arose after a heated semantic argument of "*Was ist norisch?*" (Bittner, 1892), when it was realized that most of the strata that had defined a "pre-Carnian" Norian Stage (Mojsisovics, 1869) were actually deposited *after* the Carnian (Mojsisovics, 1893). This debate and the emergence of the Ladinian Stage split the Vienna geological establishment (vividly reviewed by Tozer, 1984). The Ladinian, named after the Ladini inhabitants of the Dolomites region of northern Italy, encompassed the Wengen and Buchenstein beds (Bittner, 1892).

This historical major revision and even partial inversion of upper Triassic stratigraphy, coupled with uncertainties about correlation potentials and definition of ammonoid zones, contributed to discussions in assigning the basal limit of the Ladinian Stage (e.g., Gaetani, 1993; Brack and Rieber, 1994; Mietto and Manfrin, 1995; Brack *et al.*, 1995; Vörös *et al.*, 1996; Muttoni *et al.*, 1996a; Orchard and Tozer, 1997; Pálfy and Vörös, 1998). The ammonoid contenders for the primary correlation markers were distributed over at least two zones; including the lowest occurrence of representatives of the *Kellnerites* genus, of the *Nevadites* genus, of the *Eoprotrachyceras* genus, of the *Reitziites reitzi* species, and of the *Aplococeras avisianum* species. In addition, the lowest occurrence of the *Budurovignathus* conodont genus was considered.

The base of the *Eoprotrachyceras curionii* Zone (lowest occurrence of *Eoprotrachyceras* ammonoid genus, which is the onset of the Trachyceratidae ammonoid family) was eventually preferred. The Bagolino section (eastern Lombardian Alps, Province of Brescia, Northern Italy) was selected for its multiple stratigraphic records, including bracketing of the boundary interval by dated volcanic ashes (Brack *et al.*, 2005). The Ladinian GSSP at Bagolino is located at the top of a distinct 15–20 cm thick interval of limestone nodules in a shaly matrix ("Chiesense groove"), located at approximately 5 m above the base of the Buchenstein Beds (Figure 25.3). The Ladinian GSSP site is accessible through a geological pathway with explanatory notes and ammonoid casts (Brack, 2010). The *Nevadites secedensis* ammonoid zone of the lowermost Buchenstein Beds, which was historically assigned as Ladinian (e.g., Bittner, 1892), has now become the uppermost zone of the Anisian. Secondary global markers include the lowest occurrence of the conodont *Budurovignathus praehungaricus* and a brief normal-polarity magnetozone (MT8n of Hounslow and Muttoni, 2010) within the uppermost Anisian. The bracketing U-Pb dated volcanic ashes indicate a boundary age of approximately 241 Ma (Brack *et al.*, 2005).

Ladinian Substages

Mojsisovics *et al.* (1895) divided the Ladinian into two substages – Lower or Fassanian (named after Val di Fassa in northern Italy, where it was equated to the Buchenstein Beds and Marmolada Limestone), and Upper or Longobardian (named after the Langobard people of northern Italy, and spanning the Wengen Beds). The substage boundary is approximately at the base of the *Protrachyceras longobardicum* ammonoid zone in the Alpine zonation or the base of *Meginoceras meginae* ammonoid zone in the Canadian zonation.

25.1.4. Subdivisions of the Upper Triassic

The Upper Triassic consists of three stages – Carnian, Norian and Rhaetian – that were originally defined by characteristic ammonoids (Mojsisovics, 1869). However, these units were originally recognized in different locations in the northern Alps of Austria with uncertain stratigraphic relationships. Indeed, until 1892, Norian units were considered to underlie the Carnian. It was only after resolving this major geological controversy that the name "Norian" applied to the same units after recognition that they were younger than Carnian (reviewed in Tozer, 1984).

25.1.4.1. Carnian

History, Revised Definition, and Boundary Stratotype Candidates

The Carnian stage, either named after localities in the Kärnten (Carinthia) region of Austria, or after the nearby Carnian Alps, was originally applied to Hallsatt Limestone beds bearing ammonoids of *Trachyceras* and *Tropites* (Mojsisovics, 1869, p. 127). The first occurrence of the ammonoid *Trachyceras* (= base of *Trachyceras aon* Zone in Tethys or *Trachyceras desatoyense* Zone in Canada) was the traditional base, although it appears that a *Trachyceras* datum would be

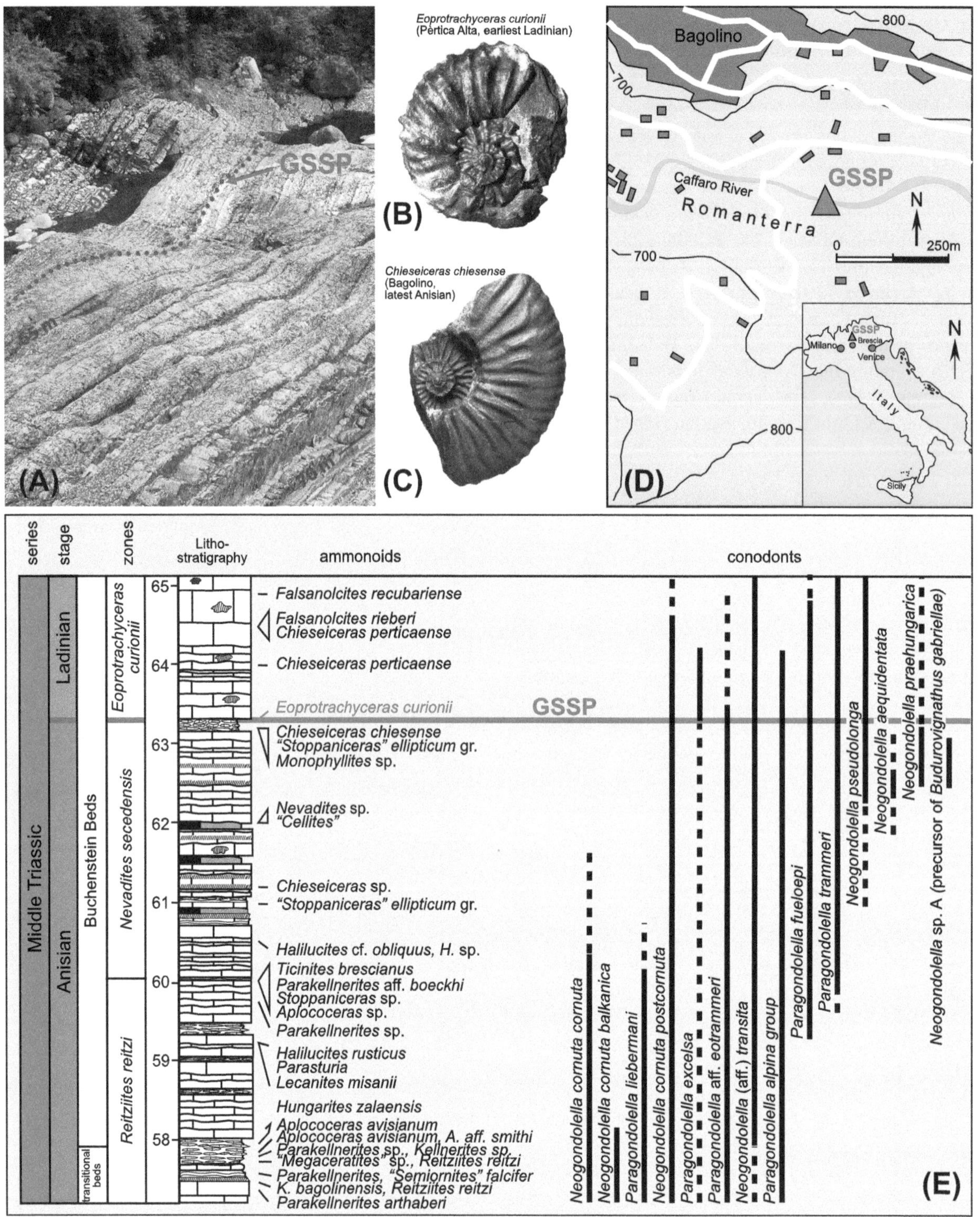

FIGURE 25.3 **GSSP for base of the Ladinian Stage at Bagolino, Italy.** The GSSP level coincides with the lowest occurrence of *Eoprotrachyceras* ammonoid genus (base of *Eoprotrachyceras curionii* Zone). *Photos provided by Peter Brack.*

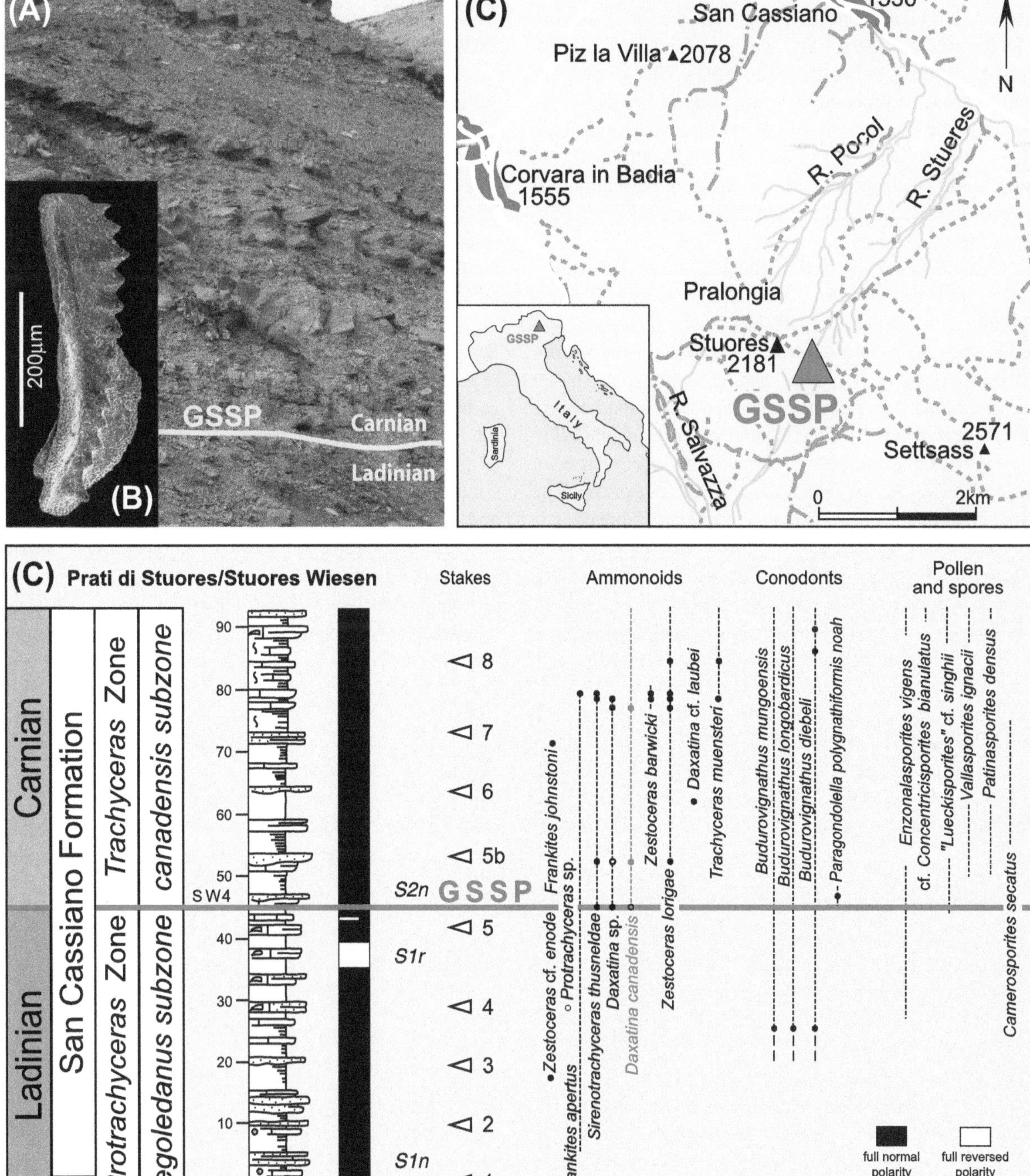

FIGURE 25.4 **GSSP for base of the Carnian Stage at Prati di Stuores/Stuores Wiesen (Dolomites, northern Italy).** The GSSP level coincides with the lowest occurrence of the cosmopolitan ammonoid *Daxatina* (base of *Daxatina canadensis* Subzone, lowest subzone of *Trachyceras* Zone), and is near the lowest occurrence of conodont *Paragondolella polygnathiformis* (inset "B"). *Photos of outcrop and conodont provided by Manuel Rigo.*

asynchronous and not cosmopolitan (e.g., Mietto and Manfrin, 1999). Mojsisovics *et al.* (1895) included the St. Cassian Beds of northern Italy in a revised Carnian subdivision; therefore the level with lowest occurrence of the cosmopolitan ammonoid *Daxatina* at the Prati di Stuores type locality in the Dolomites (northern Italy) was proposed for the base-Carnian GSSP (Broglio Loriga *et al.*, 1998). This section has relatively rapid sedimentation and proved suitable for multiple types of stratigraphy and therefore it was ratified ten years later (2008). Two other reference sections with multiple biostratigraphic successions are in Spiti in Himalaya of northwest India (Balini *et al.*, 1998, 2001) and the New Pass section of Nevada (USA).

The Carnian GSSP at Prati di Stuores/Stuores Wiesen is 45 m above the base of the St. Cassian (San Cassiano) Formation (Figure 25.4) (Mietto *et al.*, in press). This level was selected to coincide with the lowest occurrence of the ammonoid *Daxatina* (base of *Daxatina canadensis* Subzone, lowest subzone of a broad *Trachyceras* Zone). Secondary markers are the lowest occurrences of the conodont *"Paragondolella" polygnathiformis* and the palynomorphs *Vallasporites ignacii* and *Patinasporites densus* (Mietto *et al.*, 2007, in press). The evolutionary transition from *Daonella* to *Halobia* bivalves is probably near this level (McRoberts, 2010). The Carnian GSSP is just above the base of a normal-polarity magnetic magnetozone (S2n in the local scale of Borglio Loriga *et al.*, 1999; or UT1n in the synthesis scale of Hounslow and Muttoni, 2010), and lies just above an interpreted maximum flooding surface within Sequence Lad 3 of Hardenbol *et al.* (1998).

The ratified placement of the base-Carnian at the appearance of *Daxatina* ammonoids, rather than the *Trachyceras aon* ammonoid, implies that the Carnian now begins in the middle of the classical "uppermost Ladinian" *Frankites regoledanus* ammonoid zone.

Carnian Substages and Wet Intermezzo

Mojsisovics *et al.* (1895) subdivided the Carnian into three substages (Cordevolian, Julian and Tuvalian) corresponding to his three ammonoid zones. Cordevolian (his *Trachyceras aon* Zone), from the St. Cassian Beds, was named after the Cordevol people who lived in this area of northern Italy. Julian (*Trachyceras aonoides* Zone) was based on the Raibl Formation of the Julian Alps in southern Austria. The Tuvalian (*Tropites subbullatus* Zone) was named after the Tuval mountains, the Roman term for the region between Berchtesgaden and Hallein near Salzburg, Austria. Mojsisovics' trio of ammonoid zones were later split into additional zones; but his main divisions can be correlated among regions.

Stratigraphers often combine the Cordevolian and Julian into a single Lower Carnian, with the Lower/Upper Carnian boundary traditionally assigned as the first occurrence of *Tropites* ammonoids (base of the *Tropites subbullatus* ammonoid zone of Tethys and *Tropites dilleri* Zone of Canada). The ammonoid change at this substage boundary is more significant than at the bases of either the Carnian or the Norian stages (Tozer, 1984), conodont diversity may have been reduced to two genera (Mazza *et al.*, 2010), and there were major changes in radiolarians and other faunal groups (Kozur and Bachmann, 2010).

An unusual climate–oceanographic event in the latest Julian has various regional names – e.g., "Reingraben turnover" (Schlager and Schöllnberger, 1974), "Raibl Event", "Carnian pluvial episode" (Simms and Ruffell, 1989), and "Middle Carnian Wet Intermezzo" (Kozur and Bachmann, 2010). This episode was initially recognized within the Alpine region by the termination of the prograding reef-building of the earlier Carnian by terrigenous clastics (Raibl Formation) and within the Germanic Basin by an influx of fluvial to brackish-water sands (Stuttgart Formation or Schilfsandstein). Black shale facies developed in restricted basins (e.g., Hornung and Brandner, 2005), and an absence of carbonate components in adjacent oceanic basins has been interpreted as a rise in carbonate compensation depth (CCD) (Rigo *et al.*, 2007). There is growing evidence that this simultaneous climate–oceanographic excursion, biotic crisis and anomalous facies package can be recognized globally (e.g., Simms and Ruffel, 1989; L. Krystyn and H. Kozur, pers. comm. to J. Ogg, 2010). The triggers for both the onset of this "Carnian pluvial episode" and its sudden termination could be a combination of paleogeographic and paleoceanographic factors (e.g., Kozur and Bachmann, 2010). The eruption of the Wrangellia flood basalts may also have had an effect (Dal Corso *et al.*, 2012). The distinct fossil assemblages within this "wet intermezzo" interval provide an important means of calibration among terrestrial settings (conchostracans, pollen, ostracod and tetrapod biostratigraphy) from the southwest USA to the Germanic Basin and into marine settings (ammonoid, conodont, ostracod and bivalve biostratigraphy) (Roghi *et al.*, 2010; Kozur and Weems, 2010). The abrupt end to this "wet intermezzo" interval in the Germanic-Alpine region coincides with the Lower/Middle Carnian substage boundary as assigned by ammonoid biostratigraphy.

25.1.4.2. Norian

History, Revised Definition, and Boundary Stratotype Candidates

Norian derives its name from the Roman province of Noria, south of the Danube and including the area of Hallstatt, Austria (Mojsisovics, 1869). The stratigraphic extent of strata assigned as "Norian" have a contorted history (Tozer, 1984).

Ammonoid successions in Nevada and British Columbia led to a proposal that the base of the Norian be assigned to the base of the *Stikinoceras kerri* ammonoid zone, which overlies the *Klamathites macrolobatus* Zone (Silberling and Tozer, 1968). This level is approximately coeval with a Tethyan placement between the *Anatropites* and *Guembelites jandianus* ammonoid zones (Krystyn, 1980; Orchard *et al.*, 2000). Examinations of conodont lineages have indicated that the first appearance of *Metapolygnathus* ex gr. *M. echinatus* is at approximately the beginning of the *Stikinoceras kerri* ammonoid zone and coincides with a major faunal turnover (Orchard, 2010). It is approximately coincident with the FAD of the widespread *Halobia austriaca* bivalve; therefore, this level has become favored for assigning the base of the Norian.

The leading candidates for the Norian GSSP (based on the FAD of *Metapolygnathus echinatus* conodont) are Pizzo Mondello in Sicily (Muttoni *et al.*, 2001; Nicora *et al.*, 2007) or Black Bear Ridge on Williston Lake of northeast British Columbia (Orchard *et al.*, 2001; Orchard, 2007b; McRoberts, 2007; Orchard *et al.*, in prep.). The section at Black Bear Ridge did not yield a useful magnetostratigraphy; whereas the recommended Norian GSSP level at Pizzo Mondello coincides with the top of a narrow reversed-polarity magnetozone (magnetozone "PM4r" in local terminology; or "UT12r" in the synthesis of Hounslow and Muttoni, 2010) and is just above a positive shift in $\delta^{13}C$ (Nicora *et al.*, 2007).

Norian Substages

The Norian is traditionally subdivided into three substages, following Mojsisovics *et al.* (1895). The boundary between the lower Norian (or "Lacian", after the Roman name for the Salzkammergut region of the northern Austrian Alps) and middle Norian (or "Alaunian", named for the Alauns, who lived in the Hallein region of Austria during Roman times) is the base of the Tethyan *Cyrtopleurites bicrenatus* ammonoid zone. The base of the upper Norian (or "Sevatian", after the Celtic tribe who lived between the Inn and Enns rivers of Austria) is generally assigned as the base of the North American *Gnomohalorites cordilleranus* ammonoid zone or the Tethyan *Sagenites quinquepunctatus* ammonoid zone; however, there has not been a consistent usage of this Sevatian substage and some include the underlying *Halorites macer* ammonoid zone within it (e.g., Kozur, 1999).

25.1.4.3. Rhaetian

The Rhaetian was the first Triassic stage to be established, when Carl Wilhelm Ritter von Gümbel (1861) applied the term to strata containing the pteriid bivalve *Rhaetavicula contorta*, such as the Kössen Beds of Austria. This distinction bivalve is found in shallow-marine facies from the western Tethys and across northwestern Europe. His "Rhätische Gebilde" name was derived from either the Rhätische Alpen or the Roman province of Rhaetium. For a while, it appeared that the Rhaetian interval would be incorporated into the Jurassic (and perhaps renamed as a "Bavarian Stage") or incorporated into the Norian Stage (reviewed in Lucas, 2010a). For example, the Rhaetian was eliminated in some Triassic time scales (e.g., Zapfe, 1974; Palmer, 1983; Tozer, 1984, 1990). In 1991, the Subcommission on Triassic Stratigraphy decided to retain the Rhaetian as an independent stage. Many options were considered for the primary biostratigraphic marker for the lower boundary.

In 2010, the base of the Rhaetian was placed by the Task Group at the lowest occurrence of the conodont *Misikella posthernsteini* (Krystyn, 2010). This conodont is a phylogenetic descendent of *Misikella hernsteini*, but is very rare at the beginning of its range. Therefore, several secondary markers should also be employed to assign the base of the Rhaetian (Krystyn, 2010) including:

(1) Lowest occurrence of conodont *Epigondolella mosheri* (morphotype B *sensu* Orchard),
(2) Lowest occurrence of ammonoid *Paracochloceras suessi* and the closely allied genus Cochloceras and other taxa,
(3) Disappearance of ammonoid genus *Metasibirites*,
(4) Lowest occurrence of radiolarian *Proparvicingula moniliformis* and other species (Carter and Orchard, 2007),
(5) Disappearance of *Monotis* bivalves, except for continuation by dwarf *Monotis* species in parts of the Tethys (McRoberts *et al.*, 2008), and
(6) Just below is a prominent change from an extended normal-polarity magnetozone upward into a reversed-polarity magnetozone (UT23n to UT23r in the tentative Norian–Rhaetian composite scale of Hounslow and Muttoni, 2010).

The leading candidate for the Rhaetian GSSP is the Steinbergkogel section near Hallstatt, Austria (Krystyn *et al.*, 2007c,d). The published magnetostratigraphy from the condensed interval spanning the proposed boundary has been verified; but correlation to the cycle-scaled magnetic polarity scale from non-marine strata (Newark group) is disputed (e.g., Gallet *et al.*, 2007; Muttoni *et al.*, 2010; Hounslow and Muttoni, 2010; Hüsing *et al.*, 2011). Other important reference sections are in British Columbia, Canada and in Turkey.

25.2. TRIASSIC STRATIGRAPHY

Ammonoids dominate the historical zonation of the Triassic, but conodonts have become the major tool for global correlation. Thin-shelled bivalves (e.g., *Daonella*, *Halobia*, etc.) provide important regional markers. During much of the Triassic, the sedimentary record across the Pangea supercontinent was dominated by terrestrial deposits, therefore widespread conchostracan, tetrapod and plant remains are important for global correlation.

Other biostratigraphic, magnetostratigraphic, chemostratigraphic and other events are typically calibrated to these standard ammonoid or conodont zones. Extensive compilations and inter-correlation of Triassic stratigraphy of European basins were coordinated by Hardenbol *et al.* (1998), and a suite of detailed Triassic stratigraphic scales are in *The Triassic Timescale* (Lucas, 2010a).

25.2.1. Marine Biostratigraphy

25.2.1.1. Ammonoids

The ammonoid successions of the Alps and Canada have historically served as global primary standards for the Triassic (reviews in Tozer, 1967, 1984; Balini *et al.*, 2010). Ammonoids were nearly extinguished at the end of the Permian, with termination of three of the four major Permian clades. However, the rare surviving ammonoids diversified much faster than other marine groups following this catastrophe (Brayard *et al.*, 2009a; Marshall and Jacobs, 2009). The entire Triassic ceratitid group of Ammonoidea is usually considered to have been derived from the morphologically simple *Ophiceras* genus survivor. This rapid surge in diversity was interrupted by major waves of extinction at the end of the Smithian substage (mid-Olenekian), base of Anisian and end of Julian substage (mid-Carnian) (e.g., Brayard *et al.*, 2009a, b). Peak diversity was attained in middle Norian. A decline in diversity through the latest Triassic culminated in apparently only a single genus, *Psiloceras*, surviving the end-Triassic mass extinctions and rapidly evolving to conquer the Jurassic seas.

Despite their historical importance in subdividing the Triassic, there is not yet a standardized ammonoid zonation (or nomenclature) for the alpine regions. For example, Mietto and Manfrin (1995) proposed a generalized standard for the Middle Triassic of the Tethyan realm that utilized first appearances of widespread genera to define zones and major species to define subzones. However, this zonal scheme was immediately rejected by some alpine workers (e.g., critiques by Brack and Rieber, 1996, and Vörös *et al.*, 1996). Contributing to this situation are the lack of consensus definitions of species and genera, historical confusions with taxonomy and relative chronostratigraphic placement of taxa, distinct latitudinal and endemic assemblages, use of assemblage or association zones, omission by many authors to provide clarity in their definition of zonal boundaries, and a current reduction in the number of active ammonoid specialists (Balini *et al.*, 2010).

A selection of Triassic zonations and their main index ammonoids are given by Balini *et al.* (2010). The diagrams in this chapter are a generalized version for the Tethyan and North American realms derived from different sources (e.g., Kozur, 2003 and pers. comm., 2010; Balini *et al.*, 2010; McRoberts, 2010). A sample suite of Tethyan zones are summarized in Table 25.3.

25.2.1.2. Conodonts

Conodonts are phosphatic feeding apparatuses of an enigmatic pelagic swimmer. The conodont taxa are based on variability of these jaw-like elements, and these evolving features enable widespread correlation of Triassic strata. After surviving the end-Permian mass extinctions and diversifying explosively in the early Olenekian (Orchard, 2007a), the conodonts mysteriously vanished at the end of the Triassic. Compilations of the calibrations among conodont zones and ammonoids have been proposed for several realms, including Canada (e.g., Orchard and Tozer, 1997), the Tethys region (e.g., Muttoni *et al.*, 1998; Krystyn *et al.*, 2002; Kozur, 2003) and European basins (Vrielynck, 1998). Even though there are persistent problems due to the lack of a consensus on standardized taxonomy/nomenclature, paleogeographic variability, artifacts from preservation and other factors, the conodonts have proven to be the most important means of inter-regional correlation. Orchard (2010) compiled ranges of major conodont taxa relative to regional ammonoid zones for the intervals spanning Triassic stage boundaries.

25.2.1.3. Bivalves

The end-Permian mass extinctions terminated the dominance by the communities of brachiopods, crinoids (and other pelmatozoan echinoderms) and bryozoans that had been typical of Paleozoic marine sea beds. Bivalve and gastropod mollusk communities are typical of Early Triassic shelves, with a later increase in the importance of scleractinian corals. Among the Triassic bivalves, a succession of pelagic forms of thin-shelled "flat clams" (pectinacean pteriid bivalves) were largely restricted to deeper water settings and tend to occur in high densities on certain bedding planes (Hallam, 1981). These genera (*Claria*, *Enteropleura*, *Daonella*, *Halobia*, *Monotis*, etc.), which have no modern counterparts, are valuable for global Triassic correlations because their species have both widespread distributions and a mean duration of only 1–2 myr. Their Jurassic and Cretaceous relatives (*Bositra*, *Buchia*, inoceramids) occupied the same settings. The first comprehensive summary of the ranges of these Triassic pelagic bivalves relative to ammonoid zones is by McRoberts (2010), who defined 30 discrete zones and regional variants.

25.2.1.4. Radiolarians

The sand-sized opaline skeletons of radiolarians are typically found in Triassic deep-marine facies. After the end-Permian, the most severe extinction event for radiolarians, diversity slowly increased through the Early Triassic, then surged in the Anisian to reach a maximum in early Carnian. Diversity declined through the Late Triassic until the end-Triassic mass extinctions again decimated the radiolarian genera (O'Dogherty *et al.*, 2010).

Radiolarian datums and zonations have been compiled for different regions, including alpine exposures (e.g., De Wever, 1982, 1998; Kozur and Mostler, 1994; Kozur, 2003), Japan (e.g., Sugiyama, 1997) and western North America (e.g., Blome, 1984; Carter, 1993; Carter and Orchard, 2007). A correlation among these regional zonations and a summary for the ranges of the main 281 genera (including schematic images) relative to Triassic substages has been compiled by O'Dogherty *et al.* (2010).

25.2.1.5. Other Microfossils

Except for radiolarians, marine microfossil biostratigraphy has not yet been developed as a widespread correlation tool within the Triassic. In contrast to Permian and Jurassic syntheses, the benthic foraminifer stratigraphy of the Triassic has not been compiled on a global scale. Regional scales include a stratigraphic summary of larger benthic foraminifera of the Tethyan realm (Peybernes, 1998) and zonations proposed for the Caucasus (e.g., Vuks, 2000, 2007).

Calcareous nannofossils are only known from Carnian and younger strata, and the first "real coccoliths" appear in the Norian (von Salis, 1998; Bown, 1998).

Records of dinoflagellates are rare, and the oldest representative of this group may be Middle Triassic (Hochuli, 1998). However, a latest Triassic species, *Rhaetogonyaulax rhaetica*, is an important means for correlating Rhaetian strata.

25.2.1.6. Reefs

Although not biostratigraphy, *per se*, there are broad trends in global shallow-water carbonate systems during the Triassic. The metazoan communities that built Permian reefs were completely destroyed in the end-Permian mass extinction, and a "reef gap" ensued that spanned much of the Early Triassic. Simultaneously, there was an Early Triassic "chert gap" following the termination of the extensive Permian chert accumulations (Beauchamp and Baud, 2002). The accompanying extinction of many benthic grazers probably contributed to the distinctive microbial-dominated carbonates (stromatolites, wavy-laminated micrites, oolites) that are characteristic of the earliest Triassic (e.g., Flügel, 2002). Elevated carbon dioxide levels and/or a pulse of carbonate saturation following CO_2-induced dissolution in an ocean that lacked deep-water carbonate buffering may have contributed to this latest Permian and earliest Triassic interval of microbial and oolitic limestones (e.g., Payne *et al.*, 2007, 2009; Xie *et al.*, 2010). This Early Triassic "reef gap" was mainly caused by a lack of reef-building biota, rather than a crisis in carbonate production/accumulation (Preto *et al.*, 2010). Microbial-sponge and *Tubiphytes* reef communities with a modest contribution from the oldest crown-type scleractinian corals are characteristic of the middle and late Anisian (Kiessling, 2010). Growth rates for some late Anisian through Ladinian reefs were phenomenal, with radio-isotopic dates suggesting up to 500 m/myr for some platforms in the Dolomites of northern Italy (e.g., Brack *et al.*, 2007; Meyers, 2008). The Ladinian—earliest Carnian peak of these microbial-sponge reefs was terminated by the events associated with the mid-Carnian pluvial event ("wet intermezzo") (e.g., Stefani *et al.*, 2010). Scleractinian corals contributed to the second phase of reef expansion, which peaked in the late Norian (Kiesling, 2010). A rapid sea-level fall near the time of the Norian/Rhaetian boundary contributed to the end-Norian cessation of the spectacular Dolomia Principale/Hauptdolomit platforms of the Alps and other regions (e.g., Berra *et al.*, 2010). The end-Rhaetian wave of extinctions led to a reef crisis that mirrored the end-Permian episode, and reefs were again rare through much of the Early and Middle Jurassic (Kiessling, 2010).

25.2.1.7. Marine Reptiles

Beginning in earliest Triassic, the seas became inhabited by large marine reptiles. Ichthyopterygia ("fish flippers", or Ichthyosaurs) with their dolphin-like morphology and Sauropterygia ("lizard flippers") that retained a more crocodile-like form rapidly diversified during the Early Triassic (e.g., Motani, 2010). The adaptation of the ichthyosaurs to the open seas, perhaps including a warm-blooded anatomy, enabled their expansion in the Middle and Late Triassic as coastal habitats for other marine reptiles declined (Motani, 2010). Most of the Sauropterygia clade vanished during the Late Triassic, although the long-necked Jurassic—Cretaceous plesiosaurs may represent an emergence from a distant branch.

25.2.2. Terrestrial Biostratigraphy

The terrestrial successions of tetrapods (amphibians, reptiles, earliest mammals), plants (spores/pollen, macroflora) and lacustrine organisms (ostracods, and especially conchostracans) are broadly inter-correlated in different regions of Pangea. The calibration of the evolving terrestrial ecosystems to marine-based Triassic stages and substages is fairly well established for the Lower Triassic via a distinctive cycle-scaled magnetostratigraphy in the Germanic Basin, but ties are much less certain and controversial through the Middle and Upper Triassic. An unambiguous correlation of the cycle-scaled magnetostratigraphy of the Carnian—Norian lacustrine-rich strata of the Newark Group from the rift valleys of easternmost North America to the magnetostratigraphy of Tethyan marine zones would enable a precise Late Triassic time scale. However, the inter-regional calibration of the pollen—tetrapod—conchostracan occurrences within those Newark beds is disputed (e.g., Kozur and Weems, 2010; Irmis *et al.*, 2010; Lucas, 2010c,d).

Other non-marine biostratigraphy, including charophyte cysts (gyrogonites or calcified spores of charophyte green algae), macrofossil plant assemblages, ostracods, bivalves and fishes, can be useful on a regional scale, but have not yet been developed for use in global correlations (e.g., review by Lucas, 2010c).

25.2.2.1. Tetrapods and Dinosaurs

The Mesozoic is popularly known as the "Age of Dinosaurs", but these famed reptiles did not diverge from other tetrapods until the mid-Triassic. The suite of proposed global zones for tetrapod stratigraphy for Pangea is largely based on widespread occurrences of select semi-aquatic tetrapods (both reptiles and amphibians) whose thick skulls, body armor, or other resistant body parts were commonly preserved in the sedimentary record. The Pangean configuration enabled many of these forms to spread across most of the world's land area, therefore the lowest occurrences of distinctive widespread genera enable subdivision of the Triassic into eight "land-vertebrate faunachrons" (LVFs) (Lucas, 1998, 1999, 2010d and references therein). These "LVFs" have an inter-continental correlation potential that is approximately equivalent to standard Triassic stages (Figure 25.2 = NON-MARINE scales); although the placement relative to marine-based substages has not yet been firmly established for all LVFs (e.g., Irmis *et al.*, 2010). When skeletal material is not available, distinctive footprints of tetrapods are a secondary, but less precise, method of correlation (Klein and Lucas, 2010).

The predominant terrestrial or semi-aquatic forms in the Early Triassic were therapsids, a major lineage of synapsids ("fused arch") that are considered to be "mammal-like reptiles" (e.g., Fraser, 2006). Three groups of therapsids survived the end-Permian mass extinction – cynodonts ("dog-toothed"), dicynodonts ("two dog-toothed", with two tusks in the upper jaw) and large-skulled therocephalian ("beast head") carnivores. The therapsids diminished in importance through the Triassic, but true mammals, of which the oldest record is a mouse-sized *Adelobasileus* of Late Triassic (e.g., Lucas, 2008), are interpreted as being derived from this group. The type-area for the Early Triassic LVFs is the Karoo Basin in South Africa. The Lootsbergian LVF begins in latest Permian with the first appearance of *Lystrosaurus*, a squat, dog-sized, dicynodont herbivore that may have been semi-aquatic. The extinction of the *Dicynodon* dicynodont is approximately at the end of the Permian (Lucas, 2009). The lowest occurrence of the robustly built *Cynognathus* cynodont defines the base of the Nonesian LVF, which is approximately equivalent to the marine Olenekian Stage.

The Middle Triassic LVFs (Perokvan and Berdyankian) are derived from the lowest occurrences of massive temnospondyl amphibians (*Eocyclotosaurus* and *Mastodonsaurus giganteus*, respectively) in the Ural foreland basin in Russia.

Diapsid ("two arches") reptiles had diverged into different major orders during the Early Triassic, and its Archosauria ("ruling reptiles") group (dinosaurs, pterosaurs, crocodiles, etc.) would dominate the terrestrial and aerial ecosystems of the Late Triassic through Cretaceous. The first occurrences of types of crocodile-like phytosaurs and of large armored herbivore aetosaurs, which resembled elongate armadillos with stout claws, define most of the Late Triassic LVFs. The fossil-rich deposits of the Chinle Basin of southwest USA are the main reference sections for four Late Triassic LVFs. The Otischalkian and Adamanian LVFs of Carnian are defined by FADs of phytosaurs *Parasuchus* (= *Paleorhinus*) and *Rutiodon*, respectively, and the late Carnian and early Norian-aged Revueltian LVF begins with the aetosaur *Typothorax coccinarium*. The Apachean LVF of approximately latest Norian and Rhaetian age is defined by the FAD of the phytosaur *Redondasaurus* (e.g., Lucas, 2010d). The Jurassic begins with the FAD of the crocodylomorph *Protosuchus*, defining the base of the Wassonian LVF.

Within this tetrapod-LVF framework, the main groups of dinosaurs seem to have radiated during the late Middle Triassic (e.g., Heckert and Lucas, 1999; Fraser, 2006). The major dinosaur lineages (Sauropodomorpha, Theropoda and Ornithischia divisions) became established during the early Carnian, as indicated by the Ischigualasto assemblage of Argentina and of other future continents (e.g., summary table in Nesbitt *et al.*, 2009). Dinosaurs became dominant land reptiles in the Norian.

25.2.2.2. Conchostracans

Conchostracans or clam shrimps are brackish-to-freshwater crustaceans (Class Branchiopoda, Order Conchostraca) with a chitinous bivalve carapace of two lateral valves that are typically 2 to 40 mm in length. Their tiny drought-resistant eggs were easily dispersed by wind and water, and rapidly hatched upon exposure to suitable environments. These characteristics and their brief life cycle of only one to three weeks enabled conchostracans to become widespread in lakes and temporary pools throughout Pangea. Their distinctive carapaces were preserved in lacustrine, salt flat and floodplain deposits. Even though conchostracans are present in pre-Permian through modern sediments, their potential for biostratigraphy and inter-regional correlation has only been fully developed in Triassic strata (e.g., Kozur and Weems, 2010, and references therein). The application of conchostracan biostratigraphy for other time intervals awaits careful studies of taxonomy, recognition of distinctive taxa, and calibration to other biostratigraphic scales.

Approximately 30 conchostracans zones and regional variants have been defined within the Triassic. These zones have been recognized and inter-correlated within the Germanic Basin, through the southwest USA, in the Newark Supergroup of easternmost North America and within other regions (summarized in Kozur and Weems, 2010; see also Kozur and Bachmann, 2008). Correlations to marine-based Triassic substages are constrained by interbedded ostracod zones, palynology zones, distinctive facies shifts (e.g., the mid-Carnian "wet intermezzo") and cycle-scaled magnetic magnetozones. The zonal detail of conchostracans enables an important calibration of the tetrapod "land-vertebrate faunachrons" (LVFs) to marine substages and aids in projecting magnetostratigraphy and radio-isotopic ages between terrestrial and marine strata.

25.2.2.3. Plants, Pollen and Spores

Spores and pollen are important for correlation of marine and terrestrial strata. However, most taxa have relatively long ranges, and changes in assemblages may indicate local climatic-ecosystem shifts rather than a useful temporal marker. Selected compilations of Triassic palynology and plant ecosystem evolution are those of Wing and Sues (1992), Warrington (2002) and Traverse (2007). Palynoflora zonations for Triassic strata have been compiled for the Alpine and Germanic regions (e.g., Visscher and Brügman, 1981; Visscher *et al.*, 1994; Hochuli, 1998; Herngreen, 2005; Kürschner and Herngreen, 2010), Australia (Helby *et al.*, 1987), southwest United States (Litwin *et al.*, 1991; Cornet, 1993), Newark Basin of eastern United States (e.g., Cornet, 1977; Cornet and Olsen, 1985), and Arctic (Van Veen *et al.*, 1998). However, only the Upper Triassic and lowest Jurassic have a detailed inter-regional compilation among different continents (Cirilli, 2010).

Regional diversity of palynomorphs declined at the time of the end-Permian mass extinctions of shallow-marine faunas (Kürschner and Herngreen, 2010). Latest Permian plant ecosystems may have undergone a short-lived explosion in abundance in lycopods that interrupted the dominance by gymnosperms to create a c. 10-kyr "spore-spike" within the major negative 13_C isotope decline, but the plant communities rapidly recovered before the main catastrophic marine extinction (Hochuli *et al.*, 2010). Widespread marine clastics of lower Triassic (Olenekian to lower Induan) record a uniquely cosmopolitan "acritarch spike" assemblage of lycopsid spores, small acanthomorph acritarchs and *Lunatisporites* coniferalean pollen (e.g., Balme and Foster, 1996). Diversity of both pollen and spores peaked in the Germanic Basin during Ladinian–Carnian (Kürschner and Herngreen, 2010). Globally during this time, there is a rapid diversification of Circumpolloid genera. This broad episode was followed, at least in the Tethyan realm, by a gradual change in palynofloral assemblages without abrupt changes from Carnian into earliest Hettangian (e.g., Cirilli, 2010).

Even though the palynology shifts are generally gradual and mainly important at regional scales, the trends can help identify major hiatuses and correlations in terrestrial records. For example, the strata slightly below flood basalts in the Newark rift basins in the eastern North America display a relatively sharp transition from diverse assemblages of monosaccate and bisaccate pollen to an overlying assemblage containing 60–90% *Corollina meyeriana* spores, and this level was considered as a regional marker for the Triassic/Jurassic boundary (e.g., Cornet, 1977, pp. 175–184; Cornet and Olsen, 1985; Fowell and Olsen, 1993, 1995). However, similar palynological changes are recorded near the base of the typical Rhaetian of Europe (e.g., Schuurman, 1979; Orbell, 1983; Van Veen, 1995), therefore some palynologists have interpreted this level to be a major hiatus that includes most of the Rhaetian stage (e.g., Cirilli *et al.*, 2009), as is also indicated by the conchostracan assemblages (Kozur and Weems, 2005, 2010). This interpretation of the absence of the uppermost Norian and most of the Rhaetian between the radio-isotopic-dated flood basalts and the underlying cycle-scaled magnetozones of the Newark Supergroup lacustrine strata would have a major implication for the Late Triassic age scale, as will be discussed in Section 25.3.

25.2.3. Physical Stratigraphy

25.2.3.1. Magnetostratigraphy

The compilation of the magnetic polarity time scale for the Triassic is better developed than for the Jurassic or for any of the Paleozoic systems. A concentrated effort by paleomagnetists working closely with biostratigraphers and cyclostratigraphers during the 1990s and the first decade of the 21st century has revealed approximately 50 main magnetozones and twice as many lesser polarity subzones (Hounslow and Muttoni, 2010). The composite polarity scale from marine strata has been calibrated to ammonoid and conodont datums in different regions. A parallel polarity scale from terrestrial settings has been partly correlated to conchostracan zones and scaled with cycle-stratigraphy. Magnetic reversals are globally synchronous; therefore, the placement of biostratigraphic datums relative to distinctive polarity patterns has been used to determine diachroneity or local distortions in relative timing of markers and a polarity boundary may potentially be used as the primary marker for at least one Triassic GSSP (e.g., GSSP for Anisian by Hounslow *et al.*, 2007). Within Triassic substages, the average of about four major and five minor polarity intervals commonly display a characteristic pattern. However, the sheer abundance of magnetic polarity chrons and lack of broad fingerprints in the patterns at the stage level implies that utilization of Triassic magnetostratigraphy for inter-regional or inter-facies correlations requires adequate biostratigraphic constraints.

The total of the 133 validated magnetozones in the Triassic have a mean reversal rate of 2.6/myr, which is similar to the average Cenozoic rate of geomagnetic reversals (Hounslow and Muttoni, 2010). The apparent reversal rate of the Early and Middle Triassic (4/myr) is twice the average rate within the Late Triassic.

Each magnetostratigraphy study employed a different system for labeling the observed magnetozones. The verification of these studies by demonstrating reproducibility within the biostratigraphic constraints among different regions enables a nomenclature for the main and minor polarity chrons. In the milestone synthesis by Hounslow and Muttoni (2010), the subjective groupings into main magnetozones are systematically numbered within each Triassic series (e.g., MT7 is the seventh "main" polarity episode, although it may contain several brief polarity subchrons, with a normal-polarity-dominated portion followed by a reversed-polarity-dominated portion). An alternative would be to label each magnetozone according to the Triassic subzone (e.g., Illy-N2; for the second cluster of normal-polarity in the Illyrian substage of Anisian; which is the same as "MT7n", thereby enabling a user to immediately know the approximate age relationship (e.g., suggested scales in Ogg *et al.*, 2008). This "stage-abbreviation-then-number" system is the same nomenclature philosophy as is used for Phanerozoic sequences (e.g., Hardenbol *et al.*, 1998). In the summary figures, both options are shown. In any case, there is a caveat that portions of these composite polarity patterns are not well calibrated to boundaries of ammonoid (or conodont) zones and the scaling of the patterns within zones is probably distorted by variable sedimentation rates in the reference sections.

The following summary is mainly from Hounslow and Muttoni (2010), and only a selection of their main reference sections for bio-magnetic calibrations and conclusions are given here. In addition to placing the generalized patterns onto the relevant biostratigraphic scales, they standardized the stratigraphy of the different regional sections onto a common Tethyan ammonoid-zone scale.

Early Triassic Magnetic Polarity Scales

The Induan stage has two pairs and the Olenekian has seven pairs of polarity chrons that are correlated to Boreal ammonoid and conodont datums in the Arctic (e.g., Ogg and Steiner, 1991; Hounslow *et al.*, 2008a,b) and to conodont ranges in several Tethyan sections in China (e.g., Steiner *et al.*, 1989; Heller *et al.*, 1995; Glen *et al.*, 2009), Iran and Italy. The Permian–Triassic boundary is near the base of a relatively long normal-magnetozone ("LT1n" or "Gries-N1"; Steiner, 2006). This feature can be identified in the composite magnetostratigraphy from terrestrial deposits within the Germanic Basin (e.g., Nawrocki, 1997; Szurlies, 2004, 2007), implying that the base-Triassic is approximately the base of the Buntsandstein Formation.

The correlation of the marine-based magnetic polarity scale to this Germanic Basin polarity pattern is well established. There is a distinctive dominance of reversed polarity in the upper Induan and lowermost Olenekian followed by predominantly normal polarity in upper Lower Triassic in both scales; which enables a one-to-one correlation of the individual cycle-scaled Buntsandstein magnetic polarity chrons to those in the marine composite (e.g., Szurlies, 2007; Hounslow and Muttoni, 2010). The same change in polarity dominance and other features are tentatively used to assign geologic stages/substages to terrestrial deposits in the Karoo Basin, southwestern USA and Russia (Hounslow and Muttoni, 2010: fig. 4). Therefore, the proposed astronomical cyclicity can be projected via these bio-magnetostratigraphic correlations to estimate the time spans for Lower Triassic stages and substages. The resulting stage durations generally agree with the estimates derived from radio-isotopic ages (e.g., Menning *et al.*, 2005; Kozur and Bachmann, 2005, 2008; but see Galfetti *et al.*, 2007b for an alternate conclusion in which the Spathian substage is significantly longer than the combined Griesbachian, Dienerian and Smithian stages, a conclusion that is difficult to reconcile with the cycle-magnetostratigraphic model).

Middle Triassic Magnetic Polarity Scale

The seven main polarity pairs of the Anisian display dominance by normal polarity in the lower substages (Aegean, Bithynian) followed by a relative dominance by reversed polarity through the lowermost Ladinian. This pattern is derived from sections containing Boreal ammonoid zones in Spitsbergen (arctic Norway; Hounslow *et al.*, 2008a,b) and from those with conodont ranges in China (e.g., Lehrmann *et al.*, 2006); Albania (e.g., Muttoni *et al.*, 1996b), Romania (e.g., Gradinaru *et al.*, 2007), Austria (Muttoni *et al.*, 1996a) and other regions. The distinctive switch in dominance of polarity is the first-order constraint on projecting ages onto terrestrial facies in the Germanic Basin (e.g., Nawrocki and Szulc, 2000; Szurlies, 2007), the Catalan and Iberian basins of Spain (e.g., Dinarés-Turell *et al.*, 2005), England and China (Hounslow and Muttoni, 2010: fig. 4).

The main reference sections for the eight main polarity pairs of the Ladinian are in the alpine region of Italy and Austria (Seceda, Mayerling, Stuores sections; e.g., Gallet *et al.*, 1998; Broglio Loriga *et al.*, 1999; Muttoni *et al.*, 2004a).

Late Triassic Magnetic Polarity Scales

The alpine Mayerling and Stuores sections, plus Bolücektasi Tepe in Turkey (Gallet *et al.*, 1992), are the main reference sections for the Julian substage (lower Carnian). There are no adequate biostratigraphic controls on magnetostratigraphic studies that might span the middle Carnian (lower two ammonoid zones of Tuvalian substage); therefore Hounslow and Muttoni (2010) assigned a "holding" nomenclature of "UT5–UT9" with a temporary schematic pattern similar to zones "E2–E5" of the terrestrial deposits of the lower Newark supergroup (Kent *et al.*, 1995). As of the year 2010, this was the only significant gap in the Triassic composite magnetic polarity time scale.

The upper Carnian through Norian pattern is mainly derived from the GSSP candidate for base-Norian at Pizzo Mondello in Sicily (Muttoni *et al.*, 2001, 2004b) and the Silickà Brezovà section of Turkey (Channel *et al.*, 2003). The uppermost Norian to lowermost Rhaetian is calibrated to conodont datums in Austria (Scheiblkogel: Gallet *et al.*, 1998; and the candidate for the Rhaetian GSSP at Steinbergkogel: Krystyn *et al.*, 2007c,d, and Hüsing *et al.*, 2011). A pair of thick overlapping sections in the southern Alps (Brumano and Italcementi Quarry: Muttoni *et al.*, 2010) spans the middle Rhaetian through lowermost Hettangian, although minor faults complicate the stratigraphy.

In general, the eight main magnetozones (UT13–UT20) of the lower Norian (Lacian and Alaunian substages) have a dominance by reversed polarity; whereas the three main polarity Zones of the upper Norian (Sevalian) have approximately equal proportions of normal and reversed polarity. The basal Rhaetian is mainly reversed ("UT23r–UT24");

followed by a prevalence of normal polarity that continues into the Hettangian.

During the Late Triassic, a series of rift valleys along the western margin of the future Central Atlantic accumulated very thick successions of lacustrine deposits that recorded climatic responses to Milankovitch orbital cycles. Drilling of these Newark basin strata has yielded a complete 30 myr cycle-scaled pattern of the magnetic reversal history during the Late Triassic (Kent *et al.*, 1995). A few meters above the brief reversed-polarity E23r zone is a palynological turnover event, and a few centimeters higher the lacustrine deposits are overlain by the Orange Mountain basalts, part of the onset of a regional Central Atlantic Magmatic Province dated at approximately 201.5 Ma (e.g., Mundil *et al.*, 2010). Therefore, Kent and Olsen (1999 with 2002 online update) assigned an age of 202 Ma to the top of E23 and tuned the cyclic stratigraphy using the 404-kyr eccentricity cycle and a 1.75-myr-long modulating cycle to project the ages of the late Triassic polarity pattern. Based on palynology (e.g., Cornet, 1977), the base of the Norian was tentatively assigned to Newark magnetozone E13 at the base of the Passiac Formation (Olsen *et al.*, 1996), but projected to a much lower level by other interpretations (e.g., option 2 of Muttoni *et al.*, 2004b). However, there is the problem of the palynological turnover event below the basalts, which was originally considered to mark the Rhaetian/Hettangian boundary and hence the Triassic/Jurassic boundary (e.g., Kent and Olsen, 1999); but other palynology and conchostracan investigations interpret this as a major hiatus at the Norian/Rhaetian boundary with the majority of the Rhaetian being absent (e.g., Kozur and Weems, 2005; Cirilli *et al.*, 2009).

Depending on the choice of interpreted placement of the Carnian/Norian boundary and the continuity of the uppermost Norian–Rhaetian for the Newark succession, there are many published versions of how the cycle-scaled Newark polarity pattern might correlate to the upper Carnian, Norian and Rhaetian magnetostratigraphy from marine sections and composites (e.g., options 1 and 2 of Muttoni *et al.*, 2004b; Ogg, 2004; Ogg *et al.*, 2008; options A, B and C of Hounslow and Muttoni, 2010; Gallet *et al.*, 2007; Muttoni *et al.*, 2010; Lucas *et al.*, submitted; etc.). Radio-isotopic age control on the Newark lacustrine beds or on Norian and Rhaetian magnetozones or identification of unambiguous markers in the terrestrial and marine facies is required. Once there is an established correlation of the terrestrial to the marine polarity patterns, then the extensive cycle-scaled Newark suite will provide a precise duration and timing for most of the events in the Late Triassic.

25.2.3.2. Chemical Stratigraphy

Major excursions in carbon, sulfur and strontium isotopes in uppermost Permian and lower Triassic and in the uppermost Rhaetian through early Hettangian strata are well documented. The recognition of global excursions in carbon and strontium isotopes during the majority of the Middle and Upper Triassic is relatively less developed than in adjacent geological periods owing to a dearth of extended sections with high-resolution biostratigraphy and unambiguous global correlation. Comprehensive reviews are by Tanner (2010a) and within the geochemical stratigraphy chapters of this volume.

Carbon, Oxygen and Sulfur Stable Isotopes

The main set of end-Permian marine mass extinctions occurs abruptly within a broader major negative excursion in carbon isotopes in both marine carbonates and organic carbon. High-resolution studies indicate that the stepwise downward trend in negative carbon isotopes was interrupted by a minor positive isotope excursion that approximately coincided with this main marine extinction level, followed by the minimum 13_C values immediately above the basal Triassic (e.g., Korte and Kozur, 2010; Hermann *et al.*, 2010; Luo *et al.*, 2011). This pronounced trend toward negative carbon isotope values may have been caused by a combination of decreased marine productivity and influx of light carbon from volcanic, soil-carbon or methane sources (e.g., Holser and Magaritz, 1987; Baud *et al.*, 1989, 1996; Krull and Retallack, 2000; Yin *et al.*, 2001; Sephton *et al.*, 2002a; Erwin *et al.*, 2002; Korte *et al.*, 2010; Horacek *et al.*, 2010a).

Three positive peaks in 13_C occur in the lower Triassic: the lower-middle of the Smithian substage, the base of the Spathian substage, and a broader peak in late Spathian through lower Anisian. The carbon isotope values never return to the relatively heavy δ^{13}C of $+4‰$ that characterizes the Late Permian (Figure 25.4; Chapter 11). The measured magnitude of these shifts vary among regional syntheses (e.g., the peak of the middle Smithian excursion is 7‰ in the compilation of Payne *et al.* (2004), but 2–3‰ in the compilations of Korte *et al.*, (2005) and Galfetti *et al.* (2007b, 2008)). The curve in Figure 25.4 is a generalized synthesis of the main trends from these studies. An abrupt drop in ammonoid and conodont diversity is associated with the onset of the positive 13_C excursion at the base of the Spathian substage (Galfetti *et al.*, 2007a). A hypothesis for the origin of these relatively high-amplitude excursions is that there were multiple pulses of carbon release during phases of eruption and intrusion of the Siberian Traps flood basalts (e.g., Payne and Kump, 2008).

Carbon isotopes generally remain in the +1.5 to +2.5‰ range throughout most of the Middle Triassic, then rise to a broad δ^{13}C peak of c. +3.5‰ upper Carnian through lower Norian (Veizer *et al.*, 1999; Hayes *et al.*, 1999; Korte *et al.*, 2005). A brief, but significant, negative carbon isotope excursion is reported from the mid-Carnian (Dal Corso *et al.*, 2012). A brief positive δ^{13}C peak in organic in organic carbon components at the Norian–Rhaetian boundary has been tentatively attributed to widespread oceanic stagnation coincident with extinction of deep-water invertebrate fauna (Sephton *et al.*, 2002b), but this requires verification in additional sections. The late Rhaetian (end-Triassic) mass extinctions coincide with a negative carbon isotope excursion, which, like the end-Permian event, may be linked to widespread volcanism, oceanic productivity collapse and release of methane (e.g., Pálfy *et al.*, 2001; Ward *et al.*, 2001; Hesselbo *et al.*, 2002; Ruhl *et al.*, 2009). However, the interpretation of these carbon isotope records may be distorted by facies variations in some sections.

Oxygen isotope trends and other paleoenvironmental evidence indicate a general progressive cooling from Late Permian through the Triassic. The global ^{18}O pattern suggests a total cooling of about 4°C in tropical seas (Veizer *et al.*, 2000; Wallman, 2001; Korte *et al.*, 2005).

The uppermost-Permian through Lower Triassic is characterized by two extraordinary high-amplitude excursions in sulfur isotopes. Evaporite and carbonate-associated sulfates display low δ^{34}S values of c. 10‰ prior to the end-Permian mass extinction, followed by a brief excursion to approximately 25–30‰ near the end-Permian and a second broader peak to similar values near the Induan–Olenekian boundary (Dienerian–Smithian substage boundary) (e.g., Kampschulte and Strauss, 2004; Newton *et al.*, 2004; Horacek *et al.*, 2010a,b; Luo *et al.*, 2010). One interpretation for the rate and magnitude of these excursions is that sequestered hydrogen sulfide in a latest Permian anoxic ocean was released by oceanic overturning (e.g., review in Payne and Gray, Chapter 9 of this volume); and that the oceanic sulfate reservoir was anomalously low, perhaps less than 15% of modern size (Luo *et al.*, 2010). Sulfur isotope stratigraphy within the rest of the Triassic is currently at a low-precision reconnaissance status.

Strontium and Osmium Isotope Ratios

The curve of marine ^{87}Sr/^{86}Sr through the Triassic (Figure 25.4) is dominated by a broad trough bounded by twin peaks in the late-Early and latest Triassic (e.g., review in McArthur *et al.*, Chapter 7 of this volume). Values of ^{87}Sr/^{86}Sr show a sharp rise through the latest Permian into a late-Olenekian maximum (0.7082), a minimum (0.7076) in early Carnian, a second peak (c. 0.7080) in late Norian, followed by a rapid decline through the Rhaetian (to c. 0.707 65) that ended upon eruption of the end-Triassic Central Atlantic Magmatic Province (CAMP) (Koepnick *et al.*, 1990; Korte *et al.* 2003; Cohen and Coe, 2007).

The major rise in marine ^{87}Sr/^{86}Sr during the Late Permian and Early Triassic may be a long-term response to increased continental weathering following the glacial climates of the early Permian (Martin and Macdougall, 1995). The abrupt initiation of the end-Triassic downturn in marine ^{87}Sr/^{86}Sr coincided with major flood basalt outpouring along the future Central Atlantic seaway, followed by the prolonged decline in ^{87}Sr/^{86}Sr to the end of the Pliensbachian (Jones and Jenkyns, 2001; McArthur *et al.*, 2001; Jenkyns *et al.*, 2002). Except for the broad trough of Carnian and Early Norian, the rate of change of ^{87}Sr/^{86}Sr through the Triassic offers some promise for global correlation at high resolution.

Osmium isotope stratigraphy has not yet been systematically compiled for the Triassic. Analysis of black shales of latest Anisian from Svalbard yielded as initial ^{187}Os/^{188}Os ratio of 0.83 ± 0.03, the highest recorded ratio for global seawater between the earliest Cambrian and the late Early Jurassic (Xu *et al.*, 2009). This peak remains to be verified and delimited in additional sections. Mirroring the strontium curve of late Triassic, a plunge in initial ^{187}Os/^{188}Os ratio from a late Norian peak of 0.75 is interrupted by a brief excursion to high (radiogenic) ratios upon eruption of the CAMP (e.g., Cohen and Coe, 2007; Peucker-Ehrenbrink and Ravizza, Chapter 8 of this volume).

25.2.3.3. Cyclostratigraphy

The monsoon-dominated climate of the Pangea megacontinent was sensitive to Milankovitch cycles, especially the precession–eccentricity components of these orbital–climate oscillations. The interpretations and controversies concerning these Triassic cyclic deposits are critically examined by Tanner (2010b). Extended and quasi-continuous deposits of continental facies which have excellent magnetostratigraphy in central Europe and eastern North America are the basis of cycle-scaled polarity patterns for the Early and the Late Triassic. In theory, these successions should be the Rosetta stone to project cycle-scaled durations onto marine sequences for a precise relative time scale, similar to what has been developed for the Cenozoic. In practice, there is a lack of a unique pattern match for correlation of these extended intervals of cycle-scaled magnetostratigraphies with marine-based composite polarity patterns.

Variations in clastic input into the Buntsandstein basins of central Europe during the Early Triassic provide a detailed regional stratigraphy that is applicable to surface exposures and downhole logs (e.g., reviews in Röhling, 1991; Bachmann and Kozur, 2004; Szurlies, 2004; Menning *et al.*, 2005; Feist-Burkhardt *et al.*, 2008). The cycles, spanning about 10–20 m with sandstones fining upward into more clay-rich sediments, are generally interpreted as oscillations between more arid and more humid conditions. Constraints from terrestrial biostratigraphy (conchostracan, pollen spores) combined with radio-isotope ages on the span of the Early Triassic indicate that the depositional sequences appear to coincide with the 100-kyr short eccentricity cycle (e.g., Bachmann and Kozur, 2004; Menning *et al.*, 2005). However, the expected 400-kyr-long eccentricity has not been unambiguously resolved. The magnetostratigraphy from the Buntsandstein, especially within the lower portion which has relatively longer duration polarity zones and biostratigraphic constraints, is fairly well correlated to the Early Triassic composite (Szurlies, 2007; Hounslow and Muttoni, 2010). Even though a monotonic 100-kyr periodicity is not expected for short eccentricity and there is a possibility of "missing beats" at possible exposure horizons within this

Buntsandstein succession, the projected cycle-scaling of the marine zonation and associated Early Triassic substages via this magnetostratigraphy is a close fit to radio-isotopic ages and was used here for detailed scaling of the Early Triassic and early Anisian.

Interbedded marls and limestones of shallow-marine origin spanning the Permian–Triassic boundary interval in the Austrian Alps display cycles with ratios matching Milankovitch periodicities, and have been interpreted to imply that the latest Permian extinction and negative carbon isotope spike spanned less than 30 kyr (Rampino *et al.*, 2000, 2002).

The Latemar massif in the Italian Dolomites was an atoll-like feature with a core of flat-lying Anisian and Ladinian platform carbonates. Oscillations in sea level were created over 500 thin depositional cycles (Goldhammer *et al.*, 1987). Stacking patterns and spectral analysis of the sea-level oscillations had been interpreted as representing precession modulated by short-term (100 kyr) eccentricity, therefore yielding an implication that the Latemar deposit spans approximately 10 myr (Goldhammer *et al.*, 1990; Hinnov and Goldhammer, 1991). In contrast, U-Pb ages from coeval tuff-bearing basinal deposits appear to constrain the Latemar platform to span only 2 to 4 myr (e.g., Brack *et al.*, 1996, 1997; Mundil *et al.*, 1996; Hardie and Hinnov, 1997; and extended review in Tanner, 2010a). A possible solution to this disparity is that an extremely rapid rate of platform construction (c. 500 m/myr or greater) enabled recording of sub-Milankovitch sea-level oscillations with misleading similarity in ratios to precession–eccentricity (e.g., Kent *et al.*, 2004, 2006; Hinnov, 2006; Meyers, 2008). This debate demonstrates that any cyclestratigraphic analysis based on a single section requires verification from other independent basins and facies.

Radiolarian-rich pelagic chert successions from Japan spanning the Middle Triassic are characterized by ribbon bedding. These chert-clay couplets have been interpreted as productivity fluctuations induced by 20-kyr precession cycles, and the longer-term trends in bed thickness correspond to 100-kyr and 405-kyr eccentricity cycles (Ikeda *et al.*, 2010). These cyclostratigraphic interpretations, the tentative correlation of radiolarian taxa to geologic stages, a potential long-term modulation of c. 3.6 myr, and the continuity of the bedded-chert sections await further verification.

Studies of similar oscillating Lofer facies within upper Triassic platform carbonates of the Austrian Alps played an important role in developing fundamental concepts of cyclostratigraphy (e.g., Fischer, 1964), but the reality of regular cyclicity in these deposits has also been debated (e.g., Satterley, 1996, versus Schwarzacher, 2005, and Cozzi *et al.*, 2005; reviewed in Tanner, 2010b).

During the late-Middle Triassic through Early Jurassic, a set of rift basins formed as Pangea underwent an initial phase of breakup. The thick Newark Group of lacustrine sediments from these tropical basins are characterized by oscillations between semi-stagnant deep lakes and arid playas as the intensity of monsoonal rains responded to Earth's precession modulated by short-term (c. 100 kyr) and long-term (c. 400 kyr) eccentricity cycles. Spectral analysis of sediment facies successions in a series of deep-drilling cores enabled compilation of a cycle-scaled stratigraphic record, including a detailed polarity pattern that is unprecedented in its 30-myr temporal span (e.g., Kent *et al.*, 1995; Olsen *et al.*, 1996; Kent and Olsen, 1999). As discussed previously, the comparison of the cycle-scaled terrestrial polarity signature to the unscaled marine magnetostratigraphy does not always provide a unique match. However, uppermost Triassic lacustrine deposits with alternating red-to-green coloration at St. Audrie's Bay have yielded both a magnetostratigraphy (Hounslow *et al.*, 2004) and an interpreted 3.7-myr cyclostratigraphy (Kemp and Coe, 2007) that have a polarity scaling resembling the upper Newark interval of polarity zones E19n-E16n.

25.2.3.4. Sequence Stratigraphy

Triassic sea-level trends and sequences of different relative magnitudes have been compiled for Boreal basins (e.g., Embry, 1988; Mørk *et al.*, 1989; Skjold *et al.*, 1998), the classic Germanic Trias (e.g., Aigner and Bachmann, 1992; Geluk and Röhling, 1997), the Dolomites and Italian Alps regions (e.g., De Zanche *et al.*, 1993; Gaetani *et al.*, 1998; Gianolla *et al.*, 1998) and other regions. Some of these sea-level trends appear to correlate on an inter-basin to global scale (e.g., Haq *et al.*, 1988; Hallam, 1992; Embry, 1997; Gianolla and Jacquin, 1998), The significant disparity in some proposed major global features for the Triassic (e.g., Hardenbol *et al.*, 1998, compared to Haq and Al-Qahtani, 2005, or to Simmons *et al.*, 2007) is difficult to resolve, because the supporting details of reference sections and biostratigraphic control are generally not adequately published.

The major sequences in Figure 25.7 are mainly based on a widely used compilation and systematic numbering system by Jacquin and Vail (1998). In this compilation, the main first-order Triassic sea-level trend is dominated by a single cycle – a progressive transgression that began in the latest Permian, peaks in the Anisian–Ladinian boundary interval, followed by a regression to the Late Norian (sequence boundary "*No2*"). The second major transgression–regression cycle beginning in the Rhaetian reaches a peak in the late Jurassic. Superimposed on these main cycles are at least four second-order facies cycles (major sequence boundaries *Ol4* in uppermost Olenekian, *Lad3* in uppermost Ladinian, *Car3* in mid-Carnian, and *No2* in uppermost Norian). These, in turn, have up to 23 proposed third-order sequences (Gianolla and Jacquin, 1998). In contrast, Simmons *et al.* (2007) have published a general Triassic scale with 7 main sequences (*Tr10* to *Tr80*). The long-term eustatic sea-level curve of Haq and Al-Qahtani (2005) displays a late Ladinian through early Carnian transgression between a general Induan through early

Ladinian lowstand to a general late Carnian through mid-Rhaetian highstand. A schematic intermediate-term sea-level curve can be constructed by offsetting the amplitudes of the Triassic sequences of Hardenbol *et al.* (1998) from the long-term Triassic curve of Haq and Al-Qahtani (2005) following advice of Bilal Haq (pers. comm., 2006; and available as a column within the TimeScale Creator visualization suite, *www.tscreator.org*). These proposed Triassic suites of sequences and sea-level trends await published documentation of the different compilations and a community consensus on the regional versus global components.

25.2.3.5. Other Major Stratigraphic Events

Large Igneous Provinces

The Triassic is delimited by two major volcanic provinces: the Siberian Traps at the base and the Central Atlantic Magmatic Province at the top.

The Siberian Traps, exposed mainly on the Siberian craton, was one of the most voluminous volcanic eruption provinces of the Phanerozoic, with an estimated volume of over 2 million cubic kilometers of basalt flows and volcaniclastic rocks (e.g., Large Igneous Provinces Commission, 2008; and review by Reichow *et al.*, 2004). The volcanic province on the Siberian craton is generally subdivided into four distinct geographic regions: Noril'sk, Putorana, Nizhnaya Tunguska and Maimecha-Kotuy. The main pulses of the voluminous Siberian Traps flood basalts are approximately coeval with the latest Permian mass extinctions, and the waning stages of this volcanic activity continued into the earliest Triassic (e.g., Renne *et al.*, 1995; Erwin *et al.*, 2002 and references therein; Reichow *et al.*, 2009).

The Wrangellia terrane, which is accreted to British Columbia and Alaska, contains a major episode of tholeiitic flood volcanism in submarine and subaerial environments. Radiometric dating (c. 227 and 232 Ma; see Table 25.1) and the overlying fossiliferous strata indicate an eruption during the Carnian. The original volcanic volume is estimated as 1 million cubic kilometers (e.g., Greene *et al.*, 2008, 2010, 2011; Large Igneous Provinces Commission, 2008). The Wrangellia eruptions may have caused a major negative excursion in carbon isotopes in the mid-Carnian (Dal Corso *et al.*, 2012).

The Central Atlantic Magmatic Province (CAMP) has ages clustering at c. 201 Ma just prior to Triassic–Jurassic boundary and is considered to have been a major causal factor in the end-Triassic extinctions (e.g., Marzoli *et al.*, 1999; Jourdan *et al.*, 2009a; Schoene *et al.*, 2010). The total extrusive volume may have been even greater than the Siberian Traps.

Major Bolide Impacts

The 100-km diameter Manicouagan impact structure of Quebec is the only Phanerozoic impact structure that has melt rock directly dated by modern U/Pb methods (214.56 ± 0.05 Ma; Ramezani *et al.*, 2005, and Hodych and Dunning, 1992, as re-evaluated by Jourdan *et al.*, 2009b; see Table 25.1). This impact event and associated environmental catastrophe may have contributed to the large-scale turnover of continental tetrapods during mid-Norian, during which dinosaurs attained dominance over competing families (e.g., Benton, 1986, 1993; Lucas, 2010d). However, no record of this impact or major environmental catastrophe has yet been discovered within the extensive lacustrine deposits of the Newark Group of northeastern North America.

Two significant impacts, with craters of c. 40 km diameter, are assigned to the Triassic. The Araguainha crater in Brazil has a reported Early Triassic age of 244.4 ± 3.25 Ma, and the Saint Martin crater in Manitoba, Canada, has an estimated mid-Triassic age of 220 ± 32 Ma (Earth Impact Database, 2010).

25.3. TRIASSIC TIME SCALE

After the publication of GTS2004, the Triassic was the focus of extensive sampling and application of ultra-high-resolution methods by different geochronology laboratories. Triassic and uppermost Permian strata enabled testing enhanced techniques for single-zircon dating, provided sets for intercalibration of U-Pb and Ar-Ar methods and the standardization of procedures among laboratories. The new age sets have replaced or called into question nearly all of the Triassic radiogenic isotope ages published before 2004 (reviewed by Mundil *et al.*, 2010). The initiation and termination of the Triassic Period have been constrained by a remarkably extensive suite of ages, implying a span from 252.2 to 201.3 Ma (51 myr).

25.3.1. Constraints from Radio-Isotopic Dates

As of the time of this compilation (early 2011), approximately 30 radio-isotopic dates constrain the Triassic numerical time scale (see Chapter 6 of this volume). An extensive set of ^{206}Pb/^{238}U CA-TIMS dates from analyses of individual zircons treated by annealing followed by chemical abrasion have replaced the dates derived from multi-grain analyses and K-Ar methods. An extensive evaluation of these earlier published dates has been compiled by Mundil *et al.* (2010), and their selection of preferred Triassic dates from marine strata and estimates for stage-boundary ages has been largely followed in the composite Phanerozoic radio-isotopic age constraint table (see radio-isotopic dating chapter (Chapter 6), this volume). Some additional newly published dates (e.g., Schoene *et al.*, 2010; Bowring in Shen *et al.*, 2010; Shen *et al.*, 2011) are included, along with some additional secondary guide ages that require future re-evaluation. A

TABLE 25.1 Selected Triassic Radio-isotopic Dates – Terrestrial strata or Flood basalts

NOTE: This is only a subset of the published age suites for most of these episodes. See the selected references for reviews.

Event	Primary controls (*)	Source (location)	Age for GTS2012; *uncertainty (95%) in myr*	Annotation; Dated Material; Method	Locality; formation; biostratigraphy	Primary Reference	Review Reference *(and evaluation comments from references)*
Central Atlantic Magmatic Province (CAMP; end-Triassic)							
	X	North Mtn. basalt; Fundy	201.38 ±0.02	NMB-03-1 re-dated; Zircon from basalt; 206Pb/238U ID-TIMS; Annealing/ chemical abrasion, single grain	Fundy Basin, Nova Scotia, Canada. Pegmatitic lenses within the lowest basalt flow of the North Mountain basalt (often regarded as the oldest CAMP basalt flow in North America); same locality as Hodych and Dunning (1992).	Schoene *et al.*, 2010.	See also: Schaltegger *et al.*, 2008.
	X	CAMP peak	201.09 ±0.7	NEW-CUL sets; Biotite and plagioclase from ten sill and lava flow samples; Ar/Ar	Composite from Newark and Culpeper basins	Marzoli *et al.*, 2011.	
	X	North Mtn. basalt; Fundy	201.33 ±0.9	NS3; Plagioclase from basalt; Ar/Ar	Fundy Basin, Nova Scotia, Canada. East Ferry Member of North Mountain basalt (often regarded as the oldest CAMP basalt flow in North America).	Jourdan *et al.*, 2009a.	
	X	North Mtn. basalt; Fundy	200.62 ±0.8	NS19, NS21; Plagioclase from pegamitite; Ar/Ar	Fundy Basin, Nova Scotia, Canada. Pegmatitic lenses within East Ferry Member of North Mountain basalt (often regarded as the oldest CAMP basalt flow in North America).	Jourdan *et al.*, 2009a.	
	secondary guide	North Mtn. basalt; Fundy	201.7 +1.4/-1.1	PSM6, GRAD285,286; Zircon from basalt; 206Pb/ 238U	Fundy Basin, Nova Scotia, Canada. North Mountain basalt (often regarded as the oldest CAMP basalt flow in North America). Base of basalt flows are 20m above a turnover in palynology and vertebrate assemblages that was tentatively considered equivalent to Triassic-Jurassic boundary.	Hodych and Dunning, 1992.	Pálfy *et al.*, 2000a.
	secondary guide	Palisades; Newark Basin	200.9 ±1	PSM8; Zircon and baddeleyite from basalt; U-Pb	Newark Basin, eastern U.S. Palisades sill. Palisades and Gettysburg sills are considered to be feeders for regional basalt that immediately overlies a turnover in palynology and vertebrate assemblages that was tentatively considered equivalent to Triassic-Jurassic boundary.	Dunning and Hodych, 1990.	Pálfy et al., 2000a.
	secondary guide	Gettysburg sill; Newark Basin	201.3 ±1	PSM7; Zircon from basalt; U-Pb	Newark Basin, eastern U.S. Gettysburg sill. [See above note on Palisades]	Dunning and Hodych, 1990.	Pálfy *et al.*, 2000a.
Manicouagan Impact Crater (mid-Norian)							
	X	Melt rock	214.56 ±0.05	Zircons from melt rock; 206Pb/238U and ArAr	Manicouagan crater in east-central Quebec, Canada. "Melt rock sheet (ca. 230m) underlying the central island of the 65-km-diameter Manicouagan crater."	Hodych and Dunning, 1992.	Jourdan *et al.*, 2009b.
	X	Melt rock	215.5 (*± not given in reference)*	Zircons from melt rock; 206Pb/238U	Manicouagan crater in east-central Quebec, Canada. "Melt rock sheet (ca. 230m) underlying the central island of the 65-km-diameter Manicouagan crater."	Ramezani *et al.*, 2005.	

(Continued)

TABLE 25.1 Selected Triassic Radio-isotopic Dates – Terrestrial strata or Flood basalts—cont'd

NOTE: This is only a subset of the published age suites for most of these episodes. See the selected references for reviews.

Event	Primary controls (*)	Source (location)	Age for GTS2012; *uncertainty (95%) in myr*	Annotation; Dated Material; Method	Locality; formation; biostratigraphy	Primary Reference	Review Reference *(and evaluation comments from references)*
Wrangellia Large Igneous Province (mid-Carnian)							
	X	Vancouver	232.2 ±1	207Pb/206Pb	Vancouver Island, Canada. Large igneous province (LIP; >1 million km2) of Wrangellia. Potential causation factor in "Carnian pluvial event" just below onset of M. nodosus conodont in Apennines (Furin et al., 2006; "This apparent coincidence needs to be further tested with new radiometric dates for this LIP and of the Carnian crisis."	Mortensen and Hulbert, 1992.	Furin *et al.*, 2006.
	X	Vancouver	227 ±3	207Pb/206Pb	Vancouver Island, Canada. Large igneous province (LIP; >1 million km2) of Wrangellia. Potential causation factor in "Carnian pluvial event" just below onset of M. nodosus conodont in Apennines (Furin et al., 2006: "This apparent coincidence needs to be further tested with new radiometric dates for this LIP and of the Carnian crisis."	Parrish and McNicoll, 1992.	Furin *et al.*, 2006.
Selected Constraints on Terrestrial Vertebrate Megafauna (Carnian-Norian)							
	x	Petrified Forest Fm.; AZ	201 *(± not given in reference)*	Black Forest (Ramezani); Redeposited zircons; 206Pb/238U. U-Pb ID-TIMS single-zircon	Petrified Forest National Park, Arizona, southwest USA, Black Forest Bed of the Petrified Forest Formation. Lower-mid Revueltian of tetrapod zonation. *"dinosaurs … were still not a dominant component of the ecosystem".*	Ramezani *et al.*, 2009.	*"Results indicate that the lower two-thirds of the Petrified Forest section was deposited within a period of ca. 10 Myr entirely within the Norian Stage of Late Triassic."*
	x	Petrified Forest Fm.; AZ	211 ±0.7	Black Forest (Heckert); Redeposited zircons; U-Pb. air-abraded ID-TIMS	Petrified Forest National Park, Arizona, southwest USA, Black Forest Bed of the Petrified Forest Formation. Lower-mid Revueltian of tetrapod zonation. *"dinosaurs … were still not a dominant component of the ecosystem".*	Heckert *et al.*, 2009.	*"consistent with the approximately 218 Ma Carnian-Norian boundary in the Newark Supergroup, not a "long Norian" extending to approximately 228 Ma as recently proposed"*
	x	Petrified Forest Fm.; AZ	213.15 *(± not given in reference)*	Flattops 1; Redeposited zircons; 206Pb/238U. ID-TIMS single-zircon	Petrified Forest National Park, Arizona, southwest USA. Flattops 1 sandy interval, lower Petrified Desert formation. Projected to be lower part of Revueltian of tetrapod zonaton.	Ramezani *et al.*, 2009.	*"Adamanian-Revueltian faunal turn-over occurred between 219.37 and 213.15 Ma.*
	x	Petrified Forest Fm.; AZ	219.37 *(± not given in reference)*	Sonsela Sst; Redeposited zircons; 206Pb/238U. ID-TIMS single-zircon	Petrified Forest National Park, Arizona, southwest USA. Just below Sonsela Sandstone, in middle of Sonsela member. Upper-middle Adamanian of tetrapod zonation (Lucas, 2010)	Ramezani *et al.*, 2009.	
	x	Petrified Forest Fm.; AZ	223.1 *(± not given in reference)*	Blue Mesa (Ramezani); Redeposited zircons; 206Pb/238U. ID-TIMS single-zircon	Petrified Forest National Park, Arizona, southwest USA. Upper part of Blue Mesa member. Middle of Adamanian of tetrapod zonation (Lucas, 2010)	Ramezani *et al.*, 2009.	

x	Petrified Forest Fm.; AZ	220.9 ±0.6	Blue Mesa (Heckert); Redeposited zircons; U-Pb. air-abraded ID-TIMS	Six Mile Canyon area, western New Mexico, southwest USA. *"A volcanic-rich litharenite at the base of the Blue Mesa Member of the Petrified Forest Formation"*. Middle of lower Adamanian of tetrapod zonation (Lucas, 2010). *"Palynostratigraphic constraints supported by vertebrate and conchostracan biostratigraphy assign a Carnian age to the Blue Mesa Member, which is consistent with the approximately 218 Ma Carnian-Norian"* boundary in the Newark Supergroup (Heckert et al., 2009)"	Heckert *et al.*, 2009	*"Comparison to the astronomically-calibrated Late Triassic polarity timescale indicates that the Black Forest Bed likely corresponds to the normal chron E16"*
x	Petrified Forest Fm.; AZ	219.2 ±0.7	Blue Mesa (Mundil); Redeposited zircons; 206Pb/238U. CA-TIMS	Six Mile Canyon area, western New Mexico, southwest USA. Tuffaceous sandstone at base of Blue Mesa Member, Petrified Forest Formation, Chinle Group. Middle of lower Adamanian of tetrapod zonation (Lucas, 2010).	Mundil and Irmis, 2008	
x	Ischigualasto; Argentina	230.59 ±0.3	Ischigualasto; Sanidine from bentonite; $^{40}Ar/^{39}Ar$	northwest Argentina. Ischigualasto Formation, Herr Toba bentonite. Earliest dinosaurs are included in this extensive tetrapod assemblage. Ischigualasto Fm has fauna considered equivalent to Adamanian of southwest USA (assigned to late Carnian by Lucas, 2010). The assemblage *"slightly overlaps and mostly overlies the Herr Toba bentonite"* (Lucas, 2010).	Rogers *et al.*, 1993.	Furin *et al.*, 2006.
Italian volcanic center (Ladinian-Anisian)						
secondary guide	Predazzo; n. Italy	237.3 +0.4/-1.0	U-Pb	Dolomites, Italy. Predazzo volcanics. Pálfy *et al.* (2003) "tentatively assign to the *Regoledanus* ammonite zone" (highest zone of Ladinian). They schematically show the biostratigraphic uncertainty as spanning that entire zone.	Brack *et al.*. 1997.	
Siberian Traps (Permian-Triassic boundary interval)						
x	Noril'sk	250.2 ±0.6	*Ar-Ar*	Noril'sk region, Russia. Bracketing age for upper part of 3.5 km succession (part of an extensive suite, and review).	Reichow *et al.*, 2009.	
x	Noril'sk	251.81 ±1.1	*Ar-Ar*	Noril'sk region, Russia. Bracketing age for lower part of 3.5 km succession (part of an extensive suite, and review).	Reichow *et al.*, 2009.	
x	Upper Traps	251.1 ±0.3	*Upper Traps (Kamo'03); Zircons from volcanics; U-Pb*	Maymecha–Kotuy area (northeastern part of the Siberian flood-volcanic province), Russia. Lavas of the uppermost unit of volcanic sequence – Delkansky Suite (trachydacitic and trachyrhyodacitic samples). "*This is a maximum age because 1400 m of high-Ti, Mg-rich lava of the Meymechinsky Suite lies above the Delkansky Suite, and may represent a few hundred thousand years of eruptive activity.*"	Kamo *et al.*, 2003.	
x	Lower Traps	251.7 ±0.4	*Lower Traps (Kamo'03); Perovskite; U-Pb*	Maymecha–Kotuy area (northeastern part of the Siberian flood-volcanic province), Russia. Lavas of the uppermost unit of volcanic sequence – 350 m thick, melilite-bearing, mela-nephelinitic to limburgitic rocks of the Arydzhansky Suite.	Kamo *et al*., 2003.	
secondary guide	Noril'sk	251.2 ±0.3	*Traps (Kamo'96); Zircon and Badellyite; U-Pb*	Noril'sk-1 gabroic intrusion. Siberian Traps. Age is from Noril'sk-1 gabbroic intrusion, which cuts lower third of the Traps.	Kamo *et al*., 1996.	Bowring *et al.*, 1998.
secondary guide	Noril'sk	250.1 ±1.5	*Traps (Renne'95); Biotite; Ar-Ar*	Noril'sk-1 gabroic intrusion. Siberian Traps. Age is from Noril'sk-1 gabbroic intrusion, which cuts lower third of the Traps. [Ar-Ar on bulk Hornblende sample yielded isochron of 249.3 ±1.6 Ma.]	Renne *et al.*, 1995.	
secondary guide	Traps	251.5 ±1.6	GRAD317, 318; Hornblende, Whole Rock; Ar-Ar	Siberian Traps [Increased uncertainty on recalculated age *"incoporated uncertainty in the standard's age"* (Renne *et al.*, 1995). Internal error was only 0.30 myr.]	Renne and Basu, 1991.	Renne *et al.*, 1995.

selected suite of radio-isotopic dates from major flood basalt episodes and from strata that constrain terrestrial vertebrate evolution are also compiled (Table 25.1).

Nearly all the ages from marine strata having adequate biostratigraphic control are clustered within the early half of the Triassic (c. 255 to 239 Ma) and within a brief interval spanning the Triassic–Jurassic boundary (c. 202–200 Ma). In particular, there is a 30-myr gap in reliable dates encompassing the upper half of the Triassic. Therefore, this age gap remains an interval of controversy in correlation of terrestrial zonations (e.g., dinosaur evolution) to marine stages.

The Permian–Triassic boundary is well constrained as approximately 252.2 Ma (U-Pb, TIMS) from sets of samples from the Induan GSSP at Meishan, Zhejiang Province and the Shangsi section from Sichuan Province in China (between samples of Bowring re-dated MAW-b25 and Mundil SH32(29)). Shen *et al.* (2010 2011) estimate the P/T boundary (U-Pb, TIMS) at 252.16 Ma. However, the Ar-Ar age estimate is slightly younger – Reichow *et al.* (2009) obtained an age from sanidines in volcanic-ash Bed 28 about 8 cm above the GSSP at Meishan of 248.25 ± 0.14 Ma (based on a monitor standard FCs of 28.02 Ma) which converts to 249.85 ± 0.14 Ma using the revised FCs monitor standard of 28.201 Ma. Renne *et al.* (2010) suggest a "best-fit" (U-Pb, plus recalculated Ar-Ar using a slightly older FCs of 28.30 Ma) of 252.3 ± 0.2 Ma. The FCs standard of 28.201 Ma is used in other GTS2012 radio-isotopic tables, therefore we adopt the P/T age of 252.16 Ma, with an estimated uncertainty of ± 0.2.

At first glance, the U-Pb TIMS age sets would seem to imply that the onset of the eruption of the main basalt complex of the Siberian Traps of Russia (Table 25.1) slightly post-dates the mass extinction; which would contradict the presumed volcanic-induced cause for the end-Permian extinctions. However, a comparison of ages derived by the Ar-Ar methods from the Meishan GSSP and the Siberian Traps indicates an overlap with the main mass extinction (Reichow *et al.*, 2009). This apparent discrepancy between the implications of the independent U-Pb and Ar-Ar suites requires further study (Burgess and Bowring, 2010).

The base of the Olenekian Stage, or rather a leading candidate level for placing a future GSSP, may be slightly older than sample CHIN-40 that yielded 251.2 ± 0.2 Ma (Galfetti *et al.*, 2007b).

The base of the Anisian Stage, as potentially recognized by conodont zonation in southern China, was estimated as 247.2 ± 0.1 Ma (between samples PGD Tuff-3 and Tuff-2) by Lehrmann *et al.* (2006). However, U-Pb dates from a nearby ammonoid-bearing basinal section suggest that the ammonoid-defined (top of *haugi* Zone) placement for the base-Anisian maybe be slightly younger than 246.9 Ma (Ovtcharova *et al.*, 2010).

The base of the Ladinian is constrained to be between 239.3 ± 0.2 Ma (sample FP2 of mid-Ladinian; Brühwiler *et al.*, 2007) and 242.1 ± 0.6 Ma (sample Mundil MSG.09 at the Grenzbitumen horizon near base of the uppermost Anisian *Nevadites secedensis* ammonoid zone; Mundil *et al.*, 2010). Mundil *et al.* (2010) estimate the Anisian–Carnian boundary as 242.0 Ma. However, this boundary age would truncate the *Nevadites secedensis* ammonoid zone, and an Ar-Ar measurement on sanadines in MSG.09 from the same Grenzbitumen horizon by Mundil *et al.* (2010) yielded 240.95 ± 0.5 Ma (after adjusting the original 239.5 ± 0.5 Ma to an FCs monitor standard of 28.201 Ma). Therefore, pending further work to converge the Ar-Ar and U-Pb ages on the Grenzbitumen horizon, the base-Ladinian is estimated as c. 241.5 ± 1 Ma.

Spanning the Ladinian–Carnian boundary is a gap of about 9 myr in radio-isotopic dates that satisfy both analytical criteria (e.g., suite accepted by Mundil *et al.*, (2010), for their table 3) and biostratigraphic control. However, a U-Pb date of 237.3 Ma (+0.4/−1.0) from the Predazzo granites of northern Italy (Brack *et al.*, 1997) was tentatively correlated to the *Frankites regoledanus* ammonite zone (highest zone of Ladinian) by Pálfy *et al.* (2003). A new high-precision age of 237.85 ± 0.05 Ma has been obtained from a set of seven CA-TIMS-processed single zircons from a volcanic ash in the Alpe di Siusi/Seiser Alm section of northern Italy within the Upper Ladinian, but it has an imprecise correlation to ammonite zones (Mietto *et al.*, in press). A volcanic ash from the southern Appenines yielding a date of 230.9 ± 0.06 Ma (sample Aglianico; Furin *et al.*, 2006) was tentatively assigned as mid-upper Carnian, and may be near the base of the Tuvalian substage (H. Kozur, pers. comm., 2010). The date from this horizon is similar to Ar-Ar ages from strata in Argentina containing some of the earliest dinosaurs (Ischigualasto Formation). If one uses the Predazzo granite as a constraint within the latest Ladinian ammonite zone, then a provisional assignment of 237 ± 1 Ma is assigned to the Ladinian–Carnian boundary.

Although no reliable Norian or early Rhaetian radio-isotopic ages have been published, there are suggestions that the Norian may extend to 225 Ma or older. Suites of redeposited zircons within sediments of the southwestern USA that have yield Adamanian and Revueltian tetrapods and early dinosaurs have enabled a temporal framework for that interval of tetrapod evolution (Table 25.1). The array of dated horizons suggests that the "Adamanian-Revueltian faunal turn–over occurred between 219.37 and 213.15 Ma" (Ramezani *et al.*, 2009). However, the correlation of the Norian–Carnian stage boundary interval to these deposits is controversial. If the base-Norian is close to 225 Ma, then:

> *'A mid-Norian age for the Adamanian to Revueltian land vertebrate faunachron boundary, as suggested by the revised Late Triassic timescale, is no longer compatible with the idea that the faunachron boundary is coincident with the Carnian-Norian Stage boundary.'*
>
> (Ramezani *et al.*, 2010)

Alternatively, palynology, vertebrate and conchostracan biostratigraphy correlations from the southwestern USA to the Germanic Basin and European stages are interpreted to imply that the Adamanian is mainly Carnian, and that the redeposited zircon dates are:

'consistent with the approximately 218 Ma Carnian-Norian boundary in the Newark Supergroup, not a "long Norian" extending to approximately 228 Ma as recently proposed.'

(Heckert *et al.*, 2009)

Resolution of this ten-million-year divergence in opinion on the placement of the Carnian–Norian boundary requires future acquisition of radio-isotopic dates on marine-zoned lowermost Norian.

There are no constraints from radio-isotopic dates on the Norian–Rhaetian boundary; nor estimates from cycle stratigraphy on the possible durations of the Rhaetian or the Norian stages.

In contrast, the Triassic–Jurassic boundary has a precise radio-isotopic age. Schoene *et al.* (2010) project that the T/J boundary age is 201.31 Ma (± 0.18/0.38/0.43) based on constraints from ammonoid-bearing strata in Peru (LM4-90 and LM4-100/101) and a similar age from the former GSSP candidate section in New York Canyon, Nevada. They conclude that initiation of the main phase of the Central Atlantic Volcanic Province (CAMP, Table 25.1) preceded the Triassic–Jurassic boundary by only c. 70 kyr (or a maximum of 290 kyr if the extremes on the uncertainties are applied). Radio-isotopic ages from CAMP flows in North America and Morocco and cyclostratigraphy of intervening periods of sedimentation within the flow succession indicate that the main phase of eruptions was a brief peak spanning c. 600 kyr (e.g., Whiteside *et al.*, 2007; Jourdan *et al.*, 2009a; Marzoli *et al.*, 2011). A proposed significant age difference between these major volcanic eruptions and the marine extinctions (e.g., Pálfy *et al.*, 2000a,b) is now considered to have been an artifact from those multi-grain zircon analysis techniques (reviewed in Mundil *et al.*, 2010).

25.3.2. Early and Middle Triassic Scaling

The primary method for assigning numerical ages to the Early Triassic is the cycle-scaled magnetostratigraphy from the Germanic Basin (Figure 25.5). The following assumptions are made:

(a) The base of the Triassic is 252.16 Ma (Shen *et al.*, 2010, 2011). The uncertainty of 0.2 myr on this U-Pb isotopic age would apply to all numerical ages, but not to the cycle-derived durations (e.g., if the base-Triassic is shifted older by 0.1 myr, then the base-Olenekian and base-Anisian must also be shifted older by the same 0.1 myr).

(b) The Germanic Basin cycles are a uniform 100-kyr Milankovitch orbital-climate signal with the base of the Triassic at the base of Calvorde cycle s1.2 and the base of the Anisian at the base of Röt cycle s7.1 (Bachmann and Kozur, 2004; Kozur and Bachmann, 2005; Menning *et al.*, 2005). Therefore, the 51 cycles span 5.1 myr, which is identical to the span of 5.2 myr for the combined Induan and Olenekian stages derived from radio-isotopic dates (c. 252.16 to 247 Ma).

(c) The magnetostratigraphy from these cycle-scaled deposits in the Germanic Basin (Szurlies, 2004, 2007) is correlated to the ammonoid-zoned Early Triassic bio-magnetostratigraphy of the Boreal realm according to Hounslow and Muttoni (2010: fig. 4). Their correlation utilizes the main trends in polarity patterns, although there may be alternative correlations of the details of the meter-scaled Boreal polarity pattern to the cycle-scaled Germanic pattern.

(d) Assigned regional working definitions of stage and sub-stage boundaries are according to their traditional placement relative to Boreal ammonoid zones; with a mean position of each ammonoid zone relative to magnetozones taken as the approximate middle of the uncertainty interval in the summary diagrams of Hounslow and Muttoni (2010). Some of these substage boundary assignments may not correspond to future GSSPs or to current working definitions in other regions (e.g., in China where the majority of radio-isotopic dates originate).

(e) Conodont zones and other stratigraphic scales are placed relative to Boreal ammonoid zones according to selected correlation diagrams of other paleontologists (e.g., charts by Hardenbol *et al.*, 1998; Hounslow *et al.*, 2008b; etc.).

These assumptions yield a suite of numerical age assignments for placement of Lower Triassic stages as currently used in the Boreal faunal realm:

Dienerian substage base (base of *Proptychites candidus* ammonoid zone) is estimated by Hounslow and Muttoni (2010) as 25% up in magnetozone LT2n (= magnetozone CG5n of Szurlies, 2007). This has a projected age from the cycle-stratigraphy of 250.9 Ma.

Olenekian stage boundary (base of Smithian substage; base of *Hedenstroemia hedenstroemi* ammonoid zone) is near the base of magnetozone LT3n, which they correlate to magnetozone CG6n of Szurlies (2007). This yields an estimated numerical age of 250.0 Ma. This projected age is significantly younger than a radio-isotope-derived estimate from China of c. 251.2 Ma ± 0.2 (Galfetti *et al.*, 2007b), which may indicate different placements of the base-Olenekian, a miscorrelation between Boreal and Germanic magnetostratigraphy, or a distortion in the cycle stratigraphy.

Spathian substage base (base of *Bajarunia euomphala* ammonoid zone in Svalbard or Siberia zonation) is approximately 70% up in Hounslow and Muttoni's

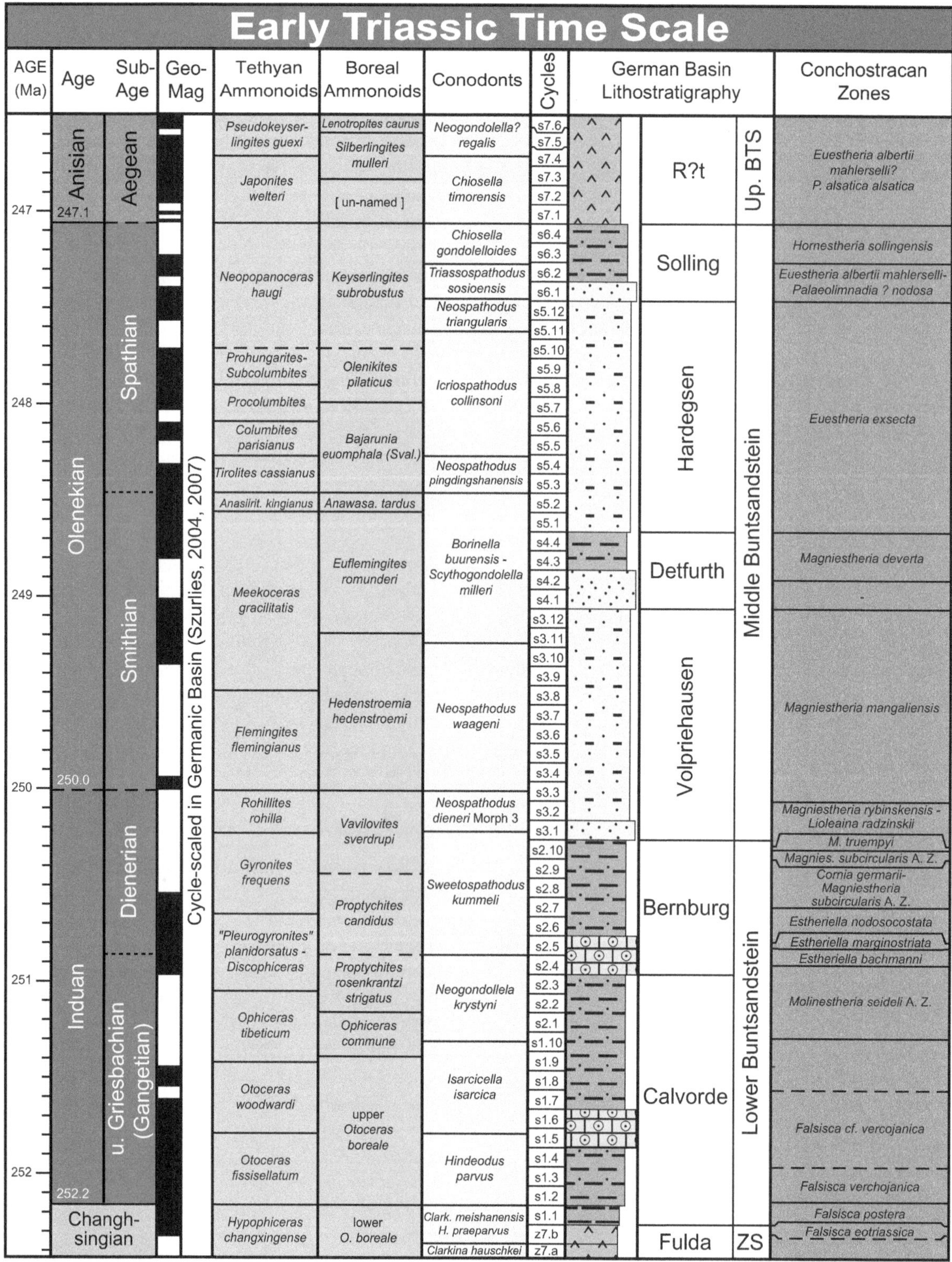

FIGURE 25.5 **Early Triassic time scale with magnetic polarity chrons, selected biostratigraphic zonations and Germanic Basin cyclostratigraphy and main lithostratigraphic units.** Potential definitions of the Olenekian and Anisian stage boundaries are indicated by dashed lines; but the final decisions will be made by the International Commission on Stratigraphy. Germanic-basin cycles are interpreted as eccentricity-induced 100-kyr climatic oscillations (e.g., Szurlies, 2004, 2007; Menning et al., 2005; Kozur and Bachmann, 2005, 2008). Magnetostratigraphy correlated to that cycle-scale is modified from Szurlies (2007) and Hounslow and Muttoni (2010). A selection of marine biostratigraphy relative to the magnetostratigraphy and/or adjacent zonations is represented by generalized ammonoid zones for the Tethyan and Boreal (western North America) realms (e.g., Kozur, 2003 and pers. comm., 2010; Balini et al., 2010; McRoberts, 2010) and conodont zonation for Tethyan realm (Kozur, 2003; and pers. comm., 2010). Terrestrial biostratigraphy, which is calibrated to the Germanic-basin cycles, is represented by conchostracan zones calibrated to the Germanic-basin cycles (Kozur and Weems, 2010, and Kozur, pers. comm., 2010).

normal-polarity-dominated interval of LT5n-7n which they correlate to magnetozone CG8n of the Germanic Basin. The implied base-Spathian age is 248.5 Ma.

The base of the Anisian is assigned as the base of Röt Formation of Germany and the base of magnetozone MT1n of Hounslow and Muttoni (2010), which is upper part of magnetozone CG10r of Szurlies (2007). The cycle-stratigraphy age for the base of Anisian (base of Middle Triassic) is 247.06 Ma.

The base of the Bithyrian substage of the lower Anisian is estimated by Hounslow and Muttoni (2010) as approximately the base of polarity subzone MT3r.3. This subzone spans the upper third of CG11r of Szurlies (2007) in the uppermost Rot Formation, therefore implying a cycle-stratigraphy age of 246.36 Ma. The implied short duration of the underlying Aegean substage is consistent with its compact thickness relative to the Germanic Basin cycle-stratigraphy (Szurlies, 2007) and a radio-isotopic age of 246.3 ± 0.07 Ma in early Anisian at Guizhou, China (sample GDGB-0, Ramezani *et al.*, 2007).

There is no verified cycle stratigraphy calibrated to ammonoid/conodont biostratigraphy for the middle and upper Anisian or Ladinian. Therefore, until additional constraints become accepted, a schematic display of ammonite zones within each interval was incorporated, in which the relative duration of ammonite zones was apportioned according to their relative number of ammonite subzones or allocating 1.5 "subzonal units" for undivided zones. The Tethyan ammonite zonal scheme was selected as the standard; therefore all other apparent ages for other biostratigraphic, magnetostratigraphic or chemostratigraphic scales are according to the estimated correlations to this Tethyan scale. This scaling was done for the Bithyrian, Pelsonian and Illyrian substages relative to the U-Pb-derived age of 242.1 Ma for the base of the uppermost Anisian *Nevadites secedensis* ammonoid zone. A similar scaling was applied for the Ladinian from the 241 ± 1 Ma for the lower boundary relative to the interim assigned age of 237 Ma for the base-Carnian.

This Anisian through Carnian interval awaits a CONOP-type compilation of multiple biostratigraphic sections to achieve a more realistic scaling of biozones within each stage.

25.3.3. Late Triassic Scaling

The relative durations of the Carnian, Norian and Rhaetian stages and the assignment of numerical ages to events within them has been a debated topic with extreme divergence of models. When this chapter was prepared (March, 2011), there was a lack of any reliable radio-isotopic dates within the latest Carnian through early Rhaetian constrained by marine-based biostratigraphy. There is a high-resolution magnetostratigraphy scaled to Milankovitch cycles from lacustrine strata in the Newark Basin that spanned much of the Norian and Carnian (and maybe Rhaetian). However, this pattern lacked an adequate "fingerprint" to correlate with compilations of magnetostratigraphy of marine-zoned strata (e.g., two options of Muttoni *et al.*, 2004b, and in Ogg *et al.*, 2008; three options discussed in detail within Hounslow and Muttoni, 2010).

The "puzzle" of how to correlate the polarity zones is illustrated in Figure 25.6(a), which assumes an arbitrary Rhaetian duration of 3 myr and assigned the Carnian–Norian boundary as 223 Ma. In this initial scale, the ammonoid zones (and their magnetic polarity patterns) within each stage have been uniformly scaled by allocating equal durations for each ammonoid subzone and 1.5 subzonal equivalents for those ammonoid zones that have no subzonal divisions. It is obvious that there are many options for visually correlating between the Newark cycle-magnetostratigraphy and the Late Triassic outcrop-based polarity patterns. Once such a correlation is attained, then the duration of each Late Triassic ammonoid zone would be known, plus estimates of sediment accumulation rate changes within the reference sections.

However, in addition to ambiguous matching of magnetic polarity patterns, there are at least four main factors in inter-regional correlation, each of which is disputed:

(1) Whether the published correlation of terrestrial-based biostratigraphy (conchostracans, palynology, tetrapods) from the Germanic Basin and the southwest USA to the stratigraphic sequences in the Newark basins and to marine-based stages and substages is valid.
(2) Whether the Newark lacustrine cycle succession (the standard for the scaling of magnetostratigraphy) is continuous, and whether the overlying basalts dated at 201 Ma are conformably overlying the highest lacustrine deposits without a significant break in deposition. These factors determine whether numerical ages can be reliably assigned to the magnetozones. For example, some paleontologists have interpreted a major stratigraphic hiatus omitted at least part of the Rhaetian stage below the basalts (e.g., Cirilli *et al.*, 2009; Kozur and Weems, 2005), and a comparison of cycle-magnetostratigraphy below the CAMP volcanics in Morocco suggested a c. 1-myr hiatus that shortened a major normal-magnetozone (E22n) in the Newark reference succession about 1.8 myr below its CAMP volcanics (Deenen *et al.*, 2010a; Deenen, 2010, p. 60).
(3) Deciding on the temporal proximity of the upper-mid Carnian U-Pb date of 230.9 ± 0.1 Ma (Aglianico date in Phanerozoic table in Chapter 6 of this volume) relative to the base of the Norian, and whether a poorly documented date of 225 ± 3 Ma (Gehrels *et al.*, 1987) is a reliable constraint on base of the early Norian.
(4) Selecting an appropriate correlation of Rhaetian magnetostratigraphy to the Newark succession, thereby constraining the upper limit of the Norian relative to Newark cyclostratigraphy (e.g., Hüsing *et al.*, 2011, compared to other options in Hounslow and Muttoni, 2010).

Therefore, the two end-members that represent current (August, 2011) published views will be summarized and diagrammed in this chapter.

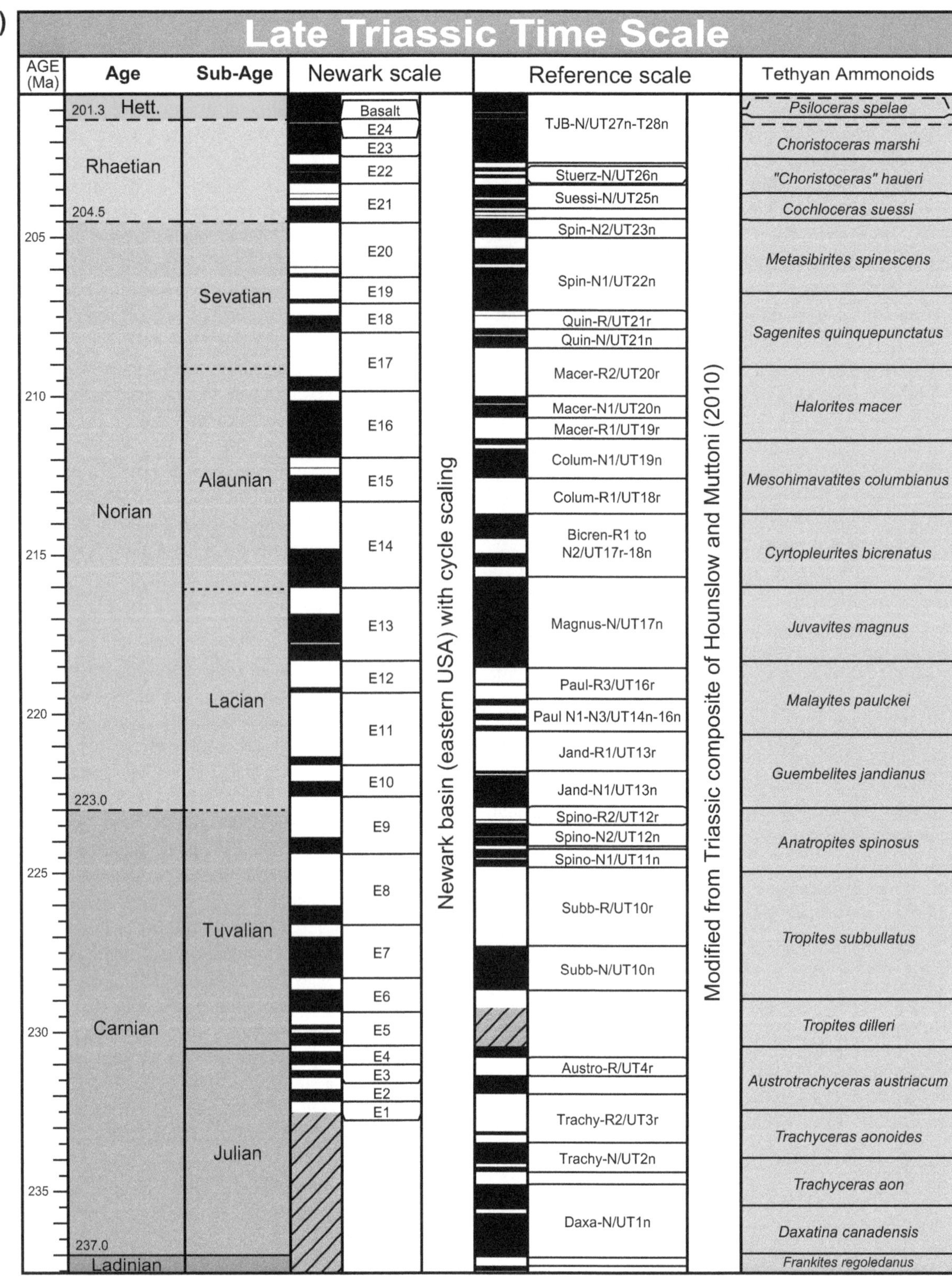

FIGURE 25.6 **The Late Triassic magnetic correlation puzzle.** (a) The ambiguity in correlating Upper Triassic magnetic polarity scales derived from two independent sources. Lacustrine basins of the Newark Supergroup (eastern USA) have yielded a high-resolution magnetic polarity pattern with durations derived from Milankovitch cycles (Kent et al., 1995), but lack direct calibrations to marine stratigraphy. Magnetostratigraphy from ammonite- or conodont-zoned marine sections have been assembled into a composite bio-magnetostratigraphy scale (modified from Hounslow and Muttoni (2010) by adding stage-based abbreviations for major polarity zones to their "UT" numbering), but lack cycle stratigraphy to estimate elapsed durations. This initial bio-magnetostratigraphy reference pattern is scaled by assigning equal durations to ammonite subzones. Several possible correlations are possible due to: (1) the lack of a distinctive "fingerprint" to correlate cycle-scaled polarity zones to marine sections that have uncertain continuity in sedimentation rates, (2) uncertainties in the correlation of terrestrial biostratigraphy (Newark) to marine biostratigraphy, and (3) possible significant gaps in both records. (b) Examples of two suggested end-member correlations are shown: **(Left)** a "Long-Tuvalian" and absence of Rhaetian in the Newark

(b)

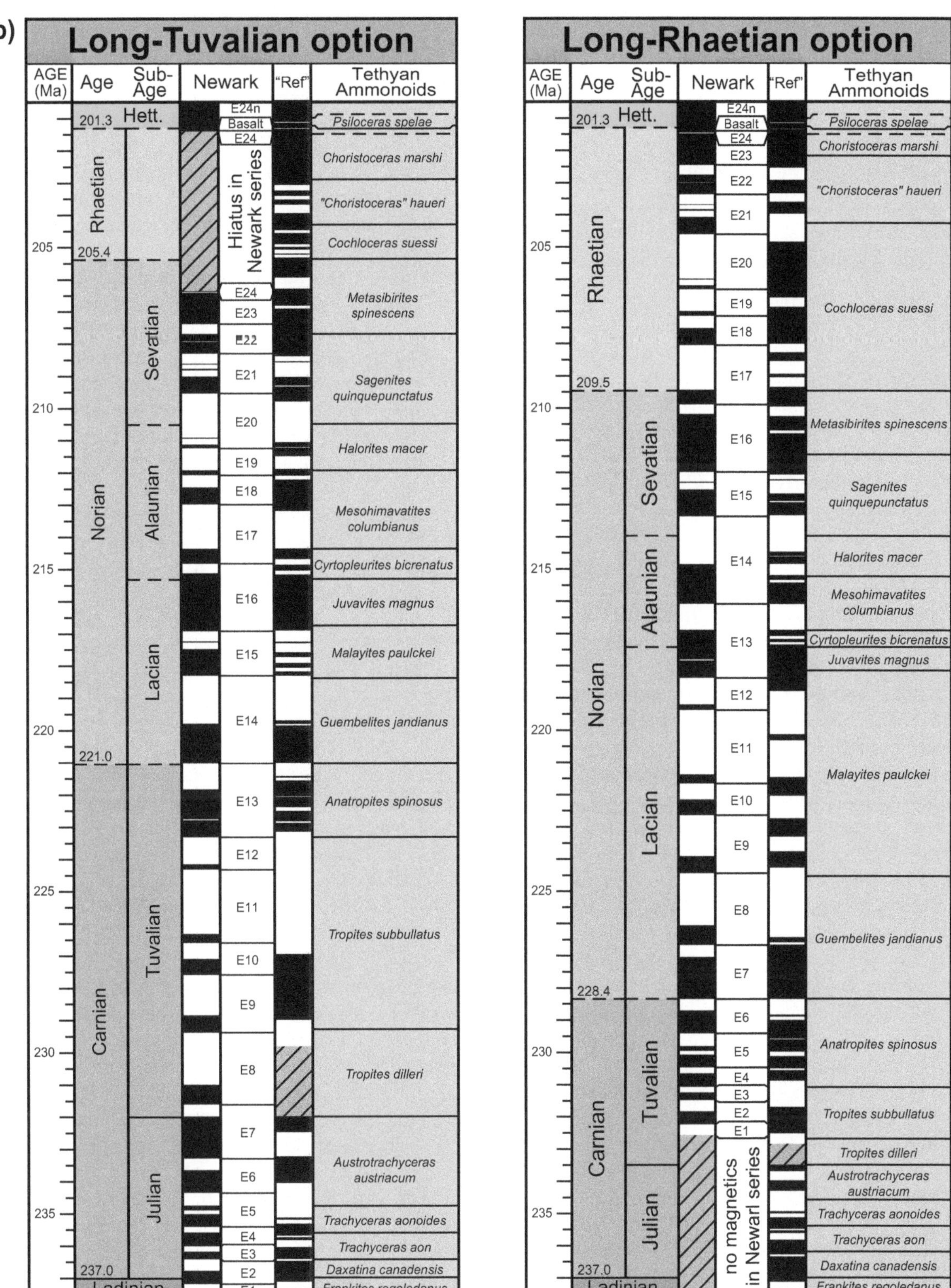

FIGURE 25.6 (*Continued*).

---

cycle-magnetostratigraphy (modified after Lucas et al., submitted, and others), and (**Right**) a "Long-Rhaetian" spanning the upper Newark cycle-magnetostratigraphy that would imply a short-duration Carnian (modified after Muttoni et al., 2010; Hüsing et al., 2011; and others). In each case, the ages for the top of the Triassic (201.3 Ma) and base of the Carnian (237.0 Ma) are the same. See text for details. In each option, the potential definitions of the Norian and Rhaetian stage boundaries are indicated by dashed lines; the final decisions will be made by the International Commission on Stratigraphy. For GTS2012, the numerical age scale of the second option ("Long-Rhaetian") was selected for scaling the upper Triassic in other diagrams.

Option #1 – Long-Duration Tuvalian Substage and Absence of Rhaetian in Newark Cycle-Magnetostratigraphy (Left Column in Figure 25.6(b)

One fit of this model (modified after Lucas *et al.*, submitted, and Kozur and Weems, 2010) incorporates six main paleontological and stratigraphic conclusions:

(1) The upper-mid Carnian U-Pb date of 230.9 ± 0.1 Ma is above the Carnian pluvial event and just below lowest occurrence of the conodont *P. carpathicus* (same as *M. carpathicus*, because genus assignment of *carpathicus* is not yet established). This *M. carpathicus* Zone begins below the base of the *T. subbullatus* ammonoid zone, therefore this radio-isotopic age is assigned to the middle of the underlying *T. dilleri* ammonoid zone of lowest Tuvalian substage.

(2) The upper Stockton Formation of the Newark Group has conchostracans of the lower Tuvalian *Gregoriusella* n. sp. Zone (Kozur and Weems, 2010). Therefore, Newark magnetozones E8 or E7 are in the lowermost Tuvalian. The estimated age of c. 237.0 Ma for the base of the Carnian implies that the four main magnetozone pairs and three subzones (UT1n-UT4r of Hounslow and Muttoni, 2010) are compacted into approximately 5 myr, therefore should correlate to a relatively high-frequency interval in the older portion of the Newark magnetic polarity pattern, hence potentially within the E1-E6 interval.

(3) Newark magnetozone E11r (a relatively long reversed-polarity interval in the middle of Lockatong Formation) correlates with the upper-Lower Tuvalian (upper mid-Carnian). Common taxa indicate that the conchostracan *Laxitextella seegisi* Zone of the Germanic Basin is coeval with the *Howellisaura princetonensis* conchostracan zone of the Newark succession in this interval and with the *Anyuanestheria wingatella* conchostracan zone of southwest USA within the lower Adamanian land vertebrate faunal chron (e.g., Kozur and Weems, 2010). This zone in the Lehrberg Beds of the Weser Formation in the Germanic Basin (following the Stuttgart Formation of Schilfsandstein deposited during the "mid-Carnian pluvial episode") contains ostracod *Simeonella nostorica*, which is common in the marine lower Tuvalian of Hungary and Austria (Kozur and Weems, 2010, p. 332). Therefore, Newark magnetozone E11r is correlated with the *T. subbullatus* ammonoid zone and its relatively long reversed-magnetozone "Subb-R" ("UT10r" of Hounslow and Muttoni, 2010).

(4) The *Tropites subbullatus* ammonoid zone of the Tuvalian usually encompasses a relatively thicker lithostratigraphic interval than most Carnian–Norian ammonoid zones; therefore, it conceivably spans a correspondingly greater interval of time (H. Kozur, pers. comm., 2010).

(5) Palynostratigraphy had placed the Carnian–Norian boundary in the Newark Supergroup succession near the base of the Passaic Formation (e.g., Cornet and Olsen, 1985) or within Newark magnetozone E13 (as used in Kent and Olsen, 1999). Magnetostratigraphy of the proposed base-Norian GSSP at Pizzo Mondello (Muttoni *et al.*, 2001, 2004a) places the boundary horizon at the top of a relatively narrow reversed-magnetozone (PM4r) between two relatively longer normal-magnetozones. One interpretation is that this Carnian–Norian boundary level corresponds to the top of Newark magnetozone E13r. (NOTE: In contrast, Muttoni *et al.* (2004a) prefer a correlation to the top of magnetozone E7; which is c. 10-myr older.)

(6) The Rhaetian stage is not represented in the uppermost Newark succession. In sections of the upper Passaic Formation which have yielded conchostracans, the fauna represent the *Shipingia olseni* Zone of uppermost Norian (no well-dated Rhaetian section in any part of the world has yielded any *Shipingia*) overlain locally by a very short interval of the uppermost *E. brodieana* Zone of latest Rhaetian (Kozur and Weems, 2005, 2010). The uppermost Sevatian, lower Rhaetian and part of the upper Rhaetian are not present. Therefore, the numerical ages of the magnetozones of the underlying Newark Supergroup cannot be assigned using cycle-stratigraphy "downward" from the overlying basalts of 201 Ma.

These constraints imply a "Long Tuvalian" with Newark magnetozone E7 near the base of the Tuvalian, the 230.9 Ma radio-isotopic age in the middle of the *T. dilleri* ammonoid zone, the Newark magnetozone E11r in the middle of the *T. subbullatus* ammonoid zone, and the top of Newark magnetozone E13r (c. 4 myr later) marking the base of the Norian.

A conservative estimate is to assign the base of the Tuvalian (base of the *T. dilleri* ammonoid zone) as 232 Ma and to position this boundary at the base of E7r. Then, assuming that the Newark cycle-magnetic sequence is complete, an age can be assigned to each magnetozone progressively upward and downward from this E7r control age. This implies that the age of the top of magnetozone E13r, which was interpreted as the Carnian–Norian boundary (item #5 above), is c. 221.9 Ma. The uppermost portion of the preserved Newark magnetostratigraphy, the lower portion of magnetozone E24n of the uppermost Norian, is projected as 206.5 Ma. The Norian–Rhaetian stage boundary would be slightly younger than this level, hence about 205.5 Ma. The age of the Triassic–Jurassic boundary at 201.3 Ma would imply that the Rhaetian was 4 myr in duration, which would be essentially the time-span for the interpreted hiatus in the Newark succession.

If the base of the Carnian is assigned as exactly 237 Ma, then the age assignments to Newark magnetozones imply that the basal Carnian reversed-polarity zone Daxa-R corresponds to Newark magnetozone E2r.

One can propose a suite of possible correlations of the magnetostratigraphy patterns of the other ammonoid zones of the Carnian and Norian to this model of the age-assigned Newark magnetic pattern. One set of possible Newark-to-ammonoid zone correlations has partially used the relative number of ammonoid subzones per zone as an approximate guide to relative durations for adjacent zones. The scaling of polarity patterns within individual ammonite zones is according to magnetostratigraphy reference sections (e.g., Hounslow and Muttoni, 2010), and only the placement of ammonite zonal boundaries have been adjusted for the potential fit to the Newark cycle-scaled polarity pattern. There is the caveat that portions of these composite polarity patterns are not well calibrated to ammonoid (or conodont) zones. Such a model could be tested in different ways, such as obtaining cycle-stratigraphy from the ammonite-zoned reference sections or clarifying the magnetostratigraphy within middle Carnian ammonoid zones (e.g., *T. dilleri*) to ascertain the predicted presence of a long reversed-polarity interval (E8r) from the Newark succession.

In the illustrated Option 1 model, the 36-myr span of the Late Triassic is allocated to equal ~16-myr durations for the Carnian and Norian, and a 4-myr "short" Rhaetian.

Option #2 – Short-Duration Carnian and Presence of a Long Rhaetian in Newark Cycle-Magnetostratigraphy (Right Column in Figure 25.6(b)

At the other end of the spectrum of models, the suite of biostratigraphic interpretations is relaxed in favor of a hiatus-free Newark succession that is continuous into the CAMP basalts of 201.4 Ma. This assumption is combined with additional interpretations from magnetostratigraphy patterns and radio-isotopic ages:

(1) The ages of all Newark magnetozones are computed downward from the CAMP age of 201.4 Ma.
(2) The Rhaetian is fully present in the uppermost Newark magnetostratigraphy. There is no lithostratigraphic evidence for the interpreted major biostratigraphic gap in the uppermost Newark lacustrine cycles (P. Olsen, pers. comm., 2010), and studies of cycle-magnetostratigraphy of coeval deposits in other basins indicate a nearly identical time span (within 20 kyr) for the uppermost magnetozones to the onset of CAMP volcanics (Deenen, 2010; Deenen *et al.*, 2010b). Therefore, the narrow UT27r of the latest Hettangian in the composite by Hounslow and Muttoni (2010) is correlated with magnetozone E23r of the uppermost Newark. The underlying upper and middle Rhaetian magnetozones (BIT5r to BIT1r/UT24r) are progressively correlated to Newark magnetozones through E22r (Muttoni *et al.*, 2010).
(3) The cluster of reversed-polarity-dominated magnetozones in the lower Rhaetian is underlain by a relatively thick normal-polarity magnetozone UT22n. This may imply that this lowermost Rhaetian interval is interpreted as condensed in the Austrian reference section; therefore this suite corresponds to the expanded set of reversed-polarity dominated E20r-E17r underlain by normal-polarity magnetozone E16n-E17n in the Newark succession (Option A of Hounslow and Muttoni, 2010, or similar model by Hüsing *et al.*, 2011). The Norian–Rhaetian boundary is assigned to just above the base of E17r with an age of 209.5 Ma. A similar "8-myr Rhaetian" model was proposed by McArthur (2008, 2010) upon applying a uniform rate of change to the declining $^{87}Sr/^{86}Sr$ ratios from latest Norian through Sinemurian.
(4) The polarity change at the Carnian–Norian boundary at the candidate GSSP of Pizzo Mondello corresponds to the base of E7n (Muttoni *et al.*, 2004a) with an age projected from the Newark cycle-magnetostratigraphy of c. 228.5 Ma. This is consistent with reported ages of c. 225 Ma for lower Norian strata (e.g., Gehrels *et al.*, 1987).
(5) Given these correlations for the top and base of the Norian, then the general trends in polarity dominance would project the base of the Alaunian stage (base of UT17r, above a relatively thick normal-polarity Magnus-N/UT17n) to the base of Newark magnetozone E13r, and the base of the Sevatian (base of Quin-N/UT21n) to perhaps the base of Newark magnetozone E15n (Option A of Hounslow and Muttoni, 2010). One problem is that the outcrop-derived magnetostratigraphy within the mid-Norian Alaunian substage has more interpreted magnetozones than the Newark cycle-magnetostratigraphy; and this apparent inconsistency was noted by Hounslow and Muttoni (2010) in their similar Option A for Norian correlations.
(6) Correlations within the Carnian proceed downward from the assigned base-Norian at E7r. This would imply that the *A. spinosus* ammonite zone may be slightly longer than the preceding *T. subbullatus* Zone, and that the apparently simple polarity pattern of that *T. subbullatus* Zone requires further resolution. The duration of the Carnian is much shorter than the Norian, with the Julian–Tuvalian substage boundary projected as approximately midway at 233.5 Ma.

In this model, the 36-myr span of the Late Triassic is apportioned among a "long" 8-myr Rhaetian, a 19-myr Norian, and a ~10-myr Carnian.

We selected the second option (short-duration Carnian and long Rhaetian) for the Triassic summary figures (Figure 27.7; Tables 27.2 and 27.3). This choice was partly influenced by reported radio-isotopic dates of 223–225 Ma (223.81 ± 0.78 Ma and 224.52 ± 0.22 Ma) derived by U-Pb CA-TIMS methods from single zircons from volcanic tuffs in the Nicola Group of British Columbia that are constrained by conodont assemblages to bracket the Lower/Middle Triassic boundary interval as used in North America (Diakow *et al.*, 2011). If confirmed, then these dates would be inconsistent with the "long-Carnian" option.

TABLE 25.2 Comparison of Triassic Age Model of GTS2012 (Dual Versions) with Selected Previous Publications

	GTS2004		Concise GTS2008 Alternate (Modified from Brack *et al.*, 2005)		Mundil *et al.* (2010)		GTS2012 Option 1 (Long Carnian)		GTS2012 Option 2 (Long Norian-Rhaet)		Derivation and Uncertainty (95% Confidence) in GTS2012
	Base	Duration	Base	Duration	Base	Duration	Base	Duration	Base	Duration	
Jurassic (Hettangian)	*199.6*		*200.9*		*201.5*		201.3		**201.3**		Age assigned from bounding radio-isotopic dates; therefore, uncertainty = ± 0.2 myr.
Rhaetian	*203.6*	*4*	*204.5*	*3.6*	x		205.4	4.1	**209.5**	8.2	***Dashed* – *See text for options*.** *Basal placement is also awaiting GSSP decision. Not possible to estimate uncertainty, but probably ± 1 myr under the assumptions of each model.*
Norian	*216.5*	*12.9*	*228.8*	*24.3*	<230		221.0	15.7	**228.4**	18.9	***Dashed* – *See text for options*.** *Basal placement is also awaiting GSSP decision. Not possible to estimate uncertainty, but probably ± 2 myr under the assumptions of each model.*
Carnian	*228.7*	*12.2*	*236.8*	*8*	<236		237.0	16.0	**237.0**	8.7	Age estimated from radio-isotopic dates immediately below base; therefore, uncertainty = ± 1 myr.
Ladinian	*237*	*8.3*	*240.5*	*3.7*	242		241.5	4.5	**241.5**	4.5	Age assigned from bounding radio-isotopic dates; therefore, uncertainty = ± 1 myr.
Anisian	*245.9*	*8.9*	*247.4*	*6.9*	247.2	*5.2*	247.1	5.6	**247.1**	5.6	**Dashed** – Basal-age derived from cycles in Germanic Basin from base-Triassic; consistent with radio-isotopic constraints, therefore uncertainty = ± 0.2 myr (or ± 0.5 with external errors).
Olenekian	*249.5*	*3.6*	*251*	*3.6*	251.3	*4.1*	250.0	2.9	**250.0**	2.9	**Dashed** – Basal-age derived from cycles in Germanic Basin from base-Triassic; Basal placement is awaiting GSSP decision. Uncertainty estimated as ± 0.5 myr.
Induan	*251*	*1.5*	*252.5*	*1.5*	252.3	*1*	252.2	2.2	**252.2**	2.2	Age assigned from bounding radio-isotopic dates; therefore, uncertainty = ± 0.2 (or ± 0.5 myr with external errors).
Permian (Changhsingian)											

Values are rounded to nearest 0.1 myr. "Option 2" of GTS2012 is used for summary figures and other tables. The term 'Dashed' under the 'Derivation' column indicates either that there is not yet a GSSP level to define that stage, and/or the assigned age depends on different views of stratigraphers on correlations of potential stage boundaries to reference Newark Basin magnetostratigraphy.

TABLE 25.3 Interpolated Numerical Ages for Tethyan Ammonoid Zones

Stage/Age	Substage/Sub-Age	Tethyan Zone	Basal Age (Ma)	Notes on Calibration of Zone Base and Selected Zonal Usages
JURASSIC (Hettangian) base	Early Hettangian	*Psiloceras spelae*	201.30	Base of Jurassic (by definition)
		TOP of Zone	201.50	Assigned as arbitrary 0.1 myr below top of *C. crickmayi* (N. Amer.) following schematic of Guex *et al.* (EPSL, 2004, fig. 4).
		Choristoceras marshi	202.16	Base at 30% in Newark Chron E23n. Subzones of *Chor. marshi* Zone are assigned as separate zones in some schemes – e.g., Kozur (2003, table 3; and in Bachmann-Kozur (2004) divided into an upper *Chor. marshi* subzone and a lower *Chor. ammonitiforme* subzone.
		"Choristoceras" haueri	204.26	Base at approximately 90% up in combined *M. hernsteini–M.posthernsteini* conodont zone. Kozur (2006, pers. comm. to J. Ogg): "*Choristoceras haueri* Zone is junior synonym of the later *Vandites stuerzenbaumi* Zone." However, his scheme (2003; and 2006 sent to J. Ogg) has an upper substage of *Vandites stuerzenbaumi.*
Rhaetian base		*Cochloceras suessi*	209.46	Derived from FAD of *M.posthernsteini* (assigned as lowest part of Newark Chron E17r). Onset of *Cochloceras* genus is coincident with FAD of conodont *M. posthernsteini* (base of Rhaetian).
		Metasibirites spinescens	211.45	Approximately 30% up in Newark Chron E16n. Formerly considered completely Norian as a *Sagenites reticulatus Zone;* but revised with Rhaetian GSSP "cutting former zone" by Krystyn (pers. comm. L. Krystyn) Begins in latest late Norian = relatively brief). Krystyn's version would continue into Rhaetian (instead of a *Cochloceras suessi* Zone).
	Sevatian	*Sagenites quinquepunctatus*	213.97	60% up in Newark Chron E14r. Was a subzone of a larger *R. suessi* Zone, but following Kozur' 2003, it is shown here as a separate zone. (NOTE: The zonal/subzonal division for *Sagenites quinquepunctatus* through Rhaetian interval varies among authors (e.g., SEPM/GTS2004 versus Kozur'2003).)
		Halorites macer	215.23	70% up in Newark Chron E14n. Kozur (2003, and pers. comm. J.Ogg, 2006) puts this ammonite zone into Sevatian, rather than Alaunian; and correlates with base of *M. bidentata* conodont zone (his expanded version of that zone). GTS2004 has this zone with 2 subzones (lower *Amarassites semiplicatus*, upper "unnamed"; but these are not indicated in Kozur's diagrams => omitted here => relative age span is different.
		Mesohimavatites columbianus	216.90	Base at base of Newark Chron E13r. Zone was called *Himavites hogarti* in GTS2004 (using SEPM'98), but Kozur (2003; and Bachmann and Kozur, 2004) calls this *Mesohimavatites columbianus* (projecting that zone from British Columbia), with no indicated subzones. However, McRoberts (2010) (bivalve chart) has *Hima. hogarti* and an underlying *Hima. watsoni* Zone. Confusing lack of "standard scale".
	Alaunian	*Cyrtopleurites bicrenatus*	217.42	65% up in combined Newark Chron E13n.
		Juvavites magnus	218.16	15% up in combined Newark Chron E13n.
		Malayites paulckei	224.53	95% up in Chron E8r. Subzones of *Malayites* (based on Krystyn *et al.*, fig. 1) are given equal duration.
Norian base	Lacian	*Guembelites jandianus*	228.35	Base of Newark Chron E7n. Kozur uses *S. kerri* (N.Amer. zone) for this "standard zone".
		Anatropites spinosus	231.08	Base of Newark Chron E4n. (see note below).
		Tropites subbullatus	232.69	Uncertain (no magnetostratigraphy); therefore proportioned from equal subzones (two subzonal units in relative duration). Kozur's 2003 "standard" uses N. Amer. zones of *K. macrolobatus* and *T. welleri* for middle-upper Tuvalian.
	Tuvalian	*Tropites dilleri*	233.50	Uncertain (no magnetostratigraphy). No subzonal divisions => scaled as "1.5 subzonal units" in relative duration.
		Austrotrachyceras austriacum	234.58	Two subzonal units in relative duration.

(Continued)

TABLE 25.3 Interpolated Numerical Ages for Tethyan Ammonoid Zones—cont'd

Stage/Age	Substage/Sub-Age	Tethyan Zone	Basal Age (Ma)	Notes on Calibration of Zone Base and Selected Zonal Usages
	Julian	*Trachyceras aonoides*	235.39	No subzonal divisions => scaled as "1.5 subzonal units" in relative duration. (see note below).
		Trachyceras aon	236.19	No subzonal divisions => scaled as "1.5 subzonal units" in relative duration. *T. aonoides* and *T. aon* (and even *D. canadensis*) are combined into a general *Trachyceras* zone in some zonal schemes.
Carnian base	*"Cordevolian"*	*Daxatina canadensis*	237.00	Base of Carnian – main marker at GSSP. Fixed at 237 Ma. No subzonal divisions => scaled as "1.5 subzonal units" in relative duration.
		Frankites regoledanus	237.90	No subzonal divisions => scaled as "1.5 subzonal units" in relative duration.
		Protrachyceras neumayri	238.50	One subzonal unit in relative duration. Zone status in Balini (2010) (used here). (see note below).
	Longobardian	*Protrachyceras longobardicum*	239.10	One subzonal unit in relative duration. Zone status in Balini (2010) (used here). Kozur (2003) shows an undivided *"Protrachy. archelaus"* Zone with *P.neumayri–P.longobardicum* subzones.
		"Eotrachyceras" gredleri	239.70	One subzonal unit in relative duration. Zone is not used on some charts; but is on Balini (2010) and McRoberts (2010).
		Protrachyceras margaritosum	240.30	One subzonal unit in relative duration. Zone is merged with *P. gredleri* for a united *"P. gredleri"* on some charts; but is on Balini (2010) and McRoberts (2010).
Ladinian base	Fassanian	*Eoprotrachyceras curionii*	241.50	GTS2004 – Radiometric age of 241.5 (± 1.5 myr) = control on scaling of Anisian –Ladinian. *E. curionii* spans two subzonal units in relative duration.
		Nevadites secedensis	242.10	Radiometric age of 242 (± 1 myr). *Nevadites secedensis* Zone is called *Nevadites* Zone by some authors, after the FAD of the first species of this genus.
		Reitziites reitzi	242.57	Zone is called *"Hungarites"* Zone by some authors. Two subzonal units in relative duration.
		Kellnerites felsoeoersensis	243.05	Two subzonal units in relative duration.
	Illyrian	*Paraceratites trinodosus*	243.99	Four subzonal units in relative duration. Zone has alternative zone/subzone divisions (e.g., Table 2 in Kozur (2003); Bachmann-Kozur (2004) is used here => middle subzone is "two" subzonal units).
		Schreyerites binodosus	244.47	Two subzonal units in relative duration.
	Pelsonian	*Balatonites balatonicus*	244.94	Two subzonal units in relative duration.
		Acrocordiceras ismidicus	245.30	1.5 subzonal units in relative duration (no subzone divisions). Was called *"Aghdarbandites" ismidicus* in Kozur (2003); but McRoberts (2010) uses *"Acrochodiceras"* (same genus as coeval British Columbia zone)
		Nicomedites osmani	245.65	1.5 subzonal units in relative duration (no subzone divisions).
		Lenotropites caurus	246.01	1.5 subzonal units in relative duration (no subzone divisions).
	Bithyrian	*Silberlingites mulleri*	246.36	Base of Bithyrian = set as base of Chron MT3r.3 (1/3rd up in CG11r of Szurlies (2007) = uppermost Rot. 1.5 subzonal units in relative duration (no subzone divisions). Applied Bithynian zonation of Kozur (2003) which has the lower two zones of Boreal scale, then elevates subzones of former *Kocelia* super-zone for upper two divisions.
		Pseudokeyserlingites guexi	246.71	The two Aegean zones are given equal durations. *Paracrochordiceras* has equivalent of 5 subzones in Boreal realm.

Anisian base	Aegean	*Japonites welteri*	247.06	Base-Anisian is not yet defined. Base-Anisian set here using Menning cycle-strat as base of Rot (upper unit of Bundsandstein). Cycles and radio-isotope dating both suggest this age should be ~247 Ma. Zonation of Kozur (2003) used for Aegean zones.
		Neopopanoceras haugi	247.71	Equivalent to *K. subrobustus* zone of Boreal (Orchard, May 2007, pers. comm.). That zone spans Chrons LT6r through mid-LT9r in Spitsbergen, based on fig. 3 of Hounslow-Muttoni (2010)) => relatively long and would eliminate most of "Lower" Spathian (as they had to do on their diagram). But, if use Canada Arctic magnetostratigraphy, then it begins later at base-LT8r (dashed).
		Prohungarites-Subcolumbites	247.90	Four Spathian zones below *N. haugi* given equal durations. Approximately equivalent to base-*Tozericeras pakistanum* in Himalayas.
		Procolumbites	248.09	Four Spathian zones below *N. haugi* given equal durations.
		Columbites parisianus	248.27	Four Spathian zones below *N. haugi* given equal durations.
	Spathian	*Tirolites cassianus*	248.46	Base of Spathian in Boreal realm is about 70% up in Chron LT5n-7n of Hounslow-Muttoni (2010) (= a single CG8n). Szurlies (2004) has the conchostracan-correlated base-Spathian at about 75% up in his Chron CG7n (which may be a small gap) = base of Detfurth cycles in Germanic system. Four Spathian zones below *N. haugi* given equal durations for now.
		Anasiirites kingianus	248.56	Set as base of *A. tardus* (Boreal), which is middle of Chron LT6 of Hounslow-Muttoni (2010); which would be approximately middle of Szurlies (2007) Chron CG8n. Very brief in Hounslow-Muttoni (2010) composite.
		Meekoceras gracilitatis	249.49	One third up in Smithian substage.
Olenekian base	Smithian	*Flemingites flemingianus*	250.01	Near the base of Chron LT3n of Hounslow-Muttoni (2010), which they correlate to magnetozone CG6n of Szulies (2007).
		Rohillites rohilla	250.23	Short zone that was assigned to "*Flemingites*" genus in basal Smithian, but now is assigned as *Rohillites* due to GSSP usage => assigned as 2/3rds up between *G. frequens* and top of Dienerian.
	Dienerian	*Gyronites frequens*	250.65	Bachmann and Kozur (2004) display base as about 1/4 up in Dienerian.
		"Pleurogyronites" planidorsatus – Discophiceras	251.05	Zone in Bachmann and Kozur (2004) shown as spanning uppermost "Gangetian" (uppermost Griesbachian) and lowermost Dienerian => spans the substage boundary (put at approximate middle of zone).
		Ophiceras tibeticum	251.42	Revised (Kozur (2004)) zonation for Griesbachian substage has uncertain correlations, therefore each zone (3.5) is dashed as equal duration.
		Otoceras woodwardi	251.79	(see above note).
TRIASSIC base	Griesbachian (pars.)	*Otoceras fissisellatum*	252.16	Radiometric age on GSSP from Chen *et al.*, (2010) is 252.16 Ma. Revised (Kozur 2004) zonation for Griesbachian substage has uncertain correlations, therefore each zone (3.5) is dashed as equal duration.

Ages of Carnian through Rhaetian zones are assigned through their calibration to the bio-magnetostratigraphic scale (e.g., Hounslow and Muttoni, 2010) with its estimated equivalence to the Milankovitch-cycle-scaled magnetic polarity time scale from the Newark Basin of eastern USA using the "Long Rhaetian" option (see sections 25.2.3 and 25.3.3). Placement of some stage boundaries is approximate or they are working definitions pending decision of the International Commission on Stratigraphy. Two-decimal ages are given to show relative durations among the zones; and uncertainties are equal to or greater than the uncertainties on the boundaries of its stage (Table 25.2). Ammonoid zonations vary among authors, and this scale is based mainly on Kozur (e.g., Kozur, 2003; Bachmann and Kozur, 2004; and pers. comm. to J. Ogg), Balini *et al.* (2010), and McRoberts (2010). The column of notes should be considered as informal remarks on calibrations and different zonal usages. Details for subzone age models and for calibrations of other biostratigraphic zonations can be found in pop-up windows for item within the TimeScale Creator database; *www.tscreator.org*.

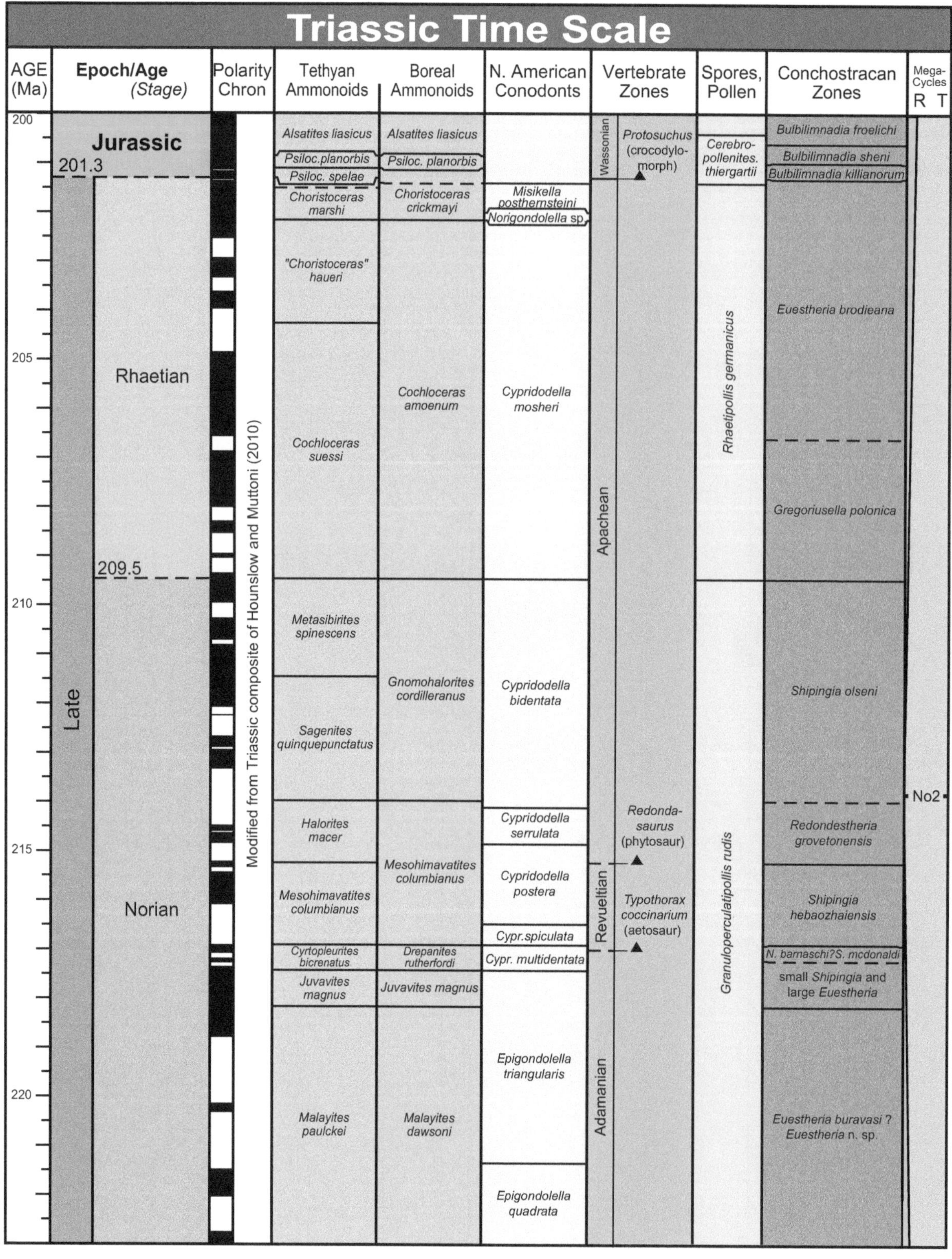
Triassic Time Scale
AGE (Ma)
Epoch/Age (Stage)
Polarity Chron
Tethyan Ammonoids
Boreal Ammonoids
N. American Conodonts
Vertebrate Zones
Spores, Pollen
Conchostracan Zones
Mega-Cycles
R T
200
205
210
215
220
Jurassic
201.3
Rhaetian
209.5
Late
Norian
Modified from Triassic composite of Hounslow and Muttoni (2010)
Alsatites liasicus
Psiloc.planorbis
Psiloc. spelae
Choristoceras marshi
"Choristoceras" haueri
Cochloceras suessi
Metasibirites spinescens
Sagenites quinquepunctatus
Halorites macer
Mesohimavatites columbianus
Cyrtopleurites bicrenatus
Juvavites magnus
Malayites paulckei
Alsatites liasicus
Psiloc. planorbis
Choristoceras crickmayi
Cochloceras amoenum
Gnomohalorites cordilleranus
Mesohimavatites columbianus
Drepanites rutherfordi
Juvavites magnus
Malayites dawsoni
Misikella posthernsteini
Norigondolella sp.
Cypridodella mosheri
Cypridodella bidentata
Cypridodella serrulata
Cypridodella postera
Cypr.spiculata
Cypr. multidentata
Epigondolella triangularis
Epigondolella quadrata
Wassonian
Apachean
Revueltian
Adamanian
Protosuchus (crocodylo-morph)
Redonda-saurus (phytosaur)
Typothorax coccinarium (aetosaur)
Cerebro-pollenites. thiergartii
Rhaetipollis germanicus
Granuloperculatipollis rudis
Bulbilimnadia froelichi
Bulbilimnadia sheni
Bulbilimnadia killianorum
Euestheria brodieana
Gregoriusella polonica
Shipingia olseni
Redondestheria grovetonensis
Shipingia hebaozhaiensis
N. bamaschi?S. mcdonaldi
small Shipingia and large Euestheria
Euestheria buravasi ? Euestheria n. sp.
No2

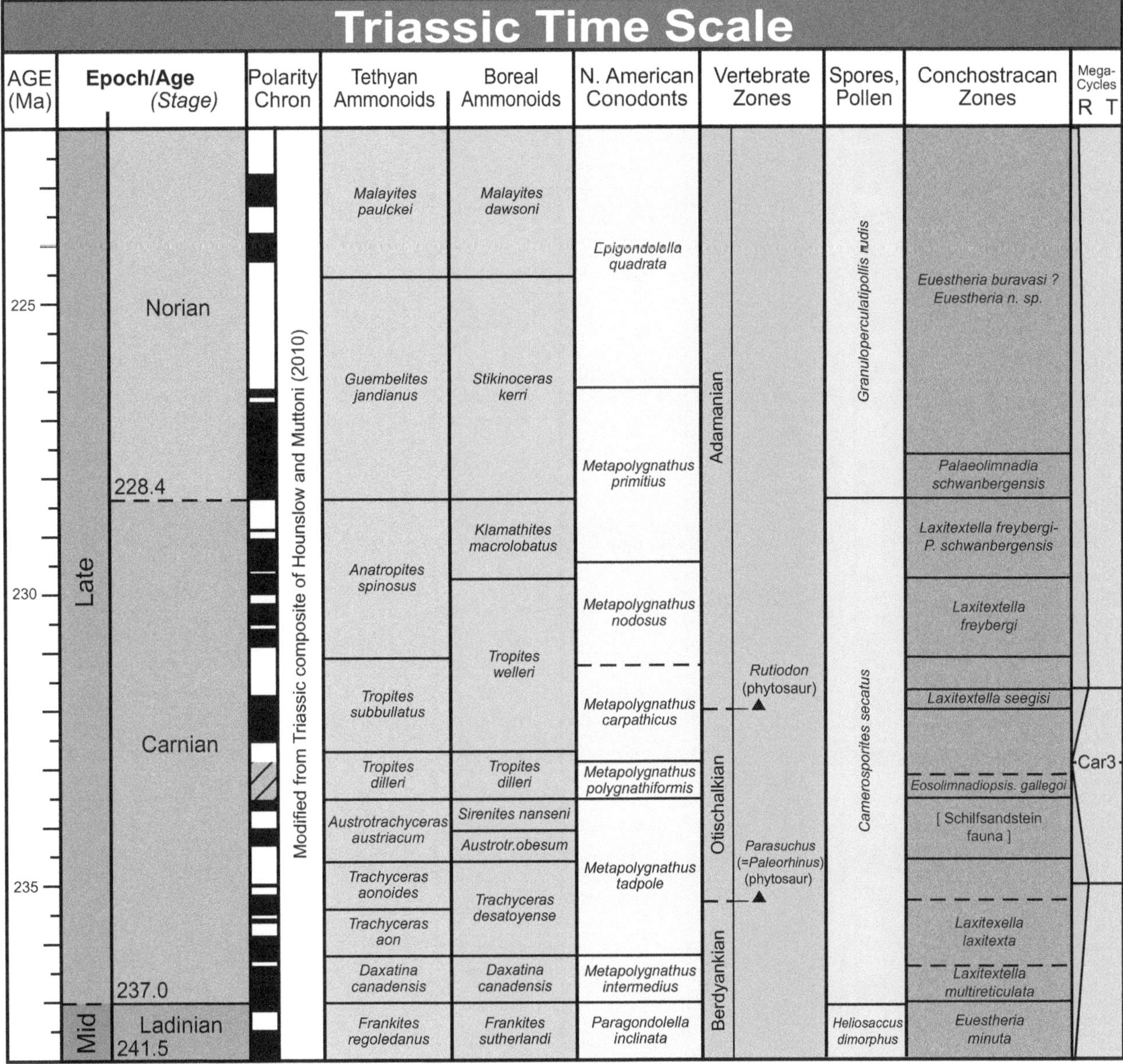

FIGURE 25.7 (*Continued*).

FIGURE 25.7 **Summary of numerical ages of epoch/series and age/stage boundaries of the Triassic with selected marine biostratigraphic zonations and principle trends in sea level.** ("Age" is the term for the time equivalent of the rock-record "stage".) Potential definitions of the Olenekian, Anisian, Norian and Rhaetian are indicated by dashed lines; the final decisions will be made by the International Commission on Stratigraphy. For GTS2012, the numerical age scale of the second option ("Long-Rhaetian") was selected for scaling the upper Triassic. Magnetic polarity pattern is modified from Hounslow and Muttoni (2010). Marine biostratigraphy columns are representative ammonoid zones for the Tethyan and Boreal (western North America) realms (e.g., Kozur, 2003 and pers. comm., 2010; Balini et al., 2010; McRoberts, 2010) and conodont zonation for North American realm (e.g., Orchard and Tozer, 1997; Orchard, 2010). Terrestrial biostratigraphy is represented by tetrapod "faunachrons" with defining first appearances (Lucas, 2010d), generalized spore–pollen zones (Kürschner and Herngreen, 2010) and Conchostracan zones (Kozur and Weems, 2010; Kozur, pers. comm., 2010). The major sequences are from Jacquin and Vail (1998, as inter-calibrated in Hardenbol et al., 1998). For details in Early Triassic, see expanded scale in **Figure 25.5**. Additional Triassic zonations, geochemical trends, sea-level curves, etc. are compiled in the internal data sets within TimeScale Creator (*www.tscreator.org*).

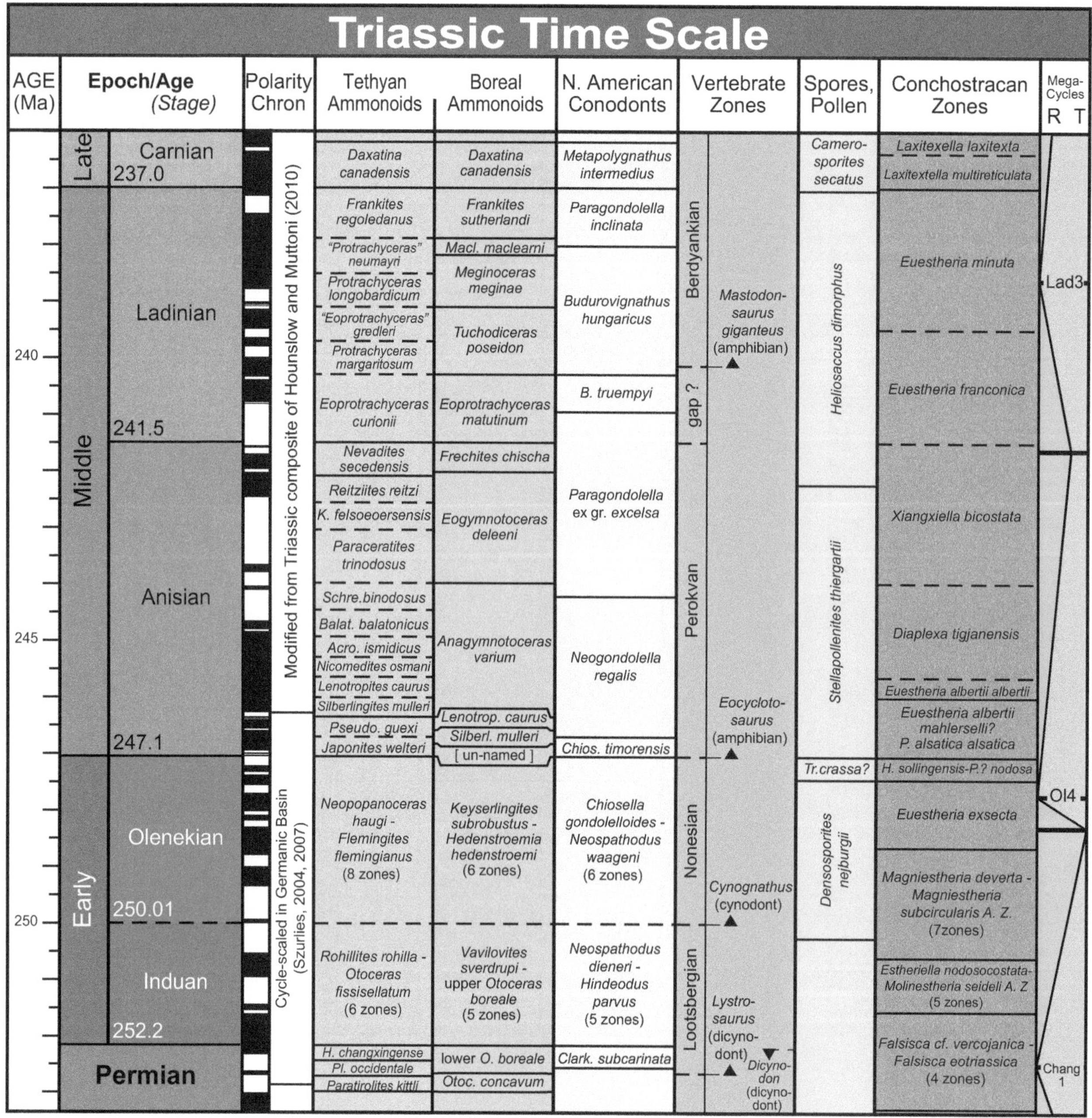

FIGURE 25.7 (*Continued*).

25.3.4. Summary

During the past decade, Triassic workers have defined most of the stages, greatly enhanced the inter-correlation of biostratigraphic zones, enabled compilation of a nearly complete magnetic polarity pattern calibrated to marine biostratigraphic datums, discovered major excursions in stable isotopes (especially carbon isotope excursions within the lower Triassic), and achieved or rejected cycle-stratigraphic scaling of several intervals. A generalized synthesis of selected Triassic stratigraphic scales is compiled in Figure 25.7; and additional geochemical trends are summarized in the geochemical chapters of this volume.

Extensive radio-isotopic dating with advanced techniques have replaced nearly the entire radiometric data set used in GTS2004 and established well-constrained numerical ages for the bases of the Early Triassic (Induan Stage), of both Middle Triassic stages (Anisian, Ladinian) and top of the Triassic (base of Hettangian Stage). There are lingering major uncertainties on the numerical ages and durations of the Upper Triassic stages. Establishing a robust Late Triassic time scale requires definitive radio-isotopic ages and cyclostratigraphy on marine sections that have standard biostratigraphy.

TABLE 25.4 GSSPs of the Triassic Stages, with Location and Primary Correlation Criteria (Status Jan. 2012)

Stage	GSSP Location	Latitude, Longitude	Boundary Level	Correlation Events	Reference
Rhaetian	*Key sections in Austria, British Columbia (Canada), and Turkey*			*Near FADs of the conodont* Misikella posthernsteini, the *ammonite* Paracochloceras, *and radiolarian* Proparvicingula moniliformis	
Norian	*Candidates are Black Bear Ridge in British Columbia (Canada) and Pizzo Mondello, Sicily (Italy)*			*Base of* Stikinoceras kerri *ammonoid zone and near FAD of* Metapolygnathus echinatus	
Carnian	Section at Prati di Stuores, Dolomites, Italy	46°31′37″N 11°55′49″E	GSSP is base of marly limestone bed SW4, 45m from base of San Cassiano Formation	FADs of the ammonoid *Daxatina*, the conodont *"Paragondolella" polygnathiformis* and *Halobia* bivalves. Just above base of S2n magnetic polarity zone and above the maximum flooding surface of Sequence Lad 3 of Hardenbol *et al* (1998)	Episodes 35, 2012 *(in press)*
Ladinian	Bagolino, Province of Brescia, Northern Italy	45°49′09.5″N 10°28′15.5″E	Base of a 15–20 cm thick limestone bed overlying a distinctive groove ("Chiesense groove") of limestone nodules in a shaly matrix, located about 5m above the base of the Buchenstein Beds	FAD of ammonite *Eoprotrachyceras curionii* (base of the *E. curionii* Zone). FAD of conodont *Budurovignathus praehungaricus* is in the uppermost Anisian	Episodes 28/4, 2005
Anisian	*Candidate section at Desli Caira (Dobrogea, Romania); significant sections in Guizhou Province (China) and South Primorye (Russia)*			*FAD of conodont* Chiosella timorensis *or Magnetic – base of magnetic normal-polarity chronozone MT1n*	
Olenekian	*Candidate GSSPs Chaohu, China and Mud (Muth) village, Spiti valley, India*			*FAD of conodont* Neospathodus waageni, *just above base of* Rohillites rohilla *ammonite zone, and below lowest occurrence of* Flemingites and Euflemingites *ammonite genera. Within a prominent positive Carbon-13 peak, and just above widely recognizable sequence boundary*	
Induan *(base Triassic)*	Meishan, Zhejiang Province, China	31° 4′47.28″N 119°42′20.90″E	Base of Bed 27c in the Meishan Section	FAD of conodont *Hindeodus parvus*	Episodes 24/2, 2001

*according to Google Earth

ACKNOWLEDGMENTS

The compilation of the 2011 status of the Triassic chronostratigraphy and understanding the current disputes on correlations and the numerical age scalings were greatly aided by discussions with several colleagues, including in alphabetical order, Marco Balini, Mark Hounslow, Heinz Kozur, Leopold Krystyn, Spencer Lucas (who also provided preprints of several articles), Manfred Menning, Michael Orchard (previous chair of the Triassic subcommission), Paul Olsen and Lawrence Tanner. None of them will entirely agree with the brief summaries, with the selected genera-species nomenclature and zonations used in the figures, or with either of the dual possibilities for numerical age scaling, but all will agree that future research into the Triassic correlations and acquisition of numerical ages calibrated to standard biostratigraphy will lead to inevitable surprises. Additional or alternative zonal schemes are available through the *TimeScale Creator* visualization datapacks (*www.tscreator.org*). Early drafts of this Triassic synthesis were reviewed by Mike Orchard, who also provided in-press radio-isotopic dates supporting the "short Carnian" scaling. Gabi Ogg prepared all of the figures. Further advances in formalizing GSSPs, zonal schemes, inter-regional correlations, and eventual consensus on the best numerical age model interpolations will be found in the *Albertiana* newsletters of the Subcommission on Triassic Stratigraphy.

The compilation and application of cycle stratigraphy to enhance the Mesozoic time scale was partially supported by U.S. National Science Foundation grant EAR-0718905 (J. Ogg, L. Hinnov).

REFERENCES

Aigner, T., Bachmann, G.H., 1992. Sequence-stratigraphic framework of the German Triassic. Sedimentary Geology 80, 115–135.

Assereto, R., 1974. Aegean and Bithynian: Proposal for two new Anisian substages. Schriftenreihe der Erdwissenschaftlichen Kommissionen, Österreichische Akademie der Wissenschaften 2, 23–39.

Bachmann, G.H., Kozur, H.W., 2004. The Germanic Triassic: Correlations with the international scale, numerical ages and Milankovitch cyclicity. Hallesches Jahrbuch fur Geowissenschaften B26, 17–62.

Balini, M., Drystyn, L., Torti, V., 1998. In search of the Ladinian/Carnian boundary: Perspectives from Spiti (Tethys Himalaya). Albertiana 21, 26–32.

Balini, M., Krystyn, L., Nicora, A., Torti, V., 2001. The Ladinian-Carnian boundary succession in Spiti (Tethys Himalaya) and its bearing to the definition of the GSSP for the Carnian stage (Upper Triassic). Journal of Asian Earth Sciences 19 (3A), 3–4.

Balini, M., Lucas, S.G., Jenks, J.F., Spielmann, J.A., 2010. Triassic ammonoid biostratigraphy: An overview. In: Lucas, S.G. (Ed.), The Triassic Timescale. Geological Society Special Publication, 334, pp. 221–262.

Balme, B.E., Foster, C.B., 1996. Triassic, Explanatory Notes on Biostratigraphic Charts. In: Young, G.C., Laurie, J.R. (Eds.), Australian Phanerozoic Timescales. Oxford University Press, Chart 7, Oxford.

Baud, A., 2001. The new GSSP, base of the Triassic: Some consequences. Albertiana 26, 4–6.

Baud, A., Beauchamp, B., 2001. Proposals for the redefinition of the Griesbachian substage and for the base of the Triassic in the Arctic regions. In: Yan, J., Peng, Y. (Eds.), Proceedings of the International Symposium on the Global Stratotype of the Permian-Triassic Boundary and the Paleozoic-Mesozoic Events. University of Geosciences Press, Changxing (China), pp. 26–28.

Baud, A., Magaritz, M., Holser, W.T., 1989. Permian-Triassic of the Tethys: Carbon isotope studies. Sonderdruck aus Geologische Rundschau 78, 649–677.

Baud, A., Atudorei, V., Sharp, Z., 1996. Late Permian and early Triassic evolution of the northern Indian margin: Carbon isotope and sequence stratigraphy. Geodinamica Acta 9, 57–77.

Benton, M.J., 1986. The Late Triassic tetrapod extinction events. In: Padian, K. (Ed.), The Beginning of the Age of Dinosaurs. Faunal Change across the Triassic-Jurassic Boundary. Cambridge University Press, Cambridge, pp. 303–320.

Benton, M.J., 1993. Late Triassic extinctions and the origin of the dinosaurs. Science 260, 769–770.

Berra, F., Jadoul, F., Anelli, A., 2010. Environmental control on the end of the Dolomia Principale/Hauptdolomit depositional system in the central Alps: Coupling sea-level and climatic changes. In: Kustatscher, E., Preto, N., Wignall, P. (Eds.), Triassic climates. Palaeogeography, Palaeoclimatology, Palaeoecology, 290, pp. 138–150.

Beauchamp, B., Baud, A., 2002. Growth and demise of Permian biogenic chert along northwest Pangea: Evidence for end-Permain collapse of thermohaline circulation. Palaeogeography, Palaeoclimatology, Palaeoecology 184, 37–63.

Bittner, A., 1892. Was ist norisch? Jahrbuch Geologischen Reichsanstalt 42, 387–396.

Blome, C.D., 1984. Upper Triassic Radiolaria and radiolarian zonation from western North America. Bulletin of American Paleontology 85, 88.

Bown, P.R. (Ed.), 1998. Calcareous Nanofossils Biostratigraphy. Chapman and Hall, London, p. 314.

Bowring, S.A., Erwin, D.H., Jin, Y.G., Martin, M.W., Davidek, K., Wang, W., 1998. U/Pb zircon geochronology and tempo of the end-Permian mass extinction. Science 280, 1039–1045.

Brack, P., 2010. The "golden spike" for the Ladinian is set! Albertiana 38, 8–10.

Brack, P., Rieber, H., 1994. The Anisian/Ladinian boundary: Retrospective and new constraints. Albertiana 13, 25–36.

Brack, P., Rieber, H., Mundil, R., 1995. The Anisian/Ladinian boundary interval at Bagolino (Southern Alps, Italy): I. Summary and new results on ammonoid horizons and radiometric dating. Albertiana 15, 45–56.

Brack, P., Rieber, H., 1996. The new 'High-resolution Middle Triassic ammonoid standard scale' proposed by Triassic researchers from Padova – a discussion of the Anisian/Ladinian boundary interval. Albertiana 17, 42–50.

Brack, P., Mundil, R., Oberli, F., Meier, M., Rieber, H., 1996. Biostratigraphic and radiometric age data question the Milankovitch characteristics of the Latemar cycles (Southern Alps, Italy). Geology 24, 371–375.

Brack, P., Mundil, R., Oberli, F., Meier, M., Rieber, H., 1997. Biostratigraphic and radiometric age data question the Milankovitch characteristics of the Latemar cycles (Southern Alps, Italy) – Reply. Geology 25, 471–472.

Brack, P., Rieber, H., Nicora, A., Mundil, R., 2005. The Global boundary Stratotype Section and Point (GSSP) of the Ladinian Stage (Middle Triassic) at Bagolino (Southern Alps, Northern Italy) and its implications for the Triassic time scale. Episodes 28 (4), 233–244.

Brack, P., Rieber, H., Mundil, R., Blendinger, W., Maurer, F., 2007. Geometry and chronology of growth and drowning of Middle Triassic carbonate platforms (Cernera and Bivera/Clapsavon) in the Southern Alps (northern Italy). Swiss Journal of Geosciences 10, 327–347.

Brayard, A., Escarguel, G., Bucher, H., Brühwiler, 2009a. Smithian and Spathian (Early Triassic) ammonoid assemblages from terranes: Paleoceanographic and paleogeographic implications. In: Metcalfe, I., Isozaki, Y. (Eds.), End-Permian Mass Extinction: Events & Processes, Age & Timescale, Causative Mechanism(s) & Recovery. Journal of Asian Earth Sciences, 36(6), pp. 420–433.

Brayard, A., Escarguel, G., Bucher, H., Monnet, C., Brühwiler, Goudemand, N., Galfetti, T., Guex, J., 2009b. Good genes and good luck: Ammonoid diversity and the end-Permian mass extinction. Science 325, 1118–1121.

Broglio Loriga, C., Cassinis, G., 1992. The Permo-Triassic boundary in the Southern Alps (Italy) and in adjacent Periadriatic regions. In: Sweet, W.C., Yang, Z., Dickins, J.M., Yin, H. (Eds.), Permo-Triassic Events in the Eastern Tethys. Stratigraphy, Classification, and Relations with the Western Tethys. Cambridge University Press, Cambridge, pp. 78–97.

Broglio Loriga, C., Cirilli, S., De Zanche, V., di Bari, D., Gianolla, P., Laghi, G.R., Lowrie, W., Manfrin, S., Mastandrea, A., Mietto, P., Muttoni, G., Neri, C., Posenato, R., Rechichi, M., Rettori, R., Rohgi, G., 1998. A GSSP candidate for the Ladinian-Carnian boundary: The Prati di Stuores/Stuores Wiesen section (Dolomites, Italy). Albertiana 21, 2–18.

Broglio Loriga, C., Cirilli, S., De Zanche, V., di Bari, D., Gianolla, P., Laghi, G.R., Lowrie, W., Manfrin, S., Mastandrea, A., Mietto, P., Muttoni, G., Neri, C., Posenato, R., Rechichi, M., Rettori, R., Rohgi, G., 1999. The Prati di Stuores/Stuores Wiesen section (Dolomites, Italy): A candidate Global Stratotype Section and Point for the base of the Carnian stage. Rivista Italiana di Paleontologia e Stratigrafia 105, 37–78.

Brühwiler, T., Hochuli, P., Mundil, R., Schatz, W., Brack, P., 2007. Bio- and chronostratigraphy of the Middle Triassic Reifling Formation of the westernmost Northern Calcareous Alps. Swiss Journal of Geosciences 100, 443–455.

Burgess, S.D., Bowring, S.A., 2010. Evaluating a link between eruption of the Siberian Traps and the end-Permian Mass Extinction with high-precision geochronology. Geophysical Research Abstracts 12, EGU2010-11490-2.

Cao, C.Q., Love, G.D., Hays, L.E., Wang, W., Shen, S.Z., Summons, R.E., 2009. Biogeochemical evidence for euxinic oceans and ecological disturbance presaging the end-Permian mass extinction event. Earth and Planetary Science Letters 281, 188–201.

Carter, E.S., 1993. Biochronology and paleontology of uppermost Triassic (Rhaetian) radiolarians, Queen Charlotte Islands, British Columbia, Canada. Mémoires Géologie (Lausanne) 11, 175.

Carter, E.S., Orchard, M.J., 2007. Radiolarian – conodont – ammonoid intercalibration around the Norian-Rhaetian Boundary and implications for trans-Panthalassan correlation. Albertiana 36, 149–163.

Channell, J.E.T., Kozur, H.W., Sievers, T., Mocl, R., Aubrecht, R., Sykora, M., 2003. Carnian-Norian biomagnetostratigraphy at Silickà Brezovà (Slovakia): Correlation to other Tethyan sections and to the Newark Basin. Palaeogeography, Palaeoclimatology, Palaecology 191, 65–109.

Chinese Triassic Working Group, 2007. Final report of the GSSP candidate for the I/O boundary at West Pingdingshan Section in Chaohu, Southeastern China. Albertiana 36, 10–21.

Cirilli, S., 2010. Upper Triassic-lowermost Jurassic palynology and palynostratigraphy: A review. In: Lucas, S.G. (Ed.), The Triassic Timescale. Geological Society, Special Publication, 334, pp. 285–314.

Cirilli, S.A., Marzoli, L., Tanner, H., Bertrand, N., Buratti, F., Jourdan, G., Bellieni, D, Kontak, Renne, P.R., 2009. Latest Triassic onset of the Central Atlantic Magmatic Province (CAMP) volcanism in the Fundy Basin (Nova Scotia): New stratigraphic constraints. Earth and Planetary Science Letters 286, 514–525.

Cohen, A.S., Coe, A.L., 2007. The impact of the Central Atlantic Magmatic Province on climate and on the Sr- and Os-isotope evolution of seawater. Palaeogeography, Palaeoclimatology, Palaeoecology 244, 374–390.

Cornet, B., 1977. The Palynology and Age of the Newark Supergroup. PhD thesis. Pennsylvania State University, College Park, Pennsylvania. p. 505.

Cornet, B., 1993. Applications and limitations of palynology in age, climatic, and paleoenvironmental analyzes of Triassic sequences in North America. In: Lucas, S.G., Morales, M. (Eds.), The Nonmarine Triassic. New Mexico Museum of Natural History & Science Bulletin, 3, pp. 75–93.

Cornet, B., Olsen, P.E., 1985. A summary of the biostratigraphy of the Newark Supergroup of Eastern North America with comments on early Mesozoic provinciality. In: Weber, R. (Ed.), III Congreso Latinoamericano de Paleontologia Mexico, Simposio Sobre Floras del Triasico Tardio, su Fitogeografia y Paleoecologia, Memoria. Universidad Nacional Autónoma de México, Mexico, pp. 67–81.

Cozzi, A., Hinnov, L.A., Hardie, L.A., 2005. Orbitally forced Lofer cycles in the Dachstein Limestone of the Julian Alps (northeastern Italy). Geology 33, 789–792.

Dal Corso, J., Mietto, P., Newton, R.J., Pancost, R.D., Preto, N., Roghi, Wignall, P.B., 2012. Discovery of a major negative δ13C spike in the Carnian (Late Triassic) linked to eruption of Wrangellia flood basalts. Geology 40, 79–82.

Deenen, M.H.L., 2010. A New Chronology for the Late Triassic to Early Jurassic. PhD thesis. Faculty of Geosciences, Department of Earth Sciences, Utrecht University. Geologica Ultraiectina, 323.

Deenen, M.H.L., Langereis, C.G., Krijgsman, W., El Hachimi, H., El Hassane, C., 2010a. Paleomagnetic research in the Argana basin, Morocco: Trans-Atlantic correlation of CAMP volcanism and implications for the late Triassic geomagnetic polarity time scale. In: Deenen, M.H.L. (Ed.), A New Chronology for the Late Triassic to Early Jurassic. Geologica Ultraiectina, 323, pp. 43–64.

Deenen, M.H.L., Ruhl, M., Bonis, N.R., Krijgsman, W., Kuerschner, W.M., Reitsma, M., van Bergen, M.J., 2010b. A new chronology for the end-Triassic mass extinction. Earth and Planetary Science Letters 291, 113–125.

De Wever, P., 1982. Radiolaires du Trias et du Lias de la Téthys: Systématique, stratigraphy. Societé géologique du Nord, Lille, Publication 7, 599.

De Wever, P., 1998. Radiolarians. Column for Triassic chart of Mesozoic and Cenozoic sequence chronostratigraphic framework of European basins, by Hardenbol, J., Thierry, J., Farley, M.B., Jacquin, Th., de Graciansky, P.-C., and Vail, P.R. (coordinators). In: de Graciansky, P.-C., Hardenbol, J., Jacquin, Th., Vail, P.R. (Eds.), Mesozoic-Cenozoic Sequence Stratigraphy of European Basins. SEPM Special Publication, 60 Chart 8.

De Zanche, V., Gianolla, P., Mietto, P., Siorpaes, C., Vail, P.R., 1993. Triassic sequence stratigraphy in the Dolomites. Memorie di Scienze Geologische 45, 1–27.

Diakow, L., Orchard, M.J., Friedman, R., 2011. Absolute ages for the Norian Stage: A contribution from southern British Columbia, Canada. 21st Canadian Paleontological Conference. University of British Columbia. 19–20 Aug 2011. Abstract sent by M.J. Orchard to J. Ogg. [July 2011].

Diener, C., 1912. The Trias of the Himalayas. Geological Survey of India Memoirs 36 (3), 159.

Dinarès-Turell, J., Diez, J.D., Rey, D., Arnal, I., 2005. 'Buntsandstein' magnetostratigraphy and biostratigraphic reappraisal from eastern Iberia: Early and Middle Triassic stage boundary definitions through correlation to Tethyan sections. Palaeogeography, Palaeoclimatology, Palaeoecology 229, 158–177.

Dunning, G.R., Hodych, J.P., 1990. U/Pb zircon and baddeleyite ages for the Palisades and Gettysburg sills of the northeastern United States: Implications for the age of the Triassic/Jurassic boundary. Geology 18, 795–798.

Earth Impact Database, 2010. Compiled by Planetary and Space Science Centre. University of New Brunswick. Available at: www.passc.net/EarthImpactDatabase/.

Embry, A.F., 1988. Triassic sea-level changes: Evidence from the Canadian Arctic Archipelago. In: Wilgus, C. (Ed.), Sea Level Changes—An Integrated Approach. SEPM Special Publication, 42, pp. 249—259.

Embry, A.F., 1997. Global sequence boundaries of the Triassic and their identification in the Western Canada Sedimentary Basin. Bulletin of Canadian Petroleum Geology 45, 415—533.

Erwin, D.H., 1993. The Great Paleozoic Crisis: Life and Death in the Permian. Columbia University Press, New York, p. 327.

Erwin, D.H., 2006. Extinction: How Life on Earth Nearly Ended 250 Million Years Ago. Princeton University Press, Princeton, p. 320.

Erwin, D.H., Bowring, S.A., Yugan, J., 2002. End-Permian mass extinctions: A review. In: Koeberl, C., MacLeod, K.G. (Eds.), Catastrophic Events and Mass Extinctions: Impacts and Beyond. Geological Society of America Special Paper, 356, pp. 363—383.

Feist-Burkhardt, S., Götz, A.E., Szulc, J., Borkhataria, R., Geluk, M., Haas, J., Hornung, J., Jordan, P., Kempf, O., Michalik, J., Nawrocki, J., Reinhardt, L., Ricken, W., Röhling, H.-G., Rüffer, T., Török, A., Zühlke, R., 2008. Triassic. In: McCann, T. (Ed.), The Geology of Central Europe. Vol. 2: Mesozoic and Cenozoic. The Geological Society, London, pp. 749—821.

Fischer, A.G., 1964. The Lofer cyclothems of the Alpine Triassic. In: Merriam, D.F. (Ed.), Symposium on Cyclic Sedimentation. Kansas Geological Survey Bulletin, 169, pp. 107—149.

Flügel, E., 2002. Triassic reef patterns. In: Kiessling, W., Flügel, E., Golonka, J. (Eds.), Phanerozoic Reef Patterns. SEPM Special Publication, 72, pp. 691—733.

Fowell, S.J., Olsen, P.E., 1993. Time calibration of Triassic/Jurassic microfloral turnover, eastern North America. Tectonophysics 222, 361—369.

Fowell, S.J., Olsen, P.E., 1995. Time calibration of Triassic/Jurassic microfloral turnover, eastern North America — Reply. Tectonophysics 245, 96—99.

Fraser, N.D., 2006. Dawn of the Dinosaurs: Life in the Triassic. Indiana University Press, Bloomington, p. 310.

Furin, S., Preto, N., Rigo, M., Roghi, G., Gianolla, P., Crowley, J.L., Bowring, S.A., 2006. High-precision U-Pb zircon age from the Triassic of Italy: Implications for the Triassic time scale and the Carnian origin of calcareous nannoplankton and dinosaurs. Geology 34, 1009—1012.

Gaetani, M., 1993. Anisian/Ladinian boundary field workshop, Southern Alps — Balaton Highlands. 27 June — 4 July 1993. Albertiana 12, 5—9.

Gaetani, M., Gnaccolini, M., Jadoul, Fl., Garzanti, E., 1998. Multiorder sequence stratigraphy in the Triassic system of the western Southern Alps. In: de Graciansky, P.-C., Hardenbol, J., Jacquin, T., Vail, P.R. (Eds.), Mesozoic-Cenozoic Sequence Stratigraphy of European Basins. SEPM Special Publication, 60, pp. 701—717.

Galfetti, T., Hochuli, P.A., Brayard, A., Bucher, H., Weissert, H., Vigran, J.O., 2007a. The Smithian/Spathian boundary event: A global climatic change in the wake of the end-Permian biotic crisis. Evidence from palynology, ammonoids and stable isotopes. Geology 35, 291—294.

Galfetti, T., Bucher, H., Ovtcharova, M., Schaltegger, U., Brayard, A., Brühwiler, T., Goudemand, N., Weissert, H., Hochuli, P.A., Cordey, F., Guodun, K.A., 2007b. Timing of the Early Triassic carbon cycle perturbations inferred from new U-Pb ages and ammonoid biochronozones. Earth and Planetary Science Letters 258, 593—604.

Galfetti, T., Bucher, H., Brayard, A., Hochuli, P.A., Weissert, H., Guodun, K., Atudorei, V., Guex, J., 2007c. Late Early Triassic climate change: Insights from carbonate carbon isotopes, sedimentary evolution and ammonoid paleobiogeography. Palaeogeography, Palaeoclimatology, Palaeoecology 243, 394—411.

Galfetti, T., Bucher, H., Martini, R., Hochuli, P.A., Weissert, H., Crasquin-Soleua, S., Brayard, A., Goudemand, Brühwiler, N., Kuang, G., 2008. Evolution of Early Triassic outer platform paleoenvironments in the Nanpanjiang Basin (South China) and their significance for the biotic recovery. Sedimentary Geology 204, 36—60.

Gallet, Y., Besse, J., Krystyn, L., Marcoux, J., Théveniaut, H., 1992. Magnetostratigraphy of the late Triassic Bolücektasi Tepe section (southwestern Turkey): Implications for changes in magnetic reversal frequency. Physics of the Earth and Planetary Interiors 73, 85—108.

Gallet, Y., Krystyn, L., Besse, J., 1998. Upper Anisian to Lower Carnian magnetostratigraphy from the Northern Calcareous Alps (Austria). Journal of Geophysical Research 103, 605—621.

Gallet, Y., Krystyn, L., Marcoux, J., Besse, J., 2007. New constraints on the End-Triassic (Upper Norian—Rhaetian) magnetostratigraphy. Earth and Planetary Science Letters 255, 458—470.

Gehrels, G.E., Saleeby, J.B., Berg, H.C., 1987. Geology of Annette, Gravina, and Duke islands, southeastern Alaska. Canadian Journal of Earth Sciences 24, 866—881.

Geluk, M.C., Röhling, H.-G., 1997. High-resolution sequence stratigraphy of the Lower Triassic 'Buntsandstein' in the Netherlands and northwestern Germany. Geologie en Mijnbouw 76, 227—246.

Gianolla, P., Jacquin, T., 1998. Triassic sequence stratigraphic framework of western European basins. In: de Graciansky, P.-C., Hardenbol, J., Jacquin, Th., Vail, P.R. (Eds.), Mesozoic-Cenozoic Sequence Stratigraphy of European Basins. SEPM Special Publication, 60, pp. 643—650.

Gianolla, P., De Zanche, V., Mietto, P., 1998. Triassic sequence stratigraphy in the Southern Alps (northern Italy): Definition of sequences and basin evolution. In: de Graciansky, P.-C., Hardenbol, J., Jacquin, Th., Vail, P.R. (Eds.), Mesozoic-Cenozoic Sequence Stratigraphy of European Basins. SEPM Special Publication, 60, pp. 719—747.

Glen, J.M.G., Nomade, S., Lyons, J.J., Metcalfe, I., Mundil, R., Renne, J.R., 2009. Magnetostratigraphic correlations of Permian-Triassic marine-to-terrestrial sections from China. In: Metcalfe, I., Isozaki, Y. (Eds.), End-Permian Mass Extinction: Events & Processes, Age & Timescale, Causative Mechanism(s) & Recovery. Journal of Asian Earth Sciences, 36(6), pp. 520—540.

Goldhammer, R.K., Dunn, P.A., Hardie, L.A., 1987. High frequency glacio-eustatic oscillations with Milankovitch characteristics recorded in northern Italy. American Journal of Science 287, 853—892.

Goldhammer, R.K., Dunn, P.A., Hardie, L.A., 1990. Depositional cycles, composite sea level changes, cycle stacking patterns, and the hierarchy of stratigraphic forcing: Examples from the Alpine Triassic platform carbonates. Geological Society of America Bulletin 102 (535), 562.

Goudemand, N., Orchard, M., Bucher, H., Brayard, A., Brühwiler, T., Galfetti, T., Hochuli, P.A., Hermann, E., Ware, D., 2008. Smithian-Spathian boundary: The biggest crisis in Triassic conodont history. Abstracts with Program, Geological Society of America 40 (6), 505.

Gradinaru, E., Orchard, M.J., Nicora, A., Gallet, Y., Besse, J., Krystyn, L., Sobolev, E.S., Atudorei, N.-V., Ivanova, D., 2007. The Global Boundary Stratotype Section and Point (GSSP) for the base of the Anisian Stage: Desli Caira Hill, North Dobrogea, Romania. Albertiana 36, 54—71.

Greene, A., Scoates, J., Weis, D., 2008. The accreted Wrangellia oceanic plateau in Alaska, Yukon, and British Columbia. Internet article from Large Igneous Provinces Commission. Large Igneous Province of the Month. Available at: www.largeigneousprovinces.org/ December 2008.

Greene, A.R., Scoates, J.S., Weis, D., Katvala, E.C., Israel, S., Nixon, G.T., 2010. The architecture of oceanic plateaus revealed by the volcanic stratigraphy of the accreted Wrangellia oceanic plateau. Geosphere 6, 47–73.

Greene, A., Scoates, J., Weis, D., 2011. The accreted Wrangellia oceanic plateau in Alaska, Yukon, and British Columbia. Available at: wwweos.ubc.ca/research/wrangellia [Last update was 2010 when viewed Feb., 2011].

Griesbach, C.L., 1880. Paleontological notes on the Lower Trias on the Himalayas. Records of the Geological Survey of India 13 (2), 94–113.

Hallam, A., 1981. The end Triassic bivalve extinction event. Palaeogeography, Palaeoclimatology, Palaeoecology 35, 1–44.

Hallam, A., 1992. Phanerozoic Sea-Level Changes. Columbia University Press, New York, p. 266.

Hallam, A., Wignall, P.B., 1999. Mass extinctions and sea-level changes. Earth-Science Reviews 48, 217–250.

Haq, B.U., Al-Qahtani, A.M., 2005. Phanerozoic cycles of sea-level change on the Arabian Platform. GeoArabia 10 (2), 127–160.

Haq, B.U., Hardenbol, J., Vail, P.R., 1988. Mesozoic and Cenozoic chronostratigraphy and eustatic cycles. In: Wilgus, C.K., Hastings, B.S., Kendall, G.St.C., Posamentier, H.W., Ross, C.A., van Wagoner, J.C. (Eds.), Sea-Level Changes: An Integrated Approach. SEPM Special Publication, 42, pp. 71–108.

Hardenbol, J., Thierry, J., Farley, M.B., Jacquin, Th., de Graciansky, P.-C., Vail, P.R., (with numerous contributors), 1998. Mesozoic and Cenozoic sequence chronostratigraphic framework of European basins. In: de Graciansky, P.-C., Hardenbol, J., Jacquin, Th., Vail, P.R. (Eds.), Mesozoic-Cenozoic Sequence Stratigraphy of European Basins. SEPM Special Publication, 60, pp. 3–13, 763–781.

Hardie, L.A., Hinnov, L.A., 1997. Biostratigraphic and radiometric age data question the Milankovitch characteristics of the Latemar cycles (Southern Alps, Italy) – Comment. Geology 25, 470–471.

Hayes, J.M., Strauss, H., Kaufman, 1999. The abundance of ^{13}C in marine organic matter and isotopic fractionation in the global biogeochemical cycle of carbon during the past 800 Ma. Chemical Geology 161, 103–125.

Heckert, A.B., Lucas, S.G., 1999. Global correlation and chronology of Triassic theropods (Archosauria: Dinosauria). Albertiana 23, 22–35.

Heckert, A.B., Lucas, S.G., Dickinson, W.R., Mortensen, J.K., 2009. New ID-TIMS U-Pb ages for Chinle Group strata (Upper Triassic) in New Mexico and Arizona, correlation to the Newark Supergroup, and implications for the "long Norian". Geological Society of America Abstracts with Programs 41 (7), 123.

Helby, R., Morgan, R., Partridge, A.D., 1987. A palynological zonation of the Australian Mesozoic. In: Jell, P.A. (Ed.), Studies in Australian Mesozoic Palynology. Memoir Association Australaisian Paleontologists, 4, pp. 1–94.

Heller, F., Haihong, C., Dobson, J., Haag, M., 1995. Permian-Triassic magnetostratigraphy – new results from south China. Physics of the Earth and Planetary Interiors 89, 281–295.

Hermann, E., Hochuli, P.A., Bucher, H., Vigran, J.O., Weissert, H., Bernasconi, S.M., 2010. A close-up view of the Permian-Triassic boundary based on expanded organic carbon isotope records from Norway (Trøndelag and Finnmark Platform). Global and Planetary Change 74, 156–167.

Herngreen, G.F.W., 2005. Triassic sporomorphs of NW Europe: Taxonomy, morphology and ranges of marker species with remarks on botanical relationship and ecology and comparison with ranges in the Alpine Triassic. Kenniscentrum Biogeology, Nederlands Instituut voor Toegepaste Geowetenschappen TNO, Utrecht, TNO report NITG 04-176-C.

Hesselbo, S.P., Robinson, S.A., Surlyk, F., Piasecki, S., 2002. Terrestrial and marine extinctions at the Triassic-Jurassic boundary synchronized with major carbon-cycle perturbation: A link to initiation of massive volcanism? Geology 30, 251–254.

Hinnov, L.A., 2006. Discussion of "Magnetostratigraphic confirmation of a much faster tempo for sea-level change for the Middle Triassic Latemar platform carbonates" by D.V. Kent, G. Muttoni and P. Brack [Earth Planet. Sci. Lett. 228 (2004), 369–377]. Earth and Planetary Science Letters 243, 841–846.

Hinnov, L.A., Goldhammer, R.K., 1991. Spectral analysis of the Middle Triassic Latemar Limestone. Journal of Sedimentary Petrology 61, 1173–1193.

Hochuli, P., 1998. Dinoflagellate cysts and spore pollen. Columns for Triassic chart of Mesozoic and Cenozoic sequence chronostratigraphic framework of European basins, by Hardenbol, J., Thierry, J., Farley, M.B., Jacquin, Th., de Graciansky, P.-C., and Vail, P.R. (coordinators). In: de Graciansky, P.-C., Hardenbol, J., Jacquin, Th., Vail, P.R. (Eds.), Mesozoic-Cenozoic Sequence Stratigraphy of European Basins. SEPM Special Publication, 60 Chart 8.

Hochuli, P.A., Hermann, E., Vigran, J.O., Bucher, H., Weissert, W., 2010. Rapid demise and recovery of plant ecosystems across the end-Permian extinction event. Global and Planetary Change 74, 144–155.

Hodych, J.P., Dunning, G.R., 1992. Did the Manicouagan impact trigger end-of-Triassic mass extinction? Geology 20, 51–54.

Holser, W.T., Magaritz, M., 1987. Events near the Permian-Triassic boundary. Modern Geology 11, 155–180.

Horacek, M., Brandner, R., Richoz, S., Povoden-Karadeniz, E., 2010a. Lower Triassic sulphur isotope curve of marine sulphates from the Dolomites, N-Italy. Palaeogeography, Palaeoclimatology, Palaeoecology 290, 65–70.

Horacek, M., Povoden, E., Richoz, S., Brandner, R., 2010b. High-resolution carbon isotope changes, litho- and magnetostratigraphy across Permian-Triassic boundary sections in the Dolomites, N-Italy. New constraints for global correlation. Palaeogeography, Palaeoclimatology, Palaeoecology 290, 58–64.

Hornung, T., Brandner, R., 2005. Biochronostratigraphy of the Reingraben Turnover (Hallstatt Facies Belt): Local black shale events controlled by regional tectonics, climatic change and plate tectonics. Facies 51, 460–479.

Hounslow, M.K., Muttoni, G., 2010. The geomagnetic polarity timescale for the Triassic: Linkage to stage boundary definitions. In: Lucas, S.G. (Ed.), The Triassic Timescale. Geological Society, Special Publication, 334, pp. 61–102.

Hounslow, M.W., Posen, P.E., Warrington, G., 2004. Magnetostratigraphy and biostratigraphy of the Upper Triassic and lowermost Jurassic succession, St. Audrie's Bay, UK. Palaeogeography, Palaeoclimatology, Palaeoecology 213, 331–358.

Hounslow, M.K., Szurlies, M., Muttoni, G., Nawrocki, J., 2007. The magnetostratigraphy of the Olenekian-Anisian boundary and a proposal to define the base of the Anisian using a magnetozone datum. Albertiana 36, 72–77.

Hounslow, M.K., Peters, C., Mørk, A., Weitschat, W., Vigran, J.O., 2008a. Biomagnetostratigraphy of the Vikinghøgda Formation, Svalbard (arctic Norway) and the geomagnetic polarity timescale for the Lower Triassic. Geological Society of America Bulletin 120, 1305–1325.

Hounslow, M.K., Hu, M., Mørk, A., Weitschat, W., Vigran, J.O., Karloukovski, V., Orchard, M.J., 2008b. Intercalibration of Boreal and Tethyan timescales: The magneto-biostratigraphy of the Middle Triassic and the latest Early Triassic, central Spitsbergen (arctic Norway). Polar Research 27, 469–490.

Hüsing, S.K., Deenen, M.H.L., Koopmans, J.G., Krijgsman, W., 2011. Magnetostratigraphic dating of the proposed Rhaetian GSSP at Steinbergkogel (Upper Triassic, Austria): Implications for the Late Triassic time scale. Earth and Planetary Science Letters 302, 203–216.

Ikeda, M., Tada, R., Sakuma, H., 2010. Astronomical cycle origin of bedded chert: A middle Triassic bedded chert sequence, Inuyama, Japan. Earth and Planetary Science Letters 297, 369–378.

Irmis, R.B., Martz, J.W., Parker, W.G., Nesbitt, S.J., 2010. Re-evaluating the correlation between Late Triassic terrestrial vertebrate biostratigraphy and the GSSP-defined marine stages. Albertiana 38, 40–53.

Isozaki, Y., 2009. Integrated "plume winter" scenario for the double-phased extinction during the Paleozoic–Mesozoic transition: The G-LB and P-TB events from a Panthalassan perspective. In: Metcalfe, I., Isozaki, Y. (Eds.), End-Permian Mass Extinction: Events & Processes, Age & Timescale, Causative Mechanism(s) & Recovery. Journal of Asian Earth Sciences, 36(6), pp. 459–480.

Jacquin, Th., Vail, P.R., (coordinators), 1998. Sequence chronostratigraphy. Columns for Triassic chart, Mesozoic and Cenozoic sequence chronostratigraphic framework of European basins by Hardenbol, J., Thierry, J., Farley, M.B., Jacquin, Th., de Graciansky, P.-C., and Vail, P.R. In: de Graciansky, P.-C., Hardenbol, J., Jacquin, Th., Vail, P.R. (Eds.), Mesozoic-Cenozoic Sequence Stratigraphy of European Basins. SEPM Special Publication, 60 Chart 8.

Jenkyns, H.C., Jones, C.E., Gröcke, D.R., Hesselbo, S.P., Parkinson, D.N., 2002. Chemostratigraphy of the Jurassic system: Applications, limitations and implications for palaeoceanography. Journal of the Geological Society 159, 351–378.

Jones, C.E., Jenkyns, H.C., 2001. Seawater strontium isotopes, oceanic anoxic events, and seafloor hydrothermal activity in the Jurassic and Cretaceous. American Journal of Science 301, 112–149.

Jourdan, F., Marzoli, A., Bertrand, H., Cirilli, S., Tanner, L.H., Kontak, D.J., McHone, G., Renne, P.R., Bellieni, G., 2009a. $^{40}Ar/^{39}Ar$ ages of CAMP in North America: Implications for the Triassic-Jurassic boundary and the ^{40}K decay constant bias. Lithos 110, 167–180.

Jourdan, F., Renne, P.R., Reimold, W.U., 2009b. An appraisal of the ages of terrestrial impact structures. Earth and Planetary Science Letters 286, 1–13.

Kamo, S.L., Czamanske, G.K., Drogh, T.E., 1996. A minimum U-Pb age for Siberian flood-basalt volcanism. Geochimica et Cosmochimica Acta 60, 3505.

Kamo, S.L., Czamanske, G.K., Amelin, Y., Fedorenko, V.A., Davis, D.W., Trofimov, V.R., 2003. Rapid eruption of Siberian flood-volcanic rocks and evidence for coincidence with the Permian-Triassic boundary and mass extinction at 251 Ma. Earth and Planetary Science Letters 214, 75–91.

Kampschulte, A., Strauss, H., 2004. The sulfur isotopic evolution of Phanerozoic sea water based on the analysis of structurally substituted sulfate in carbonates. Chemical Geology 204, 255–286.

Kemp, D.B., Coe, A.L., 2007. A nonmarine record of eccentricity forcing through the Upper Triassic of southwest England and its correlation with the Newark Basin astronomically calibrated geomagnetic polarity time scale from North America. Geology 35, 991–994.

Kent, D.V., Olsen, P.E., 1999. Astronomically tuned geomagnetic polarity timescale for the Late Triassic. Journal of Geophysical Research 104, 12831–12841. Web page update (2002) posted at Newark Basin Coring Project website: www.ldeo.columbia.edu/~polsen/nbcp/nbcp.timescale.htm [Accessed 3 July 2010].

Kent, D.V., Olsen, P.E., Witte, W.K., 1995. Late Triassic-earliest Jurassic geomagnetic polarity sequence and paleolatitudes from drill cores in the Newark rift basin, eastern North America. Journal of Geophysical Research 100, 14965–14998.

Kent, D.V., Muttoni, G., Brack, P., 2004. Magnetostratigraphic confirmation of a much faster tempo for sea-level change for the Middle Triassic Latemar platform carbonates. Earth and Planetary Science Letters 228, 369–377.

Kent, D.V., Muttoni, G., Brack, P., 2006. Reply to Discussion of "Magnetostratigraphic confirmation of a much faster tempo for sea-level change for the Middle Triassic Latemar platform carbonates" by L. Hinnov. Earth and Planetary Science Letters 243, 847–850.

Kiessling, W., 2010. Reef expansion during the Triassic: Spread of photosymbiosis balancing Climatic cooling. In: Kustatscher, E., Preto, N., Wignall, P. (Eds.), Triassic Climates. Palaeogeography, Palaeoclimatology, Palaeoecology, 290, pp. 11–19.

Kiparisova, L.D., Popov, Yu.N, 1956. Subdivision of the Lower series of the Triassic system into stages. Doklady Academy Sciences U.S.S.R 109, 842–845 [In Russian].

Kiparisova, L.D., Popov, Yu.N, 1964. The project of subdivision of the Lower Triassic into stages. XXII International Geological Congress, Reports of Soviet geologists, Problem 16a, 91–99 [In Russian].

Klein, H., Lucas, S.G., 2010. Tetrapod footprints – their use in biostratigraphy and biochronology of the Triassic. In: Lucas, S.G. (Ed.), The Triassic Timescale. Geological Society, Special Publication, 334, pp. 419–446.

Knoll, A.H., Bambach, R.K., Payne, J.L., Pruss, S., Fischer, W.W., 2007. Paleophysiology and end-Permian mass extinction. Earth and Planetary Science Letters 256, 295–313.

Koepnick, R.B., Denison, R.E., Burke, W.H., Hetherington, E.A., Dahl, D.A., 1990. Construction of the Triassic and Jurassic portion of the Phanerozoic curve of seawater $^{87}Sr/^{86}Sr$. Chemical Geology 80, 327–349.

Korte, C., Kozur, H.W., 2010. Carbon-isotope stratigraphy across the Permian-Triassic boundary: A review. Journal of Asian Earth Sciences 39, 215–235.

Korte, C., Kozur, H.W., Bruckschen, P., Veizer, J., 2003. Strontium isotope evolution of late Permian and Triassic seawater. Geochimica et Cosmochimica Acta 67, 47–62.

Korte, C., Kozur, H.W., Veizer, J., 2005. $\partial^{13}C$ and $\partial^{18}O$ values of Triassic brachiopods and carbonate rocks as proxies for coeval seawater and palaeotemperature. Palaeogeography, Palaeoclimatology, Palaeoecology 226, 287–306.

Korte, C., Pande, P., Kalia, P., Kozur, H.W., Joachimski, M.M., Oberhänsli, 2010. Massive volcanism at the Permian-Triassic boundary and its impact on the isotopic composition of the ocean and atmosphere. Journal of Asian Earth Sciences 37, 293–311.

Kozur, H.W., 1998. Some aspects of the Permian-Triassic boundary (PTB) and of the possible causes for the biotic crisis around this boundary. Palaeogeography, Palaeoclimatology, Palaeoecology 143, 227–272.

Kozur, H.W., 1999. Remarks on the position of the Norian-Rhaetian boundary. Proceedings of the Epicontinental Triassic International Symposium. Halle (Germany), 21–23. Sept., 1998. Zentralblatt für Geologie und Paläontologie, I(7–8): 523–535.

Kozur, H.W., 2003. Integrated ammonoid, conodont and radiolarian zonation of the Triassic. Hallesches Jahrbuch fur Geowissenschaften B25, 49–79.

Kozur, H.W., 2007. Biostratigraphy and event stratigraphy in Iran around the Permian–Triassic boundary (PTB): Implications for the causes of the PTB biotic crisis. Global and Planetary Change 55, 155–176.

Kozur, H.W., Mostler, H., 1994. Anisian to Middle Carnian radiolarian zonation and description of some stratigraphically important radiolarians. Geologisch-Paläontologische Mitteilungen Innsbruck, Sonderbd 3, 39–255.

Kozur, H.W., Bachmann, G.H., 2005. Correlation of the Germanic Triassic with the international scale. Albertiana 32, 21–35.

Kozur, H.W., Weems, R.E., 2005. Conchostracan evidence for a late Rhaetian to early Hettangian age for the CAMP volcanic event in the Newark Supergroup, and a Sevatian (late Norian) age for the immediately underlying beds. Hallesches Jahrbuch für Geowissenschaften, Reihe B: Geologie, Paläontologie, Mineralogie 27, 21–51.

Kozur, H.W., Bachmann, G.H., 2008. Updated correlation of the Germanic Triassic with the Tethyan scale and assigned numeric ages. Berichte der Geologischen Bundesanstalt 76, 53–58.

Kozur, H.W., Bachmann, G.H., 2010. The Middle Carnian Wet Intermezzo of the Stuttgart Formation (Schilfsandstein), Germanic Basin. In: Kustatscher, E., Preto, N., Wignall, P. (Eds.), Triassic Climates. Palaeogeography, Palaeoclimatology, Palaeoecology, 290, pp. 107–119.

Kozur, H.W., Weems, R.E., 2010. The biostratigraphic importance of conchostracans in the continental Triassic of the northern hemisphere. In: Lucas, S.G. (Ed.), The Triassic Timescale. Geological Society, Special Publication, 334, pp. 315–417.

Krull, E.S., Retallack, G.J., 2000. $\partial^{13}C$ depth profiles from paleosols across the Permian-Triassic boundary: Evidence for methane release. Geological Society of America Bulletin 112, 1459–1472.

Krystyn, L., 1980. Triassic conodont localities of the Salzkammergut region (northern Calcareous Alps). In: Schonlaub, H.P. (Ed.), Second European Conodont Symposium ECOS II: Guidebook and Abstracts. Abhandlungen der Geologischen Bundesanstalt, 35, pp. 61–98.

Krystyn, L., 2010. Decision report on the defining event for the base of the Rhaetian stage. Albertiana 38, 11–12.

Krystyn, L., Orchard, M.J., 1996. Lowermost Triassic ammonoid and conodont biostratigraphy of Spiti, India. Albertiana 17, 10–21.

Krystyn, L., Gallet, Y., Besse, J., Marcoux, J., 2002. Integrated Upper Carnian to Lower Norian biochronology and implications for the Upper Triassic magnetic polarity time scale. Earth and Planetary Science Letters 203, 343–351.

Krystyn, L., Richoz, S., Bhargava, O.N., 2007a. The Induan-Olenekian Boundary (IOB) in Mud – an update of the candidate GSSP section M04. Albertiana 36, 33–49.

Krystyn, L., Bhargava, O.N., Richoz, S., 2007b. A candidate GSSP for the base of the Olenekian Stage: Mud at Pin Valley, district Lahul & Spiti, Himachal Pradesh (Western Himalaya, India). Albertiana 35, 5–29.

Krystyn, L., Boquerel, H., Kuerschner, W., Richoz, S., Gallet, Y., 2007c. Proposal for a candidate GSSP for the base of the Rhaetian Stage. New Mexico Museum of Natural History and Science Bulletin 41, 189–199.

Krystyn, L., Richoz, S., Gallet, Y., Bouquerel, H., Kürschner, W.M., Spötl, C., 2007d. Updated bio- and magnetostratigraphy from the Steinbergkogel (Austria), candidate GSSP for the base of the Rhaetian stage. Albertiana 36, 164–173.

Kürschner, W.M., Herngreen, G.F.W., 2010. Triassic palynology of central and northwestern Europe: A review of palynofloral diversity patterns and biostratigraphic subdivisions. In: Lucas, S.G. (Ed.), The Triassic Timescale. Geological Society, Special Publication, 334, pp. 263–283.

Large Igneous Provinces Commission (International Association of Volcanology and Chemistry of the Earth's Interior), 2011. LIP record. Internet database available at. www.largeigneousprovinces.org [Last update was Feb. 2008, when viewed Feb. 2011).

Lehrmann, D.J., Ramezani, J., Bowring, S.A., Martin, M.W., Montgomery, P., Enos, P., Payne, J.L., Orchard, J.J., Wang, H., Wei, J., 2006. Timing of recovery from the end-Permian extinction: Geochronologic and biostratigraphic constraints from south China. Geology 34, 1053–1056.

Litwin, R.J., Traverse, A., Ash, S.R., 1991. Preliminary palynological zonation of the Chinle Formation, southwestern U.S.A., and its correlation to the Newark Supergroup (eastern U.S.A.). Review of Palaeobotany and Palynology 68, 269–287.

Lucas, S.G., 1998. Global Triassic tetrapod biostratigraphy and biochronology. Palaeogeography, Palaeoclimatology, Palaeoecology 143, 345–382.

Lucas, S.G., 1999. A tetrapod-based Triassic timescale. Albertiana 22, 31–40.

Lucas, S.G., 2008. Triassic New Mexico: Dawn of the Dinosaurs. New Mexico Museum of Natural History and Science, Albuquerque, p. 48.

Lucas, S.G., 2009. Timing and magnitude of tetrapod extinctions across the Permo-Triassic boundary. In: Metcalfe, I., Isozaki, Y. (Eds.), End-Permian Mass Extinction: Events & Processes, Age & Timescale, Causative Mechanism(s) & Recovery. Journal of Asian Earth Sciences, 36(6), pp. 491–502.

Lucas, S.G. (Ed.), 2010a. The Triassic Timescale. Geological Society London, Special Publication, 334, p. 500.

Lucas, S.G., 2010b. The Triassic chronostratigraphy scale: History and status. In: Lucas, S.G. (Ed.), The Triassic Timescale. Geological Society Special Publication, 334, pp. 447–500.

Lucas, S.G., 2010c. The Triassic timescale: An introduction. In: Lucas, S.G. (Ed.), The Triassic Timescale. Geological Society, Special Publication, 334, pp. 1–16.

Lucas, S.G., 2010d. The Triassic timescale based on nonmarine tetrapod biostratigraphy and biochronology. In: Lucas, S.G. (Ed.), The Triassic Timescale. Geological Society, Special Publication, 334, pp. 17–39.

Lucas, S.G., Tanner, L.H., Kozur, H.W., Weems, R.E., and Heckert, A.B., submitted, The Late Triassic timescale: age and correlation of the Carnian-Norian boundary.

Luo, G.M., Kump, L.R., Wang, Y.B., Tong, J.N., Arthur, M.A., Yang, H., Huang, J.H., Yin, H.F., Xie, S.C., 2010. Isotopic evidence for an anomalously low oceanic sulfate concentration following end-Permian mass extinction. Earth and Planetary Science Letters 300, 101–111.

Luo, G.M., Wang, Y.B., Yang, H., Algeo, T.J., Kump, L.R., Huang, J.H., Xie, S.C., 2011. Stepwise and large-magnitude negative shift in $\delta^{13}C_{carb}$ preceded the main marine mass extinction of the Permian-Triassic crisis interval. Palaeogeography, Palaeoclimatology, Palaeoecology 299, 70–82.

Marshall, C.R., Jacobs, D.K., 2009. Flourishing after the end-Permian mass extinction. Science 325, 1079–1080.

Martin, E.E., Macdougall, J.D., 1995. Sr and Nd isotopes at the Permian/Triassic boundary: A record of climate change. Chemical Geology 125, 73–95.

Marzoli, A., Renne, P.R., Piccirillo, E.M., Ernesto, M., Bellieni, G., DeMin, A., 1999. Extensive 200-million-year-old continental flood basalts of the Central Atlantic Magmatic Province. Science 284, 616–618.

Marzoli, A., Jourdan, F., Puffer, J.H., Cuppone, T., Tanner, L.H., Weems, R.E., Bertrand, H., Cirilli, S., Bellieni, G., De Min, A., 2011. Timing and duration of the Central Atlantic magmatic province in the Newark and Culpeper basins, eastern U.S.A. Lithos 122, 175–188.

Mazza, M., Furin, S., Spöti, C., Rigo, M., 2010. Generic turnovers of Carnian/Norian conodonts: Climatic control or competition? In: Kustatscher, E., Preto, N., Wignall, P. (Eds.), Triassic Climates. Palaeogeography, Palaeoclimatology, Palaeoecology, 290, pp. 120–137.

McArthur, J.M., 2008. Comment on 'The impact of the Central Atlantic Magmatic Province on climate and on the Sr- and Os-isotope evolution of seawater' by Cohen, A.S. and Coe, A.L. 2007. [Palaeogeography, Palaeoclimatology, Palaeoecology 244] Palaeogeography, Palaeoclimatology, Palaeoecology, 263: 146–149374–390.

McArthur, J.M., 2010. Correlation and dating with strontium-isotope stratigraphy. In: Whitaker, J.E., Hart, M.B. (Eds.), Micropalaeontology, Sedimentary Environments and Stratigraphy: A Tribute to Dennis Curry (1912–2001). The Micropalaeontological Society Special Publication, TSM004, pp. 133–145.

McArthur, J.M., Howarth, R.J., Bailey, T.R., 2001. Strontium isotope stratigraphy: LOWESS Version 3: Best fit to the marine Sr-isotope curve for 0–509 Ma and accompanying look-up table for deriving numerical age. Journal of Geology 109, 155–170.

McRoberts, C.A., 2007. The halobid bivalve succession across a potential Carnian/Norian GSSP at Black Bear Ridge, Williston Lake, northeast British Columbia, Canada. Albertiana 36, 142–145.

McRoberts, C.A., 2010. Biochronology of Triassic pelagic bivalves. In: Lucas, S.G. (Ed.), The Triassic Timescale. Geological Society Special Publication, 334, pp. 201–219.

McRoberts, C.A., Krystyn, L., Shea, A., 2008. Rhaetian (Late Triassic *Monotis* (Bivalvia: Pectinacea) from the Northern Calcareous Alps (Austria) and the end-Norian crisis in pelagic faunas. Journal of Paleontology 51, 721–735.

Menning, M., Gast, R., Hagdorn, H., Kading, K.-C., Simon, T., Szurlies, M., Nitsch, E., 2005. Zeitskala für Perm und Trias in der Stratigraphischen Tabelle von Deutschland 2002, zyklostratigraphische Kalibrierung der höheren Dyas und Germanischen Trias und das Alter der Stufen Roadium bis Rhaetium 2005. In: Menning, M., Hendrich, A. (Eds.), Erläuterungen zur Stratigraphischen Tabelle von Deutschland. Newsletters on Stratigraphy, 41(1/3), pp. 173–210.

Metcalfe, I., Nicoll, R.S., Black, L.P., Mundil, R., Renne, P., Jagodzinski, E.A., Wang, C., 1999. Isotope geochronology of the Permian-Triassic boundary and mass extinction in South China. In: Yin, H., Tong, J. (Eds.), Pangea and the Paleozoic-Mesozoic Transition. China University of Geosciences Press, Wuhan, pp. 134–137.

Metcalfe, I., Isozaki, Y. (Eds.), 2009a. End-Permian Mass Extinction: Events & Processes, Age & Timescale, Causative Mechanism(s) & Recovery. Journal of Asian Earth Sciences, 36(6), pp. 407–540.

Metcalfe, I., Isozaki, Y., 2009b. Current perspectives on the Permian-Triassic boundary and end-Permian mass extinction: Preface. In: Metcalfe, I., Isozaki, Y. (Eds.), End-Permian Mass Extinction: Events & Processes, Age & Timescale, Causative Mechanism(s) & Recovery. Journal of Asian Earth Sciences, 36(6), pp. 407–412.

Metcalfe, I., Foster, C.B., Afonin, S.A., Nicoll, R.S., Wang, X., Lucas, S.G., 2009c. Stratigraphy, biostratigraphy and C-isotopes of the Permian–-Triassic non-marine sequence at Dalongkou and Lucaogou, Xinjiang Province, China. In: Metcalfe, I., Isozaki, Y. (Eds.), End-Permian Mass Extinction: Events & Processes, Age & Timescale, Causative Mechanism(s) & Recovery. Journal of Asian Earth Sciences, 36(6), pp. 407–412.

Meyers, S.R., 2008. Resolving Milankovitchian controversies: The Triassic Latemar Limestone and the Eocene Green River Formation. Geology 36, 319–322.

Mietto, P., Manfrin, S., 1995. A high resolution Middle Triassic ammonoid standard scale in the Tethys Realm: A preliminary report. Bulletin de la Société géologique de France 166 (5), 539–563.

Mietto, P., Manfrin, S., 1999. A debate on the Ladinian-Carnian boundary. Albertiana 22, 23–27.

Mietto, P., Andreetta, R., Broglio Loriga, C., Buratti, N., Cirilli, S., De Zanche, V., Furin, S., Gianolla, P., Manfrin, S., Muttoni, G., Neri, C., Nicora, A., Posenato, R., Preto, N., Rigo, M., Roghi, G., Spötl, C., 2007. A candidate of the Global Boundary Stratotype Section and Point for the base of the Carnian Stage (Upper Triassic): GSSP at the base of the *canadensis* Subzone (FAD of *Daxatina*) in the Prati di Stuores/Stuores Wiesen section (Southern Alps, NE Italy). Albertiana 36, 78–97.

Mietto, P., Manfrin, S., Preto, N., Rigo, M., Roghi, G., Furin, S., Gianolla, P., Posenato, R., Muttoni, G., Nicora, A., Buratti, N., Cirilli, S., Spötl, C., Bowring, S.A., and Ramezani, J., *in press*, A candidate of the Global Boundary Stratotype Section and Point for the base of the Carnian Stage (Upper Triassic): GSSP at the base of the *canadensis* Subzone (FAD of *Daxatina*) in the Prati di Stuores/Stuores Wiesen section (Southern Alps, NE Italy). *Episodes*.

Mojsisovics, E., von, 1869. Über die Gliederung der oberen Triasbildungen der östlichen Alpen. Jahrbuch Geologischen Reichsanstalt 19, 91–150.

Mojsisovics, E., von, 1893. Faunistische Ergebnisse aus der Untersuchung der Ammoneen-faunen der Mediterranen Trias. Abhandlungen der Geologischen Reichsanstalt 6, 810.

Mojsisovics, E., vonWaagen, W., Diener, C., 1895. Entwurf einer Gliederung der pelagischen Sedimente des Trias-Systems. Sitzungberichte Akademie Wissenschaften Wien 104, 1271–1302.

Mørk, A., Embry, A.F., Weitschat, W., 1989. Triassic transgressive-regressive cycles in the Sverdrup Basin, Svalbard and the Barents Shelf. In: Collinson, J.D. (Ed.), Correlation in Hydrocarbon Exploration. Graham and Trotman, London, pp. 113–130.

Mortensen, J.K., Hulbert, L.J., 1992. A U-Pb zircon age for a Maple Creek gabbro sill, Tatamagouche Creek area, southwest Yukon Territory. In: Radiogenic Age and Isotopic Studies, Report 5: Geological Survey of Canada, Paper, 91–2, pp. 175–179.

Motani, R., 2010. Warm blooded sea dragons? Science 328, 1361–1362.

Mundil, R., Brack, P., Meier, M., Rieber, H., Oberli, F., 1996. High resolution U-Pb dating of Middle Triassic volcaniclastics: Time-scale calibration and verification of tuning parameters for carbonate sediments. Earth and Planetary Science Letters 141, 137–151.

Mundil, R., Metcalfe, I., Ludwig, K.R., Renne, P.R., Oberli, F., Nicoll, R.S., 2001. Timing of the Permian-Triassic biotic crisis: Implications from new zircon U/Pb age data (and their limitations). Earth and Planetary Science Letters 187, 131–145.

Mundil, R., Ludwig, K.R., Metcalfe, I., Renne, P.R., 2004. Age and timing of the Permian mass extinctions: U/Pb dating of closed-system zircons. Science 305, 1760–1763.

Mundil, R., Irmis, R., 2008. New U-Pb age constraints for terrestrial sediments in the Late Triassic: Implications for faunal evolution and correlations with marine environments 33rd International Geological Congress, Oslo 2008 (abstract): Online at: http://www.cprm.gov.br/33IGC/1342538.html.

Mundil, R., Pálfy, J., Renne, P.R., Brack, P., 2010. The Triassic timescale: New constraints and a review of geochronological data. In: Lucas, S.G. (Ed.), The Triassic Timescale. Geological Society, Special Publication, 334, pp. 41–60.

Muttoni, G., Kent, D.V., Nicora, A., Rieber, H., Brack, P., 1996a. Magneto-biostratigraphy of the 'Buchenstein Beds' at Frötschbach (western Dolomites, Italy). Albertiana 17, 51–56.

Muttoni, G., Kent, D.V., Meço, S., Nicora, A., Gaetani, M., Balini, M., Germani, D., Rettori, R., 1996b. Magnetobiostratigraphy of the Spathian to Anisian (Lower to Middle Triassic) Kçira section, Albania. Geophysical Journal International 127, 503–514.

Muttoni, G., Kent, D.V., Meço, S., Balini, M., Nicora, A., Rettori, R., Gaetani, M., Krystyn, L., 1998. Towards a better definition of the Middle Triassic magnetostratigraphy and biostratigraphy in the Tethyan realm. Earth and Planetary Science Letters 164, 285–302.

Muttoni, G., Kent, D.V., DiStefano, P., Gullo, M., Nicora, A., Tait, J., Lowrie, W., 2001. Magnetostratigraphy and biostratigraphy of the Carnian/Norian boundary interval from the Pizzo Mondello section (Sicani Mountains, Sicily). Palaeogeography, Palaeoclimatology, Palaeoecology 166, 383–399.

Muttoni, G., Nicora, A., Brack, P., Kent, D.V., 2004a. Integrated Anisian-Ladinian boundary chronology. Palaeogeography, Palaeoclimatology, Palaeoecology 208, 85–102.

Muttoni, G., Kent, D.V., Olsen, P.E., DiStefano, P., Lowrie, W., Bernasconi, S.M., Hernandez, F.M., 2004b. Tethyan magneto-stratigraphy from Pizzo Mondello (Sicily) and correlation to the Late Triassic Newark astrochronological polarity time scale. Geological Society of America Bulletin 116, 1043–1058.

Muttoni, G., Kent, D.V., Flavio, J., Olsen, P., Rigo, M., Galli, M.T., Nicora, A., 2010. Rhaetian magnetostratigraphy from the Southern Alps (Italy): Constraints on Triassic chronology. Palaeogeography, Palaeoclimatology, Palaeoecology 285, 1–16.

Nawrocki, J., 1997. Permian to Early Triassic magnetostratigraphy from the Central European Basin in Poland: Implications on regional and worldwide correlations. Earth and Planetary Science Letters 152, 37–58.

Nawrocki, J., Szulc, J., 2000. The Middle Triassic magnetostratigraphy from the Peri-Tethys basin in Poland. Earth and Planetary Science Letters 182, 77–92.

Nesbitt, S.J., Smith, N.D., Irmis, R.B., Turner, A.H., Downs, A., Norell, M.A., 2009. A complete skeleton of a Late Triassic Saurischian and the early evolution of dinosaurs. Science 326, 1530–1533.

Newell, N.D., 1994. Is there a precise Permian-Triassic boundary? Permophiles 24, 46–48.

Newton, R.J., Pevitt, E.L., Wignall, P.B., Bottrell, S.H., 2004. Large shifts in the isotopic composition of seawater sulphate across the Permo-Triassic boundary in northern Italy. Earth and Planetary Science Letters 218, 331–345.

Nicora, A., Balini, M., Bellanea, A., Bowring, S.A., Di Stefano, P., Dumitrica, P., Guaiumi, C., Gullo, M., Hungerbuehler, A., Levera, M., Mazza, M., McRoberts, C.A., Muttoni, G., Preto, N., Rigo, M., 2007. The Carnian/Norian boundary interval at Pizzo Mondello (Sicani Mountains, Sicily) and its bearing for the definition of the GSSP of the Norian Stage. Albertiana 36, 102–129.

O'Dogherty, L., Carter, E.S., Gorican, S., Dumitrica, P., 2010. Triassic radiolarian biostratigraphy. In: Lucas, S.G. (Ed.), The Triassic Timescale. Geological Society, Special Publication, 334, pp. 163–200.

Ogg, J.G., 2004. The Triassic Period. In: Gradstein, F.M., Ogg, J.G., Smith, A.L. (Eds.), A Geologic Time Scale 2004. Cambridge University Press, Cambridge, pp. 271–306.

Ogg, J.G., Steiner, M.B., 1991. Early Triassic magnetic polarity time scale – integration of magnetostratigraphy, ammonite zonation and sequence stratigraphy from stratotype sections (Canadian Arctic Archipelago). Earth and Planetary Science Letters 107, 69–89.

Ogg, J.G., Ogg, G.M., Gradstein, F.M., 2008. Triassic Period. In: Ogg, J.G., Ogg, G.M., Gradstein, F.M. (Eds.), Concise Geologic Time Scale. Cambridge University Press, Cambridge, pp. 95–106.

Olsen, P.E., Kent, D.V., Cornet, B., Witte, W.K., Schlische, R.W., 1996. High-resolution stratigraphy of the Newark rift basin (early Mesozoic, eastern North America). Geological Society of America Bulletin 108, 40–77.

Orbell, G., 1983. Palynology of the British Rhaetian. Bulletin of the Geological Survey of Great Britain 44, 1–44.

Orchard, M.J., 2007a. Conodont diversity and evolution through the latest Permian and Early Triassic upheavals. Palaeogeography, Palaeoclimatology, Palaeoecology 252, 93–117.

Orchard, M.J., 2007b. A proposed Carnian-Norian Boundary GSSP at Black Bear Ridge, northeast British Columbia, and a new conodont framework for the boundary interval. Albertiana 36, 130–141.

Orchard, M.J., 2010. Triassic conodonts and their role in stage boundary definitions. In: Lucas, S.G. (Ed.), The Triassic Timescale. Geological Society, Special Publication, 334, pp. 139–161.

Orchard, M.J., Tozer, E.T., 1997. Triassic conodont biochronology, its intercalibration with the ammonoid standard, and a biostratigraphic summary for the Western Canada Sedimentary Basin. Bulletin of Canadian Petroleum Geology 45, 675–692.

Orchard, M.J., Carter, E.S., Tozer, E.T., 2000. Fossil data and their bearing on defining a Carnian-Norian (upper Triassic) boundary in western Canada. Albertiana 24, 43–50.

Orchard, M.J., Zonneveld, J.P., Johns, M.J., McRoberts, C.A., Sandy, M.R., Tozer, E.T., Carrelli, G.G., 2001. Fossil succession and sequence stratigraphy of the Upper Triassic of Black Bear Ridge, northeast British Columbia, a GSSP prospect for the Carnian-Norian boundary. Albertiana 25, 10–22.

Ovtcharova, M., Bucher, H., Goudemand, N., Schaltegger, U., Brayard, A., Galfetti, T., 2010. New U/Pb ages from Nanpanjiang Basin (South China): Implications for the age and definition of the Early-Middle Triassic boundary. Geophysical Research Abstracts 12, EGU2010-12505-3.

Pálfy, J., Vörös, A., 1998. Quantitative ammonoid biochronological assessment of the Anisian-Ladinian (Middle Triassic) stage boundary proposals. Albertiana 21, 19–26.

Pálfy, J., Smith, P.L., Mortensen, J.K., 2000a. A U-Pb and $^{40}Ar/^{39}Ar$ time scale for the Jurassic. Canadian Journal of Earth Sciences 37, 923–944.

Pálfy, J., Mortensen, J.K., Carter, E.S., Smith, P.L., Friedman, R.M., Tipper, H.W., 2000b. Timing the end-Triassic mass extinction: First on land, then in the sea? Geology 28, 39–42.

Pálfy, J., Demeny, A., Haas, J., Hetenyi, M., Orchard, M.J., Veto, I., 2001. Carbon isotope anomaly and other geochemical changes at the Triassic-Jurassic boundary from a marine section in Hungary. Geology 29, 1047–1050.

Pálfy, J., Parrish, R.R., David, K., Voros, A., 2003. Mid-Triassic integrated U/Pb geochronology and ammonoid biochronology from the Balaton Highland (Hungary). Journal of the Geological Society 160, 271–284.

Palmer, A.R., 1983. Geologic Time Scale, Decade of North American Geology (DNAG). Geological Society of America, Boulder.

Parrish, R.R., McNicoll, V.J., 1992. U/Pb age determinations from the southern Vancouver Island area, British Columbia. In: Radiogenic Age and Isotopic Studies: Geological Survey of Canada, Paper, 91-2, 79–86.

Paull, R.K., 1997. Observations on the Induan-Olenekian boundary based on conodont biostratigraphic studies in the Cordillera of the western United States. Albertiana 20, 31–32.

Payne, J.L., Kump, L.R., 2007. Evidence for recurrent Early Triassic massive volcanism from quantitative interpretation of carbon isotope fluctuations. Earth and Planetary Science Letters 256, 264–277.

Payne, J.L., Lehrmann, D.J., Wei, J., Orchard, M.J., Schrag, D.P., Knoll, A.H., 2004. Large perturbations of the carbon cycle during recovery from the end-Permian extinction. Science 305, 506–509.

Payne, J.L., Lehrmann, D.J., Follet, D., Seibel, M., Kump, L.R., Riccardi, A., Altiner, D., Sano, H., Wei, J.-Y., 2009. Erosional truncation of uppermost Permian shallow-marine carbonates and implications for Permian-Triassic boundary events: Reply. Geological Society of America Bulletin 121, 957–959.

Peybernes, B., 1998. Larger benthic foraminifer. Column for Triassic chart of Mesozoic and Cenozoic sequence chronostratigraphic framework of European basins, by Hardenbol, J., Thierry, J., Farley, M.B., Jacquin, Th., de Graciansky, P.-C., and Vail, P.R. (coordinators). In: de Graciansky, P.-C., Hardenbol, J., Jacquin, Th., Vail, P.R. (Eds.), Mesozoic-Cenozoic Sequence Stratigraphy of European Basins. SEPM Special Publication, 60 Chart 8.

Phillips, J., 1840. Palaeozoic Series. In: Long, G. (Ed.), The Penny Cyclopaedia of the Society for the Diffusion of Useful Knowledge, vol. 17. Charles Knight, London, pp. 153–154.

Phillips, J., 1841. Figures and Descriptions of the Palaeozoic Fossils of Cornwall, Devon and east Somerset. Longman, Brown, Green and Longmans, London.

Pia, J., 1930. Grundbegriffe der Stratigraphie mit ausführlicher anwendung auf die Europäische Mitteltrias. Deuticke, Leipzig and Wien.

Preto, N., Kustatscher, E., Wignall, P.B., 2010. Triassic climates — state of the art and perspectives. In: Kustatscher, E., Preto, N., Wignall, P. (Eds.), Triassic Climates. Palaeogeography, Palaeoclimatology, Palaeoecology, 290, pp. 1–10.

Ramezani, J., Bowring, S.A., Pringle, M.S., Winslow III, F.D., Rasbury, E.T., 2005. The Manicouagan impact melt rock: A proposed standard for the intercalibration of U-Pb and ^{40}Ar/^{39}Ar isotopic systems. Geochimica et Cosmochimica Acta Suppliment 69(10), Supplement 1, Goldschmidt Conference Abstracts 2005, p. A321.

Ramezani, J., Bowring, S.A., Martin, M.W., Lehrmann, D.J., Montgomery, P., Enos, P., Payne, J.L., Orchard, M.J., Wang, H.M., Wei, J., 2007. Reply: Timing of recovery from the end-Permian extinction: Geochronologic and biostratigraphic constraints from south China: Comment. Geology Online Forum, e137.

Ramezani, J., Bowring, S.A., Fastovsky, D.E., Hoke, G.D., 2009. U-Pb-ID-TIMS geochronology of the Late Triassic Chinle Formation, Petrified Forest National Park, Arizona. Geological Society of America Abstracts with Programs 41 (7), 421 [plus copy of summary slide sent to H. Kozur].

Ramezani, J., Bowring, S.A., Fastovsky, D.E., Hoke, G.D., 2010. Depositional history of the Late Triassic Chinle fluvial system at the Petrified Forest National Park: U-Pb geochronology, regional correlation and insights into early dinosaur evolution. American Geophysical Union Fall Meeting 2010, abstract #V31A-2313.

Rampino, M.R., Prokoph, A., Adler, A.C., 2000. Tempo of the end-Permian event: High-resolution cyclostratigraphy at the Permian-Triassic Boundary. Geology 28, 643–646.

Rampino, M.R., Prokoph, A., Adler, A.C., Schwindt, D.M., 2002. Abruptness of the end-Permian mass extinction as determined from biostratigraphic and cyclostratigraphic analysis of European western Tethyan sections. In: Koeberl, C., MacLeod, K.G. (Eds.), Catastrophic Events and Mass Extinctions: Impacts and Beyond. Geological Society of America Special Paper, 356, pp. 415–427.

Reichow, M.K., Saunders, A.D., Ivanov, A.V., Puchkov, V.N., 2004. The Siberian large igneous province. Internet article from Large Igneous Provinces Commission. Large Igneous Province of the Month. Available at: www.largeigneousprovinces.org/ March 2004.

Reichow, M.K., Pringle, M.S., Al'Mukhamedov, A.I., Allen, M.B., Andreichev, V.L., Buslov, M.M., Davies, C.E., Fedoseev, G.S., Fitton, J.G., Inger, S., Medvedev, A.Ya., Mitchell, C., Puchkov, V.N., Safonova, I.Yu., Scott, R.A., Sauders, A.D., 2009. The timing and extent of the eruption of the Siberian Traps large igneous province: Implications for the end-Permian environmental crisis. Earth and Planetary Science Letters 27, 9–20.

Renne, P.R., Basu, A.R., 1991. Rapid eruption of the Siberian traps flood basalts at the Permo-Triassic boundary. Science 253, 176–179.

Renne, P.R., Zichao, Zhang, Richards, M.A., Black, M.T., Basu, A.R., 1995. Synchrony and causal relations between Permian-Triassic boundary crisis and Siberian flood volcanism. Science 269, 1413–1416.

Renne, P.R., Mundil, M., Balco, G., Min, K., Ludwig, K.R., 2010. Joint determination of ^{40}K decay constants and ^{40}Ar*/^{40}K for the Fish Canyon sanidine standard, and improved accuracy for ^{40}Ar/^{39}Ar geochronology. Geochimica et Cosmochimica Acta 74, 5349–5367.

Rigo, M., Preto, N., Roghi, G., Tateo, F., Mietto, P., 2007. A rise in the carbonate compensation depth of western Tethys in the Carnian (Late Triassic): Deep-water evidence for the Carnian pluvial event. Palaeogeography, Palaeoclimatology, Palaeoecology 246, 188–205.

Rogers, R.R., Swisher III, C.C., Sereno, P.C., Monetta, A.M., Forster, C.A., Martinez, R.N., 1993. The Ischigualasto tetrapod assemblage (Late Triassic, Argentina) and ^{40}Ar/^{39}Ar dating of dinosaur origins. Science 260, 794–797.

Roghi, G., Gianolla, P., Minarelli, L., Pilati, C., Preto, N., 2010. Palynological correlation of Carnian humid pulses throughout western Tethys. In: Kustatscher, E., Preto, N., Wignall, P. (Eds.), Triassic Climates. Palaeogeography, Palaeoclimatology, Palaeoecology, 290, pp. 89–106.

Röhling, H.-G., 1991. A lithostratigraphic subdivision of the Lower Triassic in the northwest German Lowland and the German Sector of the North Sea, based on gamma-ray and sonic logs. Geologisches Jahrbuch A 119, 3–24.

Ruhl, M., Kuerschner, W.M., Krystyn, L., 2009. Triassic-Jurassic organic carbon isotope stratigraphy of key sections in the western Tethys realm (Austria). Earth Planetary Science Letters 281, 169–187.

Satterley, A.K., 1996. The interpretation of cyclic successions of the Middle and Upper Triassic of the Northern and Southern Alps. Earth-Science Reviews 40, 181–207.

Schaltegger, U., Guex, J., Bartolini, A., Schoene, B., Ovtcharova, M., 2008. Precise U-Pb age constraints for end-Triassic mass extinction, its correlation to volcanism and Hettangian post-extinction recovery. Earth and Planetary Science Letters 267, 266–275.

Schlager, W., Schöllnberger, W., 1974. Das Prinzip stratigraphischer Wenden in der Schichtenfolge der Nördlichen Kalkalpen. Mitteilungen der Österreichischen Geologischen Gesellschaft, Wien 66/67, 165–193.

Schoene, B., Guex, J., Bartolini, A., Schaltegger, U., Blackburn, T.J., 2010. Correlating the end-Triassic mass extinction and flood basalt volcanism at the 100 ka level. Geology 38, 387–390, and 13 page supplement.

Schuurman, W.M.N., 1979. Aspects of Late Triassic palynology. 3. Palynology of the latest Triassic and earliest Jurassic deposits of the Northern Limestone Alps in Austria and southern Germany, with special reference to a palynological characterization of the Rhaetian stage in Europe. Review of Palaeobotany and Palynology 27, 53–75.

Schwarzacher, W., 2005. The stratification and cyclicity of the Dachstein Limestone in Lofer, Leogang and Steinernes Meer (Northern Calcareous Alps, Austria). Sedimentary Geology 181, 93–106.

Sephton, M.A., Looy, C.V., Veefkind, R.J., Brinkhuis, H., De Leeuw, J.W., Visscher, H., 2002a. Synchronous record of $\partial^{13}C$ shifts in the oceans and atmosphere at the end of the Permian. In: Koeberl, C., MacLeod, K.G. (Eds.), Catastrophic Events and Mass Extinctions: Impacts and Beyond. Geological Society of America Special Paper, 356, pp. 455–462.

Sephton, M.A., Amor, K., Franchi, I.A., Wignall, P.B., Newton, R., Zonneveld, J.-P., 2002b. Carbon and nitrogen isotope disturbances and an end-Norian (Late Triassic) extinction event. Geology 30, 1119–1122.

Shen, S.-Z., Henderson, C.M., Bowring, S.A., Cau, C.-Q., Wang, Y., Wang, W., Zhang, H., Zhang, Y.-C., Mu, L., 2010. High-resolution Lopingian (Late Permian) timescale of South China. Geological Journal 45, 122–134.

Shen, S.Z., Crowley, J.L., Wang, Y., Bowring, S.A., Erwin, D.H., Sadler, P.M., Cao, C.Q., Rothman, D.H., Henderson, C.M., Ramezani, J., Zhang, H., Shen, Y., Wang, X.-D., Wang, W., Mu, L., Li, W.-Z., Tang, Y.G., Liu, X.L., Liu, L.J., Zeng, Y., Jiang, Y.F., Jin, Y.G., 2011. Calibrating the end-Permian mass extinction. Science 334, 1367–1372.

Silberling, N.J., Tozer, E.T., 1968. Biostratigraphic classification of the marine Triassic in North America. Geological Society of America Special Paper 10, 1–63.

Simmons, M.D., Sharland, P.R., Casey, D.M., Davies, R.B., Sutcliffe, O.E., 2007. Arabian Plate sequence stratigraphy: Potential implications for global chronostratigraphy. GeoArabia 12 (4), 101–130.

Simms, M.J., Ruffel, A.H., 1989. Synchroneity of climate change and extinctions in the Late Triassic. Geology 17, 265–268.

Skjold, L.J., Van Veen, P.M., Kristensen, S.-E., Rasmussen, A.R., 1998. Triassic sequence stratigraphy of the southwestern Barents Sea. In: de Graciansky, P.-C., Hardenbol, J., Jacquin, Th., Vail, P.R. (Eds.), Mesozoic-Cenozoic Sequence Stratigraphy of European Basins. SEPM Special Publication, 60, pp. 651–666.

Stefani, M., Furin, S., Gianolla, P., 2010. The changing climate framework and depositional dynamics of Triassic carbonate platforms from the Dolomites. In: Kustatscher, E., Preto, N., Wignall, P. (Eds.), Triassic Climates. Palaeogeography, Palaeoclimatology, Palaeoecology, 290, pp. 43–57.

Steiner, M.B., Ogg, J.G., Zhang, Z., Sun, S., 1989. The Late Permian/Early Triassic magnetic polarity time scale and plate motions of south China. Journal of Geophysical Research 94, 7343–7363.

Steiner, M.B., 2006. The magnetic polarity time scale across the Permian–Triassic boundary. In: Lucas, S.G., Cassinis, G., Schneider, J.W. (Eds.), Non-Marine Permian Biostratigraphy and Biochronology. Geological Society Special Publications, 265, pp. 15–38.

Sugiyama, K., 1997. Triassic and Lower Jurassic radiolarian biostratigraphy in the siliceous claystone and bedded chert units of the southeastern Mino Terrane, Central Japan. Bulletin of the Mizunami Fossil Museum 24, 79–193.

Sun, Z., Hounslow, M.W., Pei, J., Zhao, L., Tong, J., Ogg, J.G., 2007. Magnetostratigraphy of the West Pingdingshan section, Chaohu, Anhui Province: Relevance for base Olenekian GSSP selection. Albertiana 36, 22–32.

Szurlies, M., 2004. Magnetostratigraphy: the key to global correlation of the classic Germanic Trias – case study Volpriehausen Formation (Middle Buntsandstein), Central Germany. Earth and Planetary Science Letters 227, 395–410.

Szurlies, M., 2007. Latest Permian to Middle Triassic cyclo-magnetostratigraphy from the Central European Basin, Germany: Implications for the geomagnetic polarity timescale. Earth and Planetary Science Letters 261, 602–619.

Tanner, L.H., 2010a. The Triassic isotope record. In: Lucas, S.G. (Ed.), The Triassic Timescale. Geological Society, Special Publication, 334, pp. 103–118.

Tanner, L.H., 2010b. Cyclostratigraphy record of the Triassic: A critical examination. In: Lucas, S.G. (Ed.), The Triassic Timescale. Geological Society, Special Publication, 334, pp. 119–137.

Tong, J., Zakharov, Y.D., Orchard, M.J., Yin, H., Hansen, H.J., 2004. Proposal of the Chaohu section as the GSSP candidate of the I/O boundary. Albertiana 29, 13–28.

Tozer, E.T., 1965. Lower Triassic stages and ammonoid zones of Arctic Canada. Geological Survey of Canada Paper 65–12, 1–14.

Tozer, E.T., 1967. A standard for Triassic time. Geological Survey of Canada Bulletin 156, 104.

Tozer, E.T., 1984. The Trias and its Ammonites: The evolution of a time scale. Geological Survey of Canada Miscellaneous Report 35, 171.

Tozer, E.T., 1986. Definition of the Permian-Triassic (P-T) boundary: The question of the age of the *Otoceras* beds. Memorie della Societa Geologica Italiana 34, 291–301.

Tozer, E.T., 1990. How many Rhaetians? Albertiana 8, 10–13.

Tozer, E.T., 1994. Age and correlation of the *Otoceras* beds at the Permian-Triassic Boundary. Albertiana 14, 31–37.

Traverse, A., 2007. Paleopalynology, second ed. Springer, Dordrecht. p. 813.

Van Veen, P.M., 1995. Time calibration of Triassic/Jurassic microfloral turnover, eastern North America – Comment. Tectonophysics 245, 93–95.

Van Veen, P., Hochuli, P.A., Vigran, J.O., 1998. Arctic spores/pollen. Column for Triassic chart of Mesozoic and Cenozoic sequence chronostratigraphic framework of European basins, by Hardenbol, J., Thierry, J., Farley, M.B., Jacquin, Th., de Graciansky, P.-C., and Vail, P.R. (coordinators). In: de Graciansky, P.-C., Hardenbol, J., Jacquin, Th., Vail, P.R. (Eds.), Mesozoic-Cenozoic Sequence Stratigraphy of European Basins. SEPM Special Publication, 60 Chart 8.

Veizer, J., Ala, D., Azmy, K., Bruckschen, P., Buhl, D., Bruhn, F., Carden, G.A.F., Diener, A., Ebneth, S., Godderis, Y., Jasper, T., Korte, C., Pawellek, F., Podlaha, O.G., Strauss, H., 1999. $^{87}Sr/^{86}Sr$, $\partial^{13}C$ and $\partial^{18}O$ evolution of Phanerozoic seawater. Chemical Geology 161, 59–88. Datasets available at: http://www.science.uottawa.ca/geology/isotope_data.

Veizer, J., Godderis, Y., François, L.M., 2000. Evidence for decoupling of atmospheric CO_2 and global climate during the Phanerozoic eon. Nature 408, 698–701.

Visscher, H., 1992. The new STS Triassic stage nomenclature. Albertiana 10, 1–2.

Visscher, H., Brugman, W.A., 1981. Ranges of selected palynomorphs in the Alpine Triassic of Europe. Review of Palaeobotany and Palynology 34, 115–128.

Visscher, H., Van Houte, M., Brugman, W.A., Poort, R.J., 1994. Rejection of a Carnian (Late Triassic) 'pluvial event' in Europe. Reviews of Palaeobotany and Palynology 83, 217–226.

von Alberti, F.A., 1834. Beitrag zu einer Monographie des Bunter Sandsteins, Muschelkalks und Keupers und die Verbindung dieser Gebilde zu einer Formation. Verlag der J.G, Stuttgart and Tübingen, Cottaíshen Buchhandlung, p. 326.

von Gümbel, C.W., 1861. Geognostische Beschreibung des bayerischen Alpengebirges und seines Vorlands. Perthes, Gotha, p. 950.

von Salis, K., 1998. Calcareous nannofossils. Column for Triassic chart of Mesozoic and Cenozoic sequence chronostratigraphic framework of European basins, by Hardenbol, J., Thierry, J., Farley, M.B., Jacquin, Th., de Graciansky, P.-C., and Vail, P.R. (coordinators). In: de Graciansky, P.-C., Hardenbol, J., Jacquin, Th., Vail, P.R. (Eds.), Mesozoic-Cenozoic Sequence Stratigraphy of European Basins. SEPM Special Publication, 60 Chart 8.

Vörös, A., Szabó, I., Kovács, S., Dosztály, L., Budai, T., 1996. The Felsöörs section: A possible stratotype for the base of the Ladinian Stage. Albertiana 17, 25–40.

Vrielynck, B., 1998. Conodonts. Column for Triassic chart of Mesozoic and Cenozoic sequence chronostratigraphic framework of European basins, by Hardenbol, J., Thierry, J., Farley, M.B., Jacquin, Th., de Graciansky, P.-C., and Vail, P.R. (coordinators). In: de Graciansky, P.-C., Hardenbol, J., Jacquin, Th., Vail, P.R. (Eds.), Mesozoic-Cenozoic Sequence Stratigraphy of European Basins. SEPM Special Publication, 60 Chart 8.

Vuks, V.J., 2000. Triassic foraminifers of the Crimea, Caucasus, Mangyshlak and Pamirs (biostratigraphy and correlation). Zentralblatt für Geologie und Paläontologie I (11–12), 1353–1365.

Vuks, V.J., 2007. New data on the Late Triassic (Norian-Rhaetian) foraminiferans of the western Pre-caucasus (Russia). New Mexico Museum of Natural History and Science Bulletin 41, 411–412.

Waagen, W., Diener, C., 1895. Untere Trias. In: Mojsisovics, E., von Waagen, W., Diener, C. (Eds.), Entwurf einer Gliederung der pelagischen Sedimente des Trias-Systems. Sitzungberichte Akademie Wissenschaften Wien, 104, pp. 1271–1302.

Wallman, K., 2001. The geological carbon cycle and the evolution of marine $\partial^{18}O$ values. Geochimica et Cosmochimica Acta 65, 2469–2485.

Ward, P.D., Haggart, J.W., Carter, E.S., Wilbur, D., Tipper, H.W., Evans, T., 2001. Sudden productivity collapse associated with the Triassic-Jurassic boundary mass extinction. Science 292, 1148–1151.

Warrington, G., 2002. Triassic spores and pollen. In: Jansonius, J., McGregor, D.C. (Eds.), Palynology: Principles and Applications. Second ed. American Association of Stratigraphic Palynologists Foundation, 2, pp. 755–766.

Whiteside, J.H., Olsen, P.E., Kent, D.V., Fowell, S.J., Et-Touhami, E., 2007. Synchrony between the Central Atlantic magmatic province and the Triassic-Jurassic mass-extinction event? Palaeogeography, Palaeoclimatology, Palaeoecology 244, 345–367.

Wignall, P.B., 2001. Large igneous provinces and mass extinctions. Earth-Science Reviews 53, 1–33.

Wignall, P.B., 2007. The End-Permian mass extinctions – how bad did it get? Geobiology 5, 303–309.

Wignall, P.B., Hallam, A., 1992. Anoxia as a cause of the Permo-Triassic mass extinction: Facies evidence from northern Italy and the western United States. Palaeogeography, Palaeoclimatology, Palaeoecology 93, 21–46.

Wing, S.L., Sues, H.-D., 1992. Mesozoic and early Cenozoic terrestrial ecosystems. In: Behrensmeyer, A.K., Damuth, J.D., DiMichele, W.A., Potts, R., Sues, H.-D., Wing, S.L. (Eds.), Terrestrial Ecosystems through Time: Evolutionary Paleoecology of Terrestrial Plants and Animals. University of Chicago Press, Chicago, pp. 324–416.

Xie, S.C., Pancost, R.D., Wang, Y.B., Yang, H., Wignall, P.B., Luo, G.M., Jia, C.L., Chen, L., 2010. Cyanobacterial blooms tied to volcanism during the 5 m.y. Permo-Triassic biotic crisis. Geology 38, 447–450.

Xu, G., Hannah, J.L., Stein, H.J., Bingen, B., Yang, G., Zimmerman, A., Weitschat, W., Mork, A., Weiss, H.M., 2009. Re-Os geochronology of Arctic black shales to evaluate the Anisian-Ladinian boundary and global faunal correlations. Earth and Planetary Science Letters 288, 581–587.

Yin, H., Zhang, K., Tong, J., Yang, Z., Wu, S., 2001. The Global Stratotype Section and Point (GSSP) of the Permian-Triassic boundary. Episodes 24, 102–114.

Yin, H., Tong, J., Zhang, K., 2005. A review of the global stratotype section and point of the Permian-Triassic boundary. Acta Geologica Sinica 79, 715–728.

Yin, H., Feng, Q., Xie, S., Yu, J., He, W., Liang, H., Lai, X., Huang, X., 2007. Recent achievements on the research of the Paleozoic-Mesozoic transitional period in South China. Frontiers in Earth Science of China 1, 129–141.

Zakharov, Y.D., 1994. Proposals on revision of the Siberian standard for the Lower Triassic and candidate stratotype section and point for the Induan-Olenekian boundary. Albertiana 14, 44–51.

Zakharov, Y.D., Shigata, Y., Popov, A.M., Sokarev, A.N., Buryi, G.I., Golozubov, V.V., Panasenko, E.S., Dorukhovskaya, E.A., 2000. The candidates of global stratotype of the boundary of the Induan and Olenekian stages of the Lower Triassic in Southern Primorye. Albertiana 24, 12–26.

Zakharov, Y.D., Shigeta, Y., Popov, A.M., Buryi, G.E., Oleinikov, A.V., Dorukhovskaya, E.A., Mikhalik, T.M., 2002. Triassic ammonoid succession in South Primorye: 1. Lower Olenekian *Hedenstroemia bosphorensis* and *Anasibirites nevolini* Zones. Albertiana 27, 42–64.

Zakharov, Y.D., Shigeta, Y., Igo, H., 2009. Correlation of the Induan-Olenekian boundary beds in the Tethys and Boreal realm: Evidence from conodont and ammonoid fossils. Albertiana 37, 20–27.

Zapfe, H., 1974. Die Stratigraphie der Alpin-Mediterranen Trias. Schriftenreihe der Erdwissenschaftlichen Kommissionen, Österreichische Akademie der Wissenschaften 2, 137–144.

J.G. Ogg and L.A. Hinnov
Contributor: C. Huang

Chapter 26

Jurassic

Abstract: Ammonites exploded in diversity after the end-Triassic near-extinction of their precursor taxa and have been the main tool for global correlation for nearly two centuries. The Pangea super-continent split during mid-Jurassic into Laurasia and Gondwana separated by a Tethys-Atlantic seaway. Late Jurassic organic-rich shales would later provide nearly half of our oil-gas source rocks. Dinosaurs dominated the land surface, and their cousins took to the air.

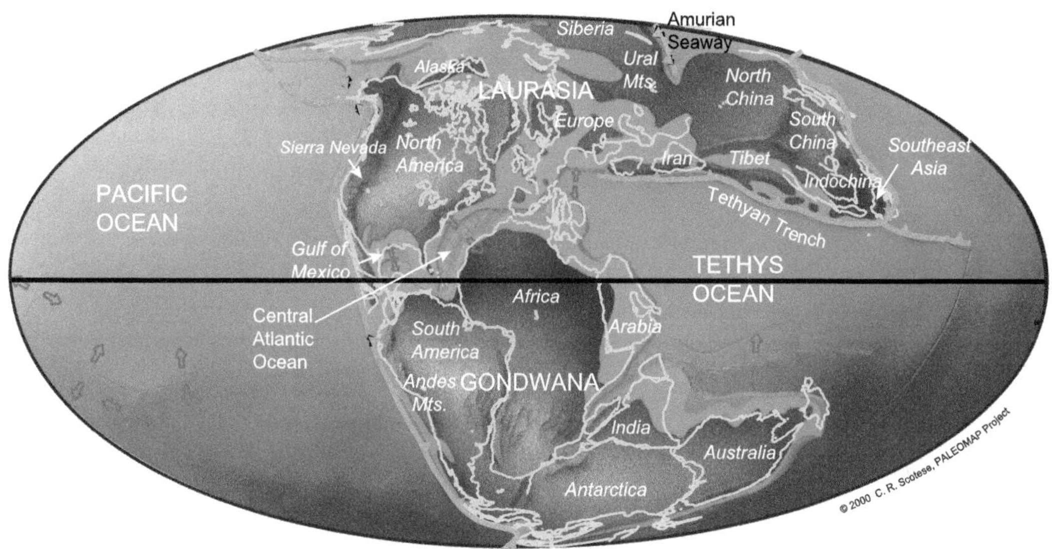

Chapter Outline

The Geologic Time Scale 2012 . DOI: 10.1016/B978-0-444-59425-9.00026-3

26.1. HISTORY AND SUBDIVISIONS

26.1.1. Overview of Jurassic

The immense wealth of fossils, particularly the ammonites with their diversity of ornamented spirals, in the Jurassic strata of Britain and northwest Europe were a magnet for innovative geologists. Modern concepts of biostratigraphy, chronostratigraphy, correlation, sequence stratigraphy and paleogeography grew out of their studies. Among recent excellent reviews of Jurassic stratigraphy and environments are those by Hallam (1975), Page (2005) and a special Jurassic thematic issue of *Proceedings of the Geologists' Association* (2008, Vol. 119, Issue 1).

The term "Jura Kalkstein" was applied by Alexander von Humboldt (1799) to a series of carbonate shelf deposits exposed in the mountainous Jura region of northernmost Switzerland. Even though he recognized that these strata were distinct from the German Muschelkalk (middle Triassic), he erroneously considered his unit to be older. Alexander Brongniart (1829) coined the term "Terrains Jurassiques" when correlating this "Jura Kalkstein" to the Lower Oolite Series (now assigned to Middle Jurassic) of the British succession. Leopold von Buch established a three-fold subdivision for the Jurassic. The basic framework of von Buch has been retained as the three Jurassic series, although the nomenclature has evolved (Black–Brown–White, Lias–Dogger–Malm, and currently Lower–Middle–Upper).

Alcide d'Orbigny (1842–1851, 1852) grouped the Jurassic ammonite and other fossil assemblages of France and England into 10 main divisions, which he termed "étages" (stages). Seven of d'Orbigny's stages are used today, but none of them has retained its original meaning. Simultaneously, Quenstedt (1858) subdivided each of the three Jurassic series of von Buch of the Swabian Alb of southwestern Germany into six lithostratigraphic subdivisions characterized by ammonites and other fossils (Geyer and Gwinner, 1979). Quenstedt had denoted his subdivisions by Greek letters (*alpha* to *zeta*). Alfred Oppel, his pupil, was the first to successfully correlate Jurassic units among England, France and southwestern Germany. Oppel (1856–1858) modified d'Orbigny's stage framework and further subdivided the Jurassic into biostratigraphic zones. His basic philosophy of defining the bases of Jurassic stages and substages according to ammonite zones in marginal-marine sections was retained at the Colloque du Jurassique à Luxembourg 1962 (Maubeuge, 1964; see also Morton, 1974) during which the majority of the current suite of eleven Jurassic stages were formalized.

Establishing the precise limits of those stages with GSSPs has been more difficult and is not yet complete. The traditional subdivision of the Jurassic was mainly based on shallow-marine deposits of the northwest European basin (England to southwest Germany). Hiatuses or other complications commonly punctuate these "stratotypes" and their boundaries, and many of the ammonite taxa represent only certain paleoclimate realms in the Jurassic paleogeography. In particular, a pronounced provinciality during the Late Jurassic has inhibited efforts to assign GSSPs. Establishing reliable high-resolution correlations to tropical (Tethyan), Pacific, deep-sea, terrestrial and other settings has commonly remained tenuous. Nevertheless, these European basins contain all of the ratified GSSP sites and nearly all of the proposed ones for the chronostratigraphic framework of the Jurassic (Figure 26.1).

Detailed reviews of the history, subdivisions, biostratigraphic zonations, and correlation of individual Jurassic stages are compiled in several sources, including Arkell

System	Series	Stage	Boundary Horizons (GSSPs)
Creta-ceous	Lower	Berriasian	
Jurassic	Upper	Tithonian	
		Kimmeridgian	
		Oxfordian	
	Middle	Callovian	
		Bathonian	FAD of *Gonolkites convergens*
		Bajocian	FAD of *Hyperlioceras mundum*
		Aalenian	FAD of *Leioceras opalinum*
	Lower	Toarcian	
		Pliensbachian	FAD of *Bifericeras donovani*
		Sinemurian	FAD of *Vermiceras quantoxense*
		Hettangian	FAD of *Psiloceras spelae*
Triassic	Upper	Rhaetian	

FIGURE 26.1 **Summary of Jurassic stage nomenclature and status.** Stages with ratified GSSPs are indicated by the first-appearance datum (FAD) of the main correlation marker, which are ammonite taxa.

(1933, 1956), Morton (1974, 2008), Cope *et al.* (1980a,b), Harland *et al.* (1982, 1990), Krymholts *et al.* (1982), Burger (1995) and Groupe Français d'Étude du Jurassique (1997). An extensive on-line library of Jurassic stratigraphy articles with an emphasis on Eurasia is maintained at *www.jurassic.ru/epubl.htm*. The International Commission on Jurassic Stratigraphy website is *http://jurassic.earth.ox.ac.uk/*.

26.1.2. Subdivisions of the Lower Jurassic

A marine transgression in northwest Europe during the latest Triassic and Early Jurassic deposited widespread clay-rich calcareous deposits. These distinctive strata in southwest Germany were called the Black Jurassic ("*Schwarzer Jura*") by von Buch (1839), and named "Lias" in southern England by Conybeare and Phillips (1822). This series was subdivided into three stages (Sinemurian, Liasian, and Toarcian) by d'Orbigny (1842–1851, 1852). Oppel (1856–1858) replaced the Liasian with a Pliensbachian Stage, and Renevier (1864) separated the lower Sinemurian as a distinct Hettangian Stage. Widespread hiatuses or condensation horizons mark the bases of the classical Sinemurian, Pliensbachian and Toarcian stages.

26.1.2.1. Triassic–Jurassic Boundary

The original Sinemurian Stage of d'Orbigny (1842–51, 1852) extended to the base of the Jurassic. Indeed, the original extent of the Lower Jurassic tentatively included beds that are now assigned to the Rhaetian of uppermost Triassic (Bonebed of southwest Germany, portions of Penarth Beds in England, Rhätische Gruppe of German and Austrian Alb, etc.). Oppel (1856–1858) redefined the base of the Jurassic as the lowest ammonite assemblage characterized by *Psiloceras planorbis* species that overlay the Bonebed. He referred to characteristic coastal sections in southern England including Lyme Regis in Dorset and Watchet in Somerset.

The end-Triassic mass extinction terminated many groups of marine life, including the conodonts, whose distinctive phosphatic jaw elements constitute a primary zonation for much of the Paleozoic and Triassic, and the majority of ammonoids. Indeed, in the few regions with continuous deposition there is an interval devoid of either typical latest-Triassic taxa (e.g., conodonts or *Choristoceras* ammonoids) or earliest-Jurassic forms (e.g., *Psiloceras* ammonites). A sea-level fall produced extended gaps in many shallow-marine sections; therefore, the boundary between upper Triassic and the overlying lower Jurassic was commonly a sequence boundary and hiatus (Hesselbo and Jenkyns, 1998; Hallam and Wignall, 1999). Ammonite diversity was very low in late Rhaetian time (*Choristoceras marshi* zone), and the Hettangian genus *Psiloceras* must be derived from the Triassic genera of the family Discophyllitidae, which lived mainly in the open sea (Hillebrandt, 1997). The informal base of the Jurassic (Hettangian Stage) in southern England was the initial influx of these *Psiloceras* ammonites (*Psiloceras planorbis* species) during the early stages of the transgression; but this local migration event was deemed unsuitable for an international definition. However, species within the genus *Psiloceras* are ubiquitous from the eastern Pacific and Tethys to the European Boreal province; therefore, one option was to utilize the earliest forms for a base-Jurassic definition. However, the earliest form, *Psiloceras spelae*, first occurs during a global sea-level fall that caused an extended gap in shallow seas, therefore this taxa is only found in rare complete sections (e.g., Peru, Alps, Nevada) (Bloos, 2008; Hillebrandt and Krystyn, 2009).

Other non-ammonoid possibilities that were considered for defining a precise base-Jurassic level that could be globally correlated included a major turnover of siliceous radiolarian microfossils or a negative excursion in carbon isotopes. After a major international effort to correlate environmental and biostratigraphic events associated with the end-Triassic extinctions and the extensive eruption of the Central Atlantic Magmatic Province at ~201 Ma (e.g., review by Hesselbo *et al.*, 2007a), it was decided to utilize the earliest forms of *Psiloceras* ammonites. The base of the Jurassic is thus placed at the initial stages of biological recovery from the end-Triassic extinction. High-precision U-Pb radio-isotopic dating demonstrated a coincidence between the onset of the Central Atlantic Magmatic Province eruptions and the end-Triassic mass extinctions; and a delay of only about 100 kyr until the recovery at the beginning of the Jurassic (e.g., Schoene *et al.*, 2010).

The GSSP for the base of the Jurassic (base of Hettangian Stage) was ratified in 2010 as the Kuhjoch section within the Northern Calcareous Alps of Austria (Figure 26.2). The GSSP level, 5.80 m above the base of the Tiefengraben Member of the Kendelbach Formation, corresponds to the local lowest occurrence of the ammonite *Psiloceras spelae* (new subsp. *tirolicum* Hillebrandt and Krystyn). Other markers include the lowest occurrences of the widely distributed continental palynomorph *Cerebropollenites thiergartii* (Kuerschner *et al.*, 2007), of the aragonitic foraminifer *Praegubkinella turgescens* and of the ostracod *Cytherelloidea buisensis* (Hillebrandt *et al.*, 2008). The $\delta^{13}C_{org}$ record shows an initial negative excursion near the boundary between the underlying Koessen and the Kendelbach formations, and a shift to more positive $\delta^{13}C_{org}$ at the GSSP level; and this carbon isotope signature provides a primary method for high-resolution correlation to other sections (e.g., Ruhl *et al.*, 2009; Deenen *et al.*, 2010). This Austrian section has some problems, including limited exposed stratigraphic extent, ammonite species are sparse and ranges are difficult to delimit, lack of magnetostratigraphy zones, no identified cyclic sediments to provide estimates of elapsed durations among events, and a possible tectonic break (thrust fault) about 4 m below the GSSP that may have truncated part of the palynology record (e.g., Bonis *et al.*, 2009; Warrington, 2010). Nevertheless, compared to other candidates, the Kuhjoch section appeared to represent the best available uppermost Triassic and lowermost Jurassic

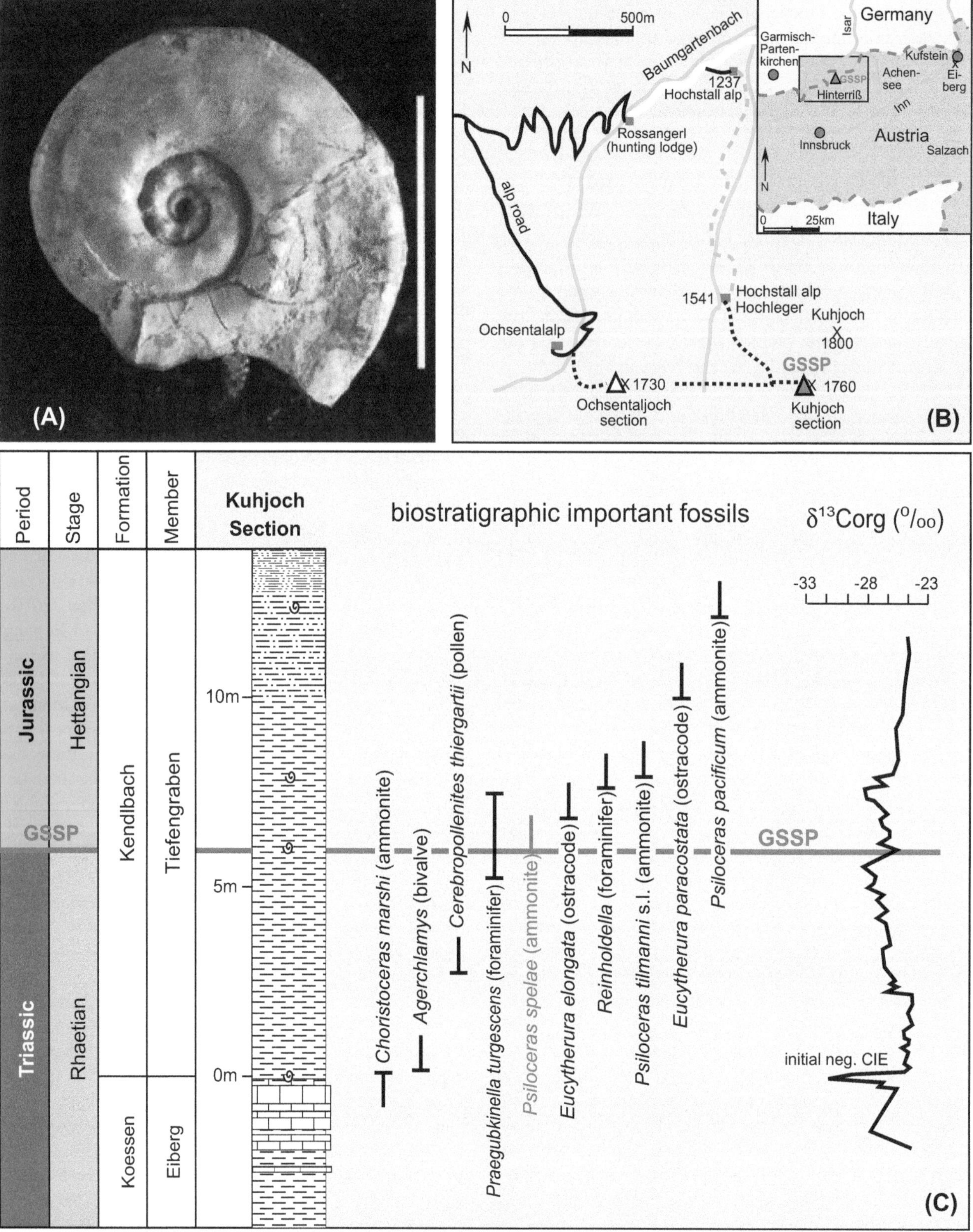

FIGURE 26.2 **GSSP for base of Jurassic (base of Hettangian Stage) at Kuhjoch section, Northern Calcareous Alps, Austria.** The GSSP level coincides with the lowest occurrence of ammonite *Psiloceras spelae* (new subsp. *tirolicum* Hillebrandt and Krystyn), marking the beginning of a new marine ecosystem after the major end-Triassic extinctions. The brief negative-excursion in carbon isotopes is approximately simultaneous with those end-Triassic extinctions and main eruption phase of the Central Atlantic Magmatic Province. Photo provided by Axel von Hillebrandt.

succession with multiple biostratigraphy and carbon isotope signatures.

26.1.2.2. Hettangian

Renevier (1864) proposed the Hettangian Stage to encompass the *Psiloceras planorbis* and *Schlotheimia angulata* ammonite zones as interpreted by Oppel. The stage was named after a quarry near Hettange-Grande village in Lorraine (northeastern France), 22 kilometers south of Luxembourg, although the strata in this locality are primarily sandstone with no fossils in the lower part.

The assignment of the base-Jurassic GSSP is only slightly older than the informal base of the Hettangian Stage placed at the initial entry of *Psiloceras planorbis* ammonites into southern England. Cyclostratigraphy of the main Hettangian reference sections in Britain indicate that this downward extension, as measured relative to carbon isotope signatures near the Triassic–Jurassic boundary, is on the order of 100 kyr (Ruhl *et al.*, 2010). This cyclostratigraphy indicates that the Hettangian spans about 2 myr in the British sections; a duration consistent with radio-isotopic dating constraints from Peru (Schaltegger *et al.*, 2008; Schoene *et al.*, 2010).

There are no accepted groupings into substages of the three Hettangian ammonite zones (*Psiloceras planorbis*, *Alsatites liasicus* and *Schlotheimia angulata*).

26.1.2.3. Sinemurian

History, Definition and Boundary Stratotype

The Sinemurian Stage was named by d'Orbigny (1842–51, 1852) after the town of Semur-en-Auxois (*Sinemurum Briennense castrum* in Latin) in the Cote d'Or department in the Burgundy region of east-central France. After the establishment of the Hettangian Stage incorporated its former lower ammonite zones (Renevier, 1864), the base of the revised Sinemurian was assigned to the proliferation of the Arietitidae ammonite group, particularly the lowest occurrence of the early genera *Vermiceras* and *Metophioceras* (base of *Metophioceras conybeari* Subzone of the *Arietites bucklandi* Zone). However, that informal stage boundary was not defined by a generally accepted species or assemblage (Sinemurian Boundary Working Group, 2000), and a stratigraphic gap exists between the Hettangian and Sinemurian throughout most of northwest Europe.

Sedimentation was continuous across the boundary interval in rapidly subsiding troughs in western Britain. Therefore, the boundary GSSP was placed in interbedded limestone and claystone at a coastal exposure near East Quantoxhead in Kilve, Somerset, England (Page *et al.*, 2000; Sinemurian Boundary Working Group, 2000; Bloos and Page, 2002) (Figure 25.3). The GSSP at 0.9 m above the base of Bed 145 coincides with the lowest occurrence of arietitid ammonite genera *Vermiceras* and *Metophioceras*. This level is near the highest occurrence of the ammonite genus *Schlotheimia* that is characteristic of the uppermost Hettangian. This turnover of ammonite genera is a global event for correlating the boundary interval (Sinemurian Boundary Working Group, 2000; Bloos and Page, 2002).

Upper Sinemurian Substages

The Sinemurian Stage has two substages. The Colloque du Jurassique à Luxembourg 1962 (Maubeuge, 1964) assigned the base of an upper stage, called Lotharingian (named by Haug (1910) after the Lorraine region of France) to the base of the *Caenisites turneri* ammonite zone. However, current usage follows Oppel (1856–1858) in assigning the base of the Lotharingian substage as the base of the overlying *Asteroceras obtusum* Zone (e.g., Krymholts *et al.*, 1982; Groupe Français d'Étude du Jurassique, 1997). The lower substage does not have a secondary name.

26.1.2.4. Pliensbachian

History, Definition and Boundary Stratotype

The Pliensbachian Stage was proposed by Oppel (1856–1858) to replace the Liasian stage of d'Orbigny, which lacked a geographic base. The stage was named after the outcrops along the Pliensbach stream near the village of Pliensbach (Geppingen, 35 km southeast of Stuttgart) in the Baden-Württemberg district of Germany. Even though this section has an unconformity to the underlying Sinemurian, the lowest ammonite subzone in this section (*Phricodoceras taylori* Subzone of the *Uptonia jamesoni* Zone) was used as the base of the Pliensbachian Stage (e.g., Dean *et al.*, 1961; Meister, 1999a).

At this level, the Psiloceratoidea ammonites, which dominated the lower stages of Hettangian and Sinemurian disappear, and the Eoderoceratoidea superfamily diversifies to dominate the NE European fauna of the Pliensbachian Stage (Meister *et al.*, 2003). This faunal event occurs globally. However, a sedimentary or paleontological gap between the Pliensbachian and Sinemurian sequences is a common feature. Of 27 regions considered by the Pliensbachian boundary working group, only a single candidate in North Yorkshire, England, was satisfactory for a potential GSSP (Meister, 1999a,b; Meister *et al.*, 2003, 2006).

At the coastal section of Wine Haven at Robin Hood's Bay, Yorkshire, the GSSP in the "Pyritous Shales Member" is within a claystone bed, 6 cm above a nodule layer (Figure 26.4). This level coincides with the lowest ammonite occurrences of *Bifericeras donovani* species and *Apoderoceras* genera (Meister *et al.*, 2006). Secondary markers include a brief reversed-polarity magnetozone within the uppermost Sinemurian and a sharply defined negative carbon isotope excursion peaking in the lowermost Pliensbachian (Meister *et al.*, 2006; Korte and Hesselbo, 2011). An auxiliary reference section at Aselfingen in the historical area in southwest Germany is in condensed limestone and clay with rare ammonites, but allows calibration of secondary markers of ostracods and dinocysts (Meister, 1999a).

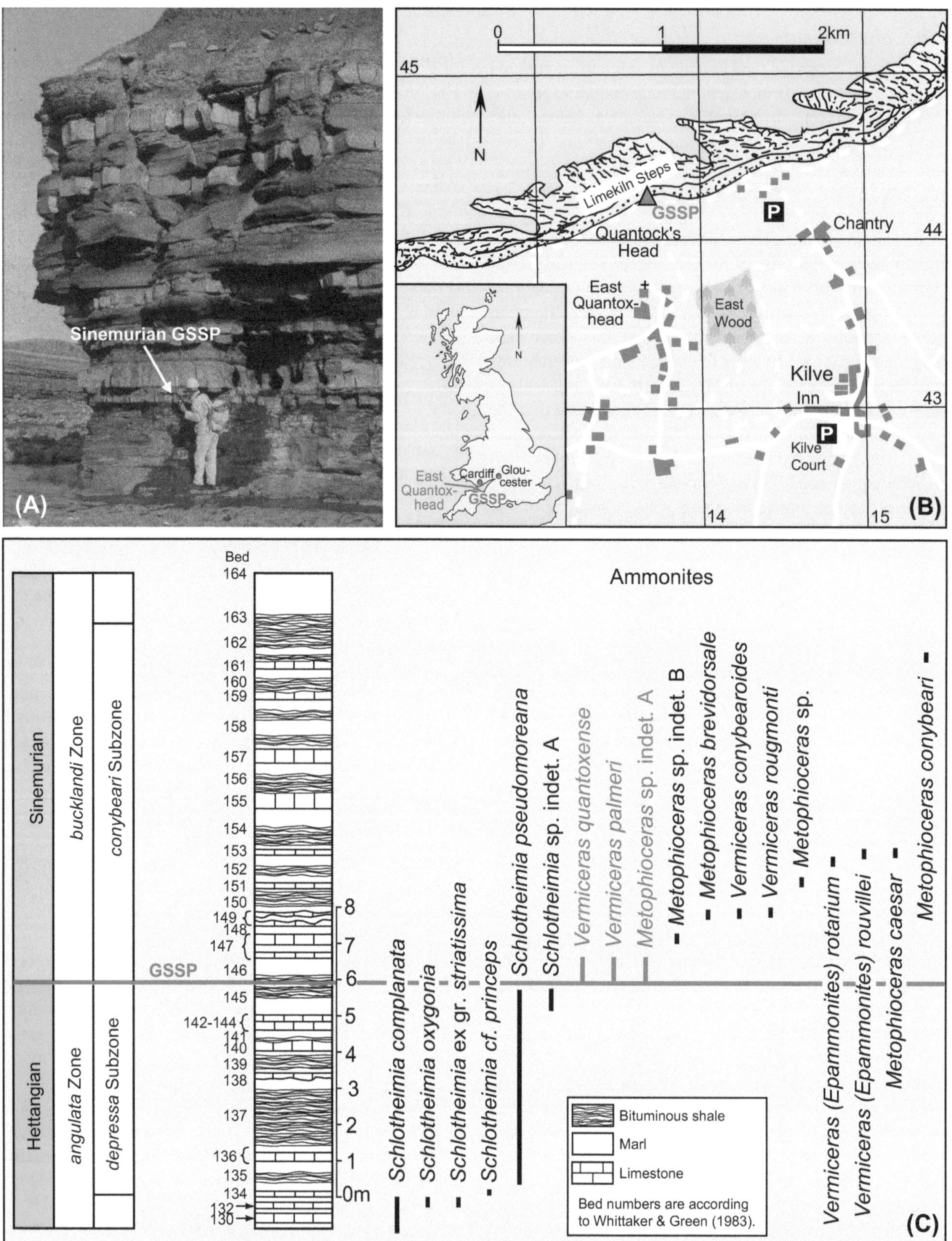

FIGURE 26.3 **GSSP for base of the Sinemurian Stage at East Quantoxhead section, Somerset, southwest England.** The GSSP level coincides with the lowest occurrence of *Vermiceras quantoxense* ammonite (base of *Metophioceras conybeari* Subzone of the *Arietites bucklandi* Zone). *Photo provided by Kevin Page.*

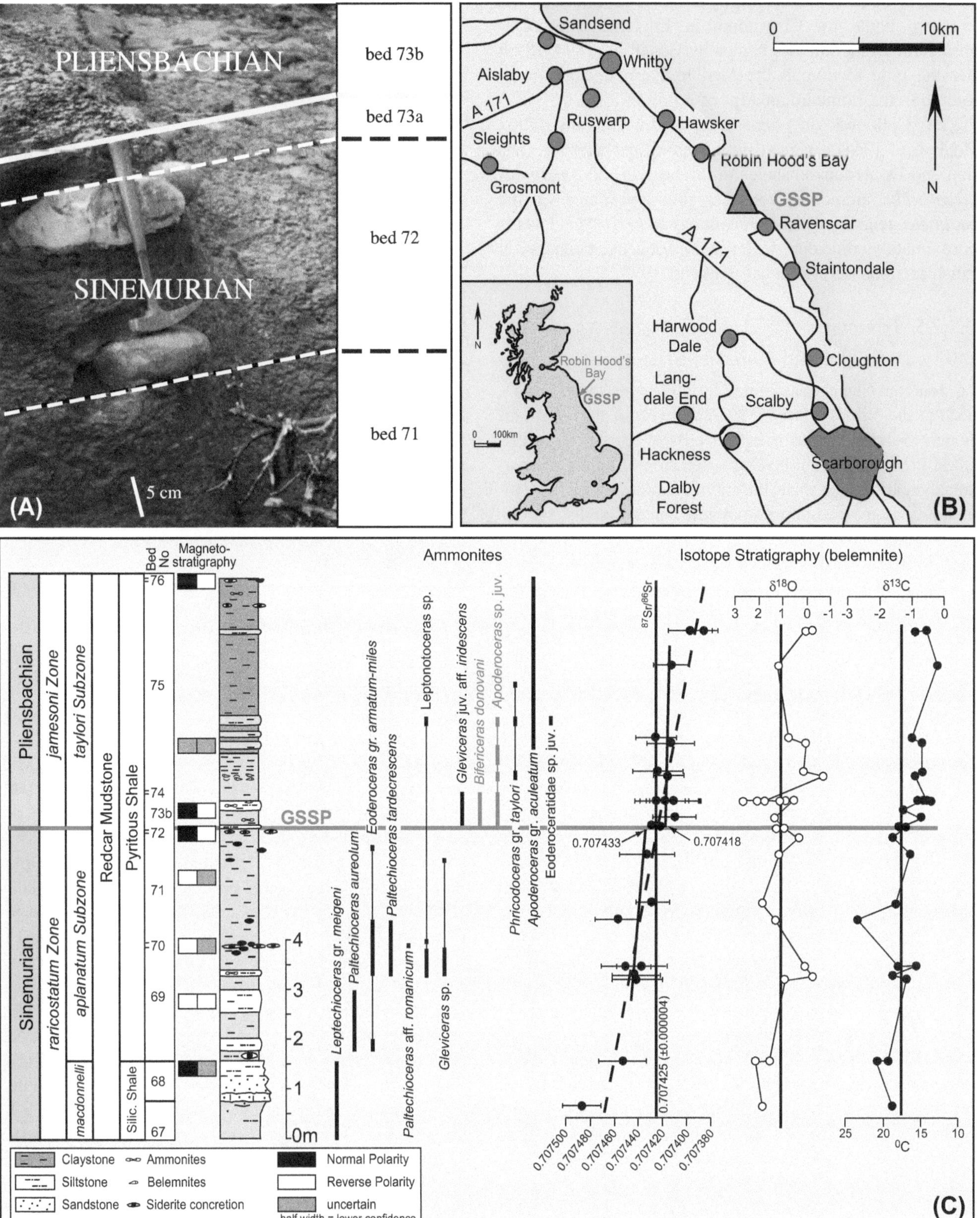

FIGURE 26.4 **GSSP for base of the Pliensbachian Stage at Wine Haven section, Yorkshire, east England.** The GSSP level coincides with the lowest occurrence of ammonite *Bifericeras donovani* (base of *Uptonia jamesoni* Zone). *Photo provided by Kevin Page.*

Upper Pliensbachian Substages

The Pliensbachian Stage has two substages. The lower substage of Carixian was named by Lang (1913) after Carixia, the Latin name for Charmouth in England. The upper substage of Domerian was named by Bonarelli (1894, 1895) after the type section in the Medolo Formation at Monte Domaro in the Lombardian Alps of northern Italy.

The Colloque du Jurassique à Luxembourg 1962 (Maubeuge, 1964) assigned the boundary between the Carixian and Domerian substages to the base of the *Amaltheus margaritatus* ammonite zone, at the appearance of the *Amaltheus* genera (typically *Amaltheus stokesi*). This level is approximately coincident with a major sequence boundary in British sections (Hesselbo and Jenkyns, 1998).

26.1.2.5. Toarcian

History, Definition and Boundary Stratotype

The Toarcian Stage was defined by d'Orbigny (1842–1851, 1852) at the Vrines quarry, 2 km northwest of the village of Thouars (Toarcium in Latin) in the Deux-Sèvres region of west-central France. The thin-bedded succession of blue-gray marl and clayey limestone spans the entire Toarcian with 27 ammonite horizons grouped into 8 ammonite zones (Gabilly, 1976).

The Toarcian–Pliensbachian boundary interval is the first phase of a major extinction event in western Europe among rhynchonellid brachiopods, ostracod faunas, benthic Foraminifera and bivalves, and marks a turnover in ammonites and belemnites. The base of the Toarcian is marked by a massive surge of Dactylioceratide (*Eodactylites*) ammonites and extinction of the boreal amaltheid family. The magnitude of the base-Toarcian extinction event appears to be a phenomenon of regional extent rather than global (Hallam, 1986). This episode is followed approximately one ammonite zone higher by a second extinction phase that coincides with a widespread anoxic event and a pronounced extinction among benthic invertebrates, such as bivalves and brachiopods (e.g., review in Wignall and Bond, 2008). Seawater strontium isotope ratios, which had been declining since the Hettangian, reach a minimum within this boundary interval.

At Thouars and throughout northwest Europe, the base of the Toarcian strata is an important maximum flooding surface above a major sequence boundary and associated condensation or gaps in the *Dactylioceras tenuicostatum* ammonite zone. This widespread hiatus necessitated selection of candidate GSSPs in the Mediterranean region where gaps in the succession are less pronounced (Elmi, 1999). The selected primary marker of the Toarcian GSSP was the lowest occurrence of a diversified *Eodactylites* ammonite fauna (*Odactylites simplex* horizon, *sensu* Goy *et al.*, 1996).

The official candidate is at Ponto do Trovao section, Peniche, western Portugal. A GSSP level at the base of Bed 15e was approved at the Jurassic Congress in 2006 (Elmi, 2006), but Serge Elmi's death delayed submission of a formal proposal for ICS/IUGS ratification. The candidate GSSP level coincides with the lowest occurrence of the ammonite *Eodactylites polymorphum.* This Tethyan *Eodactylites* fauna interval correlates with the NW European *Protogrammoceras paltus* horizon and is succeeded by an "English" *Orthodactylites* succession (Elmi, 2003). A proposed complementary section spanning the uppermost Pliensbachian and lowermost Toarcian ammonite zones at Almonacid de la Cuba in the Iberian Range of central-eastern Spain has detailed magnetostratigraphy and isotope stratigraphy (carbon, oxygen, strontium) (Comas-Rengifo *et al.*, 2010a,b).

Upper Toarcian Substages

There is no agreement on the number of substages of the Toarcian. A binary subdivision following the division by Oppel (1856–1858) places a substage boundary at the base of the *Haugia variabilis* ammonite zone, at the appearance of abundant Phymatoceratinae group of ammonites, particularly the *Haugia* genus (e.g., Krymholts *et al.*, 1982; Burger, 1995). An alternate three-substage division (e.g., Groupe Français d'Étude du Jurassique, 1997) groups the *Haugia variabilis* and underlying *Hildoceras bifrons* zones into a Middle Toarcian, and places the limit of an Upper Toarcian at the base of the *Grammoceras thouarsense* Zone.

26.1.3. Subdivisions of the Middle Jurassic

Black Lower Jurassic clays ("*Schwarzer Jura*") in southwestern Germany are overlain by strata containing clayey sandstone and brown-weathering ferruginous oolite. This lithologic change to the Brown Jurassic ("*Brauner Jura*") of Quenstedt (1858) has been retained as the base of the Middle Jurassic (base of Aalenian). The Middle Jurassic in southern England is characterized by shallow-marine carbonates of the Lower Oolite group of Conybeare and Phillips (1822), which comprised an expanded "Bathonian" stage of d'Omalius d'Halloy (1843). The lower portion of d'Omalius d'Halloy's Bathonian was classified as a separate "Bajocian" stage by d'Orbigny (1842–1851, 1852). In turn, Mayer-Eymar (1864) further removed the lower portion of d'Orbigny's Bajocian into a distinct "Aalenian" Stage.

Oppel (1856–1858) assigned the upper limit of his Middle Jurassic series or "Dogger" at the base of the Kellaway Rock of England, or the base of the Callovian Stage of d'Orbigny (1842–1851, 1852). This Callovian stage was shifted into the Middle Jurassic series by the Colloque du Jurassique à Luxembourg 1962 (Maubeuge, 1964) as preferred by Arkell (1956).

The bases of the Aalenian through Bathonian stages have been marked by GSSPs in expanded sections of rhythmic alternations of limestone and marl. The placement of a base for the Callovian Stage has been hindered by a ubiquitous condensation or hiatuses in strata of northwest Europe.

26.1.3.1. *Aalenian*

History, Definition and Boundary Stratotype

The Aalenian Stage was proposed by Karl Mayer-Eymar (1864, and redefined in 1874) for the lowest part of the "*Brauner Jura*" in the vicinity of Aalen at the northeastern margin of the Swabian Alb (Baden-Württemberg, southwestern Germany) where iron ore was mined from the associated ferruginous oolite sandstones (Dietl and Etzold, 1977; Rieber, 1984). His Aalenian stratotype truncated the Bajocian Stage of d'Orbigny (1842–1851, 1852) at the base of the *Sonninia sowerbyi* ammonite zone.

The biostratigraphic placement of the base of the Middle Jurassic was the evolution of the ammonite subfamilies Grammoceratinae and Leioceratinae, especially the first occurrence of species of the genus *Leioceras*, which evolved from *Pleydellia*. The Aalenian GSSP in the Fuentelsaz section in the Iberian Chain of Spain corresponds to this ammonite marker (Goy *et al.*, 1994, 1996; Cresta *et al.*, 2001). The base of marl Bed FZ107 in the Fuentelsaz section of alternating marl and limestone coincides with the first occurrence of *Leioceras opalinum* (base of *Leioceras opalinum* Zone) (Figure 26.5). The magnetostratigraphy of Fuentelsaz correlates to a composite magnetic pattern derived from other sections in Europe. A secondary reference section for basal Aalenian is at Wittnau near Freiburg in south Germany (Ohmert, 1996).

Middle and Upper Aalenian Substages

The four ammonite zones of the Aalenian are grouped into three substages; the Lower Aalenian is the *Leioceras opalinum* Zone, the Middle Aalenian comprises the *Ludwigia murchisonae* and *Brasilia bradfordensis* zones, and the Upper Aalenian is the *Graphoceras concavum* Zone.

26.1.3.2. *Bajocian*

History, Definition and Boundary Stratotype

The Bajocian Stage was named by d'Orbigny (1842–1851, 1852) after the town of Bayeux in Normandy (Bajoce in Latin). The abandoned quarries are now overgrown, and the nearby coastal cliff section of Les Hachettes indicates that most of the early Bajocian is absent at a hiatus and erosional surface. Indeed, most of the late Bajocian is largely condensed into a 15-cm-thick layer (Rioult, 1964). Ammonite lists of d'Orbigny indicate that he erroneously assigned species of upper Toarcian to lower Bajocian and vice versa. This confusion was one reason for Mayer-Eymar (1864) to distinguish a new Aalenian Stage between the Toarcian and Bajocian.

The Colloque du Jurassique à Luxembourg 1962 (Maubeuge, 1964) defined the Bajocian Stage to begin at the base of the *Sonninia sowerbyi* ammonite zone and extend to the top of the *Parkinsonia parkinsoni* Zone. However, the holotype of the *Sonninia sowerbyi* index species was later discovered to be a nucleus of a large Sonniniidae (*Papilliceras*) from the overlying *Otoites sauzei* Zone (Westermann and Riccardi, 1972). Therefore, the basal ammonite zone of the Bajocian was redefined to be the *Hyperlioceras discites* Zone, with the zonal base marked by the lowest occurrence of the ammonite genus *Hyperlioceras* (*Toxolioceras*), which evolved from *Graphoceras* (both in ammonite family Graphoceratidae).

Two sections recorded this ammonite datum with supplementary biostratigraphic and magnetostratigraphic data – Murtinheira at Cabo Mondego in Portugal (selected for the GSSP) and Bearreraig Bay on the Isle of Skye in Scotland (selected as an Auxiliary Stratotype Point) (Pavia and Enay, 1997). The GSSP of Cabo Mondego is at the base of Bed AB11 in a rhythmic alternation of gray limestone and marl (Henriques, 1992; Henriques *et al.*, 1994; Pavia and Enay, 1997) (Figure 26.6). The GSSP level is just below the base of a normal-polarity magnetozone.

Upper Bajocian Substage

The base of the Upper Bajocian is the base of the *Strenoceras (Strenoceras) niortense* ammonite zone. In older literature, the base was assigned as the base of a *Strenoceras subfurcatum* zone, until it was recognized by Dietl (1981) that the holotype of the index species belongs to *Garantiana* and had originated from the overlying zone, therefore this *subfurcatum* zone became invalid. The base of the Upper Bajocian corresponds to a major turnover of ammonite genera.

26.1.3.3. *Bathonian*

History, Definition and Boundary Stratotype

The Bathonian Stage was named by Jean Julien d'Omalius d'Halloy (1843) after the town of Bath (Bathonium in Latin) in southwest England, where oolitic limestones are exposed in a number of quarries. But this stratotype succession is incomplete and lacks adequate characterization by ammonites (Torrens, 1965). The lower half of this original Bathonian was reclassified in the system of d'Orbigny (1842–1851, 1852) as a Bajocian Stage exposed in Normandy, but d'Orbigny did not specify a revised stratotype for the truncated Bathonian, nor provide an unambiguous lower boundary. Indeed, d'Orbigny's original description suggests that he had included the equivalent of the present "Lower Bathonian" substage within his Bajocian stratotype (Rioult, 1964). A century of confusion ended when the base of the Bathonian Stage was defined by the Colloque du Jurassique à Luxembourg 1962 (Maubeuge, 1964) as the base of the *Zigzagiceras zigzag* ammonite zone.

The basal Bathonian is well developed in cyclic alternations of marl and limestone in southeastern France. The Bathonian GSSP was placed at the base of limestone bed RB071 at Ravin du Bès, Bas Auran near Digne, Basses-Alpes, France (Innocenti *et al.*, 1988; Fernández-López, 2007, Fernández-López *et al.*, 2009) (Figure 26.7). This

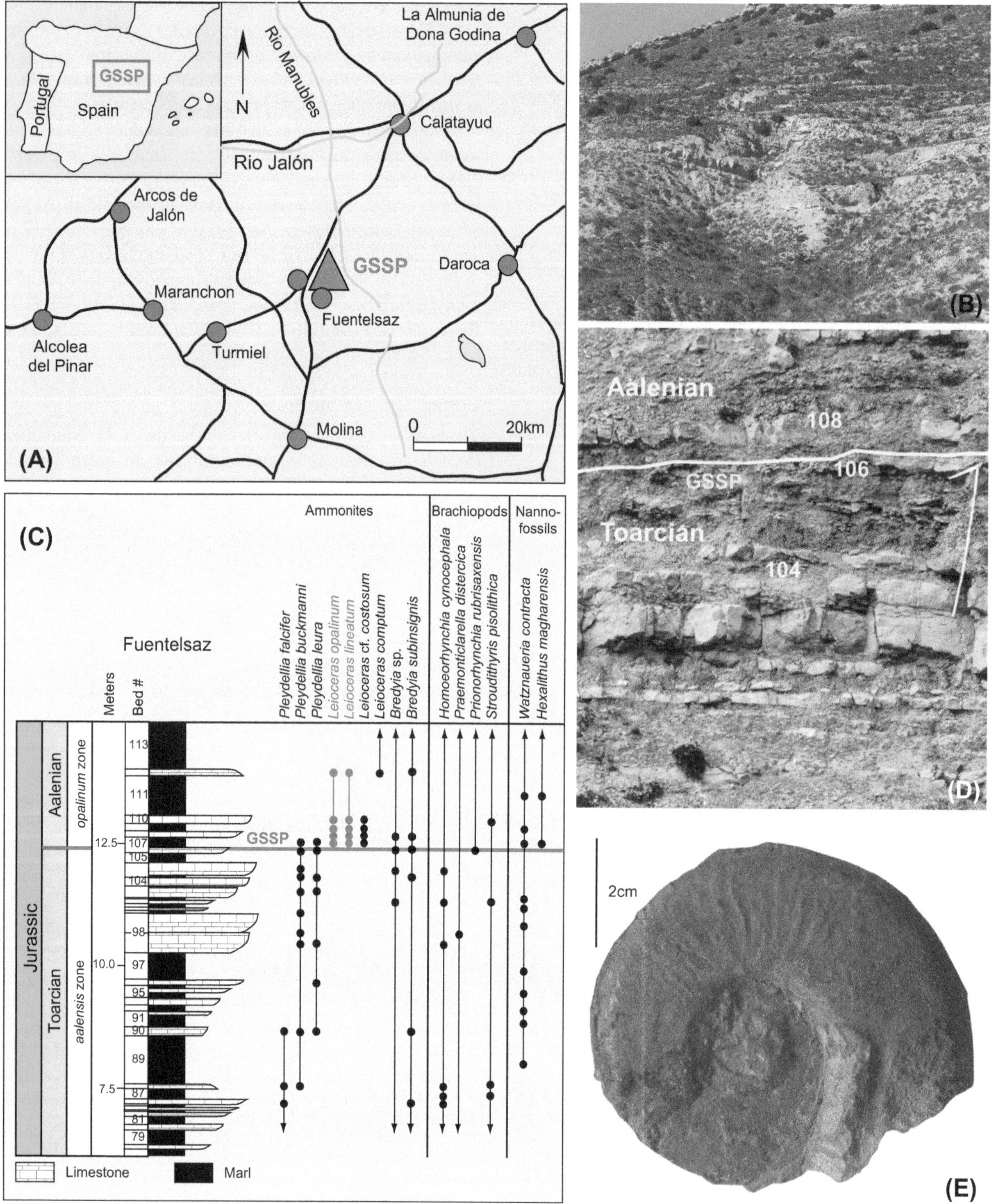

FIGURE 26.5 **GSSP for base of the Aalenian stage at Fuentelsaz section, Castilian Branch of the Iberian Range, Spain.** The GSSP level coincides with the lowest occurrence of ammonite *Leioceras opalinum* (base of *Leioceras opalinum* Zone). *Photos provided by Maria Soledad Ureta.*

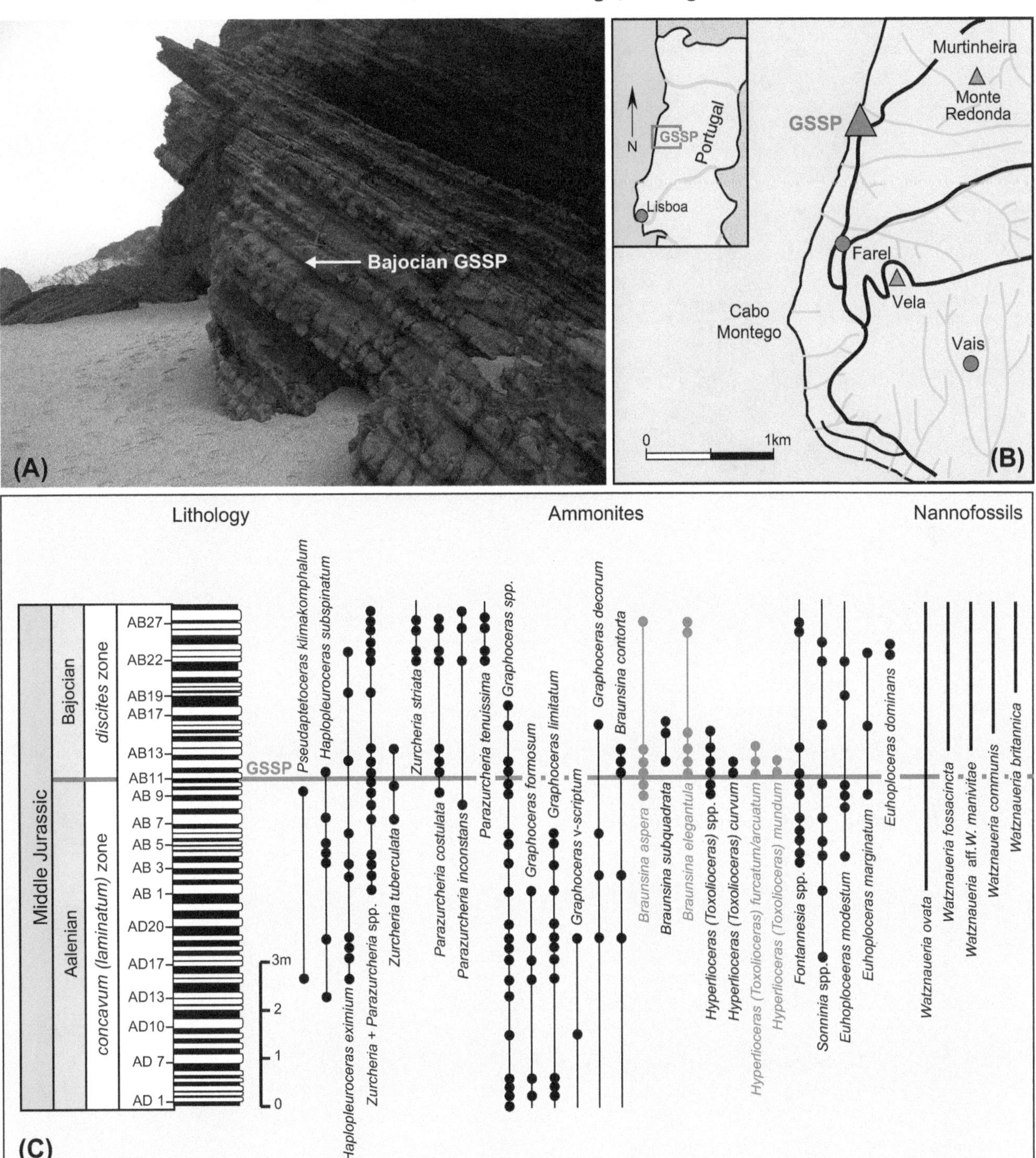

FIGURE 26.6 **GSSP for base of the Bajocian Stage at Cabo Mondego on the Atlantic coast of Portugal.** The GSSP level coincides with the lowest occurrence of ammonite *Hyperlioceras* (base of *Hyperlioceras discites* Zone). *Photo provided by Maria Helena Henriques.*

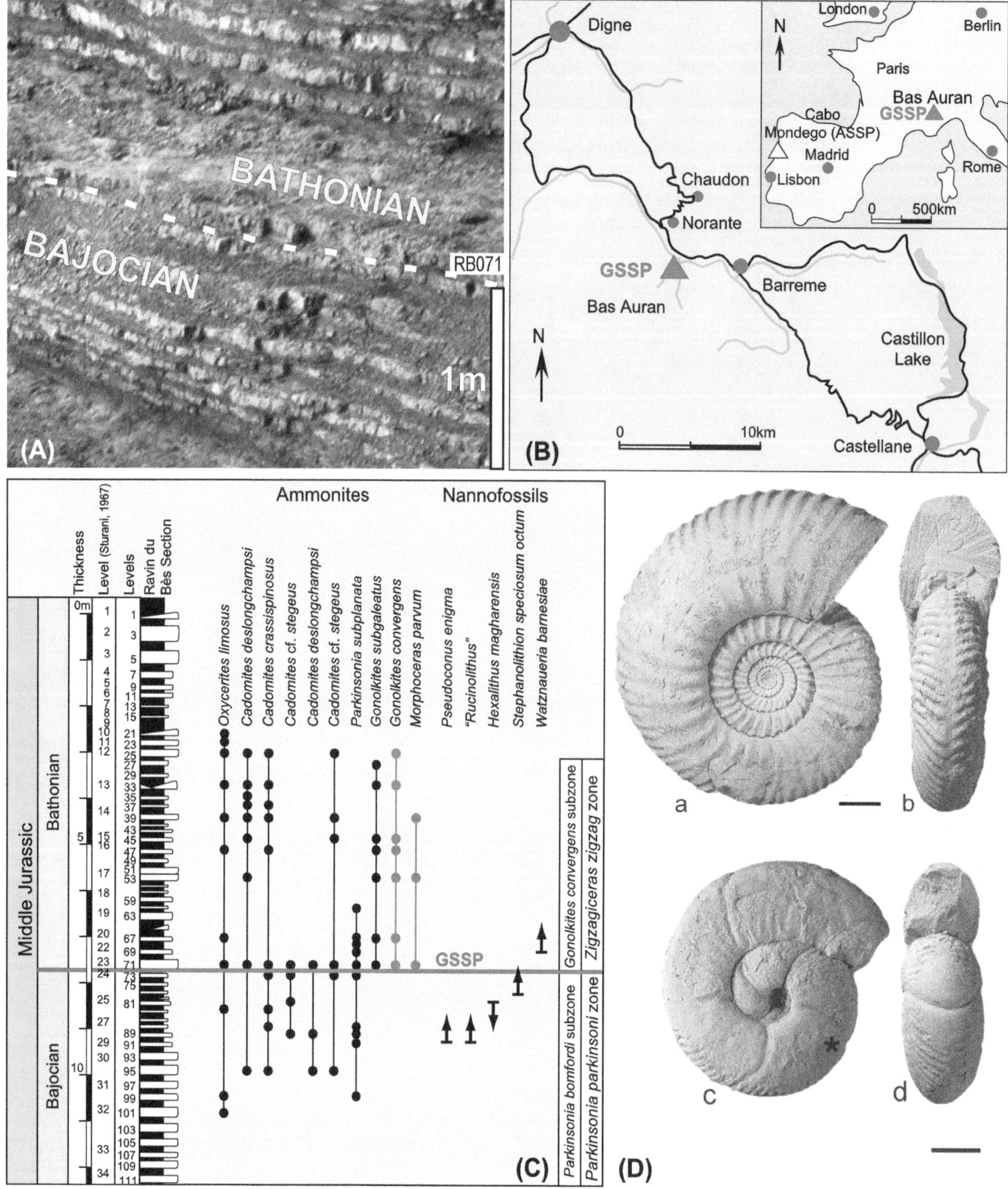

FIGURE 26.7 **GSSP for base of the Bathonian Stage at Ravin du Bès, Bas-Auran area, Alpes de Haute Provence, France.** The GSSP level coincides with the lowest occurrence of ammonite *Gonolkites convergens* (base of *Zigzagiceras zigzag* Zone). *Photo provided by Sixto Fernández-López.*

level coincides with the lowest occurrence of ammonites *Gonolkites convergens* (base of *Zigzagiceras zigzag* Zone) and *Morphoceras parvum*. The GSSP level is just below the local lowest occurrence of calcareous nannofossil *Watznaueria barnesiae* (base of zone NJT11).

An auxiliary section for the Bathonian Stage at Cabo Mondego, near the Bajocian GSSP site, provides complementary data about the biochronostratigraphic subdivision and ammonite succession of the Sub-Mediterranean *parvum* Subzone and the Northwest European *convergens* Subzone. Neither section has yielded an unambiguous cyclostratigraphy or magnetostratigraphy.

Middle and Upper Bathonian Substages

The Bathonian is generally divided into three substages, with the base of the Middle Bathonian placed at the base of the *Procentes progracilis* ammonite zone.

A divergence of ammonite assemblages in the upper Middle Bathonian has resulted in different basal-definitions for an Upper Bathonian substage in each province. In the Sub-Mediterranean province (Tethyan province), a Middle/Upper Bathonian boundary is assigned to the base of the *Hecticoceras (Prohecticoceras) retrocostatum* Zone. In the Northwest European province (Boreal province), a substage boundary is commonly assigned to the base of the *Procerites (Procerites) hodsoni* Zone, which is a significantly older level (Groupe Français d'Étude du Jurassique, 1997).

26.1.3.4. Callovian

History, Definition and Boundary Stratotype

The Callovian Stage was named by d'Orbigny (1842–1851, 1852) after Kellaways (various spellings in the geological literature) in Wiltshire, 3 km northeast of Chippenham, England, because he considered "*Calloviensis*" to be a derivative of the place name. The "Kelloways Stone" contains abundant cephalopods, including *Ammonites calloviensis* (called *Sigaloceras calloviensis* in current taxonomy). Oppel (1856–1858) placed the base of his "Kelloway gruppe" at the base of the *Macrocephalites macrocephalus* Zone, or at the lithologic contact of the Upper Cornbrash to the underlying Forest Marble Formation (currently the upper part of the *Clydoniceras discus* Subzone of uppermost Bathonian). At this level, ammonites of the genus *Macrocephalites* replace *Clydoniceras*, but the Upper Cornbrash is a condensed deposit "*representing but a fraction of the time-intervals involved*" (Cope *et al.*, 1980).

Callomon (1964, 1999) noted that the base of the *Macrocephalites macrocephalus* Subzone in "standard chronostratigraphy" was initially defined as the base of Bed 4 by Arkell (1954) at the Sutton Bingham section near Yeovil, Somerset, England, therefore this level served as the de facto GSSP for the base of the Callovian Stage. However, the lowest occurrence of *Macrocephalites* genus was later discovered to be in strata equivalent to the Upper Bathonian (Dietl, 1981; Dietl and Callomon, 1988). Therefore, the "standard" *Macrocephalites macrocephalus* Zone was abandoned, and the base of the Callovian was assigned to the lowest occurrence of the genus *Kepplerites* (Kosmoceratidae), which defined a basal horizon of *Kepplerites (Kepplerites) keppleri* (base of *K. keppleri* Subzone of *Macrocephalites herveyi* Zone) in the Sub-Boreal province (Great Britain to southwest Germany). The uppermost Bathonian is the *hockstetteri* horizon (var. *hockstetteri* of *Clydoniceras discus*) of the *C. discus* Subzone, *C. discus* Zone.

The boundary between the uppermost Bathonian and basal Callovian is rarely preserved. A proposed GSSP with an apparently complete boundary at the resolution level of ammonite successions is in the Albstadt district of the Swabian Alb in the Baden-Württemberg region of southwest Germany (Dietl, 1994; Callomon and Dietl, 2000). The Macrocephalen-Oolith Formation (Unit ε of the Brown Jura facies) is a condensed facies of ferruginous-oolite-bearing clay to marly limestone, and the compact Bathonian–Callovian boundary interval is bounded by unconformities and may contain minor hiatuses. The magnetostratigraphy from this suggested GSSP has not yet been verified in other sections. This potential GSSP at base of Bed 6a in the Albstadt-Pfeffingen section has not yet been adopted by the International Stratigraphic Commission. A more expanded section in the Novgorod region of Russia is an additional candidate (Guzhikov *et al.*, 2010).

In the Sub-Mediterranean province (southern Paris Basin to North Africa and Italy, the basal Callovian zone is the *Bullatimorphites (Kheraiceras) bullatus* Zone defined by the range of the index species (Groupe Français d'Étude du Jurassique, 1997). Strong ammonite biogeographic differences required these two regions to have distinct and poorly correlated zonations up to the middle of the Callovian.

Middle and Upper Callovian Substages

The Callovian Stage is generally divided into three substages. The substage boundaries correspond to two important changes in ammonite fauna, but ammonite provincialism and utilization of different faunal successions led to different placements within each province that do not necessarily coincide (Groupe Français d'Étude du Jurassique, 1997).

In the Sub-Boreal province, the Lower/Middle Callovian substage boundary is placed at the base of the *Kosmoceras (Zugokosmoceras) jason* Zone (base of *Kosmoceras (Zugokosmoceras) medea* Subzone), above the *Sigaloceras (Sigaloceras) calloviense* Zone (*Sigaloceras (Catasigaloceras) enodatum* Subzone). In the Sub-Mediterranean province, the Lower/Middle Callovian boundary is placed at the base of the *Reineckeia anceps* Zone (*Reineckeia stuebeli* Subzone) above the *Macrocephalites (Dolikephalites) gracilis* Zone (*Indosphinctes patina* Subzone). These two levels are considered approximately coeval.

The Middle/Upper Callovian substage boundary in the Sub-Boreal province is assigned to the base of the *Peltoceras (Peltoceras) athleta* Zone (*Kosmoceras (Lobokosmoceras) phaeinum* Subzone) above the *Erymnoceras coronatum* Zone (*Kosmoceras (Zugokosmoceras) grossouvrei* Subzone). The Middle/Upper Callovian boundary in the Sub-Mediterranean province is assigned to the base of the *Peltoceras (Peltoceras)* athleta Zone (*Hecticoceras (Orbignyiceras) trezeense* Subzone) above the *Erymnoceras coronatum* Zone (*Rehmannia (Loczyceras) rota* Subzone), or approximately a subzone higher than in the Sub-Boreal province (Groupe Français d'Étude du Jurassique, 1997).

26.1.4. Subdivisions of the Upper Jurassic

The brownish-weathering deposits "*Braunner Jura*" of the Middle Jurassic in southwestern Germany are overlain by units dominated by calcareous claystone and limestone. The base of the current Upper Jurassic (base of Oxfordian) coincides approximately with this lithologic change to the White Jurassic ("*Weisser Jura*") of Quenstedt (1858). This Upper Jurassic interval is approximately equivalent to the Middle and Upper Oolite group of Conybeare and Phillips (1822) in England; which is also known as 'Malm' from the local name for the clayey oolitic limestone. Both the White Jurassic of southwest Germany and the English "Malm" strata undergo a shallowing-upward trend in the latest Jurassic that terminates in non-marine facies or erosional truncation.

d'Orbigny (1842–1851, 1852) named four stages (Oxfordian, Corallian, Kimmeridgian, and Portlandian) after reference sections in southern England, and he designated the base of the Cretaceous as a "Purbeck stage" followed by a "Neocomian stage". Oppel (1856–1858) eliminated d'Orbigny's Corallian and Portlandian stages to extend the Kimmeridgian stage to the base of the Purbeckian (also considered to be Cretaceous). Oppel had left an interval "unassigned" between his Oxfordian and Kimmeridgian groups (his *Diceras arietina* Zone, approximately equivalent to the Upper Calcareous Grit Formation of Dorset, England). Later, Oppel (1865) created a new uppermost Jurassic stage, the "Tithonian", in the Mediterranean region that encompassed the upper part of his previous "Kimmeridgian group" and extended through the former Purbeckian to the base of the Neocomian stage. However, Oppel did not specify the limits or reference sections for this Tithonian stage concept. This confusing situation was further distorted when the "Berriasian" Stage of Coquand (1871) came into common use to designate the lowermost Cretaceous, even though it overlapped with the original concept of Oppel's Tithonian Stage.

The combination of (1) shuffling of Upper Jurassic stage nomenclature, (2) imprecise definitions, (3) a pronounced faunal provincialism during the majority of the Upper Jurassic that precluded precise correlation even within northwest Europe, and (4) widespread hiatuses in the reference sections resulted in a proliferation of regional stage and substage nomenclature. Finally, after a century of debate, the Colloque du Jurassique à Luxembourg 1962 (Maubeuge, 1964) voted to "*return to the original sense of this stage [Oxfordian] as defined by A. d'Orbigny and given precision by W.J. Arkell*" and to discontinue the regional usage of a "Purbeckian" stage, because it was primarily a local facies. However, the controversy over other Upper Jurassic stage definitions and the placement of the Jurassic/Cretaceous boundary led the Collogue to "refer the question back for consultation among interested specialists". These issues have yet to be satisfactorily resolved.

During the 1980s and 1990s, the International Subcommission on Jurassic Stratigraphy decided that the Upper Jurassic consists of three stages: Oxfordian, Kimmeridgian and Tithonian. Through a fortunate episode of biogeography, an inter-regional biostratigraphic definition of the base of the Oxfordian is well established with ammonites. However, it has proven difficult to correlate potential definitions for the bases of the Kimmeridgian and Tithonian stages, and long-held traditions of regional equivalence have proven to be erroneous.

26.1.4.1. Oxfordian

History, Definition and Boundary Stratotype

The Oxfordian Stage of d'Orbigny (1842–1851, 1852), named after the town of Oxford in south-central England with reference to the Oxford Clay Formation, was overlain by his "Coralline Stage". Oppel (1856–1858) incorporated the majority of that "Corallian Stage" into an expanded "Oxfordian group". In an error of correlation, he simultaneously considered the base of his Oxfordian to be the lithologic base of the Oxford Clay Formation onto the Kelloway Rock in Yorkshire (now considered to be approximately the Lower/Middle Callovian boundary) and the biostratigraphic top of the *Peltoceras athleta* ammonite zone (now considered to be the middle of Upper Callovian). He also left "unassigned" a suite of strata between his "Oxfordian" and the overlying Kimmeridgian Stage.

Ammonites across the Callovian/Oxfordian boundary interval were studied by Arkell (1939, 1946), who placed this boundary at the lowest occurrence of *Vertumniceras mariae* (now placed in the *Quenstedtoceras* genus) above *Quenstedtoceras lamberti*, or essentially at the base of the Oxford Clay Formation. This is similar to the historical usage in southwest Germany, where the Upper or "White" Jurassic begins just above the Lamberti Knollen bed. The Colloque du Jurassique à Luxembourg 1962 (Maubeuge, 1964) selected Arkell's biostratigraphic definition for the base of the Oxfordian Stage. This Colloque also assigned the upper limit of the Oxfordian as the top of the *Ringsteadia pseudocordata* ammonite zone in the Boreal province.

The main biostratigraphic event is the "Boreal Spread", an extensive expansion of the Boreal Cardioceratidae ammonites from their Arctic province across Europe and mixing with Mediterranean-province fauna rich in Phylloceratidae ammonites (Page *et al.*, 2010a). The current candidate for the Oxfordian GSSP is a coastal section at Redcliff Point near Weymouth, Dorset, southern England (Page *et al.*, 2010b). The GSSP in this clay-rich section would be placed at the lowest occurrence of ammonite *Cardioceras redcliffense*, which would define the base of the *Cardioceras scarburgense* Subzone of the *Quenstedtoceras mariae* Zone. This level is within the youngest part of a normal-polarity magnetozone correlated to marine magnetic anomaly M37n (Ogg *et al.*, 2010a, 2011). Recognition of this level with other biostratigraphic groups is not yet established, but the level is immediately followed by a widespread surge in primitive planktonic foraminifers (Oxford *et al.*, 2002). One concern is that the exposure is in an active landslip that lacks any lithological marker near the boundary; therefore, there may be cryptic slip planes related to the landslip that might cut out or repeat strata (S. Hesselbo, pers. comm. to J. Ogg, 2011).

The ammonite biostratigraphy across the *Q. lamberti* to *Q. mariae* transition interval has been studied in expanded dark claystone successions in southeast France (e.g., Fortwengler and Marchand, 1994; Page *et al.*, 2010b) where two sections had been recommended as complimentary basal-Oxfordian GSSPs (Melendez, 1999). Even though the same *Cardioceras redcliffense* horizon was identified, these French sections have not yet proved suitable for other stratigraphic correlation methods except for dinoflagellates, therefore this GSSP candidate may be considered as an auxiliary reference section (e.g., Melendez, 2002, 2003, 2006).

Middle and Upper Oxfordian Substages

The base of a Middle Oxfordian substage is commonly placed at base of the ammonite *Perisphinctes (Arisphinctes) plicatilis* Zone of the Tethyan and Sub-Boreal faunal provinces, which is coeval with the base of the *Cardioceras densiplicatus* Zone of the Boreal province.

Beginning with the Middle Oxfordian, faunal differentiation among faunal realms became more pronounced and has inhibited standardization and correlation of regional ammonite zones. In addition, even though regional zonal nomenclatures have commonly remained constant, the assigned biostratigraphic boundaries of these "standard" zonal units have been undergoing redefinitions (e.g., Głowniak *et al.*, 1997, 2008; Groupe Français d'Étude du Jurassique, 1997; among others).

In the Sub-Mediterranean province (Tethyan province), the base of an Upper Oxfordian substage is commonly assigned to the base of the *Perisphinctes (Dichotomoceras) bifurcatus* Zone; although the definition of the zonal base is not always consistent among regional studies. In the Sub-Boreal province, the Middle/Upper substage boundary is assigned as the base of the regional *Perisphinctes cautisnigrae* Zone, and the Boreal province uses the base of the regional *Amoeboceras serratum* Zone. Cross-regional ammonite markers and magnetostratigraphic studies suggest that these three levels are within about 250 kyr of each other (base of *A. serratum* is youngest; base of *P. cautisnigrae* is oldest).

26.1.4.2. Kimmeridgian

History, Definition and Boundary Stratotype

The Kimmeridgian Stage was named by d'Orbigny (1842–1851, 1852) after the coastal village of Kimmeridge in Dorset, England, where the spectacular cliffs of black Kimmeridge Clay expose a continuous record of that interval. Oppel (1856–1858) expanded the Kimmeridgian downward by incorporating a portion of d'Orbigny's former "Corallian stage", but, rather than assign a boundary between the Oxfordian and Kimmeridgian, he left the intervening Upper Calcareous Grit Formation "unassigned". Oppel initially indicated that the Kimmeridgian "group" would continue upward to the base of the Purbeck (his base of Cretaceous), but later, Oppel (1865) inserted a Tithonian Stage as the uppermost Jurassic stage. Therefore, neither of the boundaries of the Kimmeridgian Stage was adequately defined.

The Oxfordian/Kimmeridgian boundary was defined by Salfeld (1914) after studying the Perisphinctidae ammonite succession from the boundary interval. He proposed a *Ringsteadia anglica* Zone (now called *Ringsteadia pseudocordata* Zone) in the uppermost Oxfordian and that the base of the Kimmeridgian would correspond to the appearance of *Pictonia.* The Colloque du Jurassique à Luxembourg 1962 (Maubeuge, 1964) fixed the base of the Kimmeridgian as the base of the *Pictonia baylei* Zone in the Sub-Boreal province. This level is considered equivalent to the base of the *Amoeboceras bauhini* Zone of the Boreal realm, although the precise definitions of the corresponding zonal boundaries are disputed. However, due to faunal provincialism that began in the middle Oxfordian, the ammonite zonations of Great Britain (Boreal and Sub-Boreal provinces) could not be correlated to the Sub-Mediterranean province (Tethyan province). The Colloque du Jurassique à Luxembourg 1962 (Maubeuge, 1964) indicated that this level was equivalent to the base of the *Sutneria platynota* Zone of the Sub-Mediterranean province.

However, this presumed equivalence was later demonstrated to be incorrect from comparisons of dinoflagellate cyst assemblages (Brenner, 1988; Poulson, in Melendez and Atrops, 1999) and rare incursions of ammonites from the Boreal province into the Sub-Mediterranean successions in Poland and in the Swabian Alb (Atrops *et al.*, 1993; Matyja and Wierzbowski, 1997, 2002, 2006a,b; Schweigert and Callomon, 1997). These constraints on the correlation among faunal provinces near the Oxfordian-Kimmeridgian boundary imply that much of the upper *Epipeltoceras bimammatum* Zone and *Idoceras planula* Zone of the Sub-Mediterranean Province of the Tethyan province, which had constituted the entire upper half of that regional "Upper Oxfordian", should be coeval with the lowest Kimmeridgian (*Amoeboceras bauhini* Zone and *Amoeboceras bayi* horizon of the *Amoebites kitchini* Zone) of the Boreal province (Matyja *et al.*, 2006). This offset in the regional "Oxfordian/Kimmeridgian" boundary placements by one-and-a-half ammonite zones (c. 1.3 myr) has been demonstrated by comparing magnetostratigraphy of Great Britain (Sub-Boreal and Boreal provinces) with different regions in the Sub-Mediterranean province (Przybylski *et al.*, 2010a). The temporal offset had created a dilemma in selecting a GSSP for the base of the Kimmeridgian Stage, because an initial choice must be made between faunal provinces and the corresponding definition of the Oxfordian/Kimmeridgian boundary (reviewed in Atrops, 1999; Melendez and Atrops, 1999; and Wierzbowski, 2001, 2002, 2010). It was decided by the Oxfordian–Kimmeridgian boundary working group to retain a definition similar to Salfeld (1914) of the appearance of *Pictonia* in the Sub-Boreal province.

An intertidal coastal section in dark-gray claystone at Flodigarry, Staffin Bay, Isle of Skye (northwest Scotland) has yielded detailed Sub-Boreal and Boreal ammonite successions and a magnetostratigraphy (e.g., Sykes and Callomon, 1979; Wierzbowski *et al.*, 2006; Matyja *et al.*, 2006; Przybylski *et al.*, 2010a). Even though this Flodigarry section was accepted as the GSSP location in 2007, it took additional years to agree on the precise *Pictonia* ammonite horizon (a new *Pictonia flodigarriensis* taxon or the previous usage of *Pictonia densicostata*). The impasse will probably be decided based upon which level has wider correlation potential (Wierzbowski, 2010). In GTS2012, we have utilized the proposed *Pictonia flodigarriensis* level, which corresponds almost exactly to the base of a reversed-polarity interval that is correlated as magnetic polarity chron M26r.

Upper Kimmeridgian Substage

In the Sub-Boreal province, the base of an Upper Kimmeridgian substage is the ammonite *Aulacostephanoides mutabilis* Zone, whereas in the Sub-Mediterranean province, it is typically assigned to the base of the *Aspidoceras acanthicum* Zone. Magnetostratigraphy suggests that the Sub-Mediterranean placement is about 0.3 myr younger than the Sub-Boreal assignment (Ogg *et al.*, 2010a,b). An alternate assignment in the Boreal province, the base of ammonite *Aulacostephanus eudoxus* zone, is one ammonite zone higher than the Sub-Boreal substage placement, or about 0.8 myr younger.

26.1.4.3. Tithonian

History, Definition and Boundary Stratotype

In an enlightened departure from the lithology-based stratotype concept, Oppel (1865) defined the Tithonian Stage solely on biostratigraphy. In mythology, Tithon is the spouse of Eos (Aurora), goddess of dawn, and Oppel used this name in a poetic allusion to the dawn of the Cretaceous. He referenced sections in western Europe, from Poland to Austria, for the characteristics of the Tithonian.

The base of Oppel's Tithonian was placed at the top of the Kimmeridgian *Aulacostephanus eudoxus* ammonite zone, which can be recognized in both the Sub-Boreal and Sub-Mediterranean faunal provinces. Later, Neumayr (1873) established the *Hybonoticeras beckeri* Zone above the *A. eudoxus* Zone in the Sub-Mediterranean province and assigned it to the Kimmeridgian Stage.

Neumayr's revised placement of the Tithonian–Kimmeridgian boundary corresponded closely with the boundary between the "Portlandian" and Kimmeridgian stages as initially assigned by Alcide d'Orbigny (1842–1851, 1852), who had assigned "des *Ammonites giganteus* et *Irius*" as Portlandian index fossils. However, d'Orbigny did not visit England, and he inadvertently combined fossil assemblages from outcrops at Bologna in Italy with a name derived from a "type section" on Portland Island in England. The "*Ammonite irius*" is one representative of the *Gravesia* genus, which has a lowest occurrence in the basal *Hybonoticeras hybonotum* Zone of the revised Tithonian. Accordingly, the *Gravesia gravesiana* ammonite zone was assigned as the basal zone of the British Portlandian [Tithonian] Stage by Salfield (1913). Following Salfield's oral presentation to the Geological Society of London, it was noted that the chronostratigraphic term "Kimmeridgian" only partially encompassed the "Kimmeridgian Clay" Formation; therefore it was recommended that Salfield "*should invent a dual nomenclature – one for the stratigraphical and another for the zoological sequence*" and replace "Kimmeridgian Stage" by a new name (in Salfield, 1913). Unfortunately, this enlightened recommendation was not pursued, and a confusing equivalence of a "Kimmeridgian Stage" with the "Kimmeridge Clay" Formation and associated lifting the base of d'Orbigny's "Portlandian" stage became common usage in England, but a lower Tithonian–Kimmeridgian boundary was used elsewhere in Europe. In Britain, the Kimmeridge Clay Formation was arbitrarily subdivided into a lower and upper member at the approximate Kimmeridgian–Tithonian boundary level at the lowest occurrence of *Gravesia gravesiana* at the "Maple Ledge" bed (reviewed in Cox and Gallois, 1981). The "Tithonian" was formally adopted as the name of the uppermost stage of the Jurassic by a vote of the International Commission on Stratigraphy in September, 1990.

The Second Colloquium on the Jurassic System, held in Luxembourg in 1967, recommended that the top of the Kimmeridgian be assigned to the base of the *Gravesia gravesiana* Zone (Resolution du deuxième Colloque du Jurassique à Luxembourg, 1970). However, Cope (1967) subdivided the lowermost Tithonian portion of the Upper Kimmeridge Clay into several ammonite zones based on successive species of his reconstituted *Pectinatites* genus and abandoned Salfield's two *Gravesia* zones. Cope raised the upper limit of the uppermost Kimmeridgian *Aulacostephanus autissiodorensis* Zone to the base of his new *Pectinatites (Virgatosphinctoides) elegans* Zone, thereby effectively lifting the associated biostratigraphic division between Lower and Upper Kimmeridge Clay. Cox and Gallois (1981) note that the top of the international Kimmeridgian Stage now falls within the middle of Cope's expanded *A. autissiodorensis* Zone in the Sub-Boreal province, therefore, they suggested reinstating a truncated *Gravesia gravesiana* Zone below the *P. (V.) elegans* Zone.

In addition to this complex history of changing concepts, the extreme provincialism of macrofossils and microfossils has proven a daunting challenge to find a suitable GSSP definition that is amenable to global correlation. The Tithonian–Kimmeridgian boundary interval in the Tethyan faunal realm is marked by the simultaneous lowest occurrence of the ammonites *Hybonoticeras* aff. *hybonotum* and *Glochiceras lithographicum* (base of *H. hybonotum* Zone), immediately followed by the lowest occurrence of the *Gravesia* genus. Candidate sections in southeast France for the GSSP include a thick pelagic-limestone succession just below a castle on Crussol mountain overlooking the Rhône river just west of Valence (Atrops, 1982, 1994) and at a quarry within a military training area at Canjuers (Var district) (Atrops *et al.*, in prep.). The base of the *H. hybonotum* Zone at Crussol is at the base of normal-polarity zone M22An (Ogg *et al.*, 2010b). The Canjuers quarry did not yield a magnetostratigraphy, but its ammonite succession is better established. Other potential GSSP sections include the Swabian region of southern Germany (Olóriz and Schweigert, 2010).

Upper Tithonian Substage

In the Tethyan faunal domain, the base of an Upper Tithonian substage is informally assigned to a major turnover in ammonite assemblages at the base of the *Micracanthoceras microcanthum* Zone. This level is approximately where calpionellid microfossils become important in the biostratigraphic correlation of pelagic limestone and is at the base of normal-polarity chron M20n.

Bolonian-Portland and Volgian Regional Stages of Europe

The century of controversy over the subdivision and nomenclature for the uppermost Jurassic and lowermost Cretaceous stages, coupled with markedly distinct ammonite assemblages in different regions of Europe, led to extensive usage of regional stages.

The Volgian Stage in western Russia was established by Nikitin (1881), capped by a Ryazanian Horizon (Bogoslovsky, 1897), and later extended downward (Resolution of the All-Union Meeting on the unified scheme of the Mesozoic stratigraphy for the Russian Platform, 1955) (reviewed in Krymholts *et al.*, 1982). The Volgian is zoned by ammonite assemblages that are extensively distributed in the Boreal faunal realm, therefore it became the standard in the northern high latitudes outside of Britain. However, the zonations are different among the Russian Platform, northern Siberia, subpolar Urals, and Spitsbergen regions (e.g., Rogov and Zakharov, 2009). In 1996, the Russian Interdepartmental Stratigraphic Committee resolved to equate the Lower and Middle Volgian (*Ilowaiskya klimovi* through *Epivirgatites nikitini* ammonite zones of the Russian Platform) to the Tithonian Stage, assign the Upper Volgian (*Kachpurites fulgens* through *Craspedites nodiger* ammonite zones) to the lowermost Cretaceous, and use only Tithonian and Berriasian as chronostratigraphic units in the Russian geological time scale (Rostovtsev and Prozorowsky, 1997). In the Russian Platform region of the Boreal realm, the boundary between Lower and Middle "Volgian" regional substages is assigned at the base of the *Dorsoplanites panderi* ammonite zone.

The "Bolonian" and "Portlandian" have been promoted as "secondary standard stages" for usage in southern English regional geology, especially in Dorset (e.g., Cope 1993, 1995, 2003; Taylor *et al.*, 2001), but have not been widely adopted. In the Sub-Boreal faunal province in Britain, the base of the "Portlandian" regional stage, named by d'Orbigny (1842–1851) after Isle of Portland in Dorset, is placed at the base of the *Progalbanites albani* ammonite zone at the base of the Portland Sand Formation. The "Bolonian" was named by Blake (1881) after the Boulogne-sur-Mer section at Pas-de-Calais, northwest France. The "Bolonian" of Britain is equivalent to the upper Kimmeridgian Clay Formation between the Tithonian–Kimmeridgian boundary and the Portland Sand, or *Pectinatites (Virgatosphinctoides) elegans* through *Virgatopaviovia fittoni* ammonite zones. The overlying "Portlandian" is equivalent to the Portland Formation in Dorset, or *Progalbanites albani* through approximately the *Titanites anguiformis* ammonite zones (usage varies).

The approximate correlation and scaling of ammonite zonations among the Boreal (Volgian), Sub-Boreal (Bolonian, regional upper Kimmeridgian) and Tethyan (Tithonian) stages has been only partially resolved through applying magnetostratigraphy (e.g., Houša *et al.*, 2007; Ogg *et al.*, 2010b) and detailed inter-correlation webs of macrofossil biostratigraphy (e.g., Rogov and Zakharov, 2009; Rogov, 2010a,b). The base of the Volgian in Russia appears to coincide exactly with the base of the Tithonian in western Europe (Rogov and Price, 2010).

Top of Tithonian (Base of Cretaceous)

In contrast to most geological systems, there are nearly no "global events" within the uppermost Jurassic and lowermost Cretaceous. Indeed, it has been nearly impossible to find any significant widespread biostratigraphic, geochemical or other marker for inter-regional correlation within this interval. As a result, the span of the Tithonian varies among regions, among usage of different paleontological markers (ammonites, calpionellids, nannofossils) and among secondary stratigraphic correlation methods (e.g., magnetostratigraphy). The Tethyan-based ammonite definition for the base of the Cretaceous as the base of the *Berriasella jacobi* Zone appears to fall within the middle of relatively long ammonite zones in both the Sub-Boreal (*Subcraspedites preplicomphalus* Zone) and Boreal (*Craspedites nodiger* Zone) provinces. The base of reversed-polarity chron M18r was suggested as a convenient global correlation horizon near the clustering of possible biostratigraphic-based boundaries (Ogg and Lowrie, 1986). The recognition of this magnetozone in Sub-Boreal and Boreal realms (e.g., Ogg *et al.*, 1994; Houša *et al.*, 2007) has enabled inter-regional correlations; therefore, by default, this base-M18r level is gaining support as a global marker for a Jurassic/Cretaceous boundary (W. Wimbledon, chair of Jurassic–Cretaceous boundary working group, pers. comm., 2010). Another alternative is to make a more dramatic shifting of the "Tethyan" Jurassic–Cretaceous boundary, and utilize more major evolutionary–environmental events that occur in the mid-Tithonian or upper Berriasian (e.g., shift that uncertain boundary by a half-stage or more). Less dramatic would be adopting the base of Chron M17r, which is about 1 myr after top-M18r, but corresponds relatively close to the Boreal/Sub-Boreal usage of Ryazanian/Volgian and Purbeckian/Bolonian regional stage boundaries (e.g., Figure 26.8).

In GTS2012, we utilize the base of chron M18r for assigning the numerical age to the top of the Jurassic.

26.2. JURASSIC STRATIGRAPHY

The ammonite successions of Europe have historically served as global primary standard for the Jurassic. Other biostratigraphic, magnetostratigraphic, chemostratigraphic and other events are typically calibrated to these standard European ammonite zones. An extensive compilation and inter-correlation of Jurassic stratigraphy of European basins was coordinated by Hardenbol *et al.* (1998), but later workers have greatly enhanced or revised portions of these comprehensive chart series. Correlation outside of these European basins, especially to terrestrial realms, remains relatively uncertain. Due to pronounced climate–oceanic provincialism during the Late Jurassic, this interval has been the most challenging for global correlations (e.g., summary by Zeiss, 2003).

26.2.1. Marine Biostratigraphy

26.2.1.1. Ammonites

Alfred Oppel (1856–1858) developed the concept of a biostratigraphic zone, and used ammonites to define two-thirds of his 33 Jurassic zones. Jurassic ammonite zonations have undergone constant revision since Oppel, and the Jurassic is currently subdivided into 70 to 80 zones and typically 160 to 170 subzones in each faunal realm. Reviews of the development, definitions and inter-correlation of European ammonite zonations are in Thierry (1998, including correlation charts), Krymholts *et al.* (1982), and Groupe Français d'Étude du Jurassique (1997). The precision of correlation of the northwest European standard to the regional ammonite zones of western North America (e.g., Pálfy *et al.*, 2000a) and of Russia (e.g., Zakharov *et al.*, 1997; Rogov, 2004, 2010a,b; Rogov and Zakharov, 2009) and of other regions depends on the varying provinciality of genera and index species. Globally, up to seven suborders of Order Ammonoidea are recognized; and these range through up to 20 biogeographic provinces and subprovinces (Page, 2008).

Only a single genus, the *Psiloceras* of basal Jurassic, appears to have survived the end-Triassic mass extinction of the ammonoids. This single form, indeed probably only a single *P. spelae* species, rapidly diversified into over twenty ammonite genera during the Hettangian (Guex, 1995; Guex *et al.*, 2004). A multi-stage "Early Toarcian crisis" caused a lesser bottleneck, followed by a rapid recovery (e.g., Dera *et al.*, 2010). In the Bajocian, the first Perisphinctoidea evolved and became the globally most important ammonites of the Late Jurassic (Page, 2008).

Belemnites, the cousins of the ammonites, have local Jurassic zonations that provide correlation to the stage or substage level (e.g., Combemorel, 1998).

Nomenclature for many of the Jurassic ammonite zones can be misleading. According to the International Stratigraphic Code, biostratigraphic zones are named according to the genus and species of associated taxa (e.g., *Kotetishvilia compressissima* Zone begins with the lowest occurrence of that ammonite taxon) combined with the appropriate term for the kind of biostratigraphic unit. This system is followed for most Paleozoic, Triassic and Cretaceous ammonoid zonations. In contrast, some Jurassic ammonite specialists in northwest Europe advocate and apply "Standard Chronozones", in which the nomenclature is not directly associated with the temporal range of the name-giving species (e.g., Callomon, 1985, 1995). Therefore, a standard Eudoxus Zone (capitalized, non-italics, no genus designation) of the Kimmeridgian in southern England has a base defined at bed E1 at a quarry near Westbury, Wiltshire (Birkelund *et al.*, 1983), and continues to the base of the Autissiodorensis Zone, which was assigned as the top of Flats Stone Band bed at beach exposures of the Kimmeridge Clay near the village of Kimmeridge, Dorset (Cox and Gallois, 1981; tables in Cox, 1990). In some cases, the "name-species" may not appear until quite high in its "Standard Chronozone" – e.g., the lowest occurrence of ammonite *Cardioceras cordatum* is in the upper third of the "Cordatum Zone". Unfortunately, the basal limits of many of these "standard chronozones" remain unstandardized among regions and specialists and have changed through the decades; which often creates correlation

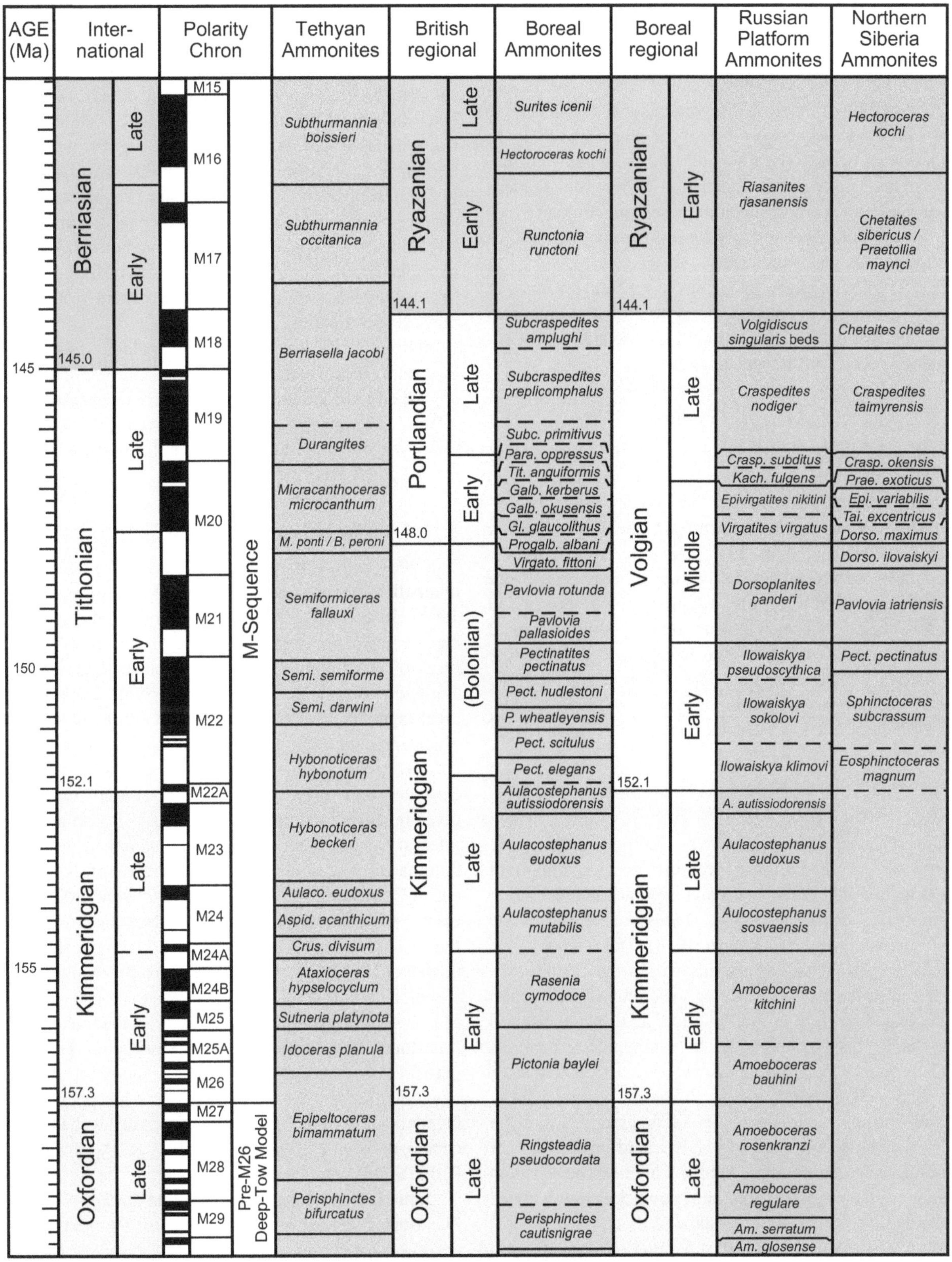

AGE (Ma): 145; 150; 155

International: Berriasian (Late; Early); 145.0; Tithonian (Late; Early); 152.1; Kimmeridgian (Late; Early); 157.3; Oxfordian (Late)

Polarity Chron: M15; M16; M17; M18; M19; M20; M21; M22; M22A; M23; M24; M24A; M24B; M25; M25A; M26; M27; M28; M29 — M-Sequence; Pre-M26 Deep-Tow Model

Tethyan Ammonites: *Subthurmannia boissieri*; *Subthurmannia occitanica*; *Berriasella jacobi*; *Durangites*; *Micracanthoceras microcanthum*; *M. ponti / B. peroni*; *Semiformiceras fallauxi*; *Semi. semiforme*; *Semi. darwini*; *Hybonoticeras hybonotum*; *Hybonoticeras beckeri*; *Aulaco. eudoxus*; *Aspid. acanthicum*; *Crus. divisum*; *Ataxioceras hypselocyclum*; *Sutneria platynota*; *Idoceras planula*; *Epipeltoceras bimammatum*; *Perisphinctes bifurcatus*

British regional: Ryazanian (Late; Early); 144.1; Portlandian (Late; Early); 148.0; Kimmeridgian ((Bolonian); Late; Early); 157.3; Oxfordian (Late)

Boreal Ammonites: *Surites icenii*; *Hectoroceras kochi*; *Runctonia runctoni*; *Subcraspedites amplughi*; *Subcraspedites preplicomphalus*; *Subc. primitivus*; *Para. oppressus*; *Tit. anguiformis*; *Galb. kerberus*; *Galb. okusensis*; *Gl. glaucolithus*; *Progalb. albani*; *Virgato. fittoni*; *Pavlovia rotunda*; *Pavlovia pallasioides*; *Pectinatites pectinatus*; *Pect. hudlestoni*; *P. wheatleyensis*; *Pect. scitulus*; *Pect. elegans*; *Aulacostephanus autissiodorensis*; *Aulacostephanus eudoxus*; *Aulacostephanus mutabilis*; *Rasenia cymodoce*; *Pictonia baylei*; *Ringsteadia pseudocordata*; *Perisphinctes cautisnigrae*

Boreal regional: Ryazanian (Early); 144.1; Volgian (Late; Middle; Early); 152.1; Kimmeridgian (Late; Early); 157.3; Oxfordian (Late)

Russian Platform Ammonites: *Riasanites rjasanensis*; *Volgidiscus singularis* beds; *Craspedites nodiger*; *Crasp. subditus*; *Kach. fulgens*; *Epivirgatites nikitini*; *Virgatites virgatus*; *Dorsoplanites panderi*; *Ilowaiskya pseudoscythica*; *Ilowaiskya sokolovi*; *Ilowaiskya klimovi*; *A. autissiodorensis*; *Aulacostephanus eudoxus*; *Aulocostephanus sosvaensis*; *Amoeboceras kitchini*; *Amoeboceras bauhini*; *Amoeboceras rosenkranzi*; *Amoeboceras regulare*; *Am. serratum*; *Am. glosense*

Northern Siberia Ammonites: *Hectoroceras kochi*; *Chetaites sibericus / Praetollia maynci*; *Chetaites chetae*; *Craspedites taimyrensis*; *Crasp. okensis*; *Prae. exoticus*; *Epi. variabilis*; *Tai. excentricus*; *Dorso. maximus*; *Dorso. ilovaiskyi*; *Pavlovia iatriensis*; *Pect. pectinatus*; *Sphinctoceras subcrassum*; *Eosphinctoceras magnum*

FIGURE 26.8 **Inter-regional correlation of Late Jurassic and earliest Cretaceous ammonite zones and regional stages.** Magnetostratigraphic correlations to the marine magnetic anomaly M-sequence and cycle stratigraphy provide the reference numerical scale and absolute durations for Tethyan ammonite zones (e.g., Galbrun, 1984; Ogg et al., 1984, 2010a; Boulila et al., 2008b, 2010a,b; Pruner et al., 2010). Sub-Boreal ammonite calibrations to Tethyan zone and/or M-sequence incorporates magnetostratigraphy (e.g., Ogg et al., 1994; Przybylski et al., 2010a,b), cycle stratigraphy for scaling of "Bolonian" succession (e.g., Huang et al., 2010b), and biostratigraphic correlations (e.g., Wierzbowski et al., 2006; Główniak et al., 2008; Wimbledon et al., 2011). Boreal calibrations of Russian Platform to Northern Siberia and to Sub-Boreal zones are mainly biostratigraphic (e.g., Rogov and Zakharov, 2009; Rogov, 2010b; Rogov and Price, 2010; Harding et al., 2011) supplemented with magnetostratigraphy for the Late Volgian of Northern Siberia (Houša et al., 2007). However, the reference sections for Boreal zonations and for the Portlandian–Ryazanian of the Sub-Boreal realm commonly have ammonite levels separated by stratigraphic gaps, therefore this lack of continuity precludes exact inter-regional correlations (e.g., charts in Harding et al., 2011). In addition, some ammonite zones are poorly defined (e.g., Cope, 2007, recommended abandoning the *Paracraspedites oppressus* Zone of England). *This summary diagram is modified from Ogg et al. (2010b) with recommendations by William Wimbledon and Mikhail Rogov.*

problems for non-ammonite workers. For example, the base of the Kimmeridgian is accepted as the base of the Baylei Zone, but, as of 2010, there is disagreement whether this zone begins with the lowest occurrence of the ammonite *Pictonia flodigarriensis* or the slightly later *Pictonia densicostata*. Both systems – a regional designation of Jurassic standard chronozones versus a nomenclature based on ammonite biozones – have defenders (mainly Jurassic ammonite specialists in the former case) and critics.

For clarity in our charts and tables (e.g., Tethyan zones in Table 26.3), we have included the genera of the ammonite "index" species, but with a caution that these are not always the "guide" species of the named zone.

26.2.1.2. Fish and Marine Reptiles

Ray-finned fish of the primitive teleost group, to which most living bony fish belong, were on the rise, but there is no established zonation. The earliest large filter feeders, pachycormids up to 9 m long, are known from the Middle Jurassic into the Cretaceous (Calvin, 2010; Friedman *et al.*, 2010).

The most famous dwellers of the Jurassic seas were abundant large marine reptiles (Sauropterygia) comprised of long-necked plesiosaurs ("near lizard"), shorter-necked pliosaurs ("more lizard") and the streamlined dolphin-like ichthyosaurs ("fish lizard"). The Callovian and Late Jurassic assemblages included the pliosaur, *Liopleurodon*, the world's largest known carnivore of over 15 m in length (Page, 2005).

26.2.1.3. Other Marine Macrofauna

Brachiopod zonations for northwest Europe and for the northern part of the Tethyan province provide important markers within individual basins and approach the resolution of ammonite zones in some stages of the Jurassic (e.g., Alméras *et al.*, 1997; Laurin, 1998). The *Buchia* genus of bivalves is used for correlations within the Oxfordian–Hauterivian strata, especially within the circum-Boreal realm of Eurasia and western North America, and several buchiid zones have been calibrated to ammonite zonations (e.g., Jeletzky, 1965; Surlyk and Zakharov, 1982; Zakharov, 1987; Sha and Fürsich, 1994; Zakharov *et al.*, 1997). However, the correlation potential of brachiopods and the slower-evolving successions of bivalves and gastropods are compromised by their benthic habits, which can be reflected in ecological-facies associations and provincialism (reviewed in Hallam, 1976; Cope *et al.*, 1980b).

Ostracodes are small (0.2 to 1.5 mm) crustaceans with calcified bivalved shells. They are a major constituent of shallow marine and brackish benthic faunas, and their evolving assemblages enable a biostratigraphic resolution approaching that provided by ammonite zones, especially within portions of the Lower and Middle Jurassic (e.g., reviews by Cox, 1990; Colin, 1998).

26.2.1.4. Foraminifers and Calpionellids

Jurassic benthic Foraminifera datums and zonations have been developed for both calcareous and agglutinated forms. Compilations are available for the British and North Sea region (e.g., Copestake *et al.*, 1989), for larger benthic Foraminifera in the Tethyan domain (Peyberne, 1998), and smaller benthic Foraminifera in European basins (Ruget and Nicollin, 1998). The earliest planktonic Foraminifera genus is meroplanktonic *Conoglobigerina* (only in Bajocian) and the first fully planktonic *Globuligerina* occurs from Bathonian onward (Simmons *et al.*, 1997).

Calpionellids are vase-shaped pelagic microfossils of uncertain origin, which appeared in the late Tithonian and continued until the middle of the Early Cretaceous (Remane, 1985). They provide important correlation markers, especially in pelagic carbonates of the Tethyan-Atlantic seaway (reviewed by Remane, 1998).

26.2.1.5. Calcareous Nannofossils

The Mesozoic history of nannofossils consists of distinct diversification episodes (summarized in de Kaenel *et al.*, 1996; Bown *et al.*, 2004; Erba, 2006). The earliest known calcareous nannofossils appeared in the mid-Triassic; but most of the initial robust types vanished during the end-Triassic extinctions. A single surviving coccolithophorid algal species diversified near the Hettangian/Sinemurian boundary; and the major radiation of Jurassic placolith coccoliths (plates from coccolithophores) took place during Pliensbachian to early Toarcian. A second major pulse of diversification within the Aalenian–Bajocian boundary interval included the genus *Watznaueria*, which would dominate most assemblages during the rest of the Jurassic and entire Cretaceous (e.g., Erba, 1990; Olivero, 2008). A major overturn of Tethyan nannofossil assemblages took place in the late Tithonian. The Jurassic closed with the appearance of nannoconids, a heavily calcified form, which are often regarded as the first effective carbonate-producers in the Mesozoic oceans and contributed to the nannofossil-rich limestones that characterize the Cretaceous pelagic realm (Erba, 2006; Casellato *et al.*, 2008). The onset of calcareous nannoplankton, their size and general morphology and the diversification trends may be responses to the chemistry of the ocean–atmosphere system, especially pCO_2 and Mg/Ca ratio (Erba, 2006).

Jurassic nannofossil zonations and markers in the Boreal/Sub-Boreal realm are calibrated to ammonite zones in northwestern Europe (e.g., Bown *et al.*, 1988; Bown, 1998; von Salis *et al.*, 1998). Nannofossil datums in the Tethyan/Sub-Mediterranean realm are partially calibrated to ammonite zones (e.g., de Kaenel *et al.*, 1996; Mattioli and Erba, 1999; Mailliot *et al.*, 2006) and to late Jurassic magnetic polarity zones (e.g., Bralower *et al.*, 1989; Channell *et al.*, 2010; Casellato, 2010).

26.2.1.6. Dinoflagellate Cysts

Organic-walled cysts of dinoflagellates are an important correlation tool for the North Sea, where their datums are correlated to ammonite zones of the Boreal realm (e.g., Woollam and Riding, 1983; Riding and Ioannides, 1996; Ioannides *et al.*, 1998). Several of these markers also occur in the Tethyan realm, but the ranges and correlation to ammonite zones are not as well established (Habib and Drugg, 1983; Ioannides *et al.*, 1998). Independent dinocyst zonations have been developed for the Jurassic of Australia (Helby *et al.*, 1987), for the upper Jurassic of New Zealand (Wilson, 1984) and for other basins.

26.2.1.7. Radiolarians

Siliceous radiolarians are a major component of Jurassic pelagic sediments deposited under high productivity conditions, but their tests are rarely preserved jointly with aragonitic ammonite shells. Detailed radiolarian zonations for the Middle and Late Jurassic have been developed for the western margin of North America (e.g., Pessagno *et al.*, 1993, 2009), for Japan (Matsuoka and Yao, 1986; Matsuoka, 1992), for the former Tethyan seaways exposed in Europe (Baumgartner, 1987; INTERRAD Jurassic-Cretaceous Working Group, 1995; De Wever, 1998), and for the North Sea (Dyer and Copestake, 1989; Dyer, 1998).

These zonations can be partially correlated to each other; however, the calibration of the radiolarian assemblages to standard geological stages and to the reference ammonite scales of Europe has been challenging and controversial (e.g., Pessagno and Meyerhoff Hull, 1996). An example is the divergent correlations for the radiolarian assemblage overlying basalt at Ocean Drilling Program (ODP) Site 801, which provided a key age control on the Callovian–Oxfordian portion of the marine magnetic anomaly M-sequence and global spreading rates. The radiolarian assemblages from the basal and inter-pillow sediments were originally interpreted as "late Bathonian to early Callovian" (Shipboard Scientific Party, 1990; Matsuoka, 1992; reviewed in Ogg *et al.*, 1992), but were re-interpreted as equivalent to "middle Oxfordian" of western North America by Pessagno and Meyerhoff Hull (1996). In contrast, the same radiolarian assemblages were assigned to "Bajocian" (Bartolini *et al.*, 2000; Bartolini and Larson, 2001) based partly on the zonal calibrations developed by the INTERRAD Jurassic-Cretaceous Working Group (1995) for the Mediterranean region. Possible contributing factors to this divergence are diachroneity of radiolarian or ammonite datums and ranges among basins, errors in taxonomy assignments, imprecise correlation of radiolarian markers to regional ammonite stratigraphy, and miscorrelation of ammonite assemblages among paleogeographic provinces (Pessagno and Meyerhoff Hull, 1996; Kiessling *et al.*, 1999; reviewed in Pessagno *et al.*, 2009).

26.2.2. Terrestrial Biostratigraphy

26.2.2.1. Dinosaurs

Dinosaurs are the most famous Jurassic fauna and the largest land animals that ever lived. Even though they dominated the terrestrial herbivores and carnivores, the biostratigraphic ranges of these enormous creatures are not well established relative to marine-based stages. A set of six generalized "land-vertebrate faunachrons" (LVFs) can be partially correlated across Pangea (Lucas, 2009).

Crocodile-like crurotarsan archosaurs were the main competitors of dinosaurs during the Late Triassic, but their mass extinction in latest Triassic enabled the dinosaurs to flourish and diversify. This crurotarsan extinction either coincided with the major mass extinction of marine life induced by environmental impacts of the CAMP flood basalt eruptions (e.g., Whiteside *et al.*, 2010) or began slightly earlier (Lucas *et al.*, 2011). The Jurassic begins with the appearance of the crocodylomorph *Protosuchus*, defining the base of the Wassonian LVF. The Dawan LVF begins with the appearance of theropod *Megapnosaurus* ("*Syntarsus*"), a 3-m long bipedal dinosaur, in approximately mid-Sinemurian.

No dinosaur genera are shared among any of the continents through most of the Middle Jurassic, despite the presumed high mobility of the enormous sauropod dinosaurs (Lucas, 2009). A vaguely calibrated Dashanpuan LVF of possible Bajocian age and a Tuojiangian LVF of possible Bathonian–Callovian age are characterized by stegosaur and sauropod assemblages.

The sparse record of dinosaurs in Lower and Middle Jurassic strata contrasts with the rich fossil deposits of the Upper Jurassic, such as the Morrison Formation of western North America. This Kimmeridgian–Tithonian interval included the carnivores of *Allosaurus* and *Ceratosaurus*, the huge sauropod *Brachiosaurus*, and moderate-sized ornithopids (iguanodontids and hypsilophodontids). The Comobluffian LVF, named from a Morrison excavation site, is coeval with the Ningjiangouan LVF of China (Lucas, 2009). These deposits can be partially correlated to the Kimmeridgian stage through magnetostratigraphy (e.g., Steiner *et al.*, 1987), which is consistent with U-Pb ages of 152–156 Ma (Trujillo *et al.*, 2006, 2008).

Flying over this Jurassic park were pterosaurs, which had evolved during the Late Triassic. The first bird, the renowned *Archaeopteryx* fossils from lagoonal settings in southern Germany, appeared in the Tithonian. Theropod ancestors to the future birds are known from the Oxfordian (Hu *et al.*, 2009; Stone, 2010; Choiniere *et al.*, 2010).

26.2.2.2. Plants, Pollen and Spores

Terrestrial ecosystem stratigraphy has uncertain correlation to the Jurassic marine record. Sporomorph records suggest that the land plant ecosystems did not change significantly during the interval of end-Triassic mass extinctions in the marine realm, and paleobotanists often combine the Rhaetian flora with Early Jurassic ones (e.g., Dobruskina, 1994; van Konijnenburg-van Cittert, 2008). However, some regional macroplant records indicate up to an 85% decline in species richness on the local scale, partly because some plant types are not well represented by the sporomorph record (Mander *et al.*, 2010). Cycads, ginkgos, and conifers have diverse assemblages that respond to local climate (e.g., review by van Konijnenburg-van Cittert, 2008, for UK), but only a few regional spore–pollen zones/ranges have been published for selected time intervals (e.g., Srivastava, 1987, for Normandy–Germany; Morris *et al.*, 2009, for offshore Norway). Transitional forms between gymnosperms and the angiosperms plants which diversify in the Cretaceous, are interpreted from floral assemblages from North China that are below an ash bed dated as ~162 Ma, hence are early Kimmeridgian or older (Chang *et al.*, 2009; see Table 26.1).

26.2.3. Physical Stratigraphy

26.2.3.1. Magnetostratigraphy

The details and calibrations of the Jurassic magnetic polarity time scale have been considerably enhanced relative to the schematic patterns given in GTS2004. The Jurassic magnetic polarity patterns are primarily compiled from ammonite-bearing sections in Europe, with the Sinemurian and the Callovian stages remaining the main uncertain intervals.

The Late Jurassic M-sequence of marine magnetic anomalies has been correlated via magnetostratigraphy to ammonite zones. Polarity chrons prior to the Callovian do not have a corresponding marine magnetic anomaly sequence to provide an independent nomenclature or scaling system. Incorporating abbreviations derived from the corresponding ammonite zone or stage is one option (e.g., Ogg, 1995); and a similar philosophy was applied at the series-level for the Triassic compilation by Hounslow and Muttoni (2010). However, until the polarity pattern spanning each stage has been adequately verified, a standardized nomenclature system is premature.

Early Jurassic through Aalenian Magnetic Polarity Scale

The Hettangian stage is dominated by normal-polarity in the Paris Basin (Yang *et al.*, 1996), England (Hounslow *et al.*, 2004) and in the Hartford Basin of eastern North America (Kent *et al.*, 1995; Kent and Olsen, 2008) with only three brief well-documented reversed-polarity zones ("H24r–H26r" of Kent and Olsen, 2008). Cycle stratigraphy of the Hartford Basin sequence enables assignment of the polarity ages relative to the radio-isotope-dated CAMP basalts (Kent and Olsen, 2008), which suggests that the uppermost reversed-polarity zone "H26r" is at the base of Sinemurian.

The Sinemurian in outcrop sections is dominated by reversed-polarity with relatively brief normal-polarity zones (e.g., Channell *et al.*, 1984; Steiner and Ogg, 1988). A detailed magnetostratigraphic study of a Paris Basin core was interpreted as indicating numerous additional brief polarity changes (Yang *et al.*, 1996); but the absence of such features in outcrop studies suggests that most of these were single-sample anomalies or possible other artifacts. The approximate ammonite-zoned stratigraphy of that borehole provides better biostratigraphic constraints on the main features of the Sinemurian polarity pattern, and a generalized column from that study is used for the GTS2012 scale.

The lower Pliensbachian (Carixian substage) is also dominated by reversed polarity. The main polarity patterns are consistent between the major studies in southern Switzerland (Horner, 1983; Heller, 1983) and a Paris Basin borehole (Moreau *et al.*, 2002). However, a detailed study in central Italy (Speranza and Parisi, 2007) is difficult to correlate to the first two compilations, and this contrast may be a combination of different ammonite subzonal schemes, hiatuses and redeposited intervals, variable sedimentation rates and other factors. Therefore, the polarity patterns within each ammonite zone according to the synthesis by Moreau *et al.* (2002) are used in GTS2012, based on the mid-points of their reinterpreted ammonite zone boundaries. Additional sections are required to determine a reproducible bio-magnetostratigraphy. Speranza and Parisi (2007) have interpreted two unusually long-duration intervals of transitional or intermediate polarity during earliest Pliensbachian, but these await confirmation.

The upper Pliensbachian (Domerian substage) through Aalenian polarity pattern is primarily from a detailed study in southern Switzerland (Horner, 1983; Horner and Heller, 1983). This pattern has been enhanced and calibrated to other regional ammonite zones in several other study areas. The upper Pliensbachian bio-magnetostratigraphy has been revised using a borehole in the Paris Basin and reanalysis of the Switzerland ammonite datums (Moreau *et al.*, 2002). Details across the Pliensbachian–Toarcian boundary in southern Spain verified the main features of the southern Switzerland pattern, and enabled correlation at a subzonal level (Comas-Rengifo *et al.*, 2010b).

The Toarcian bio-magnetostratigraphy composite is also based on the southern Switzerland suite (Horner, 1983; Horner and Heller, 1983) with enhanced biostratigraphy from the relatively condensed type Toarcian section at Thouars in France (Galbrun, 1988, revised using the inter-province ammonite zone correlation by Thierry, 1998) and the expanded lower and middle Toarcian in Spain (Osete *et al.*, 2007; Comas-Rengifo *et al.*, 2010b). The proportional scaling of magnetic zones within ammonite subzones in the lower

TABLE 26.1 Selected Radio-Isotopic Ages from Terrestrial and Flood-Basalt Deposits of Uppermost Triassic, Jurassic and Lowermost Cretaceous

Selected Jurassic Radio-isotopic Dates – Terrestrial strata or Flood basalts

NOTE: This is only a subset of the published age suites for most of these episodes. See the selected references for reviews.

EVENT	Primary controls (*)	Source (location)	Age for GTS2012; uncertainty (95%) in myr	Annotation; Dated Material; Method	**Locality; formation; biostratigraphy**	Primary Reference	Review Reference (and evaluation comments from references)
Morokweng crater							
	X	Morokweng	145.2 ±0.8	Recrystallized zircons; U-Pb	Morokweng crater (70 km diameter); Kalahari desert of South Africa	Hart *et al.*, 1997; Koeberl *et al.*, 1997	Jourdan et al., 2009b, evaluation: "Zircon age is indistinguishable from biotite $^{40}Ar/^{39}Ar$ age at 143 ± 4 Ma."
Morrison Formation (famous dinosaur deposits, usually considered to be Kimmeridgian)							
	X	upper Morrison	152.37 ±0.62	Trujullo'06; zircons from bentonite; U-Pb SHRIMP	south of Laramie; southeastern Wyoming. upper Morrison Formation, smectitic mudstone. Dinosaurs – Apatosaurus, Allosaurus, Camarasaurus; etc.	Trujillo *et al.*, 2008.	
	X	upper Morrison	156.3 ±2	Trujullo'06; zircons from bentonite; U-Pb SHRIMP	Ninemile Hill, Medicine Bow, southeastern Wyoming. upper third of Morrison Formation, smectitic mudstone. Same level as dinosaur Quarry Nine, Como Bluff, Wyoming. Magnetostrat by Steiner would suggest Kimmeridgian (M22-M24 interval).	Trujillo *et al.*, 2006.	
	secondary guide	upper Morrison	149 ±1	PSM55; sanidines from bentonite; Ar-Ar, Laser fusion	Brushy Basin Member 6, upper Morrison Formation. Set of 6 stratigraphic intervals in upper part of Brushy Basin member above a change in clay mineralogy, considered to be a regional hiatus. Ostracod and charophyte assemblages suggest Kimmeridgian in lower part to Tithonian in upper part.	Kowallis *et al.*, 1998.	Pálfy *et al.*, 2000a. The suite of 6 ages from the Brushy Basin Member are stratigraphically consistent. Pálfy *et al.* (2000a) increase all Ar-Ar uncertainties by an additional 2.5 myr to indicate 'external errors' due to calibration of radiometric decay constants and other factors.
	secondary guide	upper Morrison	149.9 ±.8	PSM54; sanidines from bentonite; Ar-Ar, Laser fusion	Brushy Basin Member 5, upper Morrison Formation	Kowallis *et al.*, 1998.	Pálfy *et al.*, 2000a.
	secondary guide	upper Morrison	150.2 ±1	PSM53; sanidines from bentonite; Ar-Ar, Laser fusion	Brushy Basin Member 4, upper Morrison Formation	Kowallis *et al.*, 1998.	Pálfy *et al.*, 2000a.
	secondary guide	upper Morrison	150.2 ±1.1	PSM52; sanidines from bentonite; Ar-Ar, Laser fusion	Brushy Basin Member 3, upper Morrison Formation	Kowallis *et al.*, 1998.	Pálfy *et al.*, 2000a.
	secondary guide	upper Morrison	151.1 ±1	PSM51; sanidines from bentonite; Ar-Ar, Laser fusion	Brushy Basin Member 2, upper Morrison Formation	Kowallis *et al.*, 1998.	Pálfy *et al.*, 2000a.
	secondary guide	upper Morrison	151.2 ±.5	PSM50; sanidines from bentonite; Ar-Ar, Laser fusion	Brushy Basin Member 1, upper Morrison Formation	Kowallis *et al.*, 1998.	Pálfy *et al.*, 2000a.
N.China (angiosperm precursors)							
	secondary guide	Lanqi Formation	161.7 ±0.4	LQ07-1; sanidines from bentonite; Ar-Ar, Laser fusion	Yujiagou village of Beipaio City, Lioining, North China. Lanqi Formation (basal), above Haifanggou Formation bearing plant assemblages that include interpreted precursors to angiosperms.	Chang et al., 2009.	

(Continued)

TABLE 26.1 Selected Radio-Isotopic Ages from Terrestrial and Flood-Basalt Deposits of Uppermost Triassic, Jurassic and Lowermost Cretaceous—cont'd

Selected Jurassic Radio-isotopic Dates – Terrestrial strata or Flood basalts							
Bajocian							
	secondary guide	mid-Bajoc., Utah	167.3 ±0.8	PSM37; sanidines from bentonite; Ar-Ar	Gunlock, Carmel Formation, Utah, USA. Bivalve correlation of Carmel Formation to ammonite-bearing Twin Creek Formation indicates an age of late Early to early Late Bajocian (according to Imlay, 1967, 1980)	Kowallis *et al.*, 1993	Pálfy *et al.*, 2000a. The suite of 6 ages from the Brushy Basin Member are stratigraphically consistent. Pálfy *et al.* (2000a) increase all Ar-Ar uncertainties by an additional 2.5 myr to indicate 'external errors' due to calibration of radiometric decay constants and other factors. Age is slightly older than with previous K-Ar and Rb-Sr methods.
Karoo Volcanic Province (early Toarcian)							
	X	Karoo sills	182.7 ±0.4	zircons from dolerites; U-Pb	Karoo Basin, South Africa. "sampled 48 pegmatites found in thick sill intrusions from different parts of the basin, separated by as much as 1000 km. Thirteen contained fresh euhedral and inclusion-free zircons suited for high resolution dating." (no details on localities within the abstract). Usually considered as the cause for onset of a negative carbon-isotope excursion and widespread organic-rich sediments of mid-lower Toarcian.	Polteau *et al.*, 2010.	Mean of 13 sites; which had range of 182.3 to 183.0 Ma (all overlapped within the average uncertainty of 0.4 myr).
	X	Karoo sills	182.5 ±0.4	G39974-596m; zircons from dolerites; U-Pb ID-TIMS (multi-grain)	Borehole G39974, Calvinia area, western Karoo Basin, South Africa. Ecca Group, pegmatite interval in lowermost sill in borehole; below Prince Albert formation, 596m level in borehole.	Svensen *et al.*, 2007.	"The five analyses plot on or slightly to the right of the Concordia curve but have uniform 206Pb/238U ages of 182.7 to 182.2 Ma, which are reflected in the Concordia age of 182.5±0.4 Ma."
	X	Karoo lavas	183.4 ±1.6	plagioclase from basalt; Ar-Ar	Lesotho, South Africa. Lesotho basalts – from the main Normal-polarity upper interval (NOTE: mean age is anomalously younger than mean age of underlying reversed-polarity flows).	Jourdan *et al.*, 2007.	
	X	Karoo lavas	182.7 ±0.7	plagioclase from basalt; Ar-Ar	Lesotho, South Africa. Lesotho basalts (different locations within lava sequence). Mean of 4 ages from lower Reversed polarity suite, and 1 age from overlying normal-polarity suite (Fig. 3 in Jourdan *et al.*, 2007).	Jourdan *et al.*, 2007.	
	X	Karoo lavas	179.2 ±1.8	NN13-NN45; basalt; K-Ar (not Ar-Ar) Cassignol-Gillot technique	Naude's Nek in South Africa, near the southern border of Lesotho. upper 650 m (normal magnetic polarity) of the traps of Lesotho lava pile, the main remnant of the Karoo traps. Usually considered as the cause for onset of a negative carbon-isotope excursion and widespread organic-rich sediments of mid-lower Toarcian.	Moulin *et al.*, 2011.	
	X	Karoo lavas	184.8 ±2.6	NN57; basalt; K-Ar (not Ar-Ar) Cassignol-Gillot technique	Naude's Nek in South Africa, near the southern border of Lesotho. lower 115 m (reversed magnetic polarity) of the traps of Lesotho lava pile, the main remnant of the Karoo traps. Possible cause for onset of a negative carbon-isotope excursion of basal Toarcian.	Moulin *et al.*, 2011.	
end-Rhaetian CAMP basalts – See Triassic table							

Argon-argon ages have been adjusted to a FCs monitor standard of 28.201 Ma. There are many other studies of these episodes or deposits; see references for reviews and syntheses. The 'Annotation' column is the orginal sample identification in the publication or a shortened designation.

and middle Toarcian synthesis by Osete *et al.* (2007) was used for the GTS2012 chart.

Details on the magnetostratigraphy at the Toarcian–Aalenian boundary and Aalenian–Bajocian boundary come from the respective GSSPs at Fuentelsaz in Spain (Goy *et al.*, 1996; Cresta *et al.*, 2001) and Capo Mondego in Portugal (Henriques *et al.*, 1994; Pavia and Enay, 1997).

M-Sequence of Marine Magnetic Anomalies and Bajocian through Late Jurassic Magnetic Polarity Scale

The M-sequence of marine magnetic anomalies provides a reference pattern and chron nomenclature for the magnetostratigraphy from Upper Jurassic through Lower Cretaceous fossiliferous sections. The most commonly used M-sequence is derived from the Pacific spreading centers as a combination of the Hawaiian and Japanese magnetic lineations with a slowly varying spreading rate (see Chapter 5 of this volume). The oldest magnetic anomaly that is documented in all ocean basins is M25.

Numerous high-resolution magnetostratigraphy studies of Kimmeridgian and Tithonian strata have calibrated ammonites, calpionellids and calcareous nannofossils from the Sub-Mediterranean faunal province and DSDP cores from the central Atlantic to polarity chrons M25 through M18 (e.g., Ogg *et al.*, 1984, 1991; Lowrie and Ogg, 1986; Ogg, 1988; Bralower *et al.*, 1989; Speranza *et al.*, 2005; Channell *et al.*, 2010; Pruner *et al.*, 2010; Przybylski *et al.*, 2010a). These calibrations constrain the relative duration of each ammonite zone within the Kimmeridgian and Tithonian stages. In turn, orbital-climate cycles recorded in Kimmeridgian and Tithonian strata enable computation of actual durations for some of the zones and polarity chrons and, thereby a determination of oceanic spreading rates for the associated marine magnetic anomalies (e.g., Ogg *et al.*, 2010b; see Chapter 5 of this volume). The succession and relative widths of the modeled marine magnetic anomalies from M18r (base of Cretaceous) through M25r of the "classic" Hawaiian magnetic lineation M-sequence (e.g., Larson and Hilde, 1975) have been an excellent match to these cycle-magnetic composites, whereas a new model that averaged three Pacific lineation sets (Tominaga and Sager, 2010) seems to be slightly distorted. Calibration of Boreal sections to this magnetic time scale has been partially achieved for the Kimmeridgian and Tithonian (e.g., Ogg *et al.*, 1994, 2010b; Houša *et al.*, 2007; Przybylski *et al.*, 2010a). Reversed-polarity Chron M25r is early Kimmeridgian.

Marine magnetic anomalies older than Chron M25 are both closer spaced and lower in amplitude. Magnetic profiles over pre-M26 oceanic crust in the Pacific (Cande *et al.*, 1978; Handschumacher *et al.*, 1988) have been supported and extended by deep-tow surveys (Sager *et al.*, 1998; Tivey *et al.*, 2006; Tominaga *et al.*, 2008) indicating a possible set of between 50 to 100 polarity chrons (semi-arbitraily grouped into clusters with a nomenclature of "M26" through "M44"). Pacific IODP Site 801 is within the marine anomaly cluster "M42", and the interpreted Bajocian–Bathonian age of Site 801 implies that marine magnetic anomalies M26 through M44 should span the Oxfordian, Callovian, Bathonian and potentially the Bajocian stage.

However, unlike the standard sea-surface polarity model of M0r through M25r, the interpretation of the deep-tow surveys is more ambiguous. A direct modeling of the deep-tow signals emphasizes the narrow paleomagnetic intensity fluctuations near the oceanic crust; thereby creating a model of numerous short-duration, low-amplitude polarity intervals. Alternatively, if the deep-tow data are projected to a mid-depth level, then close-spaced narrow-width fluctuations are diminished and the wider-width features are emphasized. Therefore, both geophysics models for pre-M25r marine magnetic anomaly interpretations are published, with the hypothesis that the actual reversal history of the Earth's geomagnetic field is perhaps between these alternatives (e.g., Tominaga *et al.*, 2008). Indeed, the outcrop-based magnetostratigraphy for the Oxfordian through Bajocian seems consistent with an intermediate model for such pre-M25 marine magnetic anomalies (Figure 26.9).

Magnetostratigraphic studies in Oxfordian strata with Sub-Mediterranean ammonite zonation in Spain and Poland yielded a frequency of reversal patterns that seemed consistent with the extended Pacific model (e.g., Steiner *et al.*, 1986; Juárez *et al.*, 1994, 1995; Ogg and Gutowski, 1996), but correlations were ambiguous. Scaling of Oxfordian ammonite zones and subzones by cycle stratigraphy (Boulila *et al.*, 2008a, 2010a) enabled a better approximation of the durations of the polarity zones observed within each of those subzones. Therefore, a composite magnetostratigraphy compiled from approximately 20 Oxfordian sections in Europe with partial cycle-scaling verified the main features of the deep-tow pattern from M26 through M37 (Przybylski *et al.*, 2010b; Ogg *et al.*, 2010a). Chron M37n of that marine magnetic anomaly model was correlated to the Callovian–Oxfordian boundary.

Magnetostratigraphy investigations of Callovian strata in England, France and Poland are generally dominated by normal polarity with relatively few reversed-polarity intervals (e.g., Ogg *et al.*, 1990; Channell *et al.*, 1990; Belkaaloul *et al.*, 1995, 1997; Callomon and Dietl, 2000; Guzhikova *et al.*, 2010; and unpublished sections of Ogg *et al.*). These reversed-polarity clusters are mainly in the Bathonian–Callovian boundary interval, within the *Macrocephalites gracilis* ammonite zone, two in the *Erymnoceras coronatum* Zone, and in the latest Callovian. Similar normal-polarity dominance is interpreted for marine magnetic anomaly clusters M37–upper M39; and the reversed-polarity clusters are correlated to similar features in the mid-depth anomaly model (Figure 26.9). The implied span of Callovian-aged Pacific crust encompassed by anomalies M37–M39 is approximately half of the span of the Oxfordian-aged crust (intervals of anomalies M26–M37).

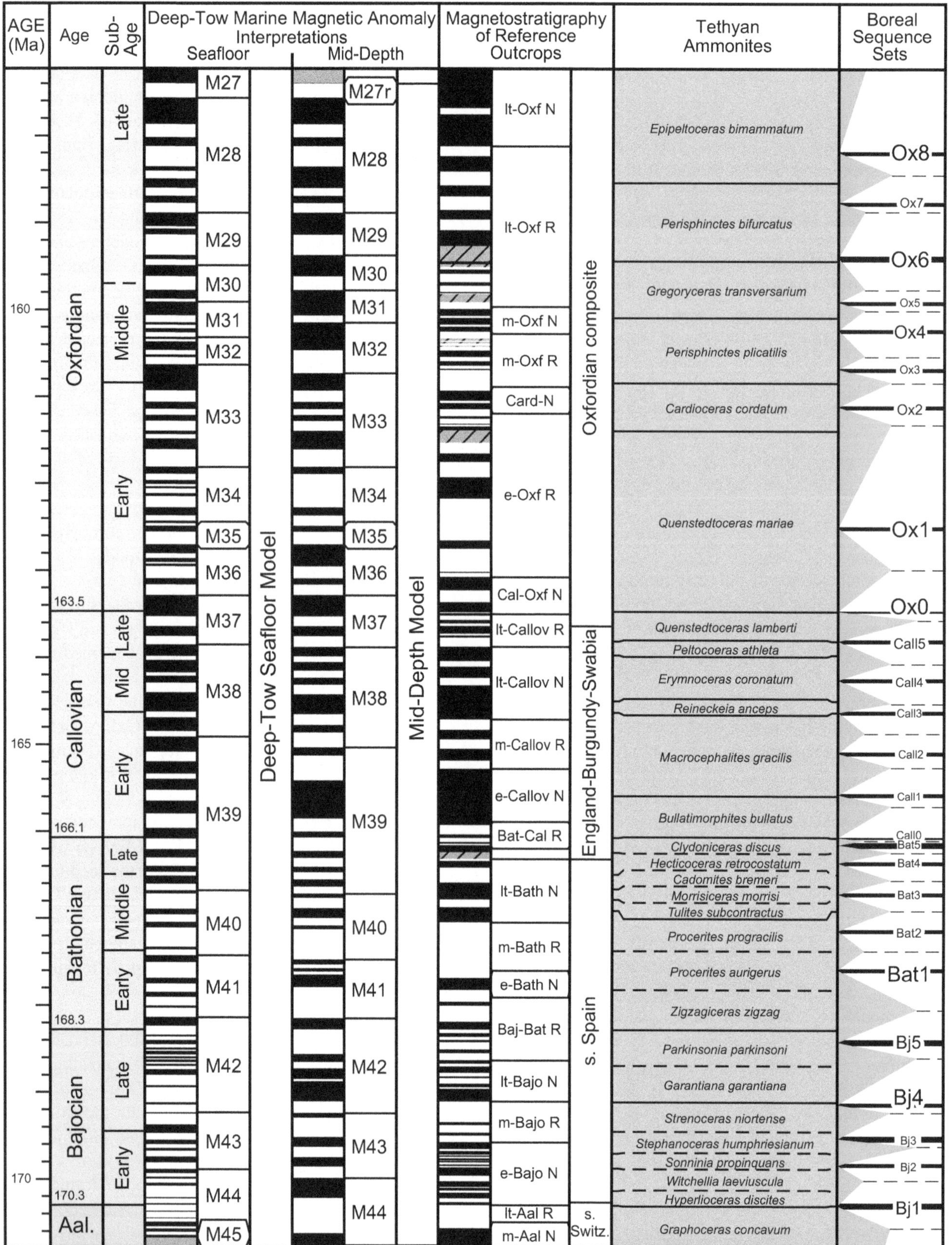

FIGURE 26.9 **Magnetostratigraphic correlations to possible geomagnetic polarity time scales as interpreted from deep-tow marine magnetic anomalies (Tominaga *et al.*, 2008)**; see Chapter 5 of this volume for age model. The deep-tow data can be interpreted from the records near the seafloor as a series of relatively high-frequency events or by projecting those signals to a mid-depth level to enhance the higher-amplitude and more major events. Magnetostratigraphic records from a composite of ammonite-zoned land outcrops (see text for sources, with suggested nomenclature of main polarity features within each stage) suggest that the mid-depth model may be more appropriate. The suggested correlation enables assignment of Tethyan ammonite zones to the extended M-sequence age model (see Table 26.3). Bajocian–Bathonian–Callovian sequences were calibrated to these ammonite zones by Hardenbol *et al.* (1998). The GTS2012 age model suggests that many of these fine-scale sequences may coincide with orbital-climate cyclic caused by the 405-kyr eccentricity oscillations. The age model and correlation require additional cycle and magnetostratigraphy studies of Callovian through Bajocian strata.

A suite of Bathonian and Bajocian ammonite-rich sections in Spain display a complex magnetic polarity pattern of frequent reversals within a reversed-polarity dominance (Steiner *et al.*, 1987). These features are consistent with the interpreted high-frequency and reversed-polarity bias of pre-M39n deep-tow marine magnetic anomalies (Figure 26.9). ODP Site 801 is on deep-tow marine magnetic anomaly M42; but the relationship of its radiolarian assemblages to geologic stages is disputed (see discussion on Jurassic radiolarians above). However, the model that the lower units of oceanic crust at Site 801 formed during late Bajocian is consistent with the general match of magnetostratigraphy to trends of marine magnetic anomalies. As with the Oxfordian, a reliable correlation to the deep-tow record will require extensive cycle tratigraphy or other means of adjusting the Callovian–Bajocian magnetostratigraphy records for variable sedimentation rates and hiatuses. Once this is achieved, then the ages and spreading rates for the earliest history of the Pacific plate can be determined. The suggested correlation in GTS2012 is based on a visual fit of the main normal- or reversed-polarity trends of the Bathonian–Bajocian magnetostratigraphy to the mid-depth deep-tow model. This tentative correlation awaits additional deep-tow marine magnetic surveys to verify the Pacific pattern (cruise by M. Tominaga *et al.* is scheduled for Nov–Dec 2011) and further land-based magnetostratigraphy. At this point, one can only conclude that the main trends in relative polarity dominance appear to be consistent with the mid-depth deep-tow marine magnetic surveys of Pacific crust of this same age span (Figure 26.9).

26.2.3.2. Geochemical Stratigraphy

The main geochemical anomalies of the Jurassic which are important for inter-regional correlation are at the Triassic–Jurassic boundary interval and within the lowest Toarcian. A comprehensive compilation of Jurassic chemostratigraphy trends and excursions by Jenkyns *et al.* (2002) was enhanced by a detailed synthesis of the geochemical signatures associated with Jurassic–Cretaceous anoxic events (Jenkyns, 2010). Aspects of Jurassic trends for these different geochemical systems are also within the relevant geochemical chapters in this volume. The wealth of detailed studies and models on these Jurassic geochemical events and trends means that only a very small subset of representative articles are cited below.

Carbon Stable Isotopes

The end-Triassic mass extinctions coincide with a negative carbon isotope excursion that may be linked to widespread volcanism of the massive CAMP flood basalts, oceanic productivity collapse and release of methane (e.g., Pálfy *et al.*, 2001; Hesselbo *et al.*, 2002, 2007a; Ruhl *et al.*, 2009; see Chapter 25 of this volume). The brief latest Rhaetian "initial" negative excursion is followed by an earliest Hettangian "main" excursion. Carbon-13 rises to a peak between +5 and +6‰ in the mid-Hettangian, followed by a stabilization at about +5‰ during the late Hettangian through Sinemurian (e.g., Williford *et al.*, 2007; van de Schootbrugge *et al.*, 2008). The positive carbon isotope excursion of the mid-Hettangian after the end-Triassic extinctions may reflect decreased skeletal carbonate production accompanied by increased organic carbon burial and elevated atmospheric carbon dioxide levels (van de Schootbrugge *et al.*, 2008).

The lower Toarcian exhibits similar but even larger geochemical excursions. A significant negative excursion in carbon-13 beginning at the Pliensbachian–Toarcian boundary spans nearly the lower half of the basal ammonite subzone (Bodin *et al.*, 2010). Due to common condensation in the boundary interval, this basal episode is not as well-studied as the first phase of the major "Toarcian Oceanic Anoxic Event" (T-OAE) that follows it.

This T-OAE event is heralded by a major negative carbon isotope excursion (c. −5 to −7‰ shift in organic-hosted C-13), which is the largest excursion in the Mesozoic, if not the whole Phanerozoic. The onset of this excursion at the base of the *Harpoceras falciferum* Zone (second ammonite zone of Toarcian; equivalent to *Harpoceras serpentinum* in Tethyan realm) is commonly thought to coincide with the initial phase of eruptions of the large Karoo–Ferrar igneous province in Africa–Antarctica at c. 183 Ma (e.g., Pálfy and Smith, 2000; Kerr, 2000; Kemp *et al.*, 2005); however, direct U-Pb dates within the T-OAE of c. 182 Ma suggest that an initial Karoo–Ferrar eruption may have caused the basal Toarcian carbon isotope excursion, and later eruptive phases led to the T-OAE suite. This carbon isotope excursion marks the beginning of widespread occurrence of organic carbon-rich marine deposits, such as the Posidonienschiefer of Germany and Jet Rock of England. The burial of these organics contributed to the termination of the negative isotope excursion and the following broader positive isotope peak in carbon-13. A major transgression during the upper *Dactylioceras tenuicostatum* and lower *H. falciferum* zones contributed to this episode (e.g., Hallam, 1981; Jenkyns, 1988), and negative steps in the carbon isotope excursion seem to coincide with flooding surfaces in the sea-level rise (Hesselbo and Pienkowski, 2011). The oxygen-poor conditions contributed to the bivalve extinction pulses during the earliest Toarcian (e.g., Wignall and Bond, 2008). As with the end-Triassic episode, this Toarcian event is interpreted as the cascading effects from volcanic-induced global warming, with the periodic step-wise rapid surges in the initial C-13 decline suggesting contributions from destabilization of submarine methane hydrate with its ultra-negative C-13 ratios (e.g., Hesselbo *et al.*, 2000, 2007b; Kemp *et al.*, 2005; Cohen *et al.*, 2007). The estimated duration for this C-13 negative excursion depends on the assignment of Milankovitch periodicity to the observed facies and geochemical oscillations (e.g., Cohen *et al.*, 2007; Kemp *et al.*, 2011; contrasted with Suan *et al.*, 2008). The greenhouse warming and associated enhanced hydrological cycle caused amplified weathering and nutrient influx from continents and the stratification and enhanced burial of stimulated organic productivity in semi-restricted regional basins and shelves, especially in the "Western Europe Euxinic Basin" (e.g., Gómez and Goy, 2011). These changes in environmental and sediment–water interface conditions caused distortions in carbon, sulfur and strontium isotope ratios (e.g., synthesis in Jenkyns, 2010).

A negative excursion in carbon-13 is reported near the Sinemurian–Pliensbachian boundary (Jenkyns *et al.*, 2002; Korte and Hesselbo, 2011). Positive, isotopically heavy, carbon isotope excursions occur within Pliensbachian, in the lower Aalenian and in the lower Bajocian; and isotopically light values characterize the Aalenian–Bajocian and Bajocian–Bathonian boundary intervals (e.g., Bartolini *et al.*, 1996, 1999; Jenkyns *et al.*, 2002; Morettini *et al.*, 2002, Sandoval *et al.*, 2008).

The Oxfordian contains significant positive and negative excursions in C-13 (e.g., Padden *et al.*, 2001, 2002; Jenkyns *et al.*, 2002; Gröcke *et al.*, 2003; Weissert and Erba, 2004), of which a relatively brief pronounced positive peak within the upper *Perisphinctes plicatilis* ammonite zone might serve as a useful global correlation horizon (e.g., Pearce *et al.*, 2005; Głowniak and Wierzbowski, 2007). Smaller changes are recognized throughout the Kimmeridgian–Tithonian (e.g., Bartolini *et al.*, 1999); however, these features have not yet been calibrated to ammonite zones. The Jurassic–Cretaceous boundary interval has generally low C-13 values, but lacks any carbon isotope excursions to aid in global correlation of the system boundary (Weissert and Channell, 1989; Jenkyns *et al.*, 2002; Zák *et al.*, 2011).

Oxygen Stable Isotopes and Other Climate Proxies

The Jurassic is commonly considered as an interval of sustained warmth without any well-documented glacial deposits. Oxygen isotope records of oceanic temperature trends are patchy and heavily biased toward records from Europe and Russia (e.g., Veizer *et al.*, 1999, 2000; Jenkyns *et al.*, 2002; Zakharov *et al.*, 2006). However, occurrences of glendonites, a form of carbonate concretion within claystone or other terrigenous clastics, are generally considered as an indicator for cold to near-freezing conditions. Glendonite occurrences are observed in high-latitude settings during a few Jurassic intervals (Rogov and Zakharov, 2010). From these proxies, it appears that the average climates were an overall warm period (lighter O-18 values) from the Hettangian through a peak in the Toarcian, but with a possible cold (glendonite-rich) interval during the latest Pliensbachian and an early Toarcian greenhouse episode coinciding with a major carbon isotope excursion (e.g., Korte and Hesselbo, 2011). Cooler temperatures in the Aalenian culminated in a peak in glendonite abundance in the Bajocian. There are only rare reports of glendonites in the Bathonian through lower Callovian (Rogov and Zakharov, 2010). No cold-climate markers are found to support a postulated "cold snap" and glacial-induced sea-level regression at the Callovian–Oxfordian boundary (e.g., Dromart *et al.*, 2003), although the middle Callovian may have been an intermediate climate optimum followed by a relative cooling (e.g., Cecca *et al.*, 2005; Zakharov *et al.*, 2006). No glendonites are recorded from the Upper Jurassic, during which there is a trend toward more equable climates – the Boreal/Tethyan temperature difference of c. 7–9°C in the middle Oxfordian decreases toward the end-Jurassic (Zák *et al.*, 2011). Greenhouse conditions reached a maximum during the Kimmeridgian (e.g., Frakes, 1979; Valdes and Sellwood, 1992; Frakes *et al.*, 1992; Zahkarov *et al.*, 2006) coinciding with a peak in global sea level. In Britain, the humid Kimmeridgian was followed by aridity through the end of the Jurassic (e.g., Hesselbo *et al.*, 2009). These general temperature trends are consistent with other pale-oclimate indicators (Jenkyns *et al.*, 2002).

Modeling of Jurassic climates suggests that atmospheric carbon dioxide levels were at least four times the present level (reviewed by Sellwood and Valdes, 2008). An elevated carbon dioxide level may have played a role in the relatively shallow carbonate-compensation depth (CCD) in the oceans (e.g., Ogg *et al.*, 1992; Ogg, 2007). An explosion in productivity by calcareous plankton during the Tithonian, especially the robust nannoconid types of nannoplankton, contributed to the lowering of the CCD and onset of the chalk ("*creta*") deposits that characterize the lower Cretaceous in all ocean basins. A decrease in pCO_2 during the Tithonian may have been a major factor in this evolutionary surge of calcified plankton (e.g., Erba, 2006; Casellato *et al.*, 2008).

Strontium and Osmium Isotope Ratios

Marine $^{87}Sr/^{86}Sr$ through the Jurassic progressively decreased from its end-Norian peak (0.707 95) to an intermediate low (0.707 08) at the end of the Pliensbachian, indicating a minimum in contribution of radiogenic Sr-87 from weathering of continental crust or ancient sediments relative to non-radiogenic Sr-87 from hydrothermal alteration of mid-ocean ridge basalts (Jones *et al.*, 1994a,b; McArthur *et al.*, 2000, 2001; Jones and Jenkyns, 2001; McArthur and Howarth, 2004; Chapter 7 of this volume). The initial stages of the downturn in marine $^{87}Sr/^{86}Sr$ from the latest Norian peak (0.707 95) spanned the major flood basalt outpouring of the "Central Atlantic Magmatic Province" along the future Central Atlantic seaway in the latest Rhaetian, but the interpretation of the impact of these CAMP eruptions and major environmental shifts upon the rate of change in the oceanic strontium ratios depends on estimates for the duration of the Rhaetian and Hettangian stages and the filtering of Sr ratios derived from different fossil types (Cohen and Coe, 2007;

McArthur, 2008, 2010). Similarly, the effects of the early Toarcian Karoo–Ferrar volcanic province on the rate and timing of the reversal in strontium isotope trends depends on estimates of the elapsed duration of the earliest Toarcian ammonite zone (e.g., Kemp *et al.*, 2005; Cohen and Coe, 2007; McArthur *et al.*, 2008; Jenkyns, 2010). In contrast, the short residence time of osmium enabled recording of pronounced excursions in response to the chemical weathering of the fresh basalts of the end-Triassic and early Toarcian flood basalts (e.g., Cohen *et al.*, 2004; Cohen and Coe, 2007), although estimates of the magnitude of the global signal can be distorted by basinal effects.

By assuming that this decrease was linear through the Hettangian, Sinemurian and Pliensbachian stages, and scaling the slope with cycle stratigraphy in the lower Pliensbachian, Weedon and Jenkyns (1999) estimated that the minimum duration of these three stages are 2.86, 7.62 and 6.67 myr, respectively, for a total of 17.15 myr. A similar calculation using an expanded Lower Jurassic database (McArthur *et al.*, 2001; also used in GTS2004) yields 3.10, 6.90, and 6.60 myr, respectively, for a total of 16.90 myr. However, a much shorter Hettangian duration is derived if oyster-derived values are largely removed (McArthur, 2008). Fitting this trend (without oyster-derived points) through the Rhaetian would imply an estimated duration of up to 8 myr for that Rhaetian stage (McArthur, 2010). In GTS2012, a similar linear fit to the decreasing Sr curve was used to scale the relative durations of the individual Sinemurian through middle Toarcian ammonite zones.

Strontium isotope ratios progressively rose during the Toarcian and crested with a sustained plateau (0.707 30) through the Aalenian. Strontium isotope ratios again decreased through the Bajocian to middle Callovian, with a possible shoulder spanning the Bajocian–Bathonian boundary (Jones *et al.*, 1994b).

During the early Oxfordian, marine $^{87}Sr/^{86}Sr$ reached its lowest ratio (0.706 86) throughout the entire Phanerozoic (McArthur *et al.*, 2001). This pronounced episode may indicate a major pulse of seafloor hydrothermal activity (Jones *et al.*, 1994b; Jones and Jenkyns, 2001), which is supported by the formation and expansion of new spreading centers as Pangea began its breakup, and by interpretations of other geochemical, deep-sea sediment and spreading-rate evidence (e.g., Ogg *et al.*, 1992). The only Phanerozoic minimum that approached this ultra-low $^{87}Sr/^{86}Sr$ ratio was at the end of the Middle Permian.

After the middle Oxfordian, the strontium isotope ratio began a long-term increase that peaked in the Barremian of Early Cretaceous during the ammonite *P. elegans* Zone. The limited data set for seawater Os isotopic composition also shows a rapid change from low $^{187}Os/^{188}Os$ (c. 0.25) in Callovian to more radiogenic values (c. 0.6) in Kimmeridgian–Tithonian, which may suggests progressive radiogenic crustal material into the late Jurassic ocean (Cohen *et al.*, 1999; Selby, 2007).

Except within the plateaus during the Aalenian and the Oxfordian, the rapidly changing ratios of marine $^{87}Sr/^{86}Sr$ enable global correlation at high resolution (McArthur *et al.*, 2001).

26.2.3.3. Cyclostratigraphy

Contributed by Chunju Huang

Astronomically forced cyclostratigraphy has become an important tool in measuring Jurassic geologic time and establishing "floating" astronomical time scales. The signatures of orbital eccentricity forcing of the precession index dominates much of Jurassic cyclostratigraphy; therefore, most Jurassic cyclostratigraphic scales have a resolving power of the long-eccentricity cycle (0.4 myr) or short-eccentricity (0.1 myr). Obliquity (~37-kyr period) forcing had a lesser, although still important, role.

Early and Middle Jurassic Cyclostratigraphy

Hettangian Obliquity-dominated cyclicity in the Blue Lias Formation of southern England yields a minimum duration of 1.29 myr for the Hettangian Stage, but this estimate incorporates known stratigraphic breaks (Weedon, 1985; Weedon *et al.*, 1999). A maximum of 2.4 myr duration for the Hettangian Stage is suggested from tuning of lacustrine sequences in eastern North America to long-period eccentricity signals (Kent and Olsen, 2008). Cycle stratigraphy of the relatively complete St. Audrie's Bay (UK) section yielded a total Hettangian span of 1.8 myr with estimates for each ammonite zone (*P. planorbis* ~0.25 myr; *A. liasicus* ~0.75 myr; *S. angulata* ~0.8 myr) (Ruhl *et al.*, 2010). These durations, plus an adjustment for the pre-*planorbis* interval to the base-Jurassic, are consistent with radio-isotopic ages for the limits of the Hettangian, and are significantly shorter than the 3.1 myr total duration in GTS2004 that had been based on the assumption of a linear strontium isotope trend (McArthur *et al.*, 2001). However, re-evaluation of the Sr isotope data sets indicates that a 2-myr-duration Hettangian is consistent (McArthur *et al.*, 2008).

Sinemurian Within the Blue Lias Formation at Lyme Regis (southern England), a 0.34 myr span was interpreted for the *A. bucklandi* Zone of lowermost Sinemurian (Weedon, 1985; Weedon *et al.*, 1999). However, Ruhl *et al.* (2010) obtained 600 kyr for only the lower part of the *C. bucklandi* Zone at St. Audrie's Bay, and a duration of 0.73 kyr was obtained for this ammonite zone upon re-study of Weedon's 1985 magnetic susceptibility data (Huang *et al.*, 2010a). The other five standard ammonite zones of the Sinemurian have not yet been astronomically calibrated.

Pliensbachian A minimum duration of 6.67 myr for the Pliensbachian is estimated from precession-dominated cyclicity in the Belemnite Marls of southern England (Weedon and Jenkyns, 1999), combined with linear strontium isotope trends and cycle-stratigraphic data from Robin Hood's Bay in northeast England (van Buchem *et al.*, 1994) and Breggia Gorge in southern Switzerland (Weedon, 1989). This is consistent with the minimum estimate of 5 myr derived from precession-dominated strata in northern Italy (Hinnov and Park, 1999). Two of the ammonite zones at Breggia Gorge have spans estimated from cycles: *P. davoei* (~0.40 myr) of middle Pliensbachian, and the *P. spinatum* (~0.89 myr) of uppermost Pliensbachian (Weedon, 1989). The combined *A. margaritatus* and *P. spinatum* zones span an estimated 3 myr according to interpreted cycles within the Sancerre core of France (Huret *et al.*, 2008, in prep.).

Toarcian The facies and high-resolution geochemical records from outcrops spanning the lower Toarcian display definite oscillations, especially through the major carbon isotope excursion and widespread organic-enriched "oceanic anoxic event" deposits. However, there is disagreement on whether the dominant cycles are a product of 20-kyr precession, 40-kyr obliquity, or 100-kyr precession (e.g., Kemp *et al.*, 2011; Cohen *et al.*, 2007; Suan *et al.*, 2008). If precession cycles are the main feature, then the onset and the main episode of C-13 depletion were brief episodes (c. 80 kyr and 120 kyr, respectively) that are comparable to the signature and span of the PETM (Paleocene–Eocene Thermal Maximum), therefore suggesting a similar ocean–atmosphere–carbon cycle response that may have involved methane hydrate releases (Cohen *et al.*, 2007). However, if the main cyclicity is a product of eccentricity, then these events were more extended (c. 150 kyr and 450 kyr, respectively), therefore requiring sustained inputs of light carbon (e.g., Suan *et al.*, 2008).

This ambiguity in cycle-stratigraphic analyses includes the remainder of the Toarcian. A duration of 0.8 myr for the basal *D. tenuicostatum* ammonite zone (or coeval *D. polymorphum* Zone, depending on region) and of 1.2 myr for the overlying zone (*H. levisoni, H. serpentinum*, or *H. falciferum*, depending on region) is estimated from a 100-kyr eccentricity interpretation applied to a 38-m-thick section at Peniche, Portugal (Suan *et al.*, 2008). However, a condensation of the basal meters of this Peniche outcrop is suggested by the carbon isotope patterns within a more expanded 220 m-thick coeval section in Morocco (Bodin *et al.*, 2010).

Carbonate cycles interpreted as obliquity-dominated oscillations within the Sogno Formation in northern Italy yielded an 11.37 ± 0.05 myr duration for the combined Toarcian and Aalenian stages (Hinnov and Park, 1999; Hinnov *et al.*, 1999). Huret *et al.* (2008) estimated a 7.3 myr minimum duration for the Toarcian as recorded in the Sancerre core (France). The magnetic susceptibility data from this Sancerre core were further analyzed and retuned on 405-kyr long-eccentricity signals (Huang *et al.*, 2010a; Boulila *et al.*, in press) resulting in a minimum total Toarcian span of 8.3 myr. The duration of each zone was estimated as *D. tenuicostatum* (minimum of 1 myr), *H. serpentinum* (1.08 myr), *H. bifrons* (2.12 myr), *H. variabilis* (3.01 myr), *G. thouarsense* (0.26 myr), *P. dispansum* (0.26 myr), *D. pseudoradiosa* (0.43 myr) and *P. aalensis* (0.13 myr); although the boundaries of some of these ammonite zones were poorly delimited in the cored material. A modified version of these cycle durations is used in GTS2012.

Aalenian The Aalenian strata of Sogno (Italy) display precession–eccentricity patterns. Tuning the longer period to 405 kyr yields the total duration of 3.85 myr (Huret *et al.*, 2008) used in GTS2012. The estimated durations for the ammonite zones are: *L. opalinum* (2.0 myr), *L. murchisonae* (0.8 myr), *B. bradfordensis* (0.4 myr), and *G. concavum* (0.43 myr); although placements of some of these ammonite zonal boundaries are indirectly assigned according to calibrated nannofossil datums.

Bajocian–Callovian Cycle stratigraphy has been interpreted only for short intervals within the Bajocian, Bathonian and Callovian stages. The Gnaszyn Formation (Poland) suggests the Lower Bathonian spans a minimal of 2 myr based on tuning to cycles interpreted as 100 kyr (Ziólkowski and Hinnov, 2009). Potential durations from the well-defined lithologic oscillations across the Bathonian GSSP Ravin du Bès section, SE France (Fernández-López *et al.*, 2009) suggest ammonite zone durations for *P. parkinsoni* (~0.6 myr) of uppermost Bajocian; and *Z. zigzag* (~0.6 myr) and *P. aurigerus* (~0.2 myr) of lower Bathonian.

Analysis of boreholes to test the feasibility of radioactive-waste storage into upper Callovian clay-rich deposits in the Paris Basin indicated that the Upper Callovian spanned 0.9 ± 0.1 myr and the middle Callovian *E. coronatum* Zone spanned a similar 0.9 ± 0.1 myr (Huret, 2006; Lefranc *et al.*, 2008).

Late Jurassic Cyclostratigraphy

The majority of the Late Jurassic has been scaled by cycle stratigraphy, and the Boreal and Tethyan data sets have been merged using magnetostratigraphic inter-correlations. The intervals of cycle-scaled magnetozones were correlated to the M-sequence of marine magnetic anomalies, thereby enabling a spreading rate model. A linear fit to these cycle-determined spreading rates was projected to the pre-Oxfordian deep-tow record of marine magnetic anomalies, which, in turn, was used as a reference magnetic polarity scale to correlate the main features of the Bajocian through Callovian magnetostratigraphy. These predictions will be tested by further cycle-stratigraphic studies of Middle and Late Jurassic strata.

Oxfordian Cycle stratigraphy of the ammonite-zoned Lower to Middle Oxfordian Terres Noires Formation in the Vocontian Basin (SE France) indicated a dominant 405-kyr signal, and a total span for the lowermost four ammonite zones of 4.1 myr (Boulila *et al.*, 2008a, 2010a). Even though the total duration is similar to the estimate in GTS2004, the distribution of time among ammonite zones and subzones is considerably different. For example, the basal *Q. mariae* ammonite zone occupies nearly 2.1 myr, in contrast to the 0.6 myr estimated in GTS2004; whereas the *G. transversarium* Zone spans 0.65 myr, only half of its estimated duration in GTS2004. The long span for the *Q. mariae* ammonite zone is verified by independent cycle stratigraphy analysis of French boreholes (Lefranc *et al.*, 2008).

Middle Oxfordian through Lower Kimmeridgian carbonate platform successions with rare ammonite levels in the Swiss and French Jura Mountains record major depositional sequences in response to sea-level changes (e.g., Gygi *et al.*, 1998; Gygi, 2000). Small-scale oscillations within these sequences were interpreted as a record of short-term 100-kyr eccentricity orbital-climate cycles to give an estimate of durations of the associated ammonite zones (summarized in Strasser, 2007), although it is probable that there are "missing beats" at some emergent surfaces of sequence boundaries.

Kimmeridgian–Tithonian Cycle stratigraphy at La Méouge (France) yielded durations for the lower Kimmeridgian ammonite zones (Boulila *et al.*, 2008b, 2010b) that were quite similar to magnetostratigraphy-derived estimates in GTS2004, indicating that the calibration of polarity zones and the associated ammonite zones to the M-sequence magnetic anomalies and the spreading model for those Pacific Hawaiian lineations was reliable.

Lithologic and magnetic-susceptibility variations within the Kimmeridge Clay Formation of Dorset (southern England) appear to be associated with obliquity, with perhaps minor contributions from precession (e.g., Waterhouse, 1995; Weedon *et al.*, 1999). Condensation within the lower two ammonite zones precluded deriving a cycle stratigraphy, but early studies of portions of the main Kimmeridge Clay (e.g., Weedon *et al.*, 1999) demonstrated the importance of obtaining a complete record. Therefore, the Kimmeridge Clay was continuously cored with calibrations to individual ammonite zones (e.g., Cope, 2009), and a record of its cyclicity was derived from multiple proxies (Morgans-Bell *et al.*, 2001; Weedon *et al.*, 2004). This data set was further refined and tuned to 405-kyr and 100-kyr eccentricity signals (Huang *et al.*, 2010b). The middle part of *A. autissiodorensis* ammonite zone through the top of *V. fittoni* Zone spans 3.72 myr. This duration incorporates a refinement based on correlation to French magnetostratigraphy sections, which implies that there are three 405-kyr cycles within an ambiguous cycle stratigraphy interval (either two or three 405-kyr cycles) of the *Pect. elegans–Pect. wheatleyensis* ammonite zones in the Kimmeridgian Clay (Figure 2 of Huang *et al.*, 2010b; Ogg *et al.*, 2010b). This total span is similar to the 3.91 myr estimate based only on 38-kyr obliquity tuning (Weedon *et al.*, 2004).

The lower part of the Volgian Hekkingen Formation (Greenland–Norwegian Seaway) (Swientek, 2002) overlaps with a portion of the astronomically-forced cycles in the Kimmeridge Clay. Upon incorporating potential Sub-Boreal-to-Boreal correlations of ammonite zones, the composite cyclicity of the two formations indicates a duration of 6.03 myr for the Lower and Middle Volgian, and possibly a 5.2 myr duration for the Upper Volgian (Huang *et al.*, 2010c).

26.2.3.4. Sequence Stratigraphy

Jurassic sea-level trends have been compiled for various intervals in different basins. Examples include: Britain and North Sea (e.g., Partington *et al.*, 1993; Coe, 1995; Hesselbo and Jenkyns, 1998; Taylor *et al.*, 2001; Hesselbo, 2008), Switzerland–France (e.g., Gygi *et al.*, 1998; Gygi, 2000; Strasser, 2007; Colombié and Rameil, 2007), Russia (Sahagian *et al.*, 1996) and Arabia (Sharland *et al.*, 2001, 2004; Al-Husseini and Matthews, 2006). The regional trends have been compared on a global scale to construct global sequence-stratigraphy scales and eustatic curves (e.g., Arkell, 1956; Hallam, 1978, 1981, 1988, 2001; Haq *et al.*, 1988; Hardenbol *et al.*, 1998).

The main Jurassic sea-level trend is a progressive rise from the latest Triassic until the late Kimmeridgian (peaking in the *A. eudoxus* ammonite zone). A major sea-level fall trend through the Tithonian reaches a minimum in the late Berriasian. Superimposed on this main cycle are several major sequences (e.g., Figure 26.9).

Assignments of even fine-scale sequences depend on interpretation models for the response of sediment facies (other than obvious emergent soils or flooding surfaces) to relative sea-level changes. Therefore interpretations vary among stratigraphers in assigning small-scale sequences within a given region (e.g., comparison charts within Hesselbo and Jenkyns, 1998; Newell, 2000; Taylor *et al.*, 2001; Hesselbo, 2008; and Simmons, Chapter 13 of this volume). For example, the same relative enrichment in carbonate within Kimmeridgian basinal successions can be interpreted as a sea-level maximum-flooding episode (e.g., Boulila *et al.*, 2010b, 2011) or as a sea-level lowstand (e.g., Mattioli *et al.*, 2011) depending on whether the carbonate is considered to be mainly pelagic or mainly export wafted from adjacent carbonate platforms.

Many of these fine-scale sequences appear to correspond to 405-kyr long-eccentricity-induced orbital-climate cycles. For example, the main sequences interpreted from carbonate-clay changes in the Lower Oxfordian of southern France and from sand-influxes and hiatuses in the lower-to-middle Oxfordian of the Dorset coast (England) represent these

405-kyr orbital-climate oscillations (Boulila *et al.*, 2010a; Ogg *et al.*, 2010a). In this particular case, a lowstand exposure in Dorset corresponds to an episode of carbonate-enrichment in the basinal successions of SE France. A 405-kyr periodicity is indicted when the fine-scale Bajocian–Bathonian–Callovian sequences of Hardenbol *et al.* (1998) are plotted on the GTS2012 timescale for the associated ammonite zones (Figure 26.9). It remains speculative whether these periodic fine-scale sequences represent major eustatic sea-level oscillations (e.g., storage of water in high-latitude or high-altitude glaciers) or if they are mainly advances/retreats of local coastlines in response to 405-kyr climate cycles altering regional weathering styles and runoff (and oceanic productivity). High-resolution correlation and systematic interpretation of such sequences are required among different paleogeographic realms.

26.2.3.5. Other Major Stratigraphic Events

Large Igneous Provinces

Central Atlantic Magmatic Province (CAMP) The Central Atlantic Magmatic Province may be the largest known large igneous province of flood basalts and sills (e.g., Marzoli *et al.*, 1999; Hames *et al.*, 2000). The eruptive centers of this c. 10 million km^2 volcanic outpouring extended from northern South America across eastern North America and western Sahara to Spain. The recalibrated radio-isotopic dates center on ~201 Ma, just prior to the base of the Jurassic (Table 26.1). Therefore, the massive CAMP eruptions are considered to be major causal factors in the end-Triassic mass extinctions (e.g., Olsen *et al.*, 2003; Hesselbo *et al.*, 2002, 2007a; Wignall and Bond, 2008; Jourdan *et al.*, 2009a; Marzoli *et al.*, 2011).

Karoo–Ferrar and the GTS2012 Age Model for Basal-Toarcian The Karoo basalts in South Africa and the Ferrar volcanics in Antarctica are part of a volcanic province with a total original volume of ~5 million km^3 (White, 1997). The numerous radio-isotopic dates span from about 185 to 179 Ma (e.g., selected suite in Table 26.1; reviewed in Jourdan *et al.*, 2007; Moulin *et al.*, 2011).

The lower 100 m of the Karoo basalts in the Lesotho region are reversed magnetic polarity. The main overlying suite is entirely normal polarity, and a possible hiatus between these polarity units is indicated by intervening aeolian deposits and the possible clustering of their radio-isotopic ages (Prévot *et al.*, 2003; Moulin *et al.*, 2011). Ages for the lower reversed-magnetized Karoo episode and earliest sills average at approximately 182.7 ± 0.7 Ma (e.g., Jourdan *et al.*, 2007; Svensen *et al.*, 2007). The initiation and age span of the overlying main normal-polarity Karoo–Ferrar suite is less constrained (e.g., 183.4 ± 1.6 Ma is reported in Jourdan *et al.*, 2007, compared to mean of 179.2 ± 1.8 Ma as computed by Moulin *et al.*, 2011).

The eruption of the Karoo–Ferrar large igneous province is commonly considered as the primary trigger for the early Toarcian "Oceanic Anoxic Event" suite of organic-rich strata, major geochemical anomalies (e.g., the largest negative excursion in carbon-13 during the Mesozoic), and faunal extinctions (e.g., Jones and Jenkyns, 2001; Wignall, 2001; Pálfy *et al.*, 2002; Wignall and Bond, 2008). The T-OAE began at the base of the *H. serpentinum* ammonite zone (or coeval *H. levisoni* or *H. falciferum* Zone), which is 0.8 myr above the Pliensbachian/Toarcian boundary according to cycle stratigraphy of Suan *et al.* (2008) or perhaps 1.2 myr after including an adjustment for basal condensation (e.g., Bodin *et al.*, 2010). However, the magnetostratigraphy of the lower *H. serpentinum* ammonite zone and the upper underlying *D. tenuicostatum* Zone is normal polarity; whereas only the lowermost part of the *D. tenuicostatum* Zone is reversed polarity (e.g., Comas-Rengifo *et al.*, 2010a,b). Therefore, it is likely that the eruption of the lower reversed-polarity Karoo basalts was associated with the basal-Toarcian negative carbon isotope excursion, hence an age of 182.7 ± 0.7 Ma is assigned to this boundary level.

Tuffs from within a set of deposits with a carbon isotope negative excursion that had been interpreted as the early Toarcian T-OAE episode yielded U-Pb dates of 180.6 ± 0.4 Ma and 181.4 ± 0.2 Ma (Mazzini *et al.*, 2010). However, even though these dates are consistent with the mean age of 179.2 ± 1.8 Ma for the upper normal-polarity basalts of the Karoo province (Moulin *et al.*, 2011), there is stratigraphic evidence suggesting that these beds are reworked deposits of mid-Toarcian (*H. bifrons* ammonite zone) that yielded fortuitous clusters of zircon dates (S. Hesselbo and A. Al-Suwaidi, pers. comm. to J. Ogg, and in prep.).

Merging the cycle stratigraphy constraints on durations with the radio-isotopic dates suggests a possible age model in which the initial eruption of the reversed-polarity Karoo basalts at ~182.7 Ma caused the basal "boundary" carbon isotope excursion. Then, after a relative quiescence, the second and main eruptive phase began ~1 myr later at ~181.7 Ma to initiate the "main T-OAE" carbon isotope excursion with a minimum in C-13 that lasted through ~180.3 Ma. For lack of direct radio-isotopic ages on the base-Toarcian GSSP level, this assumption that the LIP caused the carbon isotope excursion at the Pliensbachian–Toarcian boundary is the age-model adopted in GTS2012.

Major Bolide Impacts

A modest iridium anomaly has been reported from near the palynology-defined Triassic–Jurassic boundary in the eastern United States; and associated features, such as the fern spike and apparent suddenness of the terrestrial extinctions, had suggested a possible impact relationship (Olsen *et al.*, 2003). However, extensive follow-up research has failed to support this hypothesis.

Puchezh-Katunki crater in Volga Federal District of Russia has a diameter of 80 km (Earth Impact Database, 2010). Its age, based on pollen, implies a relatively large impact during the Bajocian, but there are no reliable radio-isotopic dates available (reviewed by Jourdan *et al.*, 2009b)

Morokweng crater in the Kalahari desert of South Africa has a similar diameter (70 km) (Earth Impact Database, 2010) with a "recommended" age from U-Pb and Ar-Ar analyses of 145.2 ± 0.8 Ma (Jourdan *et al.*, 2009b). This implies an impact event very close to the Jurassic/Cretaceous boundary.

26.3. JURASSIC TIME SCALE

The primary reference scales for most stage boundaries and other events in Jurassic stratigraphy are the ammonite zones for the Tethyan and Sub-Boreal faunal realms (Figure 26.10). The numerical age scale for these ammonite zones requires integration of several constraints, in approximate order of importance:

(a) Selected radio-isotopic dates, especially for Hettangian, Toarcian and Berriasian.
(b) Durations of ammonite zones derived from cycle stratigraphy, especially in Hettangian, Toarcian–Aalenian, and Oxfordian–Tithonian.
(c) Calibrations of ammonite zones via magnetostratigraphy to a spreading rate model calibrated to numerical ages for the early Pacific plate; especially in Oxfordian through Tithonian, but also a preliminary correlation for Bajocian through Callovian.
(d) Proportional scaling of ammonite zones/subzones derived from assumptions of geochemical trends (e.g., linear trends in strontium isotope variation in Sinemurian–Pliensbachian)
(e) Proportional scaling of ammonite zones according to relative numbers of subzones in the intervals that do not have adequate constraints by one of the above methods.

In turn, the assignment of numerical ages for most of the other Jurassic stratigraphic events is according to their calibrations to this age model for the ammonite zonations.

The assumptions, scaling methods and derived numerical age for each Jurassic stage are summarized in Table 26.2, and each Tethyan ammonite zone in Table 26.3.

26.3.1. Constraints from Radio-Isotopic Dates

It is ironic that the radio-isotopic database constraining the Jurassic time scale in GTS2012 has fewer "control dates" than were used in GTS2004. In turn, GTS2004 largely followed the analysis of Pálfy *et al.* (2000a), who had compiled a database of 55 latest Triassic through Tithonian ages derived from U-Pb and $^{40}Ar/^{39}Ar$ methods, of which only 12 were from publications of pre-1995 vintage. That compilation had included recalibration of North American ammonite zones or specific ammonite datums to the standard Northwest European zones and associated definition of geological stages (e.g., Pálfy *et al.*, (1997), observed that there were approximately 25% erroneous identifications of Toarcian ammonites at the species level in other reference biostratigraphic sections).

The more restrictive GTS2012 criteria for incorporation of radio-isotopic dates from U-Pb methods require only single-zircon analyses (no multi-grain) by CA-TIMS which must follow treatments of annealing followed by chemical abrasion; whereas, nearly all U-Pb ages in GTS2004 were multi-grain zircon analyses without this type of treatment. In striking contrast to post-2003 radio-isotopic studies of Triassic strata, only a few U-Pb dates for the Jurassic have been acquired by these methods. Unfortunately, nearly all of the U-Pb dates compiled by Pálfy *et al.* (2000a) had utilized multi-grain analyses without these pre-processing techniques. However, many of these Jurassic radio-isotopic dates are retained as "secondary guides" in GTS2012, because the values still provide an indication of the approximate age (e.g., comparisons of U-Pb dating from basal-Triassic indicate that earlier methods are usually within 2 myr of the newer analysis results). Additional U-Pb dates for Pliensbachian and Bathonian are as cited by Pálfy (2008).

Dates derived from $^{40}Ar/^{39}Ar$ methods reported in GTS2004 are recalculated to the monitor standard FCs value of 28.20 Ma (see discussions and tables in Chapter 6 of this volume). This shift from the GTS2004 standard of 28.02 Ma added ~1 myr to each value published before 2009.

The Triassic–Jurassic boundary has a precise radio-isotopic age. Based on constraints from dated ammonoid-bearing strata in Peru (LM4-90 and LM4-100/101) and a similar age from the former GSSP candidate section in New York Canyon, Nevada, Schoene *et al.* (2010) project the T/J boundary age to be 201.31 Ma (± 0.18/0.38/0.43). The onset of the main phase of the Central Atlantic magmatic Province (CAMP) preceded the Triassic–Jurassic boundary by only c. 70 kyr (or a maximum of 290 kyr if the extremes on the uncertainties are applied). An apparent age difference between these major volcanic eruptions and the end-Triassic marine extinctions in GTS2004 (e.g., Pálfy *et al.*, 2000b) was an artifact from multi-grain zircon analysis techniques (reviewed in Mundil *et al.*, 2010).

The Hettangian–Sinemurian boundary is constrained by U-Pb ages of 199.5 ± 0.2 Ma from boundary interval Peru (Schaltegger *et al.*, 2008) and 198.0 ± 0.6 Ma in early Sinemurian from Hungary (reported in Pálfy and Mundil, 2006).

There are no U-Pb ages from marine strata of the middle Sinemurian through Bathonian that meet the restrictive GTS2012 criteria – all dates from biostratigraphic-constrained

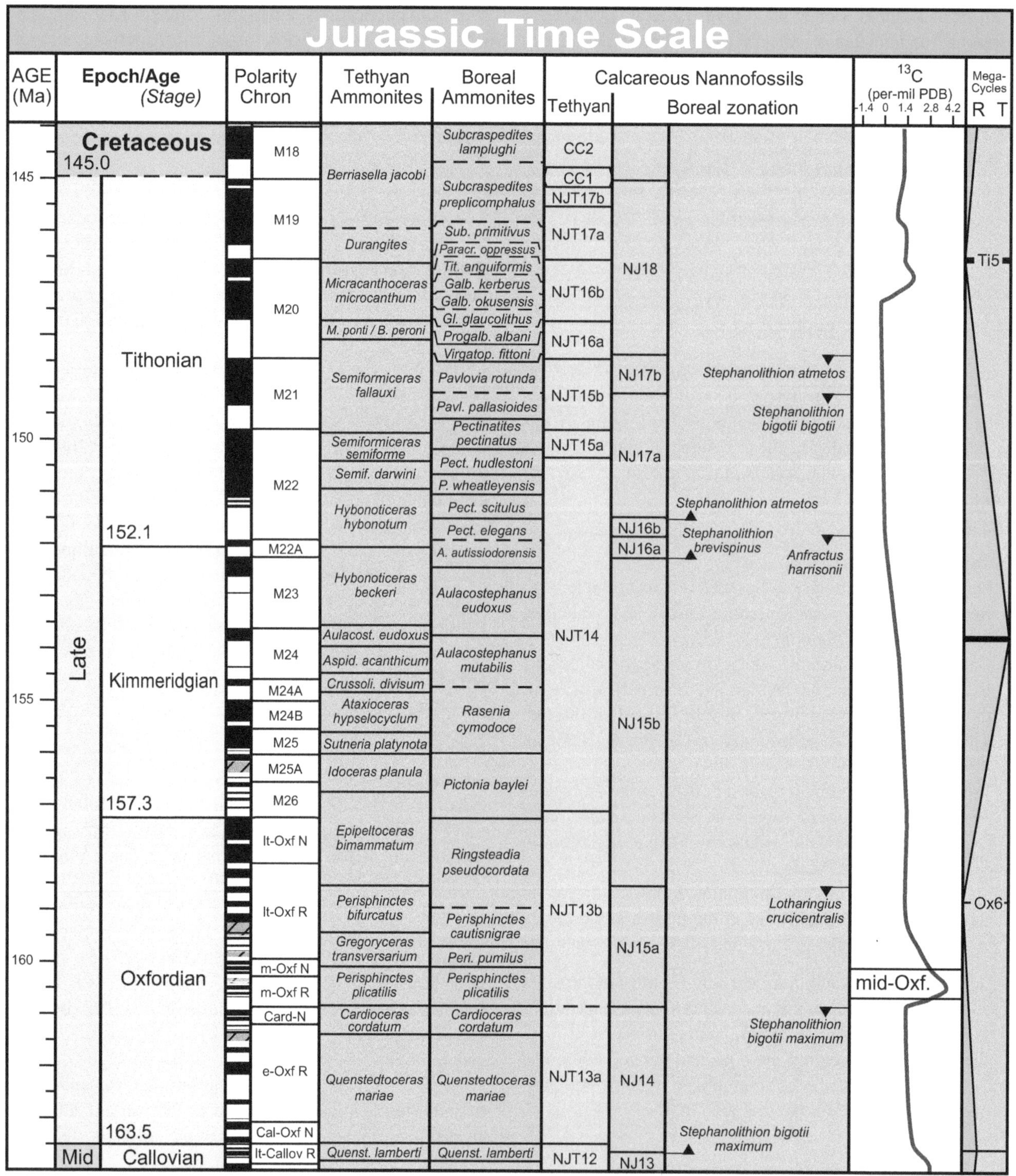

FIGURE 26.10 **Summary of numerical ages of epoch/series and age/stage boundaries of the Jurassic with selected marine biostratigraphic zonations and principal trends in carbon isotopes and sea level.** ("Age" is the term for the time equivalent of the rock-record "stage".) Potential definitions of Oxfordian, Kimmeridgian, Tithonian and Berriasian are indicated; the final decisions will be made by the International Commission on Stratigraphy. Marine biostratigraphy columns are representative ammonoid zones for the Tethyan (Sub-Mediterranean province) and Boreal (Sub-Boreal for Late Jurassic) realms. For detailed ammonite zonations in Late Jurassic, see expanded scale in Figure 26.8. Calcareous nannofossil zones and Boreal zone-markers are a composite from Casellato (2010), Mattioli and Erba (1999) and Bown (1998). Composite carbon-13 trends are generalized from Jenkyns *et al.* (2002), with additional details from Głowniak and Wierzbowski (2007) for middle Oxfordian, Kemp *et al.* (2005) and Gómez and Goy (2011) for early Toarcian, and Pálfy *et al.* (2001) for the Triassic–Jurassic boundary interval. The mega-cycles of sea-level trends are from Hardenbol *et al.* (1998). Additional Jurassic zonations, geochemical trends, sea-level curves, etc. are compiled in the internal data sets within TimeScale Creator (*www.tscreator.org*).

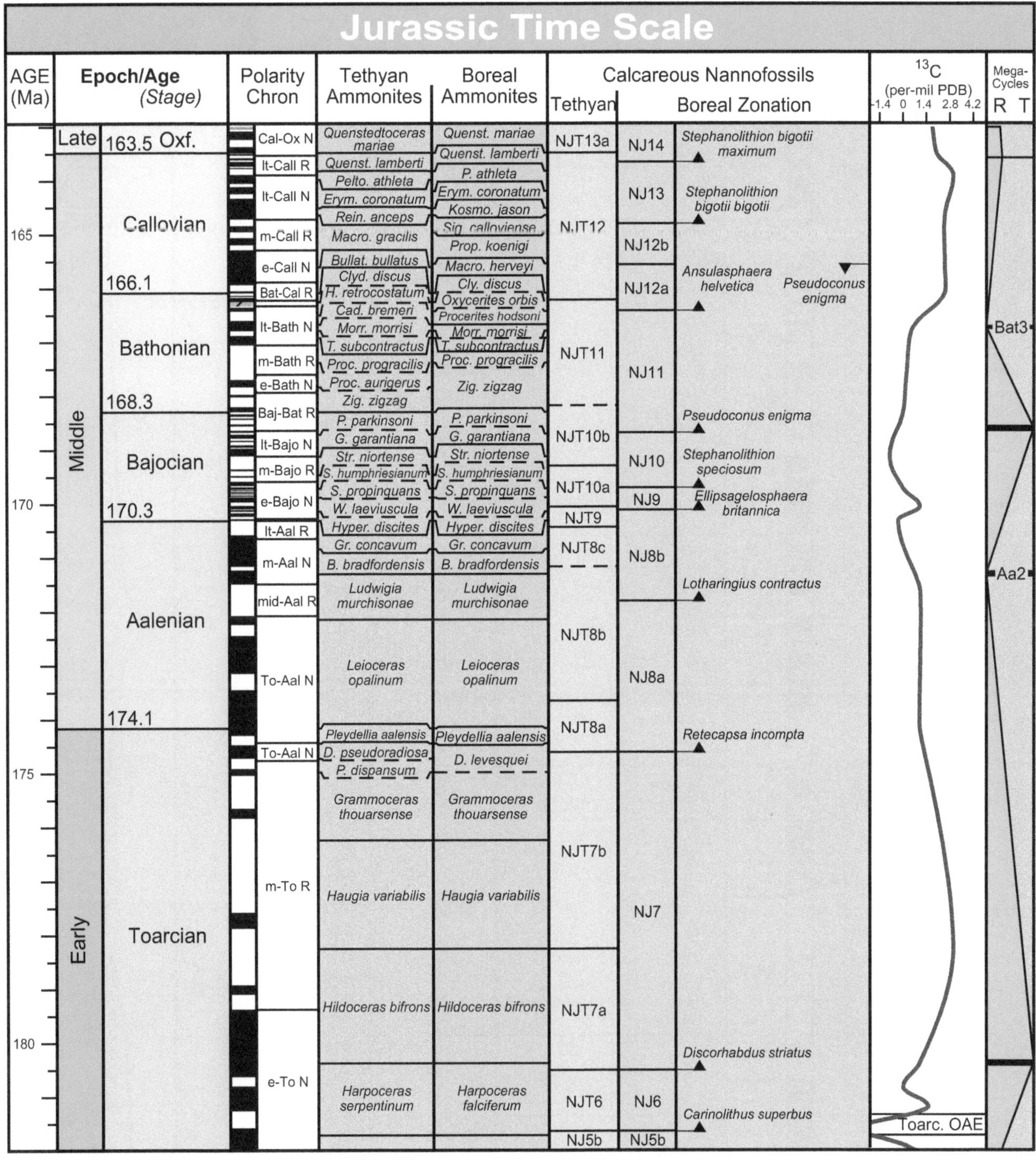

FIGURE 26.10 (*Continued*).

intervals are multi-grain zircon analyses or have biostratigraphic uncertainties that exceed one ammonite zone. However, these lower-quality dates (Appendix 2) suggest the base-Pliensbachian is ~190 Ma, the base-Toarcian is ~183 Ma, and the base-Aalenian is ~176 Ma. The Bajocian/Bathonian boundary interval is ~169 Ma, based on $^{40}Ar/^{39}Ar$ dating of lower volcanic units in ODP Site 801 (Koppers *et al.*, 2003); although there is a large uncertainty on the biostratigraphic age of these basalts.

The Bathonian−Callovian boundary is ~165 Ma, based on a U-Pb date of 164.6 ± 0.2 Ma from an ash bed in Argentina with ammonites of possible early Callovian age (Kamo and Riccardi, 2009). Earlier analyses of multi-grain zircon suites from this same ash bed had yielded 161.0 ± 0.5

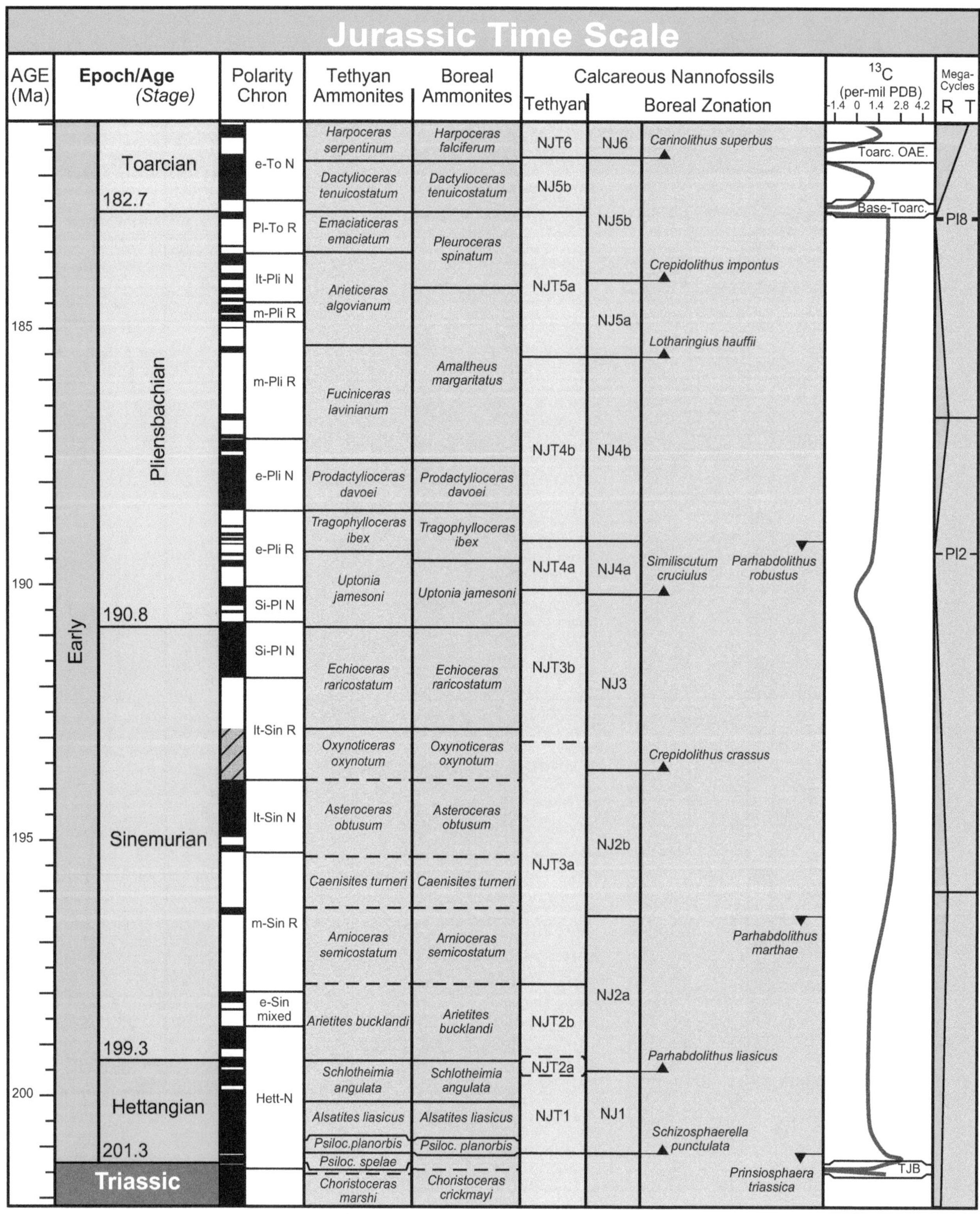

FIGURE 26.10 (*Continued*).

TABLE 26.2 Summary of the Derivation of the Jurassic Numerical Age Models for Stages and for the Primary Scale of Tethyan Ammonite Zones (Tabulated in Table 26.3)

Stage	Spline-Fit Age; (Ma)	*Spline-Fit Uncertainty (2-Sigma)*	Revised Age (Ma)	*Estimated Uncertainty (2-Sigma)*	*Duration (myr)*	Calibration (Brief)	Stage Primary Marker and GSSP or Working Definition; Calibration	Method to Compute Basal Age of Stage	Method for Scaling Ammonite Zones within Stage
CRETACEOUS (Berriasian)	144.6	*1.5*	**145.0**	*0.8*		Base of Chron M18r	Not yet defined. Set as Base of Chron M18r.	Cycle-scaled spreading rate model for Pacific M-sequence.	
Tithonian	150.8	*1.4*	**152.1**	*0.9*	*7.1*	Base of Chron M22An	Not yet defined. Set as Base of Chron M22An, which is nearly coeval with base of *H. hybonotum* ammonite zone at Crussol Mountain, France.	[*same*]	Tethyan zones have magnetostratigraphic placements; Sub-Boreal zones (England) have a mixture of cycle-durations and magnetostratigraphic placements.
Kimmeridgian	156.0	*1.7*	**157.3**	*1.0*	*5.2*	Base of Chron M26r	Base of *P. baylei* ammonite zone (Isle of Skye), using lowest occurrence of *P. flodigarriensis* ammonite (not yet ratified) which is essentially coeval with Base of Chron M26r at the GSSP.	[*same*]	Tethyan and Sub-Boreal zones have magnetostratigraphic placements; and some of their cycle-derived durations are constraints on the Pacific M-sequence spreading model.
Oxfordian	162.5	*1.1*	**163.5**	*1.1*	*6.2*	25% up in Chron M37n.1n	Base of *Q. mariae* ammonite zone. GSSP is not yet decided, but this level at Redcliff (England) candidate has been correlated to 25% up in Chron M37n.1n of deep-tow extension to M-sequence.	Cycle-scaled spreading rate model for Pacific M-Sequence (deep-tow extension).	Tethyan and Sub-Boreal zones and subzones have magnetostratigraphic placements; and some of their cycle-derived durations are constraints on the Pacific M-sequence spreading model.
Callovian	165.8	*0.9*	**166.1**	*1.2*	*2.6*	Base of Chron M39n.3n (mid-depth)	Base of *B. bullatus* ammonite zone. GSSP is not yet decided. In magnetostratigraphy from the Albstadt-Pfeffingen GSSP candidate (Swabia, Germany), this level is the base of a brief normal-polarity zone, which is interpreted to be the base of marine magnetic anomaly M39n.3n in the deep-tow extension (mid-depth projection) to M-sequence.	[*same*]	Most Tethyan and Sub-Boreal zones (and some subzones) have magnetostratigraphy with possible correlations to the mid-depth deep-tow extension to the Pacific M-sequence spreading model.

(*Continued*)

TABLE 26.2 **Summary of the Derivation of the Jurassic Numerical Age Models for Stages and for the Primary Scale of Tethyan Ammonite Zones (Tabulated in Table 26.3)—cont'd**

Stage	Spline-Fit Age; (Ma)	*Spline-Fit Uncertainty (2-Sigma)*	Revised Age (Ma)	*Estimated Uncertainty (2-Sigma)*	*Duration (myr)*	Calibration (Brief)	Stage Primary Marker and GSSP or Working Definition; Calibration	Method to Compute Basal Age of Stage	Method for Scaling Ammonite Zones within Stage
Bathonian	168.6	*0.7*	**168.3**	*1.3*	*2.2*	Base of Chron M42n.1n (mid-depth)	Base of *Z. zigzag* ammonite zone at Ravin du Bès, Bas Auran (France). In magnetostratigraphy of Spain, this level is the base of a normal-polarity zone, which is interpreted to be the base of marine magnetic anomaly M42n.1n in the deep-tow extension (mid-depth projection) to M-sequence.	[*same*]	Most Tethyan zones have magnetostratigraphy (Spain) with possible correlations to the mid-depth deep-tow extension to the Pacific M-sequence spreading model. Subzones within each zone dashed. Lower Bathonian zones scaled proportionally to the relative number of subzones, due to uncertainty in limits for *P. aurigerus* Zone in reference section.
Bajocian	171.3	*0.9*	**170.3**	*1.4*	*2.0*	80% up in Chron M44n.1r (mid-depth)	Base of *H. discites* ammonite zone at Cabo Mondego (Portugal). In combined magnetostratigraphy of Spain and Switzerland, this level is in the uppermost part (c. 80% up) of a reversed-polarity-dominated zone, interpreted to be marine magnetic anomaly M44n.1r in the deep-tow extension (mid-depth projection) to M-sequence.	[*same*]	Most Tethyan zones have magnetostratigraphy (Spain) with possible correlations to the mid-depth deep-tow extension to the Pacific M-sequence spreading model. However, portions of uppermost Bathonian and the Middle and Lower Bathonian scaled proportionally to the relative number of subzones, due to ambiguities in deep-tow models.
Aalenian	176.2	*1.8*	**174.1**	*1.0*	*3.8*	*3.85 myr before base of Bajocian*	Base of *L. opalinum* ammonite zone at Fuentelsaz (Spain).	Durations of Toarcian and Aalenian ammonite zones from cycle stratigraphy.	Durations of Aalenian ammonite zones from cycle stratigraphy (France, Italy).
Toarcian	183.4	*0.6*	**182.7**	*0.7*	*8.6*	*U-Pb age on initial reversed-polarity phase of Karoo basalts.*	Base of *D. tenuicostatum* ammonite zone; but GSSP not yet decided (probably in either Portugal or Spain).	High-resolution U-Pb radio-isotopic age.	Base of Toarcian is onset of a major carbon isotope excursion that is assumed to be the initial reversed-polarity eruptive phase of Karoo igneous province, dated as ~182.7 ±0.7 Ma. Durations of Toarcian ammonite zones from cycle-stratigraphy (France for *H. serpentinum* upward; Portugal for *D. tenuicostatum*).

Pliensbachian	190.2	*1.0*	**190.8**	*1.0*	*8.1*	*51% of combined Sinemurian-Pliensbachian span*	Base of *U. jamesoni* ammonite zone at Robin Hood's Bay (England).	Linear trend in strontium isotopes from base-Sinemurian through the Pliensbachiann to the base of Toarcian.	Linear trend in strontium isotopes for upper Pliensbachian zones (Sub-Boreal). Lower Pliensbachian zones scaled proportional to the relative number of subzones, due to uncertain zonation in this interval in the strontium isotope reference sections.
Sinemurian	199.2	*0.5*	**199.3**	*0.3*	*8.5*	*Cycle strat from base = U-Pb radio-isotope age*	Base of *A. bucklandi* ammonite zone at East Quantoxhead (England).	High-resolution U-Pb radio-isotopic ages.	Proportional to the relative number of subzones, due to uncertain zonation between stage limits in the strontium isotope reference sections.
Hettangian	201.3	*0.2*	**201.3**	*0.2*	*2.0*	*U-Pb radio-isotope age*	Lowest occurrence of the ammonite *Psiloceras spelae* at Kuhjoch (Austria).	Bracketed by high-resolution U-Pb radio-isotopic ages.	Cycle stratigraphy (England).

Values are rounded to nearest 0.1 myr. Spline fit through the database of Jurassic—Cretaceous radio-isotopic ages (Appendix 2) was run, then the final age model incorporated cycle stratigraphic constraints on stage and zone durations, segments with linear trends in strontium isotopes (Sinemurian—Pliensbachian), removing anomalous jumps in Pacific spreading rates (late Jurassic through early Cretaceous) and other stratigraphic or radio-isotopic constraints (see text).

TABLE 26.3 Interpolated Numerical Ages for Tethyan Ammonite Zones

Stage/Age	Tethyan Zone	Basal Age (Ma)	Notes on Calibration and Zonation
CRETACEOUS basal working definition (base-M18r) is in lower part of this zone	*Berriasella jacobi*	145.95	Dash at about Chron M19n.2n.3; due to probable distortions in sedimentation in the compact reference sections – at Puerto Escano (Pruner *et al.*, 2010), base is about M19n.2n.1. At Sierra Gorda (Ogg *et al.*, 1984), this zone base is M19n.2n.55 (+/–.05); Uncertainty in magnetostratigraphic correlation to top of Tithonian *Durangites* Zone is ±.05 of polarity chron M19n.2n = approx. 0.04 myr. Base-Berriasian, using base-M18r, is WITHIN this zone.
	Durangites	146.61	Placed at Chron M20n.1n.8. At Puerto Escano (Pruner *et al.*, 2010), is about M20n.1n.65. Higher at Sierra Gorda (Ogg *et al.*, 1984) at M19r.1 (+/–.2); which is only about 0.1 myr higher. Uncertainty in magnetostratigraphic correlation to base of Tithonian *Durangites* Zone is ±.2 of polarity chron M19r = approx. 0.08 myr.
	Micracanthoceras microcanthum	147.72	Base Chron M20n (+/–.1) at both Sierra Gorda (Ogg *et al.*, 1984) and Puerto Escano (Pruner *et al.*, 2010). The zones of *P. transitorius* and *"Simplisphinctes"* are combined into a *Microcanthum* Zone in some schemes (used here; following diagarm from F. Oloriz); but separated as zones in the main reference section. Base of Calpionellid Zone A1 is just above base of *Microcanthum* Zone.
	Micracanthoceras ponti / Burckhardticeras peroni	148.08	Chron M20r.5 (+/–.1).
	Semiformiceras fallauxi	149.87	Chron M22n.95 (+/–.05).
	Semiformiceras semiforme	150.41	Chron M22n.6 (+/–.1); *Semiformiceras semiforme* (= zone of *Haploceras (Volanites) verruciferum*).
	Semiformiceras darwini	150.94	Chron M22n.25 (+/–.05); *Semiformiceras darwini* (= zone of *Virgatosimoceras albertinum*); subzones arbitrarily given equal durations.
Tithonian base	*Hybonoticeras hybonotum*	152.06	Base Chron M22An (+/–.1).
	Hybonoticeras beckeri	153.55	Chron M23r.2r.1 (+/–.05) Subzones given equal duration.
	Aulacostephanus eudoxus	153.96	Full name = *Aulacostephanus (Pseudomutabilis) eudoxus*. Base = Chron M24r.1r.8 (+/–.1).
	Aspidoceras acanthicum	154.47	Base = Chron M24r.2r.6 (+/–.1).
	Crussoliceras divisum	154.84	Assigned as Chron M24Ar.6 (+/–.1).
	Ataxioceras hypselocyclum	155.60	Full name of Zone = *Ataxioceras (A.) hypselocyclum*. Assigned as Chron M25n.8 (+/–.1).
	Sutneria platynota	156.02	Chron M25r.1 (+/–.1). Cycles => 0.40 myr duration. French sections (e.g., Crussol) indicate a 1:2:3 relative duration of the subzones.
	Idoceras planula	156.75	Base = base Chron M26n.1r. Duration from Cycles (Strasser, 2007) was 0.3 to 0.4 myr, but there is apparently a 0.4 myr "missed beat" => 0.75 myr total fits magnetostratigraphy.
Kimmeridgian base (base of a Boreal ammonite zone) is in middle of this Tethyan zone	*Epipeltoceras bimammatum*	158.54	Base = Chron M28Ar.25. Span of full *Bimammatum* Zone by Strasser (2007) had as 1.2 myr; but there is apparently a missing 0.4 myr "missed beat"; plus another 0.1 myr for entire *Bimammatum zone* scaling.
	Perisphinctes bifurcatus	159.44	Full name of zone = *Perisphinctes (Dichotomoceras) bifurcatus*. Base = Chron M29r.5 (approx. due to hiatus in Aguilon, Spain, magnetostratigraphy reference section). Only 1.5 cycles of 100 kyr known in lower part of zone. Total duration for *Bifurcatus* zone of 0.9 myr, based partly on scaling relative to M-sequence.
	Gregoryceras transversarium	160.09	Base = approx. base Chron M30An. Cycle duration = 0.65 (Boulila *et al.*, 2010).
	Perisphinctes plicatilis	160.84	Full name = *Perisphinctes (Arisphinctes) plicatilis*. Approx. base of Chron M32r on short-wavelength deeptow = average of Spain and Dorset.
	Cardioceras cordatum	161.39	Full name = *Cardioceras (C.) cordatum*. Base = approx. Chron M33Bn.5.
Oxfordian base	*Quenstedtoceras mariae*	163.47	Base of Oxfordian is assigned here as M37n.1n.25. Full name = *Quenstedtoceras (Q.) mariae*. Cycle duration (Boulila *et al.*, 2008a) is ~2.1 myr.
	Quenstedtoceras lamberti	163.81	Full name of zone = *Quenstedtoceras (Lamberticeras) lamberti*. Base had been Chron M37n.2n.3 (+/–.2); but is in R-zone in England. Cycle strat (Huret *et al.*, 2006) implies *Lamberti* is about 0.76 myr duration.
	Peltocoeras athleta	163.97	Base = within lower Chron M38n.1n in mid-depth deep-tow model. Cycle-strat implies very brief zone (0.1 myr in Huret *et al.*, 2006).

	Erymnoceras coronatum	164.50	Chron M38n.4n.7 in mid-depth deep-tow model.
	Reineckeia anceps	164.63	Chron M38n.4n.1 in mid-depth deep-tow model. Implied short duration fits cycles (0.2 myr) of Huret *et al.* (2006).
	Macrocephalites gracilis	165.59	Full zone name = *Macrocephalites (Dolikephalites) gracilis.* Base = Chron M39n.2n.6 in mid-depth deep-tow model.
Callovian base	*Bullatimorphites bullatus*	166.07	Full name = *Bullatimorphites (Kheraiceras) bullatus.* Base = base Chron M39.3n in mid-depth deep-tow model (assumed small-N at Albstadt proposed GSSP for base-Callovian is this level).
	Clydoniceras discus	166.24	Full zone name = *Clydoniceras (C.) discus* (Late-Middle Bathonian ammonite subzones (or 1.5 s.z. if single-subzone zone = total of 10) given equal duration from base-*subcontractus* to base-Callovian).
	Hecticoceras retrocostatum	166.49	Full zone name = *Hecticoceras (Prohecticoceras) retrocostatum.* (See note on *C. discus* for scaling method.)
	Cadomites bremeri	166.66	Full zone name = *Cadomites (C.) bremeri.* (See note on *C. discus* for scaling method.)
	Morrisiceras morrisi	166.78	Full zone name = *Morrisiceras (M.) morrisi.* M. morrisi Zone given a single-zone (1.5 s.z.) duration. (See note on *C. discus* for scaling method.)
	Tulites subcontractus	166.91	Full name = *Tulites (T.) subcontractus.* Base placed as Chron M40n.2n.8, which equates magnetostratigraphy-zone "Subc-R" with Chron M40n.1r. Zone given 1.5 s.z. duration. (See note on *C. discus* for scaling method.)
	Procerites progracilis	167.37	Full zone name = *Procerites (P.) progracilis.* (Early Bathonian ammonite subzones (total of 6) given equal duration from base-*subcontractus* to base-Bathonian.)
	Procerites aurigerus	167.82	Full name = *Procerites (Siemiradzkia) aurigerus.* Base may be at base of Chron M41n.3n; but magnetostrat had undifferentiated *Zigzag* and *Aurigenus* zones => subzone scaling was done. (See note on *P. progracilis* for scaling method.)
Bathonian base	*Zigzagiceras zigzag*	168.28	Full name = *Zigzagiceras (Z.) zigzag.* Assigned as Chron Base M42n.1n in mid-depth deep-tow model.
	Parkinsonia parkinsoni	168.69	Full name = *Parkinsonia (P.) parkinsoni.* (Late Bajocian ammonite subzones (6 of them in *Parkisoni* and *Garantiana* zones) are given equal duration.)
	Garantiana garantiana	169.11	Full zone name = *Garantiana (G.) garantiana.* Calibrated as Base of Chron M42n.4n in mid-depth deep-tow model.
	Strenoceras niortense	169.45	Full zone name = *Strenoceras (S.) niortense.* (Early-Middle Bajocian ammonite subzones (14 of them) are given equal duration.)
	Stephanoceras humphriesianum	169.70	Full zone name = *Stephanoceras (S.) humphriesianum.* (See note on *S. niortense* for scaling method.)
	Sonninia propinquans	169.87	Full subzone name = *Emileia (Otoites) sauzei.* (See Note on *S. niortense* for scaling method.)
	Witchellia laeviuscula	170.13	Full subzone name = *Euhoploceras (Fissilobiceras) ovalis.* (See note at S. niortense for scaling method.)
Bajocian base	*Hyperlioceras discites*	170.30	Full zone name = *Hyperlioceras (H.) discites.* Assigned as Chron M44n.1r.8 of mid-depth deep-towmodel (uppermost-Aalenian long-reversed-polarity zone is interpreted as this Chron, and the onset of the long-normal-polarity zone "Disc-N2" interpreted as Chron M44n.1n).
	Graphoceras concavum	170.83	Duration was interpreted as 0.43 myr by Huret *et al.*, 2008; but shifted here by 0.1 myr to partly adjust its magnetostratigraphy (which, also, might have uncertainties) to fit base of Chron M44n.1r in deep-tow model.
	Brasilia bradfordensis *Ludwigia murchisonae*	171.29 172.13	Duration from cycle stratigraphy of Huret *et al.* (2008) Genus is *"Ludwigia"* in some schemes. *Ludwigia murchisonae* Zone of S.Switzerland also includes a *bradfordensis* Zone (This follows Arkell's usage). Duration from cycle stratigraphy of Huret *et al.* (2008).
Aalenian base	*Leioceras opalinum*	174.15	Duration from cycle stratigraphy of Huret *et al.* (2008).
	Pleydellia aalensis	174.43	Duration from Chunju Huang (pers. comm., 10 Jan 2010) restudy of cycle stratigraphy by Huret *et al.* (2008) who has slightly shorter duration. However, for *P. aalenis* and *D. pseudoradiosa* an adjustment of 0.15 myr was made (shifted from *pseudoradiosa* to *aalensis*). Dash base – exact position of zonal limits with respect to cycles is not precise.
	Dumortieria pseudoradiosa	174.71	Cycle strat duration (Huret *et al.*, 2008; as revised by Chunju Huang). Dash base – exact position of zonal limits with respect to cycles is not precise.
	Phlyseogrammoceras dispansum	174.97	Cycle strat duration (Huret *et al.*, 2008; as revised by Chunju Huang). Dash base – exact position of zonal limits with respect to cycles is not precise.
	Grammoceras thouarsense	176.23	Cycle strat duration (Huret *et al.*, 2008; as revised by Chunju Huang). Dash base – exact position of zonal limits with respect to cycles is not precise.
	Haugia variabilis	178.24	Cycle strat duration (Huret *et al.*, 2008; as revised by Chunju Huang). However, this led to a pronounced distortion in Sr isotope trend; therefore a portion is shifted to the overlying *G. thouarsense* Zone (of which its zonal base is not well established relative to the cycle strat).

(Continued)

TABLE 26.3 Interpolated Numerical Ages for Tethyan Ammonite Zones—cont'd

Stage/Age	Tethyan Zone	Basal Age (Ma)	Notes on Calibration and Zonation
	Hildoceras bifrons	180.36	Cycle strat duration (Huret *et al.*, 2008; as revised by Chunju Huang).
	Harpoceras serpentinum	181.70	Onset of main OAE = base of *H. serpentinum* (see Hesselbo *et al.*, 2007, about problems with inter-regional correlation; this is their preferred assignment). The basal age derived from cycle duration of underlying zone determines the age assignments of all overlying Toarcian–Aalenian cycle-scaled-duration zones.
Toarcian base	*Dactylioceras tenuicostatum*	182.70	Full name = *Dactylioceras (Orthodactylites) tenuicostatum*. Age is 182.7 ± 0.7 Ma = Karoo lavas (same dates using mean Ar-Ar converted (Jourdan *et al.*, 2007) with ±0.7 and U-Pb (EGU abstract by Moulin *et al.*, 2010) with ±0.4)). Eruption of these reversed-polarity basalts is assumed to cause the LOWER C-13 excursion, which is in a reversed-polarity interval. Duration of zone was estimated as 0.3 myr from linear Sr isotope trend; but Os isotopes (e.g., Cohen *et al.*, 2007) suggest a longer span; and cycle strat (Suan *et al.*, 2008) yields c. 0.8 myr (Chunju Huang, 2010, pers. comm., computed as 1.0 myr). Used Suan *et al.*'s published (2008) 0.8 myr used here, plus 0.2 myr for condensed boundary (Bodin *et al.*, 2011).
	Emaciaticeras emaciatum	183.49	Scaling from linear Sr trends in England to Sub-Boreal zones; and calibration to Tethyan zone.
	Arieticeras algovianum	185.31	Scaling from linear Sr trends in England to Sub-Boreal zones; and calibration to Tethyan zone.
	Fuciniceras lavinianum	187.56	Relatively long duration, due to linear Sr-trend scaling in England implying an expanded *A. margaritatus* Zone.
	Prodactylioceras davoei	188.54	Lower Pliensbachian ammonite subzones are given equal duration.
	Tragophylloceras ibex	189.35	Scaling from linear Sr trends in England to Sub-Boreal zones; and calibration to Tethyan zone. *Ibex* Zone in Tethyan realm doesnot extend as old as Boreal *Ibex*, according to SEPM chart (Hardenbol *et al.*, 1998).
Pliensbachian base	*Uptonia jamesoni*	190.82	Scaling from linear Sr trends in England.
	Echioceras raricostatum	192.81	Scaling from linear Sr trends in England.
	Oxynoticeras oxynotum	193.81	Sinemurian ammonite subzones (17 of them) are given equal duration.
	Asteroceras obtusum	195.31	Sinemurian ammonite subzones (17 of them) are given equal duration.
	Caenisites turneri	196.31	Sinemurian ammonite subzones (17 of them) are given equal duration.
	Arnioceras semicostatum	197.80	Full subzone name = *Coroniceras (Paracoroniceras) charlesi*.
Sinemurian base	*Arietites bucklandi*	199.30	Basal age from U-Pb dating of base-Sinemurian interval.
	Schlotheimia angulata	200.10	Duration from cycle stratigraphy in England (Rohl *et al.*, 2010).
	Alsatites liasicus	200.85	Duration from cycle stratigraphy in England (Rohl *et al.*, 2010).
	Psiloceras planorbis	201.10	Former base of Jurassic in British stratigraphy. Assigned as 0.1 myr above base of *P. pacificum* (hence 0.2 above base of *P. tilmanni*), based on Kozur's and Lucas's schematic diagrams of Triassic–Jurassic boundary events. Duration from cycle stratigraphy in England (Rohl *et al.*, 2010).
JURASSIC (Hettangian) base	*Psiloceras spelae*	201.30	Basal "new" zone (may be called "*Psiloceras tilmanni*") not present in traditional England zonation, and the Tethyan zonation lacks standardization here. Appears to have a brief duration (0.2 myr assigned here).

Placement of some stage boundaries are approximate or are working definitions pending decision of the International Commission on Stratigraphy. Usage of zones/subzones varies among Tethyan regions, and this suite is mainly for the Sub-Mediterranean province. Two-decimal ages are given to show relative durations among the zones; and uncertainties on those ages are equal to or greater than the uncertainties on the boundaries of its stage (Table 26.4). The column of notes are informal remarks on calibrations and different zonal usages. (Details for subzone age models; and for calibrations of other biostratigraphic zonations can be found in pop-up windows for those items within the TimeScale Creator database; *www.tscreator.org*)

Ma (Odin *et al.*, 1992); which emphasizes the importance of the new standards for zircon processing and single-grain analysis techniques.

There are no dates from the Late Jurassic that meet GTS2012 standards, and the secondary guides are contradictory in some intervals. The suites suggest that the base-Oxfordian is ~162 Ma and the base-Kimmeridgian is ~156 Ma, but with large uncertainties. A date of 154.1 ± 2.2 Ma (Selby, 2007) obtained by a novel $^{187}Re/^{188}Os$ method from black shales just above the proposed base-Kimmeridgian GSSP on the Isle of Skye awaits confirmation by other techniques in coeval sections.

The Jurassic–Cretaceous boundary, using the working definition of base of magnetic polarity chron M18r, is constrained by a recalibrated $^{40}Ar/^{39}Ar$ date of 146.5 ± 1.6 Ma on reversed-polarity sills intruding earliest Berriasian sediments on the Shatsky Rise of the Pacific (Mahoney *et al.*, 2005).

The majority of this limited data set of radio-isotopic dates is consistent with the Jurassic age model of GTS2012; but definitive dating is urgently needed on strata having cosmopolitan biostratigraphic markers and employing modern zircon processing.

26.3.2. Direct Spline-Fitting of Radio-Isotopic Age Suite

As a first approximation, an initial spline-fit model was performed of all Jurassic–Cretaceous radio-isotopic ages (Appendix 2) versus their estimated placements (with uncertainties) on the primary Tethyan ammonite zonation. In the first run with a smoothing factor of 1.5, there were 11 radio-isotopic dates, mainly in the Jurassic, that did not pass the chi-squared test. This reflects the inconsistency among some of these published ages or their estimates biostratigraphic assignments. Therefore, a second run relaxed the smoothing factor to 0.975.

However, the parameters for this combined Jurassic–Cretaceous run resulted in several high-precision Late Cretaceous dates no longer falling within their known ammonite zones. Therefore, a pair of second spline-fits were made with the Late Cretaceous segment using a different smoothing factor to the Jurassic and Early Cretaceous segment. The computed Jurassic stage boundaries are indicated in the spline columns of Table 26.2. This second spline-fit mainly altered the projected age from the first spline-fit for the base-Aalenian (175.6 versus 176.2 Ma) and caused minor 0.2 myr shifts to some other levels. One surprising implication from the radio-isotopic database is that the durations of the Callovian, Bathonian and Bajocian stages are relatively short (less than ~2.5 myr each); a conclusion that is supported by the Callovian-cycle stratigraphy estimates, but is considerably less than the average of ~3.5 myr for these three stages in GTS2004.

The Jurassic radio-isotopic set for this spline had been augmented by including one date from the lower Karoo basalts (182.7 ± 0.7 Ma; Jourdan *et al.*, 2007) assigned to the Pliensbachian–Toarcian carbon isotope excursion interval.

However, the Jurassic scale based only on this fit to radio-isotopic ages (which included dates derived from multi-zircon methods) had inconsistencies with other constraints. Some of the main problems were:

(1) Toarcian and Aalenian durations from cycle stratigraphy had been estimated as ~8.3 (or ~8.6 after adjusting for base-Toarcian condensation) and ~3.9 myr; but the Spline #2 fit yielded 7.2 and 4.9 myr, respectively. This is probably reflecting that there is only a single Aalenian radio-isotopic date (with a biostratigraphic uncertainty greater than the entire lower Aalenian interval).
(2) The spline-fit Tithonian yields a span for that stage of 6.3 myr. When incorporating the magnetostratigraphic correlations to M-sequence, it would require that the Hawaiian marine magnetic anomaly reference scale had a spreading rate of 28.3 km/myr in Kimmeridgian, surging to 31.3 km/myr in Tithonian, then returning to 27.7 km/myr in Berriasian. This suggests that the spline-fit Tithonian is 10% "too long" in duration as estimated only from the spline-fit. In contrast, the Site 801 age constraint of 168.7 ± 1.7 Ma for the age of marine magnetic anomaly M42n.4r that is near the Bajocian–Bathonian boundary would require a dramatic 40% increase in the spreading rates of the Pacific synthetic Hawaiian reference profile from a steady 22 km/myr in Bajocian–Callovian to a steady 28 km/myr in Oxfordian–Kimmeridgian.
(3) The spline-fit durations for Sinemurian and Pliensbachian would require that the strontium isotope trend change to a lesser slope as it passes from Hettangian to Sinemurian, then dramatically surge to a very high slope in the Pliensbachian, before leveling out in the lower Toarcian. Alternatively, considering that there are no late Sinemurian or early Pliensbachian radio-isotopic age constraints, then the durations of these stages could be derived by assuming a linear strontium isotope trend, as suggested by McArthur *et al.* (2001).

Therefore, the pure spline-fit to the limited Jurassic suite of radio-isotopic age constraints was modified to incorporate stage (and zone) durations obtained from cycle stratigraphy, use of stable geochemical trends (Sinemurian–Pliensbachian), and incorporation of Pacific spreading rates that avoid stage-to-stage discontinuities.

26.3.3. Jurassic Numerical Age Model

The dearth of high-precision radio-isotopic ages with unambiguous precise biostratigraphic calibrations requires portions of the Jurassic numerical age scale to be interpolated from matching magnetostratigraphy from cycle-derived durations for stage/zones, from matching magnetostratigraphy of outcrops to the spreading rate model for the M-sequence of marine magnetic anomalies, from assumptions of linear trends in Sr isotopic ratios, and from relative numbers of ammonite subzones.

The main radio-isotopic controls are at the basal limits of the Hettangian and the Sinemurian stages; an assumption that the well-dated onset of the Karoo–Ferrar large igneous province was the primary cause of the initiation of the major carbon isotope excursion in the basal Toarcian, and the consistency of the cycle stratigraphy-scaled M-sequence pattern of Middle Jurassic through Early Cretaceous with secondary guides from radio-isotopic dates. This Jurassic age model requires testing by acquiring radio-isotopic dates with precise biostratigraphic placements, verification from independent cycle stratigraphy analyses, and obtaining a reliable pre-Oxfordian marine magnetic anomaly reference scale for calibrating the Bajocian through Callovian magnetostratigraphy.

26.3.3.1. *Early Jurassic Scaling*

Hettangian

The Triassic/Jurassic boundary age is 201.3 ± 0.2 Ma according to bracketing U-Pb dates (Schoene *et al.*, 2010). This is currently the only well-documented U-Pb zircon age with narrow 95% confidence limits (2-sigma) that directly constrains a stage boundary within the Jurassic! Each of the Hettangian ammonite zones of Britain is cycle-scaled for a total of 1.8 myr (Ruhl *et al.*, 2010). An additional 0.2 myr was added to the basal zone to compensate for the correlation of the ammonite-barren pre-*planorbis* interval of Britain to the first occurrence of the precursor species to *P. planorbis* that marks the base-Jurassic GSSP.

Sinemurian

The Hettangian cycle stratigraphy projects the base-Sinemurian as 199.3 Ma, which is consistent with the 199.5 ± 0.2 Ma for this boundary interval in Peru (Schaltegger *et al.*, 2008). An uncertainty of 0.3 myr is estimated for this 199.3 Ma boundary age.

As in GTS2004, the Sinemurian/Pliensbachian boundary is extrapolated by assuming a linear decline in the cycle-scaled Sr trend from base-Sinemurian through base-Toarcian (LOWESS fit of McArthur *et al.*, 2001). At the Pliensbachian/Toarcian boundary, the strontium isotopes reach a minimum followed by a sharp upward increase in earliest Toarcian, which perhaps may be related to the initial eruptive phase of the Karoo flood basalts. Based on this assumption of a linear Sr isotope trend, the Sr isotope values at the Sinemurian/Pliensbachian boundary imply an age assignment that is 51% between base-Sinemurian (199.3 Ma) and base-Toarcian (182.5 Ma), therefore at 190.7 Ma.

The Sinemurian ammonite zones are scaled proportionally to the relative number of component subzones (total of 17 subzones distributed within the 6 ammonite zones). This assumption of equal subzonal units was necessitated because of uncertainties in correlation of the individual zonal boundaries to the main strontium isotope trend line and the lack of verified cycle durations for any of the zones.

Pliensbachian

As explained above, the Sinemurian/Pliensbachian boundary was extrapolated from an assumption of linear Sr isotope trends as 190.8 Ma. The uncertainty on this extrapolation is difficult to estimate; therefore an arbitrary ± 1 myr is indicated.

The relative durations of the upper Pliensbachian zones (Sub-Boreal realm) are scaled according to their correlations to the linear-decay strontiumisotope trend. The lower Pliensbachian zones are scaled proportionally to the relative number of subzones, due to uncertainties in zonal correlation to main strontium isotope trend line. Tethyan ammonite zones are correlated to Sub-Boreal using published charts (e.g., Groupe Français d'Étude du Jurassique, 1997; Hardenbol *et al.*, 1998).

Toarcian

The base-Toarcian is marked by the onset of a major negative excursion in carbon isotopes, and it is suggested that the basal-Toarcian of reversed-polarity magnetization is coeval with the reversed-polarity initial eruptive episode of Karoo–Ferrar flood basalts. A mean age of 182.7 ± 0.7 Ma for these basalts and earliest sills is therefore assigned as the onset of the base-Toarcian carbon isotope excursion (see above discussions under Large Igneous Provinces).

Toarcian ammonite zones of France have durations from cycle stratigraphy (e.g., Suan *et al.*, 2008; Huang *et al.*, 2010a; Boulila *et al.*, in press). An additional 0.2 myr was added to the basal ammonite zone to compensate for the condensation of the basal Toarcian carbon isotope excursion in those reference sections as compared to an expanded 220-m-thick coeval section in Morocco (Bodin *et al.*, 2010). Some ammonite zones within the middle and upper Toarcian have uncertain zonal boundaries relative to the detailed cycle stratigraphy, therefore their limits were partially adjusted according to relative thicknesses and numbers of component subzones in other reference sections (e.g., an adjustment of 0.15 myr was added to *P. aalensis* from the adjacent *D. pseudoradiosa*).

26.3.3.2. *Middle Jurassic Scaling*

Aalenian

The Toarcian spans ~8.5 myr – 8.3 myr from the cycle stratigraphy in reference sections, plus a minor compensation for the condensed basal portion in those sections. This projects the base-Aalenian as 174.1 Ma. The uncertainty on this age is estimated as 0.8 myr, based on the 0.5 myr uncertainty on base-Toarcian plus an additional allowance for cycle stratigraphy values that await verification in additional reference sections.

The Aalenian ammonite zones are scaled from cycle stratigraphy, and the stage spans a total of 3.85 myr (Huret *et al.*, 2008). Therefore, the Aalenian/Bajocian boundary is 170.3 Ma.

Bajocian

From uppermost Aalenian through the Barremian Stage of the Cretaceous, the ages of stage boundaries and most ammonite zones are derived from magnetostratigraphic calibration to the M-sequence of Pacific marine magnetic anomalies. In turn, the assigned numerical ages are from the spreading rate model for that M-sequence. As explained in the Geomagnetic Polarity Timescale chapter (Chapter 5 of this volume), that spreading rate model is a linear fit to a suite of six intervals of cycle-scaled durations for clusters of polarity chrons and the associated magnetic anomalies between early Oxfordian to Barremian. The linear fit implies that the spreading rate had a gradual slowing from early Oxfordian through the Barremian. This spreading rate model from base-Aptian (base of Chron M0r, assigned as 126 Ma) back through base-Oxfordian (Chron M37n.1n.25) is projected for the earlier pre-M37 set of magnetic anomalies obtained from deep-tow surveys.

Results from drilling of anomaly M42n.4r at ODP Site 801 suggest that this crustal interval formed near the Bathonian–Bajocian boundary and has a U-Pb age of 168.4 ± 1.7 Ma (Koppers *et al.*, 2003). This is an excellent fit to the linear-fit spreading rate model for the M-sequence, which projects an age of 168.5 Ma for this same anomaly.

The Aalenian/Bajocian boundary is approximately 80% up within a relatively long reversed-polarity magnetozone (Henriques *et al.*, 1994; Pavia and Enay, 1997). Based on the estimated 170.3 Ma age of this boundary derived from cycle stratigraphy relative to the base-Toarcian (182.7 Ma) and the projected M-sequence of marine magnetic anomalies, this reversed-polarity magnetozone is correlated to polarity chron M44n.1r.

The magnetostratigraphy of Bajocian reference sections (Steiner *et al.*, 1987) is used to partially constrain the age model for the Bajocian. For example, the base of ammonite zone *G. garantiana* is at the beginning of a relatively long normal-polarity magnetozone correlated to Chron M42n.4n, therefore it is assigned an age of 169.1 Ma according to the M-sequence age model (Figure 26.9). Where the magnetostratigraphic correlations or limits of ammonite boundaries are less precise, then the relative durations of Bajocian ammonite zones are apportioned according to relative numbers of component subzones.

Bathonian

The Bajocian/Bathonian boundary is at the base of a normal-polarity magnetozone that is tentatively correlated with Chron M42n.1n. This has a projected age of 168.3 Ma according to the M-sequence age model, and is supported by the 168.7 ±1.7 Ma date for the older marine magnetic anomaly M42n.4r as drilled at ODP Site 801 (adjusted from Koppers *et al.*, 2003).

As with the Bajocian, several of the Bathonian ammonite zones are correlated via magnetostratigraphy to the M-sequence polarity pattern and therefore have extrapolated numerical ages. Other intervals necessitated scaling according to relative numbers of component ammonite subzones. These extrapolated ages fit well with the 167–163 Ma range of secondary radio-isotopic age guides.

Callovian

The base of the Callovian is within a normal-polarity-dominated interval that is tentatively correlated with Chron M39n.3n of the M-sequence pattern and hence has a projected age of 166.1 Ma (with ~1 myr uncertainty). This is consistent with the 164.6 ± 0.2 Ma age on early Callovian from Argentina (Kamo and Riccardi, 2009), although the ammonites from that section lack cosmopolitan markers.

Each Callovian ammonite zone has tentative correlations to the M-sequence. Some of these correlations incorporate estimated durations from cycle stratigraphy of those zones (e.g., according to Huret *et al.* (2008), the *R. anceps* Zone spans only about 0.2 myr and *P. athleta* only about 0.1 myr).

This suite of correlations of magnetostratigraphy to the M-sequence model, although inevitably subject to possible alternative matches, implies that the three stages – Bajocian, Bathonian and Callovian – had approximately equal durations of only 2.0 to 2.6 myr. These relatively brief stages are consistent with the spline-fit of the few radio-isotopic age constraints (Appendix 2; and Table 26.2) and with the cycle stratigraphy analyses of Callovian ammonite zones in boreholes (Huret *et al.*, 2006; Lefranc *et al.*, 2008).

The uncertainties on the extrapolated boundary ages are based on the 0.3 myr uncertainty on the base of Chron M0r that is the age control for the M-sequence and the statistical uncertainties on the spreading rate line-fit slope (see Chapter 5 of this volume). The total uncertainty is estimated as ± 0.8 myr at base-Cretaceous (M18r) that progressively increases to ± 1.3 myr at base-Bathonian (M42n), assuming that the magnetostratigraphic correlations are accurate.

The projection of the Pacific spreading rate model to pre-Oxfordian marine magnetic anomalies interpreted from deep-tow data requires verification of the polarity signatures, cycle-stratigraphic work on Bajocian–Bathonian–Callovian zones/chrons, and additional direct radio-isotopic dating. If the magnetostratigraphic correlations to the M-sequence are supported by future studies, then the uncertainty on the duration of each stage is probably only about 10% of its total current estimated span, based on the cycle-scaled spreading rates in Kimmeridgian–Oxfordian and the consistency of the base-Bajocian age "projected downward from base-Oxfordian" with the cycle-scaled span of the Toarcian and Aalenian that gives the same base-Bajocian age "projected upward from base-Toarcian". At this time, the age model used in GTS2012 for the Middle Jurassic should be regarded as a hypothesis awaiting testing.

26.3.3.3. Late Jurassic Scaling

The progressive efforts of numerous investigators have compiled magnetostratigraphy patterns from nearly every ammonite zone of the Sub-Boreal and Tethyan realms, and correlated these to the M-sequence of marine magnetic anomalies. Several of the Oxfordian and Kimmeridgian ammonite zones have durations computed from cycle stratigraphy (e.g., Boulila *et al.*, 2008a,b, 2010a,b), which, in turn, are important controls on the Pacific spreading rate model for that M-sequence (see Chapter 5). The magnetostratigraphy of cycle-scaled Sub-Boreal sections of Tithonian age are consistent with this model.

Oxfordian

The Callovian/Oxfordian boundary at 25% up in Chron M37n.1n has a projected M-sequence age of 163.5 Ma. This is older than "secondary guide" ages in the Oxfordian; but none of those dates were obtained from single-zircon analyses or ODP basalts with unambiguous ages.

Each Oxfordian ammonite zone has its age derived from a merger of it cycle-derived and magnetostratigraphic correlation to the M-sequence (e.g., Boulila *et al.*, 2010a; Ogg *et al.*, 2010a; Przybylski *et al.*, 2010a). The Oxfordian spans 6.0 myr, of which 2.0 myr is within the basal *Q. mariae* ammonite zone according to cycle stratigraphy from SE France (Boulila *et al.*, 2008a). This is a major change from GTS2004, in which that ammonite zone was estimated from condensed sections in Britain as spanning only about 0.6 myr.

Kimmeridgian

The base of the Kimmeridgian, using the Boreal/Sub-Boreal definition of the base of the *P. baylei* ammonite zone corresponds to the base of Chron M26r with a projected age of 157.25 Ma. This is within the uncertainty on the recalibrated $^{40}Ar/^{39}Ar$ date (156.3 $\pm$ 3.4 Ma) obtained from within oceanic basalts drilled at Site 765 on marine anomaly M26r (Ludden, 1992).

Each Kimmeridgian ammonite zone has its age derived from magnetostratigraphic correlation to the M-sequence. The implied durations of zones are generally consistent with the cycle-derived durations from cycle stratigraphy (e.g., Boulila *et al.*, 2008b, 2010b; Strasser, 2007), although the span of a couple of the ammonite zones implied "missing cycle beats" at sequence boundaries. For example, the *Idoceras planula* Zone of basal Kimmeridgian is estimated as ~0.35 myr by Strasser (2007), but the magnetostratigraphy implies a duration of 0.75 myr, implying that a full 400-kyr-long eccentricity cycle is probably missing at the regional Oxf-8 sequence boundary below its transgressive base.

Tithonian

The base of the Tithonian is assigned here as the base of Chron M22An, which is essentially coeval with the historical boundary in Tethyan ammonite stratigraphy. The corresponding M-sequence age is 152.1 Ma.

Each Tithonian ammonite zone of the Sub-Mediterranean faunal realm has its age derived from magnetostratigraphic correlation to the M-sequence (Figure 26.8). The ages of Sub-Boreal/Boreal zones are estimated from a combination of rare correlation levels to the Sub-Mediterranean succession, magnetostratigraphy (e.g., Ogg *et al.*, 1994; Houša *et al.*, 2007), and scaling from cycle stratigraphy (e.g., Huang *et al.*, 2010b).

A working definition for the Jurassic/Cretaceous boundary is the base of Chron M18r, which we have adopted for the GTS2012 time scale summary. The M-sequence model projects this age as 145.0 Ma, which matches a recalibrated $^{40}Ar/^{39}Ar$ date of 145.5 $\pm$ 0.8 Ma from reversed-magnetized (interpreted as magnetozone M18r) sills intruded into earliest Berriasian pelagic sediments drilled at ODP Site 1213B on Shatsky Rise (Mahoney *et al.*, 2005).

As with the Middle Jurassic ages projected from the M-sequence calibrations, the uncertainties on these extrapolated boundary ages are also estimated as $\pm$ 1 myr. The crucial factor is the reliability of the age-model for the M-sequence. Even though it is derived from cycle-derived durations for intervals of magnetozones within the early Cretaceous (three intervals) and late Jurassic (three intervals) and is consistent with radio-isotopic dates within the Early Cretaceous, the numerical ages of pre-Berriasian marine magnetic anomalies ultimately hinge on a fairly simplistic spreading rate model of a gradual increase in Pacific spreading from Bajocian through Barremian.

26.3.4. Summary

This Jurassic compilation should be considered as a work in progress. The combination of extensive cycle stratigraphy studies and enhanced methods of radio-isotopic dating has greatly refined the numerical age scale for the Jurassic, but several intervals lack adequate constraints.

In general, the assigned or extrapolated ages have shifted to slightly older values relative to the estimates in GTS2004 (Table 26.4). The extension of the M-sequence by deep-tow magnetometer surveys into Middle Jurassic crust in the Pacific enabled tentative correlation of the magnetostratigraphy of the Callovian through Bajocian. The cycle-scaling of the M-sequence spreading rate model has greatly reduced the

TABLE 26.4 **Jurassic Age Model of GTS2012 and Comparison with GTS2004 (Gradstein *et al.*, 2004).**

	GTS2004			GTS2012			
	Base	Est. Uncertainty	Duration	Base	Est. Uncertainty	Duration	Derivation in GTS2012; and Differences from GTS2004
Cretaceous (Berriasian)	145.5	4.0		145.0	0.8		Base of Chron M18r used to assign base-Cretaceous, instead of mid-Chron M19n
Tithonian	150.8	4.0	5.3	152.1	0.9	7.1	Ages of M-sequence; hence Tith–Oxf shifted older than GTS2004 – partly due to Ar-Ar monitor change, partly due to new spreading rate model
Kimmeridgian	155.7	4.0	4.9	157.3	1.0	5.2	"
Oxfordian	161.2	4.0	5.5	163.5	1.1	6.2	"
Callovian	164.7	4.0	3.5	166.1	1.2	2.6	Ages of Callovian–Bathonian–Bajocian derived from radio-isotope age constraints and M-sequence spreading rate model (deep-tow extension) => much shorter stage durations than in GTS2004.
Bathonian	167.7	3.5	3	168.3	1.3	2.2	"
Bajocian	171.6	3.0	3.9	170.3	1.4	2.0	"
Aalenian	175.6	2.0	4	174.1	1.0	3.8	
Toarcian	183.0	1.5	7.4	182.7	0.7	8.6	Duration of Toarcian from cycle-stratigraphy is 1 myr longer than estimate in GTS2004
Pliensbachian	189.6	1.5	6.6	190.8	1.0	8.1	Durations of Pliensbachian and Sinemurian increased due to older age for base-Sinemurian
Sinemurian	196.5	1.0	6.9	199.3	0.3	8.5	Radio-isotopic U-Pb age for base-Sinemurian
Hettangian	199.6	0.6	3.1	201.3	0.2	2.0	Age assigned from bounding U-Pb radio-isotopic dates from single zircons. Duration of Hettangian from cycle stratigraphy.
Triassic (Rhaetian)							

Values are rounded to nearest 0.1 myr.

estimated statistical uncertainties on Middle and Late Jurassic ages relative to the estimates in GTS2004; although the magnetostratigraphic calibration of the Middle Jurassic awaits an improved pre-M29 marine magnetic model, confirmation of the polarity pattern for each ammonite zone, and additional cycle stratigraphy analyses of the durations of the component ammonite zones in multiple independent sections.

The greatest relative change in the numerical age scale relative to GTS2004 is in the earliest Jurassic. In this interval, the dates derived from multi-zircon methods (e.g., Pálfy *et al.*, 2000b) have been replaced with significantly more precise dates (and older values) derived from enhanced processing for single-zircon analyses. Similar high-resolution redating of horizons and the acquisition of new radio-isotopic ages is required for the majority of the Jurassic.

When cycle-stratigraphic durations of ammonite zones have been verified in additional sections, then the Jurassic will have a very stable scaling for use in global correlation.

TABLE 26.5 GSSPs of the Jurassic Stages, with Location and Primary Correlation Criteria (Status Jan. 2012)

Stage	GSSP Location	Latitude, Longitude	Boundary Level	Correlation Events	Reference
Tithonian	*Candidates are Mt. Crussol or Canjuers (SE France) and Swabia, Germany*			*Near base of* Hybonoticeras hybonotum *ammonite zone and lowest occurrence of* Gravesia *genus, and the base of magnetic polarity chronozone M22An*	
Kimmeridgian	*Candidate is Flodigarry (Isle of Skye, NW Scotland)*			*Ammonite, near base of* Pictonia flodigarriensis *ammonite zone of Boreal realm, and the base of magnetic polarity chronozone M26r*	
Oxfordian	*Candidate is at Redcliff Point (Dorset, SW England)*			*Ammonite* Cardioceras redcliffense *Horizon at base of the* Cardioceras scarburgense *Subzone* (Quenstedtoceras mariae *Zone)*	
Callovian	*Candidates are Pfeffingen, Swabian Alb, SW Germany, a nd Novgorod region, Russia*			*Ammonite, FAD of the genus* Kepplerites (Kosmoceratidae) *(defines base of* Macrocephalites herveyi *Zone in Sub-Boreal province of Great Britain to southwest Germany)*	
Bathonian	Ravin du Bès, Bas-Auran area, Alpes de Haute Provence, France	43°57′38″N 6°18′55″E*	Base of limestone bed RB071	Ammonite, FAD of *Gonolkites convergens* (defines base of *Zigzagiceras zigzag* Zone) and *Morphoceras parvum*	Episodes 32/4, 2009
Bajocian	Murtinheira Section, Cabo Mondego, Portugal	40°11′57″N 8°54′15″W*	Base of Bed AB11 of the Murtinheira Section	Ammonite FAD *Hyperlioceras mundum, H. furcatum, Braunsina aspera, B. elegantula*	Episodes 20/1, 1997
Aalenian	Fuentelsaz, Spain	41°10′15″N 1°50′W	Base of Bed FZ 107 in Fuentelsaz Section	Ammonite FAD *Leioceras opalinum* and *Leioceras lineatum*	Episodes 24/3, 2001
Toarcian	*Peniche (Portugal)*			*Ammonite, near FAD of a diversified* Eodactylites *ammonite fauna; correlates with the NW European* P. paltus *Horizon*	
Pliensbachian	Robin Hood's Bay, Yorkshire Coast, England	54°24′25″N 0°29′51″W	Base of Bed 73b at Wine Haven, Robin Hood's Bay	Ammonite association of *Bifericeras donovani* and *Apoderoceras* sp	Episodes 29/2, 2006
Sinemurian	East Quantoxhead, SW England	51°11′27.3″N 3°14′11.2″W*	0.90m above the base of Bed 145	Ammonite FAD *Vermiceras quantoxense, Vermiceras palmeri*	Episodes 25/1, 2002
Hettangian (base Jurassic)	Kuhjoch Section, Karwendel Mountains, Austria	47°29′02″N 11°31′50″E	5.80m above the base of the Tiefengraben Member of the Kendelbach Formation	FAD of *Psiloceras spelae*	

**according to Google Earth*

ACKNOWLEDGMENTS

Our colleagues within the Jurassic Subcommission of the ICS and the participants at the Jurassic Congress meetings and other venues have unselfishly shared their data, preliminary results and ideas. In particular, we wish to thank Slah Boulila, and Bruno Galbrun, among others, for enabling the cycle stratigraphy scaling of zones, stages and spreading rates. József Pálfy, Nicol Morton, Angela Coe, Ewa Glowniak, Steve Hesselbo, Paul Wignall and Mikhail Rogov provided insights and reviews of inter-regional correlations, geochemical trends and age constraints. Undoubtedly, most of these groups will consider this set of syntheses to be too brief in covering their important specialties. Steve Hesselbo was outstanding in giving an extensive review of an early version of this chapter, and Gabi Ogg prepared the figures.

Additional or alternative zonal schemes are available through the *TimeScale Creator* visualization datapacks (*www.tscreator.org*). Documentation for post-2011 advances in formalizing GSSPs, zonal schemes, inter-regional correlations, and eventual consensus on the best numerical scalings will be found in the newsletters of the Subcommission on Jurassic Stratigraphy.

The compilation and application of cycle stratigraphy to enhance the Mesozoic time scale was partially supported by U.S. National Science Foundation grant EAR-071895.

REFERENCES

Al-Husseini, M., Matthews, R.K., 2006. Stratigraphic Note: Orbital calibration of the Arabian Jurassic second-order sequence stratigraphy. GeoArabia 11, 161–170.

Alméras, Y., Boullier, A., Laurin, B., 1997. Brachiopodes. In: Groupe Français d'Étude du Jurassique (Ed.), Biostratigraphie du Jurassique Ouest-Européen et Méditerranéen: Zonations Parallèles et Distribution des Invertébrés et Microfossiles. Bulletin des Centres de Recherches Exploration- Production Elf-Aquitaine, Mémoire, 17, pp. 169–195.

Arkell, W.J., 1933. The Jurassic System in Great Britain. Clarendon Press, Oxford, p. 681.

Arkell, W.J., 1939. The ammonite succession at the Woodham Brick Company's pit, Akeman Street Station, Buckinghamshire, and its bearing on the classification of the Oxford Clay. Quarterly Journal of the Geological Society 95, 135–221.

Arkell, W.J., 1946. Standard of the European Jurassic. Geological Society of America Bulletin 57, 1–34.

Arkel, W.J., 1954. Three complete sections of the Cornbrash. Proceedings of the Geological Association London 65, 115.

Arkell, W.J., 1956. Jurassic Geology of the World. Oliver and Boyd, Edinburgh, pp. 806.

Atrops, F., 1982. La sous-famille des Ataxioceratinae (Ammonitina) dans le Kimméridgien inférieur du Sud-Est de la France: Systématique, évolution, chronostratigraphie des genres *Orthosphinctes* et *Ataxioceras*. Documents des Laboratoires de Géologie Lyon 83, p. 463.

Atrops, F., 1994. Upper Oxfordian to Lower Kimmeridgian ammonite successions and biostratigraphy of the Crussol and Châteauneuf d'Oze sections. In: Atrops, F., Fortwengler, D., Marchand, D., Melendez, G. (Eds.), IV Oxfordian & Kimmeridgian Working Group Meeting, Lyon and SE France Basin, Guide Book and Abstracts, pp. 50–60 and 106–111.

Atrops, F., 1999. Report of the Oxfordian-Kimmeridgian Boundary Working Group. International Subcommission on Jurassic Stratigraphy Newsletter 27, 34.

Atrops, F., Gygi, R., Matyja, B.A., Wierzbowski, A., 1993. The *Amoeboceras* faunas in the Middle Oxfordian – Lowermost Kimmeridgian, Submediterranean succession, and their correlation value. Acta Geologica Polonica 43, 213–227.

Bartolini, A., Larson, R.L., 2001. The Pacific microplate and the Pangea supercontinent in the Early to Middle Jurassic. Geology 29, 735–738.

Bartolini, A., Baumgartner, P.O., Hunziker, J., 1996. Middle and Late Jurassic carbon stable-isotope stratigraphy and radiolarite sedimentation of the Umbria-Marche Basin (Central Italy). Eclogae Geologicae Helvetiae 89, 811–844.

Bartolini, A., Baumgartner, P.O., Guex, J., 1999. Middle and Late Jurassic radiolarian palaeoecology versus carbon-isotope stratigraphy. Palaeogeography, Palaeoclimatology, Palaeoecology 145, 43–60.

Bartolini, A., Larson, R., Baumgartner, P.O., the ODP Leg 185 Scientific Party, 2000. Bajocian radiolarian age of the oldest oceanic crust in situ (Pigafetta Basin, western Pacific, ODP Site 801, Leg 185). European ODP Forum, La Grande-Motte, France, abstracts, 32.

Baumgartner, P.O., 1987. Age and genesis of Tethyan Jurassic radiolarites. Eclogae Geologicae Helvetiae 80, 831–879.

Belkaaloul, K.N., Aissaoui, D.M., Rebelle, M., Sambet, G., 1995. Magnetostratigraphic correlation of the Jurassic carbonates from the Paris Basin: Implications for petroleum exploration. Journal of the Geological Society 95, 173–186.

Belkaaloul, K.N., Aissaoui, D.M., Rebelle, M., Sambet, G., 1997. Resolving sedimentological uncertainties using magnetostratigraphic correlation: An example from the Middle Jurassic of Burgundy, France. Journal of Sedimentary Research 67, 676–685.

Birkelund, T., Callomon, J.H., Clausen, C.K., Nøhr Hansen, H., Salinas, I., 1983. The Lower Kimmeridge Clay at Westbury, Wiltshire, England. Proceedings of the Geological Association 94, 289–309.

Blake, J.F., 1881. On correlation of the Kimmeridge and Portland rocks of England with those of the continent. Part 1. The Paris Basin. Quarterly Journal of the Geological Society 37, 497–587.

Bloos, G., 2008. Comment on the T-J boundary level at the FAD of the earliest psiloceratid. *P. spelae* [Unpublished document accompanying ICS voting package on base-Jurassic GSSP from the secretary of Triassic/Jurassic Boundary Working Group].

Bloos, G., Page, K.N., 2002. Global stratotype section and point for base of Sinemurian Stage (Lower Jurassic). Episodes 25, 22–28.

Bodin, S., Mattioli, E., Fröhlich, S., Marshall, J.D., Boutib, L., Lahsini, S., Redfern, J., 2010. Toarcian carbon isotope shifts and nutrient changes from the Northern margin of Gondwana (High Atlas, Morocco, Jurassic): Palaeoenvironmental implications. Palaeogeography, Palaeoclimatology, Palaeoecology 297, 377–390.

Bogoslovsky, N.L., 1897. The Ryazanian horizon: Fauna, stratigraphic relationships, and probable age. Materials on Geology of Russia (St. Petersburg) 18, 157 [In German].

Bonarelli, G., 1894. Contribuzione alla conoscenza del Giura-Lias Lombardo. Atti della Regia Accademia delle Scienze di Torino 30, 63–78.

Bonarelli, G., 1895. Fossili domeriani della Brianza. Rendiconti del Reale Istituto. Lombardo di Scienze e Lettere (Milano), serie 2. 28, 326–347.

Bonis, N.R., Kürschner, W.M., Krystyn, L., 2009. A detailed palynological study of the Triassic-Jurassic transition in key sections of the Eiberg Basin (Northern Calcareous Alps, Austria). Review of Palaeobotany and Palynology 156, 376–400.

Boulila, S., Hinnov, L.A., Collin, P.Y., Huret, E., Galbrun, B., Fortwengler, D., 2008a. Astronomical calibration of the Lower Oxfordian (Terres Noires, Vocontian Basin, France): Consequences of revising

Late Jurassic time scale. Earth and Planetary Science Letters 276, 40–51.

Boulila, S., Galbrun, B., Hinnov, L., Collin, P.Y., 2008b. Orbital calibration of the Early Kimmeridgian (Southeastern France): Implications for geochronology and sequence stratigraphy. Terra Nova 20, 455–462.

Boulila, S., Galbrun, B., Hinnov, L.A., Collin, P.-Y., Ogg, J.G., Fortwengler, D., Marchand, D., 2010a. Milankovitch and sub-Milankovitch forcing of the Oxfordian (Late Jurassic) Terres Noires Formation (SE France) and global implications. Basin Research 22, 712–732.

Boulila, S., de Rafélis, M., Hinnov, L.A., Gardin, S., Galbrun, B., Collin, P.-Y., 2010b. Orbitally forced climate and sea-level changes in the Paleoceanic Tethyan domain (marl-limestone alternations, Lower Kimmeridgian, SE France). Palaeogeography, Palaeoclimatology, Palaeoecology 292, 57–70.

Boulila, S., Gardin, S., de Rafélis, M., Hinnov, L.A., Galbrun, B., Collin, P.-Y., 2011. Reply to the comment on "Orbitally forced climate and sea-level changes in the Paleoceanic Tethyan domain (marl-limestone alternations, Lower Kimmeridgian, SE France)." by S. Boulila, M. de Rafélis, L.A. Hinnov, S. Gardin, B. Galbrun, and P.-Y. Collin [Palaeogeography, Palaeoclimatology, Palaeoecology 292 (2010) 57–70]. Palaeogeography, Palaeoclimatology, Palaeoecology 306, 252–257.

Boulila, S., Galbrun, B., Huret, E., Hinnov, L.A., Rouget, I., Gardin, S., Bartoloni, A., and Huang, C., in press, Astronomical calibration of the Toarcian stage in the Paris Basin and implications for the CIE duration. Geology.

Bown, P.R., Cooper, M.K.E., Lord, A.R., 1988. A calcareous nannofossil biozonation for the early to mid Mesozoic. Newsletters on Stratigraphy 20, 91–114.

Bown, P.R. (Ed.), 1998. Calcareous Nannofossil Biostratigraphy. Chapman & Hall, London, p. 328.

Bown, P.R., Lees, J.A., Young, J.R., 2004. Calcareous nannoplankton evolution and diversity through time. In: Thierstein, H.R., Young, J.R. (Eds.), Coccolithophores: From Molecular Processes to Global Impact. Springer-Verlag, Berlin, Berlin, pp. 481–508.

Bralower, T.J., Monechi, S., Thierstein, H.R., 1989. Calcareous nannofossil zonation of the Jurassic-Cretaceous boundary interval and correlation with the geomagnetic polarity timescale. Marine Micropaleontology 14, 153–235.

Brenner, W., 1988. Dinoflagellaten aus dem Unteren Malm (Oberer Jura) von Süddeutschland; Morphologie, Ökologie, Stratigraphie. Tübinger Mikropaläontologische Mitteilungen 6, 116.

Brongniart, A., 1829. Tableau des Terrains qui Composent L'écorce du Globe ou Essai sur la Structure de la Partie Connue de la Terre. F.G. Levrault, Paris, pp. 435.

Burger, D., 1995. Timescales 8. Jurassic. Calibration and development correlation charts and explanatory notes. Australian Geological Survey Organization Record 37, 1–30.

Callomon, J.H., 1964. Notes on the Callovian and Oxfordian Stages. In: Mautbeuge, P.L. (Ed.), Colloque du Jurassique à Luxembourg 1962. Publication de l'Institut Grand-Ducal, Section des Sciences Naturelles, Physiques et Mathématiques, Luxembourg, pp. 269–291.

Callomon, J.H., 1985. Biostratigraphy, chronostratigraphy and all that—again. In: Michelsen, O., Zeiss, A. (Eds.), International Symposium on Jurassic Stratigraphy, Erlangen, 1984, Vol. 3. Geological Survey of Denmark, Copenhagen, pp. 611–624.

Callomon, J.H., 1995. Time from fossils: S. S. Buckman and Jurassic high-resolution geochronology. In: Le Bas, M.J. (Ed.), Milestones in Geology. Geological Society, Memoir, 16, pp. 127–150.

Callomon, J.H., 1999. Report of the Bathonian – Callovian Boundary Working Group. International Subcommission on Jurassic Stratigraphy Newsletter 26, 53–60.

Callomon, J.H., Dietl, G., 2000. On the proposed basal boundary stratotype (GSSP) of the Middle Jurassic Callovian Stage. In: Hall, R.L., Smith, P.L. (Eds.), Advances in Jurassic Research. GeoResearch Forum, 6, pp. 41–54.

Calvin, L., 2010. On giant filter feeders. Science 327, 968–969.

Cande, S.C., Larson, R.L., LaBrecque, J.L., 1978. Magnetic lineations in the Pacific Jurassic Quiet Zone. Earth and Planetary Science Letters 41, 434–440.

Casellato, C.E., 2010. Calcareous nannofossil biostratigraphy of upper Callovian-lower Berriasian successions from the Southern Alps, North Italy. Rivista Italiana di Paleontologia e Stratigrafia 116, 357–404.

Casellato, C.E., Andreini, G., Erba, E., Parisi, G., 2008. Calcareous Nannofossil and Calpionellid calcification events across Tithonian – Berriasian time interval and low latitudes paleoceanographic implications. Journal of Nannoplankton Research, 12nd INA Conference, 33.

Cecca, F., Martin Garin, B., Marchand, D., Lathuiliere, B., Bartolini, A., 2005. Paleoclimatic control of biogeographic and sedimentary events in Tethyan and peri-Tethyan areas during the Oxfordian (Late Jurassic). Palaeogeography, Palaeoclimatology, Palaeoecology 222, 10–32.

Chang, S.-C., Zhang, H., Renne, P.R., Fang, Y., 2009. High-precision $^{40}Ar/^{39}Ar$ age constraints on the basal Lanqi Formation and its implications for the origin of angiosperm plants. Earth and Planetary Science Letters 30, 212–221.

Channell, J.E.T., Lowrie, W., Pialli, P., Venturi, F., 1984. Jurassic magnetic stratigraphy from Umbrian (Italian) land sections. Earth and Planetary Science letters 68, 309–325.

Channell, J.E.T., Massari, F., Benetti, A., Pezzoni, N., 1990. Magnetostratigraphy and biostratigraphy of Callovian-Oxfordian limestones from the Trento Plateau (Monti Lessini, Northern Italy). Palaeogeography, Palaeoclimatology, Palaeoecology 79, 289–303.

Channell, J.E.T., Casellato, C.E., Muttoni, G., Erba, E., 2010. Magnetostratigraphy, nannofossil stratigraphy and apparent polar wander for Adria-Africa in the Jurassic–Cretaceous boundary interval. Palaeogeography, Palaeoclimatology, Palaeoecology 293, 51–75.

Choiniere, J.N., Xu, X., Clark, J.M., Forster, C.A., Guo, Y., Han, F., 2010. A basal alvarezsauroid theropod from the early Late Jurassic of Xinjiang, China. Science 327, 571–574.

Coe, A.L., 1995. A comparison of the Oxfordian successions of Dorset, Oxfordshire, and Yorkshire. In: Taylor, P.D. (Ed.), Field Geology of the British Jurassic. Geological Society, London, pp. 151–172.

Cohen, A.S., Coe, A.L., 2007. The impact of the Central Atlantic Magmatic Province on climate and on the Sr- and Os-isotope evolution of seawater. Palaeogeography, Palaeoclimatology, Palaeoecology 244, 374–390.

Cohen, A.S., Coe, A.L., Bartlett, J.M., Hawkesworth, C.J., 1999. Precise Re-Os ages of organic-rich mudrocks and the Os isotope composition of Jurassic seawater. Earth and Planetary Science Letters 167, 150–173.

Cohen, A.S., Coe, A.L., Harding, S.M., Schwark, L., 2004. Osmium isotope evidence for the regulation of atmospheric CO_2 by continental weathering. Geology 32, 157–160.

Cohen, A.S., Coe, A.L., Kemp, D.B., 2007. The Late Palaeocene-Early Eocene and Toarcian (Early Jurassic) carbon isotope excursions: A comparison of their time scales, associated environmental changes, causes and consequences. Journal of the Geological Society 164, 1093–1108.

Collin, J.-P., 1998. Ostracodes. Columns for Jurassic chart of Mesozoic and Cenozoic sequence chronostratigraphic framework of European basins, by Hardenbol, J., Thierry, J., Farley, M.B., Jacquin, Th., de Graciansky, P.-C., and Vail, P.R. (coordinators). In: de Graciansky, P.-C., Hardenbol, J., Jacquin, Th., Vail, P.R. (Eds.), Mesozoic–Cenozoic Sequence Stratigraphy of European Basins. SEPM Special Publication, 60 Chart 7.

Colombié, C., Rameil, N., 2007. Tethyan-to-boreal correlation in the Kimmeridgian using high-resolution sequence stratigraphy (Vocontian Basin, Swiss Jura, Boulonnais, Dorset). International Journal of Earth Science (*Geol. Rundsch.*) 96, 567–591.

Comas-Rengifo, M.J., Gómez, J.J., Goy, A., Osete, M.L., Palencia-Ortas, A., 2010a. The base of the Toarcian (Early Jurassic) in the Almonacid de la Cuba section (Spain): Ammonite biostratigraphy, magnetostratigraphy and isotope stratigraphy. Episodes 33, 15–22.

Comas-Rengifo, M.J., Arias, C., Gómez, J.J., Goy, A., Herrero, C., Osete, M.L., Palencia, A., 2010b. A complementary section for the proposed Toarcian (Lower Jurassic) global stratotype: The Almonacid De La Cuba Section (Spain). Stratigraphy and Geological Correlation 18, 133–152.

Combemorel, R., 1998. Belemnites. Columns for Jurassic chart of Mesozoic and Cenozoic sequence chronostratigraphic framework of European basins, by Hardenbol, J., Thierry, J., Farley, M.B., Jacquin, Th., de Graciansky, P.-C., and Vail, P.R. (coordinators). In: de Graciansky, P.-C., Hardenbol, J., Jacquin, Th., Vail, P.R. (Eds.), Mesozoic-Cenozoic Sequence Stratigraphy of European Basins. SEPM Special Publication, 60 Chart 7.

Conybeare, W.D., Phillips, W., 1822. Outlines of the Geology of England and Wales, with an Introduction Compendium of the General Principles of that Science, and Comparative Views of the Structure of Foreign Countries, Part 1. William Phillips, London, p. 470.

Cope, J.C.W., 1967. The paleontology and stratigraphy of the lower part of the Upper Kimmeridge Clay of Dorset. Bulletin of the British Museum. Natural History, Geology Series 15, 3–79.

Cope, J.C.W., Getty, T.A., Howarth, M.K., Morton, N., Torrens, H.S., 1980a. A correlation of Jurassic rocks in the British Isles. Part One: Introduction and Lower Jurassic. Geological Society, Special Report 14, 73.

Cope, J.C.W., Duff, K.L., Parsons, C.F., Torrens, H.S., Wimbledon, W.A., Wright, J.K., 1980b. A correlation of Jurassic rocks in the British Isles. Part Two: Middle and Upper Jurassic. Geological Society, Special Report 15, p. 109.

Cope, J.C.W., 1993. The Bolonian Stage: An old answer to an old problem. Newsletters on Stratigraphy 28, 151–156.

Cope, J.C.W., 1995. Towards a unified Kimmeridgian Stage. Petroleum Geoscience 1, 351–354.

Cope, J.C.W., 2003. Latest Jurassic stage nomenclature. International Subcommission on Jurassic Stratigraphy Newsletter 30, 27–29.

Cope, J.C.W., 2007. Drawing the line: The history of the Jurassic–Cretaceous boundary. Proceedings of the Geologists' Association 119, 105–117.

Cope, J.C.W., 2009. Correlation problems in the Kimmeridge Clay Formation (Upper Jurassic, UK): Lithostratigraphy versus biostratigraphy and chronostratigraphy. Geological Magazine 146, 266–275.

Copestake, P., Johnson, B., Morris, P.H., Coleman, B.E., Shipp, D.J., 1989. Jurassic. In: Jenkyns, D.G., Murray, J.W. (Eds.), Stratigraphical Atlas of Fossil Foraminifera. The British Micropalaeontological Society, Chichester, pp. 125–272.

Coquand, H., 1871. Sur le Klippenkalk du département du Var et des Alpes-Maritimes. Bulletin de la Société Géologique de France 28, 232–233.

Cox, B.M., 1990. A review of Jurassic chronostratigraphy and age indicators for the UK. In: Hardman, R.F.P., Brooks, J.R.V. (Eds.), Tectonic Events Responsible for Britain's Oil and Gas Reserves. Geological Society Special Publication, 55, pp. 169–190.

Cox, B.M., Gallois, R.W., 1981. The stratigraphy of the Kimmeridge clay of the Dorset type area and its correlation with some other Kimmeridgian sequences. Report – Natural Environment Research Council, Institute of Geological Sciences 80–4, 44.

Cresta, S., Goy, A., Ureta, S., Arias, C., Barrón, E., Bernard, J., Canales, M.L., Garcí a-Joral, F., Garcí a-Romero, E., Gialanella, P.R., Gomes, J.J., González, J.A., Herrero, C., Marínez, G., Osete, M.L., Perilli, N., Villalaín, J.J., 2001. The global stratotype section and point (GSSP) of the Toarcian-Aalenian boundary (Lower-Middle Jurassic). Episodes 24, 166–175.

de Kaenel, E., Bergen, J.A., von Salis Perch-Nielsen, K., 1996. Jurassic calcareous nannofossil biostratigraphy of western Europe. Compilation of recent studies and calibration of bioevents. Bulletin de la Société géologique de France 167, 15–28.

De Wever, P., (coordinator), 1998. Radiolarians. Columns for Jurassic chart of Mesozoic and Cenozoic sequence chronostratigraphic framework of European basins, by Hardenbol, J., Thierry, J., Farley, M.B., Jacquin, Th., de Graciansky, P.-C., and Vail, P.R. (coordinators). In: de Graciansky, P.-C., Hardenbol, J., Jacquin, Th., Vail, P.R. (Eds.), Mesozoic-Cenozoic Sequence Stratigraphy of European Basins. SEPM Special Publication, 60 Chart 7.

Dean, W.T., Donovan, D.T., Howarth, M.K., 1961. The Liassic ammonite zones and subzones of the North West European Province. Bulletin of the British Museum of Natural History, Geology 4, 435–505.

Deenen, M.H.L., Ruhl, M., Bonis, N.R., Krijgsman, W., Kuerschner, W.M., Reitsma, M., van Bergen, M.J., 2010. A new chronology for the end-Triassic mass extinction. Earth and Planetary Science Letters 291, 113–125.

Dera, G., Neige, P., Dommergues, J.-L., Fara, E., Laffont, R., Pellenard, P., 2010. High-resolution dynamics of Early Jurassic marine extinctions: The case of Pliensbachian-Toarcian ammonites (Caphalopoda). Journal of the Geological Society 167, 21–33.

Dietl, G., 1981. Zur systematischen Stellung von *Ammonites subfurcatus* Zieten und deren Bedeutung für *subfurcatum*-Zone (Bajocium, Mittlerer Jura). Stuttgarter Beiträge zur Naturkunde, Serie B (Geologie und Paläontologie) 81, 1–11.

Dietl, G., 1994. Der hochstetteri-Horizont – ein Ammonitenfaunen-Horizont (Discus-Zone, Ober-Bathonium, Dogger) aus dem Schwäbischen Jura. Stuttgarter Beiträge zur Naturkunde, Serie B (Geologie und Paläontologie) 202, 1–39.

Dietl, G., Etzold, A., 1977. The Aalenian at the Type Locality. Stuttgarter Beiträge zur Naturkunde, Serie B (Geologie und Paläontologie) 30, 13.

Dietl, G., Callomon, J.H., 1988. Der Orbis-Oolith (Ober-Bathonium, Mittl. Jura) von Sengenthal/Opf. Fränk. Alb, und siene Bedeutung für Korrelation und Gliederung der Orbis-zone. Stuttgarter Beiträge zur Naturkunde, Serie B (Geologie und Paläontologie) 142, 1–31.

Dobruskina, I.A., 1994. Triassic floras of Eurasia. Österreichische Akademie der Wissenschaften, Schriftenreihe der Erdwissenschaftlichen Kommissionen 10, 422.

d'Omalius d'Halloy, J.J., 1843. Précis élémentaire de Géologie. Bruxelles-Paris, Paris, p. 636.

d'Orbigny, A., 1842–1851. Paléontologie Français. Description zoologique et géologique de tous les animaux mollusques et rayonnés fossiles de France. Terrains oolitiques ou jurassiques. I. Céphalopodes. Masson, Paris. p. 642.

d'Orbigny, A., 1852. Cours élémentaire de paléontologie et de géologie stratigraphique, Vol. 2. Masson, Paris. 383–847.

Dromart, G., Garcia, J.-P., Picard, S., Atrops, F., Lécuyer, C., Sheppard, S.M.F., 2003. Ice age at the Middle-Late Jurassic transition? Earth and Planetary Science Letters 213, 205–220.

Dyer, R., Copestake, P., 1989. A review of Late Jurassic to earliest Cretaceous radiolaria and their biostratigraphic potential to petroleum exploration in the North Sea. In: Batten, D.J., Keen, M.C. (Eds.), Northwest European Micropalaeontology and Palynology. British Micropalaeontological Society, Chichester, pp. 214–235.

Dyer, R., 1998. Radiolarians: North Sea Central and Viking Graben. Columns for Jurassic chart of Mesozoic and Cenozoic sequence chronostratigraphic framework of European basins, by Hardenbol, J., Thierry, J., Farley, M.B., Jacquin, Th., de Graciansky, P.-C., and Vail, P.R. (coordinators). In: de Graciansky, P.-C., Hardenbol, J., Jacquin, Th., Vail, P.R. (Eds.), Mesozoic-Cenozoic Sequence Stratigraphy of European Basins. SEPM Special Publication, 60 Chart 7.

Earth Impact Database, maintained by the Planetary and Space Science Centre, University of New Brunswick, Fredericton, New Brunswick, Canada. Viewed 2010. http://www.passc.net/*EarthImpactDatabase*/.

Elmi, S., (convenor), 1999. Report of the Pliensbachian-Toarcian Boundary Working Group. International Subcommission on Jurassic Stratigraphy Newsletter 26, 43–46.

Elmi, S., (convenor), 2003. Toarcian working group. International Subcommission on Jurassic Stratigraphy Newsletter 30, 14.

Elmi, S., 2006. Pliensbachian/Toarcian boundary: The proposed GSSP at Peniche (Portugal). Volumina Jurassica 4, 5–16.

Erba, E., 1990. Calcareous nannofossil biostratigraphy of some Bajocian sections from the Digne area (SE France). Memorie Descrittive della Carta Geologica d'Italia, XL, 237–255.

Erba, E., 2006. The first 150 million years history of calcareous nannoplankton: Biosphere-geosphere interactions. Palaeogeography, Palaeoclimatology, Palaeoecology 232, 237–250.

Fernández-López, S.R., 2007. Ammonoid taphonomy, palaeoenvironments and sequence stratigraphy at the Bajocian/Bathonian boundary on the Bas Auran area (Subalpine Basin, south-eastern France). Lethaia 40, 377–391.

Fernández-López, S.R., Pavia, G., Erba, E., Guiomar, M., Henriques, M.H., Lanza, R., Mangold, C., Morton, N., Olivero, D., Tiraboschi, D., 2009. The Global Boundary Stratotype Section and Point (GSSP) for base of the Bathonian Stage (Middle Jurassic), Ravin du Bès Section, SE France. Episodes 32, 222–248.

Fortwengler, D., Marchand, D., 1994. Nouvelles unites biochronologiques de la zone à Mariae (Oxfordien inférieur). Geobios, Mémoire Spécial 17, 203–209.

Frakes, L.A., 1979. Climates through Geologic Time. Elsevier, Amsterdam, p. 322.

Frakes, L.A., Francis, J.E., Syktus, J.I., 1992. Climate Modes of the Phanerozoic. Cambridge University Press, Cambridge, p. 286.

Friedman, M., Shimada, K., Martin, L.D., Everhart, M.J., Liston, J., Maltese, A., Triebold, 2010. 100-million-year dynasty of giant planktivorous bony fishes in the Mesozoic seas. Science 327, 990–993.

Gabilly, J., 1976. Le Toarcien à Thouars et dans le centre-ouest de la France: biostratigraphie, evolution de la faune (Harpoceratinae-Hildoceratinae). In: Les Stratotypes Français, 3. Centre National Recherche Scientifique, Paris, p. 217.

Galbrun, B., Gabilly, J., Rasplus, L., 1988. Magnetostratigraphy of the Toarcian stratotype sections at Thouars and Airvault (Deux-Sevres, France). Earth and Planetary Science Letters 87, 453–462.

Geyer, O.F., Gwinner, M.P., 1979. Die Schwäbische Alb und ihr Vorland. Sammlung Geologischer Führer 67, 271.

Głowniak, E., 1997. Middle Oxfordian ammonites. International Subcommission on Jurassic Stratigraphy Newsletter 25, 45–46.

Głowniak, E., Wierzbowski, A., 2007. Comment on mid-Oxfordian (Late Jurassic) positive carbon-isotope excursion recognized from fossil wood in the British Isles by C.R. Pearce, S.P. Hesselbo, A.L. Coe, Palaeogeography, Palaeoclimatology, Palaeoecology, 221: 343–357. Palaeogeography, Palaeoclimatology, Palaeoecology 248, 252–254.

Głowniak, E., Matyja, B.A., Wierzbowski, A., 2008. Upgraded subdivision of the Oxfordian and Kimmeridgian as a consequence of recent correlations. In: The 5th International Symposium IGCP 506, Hammamet – Tunisia, March 28–31, 2008, Abstract Volume, pp. 34–35.

Gómez, J.J., Goy, A., 2011. Warming-driven mass extinction in the Early Toarcian (Early Jurassic) of northern and central Spain, correlation with other time-equivalent European sections. Palaeogeography, Palaeoclimatology, Palaeoecology 306, 176–195.

Goy, A., Ureta, M.S., Arias, C., Canales, M.L., Garcia Joral, F., Herrero, C., Martinez, G., Perilli, N., 1994. The Fuentelsaz section (Iberian Range, Spain), a possible stratotype for the base of the Aalenian Stage. Miscellanea del Servizio Geologico Nazionale 5, 1–31.

Goy, A., Ureta, M.S., Arias, C., Canales, M.L., Garcia Joral, F., Herrero, C., Martinez, G., Perilli, N., 1996. Die Toarcium/Aalenium-Grenze im Profil Fuentelsaz (Iberische Ketten, Spanien). In: Ohmert, W. (Ed.), Die Grenzziehung Unter-/Mitteljura (Toarciuim/Aalenium) bei Wittnau und Fuentelsaz. Beispiele interdisziplinarer geowissenschaftlicher Zusammenarbeit. Geologisches Landesamt Baden-Wurttemberg, Informationen, 8, pp. 43–52.

Gröcke, D.R., Price, G.D., Ruffell, A.H., Mutterlose, J., Baraboshkin, E., 2003. Isotopic evidence for Late Jurassic-Early Cretaceous climate change. Palaeogeography, Palaeoclimatology, Palaeoecology 202, 97–118.

Groupe Français d'Étude du Jurassique (coordinated by Cariou, E., and Hantzpergue, P.), 1997. Biostratigraphie du Jurassique ouest-européen et méditerranéen: zonations parallèles et distribution des invertébrés et microfossiles. Bulletin des Centres de Recherches Exploration-Production Elf-Aquitaine, Mémoire 17, 422.

Guex, J., 1995. Ammonites hettangiennes de la Gabbs Valley Range (Nevada). Mémoires Géologie (Lausanne) 27, 130.

Guex, J., Bartolini, A., Atudorei, V., Taylor, D., 2004. High-resolution ammonite and carbon isotope stratigraphy across the Triassic - Jurassic boundary at New York Canyon. Earth and Planetary Science Letters 225, 29–41.

Guzhikova, A.Yu., Pimenova, M.V., Malenkinab, S.Yu., Manikina, A.G., Astarkina, S.V., 2010. Paleomagnetic, petromagnetic, and terrigenous-mineralogical studies of Upper Bathonian-Lower Callovian sediments in the Prosek Section, Nizhni Novgorod Region. Stratigraphy and Geological Correlation 18, 42–62.

Gygi, R.A., 2000. Integrated stratigraphy of the Oxfordian and Kimmeridgian (Late Jurassic) in northern Switzerland and adjacent southern Germany. Memoires of the Swiss Academy of Sciences 104, p. 152.

Gygi, R.A., Coe, A.L., Vail, P.R., 1998. Sequence stratigraphy of the Oxfordian and Kimmeridgian stages (Late Jurassic) in northern Switzerland. In: de Graciansky, P.-C., Hardenbol, J., Jacquin, Th., Vail, P.R. (Eds.), Mesozoic-Cenozoic Sequence Stratigraphy of European Basins. SEPM Special Publication, 60, pp. 527–544.

Habib, D., Drugg, W.S., 1983. Dinoflagellate age of Middle Jurassic-Early Cretaceous sediments in the Blake-Bahama Basin. Initial Reports of the Deep Sea Drilling Project 76, 623–638.

Hallam, A., 1975. Jurassic Environments. Cambridge University Press, Cambridge, p. 269.

Hallam, A., 1976. Stratigraphic distribution and ecology of European Jurassic bivalves. Lethaia 9, 245–259.

Hallam, A., 1978. Eustatic cycles in the Jurassic. Palaeogeography, Palaeoclimatology, Palaeoecology 23, 1–32.

Hallam, A., 1981. A revised sea-level curve for the Early Jurassic. Journal of the Geological Society 138, 735–743.

Hallam, A., 1986. The Pliensbachian and Tithonian extinction events. Nature 319, 765–768.

Hallam, A., 1988. A re-evaluation of the Jurassic eustasy in the light of new data and the revised Exxon curve. In: Wilgus, C.K., Hastings, B.S., Kendall, G.St.C., Posamentier, H.W., Ross, C.A., van Wagoner, J.C. (Eds.), Sea-level Changes: An Integrated Approach. Special Publication, 42, pp. 261–273.

Hallam, A., Wignall, P.B., 1999. Mass extinctions and sea-level changes. Earth-Science Reviews 48, 217–250.

Hallam, A., 2001. A review of the broad pattern of Jurassic sea-level changes and their possible causes in the light of current knowledge. Palaeogeography, Palaeoclimatology, Palaeoecology 167, 23–37.

Hames, W.E., Renne, P.R., Ruppel, C., 2000. New evidence for geologically instantaneous emplacement of earliest Jurassic Central Atlantic magmatic province basalts on the North American margin. Geology 28, 859–862.

Handschumacher, D.W., Sager, W.W., Hilde, T.W.C., Bracey, D.R., 1988. Pre-Cretaceous evolution of the Pacific plate and extension of the geomagnetic polarity reversal time scale with implications for the origin of the Jurassic "Quiet Zone". Tectonophysics 155, 365–380.

Haq, B.U., Hardenbol, J., Vail, P.R., 1988. Mesozoic and Cenozoic chronostratigraphy and eustatic cycles. In: Wilgus, C.K., Hastings, B.S., Kendall, G.St.C., Posamentier, H.W., Ross, C.A., van Wagoner, J.C. (Eds.), Sea-level Changes: An Integrated Approach. SEPM Special Publication, 42, pp. 71–108.

Hardenbol, J., Thierry, J., Farley, M.B., Jacquin, Th., de Graciansky, P.-C., Vail, P.R., (with numerous contributors), 1998. Mesozoic and Cenozoic sequence chronostratigraphic framework of European basins. In: de Graciansky, P.-C., Hardenbol, J., Jacquin, Th., Vail, P.R. (Eds.), Mesozoic-Cenozoic Sequence Stratigraphy of European Basins. SEPM Special Publication, 60, pp. 3–13 and 763–781.

Harding, I.C., Smith, G.A., Riding, J.B., and Wimbledon, W.A.P, 2011. Inter-regional correlation of Jurassic/Cretaceous boundary strata based on the Tithonian-Valanginian dinoflagellate cyst biostratigraphy of the Volga Basin, western Russia. Review of Palaeobotany and Palynology 167, 82–116.

Harland, W.B., Cox, A.V., Llewellyn, P.G., Pickton, C.A.G., Smith, A.G., Walters, R., 1982. A Geologic Time Scale. Cambridge University Press, Cambridge, p. 131.

Harland, W.B., Armstrong, R.L., Cox, A.V., Craig, L.E., Smith, A.G., Smith, D.G., 1990. A Geologic Time Scale 1989. Cambridge University Press, Cambridge, p. 263.

Hart, R.J., Andreoli, M.A.G., Tredoux, M., Moser, D., Ashwal, L.D., Eide, E.A., Webb, S.J., Brandt, D., 1997. Late Jurassic age for the Morokweng impact structure, southern Africa. Earth and Planetary Science Letters 147, 25–35.

Haug, E., 1910. Traité de Géologie. II. Les périodes géologiques. Fasicule 2, Jurassique et Crétacé. Armand Collin, Paris, pp. 929–1396.

Helby, R., Morgan, R., Partridge, A.D., 1987. A palynological zonation of the Australian Mesozoic. Association of Australasian Palaeontologists Memoir 4, p. 94.

Henriques, M.H., 1992. Biostratigrafia e paleontologia (Ammonoidea) do Aaleniano em Portugal (Sector Setentrional da Bacia Lusitaniana). Unpublished Ph.D Thesis, Centro de Geociências da Universidade de Coimbra, Instituto Nacional de Investigacao Cientifica, 301.

Henriques, M.H., Gardin, S., Gomes, C.R., Soares, A.F., Rocha, R.B., Marques, J.F., Lapa, M.R., Montenegro, J.D., 1994. The Aalenian-Bajocian boundary at Cabo Mondego (Portugal). In: Cresta, S., G. Pavia, G. (Eds.), Proceedings of the 3rd International Meeting on Aalenian and Bajocian Stratigraphy. Miscellanea del Servizio Geologico Nazionale, 5, pp. 63–77.

Hesselbo, S.P., Jenkyns, H.C., 1998. British Lower Jurassic sequence stratigraphy. In: de Graciansky, P.-C., Hardenbol, J., Jacquin, Th., Vail, P.R. (Eds.), Mesozoic-Cenozoic Sequence Stratigraphy of European Basins. SEPM Special Publication, 60, pp. 562–581.

Hesselbo, S.P., Gröcke, D.R., Jenkyns, H.C., Bjerrum, C.J., Farrimond, P., Morgans, B.H.S., Green, O.R., 2000. Massive dissociation of gas hydrate during a Jurassic oceanic anoxic event. Nature 406, 392–395.

Hesselbo, S.P., Robinson, S.A., Surlyk, F., Piasecki, S., 2002. Terrestrial and marine extinctions at the Triassic-Jurassic boundary synchronized with major carbon-cycle perturbation: A link to initiation of massive volcanism? Geology 30, 251–254.

Hesselbo, S.P., McRoberts, C.A., Pálfy, J., 2007a. Triassic-Jurassic boundary events: Problems, progress, possibilities. Palaeogeography, Palaeoclimatology, Palaeoecology 244, 1–10.

Hesselbo, S.P., Jenkyns, H.C., Duarte, L.V., Oliveira, L.C.V., 2007b. Carbon isotope record of the Early Jurassic (Toarcian) Oceanic Anoxic Event from fossil wood and marine carbonate (Lusitanian Basin, Portugal). Earth and Planetary Science Letters 253, 455–470.

Hesselbo, S.P., 2008. Sequence stratigraphy and inferred relative sea-level change from the onshore British Jurassic. Proceedings of the Geologists' Association 119, 19–34.

Hesselbo, S.P., Deconinck, J.-F., Huggett, J.M., Morgans-Bell, H.S., 2009. Late Jurassic palaeoclimatic change from clay mineralogy and gamma-ray spectrometry of the Kimmeridge Clay, Dorset, UK. Journal of the Geological Society 166, 1123–1133.

Hesselbo, S.P., Pienkowski, G., 2011. Stepwise atmospheric carbon-isotope excursion during the Toarcian Oceanic Anoxic Event (Early Jurassic, Polish Basin). Earth and Planetary Science Letters 301, 365–373.

Hillebrandt, A. von, 1997. Proposal for the Utcubamba Valley sections in northern Peru. International Subcommission on Jurassic Stratigraphy Newsletter 24, 21–25.

Hillebrandt, A. von, Krystyn, L., 2009. On the oldest Jurassic ammonites of Europe (Northern Calcareous Alps, Austria) and their global significance. Neues Jahrbuch für Geologie und Paläontologie Abhandlungen 253, 163–195.

Hillebrandt, A. von, Krystyn, L., Kürschner, W.M., with contributions by Bown, P.R., McRoberts, Ch., Kment, K., Ruhl, M., Simms, M., Tomasovych, A., and Urlichs, M., 2008. A candidate GSSP for the base of the Jurassic in the Northern Calcareous Alps (Kuhjoch section; Karwendel Mountains, Tyrol, Austria). International Subcommission on Jurassic Stratigraphy, Triassic/Jurassic Boundary Working Group Ballot 2008, 44.

Hinnov, L.A., Park, J.J., 1999. Strategies for assessing Early-Middle (Pliensbachian-Aalenian) Jurassic cyclochronologies. Philosophical Transactions of the Royal Society, Series A 357, 1831–1859.

Hinnov, L.A., Park, J.J., Erba, E., 1999. Lower-Middle Jurassic rhythmites from the Lombard Basin, Italy: A record of orbitally-forced cycles modulated by long-term secular environmental changes in West Tethys. In: Hall, R.L., Smith, P.L. (Eds.), Advances in Jurassic Research 2000; Proceedings of the Fifth International Symposium on the Jurassic System, 1998. GeoResearch Forum, 6, pp. 437–453.

Horner, F., 1983. Palaeomagnetismus von Karbonatsedimenten der Suedlichen Tethys: Implikationen für die Polaritaet des Erdmagnetfeldes im untern Jura und für die tektonik der Ionischen zone Grieschenlands. PhD. thesis, ETH Zürich, 139.

Horner, F., Heller, F., 1983. Lower Jurassic magnetostratigraphy at the Breggia Gorge (Ticino, Switzerland) and Alpi Turati (Como, Italy). Geophysics Journal of the Royal Astronomical Society 73, 705–718.

Hounslow, M.K., Muttoni, G., 2010. The geomagnetic polarity timescale for the Triassic: Linkage to stage boundary definitions. In: Lucas, S.G. (Ed.), The Triassic Timescale. Geological Society, Special Publication, 334, pp. 61–102.

Hounslow, M.W., Posen, P.E., Warrington, G., 2004. Magnetostratigraphy and biostratigraphy of the Upper Triassic and lowermost Jurassic succession, St. Audrie's Bay, UK. Palaeogeography, Palaeoclimatology, Palaeoecology 213, 331–358.

Houša, V., Pruner, P., Zakharov, V.A., Kostak, M., Chadima, M., Rogov, M.A., Slechta, S., Mazuch, M., 2007. Boreal-Tethyan correlation of the Jurassic-Cretaceous boundary interval by magneto- and biostratigraphy. Stratigraphy and Geological Correlation 15 (3), 297–309.

Hu, D., Hou, L., Zhang, L., Xu, X., 2009. A pre-*Archaeopteryx* troodontid theropod from China with long feathers on the metatarsus. Nature 461, 640–643.

Huang, C., Hinnov, L.A., Ogg, J.G., Galbrun, B., Boulila, S., Huret, E., 2010a. Astronomical calibration of the Jurassic time scale. Earth Science Frontiers 17, 108–109.

Huang, C., Hesselbo, S.P., Hinnov, L.A., 2010b. Astrochronology of the late Jurassic Kimmeridgian Clay (Dorset, England) and implications for Earth system processes. Earth and Planetary Science Letters 289, 242–255.

Huang, C., Hinnov, L.A., Swientek, O., Smelnor, M., 2010c. Astronomical tuning of Late Jurassic-early Cretaceous sediments (Volgian-Ryazanian stages), Greenland-Norwegian Seaway. AAPG Annual Convention, New Orleans, LA, abstract.

Huret, E., 2006. Analyse cyclostratigraphique des variations de la susceptibilité magnétique des argilites callovo-oxfordiennes de l'Est du Bassin de Paris: Application à la recherche de hiatus sédimentaires. Ph.D thesis, Université Pierre et Marie Curie, Paris, p. 321.

Huret, E., Hinnov, L.A., Galbrun, B., Collin, P.-Y., Gardin, S., Rouget, I., 2008. Astronomical calibration and correlation of the Lower Jurassic, Paris and Lombard basins (Tethys). 33rd International Geological Congress, Oslo, Norway, abstract.

Ioannides, I., Riding, J., Stover, L.E., Monteil, E., 1998. Dinoflagellate Cysts. Columns for Jurassic chart of Mesozoic and Cenozoic sequence chronostratigraphic framework of European basins, by Hardenbol, J., Thierry, J., Farley, M.B., Jacquin, Th., de Graciansky, P.-C., and Vail, P.R. (coordinators). In: de Graciansky, P.-C., Hardenbol, J., Jacquin, Th., Vail, P.R. (Eds.), Mesozoic-Cenozoic Sequence Stratigraphy of European Basins. SEPM Special Publication, 60 Chart 7.

Innocenti, M., Mangold, C., Pavia, G., Torrens, H.S., 1988. A proposal for the formal ratification of the boundary stratotype of the Bathonian stage based on a Bas Auran section (S.E. France). In: Rocha, R.B., Soares, A.F. (Eds.), 2nd International Symposium on Jurassic Stratigraphy, 1, pp. 333–346.

INTERRAD Jurassic-Cretaceous Working Group, 1995. Middle Jurassic to Lower Cretaceous Radiolaria of Tethys: Occurrences, systematics, biochronology. Mémoires de Géologie Lausanne 23, 1172.

Jenkyns, H.C., 1988. The early Toarcian (Jurassic) anoxic event: Stratigraphic, sedimentary and geochemical evidence. American Journal of Science 288, 101–151.

Jenkyns, H.C., 2010. Geochemistry of oceanic anoxic events. Geochemistry, Geophysics, Geosystems 11, Q03004. doi: 10.1029/2009GC002788.

Jenkyns, H.C., Jones, C.E., Gröcke, D.R., Hesselbo, S.P., Parkinson, D.N., 2002. Chemostratigraphy of the Jurassic System: Applications, limitations and implications for palaeoceanography. Journal of the Geological Society 159, 351–378.

Jones, C.E., Jenkyns, H.C., 2001. Seawater strontium isotopes, oceanic anoxic events, and seafloor hydrothermal activity in the Jurassic and Cretaceous. American Journal of Science 301, 112–149.

Jones, C.E., Jenkyns, H.C., Hesselbo, S.P., 1994a. Strontium isotopes in Early Jurassic seawater. Geochimica et Cosmochimica Acta 58, 1285–1301.

Jones, C.E., Jenkyns, H.C., Coe, A.L., Hesselbo, S.P., 1994b. Strontium isotopic variations in Jurassic and Cretaceous seawater. Geochimica et Cosmochimica Acta 58, 3061–3074.

Jourdan, F., Féraud, G., Bertrand, H., Watkeys, M.K., Renne, P.R., 2007. Distinct brief major events in the Karoo large igneous province clarified by new $^{40}Ar/^{39}Ar$ ages on the Lesotho basalts. Lithos 98, 195–209.

Jourdan, F., Marzoli, A., Bertrand, H., Cirilli, S., Tanner, L., Kontak, D.J., McHone, G., Renne, P.R., Bellieni, G., 2009a. $^{40}Ar/^{39}Ar$ ages of CAMP in North America: Implications for the Triassic-Jurassic boundary and the ^{40}K decay constant bias. Lithos 110, 167–180.

Jourdan, F., Renne, P.R., Reimold, W.U., 2009b. An appraisal of the ages of terrestrial impact structures. Earth and Planetary Science Letters 286, 1–13.

Juárez, M.T., Osete, M.L., Melendez, G., Langereis, C.G., Zijderveld, J.D.A., 1994. Oxfordian magnetostratigraphy of Aguilón and Tosos sections (Iberian Range, Spain) and evidence of a pre-Oligocene overprint. Physics of the Earth and Planetary Interiors 85, 195–211.

Juárez, M.T., Osete, M.L., Melendez, G., Lowrie, W., 1995. Oxfordian magnetostratigraphy in the Iberian Range. Geophysical Research Letters 22, 2889–2892.

Kamo, S.L., Riccardi, A.C., 2009. A new U-Pb zircon age for an ash layer at the Bathonian-Callovian boundary, Argentina. GFF 131, 177–182.

Kemp, D.B., Coe, A.L., Cohen, A.S., L.Schwark, L., 2005. Astronomical pacing of methane release in the Early Jurassic period. Nature 437, 396–399.

Kemp, D.B., Coe, A.L., Cohen, A.S., Weedon, G.P., 2011. Astronomical forcing and chronology of the early Toarcian (Early Jurassic) Oceanic Anoxic Event in Yorkshire, UK. Paleoceanography 26, PA4210. doi:10.1029/2011PA002122.

Kent, D.V., Olsen, P.E., Witte, W.K., 1995. Late Triassic-Early Jurassic geomagnetic polarity sequence and paleolatitudes from drill cores in the Newark rift basin, eastern North America. Journal of Geophysical Research 100, 14965–14998.

Kent, D.V., Olsen, P.E., 2008. Early Jurassic magnetostratigraphy and paleolatitudes from the Hartford continental rift basin (eastern North America): Testing for polarity bias and abrupt polar wander in association with the central Atlantic Magmatic province. Journal of Geophysical Research 113, B06105. doi: 10.1029/2007JB005407.

Kerr, R.A., 2000. Did volcanoes drive ancient extinctions? Science 289, 1130–1131.

Kiessling, W., Scasso, R., Zeiss, A., Riccardi, A., Medina, F., 1999. Combined Radiolaria-ammonite stratigraphy for the Late Jurassic of the Antarctic Peninsula: Implications for radiolarian stratigraphy. Geodiversitas 21, 687–713.

Koeberl, C., Armstrong, R.A., Reimold, W.U., 1997. Morokweng, South Africa: A large impact structure of Jurassic-Cretaceous Boundary age. Geology 25, 731–734.

Koppers, A.A.P., Staudigel, H., Duncan, R.A., 2003. High resolution $^{40}Ar/^{39}Ar$ dating of the oldest oceanic basement basalts in the western Pacific basin. Geochemistry, Geophysics, Geosystems 4, 8914. doi: 10.1029/2003GC000574.

Kowallis, B.J., Christiansen, E.H., Deino, A.L., et al., 1998. The age of the Morrison Formation. Modern Geology 22, 235–260.

Korte, C., Hesselbo, S.P., 2011. Shallow marine carbon and oxygen isotope and elemental records indicate icehouse-greenhouse cycles during the Early Jurassic. Paleoceanography 26, PA429.

Krymholts, G.Ya., Mesezhnikov, M.S., Westermann, G.E.G. (Eds.), 1982. Zony iurskoi sistemy v SSSR. Interdepartmental Stratigraphic Committee of the USSR Transactions, 10 [In Russian. English translation by Vassiljeva, T.I., 1988, The Jurassic Ammonite Zones of the Soviet Union. Geological Society of America, Special Paper, 223, pp. 116].

Kuerschner, W.M., Bonis, N.R., Krystyn, L., 2007. Carbon-isotope stratigraphy and palynostratigraphy of the Triassic-Jurassic transition in the Tiefengraben section – Northern Calcareous Alps (Austria). Palaeogeography, Palaeoclimatology, Palaeoecology 244, 257–280.

Lang, W.D., 1913. The Lower Pliensbachian – "Carixian" – of Charmouth. Geological Magazine 5, 401–412.

Larson, R.L., Hilde, T.W.C., 1975. A revised time scale of magnetic reversals for the Early Cretaceous and Late Jurassic. Journal of Geophysical Research 80, 2586–2594.

Laurin, B., 1998. Brachiopods. Columns for Jurassic chart of Mesozoic and Cenozoic sequence chronostratigraphic framework of European basins, by Hardenbol, J., Thierry, J., Farley, M.B., Jacquin, Th., de Graciansky, P.-C., and Vail, P.R. (coordinators). In: de Graciansky, P.-C., Hardenbol, J., Jacquin, Th., Vail, P.R. (Eds.), Mesozoic-Cenozoic Sequence Stratigraphy of European Basins. SEPM Special Publication, 60 Chart 7.

Lefranc, M., Beaudoin, B., Chilès, J.P., Guillemot, D., Ravenne, C., Trouiller, A., 2008. Geostatistical characterization of Callovo-Oxfordian clay variability from high-resolution log data. Physics and Chemistry of the Earth 33, S2–S13.

Lowrie, W., Ogg, J.G., 1986. A magnetic polarity time scale for the Early Cretaceous and Late Jurassic. Earth and Planetary Science Letters 76, 341–349.

Lucas, S.G., 2009. Global Jurassic tetrapod biochronology. Volumina Jurassica VI, 99–108.

Lucas, S.G., Tanner, L.H., Donohoo-Hurley, L.L., Geissman, J.W., Kozur, H.W., Heckert, A.B., Weems, R.E., 2011. Position of the Triassic-Jurassic boundary and timing of the end-Triassic extinctions on land: Data from the Moenave Formation on the southern Colorado Plateau, USA. Palaeogeography, Palaeoclimatology, Palaeoecology 302, 194–205.

Ludden, J., 1992. Radiometric age determinations for basement from Sites 765 and 766, Argo Abyssal Plain and Northwestern Australia. Proceedings of the Ocean Drilling Program, Scientific Results 123, 557–559.

Mahoney, J.J., Duncan, R.A., Tejada, M.L.G., Sager, W.W., Bralower, T.J., 2005. Jurassic-Cretaceous boundary age and mid-ocean-ridge-type mantle source for Shatsky Rise. Geology 33, 185–188.

Mailliot, S., Mattioli, E., Guex, J., Pittet, B., 2006. The Early Toarcian anoxia, a synchronous event in the Western Tethys? An approach by quantitative biochronology (Unitary Associations), applied on calcareous nannofossils. Palaeogeography, Palaeoclimatology, Palaeoecology 240, 562–586.

Mander, L., Kürschner, W.M., McElwain, J.C., 2010. An explanation for conflicting records of Triassic-Jurassic plant diversity. Proceedings of the National Academy of Science 107, 15351–15356 [www.pnas.org/cgi/doi/10.1073/pnas.1004207107].

Marzoli, A., Renne, P.R., Piccirillo, E.M., Ernesto, M., Bellieni, G., DeMin, A., 1999. Extensive 200-million-year-old continental flood basalts of the Central Atlantic Magmatic Province. Science 284, 616–618.

Marzoli, A., Jourdan, F., Puffer, J.H., Cuppone, T., Tanner, L.H., Weems, R.E., Bertrand, H., Cirilli, S., Bellieni, G., De Min, A., 2011. Timing and duration of the Central Atlantic magmatic province in the Newark and Culpeper basins, eastern U.S.A. Lithos 122, 175–188.

Matsuoka, A., 1992. Jurassic and Early Cretaceous radiolarians from Leg 129, Sites 800 and 801, Western Pacific Ocean. Proceedings of the Ocean Drilling Program, Scientific Results 129, 203–211.

Matsuoka, A., Yao, A., 1986. A newly proposed radiolarian zonation for the Jurassic of Japan. Marine Micropaleontology 11, 91–105.

Mattioli, E., Erba, E., 1999. Synthesis of calcareous nannofossil events in tethyan Lower and Middle Jurassic successions. Rivista Italiana di Paleontologia e Stratigrafia 105, 343–376.

Mattioli, E., Colombié, C., Giraud, F., Olivier, N., Pittet, B., 2011. Comment on "Orbitally forced climate and sea-level changes in the Paleoceanic Tethyan domain (marl-limestone alternations, Lower Kimmeridgian, SE France)." by S. Boulila, M. de Rafélis, L.A. Hinnov, S. Gardin, B. Galbrun, and P.-Y. Collin [Palaeogeography, Palaeoclimatology, Palaeoecology 292 (2010) 57–70]. Palaeogeography, Palaeoclimatology, Palaeoecology 306, 249–251.

Matyja, B.A., Wierzbowski, A., 1997. The quest for a unified Oxfordian/Kimmeridgian boundary: Implications of the ammonite succession at the turn of the Bimammatum and Planula Zones in the Wielun Upland, Central Poland. Acta Geologica Polonica 47, 77–105.

Matyja, B.A., Wierzbowski, A., 2002. Boreal and Subboreal ammonites in the Submediterranean uppermost Oxfordian in the Bielawy section (northern Poland) and their correlation value. Acta Geologica Polonica 52, 411–421.

Matyja, B.A., Wierzbowski, A., 2006a. , Syborowa Góra (Upper Oxfordian: Amoeboceras layer of lower Bimammatum Zone). In: Wierzbowski, A., Aubrecht, R., Golonka, J., Gutowski, J., Krobicki, M., Matyja, B.A., Pieńkowski, G., Uchman, A. (Eds.), Jurassic of Poland and Adjacent Slovakian Carpathians, Field Guide – 7th International Congress on the Jurassic System, Kraków, September 6–18, 2006. Polish Geological Institute, Warsaw, pp. 141–143.

Matyja, B.A., Wierzbowski, A., 2006b. Quarries at Raciszyn and Lisowice, Upper Oxfordian ammonite succession (upper Bimammatum to Planula zones). In: Wierzbowski, A., Aubrecht, R., Golonka, J., Gutowski, J., Krobicki, M., Matyja, B.A., Pieńkowski, G., Uchman, A. (Eds.), Jurassic of Poland and Adjacent Slovakian Carpathians, Field Guide – 7th International Congress on the Jurassic System, Kraków, September 6–18, 2006. Polish Geological Institute, Warsaw, pp. 163–165.

Matyja, B.A., Wierzbowski, A., Wright, J.K., 2006. The Subboreal/Boreal ammonite succession at the Oxfordian/Kimmeridgian boundary at Flodigarry, Staffin Bay (Isle of Skye), Scotland. Transactions of the Royal Society of Edinburgh, Earth Science 96, 309–318.

Maubeuge, P.L. (Ed.), 1964. Colloque du Jurassique à Luxembourg 1962. Publication de l'Institut Grand-Ducal, Section des Sciences Naturelles, Physiques et Mathématiques, Luxembourg, p. 948.

Mayer-Eymar, C., 1864. Tableau Synchronistique des Terrains Jurassiques. J. Höfer, Zurich.

Mazzini, A., Svensen, H., Leanza, H.A., Corfu, F., Planke, S., 2010. Early Jurassic shale chemostratigraphy and U-Pb ages from the Neuquén Basin (Argentina): Implications for the Toarcian Oceanic Anoxic Event. Earth and Planetary Science Letters 297, 633–645.

McArthur, J.M., Donovan, D.T., Thirlwall, M.F., Fouke, B.W., Mattey, D., 2000. Strontium isotope profile of the early Toarcian (Jurassic) Oceanic Anoxic Event, the duration of ammonite biozones, and belemnite paleotemperatures. Earth and Planetary Science Letters 179, 269–285.

McArthur, J.M., Howarth, R.J., Bailey, T.R., 2001. Strontium isotope stratigraphy: LOWESS Version 3: Best fit to the marine Sr-isotope curve for 0–509 Ma and accompanying look-up table for deriving numerical age. Journal of Geology 109, 155–170.

McArthur, J.M., Howarth, R.J., 2004. Sr-isotope stratigraphy: The Phanerozoic $^{87}Sr/^{86}Sr$-curve and explanatory notes. In: Gradstein, F.M., Ogg, J.G., Smith, A.G. (Eds.), A Geologic Time Scale 2004. Cambridge University Press, Cambridge, pp. 96–105.

McArthur, J.M., Algeo, T.J., van de Schootbrugge, B., Li, Q., Howarth, R.J., 2008. Basinal restriction, black shales, Re-Os dating, and the early Toarcian (Jurassic) oceanic anoxic event. Paleoceanography 23, PA4217. doi: 10.1029/2008PA001607.

McArthur, J.M., 2008. Comment on 'The impact of the Central Atlantic Magmatic Province on climate and on the Sr- and Os-isotope evolution of seawater' by Cohen, A. S. and Coe, A. L. 2007, Palaeogeography, Palaeoclimatology, Palaeoecology, 244, 374–390. Palaeogeography, Palaeoclimatology, Palaeoecology 263, 146–149.

McArthur, J.M., 2010. Correlation and dating with strontium-isotope stratigraphy. In: Whittaker J.E., Hart M.B. (Eds.), Micropalaeontology, Sedimentary Environments and Stratigraphy: A Tribute to Dennis Curry (1912–2001). The Micropalaeontological Society, Special Publications, 4, pp. 133–145.

Meister, C., 1999a. Report of the Sinemurian-Pliensbachian Boundary Working Group. International Subcommission on Jurassic Stratigraphy Newsletter 26, 33–42.

Meister, C., 1999b. Report of the Sinemurian-Pliensbachian Boundary Working Group. International Subcommission on Jurassic Stratigraphy Newsletter 27, 25–26.

Meister, C., Blau, J., Dommergues, J.-L., Feistburkhardt, S., Hart, M., Hesselbo, S.P., Hylton, M., Page, K., Price, G., 2003. A proposal for the Global Boundary Stratotype Section and Point (GSSP) for the base of the Pliensbachian Stage (Lower Jurassic). Elogae Geologicae Helvetiae 96, 275–298.

Meister, C., Aberhan, M., Blau, J., Dommergues, J.-L., Feist-Burkhardt, S., Hailwood, E.A., Hart, M., Hesselbo, S.P., Hounslow, M.W., Hylton, M., Morton, N., Page, K., Price, G.D., 2006. The Global Boundary Stratotype Section and Point (GSSP) for the base of the Pliensbachian Stage (Lower Jurassic), Wine Haven, Yorkshire, UK. Episodes 29, 93–106.

Melendez, G., 1999. Report of the Callovian – Oxfordian Boundary Working Group. International Subcommission on Jurassic Stratigraphy Newsletter 26, 61–67.

Melendez, G., Atrops, F., 1999. Report of the Oxfordian-Kimmeridgian Boundary Working Group. International Subcommission on Jurassic Stratigraphy Newsletter 26, 67–74.

Melendez, G., 2002. Oxfordian Working Group. International Subcommission on Jurassic Stratigraphy Newsletter 29, 10.

Melendez, G., (convenor), 2003. Oxfordian Working Group. International Subcommission on Jurassic Stratigraphy Newsletter 30, 16–18.

Melendez, G., 2006. Oxfordian Working Group [with contributions by Page, K.N., Atrops, F. and Rogov, M.]. International Subcommission on Jurassic Stratigraphy Newsletter 33, 16–19.

Moreau, M.-G., Bucher, H., Bodergat, A.-M., Guex, J., 2002. Pliensbachian magnetostratigraphy: New data from Paris Basin (France). Earth and Planetary Science Letters 203, 755–767.

Morettini, E., Santantonio, M., Bartolini, A., Cecca, F., Baumgartner, P.O., Hunziker, J.C., 2002. Carbon isotope stratigraphy and carbonate production during the Early-Middle Jurassic: Examples from the Umbria-Marche-Sabina Apennines (central Italy). Palaeogeography, Palaeoclimatology, Palaeoecology 184, 251–273.

Morgans-Bell, H.S., Coe, A.L., Hesselbo, S.P., Jenkyns, H.C., Weedon, G.P., Marshall, J.E.A., Tyson, R.V., Williams, C.J., 2001. Integrated stratigraphy of the Kimmeridge Clay Formation (Upper Jurassic) based on exposures and boreholes in south Dorset, UK. Geological Magazine 138, 511–539.

Morris, P.H., Cullum, A., Pearce, M.A., Batten, D.J., 2009. Megaspore assemblages from the Are Formation (Rhaetian-Pliensbachian) offshore mid-Norway, and their value as field and regional stratigraphic markers. Journal of Micropalaeontology 28, 161–181.

Morton, N. (Ed.), 1974. The definition of standard Jurassic Stages. Colloque du Jurassique à Luxembourg 1967; Generalities, methods. Mémoires du Bureau de Recherches Géologiques et Minières, 75, pp. 83–93.

Morton, N., 2008. The International Subcommission on Jurassic Stratigraphy. Proceedings of the Geologists' Association 119, 97–103.

Moulin, M., Fluteau, F., Courtillot, V., Marsh, J., Delpech, G., Quidelleur, X., 2011. An attempt to constrain the age, duration, and eruptive history of the Karoo flood basalt: Naude's Nek section (South Africa). Journal of Geophysical Research 116 B07403. doi: 10.1029/2011JB008210.

Mundil, R., Pálfy, J., Renne, P.R., Brack, P., 2010. The Triassic timescale: New constraints and a review of geochronological data. In: Lucas, S.G. (Ed.), The Triassic Timescale. Geological Society, Special Publication, 334, pp. 41–60.

Newell, A.J., 2000. Fault activity and sedimentation in marine rift basin (Upper Jurassic, Wessex Basin, UK). Journal of the Geological Society 157, 83–92.

Neumayr, M., 1873. Die Fauna des Schichten mit Aspidoceras acanthicum. Abhandlungen der Kais.-Königl. Geologischen Reichsanstalt 5 (6), 141–257.

Nikitin, S.N., 1881. Jurassic formations between Rogbinsk, Mologa, and Myshkin. Materials on the Geology of Russia 10, 194 [In Russian].

Ohmert, W.(Ed.). Die Grenzziehung Unter-/Mitteljura (Toarciuim/Aalenium) bei Wittnau und Fuentelsaz. Beispiele interdisziplinarer geowissenschaftlicher Zusammenarbeit. Geologisches Landesamt Baden-Württemberg, Informationen, 8. p. 53.

Odin, G.S., Baadsgaard, H., Hurford, A.J., Riccardi, A.C., 1992. U-Pb & fission track geochronology of Bathonian–Callovian 'tuffs' from Argentina. In: Odin, G.S. (Ed.), Phanerozoic Time Scale. Bulletin de Liaison et Information, IUGS Subcommission on Geochronology, 11, pp. 11–17.

Ogg, J.G., 1988. Early Cretaceous and Tithonian magnetostratigraphy of the Galicia margin (Ocean Drilling Program Leg 103). Proceedings of the Ocean Drilling Program, Scientific Results 103, 659–682.

Ogg, J.G., 1995. Magnetic polarity time scale of the Phanerozoic. In: Ahrens, T.J. (Ed.), Global Earth Physics: A Handbook of Physics Constants. American Geophysical Union Reference Shelf, 1, pp. 240–270.

Ogg, J.G., Lowrie, W., 1986. Magnetostratigraphy of the Jurassic-Cretaceous boundary. Geology 14, 547–550.

Ogg, J.G., Gutowski, J., 1996. Oxfordian and lower Kimmeridgian magnetic polarity time scale. In: Riccardi, A.C. (Ed.), Advances in Jurassic Research. GeoResearch Forum, 1–2, pp. 406–414.

Ogg, J.G., Steiner, M.B., Oloriz, F., Tavera, J.M., 1984. Jurassic magnetostratigraphy, 1. Kimmeridgian-Tithonian of Sierra Gorda and Carcabuey, southern Spain. Earth and Planetary Science Letters 71, 147–162.

Ogg, J.G., Wieczorek, J., Steiner, M.B., Hoffmann, M., 1990. Jurassic magnetostratigraphy, 4. Early Callovian through Middle Oxfordian of Krakow Uplands (Poland). Earth and Planetary Science Letters 104, 289–303.

Ogg, J.G., Hasenyager, R.W., Wimbledon, W.A., Channell, J.E.T., Bralower, T.J., 1991. Magnetostratigraphy of the Jurassic-Cretaceous boundary interval –Tethyan and English faunal realms. Cretaceous Research 12, 455–482.

Ogg, J.G., Karl, S.M., Behl, R.J., 1992. Jurassic through Early Cretaceous sedimentation history of the central Equatorial Pacific and of Sites 800 and 801. Proceedings of the Ocean Drilling Program, Scientific Results 129, 571–613.

Ogg, J.G., Hasenyager II, R.W., Wimbledon, W.A., 1994. Jurassic-Cretaceous boundary: Portland-Purbeck magnetostratigraphy and possible correlation to the Tethyan faunal realm. Géobios 17, 519–527.

Ogg, J.G., 2007. Paleoenvironments of the Jurassic and Cretaceous oceans: Selected highlights. American Geophysical Union, Fall Meeting 2007, abstract T12A–07 .

Ogg, J.G., Coe, A.L., Przybylski, P.A., Wright, J.K., 2010a. Oxfordian magnetostratigraphy of Britain and its correlation to Tethyan regions and Pacific marine magnetic anomalies. Earth and Planetary Science Letters 289, 433–448.

Ogg, J.G., Hinnov, L.A., Huang, C., Przybylski, P.A., 2010b. Late Jurassic time scale: Integration of ammonite zones, magnetostratigraphy, astronomical tuning and sequence interpretation for Tethyan, Sub-boreal and Boreal realms. Earth Science Frontiers 17, 81–82.

Ogg, J.G., Coe, A.L., Przybylski, P.A., Wright, J.K., 2011. Magnetostratigraphy of the Redcliff Section. [submitted to Oxfordian Working Group for base-Oxfordian GSSP voting documentation coordinated by Guillermo Melendez, chair], Dorset.

Olivero, D., (with contribution of E. Mattioli), 2008. The Aalenian-Bajocian (Middle Jurassic) of the Digne area. In Mattioli, E. (special ed), Guidebook for the Post-Congress Fieldtrip in the Vocontian Basin, SE France (September 11–13, 2008). 12th Meeting of the International Nannoplankton Association (Lyon, September 7–10, 2008). Available at http://paleopolis.rediris.es/cg/CG2008_BOOK_01/index.htm.

Olóriz, F., Schweigert, G., 2010. Working Group on the Kimmeridgian-Tithonian boundary. International Commission on Jurassic Stratigraphy Newsletter 36, 5–12.

Olsen, P.E., Kent, D.V., Et Touhami, M., Puffer, J.H., 2003. Cyclo-, magneto-, and bio-stratigraphic constraints on the duration of the CAMP event and its relationship to the Triassic-Jurassic boundary. American Geophysical Union Geophysical Monograph 136, 7–32.

Oppel, C.A., 1856–1858. Die Juraformation Englands, Frankreichs und des Südwestlichen Deutschlands. Württemberger Naturforschende Jahreshefte 12–14, 857.

Oppel, C.A., 1865. Die Tithonische Etage. Zeitschrift der Deutschen Geologischen Gesellschaft, Jahrgang 17, 535–558.

Osete, M.L., Gialanella, P.R., Gómez, J.J., Villalaiín, J.J., Goy, A., Heller, F., 2007. Magnetostratigraphy of early-middle Toarcian expanded sections from the Iberian Range (Central Spain). Earth and Planetary Science Letters 259, 319–332.

Oxford, M.J., Gregory, F.J., Hart, M.B., Henderson, A.S., Simmons, M.D., Watkinson, M.P., 2002. Jurassic planktonic foraminifera from the United Kingdom. Terra Nova 14, 205–209.

Padden, M., Weissert, H., de Rafelis, M., 2001. Evidence for Late Jurassic release of methane from gas hydrate. Geology 29, 223–226.

Padden, M., Weissert, H., Funk, H., Schneider, S., Gansner, C., 2002. Late Jurassic lithological evolution and carbon-isotope stratigraphy of the western Tethys. Eclogae Geologicae Helvetiae 95, 333–346.

Page, K.N., Bloos, G., Bessa, J.L., Fitzpatrick, M., Hesselbo, S., Hylton, M., Morrise, A., Randall, D.E., 2000. East Quantoxhead, Somerset: A candidate Global Stratotype Section and Point for the base of the Sinemurian Stage (Lower Jurassic). GeoResearch Forum 6, 163–171.

Page, K.N., 2005. Jurassic. In: Selley, R.C., Cocks, L.R.M., Plimer, I.R. (Eds.), vols 1–5. Encyclopedia of Geology. Elsevier, Amsterdam, pp. 353–360.

Page, K.N., 2008. The evolution and geography of Jurassic ammonoids. Proceedings of the Geologists' Association 119, 35–57.

Page, K.N., Melendez, G., Wright, J.K., 2010a. The ammonite faunas of the Callovian-Oxfordian boundary interval in Europe and their relevance to the establishment of an Oxfordian GSSP. Volumina Jurassica 7, 89–99.

Page, K.N., Melendez, G., Hart, M.B., Price, G., Wright, J.K., Bown, P., Bello, J., 2010b. Integrated stratigraphical study of the candidate Oxfordian Global Stratotype Section and Point (GSSP) at Redcliff Point, Weymouth, Dorset, UK. Volumina Jurassica 7, 101–111.

Pálfy, J., Parrish, R.R., Smith, P.L., 1997. A U-Pb age from the Toarcian (Lower Jurassic) and its use for time scale calibration through error analysis of biochronologic dating. Earth and Planetary Science Letters 146, 659–675.

Pálfy, J., Smith, P.L., 2000. Synchrony between Early Jurassic extinction, oceanic anoxic event, and the Karoo-Ferrar flood basalt volcanism. Geology 28, 747–750.

Pálfy, J., Smith, P.L., Mortensen, J.K., 2000a. A U-Pb and $^{40}Ar/^{39}Ar$ time scale for the Jurassic. Canadian Journal of Earth Science 37, 923–944.

Pálfy, J., Mortensen, J.K., Carter, E.S., Smith, P.L., Friedman, R.M., Tipper, H.W., 2000b. Timing the end-Triassic mass extinction: First on land, then in the sea? Geology 28, 39–42.

Pálfy, J., Demeny, A., Haas, J., Hetenyi, M., Orchard, M.J., Veto, I., 2001. Carbon isotope anomaly and other geochemical changes at the Triassic-Jurassic boundary from a marine section in Hungary. Geology 29, 1047–1050.

Pálfy, J., Smith, P.L., Mortensen, J.K., 2002. Dating the end-Triassic and Early Jurassic mass extinctions, correlative large igneous provinces, and isotopic events. In: Koeberl, C., MacLeod, K.G. (Eds.), Catastrophic Events and Mass Extinctions: Impacts and Beyond. Geological Society of America Special Paper, 356, pp. 523–532.

Pálfy, J., Mundil, R., 2006. The age of the Triassic/Jurassic boundary: New data and their implications for the extinction and recovery. Volumina Jurassica IV, 294. [Available at http://voluminajurassica.org/pdf/volumen_IV_session_IX.pdf]

Pálfy, J., 2008. The quest for refined calibration of the Jurassic time-scale. Proceedings of the Geologists' Association 119, 85–95.

Partington, M.A., Copestake, P., Mitchener, B.C., Underhill, J.R., 1993. Biostratigraphic calibration of genetic stratigraphic sequences in the Jurassic-lowermost Cretaceous (Hettangian to Ryazanian) of the North Sea and adjacent areas. In: Parker, J.R. (Ed.), Petroleum Geology of Northwest Europe: Proceedings of the 4th Conference. The Geological Society, London, pp. 371–386.

Pavia, G., Enay, R., 1997. Definition of the Aalenian-Bajocian Stage boundary. Episodes 20, 16–20.

Pearce, C.R., Hesselbo, S.P., Coe, A.L., 2005. The mid-Oxfordian (Late Jurassic) positive carbon-isotope excursion recognized from fossil wood in the British Isles. Palaeogeography, Palaeoclimatology, Palaeoecology 221, 343–357.

Pessagno Jr., E.A., Blome, C.D., Hull, D.M., Six, W.M., 1993. Jurassic Radiolaria from the Josephine ophiolite and overlying strata, Smith River subterrane (Klamath Mountains), northwestern Californian and southwestern Oregon. Micropaleontology 39, 93–166.

Pessagno Jr., E.A., Meyerhoff Hull, D., 1996. Once upon a time in the Pacific: Chronostratigraphic misinterpretation of basal strata at ODP Site 801 (Central Pacific) and its impact on geochronology and plate tectonics models. GeoResearch Forum 1–2, 79–92.

Pessagno Jr., E.A., Cantú-Chapa, A., Mattinson, J.M., Meng, X., Kariminia, S.M., 2009. The Jurassic-Cretaceous boundary: New data from North America and the Caribbean. Stratigraphy 6, 185–262.

Peybernes, B., 1998. Larger benthic foraminifera. Columns for Jurassic chart of Mesozoic and Cenozoic sequence chronostratigraphic framework of European basins, by Hardenbol, J., Thierry, J., Farley, M.B., Jacquin, Th., de Graciansky, P.-C., and Vail, P.R. (coordinators). In: de Graciansky, P.-C., Hardenbol, J., Jacquin, Th., Vail, P.R. (Eds.), Mesozoic-Cenozoic Sequence Stratigraphy of European Basins. SEPM Special Publication, 60 Chart 7.

Polteau, S., Corfu, F., Svensen, H., Planke, S., 2010. Rapid emplacement of the Karoo Basin sill complex during the Toarcian revealed by U-Pb dating of zircons. Geophysical Research Abstracts 12, EGU2010-13234-1. [on-line at http://meetingorganizer.copernicus.org/EGU2010/EGU2010-132341.pdf]

Prévot, M., Roberts, N., Thompson, J., Faynot, L., Perrin, M., Camps, P., 2003. Revisiting the Jurassic magnetic reversal recorded in the Lesotho Basalt (Southern Africa). Geophysical Journal International 155, 367–378.

Pruner, P., Houša, V., Olóriz, F., Koštak, M., Krs, M., Man, O., Schnable, P., Venhodová, D., Tavera, J.M., Mazuch, M., 2010. High-resolution magnetostratigraphy and the biostratigraphic zonation of the Jurassic/Cretaceous boundary strata in the Puerto Escaño section (southern Spain). Cretaceous Research 31, 192–206.

Przybylski, P.A., Ogg, J.G., Wierzbowski, A., Coe, A.L., Hounslow, M.W., Wright, J.K., Atrops, F., Settles, E., 2010a. Magnetostratigraphic correlation of the Oxfordian-Kimmeridgian Boundary. Earth and Planetary Science Letters 289, 256–272.

Przybylski, P.A., Głowniak, E., Ogg, J.G., Ziółkowski, P., Sidorczuk, M., Gutowski, J., Lewandowski, M., 2010b. Oxfordian magnetostratigraphy of Poland and its Sub-Mediterranean correlations. Earth and Planetary Science Letters 289, 417–432.

Quenstedt, F., 1858. Der Jura. Tübingen: H. Laupp, 103.

Remane, J., 1985. Calpionellids. In: Bolli, H.M., Saunders, J.B., Perch Nielsen, K. (Eds.), Plankton Stratigraphy. Cambridge University Press, Cambridge, pp. 555–572.

Remane, J., 1998. Calpionellids. Columns for Jurassic chart of Mesozoic and Cenozoic sequence chronostratigraphic framework of European basins, by Hardenbol, J., Thierry, J., Farley, M.B., Jacquin, Th., de Graciansky, P.-C., and Vail, P.R. (coordinators). In: de Graciansky, P.-C., Hardenbol, J., Jacquin, Th., Vail, P.R. (Eds.), Mesozoic-Cenozoic Sequence Stratigraphy of European Basins. SEPM Special Publication, 60 Chart 7.

Renevier, E., 1864. Notices géologique et paléontologiques sur les Alpes Vaudoises, et les régions environnantes. I. Infralias et Zone à Avicula contorta (Ét. Rhaetien) des Alpes Vaudoises. Bulletin de la Société Vaudoise des Sciences Naturelles 8, 39–97.

Resolution of the All-Union Meeting on the unified scheme of the Mesozoic stratigraphy for the Russian Platform, 1955. Gostoptechnizdat, Leningrad, p. 30 [In Russian].

Resolution du deuxième Colloque du Jurassique à Luxembourg 1967, 1970. Vagner, Nancy, p. 38.

Riding, J.B., Ioannides, N.S., 1996. Jurassic dinoflagellate cysts. Bulletin de la Société géologique de France 167, 3–14.

Rieber, H., 1984. Report of the Aalenian Working Group. Aalenian, present status and open problems. Proceedings of 1st International Symposium on Jurassic Stratigraphy 1, 45–54.

Rioult, M., 1964. Le stratotype du Bajocien. In: Maubeuge, P.L. (Ed.), Colloque du Jurassique à Luxembourg 1962. Publication de l'Institut Grand-Ducal, Section des Sciences Naturelles, Physiques et Mathématiques, Luxembourg, pp. 239–258.

Rogov, M.A., 2004. Ammonite-based correlation of the Lower and Middle (*Pandieri* Zone) Volgian substages with the Tithonian Stage. Stratigraphy and Geological Correlation 12 (7), 35–57.

Rogov, M.A., Zakharov, V.A., 2009. Ammonite- and bivalve-based biostratigraphy and Panboreal correlation of the Volgian Stage. Science in China Series D: Earth Sciences 52, 1890–1909.

Rogov, M.A., 2010a. New data on ammonites and stratigraphy of the Volgian Stage of Spitzbergen. Stratigraphy and Geological Correlation 18, 505–531.

Rogov, M.A., 2010b. A precise ammonite biostratigraphy through the Kimmeridgian-Volgian boundary beds in the Gorodischi section (Middle Volga area, Russia), and the base of the Volgian Stage in its type area. Volumina Jurassica 8, 103–130.

Rogov, M.A., Price, G.D., 2010. New stratigraphic and isotope data on the Kimmeridgian-Volgian boundary beds of the Subpolar Urals, Western Siberia. Geological Quarterly 54, 33–40.

Rogov, M.A., Zakharov, V.A., 2010. Jurassic and Lower Cretaceous glendonite occurrences and their implication for Arctic paleoclimate reconstructions and stratigraphy. Earth Science Frontiers 17, 345–347.

Rostovtsev, K.O., Prozorowsky, V.A., 1997. Information on resolutions of standing commissions of the Interdepartmental Stratigraphic Committee (ISC) on the Jurassic and Cretaceous systems. International Subcommission on Jurassic Stratigraphy Newsletter 24, 48–49.

Ruget, C., Nicollin, J.-P., 1998. Smaller benthic foraminifera. Columns for Jurassic chart of Mesozoic and Cenozoic sequence chronostratigraphic framework of European basins, by Hardenbol, J., Thierry, J., Farley, M.B., Jacquin, Th., de Graciansky, P.-C., and Vail, P.R. (coordinators). In: de Graciansky, P.-C., Hardenbol, J., Jacquin, Th., Vail, P.R. (Eds.), Mesozoic-Cenozoic Sequence Stratigraphy of European Basins. SEPM Special Publication, 60 Chart 7.

Ruhl, M., Kürschner, W.M., Krystyn, L., 2009. Triassic-Jurassic organic carbon isotope stratigraphy of key sections in the western Tethys realm (Austria). Earth and Planetary Science Letters 281, 169–187.

Ruhl, M., Deenen, M.H.L., Abels, H.A., Bonis, N.R., Krijgsman, W., Kürschner, W.M., 2010. Astronomical constraints on the duration of the early Jurassic Hettangian stage and recovery rates following the end-Triassic mass extinction (St Audrie's Bay/East Quantoxhead, UK). Earth and Planetary Science Letters 295, 262–276.

Sager, W.W., Weiss, C.J., Tivey, M.A., Johnson, H.P., 1998. Geomagnetic polarity reversal model of deep-tow profiles from the Pacific Jurassic Quiet Zone. Journal of Geophysical Research — Solid Earth 103, 5269—5286.

Sahagian, D., Pinous, O., Olferiev, A., Zakharov, V., 1996. Eustatic curve for the Middle Jurassic-Cretaceous based on Russian Platform and Siberian stratigraphy: Zonal resolution. American Association of Petroleum Geologists Bulletin 80, 1433—1458.

Salfeld, H., 1913. Certain Upper Jurassic strata of England. The Quarterly Journal of the Geological Society 69, 423—430.

Salfeld, H., 1914. Die Gliederung des Oberen Jura in Nordwest Europa. Neues Jahrbuch für Mineralogie, Geologie und Palaeontologie 32, 125—246.

Sandoval, J., O'Dogherty, L., Aguado, R., Bartolini, A., Bruchez, S., Bill, M., 2008. Aalenian carbon-isotope stratigraphy: Calibration with ammonite, radiolarian and nannofossil events in the Western Tethys. Palaeogeography, Palaeoclimatology, Palaeoecology 267, 115—137.

Schaltegger, U., Guex, J., Bartolini, A., Schoene, B., Ovtcharova, M., 2008. Precise U-Pb age constraints for end-Triassic mass extinction, its correlation to volcanism and Hettangian post-extinction recovery. Earth and Planetary Science Letters 267, 266—275.

Schoene, B., Guex, J., Bartolini, A., Schaltegger, U., Blackburn, T.J., 2010. Correlating the end-Triassic mass extinction and flood basalt volcanism at the 100 ka level. Geology 38, 387—390.

Schweigert, G., Callomon, W.J., 1997. Der bauhini-Faunenhorizont unde seine Bedeutung für die Korrelation zwischen tethyalem und subborealem Oberjura. Stuttgarter Beiträge zur Naturkunde, Serie B (Geologie und Paläontologie) 247, 1—69.

Selby, D., 2007. Direct rhenium-osmium age of the Oxfordian-Kimmeridgian boundary, Staffin Bay, Isle of Skye, UK and the Late Jurassic geologic timescale. Norwegian Journal of Geology 87, 291—299.

Sellwood, B.W., Valdes, P.J., 2008. Jurassic climates. Proceedings of the Geologists' Association 119, 5—17.

Sha, J., Fürsich, F.T., 1994. Bivalve faunas of eastern Heilongjiang, northeastern China. II. The Late Jurassic and Early Cretaceous buchiid fauna. Berigeria 12, 3—93.

Sharland, P.R., Archer, R., Casey, D.M., Davies, R.B., Hall, S.H., Heward, A.P., Horbury, A.D., Simmons, M.D., 2001. Arabian Plate Sequence Stratigraphy. GeoArabia Special Publication 2, 371.

Sharland, P.R., Casey, D.M., Davies, R.B., Simmons, M.D., Sutcliffe, O.E., 2004. Arabian Plate Sequence Stratigraphy — revisions to SP2. GeoArabia 9, 199—214.

Shipboard Scientific Party, 1990. Site 801: Pigafetta Basin, western Pacific. Proceedings of the Ocean Drilling Program, Initial Reports 129, 91—170.

Simmons, M.D., BouDagher-Fadel, M.K., Banner, F.T., Whittaker, J.E., 1997. The Jurassic Favusellacea, the earliest Globigerinina. In: BouDagher-Fadel, M.K., Banner, F.T., Whittaker, J.E. (Eds.), The Early Evolutionary History of Planktonic Foraminifera. Chapman and Hall, London, pp. 17—52.

Sinemurian Boundary Working Group (G. Bloos, coordinator), 2000. Submission of East Quantoxhead (West Somerset, SW England) as the GSSP for the base of the Sinemurian Stage. International Commission on Stratigraphy Document, 14.

Speranza, F., Parisi, G., 2007. High-resolution magnetic stratigraphy at Bosso Stirpeto (Marche, Italy): Anomalous geomagnetic field behavior during early Pliensbachian (early Jurassic) times? Earth and Planetary Science Letters 256, 344—359.

Speranza, F., Satolli, S., Mattioli, E., Calamita, F., 2005. Magnetic stratigraphy of Kimmeridgian-Aptian sections from Umbria-Marche (Italy): New details on the M-polarity sequence. Journal of Geophysical Research 110, B12109. doi: 10.1029/2005JB003884.

Srivastava, S.K., 1987. Jurassic spore-pollen assemblages from Normandy (France) and Germany. Geobios 20, 5—79.

Svensen, H., Planke, S., Chevallier, L., Malthe-Sørenssen, A., Corfu, F., Jamtveit, B., 2007. Hydrothermal venting of greenhouse gases triggering Early Jurassic global warming. Earth and Planetary Science Letters 256, 554—566.

Steiner, M.B., Ogg, J.G., Melendez, G., Sequieros, L., 1986. Jurassic magnetostratigraphy, 2. Middle-Late Oxfordian of Aguilon, Iberian Cordillera, northern Spain. Earth and Planetary Science Letters 76, 151—166.

Steiner, M.B., Ogg, J.G., J. Sandoval, J., 1987. Jurassic magnetostratigraphy, 3. Bajocian-Bathonian of Carcabuey, Sierra Harana and Campillo de Arenas, (Subbetic Cordillera, southern Spain). Earth and Planetary Science Letters 82, 357—372.

Steiner, M.B., Ogg, J.G., 1988. Early and Middle Jurassic magnetic polarity time scale. In: Rocha, R.B., Soares, A.F. (Eds.), 2nd International Symposium on Jurassic Stratigraphy. Instituto Nacional de Investigação Científica, Lisbon, pp. 1097—1111.

Stone, R., 2010. Bird-dinosaur link firmed up, and in brilliant color. Science 327, 570—571.

Strasser, A., 2007. Astronomical time scale for the Middle Oxfordian to Late Kimmeridgian in the Swiss and French Jura Mountains. Swiss Journal of Geosciences 100, 407—429.

Suan, G., Pittet, B., Bour, I., Mattioli, E., Duarte, L.V., Mailliot, S., 2008. Duration of the Early Toarcian carbon isotope excursion deduced from spectral analysis—Consequence for its possible causes. Earth and Planetary Science Letters 267, 666—679.

Surlyk, F., Zakharov, V.A., 1982. Buchiid bivalves from the Upper Jurassic and Lower Cretaceous of East Greenland. Palaeontology 25, 727—753.

Svensen, H., Planke, S., Chevallier, L., Malthe-Sørenssen, A., Corfu, F., Jamtveit, B., 2007. Hydrothermal venting of greenhouse gases triggering Early Jurassic global warming. Earth and Planetary Science Letters 256, 554—566.

Swientek, O., 2002. The Greenland Norwegian Seaway: Climatic and Cyclic Evolution of Late Jurassic-Early Cretaceous Sediments. Inaugural dissertation. University of Köln, p. 148.

Sykes, R.M., Callomon, J.H., 1979. The *Amoeboceras* zonation of the Boreal Upper Oxfordian. Palaeontology 22, 839—903.

Taylor, S.P., Sellwood, B., Gallois, R., Chambers, M.H., 2001. A sequence stratigraphy of Kimmeridgian and Bolonian stages, Wessex-Weald Basin. Journal of the Geological Society 158, 179—192.

Thierry, J., 1998. Ammonites. Columns for Jurassic chart of Mesozoic and Cenozoic sequence chronostratigraphic framework of European basins, by Hardenbol, J., Thierry, J., Farley, M.B., Jacquin, Th., de Graciansky, P.-C., and Vail, P.R. (coordinators). In: de Graciansky, P.-C., Hardenbol, J., Jacquin, Th., Vail, P.R. (Eds.), Mesozoic-Cenozoic Sequence Stratigraphy of European Basins. SEPM Special Publication, 60, Chart 7.

Tivey, M.A., Sager, W.W., Lee, S.-M., Tominaga, M., 2006. Origin of the Pacific Jurassic Quiet Zone. Geology 34, 789—792.

Tominaga, M., Sager, W.W., 2010. Revised Pacific M-anomaly geomagnetic polarity timescale. Geophysical Journal International 182, 203—232.

Tominaga, M., Sager, W.W., Tivey, M.A., Lee, S.-M., 2008. Deep-tow magnetic anomaly study of the Pacific Jurassic Quiet Zone and implications for the geomagnetic polarity reversal time scale and geomagnetic field behavior. Journal of Geophysical Research 113, B07110. doi: 10.1029/2007JB005527.

Torrens, H.S., 1965. Revised zonal scheme for the Bathonian stage of Europe. Reports of the Seventh Carpato-Balkan Geological Association Congress 2, 47–55.

Trujillo, K.C., Chamberlain, K.R., Strickland, A., 2006. Oxfordian U/Pb ages from SHRIMP analysis for the Upper Jurassic Morrison Formation of southeastern Wyoming with implications for biostratigraphic correlations. Geological Society of America Abstracts with Programs 38 (6), 7 [Geol. Soc. Amer. Rocky Mountain Section annual meeting (17–19 May 2006) abstract 4–5 (gsa.confex.com/gsa/2006RM/finalprogram/abstract_104814.htm)]

Trujillo, K.C., Chamberlain, K.R., Bilbey, S.A., 2008. U/Pb age for a pipeline dinosaur site; new techniques and stratigraphic challenges. Geological Society of America Abstracts with Programs 40 (1), 42.

Valdes, P.J., Sellwood, B.W., 1992. A palaeoclimate model for the Kimmeridgian. Palaeogeography, Palaeoclimatology, Palaeoecology 95, 47–72.

Van Buchem, F.S.P., McCave, I.N., Weedon, G.P., 1994. Orbitally induced small-scale cyclicity in a siliciclastic epicontinental setting (Lower Lias, Yorkshire, UK). In: Boer, P.L., Smith, D.G. (Eds.), Orbital Forcing and Cyclic Sequences. International Association of Sedimentologists Special Publication, 19, pp. 345–366.

van de Schootbrugge, B., Payne, J.L., Tomasovych, A., Pross, J., Fiebig, J., Benbrahim, M., Follmi, K.B., Quan, T.M., 2008. Carbon cycle perturbation and stabilization in the wake of the Triassic-Jurassic boundary mass-extinction event. Geochemistry, Geophysics, Geosystems 9, Q04028. doi: 10.1029/2007GC001914.

van Konijnenburg-van Cittert, J.H.A., 2008. The Jurassic fossil plant record of the UK area. Proceedings of the Geologists' Association 119, 59–72.

Veizer, J., Ala, D., Azmy, K., Bruckschen, P., Buhl, D., Bruhn, F., Carden, G.A.F., Diener, A., Ebneth, S., Godderis, Y., Jasper, T., Korte, C., Pawellek, F., Podlaha, O.G., Strauss, H., 1999. $^{87}Sr/^{86}Sr$, $\partial^{13}C$ and $\partial^{18}O$ evolution of Phanerozoic seawater. Chemical Geology 161, 59–88. Datasets posted at http://www.science.uottawa.ca/geology/isotope_data/.

Veizer, J., Godderis, Y., François, L.M., 2000. Evidence for decoupling of atmospheric CO_2 and global climate during the Phanerozoic eon. Nature 408, 698–701.

von Buch, L., 1839. Über den Jura in Deutschland. Der Königlich Preussischen Akademie der Wissenschaften, Berlin, p. 87.

von Humboldt, F.W.H.A., 1799. Über die Unterirdischen Gasarten und die Mittel ihren Nachtheil zu Vermindern. Ein Beitrag zur Physik der Praktischen Bergbaukunde. Braunschweig, Wiewag, p. 384.

von Salis, K., Bergen, J., De Kaenel, E., 1998. Calcareous nannofossils. Columns for Jurassic chart of Mesozoic and Cenozoic sequence chronostratigraphic framework of European basins, by Hardenbol, J., Thierry, J., Farley, M.B., Jacquin, Th., de Graciansky, P.-C., and Vail, P.R. (coordinators). In: de Graciansky, P.-C., Hardenbol, J., Jacquin, Th., Vail, P.R. (Eds.), Mesozoic-Cenozoic Sequence Stratigraphy of European Basins. SEPM Special Publication, 60 Chart 7.

Warrington, G., 2010. The Hettangian GSSP. International Commission on Jurassic Stratigraphy Newsletter 36, 5–12.

Waterhouse, H.K., 1995. High-resolution palynofacies investigation of Kimmeridgian sedimentary rocks. In: House, M.R., Gale, A.S. (Eds.), Orbital Forcing Timescales and Cyclostratigraphy. Geological Society Special Publication, 85, pp. 75–114.

Weedon, G.P., 1985. Hemipelagic shelf sedimentation and climatic cycles: The basal Jurassic (Blue Lias) of South Britain. Earth and Planetary Science Letters 76, 321–335.

Weedon, G.P., 1989. The detection and illustration of regular sedimentary cycles using Walsh power spectra and filtering, with examples from the Lias of Switzerland. Journal of the Geological Society 146, 133–144.

Weedon, G.P., Jenkyns, H.C., 1999. Cyclostratigraphy and the Early Jurassic timescale: Data from the Belemnite Marls, southern England. Geological Society of America Bulletin 111, 1823–1840.

Weedon, G.P., Jenkyns, H.C., Coe, A.L., Hesselbo, S.P., 1999. Astronomical calibration of the Jurassic time-scale from cyclostratigraphy in British mudrock formations. Philosophical Transactions of the Royal Society, Series A 357, 1787–1813.

Weedon, G.P., Coe, A.L., Gallois, R.W., 2004. Cyclostratigraphy, orbital tuning and inferred productivity for the type Kimmeridge Clay (Late Jurassic), Southern England. Journal of the Geological Society 161, 655–666.

Weissert, H., Channell, J.E.T., 1989. Tethyan carbonate carbon isotope stratigraphy across the Jurassic-Cretaceous boundary: An indicator of decelerated global carbon cycling? Paleoceanography 4, 483–494.

Weissert, H., Erba, E., 2004. Volcanism, CO_2 and palaeoclimate: A Late Jurassic-Early Cretaceous carbon and oxygen isotope record. Journal of the Geological Society 161, 695–702.

Westerman, G.E.G., Riccardi, A.C., 1972. Middle Jurassic ammonite fauna and biochronology of the Argentine-Chilean Andes. 1. Hildoceratacea. Palaeontolographica, A 140, 1–116.

White, R.S., 1997. Mantle plume origin for the Karoo and Ventersdorp flood basalts, South Africa. In: Hatton, C.J. (Ed.), Special Issue on the Proceedings of the Plumes, Plates and Mineralisation '97 Symposium. South African Journal of Geology, 100, pp. 271–282.

Whiteside, J.H., Olsen, P.E., Eglinton, T.I., Montlućon, D., Brookfield, M.E., Sambrotto, R.N., 2010. Compound-specific carbon isotopes from Earth's largest flood basalt province directly link eruptions to the end-Triassic mass extinction. Proceedings of the National Academy of Sciences 107, 6721–6725.

Wierzbowski, A., 2001. Kimmeridgian Working Group. International Subcommission on Jurassic Stratigraphy Newsletter 28, 15–16.

Wierzbowski, A., 2002. Kimmeridgian Working Group. International Subcommission on Jurassic Stratigraphy Newsletter 29, 18–19.

Wierzbowski, A., Coe, A.L., Hounslow, M.W., Matyja, B.A., Ogg, J.G., Page, K.N., Wierzbowski, H., Wright, J.K., 2006. A potential stratotype for the Oxfordian/Kimmeridgian boundary: Staffin Bay, Isle of Skye, U.K. Volumina Jurassica 4, 17–33.

Wierzbowski, A., 2010. Kimmeridgian. International Commission on Jurassic Stratigraphy Newsletter 36, 5–12.

Wignall, P.B., 2001. Large igneous provinces and mass extinctions. Earth-Science Reviews 53, 1–33.

Wignall, P.B., Bond, D.P.G., 2008. The end-Triassic and Early Jurassic mass extinction records in the British Isles. Proceedings of the Geologists' Association 119, 73–84.

Williford, K.H., Ward, P.D., Garrison, G.H., R.Buick, R., 2007. An extended organic carbon-isotope record across the Triassic-Jurassic boundary in the Queen Charlotte Islands, British Columbia, Canada. Palaeogeography, Palaeoclimatology, Palaeoecology 244, 290–296.

Wilson, G.J., 1984. New Zealand Late Jurassic to Eocene dinoflagellate biostratigraphy. Newsletters on Stratigraphy 13, 104–117.

Wimbledon, W.A.P., Casellato, C.E., Reháková, D., Bulot, L.G., Erba, E., Gardin, S., Verreussel, R.M.C.H., Munsterman, D.K., Hunt, C.O., 2011. Fixing a basal Berriasian and Jurassic-Cretaceous (J-K) boundary – is there perhaps some light at the end of the tunnel? Rivista Italiana di Paleontologia e Stratigrafia 117, 295–307.

Woollam, R., Riding, J.B., 1983. Dinoflagellate cyst zonation of the English Jurassic. Report of the Institute of Geological Sciences 83, 41.

Yang, Z., Moreau, M.-G., Bucher, H., Dommergues, J.-L., Trouiller, A., 1996. Hettangian and Sinemurian magnetostratigraphy from the Paris Basin. Journal of Geophysical Research 101, 8025–8042.

Zák, K., Košták, M., Man, O., Zakharov, V.A., Rogov, M.A., Pruner, P., Rohovec, J., Dzyuba, O.S., Mazuch, M., 2011. Comparison of carbonate C and O stable isotope records across the Jurassic/Cretaceous boundary in the Tethyan and Boreal Realms. Palaeogeography, Palaeoclimatology, Palaeoecology 299, 83–96.

Zakharov, V.A., 1987. The Bivalve *Buchia* and the Jurassic-Cretaceous boundary in the Boreal Province. Cretaceous Research 8, 141–153.

Zakharov, V.A., Bogomolov, Yu.I., Il'ina, V.I., Konstantinov, A.G., Kurushin, N.I., Lebedeva, N.K., Meledina, S.V., Nikitenko, B.L., Sobolev, E.S., Shurygin, B.N., 1997. Boreal zonal standard and biostratigraphy of the Siberian Mesozoic. Russian Geology and Geophysics 38, 965–993.

Zakharov, Y.D., Smyshlyaeva, O.P., Shigeta, Y., Popov, A.M., Zonova, T.D., 2006. New data on isotopic composition of Jurassic-Early Cretaceous cephalopods. Progress in Natural Science 16 (special issue), 50–67.

Zeiss, A., 2003. The Upper Jurassic of Europe: Its subdivision and correlation. Geological Survey of Denmark and Greenland Bulletin 1, 75–114.

Ziółkowski, P., Hinnov, L.A., 2009. Cyclostratigraphy of Bathonian using magnetic susceptibility – preliminary report. In: Krobicki, M. (Ed.), Jurassica VIII, Vrsatec 09–11.10.2009. Geologia 35 (3/1), 115 [In Polish].

J.G. Ogg and L.A. Hinnov
Contributor: C. Huang

Chapter 27

Cretaceous

Abstract: The breakup of the former Pangea supercontinent culminated in the modern drifting continents. An explosion of calcareous nannoplankton and foraminifers in the warm seas created massive chalk deposits. A surge in undersea volcanic activity enhanced super-greenhouse conditions in the middle Cretaceous. Angiosperm plants bloomed on the dinosaur-dominated land during late Cretaceous. The Cretaceous dramatically ended with an asteroid impact.

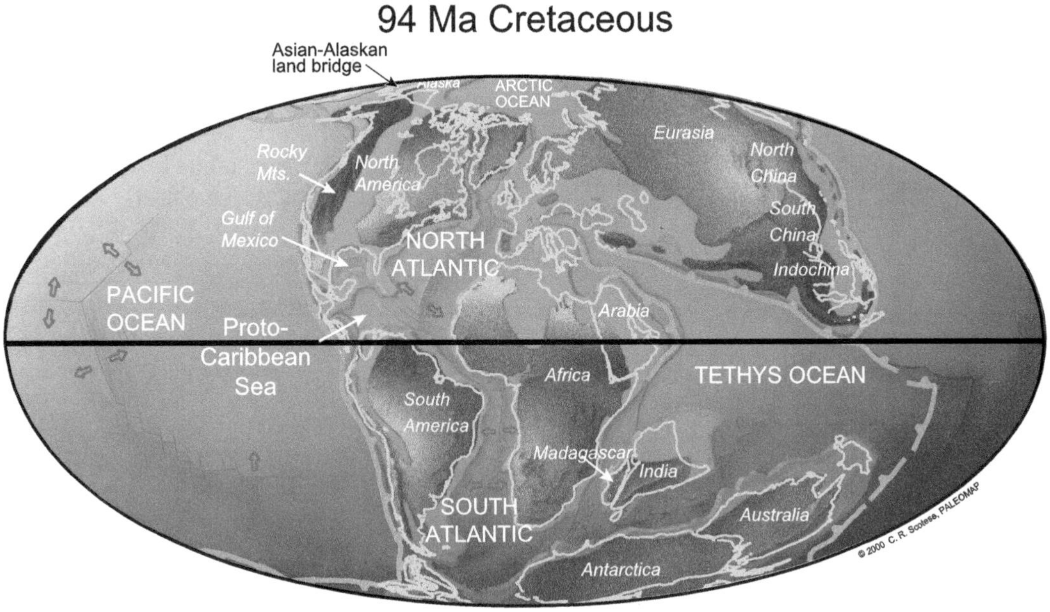

Chapter Outline

The Geologic Time Scale 2012. DOI: 10.1016/B978-0-444-59425-9.00027-5

27.1. HISTORY AND SUBDIVISIONS

27.1.1. Overview of the Cretaceous

The "Terrain Crétacé" was established by J.G.J. d'Omalius d'Halloy (1822) to encompass the Chalk (*creta* in Latin) that characterizes a major unit of strata in the Paris Basin and across much of Europe. He defined the "Terrain" in 1823 to include "the formation of the chalk, with its tufas, its sands and its clays".

William Smith had already mapped four stratigraphic units between the "lower clay" (Eocene) and the "Portland Stone" (Jurassic). His units were grouped by Conybeare and Phillips (1822) into two main divisions – an upper Chalk and the formations below. This two-fold division, adopted in England and France at an early stage, has persisted as the two Cretaceous series and epochs.

Alcide d'Orbigny (1840) divided the Cretaceous fossil assemblages of France into five divisions, which he termed "étages" (stages): Neocomian, Aptian, Albian, Turonian and Senonian. He later added a "Urgonian" (between the Neocomian and Aptian) and a "Cenomanian" (between Albian and Turonian). The term "Neocomian" had been coined by Thurmann in 1836 for the strata in the vicinity of Neuchâtel, Switzerland. This Neocomian has been subdivided into the three stages of Berriasian, Valanginian and Hauterivian following the recommendation of Barbier, Debelmas and Thieuloy in the Colloque sur le Crétacé inférieur (Barbier and Thieuloy, 1965). d'Orbigny's "Urgonian" stage has been replaced by a "Barremian" stage and an expanded Aptian stage. The "Senonian" (named by d'Orbigny after the Sénones people of the Burgundy region, France) has been divided into four stages: Coniacian, Santonian, Campanian and Maastrichtian.

The uppermost units of the chalks of the Danish coast, which had traditionally been included within the Cretaceous, are now classified as a Danian Stage of the lowermost Cenozoic. The termination of the Cretaceous is now defined by a major mass extinction and impact-generated iridium anomaly horizon at the base of this Danian Stage.

The boundaries of the twelve historical Cretaceous stages were primarily defined by ammonites in France and the Netherlands (Birkelund *et al.*, 1984; Kennedy, 1984). Refined recognition or proposed definitions of the basal boundaries of these stages have other global criteria, including geomagnetic reversals, carbon isotope excursions and microfossil datums (see summary table at end of this chapter).

Historical usage, coupled with the expertise of members of the various boundary working groups, has dictated a preferential selection of boundary stratotypes (GSSPs) for most stages and substages to be within western European basins. However, a major problem in Cretaceous chronostratigraphy is how to correlate these regional biostratigraphic events and associated stage boundary definitions to other paleogeographic and paleoceanographic realms. At the time of preparation of this book (Sept, 2011), only three of the twelve Cretaceous stages had ratified GSSPs (Cenomanian, Turonian, Maastrichtian) (Figure 27.1).

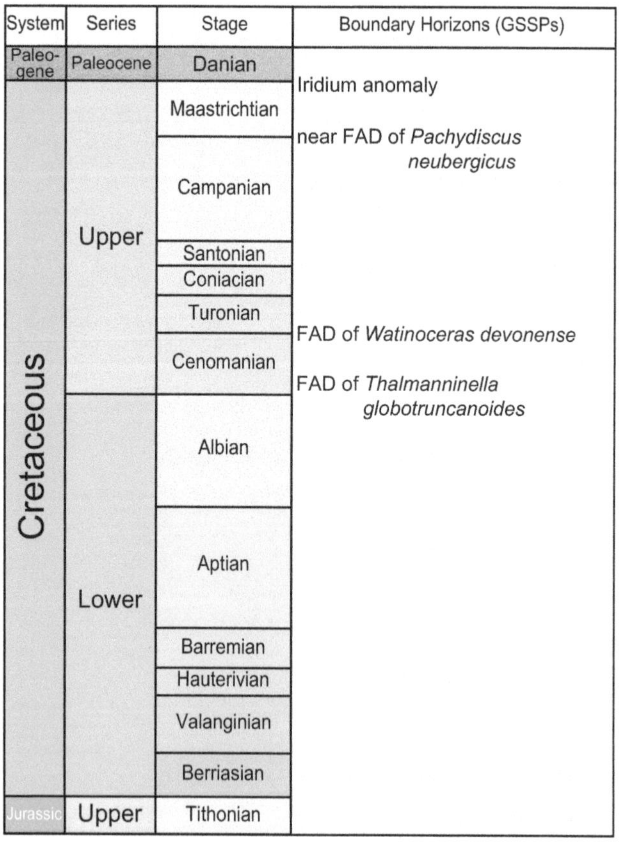

System	Series	Stage	Boundary Horizons (GSSPs)
Paleo-gene	Paleocene	Danian	
Cretaceous	Upper	Maastrichtian	Iridium anomaly
		Campanian	near FAD of *Pachydiscus neubergicus*
		Santonian	
		Coniacian	
		Turonian	
		Cenomanian	FAD of *Watinoceras devonense*
	Lower	Albian	FAD of *Thalmanninella globotruncanoides*
		Aptian	
		Barremian	
		Hauterivian	
		Valanginian	
		Berriasian	
Jurassic	Upper	Tithonian	

FIGURE 27.1 **Summary of Cretaceous stage nomenclature and status.** Stages with ratified GSSPs are indicated by the first-appearance datum (FAD) of the main correlation marker (planktonic foraminifer, ammonite, etc.) or other primary correlation level.

27.1.2. Subdivisions of the Lower Cretaceous

As of late 2011, none of the six Lower Cretaceous stages (Berriasian through Albian) have GSSP-based definitions.

The subdivisions of the Berriasian through Aptian stages of the Lower Cretaceous were originally derived from exposures in southeast France and adjacent northwest Switzerland. In this region, the paleogeographic Vocontian trough and adjacent margins preserved a nearly continuous record, generally in limestone–marl successions. The fauna and microflora thriving in the tropical ocean of this Tethyan realm did not extend into the colder Boreal realm of northwest Europe and other northern regions; therefore, the Boreal equivalents of most of these Tethyan-defined stages remain uncertain.

Traditionally, the primary markers that were considered for defining stage and substage boundaries of the Berriasian through Barremian were the lowest or highest occurrence of species of ammonoids. However, in an effort to achieve global correlation potential, these macrofossil datums have been replaced or enhanced by magnetostratigraphic, geochemical or microfossil events. For example, proposed bases of the Berriasian Stage (base of Cretaceous) and the Aptian Stage are magnetic reversals, and a possible correlation marker for the base of the Albian Stage is a widespread organic-enrichment episode.

27.1.2.1. Jurassic–Cretaceous Boundary and Berriasian

Tithonian–Berriasian (J/K) Boundary Definition

The Cretaceous is the only Phanerozoic system that does not yet have an accepted global boundary definition, despite over a dozen international conferences and working group meetings having been dedicated to the issue since 1974 (e.g., reviews by Zakharov *et al.*, 1996; Cope, 2007; Wimbledon *et al.*, 2011). The difficulties in assigning a global Jurassic/Cretaceous boundary are the combined product of historical usage, of the lack of any major faunal change between the latest Jurassic and earliest Cretaceous, of pronounced provincialism of marine fauna and flora, of the long-term "Purbeckian regression" of global sea level that caused hiatuses or condensations in the main reference sections of England and Russia, and of persistent debates on which stratigraphic markers might provide the most useful global correlation levels.

The Tithonian Stage of latest Jurassic was defined by A. Oppel (1865) to include all deposits in the Mediterranean area that lie between a restricted Kimmeridgian and "Valanginian", but no representative section or upper limit was designated. A "Berriasian" was proposed for a limestone succession near the village of Berrias (Ardèche Département, southeast France) (Coquand, 1871; reviewed in Rawson, 1983). Originally conceived as a subdivision of Valanginian, this interval was subsequently often referred to as "Infra-Valanginian" until the concept and name of "Berriasian" was eventually brought back into use. To some degree, this "Berriasian" Stage overlapped with the original concept of the Tithonian Stage. The historical lower boundary of the Berriasian lacks any significant faunal change; indeed the basal part of the Berrias stratotype lacks any diagnostic ammonites (Cope, 2007). The Colloque sur la limite Jurassique-Crétacé (1975) voted to define the base of the Berriasian as the base of the *Berriasella jacobi* ammonite subzone, which was a shift downward from a previous vote (1963) to use the base of the *Pseudosubplanites grandis* Subzone (reviewed in Wimbledon *et al.*, 2011).

Some workers have suggested assigning at least a portion of this historical Berriasian Stage into the Jurassic, and assign the Jurassic/Cretaceous boundary to the base of the "middle Berriasian" at the beginning of the *Subthurmannia occitanica* Zone (*Subthurmannia subalpina* Subzone) (Remane, 1991) or even raised to the beginning of the current Valanginian Stage (e.g., Rawson, 1990, as cited in Zakharov *et al.*, 1996). The ammonite turnover at the base of the *Subthurmannia occitanica* Zone corresponds to the effective base of the ammonite zonation at the historical Berrias stratotype:

> '*...because of the virtual ammonites and the frequency of intervals with reworked sediments below that zone.*'
>
> (Hoedemaeker *et al.*, 2003)

This level also has an advantage in that it can be correlated to the Sub-Boreal realm, where it is approximately coincident with the base of the *Runctonia runctoni* ammonite zone, it is close to the Boreal regional Volgian/Ryazanian stage boundary, and it coincides with a widespread depositional sequence boundary of possible eustatic significance (reviewed in Hoedemaeker and Leereveld, 1995).

The current Berriasian Working Group of the International Commission on Cretaceous Stratigraphy has worked to integrate regional ammonite zonations, calpionellid zones, calcareous nannofossil datums, palynomorphs and magnetostratigraphy (Wimbledon *et al.*, 2011; summarized in Figure 27.2). Its emphasis is on identifying a GSSP level near markers of both regional and global significance.

(a) Magnetostratigraphy has proven to be a reliable method for placing biostratigraphic events into a common framework (e.g., summaries in Ogg and Lowrie, 1986; Speranza *et al.*, 2005; Channell *et al.*, 2010), correlating stratigraphic scales of Tethyan (e.g., Pruner *et al.*, 2010) and Boreal (e.g., Houša *et al.*, 2007) regions, and comparing marine and non-marine facies (e.g., Purbeck facies of England by Ogg *et al.*, 1991, 1994).

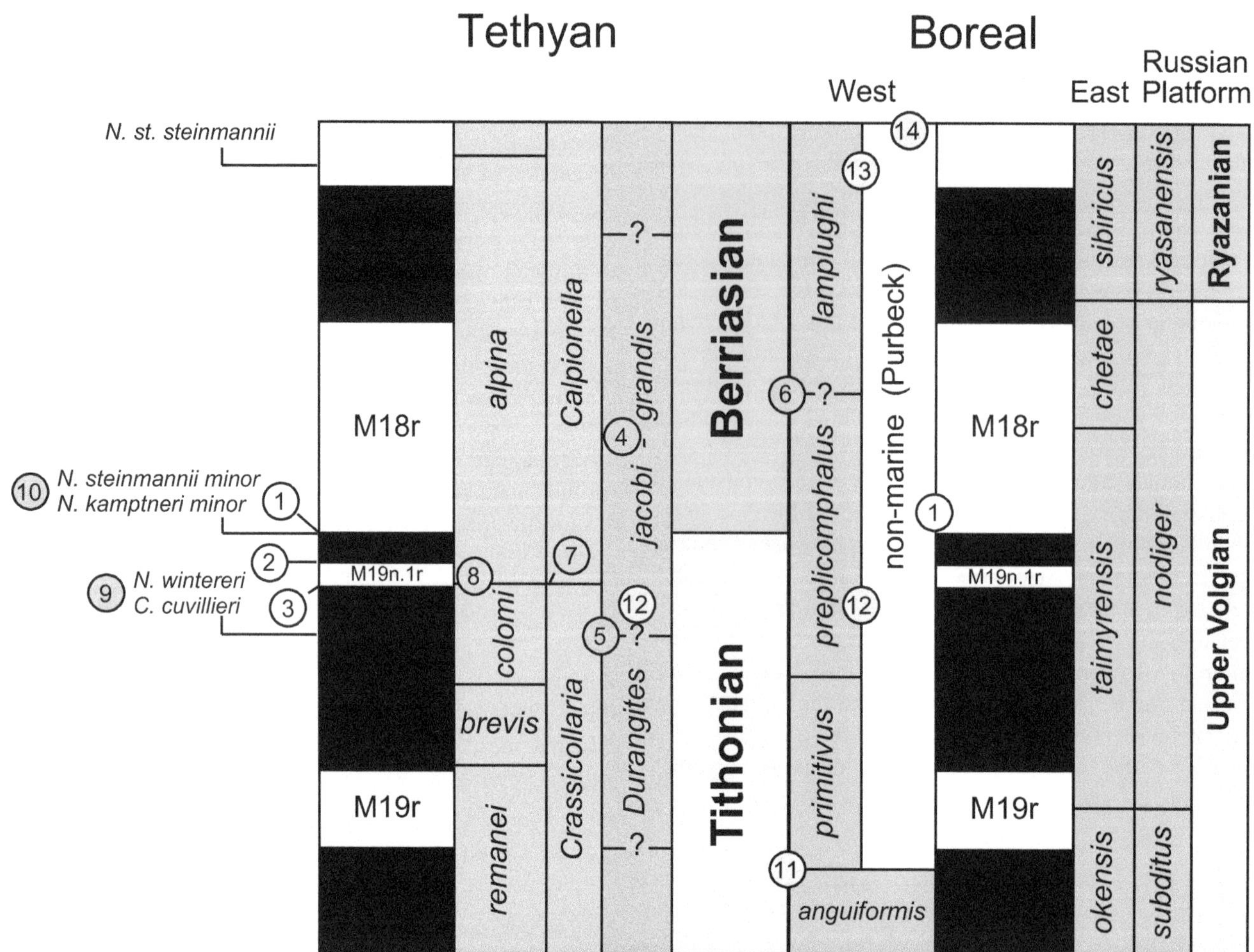

FIGURE 27.2 **Relative placement of selected markers under consideration for Jurassic–Cretaceous boundary or useful for correlations, and inter-regional comparison of zonations.** Magnetic polarity zones: (1) Base of Chron M18r (*placement used in GTS2012*), (2) Base of Chron M19n.1n, (3) Base of Chron M19n.1r. Ammonites: (4) Base of *Pseudosubplanites grandis* Subzone (1965 colloquium decision on J–K boundary), (5) Base of *Berriasella jacobi* Subzone (1975 colloquium decision on J–K boundary), (6) base *Subcraspedites lamplughi* Zone. Calpionellids: (7) base *Calpionella* biozone (biozone A/B boundary), (8) "explosion" of small, globular *Calpionella alpina*. Calcareous nannofossils: (9) FAD of *Nannoconus wintereri* and *Cruciellipsis cuvillieri*, (10) FAD of *Nannoconus steinmannii minor* and *N. kamptneri minor*. Palynomorphs: (11) LAD of *Dichadogonyaulax pannea*, *Egmontodinium polyplacophorum* etc. in the late Portlandian extinction, (12) FAD of *Warrenia californica*, *Dichadogonyaulax bensonii* and *Ampiculatisporis verbitskayae*, (13) FAD of *Matonisporites elegans* and *Aequitriradites spinulosum*, (14) *Classopolis* collapse (*modified from Wimbledon et al., 2011*).

'Magnetostratigraphers, pragmatically, have seen magnetozone M18r as a workable J/K boundary indicator, being the first long, reversed interval (after the very long M19n) as one works up-sequence from the Jurassic. ... The base of M18r has been chosen as a significant event, in preference to the short magnetic intervals below (M19n.1r, M19n.1n) ...'

(Wimbledon *et al.*, 2011)

In GTS2012, we have selected this working definition of the base of Chron M18r as the Jurassic/Cretaceous boundary.

(b) Ammonites will always be a primary method of approximate correlation of the Jurassic/Cretaceous boundary interval within different regional basins. But, as indicated in Figure 27.2, it appears that there are no ammonite-zone boundaries that are synchronous among the main European regions within the basal Cretaceous transition interval.

(c) Calpionellid microfossils, especially the "explosion" of small, globular *Calpionella alpina* that marks the base of the Calpionellid Zone B, are very useful for deep-shelf to pelagic limestone facies. In contrast to the endemic ammonite assemblages, *Calpionella alpina* occurs from Mexico to Tibet (reviewed in Wimbledon *et al.*, 2011).

(d) Calcareous nannoplankton underwent a rapid diversification within this interval. Several main and secondary events, especially within the nannoconnid lineage of heavily calcified forms, enable a high-resolution subdivision (e.g., Casellato, 2010). Indeed, this surge in the evolution and abundance of nannoplankton had a major role in the deposition of the chalk "creta" that characterizes the Cretaceous (e.g., reviews in Roth, 1989; Erba, 2006; and Casellato *et al.*, 2008).

In a suggested boundary stratotype at the Puerto Escano section (Cordoba province, Spain), the pelagic limestone facies contains magnetostratigraphy, ammonoids, calpionellids and calcareous nannoplankton (Tavera *et al.*, 1994; Pruner *et al.*, 2010).

Portlandian–Purbeckian and Volgian–Ryazanian

Alcide d'Orbigny (1849–50) used the facies of Portland Limestone and non-marine Purbeck facies of England and northern France as the Portlandian and Purbeckian stages, but this usage has been largely discontinued (reviewed in Cope, 2007). The Berriasian stage, nearly regardless of its eventual basal definition, encompasses at least half of the Purbeck succession based on correlations using magnetostratigraphy (e.g., Figure 27.2, and Ogg *et al.*, 1994) and palynology. However, the succession of ammonites endemic to the Boreal realm required different zonations, none of which could be reliably correlated to the Berriasian of the Tethyan realm (see Figure 26.8 in the Jurassic chapter (26) this volume). A regional Volgian stage was defined by Nikitin (1881) and later extended downward (Gerasimov and Michailov, 1966) to approximately coincide with the base of the Tithonian (reviewed in Cope, 2007). On the Russian Platform and in Siberia, the Volgian is followed by a Ryazanian regional stage (Sazonov, 1951), and use of "Volgian–Ryazanian" had been applied to the stratigraphy of other North Sea and high-latitude regions. The base of the regional Ryazanian Stage is assigned as the *Praetollia maynci* or *Chetaites sibiricus* ammonite zone (Russian Platform versus Siberian zonations). This level was thought to correlate to the base of the *Ructonia ructoni* ammonite zone of eastern England, but these levels may not be equivalent (Hancock, 1991; Hardenbol *et al.*, 1998). Magnetostratigraphy from Siberian sections suggests that the base of the Ryazanian Stage is within the middle to upper part of Chron M18n (Houša *et al.*, 2007).

The top of the regional Ryazanian is probably close to the base of the Valanginian Stage, as revised by Hoedemaeker *et al.* (2003).

Berriasian Substages

The three main ammonite zones of the Berriasian Stage in the Tethyan region are generally used as informal substages (e.g., Reboulet *et al.*, 2011). No boundary stratotypes for these substages have yet been recommended. The base of the Middle Berriasian in the Mediterranean region (base of *Subthurmannia occitanica* Zone) correlates to the lower portion of polarity Chron M17r.

The base of the Upper Berriasian substage in the Mediterranean region is currently placed at the base of the *Malbosiceras paramimounum* ammonite subzone (base of *Subthurmannia boissieri* Zone) (Zakharov *et al.*, 1996). This level is close to the base of Calpionellid Zone D, and is in the middle of magnetic polarity zone M16r (M16r.5) (Galbrun, 1984). Magnetostratigraphic correlation to southern England tentatively projects this level to be within the *Runctonia runctoni* ammonite zone of the Boreal realm (Ogg *et al.*, 1991).

27.1.2.2. Valanginian

History, Definition and Boundary Stratotype

The original type section of the Valanginian included all post-Jurassic strata to the base of the "Marnes de Hauterive" in shallow-marine facies at Seyon Gorge near Valangin (Neuchâtel, Switzerland) (Desor, 1854; Desor and Greesly, 1859; reviewed in Rawson, 1983). Because this section is poor in ammonoids, a pair of "hypostratotypes" at Angles (Alpes-de-Haute-Provence province) and Barret-le-Bas (Hautes-Alpes province) in southeast France have served as reference sections (Busnardo *et al.*, 1979). The base of the Valanginian was provisionally placed at the base of the *Thurmanniceras otopeta* ammonite zone (Busnardo *et al.*, 1979; Birkelund *et al.*, 1984).

Taxonomic and correlation problems with the ammonoid successions (e.g., Bulot *et al.*, 1993) led the Valanginian Working Group (Hoedemaeker *et al.*, 2003) to recommend moving the base of the Valanginian Stage upward to the base of Calpionellid Zone E, defined by the first occurrence of *Calpionellites darderi* (Bulot *et al.*, 1996; Aguado *et al.*, 2000). This transferred the *T. otopeta* ammonite zone into the Berriasian Stage. This level can be traced from France to Mexico, is associated with other bioevents, and coincides with rare migrations of cold-water taxa into the Tethyan realm (reviewed in Aguado *et al.*, 2000). Magnetostratigraphic–calpionellid studies in Italy and Spain have indicated an apparent 300 000-year variability or diachroneity of the base of Calpionellid Zone E between the middle of magnetic polarity Chron M14r to earliest part of polarity Chron M14n (Channell and Grandesso, 1987; Channell *et al.*, 1987; Ogg *et al.*, 1988). One candidate GSSP at Montbrun-les-Bains in southeast France is within alternating limestone and marl with well-preserved ammonoid and calpionellid successions, where the proposed boundary level is followed by the lowest occurrence of ammonite *Thurmanniceras pertransiens* (Blanc *et al.*, 1994). However, this French section was unsuitable for magnetostratigraphy. Another candidate GSSP section at Barranco de Cañada Luenga, south of Cehegín in southeast Spain, has integrated ammonoid, nannofossil and magnetostratigraphy, but poorer preservation of calpionellids (Aguado *et al.*, 2000).

Correlation of the base of Valanginian from the Tethyan realm into the Boreal realm is not precise. However, episodes of ammonoid migration and comparisons of nannofossil and dinoflagellate cysts suggest a correspondence near the base of the regional usage of "Valanginian" in northwest Germany and England (Bulot *et al.*, 1996; Leereveld, 1995).

Upper Valanginian Substage

The base of the Upper Valanginian is traditionally placed at the base of the distinctive *Saynoceras verrucosum* ammonite zone, which can also be recognized in the West European province of the Boreal realm. Magnetostratigraphic, chemostratigraphic and microfossil markers (e.g., Channell *et al.*, 1993) await calibration with this ammonoid datum. Potential boundary stratotypes in southeast France (section at Vergol) and in the Betic Cordillera of Spain are being considered (Bulot *et al.*, 1996; Subcommission on Cretaceous Stratigraphy, 2009). The base of the Upper Valanginian is marked by the onset of a major sea-level transgression, and the following pronounced sea-level highstand is associated with the greatest ammonoid diversity of the Valanginian in the Mediterranean faunal province (Hoedemaeker and Leereveld, 1995; Reboulet, 1996; Reboulet and Atrops, 1999).

27.1.2.3. Hauterivian

History, Definition and Boundary Stratotype

The original Hauterivian Stage in the area of Hauterive (Neuchâtel, Switzerland) encompassed "marnes à *Astieria*" (later transferred to Valanginian), "marnes bleues d'Hauterive" and "Pierre Jaune de Neuchâtel" formations (Renevier, 1874; reviewed in Rawson, 1983). The base of the Hauterivian in the Tethyan realm is traditionally recognized by the lowest occurrence of the ammonite species *Acanthodiscus radiatus*. Reboulet (1996) suggested that the base of the Radiatus zone should be defined by the first occurrence of the ammonite genus *Acanthodiscus* (Reboulet and Atrops, 1999).

Expanded sections of ammonoid-rich, limestone–marl alternations in southeast France have served as the reference for this stage. The section at the village of La Charce in this region is the proposed global boundary stratotype for both the base of the Hauterivian Stage and base of the Upper substage (Thieuloy, 1977; Mutterlose *et al.*, 1996). Inter-realm exchanges of dinoflagellate cyst and ammonoid assemblages suggest that the traditional Boreal usage of "Hauterivian" (base of *Endemoceras amblygonium* ammonite zone) is nearly coeval with this candidate GSSP level in the Tethyan realm (Leereveld, 1995; Mutterlose *et al.*, 1996). Magnetostratigraphic sections in Italy were originally interpreted so that the base of the Hauterivian was nearly coincident with the beginning of magnetic polarity Chron M11n (Channell *et al.*, 1995a). This interpretation was later revised to assign the base of the Hauterivian to be near the base of the next higher Chron M10n (e.g., Weissert *et al.*, 1998). This magnetostratigraphic placement was supported by later studies (e.g., McArthur *et al.*, 2007), and is consistent with the cycle-scaled duration of the Valanginian.

Upper Hauterivian Substage

The traditional base of the Upper Hauterivian substage in the Tethyan realm was the highest occurrence of Neocomitinae ammonoids and lowest occurrence of the ammonite *Subsaynella sayni*, but these events cannot be recognized in the Boreal realm. Therefore, it was recommended that the highest occurrence of the calcareous nannofossil *Cruciellipsis cuvillieri* should serve as the global marker for the base of the Upper Hauterivian (Mutterlose *et al.*, 1996). The highest occurrence of *C. cuvillieri* has been reported from the middle of the *Subsaynella sayni* ammonite zone (Bergen, 1994) and is near the base of magnetic polarity zone M8r in Italy (Channell *et al.*, 1993, 1995a). The section at La Charce that is being considered as the Hauterivian GSSP could also be a candidate for the Lower/Upper substage boundary section (Subcommission on Cretaceous Stratigraphy, 2009).

27.1.2.4. Barremian

History, Definition and Boundary Stratotype

The original Barremian concept of Coquand (1861) with a type locality near Barrême (Alpes-de-Haute-Provence, southeast France) was a vague assignment to an assemblage of belemnites and ammonoids that also encompassed the current Upper Hauterivian substage. Busnardo (1965) redefined the Barremian Stage using the roadside exposures of Angles in southeast France as the type locality.

The base of the Barremian is marked by the lowest occurrence of the *Taveraidiscus hugii* auctorum (formerly called "*Spitidiscus*" *hugii*) (Hoedemaeker *et al.*, 2003) and the *Avramidiscus vandeckii* ammonite group. A candidate for the global boundary stratotype is at Río Argos near Caravaca, Spain (Company *et al.*, 1995, Rawson *et al.*, 1996a). The boundary interval in Italy falls within the uppermost part of magnetic polarity zone M4n (at approximately Chron M4n.8 = "M5n.8") (Bartolocci *et al.*, 1992; Channell *et al.*, 1995a). Correlation to the Boreal realm by dinoflagellate cysts indicates that this base of the Barremian Stage is approximately at the base of the "*Hoplocrioceras*" *rarocinctum* ammonite zone (Leereveld, 1995; Rawson *et al.*, 1996a).

Upper Barremian Substage

The boundary between the Lower and Upper Barremian is placed at the lowest occurrence of the ammonite *Ancyloceras vandenheckei* (Rawson *et al.*, 1996a; Reboulet *et al.*, 2011). The proposed substage GSSP is near Caravaca, Spain, (Company *et al.*, 1995). This substage boundary level is within the uppermost part of magnetic polarity zone M3r (at approximately Chron M3r.8) in Italy (Bartolocci *et al.*, 1992; Channell *et al.*, 1995a) and is correlated to the lower part of the *Paracrioceras elegans* ammonite zone of the Boreal realm (Leereveld, 1995; Rawson *et al.*, 1996a).

27.1.2.5. Aptian

History, Definition and Boundary Stratotype

The Aptian Stage was a vague designation by Alcide d'Orbigny (1840) for strata containing "Upper Neocomian" fauna and named after the village of Apt (Vaucluse province, southeast France). The extent and nomenclature of the Aptian and underlying stages have undergone major revisions (vividly reviewed by Moullade *et al.*, 2011). The French sections are poor in ammonoids; therefore, the classical marker for the base of the Aptian was the lowest occurrence of the deshayesitid ammonite *Prodeshayesites* in northwest Europe (Rawson, 1983; Moullade *et al.*, 1998a,b; Hoedemaeker *et al.*, 2003). However, the local lowest occurrence of this ammonite genus is commonly associated with a major progressive transgression in earliest Aptian, and virtually no ammonoid-bearing section represents a continuous and complete Barremian–Aptian boundary interval (Erba *et al.*, 1996). Therefore, the proposed primary marker for base of the Aptian Stage is the beginning of magnetic polarity Chron M0r. A quite different boundary concept was proposed by Moullade *et al.* (2011), who on historical grounds recommended converting the lower three ammonite zones of the current Aptian into a Bedoulian Stage, and beginning the Aptian at the base of the *Dufrenoyia furcata* Zone (marked by lowest occurrence of *Dufrenoyia* ammonite genus) with a GSSP near Roquefort-La Bédoule in southeastern France. Reboulet *et al.* (2011) reject the usage of Bedoulian (see below under "Substages of Aptian"). The GTS2012 scale retains the current Aptian/Barremian usage with its working boundary definition as the base of Chron M0r.

The proposed global boundary stratotype in pelagic limestone at Gorgo a Cerbara in central Italy has an integrated stratigraphy of paleomagnetics, biostratigraphy (calcareous nannofossils, planktonic foraminifers, radiolarians, dinoflagellates) carbon isotope chemostratigraphy, and cycle stratigraphy (Erba *et al.*, 1996, 1999; Channell *et al.*, 2000). Cyclic stratigraphy implies that the lowest occurrence of a *Prodeshayesites* ammonite in uppermost polarity zone M0r occurs about 300 000 years after the boundary (Herbert, 1992; Channell *et al.*, 1995a; Erba *et al.*, 1996). An important event about 1 myr after the boundary is Oceanic Anoxic Event "OAE1a", marked by widespread organic-rich shale (e.g., Selli level in Italy, Goguel level in southeast France, Fischschiefer in northwestern Germany) and the beginning of a positive carbon-13 isotopic excursion (Weissert and Lini, 1991; Mutterlose and Böckel, 1998; Li *et al.*, 2008). This carbon isotope signature provides a correlation between magnetic Chron M0r and the basal Aptian *Deshayesites oglanlensis* ammonite zone in southeastern France (Subcommission on Cretaceous Stratigraphy, 2009).

Substages of the Aptian

Two substages are recommended for subdividing the Aptian Stage. The Lower/Upper substage boundary is at the base of the *Epicheloniceras martinoides* ammonite zone in the standard Tethyan zonation which is approximately equivalent to the base of the *Tropaeum bowerbanki* ammonite zone of the Boreal zonations (Reboulet *et al.*, 2011). At this level, there is an important change in ammonite fauna in both the Tethyan and Sub-Boreal realms.

This level coincides with a traditional boundary between a lower "Bedoulian" and an upper "Gargasian" stage, which were based on sections at Cassis-La Bédoule (Bouches du Rhône province; near Marseilles) and at Gargas (near Apt) in southern France, although historical usage of "Bedoulian" was only for the lower portion of this Lower Aptian (Moullade *et al.*, 2011). A comprehensive synthesis has been compiled for the integrated stratigraphy of the type Bedoulian (Moullade *et al.*, 1998a,b; Ropolo *et al.*, 1998) and type Gargasian (Moullade and Tronchetti, 2004; Moullade *et al.*, 2004, 2005; and later articles in *Notebooks on Geology*). The lowest occurrence of the planktonic foraminifer *Praehedbergella luterbacheri* is just above the substage boundary followed by the lowest *Globigerinelloides ferreolensis* (Moullade *et al.*, 2005), and the lowest calcareous nannofossil *Eprolithus floralis* (base of Zone NC7) slightly precedes the boundary (Moullade *et al.*, 1998b).

In France, an additional uppermost substage of "Clansayesian" was added when Breistroffer (1947) moved the thin "Clansayes" horizon from Albian into the underlying Aptian. However, this substage was not considered useful because:

(a) The reference sections at Gargas (near Apt) and Clansayes are not suitable for correlation purposes (Rawson, 1983);
(b) The rationale for a separate "Clansayesian" substage is questioned (e.g., Owen, 1996a; Casey *et al.*, 1998); and
(c) The Aptian/Albian boundary has not yet been defined and could potentially include a large part of the Clansayesian interval.

The IUGS Lower Cretaceous Ammonite Working Group or "Kilian Group" voted to:

'abandon the terms Bedoulian, Gargasian and Clansayesian as they are not recognized internationally, but mainly used in France … Moreover, the type sections of these French substages do not offer good prospects (low number and/or bad preservation of ammonoids) …'

(reported in Reboulet *et al.*, 2011)

27.1.2.6. Albian

History, Definition and Boundary Stratotype

The Albian Stage was named after the Aube region (Roman name was *Alba*) in northeast France (d'Orbigny, 1842). The placement of the Aptian/Albian boundary has a convoluted history and is currently undecided. The main sets of proposed microfossil or macrofossil markers are near one of three organic-rich episodes in Europe, each of which have been calibrated to cycle stratigraphy. The working version used in GTS2012 is near the middle of these three episodes.

Prior to 1947, the base of the Albian was assigned as the base of the *Nolaniceras nolani* ammonite zone. Breistroffer (1947) shifted this zone and the overlying *Hypacanthoplites jacobi* ammonite zone into an expanded uppermost Aptian, thereby placing the base of the Albian in the Northwest European faunal province of the Boreal realm at the base of the *Leymeriella tardefurcata* ammonite zone with a basal level characterized by *Leymeriella schrammeni*. However, this boundary interval in western Europe is marked by endemic ammonites, and the occurrences of the earliest *Leymeriella* ammonite species (e.g., *L. schrammeni*) appear to be restricted to northern Germany (Casey, 1996; Hart *et al*., 1996; Mutterlose *et al*., 2003). This choice for an Aptian−Albian boundary interval also coincides with a widespread hiatus associated with a major sequence boundary; therefore, the *L. schrammeni* ammonite subzone is represented by sediments in only isolated localities (e.g., the Harz foredeep of north Germany and in north Greenland) (H. Owen, pers. comm., 1999). An important section associated with the lowest occurrence of the earliest "anterior" form of *L. schrammeni* is at Vörhum, near Hannover in northern Germany (Hart *et al*., 1996; Mutterlose *et al*., 2003). However, Mutterlose *et al*. (2003) conclude that defining "the base of the Albian on these criteria cannot be used in a global context", therefore an alternate biostratigraphic or non-biostratigraphic marker must be identified. In southeast England, the base of the Albian was placed above a transgressive condensation horizon at the lowest occurrence of ammonite *Farnhamia farnhamensis* (Casey, 1961; Hesselbo *et al*., 1990; Hart *et al*., 1996). Whether the *F. farnhamensis* ammonite subzone is equivalent to the *L. schrammeni* Subzone of northern Germany is debated (Owen, 1996a, b; Casey, 1996). As a result, ammonite-based definitions for the base of the Albian Stage and its component substages are controversial (reviewed in Hancock, 2001, and in Owen, 2002).

The Aptian−Albian boundary interval in deep-water facies of the Tethyan realm is commonly in clay-rich facies punctuated by four main widespread organic-rich horizons and associated carbon isotope excursions − the "Jacob", "Kilian", "Paquier" (called "Urbino" in Italy) and "Leenhardt" levels in reference sections of southeastern France. An array of options are being considered as potential definitions for the base of the Albian (summarized from Hart *et al*., 1996; Mutterlose *et al*., 2003; Gale *et al*., 2011; Petrizzo *et al*., in press). The options are listed in approximate descending stratigraphic order:

(a) Base of the *Lyelliceras lyelli* ammonite subzone (base of *Hoplites dentatus* Zone) of the Sub-Boreal realm, which has been traditionally assigned as the base of the Middle Albian substage (Hancock, 2001; Owen, 2002). This relatively extreme suggestion was to resolve the ambiguous global correlation using ammonites near the traditional Aptian/Albian boundary, and would shorten the Albian by ~3 million years.

(b) Lowest occurrence of ammonite *Douvilleiceras* ex. gp. *mammillatum* (which is the top of the *Leymeriella tardefurcata* Zone). This may be near the "Leenhardt" organic-rich level (Gale *et al*., 2011).

(c) "Paquier/Urbino" organic-rich event (coinciding with "oceanic anoxic event" OAE 1b). The beginning of this event is nearly coincident with the lowest occurrence of ammonite *Leymeriella tardefurcata* (Kennedy *et al*., 2000a), but may be difficult to reliably correlate outside of southeastern France. This was advocated by Gale *et al*. (2011; Figure 60); and, according to their cycle stratigraphy correlations, would shorten the traditional "Early Albian" to less than 1 million years. However, a proposal in 2007 to place the Albian GSSP in the Tartonne section at the lowest occurrence of ammonite *Leymeriella tardefurcata* and the nearly coincident base of the Paquier organic-rich episode did not receive a quorum vote from the Albian working group. Rejection reasons included the perceived low stratigraphic value of ancillary biostratigraphic markers (ammonites, foraminifers, etc.) (Subcommission for Cretaceous Stratigraphy, 2009).

(d) Highest occurrence of ammonite *Hypacanthoplites jacobi*.

(e) Lowest occurrence of the earliest "anterior" form of *Leymeriella schrammeni*, with a reference section at Vörhum in northern Germany (Hart *et al*., 1996; Mutterlose *et al*., 2003). A volcanic ash just above this level was dated by U-Pb as 113.1 ± 0.3 Ma (Selby *et al*., 2009). The *L. schrammeni* zone of the Boreal realm precedes the regional *Leymeriella tardefurcata* Zone, and overlaps with the upper extend of *Hypacanthoplites jacobi* (Mutterlose *et al*., 2003).

(f) Microfossil datums near the "Kilian" organic-rich bed:

  (i) The lowest occurrence of calcareous nannofossil *Praediscosphaera columnata* (subcirculular form; called *P. cretacea* in some earlier studies), which is the base of nannofossil zone NC8, may be just above this Kilian level according to calibrations to an Italian cycle stratigraphy reference section (e.g., Fiet *et al*., 2001; Grippo *et al*., 2004; Huang *et al*., 2010). This nannofossil event has been a candidate marker for global boundary correlation (e.g., Hart *et al*., 1996; Owen, 2002), and may occur near the lowest occurrence of the ammonite *L. schrammeni* in Germany (Mutterlose *et al*.,

2003). However, the evolution from elliptical to round is a transition, so it is difficult to unambiguously differentiate the "subcircular" *P. columnata* from its ancestors or descendants (e.g., Kennedy *et al.*, 2000a; Mutterlose *et al.*, 2003; P. Bown, pers. comm., 2011).

(ii) An abrupt turnover in planktonic foraminifers begins with the extinction of *Paraticinella eubejaouaensis* planktonic foraminifer (a commonly used base-Albian working definition among microfossil workers) just below the Kilian level in southeastern France (e.g., Petrizzo *et al.*, in press). The lowest occurrence of the planktonic foraminifer *Microhedbergella miniglobularis* is within the Kilian level, is followed by the lowest occurrence of *Mi. renilaevis* in the middle of the Kilian episode, then the lowest occurrence of *Mi. rischi* (Huber and Leckie, 2011; Huber *et al.*, 2011; Petrizzo *et al.*, in press). This succession of close-spaced foraminifer events is also observed in drill sites within the southern and northern Atlantic and within the Indian Ocean.

(g) "Jacob" organic-rich shale and coeval carbon-13 isotopic excursion; which, according to cycle stratigraphy (Huang *et al.*, 2010) is about 2 myr before the FAD of the subcircular *Praediscosphaera columnata* nannofossil.

(h) base of *Hypacanthoplites jacobi* ammonite zone (first occurrence of *Hypacanthoplites* ammonite genus). This proposal would shift the informal "Clansayesian" interval, that Breistroffer (1947) had assigned as uppermost Aptian, back into the lowermost Albian to restore the historical usage of Albian (Moullade *et al.*, 2011).

In GTS2004 (Gradstein *et al.*, 2004), the base of the Albian Stage was assigned to the lowest occurrence of the calcareous nannofossil *Praediscosphaera columnata* (subcircular form). This Aptian/Albian boundary assignment (base of nannofossil zone NC8) is retained in GTS2012. Despite problems in precision taxonomy caused by the evolutionary succession in morphology, the datum can be put into a temporal context based on cycle stratigraphy spanning the entire Albian–Aptian (e.g., Huang *et al.*, 2010), is near the "Kilian" carbon isotope excursion and the base of the *Microhedbergella rischi* foraminifer zone, can be applied to global oceanic sediments, and is approximately dated by a precise U-Pb age in Germany (Selby *et al.*, 2009). It is possible that one of the events in the turnover of planktonic foraminifer species in this same "Kilian" interval will be utilized as the main base-Albian reference marker (e.g., Huber *et al.*, 2011; Petrizzo *et al.*, in press).

However, considering the range of current proposals, the Aptian–Albian boundary may eventually be ratified with a definition that is much older by ~2 myr, or younger by ~2 myr.

Substages of the Albian

The traditional base of the Middle Albian is placed at the lowest occurrence of the ammonite *Lyelliceras lyelli* (base of *Hoplites dentatus* Zone) (Gale *et al.*, 2011). The appearance of this ammonite in the European marginal basins is a temporary incursion of a more southern and cosmopolitan form (Owen, 1996a). A proposed boundary stratotype lies within dark gray clay-rich strata near St-Dizier, northern France (Hart *et al.*, 1996; Hancock, 2001).

The base of the Upper Albian is assigned to the lowest occurrence of the ammonite *Dipoloceras cristatum* (Hart *et al.*, 1996; Gale *et al.*, 2011). This event commonly coincides with a transgression following a major sequence boundary (e.g., Hesselbo *et al.*, 1990; Amédro, 1992; Hardenbol *et al.*, 1998), thereby causing the Middle/Upper Albian substage boundary interval to be quite condensed and probably incomplete at the key sections along the English Channel at Wissant (Pas-de-Calais province, northwest France) and at Folkstone (Kent, United Kingdom) (Hart *et al.*, 1996). Therefore, an expanded basinal clay-rich section at Col de Palluel in southeast France was extensively studied as a substage GSSP candidate and tied to cyclostratigraphy (Gale *et al.*, 2011).

27.1.3. Subdivisions of the Upper Cretaceous

The majority of the Late Cretaceous subdivisions were derived from facies successions in marginal marine to chalk facies within eastern France and the Netherlands. The original Turonian and Senonian stages of d'Orbigny (1847) were progressively subdivided into the current six stages. However, few of the classical stratotypes are suitable for placing the limits of the stages.

A diverse array of primary markers are used or under consideration for defining stage and substage boundaries, including ammonoids, inoceramid bivalves, planktonic foraminifers, pelagic crinoids and magnetic polarity chrons.

27.1.3.1. Cenomanian

History, Definition and Boundary Stratotype

In 1847, d'Orbigny converted the lower portion of his original Turonian into a Cenomanian Stage and assigned the type region as the vicinity of the former Roman town of Cenomanum, now called Le Mans (Sarthe region, northern France) (d'Orbigny, 1847, p. 270). The conventional ammonoid markers for the base of the Cenomanian were the lowest occurrence of the acanthoceratid genus *Mantelliceras*, which was derived from the genus *Stoliczkaia* of

Upper Albian, or the lowest occurrence of the genus *Neostlingoceras*.

Because these ammonoid groups are relatively rare in many regions (Hancock, 1991), the Cenomanian Working Group selected the lowest occurrence of the planktonic foraminifer, *"Rotalipora" globotruncanoides* (= *R. brotzeni* of some studies and now classified as *Thalmanninella globotruncanoides*), as the basal boundary criterion for the Cenomanian Stage. This foraminifer level is slightly lower than the lowest occurrence of the Cenomanian ammonoid marker of *Mantelliceras mantelli*. The Mont Risou section in southeast France was chosen as the GSSP section (Tröger and Kennedy, 1996; Gale *et al.*, 1996) (Figure 27.3) and ratified in 2002 (Kennedy *et al.*, 2004). In many regions, the Albian–Cenomanian boundary interval is coincident with a widespread hiatus and condensation associated with a major sequence boundary (e.g., Tröger and Kennedy, 1996; Hardenbol *et al.*, 1998).

Substages of the Cenomanian

The sudden entry of the ammonite genera *Cunningtoniceras* and *Acanthoceras s.s.* is a major biostratigraphic event in Europe. The base of the Middle Cenomanian is currently placed at the lowest occurrence of the ammonite *Cunningtoniceras inerme* at the Southerham Gray Quarry in the Sussex province of England) (Tröger and Kennedy, 1996). The lowest occurrences of ammonite *Acanthoceras rhotomagense* and the beginning of a positive carbon-13 isotope excursion are approximately 5 couplets higher (~100 kyr) (Gale, 1995). This Lower/Middle Cenomanian boundary interval is missing over large regions due to its coincident with a major sequence boundary (e.g., Hardenbol *et al.*, 1998).

The replacement of *Acanthoceras* ammonites by the *Calycoceras* genus is commonly used to mark the base of the Upper Cenomanian (Hancock, 1991). A marker for the base of the Upper Cenomanian has not yet been selected, but the placement will probably be near the limits of, or within, the *Acanthoceras jukesbrownei* ammonite zone (Tröger and Kennedy, 1996). This zone is coeval with the *Dunveganoceras pondi* Zone of the Western Interior, the base of which is used by Cobban *et al.* (2006) for their Upper/Middle substage boundary.

27.1.3.2. Turonian

History, Definition and Boundary Stratotype

The concept of the Turonian Stage has undergone continual redefinition (reviewed in Bengtson *et al.*, 1996). The Turonian proposed by d'Orbigny in 1842 was later divided by him (d'Orbigny, 1847) into a lower Cenomanian Stage and an upper Turonian Stage. The name is derived from the Tours or Touraine region of France (Turones and Turonia of the Romans), and d'Orbigny (1852) clarified his later definition by selecting a type region lying between Saumur (on the Loire river) and Montrichard (on the Cher river). In this region, the lowest Turonian formation contains the ammonite *Mammites nodosoides*, therefore its lowest occurrence was considered the marker for the base of the Turonian Stage (e.g., Harland *et al.*, 1990). Below this level is a worldwide oceanic anoxic event (OAE 2), a major positive excursion in carbon-13 isotopes, and a mass extinction of over half of ammonoid and brachiopod genera (e.g., Schlanger *et al.*, 1987; Kerr, 1998).

After considering several potential placements, the Turonian Working Group placed the base of the Turonian at the lowest occurrence of the ammonite *Watinoceras devonense* (two ammonite zones below the *Mammites nodosoides* Zone) near the global oceanic anoxic event (Bengtson *et al.*, 1996). The global boundary stratotype section and point is at Rock Canyon Anticline, east of Pueblo (Colorado, west-central USA) (Kennedy and Cobban, 1991; Bengtson *et al.*, 1996; Kennedy *et al.*, 2000b, 2005) (Figure 27.4) and was ratified in 2003. The maximum major carbon isotope peak associated with the oceanic anoxic event occurs 0.5 m above the boundary. The age of the Cenomanian–Turonian boundary is well constrained by $^{40}Ar/^{39}Ar$ ages from bentonites as 94 Ma (Obradovich, 1993; Meyers *et al.*, 2010).

Substages of the Turonian

The base of the Middle Turonian is marked by the lowest occurrence of ammonite *Collignoniceras woollgari*. The candidate for the GSSP is the base of Bed 120 in the Rock Canyon Anticline section, approximately 5 m above the GSSP defining the base of the Turonian Stage in the same section (Bengtson *et al.*, 1996; Kennedy *et al.*, 2000b).

The base of the Upper Turonian is not yet formalized, but potential datums are the lowest occurrences of the ammonite *Subprionocyclus neptuni* in the Sub-Boreal realm (e.g., Germany, England), of the ammonite *Romaniceras deverianum* in the Tethys realm (e.g., southern France, Spain), of the ammonite *Scaphites whitfieldi* in the North American Western Interior (as used by Cobban *et al.*, 2006, and GTS2012) or of an inoceramid bivalve, *Inoceramus perplexus* (= *Mytiloides costellatus* in some studies) (Bengtson *et al.*, 1996; Wiese and Kaplan, 2001).

27.1.3.3. Coniacian

History, Definition and Boundary Stratotype

Coquand (1857a,b) defined the Coniacian Stage with the type locality at Richemont Seminary near Cognac (Charente province, northern part of the Aquitaine Basin, western France). In this region, basal Coniacian glauconitic sands overlie Turonian rudistid-bearing limestones. The entry of ammonoid *Forresteria (Harleites) petrocoriensis* was used to mark the base of the Coniacian Stage. However, there can be problems in identifying this species, and ammonoids can be rare or absent in important Coniacian sections (Hancock,

Base of the Cenomanian Stage of the Cretaceous System, Mont Risou, Hautes-Alpes, France.

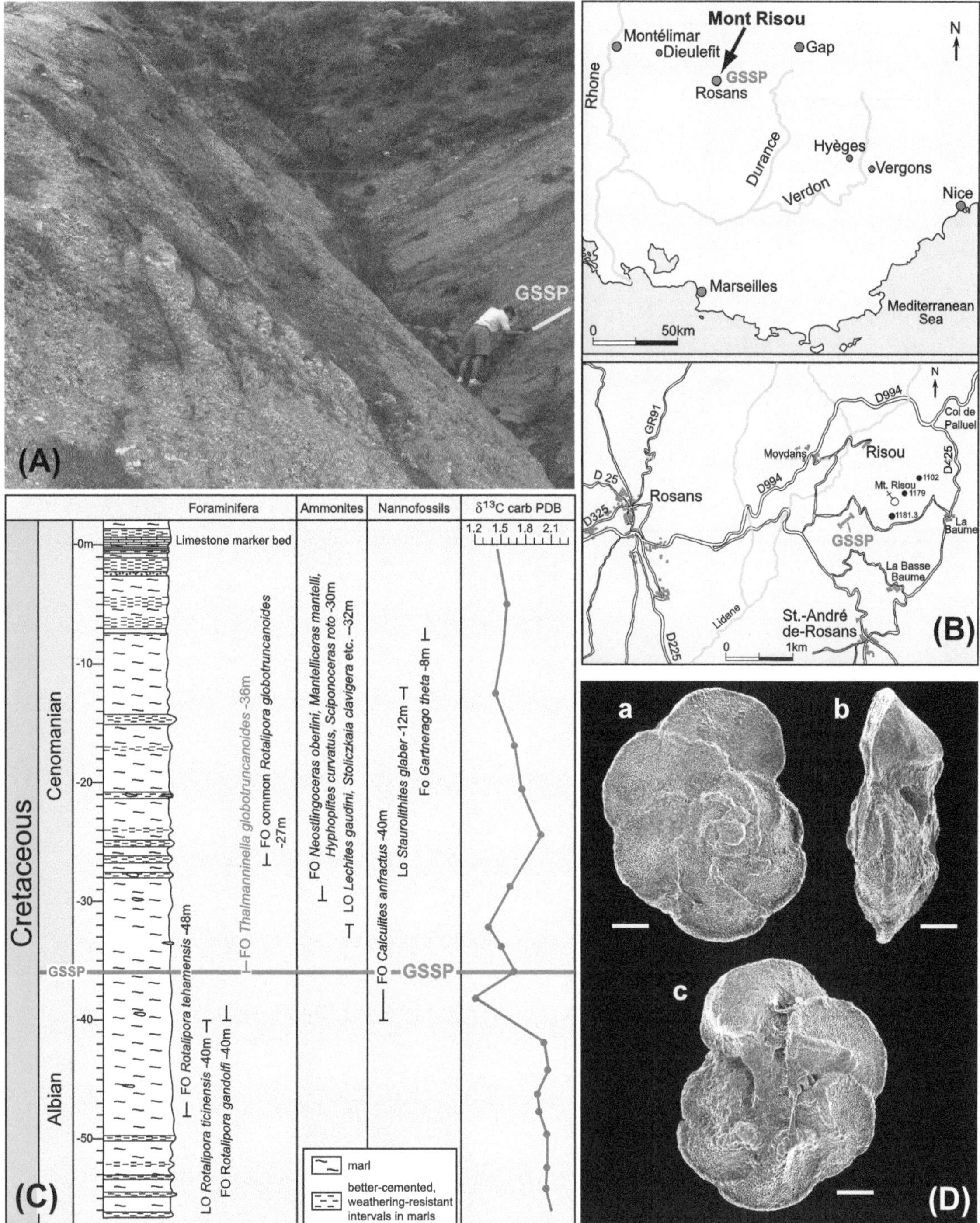

FIGURE 27.3 **GSSP for base of the Cenomanian Stage at Mont Risou, southeastern France.** The GSSP level coincides with the lowest occurrence of the planktonic foraminifera *Rotalipora globotruncanoides* (now classified as "*Thalmanninella globotruncanoides*"). Photo of the marker foraminifer (provided by Atsushi Ando) is the metatype of *globotruncanoides* from the type locality (Ando and Huber, 2007).

1991; Kauffman *et al.*, 1996). Therefore, the Coniacian Working Group redefined the Coniacian Stage and its substage boundaries using lowest occurrences of widespread inoceramid bivalves that lived in the chalk seas.

The proposed marker for the base of the Coniacian is the lowest occurrence of inoceramid bivalve *Cremnoceramus deformis erectus* (= *C. rotundatus* (*sensu* Tröger *non* Fiege)), which is above the lowest occurrence of *F. (H.) petrocoriensis*

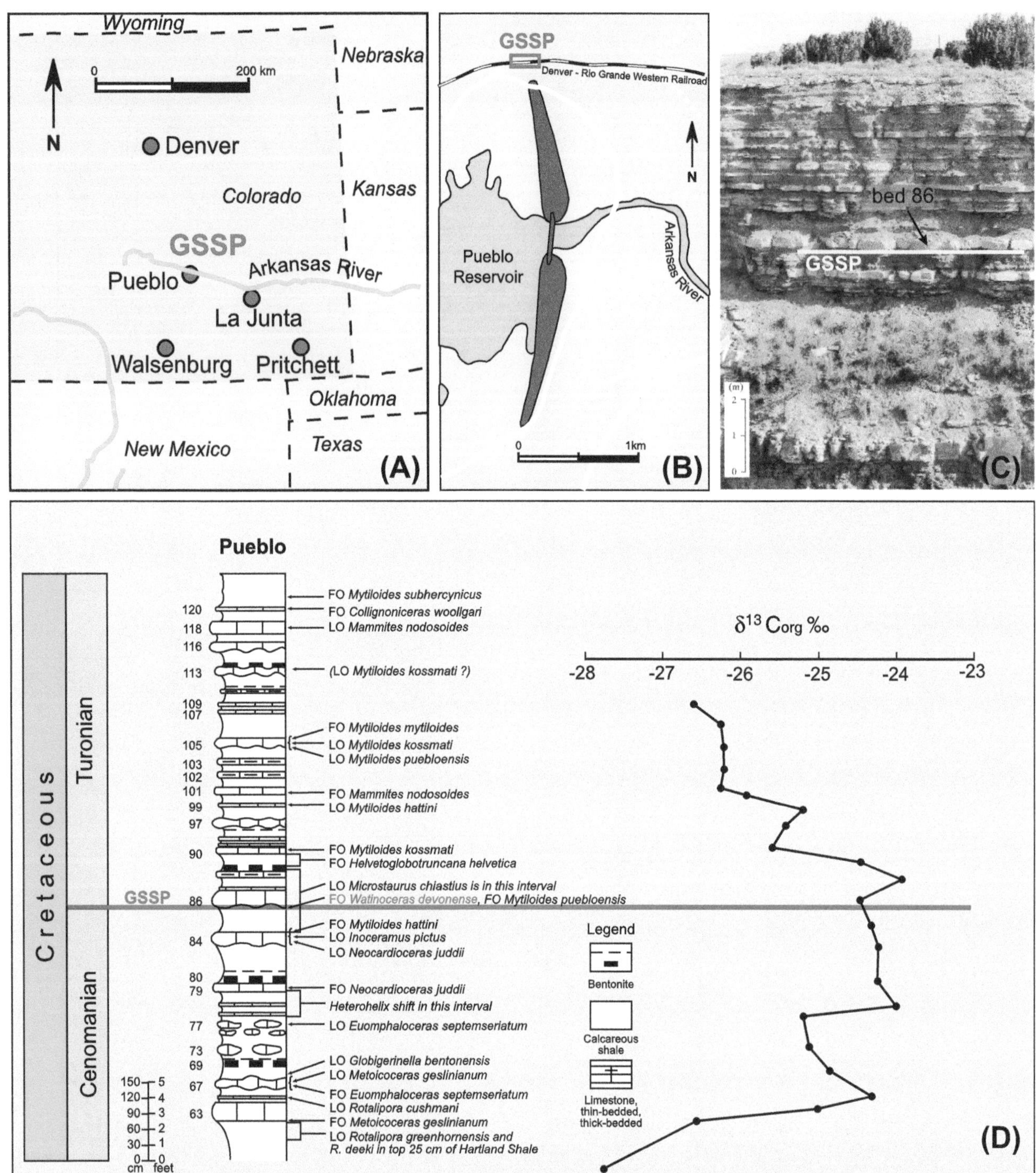

FIGURE 27.4 GSSP for base of the Turonian Stage near Pueblo, Colorado, USA. The GSSP level coincides with the lowest occurrence of the ammonite *Watinoceras devonense. Photo provided by Jim Kennedy.*

in Europe. This inoceramid is present throughout the Europe–American biogeographic province and in the Tethyan realm, and can be easily correlated to the North Pacific and southern hemisphere (Walaszczyk *et al.*, 2010).

A candidate boundary stratotype is the active Salzgitter-Salder Quarry, southwest of Hannover (Lower Saxony province, northern Germany) with an extensive upper Turonian through lower Coniacian succession with macrofossils, microfossils and geochemistry (Kauffman *et al.*, 1996; Walaszczyk and Wood, 1998; Walaszczyk *et al.*, 2010). However, at Salzgitter-Salder, the boundary interval is condensed and may contain a minor hiatus in uppermost Turonian. The most complete succession across the actual boundary is at Słupia Nadbrzeżna in central Poland (Walaszczyk and Wood, 1998; Walaszczyk *et al.*, 2010), but it is limited to only a very brief interval spanning the latest Turonian and earliest Coniacian and is poorly exposed. Therefore, a composite GSSP is being proposed that will use both of these reference sections.

The lowest occurrence of inoceramid bivalve *Cremnoceramus deformis erectus* correlates approximately to the base of the *Scaphites preventricosus* ammonite zone of the North American Western Interior basin (Cobban *et al.*, 2006). Bentonites from this ammonite zone and the uppermost Turonian zone have yielded ^{40}Ar/^{39}Ar ages indicating a boundary age of 90 Ma (Obradovich, 1993; Siewert, 2011; Siewert *et al.*, in press).

Substages of the Coniacian

The base of the Middle Coniacian is placed at the lowest occurrence of the inoceramid bivalve genus *Volviceramus*, which is at or near the lowest occurrence of ammonoid *Peroniceras (Peroniceras) tridorsatum* (Kauffman *et al.*, 1996). Potential boundary stratotypes are near Dallas-Fort Worth (Texas, southern USA) or Seaford Head in southern England. The base of the Upper Coniacian is placed at the lowest occurrence of the inoceramid bivalve *Magadiceramus subquadratus* (Kauffman *et al.*, 1996). No stratotypes have yet been proposed for this substage boundary.

27.1.3.4. Santonian

History, Definition and Boundary Stratotype

The Santonian Stage was named after Saintes (Cognac, southwest France) by Coquand (1857b), who placed the lower boundary at a glauconitic hardground capping Santonian chalk.

The sudden turnover from *Magadiceramus* inoceramid bivalves to the lowest occurrence of the widespread inoceramid *Cladoceramus* (=*Platyceramus*) *undulatoplicatus* has been selected as the marker for the base of the Santonian (Lamolda *et al.*, 1996). No other significant calcareous microfossil or nannofossil datums occur near this level – significantly below are the lowest *Lucianorhabdus cayeuxii* nannoplankton (base of Zone UC11c) and lowest rare *Dicarinella asymetrica* planktonic foraminifers (base of *D. asymetrica* Zone) (Lamolda *et al.*, 2011; Gale *et al.*, 2007). (The FAD of *Dicarinella asymetrica* planktonic foraminifers had been a competing candidate marker for the boundary.) There are lesser and some newly described calcareous nannofossils that may provide correlation into other facies (Blair and Watkins, 2009).

The leading GSSP candidate is a working quarry to the south of Olazagutia in the Navarra region of Spain (D'Hondt *et al.*, 2007, special issue of *Cretaceous Research*; Lamolda *et al.*, 2011). The Olazagutia section is not ideal – the biostratigraphic record may be incomplete and the abandoned quarry wall might be unsuitable for future sampling. An alternative candidate is Ten Mile Creek, Dallas County, Texas (Gale *et al.*, 2007), where portions of the record are a composite of overlapping sections. The Olazagutia section was approved as the GSSP candidate by the Subcommission on Cretaceous Stratigraphy, and submitted to the ICS in early 2012.

The lowest occurrence of the inoceramid *Cladoceramus undulatoplicatus* boundary marker is just below the base of the *Clioscaphites saxitonianus* ammonite zone of the North American Western Interior (Walaszczyk and Cobban, 2007). A set of ^{40}Ar/^{39}Ar ages from bracketing bentonites indicate a boundary age near 86 Ma (Obradovich, 1993; Siewert, 2011; Siewert *et al.*, in press).

Substages of the Santonian

The traditional Santonian has three substages, but no markers for boundary stratotypes have yet been formalized. A possible datum for the base of the Middle Santonian is the extinction of the same *Cladoceramus undulatoplicatus* inoceramid bivalve that marks the Coniacian–Santonian boundary (Lamolda *et al.*, 1996).

The lowest occurrence of stemless crinoid *Uintacrinus socialis* is commonly used to place the base of the Upper Santonian, and this level is near the FAD of nannofossil *Arkhangelskiella cymbifomis* (Lamolda *et al.*, 1996; Dhont *et al.*, 2007). Magnetostratigraphy of chalk successions in southern England suggested that the base of this *Uintacrinus socialis* Zone is at or just below the base of reversed-polarity Chron C33r (Montgomery *et al.*, 1998), but this has not yet been confirmed in other sections. If the magnetostratigraphy is supported and the base of Chron C33r is selected as the main base-Campanian marker, then the traditional "Upper Santonian" in the English Chalk would be reassigned as lowermost Campanian (A. Gale, pers. comm., 2011).

27.1.3.5. Campanian

History, Definition and Boundary Stratotype

The Campanian Stage of Coquand (1857b) was named after the hillside exposures of Grande Champagne near Aubeterre-sur-Dronne (45 km west of Périgueux, northern Aquitaine province, France), but the bulk of the type "Campanian" at Aubeterre is now classified as Maastrichtian (e.g., van Hinte, 1965; Séronie-Vivien, 1972).

The lower part of the type section had no obvious base to the shallow-water limestone formation. The base of the Campanian was placed at the lowest occurrence of ammonoid *Placenticeras bidorsatum* by de Grossouvre (1901), but this extremely rare species is not a practical marker (reviewed in Hancock and Gale, 1996). In contrast, stemless crinoids of the *Uintacrinus* and *Marsupites* genera, with their planktonic or benthic habitat, have a near-global distribution in shelf chalks (Gale *et al.*, 1995). Therefore, the extinction of crinoid *Marsupites testudinarius* has been a provisional boundary marker for the base of the Campanian Stage (Hancock and Gale, 1996), although the occurrence of this taxon is restricted to certain environments. The base of the traditional Campanian is at or within the lower portion of reversed-polarity Chron C33r (e.g., Montgomery *et al.*, 1998). Therefore, the Campanian Working Group is considering using the beginning of Chron C33r as the primary boundary definition, thereby enabling global recognition in pelagic, continental and other non-shallow-marine settings (A. Gale, pers. comm. and discussions at NERC Workshop on Late Cretaceous, University College London, June 2011). A side effect might be to shift the traditional "upper Santonian" of the English Chalk succession to the lowermost Campanian.

Potential boundary stratotypes might be the Waxahachie dam spillway section (north central Texas) if the crinoid marker is selected (Gale *et al.*, 2008), or, if the base of C33r is chosen, perhaps the Gubbio section in Italy with its classic compilation of bio-magnetostratigraphy (e.g., Alvarez *et al.*, 1997; Lowrie and Alvarez, 1977).

The base-Campanian is generally correlated to the base of the *Scaphites leei III* ammonite zone of the North American Western Interior (e.g., Cobban *et al.*, 2006). The age of the base of this ammonite zone is constrained by $^{40}Ar/^{39}Ar$ dates to be between 83 and 84 Ma (Obradovich, 1993; Siewert, 2011; Siewert *et al.*, in press).

Substages of the Campanian

The Campanian is generally subdivided into Lower, Middle and Upper substages of approximately equal duration, but there are not yet any formal recommendations for primary markers or boundary stratotypes for substages. In the North American Western Interior, these substage bases are informally placed by Cobban (1993; and Cobban *et al.*, 2006) as the lowest occurrences of ammonoids *Baculites obtusus* and *Didymoceras nebrascense*, respectively. In northwest Europe, the base of the belemnite *Belemnitella mucronata* is often used for a single Upper/Lower Campanian division marker, and this level projects slightly below the "Middle/Lower" Campanian boundary in North American usage (A. Gale, Aug. 2011 correlation chart to J. Ogg).

27.1.3.6. Maastrichtian

History, Definition and Boundary Stratotype

The Maastrichtian Stage was introduced by Dumont (1849) for the "Calcaire de Maastricht" with a type locality at the town of Maastricht (southern Netherlands, near the border with Belgium). The stratotype was fixed by the Comité d'étude du Maastrichtian as the section of the Tuffeau de Maastricht exposed in the ENCI company quarry at St. Pietersberg on the outskirts of Maastricht, but this local detrital-carbonate facies would correspond only to part of the upper Maastrichtian in current usage (reviewed in Rawson *et al.*, 1978, and Odin, 2001). A revised concept of the Maastrichtian Stage was based on belemnites in the white chalk facies. Accordingly, the base of the stage was assigned to the lowest occurrence of belemnite *Belemnella lanceolata*, with a reference section in the chalk quarry at Kronsmoor (50 km northwest of Hamburg, north Germany (e.g., Birkelund *et al.*, 1984; Schönfeld *et al.*, 1996). The lowest occurrence of ammonoid *Hoploscaphites constrictus* above this level provided a secondary marker. Comparison of strontium isotope stratigraphy and indirect correlations by ammonoids indicate that this level is approximately equivalent to the base of the *Baculites eliasi* ammonoid zone of the North American Western Interior (Landman and Waage, 1993; McArthur *et al.*, 1992).

However, the belemnite *Belemnella lanceolata* is not a useful marker into the Tethyan faunal realm, whereas the ammonoid *Pachydiscus neubergicus* has a much wider geographical distribution (reviewed in Hancock, 1991). Therefore, in a mixed decision, the Maastrichtian Working Group recommended the base of the Maastrichtian to be assigned as the lowest occurrence of ammonoid *Pachydiscus neubergicus* (Odin *et al.*, 1996).

The ratified GSSP boundary is in an abandoned quarry near the village of Tercis les Bains in southwest France, at 90 cm below a coincident lowest occurrence of *Pachydiscus neubergicus* and *Hoploscaphites constrictus* ammonoids (Odin, 1996, 2001; Odin and Lamaurelle, 2001) (Figure 27.5). The GSSP level was selected as arithmetic mean of 12 biohorizons with correlation potential, including ammonoids, dinoflagellate cysts, planktonic and benthic foraminifers, inoceramid bivalves and calcareous nannofossils (Odin and Lamaurelle, 2001). The history, stratigraphy, paleontology and inter-continental correlations are extensively compiled in a special volume (Odin, 2001). The weakly magnetized limestones at the Tercis stratotype did not yield a magnetostratigraphy above the upper-middle Campanian. A complex discussion of debated correlations of the Tercis GSSP to sections having magnetostratigraphy, geochemistry, North American ammonoid zones, strontium isotope curves, radio-isotopic-dated bentonites, and other markers was presented in GTS2004. Most of those uncertainties have now been resolved through a combination of

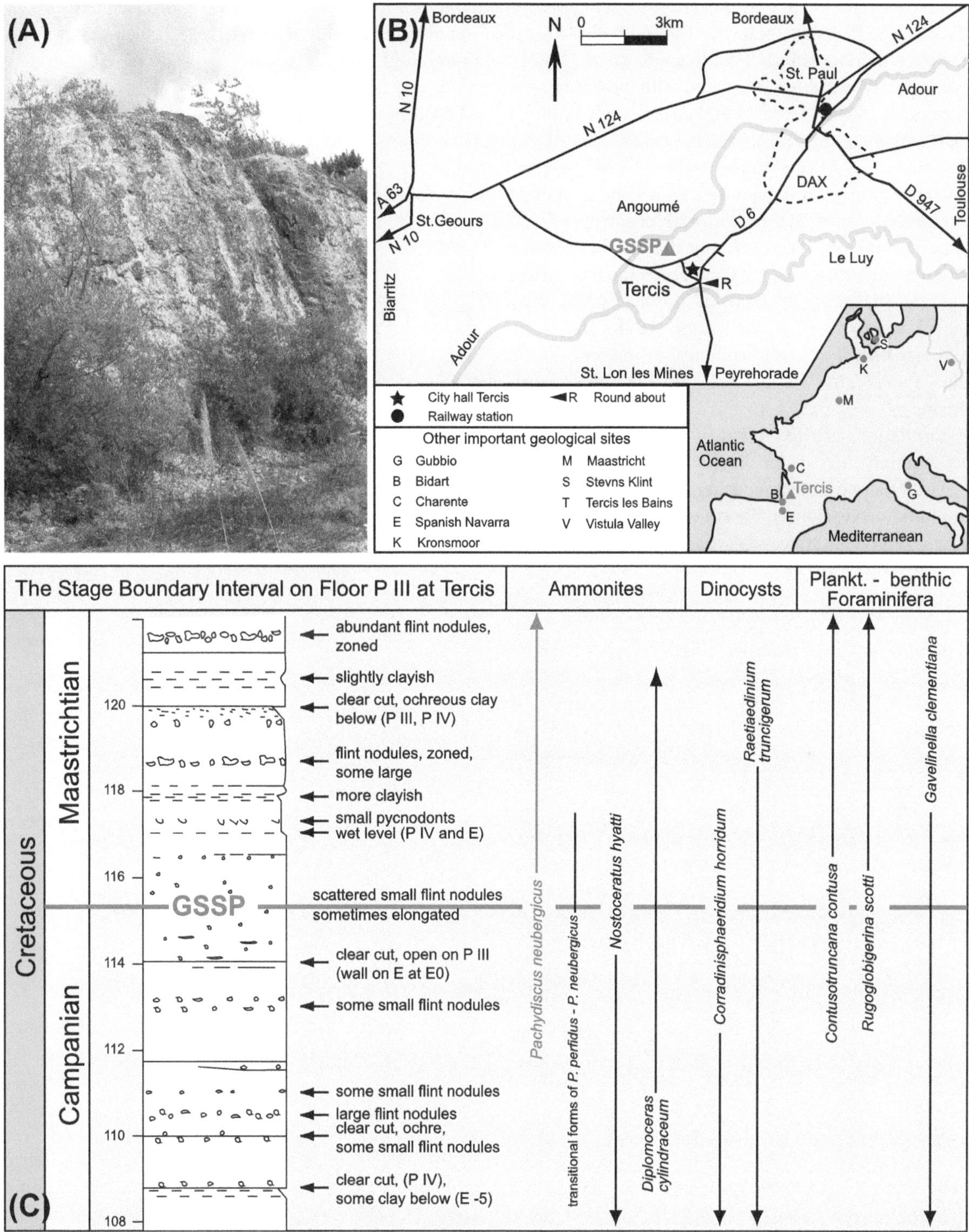

FIGURE 27.5 **GSSP for base of the Maastrichtian Stage at Tercis, France,** is 90 cm below the lowest occurrence of the ammonoid *Pachydiscus neubergicus. Photos provided by Andy Gale.*

carbon isotope stratigraphy studies, inoceramid correlations, cycle-stratigraphic scaling of C-sequence chrons, and revised spline-fit to recalibrated $^{40}Ar/^{39}Ar$ dates.

The Tercis GSSP is just below a bed with the lowest *Endocostae typica* inoceramids (Walaszczyk *et al.*, 2002). The base of the *E. typica* Zone is equated with the base of the *Baculites baculus* Zone in the North American ammonite succession (Cobban *et al.*, 2006). The age for the base of this ammonite zone is 72.1 Ma according to the GTS2012 spline-fit of dates from bracketing bentonites, as will be explained below. Interpretations of the macrofossil occurrences and microfossil stratigraphy of Tercis imply that the base of the Maastrichtian Stage projects approximately as the base of the *Belemnella obtusa* belemnite zone of northwest Germany (e.g., Hancock *et al.*, 1993; Burnett *et al.*, 1998); therefore, one zone higher than the "traditional" use of the base of the *Belemnella lanceolata* Zone for placing the base-Maastrichtian.

These correlations are supported by a distinctive carbon isotope stratigraphy across the Tercis GSSP compared to other Maastrichtian–Campanian reference sections, especially the chalk successions of northern Germany and the magnetostratigraphy-calibrated curve from Gubbio, Italy (Voigt *et al.*, 2010, and in prep.). The GSSP level is just below a minimum in the carbon isotope curve seen in all sections, implying that the base-Maastrichtian correlates to the base of the *Belemnella obtusa* belemnite zone in north Germany, to approximately the base of nannofossil zone UC17 (highest occurrence of *Broinsonia parca constricta*; although this datum displays diachroneity among the studied sections), and approximately 7/8ths up in magnetic polarity Chron C32n.2n (~C32n.2n.88). According to the GTS2012 age model for the C-sequence, which incorporates cycle-scaled polarity zone durations for the Maastrichtian set (Husson *et al.*, 2011), the projected age for C32n.2n.88 is 72.1 Ma. This is exactly the same as the GTS2012 spline-fit age of 72.1 Ma for the base of the *Baculites baculus* Zone in the North American ammonite succession. (Note: the initial GSSP summary by Odin (2001, pp. 775–782) had a slightly lower set of proposed correlations.)

This agreement between the independent sets of correlations implies that the base-Maastrichtian is essentially equivalent to the base of the *Belemnella obtusa* belemnite zone, the base of the *Endocostae typica* inoceramid zone, the base of the *Baculites baculus* ammonite zone of North American, and approximately to the base of nannofossil zone UC17. The magnetostratigraphic placement is Chron C32n.2n.88, and the GTS2012 age for all of these base-Maastrichtian levels is approximately 72.1 Ma.

Upper Maastrichtian Substage

The Maastrichtian is commonly divided into two substages, although there is no agreement on the boundary criterion for the base of the Upper Maastrichtian (Odin *et al.*, 1996; Odin, 2001). One potential marker is the lowest occurrence of ammonoid *Pachydiscus fresvillensis* with a possible stratotype at Zumaya in northern Spain (Kennedy *et al.*, 1995, unpublished; Odin *et al.*, 1996). The lowest occurrence of ammonoid *Hoploscaphites birkelundi* (formerly "*H.* aff. *nicolleti*) is an informal marker for the base of the Upper Maastrichtian in the North American Western Interior (Landman and Waage, 1993; Cobban, 1993; Cobban *et al.*, 2006). Alternative criteria include a magnetic polarity reversal (e.g., base of Chron C31n), the extinction of rudistid reefs, or the extinction of the majority of inoceramid bivalves.

27.2. CRETACEOUS STRATIGRAPHY

Similarly to the Jurassic, the ammonite and other macrofossil zones of European and North American basins had provided the traditional standards for subdividing Cretaceous stages. With the advent of petroleum and scientific drilling, microfossil datums from pelagic successions are commonly used for global correlations, especially when augmented by magnetostratigraphy. Cycle stratigraphy has now enabled scaling of many of these zonations, and detailed carbon isotope curves are becoming a major method for inter-regional correlation.

27.2.1. Marine Biostratigraphy

Ammonoids dominate the historical zonation of the Lower Cretaceous of Europe. Belemnites, inoceramid bivalves and pelagic or benthic crinoids (e.g., *Marsupites*) provide important macrofossil horizons within the Upper Cretaceous of northwest Europe, and buchiid brachiopods are used for correlation within the Lower Cretaceous of the Boreal realm. Important microfossil biostratigraphic zonations include planktonic foraminifers, calcareous nannoplankton, dinoflagellate cysts, and calpionellids.

27.2.1.1. Ammonites

Extreme faunal provincialism necessitated the establishment of different regional scales throughout most of the Cretaceous, and these regional scales were commonly non-standardized among publications. To partially rectify this situation, the grouping of ammonoid datums into zones and subzones has undergone significant revisions since 1990. Only a few of the Cretaceous ammonoid zones compiled by Hancock (1991) are currently used by the various Cretaceous working groups (e.g., Rawson *et al.*, 1996b; Reboulet *et al.*, 2011). The relative grouping into zones also varies among regions and stages. For example, many of the high-resolution "zones" of the North American Western Interior would be classified as "horizons" in the broader zonal schemes used in Europe.

In particular, the IUGS Lower Cretaceous Working Group ("Kilian Group") has been striving to develop a systematic zonation for the Tethyan region that can be applied over large regions (Hoedemaeker *et al.*, 1993; Reboulet *et al.*, 2006, 2009, 2011). In particular, the new schemes have a more logical nomenclature in which most of the revised zones begin with the first occurrence of the index species. This is in contrast to the irregular association of zonal spans with their "index" taxa in the "traditional" schemes for the Cretaceous and still used for much of the Jurassic. As a result, the current (Aug. 2011) Tethyan zonation for Berriasian through Albian (Figure 27.6) has only a few zonal names in common with the Lower Cretaceous zonation in Hardenbol *et al.* (1998) or even in GTS2004. For example, the uppermost Albian "classic *Stoliczkaia dispar* Zone", which had varying definitions among different authors, is replaced by elevating its subzones to zone status (summarized by Kennedy in Gale *et al.*, 2011). These zones are cross-calibrated to microfossil datums and cycle stratigraphy in southeastern France (Gale *et al.*, 2011). The "Kilian Group" is now undertaking the challenging task of correlating the revised Tethyan scales to the Boreal ammonite zonations.

This revision of zone definitions, taxonomic changes in genera assignment of index species, and extensive high-resolution subzonal divisions will simplify and standardize correlations, but will undoubtedly lead to confusion when trying to decipher zonal scales in older literature. Another problem is that details of reference sections for these revised and enhanced zonations have not yet been published; therefore, it is uncertain how to correlate other microfossil, geochemical and paleomagnetic stratigraphy to the zones. Estimates of zonal durations with the revised definitions also require precise placement onto the existing reference sections that had been used for strontium isotope trends and cycle stratigraphy. For GTS2012, estimates of the placement of these renamed/redefined zones relative to the former "inter-calibrated" ammonite zones were used to make an approximate age model.

A detailed ammonite zonation for the Cenomanian through lower Maastrichtian of the North American Western Interior has been standardized by Cobban (1993) and Cobban *et al.* (2006). The co-occurrence of similar inoceramid taxa in North America and in Europe has enabled correlation of most European ammonite zones (and other marine macrofossils) to this North American scale (summarized in Figure 27.6).

27.2.1.2. Other Marine Macrofauna

Upper Cretaceous chalk-rich successions of northwest Europe have a variety of markers and zonations based on belemnites, pelagic crinoids and other echinoderms (e.g., belemnite zonations summarized by Combemorel and Christensen, 1998). Inoceramid bivalves provide an important tool for correlation within Late Cretaceous basins, especially within the North American Western Interior (e.g., D'Hondt, 1998; Cobban *et al.*, 2006). Rudist bivalves enable correlation of shallow-water carbonate platforms (reviewed by Masse and Philip, 1998). Ostracode datums and associated zones are correlated to Boreal and Tethyan ammonoid zones within the lower Cretaceous, and to Boreal belemnite–echinoderm zonations within the upper Cretaceous (reviewed by Colin and Babinot, 1998).

27.2.1.3. Foraminifera and Calpionellids

The remarkable evolution, diversification and global spread of planktonic foraminifera in mid Cretaceous provide a series of high-resolution global markers from Aptian through Maastrichtian (e.g., Robaszynski and Caron, 1995; Bralower *et al.*, 1995, 1997; Premoli Silva and Sliter, 1999). There are only relatively broad and geographically restricted biozones in Berriasian through Barremian. The taxonomy of this group is currently under revision using a more systematic and evolutionary framework (e.g., Mesozoic Planktonic Foraminiferal Working Group, 2006). The Early Cretaceous zonal scale in Figure 27.6 uses the synthesis by Robaszynski (in Hardenbol *et al.*, 1998) with selected modifications (e.g., Erba *et al.*, 1999; composite scales by Shipboard Scientific Party, 2004). The Aptian–Albian zonation includes various revisions (Erba *et al.*, 1999; Leckie *et al.*, 2002; Huber and Leckie, 2011; Petrizzo *et al.*, in press; and B. Huber, pers. comm., 2011). The revised Late Cretaceous foraminifer zonation and its correlation to geologic stages and nannofossil datums is partially from a Late Cretaceous microfossil workshop at University College London (June 2011), which incorporated revised ranges, taxonomy revisions and zonation recommendations by Petrizzo *et al.* (2011) and others.

A detailed suite of correlations of smaller and larger benthic foraminifera datums to ammonoid zones is partially established (e.g., Magniez-Jannin, 1995; Arnaud-Vanneau and Bilotte, 1998), although these zonations are mainly applicable to European basins.

Calpionellids are enigmatic pelagic microfossils with distinctive vase-shaped tests in thin-section. Calpionellids appeared in the Tithonian and vanished in the latest Valanginian or earliest Hauterivian (Remane, 1985), and their abundance in carbonate-rich shelf to basinal settings within the Tethyan realm enables biostratigraphic correlation prior to the increase in the diversity of planktonic foraminifera. Six standard zones (Allemann *et al.*, 1971) with finer subdivisions (e.g., Remane *et al.*, 1986) provide the basic framework for interregional correlation (summarized by Remane, 1998). There are variations upon this basic framework (e.g., Pruner *et al.*, 2010), but a nomenclature using lettered zones is a common system (Figure 27.6).

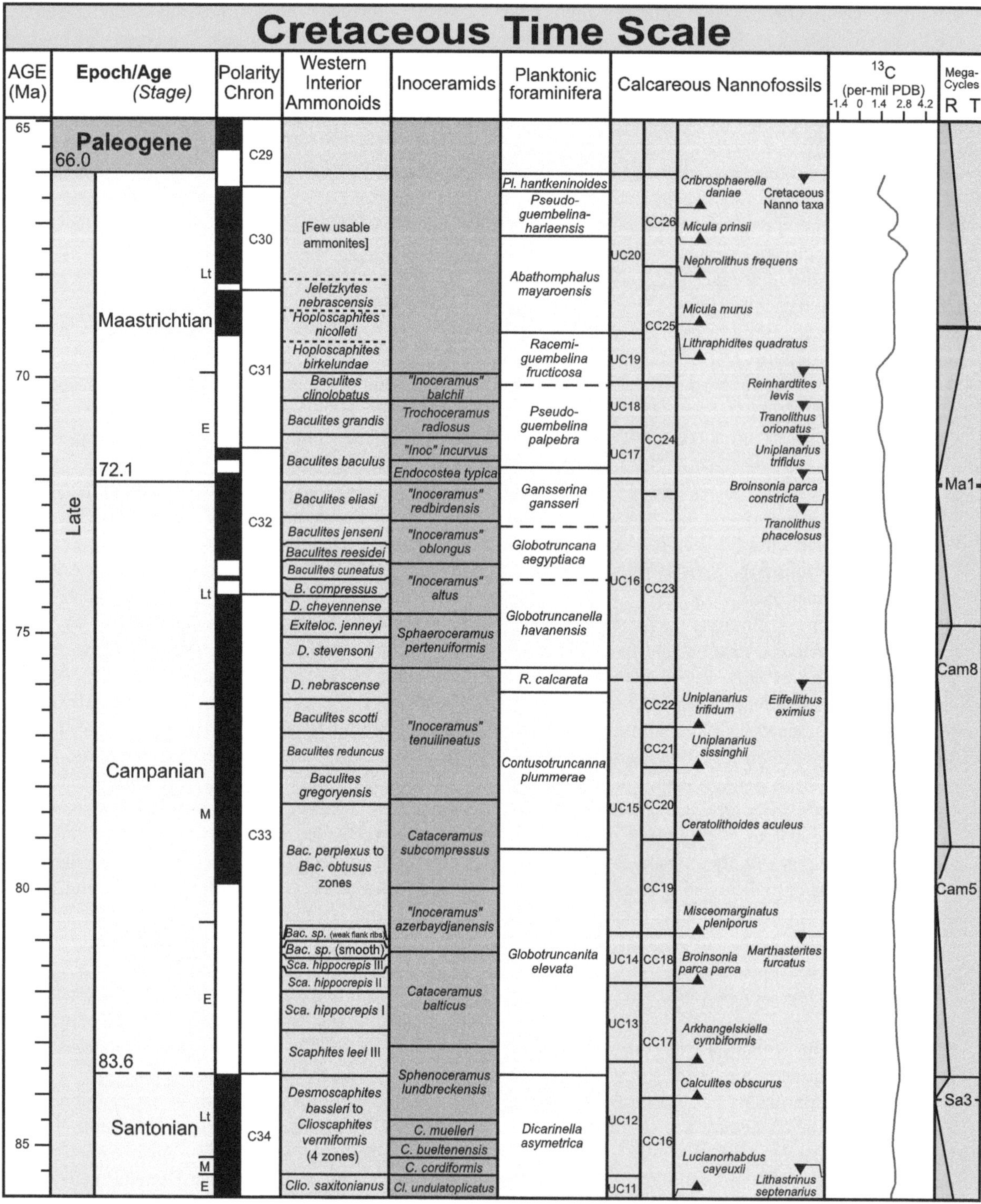

FIGURE 27.6 **Summary of numerical ages of epoch/series and age/stage boundaries of the Cretaceous with selected marine biostratigraphic zonations and principal trends in carbon isotopes and sea level.** ("Age" is the term for the time equivalent of the rock-record "stage".) Potential definitions for Cretaceous stages are indicated; the final decisions will be made by the International Commission on Stratigraphy. Selected marine macrofossil biostratigraphy columns for Early Cretaceous are ammonoid zones for the Tethyan realm (Sub-Mediterranean province; Reboulet *et al.*, 2011) and Boreal realm (Sub-Boreal province based mainly on Thierry, 1998; with lower Albian from Mutterlose *et al.*, 2003). For detailed regional ammonite zonations in earliest Cretaceous, see expanded scale in Figure 26.8 in Jurassic chapter (this volume). Marine macrofossil zones for Late Cretaceous are ammonoids of the Western Interior of USA (a full list of zone names is in Table 27.3) and inoceramids of North America and Europe (Cobban *et al.*, 2006; with partial modification by A. Gale, pers. comm., 2010). Selected microfossil zones are planktonic foraminifers (composite from Robaszynski, 1998; Huber *et al.*, 2008; Huber and Leckie, 2011; Petrizzo *et al.*, 2011; and other sources, including B. Huber and M.R. Petrizzo, pers. comm., 2011) and early Cretaceous calpionellid zones (Remane, 1998). Early Cretaceous calcareous nannofossil zones are Boreal (BC; Bown *et al.*, 1998) and Tethyan (CC with selected zone/subzone markers; with calibrations compiled from Bergen, 1994; Bralower *et al.*, 1995; and other sources, including J. Bergen, P. Bown and E. de Kanael, pers. comm., 2007–2010).

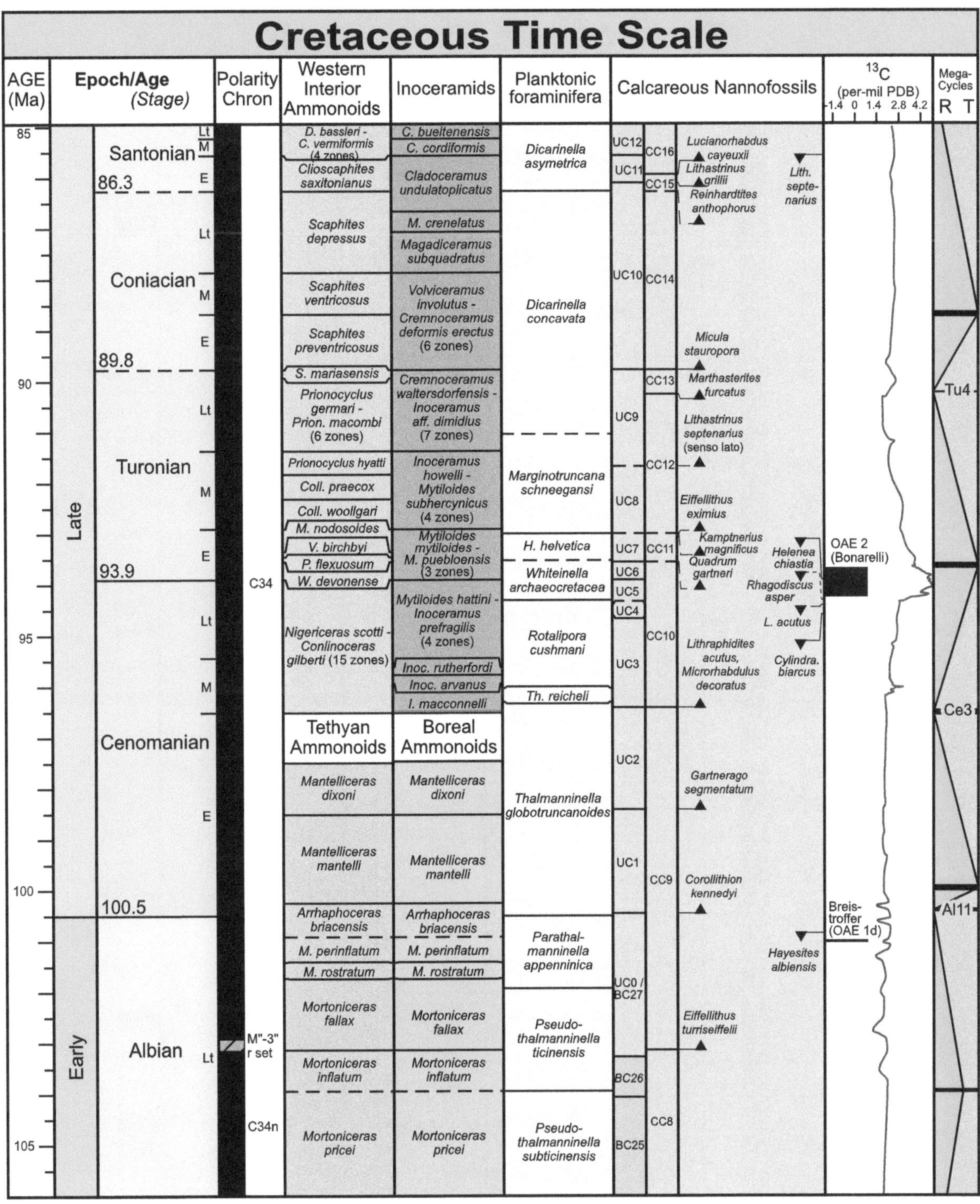

FIGURE 27.6 *(Continued)*.

Late Cretaceous UC zones are relatively cosmopolitan; and the calibration of the main markers was provided by Jackie Lees and the Late Cretaceous microfossil workshop (July, 2010). The ^{13}C curve (with widespread anoxic events) is a synthesis of generalized trends with relative magnitudes: Berriasian through middle-late Aptian from Föllmi et al. (2006) with early Aptian modifications from Renard *et al.* (2005), the late Aptian to early Albian generalized from Herrle *et al.* (2004), middle and late Albian from Gale *et al.* (2011), Cenomanian through Campanian from Jarvis *et al.* (2006), early Maastrichtian from Voigt *et al.* (2010, and submitted) and late Maastrichtian from Thibault et al. (in press); although values of the observed ^{13}C measurements are usually systematically offset among different regions. The mega-cycles of sea-level trends are from Jacquin and de Graciansky (1998) and Hardenbol *et al.* (1998). Additional Cretaceous zonations, geochemical trends, sea-level curves, etc. are compiled in the internal data sets within TimeScale Creator (*www.tscreator.org*).

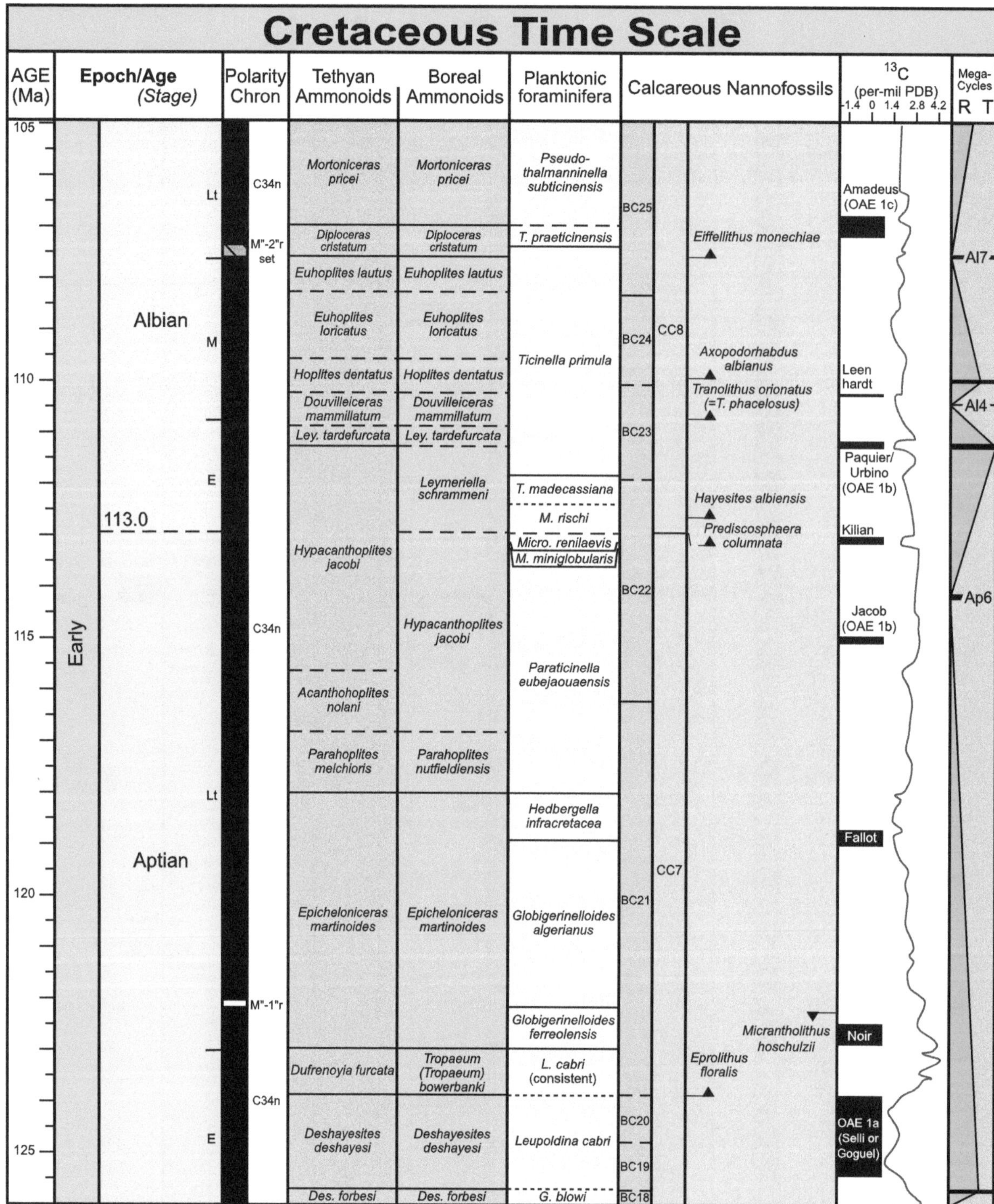

FIGURE 27.6 (*Continued*).

27.2.1.4. Calcareous Nannofossils

The Cretaceous period was named for the immense chalk formations that blanket much of northwestern Europe, and the main components of this chalk are calcareous nannofossils. Following their rapid surge in abundance at the end of the Jurassic (e.g., Caselleto *et al*., 2008), calcareous nannofossils remained ubiquitous throughout the Cretaceous and Cenozoic in all oceanic settings and sediments above the carbonate dissolution depth.

Calibrations of major calcareous nannofossil datums to ammonoid zones or magnetic polarity zones have been

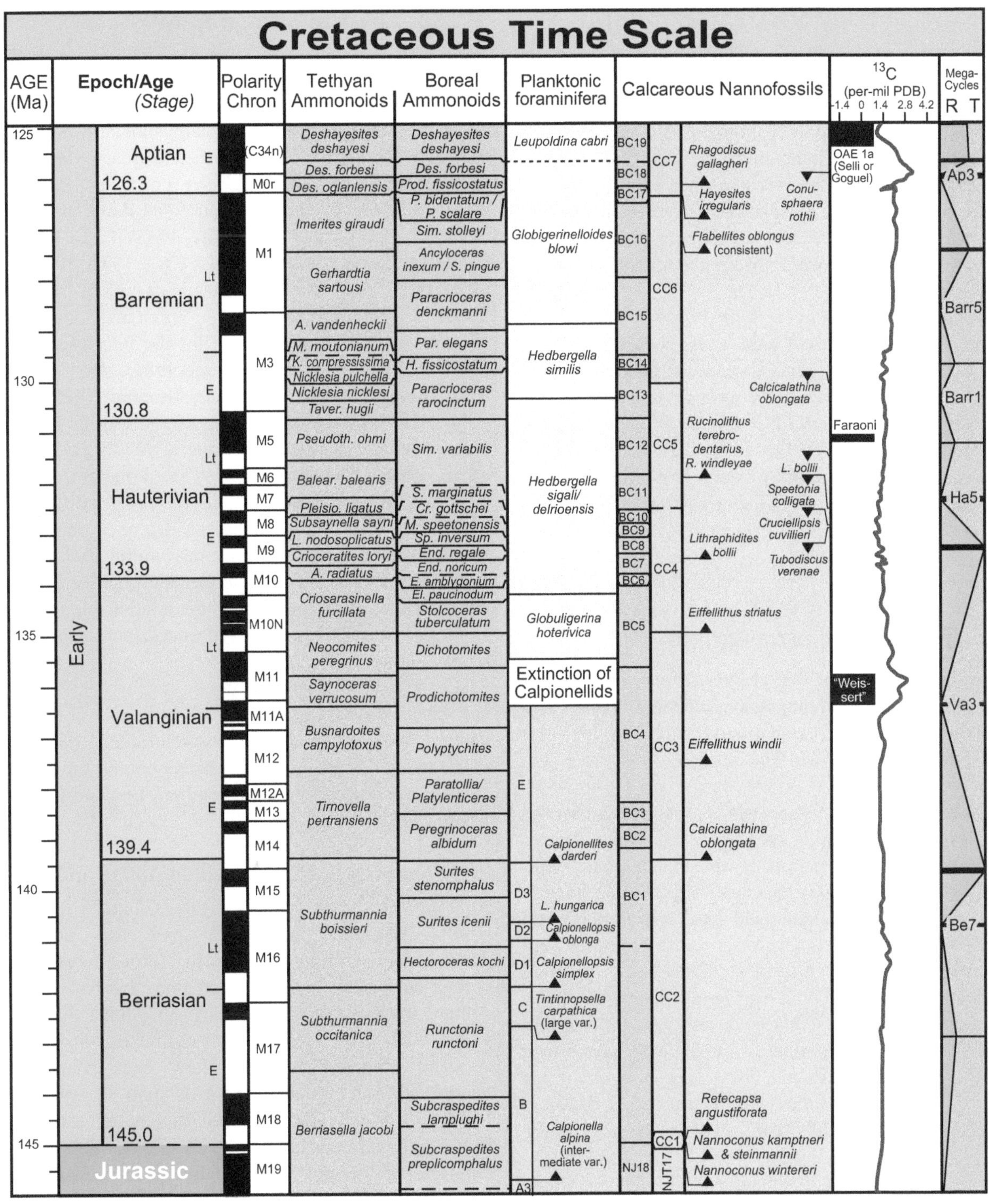

FIGURE 27.6 (Continued).

established for several intervals in the Tethyan and Boreal realms (e.g., Lower Cretaceous of Tethyan realm by Bergen, 1994: Bralower *et al.*, 1995; Erba *et al.*, 1999; Channell *et al.*, 2000, 2010; Lower Cretaceous of Sub-Boreal-Boreal realm by Bown, 1998; Upper Cretaceous by Burnett *et al.*, 1998; von Salis, 1998; Huber *et al.*, 2008; and the Late Cretaceous microfossil workshop at University College London, June 2011). The calibrations for the zonations in Figure 27.6 are a synthesis of selected zonal scales and markers.

27.2.1.5. Organic and Siliceous Microfossils

Organic-walled cysts of dinoflagellates have been correlated directly to ammonoid zones in the Tethyan realm for Berriasian through Turonian, and in the Boreal realm for Berriasian through Aptian (compiled by Monteil and Foucher, 1998). In some intervals, these widespread dinoflagellate cysts have helped to resolve uncertainties in interregional correlations (e.g., earliest Cretaceous between Sub-Boreal and Boreal realms by Harding *et al.*, 2011; Barremian–Aptian between Tethyan and Austral realms by Oosting *et al.*, 2006).

Siliceous radiolarians (pelagic sediments), charophytes (brackish-water algae tests) and calcareous algae have a relatively low resolution set of datums and zones compared to other Cretaceous microfossil groups (e.g., respective syntheses by De Wever, 1998; Riveline, 1998; and Masse, 1998). Siliceous diatoms evolved in the Jurassic, but did not undergo a major evolutionary radiation until the mid-Cretaceous, especially after the Cenomanian–Turonian boundary (e.g., Round *et al.*, 1990; Sinninghe Damsté *et al.*, 2004).

27.2.2. Terrestrial Biostratigraphy

It is beyond the scope of this chapter to make a thorough review of terrestrial biostratigraphy, and only a few highlights will be mentioned. Authoritative books on this topic include Lucas *et al.* (1998), Woodburne (2004), and Kemp (2005).

Dinosaurs, the most renowned group of Cretaceous vertebrates, provide only a broad biostratigraphy (Lucas, 1997). Early Cretaceous sauropods were smaller, but ornithopods (such as *Iguanodon*) were larger than their Jurassic cousins. Stegosaurs, iguanodontid and hypsilophodontid ornithopods and sauropods (except in South America) were nearly extinct by the end of the Early Cretaceous. The rapid diversification of angiosperms (flowering plants) displaced gymnosperms in the mid-Cretaceous, and this was probably a major factor in the evolution of the suite of hadrosaurid ornithopods, ceratopsians, and ankylosaur browsers. This suite and their tyrannosaurid and coelurosaurian theropod predators were dramatically terminated at the end of the Cretaceous.

The Yixian Formation of Aptian age (Table 27.1) in China is famous for the well-preserved Jehol Biota (plants, birds, terrestrial life). Radio-isotope dating has yielded an early Aptian age for the main Yixian Formation fossil localities (Table 27.1). The Jehol Biota and other localities indicate that placental and marsupial mammals appeared near the end of the Early Cretaceous. The North American mammal age "Judithian" began in the early Campanian (e.g., Jinnah *et al.*, 2009), and the earliest NALMA of "Aquilan" is possibly of Santonian age.

27.2.3. Physical Stratigraphy

27.2.3.1. Magnetostratigraphy

Cretaceous Portion of M-Sequence

The M-sequence of marine magnetic anomalies formed from the Late Jurassic to the earliest Aptian. Several biomagnetostratigraphic studies have correlated Early Cretaceous calpionellid, calcareous microfossil and dinoflagellate datums to the M-sequence polarity chrons (e.g., Channell and Gradesso, 1987; Channell *et al.*, 1987, 1993, 1995b, 2000, 2010; Ogg, 1987, 1988; Speranza *et al.*, 2005; Pruner *et al.*, 2010). Correlation of Tethyan ammonite zones to the M-sequence has been achieved for the Berriasian spanning Chrons M18 to M15 (Galbrun, 1984), the Berriasian–Valanginian boundary interval spanning M15 to M13 (Ogg *et al.*, 1988; Aguado *et al.*, 2000), and the Hauterivian and Barremian spanning portions of M10N to M1 (Cecca *et al.*, 1994; Channell *et al.*, 1995a). Polarity zone M0r is a primary marker associated with the proposed GSSP at the base of the Aptian. When coupled with a spreading rate model for the Pacific magnetic lineations within each individual stage (see the chapter on geomagnetic polarity time scale), these correlations also constrain the relative duration of each ammonite zone within the Berriasian, Hauterivian and Barremian stages.

Correlation of Boreal ammonite zones to the M-sequence has been directly achieved only for the equivalent of uppermost Tithonian and Berriasian in Siberia (Houša *et al.*, 2007) and indirectly for the equivalent of the Berriasian Stage in the Purbeck beds of southern England (Ogg *et al.*, 1991, 1994).

Reported Brief Subchrons Within Aptian and Albian

An extended normal-polarity Chron C34n, or "Cretaceous Normal Polarity Superchron", spans the early Aptian through to the middle Santonian. Brief reversed-polarity chrons have been reported from three intervals – middle Aptian, middle Albian, and mid-late Albian – especially within drilling cores of deep-sea sediments. However, none of these proposed subchrons have been unambiguously interpreted from marine magnetic anomaly surveys, nor have M"-2r" or M"-3r" been verified in outcrop sections. The following summary is revised from GTS2004; and the reported possible placement relative to microfossil zones is illustrated in Figure 27.6.

Ryan *et al.* (1978) proposed a negative numbering for these three "pre-M0r" reversed-polarity events or clusters of events:

(1) M"-1r" in middle Aptian with a biostratigraphic age near the base of the *Globigerinelloides algerianus* planktonic foraminifer zone (Pechersky and Khramov, 1973; Jarrard, 1974; VandenBerg *et al.*, 1978; Keating and Helsley, 1978a,b,c; Hailwood, 1979, VandenBerg and Wonders,

TABLE 27.1 Selected Radio-Isotopic Ages from Terrestrial and Flood-Basalt Deposits of Uppermost Jurassic, Cretaceous and Lowermost Paleogene

Note: This is only a subset of the published age suites for most of these episodes. See the selected references for reviews.

Event	Source (Biostrat; Location)	Age for GTS2012	Uncertainty (95%)	Annotation; Dated Material; Method	Locality; Formation; Biostratigraphy	Primary Reference [Review Reference]
Jehol Biota (Barremian–Aptian)						
	Yixian	121.96	0.50	Mt9, Mt14, ZCZ28; Andesite (bulk matrix); Ar-Ar step-heating; weighted mean of 3 flows	Mashenmiao-Zhuanchengzi section (121°E, 41.5°N), Yixian Basin, 17 km southwest of the Yixian City, in western Liaoning, northeast China. Lower part of Yixian Formation of intercalated volcanic and lacustrine sedimentary rocks. Paleomagnetic behavior of all 12 sampled levels in volcanics are interpreted as a primary reversed-polarity; therefore Chron M0r was proposed, but age indicates M"-1r" in the GTS2012 scale.	He *et al.*, 2008.
	Yixian	125.38	0.10	P4T-1, P1T-2; Sanidine separates from tuffs; Ar-Ar, Laser fusion; Single crystals, mean age	Two sites in main fossil-bearing region of Sihetun, western Liaoning, northeast China. Tuffs interbedded directly in the fossiliferous horizons (Bed 6) of the lower Yixian Formation. P4T-1 (Jianshangou section) is 50cm above the type specimen of primitive mammal *Zhangheotherium quinquecuspidens*. PIT-2 (Sihetun section) is 3.4 m above the layer bearing the type-specimen of primitive bird *Confuciusornis sanctu*.	Swisher *et al.*, 1999.
	Yixian	124.88	0.16	YX07-4; Sanidine separates from tuff; Ar-Ar; mean plateau age	Site in main fossil-bearing region of Sihetun, western Liaoning, northeast China. Tuff interbedded directly in the fossiliferous horizons (Bed 6) of the lower Yixian Formation. Same level as Swisher's P1T-2.	Chang *et al.*, 2009. [Renne *et al.*, 2010.]
	Yixian	125.65	0.17	P1T-2; Zircons from tuff; $^{206}Pb/^{238}U$; weighted mean from 13 zircones	Site in main fossil-bearing region of Sihetun, western Liaoning, northeast China. Tuff interbedded directly in the fossiliferous horizons (Bed 6) of the lower Yixian Formation. Same level as P1T-2 of Swisher *et al.* (1999) and YX07-4 of Chang *et al.* (2009).	Renne *et al.*, 2010.
Paraná-Etendeka flood basalts (Valanginian–Hauterivian)						
	Parana, Uruguay	133.06	0.26	PR94-7; Zircons from tuff; $^{206}Pb/^{238}U$; weighted mean from 13 zircons	4 km west of Lascano, SE Uruguay. Arequita Formation. Sample is a quartz–plagioclase–sanidine phyric rhyolitie ash-flow tuff (ignimbrite), taken about 20m up in the 50m thick tuff. Late-stage eruption of the voluminous flood volcanic province.	Renne *et al.*, 2010.
		134.30	0.80	OU-99 Dacite; Baddeleyite/ zircon; $^{206}Pb/^{238}U$	Ourinhos (northern Paraná Basin), South Brazil. Felsic volcanic rock (hypocrystalline Chapecó-type dacite), considered to be regional onset of flood basalt volcanism in northern and western portions of province.	Janasi *et al.*, 2011.

(*Continued*)

TABLE 27.1 Selected Radio-Isotopic Ages from Terrestrial and Flood-Basalt Deposits of Uppermost Jurassic, Cretaceous and Lowermost Paleogene—cont'd

Note: This is only a subset of the published age suites for most of these episodes. See the selected references for reviews.

Event	Source (Biostrat; Location)	Age for GTS2012	Uncertainty (95%)	Annotation; Dated Material; Method	Locality; Formation; Biostratigraphy	Primary Reference [Review Reference]
	Overview	Main pulse c. 135–133 Ma		Summary; Sanidines from alkaline igneous; Ar-Ar	Regional overview; plus suites of ages from Paraguay (western margin of igneous province) —Main pulse c. 133–135 Ma; but with precursor phases and late stages of activity.	Extensive review of Ar-Ar ages for Paraná —Etendeka flood basalts in Gibson *et al.*, 2006; with summary in Gibson, 2006.

Radio-isotopic ages with the $^{40}Ar/^{39}Ar$ method have been adjusted to a FCs monitor standard of 28.201 Ma. There are many other studies of these episodes or deposits; see references for reviews and syntheses. The 'Annotation' column is the original sample identification in the publication or a shortened designation.

1980; Lowrie *et al.*, 1980; Tarduno *et al.*, 1989; Ogg *et al.*, 1992). This subchron has also been called the "ISEA" event from an Italian outcrop sample code (Tarduno *et al.*, 1989), and has an estimated duration of less than 100 000 years (Tarduno, 1990). Based on cycle stratigraphy, the base of this foraminifer zone is at approximately 122 Ma in the GTS2012 scale. Andesite volcanic beds in the lower Yixian Formation of China, which are interpreted as eruptions during a reversed-polarity episode, have a similar date of 122.0 ± 0.5 Ma (He *et al.*, 2008).

(2) M"-2r" set of Middle Albian events near the boundary of the *Biticinella breggiensis* and *Ticinella primula* planktonic foraminifer zones (Keating and Helsley, 1978a; Jarrard, 1974; VandenBerg and Wonders, 1980; Tarduno *et al.*, 1992; Shipboard Scientific Party, 1998; Ogg and Bardot, 2001).

(3) M"-3r" set in Late Albian (Green and Brecher, 1974; Jarrard, 1974; Hailwood, 1979), which may occur at the end of the *Praediscosphaera cretacea* or within the *Eiffellithus turriseiffeli* nannoplankton zones (Tarduno *et al.*, 1992).

Another reversed-polarity event, possibly near the Aptian–Albian boundary, has been reported within basalt flows with a radiometric age of 113.3 ±1.6 Ma (Gilder *et al.*, 2003).

Cretaceous Portion of C-Sequence

Polarity Chrons C33r through lower C29r have been correlated to microfossil and nannofossil datums from basal Cenomanian to the base of the Cenozoic (e.g., Alvarez *et al.*, 1977; Lowrie and Alvarez, 1977; Huber *et al.*, 2011). This polarity time scale has been calibrated in the western interior seaway of North America to regional ammonoid zones and an array of $^{40}Ar/^{39}Ar$ dates from bentonites (e.g., Obradovich, 1993; Hicks and Obradovich, 1995; Hicks *et al.*, 1995, 1999; Lerbekmo and Braman, 2002), and Chrons C32 through C29 have been scaled from cycle stratigraphy in ocean drilling cores (Herbert *et al.*, 1995; Husson *et al.*, 2011) (see tables in Chapters 5 and 28). The base of Chron C33r was reported as being within the regional "upper Santonian" of England (Montgomery *et al.*, 1998), but this polarity reversal is now being considered to mark the base of the Cenomanian (A. Gale, pers. comm., 2011). The ages on the Campanian–Maastrichtian portion of the C-sequence constrain the synthetic age-distance model for the magnetic anomalies of the South Atlantic (revised from Cande and Kent, 1992, 1995; see Chapter 5).

27.2.3.2. Geochemical Stratigraphy

Carbon Stable Isotopes and Carbon-Enrichment Episodes

Several significant excursions in the carbon cycle punctuate the Cretaceous stratigraphic record (Figure 27.6), of which the early Aptian "Selli" and latest Cenomanian "Bonarelli" events are the most significant global events. Most of the major positive excursions (>1.5 ‰) in carbon-13 are associated with widespread organic-rich sediments and drowning of carbonate platforms, and appear to be preceded by or coincide with the eruption of major flood basalt provinces which provided a source of the excess carbon (e.g., reviews and syntheses by Weissert *et al.*, 1998; Jenkyns, 1999, 2010;

Larson and Erba, 1999; Erba, 2004; Wortmann *et al.*, 2004; see also Chapter 11, this volume). The middle Cretaceous concentrations of black shale deposits, which are the source rocks for over a quarter of current oil–gas deposits, were originally considered as indications of widespread oceanic anoxic events ("OAE") (e.g., Schlanger and Jenkyns, 1976). Enhanced oceanic productivity, perhaps in response to greenhouse conditions, also played a role (e.g., Hochuli *et al.*, 1999; Leckie *et al.*, 2002; Jenkyns, 2010). The nomenclature for the main events is partly derived from distinctive European horizons and an "OAE" numbering.

(1) A Late Valanginian positive C-13 excursion of ~2 ‰ has an onset in the mid-Valanginian (base of *Say. verrucosum* ammonite zone), peaks in the following *Neocomites peregrinus* ammonite zone, and returns to background levels at the beginning of the Hauterivian (Lini *et al.*, 1992; Channell *et al.*, 1993; Weissert *et al.*, 1998). The excursion approximately coincides with the early phases of eruption of the Paraná–Etendeka large igneous province of South America and Namibia (Erba *et al.*, 2004). The onset of this excursion, informally called the "Weissert" episode after one of its discoverers, is heralded by four thin organic carbon-rich layers ("Barrande" layers B1 to B4) in the Vocontian basin of southeastern France (Reboulet, 2001, Reboulet *et al.*, 2003).

(2) A latest Hauterivian ("Faraoni") organic-enrichment episode during the *Pseudothurmannia catulloi* ammonite subzone (middle of *Ps. ohmi* Zone) is documented by a pair of organic-rich sediments in Mediterranean and Atlantic pelagic sections (Bodin *et al.*, 2006). It coincides with only a relatively minor positive C-13 excursion (e.g., Baudin *et al.*, 1997, 1999; Coccioni, 2003; Föllmi *et al.*, 2006). This level coincides with a maximum flooding interval and regional drowning of platforms, and occurs 0.5 myr before the base of the Barremian (Bodin *et al.*, 2006).

(3) An Early Aptian excursion ("OAE1a") with a complex C-13 signature that begins in the *Desh. deshayesi* ammonite zone has been extensively studied. The organic-rich "Selli" event (called "Goguel" level in SE France) in the lower portion of the *Leopoldina cabri* foraminifer zone is preceded by a negative excursion in C-13 and followed by a major positive C-13 excursion (e.g., Weissert and Lini, 1991; Weissert *et al.*, 1998; Mutterlose and Böckel, 1998; Moullade *et al.*, 1998b; Hochuli *et al.*, 1999; Larson and Erba, 1999; Leckie *et al.*, 2002; Renard *et al.*, 2005). A global "nannoconid crisis" of a major decrease of these rock-forming calcareous nannofossils precedes and continues through the main episode of black shales (e.g., Erba, 1994, 2004). The beginning of this OAE1a Selli event appears to be synchronous with the main eruptive phase of the massive Ontong Java Plateau flood basalts in the eastern equatorial Pacific. The major positive isotope excursion continues into the early Late Aptian. A relatively minor organic-rich "Noir" bed at the base of the Upper Aptian *Epicheloniceras martinoides* ammonite zone in the reference sections in southeastern France is at the top of the main positive isotope peak, and another minor "Fallot" organic-rich level occurs at a minimum in C-13 near the top of that ammonite zone (Föllmi *et al.*, 2006).

(4) A complex multi-phase set of oscillating positive and negative excursions spanning the Aptian–Albian boundary interval contains four widespread organic-rich layers in the reference sections of southeastern France – the "Jacob", "Kilian", "Paquier" (called "Urbino" in Italy) and "Leenhardt" levels (Bréheret, 1988; Weissert and Bréheret, 1991; Bralower *et al.*, 1993; Weissert *et al.*, 1998; Leckie *et al.*, 2002; Föllmi *et al.*, 2006; Gale *et al.*, 2011). The designation of "OAE1b" has been applied either to all or to only part of this interval, especially the "Jacob" level. Cycle stratigraphy studies have quantified the placement of the Jacob, Kilian and Paquier/Urbino events as approximately 36, 31 and 26.5 long-eccentricity 405-kyr cycles prior to the base-Cenomanian, respectively (e.g., Fiet *et al.*, 2001; Grippo *et al.*, 2004; Huang *et al.*, 2010; Gale *et al.*, 2011; see Figure 27.6).

(5) An "OAE1c" middle Albian excursion (called "Jassines" event in Gale *et al.*, 2011) is associated with an organic-rich "Amadeus" layer (after Mozart) or "Toolebuc" level (e.g., Leckie *et al.*, 2002; Coccioni, 2003).

(6) An "OAE1d" latest Albian excursion (e.g., Erbacher *et al.*, 1996; Leckie *et al.*, 2002) or the series of positive isotope excursions called the "Albian/Cenomanian Boundary Event" (Gale *et al.*, 2011) coincides with a set of organic-rich "Breistroffer" or "Pialli" layers.

(7) The major "OAE2" Cenomanian–Turonian boundary excursion spans the *Meto. geslinianum* to *Watin. devonense* ammonite zones, with the peak in uppermost Cenomanian (e.g., Schlanger *et al.*, 1987; Jenkyns *et al.*, 1994; Kerr, 1998; Jenkyns, 1999, 2010; Jarvis *et al.*, 2011). The associated organic-rich level is named "Bonarelli" after its occurrence in central Italy.

In addition to these major excursions, there are a multitude of lesser, but commonly named, isotopic events that enable high-resolution correlations within each Cretaceous stage (e.g., syntheses by Erba, 2004; Herrle *et al.*, 2004; Föllmi *et al.*, 2006; Jarvis *et al.*, 2006; Voigt *et al.*, 2010). A selection of these syntheses that are tied to ammonite zones of southeastern France (Early Cretaceous) and of NW Europe (Late Cretaceous) have been merged into a synthetic curve of the major trends in Figure 27.6. The selected sets are the Berriasian through middle-late Aptian from Föllmi *et al.* (2006; but inserting the isotope–ammonite calibration within the Early Aptian OAE1a in *Desh. deshayesi* ammonite zone from Renard *et al.*, (2005)), the late Aptian to early Albian from

Herrle *et al.* (2004), middle and late Albian from Gale *et al.* (2011), Cenomanian through Campanian from Jarvis *et al.* (2006) and Maastrichtian from Voigt *et al.* (2010; and in prep.).

Oxygen Stable Isotopes and Other Climate Proxies

Oxygen isotope records of oceanic temperature trends through the Cretaceous are patchy; and many of these records may be overprinted to create a false impression of cooler values (e.g., Pearson *et al.*, 2001; Wilson *et al.*, 2002; see also Chapter 10, this volume).

Some isotopic and paleontological studies have suggested that the overall warm Early Cretaceous was punctuated by cold spells during the Berriasian–Valanginian boundary interval, during earliest Late Valanginian and during the earliest Aptian (e.g., Weissert and Lini, 1991; Hochuli *et al.*, 1999; Miller *et al.*, 1999; Pucéat *et al.*, 2003). Paleontologic and isotopic evidence indicate that globally averaged surface temperatures during the middle part of the Cretaceous were more than 10°C higher than today (e.g., reviews by Barron, 1983; Huber *et al.*, 1995, 2002; Bice *et al.*, 2002). The general model for the middle through late Cretaceous, from isotopic records from DSDP-ODP sites, uplifted chalk and fish tooth enamels, is that there was an overall warming trend from the early Aptian to a peak in the early Turonian, followed by a gradual cooling trend through the end of the Maastrichtian (e.g., Douglas and Savin, 1975; Jenkyns *et al.*, 1994; Clarke and Jenkyns, 1999; Wilson *et al.*, 2002; Pucéat *et al.*, 2003; Miller *et al.*, 2004). However, even this "supergreenhouse" interval may have experienced cooling to produce glaciations that affected sea levels (e.g., Miller *et al.*, 1999, 2003, 2004; Bornemann *et al.*, 2008).

Strontium and Osmium Isotope Ratios

The marine $^{87}Sr/^{86}Sr$ record displays a progressive rise from the Berriasian to a maximum of 0.707 493 in the Barremian *P. elegans* ammonite zone (see Chapter 7, this volume). If one assumes that this was a quasi-linear trend through the Valanginian and Hauterivian, then the relative duration of each ammonite zone and its subzones can be estimated from the reference sections in southeastern France (McArthur *et al.*, 2007).

The strontium isotope ratio decreases to a pronounced minimum of 0.707 220 just before the Aptian/Albian boundary, rises sharply in the Early Albian, flattens to a broad maximum through the mid-to-late Albian, then declines to a late Turonian minimum of approximately 0.707 275. From this Turonian minimum, it rises to 0.707 830 at the end of the Cretaceous.

These $^{87}Sr/^{86}Sr$ trends enable global correlation and relative dating (e.g., McArthur *et al.*, 1993), except on the cusps of reversals (Aptian/Albian boundary, Late Turonian, middle Barremian (*sensu lato*) or where the curve is relatively flat (e.g., mid-to-late Albian). For example, the age of the Campanian–Maastrichtian was calibrated in GTS2004, using $^{87}Sr/^{86}Sr$ curve correlations between Kronsmoor, Germany and the North American Western Interior (Chapter on Cretaceous in GTS2004).

27.2.3.3. Cyclostratigraphy

Contributed by Chunju Huang

Nearly the entire Cretaceous has been scaled using astronomically-forced cycle stratigraphy, especially using the long-eccentricity 405-kyr and short-eccentricity 100-kyr cycles. A precise "tie to Present" that utilizes the stable 405-kyr signal awaits unambiguous and continuous scaling of the Paleogene, and a reliable model of how different sediment regimes respond to maximum and to minimum intervals of this eccentricity oscillation.

This review will highlight only a few cycle stratigraphy studies, especially those which are directly used in scaling the GTS2012 Cretaceous bio-magnetostratigraphy.

Early Cretaceous Cyclostratigraphy

Berriasian–Valanginian A cycle stratigraphy in southeastern Spain from the lowest occurrence of *Calpionellopsis simplex* to the lowest occurrence of *L. hungarica* implies that the span of Calpionellid Zones D1 and D2 is about 1.3 myr (Sprenger and ten Kate, 1993). The corresponding span in magnetic polarity Chrons M16r and M16n is estimated from a composite of calpionellid-bearing sections in southern Europe (e.g., Ogg *et al.*, 1988; Channell and Grandesso, 1987), and was used as a constraint on the M-sequence spreading rate model (see Chapter 5 on the Geomagnetic Time Scale, this volume).

Sections in the middle Berriasian in the Swiss and French Jura suggest durations for the subzones of *Berr. privasensis* (300–400 kyr), *Dal. dalmasi* (500–600 kyr) and *Malbo. paramimounum* (900–1000 kyr) (Strasser *et al.*, 2004). The implied 1.8 myr span is nearly identical to the 1.7 myr span in the GTS2012 scale based on magnetostratigraphic correlation of these zones (e.g., Galbrun, 1984) to the M-sequence age model.

For regional stages in the Boreal realm, Huang *et al.* (in prep.) estimate a duration of ~5.2 myr (almost thirteen 405-kyr cycles) for Upper Volgian, ~1.8 myr for Ryazanian and ~5.3 myr (thirteen 405-kyr cycles) for Valanginian Stage based on 405-kyr tuning from high-resolution grayscale logging of cores from the Greenland–Norwegian Seaway. However, there are only indirect biostratigraphic constraints on the placement of these regional stage boundaries.

In ammonite-rich pelagic carbonates of the Vocontian basin in southeast France, the uppermost Berriasian *Thurmanniceras otopeta* Subzone through lower Valanginian *Busnardoites campylotoxus* ammonite zone spans an estimated minimum of 3.15 myr, and the upper Valanginian

Sayno. verrucosum ammonite zone through *Teschenites callidiscus* Subzone (uppermost Valanginian) spans an estimated 2.75 myr. Therefore, the Valanginian portion of this interval would have a minimum duration of ~5.5 myr (Huang *et al.*, 1993, and similar results by Giraud, 1995, and by Giraud *et al.*, 1995). After adjusting for the current placement of the Valanginian boundaries, these results are consistent with the 5.5-myr Valanginian span derived from magnetostratigraphic correlation to the M-sequence age model.

There are other cycle stratigraphy analyses of portions of the Valanginian (e.g., Sprovieri *et al.*, 2006; Amodio *et al.*, 2008), but with less precise biostratigraphic control.

Hauterivian–Barremian Spectral analysis of high-resolution carbon-13 records from the uppermost Valanginian (Chron M11n) through lower Hauterivian (Chron M7n) magnetostratigraphic reference sections in Italy yielded precession, short-term and long-term eccentricity cycles (Sprovieri *et al.*, 2006). The cycle-derived span of 3.58 myr for this chron interval was used to scale the M-sequence spreading rate model (See Chapter 5 on the Geomagnetic Time Scale, this volume).

The lower Hauterivian *Acanthodiscus radiatus* through *Lyticoceras nodosoplicatum* ammonite zones span an estimated minimum of 2.40 myr, and the upper Hauterivian *Subsaynella sayni* ammonite zone to the lowest occurrence of *C. emerci* (possible base of Barremian Stage) spans an estimated 2.93 myr, for a total Hauterivian minimum duration of 5.3 myr in southeast France (Huang *et al.*, 1993, and similar results by Giraud, 1995, and Bodin *et al.*, 2006). Huang *et al.* (1993) estimated a duration of 5.3 myr for the Hauterivian Stage from the Vocontian Basin (southeast France) and coeval strata in the western central Atlantic Ocean (DSDP sites 391C, 534A, 603B). This was confirmed by Fiet *et al.* (2006) who applied more rigorous analyses to the same Vocontian Basin sections to obtain 5.3 ± 0.4 myr. The shorter (3.9 myr) duration in GTS2012 probably indicates a difference in placement of the stage boundaries, and is consistent with the 3.5 myr duration for the Hauterivian from cycle stratigraphy analysis of Italian magnetostratigraphy sections (Sprovieri *et al.*, 2006). Durations for uppermost Hauterivian ammonite zones were estimated from cycle stratigraphy by Bodin *et al.* (2006). Their duration estimates are considerably greater than the scalings derived by McArthur *et al.* (2007) based on a model of linear trends in Hauterivian strontium isotope curves; but that portion of the strontium isotope database was relatively poorly constrained.

The Barremian has been analyzed by several teams using magnetostratigraphic reference sections in Italy. Chron M3r to the base of Chron M0r, encompassing most of the Barremian interval, spans 4.3 myr according to Sprovieri *et al.* (2006). Their result is similar to estimates in nearby sections (e.g., Herbert, 1992; Fiet and Gorin, 2000; Wissler *et al.*, 2004; although Bodin *et al.*, 2006, estimated ~4.5 myr), and is used for the M-sequence spreading rate model of GTS2012. Durations of individual Barremian ammonite zones were estimated from cycle stratigraphy by Bodin *et al.*, 2006. Even though some of their Tethyan zonal boundaries are different than those that were later recommended by the "Kilian Group" (e.g., Reboulet *et al.*, 2006, 2009, 2011) and the schematic sections did not indicate the ammonite biostratigraphy, the durations are partially incorporated into the GTS2012 age model, especially in the uppermost Hauterivian and lowermost Barremian.

Aptian–Albian An international effort has yielded a cycle stratigraphy for the entire Aptian–Albian interval and its planktonic foraminifer and calcareous nannofossil datums (e.g., Erba *et al.*, 1999; Fiet, 2000; Fiet *et al.*, 2001; Grippo *et al.*, 2004; Huang *et al.*, 2010; Gale *et al.*, 2011). There is a total of 62.5 long-eccentricity cycles of 405 kyr from the base of Cenomanian downwards to the top of Chron M0r (Huang *et al.*, 2010). If the Aptian–Albian boundary is placed at the first occurrence (FAD) nannofossil *Predicosphaera columnata* (subcircular), then the Albian spans 12.45 myr (Grippo *et al.*, 2004; Huang *et al.*, 2010) and the Aptian down to the top of Chron M0r spans 12.8 myr (Huang *et al.*, 2010). Chron M0r spans 490 ± 0.25 kyr in cycle stratigraphy of the Cismon APTICORE of Italy (Malinverno *et al.*, 2010), therefore, the entire Aptian stage spans 13.3 myr.

The early Aptian organic-enrichment episode (OAE1a or "Selli") spans 1.11 ± 0.11 myr according to cycle stratigraphy (e.g., Malinverno *et al.*, 2010); and precession cycles in a section at Marcoulline, France, indicate that the *Globigerinelloides ferreolensis* foraminifer zone spans 760 kyr (Kuhnt and Moullade, 2007). This listing is only a selected subset of the extensive array of studies in these intervals.

Late Cretaceous Cyclostratigraphy

Cenomanian–Turonian A cycle stratigraphy has been constructed for most of the uppermost Albian, Cenomanian and lower Turonian in western Europe (Gale, 1995; Gale *et al.*, 1999), although there are potentially some missed-beats in the succession of the interpreted 212 precession-couplets, which have not been analyzed by spectral-analysis techniques. According to the couplet-counting, the duration of the Middle and Late Cenomanian (*Cunningtoniceras inerme* through *N. juddii* ammonite zones) is 3.0 myr (Gale *et al.*, 1999). The Early Cenomanian (*Mantelliceras mantelli* and *Mant. dixoni* ammonite zones) has a partial cycle stratigraphy, but indicates a minimum span of about 1.5 myr. In the Vocontian Basin of SE France, there are 22 couplets, considered to be precession cycles, within the

Neostlingoceras carcitanense Subzone in outcrops of apparent continuity down through the uppermost Albian (Gale, 1995) implying that the *N. carcitanense* Subzone spans at least 0.44 myr. Herbert *et al.* (1995) estimated the duration for the Cenomanian Stage as 6 ± 0.5 myr based on precessional chronology and carbonate cycles at the South Atlantic DSDP sites 356, 357, 516F, 525A, 527, 528, 529, and at Zumaya (Spain) and Italian pelagic series.

Gale *et al.* (2002) interpreted 12 sea-level sequences in the Cenomanian as being driven by 405-kyr long-eccentricity cycles through the high-resolution ammonite correlation of marine successions in southeast India and northwest Europe to demonstrate that these sea-level changes are globally synchronous.

The organic-rich OAE2 "Bonarelli" level at Furlo of the Scaglia Bianca Formation in the Umbria-Marches region of Italy was estimated to have a cycle-derived duration of ~1.8 myr (Mitchell *et al.*, 2008). In contrast, the OAE2 organic-rich interval at Wunstorf in northern Germany has an eccentricity–precession-derived duration of 430–445 kyr (Voigt *et al.*, 2008). Durations of each ammonite zone have been derived from a cyclic-bedded succession across the Cenomanian–Turonian boundary in the Western Interior Basin of North America and used with interbedded radio-isotope-dated bentonites to obtain a precise age for that stage boundary (Meyers *et al.*, 2010, 2012). There are numerous other cycle stratigraphy analyses of individual ammonite zones and other intervals within the Cenomanian–Turonian (e.g., Sageman *et al.*, 2006; Kuhnt *et al.*, 2009).

Coniacian–Santonian Borehole-resistivity and lithostratigraphy of the Niobrara Formation of the Western Interior Basin tied to North American ammonite zones in coeval outcrops has been tuned to short- and long-eccentricity cycles (Locklair and Sageman, 2008). The regional usage of the Coniacian stage spans 3.4 ± 0.13 myr, and the Santonian spans 2.39 ± 0.15 myr. The Coniacian chalks in the Anglo-Paris Basin display eight depositional cycles. If these correspond to a 405-kyr eccentricity rhythm, then the Coniacian stage (as used in England) spans 3.3 myr (Grant *et al.*, 1999), similar to the North American estimate.

Campanian–Maastrichtian The long span of the Campanian, coupled with imprecise microfossil biostratigraphy, hiatuses from sea-level changes, and a widespread low-carbonate facies within the lower Campanian in oceanic cores has precluded a full cycle stratigraphy for this interval. The problems can be seen in the divergent "minimum" cycle-derived durations for this stage, which, according to radio-isotopic dating, should span at least 11 myr. The "middle substage" in marl–limestone beds from northern Germany spans a minimum of 3.6 myr (Niebuhr *et al.*, 2005); and interpreted long-eccentricity in chalk of Lägerdorf-Kronsmoor of northern Germany suggests a total Campanian span of 6.2 myr (Voigt and Schönfeld, 2010). Estimates of ~10 myr were obtained by counting interpreted 100-kyr short-eccentricity and precession cycles in central Tunisia (Hennebert *et al.*, 2009).

In contrast, a reproducible and tuned cycle stratigraphy calibrated to C-sequence Chrons C29r through C32r and to microfossil datums has been obtained from analysis of magnetic susceptibility and grayness-scanning from ODP (Ocean Drilling Program) Holes 1258A (Leg 207, equatorial Atlantic), 1267B (Leg 208, South Atlantic), 762C (Leg 122, Indian Ocean) and DSDP (Deep Sea Drilling Program) Hole 525A (Leg 74, South Atlantic) (Husson *et al.*, 2011). This 8 myr record includes detailed durations for each polarity chron and subchron. For example, this composite cycle stratigraphy places the Cenozoic ("K/T" boundary) at 0.30 ± 0.02 myr above the beginning of Chron C29r.

27.2.3.4. Sequence Stratigraphy

Cretaceous marginal-marine to deep shelf successions in Europe and North America record an abundance of basinal and regional transgressions and regressions. At the largest scale, the Cretaceous strata encompass a single transgressive–regressive cycle (the "North Atlantic" cycle of Jacquin and de Graciansky, 1998). The lower boundary is a widespread unconformity during Late Berriasian, the transgression peaked in the Early Turonian, and average sea levels continued to decrease into the Paleocene.

The common features from an extensive suite of compilations edited by de Graciansky *et al.* (1998) were assembled in a comprehensive synthesis and systematically numbered from the base of each stage (Hardenbol *et al.*, 1998). Coeval emergent horizons recorded by Aptian–Albian seamount carbonate platforms in the central Pacific Ocean imply that some of these depositional sequences reflect global eustatic sea-level oscillations (Röhl and Ogg, 1996). Detailed analysis of facies successions and biostratigraphic constraints across the Arabian Plate yielded a detailed sequence stratigraphy for the Cretaceous (e.g., Sharland *et al.*, 2001, 2004; Simmons *et al.*, 2007; van Buchem *et al.*, 2011), in which many of the main features appear to coincide with the European–American-derived sequence scale. The magnitude and cause of these sea-level changes during the presumed "super-greenhouse" of the mid-Cretaceous remain controversial (e.g., Miller *et al.*, 2003, 2004) and some features require additional verification (for example, the mid-Valanginian "major sea-level lowstand excursion" may be partially an artifact of the selected reference sections).

Details of the Cretaceous sequence stratigraphy are continually undergoing refinement, but the major global oscillations have probably been identified (see also Chapter 13 on sequence stratigraphy, this volume). The main large-scale deepening and shallowing trends from the SEPM'98 sequence stratigraphy charts (Hardenbol *et al.*, 1998) are summarized on Figure 27.6.

27.2.3.5. Other Major Stratigraphic Events

Large Igneous Provinces

At least five major large igneous provinces formed during the Cretaceous. Most of these appear to be associated with major distortions in the global carbon budget as indicated by excursions in carbon isotopes, widespread organic-rich shales or "OAE" episodes, increased oceanic carbonate dissolution, and other changes in climate and oceanic chemistry (e.g., Larson and Erba, 1999; Jones and Jenkyns, 2001; Bice *et al.*, 2002). Age constraints and possible feedbacks from these episodes are reviewed by Wignall (2001) and Courtillot and Renne (2003); and selected ages are summarized in Table 27.1.

(1) **Paraná–Etendeka, *~136–133 Ma*.** The early stages of rifting of the South Atlantic were accompanied by extrusion of a large igneous province onto South America (Paraná flood basalts) and smaller fields in Namibia (Etendeka Traps). The full suite of volcanic activity may have had multiple pulses that spanned over 10 myr, but the main pulse of tholeiitic volcanism appears to be between 135 and 133 Ma (e.g., Stewart *et al.*, 1996; Gibson, 2006; Gibson *et al.*, 2006; dates are after adjustment to a $^{40}Ar/^{39}Ar$ FCs monitor age of 28.20 Ma). The central Paraná volcanic suite spans a minimum of a triplet of polarity zones (normal–reversed–normal) (Mena *et al.*, 2006). The onset of the northern and western Paraná volcanics from U-Pb dating of baddeleyite/zircon is 134.3 ± 0.8 Ma (Janasi *et al.*, 2011). The onset and main eruptive phase of the Paraná–Etendeka flood basalts approximately coincides with the late Valanginian "Weissert" C-13 positive excursion (e.g., Weissert *et al.*, 1998; Erba *et al.*, 2004), which has a GTS2012 age assignment of ~136 ± 1 Ma.

(2) **Ontong Java Plateau–Manihiki Plateau, *~125–123 Ma*.** During the middle of the early Aptian, the largest series of volcanic eruptions of the past quarter-billion years built the Ontong Java Plateau and Manihiki Plateau in the western equatorial Pacific. A series of deep-sea drilling legs documented that the multi-kilometer-thick series of volcanic flows forming the Ontong Java Plateau occurred during a short time span at ~125–122 Ma (e.g., Mahoney *et al.*, 1993; Chambers *et al.*, 2004). A controversial theory is that its initiation may have been caused by a bolide impact (Ingle and Coffin, 2004). The upper portions of the basalt flows are interbedded with pelagic sediments of the lower portion of the *Leopoldina cabri* foraminifer zone (Mahoney *et al.*, 2001; Sikora and Bergen, 2004). A cascade of environmental effects from the eruption of the Ontong Java Plateau is the suspected culprit for the organic-rich deposits associated with the Early Aptian "OAE1a" or "Selli" episode which has an estimated age of 125–124 Ma and was followed by a large positive carbon isotope excursion (e.g., Larson, 1991; Tarduno *et al.*, 1991; Larson and Erba, 1999).

(3) **Kerguelen Plateau–Rajmahal Traps, *~118 Ma*.** The Kerguelen plateau in the southern Indian Ocean is the second largest oceanic plateau after the Ontong Java Plateau. The peak of construction of the southern and largest portion may have been simultaneous with the eruption of the formerly adjacent Rajmahal Traps of eastern India at ~118 Ma (reviewed in Wignall, 2001, and in Courtillot and Renne, 2003). This episode probably contributed to the broad carbon isotope excursion that characterizes the late Aptian. A second eruptive episode at ~83 Ma enlarged the Kerguelen Plateau and constructed the Broken Ridge (Courtillot and Renne, 2003).

(4) **Caribbean–Columbian Province, *~90 Ma*.** The Caribbean–Columbian volcanic province is a large Pacific Ocean plateau that was later emplaced between the North and South American plates. Its eruption may have contributed to the end-Cenomanian OAE2 carbon isotope excursion (e.g., Kerr, 1998); however, its apparent average age of 89.5 ± 0.3 Ma, coincides with the Turonian–Coniacian boundary (Courtillot and Renne, 2003).

(5) **Deccan Traps, *~66–65 Ma*.** The Deccan Traps cover most of central India. Their eruption peak at ~66 Ma coincides with the catastrophic termination of the Cretaceous (Courtillot and Renne, 2003).

In addition to these large igneous provinces, pulses of large-scale volcanism constructed the Shatsky Rise in the central Pacific during the earliest Cretaceous (Mahoney *et al.*, 2005), the Madagascar Traps at ~87 Ma (Bryan and Ernst, 2008), and the Sierra Leone Rise in the central Atlantic at about 73 Ma (Ernst and Buchan, 2001; Large Igneous Provinces Commission, 2011).

Major Bolide Impacts

Five impact craters with diameters greater than 40 km are currently documented from the Cretaceous (details from Joudan *et al.*, 2009, and Earth Impact Database, 2010; unless otherwise noted).

(1) ***Morokweng*** crater (~70 km; or perhaps originally over 100 km) in the Kalahari desert of South Africa has a "recommended" age from U-Pb and $^{40}Ar/^{39}Ar$ analyses of 145.2 ± 0.8 Ma (Koeberl *et al.*, 1997; Reimond *et al.*, 1999; Jourdan *et al.*, 2009). This coincides with the Jurassic–Cretaceous boundary interval, although there is

no unambiguous evidence of a wave of extinctions of this age.

(2) ***Mjølnir*** crater (~40 km) offshore northern Norway has an estimated age of c. 144 ± 3 Ma based on biostratigraphy of regional coring that indicates the impact was near the Volgian–Ryazanian boundary interval (Smelror *et al.*, 2001; Tsikalas, 2005).

(3) ***Tookoonoka*** crater (~50 km) in west Queensland, Australia, has a poorly constrained age estimated as 128 ± 5 Ma.

(4) ***Kara*** crater (~65 km) in the northern Urals of Russia is dated as 70.3 ± 2.2 Ma.

(5) the immense ***Chicxulub*** crater (~170 km) in Yucatan, Mexico, that dramatically terminated the Mesozoic Era at 66.0 Ma.

27.3. CRETACEOUS TIME SCALE

The numerical time scale for the Cretaceous consists of three main "primary scales":

(1) Tethyan ammonite zones of Berriasian through Barremian are calibrated to the M-sequence age model (Chapter 5, this volume),

(2) microfossil zones in Aptian–Albian are calibrated to cycle stratigraphy that is constrained by radio-isotopic dates, and

(3) North American ammonite zones of Cenomanian through early Maastrichtian that have an abundance of interbedded bentonites with radio-isotopic ages are scaled with cycle stratigraphy and a spline fit.

Most other Cretaceous events are assigned ages according to their calibration to these primary biostratigraphic scales or via direct calibrations to the M-sequence or C-sequence chrons.

27.3.1. Constraints from Radio-Isotopic Dates

Compared with any other Phanerozoic interval, the Late Cretaceous has the highest concentration of radio-isotopic dates derived from bentonites interbedded with fossiliferous limestone (see Figure 1.5 of Chapter 1 and Appendix 2, this volume). In contrast, the Early Cretaceous has only a few well-constrained radio-isotopic ages; and a selected subset of the main constraints is briefly reviewed here.

The Tithonian–Berriasian boundary (base of Cretaceous) is constrained by a ^{40}Ar/^{39}Ar date of 145.5 ± 0.8 Ma on reversed-polarity sills intruding earliest Berriasian sediments on the Shatsky Rise of the Pacific (Mahoney *et al.*, 2005). A Berriasian–Valanginian age older than ~138 Ma is supported by U-Pb dates from calcareous nannofossil-zoned sediments in the Great Valley of California (Shimokawa, 2010).

Deposits in the Neuquén Basin of Argentina that are correlated to the basal ammonite zone of the Upper Hauterivian substage yielded a U-Pb SHRIMP age of 132.7 ± 1.3 Ma (Aguirre Urreta *et al.*, 2008). Basalts from Resolution Guyot in the Pacific with reversed-polarity magnetization interpreted as Chron M5r or M3r of Barremian yielded a ^{40}Ar/^{39}Ar age from whole rock incremental heating of 128.4 ± 2.1 Ma (Pringle and Duncan, 1995a); but this method and its biostratigraphic constraints are not considered to be conclusive.

Several dates indicate that the Aptian begins at approximately 126 Ma. The basaltic basement on the Ontong Java Plateau, which is interbedded and overlain by limestone assigned to the *Leopoldina cabri* foraminifer zone (Mahoney *et al.*, 2001; Sikora and Bergen, 2004), yielded an average ^{40}Ar/^{39}Ar age of 124.3 ± 1.8 Ma (Chambers *et al.*, 2004). As noted above, the eruption of this massive igneous province is considered to be a causal factor in the OAE1a or "Selli" episode of carbon-rich oceanic sediment accumulation and carbon isotope excursion that begins at about 0.5 myr after the end of Chron M0r (hence about 0.9 myr after the beginning of the Aptian) and spans about 1.1 myr (Larson and Erba, 1999; Malinverno *et al.*, 2010). Dating of reversed-polarity basalts at MIT Guyot in the Pacific yielded an age of 125.4 ± 0.2 Ma (Pringle and Duncan, 1995b; as recomputed to FCs of 28.201 Ma). The overlying basalt is a transition into a well-developed soil that is overlain by transgressive marine sediments containing early Aptian nannofossils, and then capped by a thick shallow-water carbonate platform. Initially, the reversed-polarity zone had been interpreted to be the uppermost portion of polarity Chron M1r of middle Barremian, but the nannofossil criteria and the character of the carbon isotope values from the overlying sediments are consistent with interpreting this zone as uppermost Chron M0r of basal Aptian (Pringle *et al.*, 2003). Chron M0r spans 0.4 myr (e.g., Herbert *et al.*, 1995), therefore its base, which is a proposed marker for the base of the Aptian Stage, is at approximately 126.0 Ma. This is consistent with a U-Pb date of 124.07 ± 0.17 from calcareous nannofossil-zoned sediments of Early Aptian in the Great Valley of California (Shimokawa, 2010) and an Ar-Ar date of 124.3 ± 1.8 Ma from basalts of Ontong Java Plateau (Chambers *et al.*, 2004) that are considered to be coeval with the onset of Early Aptian OAE1a anoxic horizon (Table 27.1).

In contrast, a volcanic episode of reversed polarity within continental deposits of the Yixian Formation of China yielded a 122.0 ± 0.5 Ma date from the ^{40}Ar/^{39}Ar step-heating method (He *et al.*, 2008). He *et al.* (2008) had interpreted this reversed-polarity event to be during Chron M0r, thereby implying that the base of the Aptian was younger than 123 Ma. Assuming that the dating and interpretation of the magnetic polarity are reliable, another alternative is that this eruption captured the brief M"-1r" or ISEA event of mid-

Early Aptian (c. 122 Ma; see summary in Magnetostratigraphy section of this chapter).

A volcanic ash within the basal Albian strata in northwest Germany yielded a U-Pb age of 113.1 ± 0.3 Ma (Selby *et al.*, 2009). The top of the basal Cenomanian ammonite subzone has a $^{40}Ar/^{39}Ar$ date of 99.8 ± 0.4 Ma (Obradovich *et al.*, 2002).

From the lowest Cenomanian through lower Maastrichtian, there are nearly 45 radio-isotopic-dated horizons from the Western Interior Basin of North America (detailed in Appendix 2). A Late Cretaceous time scale for North American ammonoid zones was initially calibrated by Obradovich (1993) from his extensive suite of $^{40}Ar/^{39}Ar$ dates on bentonites using a multi-grain analysis of sanidine grains at the USGS lab. He added additional dates from sections calibrated to Campanian–Maastrichtian magnetostratigraphy (Hicks *et al.*, 1995, 1999), then summarized and enhanced the sets using a refined ammonite zonation for the Western Interior Basin (Cobban *et al.*, 2006). The collections of sanidine grains from Obradovich's separates have been undergoing single-crystal analyses to narrow the uncertainties at the University of Wisconsin at Madison; and this group has analyzed additional horizons and applied U-Pb dating to zircons from the same levels (e.g., Siewert, 2011; Siewert *et al.*, in press;* Meyers *et al.*, 2012). Supporting U-Pb and Ar-Ar results have been contributed by Quidelleur *et al.* (2011) from Japanese sections, including verification that the base of the Cenomanian is at approximately 100.0 Ma.

The youngest ammonite-zoned age from the Western Interior suite is ~69.9 Ma in the earliest Maastrichtian. The Maastrichtian–Danian boundary (base of Cenozoic) is well-dated as ~66.0 Ma (see Chapter 28 on the Paleogene, this volume).

27.3.2. Direct Spline-Fitting of Radio-Isotopic Age Suite

An initial spline fit incorporated the majority of the Cretaceous and Jurassic suite of radio-isotopic ages (Appendix 2) and their bio-chronostratigraphic assignments with uncertainties to a merged scale of North American Western Interior ammonite zones (Late Cretaceous) and Tethyan ammonite zones (basal Cenomanian through Jurassic). Some Cretaceous radio-isotope dates were omitted that were difficult to correlate to the primary ammonite zonal scale (e.g., the Kneehills Tuff within *Triceratops* dinosaur beds). The spline-fit processed by Øyvind Hammer used the methods described in Chapter 14 on Geomathematics (this volume). The first spline run with a smoothing factor of 1.5 indicated that 11 dates did not pass the chi-squared distribution test (mainly in the Jurassic); therefore a second spline run relaxed the smoothing factor to 0.975.

However, the relaxed fit for this spline did not fully utilize the detailed high-precision radio-isotope dates with excellent ammonite zone placements for the Late Cretaceous. As a result, many of those dates were no longer being assigned by that spline-fit into the computed age ranges for those zones. Therefore, a second hybrid spline was constructed – in which the Late Cretaceous had a stricter spline-fit model, and the Early Cretaceous through Jurassic had a less restrictive smoothing factor. The resulting estimates for Cretaceous stage boundaries are listed under "Spline" in Table 27.2. The main change was for base of Hauterivian, which differed by 0.4 myr between these two spline-fit versions.

This Spline #2 also yielded a set of smoothly varying durations for the ammonite zones. However, in many intervals of the Cretaceous, this smoothly varying scaling can be enhanced by using the durations for individual ammonite zones based on cycle stratigraphy. For example, the cycle-scaled duration of the *Pseudaspidoceras flexuosum* ammonite zone is 0.45 myr (Meyers *et al.*, 2012), but the spline-fit projects it as 0.84 myr, twice the actual duration. In other intervals, improved scalings of ammonite zones can incorporate calibrations to the Pacific M-sequence, relative scalings from linear strontium isotope trends (e.g., McArthur *et al.*, 2007) or correlations to cycle-scaled microfossil zones (e.g., Grippo *et al.*, 2004; Huang *et al.*, 2010; Gale *et al.*, 2011).

The spline-fit was therefore enhanced by incorporating these other sets of stratigraphic studies or integrated stratigraphy (e.g., the Bayesian statistical method by Meyers *et al.*, 2012, that optimizes the merger of radio-isotopic dates with the cycle stratigraphy in the Turonian–Cenomanian boundary interval). In most cases, the modification of the ages for those stage boundaries based on the enhanced calibrations of the ammonite zones is less than 0.3 myr from one of the spline-fit versions.

27.3.3. Cretaceous Numerical Age Model

The primary standards for Cretaceous calibrations are the ammonite zones of the Tethyan realm (Berriasian through Albian) and North American Western Interior (Cenomanian through mid-Maastrichtian) (Table 27.3). The age model for the Berriasian through Barremian ammonite zones is mainly derived from correlations to the M-sequence of marine magnetic anomalies (see Geomagnetic Time Scale, Chapter 5) plus intervals with cycle stratigraphy and/or linearization of strontium isotope trends (Table 27.3). The Aptian–Albian ammonite zones are scaled according to their correlation with microfossil and nannofossil datums, which, in turn, are scaled

* Sarah Siewert et al. had kindly provided us with their first manuscript submission. However, their resubmission as GTS2012 was going to press had applied slightly different statistical methods, therefore some computed dates shifted by up to 0.15 myr.

TABLE 27.2 **Summary of the Derivation of the Cretaceous Numerical Age Model for Stages and the Main Ammonite Zonations**

Stage	Spline-Fit Age; (Ma)	Spline-Fit Uncertainty (2-Sigma)	Revised Age (Ma)	Estimated Uncertainty (2-Sigma)	Duration (myr)	Calibration (Brief)	Stage Primary Marker and GSSP or Working Definition; Calibration	Method to Compute Basal Age of Stage	Method for Scaling Ammonite Zones within Stage
PALEOGENE (Danian)	66.0	0.1	**66.0**	0.1		Ar-Ar radio-isotope age	Iridium anomaly caused by bolide impact.	Maastrichtian/ Paleocene boundary = 66.0 Ma (radiometrics in GTS2012) = Chron C29r.5 (Husson et al., 2011, cycle stratigraphy).	
Maastrichtian	72.1	0.2	**72.1**	0.2	6.1	Bracketing radio-isotopic ages	GSSP at Tercis, France, is 90 cm below the lowest occurrences of the ammonoids *Pachydiscus neubergicus* and *Hoploscaphites constrictus*. Carbon-13 curves imply that GSSP level is about 85% up in Chron C32n.2n.88.	Equivalence with base of *Baculites baculus* ammonite zone of Western Interior which is bracketed by Ar-Ar dates. This age is same as cycle-scaled C32n.2n.88 relative to base-Cenozoic.	Spline-fit of Western Interior zones relative to Ar-Ar dates from interbedded ash beds for Lower Maastrichtian. Foraminifer and nannofossil datums for Maastrichtian are calibrated to C-sequence magnetostratigraphy.
Campanian	83.8	0.2	**83.6**	0.3	11.6	Bracketing radio-isotopic ages	GSSP is not yet decided. Set here as base of Chron 33r. Alternate base-Campanian is the extinction of crinoid *Marsupites testudinarius*.	Assumed equivalence with base of *Scaphites leei III* ammonite zone of Western Interior which is bracketed by Ar-Ar dates.	Spline-fit of Western Interior zones relative to Ar-Ar dates from interbedded ash beds. Foraminifer and nannofossil datums for upper Campanian are calibrated to C-sequence magnetostratigraphy.
Santonian	86.2	0.2	**86.3**	0.5	2.6	Bracketing radio-isotopic ages	Lowest occurrence of inoceramid *Cladoceramus undulatoplicatus*.	Assumed to be equivalent to the base of the *Clioscaphites saxitonianus* ammonite zone of the Western Interior, which is bracketed by Ar-Ar and U-Pb dates.	Mainly from cycle stratigraphy durations for these Western Interior zones, plus spline-fit of zones relative to Ar-Ar dates from interbedded ash beds.
Coniacian	89.7	0.1	**89.8**	0.4	3.5	Bracketing radio-isotopic ages	Current working placement is the lowest occurrence of the *Cremnoceramus rotundatus* inoceramid bivalve.	Assumed to be equivalent to the base of the *Scaphites preventricosus* ammonite zone of the Western Interior, which is bracketed by Ar-Ar and U-Pb dates.	Mainly from cycle stratigraphy durations for these Western Interior zones, plus spline-fit of zones relative to Ar-Ar dates from interbedded ash beds.

TABLE 27.2 Summary of the Derivation of the Cretaceous Numerical Age Model for Stages and the Main Ammonite Zonations—cont'd

Stage	Spline-Fit Age; (Ma)	Spline-Fit Uncertainty (2-Sigma)	Revised Age (Ma)	Estimated Uncertainty (2-Sigma)	Duration (myr)	Calibration (Brief)	Stage Primary Marker and GSSP or Working Definition; Calibration	Method to Compute Basal Age of Stage	Method for Scaling Ammonite Zones within Stage
Turonian	94.0	0.2	**93.9**	0.2	4.1	Bracketing radio-isotopic ages	GSSP at Pueblo, Colorado, coincides with lowest occurrence of the ammonite *Watinoceras devonense* and is in the middle of a global positive excursion in carbon-13 isotopes.	GSSP level is bracketed by widespread bentonites that have yielded age of 93.79 to 93.96 Ma; and cycle-extrapolated as 93.90 ± 0.16 myr (Meyers *et al.*, 2010).	Mainly from cycle stratigraphy durations for these Western Interior zones.
Cenomanian	100.3	0.3	**100.5**	0.4	6.6	Bracketing radio-isotopic ages	GSSP at Mt. Risou in SE France coincides with the first appearance datum (FAD) of foraminifer *Rotalipora globotruncanoides*, which is 6 m lower than base of *M. mantelli* ammonite zone at this section.	The uppermost part of the *N. caractanense* Subzone (basal subzone of Cenomanian) is ~99.86 ± 0.37 Ma (Obradovich *et al.*, 2002), and cycle stratigraphy scaling of its duration (0.44 myr) = 100.3 Ma, and a slight 0.25 myr offset to the GSSP marker below implies the base of the Cenomanian is 100.5 ± 0.4 Ma.	Spline-fit of Western Interior zones relative to Ar-Ar dates from interbedded ash beds, plus cycle-stratigraphy durations for some zones.
Albian	113.1	0.2	**113.0**	*0.4*	12.5	12.5 myr below base of Cenomanian	Working definition for GTS2012 is the first appearance datum (FAD) of *Praediscosphaera columnata* (subcircular). GSSP not yet decided.	Total of 30.75 cycles of 405 kyr from base of Cenomanian to FAD of *P. columnata* in Piobbico core.	Ammonite zones of the Tethyan realm have been correlated to most of the nannofossil and foraminifer datums that are in the Albian cycle stratigraphy scale.
Aptian	125.9	0.5	**126.3**	0.4	13.3	25.8 myr below base of Cenomanian is the base of Chron M0r	Working definition for GTS2012 is the base of Chron M0r. GSSP not yet decided.	Total of 62.5 cycles of 405 kyr from base of Cenomanian to top of M0r (hence, 25.31 myr), plus 0.49 myr for cycle duration of M0r = total of 25.8 myr before base-Cenomanian.	Ammonite zones of the Tethyan realm have been correlated to most of the nannofossil and foraminifer datums that are in the Aptian cycle stratigraphy scale.

(*Continued*)

TABLE 27.2 Summary of the Derivation of the Cretaceous Numerical Age Model for Stages and the Main Ammonite Zonations—cont'd

Stage	Spline-Fit Age; (Ma)	Spline-Fit Uncertainty (2-Sigma)	Revised Age (Ma)	Estimated Uncertainty (2-Sigma)	Duration (myr)	Calibration (Brief)	Stage Primary Marker and GSSP or Working Definition; Calibration	Method to Compute Basal Age of Stage	Method for Scaling Ammonite Zones within Stage
Barremian	130.5	0.5	**130.8**	0.5	4.5	Chron M5n.8	Base of *Spitidiscus hugii* ammonite zone; assigned as 80% up in Chron M5n. GSSP not yet decided.	Chron M5n.8 from cycle-scaled spreading rate model for Pacific M-sequence.	Base of the Upper Barremian (base of *Ancyloceras vandenheckii* Zone) is approximately 90% up in Chron M3r; and zones above/below are scaled according to cycle stratigraphy durations (e.g., Bodin *et al.*, 2006) or using relative numbers of subzones.
Hauterivian	133.5	0.5	**133.9**	0.6	3.1	Base of Chron M10n	Base of *Acanthodiscus radiatus* ammonite zone of Tethyan realm is near the beginning of Chron M10n (e.g., Weissert and others), as supported by later studies (e.g, McArthur *et al.*, 2007), and consistent with the cycle-scaled duration of the Valanginian. Assigned as base M10n. GSSP not yet decided.	Base of Chron M10n from cycle-scaled spreading rate model for M-sequence of marine magnetic anomalies in Pacific.	Late Hauterivian zones scaled according to cycle stratigraphy durations (e.g., Bodin *et al.*, 2006); and Early Hauterivian using relative duration of each ammonite subzone derived from the assumption by McArthur *et al.* (2007) of a linear trend in strontium isotope values.
Valanginian	139.0	0.8	**139.4**	0.7	5.5	Chron M14r.3	Base of *Tirnovella pertransiens* ammonite zone in Tethyan realm, which is 30% up in Chron M14r. GSSP not yet decided.	Chron M14r.3 from cycle-scaled spreading rate model for Pacific M-sequence.	Relative duration of each ammonite subzone within the stage is derived from the assumption by McArthur *et al.* (2007) of a linear trend in strontium isotope values.
Berriasian	144.6	1.5	**145.0**	0.8	5.6	Base of Chron M18r	Not yet defined. Set as Base of Chron M18r for GTS2012, but may eventually be assigned to a calpionellid, nannofossil or ammonite datum.	Base of Chron M18r from cycle-scaled spreading rate model for Pacific M-sequence.	Correlation of Tethyan ammonite zones to M-sequence magnetostratigraphy.

Values are rounded to nearest 0.1 myr. A spline fit through the database of Jurassic—Cretaceous radio-isotopic ages (Appendix 2) was run, then the final age model incorporated cycle stratigraphic constraints on stage and zone durations, segments with linear trends in strontium isotopes (Sinemurian—Pliensbachian), removing anomalous jumps in Pacific spreading rates (late Jurassic through early Cretaceous) and other stratigraphic or radio-isotopic constraints (see text).

TABLE 27.3 Interpolated Numerical Ages for Ammonite Zones of the Early Cretaceous Tethyan Faunal Realm and the Late Cretaceous Western Interior Realm

Stage/Age	Tethyan Zone	Basal Age (Ma)	Notes on Calibration and Zonal Usages
PALEOGENE base	TOP	66.00	
	[Few usable ammonites]	68.08	DOTTED —No constraints.
	Jeletzkytes nebrascensis	68.69	DOTTED —Used average durations of *H. birkelundae* through *J. nebrascensis* from spline-fit (0.61 myr).
	Hoploscaphites nicolleti	69.30	DOTTED —Used average durations of *H. birkelundae* through *J. nebrascensis* from spline-fit (0.61 myr).
	Hoploscaphites birkelundae	69.91	Base is uppermost age constraint on Cretaceous spline (base constrained by 70.08 ± 0.37 Ma in underlying uppermost *B. clinolobatus*). Used average durations of *H. birkelundae* through *J. nebrascensis* from spline-fit (0.61 myr). Cobban *et al.* (2006) spelled it as *"birkelundae"* instead of their previous *"birkelundi"*.
	Baculites clinolobatus	70.44	Top of zone contains bentonite dated as 70.08 ± 0.37 Ma. 0.53 myr duration from spline-fit of dated interbedded bentonites in W. Interior.
	Baculites grandis	71.13	Zone contains bentonite dated as 70.66 ± 0.65 Ma. 0.69 myr duration from spline-fit of dated interbedded bentonites in W. Interior.
Maastrichian base equivalent	*Baculites baculus*	72.05	Duration = 0.92 myr from spline-fit of dated interbedded bentonites in W. Interior.
	Baculites eliasi	72.74	Zone contains bentonite dated as 72.50 ± 0.31 Ma. Duration = 0.69 myr (spline-fit of dated interbedded bentonites in W. Interior).
	Baculites jenseni	73.27	Duration = 0.53 myr (spline-fit of dated interbedded bentonites in W. Interior).
	Baculites reesidei	73.63	Upper half of this zone contains bentonite dated as 73.41 Ma (± 0.47). Duration = 0.36 myr (spline-fit of dated interbedded bentonites in W. Interior).
	Baculites cuneatus	73.91	Duration = 0.28 myr (spline-fit of dated interbedded bentonites in W. Interior).
	Baculites compressus	74.21	Contains bentonite dated as 74.04 Ma (± 0.39). Duration = 0.30 myr (spline-fit of dated interbedded bentonites in W. Interior).
	Didymoceras cheyennense	74.60	Duration = 0.39 myr (spline-fit of dated interbedded bentonites in W. Interior).
	Exiteloceras jenneyi	75.08	Contains bentonite dated as 74.85 Ma (± 0.43). Duration = 0.48 myr (spline-fit of dated interbedded bentonites in W. Interior).
	Didymoceras stevensoni	75.64	Duration = 0.56 myr (spline-fit of dated interbedded bentonites in W. Interior).
	Didymoceras nebrascense	76.27	Duration = 0.63 myr (spline-fit of dated interbedded bentonites in W. Interior).
	Baculites scotti	76.94	Contains bentonite dated as 76.62 Ma (± 0.51). Duration = 0.67 myr (spline-fit of dated interbedded bentonites in W. Interior).

(Continued)

TABLE 27.3 **Interpolated Numerical Ages for Ammonite Zones of the Early Cretaceous Tethyan Faunal Realm and the Late Cretaceous Western Interior Realm—cont'd**

Stage/Age	Tethyan Zone	Basal Age (Ma)	Notes on Calibration and Zonal Usages
	Baculites reduncus	77.63	Zone in Cobban *et al.* (2006); not in GTS2004 that used an earlier Cobban zonation. Duration = 0.69 myr (spline-fit of dated interbedded bentonites in W. Interior).
	Baculites gregoryensis	78.34	Duration = 0.71 myr (spline-fit of dated interbedded bentonites in W. Interior).
	Baculites perplexus	79.01	Duration = 0.67 myr (spline-fit of dated interbedded bentonites in W. Interior).
	Baculites sp. (smooth)	79.64	Duration = 0.63 myr (spline-fit of dated interbedded bentonites in W. Interior).
	Baculites asperiformis	80.21	Duration = 0.57 myr (spline-fit of dated interbedded bentonites in W. Interior). Reported to span Chron C33r/C33n polarity reversal (Leberkmo, 1989; who reports a 79.92 ± 0.6 Ma age).
	Baculites maclearni	80.67	Duration = 0.46 myr (spline-fit of dated interbedded bentonites in W. Interior).
	Baculites obtusus	80.97	Contains bentonites dated as 80.62 Ma (± 0.40) and 81.30 ± 0.55 Ma. Duration = 0.30 myr (spline-fit of dated interbedded bentonites in W. Interior).
	Baculites sp. (weak flank ribs)	81.13	Duration = 0.16 myr (spline-fit of dated interbedded bentonites in W. Interior).
	Baculites sp. (smooth)	81.28	Duration = 0.15 myr (spline-fit of dated interbedded bentonites in W. Interior).
	Scaphites hippocrepis III	81.53	Duration = 0.25 myr (spline-fit of dated interbedded bentonites in W. Interior).
	Scaphites hippocrepis II	82.00	Contains bentonites dated as 81.71 Ma (± 0.51) and 82.29 ± 0.34 within lower part of zone. Duration = 0.47 myr (spline-fit of dated interbedded bentonites in W. Interior).
	Scaphites hippocrepis I	82.70	Duration = 0.76 myr (spline-fit of dated interbedded bentonites in W. Interior); but reduced to 0.7 to fit base-Campanian extrapolated age.
Campanian base equivalent	*Scaphites leei III*	83.64	Age of base is extrapolated as 83.64 ± 0.23 Ma by (Siewert, 2011; or 83.75 ± 0.11 Ma by Siewert *et al.*, in press). Duration was estimated as 0.9 myr (spline-fit of dated interbedded bentonites in W. Interior).
	Desmoscaphites bassleri	84.08	Cycle strat duration = 0.44 myr (Siewert *et al.*, in press). Contains bentonites dated as 84.33 ± 0.27 and 84.43 ± 0.15 Ma.
	Desmoscaphites erdmanni	84.52	Cycle strat duration = 0.44 myr (Siewert *et al.*, in press). Contains bentonite dated as 84.57 ± 0.35 Ma in upper part.
	Clioscaphites choteauensis	85.23	Base-age from cumulative cycles from base-Santonian; and top-age from cumulative cycles from base-Campanian (Siewert *et al.*, in press).
	Clioscaphites vermiformis	85.56	Cycle strat duration = 0.33 myr (Siewert *et al.*, in press). Contains bentonite dated as 85.55 ± 0.25 Ma in middle of zone.

TABLE 27.3 Interpolated Numerical Ages for Ammonite Zones of the Early Cretaceous Tethyan Faunal Realm and the Late Cretaceous Western Interior Realm—cont'd

Stage/Age	Tethyan Zone	Basal Age (Ma)	Notes on Calibration and Zonal Usages
Santonian base equivalent	*Clioscaphites saxitonianus*	86.26	Age is extrapolated as 86.26 ± 0.45 Ma by Siewert, 2011; or 86.35 ± 0.11 Ma by Siewert *et al.*, in press. Cycle strat duration of zone = 0.70 myr (Siewert *et al.*, in press). Base-Santonian is lowest occurrence of the widespread inoceramid bivalve *Cladoceramus undulatoplicatus.* Following chart in Cobban *et al.* (2006), this level is equated to the base of the *C. saxitonianus* ammonite zone of Western Interior; but Gale (chart to J. Ogg, Aug 2011) projects as uppermost *S. depressus* Zone.
	Scaphites depressus	87.86	Cycle strat duration = 1.60 myr (Siewert *et al.*, in press). Contains bentonites dated as 86.98 ± 0.24 and 87.11 ± 0.15 Ma in its lower part. Sometimes a hybrid zone of *Scaphites depressus–Protexanites bourgeoisianus*
	Scaphites ventricosus	88.77	Base-age from cumulative cycles from base-Coniacian; and top age from cumulative cycles downward from base-Santonian (Siewert *et al.*, in press).
Coniacian base equivalent	*Scaphites preventricosus*	89.77	Cycle strat duration = 1.0 myr (Siewert *et al.*, in press). Contains bentonite dated as 89.09 ± 0.33 Ma in upper part, and 89.30 ± 0.27 and 89.37 ± 0.15 Ma in lower part. Formerly "*Forresteria allaudi–Scaphites preventricosus*" Zone; but called *S. preventricosus* in Cobban *et al.* (2006).
	Scaphites mariasensis	89.87	Duration = 0.10 myr (spline-fit of dated interbedded bentonites in W. Interior). Formerly *Forresteria peruana*; but now *Scaphites mariasensis* in Cobban *et al.* (2006).
	Prionocyclus germari	89.98	Duration = 0.11 myr (spline-fit of dated interbedded bentonites in W. Interior).
	Scaphites nigricollensis	90.24	Contains bentonite dated as 89.86 ± 0.26 Ma. Duration = 0.26 myr (spline-fit of dated interbedded bentonites in W. Interior).
	Scaphites whitfieldi	90.65	Duration = 0.41 myr (spline-fit of dated interbedded bentonites in W. Interior).
	Scaphites ferronensis	91.08	Duration = 0.43 myr (spline-fit of dated interbedded bentonites in W. Interior).
	Scaphites warreni	91.34	Duration = 0.26 myr (spline-fit of dated interbedded bentonites in W. Interior).
	Prionocyclus macombi	91.41	Contains bentonites dated as 91.07 ± 0.28 and 91.37 ± 0.16 in uppermost and 91.30 ± 0.27 in lower part of this zone. Duration is only 0.07 myr from spline-fit of dated interbedded bentonites in W. Interior; which may indicate one or more of these levels require re-dating.
	Prionocyclus hyatti	91.60	Contains bentonite dated as 91.15 ± 0.26 Ma. Duration = 0.19 myr (spline-fit of dated interbedded bentonites in W. Interior).
	Collignoniceras praecox	92.08	Duration = 0.48 myr (spline-fit of dated interbedded bentonites in W. Interior). Formerly called *Prionocyclus percarinatus*; but renamed in Cobban *et al.* (2006)

(*Continued*)

TABLE 27.3 **Interpolated Numerical Ages for Ammonite Zones of the Early Cretaceous Tethyan Faunal Realm and the Late Cretaceous Western Interior Realm—cont'd**

Stage/Age	Tethyan Zone	Basal Age (Ma)	Notes on Calibration and Zonal Usages
	Collignoniceras woollgari	92.90	Basal age from cumulative cycles from base-Turonian; upper age from cumulative spline-fit durations from base-Coniacian.
	Mammites nodosoides	93.35	Cycle strat duration = 0.45 myr (Meyers *et al.*, 2012).
	Vascoceras birchbyi	93.45	Cycle strat duration = 0.1 myr (Meyers *et al.*, 2012).
	Pseudaspidoceras flexuosum	93.55	Cycle strat duration = 0.1 myr (Meyers *et al.*, 2012). Contains bentonite dated as 93.67 ± 0.26 Ma in top part of this zone.
Turonian base	*Watinoceras devonense*	93.90	The base of the Turonian Stage is at the base of Bed 86 of the Bridge Creek Limestone Member, west of Pueblo, Colorado, USA. The GSSP coincides with the first occurrence of the ammonite *Watinoceras devonense*, is in the middle of a global positive excursion in carbon-13 isotopes, and is bracketed by widespread bentonites that have yielded age of 93.79 to 93.96 Ma; and cycle-extrapolated as 93.90 ± 0.15 myr (Meyers *et al.*, 2010, 2012). Cycle strat duration = 0.35 myr (Meyers *et al.*, 2012); contains bentonite dated as 93.79 ± 0.26 Ma in upper part of this zone.
	Nigericeras scotti	93.98	Cycle strat duration = 0.08 myr (Meyers *et al.*, 2012).
	Neocardioceras juddii	94.15	Cycle strat duration = 0.17 myr (Meyers *et al.*, 2012). Contains bentonite set dated as 94.10, 94.01 and 94.29 Ma (all about ± 0.2) are in upper part of this zone.
	Burroceras clydense	94.27	Duration = 0.11 myr (spline-fit of dated interbedded bentonites in W. Interior).
	Sciponoceras gracile (Euomphaloceras septemseriatum)	94.39	*Sciponoceras gracile* was subdivided into two zones in Cobban et al. (2006). Duration of upper = 0.12 myr (spline-fit of dated interbedded bentonites in W. Interior). Contains bentonite dated as 94.20 ± 0.28 Ma.
	Sciponoceras gracile (Vascoceras diartianum)	94.57	*Sciponoceras gracile* was subdivided into two zones in Cobban et al. (2006). Duration of lower = 0.21 myr (spline-fit of dated interbedded bentonites in W. Interior). Contains bentonite dated as 94.43 ± 0.29 Ma.
	Dunveganoceras conditum	94.78	Duration = 0.21 myr (spline-fit of dated interbedded bentonites in W. Interior).
	Dunveganoceras albertense	95.01	Duration = 0.23 myr (spline-fit of dated interbedded bentonites in W. Interior).
	Dunveganoceras problematicum	95.24	Duration = 0.23 myr (spline-fit of dated interbedded bentonites in W. Interior).
	Dunveganoceras pondi	95.47	Formerly called *Calycoceras canitaurinum–Dunveganoceras pondi*. Duration = 0.23 myr (spline-fit of dated interbedded bentonites in W. Interior). Contains bentonite dated as 95.32 ± 0.61 Ma.
	Plesiacanthoceras wyomingsense	95.67	Duration = 0.20 myr (spline-fit of dated interbedded bentonites in W. Interior).
	Acanthoceras amphibolum	95.81	Duration = 0.14 myr (spline-fit of dated interbedded bentonites in W. Interior). Contains bentonites dated as 95.53 ± 0.16 and 95.87 ± 0.1 Ma.
	Acanthoceras bellense	95.90	Duration = 0.09 myr (spline-fit of dated interbedded bentonites in W. Interior).

TABLE 27.3 Interpolated Numerical Ages for Ammonite Zones of the Early Cretaceous Tethyan Faunal Realm and the Late Cretaceous Western Interior Realm—cont'd

Stage/Age	Tethyan Zone	Basal Age (Ma)	Notes on Calibration and Zonal Usages
	Acanthoceras muldoonense	95.98	Duration = 0.08 myr (spline-fit of dated interbedded bentonites in W. Interior).
	Acanthoceras granerosense	96.08	Duration = 0.10 myr (spline-fit of dated interbedded bentonites in W. Interior).
	Conlinoceras tarrantense–Conlinoceras gilberti	96.24	Duration = 0.16 myr (spline-fit of dated interbedded bentonites in W. Interior). Contains bentonite dated as 96.12 ± 0.31 Ma.
	[gap in N.Amer. zonation]	97.26	Duration = 0.43 myr (spline-fit of dated interbedded bentonites in W. Interior).
	Neogastroplites maclearni	97.76	Duration = 0.50 myr (spline-fit of dated interbedded bentonites in W. Interior).
	Neogastroplites americanus	98.19	Duration = 0.43 myr (spline-fit of dated interbedded bentonites in W. Interior).
	Neogastroplites muelleri	98.75	Duration = 0.56 myr (spline-fit of dated interbedded bentonites in W. Interior).
	Neogastroplites cornutus	99.17	Duration = 0.28 myr (spline-fit of dated interbedded bentonites in W. Interior); changes to 0.35 to fit overlying adjustment. Contains bentonite dated as 99.25 ± 0.41 Ma.
BASE of N. Amer. ammonite zonation	*Neogastroplites haasi* (N. American zonation)	99.81	Contains bentonites dated as 99.24 ± 0.7 and 99.46 ± 0.59 Ma; which is consistent with dated bentonites from *M. saxbii* Subzone.
Cenomanian base is about 0.2 myr below	*Mantelliceras mantelli* (Tethyan zonation)	100.25	The upper *Mantelliceras saxbii* s.z. (Europe) contains bentonite dated as 99.70 ± 0.38 Ma near base. The top of *N. carcitanense* Subzone (lower subzone) contains bentonite dated as 99.89 ± 0.37 near top, and cycle stratigraphy scaling of that subzone duration (0.44 myr) and a slight offset to the GSSP marker below implies the base of the Cenomanian is ~100.25 Ma.
	Arrhaphoceras briacensis	100.91	Full name = *Arrhaphoceras (Praeschloenbachia) briacensis*. "introduced by Scholz (1973) for an uppermost Albian interval in which *Mortoniceras* and *Ostlingoceras (O.) puzosianum* were absent, and *Hyphoplites* had appeared." (Kennedy, in Gale *et al.*, 2011). Spans Albian/Cenomanian boundary. Projected as 1 cycle (405 kyr) below base-Cenomanian at Col de Palluel (Gale *et al.*, 2011).
	Mortoniceras perinflatum	101.41	Full name = *Mortoniceras (Subschloenbachia) perinflatum*. Begins at FAD of index species; which is projected as 2 1/4 cycles (of 405 kyr) below base-Cenomanian at Col de Palluel (Gale *et al.*, 2011).
	Mortoniceras rostratum	101.72	Full name = *Mortoniceras (Subschloenbachia) rostratum.* Short zone defined by FAD of index species, which begins 3 cycles (of 405 kyr) below base-Cenomanian at Col de Palluel (Gale *et al.*, 2011); or just above base of *P. buxtorfi* foram zone. GTS2004: Base of *S. dispar* correlated to base of *R. appenninica* foraminifer zone (cycle-scaled as 100.9 Ma).
	Mortoniceras fallax	103.13	Base (poorly delimited) is about 6 1/2 cycles (of 405 kyr) below base-Cenomanian at Col de Palluel, and coincides with FAD of *Turriseffeli* nannofossil (Gale *et al.*, 2011).

(Continued)

TABLE 27.3 **Interpolated Numerical Ages for Ammonite Zones of the Early Cretaceous Tethyan Faunal Realm and the Late Cretaceous Western Interior Realm—cont'd**

Stage/Age	Tethyan Zone	Basal Age (Ma)	Notes on Calibration and Zonal Usages
	Mortoniceras inflatum	103.94	Dashed (see last sentence). Base is FAD of index species = 8 1/2 cycles (of 405 kyr) below base-Cenomanian at Col de Palluel, and is at base of *T. ticinensis* Foram zone (Gale *et al.*, 2011) BUT, there may be a hiatus in their section at this point.
	Mortoniceras pricei	106.98	Base is FAD of index species above LAD of *D.cristatum* (underlying zone index). Base is 16 cycles (of 405 kyr) below base-Cenomanian at Col de Palluel (Gale *et al.*, 2011).
	Diploceras cristatum	107.59	FAD of *D. cristatum* defines base of Upper Albian; and "traditionally" was a subzone in an expanded *M. inflatum,* but raised to zone status (e.g., review by Kennedy in Gale *et al.*, 2011). Base is 17 1/2 cycles (of 405 kyr) below base-Cenomanian at Coll de Palluel, and is just above base of *T. praeticinensis* Foram zone (Gale *et al.*, 2011). In GTS2004, the base of extended *M. inflatum* (with *D. cristatum* as subzone) was assigned as FAD of foraminifer *B. breggiensis* (base of *T. praeticinensis* Subzone in Leckie et al., 2002), which is essentially same correlation.
	Euhoplites lautus	108.25	Base of *lautus/nitidus.* No direct assignment to Foram zones. Equal subzones in Middle Albian of Boreal zones used for scaling.
	Euhoplites loricatus	109.56	Base of *loricatus/intermedius.* No direct assignment to Foram zones. Equal-subzones in Middle Albian of Boreal zones used for scaling.
	Hoplites dentatus	110.22	Base of Upper Albian = Base of *dentatus/lyelli* Zone/ Subzone (full zone name = *Hoplites (Hoplites) dentatus*). No direct assignment to Foram zones; but Gale *et al.* (2011) assign (dashed) at 25 1/4 cycles (of 405 kyr) below base-Cenomanian. Duration scaled according to its 2 subzones.
	Douvilleiceras mammillatum	110.87	Base of *mammillatum/perinflata.* There is a problem of correlation to Tethys; summarized in Reboulet *et al.* (2011; Kilian Group report). Herrle and Mutterlose (2003) dash in their composite section the zone base at 15 m above Paquier organic-rich layer, compared to 90 m span between Kilian and Paquier => considering these two levels are 4.6 cycles of 0.405 myr apart, a constant accumulation rate assigns the base-*mammillatum* as approximately 1 cycle (0.405 myr) above Paquier (used here).
	Leymeriella tardefurcata	111.27	Tethyan version correlates with base of Paquier organic-rich layer, which is 26.6 cycles below base-Cenomanian (Huang *et al.*, 2010; Gale *et al.*, 2011). NOTE: In N. Germany regional zonation, there is another ammonite zone of lowest Albian below *L. tardefurcata.*
Albian base in GTS2012 (working version) is in upper part of this Tethyan zone	*Hypacanthoplites jacobi*	115.64	Base of *H. jacobi* is dashed as 1/2 up in *T. bejaouensis* Foram zone (on composite meters) by Herrle *et al.* (2004) in SE France composite. It is a very long zone in SE France sections and crosses the base-Albian working definition used in GTS2012.

TABLE 27.3 Interpolated Numerical Ages for Ammonite Zones of the Early Cretaceous Tethyan Faunal Realm and the Late Cretaceous Western Interior Realm—cont'd

Stage/Age	Tethyan Zone	Basal Age (Ma)	Notes on Calibration and Zonal Usages
	Acanthohoplites nolani	116.83	Zone is also called "*Nolaniceras nolani*". Base of *Nolani* (hence, top of *P. nutfieldiensis* Zone) is dashed as 1/4th up in *T. bejaouensis* Foram zone (on composite meters) by Herrle *et al.* (2004) in SE France composite.
	Parahoplites melchioris	118.02	Base of *P. melchioris* or "*Parahoplites nutfieldiensis*" ammonite zone is at base of *T. bejaouaensis* Foram zone in SE France (summarized in Herrle *et al.*, 2004); and that Foram zone is cycle-scaled relative to Apt–Alb boundary (Huang *et al.*, 2010) = scale for GTS2012.
	Epicheloniceras martinoides	122.98	Base of *E.martinoides* ammonite zone (full name = *Cheloniceras (Epicheloniceras) martinoides*; called *Epicheloniceras subnodosocostatum* on SEPM chart) is at base of *G. ferreolensis* Foram zone in SE France (summarized in Herrle *et al.*, 2004); and that Foram zone is cycle-scaled relative to Apt–Alb boundary (Huang *et al.*, 2010). Equivalent to *Cheloniceras (Epicheloniceras) martinoides* Zone.
	Dufrenoyia furcata	123.88	Base of *D. furcata* Zone = base of nannofossil *E. floralis* in SE France (Moullade *et al.*, 1998); therefore used Aptian cycle-placement of this event relative to Apt–Alb boundary (Huang *et al.*, 2010).
	Deshayesites deshayesi	125.71	Base of *D. deshayesi* = FAD of foraminifer *R. angustus* in SE France; therefore used Aptian cycle-placement of this event relative to Apt–Alb boundary (Huang *et al.*, 2010). *D. deshayesi* to *D. oglanlensis* (3 zones) assigned equal duration.
	Deshayesites forbesi	126.01	Also called *Deshayesites weissi* in some zonations. DASH – midway between base of *D. deshayesi* and *D. oglanlensis* Zone an "unnamed" subzone was added to the base when an "upper" subzone was created (Kilian group; 2011).
Aptian base equivalent	*Deshayesites oglanlensis*	126.30	Working version for base of Aptian is base of polarity Chron M0r; and the base of *D. oglanlensis* is assumed to be close to this level. *D. tuarkyricus* Zone is used as basal Aptian in some charts; but the *D. tuarkyricus* index species is only found in Turkmenistan; therefore, IUGS Lower Cretaceous group proposes *D. oglanlensis*. Zone is approximately equivalent to *Prodeshayesites fissicostatus* of Boreal realm.
	Imerites giraudi	127.47	Barremian tethyan ammonite zone '*I. giraudi*' replaces '*Heroceras astieri*' in some other zonal scales. Was a brief zone (in GTS2008) but that former range was demoted to a subzone (Killian group; 2009/2011), and the zone name extended upward to encompass the previous two "overlying zones" as subzones. Upper Barremian zones assigned equal duration until cycle stratigraphy or magnetostratigraphy calibrations are accomplished.
	Gerhardtia sartousi	128.63	Barremian tethyan ammonite zone '*H. sartousi*' (or '*sartousiana*') is a subzone of the '*H. feraudianus*' zone in some other zonal scales.

(Continued)

TABLE 27.3 **Interpolated Numerical Ages for Ammonite Zones of the Early Cretaceous Tethyan Faunal Realm and the Late Cretaceous Western Interior Realm—cont'd**

Stage/Age	Tethyan Zone	Basal Age (Ma)	Notes on Calibration and Zonal Usages
	Ancyloceras vandenheckii	129.41	Approximately Chron M3r.8 in Italy (Bartolocci *et al.*, 1992; Channell *et al.*, 1995a). Barremian tethyan ammonite zone '*A. vandenheckii*' has replaced the lower portion of the '*Emiriceras*' *barremense*' of other zonal scales.
	Moutoniceras moutonianum	129.60	Was named "*Coronites darsi*" in GTS2008, which is now used the basal horizon in this new zone. Zones of Lower Barremian were extensively revised (Kilian Group, 2002). In particular, the former "*H. caillaudianus*" was subdivided into 3 zones. Upper 3 zones of Early Barremian given equal durations.
	Kotetishvilia compressissima	129.78	Former base of *Holcodiscus caillaudianus* Zone. Zones of Lower Barremian were extensively revised (Kilian Group, 2002). Formerly "*Pulchellia*" *compressissima*". Barremian tethyan ammonite zones of '*Holcodiscus caillaudianus*' (formerly spelled '*H. callaudi*') and '*N. nicklesi*' were incorporated as subzones in a single zone '*Pulchellia compressissima*' in earlier time scales. Upper 3 zones of Early Barremian given equal durations.
	Nicklesia pulchella	129.97	*N. pulchella* was elevated to zone status by Killian Group (2002); was formerly a horizon. Upper 3 zones of Early Barremian given equal durations.
	Nicklesia nicklesi	130.37	Cycle stratigraphy (Bodin *et al.*, 2006) suggests a duration of about 0.4 myr. Barremian Tethyan ammonite zone *N. nicklesi* had incorporated a former separate zone '*Niclesia pulchella*' of SE France as its upper horizon (Hoedemaeker *et al.*, 1993). This former zone only occupied a few beds in the Angles section, and could not be recognized without the index species.
Barremian base	*Taveraidiscus hugii*	130.77	Chron M5n.8. Full name = *Taveraidiscus hugii auctorum.* Formerly called *Spitidiscus hugii.* Cycle stratigraphy (Bodin *et al.*, 2006) suggests a duration of about 0.4 myr. New subzone at base (Reboullet *et al.*, 2009).
	Pseudothurmannia ohmi	131.57	Cycle stratigraphy (Bodin *et al.*, 2006) suggests a duration of about 0.8 myr. Formerly called *Pseudothurmannia angulicostata auctorum.*
	Balearites balearis	132.37	Cycle stratigraphy (Bodin *et al.*, 2006) suggests a duration of about 0.8 myr. Genus was formerly *Pseudothurmannia.*
	Pleisiospitidiscus ligatus	132.68	Duration = 21% of pre-*balearis* Hauterivian (McArthur *et al.*, 2007). Hauterivian tethyan ammonite zone "*Plesiospitidiscus ligatus*" is sometimes in quotes, because the index species is not a true '*P. ligatus*', but 'sed subseq. litt.'
	Subsaynella sayni	133.09	Duration = 27% of pre-*balearis* Hauterivian (McArthur *et al.*, 2007).
	Lyticoceras nodosoplicatus	133.35	Duration = 17% of pre-*balearis* Hauterivian (McArthur *et al.*, 2007).
	Crioceratites loryi	133.59	Duration = 16% of pre-*balearis* Hauterivian (McArthur *et al.*, 2007).

TABLE 27.3 **Interpolated Numerical Ages for Ammonite Zones of the Early Cretaceous Tethyan Faunal Realm and the Late Cretaceous Western Interior Realm—cont'd**

Stage/Age	Tethyan Zone	Basal Age (Ma)	Notes on Calibration and Zonal Usages
Hauterivian base	*Acanthodiscus radiatus*	133.88	Base of Chron M10n (GTS2004 had as base Chron M11n, using an earlier published correlation). *A. radiatus* duration is 19% of duration of pre-*balearis* Hauterivian (McArthur *et al.*, 2007).
	Criosarasinella furcillata	134.98	*C. furcillata* duration is 20% of duration of Valanginian (McArthur *et al.*, 2007). Former "*Neocomites (Teschenites) pachydicranus*" Zone has been completely replaced in new zonal scheme by Kilian Group (2002). New upper zone has 2 subzones (*Teschenites callidiscus* and *C. furcillata*).
	Neocomites peregrinus	135.81	*N. peregrinus* duration is 15% of duration of Valanginian (McArthur *et al.*, 2007).
	Saynoceras verrucosum	136.41	*S. verrucosum* duration is only 11% of duration of Valanginian (McArthur *et al.*, 2007).
	Busnardoites campylotoxus	137.68	*B. campylotoxus* duration is 23% of duration of Valanginian (McArthur *et al.*, 2007).
Valanginian base	*Tirnovella pertransiens*	139.39	Base of Chron M14r.3. *T. pertransiens* duration is 31% of duration of Valanginian (McArthur *et al.*, 2007). Name "*Thurmanniceras*" *pertransiens* has also been used for this ammonite.
	Subthurmannia boissieri	141.93	Chron M16r.5. "*Fauriella*" and "*Berriasella*" have also been used as genus names for this species "*boissieri*".
	Subthurmannia occitanica	143.57	Chron M17r.3. "*Neocomites*" and "*Tirnovella*" *subalpina* have also been used for this ammonite.
Berriasian working definition (base-M18r) is in lower part of this zone	*Berriasella jacobi*	145.95	About Chron M19n.2n.3; uncertain due to probable distortions in sedimentation in the compact reference sections. At Puerto Escano (Pruner *et al.*, 2010) base is about M19n.2n.1. At Sierra Gorda (Ogg *et al.*, 1984), zone base is M19n.2n.55 (+/-.05). A potential definition for base-Berriasian. Base of Chron M18r (working version for boundary in GTS2012) is within this zone.

NOTE: placement of some stage boundaries are approximate or are working definitions pending decision of the International Commission on Stratigraphy. Usage of zones/subzones varies among Tethyan regions, and this suite is mainly for the Sub-Mediterranean province following recommendations of the IUGS Lower Cretaceous Ammonite Working Group or "Kilian Group" (e.g., Reboulet *et al.*, 2011). Western Interior zones are from Cobban *et al.* (2006). Two-decimal ages are given to show relative durations among the zones; and uncertainties are equal to or greater than the uncertainties on the boundaries of its stage (Table 27.4). The column of notes includes informal remarks on calibrations and different zonal usages. Details for subzone age models and for calibrations of other biostratigraphic zonations can be found in pop-up windows for item within the TimeScale Creator database; *www.tscreator.org*.

by cycle stratigraphy. A spline-fit (Spline #2, as explained above) of numerous radio-isotopic dates from volcanic ash horizons interbedded with Western Interior ammonites with adjustments for cycle stratigraphy of some intervals provides a high-resolution numerical scale for the Cenomanian through early Maastrichtian. The late Maastrichtian correlations rely on microfossil datums calibrated to a spline- and cycle-fit of C-sequence marine magnetic anomalies.

Only three Cretaceous stages have ratified GSSPs, therefore the numerical ages for the other stages are based on the ages for selected working definitions (e.g., base of Aptian placed at base of magnetic polarity Chron M0r; base of Albian assigned to the lowest occurrence of a nannofossil in the cycle-scaled reference section) (Table 27.2).

27.3.3.1. Early Cretaceous Scaling

Berriasian Through Barremian

An extensive suite of studies by numerous magnetostratigraphy groups have calibrated nearly every ammonite zone, microfossil zone and major geochemical excursion from the Tethyan realm to the M-sequence of marine magnetic anomalies. Cycle stratigraphy has provided constraints on the duration of suites of these M-sequence chrons and of several ammonite zones. In turn, these cycle-duration segments, coupled with radio-isotopic dating of Chron M0r, enable assignment of numerical ages to each polarity chron (see Chapter 5 on the geomagnetic time scale, this volume), and therefore to all biostratigraphic datums that are calibrated to that M-sequence. Uncertainties are derived from the variance on the M-sequence line-fit model plus the uncertainty on the radio-isotopic date near the base of Chron M0r.

The base of the Berriasian is assigned in GTS2012 to the base of Chron 18r, which has an age of 145.0 ± 0.8 Ma from the M-sequence model. This is consistent with a recalibrated $^{40}Ar/^{39}Ar$ date of 145.5 ± 0.8 Ma from reversed-magnetized (interpreted as magnetozone M18r) sills intruded into earliest Berriasian pelagic sediments drilled at ODP Site 1213B on Shatsky Rise (Mahoney *et al.*, 2005). Each Berriasian ammonite zone is calibrated to the M-sequence.

The base of the Valanginian (base of *Tirnovella pertransiens* ammonite zone in the Tethyan realm) is assigned here as Chron M14r.3, which has an age of 139.4 ± 0.7 Ma. Magnetostratigraphy within ammonite-zoned sections spanning the bulk of the Valanginian (e.g., southeastern France) has been largely unsuccessful. However, ammonite-zoned sections have yielded a pronounced trend in strontium isotopes (McArthur *et al.*, 2007). Therefore, the relative duration of each ammonite subzone within the stage is derived from the assumption by McArthur *et al.* (2007) of a linear trend in those isotope values.

The base of the Hauterivian (base of *Acanthodiscus radiatus* ammonite zone of the Tethyan realm) is near the beginning of Chron M10n (e.g., Weissert *et al.*, 1998; McArthur *et al.*, 2007), which has a model age of 133.9 ± 0.6 Ma. The implied 5.5-myr duration of the Valanginian is consistent with cycle stratigraphy of that stage. The internal scaling of the lower and middle Hauterivian is similar to the Valanginian, in which the proportional duration of each ammonite subzone relies on the estimates for that subzone from strontium isotope trends (McArthur *et al.*, 2007). The uppermost ammonite zones (*B. balearis* and *P. ohmi*) are each about 0.8 myr in duration, based on the generalized estimates from cycle stratigraphy by Bodin *et al.* (2006).

The base of the Barremian (base of *Spitidiscus hugii* ammonite zone) is approximately 80% up in Chron M5n, which implies an age of 130.8 ± 0.5 Ma. The base of the Upper Barremian (base of *Ancyloceras vandenheckii* Zone) is approximately 90% up in Chron M3r; and the base of the Aptian is assigned as the base of Chron M0r. Between these three calibrated levels, the suites of ammonite zones, which have been extensively redefined and renamed since GTS2004 (e.g., Reboulet *et al.*, 2006, 2009, 2011), are proportionally scaled according the relative numbers of subzones (Upper Barremian), as equal zones (Lower Barremian above *N. nicklesi* Zone) and according to generalized cycle stratigraphy estimates by Bodin *et al.* (2006) assigning 0.4 myr durations for both the *T. hugii* and *N. nicklesi* zones. The Barremian awaits a more precise method of scaling these newly defined sets of Tethyan ammonite zones and their correlations to other zonations.

It is anticipated that direct calibrations of each Valanginian–Hauterivian–Barremian ammonite zone to the M-sequence will be acquired and verified in the future. This will require sections in which the ammonite zone boundaries are precisely delimited by ranges of ammonites; a feature that is sometimes only estimated in the reference sections currently used by these different studies. Therefore, the age model for scaling events within the Valanginian, Hauterivian and Barremian stages should be considered as a temporary model that awaits detailed enhancement.

Aptian and Albian

The relative placement and duration of nannofossil and foraminifer datums and zones through the entire Aptian and Albian has been calibrated by a series of cycle stratigraphy studies in Italian reference sections and drilling projects (e.g., Erba *et al.*, 1999; Fiet, 2000; Fiet *et al.*, 2001; Grippo *et al.*, 2004; Huang *et al.*, 2010; Gale *et al.*, 2011).

Ammonite zones of the Tethyan realm have been correlated to most of these same nannofossil and foraminifer datums from reference sections in southeastern France, including outcrops that have been suggested as possible stratotypes for substage boundaries – Lower and Middle Aptian, including Bedoulian and Gargasian historical stratotypes (e.g., Moullade *et al.*, 1998b, 2005; Renard *et al.*, 2005), Middle Aptian through Lower Albian (synthesis in Herrle *et al.*, 2004), and Middle Albian through basal Cenomanian (Gale *et al.*, 2011). A portion of the middle-Lower Aptian (*D. deshayesi* ammonite zone and adjacent intervals) of the GTS2012 composite ammonite–microfossil scale uses indirect inter-calibrations using the distinctive carbon isotope trends associated with organic-enrichment episode OAE1a in ammonite-bearing (Renard *et al.*, 2005) and foraminifer–nannofossil–bearing sections (synthesis in Erba, 2004). The bases of three ammonite zones in the lower and middle Albian (*Douvilleiceras mammillatum*, *Euhoplites loricatus*, and *Euhoplites lautus*) are lacking a direct inter-calibration to these foraminifer–nannofossil datums in the cycle stratigraphy reference sections; therefore, these were proportionally scaled assuming equal durations of their component subzones. The ammonite–foraminifer correlations to cycle stratigraphy imply that the middle Aptian ammonite *Epicheloniceras martinoides* Zone Spans nearly one third of the Aptian, which is consistent with its thickness relative to foraminifer zones and to adjacent ammonite zones in reference sections. The correlations imply that the organic-rich Jacob (OAE 1b) event in the lower part of the

Hypacanthoplites jacobi Zone in southeastern France reference sections (Herrle *et al.*, 2004) is 2 myr before the working definition for the base of the Albian (FAD of nannofossil *P. columnata*), which verifies the conjectured placement of this episode in the cycle stratigraphy records in Italian sections (e.g., Huang *et al.*, 2010).

Most of these ammonite zones and subzones are relatively cosmopolitan, and the majority of sea-level oscillations on European margins, microfossil biostratigraphy, geochemical trends, carbon-burial episodes and other datums are calibrated to these ammonite zones/subzones (e.g., charts in Hardenbol *et al.*, 1998; Herrle *et al.*, 2004; Föllmi *et al.*, 2006).

Assignment of numerical ages onto this cycle-scaled Aptian–Albian integrated stratigraphy requires at least one radio-isotopic date. There are three dated horizons that are most relevant and have relatively high precision relative to the cycle stratigraphy events (details in Appendix 2, this volume).

(1) Base of Cenomanian = 100.5 Ma (± 0.4) Obradovich *et al.* (2002) dated two horizons that assign the age for the top of the Albian. An ^{40}Ar-^{39}Ar age of 99.89 ± 0.37 Ma (recalibrated to GTS2012 monitor FCs age of 28.201 Ma) from the uppermost part of the *Graysonites wooldridgei* ammonite zone, which is equivalent to European subzone of *Neostlingoceras carcitanense*, the lowermost subzone of the *Mantelliceras mantelli* Zone, or the basal ammonite subzone of the Cenomanian. The overlying lowermost part of the *Mantelliceras saxbii* ammonite subzone yielded a date of 99.70 ± 0.38 Ma. Therefore, the age for the top of the *N. carcitanense* Subzone is ~99.8 Ma (± 0.4 myr). In the Vocontian Basin of SE France, there are 22 couplets within the *Neostlingoceras carcitanense* Subzone in outcrops of apparent continuity down through the uppermost Albian (Gale, 1995). If, following Gale (1995), we assign these couplets to be precession cycles (20 kyr), then the *N. carcitanense* Subzone spans at least 0.44 myr; therefore, its basal age is projected as ~100.25 Ma (± 0.4). At the Risou GSSP section, the base of the *N. carcitanense* Subzone is 6 m above the GSSP (base of foraminifer *Thalmanninella globotruncanoides*), and the relative spacing of the immediately underlying foraminifer datums of uppermost Albian relative to their cycle-scaling in Italy (Grippo *et al.*, 2004; Gale *et al.*, 2011) implies a rapid sedimentation rate; therefore, it is estimated that the Cenomanian GSSP is ~0.25 myr below the base of the *N. carcitanense* Subzone, or 100.5 (± 0.4) Ma.

This age for the basal Cenomanian ammonite zone is supported by Ar-Ar and U-Pb dates from Japan (100.83 ± 1.00 Ma from Ar-Ar analyses of biotite separates from within the *M. saxbii* ammonite zone; and 99.7 ± 1.1 Ma from LA-ICPMS U-Pb method on 12 zircon grains from a level nearly coeval with the local FAD of *Thalmanninella globotruncanoides* (primary marker of Albian–Cenomanian boundary)).

The dating of this *Mantelliceras mantelli* basal-Cenomanian ammonite zone in Japan has been questioned by methods that compare North American Western Interior and European successions of dinoflagellate cysts and other events (e.g., graphical correlation of dinoflagellate cysts by Oboh-Ikuenobe *et al.*, 2007; macrofossils by Scott, 2007, 2009; Scott *et al.*, 2009). These studies prefer a correlation of the base-Cenomanian with the Clay Spur Bentonite at the top of the Mowry Shale, which has a recalibrated age of 97.88 ± 0.69 Ma (Appendix 2 of this book). This would imply that the *Neogastroplites haasi* through *Neo. muelleri*, and perhaps even *Neo. americanus* and *Neo. maclearni* ammonite zones of the Western Interior are Albian, rather than the Cenomanian assignment for these zones as implied by the Obradovich *et al.* (2002) and Quidelleur *et al.* (2011) radio-isotopic dates from the *Mantelliceras mantelli* Zone in Japan. This interregional correlation problem and implications for timing of sea-level variations (e.g., Oboh-Ikuenobe *et al.*, 2008) and other Albian–Cenomanian boundary correlations needs to be explored further. It is noted that graphic correlation is sensitive to sedimentation rate changes between sections that may distort high-resolution correlations (F. Gradstein, pers. comm., 2011). GTS2012 utilizes the current direct radio-isotopic dating of the *Mantelliceras mantelli* Zone and its subzones and therefore extrapolates the base-Cenomanian as 100.5 Ma.

(2) Base of Albian = 113 Ma The Albian, using a working definition for its base as the FAD of nannofossil *P. columnata* (subcircular), spans 30.75 long-eccentricity cycles of 405 kyr, for a total duration of 12.45 myr (Grippo *et al.*, 2004; Huang *et al.*, 2010). Using the base-Cenomanian age of 100.5 ± 0.4 Ma implies a base-Albian age of 112.95 Ma (± 0.4). This is supported by a U-Pb age of 113.08 ± 0.14 Ma from zircons in a tuff located 65 cm above a local working definition for the base-Albian (first occurrence of the ammonite *Leymeriella (Proleymeriella) schrammeni anterior*) at Vöhrum clay pit near Hannover, Germany (Selby *et al.*, 2009). The FAD of nannofossil *P. columnata* (subcircular) is approximately near this dated level (Mutterlose *et al.*, 2003).

(3) Chron M0r, near base of Aptian = 126 Ma There is a total of 62.5 long-eccentricity cycles of 405 kyr from base of Cenomanian to top of Chron M0r (Huang *et al.*, 2010). Therefore, the top of M0r is 25.3 myr before the base-Cenomanian (100.5 ± 0.4 Ma), which would imply an age of 125.8 ± 0.4 Ma. This overlaps with an ^{40}Ar-^{39}Ar age of 125.4 ± 0.2 Ma (recalibrated to FCs of 28.201 Ma) from reversed-polarity basalts of MIT Guyot in the Pacific interpreted as being within polarity zone M0r (Pringle *et al.*, 2003; see details in Chapter 5 on the Geomagnetic Time Scale, this volume).

TABLE 27.4 Cretaceous Age Model of GTS2012

	GTS2008/GTS2004			GTS2012			Derivation in GTS2012; and Differences from GTS2008 (and GTS2004)
	Base	Est. Uncertainty	Duration	Base	Est. Uncertainty	Duration	
PALEOGENE (Danian)	65.5	0.3		66.0	0.1		Radio-isotopic dates with revised Ar-Ar monitor standard
Maastrichtian	70.6	0.6	5.1	72.1	0.2	6.1	Cycle stratigraphy calibration of C-sequence; and correlation via Carbon-13
Campanian	83.5	0.7	12.9	83.6	0.3	11.6	Radio-isotopic dates with revised Ar-Ar monitor standard
Santonian	85.8	0.7	2.3	86.3	0.5	2.6	Radio-isotopic dates with revised Ar-Ar monitor standard and cycle stratigraphy
Coniacian	88.8	1.0	3	89.8	0.4	3.5	Radio-isotopic dates with revised Ar-Ar monitor standard and cycle stratigraphy *[Note: GTS2004 had a different correlation to N. Amer. ammonite zones that was revised in GTS2008 - the value shown here]*
Turonian	93.6	0.8	4.8	93.9	0.2	4.1	Radio-isotopic dates with revised Ar-Ar monitor standard and cycle stratigraphy
Cenomanian	99.6	0.9	6	100.5	0.4	6.6	Radio-isotopic dates with revised Ar-Ar monitor standard
Albian	112.0	1.0	12.4	113.0	0.4	12.5	Cycle stratigraphy; agrees with radio-isotopic U-Pb date
Aptian	125.0	1.0	13	126.3	0.4	13.3	Cycle stratigraphy; agrees with radio-isotopic dates with revised Ar-Ar monitor standard
Barremian	130.0	1.5	5	130.8	0.5	4.5	Cycle stratigraphy calibration of M-sequence
Hauterivian	133.9	2.0	3.9	133.9	0.6	3.1	Cycle stratigraphy calibration of M-sequence *[Note: GTS2004 had an earlier published calibration to M-sequence that was revised in GTS2008 - the value shown here]*
Valanginian	140.2	3.0	6.3	139.4	0.7	5.5	Cycle stratigraphy calibration of M-sequence
Berriasian	145.5	4.0	5.3	145.0	0.8	5.6	Cycle stratigraphy calibration of M-sequence. *[Note: GTS2004/ GTS2008 had used base of a Tethyan ammonite zone; GTS2012 uses a slightly higher magnetic chron.]*
JURASSIC (Tithonian)							

Cretaceous age model of GTS2012 and comparison with *The Concise Geologic Time Scale* (GTS2008; Ogg *et al.*, 2008) and GTS2004 (Gradstein *et al.*, 2004). Values are rounded to nearest 0.1 myr.

TABLE 27.5 GSSPs of the Cretaceous Stages, with Location and Primary Correlation Criteria (Status Jan. 2012)

Stage	GSSP Location	Latitude, Longitude	Boundary Level	Correlation Events	Reference
Maastrichtian	Tercis les Bains, Landes, France	43°40′46.1″N 1°06′47.9″W*	Level 115.2 on platform IV of the geological site at Tercis les Bains	Mean of 12 biostratigraphic criteria of equal importance. Closely above is FAD of ammonite *Pachydiscus neubergicus*	Episodes 24/4, 2001
Campanian	*Candidates are in Italy and in Texas*			*Either Crinoid, extinction of* Marsupites testudinarius; *or base of magnetic polarity chronozone C33r*	
Santonian	*Leading candidate is Olazagutia (Spain)*			*Inoceramid bivalve, FAD of* Cladoceramus undulatoplicatus	
Coniacian	*Leading candidates are in Poland (Slupia Nadbrzena) and Germany (Salzgitter-Salder Quarry)*			*Inoceramid bivalve, FAD of* Cremnoceramus deformis erectus	
Turonian	Pueblo, Colorado, USA	38°16′56″N 104°43′39″W*	Base of Bed 86 of the Bridge Creek Limestone Member	Ammonite FAD *Watinoceras devonense*	Episodes 28/2, 2005
Cenomanian	Mount Risou, Hautes-Alpes, France	44°23′33″N 5°30′43″E	36 m below the top of the Marnes Bleues Formation on the south side of Mont Risou	Planktonic foraminifer FAD *Thalmanninella globotruncanoides*	Episodes 27/1, 2004
Albian	Southeastern France			*Candidates include: (1) calcareous nannofossil, FAD of* Praediscosphaera columnata, *(2) carbon isotope excursion (black shale episode), (3) planktonic foraminfer or ammonite markers*	
Aptian	*Candidate is Gorgo a Cerbara, Piobbico, Umbria-Marche, central Italy*			*Base of magnetic polarity chronozone M0r; near base of* Paradeshayesites oglanlensis *ammonite zone*	

(Continued)

TABLE 27.5 GSSPs of the Cretaceous Stages, with Location and Primary Correlation Criteria (Status Jan. 2012)—cont'd

Stage	GSSP Location	Latitude, Longitude	Boundary Level	Correlation Events	Reference
Barremian	*Candidate is Río Argos near Caravaca, Murcia province, Spain*			*Ammonite, FAD of* Taveraidiscus hugii *auctorum* – Avramidiscus vandeckii *group*	
Hauterivian	*Candidate is La Charce village, Drôme province, southeast France*			*Ammonite, FAD of genus* Acanthodiscus *(especially* A. radiatus*)*	
Valanginian	*Candidates are near Montbrun-les-Bains (Drôme province, SE France) and Cañada Luenga (Betic Cordillera, S. Spain)*			*Calpionellid, FAD of* Calpionellites darderi *(base of Calpionellid Zone E); followed by the FAD of ammonite* "Thurmanniceras" pertransiens	
Berriasian (base Cretaceous)				*Candidates include base of magnetic polarity chronozone M18r, or near FAD of ammonite* Berriasella jacobi	

*according to Google Earth

The duration of Chron M0r from cycle stratigraphy is 0.5 myr, therefore the Aptian begins 25.8 myr below the base of the Cenomanian (Huang *et al.*, 2010), or at 126.3 Ma (± 0.4) using the 100.5 ± 0.4 Ma value. Therefore, the beginning of Chron M0r was set at 126.3 Ma (± 0.4) for the age model applied to the M-sequence polarity time scale (see Chapter 5 on the Geomagnetic Time Scale, this volume). This 126 Ma age for the base of the Aptian is supported by a U-Pb date of 124.07 ± 0.24 Ma from the early Aptian *Chiastozygus litterarius* biozone (NC6; subzone 6b) (Shimokawa, 2010) and Ar-Ar dating of 124.3 ± 1.8 Ma on lower Aptian basalts of the Ontong Java Plateau (Chambers *et al.*, 2004).

27.3.3.2. Late Cretaceous Scaling

Ash beds within Cenomanian through lower Maastrichtian ammonite zones of the North American Western Interior have been systematically collected and dated (Obradovich, 1993; Hicks *et al.*, 1995; Hicks and Obradovich, 1995; Hicks *et al.*, 1999; Cobban *et al.*, 2006; Meyers *et al.*, 2010, 2012; Siewert, 2011; Siewert *et al.*, in press). There are nearly 45 radio-isotopic dates from sanidine feldspar crystals ($^{40}Ar/^{39}Ar$) or zircons (U-Pb); therefore, this interval is the best-dated succession in the entire Phanerozoic. In addition, many of the ammonite zones have cycle-scaled durations (e.g., Locklair and Sageman, 2008; Meyers *et al.*, 2012; Siewert *et al.*, in press).

In GTS2004, the Late Cretaceous time scale was based on a spline-fit of the existing database of $^{40}Ar/^{39}Ar$ dates to estimate the ages of the full suite of Western Interior ammonite zones. As explained in the previous section, we applied a similar spline-fit to a revised ammonite zonation (Cobban *et al.*, 2006) with an expanded and recalibrated set of radio-isotopic ages (Appendix 2). The spline results were adjusted to incorporate constraints of cycle-scaled durations for a subset of those ammonite zones. The fit to radio-isotope dates implied that ammonite zones in some intervals are very brief, followed by intervals with relatively long-duration zones; these projections require testing with cycle stratigraphy and independent radio-isotopic dating.

Ages for other stratigraphic scales are assigned based on correlations to this primary age-scale of Western Interior ammonite zones (Table 27.3). In particular, only one of the Late Cretaceous stages (the Cenomanian/Turonian boundary) is defined relative to one of these Western Interior ammonite zones. The possible definitions being considered for the other Late Cretaceous stages or the estimated placement of stage boundaries relative to these ammonite zones are discussed in the first part of this chapter. For the "working scale" of Late Cretaceous of GTS2012, we have generally used the placements for geologic stages by Cobban *et al.* (2006). The numerical ages from the spline-fits for these stage boundaries are nearly identical to those proposed by Meyers *et al.* (2010, 2012) for the base-Turonian and by Siewert *et al.* (in press) for the base-Coniacian, base-Santonian and base-Campanian; therefore, we used their published detailed analyses for the ages of each of those boundaries. However, a caveat: the eventual GSSP definition for some of these stages may be chosen at levels that could be considerably different from these working definitions.

Correlation of other stratigraphic scales (e.g., European-based macrofossil and microfossil zones) to the Western Interior ammonites is based partly on shared inoceramid and calcareous nannofossil taxa, some common ammonites, strontium isotope trends, carbon-13 trends, magnetostratigraphy and other methods (e.g., Cobban, 1993; McArthur *et al.*, 1993, 1994; Hardenbol *et al.*, 1998).

We have assigned the base-Campanian to be coeval with the base of Chron C33r, which puts a constraint on the earliest portion of the C-sequence. The duration of all Maastrichtian and upper Campanian chrons are known from cycle stratigraphy (Husson *et al.*, 2011; see Chapter 5, this volume). Most foraminifer and nannofossil datums are correlated to those polarity chrons (e.g., Burnett *et al.*, 1998; Huber *et al.*, 2008), and therefore can be assigned numerical ages from the C-sequence age model. Carbon isotope curves from Campanian–Maastrichtian magnetostratigraphy reference sections and coeval macrofossil-zoned sections provide a similar set of numerical age assignments (e.g., Voigt *et al.*, 2010, in press). As noted above for the Maastrichtian GSSP, these different correlation methods converged on its correlation to the base of the *Baculites baculus* ammonite zone of North America, approximately to the base of nannofossil zone UC17, a magnetostratigraphic placement at Chron C32n.2n.88, and a numerical age of 72.1 Ma (± 0.2).

The end of the Cretaceous occurred at 66.0 Ma (± 0.1), based on radio-isotopic dating and the placement of this mass extinction level relative to the C-sequence polarity time scale (See Chapter 28 on the Paleogene, this volume).

27.3.4. Summary

An integrated scaling for a selection of Cretaceous stratigraphy is illustrated in Figure 27.6. A comparison of the projected ages and durations of the geologic stages to those of *The Concise Geologic Time Scale* (GTS2008; Ogg *et al.*, 2008) is compiled in Table 27.4. The durations of stages range from over 10 myr for Aptian, Albian and Campanian to less than 3 myr for the Santonian. The differences in the age models for GTS2008 and GTS2012 are relatively minor, mainly a ~1 myr shift to older ages due to the revised Ar-Ar monitor standard. This general stability suggests that the global efforts to integrate cycle stratigraphy with high-precision radio-isotope dates and improved inter-zonal calibrations are converging on a reliable Cretaceous time scale.

However, several of these stages are not yet defined by GSSPs, and those decisions may select different horizons for global correlations than are used as the working definitions for these stages in GTS2012. The numerical age model for stratigraphic events might be relatively stable, but the future assignment of those events into chronostratigraphic stages may shift when the GSSPs are eventually decided.

The next major improvement in the Cretaceous numerical age model will occur when the floating intervals of cycle stratigraphy are unambiguously tied to the Present by using the stability of the 405-kyr long-eccentricity cycle. At least four factors precluded accomplishing this achievement for GTS2012:

(a) The Eocene gap in astronomical-cycle chronostratigraphy;
(b) The inherent uncertainties on the radio-isotopic age constraints which equal or exceed one eccentricity cycle;
(c) Gaps in the composite records to enable splicing among basins; and
(d) Uncertainties on which component of the recognized facies alternations in the deep basins (e.g., carbonate-rich versus clay-rich intervals) are indicative of the minimum or the maximum phases on that eccentricity orbital-climate curve.

When these uncertainties are overcome, then the precision of the age assignments for the datums calibrated to those Cretaceous cycles will become less than 0.1 myr. An even more critical hurdle is the lack of verified inter-calibration of the different stratigraphic scales among regions (e.g., complications of faunal provincialism during Early Cretaceous, and how to correlate into terrestrial realms) and among methods (e.g., direct calibration of microfossil zonations to the dated Western Interior ammonite zones). Chronostratigraphic databases, such as the "CRET1" set of ranges of fossil taxa in selected worldwide sections (Scott, 2011), will aid in achieving this goal. The next generation of Cretaceous time scales will follow the lead of the Paleozoic (see Chapters 20, 21 and 23 of this book) in uniting statistical analyses of biostratigraphic data with cycle stratigraphy, radio-isotopic dates and terrestrial records.

In many respects, these correlation puzzles are reasons why so few of the Cretaceous stages have been assigned stable definitions based on globally useful GSSPs.

ACKNOWLEDGMENTS

Our colleagues within the Cretaceous Subcommission of the ICS, participants at the Late Cretaceous microfossil workshop at University College London, and many other experts have unselfishly shared their data, preliminary results and ideas. In particular, we wish to thank the radio-isotope group at the University of Wisconsin at Madison (Brad Singer, Sarah Siewert, Steve Meyers) for providing their extensive data sets and calibration details, John Obradovich for discussions on Late Cretaceous numerical time scales, the cycle stratigraphy teams (Linda Hinnov, Chunfu Zhang, Bruno Galbrun, Dorothée Husson, Nicolas Thibault, Robert Locklair) for hosting working sessions with their groups and gladly giving us advance copies of their forthcoming publications to establish the Cretaceous cycle stratigraphy, and the microfossil gurus (Jackie Lees, Paul Bown, Jim Bergen, Brian Huber, Maria Rose Petrizzo, Dave Watkins, Elisabeth Erba, Isabella Primola Silva, and others) for helping prepare, review and update these data sets and clarifying calibrations. Andy Gale, Silke Voigt, Ian Jarvis and others provided insights and in-press data sets for inter-regional calibrations and geochemical trends. Undoubtedly, most of these groups will consider this set of syntheses to be too brief in covering their important specialties and will emphasize that this Cretaceous chapter is only an inadequate synopsis of the current (2011) status of this rapidly progressing effort. Øyvind Hammer and Felix Gradstein assisted with spline fittings of the data, and Gabi Ogg patiently prepared all the figures.

Additional or alternative zonal schemes for the Cretaceous are available through the *TimeScale Creator* visualization datapacks (*www.tscreator.org*).

The compilation and application of cycle stratigraphy to enhance the Mesozoic time scale was partially supported by U.S. National Science Foundation grant EAR-071895.

REFERENCES

Aguado, R., Company, M., Tavera, J.M., 2000. The Berriasian/Valanginian boundary in the Mediterranean region: New data from the Caravaca and Cehegin sections, SE Spain. Cretaceous Research 21, 1–21.

Aguirre Urreta, M.B., Pazos, P.J., Lazo, D.G., Fanning, C.M., Litvak, V.D., 2008. First U/Pb SHRIMP age of the Hauterivian stage, Neuquen Basin, Argentina. Journal of South American Earth Sciences 26, 91–99.

Allemann, F., Catalano, R., Fares, F., Remane, J., 1971. Standard calpionellid zonation (Upper Tithonian-Valanginian) of the Western Mediterranean province. Proceedings II Planktonic Conference, Roma 1971, 1337–1340.

Alvarez, W., Arthur, M.A., Fischer, A.G., Lowrie, W., Napoleone, G., Premoli Silva, I., Roggenthen, M.W., 1977. Upper Cretaceous-Palaeocene magnetic stratigraphy at Gubbio, Italy – V. Type section for the Late Cretaceous-Palaeocene geomagnetic reversal time scale. Geological Society of America Bulletin 88, 383–389.

Amédro, F., 1992. L'Albien du bassin Anglo-Parisien: Ammonites, zonation phylétique, séquences. Bulletin des Centres de Recherches Exploration-Production Elf-Aquitaine 16, 187–233.

Amodio, S., Ferreri, V., D'Argenio, B., Weissert, H., Sprovieri, M., 2008. Carbon-isotope stratigraphy and cyclostratigraphy of shallow-marine carbonates: The case of San Lorenzello, Lower Cretaceous of southern Italy. Cretaceous Research 29, 803–813.

Ando, A., Huber, B.T., 2007. Taxonomic revision of the Late Cenomanian planktonic foraminifera *Rotalipora greenhornensis* (Morow, 1934). Journal of Foraminiferal Research 37, 160–174.

Arnaud-Vanneau, A., Bilotte, M., 1998. Larger benthic foraminifera. Columns for Jurassic chart of Mesozoic and Cenozoic sequence chronostratigraphic framework of European basins, by Hardenbol, J., Thierry, J., Farley, M.B., Jacquin, Th., de Graciansky, P.-C., and Vail, P.R. (coordinators). In: de Graciansky, P.-C., Hardenbol, J., Jacquin, Th., Vail, P.R. (Eds.), Mesozoic-Cenozoic Sequence Stratigraphy of European Basins. SEPM Special Publication, 60, pp. 763–781.

Barbier, R., Thieuloy, J.P., 1965. Étage Valanginien. Mémoires du Bureau de Recherches Géologiques et Minières 34, 79–84.

Barron, E.J., 1983. A warm, equable Cretaceous: The nature of the problem. Earth-Science Reviews 19, 305–338.

Bartolocci, P., Beraldini, M., Cecca, F., Faraoni, P., Marini, A., Pallini, G., 1992. Preliminary results on correlation between Barremian ammonites and magnetic stratigraphy in Umbria-Marche Apennines (Central Italy). Palaeopelagos 2, 63–68.

Baudin, F., Faraoni, P., Marini, A., Pallini, G., 1997. Organic matter characterization of the "Faraoni Level" from Northern Italy (Lessini Mountains and Trento Plateau): Comparison with that from Umbria-Marche Appennines. Palaeopelagos 7, 41–51.

Baudin, R., Bulot, L.G., Cecca, F., Coccioni, R., Gardin, S., Renard, M., 1999. Un équivalent du "Niveau Faraoni" dans le Bassin du Sud-Est de la France, indice possible d'un événement anoxique fini-hauterivien étendu à la téthys méditerranéenne. Bulletin de la Société géologique de France 170, 487–498.

Bengtson, P., Cobban, W.A., Dodsworth, P., Gale, A.S., Kennedy, W.J., Lamolda, M.A., Matsumoto, T., Reyment, R.A., Siebertz, E., Tröger, E., 1996. The Turonian stage and substage boundaries. Bulletin de l'Institut Royal des Sciences Naturelles de Belgique, Sciences de la Terre 66 (Suppl.), 69–79.

Bergen, J.A., 1994. Berriasian to early Aptian calcareous nannofossils from the Vocontian trough (SE France) and Deep Sea Drilling Site 534: New nannofossil taxa and a summary of low-latitude biostratigraphic events. Journal of Nannoplankton Research 16, 59–69.

Bice, K.L., Bralower, T.J., Duncan, R.A., Huber, B.T., Leckie, R.M., Sageman, B.B., 2002. Cretaceous Climate-Ocean Dynamics: Future Directions for IODP. A JOI/USSSP and NSF Sponsored Workshop Available online: http://www.whoi.edu/ccod/CCOD_report.html.

Birkelund, T., Hancock, J.M., Hart, M.B., Rawson, P.F., Remane, J., Robaszynski, R., Schmid, F., Surlyk, F., 1984. Cretaceous stage boundaries – proposals. Geological Society of Denmark Bulletin 33, 3–20.

Blair, S.A., Watkins, D.K., 2009. High-resolution calcareous nannofossil biostratigraphy for the Coniacian/Santonian Stage boundary, Western Interior Basin. Cretaceous Research 30, 367–384.

Blanc, E., Bulot, L.G., Paicheler, J.-C., 1994. La coupe de référence de Monbrun-les-Bains (Drôme, SE France): Un stratotype potentiel pour la limite Berriasien-Valanginien. Comptes Rendus de l'Académie des Sciences, Paris, Série II. Sciences de la Terre et des Planetes 318, 101–108.

Bodin, S., Godet, A., Föllme, K.B., Vermeulen, J., Arnaud, H., Strasser, A., Fiet, N., Adatte, T., 2006. The late Hauterivian Faraoni oceanic anoxic event in the western Tethys: Evidence from phosphorus burial rates. Palaeogeography, Palaeoclimatology, Palaeoecology 235, 245–264.

Bornemann, A., Norris, R.D., Friedrich, O., Beckmann, B., Schouten, S., Sinninghe Damsté, J., Vogel, J., Hofmann, P., Wagner, T., 2008. Isotopic evidence for glaciation during the Cretaceous supergreenhouse. Science 319, 189–192.

Bown, P.R., Rutledge, D.C., Crux, J.A., Gallagher, L.T., 1998. Lower Cretaceous. In: Bown, P.R. (Ed.), Calcareous Nannofossil Biostratigraphy. British Micropalaeontology Society Publication Series. Chapman & Hall, London, pp. 86–131.

Bralower, T.J., Sliter, W.V., Arthur, M.A., Leckie, R.M., Allard, D.J., Schlanger, S.O., 1993. Dysoxic/anoxic episodes in the Aptian-Albian (Early Cretaceous). In: Pringle, M.S., Sager, W.W., Sliter, W.V., Stein, S. (Eds.), The Mesozoic Pacific: Geology, Tectonics, and Volcanism. American Geophysical Union Geophysics Monograph, 77, pp. 5–37.

Bralower, T.J., Leckie, R.M., Slliter, W.V., Thierstein, H.R., 1995. An integrated Cretaceous microfossil biostratigraphy. In: Berggren, W.A., Kent, D.V., Hardenbol, J. (Eds.), Geochronology, Time Scales and Global Stratigraphic Correlations. SEPM Special Publication, 54, pp. 65–79.

Bralower, T.J., Fullagar, P.D., Paull, C.K., Dwyer, G.S., Leckie, R.M., 1997. Mid-Cretaceous strontium-isotope stratigraphy of deep-sea sections. Geological Society of America Bulletin 109, 1421–1442.

Bréheret, J.-G., 1988. Épisodes de sedimentation riche en matiere organique dans les marnes bleues d'âge aptien et albien de la partie pélagique du bassin vocontien. Bulletin de la Société géologique de France 8 (IV), 349–356.

Breistroffer, M., 1947. Sur les zones d'ammonites dans l'Albien de France et d'Angleterre. Travaux du Laboratoire de Géologie de la Faculté des Sciences de l'Université de Grenoble 26, 17–104.

Bryan, S.E., Ernst, R.E., 2008. Revised definition of Large Igneous Provinces (LIPS). Earth-Science Reviews 86, 175–202.

Bulot, L.G., Blanc, E., Thieuloy, J.-P., Remane, J., 1993. La limite Berriasien-Valanginien dans le Sud-Est de la France: Données biostratigraphiques nouvelles. Comptes Rendus de l'Académie des Sciences, Paris, Série II. Sciences de la Terre et des Planetes 317, 387–394.

Bulot, L.G., Blanc, E., Company, M., Gardin, S., Hennig, S., Hoedemaeker, Ph.J., Leereveld, H., Magniez-Jannin, F., Mutterlose, J., Pop, G., Rawson, P.F., 1996. The Valanginian Stage. Bulletin de l'Institut Royal des Sciences Naturelles de Belgique, Sciences de la Terre 66 (Suppl.), 11–18.

Burnett, J.A., Gallagher, L.T., Hampton, M.J., 1998. Upper Cretaceous. In: Bown, P.R. (Ed.), Calcareous Nannofossil Biostratigraphy. British Micropalaeontological Society Publication Series. Chapman & Hall, London, pp. 132–199.

Busnardo, R., 1965. Le Stratotype du Barrémien. 1. – Lithologie et Macrofaune, and Rapport sur l'étage Barrémien. Mémoires du Bureau de Recherches Géologiques et Minières 34, 101–116 and 161–169.

Busnardo, R., Thieuloy, J.-P., Moullade, M., 1979. Hypostratotype Mesogéen de l'Étage Valanginien (sud-est de la France). Les Stratotypes Français 6, p. 143.

Cande, S.C., Kent, D.V., 1992. A new geomagnetic polarity time scale for the Late Cretaceous and Cenozoic. Journal of Geophysical Research 97, 13917–13951.

Cande, S.C., Kent, D.V., 1995. Revised calibration of the geomagnetic polarity timescale for the Late Cretaceous and Cenozoic. Journal of Geophysical Research 100, 6093–6095.

Casellato, C.E., 2010. Calcareous nannofossil biostratigraphy of upper Callovian-lower Berriasian successions from the Southern Alps, North Italy. Rivista Italiana di Paleontologia e Stratigrafia 116, 357–404.

Casellato, C.E., Andreini, G., Erba, E., Parisi, G., 2008. Calcareous Nannofossil and Calpionellid calcification events across Tithonian – Berriasian time interval and low latitudes paleoceanographic implications. 12nd INA Conference Lyon, France, 7–11 Sept 2008. Journal of Nannoplankton Research, Special Issue, 33.

Casey, R., 1961. The stratigraphical palaeontologiy of the Lower Greensand. Palaeontology 3, 487–621.

Casey, R., 1996. Lower Greensand ammonites and ammonite zonation. Proceedings of the Geologists' Association 107, 69–76.

Casey, R., Boyliss, H.M., Simpson, M.E., 1998. Observations on the lithostratigraphy and ammonite succession of the Aptian (Lower Cretaceous) Lower Greensand of Chale Bay, Isle of Wight, UK. Cretaceous Research 19, 511–535.

Cecca, F., Pallini, G., Erba, E., Premoli-Silva, I., Coccioni, R., 1994. Hauterivian-Barremian chronostratigraphy based on ammonites, nannofossils, planktonic foraminifra and magnetic chrons from the Mediterranean domain. Cretaceous Research 15, 457–467.

Chambers, L.M., Pringle, M.S., Fitton, J.G., 2004. Phreatomagmatic eruptions on the Ontong Java Plateau: An Aptian $^{40}Ar/^{39}Ar$ age for volcaniclastic rocks at ODP Site 1184. In: Fitton, J.G., Mahoney, J.J., Wallace, P.J., Sanders, A.D. (Eds.), Origin and Evolution of the Ontong Java Plateau. Geological Society Special Publications, 229, pp. 325–331.

Chang, S.-C., Zhang, H., Renne, P.R., Fang, Y., 2009. High-precision $^{40}Ar/^{39}Ar$ age of the Jehol Biota. Palaeogeography, Palaeoclimatology, Palaeoecology 280, 94–104.

Channell, J.E.T., Grandesso, P., 1987. A revised correlation of magnetozones and calpionellid zones based on data from Italian pelagic limestone sections. Earth and Planetary Science Letters 85, 222–240.

Channell, J.E.T., Bralower, T.J., Grandesso, P., 1987. Biostratigraphic correlation of M-sequence polarity chrons M1 to M22 at Capriolo and Xausa (S. Alps, Italy). Earth and Planetary Science Letters 85, 203–221.

Channell, J.E.T., Erba, E., Lini, A., 1993. Magnetostratigraphic calibration of the Late Valanginian carbon isotope event in pelagic limestones from Northern Italy and Switzerland. Earth and Planetary Science Letters 118, 145–166.

Channell, J.E.T., Cecca, F., Erba, E., 1995a. Correlations of Hauterivian and Barremian (Early Cretaceous) stage boundaries to polarity chrons. Earth and Planetary Science Letters 134, 125–140.

Channell, J.E.T., Erba, E., Nakanishi, M., Tamaki, K., 1995b. Late Jurassic-Early Cretaceous time scales and oceanic magnetic anomaly block models. In: Berggren, W.A., Kent, D.V., Hardenbol, J. (Eds.), Geochronology, Time Scales and Global Stratigraphic Correlations. SEPM Special Publication, 54, pp. 51–63.

Channell, J.E.T., Erba, E., Muttoni, G., Tremolada, F., 2000. Early Cretaceous magnetic stratigraphy in the APTICORE drill core and adjacent outcrop at Cismon (Southern Alps, Italy), and correlation to the proposed Barremian-Aptian boundary stratotype. Geological Society of America Bulletin 112, 1430–1443.

Channell, J.E.T., Casellato, C.E., Muttoni, G., Erba, E., 2010. Magnetostratigraphy, nannofossil stratigraphy and apparent polar wander for Adria-Africa in the Jurassic-Cretaceous boundary interval. Palaeogeography, Palaeoclimatology, Palaeoecology 293, 51–75.

Clarke, L.J., Jenkyns, H.C., 1999. New oxygen evidence for long-term Cretaceous climatic change in the Southern Hemisphere. Geology 27, 699–702.

Cobban, W.A., 1993. Diversity and distribution of Late Cretaceous ammonites, Western Interior, United States. In: Caldwell, W.G.E., Kauffman, E.G. (Eds.), Evolution of the Western Interior Basin. Geological Association of Canada, Special Paper, 39, pp. 435–451.

Cobban, W.A., Walaszczyk, I., Obradovich, J.D., McKinney, K.C., 2006. A USGS Zonal Table for the Upper Cretaceous Middle Cenomanian-Maastrichtian of the Western Interior of the United States Based on Ammonites, Inoceramids, and Radiometric Ages. U.S. Geological Survey Open-File Report 2006–1250, 45.

Coccioni, R., 2003. Cretaceous anoxic events: The Italian record. Abstract volume of the Séance spécialisée de la Société géologique de France. Paléocéanographie du Mésozoïque en réponse aux forçages de la paléogéographie et du paléoclimat, 10–11 June 2003, Paris, p. 10.

Colin, J.-P., Babinot, J.F., 1998. Ostracodes. Columns for Jurassic chart of Mesozoic and Cenozoic sequence chronostratigraphic framework of European basins, by Hardenbol, J., Thierry, J., Farley, M.B., Jacquin, Th., de Graciansky, P.-C., and Vail, P.R. (coordinators). In: de Graciansky, P.-C., Hardenbol, J., Jacquin, Th., Vail, P.R. (Eds.), Mesozoic-Cenozoic Sequence Stratigraphy of European Basins. SEPM Special Publication, 60, pp. 763–781, and chart supplements.

Colloque sur la limite Jurassique-Crétacé, 1975. Mémoires du Bureau de Recherches Géologiques et Minières 86, 386–393 (summary).

Combemorel, R., Christen, W.K., 1998. Belemnites. Columns for Jurassic chart of Mesozoic and Cenozoic sequence chronostratigraphic framework of European basins, by Hardenbol, J., Thierry, J., Farley, M.B., Jacquin, Th., de Graciansky, P.-C., and Vail, P.R. (coordinators). In: de Graciansky, P.-C., Hardenbol, J., Jacquin, Th., Vail, P.R. (Eds.), Mesozoic-Cenozoic Sequence Stratigraphy of European Basins. SEPM Special Publication, 60, pp. 763–781.

Company, M., Sandoval, J., Tavera, J.M., 1995. Lower Barremian ammonite biostratigraphy in the Subbetic Domain (Betic Cordillera, southern Spain). Cretaceous Research 16, 243–256.

Conybeare, W.D., Phillips, W., 1822. Outlines of the Geology of England and Wales, with an Introduction Compendium of the General Principles of that Science, and Comparative Views of the Structure of Foreign Countries. Part 1. William Phillips, London, p. 470.

Cope, J.C.W., 2007. Drawing the line: the history of the Jurassic-Cretaceous boundary. Proceedings of the Geologists' Association 119, 105–117.

Coquand, H., 1857a. Notice sur la formation crétacée du départment de la Charente. Bulletin de la Société géologique de France, Série 2 (14), 55–98.

Coquand, H., 1857b. Position des *Ostrea columba* et *biauriculata* dans le groupe de la craie inférieure. Bulletin de la Société géologique de France, Série 2 (14), 745–766.

Coquand, H., 1861. Sur la convenance d'établir dans le groupe inférieur de la formation crétacée un nouvel étage entre le Néocomien proprement dit (couches à *Toxaster complanatus* et à *Ostrea couloni*) et le Néocomien supérieur (étage Urgonien de d'Orbigny). Mémoires de la Société d'Emulation de Provence 1, 127–139.

Coquand, H., 1871. Sur le Klippenkalk du département du Var et des Alpes-Maritimes. Bulletin de la Société géologique de France 28, 232–233.

Courtillot, V.E., Renne, P.R., 2003. On the ages of flood basalt events. Comptes Rendus Geoscience 335, 113–140.

de Graciansky, P.-C., Hardenbol, J., Jacquin, T., and Vail, P. R. (Eds), Mesozoic-Cenozoic Sequence Stratigraphy of European Basins. SEPM Special Publication, 60, p. 786.

De Grossouvre, A., 1901. Recherches sur la Craie supérieure. Partie 1: Stratigraphie générale. Mémoires pour servir à l'explication de la carte géologique détaillée de la France. Imprimerie Nationale, Paris, p. 1013.

Desor, E., 1854. Quelques mots sur l'étage inférieur du groupe néocomien (étage Valanginien). Bulletin de la Société des Sciences Naturelles de Neuchâtel 3, 172–180.

Desor, E., Gressly, A., 1859. Études géologiques sur le Jura neuchâtelois. Bulletin de la Société des Sciences Naturelles de Neuchâtel 4, 1–159.

De Wever, P., 1998. Radiolarians Columns for Jurassic chart of Mesozoic and Cenozoic sequence chronostratigraphic framework of European basins, by Hardenbol, J., Thierry, J., Farley, M.B., Jacquin, Th., de Graciansky, P.-C., and Vail, P.R. (coordinators). In: de Graciansky, P.-C., Hardenbol, J., Jacquin, Th., Vail, P.R. (Eds.), Mesozoic-Cenozoic Sequence Stratigraphy of European Basins. SEPM Special Publication, 60, pp. 763–781.

D'Hondt, A.V., Lamolda, M.A., Pons, J.M., (coordinators), 2007. Stratigraphy of the Coniacian-Santonian transition, Meeting organised by the Santonian working group of the Subcommission on Cretaceous Stratigraphy. Cretaceous Research 28 (1), 142.

D'Hondt, A.V., 1998. Inoceramids. Columns for Jurassic chart of Mesozoic and Cenozoic sequence chronostratigraphic framework of European basins, by Hardenbol, J., Thierry, J., Farley, M.B., Jacquin, Th., de Graciansky, P.-C., and Vail, P.R. (coordinators). In: de Graciansky, P.-C., Hardenbol, J., Jacquin, Th., Vail, P.R. (Eds.), Mesozoic-Cenozoic Sequence Stratigraphy of European Basins. SEPM Special Publication, 60, pp. 763–781.

d'Omalius d'Halloy, J.G.J., 1822. Observations sur un essai de cartes géologiques de la France, des Pays-Bas, et des contrées voisines. Annales de Mines 7, 353–376.

d'Orbigny, A., 1840. Paléontologie française. Terrains crétacés. 1. Céphalopodes. A. d'Orbigny, Paris, p. 622.

d'Orbigny, A., 1842. Paléontologie française. Terrains crétacés. 2. Gastropodes. Masson, Paris, p. 456.

d'Orbigny, A., 1847. Paléontologie française. Terrains crétacés. 4. Brachiopodes. Masson, Paris, p. 390.

d'Orbigny, A., 1849–1850. Paléontologie française. Terrains jurassique. 1. Céphalopodes. Bertrand, Paris.

d'Orbigny, A., 1852. Cours élémentaire de paléontologie et de géologie stratigraphique, vol. 2. Masson, Paris, pp. 383–847.

Douglas, R.G., Savin, S.M., 1975. Oxygen and carbon isotope analyses of Tertiary and Cretaceous microfossils from Shatsky Rise and other sites in the North Pacific Ocean. Initial Reports of the Deep Sea Drilling Project 32, 509–520.

Dumont, A., 1849. Rapport sur la carte géologique du Royaume. Bulletin de l'Académie royal des Sciences, des Lettres et des Beaux-Arts de Belgique 16, 351–373.

Earth Impact Database. Planetary and Space Science Centre, University of New Brunswick, Fredericton, New Brunswick, Canada. Website: http://www.passc.net/EarthImpactDatabase/. Viewed 2010.

Erba, E., 1994. Nannofossils and superplumes: The Early Aptian nannoconid crisis. Paleoceanography 9, 483–501.

Erba, E., 2004. Calcareous nannofossils and Mesozoic oceanic anoxic events. Marine Micropaleontology 52, 85–106.

Erba, E., 2006. The first 150 million years history of calcareous nannoplankton: Biosphere–geosphere interactions. Palaeogeography, Palaeoclimatology, Palaeoecology 232, 237–250.

Erba, E., Aguaro, R., Avram, E., Barboschkin, E.J., Bergen, J., Bralower, T.J., Cecca, F., Channell, J.E.T., Coccioni, R., Company, M., Delanoy, G., Erbacher, J., Herbert, T.D., Hoedemaeker, P., Kakabadze, M., Leereveld, H., Lini, A., Mikhailova, I.A., Mutterlose, J., Ogg, J.G., Premoli Silva, I., Rawson, P.F., von Salis, K., Weissert, H.,

1996. The Aptian Stage. Bulletin de l'Institut Royal des Sciences Naturelles de Belgique, Sciences de la Terre 66 (Suppl.), 31–43.

Erba, E., Channell, J.E.T., Claps, M., Jones, C., Larson, R., Opdyke, B., Premoli Silva, I., Riva, A., Salvini, G., Torricelli, S., 1999. Integrated stratigraphy of the Cismon Apticore (southern Alps, Italy): A "reference section" for the Barremian-Aptian interval at low latitudes. Journal of Foraminiferal Research 29, 371–391.

Erba, E., Bartolini, A., Larson, R.I., 2004. Valanginian Weissert oceanic event. Geology 32, 149–152.

Erbacher, J., Thurow, J., Littke, R., 1996. Evolution patterns of radiolaria and organic matter variations: A new approach to identify sea-level changes in mid-Cretaceous pelagic environments. Geology 24, 499–502.

Ernst, R.E., Buchan, K.L., 2001. Large mafic magmatic events through time and links to mantle plume-heads. In: Ernst, R.E., Buchan, K.L. (Eds.), Mantle Plumes: Their identification through time. Geological Society of America Special Paper, 352, pp. 483–575.

Fiet, N., 2000. Calibrage temporal de l'Aptien et des sous-étages associés par une approche cyclostratigraphique appliquée à la série pélagique de Marches-Ombrie (Italie centrale). Bulletin de la Société géologique de France 171, 103–113.

Fiet, N., Gorin, G., 2000. Lithological expression of Milankovitch cyclicity in carbonate-dominated, pelagic, Barremian deposits in central Italy. Cretaceous Research 21, 457–467.

Fiet, N., Beaudoin, B., Parize, O., 2001. Lithostratigraphic analysis of Milankovitch cyclicity in pelagic Albian deposits of central Italy: Implications for the duration of the stage and substages. Cretaceous Research 22, 265–275.

Fiet, N., Quidelleur, X., Parize, O., Bulot, L.G., Gillot, P.Y., 2006. Lower Cretaceous stage durations combining radiometric data and orbital chronology: Towards a more stable relative time scale? Earth and Planetary Science Letters 246, 407–417.

Föllmi, K.B., Godet, A., Bodin, S., Linder, P., 2006. Interactions between environmental change and shallow water carbonate buildup along the northern Tethyan margin and their impact on the Early Cretaceous carbon isotope record. Paleoceanography 21 PA4211, doi: 10.1029/2006PA001313.

Galbrun, B., 1984. Magnétostratigraphie de la limitè Jurassique–Crétacé. Proposition d'une échelle de polarité à partir du stratotype du Berriasien (Berrias, Ardèche, France) et la Sierra de Lugar (Province de Murcie, Espagne). Mémoires des Sciences della Terre 38, 95.

Gale, A.S., 1995. Cyclostratigraphy and correlation of the Cenomanian Stage in Western Europe. In: House, M.R., Gale, A.S. (Eds.), Orbital Forcing Timescales and Cyclostratigraphy. Geological Society Special Publication, 85, pp. 177–197.

Gale, A.S., Montgomery, P., Kennedy, W.J., Hancock, J.M., Burnett, J.A., McArthur, J.M., 1995. Definition and global correlation of the Santonian-Campanian boundary. Terra Nova 7, 611–622.

Gale, A.S., Kennedy, W.J., Burnett, J.A., Caron, M., Kidd, B.E., 1996. The Late Albian to Early Cenomanian succession at Mont Risou near Rosans (Drôme, SE France): An integrated study (ammonites, inoceramids, planktonic foraminifera, nannofossils, oxygen and carbon isotopes). Cretaceous Research 17, 515–606.

Gale, A.S., Young, J.R., Shackleton, N.J., Crowhurst, S.J., Wray, D.S., 1999. Orbital tuning of Cenomanian marly chalk successions: Towards a Milankovitch time-scale for the Late Cretaceous. Philosophical Transactions of the Royal Society, series A 357, 1815–1829.

Gale, A.S., Hardenbol, J., Hathway, B., Kennedy, W.J., Young, J.R., Phansalkar, V., 2002. Global correlation of Cenomanian (Upper Cretaceous) sequences: Evidence for Milankovitch control on sea level. Geology 30, 291–294.

Gale, A.S., Kennedy, J.W., Lees, J.A., Petrizzo, M.R., Walaszczyk, I., 2007. An integrated study (inoceramid bivalves, ammonites, calcareous nannofossils, planktonic foraminifera, stable carbon isotopes) of the Ten Mile Creek section, Lancaster, Dallas County, north Texas, a candidate Global boundary Stratotype Section and Point for the base of the Santonian Stage. Acta Geologica Polonica 57, 113–160.

Gale, A.S., Hancock, J.M., Kennedy, W.J., Petrizzo, M.R., Lees, J.A., Walaszczyk, I., Wray, D.S., 2008. An integrated study (geochemistry, stable oxygen and carbon isotopes, nannofossils, planktonic foraminifera, inoceramid bivalves, ammonites and crinoids) of the Waxahachie Dam Spillway section, north Texas, a possible boundary stratotype for the base of the Campanian Stage. Cretaceous Research 29, 131–167.

Gale, A.S., Bown, P., Caron, M., Crampton, J., Crowhurst, S.J., Kennedy, W.J., Petrizzo, M.R., Wray, D.S., 2011. The uppermost Middle and Upper Albian succession at the Col de Palluel, Hautes-Alpes, France: An integrated study (ammonites, inoceramid bivalves, planktonic foraminifera, nannofossils, geochemistry, stable oxygen and carbon isotopes, cyclostratigraphy). Cretaceous Research 32, 59–130.

Gerasimov, P.A., Mikhailov, N.R., 1966. Volgian Stage and general stratigraphic scale of the Upper Series of the Jurassic System. Isvestia Akademia Nauka, S.S.S.R., Geology Series 2, 118–138 [In Russian].

Gibson, F., 2006. Timescales of plume-lithosphere interactions in LIPs: $^{40}Ar/^{39}Ar$ geochronology of alkaline igneous rocks from the Paraná-Etendeka large igneous province. Available online: www.largeigneousprovinces.org/06nov.html.

Gibson, S.A., Thompson, R.N., Day, J.A., 2006. Timescales and mechanisms of plume-lithosphere interactions: $^{40}Ar/^{39}Ar$ geochronology and geochemistry of alkaline igneous rocks from the Paraná-Etendeka large igneous province. Earth and Planetary Science Letters 256, 1–17.

Gilder, S., Chen, Y., Cogne, J.P., Tan, X.D., Courtillot, V., Sun, D.J., Li, Y.G., 2003. Paleomagnetism of Upper Jurassic to Lower Cretaceous volcanic and sedimentary rocks from the western Tarim basin and implications for inclination shallowing and absolute dating of the M-0 (ISEA?) chron. Earth and Planetary Science Letters 206, 587–600.

Giraud, F., 1995. Recherche des périodicités astronomiques et des fluctuations du niveau marin à partir de l'étude du signal carbonaté des séries pélagiques alternantes. Documents des Laboratoires de Géologie Lyon 134, 279.

Giraud, F., Beaufort, L., Cotillon, P., 1995. Periodicities of carbonate cycles in the Valanginian of the Vocontian Trough: A strong obliquity control. In: House, M.R., Gale, A.S. (Eds.), Orbital Forcing Timescales and Cyclostratigraphy. Geological Society Special Publication, 85, pp. 143–164.

Gradstein, F.M., Ogg, J.G., Smith, A.G., (coordinators), 2004. A Geologic Time Scale 2004. Cambridge University Press, Cambridge, 589 p.

Grant, S.F., Coe, A.L., Armstrong, H.A., 1999. Sequence stratigraphy of the Coniacian succession of the Anglo-Paris Basin. Geological Magazine 136, 17–38.

Green, K.A., Brecher, A., 1974. Preliminary paleomagnetic results for sediments from Site 263, Leg 27. Initial Reports, Deep Sea Drilling Project 27, 405–413.

Grippo, A., Fischer, A.G., Hinnov, L.A., Herbert, T.M., Premoli Silva, I., 2004. Cyclostratigraphy and chronology of the Albian stage (Piobbico core, Italy). In: D'Argenio, B., Fischer, A.G., Premoli Silva, I.,

Weissert, H., Ferreri, V. (Eds.), Cyclostratigraphy: Approaches and Case Histories. SEPM Special Publication, 81, pp. 57–81.

Hailwood, E.A., 1979. Paleomagnetism of Late Mesozoic to Holocene sediments from the Bay of Biscay and Rockall Plateau, drilled on IPOD Leg 48. Initial Reports, Deep Sea Drilling Project 48, 305–339.

Hancock, J.M., 1991. Ammonite scales for the Cretaceous System. Cretaceous Research 12, 259–291.

Hancock, J.M., 2001. A proposal for a new position for the Aptian/Albian boundary. Cretaceous Research 22, 677–683.

Hancock, J.M., Gale, A.S., 1996. The Campanian Stage. Bulletin de l'Institut Royal des Sciences Naturelles de Belgique, Sciences de la Terre 66 (Suppl.), 103–109.

Hancock, J.M., Peake, N.B., Burnett, J., Dhondt, A.V., Kennedy, W.J., Stokes, R., 1993. High Cretaceous biostratigraphy at Tercis, SW France. Bulletin de l'Institut Royal des Sciences Naturelles de Belgique, Sciences de la Terre 63, 133–148.

Hardenbol, J., Thierry, J., Farley, M.B., Jacquin, Th., de Graciansky, P.-C., Vail, P.R., (with numerous contributors), 1998. Mesozoic and Cenozoic sequence chronostratigraphic framework of European basins. In: de Graciansky, P.-C., Hardenbol, J., Jacquin, Th., Vail, P.R. (Eds.), Mesozoic-Cenozoic Sequence Stratigraphy of European Basins. SEPM Special Publication, 60, pp. 3–13, 763–781, and chart supplements.

Harding, I.C., Smith, G.A., Riding, J.B., Wimbledon, W.A.P., 2011. Interregional correlation of Jurassic/Cretaceous boundary strata based on the Tithonian-Valanginian dinoflagellate cyst biostratigraphy of the Volga Basin, western Russia. Review of Palaeobotany and Palynology 167 (1–2), 82–116.

Harland, W.B., Armstrong, R.L., Cox, A.V., Craig, L.E., Smith, A.G., Smith, D.G., 1990. A Geologic Time Scale 1989. Cambridge University Press, Cambridge, p. 263.

Hart, M., Amédro, F., Owen, H., 1996. The Albian stage and substage boundaries. Bulletin de l'Institut Royal des Sciences Naturelles de Belgique, Sciences de la Terre 66 (Suppl.), 45–56.

He, H.Y., Pan, Y.X., Tauxe, L., Qin, H.F., Zhu, R.X., 2008. Toward age determination of the M0r (Barremian-Aptian boundary) of the Early Cretaceous. Physics of the Earth and Planetary Interiors 169, 41–48.

Hennebert, M., Robaszynski, F., Goolaerts, S., 2009. Cyclostratigraphy and chronometric scale in the Campanian – Lower Maastrichtian– the Abiod Formation at Ellès, central Tunisia. Cretaceous Research 30, 325–338.

Herbert, T.D., 1992. Paleomagnetic calibration of Milankovitch cyclicity in Lower Cretaceous sediments. Earth and Planetary Science Letters 112, 15–28.

Herbert, T.D., D'Hondt, S.L., Premoli-Silva, I., Erba, E., Fischer, A.G., 1995. Orbital chronology of Cretaceous-Early Palaeocene marine sediments. In: Berggren, W.A., Kent, D.V., Hardenbol, J. (Eds.), Geochronology, Time Scales and Global Stratigraphic Correlations. SEPM Special Publication, 54, pp. 81–94.

Herrle, J.O., Mutterlose, J., 2003. Calcareous nannofossils from the Aptian-Lower Albian of southeast France: Palaeoecological and biostratigraphic implications. Cretaceous Research 24, 1–22.

Herrle, J.O., Kössler, P., Friedrich, O., Erlenkeuser, H., Hemleben, C., 2004. High-resolution carbon isotope records of the Aptian to Lower Albian from SE France and the Mazagan Plateau (DSDP Site 545): A stratigraphic tool for paleoceanographic and paleobiologic reconstruction. Earth and Planetary Science Letters 218, 149–161.

Hesselbo, S.P., Coe, A.L., Jenkyns, H.C., 1990. Recognition and documentation of depositional sequences from outcrop: An example from the Aptian and Albian on the eastern margin of the Wessex Basin. Journal of the Geological Society 147, 549–559.

Hicks, J.F., Obradovich, J.D., 1995. Isotopic age calibration of the GRTS from C33N to C31N: Late Cretaceous Pierre Shale, Red Bird section, Wyoming, USA. Geological Society of America, Abstracts with Programs 27, A–174.

Hicks, J.F., Obradovich, J.D., Tauxe, L., 1995. A new calibration point for the Late Cretaceous time scale: the $^{40}Ar/^{39}Ar$ isotopic age of the C33r/C33n geomagnetic reversal from the Judith River Formation (Upper Cretaceous), Elk Basin, Wyoming, USA. Journal of Geology 103, 243–256.

Hicks, J.F., Obradovich, J.D., Tauxe, L., 1999. Magnetostratigraphy, isotopic age calibration and intercontinental correlation of the Red Bird section of the Pierre Shale, Niobrara County, Wyoming, USA. Cretaceous Research 20, 1–27.

Hochuli, P.A., Menegatti, A.P., Weissert, H., Riva, A., Erba, E., Premoli Silva, I., 1999. Episodes of high productivity and cooling in the early Aptian Alpine Tethys. Geology 27, 657–660.

Hoedemaeker, P.J., Leereveld, H., 1995. Biostratigraphy and sequence stratigraphy of the Berriasian-lowest Aptian (Lower Cretaceous) of the Río Argos succession, Caravaca, SE Spain. Cretaceous Research 16, 195–230.

Hoedemaeker, P.J., Company, M.R., Aguirre Urreta, M.B., Avram, E., Bogdanova, T.N., Bujtor, L., Bulot, L., Cecca, F., Delanoy, G., Ettachfini, M., Memmi, L., Owen, H.G., Rawson, P.F., Sandoval, J., Tavera, J.M., Thieuloy, J.-P., Tovbina, S.Z., Vasicek, Z., 1993. Ammonite zonation for the Lower Cretaceous of the Mediterranean region; basis for the stratigraphic correlation within IGCP, Project 262. Revista Española de Paleontología 8, 117–120.

Hoedemaeker, P.J., Reboulet, S., Aguirre-Urreta, M.B., Alsen, P., Atrops, F., Barragan, R., Company, M., González Arreola, C., Klein, J., Lukeneder, A., Ploch, I., Raisossadat, N., Rawson, P.F., Ropolo, P., Vasicek, Z., Vermeulen, J., Wippich, M.G.E., 2003. Report on the 1st International Workshop of the IUGS Lower Cretaceous Ammonite Working Group, the 'Kilian Group' (Lyon, 11 July 2002). Cretaceous Research, 24 89–94.

Houša, V., Pruner, P., Zakharov, V.A., Kostak, M., Chadima, M., Rogov, M.A., Slechta, S., Mazuch, M., 2007. Boreal-Tethyan correlation of the Jurassic–Cretaceous boundary interval by magneto- and biostratigraphy. Stratigraphy and Geological Correlation 15 (3), 297–309.

Huang, C., Hinnov, L.A., Fischer, A.G., Grippo, A., Herbert, T., 2010. Astronomical tuning of the Aptian Stage from Italian reference sections. Geology 38, 899–902.

Huang, Z., Ogg, J.G., Gradstein, F.M., 1993. A quantitative study of Lower Cretaceous cyclic sequences from the Atlantic Ocean and the Vocontian Basin (SE France). Paleoceanography 8, 275–291.

Huber, B.T., Leckie, R.M., 2011. Planktic foraminiferal species turnover across deep-sea Aptian/Albian boundary sections. Journal of Foraminiferal Research 41, 53–95.

Huber, B.T., Hodell, D.A., Hamilton, C.P., 1995. Middle-Late Cretaceous climate of the southern high latitudes: Stable isotopic evidence for minimal equator-to-pole thermal gradients. Geological Society of America Bulletin 107, 1164–1191.

Huber, B.T., Norris, R.D., MacLeod, K.G., 2002. Deep sea paleotemperature record of extreme warmth during the Cretaceous. Geology 30, 123–126.

Huber, B.T., MacLeod, K.G., Tur, N.A., 2008. Chronostratigraphic framework for Upper Campanian-Maastrichtian sediments on the Blake Nose (subtropical North Atlantic). Journal of Foraminiferal Research 38, 162–182.

Huber, B.T., Petrizzo, M.R., Gale, A.S., Barchetta, A., 2011. New criteria for defining the base of the Albian Stage. Geological Society of America Annual Meeting, Minneapolis. 9–12 October 2011, Abstract, 267–4. Available online: gsa.confex.com/gsa/2011AM/finalprogram/abstract_196763.htm.

Husson, D., Galbrun, B., Laskar, J., Hinnov, L.A., Thibault, N., Gardin, S., Locklair, R.E., 2011. Astronomical calibration of the Maastrichtian. Earth and Planetary Science Letters 305, 328–340.

Ingle, S., Coffin, M.F., 2004. Impact origin for the greater Ontong Java Plateau? Earth and Planetary Science Letters 218, 123–134.

Jacquin, T., de Graciansky, P.-C., 1998. Major transgressive-regressive cycles: The stratigraphic signature of European basin development. In: de Graciansky, P.-C., Hardenbol, J., Jacquin, Th., Vail, P.R. (Eds.), Mesozoic–Cenozoic Sequence Stratigraphy of European Basins. SEPM Special Publication, 60, pp. 15–29.

Janasi, V.A., Freitas, V.A., Heaman, L.H., 2011. The onset of flood basalt volcanism, Northern Paraná Basin, Brazil: A precise U–Pb baddeleyite/zircon age for a Chapecó-type dacite. Earth and Planetary Science Letters 302, 147–153.

Jarrard, R.D., 1974. Paleomagnetism of some Leg 27 sediment cores. Initial Reports, Deep Sea Drilling Project 27, 415–423.

Jarvis, I., Gale, A.S., Jenkyns, H.C., Pearce, M.A., 2006. Secular variation in Late Cretaceous carbon isotopes: A new $\delta^{13}C$ carbonate reference curve for the Cenomanian-Campanian (99.6–70.6 Ma). Geological Magazine 143, 561–608.

Jarvis, I., Lignum, J.S., Gröcke, D.R., Jenkyns, H.C., Pearce, M.A., 2011. Black shale deposition, atmospheric CO_2 drawdown, and cooling during the Cenomanian-Turonian Oceanic Anoxic Event. Paleoceanography 26 PA3201, doi: 10.1029/2010PA002081.

Jenkyns, H.C., 1999. Mesozoic anoxic events and palaeoclimate. Zentralblatt für Geologie und Paläontologie, Teil I, H 7–9, 943–949.

Jenkyns, H.C., 2010. Geochemistry of oceanic anoxic events. Geochemistry, Geophysics, Geosystems 11 Q03004. doi: 10.1029/2009GC002788.

Jenkyns, H.C., Gale, A.S., Corfield, R.M., 1994. Carbon- and oxygen-isotope stratigraphy of the English Chalk and the Italian Scaglia and its palaeoclimatic significance. Geological Magazine 131, 1–34.

Jinnah, Z.A., Roberts, E.M., Deino, A.L., Larsen, J.S., Link, P.K., Fanning, C.M., 2009. New $^{40}Ar/^{39}Ar$ and detrital zircon U-Pb ages for the Upper Cretaceous Wahweap and Kaiparowits formations on the Kaiparowits Plateau, Utah: Implications for regional correlation, provenance, and biostratigraphy. Cretaceous Research 30, 287–299.

Jones, C.E., Jenkyns, H.C., 2001. Seawater strontium isotopes, oceanic anoxic events, and seafloor hydrothermal activity in the Jurassic and Cretaceous. American Journal of Science 301, 112–149.

Jourdan, F., Renne, P.R., Reimold, W.U., 2009. An appraisal of the ages of terrestrial impact structures. Earth and Planetary Science Letters 286, 1–13.

Kauffman, E.J., Kennedy, W.J., Wood, C.J., 1996. The Coniacian stage and substage boundaries. Bulletin de l'Institut Royal des Sciences Naturelles de Belgique, Sciences de la Terre 66 (Suppl.), 81–94.

Keating, B.H., Helsley, C.E., 1978a. Magnetostratigraphic studies of Cretaceous age sediments from Sites 361, 363, and 364. Initial Reports, Deep Sea Drilling Project 40, 459–467.

Keating, B.H., Helsley, C.E., 1978b. Magnetostratigraphic studies of Cretaceous sediments from DSDP Site 369. Initial Reports, Deep Sea Drilling Project 41 (Suppl.), 983–986.

Keating, B.H., Helsley, C.E., 1978c. Paleomagnetic results from DSDP Hole 391C and the magnetostratigraphy of Cretaceous sediments from the Atlantic Ocean floor. Initial Reports, Deep Sea Drilling Project 44, 523–528.

Kemp, T.S., 2005. The Origin and Evolution of Mammals. Oxford University Press, Oxford, p. 342.

Kennedy, W.J., 1984. Ammonite faunas and the 'standard zones' of the Cenomanian to Maastrichtian Stages in their type areas, with some proposals for the definition of the stage boundaries by ammonites. Geological Society of Denmark Bulletin 33, 147–161.

Kennedy, W.J., Cobban, W.A., 1991. Stratigraphy and inter-regional correlation of the Cenomanian-Turonian transition in the Western Interior of the United States near Pueblo, Colorado. A potential boundary stratotype for the base of the Turonian Stage. Newsletters on Stratigraphy 24, 1–33.

Kennedy, W.J., Christensen, W.K., Hancock, J.M., 1995. Defining the base of the Maastrichtian and its substages. Internal report for the Maastrichtian Working Group. Second International Symposium on Cretaceous Boundaries, Brussels, 8–16 September 1995, p. 13 [Unpublished].

Kennedy, W.J., Gale, A.S., Bown, P.R., Caron, M., Davey, R.J., Gröcke, D., Wray, D.S., 2000a. Integrated stratigraphy across the Aptian-Albian boundary in the Marnes Bleues at the Col de Pré-Guittard, Arnayon (Brôme), and at Tartonne (Alpes-de-Haute-Provence), France: A candidate Global Boundary Stratotype Section and Boundary Point for the base of the Albian Stage. Cretaceous Research 21, 591–720.

Kennedy, W.J., Walaszczyk, I., Cobban, W.A., 2000b. Pueblo, Colorado, USA, candidate global boundary stratotype section and point for the base of the Turonian Stage of the Cretaceous and for the Middle Turonian substage, with a revision of the Inoceramidae (Bivalve). Acta Geologica Polonica 50, 295–334.

Kennedy, W.J., Gale, A.S., Lees, J.A., Caron, M., 2004. Definition of a Global Boundary Stratotype Section and Point (GSSP) for the base of the Cenomanian Stage, Mont Risou, Hautes-Alpes, France. Episodes 27, 21–32.

Kennedy, W.J., Walaszczyk, I., Cobban, W.A., 2005. The Global Boundary Stratotype Section and Point for the base of the Turonian Stage of the Cretaceous: Pueblo, Colorado, U.S.A. Episodes 28 (2), 93–104.

Kerr, A.C., 1998. Oceanic plateau formation: A cause of mass extinction and black shale deposition around the Cenomanian-Turonian boundary? Journal of the Geological Society 155, 619–626.

Koeberl, C., Armstrong, R.A., Reimold, W.U., 1997. Morokweng, South Africa: A large impact structure of Jurassic-Cretaceous boundary age. Geology 25, 731–734.

Kuhnt, W., Moullade, M., 2007. The Gargasian (middle Aptian) of La Marcouline section at Cassis-La Bédoule (SE France): Stable isotope record and orbital cyclicity. Carnets de Géologie 2007 (02), 1–9.

Kuhnt, W., Holbourn, A., Gale, A., Chellai, E.H., Kennedy, W.J., 2009. Cenomanian sequence stratigraphy and sea-level fluctuations in the Tarfaya Basin (SW Morocco). Geological Society of America Bulletin 121, 1695–1710.

Lamolda, M.A., Hancock, J.M., with contributions by Burnett, J.A., Collom, C.J., Christensen, W.K., Dhondt, A.V., Gardin, S., Gräfe, K.-U., Ion, J., Kauffman, E.G., Kennedy, W.J., Kopaevich, L.F., Lopez, G., Matsumoto, T., Mortimore, R., Premoli Silva, I., Robaszynski, F., Salij, J.,

Summesberger, H., Toshimitsu, S., Tröger, K.-A., Wagreich, M., Wood, C.J., and Yazykova, E.A., 1996. The Santonian Stage and substages. Bulletin de l'Institut Royal des Sciences Naturelles de Belgique, Sciences de la Terre 66 (Suppl.), 95–102.

Lamolda, M.A., the Santonian Working Group, 2011. The "Cantera de Margas" section, Olazagutia, northern Spain. A candidate GSSP for the base of the Santonian Stage: Its stratigraphy across the Coniacian-Santonian transition [Documentation submitted to International Commission on Stratigraphy for consideration as a GSSP, draft of April 2011].

Landman, N.H., Waage, K.M., 1993. Scaphitid ammonites of the Upper Cretaceous (Maastrichtian) Fox Hills Formation in South Dakota and Wyoming. Bulletin of the American Museum of Natural History 215, 257.

Large Igneous Provinces Commission (of the International Association of Volcanology and Chemistry of the Earth's Interior), 2011. LIP record. Available at. http://largeigneousprovinces.org/record [and subpages].

Larson, R.L., 1991. Latest pulse of the Earth: Evidence for a mid Cretaceous super plume. Geology 19, 547–550.

Larson, R.L., Erba, E., 1999. Onset of the mid-Cretaceous greenhouse in the Barremian-Aptian: Igneous events and the biological, sedimentary, and geochemical responses. Paleoceanography 14, 663–678.

Leckie, R.M., Bralower, T.J., Cashman, R., 2002. Oceanic anoxic events and plankton evolution: Biotic response to tectonic forcing during the mid-Cretaceous. Paleoceanography 17, 1041. doi: 10.1029/2001PA000623.

Leereveld, H., 1995. Dinoflagellate cysts from the Lower Cretaceous Río Argos succession (SE Spain). LPP Contributions Series 2, 175.

Lerbekmo, J.F., Braman, D.R., 2002. Magnetostratigraphic and biostratigraphic correlation of late Campanian and Maastrichtian marine and continental strata from the Red Deer River Valley to the Cypress Hills, Alberta, Canada. Canadian Journal of Earth Sciences 39, 539–557.

Li, Y.-X., Bralower, T.J., Montañez, I.P., Osleger, D.A., Arthur, M.A., Bice, D.M., Herbert, T.D., Erba, E., Premoli Silva, I., 2008. Toward an orbital chronology for the early Aptian Oceanic Anoxic Event (OAE1a, ~120 Ma). Earth and Planetary Science Letters 271, 88–100.

Lini, A., Weissert, H., Erba, E., 1992. The Valanginian carbon isotope event: A first episode of greenhouse climate conditions during the Cretaceous. Terra Nova 4, 374–384.

Locklair, R.E., Sageman, B.B., 2008. Cyclostratigraphy of the Upper Cretaceous Niobrara Formation, Western Interior, U.S.A.: A Coniacian-Santonian orbital timescale. Earth and Planetary Science Letters 269, 539–552.

Lowrie, W., Alvarez, W., 1977. Late Cretaceous geomagnetic polarity sequence: Detailed rock and palaeomagnetic studies of the Scaglia Rossa limestone at Gubbio, Italy. Geophyics Journal of the Royal Astronomical Society 51, 561–581.

Lowrie, W., Alvarez, W., Premoli–Silva, I., Monechi, S., 1980. Lower Cretaceous magnetic stratigraphy in Umbrian pelagic carbonate rocks. Geophysical Journal, Royal Astronomical Society 60, 263–281.

Lucas, S.G., 1997. Dinosaurs, the Textbook, second ed. Wm. C. Brown Publishers, Dubuque, Iowa, p. 336.

Lucas, S.G., Kirkland, J.I., aand Estep, J.W. (Eds.), 1998. Lower and Middle Cretaceous Terrestrial Ecosystems. New Mexico Museum of Natural History and Science Bulletin, 14, p. 330.

Magniez-Jannin, F., 1995. Cretaceous stratigraphic scales based on benthic foraminifera in West Europe (biochronohorizons). Bulletin de la Société géologique de France 166, 565–572.

Mahoney, J.J., Storey, M., Duncan, R.A., Spencer, K.J., Pringle, M.S., 1993. Geochemistry and age of the Ontong Java Plateau. In: Pringle, M.S. (Ed.), The Mesozoic Pacific: Geology, Tectonics, and Volcanism. American Geophysical Union Geophysical Monographs, 77, pp. 233–262.

Mahoney, J.J., et al., 2001. Proceedings of the Ocean Drilling Program, Initial Reports, 192 Available online. http://www-odp.tamu.edu/publications/192_IR/192ir.htm.

Mahoney, J.J., Duncan, R.A., Tejada, M.L.G., Sager, W.W., Bralower, T.J., 2005. Jurassic-Cretaceous boundary age and mid-ocean-ridge-type mantle source for Shatsky Rise. Geology 33, 185–188.

Malinverno, A., Erba, E., Herbert, T.D., 2010. Orbital tuning as an inverse problem: Chronology of the early Aptian oceanic anoxic event 1a (Selli Level) in the Cismon APTICORE. Paleoceanography 25 PA2203. doi: 10.1029/2009PA001769.

Masse, J.-P., 1998. Calcareous algae. Columns for Jurassic chart of Mesozoic and Cenozoic sequence chronostratigraphic framework of European basins, by Hardenbol, J., Thierry, J., Farley, M.B., Jacquin, Th., de Graciansky, P.-C., and Vail, P.R. (coordinators). In: de Graciansky, P.-C., Hardenbol, J., Jacquin, Th., Vail, P.R. (Eds.), Mesozoic-Cenozoic Sequence Stratigraphy of European Basins. SEPM Special Publication, 60, pp. 763–781.

Masse, J.-P., Philip, J., 1998. Rudists. Columns for Jurassic chart of Mesozoic and Cenozoic sequence chronostratigraphic framework of European basins, by Hardenbol, J., Thierry, J., Farley, M.B., Jacquin, Th., de Graciansky, P.-C., and Vail, P.R. (coordinators). In: de Graciansky, P.-C., Hardenbol, J., Jacquin, Th., Vail, P.R. (Eds.), Mesozoic-Cenozoic Sequence Stratigraphy of European Basins. SEPM Special Publication, 60, pp. 763–781.

McArthur, J.M., Kennedy, W.J., Gale, A.S., Thirlwall, M.F., Chen, M., Burnett, J., Hancock, J.M., 1992. Strontium isotope stratigraphy in the late Cretaceous: International correlation of the Campanian/Maastrichtian boundary. Terra Nova 4, 332–345.

McArthur, J.M., Thirlwall, M.F., Chen, M., Gale, A.S., Kennedy, W.J., 1993. Strontium isotope stratigraphy in the late Cretaceous: Numerical calibration of the Sr isotope curve, and international correlation for the Campanian. Paleoceanography 8, 859–873.

McArthur, J.M., Kennedy, W.J., Chen, M., Thirlwall, M.F., Gale, A.S., 1994. Strontium isotope stratigraphy for the Late Cretaceous: Direct numerical age calibration of the Sr-isotope curve for the U.S. Western Interior Seaway. Palaeogeography, Palaeoclimatology, Palaeoecology 108, 95–119.

McArthur, J.M., Janssen, N.M.M., Reboulet, S., Leng, M.J., Thirlwall, M.F., van de Schootbrugge, B., 2007. Palaeotemperatures, polar ice-volume, and isotope stratigraphy (Mg/Ca, $\delta^{18}O$, $\delta^{13}C$, $^{87}Sr/^{86}Sr$): The Early Cretaceous (Berriasian, Valanginian, Hauterivian). Palaeogeography, Palaeoclimatology, Palaeoecology 248, 391–430.

Mena, M., Orgeira, M.J., Lagorioi, S., 2006. Paleomagnetism, rock-magnetism and geochemical aspects of early Cretaceous basalts of the Paraná Magmatic Province, Misiones, Argentina. Earth Planets Space 58, 1283–1293.

Mesozoic Planktonic Foraminiferal Working Group (Huber, B.T., coordinator), 2006. Mesozoic Planktonic Foraminiferal Taxonomic Dictionary Available at. www.chronos.org.

Meyers, S.R., Siewert, S.E., Singer, B.S., Sageman, B.B., Condon, D., Obradovich, J.D., Jicha, B.R., Sawyer, D.A., 2010. Reducing error bars through the intercalibration of radioisotopic and astrochronologic time scales for the Cenomanian/Turonian Boundary Interval, Western Interior Basin, USA. American Geological Union Fall Meeting, San Francisco, CA. 13–17 Dec 2010, Abstract. Available online. http://www.agu.org/meetings/fm10/waisfm10.html.

Meyers, S.R., Siewert, S.E., Singer, B.S., Sageman, B.B., Condon, D., Obradovich, J.D., Jicha, B.R., and Sawyer, D.A., 2012. Intercalibration of radioisotopic and astrochronologic time scales for the Cenomanian-Turonian Boundary interval, Western Interior Basin, USA. Geology 40, 7–10.

Miller, K.G., Barrera, E., Olsson, R.K., Sugarman, P.J., Savin, S.M., 1999. Does ice drive early Maastrichtian eustasy? Geology 27, 783–786.

Miller, K.G., Sugarman, P.J., Browning, J.V., Kominz, M.A., Hernàndez, J.C., Olsson, R.K., Wright, J.D., Feigenson, M.D., Van Sickel, W., 2003. Late Cretaceous chronology of large, rapid sea-level changes: Glacioeustasy during the greenhouse world. Geology 31, 585–588.

Miller, K.G., Sugarman, P.J., Browning, J.V., Kominz, M.A., Olsson, R.K., Feigenson, M.D., Hernàndez, J.C., 2004. Upper Cretaceous sequences and sea-level history, New Jersey coastal plain. Geological Society of America Bulletin 116, 368–393.

Mitchell, R.N., Bice, D.M., Montanari, A., Cleaveland, L.C., Christianson, K.T., Coccioni, R., Hinnov, L.A., 2008. Oceanic anoxic cycles? Orbital prelude to the Bonarelli Level (OAE 2). Earth and Planetary Science Letters 267, 1–16.

Monteil, E., and Foucher, J.-C., Dinoflagellate cysts. Columns for Jurassic chart of Mesozoic and Cenozoic sequence chronostratigraphic framework of European basins, by Hardenbol, J., Thierry, J., Farley, M.B., Jacquin, Th., de Graciansky, P.-C., and Vail, P.R. (coordinators). In: de Graciansky, P.-C., Hardenbol, J., Jacquin, Th., and Vail, P.R. (Eds.), Mesozoic-Cenozoic Sequence Stratigraphy of European Basins. SEPM Special Publication, 60, pp. 763–781.

Montgomery, P., Hailwood, E.A., Gale, A.S., Burnett, J.A., 1998. The magnetostratigraphy of Coniacian-Late Campanian chalk sequences in southern England. Earth and Planetary Science Letters 156, 209–224.

Moullade, M., Tronchetti, G., 2004. The Gargasian (Middle Aptian) substage in the Aptian historical stratotypes (SE France): General introduction. Carnets de Géologie / Notebooks on Geology, Letter 2004/01, 3. Available online. http://paleopolis.rediris.es/cg/CG2004_L01_MM-GT/index.html.

Moullade, M., Tronchetti, G., Masse, J.-P., 1998a. Le stratotype historique de l'Aptien inférieur (Bédoulien) dans la région de Cassis-La Bédoule (S.E. France). Géologie Méditerranéenne XXV (3–4), 298.

Moullade, M., Masse, J.-P., Tronchetti, G., Kuhnt, W., Ropolo, P., Bergen, J.A., Masure, E., Renard, M., 1998b. Le stratotype historique de l'Aptien (région de Cassis-La Bédoule, SE France): Synthèse stratigraphique. Géologie Méditerranéenne XXV (3–4), 289–298.

Moullade, M., Tronchetti, G., Kuhnt, W., Renard, M., Bellier, J.-P., 2004. The Gargasian (Middle Aptian) of Cassis-La Bédoule (Lower Aptian historical stratotype, SE France): Geographic location and lithostratigraphic correlations. Carnets de Géologie / Notebooks on Geology, Letter 2004/02, 24. Available online. http://paleopolis.rediris.es/cg/CG2004_L02_MM-etal/index.html.

Moullade, M., Tronchetti, G., Bellier, J.-P., 2005. The Gargasian (Middle Aptian) strata from Cassis-La Bédoule (Lower Aptian historical stratotype, SE France): Planktonic and benthic foraminiferal assemblages and biostratigraphy. Carnets de Géologie / Notebooks on Geology, Article 2005/02, 20. Available online. http://paleopolis.rediris.es/cg/CG2004_A02/index.html.

Moullade, M., Granier, B., Tronchetti, G., 2011. The Aptian Stage: Back to fundamentals. Episodes 34, 148–156.

Mutterlose, J., Böckel, B., 1998. The Barremian-Aptian interval in NW Germany: A review. Cretaceous Research 19, 539–568.

Mutterlose, J., Autran, G., Baraboschkin, E.J., Cecca, F., Erba, E., Gardin, S., Herngreen, W., Hoedemaeker, Ph.J., Kakabadze, M., Klein, J., Leereveld, H., Rawson, P.F., Ropolo, P., Vasicek, Z., von Salis, K., 1996. The Hauterivian Stage. Bulletin de l'Institut Royal des Sciences Naturelles de Belgique, Sciences de la Terre 66 (Suppl.), 19–24.

Mutterlose, J., Bornemann, A., Luppold, F.W., Owen, H.G., Ruffell, A., Weiss, W., Wray, D.S., 2003. The Vöhrum section (northwest Germany) and the Aptian/Albian boundary. Cretaceous Research 24, 203–252.

Niebuhr, B., 2005. Geochemistry and time-series analyses of orbitally forced Upper Cretaceous marl-limestone rhythmites (Lehrte West Syncline, northern Germany). Geological Magazine 142, 31–55.

Nikitin, S.N., 1881. Jurassic Deposits between Rybinsk, Mologa, and Myshkin. Materialy dlya geologii Rossii X, 201–331 [In Russian].

Oboh-Ikuenobe, F.E., Benson, D.G., Scott, R.W., Holbrook, J.M., Evetts, M.J., Erbacher, J., 2007. Re-evaluation of the Albian-Cenomanian boundary in the U.S. Western Interior based on dinoflagellate cysts. Review of Palaeobotany and Palynology 144, 77–97.

Oboh-Ikuenobe, F.E., Holbrook, J.M., Scott, R.W., Akins, S.L., Evetts, M.J., Benson, D.G., Pratt, L.M., 2008. Anatomy of epicontinetal flooding: Late Albian-early Cenomanian of the southern U.S. Western Interior Basin. In: Pratt, B.R., Homden, C. (Eds.), Dynamics of Epeiric Seas. Geological Association of Canada Special Publication, 48, pp. 201–227.

Obradovich, J.D., 1993. A Cretaceous time scale. In: Caldwell, W.G.E., Kauffman, E.G. (Eds.), Evolution of the Western Interior Basin. Geological Association of Canada, Special Paper, 39, pp. 379–396.

Obradovich, J.D., Matsumoto, T., Nishida, T., Inoue, Y., 2002. Integrated biostratigraphic and radiometric scale on the Lower Cenomanian (Cretaceous) of Hokkaido, Japan. Proceedings of the Japan Academy, Series B-Physical and Biological Sciences 78 (6), 149–153.

Odin, G.S., 1996. Le site de Tercis (Landes). Observations stratigraphiques sur le Maastrichtien. Arguments pour la localisation et la corrélation du Point Stratotype Global de la limite Campanien – Maastrichtien. Bulletin de la Société géologique de France 167, 637–643.

Odin, G.S. (Ed.), 2001. The Campanian - Maastrichtian Boundary: Characterisation at Tercis les Bains (France) and correlation with Europe and other Continents. Developments in Palaeontology and Stratigraphy, vol. 19. Elsevier, Amsterdam, p. 910.

Odin, G.S., Lamaurelle, M.A., 2001. The global Campanian-Maastrichtian stage boundary. Episodes 24, 229–238.

Odin, G.S., Hancock, J.M., Antonescu, E., Bonnemaison, M., Caron, M., Cobban, W.A., Dhondt, A., Gaspard, D., Ion, J., Jagt, J.W.M., Kennedy, W.J., Melinte, M., Néraudeau, D., von Salis, K., Ward, P.D., 1996. Definition of a Global Boundary Stratotype Section and Point for the Campanian/Maastrichtian boundary. Bulletin de l'Institut Royal des Sciences Naturelles de Belgique, Sciences de la Terre 66 (Suppl.), 111–117.

Ogg, J.G., 1987. Early Cretaceous magnetic polarity time scale and the magnetostratigraphy of DSDP Sites 603 and 534, western Central Atlantic. Initial Reports of the Deep Sea Drilling Project 93, 849–888.

Ogg, J.G., 1988. Early Cretaceous and Tithonian magnetostratigraphy of the Galicia margin (Ocean Drilling Program Leg 103). Proceedings of the Ocean Drilling Program, Scientific Results 103, 659–682.

Ogg, J.G., Lowrie, W., 1986. Magnetostratigraphy of the Jurassic-Cretaceous boundary. Geology 14, 547–550.

Ogg, J.G., Bardot, L., 2001. Aptian through Eocene magnetostratigraphic correlation of the Blake Nose Transect (Leg 171B), Florida Continental

Margin. Proceedings of the Ocean Drilling Program, Scientific Results 171B, 59. Available online. http://www-odp.tamu.edu/ publications/ 171B_SR/ chap_09/ chap_09.htm.

Ogg, J.G., Steiner, M.B., Oloriz, F., Tavera, J.M., 1984. Jurassic magnetostratigraphy, 1. Kimmeridgian-Tithonian of Sierra Gorda and Carcabuey, southern Spain. Earth and Planetary Science Letters 71, 147–162.

Ogg, J.G., Steiner, M.B., Company, M., Tavera, J.M., 1988. Magnetostratigraphy across the Berriasian–Valanginian stage boundary (Early Cretaceous) at Cehegin (Murcia Province, southern Spain). Earth and Planetary Science Letters 87, 205–215.

Ogg, J.G., Hasenyager II, R.W., Wimbledon, W.A., Channell, J.E.T., Bralower, T.J., 1991. Magnetostratigraphy of the Jurassic-Cretaceous boundary interval – Tethyan and English faunal realms. Cretaceous Research 12, 455–482.

Ogg, J.G., Kodama, K., Wallick, B.P., 1992. Lower Cretaceous magnetostratigraphy and paleolatitudes off northwest Australia, ODP Site 765 and DSDP Site 261, Argo Abyssal Plain, and ODP Site 766, Gascoyne Abyssal Plain. Proceedings of the Ocean Drilling Program, Scientific Results 123, 523–548.

Ogg, J.G., Hasenyager II, , RW., Wimbledon, W.A., 1994. Jurassic-Cretaceous boundary: Portland-Purbeck magnetostratigraphy and possible correlation to the Tethyan faunal realm. Géobios, Mémoire Spécial 17, 519–527.

Ogg, J.G., Ogg, G., Gradstein, F.M., 2008. The Concise Geologic Time Scale. Cambridge University Press, Cambridge, p. 177.

Oppel, C.A., 1865. Die Tithonische Etage. Zeitschrift der Deutschen Geologischen Gesellschaft, Jahrgang 17, 535–558.

Oosting, A.M., Leereveld, H., Dickens, G.R., Henderson, R.A., Brinkhuis, 2006. Correlation of Barremian-Aptian (mid-Cretaceous) dinoflagellate cyst assemblages between the Tethyan and Austral realms. Cretaceous Research 27, 762–813.

Owen, H.G., 1996a. Boreal and Tethyan late Aptian to late Albian ammonite zonation and palaeobiogeography. In: Spaeth, C. (Ed.), Jost Wiedmann Memorial Volume, Proceedings of the 4th International Cretaceous Symposium, Hamburg, 1992. Mitteilung aus dem Geologisch-Paläontologischen Institut der Universität Hamburg, 77, pp. 461–481.

Owen, H.G., 1996b. "Uppermost Wealden facies and Lower Greensand Group (Lower Cretaceous) in Dorset, southern England: Correlation and palaeoenvironment" by Ruffell & Batten (1994) and "The Sandgate Formation of the M20 Motorway near Ashford, Kent and its correlation" by Ruffell & Owen (1995)": reply. Proceedings of the Geologists' Association 107, 74–76.

Owen, H.G., 2002. The base of the Albian Stage; comments on recent proposals. Cretaceous Research 23, 1–13.

Pearson, P.N., Ditchfield, P.W., Singano, J., Harcourt-Brown, K.G., Nicholas, C.J., Olsson, R.K., Shackleton, N.J., Hall, M.A., 2001. Warm tropical sea surface temperatures in the Late Cretaceous and Eocene epochs. Nature 413, 481–487.

Pechersky, D.M., Khramov, A.N., 1973. Mesozoic paleomagnetic scale of the U.S.S.R. Nature 244, 499–501.

Petrizzo, M.R., Falzoni, F., Premoli Silva, I., 2011. Identification of the base of the lower-to-middle Campanian *Globotruncana ventricosa* Zone: Comments on reliability and global correlations. Cretaceous Research 32, 387–405.

Petrizzo, M.R., Huber, B.T., Gale, A.S., Barchetta, A., Jenkyns, H.C., in press, Abrupt planktic foraminiferal turnover across the Niveau Kilian at Col de Pré-Guittard (Vocontian Basin, southeast France): New criteria for defining the Aptian/Albian boundary. Newsletters on Stratigraphy.

Premoli Silva, I., Sliter, W.V., 1999. Cretaceous paleoceanography: Evidence from planktonic foaminiferal evolution. In: Barrera, E., Johnson, C.C. (Eds.), Evolution of the Cretaceous Ocean-Climate System. Geological Society of America Special Paper, 332, pp. 301–328.

Pringle, M.S., Duncan, R.A., 1995a. Radiometric ages of basaltic lavas recovered Sites 865, 866, and 869. Proceedings of the Ocean Drilling Program, Scientific Results 143, 277–283.

Pringle, M.S., Duncan, R.A., 1995b. Radiometric ages of basaltic lavas recovered at Lo-En, Wodejebato, MIT, and Takuyo-Daisan Guyots, northwestern Pacific Ocean. Proceedings of the Ocean Drilling Program, Scientific Results 144, 547–557.

Pringle, M.S., Chambers, L., Ogg, J.G., 2003. Synchronicity of volcanism on Ontong Java and Manihiki plateaux with global oceanographic events? American Geophysical Union and European Union of Geophysics Conference, Nice, France. May 2003, Abstract.

Pruner, P., Houša, V., Olóriz, F., Koštak, M., Krs, M., Man, O., Schnable, P., Venhodová, D., Tavera, J.M., Mazuch, M., 2010. High-resolution magnetostratigraphy and the biostratigraphic zonation of the Jurassic/Cretaceous boundary strata in the Puerto Escaño section (southern Spain). Cretaceous Research 31, 192–206.

Pucéat, E., Lécuyer, C., Sheppard, S.M.F., Dromart, G., Reboulet, S., Grandjean, P., 2003. Thermal evolution of Cretaceous Tethyan marine waters inferred from oxygen isotope composition of fish tooth enamels. Paleoceanography 18, 1029. doi: 10.1029/2002PA00823.

Quidelleur, X., Paquette, J.L., Fiet, N., Takashima, R., Tiepolo, M., Desmares, D., Nishi, H., Grosheny, D., 2011. New U-Pb (ID-TIMS and LA-ICPMS) and $^{40}Ar/^{39}Ar$ geochronological constraints of the Cretaceous geologic time scale calibration from Hokkaido (Japan). Chemical Geology 286, 72–83.

Rawson, P.F., 1983. The Valanginian to Aptian stages – current definitions and outstanding problems. Zitteliania 10, 493–500.

Rawson, P.F., 1990. Event stratigraphy and the Jurassic-Cretaceous boundary. Transactions of the Institute of Geology and Geophysics, Academy Sciences USSR, Siberian Branch 699, 48–52 [In Russian with English summary].

Rawson, P.F., Curry, D., Dilley, F.C., Hancock, J.M., Kennedy, W.J., Neale, J.W., Wood, C.J., Worssam, B.C., 1978. A correlation of Cretaceous rocks in the British Isles. Geological Society Special Report 9 p. 70.

Rawson, P.F., Avram, E., Baraboschkin, E.J., Cecca, F., Company, M., Delanoy, G., Hoedemaeker, Ph.J., Kakabadze, M., Kotetishvili, E., Leereveld, H., Mutterlose, J., von Salis, K., Sandoval, J., Tavera, J.M., Vasicek, Z., 1996a. The Barremian Stage. Bulletin de l'Institut Royal des Sciences Naturelles de Belgique, Sciences de la Terre 66 (Suppl.), 25–30.

Rawson, P.F., Dhondt, A.V., Hancock, J.M., Kennedy, W.J. (Eds.), 1996b. Proceedings of the Second International Symposium on Cretaceous Stage Boundaries, Brussels, 8–16 September 1995. Bulletin de l'Institut Royal des Sciences Naturelles de Belgique, Sciences de la Terre 66 (Suppl.), 117.

Reboulet, S., 1996. L'évolution des ammonites du Valanginien-Hauterivien inférieur du bassin vocontien et de la plate-forme provençale (sud-est de la France): relations avec la stratigraphie séquentielle et implications biostratigraphiques. Documents des Laboratoires de Géologie Lyon 137, 370.

Reboulet, S., 2001. Limiting factors on shell growth, mode of life and segregation of Valanginian ammonoid populations: Evidence from adult-size variations. Geobios 34, 423–435.

Reboulet, S., Atrops, F., 1999. Comments and proposals about the Valanginian-Lower Hauterivian ammonite zonation of south-eastern France. Eclogae Geologicae Helvetiae 92, 183–197.

Reboulet, S., Mattioli, E., Pittet, B., Baudin, B., Olivero, D., Proux, O., 2003. Ammonoid and nannoplankton abundance in Valanginian (early Cretaceous) limestone-marl alternations from the southeast France Basin: Carbonate dilution or productivity? Palaeogeography, Palaeoclimatology, Palaeoecology 201, 113–139.

Reboulet, S., Hoedemaeker, P.J., Aguirre-Urreta, M.B., Alsen, P., Atrops, F., Baraboshkin, E.Y., Company, M., Delanoy, G., Dutour, Y., Klein, J., Latil, J.L., Lukeneder, A., Mitta, V., Mourgues, F.A., Ploch, I., Raisossadat, N., Ropolo, P., Sandoval, J., Tavera, J.M., Vasicek, Z., Vermeulen, J., 2006. Report on the 2nd International Meeting of the IUGS Lower Cretaceous Ammonite Working Group, the "Kilian Group", Neuchâtel, Switzerland, 8 September 2005. Cretaceous Research 27, 712–715.

Reboulet, S., Klein, J., Barragán, R., Company, M., González-Arreola, C., Lukeneder, A., Raisossadat, S.N., Sandoval, J., Szives, O., Tavera, J.M., Vasícek, Z., Vermeulen, J., 2009. Report on the 3rd International Meeting of the IUGS Lower Cretaceous Ammonite Working Group, the "Kilian Group", Vienna, Austria, 15th April 2008. Cretaceous Research 30, 496–502.

Reboulet, S., Rawson, P.F., Moreno-Bedmar, J.A., Aguirre-Urreta, M.B., Barragán, R., Bogomolov, Y., Company, M., González-Arreola, C., Stoyanova, V.I., Lukeneder, A., Matrion, B., Mitta, V., Randrianaly, H., Vasicek, Z., Baraboshkin, E.J., Bert, D., Bersac, S., Bogdanova, T.N., Bulot, L.G., Latil, J.-L., Mikhailova, I.A., Ropolo, P., Szives, O., 2011. Report on the 4th International Meeting of the IUGS Lower Cretaceous Ammonite Working Group, the "Kilian Group", Dijon, France, 30th August, 2010. Cretaceous Research 32, 786–793.

Reimond, W.U., Koeberl, C., Brandstäter, F., Kruger, F.J., Armstrong, R.A., Bootsman, C., 1999. Morokweng impact structure, South Africa: Geologic, petrographic, and isotopic results, and implications for the size of the structure. In: Dressler, B.O., Sharpton, V.L. (Eds.), Large Meteorite Impacts and Planetary Evolution II. Geological Society of America Special Paper, 339, pp. 61–90.

Remane, J., 1985. Calpionellids. In: Bolli, H.M., Saunders, J.B., Perch Nielsen, K. (Eds.), Plankton Stratigraphy. Cambridge University Press, Cambridge, pp. 555–572.

Remane, J., 1991. The Jurassic-Cretaceous boundary: Problems of definition and procedure. Cretaceous Research 12, 447–453.

Remane, J., 1998. Calpionellids. Columns for Jurassic chart of Mesozoic and Cenozoic sequence chronostratigraphic framework of European basins, by Hardenbol, J., Thierry, J., Farley, M.B., Jacquin, Th., de Graciansky, P.-C., and Vail, P.R. (coordinators). In: de Graciansky, P.-C., Hardenbol, J., Jacquin, Th., Vail, P.R. (Eds.), Mesozoic-Cenozoic Sequence Stratigraphy of European Basins. SEPM Special Publication, 60, pp. 763–781.

Remane, J., Bakalova-Ivanova, D., Borza, K., Knauer, J., Nagy, I., Pop, G., Tardi-Filacz, E., 1986. Agreement on the subdivision of the Standard Calpionellid Zones defined at the IInd Planktonic Conference Roma 1971. Acta Geologica Hungarica 29, 5–14.

Renard, M., de Rafélis, M., Emmanuel, L., Moullade, M., Masse, J.-P., Kuhnt, W., Bergen, J.A., Tronchetti, G., 2005. Early Aptian $\partial^{13}C$ and manganese anomalies from the historical Cassis-La Bédoule stratotype sections (S.E. France): Relationship with a methane hydrate dissociation event and stratigraphic implications. Carnets de Géologie / Notebooks on Geology, Article 2005/04, 18. Available at. http://paleopolis.rediris.es/cg/CG2005_A04/index.html.

Renevier, E., 1874. Tableau des terrains sédimentaires. Bulletin de la Société vaudoise des Sciences naturelles 13, 218–252.

Renne, P.R., Mundil, M., Balco, G., Min, K., Ludwig, K.R., 2010. Joint determination of ^{40}K decay constants and $^{40}Ar^{*}/^{40}K$ for the Fish Canyon sanidine standard, and improved accuracy for $^{40}Ar/^{39}Ar$ geochronology. Geochimica et Cosmochimica Acta 74, 5349–5367.

Riveline, J., 1998. Charophytes. Columns for Jurassic chart of Mesozoic and Cenozoic sequence chronostratigraphic framework of European basins, by Hardenbol, J., Thierry, J., Farley, M.B., Jacquin, Th., de Graciansky, P.-C., and Vail, P.R. (coordinators). In: de Graciansky, P.-C., Hardenbol, J., Jacquin, Th., Vail, P.R. (Eds.), Mesozoic-Cenozoic Sequence Stratigraphy of European Basins. SEPM Special Publication, 60, pp. 763–781.

Robaszynski, F., Caron, M., 1995. Foraminiféres planctoniques du crétacé: commentaire de la zonation Europe-Mediterranée. Bulletin de la Société géologique de France 166, 681–692.

Robaszynski, F., 1998. Planktonic foraminifera. Columns for Jurassic chart of Mesozoic and Cenozoic sequence chronostratigraphic framework of European basins, by Hardenbol, J., Thierry, J., Farley, M.B., Jacquin, Th., de Graciansky, P.-C., and Vail, P.R. (coordinators). In: de Graciansky, P.-C., Hardenbol, J., Jacquin, Th., Vail, P.R. (Eds.), Mesozoic-Cenozoic Sequence Stratigraphy of European Basins. SEPM Special Publication, 60, pp. 763–781.

Röhl, U., Ogg, J.G., 1996. Aptian-Albian sea level history from guyots in the western Pacific. Paleoceanography 11, 595–624. 1996.

Ropolo, P., Conte, G., Gonnet, R., Masse, J.-P., Moullade, M., 1998. Les faunes d'ammonites du Barrémien supérieur/Aptien inférieur (Bédoulien) dans la région stratotypique de Cassis-La Bédoule (S.E. France): état des connaissances et propositions pour une zonation par ammonites du Bédoulien type. Géologie Méditerranéene 25, 167–175.

Roth, P.H., 1989. Ocean circulation and calcareous nannoplankton evolution during the Jurassic and Cretaceous. Palaeogeography, Palaeoclimatology, Palaeoecology 74, 111–126.

Round, F.E., Crawford, R.M., Mann, D.G., 1990. The Diatoms: Biology and Morphology of the Genera. Cambridge University Press, Cambridge, p. 747.

Ryan, W.B.F., Bolli, H.M., Foss, G.N., Natland, J.H., Hottman, W.E., Foresman, J.B., 1978. Objectives, principal results, operations, and explanatory notes of Leg 40, South Atlantic. Initial Reports, Deep Sea Drilling Project 40, 5–20.

Sageman, B.B., Meyers, S.R., Arthur, M.A., 2006. Orbital time scale and new C-isotope record for Cenomanian-Turonian boundary stratotype. Geology 34, 125–128.

Sazonov, N.T., 1951. On some little-known ammonites of the Lower Cretaceous. Byulleten' Moskovskogo Obshchestva Ispytatelei Prirody 56, 1–176 [In Russian].

Schlanger, S.O., Jenkyns, H.C., 1976. Cretaceous oceanic anoxic events: Causes and consequences. Geology en Mijnbouw 55, 179–184.

Schlanger, S.O., Arthur, M.A., Jenkyns, H.C., Scholle, P.A., 1987. The Cenomanian-Turonian oceanic anoxic event, I. Stratigraphy and distribution of organic carbon-rich beds and the marine ^{13}C excursion. In: Brooks, J., Fleet, A.J. (Eds.), Marine Petroleum Source Rocks. Geological Society Special Publication, 26, pp. 371–399.

Schönfeld, J., Schulz, M.-G., McArthur, J.M., Burnett, J., Gale, A., Hambach, U., Hansen, H.J., Kennedy, W.J., Rasmussen, K.L., Thirlwall, M.F., Wray, D.S., 1996. New results on biostratigraphy, geochemistry and correlation from the standard section for the Upper Cretaceous white chalk of northern Germany (Lägerdorf – Kronsmoor – Hemmoor). In: Spaeth, C. (Ed.), Jost Wiedmann Memorial Volume;

Proceedings of the 4th International Cretaceous Symposium, Hamburg, 1992. Mitteilung aus dem Geologisch–Paläontologischen Institut der Universität Hamburg, 77, pp. 545–575.

Scott, R.W., 2007. Calibration of the Albian/Cenomanian boundary by ammonite biostratigraphy: U.S. Western Interior. Acta Geologica Sinica 81, 940–948.

Scott, R.W., 2009. Uppermost Albian biostratigraphy and chronostratigraphy. Carnets de Géologie/Notebooks on Geology, Article 2009 (03), 15.

Scott, R.W., 2011. CRET1 Chronostratigraphic Database Available online. http://precisionstratigraphy.com.

Scott, R.W., Oboh-Ikuenobe, F.E., Benson, D.G., Holbrook, J.M., 2009. Numerical age calibration of the Albian/Cenomanian boundary. Stratigraphy 6, 17–32.

Selby, D., Mutterlose, J., Condon, D.J., 2009. U-Pb and Re-Os Geochronology of the Aptian/Albian and Cenomanian/Turonian stage boundaries: Implications for timescale calibration, osmium isotope seawater composition and Re-Os systematics in organic-rich sediments. Chemical Geology 265, 394–409.

Séronie-Vivien, M., 1972. Contribution à l'étude de Sénonien en Aquitaine septentrionale, ses stratotypes: Coniacien, Santonien, Campanien. Les Stratotypes Français 2, 195.

Sharland, P.R., Archer, R., Casey, D.M., Davies, R.B., Hall, S.H., Heward, A.P., Horbury, A.D., Simmons, M.D., 2001. Arabian Plate Sequence Stratigraphy. GeoArabia Special Publication 2, 372.

Sharland, P.R., Casey, D.M., Davies, R.B., Simmons, M.D., Sutcliffe, O.E., 2004. Arabian Plate sequence stratigraphy. GeoArabia: Middle East Petroleum Geosciences 9 (2), 199–214.

Shimokawa, A., 2010. Zircon U-Pb Geochronology of the Great Valley Group: Recalibrating the Lower Cretaceous Time Scale. M.S. thesis, University of North Carolina at Chapel Hill, 46 [See also: Shimokawa, A., Coleman, D.S., and Bralower, T.J., 2010, Recalibrating the Lower Cretaceous time scale with U-Pb zircon ages from the Great Valley Group. Geological Society of America Annual Meeting, Denver, 31 Oct – 3 Nov 2010, Abstract, 160–7.].

Shipboard Scientific Party, 1998. Site 1049: Paleomagnetism section (authored by Ogg, J.G., Bardot, L., and Foster, J). Proceedings Ocean Drilling Program, Initial Reports 171B, 70–75.

Shipboard Scientific Party, 2004. Explanatory notes: Biostratigraphy. Proceedings of the Ocean Drilling Program, Initial Reports 207 Available online. http://www-odp.tamu.edu/ publications/ 207_IR.

Sikora, P., Bergen, J., 2004. Lower Cretaceous Biostratigraphy of Ontong Java Sites from DSDP Leg 30 and ODP Leg 192. In: Fitton, J.G., Mahoney, J.J., Wallace, P.J., Sanders, A.D. (Eds.), Origin and Evolution of the Ontong Java Plateau. Geological Society Special Publication, 229, pp. 83–111.

Siewert, S.E., 2011. Integrating $^{40}Ar/^{39}Ar$, U-Pb and astronomical clocks in the Cretaceous Niobrara Formation. M.S. thesis, University of Wisconsin at Madison. 74.

Siewert, S.E., Singer, B.S., Condon, D., Meyers, S.R., Sageman, B.B., Jicha, B.R., Obradovich, J.D., and Sawyer, D.A., in press, Integrating $^{40}Ar/^{39}Ar$, U-Pb, and astronomical clocks in the Cretaceous Niobrara Formation, Western Interior Basin, USA. Geological Society of America Bulletin.

Simmons, M.D., Sharland, P.R., Casey, D.M., Davies, R.B., Sutcliffe, O.E., 2007. Arabian Plate sequence stratigraphy: Potential implications for global chronostratigraphy. GeoArabia: Middle East Petroleum Geosciences 12 (4), 101–130.

Sinninghe Damsté, J.S., Muyzer, G., Abbas, B., Rampen, S.W., Massé, G., Allard, W.G., Belt, S.T., Robert,-M., J., Rowland, S.J., Moldowan, J.M., Barbanti, S.M., Fago, F.J., Denisevich, P., Dahl, J., Trindad, L.A.F., Schouten, S., 2004. The rise of rhizosolenid diatoms. Science 304, 584–587.

Smelror, M., Kelley, S.R.A., Dypvik, H., Mørk, A., Nagy, J., Tsikalas, F., 2001. Mjølnir (Barents Sea) meteorite impact ejecta offers a Volgian-Ryazanian boundary marker. Newsletters on Stratigraphy 38, 129–140.

Speranza, F., Satolli, S., Mattioli, E., Calamita, F., 2005. Magnetic stratigraphy of Kimmeridgian–Aptian sections from Umbria-Marche (Italy): New details on the M-polarity sequence. Journal of Geophysical Research 110 B12109, doi: 10.1029/2005JB003884.

Sprenger, A., ten Kate, W.G., 1993. Orbital forcing of calcilutite-marl cycles in southeast Spain and an estimate for the duration of the Berriasian stage. Geological Society of America Bulletin 105, 807–818.

Sprovieri, M., Coccioni, R., Lirer, F., Pelosi, N., Lozar, F., 2006. Orbital tuning of a lower Cretaceous composite record (Maiolica Formation, central Italy). Paleoceanography 21 PA4212. doi: 10.1029/ 2005PA001224.

Stewart, K., Turner, S., Kelley, S., Hawkesworth, C., Kirstein, L., Mantovani, M., 1996. 3-D ^{40}Ar-^{39}Ar geochronology in the Parana continental flood basalt province. Earth and Planetary Science Letters 143, 95–109.

Strasser, A., Hillgartner, H., Pasquier, J.B., 2004. Cyclostratigraphic timing of sedimentary processes: An example from the Berriasian of the Swiss and French Jura Mountains. In: D'Argenio, B., Fischer, A.G., Premoli Silva, I., Weissert, H., Ferreri, V. (Eds.), Cyclostratigraphy: Approaches and Case Histories. SEPM Special Publication, 81, pp. 135–151.

Subcommission on Cretaceous Stratigraphy, 2009. Annual Report 2009, 12 Available online: http://www2.mnhn.fr/hdt203/media/ISCS/ICS2009_ Report_Creta.pdf.

Swisher III, C.C., Wang, Y.Q., Wang, X.L., Xu, X., Wang, Y., 1999. Cretaceous age for the feathered dinosaurs of Liaoning, China. Nature 400, 58–61.

Tarduno, J.A., 1990. A brief reversed polarity interval during the Cretaceous Normal Polarity Superchron. Geology 18, 638–686.

Tarduno, J.A., Sliter, W.V., Bralower, T.J., McWilliams, M., Premoli-Silva, I., Ogg, J.G., 1989. M-sequence reversals recorded in DSDP Sediment Cores from the Western Mid-Pacific Mountains and Magellan Rise. Geological Society of America Bulletin 101, 1306–1319.

Tarduno, J.A., Sliter, W.V., Kroenke, L., Leckie, R.M., Mayer, H., Mahoney, J.J., Musgrave, R., Storey, M., Winterer, E.L., 1991. Rapid formation of Ontong Java Plateau by Aptian mantle volcanism. Science 254, 399–403.

Tarduno, J.A., Lowrie, W., Sliter, W.V., Bralower, T.J., Heller, F., 1992. Reversed polarity characteristic magnetizations in the Albian Contessa Section, Umbrian Apennines, Italy: Implications for the existence of a Mid-Cretaceous Mixed Polarity Interval. Journal of Geophysical Research 97 (B1), 241–271.

Tavera, J.M., Aguado, R., Company, M., Oloriz, F., 1994. Integrated biostratigraphy of the Durangites and Jacobi zones (J/K boundary) at the Puerto Escano section in southern Spain (Province of Cordoba). In: Cariou, E., Hantzpergue, P. (Eds.), Third International Symposium on Jurassic Stratigraphy, Poitiers, France, 22–29 Sept 1991. Geobios, Mémoire Special 17 (1), 469–476.

Thibault, N., Husson, D., Harlou, R., Gardin, S., Galbrun, B., Huret, E., and Minoletti, F., in press, Astronomical calibration of upper Campanian-Maastrichtian carbon isotope events and calcareous plankton biostratigraphy in the Indian Ocean (ODP Hole 762C): implication for the age of the Campanian-Maastrichtian boundary. *Palaeogeography, Palaeoclimatology, Palaeoecology.*

Thierry, J., 1998. Ammonites. Columns for Jurassic chart of Mesozoic and Cenozoic sequence chronostratigraphic framework of European basins, by Hardenbol, J., Thierry, J., Farley, M.B., Jacquin, Th., de Graciansky, P.-C., and Vail, P.R. (coordinators). In: de Graciansky, P.-C., Hardenbol, J., Jacquin, Th., Vail, P.R. (Eds.), Mesozoic-Cenozoic Sequence Stratigraphy of European Basins. SEPM Special Publication, 60, pp. 776–777.

Thieuloy, J.-P., 1977. La zone à Callidiscus du Valanginien supérieur vocontien (Sud-Est de la France). Lithostratigraphie, ammonitofaune, limite Valanginien-Hauterivien, corrélations. Géologie Alpine 53, 83–143.

Thurmann, J., 1836. Lettre à ME de Beaumont. Bulletin de la Société géologique de France, Serie 1 7, 207–211.

Tröger, K.-A., Kennedy, W.J., 1996. The Cenomanian Stage. Bulletin de l'Institut Royal des Sciences Naturelles de Belgique, Sciences de la Terre 66 (Suppl.), 57–68.

Tsikalas, F., 2005. Mjølnir impact crater. Geophysics Research Group, Department of Geology, University of Oslo Website. http://folk.uio.no/ftsikala/mjolnir/.

van Buchem, F.S.P., Simmons, M.D., Droste, H.J., Davies, R.B., 2011. Late Aptian to Turonian stratigraphy of the eastern Arabian Plate – depositional sequences and lithostratigraphic nomenclature. Petroleum Geosciences 17, 211–222.

VandenBerg, J., Wonders, A.A.H., 1980. Paleomagnetism of Late Mesozoic pelagic limestones from the Southern Alps. Journal of Geophysical Research 85, 3623–3627.

VandenBerg, J., Klootwijk, C.T., Wonders, A.A.H., 1978. Late Mesozoic and Cenozoic movements of the Italian peninsula: Further paleomagnetic data from the Umbrian sequence. Geological Society of America Bulletin 89, 133–150.

van Hinte, J.E., 1965. The type Campanian and its planktonic Foraminifera. Proceedings of the Koninklijke Nederlandse Akademie van Wetenschappen, Series B 68, 8–28.

Voigt, S., Schönfeld, J., 2010. Cyclostratigraphy of the reference section for the Cretaceous white chalk of northern Germany, Lägerdorf-Kronsmoor– A late Campanian-early Maastrichtian orbital time scale. Palaeogeography, Palaeoclimatology, Palaeoecology 287, 67–80.

Voigt, S., Erbacher, J., Mutterlose, J., Weiss, W., Westerhold, T., Wiese, F., Wilmsen, M., Wonik, T., 2008. The Cenomanian – Turonian of the Wunstorf section - (North Germany): Global stratigraphic reference section and new orbital time scale for Oceanic Anoxic Event 2. Newsletters on Stratigraphy 43, 65–89.

Voigt, S., Friedrich, O., Norris, R.D., Schönfeld, J., 2010. Campanian - Maastrichtian carbon isotope stratigraphy: Shelf–ocean correlation between the European shelf sea and the tropical Pacific Ocean. Newsletters on Stratigraphy 44, 57–72.

von Salis, K., 1998. Calcareous nannofossils. Columns for Jurassic chart of Mesozoic and Cenozoic sequence chronostratigraphic framework of European basins, by Hardenbol, J., Thierry, J., Farley, M.B., Jacquin, Th., de Graciansky, P.-C., and Vail, P.R. (coordinators). In: de Graciansky, P.-C., Hardenbol, J., Jacquin, Th., Vail, P.R. (Eds.), Mesozoic-Cenozoic Sequence Stratigraphy of European Basins. SEPM Special Publication, 60, pp. 763–781.

Walaszczyk, I., Wood, C.J., 1998. Inoceramid and biostratigraphy at the Turonian/Coniacian boundary; based on the Salzgitter-Salder Quarry, Lower Saxony, Germany, and the Słupia Nadbrzeżna section, Central Poland. Acta Geologica Polonica 48, 395–434.

Walaszczyk, I., Cobban, W.A., 2007. Inoceramid fauna and biostratigraphy of the upper Middle Coniacian-lower Middle Santonian of the Pueblo Section (SE Colorado, US Western Interior). Cretaceous Research 28, 132–142.

Walaszczyk, I., Odin, G.S., Dhondt, A.V., 2002. Inoceramids from the Upper Campanian and Lower Maastrichtian of the Tercis section (SW France), the Global Stratotype Section and Point for the Campanian - Maastrichtian boundary; taxonomy, biostratigraphy and correlation potential. Acta Geologica Polonica 52, 269–305.

Walaszczyk, I., Wood, C.J., Lees, J.A., Peryt, D., Voigt, S., Wiese, F., 2010. The Salzgitter-Salder Quarry (Lower Saxony, Germany) and Słupia Nadbrzeżna river cliff section (central Poland): A proposed candidate composite Global Boundary Stratotype Section and Point for the base of the Coniacian Stage (Upper Cretaceous). Acta Geologica Polonica 60, 445–477.

Weissert, H., Lini, A., 1991. Ice Age interludes during the time of Cretaceous greenhouse climate? In: Muller, D.W., McKenzie, J.A., Weissert, H. (Eds.), Controversies in Modern Geology. Academic Press, London, pp. 173–191.

Weissert, H., Bréheret, J.G., 1991. A carbonate-isotope record from Aptian-Albian sediments of the Vocontian Troght (SE France). Bulletin de la Société Géologique de France 162, 1133–1140.

Weissert, H., Lini, A., Föllmi, K.B., Kuhn, O., 1998. Correlation of Early Cretaceous carbon isotope stratigraphy and platform drowning events: A possible link? Palaeogeography, Palaeoclimatology, Palaeoecology 137, 189–203.

Wiese, F., Kaplan, U., 2001. The potential of the Lengerich section (Münster Basin, northern Germany) as a possible candidate Global boundary Stratotype Section and Point (GSSP) for the Middle/Upper Turonian boundary. Cretaceous Research 22, 549–563.

Wignall, P.B., 2001. Large igneous provinces and mass extinctions. Earth-Science Reviews 53, 1–33.

Wilson, P.A., Norris, R.D., Cooper, M.J., 2002. Testing the Cretaceous greenhouse hypothesis using glassy foraminiferal calcite from the core of the Turonian tropics on Demerara Rise. Geology 30, 607–610.

Wimbledon, W.A.P., Casellato, C.E., Reháková, D., Bulot, L.G., Erba, E., Gardin, S., Verreussel, R.M.C.H., Munsterman, D.K., Hunt, C.O., 2011. Fixing a basal Berriasian and Jurassic-Cretaceous (J-K) boundary – is there perhaps some light at the end of the tunnel? Rivista Italiana di Paleontologia e Stratigrafia 117, 295–307.

Wissler, L., Weissert, H., Buonocunto, F.P., Ferreri, V., D'Argenio, B., 2004. Calibration of the Early Cretaceous time scale; a combined chemostratigraphic and cyclostratigraphic approach to the Barremian-Aptian interval, Campania Apennines and Aouthern Alps (Italy). In: D'Argenio, B., Fischer, A.G., Premoli Silva, I., Weissert, H., Ferreri, V. (Eds.), Cyclostratigraphy: Approaches and Case Histories. SEPM Special Publication, 81, pp. 123–133.

Woodburne, M.O., 2004. Late Cretaceous and Cenozoic Mammals of North America: Biostratigraphy and Geochronology. Columbia University Press, New York, p. 376.

Wortmann, U.G., Herrle, J.O., Weissert, H., 2004. Altered carbon cycling and coupled changes in Early Cretaceous weathering patterns: Evidence from integrated carbon isotope and sandstone records of the western Tethys. Earth and Planetary Science Letters 220, 69–82.

Zakharov, V.A., Bown, P., Rawson, P.F., 1996. The Berriasian Stage and the Jurassic-Cretaceous boundary. Bulletin de l'Institut Royal des Sciences Naturelles de Belgique, Sciences de la Terre 66 (Suppl.), 7–10.

Chapter 28

N. Vandenberghe, F.J. Hilgen, R.P. Speijer
With contributions by J. G. Ogg, F. M. Gradstein and
O. Hammer on the magnetochronology, biochronology
and the time scale, and by C. J. Hollis and J. J. Hooker on the biostratigraphy

The Paleogene Period

Abstract: All Paleocene stages (i.e., Danian, Selandian and Thanetian) have formally ratified definitions, and so have the Ypresian and Lutetian Stages in the Eocene, and the Rupelian Stage in the Oligocene. The Bartonian, Priabonian and Chattian Stages are not yet formally defined. After the global catastrophe and biotic crisis at the Cretaceous–Paleogene boundary, stratigraphically important marine microfossils started new evolutionary trends, and on land the now flourishing mammals offer a potential for stratigraphic zonation. During the Paleogene the global climate, being warm until the late Eocene, shows a significant cooling trend culminating in a major cooling event in the beginning of the Oligocene, preparing the conditions for modern life and climate. Orbitally tuned cyclic sedimentation series, calibrated to the geomagnetic polarity and biostratigraphic scales, have considerably improved the resolution of the Paleogene time scale.

50.2 Ma Paleogene

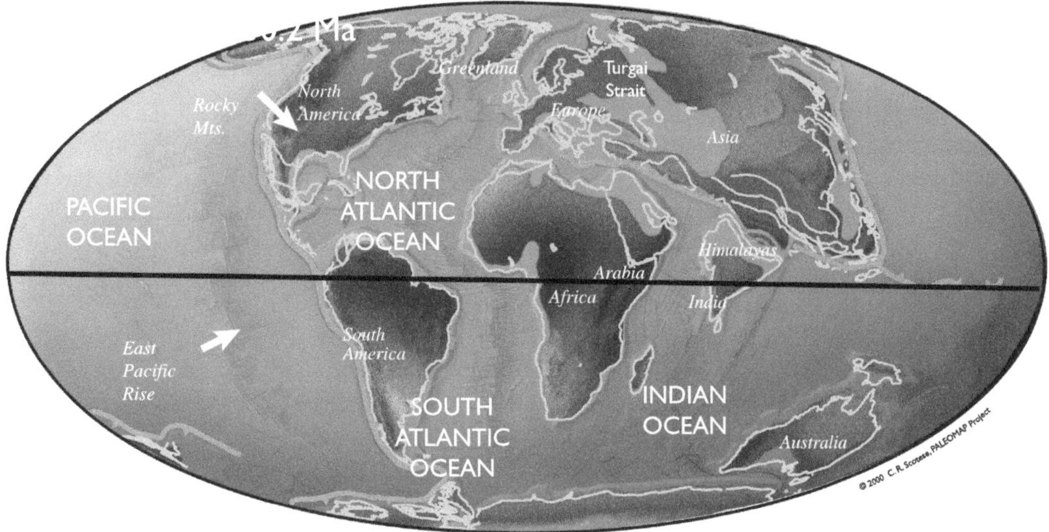

Chapter Outline

The Geologic Time Scale 2012. DOI: 10.1016/B978-0-444-59425-9.00028-7

28.1. HISTORY AND SUBDIVISIONS

28.1.1. Overview of the Paleogene

The Cenozoic (originally Cainozoic) Era (Phillips, 1841) derives its name from the relatively new (kainos) biota, compared to the Mesozoic Era. The Cenozoic Era is subdivided into the Paleogene (*palaios* = old, *genos* = birth), Neogene and Quaternary periods. The term Tertiary (Arduino, 1759) has been traditionally used to indicate the time interval between the Mesozoic and the Quaternary and therefore comprises the Paleogene and the Neogene. The use of the term Tertiary is discouraged, since the term is equally as antiquated as Primary and Secondary, and the latter have fallen into disuse in the 20th century. However, the term Tertiary still has widespread use, in particular since the ratification of the term Quaternary as a period (Head *et al.*, 2008).

The Paleogene System is subdivided into three series (Figure 28.1): the Paleocene, Eocene, and Oligocene, referring again to the evolution of the biota (eos = dawn and oligos = little). In his original subdivision of the Tertiary, Lyell (1833) introduced the term Eocene to refer to the older part of the Tertiary in the classical European Cenozoic basins, in which he recognized less than 3% of extant mollusk species. Later, Beyrich (1854), working mainly in northern Germany, separated the Oligocene from the Eocene. Naumann (1866) combined the Eocene and Oligocene in his "Paleogen Stufe", as opposed to the "Neogen Stufe" of Hörnes (1853) which included not only the Miocene and the Pliocene, but also fauna of the Pleistocene. Finally, Schimper (1874), added the Paleocene based on paleobotanic studies in the Paris Basin and other basins of West Europe, although this met with opposition for a long time (e.g., Mangin, 1957).

Larger foraminifera, and particularly the nummulites, have played an important role in the development of Paleogene biostratigraphy, and French-speaking stratigraphers such as Renevier (1873), Haug (1908–11), and Gignoux (1950) have used the term "Nummulitique" as an equivalent of Paleogene, but this term has now disappeared from stratigraphic practice.

The three Paleogene series have been formally subdivided into nine stages (Figure 28.1) as decided by the International Subcommission on Paleogene Stratigraphy at the 1989 International Geological Congress in Washington (Jenkins and Luterbacher, 1992). The Paleocene Series is further subdivided into the Danian, Selandian and Thanetian stages, the Eocene Series into the Ypresian, Lutetian, Bartonian and Priabonian stages and the Oligocene Series into the Rupelian and Chattian stages, all stages being listed successively from old to young. All the unit stratotypes of these stages, as well as several other now disused Paleogene stages, were historically defined in the mid 19th and early 20th century in the North Sea Basin and its southern part, the Paris Basin (see Pomerol, 1981a).

In the last three decades, the practice of Paleogene stratigraphy has been based on classical Berggren *et al.* (1985, 1995) magnetobiochronology, emphasizing calcareous microfossils, and has recently been revised by Berggren and Pearson (2005) and Wade *et al.* (2011). Organic-walled dinoflagellate cysts (dinocysts) have also been recognized as a valuable additional tool in Paleogene stratigraphy, especially in shallow basins of the northern hemisphere.

Originally, the definition and characterization of the unit stratotypes were needed to reconstruct the fine lines of Paleogene historical geological evolution. Developing a high-resolution time scale for the evolution of our planet, however, requires precise chronostratigraphy and geochronology of well defined levels that can be correlated between areas. Therefore, since the 1980s the GSSP concept has dominated Paleogene stratigraphic research in the attempts that have been made to precisely define the boundary stratotypes of the stages at particular physical levels in well-documented sections. A major concern in the definition of the Paleogene GSSPs has been that the geographic locations of the traditional unit stratotypes did not prove suitable for defining the boundaries of the stratotypes. As stratigraphic

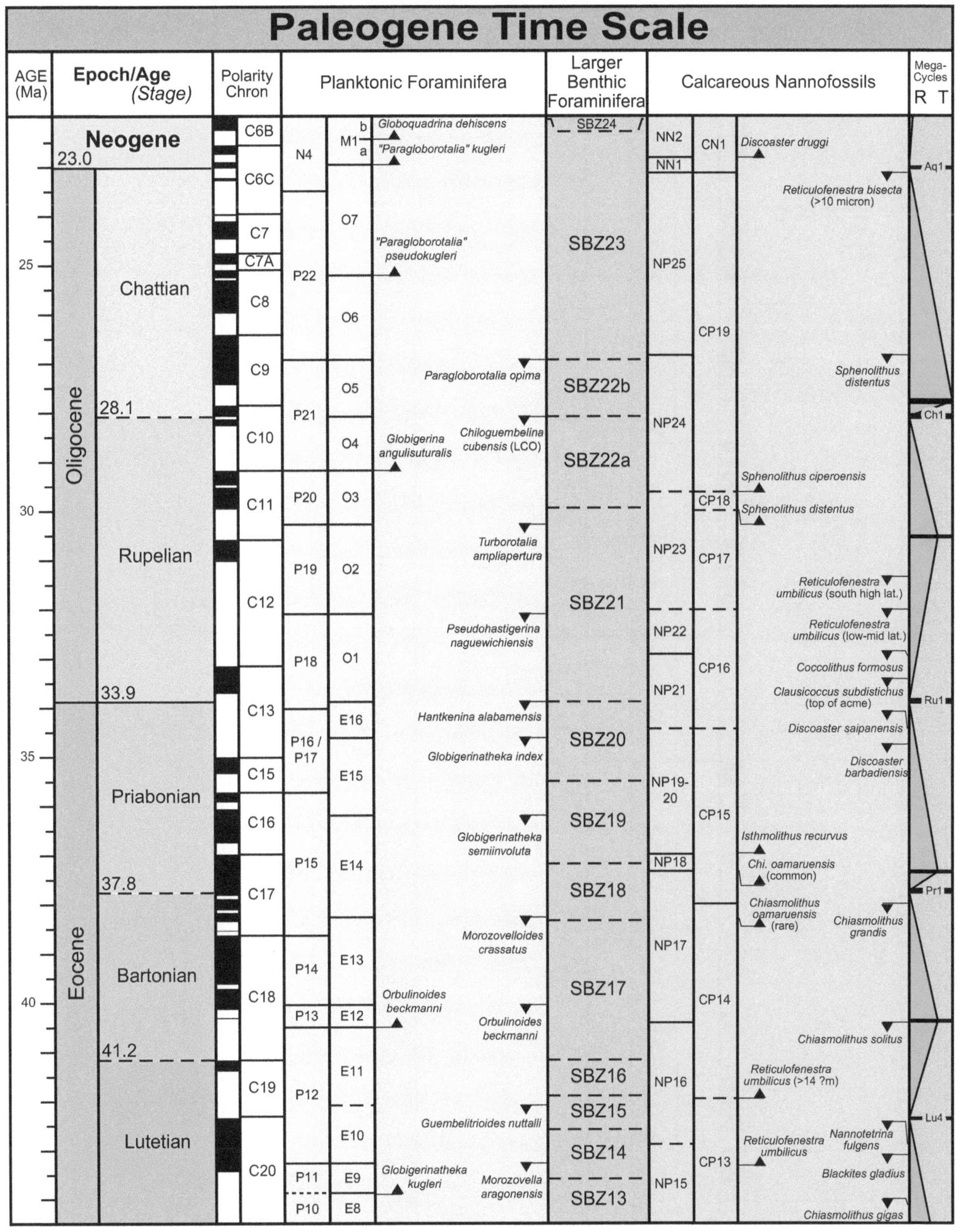

FIGURE 28.1 **Paleogene stratigraphic subdivisions, geomagnetic polarity scale, zonations of planktonic foraminifera, larger benthic foraminifera and calcareous nannofossils, and main trends in eustatic sea level**. Planktonic foraminiferal stratigraphy is modified from Wade *et al.* (2011), Tethyan zonation of larger benthic foraminifera is modified from Serra-Kiel *et al.* (1998), and the calcareous nannofossil stratigraphy is modified from tables in the Pacific Equatorial Age Transect (PEAT) program (Pälike *et al.*, 2010) with assistance of Paul Bown. The main Paleogene transgressive–regressive trends are modified from Hardenbol *et al.* (1998).

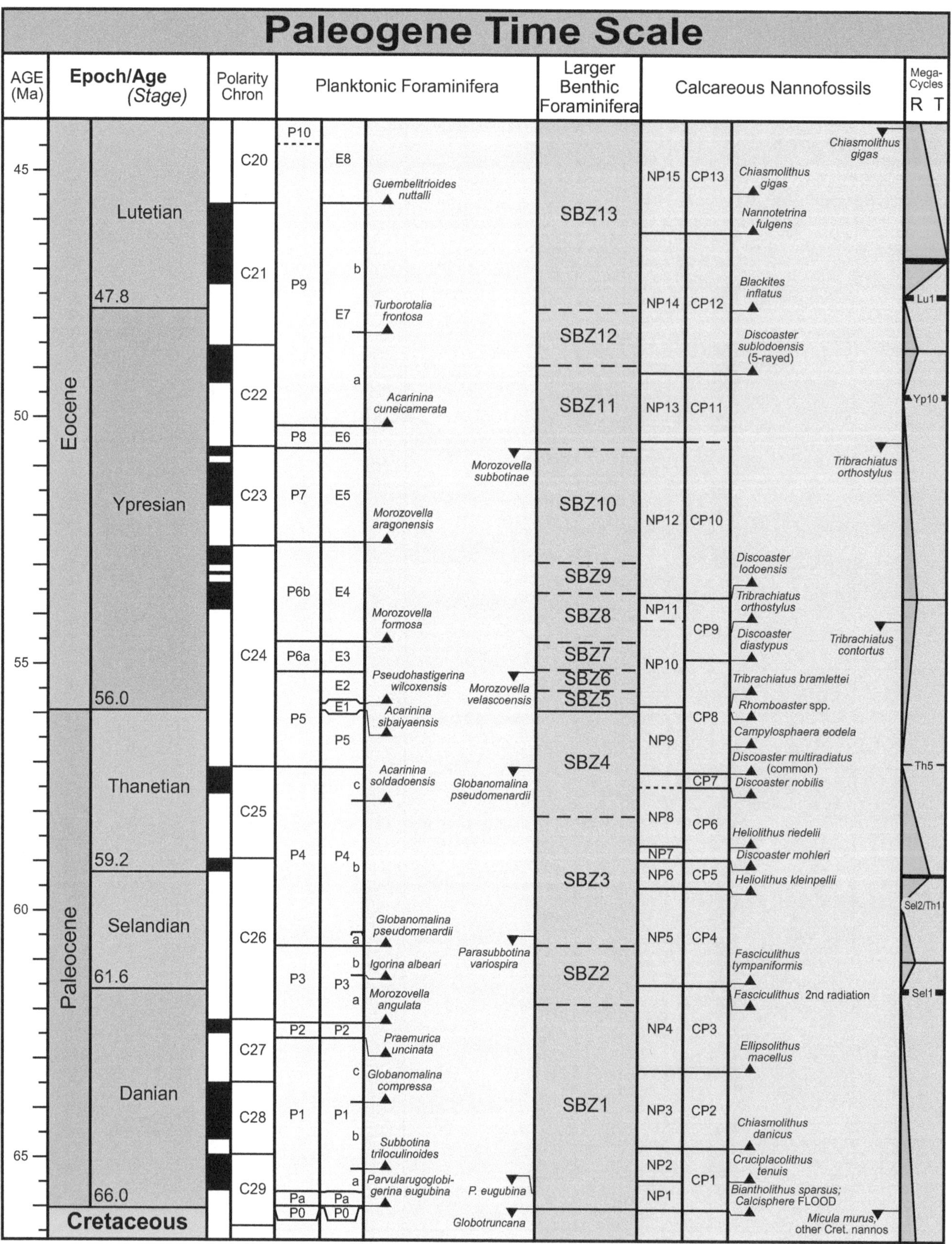

FIGURE 28.1 *(Continued).*

communication requires stability and coherence over time in the definitions of the published terminology, boundary stratotypes and GSSPs have been sought that reasonably respect the stage definitions as they were originally defined in the historical areas.

As of late 2011, the global stratotype sections and points (GSSP) defining the bases of the Paleogene series have been formally ratified: the base of the Danian Stage as the Cretaceous Paleogene boundary, the base-Ypresian as the Paleocene–Eocene boundary, and the base-Rupelian as the Eocene–Oligocene boundary. In addition, the GSSPs of the Selandian, Thanetian and Lutetian stages have been ratified. GSSP proposals for Priabonian and Chattian are being prepared for submission. Although the original Bartonian sections in England are well studied, a suitable GSSP for this stage is yet to be found.

28.1.2. Definition of the Base of the Danian

According to stratigraphic rules, the definition of the lowermost stage of the Paleogene, the Danian, also defines the bases of the Paleocene Series, the Paleogene System, of the Tertiary and of the Cenozoic Erathem. Therefore the base of the Danian also represents the much-discussed Cretaceous–Paleogene (K/Pg) or Cretaceous–Tertiary (K/T) boundary level.

28.1.2.1. The Danian Stage

The Danian Stage, named after its type area in Denmark, was introduced by Desor (1847) following his studies of echinoids within the *Cerithium* and the Bryozoan Limestones at Stevns Klint and Fakse on Seland (for a history and modern definition, see Floris and Thomsen, 1981). The Danian in the type area corresponds to the interval between the top of the Maastrichtian Chalk and the basal conglomerate of the Selandian. Its range there spans the calcareous nannoplankton Zones NP1–NP4 (Martini, 1971), planktonic foraminifera Zones P0–P2 (*Guembelitria cretacea* Zone to *Praemurica uncinata* Zone; Berggren and Pearson, 2005; Wade *et al.*, 2011), and includes dinoflagellate cyst Viborg Zone 1 (Heilmann-Clausen, 1988), all of definite Paleogene character (e.g., Michelsen *et al.*, 1998, and references therein). Before they were recognized as being of Paleogene age, however, these Danian strata were originally considered to be the youngest stage of the Cretaceous (Desor, 1847); consequently a supposedly younger stage at the base of the Paleogene was introduced by Dewalque (1868): the Montian (after Mons, Belgium). The Montian, however, has lost its significance, as in its type section it represents only the later part of the Danian Stage in its present definition, and also because of the poor stratotype conditions (Robaszynski, 1981; De Geyter *et al.*, 2006).

28.1.2.2. The Danian GSSP

The base-Danian GSSP (Figure 28.2) has been fixed in a section of Oued Djerfane, 8 km west of El Kef (168 km southwest of Tunis), in Tunisia (coordinates N036°09′13.2″, E008°38′54.8″) and ratified by IUGS in 1991. Details of the GSSP section and a summary of the studies since the original definition have been published by Molina *et al.* (2006).

The GSSP level has been defined at the rusty colored base of a 50 cm thick boundary clay. A similar layer occurs in many K/Pg sections worldwide and it includes an iridium anomaly, microtektites, Ni-rich spinel crystals and shocked quartz. Since the seminal paper by Alvarez *et al.* (1980) on the events at the K/Pg boundary, their causes, the impact of an extraterrestrial body, inter-relationships and consequences have been much debated (e.g., Schulte *et al.*, 2010). The bulk of this layer was deposited during a few days and the level is considered isochronous all over the world, in marine as well as in continental sections. It corresponds to a drastic change in the marine plankton, the extinction of the ammonites and the demise of the dinosaurs, and is accompanied by biotic turnover/crises in many other groups of organisms, all of which can be used for correlation purposes (Molina *et al.*, 2006). The GSSP level represents the moment of the extraterrestrial impact, implying that all sediments generated by the impact are already Danian in age. The Cretaceous–Paleogene boundary is also marked by a 2–3‰ negative carbon isotopic excursion in calcite and organic matter that formed in ocean surface waters.

Concerns about the possible degradation of the El Kef GSSP section have led Molina *et al.* (2009a) to propose several well-studied auxiliary sections at varying distances from the original meteorite impact (Chicxulub, Yucatan Mexico). The sections are located in Mexico (Bochil and Mulato), Spain and southwest France (Caravaca, Zumaia and Bidart) and near the GSSP itself in Tunisia (Aïn Settara and Ellès). They can be correlated to the GSSP at El Kef using the geochemical and mineralogical impact evidence, and by using biozonations in particular those of planktonic foraminifera, calcareous nannofossils and dinoflagellate cysts (Figure 3 in Molina *et al.*, 2009a).

Using a combination of $^{40}Ar/^{39}Ar$ and astronomical ages at the classical Zumaia section, Kuiper *et al.* (2008) have proposed an age of ~65.95 Ma for the Cretaceous/ Paleogene boundary. The most reliable estimate of the position of the K/Pg boundary with respect to the magnetostratigraphic scale is based on astronomical data and situates the GSSP at C29r.5 (Westerhold *et al.*, 2008).

28.1.3. Definition of the Base of the Selandian

28.1.3.1. The Selandian Stage

The division of the Paleocene Series into three stages, Danian, Selandian and Thanetian, was decided at the IGC in

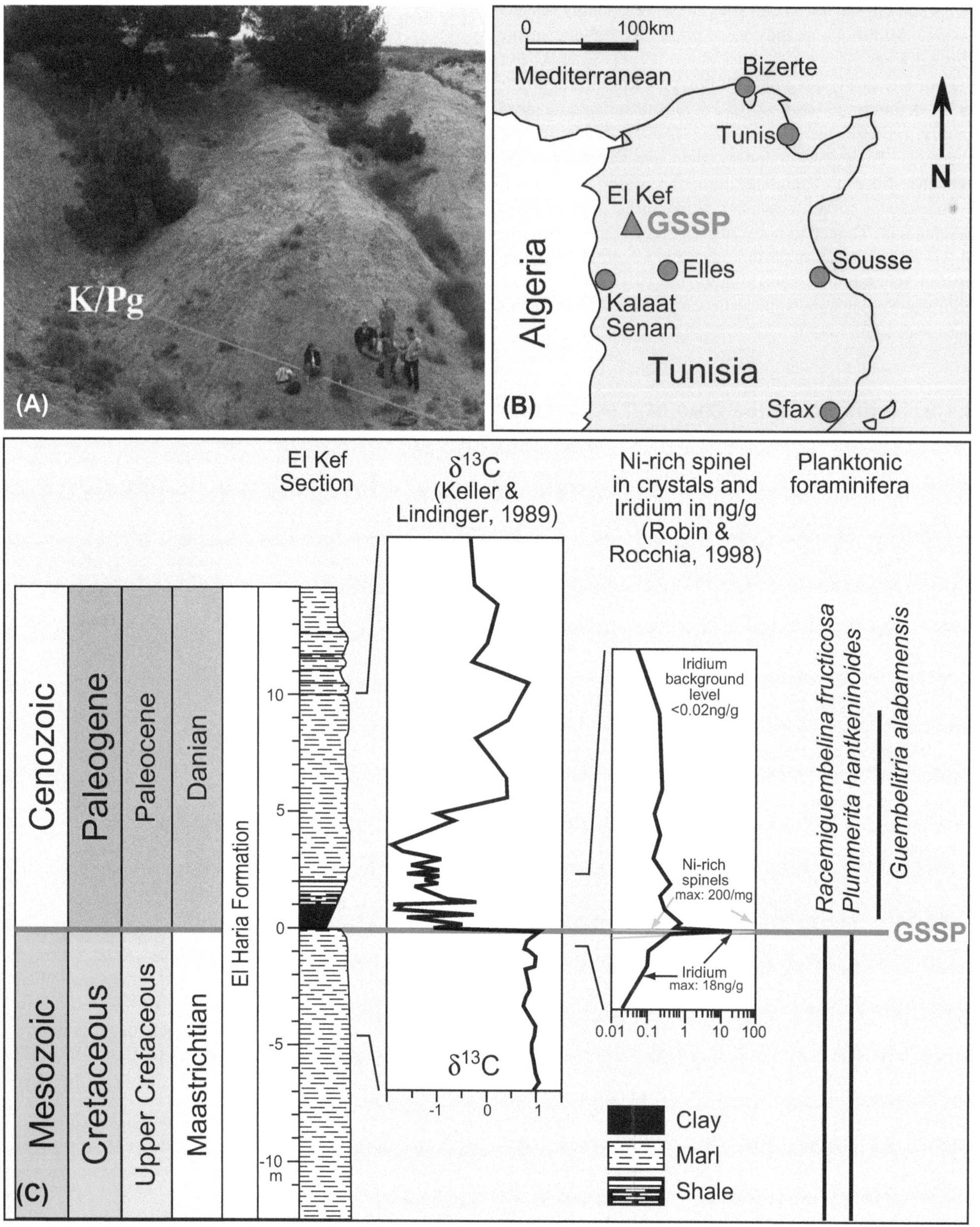

FIGURE 28.2 **The Danian GSSP located at El Kef in Tunisia (B).** Photograph (A) in Ogg *et al.* (2008) and stratigraphic data (C) from Molina *et al.* (2006).

Washington (Jenkins and Luterbacher, 1992). Before this threefold subdivision of the Paleocene became official in 1989, many stratigraphers used a twofold subdivision, and the present Selandian Stage was in fact included in the younger Thanetian Stage (Harland *et al.*, 1990; for a review see Bignot *et al.*, 1997).

The Selandian Stage was originally introduced by Rosenkrantz (1924) based on a succession composed of conglomerates, greensand (Lellinge Greensand), marls (Kerteminde Marl), and clays that unconformably overlie the upper Danian limestones and underlie the ash-bearing Mo-Clay (upper Paleocene to lower Eocene, see Hansen 1979) in Denmark, and is typified in the Copenhagen area (e.g., Perch-Nielsen and Hansen, 1981; Berggren, 1994). Towards the top of this sedimentary sequence a considerable overlap exists with the original Thanetian type section (see Bignot *et al.*, 1997) and in the international threefold subdivision of the Paleocene only the lower part is referred to as Selandian; in its type area this Selandian contains characteristic dinoflagellate cyst assemblages (Zones Viborg 2 and 3 *pro parte* of Heilmann-Clausen, 1985, 1988; see also Figure 23 in Michelsen *et al.*, 1998) and calcareous nannoplankton (Zones NP4 *pro parte*, NP5–NP6 of Martini, 1971; e.g., Thomsen, 1994). The Danian–Selandian boundary in the North Sea area ends a long lasting period of marine carbonate sedimentation; this explains why it was common until the mid 20th century to consider the top of the Danian chalks as the Cretaceous/Tertiary boundary.

The now disused Heersian Stage, introduced by Dumont (1851) and named after the Heers village 20 km northwest of Liège in Belgium, is equivalent to the middle and upper Selandian (De Geyter *et al.*, 2006).

28.1.3.2. The GSSP of the Selandian

The Selandian GSSP was ratified in 2008, as extensively documented by the Paleocene Working Group (Schmitz *et al.*, 2008, 2011; Figure 28.3). The GSSP of the Selandian is placed in the section exposed at Itzurun Beach in Zumaia, Basque Country, northern Spain (coordinates 43° 17′57.1″N, 2° 15′39.6″W). The Zumaia section is part of a continuous lower Santonian to lower Eocene sea-cliff outcropping along the coast of the Gipuzkoa province halfway between Bilbao and San Sebastian. The precise GSSP level is defined at the base of the red marls of the Itzurun Formation, and overlies the uppermost limestone bed of the limestone–marl couplets in the upper part of the Aitzgorri Limestone Formation. The sediments are deep marine deposits, and sedimentation is thought to be continuous as bed-by-bed correlation is possible across the whole outcrop. The change from the Aitzgorri Limestone Formation to the Itzurun Formation marls is considered the deep water correlative conformity of the major Se1 sequence boundary and sea-level low as described in Hardenbol *et al.* (1998). Detailed stratigraphic studies show that this lithologic change at the Selandian GSSP is time equivalent to the facies shift from Danian limestones to Selandian detrital sediments in the type area of Denmark, suggesting therefore the effect of the same sea-level event (Schmitz *et al.*, 1998).

The GSSP occurs approximately at the top of the lower third of Chron C26r. The age of the boundary can be further estimated using astrocyclicity (Dinarès-Turell *et al.*, 2007; Schmitz *et al.*, 2008) and the GSSP is reported at 32 precession cycles above the top of C27n in the Zumaia section. More recently, in a comparative study between the Bjala section in Bulgaria and the Zumaia section, Dinarès-Turell *et al.* (2010) have refined the placement of the chron C27n at Zumaia, and report the Selandian GSSP to occur at 30 precession cycles above the top of C27n. A global marine correlation event of the GSSP level is the second radiation of the calcareous nannofossil *Fasciculithus* (base Zone NP5), that is situated just above the base of the Selandian. For regional correlations the end of the acme of the Braarudosphaeraceae nannofossil family can also be used (Steurbaut and Sztrákos, 2008; Bernaola *et al.*, 2009). Several foraminiferal events around the boundary can be used to bracket the GSSP level. However, it should be noted that the now defined GSSP of the Selandian is distinctly younger than a previous informal working definition, as used in GTS2004, which placed the Danian–Selandian boundary at the lowest occurrence of the planktonic foraminifer *Morozovella angulata* at the base of Zone P3. The base of the Selandian is also about 600 kyr younger than the latest Danian Event (Bornemann *et al.*, 2009), a distinctly developed stratigraphic level observed in Egypt and Tunisia, and until 2007 considered as an alternative for the placement of the GSSP of the Selandian (Guasti *et al.*, 2006; Schmitz *et al.*, 2008; Sprong *et al.*, 2009).

28.1.4. Definition of the Base of the Thanetian

28.1.4.1. The Thanetian Stage

The Thanetian is the youngest stage of the Paleocene. The name of this stage was first used by Renevier (1873), but its meaning was subsequently narrowed by Dollfus (1880), who included only the Thanet Sands, the type strata on the Isle of Thanet in southeast England. Since then the Thanetian has always been used *sensu* Dollfus (Curry, 1981). These strata contain calcareous nannoplankton Zones NP6 *pro parte* up to NP9 *pro parte* (Martini, 1971) and dinoflagellate cyst Zones Viborg 3 *pro parte* to 6 (Heilmann-Clausen, 1985, 1988) and D4b–D5a (Costa and Manum, 1988). The Thanet Sands span polarity Chrons C26n, C25 and the lower part of C24r (Ali and Jolley, 1996).

The Landenian has been used as an approximately equivalent alternative stage name in the past. This is named

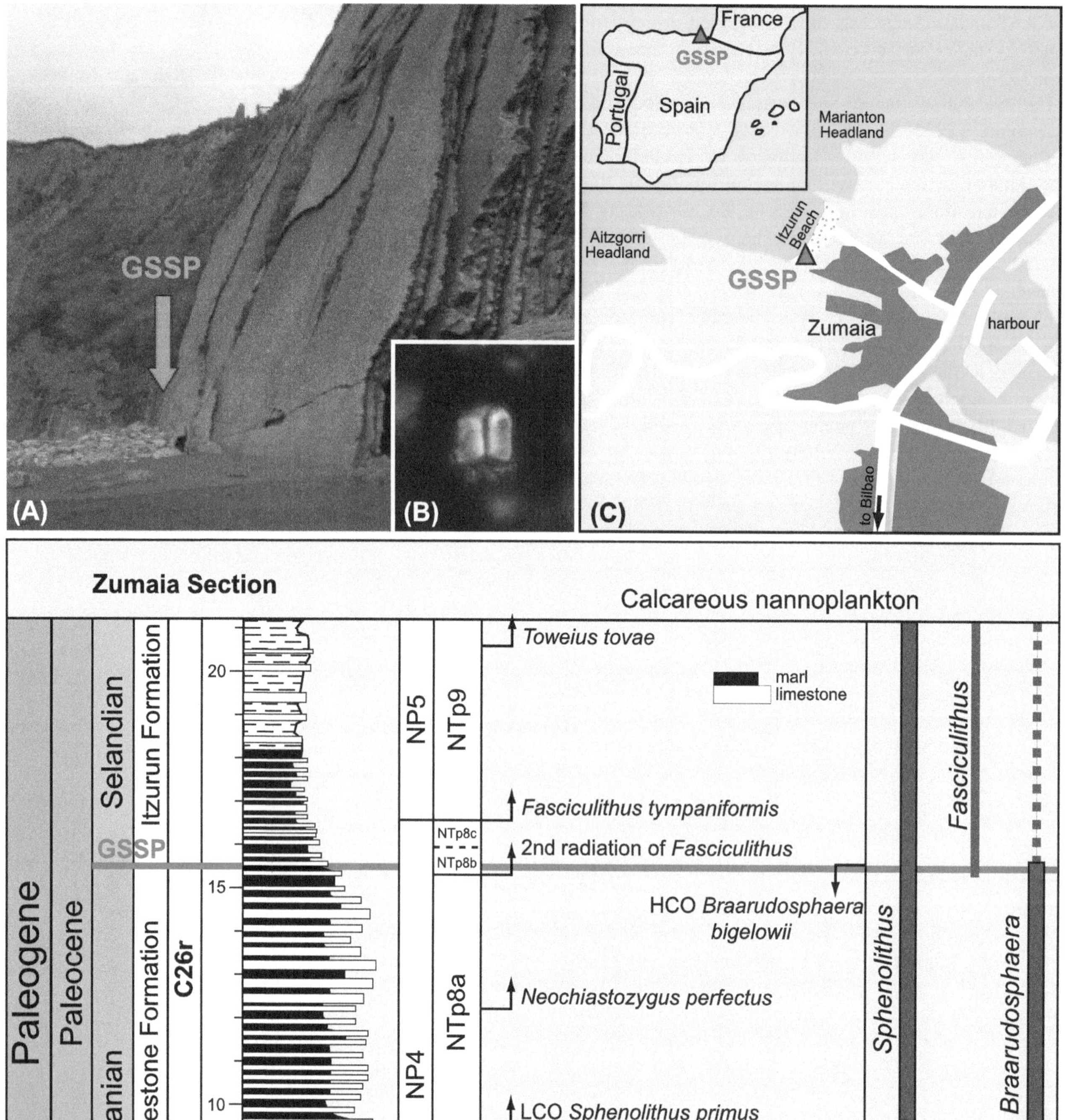

FIGURE 28.3 **The Selandian GSSP located at Zumaia, Basque Country, Spain (C).** The photograph (A) and the stratigraphic data (D) are based on Schmitz *et al.* (2008, 2011); the image of *Fasciculithus tympaniformis* (B) courtesy of Simonetta Monechi.

after a town in central Belgium, and was introduced by Dumont (1839, 1849) (Curry, 1981; De Geyter *et al.*, 2006).

28.1.4.2. The GSSP of the Thanetian

The Thanetian GSSP proposal was ratified in 2008, as extensively documented by the Paleocene Working Group (Schmitz *et al.*, 2008, 2011; Figure 28.4). The GSSP of the Thanetian is placed in the section exposed at Itzurun Beach in Zumaia, Basque Country, northern Spain (coordinates 43° 17′58.4″N, 2° 15′39.1″W), the same section in which the Selandian GSSP has been defined (see Section 28.1.3.2).

The precise level of the Thanetian GSSP in the Zumaia section occurs 8 precession cycles, or 2.8 m, above the base of the core of a distinct clay rich level, about 1 m thick and with a reduced carbonate content and an increased magnetic susceptibility. The GSSP level corresponds to the base of magnetochron C26n, or the C26r/C26n reversal.

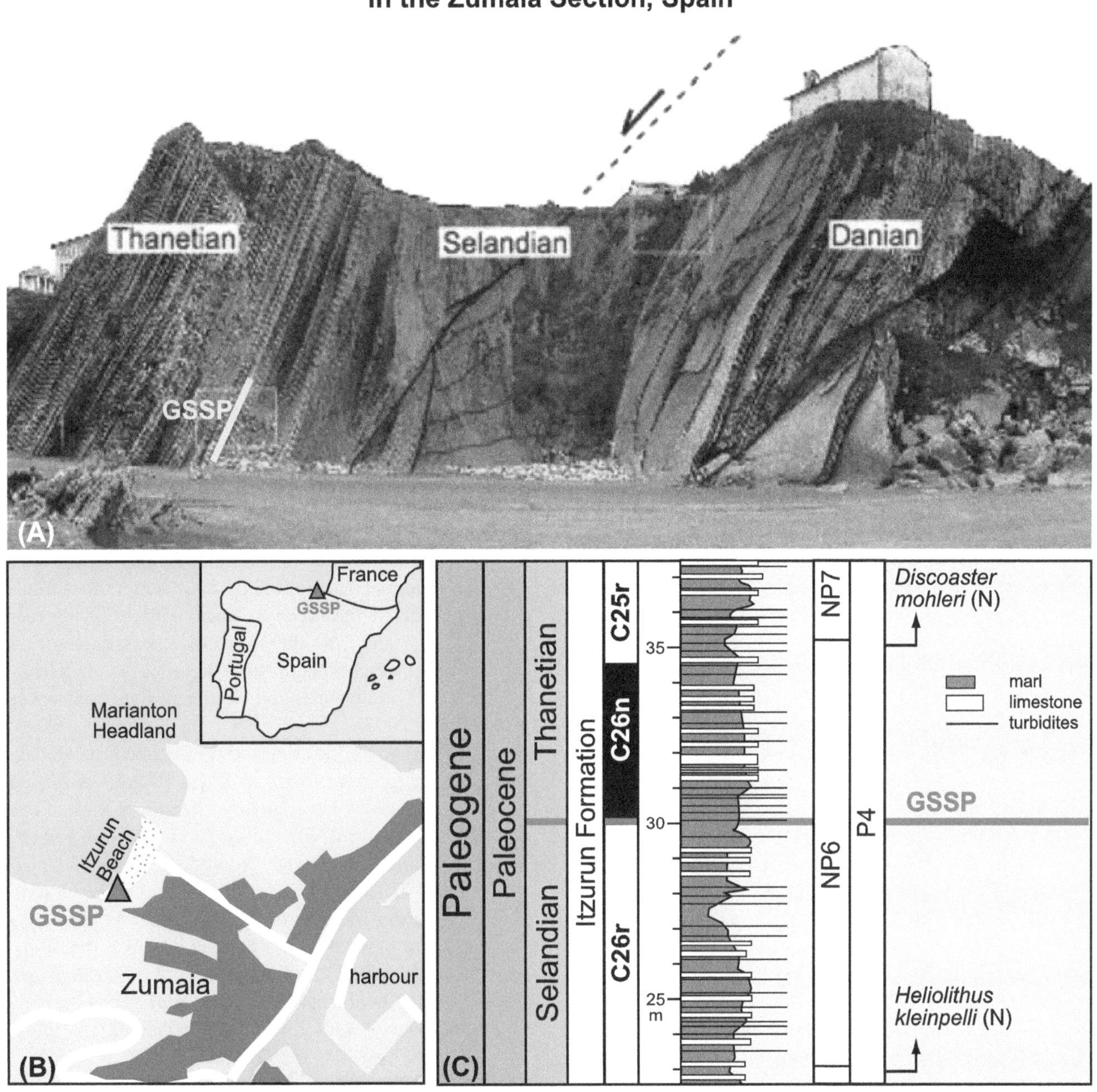

FIGURE 28.4 **The Thanetian GSSP located at Zumaia, Basque Country, Spain (B).** The photograph (A) and the stratigraphic data (C) are based on Schmitz *et al.* (2008, 2011).

The GSSP level occurs c. 6.5m above the base of Member B of the Itzurun Formation; this Member B contains indurated limestone beds, in contrast to the underlying Member A which only consists of marls. The distinct clay horizon is characterized by a marked change in calcareous nannofossil and foraminifera content, known as the short-lived Mid-Paleocene Biotic Event (MPBE) (Bernaola *et al.*, 2007). The MPBE is located c. 4.5 m above the first occurrence of *H. kleinpelli*, the Zone NP6 marker, and within planktonic foraminifera Zone P4. A major increase in the abundance of the dinoflagellate *Alisocysta gippingensis* is considered a useful event for recognizing the base of the Thanetian within the North Sea Basin, while the last occurrence of *Palaeoperidinium pyrophorum* and *Palaeocystodinium australinum/bulliforme* are late Selandian events that are useful for inter-regional correlations (Schmitz *et al.*, 2008).

28.1.5. Definition of the Base of the Ypresian

The Ypresian is the lowermost stage of the Eocene Series and accordingly, the base of the Ypresian also defines the base of the Eocene (Salvador, 1994). It is succeeded successively by the overlying Eocene stages of the Lutetian, Bartonian and Priabonian.

28.1.5.1. The Ypresian Stage

The Ypresian Stage was originally introduced by Dumont (1849) to include the clayey to fine-sandy shelf-facies strata lying between the terrestrial to marginally marine Landenian (see section 28.1.4.1) and the marine Brusselian (see section 28.1.6.1) in western Belgium. Dumont (1851) later assigned the upper, sandier part of this stage to a separate Paniselian Stage (Willems *et al.*, 1981). The Paniselian (named after the Mont-Panisel hill near Mons in southern Belgium), however, is a poorly defined term and became obsolete as a chronostratigraphic term (De Geyter *et al.*, 2006), leaving the Ypresian as it was originally defined by Dumont (1849). The term Ypresian is derived from the town of Ieper (Dutch spelling) or Ypres (French translation), located in western Belgium.

The Ypresian sediments in the type area are well constrained by dinocyst and calcareous nannoplankton associations, by sequence-stratigraphic analysis and magnetostratigraphy (e.g., Ali *et al.* 1993; Vandenberghe *et al.*, 1998, 2004; Steurbaut, 2006).

Cuisian and Ilerdian are regional stage names that have been used in parallel with the Ypresian (Pomerol, 1981a). The Cuisian has been defined and used in the Paris Basin, but it was recommended to refrain from applying the term by Bignot (1981). In the Tethyan realm, the Ilerdian is a regional Mediterranean stage (Hottinger and Schaub, 1960; Plaziat, 1981) corresponding to an important phase in the evolution of the larger foraminifera which is not represented in northwest European basins. Based on integrated stratigraphic studies, the base of the Ilerdian was recently correlated to the base of the Ypresian (Pujalte *et al.*, 2009; Scheibner and Speijer, 2009). In a larger foraminifera zonation, the position of the Ilerdian and the Cuisian occurs in between Thanetian and Lutetian (Figure 28.8).

28.1.5.2. The GSSP of the Ypresian

The events taking place in the Paleocene–Eocene boundary interval, their inter-relations, and possible causes have been intensively studied by the ISPS Paleocene–Eocene Boundary Working Group (e.g., reviews in Aubry *et al.*, 1998, 2003, 2007; Schmitz *et al.*, 2001). Although several biostratigraphic events occur in this boundary interval, a geochemical event, namely the onset of a pronounced negative carbon isotope excursion (CIE), has been selected as the best criterion for the recognition and correlation of the Paleocene–Eocene boundary (Figure 28.5).

This event at the base of the Eocene was estimated to be 0.94 myr younger than the base of magnetochron C24r by cyclostratigraphy (Norris and Röhl, 1999). In a more recent cyclostratigraphic approach, Westerhold *et al.* (2008) have found the base PETM at C24r.36. The carbon isotope signal has the advantage of being observable in both marine and terrestrial successions (Stott *et al.*, 1996). The carbon isotope excursion is the expression of the Paleocene–Eocene thermal maximum (PETM), a major global warming event traditionally related to a sudden release of methane from seafloor clathrates (e.g., Dickens *et al.*, 1997). The CIE coincides with major paleontological events: a deep-sea benthic-foraminiferal extinction event, diversifications in planktonic foraminifera, calcareous nannofossils and larger foraminifera, a global acme of the dinocyst *Apectodinium* in the marine environment, and in the terrestrial environment with a mammal dispersal event in North America which defines the base of the Wasatchian Land Mammal Age (references in Aubry *et al.*, 2007).

The GSSP of the Ypresian and the Eocene is located in the Dababiya section near Luxor in Egypt (Dupuis *et al.*, 2003; Aubry *et al.*, 2007). The Dababiya section is located on the east bank of the upper Nile Valley, 23 km south of Luxor and 30 km north of Esna (coordinates 25°30′04.7″N, 32° 31′19.14″E). The GSSP level is located at the base of the Dababiya Quarry Member of the Esna Shale Formation. It is defined at the base of a thin dark gray clayey horizon, the Dababiya Quarry Bed 1, which underlies the 2-m-thick, partly phosphatic and laminated Dababiya Quarry beds 2 to 5. The Dababiya Quarry Beds distinctly weather out in the field making it easy to locate the GSSP level at its base.

This GSSP definition defines the base of the Ypresian at a slightly older age than the base of the traditional and historical Ypresian type section in the Ieper area in Belgium.

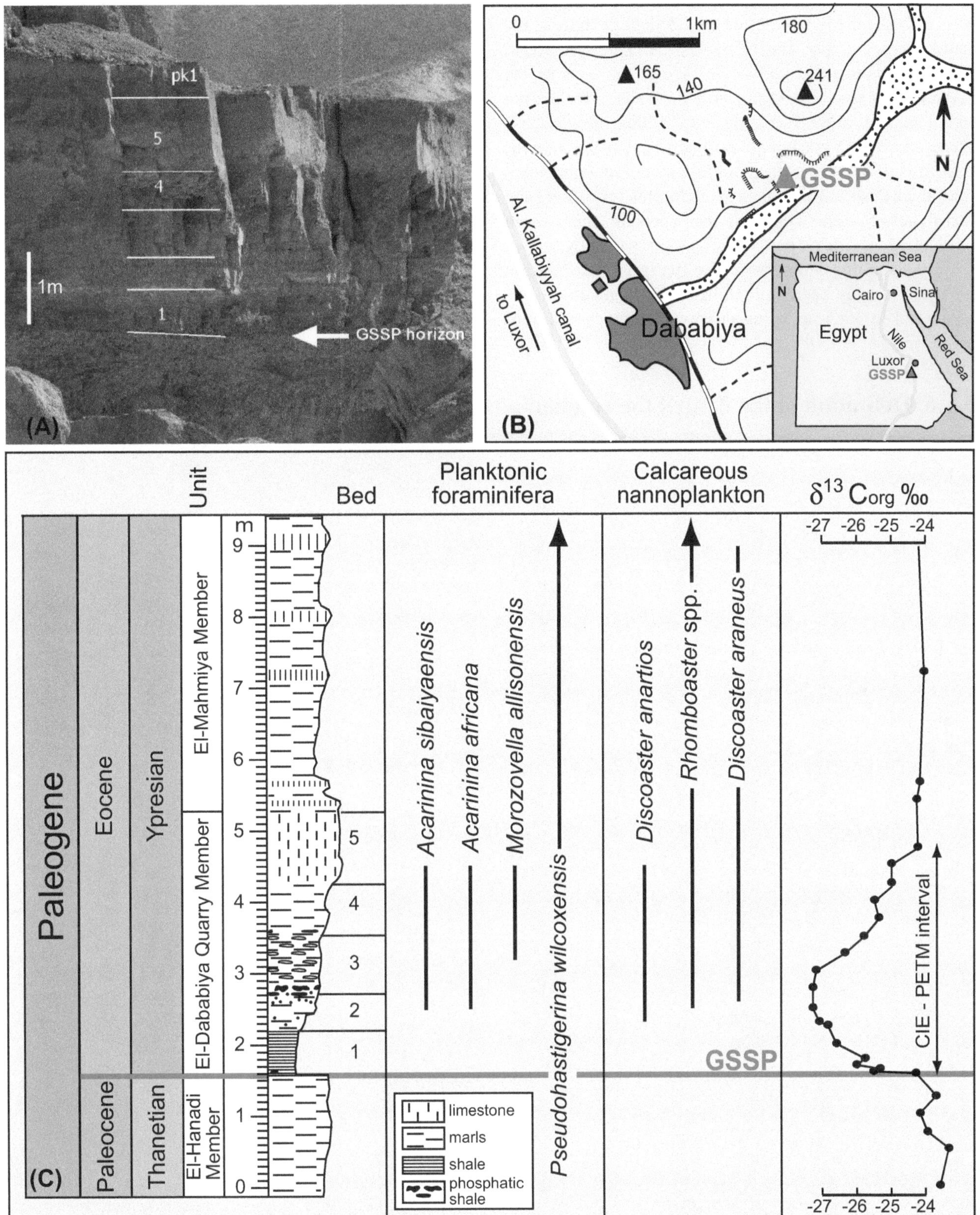

FIGURE 28.5 **The Ypresian GSSP located at Dababiya, Egypt (B).** The photograph (A) and the stratigraphic data (C) are based on Aubry *et al.* (2007).

In fact, the CIE-defined base of the Ypresian occurs in the historical type area almost at the base of the fluvio-lagoonal upper part of the Landenian (see Section 28.1.4.1; Steurbaut, 2006). In 2003, the IUGS ratified the GSSP of the base of the Eocene as being also the GSSP of the Ypresian Stage, thereby attributing a slightly older age to the base of the Ypresian Stage; moving it from 54.8 Ma (see also Hardenbol *et al.*, 1998) to 55.80 ± 0.2 Ma in GTS2004. Hence, base Eocene and base Ypresian are ratified together in one definition. It is therefore not meaningful and not recommended to introduce an additional stage name for this very short time interval between the GSSP of the Ypresian and its old historical base. The "Sparnacian", as discussed by Aubry *et al.* (2003) in this context (see also Berggren and Pearson, 2005), had already been omitted from the Paleogene stages by Pomerol in 1981, as being of less importance and non-marine.

28.1.6. Definition of the Base of the Lutetian

28.1.6.1. The Lutetian Stage

The Lutetian was defined in the Paris Basin, Lutetia being the Roman name for Paris. According to its author, De Lapparent (1883), the Lutetian Stage (early-Middle Eocene) is typified by the "Calcaire grossier" of the Paris Basin. As no stratotype section was indicated by De Lapparent, a neostratotype was selected by Blondeau (1981) near Creil, approximately 50 km north of Paris, at St. Leu d'Esserent and St. Vaast-les-Mello. The type-Lutetian contains typical larger foraminifera (*Nummulites laevigatus*, *Orbitolites complanatus*) and, to a lesser extent, palynomorphs and calcareous nannoplankton. Traditionally in open marine sequences, the base of the Lutetian was recognized by the lowest occurrence of the planktonic foraminifer *Hantkenina nuttalli* (or the former P9/P10 zonal boundary: e.g., Berggren *et al.*, 1995; GTS2004, p. 386). However, it was recently shown convincingly that this species does not appear at the base of C21r, but at the base of C20r, i.e. about 3 myr later (Wade *et al.* 2011). Another traditional criterion used to approach the base of the Lutetian was the C22n/C21r boundary (Molina *et al.*, 2011). Currently, the base of the Lutetian is considered to be slightly younger than the first appearance of *Turborotalia frontosa.* Blondeau *et al.* (1965) have discussed the correlation of the Lutetian with other, now obsolete, stage names in the Belgian Basin, such as Brusselian, Ledian, Wemmelian (see also De Geyter *et al.*, 2006) and the Bracklesham Beds of the Hampshire Basin in England.

28.1.6.2. The GSSP of the Lutetian

A Lutetian GSSP was ratified in 2011 (Molina *et al.*, 2011; Figure 28.6). Due to the presence of an important global sea-level drop near the Ypresian–Lutetian boundary, a continuous boundary stratotype had to be sought in deep-water deposits outside the historical Paris Basin. The Agost (Molina *et al.*, 2000) and Gorrondatxe (Molina *et al.*, 2009b) sections in Spain have been extensively studied for this purpose. The GSSP for the Lutetian is proposed in the cliffs of the Gorrondatxe (or Azkorri) beach section near Getxo village in the western Pyrenees (province of Biscay, Basque Country, North Spain) at coordinates 43° 22′ 46.47″ N and 3° 00′ 51.61″ W. The Gorrondatxe section is mostly composed of hemipelagic marls and limestones; thin-bedded siliciclastic turbidites occur but are distal and do not scour the hemipelagic marls. The GSSP level is at 167.85 m of the Gorrondatxe section in a dark marly level occurring in the section 2.15 m below a prominent 15-cm-thick turbidite at meter 170. This marly level coincides with a global maximal flooding event (Molina *et al.*, 2011). The GSSP level can be correlated with the lowest occurrence of the nannofossil *Blackites inflatus,* corresponding to the base of Subzone CP12b within Zone NP14.

The calcareous nannofossil *Blackites inflatus* was found to almost coincide with the base of the original Lutetian stratotype. Counting 39 precession cycles between the base of C21r and the lowest occurrence of *Blackites inflatus* determines an age about 819 kyr younger than the base of this magnetochron boundary (Molina *et al.* 2011). This gives an age of 47.78 Ma for the GSSP, based on the GTS2004 age of base C21r; the GSSP is also estimated 507 kyr older than the C21r/C21n boundary (Molina *et al.* 2011) leading to the GSSP position at chron C21r.6. This GSSP level is close to the base of the original Lutetian stratotype, and also coincides with the base of the Earnley Formation in the Hampshire Basin in the UK.

28.1.7. Definition of the Base of the Bartonian

28.1.7.1. The Bartonian Stage

The Bartonian (Mayer-Eymar, 1858) refers to the marine clays and sands of the Barton Beds of the Hampshire Basin, central southern England, exposed in the coastal section between Highcliff and Barton-on-Sea (Curry, 1981). These Barton Beds contain a rich and diversified dinoflagellate cyst assemblage including *Rhombodinium draco* and *Rhombodinium porosum*. The planktonic microfossils in the Bartonian type area, NP17 and P13–14 (or E12–E13 of Berggren and Pearson (2005) and Wade *et al.* (2011)), are generally considered indicative of a middle Eocene age for the Bartonian, in contrast to the late Eocene Priabonian containing *Isthmolithus recurvus* (Hardenbol and Berggren, 1978).

In the late 19th century, the base of the Bartonian in the type area was taken at the base of a horizon rich in *Nummulites prestwichianus* (Curry, 1981). Hooker (1986) shifted the base of the Barton Clay Formation downwards to a major lithological boundary, namely a thin glauconitic and pebbly omission surface. In fact this shift means the return to the

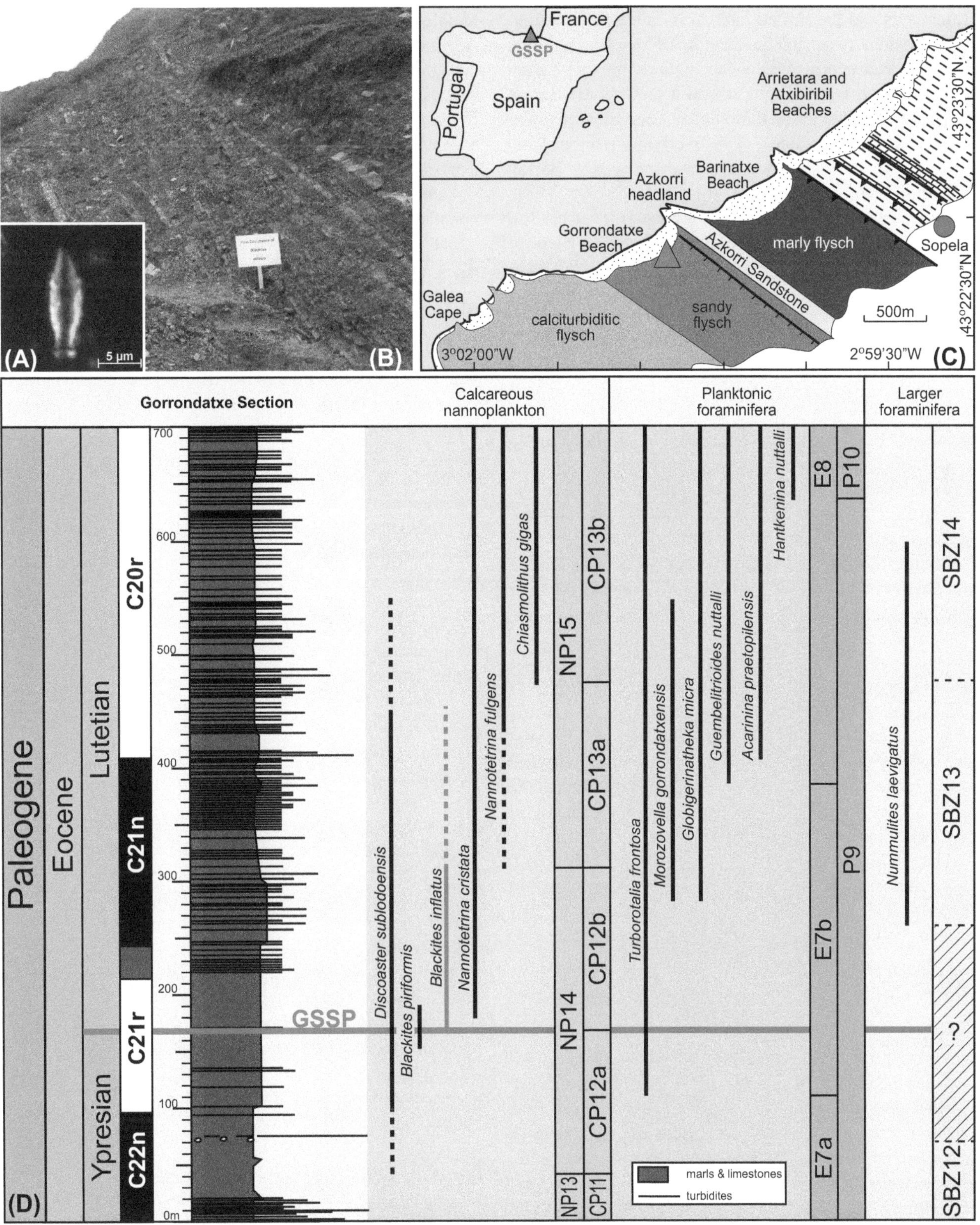

FIGURE 28.6 **The Lutetian GSSP located at Gorrondatxe, Basque Country, Spain (C).** Photograph (B) courtesy of E. Molina. Stratigraphical data and section (D) based on Molina *et al.* (2011). Image of *Blackites inflatus* (A) by courtesy of Simonetta Monechi.

original definition of the Barton Beds by Prestwich in 1857 (Curry, 1981). In the type section, the thickness involved in the shift is minimal, but laterally the thickness involved in the shift increases and the shifted interval is called the Huntingbridge Division, occurring in the top of the Bracklesham Beds. The Huntingbridge Division, underlying the Barton Clay, which itself is composed of Lower and Middle Barton Beds, is now included in the Barton Clay Formation, together with the Barton Clay Beds. The overlying Becton Sand Formation corresponds to the Upper Barton Beds or Barton Sand (King, 2006).

Cavelier and Pomerol (1986), following Hooker (1986), proposed placing the base of the standard Bartonian Stage at the base of the Huntingbridge Beds in the top of the Bracklesham Beds in England. In a review of data on the type Bartonian, these authors report that the Lower Barton Beds correspond to the *Rhombodinium draco* Zone (W10) (D10b in Figure 28.9), and the base of the Bartonian is situated in the upper part of NP16. These authors also report radiometric ages of 38.9 ± 1.8 and 39.6 ± 1.8 Ma for the Lower Barton Beds, and 39.1 ± 1.5 Ma for the Huntingbridge Beds in the type area. This redefinition explains why the base of the Barton Clay Formation in recent British stratigraphic tables, such as in King (2006) and Dawber *et al.* (2011), is put in the upper Lutetian rather than in the Bartonian.

Although the Barton Beds have yielded a rich marine fauna, they contain relatively few species suitable for long distance correlation. It has been suggested that the base of the Bartonian occurs near the common occurrence of the calcareous nannofossil *Reticulofenestra reticulata* (misquoted in Ogg *et al.*, 2008, p. 131) and close to the C19n/C18r magnetochron boundary.

The upper limit of the Bartonian in its historical type area remains poorly defined, since it corresponds to a widespread sea-level lowstand and therefore it is frequently marked by a hiatus separating it from the Priabonian. In the past in France, also poorly defined regional stages such as the Auversian, a substage of the Lower Bartonian (Pomerol, 1981b), Marinesian (Schuler *et al.*, 1992), and the Biaritzian have been introduced, which in part overlap with the Bartonian (Pomerol, 1981c; Cavelier and Pomerol, 1986).

28.1.7.2. The Criterion for the Base of the Bartonian

At the present time no formal proposal has been made for a Bartonian GSSP. A well-documented section, unfortunately also with some shortcomings, has been published. This is near the classical Contessa Highway in the Umbrian Apennines in Italy containing Lutetian and Bartonian strata (Jovane *et al.*, 2007). Several distinctive lithologic marker beds occur near the C19n/C18r magnetochron boundary and also foraminiferal and nannofossil events occur around that boundary that could help correlation. A clear oxygen and carbon isotope excursion starts about 6 m above the C19n/C18r magnetochron boundary, within C18n.2n and is accompanied by a significant planktonic foraminifera turnover. This is comparable to the middle Eocene climatic optimum as observed in the Indian–Atlantic sector of the Southern Ocean (Bohaty and Zachos, 2003) and is also identified in the Barton Clay Formation of the Hampshire Basin in the UK (Dawber *et al.*, 2011).

In the absence of a formal Bartonian GSSP, it seems reasonable, from the data in the Hampshire Basin and the Contessa Highway section, to keep the base Bartonian as it was defined in the previous GTS, namely at the C19n/C18r boundary.

28.1.8. Definition of the Base of the Priabonian

28.1.8.1. The Priabonian Stage

The Upper Eocene Priabonian Stage (Munier Chalmas and De Lapparent, 1893) has its historical type section in Priabona, northern Italy (Roveda, 1961; Hardenbol, 1968; Barbin, 1988). Hardenbol (1968) discusses the historical evolution of the usage of the term Priabonian since its introduction by Suess (1868). In the propositions at the end of the Colloque sur l'Eocène (BRGM, 1968), several parastratotype sections were designated in the same area in addition to the type section.

Bryozoan marls and limestones, as well as small *Nummulites* beds are included in the shallow water facies of the Priabonian at the type locality (Barbin, 1988). The Priabonian in the type section contains rare planktonic foraminifera, such as *Turborotalia cerroazulensis,* assigned to upper Zone E15 and E16, and spans the calcareous nannofossil Zones NP19/20 with *Isthmolithus recurvus* (Roth *et al.*, 1971); this age estimate is confirmed by the recognition of the dinoflagellate cyst *Melitasphaeridium pseudorecurvatum* Zone in the historical type section (Brinkhuis and Biffi, 1993).

The Ludian is a regional stage name used in the Paris Basin that can be attributed to the Priabonian (Schuler *et al.*, 1992).

At the Bartonian–Priabonian boundary in Western Europe, uplift by the Pyrenean tectonic phase can be recognized in many regions. Another particularity of the Priabonian is the occurrence of at least two small impact-related iridium anomalies reported from several sections (Coccioni *et al.*, 2000; Vonhof *et al.*, 2000). This, apparently general, increase in flux of extraterrestrial material during the late Eocene has been related to the massive impact at Chesapeake Bay along the US Mid Atlantic coast dated at about 35.5 Ma (Gohn *et al.*, 2009).

28.1.8.2. The Criterion for the Base of the Priabonian

At the present time no formal proposal has been made for a Priabonian GSSP. The Bartonian/Priabonian boundary has been placed at the lowest occurrence of *Chiasmolithus oamaruensis*, marker of the base of NP18, in Berggren *et al.* (1985, 1995).

In the GTS2004 and in Ogg *et al.* (2008), this base Priabonian criterion was replaced by the slightly older base of magnetochron C17n.1n. This criterion change was inspired by doubt on the clear-cut nature of the nannoplankton event, and by the possibility of recognizing the magnetochron boundary in non-marine facies (H.P. Luterbacher, pers. comm., 2010).

Agnini *et al.* (2011) have presented an integrated biomagnetostratigraphy of the Alano di Piave section in the Venetian southern Alps of Italy in which they propose the base of the Tiziano volcanic crystal tuff bed as the Priabonian GSSP. The section is about 50 km away from the historical Priabona section, and is close to Possagno (45°54′51.10″N and 11°55′4.87″E), one of the deep-water parastratotype sections chosen during the Colloque sur l'Eocène (BRGM, 1968) but unfortunately itself poorly outcropping. Calcareous microfossils and palynomorphs are abundant in the hemipelagic marls of the section. Paleomagnetic signals and calcareous biohorizons, based on their reliable intercalibration at ODP Site 1052, show that the section spans the interval from the upper part of chron C18r to the basal part of chron C16r. The boundaries between the different chrons can be well identified, allowing good age control in the section. The suggested Tiziano bed is immediately above the lowest occurrence of *Chiasmolithus oamaruensis*. However, according to the analysis in the Alano section, this seems to occur in chron 17n.2n, and not as commonly assumed in chron 17n.1n. The age estimate of the Tiziano bed, based on different time scales, varies between 37.537 (GTS2004) and 37.833 Ma (Cande and Kent, 1995); several micropaleontological correlation criteria are available around the Tiziano bed (Agnini *et al.*, 2011). The historical Priabonian section that contains *I. recurvus* at its base starts somewhat higher than the Tiziano bed, as in the Mediterranean *I. recurvus* becomes common higher in C16n (Figure 4 in Coccioni *et al.*, 1988).

Although the Alano section is a good candidate for a Priabonian GSSP, in the absence of a formal Priabonian GSSP it seems reasonable for practical reasons to keep the base Priabonian as it was defined in the previous GTS, being the base of magnetochron C17n.1n.

28.1.9. Definition of the Base of the Rupelian

A threefold subdivision of the Oligocene was introduced by Beyrich (1854). There has been some stratigraphic confusion related to the Oligocene or Eocene nature of some stages, especially relating to stages such as the Latdorfian and Tongrian (see e.g., table in Cavelier and Pomerol, 1983; table 1 in Schuler *et al.*, 1992).

A pragmatic chronostratigraphic scheme for the standardization of the Oligocene in two stages was provided by Hardenbol and Berggren (1978), who distinguished two easily recognizable lithostratigraphic units in the historical Oligocene stratotype area of northwest Europe:

(i) A lower open marine, clayey unit, which includes the typical Rupelian type area sediments; and
(ii) An upper, predominantly shallow-marine, sandy unit, which incorporates the type section of the Chattian.

This proposal has been followed ever since, and the twofold subdivision became official in the IGC of Washington in 1989 (Jenkins and Luterbacher, 1992). According to stratigraphic rules (Salvador, 1994), the base of the Rupelian Stage equals the base of the Oligocene Series.

28.1.9.1. The Rupelian Stage

The name Rupelian was introduced by Dumont (1849) for the Boom Clay deposits exposed in a series of brick yards along the rivers Rupel and Scheldt in Belgium. This clay deposit constituted the upper Rupelian of Dumont, while a lower Rupelian part consisting of a clay-like deposit, the *Nucula comta* clay, between two sand deposits, was exposed in northeast Belgium. In its outcrop area, the Boom Clay, is substantially eroded but it becomes thicker and stratigraphically more complete in the subsurface of North Belgium (Vandenberghe *et al.*, 2001; Van Simaeys and Vandenberghe, 2006). These Rupelian clays form a characteristic and continuous deposit in a large part of the North Sea area. Distinct layering in the clay reflects mainly 41 kyr obliquity cycles (Abels *et al.*, 2007). Several characteristic layers of septaria (carbonate concretions) systematically occur in the clay, explaining the name Septarienton that is used in Germany. Many macro- and microfossil groups have been studied in the stratotype area (references in Vandenberghe, 1981; Van Simaeys and Vandenberghe, 2006). Calcareous nannoplankton Zones NP23 and NP24 and planktonic foraminifera Zones O1 to O4 (Berggren and Pearson 2005; Wade *et al.*, 2011) were recognized in the clay, and a benthic foraminiferal and a dinoflagellate cyst zonation have been established in the Rupelian stratotype (references in Van Simaeys and Vandenberghe, 2006).

Within the concept of a threefold subdivision of the Oligocene by Beyrich (1854), the historical stratotype as defined by Dumont in 1849 was considered as the Middle Oligocene, between the Lower Oligocene Latdorfian (Ritzkowski, 1981a) and the Upper Oligocene Chattian (Ritzkowski, 1981b). However, the twofold subdivision of the Oligocene advocated by Hardenbol and Berggren (1978) became the standard practice. The Upper Oligocene Chattian remained unchanged. The newly introduced Lower Oligocene, after the definition of the base-Oligocene GSSP (see Section 28.1.9.2.) and more detailed micropaleontologic studies (Stover and Hardenbol, 1993; Brinkhuis and Visscher, 1995), appeared to encompass the approximately correlative regional stages Latdorfian (Germany) and Tongrian (Belgium), as well as the historical Rupelian Stage. However, the name Rupelian is now used for the whole newly defined Lower Oligocene. Consequently this base Rupelian is older than the base of the historical Rupelian stratotype and a lowermost Rupelian sedimentary sequence, below the Boom Clay, containing calcareous nannoplankton Zone NP21 occurs at the very base of the Lower Oligocene Rupelian (Vandenberghe *et al.*, 2003, 2004).

Two events which are usually associated with the start of the Oligocene in fact occur between this lowermost Rupelian sequence (NP21) and the overlying sequences containing the Boom Clay and starting in calcareous nannoplankton Zone NP22. These events are the Oi-1 most significant oxygen isotope cooling event (Zachos *et al.*, 1996; De Man *et al.*, 2004a), and the "Grande Coupure" in the mammal record, separating Eocene from Oligocene vertebrate fauna after new land bridges had been created (Woodburne and Swisher, 1995; Vandenberghe *et al.*, 2003). The considerable faunal and floral changes at the Eocene–Oligocene transition have been related to widespread cooling (e.g., Premoli Silva *et al.*, 1988; Vonhof *et al.*, 2000).

Before the Rupelian was officially adopted by the ICS as the Lower Oligocene stage name, the "Stampian" has been used in the Paris Basin for the Lower Oligocene; it consists mainly of fine marine sands with aeolian sands at the top, and includes continental "Sannoisian" facies at the base and Etampes limestone at the top (Cavelier, 1981; Cavelier and Pomerol, 1977, 1983; Plaziat *et al.*, 2009).

28.1.9.2. The GSSP of the Rupelian

The GSSP of the Rupelian Stage and the Oligocene Series has been defined in the Massignano section, 10 km southeast of Ancona, on the Adriatic coast of northern Italy (Premoli Silva *et al.*, 1988; Premoli Silva and Jenkins, 1993) (Figure 28.7). The GSSP was defined in an abandoned quarry on the east side of the Ancona–Sirolo road near Massignano; at coordinates 43°32′58.2″N and 13°36′03.8″E. The Massignano section covers a continuous, 23 m, Upper Eocene to Lower Oligocene sequence of open marine marls and calcareous marls. The GSSP is defined at the base of a greenish-gray 0.5-m-thick marl bed, 19 m above the base of the section. The key marker of the GSSP is the extinction of the hantkeninid planktonic foraminifera (Figure 28.7(b)) at the top of Zone E16, which lies within nannofossil Zone NP21, and magnetic polarity Chron 13r.1r. However the normal-polarity subzone C13r.2n observed in the Massignano section is not present in marine magnetic surveys, which is the basis of the C-sequence km model and therefore the GSSP position projected to the magnetic polarity scale is better expressed as C13r.86, as was already reported in GTS2004. K-Ar and Ar-Ar dating of biotite grains from 4.3 m below the base of the GSSP give an age of 34.6 ± 0.3 Ma, suggesting a numerical age of 34 Ma for the GSSP itself (Premoli Silva and Jenkins, 1993). The selection of this GSSP for the base of the Oligocene honors the original meaning of the Oligocene with a threefold subdivision as defined by Beyrich (1854) who considered the Latdorfian transgression as the start of the Oligocene (see also Brabb, 1968).

28.1.10. The Definition of the Base of the Chattian

28.1.10.1. The Chattian Stage

The upper Oligocene Chattian was introduced and defined with the "Kasseler Meeressande" as the name-giving strata in Hessen in Germany by Fuchs (1894), who recognized their similarity to the Doberg strata near Bünde (Westphalia, Germany). Goerges (1957) selected the section at Doberg as the stratotype, and Ritzkowski (1981b) presented a detailed drawing of this section. The stratigraphy of key Chattian sections in the North Sea Basin is discussed by Van Simaeys *et al.* (2004).

The most distinct biostratigraphic marker characterizing the base of the Chattian in the North Sea Basin is the record of the *Asterigerina guerichi guerichi* benthic foraminifer bloom event, known as the *Asterigerina* Horizon. It is a biohorizon recognized over the entire North Sea Basin. In shallow marine facies it is a suitable regional stratigraphic correlation marker to define the base of the Chattian in the North Sea area. Other benthic foraminiferal bioevents also characterize the Rupelian–Chattian boundary in the North Sea Basin (De Man *et al.*, 2004b).

The *Asterigerina* Horizon falls within NP24, adopting the lowest occurrence (LO) of *Helicosphaera recta* as the substitute marker of the base of Zone NP24 (Van Simaeys *et al.*, 2004). The lower unit A of the Chattian is within Zone NP24 according to a classical threefold subdivision based on pectinid ranges (Anderson, 1961); the middle and the upper subdivisions, B and C, are situated in Zone NP25, defined in the North Sea Basin by the lowest occurrence of *Pontosphaera enormis*. Chattian A and B make up the Eochattian as defined by Hubach (1957) and Chattian C corresponds to the Neochattian of this author.

Dinocyst assemblages are diverse and well preserved in both the Rupelian and the Chattian of the North Sea (Köthe, 1990). The lowest occurrence of *Artemisiocysta cladodichotoma* coincides with the onset of the *Asterigerina guerichi guerichi* bloom marking the base of the Chattian in the North Sea Basin. The Rupelian–Chattian boundary in its historical North Sea type area falls in a short hiatus associated with the *Svalbardella* event known to have occurred in the middle and upper part of chron C9n (Van Simaeys *et al.*, 2005a,b); the migration of the high-latitude dinocyst *Svalbardella cooksoniae* into low latitudes reflects the influence of anomalously cool surface waters. Its occurrence in Chron C9 relates this

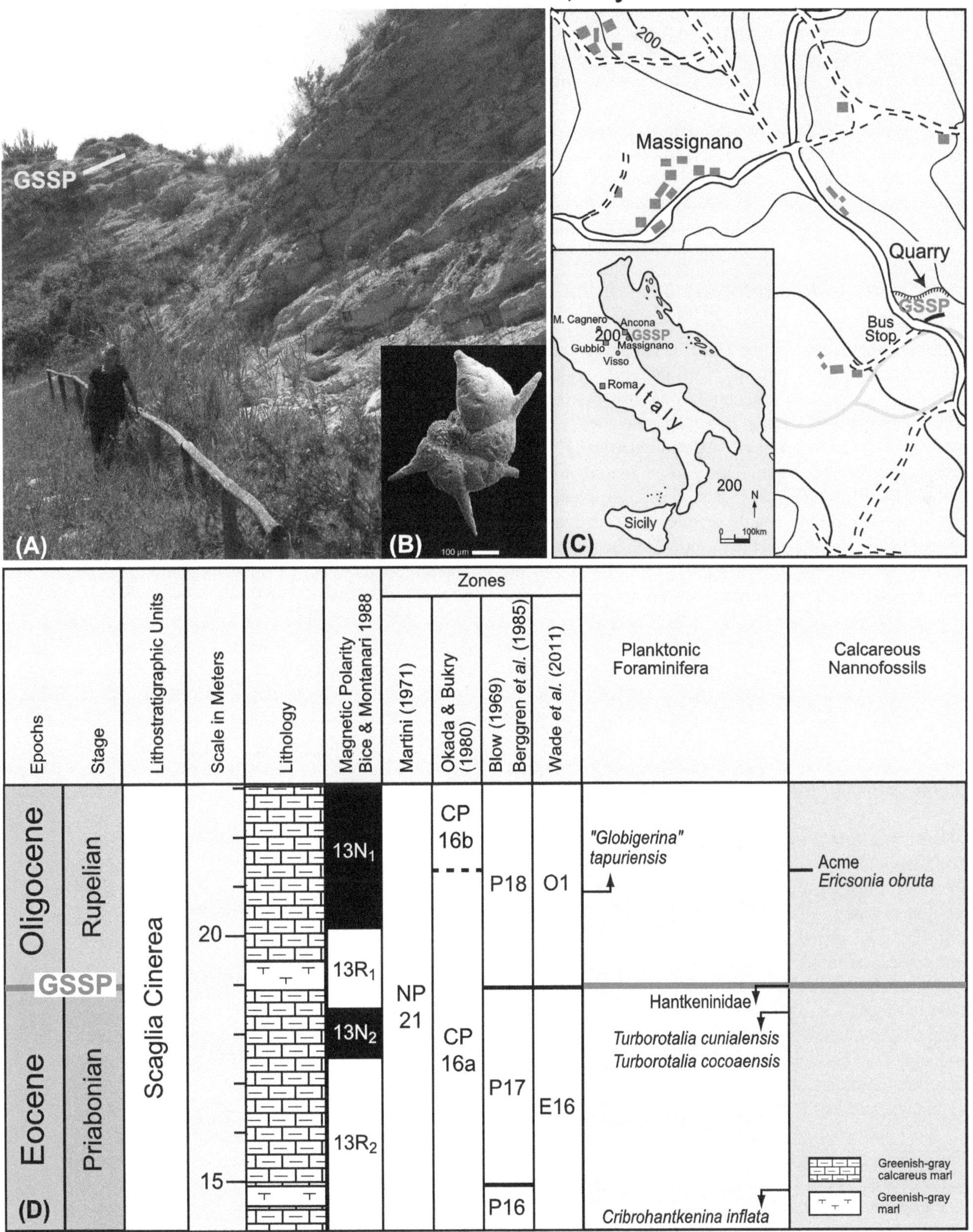

FIGURE 28.7 **The Rupelian GSSP located at Massignano, Ancona, Italy (map in C).** Photograph of the outcrop (A) courtesy of E. Molina. *Hantkenina alabamensis* in (B) courtesy of B. Huber. Stratigraphic data (D) are based on Premoli Silva and Jenkins (1993).

event to the important mid-Oligocene cooling event observed in deep-sea benthic foraminiferal oxygen isotope records described by Miller *et al.* (1998) (see Section 28.3.2); it is thought that a significant sea-level lowering accompanied this cold climatic spell and caused the general hiatus in the historical Oligocene type areas, making the areas unsuitable for the definition of a GSSP.

The presence of the larger foraminifer *Miogypsina septentrionalis* in the middle part of the Doberg Chattian reference section (Anderson *et al.*, 1971) and the abundant *Asterigerina guerichi guerichi* (De Man and Van Simaeys, 2004) are indicators of tropical to subtropical conditions.

28.1.10.2. The Criterion for the Base of the Chattian

No formal proposition for the GSSP Chattian has yet been made. Chronostratigraphically meaningful calcareous plankton are sparse in the generally marginal marine sedimentary environment of the historical reference sections, paleomagnetic signals are poor, and radiometrically datable volcanic horizons are absent. Therefore, a continuous, well-calibrated base Chattian GSSP section offering good correlation potential, has to be sought outside the North Sea Basin.

Three pelagic sections in the Umbria-Marche northeastern Apennines of Italy provide continuous information on the Rupelian to Chattian transition: Pieve d'Accinelli, Monte Cagnero and Contessa (Coccioni *et al.*, 2008). The base Chattian marker in the sections figured by these authors occurs at the highest or last common occurrence of *Chiloguembelina cubensis* or between planktonic foraminiferal Zones O4 and O5 (Berggren and Pearson, 2005; i.e. subzones P21a and P21b of Berggren *et al.*, 1995); based on this criterion the level at meter 188 in the Monte Cagnero section is a potential GSSP for the Chattian. The highest common occurrence of *C. cubensis* is considered by Coccioni *et al.* (2008) to be a robust event, contrary to the analysis of Van Simaeys *et al.* (2004) indicating this event to be diachronous among basins and most likely controlled by paleolatitude and paleobiogeography. Also the appearance of the colder-water immigrant dinoflagellate *Distatodinium biffi* occurs in these Mediterranean sections very close to the last occurrence of *Chiloguembelina cubensis*. In the North Sea *D. biffi* occurs in the top of the Rupelian (Van Simaeys *et al.*, 2005a), but this apparent contradiction is solved in Pross *et al.* (2010) by introducing *D. biffi s.s* and *s.l*, with *D. biffi s.s* occurring in the base of the Chattian.

The suggested GSSP level is within calcareous nannofossil Zone NP24 and in the upper part of Chron 10n. Volcano-clastic biotite-rich layers were found in all three sections and interpolation between these levels suggests an age of 28.3 ± 0.2 Ma for the Rupelian–Chattian boundary at this location (Coccioni *et al.*, 2008). Sr isotope and glauconite dates of the historical Rupelian and Chattian stratotype areas bracket the Rupelian–Chattian boundary between about 29 Ma and about 27 Ma (De Man *et al.*, 2010), therefore the proposed GSSP level at Monte Cagnero is consistent with the original position of the historical stratotypes. At Monte Cagnero a choice could be made amongst several criteria to identify the Rupelian–Chattian boundary, including: last occurrence of *Chiloguembelina cubensis*, last common occurrence of *Sphenolithus distentus*, *Svalbardella* abundance, a glacio-climatic oxygen isotope excursion or base of Chronozone 10n. This section at Monte Cagnero is therefore a good candidate for a GSSP of the Chattian. However, as no formal proposition for the GSSP has yet been made, and as in the previous GTS Luterbacher *et al.* (2004) also placed the Rupelian–Chattian boundary within Chron 10n, at the base of C10n.1n near the position of the last common occurrence of *Chiloguembelina cubensis*, between P21a and P21b, it seems practical to keep this position as the base of the Chattian.

The Paleogene–Neogene boundary (base of the Aquitanian Stage) coincides with the magnetic reversal from polarity Chron C6Cn.2r to C6Cn.2n in the section at Lemme-Carrosio, northern Italy (Steininger *et al.*, 1997), orbitally tuned at 23.03 ± 0.1 Ma (see Chapter 29).

28.2. PALEOGENE BIOSTRATIGRAPHY

The biostratigraphy of Paleogene marine successions is well established, using several microfossil groups. Calcareous nannofossils and planktonic foraminifera are the two most widely used groups for biostratigraphic purposes in open marine deposits, especially in low- and mid-latitude settings where deposition occurs above the carbonate compensation depth (CCD). In areas where calcareous remains do not preserve well (below CCD or in restricted basins) alternative biostratigraphic schemes have been established. These are based on organic microfossils, notably dinocysts, or siliceous microfossils such as radiolarians. In siliciclastic neritic facies, dinoflagellates are also particularly useful; information concerning Paleogene dinocyst distribution is most comprehensive for the mid latitudes of the northern hemisphere. Larger foraminifera are a key stratigraphic tool in carbonate-dominated shallow-marine low-to-middle latitudes. Calibration between various biostratigraphic schemes can be achieved in areas where pelagic and continental margin taxa co-occur. The Paleogene is also marked by rapid diversification of land mammals, yielding detailed terrestrial zonations.

Figure 28.1 shows the correlation of Paleogene stages, magnetostratigraphy, and standard marine zonations based on lower to mid latitude planktonic foraminifera, calcareous nannoplankton and larger foraminifera. Additional regional biozonal schemes are provided for radiolarians and dinoflagellate cysts (Figure 28.9). The intricate terrestrial zonations for North America and Europe with mammalian fossils and the calibration of the regional terrestrial "stages" with the Paleogene standard stages are shown in Figure 28.10.

28.2.1. Foraminifera

28.2.1.1. Planktonic Foraminifera

The application of planktonic foraminifera to Paleogene stratigraphy (Figure 28.1) has been primarily developed in two areas: the southern part of central Asia (e.g., Subbotina, 1953) and the Caribbean (e.g. Bolli, 1957a,b). In the late 1950s and 1960s, planktonic foraminifera became one of the main biostratigraphic tools, and assumed even greater importance with the advent of scientific ocean drilling. Paleogene planktonic foraminifera and associated zonations have been discussed by many authors (e.g., Bolli, 1966; Blow, 1979; Bolli and Saunders, 1985; Toumarkine and Luterbacher, 1985; Berggren *et al.*, 1985, 1995; Berggren and Miller, 1988; Olsson *et al.*, 1999; Berggren and Pearson, 2005; Pearson *et al.*, 2006; Wade *et al.*, 2011). The planktonic foraminiferal zonation used here is based on the most recent update of Wade *et al.* (2011). Planktonic foraminifera can be best used in open-marine deposits of the low-to-middle latitudes. Important marker species are generally absent in higher latitudes and zonations proposed for high latitudes are therefore considerably less detailed (Jenkins, 1985).

Estimates of the number of planktonic foraminifera species surviving the Cretaceous–Paleogene boundary mass extinction vary between authors, depending on the amount of taxa in basal Danian beds considered as either *in situ* or reworked (e.g., Olsson, 1970; Keller, 1988; Olsson and Liu, 1993; Arz *et al.*, 1999). The proposed survivors are small opportunistic species belonging to the genera *Heterohelix, Guembelitria,* and *Hedbergella*. The recovery of planktonic foraminifera during the Paleocene led to a high degree of diversification, thereby allowing a detailed biostratigraphic subdivision. Olsson *et al.* (1999) provided a state-of-the-art atlas of the taxonomy, phylogeny and biostratigraphy of Paleocene planktonic foraminifera. Several new genera developed during the early Paleocene, in particular forms with spinose tests (*Subbotina* and *Parasubbotina*), muricae (*Praemurica*), and smooth surfaces (*Globanomalina*) (Olsson *et al.*, 1999). The late early Paleocene (Zones P2–P3; ~61–62 Ma) is marked by the appearance and first diversification of the symbiont bearing genera *Acarinina*, *Igorina* and *Morozovella* (Norris, 1996; Quillévéré *et al.*, 2001). During the late Paleocene to early Eocene, tropical to subtropical open-marine assemblages are usually dominated by representatives of *Morozovella* and *Acarinina*. These genera provide numerous biozonal markers starting at the base of Zone P3a (*Morozovella angulata*; at ~62 Ma) up to middle Eocene Zone E10 (*Acarinina topilensis*; at ~43 Ma).

Members of the Paleogene Planktonic Foraminiferal Working Group recently published a comprehensive and excellently documented atlas of Eocene planktonic foraminifera (Pearson *et al.*, 2006). The atlas provides up-to-date views on their taxonomy, phylogeny and biostratigraphy. The basal Eocene (during the PETM) is characterized by an increase or even proliferation of *Acarinina*, both in continental margin basins (e.g., Molina *et al.*, 1999; Speijer *et al.*, 2000; Guasti and Speijer, 2007) and in the open ocean (e.g., Kelly *et al.*, 1996, 1998; Pardo *et al.*, 1997). Some taxa (*M. allisonensis*, *A. sibaiyaensis*, *A. africana* and *A. multicamerata*) flourished during, and are largely restricted to, the PETM (Kelly *et al.*, 1996; Berggren and Ouda, 2003; Guasti and Speijer, 2008) and allow a threefold subdivision of the former upper Paleocene to lower Eocene Zone P5 into three distinct zones (P5, E1, E2; Pardo *et al.*, 1999; Speijer *et al.*, 2000; Berggren and Pearson, 2005).

The early Eocene also witnessed the first appearance of new genera with a distinct morphology, such as *Catapsydrax*, *Globoturborotalita*, *Jenkinsella* and *Pseudohastigerina*. These were joined in the middle Eocene by e.g., *Clavigerinella, Globigerinatheka, Hantkenina, Morozovelloides and Turborotalia* (Pearson *et al.*, 2006). These genera provide a range of distinct middle to upper Eocene biostratigraphic markers, some of which have very short stratigraphic ranges (e.g., *Orbulinoides beckmanni*). Maximum diversity in Eocene planktonic foraminifera was reached during the early part of the middle Eocene, prior to the extinction of *Morozovella*, *Globanomalina* and *Igorina*. From then on, pulses of extinction led to generally less diverse planktonic assemblages and a series of last appearances constitute the majority of middle to upper Eocene biostratigraphic events (Pearson *et al.*, 2006).

The Eocene–Oligocene boundary is generally considered to correspond to the extinction of the last representatives of the genera *Hantkenina* and *Cribrohantkenina* (e.g., Wade and Pearson, 2008), although some studies suggest that this extinction is diachronous (e.g., van Mourik and Brinkhuis, 2005). The early Oligocene is generally dominated by small-sized low-diversity globigerinid assemblages, but the diversity and average size of the assemblages gradually recovered somewhat during late Oligocene, when non-carinate *Globorotalia* started to diversify (Bolli and Saunders, 1985).

Clearly, the numerous changes proposed in the foraminiferal biostratigraphy of the Oligocene (and to a lesser extent the Eocene) since GTS2004 indicate that it is not yet as well constrained as the biostratigraphy of the Paleocene. The reliance on numerous last appearances makes the middle Eocene and Oligocene biozonation scheme also more vulnerable to reworking.

28.2.1.2. Larger Benthic Foraminifera

Many of the west European centers in which the science of stratigraphy evolved during the 18th and 19th centuries are located on Paleogene strata rich in larger foraminifera. Hence, larger foraminifera played a key role in the development of the Paleogene stratigraphy, as demonstrated by the now abandoned term "Nummulitique", which corresponds to the "Paleogene". Larger foraminifera continue to be a decisive stratigraphic tool in shallow-marine low-to-middle latitude areas (Figures 28.1, 28.8). Zones for larger foraminifera are ideally based on successions of biometric populations within phylogenetic lineages; species are essentially morphometric units. During the second half of the 20th century, several monographs on larger foraminifera groups have considerably increased their stratigraphic usefulness (e.g., Hottinger, 1960, 1977; Less, 1987; Schaub, 1981).

The Danian and early Selandian larger foraminiferal assemblages are poorly differentiated and consist mainly of a few rotaliids. They mark the start of the recovery from the extinction of virtually all larger foraminiferal genera at the K/Pg boundary.

Within the Thanetian, the rapid radiations of the alveolinids and orthophragminids followed by that of the nummulitids at the onset of the Ypresian gave origin to several phylogenetic lineages within each group which allow a well-defined zonation of the upper Thanetian, Ypresian, Lutetian and lower Bartonian.

A major change took place within the Bartonian, with the extinction of *Alveolina* and large *Nummulites*. The younger part of the Bartonian and the Priabonian are characterized by relatively small nummulitids and other rotaliids and by orthophragminids. Within the larger foraminiferal assemblages, the Eocene−Oligocene boundary is not marked by drastic changes. The subdivisions of the Oligocene based on larger foraminifera are mainly based on lepidocyclinids, miogypsinids, and nummulitids.

A general larger foraminiferal zonation of the Paleocene and Eocene of the Tethyan area (between the Pyrenees and India) was proposed by Serra-Kiel *et al.* (1998) as one of the results of IGCP Project 286 *Early Paleogene Benthos* and provides a widely used reference framework. Cahuzac and Poignant (1997) proposed a similar larger foraminiferal zonation for the Oligocene−Miocene of the west European basins. The "letter-stage" subdivision of the Indo-Pacific Cenozoic (Leupold and van der Vlerk, 1936; Adams, 1970; Chaproniere, 1984; Boudagher-Fadel and Banner, 1999) is based on larger foraminifera. These are also largely used in the western hemisphere (Barker and Grimsdale, 1936; Butterlin, 1988; Caudri, 1996).

The Tethyan zonations proposed by Serra-Kiel *et al.* (1998) and the Oligocene part of the zonation by Cahuzac and Poignant (1997) are shown in Figure 28.8. The correlation between the planktonic zonations and those based on larger foraminifera is partly based on direct correlation (e.g., Molina *et al.*, 2003) and partly through correlation with magnetostratigraphic data (mostly from Spain: e.g., Burbank *et al.*, 1992; Molina *et al.*, 1992; Serra-Kiel *et al.*, 1994) and calibration with the integrated time scale of Berggren *et al.* (1995). It should be noted, however, that larger foraminiferal successions are generally discontinuous and that it is not possible to accurately pinpoint the positions of biozonal boundaries in the time scale with great confidence. This is particularly true for the Paleocene and the Oligocene.

For the Paleocene and Eocene, we largely followed the correlations between the Shallow Benthic Zonation (SBZ) scheme and the magnetostratigraphy and the partially indirect correlation with the oceanic planktonic schemes as indicated by Serra-Kiel *et al.* (1998), and modified this scheme to the updated magnetobiostratigraphic schemes. Around the Thanetian/Ypresian and Ypresian/Lutetian boundaries, new data led to improvements in the correlation scheme. Specifically, Serra-Kiel *et al.* (1998) correlated the basal Ilerdian SBZ5 with the upper Thanetian. It is now well established that this zone corresponds with the basal Ypresian and thus that the classical Ilerdian stage correlates entirely with the lower Ypresian (Scheibner and Speijer, 2009; Pujalte *et al.*, 2009).

28.2.1.3. Smaller Benthic Foraminifera

Smaller benthic foraminifera can be useful for correlation of both shallow- and deep-water Paleogene sedimentary sections. In the following text section we address calcareous and agglutinated benthics separately.

Agglutinated Foraminifera

The majority of Paleogene agglutinated taxa occupy deeper marine habitats on fine-grained siliciclastic wedges off continental margins. Hence, deeper marine shales in many petroleum basins are rich in deep-water agglutinated foraminifera (DWAF); the latter is particularly true for Paleogene sediments in basins around the North Atlantic, the Carpathians (flysch sediments), the Caribbean and New Zealand (e.g., Gradstein *et al.*, 1994; Hornibrook *et al.*, 1989). In general fine-grained, high-latitude sediments, poor in carbonate harbor diversified agglutinated benthic assemblages.

There is no major taxonomic turnover of DWAF at the K/Pg boundary, or in the Paleogene itself, although the PETM is marked by distinct faunal shifts, leading to temporary widespread blooms of *Repmanina* (*Glomospira*) *charoides* (e.g., Kaminski *et al.*, 1994; Galeotti *et al.*, 2004). Paleoenvironmental changes in basins, often the result of shallowing due to sediment infill, basin uplift, or the return of carbonate facies truncate the stratigraphic range of various taxa. Well-known examples of such regional stratigraphic truncations are the late Paleocene disappearance of the Lizard Springs assemblage of Trinidad and the middle-late Eocene disappearance of flysch-type agglutinated assemblages in the Carpathians.

There are over 150 stratigraphically useful cosmopolitan Paleogene deeper-water agglutinated benthic foraminifera, and many taxa have stratigraphic ranges that vary slightly or even markedly from basin to basin. However, correlation of wells within a basin can be accomplished with local zonations (e.g., Hornibrook *et al.*, 1989; Gradstein *et al.*, 1994; Kaminski and Gradstein, 2005). These zonations are particularly useful in regions where calcareous planktonic

Epoch	Stage	Substage	Zonation	Larger Benthic Foraminifera
	Aquitanian		SBZ 24	*Miogypsina gr. gunteri / tani*
Oligocene	Chattian		SBZ 23	*Miogypsinoides, Lepidocyclinids, Nummulites bouillci*
Oligocene	Chattian		SBZ 22 b	*Lepidocyclinids, Nummulites vascus* — *Cycloclypeus*
Oligocene	Rupelian		SBZ 22 a	*N. fichteli, N. bouillei* — *Bullalveolina*
Oligocene	Rupelian		? SBZ 21	*Nummulites vascus, N. fichteli*
Eocene	Priabonian		SBZ 20	*Nummulites retiatus, Heterostegina gracilis*
Eocene	Priabonian		SBZ 19	*Nummulites fabianii, N. garnieri, Discocyclina pratti minor*
Eocene	Bartonian		SBZ 18	*Nummulites biedai, N. cyrenaicus*
Eocene	Bartonian		SBZ 17	*Alveolina elongata, A. fragilis, A. fusiformis, Discocyclina pulcra baconica Nummulites perforatus, N. brogniarti, N. biarritzensis*
Eocene	Lutetian		SBZ 16	*Nummulites herbi, N. aturicus, Assilina gigantea, Discocyclina pulcra balatonica*
Eocene	Lutetian		SBZ 15	*Alveolina prorrecta, Nummulites millecaput, N. travertensis*
Eocene	Lutetian		SBZ 14	*Alveolina munieri, Nummulites beneharnensis, N. boussaci, Assilina spira spira*
Eocene	Lutetian		SBZ 13	*Alveolina stipes, Nummulites laevigatus, N. uranensis*
Eocene	Ypresian	Cuisian	SBZ 12	*Alveolina violae, N. manfredi, N. campesinus, N. caupennensis, Assilina major, A. cuvillieri*
Eocene	Ypresian	Cuisian	SBZ 11	*Alveolina cremae, A. dainellii, Nummulites praelaevigatus, N. nitidus, N. archiaci, Assilina laxispira*
Eocene	Ypresian	Cuisian	SBZ 10	*Alveolina schwageri, A. indicatrix, Nummulites burdigalensis burdigalensis, N. planulatus, Assilina placentula, Discocyclina archiaci archiaci*
Eocene	Ypresian	Ilerdian	SBZ 9	*Alveolina trempina, Nummulites involutus, Assilina adrianensis*
Eocene	Ypresian	Ilerdian	SBZ 8	*Alveolina corbarica, Nummulites exilis, N. atacicus, Assilina leymeriei*
Eocene	Ypresian	Ilerdian	SBZ 7	*Alveolina moussoulensis, Nummulites praecursor, N. carcasonensis*
Eocene	Ypresian	Ilerdian	SBZ 6	*Alveolina ellipsoidalis, A. pasticillata, Nummulites minervensis*
Eocene	Ypresian	Ilerdian	SBZ 5	*Orbitolites gracilis, Alveolina vredenburgi, Nummulites gamardensis*
Paleocene	Thanetian		SBZ 4	*Glomalveolina levis, Nummulites catari, Assilina yvettae*
Paleocene	Selandian		SBZ 3	*Glomalveolina primaeva, Fallotella alavensis, Miscellanea yvettae*
Paleocene	Selandian		SBZ 2	*Miscellanea globularis, Ornatononion minutus, Paralockhartia eos, Lockhartia akbari*
Paleocene	Danian		SBZ 1	*Bangiana hanseni, Laffitteina bibensis*

FIGURE 28.8 **Paleogene zonation of larger benthic foraminifera with selected taxa.**

microfossils are generally rare and only present in narrow stratigraphic intervals. For example, in exploration wells in deep-marine sediments in the central and northern North Sea, *Adercotryma agterbergii* is a useful marker for the lower Oligocene (Rupelian), *Ammomarginulina aubertae* for the middle Eocene, *Reticulophragmium intermedium* for the lower Eocene (approx. Ypresian), *R. pauperum* and *Rzehakina minima* for the upper Paleocene (Thanetian), and *Ammoanita ingerlisae* for the middle Paleocene (approx. Selandian).

Smaller Calcareous Benthic Foraminifera

Just as investigations of the agglutinated taxa received a boost in the late 1970s and 1980s from petroleum exploration in North Atlantic and other continental margin basins and deep-sea scientific drilling, the study of calcareous smaller benthic foraminifera also advanced significantly. Remarkably diverse bathyal–abyssal Paleogene assemblages were discovered that turned out to have some stratigraphic potential, although rare or patchy single taxa distributions require the use of zonal assemblages. The assemblages at the same time track important paleoceanographic changes. The most important turnover of the calcareous benthic fauna took place during the Paleocene–Eocene thermal maximum (see Section 28.3.2.). At that time, benthic deep-sea fauna at many sites experienced a severe reduction in diversity and composition with a ~40% extinction of many Velasco-type taxa (e.g., Thomas and Shackleton, 1996; Thomas, 2007). Well known deep-sea taxa that became extinct are *Stensioeina beccariiformis, Angulogavelinella avnimelechi, Cibicidoides velascoensis*, *Bulimina velascoensis*, *Pullenia coryelli* and *Aragonia velascoensis*. At shallower depths extinction during the PETM was less severe, although marked faunal turnovers are observed in shallow continental margin sequences (e.g., Speijer *et al.*, 1996; Stassen *et al.*, 2012).

The genus *Turrilina* with the species *T. brevispira* and *T. robertsi* in the uppermost Paleocene through Eocene (planktonic foraminifer Zones P5–E14), the species *T. alsatica* in the lower Oligocene (Rupelian), and the related *Rotaliatina bulimoides* in middle Eocene through lower Oligocene are examples of taxa that have global correlation potential in deep neritic and bathyal environments, particularly in mid-high latitudes with shale facies. Similarly, the extinction of *Nuttallides truempyi* during a long-term turnover at the end of the Eocene (Thomas, 2007) is a useful bathyal–abyssal event, as is the appearance of the genus *Siphonina* in the late Eocene, and of *Sphaeroidina* in the early Oligocene.

28.2.2. Calcareous Nannoplankton

The recognition of calcareous nannofossils as a useful tool for biostratigraphic correlations is generally credited to Bramlette and co-workers (Bramlette and Riedel, 1954; Bramlette and Sullivan, 1961; Bramlette and Wilcoxon, 1967). Following these pioneer efforts, intensive taxonomic and biostratigraphic studies have been carried out which formed the basis for the presently existing calcareous nannofossil biozonal schemes for the Mesozoic and Cenozoic.

28.2.2.1. Development of Zonations

In the middle of the last century, several authors described species that are important for the correlation of Paleogene marine sediments in Europe and North America (Bramlette and Riedel, 1954; Deflandre and Fert, 1954; Martini, 1958, 1959a,b; Stradner, 1958, 1959a,b; Brönnimann and Stradner, 1960). In the 1960s, significant contributions to Paleogene zonations were made (Bramlette and Sullivan, 1961; Bystricka, 1963, 1965; Hay, 1964; Sullivan, 1964, 1965; Bramlette and Wilcoxon, 1967; Hay and Mohler, 1967; Hay *et al.*, 1967). But it was only in the 1970s with the onset of scientific ocean drilling that calcareous nannofossils became one of the most important biostratigraphic microfossil groups.

The two most widely used Paleogene zonal schemes are the standard NP-zonation of Martini (1971), and the CP-zonation of Bukry (1973, 1975; emended by Okada and Bukry, 1980) (Figure 28.1). Martini's zonation relied on studies of land sequences from largely temperate areas, whereas Bukry's zonation was developed in low-latitude oceanic sections.

High-resolution studies (e.g., Romein, 1979; Perch-Nielsen, 1985; Varol, 1989; Aubry, 1996) redefined and subdivided these zones. For example, a fourfold subdivision was proposed for Zone NP10, and Zones NP19 and NP20 were combined because *Sphenolithus pseudoradians* is an unreliable marker at the base of Zone NP20. Aubry and Villa (1996) emended Zone NP25 in the late Oligocene with a threefold subdivision that accounts for the definition of the Chattian–Aquitanian boundary (Oligocene–Miocene boundary). Fornaciari and Rio (1996) proposed a twofold subdivision of Zone NP25 for the Mediterranean region, and redefined the top of the zone with the last common occurrence (LCO) of *Reticulofenestra bisecta*, because the zonal marker species *H. recta* extends well into the Miocene (Rio *et al.*, 1990). The LCO of *R. bisecta* is very close to the Oligocene–Miocene boundary (Berggren *et al.*, 1995).

The detailed scheme of calcareous nannofossil zones and key events in Figure 28.1 is largely based on the magnetobiostratigraphic correlations of the Pacific Equatorial Age Transect (PEAT) program (Pälike *et al.*, 2010).

Calcareous nannoplankton responds quickly to changes in the thermal structure of oceanic water masses. As a consequence, the presence and stratigraphic ranges of marker species strongly differ from low to high latitudes. Therefore, the "standard zonal schemes" are only partially applicable to sedimentary sequences of the North Sea and the sub-Antarctic South Atlantic and modified Paleogene zonations have been proposed for these areas (Wise, 1983; van Heck and Prins, 1987; Varol, 1989, 1998; Wei and Wise, 1990a,b; Crux, 1991; Wei and Pospichal, 1991; Steurbaut, 1998).

28.2.2.2. Evolutionary Trends

Following the biotic crisis at the K/Pg boundary, calcareous nannoplankton underwent a major diversification in the early-middle Paleocene giving rise to several key genera for biostratigraphy (e.g., *Fasciculithus, Chiasmolithus, Sphenolithus, Discoaster*, and *Helicosphaera*). A major taxonomic turnover took place at the Paleocene–Eocene boundary interval in response to the Paleocene–Eocene thermal maximum (PETM). The *Rhomboaster–Tribrachiatus* lineage and the *Discoaster* radiation characterize the zones at the Paleocene–Eocene transition (Angori and Monechi, 1996; Aubry, 1996, 1998a).

While the early Paleogene calcareous nannoplankton evolution reflects an increasing temperature trend and oligotrophic conditions, the late Paleogene calcareous nannoplankton evolution is influenced by climatic deterioration and eutrophication. A progressive decline in diversity with a low rate of evolution characterizes this interval (Aubry, 1992, 1998b). A pronounced taxonomic turnover occurred near the middle-late Eocene boundary. A sharp impoverishment in species of the genus *Discoaster* took place prior to the Eocene–Oligocene boundary with the extinction of the last two representatives of the rosette-shaped discoasters *D. barbadiensis* and *D. saipanensis* (Monechi, 1986; Nocchi *et al.*, 1988).

A large number of nannofossil extinctions took place in the early Oligocene. This decreasing diversity trend continued through the Oligocene, in particular at high latitudes; it is actually reversed at low latitudes with a radiation of the genera *Sphenolithus* and *Helicosphaera*. A succession of several last occurrences characterizes the Oligocene–Miocene transition.

Recent onshore drilling in Tanzania revealed extremely diverse tropical Paleogene nannofossil assemblages with numerous newly described species and genera (Bown, 2005, Bown and Dunkley-Jones, 2006). These data suggest that in oceanic sediments the tiniest and most fragile species are generally selectively removed from the assemblage during and after deposition. This also means that in an absolute sense Paleogene nannofossil diversity might be significantly underestimated.

28.2.3. Radiolaria

(Revised by C.J. Hollis)

Investigations contributing to a detailed Cenozoic radiolarian biostratigraphy have been carried out largely on deep-sea sediment samples in which calcareous microfossils co-occur, or on cores with magnetostratigraphy or other methods for estimating sediment age. This calibration is essential because radiolarians are commonly absent in stage stratotypes. For this reason, a two-step correlation via calcareous nannofossils is usually unavoidable. The latter group can then be tied to the geomagnetic polarity time scale (GPTS), epoch/series boundaries, and numerical age estimates, thus allowing for accurate correlation of zonations across paleobiogeographic boundaries.

Riedel (1957) was the first to realize the potential of radiolarians for stratigraphic purposes. The Cenozoic zonation of Riedel and Sanfilippo (1970, 1971) has received only minor modifications and additions (Moore, 1971; Nigrini, 1971, 1974; Foreman, 1973; Maurasse and Glass, 1976; Saunders *et al.*, 1985) and is the standard scheme for low-latitude regions (Sanfilippo *et al.*, 1985).

A late Eocene to Pleistocene northern high-latitude zonation was developed for the Norwegian and Greenland Seas by Bjørklund (1976). Independent zonations for the Antarctic sediments have been established by Petrushevskaya (1975), Takemura (1992), and Abelmann (1990) who proposed two zones for the upper Oligocene. Takemura and Ling (1997) extended the Paleogene high-latitude Southern Ocean zonation into the late Eocene mainly based on material from ODP Legs 114 and 120.

Foreman (1973) created the first lower Paleogene low-latitude radiolarian zonation based on material from DSDP Leg 10 in the Gulf of Mexico. The lowermost upper Paleocene *Bekoma campechensis* Zone was introduced by Nishimura (1987), who subsequently (1992) subdivided this zone into three subzones. Hollis (1993, 1997) established the first complete Paleocene zonation based on onshore Cretaceous–Paleogene sections in New Zealand and DSDP site 208, Lord Howe Rise. This zonation may have wider application to mid to high latitudes. However, radiolarian-bearing lower Paleocene records have been encountered only infrequently beyond the Southwest Pacific (e.g., Hollis and Kimura, 2001). Kozlova (1999) established a boreal zonation for the lower Paleogene based on previous biostratigraphic studies by herself and other Russian micropaleontologists of Paleocene and Eocene strata in Russia and adjoining regions.

For the Paleogene record, where magnetostratigraphic data are not available for most radiolarian sequences, Sanfilippo and Nigrini (1998) used a combination of data from previous literature and an unpublished integrated compilation chart based on data from DSDP/ODP Legs 1–135 to construct a composite chronology of radiolarian zonal boundary events tied to numerical ages. Sanfilippo and Nigrini (1998) also introduced code numbers for the radiolarian zonation for the low-latitude Pacific, Indian and Atlantic oceans (RP1–RP22 for the Paleogene and RN1–RN17 for the Neogene). This low-latitude zonation and the mid-latitude South Pacific Paleogene radiolarian zonations (Hollis, 1993, 1997, 2002; Hollis *et al.*, 1997, 2005), which are both calibrated to planktonic foraminiferal and calcareous nannofossil zones, are shown in Figure 28.9. A limited number of corrections have been made compared to GTS2004.

Study of a nearly complete radiolarian record for the upper Lower Paleocene to upper Middle Eocene (Sanfilippo and Blome, 2001) in ODP Hole 1051A (Leg 171B), western North Atlantic, resulted in a well-resolved succession of 200 bioevents for low-latitude zones RP6 to RP16. Comparison of the stratigraphic ranges of species from Hole 1051A with those from the tropics indicates that a high proportion of these species have diachronous first and/or last occurrences. Similarly, delayed first occurrences and early last occurrences are a feature of radiolarian index species in the South Pacific Paleogene. As an example, nine species have first occurrences at the Paleocene–Eocene boundary in New Zealand (Hollis, 2006), whereas five of these species first occur in the late Paleocene in low-latitudes. This appears to reflect the poleward expansion of the ranges of warm-water radiolarians at times of pronounced global warmth such as the Paleocene–Eocene thermal maximum (PETM). Two well-known radiolarian index species have first occurrences at the onset of the PETM: *Podocyrtis papalis* and *Phormocyrtis turgida* (Sanfilippo and Blome, 2001; Hollis, 2006).

28.2.4. Dinoflagellate Cysts

Applications of fossil dinocysts in global Paleogene biostratigraphy and paleoecology have been reviewed in detail in several papers, including Williams and Bujak (1985), Powell (1992), Stover *et al.* (1996), Williams *et al.* (1998, 2004), and Fensome and Williams (2004). These compilations indicate that the stratigraphic range of a given Paleogene dinocyst is rarely synchronous worldwide. Many authors have demonstrated climatic and environmental control on the stratigraphic distribution of taxa in the Paleogene (e.g., Brinkhuis, 1994; Wilpshaar *et al.*, 1996; Bujak and Brinkhuis, 1998; Crouch *et al.*, 2001). Several dinocyst papers also deal with the differentiation of northern and southern hemisphere assemblages and/or endemic Antarctic assemblages (e.g., Wilson *et al.*, 1998; Wrenn and Hart, 1988; Truswell, 1997; Guerstein *et al.*, 2002; Brinkhuis *et al.*, 2003a,b).

Williams *et al.* (2004) recognized the need to accommodate both latitudinal and hemispherical control of dinocyst assemblages in Paleogene distribution charts. Accordingly, these authors give ranges for low, mid, and high latitudes in both northern and southern hemispheres, and the contribution also updates data presented in Williams *et al.* (1998).

Information concerning Paleogene dinocyst distribution is most comprehensive for the mid latitudes of the northern hemisphere. This is a reflection of the more intense study of assemblages from these regions, notably from northwest Europe (particularly the greater North Sea Basin). Although first-order calibration of dinocyst events against magnetostratigraphy is largely absent, reliable age control is possible through published studies on the biostratigraphy of type sections of the Paleogene stages. This is supplemented by large volumes of largely unpublished subsurface data. The succession of dinocyst events offshore of northwest Europe is becoming better documented in the public domain (e.g., Harland *et al.*, 1992; Mudge and Bujak, 1996; Neal, 1996; Powell *et al.*, 1996; Mangerud *et al.*, 1999). An overview of Paleogene index dinocyst events for northwest Europe is presented in Figure 28.9.

Recent high-latitude ocean drilling (ODP leg 189, Tasman Plateau; IODP 302, Lomonosov Ridge, Arctic Sea) has been providing a wealth of new high-latitude paleoenvironmental and stratigraphic data, especially on the early to middle Eocene (e.g., Sluijs *et al.*, 2006, 2009, 2011; Warnaar *et al.*, 2009), leading to a better global coverage of dinocyst distribution. A compilation of biostratigraphic ranges for five different latitudinal zones (low latitude and high and mid latitude for each hemisphere) is provided by Brinkhuis *et al.* (2009).

28.2.5. Mammals

(Revised by J.J. Hooker)

During the Paleogene, the various continental masses had distinctive land mammal faunas. These exhibit rapid evolution and have been much used for the correlation of nonmarine strata. Inter-continental correlation, however, has often proved problematic owing to endemism, except during geologically brief periods of faunal inter-change facilitated by paleogeographic features such as land bridges.

Because mammals are generally rarer as fossils than are invertebrates or microbiota, and because of the often laterally discontinuous nature of continental strata, occurrences may be in isolated exposures whose superpositional relationships are unknown. A notable exception, however, is the stratigraphically and geographically extensive sequences in western North America. Solutions to the problems of correlating isolated mammalian fauna have varied.

Series of broad biochronological–biostratigraphic units known as "land mammal ages" have been widely applied, with a separate series in each continent for North America (NALMA), Europe (ELMA), Asia (ALMA), and South America (SALMA). These can stand independently when correlation to standard marine biostratigraphies and to global chronostratigraphy–geochronology is uncertain.

Owing to endemism, smaller biostratigraphic–biochronological units vary from having continent-wide applicability to only local use. These may be conventional biozones or, commonly in Europe, reference levels (MP). Reference levels purport to order superpositionally isolated fauna according to evolutionary grade, avoiding the problem of fixing boundaries (Schmidt-Kittler, 1987). In practice, because of referral of a fauna other than the reference fauna to a given reference level, a temporal range is spanned and

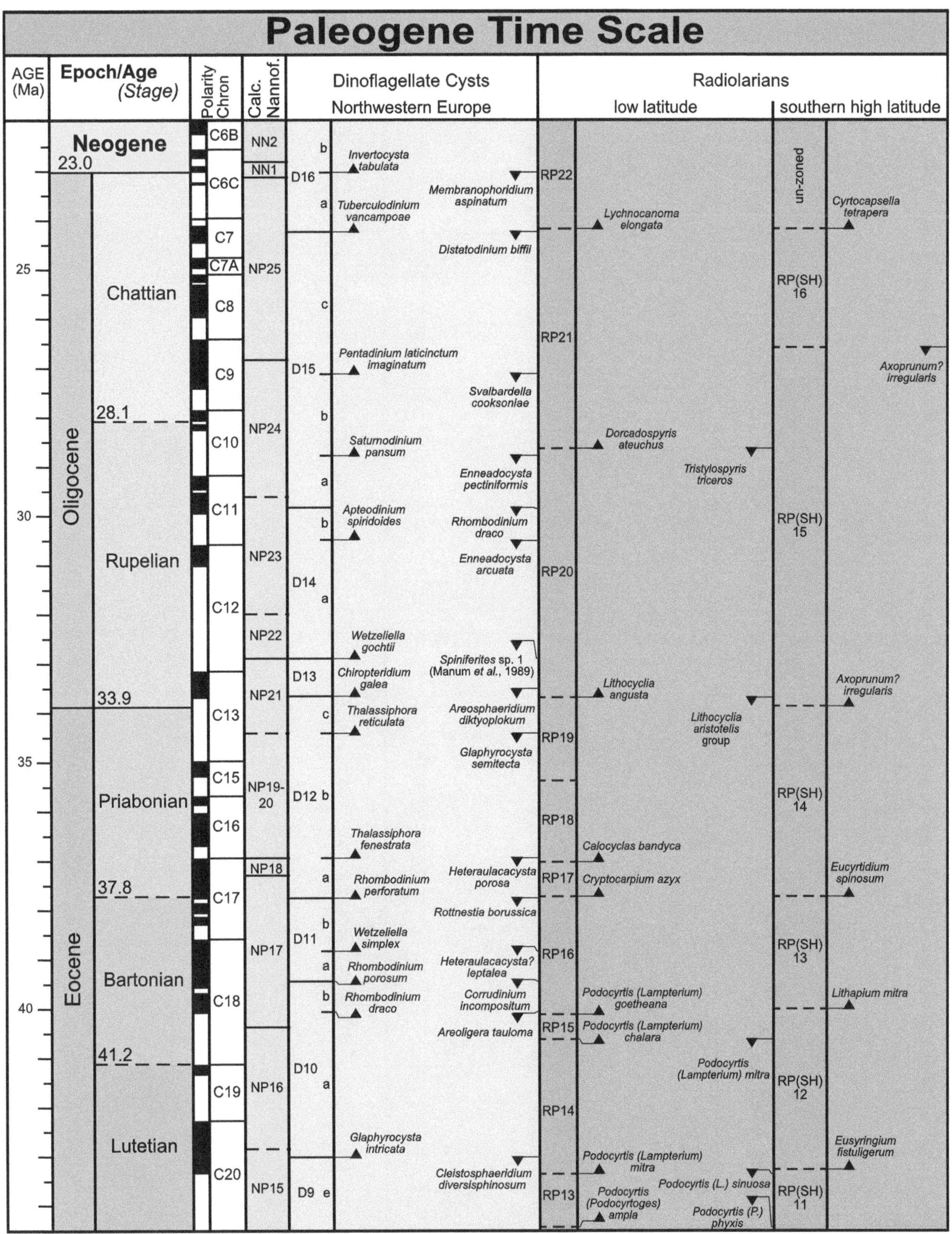

FIGURE 28.9 **Paleogene dinoflagellate cyst zonation and datums and radiolarian zonation, with their estimated correlation to magnetostratigraphy and calcareous nannoplankton zones.** Dinoflagellate stratigraphy was compiled by A. J. Powell and H. Brinkhuis.

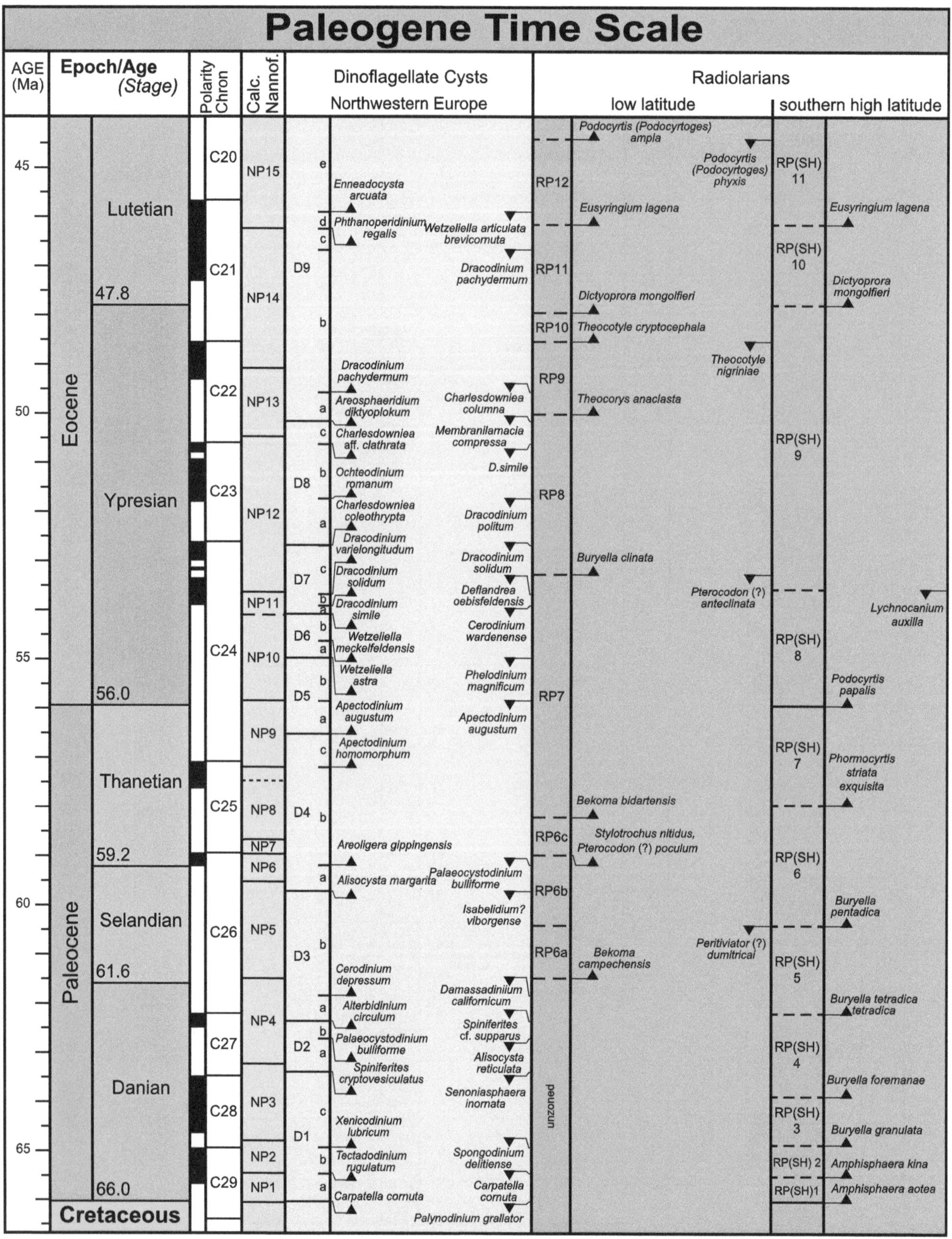

FIGURE 28.9 (*Continued*).

a reference level is thus used in much the same way as a standard assemblage biozone (Aguilar *et al.*, 1997). An alternative solution to the isolation problem is the application of parsimony to concurrence (Alroy, 1992).

Calibration of mammalian biostratigraphic–biochronological systems to the geomagnetic polarity time scale is most extensively documented in North America and Europe (Aguilar *et al.*, 1997; Janis, 1998; Secord, 2008; Smith *et al.*, 2008; Gunnell *et al.*, 2009) and recently also in Asia (Ting *et al.*, 2011). It has been achieved either directly or via links with other biostratigraphies, through intercalation with marine strata or through co-occurrence of mammalian and other zonal indicators in paralic facies (e.g., Smith and Smith, 2003; Hooker and Weidmann, 2007). Radiometric dating is helping parts of the European, central Asian, South American, and Australasian sequences (Flynn and Swisher, 1995; Aguilar *et al.*, 1997; Kay *et al.*, 1999; Franzen, 2005), which are otherwise still proving difficult to correlate globally. The broadly construed Casamayoran unit was until recently thought to be early Eocene in age. One of its two component SALMAs, the Barrancan, has now been radiometrically dated, which shifts it to late Eocene, thereby shifting the Mustersan (Kay *et al.*, 1999). No LMA or zonal system currently exists in Africa, but magnetostratigraphic and sequence stratigraphic studies are improving calibration (Gingerich, 1992; Kappelman *et al.*, 1992; Gheerbrant *et al.*, 1998).

Three major faunal turnovers occurred in the Paleogene, at or close to epoch/series boundaries. The first, at the beginning of the Paleogene, set the scene for rapid evolutionary radiation and continental endemism following the major tetrapod extinctions. Most of this Paleocene radiation is recorded in the first one and a half million years of the epoch in North America and is represented by the Puercan NALMA. Correlation of the Puercan to continental strata elsewhere in the world is difficult.

The second major faunal turnover at the Paleocene–Eocene boundary coincides with the carbon isotope excursion associated with the PETM. The turnover involved similar innovations in all three northern hemisphere continents, and is known as the Mammalian Dispersal Event (MDE). This marks the first appearance of many modern mammalian orders, especially primates, bats, artiodactyls, perissodactyls, and proboscideans. The PETM and MDE coincide with the beginning of the Wasatchian NALMA (Bowen *et al.*, 2001) and probably also of the Neustrian ELMA (Hooker, 1998). The end of the Wasatchian has been recently found to be older than previously thought thanks to magnetostratigraphic studies (Clyde *et al.*, 2001).

The third major faunal turnover, at the end of the Headonian, in the earliest Oligocene, was less widespread, but well represented in Europe where it is known as the Grande Coupure. It marks the extinction of many endemic European taxa and the incoming of new ones from Asia. A supposedly contemporaneous major turnover in central Asia, termed the Mongolian Remodelling (Meng and McKenna, 1998) is now better dated radiometrically and magnetostratigraphically (Kraatz and Geisler, 2010) and may be younger than the Grande Coupure. A less major turnover, but with some elements similar to the Grande Coupure, occurs slightly earlier at the beginning of the Ergilian (Dashzeveg, 1993).

Figure 28.10 shows the current state of knowledge on calibration of LMAs and relevant biozones (concurrent range zones, CRZ, and interval zones, IZ).

28.3. PHYSICAL STRATIGRAPHY

28.3.1. Magnetostratigraphy

The reversal history of the Earth's magnetic field in the form of the GPTS has been the backbone of the Cenozoic and, hence, the Paleogene time scale ever since it was constructed for the first time (Heirtzler *et al.*, 1968). Until 1990, the GPTS was established by applying a limited number of radio-isotopically dated age calibration points in seafloor anomaly profiles; the combination of radio-isotopic age determinations and polarities of lava samples was used to construct the GPTS for the last 5 million years (e.g., Mankinen and Dalrymple, 1979).

The correlation of magnetostratigraphic records of land-based sections and deep-sea cores to the GPTS provides numerical ages for bioevents and chronostratigraphic boundaries, and is used to construct integrated magneto-biochronostratigraphic scales. The role of magnetostratigraphy has changed with the advent of astronomical dating; it now directly provides numerical ages for reversal boundaries through (linear) interpolation of the sedimentation rate between astronomically dated calibration points. This has become a standard approach for the Neogene and is now being extended into the Paleogene and even already into the Cretaceous. As a consequence, magnetostratigraphy of cyclic deep-marine carbonate successions has become more important for time scale construction itself. In this respect, ODP Site 1218 in the equatorial Pacific provides an outstanding example, offering an excellent high-resolution magnetobiostratigraphy and an equally high-resolution and high-quality stable isotope record for the entire Oligocene and Lower Miocene, which in combination with other proxy records has been tuned to astronomical target curves (Pälike *et al.*, 2006). The classical magnetostratigraphy of the Contessa Highway section has been considerably improved for the Middle Eocene to investigate the Middle Eocene Climatic Optimum (Jovane *et al.*, 2007). The K/Pg boundary section of Zumaia provided a reliable magnetostratigraphy for the entire Paleocene (Dinares-Turell *et al.*, 2003, 2007), and has further been improved using an integrated stratigraphic approach (Dinares-Turell *et al.*, 2010). All the above

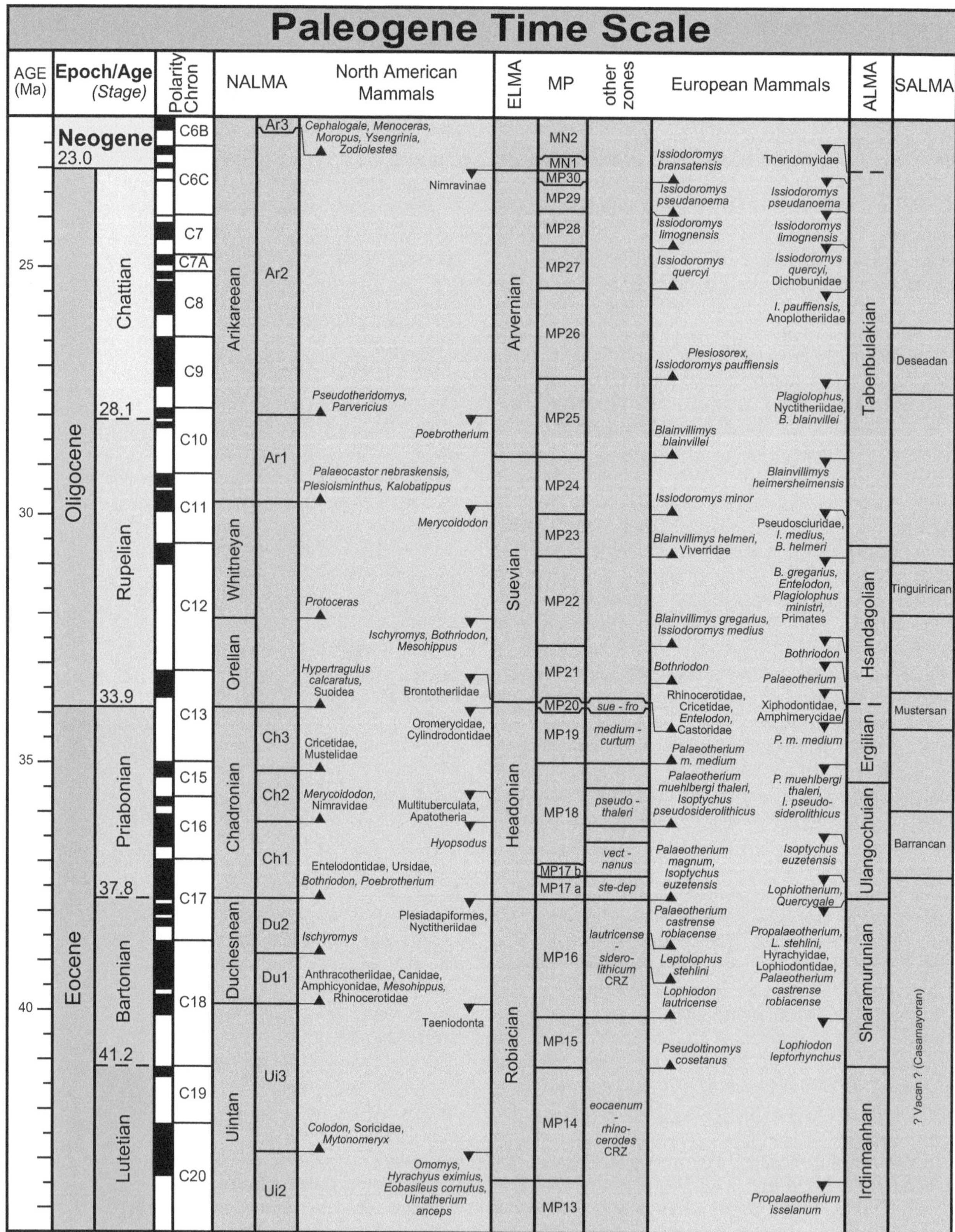

FIGURE 28.10 **Mammalian zonations and biostratigraphy and events of the Paleogene.** NALMA, North American Land Mammal Ages; SALMA, South American Land Mammal Ages; ALMA, Asian Land Mammal Ages; ELMA, European Land Mammal Ages; MP, European Reference Levels. The sequence of NALMAs and their subdivision follows Woodburne (1987) with calibrations to the geomagnetic polarity time scale by Butler *et al.* (1987), Prothero and Emry (1996), Williamson (1996), Clyde *et al.* (2001), Clemens (2002), Secord (2008), Smith *et al.* (2008) and Gunnell *et al.* (2009). The sequence of ELMAs follows Fahlbusch (1976) with some modifications in the Eocene (see Savage and Russell, 1977; Franzen and Haubold, 1987; Hooker 1998). They are divided into MP reference levels (Schmidt-Kittler, 1987; Aguilar *et al.*, 1997) and partly into biozones (Hooker, 1986, 1987, 1996). Several Paleocene localities with significant mammals are tabulated as local fauna (LF): Fontllonga 3 (Peláez-Campomanes *et al.*, 2000); Hainin (Sigé and Marandat, 1997; Steurbaut, 1998), Menat (Gingerich, 1976), and Walbeck (references in Hooker, 1991). Calibration to the GPTS varies from direct (Engesser and Mödden, 1997; Legendre and

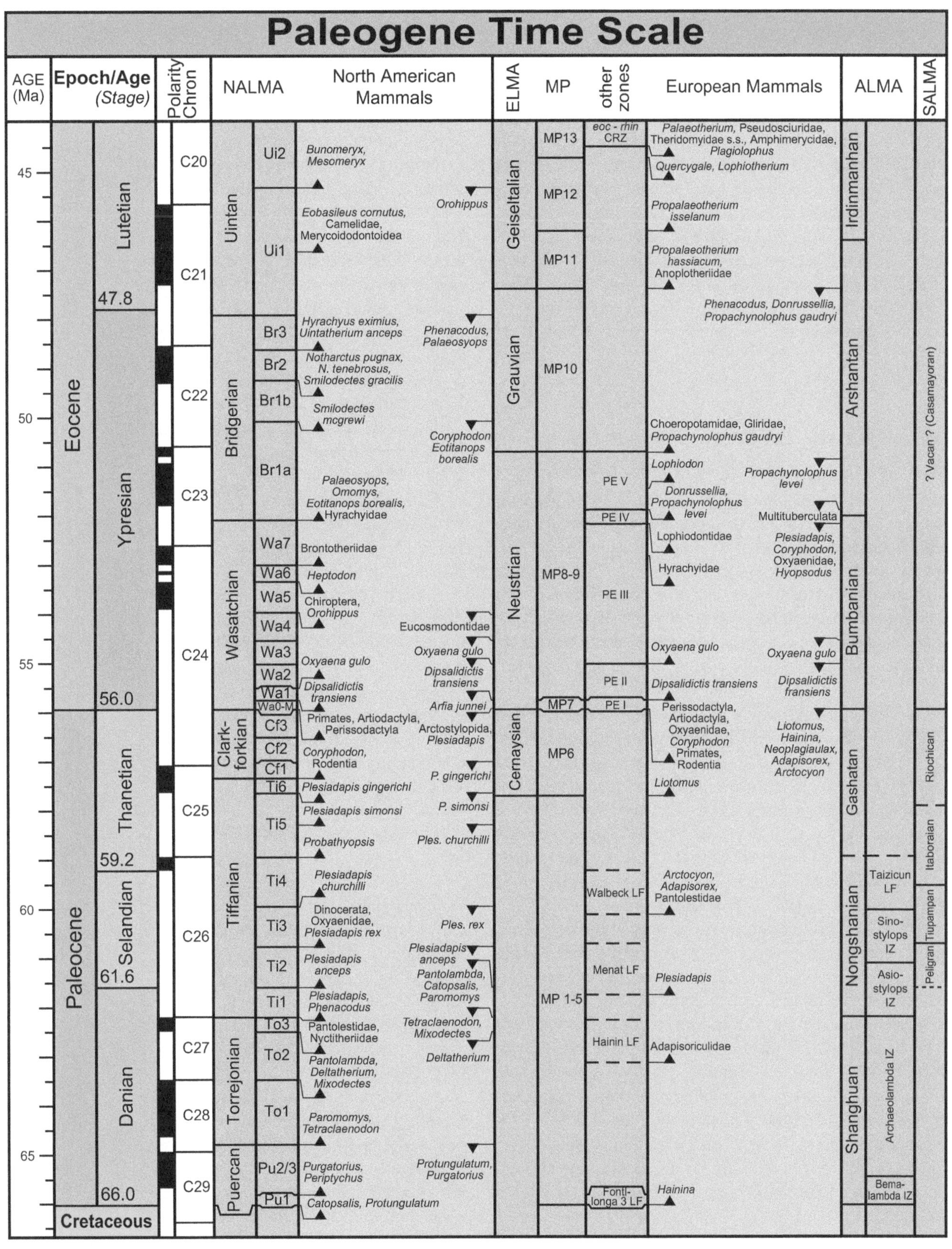

Levêque, 1997; Hooker, 1998; López-Martínez and Peláez-Campomanes, 1999; Hooker and Millbank, 2001) to indirect (Hooker, 1986; Franzen and Haubold, 1987; Steurbaut, 1992; Mertz *et al.*, 2000). Calibration of ALMAs and their subdivisions essentially follows Holroyd and Ciochon (1994), Meng and McKenna (1998) and Ting *et al.* (2011). That of SALMAs follows Flynn and Swisher (1995) and Kay *et al.* (1999). In the case of Europe and North America, selected first and last appearances are given. They are intended to reflect a balance between those recording important faunal changes, used in biostratigraphy and inter-continental correlation, and those highlighting inter-continental diachronism. Sources additional to the above references are Köhler and Moyá-Solá (1999) and Stucky (1992).

magnetostratigraphic records provide the basis for attempts to construct an astronomical – polarity – time scale for the Paleogene (see Section 28.4).

In addition, magnetostratigraphic records are still extremely useful for generating magnetostratigraphic age models for sedimentary successions based on their correlation to the GPTS (or APTS in case reversal ages are based on tuning). Paleogene examples are the magnetostratigraphic framework established for the Bighorn Basin in North America to unravel faunal, climate and basin evolution (Clyde *et al.*, 2007), for the Tibetan Plateau to constrain climate and uplift history (Dupont-Nivet *et al.*, 2007; Wang *et al.*, 2008), for Cenozoic ocean drilling cores to constrain planktonic foraminiferal biochronology (Wade *et al.*, 2011), and for Antarctic cores to reconstruct glaciation history (Florindo *et al.*, 2005).

28.3.2. Climatic Evolution and Events

Carbon and oxygen stable isotope analyses are routinely performed in many stratigraphic studies of Paleogene sequences. The data reflect respectively biological productivity, changes in the carbon cycle and climatic changes, all related to each other (see Chapter 11).

An integral perspective on the climate evolution during the Paleogene has been possible through the study of oxygen isotopes of deep-sea benthics in ocean drill cores; the data are represented on classical curves by Miller *et al.* (1987) and Zachos *et al.* (1993, 1999) (Figure 28.11).

After the climatic optimum in the Late Cretaceous and a relative cooling during the Maastrichtian, the isotope shift to lighter oxygen points to a general warming during the Paleocene and the lower Eocene where the warmest times of the whole Paleogene are recorded between about 50 and 52 Ma ago. From this Early Eocene Climatic Optimum (EECO) onwards, the oxygen becomes heavier, pointing to a cooling of the climate. At the Eocene–Oligocene transition a major jump to an even cooler climate is observed, which recovers to late Eocene temperatures only at the end of the Oligocene, before renewed cooling takes place in the beginning of the Neogene. These long term trends are generally considered to be related to important tectonic phenomena. The North Atlantic rifting and associated volcanism at the end of the Paleocene and the beginning of the Eocene could explain the early Eocene warm times, whilst the reduction in seafloor spreading rates during the later Eocene is associated with the cooling; the plate evolution with the opening of the Tasmania and Drake passages and the installation of a circum Antarctic ocean current is commonly associated with the sudden cooling in the beginning of the Oligocene (see Zachos *et al.*, 2001).

Superposed on this general climatic trend are several short-lived events, taking only some tens to a few hundred thousand years, and observed in different paleogeographic settings. As this type of research is rapidly developing, the nomenclature and coding in use for these events have not settled. The events recognized are discussed from old to young (Figure 28.11). The Dan-C2 is a sequence of two short events observed in both carbon and oxygen isotopes and is thought to represent a redistribution of existing carbon in the biosphere at about 65.2 Ma (Quillévéré *et al.*, 2008). Slightly preceding the Selandian, the Latest Danian Event (LDE) is recognized at about 62 Ma in the Tethys, Atlantic and Pacific; it is situated within Zone P3, in and close to the base of C26r (Bornemann *et al.*, 2009; Westerhold *et al.*, 2011). The LDE appears to correlate to the mammal transition from Torrejonian to Tiffanian in North America and from Shanghuan to Nongshanian in Asia (Clyde *et al.*, 2008, 2010). A thin clay-rich dissolution layer at Shasky Rise was described as the Mid-Paleocene Biotic Event (MPBE) or Early Latest Paleocene Event (ELPE) (Bralower *et al.*, 2002; Petrizzo, 2005). The event has also been reported from the Zumaia section (Bernaola *et al.*, 2007). The position of the event is situated close to the top of C26r (Bernaola *et al.*, 2007; Westerhold *et al.*, 2011). According to Westerhold *et al.* (2011) data are still too sparse to conclude that the ELPE was a global hyperthermal-type event (Figure 28.11).

Definitely the most discussed and most widely recognized event is the Paleocene–Eocene Thermal Maximum (PETM) at the base of the Eocene. Albeit short lived, it is characterized by the highest global temperatures of the Cenozoic. The onset of a 2–3‰ negative carbon isotopic excursion (CIE) marks the Paleocene–Eocene boundary (Kennett and Stott, 1991; Schmitz *et al.*, 2001). The isotopic anomaly developed within a few thousand years in both the deep and the shallow ocean indicating that the entire oceanic carbon reservoir experienced a rapid change in isotopic composition. It took ~200 kyr before $\delta^{13}C$ returned to its background levels. The oceanic perturbation is also shown by benthic fauna turnovers and dissolution phenomena leading to changing carbonate compensation depth, phenomena also observed with other hyperthermal events. The $\delta^{13}C$ anomaly has been used for marine–terrestrial correlations (Koch *et al.*, 1992). This study showed that the Clarkforkian–Wasatchian land-mammal turnover and the major extinction event among benthic foraminifera in the deep sea at ~56 Ma are more or less coeval, triggered by the same global environmental perturbation.

Another hyperthermal event following the PETM, is the Eocene Thermal Maximum 2 (ETM2, including the ELMO bed at Walvis Ridge) or the $\delta^{13}C$ excursion H1. It is followed after about 100 kyr by another isotopic excursion, known as H2, occurring just below the top of C24r. The ETM2 occurs about 2 myr after the PETM and it has similar isotopic and microfaunal characteristics, although they are less pronounced (Lourens *et al.*, 2005; D'haenens *et al.*, 2012). Lourens *et al.* (2005) suggested that both the PETM and

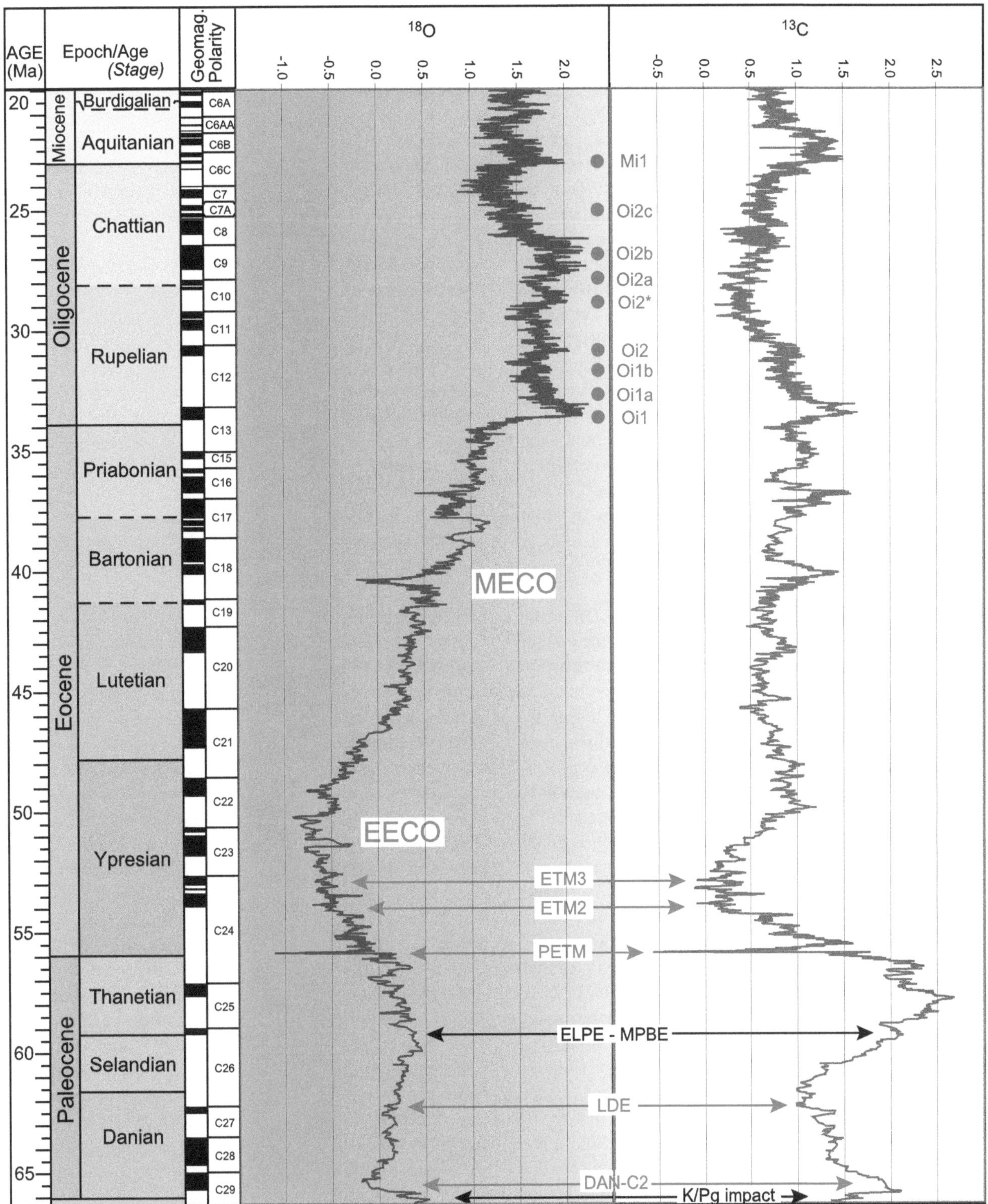

FIGURE 28.11 **Position of main Paleogene climatic events against the generalized oxygen and carbon isotope curves for the Paleogene.** Oxygen and carbon isotopic curves are from Cramer *et al.* (2009); original carbon and oxygen isotope values are recalibrated to GTS2012 and the curves shown are obtained by a 9-point moving window. The references for the positioning of the climatic events are discussed in the text (Section 28.3.2).

ELMO are paced by astronomical cyclicity, although this is controversial (Westerhold *et al.*, 2007).

ETM3, also known as the K or X event, has been recognized in C24n.1n (Agnini *et al.*, 2009). The H and K codes originate from an analysis by Cramer *et al.* (2003), who recognized almost 20 rapid negative $\delta^{13}C$ excursions, labeled A to L in ocean drilling cores within the interval of C24n–C25n time. These authors have argued that the negative excursions were related to maxima in eccentricity, but also observed that the CIE at the PETM occurred at a minimum in the eccentricity cycle and therefore suggest a probably more stochastical than orbital control. However, Lourens *et al.* (2005) do link the PETM to a 405-kyr eccentricity maximum and further claimed that the PETM and ETM2 events are paced by the very long 2.0-myr eccentricity cycle, although this relation is not generally accepted (see also Westerhold *et al.*, 2007).

A last supra-regionally recognized Eocene thermal event is the Mid Eocene Climatic Optimum (MECO) having typical hyperthermal properties regarding carbon and oxygen isotopes, oscillation in CCD, and a short duration of about 500 kyr; it occurs in the top of C18r or in C18n.2n. at 40 to 40.5 Ma during the Bartonian (Bohaty *et al.*, 2009; Luciani *et al.*, 2010).

The remarkable cooling at the beginning of the Oligocene is something already noted by earlier workers in the oxygen isotopes of coastal shell material from the North Sea area (e.g., Buchardt, 1978). The global nature of this cooling drop and the existence of several other cooling events during the Oligocene have been recognized and coded by Miller *et al.* (1991). It is generally recognized that the Oligocene cooling – the transition from greenhouse to icehouse – is associated with the buildup of permanent ice masses on Antarctica. The subsequent recognition of Oligocene cooling events in many other sections has led to some confusing coding, and the one by Pekar *et al.* (2002), amended by Wade and Pälike (2004), and the timing estimates for the events made by these authors, is followed here. The oldest event, bringing the major cooling, is recognized by many authors to occur a few 100 kyr later than the GSSP Rupelian, and is coded Oi1 and estimated to occur at 33.5 Ma. The Oi1a, Oi1b, Oi2, Oi2* oxygen isotope events are Rupelian events in C12, C11 and C10 magnetochrons. The cooling event at the Rupelian–Chattian boundary is coded Oi2a and occurs in the base of C9r. Chattian events, coded Oi2b and Oi2c, occur respectively in the upper part of C9n and in magnetochron C7. The cooling event at the Oligocene–Miocene boundary is coded Mi 1. All these Oligocene cooling events have been shown to be related to 405-kyr, 127-kyr, 96-kyr eccentricity cycles and 1.2-myr obliquity cycles (Pälike *et al.*, 2006).

This succession of cooling events with warmer times is thought to reflect the waxing and waning of an ice mass on Antarctica. Consequently this must have also had an influence on eustacy. Eustasy estimates for the Oligocene Oi events are based on Pleistocene calibration and sea-level changes associated with these events are estimated to be in the order of 50 to 65 m (Wade and Pälike, 2004). Oxygen fluctuations during the earlier Paleogene are more difficult to approach in terms of eustasy as the pre-Oligocene glacio-eustatic origin of the oxygen excursions is more debatable given the more controversial presence of continental ice volumes. However the correlation of sea-level variations between tectonically independent basins, as derived from sequence stratigraphic analysis, and the correlation of such sea-level fluctuations with isotopic data have been put forward as strong arguments for the reality of glacio-eustacy also in the pre-Oligocene (Abreu and Haddad, 1998). Abreu and Anderson (1998) have demonstrated this for the Middle and Upper Eocene and Abreu and Haddad (1998) have even suggested the presence of polar ice as early as the Aptian, based on the similarity between Cretaceous and Eocene isotope values.

Logically, the eustatic sea-level cycles should leave their impact on the sedimentary architecture of strata, especially on the shelves. This architectural pattern was discovered on seismic sections and described in the classical AAPG Memoir 26 edited by Payton (1977). Vail *et al.* (1977) concluded from the interbasinal comparison that the sedimentary sequences, with a duration of about 1 to 2 myr, commonly referred to as third order, could be correlated worldwide and therefore had to be eustatic; these sequences formed the basis for the construction of a global sea-level curve. It was already suggested from the beginning that these third-order sequences were of glacio-eustatic origin, inferred from the rate of change of sea level. The idea was complemented by field data and a comprehensive sea-level chart, coded for the sequence boundaries, was published by Haq *et al.* (1987), and followed later by a new study on European basins (Hardenbol *et al.*, 1998). However seismic resolution was limited at that time. The present recognition that Milankovitch steering of the sedimentary record has occurred generally and produces analogous sedimentary patterns on different scales smaller than the third-order scale, the duration of which is even approached by the long-duration obliquity of 1.2 myr (see also Boulila *et al.*, 2011), made it clear that the published charts are not standard eustasy charts. Nevertheless, some of the recognized fluctuations are strongly suspected to have a eustatic origin (see also Chapter 13); interbasinal comparison of sea-level signals remains an essential part of the demonstration of eustasy for which, however, Milankovitch scale stratigraphic resolution is required.

The major transgressive–regressive cycles recognized in the Paleogene have been represented on Figure 28.1; this curve is based on a combination of the major T-R facies cycles and major sequence boundaries identified in Hardenbol *et al.* (1998), and adding the major cooling event in the earliest Oligocene.

28.4. PALEOGENE TIME SCALE

The Paleogene time scale can be constructed in two fundamentally different ways. The traditional way is to use the geomagnetic polarity time scale (GPTS) as backbone; the GPTS is constructed using seafloor anomaly profiles in combination with a limited number of radio-isotopically dated age control points. The alternative method is astronomical dating; this provides direct numerical ages for magnetic reversals and bio-events by tuning cyclic sedimentary successions to astronomical target curves. Both methods will be presented independently from one another, although astronomical dating critically depends on selected radio-isotopic ages for first-order tuning to 405-kyr eccentricity. Subsequently, it will be discussed which age model is preferred to construct the Paleogene time scale in GTS2012.

28.4.1. Radio-Isotopic Age Model

Due to progress with cyclostratigraphy and astrochronology, the importance of radio-isotopic dating in constructing the Paleogene time scale is decreasing, but it is still required for scaling of selected intervals, and for providing anchor points, like at the Cretaceous–Paleogene and Paleocene–Eocene boundaries. The radio-isotopic ages discussed in this chapter (Table 28.1) are directly linked via magnetostratigraphy to the GPTS and the South Atlantic seafloor spreading profile, and can thus be directly used to calibrate the Paleogene time scale (see Figure 28.12a). All ages are based on carefully calibrated high-temperature radio-isotopic dating. This also holds for a limited number of ages which are not directly calculated from a radio-isotopic age, but are an offset from a radio-isotopic age tie-point; the duration of this offset is calculated with the help of cyclostratigraphy. The age of the FCT (Fish Canyon tuff) ^{40}Ar/^{39}Ar monitor in GTS2012 is 28.201 Ma (was 28.02 Ma in GTS2004; see Chapter 6). The new FCT monitor age makes all ^{40}Ar/^{39}Ar ages 1.0064× or 0.64 % older than in GTS2004. All ^{40}Ar/^{39}Ar ages reported below are thus reported relative to the new FCT age.

Recalculation of ^{40}Ar/^{39}Ar ages close to the Cretaceous–Paleogene boundary yields age estimates close to ~66.0 Ma (Kuiper *et al.*, 2008; see also Table 28.1), instead of 65.5 ± 0.3 Ma in GTS2004. We further incorporated recalculated single crystal ^{40}Ar/^{39}Ar sanidine ages of Swisher *et al.* (1992) from the K/Pg boundary up to base C28n. The Belt ash with a recalculated multicrystal sanidine age of 59.39 ± 0.30 Ma is assigned a position of C26r.85 based on the magnetostratigraphy of the Southeast Polecat Bench section (Secord *et al.*, 2006).

Before 2011, there was no direct radio-isotopic age for the Paleocene/Eocene boundary, as defined by the δ^{13}C excursion within the lower part of C24r. However, Charles *et al.* (2011) have now documented a single crystal U/Pb zircon age for an ash bed in the lower part of the PETM in two cyclic sections from Spitsbergen. The radiometric data were reprocessed by Mark Schmitz for error analysis, yielding an age of 55.785 ± 0.034/0.075 Ma (without/with decay constant error, using a ET535 spike calibration error of ± 0.05% to be consistent with the rest of GTS2012) for the ash bed. Using cycle scaling for two different sections which contain the ash bed, Charles *et al.* (2011) arrive at an age of 55.846 ± 0.118 Ma (55.728 to 55.964) for the base of the PETM and, hence, base Eocene. Danish ash -17 has been dated at DSDP site 550 by single crystal ^{40}Ar/^{39}Ar sanidine dating, yielding a recalculated age of 55.48 Ma that is indistinguishable from the ^{40}Ar/^{39}Ar age of the supposedly correlative Skraenterne Formation tuff on Greenland (Storey *et al.*, 2007). This ash is placed in the middle of C24r (C24r.5) based on the magnetostratigraphy of DSDP site 550 (Storey *et al.*, 2007). Note, however, that this position is not without uncertainty as a hiatus is supposedly present between the PETM and the -17 ash bed (see Westerhold and Röhl, 2009; Hilgen *et al.*, 2010). ^{40}Ar/^{39}Ar dating initially yielded an age of 52.8 ± 0.3 Ma for the Willwood ash in Wyoming (USA), which magnetostratigraphically coincides with base C24n.1n (Wing *et al.*, 1991; Tauxe *et al.*, 1994). This age was obtained relative to MMhb-1 (McClure Mountain hornblende) at 520.4 and is recalculated to 53.45 ± 0.3 Ma. The ash was redated by Smith *et al.* (2004), yielding a significantly younger recalculated age of 52.93 Ma ± 0.23 Ma. It is this age that we selected for the Willwood ash and, hence, base C24n.1n, as Smith *et al.* (2004) used single to three crystal aliquots instead of multi-crystal aliquots for dating.

An important radio-isotopic data set that is partly magnetostratigraphically constrained comes from the Lower Eocene of the Greater Green River and surrounding basins in North America (e.g., Smith *et al.*, 2008). The calibration of the magnetostratigraphy to the GPTS is based in particular on Clyde *et al.* (1997, 2001), who initially provided two possible correlations, but later favored the older correlation. Comparing the radio-isotopic ages of the intercalated tuffs with the GPTS (i.e. CK95: Cande and Kent, 1995), Smith *et al.* (2003) initially started from the younger correlation, and assumed that C23n was not expressed due to weak natural remanent magnetization intensities and erratic demagnetization behavior. They switched to the older option of Clyde *et al.* (2001; see also Clyde *et al.*, 2004; Machlus *et al.*, 2004) in the Smith *et al.* (2008) paper. In this paper they used single and multicrystal ^{40}Ar/^{39}Ar sanidine ages from seven magnetostratigraphically constrained ash beds to revise the GPTS of Cande and Kent (1995) for the interval from C24 up to C20; the resultant reversal ages were significantly younger by 1 to 1.5 million years, but the discrepancies were already reduced with respect to an initial attempt by Machlus *et al.* (2004) to revise the polarity time scale.

Recently, Smith *et al.* (2010) published mutually consistent ^{40}Ar/^{39}Ar and U/Pb ages for two ash layers, using the FCs age of 28.201 Ma of Kuiper *et al.* (2008) to calculate the

TABLE 28.1 **Paleogene Radiometric and Cyclostratigraphic Ages for the C-Sequence Polarity Chrons**

Placement in Polarity Chron	Distance (km) in S. Atlantic Profile	Age (Ma)	Type	Dated Unit	Locality; Formation; Calibration	Geological Age	Source or Review References
C5Br (base)	318.4 ± 1.4	15.97			Base-Langhian (assigned chron).	early Neogene	see Chapter 29 (Neogene)
C6An.1r (base)	434.18 ± 0.3	20.34			Aquitanian/Burdigalian boundary (assigned chron). Proportional-placement in chron implies "0.0" km uncertainty (if this is precise chron-age definition); plus an additional 4% uncertainty on 6.4km anomaly width (= 0.3 km).	early Neogene	see Chapter 29 (Neogene)
C6Cn.2n (base)	501.55 ± 0.1	23.03			Oligocene/Miocene boundary (assigned chron). Proportional-placement in "lower" chron implies "0.0" km uncertainty (if this is definition); plus an additional 4% uncertainty on 3km width (0.1 km).	base of Neogene	see Chapter 29 (Neogene)
Uppermost C9n (assumed C9n.95 ± 0.05)	584.53 ± 2.8	27.0 ± 0.1	$^{40}Ar/^{39}Ar$	MCA98-7	208.3 mab, Monte Cagnero section, northeastern Apennines, Italy. Scaglia Cinerea Formation. Proportional-placement in "uppermost" chron C9n (5% of width of 24.7 km) implies "1.3" km uncertainty; plus an additional 6% uncertainty on 24.7 km width (1.5 km) = 2.8 km.	early Chattian	Coccioni *et al.*, 2008
Uppermost C12r (assumed C12r.85 ± 0.1)	694.9 ± 7.3	31.8 ± 0.2	$^{40}Ar/^{39}Ar$	MCA/84-3	145.5 mab, Monte Cagnero section, northeastern Apennines, Italy. Scaglia Cinerea Formation. Proportional-placement in uppermost chron C12r (10% of width of 56.1 km) implies "5.6" km uncertainty; plus an additional 3% uncertainty on 56.1 km width (1.7 km) = 7.3 km.	late Rupelian	Coccioni *et al.*, 2008
C13r.4 ± 0.05	772.82 ± 2.0	34.4 ± 0.2	$^{40}Ar/^{39}Ar$	MAS/86-14.7	Massignano, near Ancona, northeastern Apennines, Italy. Diagram in Brown *et al.*, 2009 indicates 14.7m ash is about 3.5-m up in 9-m C13r; therefore C13r.4 ± 0.05 if constant sedimentation rate. Uncertainty on exact position, plus 2% uncertainty on 29 km width of C13r = 2.0 km.	latest Eocene	Odin *et al.*, 1991; Hilgen and Kuiper, 2009; Brown *et al.*, 2009
C13r.14 ± 0.03	780.3 ± 1.5	35.2 ± 0.2	$^{40}Ar/^{39}Ar$	MAS/86-12.7	Massignano, near Ancona, northeastern Apennines, Italy. Diagram in Brown *et al.*, 2009 indicates 14.7m ash is about 1.3-m up in 9-m C13r; therefore C13r.14 ± 0.03 if constant sedimentation rate. Uncertainty on exact position, plus 2% uncertainty on 29 km width of C13r = 1.4 km.	latest Eocene	Odin *et al.*, 1991; Hilgen and Kuiper, 2009; Brown *et al.*, 2009
C20n.1 ± 0.1	1003.22 ± 3.5	43.35 ± 0.5	$^{40}Ar/^{39}Ar$	Mission Valley ash	California. Mission Valley Formation. Proportional-placement in chron implies 2.8 km uncertainty; plus an additional 3% uncertainty on width (0.7 km).	middle Eocene	Prothero and Emry, 1996; Walsh *et al.*, 1996; Smith *et al.*, 2010.

C21n.75 ± 0.1	1068.89 ± 4.6	46.24 ± 0.5	$^{40}Ar/^{39}Ar$	DSDP Hole 516F	^{40}Ar-^{39}Ar dating of magnetostratigraphy in DSDP Hole 516F. Proportional-placement in chron implies 3.4 km uncertainty; plus an additional 4% uncertainty on width (1.2 km).	middle Eocene (earliest Lutetian)	Swisher and Montanari, in prep, cited in Berggren *et al.* 1995 (postscript); Smith *et al.*, 2010.
C21r.1 ± 0.1	1115.27 ± 3.8	48.41 ± 0.21	$^{40}Ar/^{39}Ar$	Blue Point Marker ash	Absaroka volcanic province, western USA. Overlies Aycross Formation and Trout Peak Trachyandesite. Proportional-placement in chron implies 2.3 km uncertainty; plus an additional 3% uncertainty on width (1.5 km).	early Eocene (latest Ypresian)	Hiza, 1999; Flynn, 1986 and references therein; Smith *et al.*, 2010.
C22n.45 ± 0.1	1124.83 ± 2.5	48.96 ± 0.33	$^{40}Ar/^{39}Ar$	Continental tuff	Bridger Basin, western USA. Bridger Formation. Proportional-placement in chron implies 1.3 km uncertainty; plus an additional 4% uncertainty on width (1.2 km).	early Eocene (Ypresian)	Smith *et al.*, 2008, 2010; Clyde *et al.*, 2001
C24n.1n (base)	1184.03 ± 0.5	52.93 ± 0.23	$^{40}Ar/^{39}Ar$	Willwood ash	Bighorn Basin, western USA. Willwood Formation. Proportional-placement in chron (c. ±5% assumed) implies 0.25 km uncertainty; plus an additional 5% uncertainty on width (0.25 km).	early Eocene (early Ypresian)	Wing *et al.*, 1991; Smith *et al.*, 2004, 2010; Clyde *et al.*, 1994; Tauxe *et al.*, 1994
C24r.5 ± 0.1	1214.93 ± 5.65	55.48 ± 0.12	$^{40}Ar/^{39}Ar$	Ash-17	DSDP Site 550, north Atlantic. Proportional-placement of PETM placement within chron (C24r.5 ± 10% approx.) = ±3.9 km; plus 4.5% uncertainty on chron-width (= 1.75 km) = 5.65 km total uncertainty.	earliest Eocene	Storey *et al.*, 2007
C24r.375 ± 0.05	1219.83 ± 3.71	55.785 ± 0.075	$^{206}Pb/^{238}U$	SB01-01 bentonite	Longyearbyen section, Spitsbergen, Svalbard Archipelago. Bentonite is 10.90 m above the base of the Gilsonryggen Member, Frysjaodden Formation, within the PETM carbon isotope excursion. Assigned as 0.045 myr above PETM (assigned as C24r.36) => C24r.375. Uncertainty of placement within chron (same as PETM, or 5% approx.) = ±1.95 km; plus 4.5% uncertainty on chron-width (= 1.75 km).	earliest Eocene	Charles *et al.*, 2011
C24r.36 ± 0.05	1220.41 ± 3.71	55.85 ± 0.10	*CYCLES*		PETM relative to $^{206}Pb/^{238}U$-dated bentonite (see above). Proportional-placement of PETM placement within chron (C24r.36 ± 5% approx.) = ±1.95 km; plus 4.5% uncertainty on chron-width (= 1.75 km).	Paleocene–Eocene boundary	Charles *et al.*, 2011
C24r (base)	1234.51 ± 1.75	56.95 ± 0.12	*CYCLES*		P/E boundary (= C-spike) is c. 1.1 myr after the end of Chron C25n, added to PETM age above. Km-uncertainty assumed as half of the uncertainty on total width of C24r, or 4.5 % of 39.14 km = 1.75 km.	late Paleocene	Westerhold *et al.*, 2008

(Continued)

TABLE 28.1 Paleogene Radiometric and Cyclostratigraphic Ages for the C-Sequence Polarity Chrons—cont'd

Placement in Polarity Chron	Distance (km) in S. Atlantic Profile	Age (Ma)	Type	Dated Unit	Locality; Formation; Calibration	Geological Age	Source or Review References
C26n (base)	1262.74 ± 2.53	59.18 ± 0.30	*CYCLES*		Duration of C26r is 2.93 kyr. Cyclo-magnetostratigraphy (central Atlantic & Caribbean). Uncertainty is km-uncertainty on K/P boundary (1.3) plus 3% on 41.0 km width of C26r.	early to late Paleocene boundary	Westerhold *et al.*, 2008; Hilgen, 2010 (averaged)
C26r.9 ± 0.05	1266.8 ± 4.1	59.39 ± 0.3	$^{40}Ar/^{39}Ar$	Belt Ash	Southeast Polecat Bench section, northern Bighorn Basin, Wyoming, USA. Silver Coulee beds, Fort Union Formation. About 85% up in Chron C26r according to published magstrat summary diagram = C26r.85 ± 0.05, or 1266.85 ± 2.05 km; plus c. 5% uncertainty on 41km width = total of 4.1 km uncertainty.	late Paleocene	Secord *et al.* (2006)
C26r (base)	1303.81 ± 2.53	62.11 ± 0.3	*CYCLES*		Duration of C27n–C27r is 1.29 kyr. Cyclo-magnetostratigraphy (central Atlantic & Caribbean). Uncertainty is km-uncertainty on K/P boundary (1.3) plus 3% on 41.0 km width of C26r.	early to late Paleocene boundary	Westerhold *et al.*, 2008; Hilgen, 2010 (averaged)
C27r (base)	1325.71 ± 2.15	63.40 ± 0.3	*CYCLES*		Interpolated duration of C28n–C28r of 1.46 myr (Hilgen, 2010 reinterpretation of Westerhold *et al.*, 2008). Cyclo-magnetostratigraphy (central Atlantic & Caribbean). Uncertainty is km-uncertainty on K/P boundary (1.3) plus 5% on 17.0 km width of C27r.	mid-Danian	Westerhold *et al.*, 2008; Hilgen, 2010
C28r.9 ± 0.1	1342.5 ± 1.3	64.73 ± 0.12	$^{40}Ar/^{39}Ar$	U-Coal; Swisher'93	Montana. "Rocks immediately above the W-Coal through the U-Coal, which include the Farrand Channel, are in a zone of reversed polarity, correlated here with chron C28r. ..." (Swisher *et al.*, 1993). ^{40}Ar-^{39}Ar age on U-Coal, at C28n/C28r transition (5% assumed). Proportional-placement in chron implies 0.3 km uncertainty; plus an additional 10% uncertainty on width (0.5 km).	early Danian	Swisher *et al.*, 1993.
C28r (base)	1347.03 ± 1.55	64.86 ± 0.3	*CYCLES*		Duration of C29n is 720 kyr ± 0.03. Cyclo-magnetostratigraphy (south Atlantic & Spain). Uncertainty is km-uncertainty on K/P boundary (1.3) plus 5% on 5.0 km width of C28r.	early Danian	Westerhold *et al.*, 2008; Hilgen, 2010 (averaged)
C29n.9	1348.19 ± 1.8	64.94 ± 0.12	$^{40}Ar/^{39}Ar$	W-Coal; Swisher'93	Uppermost-C29n Montana. "Rocks immediately above the W-Coal ... are in a zone of reversed polarity, correlated here with chron C28r. ..." (Swisher *et al.*, 1993). ^{40}Ar-^{39}Ar age on W-Coal, at about 90% up (±5% assumed) in polarity zone C29r. Proportional-placement in chron implies 0.6 km uncertainty; plus an additional 10% uncertainty on width (1.2 km).	early Danian	Swisher *et al.*, 1993.

C29n (base)	1358.66 ± 1.3	65.58 ± 0.3	*CYCLES*		Base of C29n is 420 kyr above K/P boundary (total duration of C29r = 720 kyr - 300 kyr for K/P to base-C29r). Cyclo-magnetostratigraphy (south Atlantic & Spain). Uncertainty is km-uncertainty on K/P boundary.	earliest Danian	Westerhold *et al.*, 2008; Husson *et al.*, 2011
C29r.8	1361.3 ± 1.3	65.611 ± 0.15	$^{40}Ar/^{39}Ar$	HFZ-Coal; Swisher'93	Upper-C29r Montana. "Rocks up to midway between the HFZ- and Y-coals are contained in a zone of reversed polarity, correlated here with chron C29r. ." (from Swisher'93). ^{40}Ar-^{39}Ar age on HFZ-Coal, at about 80% up (±5% assumed) in polarity zone C29r. Proportional-placement in chron implies 0.5 km uncertainty; plus an additional 5% uncertainty on width (0.8 km).	earliest Danian	Swisher *et al.*, 1993.
C29r.57	1364.33 ± 1.3	65.84 ± 0.12	$^{40}Ar/^{39}Ar$	Z-Coal; Swisher'93	Hell Creek, Garfield County, Montana. Z-coal bentonite; Hell Creek bentonite (lower of two volcanic ash beds in 1-m-thick Lower Z coal seam at base of Fort Union Fm. Assumed to be only 50 kyr after base-Cenozoic (which is assigned as Chron C29r.5). A 5% uncertainty in position in chron, plus the 5% uncertainty on 13.1 km width of chron.	just above base of Cenozoic	Swisher *et al.*, 1993.
C29r.57	1364.33 ± 1.3	66.28 ± 0.18	$^{40}Ar/^{39}Ar$	Z-coal bentonite	Hell Creek, Garfield County, Montana. Z-coal bentonite; Hell Creek bentonite (lower of two volcanic ash beds in 1-m-thick Lower Z coal seam at base of Fort Union Fm. Assumed to be only 50 kyr after base-Cenozoic (which is assigned as Chron C29r.5). A 5% uncertainty in position in chron, plus the 5% uncertainty on 13.1 km width of chron.	just above base of Cenozoic	Swisher *et al.*, 1993.
C29r.5	1365.25 ± 1.3	65.99 ± 0.12	$^{40}Ar/^{39}Ar$	IrZ-coal bentonite	Ir Hill (named after iridium anomaly in lowest lignite) - Hauso Flats, Garfield County; eastern Montana. 2-mm bentonite in middle of 4-cm basal lignite (IrZ after a small iridium anomaly in its lower half) at base of Tullock Formation. The Tullock Formation overlies the dinosaur-bearing Hell Creek Formation with the highest in-situ dinosaur 3.3 m below. Assumed to be exactly base-Cenozoic. A 5% uncertainty in position in chron, plus the 5% uncertainty on width (13 km) of chron.	base of Cenozoic	Swisher, *et al.* 1993; Kuiper *et al.* 2008
C29r (base)	1371.84 ± 2.24	66.30 ± 0.3	*CYCLES*		Cyclo-magnetostratigraphy (ODP site 1267B as primary reference). 300 ± 20 kyr relative to K/T boundary. Km-uncertainty assumed as half the uncertainty on total width of C28–C29.	latest Maastrichtian	Husson *et al.*, 2011

(*Continued*)

TABLE 28.1 Paleogene Radiometric and Cyclostratigraphic Ages for the C-Sequence Polarity Chrons—cont'd

Placement in Polarity Chron	Distance (km) in S. Atlantic Profile	Age (Ma)	Type	Dated Unit	Locality; Formation; Calibration	Geological Age	Source or Review References
C30n (base)	1407.22 ± 2.87	68.20 ± 0.3	*CYCLES*		Cyclo-magnetostratigraphy (ODP site 1267B as primary reference). 2.2 ± 0.02 myr relative to K/T boundary. Km-uncertainty assumed as half the uncertainty on total width of C30–C31n.	Late Maastrichtian	Husson *et al.*, 2011
C31n (base)	1429.14 ± 2.87	69.22 ± 0.2	*CYCLES*		Cyclo-magnetostratigraphy (ODP site 762C as primary reference). 3.22 ± 0.01 myr relative to K/T boundary. Km-uncertainty assumed as half the uncertainty on total width of C30–C31n.	Late Maastrichtian	Husson *et al.*, 2011
C31n (base)	1481.12 ± 1.72	71.40 ± 0.3	*CYCLES*		Cyclo-magnetostratigraphy (ODP site 762C as primary reference). 5.4 ± 0.02 myr relative to K/T boundary. Km-uncertainty assumed as half the uncertainty on total width of C31r.	base Late Maastrichtian	Husson *et al.*, 2011
C32n (base)	1531.81 ± 2.87	73.60 ± 0.3	*CYCLES*		Cyclo-magnetostratigraphy (ODP site 762C as primary reference). 7.6 ± 0.02 myr relative to K/T boundary. Km-uncertainty assumed as half the uncertainty on total width of C32.	early Maastrichtian	Husson *et al.*, 2011
C32r (base)	1549.41 ± 2.87	74.10 ± 0.3	*CYCLES*		Cyclo-magnetostratigraphy (ODP site 762C as primary reference). 8.1 ± 0.02 myr relative to K/T boundary. Km-uncertainty assumed as half of the uncertainty on total width of C32.	early Maastrichtian	Husson *et al.*, 2011
C33n (base)	1723.76 ± 2.88	79.84 ± 0.5	$^{40}Ar/^{39}Ar$		US western interior (Elk Basin ^{40}Ar-^{39}Ar ages and magnetostratigraphy). Km-uncertainty assumed as half of the uncertainty on total width of C33n.	mid-Campanian	Hicks *et al.*, 1995
C33r (base)	1862.32 ± 2.91	83.6 ± 0.5	$^{40}Ar/^{39}Ar$		^{40}Ar-^{39}Ar age constraints on biostratigraphy (84.35 ± 0.27 in latest-Santonian *Desmoscaphites bassleri* ammonite zone) associated with magnetostratigraphy. Km-uncertainty assumed as half of the uncertainty on total width of C33r.	~ base Campanian *(was uppermost Santonian in GTS2004)*	Obradovich, 1993

Selected Paleogene radio-isotopic dates and selected cyclostratigraphy durations for C-sequence polarity chrons C5Br through C33r that were used for the cubic spline-fit in Figure 28.12a. Details on radio-isotopic dates are given in Appendix 2. All Ar-Ar ages are recalculated to FCs of 28.201 Ma. Chron-events are assigned positions (km) in the South Atlantic seafloor spreading profile of Cande and Kent (1992) based on proportion position in chron, rather than attempting to adjust for trends in spreading rate within the anomaly. Uncertainties on km-placement partially incorporate the uncertainty on magnetic-anomaly widths from tables in Cande and Kent (1992).

$^{40}Ar/^{39}Ar$ ages. In addition, ages proved older than previously published ages for the same ash layers. These older $^{40}Ar/^{39}Ar$ ages resulted from eliminating samples with low radiogenic $^{40}Ar^*$ that produce a bias towards young ages. Unfortunately, the position of these two ash beds is magnetostratigraphically not constrained. Here we select the original tie points used by Smith *et al.* (2008) for the revision of the GPTS, recalculated to 28.201 Ma and after (minor) removal of $^{40}Ar/^{39}Ar$ dates with low $^{40}Ar^*$ (M.E. Smith, pers. comm., 2011; see Table 28.2). However, we excluded the LT and 6th tuff as their position relative to the magnetostratigraphy and thus the chron boundaries remains uncertain (M.E. Smith, pers. comm., 2011). These two tie points are responsible for the large discrepancy between the revised GPTS of Smith *et al.* (2008) and CK95 (Smith *et al.*, 2008). Moreover, inclusion of these tie points would result in an aberrant increase in spreading rates followed by a decrease which are difficult to explain. Following the selection of tie points by Smith *et al.* (2008), we used $^{40}Ar/^{39}Ar$ ages of 46.24 $\pm$ 0.5 Ma for an ash bed from DSDP Site 516F approximately at the magnetostratigraphic level of C21n(0.67) (Berggren *et al.*, 1995 (postscript)) and of 43.35 $\pm$ 0.5 Ma for the Mission Valley ash bed in California (Prothero and Emry, 1996; Walsh *et al.*, 1996).

Numerous radio-isotopic ages are available for the interval around the E−O boundary, both from the deep-marine succession in Italy and the continental succession in North America. Single crystal $^{40}Ar/^{39}Ar$ anorthoclase ages of Swisher and Prothero (1990) and Prothero and Swisher (1992) for the interval between C15/16 and C11 in North America are not included as they are inconsistent (i.e. ~300−400 kyr older) with ages for some of the same ash beds published in an abstract by Obradovich *et al.* (1995); in addition the calibration of the magnetostratigraphy to the GPTS remains problematic (see Hilgen and Kuiper, 2009 for details). The Italian dates are mainly old K-Ar and some $^{40}Ar/^{39}Ar$ ages of biotite and should thus preferably not be used for time scale construction. Nevertheless, in the absence of more reliable sanidine dates, we included $^{40}Ar/^{39}Ar$ biotite ages of two ash layers (Odin *et al.*, 1991) that occur in the reversed polarity interval of C13r in the E/O boundary stratotype section of Massignano (e.g., see Brown *et al.*, 2009). It is important to note that the short normal polarity interval directly above the 18 m level corresponds to C13n in Odin *et al.* (1991) and that the ages given in Brown *et al.* (2009) do not represent $^{40}Ar/^{39}Ar$ ages, but are based on all the available age data in 1993 (Montanari *et al.*, 1993: i.e. K/Ar, $^{40}Ar/^{39}Ar$, U/Pb and Rb/Sr). Based on the magnetostratigraphy, these two ash beds are assigned positions C13r.14 and C13r.4. Earliest Chattian tuffs in Italy assigned to the base of polarity Chron C9n have an age of 28.28 $\pm$ 0.30 Ma (Odin *et al.* 1991; Cande and Kent, 1992; Wei, 1995). This is K-Ar on a single biotite separate, with no standards mentioned. Presumably, based upon related literature, this age was calculated with respect to FCT biotite at 27.84 Ma. This is a low confidence biotite result and hence is discarded here. Recently, $^{40}Ar/^{39}Ar$ biotite ages were published from three sections in Northern Italy in a proposal for defining the Rupelian/Chattian boundary (Coccioni *et al.*, 2008). In the absence of more reliable sanidine ages, we selected recalculated biotite ages of 31.8 $\pm$ 0.2 and 27.0 $\pm$ 0.1 Ma of two ash beds, which occur at 145.6 and 208 m in the Monte Cagnero section; these ash beds are assigned a position of C12r.85 and C9n.95, respectively. We refrained from incorporating the dataset of McIntosh et al. (1992) as the combined single crystal $^{40}Ar/^{39}Ar$ sanidine ages and polarity directions come from the discontinuous ignimbrite succession of the Eocene-Oligocene Mogollon-Datil volcanic field in New Mexico. However, this is arguable as this study provides by far the best radioisotopic data at present available for the interval across the Eocene-Oligocene boundary.

The Oligocene/Miocene boundary as defined by its GSSP in the Carrosio/Lemme section (Steininger *et al.*, 1997) has not been directly dated, but is associated with base C6n.2n with an orbitally tuned age of 23.03 Ma and an uncertainty of 0.1 kyr that is also used in the Neogene time scale. This age is included for consistency with that time scale. For details see Chapter 29 on the Neogene Period.

28.4.2. Astronomical Age Model

As far as numerical calibration is concerned, the Paleogene time scale, like its predecessor in GTS2004 (Luterbacher *et al.*, 2004), remains a time scale in transition towards one that is fully underlain by astronomical tuning. As yet, this has not been achieved, resulting in a time scale of which the age model is only partially tuning based. Moreover, existing literature reveals uncertainties and discrepancies for the tuning of parts of the Paleogene. At present, not all these uncertainties and discrepancies have been solved in a satisfactory way.

The Paleogene time scale in GTS2004 (Luterbacher *et al.*, 2004) was constructed based on an integration of well constrained radio-isotopic ages, the GPTS and cyclostratigraphy. However, at present, more stringent quality criteria are applied as far as geochronology is concerned (i.e. preferably single crystal $^{40}Ar/^{39}Ar$ sanidine and U/Pb zircon dates) and, in 2004, Paleogene cyclostratigraphy was still in its infancy. The cyclostratigraphy was especially based on cycle counts in old DSDP holes for Chron 29 (e.g., Herbert, 1999) and in ODP holes for the older part of the Paleogene (e.g., Röhl *et al.*, 2001, 2003). These studies focused on duration of magnetochrons and, apart from a premature effort of Ten Kate and Sprenger (1992), did not attempt to include a tuning as no reliable astronomical solution was available for this interval at that time. These studies have largely been superseded by more recent cyclostratigraphic studies, which aim to tune the entire Paleogene on at least the 405-kyr eccentricity scale, using new astronomical solutions (La2004: Laskar *et al.*, 2004; La2010: Laskar *et al.*, 2011). These studies are used here in constructing an astronomical time scale for the Paleogene.

TABLE 28.2 Comparison of Estimated Ages for Paleogene C-Sequence Polarity Chrons from Cyclostratigraphy

Polarity Chron	Pälike *et al.*, 2006		Pälike *et al.* 2001	Brown *et al.* 2009 Hyland *et al.* 2009 Jovane *et al.* 2010	Westerhold *et al.* 2007, 2008 Westerhold & Röhl 2009			Hilgen *et al.* 2010	Smith *et al.*		GTS2012
	1218 Man	1218 Auto	1052	Italy	option 1	option 2	option 3		2008	2010	Astro-Age Model
C6Cn.2n	23.024	23.026									23.024
C6Cn.2r	23.233	23.278									23.233
C6Cn.3n	23.295	23.340									23.295
C6Cr	23.962	24.022									23.962
C7n.1n	24.000	24.062									24.000
C7n.1r	24.109	24.147									24.109
C7n.2n	24.474	24.459									24.474
C7r	24.761	24.756									24.761
C7An	24.984	24.984									24.984
C7Ar	25.099	25.110									25.099
C8n.1n	25.264	25.248									25.264
C8n.1r	25.304	25.306									25.304
C8n.2n	25.987	26.032									25.987
C8r	26.420	26.508									26.420
C9n	27.439	27.412									27.439
C9r	27.859	27.886									27.859
C10n.1n	28.087	28.126									28.087
C10n.1r	28.141	28.164									28.141
C10n.2n	28.278	28.318									28.278
C10r	29.183	29.166									29.183
C11n.1n	29.477	29.467									29.477
C11n.1r	29.527	29.536									29.527
C11n.2n	29.970	29.957									29.970
C11r	30.591	30.617									30.591

C12n	31.034	31.021		31.60							31.034
C12r	33.157	33.232		33.23							33.157
C13n	33.705	33.705		33.78							33.705
Olig/Eoc = 86% up in C13r	*33.79*	*33.79*		33.93							
C13r	*35.126*	*35.126*		34.73							*34.999*
C15n	*35.254*	*35.254*		34.97							*35.294*
C15r	*35.328*	*35.328*	35.185	35.14							*35.706*
C16n.1n	*35.554*	*35.554*	35.524								*35.892*
C16n.1r	*35.643*	*35.643*	35.605								*36.051*
C16n.2n	*36.355*	*36.355*	36.051								*36.700*
C16r	*36.668*	*36.668*	36.404								*36.969*
C17n.1n	*37.520*	*37.520*	37.300								*37.785*
C17n.1r	*37.656*	*37.656*	37.399								*37.908*
C17n.2n	*37.907*	*37.907*	37.618								*38.135*
C17n.2r	*37.956*	*37.956*	37.692								*38.202*
C17n.3n	*38.159*	*38.159*	37.897								*38.380*
C17r	*38.449*	*38.449*	38.186								*38.668*
C18n.1n	*39.554*	*39.554*	39.441								*39.686*
C18n.1r	*39.602*	*39.602*	39.486								*39.756*
C18n.2n	*40.084*	*40.084*	39.828	40.12							*40.201*
C18r	*41.358*	*41.358*		41.25							*41.204*
C19n	*41.510*	*41.510*		41.51							*41.438*
C19r				42.54					42.12	42.37	*42.351*
C20n				43.79					43.16	43.42	*43.505*
C20r				46.31					45.71	45.99	*45.942*
C21n					47.396	47.791	48.191	47.837	46.52	46.80	47.837
C21r					48.661	49.056	49.456	49.102	48.30	48.59	49.102
C22n					49.299	49.694	50.094	49.740	49.07	49.37	49.740
C22r					50.768	51.163	51.563	51.209	49.71	50.01	51.209

(Continued)

TABLE 28.2 Comparison of Estimated Ages for Paleogene C-Sequence Polarity Chrons from Cyclostratigraphy—cont'd

Polarity Chron	Pälike *et al.*, 2006		Pälike *et al.* 2001	Brown *et al.* 2009 Hyland *et al.* 2009 Jovane *et al.* 2010	Westerhold *et al.* 2007, 2008 Westerhold & Röhl 2009			Hilgen *et al.* 2010	Smith *et al.*		GTS2012
	1218 Man	1218 Auto	1052	Italy	option 1	option 2	option 3		2008	2010	Astro-Age Model
C23n.1n					50.992	51.387	51.787	51.433	49.88	50.18	51.433
C23n.1r					51.064	51.459	51.859	51.505	50.04	50.34	51.505
C23n.2n					51.443	51.838	52.238	51.884	50.50	50.80	51.884
C23r					52.243	52.638	53.038	52.684	52.23	52.54	52.684
C24n.1n					52.633	53.028	53.428	53.074	52.59	52.91	53.074
C24n.1r					52.758	53.153	53.553	53.199	52.71	53.03	53.199
C24n.2n					52.833	53.228	53.628	53.274	52.75	53.07	53.274
C24n.2r					52.975	53.370	53.770	53.416	52.89	53.21	53.416
C24n.3n					53.542	53.937	54.337	53.983	53.44	53.76	53.983
Eoc/Pal = ~36% up in C24r					*55.53*	*55.93*	*56.33*	*56.00*			
C24r					56.660	57.055	*57.455*	57.101			57.101
C25n					57.197	57.612	*58.012*	57.656			57.656
C25r					58.551	58.935	*59.335*	58.959			58.959
C26n					58.878	59.273	*59.673*	59.237			59.237
C26r					61.774	62.154	*62.554*	62.221			62.221
C27n					62.098	62.468	*62.868*	62.517			62.517
C27r					63.079	63.463	*63.863*	63.494			63.494
C28n					64.028	64.385	*64.785*	64.667			64.667
C28r					64.205	64.572	*64.972*	64.958			64.958
C29n					64.912	65.282	*65.682*	65.688			65.688
Pg/C = ~50% up in C29r					*65.28*	*65.68*	*66.08*	*65.95*			*65.950*
C29r					65.625	66.007					

Age estimates based on cyclostratigraphy and correlations to astronomical target curves. The three Paleocene options of Westerhold *et al.* (2008) have been extended up to C21n using astronomical durations of (1) 3118 kyr for C24r (Westerhold *et al.*, 2007) and (2) based on a tuning to stable 405-kyr long eccentricity for C24n.3n to C21n of Westerhold and Röhl (2009; see their Table 1). In addition, chron age estimates of Smith *et al.* (2008, 2010) are shown based on a combination of $^{40}Ar/^{39}Ar$ ages and magnetostratigraphy from the greater Green River basin. The GTS2012 astronomical age model (right-most column) is based on ODP Site 1218 (manual fit of Pälike *et al.*, 2006) for the upper Paleogene (chrons C6C-C13n) and on Hilgen *et al.* (2010) combined with the extended option 2 of Westerhold *et al.* (2008) for the lower Paleogene (chrons C24-C21n) placed on top of C25n with an age of 57.101 Ma. The model is completed by applying a 6th order polynomial fit for chrons C13r through C23r to bridge the gap between the astronomical age models above and below.

The age calibration of the Neogene in GTS2012 is based almost entirely on astronomical tuning (ATNTS2012). This astronomical time scale is rapidly extended towards older time intervals. The entire Oligocene was tuned in a single effort, using climatic proxy records of ODP Site 1218 in the Pacific with its excellent magneto-bio-cyclostratigraphy (Pälike *et al.*, 2006). Despite the advancement, seafloor spreading rates suggest that minor problems may still exist in the tuning of the O–M boundary interval and the oldest part of the Oligocene (D. Wilson, pers. comm., 2011). Indeed, recent attempts to tune the E–O boundary interval in land-based marine sections in Italy arrived at slightly older ages for this boundary (33.91–33.95 Ma) and the C13n reversal boundaries (Brown *et al.*, 2009; Hyland *et al.*, 2009), while Jovane *et al.* (2006) obtained a slightly younger age. In addition, an attempt has been made to tune the Upper Eocene in ODP core 1052 (Pälike *et al.*, 2001), while the Middle Eocene was recently tuned in the Contessa section by Jovane *et al.* (2010). However, it remains unclear whether the first-order tuning to 405-kyr eccentricity of Jovane *et al.* (2010) is correct as their reversal ages remain very close to the ages of CK95 which they used to establish their initial tuning. Overall, the Eocene proved particularly difficult to tune, because the shallow position of the calcite compensation depth (CCD) at the time prevented the deposition of cyclic carbonate-rich sediments in the marine realm most suitable for tuning. Recently, IODP Leg 320 focused timely on this time interval (Pälike *et al.*, 2008), and its cores may prove instrumental in constructing a reliable tuning for the Eocene. Nevertheless, several attempts have been made to establish a tuning for the older part of the Eocene; these attempts are linked to the tuning of the entire Paleocene down to the K/Pg boundary.

Recognizing 24 × 405-kyr eccentricity related cycles, Westerhold *et al.* (2008) tuned the entire Paleocene but, due to the Eocene gap in the astronomical time scale, had to present three different options, each with a 405-kyr offset (Table 28.2). Kuiper *et al.* (2008) used an astronomically calibrated age of 28.201 ± 0.046 Ma for the FCs dating standard to recalculate $^{40}Ar/^{39}Ar$ ages of ash beds found directly above the K/Pg boundary in continental successions of North America; these ages were subsequently used to constrain the tuning of the older part of the Paleocene in the K/Pg boundary section at Zumaia (Spain) on the ~405-kyr eccentricity scale. The resulting age of ~65.95 Ma for the boundary proved incompatible with the two main tuning options in Westerhold *et al.* (2008), but was consistent with the third that had been added for consistency with the astronomically calibrated age of the FCs standard.

A critical evaluation of the cyclostratigraphic interpretation of Westerhold *et al.* (2008) indicated that 25 rather than 24 × 405-kyr cycles are present in the Paleocene (Hilgen *et al.*, 2010). As a result, Hilgen *et al.* (2010) retuned the Paleocene on the 405-kyr eccentricity scale, starting from an age 65.95 Ma for the K/Pg boundary and assuming that 25 × 405-kyr cycles are present (Table 28.2). This resulted in an age of ~56.0 Ma for the base of the PETM and, hence, the Paleocene–Eocene boundary (Hilgen *et al.*, 2010; 55.93 Ma of Westerhold *et al.*, 2008, 2009). A tuning to short (~100 kyr) eccentricity is unrealistic in view of the limited reliability of the astronomical solution in this interval as far as the short eccentricity cycle is concerned. The tuning can be extended upwards (i.e. up to base C21n) by incorporating the cyclostratigraphic interpretation and tuning of proxy records from ODP sites 1258 (Leg 207, Demerara Rise) and 1262 (Leg 208, Walvis Ridge) (Lourens *et al.*, 2005; Westerhold *et al.*, 2007, 2008; Westerhold and Röhl, 2009). Note that option 2 of Westerhold *et al.* (2007) and Westerhold and Röhl (2009) is followed for this interval as we distinguish 25 instead of 24 × ~405-kyr cycles in the Paleocene (see caption to Table 28.2). In this way an astronomical-tuned time scale is constructed for the entire Paleocene and Early Eocene, covering the interval from ~66 to 47 Ma (Table 28.2). We did not incorporate tuned reversal ages for the Middle–Late Eocene as it is not clear whether published tunings are correct. For the interval between 47 and 34 Ma, we preferred to calculate ages for magnetic reversal boundaries using the synthetic anomaly profile of Cande and Kent (1992) and an interpolation between the nearest older and younger astronomically dated reversal boundary (i.e. C21n(o) at ~47.8 Ma, and C13n(o) at 33.705 Ma, using a 6-order polynomial interpolation; Table 28.3: "Astronomic age model with 6-order polynomial fit in Eocene").

28.4.3. Age Model Comparison and Final Time Scale

Above, the Paleogene time scale is constructed in two different ways, *yielding partly diverging results.* Firstly, we created a Paleogene time scale from the integration of well-constrained radiogenic isotope age dates, the geomagnetic polarity time scale (GPTS) and the magnetochronology of the South Atlantic seafloor spreading profile. Carefully selected cyclostratigraphic arguments were added to the magnetochronology. This data set is detailed in Section 28.4.1. Secondly, cyclostratigraphy and astrochronology are used, using a 6-order polynomial fit for bridging the cyclostratigraphic gap in the Eocene. The resulting chron ages for this combined astrocyclic and best fit model are in Table 28.3 (column with astronomical age model with 6-order polynomial fit in Eocene). This data set is the one that was detailed in Section 28.4.2. Both age models are used to calculate South Atlantic seafloor spreading history and ages for Paleogene GSSPs, as the position of the latter relative to chron boundaries is known.

Figure 28.12A displays the cubic spline for the Paleogene radiometric GTS solution, using the dates and km interpolation in Table 28.1. The initial smoothing factor was

TABLE 28.3 Numerical Age Models for Paleogene C-Sequence Polarity Chrons

Marine Magnetic Anomaly Profile		CK92/95	GTS2004	Radio-Isotopic Age Model				Astronomic Age Model with 6-order Polynomial Fit in Eocene			Combined Age Model	
Polarity Chron	Distance, Old End (km)	Age (Ma)	Age (Ma)	Age, Old End (Ma)	*2 Sigma (myr)*	Duration (myr)	*2 Sigma (myr)*	Age (Ma)	Duration (myr)	*Δ Astro-Radio (myr)*	Age, Old End (Ma)	Duration (myr)
C6Cn.2n	501.55	23.8	23.03	23.03	*0.04*	0.13	*0.00*	23.024	–	*−0.007*	23.03	0.13
C6Cn.2r	506.47	23.999	23.249	23.25	*0.04*	0.22	*0.01*	23.233	0.209	*−0.017*	23.23	0.20
C6Cn.3n	509.41	24.118	23.375	23.38	*0.04*	0.13	*0.00*	23.295	0.062	*−0.088*	23.30	0.06
C6Cr	524.64	24.73	24.044	24.09	*0.05*	0.71	*0.02*	23.962	0.667	*−0.127*	23.96	0.67
C7n.1n	525.92	24.781	24.102	24.15	*0.05*	0.06	*0.00*	24.000	0.038	*−0.150*	24.00	0.04
C7n.1r	527.29	24.835	24.163	24.21	*0.05*	0.07	*0.00*	24.109	0.109	*−0.106*	24.11	0.11
C7n.2n	536.04	25.183	24.556	24.63	*0.05*	0.42	*0.01*	24.474	0.365	*−0.160*	24.47	0.36
C7r	543.97	25.496	24.915	25.02	*0.06*	0.38	*0.01*	24.761	0.287	*−0.258*	24.76	0.29
C7An	547.82	25.648	25.091	25.21	*0.07*	0.19	*0.01*	24.984	0.223	*−0.222*	24.98	0.22
C7Ar	552.30	25.823	25.295	25.43	*0.07*	0.22	*0.01*	25.099	0.115	*−0.327*	25.10	0.11
C8n.1n	555.55	25.951	25.444	25.58	*0.08*	0.16	*0.01*	25.264	0.165	*−0.321*	25.26	0.16
C8n.1r	556.60	25.992	25.492	25.64	*0.08*	0.05	*0.00*	25.304	0.040	*−0.332*	25.30	0.04
C8n.2n	571.04	26.554	26.154	26.34	*0.09*	0.71	*0.02*	25.987	0.683	*−0.356*	25.99	0.68
C8r	583.30	27.027	26.714	26.94	*0.10*	0.59	*0.02*	26.420	0.433	*−0.516*	26.42	0.43
C9n	607.96	27.972	27.826	28.09	*0.12*	1.16	*0.05*	27.439	1.019	*−0.655*	27.44	1.02
C9r	616.12	28.283	28.186	28.47	*0.12*	0.37	*0.02*	27.859	0.420	*−0.607*	27.86	0.42
C10n.1n	622.16	28.512	28.45	28.74	*0.13*	0.27	*0.01*	28.087	0.228	*−0.652*	28.09	0.23
C10n.1r	623.90	28.578	28.525	28.82	*0.13*	0.08	*0.00*	28.141	0.054	*−0.675*	28.14	0.05
C10n.2n	628.29	28.745	28.715	29.01	*0.14*	0.20	*0.01*	28.278	0.137	*−0.734*	28.28	0.14
C10r	645.65	29.401	29.451	29.77	*0.16*	0.76	*0.04*	29.183	0.905	*−0.587*	29.18	0.91
C11n.1n	652.56	29.662	29.74	30.06	*0.17*	0.30	*0.02*	29.477	0.294	*−0.588*	29.48	0.29
C11n.1r	655.31	29.765	29.853	30.18	*0.17*	0.12	*0.01*	29.527	0.050	*−0.654*	29.53	0.05
C11n.2n	664.15	30.098	30.217	30.55	*0.18*	0.37	*0.02*	29.970	0.443	*−0.581*	29.97	0.44

C11r	674.26	30.479	30.627	30.97	*0.19*	0.42	*0.02*	30.591	0.621	*−0.377*	30.59	0.62
C12n	686.50	30.939	31.116	31.46	*0.19*	0.49	*0.03*	31.034	0.443	*−0.428*	31.03	0.44
C12r	742.63	33.058	33.266	33.61	*0.26*	2.15	*0.26*	33.157	2.123	*−0.450*	33.16	2.12
C13n	755.44	33.545	33.738	34.08	*0.31*	0.47	*0.07*	33.705	0.548	*−0.374*	33.71	0.55
C13r	784.40	34.655	34.782	35.14	*0.42*	1.06	*0.14*	*34.999*	1.294	*−0.143*	35.00	1.29
C15n	791.78	34.94	35.043	35.41	*0.44*	0.27	*0.03*	*35.294*	0.295	*−0.117*	35.29	0.30
C15r	802.15	35.343	35.404	35.79	*0.45*	0.38	*0.03*	*35.706*	0.411	*−0.084*	35.71	0.41
C16n.1n	806.87	35.526	35.567	35.96	*0.46*	0.17	*0.01*	*35.892*	0.186	*−0.070*	35.89	0.19
C16n.1r	810.93	35.685	35.707	36.11	*0.47*	0.15	*0.01*	*36.051*	0.159	*−0.058*	36.05	0.16
C16n.2n	827.67	36.341	36.276	36.72	*0.49*	0.61	*0.05*	*36.700*	0.649	*−0.017*	36.70	0.65
C16r	834.68	36.618	36.512	36.97	*0.50*	0.25	*0.02*	*36.969*	0.269	*−0.002*	36.97	0.27
C17n.1n	856.19	37.473	37.235	37.75	*0.52*	0.78	*0.07*	*37.785*	0.816	*0.032*	37.75	0.78
C17n.1r	859.46	37.604	37.345	37.87	*0.52*	0.12	*0.01*	*37.908*	0.123	*0.036*	37.87	0.12
C17n.2n	865.54	37.848	37.549	38.09	*0.51*	0.22	*0.02*	*38.135*	0.228	*0.042*	38.09	0.22
C17n.2r	867.33	37.92	37.61	38.16	*0.52*	0.07	*0.01*	*38.202*	0.067	*0.044*	38.16	0.07
C17n.3n	872.10	38.113	37.771	38.33	*0.52*	0.17	*0.02*	*38.380*	0.178	*0.047*	38.33	0.17
C17r	879.83	38.426	38.032	38.62	*0.52*	0.28	*0.03*	*38.668*	0.288	*0.052*	38.62	0.28
C18n.1n	907.31	39.552	38.975	39.63	*0.51*	1.01	*0.11*	*39.686*	1.018	*0.058*	39.63	1.01
C18n.1r	909.21	39.631	39.041	39.70	*0.51*	0.07	*0.01*	*39.756*	0.070	*0.058*	39.70	0.07
C18n.2n	921.21	40.13	39.464	40.14	*0.50*	0.45	*0.04*	*40.201*	0.445	*0.056*	40.14	0.45
C18r	947.96	41.257	40.439	41.15	*0.50*	1.01	*0.09*	*41.204*	1.003	*0.049*	41.15	1.01
C19n	954.12	41.521	40.671	41.39	*0.50*	0.24	*0.02*	*41.438*	0.234	*0.048*	41.39	0.24
C19r	977.65	42.536	41.59	42.30	*0.48*	0.91	*0.08*	*42.351*	0.913	*0.049*	42.30	0.91
C20n	1006.06	43.789	42.774	43.43	*0.42*	1.13	*0.11*	*43.505*	1.154	*0.073*	43.43	1.13
C20r	1060.24	46.264	45.346	45.72	*0.28*	2.29	*0.27*	*45.942*	2.438	*0.218*	45.72	2.29
C21n	1094.71	47.906	47.235	47.35	*0.20*	1.62	*0.18*	47.837	1.895	*0.488*	47.35	1.62
C21r	1117.55	49.037	48.599	48.57	*0.20*	1.22	*0.12*	49.102	1.265	*0.536*	48.57	1.22
C22n	1130.78	49.714	49.427	49.34	*0.22*	0.78	*0.06*	49.740	0.638	*0.396*	49.34	0.78
C22r	1150.83	50.778	50.73	50.63	*0.22*	1.28	*0.07*	51.209	1.469	*0.581*	50.63	1.28
C23n.1n	1153.90	50.946	50.932	50.83	*0.22*	0.21	*0.01*	51.433	0.224	*0.598*	50.83	0.21
C23n.1r	1155.75	51.047	51.057	50.96	*0.22*	0.13	*0.01*	51.505	0.072	*0.544*	50.96	013

(Continued)

TABLE 28.3 Numerical Age Models for Paleogene C-Sequence Polarity Chrons—cont'd

Marine Magnetic Anomaly Profile		CK92/95	GTS2004	Radio-Isotopic Age Model				Astronomic Age Model with 6-order Polynomial Fit in Eocene			Combined Age Model	
Polarity Chron	Distance, Old End (km)	Age (Ma)	Age (Ma)	Age, Old End (Ma)	*2 Sigma (myr)*	Duration (myr)	*2 Sigma (myr)*	Age (Ma)	Duration (myr)	*Δ Astro-Radio (myr)*	Age, Old End (Ma)	Duration (myr)
C23n.2n	1168.20	51.743	51.901	51.83	*0.21*	0.87	*0.04*	51.884	0.379	*0.051*	51.83	0.87
C23r	1178.96	52.364	52.648	52.62	*0.18*	0.79	*0.04*	52.684	0.800	*0.064*	52.62	0.79
C24n.1n	1184.03	52.663	53.004	53.00	*0.16*	0.38	*0.02*	53.074	0.390	*0.073*	53.07	0.45
C24n.1r	1185.61	52.757	53.116	53.12	*0.15*	0.12	*0.01*	53.199	0.125	*0.078*	53.20	0.13
C24n.2n	1186.34	52.801	53.167	53.18	*0.15*	0.06	*0.00*	53.274	0.075	*0.098*	53.27	0.08
C24n.2r	1188.05	52.903	53.286	53.31	*0.15*	0.13	*0.01*	53.416	0.142	*0.109*	53.42	0.14
C24n.3n	1195.35	53.347	53.808	53.87	*0.11*	0.56	*0.04*	53.983	0.567	*0.112*	53.98	0.57
C24r	1234.51	55.904	56.665	56.94	*0.06*	3.07	*0.15*	57.101	3.118	*0.158*	57.10	3.12
C25n	1241.50	56.391	57.18	57.49	*0.08*	0.55	*0.02*	57.656	0.555	*0.168*	57.66	0.56
C25r	1257.81	57.554	58.379	58.74	*0.11*	1.25	*0.06*	58.959	1.303	*0.217*	58.96	1.30
C26n	1262.74	57.911	58.737	59.11	*0.12*	0.37	*0.02*	59.237	0.278	*0.122*	59.24	0.28
C26r	1303.81	60.92	61.65	62.07	*0.13*	2.95	*0.11*	62.221	2.984	*0.155*	62.22	2.98
C27n	1308.70	61.276	61.983	62.40	*0.12*	0.33	*0.01*	62.517	0.296	*0.119*	62.52	0.30
C27r	1325.71	62.499	63.104	63.52	*0.09*	1.13	*0.05*	63.494	0.977	*−0.030*	63.49	0.98
C28n	1341.99	63.634	64.128	64.56	*0.04*	1.03	*0.06*	64.667	1.173	*0.108*	64.67	1.17
C28r	1347.03	63.976	64.432	64.87	*0.04*	0.31	*0.02*	64.958	0.291	*0.090*	64.96	0.29
C29n	1358.66	64.745	65.118	65.56	*0.04*	0.69	*0.04*	65.688	0.730	*0.127*	65.69	0.73

Radioisotopic, astronomical (middle columns) and combined (right columns) age models for Paleogene chron boundaries as discussed in the text. Also shown are the differences between the astronomical and radioisotopic age models and the positions of chron boundaries in the marine magnetic anomaly profile and their age in CK92/95 and GTS2004.

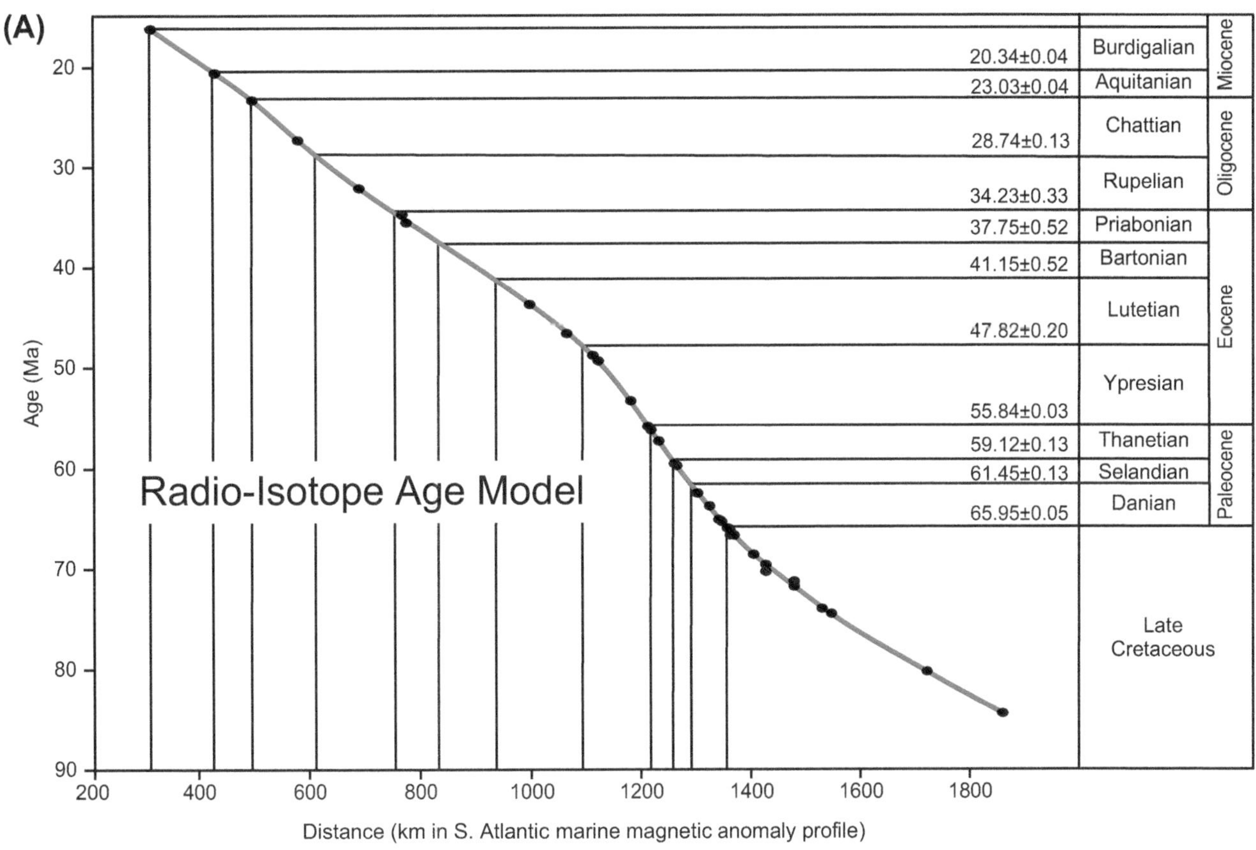

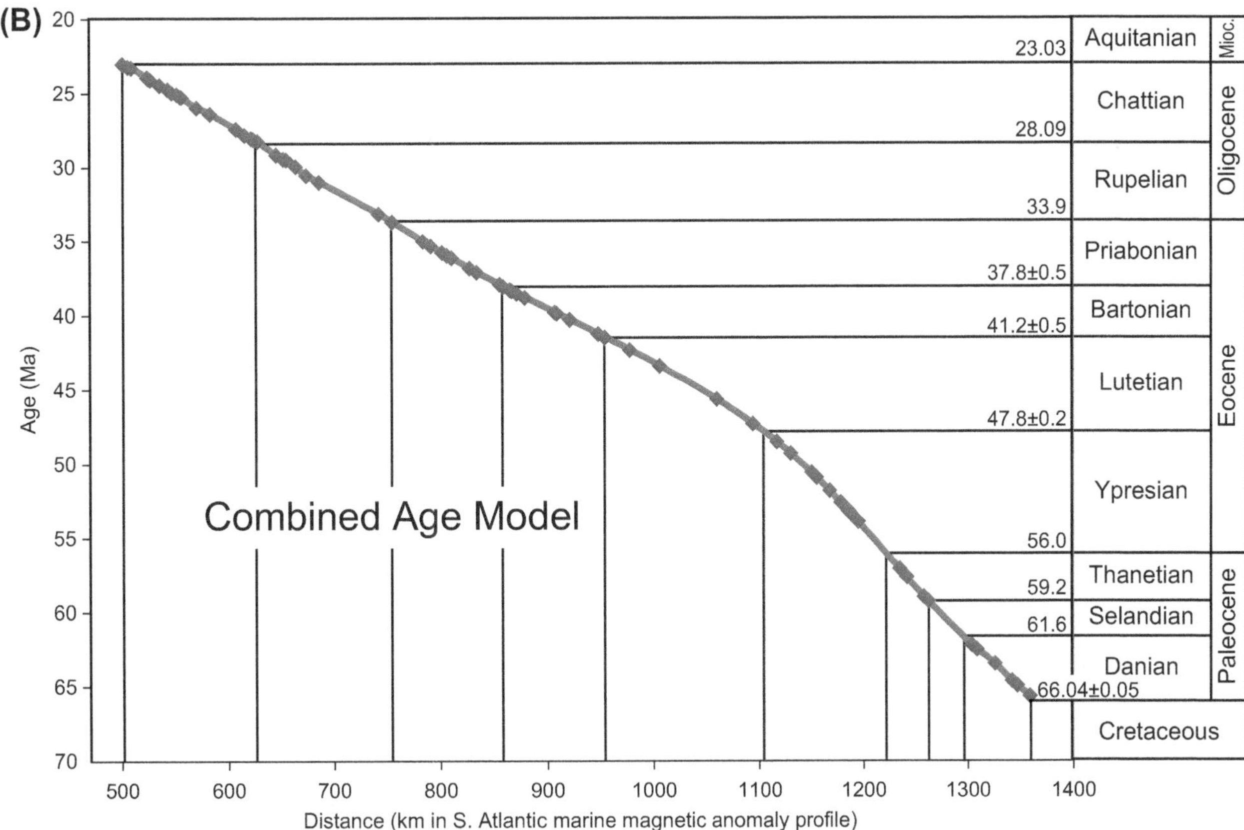

FIGURE 28.12 Age models of C-sequence marine magnetic anomalies in the South Atlantic for two Paleogene GTS solutions. (A) Spline fit of selected Paleogene radio-isotopic dates for chron age-distances and selected cyclostratigraphy durations (Table 28.1); the interpolated chron ages are tabulated under "Radio-Isotopic Age Model" columns in Table 28.3. (B) Chron age-distances (diamonds) from the "Combined Age Model" (rightmost column in Table 28.3) that incorporated "Astro-Age Model" (right column of Table 28.2) for Chrons C6C through C13n, a bridging polynomial for Chron C15n, the spline-fit to radio-isotope dates for Chrons C15r-C23r (as shown in the upper "A" panel, and the Astro-Age Model for Chrons C24n-C29r).

1.125. Five points did not pass the chi-squared test at $p>0.05$: 767.02 km, 772.82 km, 1150.83 km, 1214.93 km and 1222.82 km. Hence the uncertainties were increased for these data points. The final smoothing factor is 0.475. The complete C-sequence results are tabulated in Table 28.3 (middle part) and the stage boundary ages in Table 28.4 (middle part under "radio-isotope age model").

The GTS2012 orbitally tuned cycle ages for the C-sequence are shown in Table 28.3 (right side), and the orbitally tuned cycle ages for the Paleogene stage boundaries in Table 28.4. Comparison of the chron ages reveals that the two age models are remarkably consistent, even though significant discrepancies are found around 28 and 50 Ma, with differences exceeding 0.5 myr (Table 28.3, column with delta astro-radio). Around 50 Ma, the astronomically derived ages are older, while they are younger around 28 Ma. A smaller discrepancy with differences of up to 200 kyr is found around 60 Ma, with astronomical ages being older (Table 28.3, column with Δ astro-radio).

Problems that exist in the radio-isotopic age model include the use of less suitable minerals (biotite) and multi-crystal – instead of single crystal – dating for some of the tie points, and also the link to the magnetostratigraphy and the calibration of this magnetostratigraphy to the polarity time scale. Biotite ages had to be included around the E/O boundary in absence of more reliable single crystal $^{40}Ar/^{39}Ar$ sanidine or U/Pb zircon ages, but are in fact unsuitable for high-resolution time scale work. A recent study revealed that their ages can be hundreds of thousands years too old at the young end of the time scale (Hora *et al.*, 2010). Uncertainty in the placement relative to the magnetostratigraphy and, hence, in the chronozonal assignment is the reason for removing two data points from the original set of seven tie points selected by Smith *et al.* (2008) to revise the GPTS. However, also for some of the selected tie points this chron position is not without uncertainty. For instance, several alternative magnetostratigraphic interpretations/ records have been published for the Massignano section in Italy; this section provides two biotite dates of Late Eocene age that are included as tie points in radio-isotopic age model. Uncertainties also exist for geological reasons in the chron placement of some of the older ashes (Belt, -17) as we will see when the main discrepancies between the two age models are discussed.

The astronomical age model is also uncertain for several reasons. In the first place, it can be questioned whether the tuning to ~405-kyr eccentricity of the Paleocene (and early Eocene), although apparently constrained by recalculated single crystal $^{40}Ar/^{39}Ar$ sanidine dates (of Swisher *et al.*, 1993), is correct. The recalculated ages were not part of an interlaboratory assessment; they have only been analyzed in a single laboratory, but are consistent with $^{40}Ar/^{39}Ar$ ages of K/Pg boundary tektites of Haiti (Izett *et al.*, 1991). Results from an interlaboratory comparison project revealed discrepancies of up to 1–1.5% (Heizler *et al.*, 2008). This is pertinent because Swisher and Prothero (1990) published much older single crystal $^{40}Ar/^{39}Ar$ sanidine ages than Obradovich and others (1995; see Hilgen and Kuiper, 2009) for the same ash layers of late Eocene age in North America, while single crystal ages of Swisher *et al.* (1993) were used to constrain the tuning of the K/Pg boundary in the Zumaia section (Kuiper *et al.*, 2008). However, the Eocene discrepancy might have a geological origin (contamination, Sahy *et al.*, 2011) and, thus, have no bearing on the reliability of the K/Pg boundary ages. Nevertheless, an uncertainty in the tuning of one cycle to 405-kyr eccentricity can not be excluded until additional state-of-the-art single crystal $^{40}Ar/^{39}Ar$ sanidine and U/Pb zircon ages have been published, preferably from different laboratories and from different key stratigraphic levels, such as the K/Pg, P/E and E/O boundaries.

As far as the tuning uncertainty is concerned, a 405-kyr younger tuning is more likely than a 405-kyr older tuning (see e.g., Hilgen *et al.*, 2010). Such a younger tuning would reduce discrepancies with the radio-isotopic ages from North America (Smith *et al.*, 2008), which are responsible for the discrepancies around 50 Ma, and restore the inferred relation between the obliquity dominated interval at ODP site 1258 and the supposedly correlative 2.4 myr eccentricity minimum as inferred by Westerhold and Röhl (2009). However, such a younger tuning would at the same time result in discrepancies with the radio-isotopic age model in the interval between 66 and 52 Ma, and a shift towards younger ages is not corroborated by published radio-isotopic ages. Our age model is consistent with single crystal $^{40}Ar/^{39}Ar$ sanidine ages of ash beds of early Paleocene age from the continental succession of North America published two decades ago (Swisher *et al.*, 1993), as these were used to establish the initial tuning (Kuiper *et al.*, 2008). U/Pb ages of 66.1 Ma for the K/Pg boundary and of 64.7 Ma for C28r have been presented in abstracts (e.g., Bowring *et al.*, 2008) and are consistent with the astronomical time scale presented in Section 28.4.2 as well. The same holds for the U/Pb zircon age of 55.785 ± 0.034 Ma for the ash bed intercalated in the lower part of the PETM on Spitsbergen (Charles *et al.*, 2011). Combined with cyclostratigraphy, this resulted in an age of 55.84 for the PETM base, which is, within error, compatible with the astronomical age of 55.93 Ma. This is encouraging as the youngest U/Pb ID-TIMS age of an ash bed, following new preparation techniques, might be very close to the eruption age (Schaltegger *et al.*, 2009). Also the $^{40}Ar/^{39}Ar$ sanidine age of 52.93 ± 0.12 Ma for the Willwood ash in North America (Smith *et al.*, 2004) is consistent with the astronomical age of 53.021 Ma for the base C24n.1n. Up to now, there are no published radio-isotope data for the interval between the K/Pg boundary and C24n.1n that unambiguously support

TABLE 28.4 Summary of Paleogene Stage Boundary Calibrations to C-Sequence Polarity Chrons with Corresponding Ages and Durations

		Calibration	GTS2004				Radio-Isotopic Age Model		Astronomical Age Model		Combined Age Model = Final GTS2012 Version		
Epoch	Age/Stage	Base of Stage –Chron Assignment or Cycle-Offset from Chron Boundary	(Ma)	Distance (km) in S. Atlantic Profile	*Uncertainty (km)*	Age of Base (Ma)	*2 Sigma*	Duration	*Age*	*Duration*	Age of Base	2 *Sigma*	Duration
	Langhian	Base of Chron C5Br.	16.0	318.39	*0.46*	15.97	*0.04*		15.97		**15.97**		
	Burdigalian	Assigned here as base of Chron C6An.1r; but another option is base of Chron C6An.1n.	20.4	434.18	*0.27*	20.34	*0.04*	4.36	20.34	4.36	**20.44**		4.47
Miocene	Aquitanian	Base of Chron C6Cn.2n.	23.0	501.55	*0.12*	23.03	*0.04*	2.70	23.03	2.70	**23.03**		2.59
	Chattian	Assigned here as base of Chron C10n.1n. [*NOTE: Base of Chattian is potentially at ~70% up in "undifferentiated Chron C10n" in candidate GSSP in Italy (Coccioni et al., 2008), which would project as equivalent to C10n.1n.4.*]	28.4	622.16	*0.36*	28.74	*0.13*	5.71	28.09	5.06	**28.09**		5.06
Oligocene	Rupelian	Chron C13r.86; this chron may include a brief normal-polarity cryptochron near the top of Eocene.	33.9	759.49	*2.03*	34.23	*0.33*	5.49	33.89	5.80	**33.9**		5.8
	Priabonian	Assigned here as base of Chron C17n.1n.	37.2	856.19	*0.65*	37.75	*0.52*	3.53	37.83	3.94	**37.8**	*0.5*	3.9
	Bartonian	Assigned here as base of Chron C18r; but potential GSSP may be as low as mid-C19r.	40.4	947.96	*0.53*	41.15	*0.52*	3.40	41.22	3.40	**41.2**	*0.5*	3.4
	Lutetian	Approximately C21r.6. [*NOTE: another estimate for GSSP is 39 precession cycles (0.80 myr if 20.5-kyr cycles) above base of Chron C21r*].	48.6	1103.85	*2.51*	47.82	*0.20*	6.67	48.29	7.07	**47.8**	*0.2*	6.7

(Continued)

TABLE 28.4 Summary of Paleogene Stage Boundary Calibrations to C-Sequence Polarity Chrons with Corresponding Ages and Durations—cont'd

		Calibration	GTS2004			Radio-Isotopic Age Model			Astronomical Age Model		Combined Age Model = Final GTS2012 Version		
Epoch	Age/Stage	Base of Stage –Chron Assignment or Cycle-Offset from Chron Boundary	(Ma)	Distance (km) in S. Atlantic Profile	*Uncertainty (km)*	Age of Base (Ma)	*2 Sigma*	Duration	*Age*	*Duration*	Age of Base	2 *Sigma*	Duration
Eocene	Ypresian	This PETM isotope excursion begins at c. 1.4 myr above the base of Chron C24r, or at approximately C24r.36 (Westerhold *et al.*, 2008). An U-Pb age of 55.79 Ma is about 0.05 myr above the base of the PETM	55.8	1220.41	*1.96*	55.84	*0.03*	8.02	55.93	7.64	**56.0**		8.1
	Thanetian	Base of Chron C26n.	58.7	1262.74	*0.15*	59.12	*0.13*	3.28	59.24	3.31	**59.2**		3.3
	Selandian	The isotope shift begins at 30 precession cycles (0.62 myr if 20.5-kyr cycles) above base of C26r at GSSP.	61.7	1303.81 km to reference Chron; then subtract 0.62 myr	*1.23*	61.45	*0.13*	2.33	61.60	2.36	**61.6**		2.4
Paleocene	Danian	Mesozoic/Cenozoic boundary is approximately Chron C29r.5. Chron C29r.5 in total C29r span of c. 710 kyr, according to cycles (Husson *et al.*, 2011; Thibault *et al.*, submitted)	65.5	1365.30	*1.30*	65.95	*0.05*	4.50	65.95	4.35	**66.04**	*0.05*	4.4
	Maastrichtian	Base of Maastrichtian is approximately at Chron C32n.2n.9; or at 72.05 Ma from spline-fit to ammonite zones (see Cretaceous chapter).	70.6	1497.70	*0.88*	72.10	*0.19*	6.15			**72.1**	*0.2*	6.1
		NOTE: Km-uncertainty on fixed chron-bases (definitions) uses half of the percent-error on mean width in Cande-Kent table. For estimated placements (e.g., C29r.5), a ±0.05 chron uncertainty was assumed.											

The computed ages from the two C-sequence models in Table 28.3 (spline-fit to all values in Table 28.1, and a selected composite of cyclostratigraphy and spline-fit) are compared to GTS2004. The full spline-fit also provided statistics on uncertainties and only a few of these are estimated for the cyclostratigraphy based model. The final Paleogene numerical time scale for GTS2012 is the cyclostratigraphy based model with a portion of the Eocene radio-isotope spline of Figure 28.12a. Durations in rightmost column have been rounded.

a 405-kyr younger tuning. The radio-isotopic ages of the Belt and -17 ash beds apparently result in chron boundary ages that are 100–200 kyr younger than the astronomically derived age estimates, but there is a more logical geological explanation for this offset (see below).

Another uncertainty that remains in the astronomical age model is the age of the FCs dating standard as recalculated $^{40}Ar/^{39}Ar$ ages were used to constrain the first-order tuning to 405-kyr eccentricity (Kuiper *et al.*, 2008; Hilgen *et al.*, 2010). However, recently published ages for this standard vary between 27.93 and 28.305 ± 0.036 Ma (Channell *et al.*, 2010; Renne *et al.*, 2010), and include the astronomically calibrated age of 28.201 ± 0.046 Ma of Kuiper *et al.* (2008) applied throughout GTS2012. The latter age was employed to (re) calculate single crystal $^{40}Ar/^{39}Ar$ sanidine ages in order to constrain the astronomical tuning of the Paleocene on the 405-kyr cycle scale (Kuiper *et al.*, 2008; Hilgen *et al.*, 2010). The age has been independently confirmed by new sanidine ages of an ash layer in a well-tuned section of the Mediterranean Neogene (Rivera *et al.*, 2011). Application of the age also results in consistency between $^{40}Ar/^{39}Ar$ and U/Pb ages for the Bishop tuff, where other ages result in a less coherent picture (Rivera *et al.*, 2011). Together, this suggests that the age of 28.305 Ma of Renne *et al.* (2010) based on Ar/Ar-U/Pb age pairs is too old and that the age of 27.93 Ma of Channell *et al.* (2010) is too young. However, a different FCs age does not explain the age discrepancy between the astronomical and radio-isotopic time scale as it would similarly affect both scales.

Another uncertainty in the astronomical age model is that one 405-kyr cycle may have been overlooked in the cyclostratigraphic studies of the deep-sea cores. As such, discussion is still ongoing whether the Paleocene contains 24 or 25 × 405-kyr cycles (Westerhold *et al.*, 2008; Hilgen *et al.*, 2010). Evidently, this has consequences for the astronomical age model presented here (Section 28.4.2), as we assume that 25 × 405-kyr cycles are present. A missing 405-kyr cycle might also be responsible for the abrupt beginning of the discrepancy around 50 Ma. This will be outlined below, as we will next discuss the main misfits between the radiometric and astronomical models around 60, 50 and 28 Ma from old to young.

28.4.3.1. The 60 Ma Discrepancy

The discrepancy between 63 and 57 Ma with astronomical ages being 100 to 200 kyr older is most likely related to the inclusion of the Belt ash and ash -17 as tie points in the radio-isotopic age model. The chron position of the Belt ash is given as C26r.85 on the basis of the magnetostratigraphy of the Southeast Polecat Bench section, but a hiatus is inferred at the base of the reversed of C26r, while in addition the sedimentation rate is assumed to increase (Secord *et al.*, 2006). As a consequence the position of the Belt ash is probably higher up in C26r, which would explain the younger radio-isotopic ages for the chron boundaries in this interval. Downward extrapolation of the C26n sedimentation rate in the Polecat Bench section, starting from the astronomical ages of the chron boundaries, yields an age of 59.35 Ma for the Belt ash, which is in perfect agreement with its $^{40}Ar/^{39}Ar$ age of 59.39 ± 0.3 Ma. However, this would correspond to a position at 0.963 within C26r rather than the C26r.9 incorporated in Table 28.1 on the basis of its magnetostratigraphic position. This seemingly small difference is largely responsible for the ~200 kyr as C26r is almost 3 million years long. The same reasoning applies to ash -17, as a hiatus is inferred on the basis of the biostratigraphy between ash -17 and the PETM (see Westerhold *et al.*, 2008; Hilgen *et al.*, 2010); again this would result in radio-isotopic ages of chron boundaries that are too young.

28.4.3.2. The 50 Ma Discrepancy

Around 50 Ma, the astronomical age model point reveals a duration of 380 kyr for C23n.2n (and a total duration of 708 kyr for C23n) as compared to 870 kyr (and 1200) in the radio-isotopic age model, which is more than twice as long. The astronomical age model gives a duration of 800 kyr for C23r, which is consistent with a 790 kyr duration according to the radio-isotopic age model. The 400–500 kyr difference remains essentially the same from C23n up to C21r, and is largely balanced by the longer astronomical duration for C20r and C21r between ~47.5 and ~43.5 Ma. The comparison shows that the problem resides in the short astronomical duration of C23n.2n (or C23n as the reversal of C23n.1r is not well resolved at ODP Sit 1258); this duration shows a 1:1 ratio to the duration of C23r. This is in sharp contrast with the observed 1:2 width ratio of C23n.2n and C23r in the synthetic anomaly profile of the southern Atlantic of Cande and Kent (1992) (note that this ratio is in fact based on the incorporation of data from Indian Ocean profiles). The same ratio is found in anomaly profiles from different ocean basins (e.g., Heirtzler *et al.*, 1968; LaBrecque *et al.*, 1977).

Excluding an abrupt global increase (doubling) of spreading rates in C23n.2n, several options remain to explain the discrepancy around 50 Ma. The first option is that a 405-kyr eccentricity related cycle has been missed by Westerhold and Röhl (2009) in the interval of ODP Site 1258 that corresponds to C23n.2n. Checking the detailed patterns shows that the cyclostratigraphic interpretation is not straightforward and that a missing cycle cannot be excluded. Evidently, independent confirmation of the cycle pattern from other cyclic deep-marine records is needed to solve this problem. Another explanation would be a fault or hiatus of 400–500 kyr, or – coincidentally – one 405-kyr cycle, in the C23n interval. This was originally not precluded by Suganuma and Ogg (2006), although the improved splice and nannofossil biostratigraphy helped to solve some of these issues (Westerhold and Röhl, 2006). Also the magnetostratigraphic interpretation is not without difficulty, but additional land-based measurements proved very useful in this respect (Suganuma and Ogg, 2006; Westerhold and Röhl, 2009). In fact, the total astronomical duration for the C24n.3n(o) to C21r(y) interval is roughly 400-kyr shorter (note that the identification of base C24 and C22r(y) to C21n at site 1258 is not disputed) than the duration according to the radio-isotopic age model. Combined this makes a fault/hiatus of approximately 400-kyr, or a missing 405-kyr cycle the most likely explanation for the observed discrepancies between the two age models around 50 Ma. This would imply that the radio-isotopic age model is better in this interval. The fact that the discrepancy is again compensated between C21n(y) and C20n(o); i.e. over 4 myr, is most likely a consequence of the spline, as it starts directly above the last directly astronomically dated tie point, where the interpolation using the CK95 anomaly profile starts in the astronomical age model (see Section 28.4.2).

28.4.3.3. The 28 Ma Discrepancy

The major discrepancy between the two age models around 28 Ma is most likely related to the incorporation of the $^{40}Ar/^{39}Ar$ biotite ages of the two ash layers in the Monte Cagnero section as tie points in the radio-isotopic age model (see Section 28.4.1; Coccioni *et al.*, 2008). In fact, biotite ages are considered unsuitable for high-resolution time scale studies and they were only used in the absence of more reliable sanidine ages. The problem of biotite ages is nicely demonstrated by the recent study of Hora *et al.* (2010), indicating that offsets to older ages in the order of hundreds of thousands of years might be anticipated at the younger end of the geological time scale. Nevertheless, the observed discrepancy seems to be too large to be explained in this way. The discrepancy would have been significantly reduced - to ~200 kyr - if we would have included the single crystal $^{40}Ar/^{39}Ar$ sanidine ages of McIntosh *et al.* (1992; see also Hilgen and Kuiper, 2009) in our radioisotopic age model instead of the biotite ages of Coccioni et al. (2010).

The fact that Oligocene spreading rates vary more smoothly according to the radio-isotopic age model results from the limited number of tie points applied while astronomical dating attempts to date every single magnetic reversal boundary. The more jumpy spreading rates resulting from the astronomical age model are probably due to small uncertainties in the tuning or in locating the exact reversal position in the magnetostratigraphy, although latter uncertainties are very small at ODP Site 1218 used to establish the astronomical polarity time scale for the entire Oligocene (Lanci *et al.*, 2005; Pälike *et al.*, 2006). Nevertheless independent testing of the tuning in other deep-sea cores such as those recently recovered during IODP legs 320/321 (Pälike *et al.*, 2008) is needed as – minor – uncertainties may still exist in the tuning around the E–O and O–M boundaries.

28.4.3.4. Paleogene Time Scale

Finally, we have to select the age model that we are going to use for building the Paleogene time scale. Evidently, neither of the current two radio-isotopic and astronomical age models is better over the entire range of the Paleogene. The astronomical age model can be held responsible for the discrepancies around 50 Ma, even though the exact cause of the discrepancy remains uncertain (i.e. hiatus, missing cycle). On the other hand, the radio-isotopic age model is most likely responsible for the discrepancies observed around 60 and 28 Ma. Apparently, the best solution would be a combination of both age models. For this purpose, we selected the astronomically derived ages for the chron boundaries between 66 and 53 Ma, and between 37 and 23 Ma. The ages for the chron boundaries in between then come from the radio-isotopic age model. The resulting chron boundary ages are given in Table 28.3 (combined age model column), and it is this polarity time scale that forms the backbone of our integrated Paleogene time scale (Table 28.4, right column). In fact, such a hybrid time scale nicely demonstrates that the Paleogene time scale is in a state of transition towards a time scale that is fully underlain by astronomical dating.

The preferred Paleogene geochronology for GTS2012 is thus tabulated in Table 28.4 (under "Combined age model = final GTS2012 version"). It shows the age of the bases for the nine Paleogene stages, the associated 2-sigma uncertainty as far as derived from direct radiometric or cubic spline interpolations, and the duration in millions of years for each stage. Durations are listed with rounded off decimal fractions. The base Danian and base Paleocene is 66.04 ± 0.05 Ma; the stage lasted 4.4 myr. The Selandian had a duration of 2.4 myr, and its base is 61.6 Ma. Base Thanetian is 59.2 Ma and the stage lasted 3.3 myr. The base Ypresian and base Eocene is 56.0 Ma; the stage is the longest in the Paleogene with 8.1 myr. Base Lutetian is 47.8 ± 0.2 Ma and the stage is 6.7 myr long. Base Bartonian is 41.2 ± 0.5 Ma and Bartonian lasted 3.3 myr. Priabonian started at 37.8 ± 0.5 Ma and lasted 3.9 myr. The Oligocene Epoch and Rupelian Stage started at 33.9 Ma, and the stage lasted 5.8 myr. The final stage of the Oligocene, the Chattian, started at 28.09 Ma and lasted 5.06 myr. Orbital tuning of more Paleogene sections, resolving discrepancies in geomagnetic polarity assignments for selected Eocene strata, and formal definition of the Bartonian, Priabonian and Chattian stage boundaries will lead to further refinement of Paleogene geochronology.

TABLE 28.5 GSSPs of the Paleogene Stages, with Location and Primary Correlation Criteria (Status Jan. 2012)

Stage	GSSP Location	Latitude, Longitude	Boundary Level	Correlation Events	Reference
Chattian	*Not yet defined*			*Provisional: base of magnetic polarity chronozone C10n.1n*	
Rupelian	Massignano, near Ancona, Italy	43° 31′58.2″N 13° 36′03.8″E*	Base of a 0.5m thick, greenish-gray marl bed, 19 m above base of section	Foraminifer LADs of *Hantkenina* and *Cribrohantkenina*	Episodes 16/3 1993
Priabonian	*Not yet defined*			*Provisional: base of magnetic polarity chronozone C17n.1n*	
Bartonian	*Not yet defined*			*Provisional: base of magnetic polarity chronozone C18r*	
Lutetian	Gorrondatxe beach section near Getxo village, W Pyrenees, Spain	43° 22′ 46.47″ N 3° 00′ 51.61″ W	167.85m above the base of the Gorrondatxe section in a dark marly level	Calcareous nannofossil FAD of *Blackites inflatus*	Episodes 34/2, 2011
Ypresian	Dababiya, near Luxor, Egypt	25° 30′04.70″ N* 32° 31′19.14″ E*	Base of Bed 1 in DBH subsection	Base of the Carbon Isotope Excursion (CIE)	Episodes 30/4 2007
Thanetian	Zumaia section, northern Spain	43° 17′ 58.4″N* 02° 15′ 39.1″W*	6.5m above the base of Member B of the Itzurun Formation	Magnetic polarity chronozone, base of C26n	Episodes 35, 2012.
Selandian	Zumaia section, northern Spain	43° 17′ 57.1″N* 02° 15′ 39.6″W*	Base of the red marls of the Itzurun Formation	Major Se1 sequence boundary and sea-level low; second radiation of the calcareous nannofossil *Fasciculithus* (base Zone NP5)	Episodes 35, 2012.
Danian *(base Paleogene)*	Oued Djerfane, west of El Kef, Tunisia	36° 09′13.2″N 8° 38′54.8″E	Reddish layer at the base of the 50 cm thick, dark boundary clay	Iridium geochemical anomaly. Associated with a major extinction horizon (foraminifers, calcareous nannofossils, dinosaurs, etc.)	Episodes 29/4, 2006.

**according to Google Earth*

ACKNOWLEDGMENTS

This chapter is a major revision and update of Chapter 20 on the Paleogene Period in GTS2004 by H.P. Luterbacher, J.R. Ali, H. Brinkhuis, F.M. Gradstein, J.J. Hooker, S. Monechi, J.G. Ogg, J. Powell, U. Röhl, A. Sanfilippo, and B. Schmitz.

REFERENCES

Abelmann, A., 1990. Oligocene to middle Miocene radiolarian stratigraphy of southern high latitudes from Leg 113, Sites 689 and 690, Maud Rise. Proceedings of the Ocean Drilling Program, Scientific Results 113, 675–708.

Abels, H.A., Van Simaeys, S., Hilgen, F.J., De Man, E., Vandenberghe, N., 2007. Obliquity dominated glacio-eustatic sea-level change in the early Oligocene: Evidence from the shallow marine siliciclastic Rupelian stratotype (Boom Formation, Belgium). Terra Nova 19, 65–73.

Abreu, V.S., Anderson, J.B., 1998. Glacial eustacy during the Cenozoic: Sequence stratigraphic implications. American Association Petroleum Geologists Bulletin 82, 1385–1400.

Abreu, V.S., Haddad, G.A., 1998. Glacioeustacy: The mechanism linking isotope event and sequence stratigraphy. In: de Graciansky, P.-C., Hardenbol, J., Jacquin, T., Vail, P.R. (Eds.), Mesozoic and Cenozoic Sequence Stratigraphy of European Basins. SEPM Special Publication, 60, pp. 245–259.

Adams, C.G., 1970. A reconsideration of the East Indian letter classification of the Tertiary. Bulletin of the British Museum of Natural History (Geology) 19, 87–137.

Agnini, C., Macri, P., Backman, J., Brinkhuis, H., Fornaciari, E., Giusberti, L., Luciani, V., Rio, D., Sluijs, A., Speranza, F., 2009. An early Eocene carbon cycle perturbation at ~52.5 Ma in the Southern Alps: Chronology and biotic response. Paleoceanography 24 PA2209. doi: 10.1029/2008PA001649.

Agnini, C., Foranciari, E., Giusberti, L., Grandesso, P., Lanci, L., Luciani, V., Muttoni, G., Pälike, H., Rio, D., Spofforth, D.J.A., Stefani, C., 2011. Integrated bio-magnetostratigraphy of the Alano section (NE Italy): A proposal for defining the middle-late Eocene boundary. Geological Society of America Bulletin 123, 841–872.

Aguilar, J.-P., Legendre, S., Michaux, J., 1997. Actes du Congrès Biochrom'97 Montpellier, 14–17 Avril. Biochronologie mammalienne du Cénozoïque en Europe et domaines reliés. Mémoires et Travaux de l'Institut de Montpellier de l'École Pratique des Hautes Études 21, 1–181.

Ali, J.R., King, C., Hailwood, E.A., 1993. Magnetostratigaphic calibration of early Eocene depositional sequences in the southern North sea Basin. In: Hailwood, E.A., Kidd, R.B. (Eds.), High Resolution Stratigraphy. Special Publication, n°70. Geological Society, London, pp. 99–125.

Ali, J.R., Jolley, D.W., 1996. Chronostratigraphic framework for the Thanetian and lower Ypresiand eposits of SE England. Geological Society Special Publication 101, 129–144.

Alroy, J., 1992. Conjunction among taxonomic distributions and the Miocene mammalian biochronology of the great-plains. Paleobiology 18, 326–343.

Alvarez, L.W., Alvarez, W., Asaro, E., Michel, H.V., 1980. Extraterrestrial cause for the Cretaceous-Tertiary extinction. Science 208, 1095–1108.

Anderson, A.J., 1961. Gliederung und paläogeographische Entwicklung der Chattischen Stufe (Oberolkigozän) im Nordseebecken. Meyniana 10, 118–146.

Anderson, A.J., Hinsch, W., Martini, E., Müller, C., Ritzkowski, S., 1971. Chattian. Giornale di Geologia 37, 69–79.

Angori, E., Monechi, S., 1996. High-resolution nannofossil biostratigraphy across the Paleocene-Eocene boundary at Caravaca (southern Spain). Israel Journal of Earth Sciences 44, 207–216.

Arduino, G., 1759. Lettera seconda sopre varie osservazioni fatti in diversi parti del territorio di Vicenza, ed altrove, appartenenti alla teoria terrestre, ed alla mineralogía. Venezia.

Arz, J.A., Arenillas, I., Molina, E., Dupuis, C., 1999. Los efectos tafonómico y "Signor-Lipps" sobre la extinción en masa de foraminíferos planctónicos en el límite Cretácico/Terciario de Elles (Tunicia). Revista de la Sociedad Geológica de España 12, 251–268.

Aubry, M.-P., 1992. Late Paleogene calcareious nannoplankton evolution: a tale of climatic amelioration. In: Prothero, D.R., Berggren, W.R. (Eds.), Eocene-Oligocene Climatic and Biotic Evolution. Princeton University Press, Princeton, N.J, pp. 272–309.

Aubry, M.-P., 1996a. Towards an Upper Paleocene-Lower Eocene high resolution stratigraphy based on calcareous nannofossil stratigraphy. Israel Journal of Earth Sciences 44, 239–253.

Aubry, M.-P., 1998a. The upper Paleocene-lower Eocene stratigraphic puzzle. Strata 6, 7–9.

Aubry, M.-P., 1998b. Early Paleogene calcareous nannplankton evolution. A tale of climatic amelioration. In: Aubry, M.-P., Lucas, S.G., Berggren, W.A. (Eds.), Late Paleocene–Early Eocene Climatic and Biotic Events in the Marine and Terrestrial Records. Columbia University Press, New York, pp. 158–203.

Aubry, M.-P., Villa, G., 1996. Calcareous nannplankton stratigraphy of the Lemme-Carrosio Paleogene/Neogene Global Stratotype Section and Point. Giornale di Geologia, series 3 58 (1–2), 51–69.

Aubry, M.-P., Lucas, S.G., Berggren, W.A. (Eds.), 1998. Late Paleocene-Early Eocene Climatic and Biotic Events in the Marine and Terrestrial Records. Columbia University Press, New York p. 513.

Aubry, M.-P., Berggren, W.A., Van Couvering, J.A., Ali, J., Brinkhuis, H., Cramer, B., Kent, D.V., Swisher III, C.C., Dupuis, C., Gingerich, P.D., Heilmann-Clausen, C., King, C., Ward, D.J., Knox, R.W., Ouda, K., Stott, L.D., Thiry, M., 2003. Chronostratigraphic terminology at the Paleocene/Eocene boundary. In: Wing, S.L., Gingerich, P.D., Schmitz, B., Thomas, E. (Eds.), Causes and Consequences of Globally Warm Climates in the Early Paleogene. Geological Society of America Special Paper, 369, pp. 551–566.

Aubry, M.-P., Ouda, K., Dupuis, C., Berggren, W.A., Van Couvering, J.A., the Members of the Working Group on the Paleocene/Eocene Boundary, 2007. The Global Standard Stratotype-section and Point (GSSP) for the base of the Eocene Series in the Dababiya section (Egypt). Episodes 30, 271–286.

Barbin, V., 1988. The Eocene-Oligocene transition in shallow-water environment: The Priabonian stage type area (Vicentin, northern Italy). In International Subcommission on Paleogene Stratigraphy. Proceedings of the Eocene-Oligocene Boundary Meeting, Ancona, Oct. 1987, pp. 163–171.

Barker, R.W., Grimsdale, T.F., 1936. A contribution to the phylogeny of the orbitoidal foraminifera, with description of new forms from the Eocene of Mexico. Journal of Paleontology 10, 231–247.

Berggren, W.A., 1994. In defense of the Selandian Age/Stage. GFF 116, 44–46.

Berggren, W.A., Miller, K.G., 1988. Paleogene tropical planktonic foraminiferal biostratigraphy and magnetobiochronology. Micropaleontology 34, 362–380.

Berggren, W.A., Ouda, K., 2003. Upper Paleocene-lower Eocene planktonic Foraminifera biostratigraphy of the Qreiya (Gebel Abu Had) section, upper Nile Valley (Egypt). In: Ouda, K., Aubry, M.-P. (Eds.), The Upper Paleocene-Lower Eocene of the Upper Nile Valley; Part 1, Stratigraphy. Micropaleontology 49 (Suppl. 1), 105–122.

Berggren, W.A., Pearson, P.N., 2005. A revised tropical to subtropical Paleogene planktonic foraminiferal zonation. Journal of Foraminiferal Research 35, 279–298.

Berggren, W.A., Kent, D.V., Flynn, J.J., 1985. Paleogene geochronology and chronostratigraphy. In: Snelling, N.J. (Ed.), The Chronology of the Geological Record. Geological Society Memoir, 10, pp. 141–195.

Berggren, W.A., Kent, D.V., Swisher III, C.C., Aubry, M.-P., 1995. A revised Cenozoic geochronology and chronostratigraphy. In: Berggren, W.A., Kent, D.V., Hardenbol, J. (Eds.), Geochronology, Time Scales and Global Stratigraphic Correlations SEPM Special Publication, 54, pp. 129–212.

Bernaola, G., Baceta, J.I., Orue-Etxebarria, X., Alegret, L., Martín-Rubio, M., Arostegui, J., Dinarès-Turell, J., 2007. Evidence of an abrupt environmental disruption during the mid-Paleocene biotic event (Zumaia section, Western Pyrenees). Geological Society of America Bulletin 119, 785–795.

Bernaola, G., Martín-Rubio, M., Baceta, J.I., 2009. New high resolution calcareous nannofossil analysis across the Danian/Selandian transition at the Zumaia section: Comparison with South Tethys and Danish sections. Geologica Acta 7, 79–92.

Beyrich, E., 1854. Über die Stellungder hessischen Tertiärbilldungen. Berichte der Verhandlungen der königlichen. Preussischen Akademie der Wissenschaften, Akademie der Wissenschaften zu Berlin 1854, 640–666.

Bignot, G., 1981. Cuisian. In: Pomerol, C. (Ed.), Stratotypes of Paleogene Stages. Bulletin d'information des géologues du Bassin de Paris, Mémoire hors série, 2, pp. 63–75.

Bignot, G., Curry, D., Pomerol, C., 1997. The resistible rise of the Selandian. Neues Jahrbuch für Geologie und Paläontologie Monatshefte 2, 114–128.

Bjørklund, K.R., 1976. Radiolaria from the Norwegian Sea, Leg 38 of the Deep Sea Drilling Project. Initial Reports of the Deep Sea Drilling Project 38, 1101–1168.

Blondeau, A., 1981. Lutetian. In: Pomerol, C. (Ed.), Stratotypes of Paleogene Stages. Bulletin d'information des géologues du Bassin de Paris, Mémoire hors série, 2, pp. 167–180.

Blondeau, A., Cavelier, C., Feugueur, L., Pomerol, C., 1965. Stratigraphie du Paléogène du Bassin de Paris en relation avec les bassins avoisinants. Bulletin de la Société Géologique de France 7, 200–221.

Blow, W.H., 1979. The Cainozoic Foraminifera, Vols.1–3. E.J. Brill, Leiden, p. 1413.

Bohaty, S.M., Zachos, J.C., 2003. Significant Southern Ocean warming event in the late middle Eocene. Geology 31, 1017–1020.

Bohaty, S.M., Zachos, J.C., Florindo, F., Delaney, M.L., 2009. Coupled greenhouse warming and deep-sea acidification in the Middel Eocene. Paleoceanography 24 PA2207. doi: 10.1029/2008PA001676.

Bolli, H.M., 1957a. The genera *Globigerina* and *Globorotalia* in the Paleocene – Lower Eocene Lizard Springs Formation of Trinidad, B.W.I. Bulletin of the United States National Museum 215, 61–82.

Bolli, H.M., 1957b. Planktonic foraminifera from the Eocene Navet and San Fernando Formations of Trinidad, B.W.I. Bulletin of the United States National Museum 215, 155–172.

Bolli, H.M., 1966. Zonation of Cretaceous to Pliocene marine sediments based on planktonic foraminifera. Boletín Informativo de la Asociación Venezolana de Geología, Minería y Petróleo 9, 3–32.

Bolli, H.M., Saunders, J.B., 1985. Oligocene to Holocene low latitude planktic foraminifera. In: Bolli, H.M., Saunders, J.B., Perch-Nielsen, K. (Eds.), Plankton Stratigraphy. Cambridge University Press, Cambridge, pp. 155–262.

Bornemann, A., Schulte, P., Sprong, J., Steurbaut, E., Youssef, M., Speijer, R.P., 2009. Latest Danian carbon isotope anomaly and associated environmental change in the southern Tethys (Nile basin, Egypt). Journal of the Geological Society 166, 1135–1142.

Boudagher-Fadel, M.K., Banner, F.T., 1999. Revision of the stratigraphic significance of the Oligocene-Miocene "letter-stages". Revue de Micropaléontologie 42, 93–97.

Boulila, S., Galbrun, B., Miller, K.G., Pekar, S.F., Browning, J.V., Laskar, J., Wright, J.D., 2011. On the origin of Cenozoic and Mesozoic "third-order" eustatic sequences. Earth-Science Reviews 109, 94–112. doi: 10.1016/j.earscirev.2011.09.003.

Bowen, G.J., Koch, P.L., Gingerich, P.D., Norris, R.D., Bains, S., Corfield, R.M., 2001. Refined isotope stratigraphy across the continental Paleocene-Eocene boundary at Polecat Bench in the northern Bighorn Basin. University of Michigan Papers on Paleontology 33, 73–88.

Bown, P.R., 2005. Palaeogene calcareous nannofossils from the Kilwa and Lindi areas of coastal Tanzania (Tanzania Drilling Project 2003–4). Journal of Nannoplankton Research 27, 21–95.

Bown, P.R., Dunkley Jones, T., 2006. New Palaeogene calcareous nannofossil taxa from coastal Tanzania: Tanzania Drilling Project Sites 11–14. Journal of Nannoplankton Research 28, 17–34.

Bowring, S.A., Johnson, K.R., Clyde, W.C., Ramezani, J., Miller, I., Peppe, D., 2008. A Paleocene Timescale for the Rocky Mountains: Status and potential. Geological Society of America Abstracts with Programs 40 (6), 322.

Brabb, E.E., 1968. Comparison of the Belgian and German Oligocene sequences for the purpose of selecting a stratotype. Bureau de Recherches Géologiques et Minières Mémoire 69, 77–82.

Bralower, T.J., Premoli-Silva, I., Malone, M.J., et al., 2002. Proceedings of the Ocean Drilling Program, Initial Reports, Leg 198. Available at: http://www.-odp.tamu.edu/publications/198_IR/198ir.htm.

Bramlette, M.N., Riedel, W.R., 1954. Stratigraphic value of discoasters and some other microfossils related to Recent coccolithophores. Journal of Paleontology 28, 385–403.

Bramlette, M.N., Sullivan, F.R., 1961. Coccolithophorids and related nannoplankton of the early Tertiary in California. Micropaleontology 7, 129–188.

Bramlette, M.N., Wilcoxon, J.A., 1967. Middle Tertiary calcareous nannoplankton of the Cipero Section, Trinidad, W.I. Tulane Studies in Geology and Paleontology 5, 93–131.

BRGM, 1968. Colloque sur l'Eocène. 3 vols. Mémoires Bureau de Recherche Géologique et Minière 58 (I) 59(II), 69(III).

Brinkhuis, H., 1994. Late Eocene to Early Oligocene dinoflagellate cysts from the Priabonian type-area (northeast Italy); biostratigraphy and palaeoenvironmental interpretation. Palaeogeography, Palaeoclimatology. Palaeoecology 107, 121–163.

Brinkhuis, H., Biffi, U., 1993. Dinoflagellate cyst stratigraphy of the Eocene Oligocene transition in Central Italy. Marine Micropaleontology 22, 131–183.

Brinkhuis, H., Visscher, H., 1995. The upper boundary of the Eocene Series: A reappraisal based on dinoflagellate cyst biostratigraphy and sequence stratigraphy. SEPM Special Publication 54, 295–304.

Brinkhuis, H., Munsterman, D.K., Sengers, S., Sluijs, A., Warnaar, J., Williams, G.L., et al., 2003a. Late Eocene to Quaternary dinoflagellate cysts from ODP Site 1168, Off western Tasmania. In: Exon, et al. (Eds.), Scientific Results ODP Leg 189, College Station. Available at. http://www-odp.tamu.edu/publications/189_SR/105/105.htm.

Brinkhuis, H., Sengers, S., Sluijs, A., Warnaar, J., Williams, G.L., 2003b. Latest Cretaceous to earliest Oligocene, and Quaternary dinoflagellate cysts from ODP Site 1172, East Tasman Plateau. In: Exon, et al. (Eds.), Scientific Results ODP Leg 189, College Station. Available at. http://www-odp.tamu.edu/publications/189_SR/106/106.htm.

Brinkhuis, H., Head, M.J., Pearce, M.A., Riding, J., Pross, J., Fensome, R.E., Williams, G.L., 2009. Mesozoic-Cenozoic dinoflagellate cyst biostratigraphy. Short Course, Urbino 2009. LPP Foundation Publications, Utrecht.

Brönnimann, P., Stradner, H., 1960. Die Foraminiferen- und Discosasteriden-Zonen von Kuba und ihre interkontinentale Korrelation. Erdölzeitschrift 76, 364–369.

Brown, R.E., Koeberl, C., Montanari, A., Bice, D.M., 2009. Evidence for a change in Milankovitch forcing caused by extraterrestrial events at Massignano, Italy, Eocene-Oligocene boundary GSSP. Geological Society of America Special Papers 452, 119–137.

Buchardt, B., 1978. Oxygen isotope palaeotemperatures from the tertiary period in the North Sea area. Nature 275, 121–123.

Bujak, J.P., Brinkhuis, H., 1998. Global warming and dinocyst changes across the Paleocene/Eocene boundary. In: Aubry, M.-P., Lucas, S.G.,

Berggren, W.A. (Eds.), Late Paleocene—Early Eocene Climatic and Biotic Events in the Marine and Terrestrial Records. Columbia University Press, New York, pp. 277—295.

Bukry, D., 1973. Low latitude coccolith biostratigraphic zonation. Initial Reports of the Deep Sea Drilling Project 15, 658—677.

Bukry, D., 1975. Coccolith and Silicoflagellate stratigraphy, northwestern Pacific Ocean. Initial Reports of the Deep Sea Drilling Project 32, 67—701.

Burbank, D.W., Puigdefàbregas, C., Muñoz, J.A., 1992. The chronology of the Eocene tectonic and stratigraphic development of the eastern Pyrenean foreland basin, northeast Spain. Geological Society of America Bulletin 104, 1101—1120.

Butler, R.F., Krause, D.W., Gingerich, P.D., 1987. Magnetic polarity stratigraphy and biostratigraphy of middle—late Paleocene continental deposits of south-central Montana. Journal of Geology 95, 647—657.

Butterlin, J., 1988. Origine et évolution des Lépidocyclines de la région Caraïbes, comparaison et relations avec les Lépidocyclines des autres régions du monde. Revue de Paléobiologie, volume spécial 2, 623—629.

Bystricka, H., 1963. Die unter-eozänen Coccolithophoriden (Flagellata) des Myjavaer Paleogens. Geologiske Sbornik (Bratislava) 14, 269—281.

Bystricka, H., 1965. Der stratigraphische Wert von Discoasteriden im Palaeogen der Slowakei. Geologiske Sbornik (Bratislava) 16, 7—10.

Cahuzac, B., Poignant, A., 1997. Essai de biozonation de l'Oligo-Miocène dans les bassins européens à l'aide des grands foraminifères néritiques. Bulletin de la Société Géologique de France 168, 155—169.

Cande, S.C., Kent, D.V., 1992. A new geomagnetic polarity time scale for the Late Cretaceous and Cenozoic. Journal of Geophysical Research 97, 13917—13951.

Cande, S.C., Kent, D.V., 1995. Revised calibrations of the geomagnetic polarity timescale for the Late Cretaceous and Cenozoic. Journal of Geophysical Research 100, 6093—6095.

Caudri, C.M.B., 1996. The larger foraminifera of Trinidad. Eclogae Geologicae Helvetica 89, 1137—1309.

Cavelier, C., 1981. Stampian. In: Pomerol, C. (Ed.), Stratotypes of Paleogene Stages. Bulletin d'information des géologues du Bassin de Paris, Mémoire hors série, 2, pp. 231—254.

Cavelier, C., Pomerol, C., 1977. Proposition d'une échelle stratigraphique standard pour le Paléogène. Newsletters on Stratigraphy 6, 56—65.

Cavelier, C., Pomerol, C., 1983. Echelle de corrélation stratigraphique du Paléogène. Stratotypes, étages standards, biozones, chimiozones et anomalies magnétiques. Géologie de la France 3, 261—262.

Cavelier, C., Pomerol, C., 1986. Stratigraphy of the Paleogene. Bulletin de la Société Géologique de France 8, 255—265.

Channell, J.E.T., Hodell, D.A., Singer, B.S., Xuan, C., 2010. Reconciling astrochronological and $^{40}Ar/^{39}Ar$ ages for the Matuyama-Brunhes boundary and late Matuyama Chron. Geochemistry, Geophysics, Geosystems 11 Q0AA12. doi: 10.1029/2010GC003203.

Chaproniere, C.G.H., 1984. Oligocene and Miocene larger Foraminiferida from Australia and New Zealand. Bulletin of the Bureau of Mineral Resources of Australia 188, 1—98.

Charles, A.J., Condon, D.J., Harding, I.C., Pälike, H., Marshall, J.E.A., Cui, Y., Kump, L., Croudace, I.W., 2011. Constraints on the numerical age of the Paleocene/Eocene boundary. Geochemistry, Geophysics, Geosystems 12 Q0AA17. doi: 10.1029/2010GC003426.

Clemens, W.A., 2002. Evolution of mammalian fauna across the Cretaceous-Tertiary boundary in northeastern Montana and other areas of the Western Interior. Geological Society of America Special Paper 361, 217—245.

Clyde, W.C., Stamatakos, J., Gingerich, P.D., 1994. Chronology of the Wasatchian land-mammal age: Magnetostratigraphic results from the McCullough Peaks section, northern Bighorn Basin, Wyoming. Journal of Geology 102, 367—377.

Clyde, W.C., Zonneveld, J.-P., Stamatakos, J., Gunnell, G., Bartels, W.S., 1997. Magnetostratigraphy across the Wasatchian/Bridgerian NALMA boundary (Early to Middle Eocene) in the western Green River Basin, Wyoming. Journal of Geology 105, 657—669.

Clyde, W.C., Sheldon, N.D., Koch, P.L., Gunnell, G.F., Bartels, W.S., 2001. Linking the Wasatchian/Bridgerian boundary to the Cenozoic Climate Optimum: New magnetostratigraphic and isotopic results from South Pass, Wyoming. Palaeogeography, Palaeclimatology, Palaeoecology 167, 175—199.

Clyde, W.C., Bartels, W.S., Gunnell, G.F., Zonneveld, J.-P., 2004. Discussion: $^{40}Ar/^{39}Ar$ geochronology of the Eocene Green River Formation, Wyoming. Geological Society of America Bulletin 116, 251—252.

Clyde, W.C., Hamzi, W., Finarelli, J.A., Wing, S.L., Schankler, D., Chew, A., 2007. Basin-wide magnetostratigraphic framework for the Bighorn Basin, Wyoming. Geological Society of America Bulletin 119, 848—859.

Clyde, W.C., Tong, Y., Snell, K.E., Bowen, G.J., Ting, S., Koch, P.L., Li, Q., Wang, Y., Meng, J., 2008. An integrated stratigraphic record from the Paleocene of the Chijiang Basin, Jiangxi Province (China): Implications for mammalian turnover and Asian block rotations. Earth and Planetary Science Letters 269, 553—563.

Clyde, W.C., Ting, S., Snell, K.E., Bowen, G.J., Tong, Y., Koch, P.L., Li, Q., Wang, Y., 2010. New paleomagnetic and stable-isotope results from the Nanxiong Basin, China: Implications for the K/T Boundary and the timing of Paleocene mammalian turnover. Journal of Geology 118, 131—143.

Coccioni, R., Monaco, P., Monechi, S., Nocchhi, M., Parisi, G., 1988. Biostratigraphy of the Eocene-Oligocene boundary at Macignano (Ancona, Italy). In: Premoli Silva, I., Coccioni, R., Montanari, A. (Eds.), The Eocene-Oligocene Boundary in the Marche-Umbria Basin (Italy). International Union of Geological Sciences, International Subcommission on Paleogene Stratigraphy. Eocene-Oligocene Boundary Meeting Report, Ancona, Italy, pp. 59—80.

Coccioni, R., Basso, D., Brinkhuis, H., Galeotti, S., Gardin, S., Monechi, S., Spezzaferri, S., 2000. Marine biotic signals across a late Eocene impact layer at Masssignano, Italy. Terra Nova 12, 258—263.

Coccioni, R., Marsili, A., Montanari, A., Bellanca, A., Neri, R., Bice, D.M., Brinkhuis, H., Church, N., Macalady, A., McDaniel, A., Deino, A., Lirer, F., Sprovieri, M., Maiorano, P., Monechi, S., Nini, C., Nocchi, M., Pross, J., Rochette, P., Sagnotti, L., Tateo, F., Touchard, Y., Van Simaeys, S., Williams, G.L., 2008. Integrated stratigraphy of the Oligocene pelagic sequence in the Umbria-Marche basin (northeastern Apennines, Italy): A potential Global Stratotype Section and Point (GSSP) for the Rupelian/Chattian boundary. Geological Society of America Bulletin 120, 487—511.

Costa, L.L., Manum, S.B., 1988. Dinoflagellate cysts: The description of the international zonation of the Paleogene (D1 — D15) and the Miocene (D16 — D20). Geologisches Jahrbuch A100, 321—330.

Cramer, B.S., Toggweiler, J.R., Wright, J.D., Katz, M.E., Miller, K.G., 2009. Ocean overturning since the Late Cretaceous: Inferences from a new benthic foraminiferal isotope compilation. Paleoceanography 24 PA4216. doi: 10.1029/2008PA001683.

Cramer, B.S., Wright, J.D., Kent, D.V., Aubry, M.P., 2003. Orbital climate forcing of $\delta^{13}C$ excursions in the late Paleocene-early Eocene (chrons

C24n–C25n). Paleoceanography 18, PA1097. doi: 10.1029/2003PA000909.
Crouch, E.M., Heilmann-Clausen, C., Brinkhuis, H., Morgans, H.E.G., Rogers, K.M., Egger, H., Schmitz, B., 2001. Global dinoflagellate event associated with the late Paleocene thermal maximum. Geology 29, 315–318.
Crux, J., 1991. Calcareous nannoplankton recovered by Leg 114 in the Subantarctic south Atlantic Ocean. Proceedings of the Ocean Drilling Project, Scientific Results 114, 155–177.
Curry, D., 1981. Thanetian. In: Pomerol, C. (Ed.), Stratotypes of Paleogene Stages. Bulletin d'information des géologues du Bassin de Paris, Mémoire hors série, 2, pp. 255–265.
Dashzeveg, D., 1993. Asynchronism of the main mammalian events near the Eocene-Oligocene boundary. Tertiary Research 14, 141–149.
Dawber, C.F., Tripati, A.K., Gale, A.S., MacNiocaill, C., Hesselbo, S.P., 2011. Glacioeustacy during the middle Eocene? Insights from the stratigraphy of the Hampshire Basin, UK. Palaeogeography, Palaeoclimatology, Palaeoecology 300, 84–100.
Deflandre, G., Fert, C., 1954. Observations sur les coccolithophoridés actuels et fossiles en microscopie ordinaire et éléctronique. Annales de Paléontologie 40, 115–176.
De Geyter, G., De Man, E., Herman, J., Jacobs, P., Moorkens, T., Steurbaut, E., Vandenberghe, N., 2006. Disused Paleogene regional stages from Belgium: Montian, Heersian, Landenian, Paniselian, Bruxellian, Laekenian, Ledian, Wemmelian and Tongrian. Geologica Belgica 9, 203–213.
De Lapparent, A., 1883. Traité de Géologie. Savy, Paris, p. 1280.
De Man, E., Van Simaeys, S., 2004. Late Oligocene Warming Event in the southern North Sea Basin benthic Foraminifera as paleotemperature proxies. Netherlands Journal of Geosciences 83, 227–239.
De Man, E., Ivany, L., Vandenberghe, N., 2004. Stable oxygen isotope record of the Eocene-Oligocene transition in the southern North Sea Basin: Positioning the Oi-1 event. Netherlands Journal of Geosciences 83, 193–197.
De Man, E., Van Simaeys, S., De Meuter, F., King, C., Steurbaut, E., 2004. Oligocene benthic foraminiferal zonation for the southern North Sea Basin. Bulletin Koninklijk Belgisch Instituut voor Natuurwetenschappen Aardwetenschappen 74, 177–195.
De Man, E., Van Simaeys, S., Vandenberghe, N., Harris, W.B., Wampler, J.M., 2010. On the nature and chronostratigraphic position of the boundary between the historical Rupelian and Chattian unit stratotypes in the southern North Sea Basin. Episodes 33, 3–13.
Desor, F., 1847. Sur le terrain danien, nouvel étage de la Craie. Bulletin de la Société géologique de France 4, 179–182.
Dewalque, C., 1868. Prodrome d'une description geologique de la Belgique. Librairie Polytechnique De Decq, Bruxelles and Liege. p. 442.
D'haenens, S., Bornemann, A., Stassen, P., and Speijer, R.P., 2012. Multiple early Eocene benthic foraminiferal assemblage and $\delta^{13}C$ fluctuations at DSDP Site 401 (Bay of Biscay—NE Atlantic). Marine Micropaleontology 88–89, 15–35.
Dickens, G.R., Castillo, M.M., Walter, J.C.G., 1997. A blast of gas in the latest Paleocene: Simulating first-order effects of massive dissociation of oceanic methane hydrate. Geology 25, 259–262.
Dinarès-Turell, J., Baceta, J.I., Bernaola, G., Orue-Extebarria, X., Pujalte, V., 2007. Closing the Mid-Palaeocene gap: Toward a complete astronomically tuned Palaeocene Epoch and Selandian and Thanetian GSSPs at Zumaia (Basque basin, W Pyrenees). Earth and Planetary Science Letters 262, 450–467.
Dinarès-Turell, J., Stoykova, K., Baceta, J.L., Ivanov, M., Pujalte, V., 2010. High-resolution intra- and interbasinal correlation of the Danian-Selandian transition (Early Paleocene): The Bjala section (Bulgaria) and the Selandian GSSP at Zuamaia (Spain). Palaeogeography, Palaeoclimatology, Palaeoecology 297, 511–533.
Dollfus, G.F., 1880. Essai sur l'étendue des terrains tertiaires dans le bassin anglo-parisien. Bulletin de la Société géologique de Normandie 6, 584–605.
Dumont, A., 1839. Rapport sur les travaux de la carte géologique pendant l'année 1839. Bulletin de l'Académie Royale de Bruxelles 6, 464–485.
Dumont, A., 1849. Rapport sur la carte géologique de la Belgigue. Bulletin de l'Académie royales des Sciences et des Lettres de la Belgique 16, 351–373.
Dumont, A., 1851. Sur la position géologique de l'argile rupelienne et sur le synchronisme des formations tertiaires de la Belgique, de l'Angleterre et du Nord de la France. Bulletin de l'Académie royales des Sciences et des Lettres de la Belgique 18, 179–195.
Dupont-Nivet, G., Krijgsman, W., Langereis, C.G., Abels, H.A., Dai, S., Fang, X., 2007. Tibetan plateau aridification linked to global cooling at the Eocene-Oligocene transition. Nature 445, 635–638.
Dupuis, C., Aubry, M.-P., Steurbaut, E., Berggren, W.A., Ouda, K., Magioncalda, R., Cramer, B.S., Kent, D.V., Speijer, R.P., Heilmann-Clausen, C., 2003. The Dababiya Quarry Section: Lithostratigraphy, clay mineralogy, geochemistry and paleontology. Micropaleontology 49, 41–59.
Engesser, B., Mödden, C., 1997. A new version of the biozonation of the Lower Freshwater Molasse (Oligocene and Agenian) of Switzerland and Savoy on the basis of fossil mammals. In: Aguilar, J.-P., Legendre, S., Michaux, J. (Eds.), Actes du Congrès BiochroM'97. Mémoires et Travaux de l'Ecole Pratique des Hautes Etudes, Institut de Montpellier, 2, pp. 475–499.
Fahlbusch, V., 1976. Report on the International Symposium on Mammalian Stratigraphy of the European Tertiary, München, April 11–14, 1975. Newsletters on Stratigraphy 5, 160–167.
Fensome, R.A., Williams, G.L., 2004. The Lentin and Williams Index of Fossil Dinoflagellates. 2004 Edition. American Association of Stratigraphic Palynologists (AASP), Houston, p. 909.
Florindo, F., Wilson, G.S., Roberts, A.P., Sagnotti, L., Verosub, K.L., 2005. Magnetostratigraphic chronology of a late Eocene to early Miocene glacimarine succession from the Victoria Land Basin, Ross Sea, Antarctica. Global and Planetary Change 45 (special issue 1–3), 207–236.
Floris, S., Thomsen, E., 1981. Danian. In: Pomerol, C. (Ed.), Stratotypes of Paleogene Stages. Bulletin d'information des géologues du Bassin de Paris, Mémoire hors série 2, pp. 77–81.
Flynn, J.J., 1986. Global stratigraphic correlation of Mesoic-Cenozoic sediments of oceans and continents. Palaeogeography, Palaeoclimatology, Palaeoecology 55, 335–406.
Flynn, J.J., Swisher, C.C., 1995. Cenozoic South American Mammal Land Ages: Correlations to global geochronologies. SEPM Special Publication 54, 317–333.
Foreman, H., 1973. Radiolaria of Leg 10 with systematics and ranges for the families Amphipyndacidae, Artostrobiidae and Theoperidae. Initial Reports of the Deep Sea Drilling Project 10, 407–474.

Fornaciari, E., Rio, D., 1996. Latest Oligocene to early Middle Miocene quantitative calcareous nannofossil biostratigraphy in the Mediterranean region. Micropaleontology 42, 1–36.

Franzen, J.L., 2005. The implications of numerical dating of the Messel fossil deposit (Eocene, Germany) for mammalian biochronology. Annales de Paléontologie 91, 329–335.

Franzen, J.L., Haubold, H., 1987. The biostratigraphic and palaeoecologic significance of the middle Eocene locality Geiseltal near Halle (German Democratic Republic). Münchner Geowissenschaftliche Abhandlungen A 10, 93–100.

Fuchs, T., 1894. Tertaerfossilien aus den kohlenführenden Miocaenablagerungen der Umgebung von Krapina und Radobog und über die Stellung der sogenannten Stufe. Mitteilungen und Jahrbuch der königlichen Ungarischen Geologischen Anstalt 10, 161–175.

Galeotti, S., Kaminski, M.A., Coccioni, R., Speijer, R.P., 2004. High-resolution deep-water agglutinated foraminiferal record across the Paleocene/ Eocene transition in the Contessa Road section (central Italy). In: Bubik, M., Kaminski, M.A. (Eds.), Proceedings of the Sixth International Workshop on Agglutinated Foraminifera. Grzybowski Foundation Special Publication, 8, pp. 83–103.

Gheerbrant, E., Sudre, J., Sen, S., Abrial, C., Marandat, B., Sigé, B., Vianey-Liaud, M., 1998. Nouvelles données sur les mammifères du Thanétien et de l'Yprésien du Bassin d'Ouarzazate (Maroc) et leur contexte stratigraphique. Palaeovertebrata 27, 155–202.

Gignoux, M., 1950. Géologie stratigraphique, fourth ed. Masson, Paris. p. 735.

Gingerich, P.D., 1976. Cranial anatomy and evolution of early Tertiary Plesiadapidae (Mammalia, Primates). University of Michigan Papers on Paleontology 15 141.

Gingerich, P.D., 1992. Marine mammals (Cetacea and Sirenia) from the Eocene of Gebel Mokattam and Fayum, Egypt: Stratigraphy, age, and paleoenvironments. University of Michigan Papers on Paleontology 30, 1–84.

Goerges, J., 1957. Die Mollusken der oberoligozänen Schichten des Doberges bei Bünde in Westfalen. Paläontologische Zeitschrift 31, 116–134.

Gohn, G.S., Koeberl, C., Miller, K.G., Reimold, W.U., 2009. The ICDP-USGS Deep Drilling Project in the Chesapeake Bay impact structure: Results from the Eyreville core holes. Geological Society of America, Special Paper 458, 977.

Gradstein, F.M., Kaminski, M.A., Berggren, W.A., Kristiansen, I.L., D'Ioro, M.A., 1994. Cenozoic biostratigraphy of the North Sea and Labrador Shelf. Micropaleontology 40 (suppl.), 1–152.

Guasti, E., Speijer, R.P., 2007. The Paleocene-Eocene thermal maximum in Egypt and Jordan; an overview of the planktic foraminiferal record. In: Monechi, S., Coccioni, R., Rampino, M.R. (Eds.), Large Ecosystem Perturbations; Causes and Consequences. Geological Society of America Special Paper, 424, pp. 53–67.

Guasti, E., Speijer, R.P., 2008. *Acarinina multicamerata* n. sp. (Foraminifera): A new marker for the Paleocene-Eocene Thermal Maximum. Journal of Micropalaeontology 27, 5–12.

Guasti, E., Speijer, R.P., Brinkhuis, H., Smit, J., Steurbaut, E., 2006. Paleoenvironmental change at the Danian-Selandian transition in Tunisia; Foraminifera, organic-walled dinoflagellate cyst and calcareous nannofossil records. Marine Micropaleontology 59, 210–229.

Guerstein, G.R., Chiesa, J.O., Guler, M.V., Camacho, H.H., 2002. Bioestratigrafia Basada en Quiestes de Dinoflagelados de la Formacion Cabo Pena (Eoceno terminal-Oligoceno temprano), Tierra del Fuego, Argentina. Revista Española de Micropaleontologia 34, 105–116.

Gunnell, G.F., Murphey, P.C., Stucky, R.K., Townsend, K.E.B., Robinson, P., Zonneveld, J.-P., Bartels, W.S., 2009. Biostratigraphy and biochronology of the latest Wasatchian, Bridgerian, and Uintan North American Land Mammal "Ages". In: Albright III, L.B. (Ed.), Papers on Geology, Vertebrate Paleontology, and Biostratigraphy in honor of Michael O. Woodburne. Museum of Northern Arizona Bulletin, 65, pp. 279–330.

Hansen, J.M., 1979. Age of the Mo-Clay Formation. Bulletin Geological Society of Denmark 27, 89–91.

Haq, B.U., Hardenbol, J., Vail, P.R., 1987. Chronology of fluctuating sea levels since the Triassic. Science 235, 1156–1167.

Hardenbol, J., 1968. The Priabon type section (Italy). Mémoires du Bureau de Recherches géologiques et minières 58, 629–635.

Hardenbol, J., Berggren, W.A., 1978. A new Paleogene numerical time scale. In: Cohee, G.V., Glaessner, M.F., Hedberg, H.D. (Eds.), Contributions to the Geologic Time Scale. American Association of Petroleum Geologists Studies in Geology, 6, pp. 213–234.

Hardenbol, J., Thierry, J., Farley, M.B., Jacquin, T., de Graciansky, P.-C., Vail, P.R., 1998. Mesozoic and Cenozoic sequence chronostratigraphic framework of European Basins. In: de Graciansky, P.-C., Hardenbol, J., Jacquin, T., Vail, P.R. (Eds.), Mesozoic and Cenozoic Sequence Stratigraphy of European Basins. SEPM Special Publication, 60, pp. 3–14.

Harland, W.B., Armstrong, R.L., Cox, A.V., Craig, L.A., Smith, A.G., Smith, D.G., 1990. A Geologic Time Scale 1989. Cambridge University Press, Cambridge, p. 263.

Harland, W.B., Hine, N.M., Wilkinson, I.P., 1992. Paleogene biostratigraphic markers. In: Knox, R.W.O.'B., Holloway, S. (Eds.), Paleogene of the Central and Northern North-Sea. British Geological Survey, Nottingham, pp. A1–A5.

Haug, E., 1908–1911. Les Périodes géologiques. In: Traité de Géologie, Vol. II. Armand Collin, Paris. in 3 parts 1766–1776.

Hay, W.W., 1964. Utilisation stratigraphique des discoasteridés pour la zonation du Paléocène et de l'Eocène inférieur. Mémoires du Bureau de Recherches géologiques et minières 28, 885–889.

Hay, W.W., Mohler, H.P., 1967. Calcareous nannoplankton from early Tertiary rocks at Pont Labau, France, and Paleocene-Eocene correlations. Journal of Paleontology 41, 1505–1541.

Hay, W.W., Mohler, H.P., Roth, H.P., Schmidt, R.R., Boudreaux, J.E., 1967. Calcareous nannoplankton zonation of the Cenozoic of the Gulf Coast and Caribbean-Antillan area, and trans-oceanic correlation. Transactions of the Gulf-Coast Association of Geological Societies 17, 428–480.

Head, M.J., Gibbard, P., Salvador, A., 2008. The Tertiary: A proposal for its formal definition. Episodes 31, 248–250.

Heilmann-Clausen, C., 1985. Dinoflagellate stratigraphy of the uppermost Danian to Ypresian in the Viborg 1 borehole, central Jylland, Denmark. Danske Geologiske Undersøgelse A7, 1–89.

Heilmann-Clausen, C., 1988. The Danish Sub-basin: Paleogene dinoflagellates. Neues Jahrbuch für Geologie und Paläontologie Abhandlungen A101, 339–343.

Heirtzler, J.R., Dickson, G.O., Herron, E.M., Pitman, W.C., Le Pichon, X., 1968. Marine magnetic anomalies, geomagnetic field reversals, and motions of the ocean floor and continents. Journal of Geophysical Research 73, 2119–2139.

Heizler, M.T., Parson, I., Sanders, R.E., Heizler, L.L., 2008. K-Feldspar Microtexture and Argon Transport. Geological Society of America Abstracts with Programs 40 (6), 310.

Herbert, T.D., 1999. Toward a composite orbital chronology for the Late Cretaceous and Early Paleocene GPTS 357, 1891–1905.

Hilgen, F.J., Kuiper, K.F., 2009. A critical evaluation of the numerical age of the Eocene-Oligocene boundary. In: Koeberl, C., Montanari, A. (Eds.), The Late Eocene Earth—hothouse, icehouse, and impacts. Geological Society of America Special Paper, 452, pp. 1–10.

Hilgen, F.J., Kuiper, K.F., Lourens, L.J., 2010. Evaluation of the astronomical time scale for the Paleocene and earliest Eocene. Earth and Planetary Science Letters 300, 139–151.

Hiza, M.M., 1999. The geochemistry and geochronology of the Eocene Absaroka volcanic province northern Wyoming and southwest Montana, USA. Ph.D. thesis, Oregon State University, 243.

Hollis, C.J., 1993. Latest Cretaceous to Late Paleocene radiolarian biostratigraphy: A new zonation from the New Zealand region. Marine Micropaleontology 21, 295–327.

Hollis, C.J., 1997. Cretaceous-Paleocene Radiolaria from eastern Marlborough, New Zealand. Institute of Geological and Nuclear Sciences, Monograph 17, 152.

Hollis, C.J., 2002. Biostratigraphy and paleoceanographic significance of Paleocene radiolarians from offshore eastern New Zealand. Marine Micropaleontology 46, 265–316.

Hollis, C.J., 2006. Radiolarian faunal change across the Paleocene-Eocene boundary at Mead Stream, New Zealand. Eclogae Geologicae Helvetiae 99, S79–S99.

Hollis, C.J., Kimura, K., 2001. A unified radiolarian zonation for the Late Cretaceous and Paleocene of Japan. Micropaleontology 47, 235–255.

Hollis, C.J., Waghorn, D.P., Strong, C.P., Crouch, E.M., 1997. Integrated Paleogene biostratigraphy of DSDP Site 277 (Leg 29), foraminifera, calcareous nannofossils, Radiolaria and palynomorphs. Institute of Geological and Nuclear Sciences, Science Report 97/07.

Hollis, C.J., Dickens, G.R., Field, B.D., Jones, C.M., Strong, C.P., 2005. The Paleocene-Eocene transition at Mead Stream, New Zealand; a southern Pacific record of early Cenozoic global change. Palaeogeography, Palaeoclimatology, Palaeoecology 215, 313–343.

Holroyd, P.A., Ciochon, R.L., 1994. Relative ages of Eocene primate-bearing deposits of Asia. In: Fleagle, J.G., Kay, R.F. (Eds.), Anthropoid Origins. Plenum Press, New York, pp. 123–141.

Hooker, J.J., 1986. Mammals from the Bartonian (middle/late Eocene) of the Hampshire Basin, southern England. Bulletin of the British Museum of Natural History (Geology) 39, 191–478.

Hooker, J.J., 1987. Mammalian faunal events in the English Hampshire Basin (late Eocene-early Oligocene) and their application to European biostratigraphy. Münchner Geowissenschaftliche Abhandlungen A 10, 109–116.

Hooker, J.J., 1991. The sequences of mammals in the Thanetain and Ypresian of the London and Belgian basins: Location of the Paleocene-Eocene boundary. Newsletters on Stratigraphy 25, 75–90.

Hooker, J.J., 1996. Mammalian biostratigraphy across the Paleocene-Eocene boundary in the Paris, London and Belgian basins. Geological Society Special Publication 101, 205–218.

Hooker, J.J., 1998. Mammalian faunal change across the Paleocene-Eocene transition in Europe. In: Aubry, M.-P., Lucas, S.G., Berggren, W.A. (Eds.), Late Paleocene – Early Eocene Climatic and Biotic Events in the Marine and Terrestrial Records. Columbia University Press, New York, pp. 428–450.

Hooker, J.J., Millbank, C., 2001. A Cernaysian mammal from the Upnor Formation (Late Paleocene, Herne Bay, UK) and its implications for correlation. Proceedings of the Geologists' Association 89, 331–338.

Hooker, J.J., Weidmann, M., 2007. A diverse rodent fauna from the middle Bartonian (Eocene) of Les Alleveys, Switzerland: Snapshot of the early theridomyid radiation. Swiss Journal of Geosciences 100, 469–493.

Hora, J.M., Singer, B.S., Jicha, B.R., Beard, B.L., Johnson, C.M., de Silva, S., Salisbury, M., 2010. Volcanic biotite-sanidine $^{40}Ar/^{39}Ar$ age discordances reflect Ar partitioning and pre-eruption closure in biotite. Geology 38, 923–926.

Hörnes, M., 1853. Mitteilung an Prof. Bronn gerichtet: Wien, 3. Okt., 1853. Neues Jahrbuch für Mineralogie, Geologie, Geognosie und Petrefaktenkunde, 806–810.

Hornibrook, N. de B., Brazier, R.C., Strong, P.C., 1989. Manual of New Zealand Permian to Pleistocene foraminiferal biostratigraphy. New Zealand Geological Survey Paleontological Bulletin 56, 175.

Hottinger, L., 1960. Recherches sur les Alvéolines du Paléocène et de l'Eocène. Mémoires Suisses de Paléontologie 75–76, 243.

Hottinger, L., 1977. Foraminifères operculiniformes. Mémoires du Museum National d'Histoire Naturelle de Paris C40, 159.

Hottinger, L., Schaub, H., 1960. Zur Stufeneinteilung de Paleocaens und des Eocaens. Einführung der Stufen Ilerdien und Biarritzien. Eclogae Geologicae Helvetica 53, 453–480.

Hubach, H., 1957. Das Oberoligozän des Dobergs bei Bünde in Westfalen. Berichte der naturhistorischen Gesellschaft 103, 5–69.

Husson, D., Galbrun, B., Laskar, J., Hinnov, L.A., Thibault, N., Gardin, S., Locklair, R.E., 2011. Astronomical calibration of the Maastrichtian. Earth and Planetary Science Letters 305, 328–340.

Hyland, E., Murphy, B., Varela, P., Marks, K., Colwell, L., Tori, F., Monechi, S., Cleaveland, L., Brinkhuis, H., van Mourik, C.A., Coccioni, R., Bice, D., Montanari, A., 2009. Integrated stratigraphic and astrochronologic calibration of the Eocene-Oligocene transition in the Monte Cagnero section (northeastern Apennines, Italy): A potential parastratotype for the Massignano global stratotype section and point (GSSP). Geological Society of America Special Paper 452, 303–322.

Izett, G.A., Dalrymple, G.B., Snee, L.W., 1991. $^{40}Ar/^{39}Ar$ age of Cretaceous-Tertiary boundary tektites from Haiti. Science 252, 1539–1542.

Janis, C.M. (Ed.), 1998. Evolution of Tertiary Mammals of North America. Terrestrial Carnivores, Unglulates and Ungulate-Like Mammals, Vol. 1. Cambridge University Press, Cambridge, p. 691.

Jenkins, D.G., 1985. Southern mid-latitude Paleocene to Holocene planktic foraminifera. In: Bolli, H.M., Saunders, J.B., Perch-Nielsen, K. (Eds.), Plankton Stratigraphy. Cambridge University Press, Cambridge, pp. 263–282.

Jenkins, D.G., Luterbacher, H.P., 1992. Paleogene stages and their boundaries: Introductory remarks. Neues Jahrbuch für Geologie und Paläontologie Abhandlungen 186, 1–5.

Jovane, L., Florindo, F., Sprovieri, M., Palike, H., 2006. Astronomic calibration of the late Eocene/early Oligocene Massignano section (central Italy). Geochemistry, Geophysiscs and Geosystems 7, Q07012. doi: 10.1029/2005GC001195.

Jovane, L., Florindo, F., Coccioni, R., Dinarès-Turell, J., Marsili, A., Monechi, S., Roberts, A.P., Sprovieri, M., 2007. The middle Eocene climatic optimum event in the Contessa Highway section, Umbrian Apennines, Italy. Geological Society of America Bulletin 119, 413–427.

Jovane, L., Sprovieri, M., Florindo, F., Coccioni, R., Marsili, A., 2010. Astronomic calibration of the middle Eocene Contessa Highway section (Gubbio, Italy). Earth and Planetary Science Letters 298, 77–88.

Kaminski, M.A., Gradstein, F.M., 2005. Atlas of Paleogene cosmopolitan deep-water agglutinated Foraminifera. Grzybowski Foundation Special Publication 10, 547.

Kaminski, M.A., Kuhnt, W., Radley, J.D., 1994. The significance of Paleocene-Eocene deep water agglutinated foraminifera from the Numidian Flysch (northern Morocco) for the paleoceanography of the Gibraltar seaway. African Micropaleontology Colloqium Angers, July 1994.

Kappelman, J., Simons, E.L., Swisher, C.C.I., 1992. New age determinations for the Eocene-Oligocene boundary sediment in the Fayum Depression, northern Egypt. Journal of Geology 100, 647–668.

Kay, R.F., Madden, R.H., Vucetich, M.G., Carlini, A.A., Mazzoni, M.M., Re, G.H., Heizler, M., Sandeman, H., 1999. Revised geochronology of the Casamayoran South American Land Mammal Age: Climatic and biotic implications. Proceedings of the National Academy of Sciences, USA 96, 13235–13240.

Keller, G., 1988. Extinction, survivorship an evolution of planktonic foraminifera across the Cretaceous-Tertiary boundary at El Kef, Tunisia. Marine Micropaleontology 13, 239–263.

Keller, G., Lindiger, M., 1989. Stable isotope, TOC and $CaCO_3$ record across the Cretaceous/Tertiary boundary at El Kef, Tunesia. Palaeogeography, Palaeoclimatology, Palaeoecology 73, 243–265.

Kelly, D.C., Bralower, T.J., Zachos, J.C., Premoli Silva, I., Thomas, E., 1996. Rapid diversification of planktonic foraminifera in the tropical Pacific (ODP Site 865) during the late Paleocene thermal maximum. Geology 24, 423–426.

Kelly, D.C., Bralower, T.J., Zachos, J.C., 1998. Evolutionary consequences of the latest Paleocene thermal maximum for tropical planktonic Foraminifera. Palaeogeography, Palaeoclimatology, Palaeoecology 141, 139–161.

Kennett, J.P., Stott, L.D., 1991. Abrupt deep sea warming, paleoceanographic changes and benthic extinctions at the end of the Paleocene. Nature 353, 225–229.

King, C., 2006. Paleogene and Neogene: Uplift and a cooling climate. In: Brenchley, P.J., Rawson, P.F. (Eds.), The Geology of England and Wales, second ed. Geological Society of London, London, pp. 395–427.

Koch, P.L., Zachos, J.C., Gingerich, P.D., 1992. Correlation between isotope records in marine and continental carbon reservoirs near the Paleocene/Eocene boundary. Nature 358, 319–322.

Köhler, M., Moyá-Solá, S., 1999. A finding of Oligocene primates on the European continent. Proceedings of the National Academy of Sciences, USA 96, 13235–13240.

Köthe, A., 1990. Paleogene dinoflagellates from northwest Germany-biostratigraphy and paleoenvironment. Geologisches Jahrbuch A118, 3–111.

Kozlova, G.E., 1999. Paleogene boreal radiolarians from Russia. VNIGRI, St Petersburg, p. 323.

Kraatz, B.P., Geisler, J.H., 2010. Eocene-Oligocene transition in central Asia and its effects on mammalian evolution. Geology 38, 111–114.

Kuiper, K.F., Deino, A., Hilgen, F.J., Krijgsman, W., Renne, P.R., Wijbrans, J.R., 2008. Synchronizing rock clocks of Earth history. Science 320, 500–504.

LaBrecque, J.L., Kent, D.V., Cande, S.C., 1977. Revised magnetic polarity time scale for Late Cretaceous and Cenozoic time. Geology 5, 330–335.

Lanci, L., Parés, J.M., Channell, J.E.T., Kent, D.V., 2005. Oligocene magnetostratigraphy from equatorial Pacific sediments (ODP Sites 1218 and 1219, Leg 199). Earth and Planetary Science Letters 237, 617–634.

Laskar, J., Robutel, P., Joutel, F., Gastineau, M., Correia, A.C.M., Levrard, B., 2004. A long term numerical solution for the insolation quantities of the Earth. Astronomy and Astrophysics 428, 261–285.

Laskar, J., Fienga, A., Gastineau, M., Manche, H., 2011. La2010: A new orbital solution for the long term motion of the Earth. Astronomy and Astrophysics 532 A89. doi: 10.1051/0004-6361/201116836.

Legendre, S., Lévêque, F., 1997. Etalonnage de l'échelle biochronologique mammalienne du Paléogène d'Europe occidentale: vers une intégration à l'échelle globale. In: Aguilar, J.-P., Legendre, S., Michaux, J. (Eds.), Actes du Congrès BiochroM'97. Mémoires et Travaux de l'Ecole Pratique des Hautes Etudes, Institut de Montpellier, 2, pp. 461–473.

Less, G., 1987. Paleontology and stratigraphy of the European Orthophragminae. Geologica Hungarica (Paleontologia) 51, 373.

Leupold, W., Van der Vlerk, I.M., 1936. De stratigraphie van Nederlandsch Oost-Indie. Leidsche Geologisch Mededelingen 5, 611–648.

López-Martínez, N., Peláez-Campomanes, P., 1999. New mammals from south-central Pyrenees (Tremp Formation, Spain) and their bearing on late Paleocene marine-continental correlations. Bulletin de la Société géologique de France 170, 681–696.

Lourens, L.J., Sluijs, A., Kroon, D., Zachos, J.C., Thomas, E., Röhl, U., Bowles, J., Raffi, I., 2005. Astronomical pacing of late Palaeocene to early Eocene global warming events. Nature 435, 1083–1087.

Luciani, V., Giusberti, L., Agnini, C., Fornaciari, E., Rio, D., Spofforth, D.J.A., Pälike, H., 2010. Ecological and evolutionary response of Tethyan planktonic foraminifera to the middle Eocene climatic optimum (MECO) from the Alano section (NE Italy). Palaegeography, Palaeoclimatology, Palaeoecology 292, 82–95.

Luterbacher, H.P., Ali, J.R., Brinkhuis, H., Gradstein, F.M., Hooker, J.J., Monechi, S., Ogg, J.G., Powell, J., Röhl, U., Sanfilippo, A., Schmitz, B., 2004. The Paleogene Period. In: Gradstein, F.M., Ogg, J.G., Smith, A.G. (Eds.), A Geologic Time Scale 2004. Cambridge University Press, Cambridge, pp. 384–408.

Lyell, C., 1833. Principles of Geology: Being an inquiry how far the former changes of the Earth's surface are referable to causes now in operation, Vol. III. John Murray, London, p. 398.

Machlus, M., Hemming, S.R., Olsen, P.E., Christie-Blick, N., 2004. Eocene calibration of geomagnetic polarity time scale reevaluated: Evidence from the Green River Formation of Wyoming. Geology 32, 137–140.

Mangerud, G., Dreyer, T., Søyseth, L., Martinsen, O., Ryseth, A., 1999. High-resolution biostratigraphy and sequence development of the Paleocene succession, Grane Field, Norway. Geological Society Special Publication 152, 167–184.

Mangin, J.-P., 1957. Remarques sur le terme Paléocène et sur la limite Crétacé-Tertiaire. Comptes rendus des séances de la Société géologique de France 14, 319–321.

Mankinen, E.A., Dalrymple, G.B., 1979. Revised geomagnetic polarity time scale for the interval 0-5 m.y. B.P. Journal of Geophysical Research 84, 615–626.

Martini, E., 1958. Discoasteriden und verwandte Formen im NW-deutschen Eozän (Coccolithophorida). 1.Taxonomische Untersuchungen. Senckenbergiana Lethaea 39, 353–388.

Martini, E., 1959a. Discoasteriden und verwandte Formen im NW-deutschen Eozän (Coccolithophorida). 2. Stratigraphische Auswertng. Senckenbergiana Lethaea 40, 137–157.

Martini, E., 1959b. Der stratigraphische Wert von Nanno-Fossilien im norwestdeutschen Tertiär. Erdöl und Kohle 12, 137–140.

Martini, E., 1971. Standard Tertiary and Quaternary calcareous nannoplankton zonation. In: Farinacci, A. (Ed.), Proceedings of the II Planktonic Conference, Roma, 1969. Tecnoscienza, Rome, pp. 739–785.

Maurasse, F., Glass, B.P., 1976. Radiolarian stratigraphy and North American microtektites in Caribbean RC9-58: implications concerning Late Eocene radiolarian chronology and the age of the Eocene-Oligocene boundary. Transactions de la VII Conférence géologique des Caraïbes, 205–212.

Mayer-Eymar, K., 1858. Versuch einer neuen Klassifikation der Tertiär-Gebilde Europa's. Verhandlungen der Schweizer Naturforschenden Gesellschaft fur die Gesammten Naturwissenschaften Trogen 42, 165–199.

McIntosh, W.C., Geissman, J.W., Chapin, C.E., Kunk, M.J., Henry, C.D., 1992. Calibration of the latest Eocene-Oligocene geomagnetic polarity time scale using $^{40}Ar/^{39}Ar$ dated ignimbrites. Geology 20, 459–463.

Meng, J., McKenna, M.C., 1998. Faunal turnovers of Paleogene mammals from the Mongolian Plateau. Nature 395, 364–367.

Mertz, D.F., Swisher, C.C., Franzen, J.L., Neuffer, F.O., Lutz, H., 2000. Numerical dating of the Eckfeld Maar fossil site, Eifel, Germany: A calibration mark for the Eocene time scale. Naturwissenschaften 87, 270–274.

Michelsen, O., Thomsen, E., Danielsen, M., Heilmann-Clausen, C., Jordt, H., Laurssen, G.V., 1998. Cenozoic sequence stratigraphy in the eastern North Sea. SEPM Special Publication 60, 91–118.

Miller, K.G., Fairbanks, R.G., Mountain, G.S., 1987. Tertiary oxygen isotope synthesis, sea-level history and continental margin erosion. Paleoceanography 2, 1–19.

Miller, K.G., Wright, J.D., Fairbanks, R.G., 1991. Unlocking the ice house: Oligocene-Miocene oxygen isotopes, eustasy and margin erosion. Journal of Geophysical Research 96, 6829–6848.

Miller, K.G., Mountain, G.S., Browning, J.V., Kominz, M., Sugerman, P.J., Christie-Blick, N., Katz, M.E., Wright, J.D., 1998. Cenozoic global sea level, sequences, and the New Jersey transect: Results from coastal plain and continental slope drilling. Reviews of Geophysics 36, 663–672.

Molina, E., Canudo, J.I., Guernet, C., McDougall, K., Ortiz, N., Pascual, J.O., Pares, J.M., Samso, J.M., Serra-Kiel, J., Tosquella, J., 1992. The stratotypic Ilerdian revisited: Integrated stratigraphy across the Paleocene/Eocene boundary. Revue de Micropaléontologie 35, 143–156.

Molina, E., Arenillas, I., Pardo, A., 1999. High resolution planktic foraminiferal biostratigraphy and correlation across the Palaeocene/Eocene boundary in the Tethys. Bulletin de la Société Géologique de France 170, 521–530.

Molina, E., Cosovic, V., Gonzalvo, C., Von Salis, K., 2000. Integrated biostratigraphy across the Ypresian/Lutetian boundary at Agost, Spain. Revue de Micropaleontologie 43, 381–391.

Molina, E., Angori, E., Arenillas, I., Brinkhuis, H., Crouch, E.M., Luterbacher, H., Monechi, S., Schmitz, B., 2003. Correlation between the Paleocene/Eocene boundary and the Ilerdian at Campo, Spain. Revue de Micropaléontologie 46, 95–109.

Molina, E., Alegret, L., Arenillas, I., Arz, J.A., Gallala, N., Hardenbol, J., von Salis, K., Steurbaut, E., Vandenberghe, N., Zaghbib-Turki, D., 2006. The Global Boundary Stratotype Section and Point for the base of the Danian Stage (Paleocene, Paleogene, "Tertiary", Cenozoic) at El Kef, Tunesia - Original definition and revision. Episodes 29, 263–273.

Molina, E., Alegret, L., Arenillas, I., Arz, J.A., Gallala, N., Grajales-Nishimura, J.M., Murillo-Muñetón, G., Zaghbib-Turki, D., 2009. The Global Boundary Stratotype Section and Point for the base of the Danian Stage (Paleocene, Paleogene, "Tertiary", Cenozoic): Auxillary sections and correlation. Episodes 32, 84–95.

Molina, E., Alegret, L., Apellaniz, E., Bernaola, G., Caballero, F., Hardenbol, J., Heilmann-Clausen, C., Larrasoaña, J.C., Luterbacher, H.P., Monechi, S., Ortiz, S., Orue-Etxebarria, X., Payros, A., Pujalte, V., Rodríguez-Tovar, F.J., Tori, F., Tosquella, J., 2009. Proposal for the Global Standard Stratotype-section and Point (GSSP) for the base of the Lutetian Stage at the Gorrondatxe section (Spain). International Subcommission on Paleogene Stratigraphy, Ypresian/Lutetian Working Group, 44.

Molina, E., Alegret, L., Apellaniz, E., Bernaola, G., Caballero, F., Dinarès-Turell, J., Hardenbol, J., Heilmann-Clausen, C., Larrasoaña, J.C., Luterbacher, H.P., Monechi, S., Ortiz, S., Orue-Etxebarria, X., Payros, A., Pujalte, V., Rodríguez-Tovar, F.J., Tori, F., Tosquella, J., Uchman, A., 2011. The Global Stratotype Section and Point (GSSP) for the base of the Lutetian Stage at the Gorrondatxe section, Spain. Episodes 34, 86–108.

Monechi, S., 1986. Calcareous nannofossil events around the Eocene-Oligocene boundary in the Umbrian Apennines (Italy). Palaeogeography, Palaeoclimatology, Palaeoecology 57, 61–69.

Montanari, A., Asaro, F., Kennett, J.P., Michel, E., 1993. Iridium anomalies of late Eocene age at Massignano (Italy), and in ODP Site 689B (Maud Rise, Antarctica). Palaios 8, 420–437.

Moore, T.C., 1971. Radiolaria. Initial Reports of the Deep Sea Drilling Project 8, 727–775.

Mudge, D.C., Bujak, J.P., 1996. An integrated stratigraphy for the Paleocene and Eocene of the North Sea. Geological Society Special Publication 101, 91–113.

Munier Chlamas, E., De Lapparent, A., 1893. Note sur la nomenclature des terrains sedimentaires. Bulletin de la Société géologique de France 3 (21), 438–493.

Naumann, C.F., 1866. Lehrbuch der Geognosie, Vol. 3. Engelmann, Leipzig.

Neal, J.E., 1996. A summary of Paleogene sequence stratigraphy in northwest Europe and the North Sea. Geological Society Special Publication 101, 15–42.

Nigrini, C., 1971. Radiolarian zones in the Quaternary of the equatorial Pacific Ocean. In: Funnell, M.B., Riedel, W.R. (Eds.), The Micropaleontology of the Oceans. Cambridge University Press, Cambridge, pp. 443–461.

Nigrini, C., 1974. Cenozoic Radiolaria from the Arabian Sea, DSDP Leg 23. Initial Reports of the Deep Sea Drilling Project 23, 1051–1121.

Nishimura, A., 1987. Cenozoic Radiolaria in the western North Atlantic, Site 603, Leg 93 of the Deep Sea Drilling Project. Initial Reports of the Deep Sea Drilling Project 93, 713–737.

Nishimura, A., 1992. Paleocene radiolarian biostratigraphy in the northwest Atlantic at Site 384, Leg 43 of the Deep Sea Drilling Project. Micropaleontology 38, 317–362.

Nocchi, M., Parisi, G., Monaco, P., Monechi, S., Madile, M., 1988. Eocene and early Oligocene micropaleontology and paleoenvironments in SE Umbria, Italy. Palaeogeography, Palaeoclimatology, Palaeoecology 67, 181–244.

Norris, R.D., 1996. Symbiosis as an evolutionary innovation in the radiation of Paleocene planktic foraminifera. Paleobiology 22, 461–480.

Norris, R.D., Röhl, U., 1999. Astronomical chronology for the Paleocene/Eocene transient global warming and carbon isotope anomaly. Nature 401, 775–778.

Obradovich, J.D., 1993. A Cretaceous time scale. In: Caldwell, W.G.E., Kauffman, E.G. (Eds.), Evolution of the Western Interior

Basin. Geological Association of Canada, Special Paper, 39, pp. 379–396.

Obradovich, J.D., Evanoff, E., Larson, E.E., 1995. Revised single-crystal laser-fusion $^{40}Ar/^{39}Ar$ 30 ages of Chadronian tuffs in the White River Formation of Wyoming. Geological Society of America Abstracts with Programs 27 (3) A–77.

Odin, G.S., Montanari, A., Deino, A., Drake, R., Guise, P.G., Kreuzer, H., Rex, D.C., 1991. Reliability of volcano-sedimentary biotite ages across the E-O boundary. Chemical Geology 86, 203–224.

Ogg, J.G., Ogg, G., Gradstein, F.M., 2008. The Concise Geologic Time Scale. Cambridge University Press, Cambridge, p. 177.

Okada, H., Bukry, D., 1980. Supplementary modification and introduction of code numbers to the low-latitude coccolith biostratigraphic zonation (Bukry, 1973; 1975). Marine Micropaleontology 51, 321–325.

Olsson, R.K., 1970. Planktonic foraminifera from the base of Tertiary, Millers Ferry, Alabama. Journal of Paleontology 44, 598–604.

Olsson, R.K., Liu, C., 1993. Controversies on the placement of Cretaceous-Paleogene boundary and the K/P mass extinction of planktonic foraminifera. Palaios 8, 127–139.

Olsson, R.K., Hemleben, C., Berggren, W.A., Huber, B.T. (Eds.), 1999. Atlas of Paleocene Planktonic foraminifera. Smithsonian Contributions to Paleobiology, 85, p. 252.

Pälike, H., Shackleton, N.J., Röhl, U., 2001. Astronomical forcing in late Eocene marine sediments. Earth and Planetary Science Letters 193, 589–602.

Pälike, H., Norris, R.D., Herrie, J.O., Wilson, P.A., Coxall, H.K., Lear, C.H., Shackleton, N.J., Tripati, A.K., Wade, B.S., 2006. The heartbeat of the Oligocene climate system. Science 414, 1894–1898.

Pälike, H., Lyle, M.W., Ahagon, N., Raffi, I., Gamage, K., Zarikian, C.A., 2008. Pacific equatorial age transect. Integrated Ocean Drilling Program Scientific Prospectus 320/321 doi: 10.2204/iodp.sp.320321.2008.

Pälike, H., Lyle, M., Nishi, H., Raffi, I., Gamage, K., Klaus, A., the Expedition 320/321 Scientists, 2010. Proceedings of the Integrated Ocean Drilling Program (IODP), 320/321. Integrated Ocean Drilling Program Management International, Tokyo. doi:10.2204/iodp.sd.9.01.2010.

Pardo, A., Keller, G., Molina, E., Canudo, J.I., 1997. Planktic foraminiferal turnover across the Paleocene-Eocene transition at DSDP Site 401, Bay of Biscay, North Atlantic. Marine Micropaleontology 29, 129–158.

Pardo, A., Keller, G., Oberhänsli, H., 1999. Paleoecologic and paleoceanographic evolution of the Tethyan realm during the Paleocene-Eocene transition. Journal of Foraminiferal Research 29, 37–57.

Payton, E. (Ed.), 1977. Seismic Stratigraphy - applications to hydrocarbon exploration. American Association Petroleum Geologists Memoir, 26, p. 516.

Pearson, P.N., Olsson, R.K., Huber, B.T., Hemleben, C., Berggren, W.A. (Eds.), 2006. Atlas of Eocene Planktonic Foraminifera. Cushman Foundation for Foraminiferal Research, Fredericksburg, p. 514.

Pekar, S.F., Christie-Blick, N., Kominz, M.A., Miller, K.G., 2002. Calibration between eustatic estimates from backstripping and oxygen isotopic records for the Oligocene. Geology 30, 903–906.

Peláez-Campomanes, P., López-Martinez, N., Álvarez-Sierra, D.M., Daams, R., 2000. The earliest mammal of the European Paleocene: The multituberculate *Hainina*. Journal of Paleontology 74, 701–711.

Perch-Nielsen, K., 1985. Cenozoic calcareous nannofossils. In: Bolli, H.M., Saunders, J.B., Perch-Nielsen, K. (Eds.), Plankton Stratigraphy. Cambridge University Press, Cambridge, pp. 427–554.

Perch-Nielsen, K., Hansen, J.M., 1981. Selandian. In: Pomerol, C. (Ed.), Stratotypes of Paleogene Stages. Bulletin d'information des géologues du Bassin de Paris, Mémoire hors série, 2, pp. 219–230.

Petrizzo, M.R., 2005. An early late Paleocene event on Shatsky Rise, northwest Pacific Ocean (ODP Leg 198): Evidence from planktonic foraminiferal assemblages. In: Bralower, T.J., Premoli Silva, I., Malone, M.J. (Eds.), Proceedings of the Ocean Drilling Program, Scientific Results, Leg 198. Available at. http://www-odp.tamu.edu/publications/198_SR/102/102.htm.

Petrushevskaya, M.G., 1975. Cenozoic radiolarians of the Antarctic, Leg 29, DSDP. Initial Reports of the Deep Sea Drilling Project 29, 541–675.

Phillips, J., 1841. Figures and descriptions of the Palaeozoic fossils of Cornwall, Devon and east Somerset. Longman, Brown, Green and Longmans, London.

Plaziat, J.-C., 1981. Ilerdian. In: Pomerol, C. (Ed.), Stratotypes of Paleogene Stages. Bulletin d'information des géologues du Bassin de Paris, Mémoire hors série, 2, pp. 103–121.

Plaziat, J.-C., Koeniguer, J.-C., Génault, B., 2009. Diversité et localisation chronologique des séismites dans les sables marins et éoliens du Stampien, au sud de Paris (entre Eure et Seine). Bulletin d'Information des géologues du Bassin de Paris 46, 3–73.

Pomerol, C. (Ed.), 1981a. Stratotypes of Paleogene Stages. Bulletin d'information des géologues du Bassin de Paris, Mémoire Hors Série, 2, p. 301.

Pomerol, C., 1981b. Auversian. In: Pomerol, C. (Ed.), Stratotypes of Paleogene Stages. Bulletin d'information des géologues du Bassin de Paris, Mémoire hors série, 2, pp. 11–22.

Pomerol, C., 1981c. Biarritzian. In: Pomerol, C. (Ed.), Stratotypes of Paleogene Stages. Bulletin d'information des géologues du Bassin de Paris, Mémoire hors série, 2, pp. 37–42.

Powell, A.J., 1992. A stratigraphic index of dinoflagellate cysts. Chapman and Hall, London, p. 290.

Powell, A.J., Brinkhuis, H., Bujak, J.P., 1996. Upper Paleocene-Lower Eocene dinoflagellate cyst sequence of southeast England. Geological Society Special Publication 101, 145–183.

Premoli Silva, I., Jenkins, D.G., 1993. Decision on the Eocene-Oligocene boundary stratotype. Episodes 16, 379–382.

Premoli Silva, I., Coccioni, R., Montanari, A. (Eds.), 1988. The Eocene-Oligocene boundary in the Marche-Umbria Basin (Italy). International Union of Geological Sciences, Eocene-Oligocene Boundary Meeting, Ancona, Italy, Oct. 1987, pp. 250.

Pross, J., Houben, A.J.P., Van Simaeys, S., Williams, G.L., Kotthoff, U., Coccioni, R., Wilpshaar, M., Brinkhuis, H., 2010. Umbria-Marche revisited. A refined magnetostratigraphic calibration of dinoflagellate cyst events for the Oligocene of the Western Tethys. Review of Palaeobotany and Palynology 158, 213–235.

Prothero, D.R., Swisher III, C.C., 1992. Magnetostratigraphy and geochronology of the terrestrial Eocene-Oligocene transition in North America. In: Prothero, D.R., Berggren, W.A. (Eds.), Eocene-Oligocene Climatic and Biotic Evolution. Princeton University Press, Princeton, pp. 46–73.

Prothero, D.R., Emry, R.J. (Eds.), 1996. The Terrestrial Eocene-Oligocene Transition in North America. Cambridge University Press, Cambridge, p. 688.

Pujalte, V., Schmitz, B., Baceta, J.I., Orue-Etxebarria, X., Bernaola, G., Dinarès-Turell, J., Payros, A., Apellaniz, E., Caballero, F., 2009. Correlation of the Thanetian-Ilerdian turnover of larger foraminifera and the Paleocene-Eocene thermal maximum confirming evidence from the Campo area (Pyrenees, Spain). Geologica Acta 7, 161–175.

Quillévéré, F., Norris, R.D., Moussa, I., Berggren, W.A., 2001. Role of photosymbiosis and biogeography in the diversification of early Paleogene acarininids (planktonic Foraminifera). Paleobiology 27, 311–326.

Quillévéré, F., Norris, R.D., Kroon, D., Wilson, P.A., 2008. Transient ocean warming and shifts in carbon reservoirs during the early Danian. Earth and Planetary Science Letters 265, 600–615.

Renevier, E., 1873. Tableau de terrains sédimentaires formés pendant les époques de la phase organique du globe terrestre. Bulletin de la Société Vaudoise des Sciences Naturelles, Laussane 12, 218–252.

Renne, P.R., Mundil, R., Balco, G., Min, K., Ludwig, K.R., 2010. Joint determination of ^{40}K decay constants and ^{40}Ar*/^{40}K for the Fish Canyon sanidine standard, and improved accuracy for ^{40}Ar/^{39}Ar geochronology. Geochimica et Cosmochimica Acta 74, 5349–5367.

Riedel, W.R., 1957. Radiolaria: A preliminary stratigraphy. In: Petterson, H. (Ed.), Reports of the Swedish Deep-Sea Expedition, 1947–1948, Vol. 6. Elanders Boktryckeri Aktiebolag, Göteborg, pp. 59–96.

Riedel, W.R., Sanfilippo, A., 1970. Radiolaria, Leg 4, Deep Sea Drilling Project. Initial Reports of the Deep Sea Drilling Project 4, 503–575.

Riedel, W.R., Sanfilippo, A., 1971. Cenozoic Radiolaria from the western tropical Pacific, Leg 7. Initial Reports of the Deep Sea Drilling Project 7, 1529–1672.

Rio, D., Fornaciari, E., Raffi, I., 1990. Late Oligocene through early Pleistocene calcareous nannofossils from the western equatorial Indian Ocean. Proceedings of the Ocean Drilling Program, Scientific Results 115, 175–235.

Ritzkowski, S., 1981a. Latdorfian. In: Pomerol, C. (Ed.), Stratotypes of Paleogene Stages. Bulletin d'information des géologues du Bassin de Paris, Mémoire hors série, 2, pp. 149–166.

Ritzkowski, S., 1981b. Chattian. In: Pomerol, C. (Ed.), Stratotypes of Paleogene Stages. Bulletin d'information des géologues du Bassin de Paris, Mémoire hors série, 2, pp. 43–61.

Rivera, T.A., Storey, M., Kuiper, K., Pälike, H., 2011. Towards an integrated geomagnetic polarity reversal timescale for the Pleistocene. American Geophysical Union Fall meeting, 5–9 Dec 2011, San Francisco, USA, Abstract V51A–2502.

Robaszynski, F., 1981. Montian. In: Pomerol, C. (Ed.), Stratotypes of Paleogene Stages. Bulletin d'information des géologues du Bassin de Paris, Mémoire hors série, 2, pp. 181–200.

Robin, E., Rocchia, R., 1998. Ni-rich spinel at the Cretaceous-Tertiary boundary of El Kef, Tunesia. Bulletin de la Société Géologique de France 169 (4), 515–526.

Röhl, U., Ogg, J.G., Geib, T., Wefer, G., 2001. Astronomical calibration of the Danian time scale. In: Norris, R.D., Kroon, D., Klaus, A. (Eds.), Western North Atlantic Palaeogene and Cretaceous Palaeoceanography. Geological Society Special Publication, 183, pp. 163–183.

Röhl, U., Norris, R.D., Ogg, J.G., 2003. Cyclostratigraphy of upper Paleocene and lower Eocene sediments at Blake Nose Site 1051 (western North Atlantic). In: Wing, S.L., Gingerich, P.D., Schmitz, B., Thomas, E. (Eds.), Causes and Consequences of Globally Warm Climates in the Early Paleogene. Geological Society of America Special Paper, 369, pp. 567–588.

Romein, A.J.T., 1979. Lineages in early Paleogene calcareous nannoplankton. Utrecht Micropaleontological Bulletin 22, 1–231.

Rosenkrantz, A., 1924. De Københavnske Grøbsandslag og deres placering i dem Danske lagrække. Meddedelinger danske geologiske Foreningen 6, 3–39.

Roth, P.H., Baumann, P., Bertolino, V., 1971. Late Eocene-Oligocene calcareous nannoplankton from central and northern Italy. Proceedings of the II Planktonic Conference, Rome 2, 1069–1097.

Roveda, V., 1961. Contributo allo studio di alcune macroforaminiferi di Priabona. Rivista Italiana di Paleontologia e Stratigrafia 67, 201–207.

Sahy, D., Fischer, A., Condon, D., Terry Jr., D.O., Hiess, J., Abels, H., Huesing, S.K., Kuipper, K., 2011. High precision radio isotopic age constraints on the Late Eocene – Early Oligocene Geomagnetic Polarity Time Scale. Annual Meeting of the Geological Society of America, 9–12 Oct. 2011, Minneapolis. Paper No. 237–2.

Salvador, A. (Ed.), 1994. International Stratigraphic Guide; A Guide to Stratigraphic Classification, Terminology, and Procedure. Geological Society of America, Boulder, p. 214.

Sanfilippo, A., Nigrini, C.A., 1998. Code numbers for Cenozoic low latitude radiolarian biostratigraphic zones and GPTS conversion tables. Marine Micropaleontology 33, 109–156.

Sanfilippo, A., Blome, C.D., 2001. Biostratigraphic implications of mid-latitude Paleocene-Eocene radiolarian faunas from Hole 1051A, Ocean Drilling Program Leg 171B, Blake Nose, western North Atlantic. Geological Society of London Special Publication 183, 185–224.

Sanfilippo, A., Westberg-Smith, M.J., Riedel, W.R., 1985. Cenozoic Radiolaria. In: Bolli, H.M., Saunders, J.B., Perch-Nielsen, K. (Eds.), Plankton Stratigraphy. Cambridge University Press, Cambridge, pp. 631–712.

Saunders, J.B., Bernoulli, D., Müller-Mertz, E., Oberhänsli, H., Perch-Nielsen, K., Riedel, W.R., Sanfilippo, A., Torrini, R., 1985. Stratigraphy of the late middle Eocene to early Oligocene in the Bath Cliff Section, Barbados, West Indies. Micropaleontology 30, 390–425.

Savage, D.E., Russell, D.E., 1977. Comments on mammalian paleontologic stratigraphy and geochronology; Eocene stages and mammal ages of Europe and North America. Géobios 10 (Suppl. 1), 47–55.

Schaltegger, U., Brack, P., Ovtcharova, M., Peytcheva, I., Schoene, B., Stracke, A., Marocchi, M., Bargossi, G.M., 2009. Zircon and titanite recording 1.5 million years of magma accretion, crystallization and initial cooling in a composite pluton (southern Adamello batholith, northern Italy). Earth and Planetary Science Letters 286, 208–218.

Schaub, H., 1981. Nummulites et Assilines de la Téthys paléogène. Mémoires Suisses de Paléontologie 236, 104–106.

Scheibner, C., Speijer, R.P., 2009. Recalibration of the shallow-benthic zonation across the Paleocene-Eocene boundary: The Egyptian record. Geologica Acta 7, 195–214.

Schimper, W.P., 1874. Traité de Paléontologie végétale III. J.P.Baillière, Paris, p. 896.

Schmidt-Kittler, N. (Ed.), 1987. International Symposium on Mammalian Biostratigraphy and Paleoecology of the European Paleogene, Mainz, February 18th–21st. Münchner Geowissenschaftliche Abhandlungen A 10, 312.

Schmitz, B., Pujalte, V., Nuñez-Betelu, K., 2001. Climate and sea-level perturbations during the initial Eocene thermal maximum: Evidence from siliciclastic units in the Basque Basin (Ermua, Zumaia and Trabakua Pass). Palaeogeography, Palaeoclimatology, Palaeoecology 165, 299–320.

Schmitz, B., Alegret, L., Apellaniz, E., Arenillas, I., Aubry, M.-P., Baceta, J.-U., Berggren, W.A., Bernaola, G., Caballero, F., Clemmensen, A., Dinarès-Turell, J., Dupuis, C., Heilmann-Clausen, C., Knox, R., Martín-

Rubio, M., Molina, E., Monechi, S., Ortiz, S., Orue-Etxebarria, X., Payros, A., Petrizzo, M.R., Pujalte, V., Speijer, R., Sprong, J., Steurbaut, E., Thomsen, E., 2008. Proposed Global Stratotype Sections and Points for the bases of the Selandian and Thanetian stages (Paleocene Series). Report of the International Subcommission on Paleogene Stratigraphy, 52. Available at. http://www.earth-prints.org/bitstream/2122/3795/1/pwg20.pdf.

Schmitz, B., Pujalte, V., Molina, E., Monechi, S., Orue-Etxebarria, X., Speijer, R.P., Alegret, L., Apellaniz, E., Arenillas, I., Aubry, M.-P., Baceta, J.I., Berggren, W.A., Bernaola, G., Caballero, F., Clemmensen, A., Dinarès-Turell, J., Dupuis, C., Heilmann-Clausen, C., Hilario Orús, A., Knox, R., Martín-Rubio, M., Ortiz, S., Payros, A., Petrizzo, M.R., von Salis, K., Sprong, J., Steurbaut, E., Thomsen, E., 2011. The Global Stratotype Sections and Points for the bases of the Selandian (Middle Paleocene) and Thanetian (Upper Paleocene) stages at Zumaia Spain. Episodes 34, 220–243.

Schuler, M., Cavelier, C., Dupuis, C., Steurbaut, E., Vandenberghe, N., 1992. The Paleogene of the Paris and Belgian basins: Standard-stages and regional stratotypes. 8th International Palynological Congress, Aix-en-Provence, 13–16th Sept. 1992, Excursion C. Cahiers de Micropaléontologie 7 (1/2), 29–92.

Schulte, P., Alegret, L., Arenillas, I., Arz, J.A., Barton, P.J., Bown, P.R., Bralower, T.J., Christeson, G.L., Claeys, P., Cockell, C.S., Collins, G.S., Deutsch, A., Goldin, T.J., Goto, K., Grajales-Nishimura, J.M., Grieve, R.A.F., Gulick, S.P.S., Johnson, K.R., Kiessling, W., Koeberl, C., Kring, D.A., MacLeod, K.G., Matsui, T., Melosh, J., Montanari, A., Morgan, J.V., Neal, C.R., Nichols, D.J., Norris, R.D., Pierazzo, E., Ravizza, G., Rebolledo-Vieyra, M., Reimold, R.U., Robin, E., Salge, T., Speijer, R.P., Swet, A.R., Urrutia-Fucugauchi, J., Vajda, V., Whalen, M.T., Willumsen, P.S., 2010. The Chicxulub asteroid impact and mass extinction at the Cretaceous-Paleogene boundary. Science 327, 1214–1218.

Secord, R., Gingerich, P.D., Smith, M.E., Clyde, W.C., Wilf, P., Singer, B.S., 2006. Geochronology and mammalian biostratigraphy of middle and upper Paleocene continental strata, Bighorn Basin, Wyoming. American Journal of Science 306, 211–245.

Secord, R., 2008. The Tiffanian Land-Mammal Age (Middle and Late Paleocene) in the northern Bighorn Basin, Wyoming. University of Michigan Papers on Paleontology 35, 192.

Serra-Kiel, J., Canudo, J.I., Dinarès, J., Molina, E., Ortiz, N., Pascual, J.O., Samso, J.M., Tosquella, J., 1994. Cronoestratigrafía de los sedimentos marinos del Terciario inferior de la Cuenca de Graus-Tremp (Zona Central Surpirenaica). Revista de la Sociedad Geologica de España 7, 273–297.

Serra-Kiel, J., Hottinger, L., Caus, E., Drobne, K., Ferràndez, C., Kumar Jauhri, A., Less, G., Pavlovec, R., Pignatti, J., Samsó, J.M., Schaub, H., Sirel, E., Strougo, A., Tambareau, Y., Tosquella, J., Zakrevskaya, E., 1998. Larger foraminiferal biostratigraphy of the Tethyan Paleocene and Eocene. Bulletin de la Société géologique de France 169, 281–299.

Sigé, B., Marandat, B., 1997. Apport à la faune du Paléocène inférieur d'Europe: un Plésiadapiforme du Montien de Hainin (Belgique). In: Aguilar, J.-P., Legendre, S., Michaux, J. (Eds.), Actes du Congrès BiochroM'97. Mémoires et Travaux de l'Ecole Pratique des Hautes Etudes, 21. Institut de Montpellier, pp. 679–686.

Sluijs, A., Schouten, S., Pagani, M., Woltering, M., Brinkhuis, H., Sinninghe Damsté, J.S., Dickens, G.R., Huber, M., Reichart, G.-J., Stein, R., Matthiessen, J., Lourens, L.J., Pedentchouk, N., Backman, J., Moran, K., the Expedition 302 Scientists, 2006. Subtropical Arctic Ocean temperatures during the Palaeocene-Eocene thermal maximum. Nature 441, 610–613.

Sluijs, A., Schouten, S., Donders, T.H., Schoon, P.L., Röhl, U., Reichart, G.-J., Sangiorgi, F., Kim, J.-H., Sinninghe Damsté, J.S., Brinkhuis, H., 2009. Warm and wet conditions in the Arctic region during Eocene Thermal maximum 2. Nature Geoscience 2, 777–780.

Sluijs, A., Bijl, P.K., Schouten, S., Röhl, U., Reichart, G.-J., Brinkhuis, H., 2011. Southern Ocean warming, sea level and hydrological change during the Paleocene-Eocene thermal maximum. Climate of the Past 7, 47–61.

Smith, M.E., Singer, B.S., Carroll, A.R., 2003. $^{40}Ar/^{39}Ar$ geochronology of the Green River Formation, Wyoming. Geological Society of America Bulletin 115, 549–565.

Smith, M.E., Singer, B.S., Carroll, A.R., 2004. Reply. Geological Society of America Bulletin 116, 253–256.

Smith, M.E., Carroll, A.R., Singer, B.S., 2008. Synoptic reconstruction of a major ancient lake system: Eocene Green River Formation, western United States. Geological Society of America Bulletin 120, 54–84.

Smith, M.E., Chamberlain, K.R., Singer, B.S., Carroll, A.R., 2010. Eocene clocks agree: Coeval $^{40}Ar/^{39}Ar$, U-Pb, and astronomical ages from the Green River Formation. Geology 38, 527–530.

Smith, T., Smith, R., 2003. Terrestrial mammals as biostratigraphic indicators in upper Paleocene-lower Eocene marine deposits of the southern North Sea Basin. In: Wing, S.L., Gingerich, P.D., Schmitz, B., Thomas, E. (Eds.), Causes and Consequences of Globally Warm Climates in the Early Paleogene. Geological Society of America Special Paper, 369, pp. 513–520.

Speijer, R.P., Schmitz, B., Aubry, M.-P., Charisi, S.D., 1996. The latest Paleocene benthic extinction event: Punctuated turnover in outer neritic foraminiferal faunas from Gebel Aweina, Egypt. In: Aubry, M.-P., Benjamini, C. (Eds.), Paleocene-Eocene Boundary Events in Space and Time. Israel Journal of Earth Sciences, 44, pp. 207–222.

Speijer, R.P., Schmitz, B., Luger, P., 2000. Stratigraphy of late Palaeocene events in the Middle East: Implications for low- to middle-latitude successions and correlations. Journal of the Geological Society 157, 37–47.

Sprong, J., Speijer, R.P., Steurbaut, E., 2009. Biostratigraphy of the Danian/Selandian transition in the southern Tethys, with special reference to the Lowest Occurrence of planktic foraminifera Igorina albeari. Geologica Acta 7, 63–77.

Stassen, P., Thomas, E., and Speijer, R.P., 2012. Restructuring outer neritic foraminiferal assemblages in the aftermath of the Paleocene-Eocene thermal maximum. Journal of Micropalaeontology 31, 89–93.

Steininger, F.F., Aubry, M.-P., Berggren, W.A., Biolzi, M., Borsetti, A.M., Cartlidge, J.E., Cati, F., Corfield, R., Gelati, R., Iaccarino, S., Napoleone, C., Ottner, R., Rögl, F., Roetzel, R., Spezzaferri, S., Tateo, F., Villas, G., Zevenboom, D., 1997. The Global Stratotype Section and Point (GSSP) for the base of the Neogene. Episodes 20, 23–28.

Steurbaut, E., 1992. Integrated stratigraphic analysis of lower Rupelian deposits (Oligocene) in the Belgian Basin. Annales de la Société Géologique de Belgique 115, 287–306.

Steurbaut, E., 1998. High-resolution holostratigraphy of middle Paleocene to early Eocene strata in Belgium and adjacent areas. Palaeontographica. Abteilung A Palaeozoologie-Stratigraphie 247, 91–156.

Steurbaut, E., 2006. Ypresian. Geologica Belgica 9, 73–93.

Steurbaut, E., Sztrákos, K., 2008. Danian/Selandian boundary criteria and North Sea Basin-Tethys correlations based on calcareous nannofossil and foraminiferal trends in SW France. Marine Micropaleontology 67, 1–29.

Storey, M., Duncan, R.A., Swisher III, C.A., 2007. Paleocene-Eocene Thermal Maximum and the opening of the northeast Atlantic. Science 316, 587–589.

Stott, L.D., Sinha, A., Thiry, M., Aubry, M.-P., Berggren, W.A., 1996. Global $\delta^{13}O$ changes across the Paleocene-Eocene boundary: Criteria for terrestrial-marine correlations. Geological Society Special Publication 101, 381–399.

Stover, L.E., Hardenbol, J., 1993. Dinoflagellates and depositional sequences in the Lower Oligocene (Rupelian) Boom Clay Formation, Belgium. Bulletin de la Société belge de géologie 102, 5–77.

Stover, L.E., Brinkhuis, H., Damassa, S.P., De Verteuil, L., Helby, R.J., Monteil, E., Partridge, A.D., Powell, A.J., Riding, J.B., Smelror, M., Williams, G.L., 1996. Mesozoic–Tertiary dinoflagellates, acritarchs and prasinophytes. In: Jansonius, J., McGregor, D.C. (Eds.), Palynology: Principles and Applications. American Association of Stratigraphic Palynologists Foundation, 2, pp. 641–750.

Stradner, H., 1958. Die fossilen Discoasteriden Österreichs: 1.Teil. Erdöl-Zeitschrift 74, 178–188.

Stradner, H., 1959a. First report on the discoasters of the Tertiary of Austria and their stratigraphic use. Proceedings of the Fifth World Petroleum Congress 1, 1080–1095.

Stradner, H., 1959b. Die fossilen Discoasteriden Österreichs: 2.Teil. Erdöl-Zeitschrift 75, 472–488.

Stucky, R.K., 1992. Mammalian faunas in North America of Bridgerian to early Arikareean "ages" (Eocene and Oligocene). In: Prothero, D.R., Berggren, W.A. (Eds.), Eocene-Oligocene Climatic and Biotic Evolution. Princeton University Press, Princeton, pp. 464–493.

Subbotina, N.N., 1953. Fossil Foraminifera of the USSR. Globigerinids, Hantkeninids and Globorotaliids. Trudy Vsesoyuznovo Nauchno-Issledovatelskoo Geologo-Razvedochnovo Insituta (VNIGRI) 76, 296 [In Russian].

Suess, E., 1868. Über die Gliederung des Vicentinischens Tertiärgebirges. Sitzungsberichte der Kaiserliche Akademie der Wissenschaften in Wien 58, 265–279.

Suganuma, Y., Ogg, J.G., 2006. Campanian through Eocene Magnetostratigraphy of Sites 1257–1261, ODP Leg 207, Demerara Rise (Western Equatorial Atlantic). In: Mosher, D.C., Erbacher, J., Malone, M.J. (Eds.), Proceedings of the Ocean Drilling Program, Scientific Results, 207. Available at. http://www-odp.tamu.edu/publications/207_SR/102/102.htm.

Sullivan, F.R., 1964. Lower Tertiary nannoplankton from the California Coast Ranges. I. Paleocene. University of California Publications in Geological Sciences 44, 163–227.

Sullivan, F.R., 1965. Lower Tertiary nannoplankton from the California Coast Ranges. II. Eocene. University of California Publications in Geological Sciences 53, 1–74.

Swisher III, C.C., Prothero, D.R., 1990. Single-crystal $^{40}Ar/^{39}Ar$ dating of the Eocene-Oligocene transition in North America. Science 249, 760–762.

Swisher III, C.C., Dingus, L., Butler, R.F., 1993. $^{40}A/^{39}Ar$ calibration of the Puercan Land Mammal Age and the age of the North American non-marine K-T boundary. Canadian Journal of Earth Sciences 30, 1981–1996.

Swisher III, C.C., Grajales-Nishimura, J.M., Montanari, A., Margolis, S.V., Claeys, P., Alvarez, W., Renne, P., Cedillo-Pardo, E., Maurrassee, F.J.-M.R., Curtis, G.H., Smit, J., McWilliams, M.O., 1992. Coeval $^{40}Ar/^{39}Ar$ ages of 65.0 million years ago from Chicxulub crater melt rock and Cretaceous-Tertiary boundary tektites. Science 257, 954–958.

Takemura, A., 1992. Radiolarian Paleogene biostratigraphy in the southern Indian Ocean, Leg 120. Proceedings of the Ocean Drilling Program, Scientific Results 120, 735–756.

Takemura, A., Ling, H.-Y., 1997. Eocene and Oligocene radiolarian biostratigraphy from the Southern Ocean: correlation of ODP Legs 114 (Atlantic Ocean) and 120 (Indian Ocean). Marine Micropaleontology 30, 97–116.

Tauxe, L., Gee, J., Gallet, Y., Pick, T., Bown, T., 1994. Magnetostratigraphy of the Willwood Formation, Bighorn Basin, Wyoming: New constraints on the location of the Paleocene/Eocene boundary. Earth and Planetary Science Letters 125, 159–172.

Ten Kate, W.G., Sprenger, A., 1992. Rhythmicity in deep-water sediments, documentation and interpretation by pattern and spectral analysis. PhD thesis, Free University, Amsterdam, 244.

Thomas, E., 2007. Cenozoic mass extinctions in the deep sea; what perturbs the largest habitat on Earth? In: Monechi, S., Coccioni, R., Rampino, M.R. (Eds.), Large Ecosystem Perturbations; Causes and Consequences. Geological Society of America, Special Paper, 424, pp. 1–23.

Thomas, E., Shackleton, N.J., 1996. The Paleocene-Eocene benthic foraminiferal extinction and stable isotope anomalies. In: Knox, R.W.O'B., Corfield, R., Dunay, R.E. (Eds.), Correlation of the Early Paleogene in Northwest Europe. Geological Society Special Publication, 101, pp. 401–441.

Thomsen, E., 1994. Calcareous nannofossil stratigraphy across the Danian-Selandian boundary in Denmark. GFF 116, 65–67.

Ting, S., Tong, Y., Clyde, W.C., Koch, P.L., Meng, J., Wang, Y., Bowen, G.J., Li, Q., Snell, K.E., 2011. Asian early Paleogene chronology and mammalian faunal turnover events. Vertebrata Pal Asiatica 49, 1–28.

Toumarkine, M., Luterbacher, H.P., 1985. Paleocene and Eocene planktic foraminifera. In: Bolli, H.M., Saunders, J.B., Perch-Nielsen, K. (Eds.), Plankton Stratigraphy. Cambridge University Press, Cambridge, pp. 87–154.

Truswell, E.M., 1997. Palynomorph assemblages from the marine Eocene sediments on the West Tasmanian continental margin and the South Tasman rise. Australian Journal of Earth Sciences 4, 633–654.

Vail, P.R., Mitchum, R.M., Thompson III, S., 1977. Seismic stratigraphy and global changes of sea level. Part 4: Global cycles of relative changes of sea level. In: Payton, C.E. (Ed.), Seismic Stratigraphy – Applications to Hydrocarbon Exploration. American Association Petroleum Geologists Memoir, 26, pp. 83–97.

van Heck, S.E., Prins, B., 1987. A refined nannoplankton zonation for the Danian of the Central North Sea. Abhandlungen der Geologischen Bundesanstalt 39, 353–388.

van Mourik, C.A., Brinkhuis, H., 2005. The Massignano Eocene-Oligocene golden spike section revisited. Stratigraphy 2, 13–30.

Van Simaeys, S., Vandenberghe, N., 2006. Rupelian. Geologica Belgica 9, 95–101.

Van Simaeys, S., De Man, E., Vandenberghe, N., Brinkhuis, H., Steurbaut, E., 2004. Stratigraphic and palaeoenvironmental analysis of the Rupelian-Chattian transition in the type region: Evidence from dinoflagellate cysts, Foraminifera and calcareous nannofossils. Palaeogeography, Palaeoclimatology, Palaeoecology 208, 31–58.

Van Simaeys, S., Munsterman, D., Brinkhuis, H., 2005. Oligocene dinoflagellate cyst biostratigraphy of the southern North Sea Basin. Review of Palaeobotany and Palynology 134, 105–128.

Van Simaeys, S., Brinkhuis, H., Pross, J., Williams, G.L., Zachos, J.C., 2005. Artic dinoflagellate migrations mark the strongest Oligocene glaciations. Geology 33, 709–712.

Vandenberghe, N., 1981. Rupelian. In: Pomerol, C. (Ed.), Stratotypes of Paleogene Stages. Bulletin d'information des géologues du Bassin de Paris, Mémoire hors série, 2, pp. 203–217.

Vandenberghe, N., Laga, P., Steurbaut, E., Hardenbol, J., Vail, P.R., 1998. Tertiary sequence stratigraphy at the southern border of the North Sea Basin in Belgium. SEPM Special Publication 60, 119–154.

Vandenberghe, N., Hager, H., van den Bosch, M., Verstraelen, A., Leroi, S., Steurbaut, E., Prüfert, J., Laga, P., 2001. Stratigraphical correlation by calibrated well logs in the Rupel Group between North Belgium, the Lower–Rhine area in Germany and Southern Limburg and the Achterhoek in The Netherlands. Aardkundige Mededelingen 11, 69–84.

Vandenberghe, N., Brinkhuis, H., Steurbaut, E., 2003. The Eocene-Oligocene boundary in the North Sea area: A sequence stratigraphic approach. In: Prothero, D.R., Ivany, L.C., Nesbitt, E.A. (Eds.), From Greenhouse to Icehouse. The Marine Eocene-Oligocene Transition. Columbia University Press, New York, pp. 419–437.

Vandenberghe, N., Van Simaeys, S., Steurbaut, E., Jagt, J.W.M., Felder, P.J., 2004. Stratigraphic architecture of the Upper Cretaceous and Cenozoic along the southern border of the North Sea Basin in Belgium. Netherlands Journal of Geosciences 83, 155–171.

Varol, O., 1989. Paleocene calcareous nannofossil biostratigraphy. In: Crux, J.A., van Heck, S. (Eds.), Nannofossils and their Applications. Ellis Horwood, Chichester, pp. 267–310.

Varol, O., 1998. Paleogene. In: Bowen, P.R. (Ed.), Calcareous Nannofossil Biostratigraphy. Kluwer, Dordrecht, pp. 200–224.

Vonhof, H.B., Smit, J., Brinkhuis, H., Montanari, A., Nederbragt, A.J., 2000. Global cooling accelerated by early late Eocene impacts? Geology 28, 687–690.

Wade, B.S., Pälike, H., 2004. Oligocene Climate dynamics. Paleoceanography 19 PA4019. doi: 10.1029/2004PA001042.

Wade, B.S., Pearson, P.N., 2008. Planktonic foraminiferal turnover, diversity fluctuations and geochemical signals across the Eocene/Oligocene boundary in Tanzania. Marine Micropaleontology 68, 244–255.

Wade, B.S., Pearson, P.N., Berggren, W.A., Pälike, H., 2011. Review and revision of Cenozoic tropical planktonic foraminiferal biostratigraphy and calibration to the geomagnetic polarity and astronomical time scale. Earth-Science Reviews 104, 111–142.

Walsh, S.L., Prothero, D.R., Lundquist, D.J., 1996. Stratigraphy and paleomagnetism of the middle Eocene Friars Formation and Poway Group, southwestern San Diego County, California. In: Prothero, D.R., Emry, R.J. (Eds.), 1996, The Terrestrial Eocene-Oligocene Transition in North America. Cambridge University Press, Cambridge, pp. 120–151.

Wang, C., Zhao, X., Liu, Z., Lippert, P.C., Graham, S.A., Coe, R.S., Yi, H., Zhu, L., Liu, S., Li, Y., 2008. Constraints on the early uplift history of the tibetan plateau. Proceedings of the National Academy of Sciences USA 105 (13), 4987–4992.

Warnaar, J., Bijl, P.K., Huber, M., Sloan, L., Brinkhuis, H., Röhl, U., Sriver, R., Visscher, H., 2009. Orbitally forced climate changes in the Tasman sector during the Middle Eocene. Palaeogeography, Palaeoclimatology, Palaeoecology 280, 361–370.

Wei, W., 1995. How many impact-generated microspherule layers in the Upper Eocene? Palaeogeography, Palaeoclimatology, Palaeoecology 114, 101–110.

Wei, W., Pospichal, J.J., 1991. Danian calcareous nannofossil succession at Site 738 in the southern Indian Ocean. Proceedings of the Ocean Drilling Program, Scientific Results 119, 495–512.

Wei, W., Wise, S.W., 1990a. Middle Eocene to Pleistocene calcareous nannofossils recovered by Ocean drilling Program Leg 113 from the Weddel Sea. Proceedings of the Ocean Drilling Program. Scientific Results 113, 639–666.

Wei, W., Wise, S.W., 1990b. Biogeographic gradients of Middle Eocene-Oligocene calcareous nannoplankton in the South Atlantic Ocean. Palaeogeography, Palaeoclimatology, Palaeoecology 79, 29–61.

Westerhold, T., Röhl, U., 2009. High resolution cyclostratigraphy of the early Eocene – new insights into the origin of the Cenozoic cooling trend. Climates of the Past 5, 309–327.

Westerhold, T., Röhl, U., Laskar, J., Raffi, I., Bowles, J., Lourens, L.J., Zachos, J.C., 2007. On the duration of magnetochrons C24r and C25n and the timing of early Eocene global warming events: Implications from the Ocean Drilling Program Leg 208 Walvis Ridge depth transect. Paleoceanography 22 PA2201, doi: 10.1029/2006PA001322.

Westerhold, T., Röhl, U., Raffi, I., Forniaciari, E., Monechi, S., Reale, V., Bowles, J., Evans, H.F., 2008. Astronomical calibration of the Paleogene time. Palaeogeography, Palaeoclimatology, Palaeoecology 257, 377–403.

Westerhold, T., Röhl, U., Donner, B., McCarren, H.K., Zachos, J.C., 2011. A complete high-resolution Paleocene benthic stable isotope record for the central Pacific (ODP Site 1209). Paleoceanography 26 PA2216. doi: 10.1029/2010PA002092.

Willems, W., Bignot, G., Moorkens, T., 1981. Ypresian. In: Pomerol, C. (Ed.), Stratotypes of Paleogene Stages. Bulletin d'information des géologues du Bassin de Paris, Mémoire hors série, 2, pp. 267–299.

Williams, G.L., Bujak, J.P., 1985. Mesozoic and Cenozoic dinoflagellates. In: Bolli, H.M., Saunders, J.B., Perch-Nielsen, K. (Eds.), Plankton Stratigraphy. Cambridge University Press, Cambridge, pp. 847–964.

Williams, G.L., Brinkhuis, H., Bujak, J.P., Damassa, S.P., Hochuli, P.A., de Verteuil, L., Zevenboom, D., 1998. Ceonoiz Era - dinoflagellates. In: de Graciansky, P.C., Hardenbol, J., Jacquin, T., Vail, P.R. (Eds.), Mesozoic and Cenozoic Sequence Stratigraphy of European Basins. SEPM Special Publication, 60, pp. 764–765. Chart 3.

Williams, G.L., Brinkhuis, H., Pearce, M.A., Fensome, R.A., Weegink, J.W., 2004. Southern Ocean and global dinoflagellate cyst events compared: Index events for the Late Cretaceous-Neogene. In: Exon, N.F., Kennett, J.P., Malone, M.J. (Eds.), Proceedings of the Ocean Drilling Program, Scientific Results, 189, pp. 1–98.

Williamson, T.E., 1996. The beginning of the Age of Mammals in the San Juan Basin, New Mexico: Biostratigraphy and evolution of Paleocene mammals of the Nacimiento Formation. Bulletin of the New Mexico Museum of Natural History and Science 8, 1–141.

Wilpshaar, M., Santarelli, A., Brinkhuis, H., Visscher, H., 1996. Dinoflagellate cysts and mid-Oligocene chronostratigraphy in the central Mediterranean region. Journal of the Geological Society 153, 553–561.

Wilson, G.S., Roberts, A.P., Verosub, K.L., Florindo, F., Sagnotti, L., 1998. Magnetobiostratigraphic chronology of the Eocene-Oligocene transition in the CIROS-1 core, Victoria Land margin, Antarctica: Implications for Antarctic glacial history. Geological Society of America Bulletin 110, 35–47.

Wing, S.L., Bown, T.M., Obradovitch, J.D., 1991. Early Eocene biotic and climatic change in interior western North America. Geology 19, 1189–1192.

Wise, S.W., 1983. Mesozoic and Cenozoic calcareous nannofossils recovered by Deep Sea Drilling Project Leg 71 in the Falkland Plateau region, Southwst Atlantic Ocean. Initial Reports of the Deep Sea Drilling Project 71, 481–550.

Woodburne, M.O., 1987. Cenozoic Mammals of North America. University of California Press, Berkeley, p. 336.

Woodburne, M.O., Swisher, C.C., 1995. Land mammal high-resolution geochronology, intercontinental overland dispersals, sea level, climate, and vicariance. In: Berggren, W.A., Kent, D.V., Hardenbol, J. (Eds.), Geochronology, Time Scales and Global Stratigraphic Correlations. SEPM Special Publication, 54, pp. 335–364.

Wrenn, J.H., Hart, G.F., 1988. Paleogene dinoflagellate cyst biostratigraphy of Seymour Island, Antarctica. In: Feldman, R.M., Woodburne, M.O. (Eds.), Geology and Paleontology of Seymour Island, Antarctic Peninsula. Geological Society of America Memoir, 169, pp. 321–447.

Zachos, J.C., Lohmann, K.C., Walker, J.C.G., Wise, S.W., 1993. Abrupt climate change and transient climates during the Paleogene: A marine perspective. Journal of Geology 101, 91–213.

Zachos, J.C., Quinn, T.M., Salamy, S., 1996. High resolution (104 yr) deep-sea foraminiferal stable isotope records of the Eocene-Oligocene climate transition. Paleoceanography 11, 251–266.

Zachos, J.C., Opdyke, B.N., Quinn, T.M., Jones, C.E., Halliday, A.N., 1999. Early Cenozoic glaciation, antarctic weathering, and seawater $^{87}Sr/^{86}Sr$: Is there a link? Chemical Geology 161, 165–180.

Zachos, J., Pagani, M., Sloan, L., Thomas, E., Billups, K., 2001. Trends, rhythms, and aberrations in global climate 65 Ma to present. Science 292, 686–693.

F.J. Hilgen, L.J. Lourens and J.A. Van Dam
With contributions by A.G. Beu, A.F. Boyes, R.A. Cooper, W. Krijgsman, J.G. Ogg, W.E. Piller and D.S. Wilson

The Neogene Period

Abstract: An Astronomically Tuned Neogene Time Scale (ATNTS2012) is presented, as an update of ATNTS2004 in GTS2004. The new scale is not fundamentally different from its predecessor and the numerical ages are identical or almost so. Astronomical tuning has in principle the potential of generating a stable Neogene time scale as a function of the accuracy of the La2004 astronomical solution used for both scales. Minor problems remain in the tuning of the Lower Miocene.

In GTS2012 we will summarize what has been modified or added since the publication of ATNTS2004 for incorporation in its successor, ATNTS2012. Mammal biostratigraphy and its chronology are elaborated, and the regional Neogene stages of the Paratethys and New Zealand are briefy discussed. To keep changes to ATNTS2004 transparent we maintain its subdivision into headings as much as possible.

14 Ma Neogene

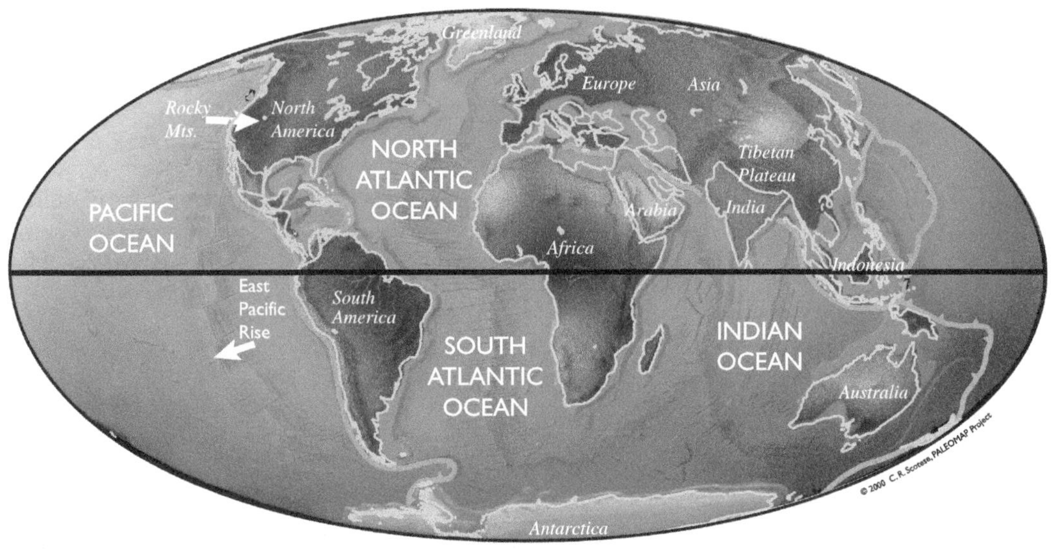

Chapter Outline

The Geologic Time Scale 2012. DOI: 10.1016/B978-0-444-59425-9.00029-9

29.1. CHRONOSTRATIGRAPHY

As in ATNTS2004, and despite the unfortunate decision by the ICS/IUGS to define a Neogene–Quaternary at 2.60 Ma coincident with the Gelasian base and Pliocene–Pleistocene boundary (Gibbard and Head, 2010; Gibbard *et al.*, 2010), we will keep the Neogene here extending up to the Recent. The Neogene thus includes the Miocene, Pliocene, Pleistocene and Holocene (Figure 29.1). The reasons for this have been outlined in Berggren (1998), Lourens *et al.* (2004), Hilgen *et al.* (2008), McGowran *et al.* (2009), Van Couvering *et al.* (2009) and Aubry *et al.* (2009). In this respect we favor an inclusive compromise solution regarding the formal definition of the Quaternary, a solution that does justice to the two main traditions (i.e. the Neogene marine and Quaternary continental) in the chronostratigraphic subdivision of the younger part of the

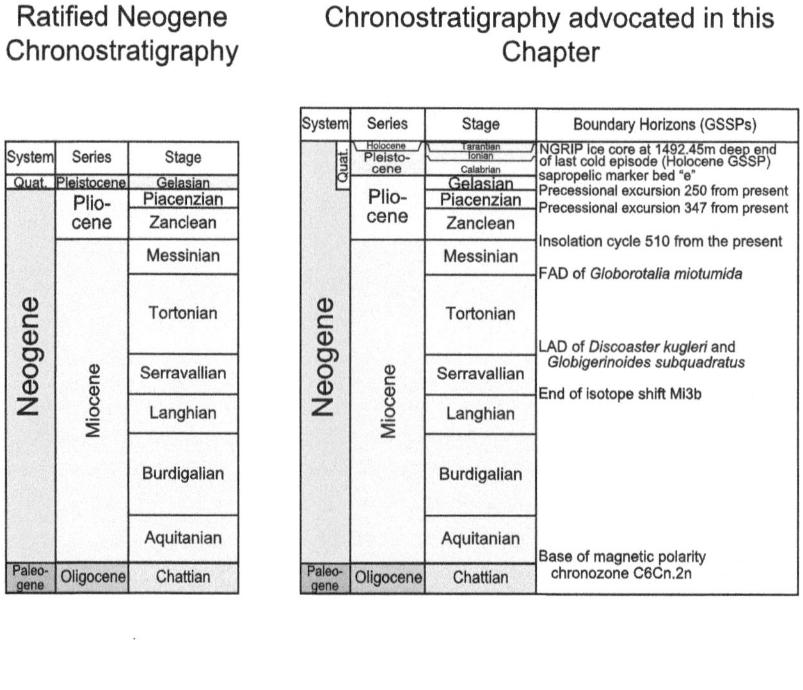

FIGURE 29.1 **Neogene stratigraphic subdivisions with summary definitions of the formalized stage boundaries.** The upper Cenozoic chronostratigraphy ratified by the IUGS in 2009 shifted the Gelasian Stage into the Pleistocene Series and a Quaternary–Neogene system boundary was defined at this level (c. 2.6 Ma). Previously, the Gelasian Stage comprised the upper Pliocene subseries; which is the chronostratigraphy favored by the authors of this chapter.

Cenozoic. Such inclusive solutions either define the Tertiary and Quaternary as sub-erathems of the Cenozoic with its Paleogene and Neogene periods (Aubry *et al.*, 2005), or define the Quaternary as a "hanging" sub-period within the Neogene (Pillans and Naish, 2004).

The base of the Quaternary is now formally defined at 2.60 Ma, coincident with the Gelasian GSSP. Nevertheless we keep the Pliocene–Pleistocene boundary at 1.81 Ma, and divide the Pliocene into an Older and Younger Pliocene Series to maintain/restore hierarchy in our preferred chronostratigraphic scheme (Lourens, 2008). We are adopting this policy because the "Gelasian shift", which results in a 44% expansion of the Pleistocene, has received much resistance concerning stability of literature and principles of chronostratigraphy. Stratigraphic communication requires stability and coherence through time in the meaning of the long published and established terminology.

An important limitation of the GSSP approach is that it focuses on stage boundaries rather than on the bodies that are defined. Hilgen *et al.* (2006) argued for a reconsideration of the unit stratotype concept for stages, albeit in a different way than anticipated by Hedberg (1976), to fit the needs of the imminent astronomical time scale (see also Aubry, 2007). These unit stratotypes cover the entire stage in a complete and continuous succession, preferably contain the formally defined stage boundaries, and their astronomical tuning and integrated stratigraphy (e.g., magnetostratigraphy, calcareous plankton biostratigraphy, isotope stratigraphy) underlies the age calibration of the standard geological time scale. Prime examples in the Neogene are the Rossello Composite section in Sicily as unit stratotype for the Zanclean and Piacenzian stages of the Pliocene, and the Monte dei Corvi section as unit stratotype for the Tortonian (Hüsing *et al.*, 2009) and Serravallian stages of the Miocene. Similarly, in the Paleogene, the Zumaia section can be considered as unit stratotype for the Danian and Selandian stages of the Paleocene (Dinarès-Turell *et al.*, 2003, 2007; Schmitz *et al.*, 2011).

Tuned ages of reversal boundaries and bioevents were incorporated in the ATNTS2004 and so remain in ATNTS2012. It is our intention that the Milankovitch cycles used for the tuning and underlying the age calibration of the time scale will be formally designated as chronozones with their own numbering system, starting from the stable 405-kyr eccentricity cycle (e.g., Pälike *et al.*, 2006a) once the tuning of the entire Cenozoic is completed (see also Hilgen *et al.*, 2006).

29.2. STAGES

29.2.1. Global Stages

In this section, we will briefly summarize the GSSPs that were already defined at the time of publication of GTS2004. We will go into more detail for GSSPs that were subsequently defined, or that are yet to be defined (see also Figure 29.1 and Table 29.1).

TABLE 29.1 ATNTS2012 Ages of Neogene Stage Boundaries and Duration of Stages

Stage/(Sub)Series	Base (Ma)	Duration (myr)
Holocene	0.0117	0.0117
Upper Pleistocene	0.1272	0.1155
Ionian	0.781	0.6538
Calabrian	1.81	1.029
Gelasian	2.59	0.78
Piacenzian	3.60	1.01
Zanclean	5.33	1.73
Messinian	7.25	1.92
Tortonian	11.63	4.38
Serravallian	13.82	2.19
Langhian	15.97	2.15
Burdigalian	20.43	4.46
Aquitanian	23.03	2.60

29.2.1.1. Miocene

GSSPs of the Miocene stages of the Messinian and Tortonian had both been formally defined at the time of publication of GTS2004. In addition, the Aquitanian GSSP was *de facto* defined by the formal definition of the Oligocene-Miocene boundary in the Lemme-C section, at least if we start from a hierarchical ordering of chronostratigraphic units. In the meantime the Serravallian GSSP has been formally designated and ratified. Progress has been made in the selection of potential boundary sections and guiding criteria for defining the Langhian GSSP. The Burdigalian GSSP remains a major concern, especially because marine sections spanning the boundary interval and potentially suitable for astronomical tuning have so far not been found in the Mediterranean.

29.2.1.1.1. Aquitanian

The Aquitanian GSSP, and hence the Paleogene/Neogene boundary, is formally defined in middle bathyal, massive and laminated siltstones with several nodule levels in the Lemme-Carrosio section in northern Italy (Steininger *et al.*, 1997; Figure 29.2). The GSSP, located at the 35 m level as measured downward from the top of the section, corresponds closely with the calcareous nannofossils *Sphenolithus delphix*

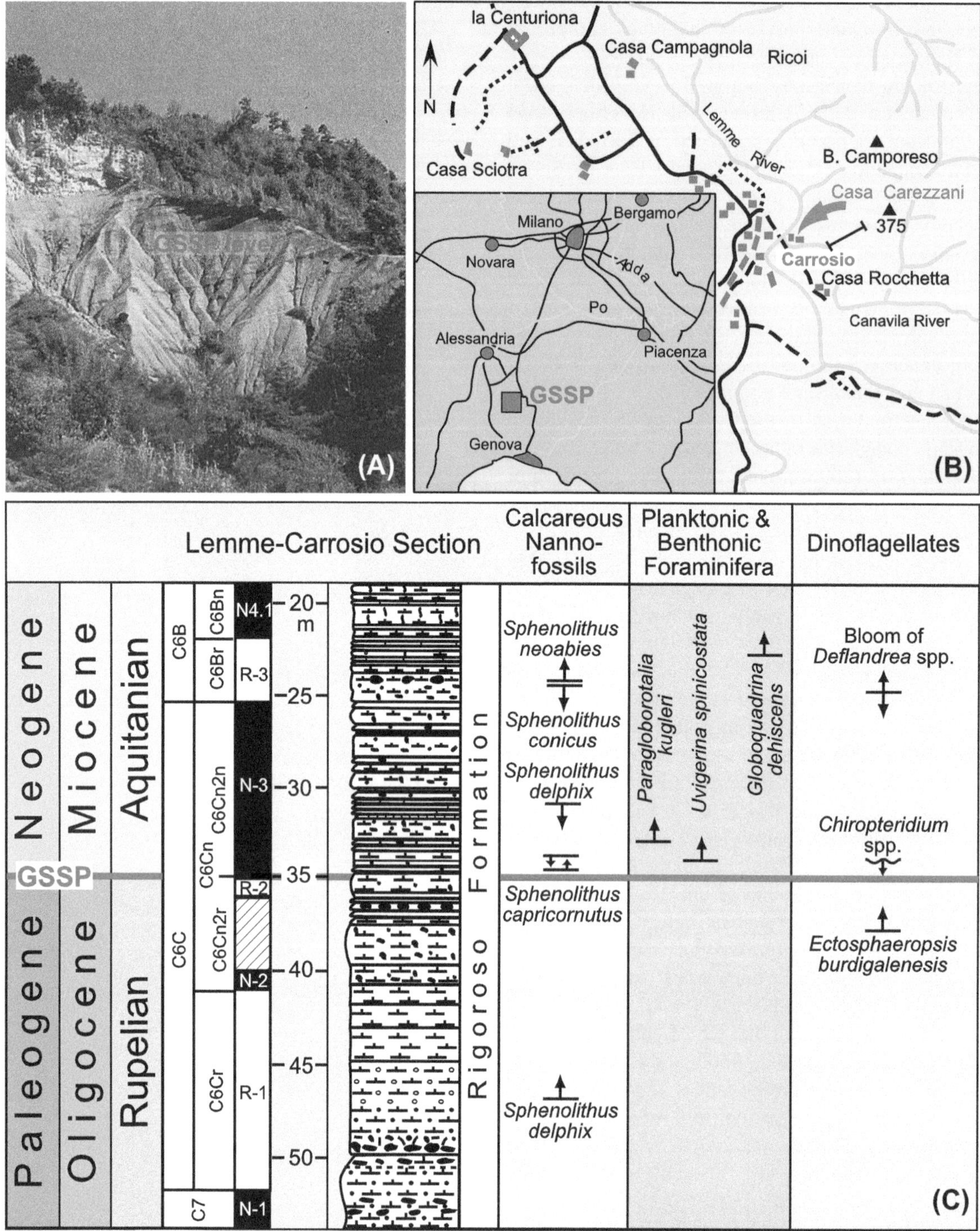

FIGURE 29.2 **Global Boundary Stratotype Section and Point (GSSP) for the Aquitanian Stage and Paleogene/Neogene boundary at Lemme-Carrosio, Italy,** showing section (A), location (B), and lithologic log and magneto-biostratigraphic data (C).

(its range) and *S. capricornutus* (FO), and with the C6Cn.2n(o) reversal boundary. The identification – and synchronicity – of these events has been confirmed by high-resolution biostratigraphic correlations to ODP site 522 in the South Atlantic Ocean (Raffi, 1999; Shackleton *et al.*, 2000). In terms of the standard, low-latitude, calcareous nannofossil zonation, the boundary falls within the older part of zones NN1 and CN1. The boundary further closely corresponds to oxygen isotope event Mi-1, and also the TB1.4 highstand of supercycle TB1 of Haq *et al.* (1987).

The poorly defined age of 23.8 Ma for the boundary in Berggren *et al.* (1995) based on the chronogram age of Harland *et al.* (1990) was replaced by the astronomical age of 22.9 ± 0.1 Ma of Shackleton *et al.* (2000). The latter age became 23.03 Ma when retuned to the La2004 solution, and it was this age, 0.77 myr younger than in the Berggren *et al.* (1995) time scale, that was incorporated in ATNTS2004 as the age for the base of the Miocene. This age is maintained in ATNTS2012 (Table 29.1).

29.2.1.1.2. Burdigalian

No consensus exists regarding the criterion and age of the Aquitanian/Burdigalian boundary. Four options exist: the *Paragloborotalia kugleri* LO (N4–N5 transition) at 21.12 Ma (Berggren *et al.*, 1995), the *Helicosphaera ampliaperta* FO (MNN2a–MNN2b transition) at 20.43 Ma (Fornaciari and Rio, 1996), the top of Chron C6An dated at 20.04 Ma (Berggren *et al.*, 1995), and the *Sphenolithus belemnos* FO (NN3–NN4 transition) dated at 19.03 Ma (Haq *et al.*, 1987) (ages according to the ATNTS2004–2012 time scale). The Aquitanian/Burdigalian boundary in ATNTS2004 was provisionally placed to coincide with the *H. ampliaperta* FO, dated astronomically at 20.43 Ma at Ceara Rise. This was chosen because the age lies close to the estimated Sr/Sr age (~20.0–20.1 Ma, CK92 recalibrated to ATNTS2004–2012) for the base of the Burdigalian in the stratotype area of Saucats, and because of the absence and presence of *H. ampliaperta* in the historical stratotypes of the Aquitanian and Burdigalian, respectively (Poignant *et al.*, 1997a,b).

No suitable sections for defining the Burdigalian GSSP have so far been identified, and it is now seriously being considered to define the boundary in an (I)ODP core in the open ocean. For this purpose ODP sites 1090 (subarctic South Atlantic) (Billups *et al.*, 2004) and 1264/1265 (Walvis Ridge) might qualify, although it will be interesting to see what comes out of the sites recently drilled during IODP Leg 321/322 in the equatorial Pacific (Pälike *et al.*, 2008).

29.2.1.1.3. Langhian

Based on marly to sandy successions exposed in (among others) the Bormida valley in the middle of the Langhe (Piedmont Basin, Italy), Pareto (1865) introduced the Langhian Stage for the middle part of the Miocene, above the now discarded Bormidian and below the Serravallian (he considered the Tortonian as belonging to the Pliocene Series). The original concept was modified by Mayer-Eymar, who in 1868 limited the term to the upper, mainly marly, part of the succession (the so-called "Pteropod Marls"). Since then the "Pteropod Marls" of the Piedmont Basin (now known as the Cessolo Formation.) have become a synonym of the Langhian, and a stratotype section was selected accordingly near the village of Cessolo in the Bormida valley (Cita and Premoli Silva, 1968). The Langhian has long been held synchronous to the Burdigalian, but it is now clear that the type Burdigalian is considerably older than the type Langhian *sensu* Mayer-Eymar (1868) (but not *sensu* Pareto, 1865).

Planktonic foraminiferal studies indicate that the first evolutionary appearance of *Praeorbulina glomerosa* occurs at the base of the Langhian and that the first evolutionary appearance of *Orbulina suturalis* (N8–9 boundary in the zonal scheme of Blow, 1969) is found in its upper part. Calcareous nannofossils (Fornaciari *et al.*, 1997) indicate that the base of the type Langhian predates the *Helicosphaera ampliaperta* LO and contains *Sphenolithus heteromorphus*, and hence falls within Zone NN4 of Martini (1971). The Langhian top lies within the *S. heteromorphus* range and thus belongs to Zone NN5 (Fornaciari *et al.*, 1997).

Lower Limit of the Langhian Stage, Base Middle Miocene The base of the Langhian, and thus the Lower–Middle Miocene boundary, is widely accepted to be approximated by the *Praeorbulina* datum (see Rio *et al.*, 1997). At this stage it is preferable to locate the Langhian GSSP in a position close to both Chron C5Cn and the *Praeorbulina* datum, in agreement with common and consolidated practice. However, the historical stratotype at Cessolo with terrigenous and turbiditic sediments in its lower part is less suitable for defining the GSSP. We have provisionally placed the Langhian GSSP to coincide with the top of C5Cn.1n, dated astronomically at 15.974 Ma in ATNTS2004 (Lourens *et al.*, 2004; Table 29.1).

Two potentially suitable sections for defining the Langhian GSSP are the downward extension of the La Vedova beach section in northern Italy (Montanari *et al.*, 1997; Mader *et al.*, 2004) and St. Peter's Pool on Malta. Both sections were presented in considerable detail at the RCMNS (Regional Committee on Mediterranean Neogene Stratigraphy) congress in Naples (2–6 September 2009), but more research is needed to examine the suitability of both sections for defining the Langhian GSSP. The study of these sections is part of the ongoing Italian research project (PRIN 2006 – prot. 2006047534 – "In search of the Global Stratotype Sections and Points of the Burdigalian and Langhian Stages and paleoceanographic implications") directed at defining the remaining GSSPs (Langhian and Burdigalian) in the Neogene.

Recently, results of integrated high-resolution magnetostratigraphic and calcareous plankton biostratigraphic studies of the La Vedova and St. Peter's Pool sections have been published (Foresi *et al.*, 2011; Turco *et al.*, 2011a). At La Vedova, the reversal boundary that corresponds to the top of C5Cn.1n is found in the interval marked by the so-called mega-beds of Montanari *et al.* (1997). However, the *Praeorbulina* datum, marked by the FO of *P. glomerosa curva* according to Turco *et al.* (2011b), is located much higher in the section, close to the C5Bn.2n/C5Br reversal boundary. These authors follow the taxonomic concept of Blow (1956, 1969) and include *P. sicanus* under the genus *Globigerinoides*. The authors also conclude that the *Praeorbulina* datum is insufficient to define the base of a chronostratigraphic unit, not only because of the controversial taxonomic concepts, but also because of its rarity and diachroneity, and its discontinuous distribution. At this stage, the best criterion to identify the base of the Langhian thus seems to be the top of Chron C5Cn. The bioevent that approximates the magnetic reversal is the LCO of the nannofossil *H. ampliaperta*, which is a reliable event in the Mediterranean, but defies exportation to open ocean sites at low-latitudes (Turco *et al.*, 2011a).

St. Peter's Pool, situated along the east coast of Malta, offers an alternative section for defining the Langhian GSSP (Foresi *et al.*, 2011). This cyclic, deep-marine section provides an excellent calcareous plankton biostratigraphy which allows a straightforward correlation to the La Vedova section (Iaccarino *et al.*, 2011). The section is easily accessible and contains the Burdigalian/Langhian boundary on the basis of (1) the historical criterion that *P. glomerosa sicana* FO (= *G. sicanus* FO of Foresi *et al.*, 2011) occurs at the base of the Langhian stratotype (Rio *et al.*, 1997) and (2) that it is close to the top of the C5Cn.1n (Lourens *et al.*, 2004), but the magnetostratigraphy is unfortunately of a rather poor quality (Mazzei *et al.*, 2009; Foresi *et al.*, 2011; Iaccarino *et al.*, 2011).

Ongoing studies focus on the cyclostratigraphy and the astronomical tuning of these sections, which is considered an important criterion for defining GSSPs in the Neogene. The younger La Vedova beach section has been studied in detail, and an astronomical tuning established (Hüsing *et al.*, 2009). Also the downward extension covering the interval for defining the GSSP looks promising from an orbital tuning perspective (Iaccarino *et al.*, 2009). A preliminary astronomical tuning and astrobiochronology have been established for the alternative St. Peter's Pool section on Malta (Lirer *et al.*, 2009). Following these studies a decision will be made regarding which section and criterion are most suitable for defining the Langhian GSSP. Evidently, both sections have their strong and weak points and are complementary to each other, with La Vedova having a higher quality magnetostratigraphy and St. Peter's Pool a better preservation of the calcareous plankton. The latter is important for biostratigraphy and stable isotopes.

29.2.1.1.4. Serravallian

The Serravallian GSSP has been formally proposed, accepted and ratified after the publication of ATNTS 2004 (Hilgen *et al.*, 2009). Until recently the *Sphenolithus heteromorphus* LO was considered the primary guiding criterion to define the Serravallian GSSP (Lourens *et al.*, 2004). However, the *S. heteromorphus* LO is demonstrably diachronous between the tropical Atlantic and the Mediterranean with astronomical ages of 13.523 and 13.654 Ma, respectively (Backman and Raffi, 1997; Abels *et al.*, 2005). Among others for this reason, the Serravallian GSSP was defined at the formation boundary (i.e. at top "Transitional Bed") between the Globigerina Limestone and Blue Clay Formation in the Ras-il-Pellegrin section on Malta (Figure 29.3), coincident with the termination of the mid-Miocene climate transition (e.g., Flower and Kennett, 1993) marked by the end of Mi-3b (Miller *et al.*, 1991, 1996) or E3 (Woodruff and Savin, 1991) oxygen isotope shift and the carbon isotope excursion CM6 (Woodruf and Savin, 1991). Although a formation boundary is not favored according to GSSP guidelines (Remane *et al.*, 1996), the level was selected because the isotope shift can readily be recognized worldwide. The abruptness and short duration (~40 kyr) of the Mi-3b isotope shift is particularly evident in the high-resolution isotope records that have recently been produced from various ODP sites in different oceanic basins (e.g., Shevenell *et al.*, 2004; Holbourn *et al.*, 2005; Tian *et al.*, 2008).

Integrated stratigraphic correlations of the Serravallian GSSP section to other sections in the Mediterranean are straightforward. For this purpose both primary calcareous plankton events, as well as secondary events such as the paracmes can be used. The *Helicosphaera walbersdorfensis* FCO is a reliable biostratigraphic marker event of regional importance for the Mediterranean middle Miocene (e.g., Fornaciari *et al.*, 1996; Rio *et al.*, 1997; Raffi *et al.*, 2003). The succession of calcareous plankton bioevents at Ras-il-Pellegrin is essentially the same as found in Monte dei Corvi (Montanari *et al.*, 1997; Hilgen *et al.*, 2003) and DSDP Site 372 (Abdul Aziz *et al.*, 2008a; Di Stefano *et al.*, 2008). The GSSP postdates the *H. walbersdorfensis* FCO and precedes the *S. heteromorphus* L(C)O. These two events respectively mark the MNN5a-b and MNN5b-6 boundaries in the standard Mediterranean zonation (Fornaciari *et al.*, 1996; Raffi *et al.*, 2003), and bracket the base of the Serravallian in its historical stratotype (Rio *et al.*, 1997).

In the open ocean, as in the Mediterranean, the GSSP precedes the NN5-6 zonal boundary of the standard low-latitude zonation of Martini (1971) and the CN4-5 zonal boundary of the Okada and Bukry (1980) zonation, both defined by the *S. heteromorphus* L(C)O. Despite its diachronous character between the Mediterranean and low-latitude tropical Atlantic, the *S. heteromorphus* LO is known to be a reliable event on a global scale, although it has rarely been calibrated directly to a reliable magnetostratigraphy and/or cyclostratigraphy. In the low-latitude open ocean the calcareous nannofossil events *S. heteromorphus* and *C. floridanus* LCOs occur above oxygen isotope event Mi-3b.

The succession of well-defined planktonic foraminiferal events around the GSSP can be used to export the boundary to other marine sections in the Mediterranean. The GSSP falls within the upper part of *Orbulina suturalis–G. peripheroronda* Zone (MMi5) of the Mediterranean zonal scheme of

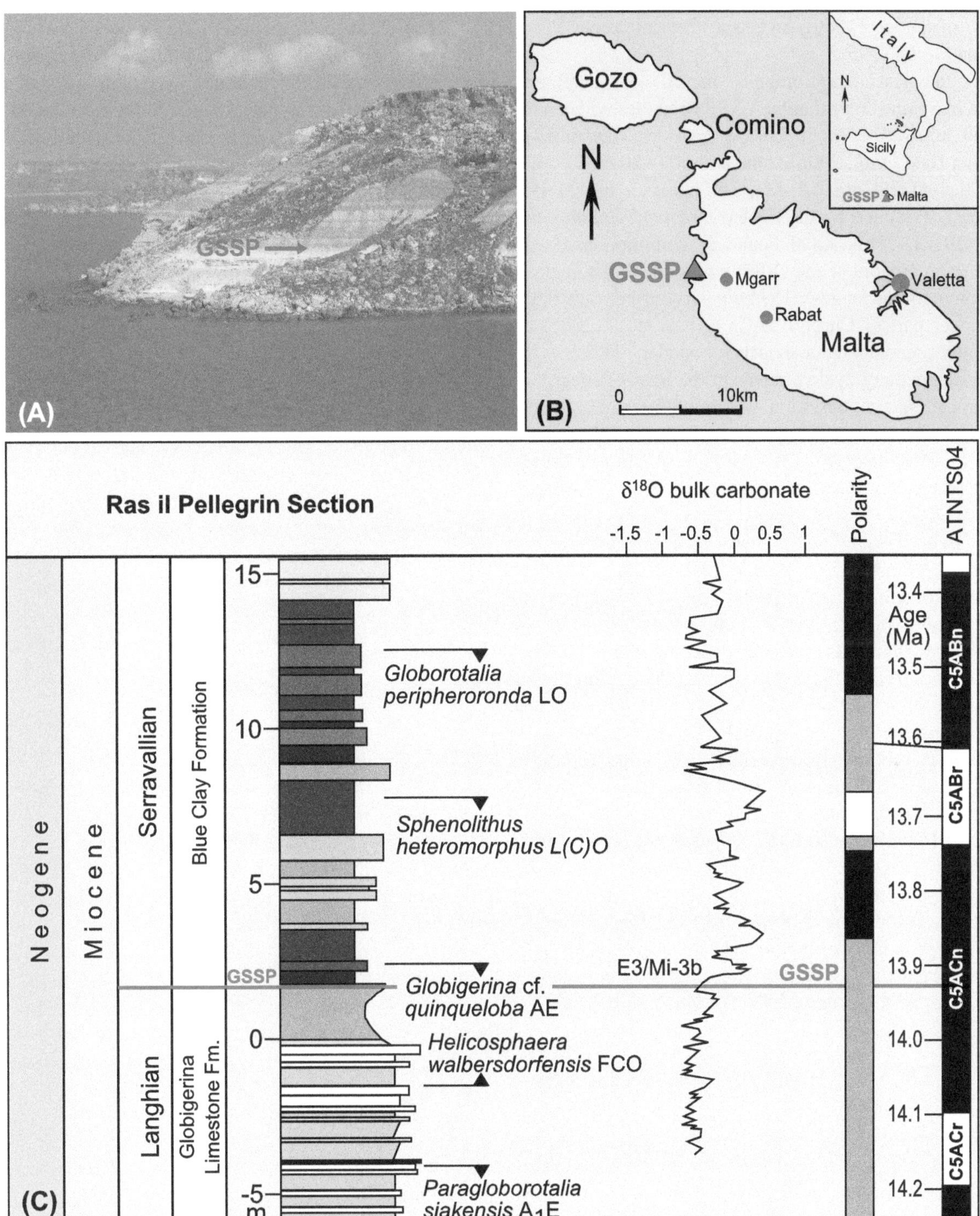

FIGURE 29.7 **Global Boundary Stratotype Section and Point (GSSP) for the Serravallian Stage at Ras-il-Pellegrin on Malta,** showing section (A), location (B), and lithologic log and magnetobiostratigraphic and oxygen isotope data (C).

Iaccarino and Salvatorini (1982) and Sprovieri *et al.* (2002), recently emended by Iaccarino *et al.* (2005). In the open ocean, the GSSP falls within standard low-latitude zones N10/11 of Blow (1969) and (sub)tropical zone M7 of Berggren *et al.* (1995).

The magnetobiostratigraphic record of continental sections in Spain shows that the GSSP coincides with a major mammal turnover event in Europe (van der Meulen *et al.*, 2005; van Dam *et al.*, 2006) at the end of a ~300 kyr cooling phase that started at 14.1 Ma, i.e. close to the MN5–6 boundary (as defined by the FHA of *Megacricetodon gersii*, section 29.3.1). This zonal boundary definition marks the middle–late Aragonian and Orleanian/Astaracian boundaries (Eurasian mammal stages). The boundary further falls within the younger part of Chron C5ACn (Abels *et al.*, 2005) and supposedly coincides with sequence boundary TB2.5.

The sedimentary cycle patterns in the Ras-il-Pellegrin are not characteristic enough to allow straightforward astronomical tuning. The tuning problems were, however, overcome by using astrochronometric ages of high-resolution calcareous plankton bioevents from the well-tuned Tremiti Islands and Monte dei Corvi sections (Hilgen *et al.*, 2003). This tuning resulted in an age of 13.82 Ma for the formation boundary between the Globigerina Limestone and Blue Clay and, hence, for the Serravallian GSSP (Table 29.1). The tuning also pointed to a particular orbital configuration, i.e. minimum eccentricity related to the 405-kyr cycle and minimum obliquity amplitude related to the 1.2-Myr cycle, at times of the major isotope shift associated with the Middle Miocene climate transition (Abels *et al.*, 2005; Holbourn *et al.*, 2005).

A high-resolution study of the Globigerina Limestone underlying the GSSP suggests that the transitional bed might be associated with a condensed interval or a hiatus (Mourik *et al.*, 2010). Similar indications for the existence of a hiatus were found for ODP Site 1146 of Holbourn *et al.* (2005, 2007). This site serves as an open ocean reference for this interval. On the other hand, the La Vedova High Cliff section in Italy is apparently continuous, but its location high in the cliffs makes it difficult to reach (Mourik *et al.*, 2010). Despite these problems, the proposed linkage between the Serravallian GSSP and the Mi-3b isotope shift remains intact.

29.2.1.1.5. Tortonian

The Tortonian GSSP has been formally defined at the midpoint of the sapropel of cycle 76 in the Monte dei Corvi section (northern Italy) astronomically dated at 11.625 Ma (Hilgen *et al.*, 2003, 2005; Hüsing *et al.*, 2007; Figure 29.4 and Table 29.1). The inferred correlation of the GSSP to Chron C5r.2n has been confirmed in the meantime (Hüsing *et al.*, 2007), thereby guaranteeing global correlation potential. The GSSP closely coincides with the calcareous nannofossil *Discoaster kugleri* LCO and thus with the MNN7b–c zonal boundary in terms of the standard Mediterranean zonation (Raffi *et al.*, 2003). It falls within Zone NN7 of the standard low-latitude zonation of Martini (1971) and in Zone CN6 of the Okada and Bukry (1980) zonation.

The GSSP corresponds to the lower part of N14 in the standard low-latitude zonation of Blow (1969), but application of the standard zonation is useless in the Mediterranean in view of the strong diachroneity of the zonal marker events. Finally the GSSP level coincides closely with oxygen isotope event Mi-5 and the associated glacio-eustatic sea-level lowstand of supercycle T3.1 and concurrent deep-sea hiatus NH4 (Hilgen *et al.*, 2005).

29.2.1.1.6. Messinian

The Messinian GSSP, ratified in 2000, is designated at the base of the reddish marl of basic cycle OA15 in section Oued Akrech, located in Morocco on the Atlantic side of the Mediterranean, with an astronomical age of 7.246 Ma (Hilgen *et al.*, 2000a,b; Figure 29.5; Table 29.1). The boundary falls in the middle of the reversed-polarity interval that corresponds to Subchron C3Br.1r and coincides closely with the first regular occurrence of the planktonic foraminiferal *Globorotalia miotumida (=conomiozea)* group. The GSSP falls within subtropical interval Zone M13B, and marks the transitional Mt9–Mt10 zonal boundary of Berggren *et al.* (1995) in terms of standard planktonic foraminiferal zonal schemes. In terms of calcareous nannofossil biostratigraphy the GSSP lies within Zones NN11b and CN9b of Martini (1971) and Okada and Bukry (1980), respectively.

29.2.1.2. Pliocene

All GSSPs of Pliocene stages have been formally defined and ratified. The base of the Zanclean and Miocene/Pliocene series boundary is formally defined at the base of the Trubi Marl Formation in the Eraclea Minoa section (Figure 29.6). This level marks the basal Pliocene flooding of the Mediterranean following the (Messinian) salinity crisis at the end of the Miocene and is positioned five precession-related cycles below the Lower Thvera magnetic reversal boundary (Figure 29.6). The definition is in line with the original distinction of Lyell (Van Couvering *et al.*, 2000). The Piacenzian GSSP is formally defined at the base of the beige marl of basic cycle 77 in the Punta Piccola partial section of the Rossello Composite, close to the Gilbert/Gauss reversal boundary (Castradori *et al.*, 1998; Figure 29.7). The Gelasian GSSP (now officially the base of the Quarternary system and of the Pleistocene series) is defined at the top of sapropel A5 (the so-called Nicola bed) in the San Nicola section, in marine isotope stage 103 close to the Gauss/Matuyama reversal (Rio *et al.*, 1998; Figure 30.3 in Chapter 30). The reader is referred to ATNTS2004 (Lourens *et al.*, 2004) and the above cited papers for further detail (see also Figure 29.1 and Table 29.1).

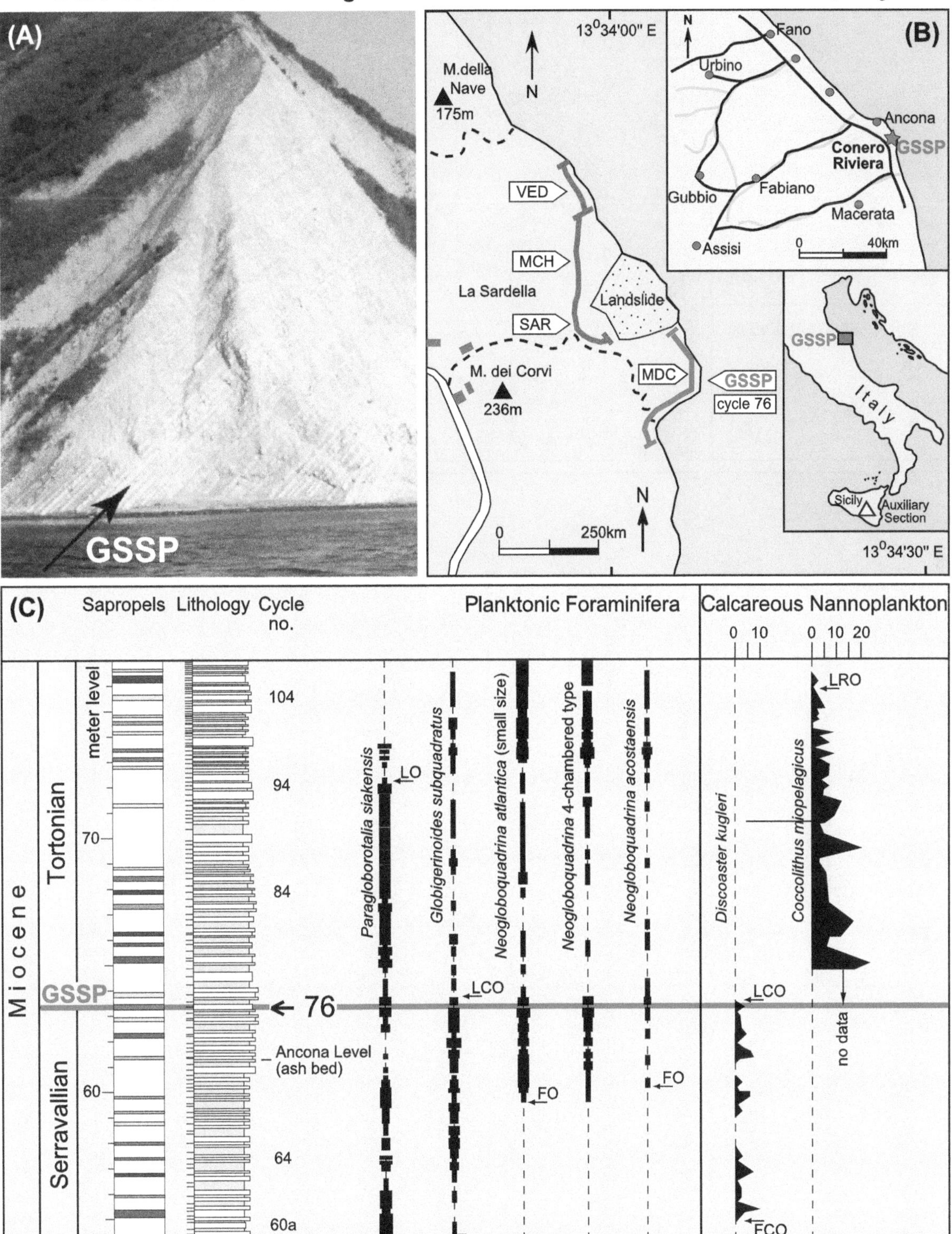

FIGURE 29.5 **Global Boundary Stratotype Section and Point (GSSP) for the Tortonian Stage at Monte dei Corvi, Italy,** showing section (A), location (B), and lithologic log and magneto-bio-cyclostratigraphic data (C).

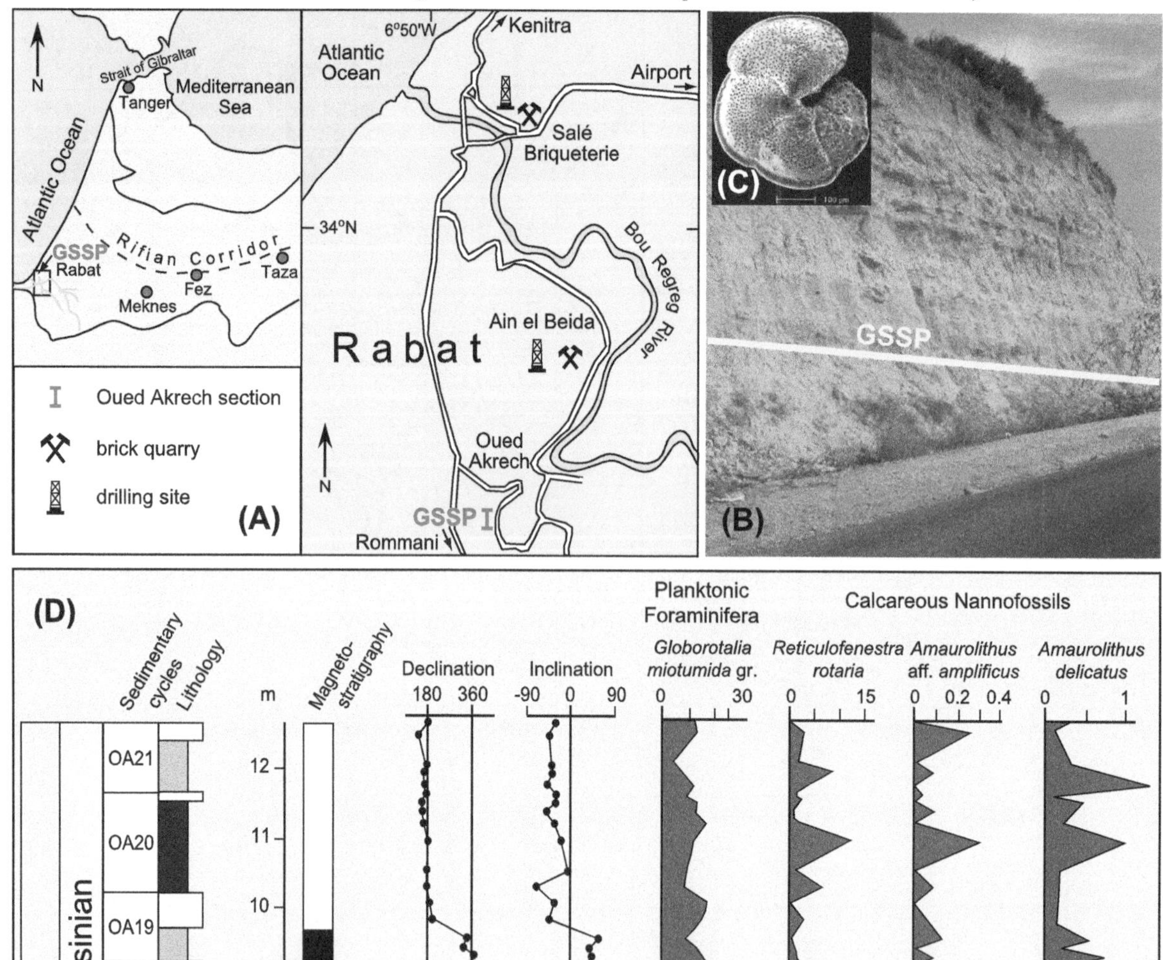

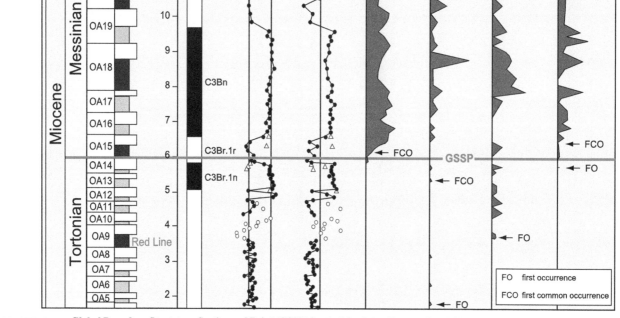

Global Boundary Stratotype Section and Point (GSSP) for the Messinian Stage at Oued Akrech, Morocco, showing location (A), section (B), the *Globorotalia miotumida* marker species (C), and lithologic log and magneto-bio-cyclostratigraphic data (D).

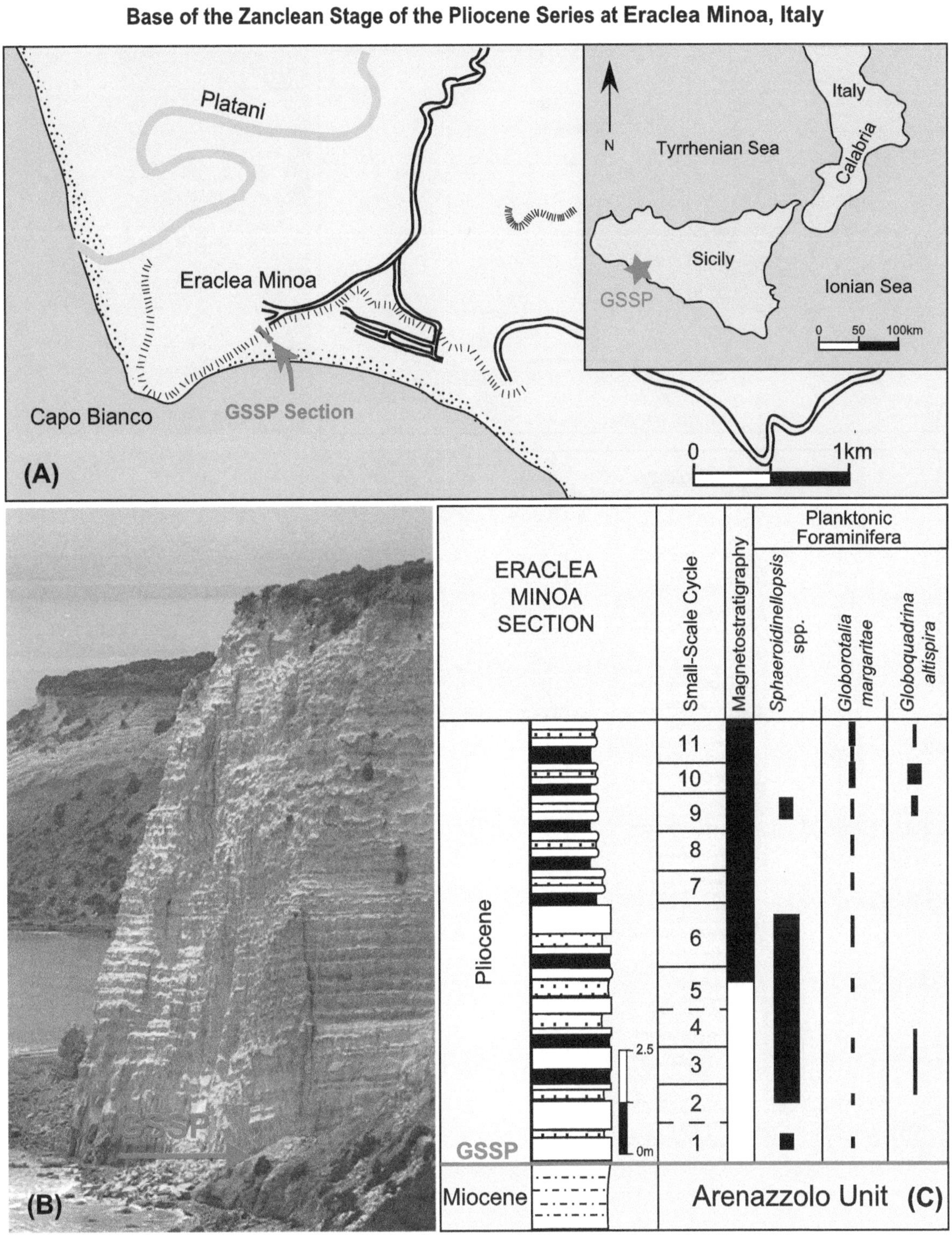

Base of the Zanclean Stage of the Pliocene Series at Eraclea Minoa, Italy

Global Boundary Stratotype Section and Point (GSSP) for the Zanclean Stage and base Pliocene at Eraclea Minoa, Italy, showing location (A), section (B), and lithologic log and magneto-bio-cyclostratigraphic data (C).

29.2.1.3. *Pleistocene*

GSSPs of Pleistocene stages have not been formally defined, except for the former Pliocene–Pleistocene boundary (Figure 30.2), which following the hierarchical ranking of the chronostratigraphic scheme should correspond to the base of the oldest Pleistocene stage, for which Cita *et al.* (2006, 2008) proposed to reintroduce the Calabrian. The notion to retain the name "Calabrian" stems from its historical use, despite the fact

Base of the Piacenzian Stage of the Pliocene Epoch at Punta Piccola, Sicily, Italy

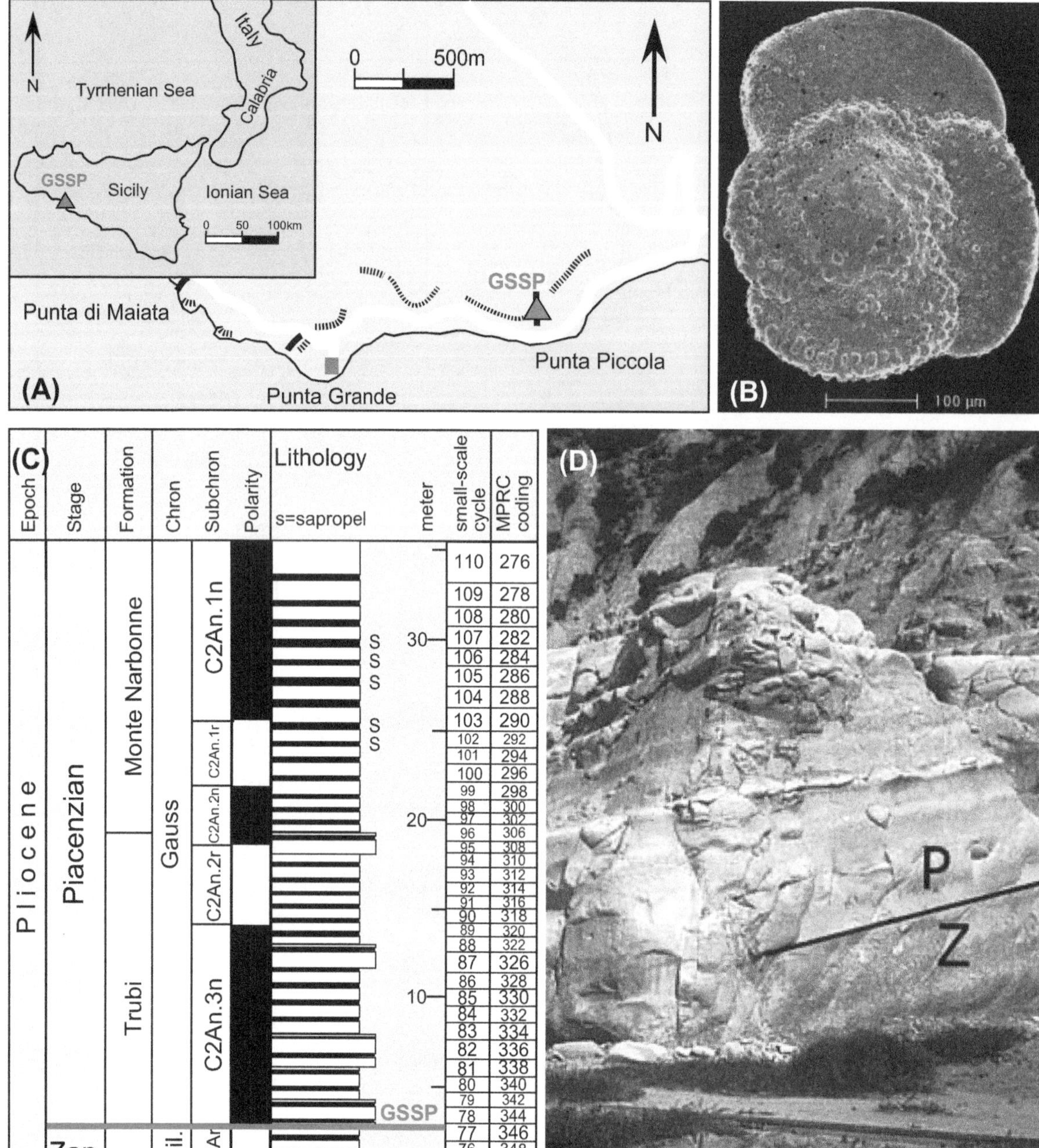

Global Boundary Stratotype Section and Point (GSSP) for the Piacenzian Stage at Punta Piccola, Italy, showing location (A), the *Globorotalia crassaformis* marker species (B), lithologic log and magneto-cyclostratigraphic data (C), and section (D).

that the historical stratotype itself proved much younger. The base of the marine claystone overlying the sapropelic marker bed "e" close to the top of the Olduvai in the Vrica section can be maintained as the GSSP (Figure 29.1; Table 29.1). (*As this volume went to press, the IUGS ratified the name "Calabrian" for this stage as defined by the GSSP at Vrica.*) A GSSP has been proposed for the Upper Pleistocene in the Amsterdam Terminal borehole at the level that marks the first signs of climate warming following the end of the Saalian Glacial (Litt and Gibbard, 2008). The proposal was accepted by the ICS, but has not (yet) been ratified by the IUGS. The Ionian GSSP for the Lower–Middle Pleistocene boundary has not been formally defined. The most likely criterion is the Brunhes–Matuyama magnetic reversal boundary (base of Chron 1n: Head *et al.*, 2008) and candidate sections are Montalbano Jonico (Ciaranfi and d'Alessandro, 2005;

Ciaranfi *et al.*, 2010); Valle di Manche (Capraro *et al.*, 2005), Italy; and Chiba, Japan. Even though the Montalbano Jonico section is suitable for astronomical tuning (Ciaranfi *et al.*, 2010) and is in many aspects judged superior for defining the GSSP, the section has been remagnetized (Sagnotti *et al.*, 2010), while the Brunhes–Matuyama is considered by many as the guiding criterion (Figure 29.1; Table 29.1).

29.2.1.4. Holocene

The GSSP of the Holocene Series has been formally defined in the Greenland ice core from NorthGRIP (NGRIP) at the 1492.45 r_m depth level (Figure 30.14 in Chapter 30) that marks the first signs of climatic warming at the end of the Younger Dryas cold phase as recorded in particular by an abrupt change in deuterium excess and accompanied by more gradual changes in $\delta^{18}O$, dust concentration, a range of chemical proxies, and annual layer thickness (Walker *et al.*, 2009). The level has been assigned an age of 11 700 yr b2k (years before 2000) based on annual layers counts, with an estimated uncertainty of 99 yr at the 2σ level (Figure 29.1; Table 29.1).

29.2.2. Regional Stages

Apart from global stages, regional stages are still widely applied as standard chronostratigraphic units on a regional scale and their linkage to the global stages is important. Below we shortly summarize the subdivision in regional stages of the Paratethys and New Zealand in the light of ATNTS2012 (Figure 29.8).

29.2.2.1. Central and Eastern Paratethys (Wout Krijgsman/Werner E. Piller)

Plate-tectonic collision of the northward moving Africa–Arabian continents with Eurasia led to the formation of the Alpine–Himalayan mountain chain during the Oligo–Miocene. These mountains subdivided the ancient Tethys Ocean into a northern Paratethys and a southern Mediterranean domain, separated by shallow sills from the Atlantic and Indian Ocean. The Paratethys was a large epicontinental sea that stretched from western Europe into central Asia at the beginning of the Oligocene (~35 Ma ago), significantly influencing the paleoclimate and paleoenvironment of Eurasia (e.g., Rögl, 1998; Popov *et al.*, 2006). A complex combination of regional tectonic uplift and glacio-eustatic sea-level lowering caused progressive sea retreat to the present-day remnants: the Black Sea and the Caspian Sea. Paratethys environments gradually evolved from open marine settings into restricted marine and, ultimately, even into lacustrine settings. Associated with this transformation, marine fauna became increasingly replaced by highly endemic, fresh to brackish water biota. The Paratethys furthermore controlled mammal migration patterns out of Africa and Asia into Europe and strongly influenced the radiation and evolution lines of endemic marine invertebrates.

Gradual growth of the Alpine–Carpathian–Dinarides orogenic system during the Miocene induced progressive restriction of the Western, Central and Eastern Paratethys. This geodynamically controlled paleogeographic and biogeographic differentiation caused major difficulties in stratigraphic correlation between the different parts of the Paratethys and the Mediterranean (Piller *et al.*, 2007) and led to the establishment of regional chronostratigraphic scales for the Paratethys, which are well documented in the series *Chronostratigraphie und Neostratotypen* (Cicha *et al.*, 1967; Steininger and Seneš, 1971; Báldi and Seneš, 1975; Papp *et al.*, 1973, 1974, 1978, 1985; Stevanović *et al.*, 1990).

These regional stages are generally defined on the basis of characteristic faunal assemblages (mainly mollusks, foraminifers, and ostracods), which are mostly endemic to the Paratethys Sea. Correlations to the standard GTS are generally strongly debated because radiometric age determinations are scarce and magnetostratigraphic studies are generally controversial. As a consequence, the ages of the Paratethyan stage boundaries can differ by more than a million years in the various geological time scales.

Here we focus on the Neogene time scale for the Central and Eastern Paratethys regions (Figure 29.8), which started to diverge during the Middle to Late Miocene by the uplift of the Carpathian orogen. In the Central Paratethys, all stratotypes correspond to periods of relative sea-level highstands, and regional stage boundaries generally correspond with erosional gaps interpreted to reflect sea-level lowstands. Because of the overall absence of absolute age control, the regional time scale is generally based on correlations to the global sea-level curves of Haq *et al.* (1987) and/or Hardenbol *et al.* (1998).

In the Central Paratethys (Figure 29.8), the Paleogene/Neogene boundary is located within the Egerian Stage. This Stage is followed by the Eggenburgian, which is marked by the first occurrence (FO) of *Oopecten gigas* (Schlotheim). Sequence stratigraphically, the Egerian/Eggenburgian boundary corresponds to sea-level lowstand Aq2 at an approximate age of 21.5 Ma. The base of the next Ottnangian Stage is defined by the FO of *Pecten hermansenni* (Dunker) and is correlated to the lowstand Bur3 at an age of 18.2 Ma. The Ottnangian/Karpatian boundary is marked by the sudden ingression of many faunal elements from the Mediterranean, and is biostratigraphically defined at the FAD of *Uvigerina graciliformis*. The base of the Karpatian is correlated to the sea-level rise at the beginning of highstand Bur4, which results in an age of ~17.3 Ma. The base of the Badenian was originally defined with the FO of *Praeorbulina* and is consequently considered to correspond to the base of the Langhian in the Mediterranean, with an age of 15.97 Ma. However, the FO of this taxon is not a suitable marker, as discussed under Langhian in Section 29.2.1. The Badenian/Sarmatian boundary corresponds to a major turnover in faunal elements, triggered by strong restriction of the connections to

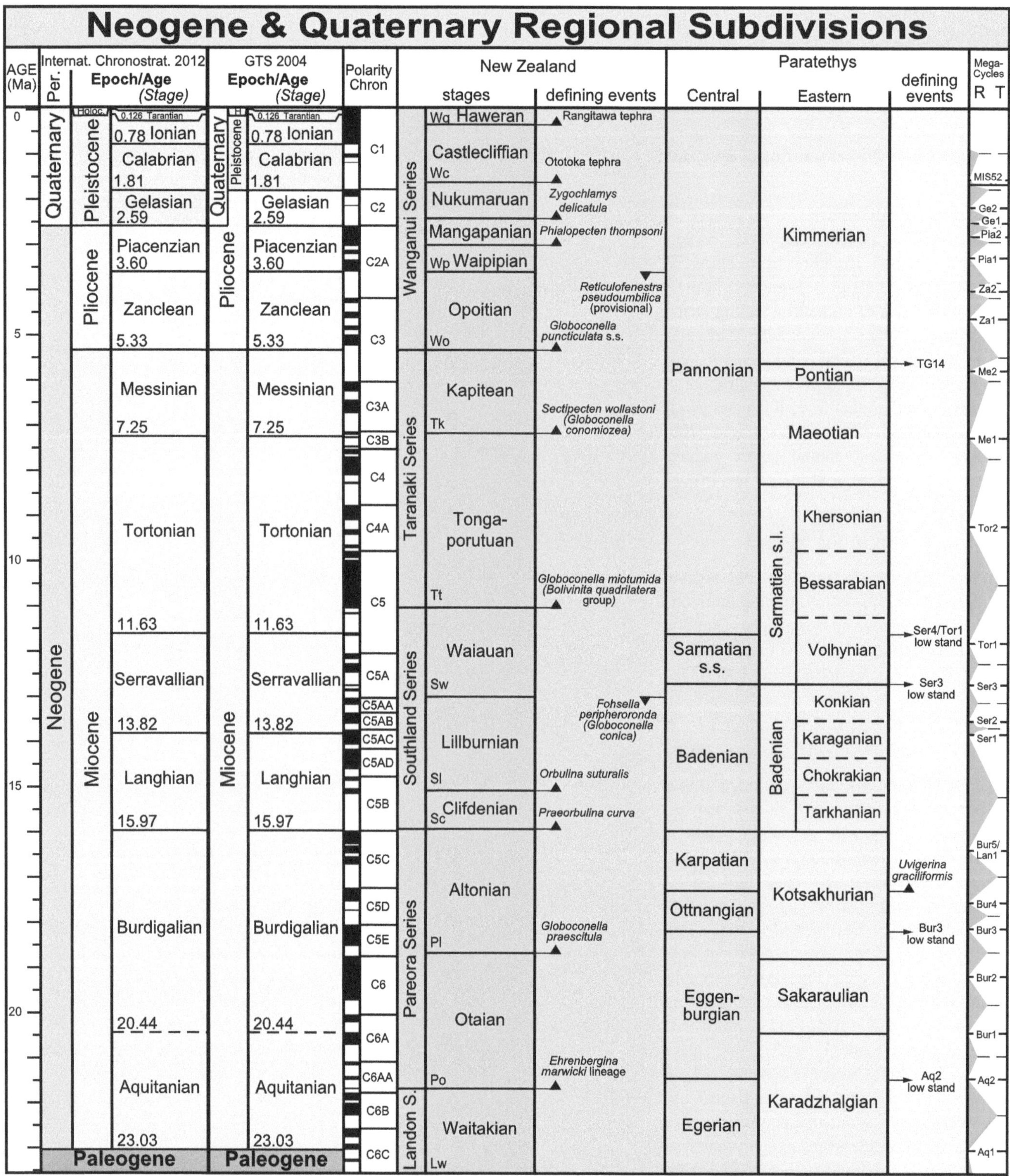

FIGURE 29.8 **Regional Neogene stages for New Zealand and the Paratethys, defining events and correlations to standard global stages.** For the Paratethys, subdivisions are shown both for the Central and Eastern Paratethys. The upper Cenozoic chronostratigraphy ratified by the IUGS in 2009 shifted the Gelasian Stage into the Pleistocene Series and a Quaternary–Neogene System boundary was defined at this level (c. 2.6 Ma). The Gelasian was formerly the upper Pliocene (as shown in GTS2004).

the open ocean. The most recent correlation is to the Ser3 lowstand at an age of 12.7 Ma. The Sarmatian/Pannonian boundary shows a marked change from fully marine to dominantly lacustrine faunal elements with a high degree of endemism, correlated to the lowstand Ser4/Tor1 at 11.60 Ma. The Late Miocene to Pliocene time scale for the Central Paratethys is still very controversial. Many authors have correlated the upper Pannonian to the Pontian Stage of the Eastern Paratethys following Stevanović *et al.* (1990). Others seriously dispute this correlation and argue that the Pontian Stage should be dismissed from the Central Paratethys time scale (Magyar *et al.*, 1999; Piller *et al.*, 2007).

In the Eastern Paratethys (Figure 29.8), the regional time scale for the Early Miocene comprises three stages: Karadzhalgian, Sakaraulian and Kotsakhurian. These stages are all poorly dated and we here follow the stratigraphic scheme of Rögl (1996) who roughly correlates the Karadzhalgian to the Aquitanian and the Sakaraulian (base 20.43 Ma) plus Kotsakhurian (base ~18.8 Ma) to the Burdigalian. The base of the Tarkhanian (15.97 Ma) is correlated to the base of the Langhian and therefore the Badenian as well. Four stages in the Eastern Paratethys form the equivalent of the Badenian stage of the Central Paratethys: the Tarkhanian, Chokrakian, Karaganian and Konkian. The base of the Sarmatian is considered to correspond to a Paratethys-wide environmental change and has consequently the same age (12.7 Ma) as in the Central Paratethys. The top of the Sarmatian is much younger in the Eastern Paratethys (boundary with Maeotian at 8.3 Ma; Vasiliev *et al.*, 2004) than in the Central Paratethys (boundary with Pannonian at 11.6 Ma), which is in serious disturbance with the stratigraphic concept of stages. Consequently, the Sarmatian in the Eastern Paratethys is commonly referred to as the Sarmatian s.l. compared to the Sarmatian s.s. of the Central Paratethys. A preferred alternative solution, already adapted in several time scales, is to promote the Eastern Paratethys substages Volhynian, Bessarabian and Khersonian to stage level. Similar problems have arisen in recent integrated stratigraphic results for the latest Miocene Pontian stage of the Eastern Paratethys. The Maeotian–Pontian boundary corresponds to a Paratethyan-wide marine flooding (Odessian Transgression) at an age of 6.05 Ma (Krijgsman *et al.*, 2010). In the Dacian basin (Romania) the Pontian extends up to 4.7 Ma where it is followed by the Dacian Stage, but in the eastern Black Sea basin (Russia) the Pontian/Kimmerian boundary is located at a much older age of 5.5 Ma (Krijgsman *et al.*, 2010). Also here consensus on a new stratigraphic scheme has to be derived in the near future.

Recently, high-resolution magnetostratigraphic studies on long and continuous sedimentary successions of the Carpathian foredeep of Romania resulted in straightforward correlations to the ATNTS and a robust chronology for the Dacian Basin (Vasiliev *et al.*, 2004, 2005). The magnetic signal of these sedimentary deposits was carried by two different populations of greigite; a biogenic component of primary origin generated by magnetotactic bacteria, and an authigenic component of secondary origin that formed by early diagenetic processes (Vasiliev *et al.*, 2007). In contrast to previous assumptions, these greigite components provided reliable records of the ancient geomagnetic field variations and can excellently be used for magnetostratigraphic dating if the proper demagnetization techniques are applied. Radioisotopic datings of volcanic ashes, based on astronomically integrated Ar/Ar standards, now also allow direct correlation to the ATNTS (Kuiper *et al.*, 2008; De Leeuw *et al.*, 2010). The combination of these new developments holds great promise that more stage boundaries will be more accurately dated in the near future.

29.2.2.2. New Zealand (Alan Beu/Roger Cooper/Andrew Boyes)

Neogene regional stages of New Zealand have been reviewed recently by Cooper (2004), integrating biostratigraphy (mollusca, foraminifera, palynology), sequence stratigraphy, tephrostratigraphy, and magnetic polarity stratigraphy. Because of New Zealand's position astride the Australia–Pacific plate boundary, Neogene marine rocks are very widespread. Some Neogene sequences in New Zealand also are extremely thick, e.g., in the Wairoa syncline, northern Hawke's Bay, the Tongaporutuan to Waipipian stages (early Late Miocene to mid-Pliocene) comprise a succession 9000 m thick. The fauna and flora also are highly endemic in this remote southwest Pacific area, although widespread planktonic groups such as dinoflagellates and calcareous nannofossils are, of course, as useful in New Zealand as elsewhere. Regional stages have seemed essential for subdividing the great thicknesses of Cenozoic rocks (Figure 29.8).

The Oligocene–Miocene boundary falls near the middle of the Waitakian regional stage, and is recognized by the highest occurrence (HO) of the planktonic foraminifer *Globigerina euapertura*. The Waitakian–Otaian boundary is defined by the lowest occurrence (LO) of the benthic foraminifer *Ehrenbergina marwicki* in the Bluecliffs section, Otaio River, and no planktonic foraminiferal criteria have been discovered at this level. Almost all younger stages of Miocene to early Pliocene age are defined at planktonic foraminiferal events. The exception is the base of the Kapitean Stage (latest Miocene), which is defined by the LO of the pectinid bivalve *Sectipecten wollastoni* at Kapitea Creek, Westland, almost coinciding with the LO of *Globoconella conomiozea*. Stratotype Sections and Points (SSPs) have not been defined for the Miocene or early Pliocene stages, as suitable localities for the designation of SSPs require further research. The base of the Opoitian Stage, which almost exactly coincides with the Miocene–Pliocene boundary, is defined by the LO of *Globoconella puncticulata* (*sensu stricto*). It is also easily identified by Mollusca, as *Cucullaea*, *Notocorbula*, *Falsicolus*, "*Lentipecten*" (new genus) and *Aturia* all become extinct in

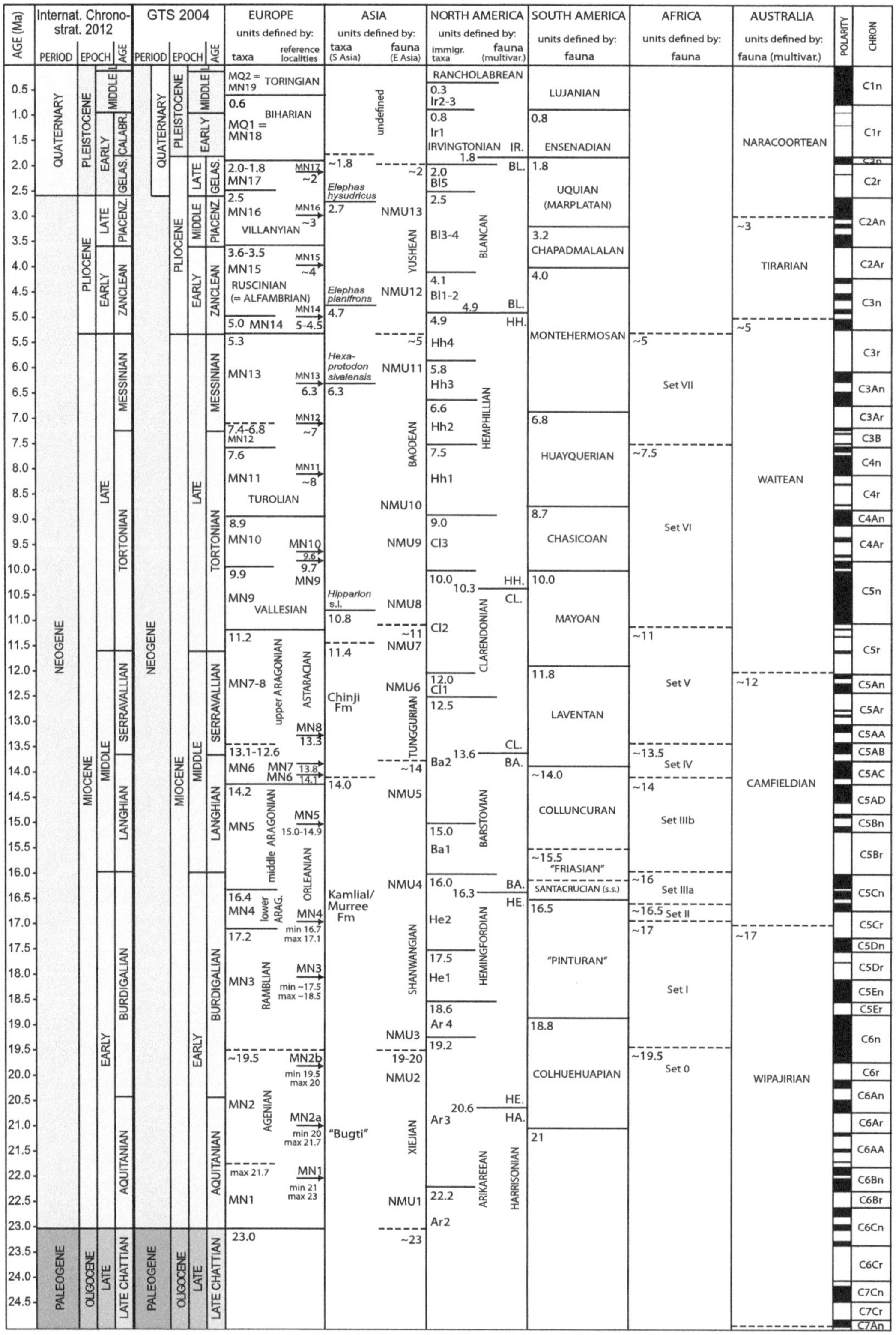

AGE (Ma): 0.5 – 24.5 (in 0.5 Ma steps)

Internat. Chrono-strat. 2012 (PERIOD | EPOCH | AGE): QUATERNARY — PLEISTOCENE (MIDDLE, EARLY: CALABR., GELAS.); NEOGENE — PLIOCENE (LATE: PIACENZ.; EARLY: ZANCLEAN); MIOCENE (LATE: MESSINIAN, TORTONIAN; MIDDLE: SERRAVALLIAN, LANGHIAN; EARLY: BURDIGALIAN, AQUITANIAN); PALEOGENE — OLIGOCENE (LATE: LATE CHATTIAN)

GTS 2004 (PERIOD | EPOCH | AGE): QUATERNARY — PLEISTOCENE (MIDDLE, EARLY); NEOGENE — PLIOCENE (LATE: GELAS., PIACENZ.; MIDDLE; EARLY: ZANCLEAN); MIOCENE (LATE: MESSINIAN, TORTONIAN; MIDDLE: SERRAVALLIAN, LANGHIAN; EARLY: BURDIGALIAN, AQUITANIAN); PALEOGENE — OLIGOCENE (LATE: LATE CHATTIAN)

EUROPE units defined by: taxa | reference localities
MQ2 = MN19, TORINGIAN; 0.6; BIHARIAN; MQ1 = MN18; 2.0-1.8 MN17, MN17 ~2; 2.5 MN16, MN16 ~3; VILLANYIAN; 3.6-3.5 MN15, MN15 ~4; RUSCINIAN (= ALFAMBRIAN); MN14; 5.0 MN14 5-4.5; 5.3; MN13, MN13 6.3; 7.4-6.8 MN12, MN12 ~7; 7.6; MN11, MN11 ~8; TUROLIAN; 8.9 MN10, MN10 9.6, 9.7; 9.9; MN9, MN9; VALLESIAN; 11.2; MN7-8; upper ARAGONIAN; ASTARACIAN; MN8 13.3; 13.1-12.6; MN6, MN7 13.8, MN6 14.1; 14.2; MN5, MN5 15.0-14.9; middle ARAGONIAN; ORLEANIAN; 16.4 MN4, lower ARAG., MN4 min 16.7 max 17.1; 17.2; MN3, RAMBLIAN, MN3 min ~17.5 max ~18.5; ~19.5, MN2b min 19.5 max 20; MN2, AGENIAN, MN2a min 20 max 21.7; max 21.7, MN1 min 21 max 23; MN1; 23.0

ASIA units defined by: taxa (S Asia) | fauna (E Asia)
undefined; ~1.8; ~2; *Elephas hysudricus* 2.7; NMU13; YUSHEAN; *Elephas planifrons* 4.7; NMU12; ~5; *Hexaprotodon sivalensis* 6.3; NMU11; BAODEAN; NMU10; NMU9; *Hipparion s.l.* 10.8; NMU8; ~11; 11.4; NMU7; Chinji Fm; NMU6; TUNGGURIAN; ~14; 14.0; NMU5; Kamlial/Murree Fm; NMU4; SHANWANGIAN; NMU3; 19-20; NMU2; "Bugti"; XIEJIAN; NMU1; ~23

NORTH AMERICA units defined by: immigr. taxa | fauna (multivar.)
RANCHOLABREAN; 0.3 Ir2-3; 0.8 Ir1; IRVINGTONIAN IR.; 1.8; BL.; 2.0 Bl5; 2.5; Bl3-4; BLANCAN; 4.1 Bl1-2; 4.9 BL.; 4.9 HH.; Hh4; 5.8 Hh3; 6.6 Hh2; HEMPHILLIAN; 7.5 Hh1; 9.0 Cl3; 10.0; 10.3 HH. CL.; Cl2; CLARENDONIAN; 12.0 Cl1; 12.5; 13.6 CL. BA.; Ba2; BARSTOVIAN; 15.0 Ba1; 16.0; 16.3 BA. HE.; He2; HEMINGFORDIAN; 17.5 He1; 18.6 Ar 4; 19.2; 20.6 HE. HA.; Ar3; ARIKAREEAN; HARRISONIAN; 22.2 Ar2

SOUTH AMERICA units defined by: fauna
LUJANIAN; 0.8; ENSENADIAN; 1.8; UQUIAN (MARPLATAN); 3.2; CHAPADMALALAN; 4.0; MONTEHERMOSAN; 6.8; HUAYQUERIAN; 8.7; CHASICOAN; 10.0; MAYOAN; 11.8; LAVENTAN; ~14.0; COLLUNCURAN; ~15.5 "FRIASIAN"; SANTACRUCIAN (s.s.); 16.5; "PINTURAN"; 18.8; COLHUEHUAPIAN; 21

AFRICA units defined by: fauna
~5; Set VII; ~7.5; Set VI; ~11; Set V; ~13.5; Set IV; ~14; Set IIIb; ~16; Set IIIa; ~16.5; Set II; ~17; Set I; ~19.5; Set 0

AUSTRALIA units defined by: fauna (multivar.)
NARACOORTEAN; ~3; TIRARIAN; ~5; WAITEAN; ~12; CAMFIELDIAN; ~17; WIPAJIRIAN

POLARITY | **CHRON**
C1n; C1r; C2n; C2r; C2An; C2Ar; C3n; C3r; C3An; C3Ar; C3B; C4n; C4r; C4An; C4Ar; C5n; C5r; C5An; C5Ar; C5AA; C5AB; C5AC; C5AD; C5Bn; C5Br; C5Cn; C5Cr; C5Dn; C5Dr; C5En; C5Er; C6n; C6r; C6An; C6Ar; C6AA; C6Bn; C6Br; C6Cn; C6Cr; C7Cn; C7Cr; C7An

FIGURE 29.9 **Neogene mammal-based chronological systems.** Europe: (left) definition of Mammal Neogene (MN) Units as Interval Biochrons based on First and Last Historical Appearances (see Table 29.2); (right) definition MN units by reference localities after De Bruijn *et al.* (1992), with estimated ages (Azanza *et al.*, 1997; Garcés *et al.*, 1997; Van Dam *et al.*, 2006; Kälin and Kempf, 2009, among others); separation of MN7 and MN8 reference localities after Kälin and Kempf (2009). Asia: (left) Late Miocene–Pliocene South Asian events (Pakistan) after Hussain *et al.* (1992), Barry *et al.* (2002, in press) and Behrensmeyer *et al.* (2007), and calibration of formations after Barry *et al.* (in press), and calibration of older units to European chronology after Van der Made (1999); (right) East Asian (Chinese) Ages, NMU units and their calibration to MN units after Qiu *et al.* (1999) and Qiu and Li (2003). North America: (left)

New Zealand at this level. The base of the Waipipian Stage provisionally is defined in deep-water rocks by the HO of the calcareous nannofossil *Reticulofenestra pseudoumbilica*. In shallow-water limestone and shell beds, the size reached in the anagenetic evolution of the pectinid bivalve *Phialopecten marwicki* is used to define the Waipipian boundary; populations in which specimens reach >110 mm in height are defined as Waipipian. Above this level, SSPs were defined for Pliocene and Pleistocene stages by Beu (compiler) in Cooper (2004), and the succession has been studied well by integrated stratigraphy (Figure 29.8). Mollusca are abundant in shallow facies, and provide the traditional and, in many areas, still the most appropriate criteria for biostratigraphy.

The base of the Mangapanian Stage (Middle Pliocene) is defined at an SSP at the base of Mangapani Shellbed in Waitotara Valley. The pectinid *Phialopecten marwicki* evolved abruptly at this horizon into *P. thompsoni*. This level, dated at 3.0 Ma, coincides approximately with HO *Truncorotalia crassaconica* and the base of the upper dextral coiling zone of *T. oceanica*. No events are recognised at the base of the Late Pliocene (Gelasian), which falls between Parihauhau and Te Rama Shellbeds in Wanganui Basin, to judge from its position immediately above the Gauss–Matuyama magnetic reversal. The base of the Nukumaruan Stage is designated at a provisional SSP at the base of Hautawa Shellbed, between the Rangitikei and Turakina valleys. At this level, the cold-water *Zygochlamys delicatula* molluscan fauna appeared abruptly in Wanganui Basin from the subantarctic, at the first sign of late Pliocene cooling in New Zealand late Neogene marine faunas. This coincides with the base of a brief overlap zone of the top of the upper dextral coiling zone of *Truncorotalia oceanica* with the HO of *T. crassula*, dated at 2.4 Ma (in MIS 97) in DSDP Site 1123 (G.H. Scott, pers. comm.). The base of the Castlecliffian Stage is defined at an SSP at the base of the Pleistocene Ototoka Tephra (1.63 Ma, in MIS 57) in the Castlecliff section, on the Wanganui coast. At this level, 15 genera of distinctive Nukumaruan Mollusca become extinct in Wanganui Basin. The base of the Haweran Stage is defined by an SSP at the base of Rangitawa Tephra (0.34 Ma, in MIS 10) in Rangitawa Stream, Rangitikei Valley.

29.3. BIOSTRATIGRAPHY

29.3.1. Vertebrates

Mammals are by far the most useful organisms to date and correlate Cenozoic terrestrial sediments. At the same time, the nature of the mammal record differs from that of the record of stratigraphically important marine organisms. Firstly, instead of being more or less continuously vertically and horizontally distributed, mammal fossils occur spot-wise in the field, albeit often in association (in "localities" or "sites"). Some types of sites, such as karstic faunas from fissure fillings, lack a stratigraphic context entirely, but may still be very useful for biochronological correlation when they contain many taxa. Secondly, the geographic domain of taxon events (appearances, disappearances) tends to be relatively small, due to strong endemism and diachroneity (e.g., Alroy, 1998; Walsh, 1998; Van Dam *et al.*, 2001; Van Dam, 2003; Van der Meulen *et al.*, 2011). Finally mammal faunas tend to contain more intrinsic temporal information than most marine faunas, which is due to rapid morphological and irreversible evolution.

The number of magnetochronologically and/or radiometrically controlled mammal-faunal sequences has steadily increased during the past three decades. This has resulted in a standard dating precision of mammal localities thus dated in the order of ~0.1 myr. The overall success in matching magnetic polarity patterns to the geomagnetic polarity time scale has also made it evident that continental sequences are more continuous than once thought. This temporal continuity has become especially clear from the results of cyclostratigraphy studies, resulting in the astrochronological calibration of continental sedimentary sequences. This new and promising approach has brought the temporal accuracy and precision of mammal sites potentially down to 0.01 myr (e.g., Abdul Aziz *et al.*, 2004; Hordijk and De Bruijn, 2009). Localities that lack first-order correlations to the numerical time scale can nevertheless be dated relatively precisely if they (and the calibrated sites to which they are tied) contain many taxa. Obviously, the precision of such correlations increases with faunal overlap, and hence with geographical proximity. The multivariate analysis of faunal content itself is very helpful for interpolation, either by using presence–absence information (e.g., Alroy, 1994, 2000), or stage of morphological evolution (such as size: Legendre and Bachelet, 1993). We refer to Walsh (1998) and Lindsay (2003) for thorough discussions on the principles of mammal-based chronology and correlation.

In order to chronologically place increasing numbers of taxa and localities, at some point chronological systems have been erected for each continent except for Antarctica (Figure 29.9). The concept of Land Mammal "Age" is central to most of these systems. The insertion of quotation marks

definition North American Land Mammal Ages (NALMAs) and subunits based on Tedford et al. (2004) and Bell et al. (2004) with modifications (see Table 29.2); (right) NALMA boundaries based on multivariate analysis of faunas (Alroy, 2000), and on calibrations (95) to the numerical time scale, (27 paleomagnetic constraints are not updated). South America: definition and chronology of South American Land Mammal Ages (SALMAs) based on Flynn and Swisher (1995) with modifications (see text). Africa: (left) faunal sets of Pickford (1981) with correlation to European chronology after Van der Made (1999). Australia: Australian Land Mammal Ages based on multivariate analysis of faunas (Megirian et al., 2010). The upper Cenozoic chronostratigraphy ratified by the IUGS in 2009 shifted the Gelasian Stage into the Pleistocene Series and a Quaternary–Neogene System boundary was defined at this level (c. 2.6 Ma). The Gelasian was formerly the upper Pliocene (as shown in GTS2004).

reflects the fact that almost none of the continental mammal-based units have been defined geochronologically, i.e. by a GSSP. One could even argue whether ages other than the contemporaneous marine ones that are included in the Standard Global Geochronologic Scale should be defined at all. Standard stages/ages are defined precisely because they are meant to define slices of time, i.e. independent of taxonomy and diachrony (see Walsh, 1998). In addition, the use of one system enhances communication between workers in the marine and continental record (Van Dam *et al.*, 2001). Although the problem is largely semantic, two solutions exist for the definition problem of continental-wide mammal-based units:

1) Defining the units bio-chronostratigraphically (on the basis of rock-based biozones) instead of geo-chronostratigraphically (e.g., Steininger, 1999), or
2) Defining them faunistically by the contents of type localities (e.g., De Bruijn *et al.*, 1992).

Whether these considerations should have consequences for the use of the stage/age extension "-an" for mammal-based units is an open question. For practical purposes these names are retained here (Figure 29.9).

In continents with a longer history of mammal paleontology, such as Europe and North America, the faunal contents and ages of most units have been established relatively well. Most recent discussions therefore focus on the exact ages of boundaries, the uncertainties of which are often related to the nature of the record, as discussed above. Sampling effects are not always appreciated, as specimen frequencies obviously affect the duration of confidence intervals preceding first and succeeding last appearances (Walsh, 1998; Van Dam, 2003).

The state of the art of the Neogene mammal-based chronological systems as used on the various continents is summarized below (Figure 29.9). Information on relevant paleomagnetic and radiometric calibrations is also provided, although no effort has been made to be complete.

29.3.1.1. Europe

With regard to European Neogene mammal chronology, two "schools of thought", the "stratigraphic school" and "faunal school", have developed side by side (see Van Dam, 2003; Lourens *et al.*, 2004). The stratigraphic approach, conforming to the strict biostratigraphic and chronostratigraphic/geochronologic principles as stated in the international stratigraphic guides, is partly rooted in well-exposed and relatively continuous series of the Iberian Peninsula and central Europe (e.g., Switzerland). In these areas "stages"/"ages" such as the Vallesian and Turolian have been defined on the basis of rock units (e.g., Crusafont and Truyols, 1960) and detailed local rodent biozonations have been constructed (e.g., Van de Weerd, 1976; Kälin and Kempf, 2009). The stratigraphic approach has also been applied at the scale of Europe as a whole (e.g., MN = Mammal Neogene Zones and Megazones in Steininger, 1999), with zones preferentially defined by one or several taxa. This resulted in proposals to base European-wide (MN) zones on the calibrations to the Geomagnetic Polarity Time Scale of first appearances of specific taxa (species or genera) in Spain and (for a large part of the Early Miocene) in Switzerland (e.g., Agustí *et al.*, 2001).

The faunal approach utilizes complete faunas as means of dividing time (biochronology *sensu* Lindsay, 2003) and has its roots in the works of Thaler (e.g., Thaler, 1965). As the temporal information of the faunas themselves is used, explicit stratigraphic support is less important. Reference localities define rather than illustrate European mammal-based units (e.g., MN units of De Bruijn *et al.*, 1992). In its extreme form, the reference localities constitute an ordinal scale, i.e. boundaries do not exist. Part of the discussion surrounding MN zones/units and their calibration stems from the fact that both approaches use the name "MN", as introduced by Mein during the early 1970s (Mein, 1976). Mein himself did not adhere to a strict stratigraphic or faunal approach; neither did he define zones formally after taxon names, nor did he follow Thaler in using reference localities to define the zones (these were added later, Fahlbusch, 1976). Mein's apparent intention to have an efficient chronologic tool rather than a formalized system explains why criteria were modified and why geographic subdivisions were applied (e.g., Mein, 1990, 1999). Similarly, "stage"/"age" names have been used in two ways, either as chronostratigraphic/geochronologic units, or as biochronological units. In both cases, the units combine several MN units (Table 29.2).

Traditionally, calibrations to the numerical time scale for European faunas and units have depended on magnetostratigraphy, although radiometric dates from volcanics in association with mammal occurrences have been compiled as well (e.g., Swisher, 1996; Kuiper *et al.*, 2004; Carbonell *et al.*, 2008; Abdul Aziz *et al.*, 2009; Wagner *et al.*, 2011). In addition, direct calibrations of mammal sites with intercalated Paratethys and open marine records have been used as well (for example, Aguilar, 1982; Rögl and Daxner-Höck, 1996; Costeur *et al.*, 2007). Quantitative estimates for ages of non-calibrated sites have been produced by multivariate analysis of faunal list information (appearance event ordination, e.g., Hernández-Fernández *et al.*, 2004; Domingo *et al.*, 2007).

Magnetostratigraphic records straddling the Oligocene–Miocene transition have been constructed for Spain and Switzerland (Levèque, 1993, Schlunegger *et al.*, 1996, Agustí *et al.*, 2001), with the latter country also being responsible for the most accurate Early Middle Miocene calibrations (Schlunegger *et al.*, 1996, Kälin and Kempf, 2009). Most important mammal-related magnetostratigraphic records for the middle Miocene have been compiled for Spain (Daams

TABLE 29.2 Defining Criteria for Neogene Mammal-based Stratigraphy/Chronology in Europe and North America

Europe			
Unit	**Option 1: Base Defined by Selected Appearance Events**	**Event Age Based On**	**Option 2: Age Based on Age Reference Locality**
(MQ2=) MN19	FHA *Arvicola*	Fejfar *et al.* (1998); Wagner et al. (2011)	Undefined
(MQ1=) MN18	FHA *Mimomys savini*	Fejfar *et al.* (1998); Maul et al. (2007)	Undefined
MN17	FHA *Mimomys pliocaenicus* (=*M. ostramosensis*)	Opdyke et al. (1997); Oms et al. (1999)	Saint Vallier
MN16	FHA *Mimomys hajnackensis* (= *M. hassiacus*)	Vangenheim et al. (2005)	Arondelli/Triversa
MN15	FHA *Mimomys*	Hordijk and de Bruijn (2009)	Perpignan (Serrat d'en Vacquer)
MN14	FHA *Promimomys*	Hordijk and de Bruijn (2009)	Podlesice
MN13	FHA *Apodemus gudrunae*	Garcés et al. (1998); Agustí et al. (2006)	El Arquillo 1
MN12	FHA *Parapodemus barbarae*	Krijgsman et al. (1996); Van Dam et al. (2001)	Los Mansuetos
MN11	FHA *Parapodemus lugdunensis*	Krijgsman et al. (1996); Van Dam et al. (2006)	Crevillente 2
MN10	FHCA *Progonomys*	Alcalá et al. (2000); Van Dam et al. (2006)	Masía del Barbo
MN9	FHA *Hippotherium*	Garcés et al. (1997b), Vasiliev et al. (2011)	Can Llobateres
MN7+8	FHA *Megacricetodon crusafonti*	Daams et al. (1999a,b)	La Grive M; Steinheim (MN7); Anwil (MN8)
MN6	FHA *Megacricetodon gersii*	Daams et al. (1999a,b)	Sansan
MN5	LHA *Ligerimys florancei*	van der Meulen et al. (2011)	Faluns Pont Levoy – Thenay
MN4	FHCA *Democricetodon*	Kempf et al. (1997); Agustí et al. (2011); Van der Meulen et al. (2011)	La Romieu
MN3	FHA *Ligerimys*	Kälin (1997); Kempf et al. (1997)	Wintershof-West
MN2	FHA *Ritteneria*	Schlunegger et al. (1996); Kälin (1997)	Montaigu (MN2a); Laugnac (MN2b)
MN1	LHA Theridomyidae	Schlunegger et al. (1996); Engesser (1997)	Paulhiac

North America	
Unit	Option 1: Base Defined by Appearance Events
Rancholabrean	FA *Bison*
Irvingtonian II	FA *Microtus meadensis*
Irvingtonian 1	FA *Microtus*
Blancan V	FA *Mictomys vetus, Plioctomys rinkeri*
Blancan III-IV	FA *Pliopotamys*
Blancan I-II	FA *Mimomys, Ophiomys, Ogmodontomys*

(Continued)

TABLE 29.2 Defining Criteria for Neogene Mammal-based Stratigraphy/Chronology in Europe and North America—cont'd

North America	
Hemphillian 4	FA *Eocoileus, Lutra, Miopetaurista, Mustela, Propliophenacomy, Sminthosinis, Trigonictis*
Hemphillian 3	FA *Agriotherium, Castor, Felis, Magalonyx, Ochotona, Plesiogulo, Pliozapus, Prosomys*
Hemphillian 2	FA *Enhydritherium, Eomellivora, Indarctos, Lutravus, Machairodus, Neotragocerus, Somocyon, Thinobadistes*
Hemphillian 1	FA *Crusafontina, Kansasimys, Lemoynea, Paramicrotoscoptes, Pliometanastes, Pliotomodon*
Clarendonian 3	FA *Actiocyon, Hoplictis, Platybelodon*
Clarendonian 2	FA *Barbourofelis*
Clarendonian 1	FA *Pseudoceras*
Barstovian 2	FA *Ramoceras*
Barstovian 1	FA *Plithocyon, Zygolophodon*
Hemingfordian 2	FA *Aphelops, Copemys, Dinogale, Eomys, Mionictis, Peraceras, Petauristodon, Plionictis, Pseudaelurus, Sthenictis, Teleoceras*
Hemingfordian 1	FA *Aletomeryx, Amphictis, Amphycyon, Antesorex, Brachyiopotherium, Craterogale, Desmatolagus, Edaphocyon, Euroxenomys, Floridaceras, Leptarctus, Miomustela, Oreolagus, Phoberocyon, Plesiosorex, Potamotherium, Ursavus*
Arikareean 4	FA *Barbouromeryx, Cynelos, Paracosoryx, Problastomeryx*
Arikareean 3	FA *Cephalogale, Menocera, Moropus, Ysengrinia, Zodiolestes*

Upper Part. Definitions of European MN units based on 1) revised selection of single-taxon appearance events and 2) reference localities (see Fig. 29.9). FHA = First Historical Appearance (in Europe), LHA = Last Historical Appearance (in Europe), FHCA = First Historical Common Appearance (in Europe). FHA estimated as midpoint between age of well-calibrated Oldest Known Record (OKR, in Europe) and age of next older well-documented and well-calibrated locality. Similar estimations apply to LHA and FHCA. FHA, LHA, OKR concepts after Walsh (1998).
Lower part. Definition of North American Land Mammal "Ages" and their subdivisions based on First Appearances (FA) of immigrant taxa (Fig. 29.9). After Tedford et al. (2004) and Bell et al. (2004), with modifications as proposed by Martin et al. (2008) and Pagnac (2009).

et al., 1999; Casanovas-Vilar *et al.*, 2008) and Switzerland (Kälin and Kempf, 2009) with additional information from southern Germany (Abdul Aziz *et al.*, 2009). Late Miocene calibrations have been documented for Spain (Garcés *et al.*, 1996, 1998, 2003; Krijgsman *et al.*, 1996a; Opdyke *et al.*, 1997; Van Dam *et al.*, 2001; Abdul Aziz *et al.*, 2004) and Eastern Europe (Vangenheim and Tesakov, 2008a,b, in press). Important latest Miocene to Pliocene magnetobiostratigraphic records are based on data from Spain (Garcés *et al.*, 1997; Opdyke *et al.*, 1997; Oms *et al.*, 1999) and from Greece, several of which are in association with radiometrically dated levels (Sen *et al.*, 1986; Swisher, 1996; Van Vugt *et al.*, 1998; Steenbrink *et al.*, 2006; Hordijk and De Bruijn, 2009). Mio–Pliocene compilations primarily focused on the Neogene of Spain can be found in Agustí *et al.* (2001) and Van Dam *et al.* (2006).

The first column in Figure 29.9 shows the mammal chronological units according to both "schools" (note that with the Neogene defined *sensu* Gradstein *et al.* (2004), the two youngest units can comfortably be numbered MN18 and 19). Both options are presented in their extreme form: by displaying boundaries as single-taxon events (thereby constituting interval zones/biochrons, left) and by displaying reference localities (right). Taxon selection (Figure 29.9) is based on earlier schemes (e.g., Fejfar *et al.* 1998; Agustí *et al.*, 2001; Kälin and Kempf, 2009; Van der Meulen *et al.*, 2011), while reference locality choice is based on De Bruijn *et al.* (1992) and Kälin and Kempf (2009). Because the systems are unambiguous in terms of their definition, both represent a good starting point for future refinement, involving replacements of some taxa by others (or possibly several taxa, e.g., disjunct Oppel zones, see Walsh, 1998), or replacements and additions of reference localities that are both species-rich and well dated. In the current sequence of reference localities, several (notably MN6–7, and MN9–10) are very close in age, which may be considered undesirable. Note that different definitions for MN units may lead to situations in which a reference locality for one MN unit correlates to another MN unit as defined by events (for example MN9–10). In general, by choosing the taxon-range approach, boundaries may change with new finds. By choosing reference localities, ages are fixed by definition, but the correlation procedure (i.e. of entire faunas) is more complex.

29.3.1.2. Asia

Because Europe and Asia basically constituted one continent during the Neogene, faunal similarities have developed between the two areas through time, and these can be used for correlation. For instance, the Neogene mammal units for Anatolia can be correlated to MN units without too much problem (De Bruijn *et al.*, in press). Various local magnetostratigraphic correlations have been realized (see Krijgsman *et al.*, 1996b; Sen, 1996; Kappelman *et al.*, 2003). With increasing distance from Europe, as in the case of the Middle East or Central Asia, the uncertainty of correlation becomes larger. Nevertheless rough correlations of mammal localities in these areas to the MN system can be made (e.g., Sotnikova *et al.*, 1997).

The Chinese Neogene mammal record has steadily grown. The faunas are mainly concentrated in the northern and northwestern part of the country (Qiu *et al.*, 1999; Qiu and Li, 2003). Although a rough correlation to the European system can be made, the establishment of an independent system appeared to be useful. Pioneered by Qiu *et al.* (1979), a system of five "ages" is now in use (see Qiu and Li, 2003). A framework of finer units at the level of, and correlated to, European MN units was presented by Qiu *et al.* (1999). Because of the paucity of volcanic materials in continental Chinese sediments, direct calibrations to the numerical time scale mostly depend on magnetostratigraphic records, which are beginning to be compiled for various basins (e.g., Flynn *et al.*, 1997a; Song *et al.*, 2003; Wang *et al.*, 2003; Zhang *et al.*, 2007). Up-to-date information on Chinese Neogene mammal chronology can be found in Wang *et al.* (in press).

The Siwaliks formations in South Asia have produced an exceptionally rich and complete Neogene mammal record, especially in its younger part (from the upper Lower Miocene onwards). The main mammal sequence originates from the Potwar plateau in Pakistan and is well calibrated to the Geomagnetic Polarity Time Scale (Barry *et al.*, 2002). Although local interval zones were defined at some point, South Asian mammal ages have never been erected, which is related to the temporal control already achieved by direct correlation to the numerical time scale, and by the relatively small size of the studied region (Flynn *et al.*, 1990). Early Miocene faunas from South Asia are best known from Baluchistan (Dera Bugti beds, Welcomme *et al.*, 1997).

29.3.1.3. North America

The first major compilation dealing with North American mammal chronology is the paper of Wood *et al.* (1941), in which 19 Cenozoic "Provincial Ages" (including one unnamed Pleistocene one) were defined as periods of time characterized by their fauna (mainly at genus level). The units were named after types (local formations or faunas) in western North America, where the record is richest. Boundaries were kept somewhat flexible in order to be modified in the future. Index fossils, additional characteristic fossils, as well as first and last appearances, accompanied these original age definitions. The loose way Wood *et al.*'s (1941) units were defined, also would characterize the MN zones that were subsequently defined in Europe (Mein, 1976). Savage (1962) introduced the name North American Land Mammal "Ages" (NALMAs), with the quotation marks indicating that these were not defined on the basis of time-stratigraphic stages.

A shift towards taxon-based definitions took place when it was proposed to use immigrants for defining the base of each age (Repenning, 1967), an approach that was used in the major reviews of Tedford *et al.* (1987, 2004). Subsequently, some of the NALMAs have been strictly defined by biostratigraphic zones or even boundary stratotypes (see Walsh, 1998; Woodburne, 2004).

Eight NALMAs are in use for the Neogene (see Figure 29.9) with the base of the Irvingtonian positioned in the latest Pliocene. Some modifications of the definition of ages have been proposed since the last major update by Tedford *et al.* (2004). Firstly, it has been suggested to redefine the base of the Late Barstovian (Ba2) by the entry of *Ramoceras* instead of that of Gompotheriidae, resulting in an age of 15.0 (Pagnac, 2009) instead of 14.8 Ma (Tedford *et al.*, 2004). Secondly, the base of the Irvingtonian is set at 2.0 Ma following the proposal of Martin *et al.* (2008) to define it by the immigration of *Microtus* as dated between 2.06 and 1.95 Ma in the Meade Basin, Kansas (for this option also see Bell *et al.*, 2004 and Woodburne, 2006). Because they were defined according to evolutionary stage and not the criteria of Woodburne (1977), Blancan 2 and 4 were not included as formal units into the recent subdivision of Bell *et al.* (2004). Here we combine them with Blancan 1 and 3, respectively.

Traditionally, calibrations of Neogene faunas to the numerical time scale in North America have depended in particular on radiometric (and fission-track) dates (Woodburne and Swisher, 1995). About a hundred Mio−Pliocene dates have been included into the compilations of Bell *et al.* (2004) and Tedford *et al.* (2004), which also contain paleomagnetic calibrations. Biochronologically important local Early Miocene magnetostratigraphic records have been produced for the Arikareean of the Great Plains (Nebraska) by MacFadden and Hunt (1998) and of the Columbian Plateau (E. Oregon) by Allbright *et al.* (2008), and for the early Hemingfordian of the California Coast (Peninsular Ranges) by Prothero and Donohoo (2001). Middle Miocene paleomagnetic calibrations have been documented for the latest Hemingfordian and Barstovian of the Southern Great Basin (New Mexico) by Barghoorn (e.g., Tedford and Barghoorn, 1993), for the Barstovian of the Northern Great Basin (California) by MacFadden *et al.* (1990a), and of the Northern Rocky Mountains (Montana, Idaho) by Burbank and Barnosky (1990), Zheng (1996) and Barnosky *et al.* (2007).

Late Miocene magnetostratigraphic records have been compiled for the Clarendonian of the Northern Great Basin (Mojave Desert, California) by Prothero and Tedford (2000) and Whistler *et al.* (2009), and for the Late Clarendonian and Hemphillian of the Southern Great Basin (New Mexico) by MacFadden (1977). Magnetostratigraphy for the latest Miocene to Pliocene (Late Hemphillian to Blancan) was performed in the Peninsular Ranges (California) by Allbright (1999) and in the Northern Great Basin (Nevada) by Lindsay *et al.* (2002). Magnetostratigraphic correlations have been made for the Plio−Pleistocene (Blancan−Irvingtonian) in a series of regions (such as California, Idaho, Arizona, New Mexico, Texas, Kansas, Florida) since the 1970s (see Bell *et al.*, 2004, for references).

Alroy (e.g., 2000) has applied multivariate analysis (appearance event ordination) of faunal contents to rank North American mammal faunas and estimate their ages using a set of calibrations. An algorithm to break the time-ordered sequence into natural units has resulted in a scheme reproducing the intervals of the successive NALMAs reasonably well, although boundary ages slightly differ. Some units, such as the long Arikareean, could house several new ages.

29.3.1.4. South America

The knowledge of South American Neogene mammal history has rapidly increased, although gaps still remain, both temporally and spatially. The richest Early Neogene records undoubtedly occur in Patagonia (e.g., Madden *et al.*, 2010), whereas Late Neogene mammal records are well exposed in the western and southern pampas region of Argentina (e.g., Alberdi *et al.*, 1995). The evidence from lower-latitude and Andean regions is more spotty, but gradually accumulating (e.g., Flynn *et al.*, 1995; MacFadden *et al.*, 1995; Kay *et al.*, 1997; Flynn, 2010). After the pioneering work of Pascual *et al.* (1965), South American Land Mammal "Ages" (SALMAs, Marshall *et al.*, 1983) for the Neogene were defined on the basis of faunal content (Figure 29.9). A series of calibrations of SALMA boundaries to the numerical time scale were compiled by Flynn and Swisher (1995), followed by successive refinements (e.g., Cione and Tonni, 2005). ^{40}Ar/^{39}Ar, K-Ar and fission track dates have proven to be useful in constraining the age of mammal faunas and various SALMAs. In addition, various important paleomagnetic records have been compiled since the 1980s.

Recent paleomagnetic and radiometric results for the Early Miocene Colhuehuapian, the "Pinturan" and Santacrucian in Patagonia (Ré *et al.*, 2010a,b) support an ~21−18.75 Ma age range of the former. The Pinturan has been proposed either as a new age or subage of the Santacrucian for the interval 18.75−16.5 Ma (Kramarz *et al.*, 2010). Additional magnetostratigraphic and radio-isotopic ages are available for the coastal and Andean Santacrucian in Patagonia (Marshall *et al.*, 1986; Fleagle *et al.*, 1995; Blisniuk *et al.*, 2005). Middle−Late Miocene calibrations have been published for the Laventan in Colombia (Flynn *et al.*, 1997b; Madden *et al.*, 1997), southern Bolivia (MacFadden *et al.*, 1990b), southern Ecuador (Hüngerbühler *et al.*, 2002), the Andean forelands of western Argentina (Jordan *et al.*, 1996; Irigoyen *et al.*, 2000), and Patagonia (Kay *et al.*, 1998).

Late Miocene paleomagnetic calibrations for the Huayquerian and Montehermosan in Bolivia and NW Argentina have been compiled by Marshall *et al.* (1983), Butler *et al.*

(1984), and MacFadden *et al.* (1990b). On the basis of $^{40}Ar/^{39}Ar$ dating of impact glass in the Buenos Aires region it has recently been suggested to place the Late Miocene Chasicoan between ~10 and 8.7 Ma (Zárate *et al.*, 2007). Various new calibrations (radiometric dates, paleomagnetism) of mammal-bearing faunas from the Late Miocene–Pliocene (Montehermosan, Chapadmalalan, Uquian = Marplatan) have been published for NW Argentina (Marshall *et al.*, 1982, Walther *et al.*, 1996), coastal Argentina (Orgeira, 1990), Bolivia (MacFadden *et al.*, 1993), and most recently, Peru (Campbell *et al.*, 2010). Finally, various magnetostratigraphic calibrations have been realized for the Lujanian (Pleistocene, e.g., Orgeira, 1990; Nabel *et al.*, 2000).

29.3.1.5. Africa

The African Neogene mammal record is unevenly distributed in time and space. For example, rich and well-dated sequences are present in Kenya and other parts of eastern Africa. The over-representation of late Neogene faunas is partly due to the intensive search for hominine fossils. No African land mammal ages have been defined, although the scheme of Pickford (1981) for eastern Africa is still considered useful (Figure 29.9). The calibrations of mammal localities to the numerical time scale in northern Africa depend on biochronological correlation to Europe, as well as on some magnetostratigraphic records (e.g., Benammi *et al.*, 1996, Benammi and Jaeger, 2001). Radiometric dates are especially important for the age calibration of eastern African faunas. Age calibrations of the South African Plio–Pleistocene hominine cave sites are sometimes controversial. Here we refer to Werdelin (2010) for an overview of the chronology of African Neogene mammal-bearing localities and sections.

29.3.1.6. Australia

Extending earlier work (e.g., Stirton *et al.*, 1968; Woodburne *et al.*, 1985), a system of Australian Land Mammal Ages has recently been proposed (Megirian *et al.*, 2010). Despite the fact that the mammal record is patchy and not well calibrated, the faunas themselves allow for subdivision in larger units. The "Ages" have been defined statistically (by constrained seriation) with five of them (see Figure 29.9) applying to the Neogene, with the Wipajirian extending backward into the Oligocene, and the Naracoortean forward into the Pleistocene.

29.3.2. Microfossil Zonations

29.3.2.1. Planktonic Foraminifera

Wade *et al.* (2011) provided an excellent and timely review of the planktonic foraminiferal biostratigraphy and chronology of the entire Cenozoic and made some important improvements compared to previously published schemes. As far as the Neogene is concerned, major revisions are made to the fohsellid lineage of the middle Miocene, while the criteria for the recognition of Zones M7, M8 and M9 have been modified with additional adjustments regarding the *Globigerinatella* lineage to Zones M2 and M3.

Ages are provided both according to the now obsolete, but still widely used, CK95 polarity time scale (Cande and Kent, 1992, 1995), and to the ATNTS2004 of Lourens *et al.* (2004). The ATNTS2004 ages for the faunal events are largely maintained in ATNTS2012 (Figure 29.10 and Appendix 3), since only minimal changes in magnetic reversal ages, used to calculate the ages of the foraminiferal event, have been introduced relative to ATNTS2004. New ages can easily be calculated, as Wade *et al.* (2011) give the position of all foraminiferal events relative to the reversal boundaries. This age determination is different from that of the calcareous nannofossil events which are based on direct tuning of the cyclic successions in which the events are found (see Section 29.3.3.2).

In the Mediterranean, calcareous plankton biostratigraphic events of Langhian age have been linked to the ATNTS2004 using the magnetobiostratigraphy of DSDP Site 372 (Abdul Aziz *et al.*, 2008a; Di Stefano *et al.*, 2008) and land-based sections such as La Vedova and St. Peter's Pool (Turco *et al.*, 2011a; Foresi *et al.*, 2011). These sections – and core – have been used to establish an integrated calcareous plankton zonal scheme for the Langhian that is directly linked to magnetostratigraphy, but that is only partially underlain by astronomical tuning (Di Stefano *et al.*, 2008). This zonal scheme represents the downward extension of a scheme that has previously been established for the Tortonian and Serravallian.

29.3.2.2. Calcareous Nannofossils

Raffi *et al.* (2006) presented an extensive review of the astrochronological ages of Recent to late Oligocene calcareous nannofossil datum events. Biohorizons included those of the widely used "standard" nannofossil zonations of Martini (1971) and Okada and Bukry (1980), as well as supplementary biohorizons defined in the literature. The biohorizons were selected on the basis of an unambiguous taxonomy of index taxa, their biostratigraphic usefulness and their presence in directly astronomically dated sections and cores. The approach means that the astrochronological age estimates are directly linked to ATNTS2012 (Figure 29.10).

A critical aspect is that the tuning of the cyclic deep-sea cores and land-based sections used for the astrobiochronology should be reliable, and consistent with the tuning of the sections/cores used to build the ATNTS2012. For instance, it is becoming clear that the tuning of ODP Leg 154 records of Shackleton and Crowhurst (1997) needs to undergo some revision. This will result in different astronomical ages for selected bioevents up to ~100 kyr (Zeeden

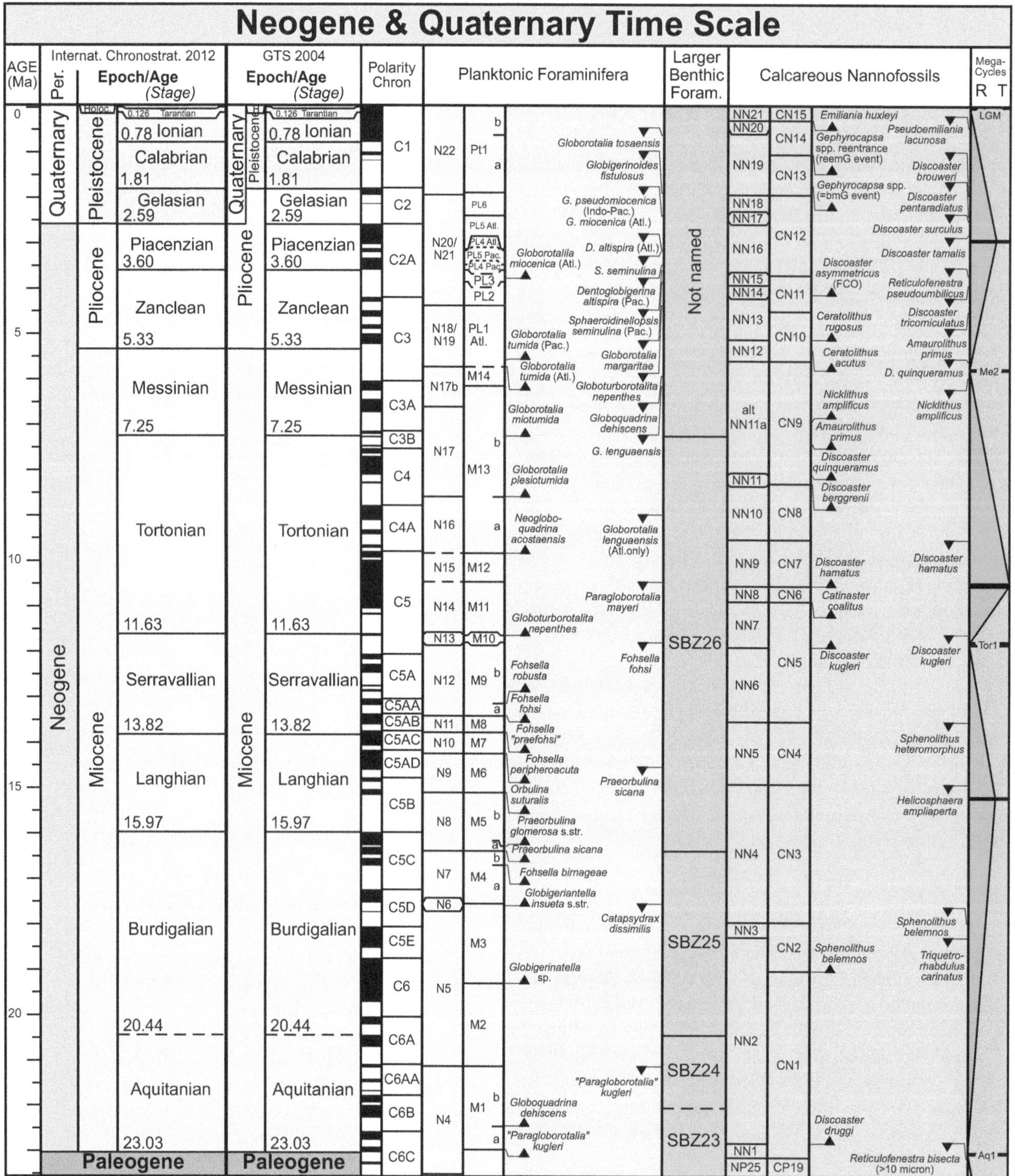

FIGURE 29.10 **Neogene stratigraphic subdivisions, geomagnetic polarity scale, standard zonations of planktonic foraminifers, larger benthic foraminifers and calcareous nannofossils, and main trends in eustatic sea level.** The upper Cenozoic chronostratigraphy ratified by the IUGS in 2009 shifted the Gelasian Stage into the Pleistocene Series and a Quaternary–Neogene System boundary was defined at this level (c. 2.6 Ma). The Gelasian was formerly the upper Pliocene (as shown in GTS2004).

et al., 2011). Similarly, a revision of the tuning of proxy records of early Miocene age may result in different ages for nannofossil events in the coming years.

Di Stefano *et al.* (2008) presented a high-resolution calcareous nannofossil biostratigraphy for the Mediterranean Langhian that is directly integrated with the planktonic foraminiferal biostratigraphy and magnetostratigraphy.

29.3.2.3. Diatoms and Radiolarians

The recovery of deep-marine sediments through deep-sea drilling led to the development of many radiolarian zonal schemes for the Neogene. However, these zonations are only applicable on a regional scale. The zonal standard developed for the equatorial Pacific provides one of the best examples (e.g., Baldauf and Iwai, 1995), partly because it is directly linked to the excellent magnetostratigraphy of ODP Leg 138 sites for the last 13 myr, thus facilitating a reliable and straightforward calibration to the APTS. The diatom biochronology of the more problematic early Miocene interval has recently been published by Barron (2006). Middle to late Miocene diatom zones have been established in the Oga Peninsula section, an onshore succession regarded as the Neogene reference section for northeast Japan (Koizumi *et al.*, 2009). Cody *et al.* (2008) developed a new chronostratigraphic framework for the Southern Ocean by integrating diatom biostratigraphy, magnetostratigraphy and tephrostratigraphy from 32 Neogene sections around the Southern Ocean and Antarctic continental margin, and used the Constrained Optimization technique to establish the best possible correlations having a temporal significance (see also Chapter 3, Figs. 3.4, 3.5 and text). Such a framework is required to study the Neogene history of the Antarctic ice sheet and Southern Ocean in detail.

As for the diatoms, many regional radiolarian zonal schemes have been developed over the last decades. Sanfilippo and Negrini (1998) merged the existing low-latitude zonations for different oceanic basins into a single standard low-latitude zonal scheme, introducing a formal code for the resulting radiolarian zones (RN1−17 for the Neogene). No major improvements have been made since the publication of the ATNTS2004 (see Figure 29.11). The development of the radiolarian biostratigraphy and chronology for the Late Neogene was briefly reviewed by Haslett (2004). Wang and Chen (2005) used the radiolarian species *Cycladophora davisiana* as a high-resolution stratigraphic tool and proxy of the glacial Subarctic Pacific Intermediate Water in the Bering Sea during the late Quaternary.

29.3.2.4. Dinoflagellates (see also Figure 29.11)

Dybkjaer and Piasecki (2010) presented a new dinocyst zonation for the eastern North Sea which reconsiders and redefines existing zonations. This zonation is important as Neogene biostratigraphy in the North Sea Basin has been problematic as a consequence of the occasionally limited connection with the North Atlantic Ocean. This resulted in the absence of many key marker species of calcareous plankton, especially those important for recognizing chronostratigraphic boundaries. This problem has apparently been solved by the new dinocyst stratigraphy, because stratigraphically important taxa do occur in the North Sea, also in marginal marine settings.

A southern North Sea Miocene dinoflagellate cyst zonation was published by Munsterman and Brinkhuis (2004). Application of the dinocyst biochronology that came out of detailed correlations to the North Atlantic and Mediterranean resulted in the first detailed age model for the Miocene in the subsurface of the Netherlands.

Dinoflagellate biostratigraphy was further successfully applied to get a better age control on the Badenian stratotype (Rögl *et al.*, 2008) and to correlate regional stages independently to global chronostratigraphy and the paleomagnetic time scale in the southern North Sea (Kuhlmann *et al.*, 2006). A Pliocene and Pleistocene dinoflagellate cyst and acritarch zonation was further published for the eastern North Atlantic (De Schepper and Head, 2009) which reaches a level of detail that surpasses earlier studies. The dinoflagellate events and zonation for northwestern Europe are shown in Figure 29.11.

Dinoflagellates were also integrated in a new multi-group microfossil zonation that comprises seven Nordic Pliocene (NP) zones, with zonal duration ranging from 300 000 to 600 000 years (Anthonissen, 2009). Figure 3.11 of Chapter 3 shows the new, multi-group Nordic Miocene biochronology and its direct and partly causal linkage to the oxygen isotope (climate) record, paleoceanography and higher order sequences. Microfossils are mainly planktonic and benthic foraminifers and bolboformids (70 events), dinoflagellates and acritarchs (16 events), and marine diatoms (6 events). The zonation and event stratigraphy for this Nordic record can now be correlated to the Mediterranean stages. As a result, standard stages can be assigned in the North Sea, instead of just Lower, Middle or Upper Miocene assignments.

29.4. EVENT STRATIGRAPHY

29.4.1. Polarity Sequences

29.4.1.1. Anomaly Profiles and Magnetostratigraphy (with Doug S. Wilson)

The GPTS was traditionally based on seafloor anomaly profiles combined with a limited number of radio-isotopically dated tie points (Cande and Kent, 1995). But, for the Neogene, this approach has been replaced by astronomical dating of sections having a reliable magnetostratigraphy with cycle tuning producing astronomical ages for magnetic reversal boundaries at the same time. In ATNTS2004, uninterrupted astronomical ages for polarity reversals only go back to 12 Ma, with reliable astronomical reversal ages only

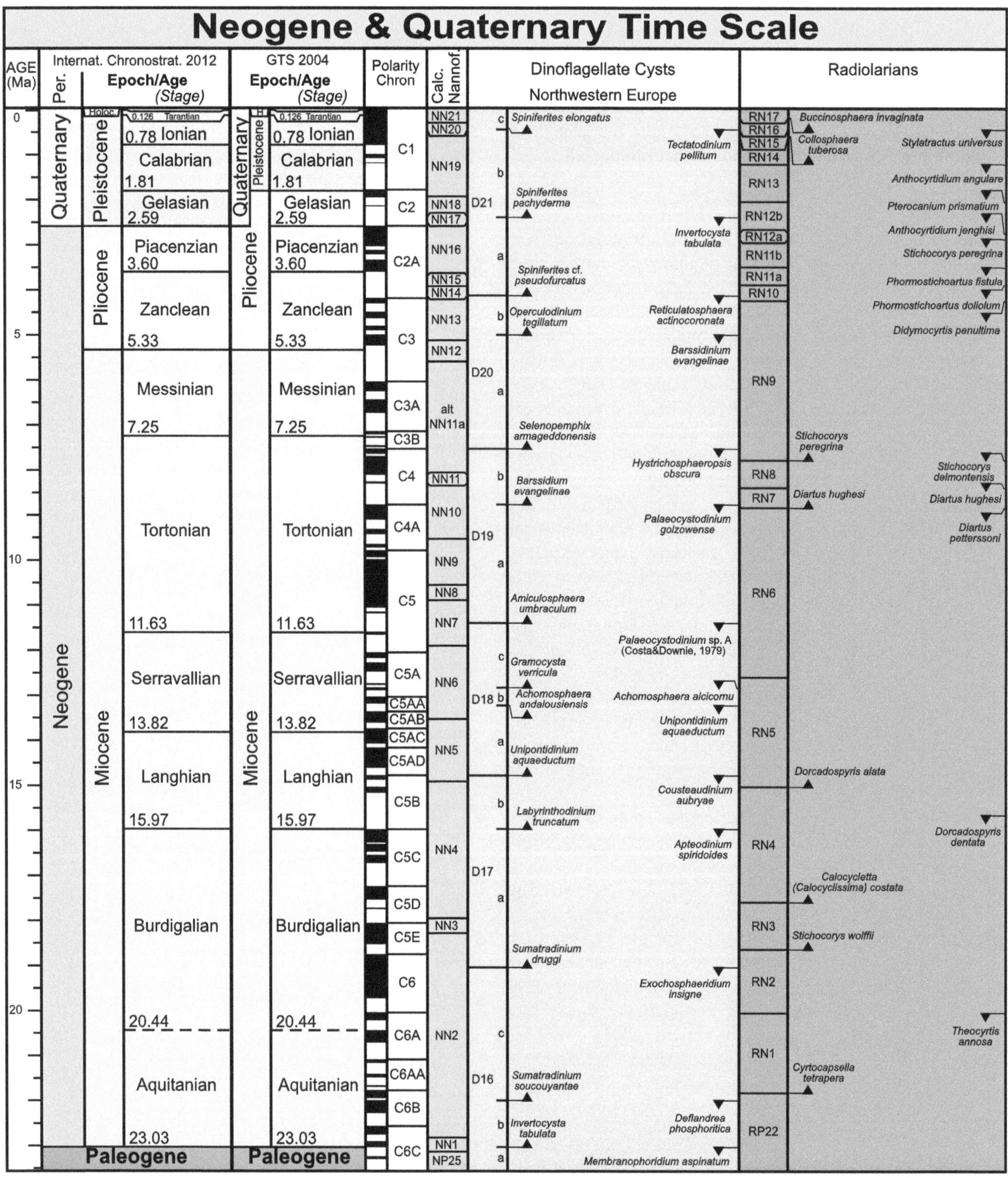

FIGURE 29.11 **Neogene dinoflagellate cyst zonation for northwestern Europe, and datums and radiolarian zonation for the low latitudes oceanic basins, with their estimated correlation to magnetostratigraphy and calcareous nannoplankton zones.** The upper Cenozoic chronostratigraphy ratified by the IUGS in 2009 shifted the Gelasian Stage into the Pleistocene Series and a Quaternary–Neogene System boundary was defined at this level (c. 2.6 Ma). The Gelasian was formerly the upper Pliocene (as shown in GTS2004).

TABLE 29.3 **ATNTS2012 Chron Ages and Comparison with ATNTS2004**

	Chron	ATNTS2012	ATNTS2004	Δ	Husing	Wilson	CK95	L&R 2005	Billups	Palike et al., 2006	
		Final			Mourik	This Study		CK95mod	1090	1218 Man	1218 Aut
T	C1n	0.000	0.000	0.000			0.000	0.000			
B	C1n	0.781	0.781	0.000			0.780	0.780			
T	C1r.1n	0.988	0.988	0.000			0.990	0.991			
B	C1r.1n	1.072	1.072	0.000			1.070	1.075			
T	C2n	1.778	1.778	0.000			1.770	1.781			
B	C2n	1.945	1.945	0.000			1.950	1.968			
T	C2An.1n	2.581	2.581	0.000			2.581	2.608			
B	C2An.1n	3.032	3.032	0.000			3.040	3.045			
T	C2An.2n	3.116	3.116	0.000			3.110	3.127			
B	C2An.2n	3.207	3.207	0.000			3.220	3.210			
T	C2An.3n	3.330	3.330	0.000			3.330	3.319			
B	C2An.3n	3.596	3.596	0.000			3.580	3.588			
T	C3n.1n	4.187	4.187	0.000			4.180	4.184			
B	C3n.1n	4.300	4.300	0.000			4.290	4.306			
T	C3n.2n	4.493	4.493	0.000			4.480	4.478			
B	C3n.2n	4.631	4.631	0.000			4.620	4.642			
T	C3n.3n	4.799	4.799	0.000			4.800	4.807			
B	C3n.3n	4.896	4.896	0.000			4.890	4.898			
T	C3n.4n	4.997	4.997	0.000			4.980	4.989			
B	C3n.4n	5.235	5.235	0.000			5.230	5.254			
T	C3An.1n	6.033	6.033	0.000			5.894				
B	C3An.1n	6.252	6.252	0.000			6.137				
T	C3An.2n	6.436	6.436	0.000			6.269				
B	C3An.2n	6.733	6.733	0.000			6.567				
T	C3Bn	7.140	7.140	0.000			6.935				
B	C3Bn	7.212	7.212	0.000			7.091				
T	C3Br.1n	7.251	7.251	0.000			7.135				
B	C3Br.1n	7.285	7.285	0.000			7.170				
T	C3Br.2n	7.454	7.454	0.000			7.341				
B	C4n.1n	7.642	7.642	0.000			7.562				
T	C4n.2n	7.695	7.695	0.000			7.650				
B	C4n.2n	8.108	8.108	0.000			8.072				
T	C4r.1n	8.254	8.254	0.000			8.225				
B	C4r.1n	8.300	8.300	0.000			8.257				
T	C4An	8.771	8.769	0.002	8.771		8.699				

(Continued)

TABLE 29.3 ATNTS2012 Chron Ages and Comparison with ATNTS2004—cont'd

	Chron	ATNTS2012	ATNTS2004	Δ	Husing	Wilson	CK95	L&R 2005	Billups	Palike et al., 2006	
		Final			Mourik	This Study		CK95mod	1090	1218 Man	1218 Aut
B	C4An	9.105	9.098	0.007	9.105		9.025				
T	C4Ar.1n	9.311	9.312	−0.001	9.311		9.230				
B	C4Ar.1n	9.426	9.409	0.017	9.426		9.308				
T	C4Ar.2n	9.647	9.656	−0.009	9.647		9.580				
B	C4Ar.2n	9.721	9.717	0.004	9.721		9.642				
T	C5n.1n	9.786	9.779	0.007	9.786		9.740				
B	C5n.1n	9.937			9.937						
T	C5n.2n	9.984			9.984						
B	C5n.2n	11.056	11.040	0.016	11.056		10.949				
T	C5r.1n	11.146	11.118	0.028	11.146		11.052				
B	C5r.1n	11.188	11.154	0.034	11.188		11.099				
T	C5r.2r-1n	11.263			11.263						
B	C5r.2r-1n	11.308			11.308						
T	C5r.2n	11.592	11.554	0.038	11.592		11.476				
B	C5r.2n	11.657	11.614	0.043	11.657		11.531				
T	C5An.1n	*12.049*	12.014	0.035		12.049	11.935				
B	C5An.1n	*12.174*	12.116	0.058		12.174	12.078				
T	C5An.2n	*12.272*	12.207	0.065		12.272	12.184				
B	C5An.2n	*12.474*	12.415	0.059		12.474	12.401				
T	C5Ar.1n	*12.735*	12.730	0.005		12.735	12.678				
B	C5Ar.1n	*12.770*	12.765	0.005		12.770	12.708				
T	C5Ar.2n	*12.829*	12.820	0.009		12.829	12.775				
B	C5Ar.2n	*12.887*	12.878	0.009		12.887	12.819				
T	C5AAn	*13.032*	13.015	0.017		13.032	12.991				
B	C5AAn	*13.183*	13.183	0.000		13.183	13.139				
T	C5ABn	*13.363*	13.369	−0.006		13.363	13.302				
B	C5ABn	13.608	13.605	0.003	13.608		13.510				
T	C5ACn	13.739	13.734	0.005	13.739		13.703				
B	C5ACn	14.070	14.095	−0.025	14.070		14.076				
T	C5ADn	14.163	14.194	−0.031	14.163		14.178				
B	C5ADn	14.609	14.581	0.028	14.609		14.612				
T	C5Bn.1n	14.775	14.784	−0.009	14.775		14.800				
B	C5Bn.1n	14.870	14.877	−0.007	14.870		14.888				
T	C5Bn.2n	15.032	15.032	0.000	(15.000)		15.034				
B	C5Bn.2n	15.160	15.160	0.000	(15.215)		15.155				
T	C5Cn.1n	15.974	15.974	0.000			16.014	15.914	15.898	*15.898*	*15.899*

TABLE 29.3 ATNTS2012 Chron Ages and Comparison with ATNTS2004—cont'd

	Chron	ATNTS2012	ATNTS2004	Δ	Husing	Wilson	CK95	L&R 2005	Billups	Palike et al., 2006	
		Final			Mourik	This Study		CK95mod	1090	1218 Man	1218 Aut
B	C5Cn.1n	16.268	16.268	0.000			16.293	16.167	16.161	*16.161*	*16.162*
T	C5Cn.2n	16.303	16.303	0.000			16.327	16.197	16.255	*16.255*	*16.256*
B	C5Cn.2n	16.472	16.472	0.000			16.488	16.343	16.318	*16.318*	*16.319*
T	C5Cn.3n	16.543	16.543	0.000			16.556	16.404	16.405	*16.405*	*16.406*
B	C5Cn.3n	16.721	16.721	0.000			16.726	16.557	16.498	*16.498*	*16.499*
T	C5Dn	17.235	17.235	0.000			17.277	17.052	17.003	*17.003*	*17.004*
B	C5Dn	17.533	17.533	0.000			17.615	17.355	17.327	*17.327*	*17.328*
T	C5Dr.1n	17.717	17.717	0.000			17.825	17.530	17.511	*17.511*	*17.512*
B	C5Dr.1n	17.740	17.740	0.000			17.853	17.579	17.550	*17.550*	*17.551*
T	C5En	18.056	18.056	0.000			18.281	17.950	17.948	*17.947*	*17.948*
B	C5En	18.524	18.524	0.000			18.781	18.396	18.431	*18.431*	*18.432*
T	C6n	18.748	18.748	0.000			19.048	18.634	18.614	*18.615*	*18.616*
B	C6n	19.722	19.722	0.000			20.131	19.606	19.599	*19.598*	*19.599*
T	C6An.1n	20.040	20.040	0.000			20.518	19.955	19.908	20.000	20.001
B	C6An.1n	20.213	20.213	0.000			20.725	20.144	20.185	20.226	20.227
T	C6An.2n	20.439	20.439	0.000			20.996	20.390	20.420	20.425	20.425
B	C6An.2n	20.709	20.709	0.000			21.320	20.687	20.720	20.651	20.652
T	C6AAn	21.083	21.083	0.000			21.768	21.099	21.150	21.114	21.114
B	C6AAn	21.159	21.159	0.000			21.859	21.183	21.191	21.196	21.197
T	C6AAr.1n	21.403	21.403	0.000			22.151	21.455	21.457	21.506	21.507
B	C6AAr.1n	21.483	21.483	0.000			22.248	21.546	21.542	21.627	21.636
T	C6AAr.2n	21.659	21.659	0.000			22.459	21.743	21.737	21.744	21.743
B	C6AAr.2n	21.688	21.688	0.000			22.493	21.776	21.780	21.783	21.780
T	C6Bn.1n	21.767	21.767	0.000			22.588	21.865	21.847	21.861	21.853
B	C6Bn.1n	21.936	21.936	0.000			22.750	22.019	21.991	22.010	21.998
T	C6Bn.2n	21.992	21.992	0.000			22.804	22.070	22.034	22.056	22.062
B	C6Bn.2n	22.268	22.268	0.000			23.069	22.323	22.291	22.318	22.299
T	C6Cn.1n	22.564	22.564	0.000			23.353	22.596	22.593	22.595	22.588
B	C6Cn.1n	22.754	22.754	0.000			23.535	22.772	22.772	22.689	22.685
T	C6Cn.2n	22.902	22.902	0.000			23.677	22.911	22.931	22.852	22.854
B	C6Cn.2n	23.030	23.030	0.000			23.800	23.031	23.033	23.024	23.026

Also shown are ages according to CK95 (Cande and Kent, 1995) and astronomical reversal ages according to Hüsing *et al.* (2009), Mourik *et al.* (2010), Lisiecki and Raymo (2005), Billups *et al.* (2004) and Pälike *et al.* (2006, both manual and automatic). Finally, modified CK95 ages according to Billups *et al.* (2004) and ages of "missing" reversals based on seafloor spreading rates (Wilson, this chapter) are shown.

being available for very short intervals around 16 and 23 Ma (Lourens *et al.*, 2004; Table 29.3).

The tuning of the Lower–Middle Miocene (and Oligocene) was initially based on ODP Leg 154 (Ceara Rise, equatorial Atlantic) sites lacking a magnetostratigraphy, using especially but not exclusively high-resolution obliquity dominated records of magnetic susceptibility (Shackleton *et al.*, 1999). Shackleton *et al.* (2000) also tuned a high-quality stable isotope record from Ceara Rise to arrive at an age of 23.03 Ma (retuned to La2004 of Laskar *et al.*, 2004, see Lourens *et al.*, 2004, Pälike *et al.*, 2006b) for the Oligocene–Miocene boundary.

At Ceara Rise, the interval between 14 and 18 Ma has proved especially problematic due to lack of data. Moreover no magnetostratigraphy is available as an independent check on the continuity of the succession. As a consequence, ages for chron boundaries in ATNTS2004 (Lourens *et al.*, 2004) were obtained indirectly, namely by using seafloor anomaly profiles of the Antarctic–Australian plate pair and the valid assumption of near-constant spreading rates for this plate pair, using the few astronomically dated reversals in this interval as constraint. A spreading rate of 67.7 mm/yr provided a good fit with the astronomical ages for C5n.2n(o)–C5Ar.2n(o) (Abdul Aziz *et al.*, 2003a) and C5Bn.2n(o) and C5Cn.1n(y) (Shackleton *et al.*, 2001). Extrapolating this rate to the beginning of the Neogene predicted an age of 23.18 for C6Cn.2n(o). To match the tuned age of 23.03 Ma, a minor rate change at C5En(o) (18.52 Ma) was introduced with a rate of 69.9 mm/yr prior to the change. The resultant ages for the reversal boundaries were added to the list of astronomically dated reversals to complete the dating of all (sub)chrons of the last 23 myr (Table 29.3).

The lack of reliable magnetostratigraphic data of tuned Lower to Middle Miocene sections was apparently solved with the tuning of the benthic $\delta^{18}O$ record of ODP Site 1090, having a magnetostratigraphy that could unequivocally be calibrated to the GPTS (Billups *et al.*, 2004). The final tuning was essentially consistent with that proposed by Shackleton *et al.* (1999, 2000), providing independent evidence for the revised younger age of ~23.0 Ma for the Oligocene–Miocene boundary. In addition the tuning provided astronomical ages for polarity Chrons C7n through C5Cn.1n between 24.1 and 15.9 Ma (Table 29.3). The new ages are generally consistent within one obliquity cycle with those obtained by rescaling the Cande and Kent (1995) time scale using the new age of the O–M boundary (23.03 Ma) and the same middle Miocene control point (14.8 Ma) used by Cande and Kent (Table 29.3). Remarkably enough, differences with the ATNTS2004 reversal ages are significantly larger, ranging from +232 kyr to −92 kyr.

In principle the astronomical reversal ages of Billups *et al.* (2004) should replace the existing ages in ATNTS2004 based on interpolation of seafloor spreading rates (see above). However, application of the same seafloor spreading rate approach used in ATNTS2004 shows that the ages of Billups *et al.* (2004) do not resolve existing flaws in the CK95 scale and are also not consistent with inferred near-constant or constantly changing seafloor spreading rates in the Pacific; it is therefore unlikely that these ages are correct. This observation has important consequences, because the proxy records used by Billups *et al.* (2004) to establish their astronomical tuning are correlated in detail to the tuned records of Ceara Rise, on which also the tuning of the O/M boundary interval is based. In fact, following the construction of the tuned ODP Site 1090 age model, these correlations were considered independent evidence that the Ceara Rise tuning, and hence the astronomical age for the O/M boundary, were correct.

The same spreading rate argument also holds for tuned reversal ages published by Pälike *et al.* (2006a), which reveal differences of up to ~100 kyr with the ages of Billups *et al.* (2004) (Table 29.3). Proxy records of ODP site 1264 in the southern Atlantic look promising to solve this issue (Liebrand *et al.*, 2011), but have not yet been tuned. However, they have been correlated in great detail to the records from ODP sites 926/929 and 1090 used by Billups *et al.* (2004).

In view of these uncertainties we decided not to incorporate the reversal ages of Billups *et al.* (2004) or Pälike *et al.* (2006a) in the ATNTS2012, but to stick to the ATNTS2004 ages until the problems encountered in the tuning of the Lower Miocene have been solved (Table 29.3). Improved astronomical ages for this time interval may come from the ODP leg 208 sites (Liebrand *et al.*, 2011) and the recently drilled IODP leg 321/322 sites (Pälike *et al.*, 2008). This conservative approach is justified by the same seafloor spreading rate approach which demonstrates that in principle only minor adjustments of the ATNTS2004 reversal ages are to be expected. This statement is made after the new ages for reversal boundaries in the 10–15 Ma interval were included in the analysis. Most likely the age of the O/M boundary will have to undergo only a minor change, in the order of ~100 kyr or less.

Further progress came from a much improved magnetostratigraphy and slightly revised tuning of the marine Monte dei Corvi section in northern Italy (Hüsing *et al.*, 2007). This study resulted in accurate astronomical ages for reversal boundaries in the interval between 12.5 and 8.5 Ma, partly replacing ages that were based on the tuning of the continental Orera section in Spain (Abdul-Aziz *et al.*, 2003a). The Monte dei Corvi beach section of Montanari *et al.* (1997) and Hilgen *et al.* (2003) was recently extended downwards to include the La Vedova section covering the interval from 14.2 to 15.3 Ma. The integrated stratigraphic study of this section resulted in a coherent tuning providing reliable astronomical ages for bioevents and reversal boundaries in this Middle Miocene time interval (Hüsing *et al.*, 2009; Table 29.3). The La Vedova beach section has been extended upwards in the high cliffs to close the approximately one million-year gap with the Monte dei Corvi section that results from the landslide that covered part of the succession along the beach. This La Vedova High Cliff section is important as it covers the interval of the Mi-3b isotope shift that marks the Langhian–Serravallian boundary and reveals the imprint of the increase in the East Antarctic ice sheet and associated glacio-eustatic sea-level fall in the Mediterranean (Mourik *et al.*, 2010; Table 29.3).

Unfortunately, the Monte dei Corvi–La Vedova composite still contains a missing interval between 13.4 and 13.6 Ma that is covered by the landslide, although this interval has now been exposed along the beach after storms. In addition, the paleomagnetic signal for the interval between 12 and 13.4 Ma is too poor to obtain a reliable magnetostratigraphy (Hilgen *et al.*, 2003; Hüsing *et al.*, 2007). Chron boundary ages for this missing interval were obtained from seafloor spreading rate history of five plate pairs in the Pacific; they are based on assuming constant rate for both Cocos–Pacific and Nazca–Pacific for C5r.2n–C5ABn(b), and taking a weighted average of the resulting ages (Figure 29.12; Table 29.3). Basically, the older ages for C5r relative to ATNTS2004 propagate to C5An but not older. Only for C5An.2n(b) does the difference between using only Cocos–Pacific or Nazca–Pacific exceed 0.015 Myr.

29.4.1.2. Excursions and Cryptochrons

Significant progress was also made in the detection and astronomical dating of so-called magnetic excursions and

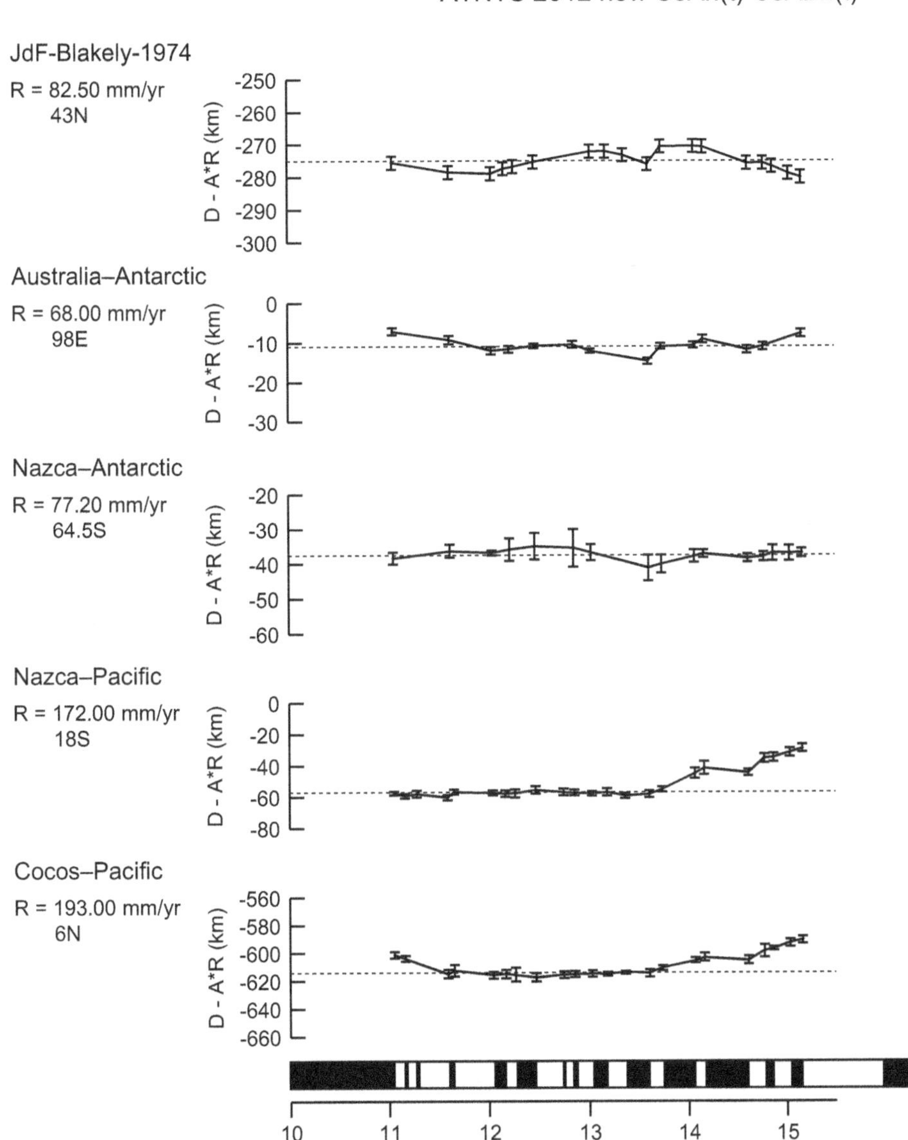

FIGURE 29.12 **Reduced distance tests for constant spreading rate to determine ages of C5An(t) down to C5ABn(t).** Observed spreading distance (D) is plotted against age (A), after subtracting predicted distance according to a constant spreading rate (R) model. Distance scales are plotted inversely to the spreading rate so that for plate pairs spreading constantly at the reduction rate, time scale errors will plot as uniform vertical departures from the reduction line (0.1-myr scale bar). Ages for C5An(t) through C5ABn(t) are based on assuming constant rate for both Cocos–Pacific and Nazca–Pacific for C5r.2n-C5ABn(b), and taking a weighted average of the resulting ages. Basically, older ages for C5r relative to ATNTS2004 propagate to C5An but not older. Only for C5An.2n(b) does the difference between using only Cocos–Pacific and only Nazca–Pacific exceed 0.015 myr.

cryptochrons. An excellent summary of the variability of the Earth's magnetic field as recorded in ODP cores is provided by Lund *et al.* (2006). As before, special attention is paid to the occurrence of excursions in the Brunhes Chron (e.g., Thouveny *et al.*, 2004; Channell, 2006) and their link to relative paleo-intensity records, such as the Sint2000 of Valet *et al.* (2005). This record provides an excellent means for high-resolution correlations between cores, and between cores and other proxies of the strength of the Earth's magnetic field as recorded in sediment and ice cores. It was extended back to 3 Ma, using cores from the equatorial Pacific (Yamazaki and Oda, 2005). A particularly interesting question that arises is the potential link with orbital control and climate (Courtillot *et al.*, 2007; Xuan and Channell, 2008; Thouveny *et al.*, 2008). New K/Ar and $^{40}Ar/^{39}Ar$ ages of the Laschamp event (Guillou *et al.*, 2004) are now consistent with a new Greenland ice core chronology based on multi-parameter annual layer counting (Svensson *et al.*, 2006). The existence of cryptochron C2r.2r-1 was confirmed in lava sequences on Hawaii and Cape Verde islands with apparently inconsistent $^{40}Ar/^{39}Ar$ ages of 2.514 ± 0.039 and 2.40 ± 0.03 Ma, respectively (Herrero-Bervera *et al.*, 2007; Knudsen *et al.*, 2009). Note that these ages were calculated respectively relative to 1.194 Ma for Alder Creek sanidine and 28.04 Ma for FCT-3 biotite.

Three cryptochrons labeled C5n.2n-1 to C5n.2n-3 are present in the normal polarity interval of C5n.2n (Evans *et al.*, 2004), but do not correspond with the cryptochrons previously identified in NE Pacific marine anomaly profiles; these excursions have now been dated astronomically at 10.154–10.157, 10.311 ± 2 and 10.826–10.829 Ma (Evans *et al.*, 2007). These ages are in good agreement with ages found for C5n.2n-2 and C5n.2n-3 at Monte dei Corvi, where C5n.2n-1 was not unambiguously identified, most likely as a consequence of the somewhat lower sampling resolution in this interval (Hüsing, 2008). These astronomical ages are in good agreement with the astronomical ages for the excursions reported from ODP Site 1092. Another excursion, labeled C4r.2r-1, was recorded in sediments off the Antarctic Peninsula (Acton *et al.*, 2006), but had previously already been detected and astronomically dated in the Mediterranean (Hilgen *et al.*, 1995; Krijgsman *et al.*, 1995). Besides their importance for unraveling the behavior of the Earth's geodynamo, such excursions and cryptochrons can be useful for increasing the temporal resolution of magnetostratigraphic correlations and dating, although care should be taken to ensure that the signal has a primary character.

29.4.1.3. Magnetostratigraphy

Magnetostratigraphy remains a primary tool to use the ATNTS for establishing initial age models or to date successions lacking a distinct cyclicity. In particular it has successfully been applied to date sedimentary successions in China (e.g., Tibetan and Loess Plateaus), which is critical in reconstructing late Cenozoic deformation and uplift history (e.g., Fang *et al.*, 2005) and the initiation and intensification of the Asian Monsoon (e.g., Nie *et al.*, 2008).

Magnetostratigraphy continues to play a crucial role in the study of our ancestors. Prime examples are the application of magnetostratigraphy in the Orche Basin to date Early Pleistocene faunas and hominid occupation sites in Spain (Scott *et al.*, 2007) and in the classical Koobi Fora Formation in Kenya to solve the problem encountered in locating the position of the Upper Olduvai reversal boundary (Lepre and Kent, 2010). Muttoni *et al.* (2010) present a critical evaluation of the magnetostratigraphic and radio-isotopic age constraints on key sites in southern Europe and propose a window between the Jaramillo Subchron and the Brunhes–Matuyama boundary (i.e. Subchron C1r.1r, 0.99–0.78 Ma) for the early presence of hominids in the area. They argue that this is related to the so-called mid-Pleistocene climate transition, and in particular to MIS22, the first prominent cold glacial of the Pleistocene characterized by a marked aridification of North Africa and Eastern Europe.

Magnetostratigraphy further played a crucial role in solving problems related to regional chronostratigraphic scales and the dating of such scales. A prime example is the progress made in establishing a reliable and consistent chronostratigraphic framework for the Paratethys that is accurately dated (e.g., Vasiliev *et al.*, 2004).

29.4.2. Sequence Stratigraphy (see also Section 29.6.1.2)

The Phanerozoic record of global sea-level change was reviewed by Miller *et al.* (2005). The record of the last 100 myr from the US mid-Atlantic coastal plain was presented in detail by Browning *et al.* (2008). From the Oligocene to the early Pliocene, sea-level changes were in the order of 30 to 60 m, mostly governed by growth and decay of the Antarctic ice sheet (Miller *et al.*, 2005). A middle Miocene $\delta^{18}O$ increase is associated with deep-water cooling and two ice-growth events resulting in the development of a permanent East Antarctic ice sheet. Northern hemisphere ice sheets have existed since at least the middle Miocene, but large ice sheets and changes in ice volume giving rise to sea-level changes of 60 to 120 m only began during the late Pliocene to Holocene (2.5 to 0 Ma).

Neogene sequence stratigraphy on the US mid-Atlantic coastal plain is dominated by two major depositional phases (Browning *et al.*, 2008). The first phase in the late early to middle Miocene is characterized by a prograding shelf under a strong wave-dominated deltaic influence associated with marked increases in sediment supply and in accommodation due to local sediment loading. The second phase occurred during the past 10 myr and is characterized by low accommodation. During this phase, coastal systems were eroded as a consequence of a long-term sea-level lowstand and low rates of sediment supply due to bypassing (Browning *et al.*, 2008). In more detail, eight Miocene sequences were distinguished and correlated regionally (Browning *et al.*, 2006). These sequences can be correlated to global $\delta^{18}O$ increases, suggesting a glacio-eustatic origin.

Another classical region for sequence stratigraphy is New Zealand. It is famous for the detailed study of higher-order, in particular sixth order, stratigraphic sequences of late Pliocene and Pleistocene age (Saul *et al.*, 1999). Sixth-order sequences have now also been described from the Middle Pliocene and Upper Miocene (Hendy and Kamp, 2004; Naish *et al.*, 2005) and are similarly related to mainly obliquity controlled glacio-eustatic sea-level changes, while four second-order sequences have been identified in the Neogene (Kamp *et al.*, 2004). Sequence stratigraphy has also been successfully applied in the Paratethys (Strauss *et al.*, 2006; Schreilechner and Sachenhofer, 2007; see also Piller *et al.*, 2007).

29.4.3. Volcanic Ash Layers and Tephrochronologies

Tephrostratigraphy, i.e. the stratigraphic application of pyroclastic layers, in particular volcanic ash beds, as a tool for correlating and dating sedimentary successions, is becoming increasingly important, because tephras represent unique time horizons. In the Neogene, this is the case both for very young time intervals, such as the last 1000 years, the last glacial termination or the last glacial cycle, and for older time intervals.

Tephrostratigraphy was used to verify the lichonometric dating of glacial landforms of the past ~400 years to study potential linkages between glacier fluctuations and climate, such as the North Atlantic oscillations and the Little Ice Age (Bradwell *et al.*, 2006). Turney *et al.* (2004) discuss the potential of – and protocol for – developing an integrated tephrostratigraphic and tephrochronologic framework for the last glacial termination. The initiative is part of INTIMATE ("INTegration of Ice core, MArine and TErrestrial records of the Last Termination"), a core project of the INQUA Palaeoclimate Commission, that has the integration of data sets from ice core, marine and land records for the interval between the Last Glacial Maximum and the Early Holocene and the study of ice–sea–atmosphere interactions and feedbacks during the last Glacial–Interglacial transition as its objective. Application of tephrostratigraphy is considered especially important because of the difficulties encountered in radiocarbon dating in this interval. This can be circumvented using tephras intercalated in ice cores and piston cores dated independently by annual layer counts (e.g., Lowe *et al.*, 2007).

The classical tephrostratigraphy and tephrochronology developed for the last 200 kyr in the Mediterranean is based on deep-sea piston cores (Keller *et al.*, 1978) and is directly linked to both oxygen isotope and sapropel chronologies (e.g., Lourens, 2004) and pyroclastic flows onland. It is now increasingly and successfully applied in continental, especially lacustrine, settings, and is a key to linking continental and marine records in the Mediterranean in detail (e.g., Wulf *et al.*, 2004, 2008; Wagner *et al.*, 2008).

Special attention is further paid to develop geochemical criteria ("fingerprinting") to distinguish between different ash layers to enhance their time-stratigraphic application (e.g., Santacroce *et al.*, 2008) in key areas such as the Antarctic (Narcisi *et al.*, 2010). Excellent examples come from the construction and application of a tephrostratigraphy and tephrochronology for the Pacific offshore central America and New Zealand (Kutterolf *et al.*, 2008a,b; Lowe, D.J., *et al.*, 2008).

29.4.4. Tektites and Impacts

Since the publication of ATNTS2004, significant contributions have been made in the field of Neogene tektites and impacts. Remarkably, a large impact from a low density extraterrestrial object has been postulated at the base of the Younger Dryas on the basis of a marked Ir-peak and the presence of such materials as microspherules, nanodiamonds and fullerenes (Firestone *et al.*, 2007). This event would account for Younger Dryas cooling, the megafauna extinction and termination of Clovis culture. However, more recent studies cast serious doubts upon the validity of an impact event as the explanation for the Younger Dryas. A study of elements belonging to the platinum group did not produce evidence for a large chondritic body impact (Paquay *et al.*, 2009), while alleged impact-related nanodiamonds (Kennett *et al.*, 2009) most likely represent aggregates of graphene- and graphene/graphane-oxides (Daulton *et al.*, 2010) and carbon spherules are in fact fungal remains that are present in layers both below, at and above the base of the Younger Dryas (Scott *et al.*, 2010).

Evidence in the form of so-called Dakhleh Glass was found for an impact event in the western Desert of Egypt between ~100 and 200 ka (Osinski *et al.*, 2007). The glass was dated at 145 ± 19 ka (Renne *et al.*, 2010a) and the potential consequences for the Middle Stone Age occupants of Dakhleh Oasis were discussed (Smith *et al.*, 2009). Remote sensing and field studies were used to demonstrate that Bajada del Diablo represents a new and remarkable crater-strewn field in Argentina (Acevedo *et al.*, 2009).

The Australasian tektite strew field has been enlarged with tektite findings on Antarctica (Folco *et al.*, 2009), but the impact crater remains to be discovered. However, the probable location of the crater can be more narrowly defined with new findings of microtektites and unmelted ejecta in deep-sea cores (Glass and Koeberl, 2006; Prasad *et al.*, 2007).

The problem of the age of the Ries impact crater and associated Moldavite tektites has apparently been solved by new ^{40}Ar/^{39}Ar stepwise heating and total fusion experiments on tektites, resulting in an age of 14.68 ± 0.11 Ma for the event (Di Vicenzo and Skala, 2009). This age is essentially consistent with results from an integrated stratigraphic study of the Brock horizon in the Upper Freshwater Molasse of the North Alpine Foreland Basin (Abdul Aziz *et al.*, 2008b).

29.4.5. Event Stratigraphy and the Astronomical Time Scale

The importance of a high-resolution integrated and astronomically tuned time scale, such as the ATNTS2012 or its predecessor ATNTS2004, for event stratigraphy is amply demonstrated by case studies of the Mediterranean Neogene, especially those concerning the Messinian salinity crisis (MSC). The application of such time scales resulted in a much better age control and, hence, understanding of the complex combination of events that lead to the onset, culmination and termination of the MSC (e.g., Krijgsman *et al.*, 1999, 2001; Roveri and Manzi, 2006). Similarly the application of a tuned time scale resulted in a much better separation and, hence, comprehension of paleoenvironmental changes that affected the Mediterranean during the late Tortonian and early Messinian (Hüsing *et al.*, 2009).

29.5. RADIO-ISOTOPIC AGES

29.5.1. ^{40}Ar/^{39}Ar Dating

29.5.1.1. Intercalibration with Astronomical Dating

During the last 25 years, the conventional ^{40}K/^{40}Ar method has been replaced by the ^{40}Ar/^{39}Ar method for dating potassium-bearing minerals of Neogene age, because the latter is fast, precise and reproducible and allows parent (first converted to ^{39}Ar by neutron irradiation) and daughter isotope to be simultaneously measured by the same method and on the same small sample aliquot; as such it even allows single crystal dating.

Nevertheless, the full error in ^{40}Ar/^{39}Ar dating remained unacceptably large (>2.5%) compared with U/Pb and astronomical dating. This large error mainly stems from uncertainties in the primary dating standard (to which secondary standards − used in the actual dating to calculate the age of the unknown sample − are calibrated) and the decay constants. Following earlier efforts (e.g., Renne *et al.*, 1994; Hilgen *et al.*, 1997), a rigorous attempt has been made during recent years in the Mediterranean Neogene to significantly reduce this error. This was achieved by calibrating the ^{40}Ar/^{39}Ar dating method directly to the astronomical dating method by means of a direct comparison between single crystal ^{40}Ar/^{39}Ar sanidine and astronomical ages for the same ash layers intercalated in astronomically tuned open marine sections in the Melilla Basin in Morocco. The ~0.65 % difference between astronomical and ^{40}Ar/^{39}Ar ages, with the latter being consistently younger, was removed by increasing the age of the most widely used secondary dating standard, the Fish Canyon Tuff sanidine, from 28.02 ± 0.28 (Renne *et al.*, 1998) to 28.201 ± 0.046 Ma (Kuiper *et al.*, 2008). The main advantage of using this astronomical age for the FCT sanidine is that uncertainties in the decay constants and the age of the primary dating standard, which contribute most to the full error in ^{40}Ar/^{39}Ar dating ages, are eliminated (Kuiper, 2003), resulting in a reduction of the full error by an order of 10. In the meantime, this age has been independently confirmed by ^{40}Ar/^{39}Ar dating of the a1 ash bed in the well-tuned Faneromeni section (Rivera *et al.*, 2011) and by direct single zircon crystal U/Pb dating of the Fish Canyon Tuff itself (Wotzlaw *et al.*, 2011), using new sample preparation techniques (Schaltegger *et al.*, 2009). Together, this suggests that the age of 28.305 Ma based on Ar/Ar−U/Pb pairs and disregarding astronomical dating (Renne *et al.*, 2010b) might be too old. This confirmation is important because a recent study shows that the Brunhes/Matuyama boundary lies at the young end of marine isotope stage (MIS) 19 with an age of 773.2 ± 0.3 ka for the mid-point of the reversal (Channell *et al.*, 2010). This age is ~7 kyr younger than the presently accepted astrochronological age for the polarity reversal and consistent with the best ^{40}Ar/^{39}Ar constraints if an age of 27.93 Ma for FCs is applied, rather than the 28.201 Ma suggested by Kuiper *et al.* (2008). However, Rivera *et al.* (2011) show that uncertainty in the age of the B/M boundary might be the cause for the discrepant young age of the FCT sanidine.

For geological time scale work, the astronomical calibration of the ^{40}Ar/^{39}Ar dating method is critical because it guarantees that both methods will in principle produce the same age when the same event is dated. It is important to realize that this was not the case in GTS2004 when an age of 28.02 Ma was accepted for the FCT sanidine. As a consequence, the numerical age calibration of parts of the time scale were based not only on different method but on methods that yielded significantly different ages for the same event. For this reason, the age of 28.201 Ma has been adopted in GTS2012 as the age for the FCs standard and all ^{40}Ar/^{39}Ar ages have been recalculated relative to this age.

29.5.1.2. Applications

Another very important application of the astronomically dated FCs standard is that it provides tight constraints for the astronomical tuning of cyclic pre-Neogene successions down to the correlative 400-kyr eccentricity cycle (Kuiper *et al.*,

2008). The approach was made obvious using the K/Pg boundary as case history. Starting from an FCT sanidine age of 28.201 Ma, the recalibrated age of 65.97 Ma for the boundary was used to constrain the tuning of the K/Pg boundary in the Zumaia section in Spain to the nearest 400-kyr cycle. This tuning yielded an age of 65.95 Ma, i.e. significantly older than the age of 65.5 Ma adopted in GTS2004 (which was calculated relative to a FCs age of 28.02 Ma).

However, a serious drawback recently came from an interlaboratory comparison study carried within the framework of the Earthtime programme. Under controlled experimental conditions this study revealed much larger age differences between laboratories than the quoted precision (McLean *et al.*, 2008). However, the new age of ~65.95 Ma is consistent with the recent U/Pb ages for the K/Pg boundary interval (Bowring *et al.*, 2008), indicating that the intercalibration and $^{40}Ar/^{39}Ar$ ages for the boundary are both essentially correct.

Despite these ongoing uncertainties in the exact intercalibration, $^{40}Ar/^{39}Ar$ dating has been successfully applied over the last years to independently test the astronomical tuning of the Serravallian/Tortonian boundary stratotype section at Monte dei Corvi (Kuiper *et al.*, 2005) and to solve existing inconsistencies between the marine and continental record (Kuiper *et al.*, 2004).

Apart from the intercalibration issue, $^{40}Ar/^{39}Ar$ dating continues to play a crucial role in dating hominid- and artefact-bearing successions and sites. Direct $^{40}Ar/^{39}Ar$ dating of the hominid-bearing level at Dmanisi confirmed the age of ~1.8 Ma, i.e. close to the Plio–Pleistocene boundary, for the oldest human remains in Eurasia (Garcia *et al.*, 2010). $^{40}Ar/^{39}Ar$ dating further revealed that the Middle Stone Age (MSA) technology evolved asynchronously in NE Africa and indicate that the oldest known MSA consistently predates fossils that belong to the earliest *Homo sapiens* (Morgan and Renne, 2008). $^{40}Ar/^{39}Ar$ age measurements, in combination with correlations to Mediterranean sapropels, indicate an age of 195 ± 5 ka for hominins in Ethiopia, making them the earliest well-dated anatomically modern humans that have been described (McDougall *et al.*, 2008; Joordens *et al.*, 2011).

29.5.2. U/Pb Dating

The Neogene is at the young end for U/Pb dating. Most U/Pb work in the Neogene is directed at intrusion dating and reconstructing metamorphic and exhumation histories. However, it has also been applied to date volcanic activity, such as the ignimbrites and lavas of the Late Miocene Heise volcanic field, with implications for the fate of Yellowstone hotspot calderas (Bindeman *et al.*, 2007), and volcanic arc activity in the SW Pacific (Mortimer *et al.*, 2010). It has further been used to find out whether extrusion ages reflect magma generation processes at depth (Cesare *et al.*, 2009).

In sedimentary successions of the Neogene, U/Pb dating is mainly used for dating detrital zircons to assess the spatial and temporal extent of sediment transport and the composition of the hinterland. However, U/Pb dating of ash layers is also used to date Neogene successions directly and reconstruct the paleoenvironmental and paleoclimatic history of, e.g., the Central Andes (Uba *et al.*, 2009; Mulch *et al.*, 2010) and NW Argentina (Bywater-Reyes *et al.*, 2010). The U/Pb method has further been used for dating lacustrine tufa in continental successions in California (Cole *et al.*, 2005), a latest Miocene aragonitic coral in the Caribbean (Denniston *et al.*, 2008), a Miocene fossil track site in Hungary (Pálfy *et al.*, 2007) and the Sterkfontein hominin site in South Africa (Walker *et al.*, 2006; Pickering and Kramers, 2010). U/Pb dating of groundwater table related speleothems was used to constrain a young 6 Ma age for tectonically driven incision of the Grand Canyon (Karlstrom *et al.*, 2008).

Another important application of U/Pb zircon dating of Neogene ash layers lies in the direct comparison with $^{40}Ar/^{39}Ar$ and astronomical ages for the same layers. This will shed light on the intercalibration with $^{40}Ar/^{39}Ar$ dating and on zircon residence times in the magma chamber through comparison with the astronomical ages (Wotzlaw and Schaltegger, 2011).

29.5.3. ^{14}C and $^{230}Th/^{234}U$ Dating

Radiocarbon dating remains the single most powerful tool to date organic material younger than 50 ka, although dissolved inorganic carbon (DIC) is datable by this method as well. The method has undergone major innovative advancements during the last decade since the publication of GTS2004 (e.g., Bronk Ramsey *et al.*, 2010), despite the fact that it has reached a relatively mature stage. These developments include extension of the calibration further back in time using, among others, varved lacustrine successions (Staff *et al.*, 2009) and Kauri trees from New Zealand (Hogg *et al.*, 2006), and the construction of the standard IntCal09 and Marine09 calibration curves, extending the previous IntCal04 and Marine04 curves from 26 to 50 ka (Reimer *et al.*, 2009). For the latter purpose, widely different archives such as pristine corals (e.g., Fairbanks *et al.*, 2005) and planktonic foraminifera of marine successions of the Cariaco Basin and Iberian margin (e.g., Bard *et al.*, 2004; Hughen *et al.*, 2006) were employed.

Important developments were further made in analytical procedures and techniques such as the pre-treatment of bone material to eliminate contamination (Bronk Ramsey *et al.*, 2004) and the use of ultra-filtration techniques and of low blanks in mass spectrometer extraction lines to measure very small or old samples (Pigati *et al.*, 2007). Finally the application of Bayesian statistical techniques (e.g., Bronk Ramsey, 2009) allowed radiocarbon ages to be combined with evidence from other sources such as stratigraphy. Together, these developments now make it possible to directly date prehistoric rock paintings by dating the pigments themselves, resulting in the remarkable, but still disputed, old age of the Chauvet Cave paintings (e.g., Valladas *et al.*, 2005; Pettitt and Pike, 2007; Pettitt, 2008) and to tackle major scientific issues such as the Middle–Upper Paleolithic transition and associated dispersal of modern humans in Europe (e.g., Mellars, 2006; Blockley *et al.*, 2008; Jöris and Street, 2008), the human population and dispersal of the New World (e.g., Waters and Stafford, 2007; Goebel *et al.*, 2008), the regional and global significance of millennial-scale climate events (e.g., Fedele *et al.*, 2008) and for instance the impact of the Campanian ignimbrite eruption (e.g., Turney *et al.*, 2010).

U-series dating of fossil corals remained a vital tool for reconstructing Pleistocene sea-level history, for instance by revealing timing, magnitude and duration of the high-stand during the last interglacial (Dumas *et al.*, 2006; Muhs *et al.*, 2011). Further improvements were made in correcting for open system behavior (Scholz *et al.*, 2004; Thompson and Goldstein, 2005) and in extending the method further back in time (Stirling and Andersen, 2009). In fact it can now be applied to date 600-ka old samples with a 2σ uncertainty of better than 15 kyr (Andersen *et al.*, 2008). Importantly, the method can reliably be applied to resolve sea-level changes on sub-Milankovitch time scales (e.g., Thompson and Goldstein, 2005; Thomas *et al.*, 2009).

Uranium-thorium dating continues to produce excellently dated climate proxy records, in particular $\delta^{18}O$, of speleothems. High-resolution intricate records from the Sanbao and Hulu caves in China now dating back to 224 000 years played a crucial role in improving marine–continental correlations and in deciphering phase relations with respect to orbital forcing (Wang *et al.*, 2001; Cheng *et al.*, 2009).

Other methods for dating young samples, such as optically stimulated luminescence (OSL: Lian and Roberts, 2006; Wintle and Murray, 2006) and electron spin resonance (ESR) dating, have also undergone rapid developments during the last decade. This is especially the case for luminescence dating, although the relatively large uncertainty remains problematical. Nevertheless, the method is becoming more reliable – and applicable to more environments – with single crystal dating (Duller, 2008) and is applicable to grains with an age of only a few tens to hundreds of years (Wintle, 2008). Infrared stimulated luminescence (IRSL) dating of feldspars has potential for extending the method further back in time from 100 ka to ~1 Ma (Wintle, 2008; Kars *et al.*, 2008).

29.5.4. $^{87}Sr/^{86}Sr$

Strontium isotope stratigraphy remains to be used in the Neogene, also after publication of ATNTS2004 (Lourens *et al.*, 2004), as it provides a powerful tool to correlate and date marine sequences in the Cenozoic, especially when other more accurate and precise dating techniques fail. The method relies on matching the $^{87}Sr/^{86}Sr$ value of a sample to a standard curve of the $^{87}Sr/^{86}Sr$ of ocean water through time (see Chapter 7). Gleason *et al.* (2004) used magnetostratigraphy, radiolarian biostratigraphy and fish teeth Sr isotope stratigraphy to date a 16 m long piston core in the equatorial Pacific. Ages based on the three different methods are generally consistent, but the Sr isotope ages are superior where the magnetostratigraphy is problematical. They therefore conclude that cores containing red clay of eolian origin can be reliably dated by the fish teeth strontium technique. This is crucial for the numerous and otherwise nondatable red clay cores from the vast red clay province of the Pacific, where it may be the only available reliable method for dating the stratigraphic record. Sr isotope stratigraphy was further applied to date, among others, the stratigraphic succession in the Neotectonic Hatay Graben in southern Turkey (Boulton *et al.*, 2007), the interglacial events of Late Neogene age that occurred in the James Ross Island region, northern Antarctic Peninsula (Smellie *et al.*, 2006), and the history of a deep-water coral mound in the northeast Atlantic (Kano *et al.*, 2007).

However, $^{87}Sr/^{86}Sr$ isotope ratios can also be employed to reconstruct inter-connectivity between basins by testing whether they both plot on the standard open ocean curve (see curve in Chapter 7) or that one (or both) reveals a deviating pattern related to enhanced riverine influence. This has been applied to determine whether the Paratethys was disconnected from the open ocean and Mediterranean across the Miocene–Pliocene boundary (Vasiliev *et al.*, 2010). They found similar values as in the Mediterranean "Upper Evaporites/Lago Mare" facies, indicating that deviating $^{87}Sr/^{86}Sr$ values in the "Upper Evaporites/Lago Mare", i.e. at the end of the Messinian salinity crisis (see also Matano *et al.*, 2005), could have been caused by drowning of the Mediterranean by Eastern Paratethys waters. Using dinoflagellate biochronology, Munsterman and Brinkhuis (2004) were able to show that previous ages obtained by $^{87}Sr/^{86}Sr$ dating were incorrect as they are markedly affected by continental run-off.

29.6. CLIMATE CHANGE AND MILANKOVITCH CYCLES

29.6.1. Oxygen and Carbon Isotopes ($\delta^{18}O$, $\delta^{13}C$)

29.6.1.1. Oxygen ($\delta^{18}O$)

Plio–Pleistocene

A major step forward in oxygen isotope stratigraphy and chronology was the publication of a benthic stack (labeled LR04) from 57 globally distributed sites for the last 5.3 myr, thus covering the entire Pliocene and Pleistocene (Lisiecki and Raymo, 2004). The stacking is used to reduce noise thereby improving quality and allowing the identification of 24 new marine isotope stages in the early Pliocene. The LR04 age model is based on tuning the $\delta^{18}O$ stack to a simple ice model, using the June 21 insolation at 65°N as input function. This tuning was done in a conservative way by using stacked sedimentation rates as additional age model constraints. Despite this conservative tuning strategy, the LR04 benthic stack exhibits significant coherency with insolation in the obliquity band throughout the entire 5.3 myr and in the precession band for more than half of the record.

By contrast Huybers (2007) followed an alternative non-tuning strategy to establish an age model for the last 2 myr using depth in cores as a proxy for time. Oxygen isotope records tied to this age model provide a detailed record of glacial variability, indicating a dominant obliquity control over the entire interval, thus including the last ~800 000 years with its prominent 100-kyr glacial cyclicity.

Miocene

The latest Miocene is marked by a distinct glacial interval ranging from ~6.3 to 5.5 Ma, followed by a major deglaciation lasting from oxygen isotope stage TG12 to 9 in terms of standard oxygen isotope stages. Remaining problems in the tuning of this characteristic interval and its relation with the Messinian salinity crisis in the Mediterranean were solved by Van der Laan *et al.* (2005, 2006). Westerhold *et al.* (2005) presented a tuned stable isotope record for the remaining part of the Late Miocene and the Serravallian part of the Middle Miocene (i.e. from 7.35 − 13.80 Ma).

Holbourn *et al.* (2005, 2007) published tuned Middle Miocene isotope records of Pacific ODP sites 1146 and 1237 with the aim to unravel the orbital pacing of the mid-Miocene climate transition and "Monterey" carbon isotope excursion. These studies showed the very high potential for developing a standard isotope stratigraphy for this time interval and revealed marked and abrupt changes in dominant orbital frequencies from precession and eccentricity to obliquity and back that mirrors the additional control by the very long period 2.4 myr eccentricity cycle, with relatively enhanced obliquity influence during eccentricity minima. Holbourn *et al.* (2007) further state that the obliquity-paced changes favored the transition into the "Icehouse" state of the climate system.

The early Miocene interval has been covered by Billups *et al.* (2004) and Pälike *et al.* (2006a), using ODP sites 1090 from the South Atlantic and 1218 from the equatorial Pacific, respectively. These high-resolution studies are critical to test the validity of the initial oxygen isotope stratigraphy developed by Miller *et al.* (1991,1996) based on low-resolution records from the Atlantic. Unfortunately, the tuning of these records remains problematical, but this may be solved through the use of high-resolution records from ODP site 1264. These records have been correlated in great detail to the isotope records of ODP sites 1090 and 926/929 and reveal marked ~100-kyr fluctuations in $\delta^{18}O$ following peak glacial episodes (Liebrand *et al.*, 2011).

Finally, Tian *et al.* (2008) used ODP site 1148 in the South China Sea to construct a benthic stable isotope record for the entire Neogene Period, using an age model based on the astronomical tuning of natural gamma radiation and color reflectance from the same site. These records were used to distinguish between linear and non-linear responses in the climate system to orbital forcing.

Modeling Temperature and Ice Volume

Recent developments in ice-sheet modeling enabled the deep-sea benthic $\delta^{18}O$ record to be decomposed into a temperature and ice-volume component (Bintanja *et al.*, 2005a,b; Bintanja and Van de Wal, 2008; De Boer *et al.*, 2010, 2011; Lourens *et al.*, 2010). The model takes advantage of the mass conservation of $\delta^{18}O$ on a global scale, while an inverse routine guarantees that changes in ocean $\delta^{18}O$ caused by both land ice sheet growth (sea level change) and deep-ocean temperature change are in agreement with marine benthic $\delta^{18}O$ reconstructions. The ice-sheet model includes a variable isotopic sensitivity and isotopic lapse rate, the mass balance height feedback, the mass balance albedo feedback and the adjustment of the underlying bedrock.

In the early stages of development, the ice-sheet model was forced with sea level reconstructions over the past 120 000 years to derive the air temperatures at sea level in areas where the northern hemisphere (NH) ice sheets grow (40°−80°N) through an inverse routine (Bintanja *et al.*, 2005a). The outcome of that study yielded a T_{NH} reconstruction, which is highly coherent with the classical Vostok temperature record. Nevertheless an obvious limitation of this work was that global sea-level observations are limited, which motivated Bintanja *et al.* (2005b) to incorporate the marine benthic $\delta^{18}O$ record as forcing. In addition, they applied a simple parameterization to separate deep-water temperature from ice-volume changes, based on a linear relation between deep-water temperature and $\delta^{18}O$ (Duplessy *et al.* 2002), and an idealized climate model (Bintanja and Oerlemans, 1996) relating changes in deep-ocean temperature to atmospheric temperature changes. This new method was first applied to calculate temperature and sea level over the past million years (Bintanja *et al.*, 2005b), and later to explore the mechanisms of the Mid-Pleistocene Transition (Bintanja and Van de Wal, 2008), both focusing on the climate in the NH, as only the Eurasian and North American ice sheet complexes were modeled explicitly. Their temperature reconstruction was compared to alkenone-derived equatorial temperatures (Lawrence *et al.* 2006) for the past 3 myr, indicating similar strength for most of the glacials (Bintanja and Van de Wal, 2008).

The last step in the model sequence until now is the explicit inclusion of ice sheets in the southern hemisphere (SH), allowing a longer time span to be covered, since for

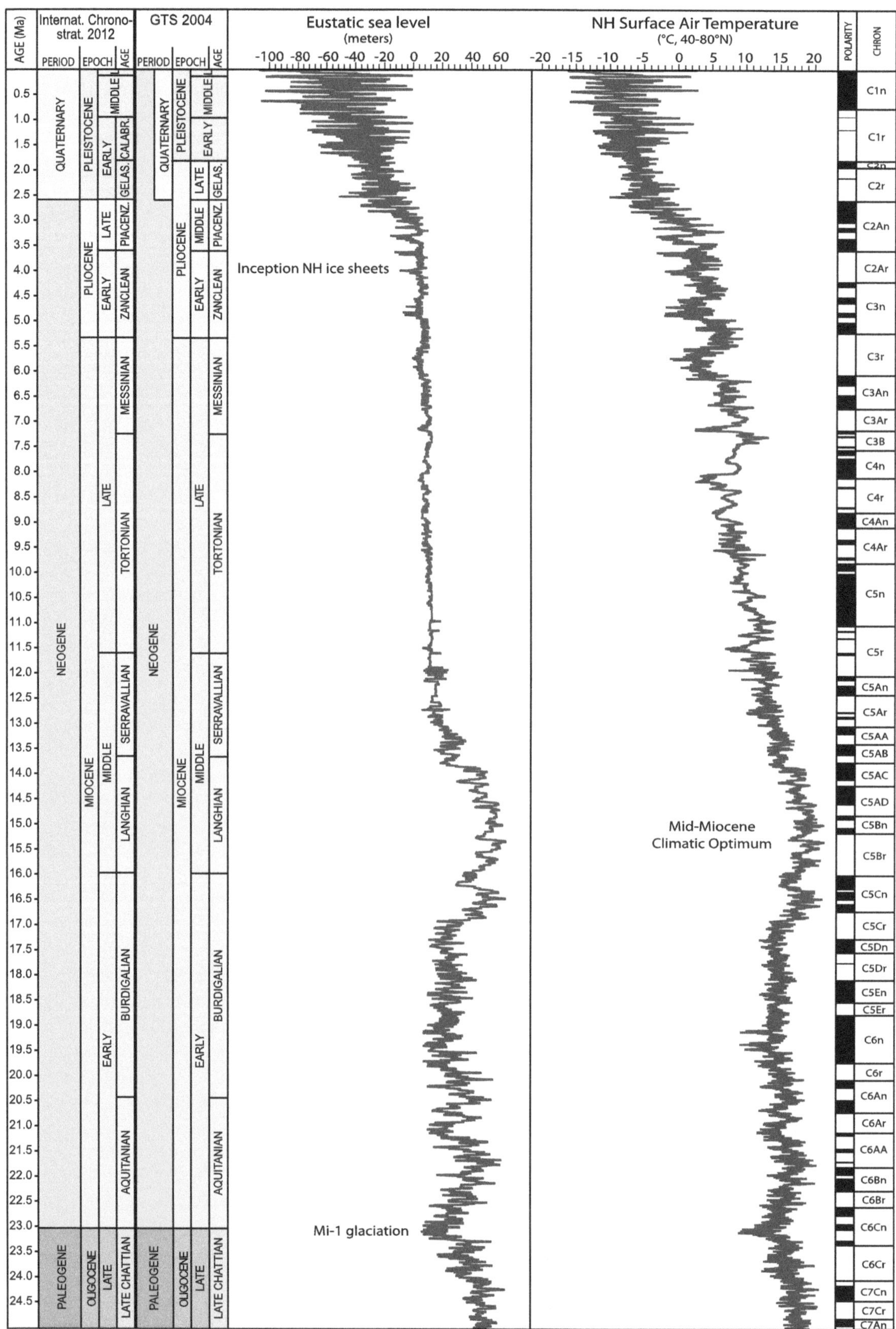

FIGURE 29.13 **Modeled sea level and northern hemisphere temperature changes during the Neogene (De Boer *et al.*, 2010).** The stacked benthic $\delta^{18}O$ record of Zachos *et al.* (2008) was smoothed, rescaled to ATNTS2004 and interpolated to obtain a continuous record with a time resolution of 100 years. The upper Cenozoic chronostratigraphy ratified by the IUGS in 2009 shifted the Gelasian Stage into the Pleistocene Series and a Quaternary–Neogene system boundary was defined at this level (c. 2.6 Ma). The Gelasian was formerly the upper Pliocene (as shown in GTS2004).

warmer conditions of the Neogene ice-volume changes are dominated by changes in the SH (De Boer *et al.*, 2010, 2011). To keep computing time manageable, this has been done at the expense of the complexity of the ice-sheet models used. Hence, five 1-D ice sheets, rather than the two 3-D ice-sheet models used previously (Bintanja *et al.*, 2005b; Bintanja and Van de Wal, 2008), were used to reconstruct temperature and sea level over the past 35 myr (see Figure 29.13). Sea level results of these late Cenozoic runs are in good agreement with the reconstructions by Miller *et al.* (2005) for the past 10 myr, as the Miller record is derived from the $\delta^{18}O$ by scaling. Large differences occur, however, for the lower part of the Neogene. Results of the deep-water temperature change since the Miocene by De Boer *et al.* (2011) are comparable to the results by Lear *et al.* (2000), although the modeled temperatures are considerably lower during the Middle Miocene than the Mg/Ca-based data.

29.6.1.2. Carbon Isotopes ($\delta^{13}C$)

Holbourn *et al.* (2007) published high-resolution tuned middle Miocene benthic isotope records of Pacific ODP sites to portray the "Monterey" carbon isotope excursion in detail. The event lasted from 16.9 to 13.5 Ma and consists of nine 400-kyr cycles, which show high coherence with long period eccentricity. Their results suggest that eccentricity played a crucial role in middle Miocene climate evolution through the modulation of long-term carbon budgets. Such long-period eccentricity related changes in $\delta^{13}C$ continue throughout the Early Miocene (e.g., Zachos *et al.*, 2001; Liebrand *et al.*, 2011) and the entire Oligocene (Pälike *et al.*, 2006b) and lag similar changes in $\delta^{18}O$ in the 400-kyr band, while they are in phase in the ~100-kyr band. This can be explained by an amplification of the longer period forcing due to the long residence time of carbon in the ocean along with a dampening effect of deep-sea sedimentary carbonate dissolution on shorter eccentricity time scales (Pälike *et al.*, 2006b). A plausible mechanism invokes continental weathering with enhanced and delayed response to 400-kyr eccentricity forcing as a consequence of the long time scale associated with weathering (Holbourn *et al.*, 2007).

In the latest Miocene, $\delta^{13}C$ records reveal a strong 41-kyr obliquity signal that is inversely related to obliquity related variations in $\delta^{18}O$, pointing to large-scale glacial-controlled variations in the deep-sea carbon reservoir (Van der Laan *et al.*, 2012).

29.6.2. Sedimentary Cycles

Apart from proxies, Milankovitch cycles are expressed as sedimentary cycles in the stratigraphic record. These cycles are well developed in deep-sea sections and cores, often in the form of marl–limestone alternations. Care should be taken that these cycles represent primary cycles and do not result from diagenetic unmixing. Many carbonate cycles reveal a diagenetic overprint, but unequivocal parameters should be used to test their primary origin (Westphal *et al.*, 2010). Nevertheless, carbonate cycles and related proxies, such as magnetic susceptibility and color, have been intensively used for the construction of the astronomical time scale (e.g., Hilgen, 1991; Tiedemann *et al.*, 1994). As an example, they were applied in the extension of the astronomical time scale for the Mediterranean Neogene back to 15 Ma (Hüsing *et al.*, 2009; Mourik *et al.*, 2010) and for the astronomical tuning of Neogene deep-sea sediments from the South China Sea (Tian *et al.*, 2008).

However, sedimentary cycles related to astronomical climate forcing are also found in continental settings. For instance, the Orera section in Spain contains a shallow lacustrine–mudflat succession, which is characterized by a remarkable regular alternation of dolomitic carbonate and mudstones that allowed astronomical tuning (Abdul Aziz *et al.*, 2003b). This tuning was used to determine astronomical ages for reversal boundaries older than 10 Ma (Abdul Aziz *et al.*, 2003a). In this way, the Mediterranean based A(P)TS was extended to ~13 Ma, although ages have now been replaced by those derived from the marine Monte dei Corvi section (Hüsing *et al.*, 2007).

Sedimentary cycles are also found in the Paratethys (see Piller *et al.*, 2007). Vasiliev *et al.* (2004) arrived at a period close to that of climatic precession for cycles deposited in lacustrine to deltaic environments during the Miocene and Pliocene in the eastern Paratethys. Sedimentary cycles expressed in spontaneous potential logs of the Central Paratethys are also astronomically controlled and were used to establish high-resolution correlations to the tuned deep-marine record of the Mediterranean (Lirer *et al.*, 2009).

29.6.3. Ice Cores

Important developments since the publication of ATNTS2004 are the completion of the drilling of the EPICA Dome C hole, thereby extending the ice core record over ~800 000 years and covering the last 8 full glacial–interglacial cycles (e.g., Loulergue *et al.*, 2008), and the construction of an independent age model, which facilitates a better – temporal – comparison with proxy records from other important climate archives. For EPICA Dome C, an age model was developed, using a snow accumulation and mechanical ice flow model, and a set of independent age markers with absolute ages that have a maximum uncertainty of 6 kyr back to 800 ka (EDC3 chronology of Parrenin *et al.*, 2007). This age model as well as other age models were used among others to reconstruct dust–climate interactions (Lambert *et al.*, 2008), moisture source and temperature based on deuterium excess (Stenni *et al.*, 2010), orbital and millennial scale changes in CO_2 and CH_4 (Lüthi *et al.*, 2008; Loulergue *et al.*, 2008), and environmental changes, using chemical proxies such as non-sea-salt calcium (Wolff *et al.*, 2010).

More accurate age models were further constructed for – parts of – Greenland ice cores by using 1) a multi-parameter counting of annual layers between 15 and 42 ka, later extended to 60 ka, thereby extending the Greenland Ice Core Chronology 2005 (GIC05 of Rasmussen *et al.*, 2006; Andersen *et al.*, 2006; Svensson *et al.*, 2006, 2008), and 2) a detailed paleoclimate comparison in combination with paired ^{14}C and Pb/Th ages on pristine corals (Shackleton *et al.*, 2004). Volcanic events were used to synchronize Greenland ice cores (Rasmussen *et al.*, 2008), while inter-hemispheric methane synchronization provides a crucial and powerful link with Antarctic ice cores (e.g., EPICA community members, 2006; Capron *et al.*, 2010; Schüpbach *et al.*, 2011). Finally, the INTIMATE group presented a revised protocol for time stratigraphic correlation in the North Atlantic region over an extended period from 8 to 30 ka (Lowe *et al.*, 2008). The NGRIP isotope record and associated GIC05 chronology are proposed as regional stratotype, Bayesian/based statistical procedures for age model construction, tephrochronology for validating regional correlations and INTCAL04 for calibration with radiocarbon dates (Lowe *et al.*, 2008), although the latter now has to be updated to INTCAL09. Clearly, such accurate high-resolution time scales are critical for understanding climate change on orbital and millennial time scales in detail (e.g., Rohling *et al.*, 2009).

29.7. ASTRONOMICALLY TUNED NEOGENE TIME SCALE – ATNTS2012

29.7.1. New Data

The cyclostratigraphy of the deep-marine Monte dei Corvi section in central Italy was slightly revised and retuned to the La2004 solution (Hüsing *et al.*, 2007). The much improved magnetostratigraphy of the lower part could be calibrated straightforwardly to the APTS and provided astronomical ages for all polarity reversals between 8.5 to 12.5 Ma. The new ages replace previous reversals ages for reversals based on the tuning of the continental Orera section in Spain (Abdul Aziz *et al.*, 2003a) and incorporated in ATNTS2004 (Lourens *et al.*, 2004). This tuning has been extended in the La Vedova Beach and High Cliff sections, covering the interval between 13.5 and 15.3 Ma (Hüsing *et al.*, 2009; Mourik *et al.*, 2010). The La Vedova composite section has further been extended along the beach but a robust tuning remains to be established for this section that is one of the candidate sections for defining the Langhian GSSP (Turco *et al.*, 2011a; Iaccarino *et al.*, 2011).

29.7.2. Solution La2004 Versus La2010

Recently a new astronomical solution, La2010, became available (Laskar *et al.*, 2011) that is reliable back to 50 Ma as far as eccentricity is concerned as compared to 40 Ma for La2004 (Laskar *et al.*, 2004). The major improvement in La2010 is a better adjustment of the parameters and initial conditions through a fit over 1 myr to a special version of the high accurate numerical ephemeris INPOP08 (Fienga *et al.*, 2009). However, we decided not to switch to the new solution because 1) the new solution solves only the orbital part used to calculate eccentricity, but not precession and obliquity, and 2) it is virtually identical to La2004 for the last 23 myr of the Neogene. The new solution will be critical in helping to solve problems encountered in the tuning of the Paleogene (see Chapter 28).

29.7.3. ATNTS2004 and 2012

Differences between ATNTS2004 and 2012 are relatively minor as can be expected from a geological time scale that is underlain by astronomical tuning. Relative to ATNTS2004, ages for chron boundaries remained the same for the last 8 myr and for the interval between 16 and 23 Ma. Maximal age differences between the two scales is ~50 kyr over the last 15 myr, and resulted from the replacement of astronomical reversal ages based on continental sections by ages based on deep marine sections (see Section 29.4.1).

29.7.4. The Age of the Paleogene/Neogene Boundary

The age of the Paleogene/Neogene boundary is kept at 23.03 Ma, i.e. the same as in ATNTS2004, although uncertainties remain in the tuning of the Early Miocene, including the boundary interval (Table 29.1). In anticipation of a final tuning, we refrained from changing the age as long as the tuning problems have not been adequately solved.

29.7.5. Incorporation of Global Chronostratigraphic Boundaries

Incorporation of the standard chronostratigraphic units, the stages, in ATNTS2012 is straightforward as most of the GSSPs have been defined in the same sections used to construct the astronomical time scale that directly underlies the age calibration of ATNTS2012 (see also Lourens *et al.*, 2004). Since ATNTS2004, the Serravallian GSSP has been added to the list of Neogene stage GSSPs that have been formally designated (Hilgen *et al.*, 2009). Significant progress has further been made in evaluating sections and guiding criteria for defining the Langhian GSSP in the near future (Iaccarino *et al.*, 2011).

GSSPs of the Neogene stages, with Location and Primary Correlation Criteria (Status Jan. 2012)

Stage	GSSP Location	Latitude, Longitude	Boundary Level	Correlation Events	Reference
Piacenzian	Punta Piccola, Sicily, Italy	37°17′20″N 13°29′36″E*	Base of the beige marl bed of small-scale carbonate cycle 77 with an age of 3.6 Ma	Precessional excursion 347 from the present with an astrochronological age estimate of 3.600 Ma	Episodes 21/2, 1998
Zanclean	Eraclea Minoa, Sicily, Italy	37°23′30″N 13°16′50″E	Base of the Trubi Formation	Insolation cycle 510 counted from the present with an age of 5.33 Ma	Episodes 23/3, 2000
Messinian	Oued Akrech, Morocco	33°56′13″N 6°48′45″W	Base of reddish layer of sedimentary cycle number 15	First regular occurrence of planktonic foraminifer *Globorotalia miotumida* and the FAD of the calcareous nannofossil *Amaurolithus delicatus*	Episodes 23/3, 2000
Tortonian	Monte dei Corvi Beach, near Ancona, Italy	43°35′12″N 13°34′10″E	Mid-point of sapropel layer of basic cycle number 76	Last Common Occurrences of the calcareous nannofossil *Discoaster kugleri* and the planktonic foraminifer *Globigerinoides subquadratus*	Episodes 28/1, 2005
Serravallian	Ras il Pellegrin section, Fomm Ir-Rih Bay, west coast of Malta	35°54′50″N 14°20′10″E	Formation boundary between the Globigerina Limestone and Blue Clay	Younger end of Mi3b oxygen isotopic event (global cooling episode)	Episodes 32/3, 2009
Langhian	*Candidate sections are La Vedova (Italy) and St. Peter's Pool (Malta)*			*Top of magnetic polarity chronozone C5Cn.1n*	
Burdigalian	*Potentially in astronomically-tuned ODP core*			*Near FAD of calcareous nannofossil* H. ampliaperta	
Aquitanian *(base Neogene)*	Lemme-Carrioso Section, Allessandria Province, Italy	44°39′32″N 08°50′11″E	35m from the top of the section	Base of magnetic polarity chronozone C6Cn.2n; FAD of calcareous nannofossil *Sphenolithus capricornutus*	Episodes 20/1, 1997

**according to Google Earth*

29.7.6. Incorporation of Zonal Schemes

Incorporation of the standard calcareous nannoplankton zonal scheme in ATNTS2012 is straightforward as the events on which the zonal scheme is based are all defined in tuned sections and cores (Raffi *et al.*, 2006). This is different from the recently published review and update of the standard planktonic foraminiferal zonation (Wade *et al.*, 2011) in which the link is mostly through magnetostratigraphy and its calibration to the polarity time scale of ATNTS2012. With respect to ATNTS2004, this will result in minor age changes up to 50 kyr for the interval between 8.5 and 15 Ma.

29.7.7. Advantages of the New Time Scale

ATNTS2012 has the same advantages as its predecessor, ATNTS2004. Work is in progress to finalize a Neogene time scale that is fully underlain by astronomical tuning and is essentially stable and will not change anymore.

ACKNOWLEDGMENTS

Everett Lindsay, Richard Madden, Maria Angeles Álvarez Sierra, Pablo Peláez-Campomanes, John Barry, Alexey Tesakov, Lutz Maul and Jan van der Made are thanked for their useful comments on the mammal biostratigraphy and chronology.

REFERENCES

Abdul Aziz, H., Hilgen, F.J., Krijgsman, W., Calvo, J.P., 2003a. An astronomical polarity time scale for the late middle Miocene based on cyclic continental sequences. Journal of Geophysical Research 108, 2159. doi: 10.1029/2002JB001818.

Abdul Aziz, H., Sanz-Rubio, E., Calvo, J.P., Hilgen, F.J., Krijgsman, W., 2003b. Palaeoenvironmental reconstruction of a middle Miocene alluvial fan to cyclic shallow lacustrine depositional system in the Calatayud Basin (NE Spain). Sedimentology 50, 211–236.

Abdul Aziz, H., Van Dam, J.A., Hilgen, F.J., Krijgsman, W., 2004. Astronomical forcing in Upper Miocene continental sequences: Implications for the Geomagnetic Polarity Time Scale. Earth and Planetary Science Letters 222, 243–258.

Abdul Aziz, H., Di Stefano, A., Foresi, L.M., Hilgen, F.J., Iaccarino, S.M., Kuiper, K.F., Lirer, F., Salvatorini, G., Turco, E., 2008a. Integrated stratigraphy and $^{40}Ar/^{39}Ar$ chronology of early Middle Miocene sediments from DSDP Leg 42A, Site 372 (Western Mediterranean). Palaeogeography, Palaeoclimatology, Palaeoecology 257, 123–138.

Abdul Aziz, H., Böhme, M., Rocholl, A., Zwing, A., Prieto, J., Wijbrans, J.R., Heissig, K., Bachtadse, V., 2008b. Integrated stratigraphy and $^{40}Ar/^{39}Ar$ chronology of the early to middle Miocene Upper Freshwater Molasse in eastern Bavaria (Germany). International Journal of Earth Sciences 97, 115–134.

Abdul Aziz, H., Böhme, M., Rocholl, A., Prieto, J., Wijbrans, J.R., Bachtadse, V., Ulbig, A., 2009. Integrated stratigraphy and $^{40}Ar/^{39}Ar$ chronology of the early to middle Miocene Upper Freshwater Molasse in western Bavaria (Germany). International Journal of Earth Science 99, 1859–1886.

Abels, H.A., Hilgen, F.J., Krijgsman, W., Kruk, R.W., Raffi, I., Turco, E., Zachariasse, W.J., 2005. Long-period orbital control on middle Miocene global cooling: Integrated stratigraphy and astronomical tuning of the Blue Clay Formation on Malta. Paleoceanography 20, PA4012. doi: 10.1029/2004PA001129.

Acevedo, R.D., Ponce, J.F., Rocca, M., Rabassa, J., Corbella, H., 2009. Bajada del Diablo impact crater-strewn field: The largest crater field in the Southern Hemisphere. Geomorphology 110, 58–67.

Acton, G., Guyodo, Y., Brachfeld, S., 2006. The nature of a cryptochron from a paleomagnetic study of Chron C4r.2r recorded in sediments off the Antarctic Peninsula. Physics of the Earth and Planetary Interiors 156, 213–222.

Aguilar, J.P., 1982. Biozonation du Miocene d'Europe occidentale a l'aide des rongeurs et correlations avec l'echelle stratigraphique marine. Comptes Rendus de l'Académie des Sciences Paris, Serie 2 294, 49–54.

Agustí, J., Cabrera, L., Garcés, M., Krijgsman, W., Oms, O., Parés, J.M., 2001. A calibrated mammal scale for the Neogene of Western Europe; State of the art. Earth-Science Reviews 52, 247–260.

Alberdi, M.T., Leone, G., Tonni, E.P. (Eds.). 1995. Evolución Biológica y Climática de la Región Pampeana Durante los Últimos Cinco Millones de Años. Monografías del Museo Nacional de Madrid 12, 1–423.

Albright III, L.B., 1999. Biostratigraphy and vertebrate paleontology of the San Timoteo Badlands, southern California. University of California Publications, Geological Sciences 144, 134.

Albright III, L.B., Woodburne, M.O., Fremd, T.J., Swisher III, C.C., MacFadden, B.J., Scott, G.R., 2008. Revised chronostratigraphy and biostratigraphy of the John Day Formation (Turtle Cove and Kimberly Members), Oregon, with implications for updated calibration of the Arikareean North American land mammal age. Journal of Geology 116, 211–237.

Alroy, J., 1994. Appearance event correlation: A new biochronological method. Paleobiology 20, 191–207.

Alroy, J., 1998. Diachrony of mammalian appearance events: Implications for biochronology. Geology 26, 23–27.

Alroy, J., 2000. New methods for quantifying macroevolutionary patterns and processes. Paleobiology 26, 707–733.

Andersen, K.K., Svensson, A., Johnsen, S.J., Rasmussen, S.O., Bigler, M., R öthlisberger, R., Ruth, U., Siggaard-Andersen, M.-L., Peder Steffensen, J., Dahl-Jensen, D., Vinther, B.M., Clausen, H.B., 2006. The Greenland Ice Core Chronology 2005, 15–42 ka. Part 1: Constructing the time scale. Quaternary Science Reviews 25, 3246–3257.

Andersen, M.B., Stirling, C.H., Potter, E.-K., Halliday, A.N., Blake, S.G., McCulloch, M.T., Ayling, B.F., O'Leary, M., 2008. High-precision U-series measurements of more than 500,000 year old fossil corals. Earth and Planetary Science Letters 265, 229–245.

Anthonissen, D.E., 2009. A new pliocene biostratigraphy for the north-eastern north Atlantic. Newsletters on Stratigraphy 43, 91–126.

Aubry, M.-P., 2007. Chronostratigraphy beyond the GSSP. Stratigraphy 4, 127–137.

Aubry, M.-P., Berggren, W.A., Van Couvering, J., McGowran, B., Pillans, B., Hilgen, F.J., 2005. Quaternary: Status, rank, definition, and survival. Episodes 28, 118–120.

Aubry, M.-P., Berggren, W.A., Van Couvering, J., McGowran, B., Hilgen, F., Steininger, F.F., Lourens, L.J., 2009. The Neogene and Quaternary: Chronostratigraphic compromise or non-overlapping magisteria? Stratigraphy 6, 1–16.

Azanza, B., Alberdi, M.T., Cerdeño, E., Prado, J.L., 1997. Biochronology from latest Miocene to Middle Pleistocene in the Western Mediterranean area. A Multivariate Approach. In: Aguilar, J.-P., Legendre, S., Michaux, J. (Eds.), Actes du Congrés Biochrom'97. Mémoire de l'École Pratique des Hautes Études, Institut de Montpellier, 21, pp. 567–574.

Backman, J., Raffi, I., 1997. Calibration of Miocene nannofossil events to orbitally-tuned cyclostratigraphies from Ceara Rise. Proceedings of the Ocean Drilling Program, Scientific Results 154, 83–99.

Baldauf, J.G., Iwai, M., 1995. Neogene diatom biostratigraphy for the eastern equatorial Pacific, Leg 138. In: Pisias, N.G., Mayer, L.A., et al. (Eds.), Proceedings of the Ocean Drilling Program, Scientific Results, 138, pp. 105–128.

Báldi, T., Seneš, J., 1975. OM-Egerien. Die Egerer, Pouzdraner, Puchkirchener Schichtengruppe und die Bretkaer Formation. Chronostratigraphie und Neostratotypen, Miozän der Zentralen Paratethys 5, 1–577.

Bard, E., Rostek, F., Ménot-Combes, G., 2004. Radiocarbon calibration beyond 20,000 ^{14}C yr B.P. by means of planktonic foraminifera of the Iberian Margin. Quaternary Research 61, 204–214.

Barnosky, A.D., Bibi, F., Hopkins, S.B., Nichols, R., 2007. Biostratigraphy and magnetostratigraphy of the mid-Miocene Railroad Canyon sequence, Montana and Idaho, and age of the mid-Tertiary unconformity west of the continental Divide. Journal of Vertebrate Paleontology 27, 204–224.

Barron, J.A., 2006. Diatom biochronology for the Early Miocene of the Equatorial Pacific. Stratigraphy 2, 281–309.

Barry, J., Morgan, M., Flynn, L., Pilbeam, D., Behrensmeyer, A.K., Raza, S.M., Khan, I.A., Badgley, C., Hicks, J., Kelley, J., 2002. Faunal and environmental change in the Late Miocene Siwaliks of Northern Pakistan. Paleobiology 28, 1–71.

Behrensmeyer, A.K., Quade, J., Cerling, T.E., Kappelman, J., Khan, I., Copeland, P., Roe, L., Hicks, J., Stubblefield, P., Willis, B., LaTorre, C., 2007. The structure and rate of late Miocene expansion of C4 plants: evidence from lateral variation in stable isotopes in paleosols of the Siwalik series, northern Pakistan. Geological Society of America Bulletin 119, 1486–1505.

Bell, C.J., Lundelius Jr., E.L., Barnosky, A.D., Graham, R.W., Lindsay, E.H., Ruez Jr., D.R., Semken Jr., H.A., Webb, S.D., Zakrzewski, R.J., 2004. The Blancan, Irvingtonian, and Rancholabrean Mammals Ages. In: Woodburne, M.O. (Ed.), Late Cretaceous and Cenozoic Mammals of North America: Biostratigraphy and Geochronology. Columbia University Press, New York, pp. 232–314.

Benammi, M., Jaeger, J.J., 2001. Magnetostratigraphy and paleontology of the continental middle Miocene of the Aït Kandoula basin (Morocco). Journal of African Earth Science 33, 335–348.

Benammi, M., Calvo, M., Prevot, M., Jaeger, J.J., 1996. Magnetostratigraphy and paleontology of Aït Kandoula basin (High Atlas, Morocco) and the African-European late Miocene terresterial fauna exchanges. Earth and Planetary Science Letters 145, 15–29.

Berggren, W.A., 1998. The Cenozoic Era: Lyellian chronostratigraphy and nomenclatural reform at the millenium. In: Blundell, J.A., Scott, A.C. (Eds.), Lyell: The Past is the Key to the Present. Geological Society Special Publication, 143, pp. 111–132.

Berggren, W.A., Kent, D.V., Swisher III, C.C., Aubry, M.-P., 1995. A revised Cenozoic geochronology and chronostratigraphy. In: Berggren, W.A., Kent, D.V., Aubry, M.-P., Hardenbol, J. (Eds.), Geochronology, Time Scales and Global Stratigraphic Correlation. SEPM Special Publication, 54, pp. 129–212.

Billups, K., Pälike, H., Channell, J.E.T., Zachos, J.C., Shackleton, N.J., 2004. Astronomic calibration of the late Oligocene through early Miocene geomagnetic polarity time scale. Earth and Planetary Science Letters 224, 33–44.

Bindeman, I.N., Watts, K.E., Schmitt, A.K., Morgan, L.A., Shanks, P.W.C., 2007. Voluminous low $\delta^{18}O$ magmas in the Late Miocene Heise volcanic field, Idaho: Implications for the fate of Yellowstone hotspot calderas. Geology 35, 1019–1022.

Bintanja, R., Oerlemans, J., 1996. The effect of reduced ocean overturning on the climate of the last glacial maximum. Climate Dynamics 12, 523–533.

Bintanja, R., Van de Wal, R.S.W., 2008. North American ice-sheet dynamics and the onset of 100,000-year glacial cycles. Nature 454, 869–872.

Bintanja, R., Van de Wal, R.S.W., Oerlemans, J., 2005a. A new method to estimate ice age temperatures. Climate Dynamics 24, 197–211.

Bintanja, R., Van de Wal, R.S.W., Oerlemans, J., 2005b. Modelled atmospheric temperatures and global sea levels over the past million years. Nature 437, 125–128.

Blisniuk, P.M., Stern, L.A., Chamberlain, C.P., Idleman, B., Zeitler, P.K., 2005. Climatic and ecologic changes during Miocene surface uplift in the southern Patagonian Andes. Earth and Planetary Science Letters 230, 125–142.

Blockley, S.P.E., Bronk Ramsey, C., Higham, T.F.G., 2008. The middle to upper paleolithic transition: Dating, stratigraphy, and isochronous markers. Journal of Human Evolution 55, 764–771.

Blow, W.H., 1956. Origin and evolution of the foraminiferal genus *Orbulina* d'Orbigny. Micropaleontology 2, 57–70.

Blow, W.H., 1969. Late middle Eocene to Recent planktonic foraminiferal biostratigraphy. In: Bronniman, P., Renz, H.H. (Eds.), Proceedings of the First International Conference on Planktonic Microfossils, Geneva, 1967, vol. 1. E.J. Brill, Leiden, pp. 199–422.

Boulton, S.J., Robertson, A.H.F., Ellam, R.M., Safak, Ü., Ünlügenç, U.C., 2007. Strontium isotopic and micropalaeontological dating used to help redefine the stratigraphy of the neotectonic Hatay Graben, southern Turkey. Turkish Journal of Earth Sciences 16, 141–179.

Bowring, S., Johnson, K.R., Clyde, W., Ramezani, J., Miller, I., Peppe, D., 2008. A Paleocene timescale for the rocky mountains: Status and potential. Geological Society of America, Abstracts with Programs 40 (6), 322.

Bradwell, T., Dugmore, A.J., Sugden, D.E., 2006. The little ice age glacier maximum in Iceland and the north Atlantic oscillation: Evidence from Lambatungnaj ökull, southeast Iceland. Boreas 35, 61–80.

Bronk Ramsey, C.B., 2009. Bayesian analysis of radiocarbon dates. Radiocarbon 51, 337–360.

Bronk Ramsey, C.B., Ditchfield, P., Humm, M., 2004. Using a gas ion source for radiocarbon AMS and GC-AMS. Radiocarbon 46, 25–32.

Bronk Ramsey, C.B., Dee, M., Lee, S., Nakagawa, T., Staff, R.A., 2010. Developments in the calibration and modeling of radiocarbon dates. Radiocarbon 52, 953–961.

Browning, J.V., Miller, K.G., McLaughlin, P.P., Kominz, M.A., Sugarman, P.J., Monteverde, D., Feigenson, M.D., Hernández, J.C., 2006. Quantification of the effects of eustasy, subsidence, and sediment supply on Miocene sequences, mid-Atlantic margin of the United States. Geological Society of America Bulletin 118, 567–588.

Browning, J.V., Miller, K.G., Sugarman, P.J., Kominz, M.A., Mclaughlin, P.P., Kulpecz, A.A., Feigenson, M.D., 2008. 100 Myr record of sequences, sedimentary facies and sea level change from Ocean Drilling Program onshore coreholes, US Mid-Atlantic coastal plain. Basin Research 20, 227–248.

Burbank, D.W., Barnosky, A.D., 1990. The magnetochronology of Barstovian mammals in southwestern Montana and implications for the initiation of Neogene crustal extension in the northern Rocky Mountains. Geological Society of America Bulletin 102, 1093–1104.

Butler, R.F., Marshall, L.G., Drake, R.E., Curtiss, G.H., 1984. Magnetic polarity stratigraphy and ^{40}K-^{40}Ar dating of Late Miocene and early Pliocene continental deposits, Catamarca Province, northwest Argentina. Journal of Geology 92, 623–636.

Bywater-Reyes, S., Carrapa, B., Clementz, M., Schoenbohm, L., 2010. Effect of late Cenozoic aridification on sedimentation in the Eastern Cordillera of northwest Argentina (Angastaco basin). Geology 38, 235–238.

Campbell, K.E., Prothero, D.R., Romero-Pittman, L., Hertel, F., Rivera, N., 2010. Amazonian magnetostratigraphy: Dating the first pulse of the Great American Faunal Interchange. Journal of South American Earth Sciences 29, 619–626.

Cande, S.C., Kent, D.V., 1992. A new geomagnetic polarity time scale for the late Cretaceous and Cenozoic. Journal of Geophysical Research 97, 13917–13951.

Cande, S.C., Kent, D.V., 1995. Revised calibration of the Geomagnetic Polarity Time Scale for the Late Cretaceous and Cenozoic. Journal of Geophysical Research 100, 6093–6095.

Capraro, L., Asioli, A., Backman, J., Bertoldi, R., Channell, J.E.T., Massari, F., Rio, D., 2005. Climatic patterns revealed by pollen and oxygen isotope records across the Brunhes-Matuyama boundary in the central Mediterranean (southern Italy). In: Head, M.J., Gibbard, P.L. (Eds.), Early-Middle Pleistocene Transitions: The Land–Ocean Evidence. Geological Society Special Publication, 247, pp. 159–182.

Capron, E., Landais, A., Lemieux-Dudon, B., Schilt, A., Masson-Delmotte, V., Buiron, D., Chappellaz, J., Dahl-Jensen, D., Johnsen, S., Leuenberger, M., Loulergue, L., Oerter, H., 2010. Synchronising EDML and NorthGRIP ice cores using δ^{18}O of atmospheric oxygen ($\delta^{18}O_{atm}$) and CH_4 measurements over MIS5 (80–123 kyr). Quaternary Science Reviews 29, 222–234.

Carbonell, E., Bermúdez De Castro, J.M., Parés, J.M., Pérez-González, A., Cuenca-Bescós, G., Ollé, A., Mosquera, M., Huguet, R., Van Der Made, J., Rosas, A., Sala, R., Vallverdú, J., García, N., Granger, D.E., Martinón-Torres, M., Rodríguez, X.P., Stock, G.M., Vergès, J.M., Allué, E., Burjachs, F., Cáceres, I., Canals, A., Benito, A., Díez, C., Lozano, M., Mateos, A., Navazo, M., Rodríguez, J., Rosell, J., Arsuaga, J.L., 2008. The first hominin of Europe. Nature 452, 465–469.

Casanovas-Vilar, I., Alba, D.M., Moyà-Solà, S., Galindo, J., Cabrera, L., Garcés, M., Furió, M., Robles, J.M., Köhler, M., Angelone, C., 2008. Biochronological, taphonomical, and paleoenvironmental background of the fossil great ape *Pierolapithecus catalaunicus* (Primates, Hominidae). Journal of Human Evolution 55, 589–603.

Castradori, D., Rio, D., Hilgen, F.J., Lourens, L.J., 1998. The Global Standard Stratotype-section and Point (GSSP) of the Piacenzian Stage (Middle Pliocene). Episodes 21, 88–93.

Cesare, B., Rubatto, D., Gómez-Pugnaire, M.T., 2009. Do extrusion ages reflect magma generation processes at depth? An example from the Neogene volcanic province of SE Spain. Contributions to Mineralogy and Petrology 157, 267–279.

Channell, J.E.T., 2006. Late Brunhes polarity excursions (Mono Lake, Laschamp, Iceland Basin and Pringle falls) recorded at ODP site 9191 (Irminger Basin). Earth and Planetary Science Letters 244, 378–393.

Channell, J.E.T., Hodell, D.A., Singer, B.S., Xuan, C., 2010. Reconciling astrochronological and ^{40}Ar/^{39}Ar ages for the Matuyama-Brunhes boundary and late Matuyama Chron. Geochemistry, Geophyphysics, Geosystems 11, Q0AA12. doi: 10.1029/2010GC003203.

Cheng, H., Edwards, R.L., Broecker, W.S., Denton, G.H., Kong, X., Wang, Y., Zhang, R., Wang, X., 2009. Ice age terminations. Science 326, 248–252.

Ciaranfi, N., d'Alessandro, A., 2005. Overview of the Montalbano Jonico area and section: A proposal for a boundary stratotype for the Lower-Middle Pleistocene, southern Italy foredeep. Quaternary International 131, 5–10.

Ciaranfi, N., Lirer, F., Lirer, L., Lourens, L.J., Maiorano, P., Marino, M., Petrosino, P., Sprovieri, M., Stefanelli, S., Brilli, M., Girone, A., Joannin, S., Pelosi, N., Vallefuoco, M., 2010. Integrated stratigraphy and astronomical tuning of lower-middle Pleistocene Montalbano Jonico section (southern Italy). Quaternary International 219, 109–120.

Cicha, I., Seneš, J., Tejkal, J., 1967. M3 (Karpatien). Die Karpatische Serie und ihr Stratotypus. Chronostratigraphie und Neostratotypen, Miozän der Zentralen Paratethys 1, 1–312.

Cione, A.L., Tonni, E.P., 2005. Bioestratigrafía basada en mamíferos del Cenozoico superior de la provincia de Buenos Aires, Argentina. In: de Barrio, R.E., Etcheverry, R.O., Caballé, M.F., Llambías, E. (Eds.), Geología y Recursos Minerales de la Provincia de Buenos Aires. Relatorio, XVI Congreso Geológico Argentino, 11, pp. 183–200.

Cita, M.B., Premoli Silva, I., 1968. Evolution of the planktonic foraminiferal assemblages in the stratigraphical interval between the type Langhian and the type Tortonian and the biozonation of the Miocene of Piedmont. Giornale di Geologia 35, 1051–1082.

Cita, M.B., Capraro, L., Ciaranfi, N., Di Stefano, E., Marino, M., Rio, D., Sprovieri, R., Vai, G.B., 2006. Calabrian and Ionian: A proposal for the definition of Mediterranean stages for the Lower and Middle Pleistocene. Episodes 29, 107–114.

Cita, M.B., Capraro, L., Ciaranfi, N., Di Stefano, E., Lirer, F., Maiorano, P., Marino, M., Raffi, I., Rio, D., Sprovieri, R., Stefanelli, S., Vai, G.B., 2008. The Calabrian Stage redefined. Episodes 31, 408–419.

Cody, R.D., Levy, R.H., Harwood, D.M., Sadler, P.M., 2008. Thinking outside the zone: High-resolution quantitative diatom biochronology for the Antarctic Neogene. Palaeogeography, Palaeoclimatology, Palaeoecology 260, 92–121.

Cole, J.M., Rasbury, E.T., Hanson, G.N., Montanez, I.P., Pedone, V.A., 2005. Using U-Pb ages of Miocene tufa for correlation in a terrestrial succession, Barstow Formation, California. Geological Society of America Bulletin 117, 276–287.

Cooper, R.A. (Ed.), 2004. The New Zealand geological timescale. Institute of Geological and Nuclear Science Monograph, 22, pp. 1–284.

Costeur, L., Legendre, S., Aguilar, J.-P., Lécuyer, C., 2007. Marine and continental synchronous climatic records: Towards a revision of the European Mid-Miocene biochronological framework. Geobios 40, 775–784.

Courtillot, V., Gallet, Y., Le Mouël, J.-L., Fluteau, F., Genevey, A., 2007. Are there connections between the Earth's magnetic field and climate? Earth and Planetary Science Letters 253, 328–339.

Crusafont, M., Truyols, J., 1960. Sobre la caracterizacion del Vallesiense. Notas y comunicaziones del Instituto Geologico y Minero de Espana 60, 109–126.

Daams, R., van der Meulen, A.J., Alvarez Sierra, M.A., Pelaez-Campomanes, P., Krijgsman, W., 1999. Aragonian stratigraphy reconsidered, and a re-evaluation of the middle Miocene mammal biochronology in Europe. Earth and Planetary Science Letters 165, 287–294.

Daulton, T.R., Pinter, N., Scott, A.C., 2010. No evidence of nanodiamonds in Younger-Dryas sediments to support an impact event. Proceedings of the National Academy of Sciences, USA 107, 16043–16047.

De Boer, B., Van de Wal, R.S.W., Bintanja, R., Lourens, L.J., Tuenter, E., 2010. Cenozoic global ice-volume and temperature simulations with 1-D ice-sheet models forced by benthic $\delta^{18}O$ records. Annals of Glaciology 51, 23–33.

De Boer, B., Van de Wal, R.S.W., Lourens, L.J., Bintanja, R., 2011. Transient nature of the Earth's climate and the implications for the interpretation of benthic $\delta^{18}O$ records. Paleogeography, Paleoecology, Paleoclimatology. doi: 10.1016/j.palaeo.2011.02.001.

De Bruijn, H., Daams, R., Daxner-H öck, G., Fahlbusch, V., Ginsburg, L., Mein, P., Morales, J., 1992. Report of the RCMNS working group on fossil mammals, Reisensburg 1990. Newsletters on Stratigraphy 26, 65–118.

De Bruijn, *et al.*, in press. In: Wang, X., Flynn, L.J., and Fortelius, M. (Eds.), *Asian Neogene Mammal Volume*. New York: Columbia University Press.

De Leeuw, A., Bukowski, K., Krijgsman, W., Kuiper, K.F., 2010. Age of the Badenian Salinity Crisis: Impact of Miocene climate variability on the Circum-Mediterranean region. Geology 38, 715–718.

Denniston, R.F., Asmerom, Y., Polyak, V.Y., McNeill, D.F., Klaus, J.S., Cole, P., Budd, A.F., 2008. Caribbean chronostratigraphy refined with U-Pb dating of a Miocene coral. Geology 36, 151–154.

De Schepper, S., Head, M.J., 2009. Pliocene and Pleistocene dinoflagellate cyst and acritarch zonation of DSDP Hole 610A, eastern North Atlantic. Palynology 33, 179–218.

Dinarès-Turell, J., Baceta, J.I., Pujalte, V., Orue-Etxebarria, X., Bernaola, G., Lorito, S., 2003. Untangling the Palaeocene climatic rhythm: An astronomically calibrated Early Palaeocene magnetostratigraphy and biostratigraphy at Zumaia (Basque Basin, northern Spain). Earth and Planetary Science Letters 216, 483–500.

Dinarès-Turell, J., Baceta, J.I., Bernaola, G., Orue-Etxebarria, X., Pujalte, V., 2007. Closing the Mid-Palaeocene gap: Toward a complete astronomically tuned Palaeocene Epoch and Selandian and Thanetian GSSPs at Zumaia (Basque Basin, W Pyrenees). Earth and Planetary Science Letters 262, 450–467.

Di Stefano, A., Foresi, L.M., Lirer, F., Iaccarino, S.M., Turco, E., Amore, F.O., Mazzei, R., Morabito, S., Salvatorini, G., Abdul Aziz, H., 2008. Calcareous plankton high resolution bio-magnetostratigraphy for the Langhian of the Mediterranean area. Rivista Italiana di Paleontologia e Stratigrafia 114, 51–76.

Di Vincenzo, G., Skála, R., 2009. ^{40}Ar-^{39}Ar laser dating of tektites from the Cheb Basin (Czech Republic): Evidence for coevality with moldavites and influence of the dating standard on the age of the Ries impact. Geochimica et Cosmochimica Acta 73, 493–513.

Domingo, M.S., Alberdi, M.T., Azanza, B., 2007. A new quantitative biochronological ordination for the Upper Neogene mammalian localities of Spain. Palaeogeography, Palaeoclimatology, Palaeoecology 255, 361–376.

Duller, G.A.T., 2008. Single-grain optical dating of Quaternary sediments: Why aliquot size matters in luminescence dating. Boreas 37, 589–612.

Dumas, B., Hoang, C.T., Raffy, J., 2006. Record of MIS 5 sea-level highstands based on U/Th dated coral terraces of Haiti. Quaternary International 145–146, 106–118.

Duplessy, J.-C., Labeyrie, L., Waelbroeck, C., 2002. Constraints on the oxygen isotopic enrichment between the Last Glacial Maximum and the Holocene: Paleoceanographic implications. Quaternary Science Reviews 21, 315–330.

Dybkjær, K., Piasecki, S., 2010. Neogene dinocyst zonation for the eastern North Sea Basin, Denmark. Review of Palaeobotany and Palynology 161, 1–29.

Epica Community Members, 2006. One-to-one coupling of glacial climate variability in Greenland and Antarctica. Nature 444, 195–198.

Evans, H.F., Westerhold, T., Channell, J.E.T., 2004. ODP Site 1092: Revised composite depth section has complications for Upper Miocene "cryptochrons". Geophysical Journal International 156, 195–199.

Evans, H.F., Westerhold, T., Paulsen, H., Channell, J.E.T., 2007. Astronomical ages for Miocene polarity chrons C4Ar-C5r (9.3–11.2 Ma), and for three excursion chrons within C5n.2n. Earth and Planetary Science Letters 256, 455–465.

Fahlbusch, V., 1976. Report on the International Symposium on Mammalian Stratigraphy of the European Tertiary (München, April 11–14, 1975). Newsletters on Stratigraphy 5, 160–167.

Fairbanks, R.G., Mortlock, R.A., Chiu, T.-C., Cao, L., Kaplan, A., Guilderson, T.P., Fairbanks, T.W., Bloom, A.L., Grootes, P.M., Nadeau, M.-J., 2005. Radiocarbon calibration curve spanning 0 to 50,000 years BP based on paired $^{230}Th/^{234}U/^{238}U$ and ^{14}C dates on pristine corals. Quaternary Science Reviews 24, 1781–1796.

Fang, X., Yan, M., Van der Voo, R., Rea, D.K., Song, C., Parés, J.M., Gao, J., Nie, J., Dai, S., 2005. Late Cenozoic deformation and uplift of the NE Tibetan Plateau: Evidence from high-resolution magnetostratigraphy of the Guide Basin, Qinghai Province, China. Geological Society of America Bulletin 117, 1208–1225.

Fedele, F.G., Giaccio, B., Hajdas, I., 2008. Timescales and cultural process at 40,000 BP in the light of the Campanian Ignimbrite eruption, Western Eurasia. Journal of Human Evolution 55, 834–857.

Fejfar, O., Heinrich, W.-D., Lindsay, E.H., 1998. Updating the Neogene rodent biochronology in Europe. Mededelingen Nederlands Instituut voor Toegepaste Geowetenschappen TNO 60, 533–554.

Fienga, A., Laskar, J., Morley, T., Manche, H., Kuchynka, P., Le Poncin-Lafitte, C., Budnik, F., Gastineau, M., Somenzi, L., 2009. INPOP08, a 4-D planetary ephemeris: From asteroid and time-scale computations to ESA Mars Express and Venus Express contributions. Astronomy and Astrophysics 507, 1675–1686.

Firestone, R.B., West, A., Kennett, J.P., Becker, L., Bunch, T.E., Revay, Z.S., Schultz, P.H., Belgya, T., Kennett, D.J., Erlandson, J.M., Dickenson, O.J., Goodyear, A.C., Harris, R.S., Howard, G.A., Kloosterman, J.B., Lechler, P., Mayewski, P.A., Montgomery, J., Poreda, R., Darrah, T., Que Hee, S.S., Smitha, A.R., Stich, A., Topping, W., Wittke, J.H., Wolbach, W.S., 2007. Evidence for an extraterrestrial impact 12,900 years ago that contributed to the megafaunal extinctions and the Younger Dryas cooling. Proceedings of the National Academy of Sciences, USA 104, 16016–16021.

Fleagle, J.G., Bown, T.M., Swisher III, C.C., Buckley, G., 1995. Age of the Pinturas and Santa Cruz Formations. Actas del VI Congreso Argentino de Paleontología y Bioestratigrafía (Trelew), 129–135.

Flower, B.P., Kennett, J.P., 1993. Middle Miocene ocean-climate transition: High-resolution oxygen and carbon isotopic records from Deep Sea Drilling Project Site 588A, southwest Pacific. Paleoceanography 8, 811–843.

Flynn, J.J., 2010. Early-Middle Cenozoic Andean mammal faunas: Integrated analyses of biochronology, geochronology, and paleoecology. American Geophysical Union, Fall Meeting 2010, Abstracts B44B–08.

Flynn, J.J., Swisher III, C.C., 1995. Cenozoic South American Land Mammal Ages: Correlation to global geochronologies. In: Berggren, W.A., Kent, D.V., Aubry, M.-P., Hardenbol, J. (Eds.),

Geochronology, Time Scales and Global Stratigraphic Correlation. SEPM Special Publication, 54, pp. 317–334.

Flynn, L.J., Wu, W.Y., Downs, W., 1997a. Dating vertebrate microfaunas in the late Neogene record of northern China. Palaeogeography, Palaeoclimatology, Palaeoecology 133, 227–242.

Flynn, J.J., Guerrero, J., Swisher III, C.C., 1997b. Geochronology of the Honda Group, Columbia. In: Kay, R.F., Madden, R.H., Cifelli, R.L., Flynn, J.J. (Eds.), Vertebrate Paleontology in the Neoptropics: The Miocene Fauna of La Venta, Columbia. Smithsonian Institution Press, Washington, D.C., pp. 44–60.

Flynn, L.J., Pilbeam, D., Jacobs, L.L., Barry, J.C., Behrensmeyer, A.K., Kappelman, J.W., 1990. The Siwaliks of Pakistan: Time and faunas in a Miocene terrestrial setting. Journal of Geology 98, 589–604.

Folco, L., D'Orazio, M., Tiepolo, M., Tonarini, S., Ottolini, L., Perchiazzi, N., Rochette, P., Glass, B.P., 2009. Transantarctic Mountain microtektites: Geochemical affinity with Australasian microtektites. Geochimica et Cosmochimica Acta 73, 3694–3722.

Foresi, L.M., Verducci, M., Baldassini, N., Lirer, F., Mazzei, R., Salvatorini, G., Ferraro, L., Da Prato, S. 2011. Integrated stratigraphy of St. Peter's Pool section (Malta): New age for the Upper Globigerina limestone Member and progress towards the Langhian GSSP. Stratigraphy 8, 125–143.

Fornaciari, E., Rio, D., 1996. Latest Oligocene to early Middle Miocene quantitative calcareous nannofossil biostratigraphy in the Mediterranean region. Micropaleontology 42, 1–36.

Fornaciari, E., Di Stefano, A., Rio, D., Negri, A., 1996. Middle Miocene quantitative calcareous nannofossil biostratigraphy in the Mediterranean region. Micropaleontology 42, 37–63.

Fornaciari, E., Iaccarino, S., Mazzei, R., Rio, D., Salvatorini, G., Bossio, A., Monteforti, B., 1997. Calcareous plankton biostratigraphy of the Langhian historical stratotype. In: Montanari, A., Odin, G.S., Coccioni, R. (Eds.), Miocene Stratigraphy: An Integrated Approach. Developments in Palaeontology and Stratigraphy, 15, pp. 89–96.

Garcés, M., Agustí, J., Cabrera, L., Parés, J.M., 1996. Magnetostratigraphy of the Vallesian (late Miocene) in the Vallès-Penedès Basin (northeast Spain). Earth and Planetary Science Letters 142, 381–396.

Garcés, M., Agustí, J., Parés, J.M., 1997. Late Pliocene continental magnetochronology from the Guadix–Baza Basin (Betic Ranges, Spain). Earth and Planetary Science Letters 146, 677–688.

Garcés, M., Krijgsman, W., Agustí, J., 1998. Chronology of the late Turolian of the Fortuna basin (SE Spain): Implications for the Messinian evolution of the eastern Betics. Earth and Planetary Science Letters 163, 69–81.

Garcés, M., Krijgsman, W., Peláez-Campomanes, P., Álvarez Sierra, M.A., Daams, R., 2003. *Hipparion* dispersal in Europe: Magnetostratigraphic constraints from the Daroca area (Spain). Coloquios de Paleontología, Volumen extraordinario 1, 171–178.

Garcia, T., Féraud, G., Falguéres, C., Lumley, H., de, Perrenoud, C., Lordkipanidze, D., 2010. Earliest human remains in Eurasia: New $^{40}Ar/^{39}Ar$ dating of the Dmanisi hominid-bearing levels, Georgia. Quaternary Geochronology 5, 443–451.

Gibbard, P.L., Head, M.J., 2010. The newly-ratified definition of the Quaternary System/Period and redefinition of the Pleistocene Series/Epoch, and comparison of proposals advanced prior to formal ratification. Episodes 33, 152–158.

Gibbard, P.L., Head, M.J., Walker, M.J.C., and the Subcommission on Quaternary Stratigraphy, 2010. Formal ratification of the Quaternary System/Period and the Pleistocene Series/Epoch with a base at 2.58 Ma. Journal of Quaternary Science 25, 96–102.

Glass, B.P., Koeberl, C., 2006. Australasian microtektites and associated impact ejecta in the South China Sea and the Middle Pleistocene supereruption of Toba. Meteoritics and Planetary Science 41, 305–326.

Gleason, J.D., Moore Jr., T.C., Johnson, T.M., Rea, D.K., Owen, R.M., Blum, J.D., Pares, J., Hovan, S.A., 2004. Age calibration of piston core EW9709-07 (equatorial central Pacific) using fish teeth Sr isotope stratigraphy. Palaeogeography, Palaeoclimatology, Palaeoecology 212, 355–366.

Goebel, T., Waters, M.R., O'Rourke, D.H., 2008. The late Pleistocene dispersal of modern humans in the Americas. Science 319, 1497–1502.

Gradstein, F.M., Ogg, J.G., Smith, A.G. (Eds.), 2004. A Geologic Time Scale 2004. Cambridge University Press, Cambridge, p. 589.

Guillou, H., Singer, B., Laj, C., Kissel, C., Scaillet, S., Jicha, B.R., 2004. On the age of the Laschamp geomagnetic excursion. Earth and Planetary Science Letters 227, 331–343.

Haq, B.U., Hardenbol, J., Vail, P.R., 1987. Chronology of fluctuating sea levels since the Triassic. Science 235, 1156–1167.

Hardenbol, J., Thierry, J., Farley, M.B., Jacquin, T., de Graciansky, P.-C., Vail, P.R., 1998. Mesozoic and Cenozoic sequence chronostratigraphic framework of European basins. In: de Graciansky, P.-C., Hardenbol, J., Jacquin, T., Vail, P.R. (Eds.), Mesozoic-Cenozoic Sequence Stratigraphy of European Basins. Society for Sedimentary Geology Special Publication, 60, pp. 3–13.

Harland, W.B., Amstrong, R., Cox, A.V., Craig, L., Smith, A., Smith, D., 1990. A Geologic Time Scale 1989. Cambridge University Press, Cambridge, p. 263.

Haslett, S.K., 2004. Late Neogene-Quaternary radiolarian biostratigraphy: A brief review. Journal of Micropalaeontology 23, 39–47.

Head, M.J., Pillans, B., Farquhar, S.A., 2008. The Early-Middle Pleistocene Transition: Characterization and proposed guide for the defining boundary. Episodes 31, 255–259.

Hedberg, H.D. (Ed.), 1976. International Stratigraphic Guide. Wiley, New York, p. 200.

Hendy, A.J.W., Kamp, P.J.J., 2004. Late Miocene to early Pliocene biofacies of Wanganui and Taranaki basins, New Zealand: Applications to paleoenvironmental and sequence stratigraphic analysis. New Zealand Journal of Geology and Geophysics 47, 769–785.

Hernández Fernández, M., Azanza, B., Álvarez Sierra, M.Á, 2004. Iberian Plio-Pleistocene biochronology: Micromammalian evidence for MNs and ELMAs calibration in southwestern Europe. Journal of Quaternary Science 19, 605–611.

Herrero-Bervera, E., Browne, E.J., Valet, J.P., Singer, B.S., Jicha, B.R., 2007. Cryptochron C2r.2r-1 recorded 2.51 Ma in the Koolau Volcano at Halawa, Oahu, Hawaii, USA: Paleomagnetic and $^{40}Ar/^{39}Ar$ evidence. Earth and Planetary Science Letters 254, 256–271.

Hilgen, F.J., 1991. Extension of the astronomically calibrated (polarity) time scale to the Miocene-Pliocene boundary. Earth and Planetary Science Letters 107, 349–368.

Hilgen, F.J., Krijgsman, W., Langereis, C.G., Lourens, L.J., Santarelli, A., Zachariasse, W.J., 1995. Extending the astronomical (polarity) time scale into the Miocene. Earth and Planetary Science Letters 136, 495–510.

Hilgen, F.J., Krijgsman, W., Wijbrans, J.R., 1997. Direct comparison of astronomical and $^{40}Ar/^{39}Ar$ ages of ash beds: Potential implications for the age of mineral dating standards. Geophysical Research Letters 24, 2043–2046.

Hilgen, F.J., Bissoli, L., Iaccarino, S., Krijgsman, W., Meijer, R., Negri, A., Villa, G., 2000a. Integrated stratigraphy and astrochronology of the

Messinian GSSP at Oued Akrech (Atlantic Morocco). Earth and Planetary Science Letters 182, 237–251.

Hilgen, F.J., Iaccarino, S., Krijgsman, W., Villa, G., Langereis, C.G., Zachariasse, W.J., 2000b. The Global Boundary Stratotype Section and Point (GSSP) of the Messinian Stage (uppermost Miocene). Episodes 23, 172–178.

Hilgen, F.J., Abdul Aziz, H., Krijgsman, W., Raffi, I., Turco, E., 2003. Integrated stratigraphy and astronomical tuning of the Serravallian and lower Tortonian at Monte dei Corvi (Middle-Upper Miocene, northern Italy). Palaeogeography, Palaeoclimatology, Palaeoecology 199, 229–264.

Hilgen, F.J., Abdul Aziz, H., Bice, D., Iaccarino, S., Krijgsman, W., Kuiper, K., Montanari, A., Raffi, I., Turco, E., Zachariasse, W.J., 2005. The Global Boundary Stratotype Section and Point (GSSP) of the Tortonian Stage (Upper Miocene) at Monte dei Corvi. Episodes 28, 6–17.

Hilgen, F.J., Brinkhuis, H., Zachariasse, W.J., 2006. Unit stratotypes for global stages: The Neogene perspective. Earth-Science Reviews 74, 113–125.

Hilgen, F.J., Aubry, M.-P., Berggren, W.A., Van Couvering, J.A., McGowran, B., Steininger, F.F., 2008. The case for the original Neogene. Newsletters in Stratigraphy 43, 23–32.

Hilgen, F.J., Abels, H.A., Iaccarino, S., Krijgsman, W., Raffi, I., Sprovieri, R., Turco, E., Zachariasse, W.J., 2009. The Global Stratotype Section and Point (GSSP) of the Serravallian Stage (Middle Miocene). Episodes 32, 152–169.

Hogg, A.G., Turney, C.S.M., Palmer, J.G., Fifield, L.K., Baillie, M.G.L., 2006. The potential for extending IntCal04 using OIS-3 New Zealand sub-fossil Kauri. PAGES News 14, 11–12.

Holbourn, A., Kuhnt, W., Schulz, M., Erlenkeuser, H., 2005. Impacts of orbital forcing and atmospheric carbon dioxide on Miocene ice-sheet expansion. Nature 438, 483–487.

Holbourn, A., Kuhnt, W., Schulz, M., Flores, J.-A., Andersen, N., 2007. Orbitally-paced climate evolution during the middle Miocene "Monterey" carbon-isotope excursion. Earth and Planetary Science Letters 261, 534–550.

Hordijk, K., De Bruijn, H., 2009. The succession of rodent faunas from the Mio/Pliocene lacustrine deposits of the Florina-Ptolemais-Servia Basin (Greece). Hellenic Journal of Geosciences 44, 21–103.

Hughen, K., Southon, J., Lehman, S., Bertrand, C., Turnbull, J., 2006. Marine-derived C-14 calibration and activity record for the past 50,000 years updated from the Cariaco Basin. Quaternary Science Reviews 25, 3216–3227.

Hungerbühler, D., Steinmann, M., Winkler, W., Seward, D., Egüez, A., Peterson, D.E., Helg, U., Hammer, C., 2002. Neogene stratigraphy and Andean geodynamics of southern Ecuador. Earth-Science Reviews 57, 75–124.

Hussain, S.T., van den Bergh, G.D., Steensma, K.J., de Visser, J.A., de Vos, J., Arif, M., Van Dam, J.A., Sondaar, P.Y., Malik, S.B., 1992. Biostratigraphy of the Plio-Pleistocene continental sediments (Upper Siwaliks) of the Mangla-Samwal Anticline, Azad Kashmir, Pakistan. Proceedings of the Koninklijke Nederlandse Academie van Wetenschappen B 95, 65–80.

Hüsing, S., 2008. Astrochronology of the Mediterranean Miocene: Linking palaeoenvironmental changes to gateway dynamics. PhD thesis, Utrecht University. Geologica Ultraiectina 295, 171.

Hüsing, S.K., Hilgen, F.J., Abdul Aziz, H., Krijgsman, W., 2007. Completing the Neogene geological time scale between 8.5 and 12.5 Ma. Earth and Planetary Science Letters 253, 340–358.

Hüsing, S.K., Kuiper, K.F., Link, W., Hilgen, F.J., Krijgsman, W., 2009. The upper Tortonian-lower Messinian at Monte dei Corvi (northern Apennines, Italy): Completing a Mediterranean reference section for the Tortonian Stage. Earth and Planetary Science Letters 282, 140–157.

Huybers, P., 2007. Glacial variability over the last two million years: An extended depth-derived age model, continuous obliquity pacing, and the Pleistocene progression. Quaternary Science Reviews 26, 37–55.

Iaccarino, S., Salvatorini, G., 1982. A framework of planktonic foraminiferal biostratigraphy for early Miocene to late Pliocene Mediterranean area. Paleontologia Stratigrafica ed Evoluzione 2, 115–125.

Iaccarino, S.M., Premoli Silva, I., Biolzi, M., Foresi, M.L., Lirer, F., Petrizzo, M.R., 2005. Practical manual of Oligocene to middle Miocene planktonic foraminifera. International School on Planktonic Foraminifera 4th Course, 14–18 February. University of Perugia Press, Perugia.

Iaccarino, S.M., Turco, E., Cascella, A., Gennari, R., Hilgen, F.J., Sagnotti, L., 2009. Integrated stratigraphy of La Vedova section (Conero Riviera, Italy), a potential candidate for the Langhian GSSP. 13th RCMNS conference Naples, Abstract volume, 20–21.

Iaccarino, S.M., Di Stefano, A., Foresi, L.M., Turco, E., Baldassini, N., Cascella, A., Da Prato, S., Ferraro, L., Gennari, R., Hilgen, F.J., Lirer, F., Maniscaldo, R., Mazzei, R., Riforgiato, F., Russo, B., Sagnotti, L., Salvatorini, G., Speranza, F., Verducci, M. 2011. High-resolution integrated stratigraphy of the Mediterranean early-middle Miocene: comparison with the Langhian historical stratotype and new perspectives for the GSSP. Stratigraphy 8, 199–215.

Irigoyen, M.V., Buchan, K.L., Brown, R.L., 2000. Magnetostratigraphy of Neogene Andean foreland-basin strata, lat. 33°S, Mendoza province, Argentina. Geological Society of America Bulletin 112, 803–816.

Joordens, J.C.A., Vonhof, H.B., Feibel, C.S., Lourens, L.J., Dupont-Nivet, G., van der Lubbe, J.H.J.L., Sier, M.J., Davies, G.R., Kroon, D., 2011. An astronomically-tuned climate framework for hominins in the Turkana Basin. Earth and Planetary Science Letters 307, 1–8.

Jordan, T.E., Tamm, V., Figueroa, G., Flemings, P.B., Richards, D., Tabbutt, K., T. Cheatham, T., 1996. Development of the Miocene Manantiales foreland basin, Principal Cordillera, San Juan, Argentina. Revista Geológica de Chile 23, 43–79.

Jöris, O., Street, M., 2008. At the end of the ^{14}C time scale–the Middle to Upper Paleolithic record of western Eurasia. Journal of Human Evolution 55, 782–802.

Kälin, D., Kempf, O., 2009. High-resolution stratigraphy from the continental record of the Middle Miocene Northern Alpine Foreland Basin of Switzerland. Neues Jahrbuch für Geologie und Paläontologie Abhandlungen 254, 177–235.

Kamp, P.J.J., Vonk, A.J., Bland, K.J., Hansen, R.J., Hendy, A.J.W., McIntyre, A.P., Ngatai, M., Cartwright, S.J., Hayton, S., Nelson, C.S., 2004. Neogene stratigraphic architecture and tectonic evolution of Wanganui, King Country, and eastern Taranaki basins, New Zealand. New Zealand Journal of Geology and Geophysics 47, 625–644.

Kano, A., Ferdelman, T.G., Williams, T., Henriet, J.P., Ishikawa, T., Kawagoe, N., Takashima, C., Kakizaki, Y., Abe, K., Sakai, S., Browing, E.L., Li, X., Andres, M.S., Bjerager, M., Cragg, B.A., De Mol, B., Dorschel, B., Foubert, A., Frank, T.D., Fuwa, Y., Gaillot, P., Gharib, J.J., Gregg, J.M., Huvenne, V.A.I., Léonide, P., Mangelsdorf, K., Monteys, X., Novosel, I., O'Donnell, R., Rüggeberg, A., Samarkin, V., Sasaki, K., Spivack, A.J., Tanaka, A., Titschack, J., van Rooij, D., Wheeler, A., 2007. Age constraints on the origin and growth history of a deep-water coral mound in the northeast Atlantic drilled during

Integrated Ocean Drilling Program Expedition 307. Geology 35, 1051–1054.

Kappelman, J., Duncan, A., Feseha, M., Lunkka, J.-P., Ekart, D., McDowell, F., Ryan, T., Swisher III, C.C., 2003. Chronology of the Sinap Formation. In: Fortelius, M., Kappelman, J., Sen, S., Bernor, R.L. (Eds.), The Geology and Paleontology of the Miocene Sinap Formation, Turkey. Columbia University Press, New York, pp. 41–68.

Karlstrom, K.E., Crow, R., Crossey, L.J., Coblentz, D., Van Wijk, J.W., 2008. Model for tectonically driven incision of the younger than 6 Ma Grand Canyon. Geology 36, 835–838.

Kars, R.H., Wallinga, J., Cohen, K.M., 2008. A new approach towards anomalous fading correction for feldspar IRSL dating – tests on samples in field saturation. Radiation Measurements 43, 786–790.

Kay, R.F., Madden, R.H., Cifelli, R.L., Flynn, J.J., 1997. Vertebrate Paleontology in the Neoptropics: The Miocene Fauna of La Venta, Columbia. Smithsonian Institution Press, Washington, D.C., p. 608.

Kay, R.F., Johnson, D., Meldrum, D.J., 1998. A new Pitheciin primate from the middle Miocene of Argentina. American Journal of Primatology 45, 317–336.

Keller, J., Ryan, W.B.F., Ninkovich, D., Altherr, R., 1978. Explosive volcanic activity in the Mediterranean over the past 200,000 yr as recorded in deep-sea sediments. Geological Society of America Bulletin 89, 591–604.

Kennett, D.J., Kennett, J.P., West, A., Mercer, C., Que Hee, S.S., Bement, L., Bunch, T.E., Sellers, M., Wolbach, W.S., 2009. Nanodiamonds in the Younger Dryas boundary sediment layer. Science 323, 94.

Knudsen, M.F., Holm, P.M., Abrahamsen, N., 2009. Paleomagnetic results from a reconnaissance study of Santiago (Cape Verde Islands): Identification of cryptochron C2r.2r-1. Physics of the Earth and Planetary Interiors 173, 279–289.

Koizumi, I., Sato, M., Matoba, Y., 2009. Age and significance of Miocene diatoms and diatomaceous sediments from northeast Japan. Palaeogeography, Palaeoclimatology, Palaeoecology 272, 85–98.

Kramarz, A.G., Vucetich, M.G., Carlini, A.A., Ciancio, M.R., Abello, M.A., Deschamps, C.M., Gelfo, J.N., 2010. A new mammal fauna at the top of the Gran Barranca sequence and its biochronological significance. In: Madden, R.H., Carlini, A.A., Vucetich, M.G., Kay, R.F. (Eds.), The Paleontology of Gran Barranca: Evolution and Environmental Change through the Middle Cenozoic of Patagonia. Cambridge University Press, Cambridge, pp. 264–277.

Krijgsman, W., Hilgen, F.J., Langereis, C.G., Santarelli, A., Zachariasse, W.J., 1995. Late Miocene magnetostratigraphy, biostratigraphy and cyclostratigraphy in the Mediterranean. Earth and Planetary Science Letters 136, 475–494.

Krijgsman, W., Garcés, M., Langereis, C.G., Daams, R., van Dam, J., van der Meulen, A.J., Agustí, J., Cabrera, L., 1996a. A new chronology for the middle to late Miocene continental record in Spain. Earth and Planetary Science Letters 142, 367–380.

Krijgsman, W., Duermeijer, C.E., Langereis, C.G., De Bruijn, H., Saraç, G., Andriessen, P.A.M., 1996b. Magnetic polarity stratigraphy of late Oligocene to middle Miocene mammal-bearing localities in Central Anatolia (Turkey). Newsletters on Stratigraphy 34, 13–29.

Krijgsman, W., Hilgen, F.J., Raffi, I., Sierro, F.J., Wilson, D.S., 1999. Chronology, causes and progression of the Messinian salinity crisis. Nature 400, 652–655.

Krijgsman, W., Fortuin, A.R., Hilgen, F.J., Sierro, F.J., 2001. Astrochronology for the Messinian Sorbas basin (SE Spain) and orbital (precessional) forcing for evaporite cyclicity. Sedimentary Geology 140, 43–60.

Krijgsman, W., Stoica, M., Vasiliev, I., Popov, V.V., 2010. Rise and fall of the Paratethys Sea during the Messinian salinity crisis. Earth and Planetary Science Letters 290, 183–191.

Kuhlmann, G., Langereis, C.G., Munsterman, D., van Leeuwen, R.-J., Verreussel, R., Meulenkamp, J.E., Wong, Th.E., 2006. Integrated chronostratigraphy of the Pliocene-Pleistocene interval and its relation to the regional stratigraphical stages in the southern North Sea region. Netherlands Journal of Geosciences 85, 19–35.

Kuiper, K., 2003. Direct intercalibration of radio-isotopic and astronomical time in the Mediterranean Neogene. PhD thesis, Utrecht University. Geologica Ultraiectina 235, p. 223.

Kuiper, K.F., Hilgen, F.J., Steenbrink, J., Wijbrans, J.R., 2004. $^{40}Ar/^{39}Ar$ ages of tephras intercalated in astronomically tuned Neogene sedimentary sequences in the eastern Mediterranean. Earth and Planetary Science Letters 22, 583–597.

Kuiper, K.F., Wijbrans, J.R., Hilgen, F.J., 2005. Radioisotopic dating of the Tortonian Global Stratotype Section and Point: Implications for intercalibration of $^{40}Ar/^{39}Ar$ and astronomical dating methods. Terra Nova 17, 385–398.

Kuiper, K.F., Deino, A., Hilgen, F.J., Krijgsman, W., Renne, P.R., Wijbrans, J.R., 2008. Synchronizing rock clocks of Earth history. Science 320, 500–504.

Kutterolf, S., Freundt, A., Schacht, U., Bürk, D., Harders, R., Mörz, T., Peréz, W., 2008a. Pacific offshore record of plinian arc volcanism in Central America: 3. Application to forearc geology. Geochemistry, Geophysics, Geosystems 9, Q02S03. doi: 10.1029/2007GC001826.

Kutterolf, S., Freundt, A., Schacht, U., Bürk, D., Harders, R., Mörz, T., Peréz, W., 2008b. Pacific offshore record of plinian arc volcanism in Central America: 1. Along arc correlations. Geochemistry, Geophysics, Geosystems 9, Q02S01. doi: 10.1029/2007 GC001631.

Lambert, F., Delmonte, B., Petit, J.R., Bigler, M., Kaufmann, P.R., Hutterli, M.A., Stocker, T.F., Ruth, U., Steffensen, J.P., Maggi, V., 2008. Dust-climate couplings over the past 800,000 years from the EPICA Dome C ice core. Nature 452, 616–619.

Laskar, J., Robutel, P., Joutel, F., Gastineau, M., Correia, A.C.M., Levrard, B., 2004. A long term numerical solution for the insolation quantities of the Earth. Astronomy & Astrophysics 428, 261–285.

Laskar, J., Fienga, A., Gastineau, M., Manche, H., 2011. La2010: A new orbital solution for the long term motion of the Earth. Astronomy & Astrophysics 532, A89.

Lawrence, K.T., Zhonghui, L., Herbert, T.D., 2006. Evolution of the eastern tropical Pacific through Plio-Pleistocene glaciation. Science 312, 79–83.

Lear, C.H., Elderfield, H., Wilson, P.A., 2000. Cenozoic deep-sea temperatures and global ice volumes from Mg/Ca in benthic formaniferal calcite. Science 287, 269–272.

Legendre, S., Bachelet, B., 1993. The numerical ages: A new method of datation applied to Paleogene mammalian localities from Southern France. Newsletters on Stratigraphy 29, 137–158.

Lepre, C.J., Kent, D.V., 2010. New magnetostratigraphy for the Olduvai Subchron in the Koobi Fora Formation, northwest Kenya, with implications for early Homo. Earth and Planetary Science Letters 290, 362–374.

Levèque, F., 1993. Correlating the Eocene-Oligocene mammalian biochronological scale from SW Europe with the marine magnetic anomaly sequence. Journal of the Geological Society 150, 661–664.

Lian, O.B., Roberts, R.G., 2006. Dating the Quaternary: Progress in luminescence dating of sediments. Quaternary Science Reviews 25, 2449–2468.

Liebrand, D., Lourens, L.J., Hodell, D.A., de Boer, B., van de Wal, R.S.W., Pälike, H., 2011. Antarctic ice sheet and oceanographic response to eccentricity forcing during the early Miocene. Climate of the Past 7, 869–880.

Lindsay, E., 2003. Chronostratigraphy, biochronology, datum events, Land Mammal Ages, stage of evolution, and appearance event ordination. Bulletin of the American Museum of Natural History 279, 212–230.

Lindsay, E., Mou, Y., Downs, W., Pedersson, J., Kelly, T., Henry, C., Trexler, J., 2002. Recognition of the Hemphillian/Blancan boundary in Nevada. Journal of Vertebrate Paleontology 22, 429–442.

Lirer, F., Di Stefano, A., Foresi, L.M., Turco, E., Iaccarino, S.M., Mazzei, R., Salvatorini, G., Baldassini, N., Da Prato, S., Verducci, M., Sprovieri, M., Pelosi, N., Vallefuoco, M., Angelino, A., Ferraro, L., 2009. Towards the Langhian astro-biochronology of Mediterranean deep marine records. 13th RCMNS conference Naples, Abstract volume, 20–21.

Lisiecki, L.E., Raymo, M.E., 2005. A Pliocene-Pleistocene stack of 57 globally distributed benthic $\delta^{18}O$ records. Paleoceanography 20, PA1003. doi: 10.1029/2004PA001071.

Litt, T., Gibbard, P.L., 2008. A proposed Global Stratotype Section and Point (GSSP) for the base of the Upper (Late) Pleistocene Subseries (Quaternary System/Period). Episodes 31, 260–261.

Loulergue, L., Schilt, A., Spahni, R., Masson-Delmotte, V., Blunier, T., Lemieux, B., Barnola, J.-M., Raynaud, D., Stocker, T.F., Chappellaz, J., 2008. Orbital and millennial-scale features of atmospheric CH_4 over the past 800,000 years. Nature 453, 383–386.

Lourens, L.J., 2004. Revised tuning of Ocean Drilling Program Site 964 and KC01B (Mediterranean) and implications for the $\delta^{18}O$, tephra, calcareous nannofossil, and geomagnetic reversal chronologies of the past 1.1 Myr. Paleoceanography 19, PA3010. doi: 10.1029/2003PA000997.

Lourens, L.J., 2008. On the Neogene-Quaternary debate. Episodes 31, 239–242.

Lourens, L.J., Hilgen, F.J., Laskar, J., Shackleton, N.J., Wilson, D., 2004. The Neogene Period. In: Gradstein, F.M., Ogg, J.G., Smith, A.G. (Eds.), A Geologic Time Scale 2004. Cambridge University Press, Cambridge, pp. 409–440.

Lourens, L.J., Becker, J., Bintanja, R., Hilgen, F.J., Tuenter, E., van de Wal, R.S.W., Ziegler, M., 2010. Linear and non-linear response of late Neogene glacial cycles to obliquity forcing and implications for the Milankovitch theory. Quaternary Science Reviews 29, 352–365.

Lowe, J.J., Blockley, S., Trincardi, F., Asioli, A., Cattaneo, A., Matthews, I.P., Pollard, M., Wulf, S., 2007. Age modelling of late Quaternary marine sequences in the Adriatic: Towards improved precision and accuracy using volcanic event stratigraphy. Continental Shelf Research 27, 560–582.

Lowe, J.J., Rasmussen, S.O., Björck, S., Hoek, W.Z., Steffensen, J.P., Walker, M.J.C., Yu, Z.C., 2008. Synchronisation of palaeoenvironmental events in the North Atlantic region during the Last Termination: A revised protocol recommended by the INTIMATE group. Quaternary Science Reviews 27, 6–17.

Lowe, D.J., Shane, P.A.R., Alloway, B.V., Newnham, R.M., 2008. Fingerprints and age models for widespread New Zealand tephra marker beds erupted since 30,000 years ago: A framework for NZ-INTIMATE. Quaternary Science Reviews 27, 95–126.

Lund, S., Stoner, J.S., Channell, J.E.T., Acton, G., 2006. A summary of Brunhes paleomagnetic field variability recorded in ocean drilling program cores. Physics of the Earth and Planetary Interiors 156, 194–204.

Lüthi, D., Le Floch, M., Bereiter, B., Blunier, T., Barnola, J.-M., Siegenthaler, U., Raynaud, D., Jouzel, J., Fischer, H., Kawamura, K., Stocker, T.F., 2008. High-resolution carbon dioxide concentration record 650,000–800,000 years before present. Nature 453, 379–382.

MacFadden, B.J., 1977. Magnetic polarity stratigraphy of the Chamita Formation stratotype (Mio-Pliocene) of north-central New Mexico. American Journal of Science 277, 769–800.

MacFadden, B.J., Hunt, R.M., 1998. Magnetic polarity stratigraphy and correlation of the Arikaree Group, Arikareean (late Oligocene-early Miocene) of northwestern Nebraska. In: Terry Jr., D.O., LaGarry, H.E., Hunt Jr., R.M. (Eds.), Depositional Environments, Lithostratigraphy, and Biostratigraphy of the White River and Arikaree groups (Late Eocene to Early Miocene, North America). Geological Society of America Special Paper, 325, pp. 143–165.

MacFadden, B.J., Swisher III, C.C., Opdyke, N.D., Woodburne, M.O., 1990a. Paleomagnetism, geochronology, and possible tectonic rotation of the middle Miocene Barstow Formation, Mojave Desert, southern California. Geological Society of America Bulletin 102, 478–493.

MacFadden, B.J., Anaya, F., Pérez, H., Naeser, C.W., Zeitler, P.K., Campbell Jr., K.E., 1990b. Late Cenozoic Paleomagnetism and chronology of Andean basins of Bolivia: Evidence for possible oroclinal bending. Journal of Geology 98, 541–555.

MacFadden, B.J., Anaya, F., Argollo, J., 1993. Magnetic polarity stratigraphy of Inchasi; a Pliocene mammal-bearing locality from the Bolivian Andes deposited just before the Great American Interchange. Earth and Planetary Science Letters 114, 229–241.

MacFadden, B.J., Anaya, F., Swisher III, C.C., 1995. Neogene paleomagnetism and oroclinal bending of the central Andes of Bolivia. Journal of Geophysical Research 100, 8153–8167.

Madden, R.H., Guerrero, J., Kay, R.F., Flynn, J.J., Swisher III, C.C., Walton, A.H., 1997. The Laventan Stage and Age. In: Kay, R.F., Madden, R.H., Cifelli, R.L., Flynn, J.J. (Eds.), Vertebrate Paleontology in the Neotropics: The Miocene Fauna of La Venta, Colombia. Smithsonian Institution Press, Washington, D.C., pp. 499–519.

Madden, R.H., Carlini, A.A., Vucetich, M.G., Kay, R.F. (Eds.). 2010. The Paleontology of Gran Barranca: Evolution and Environmental Change through the Middle Cenozoic of Patagonia. Cambridge University Press, New York, p. 458.

Mader, D., Cleaveland, L., Bice, D., Montanari, A., Koeberl, C., 2004. High-resolution cyclostratigraphic analysis of multiple climate proxies from a short Langhian pelagic succession in the Conero Riviera, Ancona (Italy). Palaeogeography, Palaeoclimatology, Palaeoecology 211, 325–344.

Magyar, I., Geary, D.H., Muller, P., 1999. Paleogeographic evolution of the Late Miocene Lake Pannon in Central Europe. Palaeogeography, Palaeoclimatology, Palaeoecology 147, 151–167.

Marshall, L.G., Butler, R.F., Drake, R.E., Curtis, G.H., 1982. Geochronology of type Uquian land mammal age, Argentina. Science 216, 986–989.

Marshall, L.G., Hoffstetter, R., Pascual, R., 1983. Mammals and stratigraphy: Geochronology of the continental mammal-bearing Tertiary of South America. Palaeovertebrata, Mémoire Extraordinaire 1983, 93.

Marshall, L.G., Drake, R.E., Curtis, G.H., Butler, R.F., Flanagan, K.M., Naeser, C.W., 1986. Geochronology of type Santacrucian (Middle Tertiary) land mammal age, Patagonia, Argentina. Journal of Geology 94, 449–457.

Martin, R.A., Pelaez-CampomanesHoney, J.G., Fox, D.L., Zakrzewski, R.J., Albright, L.B., Lindsay, E.H., Opdyke, N.D., Goodwin, H.T., 2008. Rodent community change at the Pliocene-Pleistocene transition in southwestern Kansas and identification of the Microtus immigration event on the central Great Plains. Palaeogeography, Palaeoclimatology, Palaeoecology 267, 196–207.

Martini, E., 1971. Standard Tertiary and Quaternary calcareous nannoplankton zonation. In: Farinacci, A. (Ed.), Proceedings of the II Planktonic Conference, Roma, 1970, vol. 2. Tecnoscienza, Roma, pp. 739–785.

Matano, F., Barbieri, M., Di Nocera, S., Torre, M., 2005. Stratigraphy and strontium geochemistry of Messinian evaporite-bearing successions of the southern Apennines foredeep, Italy: Implications for the Mediterranean "salinity crisis" and regional palaeogeography. Palaeogeography, Palaeoclimatology, Palaeoecology 217, 87–114.

Mayer-Eymar, K. (Ed.), 1868. Tableau Synchronistique des Terrains Tertiaires Supérieurs, fourth ed. Autographie H. Manz, Zürich.

Mazzei, R., Baldassini, N., Da Prato, S., Foresi, L.M., Lirer, F., Salvatorini, G., Verducci, M., Sprovieri, M., 2009. The St. Peter's Pool section in the Malta Island. Work in progress on the Langhian GSSP. 13th RCMNS conference Naples, Abstract volume, 22–23.

McDougall, I., Brown, F.H., Fleagle, J.G., 2008. Sapropels and the age of hominins Omo I and II, Kibish, Ethiopia. Journal of Human Evolution 55, 409–420.

McGowran, B., Berggren, B., Hilgen, F.J., Steininger, F.F., Aubry, M.-P., Lourens, L.J., van Couvering, J., 2009. Neogene and Quaternary coexisting in the geological time scale: The inclusive compromise. Earth-Science Reviews 96, 249–262.

McLean, N., Bowring, S.A., Bowring, J.F., Condon, D., Heizler, M., Parrish, R., Ramezani, J., Schoene, B., 2008. The EARTHTIME initiative: A review of accomplishments and promise. 33rd International Geological Congress, Oslo, Abstracts.

Megirian, D., Prideaux, G.J., Murray, P.F., Smit, N., 2010. An Australian land mammal age biochronological scheme. Paleobiology 36, 658–671.

Mein, P., 1976. Biozonation du Néogène méditerranéen à partir de mammifères. Proceedings of the VIth Congress of the Regional Committee on Mediterranean Neogene Stratigraphy (RCMNS), Bratislava 1975, 2. Table.

Mein, P., 1990. Updating of MN zones. In: Lindsay, E.H., Fahlbusch, V., Mein, P. (Eds.), European Neogene Mammal Chronology. Plenum Press, New York, pp. 73–90.

Mein, P., 1999. European Miocene Mammal Biochronology. In: Rössner, G.E., Heissig, K. (Eds.), The Miocene Land Mammals of Europe. Pfeil, München, pp. 25–38.

Mellars, P., 2006. A new radiocarbon revolution and the dispersal of modern humans in Eurasia. Nature 439, 931–935.

Miller, K.G., Wright, J.D., Fairbanks, R.G., 1991. Unlocking the Ice House: Oligocene-Miocene oxygen isotopes, eustacy and marginal erosion. Journal of Geophysical Research 96, 6829–6848.

Miller, K.G., Mountain, G.S., and the Leg 150 Shipboard Party Members of the New Jersey Coastal Plain Drilling Project, 1996. Drilling and dating New Jersey Oligocene-Miocene sequences: Ice volume, global sea level, and Exxon records. Science 271, 1092–1095.

Miller, K.G., Kominz, M.A., Browning, J.V., Wright, J.D., Mountain, G.S., Katz, M.E., Sugarman, P.J., Cramer, B.S., Christie-Blick, N., Pekar, S.F., 2005. The phanerozoic record of global sea-level change. Science 310, 1293–1298.

Montanari, A., Beaudoin, B., Chan, L.S., Coccioni, R., Deino, A., De Paolo, D.J., Emmanuel, L., Fornaciari, E., Kruge, M., Lundblad, S., Mozzato, C., Portier, E., Renard, M., Rio, D., Sandroni, P., Stankiewicz, A., 1997. Integrated stratigraphy of the Middle and Upper Miocene pelagic sequence of the Cònero Riviera (Marche region, Italy). In: Montanari, A., Odin, G.S., Coccioni, R. (Eds.), Miocene Stratigraphy: An Integrated Approach. Developments in Palaeontology and Stratigraphy, 15, pp. 409–450.

Morgan, L.E., Renne, P.R., 2008. Diachronous dawn of Africa's Middle Stone Age: New $^{40}Ar/^{39}Ar$ ages from the Ethiopian Rift. Geology 36, 967–970.

Mortimer, N., Gans, P.B., Palin, J.M., Meffre, S., Herzer, R.H., Skinner, D.N.B., 2010. Location and migration of Miocene-Quaternary volcanic arcs in the SW Pacific region. Journal of Volcanology and Geothermal Research 190, 1–10.

Mourik, A.A., Bijkerk, J.F., Cascella, A., Hüsing, S.K., Hilgen, F.J., Lourens, L.J., Turco, E., 2010. Astronomical tuning of the La Vedova High Cliff section (Ancona, Italy)–Implications of the Middle Miocene Climate Transition for Mediterranean sapropel formation. Earth and Planetary Science Letters 297, 249–261.

Mourik, A.A., Abels, H.A., Hilgen, F.J., Di Stefano, A., Zachariasse, W.J., 2011. Improved astronomical age constraints for the middle Miocene climate transition based on high-resolution stable isotope records from the central Mediterranean Maltese Islands. Paleoceanography 26, PA1210. doi: 10.1029/2010PA001981.

Muhs, D.R., Simmons, K.R., Schumann, R.R., Halley, R.B., 2011. Sea-level history of the past two interglacial periods: New evidence from U-series dating of reef corals from South Florida. Quaternary Science Reviews 30, 570–590.

Mulch, A., Uba, C.E., Strecker, M.R., Schoenberg, R., Chamberlain, C.P., 2010. Late Miocene climate variability and surface elevation in the Central Andes. Earth and Planetary Science Letters 290, 173–182.

Munsterman, D.K., Brinkhuis, H., 2004. A southern North Sea Miocene dinoflagellate cyst zonation. Netherlands Journal of Geosciences 83, 267–285.

Muttoni, G., Scardia, G., Kent, D.V., 2010. Human migration into Europe during the late early Pleistocene climate transition. Palaeogeography, Palaeoclimatology, Palaeoecology 296, 79–93.

Nabel, P.E., Cione, A., Tonni, E.P., 2000. Environmental changes in the Pampean area of Argentina at the Matuyama-Brunhes (C1r–C1n) Chrons boundary. Palaeogeography, Palaeoclimatology, Palaeoecology 162, 403–412.

Naish, T.R., Wehland, F., Wilson, G.S., Browne, G.H., Cook, R.A., Morgans, H.E.G., Rosenberg, M., King, P.R., Smale, D., Nelson, C.S., Kamp, P.J.J., Ricketts, B., 2005. An integrated sequence stratigraphic, palaeoenvironmental, and chronostratigraphic analysis of the Tangahoe Formation, Southern Taranaki coast, with implications for mid-Pliocene (c. 3.4–3.0 ma) glacio-eustatic sea-level changes. Journal of the Royal Society of New Zealand 35, 151–196.

Narcisi, B., Petit, J.R., Delmonte, B., 2010. Extended East Antarctic ice-core tephrostratigraphy. Quaternary Science Reviews 29, 21–27.

Nie, J., King, J.W., Fang, X., 2008. Correlation between the magnetic susceptibility record of the Chinese aeolian sequences and the marine benthic oxygen isotope record. Geochemistry, Geophysics, Geosystems 9, Q12026. doi: 10.1029/2008GC002243.

Okada, H., Bukry, D., 1980. Supplementary modification and introduction of code numbers to the low-latitude coccolith biostratigraphic zonation (Bukry, 1973; 1975). Marine Micropaleontology 51, 321–325.

Oms, O., Dinarès-Turell, J., Agustí, J., Parés, J.M., 1999. Refining the European Mammalian Biochronology: Magnetic polarity record of the Plio-Pleistocene Zújar section (Guadix-Baza Basin, SE Spain). Quaternary Research 51, 94–103.

Opdyke, N., Mein, P., Lindsay, E., Perez-Gonzales, A., Moissenet, E., Norton, V.L., 1997. Continental deposits, magnetostratigraphy and vertebrate paleontology, late Neogene of Eastern Spain. Palaeogeography, Palaeoclimatology, Palaeoecology 133, 129–148.

Orgeira, M.J., 1990. Paleomagnetism of late Cenozoic fossiliferous sediments from Barranca de los Lobos (Buenos Aires Province, Argentina). The magnetic age of the South American land-mammal ages. Physics of the Earth and Planetary Interiors 64, 121–132.

Osinski, G.R., Schwarcz, H.P., Smith, J.R., Kleindienst, M.R., Haldemann, A.F.C., Churcher, C.S., 2007. Evidence for a ~200–100 ka meteorite impact in the Western Desert of Egypt. Earth and Planetary Science Letters 253, 378–388.

Pagnac, D.C., 2009. Revised large mammal biostratigraphy and biochronology of the Barstow Formation (middle Miocene), California. PaleoBios 29, 48–49.

Pálfy, J., Mundil, R., Renne, P.R., Bernor, R.L., Kordos, L., Gasparik, M., 2007. U-Pb and ^{40}Ar/^{39}Ar dating of the Miocene fossil track site at Ipolytarnóc (Hungary) and its implications. Earth and Planetary Science Letters 258, 160–174.

Pälike, H., Norris, R.D., Herrle, J.O., Wilson, P.A., Coxall, H.K., Lear, C.H., Shackleton, N.J., Tripati, A.K., Wade, B.S., 2006a. The heartbeat of the Oligocene climate system. Science 314, 1894–1898.

Pälike, H., Frazier, J., Zachos, J.C., 2006b. Extended orbitally forced palaeoclimatic records from the equatorial Atlantic Ceara Rise. Quaternary Science Reviews 25, 3138–3149.

Pälike, H., Lyle, M.W., Ahagon, N., Raffi, I., Gamage, K., Zarikian, C.A., 2008. Pacific equatorial age transect. Integrated Ocean Drilling Program Scientific Prospectus 320/321. doi: 10.2204/iodp.sp.320321.2008.

Papp, A., Rögl, F., Seneš, J., 1973. M2–Ottnangien. Die Innviertler, Salgótarjáner, Bántapusztaer Schichtengruppe und die Rzehakia Formation. Chronostratigraphie und Neostratotypen, Miozän der Zentralen Paratethys 3, 1–841.

Papp, A., Marinescu, F., Seneš, J., 1974. M5–Sarmatien (sensu E. SUESS, 1866). Die Sarmatische Schichtengruppe und ihr Stratotypus. Chronostratigraphie und Neostratotypen, Miozän der Zentralen Paratethys 4, 1–707.

Papp, A., Cicha, I., Seneš, J., Steininger, F.F., 1978. M4-Badenien (Moravien, Wielicien, Kosovien). Chronostratigraphie und Neostratotypen, Miozän der Zentralen Paratethys 6, 1–594.

Papp, A., Jámbor, A., Steininger, F.F., 1985. M6-Pannonien (Slavonien und Serbien). Chronostratigraphie und Neostratotypen, Miozän der Zentralen Paratethys 7, 1–636.

Paquay, F.S., Goderis, S., Ravizza, G., Vanhaeck, F., Boyd, M., Surovell, T.A., Holliday, V.T., Haynes Jr., C.V., Claeys, P., 2009. Absence of geochemical evidence for an impact event at the Bølling-Allerød/Younger Dryas transition. Proceedings of the National Academy of Sciences, USA 106, 21505–21510.

Pareto, L., 1865. Note sur les subdivisions que l'on pourrait établir dans les terrains tertiaires de l'Apennin septentrionalis. Bulletin de la Société Geologique de France, série V 2 (22), 210–217.

Parrenin, F., Barnola, J.-M., Beer, J., Blunier, T., Castellano, E., Chappellaz, J., Dreyfus, G., Fischer, H., Fujita, S., Jouzel, J., Kawamura, K., Lemieux-Dudon, B., Loulergue, L., Masson-Delmotte, V., Narcisi, B., Petit, J.-R., Raisbeck, G., Raynaud, D., Ruth, U., Schwander, J., Severi, M., Spahni, R., Steffensen, J.P., Svensson, A., Udisti, R., Waelbroeck, C., Wolff, E., 2007. The EDC3 chronology for the EPICA Dome C ice core. Climate of the Past 3, 485–497.

Pascual, R., Ortega Hinojosa, E.J., Gondar, D., Tonni, E.P., 1965. Las edades del Cenozoico mamalífero de la Argentina, con especial atención a aquéllas del territorio bonarense. Comisión Investigaciones Científicas de la Provincia de Buenos Aires, Anales 6, 165–193.

Pettitt, P., 2008. Art and the Middle-to-Upper Paleolithic transition in Europe: Comments on the archaeological arguments for an early Upper Paleolithic antiquity of the Grotte Chauvet art. Journal of Human Evolution 55, 908–917.

Pettitt, P., Pike, A., 2007. Dating European Palaeolithic cave art: Progress, prospects, problems. Journal of Archaeological Method and Theory 14, 27–47.

Pickering, R., Kramers, J.D., 2010. Re-appraisal of the stratigraphy and determination of new U-Pb dates for the Sterkfontein hominin site, South Africa. Journal of Human Evolution 59, 70–86.

Pickford, M., 1981. Preliminary Miocene mammalian biostratigraphy for Western Kenya. Journal of Human Evolution 10, 73–97.

Pigati, J.S., Quade, J., Wilson, J., Jull, A.J.T., Lifton, N.A., 2007. Development of low-background vacuum extraction and graphitization systems for ^{14}C dating of old (40–60 ka) samples. Quaternary International 166, 4–14.

Pillans, B., Naish, T., 2004. Defining the Quaternary. Quaternary Science Reviews 23, 2271–2282.

Piller, W.E., Harzhauser, M., Mandic, O., 2007. Miocene Central Paratethys stratigraphy – current status and future directions. Stratigraphy 4, 151–168.

Poignant, A., Pujol, C., Ringeade, M., Londeix, L., 1997a. The Aquitanian historical stratotype. In: Montanari, A., Odin, G.S., Coccioni, R. (Eds.), Miocene Stratigraphy: An Integrated Approach. Developments in Paleontology and Stratigraphy, vol. 15. Elsevier, Amsterdam, pp. 10–16.

Poignant, A., Pujol, C., Ringeade, M., Londeix, L., 1997b. The Burdigalian historical stratotype. In: Montanari, A., Odin, G.S., Coccioni, R. (Eds.), Miocene Stratigraphy: An Integrated Approach. Developments in Paleontology and Stratigraphy, vol. 15. Elsevier, Amsterdam, pp. 17–24.

Popov, S.V., Shcherba, I.G., Ilyina, L.B., Nevesskaya, L.A., Paramonova, N.P., Khondkarian, S.O., Magyar, I., 2006. Late Miocene to Pliocene palaeogeography of the Paratethys and its relation to the Mediterranean. Palaeogeography, Palaeoclimatology, Palaeoecology 238, 91–106.

Prasad, M.S., Mahale, V.P., Kodagali, V.N., 2007. New sites of Australasian microtektites in the central Indian Ocean: Implications for the location and size of source crater. Journal of Geophysical Research 112, E06007. doi: 10.1029/2006JE002857.

Prothero, D.R., Tedford, R.H., 2000. Magnetic stratigraphy of the type Montediablan State (late Miocene), Black Hawk Ranch, Contra Costa County, California: Implications for regional correlations. PaleoBios 20, 1–10.

Prothero, D.R., Donohoo, L.L., 2001. Magnetic stratigraphy of the lower Miocene (early Hemingfordian) Sespe-Vaqueros Formations, Orange County, California. In: Prothero, D.R. (Ed.), Magnetic Stratigraphy of the Pacific Coast Cenozoic. Society for Sedimentary Geology, Pacific Section, Publication, 91, pp. 242–253.

Qiu, Z., Huang, W., Guo, Z., 1979. Hyaenidae of Qingyang (K'ingyang) Hipparion fauna. Vertebrata Palasiatica 17, 200–221 [In Chinese with English summary].

Qiu, Z., Li, C., 2003. Rodents from the Chinese Neogene: Biogeographic relationships with Europe and North America. Bulletin of the American Museum of Natural History 279, 586–602.

Qiu, Z.X., Wu, W.Y., Qiu, Z.D., 1999. Miocene mammal faunal sequence of China: Palaeozoogeography and Eurasian relationships. In: Rössner, G.E.,

Heissig, K. (Eds.), The Miocene Land Mammals of Europe. Pfeil Verlag, Munich, pp. 443–455.

Raffi, I., 1999. Precision and accuracy of nannofossil biostratigraphic correlation. Philosophical Transactions of the Royal Society, Series A 357, 1975–1993.

Raffi, I., Mozzato, C.A., Fornaciari, E., Hilgen, F.J., Rio, D., 2003. Late Miocene calcareous nannofossil biostratigraphy and astrobiochronology for the Mediterranean region. Micropaleontology 49, 1–26.

Raffi, I., Backman, J., Fornaciari, E., Pälike, H., Rio, D., Lourens, L.J., Hilgen, F.J., 2006. A review of calcareous nannofossil astrobiochronology encompassing the past 25 million years. Quaternary Science Reviews 25, 3113–3137.

Rasmussen, S.O., Andersen, K.K., Svensson, A.M., Steffensen, J.P., Vinther, B.M., Clausen, H.B., Siggaard-Andersen, M.-L., Johnsen, S.J., Larsen, L.B., Dahl-Jensen, D., Bigler, M., Röthlisberger, R., Fischer, H., Goto-Azuma, K., Hansson, M.E., Ruth, U., 2006. A new Greenland ice core chronology for the last glacial termination. Journal of Geophysical Research 111, D06102. doi: 10.1029/2005JD006079.

Rasmussen, S.O., Seierstad, I.K., Andersen, K.K., Bigler, M., Dahl-Jensen, D., Johnsen, S.J., 2008. Synchronization of the NGRIP, GRIP, and GISP2 ice cores across MIS 2 and palaeoclimatic implications. Quaternary Science Reviews 27, 18–28.

Ré, G.H., Bellosi, E.S., Heizler, M., Vilas, J.F., Madden, R.H., Carlini, A.A., Vucetich, M.G., Kay, R.F., 2010a. A geochronology for the Sarmiento Formation at Gran Barranca; Part II. Systematic Palaeontology. In: Madden, R.H., Carlini, A.A., Vucetich, M.G., Kay, R.F. (Eds.), The Paleontology of Gran Barranca: Evolution and Environmental Change through the Middle Cenozoic of Patagonia. Cambridge University Press, Cambridge, pp. 46–58.

Ré, G.H., Geuna, S.E., Vilas, J.F., 2010b. Paleomagnetism and magnetostratigraphy of the Sarmiento Formation (Eocene-Miocene) at Gran Barranca, Chubut, Argentina. In: Madden, R.H., Carlini, A.A., Vucetich, M.G., Kay, R.F. (Eds.), The Paleontology of Gran Barranca: Evolution and Environmental Change through the Middle Cenozoic of Patagonia. Cambridge University Press, Cambridge, pp. 32–45.

Reimer, P.J., Baillie, M.G.L., Bard, E., Bayliss, A., Beck, J.W., Blackwell, P.G., Ramsey, C.B., Buck, C.E., Burr, G.S., Edwards, R.L., Friedrich, M., Grootes, P.M., Guilderson, T.P., Hajdas, I., Heaton, T.J., Hogg, A.G., Hughen, K.A., Kaiser, K.F., Kromer, B., McCormac, F.G., Manning, S.W., Reimer, R.W., Richards, D.A., Southon, J.R., Talamo, S., Turney, C.S.M., van der Plicht, J., Weyhenmeyer, C.E., 2009. IntCal09 and Marine09 radiocarbon age calibration curves, 0–50,000 years CAL BP. Radiocarbon 51, 1111–1150.

Remane, J., Bassett, M.G., Cowie, J.W., Gohrbandt, K.H., Lane, H.R., Michelson, O., Naiwen, W., 1996. Revised guidelines for the establishment of global chronostratigraphic standards by the International Commission on Stratigraphy (ICS). Episodes 19, 77–81.

Renne, P.R., Deino, A.L., Walter, R.C., Turrin, B.D., Swisher III, C.C., Becker, T.A., Curtis, G.H., Sharp, W.D., Jaouni, A.R., 1994. Intercalibration of astronomical and radioisotopic time. Geology 22, 783–786.

Renne, P.R., Swisher III, C.C., Deino, A.L., Karner, D.B., Owens, T.L., DePaolo, D.J., 1998. Intercalibration of standards, absolute ages and uncertainties in ^{40}Ar/^{39}Ar dating. Chemical Geology 145, 117–152.

Renne, P.R., Schwarcz, H.P., Kleindienst, M.R., Osinski, G.R., Donovan, J.J., 2010a. Age of the Dakhleh impact event and implications for Middle Stone Age archeology in the Western Desert of Egypt. Earth and Planetary Science Letters 291, 201–206.

Renne, P.R., Mundil, R., Balco, G., Min, K., Ludwig, K.R., 2010b. Joint determination of ^{40}K decay constants and ^{40}Ar*/^{40}K for the Fish Canyon sanidine standard, and improved accuracy for ^{40}Ar/^{39}Ar geochronology. Geochimica et Cosmochimica Acta 74, 5349–5367.

Repenning, C.A., 1967. Palearctic-Nearctic mammalian dispersal in the late Cenozoic. In: Hopkins, D.M. (Ed.), The Bering Land Bridge. Stanford University Press, Stanford, pp. 288–311.

Rio, D., Cita, M.B., Iaccarino, S., Gelati, R., Gnaccolini, M., 1997. Langhian, Serravallian, and Tortonian historical stratotypes. In: Montanari, A., Odin, G.S., Coccioni, R. (Eds.), Miocene Stratigraphy: An Integrated Approach. Developments in Palaeontology and Stratigraphy, 15, pp. 57–87.

Rio, D., Sprovieri, R., Castradori, D., Di Stefano, E., 1998. The Gelasian Stage (Upper Pliocene): A new unit of the global standard chronostratigraphic scale. Episodes 21, 82–87.

Rivera, T.A., Storey, M., Zeeden, C., Hilgen, F.J., Kuiper, K. 2011. A refined astronomically calibrated ^{40}Ar/^{39}Ar age for Fish Canyon sanidine. Earth and Planetary Science Letters 311, 420–426.

Rögl, F., 1996. Stratigraphic correlation of the Paratethys Oligocene and Miocene. Mitteilungen der Gesellschaft der Geologie-und Bergbaustudenten Österreichs 41, 65–73.

Rögl, F., 1998. Palaeogeographic considerations for Mediterranean and Paratethys seaways (Oligocene to Miocene). Annalen des Naturhistorischen Museums in Wien 99, 279–310.

Rögl, F., Daxner-Höck, G., 1996. Late Miocene Paratethys correlations. In: Bernor, R.L., Fahlbusch, V., Mittmann, H.-W. (Eds.), The Evolution of Western Eurasian Neogene Mammal Faunas. Columbia University Press, New York, pp. 47–55.

Rögl, F., Ćorić, S., Harzhauser, M., Jimenez-Moreno, G., Kroh, A., Schultz, O., Wessely, G., Zorn, I., 2008. The Middle Miocene Badenian stratotype at Baden-Sooss (lower Austria). Geologica Carpathica 59, 367–374.

Rohling, E.J., Liu, Q.S., Roberts, A.P., Stanford, J.D., Rasmussen, S.O., Langen, P.L., Siddall, M., 2009. Controls on the East Asian monsoon during the last glacial cycle, based on comparison between Hulu Cave and polar ice-core records. Quaternary Science Reviews 28, 3291–3302.

Roveri, M., Manzi, V., 2006. The Messinian salinity crisis: Looking for a new paradigm? Palaeogeography, Palaeoclimatology, Palaeoecology 238, 386–398.

Sagnotti, L., Cascella, A., Ciaranfi, N., Macrì, P., Maiorano, P., Marino, M., Taddeucci, J., 2010. Rock magnetism and palaeomagnetism of the Montalbano Jonico section (Italy): Evidence for late diagenetic growth of greigite and implications for magnetostratigraphy. Geophysical Journal International 180, 1049–1066.

Santacroce, R., Cioni, R., Marianelli, P., Sbrana, A., Sulpizio, R., Zanchetta, G., Donahue, D.J., Joron, J.L., 2008. Age and whole rock-glass compositions of proximal pyroclastics from the major explosive eruptions of Somma-Vesuvius: A review as a tool for distal tephrostratigraphy. Journal of Volcanology and Geothermal Research 177, 1–18.

Saul, G., Naish, T.R., Abbott, S.T., Carter, R.M., 1999. Sedimentary cyclicity in the marine Pliocene-Pleistocene of the Wanganui Basin (New Zealand): Sequence stratigraphic motifs characteristic of the past 2.5 m.y. Geological Society of America Bulletin 111, 524–537.

Savage, D.E., 1962. Cenozoic geochronology of the fossil mammals of the Western Hemisphere. Revista Museo Argentina Ciencias Naturale "Bernardino Rivadavia," Ciencia Zoologica 8, 53–67.

Schaltegger, U., Brack, P., Ovtcharova, M., Peytcheva, I., Schoene, B., Stracke, A., Marocchi, M., Bargossi, G.M., 2009. Zircon and titanite recording 1.5 million years of magma accretion, crystallization and initial cooling in a composite pluton (southern Adamello batholith, northern Italy). Earth and Planetary Science Letters 286, 208–218.

Schlunegger, F., Burbank, D., Matter, A., Engesser, B., Mödden, C., 1996. Magnetostratigraphic calibration of the Oligocene to Middle Miocene (30–15 Ma) mammal biozones and depositional sequences of the Swiss Molasse Basin. Eclogae Geologicae Helveticae 89, 753–788.

Schmitz, B., Pujalte, V., Molina, E., *et al.*, 2011. The Global Stratotype Sections and Points for the bases of the Selandian (Middle Paleocene) and Thanetian (Upper Paleocene) stages at Zumaia, Spain. Episodes 34, 220–243.

Scholz, D., Mangini, A., Felis, T., 2004. U-series dating of diagenetically altered fossil reef corals. Earth and Planetary Science Letters 218, 163–178.

Schreilechner, M.G., Sachsenhofer, R.F., 2007. High resolution sequence stratigraphy in the eastern Styrian Basin (Miocene, Austria). Austrian Journal of Earth Sciences 100, 164–184.

Schüpbach, S., Federer, U., Bigler, M., Fischer, H., Stocker, T.F., 2011. A refined TALDICE-1a age scale from 55 to 112 ka before present for the Talos Dome ice core based on high-resolution methane measurements. Climate of the Past 7, 1175–1193.

Scott, G.R., Gibert, L., Gibert, J., 2007. Magnetostratigraphy of the Orce region (Baza basin), SE Spain: New chronologies for early Pleistocene faunas and hominid occupation sites. Quaternary Science Reviews 26, 415–435.

Scott, A.C., Pinter, N., Collinson, M.E., Hardiman, M., Anderson, R.S., Brain, A.P.R., Smith, S.Y., Marone, F., Stampanoni, M., 2010. Fungus, not comet or catastrophe, accounts for carbonaceous spherules in the Younger Dryas "impact layer". Geophysical Research Letters 37, L14302. doi: 10.1029/2010GL043345.

Sen, S., 1996. Present state of magnetostratigraphic studies in the continental Neogene of Europe and Anatolia. In: Bernor, R.L., Fahlbusch, V., Mittmann, H.-W. (Eds.), The Evolution of Western Eurasian Neogene Mammal Faunas. Columbia University Press, New York, pp. 56–63.

Sen, S., Valet, J.-P., Ioakim, C., 1986. Magnetostratigraphy of the Neogene of Kastellios Hill in Crete. Palaeogeography, Palaeoclimatology, Palaeoecology 53, 321–334.

Shackleton, N.J., Crowhurst, S., 1997. Sediment fluxes based on an orbitally tuned time scale 5 Ma to 14 Ma, Site 926. Proceedings of the Ocean Drilling Program, Scientific Results 154, 69–82.

Shackleton, N.J., Crowhurst, S.J., Weedon, G.P., Laskar, J., 1999. Astronomical calibration of Oligocene-Miocene time. Philosophical Transactions of the Royal Society, Series A 357, 1907–1929.

Shackleton, N.J., Hall, M.A., Raffi, I., Tauxe, L., Zachos, J.C., 2000. Astronomical calibration age for the Oligocene-Miocene boundary. Geology 28, 447–450.

Shackleton, N.J., Röhl, U., Raffi, I., 2001. Astronomical calibration in the Middle Miocene. EOS 82 (20) Abstract OS32A-03.

Shackleton, N.J., Fairbanks, R.G., Chiu, T.-C., Parrenin, F., 2004. Absolute calibration of the Greenland time scale: Implications for Antarctic time scales and for $\delta^{14}C$. Quaternary Science Reviews 23, 1513–1522.

Shevenell, A.E., Kennett, J.P., Lea, D.W., 2004. Middle Miocene Southern Ocean cooling and Antarctic cryosphere expansion. Science 305, 1766–1770.

Smellie, J.L., McArthur, J.M., McIntosh, W.C., Esser, R., 2006. Late Neogene interglacial events in the James Ross Island region, northern Antarctic Peninsula, dated by Ar/Ar and Sr-isotope stratigraphy. Palaeogeography, Palaeoclimatology, Palaeoecology 242, 169–187.

Smith, J.R., Kleindienst, M.R., Schwarcz, H.P., Churcher, C.S., Kieniewicz, J.M., Osinski, G.R., Haldemann, A.F.C., 2009. Potential consequences of a Mid-Pleistocene impact event for the Middle Stone Age occupants of Dakhleh Oasis, Western Desert, Egypt. Quaternary International 195, 138–149.

Song, C., Fang, X., Gao, J., Nie, J., Yan, M., Xu, X., Sun, D., 2003. Magnetostratigraphy of Late Cenozoic fossil mammals in the northeastern margin of the Tibetan Plateau. Chinese Science Bulletin 48, 188–193.

Sotnikova, M.V., Dodonov, A.E., Pen'kov, A.V., 1997. Upper Cenozoic biomagnetic stratigraphy of Central Asian mammalian localities. Palaeogeography, Palaeoclimatology, Palaeoecology 133, 243–258.

Sprovieri, R., Bonomo, S., Caruso, A., Di Stefano, A., Di Stefano, E., Foresi, L.M., Iaccarino, S.M., Lirer, F., Mazzei, R., Salvatorini, G., 2002. An integrated calcareous plankton biostratigraphic scheme and biochronology for the Mediterranean Middle Miocene. In: Iaccarino, S. (Ed.), Integrated Stratigraphy and Paleoceanography of the Mediterranean Middle Miocene. Rivista Italiana di Paleontologia e Stratigrafia, 108, pp. 337–353.

Staff, R.A., Bronk Ramsey, C., Nakagawa, T., 2010. A re-analysis of the Lake Suigetsu terrestrial radiocarbon calibration dataset. Nuclear Instruments and Methods in Physics Research, Section B: Beam Interactions with Materials and Atoms 268, 960–965.

Steenbrink, J., Hilgen, F.J., Krijgsman, W., Wijbrans, J.R., Meulenkamp, J.E., 2006. Late Miocene to early Pliocene depositional history of the intramontane Florina-Ptolemais-Servia Basin, NW Greece: Interplay between orbital forcing and tectonics. Palaeogeography, Palaeoclimatology, Palaeoecology 238, 151–178.

Steiniger, F.F., 1999. Chronostratigraphy, geochronology, and biochronology of the Miocene "European Land Mammal Mega-Zones (ELMMZ) and the Miocene Mammal-Zones (MN-Zones)." In: Rössner, G.E., Heissig, K. (Eds.), The Miocene Land Mammals of Europe. Pfeil Verlag, München, pp. 9–24.

Steininger, F.F., Seneš, J., 1971. M1 (Eggenburgien). Die Eggenburger Schichten gruppe und ihr Stratotypus. Chronostratigraphie und Neostratotypen, Miozän der Zentralen Paratethys 2, 1–827.

Steininger, F.F., Aubry, M.-P., Berggren, W.A., Biolzi, M., Borsetti, A.M., Cartlidge, J.E., Cati, F., Corfield, R., Gelati, R., Iaccarino, S., Napoleone, C., Ottner, F., Rogl, F., Roetzel, R., Spezzaferri, S., Tateo, F., Villa, G., Zvenboom, D., 1997. The Global Stratotype Section and Point (GSSP) for the base of the Neogene. Episodes 20, 23–28.

Stenni, B., Masson-Delmotte, V., Selmo, E., Oerter, H., Meyer, H., Röthlisberger, R., Jouzel, J., Cattani, O., Falourd, S., Fischer, H., Hoffmann, G., Iacumin, P., Johnsen, S.J., Minster, B., Udisti, R., 2010. The deuterium excess records of EPICA Dome C and Dronning Maud Land ice cores (East Antarctica). Quaternary Science Reviews 29, 146–159.

Stevanović, P., Nevesskaja, L.A., Marinescu, F., Sokac, A., Jámbor, A., 1990. Pl1 – Pontien (sensu F. Le Play, N. P. Barbot De Marny, N. I. Andrusov). Chronostratigraphie und Neostratotypen. Neogen der Westlichen (»Zentrale«) Paratethys 8, 1–952.

Stirling, C.H., Andersen, M.B., 2009. Uranium-series dating of fossil coral reefs: Extending the sea-level record beyond the last glacial cycle. Earth and Planetary Science Letters 284, 269–283.

Stirton, R.A., Tedford, R.H., Woodburne, M.O., 1968. Australian Tertiary deposits containing terrestrial mammals. University of California Publications in Geological Sciences 77, 1–30.

Strauss, P., Harzhauser, M., Hinsch, R., Wagreich, M., 2006. Sequence stratigraphy in a classic pull-apart basin (Neogene, Vienna Basin). A 3D seismic based integrated approach. Geologica Carpathica 57, 185–197.

Svensson, A., Andersen, K.K., Bigler, M., Clausen, H.B., Dahl-Jensen, D., Davies, S.M., Johnsen, S.J., Muscheler, R., Rasmussen, S.O., R öthlisberger, R., Peder Steffensen, J., Vinther, B.M., 2006. The Greenland Ice Core Chronology 2005, 15–42 ka. Part 2: Comparison to other records. Quaternary Science Reviews 25, 3258–3267.

Svensson, A., Andersen, K.K., Bigler, M., Clausen, H.B., Dahl-Jensen, D., Davies, S.M., Johnsen, S.J., Muscheler, R., Parrenin, F., Rasmussen, S.O., Röthlisberger, R., Seierstad, I., Steffensen, J.P., Vinther, B.M., 2008. A 60 000 year Greenland stratigraphic ice core chronology. Climate of the Past 4, 47–57.

Swisher III, C.C., 1996. $^{40}Ar/^{39}Ar$ dating of key European and West Asian Late Miocene age vertebrate localities and their contribution to Mammal Neogene (MN) Correlations. In: Bernor, R., Fahlbusch, V., Rietschel, S., Mittmann, H.-W. (Eds.), The Evolution of Western Eurasian Mammal Faunas. Columbia University Press, New York, pp. 64–77.

Tedford, R.H., Barghoorn, S., 1993. Neogene stratigraphy and mammalian biochronology of the Española Basin, northern New Mexico. New Mexico Museum of Natural History and Science Bulletin 2, 159–168.

Tedford, R.H., Galusha, T., Skinner, M.F., Taylor, B.E., Fields, R.W., Macdonald, J.R., Rensberger, J.M., Webb, S.D., Whistler, D.P., 1987. Faunal succession and biochronology of the Arikareean through Hemphillian interval (late Oligocene through earliest Pliocene epochs) in North America. In: Woodburne, M.O. (Ed.), Cenozoic Mammals of North America: Geochronology and Biostratigraphy. University of California Press, Berkeley, pp. 153–210.

Tedford, R.H., Allbright, L.B., Barnosky, A.D., Ferrusquia-Villafranca, I., Hunt Jr., R.M., Storer, J.E., Swisher III, C.R., Voorhies, M.R., Webb, S.D., Whistler, D.P., 2004. Mammalian biochronology of the Arikareean through Hemphillian interval (Late Oligocene through Early Pliocene Epochs). In: Woodburne, M.O. (Ed.), Late Cretaceous and Cenozoic Mammals of North America: Biostratigraphy and Geochronology. Columbia University Press, New York, pp. 169–231.

Thaler, L., 1965. Une échelle de zones biochronologique pour les mammifères du Tertiaire de l'Europe. Comptes Rendus Sommaires de la Société Géologique de France 4, 118.

Tiedemann, R., Sarnthein, M., Shackleton, N.J., 1994. Astronomic timescale for the Pliocene Atlantic $\delta^{18}O$ and dust flux records of Ocean Drilling Program Site 659. Paleoceanography 9, 619–638.

Thomas, A.L., Henderson, G.M., Deschamps, P., Yokoyama, Y., Mason, A.J., Bard, E., Hamelin, B., Durand, N., Camoin, G., 2009. Penultimate deglacial sea-level timing from uranium/thorium dating of Tahitian corals. Science 324, 1186–1189.

Thompson, W.G., Goldstein, S.L., 2005. Open-system coral ages reveal persistent suborbital sea-level cycles. Science 308, 401–404.

Thouveny, N., Carcaillet, J., Moreno, E., Leduc, G., Nérini, D., 2004. Geomagnetic moment variation and paleomagnetic excursions during the past 400 ka: A stacked record from sedimentary cores of the Portuguese margin. Earth and Planetary Science Letters 219, 377–396.

Thouveny, N., Bourlès, D.L., Saracco, G., Carcaillet, J.T., Bassinot, F., 2008. Paleoclimatic context of geomagnetic dipole lows and excursions in the brunhes, clue for an orbital influence on the geodynamo? Earth and Planetary Science Letters 275, 269–284.

Tian, J., Zhao, Q., Wang, P., Li, Q., Cheng, X., 2008. Astronomically modulated Neogene sediment records from the South China Sea. Paleoceanography 23, PA3210. doi: 10.1029/2007PA001552.

Turco, E., Cascella, A., Gennari, R., Hilgen, F.J., Iaccarino, S.M., Sagnotti, L., 2011a. Integrated stratigraphy of the La Vedova section (Conero Riviera, Italy) and implications for the Langhian GSSP. Stratigraphy 8, 89–110.

Turco, E., Iaccarino, S.M., Foresi, L., Salvatorini, G., Riforgiato, F., Verducci, M., 2011b. Revisiting the taxonomy of the intermediate stages in the Globigerinoides – Praeorbulina lineage. Stratigraphy 8, 163–187.

Turney, C.S.M., Lowe, J.J., Davies, S.M., Hall, V., Lowe, D.J., Wastegård, S., Hoek, W.Z., Alloway, B., 2004. Tephrochronology of last termination sequences in Europe: A protocol for improved analytical precision and robust correlation procedures (a joint SCOTAV-INTIMATE proposal). Journal of Quaternary Science 19, 111–120.

Turney, C.S.M., Fifield, L.K., Hogg, A.G., Palmer, J.G., Hughen, K., Baillie, M.G.L., Galbraith, R., Ogden, J., Lorrey, A., Tims, S.G., Jones, R.T., 2010. The potential of New Zealand kauri (*Agathis australis*) for testing the synchronicity of abrupt climate change during the Last Glacial Interval (60,000–11,700 years ago). Quaternary Science Reviews 29, 3677–3682.

Uba, C.E., Strecker, M.R., Schmitt, A.K., 2007. Increased sediment accumulation rates and climatic forcing in the Central Andes during the Late Miocene. Geology 35, 979–982.

Valet, J.-P., Meynadier, L., Guyodo, Y., 2005. Geomagnetic dipole strength and reversal rate over the past two million years. Nature 435, 802–805.

Valladas, H., Tisnérat-Laborde, N., Cachier, H., Kaltnecker, É, Arnold, M., Oberlin, C., Évin, J., 2005. Result of carbon 14 date markings performed on a charcoal of the Chauvet cave. Bulletin De La Societe Prehistorique Francaise 102, 109–113.

Van Couvering, J.A., Castradori, D., Cita, M.B., Hilgen, F.J., Rio, D., 2000. The base of the Zanclean Stage and of the Pliocene Series. Episodes 23, 179–186.

Van Couvering, J.A., Aubry, M.-P., Berggren, W.A., Gradstein, F.M., Hilgen, F.J., Kent, D.V., Lourens, L.J., McGowran, B., 2009. What, if anything, is Quaternary? Episodes 32, 125–126.

Van Dam, J.A., 2003. European mammal-based chronological systems: Past, present and future. Deinsia 10, 85–96.

Van Dam, J.A., Alcalá, L., Alonso Zarza, A.M., Calvo, J.P., Garcés, M., Krijgsman, W., 2001. The Upper Miocene mammal record from the Teruel-Alfambra region (Spain): The MN system and continental Stage/Age concepts discussed. Journal of Vertebrate Paleontology 21, 367–385.

Van Dam, J.A., Abdul Aziz, H., Álvarez Sierra, M.A., Hilgen, F.J., van den Hoek Ostende, L.W., Lourens, L., Mein, P., van der Meulen, A.J., Pelaez-Campomanes, P., 2006. Long-period astronomical forcing of mammal turnover. Nature 443, 687–691.

Van der Made, J., 1999. Intercontinental relationship Europe–Africa and the Indian Subcontinent. In: Rössner, G.E., Heissig, K. (Eds.), The Miocene Land Mammals of Europe. Pfeil Verlag, München, pp. 457–472.

Van der Laan, E., Gaboardi, S., Hilgen, F.J., Lourens, L.J., 2005. Regional climate and glacial control on high-resolution oxygen isotope records from Ain el Beida (latest Miocene, northwest Morocco): A cyclostratigraphic analysis in the depth and time domain. Paleoceanography 20, PA1001. doi: 10.1029/2003PA000995.

Van der Laan, E., Snel, E., de Kaenel, E., Hilgen, F.J., Krijgsman, W., 2006. No major deglaciation across the Miocene-Pliocene boundary: Integrated stratigraphy and astronomical tuning of the Loulja sections (Bou Regreg area, NW Morocco). Paleoceanography 21, PA3011. doi: 10.1029/2005PA001193.

Van der Laan, E., Hilgen, F.J., Lourens, L.J., de Kaenel, E., Iaccarino, S., Gaboardi, S., 2012. Astronomical forcing of Northwest African climate and glacial history during the late Messinian (6.5–5.5 Ma). Palaeogeography, Palaeoclimatology, Palaeoecology. 313–314, 107–126.

Van der Meulen, A.J., Peláez-Campomanes, P., Levin, S.A., 2005. Age structure, residents, and transients of Miocene rodent communities. The American Naturalist 165, E108–E125.

Van der Meulen, A.J., García-Paredes, I., Álvarez-Sierra, M.Á, Van den Hoek Ostende, L.W., Hordijk, K., Oliver, A., Peláez-Campomanes, P., 2011. Updated Aragonian biostratigraphy: Small mammal distribution and its implications for the Miocene European Chronology. Geologica Acta. doi: 10.101344/105.000001710.

Van de Weerd, A., 1976. Rodent faunas of the Mio-Pliocene continental sediments of the Teruel-Alfambra region, Spain. Utrecht Micropaleontological Bulletin, Special Publication 2, 217.

Vangengeim, E.A., Tesakov, A.S., 2008a. Late Sarmatian mammal localities of the Eastern Paratethys: Stratigraphic position, magnetochronology, and correlation with the European continental scale. Stratigraphy and Geological Correlation 16, 92–103.

Vangengeim, E.A., Tesakov, A.S., 2008b. Maeotian mammalian localities of Eastern Paratethys: Magnetochronology and position in European Continental Scales. Stratigraphy and Geological Correlation 16, 437–450.

Vangengeim, E.A., Tesakov, A.S., in press, Late Miocene mammal localities of Eastern Europe and Western Asia: Toward biostratigraphic synthesis. In: Wang, X., Flynn, L.J., and Fortelius, M. (Eds), *Asian Neogene Mammal Volume*. New York: Columbia University Press.

Van Vugt, N., Steenbrink, J., Langereis, C.G., Hilgen, F.J., Meulenkamp, J.E., 1998. Sedimentary cycles in the early Pliocene lacustrine sediments of Ptolemais (NW Greece) correlated to insolation and to the marine Rossello section (S. Italy). Earth and Planetary Science Letters 164, 535–551.

Vasiliev, I., Krijgsman, W., Langereis, C.G., Panaiotu, C.E., Matenco, L., Bertotti, G., 2004. Towards an astrochronological framework for the eastern Paratethys Mio-Pliocene sedimentary sequences of the Focsani basin (Romania). Earth and Planetary Science Letters 227, 231–247.

Vasiliev, I., Krijgsman, W., Stoica, M., Langereis, C.G., 2005. Mio-Pliocene magnetostratigraphy in the southern Carpathian foredeep and Mediterranean-Paratethys correlations. Terra Nova 17, 376–438.

Vasiliev, I., Dekkers, M.J., Krijgsman, W., Franke, C., Langereis, C.G., Mullender, T.A.T., 2007. Early diagenetic greigite as a recorder of the palaeomagnetic signal in Miocene-Pliocene sedimentary rocks of the Carpathian foredeep (Romania). Geophysical Journal International 171, 613–629.

Vasiliev, I., Reichart, G.-J., Davies, G.R., Krijgsman, W., Stoica, M., 2010. Strontium isotope ratios of the Eastern Paratethys during the Mio-Pliocene transition: Implications for interbasinal connectivity. Earth and Planetary Science Letters 292, 123–131.

Wade, B.S., Pearson, P.N., Berggren, W.A., Pälike, H., 2011. Review and revision of Cenozoic tropical planktonic foraminiferal biostratigraphy and calibration to the geomagnetic polarity and astronomical time scale. Earth-Science Reviews 104, 111–142.

Wagner, B., Sulpizio, R., Zanchetta, G., Wulf, S., Wessels, M., Daut, G., Nowaczyk, N., 2008. The last 40 ka tephrostratigraphic record of Lake Ohrid, Albania and Macedonia: A very distal archive for ash dispersal from Italian volcanoes. Journal of Volcanology and Geothermal Research 177, 71–80.

Wagner, G.A., Maul, L.C., Löscher, M., Schreiber, H.D., 2011. Mauer – the type site of *Homo heidelbergensis*: Palaeoenvironment and age. Quaternary Science Reviews 30, 1464–1473.

Walker, J., Cliff, R.A., Latham, A.G., 2006. U-Pb isotopic age of the StW 573 hominid from Sterkfontein, South Africa. Science 314, 1592–1594.

Walker, M., Johnsen, S., Rasmussen, S.O., Popp, T., Steffensen, J.-P., Gibbard, P., Hoek, W., Lowe, J., Andrews, J., Björck, S., Cwynar, L.C., Hughen, K., Kershaw, P., Kromer, B., Litt, T., Lowe, D.J., Nakagawa, T., Newnham, R., Schwander, J., 2009. Formal definition and dating of the GSSP (Global Stratotype Section and Point) for the base of the Holocene using the Greenland NGRIP ice core, and selected auxiliary records. Journal of Quaternary Science 24, 3–17.

Walsh, S.L., 1998. Fossil datum and paleobiological event terms, paleontostratigraphy, chronostratigraphy, and the definition of Land Mammal "Age" boundaries. Journal of Vertebrate Paleontology 18, 150–179.

Walther, A.M., Orgeira, M.J., Reguero, M.A., Verzi, D.H., Chernoglasov, C., Vilas, J.F., Sinito, A.M., Herrero Bervera, E., 1996. Nuevo estudio paleomagnético de la Formación Uquía de la provincia de Jujuy, Argentina. Revista de la Asociación Geológica Argentina 51, 355–364.

Wang, R.J., Chen, R.H., 2005. *Cycladophora davisiana* (radiolarian) in the Bering Sea during the late Quaternary: A stratigraphic tool and proxy of the glacial Subarctic Pacific Intermediate Water. Science in China Series D: Earth Sciences 48, 1698–1707.

Wang, X., Qiu, Z., Opdyke, N.D., 2003. Litho-, Bio-, and Magnetostratigraphy and Paleoenvironment of Tunggur Formation (Middle Miocene) in central Inner Mongolia, China. American Museum Novitates 3411, 1–31.

Wang, X., Flynn, L.J., Fortelius, M. (Eds.), in press, *Asian Neogene Mammal Volume*. New York: Columbia University Press.

Wang, Y.J., Cheng, H., Edwards, R.L., An, Z.S., Wu, J.Y., Shen, C.-C., Dorale, J.A., 2001. A high-resolution absolute-dated late Pleistocene monsoon record from Hulu cave, China. Science 294, 2345–2348.

Waters, M.R., Stafford Jr., T.W., 2007. Redefining the age of Clovis: Implications for the peopling of the Americas. Science 315, 1122–1126.

Welcomme, J.-L., Antoine, P.O., Duranthon, F., Mein, P., Ginsburg, L., 1997. Nouvelles découvertes de Vertébrés miocènes dans le synclinal de Dera Bugti (Balouchistan, Pakistan). Comptes Rendus de l'Académie des Sciences 325, 531–536.

Werdelin, L., 2010. Chronology of Neogene mammal localities. In: Werdelin, L., Sanders, W.J. (Eds.), Cenozoic Mammals of Africa. California University Press, Berkeley, pp. 27–44.

Westerhold, T., Bickert, T., Röhl, U., 2005. Middle to Late Miocene oxygen isotope stratigraphy of ODP Site 1085 (SE Atlantic): New constraints on Miocene climate variability and sea-level fluctuations. Palaeogeography, Palaeoclimatology, Palaeoecology 217, 205–222.

Westphal, H., Hilgen, F.J., Munnecke, A., 2010. An assessment of the suitability of individual rhythmic carbonate successions for astrochronological application. Earth-Science Reviews 99, 19–30.

Whistler, D.P., Tedford, R.H., Takeuchi, G.T., Wang, X., Tseng, Z.J., Perkins, M.E., 2009. Revised Miocene biostratigraphy and biochronology of the Dove Spring Formation, Mojave Desert, California. Museum of Northern Arizona Bulletin 65, 331–362.

Wintle, A.G., 2008. Luminescence dating: Where it has been and where it is going. Boreas 37, 471–482.

Wintle, A.G., Murray, A.S., 2006. A review of quartz optically stimulated luminescence characteristics and their relevance in single-aliquot regeneration dating protocols. Radiation Measurements 41, 369–391.

Wolff, E.W., Barbante, C., Becagli, S., Bigler, M., Boutron, C.F., Castellano, E., de Angelis, M., Federer, U., Fischer, H., Fundel, F., Hansson, M., Hutterli, M., Jonsell, U., Karlin, T., Kaufmann, P.,

Lambert, F., Littot, G.C., Mulvaney, R., Röthlisberger, R., Ruth, U., Severi, M., Siggaard-Andersen, M.L., Sime, L.C., Steffensen, J.P., Stocker, T.F., Traversi, R., Twarloh, B., Udisti, R., Wagenbach, D., Wegner, A., 2010. Changes in environment over the last 800,000 years from chemical analysis of the EPICA Dome C ice core. Quaternary Science Reviews 29, 285–295.

Wood II, H.E., Chaney, R.W., Clark, J., Colbert, E.H., Jepsen, G.L., Reeside Jr., J.B., Stock, C., 1941. Nomenclature and correlation of the North American continental Tertiary. Geological Society of America Bulletin 52, 1–48.

Woodburne, M.O., 1977. Definition and characterization in mammalian chronostratigraphy. Journal of Paleontology 51, 220–234.

Woodburne, M.O., 2006. Mammal ages. Stratigraphy 3, 229–261.

Woodburne, M.O., 2004. Principles and procedures. In: Woodburne, M.O. (Ed.), Late Cretaceous and Cenozoic Mammals of North America: Biostratigraphy and Geochronology. Columbia University Press, New York, pp. 1–20.

Woodburne, M.O., Swisher III, C.C., 1995. Land mammal high resolution geochronology, intercontinental overland dispersals, sea level, climate and vicariance. In: Berggren, W.A., Kent, D.V., Aubry, M.-P., Hardenbol, J. (Eds.), Geochronology, Time Scales and Global Stratigraphic Correlation. SEPM Special Publication, 54, pp. 335–364.

Woodburne, M.O., Tedford, R.H., Archer, M., Turnbull, W.D., Plane, M.D., Lundelius, E.L., 1985. Biochronology of the continental mammal record of Australia and New Guinea. Special Publication of the South Australian Department of Mines and Energy 5, 347–363.

Woodruff, F., Savin, S.M., 1991. Mid-Miocene isotope stratigraphy in the deep sea: High-resolution correlations, paleoclimatic cycles, and sediment preservation. Paleoceanography 6, 755–806.

Wotzlaw, J.F., Schaltegger, U., 2011. High-precision zircon U-Pb geochronology of astronomically tuned ash beds from the Mediterranean Miocene. Geophysical Research Abstracts 13 EGU2011-3442-1.

Wotzlaw, J.F., Decou, A., von Eynatten, H., Wörner, G., Frei, D., 2011. Jurassic to Palaeogene tectono-magmatic evolution of northern Chile and adjacent Bolivia from detrital zircon U-Pb geochronology and heavy mineral provenance. Terra Nova. doi: 10.1111/j.1365-3121.2011.01025.x.

Wulf, S., Kraml, M., Brauer, A., Keller, J., Negendank, J.F.W., 2004. Tephrochronology of the 100ka lacustrine sediment record of Lago Grande di Monticchio (southern Italy). Quaternary International 122 (Spec. issue 1), 7–30.

Wulf, S., Kraml, M., Keller, J., 2008. Towards a detailed distal tephrostratigraphy in the Central Mediterranean: the last 20,000 yrs record of Lago Grande di Monticchio. Journal of Volcanology and Geothermal Research 177, 118–132.

Xuan, C., Channell, J.E.T., 2008. Testing the relationship between timing of geomagnetic reversals/excursions and phase of orbital cycles using circular statistics and monte carlo simulations. Earth and Planetary Science Letters 268, 245–254.

Yamazaki, T., Oda, H., 2005. A geomagnetic paleointensity stack between 0.8 and 3.0 Ma from equatorial Pacific sediment cores. Geochemistry, Geophysics, Geosystems 6, Q11H20. doi: 10.1029/2005GC001001.

Zachos, J.C., Shackleton, N.J., Revenaugh, J.S., Pälike, H., Flower, B.P., 2001. Climate Response to Orbital Forcing Across the Oligocene-Miocene Boundary. Science 292, 274–278.

Zachos, J.C., Dickens, G.R., Zeebe, R.E., 2008. An early Cenozoic perspective on greenhouse warming and carbon-cycle dynamics. Nature 451, 279–283.

Zárate, M.A., Schultz, P., Blasi, A., Heil, C., King, J., Hames, W., 2007. Geology and geochronology of type Chasicoan (late Miocene) mammal-bearing deposits of Buenos Aires (Argentina). Journal of South American Earth Sciences 23, 81–90.

Zeeden, C., Hilgen, F.J., Lourens, L.J., Westerhold, T., Röhl, U., 2011. Revised stratigraphy and tuning of the equatorial Atlantic Ceara Rise: challenges and applications of astronomical time scales in the Miocene. EGU 2011, 3-8 April, Vienna, Austria Abstract.

Zhang, R., Yue, L.P., Wang, J.Q., 2007. Magnetostratigraphic dating of mammalian fossils in Junggar Basin, northwest China. Chinese Science Bulletin 11, 1526–1531.

Zheng, J.-Y., 1996. Magnetostratigraphy of a Miocene sedimentary sequence in Railroad Canyon, Idaho. Ph.D. dissertation. University of Pittsburgh, p. 496.

B. Pillans and P. Gibbard

The Quaternary Period

Abstract: The Quaternary Period, comprising the Holocene and Pleistocene Epochs, encompasses the last ~2.6 Ma during which time Earth's climate was strongly influenced by bi-polar glaciation and the genus *Homo* first appeared and evolved. The base of the Quaternary System/Period and Pleistocene Series/Epoch is defined by the GSSP for the Gelasian Stage at Monte San Nicola section in Italy. The base of the Holocene Series/Epoch is defined at a depth of 1492.45 m in the NGRIP ice core from Greenland, with an age based on annual layer counting, of 11 700 years b2k (before AD2000), with a 2σ uncertainty of 99 years; it is the first and only GSSP to be defined in an ice core.

21000 Years Quaternary

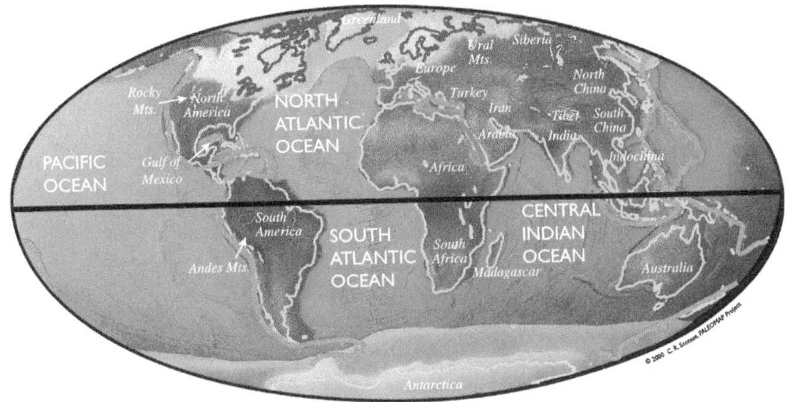

Chapter Outline

The Geologic Time Scale 2012 Set. DOI: 10.1016/B978-0-444-59425-9.00030-5

30.1. EVOLUTION OF TERMINOLOGY

The classification and interpretation of the youngest stratigraphic sequences, variously known as Pleistocene, Holocene or Quaternary, have been a matter of much debate. During the first two decades of the 19th century many of the sequences were attributed to the biblical flood (the "Diluvial" theory). This theory could account for unconsolidated sediments that rested unconformably on Tertiary rocks and capped hills, and that commonly contained exotic boulders and the remains of animals, many still extant. A flood origin for the "Diluvium" was the accepted paradigm by most eminent geologists of the time, including Buckland and Sedgwick.

During voyages of polar exploration, floating ice had frequently been seen transporting exotic materials, providing an explanation for the boulders in Diluvium, and reinforcing the Diluvial theory. This explanation led to the adoption of the term "drift" to characterize the sediments. However, geologists working in the Alps and northern Europe had been struck by the extraordinary similarity of the "drift" deposits, and their associated landforms, to those being formed by modern mountain glaciers. Several observers, such as Perraudin, Venetz-Sitten, and de Charpentier, proposed that the glaciers had formerly been more extensive, but it was the paleontologist Agassiz who first advocated that this extension represented a time that came to be termed the Ice Age by Goethe.

After having convinced Buckland and Lyell of the validity of his Glacial Theory in 1840, Agassiz's ideas became progressively accepted. The term *Drift* became established for the widespread sands, gravels, and boulder clays thought to have been deposited by glacial ice. Meanwhile, Lyell had already proposed the term *Pleistocene* in 1839 for the post-Pliocene period closest to the present. He defined this period on the basis of its molluscan faunal content, the majority of which are still extant. However, the term Quaternary (*Quaternaire* or *Tertiaire récent*) had already been proposed in 1829 by Desnoyers for marine sediments in the Seine Basin (Bourdier, 1957, p. 99). However the term had been in use from the late 18th century, originating with the Italian mining engineer, Giovanni Arduino (1714–1795), who distinguished four separate stages or "orders": Primary, Secondary, Tertiary and Quaternary, comprising the Artesine Alps, the Alpine foothills, the sub-Alpine hills and the Po Plain, respectively (Schneer, 1969; Ellenberger, 1994).

Both terms – Pleistocene and Quaternary – became synonymous with the Ice Age and also with the period during which humans evolved. However, unlike the Pleistocene concept, the span of the Quaternary included Lyell's original "Recent", later named *Holocene* by the 3rd International Geological Congress (IGC) in London in 1885. The term *Holocene* (meaning "wholly recent") refers to the percentage of living organisms and was defined by Gervais (1867–1869) "for the post-diluvial deposits approximately corresponding to the post-glacial period" (Bourdier, 1957, p. 101). The Holocene period was originally considered to represent a fifth era or *Quinquennaire* (Parandier, 1891), but this division was deemed to be "excessive"; details are given in Bourdier (1957) and de Lumley (1976).

Because the terms Primary and Secondary have been abandoned and attempts have been made to suppress Tertiary, the continued use of Quaternary has been regarded by some stratigraphers as somewhat archaic. Alternative terms, such as *Anthropogene* (extensively used in the former USSR) or *Pleistogene* (suggested by Harland *et al.*, 1990), have been proposed, but neither have found favor. Subsequently, Gradstein *et al.* (2004) did not include Quaternary as the youngest period of the Cenozoic Era. Rather, they designated the Quaternary as an informal climatostratigraphic unit, spanning:

'an interval of oscillating climatic extremes (glacial and interglacial episodes) that was initiated at about 2.6 Ma, therefore encompassing the Holocene and Pleistocene epochs and Gelasian Stage of late Pliocene.'

(Gradstein *et al.*, 2004, p. 441)

After exhaustive discussions between the International Commission on Stratigraphy (ICS) and the International Union for Quaternary Research (INQUA), a formal decision on the chronostratigraphic status of the Quaternary was reached in 2009, resulting in the ratification of the Quaternary Period/System and revised base of the Pleistocene Epoch/Series (Gibbard *et al.*, 2010).

The evolving definition of the Quaternary is summarized in Figure 30.1.

30.2. THE PLIO–PLEISTOCENE BOUNDARY AND DEFINITION OF THE QUATERNARY

In 1948, at the 18th International Geological Congress in London, agreement was reached to place the Pliocene–Pleistocene boundary:

'at the horizon of the first indication of climatic deterioration in the Italian Neogene succession.'

(King and Oakley, 1949)

Furthermore it was agreed that the boundary should be based on marine faunas and that the lower Pleistocene should include not only the marine sediments of the Italian Calabrian Stage (a stage introduced by Gignoux in 1910), but also continental deposits of the Villafranchian Stage.

The initial Calabrian boundary was thought to be marked by the first appearance of the cold-water molluscan indicators *Arctica islandica* and *Hyalinea baltica* (Sibrava, 1978), but Ruggieri and Sprovieri (1979) showed that *Hyalinea baltica* appears slightly later. Subsequently, while various sections in southern Italy competed for the position of stratotype, it

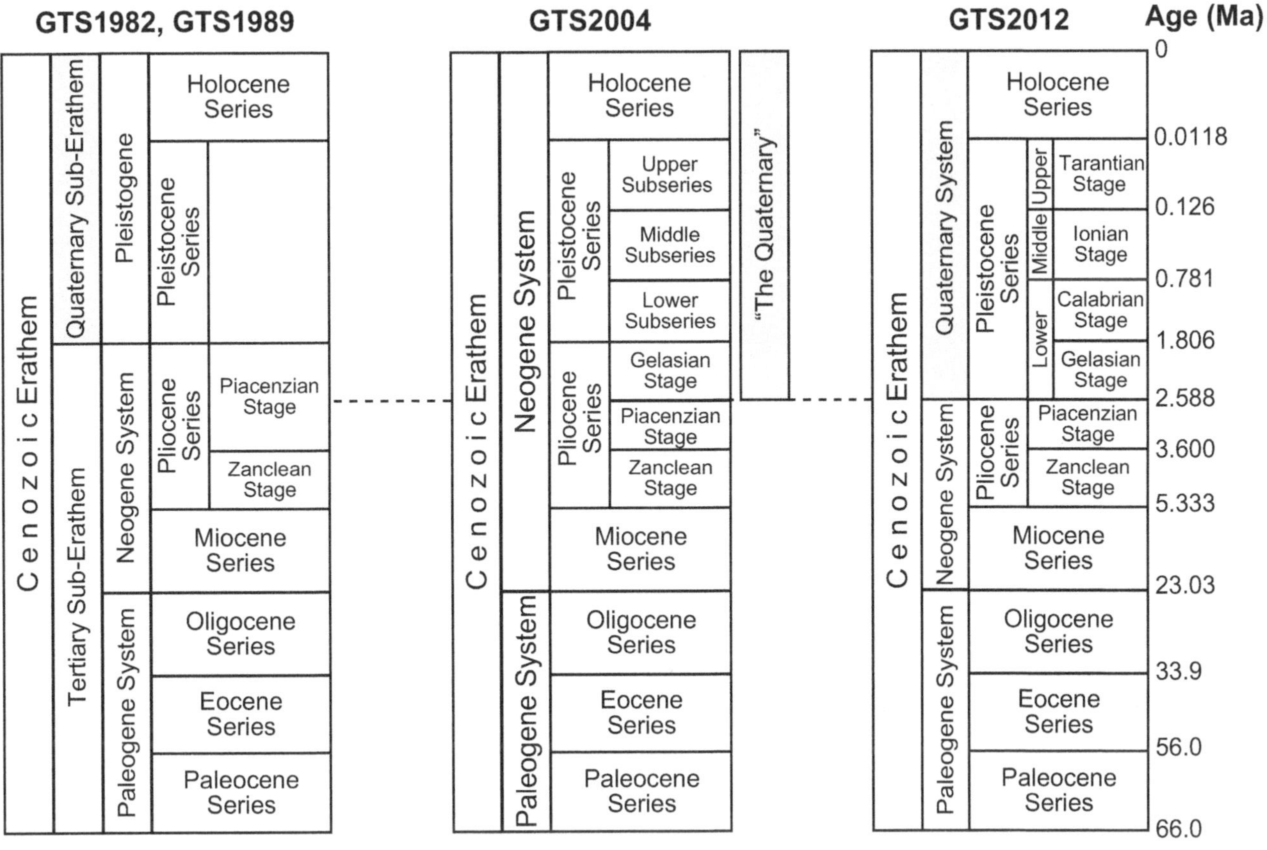

Quaternary chronostratigraphy summary *(Harland, 1990; Gradstein et al., 2004; this work).*

became clear that marine microfossils in conjunction with magnetostratigraphy were the best criteria for defining the boundary (Haq *et al.*, 1977; Backman *et al.*, 1983; Tauxe *et al.*, 1983).

After more than three decades of investigations, the GSSP for the Pliocene–Pleistocene boundary and the beginning of the Pleistocene was placed by a joint INQUA and ICS working group (IGCP Project 41) at the base of marine claystones conformably overlying bed e at the Vrica section (39° 02′ 18.61″ N, 17° 08′ 05.79″ E), approximately 4 km south of Crotone on the Marchesato Peninsula, Calabria, southern Italy (Figure 30.2; Aguirre and Pasini, 1985; Bassett, 1985). The boundary, which has an astronomically tuned age of 1.806 Ma (Lourens *et al.*, 1996), is some 3 to 6 m (representing an interval of 10–20 ky) above the top of the Olduvai normal polarity subchron (Bassett, 1985).

The magnetostratigraphy of the Vrica section has been investigated by Haq *et al.* (1977), Tauxe *et al.* (1983), Zijderveld *et al.* (1991) and, most recently, by Roberts *et al.* (2010), who showed that the complex polarity pattern at the top of the Olduvai Subchron is the result of authigenic growth of iron sulphide minerals such as greigite and pyrrhotite. However, uncertainties arising from the complex magnetostratigraphic record at Vrica result in maximum age differences of less than 50 ka for the upper Olduvai reversal (Roberts *et al.*, 2010) and hence a similar uncertainty in the boundary may be assumed.

The decision to assign the base-Pleistocene GSSP was:

> *'isolated from other more or less related problems, such as … the status of the Quaternary within the chronostratigraphic scale.'*
>
> (Aguirre and Pasini, 1985)

However, many "Quaternary" scientists, especially those working with terrestrial and climatic records, continued to favor defining the "Quaternary" as beginning significantly before the base-Pleistocene GSSP. As a result, the status and chronostratigraphic rank of Quaternary was not formally established.

The London 1948 IUGS recommendations included the notion that the base-Pleistocene boundary should be placed at the first evidence of climatic cooling. However, the Vrica GSSP boundary level is not the first severe cold-climate oscillation of the late Cenozoic. While the first evidence of continental glaciation in the northern hemisphere comes from ice-rafted debris in the Greenland Sea in the mid-Paleogene, around 44 Ma (Tripati *et al.*, 2008), it is now well established that global cooling around 2.7 to 2.6 Ma is the event that best characterizes the beginning of the Quaternary/Pleistocene ice ages (Pillans and Naish, 2004; Gibbard *et al.*, 2005; Ogg and

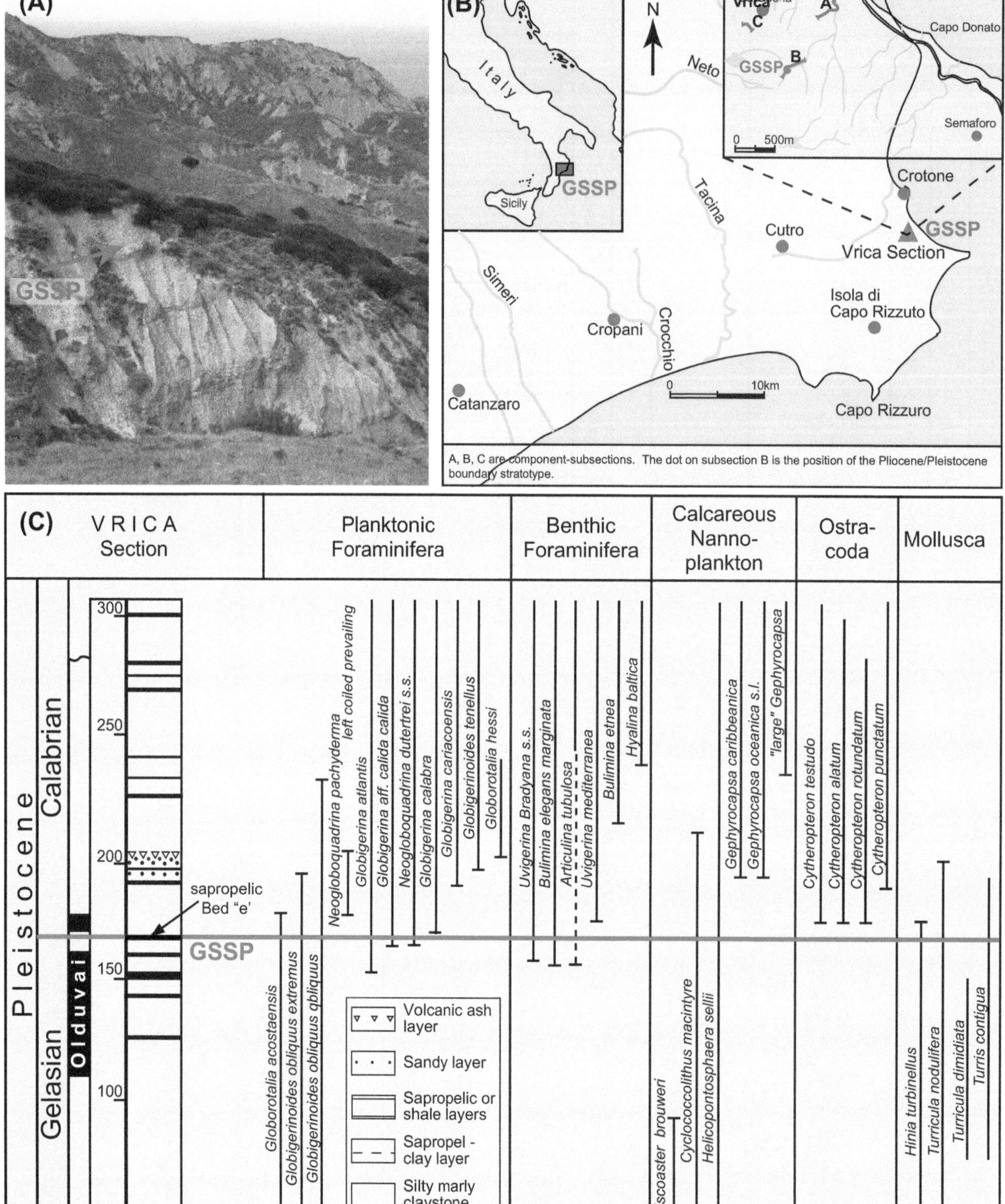

The GSSP for the Calabrian Stage in the Vrica section, Italy with photograph of the outcrop and stratigraphy. Magnetostratigraphy is from Zijderveld *et al.*, 1991.

Pillans, 2008; Sarnthein *et al.*, 2009). This major cooling takes place at a stratigraphic position equivalent to the base of the Gelasian Stage (Rio *et al.*, 1998), the GSSP for which is defined in marine sediments at Monte San Nicola in Sicily (Figure 30.3) and can be easily correlated with Marine Isotope Stage (MIS) 103 in the ocean sediments (see discussion in Suc *et al.*, 1997).

This older level corresponds to the Gauss/Matuyama magnetic epoch boundary (2.6 Ma) as well as being approximately coeval with the base of the Netherlands' terrestrial Praetiglian Stage, the base of the New Zealand Nukumaruan Stage and the base of the classic Chinese loess sequence (Figure 30.4). The event is clearly defined in the marine oxygen isotope stratigraphy and coincides with the first major influx of ice-rafted debris into the middle latitude of the North Atlantic (Shackleton *et al.*, 1984; Shackleton, 1997; Partridge, 1997a). The fossil mammalian record also shows changes that are obvious near the Gauss/Matuyama reversal marking the boundary between the Early and Middle Villafranchian (Rook and Martinez-Navarro, 2010).

In June 2009, the Executive Committee of the International Union of Geological Sciences (IUGS) formally ratified a proposal by the ICS to define the base of the Quaternary System at the GSSP for the Gelasian Stage at Monte San Nicola in Italy (Gibbard *et al.*, 2010). At the same time the base of the Pleistocene was also lowered to coincide with the Gelasian GSSP. Thus, the Gelasian Stage was transferred from the Pliocene Series to the Pleistocene Series (Figure 30.1). Alternative views on the position for the Pliocene–Pleistocene boundary have been discussed by Van Couvering (1997), Partridge (1997b) and Lourens (2008).

30.3. SUBDIVISION OF THE PLEISTOCENE

Two major types of subdivisions have been proposed for the Pleistocene Series. A standard subdivision at stage level has been advocated by workers based on sections in elevated shallow-marine sediments in Italy (Figure 30.4). Indeed, with the ratification, in 2009, of the Gelasian Stage as the basal stage of the Pleistocene, the International Stratigraphic Guide requires that the rest of the Pleistocene should be subdivided into one or more formal stages (Cita and Pillans, 2010). At the same time, however, earth scientists concerned with terrestrial and to a lesser extent shallow-marine sequences have adopted regional subdivision schemes. The regional schemes have found favor despite the difficulties of worldwide correlation. In these schemes, larger, subseries (sub-epoch) scale units have been adopted. For example, in the former USSR, and particularly in European Russia, the Pleistocene is divided into the Eopleistocene, equivalent to the Early Pleistocene subseries, and the Neopleistocene, equivalent to the Middle and Late Pleistocene subseries (Anonymous, 1982, 1984; Krasnenkov *et al.*, 1997). The most recent proposal for a revised stratigraphical scheme for the last 1 Ma in the Eastern European Plain is given by Shik *et al.* (2002).

A quasi-formal tripartite subdivision of the Pleistocene into Lower (Early), Middle and Upper (Late) has been in use since the 1930s. The first usage of the terms Lower, Middle, and Upper Pleistocene was at the second International Quaternary Association (INQUA) Congress in Leningrad in 1932 (Woldstedt, 1962), although they may have been used in a loose way before this time. Their first use in a formal sense in English was by Zeuner (1935, 1959) and Hopwood (1935) and was based on characteristic assemblages of vertebrate fossils in the European sequence.

30.3.1. Lower Pleistocene

Cita *et al.* (2008) proposed the Calabrian Stage, as the next youngest Pleistocene stage above the Gelasian, with a GSSP defined at Vrica section, former GSSP for the Pleistocene Series (Figure 30.2). Thus, the Lower Pleistocene Subseries would comprise both the Gelasian and Calabrian Stages. In November 2011, the IUGS ratified the Calabrian Stage.

30.3.2. Middle Pleistocene

Participants at the Burg Wartenstein Symposium "Stratigraphy and Patterns of Cultural Change in the Middle Pleistocene", held in Austria in 1973, recommended that:

'The beginning of the Middle Pleistocene should be so defined as to either coincide with or be linked to the Matuyama Reversed Epoch and the Brunhes Normal Epoch of palaeomagnetic chronology.'

(Butzer and Isaac, 1975, appendix 2)

A similar recommendation was made by the INQUA Commission on Stratigraphy/ICS Working Group on Major Subdivision of the Pleistocene, at the XIIth INQUA Congress in Ottawa in 1987, which placed the Lower–Middle boundary at the Brunhes–Matuyama magnetic reversal (Richmond, 1996). However, although potential GSSPs in Japan, Italy and New Zealand were discussed, no decision was reached. Pillans (2003) also advocated the Matuyama–Brunhes boundary (MBB) as the key marker, emphasizing that it constituted the most recognizable chronostratigraphic marker in weathered continental deposits.

At present the SQS Working Group on the Lower–Middle Pleistocene Boundary is considering candidate sections in Italy and Japan, having resolved at the 32nd International Geological Congress in Florence in 2004 to place the boundary as close as possible to the MBB. The age of the MBB is estimated to be 781 ka in the astronomically tuned time scale (Lourens *et al.*, 2004), though recent papers by Dreyfus *et al.* (2008), Channell *et al.* (2010) and Suganuma *et al.* (2010) support a slightly younger age of ~770 ka.

Cita *et al.* (2006) have proposed an Ionian Stage above the Calabrian Stage in Italian marine sections. The basal boundary

Base of the Gelasian Stage of the Pleistocene Series of the Quaternary System at Monte San Nicola, Italy

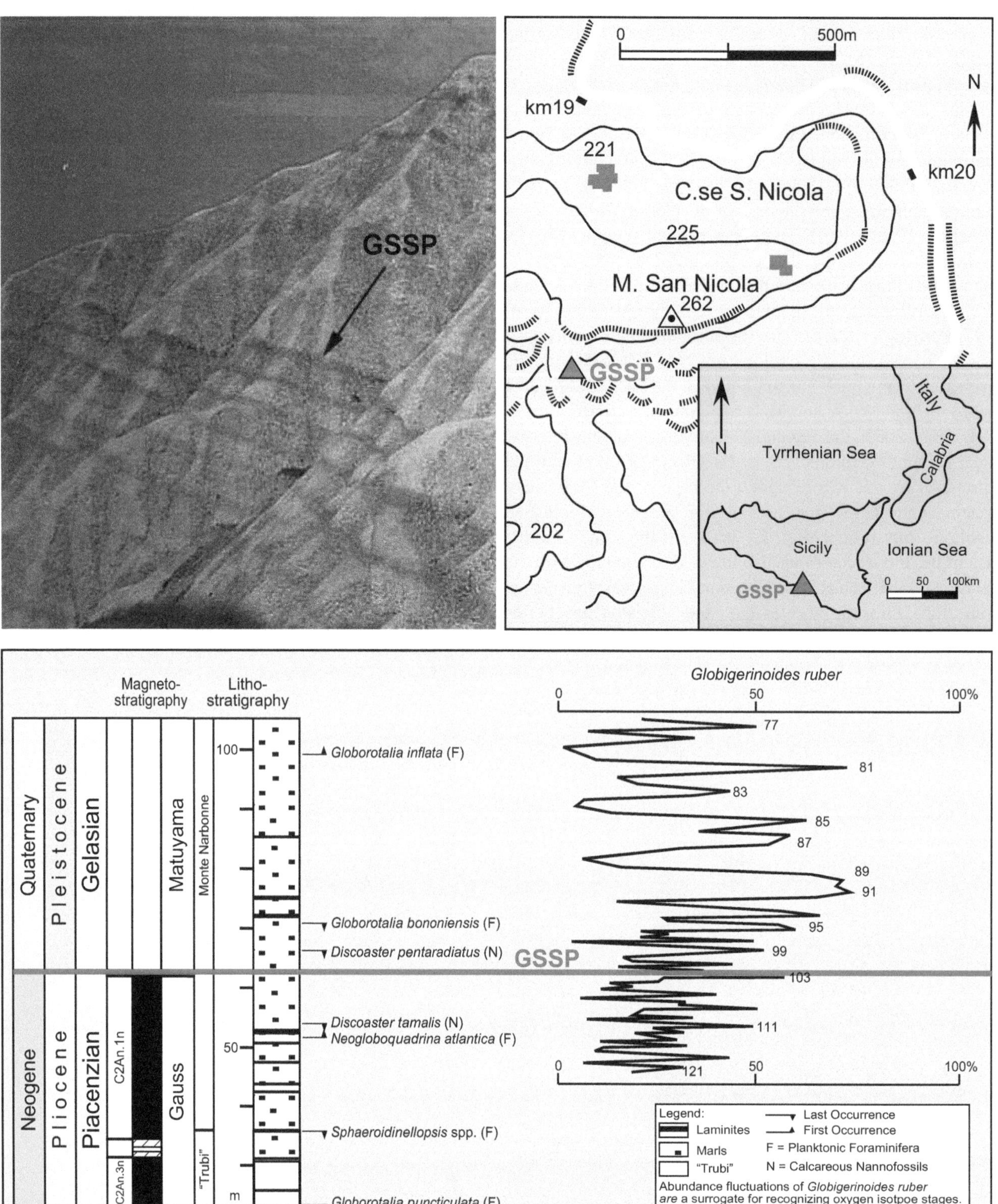

The GSSP for the Gelasian Stage is defined in marine sediments at Monte San Nicola in Sicily, and can be correlated with Marine Isotope Stage (MIS) 104 in the ocean sediments (see Section 30.2).The section is shown with a photograph and stratigraphy.

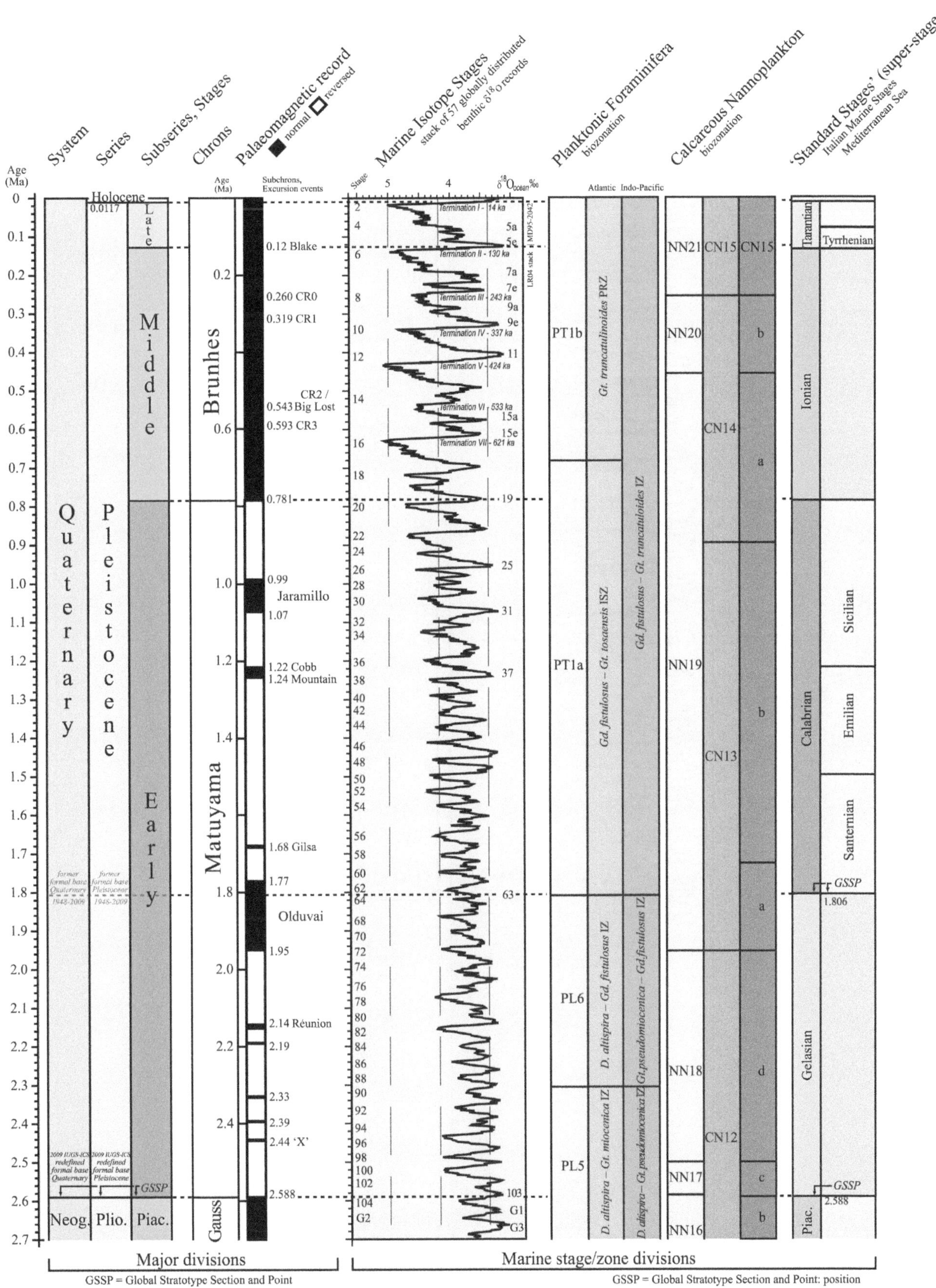
System
Series
Subseries, Stages
Chrons
Palaeomagnetic record
normal reversed
Marine Isotope Stages
stack of 57 globally distributed benthic δ18O records
Planktonic Foraminifera
biozonation
Calcareous Nannoplankton
biozonation
'Standard Stages' (super-stages)
Italian Marine Stages Mediterranean Sea
Age (Ma)
Quaternary
Pleistocene
Holocene
0.0117
Late
Middle
Early
Brunhes
Matuyama
Gauss
Neog.
Plio.
Piac.
GSSP
former formal base Quaternary 1948-2009
former formal base Pleistocene 1948-2009
2009 IUGS-ICS redefined formal base Quaternary
2009 IUGS-ICS redefined formal base Pleistocene
Subchrons, Excursion events
0.12 Blake
0.260 CR0
0.319 CR1
CR2 / 0.543 Big Lost
0.593 CR3
0.781
0.99
Jaramillo
1.07
1.22 Cobb
1.24 Mountain
1.68 Gilsa
1.77
Olduvai
1.95
2.14 Réunion
2.19
2.33
2.39
2.44 'X'
2.588
Termination I - 14 ka
Termination II - 130 ka
Termination III - 243 ka
Termination IV - 337 ka
Termination V - 424 ka
Termination VI - 533 ka
Termination VII - 621 ka
LR04 stack + MD95-2042
Atlantic Indo-Pacific
PT1b
PT1a
PL6
PL5
Gt. truncatulinoides PRZ
Gd. fistulosus – Gt. tosaensis ISZ
Gd. fistulosus – Gt. truncatuloides IZ
D. altispira – Gd. fistulosus IZ
D. altispira – Gt. miocenica IZ
Gt.pseudomiocenica – Gd.fistulosus IZ
D. altispira – Gt. pseudomiocenica IZ
NN21
NN20
NN19
NN18
NN17
NN16
CN15
CN14
CN13
CN12
Tarantian
Tyrrhenian
Ionian
Calabrian
Sicilian
Emilian
Santernian
Gelasian
Piac.
GSSP
1.806
2.588
Major divisions
GSSP = Global Stratotype Section and Point
Marine stage/zone divisions
GSSP = Global Stratotype Section and Point: position

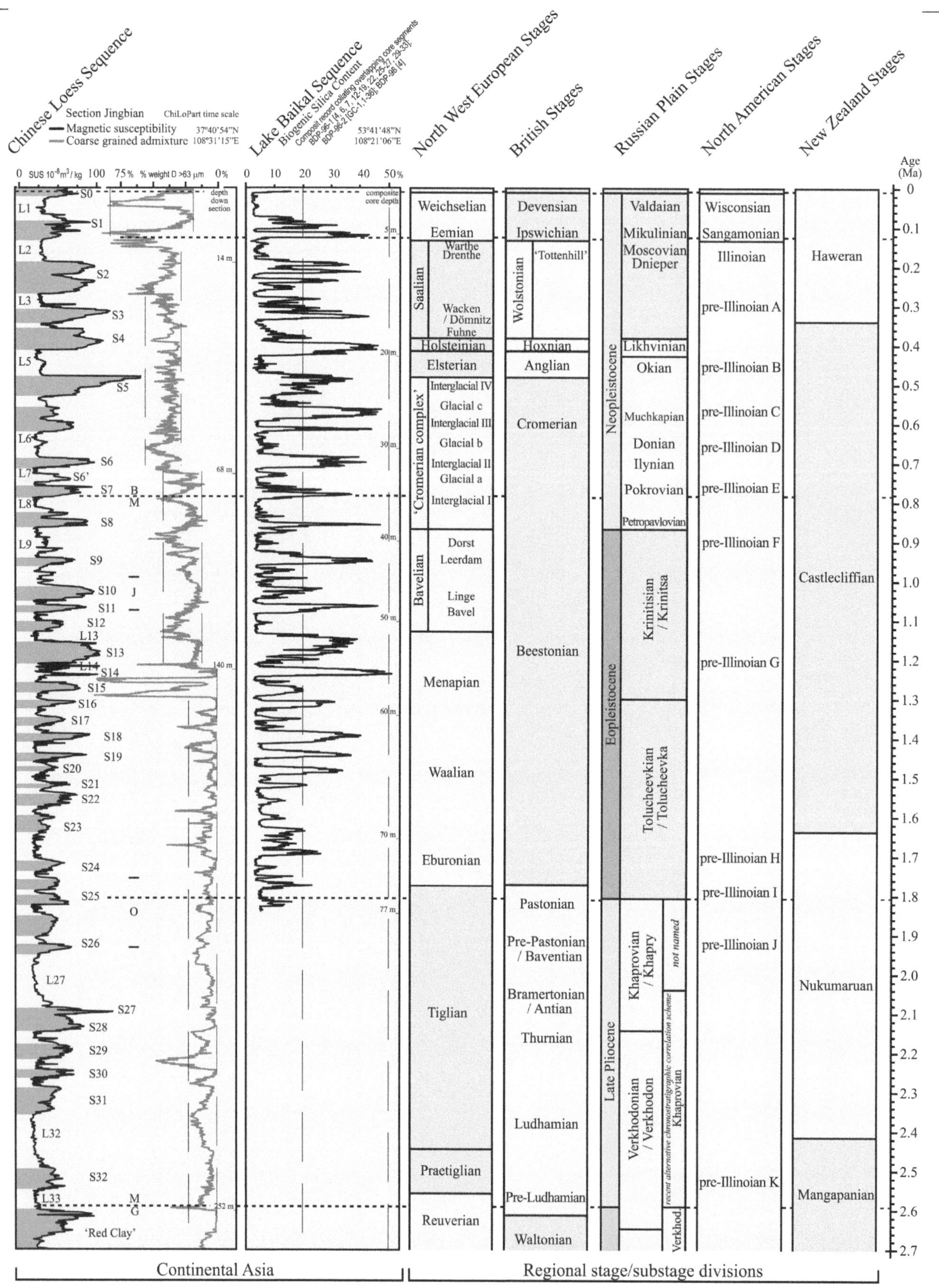
Chinese Loess Sequence
Section Jingbian
ChiLoPart time scale
Magnetic susceptibility
Coarse grained admixture
37°40'54"N
108°31'15"E
Lake Baikal Sequence
Biogenic Silica Content
53°41'48"N
108°21'06"E
North West European Stages
British Stages
Russian Plain Stages
North American Stages
New Zealand Stages
Age (Ma)
Weichselian
Eemian
Saalian
Warthe
Drenthe
Wacken / Dömnitz
Fuhne
Holsteinian
Elsterian
'Cromerian complex'
Interglacial IV
Glacial c
Interglacial III
Glacial b
Interglacial II
Glacial a
Interglacial I
Bavelian
Dorst
Leerdam
Linge
Bavel
Menapian
Waalian
Eburonian
Tiglian
Praetiglian
Reuverian
Devensian
Ipswichian
Wolstonian
'Tottenhill'
Hoxnian
Anglian
Cromerian
Beestonian
Pastonian
Pre-Pastonian / Baventian
Bramertonian / Antian
Thurnian
Ludhamian
Pre-Ludhamian
Waltonian
Neopleistocene
Eopleistocene
Late Pliocene
Valdaian
Mikulinian
Moscovian
Dnieper
Likhvinian
Okian
Muchkapian
Donian
Ilynian
Pokrovian
Petropavlovian
Krinitisian / Krinitsa
Tolucheevkian / Tolucheevka
Khaprovian / Khapry
not named
Verkhodonian / Verkhodon
recent alternative chronostratigraphic correlation scheme
Khaprovian
Verkhod.
Wisconsian
Sangamonian
Illinoian
pre-Illinoian A
pre-Illinoian B
pre-Illinoian C
pre-Illinoian D
pre-Illinoian E
pre-Illinoian F
pre-Illinoian G
pre-Illinoian H
pre-Illinoian I
pre-Illinoian J
pre-Illinoian K
Haweran
Castlecliffian
Nukumaruan
Mangapanian
'Red Clay'
Continental Asia
Regional stage/substage divisions

(*Continued*).

of the Ionian Stage is represented in two sections, Valle di Manche in Calabria (Rio *et al.*, 1996; Massari *et al.*, 2002; Capraro *et al.*, 2005) and Montalbano Jonico in the central part of the Apennine foredeep (Cita and Castradori, 1995; Ciaranfi *et al.*, 2001; Ciaranfi and D'Alessandro, 2005; Cita *et al.*, 2006; Joannin *et al.* 2008; Ciaranfi *et al.* 2010). Both sequences have been correlated using multiple parameters, including calcareous nannofossil biochronology, palynology, isotope stratigraphy and tephrostratigraphy. The Montalbano Jonico section would represent the more suitable boundary stratotype (Cita *et al.* 2006), except that attempts to obtain a magnetostratigraphy from this section have been unsuccessful (Sagnotti *et al.* 2010).

It is proposed that the Middle Pleistocene Subseries comprises a single stage – the Ionian Stage. A Working Group of the ICS Subcommission for Quaternary Stratigraphy (SQS) is currently considering three candidate sections for the GSSP – the Montalbano Jonico and Valle di Manche sections in Italy (see above) and the Chiba section in Japan (Pickering *et al.* 1999).

30.3.3. Upper Pleistocene

The boundary between the Middle and Upper Pleistocene has yet to be formally defined by the IUGS. However, as long ago as the 2nd INQUA Congress in Leningrad in 1932, a decision was made to define the boundary at the base of the Last Interglacial (Eemian Stage). More recently, the lower boundary of the Upper Pleistocene has been placed at the base of Marine Isotope Stage 5 (MIS 5), based on a proposal from the INQUA Commission on Stratigraphy (Richmond, 1996). This proposal naturally follows from the acceptance that MIS 5, substage e, is the ocean equivalent of the terrestrial northwest European Eemian Stage interglacial (Shackleton, 1977).

However, detailed pollen analyses of deep sea cores west of Portugal have shown that the base of MIS 5 is some 6 ka earlier than the base of the Eemian (Shackleton *et al.*, 2003) – see Figure 30.5. Consequently, Gibbard (2003) and Litt and Gibbard (2008) have proposed that, in keeping with the historical association of the boundary with the base of the Eemian, the GSSP should be defined in a high-resolution core sequence from the Amsterdam Terminal – which is both the parastratotype and unit-stratotype of the Eemian Stage (Van Leeuwen *et al.*, 2000).

Both the stage and the stage boundary are recognized on the basis of multidisciplinary biostratigraphy, the boundary being placed at the expansion of forest tree pollen above 50% of the total pollen assemblage, the standard practice in northwest Europe (Gibbard, 2003). In particular, the GSSP in the Amsterdam Terminal borehole is based on the steep rise of the *Betula* (birch tree) pollen curve at this point (Litt and Gibbard, 2008). A similar increase in the *Betula* curve is characteristic of many Eemian pollen diagrams across Europe and is thought to be a synchronous response to climatic amelioration, perhaps within a matter of decades as is

◀

Quaternary / Pleistocene time scale. At present the subseries (subepoch) divisions of the Pleistocene are not formalized. Series, and thereby systems, are formally defined based on Global Stratotype Section and Points (GSSP) of which two divide the Quaternary System into the Holocene and Pleistocene Series. The formal base of the Pleistocene, as ratified in 2009, coincides with a GSSP at Monte San Nicola in southern Italy, marking the base of the Gelasian Stage (Rio et al., 1994, 1998). The Gelasian GSSP at 2.58 Ma replaces the previous Pleistocene base GSSP (~1.8 Ma, defined at Vrica, southern Italy). It is based on the internationally recognised formal chronostratigraphical/geochronological subdivisions of the Quaternary System/Period; the Pleistocene and Holocene Series/Epochs, and finally the Early/Lower, Middle, Late/Upper Pleistocene Subseries/Subepochs.

The base of the Holocene is located in the North-GRIP ice core of the Greenland Ice Core Project (NGRIP: Walker et al., 2009). The deep-sea based, climatically defined stratigraphy is taken from oxygen isotope data obtained from tests of fossil benthonic (ocean-floor dwelling) foraminifera, retrieved from deep-ocean cores from 57 locations around the world. The plots depict $\delta^{18}O$ (the ratio of ^{18}O versus ^{16}O) of a stacked record as processed by Lisiecki and Raymo (2005). Their calibrated ages for the last seven major glacial terminations are included. It is plotted against the magnetostratigraphic time scale prepared and modified from Funnell (1996), supplemented with Calabrian Ridge (CR) magnetic event ages (cf. Lourens, 2004). Shifts in this ratio are a measure of global ice volume, which is dependent on global temperature and which determines global sea level. Planktonic foraminifera and calcareous nannoplankton provide an alternative biostratigraphical means of subdivision of marine sediments. The micropaleontological zonation is taken from Berggren et al. (1995). The Italian shallow marine stages are derived from Van Couvering (1997) modified by Cita et al. (2006) (cf. also Cita and Pillans, 2010). In view of their duration, covering multiple climate cycles and periods for which regional stage units of markedly shorter duration have been defined, these "standard stages" are considered as "super stages". Two plots of isotope measurements from Antarctic ice cores are shown.

For the Chinese loess deposits the chart shows the sequence of paleosols (units S0 to S32) for the Jingbian site in northern China (Ding et al., 2005). High values of magnetic susceptibility indicate repeated episodes of weathering (soil formation), predominantly in interglacials with relative strong summer monsoon. In intercalated strata (units L1 to L33; accumulated during glacials) the proportion of coarser grains (grains > 63 μm, % dry weight) is a signal of progressive desertification in Central Asia. The magnetic and grain-size data is plotted on the Chinese Loess Particle Time Scale (Ding et al., 2002). Alternating loess–paleosol sequence accumulation throughout NE China coincides with the beginning of the Pleistocene and buries the more intensively weathered Pliocene "Red Clay" Formation (An et al., 1990).

The Siberian Lake Baikal provides a bioproductivity record from the heart of the world's largest landmass, an area of extreme continental climate. High concentrations of biogenic silica indicate high aquatic production during interglacials (i.e. lake diatom blooms during ice-free summer seasons). The composite biogenic silica record from cores BDP-96-1, -96-2 and -98 is plotted on an astronomically tuned age scale (Prokopenko et al., 2006). The NW European stages are taken from Zagwijn (1992) and de Jong (1988). The British stages are taken from Mitchell et al. (1973), Gibbard et al. (1991) and Bowen (1999). The Russian Plain stages are from *Stratigraphy of the USSR: Quaternary System* (1982, 1984), Krasnenkov et al. (1997), Shik et al. (2002), Iossifova (pers. comm.) and Tesakov (pers. comm.). In addition, the Russian Pleistocene is also frequently divided into the Eopleistocene, equivalent to the Early Pleistocene Subseries, and the Neopleistocene, equivalent to the Middle and Late Pleistocene Subseries. The North American stages are taken from Richmond (unpublished). The New Zealand stages are from Pillans (1991) and Beu (2004). *(after Cohen and Gibbard, 2010)*

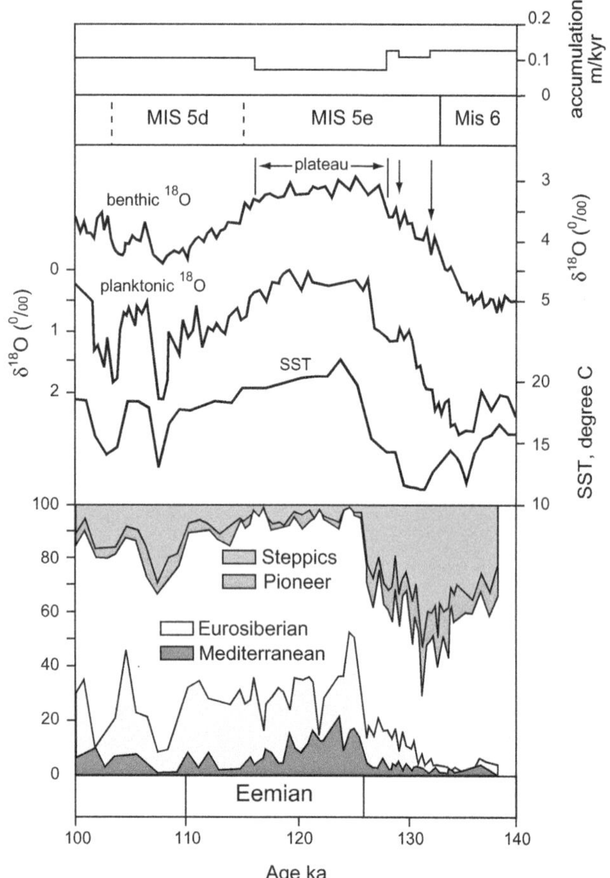

Marine and continental records of the Last (Eemian) Interglacial in core MD95-2042. *(Shackleton et al., 2003)*

observed at the transition from the Last Glacial to the Holocene. An age of 127.2 ka is estimated for the base of the Eemian from a varved-dated record at Monticchio in Italy (Brauer *et al.*, 2007), where the chronology is based on a combination of tephrochronology and annual layer counting.

Cita and Castradori (1995) proposed the Tarantian Stage for the interval corresponding to the Upper Pleistocene, i.e. from the base of the Eemian to the base of the Holocene, for marine sequences in Italy. Thus the Upper Pleistocene would comprise a single stage (Figure 30.1).

30.4. TERRESTRIAL SEQUENCES

30.4.1. Glacial Deposits and Climatostratigraphy

In contrast to the rest of the Phanerozoic, the uppermost Cenozoic has a long-established tradition of sediment sequences being divided on the basis of represented climatic changes, particularly those sequences based on glacial deposits in central Europe and mid-latitude North America. This approach was adopted for terrestrial sequences by early workers because it seemed logical to divide till (glacial diamicton) sheets and non-glacial deposits or stratigraphical sequences into *glacial (glaciation)* and *interglacial* periods, respectively (cf. West, 1968, 1977; Bowen, 1978). In other words, the divisions were fundamentally lithological. However, the overriding influence of climatic change on sedimentation and erosion has meant that, despite the enormous advances in knowledge during the last 150 years, climate-based stratigraphic classification (climatostratigraphy) has remained central to the subdivision of Quaternary sequences. Indeed, the subdivision of deep-sea sediment sequences into Marine Isotope Stages (MIS) is itself based on the same basic concept (see below).

For at least the first half of the 20th century the preferred climatostratigraphic scale for the Pleistocene was that developed for the Alps at the turn of the century by Penck and Brückner (1909–11).

For the Alps, the sequence in increasing age was:

Würm Glacial (Würmian)
Riss-Würm Interglacial
Riss Glacial (Rissian)
Mindel-Riss Interglacial
Mindel Glacial (Mindelian)
Günz-Mindel Interglacial
Günz Glacial
Donau-Günz Interglacial
Donau Glacial
?Biber Glacial

For other glaciated regions, including northern Europe, Britain, Russia, North America and New Zealand separate named glacial/interglacial sequences were also developed (see Figures 30.4 and 30.6).

Before the impact of the ocean-core isotope sequences, an attempt was made to formalize the climate-based stratigraphical terminology in the American Code of Stratigraphic Nomenclature (American Commission on Stratigraphic Nomenclature, 1961), in which so-called geologic-climate units were proposed. Here a geologic-climate unit is based on an inferred widespread climatic episode defined from a subdivision of Quaternary rocks (American Commission on Stratigraphic Nomenclature, 1961). Several synonyms for this category of units have been suggested, the most recent being climatostratigraphical units (Mangerud *et al.*, 1974) in which a hierarchy of terms is proposed. These units are neither referred to in the standard stratigraphic codes by Hedberg (1976) nor Salvador (1994), and are not followed in New Zealand, but are included in the local Norwegian Code (Nystuen, 1986). Boundaries between geologic-climate units were to be placed at those of the stratigraphic units on which they were based.

The American Commission on Stratigraphic Nomenclature (1961) defines the fundamental units of the geologic-climate classification as follows:

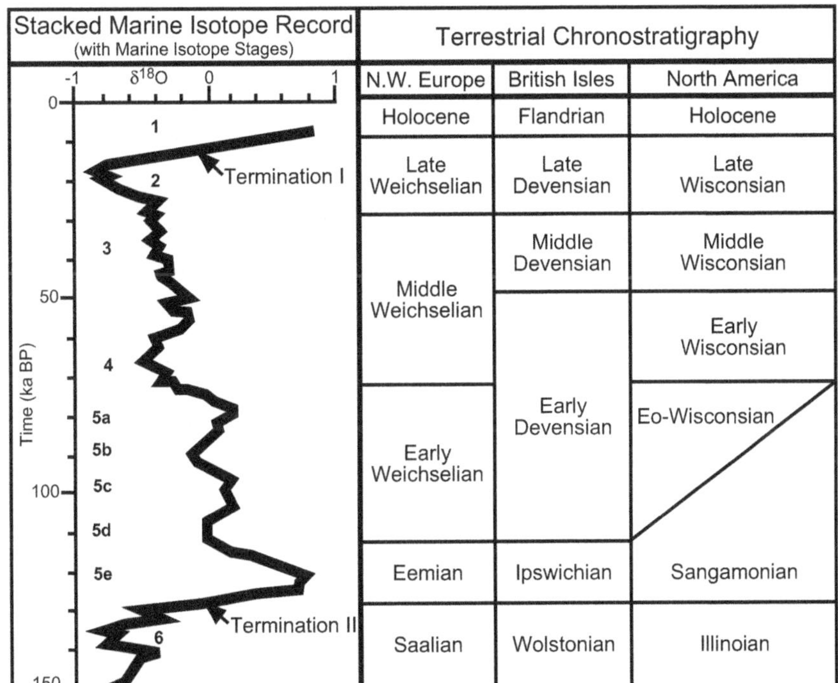

Marine and continental chronostratigraphy for the past 150 kyr. The stacked marine oxygen isotope sequence and associated stages are from Martinson et al. (1987), and the terrestrial climato- and chronostratigraphical divisions in northwest Europe and North America are modified from Lowe and Walker (1997).

A *Glaciation* is a climatic episode during which extensive glaciers developed, attained a maximum extent, and receded. A *Stadial* ("Stade") is a climatic episode, representing a subdivision of a glaciation, during which a secondary advance of glaciers took place. An *Interstadial* ("Interstade") is a climatic episode within a glaciation during which a secondary recession or standstill of glaciers took place. An *Interglacial* ("Interglaciation") is an episode during which the climate was incompatible with the wide extent of glaciers that characterize a glaciation.

In Europe, following the work of Jessen and Milthers (1928), it is customary to use the terms interglacial and interstadial to define characteristic types of non-glacial climatic conditions indicated by vegetational changes (Table 30.1); interglacial to describe a temperate period with a climatic optimum at least as warm as the present interglacial (Holocene: see below) in the same region, and interstadial to describe a period that was either too short or too cold to allow the development of temperate deciduous forest or the equivalent of interglacial-type in the same region.

In North America, mainly in the USA, the term interglaciation is occasionally used for interglacial (cf. American Commission on Stratigraphic Nomenclature, 1961). Likewise, the terms stade and interstade may be used instead of stadial and interstadial, respectively (cf. American Commission on Stratigraphic Nomenclature, 1961). The origin of

Examples of Chronostratigraphical Substage Divisions of Interglacial (Temperate) Stages and Related Cold (Glacial) Stages of the Middle and Late Pleistocene

	Chronostratigraphical Substages[c]								Vegetation Aspect	Characteristic Vegetation
Cold stage		e An		e Wo		e De			early glacial	herb-dominated
Temperate	Cromerian	Cr IV	Hoxnian	Ho IV	Ipswichian	Ip IV	Flandrian		post-temperate	birch–pine forest
stage	(~750 ka)	Cr III	(~350 ka)	Ho III	(~125 ka)	Ip III	(post-10 ka)	Fl III	late temperate	mixed deciduous–coniferous forest
		Cr II		Ho II		Ip II		Fl II	early temperate	deciduous forest
		Cr I		Ho I		Ip I		Fl I	pre-temperate	birch–pine forest
Cold stage		l Be		l An		l Wo		l De	late glacial	herb-dominated

[a]Modified after West (1968) and West & Turner (1968).
[b]For the Holocene (Flandrian), correlations with the zones of Godwin (1975) are also indicated.
[c]*e, early; l, late; Be, Beestonian; An, Anglian; Wo, Wolstonian; De, Devensian.*

these terms is not certain but the latter almost certainly derive from the French word *stade* (m), which is unfortunate since in French *stade* means (chronostratigraphical) stage (cf. Michel *et al.*, 1997), e.g., *stade isotopique marin* = marine isotope stage.

It will be readily apparent that, although in longstanding usage, the glacially based terms are very difficult to apply outside glaciated regions, i.e. most of the world. Moreover, as Suggate and West (1969) recognized, the term glaciation or glacial is particularly inappropriate since modern knowledge indicates that cold rather than glacial climates have tended to characterize the periods intervening between interglacial events. They therefore proposed that the term "cold" stage (chronostratigraphy) be adopted for "glacial" or "glaciation". Likewise, they proposed the use of the term "warm" or "temperate" stage for interglacial, both being based on regional stratotypes. The local nature of these definitions indicates that they cannot necessarily be used across great distances or between different climatic provinces (Suggate and West, 1969; Suggate, 1974; West, 1968, 1977) or indeed across the terrestrial–marine facies boundary (see below). The use of mammalian biostratigraphic data, in particular the evolution of voles, offers the possibility of long-distance correlations between local assemblages. In addition, it is worth noting that the subdivision into glacial and interglacial is mainly applied to the Middle and Late Pleistocene.

Both interglacial and glacial, or temperate and cold, stages have been subdivided into substages and zones. This is achieved in interglacial stages using paleontological, particularly vegetational, assemblages. The cyclic pattern of interglacial vegetation that typifies all known temperate events in Europe was developed as a means of subdividing, comparing, and therefore, characterizing temperate events by West (1968, 1977) and Turner and West (1968). In this scheme, temperate (interglacial) event sequences are subdivided into four substages: pre-temperate, early temperate, late temperate, and post-temperate. Finer-scale zonation schemes are also commonly in use throughout Europe and the former USSR (Table 30.1).

Late Middle- and Late Pleistocene glacial stages have been divided on various bases, but in the northern hemisphere the division is based on a combination of vegetation, lithology and occasionally pedological evidence, often resulting in an unfortunate intermixture of chrono- and climatostratigraphical terminology. Chronological control for Middle and Lower Pleistocene glacial deposits comes from tephrochronology (e.g., Boellstorff, 1978), K/Ar and $^{40}Ar/^{39}Ar$ dating of intercalated lavas (e.g., Geirsdottir and Eriksson, 1994; Singer *et al.*, 2004), magnetostratigraphy (e.g., Roy *et al.*, 2004) and cosmogenic nuclide dating (e.g., Balco *et al.*, 2005; Balco and Rovey, 2010).

The last glacial stage (Weichselian, Valdaian, Devensian, Wisconsian) has particularly been divided into three or four substages (Early Middle or Pleni-glacial, Late Weichselian, etc., and Late-glacial), using geochronology (mainly ^{14}C). Boundaries are defined at specific dates, especially in the last 30 ka (Table 30.1).

A widely used event term, the Last Glacial Maximum, frequently abbreviated to LGM, is used to refer to the maximum global ice volume during the last glacial cycle (CLIMAP, 1981). It corresponds to the trough in the marine isotope record centered on c. 18 ^{14}C ka BP (Martinson *et al.*, 1987) and the associated global eustatic sea-level low also dated to 18 ^{14}C ka BP (Yokoyama *et al.*, 2000). According to Clark *et al.* (2009), most ice sheets were at their LGM positions from 26.5 ka to 19 to 20 ka cal BP. The LGM has also been assigned chronozone status (23–19 or 24–18 ka cal BP dependent on the dating applied) by Mix *et al.* (2001) who consider the event should be centered on the calibrated date at 21 ka cal BP. (i.e. LGM *sensu stricto*). However, since the post-MIS 5e (last interglacial, Eemian Stage) last maximum glaciation occurred much earlier in some areas than in others, the term LGM should be used with caution (Ehlers and Gibbard, 2011; Ehlers *et al.*, 2011). Moreover, the chronozone has neither been formally defined nor named from a type locality. It is therefore currently of informal status. However, the SQS are investigating the possibility of formally defining this and related terms (e.g., Heinrich event) at present.

30.4.2. Loess Deposits

In contrast to continental glacial sequences, which tend to be fragmentary, thick aeolian dust (loess) deposits in Europe and Asia (particularly China) are essentially continuously deposited, much like deep-ocean sediment. In all areas, the loess deposits consist of two major stratigraphic units – loess and paleosols – representing climatically controlled variations in loess accumulation and weathering processes. In the classic sections on the Chinese Loess Plateau, 34 soil–loess couplets have been identified in loess deposits more than 150 m thick, with the Gauss/Matuyama paleomagnetic boundary (~2.6 Ma) located within loess L33 (Rutter *et al.*, 1991; Yang and Ding, 2010) – see Figure 30.7. Paleosol horizons are characterized by higher magnetic susceptibility, finer grain size and redder colors than the intervening loess horizons and can be correlated throughout the Loess Plateau.

Underlying the lowermost loess layer (L34), is the so-called Red Clay, which is still of aeolian origin, but which has been weathered much more strongly than the overlying loess. The boundary between the Red Clay and L34 is dated to ~2.8 Ma, and is characterized by a dramatic increase in grain size and loess accumulation rates, interpreted as a major climate shift from long-lasting warm humid conditions to large amplitude cold/dry and warm/humid oscillations (Yang and Ding, 2010). There was also a large increase in aridity over the dust source region.

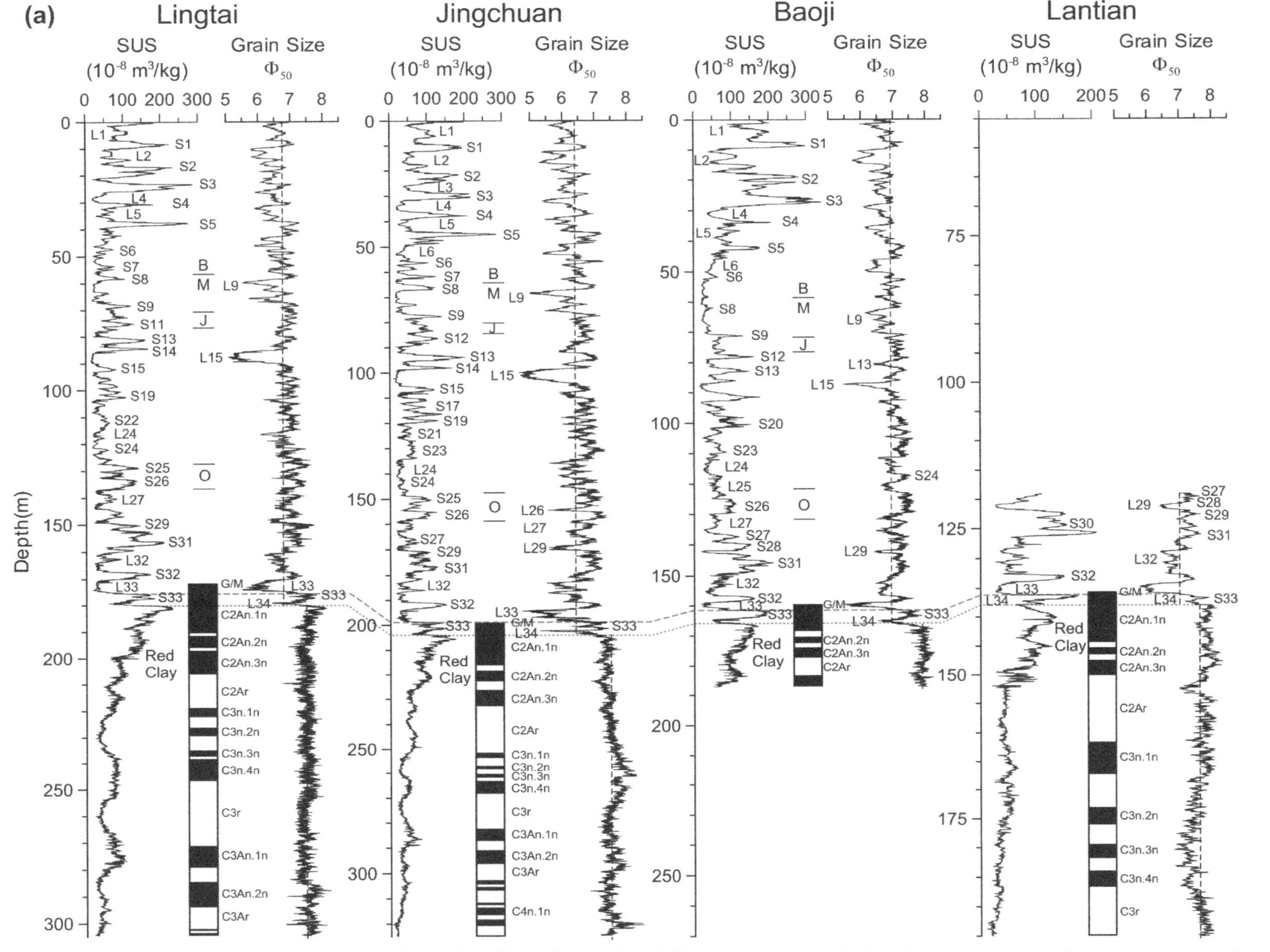

Magnetic susceptibility and grain-size variations in four Chinese loess sections. Soil horizons are characterized by high magnetic susceptibility values, which may be used to correlate between sections, in addition to magnetostratigraphy. *(Yang and Ding, 2010)*

Weinan loess from Shiling Yang. 34 soil–loess couplets have been identified, with the Gauss/Matuyama paleomagnetic boundary (~2.6 Ma) located within loess L33. Paleosol horizons (S levels) are characterized by higher magnetic susceptibility, finer grainsize and redder colors than the intervening loess (L) horizons and can be correlated throughout the Loess Plateau.

30.4.3. Ice Cores

In the past three decades, the drilling of cores into ice sheets in various parts of the world has revolutionized our records of detailed climatic change. An independent record of Late Pleistocene and Holocene climatic changes has been derived from $\delta^{16}O/\delta^{18}O$ ratios and other measurements in cores through the Greenland and Antarctic ice sheets (Johnsen *et al.*, 1972; Dansgaard *et al.*, 1993, EPICA Community Members, 2004) and from other areas, including temperate and tropical ice caps in Asia, Africa and South America (Thompson *et al.*, 1989, 1995, 2002). These have provided spectacularly unrivalled sequences which allow annual resolution of climatic events. In one case (Kilimanjaro Ice Cap) the ice cores may become the only physical record if present rates of ice cap melting continue – Thompson *et al.* (2009) have estimated that the ice cap could disappear within the next few decades.

From a stratigraphical point of view, it is the recognition of patterns of a wide range of climatically controlled parameters that provide potentially very high-resolution correlation tools in ice cores. Detailed patterns arise from determination of aerosol particles, dust, trace elements, spores, or pollen grains, etc., that have fallen onto the ice surface and become incorporated into the annual ice layers. They include, for example, dust from wind activity (e.g., Delmonte *et al.*, 2004) or volcanic eruptions (e.g., Zielinski *et al.*, 1997; Davies *et al.*, 2010). Trace gases such as carbon dioxide or methane can be trapped in air bubbles within ice crystals and provide long, continuous records of atmospheric greenhouse gas concentrations (Luthi *et al.*, 2008). In addition, naturally and artificially occurring radioactive isotopes present in the ice layers can be used to provide an independent chronology for dating the ice core sequences (e.g., Dreyfus *et al.*, 2008).

Most cores span the Holocene and provide an annually resolved sequence for the current interglacial. However, the Dome C core in Antarctica extends back to nearly 800 ka (Figure 30.8; EPICA Community Members, 2004; Luthi *et al.*, 2008), while in Greenland detailed cores from the Greenland Ice Core Project (GRIP) and the Greenland Ice Sheet Project (GISP) have been obtained that provide a sequence that extends at least as far back as the Last (Eemian) Interglacial (North Greenland Ice Core Project Members, 2004). Chronology comes from a combination of annual layer counting, ice flow modeling and independent age markers (Andersen *et al.*, 2006; Parrenin *et al.*, 2007; Svensson *et al.*, 2008). These sequences have revolutionized our understanding of patterns and rates of global climate changes, as well as the interlinking of the ocean–atmosphere–terrestrial systems (cf. Lowe and Walker, 1997, for a more detailed discussion).

The climatically determined subdivisions of ice core sequences have been used as a basis for an event-based stratigraphy at least for the end of the last glacial period, by

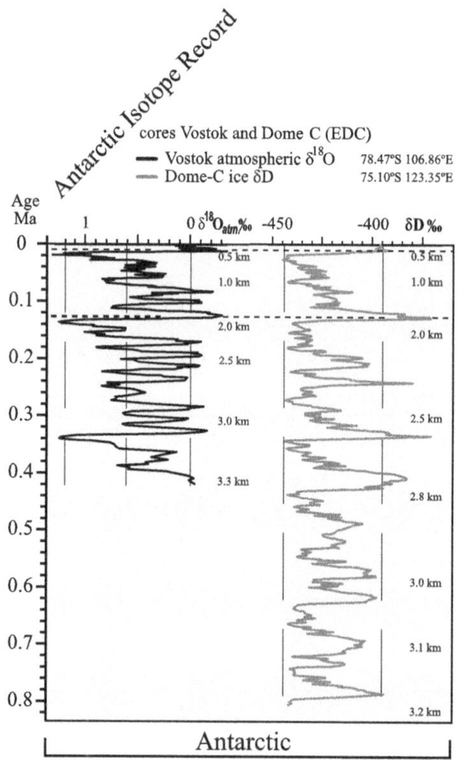

The first is the 420 ka-long plot from the Vostok core and shows atmospheric $\delta^{18}O$ (Petit et al. 1999), determined from gas bubbles in the ice. This atmospheric $\delta^{18}O$ is inversely related to $\delta^{18}O$ measurements from seawater and therefore is a measure of ice volume. It can also be used to separate ice volume and deepwater temperature effects in benthic foraminiferal $\delta^{18}O$ measurements. The deuterium measurements (δD) for the last 800 ka are from the 3.2 km deep EDC core in Dome C (EPICA Community Members, 2004; Jouzel et al., 2007). They come from samples of the ice itself and give a direct indication of Antarctic surface paleotemperature.

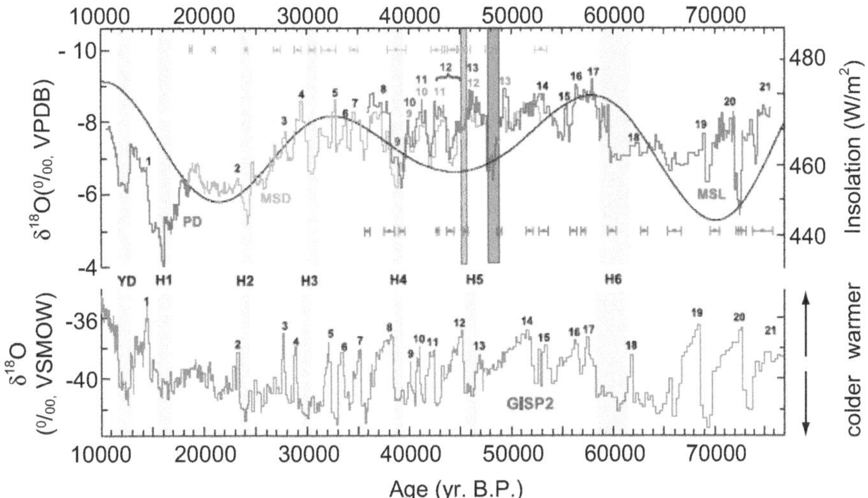

Comparison between $\delta^{18}O$ records from Hulu Cave (China) stalagmites and Greenland ice core (Wang et al., 2001), showing summer insolation at 33°N (the latitude of Hulu Cave). Cold climatic events (YD = Younger Dryas and H1–H6 = Heinrich Events) indicated by vertical bars. Interstadial (warm climate) events are numbered 1 to 21 in both records.

Björck *et al.* (1998). In this scheme warmer and cooler events are referred to as stadials or interstadials, in common with other climatostratigraphic schemes applied to the Quaternary (see also Figure 30.9).

30.4.4. Speleothems

Advances in high-precision U/Th dating methods over the past decade have permitted precise dating of carbonate deposits in caves (speleothems) such as stalagmites and stalactites. In a series of ground-breaking papers, Wang *et al.* (2001, 2004) and Cheng *et al.* (2006, 2009) have demonstrated that the paleoclimate record (principally $\delta^{18}O$) in speleothems is as detailed as that in ice cores, and with comparable dating precision (Figure 30.9).

30.4.5. Long Lake Records

Comparable evidence to that from ice cores has been retrieved in the last three decades by the recovery of sediment cores and associated studies of deep lake basins in the interiors of the continents. These investigations have produced sedimentary records of environmental changes over time spans of a few to hundreds of thousands of years, and in some rare cases of the whole Quaternary. Examples include Lake Baikal in Asia (Prokopenko *et al.*, 2006), Lake George in Australia (Singh *et al.*, 1981), Lake Malawi in Africa (Lyons *et al.*, 2007), Lake Biwa in Japan (Fuji, 1988), Tenaghi Phillipon in Greece (van der Weil and Wijmstra, 1987a,b; Tzedakis *et al.*, 2006), Lake Van in Turkey (Litt *et al.*, 2009), La Grande Pile in France (Woillard, 1978; de Beaulieu and Reille, 1992), the Great Salt Lake in the USA (Eardley *et al.*, 1973), and the Bogotá Basin in Columbia (Hooghiemstra, 1989). The long sequences in these basins preserve lithological, biological and geochemical records of past environmental changes which, together with the identification of climatic and physical events such as magnetic reversals and tephra strata, can be correlated to global, as well as regional stratigraphies (e.g., Hooghiemstra and Sarmiento, 1989). The correlation is achieved by defining individual climatic or chronostratigraphies for each basin which are then compared and correlated with external sequences, mainly the marine isotope record (e.g., Tzedakis *et al.*, 1997, 2006). For example, the Siberian Lake Baikal provides a bioproductivity record from the centre of the world's largest landmass, an area of extreme continental climate. High concentrations of biogenic silica indicate high aquatic productivity during interglacials (i.e. lake diatom blooms during ice-free summer seasons). The composite biogenic silica record from cores BDP-96-1, BDP-96-2 and BDP-98 is plotted on an astronomically tuned age scale (Prokopenko *et al.*, 2006; Prokopenko and Khursevich, 2010) – see Figure 30.4.

30.4.6. Mammal Biochronology and the Villafranchian "Stage"

The term "Villafranchian" was introduced by Pareto (1865) as a continental stage referring to fluvial and lacustrine sediments in the Villafranca d'Asti (Piedmont) region of Italy that contained mammal fossils. Gignoux (1916) proposed a correlation with the marine Calabrian Stage, which resulted in the Villafranchian being associated with the basal Pleistocene. However, the Villafranchian is now known to span the period from about 3.5 to 1.0 Ma (i.e., Pliocene–Lower Pleistocene) and is defined as a Mammal Age, or biochronological unit based on the evolution of large European mammals (Rook and Martínez-Navarro, 2010) – Figure 30.10.

CHRONOLOGICAL RANGE OF LARGE EUROPEAN MAMMALS
(newcomers from Africa in orange)
EPOCHS
AGES (Ma)
CHRON
MAMMAL AGES
FAUNAL UNITS
Local Faunas
Occurrence of genus Homo in Europe
Mid.
EARLY PLEISTOCENE
LATE PLIOCENE
1,0
1,5
2,0
2,5
3,0
3,5
BRUNHES
Jaramillo
C.M.
MATUYAMA
Olduvai
Reunion
GAUSS
Mamm.
Kaena
GILBERT
GALERIAN
Early
VILLAFRANCHIAN
Early - Middle - Late
RUSCINIAN
Atapuerca TD6
SLIVIA
Le Vallonet
COLLE CURTI
Cueva Victoria
Untermassfeld
Atapuerca TE
F.-Nueva 3
B.-Leon 5
Incarcal
PIRRO NORD
Venta Micena
Pietrafitta
FARNETA
TASSO
Dmanisi
P. Rosso
OLIVOLA
Fonelas P1
Senéze
C.S.GIACOMO
ST. VALLIER
Huélago
F.-Nueva 1
Perrier L-Et
MONTOPOLI
Villaroya
Vialette
Kvabebi
TRIVERSA
Santa Barbara
Tapirus arvernensis
Mammut borsoni
Mesopithecus monspessulanum
Sus minor
Anancus arvernensis
Acynonix pard.
Leptobos stenometopon
Stephanorhinus elatus
Pseudodama lyra
Pliocrocuta perrieri
Chasmaporthetes lunensis
Homotherium crenatidens
Canis sp.
Equus ex gr. livetzovensis
Megantereon cultridens
Mammuthus rumanus
M. meridionalis
Gazella borbonica
Eucladoceros tegulensis
Gallogoral meneghinii
Gazellospira torticornis
Equus stenonis
Stephanorhinus etruscus
Sus strozzii
Canis ex gr. etruscus
Procamptoceras brivatense
Leptobos etruscus
Eucladoceros dicranios - ctenoides
Pseudodama nestii
Pachycrocuta brevirostris
Panthera bombaszoegensis
Canis arnensis
Lycaon falconeri
Leptobos vallisarni
Praeovibos sp.
Equus stehlini
Megantereon whitei
Pseudodama eurygonos - farnetensis
Homotherium latidens
Praemegacerus oscurus
Stephanorhinus hundshaimensis
Bison spp.
Praemegaceros verticornis
Lycaon lycaonoides
Canis mosbachensis
Equus altidens
Theropithecus sp.
Hippopotamus antiquus
Sus scrofa piscus
Mammuthus trogontherii
Crocuta crocuta
Atap. TD6
Vallonet
Atap. TE
FN3
BL5
Pirro Nord
Dmanisi

Chronology of the Villafranchian Mammal Age, with range chart for selected large mammals. *(Rook and Martinez-Navarro, 2010)*

30.5. OCEAN-SEDIMENT SEQUENCES

Because the span of Quaternary time includes our own, a different order of discrimination is possible and different methods are rapidly developing. The principal development in the Pleistocene time scale depends on the regularity of the climatic cycle that was discovered around 1875 by Croll and developed, especially, by Milankovitch. This approach was not taken seriously by Quaternary geologists until Zeuner (1945), Emiliani (1955), Broecker *et al.* (1968) and Evans (1971) were among those to recalculate and relate the astronomical parameters, testing, for example, 42- and 100-ky cycles against other phenomena, such as the newly established oxygen isotope curve from the oceans. The first rigorous treatment using wide-ranging techniques was by Hays *et al.* (1976). Isotope studies from the bottom sediments of the world's oceans since then have indicated as many as 52 Quaternary glacial/interglacial cycles and have clearly shown that the continental evidence can be very fragmentary.

The marine oxygen isotope scale makes use of the fact that, when continental ice builds up as a result of global cooling and sea level is lowered, the ice is depleted in $\delta^{18}O$ relative to the ocean water, leaving the ocean water enriched in $\delta^{18}O$. The oxygen isotope composition of calcareous foraminifera and coccoliths, and of siliceous diatoms, varies in direct proportion to that of the water (cf. Shackleton and Opdyke, 1973, for discussion of the limitations of isotope stratigraphy). The 16 stages of Emiliani (1955, 1966) obtained from Caribbean and Atlantic sediment cores were extended to 22 by Shackleton and Opdyke (1973) after analysis of the V28-238 core from the equatorial Pacific. Subsequently another equatorial Pacific core V28-239 (Shackleton and Opdyke, 1976) and an Atlantic core (Van Donk, 1976) extended the reconstruction of glacial–interglacial variability through the Pliocene–Pleistocene boundary. Later developments under the flag of the Deep Sea Drilling Program resulted in the extension of the isotope record into the Early Pleistocene and Pliocene (Shackleton and Hall, 1989; Ruddiman *et al.*, 1987; Raymo *et al.*, 1989). The sequence shown in Figure 30.4 is a composite benthic $\delta^{18}O$ record from 57 globally distributed sites, containing more than 38 000 individual measurements, to maximize the signal-to-noise ratio (Lisiecki and Raymo, 2005).

As regards nomenclature, the events differentiated in isotope sequences are termed marine isotope stages (MIS); this term is preferred by paleoceanographers to the previously widely used oxygen isotope stages (OIS). This is because of the need to distinguish the isotope stages recognized from those identified from ice core or speleothem sequences (Shackleton, pers. comm.). The stages are numbered from the present day (MIS 1) backwards in time, such that cold-climate or glacial events are assigned even numbers and warm or interglacial (and interstadial) events are given odd numbers. Individual events or substages in marine isotope stages are indicated either by lower-case letters or in some cases by a decimal system, thus MIS 5 is divided into warm substages 5a, 5c, and 5e, and cold substages 5b and 5d, or 5.1, 5.3, 5.5, and 5.2 and 5.4, respectively, named from the top downwards. This apparently unconventional top-downwards nomenclature originates from Emiliani's (1955) original terminology and reflects the need to identify oscillations down cores from the ocean floor.

The biggest problem with climate-based nomenclature, like the marine isotope stratigraphy, is where the boundaries should be drawn. Ideally, the boundaries should be placed at a major climate change. However, this is problematic because of the multifactorial nature of climate. But since the events are only recognized through the responses they initiate in depositional systems and biota, a compromise must be agreed. Although there are many places at which boundaries could be drawn, in principle in ocean-sediment cores they are placed at midpoints between temperature maxima and minima. The boundary points thus defined in ocean sequences are assumed to be globally isochronous, although a drawback is that temperatures may be very locally influenced and may also show time lag. The extremely slow sedimentation rate of ocean-floor deposits and the relatively rapid mixing rate of oceanic waters argue in favor of the approach. Attempts to date these MIS boundaries are now well established (Martinson *et al.*, 1987).

30.6. LAND–SEA CORRELATION

In recent years it has become common to correlate terrestrial sequences directly with the marine isotope stages established in deep-sea cores. This arises from the need felt to correlate local sequences to a regional or global time scale, occasioned by the fragmentary and highly variable nature of many terrestrial sequences. The realization that more events are represented in the deep-sea, and indeed ice core, sequences than were recognized on land, together with the growth in geochronology, has often led to the replacement of locally established terrestrial time scales. Instead, direct correlations of terrestrial sequences to the global isotope scale are advanced, as advocated, for example, by Kukla (1977). The temptation to do this is understandable, but there are serious practical limitations to this approach (cf. Schlüchter, 1992).

In reality, there are very few means of directly and reliably correlating between the ocean and terrestrial sediment sequences. Direct correlation can be achieved using markers that are preserved in both rock sequences, such as magnetic reversals or tephra layers, and, rarely, fossil assemblages (particularly pollen). However, this is normally impossible over most of the record and in most geographical areas. Thus these correlations must rely on direct dating or less reliably on the technique of "curve matching", a widely used approach in the Quaternary. The latter can only reliably be achieved where long, continuous terrestrial sequences are available, such as long lake records (e.g., Tzedakis *et al.*, 1997), but even here it is not always straightforward (e.g., Watts *et al.*, 1995) because of overprinting by local factors. Moreover, the possibility of failure to identify "leads-and-lags" in timing by the matching of curves is very real. In discontinuous sequences which typify land and shelf environments, correlations with ocean-basin sequences are potentially unreliable, in the absence of fossil groups distributed across the facies boundaries or potentially useful markers.

Increasingly, highly characteristic events are used as a basis for correlation. This event stratigraphy (e.g., Björck *et al.*, 1998; Lowe *et al.*, 1999; Alloway *et al.*, 2007), typically using deposition of a tephra layer or magnetic reversals, can also include geological records of other potentially significant events such as floods, tectonic movements, changes of sea level and climatic oscillations. Such occurrences, often termed "sub-Milankovitch events", may be preserved in a variety of environmental settings and thus offer important potential tools for high- to very high-resolution cross-correlation.

Of particular importance are the so-called "Heinrich Layers" which represent major iceberg-rafting events in the North Atlantic Ocean (Heinrich, 1988; Bond *et al.*, 1992; Bond and Lotti, 1995). These detritus bands can potentially provide important lithostratigraphical markers for intercore correlation in ocean sediments and the impact of their accompanying sudden coolings (Heinrich Events) may be recognizable in certain sensitive terrestrial sequences such as ice cores and speleothems (Figure 30.9). Similarly the essentially time-parallel periods of abrupt climate change termed "terminations" (Broecker and Van Donk, 1970), seen in marine isotope records, can also be recognized on land as sharp changes in pollen assemblage composition or other parameters, for example, where sufficiently long and detailed sequences are available, such as in long lake cores (cf. Tzedakis *et al.*, 1997) and in speleothems (Cheng *et al.*, 2009; see Figure 30.11). However, their value for correlation may be limited in high-sedimentation-rate sequences because these "terminations" are not instantaneous but have durations of several thousand years (Broecker and Henderson, 1998). These matters essentially concern questions of resolution and scale.

Of greater concern for the development of a high-resolution marine–terrestrial correlation is that different proxies respond at different rates and in different ways to climate changes and these changes themselves may be time-transgressive. This has been forcefully demonstrated in marine cores off Portugal by Sanchez-Goñi *et al.* (1999) and Shackleton *et al.* (2003) where the MIS 6/5 boundary has been shown to have not been coeval with the Saalian–Eemian Stage boundary on land, as was previously assumed (see also Figure 30.5). The same point concerns the MIS 1–2 boundary, which pre-dates the Holocene–Pleistocene boundary by some 2000–4000 years. Thus high-resolution land sequences and low-resolution marine sequences must be correlated with an eye to the detail since it cannot be assumed that the boundaries recognized in different situations are indeed coeval.

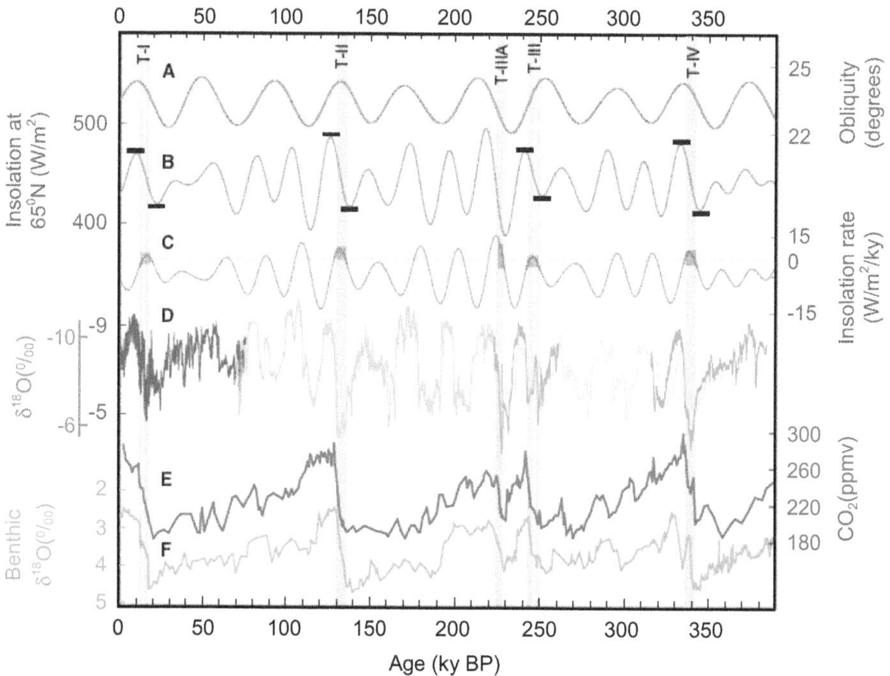

Correlations between summer insolation at 65°N, $\delta^{18}O$ in Hulu Cave stalagmites (China), Vostok CO_2 record (Antarctica), and benthic $\delta^{18}O$ in ODP 980 core (North Atlantic). *(After Cheng et al., 2009)*

Notwithstanding these problems of detail, which will no doubt be further resolved as new evidence becomes available, it is now generally possible to relate the onshore–offshore sequences fairly reliably at a coarse scale at least for the last glacial–interglacial cycle (Upper/Late Pleistocene). This was first proposed by Woillard (1978), but is now well established (e.g., Tzedakis *et al.*, 1997). Beyond, things are very much more complicated. Witness, for example, the longstanding disagreements over the nature and duration of the northwest European Saalian Stage, already referred to above (Litt and Turner, 1993). Questions of whether the Holsteinian/Hoxnian temperate (interglacial) stage relates to MIS 9 or 11 (e.g., Grün and Schwarcz, 2000; Geyh and Müller, 2005), and thus the immediately preceding Elsterian/Anglian glacial stage (cf. Zagwijn, 1992) to MIS 10 or 12 (Turner, 1998; de Beaulieu and Reille, 1995; Litt *et al.*, 2008) or even MIS 8, and precisely how many interglacial-type events occur within the Saalian, leave much potential for inaccuracy that cannot be resolved by "counting backwards" methods. In the absence of reliable dating these correlations represent little more than a matter of belief.

30.6.1. Wanganui Basin, New Zealand

An outstanding example of successful land–sea correlation comes from the shallow marine and coastal sedimentary sequences of Wanganui Basin, a back-arc basin in the North Island of New Zealand, where a cycle-by-cycle correlation between marine isotope stages and basin sediments has been achieved for the entire Quaternary (Naish *et al.*, 1998) (Figure 30.12).

Wanganui Basin is one of a small number of continental-margin basins in which sediment accumulation kept pace with subsidence throughout the Quaternary Period, resulting in a 5 km thick sequence of Neogene–Quaternary age. The basin fill consists of cyclic, unconformity-bound strata representing shelf and coastal depositional environments. Gentle regional uplift and southward migration of the basin depocenter have resulted in on-land exposure of the sequence, including a broad flight of marine terraces spanning the last 700 ka (Pillans, 1983).

Chronology comes from fission track-dated tephras (Pillans *et al.*, 2005), magnetostratigraphy (Pillans et al., 1994), biostratigraphy, amino-acid racemization (Bowen *et al.*, 1998), luminescence (Berger *et al.*, 1992) and radiocarbon dating, coupled with cyclostratigraphic correlation to the marine isotope record (Naish *et al.*, 1998; Pillans *et al.*, 2005).

Regional and global correlations of the shallow marine sequence in Wanganui Basin have been made to Taupo Volcanic Zone (Alloway *et al.*, 2005), deep-sea cores east of New Zealand (Alloway *et al.*, 2005; Pillans *et al.*, 2005), and the classic Plio–Pleistocene sections in Italy (Beu and Edwards, 1984; Naish *et al.*, 1998).

30.6.2. Ross Sea, Antarctica

In the austral summer of 2006/2007, the ANtarctic Geological DRILLing Program (ANDRILL) recovered a 1284.87-m-long core from beneath the Ross Ice Shelf at 77.89°S in Antarctica, spanning the last 13 Ma (McKay *et al.*, 2009; Naish *et al.*, 2009), the uppermost 190 m of which spans the Quaternary Period (Figure 30.13). Chronology is based on biostratigraphy, magnetostratigraphy and $^{40}Ar/^{39}Ar$ ages of interbedded volcanics. The core contains a cyclic succession of subglacial, glaciomarine and marine sediments that comprises some 58 sequences of orbital-scale duration.

The glaciomarine cycles in the core, each bounded glacial surfaces of erosion, reflect fluctuations in the extent of the Ross Ice Shelf, which is largely controlled by oscillations in the size of the West Antarctic Ice Sheet (Naish *et al.*, 2009). The Late Pliocene and Early Pleistocene cycles are characterized by subglacial diamictite (till) overlain by diatom-bearing marine mudstones and diatomites, the latter interpreted as representing open-water deposition associated with a warmer than present climate. Middle Pleistocene cycles are dominated by subglacial diamictite overlain by thin mudstones interpreted as ice-shelf deposits (McKay *et al.*, 2009).

The youngest marine diatomite in the core occurs at a depth of ~90 m, and is correlated with warmer than present conditions during MIS 31, representing partial collapse of the West Antarctic Ice Sheet (Naish *et al.*, 2009; Pollard and DeConto, 2009).

30.7. PLEISTOCENE–HOLOCENE BOUNDARY

In a previous edition of this book (Harland *et al.*, 1990) it was stated that "this boundary was thought to correspond to a climatic event around 10 000 radiocarbon years before present (BP)". At the time, the boundary was considered likely to be standardized in a varved lacustrine sequence in Sweden (cf. Mörner, 1976). It was originally proposed at the VIII INQUA Congress in Paris in 1969 and was subsequently accepted by the INQUA Holocene Commission in 1982 (Olausson, 1982). The climatic amelioration on which this boundary is identified is well established in a variety of sediments, particularly in northern Europe and North America. In Scandinavia, it corresponds to the following boundaries: European Pollen Zones III–IV, the Younger Dryas–pre-Boreal and Late Glacial–post-Glacial (Mörner, 1976; Mangerud *et al.*, 1974). However, this boundary definition was not formally ratified by the ICS. If it had been defined precisely at 10 000 (^{14}C yr BP), it would have been the first stratigraphic boundary later than the Proterozoic to be defined chronometrically.

In 2008, the Pleistocene–Holocene boundary was defined at a depth of 1492.45 m in the NGRIP ice core from Greenland, with an age, based on annual layer counting, of 11 700 years b2k (before AD2000), with a 2σ uncertainty of 99 years

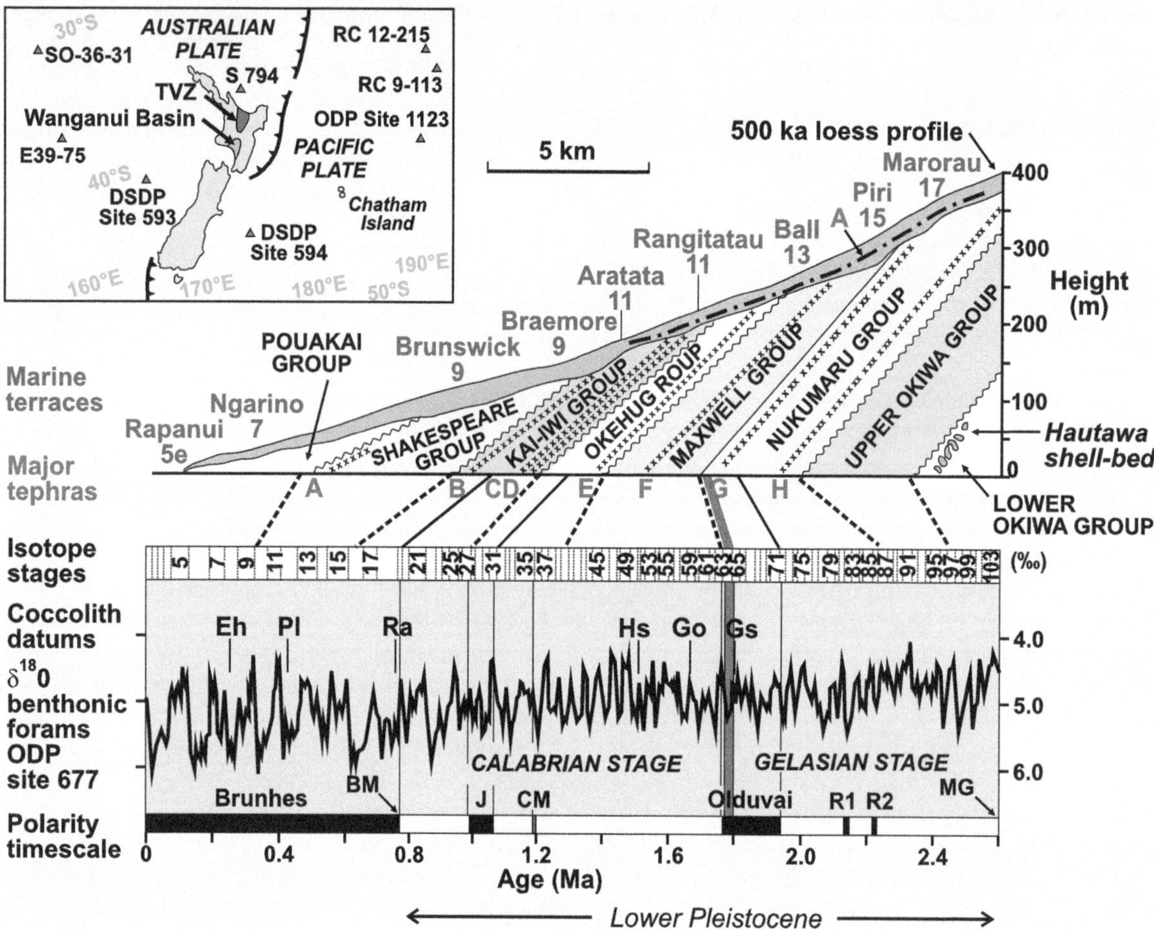

(top). Coastal outcrop of basin cycles 33 to 35 (MIS29 to 25), Okehu Stream, Castlecliff Beach, Wanganui Basin. Kaimatira Pumice Sand contains volcaniclastic sediment from the Potaka Tephra eruption (1.00 Ma) in the Taupo Volcanic Zone. (*Pillans et al., 2005*)

(bottom). Generalized cross-section of Wanganui Basin, New Zealand, showing stratigraphic relationships between marine terraces, and major lithostratigraphic units of the shallow marine basin-fill, including correlation to the marine isotope record of ODP Site 677.

Major tephras: A. Rangitawa (0.345 Ma), B. Kupe (0.63 Ma), C. Kaukatea (0.86 Ma), D. Potaka (1.00 Ma), E. Rewa (1.20 Ma), F. Pakihikura (1.58 Ma), G. Waipuru (1.79 Ma), H. Ohingaiti (Pillans et al., 2005).

Major biostratigraphic datums (coccoliths): Eh. FAD *Emiliania huxleyi*, Pl. LAD *Pseudoemiliania lacunosa*, Ra. LAD *Reticulofenestra asanoi*, Hs. LAD *Helicosphaera sellii*, Go. FAD *Geophyrocapsa oceanica*, Gs. *Geophyrocapsa sinuosa* (Naish et al., 1998).

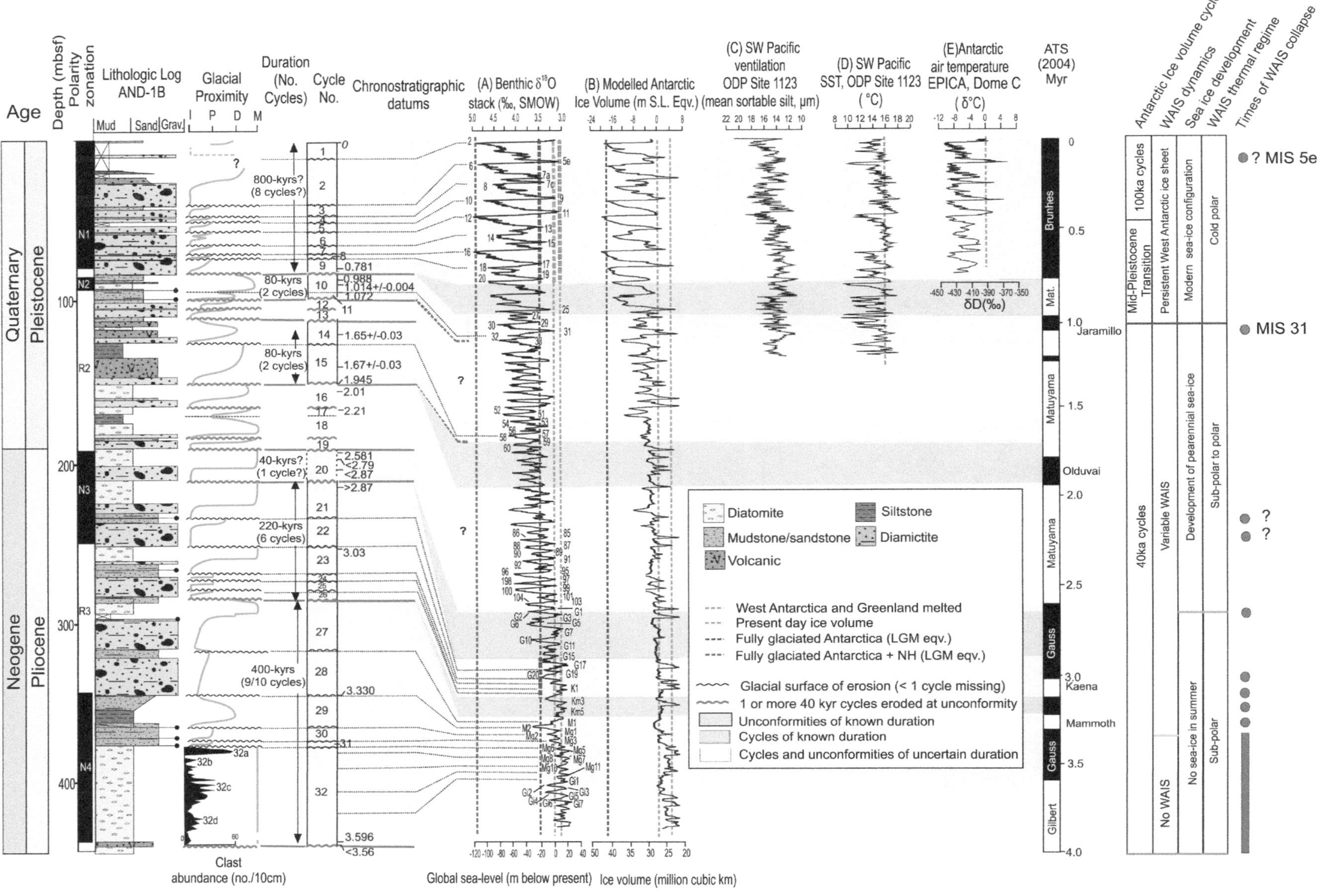

Stratigraphic and chronologic summary of the upper 450 m of the AND-1B core showing 32 sedimentary cylces of ice sheet advance—retreat—readvance during the last 4 Myr (Naish et al., 2009). Lithologies (rock-types) are plotted against depth. Glacial proximity curve tracks the relative position of the grounding line through (I) ice-contact, (P) ice-proximal, (D) ice-distal and (M) marine environments and provides a proxy for ice sheet extent (McKay et al., 2009). Cycle duration is constrained by chronostratigraphic datums coded A-Z (Table 1). The chronostratigraphy allows correlation of some cycles (blue shading) with time series of (A) stacked benthic $\delta^{18}O$ record (Lisiecki and Raymo, 2005), (B) model output Antarctic ice volume (Pollard and DeConto, 2009), (C) deep southwest Pacific Ocean ventilation (Hall et al., 2001), (D) southwest Pacific Ocean sea-surface temperatures (Crundwell et al., 2008) (E) Antarctic air temperature (Jouzel et al., 2007). Right-hand side of figure illustrates key phases in the dynamics and thermal evolution of the West Antarctic Ice Sheet during the onset and duration of the Quaternary Period. Courtesy of Tim Naish.

Pleistocene – Holocene Boundary in the NGRIP Ice Core, Greenland

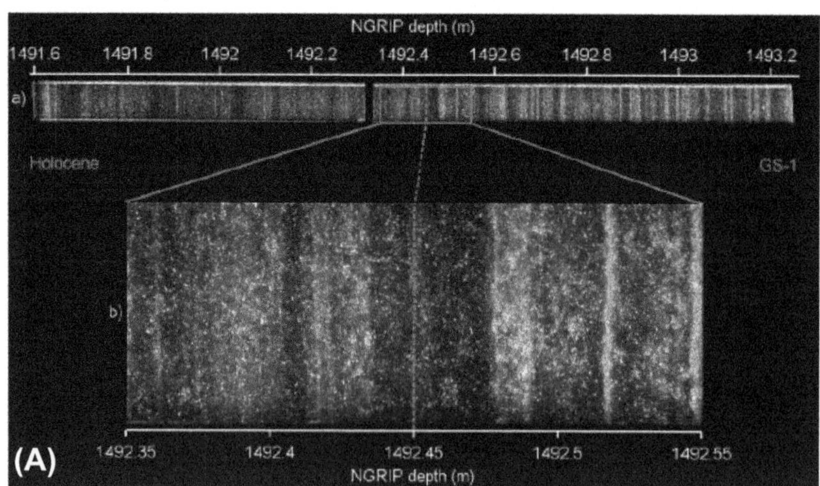

The location of the Pleistocene–Holocene boundary at 1492.45 m is shown in the enlarged lower image.

-90 -80 -70 -60 -50 -40 -30 -20 -10 0 10
80
76
Camp Century
GSSP
NGRIP
72
GISP2 GRIP
Renland
Latitude
68
continental divide
64
Dye3
60
(B)
-50 Longitude -40 -30

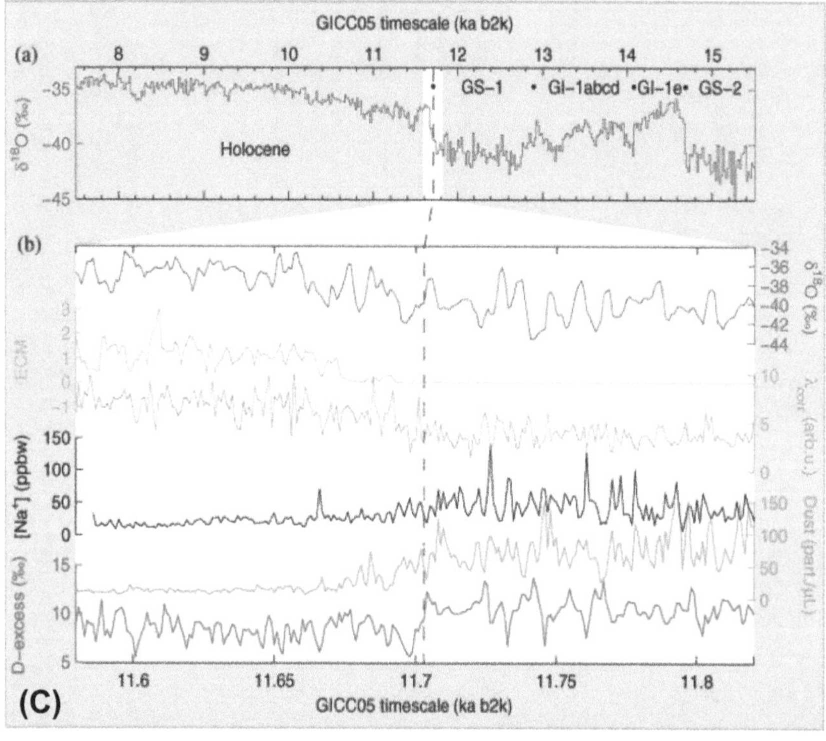

(a) The $\delta^{18}O$ record through the Last Glacial–Interglacial Transition, showing the position of the Pleistocene–Holocene boundary in the NGRIP core.

(b) High-resolution multi-parameter record across the Pleistocene–Holocene boundary: $\delta^{18}O$, electrical conductivity (ECM), annual layer thicknesses corrected for flow-induced thinning (lcorr) in arbitrary units, Na^+ concentration, dust content, and deuterium excess.

Holocene GSSP. Visual stratigraphy of the NGRIP core between 1491.6 and 1493.25 m depth obtained using a digital line scanner. In this photograph, the image is "reversed" so that clear ice shows up black, and cloudy ice (which contains impurities such as micrometer-sized dust particles) shows up white. The visual stratigraphy is essentially a seasonal signal and reveals annual banding in the ice. The position of the Pleistocene–Holocene boundary at 1492.45 m is shown in the enlarged lower image (Walker et al., 2009).

(Walker *et al.*, 2008). The boundary is most clearly marked by an abrupt decrease in deuterium excess values, indicating rapid warming within a period of 1 to 3 years and over subsequent decades. Other indicators, which change more slowly, include $\delta^{18}O$, dust concentration, annual layer thickness and a range of chemical species (Figure 30.14).

30.8. HOLOCENE SERIES

Holocene is the name for the most recent interval of Earth history and includes the present day. Although generally regarded as having begun 10 000 radiocarbon years, or the last 11 500 calibrated (i.e. calendar) years, before present (i.e.

1950), the base of the Holocene is now assigned an age of 11 700 years b2k (before AD2000) as described above. The term "Recent" as an alternative to Holocene is invalid and should not be used. Sediments accumulating or processes operating at present should be referred to as "modern" or by similar synonyms.

The term Flandrian, derived from marine transgression sediments on the Flanders coast of Belgium (de Heinzelin and Tavernier, 1957), has often been used as a synonym for Holocene (Figure 30.2). It has been adopted by authors who consider that the present (Holocene) interglacial should have the same stage status as previous interglacial events and thus be included in the Pleistocene. In this case, the latter would thus extend to the present day (cf. West, 1968, 1977, 1979; Hyvärinen, 1978). This usage, although advocated particularly in Europe, has been losing ground in the last three decades (cf. Lowe and Walker, 1997, p. 16) and is now untenable with formal ratification of the Holocene Series.

Various zonation schemes have been proposed for the Holocene (Flandrian) Epoch. The most established is that of Blytt and Sernander (cf. Lowe and Walker, 1997), which was developed for peat bogs in Scandinavia in the late 19th to earliest 20th centuries. Their terminology, based on interpreted climatic changes, comprised, in chronological order, the pre-Boreal, Boreal, Atlantic, sub-Boreal, and sub-Atlantic. This scheme was refined by the Swede von Post and others, using pollen analysis throughout Europe. Today this terminology remains in use in northern Europe, although it has been largely displaced by precise numerical chronology, particularly from ^{14}C dating, which has shown that the biostratigraphically defined zone boundaries are diachronous (cf. Godwin, 1975). An attempt to fix these boundaries to precise dates was proposed for northern Europe by Mangerud *et al.* (1974).

In prehistoric times, as well as later, climatic events have largely served to identify the divisions elaborated by modern ^{14}C, as well as other dating techniques, such as tephrochronology, and dendrochronology as well as human history. Using these techniques Holocene time can be divided into ultra-high-resolution divisions. For example, recent developments indicate that cyclic patterns of climate change of durations as short as 200 years can be differentiated and potentially used for demonstrating equivalence in peat sequences, while ice cores and annually laminated lacustrine sequences allow even more highly resolved correlations.

30.9. "ANTHROPOCENE SERIES"

A recent proposal has been suggested to establish a new series/epoch, following the Holocene, to be termed the Anthropocene (Crutzen, 2002). This term is being increasingly employed to identify the current interval of anthropogenic global environmental change, and might be proposed on stratigraphical grounds (it should not be confused with the term *Anthropogene* which was used in the former USSR as a synonym for Pleistocene). It might be adopted at formal series/epoch level for the time since the start of the Industrial Revolution or possibly earlier (cf. Ruddiman, 2003), when changes sufficient to leave a global stratigraphic signature distinct from that of the Holocene or of previous Pleistocene interglacial phases occurred. A boundary definition could be made, either using a GSSP or by adopting a numerical date (i.e. Global Standard Stratigraphic Age, GSSA). Formal adoption of this term depends on its practicality, particularly to earth scientists working on younger parts of Holocene successions (Zalasiewicz *et al.*, 2008; see also Chapter 32 of this volume). From the perspective of the far future, this datum would provide a distinctive stratigraphical boundary.

30.10. QUATERNARY DATING METHODS

There is a much wider range of geological dating techniques underpinning Quaternary chronology than for earlier periods of the geological time scale. Some of the techniques are solely applied to Quaternary deposits, while others can also be applied to older deposits. Here, we briefly summarize the principles and applications of the less familiar, but nonetheless important, Quaternary dating methods (Table 30.2). The time ranges over which each of the methods are used are shown in Figure 30.15. General references on Quaternary dating methods include Easterbrook (1988), Rutter and Cato (1995), Wagner (1998), Noller *et al.* (2000), Walker (2005) and the Encyclopedia of Quaternary Science (Elias, 2007).

RADIOCARBON or carbon-14 dating is probably the most familiar and widely used dating method in the Quaternary. Radiocarbon (^{14}C) is a cosmogenic radioactive isotope that is produced in the upper atmosphere by cosmic ray bombardment. It is then mixed through the atmosphere and the oceans and is incorporated into the bodies of animals and plants as well as non-organic carbonate deposits such as speleothems. After death of an organism or after the ^{14}C is locked into a mineral structure, it undergoes radioactive decay, with a half life of 5730 years. The method is therefore used for dating a wide range of organic and inorganic materials, such as shell, wood, charcoal, peat and coral, that are up to about 50 000 years old. Since the amount of ^{14}C in the atmosphere is not constant over time, conversion of radiocarbon age to calendar age utilizes calibration curves from dated tree rings (Reimer *et al.*, 2004).

LUMINESCENCE. Natural radiation (α, β, γ and cosmic) produces electron and hole pairs which can be trapped at defects in a crystal lattice. Luminescence is emitted when electrons liberated from the traps by heating (thermoluminescence or TL) or exposure to visible-wavelength light (optically stimulated luminescence or OSL) or infra-red light (infra-red stimulated luminescence or IRSL) recombine with trapped holes. The greater the radiation dose accrued, the greater the resulting luminescence. Age is found by calibrating the luminescence with known

Classification of Quaternary Dating Methods According to Type of Method Employed

Sidereal†	Isotopic	Radiogenic	Chemical/ Biological	Geomorphic	Correlation
Historical records	Radiocarbon	Fission-track	Amino acid racemization	Weathering rinds	Paleomagnetism
Dendrochronology	K/Ar and Ar/Ar	Luminescence	Obsidian hydration	Soil development	Stable isotopes
Varve chronology	Uranium series	Electron spin resonance	Lichenometry	Geomorphic position	Biostratigraphy
Ice cores	Cosmogenic isotopes		Soil chemistry	Rate of deposition	Lithostratigraphy
	(U+Th)/He			Rate of deformation	Orbital variations
	U/Pb				Tephrochronology
	^{210}Pb				Tektites
					Rock varnish

(modified from Colman et al., 1987)

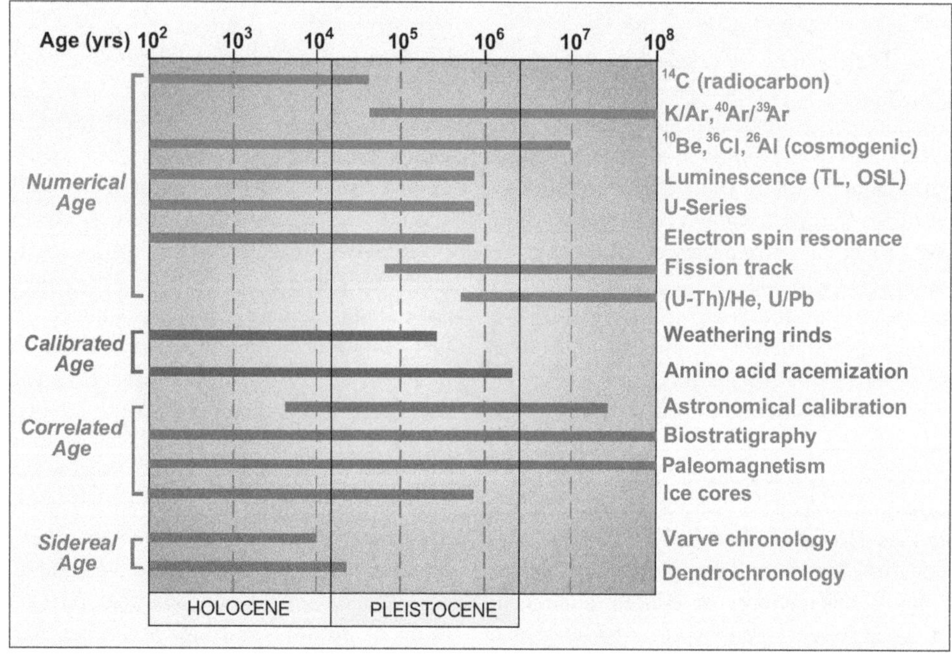

Quaternary dating methods grouped according to type of age result produced (Colman et al., 1987), showing age ranges over which each method can be applied.

radiation doses, in combination with measurements of the environmental dose rate. The luminescence method dates the last time a sample was heated or exposed to sunlight, and is especially used for dating pottery (time of heating) and aeolian deposits (last exposure to sunlight). Samples up to about 1 million years old may be dated in some circumstances (e.g., Huntley *et al.* 1993).

ELECTRON SPIN RESONANCE (ESR). As with luminescence dating, ESR dating relies on radiation producing unpaired electrons in crystal lattices, with the concentration of unpaired electrons increasing with time and radiation dose. However, in ESR dating the proportion of trapped electrons is measured directly (using an ESR spectrometer), rather than from the luminescence emitted by the electrons. The method is commonly used for dating carbonates such as corals, shells and speleothems, but arguably its most important application is the dating of fossil teeth (Grün, 2006, 2007).

COSMOGENIC NUCLIDES. Cosmic ray interactions produce ^{3}He, ^{10}Be, ^{21}Ne, ^{26}Al, and ^{36}Cl in the atmosphere and lithosphere. Accumulation of these nuclides reflects duration of cosmic ray exposure within the upper 1–2 m of Earth's surface and also varies with altitude and latitude. Useful reviews include Gosse and Phillips (2001), Blard *et al.* (2006) and Balco *et al.* (2008). The technique is very important for dating moraines and rock surfaces associated with glacial erosion (e.g., Barrows *et al.*, 2002; Balco *et al.*, 2005; Phillips *et al.*, 2009; Balco and Rovey, 2010).

URANIUM-SERIES. Based on decay or accumulation of various parent/daughter isotopes in the U-series decay chains, especially $^{234}U/^{238}U$ and $^{230}Th/^{234}U$, the method is commonly used for dating carbonates such as cave speleothems (Cheng *et al.*, 2009) and fossil corals (Stirling and Andersen, 2009). The method is generally not suitable for dating fossil mollusks because, unlike corals, mollusks do not incorporate U into their structure during growth. Rather, the uranium in mollusk shells is incorporated after death, by diagenetic processes. However, Eggins *et al.* (2005) have shown that laser ablation multi-collector ICPMS measurements can allow dating of mollusks in some circumstances. Dating of peat has also been attempted (e.g., Geyh and Müller 2005).

WEATHERING RINDS. The depth of chemical weathering on the surfaces of rocks progressively increases over time, to produce a weathering skin or rind. Has been used for dating glacial moraines and river terraces (e.g., Colman and Pierce, 1981; Knuepfer, 1988).

AMINO-ACID RACEMIZATION (AAR). Living organisms contain dominantly L-amino acids. After death, the amino acids undergo diagenetic changes, including racemization which converts L-amino-acids into D-amino-acids. The D/L ratio increases over time, at a rate dependent on temperature. Widely used for correlation and dating of coastal deposits containing molluscan shells (e.g., Bowen *et al.*, 1998). Can be used to calculate paleotemperatures when age is independently known (e.g., Miller *et al.*, 1997).

ANNUAL LAYER COUNTING (including dendrochronology, varves, corals and ice cores). Trees rings, seasonally layered sediments, ice cores and the carbonate skeletons of various fossils (corals) all contain annual layers that can provide unparalleled precision in dating Quaternary events and deposits.

GSSPs of the Quaternary Stages, with Location and Primary Correlation Criteria (Status Jan. 2012)

Stage	GSSP Location	Latitude, Longitude	Boundary Level	Correlation Events	Reference
Holocene Series	NorthGRIP ice core, central Greenland	75.10°N 42.32°W	1492.45 m depth in Borehole NGRIP2	End of the Younger Dryas Stadial, which is reflected in a shift in deuterium excess values	Journal of Quaternary Science 24/1, 2009; Episodes 31/2, 2008
Upper Pleistocene (Tarantian)	*Amsterdam-Terminal borehole, Netherlands*	*52°22′45″N 4°54′52″E*	*63.5 m below surface*	*Base of warm Marine Isotope Stage 5e, before final glacial episode of Pleistocene*	
Middle Pleistocene (Ionian)	*Candidate sections in Italy (Montalbano Jorica or Valle di Manche) and Japan (Chiba)*			*Magnetic – Matuyama/ Brunhes magnetic reversal (base of Chron 1n)*	
Calabrian	Vrica, Italy	39°02′18.61″N 17°08′05.79″E	Base of the marine claystone overlying the sapropelic marker Bed 'e'	~15 kyr after end of Oldovai normal polarity chron	Episodes 8/2, 1985
Gelasian	Monte San Nicola, Sicily, Italy	37°08′48.8″N 14°12′12.6″E*	Base of marly layer overlying sapropel MPRS 250 with an age of 2.588Ma	Corresponds to Gauss/ Matuyama magnetic epoch boundary; precessional cycle 250 from the present, Marine Isotope Stage 103, with an age of 2.588 Ma	Episodes 21/2, 1998; Episodes, 33/3, 2010; Journal of Quaternay science 25/1, 2010

**according to Google Earth*

REFERENCES

Aguirre, E., Pasini, G., 1985. The Pliocene-Pleistocene boundary. Episodes 8, 116–120.

Alloway, B.V., Pillans, B.J., Carter, L., Naish, T.R., Westgate, J.A., 2005. Onshore-offshore correlation of Pleistocene rhyolitic eruptions from New Zealand: Implications for TVZ eruptive history and paleoenvironmental construction. Quaternary Science Reviews 24, 1601–1622.

Alloway, B.V., Lowe, D.J., Barrell, D.J.A., Newnham, R.M., Almond, P.C., Augustinus, P.C., Bertler, N.A.N., Carter, L., Litchfield, N.J., McGlone, M.S., Shulmeister, J., Vandergoes, M.J., Williams, P.W., 2007. Towards a climate event stratigraphy for New Zealand over the past 30 000 years (NZ-INTIMATE project). Journal of Quaternary Science 22, 9–35.

American Commission on Stratigraphic Nomenclature, 1961. Code of stratigraphic nomenclature. Bulletin of the American Association of Petroleum Geologists 45, 645–660.

An, Z., Lui, T., Porter, S.C., Kukla, G., Wu, X., Hua, Y., 1990. The long-term paleomonsoon variation record by the loess-paleosol sequence in central China. *Quaternary International*, 7/8, 91–95.

Andersen, K.K., Svensson, A., Johnsen, S.J., Rasmussen, S.O., Bigler, M., Rothlisberger, R., Ruth, U., Siggaard-Andersen, M.-L., Steffensen, J.P., Dahl-Jensen, D., Vinther, B.M., Clausen, H.B., 2006. The Greenland Ice Core Chronology 2005, 15–42 ka. Part 1: Constructing the timescale. Quaternary Science Reviews 25, 3246–3257.

Anonymous, 1982–1984. Stratigraphy of the USSR: Quaternary System, vol. 1. Nedra, Moscow (1982), 443 pp. Vol. 2(1984), 556 pp. [In Russian].

Backman, J., Shackleton, N.J., Tauxe, L., 1983. Quantitative nannofossil correlation to open ocean deep-sea sections from Plio-Pleistocene boundary at Vrica, Italy. Nature 304, 156–158.

Balco, G., Rovey, C.W.I., 2010. Absolute chronology for major Pleistocene advances of the Laurentide Ice Sheet. Geology 38, 795–798.

Balco, G., Rovey, C.W.I., Stone, J.O.H., 2005. The first glacial maximum in North America. Science 307, 222.

Balco, G., Stone, J.O., Lifton, N.A., Dunai, T.J., 2008. A complete and easily accessible means of calculating surface exposure ages or erosion rates from ^{10}Be and ^{26}Al measurements. Quaternary Geochronology 3, 174–195.

Barrows, T.T., Stone, J.O., Fifield, L.K., Cresswell, R.G., 2002. The timing of the Last Glacial Maximum in Australia. Quaternary Science Reviews 21, 159–173.

Bassett, M.G., 1985. Towards a common language in stratigraphy. Episodes 8, 87–92.

de Beaulieu, J.-L., Reille, M., 1992. The last climatic cycle at La Grande Pile (Vosges, France): A new pollen profile. Quaternary Science Reviews 11, 431–438.

de Beaulieu, J.L., Reille, M., 1995. Pollen records from the Velay craters: A review and correlation of the Holsteinian Interglacial with isotopic stage 11. Mededelingen van de Rijks Geologische Dienst 52, 59–70.

Berger, G.W., Pillans, B.J., Palmer, A.S., 1992. Dating loess up to 800 ka by thermoluminescence. Geology 20, 403–406.

Berggren, W.S., Hilgen, F.J., Langereis, C.G., Kent, D.V., Obradovich, J.D., Raffi, I., Raymo, M.E., Shackleton, N.J., 1995. Late Neogene chronology: New perspectives in high-resolution stratigraphy. Geological Society of America Bulletin 107, 1272–1287.

Beu, A.G., 2004. Marine mollusca of oxygen isotope stages of the last 2 million years in New Zealand. Part 1: Revised generic positions and recognition of warm-water and coool-water immigrants. Journal of the Royal Society of New Zealand 34, 111–265.

Beu, A.G., Edwards, A.R., 1984. New Zealand Pleistocene and Late Pliocene glacio-eustatic cycles. Palaeogeography, Palaeoclimatology, Palaeoecology 46, 119–142.

Björck, S., Walker, M.J.C., Cwynar, L.C., Johnsen, S., Knudsen, K.L., Lowe, J.J., Wohlfarth, B., INTIMATE Members, 1998. An event stratigraphy for the Last Termination in the North Atlantic region based on the Greenland ice-core record: A proposal of the INTIMATE group. Journal of Quaternary Science 13, 283–292.

Blard, P.-H., Bourles, D., Lave, J., Pik, R., 2006. Applications of ancient cosmic-ray exposures: Theory, techniques and limitations. Quaternary Geochronology 1, 59–73.

Boellstorff, J., 1978. North American Pleistocene stages reconsidered in light of probable Pliocene-Pleistocene continental glaciation. Science 202, 305–307.

Bond, G.C., Lotti, R., 1995. Iceberg discharge into the North Atlantic on millenial time scales during the Last Glaciation. Science 267, 1005–1010.

Bond, G., Heinrich, H., Broecker, W., Labeyrie, L., McManus, J., Andrews, J., Huon, S., Jantschik, R., Clasen, S., Simet, C., Tedesco, K., Klas, M., BonanI, G., Ivy, S., 1992. Evidence for massive discharges of icebergs into the North Atlantic ocean during the last glacial period. Nature 360, 245–249.

Bourdier, F., 1957. Quaternaire. In: Pruvost, P. (Ed.), Lexique stratigraphique international. Vol. 1 Europe. Centre National de la Recherche Scientifique, Paris, pp. 99–100.

Bowen, D.Q., 1978. Quaternary Geology, first ed. Pergamon Press, Oxford, p. 221.

Bowen, D.Q., 1999. A revised correlation of the Quaternary deposits in the British Isles. Geological Society Special Report 23, 174.

Bowen, D.Q., Pillans, B., Sykes, G.A., Beu, A.G., Edwards, A.R., Kamp, P.J.J., Hull, A.G., 1998. Amino acid geochronology of Pleistocene marine sediments in the Wanganui Basin: A New Zealand framework for correlation and dating. Journal of the Geological Society 155, 439–446.

Brauer, A., Allen, J.R.M., Mingram, J., Dulski, P., Wulf, S., Huntley, B., 2007. Evidence for last interglacial chronology and environmental change from Southern Europe. Proceedings of the National Academy of Sciences, USA 104, 450–455.

Broecker, W.S., Van Donk, J., 1970. Insolation changes, ice volumes and the ^{18}O record in deep-sea cores. Review of Geophysics and Space Physics 8, 169–198.

Broecker, W.S., Henderson, G.M., 1998. The sequence of events surrounding Termination II and their implications for the cause of glacial-interglacial CO_2 changes. Paleoceanography 13, 352–364.

Broecker, W.S., Thurber, D.L., Goddard, J., Ku, T.-L., Matthews, R.K., Mesolella, K.J., 1968. Milankovitch hypothesis supported by precise dating of coral reefs and deep-sea sediments. Science 159, 297–300.

Butzer, K.W., Isaac, G.L. (Eds.), 1975. After the Australopithecines. Mouton, The Hague, p. 911.

Capraro, L., Asioli, A., Backman, J., Bertoldi, R., Channell, J.E.T., Massari, F., Rio, D., 2005. Climatic patterns revealed by pollen and oxygen isotope records across the Matuyama/Brunhes Boundary in central Mediterranean (Southern Italy). Geological Society Special Publication 247, 159–182.

Channell, J.E.T., Hodell, D.A., Singer, B.S., Xuan, C., 2010. Reconciling astrochronological and ^{40}Ar/^{39}Ar ages for the Matuyama-Brunhes boundary and late Matuyama Chron. Geochemistry Geophysics Geosystems 11 QOAA12. doi: 10.1029/2010GC003203.

Cheng, H., Edwards, R.L., Wang, Y., Kong, X., Ming, Y., Kelly, M.J., Wang, X., Gallup, C.D., Liu, W., 2006. A penultimate glacial monsoon record from Hulu Cave and two-phase glacial terminations. Geology 34, 217–220.

Cheng, H., Edwards, R.L., Broecker, W.S., Denton, G.H., Kong, X., Wang, Y., Zhang, R., Wang, X., 2009. Ice age terminations. Science 326, 248–252.

Ciaranfi, N., D'Alessandro, A., Girone, G., Maiorano, P., Marino, M., Soldani, D., Stefanelli, S., 2001. Pleistocene sections in the Montalbano Jonico area and the potential GSSP for Early-Middle Pleistocene in the Lucania Basin (Southern Italy). Memorie Scienze Geologiche 53, 67–83.

Ciaranfi, N., D'Alessandro, A., 2005. Overview of the Montalbano Jonico area and section: A proposal for a boundary stratotype for the lower-middle Pleistocene, Southern Italy Foredeep. Quaternary International 131, 5–10.

Ciaranfi, N., Lirer, F., Lirer, L., Lourens, L.J., Maiorano, P., Marino, M., Petrosini, P., Sprovieri, M., Stefanelli, S., Brilli, M., Girone, A., Joannin, S., Pelosi, N., Vallefuoco, M., 2010. Integrated stratigraphy and astronomical tuning of lower-middle Pleistocene Montalbano Jonico section (southern Italy). Quaternary International 219, 109–120.

Cita, M.B., Castradori, D., 1995. Workshop on marine sections from the Gulf of Taranto (southern Italy) usable as potential stratotypes for the GSSP of the Lower, Middle and Upper Pleistocene (Bari, Italy, September 29 to October 4, 1994). Il Quaternario 7 (2), 677–692.

Cita, M.B., Pillans, B., 2010. Global stages, regional stages or no stages in the Plio/Pleistocene? Quaternary International 219, 6–15.

Cita, M.B., Capraro, L., Ciaranfi, N., Di Stephano, E., Lirer, F., Maiorano, P., Marino, M., Raffi, I., Rio, D., Sprovieri, R., Stefanelli, S., Vai, G.B., 2008. The Calabrian Stage redefined. Episodes 31, 408–419.

Cita, M.B., Capraro, L., Ciaranfi, N., Di Stefano, E., Marino, M., Rio, D., Sprovieri, R., Vai, G.B., 2006. Calabrian and Ionian: A proposal for the definition of Mediterranean stages for Lower and Middle Pleistocene. Episodes 29 (2), 107–114.

Clark, P.U., Dyke, A.S., Shakun, J.D., Carlson, A.E., Clark, J., Wohlfarth, B., Mitrovica, J.X., Hostetler, S.W., McCabe, A.M., 2009. The Last Glacial Maximum. Science 325, 710–771.

CLIMAP, 1981. Seasonal reconstructions of the Earth's surface at the last glacial maximum. Map Series, Technical Report, MC-36. Geological Society of America, Boulder.

Cohen, K., Gibbard, P.L., 2010. Global chronostratigraphical correlation table for the last 2.7 million years. Subcommission on Quaternary Stratigraphy, Cambridge, p. 6.

Colman, S.M., Pierce, K.L., 1981. Weathering rinds on andesitic and basaltic stones as a Quaternary age indicator, Western United States. United States Geological Survey Professional Paper 1210, 41.

Colman, S.M., Pierce, K.L., Birkland, P.W., 1987. Suggested terminology for Quaternary dating methods. Quaternary Research 28, 314–319.

Crundwell, M., Scott, G., Naish, T.R., Carter, L., 2008. Glacial-interglacial ocean-climate variability spanning the Mid-Pleiocene transition in the temperate Southwest Pacific, ODP Site 1123. Palaeogeography, Palaeoclimatology, Palaeoecology 260, 202–229.

Crutzen, P.J., 2002. Geology of mankind. Nature 415, 23.

Dansgaard, W., Johnsen, S.J., Clausen, H.B., Dahl-Jensen, D., Gundestrup, N.S., Hammer, C.U., Hvidberg, C.S., Steffensen, J.P., Sveinbjörnsdottir, A.E., Jouzel, J., Bond, G., 1993. Evidence for general instability of past climate from a 250-kyr ice-core record. Nature 364, 218–220.

Davies, S.M., Wastegard, S., Abbott, P.M., Barbante, C., Bigler, M., Johnsen, S.J., Rasmussen, T.L., Steffensen, J.P., Svensson, A., 2010. Tracing volcanic records in the NGRIP ice-core and synchronising North Atlantic marine records during the last glacial period. Earth and Planetary Science Letters 294, 69–79.

Delmonte, B., Basile-Doelsch, I., Petit, J.R., Maggi, V., Revel-Rolland, M., Michard, A., Jagoutz, E., Grousset, F., 2004. Comparing the EPICA and Vostok dust records during the last 220,000 years: Stratigraphical correlation and provenance in glacial periods. Earth-Science Reviews 66, 63–87.

Ding, Z.L., Derbyshire, E., Yang, S.L., Yu, Z.W., Xiong, S.F., and Liu, T.S., 2002 Stacked 2.6-Ma grain size record from the Chinese loess based on five sections and correlation with the deep-sea $\delta^{18}O$ record. Paleoceanography, 17, 1033. doi: 10.1029/2001PA000725.

Ding, Z.L., Derbyshire, E., Yang, S.L., Sun, J.M., Liu, T.S., 2005. Stepwise expansion of desert environment across northern China in the past 3.5 Ma and implications for monsoon evolution. Earth and Planetary Science Letters 237, 45–55.

Dreyfus, G.B., Raisbeck, G.M., Parrenin, F., Jouzel, J., Guyodo, Y., Nomade, S., Mazaud, A., 2008. An ice core perspective on the age of the Matuyama-Brunhes boundary. Earth and Planetary Science Letters 274, 151–156.

Eardley, A.J., Shuey, R.T., Gvodetsky, V., Nash, W.P., Picard, M.D., Grey, D.C., Kukla, G.J., 1973. Lake cycles in the Bonneville Basin, Utah. Geological Society of America Bulletin 84, 211–215.

Easterbrook, D.J. (Ed.), 1988. Dating Quaternary sediments. Geological Society of America Special Paper, 227, p. 165.

Eggins, S.M., Grün, R., McCulloch, M.T., Pike, A.W.G., Chappell, J., Kinsley, L., Mortimer, G., Shelley, M., Murray-Wallace, C.V., Spotl, C.L.T., 2005. In situ U-series dating by laser-ablation multi-collector ICPMS: New prospects for Quaternary geochronology. Quaternary Science Reviews 24, 2523–2538.

Ehlers, J., Gibbard, P.L., 2011. Quaternary glaciation. In: Singh, V.P., Singh, P., Haritashya, U.K. (Eds.), Encyclopedia of Snow, Ice and Glaciers. Springer, Dordrecht, pp. 873–882.

Ehlers, J., Gibbard, P.L., Hughes, P.D., 2011. Introduction. In: Ehlers, J., Gibbard, P.L., Hughes, P.D. (Eds.), Quaternary Glaciations – Extent and Chronology, Part IV – A Closer Look. Elsevier, Amsterdam, pp. 1–14.

Elias, S.A. (Ed.), 2007, Encyclopedia of Quaternary Science, vol 4. Elsevier, Amsterdam.

Ellenberger, F., 1994. Histoire de la Géologie. Tome 2: Technique et documentation. Lavoisier, Paris, p. 380.

Emiliani, C., 1955. Pleistocene temperatures. Journal of Geology 63, 538–578.

Emiliani, C., 1966. Palaeotemperature analysis of Caribbean cores P6304-8 and P6304-9 and a generalized temperature curve for the past 425,000 years. Journal of Geology 74, 109–126.

EPICA Community Members, 2004. Eight glacial cycles from an Antarctic ice core. Nature 429, 623–628.

Evans, P., 1971. Towards a Pleistocene time-scale. In: Harland, W.B., Francis, H., Evans, P. (Eds.), The Phanerozoic time-scale, a supplement. Part 2. Geological Society of London, London, pp. 123–356.

Fuji, N., 1988. Palaeovegetation and palaeoclimatic changes around Lake Biwa, Japan, during the last c. 3 million years. Quaternary Science Reviews 7, 21–28.

Funnell, B.W., 1996. Plio-Pleistocene palaeogeography of the southern North Sea Basin. (3.75–0.60 Ma). Quaternary Science Reviews 15, 391–405.

Geyh, M.A., Müller, H., 2005. Numerical ^{230}Th/U dating and a palynological review of the Holsteinian/Hoxnian Interglacial. Quaternary Science Reviews 24, 1861–1872.

Geirsdottir, A., Eriksson, J., 1994. Growth of an intermittent ice sheet in Iceland during the Late Pliocene and Early Pliocene. Quaternary Research 42, 115–130.

Gervais, P., 1867–1869. Zoologie et paleontology générales. In: Nouvelles Recherches sur les Animaux Vertétebrés et Fossiles, vol 2. Bertrand, Paris, p. 263.

Gibbard, P.L., 2003. Definition of the Middle / Upper Pleistocene boundary. Global and Planetary Change 36, 201–208.

Gibbard, P.L., West, R.G., Zagwijn, W.H. (Eds.), 1991. Early and early Middle Pleistocene correlations in the southern North Sea Basin. Quaternary Science Reviews, 10, pp. 23–52.

Gibbard, P.L., Smith, A.G., Zalasiewicz, J.A., Barry, T.L., Cantrill, D., Coe, A.L., Cope, J.C.W., Gale, A.S., Gregory, F.J., Powell, J.H., Rawson, P.F., Stone, P., Waters, C.N., 2005. What status for the Quaternary? Boreas 34, 1–6.

Gibbard, P.L., Head, M.J., Walker, M., Alloway, B., Beu, A.G., Coltorti, M., Hall, V.M., Liu, J., Knudsen, K.-L., Van Kolfschoten, T., Litt, T., Marks, L., McManus, J., Partridge, T.C., Piotrowski, J.A., Pillans, B., Rousseau, D.-D., Suc, J.-P., Tesakov, A.S., Turner, C., Zazo, C., 2010. Formal ratification of the Quaternary System/Period and the Pleistocene Series/Epoch with a base at 2.588 Ma. Journal of Quaternary Science 25, 96–102.

Gignoux, M., 1916. L'étage Calabrien (Pliocène supérieur marine) sur le versant Nord-est de l'Apennin, entre le Monte Gargano et Plaisance. Bulletin de la Société Géologique de France, Series 4 (14), 324–348.

Godwin, H., 1975. The History of the British Flora, second ed. Cambridge University Press, Cambridge, p. 541.

Gosse, J.C., Phillips, F.M., 2001. Terrestrial in situ cosmogenic nuclides: Theory and application. Quaternary Science Reviews 20, 1475–1560.

Gradstein, F.M., Ogg, J.G., Smith, A.G. (Eds.), 2004. A Geologic Time Scale 2004. Cambridge University Press, Cambridge, p. 589.

Grün, R., 2006. Direct dating of human remains. Yearbook of Physical Anthropology 49, 2–48.

Grün, R., 2007. Electron spin resonance dating. In: Elias, S.A. (Ed.), Encyclopedia of Quaternary Science, vol. 2. Elsevier, Amsterdam, pp. 1505–1516.

Grün, R., Schwarcz, H.P., 2000. Revised open system U-series/ESR age calculations for teeth from Stratum C at the Hoxnian Interglacial type locality, England. Quaternary Science Reviews 19, 1151–1154.

Hall, I.R., McCave, I.N., Shackleton, N.J., Weedon, G.P., Harris, S.E., 2001. Intensified deep Pacific inflow and ventilation in Pleistocene glacial times. Nature 412, 809–812.

Haq, B.U., Berggren, W.A., Van Couvering, J.A., 1977. Corrected age of the Plio/Pleistocene boundary. Nature 269, 483–488.

Harland, W.B., Armstrong, R.L., Cox, A.V., Craig, L.E., Smith, A.G., Smith, D.G., 1990. A Geologic Time Scale 1989. Cambridge University Press, Cambridge, p. 263.

Hays, J.D., Imbrie, J., Shackleton, N.J., 1976. Variations in the Earth's orbit: Pacemaker of the ages. Science 194, 1121–1132.

Hedberg, H.D., 1976. International Stratigraphic Guide. Wiley Interscience, New York, p. 200.

Heinrich, H., 1988. Origin and consequences of cyclic ice-rafting in the Northeast Atlantic Ocean during the past 130 000 years. Quaternary Research 29, 142–152.

de Heinzelin, J., Tavernier, R., 1957. Flandrien. In: Pruvost, P. (Ed.), Lexique Stratigraphique International, vol. 1. Centre National de la Recherche Scientifique, Europe. Paris, p. 32.

Hooghiemstra, H., 1989. Quaternary and Upper-Pliocene glaciations and forest development in the tropical Andes: Evidence from a long high-resolution pollen record from the sedimentary basin of Bogotá, Columbia. Palaeogeography, palaeoclimatology, Palaeoecology 72, 11–26.

Hooghiemstra, H., Sarmiento, G., 1991. Long continental pollen record from a tropical intermontane basin: Late Pliocene and Pleistocene history from a 540-meter core. Episodes 14, 107–115.

Hopwood, A.T., 1935. Fossil elephants and Man. Proceedings of the Geologists' Association 46, 46–60.

Huntley, D.J., Hutton, J.T., Prescott, J.R., 1993. The stranded beach-dune sequence of south-east South Australia: A test of thermoluminescence dating, 0–800 ka. Quaternary Science Reviews 12, 1–20.

Hyvärinen, H., 1978. Use and definition of the term Flandrian. Boreas 7, 182.

Jessen, K., Milthers, V., 1928. Stratigraphical and palaeontological studies of interglacial freshwater deposits in Jutland and north-west Germany. Danmarks Geologisk Undersøgelse II (48), 379.

Joannin, S., Ciaranfi, N., Stefanelli, S., 2008. Vegetation changes during the late Early Pleistocene at Montalbano Jonico (Province of Matera, southern Italy) based on pollen analysis. Palaeogeography, Palaeoclimatology. Palaeoecology 270, 92–101.

Johnsen, S.J., Dansgaard, W., Clausen, H.B., Langway, C.C., 1972. Oxygen isotope profiles through Antarctic and Greenland ice sheets. Nature 235, 429–433.

de Jong, J., 1988. Climatic variability during the past three million years, as indicated by vegetational evolution in northwest Europe and with emphasis on data from The Netherlands. Philosophical Transactions of the Royal Society, Series B 318, 603–617.

Jouzel, J., Masson-Delmotte, V., Cattani, O., Dreyfus, G., Falourd, S., Hoffmann, G., Minster, B., Nouet, J., Barnola, J.M., Chappellaz, J., Fischer, H., Gallet, J.C., Johnsen, S., Leuenberger, M., Loulergue, L., Luethi, D., Oerter, H., Parrenin, F., Raisbeck, G., Raynaud, D., Schilt, A., Schwander, J., Selmo, E., Souchez, R., Spahni, R., Stauffer, B., Steffensen, J.P., Stenni, B., Stocker, T.F., Tison, J.L., Werner, M., Wolff, E.W., 2007. Orbital and millennial Antarctic climate variability over the past 800,000 years. Science 317, 793–796.

King, W.B.R., Oakley, K.P., 1949. Definition of the Pliocene-Pleistocene boundary. Nature 163, 186–187.

Knuepfer, P.L.K., 1988. Estimating ages of late Quaternary stream terraces from analysis of weathering rinds and soils. Geological Society of America Bulletin 100, 1224–1236.

Krasnenkov, R.V., Iossifova, Yu.I., Semenov, V.V., 1997. The Upper Don drainage basin – an important stratoregion for climatic stratigraphy of the early Middle Pleistocene (the early Neopleistocene) of Russia. In: Alekseev, M.N., Khoreva, I.M. (Eds.), Quaternary Geology and Paleogeography of Russia. Geosynthos, Moscow, pp. 82–96 [In Russian, abstract in English].

Kukla, G.J., 1977. Pleistocene land-sea correlations I. Europe. Earth-Science Reviews 13, 307–374.

Lisiecki, L.E., Raymo, M.E., 2005. A Pliocene-Pleistocene stack of 57 globally distributed benthic δ^{18}O records. Paleoceanography 20, PA1003. doi: 10.1029/2004PA001071.

Litt, T., Turner, C., 1993. Arbeitserbgenisse der Subkommission für Europäische Quartärstratigraphie: Die Saalesequenz in der Typusregion (Berichte der SEQS 10). Eiszeitalter und Gegenwart 43, 125–128.

Litt, T., Gibbard, P.L., 2008. A proposed Global Stratotype Section and Point (GSSP) for the base of the Upper (Late) Pleistocene Subseries (Quaternary System/Period). Episodes 31, 260–261.

Litt, T., Bettis, E.A., Bosch, A., Dodonov, A., Gibbard, P.L., Jiaqui, L., Kershaw, P., von Koenigswald, W., McManus, J., Partridge, T., Turner, C., 2008. A proposal for the Global Stratotype Section and Point (GSSP) for the Middle/Upper (Late) Pleistocene Subseries Boundary (Quaternary System/Period). Subcommission on Quaternary Stratigraphy, Cambridge, p. 35.

Litt, T., Krastel, S., Sturm, M., Kipfer, R., Örcen, S., Heumann, G., Franz, S.O., Ülgen, U.B., Niessen, F., 2009. 'PALEOVAN', International Continental Scientific Drilling Program (ICDP): Site survey results and perspectives. Quaternary Science Reviews 28, 1555–1567.

Lourens, L.J., 2004. Revised tuning of Ocean Drilling Program Site 964 and KC01B Mediterranean) and implications for the D18O, tephra, calcareous nannofossil, and geomagnetic reversal chronologies of the past 1.1 Myr. Paleoceanography 19, PA3010. doi: 10.1029/2003PA000997.

Lourens, L.J., 2008. On the Neogene-Quaternary debate. Episodes 31, 239–242.

Lourens, L.J., Antonarakou, A., Hilgen, F.J., Van Hoof, A.A.M., Vergnaud-Grazzini, C., Zachariasse, W.J., 1996. Evaluation of the Plio-Pleistocene astronomical timescale. Paleoceanography 11 (4), 391–413.

Lourens, L.J., Hilgen, F.J., Laskar, J., Shackleton, N.J., Wilson, D., 2004. The Neogene Period. In: Gradstein, F.M., Ogg, J.G., Smith, A.G. (Eds.), A Geologic Time Scale 2004. Cambridge University Press, Cambridge, pp. 409–440.

Lowe, J.J., Walker, M.J.C., 1997. Reconstructing Quaternary Environments. Longmans, London, p. 446.

Lowe, J.J., Birks, H.H., Brooks, S.J., Coope, G.R., Harkness, D.D., Mayle, F.E., Sheldrick, C., Turney, C.S.M., Walker, M.J.C., 1999. The chronology of palaeoenvironmental changes during the last Glacial – Holocene transition: Towards an event stratigraphy for the British Isles. Journal of the Geological Society 156, 397–410.

de Lumley, H., 1976. La Préhistoire Française. Tome 1. Editions CNRS, Paris, pp. 5–23.

Luthi, D., Le Floc, M., Bereiter, B., Blunier, T., Barnola, J.M., Siegenthaler, U., Raynaud, D., Jouzel, J., Fischer, H., Kawamura, K., Stocker, T.F., 2008. High-resolution carbon dioxide concentration record 650,000–800,000 years before present. Nature 453, 379–382.

Lyons, R.P., Scholz, C.A., King, J.W., Cohen, A.S., Johnson, T.C., 2007. High amplitude climate variability in a tropical rift-lake: Correlation of drillcore and seismic reflection data in Lake Malawi, East Africa. AAPG Annual Meeting, April 1–4. AAPG Search and Discover Article, Long Beach, California. 90063.

Massari, F., Rio, D., Sgavetti, M., Prosser, G., D'Alessandro, A., Asioli, A., Capraro, L., Fornaciari, E., Tateo, F., 2002. Interplay between tectonics and glacio-eustasy: Pleistocene of the Crotone Basin, Calabria (southern Italy). Bulletin of the American Geological Society 114, 1183–1209.

McKay, R., Browne, G., Carter, L., Cowan, E., Dunbar, G., Krissek, L., Naish, T., Powell, R., Reed, J., Talarico, F., Wilch, T., 2009. The stratigraphic signature of the late Cenozoic Antarctic ice sheets in the Ross Embayment. Geological Society of America Bulletin 121, 1537–1561.

Mangerud, J., Andersen, S.Th., Berglund, B.E., Donner, J.J., 1974. Quaternary stratigraphy of Norden, a proposal for terminology and classification. Boreas 3, 109–128.

Martinson, D.G., Pisias, N.G., Hays, J.D., Imbrie, J., Moore, T.C., Shackleton, N.J., 1987. Age dating and the orbital theory of the ice ages: Development of a high-resolution 0 to 300 000 year chronostratigraphy. Quaternary Research 27, 1–29.

Michel, J.-P., Fairbridge, R.W., Carpenter, M.S.N., 1997. Dictionnaire des Sciences de la Terre, third ed. Masson and J.Wiley & Sons, Paris and Chichester, p. 346.

Miller, G.H., Magee, J.W., Jull, A.J.T., 1997. Low-latitude cooling in the Southern Hemisphere from amino acids in emu eggshell. Nature 385, 241–244.

Mitchell, G.F., Penny, L.F., Shotton, F.W., West, R.G., 1973. A Correlation of Quaternary deposits in the British Isles. Geological Society Special Report 4, 99.

Mix, A.C., Bard, E., Schneider, R., 2001. Environmental processes of the ice age: Land, oceans, glaciers (EPILOG). Quaternary Science Reviews 20, 627–657.

Mörner, N.A., 1976. The Pleistocene/Holocene boundary: Proposed boundary stratotypes in Gothenburg, Sweden. Boreas 5, 193–275.

Naish, T.R., Abbott, S.T., Alloway, B.V., Beu, A.G., Carter, R.M., Edwards, A.R., Journeaux, T.D., Kamp, P.J.J., Pillans, B.J., Saul, G., Woolfe, K.J., 1998. Astronomical calibration of a Southern Hemisphere Plio-Pleistocene reference section, Wanganui Basin, New Zealand. Quaternary Science Reviews 17, 695–710.

Naish, T., Powell, R., Levy, R., Wilson, G., Scherer, R., Talarico, F., Krissek, L., Niessen, F., Pompilio, M., Wilson, T., Carter, L., DeConto, R., Huybers, P., McKay, R., Pollard, D., Ross, J., Winter, D., Barrett, P., Browne, G., Cody, R., Cowan, E., Crampton, J., Dunbar, G., Dunbar, N., Florindo, F., Gebhardt, C., Graham, I., Hannah, M., Hansaraj, D., Harwood, D., Helling, D., Henrys, S., Hinnov, L., Kuhn, G., Kyle, P., Laufer, A., Maffioli, P., Magens, D., Mandernack, K., McIntosh, W., Millan, C., Morin, R., Ohneiser, C., Paulsen, T., Persico, D., Raine, I., Reed, J., Riesselman, C., Sagnotti, L., Schmitt, D., Sjunneskog, C., Strong, P., Taviani, M., Vogel, S., Wilch, T., Williams, T., 2009. Obliquity-paced Pliocene West Antarctic ice sheet oscillations. Nature 458, 322–329.

Noller, J.S., Sowers, J.M., Lettis, W.R., 2000. Quaternary Geochronology. Methods and Applications. American Geophysical Union Reference Shelf 4, 581.

North Greenland Ice Core Project Members, 2004. High-resolution record of Northern Hemisphere climate extending into the last interglacial period. Nature 431, 147–151.

Nystuen, J.P. (Ed.), 1986. Regler og råd for navnsetting av geologiske enheter i Norge. Norsk Geologisk Tidsskrift 66 (Suppl. 1), p. 96.

Ogg, J.G., Pillans, B., 2008. Establishing Quaternary as a formal international Period/System. Episodes 31, 230–233.

Olausson, E. (Ed.), 1982. The Pleistocene/Holocene boundary in southwestern Sweden. Sveriges Geologiska Undersökning, 794, p. 288.

Parandier, H., 1891. Notice géologique et paléontologique sur la nature des terrains traversés par le chemin de fer entre Dijon at Châlons-sur-Saône. Bulletin de la Société géologique de France, series 3 19, 794–818.

Pareto, L., 1865. Note sur les subdivisions que l'on pourrait établir dans les terrains tertiaires de l'Apennin septentrional. Bulletin de la Société géologique de France 22, 210–217.

Parrenin, F., Barnola, J.M., Beer, J., Blunier, T., Castellano, E., Chappellaz, J., Dreyfus, G., Fischer, H., Fujita, S., Jouzel, J.,

Kawamura, K., Lemieux-Dudon, B., Loulergue, L., Masson-Delmotte, V., Narcisi, B., Petit, J.R., Raisbeck, G., Raynaud, D., Ruth, U., Schwander, J., Severi, M., Spahni, R., Steffensen, J.P., Svensson, A., Udisti, R., Waelbroeck, C., Wolff, E., 2007. The EDC3 chronology for the EPICA Dome C ice core. Climate of the Past 3, 485–497.

Partridge, T.C. (Ed.), 1997a. The Plio-Pleistocene boundary. Quaternary International, 40 100pp.

Partridge, T.C., 1997b. Reassessment of the position of the Plio-Pleistocene boundary: Is there a case for lowering it to the Gauss-Matuyama palaeomagnetic reversal? In: Partridge, T.C. (Ed.), The Plio-Pleistocene Boundary. Quaternary International, 40, pp. 5–10.

Penck, A., Brückner, E., 1909–1911. Die Alpen im Eiszeitalter. Taunitz, Leipzig, p. 1199.

Petit, J.R., Jouzel, J., Raynaud, D., Barkov, N.I., Barnola, J.-M., Basile, I., Bender, M., Chappellaz, J., Davis, M., Delayque, G., Delmotte, M., Kotlyakov, V.M., Legrand, M., Lipenkov, V.Y., Lorius, C., Pépin, L., Ritz, C., Saltzman, E., Stievenard, M., 1999. Climate and atmospheric history of the past 420,000 years from the Vostok ice core, Antarctica. Nature 399, 429–436.

Phillips, F.M., Zreda, M., Plummer, M.A., Elmore, D., Clark, D.H., 2009. Glacial geology and chronology of Bishop Creek and vicinity, eastern Sierra Nevada, California. Geological Society of America Bulletin 121, 1013–1033.

Pickering, K.T., Souter, C., Oba, T., Taira, A., Schaff, M., Platzman, E., 1999. Glacio-eustatic control on deep-marine clastic forearc sedimentation, Pliocene-mid Pleistocene (c. 1180–600 ka) Kazusa Group, Japan. Journal of the Geological Society 156, 125–136.

Pillans, B., 1983. Upper Quaternary marine terrace chronology and deformation, South Taranaki, New Zealand. Geology 11, 292–297.

Pillans, B., 1991. New Zealand Quaternary stratigraphy: An overview. Quaternary Science Reviews 10, 405–418.

Pillans, B., 2003. Subdividing the Pleistocene using the Matuyama-Brunhes boundary (MBB): An Australasian perspective. Quaternary Science Reviews 22, 1569–1577.

Pillans, B., Naish, T., 2004. Defining the Quaternary. Quaternary Science Reviews 23, 2271–2282.

Pillans, B.J., Roberts, A.P., Wilson, G.S., Abbott, S.T., Alloway, B.V., 1994. Magnetostratigraphic, lithostratigraphic and tephrostratigraphic constraints on Lower and Middle Pleistocene sea-level changes, Wanganui Basin, New Zealand. Earth and Planetary Science Letters 121, 81–98.

Pillans, B., Alloway, B., Naish, T., Westgate, J., Abbott, S., Palmer, A., 2005. Silicic tephras in Pleistocene shallow-marine sediments of Wanganui Basin, New Zealand. Journal of the Royal Society of New Zealand 35, 43–90.

Pollard, D., DeConto, R.M., 2009. Modelling West Antarctic ice sheet growth and collapse through the past five million years. Nature 458, 329–333.

Prokopenko, A.A., Khursevich, G.K., 2010. Plio-Pleistocene transition in the continental record from lake Baikal: Diatom biostratigraphy and age model. Quaternary International 219, 26–36.

Prokopenko, A.A., Hinnov, L.A., Williams, D.F., Kuzmin, M.I., 2006. Orbital forcing of continental climate during the Pleistocene: A complete astronomically tuned record from Lake Baikal, SE Siberia. Quaternary Science Reviews 25, 3431–3457.

Raymo, M.E., Ruddiman, W.F., Backman, J., Clement, B.M., Martinson, D.G., 1989. Late Pliocene variation in northern hemisphere ice sheets and North Atlantic deep water circulation. Paleoceanography 4 (4), 413–446.

Reimer, P.J., Baillie, M.G.L., Bard, E., 2004. IntCal04 terrestrial radiocarbon age calibration/comparison, 26–0 ka B.P. Radiocarbon 46, 1029–1058.

Richmond, G.M., 1996. The INQUA-approved provisional Lower-Middle Pleistocene boundary. In: Turner, C. (Ed.), The Early Middle Pleistocene in Europe. Balkema, Rotterdam, pp. 319–326.

Rio, D., Sprovieri, R., Di Stefano, E., 1994. The Gelasian Stage: A proposal of a new chronostratigraphic unit of the Pliocene Series. Rivista Italiano di Paleontoloqia e Stratigrafia 100, 103–124.

Rio, D., Channell, J.E.T., Massari, F., Poli, M.S., Sgavetti, M., D'Alessandro, A., Prosser, G., 1996. Reading Pleistocene eustasy in a tectonically active siliciclastic shelf setting (Crotone peninsula, southern Italy). Geology 24, 743–746.

Rio, D., Sprovieri, R., Castradori, D., Di Stefano, E., 1998. The Gelasian Stage (Upper Pliocene): A new unit of the global standard chronostratigraphic scale. Episodes 21, 82–87.

Roberts, A.P., Florindo, F., Larrasoana, J.C., O'Regan, M.A., Zhao, X., 2010. Complex polarity pattern at the former Plio-Pleistocene global stratotype section at Vrica (Italy): Remagnetization by magnetic iron sulphides. Earth and Planetary Science Letters 292, 98–111.

Rook, L., Martinez-Navarro, B., 2010. Villafranchian: The long story of a Plio-Pleistocene European large mammal biochronological unit. Quaternary International 219, 134–144.

Roy, M., Clark, P.U., Barendregt, R.W., Glasmann, J.R., Enkin, R.J., 2004. Glacial stratigraphy and paleomagnetism of late Cenozoic deposits of the north-central United States. Geological Society of America Bulletin 116, 30–41.

Ruddiman, W.F., 2003. The anthropogenic greenhouse era began thousands of years ago. Climatic Change 61, 261–293.

Ruddiman, W.F., Kidd, R.B., Thomas, E., et al., 1987. doi: 10.2973/dsdp.proc.94.1987. Initial Reports Deep Sea Drilling Project, 94. U.S. Government Printing Office, Washington.

Ruggieri, G., Sprovieri, R., 1979. Selinunitiano, nuovo superpiano per il Pleistocene inferiore. Bollettino della Sociela Geologica Italiana 96, 797–802.

Rutter, N.W., Cato, E.R., 1995. Dating methods for Quaternary deposits. Geological Association of Canada, Geotext 2, 308.

Rutter, N., Ding, Z., Evans, M.E., Liu, T.S., 1991. Baoji-type pedostratigraphic section, Loess Plateau, north-central China. Quaternary International 10, 1–22.

Sagnotti, L., Cascella, A., Ciaranfi, N., Macri, P., Maiorano, P., Marino, M., Taddeucci, J., 2010. Rock magnetism and palaeomagnetism of the Montalbano Jonico section (Italy): Evidence for late diagenetic growth of greigite and implications for magnetostratigraphy. Geophysical Journal International 180, 1049–1066.

Salvador, A., 1994. International Stratigraphic Guide, second ed. International Union of Geological Sciences and Geological Society of America, Trondheim, p. 214.

Sanchez-Goñi, M.F., Eynaud, F., Turon, J.L., Shackleton, N.J., 1999. High resolution palynological record off the Iberian margin: Direct land-sea correlation for the Last Interglacial complex. Earth and Planetary Science Letters 171, 123–137.

Sarnthein, M., Bartoli, G., Prange, M., Schmittner, A., Schneider, B., Weinelt, M., Andersen, N., Garbe-Schönberg, D., 2009. Mid-Pliocene shifts in ocean overturning ciculation and the onset of Quaternary-style climates. Climates of the Past 5, 269–283.

Schlüchter, C., 1992. Terrestrial Quaternary stratigraphy. Quaternary Science Reviews 11, 603–607.

Schneer, C.J., 1969. Introduction. In: Schneer, C.J. (Ed.), Towards a History of Geology. The Massachusetts Institute of Technology Press, Cambridge and London, pp. 1–18.

Shackleton, N.J., 1977. The oxygen isotope stratigraphic record of the Late Pleistocene. Philosophical Transactions of the Royal Society, Series B 280, 169–182.

Shackleton, N.J., 1997. The deep-sea record and the Pliocene-Pleistocene boundary. In: Partridge, T.C. (Ed.), The Plio-Pleistocene Boundary. Quaternary International, 40, pp. 33–36.

Shackleton, N.J., Opdyke, N.D., 1973. Oxygen isotope and palaeomagnetic stratigraphy of the equatorial Pacific core V28-238: Oxygen isotope temperatures and ice volumes on a 10^5 and 10^6 year scale. Quaternary Research 3, 39–55.

Shackleton, N.J., Opdyke, N.D., 1976. Oxygen isotope and palaeomagnetic stratigraphy of Equatorial Pacific core V28-239, Late Pliocene to Latest Pleistocene. In: Cline, R.M., Hays, J.D. (Eds.), Investigation of Late Quaternary Paleoceanography and Paleoclimatology. Geological Society of America Memoir, 145, pp. 449–464.

Shackleton, N.J., Hall, M.A., 1989. Stable isotope history of the Pleistocene at ODP site 677. Proceedings of the Ocean Drilling Program, Scientific Results 111, 295–316.

Shackleton, N.J., Backman, J., Zimmerman, H., Kent, D.V., Hall, A., Homrighausen, R., Huddlestun, P., Keene, J.B., Kaltenback, A.J., Krumsiek, K.A.O., Morton, A.C., Murray, J.W., Westberg-Smith, J., 1984. Oxygen-isotope calibration of the onset of ice-rafting in DSDP site 552A: History of glaciation in the North Atlantic region. Nature 307, 620–633.

Shackleton, N.J., Sanchez-Goñi, M.F., Pailler, D., Lancelot, Y., 2003. Marine Isotope Substage 5e and the Eemian Interglacial. Global and Planetary Change 36, 151–155.

Shik, S.M., Borisov, B.A., Zarrina, E.P., 2002. About the project of the interregional stratigraphic scheme of the Neopleistocene of East European Platform and improving regional stratigraphic schemes. The Third All Russian Meeting on the Quaternary research. Abstracts. Smolensk. Geological Institute RAN –Smolensky Pedagogical University, pp. 125–129 [In Russian].

Sibrava, V., 1978. Isotopic methods in Quaternary geology. In: Cohee, G.V., Glaessner, M.F., Hedberg, H.D. (Eds.), Contributions to the Geologic Time Scale. Studies in Geology, 6, pp. 165–169.

Singer, B.S., Ackert, R.P.J., Guillou, H., 2004. $^{40}Ar/^{39}Ar$ and K-Ar chronology of Pleistocene glaciations in Patagonia. Geological Society of America Bulletin 116, 434–450.

Singh, G., Opdyke, N.D., Bowler, J.M., 1981. Late Cainozoic stratigraphy, palaeomagnetic chronology and vegetational history from Lake George, N.S.W. Journal of the Geological Society of Australia 28, 435–452.

Stirling, C.H., Andersen, M.B., 2009. Uranium-series dating of fossil coral reefs: Extending the sea-level record beyond the last glacial cycle. Earth and Planetary Science Letters 284, 269–283.

Stratigraphy of the USSR: Quaternary System. 1982–1984. 2 Vol. Moscow: Nedra [In Russian].

Suc, J.P., Bertini, A., Leroy, S.A.G., Suballyova, D., 1997. Towards a lowering of the Pliocene/Pleistocene boundary to the Gauss/Matuyama Reversal. In: Partridge, T.C. (Ed.), The Plio-Pleistocene Boundary. Quaternary International, 40, pp. 37–42.

Suganuma, Y., Yokoyama, Y., Yamazaki, T., Kawamura, K., Horng, C.-S., Matsuzaki, H., 2010. ^{10}Be evidence for delayed acquisition of remanent magnetization in marine sediments: Implication for a new age for the Matuyama-Brunhes boundary. Earth and Planetary Science Letters 296, 443–450.

Suggate, R.P., 1974. When did the Last Interglacial end? Quaternary Research 4, 246–252.

Suggate, R.P., West, R.G., 1969. Stratigraphic nomenclature and subdivision in the Quaternary Working group for Stratigraphic Nomenclature, INQUA Commission for Stratigraphy (unpublished discussion document).

Svensson, A., Andersen, K.K., Bigler, M., Clausen, H.B., Dahl-Jensen, D., Davies, S.M., Johnsen, S.J., Muscheler, R., Parrenin, F., Rasmussen, S.O., Röthlisberger, R., Seierstad, I., Steffensen, J.P., Vinther, B.M., 2008. A 60 000 year Greenland stratigraphic ice core chronology. Climate of the Past 4, 47–57.

Tauxe, L., Opdyke, N.D., Pasini, G., Elmi, C., 1983. Age of the Pliocene-Pleistocene boundary in the Vrica section, southern Italy. Nature 304, 125–129.

Thompson, L.G., Mosley-Thompson, E., Davis, M.E., Bolzan, J.F., Dai, J., Yao, T., Gundestrup, N., Wu, X., Klein, L., Xie, Z., 1989. Holocene-Late Pleistocene climatic ice core records from Qinghai-Tibetan Plateau. Science 246, 474–477.

Thompson, L.G., Mosley-Thompson, E., Davis, M.E., Lin, P.-N., Henderson, K.A., Cole-Dai, J., Bolzan, J.F., Liu, K.-B., 1995. Late Glagial Stage and Holocene tropical ice core records from Huascaran, Peru. Science 269, 46–50.

Thompson, L.G., Mosley-Thompson, E., Davis, M.E., Henderson, K.A., Brecher, H.H., Zagorodnov, V.S., Mashiotta, T.A., Lin, P.-N., Mikhalenko, V.N., Hardy, D.R., Beer, J., 2002. Kilimanjaro ice core records: Evidence of Holocene climate change in tropical Africa. Science 298, 589–593.

Thompson, L.G., Brecher, H.H., Mosley-Thompson, E., Hardy, D.R., Mark, B.G., 2009. Glacier loss on Kilimanjaro continues unabated. Proceedings of the National Academy of Sciences, USA 106, 19770–19775.

Tripati, A.K., Eagle, R.A., Morton, A., Dowdeswell, J.A., Atkinson, K.L., Bahe, Y., Dawber, C.F., Khadun, E., Shaw, R.M.H., Shorttle, O., Thanabalasundaram, L., 2008. Evidence for glaciation in the Northern Hemisphere back to 44 Ma from ice-rafted debris in the Greenland Sea. Earth and Planetary Science Letters 265, 112–122.

Turner, C., 1998. Volcanic maars, long Quaternary sequences and the work of the INQUA Subcommission on European Quaternary Stratigraphy. Quaternary International 47/48, 41–49.

Turner, C., West, R.G., 1968. The subdivision and zonation of interglacial periods. Eiszeitalter und Gegenwart 19, 93–101.

Tzedakis, P.C., Andrieu, V., Beaulieu, J.L., de, Crowhurst, S., Follieri, M., Hooghiemstra, H., Magri, D., Reille, M., Sadori, L., Shackleton, N.J., Wijmstra, T.A., 1997. Comparison of terrestrial and marine records of changing climate of the last 500 000 years. Earth and Planetary Science Letters 150, 171–176.

Tzedakis, P.C., Hooghiemstra, H., Pälike, H., 2006. The last 1.35 million years at Tenaghi Philippon: Revised chronostratigraphy and long-term vegetation trends. Quaternary Science Reviews 25, 3416–3430.

Van Couvering, J., 1997. Preface, the new Pleistocene. In: Van Couvering, J. (Ed.), The Pleistocene Boundary and the Beginning of the Quaternary. Cambridge University Press, Cambridge, pp. ii–xvii.

Van Donk, J., 1976. A record of the Atlantic Ocean for the entire Pleistocene Epoch. Geological Society of America Memoir 145, 147–163.

Van Leeuwen, R.J.W., Beets, D.J., Bosch, J.H.A., Burger, A.W., Cleveringa, P., van Harten, D., Herngreen, G.F.F., Kruk, R.W., Langereis, C.G., Meijer, T., Pouwer, R., de Wolf, H., 2000. Stratigraphy and integrated facies analysis of the Saalian and Eemian sediments in the Amsterdam-Terminal borehole, the Netherlands. Netherlands Journal of Geosciences 79, 161–198.

Wagner, G.A., 1998. Age Determination of Young Rocks and Artifacts. Springer, Berlin, p. 466.

Walker, M., 2005. Quaternary Dating Methods. Wiley, Chichester, p. 286.

Walker, M., Johnsen, S., Rasmussen, S.O., Steffensen, J.P., Popp, T., Gibbard, P., Hoek, W., Lowe, J., Andrews, J., Bjorck, S., Cwynar, L., Hughen, K., Kershaw, P., Kromer, B., Litt, T., Lowe, D.J., Nakagawa, T., Newnham, R., Schwande, J., 2008. The Global Stratotype Section and Point (GSSP) for the base of the Holocene Series/Epoch (Quaternary System/Period) in the NGRIP ice core. Episodes 31, 264–267.

Walker, M., Johnsen, S., Rasmussen, S.O., Popp, T., Steffensen, J.P., Gibbard, P., Hoek, W., Lowe, J., Andrews, J., Bjorck, S., Cwynar, L., Hughen, K., Kershaw, P., Kromer, B., Litt, T., Lowe, D.J., Nakagawa, T., Newnham, R., Schwander, J., 2009. Formal definition and dating of the GSSP (Global Stratotype Section and Point) for the base of the Holocene using the Greenland NGRIP ice core, and selected auxiliary records. Journal of Quaternary Science 24, 3–17.

Wang, Y.J., Cheng, H., Edwards, R.L., An, Z.S., Wu, J.Y., Shen, C.-C., Dorale, J.A., 2001. A high-resolution absolute-dated late Pleistocene monsoon record from Hulu Cave, China. Science 294, 2345–2348.

Wang, X., Auler, A.S., Edwards, R.L., Cheng, H., Cristalli, P.S., Smart, P.L., Richards, D.A., Shen, C.-C., 2004. Wet periods in northeastern Brazil over the past 210 ka linked to distant climate anomalies. Nature 432, 740–743.

Watts, W.A., Allen, J.R.M., Huntley, B., 1995. Vegetation history and palaeoclimate of the last glacial period at Lago Grande di Monticchio, southern Italy. Quaternary Science Reviews 15, 133–153.

van der Weil, A.M., Wijmstra, T.A., 1987a. Palynology of the lower part (78–120 m) of the core Tenaghi Philippon II, Middle Pleistocene, Greece. Review of Palaeobotany and Palynology 52, 73–88.

van der Weil, A.M., Wijmstra, T.A., 1987b. Palynology of the 112.8–197.8 m interval of the core Tenaghi Philippon II, Middle Pleistocene, Greece. Review of Palaeobotany and Palynology 52, 89–117.

West, R.G., 1968. Pleistocene Geology and Biology, first ed. Longmans, London, p. 377.

West, R.G., 1977. Pleistocene Geology and Biology, second ed. Longmans, London, p. 440.

West, R.G., 1979. Further on the Flandrian. Boreas 8, 126.

West, R.G., Turner, C., 1968. The subdivision and zonation of interglacial periods. Eiszeitalter und Gegenwart 19, 93–101.

Woillard, G.M., 1978. Grande Pile peat bog: A continuous pollen record for the past 140 000 years. Quaternary Research 9, 1–21.

Woldstedt, P., 1962. Über die Bennenung einiger Unterabteilungen des Pleistozäns. Eiszeitalter und Gegenwart 3, 14–18.

Yang, S., Ding, Z., 2010. Drastic climatic shift at ~2.8 Ma as recorded in eolian deposits of China and its implications for redefining the Pliocene-Pleistocene boundary. Quaternary International 219, 37–44.

Yokoyama, Y., Lambeck, K., De Deckker, P., Johnston, P., Fifield, K., 2000. Timing of the Last Glacial Maximum from observed sea-level minima. Nature 406, 713–716.

Zagwijn, W.H., 1992. The beginning of the Ice Age in Europe and its major subdivisions. Quaternary Science Reviews 11, 583–591.

Zalasiewicz, J., Smith, A., Williams, M., Barry, T.L., Bown, P.R., Brenchley, P., Cantrill, D., Coe, A.L., Gale, A., Gibbard, P., Gregory, F.J., Hounslow, M., Knox, R., Marshall, J., Oates, M., Powell, J., Rawson, P., Stone, P., Waters, C., 2008. Are we now living in the Anthropocene? GSA Today 18, 4–8.

Zeuner, F.E., 1935. The Pleistocene chronology of central Europe. Geological Magazine 72, 350–376.

Zeuner, F.E., 1945. The Pleistocene Period: Its climate, chronology and faunal successions. Ray Society Monographs 130, 322.

Zeuner, F.E., 1959. The Pleistocene Period. Hutchinson, London, p. 447.

Zielinski, G.A., Mayewski, P.A., Meeker, D., Grönvold, K., Germani, M.S., Whitlow, S., Twickler, M.S., Taylor, K., 1997. Volcanic aerosol records and tephrachronology of the Summit, Greenland, ice cores. Journal of Geophysical Research 102, 26625–26640.

Zijderveld, J.D.A., Hilgen, F.J., Langereis, C.G., Verhallen, P.J.J.M., Zachariasse, W.J., 1991. Integrated magnetostratigraphy of the upper Pliocene-lower Pleistocene from the Monte Singa and Crotone areas in Calabria, Italy. Earth and Planetary Science Letters 107, 697–714.

J.A. Catt and M.A. Maslin

The Prehistoric Human Time Scale

Abstract: The hominin phylogenetic tree based on comparative skeletal morphology is less robust than those of most other animal groups, because hominin fossils are rare and poorly preserved. After a slow start in the Miocene and Pliocene, there was much more rapid evolution through the Pleistocene. Speciation, dispersal and extinction events often coincided in time with episodes of increased tectonic activity or more rapid climatic and vegetational change, and in space with areas of high biodiversity, such as south and east Africa and Mediterranean regions. Throughout the Pleistocene, there is much more abundant evidence for numerous hominin tool types, but at present most of the cultural changes do not seem to coincide with known evolutionary changes.

Chapter Outline

31.1. INTRODUCTION

In his classic books *The Origin of Species* (1859) and *The Descent of Man* (1871), Charles Darwin implied that modern humans (*Homo sapiens*) arose in Africa from some ancestor resembling an ape. In the mid-19th century the only known primate fossils that could support his suggestion were of the European Mid-Late Pleistocene Neanderthal man (*Homo neanderthalensis*) and the African Miocene ape *Dryopithecus*. Consequently, it was a long time before his ideas were confirmed by the discovery of other hominid fossils of intermediate date, many of them from Africa.

Thomas Huxley (1863) also suggested a phylogeny linking African apes to modern humans, based on anatomical details. In recent decades, this has been confirmed by studies of macromolecules such as proteins (Sarich and Wilson, 1967; Andrews and Cronin, 1982), which undergo more regular evolutionary changes than morphological characteristics. Subsequent genetic studies have also shown that the chimpanzees of various parts of Africa are the nearest ape-like relatives to *Homo sapiens*. Originally their divergence from an unknown common ancestor (gorilla/chimpanzee + human last common ancestor) was dated to the Mid-Miocene (~14 Ma), but the genome studies suggest it occurred in the Late Miocene (5−7 Ma) (Kumar *et al.*, 2005; Patterson *et al.*, 2006). Fossil remains from numerous African, Asian and European sites now indicate rapid subsequent (Plio−Pleistocene) evolution of the human line, with the appearance of perhaps as many as twenty morphologically distinct species ultimately leading to *Homo sapiens* in the Mid-Late Pleistocene.

Because of their genetic similarity (c. 97 %, according to Kumar *et al.*, 2005), modern humans and chimpanzees are separated from other members of the Hominidae family, such as gorillas and orangutans, as the sub-family Homininae. All are within the Primate Order of the Class Mammalia. Humans are fully bipedal, apes are not, but the tree-dwelling

The Geologic Time Scale 2012. DOI: 10.1016/B978-0-444-59425-9.00031-7

chimpanzees are occasionally bipedal. Very few chimpanzee fossils are known. The two modern African species *Pan troglodytes* (bonobo) and *Pan paniscus* (chimpanzee) are thought to have separated late in the Early Pleistocene around 1 Ma. Little is known of their earlier relatives, but it is unlikely that the African ape from which humans separated (the chimpanzee/human last common ancestor) was exactly the same as a modern chimpanzee.

When Linnaeus (1758) introduced the name *Homo sapiens*, unusually he provided no morphological description of either the genus or the species, and working definitions of both remain elusive even today (Tattershall and Schwartz, 2009). As a result, the morphological differences between modern man and earlier hominins are inadequately defined. In addition, fossil hominin remains are rare, often poorly preserved and almost always incomplete, so that the differences between them are not clearly defined, and phylogenetic linkages are correspondingly speculative. Even more than in other branches of paleontology, paleoanthropology has vacillated between the designation of new species or subspecies for almost every new discovery and the tendency to shoehorn almost any new material into a minimum of existing categories.

In what follows, summarized in Figure 31.1, we present what we believe to be the modern consensus view of hominin phylogeny, omitting most of the numerous contrary opinions for the sake of brevity. However, it is inevitable that this picture will change as further fossil remains are discovered. The main morphological feature widely taken as evidence for evolutionary development is cranial capacity, as an indicator of brain size. Even this is adequately justified only within groups representing single clades shown to be morphologically cohesive in other ways, though among the various hominins so far distinguished there is a general if somewhat irregular increase from about 300 cm^3 in the Late Pliocene to 1800 cm^3 in the Late Pleistocene, followed by a small decrease to the present. Encephalization quotient, the ratio of actual brain mass to that expected after adjustment for body size, is perhaps a more appropriate factor, as it increases more linearly, from about 2.5 in chimpanzee to 7.8 in modern humans, but it is often impossible to calculate this from incomplete fossil remains.

31.2. HOMININ PHYLOGENY AND MIGRATION EPISODES

31.2.1. Hominins Predating *Homo*

Within the last two decades, several hominin fossils post-dating divergence from chimpanzee have been discovered in the late Miocene and Pliocene deposits of Ethiopia, Kenya and Chad. They have been attributed to three different genera; *Ardipithecus* (WoldeGabriel *et al.*, 1994; White *et al.*, 1994), *Orrorin* (Senut *et al.*, 2001) and *Sahelanthropus* (Brunet *et al.*, 2002), respectively, though the last two may be variants of an early form of *Ardipithecus* known as *A. kadabba* (Haile-Selassie *et al.*, 2004). *Sahelanthropus tchadensis* and *Orrorin tugenensis* date from 6–7 Ma; both were probably bipedal and after the chimpanzee could be the earliest hominins, but conclusions are tentative because their fossil remains are sparse (Lockwood, 2008).

Much more abundant material is known for the somewhat later *Ardipithecus ramidus* from the Lower Aramis Member of the Pliocene Sagantole Formation exposed in the Middle Awash valley of the Afar Rift in Ethiopia (White *et al.*, 2009). The date of this hominin, as indicated by $^{39}Ar/^{40}Ar$ dates for volcanic tuff horizons above and below the Lower Aramis Member, is 4.4 Ma. The skeletal remains and associated paleobiological evidence suggest it was bipedal and had a small brain, similar in size (300–350 cm^3) to that of modern chimpanzees, ate a more omnivorous diet than modern (fruit-eating) chimpanzees, and clambered, often on its palms, in humid cool woodland, which was interspersed with small patches of moist grassland. So, even if bipedality originated in response to the appearance of an East African savannah habitat, as often postulated, it did not necessarily preclude a later return to the woodland environment. However, assigning an individual species to a particular habitat is often uncertain because of the problem of time-averaging within sediments (Hopley and Maslin, 2010). An earlier *Ardipithecus* from the Middle Awash valley, transitional in morphology between *A. kadabba* and *A. ramidus* is dated to 5.2–5.8 Ma (Haile-Selassie, 2001).

Probably the next youngest predecessor of *Homo* was *Australopithecus*, examples of which date from African deposits ranging in age from about 4.2 to 1 Ma (Figure 31.1). The first discovery was of a juvenile skull found at Taungs, South Africa, in 1924 and named *Australopithecus africanus* (Dart, 1925). Remains of this species have also been found at three other sites in South Africa (Sterkfontein, Makapansgat and Gladysvale), where they have been dated by the associated fauna and by electron spin resonance to 2.5–3.0 Ma. An earlier species (*Australopithecus afarensis*) is represented by a mandible fragment older than 4.15 Ma found at Tabarin, Kenya (Ward and Hill, 1987), by jaw fragments, teeth and footprints dated to 3.5–3.7 Ma at Laetoli, Tanzania (Leakey *et al.*, 1976), by a jaw and other bones found at Maka, Ethiopia and dated to 3.4–3.9 Ma (White *et al.*, 1993), and by a partial female skeleton ("Lucy"), dated to 3.2 Ma, found at Hadar in the Afar region of Ethiopia (Johanson, 2009). *A. afarensis* had anatomical features of the hand, wrist, pelvis, knee, ankle and foot indicating that it was bipedal and had abandoned a woodland habitat. A yet earlier species, *Australopithecus anamensis*, from the Early Pliocene (4.2–3.9 Ma) of the Afar region, and also known from the Lake Baringo (Bishop and Chapman, 1970; Hill, 1985) and Lake Turkana (Patterson *et al.*, 1970; Ward *et al.*, 2001) districts in Kenya, shows several skeletal characteristics transitional between *Ardipithecus ramidus* and *Australopithecus afarensis* (White *et al.*, 2009).

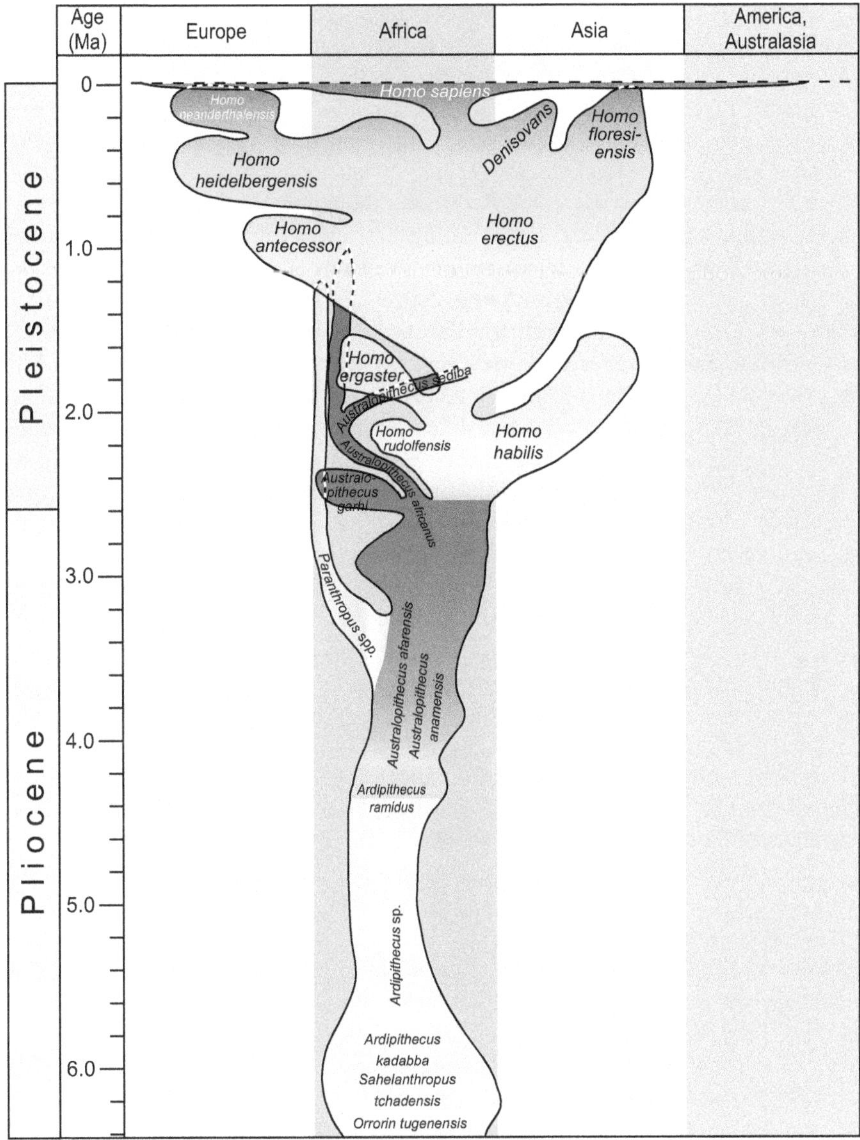

Development and distribution of the main hominin species during the later Pliocene and Pleistocene.

Later, more advanced Australopithecines, which may have been transitional to *Homo*, include *A. garhi*, from the Early Pleistocene (~2.5 Ma) at Bouri in the Middle Awash valley (Asfaw *et al.*, 1999), and *A. sediba*, two partial skeletons of which, recently discovered at Malapa Cave in the Sterkfontein area of South Africa, have been dated to 1.78–1.95 Ma (Berger *et al.*, 2010). *A. garhi* had a cranial capacity of approximately 450 cm^3.

Another group originally named *Australopithecus* (*A. aethiopicus, A. boisei* and *A. robustus*), sometimes known as the robust Australopithecines but now generally separated as the genus *Paranthropus*, have been found in Late Pliocene and Early Pleistocene deposits dating from about 3.0 Ma in various parts of Africa. They differed from the gracile Australopithecines described above in possessing heavier skulls with cranial capacities up to 530 cm^3, and probably diverged from the *Ardipithecus ramidus–Australopithecus afarensis* line in the Late Pliocene, but evolved slowly and eventually died out around 1.1 Ma. Measurements of stable carbon isotopes in Australopithecine teeth suggest that most had a diet consisting of mixed C3 (moist woodland) and C4 (dry grassland) plants (e.g., Sponheimer and Lee-Thorp, 1999).

31.2.2. Early Forms of *Homo* in the Early and Mid Pleistocene

The earliest species attributed to *Homo* date from the first part of the Early Pleistocene (2.4–1.9 Ma) mainly at East African sites, such as Olduvai Gorge in Tanzania (Leakey *et al.*, 1964), Hadar, Omo and Gona in Ethiopia, and Koobi Fora and Lokalalei in Kenya. Remains attributed to two different species, *Homo habilis* and *Homo rudolfensis*, occur at Koobi Fora, but Olduvai Gorge and other sites have only *H. habilis*. These early species of *Homo* were taller than *Australopithecus* and had generally larger cranial capacities, 500–680 cm^3 in *H. habilis* and 750 cm^3 in *H. rudolfensis*. Nevertheless, on account of the small difference in cranial capacity, Wood and Collard (1999) suggested that *H. habilis* was more closely related to the robust Australopithecines than to other species of *Homo*. About 2 Ma, *Homo habilis* may have migrated further than earlier hominins, as somewhat similar skeletal remains have been found in parts of Africa remote from the eastern rift (South Africa, Malawi, Algeria) and in the Arabian Peninsula (Dennell and Roebroeks, 2005).

Two new species appeared about 1.9 Ma in East Africa, *Homo erectus and Homo ergaster. H. ergaster* is known mainly from Africa, the best example being the almost complete skeleton of a boy dated to 1.6 Ma found near the River Nariokotome in West Turkana, Kenya (Walker and Leakey, 1993). *H. erectus* also originated in East Africa (Conroy *et al.*, 1978; Clark *et al.*, 1994), but soon dispersed more widely, probably reaching all parts of Africa except the rainforest of the Congo basin, and migrating through northeast Egypt, Israel, Syria and Turkey (the "Levant Corridor") to become established in Georgia (Gabunia *et al.*, 2000; Rightmire *et al.*, 2006), China (Tangshan Archaeological Team, 1996) and Java (Sartono, 1991) by about 1.8 Ma. In East Africa it was contemporary with *H. habilis* for several hundred thousand years, but it had a slightly bigger brain and was probably better adapted to endurance running and animal hunting (Brambel and Leibermann, 2004). The increased energy obtained from a more meat-rich diet was probably important for brain development.

By about 0.8 Ma, *H. erectus* had migrated to Israel (Saragusti and Goren-Inbar, 2001), and soon afterwards reached Europe. A mandible found at Mauer near Heidelberg in 1907 (Schoetensack, 1908; Mounier *et al.*, 2009) and dated to about 0.5 Ma has long been attributed to *Homo heidelbergensis*, though the justification for a separate species is sometimes questioned. Early Mid Pleistocene remains attributed to *H. erectus* or *H. heidelbergensis* have also been found elsewhere in Germany (Vlček, 1978) and in many other parts of Europe, including France (de Lumley and de Lumley, 1974), Greece (Poulianos, 1971), Italy (Ascenzi *et al.*, 1996), Hungary (Thoma, 1966) and southern England (Roberts and Parfitt, 1999). At the Sima de los Huesos site in the Sierra de Atapuerca, northern Spain (Arsuaga *et al.*, 1993; Falguères *et al.*, 1999; Lozano *et al.*, 2009), abundant remains of what was probably an earlier offshoot of *H. erectus*, dated to 0.8–1.2 Ma, have been named *Homo antecessor* (Figure 31.1).

There is also evidence of a remnant population of *H. erectus* persisting in various parts of Africa through the Mid Pleistocene. Perhaps the most famous is a skull with very large brow ridges found in 1921 at the Broken Hill lead–zinc mine in Rhodesia (now known as the Kabwe Mine, Zambia). Originally named *Homo rhodesiensis* (Woodward, 1921), this specimen and an associated tibia (Trinkaus, 2009) are now attributed to an African equivalent of *H. heidelbergensis*, though with some affinities to *H. erectus*. Exact dating has always been uncertain, but associated micro-mammal remains suggest it could be as recent as 200 ka (Avery, 2002).

In northern Europe, *H. heidelbergensis* often lived in caves and rock shelters during colder stages of the Mid Pleistocene, though in lower latitudes and warmer periods temporary tent-like shelters were probably built in open areas from timber and animal skins weighed down by large stones (de Lumley and Boone, 1976). Hearths at the Gesher Benot Ya'aqov site in Israel provide the earliest evidence for controlled use of fire at approximately 595 ka (Goren-Inbar *et al.*, 2004). Apart from providing warmth, illumination at night and protection from predators, fire was probably important in evolutionary development as it allowed food to be cooked (Wrangham and Carmody, 2010), making it more digestible and removing bacteria and other pathogens.

The cranial capacity of *H. erectus* ranged from about 600 cm^3 in the early part of its history to 1250 cm^3 in skulls dated to about 450 ka from the Zhoukoudian caves near Beijing (China). It seems to have become extinct by about 150 ka. But on the island of Flores (Indonesia), which has been inhabited by hominins at least periodically since 1 Ma (Brumm *et al.*, 2010), *H. erectus* probably persisted until almost 12 ka as the dwarf ("hobbit") species *Homo floresiensis*, which had a cranial capacity of only 380 cm^3 (Brown *et al.*, 2004).

Two new hominin species emerged in the later Mid Pleistocene. The populations of *Homo heidelbergensis* in Eurasia are thought to have been the immediate ancestor of *Homo neanderthalensis*, and those of *Homo erectus* remaining in East Africa the ancestor of *Homo sapiens*. The range of cranial capacity in *H. sapiens* (1350–1450 cm^3) is greater than those of later *H. erectus* and *H. heidelbergensis* (1100–1400 cm^3). But all three had smaller skulls than *H. neanderthalensis*, which had a maximum cranial capacity exceeding 1800 cm^3, though this does not necessarily imply greater intelligence.

31.2.3. *Homo Neanderthalensis* in the Mid and Late Pleistocene

H. neanderthalensis generally inhabited cooler mid-latitude regions of the northern hemisphere, such as France, Belgium, Germany, Italy, southern England, Spain and northern Arabia, though in warmer periods, such as parts of marine isotope stages (MISs) 5, 4 and 3, they also migrated temporarily across the eastern Alps and Carpathian Mountains into Poland

(Skrzypek *et al.*, 2011) and parts of Asia as far east as Uzbekistan and the Altai Mountains of southern Siberia (Krause *et al.*, 2007). The enlarged brain could have resulted from environmental conditions at these latitudes, such as the need to improve sight in low winter light levels.

The earliest putative Neanderthal remains, found at Pradayrol in France, were dated by electron spin resonance to 330 ± 5 ka, within MIS 9 (Blackwell *et al.*, 2007). The most recent, c. 40–28 ka (late MIS 3), are from final refugia to which the species had retreated in southern Iberia, notably Gorham's Cave, Gibraltar (Finlayson *et al.*, 2006). Neanderthals seem to have been the first hominin species to bury their dead. This probably explains why their fossil remains are significantly more abundant than those of earlier species, the corpses of which were probably cannibalized or left on the surface for their flesh and bones to be consumed by carnivore animals.

Neanderthals were shorter and stockier than either *H. erectus* or *H. sapiens*, with an almost triangular thorax and broad pelvis. This shape decreased surface area, and has long been interpreted as an adaptation to life in mid-latitudes during the long cold stages of the Mid and Late Pleistocene. But Stewart (2005) has questioned this, and Aiello and Wheeler (2003) have argued that such features were not necessarily an advantage in cold conditions, where the main problem is frostbite of exposed skin. *H. neanderthalensis* also had a large nose, which may have served to warm cold air before it entered the lungs. The numerous other cranial and mandibular characteristics that distinguish it from *H. sapiens* were listed by Tattershall and Schwartz (2006), and justify their separation as distinct species (Harvati *et al.*, 2004). Analyses of carbon and nitrogen isotopes in the collagen of fossil bones indicate that European Neanderthals had a very meat-rich diet, whereas contemporary *H. sapiens* was more omnivorous (Richards and Trinkaus, 2009). More speculative behavioral and social differences between the two species are discussed by Stringer (2011) in his chapters 5 and 6.

The main contraction of Neanderthal populations after c. 40 ka broadly coincides with the deteriorating and strongly oscillating climate of MIS 3. The large Campanian Ignimbrite eruption of the Phlegrean Fields (central Italy) around 39.3 ka may have combined with the extremely cold Heinrich Event 4 to produce a "volcanic winter" that initiated Neanderthal extinction (Golovanova *et al.*, 2010). At Gorham's Cave their final disappearance may coincide with the increase in global ice volume and steepening of the latitudinal temperature gradient that occurred at the MIS 3/2 boundary c. 24 ka (Tzedakis *et al.*, 2007). However, these climatic changes may not be the sole reason for their extinction, as they would also have affected *H. sapiens*, which survived them. Elsewhere in Europe the disappearance of Neanderthals seems to coincide roughly with the arrival of *H. sapiens*. The two species probably co-existed for short periods, and would have competed for decreasing food resources in the deteriorating environmental conditions. *H. sapiens* was taller, faster, better at endurance running, and thus a more efficient hunter, but also could have introduced new diseases from the tropics that proved fatal to *H. neanderthalensis* (Wolff and Greenwood, 2010).

More speculatively, Valet and Valladas (2010) noted the coincidence of the progressive decline in European Neanderthal populations with the period of weak geomagnetic field intensity between the Laschamp (40.4 ± 1.1 ka) and Mono Lake (32.4 ± 0.3 ka) excursions. At this time, western Europe would have been shielded less efficiently from high energy particles of cosmic and solar origin, so they suggested that local depletion of atmospheric ozone and increased exposure to UVB radiation could have weakened Neanderthal health, causing skin cancer, suppressed immunity to disease and eye cataracts that restricted hunting ability. In comparison, contemporary *H. sapiens* populations were more widely distributed and were less affected in areas, such as Africa, eastern Europe and Asia, where the geomagnetic field strength was not weakened as much.

Over the last 2–3 decades, some of the relationships between later hominin groups previously based on skeletal morphology have been reassessed by detailed sequencing of nuclear and mitochondrial DNA extracted from selected fossil remains; mitochondrial DNA is transmitted through the female line only, and is often more useful for evolutionary studies as it mutates more quickly than nuclear DNA. As DNA degrades more rapidly at higher temperatures and in acidic conditions, all endogenous forms are preserved better in remains found at higher latitudes, such as those inhabited by *H. neanderthalensis*, and especially from limestone caves. Comparison of DNA sequences from well-preserved fossil bones of three different *H. neanderthalensis* specimens with those from modern *H. sapiens* from five different parts of the world by Green *et al.* (2010) suggested that the two species had a common ancestor between about 825 ka and 440 ka, but (in broad agreement with the fossil record) had separated before 270 ka. It also showed that present-day populations of Eurasian areas share slightly more (1–4 %) of their genes with *H. neanderthalensis* than do the native populations of Sub-Saharan Africa, as might be expected from the known occurrence of Neanderthal fossils.

31.2.4. The Denisovans

A complete mitochondrial DNA genome obtained by Krause *et al.* (2010) from a fossil phalanx (finger bone) found in Denisova Cave in the Altai Mountains, an area known to have been inhabited by *H. neanderthalensis*, indicated that the hominin from which the phalanx came shared a much earlier (~1.04 Ma) common ancestor with *neanderthalensis* and *sapiens* than that shared by *H. neanderthalensis* and *H. sapiens* alone (466 ka). As the cave deposit in which it was found dates from between 50 ka and 30 ka, the considerably greater age of this most recent common ancestor suggests the existence of a third otherwise unknown Late Pleistocene hominin, which may have separated from *Homo erectus* populations in eastern Asia. Krause *et al.* (2010) considered it premature to distinguish this as a separate species, so they described the "Denisovans" as a "sister group" to the Neanderthals (Figure 31.1).

Further studies of the nuclear DNA extracted from the same phalanx (Reich *et al.*, 2010), including comparisons with the nuclear DNA of present-day human groups, indicated that, although there has been no Denisovan genetic contribution to the modern Mongolian and Chinese populations of the Altai region, 4−6 % of the genes of the present Melanesian population of Papua New Guinea are inherited from the Denisovans. So there was some interaction between the Denisovans and Melanesian ancestors, though probably not in southern Siberia; this implies a Denisovan migration from Siberia to Melanesia, possibly in response to deteriorating climatic conditions in MIS 3 or 2.

31.2.5. The Origin of *Homo Sapiens*

Hominins with a large proportion of anatomical features resembling those of modern *Homo sapiens* became common in East and South Africa between 195 and 100 ka, and eventually dispersed to all parts of the world other than the polar regions (Rightmire, 2001). The first evidence for emergence of *H. sapiens* in Africa during this period was provided by analyses of mitochondrial DNA and derivation of a single common ancestor for 147 modern humans distributed around the world (Cann *et al.*, 1987). More recently, it has been confirmed by further comparisons of mitochondrial, nuclear and Y-chromosome DNA sequences from modern African populations (Stringer, 2011, chapter 7), though attempts to extract uncontaminated DNA from fossil remains of *H. sapiens* have not yet been successful. Morphometric studies of hominin remains also support an African origin for early *H. sapiens*. For example, crania from the Middle Awash valley of Ethiopia, which are similar in size and general shape to modern human skulls from Oceania and Australia, have been dated to 160−154 ka (White *et al.*, 2003), and an anatomically modern child's jaw from Jebel Irhoud, Morocco, has been dated to 160 ka (Smith *et al.*, 2007).

Compared with *H. erectus*, *H. sapiens* developed reduced post-canine dentition, a smaller jaw but more prominent chin, a more barrel-shaped chest and shorter forearms, but the two were broadly similar in height. Unlike those of all earlier hominins, including chimpanzee, the skull of *H. sapiens* is more globular and has an upright forehead with almost no brow ridges. These changes may have been associated with differences in brain shape and structure of the larynx, which allowed the development of articulate speech. The brain has a large energy requirement, and for *H. sapiens* this was probably met by a more carnivorous diet than that of *H. erectus*, though lipids obtained from a seafood diet in coastal regions of Africa may also have been important in development of the *H. sapiens* brain (Parkington, 2006).

Compared with earlier hominins, and indeed almost all other mammals apart from some aquatic and subterranean species, *H. sapiens* also has a greatly decreased cover of body hair. In fact, much of the 2−3% difference between the modern human and chimpanzee genomes lie in the genes that control skin properties (e.g., scuff resistance) resulting from the change to hairlessness. This change may have originated in the need for more efficient cooling of a body that often had to expend large amounts of energy in hunting animal food (Jablonski, 2010). The ability to dissipate body heat by more efficient sweating may also have permitted enlargement of the brain, which is one of the human body's most temperature-sensitive organs, though retention of scalp hair may have been necessary to provide some insulation for the top of the head exposed directly to the sun in the open savannah environment of Africa.

The need to find new hunting areas because of a simultaneous increase in human population and decrease in prey animals, may account for an initial migration of *H. sapiens* out of Africa into southeastern Arabia by c. 125 ka (Armitage *et al.*, 2011) and into Israel between 120 and 90 ka (Valladas *et al.*, 1987; Grün and Stringer, 2000). At this time (MIS 5), the climate of north Africa and the Near East was considerably wetter than today, with numerous lakes, rivers and forest that could have assisted migration (Petraglia, 2011). A small but robust mandible and teeth dated to >100 ka from Zhiren Cave in Guizhou Province, southern China, have also been attributed to an early form of *H. sapiens* (Liu *et al.*, 2010). However, with no evidence for anatomically modern hominins between Arabia and China at this time, Dennell (2010) suggested that they could represent a late form of *H. erectus* that had become more gracile over time in an isolated situation.

The main dispersal of *H. sapiens* from Africa into the Near East and Europe through the "Levant Corridor" probably occurred much later during Greenland Interstadial (GIS) 14/13 of the ice core record (part of MIS 3). At this time (55−49 ka), migration was assisted by an increase in low latitude summer insolation, which led to northward extension of summer monsoon rainfall in Africa. Simultaneously a mild humid period in Europe resulted from resumption of North Atlantic meridional overturning (Müller *et al.*, 2011). The MIS 3 migration into Europe was probably interrupted by the brief but sharp climatic deterioration of Heinrich Event H5 c. 48 ka, but resumed when Atlantic circulation was reinstated in GIS 12 (c. 47 ka). However, GIS 12 initiated a long period of rapid climatic change, to which *H. sapiens* already in Europe adapted by innovation in technology and social organization (Gamble, 2009). This timing of the main dispersal of *H. sapiens* from Africa into Europe is supported by molecular studies, which suggest that most modern human populations had migrated out of Africa by about 60 ka (Harpending and Rogers, 2000).

From the Near East, it is likely that after about 50 ka *H. sapiens* migrated eastwards to India, southeast Asia and Indonesia, probably using coastal routes to avoid arid and often mountainous inland areas. Numerous short sea crossings across Indonesia led to Australia by about 40 ka, and a northward route through China and eastern Russia may eventually have led across the Bering Strait to North America by about 13 ka. However, these suggested routes are only weakly supported by evidence from human remains or artifacts.

In warmer regions, early *H. sapiens* sometimes constructed temporary homes using poles, large bones, skins and stones, but caves or rock shelters were the main form of habitation, especially in the increasingly cold conditions of higher latitudes in late MIS 3 and MIS 2. One of the first hominin fossils ever discovered is an ochre-stained complete skeleton of *H. sapiens* found in Paviland Cave, South Wales, by William Buckland (1823). Originally thought to be an interred Roman prostitute, it was termed the "red lady of Paviland", but the skeleton is actually that of a young man aged 26–30, which has now been radiocarbon dated to ~ 33 ka (MIS 3). The arrival of ice sheets at the height of the Last Glaciation (~ 20 ka) temporarily drove *H. sapiens* southwards out of Britain, Denmark and other north European areas for several thousand years in the middle of MIS 2, though they returned during the Allerød or Windermere Interstadial, which is equivalent to Greenland Interstadial 1 (GI-1, 14.70–12.65 ka BP) in the ice core record.

The rapid climatic amelioration at the beginning of the Holocene (11.7 ka) led to a large increase in the population of *H. sapiens* in many regions. The hunter-gatherer economy, which had persisted throughout the later Pliocene and Pleistocene, initially remained unchanged in many areas, but the need to sustain the rapidly expanding population prompted the development of agricultural techniques, such as the domestication of animals (goats, sheep, cattle) and the cultivation of arable crops (mainly cereals). Agriculture in turn led to development of farming settlements, which progressively replaced the earlier nomadic lifestyle, and sparked a rapid advancement towards modern civilization.

31.3. THE PALEOENVIRONMENTAL CONTEXT OF EARLY HOMININ EVOLUTION

The Late Miocene to Mid Pleistocene period of hominin evolution leading from the chimpanzee/human last common ancestor eventually to *Homo sapiens* was one of progressive though irregular global cooling. In the main evolutionary arena, East Africa, it was also one of massive tectonic changes leading to progressive but irregular fragmentation of the vegetation and an increase in overall aridity. The overall East African aridification is mainly indicated by the carbon isotope records of secondary carbonates in paleosols (Levin *et al.*, 2004; Wynn, 2004; Segalen *et al.*, 2007) and of *n*-alkanes in marine sediments (Feakins *et al.*, 2005), both of which suggest a progressive shift from C3 (woodland) to C4 (tropical grassland) vegetation types (Figure 31.2). The gradual drying has been attributed to rifting and uplift of the region (Sepulchre *et al.*, 2006).

However, in the Early and Mid Pleistocene, the pattern of overall aridification was punctuated by three main episodes of highly variable climate, oscillating between very wet and unusually dry, which have been dated to approximately 2.7–2.5, 1.9–1.7 and 1.1–0.9 Ma (Figure 31.2). The evidence for these is provided by the record of large, deep ephemeral lakes in the Kenyan and Ethiopian Rifts (Trauth *et al.*, 2005, 2007), by variations in the abundance of Saharan dust reaching the eastern Mediterranean (Larrasoaña *et al.*, 2003) (Figure 31.2) and by peaks in eastern Mediterranean seafloor sapropel, which probably resulted from changes in discharge of the River Nile (Lourens *et al.*, 2004). Where the lacustrine records have been investigated in detail, as for the 2.7–2.5 Ma episode in the Chemeron and Baringo Basins of the Kenya Rift (Deino *et al.*, 2006; Kingston *et al.*, 2007) and the 1.9–1.7 Ma episode in the Koobi Fora Formation of the Turkana Basin (Lepre *et al.*, 2007), separate sub-phases of higher lake levels, often indicated by diatomite beds, have been identified with a cyclicity related to the 23 ka astronomical cycle of precession. Similar precessional cyclicities have also been recognized in the Mediterranean sapropel record (Lourens *et al.*, 2004) and in the dust record of the eastern Mediterranean (Larrasoaña *et al.*, 2003), North Atlantic and Indian Ocean (de Menocal, 1995, 2004), though the dust and sapropel records are negatively correlated.

The three main episodes of intense wet–dry climatic oscillation seem to correlate with maxima in the 400 ka component of the astronomical eccentricity cycle. Although these occurred every 400 ka before 2.7 Ma, after this time the strongest maxima recur at 800 ka intervals, and the weaker intervening 400 ka maxima (at about 2.2, 1.4 and 0.6 Ma) are not so easily identified in the lake records. Because of the link with the eccentricity cycle, the three episodes correlate with important global climatic transitions (Trauth *et al.*, 2005), the 2.7–2.5 Ma episode with a considerable intensification of northern hemisphere glaciation following closure of the Straits of Panama (Haug and Tiedemann, 1998), the 1.9–1.7 Ma episode with intensification and shift in the east-west zonal atmospheric circulation known as the Walker Circulation (Ravelo *et al.*, 2004) and the 1.1–0.9 Ma episode with the change from 41 ka to 100 ka glacial–interglacial cycles, often known as the Mid Pleistocene Revolution (Berger and Jansen, 1994) (Figure 31.2). The three global climatic transitions led to polar ice expansion and compression of the Intertropical Convergence Zone, the latter influencing the strength of the southeast Asian monsoon and increasing the sensitivity of East Africa to precessional forcing of rainfall.

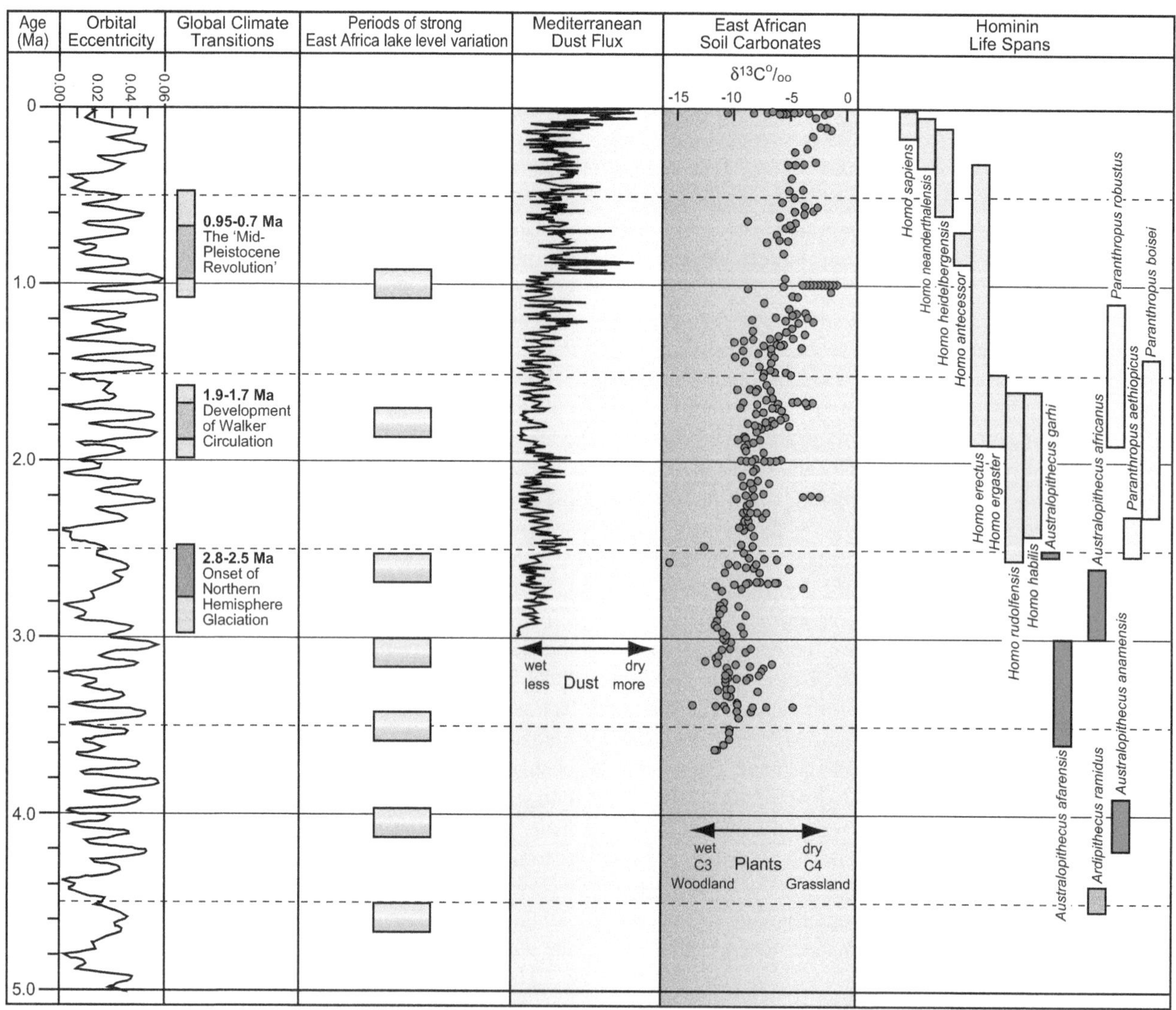

Relationship of hominin evolution to global climate transitions and indicators of African environments.

Coupled with the orographic changes resulting from uplift of the Kenyan and Ethiopian Plateaux, the increased sensitivity resulted in a rift valley climate that, by analogy with the climate and lake history of the last 150 ka, could have produced lakes up to 150 m deep with an increase in annual precipitation of only 15–30% (Bergner *et al.*, 2003).

Apart from the climatic variability in East Africa resulting from precessional cycles, even shorter range variability was likely because precessional forcing of climate is not smooth but sinusoidal (Maslin *et al.*, 2005), with short periods (<2000 years) of strong forcing separated by longer periods (>8000 years) of weaker forcing. This would have caused lakes to appear rapidly in the rift valley, probably in less than 500 years, then persist for several millennia before disappearing equally rapidly (Deino *et al.*, 2006; Kingston *et al.*, 2007). The profound effects such rapid climatic changes would have had on the vegetation of East Africa also had important implications for the speciation and dispersal of hominins in the region (Maslin and Christensen, 2007). In particular, Maslin and Trauth (2009) have pointed out that:

> In the 2.7–2.5 Ma episode, *Australopithecus africanus, A. garhi, Homo rudolfensis, H. habilis* and *Paranthropus aethiopicus* all first appeared, there was a minor migration of *H. habilis* out of Africa and *A. africanus* later became extinct.
>
> In the 1.9–1.7 Ma episode, *H. erectus, H. ergaster* and *P. robustus* first appeared, there was a major migration of *H. erectus* out of Africa and *H. habilis, H. rudolfensis* and *H. ergaster* all became extinct.
>
> Finally, in the 1.1–0.9 Ma episode, there was another migration of *H. erectus* from Africa into Europe, *H. heidelbergensis* first appeared and *P. robustus* became extinct.

Where it is possible to date them precisely, these speciation, dispersal and extinction events seem to have occurred mainly at the beginning and end of the episodes of enhanced climate variability. This supports an extreme version of the Variability Selection Hypothesis of human evolution (Potts, 1998), which has been refined as the Pulsed Climate Variability Hypothesis (Maslin and Trauth, 2009).

The episodes of enhanced climate variability also resulted in an increase in relief in many parts of the world in response to increased erosion and isostatic uplift, as indicated by river terrace sequences (Westaway *et al.*, 2009). The second episode (1.9–1.7 Ma) coincides with the migration of *H. erectus* from East Africa into Asia via the "Levant Corridor", a region that suffered especially strong erosional isostasy and possibly also tectonic uplift (Schattner, 2010) in this part of the Mid Pleistocene. The increased relief in this previously semi-arid area probably led to moister conditions with denser vegetation and more lakes, which may have facilitated the migration of *H. erectus* (King and Bailey, 2006).

31.4. HOMININ INDUSTRIES AND THE TERMINOLOGY OF PREHISTORIC PERIODS

Compared with the rare hominin skeletal remains (bones and teeth), artifacts provide a much more abundant record of human presence and the development of human activities. Based on the raw materials used and the typology of artifacts, the period from the first appearance of *Homo* early in the Pleistocene until the start of the historic period has been divided into the Old Stone Age (Paleolithic), Middle Stone Age (Mesolithic), New Stone Age (Neolithic), Bronze Age and Iron Age. By convention, the Paleolithic ends and the Mesolithic begins at the Pleistocene/Holocene boundary, so the Paleolithic is almost exactly equivalent to the Pleistocene.

Apart from the end of the Paleolithic coinciding with the close of the Pleistocene, there are no clear time lines between the various archaeological periods. Their boundaries are often irregularly diachronous because they depend upon the slow migration of tool-making techniques from region to region or on the indigenous development of techniques at different times in different places (Figure 31.3). Also each species of *Homo* did not occupy all areas continuously from the time of its first appearance. Long gaps in the human record of many regions have been attributed to factors such as climatic deterioration, depletion of food resources or changes in land distribution resulting from major modifications to sea level.

The Paleolithic is further divided into Lower, Middle and Upper. Based on the association of artifacts with human fossil remains, the Lower Paleolithic is roughly equivalent to the lifespans of *Homo habilis* and *Homo rudolfensis*, and the Middle Paleolithic to those of *Homo erectus, H. antecessor, H. ergaster, H. heidelbergensis* and *H. neanderthalensis*. The Upper Paleolithic, Mesolithic and later archaeological periods coincide with the presence of *Homo sapiens*.

31.5. EARLY AND MID PLEISTOCENE TECHNOLOGIES

31.5.1. The Lower Paleolithic

No hominid predating *Homo* produced stone tools that can definitely be identified as such in Pliocene or earlier deposits, though there is some evidence in the Late Pliocene (~ 3.4 Ma) of Ethiopia for sharp (probably natural) rock flakes having been used for butchery (McPherron *et al.*, 2010). However, both of the earliest species in the Early Pleistocene (*H. habilis* and *H. rudolfensis*) produced rudimentary stone tools of the Lower Paleolithic Oldowan culture. *H. habilis* (handy/able/skilful man) was in fact named originally because of its association with stone tools (Leakey *et al.*, 1964), which may have been essential to the meat-eating diet that facilitated brain enlargement. Oldowan tools consisted of natural stream cobbles of basalt, quartzite, chert or other hard rock which have been crudely knapped by striking them with a larger softer hammerstone, so as to remove sharp-edged flakes that could be used for cutting meat and plant material or extracting marrow from animal bones. At some sites, these "pebble-tools" are associated with the skeletal remains of Australopithecines as well as *Homo* spp, but there is no hard evidence that the former were responsible for producing tools in this way.

Three phases of the Oldowan culture can be distinguished. The Pre-Oldowan, which has been recognized at Gona, Omo and Hadar in East Africa and dates from 2.6–2.0 Ma, consists of flakes produced by simply striking cobbles with a hammerstone against an anvil stone; the unifacial flakes were used and the remaining corestones rejected. As the earliest remains of *H. habilis* are from 2.4 Ma, the earliest stone tools predate current knowledge of the species with which they are thought to be associated. In the Classic Oldowan, recognized more widely and dating from 1.9–1.6 Ma, the cobble was struck either against an anvil stone or while being held in the hand, and many of the cores were used for chopping and not simply rejected; also, there was some tendency towards characteristic shaping of flakes and cores. Finally, in the Developed Oldowan (1.6–1.5 Ma), recognized mainly at Olduvai and Koobi Fora, stones held in the hand were chipped from both sides to produce proto-bifaces with sharper and stronger cutting edges, and large numbers of flakes were skillfully fashioned into skin scrapers.

Chopper-core and simple flake industries resembling the Oldowan are also known from numerous sites outside East Africa, including raised beaches in Morocco, Spain and Portugal, high river terraces in Hungary, northern Punjab, Thailand, Malaya and Java and lake deposits in Italy, northern

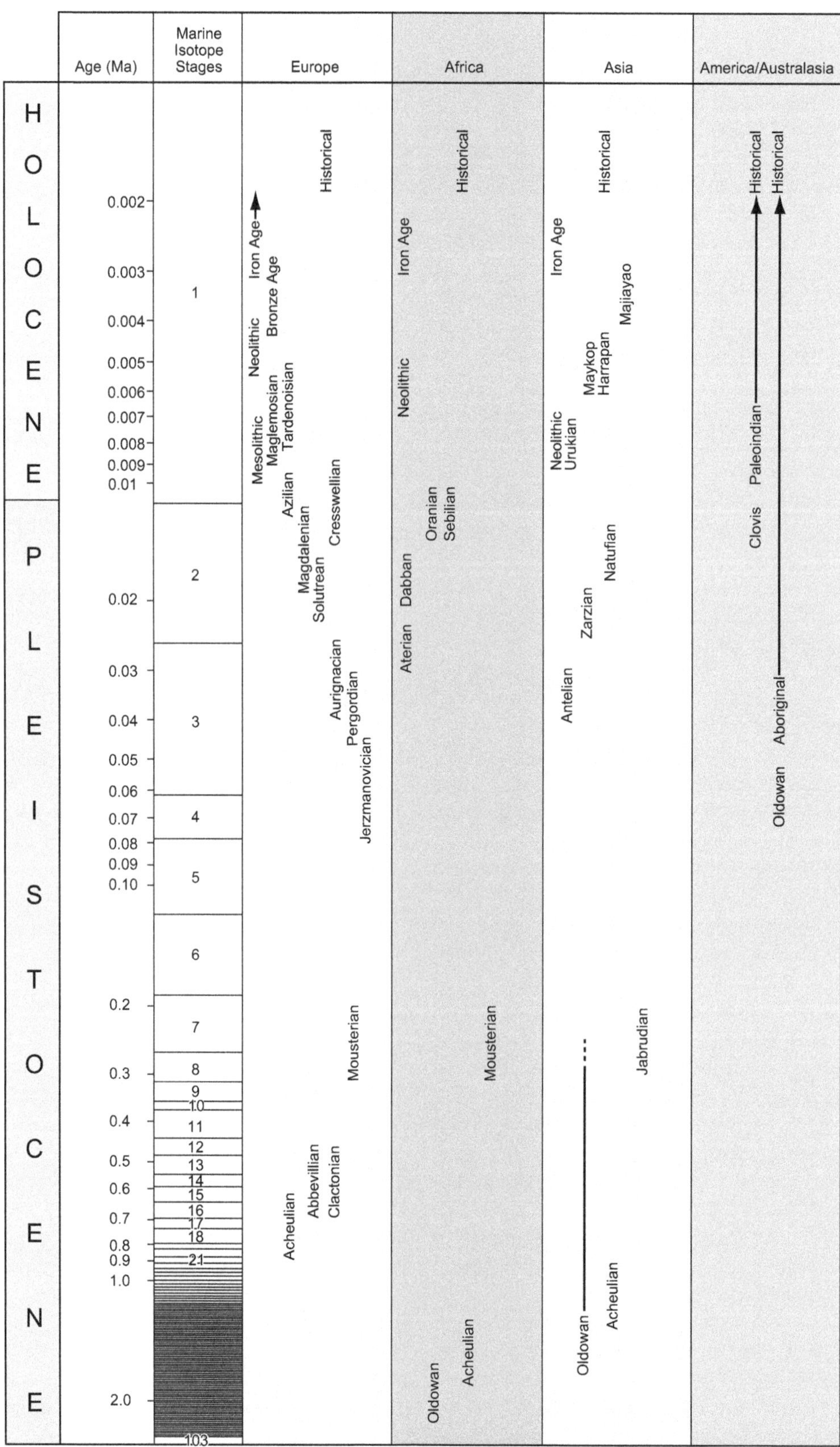

Approximate time ranges of main industries of *Homo*.

Germany and the Czech Republic. The dating of these sites is often imprecise, but many seem to be considerably younger than those in East Africa, reflecting either slow migration of the Oldowan culture away from Africa or independent emergence of similar techniques in later periods. For example, in the Ceprano basin of central Italy, Classic Oldowan pebble tools (unifacial chopping and percussion tools with cores and hammerstone flakes) are associated with a fragmentary cranium of "archaic" morphology found in lacustrine deposits now dated to MIS 11 (430−385 ka) (Manzi *et al.*, 2010).

31.5.2. The Middle Paleolithic

The more advanced Middle Paleolithic Acheulian culture, dating from about 1.9 Ma in East Africa, emerged at approximately the same time as the appearance of *H. erectus* and *H. ergaster*. Bifaces known as handaxes were more common than in the Developed Oldowan and showed a wider range of shapes, which often differ between regions, suggesting a conscious attempt repeatedly to produce tools of characteristic shape. Because of the long overlap in time between late *H. habilis* and early *H. erectus*, the Acheulian and Oldowan cultures are found together in East and South Africa between about 1.9 Ma and 1.5 Ma. Outside Africa, handaxes probably produced by *H. erectus* first appeared later in the Early Pleistocene, about 1.2 Ma in Israel (Bar-Josef and Goren-Inbar, 1993), 0.9 Ma in southern Spain (Scott and Gibert, 2009) and 0.8 Ma in southern China (Yamei *et al.*, 2000). But they are not found in other parts of Asia, where Oldowan tools persisted until about 200 ka.

Unlike the proto-bifaces of the Developed Oldowan culture, handaxes were worked symmetrically, often by hammering with bone, antler or wood as well as stone. This allowed greater control over the shape of the finished artifacts, with secondary flaking to make the handaxes thinner and retouching of flakes for further sharpening. On average, handaxes had cutting edges at least four times longer than Oldowan cores and flakes. Microwear studies suggest that they were multi-purpose tools, probably used for butchering animal carcasses, scraping hides, cutting timber and gathering plant material for bedding (Schick and Toth, 1993), and possibly also for digging for roots or soil animals. At African sites in particular, they are often associated with tools known as cleavers, which had a straight cutting edge transverse to their longest axis; these were probably also used for butchering, cutting vegetables or working timber.

The oldest tools in northern Europe, dating from 1.0−0.7 Ma and attributable to either *H. erectus* or *H. heidelbergensis*, are from Pakefield and Happisburgh at almost 53° N in eastern England (Parfitt *et al.*, 2005, 2010). The assemblages at these sites consist of cores, flakes and retouched flakes but do not include handaxes, which appeared somewhat later in northern Europe, about 0.5 Ma (Monnier, 2006). The classic Acheulian flint handaxes found widely in the early Middle Pleistocene river terrace and lacustrine deposits of Europe, including those from the historically important sites of St Acheul near Abbeville in northern France and Hoxne and Swanscombe in southern England, were probably also the work of the same species. In these areas, as elsewhere, there are also approximately contemporaneous chopper-core industries (the Abbevillian in France and Clactonian in Britain), which show some resemblance to the Classic Oldowan, and suggest continuing coexistence of tool-making traditions originating with *H. habilis* and *H. erectus* or *H. heidelbergensis*. The oldest known wooden tools, hunting spears more than 2 m long from Schoeningen in Germany, date from approximately 400 ka and were probably the work of *H. heidelbergensis* (Thieme, 1997).

Handaxe industries of the Mid-Late Pleistocene are also widespread in Africa, the Near East and India, but dating is almost always imprecise. At Sidi Abderrahman in Morocco, handaxes are associated with a mandible fragment of *H. erectus*, but where hominin remains have been found at other sites they have usually been attributed to the later *H. neanderthalensis* or even *H. sapiens*. Handaxes found on Crete (Strasser *et al.*, 2011) and other Mediterranean islands, which are surrounded by deep water and could not have been connected to Africa, the Near East or Europe, even when sea level fell eustatically during cold stages of the Mid and Late Pleistocene, suggest an ability to cross open water during the Middle Paleolithic.

The numerous stone tools found at European Neanderthal sites are mainly of the later Middle Paleolithic Mousterian industry, which is named after the type site at Le Moustier in the Dordogne region of France. Mousterian artifacts generally consist of flakes shaped into triangular knives, burins, scrapers and spearheads or toothed for use as saws, and a small number of cordate or triangular handaxes made from discoidal cores. Compared with the general purpose Acheulian handaxes, these tools were more specialized, had increased cutting edges and could be reshaped or sharpened, so that they lasted longer. Similar tools have also been found in the Near East, north Africa and India (Petraglia *et al.*, 2007), but are here thought to be at least partly the work of early *H. sapiens*. In the Near East, the industry is known as Jabrudian after a rock shelter site in Syria (Waechter, 1952).

After MIS 8, Mousterian tool production was often based on the Levallois technique (Dibble and Bar-Yosef, 1995), which involved pre-shaping a flake on the core before it was removed (Figure 31.4). Small flakes were first removed from the entire cortex of a flint nodule to produce an oval core, a striking platform was then prepared at one end and a single large flake detached by a sharp blow on the platform, thus leaving behind a "tortoise core" with a single concave surface and a multifaceted convex outer surface. Because the Levallois technique was very wasteful of raw material,

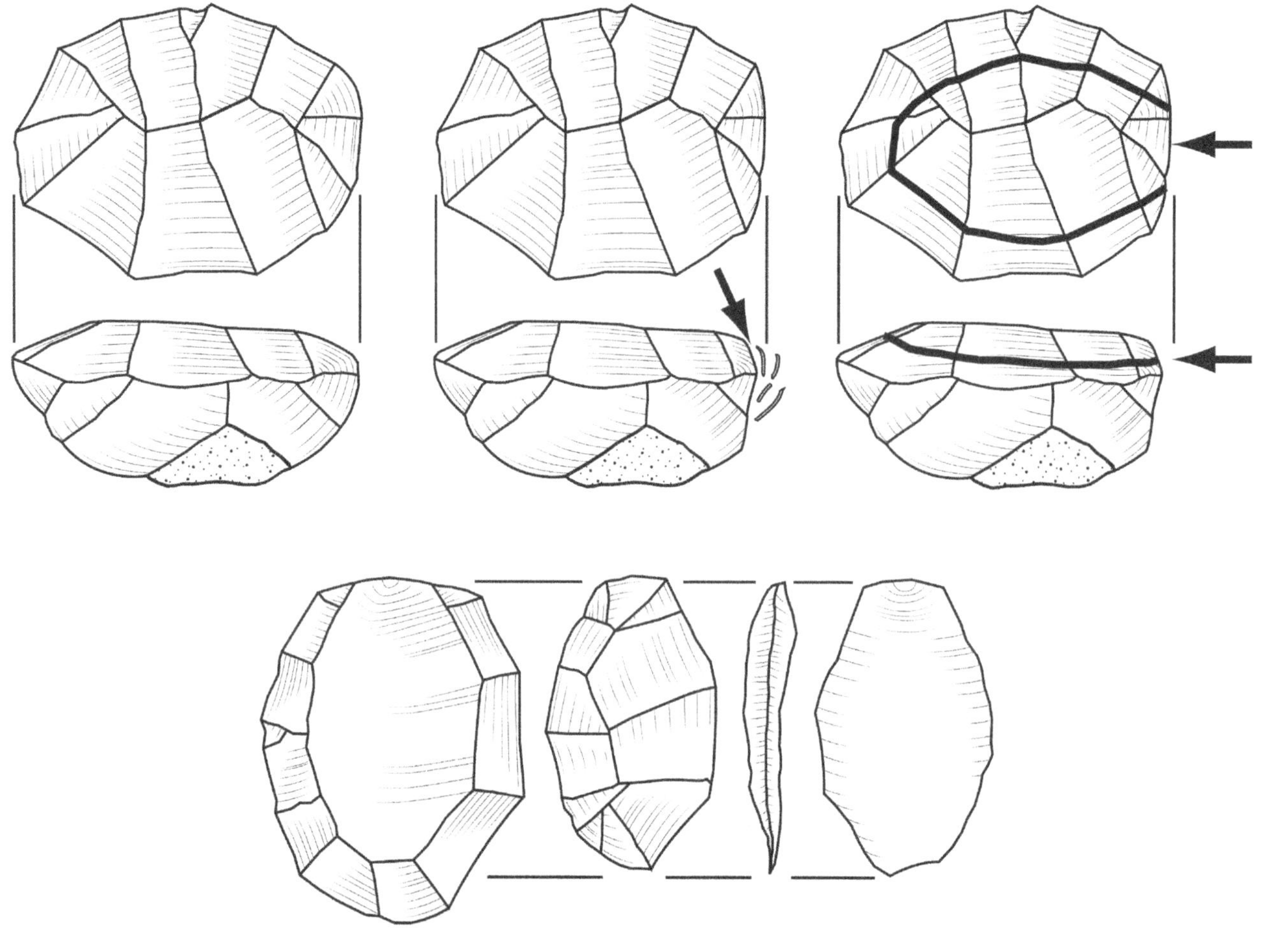

Production of Mousterian tools by the Levallois technique.

sites with large numbers of tortoise cores are mainly found close to rich sources of good quality flint, such as Baker's Hole and Purfleet in the lower Thames valley (Roe, 1981; Wymer, 1985) and near Onival on the Normandy coast (Vallin *et al.*, 2006).

A late development of the Levallois technique, dating from MIS 5 and MIS 4 in parts of southern, western and central Europe, north Africa and the Near East, involved the preparation of two flat striking platforms, one at either end of the original core, and the production of numerous elongate prismatic blades from that core. These were the precursor of the prismatic blades that dominated the subsequent Upper Paleolithic industries. Locally some late Levallois blades were modified to create crude spearheads and arrowtips.

Techniques resembling Levallois are also known from Mousterian sites in more distant areas, including South Africa (Clark, 1959, pp. 126–127), India (Wainwright and Malik, 1967), China (Chang, 1987), Crimea (Marks and Chabai, 2006) and Uzbekistan, but their dating is imprecise, so it is unclear whether they were made by migrating Neanderthals or developed independently. In many areas artifacts with features of those commonly found at Upper Paleolithic sites are curiously mixed with typical Middle Paleolithic tools, suggesting either that there were local invasions by early *H. sapiens* or local advances in human cognitive and industrial abilities.

In South Africa, Middle Paleolithic sites of various ages from 160 ka to about 80 ka, including Pinnacle Point (Marean *et al.*, 2007) and Blombos Cave (Henshilwood *et al.*, 2009), provide evidence for the exploitation of red ochre (hematite) and other pigments, probably for body decoration. Blombos Cave and other Middle Paleolithic sites in Israel and Morocco also contain numerous pierced shells of the marine gastropod *Nassarius*, which were probably strung together, again for body decoration.

31.6. THE EARLIEST TECHNOLOGIES OF *HOMO SAPIENS* – THE UPPER PALEOLITHIC

The dispersal of *H. sapiens* through the "Levant Corridor" in MIS 3 after 55 ka introduced the main wave of Upper Paleolithic culture to Europe and Russia. It spread rapidly through southern Europe, where it has long been attributed to the Cro-Magnon race, named after five skeletons found in the Abri de Crô-Magnon rock shelter, Dordogne (Lartet, 1868), but was slower to reach northern areas such as Belgium and Britain. In Russia it had reached the Altai Mountains by 43 ka (Brantingham *et al.*, 2001), the Don Valley by 41 ka (Holliday *et al.*, 2007) and the northern Urals (67° N) by 40 ka (Svendsen *et al.*, 2010).

Stone tools, forming the various Leptolithic (thin or light stone) industries, consisted of thin delicate blade-like implements fashioned from carefully prepared two-platform cores of flint or other fine hard stone. This technique utilized the raw material more economically, and allowed the production of numerous specialized forms of tool (e.g., spearheads, arrowtips, harpoons, awls, scrapers, knives) often with greatly increased length of cutting edge per unit volume of flint. Some weapons and tools (e.g., needles, picks) were also made from bone, tusk and antler. Rock painting, sculpture and the production of personal ornaments (e.g., ivory bead necklaces) flourished in some areas (Guthrie, 2005). Upper Paleolithic mural art, usually on the walls of caves, is common in France (Clottes, 2003) and Spain, but rare elsewhere. Portable art objects, such as ivory figurines, are known mainly from France, Germany, Italy and western Russia, and simple musical instruments (flutes) carved from ivory or bird bones are known from caves in southwestern Germany (Conard *et al.*, 2009). Eyed bone needles suggest production of clothing and bedding for insulation against the cold, and features of some fossil foot bones indicate the use of shoes (Trinkaus and Shang, 2008).

Most habitation sites were again in caves and rock shelters but, where these were not available, temporary structures were probably built in open country, including huts and tents of animal skin or reed stretched over a frame of timber or bone. However, these materials decompose rapidly and there is little evidence for them in the archaeological record. Locally semi-underground dwellings, known as zemlyankas in the Slovak Republic, were constructed.

The abundance of archaeologically rich Upper Paleolithic sites in France, often with well-developed stratigraphy within the range of radiocarbon dating, led to the recognition of five Upper Paleolithic stages, each with characteristic tool types, which can often be recognized elsewhere in Europe. In approximate date order, these stages are:

Perigordian (45–20 ka), which probably evolved from the local Mousterian, as at some sites there are intermediate industries stratified between them. Six substages have been recognized, the earliest of which (the Châtelperronian) is characterized by thin flakes and flint knives with curved backs, the last (Gravettian) by straight-backed knives. The Châtelperronian probably represents a transition from the Mousterian, as the two have many tool types in common.

Aurignacian (40–20 ka), which is characterized by spears and blades other than the Châtelperronian and Gravettian knives and by highly developed bone and ivory work, such as points with oval or lozenge-shaped cross-sections, and miniature female figurines or "venuses". Approximately contemporaneous with the Perigordian, the Aurignacian seems to represent a different industrial tradition introduced from the Near East, and principally associated with the Cro-Magnons. Five Aurignacian substages, each with distinct tool types, have been recognized.

Solutrean (21–19 ka), which is characterized by beautifully flaked bifacial leaf-shaped flint projectile points and fine bone artifacts such as eyed needles. The origin and fate of this short-lived industry are still uncertain.

Magdalenian (19–13 ka), which is characterized by finely worked bone and antler tools, including awls, harpoons, tridents and needles, and long flint blades used as scrapers or borers. In the southwest Paris Basin, the late Magdalenian includes a distinctive asymmetrically pointed (beak-shaped) boring tool resembling those known as "Zinken" from Upper Paleolithic (Hamburgian) sites in north Germany.

Azilian (12–11 ka), which is confined to the Pyrenean region in south-western France and northern Spain, and consists of small, less refined blades, scrapers and harpoons.

The Late Devensian (MIS 2) glaciation of Britain and Denmark, and the associated severe periglacial conditions of Holland and Belgium, divided the Upper Paleolithic of these countries into two separate parts, often known as the Early and Late Upper Paleolithic. The Early Upper Paleolithic is mainly equivalent to the Aurignacian and Gravettian, the Late to the Magdalenian. Bone and antler tools of the British Aurignacian date from as early as 40 ka, but the British Gravettian artifacts are probably younger. During both Early and Late Upper Paleolithic periods there would have been a land connection with France across the dry floor of the eastern English Channel. A land connection across the southern North Sea basin may also account for characteristic blade points from Beedings (Jacobi, 2007) and other sites in eastern England, which resemble those from Upper Paleolithic sites in Belgium, Thuringia and Poland dated to 38–37 ka.

The Later Upper Paleolithic in Britain, named Creswellian after several cave sites at Creswell Crags (Nottinghamshire), includes flint blades obliquely truncated to form points at each end, "Zinken" and barbed bone points. These are known from numerous cave sites in northern and southern parts of both England and Wales, most of which date from the Windermere Interstadial, equivalent to Interstadial 1 (GI-1, 14.7–12.65 ka) in the Greenland ice core record, though a few, some of them even open sites, are dated to the subsequent Loch Lomond Stadial (= Greenland Stadial 1 or GS-1, 12.65–11.7 ka). Cresswell Crags sites have also yielded rare examples of the earliest known cave art in Britain, including engravings of birds, bison and a horse's head.

In central and eastern Europe, industries known as Szeletian, Jerzmanovician and Bohunician, which are again transitional in type from the Mousterian to the Aurignacian, occur at cave sites in Hungary, Poland and the Czech Republic, respectively. They include numerous bifacial leaf-shaped points or "Blattspitzen". More widespread are the somewhat later Aurignacian and Gravettian artifact assemblages found in Austria, Germany, Poland, the Czech Republic, Hungary, Rumania, Bulgaria, Yugoslavia and Greece, suggesting a degree of cultural uniformity in the earlier Upper Paleolithic across most of Europe. The Gravettian site of Willendorf, in the Danube valley, Austria, is famous for a female clay statuette known as the "Venus of Willendorf". The later Magdalenian industries of north Germany (the Hamburgian), Switzerland, Poland, the Slovak Republic and western Russia are again remarkably similar to those of France; the available dates suggest a north-easterly migration of later Upper Paleolithic people during Greenland Stadial 2 (GS-2) and GI-1. Over 200 open Upper Paleolithic sites are known from the loess plains and river valleys of western Russia and Ukraine. One of the most famous is Kostenki on a low terrace of the River Don, where numerous venuses, points used as knives similar to those found at Willendorf, and the earliest known example of a hafted flint axehead have been recorded.

In western Asia, artifact assemblages with weaker affinities to the French Aurignacian are recorded from caves and rock shelters at Mount Carmel and Jabrud (the Antelian industry) and in the Zagros Mountains of Iraq and Iran (the Baradostian and later Zarzian industries). Elsewhere in Asia, possible Upper Paleolithic industries are very different from those of Europe, but are poorly dated and often mixed with earlier tool types, including Mousterian or even chopper-core tools. The earliest definite evidence for human activity in Japan includes Upper Paleolithic retouched blades and leaf points often made from obsidian, which have been found at over 10 000 sites with calibrated radiocarbon dates ranging from 23 ka to about 15 ka (Kudo, 2005). There are suggestions of Early and Middle Paleolithic industries dating from 600 ka to 35 ka on the island of Honshu and elsewhere in Japan, but their human origin and dates have been strongly disputed (Bleed, 1977; Oda and Keally, 1986), and at least some of the artifacts have recently been exposed as fakes. In the Lake Baikal region of Siberia, there are numerous rich sites showing a combination of Mousterian and Upper Paleolithic artifacts, including Mousterian points and tortoise cores and Upper Paleolithic blades, bone venuses, needles and awls, and a chisel formed from a retouched flint blade and an antler handle (Coles and Higgs, 1969).

The distinction between Middle and Upper Paleolithic industries is also blurred throughout Africa. In South Africa, leptolithic industries, generally described as Magosian, are interbedded with industries of the Middle Stone Age or transitional from it, as at Klasies River Mouth (Eastern Cape) and Howieson's Poort (Cape Province). The successions at these sites cover various short periods between 72 and 60 ka (Jacobs *et al.*, 2008). The Magosian probably developed from African Middle Stone Age traditions, because a progressive transition from one to the other can be traced at the Border Cave in KwaZulu-Natal (Butzer *et al.*, 1978), but whether this happened in more than one area and at more than one time is uncertain. In Swaziland, one of the earliest known mines was used from about 40 ka for extraction of hematite, probably for use as a ritual cosmetic (Dart and Beaumont, 1969). In North Africa, another industry transitional between the Middle and Upper Paleolithic (the Aterian) occurs in Tunisia and Morocco, and seems to cover a long period from late MIS 5 to after 18 ka. It was progressively replaced after 40 ka by a true but poorly developed Upper Paleolithic industry (the Dabban), and this in turn by the Oranian industry in northwest Africa and the Sebilian of the Nile Valley in Egypt, both of which span the period from about 14 ka to 11 ka.

The human occupation of Australia dates from at least 32 ka, and possibly as early as 62 ka, based on a range of radiocarbon and thermoluminescence dates for cremations and burials at Lake Mungo, New South Wales (Gill, 1966; Bowler *et al.*, 1970; Barbetti and Allen, 1972; Thorne *et al.*, 1999; O'Connell and Allen, 2004). In this period, the sea level was probably low enough to permit migration of *H. sapiens* from Indonesia. Other Australian occupation sites dating from MIS 4 or MIS 3 include Rottnest Island, Devil's Lair, Nauwalabila and Malakanunja (Hiscock, 2008). The earliest Australian stone industries, which continued well into the Holocene and even into historic times in Tasmania, consisted of unspecialized flakes, scrapers and chopper cores resembling some Oldowan tools, though with the notable addition at several sites of hafted and ground axeheads fashioned from fine-grained volcanic rocks and hornfels. Similar axeheads also occur in New Guinea, but are unknown from Australian sites post-dating about 12 ka, so they suggest a migration into Australia by people from New Guinea who subsequently abandoned or lost that particular manufacturing technique. They did not appear in Europe until the Neolithic (mid Holocene) period.

The earliest human artifacts found widely in North America, the Paleoindian bifacially flaked fluted projectile points of the Clovis culture (Haynes, 2002), date mainly from about 13 ka. The first of these artifacts were found in New Mexico in the 1930s and since then the origin of the culture has been hotly disputed. The commonest theory is that a few Upper Paleolithic people migrated from eastern Russia across a Beringia land bridge using Inuit methods of arctic travel, but no artifacts resembling Clovis points are known from any part of Asia. Also, the Wisconsian ice barrier formed over Alaska by the coalescence of the Cordilleran and Keewatin ice sheets probably persisted until the mid Holocene so that, unless there was an unlikely temporary "ice-free corridor" about 13 ka, any migration from Asia to North America is more likely to have occurred before the barrier was established in MIS 3 or 2. In support of this, a few pre-13 ka dates have been obtained for isolated human remains, footprints and stone tools in southeastern USA, and Central and South America, including Chile and Venezuela (Pearson and Ream, 2005). Based on the similarity of Clovis points to the Solutrean leaf-shaped points of southern France and Spain, and the fact that earlier sites dating from 14–18 ka occur mainly in southeastern USA, Stanford and Bradley (2002) proposed that the initial invasion of the Americas was from western Europe across the Atlantic, but this is accepted by few American archaeologists.

31.7. HOLOCENE TECHNOLOGIES – MESOLITHIC, NEOLITHIC, BRONZE AGE AND IRON AGE

31.7.1. Mesolithic

This period of the early Holocene is often known as the Epi-Paleolithic, because there was no real cultural change from the Upper Paleolithic and in many areas the two are difficult to separate. The term Mesolithic is used mainly in northern Europe; Epipalaeolithic is used elsewhere, often for cultures showing a partial transition to the agricultural practices which distinguish the Neolithic. In most areas the Paleolithic hunter-gatherer lifestyle continued, but at the Abu Hureyra site in Syria (Moore, 1988) and in the Epipaleolithic Natufian culture of the Levant (Munro, 2003) there is some evidence for cultivation of crops even as early as 13 ka. The rapid climatic amelioration at the beginning of the Holocene (c. 11.7 ka) resulted in a large population increase, at least in mid-latitude regions of the northern hemisphere, where Mesolithic artifacts are much more abundant than Upper Paleolithic. Initially, population would have increased by migration from areas to the south, but in Europe the widespread Mesolithic riverside and lakeside sites, which provided a range of plant and animal foods as well as supplies of fresh water and timber, suggest some preference for semi-permanent settlement over a purely nomadic lifestyle, and this could have led to a gradual increase in indigenous populations.

The Mesolithic of most European areas was characterized by the production of large numbers of very small blades (microliths), which were often mounted singly on a shaft to form projectiles such as arrows, or in groups to form saw-like tools. Other stone tools included axeheads, adzes and picks; antler and bone were also used to make spearheads, fishing hooks and barbed harpoons. Dugout canoes were used for river transport, and there is evidence for more extensive domestication of dogs than in the Upper Paleolithic. Various names are used for this type of industry (Bailey and Spikins, 2008), including Maglemosian in Denmark, Scandinavia, north Germany and Britain, Swiderian in northeast Europe and Sauveterrian in France, Belgium and Switzerland. In southwest France and northern Spain, the microlith-rich Tardenoisian industry may have developed from the Upper Paleolithic Azilian.

The frequent recovery of Mesolithic artifacts by fishing trawlers and seismic mapping of shallow seafloor areas, such as the southern North Sea (Gaffney *et al.*, 2007), suggest that much of the evidence for Mesolithic occupation and migration was lost during the early Holocene eustatic rise of sea level.

31.7.2. Neolithic

The start of the Neolithic is defined by the widespread initiation of serious farming methods, which strengthened the lifestyle based on long-term settlements, and led to early farmsteads and villages. Farming seems to have begun in the "Fertile Crescent", an arcuate region of increased rainfall extending from western Syria and northern Jordan eastwards to Iraq, and including the Tigris and Euphrates valleys. At Abu Hureyra, an 11.5 ha site in the Middle Euphrates Valley (Moore, 1988), and Tell Aswad near Damascus, the earliest evidence for settlements of semi-subterranean houses built of mudbricks and with plastered walls and floors dates from 9.5–10 ka. For food, the occupants of these villages partly used wild plants, such as einkorn wheat and vetches, but they also cultivated arable crops, including emmer wheat, barley, lentils and peas, and domesticated sheep and goats. In addition, baskets and mats were woven from straw, and bone artifacts and stone bowls were produced, the hard stone being imported from Anatolia, perhaps by trading agricultural products. Later, the small early Neolithic settlements along the Euphrates developed into larger towns of the Urukian civilization. The first of these dating from ~ 8 ka was probably the walled town of Jericho, which then had a human population exceeding 2000. About 6 ka, the much larger city of Uruk was built in southern Iraq; its buildings of kiln-fired bricks, often with internal walls adorned with mosaics, covered an area of 6 km^2 and served a population probably exceeding 40 000.

In other areas, farming began much later, for example around 6 ka in northwest Europe and the UK. It is possible that it evolved here through the creativeness of an indigenous (Mesolithic) human population in response to demands to feed the expanding population. But slow migration of farming communities from the Near East best explains a strong SE–NW gradient in modern human DNA characteristics across Europe (Evison, 1999); at no other time could this gradient have been created by mixing a sufficiently large population from the southeast with a much smaller indigenous European population. Neolithic farming resulted in the introduction of new artifacts, such as stone sickles and primitive wooden ploughs (ards), and the improvement of stone axeheads by grinding and polishing for more efficient tree-felling and working of timber. In many areas, soil cultivation on slopes resulted in widespread deposition of foot-slope colluvium and a change in river deposits from organic peat to inorganic alluvium.

Another characteristic of the Neolithic was the widespread introduction of pottery. This appeared in the Fertile Crescent around 10.5 ka, though here there was a brief Pre-Pottery Neolithic Period. It also appeared about the same time at the Nanzhuangtou site in the Yellow River valley in China, but elsewhere the first appearance of pottery coincides with the first evidence for farming; for example, in Greece, the earliest (plain) pottery dates from 8.5 ka, and in northwest Europe from 6 ka. At some sites, large amounts of pottery were produced; for example many thousand crude bevel-rimmed bowls dating from 5–6 ka have been found at the Syrian city of Tel Brak. Pottery is especially useful in Holocene archaeology, as different pot shapes, tempering methods and glazing patterns can be used for dating and reconstructing migration or trading routes. Other Neolithic inventions included stone-paved streets and canals dating from about 8.0 ka in Mesopotamia, the production of silk in China from about 5.6 ka, oval burial mounds and wetland timber trackways, such as the Sweet Track across the Somerset Levels in England, which has been dated by dendrochronology to 5.8 ka.

Early-mid Holocene coastal flooding resulting from worldwide eustatic or local isostatic changes in relative sea level was again responsible for the loss of much evidence of Neolithic civilization. Alignments of dressed stone blocks on the sea floor close to the coasts of India, China, Japan, Malta and other countries provide (often contested) evidence for drowned prehistoric cities and harbors, though dating is usually uncertain.

31.7.3. The Bronze Age

This period resulted from the discovery that ores such as malachite can easily be smelted with charcoal to produce metallic copper, from which a range of weapons (swords, daggers, spearheads and armor), tools (axeheads, adzeheads, knives), domestic ware (pans, drinking vessels) and decorated ritualistic items could be cast using moulds of clay or stone. In some areas, the earliest metal artifacts were manufactured from copper alone, which was first mined c. 6 ka in the Balkans and other Mediterranean areas. But pure copper is soft and techniques for alloying it with tin to produce the much harder bronze were developed within a few centuries. The earliest bronze artifacts thus originated in areas where both mineable copper and tin ores were located, though far-ranging trading routes were soon established to other regions, particularly for the rarer tin ore cassiterite. For example, isotopic analyses have shown that tin from Cornwall (southwest England) was traded by sea as far away as the Aegean and other eastern Mediterranean regions. The oldest surviving sea-going vessel, built of timber planks sewn together with yew string, is known from Dover in southeast England.

The earliest bronze artifacts, and therefore the local beginning of the Bronze Age, date from about 5.3 ka in Mesopotamia (the Akkadian culture) and north Caucasus (the Maykop culture), from 5.2 ka in the Aegean region, Egypt, Anatolia and Levant, from 5.0 ka in the Indus valley (the Harrapan culture), from 4.3 ka in mainland Europe, from 4.2 ka in Britain, from 4.0 ka in China (the Majiayao culture) and from as late as 2.8 ka in Korea. In Sub-Saharan Africa there is almost no evidence for prehistoric use of bronze, so that the Neolithic was followed directly by the Iron Age. Bronze Age graves, in which warrior leaders were buried individually, often with bronze, gold and pottery grave goods, occur widely, frequently beneath circular mounds. In some areas henge monuments (e.g., Stonehenge in southern England) were constructed, or at least begun, in the later Bronze Age. These earthworks provide the earliest evidence for the emergence of religious beliefs.

Superimposed on the overall warming trend of the early Holocene and gradual cooling in the later Holocene, there were several millennial-scale climatic oscillations similar to the Bond and Dansgaard–Oeschger cycles of MIS 3–2, some of which probably influenced civilization development and population migration during the Holocene. For example, beneath the late Holocene desert sands of north Africa, high resolution near-infrared satellite imagery has revealed widespread evidence of early Holocene stone circles and villages, which were probably occupied by hunter-gatherer communities living in a savannah environment, and of a later agricultural period with mudbrick towns, palaces and tombs covered by earth mounds. Mid-Holocene (~5.5 ka) aridification led to abandonment of these sites, so that occupation was restricted to areas fed by surface waters, especially the Nile valley. The ancient (Predynastic) Egyptian civilization, with evidence of advanced pottery and jewelry manufacture, architectural and artistic abilities, was established in the Nile Valley by 5 ka, and the first large temple-surrounded pyramids with causeways to the Nile were built as "Old Kingdom" royal burial sites about 4.5 ka. Later, a cool and arid event influencing southern Asia after about 4.2 ka led to partial collapse of several urban early Bronze Age civilizations, including the Akkadian of Mesopotamia, the Harrapan of the Indus Valley, the Hilmand in Afghanistan and the Hongshan of China. This climatic change also extended to north Africa, where severe drought prevented normal flooding of the Nile and terminated the Egyptian "Old Kingdom" about 4.15 ka.

The classic civilizations of the eastern Mediterranean, including that of Greece, probably originated in the later Bronze Age between 4 ka and 3 ka. One of the most advanced was the Minoan civilization of Crete, where excavations at the trading port of Knossos by Evans (1921) showed the existence of numerous high-status 2–3 storey stone buildings with windows and intricately painted plaster walls. The city of Akrotiri on the island of Santorini was another affluent Bronze Age trading port of the eastern Mediterranean at this time, but was destroyed by the paroxysmal eruption of Thera c. 1628 BC (Friedrich *et al.*, 2006), an event which was probably handed down by oral tradition and became Plato's legend of Atlantis. The earliest forms of writing, including pictograms and hieroglyphs cut into baked clay or stone tablets, originated in Egypt and Mesopotamia around 5 ka and persisted until 1.7 ka, though other writing methods were introduced in Greece about 2.9 ka.

31.7.4. The Iron Age

Bronze was progressively replaced by iron for the manufacture of artifacts from about 4.0 ka in Anatolia, 3.2 ka in Mesopotamia, parts of the Indian Subcontinent and eastern Europe, 2.8 ka in western Europe and China and 2.3 ka in Japan. Iron was eventually used more widely than bronze, principally because iron ores such as hematite and bog iron ore are much more widespread than copper and tin ores. By 2.5 ka, it was in almost universal use in Europe, Asia and Africa. The main advantage of iron over bronze is its hardness, so that tools and weapons can bear sharper and longer lasting cutting edges, though the higher melting point of iron demanded better smelting techniques. No other metals were required to produce an alloy, and most artifacts were of wrought iron, though steel containing up to 1.7 % carbon was locally produced, for example in Tanzania (Schmidt and Avery, 1978).

The Iron Age terminated in any area with the beginning of the historical period and the greatly increased production of written documents. This occurred during the Hellenic period in Greece, the spread of the Roman Empire elsewhere in Europe (e.g., AD 43 in England), the start of Confucianism in China and the beginning of Buddhism in India.

31.7.5. New World Holocene

The prehistoric paleoindian and aboriginal populations (*H. sapiens*) of America and Australia were probably direct descendants of the first (MIS 4—MIS 2) immigrants, with few or no subsequent invasions. In both continents the life-styles remained as nomadic hunter-gatherers, with the late persistence of stone tools and very little evidence for agricultural activities until c. 4.7 ka in Mesoamerica. The first pottery in America dates from c. 4.0 ka and the first metal tools from c. 2.9 ka, but neither are known from pre-colonial Australia, where organized agriculture was made difficult by climatic conditions.

31.8. CONCLUSIONS

Because hominin fossils are rare, often incomplete and poorly preserved, the current phylogenetic tree based on comparative morphology is less robust than those for most other animal groups. In part it has been over-simplified by expanding species such as *Homo erectus* to include too wide a diversity of skeletal material. The recently developed molecular and genetic methods will almost certainly provide greater clarity in the future, but they are dependent on the availability of suitably preserved material, which means the advances they bring will mainly contribute to the history of more recent species, notably *Homo heidelbergensis*, *H. neanderthalensis* and *H. sapiens*. Also, every new fossil discovery and analytical result demonstrates the amazing diversity of the hominin sub-family. Paleoanthropology is still at a very early stage in assessing this diversity, so the establishment of a clear human phylogeny is still a long way off.

At present one aspect is clear — that the evolution of hominins through the Pleistocene was especially rapid. The clear coincidence in time between hominin speciation/dispersal/extinction events and episodes of tectonic activity or rapid climatic/vegetational change demonstrates the importance of these geological processes in driving or at least strongly influencing hominin evolution and development. However, an especially puzzling aspect of hominin history is that despite the persistence of numerous species for relatively long periods during the Late Pliocene and Early and Mid Pleistocene, chimpanzee and anatomically modern humans are the only hominins that survived through the Late Pleistocene into the Holocene. Many other mammals became extinct in the latest Pleistocene and early Holocene, either because of changes in climate and vegetation (food) or through excessive hunting, but this occurred too late to explain the disappearance of *Homo erectus*, *H. heidelbergensis*, *H. neanderthalensis*, the Denisovans and others.

It would be reasonable to expect that the much more abundant evidence provided by the numerous tool types found throughout the Pleistocene would supplement evidence from the fossils in constructing a history of hominin evolution. There are certainly a few examples of biological developments apparently leading to cultural change, such as the appearance of leptolithic industries in Europe with the arrival of *Homo sapiens* and the disappearance of Mousterian tools with the extinction of *H. neanderthalensis*. However, there are also too many exceptions for the cultural evidence to be generally useful in an evolutionary context (Tattershall, 1998). The reasons for this may well emerge when the complexity of earlier parts of the human phylogenetic tree is better understood, but for the moment paleoanthropological studies and investigations of tool typology must be treated independently, both being related to the increasingly detailed Pleistocene time scale by judicious use of the growing range of available dating methods.

ACKNOWLEDGMENT

We thank Professor Chronis Tzedakis for reviewing an early version of the manuscript and making many useful suggestions for improvement.

REFERENCES

Aiello, L.C., Wheeler, P., 2003. Neanderthal thermoregulation and the glacial climate. In: van Andel, T.H., Davies, W. (Eds.), Neanderthals and Modern Humans in the European Landscape of the Last Glaciation: Archaeological Results of the Stage 3 Project. McDonald Institute for Archaeological Research, Cambridge, pp. 147–165.

Andrews, P., Cronin, J.E., 1982. The relationships of *Sivapithecus* and *Ramapithecus* and the evolution of the orang-utan. Nature 297, 541–546.

Armitage, S.J., Jasim, S.A., Marks, A.E., Parker, A.G., Usik, V.I., Uerpmann, H.-P., 2011. The southern route "out of Africa": Evidence for an early expansion of modern humans into Arabia. Science 331, 453–456.

Arsuaga, J.-L., Martínez, I., Gracia, A., Carretero, J.-M., Carbonell, E., 1993. Three new human skulls from the Sima de los Huesos Middle Palaeolithic site in Sierra de Atapuerca, Spain. Nature 362, 534–537.

Ascenzi, A., Biddittu, I., Cassoli, P.F., Segre, A.G., Segre-Naldini, E., 1996. A calvarium of late *Homo erectus* from Ceprano, Italy. Journal of Human Evolution 31, 409–423.

Asfaw, B., White, T., Lovejoy, O., Latimer, B., Simpson, S., Suwa, G., 1999. *Australopithecus garhi*: A new species of early hominid from Ethiopia. Science 284, 629–635.

Avery, D.M., 2002. Taphonomy of micromammals from cave deposits at Kabwe (Broken Hill) and Twin Rivers in central Zambia. Journal of Archaeological Science 29, 537–544.

Bailey, G., Spikins, P. (Eds.), 2008. Mesolithic Europe. Cambridge University Press, New York, p. 498.

Barbetti, M., Allen, H., 1972. Prehistoric man at Lake Mungo, Australia, by 32,000 years BP. Nature 240, 46–48.

Bar-Yosef, O., Goren-Inbar, N., 1993. The lithic assemblages of Ubeidiya: A Lower Palaeolithic site in the Jordan valley. Monographs of the Institute of Archaeology (Qedem), 34. Hebrew University of Jerusalem, Jerusalem, p. 266.

Berger, L.R., de Ruiter, D.J., Churchill, S.E., Schmid, P., Carlson, K.J., Dirks, P.H.G.M., Kibii, J.M., 2010. *Australopithecus sediba*: A new species of *Homo*-like Australopith from South Africa. Science 328, 195–204.

Berger, W.H., Jansen, E., 1994. Mid-Pleistocene climatic shift: The Nansen connection. In: Johannessen, O., Muench, R.D., Overland, J.E. (Eds.), The Polar Oceans and their Role in Shaping the Global Environment. Geophysical Monograph, 85. American Geophysical Union, Washington, pp. 295–311.

Bergner, A.G.N., Trauth, M.H., Bookhagen, B., 2003. Palaeoprecipitation estimates for the Naivasha Basin (Kenya) during the last 175 k.y. using a lake balance model. Global and Planetary Change 36, 117–136.

Bishop, W.W., Chapman, G.R., 1970. Early Pliocene sediments and fossils from the northern Kenya rift valley. Nature 226, 914–918.

Blackwell, B.A.B., Long, R.A., Montoya, A., Blickstein, J.I.B., Skinner, A.R., Séronie-Vivien, M.R., Tillier, A.M., 2007. ESR dating at Pradayrol, Lot, France: Constraining the age of the Middle Pleistocene bone-bearing layer, its Middle Palaeolithic artifacts and hominid incisor. Geological Society of America, Abstracts with Programs 39 (6), 547.

Bleed, P., 1977. Early flakes from Sozudai, Japan: Are they man-made? Science 197, 1357–1359.

Bowler, J.M., Jones, R., Allen, H., Thorne, A.G., 1970. Pleistocene human remains from Australia: A living site and human cremation from Lake Mungo, western New South Wales. World Archaeology 2, 39–60.

Brambel, D.M., Liebermann, D.E., 2004. Endurance running and the evolution of *Homo*. Nature 432, 345–352.

Brantingham, P.J., Krivoshapkin, A., Jinzeng, L., Tserendagva, Y., 2001. The initial Upper Palaeolithic in northeast Asia. Current Anthropology 42, 735–747.

Brown, P., Sutikna, T., Morwood, M.J., Soejono, R.P., Jatmiko, Wayhu Saptomo, E., Due, Rokus Awe, 2004. A new small-bodied hominin from the Late Pleistocene of Flores, Indonesia. Nature 431, 1055–1061.

Brumm, A., Jansen, G.M., Van den Bergh, G.D., Morwood, M.J., Kurniawan, I., Aziz, F., Storey, M., 2010. Hominins on Flores, Indonesia, by one million years ago. Nature 464, 748–752.

Brunet, M., Guy, F., Pilbeam, D., Mackaye, H.T., Likius, A., Ahounta, D., Beauvilain, A., Blondel, C., Bocherens, H., Boisserie, J.R., de Bonis, L., Coppens, Y., Dejax, J., Denys, C., Duringer, P., Eisenmann, V., Fanone, G., Fronty, P., Geraads, D., Lehmann, T., Lihoreau, F., Louchart, A., Mahamat, A., Merceron, G., Mouchelin, G., Otero, O., Pelaez Campomanes, P., Ponce de Leon, M., Rage, J.C., Sapanet, M., Schuster, M., Sudre, J., Tassy, P., Valentin, X., Vignaud, P., Viriot, L., Zazzo, A., Zollikofer, C., 2002. A new hominid from the Upper Miocene of Chad, central Africa. Nature 418, 145–151.

Buckland, W., 1823. Reliquiae Diluvianae; or observations on the organic remains contained in caves, fissures and diluvial gravel, and on other geological phenomena, attesting the action of an universal deluge. John Murray, London, p. 303.

Butzer, K.W., Beaumont, P.B., Vogel, J.C., 1978. Lithostratigraphy of Border Cave, KwaZulu, South Africa: A Middle Stone Age sequence beginning *c.* 195,000 b.p. Journal of Archaeological Science 5, 317–341.

Cann, R.L., Stoneking, M., Wilson, A.C., 1987. Mitchondrial DNA and human evolution. Nature 325, 31–36.

Chang, K.-C., 1987. The Archaeology of Ancient China, fourth ed. Yale University Press, New Haven, p. 483.

Clark, J.D., 1959. The Prehistory of Southern Africa. Penguin Books, London, p. 341.

Clark, J.D., de Heinzelin, J., Schick, K.D., Hart, W.K., WoldeGabriel, G., Walter, R.C., Suwa, G., Asfaw, B., Vrba, E., Haile-Selassie, Y., 1994. African *Homo erectus*: Old radiometric ages and young Oldowan assemblages in the Middle Awash Valley, Ethiopia. Science 264, 1907–1910.

Clottes, J., 2003. Return to Chauvet Cave, Excavating the Birthplace of Art: The First Full Report. Thames and Hudson, London, p. 232.

Coles, J.M., Higgs, E.S., 1969. The Archaeology of Early Man. Faber and Faber, London, p. 454.

Conard, N.J., Malina, M., Münzel, S.C., 2009. New flutes document the earliest musical tradition in southwestern Germany. Nature 460, 737–740.

Conroy, G.C., Jolly, C.J., Cramer, D., Kalb, J.E., 1978. Newly discovered fossil hominid skull from the Afar Depression, Ethiopia. Nature 276, 67–70.

Dart, R.A., 1925. *Australopithecus africanus*: The man-ape of South Africa. Nature 115, 195–199.

Dart, R.A., Beaumont, P., 1969. Evidence of iron ore mining in southern Africa in the Middle Stone Age. Current Anthropology 10, 127–128.

Darwin, C., 1859. On the Origin of Species by Means of Natural Selection, or the Preservation of Favoured Races in the Struggle for Life. John Murray, London, p. 502.

Darwin, C., 1871. The Descent of Man, and Selection in Relation to Sex, vol. 2, John Murray, London.

Deino, A.L., Kingston, J.D., Glen, J.M., Edgar, R.K., Hill, A., 2006. Precessional forcing of lacustrine sedimentation in the Late Cenozoic Chemeron Basin, Central Kenya Rift, and calibration of the Gauss/Matuyama boundary. Earth and Planetary Science Letters 247, 41–60.

Dennell, R., 2010. Early *Homo sapiens* in China. Nature 468, 512–513.
Dennell, R., Roebroeks, W.E., 2005. An Asian perpective on early human dispersal from Africa. Nature 438, 1099–1104.
Dibble, H.L., Bar-Yosef, O. (Eds.), 1995. The Definition and Interpretation of Levallois Technology. Monographs in World Archaeology, 23. Prehistory Press, Madison, p. 502.
Evans, A.J., 1921. The Palace of Minos: A Comparative Account of the Successive Stages of the Early Cretan Civilization as Illustrated by the Discoveries at Knossos, vol. 4, MacMillan & Co, London.
Evison, M.P., 1999. Perspectives on the Holocene in Britain: Human DNA. Journal of Quaternary Science 14, 615–623.
Falguères, C., Bahain, J.J., Yokoyama, Y., Arsuaga, J.L., de Castro, J.M.B., Carbonell, E., Bischoff, J.L., Dolo, J.M., 1999. Earliest humans in Europe: The age of TD6 Gran Dolina, Atapuerca, Spain. Journal of Human Evolution 37, 343–352.
Feakins, S.J., deMenocal, P.B., Eglinton, T.I., 2005. Biomarker records of Late Neogene changes in northeast African vegetation. Geology 33, 977–980.
Finlayson, C., Pacheco, F.G., Rodríguez-Vidal, J., Fa1, D.A., Gutierrez López, J.M.G., Pérez, A.S., Finlayson, G., Allue, E., Baena Preysler, J., Cáceres, I., Carrión, J.S., Fernández Jalvo, Y., Gleed-Owen, C.P., Jimenez Espejo, F.J., López, P., López Sáez, J.A., Riquelme Cantal, J.A., Sánchez Marco, A., Guzman, F.G., Brown, K., Fuentes, N., Valarino, C.A., Villalpando, A., Stringer, C.B., Martinez Ruiz, F., Sakamoto, T., 2006. Late survival of Neanderthals at the southernmost extreme of Europe. Nature 443, 850–853.
Friedrich, W.L., Kromer, B., Friedrich, M., Heinemeier, J., Pfeiffer, T., Talamo, S., 2006. Santorini eruption radiocarbon dated to 1627-1600 B.C. Science 312, 548.
Gabunia, L., Vekua, A., Lorkidpanidze, D., Swisher III, C.C., Ferring, R., Justus, A., Nioradze, M., Tvalchrelidze, M., Antón, S.C., Bosinski, G., Jöris, O., de Lumley, M.–A., Majsuradze, G., Mouskhelishvili, A., 2000. Earliest Pleistocene hominid cranial remains from Dmanisi, Republic of Georgia: Taxonomy, geological setting, and age. Science 288, 1019–1025.
Gaffney, V., Thomson, K., Fitch, S., 2007. Mapping Doggerland: The Mesolithic Landscapes of the Southern North Sea. Archaeopress, Oxford, p. 143.
Gamble, C., 2009. Human display and dispersal: A case study from biotidal Britain in the Middle and Upper Pleistocene. Evolutionary Anthropology 18, 144–156.
Gill, E.D., 1966. Provenance and age of the Keilor cranium: Oldest known human skeletal remains in Australia. Current Anthropology 7, 581–584.
Golovanova, L., Doronichev, V.B., Cleghorn, N.E., Koulkova, M.A., Sapelko, T.V., Shackley, M.S., 2010. Significance of ecological factors in the Middle to Upper Palaeolithic Transition. Current Anthropology 51, 655–691.
Goren-Inbar, N., Alperson, N., Kislev, M.E., Simchoni, O., Melamed, Y., Ben-Nun, A., Werker, E., 2004. Evidence of hominin control of fire at Gesher Benot Ya'aqov, Israel. Science 304, 725–727.
Green, R.E., Krause, J., Briggs, A.W., Maricic, T., Stenzel, U., Kircher, M., Patterson, N., Li, H., Zhai, W., Fritz, M.H., Hansen, N.F., Durand, E.Y., Malaspinas, A.S., Jensen, J.D., Marques-Bonet, T., Alkan, C., Prüfer, K., Meyer, M., Burbano, H.A., Good, J.M., Schultz, R., Aximu-Petri, A., Butthof, A., Höber, B., Höffner, B., Siegemund, M., Weihmann, A., Nusbaum, C., Lander, E.S., Russ, C., Novod, N., Affourtit, J., Egholm, M., Verna, C., Rudan, P., Brajkovic, D., Kucan, Z., Gusic, I., Doronichev, V.B., Golovanova, L.V., Lalueza-Fox, C., de la Rasilla, M., Fortea, J., Rosas, A., Schmitz, R.W., Johnson, P.L., Eichler, E.E., Falush, D., Birney, E., Mullikin, J.C., Slatkin, M., Nielsen, R., Kelso, J., Lachmann, M., Reich, D., Pääbo, S., 2010. A draft sequence of the Neanderthal genome. Science 328, 710–722.
Grün, R., Stringer, C., 2000. Tabun revisited: Revised ESR chronology and new ESR and U-series analyses of dental material from Tabun C1. Journal of Human Evolution 39, 601–612.
Guthrie, R.D., 2005. The Nature of Palaeolithic art. The University of Chicago Press, Chicago, p. 507.
Haile-Selassie, Y., 2001. Late Miocene hominids from the Middle Awash, Ethiopia. Nature 412, 178–181.
Haile-Selassie, Y., Suwa, G., White, T.D., 2004. Late Miocene teeth from Middle Awash, Ethiopia, and early hominid dental evolution. Science 303, 1503–1505.
Harpending, H., Rogers, A., 2000. Genetic perspectives on human origins and differentiation. Annual Review of Genomics and Human Genetics 1, 361–385.
Harvati, K., Frost, S.R., McNulty, K.P., 2004. Neanderthal taxonomy reconsidered: Implications of 3D primate models of intra- and inter-specific differences. Proceedings of the National Academy of Sciences, USA 101, 1147–1152.
Haug, G.H., Tiedemann, R., 1998. Effect of the formation of the Isthmus of Panama on Atlantic Ocean thermohaline circulation. Nature 393, 673–676.
Haynes, G., 2002. The Early Settlement of North America. The Clovis Era. Cambridge University Press, New York, p. 360.
Henshilwood, C.S., d'Errico, F., Watts, I., 2009. Engraved ochres from the Middle Stone Age levels at Blombos Cave, South Africa. Journal of Human Evolution 57, 27–47.
Hill, A., 1985. Early hominid from Baringo, Kenya. Nature 315, 222–224.
Hiscock, P., 2008. Archaeology of Ancient Australia. Routledge, London. p. 368.
Holliday, V.T., Hoffecker, J.F., Goldberg, P., Macphail, R.I., Forman, S.L., Anikovich, M., Sinitsyn, A., 2007. Geoarchaeology of the Kostenki-Borshchevo sites, Don River valley, Russia. Geoarchaeology 22, 181–228.
Hopley, P., Maslin, M.A., 2010. Climate-averaging of terrestrial faunas – an example from the Plio-Pleistocene of South Africa. Palaeobiology 36, 32–50.
Huxley, T.H., 1863. Evidence as to Man's Place in Nature. D. Appleton & Co, New York, p. 184.
Jablonski, N.G., 2010. The naked truth. Scientific American 302 (2), 28–35.
Jacobi, R., 2007. A collection of Early Upper Palaeolithic artifacts from Beedings, near Pulborough, West Sussex, and the context of similar finds from the British Isles. Proceedings of the Prehistoric Society 73, 229–325.
Jacobs, Z., Roberts, R.G., Galbraith, R.F., Deacon, H.J., Gün, R., Mackay, A., Mitchell, P., Vogelsang, R., Wadley, L., 2008. Ages for the Middle Stone Age of Southern Africa: Implications for human behavior and dispersal. Science 322, 733–735.
Johanson, D.C., 2009. Lucy (*Australopithecus afarensis*). In: Ruse, M., Travis, J. (Eds.), Evolution: The First Four Billion Years. Belknap Press, Cambridge, Ma, pp. 693–697.
King, G., Bailey, G., 2006. Tectonics and human evolution. Antiquity 80, 265–286.
Kingston, J.D., Deino, A.L., Edgar, R.K., Hill, A., 2007. Astronomically forced climate change in the Kenyan Rift Valley 2.7–2.55 Ma: Implications for the evolution of early hominin ecosystems. Journal of Human Evolution 53, 487–503.

Krause, J., Orlando, L., Serre, D., Viola, B., Prüfer, K., Richards, M.P., Hublin, J.-J., Hänni, C., Derevianko, A.P., Pääbo, S., 2007. Neanderthals in central Asia and Siberia. Nature 449, 902–904.

Krause, J., Fu, Q., Good, J.M., Viola, B., Shunkov, M.V., Derevianko, A.P., Pääbo, S., 2010. The complete mitochondrial DNA genome of an unknown hominin from southern Siberia. Nature 464, 894–897.

Kudo, Y., 2005. The temporal correspondence between archaeological chronology and environmental changes in the final Pleistocene in eastern Honshu Island. Daiyonki Kenkyu (Quaternary Research) 44, 51–64.

Kumar, S., Filipski, A., Swarna, V., Walker, A., Hedges, S.B., 2005. Placing confidence limits on the molecular age of the human-chimpanzee divergence. Proceedings of the National Academy of Sciences, USA 102, 18842–18847.

Larrasoaña, J.C., Roberts, A.P., Rohling, E.J., Winklhofer, M., Wehausen, R., 2003. Three million years of monsoon variability over the northern Sahara. Climate Dynamics 21, 689–698.

Lartet, L., 1868. Mémoire sur une sépulture des anciens troglodytes de Périgord. Annales des Sciences Naturelles: zoologie et paléontologie. Series 5 (10), 133–145.

Leakey, L.S.B., Tobias, P.V., Napier, J.R., 1964. A new species of the genus *Homo* from Olduvai Gorge. Nature 202, 649–650.

Leakey, M.D., Hay, R.L., Curtis, G.H., Drake, R.E., Jackes, M.K., 1976. Fossil hominids from the Laetolil Beds. Nature 262, 460–466.

Lepre, C.J., Quinn, R.L., Joordens, J.C., Swisher III, C.C., Feibel, C.S., 2007. Plio-Pleistocene facies environments from the KBS Member, Koobi Fora Formation: Implications for climate controls on the development of lake-margin hominin habitats in the northeast Turkana Basin (northwest Kenya). Journal of Human Evolution 53, 504–514.

Levin, N.E., Quade, J., Simpson, S.W., Semaw, S., Rogers, M., 2004. Isotopic evidence for Plio-Pleistocene environmental change at Gona, Ethiopia. Earth and Planetary Science Letters 219, 93–110.

Linnaeus, C., 1758. Systema Naturae, tenth ed. Salvius, Stockholm.

Liu, W., Jin, C.-Z., Zhang, Y.-Q., Cai, Y.-J., Xing, S., Wu, X.-J., Cheng, H., Edwards, R.L., Pan, W.-S., Qin, D.-G., An, Z.-S., Trinkaus, E., Wu, X.-X., 2010. Human remains from Zhirendong, South China, and modern human emergence in East Asia. Proceedings of the National Academy of Sciences, USA 107, 19201–19206.

Lockwood, C., 2008. The Human Story Where we come from and how we evolved. The Natural History Museum, London, p. 112.

Lourens, L., Hilgen, F., Shackleton, N.J., Laskar, J., Wilson, D., 2004. The Neogene Period. In: Gradstein, F., Ogg, J.G., Smith, G. (Eds.), A Geologic Time Scale 2004. Cambridge University Press, Cambridge, pp. 409–440.

Lozano, M., Mosquera, M., de Castro, J.M.B., Arsuaga, J.L., Carbonell, E., 2009. Right handedness of *Homo heidelbergensis* from Sima de los Huesos (Atapuerca, Spain) 500,000 years ago. Evolution and Human Behavior 30, 369–376.

de Lumley, H., de Lumley, M.A., 1974. Pre-Neanderthal human remains from Arago Cave in southwestern France. Yearbook of Physical Anthropology 17, 162–168.

de Lumley, H., Boone, Y., 1976. Les structures d'habitat au Paléolithique inférieur. In: de Lumley, H. (Ed.), La Préhistoire Française: Les Civilisations Paléolithiques et Mésolithiques de la France. Centre National de la Recherche Scientifique, Paris, pp. 635–643.

Manzi, G., Magri, D., Milli, S., Palombo, M.R., Margari, V., Celiberti, V., Barbieri, M., Barbieri, M., Melis, R.T., Rubini, M., Ruffo, M., Saracino, B., Tzedakis, P.C., Zarattini, A., Biddittu, I., 2010. The new chronology of the Ceprano calvarium (Italy). Journal of Human Evolution 59, 580–585.

Marean, C.W., Bar-Matthews, M., Bernatchez, J., Fisher, E., Goldberg, P., Herries, A.I.R., Jacobs, Z., Jerardino, A., Karkanas, T.M., Nilssen, P.J., Thompson, E., Watts, I., Williams, H.M., 2007. Early human use of marine resources and pigment in South Africa during the Middle Pleistocene. Nature 449, 905–908.

Marks, A.E., Chabai, V.P., 2006. Stasis and change during the Crimean Middle Palaeolithic. In: Hovers, E., Kuhn, S.L. (Eds.), Transitions before the Transition: Evolution and Stability in the Middle Palaeolithic and Middle Stone Age. Interdisciplinary Contributions to Archaeology. Springer, New York, pp. 121–135.

Maslin, M.A., Christensen, B., 2007. Tectonics, orbital forcing, global climate change, and human evolution in Africa: Introduction to the African paleoclimate special volume. Journal of Human Evolution 53, 443–464.

Maslin, M.A., Trauth, M.H., 2009. Plio-Pleistocene East African pulsed climate variability and its influence on early human evolution. In: Grine, F.E., Leakey, R.E., Fleagle, J.G. (Eds.), The First Humans – Origin and Early Evolution of the Genus Homo. Springer, New York, pp. 151–158.

Maslin, M., Mahli, Y., Phillips, O., Cowling, S., 2005. New views on an old forest: Assessing the longevity, resilience and future of the Amazon rainforest. Transactions of the Institute of British Geographers 30, 477–499.

McPherron, S.P., Alemseged, Z., Marean, C.W., Wynn, J.G., Reed, D., Geraads, D., Bobe, R., Béarat, H.A., 2010. Evidence for stone-tool-assisted consumption of animal tissues before 3.39 million years ago at Dikika, Ethiopia. Nature 466, 857–860.

de Menocal, P.B., 1995. Plio-Pleistocene African climate. Science 270, 53–59.

de Menocal, P.B., 2004. African climate change and faunal evolution during the Pliocene-Pleistocene. Earth and Planetary Science Letters 220, 3–24.

Monnier, G.F., 2006. The Lower/Middle Palaeolithic periodization in western Europe: An evaluation. Current Anthropology 47, 709–744.

Moore, A.M.T., 1988. The prehistory of Syria. Bulletin of the American Schools of Oriental Research 270, 3–12.

Mounier, A., Marchal, F., Condemi, S., 2009. Is *Homo heidelbergensis* a distinct species? New insight on the Mauer mandible. Journal of Human Evolution 56, 219–246.

Müller, U.C., Pross, J., Tzedakis, P.C., Gamble, C., Kotthoff, U., Schmiedl, G., Wulf, S., Christanis, K., 2011. The role of climate in the spread of modern humans into Europe. Quaternary Science Reviews 30, 273–279.

Munro, N.D., 2003. Small game, the Younger Dryas, and the transition to agriculture in the southern Levant. Mitteilungen der Gesellschaft für Urgeschichtel 12, 47–71.

O'Connell, J.F., Allen, J., 2004. Dating the colonization of Sahul (Pleistocene Australia-New Guinea): A review of recent research. Journal of Archaeological Science 31, 835–853.

Oda, S., Keally, C.T., 1986. A critical look at the Palaeolithic and "Lower Palaeolithic" research in Miyagi Prefecture, Japan. Jinruigaku Zasshi (Journal of the Anthropological Society of Nippon) 94, 325–361.

Parfitt, S.A., Barendregt, R.W., Breda, M., Candy, I., Collins, M.J., Coope, G.R., Durbidge, P., Field, M.H., Lee, J.R., Lister, A.M., Mutch, R., Penkman, K.E.H., Preece, R.C., Rose, J., Stringer, C.B., Symmons, R., Whittaker, J.E., Wymer, J.J., Stuart, A.J., 2005. The earliest record of human activity in northern Europe. Nature 438, 1008–1012.

Parfitt, S.A., Ashton, N.M., Lewis, S.G., Abel, R.L., Cooper, G.R., Field, M.H., Gale, R., Hoare, P.G., Larkin, N.R., Lewis, M.H., Karloukovski, V., Maher, B.A., Peglar, S.M., Preece, R.C., Whittaker, J.E., Stringer, C.B., 2010. Early Pleistocene human occupation at the edge of the boreal zone in northwest Europe. Nature 466, 229–233.

Parkington, J.E., 2006. Shorelines, Strandlopers and Shell Middens: Archaeology of the Cape Coast. Southern Cross Ventures, Cape Town, p. 127.

Patterson, B., Behrensmeyer, A.K., Sill, W.D., 1970. Geology and fauna of a new Pliocene locality in north-western Kenya. Nature 226, 918–921.

Patterson, N., Richter, D.J., Gnerre, S., Lander, E.S., Reich, D., 2006. Genetic evidence for complex speciation of humans and chimpanzees. Nature 441, 1103–1108.

Pearson, G.A., Ream, J.W., 2005. Clovis on the Caribbean coast of Venezuela. Current Research in the Pleistocene 22, 28–31.

Petraglia, M.D., 2011. Trailblazers across Arabia. Nature 470, 50–51.

Petraglia, M.D., Korisettar, R., Boivin, N., Clarkson, C., Ditchfield, P., Jones, S., Koshy, J., Mirazón Lahr, M., Oppenheimer, C., Pyle, D., Roberts, R., Schwenninger, J.-L., Arnold, L., White, K., 2007. Middle Palaeolithic assemblages from the Indian Subcontinent before and after the Toba super-eruption. Science 317, 114–116.

Potts, R., 1998. Environmental hypotheses of hominin evolution. American Journal of Physical Anthropology 107 (Suppl. 27, Yearbook of Physical Anthropology), 93–136.

Poulianos, A.N., 1971. Petralona: A Middle Pleistocene cave in Greece. Archaeology 24, 6–11.

Ravelo, A.C., Andreasen, D.H., Lyle, M., Lyle, A.O., Wara, M.W., 2004. Regional climate shifts caused by gradual global cooling in the Pliocene epoch. Nature 429, 263–267.

Reich, D., Green, R.E., Kircher, M., Krause, J., Patterson, N., Durand, E.Y., Viola, B., Briggs, A.W., Stenzel, U., Johnson, P.L.F., Maricic, T., Good, J.M., Marques-Bonet, T., Alkan, C., Fu, Q., Mallick, S., Li, H., Meyer, M., Eichler, E.E., Stoneking, M., Richards, M., Talamo, S., Shunkov, M.V., Derevianko, A.P., Hublin, J.-J., Kelso, J., Slatkin, M., Pääbo, S., 2010. Genetic history of an archaic hominin group from Denisova Cave in Siberia. Nature 468, 1053–1060.

Richards, M.P., Trinkaus, E., 2009. Isotopic evidence for the diets of European Neanderthals and early modern humans. Proceedings of the National Academy of Sciences, USA 106, 16034–16039.

Rightmire, G.P., 2001. Patterns of hominid evolution and dispersal in the Middle Pleistocene. Quaternary International 75, 77–84.

Rightmire, G.P., Lordkipanidze, D., Vekua, A., 2006. Anatomical descriptions, comparative studies and evolutionary significance of the hominin skulls from Dmanisi, Republic of Georgia. Journal of Human Evolution 50, 115–141.

Roberts, M.B., Parfitt, S.A., 1999. Boxgrove: A Middle Pleistocene Hominid Site at Eartham Quarry, Boxgrove, West Sussex. Archaeological Report 17. English Heritage, London, p. 456.

Roe, D.A., 1981. The Lower and Middle Palaeolithic Periods in Britain. Routledge and Kegan Paul Books, London, p. 340.

Saragusti, I., Goren-Inbar, N., 2001. The biface assemblage from Gesher Benot Ya'aqov, Israel: Illuminating patterns in "Out-of-Africa" dispersal. Quaternary International 75, 85–89.

Sarich, V.M., Wilson, A.C., 1967. Rates of albumin evolution in Primates. Proceedings of the National Academy of Sciences, USA 58, 142–148.

Sartono, S., 1991. A new *Homo erectus* skull from Ngawi, east Java; relation to Upper Pleistocene terraces in Java. Indo-Pacific Prehistory Association Bulletin 11, 14–22.

Schattner, U., 2010. What triggered the early-to-mid Pleistocene tectonic transition across the entire eastern Mediterranean? Earth and Planetary Science Letters 289, 539–548.

Schick, K.D., Toth, N., 1993. Making Silent Stones Speak: Human Evolution and the Dawn of Technology. Simon and Schuster, New York, p. 351.

Schmidt, P., Avery, D.H., 1978. Complex iron smelting and prehistoric culture in Tanzania. Science 201, 1085–1089.

Schoetensack, O., 1908. Der Unterkiefer des *Homo heidelbergensis* aus den Sanden von Mauer bei Heidelberg. Wilhelm Engelmann, Leipzig.

Scott, G.R., Gibert, L., 2009. The oldest hand-axes in Europe. Nature 461, 82–85.

Segalen, L., Lee-Thorp, J.A., Cerling, T., 2007. Timing of C4 grass expansion across sub-Saharan Africa. Journal of Human Evolution 53, 549–559.

Senut, B., Pickford, M., Gommery, D., Mein, P., Cheboi, K., Coppens, Y., 2001. First hominid from the Miocene (Lukeino Formation, Kenya). Comptes Rendus de l'Académie des Sciences, Series IIA: Earth and Planetary Science 332, 137–144.

Sepulchre, P., Ramstein, G., Fluteau, F., Schuster, M., Tiercelin, J.-J., Brunet, M., 2006. Tectonic uplift and eastern Africa aridification. Science 313, 1419–1423.

Skrzypek, G., Wiśniewski, A., Grierson, P.F., 2011. How cold was it for Neanderthals moving to Central Europe during warm phases of the last glaciation? Quaternary Science Reviews 30, 481–487.

Smith, T.M., Tafforeau, P., Reid, D.J., Grün, R., Eggins, S., Boutakiout, M., Hublin, J.-J., 2007. Earliest evidence of modern human life history in north African early *Homo sapiens*. Proceedings of the National Academy of Sciences, USA 104, 6128–6133.

Sponheimer, M., Lee-Thorp, J.A., 1999. Isotopic evidence for the diet of an early hominid, *Australopithecus africanus*. Science. 283, 368–370.

Stanford, D., Bradley, B., 2002. Ocean trails and prairie paths? Thoughts about Clovis origins. In: Jablonski, N.G. (Ed.), The First Americans: The Pleistocene Colonization of the New World. Memoirs of the California Academy of Sciences, 27. University of California Press, San Francisco, pp. 255–271.

Stewart, J.R., 2005. The ecology and adaptation of Neanderthals during the non-analogue environment of Oxygen Isotope Stage 3. Quaternary International 137, 35–46.

Strasser, T.F., Runnels, C., Wegmann, K., Panagopoulou, E., Mccoy, F., Digregorio, C., Karkanas, P., Thompson, N., 2011. Dating Palaeolithic sites in southwestern Crete, Greece. Journal of Quaternary Science 26, 553–560.

Stringer, C., 2011. The Origin of Our Species. Allen Lane, London, p. 333.

Svendsen, J.I., Heggen, H.P., Hufthammer, A.K., Mangerud, J., Pavlov, P., Roebroeks, W., 2010. Geo-archaeological investigations of Palaeolithic sites along the Ural Mountains – on the northern presence of humans during the last Ice Age. Quaternary Science Reviews 29, 3138–3156.

Tangshan Archaeological Team, 1996. Locality of the Nanjing Man Fossils 1993–1994. Heritage Publishing, Beijing, p. 306 [In Chinese with English abstract].

Tattershall, I., 1998. Becoming Human: Human Evolution and Human Nature. Harcourt Brace, New York, p. 272.

Tattershall, I., Schwartz, J.H., 2006. The distinctiveness and systematic context of *Homo neanderthalensis*. In: Harvati, K., Harrison, T. (Eds.),

Neanderthals Revisited: New Approaches and Perspectives. Springer, New York, pp. 9–22.

Tattershall, I., Schwartz, J.H., 2009. Evolution of the genus *Homo*. Annual Review of Earth and Planetary Sciences 37, 67–92.

Thieme, H., 1997. Lower Palaeolithic hunting spears from Germany. Nature 385, 807–810.

Thoma, A., 1966. L'occipital de l'homme Mindelien de Vertesszöllös. L'Anthropologie 70, 495–534.

Thorne, A., Grün, R., Mortimer, G., Spooner, N.A., Simpson, J.J., McCulloch, M., Taylor, M., Cumoe, D., 1999. Australia's oldest human remains: Age of the Lake Mungo 3 skeleton. Journal of Human Evolution 36, 591–612.

Trauth, M.H., Maslin, M.A., Deino, A., Strecker, M.R., 2005. Late Cenozoic moisture history of East Africa. Science 309, 2051–2053.

Trauth, M., Maslin, M., Deino, A., Strecker, M.R., Bergner, A.G.N., Duhnforth, M., 2007. High and low latitude forcing of Plio-Pleistocene East African climate and human evolution. Journal of Human Evolution 53, 475–486.

Trinkaus, E., 2009. The human tibia from Broken Hill, Kabwe, Zambia. PaleoAnthropology 2009, 145–165.

Trinkaus, E., Shang, H., 2008. Anatomical evidence for the antiquity of human footwear: Tianyuan and Sunghir. Journal of Archaeological Science 35, 1928–1933.

Tzedakis, P.C., Hughen, K.A., Cacho, I., Harvati, K., 2007. Placing late Neanderthals in a climatic context. Nature 449, 206–208.

Valet, J.-P., Valladas, H., 2010. The Laschamp-Mono Lake geomagnetic events and the extinction of Neanderthal: A causal link or a coincidence? Quaternary Science Reviews 29, 3887–3893.

Valladas, H., Reyss, J.L., Joron, J.L., Valladas, G., Bar-Yosef, O., Vandermeersch, B., 1987. Thermoluminescence dating of Mousterian "proto-Cro-Magnon" remains from Israel and the origin of modern man. Nature 331, 614–616.

Vallin, L., Masson, B., Caspar, J.-P., Depiereux, E., 2006. L'outil idéal: analyse du standard Levallois des sites moustériens d'Hermies (Nord de la France). Paléo, Revue d'Archéologie Préhistorique 18, 237–272.

Vlček, E., 1978. A new discovery of *Homo erectus* in central Europe. Journal of Human Evolution 7, 239–251.

Waechter, J.d'A., 1952. The excavation of Jabrud and its relation to the prehistory of Palestine and Syria. Annual Report of London University Institute of Archaeology 8, 10–28.

Wainwright, G.J., Malik, S.C., 1967. Recent field research on problems of archaeology and Pleistocene chronology in peninsular India. Proceedings of the Prehistoric Society 33, 132–146.

Walker, A., Leakey, R.E. (Eds.), 1993. The Nariokotome *Homo erectus* skeleton. Harvard University Press, Cambridge, Ma, p. 457.

Ward, S., Hill, A., 1987. Pliocene hominid partial mandible from Tabarin, Baringo, Kenya. American Journal of Physical Anthropology 72, 21–37.

Ward, C.V., Leakey, M.G., Walker, A., 2001. Morphology of *Australopithecus anamensis* from Kanapoi and Allia Bay, Kenya. Journal of Human Evolution 41, 255–368.

Westaway, R., Bridgland, D.R., Sinha, R., Demir, T., 2009. Fluvial sequences as evidence for landscape and climatic evolution in the late Cenozoic: A synthesis of data from IGCP 518. Global and Planetary Change 68, 237–253.

White, T.D., Suwa, G., Hart, W.K., Walter, R.C., WoldeGabriel, G., de Heinzelin, J., Clark, J.D., Asfaw, B., Vrba, E., 1993. New discoveries of *Australopithecus* at Maka in Ethiopia. Nature 366, 261–265.

White, T.D., Suwa, G., Asfaw, B., 1994. *Australopithecus ramidus*, a new species of early hominid from Aramis, Ethiopia. Nature 371, 306–312.

White, T.D., Asfaw, B., DeGusta, D., Gilbert, H., Richards, G.D., Suwa, G., Howell, F.C., 2003. Pleistocene *Homo sapiens* from Middle Awash, Ethiopia. Nature 423, 742–747.

White, T.D., Asfaw, B., Beyene, Y., Haile-Selassie, Y., Lovejoy, C.O., Suwa, G., WoldeGabriel, G., 2009. *Ardipithecus ramidus* and the palaeobiology of early hominids. Science 326, 75–86.

WoldeGabriel, G., White, T.D., Suwa, G., Renne, P., de Heinzelin, J., Hart, W.K., Heiken, G., 1994. Ecological and temporal placement of early Pliocene hominids at Aramis, Ethiopia. Nature 371, 330–333.

Wolff, H., Greenwood, A.D., 2010. Did viral disease of humans wipe out the Neanderthals? Medical Hypotheses 75, 99–105.

Wood, B., Collard, M., 1999. The human genus. Science 284, 65–71.

Woodward, A.S., 1921. A new cave man from Rhodesia, South Africa. Nature 108, 371–372.

Wrangham, R., Carmody, R., 2010. Human adaptation to the control of fire. Evolutionary Anthropology 19, 187–199.

Wymer, J.J., 1985. The Palaeolithic Sites of East Anglia. Geobooks, Norwich, p. 400.

Wynn, J.G., 2004. Influence of Plio-Pleistocene aridification on human evolution: Evidence from paleosols of the Turkana Basin, Kenya. American Journal of Physical Anthropology 123, 106–118.

Yamei, H., Potts, R., Baoyin, Y., Zhengtang, G., Deino, A., Wei, W., Clark, J., Guangmao, X., Weiwen, H., 2000. Mid-Pleistocene Acheulean-like stone technology of the Bose Basin, south China. Science 287, 1622–1626.

J. Zalasiewicz, P.J. Crutzen and W. Steffen

The Anthropocene

Abstract: He Anthropocene is a currently informal term to signify a contemporary time interval in which surface geological processes are dominated by human activities, now being studied by an ICS working group as regards potential formalization within the Geological Time Scale. Its developing stratigraphic signature includes components that are lithostratigraphic, biostratigraphic, and chemostratigraphic; and these vary from being approximately synchronous to strongly diachronous. Formalization will depend upon both scientific justification and utility to working scientists, and upon the choice of an effective boundary, whether by GSSP or GSSA.

Chapter Outline

The idea that the emergence of *Homo sapiens* – and particularly the growth of a complex cultural and material civilization – precipitated a new geological age has been mooted for well over a century. Stoppani (1873) coined the "Anthropozoic Era", noting that the arrival of humans had introduced on Earth a "new telluric force which in power and universality may be compared to the greater forces of the earth", a phenomenon described also via the "Psychozoic" of Joseph Le Conte (1879), while Marsh (1864) detailed examples of large-scale anthropogenic change across the Earth. Edouard Le Roy (1927), influenced by Vladimir Vernadsky and Teilhard de Chardin, produced the concept of the "noösphere", a concept developed further by Vernadsky himself (1945).

The ideas were in part developed philosophically, being linked with ideas such as "cephalization" (Dana, 1863), that is, of unidirectional progress culminating in the ascent and dominance of humankind. They were thus controversial, and ignored or dismissed by most mainstream geologists. Berry (1925), for instance, said that the span of human influence was tiny as regards geological time and stratal thickness, and considered Le Conte's "Psychozoic Era" as a stratigraphic irrelevance. The terms were never adopted, formally or informally, and essentially were forgotten by all except historians of the earth sciences (Grinevald, in press; see also Steffen *et al.*, 2011a).

32.1. THE ANTHROPOCENE

That situation changed when Crutzen introduced the term "Anthropocene" (Crutzen and Stoermer, 2000; Crutzen, 2002). This was at a time when the scale and significance of human influence on Earth surface processes was being increasingly accepted by mainstream scientists (including earth scientists). Hence, the term, unlike its predecessors, quickly began to be used in peer-reviewed scientific literature (e.g., Meybeck, 2003; Steffen *et al.*, 2004; Crossland, 2005; Syvitski *et al.*, 2005; Andersson *et al.*, 2006) and in formal and informal discussion, and has been widely disseminated in public media (e.g., Kolbert, 2011; Walsh, 2012). Thus, the term has now become part of the scientific (and vernacular) *lingua franca*.

However, the term currently has no formal status within the geological time scale, either as regards hierarchical level (age, epoch, period, era?) or definition, particularly as regards its beginning ("base") and where and how that might be defined. Noting, though, its increasing use, members of the Stratigraphy Commission of the Geological Society of London (Zalasiewicz *et al.*, 2008) examined the term, and considered that a reasonable case may be made for its formalization. Subsequently, an Anthropocene Working Group of the Subcommission of Quaternary Stratigraphy of the International Commission of Stratigraphy was initiated, its brief being to further examine the stratigraphic justification of

The Geologic Time Scale 2012. DOI: 10.1016/B978-0-444-59425-9.00032-9

the term and its utility for earth (and other) scientists. Subsequent publications involving the group or its members include Ellis *et al.* (2010), Zalasiewicz *et al.* (2010), a series of papers in Williams *et al.* (2011), and Steffen *et al.* (2011b).

We here discuss the Anthropocene in terms of signals preserved in recently and currently accumulating strata, as these are the signals that, in ancient strata, provide our understanding of Earth history. We note that humans typically live on land, a largely erosional realm. However, human activities also impact the sea and coastal regions, where a more continuous record is accumulating. The land also includes many sediment traps (lakes, swamps and so on) that, while geologically short-lived, are nevertheless an object of study for Quaternary and other scientists.

In this account we briefly assess the stratigraphic signature of the Anthropocene, consider possible levels for its beginning and its geological longevity, and discuss practical aspects of formalization, including its hierarchical level and the selection of a GSSP (Global Stratigraphic Section and Point) or GSSA (Global Stratigraphic Standard Age).

32.2. STRATIGRAPHIC SIGNATURE

Lithostratigraphic signals: Direct and indirect signals may be distinguished. More or less direct signals include those associated with construction, agriculture and irrigation (including the modification of rivers: Syvitski and Kettner, 2011; Merritts *et al.*, 2011). These are substantial, globally being the dominant means of surface sediment transport today (Hooke, 2000; Wilkinson, 2005; Figure 32.1), with anthropogenically modified terrain now covering >50% of the Earth's land surface (Ellis, 2011). The deposit of buildings and rubble associated with urban areas is moreover lithologically distinctive, and indeed can be mapped as a lithostratigraphic unit, such as the "Made Ground" of British Geological Survey maps (Price *et al.*, 2011).

Indirect signals include, for instance, the shifts in sedimentary facies belts associated with sea-level change (see below) and with ocean acidification. The latter will likely have profound effects on carbonate sedimentation, if current trends continue through this century (Zeebe *et al.*, 2008; Tyrrell, 2011). These include hindering the production of biogenic carbonate, especially on coral reefs and in pelagic settings (see below also), and the creation of a dissolution layer within already-deposited oceanic oozes (cf. that associated with the Paleocene–Eocene Thermal Maximum acidification event: Zachos *et al.*, 2005).

Sequence stratigraphic signals: An "Anthropocene sea-level rise effect" to date has been geologically insignificant, being of the order of half a metre (Figure 32.1). Glacioeustatic rise, however, significantly lags behind atmospheric chemistry and temperature change on human time scales, given the long response times of the major ice sheets. Geologically, in a Quaternary world with substantial polar ice, eventual sea-level change has been estimated as 10–30 m per degree centigrade (Rahmstorf, 2007), thus suggesting, if IPCC (2007) temperature predictions are broadly correct, a geologically near-future sea-level rise of a few to several tens of metres, centuries to millennia hence (see also Overpeck *et al.*, 2006). If sustained (Tyrrell *et al.*, 2007; Tyrrell, 2011) this would produce a marine transgressive signal particularly affecting shelf and coastal environments and also, via near-shore sediment storage, influencing deeper (turbidite) environments.

Biostratigraphic signals: Anthropogenic change to the Earth's biota is now pervasive, both on land (where its effects are most clearly visible and easily studied) and in the sea. Extinction rates, the most obvious indicator of change, are currently perhaps 100–1000 times background rates (Millennium Ecosystem Assessment, 2005). These very high rates are recent, and so the Earth has thus far only lost a small proportion of its biodiversity. However, many more species are classed as threatened, endangered or critically endangered (IUCN), and continuation of present trends could produce a mass extinction event comparable to the "Big Five" of Phanerozoic history in as little as a few centuries (Barnosky *et al.*, 2011; Figure 32.2).

Current biodiversity change is driven by habitat loss associated with urbanization and agriculture, the latter in itself creating a clear palynological signal (e.g., Graf and Chmura, 2007) that extends into coastal regions. In the marine realm, human predation is now extensive (Myers and Worm, 2003; Worm *et al.*, 2006), though concentrated on the top of the food chain, rather than on the protists and invertebrates that provide the bulk of the biostratigraphic signal. However, the latter are already affected by 20th century warming and acidification (Caldeira and Wickett, 2003; Orr *et al.*, 2005; Hoegh-Guldberg *et al.*, 2007) and these stressors are projected to increase throughout this century and beyond (IPCC, 2007; Wilson, 2002; Thomas *et al.*, 2004; Zeebe *et al.*, 2008). Range changes have already occurred (Edwards, 2009) and are predicted to develop on a greater scale (compare the polewards spread of the dinoflagellate *Apectodinium* at the Paleocene–Eocene Thermal Maximum: Crouch *et al.*, 2001). Extinctions are likely where tolerances are exceeded, giving rise to major biofacies transitions. For instance, experiments suggest that net coral reef loss (and hence the start of major decline of these ecosystems) takes place at carbon dioxide levels above 560 ppm (Kleypas and Yates, 2009), a level likely to be reached by the end of this century. Lacustrine ecosystems have been changed, too, often effectively irreversibly, by acid deposition (Jeffries *et al.*, 2003; Wright and Schindler, 2004).

Perhaps the most striking signal to date is that of cross-global species invasions, at a rate and on a scale unique in Earth history (Figure 32.2). Terrestrial biotas worldwide now commonly include up to a quarter of total species (and up to a half of plant species) as invasives (McNeely, 2001). Marine invasive species may be less in total, but are more rapidly

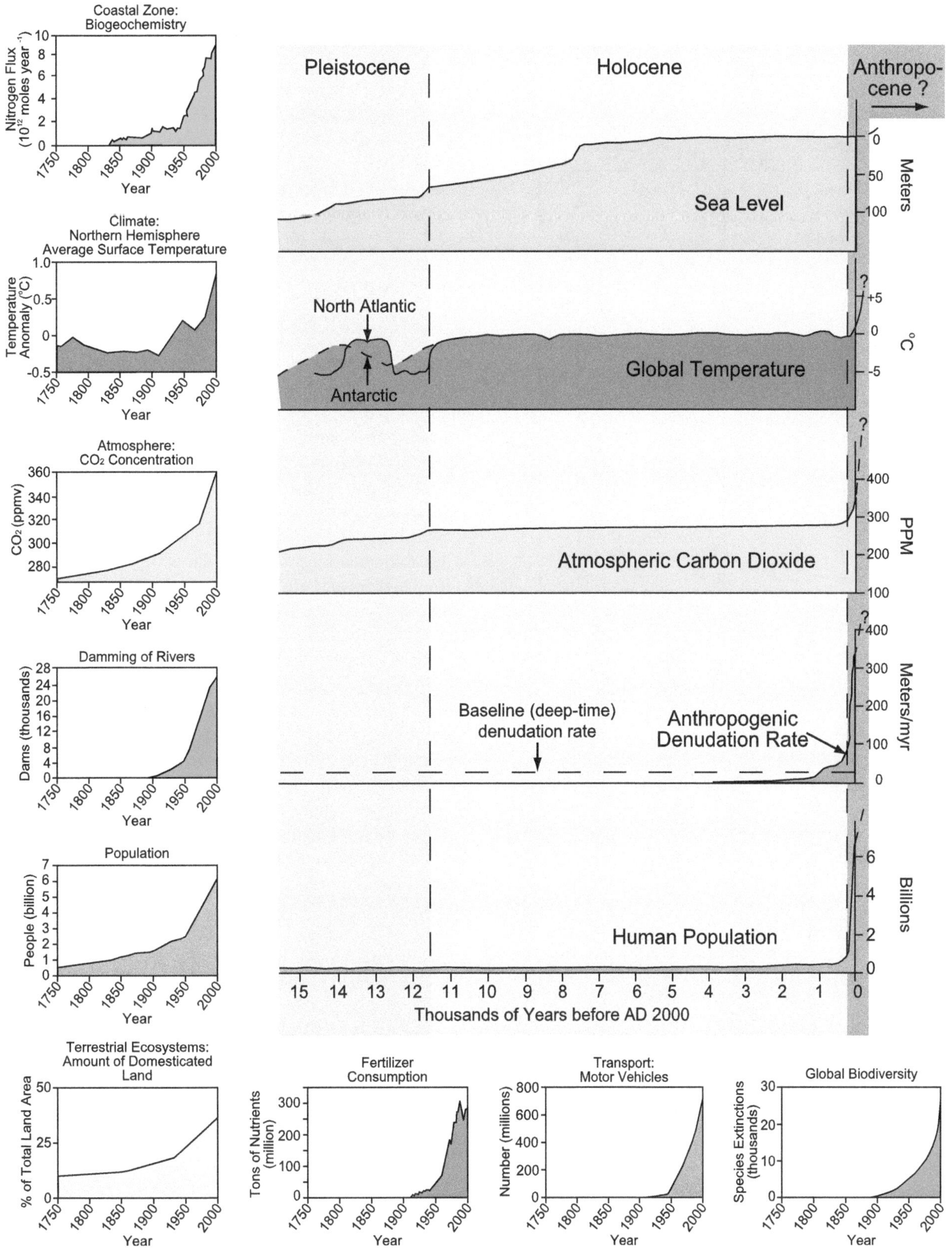

Stratigraphically significant trends over the last 15 000 years (after Zalasiewicz *et al.*, 2008) and over the past 250 years. *(After Steffen et al., 2004, 2007).*

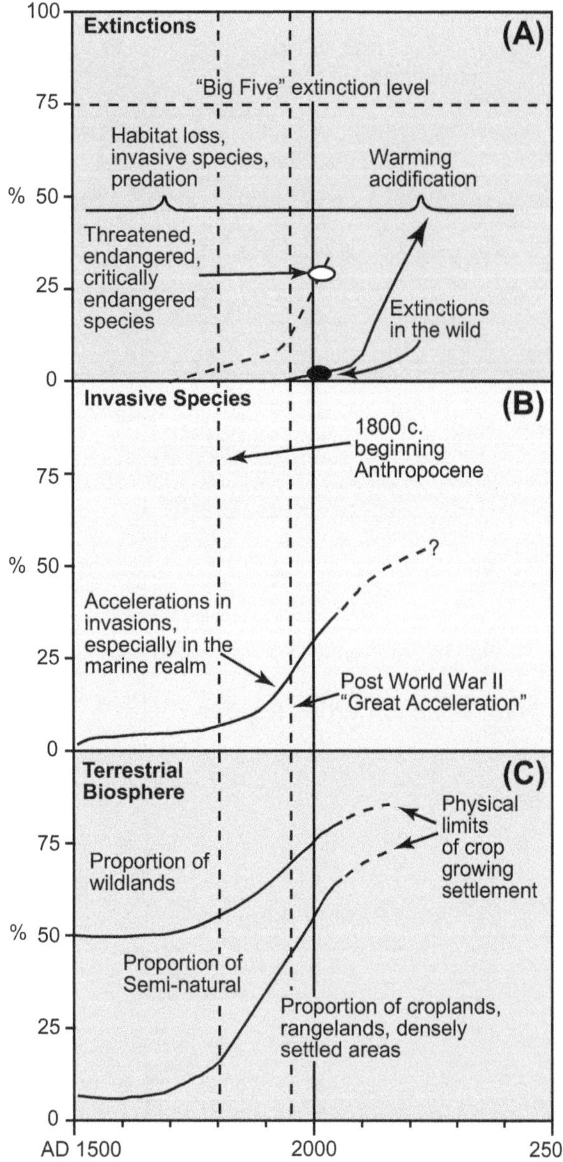

Factors within the Anthropocene biostratigraphic signal. (A) Relatively few species have been declared extinct to date, but many more are threatened; current rates may see 'Big Five' levels within centuries (Barnosky et al., 2011). (B) Species invasions have reached high levels and are ongoing. Biotic effects are effectively permanent (McNeely, 2001; SEBI, 2011. (C) Anthromes (anthropogenic biomes) have increased markedly at the expense of natural systems (data from Ellis, 2011).

growing (SEBI, 2010; see also Molnar, 2008). The numbers far exceed reported extinctions, with some 10 000 invasive species reported from Europe alone (SEBI, 2010).

Chemostratigraphic signals: These include heavy metal traces, detectable initially from Graeco-Roman smelting and subsequently from post-Industrial Revolution vehicle emissions (Dunlap *et al.*, 1999) and widespread and distinctive organic molecules (such as polyaromatic hydrocarbons) from hydrocarbon combustion (Kruge, 2008; Vane *et al.*, 2011). Isotopic signals include global shifts in carbon isotope ratios, again from fossil fuel burning, and these are already detectable in marine biogenic sediments (Al-Rousan *et al.*, 2004), as are artificial radionuclides from atomic bomb tests (see below). The approximate doubling in size of the surface nitrogen cycle by humans (Galloway *et al.*, 2008) is also likely to produce stratigraphic signals, perhaps expressed as nitrogen isotope shifts (Savard *et al.*, 2009) or as deposition of organic-rich facies linked to ongoing spread of marine "dead zones" (Diaz and Rosenberg, 2008).

32.3. BEGINNING OF THE ANTHROPOCENE?

As regards duration, there are three levels at which the beginning of a putative Anthropocene geological time interval might be placed.

Early Holocene: Ruddiman (2003) has proposed that the observed slow increase in CO_2 levels from 260 to 280 ppm over the second half of the Holocene was, firstly, caused by human agricultural (though pre-industrial) activities, and, secondly, was instrumental in maintaining interglacial conditions for a few millennia longer than in the two previous interglacials (OIS 7 and 5e; cf. Petit *et al.*, 1999). In this interpretation, the Holocene is profoundly anthropogenically influenced *per se*, and so the beginning of an "Anthropocene" thus defined may be placed somewhere in early Holocene time.

However, this thesis remains controversial (e.g., Broecker and Stocker, 2006; Stocker *et al.*, 2010). And, even if true, it represents an effective prolongation of "Holocene" conditions that would not, from a far future perspective, give rise to strata that were clearly distinguishable from those of previous Quaternary interglacials.

Beginning of Industrial Revolution: This is effectively the level chosen by Crutzen (2002), where the combination of industrialization and the acceleration of population growth created a clear step change in the human signal (Figures 32.1 and 32.2). Here the biotic (biostratigraphic), geomorphological (lithostratigraphic) and geochemical changes noted above begin to introduce a stratigraphic signal qualitatively different from that of the earlier Holocene and that of preceding interglacials. Industrialization, of course, has been diachronous globally. Significant iron-smelting took place in 11th century China, for instance, and coal-burning occurred in medieval England (Steffen *et al.*, 2007); the global phenomenon began in the UK in the late 18th century, spreading to central Europe by the mid-19th century, then to North America and beyond. However, the beginning of the 19th century may reasonably approximate the beginning of this phenomenon.

Mid-twentieth century: A significant increase in human activity took place following the Second World War, and this has been termed the "Great Acceleration" by Steffen *et al.* (2007), of population growth, industrialization, agriculture and the novel use of atomic fission in bombs, which is of stratigraphic as well as societal importance. Since 1950, for instance, world population doubled, the global economy increased some 15-fold, automobile use rose by a yet greater amount, and humans switched from being a largely rural to a largely urban species. Environmental indicators reflect this pattern (Figure 32.1), with nearly three-quarters of anthropogenic CO_2 rise, say, occurring in this interval; the rate of rise considerably exceeds that at the end of the last glacial phase (cf. Monnin *et al.*, 2001). These trends, both industrial and environmental, are currently accelerating (Steffen *et al.*, 2007).

32.4. FUTURE DURATION OF THE ANTHROPOCENE?

Part of the rationale for the Anthropocene hinges on the likely duration of anthropogenic effects, and thus on the stratigraphic thickness of strata that may come to be "Anthropocene", even if anthropogenic pressures diminished or ceased in the near future. Direct lithostratigraphic signals – e.g., the production of "urban strata" are likely to be the shortest lived, persisting for millennia only, and creating an "event bed" in terrestrial and coastal facies. Indirect signals relating to CO_2 release (warming, acidification, sea-level rise) have been projected as substantial and long-lived (Pagani *et al.*, 2010; Park and Royer, 2011), perhaps preventing glaciations for hundreds of thousands of years (Archer, 2005; Tyrrell *et al.*, 2007; Tyrrell, 2011) and halting any potential future intensification of Quaternary icehouse conditions (Crowley and Hyde, 2008). The biostratigraphical signal will be effectively permanent: even with future recovery of biodiversity, that will take place over million-year time scales by comparison with past biological crises (Hart, 1996), incoming floras and faunas will be clearly distinct from those that preceded the change.

32.5. FORMAL CONSIDERATION OF THE ANTHROPOCENE

Thus, although we are at most only two centuries into an interval that might be termed the Anthropocene, it is likely only the beginning of stratigraphically significant changes that will endure over geological time scales, modified by a variety of feedbacks, including human ones (Kellie-Smith and Cox, 2011). Some of the changes are taking place immediately – notably the direct lithostratigraphical ones and biotic phenomena such as species invasions. Others, because of lag effects, are in their early phases: atmospheric warming and direct effects thereof, and the "extinction debt" (Tilman *et al.*, 1994) where there is a time gap between habitat loss and species extinctions. Yet others, such as sea-level rise, have scarcely begun. In addition some near-future effects are effectively predictable (ocean acidification). Other effects are less predictable, particularly those that involve multiple feedback loops, such as the pattern and timing of thresholds (or "tipping points") within global warming (Schneider, 2004; Lenton *et al.*, 2008).

There clearly is more work to do in quantitatively and qualitatively comparing these anthropogenic changes with those of previous perturbations in Earth history. However, we consider that the sum total of stratigraphically relevant changes – physical, chemical, and biological – that have taken place *to date* in themselves justify the consideration of the term Anthropocene, in reflecting near-surface geological conditions distinct from those of the earlier Holocene or of previous Quaternary interglacial phases.

What remains to be resolved is whether the introduction of the term as a formal unit will serve a useful purpose for earth (and other) scientists, particularly to those working on late Quaternary strata, and especially Holocene workers. The alternative would be to continue regarding it as an informal unit. Formalization would have the advantage of helping scientific communication by precisely defining the unit as regards duration, recognition and hierarchical level. It may also be significant to the wider community (e.g., Nature, 2011; Tickell, 2011; Vidas, 2011).

There follows the question of scale; the Anthropocene, at any hierarchical level, would be thus far the briefest chronostratigraphic unit in the geological time scale. However, the scale is utilitarian, in that the Holocene (but one of many Quaternary interglacials) is currently the briefest Epoch by some orders of magnitude. Yet, because of its geological importance (literally comprising the ground under our feet), its status has never been seriously questioned. The Anthropocene is now emerging as of comparable significance.

32.6. DEFINITION

This is one of the questions being debated by the Working Group, but we consider that a potential boundary for the Anthropocene should reflect either of the early 19th century or mid-20th century candidate levels noted above. The latter includes a practical stratigraphic marker: radionucleides from atmospheric A-bomb tests detectable within strata including ice cores (Schwikowski, 2004; Turetsky *et al.*, 2004; Marshall *et al.*, 2007). But, by comparison with the newly defined Holocene boundary (Walker *et al.*, 2009), an Anthropocene boundary that more nearly reflects the beginning of global industrialization and its effects may be more appropriate. Any level selected should, at this temporal reach, offer annual resolution, as does, almost, the reference level for the beginning of the Holocene Epoch, placed in an ice core (Walker *et al.*, 2009).

The Anthropocene may be defined either by a GSSP (Global Stratigraphic Section and Point) or a GSSA (Global Standard Stratigraphic Age). The former might be placed in, for instance, lake sediment, annually layered ice/snow, or laminated marine sediment such as that of the anoxic Santa Barbara Basin (Thunnell *et al.*, 2005). Normal deep-ocean sediment would not suffice, since mixing through bioturbation would make the boundary (and indeed commonly the entire Anthropocene interval) not separately distinguishable.

Equally, for current use, a selected GSSA (say set at 1800) would be practically effective as a boundary to trace in sedimentary sections, and would be perhaps more familiar for archaeologists and other scientists.

Recognition of Anthropocene strata will be attended by all the problems of correlation typically encountered with older strata, albeit at a finer temporal scale. Thus, recognition cannot be made simply by the recognition of anthropogenic effects alone, for most of these have been diachronous across the globe. However, a combination of stratigraphic and archeological methodologies should allow reasonably effective discrimination. This would especially be the case if the boundary selected were to coincide with the beginning of the nuclear age. The incoming of ubiquitous radioactive contamination should provide an unambiguous signal in both marine and non-marine deposits, whether visibly anthopogenically influenced or not.

32.7. HIERARCHICAL LEVEL

The hierarchical scale of the Anthropocene, if formalized, should reflect the scale of environmental perturbation as recorded by stratigraphic indicators, by comparison with those seen in the deep time record. Possibilities include defining the Anthropocene as an age (or stage), that is, as a subdivision of the Holocene; this is the most conservative stance. However, we have shown here that sufficient change has been forced on the Earth system to take it beyond Holocene norms of sedimentation, biota and chemistry (see also Worldwatch Institute, 2009, pp. 20–21, table 2.1; Röckstrom *et al.*, 2009), with estimates of period- or era-scale biotic change within centuries (e.g., Barnosky *et al.*, 2011). Thus, status as an epoch seems currently appropriate and conservative. Continuation of current trends over this century may lead to its consideration as a period (or greater); we hope that effective societal mitigation will prevent this outcome.

ACKNOWLEDGMENTS

We acknowledge input and discussions from colleagues, especially those of the Stratigraphy Commission of the Geological Society of London and of the Anthropocene working Group of ICS.

REFERENCES

Al-Rousan, S., Pätzold, J., Al-Moghrabi, S., Wefer, G., 2004. Invasion of anthropogenic CO_2 recorded in planktonic foraminifera from the northern Gulf of Aquaba. International Journal of Earth Sciences 93, 1066–1076.

Andersson, A.J., Mackenzie, F.T., Lerman, A., 2006. Coastal ocean and carbonate systems in the high CO_2 world of the Anthropocene. American Journal of Science 305, 875–918.

Archer, D., 2005. Fate of fossil fuel in geologic time. Journal of Geophysical Research 110 C09S05, doi: 10.1029/2004/2004JC002625.

Barnosky, A.D., Matzke, N., Tomiya, S., Wogan, G.O.U., Swartz, B., Quental, T.B., Marshall, C., McGuire, J.L., Lindsey, E.L., Maguire, K.C., Mersey, B., Ferrer, E.A., 2011. Has the Earth's sixth mass extinction already arrived? Nature 471, 51–57.

Berry, E.W., 1925. The term Psychozoic. Science 44, 16.

Broecker, W.C., Stocker, T.F., 2006. The Holocene CO_2 rise: Anthropogenic or natural? Eos 87 (3), 27–29.

Caldeira, K., Wickett, M.E., 2003. Anthropogenic carbon and ocean pH. Nature 425, 365–365.

Crossland, C.J. (Ed.), 2005. Coastal Fluxes in the Anthropocene. Springer, Berlin, p. 231.

Crouch, E.M., Heilmann-Clausen, C., Brinkhuis, H., Morgans, H.E.G., Rogers, K.M., Egger, H., Schmitz, B., 2001. Global dinoflagellate event associated with the late Paleocene thermal maximum. Geology 29, 315–318.

Crowley, T.J., Hyde, W.T., 2008. Transient nature of late Pleistocene climate variability. Nature 456, 226–230.

Crutzen, P.J., 2002. Geology of mankind. Nature 415, 23.

Crutzen, P.J., Stoermer, E.F., 2000. The 'Anthropocene'. Global Change Newsletter 41, 17–18.

Dana, J.D., 1863. On cephalization. New Englander 22, 495–506.

Diaz, R.J., Rosenberg, R., 2008. Spreading dead zones and consequences for marine ecosystems. Science 321, 926–929.

Dunlap, C.E., Steinnes, E., Flegal, A.R., 1999. A synthesis of lead isotopes in two millenia of European air. Earth and Planetary Science Letters 167, 81–88.

Edwards, M., 2009. Sea life (pelagic and planktonic ecosystems) as an indicator of climate and global change. In: Letcher, T.P. (Ed.), Climate Change: Observed Impacts on Planet Earth. Elsevier, Amsterdam, pp. 233–251.

Ellis, E.C., 2011. Anthropogenic transformation of the terrestrial biosphere. Philosophical Transactions of the Royal Society, Series A 369, 1010–1035.

Ellis, E.C., Goldewijk, K.K., Siebert, S., Lightman, D., Ramankutty, N., 2010. Anthropogenic transformation of the biomes 1700–2000. Global Ecology and Biogeography 19, 589–606.

Galloway, J.N., Townsend, A.R., Erisman, J.W., Bekunda, M., Cai, Z., Freney, J.R., Martinelli, L.A., Seitzinger, S.P., Sutton, M.A., 2008. Transformation of the nitrogen cycle: Recent trends, questions and potential solutions. Science 320, 889–892.

Graf, M.-T., Chmura, G.L., 2007. Development of modern analogues for natural, mowed and grazed grasslands using pollen assemblages and coprophilous fungi. Review of Palaeobotany and Palynology 141, 139–149.

Grinevald, J., in press, La Biosphère de l'Anthropocène: Repères trans-disciplinaires 1824-2010, second ed. Geneva: Georg.

Hart, M.B. (Ed.), 1996. Biotic Recovery from Mass Extinction Events. Geological Society Special Publication, 102, p. 392.

Hoegh-Guldberg, O., Mumby, P.J., Hooten, A.J., Steneck, R.S., Greenfield, P., Gomez, E., Harvell, C.D., Sale, P.F., Edwards, A.J., Caldeira, K., Knowlton, N., Eakin, C.M., Iglesias-Prieto, Muthiga, N.,

Bradbury, R.H., Dubi, A., Hatziolos, M.E., 2007. Coral reefs under rapid climate change and ocean acidification. Science 318, 1737–1742.

Hooke, R., LeB, 2000. On the history of humans as geomorphic agents. Geology 28, 843–846.

IPCC (Intergovernmental Panel on Climate Change), 2007. Climate change 2007: synthesis report. Summary for policy makers. Available at: http://www.ipcc.ch/SPM2feb07.pdf.

Jeffries, D.S., Clair, T.S., Couture, S., Dillan, P.J., Dupont, J., Keller, W., McNicol, D.K., Turner, M.A., Vet, R., Veeber, R., 2003. Assessing the recovery of lakes in southeastern Canada from the effects of acidic deposition. Ambio 32, 176–182.

Kellie-Smith, O., Cox, P.M., 2011. Emergent dynamics of the climate-economy system in the Anthropocene. Philosophical Transactions of the Royal Society, Series A 369, 868–886.

Kleypas, J.A., Yates, K.K., 2009. Coral reefs and ocean acidification. Oceanography 22 (4), 108–117.

Kolbert, E., 2011. Enter the Age of Man. National Geographic 219 (3), 60–85.

Kruge, M.A., 2008. Organic chemostratigraphic markers characteristic of the (informally designated) Anthropocene Epoch. American Geophysical Union Fall Meeting Abstracts, A675.

Le Conte, J., 1879. Elements of Geology. D. Appleton & Co., New York, p. 588.

Le Roy, E., 1927. L'exigence idéaliste et le fait de l'évolution. Boivin, Paris, p. 270.

Lenton, T.M., Held, H., Kriegler, E., Hall, J.W., Lucht, W., Rahmstorf, S., Schnellhuber, H.J., 2008. Tipping points in the Earth's climate system. Proceedings of the National Academy of Sciences, USA 105, 1786–1793.

Marsh, G.M., 1864 (1965 reprint). Man and Nature: or, the Earth as Modified by Human Action. Belknap Press of Harvard University Press, Cambridge, Ma and London, p. 472.

Marshall, W.A., Gehrels, W.R., Garnett, M.H., Freeman, S.P.H.T., Maden, C., Sheng, X., 2007. The use of 'bomb spike' calibration and high-precision AMS ^{14}C analyses to date salt marsh sediments deposited during the last three centuries. Quaternary Research 68, 325–337.

McNeely, J., 2001. Invasive species: A costly catastrophe for native biodiversity. Land Use and Water Resources Research 1 (2), 1–10.

Merrits, D., Walter, R., Rahnis, M., Hartranft, J., Cox, S., Gellis, A., Potter, N., Hilgartner, W., Langland, M., Manion, L., Lippincott, C., Siddiqui, S., Rehman, Z., Scheid, C., Kratz, L., Shilling, A., Jenschke, M., Datin, C., Cranmer, E., Reed, A., Matuszewski, D., Voli, M., Ohlson, E., Neugebauer, A., Ahamed, A., Neal, C., Winter, A., Becker, C., 2011. Anthropocene streams and base-level controls from historic dams in the unglaciated mid-Atlantic region, USA. Philosophical Transactions of the Royal Society, Series A 369, 976–1009.

Meybeck, M., 2003. Global analysis of river systems: From Earth system controls to Anthropocene syndromes. Philosophical Transactions of the Royal Society, Series B 358, 1935–1955.

Millennium Ecosystem Assessment, 2005. Ecosystems and Human Well-being: Synthesis. Island Press, Washington, p. 137.

Molnar, J.L., Gamboa, R.L., Revenga, C., Spalding, M.D., 2008. Assessing the global threat of invasive species to marine biodiversity. Frontiers in Ecology and the Environment 6 (9), 485–492.

Monnin, E., Indermühle, A., Dällenbach, A., Flückiger, J., Stauffer, B., Stocker, T.S., Reynaud, D., Barnola, J.-M., 2001. Atmospheric CO_2 concentrations over the Last Glacial Termination. Science 291, 112–114.

Myers, R.A., Worm, B., 2003. Rapid worldwide depletion of predatory fish communities. Nature 423, 280–283.

Nature (editorial), 2011. The human epoch. Nature 473, 254.

Orr, J.C., Fabry, V.J., Aumont, O., Bopp, L., Doney, S.C., Feely, R.A., Gnanadesikan, A., Gruber, N., Ishida, A., Joos, F., Key, R.M., Lindsay, K., Maier-Reimer, E., Matear, R., Monfray, P., Mouchet, A., Najjar, R.G., Plattner, G.K., Rodgers, K.B., Sabine, C.L., Sarmiento, J.L., Schlitzer, R., Slater, R.D., Totterdell, I.J., Weirig, M.F., Yamanaka, Y., Yool, A., 2005. Anthropogenic ocean acidification over the twenty-first century and its impact on calcifying organisms. Nature 437, 681–686.

Overpeck, J.T., Otto-Bliesner, B.L., Miller, G.H., Muhs, D.R., Alley, R.B., Kiehl, J.T., 2006. Paleoclimatic evidence for future ice-sheet instability and rapid sea-level rise. Science 311, 1747–1750.

Pagani, M., Liu, Z., LaRiviere, J., Ravelo, A.C., 2010. High Earth-system climate sensitivity determined from Pliocene carbon dioxide concentrations. Nature Geoscience 3, 27–30.

Park, J., Royer, D.L., 2011. Geologic constraints on the glacial amplification of Phanerozoic climate sensitivity. American Journal of Science 311, 1–26.

Petit, J.R., Jouzel, J., Raynaud, D., Barkov, N.I., Barnola, J.M., Basile, I., Bender, M., Chappellaz, J., Davis, J., Delaygue, G., Delmotte, M., Kotlyakov, V.M., Legrand, M., Lipenkov, V.M., Lorius, C., Pépin, L., Ritz, C., Saltzman, E., Stievenard, M., 1999. Climate and atmospheric history of the past 420,000 years from the Vostok ice core, Antarctica. Nature 399, 429–436.

Price, S.J., Ford, J.R., Cooper, A.H., Neal, C., 2011. Humans as major geological and geomorphological agents in the Anthropocene: The significance of artificial ground in Great Britain. Philosophical Transactions of the Royal Society, Series A 369, 1056–1084.

Rahmstorf, S., 2007. A semi-empirical approach to projecting future sea-level rise. Science 315, 368–370.

Rockström, J., Steffen, W., Noone, K., Persson, Å, Chapin III, F.S., Lambin, E.F., Lenton, T.M., Scheffer, M., Folke, C., Schellnhuber, H.J., Nykvist, B., de Wit, C.A., Hughes, T., van der Leeuw, S., Rodhe, H., Sörlin, S., Snyder, P.K., Costanza, R., Svedin, U., Falkenmark, M., Karlberg, L., Corell, R.W., Fabry, V.J., Hansen, J., Walker, B., Liverman, D., Richardson, K., Crutzen, P., Foley, J.A., 2009. A safe operating space for humanity. Nature 461, 472–475.

Ruddiman, W.F., 2003. The anthropogenic Greenhouse Era began thousands of years ago. Climate Change 61, 261–293.

Savard, M.M., Bégin, C., Smirnoff, A., Marion, J., Rioux-Paquette, E., 2009. Tree-ring nitrogen isotopes reflect anthropogenic NOx emissions and climatic effects. Environmental Science and Technology 43, 604–609.

Schneider, S.H., 2004. Abrupt non-linear climate change, irreversibility and surprise. Global Environmental Change 14, 245–258.

Schwikowski, M., 2004. Reconstruction of European air pollution from Alpine ice cores. In: DeWayne Cecil, L., Green, J.R., Thompson, L.G. (Eds.), Earth Palaeoenvironments: Records Preserved in Mid- and Low-Latitude Glaciers. Developments in Palaeoenvironmental Research, 9, pp. 95–119.

SEBI (Streamlining European Biodiversity Indicators), 2010. Invasive alien species in Europe, assessment published May 2010. European Environmental Agency, Permalink: e14916abb58edb19479f670844972831.

Steffen, W., Sanderson, A., Tyson, P.D., Jaeger, J., Matson, P.A., Moore III, B., Oldfield, F., Richardson, K., Schnellnhuber, H.J.,

Turner, B.L., Wasson, R.J., 2004. Global Change and the Earth System: a Planet under Pressure. Springer, Berlin, p. 336.

Steffen, W., Crutzen, P.J., McNeill, J.R., 2007. The Anthropocene: Are humans now overwhelming the great forces of Nature? Ambio 36, 614–621.

Steffen, W., Grinevald, J., Crutzen, P., McNeill, J., 2011a. The Anthropocene: Conceptual and historical perspectives. Philosophical Transactions of the Royal Society, Series A 369, 842–867.

Steffen, W., Persson Å., Deutsch, L., Zalasiewicz, J., Williams, M., Richardson, K., Crumley, C., Crutzen, P., Folke, C., Gordon, L., Molina, M., Ramanathan, V., Rockström, J., Scheffer, M., Schellnhuber, J., Svedin, U., 2011b The Anthropocene: From global change to planetary stewardship. Ambio, 40, 739–761.

Stocker, T.S., Strassmann, K., Joos, F., 2010. Sensitivity of Holocene CO_2 and the modern carbon budget to early human land use: Analysis with a process-based model. Biogeosciences Discussions 7, 921–952.

Stoppani, A., 1873. Corsa di Geologia. Bernardoni and Brigola, Milan.

Syvitski, J.P.M., Kettner, A., 2011. Sediment flux and the Anthropocene. Philosophical Transactions of the Royal Society, Series A 369, 957–975.

Syvitski, J.P.M., Vörösmarty, C.J., Kettner, A.J., Green, P., 2005. Impact of humans on the flux of terrestrial sediment to the global coastal ocean. Science 308, 376–380.

Thomas, C.D., Cameron, A., Green, R.E., Bakkenes, M., Beaumont, L.J., Collingham, Y.C., Erasmus, B.F., De Siqueira, M.F., Grainger, A., Hannah, L., Hughes, L., Huntley, B., Van Jaarsveld, A.S., Midgley, G.F., Miles, L., Ortega–Huerta, M.A., Peterson, A.T., Phillips, O.L., Williams, S.E., 2004. Extinction risk from climate change. Nature 427, 145–148.

Thunnell, R.C., Tappa, E., Anderson, D.M., 1995. Sediment fluxes and varve formation in Santa Barbara Basin, offshore California. Geology 23, 1083–1086.

Tickell, C., 2011. Societal responses to the Anthropocene. Philosophical Transactions of the Royal Society, Series A 369, 926–932.

Tilman, D., May, R.M., Lehman, C.L., Nowak, M.A., 1994. Habitat destruction and the extinction debt. Nature 371, 65–66.

Tuterstky, M.R., Manning, W., Kelman Wieder, R., 2004. Dating recent peat deposits. Wetlands 24, 324–356.

Tyrrell, T., 2011. Anthropogenic modification of the oceans. Philosophical Transactions of the Royal Society, Series A 369, 887–908.

Tyrrell, T., Shepherd, J.S., Castle, C., 2007. The long-term legacy of fossil fuels. Tellus B 59, 664–672.

Vane, C.H., Chenery, S.R., Harrison, I., Kim, A.W., Moss-Hayes, V., Jones, D.G., 2011. Chemical signatures of the Anthropocene in the Clyde estuary, UK: Sediment-hosted $^{207}Pb/^{206}Pb$, total petroleum hydrocarbon and polychlorinated biphenyl pollution records. Philosophical Transactions of the Royal Society, Series A 369, 1085–1111.

Vernadsky, V.I., 1945 (2005 translated reprint). Some words about the Noösphere. 21st Century, Spring 2005, 16–21.

Vidas, D., 2011. The Anthropocene and the international law of the sea. Philosophical Transactions of the Royal Society, Series A 369, 909–925.

Walker, M., Johnsen, S., Rasmussen, S.O., Steffensen, J.P., Popp, T., Gibbard, P., Hoek, W., Lowe, J., Björck, S., Cwynar, L., Hughen, K., Kershaw, P., Kromer, B., Litt, T., Lowe, D.J., Nakagawa, T., Newnham, R., Schwander, J., 2009. Formal definition and dating of the GSSP (Global Stratotype Section and Point) for the base of the Holocene using the Greenland NGRIP ice core, and selected auxiliary records. Journal of Quaternary Science 24, 3–17.

Walsh, B., 2012. Nature is over. Time, 12 March 2012, 56–59.

Wilkinson, B.H., 2005. Humans as geologic agents: A deep-time perspective. Geology 33, 161–164.

Williams, M., Zalasiewicz, J., Haywood, A., Ellis, M. (Eds.), 2011. The Anthropocene: A new epoch of geological time? Philosophical Transactions of the Royal Society, Series A. 369, 833–1112.

Wilson, E.O., 2002. The Future of Life. Alfred A. Knopf, New York, p. 256.

Worldwatch Institute, 2009. State of the World 2009: Into a Warming World. W.W. Norton, New York, p. 304.

Worm, B., Barbier, E.B., Beaumont, N., Duffy, J.E., Folke, C., Halpern, B.S., Jackson, J.B.C., Lotzel, H.K., Micheli, F., Palumbi, S.R., Sala, E., Selkoe, K.A., Stachowicz, J.J., Watson, R., 2006. Impacts of biodiversity loss on ocean ecosystem services. Science 314, 787–790.

Wright, R.F., Schindler, D.W., 2004. Interaction of acid rain and global changes: Effects on terrestrial and aquatic ecosystems. Water, Air and Soil Pollution 85, 85–99.

Zachos, J.C., Röhl, U., Schellenberg, S.A., Sluijs, A., Hodell, D.A., Kelly, D.C., Thomas, E., Nicolo, M., Raffi, I., Lourens, L.J., McCarren, H., Kroon, D., 2005. Rapid acidification of the ocean during the Paleocene-Eocene thermal maximum. Science 308, 1611–1615.

Zalasiewicz, J., Williams, M., Smith, A., Barry, T.L., Bown, P.R., Rawson, P., Brenchley, P., Cantrill, D., Coe, A.E., Cope, J.C.W., Gale, A., Gibbard, P.L., Gregory, F.J., Hounslow, M., Knox, R., Powell, P., Waters, C., Marshall, J., Oates, Stone, P., 2008. Are we now living in the Anthropocene? GSA Today 18 (2), 4–8.

Zalasiewicz, J., Williams, M., Steffen, W., Crutzen, P., 2010. The New World of the Anthropocene. Environmental Science and Technology 44, 2228–2231.

Zeebe, R.E., Zachos, J.C., Caldeira, K., Tyrrell, T., 2008. Carbon emissions and acidification. Science 321, 51–52.

Appendix 1

Color Codes for Geological Timescales

CMYK Color Code according to the Commission for the Geological Map of the World (CGMW), Paris, France

Eon	Era	Period	Epoch	Age
Phanerozoic (40/0/5/0)	Cenozoic (5/0/90/0)	Quaternary (0/0/50/0)	Holocene (0/5/10/0)	(0/5/5/0)
			Pleistocene (0/5/30/0)	Upper (0/5/15/0)
				Middle (0/5/20/0)
				Lower (0/5/25/0)
				Gelasian (0/5/35/0)
		Neogene (0/10/90/0)	Pliocene (0/0/40/0)	Piacenzian (0/0/25/0)
				Zanclean (0/0/30/0)
			Miocene (0/0/100/0)	Messinian (0/0/55/0)
				Tortonian (0/0/60/0)
				Serravallian (0/0/65/0)
				Langhian (0/0/70/0)
				Burdigalian (0/0/75/0)
				Aquitanian (0/0/80/0)
		Paleogene (0/40/60/0)	Oligocene (0/25/45/0)	Chattian (0/10/30/0)
				Rupelian (0/15/35/0)
			Eocene (0/30/50/0)	Priabonian (0/20/30/0)
				Bartonian (0/25/35/0)
				Lutetian (0/30/40/0)
				Ypresian (0/35/45/0)
			Paleocene (0/35/55/0)	Thanetian (0/25/50/0)
				Selandian (0/25/55/0)
				Danian (0/30/55/0)
	Mesozoic (60/0/10/0)	Cretaceous (50/0/75/0)	Upper (35/0/75/0)	Maastrichtian (5/0/45/0)
				Campanian (10/0/50/0)
				Santonian (15/0/55/0)
				Coniacian (20/0/60/0)
				Turonian (25/0/65/0)
				Cenomanian (30/0/70/0)
			Lower (45/0/70/0)	Albian (20/0/40/0)
				Aptian (25/0/45/0)
				Barremian 30/0/50/0)
				Hauterivian (35/0/55/0)
				Valanginian (40/0/60/0)
				Berriasian (45/0/65/0)

Eon	Era	Period	Subperiod	Epoch	Age
Phanerozoic (40/0/5/0)	Mesozoic (60/0/10/0)	Jurassic (80/0/5/0)		Upper (30/0/0/0)	Tithonian (15/0/0/0)
					Kimmeridgian (20/0/0/0)
					Oxfordian (25/0/0/0)
				Middle (50/0/5/0)	Callovian (25/0/5/0)
					Bathonian (30/0/5/0)
					Bajocian (35/0/5/0)
					Aalenian (40/0/5/0)
				Lower (75/5/0/0)	Toarcian (40/5/0/0)
					Pliensbachian (50/5/0/0)
					Sinemurian (60/5/0/0)
					Hettangian (70/5/0/0)
		Triassic (50/80/0/0)		Upper (25/40/0/0)	Rhaetian (10/25/0/0)
					Norian (15/30/0/0)
					Carnian (20/35/0/0)
				Middle (30/55/0/0)	Ladinian (20/45/0/0)
					Anisian (25/50/0/0)
				Lower (40/75/0/0)	Olenekian (30/65/0/0)
					Induan (35/70/0/0)
	Paleozoic (40/10/40/0)	Permian (5/75/75/0)		Lopingian (0/35/30/0)	Changhsingian (0/25/20/0)
					Wuchiapingian (0/30/25/0)
				Guadalupian (0/55/50/0)	Capitanian (0/40/35/0)
					Wordian (0/45/40/0)
					Roadian (0/50/45/0)
				Cisuralian (5/65/60/0)	Kungurian (10/45/40/0)
					Artinskian (10/50/45/0)
					Sakmarian (10/55/50/0)
					Asselian (10/60/55/0)
		Carboniferous (60/15/30/0)	Pennsylvanian (40/10/20/0)	Upper (25/10/20/0)	Gzhelian (20/10/15/0)
					Kasimovian (25/10/15/0)
				Middle (35/10/20/0)	Moscovian (30/10/20/0)
				Lower (45/10/20/0)	Bashkirian (40/10/20/0)
			Mississippian (60/25/55/0)	Upper (30/15/55/0)	Serpukhovian (25/15/55/0)
				Middle (40/15/55/0)	Visean (35/15/55/0)
				Lower (50/15/55/0)	Tournaisian (45/15/55/0)

Eon	Era	Period	Epoch	Age
Phanerozoic (40/0/5/0)	Paleozoic (40/10/40/0)	Devonian (20/40/75/0)	Upper (5/10/35/0)	Famennian (5/5/20/0)
				Frasnian (5/5/30/0)
			Middle (5/20/55/0)	Givetian (5/10/45/0)
				Eifelian (5/15/50/0)
			Lower (10/30/65/0)	Emsian (10/15/50/0)
				Pragian (10/20/55/0)
				Lochkovian (10/25/60/0)
		Silurian (30/0/25/0)	Pridoli (10/0/10/0)	(10/0/10/0)
			Ludlow (25/0/15/0)	Ludfordian (15/0/10/0)
				Gorstian (20/0/10/0)
			Wenlock (30/0/20/0)	Homerian (20/0/15/0)
				Sheinwoodian (25/0/20/0)
			Llandovery (40/0/25/0)	Telychian (25/0/15/0)
				Aeronian (30/0/20/0)
				Rhuddanian (35/0/25/0)
		Ordovician (100/0/60/0)	Upper (50/0/40/0)	Hirnantian (35/0/30/0)
				Katian (40/0/35/0)
				Sandbian (45/0/40/0)
			Middle (70/0/50/0)	Darriwilian (55/0/35/0)
				Dapingian (60/0/40/0)
			Lower (90/0/60/0)	Floian (75/0/45/0)
				Tremadocian (80/0/50/0)
		Cambrian (50/20/65/0)	Furongian (30/0/40/0)	*Stage 10* (10/0/20/0)
				Jiangshanian (15/0/25/0)
				Paibian (20/0/30/0)
			Series 3 (35/5/45/0)	Guzhangian (20/5/30/0)
				Drumian (25/5/35/0)
				Stage 5 (30/5/40/0)
			Series 2 (40/10/50/0)	*Stage 4* (30/10/40/0)
				Stage 3 (35/10/45/0)
			Terreneuvian (45/15/55/0)	*Stage 2* (35/15/45/0)
				Fortunian (40/15/50/0)

Color composition by J.M. Pellé (BRGM, France)

Supereon	Eon	Era	Period
Precambrian (0/75/30/0)	Proterozoic (0/80/35/0)	Neoproterozoic (0/30/70/0)	Ediacaran (0/15/55/0)
			Cryogenian (0/20/60/0)
			Tonian (0/25/65/0)
		Mesoproterozoic (0/30/55/0)	Stenian (0/15/35/0)
			Ectasian (0/20/40/0)
			Calymmian (0/25/45/0)
		Paleoproterozoic (0/75/30/0)	Statherian (0/55/10/0)
			Orosirian (0/60/15/0)
			Rhyacian (0/65/20/0)
			Siderian (0/70/25/0)
	Archean (0/100/0/0)	Neoarchean (0/40/5/0)	(0/35/5/0)
		Mesoarchean (0/60/5/0)	(0/50/5/0)
		Paleoarchean (0/75/0/0)	(0/60/0/0)
		Eoarchean (10/100/0/0)	(5/90/0/0)
	Hadean (30/100/0/0)		

The CMYK color code is an additive model with percentages of Cyan, Magenta, Yellow and Black. For example: the CMYK color for Devonian (20/40/75/0) is a mixture of 20% Cyan, 40% Magenta, 75% Yellow and 0% Black. The CMYK values are the primary reference system for designating the official colors for these geological units.

FIGURE A1.1 **CMYK Color Code according to the Commission for the Geological Map of the World (CGMW), Paris, France.**

RGB Color Code according to the Commission for the Geological Map of the World (CGMW), Paris, France

Eon	Era	Period	Epoch	Stage	RGB
Phanerozoic 154/217/221	Cenozoic 242/249/29	Quaternary 249/249/127	Holocene 254/242/224		254/242/236
			Pleistocene 255/242/174	Upper	255/242/211
				Middle	255/242/199
				Lower	255/242/186
				Gelasian	255/237/179
		Neogene 255/230/25	Pliocene 255/255/153	Piacenzian	255/255/191
				Zanclean	255/255/179
			Miocene 255/255/0	Messinian	255/255/115
				Tortonian	255/255/102
				Serravallian	255/255/89
				Langhian	255/255/77
				Burdigalian	255/255/65
				Aquitanian	255/255/51
		Paleogene 253/154/82	Oligocene 253/192/122	Chattian	254/230/170
				Rupelian	254/217/154
			Eocene 253/180/108	Priabonian	253/205/161
				Bartonian	253/192/145
				Lutetian	252/180/130
				Ypresian	252/167/115
			Paleocene 253/167/95	Thanetian	253/191/111
				Selandian	254/191/101
				Danian	253/180/98
	Mesozoic 103/197/202	Cretaceous 127/198/78	Upper 166/216/74	Maastrichtian	242/250/140
				Campanian	230/244/127
				Santonian	217/239/116
				Coniacian	204/233/104
				Turonian	191/227/93
				Cenomanian	179/222/83
			Lower 140/205/87	Albian	204/234/151
				Aptian	191/228/138
				Barremian	179/223/127
				Hauterivian	166/217/117
				Valanginian	153/211/106
				Berriasian	140/205/96

Eon	Era	Period	Subperiod	Epoch	Stage	RGB
Phanerozoic 154/217/221	Mesozoic 103/197/202	Jurassic 52/178/201		Upper 179/227/238	Tithonian	217/241/247
					Kimmeridgian	204/236/244
					Oxfordian	191/231/241
				Middle 128/207/216	Callovian	191/231/229
					Bathonian	179/226/227
					Bajocian	166/221/224
					Aalenian	154/217/221
				Lower 66/174/208	Toarcian	153/206/227
					Pliensbachian	128/197/221
					Sinemurian	103/188/216
					Hettangian	78/179/211
		Triassic 129/43/146		Upper 189/140/195	Rhaetian	227/185/219
					Norian	214/170/211
					Carnian	201/155/203
				Middle 177/104/177	Ladinian	201/131/191
					Anisian	188/117/183
				Lower 152/57/153	Olenekian	176/81/165
					Induan	164/70/159
	Paleozoic 153/192/141	Permian 240/64/40		Lopingian 251/167/148	Changhsingian	252/192/178
					Wuchiapingian	252/180/162
				Guadalupian 251/116/92	Capitanian	251/154/133
					Wordian	251/141/118
					Roadian	251/128/105
				Cisuralian 239/88/69	Kungurian	227/135/118
					Artinskian	227/123/104
					Sakmarian	227/111/92
					Asselian	227/99/80
		Carboniferous 103/165/153	Pennsylvanian 153/194/181	Upper 191/208/186	Gzhelian	204/212/199
					Kasimovian	191/208/197
				Middle 166/199/183	Moscovian	199/203/185
				Lower 140/190/180	Bashkirian	153/194/181
			Mississippian 103/143/102	Upper 179/190/108	Serpukhovian	191/194/107
				Middle 153/180/108	Visean	166/185/108
				Lower 128/171/108	Tournaisian	140/176/108

Eon	Era	Period	Epoch	Stage	RGB
Phanerozoic 154/217/221	Paleozoic 153/192/141	Devonian 203/140/55	Upper 241/225/157	Famennian	242/237/197
				Frasnian	242/237/173
			Middle 241/200/104	Givetian	241/225/133
				Eifelian	241/213/118
			Lower 229/172/77	Emsian	229/208/117
				Pragian	229/196/104
				Lochkovian	229/183/90
		Silurian 179/225/182	Pridoli 230/245/225		230/245/225
			Ludlow 191/230/207	Ludfordian	217/240/223
				Gorstian	204/236/221
			Wenlock 179/225/194	Homerian	204/235/209
				Sheinwoodian	191/230/195
			Llandovery 153/215/179	Telychian	191/230/207
				Aeronian	179/225/194
				Rhuddanian	166/220/181
		Ordovician 0/146/112	Upper 127/202/147	Hirnantian	166/219/171
				Katian	153/214/159
				Sandbian	140/208/148
			Middle 77/180/126	Darriwilian	116/198/156
				Dapingian	102/192/146
			Lower 26/157/111	Floian	65/176/135
				Tremadocian	51/169/126
		Cambrian 127/160/86	Furongian 179/224/149	*Stage 10*	230/245/201
				Jiangshanian	217/240/187
				Paibian	204/235/174
			Series 3 166/207/134	Guzhangian	204/223/170
				Drumian	191/217/157
				Stage 5	179/212/146
			Series 2 153/192/120	*Stage 4*	179/202/142
				Stage 3	166/197/131
			Terreneuvian 140/176/108	*Stage 2*	166/186/128
				Fortunian	153/181/117

Color composition by J.M. Pellé (BRGM, France)

Supereon	Eon	Era	Period	RGB
Precambrian 247/67/112	Proterozoic 247/53/99	Neoproterozoic 254/179/66	Ediacaran	254/217/106
			Cryogenian	254/204/92
			Tonian	254/191/78
		Mesoproterozoic 253/180/98	Stenian	254/217/154
			Ectasian	253/204/138
			Calymmian	253/192/122
		Paleoproterozoic 247/67/112	Statherian	248/117/167
			Orosirian	247/104/152
			Rhyacian	247/91/137
			Siderian	247/79/124
	Archean 240/4/127	Neoarchean 249/155/193		250/167/200
		Mesoarchean 247/104/169		248/129/181
		Paleoarchean 244/68/159		246/104/178
		Eoarchean 218/3/127		230/29/140
	Hadean 174/2/126			

The RGB color code is an additive model of Red, Green and Blue. Each is indicated on a scale from 0 (no pigment) to 255 (saturation of this pigment). "Devonian (203/140/205)" indicates a mixture of 203 Red, 140 Green and 205 Blue.

The conversion from the reference CMYK values to the RGB codes utilizes Adobe® Illustrator® CS3's color function of "Emulate Adobe® Illustrator® 6.0" (menu Edit / Color Settings / Settings).

ATTENTION: For color conversions using a program other than Adobe® Illustrator®, it is necessary to conserve the reference CMYK, even if the resulting RGB values are slightly different.

FIGURE A1.2 **RGB Color Code according to the Commission for the Geological Map of the World (CGMW), Paris, France.**

Appendix 2

Radiometric ages used in GTS2012

M.D. SCHMITZ

No.	Period	Epoch	Age	Sample	Locality	Formation	GTS2012 Age (Ma)	± 2σ (without λ errors)	± 2σ (with λ errors)	Type
	Quaternary -- not compiled									
	Neogene -- partial list									
		Pliocene								
		Miocene								
			Messinian							
			Tortonian							
N4				GiD-3	Monte Gibliscemi section, near Mazzarino, Sicily		11.55	± 0.02	± 0.02	$^{40}Ar/^{39}Ar$
			Langhian							
N3				GiF-1	Monte Gibliscemi section, near Mazzarino, Sicily		11.97	± 0.04	± 0.04	$^{40}Ar/^{39}Ar$
			Serravallian							
N2				A2 ash layer	La Vedova section, between Ancona and Il Trave, Adriatic coast, Italy	Schlier Formation, 4.4 m above A0	14.86	± 0.02	± 0.02	$^{40}Ar/^{39}Ar$
N1				A0 ash layer	La Vedova section, between Ancona and Il Trave, Adriatic coast, Italy	Schlier Formation	14.97	± 0.02	± 0.02	$^{40}Ar/^{39}Ar$
			Burdigalian							
			Aquitanian							
	Paleogene									
		Oligocene								
			Chattian							
Pg21				MCA98-7	208.3 mab, Monte Cagnero section, northeastern Apennines, Italy	Scaglia Cinerea Formation	27.0	± 0.1	± 0.1	$^{40}Ar/^{39}Ar$
			Rupelian							
Pg20				MCA/84-3	145.5 mab, Monte Cagnero section, northeastern Apennines, Italy	Scaglia Cinerea Formation	31.8	± 0.2	± 0.2	$^{40}Ar/^{39}Ar$
		Eocene								
			Priabonian							
Pg19				MAS/86-14.7	Massignano, near Ancona, northeastern Apennines, Italy	Scaglia Cinerea Formation	34.4	± 0.2	± 0.2	$^{40}Ar/^{39}Ar$
Pg18				MAS/86-12.7	Massignano, near Ancona, northeastern Apennines, Italy	Scaglia Cinerea Formation	35.2	± 0.2	± 0.2	$^{40}Ar/^{39}Ar$
			Bartonian							
			Lutetian							
Pg17				Mission Valley ash, SDSNH Loc. 3428	La Mesa, California	terrestrial facies of Mission Valley Formation	43.35	± 0.50	± 0.50	$^{40}Ar/^{39}Ar$
Pg16				DSDP Hole 516F	DSDP Hole 516F		46.24	± 0.50	± 0.50	$^{40}Ar/^{39}Ar$
			Ypresian							
Pg15				68-0-51, 3497; Blue Point Marker ash	Two-Ocean Plateau and Irish Rock locales, Absaroka volcanic province, western USA	overlies Aycross Formation	48.41	± 0.21	± 0.21	$^{40}Ar/^{39}Ar$
Pg14				CP-1; Continental Peak tuff	Bridger Basin, western USA	Bridger Formation	48.96	± 0.28	± 0.33	$^{40}Ar/^{39}Ar$
Pg15				TR-5; Sixth tuff	Bridger Basin, western USA	Wilkins Peak Member	49.92	± 0.10	± 0.21	$^{40}Ar/^{39}Ar$
Pg14				TR-6; Layered tuff	Bridger Basin, western USA	Wilkins Peak Member	50.11	± 0.09	± 0.20	$^{40}Ar/^{39}Ar$
Pg!3				Willwood ash	Bighorn Basin, western USA	Willwood Formation	52.93	± 0.23	± 0.23	$^{40}Ar/^{39}Ar$
Pg12				Ash-17	DSDP Site 550, north Atlantic		55.48	± 0.12	± 0.12	$^{40}Ar/^{39}Ar$
Pg11				SB01-01 bentonite	Longyearbyen section, Spitsbergen, Svalbard Archipeligo	10.9 m above base of the Gilsonryggen Member, Frysjaodden Formation, within the PETM carbon isotope excursion	55.785	± 0.034	± 0.075	$^{206}Pb/^{238}U$

Primary radiometric age details	Zone assignment	Primary biostratigraphic age details	Reference
Weighted mean laser fusion age for ten (of 10) multigrain aliquots of feldspar, using FCs fluence monitor.	earliest Tortonian	at *Globigerinoides subquadratus* LCO	Kuiper et al., 2005
Weighted mean laser fusion age for nine (of 10) multigrain aliquots of feldspar,using FCs fluence monitor.	latest Serravallian	below *Discoaster kuglieri* FCO	Kuiper et al., 2005
Weighted mean laser fusion age for seven (of 8) multigrain aliquots of biotite, using FCs fluence monitor.		near Influx E of *G. dehiscens* and Acme b B of *P. siakensis*	Hüsing et al., 2010
Weighted mean laser fusion age for seven (of 9) multigrain aliquots of biotite, using FCs fluence monitor.		near Influx E of *G. dehiscens* and Acme b B of *P. siakensis*	Hüsing et al., 2010
Weighted mean of four biotite incremental heating plateau ages, originally calibrated using FCs = 27.84 Ma.		lower NP25, upper Zone O5	Coccioni et al., 2008
Weighted mean of two biotite incremental heating plateau ages, originally calibrated using FCs = 27.84 Ma.	lowermost NP23	mid *Globorotalia sellii*	Coccioni et al., 2008
One multi-grain biotite fusion age originally calibrated to LP-6 biotite = 127.7 Ma.	uppermost NP16, lower CP16a	uppermost NP16, lower CP16a	Odin et al., 1991; Hilgen and Kuiper, 2010
One multi-grain biotite fusion age originally calibrated to LP-6 biotite = 127.7 Ma.	upper NP16, upper CP15b	upper NP16, upper CP15b	Odin et al., 1991; Hilgen and Kuiper, 2010
Laser fusion single-crystal sanidine analyses, originally calibrated to TCs = 28.32 Ma; no analytical data published.	NP16, CP14a	coccoliths *Reticulofenestra umbilica* (Levin) and *Discoaster distinctus* Martini; magnetostratigraphic control	Obradovich, unpublished data cited in Walsh et al., 1996
Laser fusion of biotite and sanidine separates, no analytical data published.	NP15, CP10	NP15, CP10; magnetostratigraphic control	Swisher and Montanari, in prep, cited in Berggren et al. 1995 (postscript)
Weighted mean of multi-grain feldspar (46.8% of gas) and multi-grain hornblende (98.7% of gas) incremental heating plateau ages, originally calibrated using FCs = 27.84 Ma.		magnetostratigraphic control	Hiza, 1999; Flynn, 1986
Weighted mean age of 12 (of 16) laser fusion analyses on small multi-grain aliquots of sanidine, using FCs as fluence monitor.		magnetostratigraphic control	Smith et al., 2010, 2008; Clyde et al., 2001
Weighted mean age of 119 (of 142) incremental heating single and multi-grain biotite plateau ages; corroborated by laser fusion analyses on single and multi-grain aliquots of sanidine, using FCs as fluence monitor.		magnetostratigraphic control	Smith et al., 2010, 2008, 2006; Machlus et al., 2004; Clyde et al., 1997, 2001
Weighted mean age of 64 (of 73) laser fusion analyses on small multi-grain aliquots of sanidine, using FCs as fluence monitor.		magnetostratigraphic control	Smith et al., 2010, 2008, 2006; Clyde et al., 1997, 2001
Weighted mean age of 16 laser-fusion and 4 five-step laser incremental heating analyses on 1-3 crystal aliquots of sanidine, using FCs as fluence monitor.		Lystitean-Lostcabinian (Wa6-Wa7) North American land mammal age (NALMA) substage boundary; magnetostratigraphic control	Smith et al., 2004; Wing et al., 1991; Clyde et al., 1994, Tauxe et al., 1994
Weighted mean age of 38 laser fusion analyses on single and multi-grain aliquots of sanidine, originally calibrated to FCs = 28.02 Ma.		above the PETM carbon isotope excursion	Storey et al., 2007
Weighted mean age of five (of 13) single zircon crystal analyses, utilizing CA-TIMS and the ET535 spike (0.05% error in tracer calibration).		within the PETM carbon isotope excursion	Charles et al., 2011

No.	Period	Epoch	Age	Sample	Locality	Formation	GTS2012 Age (Ma)	± 2σ (without λ errors)	± 2σ (with λ errors)	Type
		Paleocene								
			Thanetian							
			Selandian							
Pg10				Belt Ash	Southeast Polecat Bench section, northern Bighorn Basin, Wyoming, USA	Silver Coulee beds, Fort Union Formation	59.39	± 0.30	± 0.30	$^{40}Ar/^{39}Ar$
			Danian							
Pg9				U-coal bentonite	Biscuit Butte, Garfield County, eastern Montana	U-coal, capping the Farrand Channel in the Paleocene Tullock Formation, overlying the dinosaur-bearing Hell Creek Formation	64.73	± 0.08	± 0.12	$^{40}Ar/^{39}Ar$
Pg8				W-coal bentonite	west side of road to Brownie Butte, Garfield County, eastern Montana	W-coal, capping the Garbani Channel in the Paleocene Tullock Formation, overlying the dinosaur-bearing Hell Creek Formation	64.94	± 0.04	± 0.12	$^{40}Ar/^{39}Ar$
Pg7				HFZ-coal bentonite	Iridium Hill site, Hauso Flats, Garfield county, eastern Montana	upper part of the HFZ-Coal, 18 m above the base of the Paleocene Tullock Formation	65.61	± 0.12	± 0.15	$^{40}Ar/^{39}Ar$
Pg6				Z-coal bentonite	Hell Creek State Park site, Garfield County, and McQuire Creek site, McCone County, eastern Montana	Two bentonites in the Z-Coal of the Paleocene Tullock Formation, overlying the dinosaur-bearing Hell Creek Formation	66.28	± 0.14	± 0.18	$^{40}Ar/^{39}Ar$
Pg5				Z-coal bentonite	Hell Creek State Park site, Garfield County, and McQuire Creek site, McCone County, eastern Montana	Lower of two bentonites in the Z-Coal of the Paleocene Tullock Formation, overlying the dinosaur-bearing Hell Creek Formation	65.84	± 0.06	± 0.12	$^{40}Ar/^{39}Ar$
Pg4				IrZ-coal bentonite	Iridium Hill site, Hauso Flats, Garfield county, eastern Montana.	IrZ-Coal bentonite of the Paleocene Tullock Formation, overlying the dinosaur-bearing Hell Creek Formation	65.99	± 0.08	± 0.12	$^{40}Ar/^{39}Ar$
Pg3				Haitian tektites	Beloc, Haiti	from 0.5-thick marl bed in lower part of the marine limestone Beloc Formation	65.920	± 0.14	± 0.18	$^{40}Ar/^{39}Ar$
Pg2				Haitian tektites	Beloc, Haiti	between levels f and g, in lower part of the marine limestone Beloc Formation	65.84	± 0.16	± 0.18	$^{40}Ar/^{39}Ar$
Pg1				C1 glassy melt rock	Chicxulub crater	C1 glassy melt rock	65.81	± 0.10	± 0.14	$^{40}Ar/^{39}Ar$
	Cretaceous									
		Late								
			Maastrichtian							
K71				AK-476	Strawberry Creek, Alberta	Kneehills Tuff, 18-cm bentonite, 1.8 m above the upper Kneehills Tuff zone	67.29	± 1.11	± 1.11	$^{40}Ar/^{39}Ar$
K70				91-O-14	Red Bird section, Niobrara County, Wyoming	30 cm bentonite bed from unit 112, Red Bird section (Gill & Cobban, 1966a), sec. 14, T. 38 N, R. 62 W	70.08	± 0.37	± 0.37	$^{40}Ar/^{39}Ar$
K69				92-O-32	Red Bird section, Niobrara County, Wyoming	1.40 m bentonite bed from unit 97, Red Bird section (Gill & Cobban, 1966a), sec. 14, T. 38 N, R. 62 W	70.66	± 0.65	± 0.66	$^{40}Ar/^{39}Ar$
			Campanian							
K68				unnamed bentonite	unknown	unknown	72.50	± 0.31	± 0.31	$^{40}Ar/^{39}Ar$
K67				Snakebite 1 bentonite	Cruikshank Coolee, north of Herbert, southwestern Saskatchewan	Snakebite Member of Bearpaw Formation	73.41	± 0.47	± 0.47	$^{40}Ar/^{39}Ar$
K66				90-O-15, bentonite	Big Horn County, Montana.	Bearpaw Shale -- 22-cm-thick bentonite 1.5 m above base of zone in the Bearpaw Shale, NE, NE, sec. 14, T. 1 N, R. 33 E	74.05	± 0.39	± 0.39	$^{40}Ar/^{39}Ar$
K65				93-O-16, bentonite	Montrose County, Colorado	Mancos Shale -- Bentonite bed in upper part, NE corner, SE, SE, sec. 17, T. 48 N, R. 7 W	74.85	± 0.43	± 0.43	$^{40}Ar/^{39}Ar$
K64				86-O-05, bentonite	Foreman, Little River County, Arkansas	Anonna Formation, Foreman Quarry	75.92	± 0.39	± 0.39	$^{40}Ar/^{39}Ar$
K63				92-O-13, bentonite	Rio Arriba County, New Mexico	Lewis Shale -- 15 cm bentonite (possibly equivalent to the Huerfanito bentonite marker bed in the subsurface) CW, sec. 11, T. 23 N, R. 1 W	76.62	± 0.51	± 0.51	$^{40}Ar/^{39}Ar$
K62				Judith River Bentonite	Elk Basin (northern Bighorn Basin), Wyoming	Judith River Bentonite; 22m below C33r/C33n polarity reversal	80.10	± 0.61	± 0.61	$^{40}Ar/^{39}Ar$
K61				RB92-15, bentonite	Red Bird section, Niobrara County, Wyoming	Ardmore bentonite bed in the base of the Pierre Shale at the Red Bird section, sec. 14, T. 38 N, R. 62 W	80.62	± 0.40	± 0.40	$^{40}Ar/^{39}Ar$
K60				Ardmore bentonite	Elk Basin (northern Bighorn Basin), Wyoming	Ardmore Bentonite; a 4.9 m thick bentonite near the base of the Claggett Shale	81.30	± 0.55	± 0.55	$^{40}Ar/^{39}Ar$
K59				PP-6, upper ash layer	Montagna della Maiella, Valle Tre Grotte, west of Pennapiedimonte, central Apennines, Italy	Tre Grotte Formation	81.67	± 0.21	± 0.26	$^{206}Pb/^{238}U$

Primary radiometric age details	Zone assignment	Primary biostratigraphic age details	Reference
Weighted mean age of 23 laser fusion analyses on 20 grain aliquots of sanidine, originally calibrated to TCs = 28.34 Ma.		magnetostratigraphic control	Secord et al., 2006
Weighted mean age of 10 single crystal sanidine laser fusions from two samples, originally calibrated to FCs = 27.84 Ma.		lower Puercan	Swisher et al., 1993
Weighted mean age of 35 single crystal sanidine laser fusions from two samples, originally calibrated to FCs = 27.84 Ma.		lower Puercan	Swisher et al., 1993
Weighted mean age of 10 single crystal sanidine laser fusions from two samples, originally calibrated to FCs = 27.84 Ma.		lower Puercan	Swisher et al., 1993
Weighted mean age of 33 multi-grain sanidine laser fusions from two samples, originally calibrated to FCs = 27.55 Ma.		base of the Puercan and at the K–P boundary as defined by the highest appearing local occurrences of in situ dinosaur fossils, Cretaceous pollen, and an iridium anomaly	Dalrymple et al., 1993; Izett et al., 1991; Obradovich, 1993
Weighted mean age of 33 single crystal sanidine laser fusions from two samples, originally calibrated to FCs = 27.84 Ma.		base of the Puercan and at the K–P boundary as defined by the highest appearing local occurrences of in situ dinosaur fossils, Cretaceous pollen, and an iridium anomaly	Swisher et al., 1993; Kuiper et al., 2008
Weighted mean age of 9 single crystal sanidine laser fusions, originally calibrated to FCs = 27.84 Ma.		base of the Puercan and at the K–P boundary as defined by the highest appearing local occurrences of in situ dinosaur fossils, Cretaceous pollen, and an iridium anomaly	Swisher et al., 1993; Kuiper et al., 2008
Weighted mean age of 52 laser fusions on single or several tektites, originally calibrated to FCs = 27.55 Ma.		debris from KT impact event	Dalrymple et al., 1993; Izett et al., 1991; Obradovich, 1993
Weighted mean age of 5 plateau ages on single tektites, originally calibrated to FCs = 27.84 Ma.		debris from KT impact event	Swisher et al., 1992
Weighted mean age of 5 plateau ages on glassy (andesitic) rock chips, originally calibrated to FCs = 27.84 Ma.		debris from KT impact event	Swisher et al., 1992
Multi-grain sanidine laser fusion analysis, originally using TCs = 28.32 Ma as fluence monitor; no analytical data published.		*Triceratops* dinosaur zone	Obradovich, 1993
Multi-grain sanidine laser fusion analysis, originally using TCs = 28.32 Ma as fluence monitor.	*Baculites clinolobatus* ammonite zone (uppermost)	*Top of Baculites clinolobatus* ammonite zone	Hicks et al., 1999
Multi-grain sanidine laser fusion analysis, originally using TCs = 28.32 Ma as fluence monitor.	*Baculites grandis ammonite zone.*	*Baculites grandis ammonite zone.*	Hicks et al., 1999
Multi-grain sanidine laser fusion analysis, originally using TCs = 28.32 Ma as fluence monitor; no analytical data published.	*Baculites eliasi ammonite zone.*	*Baculites eliasi ammonite zone.*	Cobban et al., 2006
Weighted mean age of 10 multi-grain sanidine laser fusion analyses, originally using FCs = 27.84 Ma as fluence monitor.	*Baculites reesidei* ammonite zone (uppermost)	top of *Baculites reesidei* ammonite zone	Baadsgaard et al., 1993
Multi-grain sanidine laser fusion analysis, originally using TCs = 28.32 Ma as fluence monitor.	*Baculites compressus* ammonite zone	*Baculites compressus* ammonite zone	Hicks et al., 1999
Multi-grain sanidine laser fusion analysis, originally using TCs = 28.32 Ma as fluence monitor.	*Exiteloceras jenneyi* ammonite zone	*Exiteloceras jenneyi* ammonite zone	Hicks et al., 1999
Multi-grain sanidine laser fusion analysis, originally using TCs = 28.32 Ma as fluence monitor; no analytical data published.		5m above lowest occurrence of foraminifer *Globotruncanita calcarata*	Obradovich, 1993
Multi-grain sanidine laser fusion analysis, originally using TCs = 28.32 Ma as fluence monitor.	*Baculites scotti* ammonite zone	*Baculites scotti* ammonite zone	Hicks et al., 1999
Multi-grain sanidine laser fusion analysis, originally using TCs = 28.32 Ma as fluence monitor.	*Baculites obtusus* to *B. perplexus* ammonite zone	*Baculites obtusus* to *B. perplexus* ammonite zone	Hicks et al., 1995
Multi-grain sanidine laser fusion analysis, originally using TCs = 28.32 Ma as fluence monitor.	*Baculites obtusus* ammonite zone	*Baculites obtusus* ammonite zone	Hicks et al., 1999
Multi-grain sanidine laser fusion analysis, originally using TCs = 28.32 Ma as fluence monitor.	*Baculites obtusus* ammonite zone	*Baculites obtusus* ammonite zone	Hicks et al., 1995
Weighted mean age of four small multi-grain zircon analyses, utilizing physical abrasion and in-house UNIGE spike (0.1% error in tracer calibration).	late Early Campanian (*Sca. hippocrepis III* to *Bac. obtusus* zone interval)	upper part of the *Aspidolithus parcus* (CC 18) calcareous nannoplankton zone; and middle of *Globotruncanita elevata* foraminifer zone	Bernoulli et al., 2004

No.	Period	Epoch	Age	Sample	Locality	Formation	GTS2012 Age (Ma)	± 2σ (without λ errors)	± 2σ (with λ errors)	Type
K58				78-O-05, bentonite	Cat Creek Oil field, Petroleum County, Montana	Eagle Sandstone, thin bentonite in lower part of Scaphites hippocrepis II ammonite zone.	81.87	± 0.15	± 0.25	$^{40}Ar/^{39}Ar$
			Santonian							
K57				TOM001	Tomiuchi section, Hokkaido, Japan	"Tomiuchibashi Tuff", Kashima Formation, Yazo Group	84.90	± 0.20	± 0.25	$^{206}Pb/^{238}U$
K56				MT08-04; 97-O-04, bentonite	East bank of McDonald Creet, Petroleum County, Montana	Telegraph Creek Formation, 10-cm bentonite	84.33	± 0.18	± 0.27	$^{40}Ar/^{39}Ar$
K55				MT08-04, bentonite	East bank of McDonald Creet, Petroleum County, Montana	Telegraph Creek Formation, 10-cm bentonite	84.43	± 0.09	± 0.15	$^{206}Pb/^{238}U$
K54				70-O-06 (OB93-12), bentonite	bentonite mine, south of Aberdeen, Monroe County, Mississippi	Bentonite in Tombigbee Sand Member of Eutaw Formation	84.70	± 0.41	± 0.41	$^{40}Ar/^{39}Ar$
K53				91-O-09, bentonite			84.57	± 0.28	± 0.35	$^{40}Ar/^{39}Ar$
K52				90-O-10, bentonite			85.55	± 0.13	± 0.25	$^{40}Ar/^{39}Ar$
K51				92-O-14, bentonite	creek bottom, 1.1 km NW of Whitewright, eastern Grayson County, Texas	Austin Chalk,91-cm bentonite. Immediately below is a 0.3-0.6-m-thick hard massive chalk with imprints and fragments of inoceramus *Cladoceramus undulatoplicatus*	85.59	± 0.28	± 0.35	$^{40}Ar/^{39}Ar$
			Coniacian							
K50				MT08-1; 91-O-08, bentonite	S. side of Yellowstone river, just south of Billings, Montana	Cody Shale, 30-cm bentonite, 30.5 m above top of Kevin bentonite group in Cody Shale	86.98	± 0.10	± 0.24	$^{40}Ar/^{39}Ar$
K49				MT08-1, bentonite	S. side of Yellowstone river, just south of Billings, Montana	Cody Shale, 30-cm bentonite, 30.5 m above top of Kevin bentonite group in Cody Shale	87.11	± 0.08	± 0.15	$^{206}Pb/^{238}U$
K48				MT08-03; 91-O-13, bentonite	Toole County, Montana	Marias River Shale, 15-cm bentonite	89.30	± 0.16	± 0.27	$^{40}Ar/^{39}Ar$
K47				MT08-03, bentonite	Toole County, Montana	Marias River Shale, 15-cm bentonite	89.37	± 0.07	± 0.15	$^{206}Pb/^{238}U$
			Turonian							
K46				MT-09-09, bentonite			89.86	± 0.13	± 0.26	$^{40}Ar/^{39}Ar$
K45				UT-08-03; SAN JUAN, bentonite	San Juan County, New Mexico	Mancos Shale, thin bentonite in Juana Lopez Member	91.07	± 0.16	± 0.28	$^{40}Ar/^{39}Ar$
K44				UT-08-03; SAN JUAN, bentonite	San Juan County, New Mexico	Mancos Shale, thin bentonite in Juana Lopez Member	91.37	± 0.08	± 0.16	$^{206}Pb/^{238}U$
K43				UT-08-05; 90-O-56, bentonite	Emery Point section, Mesa Butte quadrangle, Emery County, Utah	Ferron Sandstone, bentonite between Clausen and Washboard members	91.15	± 0.13	± 0.26	$^{40}Ar/^{39}Ar$
K42				90-O-34, bentonite	Laholi Point section, Blue Gap, Navajo Indian Reservation, Arizona	Mancos Shale, bentonite BM-17, 0.16-m-thick in lowe calcareous shale member, 25-m above base. Equated to bentonite PCB-20 at top of lower Bridge Creek Limestone Mbr, Pueblo, Colorado; and equivalent bentonite below Bed 97 in USGS #1 Portland core.	93.67	± 0.21	± 0.31	$^{40}Ar/^{39}Ar$
K41				AZLP-08-05 (same bentonite as 90-O-34)	Laholi Point section, Blue Gap, Navajo Indian Reservation, Arizona	Mancos Shale, bentonite BM-17, 0.16-m-thick in lowe calcareous shale member, 25-m above base. Equated to bentonite PCB-20 at top of lower Bridge Creek Limestone Mbr, Pueblo, Colorado; and equivalent bentonite below Bed 97 in USGS #1 Portland core.	94.09	± 0.13	± 0.19	$^{206}Pb/^{238}U$

Primary radiometric age details	Zone assignment	Primary biostratigraphic age details	Reference
Weighted mean age of 35 (of 43) laser fusions on 1-3 grain aliquots of sanidine, using FCs as fluence monitor -- as presented in Siewert (2011); but Siewert et al. (in press) add additional fusions and revise as 81.84 ±0.11/0.22 Ma.	*Scaphites hippocrepis II* ammonite zone	lower *Scaphites hippocrepis II* ammonite zone	Siewert, 2011; Obradovich, 1993
Four (of 5) multigrain chemically-abraded zircon fractions; ID TIMS.	Late Santonian to earliest Campanian?	Local *Globotruncana arca* planktonic zone. Within lower *Inoceramus japonicus* zone, which Walaszczyk and Cobban () cite as "Late Santonian evolutionary descendant of *Platyceramus ezoensis*"; althugh Quidelleur et al. (2011) show as earliest "Campanian".	Quidelleur et al., 2011
Weighted mean age of two samples, including 62 (of 78) laser fusions on 3-4 grain aliquots of sanidine, using FCs as fluence monitor -- as presented in Siewert (2011); but Siewert et al. (in press) add additional fusions and revise as 84.41 ±0.14/0.24 Ma.	*Desmoscaphites bassleri* ammonite zone	*Desmoscaphites bassleri* ammonite zone	Siewert, 2011; Obradovich, 1993
Four (of 17) youngest single grain zircon analyses yield a weighted mean 206Pb/238U age, utilizing CA-TIMS and ET535 spike (0.05% error in tracer calibration).	*Desmoscaphites bassleri* ammonite zone	*Desmoscaphites bassleri* ammonite zone	Siewert, 2011; Obradovich, 1993
Multi-grain sanidine laser fusion analysis, originally using TCs = 28.32 Ma as fluence monitor; no analytical data published.		Just below *Boehmoceras* fauna of "Uppermost Santonian"	Obradovich, 1993
Weighted mean age of 30 (of 46) laser fusions on 5-7 grain aliquots of sanidine, using FCs as fluence monitor -- as presented in Siewert (2011); but Siewert et al. (in press) revise as 84.55 ±0.24/0.37 Ma..	*Clioscaphites erdmanni* ammonite zone (upper)	Upper part of *Clioscaphites erdmanni* ammonite zone	Siewert, 2011; Obradovich, 1993
Weighted mean age of two samples, including 40 (of 46) laser fusions on 7-10 grain aliquots of sanidine, using FCs as fluence monitor -- as presented in Siewert (2011); but Siewert et al. (in press) add an additional sample and revise as 85.66 ±0.09/0.19 Ma.	*Clioscaphites vermiformis* ammonite zone (middle)	lower-Upper part of *Clioscaphites vermiformis* ammonite zone	Siewert, 2011; Obradovich, 1993
Weighted mean age of two samples, including 30 (of 48) laser fusions on 5 grain aliquots of sanidine, using FCs as fluence monitor -- as presented in Siewert (2011); but Siewert et al. (in press) revise as 85.84 ±0.24/0.37 Ma.	Top of *Clioscaphites saxitonianus* ammonite zone	Top of *Cladoceramus undulatoplicatus* inoceramus bivalve zone, equivalent to top of *Clioscaphites saxitonianus* ammonite zone	Siewert, 2011; Obradovich, 1993
Weighted mean age of three samples, including 107 (of 126) laser fusions on 4-5 grain aliquots of sanidine, using FCs as fluence monitor -- as presented in Siewert (2011); but Siewert et al. (in press) revise as 87.13 ±0.09/0.19 Ma.	middle of *Scaphites depressus* ammonite zone	*Protexanites bourgeoisianus (Scaphites depressus)* ammonite zone	Siewert, 2011; Obradovich, 1993
Seven (of 11) single grain zircon analyses yield a weighted mean 206Pb/238U age, utilizing CA-TIMS and ET535 spike (0.05% error in tracer calibration).	middle of *Scaphites depressus* ammonite zone	*Protexanites bourgeoisianus (Scaphites depressus)* ammonite zone	Siewert, 2011; Obradovich, 1993
Weighted mean age of two samples, including 76 (of 90) laser fusions on 2-10 grain aliquots of sanidine, using FCs as fluence monitor -- as presented in Siewert (2011); but Siewert et al. (in press) revise as 89.32 ±0.11/0.24 Ma.	*Scaphites preventricosus* ammonite zone (upper-middle)	*Forresteria alluaudi - Scaphites preventricosus* ammonite zone (upper ash bed)	Siewert, 2011; Obradovich, 1993
Four (of 18) single grain zircon analyses yield a weighted mean 206Pb/238U age, utilizing CA-TIMS and ET535 spike (0.05% error in tracer calibration).	*Scaphites preventricosus* ammonite zone (lower-middle)	*Forresteria alluaudi - Scaphites preventricosus* ammonite zone (lower ash bed)	Siewert, 2011; Obradovich, 1993
Weighted mean age of two samples, including 58 (of 63) laser fusions on 8->10 grain aliquots of sanidine, using FCs as fluence monitor -- as presented in Siewert (2011); but Siewert et al. (in press) revise as 89.87 ±0.10/0.18 Ma.	*Scaphites nigricollensis* ammonite zone	*Scaphites nigricollensis* ammonite zone	Siewert, 2011; Obradovich, 1993
Weighted mean age including 40 (of 45) laser fusions on single grain aliquots of sanidine, using FCs as fluence monitor -- as presented in Siewert (2011); but Siewert et al. (in press) revise as 91.24 ±0.09/0.23 Ma.	Top of *Prionocyclus macombi* ammonite zone	Top of *Prionocyclus macombi* ammonite zone; 0.9 m above *P. macombi* and 6 m below *Scaphites warreni*	Siewert, 2011; Obradovich, 1993
Five (of 11) single grain zircon analyses yield a weighted mean 206Pb/238U age, utilizing CA-TIMS and ET535 spike (0.05% error in tracer calibration).	Top of *Prionocyclus macombi* ammonite zone	Top of *Prionocyclus macombi* ammonite zone; 0.9 m above *P. macombi* and 6 m below *Scaphites warreni*	Siewert, 2011; Obradovich, 1993
Weighted mean age of two samples, including 83 (of 103) laser fusions on 1-5 grain aliquots of sanidine, using FCs as fluence monitor.	*Prionocyclus hyatti* ammonite zone	*Prionocyclus hyatti* ammonite zone	Siewert, 2011; Obradovich, 1993
Weighted mean age including 33 (of 41) laser fusions on 5 grain aliquots of sanidine, using FCs as fluence monitor.	Top of *Pseudaspidoceras flexuosum* ammonite zone	Top of *Pseudaspidoceras flexuosum* ammonite zone (zone below *V. birchbyi* Zone)	Meyers et al., 2012; Obradovich, 1993
Two youngest (of 11) single grain zircon analyses yield a weighted mean 206Pb/238U age, utilizing CA-TIMS and ET535 spike (0.05% error in tracer calibration). *[Not used in Late Cretaceous spline fit.]*	Top of *Pseudaspidoceras flexuosum* ammonite zone	Top of *Pseudaspidoceras flexuosum* ammonite zone (zone below *V. birchbyi* Zone)	Meyers et al., 2012; Obradovich, 1993

No.	Period	Epoch	Age	Sample	Locality	Formation	GTS2012 Age (Ma)	± 2σ (without λ errors)	± 2σ (with λ errors)	Type
K40				AZLP-08-04; K-07-01C; 90-O-33, bentonite	Laholi Point section, Blue Gap, Navajo Indian Reservation, Arizona; and two equivalent bentonites from other Bridge Creek limestone localities	Mancos Shale, bentonite BM-15, 0.6-m-thick in lower calcareous shale member, 6.4-m above base. Equated to bentonite PCB-17 in lower Bridge Creek Limestone Mbr, Pueblo, Colorado; equivalent bentonite below limestone Bed 90 in USGS #1 Portland core.	93.79	± 0.12	± 0.26	$^{40}Ar/^{39}Ar$
K39				AZLP-08-04	Laholi Point section, Blue Gap, Navajo Indian Reservation, Arizona; and two equivalent bentonites from other Bridge Creek limestone localities	Mancos Shale, bentonite BM-15, 0.6-m-thick in lower calcareous shale member, 6.4-m above base. Equated to bentonite PCB-17 in lower Bridge Creek Limestone Mbr, Pueblo, Colorado; equivalent bentonite below limestone Bed 90 in USGS #1 Portland core.	94.37	± 0.04	± 0.14	$^{206}Pb/^{238}U$
K38				HKt003	Hakkin section, Hokkaido, Japan	Base of Kakkin Muddysandstone Member, Yazo Group	94.30	± 0.30	± 0.35	$^{206}Pb/^{238}U$
			Cenomanian							
K37				K-07-01B; NE-08-01; 90-O-19; 90-O-49	Little Blue River, Thayer County, Nebraska; Four Corners, San Juan County, New Mexico; and one other equivalent bentonite from a Bridge Creek limestone locality	Greenhorn Limestone; HL-3 marker bed.	94.10	± 0.14	± 0.27	$^{40}Ar/^{39}Ar$
K36				NE-08-01	Little Blue River, Thayer County, Nebraska	Greenhorn Limestone	94.01	± 0.04	± 0.14	$^{206}Pb/^{238}U$
K35				Bighorn River Bentonite	Ram River below Ram Falls, Alberta	Vimy Member of Blackstone Formation; upper prominent 35-cm greenish-grey bentonite in lowest part of member.	94.29	± 0.13	± 0.17	$^{206}Pb/^{238}U$
K34				AZLP-08-02; 90-O-31, bentonite	Lohali Point section, Blue Gap, Navajo Indian Reservation, Arizona; and one equivalent bentonite from other Bridge Creek limestone locality	Mancos Shale; bentonite BM-6, 0.22 m thick in lower calcareous shale member, 6.8 m above base of Mancos Shale.	94.20	± 0.15	± 0.28	$^{40}Ar/^{39}Ar$
K33				AZLP-08-01; 90-O-30, bentonite	Lohali Point section, Blue Gap, Navajo Indian Reservation, Arizona	Mancos Shale, bentonite BM-5, 0.9-m-thick in lower calcareous shale member, 5.7-m above base of Mancos Shale.	94.43	± 0.17	± 0.29	$^{40}Ar/^{39}Ar$
K32				AZLP-08-01, bentonite	Lohali Point section, Blue Gap, Navajo Indian Reservation, Arizona	Mancos Shale, bentonite BM-5, 0.9-m-thick in lower calcareous shale member, 5.7-m above base of Mancos Shale.	94.28	± 0.08	± 0.15	$^{206}Pb/^{238}U$
K31				LP22 biotite	Lohali Point section, Blue Gap, Navajo Indian Reservation, Arizona	Bentonite BM-7; Mancos Shale	95.25	± 1.00	± 1.00	$^{40}Ar/^{39}Ar$
K30				D2315 (OB93-23)	Carter County, Wyoming	Frontier Formation	95.32	± 0.61	± 0.61	$^{40}Ar/^{39}Ar$
K29				WY-09-04; 90-O-50	Natrona County, Wyoming	Frontier Formation, Soap Creek Bentonite marker	95.53	± 0.09	± 0.25	$^{40}Ar/^{39}Ar$
K28				X bentonite	Burnt Timber Creek, Alberta	Sunkay Member of Blackstone Formation; 30-cm-thick brownish-gray bentonite enclosed by dark marine mudstone.	95.87	± 0.10	± 0.14	$^{206}Pb/^{238}U$
K27				91-O-03; bentonite	Pueblo County, Colorado	Graneros Shale; 7.5-cm-thick bentonite, approximately 0.9 m below its Thatcher Limestone Member	96.12	± 0.19	± 0.31	$^{40}Ar/^{39}Ar$
K26				Bailey Flats core (OB93-26); bentonite	Johnson County, Wyoming	Frontier Formation	96.56	± 0.45	± 0.45	$^{40}Ar/^{39}Ar$
K25				68-O-09 (OB93-27); bentonite	Casper, Wyoming	Top of Mowry Shale; Clay Spur Bentonite	97.88	± 0.69	± 0.69	$^{40}Ar/^{39}Ar$
K24				OB93-28; bentonite	Judith Basin County, Montana	Arrow Creek Member, Colorado Shale; basal 15cm of Arrow Creek bentonite	99.24	± 0.41	± 0.41	$^{40}Ar/^{39}Ar$
K23				91-O-20 (OB93-29); bentonite	Greybull, Big Horn County, Wyoming	Thermopolis Shale, 30-cm-thick bentonite, upper part of Shell Creek Shale Member	99.26	± 0.70	± 0.70	$^{40}Ar/^{39}Ar$
K22				91-O-12 (OB93-30); bentonite	Wyoming	Thermopolis Shale, bentonite 12m below top	99.46	± 0.59	± 0.59	$^{40}Ar/^{39}Ar$
K21				TNGt006; crystal tuff	Tengunosaw valley, Hokkaido, Japan	Hikagenosawa Fm.; Yazo Group	97.00	± 0.40	± 0.40	$^{206}Pb/^{238}U$ ID-TIMS
K20				R8072; crystal tuff	Hotei-zawa, Hokkaido, Japan	middle part of Yezo Group	99.70	± 0.38	± 0.38	$^{40}Ar/^{39}Ar$

Primary radiometric age details	Zone assignment	Primary biostratigraphic age details	Reference
Weighted mean age of three samples, including 93 (of 107) laser fusions on 5-7 grain aliquots of sanidine, using FCs as fluence monitor.	*Watinoceras devonense* ammonite zone (upper half)	upper part of *Watinoceras devonense* ammonite zone	Meyers et al., 2012; Obradovich, 1993
Five youngest (of 7) single grain zircon analyses yield a weighted mean 206Pb/238U age, utilizing CA-TIMS and ET535 spike (0.05% error in tracer calibration).	*Watinoceras devonense* ammonite zone (upper half)	upper part of *Watinoceras devonense* ammonite zone.	Meyers et al., 2012; Obradovich, 1993
Five (of 7) multi-grain chemically-abraded zircon fractions. ID-TIMS.	Cenomanian-Turonian boundary interval	Between LAD of *Rotalipora cushmani* and FAD of *Marginotruncana schneeganei;* therefore Cenomanian-Turonian boundary interval.	Quidelleur et al., 2011
Weighted mean age of four samples, including 103 (of 116) laser fusions on 1-3 grain aliquots of sanidine, using FCs as fluence monitor.	*Neocardioceras juddii-N. scotti* (lower-middle)	One-third up in undiffrerentiated *Neocardioceras juddii-Nigericeras scotti* ammonite zone (Meyers *et al.*, 2011 figure)	Meyers et al., 2012; Obradovich, 1993
Six youngest (of 11) single grain zircon analyses yield a weighted mean 206Pb/238U age, utilizing CA-TIMS and ET535 spike (0.05% error in tracer calibration).	*Neocardioceras juddii-N. scotti* (lower-middle)	One-third up in undiffrerentiated *Neocardioceras juddii-Nigericeras scotti* ammonite zone (Meyers *et al.*, 2011 figure)	Meyers et al., 2012; Obradovich, 1993
Six single zircon grains yielded a weighted mean 206Pb/238U age, utilizing CA-TIMS and the ET535 spike (0.05% error in tracer calibration)	*Neocardioceras juddii* ammonite zone	Correlated to "B" bentonite, or Bed 80 in Pueblo section; therefore same as other dated "*N. juddii*" bentonite horizon.	Barker et al., 2011
Weighted mean age of two samples, including 74 (of 83) laser fusions on single grain aliquots of sanidine, using FCs as fluence monitor.	Middle of combined *Vascoceras diartianum-B. clydense* interval	*Euomphaloceras septemseriatum* ammonite zone [middle of undifferentiated *S. gracile* zone (including *B. clydense*?); on Meyers *et al.*, 2011 figure]	Meyers et al., 2012; Obradovich, 1993
Weighted mean age including 42 (of 46) laser fusions on 3-6 grain aliquots of sanidine, using fluence monitor FCs = 28.201 Ma.	Middle of combined *Vascoceras diartianum-B. clydense* interval	*Vascoceras diartianum* portion of *Sciponoceras gracile* ammonite zone [middle of undifferentiated *S. gracile* zone (including *B. clydense*?); on Meyers *et al.*, 2011 figure]	Meyers et al., 2012; Obradovich, 1993
Two youngest (of 5) single grain zircon analyses yield a weighted mean 206Pb/238U age, utilizing CA-TIMS and ET535 spike (0.05% error in tracer calibration).	Middle of combined *Vascoceras diartianum-B. clydense* interval	*Vascoceras diartianum* portion of *Sciponoceras gracile* ammonite zone [middle of undifferentiated *S. gracile* zone (including *B. clydense*?); on Meyers *et al.*, 2011 figure]	Meyers et al., 2012; Obradovich, 1993
Plateau of biotite separate, originally using FCs = 28.02 Ma as fluence monitor.	Base of combined *Vascoceras diartianum-B. clydense* interval	Near base of undifferentiated *Sciponoceras gracile* ammonite zone (overlying *M. mosbyense* zone)	Quidelleur et al., 2011
Multi-grain sanidine laser fusion analysis, originally using TCs = 28.32 Ma as fluence monitor; no analytical data published.	*Dunveganoceras pondi* ammonite zone	*Dunveganoceras pondi* ammonite zone	Obradovich, 1993
Weighted mean age of three samples, including 60 (of 69) laser fusions on 1->10 grain aliquots of sanidine, using FCs as fluence monitor.	*Acanthoceras amphibolum* ammonite zone	*Acanthoceras amphibolum* ammonite zone	Siewert, 2011; Obradovich, 1993
Six single zircon grains yielded a weighted mean 206Pb/238U age, utilizing CA-TIMS and the ET535 spike (0.05% error in tracer calibration)	*Acanthoceras amphibolum* ammonite zone	Correlated to "X" bentonite of Frontier Fm in Wyoming; therefore coeval with sample 90-O-50 (*A. amphibolum* ammonite zone). Close to end of regressive phase, just prior to onset of the major transgression culminating in early Turonian most-extensive-flooding of Western Interior basin.	Barker et al., 2011
Weighted mean age of 48 (of 58) laser fusions on 5 grain aliquots of sanidine, using FCs as fluence monitor.	*Conlinoceras tarrantense - C. gilberti* ammonite zone	*Conlinoceras tarrantense (=Conlinoceras gilberti)* ammonite zone	Siewert, 2011; Obradovich, 1993
Multi-grain sanidine laser fusion analysis, originally using TCs = 28.32 Ma as fluence monitor; no analytical data published.	*N. cornutus* - base *C. gilberti* ammonite zone interval.	27m below lowest occurrence of *C. gilberti* ammonite	Obradovich, 1993
Multi-grain sanidine laser fusion analysis, originally using TCs = 28.32 Ma as fluence monitor; no analytical data published.	*N. cornutus* - base *C. gilberti* ammonite zone interval.	Top of Mowry Shale	Obradovich, 1993
Multi-grain sanidine laser fusion analysis, originally using TCs = 28.32 Ma as fluence monitor; no analytical data published.	*Neogastroplites cornutus* ammonite zone	*Neogastroplites cornutus* ammonite zone	Obradovich, 1993
Multi-grain sanidine laser fusion analysis, originally using TCs = 28.32 Ma as fluence monitor; no analytical data published.	*Neogastroplites haasi - N. cornutus* ammonite zones	0.9-4.6 m above mudstone unit containing black Mn-nodules with *Neogastroplites haasi*	Obradovich, 1993
Multi-grain sanidine laser fusion analysis, originally using TCs = 28.32 Ma as fluence monitor; no analytical data published.	*Neogastroplites haasi* ammonite zone	Similarity in age to 91-O-20 suggests the *N. haasi* zone.	Obradovich, 1993
Mean of four (of 9) multi-grain chemically abraded zircon fractions.	Early Cenomanian	"*Above FAD of Mantelliceras ammonite*" and within range of foraminifer *Thalmanninella globotruncanoides.*	Quidelleur *et al.*, 2011
Single or multi-grain sanidine laser fusion analysis, originally using TCs = 28.32 Ma as fluence monitor.	*Mantelliceras mantelli* ammonite zone (base of upper subzone)	*Mantelliceras saxbii* ammonite subzone (lower part) of *Mantelliceras mantelli* zone	Obradovich *et al.*, 2002

No.	Period	Epoch	Age	Sample	Locality	Formation	GTS2012 Age (Ma)	± 2σ (without λ errors)	± 2σ (with λ errors)	Type
K19				SK069; crystal tuff		middle part of Yezo Group	100.83	± 1.00	± 1.00	$^{40}Ar/^{39}Ar$
K18				R8943B; crystal tuff	Kyoei-Sakin-zawa, Hokkaido, Japan	middle part of Yezo Group	99.89	± 0.37	± 0.37	$^{40}Ar/^{39}Ar$
K17				TNGt005; crystal tuff	Tengunosaw valley, Hokkaido, Japan	Hikagenosawa Fm.; Yazo Group	99.70	± 1.10	± 1.10	$^{206}Pb/^{238}U$ LA-ICPMS
		Early								
			Albian							
K16				75-O-06 (OB93-31); bentonite	Peace River "Gates", British Columbia	Hulcross Formation, bentonite in its upper 50 m.	107.89	± 0.30	± 0.30	$^{40}Ar/^{39}Ar$
K15				Vohrum tuff	Vöhrum clay pit situated 30 km east of Hannover, 1.6 km SW of Vöhrum, NW Germany	Schwicheldt Ton Member, Gault Formation. 2-cm-thick tuff horizon 65 cm above the (local usage) "Aptian/Albian boundary"	113.08	± 0.07	± 0.14	$^{206}Pb/^{238}U$
			Aptian							
K14				OB93-32; bentonite	Otto gott clay pit, Sarstedt, 21 km SE of Hannover, Germany		114.84	± 1.30	± 1.30	$^{40}Ar/^{39}Ar$
K13				PA-140, tuff layer	Bano Nuevo Volcanic Complex, near Cerro Mirador, central Patagonian Cordillera, Argentina	volcanics conformably overlie unconsolidated sands of the Apeleg Formation	120.90	± 1.10	± 1.10	$^{206}Pb/^{238}U$ ion probe
K12				MC-44, MC-84, MC-86-1; lavas	Bano Nuevo Volcanic Complex, near Cerro Mirador, central Patagonian Cordillera, Argentina	volcanics conformably overlie unconsolidated sands of the Apeleg Formation	122.20	± 1.50	± 1.50	$^{40}Ar/^{39}Ar$
K11				144-878A-46M-1, 46M-2, 79R-3; hawaiites	ODP Leg 144, MIT guyot, western Pacific		122.18	± 1.43	± 1.43	$^{40}Ar/^{39}Ar$
K10				MC888; tuff	outcrops along McCarty Creek, 2 miles north of Paskenta, Tehama County, Great Valley, California, USA	Great Valley Group	124.07	± 0.16	± 0.24	$^{206}Pb/^{238}U$
K9				40R-1, 41R-1, 42R-5, 45R-1; plagioclase phenocrysts from volcaniclastic rocks	ODP Site 1184 (Leg 192), eastern salient of Ontong Java Plateau, western equatorial Pacific	Four samples from cores into subunit IIE in the lower part of the volcaniclastic succession.	124.32	± 1.80	± 1.80	$^{40}Ar/^{39}Ar$
K8				SG7, SGB25, ML475, KF36, KF53; whole rock basalts	on-shore sections of Ontong Java Plateau basalts, central Malaita, Solomon Islands; equivalent to OJP basalts drilled in ODP Site 807	Malaita Volcanic Group	125.98	± 2.86	± 2.87	$^{40}Ar/^{39}Ar$
K7				144-878A-89R-4, 91R-3; plagioclase from alkalic basalts	ODP Site 878 (Leg 144), MIT guyot, western Pacific		125.45	± 0.41	± 0.43	$^{40}Ar/^{39}Ar$
			Barremian							
			Hauterivian							
K6				Caepe Malal; tuff layer	Neuquén Basin, Argentina	Interbedded between shales of the upper member (Agua de la Mula) of the Agrio Formation	132.70	± 1.30	± 1.30	$^{206}Pb/^{238}U$ ion probe
			Valanginian							
K5				MC873A; tuff	outcrops along McCarty Creek, 2 miles north of Paskenta, Tehama County, Great Valley, California, USA	Great Valley Group	133.51	± 0.22	± 0.29	$^{206}Pb/^{238}U$
K4				XZ0506; rhyolite	Kadong section, south side of Yamzho Iyumco Lake, 28.75N, 90.70E, ca. 51km ESE of Nagarze, Tibet.	Rhyolite is from unit 13 of the upper (volcanic and volcaniclastic) part of Sangxiu Formation	136.00	± 3.00	± 3.00	$^{206}Pb/^{238}U$ ion probe

Primary radiometric age details	Zone assignment	Primary biostratigraphic age details	Reference
Mean of five multi-grain biotite separates, originally using FCs = 28.02 Ma as fluence monitor.	*Mantelliceras mantelli* ammonite zone (upper subzone)	*Above occurrence of Mantelliceras saxbii* ammonite	Quidelleur *et al.*, 2011
Single or multi-grain sanidine laser fusion analysis, originally using TCs = 28.32 Ma as fluence monitor.	*Mantelliceras mantelli* ammonite zone (top of lower subzone)	*Graysonites woodridgei* ammonite zone (equivalent to European subzone of *Neostlingoceras carcitanense*); upper part	Obradovich *et al.*, 2002
Eleven concordant spots on 12 zircon grains.	uppermost *Arrhaphoceras briacensis* ammonite zone	Nearly coeval with local FAD of pelagic forminifer *Thalmanninella globotuncanoides* (primary marker of Albian-Cenomanian boundary)	Quidelleur *et al.*, 2011
Multi-grain sanidine laser fusion analysis, originally using TCs = 28.32 Ma as fluence monitor; no analytical data published.	*Euhoplites loricatus* ammonite zone	*Pseudopulchellia pattoni* zone of North American Western Interior. Correlated to mid-Middle Albian, *Euhoplites loricatus* zone of Europe.	Obradovich, 1993
Weighted mean of five (of 7) single and 2-3 grain zircon fractions, utilizing CA-TIMS and the ET535 spike (0.05% error in tracer calibration)	*Leymeriella tardefurcata* (lower subzone)	65 cm above the first occurrence of the ammonite *Leymeriella (Proleymeriella) schrammeni anterior*	Selby et al., 2009
Multi-grain sanidine laser fusion analysis, originally using TCs = 28.32 Ma as fluence monitor; no analytical data published.	*Parahoplites melchioris* to top of *H. jacobi* zone due to uncertain zone usage	*Parahoplites nutfieldiensis* ammonite zone (but definition of this zone is not provided)	Obradovich, 1993
Weighted mean of 14 (of 17) spot analyses analyzed by SHRIMP II (ANU), using the TEMORA zircon standard.	*Deshayesites forbesi* through *E. martinoides zones* due to vague biostrat	upper beds of the Apeleg Formation have been dated as early Aptian based on the presence of the ammonite *Tropaeum* or *Australiceras* sp.	Suarez et al., 2010
Weighted mean of three hornblende 40Ar/39Ar plateau or isochron ages, originally calculated with FCs = 28.02 Ma as the fluence monitor.	*Deshayesites forbesi* through *E. martinoides zones* due to vague biostrat	upper beds of the Apeleg Formation have been dated as early Aptian based on the presence of the ammonite *Tropaeum* or *Australiceras* sp.	Suarez et al., 2010
Weighted mean of three whole rock 40Ar/39Ar isochron ages, originally calculated with TCs = 27.92 Ma as the fluence monitor.	*D. forbesi* to base of *D. furcata* ammonite zone	MIT seamount normally magnetized, upper hawaiite lavas; platform carbonates 25 m above volcanic basement are assigned to the *C. litterarius* nannofossil biozone (NC6). Implied to be younger than Chron M0r	Pringle and Duncan, 1995
Weighted mean of youngest five (of 7) 1-4 grain zircon fractions, utilizing CA-TIMS and in-house UNC spike (0.1% error in tracer calibration), with analytical error expanded in accommodate geologic scatter.	*D. forbesi* to base of *D. furcata* ammonite zone	*Chiastozygus litterarius* biozone (NC6) – Text says NC6A subzone, as supported by the first occurrence of *C. litterarius* and the last occurrence of *Conusphaera rothii* in MC887, a sample found below MC888. But, LAD of *C. rothii* = defines base subzone 6B, not 6A => sample is lower NC6B.	Shimokawa, 2010
Weighted mean of 6 (of 8) laser fusions on 4-5 grain aliquots of plagioclase, sampled from four core intervals, originally calibrated to FCs = 28.02 Ma.	*mid-D. deshayesi* to *D. furcata* ammonite zone	The oldest sediment overlying basement on the crest of the Ontong Java Plateau occurs within the upper part of the *Leupoldina cabri* planktonic foraminiferal zone.	Chambers et al., 2004
Weighted mean of 5 whole rock incremental heating plateau ages, originally calculated with FCT-3 biotite = 27.55 Ma.	*mid-D. deshayesi* to *D. furcata* ammonite zone	The oldest sediment overlying basement on the crest of the Ontong Java Plateau (ODP sites) occurs within the upper part of the *Leupoldina cabri* planktonic foraminiferal zone; but there are no biostratigraphic constaints on these Malaita exposures.	Tejada et al., 2002
Weighted average of plagioclase 40Ar/39Ar isochron ages, originally calculated with TCs = 27.92 Ma as the fluence monitor	*Deshayesites forbesi ammonite zone*	MIT seamount lower alkalic basalts at reversed-upward-to-normal polarity transition; interpreted as top of Chron M0r*.	Pringle and Duncan, 1995
Weighted mean of fifteen (of 20) spot analyses analyzed by SHRIMP II (ANU), using the TEMORA zircon standard.	*Subsaynella sayni* ammonite zone (lower)	within the *Spitidiscus riccardii* ammonoid zone, correlated with the lower *Subsaynella sayni* zone of the Tethys	Aguirre Urreta et al., 2008
Weighted mean of youngest eight (of 10) 3-5 grain zircon fractions, utilizing CA-TIMS and in-house UNC spike (0.1% error in tracer calibration), with analytical error expanded in accommodate geologic scatter.	Val/Haut boundary interval (*C. furcillata - A. radiatus* zones)	Fossils of *P. fenestrata, R. nebulosus, C. angustiforatus, Cyclagelosphaera deflandrei, Eiffellithus windii, Metadoga mercurius, R. wisei,* and *Tubodiscus verenae* place this sample in the *C. oblongata* (NK3) biozone – *R. wisei* (NK3A) subzone	Shimokawa, 2010
Weighted mean of fifteen spot analyses analyzed by SHRIMP II (Beijing), using the TEMORA zircon standard	*T. pertransiens - C. furcillata* ammonite zones	Sample is above a *Calcicalathina oblongata–Speetonia colligata* assemblage of calcareous nannofossils, but minimum age is unconstrained. Assigned here as Valanginian.	Wan et al., 2010

No.	Period	Epoch	Age	Sample	Locality	Formation	GTS2012 Age (Ma)	± 2σ (without λ errors)	± 2σ (with λ errors)	Type
K3				MC180.5; tuff	outcrops along Kelly Road, Paskenta, Tehama County, Great Valley, California, USA	Great Valley Group	137.62	± 0.07	± 0.21	$^{206}Pb/^{238}U$
			Berriasian							
K2				CA-6189; rhyolite ignimbrite	12 m thick rhyolitic ignimbrite of the Lago Norte Vocanic Complex, overlying a 50 m thick fossiliferous marine succession	Lago Norte Volcanic Complex, overlying a thin succession of Toqui Formation intercalated in the Ibáñez Formation	137.30	± 1.20	± 1.20	$^{206}Pb/^{238}U$ ion probe
K1				GC-670; tuff	Outcrops on south side of Grindstone Creek, west of State Road 306, Glen County, Great Valley, California, USA	Great Valley Group	138.46	± 0.21	± 0.29	$^{206}Pb/^{238}U$
	Jurassic									
		Late								
			Tithonian							
J19				Shatsky Rise basalt sills	ODP Hole 1213B (Leg 198) on Shatsky Rise, northwest Pacific	Lithologic Unit IV. Whole-rock = Core 28R-2 (1-6cm) and 30R-3 (1-6cm). Feldspars from Core 33R-3 (115-120cm)	146.48	± 1.62	± 1.63	$^{40}Ar/^{39}Ar$
			Kimmeridgian							
J18				Staffin black shale	Flodigarry, Staffin Bay, Isle of Skye, U.K. = proposed GSSP for base-Oxfordian.	Flodigarry Shale Member, Staffin Formation. Sampled Bed 35 at approximate proposed GSSP level is 1.30±0.1m below Concretion-rich marker Bed 36.	154.10	± 2.10	± 2.20	Re-Os
			Oxfordian							
J17				ODP 801C-15R-4 92–96, tholeiitic basalt	ODP Site 801, Pigafetta basin, western Pacific; marine magnetic anomaly M43r	Upper basalts	161.17	± 0.70	± 0.74	$^{40}Ar/^{39}Ar$
J16				ODP 801C-15R-4 92–96, oceanic tholeiitic basalt	ODP Site 801, Pigafetta basin, western Pacific; marine magnetic anomaly M43r	Upper basalts	159.86	± 3.32	± 3.33	$^{40}Ar/^{39}Ar$
		Middle								
			Callovian							
J15				boundary ash	Chacay Melehué, Neuquén Basin, western Argentina	Chacay Melehué/Los Molles Formation	164.64	± 0.20	± 0.27	$^{206}Pb/^{238}U$
			Bathonian							
			Bajocian							
J14				Samples ODP 801C-30R-4 37-41 and 801C-31R-4 138-143	ODP Site 801, Pigafetta basin, western Pacific; marine magnetic anomaly M43r	Tholeiitic basalts (ca. 750 mbsf; this unit extends from ca. 500 mbsf to base of hole at ca. 950 mbsf). Samples from oceanic thoeiitic basalt crystalline groundmass.	168.72	± 1.71	± 1.73	$^{40}Ar/^{39}Ar$
J13				GUN-F; bentonite	near top of Gunlock section, Carmel Formation, Utah, USA	Crystal Creek Member, Carmel Formation, Utah, USA	168.35	± 1.30	± 1.31	$^{40}Ar/^{39}Ar$
J12				GUN-4; bentonite	near bottom of Gunlock section, Temple Cap Formation, Utah, USA	uppermost Temple Cap Formation, Utah, USA	171.48	± 1.20	± 1.22	$^{40}Ar/^{39}Ar$
			Aalenian							
J11				Eskay Rhyolite	Eskay rhyolite east, Eskay Creek gold mine, Iskut River map area, NW British Columbia	Salmon River Formation, Hazelton Group	174.10	± 0.68	± 0.72	$^{206}Pb/^{238}U$
		Early								
			Toarcian							
J10				PCA-YR-1; volcanic ash	Yakoun River, Queen Charlotte Islands, British Columbia	Whiteaves Formation	181.40	± 0.73	± 0.78	$^{206}Pb/^{238}U$

Primary radiometric age details	Zone assignment	Primary biostratigraphic age details	Reference
Weighted mean of six oldest (of 8) 1-2 grain zircon fractions, utilizing CA-TIMS and in-house UNC spike (0.1% error in tracer calibration).	mid-lower *T. pertransiens* to mid-*B. campylotoxus* ammonite zones?	Calcareous nannofossil occurrences suggest correlation of this sample to the *C. angustiforatus* biozone – *P. fenestrata* subzone (NK-2B) at the oldest, but it is not possible to rule out a younger age with the current biostratigraphy; the sample is conservatively placed at the base of the Valanginian in the *Calcicalathina oblongata* biozone – *R. wisei* (NK-3A) subzone. [*C. oblongata* begins in lower *T. pertransiens* ammonite zone]	Shimokawa, 2010
Weighted mean of 14 (of 19) spot analyses analyzed by SHRIMP II (ANU), using the TEMORA zircon standard.	*S. boissieri* - *T. pertransiens* ammonite zone interval?	Ammonites (*Groebericeras* Leanza and *Blandfordiceras* Cossman) from the underlying Toqui Formation, the basal unit of the Coihaique Group, indicate early Berriasian and late Berriasian ages.	Suárez et al., 2009
Weighted mean of five 2-4 grain zircon fractions, utilizing CA-TIMS and in-house UNC spike (0.1% error in tracer calibration), with analytical error expanded in accommodate geologic scatter.	*S. boissieri* - *T. pertransiens* ammonite zone interval?	Sample placed in the *Cretarhabdus angustiforatus* biozone (NK-2), supported by the occurrences of *C. angustiforatus* and *Assipetra infracretacea* 1 cm above the sample, and the appearances of *Micrantholithus hoschulzii* and *Rhagodiscus nebulosus* within a meter above the sample. [their schematic figure suggests latest Berriasian or early Valanginian]	Shimokawa, 2010
Weighted mean of whole rock (middle sill), and plagioclase (lower sill) incremental heating plateau ages, originally calculated with FCT biotite = 28.03 Ma as the fluence monitor.	*Berriasella jacobi* ammonite zone?	Sills (reversely magnetized) were injected into sediments with Calcareous nannofossil zone NK1 and radiolarian zone *Pseudodictryomitra carpatica* of earliest Berriasian => emplacement of Sills must be younger than earliest Berriasian; and perhaps during Chron M18r.	Mahoney et al., 2005
Model 3 isochron age from 9 rock powder aliquots (+ 1 replicate) sampling a of 10 cm of section, dissolved via CrVI-H2SO4.	*Epipeltoceras bimammatum* ammonite zone (middle part)	"subboreal Oxfordian-Kimmeridgian (*Pseudocordata/Baylei* zones) boundary is placed between 1.65 - 1.47 m and 1.08 m below bed 36 - the last occurrence of of *Ringsteadia* and the first occurrence of Pictonia, respectively. Bed 35 is rich in *Pictonia densicostata*.	Selby, 2007
Weighted mean of 5 laser incremental heating and total fusion experiments on plagioclase and biotite separates, originally calibrated to TCs = 28.34 Ma.	*Q. mariae* - mid-*E. bimammatum* zmmonite zones	Overlain by (or, perhaps, intruded into) radiolarian claystones assigned to Oxfordian based on radiolarian calibration by Pessagno *et al*. (1996), although this age assignment is disputed (see Jurassic text).	Koppers et al., 2003
One laser incremental heating plateau age (71% of gas) on acid-leached crystalline groundmass, originally calibrated to FCT-3 biotite = 28.04 Ma. *[re-analysis by Koppers et al., 2003, is given above]*	*Q. mariae* - mid-*E. bimammatum* zmmonite zones	Overlain by (or, perhaps, intruded into) radiolarian claystones assigned to Oxfordian based on radiolarian calibration by Pessagno *et al*. (1996), although this age assignment is disputed (see Jurassic text).	Pringle et al., 1992
Weighted mean of 6 (of 9) single zircon grains, utilizing CA-TIMS and the ET535 spike (0.05% error in tracer calibration)	*Clydoniceras discus* - *R. anceps* ammonite zone interval?	Top of local *Lilloettia steinmanni* ammonite assemblage zone. "A precise age cannot be established. ... Only the overling *Hecticoceras proximum* Zone [which is above a local *Neuqueniceras bodenbenderi* assemblage zone] is chararacterised by an association of Tethyan Oppeliidae indicating the latest early Callovian (Riccardi *et al*., 1989)." Elsewhere, the *L. steinmanni* Zone is correlated with uppermost Bathonian.	Kamo and Riccardi, 2009
Weighted mean of two laser incremental heating plateau ages (41-57% of gas) on acid-leached crystalline groundmass, originally calibrated to FCT-3 biotite = 28.04 Ma.	*S. propinguans* - *P. progracilis* ammonite zone interval	Probably the lower massive tholeiitic basalt is mid-Bajocian to early Bathonian (disputed, see Jurassic text; with a much older Early Bajocian or early Aalenian proposed by Bartolini and Larson, 2001). Important constraint on Jurassic portion (Bajocian-Oxfordian) of M-sequence.	Koppers et al., 2003
Weighted mean of six single-grain sanidine laser fusion analyses, originally using FCs = 27.84 Ma as fluence monitor.	*W. laeviuscula* - *G. garantiana* ammonite zone interval	Bivalve correlation of Carmel Formation to ammonite-bearing Twin Creek Formation indicates an age of late Early to early Late Bajocian	Kowallis et al., 2001
Weighted mean of eight single-grain sanidine laser fusion analyses, originally using FCs = 27.84 Ma as fluence monitor.	*W. laeviuscula* - *G. garantiana* ammonite zone interval	Bivalve correlation of Carmel Formation to ammonite-bearing Twin Creek Formation indicates an age of late Early to early Late Bajocian	Kowallis et al., 2001
Weighted mean of 5 multi-grain zircon fractions, utilizing physical abrasion and in-house UBC spike (0.1% error in tracer calibration). Analytical error expanded in accommodate geologic scatter.	uppermost Toarcian - mid-upper Aalenian (*D. pseudoradiosa* - *B. bradfordensis* zone interval)	Flow-banded rhyolite is overlain by Upper Aalenian strata, correlated to *Erycitoides* cf. *howelli* ammonite zone of western North America zonation) [therefore, could be Lower Aalenian or older?]	Childe, 1996
Weighted mean of 6 single and multi-grain zircon fractions, utilizing physical abrasion and in-house GSC spike (0.1% error in tracer calibration).	upper *Hild. bifrons* - lower *Haugia variabilis* ammonite zone	Volcanic ash layer near base of *Crassicosta* ammonite zone (late Middle Toarcian) at type section of this regional western North American zone.	Pálfy et al., 1997

No.	Period	Epoch	Age	Sample	Locality	Formation	GTS2012 Age (Ma)	± 2σ (without λ errors)	± 2σ (with λ errors)	Type
J9				AR08AS-29C; tuff	Serrucho Creek, Neuquén Basin, Argentina	Upper tuff bed within 19.5-m-thick organic-rich shale of lower Tres Esquinas Formation (21.9 mab in measured section).	180.59	± 0.40	± 0.48	$^{206}Pb/^{238}U$
J8				AR08AS-16; tuff	Serrucho Creek, Neuquén Basin, Argentina	Lower tuff bed within 19.5-m-thick organic-rich shale of lower Tres Esquinas Formation (13.8 mab in measured section)	181.33	± 0.17	± 0.31	$^{206}Pb/^{238}U$
J7				EP-89-258; quartz monzonite pluton	Spatsizi River map area, NW British Columbia	McEwan Creek pluton	183.23	± 0.62	± 0.67	$^{206}Pb/^{238}U$
			Pliensbachian							
J6				IWE-95-111; rhyolite tuff	Skinhead Lake, NW British Columbia		184.75	± 0.60	± 0.63	$^{206}Pb/^{238}U$
J5				GGAJ-92-127; lithic crystal lapilli tuff	South side of Copper Island, Atlin Lake, NW British Columbia	Laberge Group	185.68	± 0.19	± 0.33	$^{206}Pb/^{238}U$
			Sinemurian							
J4				93JP48; rhyolite flow	southern Telkwa Range, British Columbia	Telkwa Formation, Hazelton Group	191.16	± 0.72	± 0.74	$^{206}Pb/^{238}U$
			Hettangian							
J3				L19; tuff	Utcubamba valley, northern Peru	Aramachay Formation, Pucara Group; Ash bed L-19, on fresh road section from Levanto to Maino exposing a complete deep marine sedimentary sequence,	199.53	± 0.22	± 0.30	$^{206}Pb/^{238}U$
J2				LM4-100/101; volcanic tuff	Utcubamba valley, northern Peru	Aramachay Formation, Pucara Group; approximately 5 1/2 m above sample LM4-90.	201.29	± 0.19	± 0.28	$^{206}Pb/^{238}U$
J1				NYC-N10; volcanic tuff	New York Canyon, Nevada, USA	New York Canyon section (stratigraphy, ammonites, carbon-isotopes in Guex et al., 2004); which was candidate for base-Triassic GSSP.	201.33	± 0.17	± 0.27	$^{206}Pb/^{238}U$
	Triassic									
		Late								
			Rhaetian							
T24				LM4-90; volcanic tuff	Utcubamba valley, northern Peru	Aramachay Formation, Pucara Group; Ash bed 90; approximately 2m above sample LM4-86.	201.36	± 0.16	± 0.27	$^{206}Pb/^{238}U$
T23				LM4-86; volcanic tuff	Utcubamba valley, northern Peru	Aramachay Formation, Pucara Group; Ash bed 86 at 60m level in fresh road section from Levanto to Maino exposing a complete deep marine sedimentary sequence,	201.40	± 0.20	± 0.29	$^{206}Pb/^{238}U$
			Norian							
			Carnian							
T22				Aglianico tuff bed	southern Apennines, southern Italy	"A 5-cm-thick volcanic ash bed (here named Aglianico) occurs within the hemipelagic to pelagic Calcari con Selce Formation (i.e., cherty limestones) of the Pignola 2 section". Unusual clay-bed (no carbonate) 3m below is "Carnian pluvial event" that concides with demise of rimmed carbonate platforms in the Tethys.	230.91	± 0.13	± 0.28	$^{206}Pb/^{238}U$

Primary radiometric age details	Zone assignment	Primary biostratigraphic age details	Reference
Weighted mean of 8 (of 9) multi-grain zircon fractions, utilizing either physical or chemical abrasion and in-house Oslo spike (0.1% error in tracer calibration).	*Harp, serpentinum - Haugi variabilis* ammonite zone interval?	Negative-carbon excursion used to estimate as equivalent to lower *falciferum* zone (earliest Toarcian). *[However, this may be a mid-Toarcian excursion, with reworked ashes; S. Hesselbo, 2011, written comm. to J.Ogg]*	Mazinni et al., 2010
Weighted mean of 5 multi-grain zircon fractions, utilizing either physical or chemical abrasion and in-house Oslo spike (0.1% error in tracer calibration).	*Harp, serpentinum - Hild. bifrons* ammonite zone interval?	Negative-carbon excursion used to estimate as equivalent to lower *falciferum* zone (earliest Toarcian). *[However, this may be a mid-Toarcian excursion, with reworked ashes; S. Hesselbo, 2011, written comm. to J.Ogg]*	Mazinni et al., 2010
Weighted mean of one zircon and two titanite multi-grain fractions, utilizing physical abrasion and in-house GSC spike (0.1% error in tracer calibration). Analytical error expanded in accommodate geologic scatter.	*Dact. tenuicostatum - Hild. bifrons* ammonite zone interval?	Intrudes volcanics with intercalated sediments of Early Toarcian age; pluton is perhaps co-magmatic with volcanism.	Evenchick and McNicoll, 1993
Weighted mean of 4 multi-grain zircon fractions, utilizing physical abrasion and in-house GSC spike (0.1% error in tracer calibration). Analytical error expanded in accommodate geologic scatter.	*Fuc. lavinianum* (lower half) - *Emac. emaciatum* (upper half) zones	Late Pliensbachian (*Kunae* ammonite zone of western North America)	Pálfy et al., 2000
Weighted mean of 4 multi-grain zircon fractions, utilizing physical abrasion and in-house GSC spike (0.1% error in tracer calibration).	*Fuc. lavinianum* (lower half) - *Emac. emaciatum* (upper half) zones	*Kunae* ammonite zone of western North America	Johannson and McNicoll, 1997
Weighted mean of 3 multi-grain zircon fractions, utilizing physical abrasion and in-house GSC spike (0.1% error in tracer calibration). Analytical error expanded in accommodate geologic scatter.	late Sinemurian (*O. oxynotum - E. raricostatum* zone interval?)	sample from 100 meters of epiclastic and pyroclastic strata above sediments with ammonite assemblages characteristic of Late Sinemurian *Plesechioceras? harbledownense* ammonite zone of western North America	Pálfy et al., 2000
Six (of 9) single zircon grain analyses yielded a weighted mean 206Pb/238U age, utilizing CA-TIMS and in-house UNIGE spike (0.1% error in tracer calibration)	*Schlo. angulata - Ariet. bucklandi* ammonite zone interval	Within *Badouxia canadensis* ammonite beds, with *Angulaticeras* genus about 8m below and the lowest *Vermiceras densicostatum* and *Metophioceras* occurrence about 12m above. The authors consider the Hett-Sinem boundary to be within the N. American Canadensis zone.	Schaltegger et al., 2008
Fourteen single zircon grains analyzed with CA-TIMS and ET2535 spike (0.05% tracer uncertainty) illustrate excess scatter attributed to a range of growth histories prior to eruption; the authors estimate the eruption age from youngest robust grain at 201.29 ± 0.19 Ma; weighted mean of 5 youngest grains is statistically equivalent at 201.33 ± 0.14 Ma.	*Psiloceras spelae* ammonite zone (lower half)	ca. 2.5m above lowest occurrence of Hettangian genus Psiloceras (Psiloceras spelae); which is presumed to be equivalent to the base-Jurassic GSSP level in Austria.	Schoene et al., 2010
Seventeen single zircon grains analyzed with CA-TIMS and ET2535 spike (0.05% tracer uncertainty) illustrate excess scatter attributed to a range of growth histories prior to eruption; the authors estimate the eruption age from youngest robust grain at 201.33 ± 0.19 Ma; weighted mean of 9 youngest robust grains is statistically equivalent at 201.45 ± 0.12 Ma.	*Psiloceras spelae* ammonite zone (lower half)	1.5 m above the lowest occurrence of Hettangian genus Psiloceras (Psiloceras spelae); which is presumed to be equivalent to the base-Jurassic GSSP level in Austria.	Schoene et al., 2010
Fourteen single zircon grains analyzed with CA-TIMS and ET2535 spike (0.05% tracer uncertainty) include Proterozoic xenocrysts, a cluster of 9 equivalent grains and one young analysis attributed to Pb-loss; the authors estimate the eruption age from most precise, young, robust grain at 201.36 ± 0.16 Ma; weighted mean of nine grains is statistically equivalent at 201.45 ± 0.14 Ma.	Top-Triassic ammonite zone gap	Mid-way between highest local occurrence of the topmost Triassic ammonite genus Choristoceras (Choristo crickmayi, 3m below) and lowest occurrence of Hettangian genus Psiloceras (Psiloceras spelae, about 3m higher) => middle of "extinction interval" of end-Triassic	Schoene et al., 2010
Fourteen single zircon grains analyzed with CA-TIMS and ET2535 spike (0.05% tracer uncertainty) illustrate excess scatter attributed to a range of growth histories prior to eruption; the authors estimate the eruption age from youngest robust grain at 201.32 ± 0.20 Ma; a weighted mean of youngest three robust grains is statistically equivalent at 201.43 ± 0.12 Ma.	Top-Triassic ammonite zone gap	One meter above the highest local occurrence of the topmost Triassic ammonite genus *Choristoceras (Choristo crickmayi,* in Bed 84b, 59.5 m level), and 5 m below the Hettangian genus *Psiloceras (Psiloceras spelae,* in bed 93b, 64.5 m level) and about 18 m below Psiloceras tilmanni (Bed 114-129, lowest at 77.5 m level). Tri/Jur boundary projected as about the 64 m level in section (FAD of *Psiloceras spelae*).	Schoene et al., 2010
Eight single zircon grains yield a weighted mean 206Pb/238U age utilizing CA-TIMS and in-house MIT spike. [NOTE: Unsuccessful magnetostrat, but age is consistent using correlations of M. nodosus total-range to Pizzo Mondello (with its magstrat projection within the Newark interval of E5r to E7n, hence a cycle-strat age of 228-230.5 Ma based on Kent and Olsen, 1999). They suggest this consistency indicates that the "less reliable" Newark APTS below 226.5 Ma seems fine.]	Assigned in text to either *P. carpathicus* or *E. nodosa* conodont zone of Channell et al. (2003) or to lower and middle part of the *M. nodusus* zone of Orchard (1991).	Just below FAD of *P. carpathicus* conodont and within total range of conodont *M. nodosus*, but they indicate FAD of this taxa varies among sections and workers. Mid-Upper Carnian age is suggested by "Lower Carnian conodonts (*Glad.* spp., *N. postkockeli*, *P. inclinata*, *P. tadpole*) abruptly disappear 3 m below the ash bed at a level corresponding to the Carnian pluvial event" and "Carnian-Norian boundary can be placed 6 m above the ash bed with the first occurrence of *M. communisti*." [If one uses the "more reliable" position just below FAD of P. carpathicus for the Pizzo Mondello placement, then this projects to approximately Newark chron E5n.]	Furin et al., 2006

No.	Period	Epoch	Age	Sample	Locality	Formation	GTS2012 Age (Ma)	± 2σ (without λ errors)	± 2σ (with λ errors)	Type
		Middle								
			Ladinian							
T21				Alpe di Siusi ash bed		Ash bed, Alpe di Siusi/Seiser Alm area, Dolomites, northern Italy	237.85	± 0.05	± 0.14	$^{206}Pb/^{238}U$
T20				FP2; volcaniclastic sandstone	Flexenpass, eastern Vorarlberg (N47°08'35" E10°10'01"), westernmost Northern Calcareous Alps, Austria	Graded volcaniclastic layer in the upper part (44.3 m level in section) of the Reifling Formation	239.34	± 0.28	± 0.38	$^{206}Pb/^{238}U$
T19				Litér, volcanic ash matrix of neptunian dyke	Litér dolomite quarry, Balaton Highlands, Hungary	Neptunian dyke into Tagyon Dolomite	238.58	± 0.65	± 0.70	$^{206}Pb/^{238}U$
			Anisian							
T18				MSG.09; volcanic ash	Miniera Val Porina, 820m elevation, Monte San Giorgio, Lake Lugano, Switzerland	Tuff in bituminous shales (Grenzbitumen horizon), Bed 71	242.14	± 0.45	± 0.52	$^{206}Pb/^{238}U$
T17				MSG.09; volcanic ash	Miniera Val Porina, 820m elevation, Monte San Giorgio, Lake Lugano, Switzerland	Tuff in bituminous shales (Grenzbitumen horizon), Bed 71	241.16	± 0.50	± 0.73	$^{40}Ar/^{39}Ar$
T16				Svalis Dome organic-rich shale	Svalis Dome, Barents Sea (IKU drill core 7323/07-U-09)	Near top of Steinkobbe Fm.	239.30	± 2.70	± 2.70	Re-Os
T15				LAT31; tuff	Cimon Latemar, Dolomites, N. Italy.	Lower Cyclic Facies (lower-middle)	242.65	± 0.76	± 0.80	$^{206}Pb/^{238}U$
T14				LAT30; tuff	Cimon Latemar, Dolomites, N. Italy.	Middle Tepee Facies (middle)	242.80	± 0.34	± 0.42	$^{206}Pb/^{238}U$
T13				Svalbard organic-rich shale	central Svalbard, Barents Sea	10-20 cm above base of Blanknuten Member (middle of Botneheia Fm.)	241.20		± 2.20	Re-Os
		Early								
			Olenekian							
T12				CHIN-34; volcanic ash	Jinya section, Nanpanjiang Basin, NW Guangxi, south China	Baifeng Formation (shaly base of formation)	244.60	± 0.43	± 0.50	$^{206}Pb/^{238}U$
T11				CHIN-29; volcanic ash	Jinya section, Nanpanjiang Basin, NW Guangxi, south China	Uppermost part of "Transitional beds" between Luolou and Baifeng formations	246.80	± 0.39	± 0.47	$^{206}Pb/^{238}U$
T10				GDGB-0; tuffaceous siliciclastic mudstone	Great Bank of Guizhou, Upper Guandao section, south China	7-m thick volcaniclastic tuff	246.30	± 0.14	± 0.29	$^{206}Pb/^{238}U$
T9				GDGB Tuff-110	Great Bank of Guizhou, Lower Guandao section, south China	In the "green-bean rock" interval, 260.25 m level (14.65 m above base-Anisian)	246.77	± 0.16	± 0.31	$^{206}Pb/^{238}U$
			Induan							
T8				PGD Tuff-3	Great Bank of Guizhou, Lower Guandao section	In the "green-bean rock" interval, 247.9 m level (2.3 m above their assignment of base-Anisian based on conodonts)	247.13	± 0.17	± 0.31	$^{206}Pb/^{238}U$
T7				PGD Tuff-2	Great Bank of Guizhou, Lower Guandao section	In the "green-bean rock" interval, 239.3 m level (6.3 m below their assignment of base-Anisian based on conodonts)	247.32	± 0.08	± 0.30	$^{206}Pb/^{238}U$

Primary radiometric age details	Zone assignment	Primary biostratigraphic age details	Reference
Seven single zircon grains pre-treated by the chemical abrasion (CA-TIMS)	Late Ladinian	"close to the *neumayri/regoledanus* Subzones … The presence of *Protrachyceras steinmanni* at a similar stratigraphic level in the correlative Bagolino section (Brack *et al.*, 1993) suggests that this age corresponds to the *longobardicum* Subzone." (Mietto et al., 2012, in press)	Mietto *et al.*, Episodes, 2012, in press.
Nine single zircon grain analyses yielded a weighted mean 206Pb/238U age, utilizing CA-TIMS and in-house BGC spike (0.1% error in tracer calibration)		About 2 m below level with bivalve *Daonella tyrolensis*, and about 16 m above level with *Daonella* cf. *longobardica*. The bivalve levels and ash bed markers indicate equivalence to upper part of non-ammonite interval between *Protrachyceras gredleri* and overlying *Protrachyceras archelaus* zones in Bagolino section, at the tuff where Mundil BAG.07 was dated.	Brühwiler et al., 2007
Two (of 5) multi-grain zircon analyses yielded a weighted mean 206Pb/238U age, utilizing mechanical abrasion and in-house NIGL spike (0.1% error in tracer calibration).		Redeposited tuff in neptunian dyke that contains ammonite assemblage assigned to upper *Protrachyceras gredleri* ammonite zone. Pálfy *et al.* (2003) schematically show biostratigraphic uncertainty as spanning upper-quarter of *P. gredleri* Zone through lower-quarter of *P. archelaus* Zone.	Pálfy et al., 2003
Six (of 8) single zircon grain analyses yielded a weighted mean 206Pb/238U age, utilizing CA-TIMS and in-house BGC spike (0.1% error in tracer calibration)		Daonella bivalve species are 150cm below lowest dated sample. Assigned to lowermost *Nevadites secendensis* ammonite zone by Brack *et al.*, 1996.	Mundil et al., 2010
One 5 mg sanidine laser incremental heating plateau age (97% of gas)		Daonella bivalve species are 150cm below lowest dated sample. Assigned to lowermost *Nevadites secendensis* ammonite zone by Brack *et al.*, 1996.	Mundil et al., 2010
Fit to 8 samples over a 1.23 m section.Model 3 fit (MSWD = 17) to 14 samples (8 + 6 replicates) collected over a 1.23 m interval of drillcore.		Assigned as top of *Frechites laqueatus* ammonite zone (uppermost Anisian)	Xu et al., 2009
Six (of 9) single zircon grain analyses yielded a weighted mean 206Pb/238U age, utilizing HF-leaching and in-house BGC spike (0.1% error in tracer calibration).		About 40m above base of Lower Cyclic Facies, which overlies Latemarites ammonites, a genera with range assigned by them to upper *Reitzi* Zone; therefore, sampled level might correspond to lowermost part of Nevadites secedensis ammonite zone.	Mundil et al., 2003
Ten single zircon grain analyses yielded a weighted mean 206Pb/238U age, utilizing CA-TIMS and in-house BGC spike (0.1% error in tracer calibration).		uppermost *Reitzi* zone	Brack et al., 2007
Model 3 fit (MSWD = 8.7) to 7 samples (4 + 3 replicates) collected over a 0.90 m interval of outcrop.		Assigned as between underlying *Anagymnotoceras varium* and overlying *Frechites laqueatus* ammonite zones	Xu et al., 2009
Four (of 7) single zircon grain analyses yielded a weighted mean 206Pb/238U age, utilizing CA-TIMS and in-house UNIGE spike (0.1% error in tracer calibration). Uncertainties recalculated in Galfetti et al. (2007).	*Balatonites shoshonensis*	Bracketed by layers containing an ammonoid assemblage diagnostic of the low-palolatitudinal *Balatonites shoshonensis* Zone.	Ovtcharova et al., 2006
Six (of 8) single zircon grain analyses yielded a weighted mean 206Pb/238U age, utilizing CA-TIMS and in-house UNIGE spike (0.1% error in tracer calibration). Uncertainties recalculated in Galfetti et al. (2007).	*Acrochordiceras hyatti*	Poorly preserved *Platycuccoceras*-dominated ammonoid assemblage, indicates *Acrochordiceras hyatti* ammonite zone	Ovtcharova et al., 2006
Thirteen (of 40) single zircon grain analyses yielded a weighted mean 206Pb/238U age, utilizing both physical abrasion and CA-TIMS, and in-house MIT spike (0.05% error in tracer calibration).		"a short distance above the Olenekian boundary". Nearly at base of "Bith-N" polarity zone (in nomenclature of Ogg, 2008). FAD of conodont *Chiosella timorensis* (working definition of base-Anisian) is about 3m below this ash layer, and *Ni. germanica* FAD is at about this level.	Ramezani et al., 2007
Eleven (of 26) single zircon grain analyses yielded a weighted mean 206Pb/238U age, utilizing both physical abrasion and CA-TIMS, and in-house MIT spike (0.05% error in tracer calibration).		Just above LADs of conodonts *Neogondolella regalis* and *Chiosella timorensis*, middle of range of *Neospathodus germanica*, and just above FAD of conodont *Nicoraella kockeli*. Assigned by them to lowest part of Pelsonian substage (they assign base-Bithynian to be the FAD of *Neospathodus germanica*, and base-Pelsonian to be the FAD of *Nicoraella kockeli*). ["The overlap of *N. germanicus* with *N. kockeli* is only seen in the {ammonoid} *Isculites constrictus* Subzone of the *A. hyatti* Zone." (M. Orchard, pers. commun., 2005, cited in Ovtcharova *et al.*, 2006)]	Lehrmann et al., 2006
Weighted mean of 8 (of 20) single zircon grain analyses, utilizing both physical abrasion and CA-TIMS, and in-house MIT spike (0.05% error in tracer calibration). Analytical error expanded in accommodate geologic scatter.		Just above LAD of conodont *Neospathodus symmetricus*, in uppermost part of range of *Neospathodus homeri*, in lowermost part of range of *Chiosella timorensis*, and base of range of *Neogondolella regalis*. Assigned by them as about one-third up in Aegean subzone (they assign base-Aegean, hence base-Anisian, to be the FAD of *Chiosella timorensis*). ["… cannot be precisely tied to any of the refined ammonoid zones or beds as revised by Monnet and Bucher." (Ovtcharova *et al.*, 2006)]. However, Ovtcharova *et al.* (2010) indicate that lowest occurrene of C. timorensis is within Spathian (late Early Triassic).	Lehrmann et al., 2006
Weighted mean of 11 (of 29) single zircon grain analyses, utilizing both physical abrasion and CA-TIMS, and in-house MIT spike (0.05% error in tracer calibration).		Just above PGD Tuff-1; uppermost part of range of conodont *Neospathodus triangularis*, and in upper part of range of *Neospathodus homeri* (uppermost Spathian).	Lehrmann et al., 2006

No.	Period	Epoch	Age	Sample	Locality	Formation	GTS2012 Age (Ma)	± 2σ (without λ errors)	± 2σ (with λ errors)	Type
T6				PGD Tuff-1	Great Bank of Guizhou, Lower Guandao section	In the "green-bean rock" interval, 238.8 m level (6.8 m below their assignment of base-Anisian based on conodonts)	247.38	± 0.14	± 0.30	$^{206}Pb/^{238}U$
T5				CHIN-23; volcanic ash	Jinya section, Nanpanjiang Basin, NW Guangxi, South China	Luolou Formation (upper carbonate unit)	248.12	± 0.37	± 0.45	$^{206}Pb/^{238}U$
T4				CHIN-10; volcanic ash	Jinya section, Nanpanjiang Basin, NW Guangxi, South China	Luolou Formation (upper carbonate unit)	250.55	± 0.47	± 0.54	$^{206}Pb/^{238}U$
T3				CHIN-40; volcanic ash	Jinya section, Nanpanjiang Basin, NW Guangxi, South China	Luolou Formation (lower unit of "thin-bedded, dark, laminated, suboxic limestones alternating with dark, organic-rich shales")	251.22	± 0.20	± 0.42	$^{206}Pb/^{238}U$
P18				SH32(29) bentonite	Shangsi section, north Sichuan Province, central China	Bed 29, 103.4 m level, lower Feixianguan Formation	252.50	± 0.32	± 0.42	$^{206}Pb/^{238}U$
T2b				Same bed as SH32(29); Sanidine from volcanic ash	Shangsi section, north Sichuan Province, central China.	Ash horizon, Bed 29, 103.4 m level, lowermost Feixianguan Formation	250.81	± 0.30		$^{40}Ar/^{39}Ar$
T2				AW3; bentonite	Meishan quarry Z, SE China	Bed 28, between the Permian carbonate Changhsingian Fm and the Triassic siliclastic Yinking Fm	252.48	± 0.30	± 0.47	$^{206}Pb/^{238}U$
T1				Bed 28; bentonite	Meishan quarry Z, SE China	Bed 28, between the Permian carbonate Changhsingian Fm and the Triassic siliclastic Yinking Fm	250.98	± 0.14	± 0.59	$^{40}Ar/^{39}Ar$
	Permian									
		Lopingian								
			Changhsingian							
P17				D3t - Bed 25; bentonite	Meishan section, the GSSP of the base of the Triassic; Changhsing, Zhejiang Province, S. China	Bed 25, 18 cm below GSSP Bed 27c at Meishan D section	252.41	± 0.41	± 0.49	$^{206}Pb/^{238}U$
P17b				C-2; bentonite [same as D3t]	Meishan section D, the GSSP of the base of the Triassic; Changhsing, Zhejiang Province, S. China	Bed 25 volcanic clay	251.63	± 0.20	± 0.61	$^{40}Ar/^{39}Ar$
P16				SH10(27); bentonite	Shangsi section, north Sichuan Province, central China	Bed 27, 100.05 m level, uppermost Dalong Formation	252.18	± 0.48	± 0.55	$^{206}Pb/^{238}U$
P16b				GS-1; bentonite [same as SH10(27)]	Shangsi section, Sichuan province, China	Bed 27, 100.05 m level, uppermost Dalong Formation	251.80	± 0.51	± 0.77	$^{40}Ar/^{39}Ar$
P15				SH09(25); bentonite	Shangsi section, north Sichuan Province, central China	Bed 25, 99.7 m level, upper Dalong Formation	252.53	± 0.32	± 0.42	$^{206}Pb/^{238}U$
P15b				SH-09; bentonite [same as SH09(25)]	Shangsi section, north Sichuan Province, central China.	Ash horizon, Bed 25, 99.7 m level, uppermost Dalong Formation	252.09	± 0.29	± 0.65	$^{40}Ar/^{39}Ar$
			Wuchiapingian							
P14				SH27(23); bentonite	Shangsi section, north Sichuan Province, central China	Bed 23, 97.1 m level, upper Dalong Formation	253.24	± 0.36	± 0.45	$^{206}Pb/^{238}U$
P14b				SH-27; bentonite [same as SH27(23)]	Shangsi section, north Sichuan Province, central China.	Ash horizon, Bed 23, 97.1 m level, uppermost Dalong Formation	252.57	± 0.34	± 0.67	$^{40}Ar/^{39}Ar$
P13				SH16(18); bentonite	Shangsi section, north Sichuan Province, central China	Bed 18, ~85 m level, upper-middle of Dalong Formation	253.69	± 0.32	± 0.42	$^{206}Pb/^{238}U$
P13b				SH-16; bentonite [same as SH16(18)]	Shangsi section, north Sichuan Province, central China.	Ash horizon, Bed 18, about 85 m level, upper Dalong Formation,	252.81	± 0.42	± 0.72	$^{40}Ar/^{39}Ar$
P12				SH08(15); bentonite	Shangsi section, north Sichuan Province, central China	Bed 15, 72.5 m level, lower-middle of Dalong Formation	257.30	± 0.43	± 0.51	$^{206}Pb/^{238}U$
P12b				SH-08; bentonite [same as SH08(15)]	Shangsi section, north Sichuan Province, central China	Bed 15, 72.5 m level, lower-middle of Dalong Formation	256.75	± 0.26	± 0.66	$^{40}Ar/^{39}Ar$
P11				SH03(8); bentonite	Shangsi section, north Sichuan Province, central China	Bed 8, 36.3 m level, lower-middle of Wuchiaping Formation	260.74	± 0.86	± 0.90	$^{206}Pb/^{238}U$
P10				SH01(7); bentonite	Shangsi section, north Sichuan Province, central China	Bed 7, 23.5 m level, lower-middle of Wuchiaping Formation	259.10	± 1.01	± 1.05	$^{206}Pb/^{238}U$

Primary radiometric age details	Zone assignment	Primary biostratigraphic age details	Reference
Weighted mean of 8 (of 23) single zircon grain analyses, utilizing both physical abrasion and CA-TIMS, and in-house MIT spike (0.05% error in tracer calibration).		Just below PGD Tuff-2; uppermost part of range of conodont *Neospathodus triangularis,* and in upper part of range of *Neospathodus homeri* (uppermost Spathian).	Lehrmann et al., 2006
Weighted mean of 6 (of 10) single zircon grain analyses, utilizing CA-TIMS and in-house UNIGE spike (0.1% error in tracer calibration). Uncertainties recalculated in Galfetti et al. (2007).	*Neopopanoceras haugi*	Low-paleolatitude *Neopopanoceras haugi* zone ammonite fauna (second ammonite zone down from top of Spathian) is correlated with the high-paleolatitude *Keyserlingites subrobustus* zone.	Ovtcharova et al., 2006
Weighted mean of 6 (of 8) single zircon grain analyses, utilizing CA-TIMS and in-house UNIGE spike (0.1% error in tracer calibration). Uncertainties recalculated in Galfetti et al. (2007).	*Tirolites/Columbites*	*Tirolites/Columbites* ammonite assemblage (second ammonite zone from base of Spathian)	Ovtcharova et al., 2006
Weighted mean of 6 single zircon grain analyses, utilizing CA-TIMS and in-house UNIGE spike (0.1% error in tracer calibration)		"within the "Kashmirites densistriatus beds" of early Smithian", which overlie beds with Hedenstroemia hedenstroemi ammonites that are assigned to basal-Smithian (basal Olenekian)	Galfetti et al., 2007
Twelve single zircon grain analyses yielded a weighted mean 206Pb/238U age of 252.50 ± 0.32 Ma, utilizing CA-TIMS and in-house BGC spike (0.1% error in tracer calibration); "may be compromised by re-deposited zircons"		1.1 m below local FOD of *H. parvus,* but likely within lowermost Triassic (about 3 m above extinction horizon)	Mundil et al., 2004
details not given in their source [Renne et al., 2004. The time scale of extinctions and paleoenvironmental crisis from $^{40}Ar/^{39}Ar$ dating of Permo-Triassic bentonites at Shangsi (Sichuan, China), 32nd International Geological Congress].		1.1 m below local FAD of H. parvus, but likely within lowermost Triassic (about 3 m above extinction horizon)	summarized in Mundil et al., 2010
Weighted mean of 6 single zircon grain analyses, utilizing HF leaching and in-house BGC spike (0.1% error in tracer calibration)		Bed 28 is 8cm above Permian-Triassic GSSP, the FAD of H. parvus in bed 27c	Mundil et al., 2001
Weighted mean of 6 multi-grain sanidine laser and furnace incremental heating plateau ages (91-100% gas).		Bed 28 is 8cm above Permian-Triassic GSSP, the FAD of H. parvus in bed 27c	Reichow et al., 2009
Weighted mean of 6 single zircon grain analyses, utilizing CA-TIMS and in-house BGC spike (0.1% error in tracer calibration)		Bed 25, a 5-cm volcanic clay, is the approximate culmination of the late-Permian mass extinction. This former "boundary clay" level, is 18cm below GSSP Bed 27c at Meishan D section, corresponding to first appearance of conodont *Hindeodus parvus.*	Mundil et al., 2004
5-10 mg multi-grain sanidine laser incremental heating plateau age (>99% gas)		Bed 25 at Meishan	Renne et al., 2010
Weighted mean of 5 single zircon grain analyses, utilizing CA-TIMS and in-house BGC spike (0.1% error in tracer calibration)		Uppermost Changhsingian (immediately below extinction horizon, which might correspond to Bed 25 at Meishan)	Mundil et al., 2004
Weighted mean of five multi-grain (5-10 mg) plagioclase incremental heating plateau ages (>95% gas).		Approx. 5cm below former paleontological-defined boundary. If similar stratigraphic-thickness trends as Meishan GSSP, then this is probably 10cm below revised base-Triiassic definition?	Renne et al., 2010
Weighted mean of 12 single zircon grain analyses, utilizing CA-TIMS and in-house BGC spike (0.1% error in tracer calibration)		Uppermost Changhsingian (30 cm below extinction horizon, which might correspond to Bed 25 at Meishan)	Mundil et al., 2004
Weighted mean of two multi-grain sanidine laser incremental heating plateau ages (>95% gas).		Uppermost Changhingian (30 cm below extinction horizon, which might correspond to Bed 25 at Meishan)	Renne et al., 2010
Weighted mean of 8 single zircon grain analyses, utilizing CA-TIMS and in-house BGC spike (0.1% error in tracer calibration)		Upper Changhsingian (about 3 m below top of Changhsingian; which spans 20 m in this section)	Mundil et al., 2004
Weighted mean of 9 single grain laser incremental heating plateau ages (>95% gas).		Upper Changhingian (about 3 m below top of Changhsingian; which spans 20 m in this section)	Renne et al., 2010
Weighted mean of 14 single zircon grain analyses, utilizing CA-TIMS and in-house BGC spike (0.1% error in tracer calibration)		Lowermost Changhsingian (about 5 m above base of Changhsingian; which spans 20 m in this section)	Mundil et al., 2004
One multi-grain sanidine laser incremental heating plateau age (100% of gas).		Lowermost Changhingian (about 5 m above base of Changhsingian; which spans 20 m in this section)	Renne et al., 2010
Weighted mean of 8 single zircon grain analyses, utilizing CA-TIMS and in-house BGC spike (0.1% error in tracer calibration)		Uppermost Wuchiapingian (about 7 m below base of Changhsingian; where Wuchiapingian spans about 80 m in this section)	Mundil et al., 2004
Weighted mean of four multi-grain sanidine laser incremental heating plateau ages.		Uppermost Wuchiapingian (about 7 m below base of Changhsingian; where Wuchiapingian spans about 80 m in this section)	Renne et al., 2010
Weighted mean of 5 single zircon grain analyses, utilizing CA-TIMS and in-house BGC spike (0.1% error in tracer calibration)		Lower Wuchiapingian	Mundil et al., 2004
Weighted mean of 6 single zircon grain analyses, utilizing CA-TIMS and in-house BGC spike (0.1% error in tracer calibration)		Lower Wuchiapingian	Mundil et al., 2004

No.	Period	Epoch	Age	Sample	Locality	Formation	GTS2012 Age (Ma)	± 2σ (without λ errors)	± 2σ (with λ errors)	Type
		Guadalupian								
			Capitanian							
P9				zircons	Ash bed occurs between the Hegler and Pinery limestone members of the Bell Canyon Formation; Nipple Hill, Guadalupe Mountains National Park, Texas	greyish-green bentonite	265.35	± 0.36	± 0.46	$^{206}Pb/^{238}U$
			Wordian							
			Roadian							
		Cisuralian								
			Kungurian							
			Artinskian							
P8				01DES-403; bentonite	Dal'ny Tulkas roadcut section, Southern Urals, Russia	12.5 mab of section (base of Bed 9), in Tulkas Formation	288.21	± 0.15	± 0.34	$^{206}Pb/^{238}U$
P7				DTR905; bentonite	Dal'ny Tulkas roadcut section, Southern Urals, Russia	10.5 mab of secttion (upper Bed 7), in Tulkas Formation	288.36	± 0.17	± 0.35	$^{206}Pb/^{238}U$
			Sakmarian							
P6				07DTRBed2; bentonite	Dal'ny Tulkas roadcut section, Southern Urals, Russia	4.0 mbb of section (upper Bed 2), in Tulkas Formation	290.81	± 0.17	± 0.35	$^{206}Pb/^{238}U$
P5				97USO-91.0; bentonite	Usolka (Krasnousolsky) section, Southern Urals, Russia	91.0 mab of section, in Tulkas Formation	290.50	± 0.17	± 0.35	$^{206}Pb/^{238}U$
P4				97USO-66.2; bentonite	Usolka (Krasnousolsky) section, Southern Urals, Russia	66.2 mab of section (upper Bed 28), in Tulkas Formation	291.10	± 0.18	± 0.36	$^{206}Pb/^{238}U$
			Asselian							
P3				01DES-212; bentonite	Usolka (Krasnousolsky) section, Southern Urals, Russia	41.25 mab of section (upper Bed 18), in Kurkin Formation	296.69	± 0.16	± 0.37	$^{206}Pb/^{238}U$
P2				01DES-202; bentonite	Usolka (Krasnousolsky) section, Southern Urals, Russia	36.0 mab of section (upper Bed 16), in Kurkin Formation	298.05	± 0.46	± 0.56	$^{206}Pb/^{238}U$
P1				01DES-194; bentonite	Usolka (Krasnousolsky) section, Southern Urals, Russia	32.4 mab of section (middle Bed 16), in Kurkin Formation	298.49	± 0.20	± 0.37	$^{206}Pb/^{238}U$
	Carboniferous (Pennsylvanian)									
		Late								
			Ghzelian							
Cb35				01DES-144; bentonite	Usolka (Krasnousolsky) section, Southern Urals, Russia	30.8 mab of section (lower Bed 16), in Kurkin Formation	299.22	± 0.20	± 0.37	$^{206}Pb/^{238}U$
Cb34				97USO-23.3; bentonite	Usolka (Krasnousolsky) section, Southern Urals, Russia	23.3 mab of section (lower Bed 14), in Kurkin Formation	300.22	± 0.16	± 0.35	$^{206}Pb/^{238}U$
Cb33				01DES-121; bentonite	Usolka (Krasnousolsky) section, Southern Urals, Russia	21.45 mab of section (upper Bed 12), in Kurkin Formation	301.29	± 0.16	± 0.36	$^{206}Pb/^{238}U$
Cb32				01DES-112; bentonite	Usolka (Krasnousolsky) section, Southern Urals, Russia	18.2 mab of section (middle Bed 12), in Kurkin Formation	301.82	± 0.17	± 0.36	$^{206}Pb/^{238}U$
Cb31				01DES-63; bentonite	Usolka (Krasnousolsky) section, Southern Urals, Russia	12.0 mab of section, in Kurkin Formation	303.10	± 0.16	± 0.36	$^{206}Pb/^{238}U$
Cb30				97USO-2.7; bentonite	Usolka (Krasnousolsky) section, Southern Urals, Russia	11.4 mab of section, in Kurkin Formation	303.54	± 0.18	± 0.39	$^{206}Pb/^{238}U$
			Kasimovian							
Cb29				97USO-1.2 & 01DES-63; bentonites	Usolka (Krasnousolsky) section, Southern Urals, Russia	10.65 mab of section, in Kurkin Formation	304.49	± 0.16	± 0.36	$^{206}Pb/^{238}U$
Cb28				02VD-5 & 01DES-42; bentonites	Usolka (Krasnousolsky) section, Southern Urals, Russia	9.85 mab of section, in Kurkin Formation	304.83	± 0.16	± 0.36	$^{206}Pb/^{238}U$
Cb27				01DES-31 & 08USO-8.4; bentonites	Usolka (Krasnousolsky) section, Southern Urals, Russia	7.8 mab of section, in Kurkin Formation	305.49	± 0.16	± 0.36	$^{206}Pb/^{238}U$
Cb26				01DES-371; bentonite	Dal'ny Tulkas quarry section, Southern Urals, Russia	20.15 mab of section, in Kurkin Formation	305.51	± 0.18	± 0.37	$^{206}Pb/^{238}U$
Cb25				08USO-7.09; bentonite	Usolka (Krasnousolsky) section, Southern Urals, Russia	6.5 mab of section, in Kurkin Formation	305.95	± 0.17	± 0.37	$^{206}Pb/^{238}U$
Cb24				01DES-363; bentonite	Dal'ny Tulkas quarry section, Southern Urals, Russia	19.7 mab of section, in Kurkin Formation	305.96	± 0.17	± 0.36	$^{206}Pb/^{238}U$

Primary radiometric age details	Zone assignment	Primary biostratigraphic age details	Reference
Weighted mean of 5 multigrain zircon analyses, utilizing physical abrasion and in-house MIT spike (0.1% error in tracer calibration).		20 m below the base of the *Jinogondolella postserrata* conodont zone that defines the base of the Capitanian stage of the Guadalupian Series.	Bowring et al., 1998
Weighted mean of 8 single zircon grain analyses, utilizing CA-TIMS and the EARTHTIME 535 spike.	*Sweetognathodus "whitei"* conodont zone	12.3 meters above FAD of *Sweetognathus "whitei"*; Burtsevian substage (Russian Platform) of Artinskian Stage	Schmitz and Davydov, 2012
Weighted mean of 7 single zircon grain analyses, utilizing CA-TIMS and the EARTHTIME 535 spike.	*Sweetognathodus "whitei"* conodont zone	10.3 meters above FAD of *Sweetognathus "whitei"*; Burtsevian substage (Russian Platform) of Artinskian Stage	Schmitz and Davydov, 2012
Weighted mean of 6 single zircon grain analyses, utilizing CA-TIMS and the EARTHTIME 535 spike.	*Sweetognathodus anceps* conodont zone	4.2 meters below FAD of *Sweetognathus "whitei"*; Sterlitamakian substage (Russian Platform) of Sakmarian Stage	Schmitz and Davydov, 2012
Weighted mean of 8 single zircon grain analyses, utilizing CA-TIMS and the EARTHTIME 535 spike.	*Sweetognathodus anceps* conodont zone	Sterlitamakian substage (Russian Platform) of Sakmarian Stage	Schmitz and Davydov, 2012
Weighted mean of 8 single zircon grain analyses, utilizing CA-TIMS and the EARTHTIME 535 spike.	*Sweetognathodus anceps* conodont zone	Sterlitamakian substage (Russian Platform) of Sakmarian Stage	Schmitz and Davydov, 2012
Weighted mean of 9 single zircon grain analyses, utilizing CA-TIMS and the EARTHTIME 535 spike.	*Streptognathodus fusus* conodont zone	Uskalikian substage (Russian Platform) of Asselian Stage	Schmitz and Davydov, 2012
Weighted mean of 9 single zircon grain analyses, utilizing both physical abrasion and CA-TIMS and the MIT-1L spike calibrated against the EARTHTIME gravimetric standards.	*Streptognathodus cristellaris - St. sigmoidalis* conodont zone	4.55 meters above FAD of *Streptognathodus isolatus*; Surenian substage (Russian Platform) of Asselian Stage	Ramezani et al., 2007
Weighted mean of 14 single zircon grain analyses, utilizing both physical abrasion and CA-TIMS and the MIT-1L spike calibrated against the EARTHTIME gravimetric standards.	*Streptognathodus isolatus* conodont zone	0.95 meters above FAD of *Streptognathodus isolatus;* Surenian substage (Russian Platform) of Asselian Stage	Ramezani et al., 2007
Weighted mean of 19 single zircon grain analyses, utilizing both physical abrasion and CA-TIMS and the MIT-1L spike calibrated against the EARTHTIME gravimetric standards.	*Streptognathodus waubaunsensis* conodont zone	0.65 meters below FAD of *Streptognathodus isolatus;* Noginian substage (Russian Platform) of Gzhelian stage	Ramezani et al., 2007
Weighted mean of 7 single zircon grain analyses, utilizing CA-TIMS and the EARTHTIME 535 spike.	*Streptognathodus simplex - St. bellus* conodont zone	Noginian substage (Russian Platform) of Gzhelian stage	Schmitz and Davydov, 2012
Weighted mean of 5 single zircon grain analyses, utilizing CA-TIMS and the EARTHTIME 535 spike.	*Streptognathodus virgilicus* conodont zone	Pavlovoposadian substage (Russian Platform) of Gzhelian stage	Schmitz and Davydov, 2012
Weighted mean of 5 single zircon grain analyses, utilizing CA-TIMS and the EARTHTIME 535 spike.	*Streptognathodus virgilicus* conodont zone	Pavlovoposadian substage (Russian Platform) of Gzhelian stage	Schmitz and Davydov, 2012
Weighted mean of 4 single zircon grain analyses, utilizing CA-TIMS and the EARTHTIME 535 spike.	*Streptognathodus simulator* conodont zone	Rusavkian substage (Russian Platform) of Gzhelian stage	Schmitz and Davydov, 2012
Weighted mean of 5 single zircon grain analyses, utilizing CA-TIMS and the EARTHTIME 535 spike.	*Streptognathodus firmus* conodont zone	Rusavkian substage (Russian Platform) of Gzhelian stage	Schmitz and Davydov, 2012
Two samples of same ash bed yielded identical weighted mean ages of 304.49 ± 0.16 Ma (six single grains) and 304.42 ± 0.16 Ma (five single grains), utilizing CA-TIMS and the EARTHTIME 535 spike.	*Idiognathodus toretzianus* conodont zone	Dorogomilovian substage (Russian Platform) of Kasimovian stage	Schmitz and Davydov, 2012
Two samples of same ash bed yielded identical weighted mean ages of 304.83 ± 0.16 Ma (four single grains) and 304.82 ± 0.17 Ma (three single grains), utilizing CA-TIMS and the EARTHTIME 535 spike.	*Idiognathodus sagittalis* conodont zone	Khamovnikian substage (Russian Platform) of Kasimovian stage	Schmitz and Davydov, 2012
Two samples of same ash bed yielded identical weighted mean ages of 305.49 ± 0.16 Ma (six single grains) and 305.52 ± 0.19 Ma (six single grains), utilizing CA-TIMS and the EARTHTIME 535 spike.	*Idiognathodus sagittalis* conodont zone	Khamovnikian substage (Russian Platform) of Kasimovian stage	Schmitz and Davydov, 2012
Weighted mean of 4 single zircon grain analyses, utilizing CA-TIMS and the EARTHTIME 535 spike.	*Idiognathodus sagittalis* conodont zone	Khamovnikian substage (Russian Platform) of Kasimovian stage	Schmitz and Davydov, 2012
Weighted mean of 3 single zircon grain analyses, utilizing CA-TIMS and the EARTHTIME 535 spike.	*Streptognathodus subexcelsus* conodont zone	Krevyakian substage (Russian Platform) of Kasimovian stage	Schmitz and Davydov, 2012
Weighted mean of 8 single zircon grain analyses, utilizing CA-TIMS and the EARTHTIME 535 spike.	*Streptognathodus subexcelsus* conodont zone	Krevyakian substage (Russian Platform) of Kasimovian stage	Schmitz and Davydov, 2012

No.	Period	Epoch	Age	Sample	Locality	Formation	GTS2012 Age (Ma)	± 2σ (without λ errors)	± 2σ (with λ errors)	Type
		Middle								
			Moscovian							
Cb23				01DES-362; bentonite	Dal'ny Tulkas quarry section, Southern Urals, Russia	17.4 mab of section, in Kurkin Formation	307.66	± 0.17	± 0.37	$^{206}Pb/^{238}U$
Cb22				06USO-2.0; bentonite	Usolka (Krasnousolsky) section, Southern Urals, Russia	1.4 mab of section, in Zilim Formation	308.00	± 0.18	± 0.37	$^{206}Pb/^{238}U$
Cb21				01DES-481; bentonite	Usolka (Krasnousolsky) section, Southern Urals, Russia	1.25 mab of section, in Zilim Formation	308.36	± 0.20	± 0.38	$^{206}Pb/^{238}U$
Cb20				01DES-351; bentonite	Dal'ny Tulkas quarry section, Southern Urals, Russia	12.4 mab of section, in Zilim Formation	308.50	± 0.16	± 0.36	$^{206}Pb/^{238}U$
Cb19				n1 coal; tonstein	Butovskaya Shaft	Isaevskaya Formation, C_3^1(N) lithostratigraphic index	307.26	± 0.19	± 0.38	$^{206}Pb/^{238}U$
Cb18				m3 coal; tonstein	Zasyadko Shaft	Lisitchanskaya Formation, C_2^7(M) lithostratigraphic index	310.55	± 0.19	± 0.38	$^{206}Pb/^{238}U$
Cb17				l3 coal; tonstein	Krasnolimanskaya Shaft	Almaznaya Formation, C_2^6(L) lithostratigraphic index	312.01	± 0.17	± 0.37	$^{206}Pb/^{238}U$
Cb16				l1 coal; tonstein	Kirov Shaft	Almaznaya Formation, C_2^6(L) lithostratigraphic index	312.23	± 0.18	± 0.37	$^{206}Pb/^{238}U$
Cb15				k7 coal; tonstein	Pereval'skaya Shaft	Kamenskaya Formation, C_2^5(K) lithostratigraphic index	313.16	± 0.17	± 0.37	$^{206}Pb/^{238}U$
Cb14				k3 coal; tonstein	Pereval'skaya Shaft	Kamenskaya Formation, C_2^5(K) lithostratigraphic index	314.40	± 0.17	± 0.37	$^{206}Pb/^{238}U$
		Early								
			Bashkirian							
Cb13				Bed 32; bentonite	Kljuch section, Southern Urals, Russia	Scherbakov Formation, Bed 32	317.54	± 0.17	± 0.38	$^{206}Pb/^{238}U$
Cb12				Bed 9; bentonite	Kljuch section, Southern Urals, Russia	Scherbakov Formation, Bed 9	318.63	± 0.22	± 0.40	$^{206}Pb/^{238}U$
Cb11				Bed 2; bentonite	Kljuch section, Southern Urals, Russia	Scherbakov Formation, Bed 2	319.09	± 0.18	± 0.38	$^{206}Pb/^{238}U$
	Carboniferous (Mississippian)									
		Late								
			Serpukhovian							
Cb10				Ruzeny coal (103); tonstein	Upper Silesian Basin, Ostrava, Czech Republic (ton. 106-B2 Karel, 146.92-146.94 meters, Staric2 borehole)	Hrusov Member, Ostrava Formation	328.01	± 0.18	± 0.39	$^{206}Pb/^{238}U$
Cb9				c11 coal; tonstein	Yuzhno-Donbasskaya, Shaft No 3, Ugledar	Samarskaya Formation, C_1^3(C) lithostratigraphic index, $D_1^3{}_{lower}$(=C_4) limestone	328.14	± 0.20	± 0.40	$^{206}Pb/^{238}U$
Cb8				Ludmila coal (043); tonstein	Upper Silesian Basin, Ostrava, Czech Republic	Petrkovice Member, Ostrava Formation	328.84	± 0.23	± 0.41	$^{206}Pb/^{238}U$
		Middle								
			Visean							
Cb7				Bed 8; bentonite	Verkhnyaya Kardailovka section, Southern Urals, Russia	Gusikhin Formation, Bed 21-8	333.87	± 0.18	± 0.39	$^{206}Pb/^{238}U$

Primary radiometric age details	Zone assignment	Primary biostratigraphic age details	Reference
Weighted mean of 7 single zircon grain analyses, utilizing CA-TIMS and the EARTHTIME 535 spike.	*Neognathodus roundyi - Streptognathodus cancellosus* conodont zone	Peskovian substage (Russian Platform) of Moscovian stage	Schmitz and Davydov, 2012
Weighted mean of 4 single zircon grain analyses, utilizing CA-TIMS and the EARTHTIME 535 spike.	*Neognathodus roundyi - Streptognathodus cancellosus* conodont zone	Myachkovian substage (Russian Platform) of Moscovian stage	Schmitz and Davydov, 2012
Weighted mean of 5 single zircon grain analyses, utilizing CA-TIMS and the EARTHTIME 535 spike.	*Neognathodus roundyi - Streptognathodus cancellosus* conodont zone	Myachkovian substage (Russian Platform) of Moscovian stage	Schmitz and Davydov, 2012
Five single zircon grain analyses yielded a weighted mean 206Pb/238U age of 308.50 ± 0.16 Ma, utilizing CA-TIMS and the EARTHTIME 535 spike.	*Neognathodus roundyi - Streptognathodus cancellosus* conodont zone	Myachkovian substage (Russian Platform) of Moscovian stage	Schmitz and Davydov, 2011
Weighted mean of 9 single zircon grain analyses, utilizing CA-TIMS and the EARTHTIME 535 spike.	*Neognathodus roundyi - Streptognathodus cancellosus* conodont zone	base C_3a biostratigraphic zone of Donets Basin;	Davydov et al., 2010
Weighted mean of 10 single zircon grain analyses, utilizing CA-TIMS and the EARTHTIME 535 spike.	*Idiognathodus podolskiensis* conodont zone	base C_2^mc biostratigraphic zone of Donets Basin;	Davydov et al., 2010
Weighted mean of 6 single zircon grain analyses, utilizing CA-TIMS and the EARTHTIME 535 spike.	*Streptognathodus dissectus* conodont zone	base C_2^mb biostratigraphic zone of Donets Basin;	Davydov et al., 2010
Weighted mean of 5 single zircon grain analyses, utilizing CA-TIMS and the EARTHTIME 535 spike.	*Streptognathodus dissectus* conodont zone	base C_2^mb biostratigraphic zone of Donets Basin;	Davydov et al., 2010
Weighted mean of 8 single zircon grain analyses, utilizing CA-TIMS and the EARTHTIME 535 spike.	*Neognathodus uralicus* conodont zone	base C_2^ma biostratigraphic zone of Donets Basin;	Davydov et al., 2010
Weighted mean of 7 single zircon grain analyses, utilizing CA-TIMS and the EARTHTIME 535 spike.	*Declinognathodus donetzianus* conodont zone	base C_2^ma biostratigraphic zone of Donets Basin;	Davydov et al., 2010
Weighted mean of 4 single zircon grain analyses, utilizing CA-TIMS and the EARTHTIME 535 spike.		Foraminifera correlate to upper Bashkirian Tashastian Horizon of Urals, and to just below H_6 Limestone of Donets Basin; Cheremshanian substage (Russian Platform) of Bashkirian stage	Schmitz and Davydov, 2012
Weighted mean of 5 single zircon grain analyses, utilizing CA-TIMS and the EARTHTIME 535 spike.		Foraminifera correlate to middle Bashkirian Ashyabashian Horizon of Urals, and between G_1 and G_1^2 Limestones of Donets Basin; Prikamian substage (Russian Platform) of Bashkirian stage	Schmitz and Davydov, 2012
Weighted mean of 4 single zircon grain analyses, utilizing CA-TIMS and the EARTHTIME 535 spike.		Foraminifera correlate to middle Bashkirian Ashyabashian Horizon of Urals, and between F_1 and F_1^1 Limestones of Donets Basin; Prikamian substage (Russian Platform) of Bashkirian stage	Schmitz and Davydov, 2012
Five single zircon grain analyses with a weighted mean 206Pb/238U age of 328.01 ± 0.18 Ma, utilizing CA-TIMS and the EARTHTIME 535 spike.	*Lochriea ziegleri* zone	Pendleian of lower Namurian A (lower Serpukhovian)	Gastaldo et al., 2009
Eight single zircon grain analyses with a weighted mean 206Pb/238U age of 328.14 ± 0.20 Ma, utilizing CA-TIMS and the EARTHTIME 535 spike.	*Lochriea ziegleri* zone	middle $C_1^vg_2$ biostratigraphic zone of Donets Basin; correlative with the Lochriea ziegleri zone of Urals, the Tarussian and lower Steshevian Horizons of the Russian Platform, and the *Eumorphoceras* ammonoid zone (Pendleian and Arnsbergian Stages) of the Namurian of Western Europe	Davydov et al., 2010
Five single zircon grain analyses with a weighted mean 206Pb/238U age of 328.84 ± 0.23 Ma, utilizing CA-TIMS and the EARTHTIME 535 spike.	*Lochriea ziegleri* zone	Pendleian of lower Namurian A (lower Serpukhovian)	Gastaldo et al., 2009
Four single zircon grain analyses with a weighted mean 206Pb/238U age of 333.87 ± 0.18 Ma, utilizing CA-TIMS and the EARTHTIME 535 spike. Twelve other single zircons yielded older ages indicative of reworking of the original pyroclastic deposit.	*Lochriea mononodosa* zone	1.48 meters below the FOD of *Lochriea ziegleri*, in the middle part of the *Lochriea mononodosa* zone, and *Hypergoniatites-Ferganoceras* ammonoid genozone of the Urals	Schmitz and Davydov, 2012

No.	Period	Epoch	Age	Sample	Locality	Formation	GTS2012 Age (Ma)	± 2σ (without λ errors)	± 2σ (with λ errors)	Type
Cb6				C1ve2; bentonite	Sukhaya Volnovakha, Dokuchaevsk, Tsentral'nyi rudnik, east	Styl'skaya Formation, C_1A_9 lithostratigraphic index	342.01	± 0.19	± 0.41	$^{206}Pb/^{238}U$
Cb5				C1vc; bentonite	Sukhaya Volnovakha, Dokuchaevsk, Tsentral'nyi rudnik, east	Skelevatskaya Formation, C_1A_8 lithostratigraphic index	345.00	± 0.18	± 0.41	$^{206}Pb/^{238}U$
Cb4				3/2002; bentonite	Sukhaya Volnovakha, Dokuchaevsk, Tsentral'nyi rudnik, east	Skelevatskaya Formation, C_1A_8 lithostratigraphic index	345.17	± 0.18	± 0.41	$^{206}Pb/^{238}U$
		Early								
			Tournaisian							
Cb3				5/2002; bentonite	Volnovakha River, right bank, near Businova Ravine, Donets Basin, Ukraine	Upper member, Karakubskaya Formation, C_1A_4 lithostratigraphic index	357.26	± 0.19	± 0.42	$^{206}Pb/^{238}U$
Cb2				DT-13 (α2); bentonite	Apricke section, Ruhr Basin, Nordrhein-Westfalen, Germany.	Bed 15, Hangenberg Limestone, equivalent to Bed 70 in Hasselbachtal section	358.43	± 0.19	± 0.42	$^{206}Pb/^{238}U$
Cb1				Hasselbachtel Bed 79 (α2); bentonite	Hasselbachtal section, Ruhr Basin, Nordrhein-Westfalen, Germany.	Bed 79, Hangenberg Limestone	358.71	± 0.19	± 0.42	$^{206}Pb/^{238}U$
	Devonian									
		Late								
			Famennian							
D18				DT-12 (ω3); bentonite	Apricke section, Ruhr Basin, Nordrhein-Westfalen, Germany.	Bed XX, Wocklum Limestone, equivalent to Bed xx in Hasselbachtal section	359.25	± 0.18	± 0.42	$^{206}Pb/^{238}U$
D17				black shale	Jura Creek, Alberta, Canada	Exshaw Formation	361.30	± 2.10	± 2.40	Re-Os
D16				tuff	Red Deer Creek Tuff, Alberta, Canada	Lower Member, Exshaw Formation	360.57	± 0.73	± 0.80	$^{207}Pb/^{235}U$
D15				tuff	Nordegg Tuff, Alberta, Canada	Lower Member, Exshaw Formation	363.02	± 0.73	± 0.81	$^{207}Pb/^{235}U$
D14				pumice tuff	Caldera complex, New Brunswick, Canada	Bailey Rock Rhyolite	362.87	± 0.88	± 0.96	$^{206}Pb/^{238}U$
D13				pumice tuff	Caldera complex, New Brunswick, Canada	Carrow Fm pumiceous tuff	364.08	± 2.22	± 2.25	$^{206}Pb/^{238}U$
D12				Sample WVC754; black shale (0.4-4% TOC)	Cattaraugus County, western New York, USA	Dunkirk Fm, ~6.4m above F-F boundary	367.70	± 2.50	± 2.80	Re-Os
			Frasnian							
D11				Sample WVC785; black shale (0.4-4% TOC)	Cattaraugus County, western New York, USA	Hanover Fm, ~2.9m below F-F boundary	374.20	± 4.00	± 4.20	Re-Os
D10				bentonite	Steinbruch Schmidt, Kellerwald, Germany	Kellwasser Horizons	373.68	± 1.49	± 1.54	$^{206}Pb/^{238}U$
D9				K-bentonite	Little War Gap, East Tennessee, USA	Chattanooga Shale	379.50	± 1.10	± 1.17	$^{206}Pb/^{238}U$
		Middle								
			Givetian							
			Eifelian							
D8				K-bentonite	Whyteville, Virginia, USA	Tioga ash "Middle Coarse Zone"	389.21	± 1.25	± 1.32	$^{206}Pb/^{238}U$
D7				K-bentonite	Union County, PA, USA	Tioga Ash Bed B	389.58	± 0.78	± 0.86	$^{207}Pb/^{235}U$

Primary radiometric age details	Zone assignment	Primary biostratigraphic age details	Reference
Seven single zircon grain analyses with a weighted mean 206Pb/238U age of 342.01 ± 0.19 Ma, utilizing CA-TIMS and the EARTHTIME 535 spike.		base $C_1^ve_2$ biostratigraphic zone of Donets Basin; correlative with the MFZ12 foraminiferal zone, at the base of the Holkerian stage of western Europe	Davydov et al., 2010
Six single zircon grain analyses with a weighted mean 206Pb/238U age of 345.00 ± 0.18 Ma, utilizing CA-TIMS and the EARTHTIME 535 spike.		upper C_1^vc biostratigraphic zone of Donets Basin; correlative with the base of the MFZ10 foraminiferal zone of Western Europe	Davydov et al., 2010
Six single zircon grain analyses with a weighted mean 206Pb/238U age of 345.17 ± 0.18 Ma, utilizing CA-TIMS and the EARTHTIME 535 spike.		middle C_1^vc biostratigraphic zone of Donets Basin; correlative with the base of the MFZ10 foraminiferal zone of Western Europe	Davydov et al., 2010
Eight single zircon grain analyses with a weighted mean 206Pb/238U age of 357.26 ± 0.19 Ma, utilizing CA-TIMS and the EARTHTIME 535 spike.		base of $C_1^tb_2$ biostratigraphic zone of Donets Basin; equivalent to lower Cherepetian Horizon of the eastern Russian Platform, and the MFZ3 foraminiferal zone of Western Europe	Davydov et al., 2010
Eighteen single zircon grain analyses with a weighted mean 206Pb/238U age of 358.43 ± 0.19 Ma, utilizing CA-TIMS and the EARTHTIME 535 and 2535 spikes.	Lower *Siphonodella duplicata* zone	Lower part of S.duplicata conodont zone, lower Tournaisian	Davydov et al., 2011
Nine single zircon grain analyses with a weighted mean 206Pb/238U age of 358.71 ± 0.19 Ma, utilizing CA-TIMS and the EARTHTIME 2535 spike.	Upper *Siphonodella sulcata* zone	Upper part of sulcata zone, lowermost Tournaisian	Davydov et al., 2011
Thirteen single zircon grain analyses with a weighted mean 206Pb/238U age of 359.25 ± 0.18 Ma, utilizing CA-TIMS and the EARTHTIME 535 and 2535 spikes.	Middle *Siphonodella praesulcata* zone	Middle part of *Siphonodella praesulcata* zone, uppermost Fammenian	Davydov et al., 2011
Isochron age for six samples collected over a 4m thick interval spanning the Devonian-Carboniferous boundary, processed with CrO3-H2SO4 dissolution	Middle *Palmatolepis expansa* to upper *Siphonodella duplicata* zones	Interval between the middle *Palmatolepis expansa* through upper *Siphonodella duplicata* zones	Selby and Creaser, 2005
Five monazite analysis (of six) yield a weighted mean 207Pb/235U age of 360.57 ± 0.80 Ma (recalculated using the U decay constant ratio of Mattinson, 2010).	Middle to upper *Palmatolepis expansa* zones	Nodular limestone beds below and above the Red Deer Creek Tuff are assigned to the middle and upper *Palmatolepis expansa* conodont zones, respectively.	Richards et al., 2002
Four monazite analysis yield a weighted mean 207Pb/235U age of 363.02 ± 0.81 Ma (recalculated using the U decay constant ratio of Mattinson, 2010).	Middle *Palmatolepis expansa* to lower *Siphonodella praesulcata* zones	Nodular limestone beds above and below the Nordegg Tuff are assigned to the middle *Palmatolepis expansa* through lower *Siphonodella praesulcata* conodont zones.	Richards et al., 2002
Five multigrain zircon analyses yield a weighted mean 206Pb/238U age of 362.87 ± 0.96 Ma (sample above or crosscutting spore-bearing bed)	uppermost *Palmatolepis marginifera* to upper *Palmatolepis expansa* zones	Spore-bearing horizon between the Bailey Rock rhyolite and Carrow Fm pumiceous tuff falls within the pusillites-lepidophyta spore zone (FA2d), equivalent to the uppermost *Palmatolepis marginifera* to upper *Palmatolepis expansa* conodont zones (Streel, 2000).	Tucker et al., 1998
Four multigrain zircon analyses yield a weighted mean 206Pb/238U age of 364.08 ± 2.25 Ma (95% conf. int. including geologic scatter; sample below spore-bearing bed)	uppermost *Palmatolepis marginifera* to upper *Palmatolepis expansa* zones	Spore-bearing horizon between the Bailey Rock rhyolite and Carrow Fm pumiceous tuff falls within the pusillites-lepidophyta spore zone (FA2d), equivalent to the uppermost *Palmatolepis marginifera* to upper *Palmatolepis expansa* conodont zones (Streel, 2000).	Tucker et al., 1998
Isochron age for nine samples collected from drillcore, processed with CrO3-H2SO4 dissolution	*Palmatolepis triangularis*	*Palmatolepis triangularis* conodont zone, ~6.4m above F-F boundary	Turgeon et al., 2007
Isochron age for eight samples collected from drillcore, processed with CrO3-H2SO4 dissolution	*Palmatolepis linguiformus*	*Palmatolepis linguiformus* conodont biozone, upper Frasnian stage; ~2.9m below F-F boundary	Turgeon et al., 2007
A large spread in the dates of the 24 single-zircon analyses indicates a combination of inheritance and/or Pb-loss; a cluster of 8 grains form a weighted mean 206Pb/238U age of 373.68 ± 1.54 Ma (95% conf. int. including geologic scatter).	upper *Palmatolepis rhenana*	Bed 36 bentonite occurs in the middle part of the upper *Palmatolepis rhenana* conodont Zone, Upper Frasnian.	Kaufmann et al., 2004
Nine single grain or small multigrain zircon fractions yield a weighted mean 207Pb/206Pb age of 379.52 ± 3.09 Ma (recalculated using decay constant ratio of Mattinson, 2010); two most concordant fractions give a weighted mean 206Pb/238U age of 379.50 ± 1.17 Ma.	*Palmatolepis hassi*	*Palmatolepis hassi* conodont zone, Frasnian. This age in Belpre Ash is correlated to Centre Hill ash, Frasnian.	Tucker et al., 1998; Over, 1999
Five concordant single grain or small multigrain zircon fractions in one ash bed give a weighted mean 206Pb/238U age of 389.21 ± 1.32 Ma.	*Polygnathus costatus costatus*	All Tioga ashes are between conodont bearing strata of the *Polygnathus costatus costatus* zone, middle Eifelian.	Tucker et al., 1998
Three multigrain monazite fractions yield a weighted mean 207Pb/235U age of 389.58 ± 0.86 Ma (recalculated using the U decay constant ratio of Mattinson, 2010).	*Polygnathus costatus costatus*	All Tioga ashes are between conodont bearing strata of the *Polygnathus costatus costatus* zone, middle Eifelian.	Roden et al., 1990

No.	Period	Epoch	Age	Sample	Locality	Formation	GTS2012 Age (Ma)	± 2σ (without λ errors)	± 2σ (with λ errors)	Type
		Early								
			Emsian							
D6				K-bentonite	Wetteldorf section, Germany	Heisdorf Formation	396.55	± 2.05	± 2.09	$^{206}Pb/^{238}U$
D5				Volcanic layer	Hans-Platte layer, Eschenbach quarry, Bundenbach, Germany	Lower Hunsrück Slate	407.75	± 1.08	± 1.16	$^{206}Pb/^{238}U$
D4				K-bentonite	Sprout Brook, Cherry Valley, N.Y., USA	Esopus Fm	405.44	± 1.08	± 1.16	$^{206}Pb/^{238}U$
D3				tuff	Limekilns	Winburn	409.90	± 6.60	± 6.60	$^{206}Pb/^{238}U$ ion probe
			Pragian							
			Lochkovian							
D2				volcaniclastic	type section	Turondale	413.40	± 6.60	± 6.60	$^{206}Pb/^{238}U$ ion probe
D1				K-bentonite	Kalkberg, New York State, USA	Kalkberg	415.48	± 0.97	± 2.71	$^{207}Pb/^{206}Pb$
	Silurian									
		Pridoli								
		Ludlow								
			Ludfordian							
S9				bentonite	Ludlow	Upper Whitcliffe	420.88	± 2.35	± 2.39	$^{206}Pb/^{238}U$
			Gorstian							
		Wenlock								
			Homerian							
S8				WNH15 bentonite	Lion's Mouth Cavern, Wren's Nest Hill, Dudley, England	Upper Quarried Limestone Mbr., Much Wenlock Limestone	427.77	± 0.50	± 0.68	$^{206}Pb/^{238}U$
S7				Djupvik bentonite	Djupvik 1 Locality, Gotland, Sweden	Djupvik Member, Halla Formation	428.06	± 0.48	± 0.68	$^{206}Pb/^{238}U$
S6				Grötlingbo bentonite	Hörsne 3 Locality, Gotland, Sweden	Mulde Brickclay Mbr., Halla Formation	428.47	± 0.54	± 0.72	$^{206}Pb/^{238}U$
			Sheinwoodian							
S5				bentonite	Shropsire, England	Middle Elton	431.22	± 2.62	± 3.14	$^{40}Ar/^{39}Ar$
S4				Ireviken bentonite	Ireviken 1 Locality, Gotland, Sweden	Lower Visby Formation	431.80	± 0.53	± 0.71	$^{206}Pb/^{238}U$
S3				Thin volcanic ash.	Welshpool, Wales	Buttington Shales	429.20	± 2.63	± 2.67	$^{207}Pb/^{206}Pb$
		Llandovery								
			Telychian							
S3				K-bentonite	Osmundsberg North Quarry, Siljan, Dalarna, south-central Sweden	Kallholn Shale	438.74	± 1.11	± 1.20	$^{206}Pb/^{238}U$
			Aeronian							

Primary radiometric age details	Zone assignment	Primary biostratigraphic age details	Reference
A large spread in the dates of the 19 single-zircon analyses indicates a combination of inheritance and/or Pb-loss; the 206Pb/238U age of the 11 youngest grains is 396.55 ± 2.05 Ma (95% conf. int. including geologic scatter).	uppermost *Polygnathus costatus patulus*	Uppermost part of the *Polygnathus costatus patulus* conodont zone, uppermost Emsian. The 'Hercules I' K-bentonite is situated 13 m below the formal boundary of the Lower and Middle Devonian (GSSP section).	Kaufmann et al., 2005
Ten single zircon analyses include five concordant results, which form a tightly grouped cluster with a weighted mean 206Pb/238U age of 407.75 ± 1.16 Ma.	middle to upper *Polygnathus excavatus*	Tentaculites (dacryoconarids) allow a biostratigraphic assignment to the upper part of the *Nowakia zlichovensis* dacryoconarid zone, correlated to the middle to upper part of the *Polygnathus excavatus* conodont zone, Lower Emsian (references quoted in Kaufm	Kaufmann et al., 2005
Nine concordant multigrain zircon fractions yield a weighted mean 207Pb/206Pb age of 405.34 ± 3.06 Ma (recalculated using decay constant ratio of Mattinson, 2010), and include a cluster of 5 analyses with a weighted mean 206Pb/238U age of 405.44 ± 1.16 Ma	*Polygnathus gronbergi/excavatus*	Indirect correlation with brachiopods to the conodont zones *Po.dehiscens, Po.gronbergi, and lower Po.inversus,* Lower Emsian; conodonts *Icriodus curvicauda* or *I. celtibericus* determined by José Valenzuela-Rios have been obtained from the Carlisle Center.	Tucker et al., 1998
Weighted mean of 24 zircon spots using SHRIMP II (GA), calibrated to standard QGNG		*Nowakia acuaria,* Pragian index in basal shale below age sample; sample age estimate Pragian-early Emsian	Jagodzinski and Black, 1999
Weighted mean of 27 zircon spots using SHRIMP II (GA), calibrated to standard QGNG		Replaced brachiopod fauna; its most likely extrapolated age is late Lochkovian	Jagodzinski and Black, 1999
Nine multigrain zircon fractions yield a weighted mean 207Pb/206Pb age of 415.48 ± 2.71 Ma (recalculated using the U decay constant ratio of Mattinson, 2010), including a cluster of four concordant fractions with a weighted mean 206Pb/238U age of 415.99 ± 1.93 Ma (95% conf. int. including geologic scatter).	*M. hercynicus Zone*	The *Caudicrius postwoschmidti* gp. occurs well below the bentonite, in the underlying Broncks Lake Mbr. of the true Kalkberg Fm.; this suggests a position in the upper part of the *postwoschmidti* Zone or in the upper part of the lower Lochkovian, equivalent to combined range of *M. praehercynicus + M. hercynicus nevadensis.*	Tucker et al., 1998
Four multigrain zircon fractions yield a weighted mean 206Pb/238U age of 420.88 ± 2.39 Ma (95% conf. int. including geologic scatter).	*Neocuculogr. kozlowskii* Zone to *Monogr. formosus* Zone	From regional correlations considered close to Ludlow - Pridoli boundary; here regarded as post *Saetogr. linearis* Zone to base *Neo. parultimus/ultimus* Zone	Tucker and McKerrow, 1995; Tucker et al., 1998
Six out of eleven concordant single grain zircon analyses from bentonite WNH15 yield a weighted mean 206Pb/238U age of 427.77 ± 0.68 Ma (including analytical, tracer, and decay constant errors), utilizing CA-TIMS and the EARTHTIME 535 spike.	*Uppermost Colonograptus ludensis Zone*	From detailed regional correlation this bentonite is likely only cm's below the base Ludlow GSSP. Therefore, a correlation with the uppermost part of the Colonograptus ludensis zone is assigned.	Cramer et al., 2012
Six out of nine concordant single grain zircon analyses from the lower of two bentonites separated by 16cm (Djupvik Bentonite) yield a weighted mean 206Pb/238U age of 428.06 ± 0.68 Ma (including analytical, tracer, and decay constant error), utilizing CA-TIMS and the EARTHTIME 535 spike.	*Upper part of Colonograptus praedeubeli to lower part of C.ludensis Zones*	Within the Kockelella ortus absidata conodont zone, correlated to a position somewhere from high in the Colonograptus praedeubeli/deubeli to low in the C. ludensis graptolite zones	Cramer et al., 2012
Five out of nine concordant single grain zircon analyses from the upper part of the 30-cm-thick Grötlingbo Bentonite yield a weighted mean 206Pb/238U age of 428.47 ± 0.72 Ma (including analytical, tracer, and decay error), utilizing CA-TIMS and the EARTHTIME 535 spike.	*Pristiograptus dubius parvus/Gothograptus nassa Zone*	Within the *Ozarkodina bohemica* longa conodont zone and the *Pristiograptus dubius parvus/Gothograptus nassa* graptolite zone.	Cramer et al., 2012
An 40Ar/39Ar plateau age of 425.0 ± 2.6 Ma for one biotite separate (recalculated from the data table of Kunk et al., 1985) was originally calibrated with the MMhb-1 standard age of 519.4 Ma; re-calibration with MMhb-1 of 526.04 Ma (FCs = 28.201 ± 0.046 Ma, Kuiper et al., 2008) yields an age of 431.2 ± 2.6 Ma (including estimated 0.2% fluence monitor uncertainty).	*N. nilssoni* Zone to *L. scanicus* Zone	*N. nilssoni* and *L. scanicus graptolite* Zones, Gorstian	Kunk et al., 1985
11 out of 15 concordant single grain zircon analyses from the Ireviken Bentonite yield a weighted mean 206Pb/238U age of 431.80 ± 0.71 Ma (including analytical error, tracer error, and decay constant error), utilizing CA-TIMS and the EARTHTIME 535 spike.	Middle to upper part of the *Cyrtograptus murchisoni* Zone	14 cm above Ireviken Event Datum 2 (base of the Upper *Pseudooneotodus bicornis* conodont zone), which is within cm's of the base Wenlock GSSP.	Cramer et al., 2012
Three multigrain zircon fractions yield a weighted mean 206Pb/238U age of 429.20 ± 2.67 Ma (95% conf. int. including geologic scatter).	*Monoclimacis crenulata Zone*	Within the zone of Monoclimacis crenulata	Tucker and McKerrow 1995, Tucker 1991 (USGS open file report)
Four single grain zircon analyses yield a weighted mean 206Pb/238U age of 438.74 ± 1.20 Ma (including 0.1% tracer uncertainty, Mundil et al., 2004)	*Spirograptus turriculatus* Zone	*Spirograptus turriculatus* Zone, Telychian Stage of the Llandovery Series	Bergstrom et al., 2008

No.	Period	Epoch	Age	Sample	Locality	Formation	GTS2012 Age (Ma)	± 2σ (without λ errors)	± 2σ (with λ errors)	Type
			Rhuddanian							
S2				Ash	Dobbs Linn	Birkhill Shales	439.57	± 1.24	± 1.33	$^{206}Pb/^{238}U$
S1				Ash	Esquibel Island, Alaska	Descon Formation	442.59	± 5.04	± 5.36	$^{40}Ar/^{39}Ar$
	Ordovician									
		Late								
			Hirnantian							
			Katian							
O16				Ash	Dobbs Linn (Linn Branch)	Hartfell Shales	444.88	± 1.07	± 1.17	$^{206}Pb/^{238}U$
			Sandbian							
O15				K-bentonite	Spechts Ferry Shale, Missouri	Millbrig K-bentonite, Decorah Formation	452.92	± 1.26	± 1.35	$^{206}Pb/^{238}U$
O14				MB-ST, K-bentonite	Shakertown, Kentucky	Millbrig K-bentonite, Decorah Formation	451.19	± 1.75	± 2.58	$^{40}Ar/^{39}Ar$
O13				K-bentonite	Millbrig	Diecke K-bentonite, Decorah Formation	454.50	± 1.00	± 1.10	$^{206}Pb/^{238}U$
O12				DK-LQ, K-bentonite	Lexington Quarry, Kentucky	Diecke K-bentonite, Decorah Formation	453.06	± 2.13	± 2.86	$^{40}Ar/^{39}Ar$
O11				Calcareous ash	Pont-y-ceunant, Bala, Wales	Base of Geli-grin	454.42	± 1.14	± 1.23	$^{206}Pb/^{238}U$
O10				SWE52, K-bentonite	Mossen Quarry, Kinnekulle, Sweden	Kinnekulle K-bentonite in Dalby Limestone	458.04	± 1.84	± 2.68	$^{40}Ar/^{39}Ar$
O9				K-bentonite	Mossen Quarry, Kinnekulle, Sweden	Chasmops Limestone	456.90	± 2.00	± 2.10	$^{206}Pb/^{238}U$
		Middle								
			Darriwilian							
O8				Gritty calcareous ash	Llandindrod, central Wales	Llanvirn Series	458.76	± 2.19	± 2.24	$^{206}Pb/^{238}U$
O7				ARG-1 K-bentonite	Cerro Viejo, San Juan, Argentina	Los Azules Formation	465.46	± 3.49	± 3.53	$^{206}Pb/^{238}U$
O6				Indurated bentonite	Abereiddy Bay, Wales	Lower rhyolitic tuff, Llanrian Volc Fm	462.90	± 1.23	± 1.32	$^{206}Pb/^{238}U$
O5				Ash flow	Arenig Fawr, Wales	Serv Formation	465.61	± 1.69	± 1.76	$^{206}Pb/^{238}U$
O4				Rhyolite	Central Newfoundland	Cutwell Group	469.00	± 4.00	± 6.00	$^{207}Pb/^{206}Pb$
			Dapingian							
O3				Rhyolite	Ireland	Formil Hill, Tyrone Volcanic Group	473.00	± 0.80	± 0.94	$^{206}Pb/^{238}U$
		Early								
			Floian							

Primary radiometric age details	Zone assignment	Primary biostratigraphic age details	Reference
Six multigrain zircon fractions yield a weighted mean 206Pb/238U age of 439.57 ± 0.91 Ma	*Coronogr. cyphus* Zone	Exact level uncertain, *Coronogr. cyphus* Zone assigned by Ross et al. 1982 and accepted here.	Tucker et al., 1990; Ross et al., 1982; Toghill, 1968
An 40Ar/39Ar total gas age of 436.2 ± 5 Ma for one biotite separate was originally calibrated with the MMhb-1 standard age of 519.4 Ma; re-calibration with MMhb-1 of 526.04 Ma (FCs = 28.201 ± 0.046 Ma, Kuiper et al., 2008) yields an age of 442.6 ± 5.0 Ma (including estimated 0.2% fluence monitor uncertainty).	*Corongraptus cyphus* Zone to *Spirograptus sedgwicki* Zone	4m above shale with graptolites of Coronogr. cyphus Zone age (Churkin and Carter 1970). *Co. cyphus* Zone used by Melchin *et al.* (2004) regarded as too narrow by Bergstroem et al. 2008; here given range of base *Co.cyphus* Zone to top *S.sedgwicki* Zone	Kunk et al., 1985; Churkin et al., 1971; Ross et al., 1982
Four multigrain zircon fractions yield a weighted mean 206Pb/238U age of 444.88 ± 1.17 Ma	*Paraorthograptus pacificus* Zone	Approximately 4.5m below Ordov/Sil GSSP, *Paraorthograptus pacificus* Zone	Tucker et al.,1990
Six multigrain zircon fractions yield a weighted mean 206Pb/238U age of 452.92 ± 1.35 Ma; Tucker (1992) reports a corroborating 206Pb/238U age of 453.1 ± 1.3 Ma based on five unpublished single grain zircon analyses	Upper *Diplogr. multidens* Zone	*Phragmodus undatus* Zone (lower)	Bergstrom, 1989; Huff et al., 1992; Tucker et al., 1990
Three 100% concordant 40ar/39Ar plateau spectra yield a weighted mean age of 448.0 ± 2.6 Ma (including F-value error), intercalibrated to FCs = 28.02 Ma. Re-calibration to FCs = 28.201 ± 0.046 Ma (Kuiper et al., 2008) yields an age of 450.6 ± 2.8 Ma.	Upper *Diplogr. multidens* Zone	*Phragmodus undatus* Zone (lower)	Min et al., 2001
Unpublished mean 206Pb/238U age of 454.5 ± 0.5 Ma for 5 concordant single grain analyses reported in Tucker (1992) is recalculated to 454.5 ± 1.0 Ma including 0.2% error in tracer calibration.	Upper *Diplogr. multidens* Zone	*Phragmodus undatus* Zone (lower)	Tucker, 1992; Tucker and McKerrow, 1995
Nine 100% concordant 40ar/39Ar plateau spectra yield a weighted mean age of 449.8 ± 2.4 Ma (including F-value error), intercalibrated to FCs = 28.02 Ma. Re-calibration to FCs = 28.201 ± 0.046 Ma (Kuiper et al., 2008) yields an age of 452.4 ± 2.7 Ma.	Upper *Diplogr. multidens* Zone	*Phragmodus undatus* Zone (lower)	Min et al., 2001
Four multigrain zircon fractions yield a weighted mean 206Pb/238U age of 454.42 ± 1.23 Ma; this age is corroborated by unpublished mean 206Pb/238U age of 454.8 ± 1.7 Ma for single grain analyses reported in Tucker (1992)	Close to *D. miltidens/ D. clingani* Zone boundary	Rich brachiopod, trilobite, conodont fauna. 'Longvillian' - above *D. multidens*, below *D. clingani*	Tucker et al., 1990; Tucker, 1992
Five 100% concordant 40Ar/39Ar plateau spectra yield a weighted mean age of 454.8 ± 2.6 Ma (including F-value error), intercalibrated to FCs = 28.02 Ma. Re-calibration to FCs = 28.201 ± 0.046 Ma (Kuiper et al., 2008) yields an age of 457.4 ± 2.9 Ma.	Upper *Diplograptus multidens* Zone	Upper *Amorphognathus tvaerensis* Zone, correlated by Bergstrom et al. (1995) with *Phragmodus undatus* Zone and upper *Diplogr. multidens* Zone	Min et al., 2001
Unpublished mean 206Pb/238U age of 456.9 ± 1.8 Ma for 5 concordant single grain analyses reported in Tucker & McKerrow (1995) is recalculated to 456.9 ± 2.0 Ma including 0.2% error in tracer calibration.	Upper *Diplograptus multidens* Zone	Upper *Amorphognathus tvaerensis* Zone, correlated by Bergstrom et al. (1995) with *Phragmodus undatus* Zone and upper *Diplogr. multidens* Zone	Kunk et al., 1985; Kunk and Sutter, 1984; Tucker and McKerrow, 1995; Leslie and Bergstroem, 1995
Five multigrain zircon fractions yield a weighted mean 206Pb/238U age of 458.76 ± 2.24 Ma (95% conf. int. including geologic scatter).	*Didymograptus murchisoni* Zone	*Didymograptus murchisoni* immediately below sampled ash "considered by Elles to be close to base *G. teretiusculus* Zone"	Ross et al., 1982; Tucker and McKerrow, 1995
Three multigrain zircon fractions (14 grains total) yield a weighted mean 206Pb/238U age of 465.46 ± 3.53 Ma (95% conf. int. including geologic scatter).	mid-lower *E. variabilis* Zone	10 graptolite species listed by Mitchell et al. 1998	Huff et al., 1997; Mitchell et al., 1998
Three multigrain zircon fractions yield a weighted mean 206Pb/238U age of 462.90 ± 1.32 Ma, and corroborating weighted mean 207Pb/206Pb age of 462.54 ± 3.24 Ma (recalculated using the U decay constant ratio of Mattinson, 2010).	*Didymograptus murchisoni* Zone	Immediately overlying Cyffredin Shale is of *Didymograptus murchisoni* zone age (Tucker and McKerrow 1995).	Tucker et al., 1990; Hughes et al., 1982
Two multigrain zircon fractions yield a weighted mean 206Pb/238U age of 465.61 ± 1.76 Ma.	*Didymograptus artus* Zone	Underlying mudstone contains *Didymograptus artus* Zone graptolites	Tucker et al., 1990
Three small fractions give an upper intercept age of 469 ± 4 Ma.	mid-lower *E. variabilis* Zone	Sparse fauna of mid-continental conodonts, *Histiodella holodentata*; mid-lower *E. variabilis* Zone	Dunning & Krogh 1991, Stouge, 1980
Three single zircon grains (of four) yield a weighted mean 206Pb/238U age of 473.00 ± 0.94 Ma, utilizing CA-TIMS.	*Isograptus victoriae lunatus* Zone	Immediately overlying mudstone contains *Isograptus victoriae lunatus*	Cooper et al., 2008

No.	Period	Epoch	Age	Sample	Locality	Formation	GTS2012 Age (Ma)	± 2σ (without λ errors)	± 2σ (with λ errors)	Type
			Tremodocian							
O2				Volcanic sandstone	McLeod Brook, Cape Breton Is.	Chelsey Drive Group	481.13	± 1.12	± 2.76	$^{207}Pb/^{206}Pb$
O1				Crystal-rich volcanic sandstone	Bryn-llin-fawr, N. Wales	Dolgellau Formation	486.78	± 0.53	± 2.57	$^{207}Pb/^{206}Pb$
	Cambrian									
		Furongian								
			Age 10							
C11				Crystal-rich volcanic sandstone	Ogof-ddu, Criccieth, N. Wales	Dolgellau Formation	488.71	± 1.17	± 2.78	$^{207}Pb/^{206}Pb$
			Jiangshanian							
			Paibian							
		Epoch 3								
			Guzhangian							
C10				Volcanic ash bed	Taylor Nunatak, Shackleton Glacier, Antarctica	Taylor Fm	502.10	± 2.40	± 3.50	$^{207}Pb/^{206}Pb$
			Drumian							
			Age 5							
C9				Comley ub volcanic ash bed	200 m south of Comley Quarry, Comley village, Shropshire, England	basal Quarry Ridge Grits, basalt Upper Comley Sandstone Fm	509.10	± 0.33	± 0.62	$^{206}Pb/^{238}U$
		Epoch 2								
			Age 4							
C8				SoS-56.1 volcanic ash bed	St John, New Brunswick	Hanford Brook Fm., middle Somerset St Mbr	508.05	± 1.13	± 2.75	$^{207}Pb/^{206}Pb$
			Age 3							
C7				Section Le-XI volcanic ash bed	S. Morocco	Upper Lemdad Fm.	515.56	± 1.03	± 1.16	$^{206}Pb/^{238}U$
C6				Comley lb volcanic ash bed	200 m south of Comley Quarry, Comley village, Shropshire, England	few cm below top of Green Callavia Sandstone, uppermost Lower Comley Sandstone Fm	514.45	± 0.43	± 0.69	$^{206}Pb/^{238}U$
C5				Cwm Bach 1 volcanic ash bed	Cwm Bach, near Newgale, Pembrokeshire, south Wales	closely above only known fossiliferous horizon, Caerfai Bay Shales Fm	519.30	± 0.34	± 0.64	$^{206}Pb/^{238}U$
		Terreneuvian								
			Age 2							
			Fortunian							
C4				M223 volcanic ash bed	Oud Sdas section, Anti-Atlas, Morocco	Tifnout Member, upper Adoudou Formation	525.23	± 0.28	± 0.61	$^{206}Pb/^{238}U$
C3				Volcanic ash bed	New Brunswick, Canada	Placentian Series	530.02	± 1.07	± 1.20	$^{206}Pb/^{238}U$
C2				92-N-1 volcanic ash bed	Namibia	Nomtsas Fm	538.18	± 1.11	± 1.24	$^{206}Pb/^{238}U$
C1				BB5 volcanic ash bed	Oman (3045m depth, Birba-5 well)	Ara Group, 1m above base of A4 carbonate unit	541.00	± 0.29	± 0.63	$^{206}Pb/^{238}U$
	Ediacaran									
E17				Mkz-11b volcanic ash bed	Oman (2194.4m depth, Mukhaizna-11 well)	Ara Group, 9m below top of A3 carbonate unit	542.37	± 0.28	± 0.63	$^{206}Pb/^{238}U$
E16				94-N-11 volcanic ash bed	Namibia	Urusis Fm, Upper Spitskopf Member, Schwarzrand Subgroup, Nama Group	540.61	± 0.67	± 0.88	$^{206}Pb/^{238}U$
E15				Minha-1A volcanic ash bed	Oman (3988.3m depth, Minha-1 well)	Ara Group, 3m above base of A3 carbonate unit	542.90	± 0.29	± 0.63	$^{206}Pb/^{238}U$

Primary radiometric age details	Zone assignment	Primary biostratigraphic age details	Reference
Six multigrain zircon fractions yield a weighted mean 207Pb/206Pb age of 481.13 ± 2.76 Ma (recalculated using the U decay constant ratio of Mattinson, 2010).	Late Tremadocian (Hunnebergian), Late La2 Zone	*Peltocare rotundiformis, Hunnegr.* cf. *copiosus, Adelograptus of quasimodo* type	Landing et al., 1997
Fourteen single zircon grains yield a weighted mean 207Pb/206Pb age of 486.78 ± 2.57 Ma (recalculated using the U decay constant ratio of Mattinson, 2010).	Base, *R. praeparabola* Zone, base Ordovician	Close to top *Acercare* Zone. Dated ash is 4m below appearance of *Rhabdinopora*, and 5m below *R.f. parabola*. It is therefore very close to C/O boundary	Landing et al., 2000
Nine multigrain zircon fractions yield a weighted mean 207Pb/206Pb age of 488.71 ± 2.78 Ma (recalculated using the U decay constant ratio of Mattinson, 2010).	Lower *Peltura scarabaeoides* Zone	*Peltura scarabaeoides scarabaeoides* below and *P.s. westergardi* above. = Lower Peltura scarabaeoides Zone	Davidek et al., 1998
Sample TAY-F, two multigrain zircon fractions yield a weighted mean 207Pb/206Pb age of 502.1 ± 3.5 Ma (recalculated using the U decay constant ratio of Mattinson, 2010).	Undillan stage	Trilobites in carbonate bed, 1km from dated samples. *Amphoton* cf. *oatesi, Nelsonia* cf. *schesis*, taken to indicate an Undillan, possibly late Floran, age.	Encarnacion et al., 1999
Seven single zircon grain analyses yield a weighted mean 206Pb/238U age of 509.07 ± 0.33 Ma, utilizing CA-TIMS and the EARTHTIME 535 spike.	*P.harlani* Zone, Stage 5, Series 3	*Paradoxides harlani* and other trilobites in immediately overlying beds indicate the *P. harlani* Biozone of Newfoundland	Harvey et al., 2010
Eight single zircon grains or small multigrain fractions yield a weighted mean 207Pb/206Pb age of 508.05 ± 2.75 Ma (recalculated using the U decay constant ratio of Mattinson, 2010).	*Protolenus howleyi* Zone Late Branchian, Stage 4.	*Protolenus* cf. *elegans* Matthew, *Ellipsocephalus* cf. *galeatus* Matthew associated in same bed. Suggests an age for the base of Series 3 and Stage 5 of ~507 Ma.	Landing et al., 1998
Five single zircon grains yield a weighted mean 206Pb/238U age of 515.56 ± 1.16 Ma.	*Antatlasia guttapluviae* Zone, Banian Stage; Stage 3.	*A. guttapluviae* Zone, based on detailed correlation to section Le-I, 8 km away. The trilobite, *Berabichia vertumnia*, a guide to the *A. guttapluviae* Zone, is 21m higher in sequence. Lower Botomian	Landing et al., 1998
Two single zircon grain analyses yield a weighted mean 206Pb/238U age of 514.38 ± 0.43 Ma, utilizing CA-TIMS and the EARTHTIME 535 spike.	*Callavia* Zone of upper Stage 3 of Series 2	Dated ash lies in the *Callavia* Zone of upper Stage 3 of Series 2	Harvey et al., 2010
Six single zircon grain analyses yield a weighted mean 206Pb/238U age of 519.29 ± 0.34 Ma, utilizing CA-TIMS and the EARTHTIME 535 spike.	*Fallotaspis* Zone of lower Stage 3 of Series 2	Dated ash lies in the *Fallotaspis* Zone of lower Stage 3 of Series 2	Harvey et al., 2010
Eight single zircon grains yield a weighted mean 206Pb/238U age of 525.23 ± 0.61 Ma, utilizing CA-TIMS and an EARTHTIME-calibrated spike.	Tommotian/Nemakit-Daldynian boundary; late Cordubian	Dated ash lies at the level of a major positive $\delta^{13}C$ excursion that is correlated with a global excursion at the Tommotian/Nemakit-Daldynian boundary	Maloof et al., 2005
Three multigrain fractions yield a weighted mean 206Pb/238U age of 530.02 ± 1.20 Ma.	Fortunian Stage	Middle part of trace fossil zone *Rusophycus avalonensis*, Placentian Series	Isachsen et al., 1994
Two multigrain zircon fractions (of five) yield a weighted mean 206Pb/238U age of 538.18 ± 1.24 Ma, corroborated by a weighted mean 207Pb/206Pb age of 537.15 ± 3.12 Ma (recalculated using the U decay constant ratio of Mattinson, 2010).	*Trychophycus pedum*, Fortunian Stage	*Trychophycus pedum* in lower Nomtsas Formation, which stratigraphically overlies Spitskopf Member. Regarded as a minimum age for base of Cambrian	Grotzinger et al., 1995
Sample BB5 at the Precambrian - Cambrian boundary, yields eight single zircon grain analyses with a weighted mean 206Pb/238U age of 541.00 ± 0.63 Ma, utilizing CA-TIMS and the EARTHTIME 535 spike.	Base Cambrian	Simultaneous occurrence of an extinction of Precambrian *Namacalathus* and *Cloudina* and a negative excursion in carbon isotopes taken as coincident with Cambrian/Ediacaran boundary.	Bowring et al., 2007
Sample Mkz-11b just below the Precambrian - Cambrian boundary, yields ten single zircon grain analyses with a weighted mean 206Pb/238U age of 542.37 ± 0.63 Ma, utilizing CA-TIMS and the EARTHTIME 535 spike.	*upper Ediacaran*	Associated with the highest calcified Ediacaran macrofssils (Cloudina and possibly Namacalathus). Maximum age for the top of the Ediacaran	Bowring et al., 2007
Ten multigrain zircon fractions yield a weighted mean 206Pb/238U age of 540.61 ± 0.88 Ma, corroborated by a weighted mean 207Pb/206Pb age of 540.94 ± 2.68 Ma (recalculated using the U decay constant ratio of Mattinson, 2010).	*upper Ediacaran*	Near top of the Nama Group, 130 m below youngest Ediacara-type megafossils and Cloudina. Maximum age for the top of the Ediacaran	Grotzinger et al.,1995
Sample Minha-1A yields eight single zircon grain analyses with a weighted mean 206Pb/238U age of 542.90 ± 0.63 Ma, utilizing CA-TIMS and the EARTHTIME 535 spike.	*upper Ediacaran*	Simultaneous occurrence of an extinction of Precambrian Namacalathus and Cloudina and a negative excursion in carbon isotopes.	Bowring et al., 2007

No.	Period	Epoch	Age	Sample	Locality	Formation	GTS2012 Age (Ma)	± 2σ (without λ errors)	± 2σ (with λ errors)	Type
E14				91-N-1 volcanic ash bed	Namibia	Urusis Fm, Middle Spitskopf Member, Schwarzrand Subgroup, Nama Group	542.68	± 1.25	± 2.80	$^{207}Pb/^{206}Pb$
E13				Asala-1 core 21 volcanic ash bed	Oman 3847m depth, Asala-1 well)	Ara Group, middle of A0 carbonate unit	546.72	± 0.34	± 0.66	$^{206}Pb/^{238}U$
E12				94-N-10B volcanic ash bed	Namibia	Lower Hoogland Member, Kuibis Subgroup, Nama Group	547.32	± 0.31	± 0.65	$^{206}Pb/^{238}U$
E11				JIN04-2 volcanic ash bed	Jijiawan section, Yangtze Platform, South China	top of black shale member containing Miaohe biota, uppermost Doushantuo Fm	551.09	± 0.84	± 1.02	$^{206}Pb/^{238}U$
E10				Volcanic ash bed	Zimnie Gory section, White Sea region, Russia	uppermost Ust-Pineg Fm, near base of sequence B	552.85	± 0.77	± 2.62	$^{207}Pb/^{206}Pb$
E9				CH2/95-011 volcanic ash bed	Charnwood Forest, England	near top of Beacon Hill Formation	559.30	± 7.30	± 7.30	$^{206}Pb/^{238}U$
E8				volcanic ash bed	Newfoundland, Canada	near top of Mistaken Point Formation	565.00	± 3.00	± 3.00	?
E7				volcanic ash bed	Newfoundland, Canada	Drook Formation, approximately 250 m above earliest definitive macrofossils (Thectardis) and 100 m above large Ediacaran fronds (Trepassia, Charnia)	578.80	± 0.50	± 1.00	?
E6				volcanic ash bed	Newfoundland, Canada	basal Drook Formation	582.10	± 0.50	± 1.00	?
E5				volcanic ash bed	Newfoundland, Canada	midway through Gaskiers Diamictite	582.4	± 0.50	± 1.00	?
E4				volcanic ash bed	Newfoundland, Canada	near top of Mall Bay Formation	583.7	± 0.50	± 1.00	?
E3				7527 volcanic ash bed	Wangjiagou section in the Zhangcunping area, Yichang, Hubei Province, South China	between beds 3 and 4, below erosional unconformity in middle Doushantuo Fm	614.0	± 9.00	± 9.00	$^{206}Pb/^{238}U$
E2				YG04-2 volcanic ash bed	Yichang section, Yangtzee Platform, Hubei Province, South China	2.3m above base of Doushantuo Fm, Lower Dolomite Member	632.48	± 0.84	± 1.02	$^{206}Pb/^{238}U$
E1				YG04-15 volcanic ash bed	Jijiawan section, Yangtze Platform, South China	9.5m above base of Doushantuo Fm, 5m above top of Lower Dolomite Member (Nantuo Cap Carbonate)	635.26	± 0.84	± 1.07	$^{206}Pb/^{238}U$
	Cryogenian									
Cr1				NAV.00.2B volcanic ash bed	Karibib, Namibia	Kachab dropstone interval, Ghaub Fm	635.47	± 0.82	± 1.05	$^{206}Pb/^{238}U$

Primary radiometric age details	Zone assignment	Primary biostratigraphic age details	Reference
Eight single grain and small multigrain zircon fractions yield a weighted mean 207Pb/206Pb age of 542.68 ± 2.80 Ma (recalculated using the U decay constant ratio of Mattinson, 2010).	upper Ediacaran	Lower Spitzkopf Member C-isotope plateau and Ediacara-type megafossils and Cloudina	Grotzinger et al., 1995
Sample Asala-1 core 21 yields eight single zircon grain analyses with a weighted mean 206Pb/238U age of 546.72 ± 0.66 Ma, utilizing CA-TIMS and the EARTHTIME 535 spike.	upper Ediacaran	Minimum age constraint on +4 per mil carbon isotope peak following the Shuram excursion	Bowring et al., 2007
Sample 94-N-10B yields eight single zircon grain analyses with a weighted mean 206Pb/238U age of 547.36 ± 0.65 Ma, utilizing CA-TIMS and the EARTHTIME 535 spike.	upper Ediacaran	Middle of Kuibis Subgroup positive C-isotope excursion. Ediacara-type megafossils and Cloudina	Bowring et al., 2007
Sample JIN04-02 yields two concordant (of ten total) single zircon grain analyses with a weighted mean 206Pb/238U age of 551.09 ± 1.02 Ma. A corroborating weighted mean 207Pb/206Pb age of 548.09 ± 2.61 Ma is obtained from all ten zircons (recalculated using the U decay constant ratio of Mattinson, 2010).	upper Ediacaran	Top of late acanthomorphic acritarch assemblage and Maiohe biota (minimum age for Doushantuo embryos and small bilaterians)	Condon et al., 2005
Nineteen single grain and small multigrain zircon fractions yield a weighted mean 207Pb/206Pb age of 552.85 ± 2.62 Ma (recalculated using the U decay constant ratio of Mattinson, 2010).	upper Ediacaran	Midpoint of the White Sea occurrence of Ediacaran megafossils, including Kimberella, and Dickinsonia	Martin et al., 2000
SHRIMP II ion probe analyses of the youngest 26 spots on 13 zircon grains yield a weighted mean 206Pb/238 age of 559.3 ± 7.3 Ma, (95% conf. int. including geologic scatter and an assumed 1% error in the QGNG standardization).		Mid-point of the Charnwood occurrence of Ediacaran megafossils, including Charnia	Compston et al., 1995
undocumented age reported in abstract form only		Acme of the Avalon assemblage of Ediacara-type fossils. Much quoted date available only in abstract form.	Dunning, pers. comm. in Benus, 1988
undocumented age reported in abstract form only		Minimum age of the Avalon assemblage of Ediacara-type fossils. Much quoted date available only in abstract/website form.	Bowring et al., 2002, 2003, 2006; see also <www.snowballearth.org/slides/Ch6-5.gif>
undocumented age reported in abstract form only		Maximum constraint on Gaskiers diamictite (mid-Ediacaran glaciation). Much quoted date available only in abstract and website form.	Bowring et al., 2002, 2003, 2006; see also <www.snowballearth.org/slides/Ch6-5.gif>
undocumented age reported in abstract form only		Syn-depositional constraint on Gaskiers diamictite (mid-Ediacaran glaciation). Much quoted date available only in abstract/website form.	Bowring et al., 2002, 2003, 2006; see also <www.snowballearth.org/slides/Ch6-5.gif>
undocumented age reported in abstract form only		Minimum constraint on top of the Nantuo diamictite (Late Cryogenian glaciation) and the base of the Ediacaran Period	Bowring et al., 2002, 2003, 2006; see also <www.snowballearth.org/slides/Ch6-5.gif>
SHRIMP II ion probe analyses of 18 zircon grains yield a weighted mean 206Pb/238 age of 614.0 ± 9.0 Ma, (95% conf. int. including geologic scatter and an assumed 1% error in the TEMORA standardization).			Liu et al., 2009
Sample YG04-2 yields three concordant (of nine total) single zircon grain analyses with a weighted mean 206Pb/238U age of 632.48 ± 1.02 Ma. A corroborating weighted mean 207Pb/206Pb age of 629.97 ± 2.83 Ma is obtained from all nine zircons (recalculated using the U decay constant ratio of Mattinson, 2010).	lower Ediacaran	Minimum constraint on top of the Nantuo diamictite (Late Cryogenian glaciation) and the base of the Ediacaran Period	Condon et al., 2005
Sample YG04-15 yields three concordant (of 18 total) single zircon grain analyses with a weighted mean 206Pb/238U age of 635.26 ± 1.07 Ma. A corroborating weighted mean 207Pb/206Pb age of 632.85 ± 2.82 Ma is obtained from 11 zircons (recalculated using the U decay constant ratio of Mattinson, 2010).	lower Ediacaran, 'Marinoan' glaciation	Minimum constraint on top of the Nantuo diamictite (Late Cryogenian glaciation); direct constraint on basal Ediacaran cap carbonate	Condon et al., 2005
Sample NAV.00.2B yields six concordant single zircon grain analyses with a weighted mean 206Pb/238U age of 635.47 ± 1.05 Ma.	lower Ediacaran, 'Marinoan' glaciation	ash within glacial deposits dates 'Marinoan' glaciation on Congo craton	Hoffmann et al., 2004

REFERENCES CITED

Aguirre-Urreta, M.B., Pazos, P.J., Lazo, D.G., Mark Fanning, C., Litvak, V.D., 2008. First U–Pb SHRIMP age of the Hauterivian stage, Neuquén Basin, Argentina. Journal of South American Earth Sciences 26 (1), 91–99.

Baadsgaard, H., Lerbekmo, J.F., Wijbrans, J.R., Swisher III, C.C., Fanning, M., 1993. Multimethod radiometric age for a bentonite near the top of the Baculites reesidei zone of Southwestern Saskatchewan (Campanian-Maastrichtian stage boundary?). Canadian Journal of Earth Sciences 30 (4), 769–775.

Barker, I.R., Moser, D.E., Kamo, S.L., Plint, A.G., 2011. High-precision U-Pb zircon ID-TIMS dating of two regionally extensive bentonites: Cenomanian Stage, Western Canada Foreland Basin. Canadian Journal of Earth Sciences 48 (2), 543–556.

Benus, A.P., 1988. Sedimentological context of a deep-water Ediacaran fauna (Mistaken Point, Avalon Zone, eastern Newfoundland). In: Landing, E., Narbonne, G.M., Myrow, P. (Eds.), Trace fossils, small shelly fossils and the Precambrian-Cambrian boundary, 463. State Museum and Geological Survey Bulletin, New York, pp. 8–9.

Berggren, W.A., Kent, D.V., Swisher, C.C.I., Aubry, M.-P., 1995. A revised Cenozoic geochronology and chronostratigraphy. In: Berggren, W.A., Kent, D.V., Aubry, H.P., Hardenbol, J. (Eds.), Geochronology, Time Scales and Global Stratigraphic Correlations: A Unified Temporal Framework for a Historical Geology. Society of Economic Paleontologists and Mineralogists Tulsa, pp. 129–212.

Bergstrom, S.M., 1989. Use of graphic correlation for assessing event-stratigraphic significance and trans-Atlantic relationships of ordovician K-bentonites. Proceedings of the Academy of Sciences of the Estonian SSR. Geology 38 (2), 55–59.

Bergstrom, S.M., Toprak, F.O., Huff, W.D., Mundil, R., 2008. Implications of a new, biostratigraphically well-controlled, radio-isotopic age for the lower Telychian Stage of the Llandovery Series (Lower Silurian, Sweden). Episodes 31 (3), 309.

Bernoulli, D., Schaltegger, U., Stern, W.B., Frey, M., Caron, M., Monechi, S., 2004. Volcanic ash layers in the upper cretaceous of the central apennines and a numerical age for the early campanian. International Journal of Earth Sciences 93 (3), 384–399.

Bowring, S., Erwin, D., Jin, Y., Martin, M., Davidek, K., Wang, W., 1998. U/Pb zircon geochronology and tempo of the end-Permian mass extinction. Science 280 (5366), 1039.

Bowring, S.A., Grotzinger, J.P., Condon, D.J., Ramezani, J., Newall, M.J., Allen, P.A., 2007. Geochronologic constraints on the chronostratigraphic framework of the neoproterozoic Huqf supergroup, Sultanate of Oman. American Journal of Science 307 (10), 1097–1145.

Bowring, S.A., Landing, E., Myrow, P., Ramenzani, J., Farmer, J., Blankenship, R., 2002. Geochronological constraints on terminal Neoproterozoic events and the rise of metazoans. Astrobiology 2, 457–458.

Bowring, S.A., Myrow, P.M., Landing, E., Ramezani, J., Condon, D., Hoffmann, K.H., 2003. Geochronological constraints on neoproterozoic glaciations and the rise of metazoans. Abstracts with Programs - Geological Society of America 35, 516.

Brack, P., Rieber, H., Mundil, R., Blendinger, W., Maurer, F., 2007. Geometry and chronology of growth and drowning of middle triassic carbonate platforms (Cernera and Bivera/Clapsavon) in the Southern Alps (Northern Italy). Swiss Journal of Geosciences 100 (3), 327–348.

Brühwiler, T., Hochuli, P.A., Mundil, R., Schatz, W., Brack, P., 2007. Bio- and chronostratigraphy of the middle triassic reifling formation of the westernmost Northern Calcareous Alps. Swiss Journal of Geosciences 100 (3), 443–455.

Chambers, L.M., Pringle, M.S., Fitton, J.G., 2004. Phreatomagmatic eruptions on the ontong java plateau; an Aptian $^{40}Ar/^{39}Ar$ age for volcaniclastic rocks at ODP site 1184. Geological Society Special Publication 229, 325–331.

Charles, A.J., Condon, D.J., Harding, I.C., Pälike, H., Marshall, J.E.A., Cui, Y., Kump, L., Croudace, I.W., 2011. Constraints on the numerical age of the Paleocene-Eocene boundary. Geochemistry Geophysics Geosystems 12. doi:10.1029/2010GC003426.

Childe, F., 1996. U-Pb geochronology and Nd and Pb isotope characteristics of the Au-Ag-rich Eskay Creek volcanogenic massive sulfide deposit, British Columbia. Economic Geology 91 (7), 1209–1224.

Churkin, M.J., Carter, C., Eberlein, G.D., 1971. Graptolite succession across the Ordovician-Silurian boundary in south-eastern Alaska. Quarterly Journal of the Geological Society of London 126 (pt 3), 319–330.

Clyde, W.C., Sheldon, N., Koch, P., Gunnell, G., Bartels, W., 2001a. Linking the Wasatchian/Bridgerian boundary to the cenozoic global climate optimum: new magnetostratigraphic and isotopic results from south pass, Wyoming. Palaeogeography Palaeoclimatology Palaeoecology 167 (1), 175–199.

Clyde, W.C., Sheldon, N.D., Koch, P.L., Gunnell, G.F., Bartels, W.S., 2001b. Linking the Wasatchian/Bridgerian boundary to the Cenozoic global climate optimum; new magnetostratigraphic and isotopic results from south pass, Wyoming. Palaeogeography Palaeoclimatology Palaeoecology 167 (1-2), 175–199.

Clyde, W.C., Stamatakos, J., Gingerich, P.D., 1994. Chronology of the Wasatchian land-mammal age (early Eocene); magnetostratigraphic results from the McCullough peaks section, Northern Bighorn Basin, Wyoming. Journal of Geology 102 (4), 367–377.

Clyde, W.C., Zonneveld, J.-P., Stamatakos, J., Gunnell, G.F., Bartels, W.S., 1997a. Magnetostratigraphy across the Wasatchian/Bridgerian NALMA boundary (early to middle Eocene) in the Western Green River Basin, Wyoming. Journal of Geology 105 (6), 657–669.

Clyde, W.C., Zonneveld, J.P., Stamatakos, J., Gunnell, G.F., Bartels, W.S., 1997b. Magnetostratigraphy across the Wasatchian/Bridgerian nalma Boundary (Early to Middle Eocene) in the Western Green River Basin, Wyoming. The Journal of Geology 105 (6), 657–670.

Cobban, W., Walaszczyk, I., Obradovich, J., McKinney, K., 2006. A USGS zonal table for the upper cretaceous middle Cenomanian-Maastrichtian of the Western interior of the United States based on ammonites, inoceramids, and radiometric ages. U.S. Geological Survey Open-File Report 2006-1250. 1–50.

Coccioni, R., Marsili, A., Montanari, A., Bellanca, A., Neri, R., Bice, D.M., Brinkhuis, H., Church, N., Macalady, A., Mcdaniel, A., Deino, A., Lirer, F., Sprovieri, M., Maiorano, P., Monechi, S., Nini, C., Nocchi, M., Pross, J., Rochette, P., Sagnotti, L., Tateo, F., Touchard, Y., Van Simaeys, S., Williams, G.L., 2008. Integrated stratigraphy of the oligocene pelagic sequence in the Umbria-Marche basin (Northeastern Apennines, Italy): A potential global stratotype section and point (GSSP) for the Rupelian/Chattian boundary. Geological Society of America Bulletin 120 (3-4), 487–511.

Compston, W., Wright, A.E., Toghill, P., 2002. Dating the late precambrian volcanicity of England and Wales. Journal of the Geological Society 159 (3), 323–339.

Condon, D., Zhu, M., Bowring, S., Wang, W., Yang, A., Jin, Y., 2005. U-Pb ages from the Neoproterozoic Doushantuo formation, China. Science 308 (5718), 95–98.

Cooper, M., Crowley, Q., Rushton, A., 2008. New age constraints for the Ordovician Tyrone Volcanic Group, Northern Ireland. Journal of the Geological Society 165 (1), 333.

Cramer, B.D., Condon, D.J., Soderlund, U., Marshall, C., Worton, G.J., Thomas, A.T., Calner, M., Ray, D.C., Perrier, V., Boomer, I., Patchett, P.J., Jeppsson, L., 2012. U-Pb (zircon) age constraints on the timing and duration of Wenlock (Silurian) paleocommunity collapse and recovery during the 'Big Crisis'. Geological Society of America Bulletin doi:10.1130/B30642.30641.

Dalrymple, G., Izett, G., Snee, L., Obradovich, J.D., 1993. $^{40}Ar/^{39}Ar$ age spectra and total-fusion ages of tektites from Cretaceous—Tertiary Boundary sedimentary rocks in the beloc formation, Haiti. US Geological Survey Bulletin 2065, 1–20.

Davidek, K., Landing, E., Bowring, S., Westrop, S., Rushton, A., Fortey, R., Adrain, J., 1998. New uppermost Cambrian U-Pb date from Avalonian Wales and age of the Cambrian-Ordovician boundary. Geol. Mag. 135 (3), 305–309.

Davydov, V., Crowley, J., Schmitz, M.D., Poletaev, V., 2010. High-precision U-Pb zircon age calibration of the global carboniferous time scale and Milankovitch band cyclicity in the Donets Basin, Eastern Ukraine. Geochemistry Geophysics Geosystems 11 (1), Q0AA04.

Davydov, V., Schmitz, M.D., Korn, D., 2011. The Hangenberg event was abrupt and short at the global scale; the quantitative intergration and intercalibration of biotic and geochronologic data within the Devonian-Carboniferous transition. Abstracts with Programs - Geological Society of America 43, 128.

Dunning, G.R., Krogh, T.E., 1991. Stratigraphic correlation of the appalachian ordovician using advanced U-Pb zircon geochronology techniques. In: Barnes, C.R., Williams, S.H. (Eds.), Fifth International symposium on the Ordovician system. Geological Survey of Canada Paper 90-09, Ottawa, pp. 85–92.

Encarnación, J., Rowell, A.J., Grunow, A.M., 1999. A U-Pb age for the Cambrian Taylor formation, Antarctica: implications for the Cambrian time scale. The Journal of Geology 107 (4), 497–504.

Evenchick, C., McNicoll, V.J., 1993. U-Pb age for the Jurassic McEwan Creek pluton, north-central British Columbia: regional setting and implications for the Toarcian stage boundary, Radiogenic Age and Isotopic Studies: Report 7. Geological Survey of Canada Paper. p. 91–97.

Flynn, J.J., 1986. Correlation and geochronology of middle eocene strata from the western United States. Palaeogeography Palaeoclimatology Palaeoecology 55, 335–406.

Furin, S., Preto, N., Rigo, M., Roghi, G., Gianolla, P., Crowley, J.L., Bowring, S.A., 2006. High-precision U-Pb zircon age from the Triassic of Italy: Implications for the Triassic time scale and the Carnian origin of calcareous nannoplankton and dinosaurs. Geology 34 (12), 1009.

Galfetti, T., Hochuli, P.A., Brayard, A., Bucher, H., Weissert, H., Vigran, J.O., 2007. Smithian-Spathian boundary event: Evidence for global climatic change in the wake of the end-Permian biotic crisis. Geology 35 (4), 291.

Gastaldo, R., Purkynova, E., Simunek, Z., Schmitz, M.D., 2009. Ecological Persistence in the Late Mississippian (Serpukhovian, Namurian A) Megafloral Record of the Upper Silesian Basin, Czech Republic. Palaios 24 (6), 336.

Grotzinger, J.P., Bowring, S.A., Saylor, B.Z., Kaufman, A.J., 1995. Biostratigraphic and geochronologic constraints on early animal evolution. Science 270 (5236), 598–604.

Harvey, T.H.P., Williams, M., Condon, D.J., Wilby, P.R., Siveter, D.J., Rushton, A.W.A., Leng, M.J., Gabbott, S.E., 2011. A refined chronology for the Cambrian succession of southern Britain. Journal of the Geological Society 168 (3), 705–716.

Hicks, J.F., Obradovich, J.D., Tauxe, L., 1995. A new calibration point for the Late Cretaceous time scale; the $^{40}Ar/^{39}Ar$ isotopic age of the C33r/C33n geomagnetic reversal from the Judith River Formation (Upper Cretaceous), Elk Basin, Wyoming, USA. Journal of Geology 103 (3), 243–256.

Hicks, J., Obradovich, J.D., Tauxe, L., 1999. Magnetostratigraphy, isotopic age calibration and intercontinental correlation of the Red Bird section of the Pierre Shale, Niobrara County, Wyoming, USA. Cretaceous Research 20 (1), 1–27.

Hilgen, F.J., Kuiper, K.F., Lourens, L.J., 2010. Evaluation of the astronomical time scale for the Paleocene and earliest Eocene. Earth and Planetary Science Letters 300 (1-2), 139–151.

Hiza, M.M., 1999. The geochemistry and geochronology of the Eocene Absaroka volcanic province, northern Wyoming and Southwest Montana, USA. Doctoral Thesis, Oregon State University, Corvallis.

Huff, W.D., Bergstrom, S.M., Kolata, D.R., 1992. Gigantic Ordovician volcanic ash fall in North America and Europe; biological, tectonomagmatic, and event-stratigraphic significance. Geology 20 (10), 875–878.

Huff, W.D., Davis, D.W., Bergstrom, S.M., Krekeler, M.M.P.S., Kolata, D.R., Cingolani, C.A., 1997. A biostratigraphically well-constrained K-bentonite U-Pb zircon age of the lowermost Darriwilian Stage (Middle Ordovician) from the Argentine Precordillera. Episodes 20 (1), 29–33.

Hughes, C.P., Jenkins, C.J., Rickards, R.B., 1982. Abereiddy Bay and the adjacent coast. In: Bassett, M.G. (Ed.), Geological Excursions in Dyfed, South-West Wales. National Museum of Wales, Cardiff, pp. 51–63.

Hüsing, S., Cascella, A., Hilgen, F., Krijgsman, W., Kuiper, K., Turco, E., Wilson, D., 2010. Astrochronology of the Mediterranean Langhian between 15.29 and 14.17 Ma. Earth and Planetary Science Letters 290 (3-4), 254–269.

Isachsen, C.E., Bowring, S.A., Landing, E., Samson, S.D., 1994. New constraint on the division of Cambrian time. Geology 22 (6), 496–498.

Izett, G., Dalrymple, G., Snee, L., 1991. $^{40}Ar/^{39}Ar$ age of Cretaceous-Tertiary boundary tektites from Haiti. Science 252 (5012), 1539–1542.

Jagodzinski, E.A., Black, L.P., 1999. U-Pb dating of silicic lavas, sills and syneruptive resedimented volcaniclastic deposits of the Lower Devonian Crudine Group, Hill End Trough, New South Wales. Australian Journal of Earth Sciences 46 (5), 749–764.

Johannson, G., McNicoll, V., 1997. New U–Pb data from the Laberge Group, northwest British Columbia: implications for Stikinian arc evolution and Lower Jurassic time scale calibrations. Current research, part F, Geological Survey of Canada, Paper. 121–129.

Kamo, S.L., Riccardi, A.C., 2009. A new U–Pb zircon age for an ash layer at the Bathonian–Callovian boundary, Argentina. Gff 131 (1-2), 177–182.

Kaufmann, B., Trapp, E., Mezger, K., 2004. The numerical age of the upper Frasnian (Upper Devonian) Kellwasser Horizons: A New U-Pb zircon date from Steinbruch Schmidt (Kellerwald, Germany). The Journal of Geology 112 (4), 495–501.

Kaufmann, B., Trapp, E., Mezger, K., Weddige, K., 2005. Two new Emsian (Early Devonian) U-Pb zircon ages from volcanic rocks of the Rhenish

Massif (Germany): implications for the Devonian time scale. Journal of the Geological Society 162 (2), 363–371.

Koppers, A.A.P., 2003. High-resolution $^{40}Ar/^{39}Ar$ dating of the oldest oceanic basement basalts in the western Pacific Basin. Geochemistry Geophysics Geosystems 4 (11).

Kowallis, B.J., Christiansen, E.H., Deino, A.L., Zhang, C., Everett, B.H., 2001. The record of Middle Jurassic Volcanism in the Carmel and Temple Cap Formations of southwestern Utah. Geological Society of America Bulletin 113 (3), 373–387.

Kuiper, K., Wijbrans, J., Hilgen, F., 2005. Radioisotopic dating of the tortonian global stratotype section and Point: implications for intercalibration of $^{40}Ar/^{39}Ar$ and astronomical dating methods. Terra Nova 17 (4), 385–397.

Kuiper, K.F., Deino, A., Hilgen, F.J., Krijgsman, W., Renne, P.R., Wijbrans, J.R., 2008. Synchronizing rock clocks of Earth history. Science 320 (5875), 500–504.

Kunk, M., Sutter, J., Obradovich, J., Lanphere, M., 1985. Age of biostratigraphic horizons within the Ordovician and Silurian systems. The chronology of the geological record: Geological Society of London,. Memoir 10, 89–92.

Kunk, M.J., Sutter, J.F., 1984. 40Ar/39Ar age spectrum dating of biotite from Middle Ordovician bentonites, eastern North America. In: Bruton, D.L. (Ed.), Paleontological Contributions from the University of Oslo, vol. 295. University of Oslo, Paleontologisk Museum, pp. 11–22.

Landing, E., Bowring, S., Davidek, K., Rushton, A., Fortey, R., Wimbledon, W., 2000. Cambrian-Ordovician boundary age and duration of the lowest Ordovician Tremadoc Series based on U-Pb zircon dates from Avalonian Wales. Geological Magazine 137 (5), 485–494.

Landing, E., Bowring, S.A., Davidek, K.L., Fortey, R.A., 1997a. U-Pb zircon date from Avalonian Cape Breton Island and geochronologic calibration of the early ordovician. Canadian Journal of Earth Sciences 34 (5), 724–730.

Landing, E., Bowring, S.A., Davidek, K.L., Westrop, S.R., Geyer, G., Heldmaier, W., 1998. Duration of the early Cambrian: U-Pb ages of volcanic ashes from Avalon and Gondwana. Canadian Journal of Earth Sciences 35 (4), 329–338.

Landing, E., Bowring, S.A., Fortey, R.A., Davidek, K.L., 1997b. U-Pb zircon date from Avalonian Cape Breton Island and geochronologic calibration of the early ordovician. Canadian Journal of Earth Sciences 34 (5), 724–730.

Lehrmann, D.J., Ramezani, J., Bowring, S.A., Martin, M.W., Montgomery, P., Enos, P., Payne, J.L., Orchard, M.J., Hongmei, W., Jiayong, W., 2006. Timing of recovery from the end-Permian extinction: Geochronologic and biostratigraphic constraints from south China. Geology 34 (12), 1053.

Leslie, S.A., Bergstrom, S.M., 1995. Revision of the North American late middle ordovician standard stage classification and timing of the Trenton transgression based on K-bentonite bed correlation. In: Cooper, J.D., Droser, M.L., Finney, S.F. (Eds.), Ordovician Odyssey: Short Papers for the Seventh International Symposium on the Ordovician System. Society of Economic Paleontologists and Mineralogists Special Publication 77. Pacific Section, Society for Sedimentary Geology (Society of Economic Paleontologists and Mineralogists), Tulsa, OK, pp. 49–54.

Liu, P., Yin, C., Gao, L., Tang, F., Chen, S., 2009. New material of microfossils from the Ediacaran Doushantuo Formation in the Zhangcunping area, Yichang, Hubei Province and its zircon SHRIMP U-Pb age. Chinese Science Bulletin 54 (6), 1058–1064.

Machlus, M., Hemming, S.R., Olsen, P.E., Christie-Blick, N., 2004. Eocene calibration of geomagnetic polarity time scale reevaluated; evidence from the Green River Formation of Wyoming. Geology 32 (2), 137–140.

Mahoney, J.J., Duncan, R.A., Tejada, M.L.G., Sager, W.W., Bralower, T.J., 2005. Jurassic-Cretaceous boundary age and mid-ocean-ridge–type mantle source for Shatsky Rise. Geology 33 (3), 185.

Maloof, A.C., Schrag, D.P., Crowley, J.L., Bowring, S.A., 2005. An expanded record of early Cambrian carbon cycling from the Anti-Atlas Margin, Morocco. Canadian Journal of Earth Sciences 42 (12), 2195–2216.

Martin, M., Grazhdankin, D., Bowring, S., Evans, D., Fedonkin, M., Kirschvink, J., 2000. Age of neoproterozoic bilatarian body and trace fossils, White Sea, Russia: implications for metazoan evolution. Science 288 (5467), 841.

Mazzini, A., Svensen, H., Leanza, H.A., Corfu, F., Planke, S., 2010. Early Jurassic shale chemostratigraphy and U–Pb ages from the Neuquén Basin (Argentina): implications for the Toarcian Oceanic Anoxic Event. Earth and Planetary Science Letters 297 (3-4), 633–645.

Meyers, S.R., Siewert, S.E., Singer, B.S., Sageman, B.B., Condon, D.J., Obradovich, J.D., Jicha, B.R., Sawyer, D.A., 2011. Intercalibration of radioisotopic and astrochronologic time scales for the Cenomanian-Turonian boundary interval, Western Interior Basin, USA. Geology 40 (1), 7–10.

Mietto, P., Manfrin, S., Preto, N., Rigo, M., Roghi, G., Furin, S., Gianolla, P., Posenato, R., Muttoni, G., Nicora, A., Buratti, N., Cirilli, S., Spötl, C., Bowring, S.A., Ramezani, J., in press. A candidate of the Global Boundary Stratotype Section and Point for the base of the Carnian Stage (Upper Triassic): GSSP at the base of the canadensis Subzone (FAD of Daxatina) in the Prati di Stuores/StuoresWiesen section (Southern Alps, NE Italy). Episodes 35, in press.

Min, K., Renne, P., Huff, W., 2001. $^{40}Ar/^{39}Ar$ dating of Ordovician K-bentonites in Laurentia and Baltoscandia. Earth and Planetary Science Letters 185 (1-2), 121–134.

Mitchell, C.E., Brussa, E.D., Astini, R.A., 1998. A diverse Da2 fauna preserved within an altered volcanic ash fall, eastern Precordillera, Argentina; implications for graptolite paleoecology. In: Gutierrez-Marco, J.C., Rabano, I. (Eds.), Sixth International Graptolite Conference of the GWG (IPA) and the SW Iberia Field Meeting 1998 of the International Subcommission on Silurian Stratigraphy (ICS-IUGS), Temas Geológico-Mineros 23. Instituto Technologico Geominero de Espana, Madrid, pp. 222–223.

Mundil, R., Ludwig, K., Metcalfe, I., Renne, P., 2004. Age and timing of the Permian mass extinctions: U/Pb dating of closed-system zircons. Science 305 (5691), 1760–1763.

Mundil, R., Metcalfe, I., Ludwig, K., Renne, P., Oberli, F., Nicoll, R., 2001. Timing of the Permian–Triassic biotic crisis: implications from new zircon U/Pb age data (and their limitations). Earth and Planetary Science Letters 187 (1-2), 131–145.

Mundil, R., Pálfy, J., Renne, P., Brack, P., 2010. The Triassic timescale: new constraints and a review of geochronological data. Geological Society London Special Publication 334 (1), 41.

Obradovich, J.D., 1993. A cretaceous time scale. In: Caldwell, W.G.E., Kauffman, E.G. (Eds.), Evolution of the Western Interior Basin. Geological Association of Canada Special Paper 39, pp. 379–396.

Obradovich, J.D., Matsumoto, T., Nishida, T., Inoue, Y., 2002. Integrated biostratigraphic and radiometric study on the lower cenomanian (Cretaceous) of Hokkaido, Japan. Proceedings of the Japan Academy. Series B: Physical and Biological Sciences 78 (6), 149–153.

Odin, G., Montanari, A., Deino, A., Drake, R., Guise, P., Kreuzer, H., Rex, D., 1991. Reliability of volcano-sedimentary biotite ages across the Eocene-Oligocene boundary (Apennines, Italy). Chemical Geology 86 (3), 203–224.

Over, D.J., 1999. Notes. Subcommission on Devonian Stratigraphy Newsletter 16, 13.

Ovtcharova, M., Bucher, H., Schaltegger, U., Galfetti, T., Brayard, A., Guex, J., 2006. New Early to Middle Triassic U–Pb ages from South China: Calibration with ammonoid biochronozones and implications for the timing of the Triassic biotic recovery. Earth and Planetary Science Letters 243 (3-4), 463–475.

Palfy, J., Mortensen, J., Smith, P., Friedman, R., McNicoll, V., Villeneuve, M., 2000. New U-Pb zircon ages integrated with ammonite biochronology from the Jurassic of the Canadian Cordillera. Canadian Journal of Earth Sciences 37 (4), 549–567.

Palfy, J., Parrish, R.R., Smith, P.L., 1997. A U-Pb age from the Toarcian (Lower Jurassic) and its use for time scale calibration through error analysis of biochronologic dating. Earth and Planetary Science Letters 146 (3-4), 659–675.

Pálfy, J.Ó., Parrish, R.R., David, K., Vörös, A., 2003. Mid-Triassic integrated U–Pb geochronology and ammonoid biochronology from the Balaton Highland (Hungary). Journal of the Geological Society 160 (2), 271–284.

Pringle, M., 1992. Radiometric ages of basaltic basement recovered at sites 800, 801, and 802, Leg 129, Western Pacific Ocean. Proceedings of the Ocean Drilling Project, Scientific Results 129, 389–404.

Pringle, M., Duncan, R., 1995. Radiometric ages of basement lavas recovered at Loen, Wodejebato, MIT, and Takuyo-Daisan Guyots, northwestern Pacific Ocean. Proceedings of the Ocean Drilling Program, Scientific Results 144, 547–557.

Pringle, M.S., Larson, R.L., Lancelot, Y., Fisher, A., Abrams, L., Behl, R., Busch, W.H., Cameron, G., Castillo, P.R., Covington, J.M., Duerr, G., Erba, E., Floyd, P.A., France-Lanord, C., Hauser, E.H., Karl, S.M., Karpoff, A.-M., Matsuoka, A., Molinie, A., Ogg, J.G., Salimullah, A.R.M., Steiner, M., Wallick, B.P., Wightman, W., 1990. Radiometric ages of basaltic basement recovered at Sites 800, 801, and 802, Leg 129, western Pacific Ocean. Proceedings of the Ocean Drilling Program, Scientific Results 129, 389–401.

Quidelleur, X., Paquette, J.L., Fiet, N., Takashima, R., Tiepolo, M., Desmares, D., Nishi, H., Grosheny, D., 2011. New U–Pb (ID-TIMS and LA-ICPMS) and $^{40}Ar/^{39}Ar$ geochronological constraints of the cretaceous geologic time scale calibration from Hokkaido (Japan). Chemical Geology 286 (3-4), 72–83.

Ramezani, J., Bowring, S.A., Martin, M.W., Lehrmann, D.J., Montgomery, P., Enos, P., Payne, J.L., Orchard, M.J., Hongmei, W., Jiayong, W., 2007a. Timing of recovery from the end-Permian extinction: Geochronologic and biostratigraphic constraints from south China: COMMENT AND REPLY. Geology 35 (1), e137–e138.

Ramezani, J., Schmitz, M., Davydov, V., Bowring, S., Snyder, W., Northrup, C., 2007b. High-precision U–Pb zircon age constraints on the Carboniferous–Permian boundary in the southern Urals stratotype. Earth and Planetary Science Letters 256 (1-2), 244–257.

Reichow, M.K., Pringle, M.S., al'Mukhamedov, A.I., Allen, M.B., Andreichev, V.L., Buslov, M.M., Davies, C.E., Fedoseev, G.S., Fitton, J.G., Inger, S., Medvedev, A.Y., Mitchell, C., Puchkov, V.N., Safonova, I.Y., Scott, R.A., Saunders, A.D., 2009. The timing and extent of the eruption of the Siberian Traps large igneous province: implications for the end-Permian environmental crisis. Earth and Planetary Science Letters 277 (1-2), 9–20.

Renne, P., Mundil, R., Balco, G., Min, K., Ludwig, K., 2010. Joint determination of 40K decay constants and $^{40}Ar^{*}/^{39}Ar$ for the Fish Canyon sanidine standard, and improved accuracy for $^{40}Ar/^{39}Ar$ geochronology. Geochimica et Cosmochimica Acta.

Richards, B.C., 2002. U-Pb geochronology, lithostratigraphy and biostratigraphy of tuff in the upper famennian to tournaisian exshaw formation: evidence for a mid-Paleozoic magmatic arc on the northwestern margin of North America. In: Hills, L.V., Henderson, C.M., Bamber, E.W. (Eds.), Carboniferous and Permian of the World (XIV International Congress of the Carboniferous and Permian, 19, pp. 158–207.

Roden, M.K., Parrish, R.R., Miller, D.S., 1990. The absolute age of the Eifelian Tioga ash bed, Pennsylvania. Journal of Geology 98 (2), 282–285.

Ross, R.J., Naeser Jr., C.W., Izett, G.A., Obradovich, J.D., Bassett, M.G., Hughes, C.P., Cocks, L.R.M., Dean, W.T., Ingham, J.K., Jenkins, C.J., Rickards, R.B., Sheldon, P.R., Toghill, P., Whittington, H.B., Zalasiewicz, J., 1982. Fission-track dating of British Ordovician and Silurian stratotypes. Geological Magazine 119 (2), 135–153.

Schaltegger, U., Guex, J., Bartolini, A., Schoene, B., Ovtcharova, M., 2008. Precise U-Pb age constraints for end-Triassic mass extinction, its correlation to volcanism and hettangian post-extinction recovery. Earth and Planetary Science Letters 267 (1-2), 266–275.

Schmitz, M., Davydov, V., 2012. Quantitative radiometric and biostratigraphic calibration of the Pennsylvanian–Early Permian (Cisuralian) time scale and pan-Euramerican chronostratigraphic correlation. Geological Society of America Bulletin 124 (3/4), 549–577.

Schoene, B., Guex, J., Bartolini, A., Schaltegger, U., Blackburn, T.J., 2010. Correlating the end-Triassic mass extinction and flood basalt volcanism at the 100 ka level. Geology 38 (5), 387–390.

Secord, R., Gingerich, P., Smith, M., Clyde, W., Wilf, P., Singer, B., 2006. Geochronology and mammalian biostratigraphy of middle and upper paleocene continental strata, Bighorn Basin, Wyoming. American Journal of Science 306 (4), 211.

Selby, D., 2007. Direct rhenium-osmium age of the Oxfordian-Kimmeridgian boundary, Staffin Bay, Isle of Skye, UK, and the Late Jurassic time scale. Norsk Geologisk Tidsskrift 87 (3), 291–299.

Selby, D., Creaser, R.A., 2005. Direct radiometric dating of the Devonian-Mississippian time-scale boundary using the Re-Os black shale geochronometer. Geology 33 (7), 545.

Selby, D., Mutterlose, J., Condon, D.J., 2009a. U–Pb and Re–Os geochronology of the Aptian/Albian and Cenomanian/Turonian stage boundaries: implications for timescale calibration, osmium isotope seawater composition and Re–Os systematics in organic-rich sediments. Chemical Geology 265 (3-4), 394–409.

Selby, D., Mutterlose, J., Condon, D.J., 2009b. U/Pb and Re/Os geochronology of the Aptian/Albian and Cenomanian/Turonian stage boundaries; implications for timescale calibration, osmium isotope sea water composition and Re-Os systematics in organic-rich sediments. Chemical Geology 265 (3-4), 394–409.

Shimokawa, A., 2010. Zircon U-Pb Geochronology of the Great Valley Group: Recalibrating the Lower Cretaceous Time Scale. University of North Carolina at Chapel Hill, Chapel Hill, p. 46.

Siewert, S.E., 2011. Integrating $^{40}Ar/^{39}Ar$, U-Pb, and astronomical clocks in the Cretaceous Niobrara Formation. University of Wisconsin-Madison, p. 70.

Siewert, S.E., Singler, B.S., Meyers, S.R., Sageman, B.B., Condon, D.J., Jicha, B.R., Obradovich, J.D., Sawyer, D.A., in press. Integrating

$^{40}Ar/^{39}Ar$, U-Pb, and astronomical clocks in the Cretaceous Niobrara Formation, Western Interior Basin, Geological Society of America Bulletin, USA.

Smith, M., Carroll, A., Singer, B., 2008. Synoptic reconstruction of a major ancient lake system: Eocene Green River Formation, western United States. Geological Society of America Bulletin 120 (1-2), 54.

Smith, M., Singer, B., Carroll, A., Fournelle, J., 2006. High-resolution calibration of Eocene strata: 40Ar/39Ar geochronology of biotite in the Green River Formation. Geology 34 (5), 393.

Smith, M.E., Chamberlain, K.R., Singer, B.S., Carroll, A.R., 2010. Eocene clocks agree: Coeval $^{40}Ar/^{39}Ar$, U-Pb, and astronomical ages from the Green River Formation. Geology 38 (6), 527–530.

Storey, M., Duncan, R., Iii, S.C.C., 2007. Paleocene-Eocene thermal maximum and the opening of the northeast Atlantic. Science 316 (5824), 587.

Stouge, S., 1980. Conodonts from the Davidsville Group, northeastern Newfoundland. Canadian Journal of Earth Sciences 17 (2), 268–272.

Suárez, M., De La Cruz, R., Aguirre-Urreta, B., Fanning, M., 2009. Relationship between volcanism and marine sedimentation in northern Austral (Aisén) Basin, central Patagonia: Stratigraphic, U–Pb SHRIMP and paleontologic evidence. Journal of South American Earth Sciences 27 (4), 309–325.

Suárez, M., Demant, A., De La Cruz, R., Fanning, C.M., 2010. $^{40}Ar/^{39}Ar$ and U–Pb SHRIMP dating of Aptian tuff cones in the Aisén Basin, Central Patagonian Cordillera. Journal of South American Earth Sciences 29 (3), 731–737.

Swisher III, C.C., Dingus, L., Butler, R.F., 1993. $^{40}Ar/^{39}Ar$ dating and magnetostratigraphic correlation of the terrestrial Cretaceous-Paleogene boundary and puercan mammal age, Hell Creek-Tullock formations, eastern Montana. Canadian Journal of Earth Sciences 30 (9), 1981–1996.

Swisher III, C.C., Grajales-Nishimura, J.M., Montanari, A., Margolis, S.V., Claeys, P., Alvarez, W., Renne, P., Cedillo-Pardoa, E., Maurrasse, F.J.M.R., Curtis, G.H., 1992. Coeval $^{40}Ar/^{39}Ar$ ages of 65.0 million years ago from Chicxulub crater melt rock and Cretaceous-Tertiary boundary tektites. Science 257 (5072), 954–958.

Tauxe, L., Gee, J., Gallet, Y., Pick, T., Bown, T., 1994. Magnetostratigraphy of the Willwood Formation, Bighorn Basin, Wyoming; new constraints on the location of Paleocene/Eocene boundary. Earth and Planetary Science Letters 125 (1-4), 159–172.

Tejada, M., Mahoney, J., Neal, C., Duncan, R., Petterson, M., 2002. Basement geochemistry and geochronology of Central Malaita, Solomon Islands, with implications for the origin and evolution of the ontong java plateau. Journal of Petrology 43 (3), 449–484.

Toghill, P., 1968. The graptolite assemblages and zones of the Brikhill shales (lower Silurian) at Dobb's Linn. Palaeontology 11 (Part 5), 654–668.

Tucker, R., 1991. Ordovician and Silurian stratotypes of Britain. In: Sutter, J., Zeitler, P., Tucker, R.D. (Eds.), Thermochronology: Applications to Tectonics, Petrology and Stratigraphy; Geological Society of America Short-Course Notes. US Geological Survey Open-File Report, Washington, pp. 57–58.

Tucker, R., Bradley, D., Ver Straeten, C., Harris, A., Ebert, J., McCutcheon, S., 1998. New U–Pb zircon ages and the duration and division of Devonian time. Earth and Planetary Science Letters 158 (3), 175–186.

Tucker, R., McKerrow, W., 1995. Early Paleozoic chronology: a review in light of new U-Pb zircon ages from Newfoundland and Britain. Canadian Journal of Earth Sciences 32 (4), 368–379.

Tucker, R.D., 1992. U-Pb dating of plinian-eruption ashfalls by the isotope dilution method; a reliable and precise tool for time-scale calibration and biostratigraphic correlation. Abstracts with Programs - Geological Society of America 24 (7), 198.

Tucker, R.D., Krogh, T.E., Ross, R.J.J., Williams, S.H., 1990. Time-scale calibration by high-precision U-Pb zircon dating of interstratified volcanic ashes in the Ordovician and lower Silurian stratotypes of Britain. Earth and Planetary Science Letters 100 (1-3), 51–58.

Turgeon, S.C., Creaser, R.A., Algeo, T.J., 2007. Re–Os depositional ages and seawater Os estimates for the Frasnian–Famennian boundary: implications for weathering rates, land plant evolution, and extinction mechanisms. Earth and Planetary Science Letters 261 (3-4), 649–661.

Walsh, S.L., Prothero, D.R., Lundquist, D.J., 1996. Stratigraphy and paleomagnetism of the middle Eocene Friars Formation and Poway Group, southwestern San Diego County, California. In: Prothero, D.R., Emry, R.J. (Eds.), The Terrestrial Eocene–Oligocene Transition in North America. Cambridge University Press, Cambridge, pp. 120–154.

Wan, X., Scott, R., Chen, W., Gao, L., Zhang, Y., 2010. Early cretaceous stratigraphy and SHRIMP U-Pb age constrain the Va-langinian-Hauterivian boundary in southern Tibet. Lethaia 44 (2), 231–244.

Wing, S.L., Bown, T.M., Obradovich, J.D., 1991. Early Eocene biotic and climatic change in interior western North America. Geology 19 (12), 1189–1192.

Xu, G., Hannah, J.L., Stein, H.J., Bingen, B., Yang, G., Zimmerman, A., Weitschat, W., Mørk, A., Weiss, H.M., 2009. Re–Os geochronology of Arctic black shales to evaluate the Anisian–Ladinian boundary and global faunal correlations. Earth and Planetary Science Letters 288 (3-4), 581–587.

Appendix 3

Cenozoic and Cretaceous Biochronology of Planktonic Foraminifera and Calcareous Nannofossils

D. Erik Anthonissen and James G. Ogg(compilers)

These compilations of selected planktonic foraminifer and calcareous nannofossil events and zones are based on previous detailed syntheses. The main events are tabulated only for the temperate and tropical realms. Information on the biostratigraphy of boreal, austral and other realms are given in the cited syntheses or are in the appropriate chapters in this book. Dates are revised to the GTS2012 age models, and numerous enhancements were added. The calibration-references are either to the syntheses that were used, to the original source studies (especially for Cenozoic foraminifera), or acknowledge inputs received from other specialists. The entries under comments are a very brief summary of the calibrations; therefore, the cited references should be used for details and clarifications. As with all compilations, these tables are a work in progress; and we thank the experts that have contributed to improving portions of these listings. Syntheses for siliceous, organic and other microfossil biostratigraphy are summarized in the appropriate chapters in this book, and detailed and updated versions of all stratigraphic scales can be accessed as datasets and graphics at the TimeScale Creator (www.tscreator.org) or Geologic TimeScale Foundation (http://stratigraphy.science.purdue.edu) websites.

Table Captions

APPENDIX TABLE A3.1 Cenozoic Planktonic Foraminiferal Biochronology.
The regional age-calibrations for events are mainly from Lourens *et al.* (2004; as GTS2004 appendix table A2.3 for Quaternary-Neogene) and Berggren *et al.* (1995a, b) with revisions and enhancements by Berggren and Pearson (2005) and Wade *et al.* (2011). Zonal definitions are summarized in Wade et al. (2011). The "GTS2012 scaling" column is the reported earliest first-appearance or latest last-appearance datum among the regions. Bridget Wade aided on the clarification of some zones and calibration of selected events.

APPENDIX TABLE A3.2 Cretaceous Planktonic Foraminiferal Biochronology.
The synthesis of events and zones by Robaszynski (1998, as a chart set) has been progressively enhanced by detailed studies of outcrops and ocean drilling sites and by integrated compilations prepared for ocean drilling legs. The Late Cretaceous and Aptian-Albian successions of events were enhanced during a workshop on Late Cretaceous microfossil stratigraphy (Univ. College London, June 2011, hosted by Jackie Lees and Paul Bown) and from communications and in-press publications by Brian Huber, Maria Rose Petrizzo and Paul Sikora.

APPENDIX TABLE A3.3 Quaternary-Neogene Calcareous Nannofossil Biochronology.
The regional age-calibrations of events are mainly from Lourens et al. (2004; as GTS2004 appendix table A2.2 for Quaternary-Neogene) and Berggren *et al.* (1995a, b). Raffi *et al.* (2006) placed most of these events onto a detailed isotope-stratigraphy framework. The "GTS2012 scaling" column is the reported earliest first-appearance or latest last-appearance datum among the regions.

APPENDIX TABLE A3.4 Paleogene-Cretaceous Calcareous Nannofossil Biochronology.
The syntheses of events and zones by Erba *et al.* (1995) and Von Salis (1998, as a chart set) have been progressively enhanced by detailed studies of outcrops and ocean drilling sites and by integrated compilations prepared for ocean drilling legs. The listed Paleogene events were reviewed and enhanced by Paul Bown. The Late Cretaceous events and calibrations were revised during a workshop on Late Cretaceous microfossil stratigraphy (Univ. College London, June 2011, hosted by Jackie Lees and Paul Bown), with additional communications from Dave Watkins. The UC scale of Late Cretaceous zones (Burnett *et al.*, 1998) is diagrammed in Chapter 27 (this volume). Jim Bergen aided in an earlier compilation for the Early Cretaceous events.

Table A3.1. Cenozoic Planktonic Foraminiferal Biochronology.

GTS2012 CHRONO-STRATIGRAPHY	STANDARD TROPICAL-SUBTROPICAL BIOZONE (BIOCHRON)					AGE (Ma)		CALIBRATED AGES COMPARED GLOBALLY								CALIBRATION
Age in Ma	After Blow (1969, 1979), Kennett & Srinivasan (1983), Berggren et al. (1995a), Chaisson & Pearson (1997), Pearson & Chaisson (1997)		After Berggren et al. (1995a) and Wade et al. (2011) with emended M14		BIOHORIZON (DATUM)	GTS2012 scaling		North Atlantic (incl. Gulf Mex.)	Mediterranean	South Atlantic	Indian Ocean	South Pacific	Equatorial Pacific	Southern Ocean	North Pacific	Magnetochron/ Marine Isotope Stage
STAGE (AGE)	Indo-Pacific	Atlantic	Indo-Pacific	Atlantic	T = Top/LAD; B = Base/FAD; X = Coiling change; ** = Regional age only	*This study*	Published Error	NA	M	SA	I	SP	EP	SO	NP	(various authors, see ref's)
Tarantian (Lt. Pleist.)					T Globorotalia flexuosa	**0.07**		0.07								~C1n.913 (late Brunhes); MIS 4
					** T Globigerinoides ruber rosa [Indo-Pac.]	**0.12**					0.12	0.12				~C1n.85 (late Brunhes)
0.126																
			PT1b	PT1b	B Globigerinella calida	**0.22**						0.22				C1n.715 (late Brunhes); MIS 7
					** B Globigerinoides ruber rosa [Med.]	**0.33**			0.33							C1n.57 (middle Brunhes)
					B Globorotalia flexuosa	**0.40**		0.40								C1n.486 (mid-Brunhes); MIS 11
Ionian (M. Pleist.)					B Globorotalia hirsuta	**0.45**				0.45						C1n.42 (mid-Brunhes); MIS 12
			PT1b/ PT1a	**PT1b/ PT1a**	T Globorotalia tosaensis	**0.61**						0.9	0.61			EP: C1n.165 (early Brunhes); SP: C1r.1r (latest Matuyama)
		N22			B Globorotalia hessi	**0.75**						0.75				C1n.04 (basal Brunhes)
0.781																
					** X Pulleniatina coiling change random to dextral [Pacific]	**0.80**							0.80			C1r.1r (upper)
			PT1a	PT1a	** B Globorotalia excelsa [Med.]	**1.00**			1.00							mid-C1r.1n (Middle Jaramillo)
Calabrian					T Globoturborotalita obliquus	**1.30**	± 0.1			1.30						-
					T Neogloboquadrina acostaensis	**1.58**	± 0.03			1.58						-
	N22				T Globoturborotalita apertura	**1.64**	± 0.03			1.64						-
					** B Globigerina cariacoensis [Med.]	**1.78**			1.78							Top C2n (Top Olduvai)
1.806																
			PT1a/ PL6	**PT1a/ PL6**	T Globigerinoides fistulosus	**1.88**	± 0.03			1.88			(1.77)			SA: Chron C2n (Olduvai)
		N22/ N21			** B Globorotalia truncatulinoides [S. Atl.]	**1.93**	± 0.03		2.00	**1.93**		2.58				SP: C2r/C2An (Gauss/Matuyama)
					T Globigerinoides extremus	**1.98**	± 0.03			1.98						-
					B Pulleniatina finalis	**2.04**	± 0.03			2.04						-
			PL6		** T Globorotalia exilis [Atl.]	**2.09**	± 0.02			2.09						~C2r.2n.5 (Reunion)
				PL6	** B Pulleniatina (reappearance) [Atl.]	**2.26**	± 0.03			2.26						Early Matuyama (below Reunion)
					T Globorotalia pertenuis	**2.30**				2.30			(2.60)			-
Gelasian					T Globoturborotalita woodi	**2.30**	± 0.02			2.30						-
			PL6/ PL5		** T Globorotalia pseudomiocenica [Indo-Pac.]	**2.39**							2.39			Early Matuyama (below Reunion)
				PL6/ PL5	** T Globorotalia miocenica [Atl.]	**2.39**	± 0.02			2.39						~C2r.3r.8 (Early Matuyama, below Reunion)
					T Globorotalia limbata [Atl.]	**2.39**	± 0.02			**2.39**			2.24			-
	N22/N21				** B Globorotalia truncatulinoides [Pac.]	**2.58**			2.00	1.93		**2.58**				SP: C2r/C2An (Gauss/Matuyama)
2.588																
				PL5	T Globoturborotalita decoraperta	**2.75**	± 0.03			2.75						-
					** X Globorotalia crassaformis (S coiling) [Med.]	**2.91**			2.91							C2An.1n (late Gauss)
		N21			T Globorotalia multicamerata	**2.98**	± 0.03			2.98						C2An.1r (Kaena)
					** X Globorotalia crassaformis (D coiling) [Med.]	**3.00**			3.00							C2An.1n (late Gauss)

CALIBRATION REFERENCES	COMMENTS
Joyce et al. (1990)	Calibrated in the Gulf of Mexico to ~C1n.913 = Same chron-age calibration as Berggren et al. (1995) = late Brunhes (Stage 4); 0.068 Ma
Thompson et al. (1979)	Calibrated to magnetics in the South Pacific and Indian Ocean to ~C1n.85 (0.12 Ma) in Wade et al. (2011), based on Thompson et al., 1979.
Chaproniere et al. (1994)	Calibrated in the South Pacific to C1n.715 = Same chron-age calibration as Berggren et al. (1995) = late Brunhes (Stage 7); 0.22 Ma [Chaproniere et al., 1994]
Lourens (unpublished) in Lourens et al. (2004)	Cycle-calibrated in the Eastern Mediterranean
Joyce et al. (1990)	Calibrated in the Gulf of Mexico to C1n.486 = Same chron-age calibration as Berggren et al. (1995) = mid-Brunhes (Stage 11); 0.401 Ma
Pujol & Duprat (1983)	South Atlantic calibration to C1n.42 = Same chron-age calibration as Berggren et al. (1995) = mid-Brunhes (stage 12); 0.45 Ma.
EP: Mix et al. (1995), Shipboard Scientific Party (1990); SP: Chaproniere et al. (1994)	Astronomically tuned from Equatorial Pacific ODP Legs 111 & 138. Berggren et al. (1995) had assigned as early Brunhes; 0.65 Ma (= C1n.165); and noted that Globorotalia tosaensis LAD is placed earlier, at latest Matuyama (0.9 Ma) by Caproniere et al. (1994).
Chaproniere et al. (1994)	Calibrated in the South Pacific to C1n.04 = Same chron-age calibration as Berggren et al. (1995) = basal Brunhes ; 0.75 Ma.
Pearson (1995), Wade et al. (2011)	Astronomical age by Wade et al. (2011), based on ODP Hole 144-871A (Marshall Islands) in Pearson (1995). Uppermost C1r.1r
Glaçon et al. (1990), Channel et al. (1990)	Calibrated to the middle of Jaramillo subchron in the Tyrrhenian Sea, ODP Leg 107.
Chaisson & Pearson (1997)	Astronomically tuned from South Atlantic ODP Sites 925 & 926.
Bickert et al. (1997a), Chaisson & Pearson (1997), Shipboard Scientific Party (1995)	Astronomically tuned from South Atlantic ODP Sites 925 & 926.
Bickert et al. (1997a), Chaisson & Pearson (1997), Shipboard Scientific Party (1995)	Astronomically tuned from South Atlantic ODP Sites 925 & 926.
Glaçon et al. (1990), Channel et al. (1990)	Mediterranean datum. Same chron-age calibration as Berggren et al. (1995)
SA: Chaisson & Pearson (1997). EP: Jenkins & Houghton (1989)	SA: Astronomically tuned from Leg 154 (925, 926). EP: Astronomically tuned as 1.77 Ma from Leg 111 (677) in Jenkins & Houghton (1989) but this was based on widely spaced core catcher samples according to Wade et al. (2011). These authors also note the mistake in Berggren et al. (1995b, table 6) where 1.6 Ma should read 1.77 Ma.
SA: Bickert et al. (1997a), Chaisson & Pearson (1997), Shipboard Scientific Party (1995); M: Lourens et al. (1996a), Zijderveld et al. (1991), Hilgen (1991a); SP: Dowsett (1988)	Highly diachronous marker for the base of N22 (Dowsett, 1988). The astronomically calibrated subtropical South Atlantic age is favoured here.
Chaisson & Pearson (1997)	Astronomically tuned from ODP Sites 925 & 926.
Chaisson & Pearson (1997)	Astronomically tuned from ODP Sites 925 & 926.
Berggren et al. (1985a)	Atlantic Ocean only. Astronomically tuned from ODP Sites 925 & 926. Berggren et al (1995) assign to Reunion Subchron; 2.15 Ma = C2r.2n.5.
Chaisson & Pearson (1997)	Astronomically tuned from ODP Sites 925 & 926.
SA: Bickert et al. (1997a), Chaisson & Pearson (1997), Shipboard Scientific Party (1995); EP: Keigwin (1982)	Astronomically tuned from ODP Sites 925 & 926. Wade et al. (2011) show both the 2.30 Ma age from Lourens et al. (2004) and a 2.60 Ma age according to Berggren et al. (1995b) who assigned it to top Gauss Subchron at 2.60 Ma (= C2An.1n.95) in the South Atlantic and East Equatorial Pacific based on Keigwin (1982). The better constrained astronomically calibrated age is here retained.
Chaisson & Pearson (1997)	Astronomically tuned from ODP Sites 925 & 926.
Berggren et al. (1995a)	See discussion under Zone PL5 in Berggren et al. (1995a)
Bickert et al. (1997a), Chaisson & Pearson (1997), Shipboard Scientific Party (1995). Magnetochron according to Berggren et al. (1995b)	Astronomically tuned from ODP Sites 925 & 926. Essentially same as Berggren et al (1995b), who assigned to early Matuyama (below Reunion Subchron); 2.30 Ma [a] =C2r.3r.8.
SA: Bickert et al. (1997a), Chaisson & Pearson (1997), Shipboard Scientific Party (1995); EP: Mix et al. (1995), Shipboard Scientific Party (1990), Shackleton et al. (1995b), Shackleton et al. (1995a)	Wade et al. (2011) prefer the LAD (2.39 Ma) at Atlantic Sites 925 & 926. Astronomically tuned later in the East Equatorial Pacific as 2.24 from ODP Legs 111 & 138
SA: Bickert et al. (1997a), Chaisson & Pearson (1997), Shipboard Scientific Party (1995); M: Lourens et al. (1996a), Zijderveld et al. (1991), Hilgen (1991a); SP: Dowsett (1988)	Highly diachronous marker for the base of N22 (Dowsett, 1988).
Chaisson & Pearson (1997)	Astronomically tuned from ODP Sites 925 & 926.
Langereis & Hilgen (1991), Hilgen (1991b)	Mediterranean coiling change D->S. C2An.1n.25 = Same chron-age calibration as Berggren et al. (1995) = C2An.1n (late Gauss); 2.92 Ma.
Chaisson & Pearson (1997); Berggren et al. (1995b), Berggren et al. (1985a), Pujol (1983), Berggren et al. (1983)	Astronomically tuned from ODP Sites 925 & 926. Berggren et al. (1995b) assign as C2An.1r (Kaena); 3.09 Ma [a] = C2Ar.1r.3. Berggren et al (1995b) magnetochron based on Pujol (1983) and Berggen et al. (1983), Rio Grande Rise DSDP 72, South Atlantic.
Langereis & Hilgen (1991), Hilgen (1991b)	Mediterranean coiling change. C2An.1n.1 = Same chron-age calibration as Berggren et al. (1995) = C2An.1n (late Gauss); 3.00 Ma.

GTS2012 CHRONO-STRATIGRAPHY	STANDARD TROPICAL-SUBTROPICAL BIOZONE (BIOCHRON)					AGE (Ma)		CALIBRATED AGES COMPARED GLOBALLY								CALIBRATION
Age in Ma	After Blow (1969, 1979), Kennett & Srinivasan (1983), Berggren et al. (1995a), Chaisson & Pearson (1997), Pearson & Chaisson (1997)		After Berggren et al. (1995a) and Wade et al. (2011) with emended M14		BIOHORIZON (DATUM)	GTS2012 scaling		North Atlantic (incl. Gulf Mex.)	Mediterranean	South Atlantic	Indian Ocean	South Pacific	Equatorial Pacific	Southern Ocean	North Pacific	Magnetochron/ Marine Isotope Stage
STAGE (AGE)	Indo-Pacific	Atlantic	Indo-Pacific	Atlantic	T = Top/LAD; B = Base/FAD; X = Coiling change; ** = Regional age only	*This study*	Published Error	NA	M	SA	I	SP	EP	SO	NP	(various authors, see ref's)
Piacenzian	N21	N21	PL5	**PL5/ PL4**	** T Dentoglobigerina altispira [Atl.]	**3.13**	± 0.02		3.17	**3.13**			3.47			M: C2An.2n (late Mammoth to early Kaena); SA: C2An.1r (Kaena)
				PL4/ PL3	** T Sphaeroidinellopsis seminulina [Atl.]	**3.16**	± 0.02		3.19	**3.16**			3.59			M: top Mammoth; SA: C2An.1r.0 (base Kaena)
				PL3	** X Globorotalia crassaformis (S coiling) [Med.]	**3.16**			3.16							C2An.2n (late Mammoth, early Kaena)
					*** T Globorotalia cibaoensis [S. Atl.] (unconfirmed)*	**3.23 (?)**	± 0.03			**3.23**						C3n.2n (Nunivak)
					** B Globorotalia inflata [S. Atl.]	**3.24**		2.0-2.2	2.09	**3.24**						NA: C2r.1r - C2r.2r; M: early Matuyama; SA: C2An.2r.7 (top Mammoth)
					** X Globorotalia crassaformis (D coiling) [Med.]	**3.26**			3.26							C2An.2r.6 (mid-Mammoth)
					** T Globorotalia cf. crassula [N.Atl.]	**3.29**		3.29								C2An.2r.3 (lower Mammoth)
					B Globigerinoides fistulosus	**3.33**			3.33							C2An.2r.0 (base Mammoth)
	N21/ N19-N20	**N21/ N19-N20**			B Globorotalia tosaensis	**3.35**							3.35			~C2An.3n.92
	N19-20	N19-20			** T Pulleniatina (disappearance) [Atl.]	**3.41**	± 0.03			3.41			3.45			EP: ~C2An.3n.5
			PL5/ PL4		** T Dentoglobigerina altispira [Pac.]	**3.47**				3.13			**3.47**			M: C2An.2n (late Mammoth to early Kaena); SA: C2An.1r (Kaena)
			PL4		B Globorotalia pertenuis	**3.52**	± 0.03			3.52			3.45			EP: ~C2An.3n.5
			PL4/ PL3		** T Sphaeroidinellopsis seminulina [Pac.]	**3.59**			3.17	3.16			**3.59**			M: top Mammoth; SA: C2An.1r.0 (base Kaena)
3.600																
Zanclean			PL3		** T Pulleniatina primalis [Pac.]	**3.66**							3.66			~C3Ar.9 (late Gilbert reversal)
					** B Globorotalila miocenica [Atl.]	**3.77**	± 0.02			3.77						C2An.3n
					T Globorotalia plesiotumida	**3.77**	± 0.02			3.77						-
			PL3/ PL2	**PL3/ PL2**	T Globorotalia margaritae	**3.85**	± 0.03		3.81	**3.85**			3.58			M: late Gilbert; EP: Base C2An.3n (Gauss/Gilbert)
			PL2	PL2	X Pulleniatina coiling sinistral to dextral	**4.08**	± 0.03			4.08			3.95			EP: Just above Cochiti
					** T Pulleniatina spectabilis [Pac.]	**4.21**							4.20			C3n.1n.8 (Top Cochiti)
					B Globorotalia crassaformis sensu lato	**4.31**	± 0.04	4.5	3.58	**4.31**		4.50				NA: C3n.2n; M: Base C2An.3n (Gauss/Gilbert); SP: C3n.2n.85
			PL2/ PL1	**PL2/ PL1**	T Globoturborotalita nepenthes	**4.37**	± 0.01			**4.37**			4.20			EP: C3n.1n.8 (Top Cochiti)
			PL1		** B Globorotalia exilis [Atl.]	**4.45**	± 0.04			4.45						-
					T Sphaeroidinellopsis kochi	**4.53**	± 0.17			4.53						-
					T Globorotalia cibaoensis	**4.60**		4.55		4.60						NA: C3n.2n (Nunivak)

CALIBRATION REFERENCES	COMMENTS
M: Lourens et al. (1996a), Zachariasse et al. (1989), Hilgen (1991b); SA: Chaisson & Pearson (1997), Shipboard Scientific Party (1995), Tiedemann & Franz (1997), Berggren et al. (1995b); EP: Mix et al. (1995), Shipboard Scientific Party (1990), Shackleton et al. (1995b), Shackleton et al. (1995a)	Astronomically tuned from South Atlantic ODP Sites 925 & 926. This LAD is reported as slightly earlier (3.19 Ma) in eastern Mediterranean and much earlier (3.47 Ma) at Pacific ODP Legs 111 and 138. Berggren et al. (1995) had similar assignment in the South Atlantic -- C2An.1r (Kaena); 3.09 Ma. However, 0.08 myr earlier in Mediterranean in late Mammoth to early Kaena C2An.2n; or about 3.17 Ma, equivalent to a calibration of about C2Ar.1r.3.
M: Lourens et al. (1996a), Zachariasse et al. (1989), Hilgen (1991b); SA: Chaisson & Pearson (1997), Shipboard Scientific Party (1995), Tiedemann & Franz (1997), Berggren et al. (1995b); EP: Mix et al. (1995), Shipboard Scientific Party (1990), Shackleton et al. (1995a)	Astronomically tuned as 3.14 Ma from Atlantic ODP Sites 925 & 926. Wade et al. (2011) used this 3.16 Ma age. This LAD is reported as slightly earlier (3.17 Ma) in astronomically tuned eastern Mediterranean and much earlier (3.59 Ma) at astronomically tuned Pacific ODP Legs 111 and 138. Berggren et al. (1995) assigned as base Kaena (3.12 Ma = C2An.1r.0) and noted that Pliocene foram 'Sphaeroidinellopsis seminulina' LAD occurs about 0.1 m.y. earlier (top of Mammoth, or 3.21 Ma) in the Mediterranean.
Langereis & Hilgen (1991), Hilgen (1991b)	Mediterranean coiling change. Same chron-age calibration as Berggren et al. (1995b) = C2An.2n (late Mammoth, early Kaena); 3.17 Ma = C2An.2n.5
Chaisson & Pearson (1997); Lourens et al. (2004); Wade et al. (2011)	NOTE: Wade et al. (2011) write: "The extinction of Globorotalia cibaoensis was used to subdivide Zone PL1 and had a calibration of 4.6 Ma in BKSA95. However, Chaisson and Pearson (1997; Atlantic ODP Sites 925 & 926) reported a much younger LAD for this species which was adopted by Lourens et al. (2004) to give an astronomical age on 3.23 Ma. As the much younger LAD at Ceara Rise is yet to be confirmed we use the 4.6 Ma calibration of BKSA95."
NA: Anthonissen (2008), Flower (1999), Weaver & Clement (1987); M: SA: Berggren et al. (1985a), Berggren et al. (1983), Pujol (1983); SA: M: Lourens et al. (1996a), Zijderveld et al. (1991), Hilgen (1991a)	Astronomically tuned in the eastern Mediterranean. Same chron-age calibration as Berggren et al. (1985a) in the South Atlantic as top Mammoth, 3.25 Ma (= C2An.2r.7). Berggren et al. (1995b) note that this FAD occurs about 1 myr later (early Matuyama, or about 2.09 Ma) in Mediterranean and North Atlantic. In Berggren et al.'s (1995b) table 5 the South Atlantic age was apparently accidentally deleted but should be top of Mammoth chron, or about 3.25 Ma (as in earlier draft of this paper and in Berggren et al.,1985)
Langereis & Hilgen (1991), Hilgen (1991b)	Mediterranean coiling change. Same chron-age calibration as Berggren et al. (1995) = mid-Mammoth, 3.26 Ma (= C2An.2r.6).
Weaver & Clement (1987)	North Atlantic LAD. Same chron-age calibration as Berggren et al. (1995), lower Mammoth, 3.30 Ma (= C2An.2r.3).
Hays et al. (1969)	Wade et al. (2011) astrochonology age implies same chron-age calibration as Berggren et al. (1995b), originally from Hays et al. (1969), of C2An.2r (base Mammoth) with an age of 3.33 Ma (= C2An.2r.0).
Hays et al. (1969)	Wade et al. (2011) astrochonology age implies same chron-age calibration as Berggren et al. (1995b) at just below Mammoth, [3.35 Ma in CK95 = C2An.3n.92]. Berggren et al. (1995b) magnetochron assignment according to Hays et al (1969).
Chaisson & Pearson (1997), Shipboard Scientific Party (1995), Tiedemann & Franz (1997), Saito et al. (1975), Keigwin (1982)	Astronomically tuned from South Atlantic ODP Sites 925 & 926. Berggren et al. (1995b) have similar calibration as C2An.3n (3.45 Ma [a] = C2An.3n.5). Berggren et al. (1995b) magnetochron designation was based on Saito et al. (1975) and Keigwin (1982)
Wade et al. (2011); M: Lourens et al. (1996a), Zachariasse et al. (1989), Hilgen (1991b); SA: Chaisson & Pearson (1997), Shipboard Scientific Party (1995), Tiedemann & Franz (1997), Berggren et al. (1995b); EP: Mix et al. (1995), Shipboard Scientific Party (1990), Shackleton et al. (1995b), Shackleton et al. (1995a)	Astronomically tuned LAD is reported much earlier (3.47 Ma) at Pacific ODP Legs 111 and 138 than in South Atlantic (ca. 3.13 Ma) [Wade et al. (2011)].
SA: Chaisson & Pearson (1997), Shipboard Scientific Party (1995), Tiedemann & Franz (1997); EP: Keigwin (1982)	Astronomically tuned from South Atlantic ODP Sites 925 & 926. Berggren et al. (1995) assigned as Chron C2An.3n (3.45 Ma = C2An.3n.5) in the Caribbean and East Equatorial Pacific (Keigwin, 1982).
M: Lourens et al. (1996a), Zachariasse et al. (1989), Hilgen (1991b); SA: Chaisson & Pearson (1997), Shipboard Scientific Party (1995), Tiedemann & Franz (1997), Berggren et al. (1995b); EP: Mix et al. (1995), Shipboard Scientific Party (1990), Shackleton et al. (1995a)	Base of Pacific PL4 -- This LAD is reported as much earlier (3.59 Ma) at Pacific ODP Legs 111 and 138. Wade et al. (2011) say: "The duration of Biochron PL4 is estimated to be 30 kyr in the Atlantic Ocean and 110 kyr in the Pacific Ocean." *[see S. seminulina [Atl.] notes at 3.16 Ma]*
Keigwin (1982)	Wade et al. (2011) astrochronology implies same chron-age calibration as Berggren et al. (1995) = late Gilbert reversed (3.65 Ma = Chron C3Ar.9) from the Pacific study of Keigwin (1982)
Chaisson & Pearson (1997), Shipboard Scientific Party (1995), Tiedemann & Franz (1997), Berggren et al. (1983), Pujol (1983)	Astronomically tuned from Atlantic ODP Sites 925 & 926. Berggren et al. (1995) assigned as C2An.3n (3.55 Ma = Chron C2An.3n.1). Magnetochron assignment according to Berggren et al. (1983) and Pujol (1983).
Chaisson & Pearson (1997), Shipboard Scientific Party (1995), Tiedemann & Franz (1997)	Astronomically tuned from South Atlantic ODP Sites 925 & 926.
M: Lourens et al. (1996a), Hilgen (1991b), Langereis & Hilgen (1991), Berggren et al. (1995b); SA: Chaisson & Pearson (1997), Shipboard Scientific Party (1995), Tiedemann & Franz (1997); EP: Saito et al. (1975)	Cycle-calibrated in Eastern Mediterranean as 3.81 Ma. This LAD is reported slightly earlier (3.85 Ma) at astronomically tuned South Atlantic Sites 925 & 926. This calibration was used by Wade et al. (2011) and is retained here. Berggren et al. (1995) assigned as Gauss/Gilbert boundary in the Eastern Equatorial Pacific (3.58 Ma = Base Chron C2An.3n) based on Saito et al. (1975). They reported this event to have occurred 0.2 m.y. earlier in Mediterranean (late Gilbert, or 3.79 Ma). The LCO occurs in the Mediterranean just above Cochiti Subchron (3.96 Ma).
SA: Chaisson & Pearson (1997), Shipboard Scientific Party (1995), Tiedemann & Franz (1997); EP: Saito et al. (1975), Keigwin (1982)	Astronomically tuned from South Atlantic ODP Sites 925 & 926. Berggren et al. (1995b) assigned this event to just above the Cochiti subchron in the Equatorial Pacific based on Saito et al. (1975) and Keigwin (1982).
Hays et al. (1969)	Pacific only. Same chron-age calibration as Berggren et al. (1995) as top of Cochiti Subchron, (4.20 Ma = Chron C3n.1n.8). P. spectabilis LAD near top of Cochiti subchron according to Hays et al. (1969) at Pacific sites.
NA: Anthonissen (2009a), Bylinskaya (2005), Weaver & Clement (1987); M: Hilgen (1991b), Langereis & Hilgen (1991); SP: Chaproniere et al. (1994)	Astronomically tuned from South Atlantic ODP Sites 925 & 926. Berggren et al. (1995b) assigned as Nunivak Subchron (4.50 Ma = Chron C3n.2n.85) in the South Pacific based on Chaproniere et al. (1994). Berggren et al. (1995b) reported it to occur about 1 m.y. later (Gilbert/Gauss boundary, or 3.58 Ma) in the Mediterranean based on Langereis & Hilgen (1991) and Hilgen (1991b). North Atlantic age similar to South Pacific.
SA: Chaisson & Pearson (1997), Shipboard Scientific Party (1995), Tiedemann & Franz (1997); EP: Hays et al. (1969), Saito et al. (1975), Berggren et al. (1983), Pujol (1983)	Astronomically tuned from Atlantic ODP Sites 925 & 926. Berggren et al. (1995b) assign as top Cochiti Subchron; 4.20 Ma = Chron C3n.1n.8 based on sites in the Equatorial Pacific.
Chaisson & Pearson (1997), Shipboard Scientific Party (1995), Tiedemann & Franz (1997)	Astronomically tuned from South Atlantic ODP Sites 925 & 926.
Chaisson & Pearson (1997), Shipboard Scientific Party (1995), Tiedemann & Franz (1997)	Astronomically tuned from South Atlantic ODP Sites 925 & 926. Typographic error in Lourens et al. (2004) where "Bottom" should read "Top".
Wade et al. (2011); NA: Weaver & Clement (1986); SA: Berggren et al. (1983), Poore et al. (1983)	Wade et al. (2011) write: "The extinction of Globorotalia cibaoensis was used to subdivide Zone PL1 and had a calibration of 4.6 Ma in BKSA95 based on S. Atlantic DSDP Site 519 where the the event coincided with the FAD G. crassaformis and within C3r (Gilbert lower part) at ~4.60 Ma.. However, Chaisson and Pearson (1997) reported a much younger LAD for this species which was adopted by Lourens et al. (2004) to give an astronomical age on 3.23 Ma. As the much younger LAD at Ceara Rise is yet to be confirmed we use the 4.6 Ma calibration of BKSA95." Chaisson & Pearson (1997) write: "The distinction between H. cibaoensis and H. scitula is difficult to make, which may account for the departure from the published range at Site 925." Also in the South Atlantic, Berggen (1977, in Hodell & Kennett, 1986), recorded the LAD of G. cibaoensis as coinciding with the FAD of calcareous nannofossil Ceratolithus acutus (5.35 Ma) in South Atlantic piston cores. In North Atlantic DSDP Site 609, Weaver & Clement (1986) recorded this event within C3n.2n (Nunivak) at ~4.55 Ma. For this GTS2012 table, we have therefore chosen to follow Wade et al. (2011) in retaining the older age, pending further study.

GTS2012 CHRONO-STRATIGRAPHY	STANDARD TROPICAL-SUBTROPICAL BIOZONE (BIOCHRON)					AGE (Ma)		CALIBRATED AGES COMPARED GLOBALLY								CALIBRATION
Age in Ma	After Blow (1969, 1979), Kennett & Srinivasan (1983), Berggren et al. (1995a), Chaisson & Pearson (1997), Pearson & Chaisson (1997)		After Berggren et al. (1995a) and Wade et al. (2011) with emended M14		BIOHORIZON (DATUM)	GTS2012 scaling		North Atlantic (incl. Gulf Mex.)	Mediterranean	South Atlantic	Indian Ocean	South Pacific	Equatorial Pacific	Southern Ocean	North Pacific	Magnetochron/ Marine Isotope Stage
STAGE (AGE)	Indo-Pacific	Atlantic	Indo-Pacific	Atlantic	T = Top/LAD; B = Base/FAD; X = Coiling change; ** = Regional age only	*This study*	Published Error	NA	M	SA	I	SP	EP	SO	NP	(various authors, see ref's)
				PL1	T Globigerinoides seiglei	**4.72**				5.0-5.2			4.72			SA: C3n.4n (Thvera); EP: ~C3n.2r.5 (above Sidufjall)
5.333																
	N19-20/ N18	**N19-20/ N18**			B Sphaeroidinella dehiscens sensu lato	**5.53**	± 0.04			**5.53**			5.2			EP: C3n.4n.13 (basal Thvera)
	N18/ N17b	N18	**PL1/ M14**		** B Globorotalia tumida [Pac.]	**5.57**				5.72	(5.57)	(5.57)	**5.57**			EP: ~C3r.45 (early Gilbert)
					** B Globorotalia pliozea [S. Pac.]	**5.67**						5.67				SP: ~C3r.45 (early Gilbert)
					** B Globorotalia sphericomiozea [S. Pac.]	**5.67**						5.67				SP: ~C3r.45 (early Gilbert)
		N18/ N17b	M14	**PL1/ M14**	** B Globorotalia tumida [Atl.]	**5.72**	± 0.04			**5.72**	(5.57)	(5.57)	5.57			EP: ~C3r.45 (early Gilbert)
					B Turborotalita humilis	**5.81**	± 0.17			5.81						-
					T Globoquadrina dehiscens	**5.92**					(5.92)	**5.92**	(5.5)			SP: C3r.15; EP: ~C3r.7
Messinian	N17				B Globorotalia margaritae	**6.08**	± 0.03	6.0	(5.1)	**6.08**		(6.5)				NA: ~C3An.1n.5; (M: ~C3r.8 mid-Thvera); SP: C3An.2n
		N17b	**M14/ M13b**		** T Globorotalia lenguaensis [Pac.]	**6.14**		8.58		8.97		**6.14**	6.14			SA: upper C4An; EP: C3An.1n; SP: ~C3An.1n.5
					B Globigerinoides conglobatus	**6.20**	± 0.41			6.20						C3r.25 (lowermost Gilbert)
					X Neogloboquadrina acostaensis coiling sinistral to dextral	**6.37**		**6.37**	6.35				6.20			EP: C3An.2n
				M14-M13b	** T Globorotalia miotumida (conomiozea) [temperate]	**6.52**		6.4-6.7	**6.52**							NA: C3An.2n
	N17b/ N17a	**N17b/ N17a**			B Pulleniatina primalis	**6.60**					6.60	**6.60**				SP: C3An.2n.5
					** T Globorotalia nicolae [Med.]	**6.72**			6.72							-
			M13b		X Neogloboquadrina acostaensis coiling dextral to sinistral	**6.77**						6.77				C3Ar.8
					** B Globorotalia nicolae [Med.]	**6.83**			6.83							-
					X Neogloboquadrina atlantica coiling dextral to sinistral	**6.99**		6.99								C3Ar.3
7.246	N17a	N17a														
					** B Globorotalia miotumida (conomiozea) [temperate]	**7.89**		7.7	**7.89**							NA: C4n.1r
					B Candeina nitida	**8.43**	± 0.04			**8.43**			(8.1)			EP: (C4r.1r.7)
					B Neogloboquadrina humerosa	**8.56**		(9.52)					**8.56**			EP: C4r.2r.5
	N17a/ N16	**N17a/ N16**	**M13b/ M13a**	**M13b /M13a**	B Globorotalia plesiotumida	**8.58**	± 0.03	(9.3)		**8.58**			8.3			EP: C4.2r.9
					B Globigerinoides extremus	**8.93**	± 0.03			**8.93**			8.3			EP: C4.2r.9

CALIBRATION REFERENCES	COMMENTS
SA: Keigwin (1982); EP: Berggren et al. (1983), Pujol (1983), Leonard et al. (1983)	Same chron-age calibration as Berggren et al. (1995), above Sidufjall subchron at 4.7 Ma = C3n.2r.5 in the Equatorial Pacific. LAD G. seiglei observed in Hole 502 (Caribbean) within Thvera Subchron (Keigwin, 1982)
SA: Chaisson & Pearson (1997), Shackleton & Crowhurst (1997); EP: Hays et al. (1969), Saito et al. (1975)	Astronomically tuned from South Atlantic ODP Sites 925 & 926. Berggren et al. (1995b) assign as early Gilbert reversed; 5.2 Ma = C3n.4n.13 in the Equatorial Pacific.
SA: Chaisson & Pearson (1997), Shackleton & Crowhurst (1997); I: Srinivasan & Sinha (1992); EP: Shipboard Scientific Party (1990), Shackleton et al. (1995b), Shackleton et al. (1995a), Saito et al. (1975), Keigwin (1982); SP: Hodell & Kennett (1986)	Astronomically tuned from South Atlantic ODP Sites 925 & 926, but Globorotalia tumida FAD is reported as later (5.57 Ma) at astronomically tuned Pacific ODP Legs 111 & 138. Berggren et al. (1995) dated this event in Hole 588 and found it to be synchronous in other SW Pacific sites (586B,587,590) and in Indian Ocean (Holes 114,219,237,238). This FAD is at a comparable position, but without paleomagnetic control, at Hole 806B (Ontong-Java Plateau). They assign as in Chron C3r (early Gilbert); 5.6 Ma = C3r.45
Srinivasan & Sinha (1992)	Same chron-age calibration as Berggren et al. (1995b) = Chron C3r (early Gilbert); 5.6 Ma = C3r.45 [Datum levels in C3r scaled proportionally to match Berggren's relative spacing.]. 'Globorotalia pliozea' FAD is dated in Hole 588 (SW Pacific).
Srinivasan & Sinha (1992), Hodell & Kennett (1986)	Same chron-age calibration as Berggren et al. (1995) = C3r (early Gilbert); 5.6 Ma = C3r.45 [Datum levels in C3r scaled proportionally to match Berggren's relative spacing.]. 'Globorotalia sphericomiozea' FAD is dated in Hole 588 (SW Pacific).
SA: Chaisson & Pearson (1997), Shackleton & Crowhurst (1997); I: Srinivasan & Sinha (1992); EP: Shipboard Scientific Party (1990), Shackleton et al. (1995b), Shackleton et al. (1995a), Saito et al. (1975), Keigwin (1982); SP: Hodell & Kennett (1986)	Astronomically tuned from South Atlantic ODP Sites 925 & 926. Globorotalia tumida FAD is reported as later (5.57 Ma) at astronomically tuned Pacific ODP Legs 111 & 138. Berggren et al. (1995) dated this event in Hole 588 and found it to be synchronous in other SW Pacific sites (586B,587,590) and in Indian Ocean (Holes 114,219,237,238). This FAD is at a comparable position, but without paleomagnetic control, at Hole 806B (Ontong-Java Plateau). They assign as C3r (early Gilbert); 5.6 Ma = C3r.45
Chaisson & Pearson (1997), Shackleton & Crowhurst (1997)	Astronomically tuned from South Atlantic ODP Sites 925 & 926.
EP: Hays et al. (1969), Saito et al. (1975), Berggren et al. (1983), Pujol (1983); SP: Srinivasan & Sinha, Hodell & Kennett (1986)	Astrochronology age by Wade et al. (2011) is same chron-age calibration as Berggren et al. (1995) = C3r (early Gilbert), 5.8 Ma = C3r.15. [Datum levels in C3r scaled proportionally to match Berggren's relative spacing.] LAD is calibrated in ODP Hole 588 in SW Pacific, and is synchronous in nearby Holes 586B,587,590 and in Indian Ocean. However, Berggren (1994 version of SEPM compilation) indicated that this LAD is apparently diachronous by about 1 myr from C3Bn (6.8 Ma) in subtropics to C3r.7 (5.5 Ma) in tropics. Wade et al. (2011) clarify: "However, Hodell and Kennett (1986) have shown the LAD of G. dehiscens to be diachronous, and the extinction appears to occur earlier in higher latitudes in comparison to tropical sites."
NA: Weaver & Clement (1987); M: Langereis and Hilgen (1991) SA: Chaisson & Pearson (1997), Shackleton & Crowhurst (1997); SP: Chaproniere et al. (1994)	Astronomically tuned from Atlantic ODP Sites 925 & 926. Extensive notes in Berggren et al. (1995a), who place the global FAD in C3An (mid) at 6.0 Ma (= C3An.1n.5) in the North Atlantic. The FAD is considered diachronous between the SW Pacific and Indian Ocean, observed to occur above (Ontong-Java Hole 806B) and below (Tonga Platform Hole 840) the FAD of G. tumida (Srinivasan & Sinha,1992). Calibrated to C3An.2n at South Pacific Tonga Platform. The FAD is nearly 1 myr later in the Mediterranean (about 5.32), where its inital occurrence is recorded immediately above base Zanclean in mid-Thvera [~C3r.8] and first common occurrence (FCO) only shortly thereafter at 5.07 Ma (Langereis and Hilgen, 1991).
NA: Zhang et al. (1993), Aubry (1993); SA: Shackleton & Crowhurst (1997), Turco et al. (2002), Poore et al. (1983); SP: Chaproniere et al. (1994); EP: Chaisson & Leckie (1993)	Pacific calibration: Same scaling as Berggren et al. (1995a) - LAD is directly calibrated to C3An.1n in ODP Hole 840 (Tonga Platform, SW Pacific) and Hole 806B (Ontong-Java Plateau, equatorial West Pacific). This LAD occurs just below FAD of P. primalis at ODP 840 and just above it at ODP 806B. The corresponding age is about 6.0 Ma (Cande & Kent, 1995). Chron-age assignment = C3An.1n at 6.0 Ma = C3An.1n.5. See discussion in NOTE below for older Atlantic age, which is significantly older than reported in Berggren et al. (1995a) (derived from the Tonga Plateau, SW Pacific) -- this Atlantic revision, Wade et al. (2011) argued would place this event within the M13a Subzone, "inconsistent with the established order of bioevents". Wade et al. (2011) argued that they retained the Berggren et al. (1995) age for stability, pending further investigation. In contrast to Wade et al. (2011), this study chooses to honour both calibrations and recognizes significant diachrony between the Atlantic and Pacific ages of LAD of G. lenguaensis.
Chaisson & Pearson (1997), Shackleton & Crowhurst (1997), Berggren et al. (1983), Pujol (1983), Poore et al. (1983)	Astronomically tuned from South Atlantic ODP Sites 925 & 926. Magnetochron assignment in South Atlantic DSDP Leg 72 according to Berggren et al. (1985a) references.
NA: Krijgsman et al. (unpubl.) in Lourens et al. (2004); M: Krijgsman et al. (1999), Hilgen & Krijgsman (1999), Sierro et al. (2001); EP: Srinivasan & Sinha (1992)	Astronomically tuned from the Eastern Mediterranean (6.35 Ma) and Morocco (6.37 Ma) in Lourens et al. (2004). Wade et al. (2011) astrochron age of 6.20 Ma projects about 0.3 myr "higher" than the chron-age calibration of Berggren et al. (1995b) = C3An.2n; 6.2 Ma (?6.4Ma?) = C3An.2n.5
NA: Weaver & Clement (1987), Clement & Robinson (1986); M: Krijgsman et al. (1999), Hilgen & Krijgsman (1999), Sierro et al. (2001)	Cycle-calibrated in Eastern Mediterranean. North Atlantic magnetochron designation based on DSDP Leg 94 (Hole 609, 609B).
I & SP: Srinivasan & Sinha (1992)	Wade et al. (2011) astrochronology age projects same chron-age of Berggren et al. (1995) = C3An.2n; 6.4 Ma (= C3An.2n.5). 'Pulleniatina primalis' FAD occurs at this age in the SW Pacific region (Sites 587,588,590; and at 806B, but without paleomagnetic calibration, and Indian Ocean (Sites 214,219,238). FAD is slightly above FAD of G. lenguaensis at Hole 840.
M:Krijgsman et al. (1999), Hilgen & Krijgsman (1999), Sierro et al. (2001)	Cycle-calibrated in Eastern Mediterranean
SP: Srinivasan & Sinha (1992)	Wade et al. (2011) astrochronology age projects as same with chron-age of Berggren et al. (1995a) = C3An.2r; 6.6 Ma = C3Ar.8.
M: Krijgsman et al. (1999), Hilgen & Krijgsman (1999), Sierro et al. (2001)	Cycle-calibrated in Eastern Mediterranean.
NA: Weaver & Clement (1987), Clement & Robinson (1986), Spiegler & Jansen (1989)	Wade et al. (2011) astrochronology age same as chron-age of Berggren et al. (1995a) = C3Ar; 6.8 Ma = C3Ar.3. Neogloboquadrina atlantica (D to S) was calibrated to magnetics in North Atlantic DSDP Hole 609, 611 and Norwegian Sea Hole 642.
NA: Anthonissen (2009a), Hodell et al. (2001), Weaver & Clement (1987), Clement & Robinson (1986); M: Krijgsman et al. (1994), Hilgen et al. (1995)	Cycle-calibrated in the Eastern Mediterranean. Calibrated to magnetostratigraphy in North Atlantic DSDP Leg 94 Holes 611C, 610E. Anthonissen (2009a) compared this event (according to depth in Flower, 1999) in North Atlantic Site 982 (Rockall Plateau) to the orbitally tuned isotope record of ODP Site 982 in Hodell et al. (2001), with a re-calibrated age to ATNTS04 of 7.7 Ma for the northeast Atlantic.
SA: Chaisson & Pearson (1997), Shackleton & Crowhurst (1997); EP: Chaisson & Leckie (1993), Berggren et al. (1995a)	Astronomically tuned from South Atlantic ODP Sites 925 & 926. Berggren et al. (1995) assign an age for 'Candeina nitida' FAD as recorded in Hole 806B (Ontong-Java Plateau), with correction of estimates made by Berggren et al. (1995a). See the explanantion of the mistake in the Chaisson & Leckie (1993) age model (under the remarks for FAD G. cibaoensis in Berggren et al.,1995a). Berggren et al. (1995a) assign as (C4r.1r); (8.1) = C4r.1r.7.
NA: Anthonissen (2009a), Spezzaferri (1998), Israelson & Spezzaferri (1998); EP: Ryan et al. (1974), Berggren et al. (1985a), Berggren et al. (1995a)	Wade et al. (2011) astrochronology age projects as same with chron-age for the Equatorial Pacific in Berggren et al. (1995a), based on Berggren et al. (1985a). They assign as C4r.2r; 8.5 Ma = C4r.2r.5. In the North Atlantic, the FAD of N. humerosa was recorded in ODP Hole 918D (Irminger Basin) by Spezzaferri (1998) with a strontium isotope value according to Israelson & Spezzaferri (1998), recalibrated to McArthur et al. (2001) strontium isotope age model in Anthonissen (2009a), 9.52 Ma.
NA: Zhang et al. (1993), Aubry (1993); SA: Chaisson & Pearson (1997), Shackleton & Crowhurst (1997); EP: Chaisson & Leckie (1993)	Astronomically tuned from South Atlantic ODP Sites 925 & 926. Berggren et al. (1995) note that 'Globorotalia plesiotumida' FAD is recorded in Hole 806B (Ontong-Java Plateau, Equatorial Pacific), but its FAD was difficult to define owing to questionable lower occurrences. Berggren et al. (1995) assigned the FAD based on its lowest questionable occurrence, but its effective FCO may be about 0.5 m.y. higher (coeval with FAD of Globorotalia cibaoensis). In the Gulf of Mexico, the FAD is recorded in upper part of Nannofossil Zone NN10, similar to the lower Pacific occurrences, and overlaps the upper part of Discoaster bollii range with the FAD slightly below FAD of nannofossil Minylitha convallis FAD (Zhang et al., 1993; Aubry, 1993). A similar FAD level is in Buff Bay, Jamaica (Berggren, 1993; Aubry, 1993). Berggren et al. (1995) assign the age as (C4r.2r); (8.3 Ma) = C4.2r.9 based on the Equatorial Pacific.
SA: Chaisson & Pearson (1997), Shackleton & Crowhurst (1997); EP: Chaisson & Leckie (1993), Berggren et al. (1995a)	Astronomically tuned from South Atlantic ODP Sites 925 & 926. Recorded in Hole 806B (Ontong-Java Plateau, equatorial Pacific), with correction of Chaisson & Leckie (1993) estimates made by Berggren et al. (1995a). See the explanantion of the mistake in the Chaisson & Leckie (1993) age model (under the remarks for FAD G. cibaoensis in Berggren et al.,1995a). Berggren et al. (1995a) had at (C4r.2r); (8.3 Ma) = C4.2r.9.

GTS2012 CHRONO-STRATIGRAPHY	STANDARD TROPICAL-SUBTROPICAL BIOZONE (BIOCHRON)					AGE (Ma)		CALIBRATED AGES COMPARED GLOBALLY								CALIBRATION
Age in Ma	After Blow (1969, 1979), Kennett & Srinivasan (1983), Berggren et al. (1995a), Chaisson & Pearson (1997), Pearson & Chaisson (1997)		After Berggren et al. (1995a) and Wade et al. (2011) with emended M14		BIOHORIZON (DATUM)	GTS2012 scaling		North Atlantic (incl. Gulf Mex.)	Mediterranean	South Atlantic	Indian Ocean	South Pacific	Equatorial Pacific	Southern Ocean	North Pacific	Magnetochron/ Marine Isotope Stage
STAGE (AGE)	Indo-Pacific	Atlantic	Indo-Pacific	Atlantic	T = Top/LAD; B = Base/FAD; X = Coiling change; ** = Regional age only	*This study*	Published Error	NA	M	SA	I	SP	EP	SO	NP	(various authors, see ref's)
Tortonian	N16	N16	M13a	M13a	** T Globorotalia lenguaensis [Atl.]	8.97		8.58		8.97		6.14	6.14			SA: upper C4An; EP: C3An.1n; SP: ~C3An.1n.5
					** B Neogloboquadrina pachyderma [Ind.]	9.37					9.37					I: C4Ar.1n.5
					B Globorotalia cibaoensis	9.44	± 0.05			9.44						EP: (C4n.2n)
					B Globorotalia juanai	9.69	± 0.26			9.69			(8.1)			EP: (C4r.1r.7)
	N16/ N15	**N16/ N15**	**M13a/ M12**	**M13a/ M12**	** B Neogloboquadrina acostaensis [(sub)tropical]	9.83	± 0.06	10.9	10.57	9.83						NA: C5n.2n.05
	N15	N15	M12	M12	** T Globorotalia partimlabiata [Med.]	9.94			9.94							-
					T Globorotalia challengeri	9.99				9.99						-
					** T Neogloboquadrina nympha [Ind.]	10.20					10.20					I: C5n.2n.8
	N15/ N14	**N15/ N14**	**M12/ M11**	**M12/ M11**	** T Paragloborotalia mayeri/siakensis [(sub)tropical]	10.46	± 0.02	11.40	11.19	10.46						NA: C5r.2r.2
	N14	N14	M11	M11	** B Neogloboquadrina acostaensis [temperate]	10.57		10.9	10.57	9.83						NA: C5n.2n.05
					B Globorotalia limbata	10.64	± 0.26									-
					T Cassigerinella chipolensis	10.89				10.89						-
					B Globoturborotalita apertura	11.18	± 0.13			11.18						-
					B Globorotalia challengeri	11.22				11.22						-
					B regular Globigerinoides obliquus	11.25			11.54	11.25						-
					** T Paragloborotalia mayeri/siakensis [temperate]	11.40		11.40	11.19	10.46						NA: C5r.2r.2
					B Globoturborotalita decoraperta	11.49	± 0.04			11.49						-
					T Globigerinoides subquadratus	11.54				11.54						-
11.608																
	N14/ N13	**N14/ N13**	**M11/ M10**	**M11/ M10**	B Globoturborotalita nepenthes	11.63	± 0.02	11.8		11.63						NA: C5r.3r.3
	N13	N13	M10	M10	** B Neogloboquadrina group [Med.]	11.78			11.78							-
	N13/ N12	**N13/ N12**	**M10/ M9b**	**M10/ M9b**	T Fohsella fohsi, Fohsella plexus	11.79	± 0.15	11.9		11.79						NA: top C5An.1n
					** T Globorotalia panda [Ind.]	11.93					11.93					I: C5r.3r.3
					T Clavatorella bermudezi	12.00				12.00						-
					** T Paragloborotalia mayeri sensu stricto [Med.]	12.07			12.07							-
					** T Tenuitella selleyi, T. pseudoedita, T. minutissima, T. clemenciae [Ind.]	12.37					12.37					I: C5An.2n.5
					** B Paragloborotalia partimlabiata [Med.]	12.77			12.77							-

CALIBRATION REFERENCES	COMMENTS
NA: Zhang et al. (1993), Aubry (1993); SA: Shackleton & Crowhurst (1997), Turco et al. (2002), Poore et al. (1983); SP: Chaproniere et al. (1994); EP: Chaisson & Leckie (1993)	Atlantic calibration: Astronomically tuned in South Atlantic ODP Sites 925 & 926 (8.97 Ma in Lourens et al. 2004). In Hole 519 (S. Atl.), calibrated to upper C4An (about 8.7 Ma) in Poore et al. (1983). In E68-136 (Gulf of Mexico), the LAD occurs about 10m above the FAD of G. plesiotumida (=base of N17) and within range of Minylitha convallis and above the LAD of Discoaster bollii. In contrast, in E66-73, the LAD is at the same level as FAD of G. plesiotumida (8.58 Ma), and below the FAD of M. convallis and LAD of D. bollii (Zhang et al., 1993; Aubry, 1993). This Atlantic and Gulf age is significantly older than in the SW Pacific - see discussion on Pacific placement above. This Atlantic assignment, Wade et al. (2011) argued, would place this event within the M13a Subzone, "inconsistent with the established order of bioevents" therefore they retained the SW Pacific age for stability. In contrast, this study chooses to honour both calibrations and recognizes significant diachrony between the Atlantic and Pacific ages of LAD of G. lenguaensis. The Atlantic event is herein no longer used as a zonal marker between M13b and M14 and the resulting zone in the Atlantic scheme is therefore a M14-M13b composite, pending further investigation.
I: Berggren (1992)	Same chron-age calibration as Berggren et al. (1995), calibrated in Holes 748, 751 (Kerguelen Plateau, south Indian Ocean) = Chronn C4Ar.1n; >9.2 Ma = C4Ar.1n.5.
SA: Chaisson & Pearson (1997), Shackleton & Crowhurst (1997); EP: Chaisson & Leckie (1993), Berggren et al. (1995a)	Astronomically tuned from South Atlantic ODP Sites 925 & 926. Recorded in Hole 806B (Ontong-Java Plateau, equatorial Pacific), with correction of Chaisson & Leckie (1993) estimates made by Berggren et al. (1995a). See the explanantion of the mistake in the Chaisson & Leckie (1993) age model (under the remarks for FAD G. cibaoensis in Berggren et al.,1995a). Berggren et al. (1995a) had at Chron C4n.2n; (7.8 Ma)
SA: Chaisson & Pearson (1997), Shackleton & Crowhurst (1997); EP: Chaisson & Leckie (1993), Berggren et al. (1995a)	Astronomically tuned from Atlantic ODP Sites 925 & 926. Berggren et al. (1995) assign an age for 'Globorotalia juanai' FAD as recorded in Hole 806B (Ontong-Java Plateau), with correction of estimates made by Berggren et al. (1995a). See the explanantion of the mistake in the Chaisson & Leckie (1993) age model (under the remarks for FAD G. cibaoensis in Berggren et al.,1995a). Berggren et al. (1995a) assign as Chron C4r.1r; (8.1) = C4r.1r.7.
NA: Berggren et al. (1985a), Miller et al. (1985), Miller et al. (1991); SA: Chaisson & Pearson (1997), Shackleton & Crowhurst (1997), Turco et al. (2002)	Astronomically tuned in the South Atlantic and Eastern Mediterranean. This FAD is younger (9.8 Ma) in South Atlantic Sites 925 & 926 than in cycle-calibrated Eastern Mediterranean (10.57 Ma). Wade et al. (2011) expound further: "The cyclostratraphic age of the LO of Neogloboquadrina acostaensis (9.83 Ma) is derived from Ceara Rise (Chaisson and Pearson, 1997). This calibration was adopted by Lourens et al. (2004) and is significantly younger (1.07 myr) than in Berggren et al. (1995a) (10.90 Ma) and would move this event from early Subchron C5n.2n to Subchron C5n.1n. Turco et al. (2002) noted the diachrony of the LO of Neogloboquadrina acostaensis between low latitudes and the Mediterranean. The age used in Berggren et al. (1995a) is calibrated to the magnetostratigraphy at Site 563 (Miller et al., 1985) and the discrepancy in calibrated ages may be due to further diachrony between the tropical and subtropical Atlantic Ocean, however, we note that the order of bioevents is consistent between Ceara Rise and Site 563." Berggren et al. (1995) note that 'Neogloboquadrina acostaensis' FAD is calibrated in Holes 558 and 563 and 608 in the North Atlantic at the base of chron C5n = C5n.2n; 10.9 Ma (= C5n.2n.05
M: Hilgen et al. (2005), Hilgen et al. (2000a)	Cycle-calibrated in Eastern Mediterranean.
SA: Shackleton & Crowhurst (1997), Turco et al. (2002)	Astronomically tuned from South Atlantic ODP Sites 925 & 926.
I: Berggren (1992)	Same chron-age calibration as Berggren et al. (1995a) -- Miocene foram 'Neogloboquadrina nympha' LAD is calibrated in Holes 748, 751 (Kerguelen Plateau, south Indian Ocean) = C5n.2n; 10.1 Ma = C5n.2n.8.
NA: Berggren et al. (1985a), Miller et al. (1985), Miller et al. (1991); M: Hilgen et al. (2005), Hilgen et al. (2000a); SA: Shackleton & Crowhurst (1997), Turco et al. (2002)	Here merged as P. siakensis/mayeri due to the ongoing taxonomic controversy. Turco et al. (2002): "According to Bolli and Saunders (1982, 1985), P. siakensis represents a junior synonym of P. mayeri. Other authors ... following the species concept of Blow (1969) regarded P. siakensis and P. mayeri as distinct species, but have different opinion about the distribution range and the phylogeny of P. mayeri." Astronomically tuned from South Atlantic ODP Sites 925 & 926 and cycle tuned in the Eastern Mediterranean as "Top Paragloborotalia siakensis" in Lourens et al (2004). Wade et al. (2011) clarify: "The extinction of Paragloborotalia mayeri has been recalibrated to 10.53 Ma as per Chaisson and Pearson (1997) (given as siakensis in Turco et al., 2002). This is significantly younger (870 kyr) than the reported age of 11.40 Ma in Berggren et al. (1995a). The interpolated age would place this event mid-C5n.2n rather than C5r.2r. Hilgen et al. (2000a) noted the diachrony between the tropical Atlantic Ocean and the Mediterranean; and diachrony with higher latitudes was suggested by Miller et al. (1991)." The older "Top P. mayeri" event in cycle-tuned Mediterranean (12.07 Ma in Lourens et al., 2004) is here used as a separate regional event.
NA: Berggren et al. (1985a), Miller et al. (1985), Miller et al. (1991); SA: Chaisson & Pearson (1997), Shackleton & Crowhurst (1997), Turco et al. (2002)	Astronomically tuned in the South Atlantic and Eastern Mediterranean. This FAD is younger (9.8 Ma) in South Atlantic Sites 925 & 926 than in cycle-calibrated Eastern Mediterranean (10.57 Ma). Wade et al. (2011): "The cyclostratraphic age of the LO of Neogloboquadrina acostaensis (9.83 Ma) is derived from Ceara Rise (Chaisson and Pearson, 1997). This calibration was adopted by Lourens et al. (2004) and is significantly younger (1.07 myr) than in Berggren et al. (1995a) (base C5n.2n; 10.90 Ma) and would move this event from early Subchron C5n.2n to Subchron C5n.1n. Turco et al. (2002) noted the diachrony of the LO of Neogloboquadrina acostaensis between low latitudes and the Mediterranean. The age used in Berggren et al. (1995a) is calibrated to the magnetostratigraphy at Site 563 (Miller et al., 1985) and the discrepancy in calibrated ages may be due to further diachrony between the tropical and subtropical Atlantic Ocean, however, we note that the order of bioevents is consistent between Ceara Rise and Site 563." The cycle tuned age from the Mediterranean is here preferred.
SA: Chaisson & Pearson (1997), Shackleton & Crowhurst (1997)	Astronomically tuned from South Atlantic ODP Sites 925 & 926.
SA: Shackleton & Crowhurst (1997), Turco et al. (2002)	Astronomically tuned from South Atlantic ODP Sites 925 & 926.
SA: Chaisson & Pearson (1997), Shackleton & Crowhurst (1997)	Astronomically tuned from South Atlantic ODP Sites 925 & 926.
SA: Shackleton & Crowhurst (1997), Turco et al. (2002)	Astronomically tuned from South Atlantic ODP Sites 925 & 926.
M: Hilgen et al. (2005), Hilgen et al. (2000a); SA: Shackleton & Crowhurst (1997), Turco et al. (2002)	Astronomically tuned from South Atlantic ODP Sites 925 & 926 and in the Eastern Mediterranean. The subtropical South Atlantic age is here favored.
NA: Berggren et al. (1985a), Miller et al. (1985), Miller et al. (1991); M: Hilgen et al. (2005), Hilgen et al. (2000a); SA: Chaisson & Pearson (1997), Shackleton & Crowhurst (1997), Turco et al. (2002)	Here merged as P. siakensis/mayeri due to the ongoing taxonomic controversy [See discussion above for "sub-tropical" entry.] Astronomically tuned from S. Atlantic ODP Sites 925 & 926 and cycle tuned in the Eastern Mediterranean as "Top Paragloborotalia siakensis" in Lourens et al (2004). This is about 1 myr "above" the projected chron-age (11.4 Ma = C5r.2r.2) of Berggren et al. (1995a) from N. Atlantic Sites 558, 563 and 608, where LAD of 'N. mayeri' was observed in close juxtaposition with FAD of 'N. acostaensis', essentially eliminating Zone N15 (=M12). Wade et al. (2011) clarify: "The extinction of Paragloborotalia mayeri has been recalibrated to 10.53 Ma as per Chaisson and Pearson (1997) (given as siakensis in Turco et al., 2002). This is significantly younger (870 kyr) than the reported age of 11.40 Ma in Berggren et al. (1995a). The interpolated age would place this event mid-C5n.2n rather than C5r.2r. Hilgen et al. (2000a) noted the diachrony between the tropical Atlantic Ocean and the Mediterranean; and diachrony with higher latitudes was suggested by Miller et al. (1991)." The older "Top P. mayeri" event in cycle-tuned Mediterranean (12.07 Ma in Lourens et al., 2004) is here used as a separate regional event.
SA: Chaisson & Pearson (1997), Shackleton & Crowhurst (1997)	Astronomically tuned from South Atlantic ODP Sites 925 & 926.
SA: Shackleton & Crowhurst (1997), Turco et al. (2002)	Astronomically tuned from South Atlantic ODP Sites 925 & 926. Tortonian GSSP coincides almost exactly with the Last Common Occurrences of the planktonic foraminifer Globigerinoides subquadratus in the Eastern Mediterranean.
NA: Berggren et al. (1985a), Berggren (1993), Miller et al. (1991), Miller et al. (1994); SA: Shackleton & Crowhurst (1997), Turco et al. (2002)	Astronomically tuned from South Atlantic ODP Sites 925 & 926. Berggren et al. (1995) notes that 'Globoturborotalita nepenthes' FAD occurs in North Atlantic Sites 563 and 608 and basal Buff Bay Fm., Jamaica = C5r.3r; 11.8 Ma = C5r.3r.3. The astronomically calibrated age is here favored.
M: Hilgen et al. (2005), Hilgen et al. (2000a)	Cycle-calibrated in Eastern Mediterranean.
NA: Berggren et al. (1985a); SA: Chaisson & Pearson (1997), Shackleton & Crowhurst (1997)	Astronomically tuned "F. fohsi s.l." from South Atlantic ODP Sites 925 & 926. Berggren et al. (1995a) note that Miocene foram 'Globorotalia fohsi robusta' LAD is calibrated in North Atlantic Hole 563 = C5An.1n; 11.9 Ma = top C5An.1n. Wade et al. (2011) on Fohsella lineage: "Taxonomic subdivision of this gradual chronocline into species and subspecies is inevitably subjective".
I: Berggren (1992), Wright & Miller (1992), Berggren et al. (1985a)	Berggren et al. (1995): 'Globorotalia panda' is calibrated in Hole 747A (Kerguelen Plateau, south Indian Ocean); with magnetic stratigraphy reinterpreted in Wright & Miller (1992) = C5r.3r; 11.8 Ma = C5r.3r.3.
SA: Shackleton & Crowhurst (1997), Turco et al. (2002)	Astronomically tuned from South Atlantic ODP Sites 925 & 926. See note on last common occurrence at 13.82 Ma.
M: Hilgen et al. (2005)	Cycle-calibrated in the eastern Mediterranean. See the comment for "T Paragloborotalia mayeri/siakensis"
I: Li et al. (1992)	Berggren et al. (1995) -- Miocene foraminifers 'Tenuitella selleyi' through "clemenciae" simultaneous LADs are calibrated in Hole 747A (Kerguelen Plateau, south Indian Ocean) = Chron C5An.2n (C5An.2n.5).
M: Hilgen et al. (2005)	Cycle-calibrated in Eastern Mediterranean.

GTS2012 CHRONO-STRATIGRAPHY	STANDARD TROPICAL-SUBTROPICAL BIOZONE (BIOCHRON)					AGE (Ma)		CALIBRATED AGES COMPARED GLOBALLY								CALIBRATION
Age in Ma	After Blow (1969, 1979), Kennett & Srinivasan (1983), Berggren et al. (1995a), Chaisson & Pearson (1997), Pearson & Chaisson (1997)		After Berggren et al. (1995a) and Wade et al. (2011) with emended M14		BIOHORIZON (DATUM)	GTS2012 scaling		North Atlantic (incl. Gulf Mex.)	Mediterranean	South Atlantic	Indian Ocean	South Pacific	Equatorial Pacific	Southern Ocean	North Pacific	Magnetochron/ Marine Isotope Stage
STAGE (AGE)	Indo-Pacific	Atlantic	Indo-Pacific	Atlantic	T = Top/LAD; B = Base/FAD; X = Coiling change; ** = Regional age only	*This study*	Published Error	NA	M	SA	I	SP	EP	SO	NP	(various authors, see ref's)
Serravallian					B Globorotalia lenguanensis	**12.84**	± 0.05			12.84						-
					B Sphaeroidinellopsis subdehiscens	**13.02**				13.02						
			M9b/ M9a	**M9b/ M9a**	B Fohsella robusta	**13.13**	± 0.02	12.3		**13.13**						NA: C5An.2n.5
			M9a	M9a	T Cassigerinella martinezpicoi	**13.27**				13.27						
	N12/ N11	**N12/ N11**	**M9a/M8**	**M9a/ M8**	B Fohsella fohsi	**13.41**	± 0.04	12.7		**13.41**						NA: C5Ar.2r.8
	N11	N11	M8	M8	B Neogloboquadrina nympha	**13.49**		**13.49**			13.73					NA: C5ABn.5; I: C5AB (base)
					** T Globorotalia praescitula [S. Atl.]	**13.73**				**13.73**	11.90					I: C5An.1n
	N11/ N10	**N11/ N10**	**M8/M7**	**M8/M7**	B Fohsella "praefohsi"	**13.77**		(12.7)		**13.77**						(NA: C5Ar.2r.8)
	N10	N10	M7	M7	T Fohsella peripheroronda	**13.80**		14.6		**13.80**						NA: base C5ADn
13.82																
Langhian					T regular Clavatorella bermudezi	**13.82**				13.82						-
					T Globorotalia archeomenardii	**13.87**				13.87						-
	N10/N9	**N10/N9**	**M7/M6**	**M7/M6**	B Fohsella peripheroacuta	**14.24**				**14.24**						(SA: base C5ADr)
	N9	N9	M6	M6	B Globorotalia praemenardii	**14.38**				14.38						-
					T Praeorbulina sicana	**14.53**				14.53						SA: base C5ADr
					T Globigeriantella insueta	**14.66**				14.66						
					T Praeorbulina glomerosa sensu stricto	**14.78**				14.78						SA: base C5B.1n
					T Praeorbulina circularis	**14.89**				14.89						-
	N9/N8	**N9/N8**	**M6/ M5b**	**M6/ M5b**	B Orbulina suturalis	**15.10**				15.10						SA: mid-C5Bn.2n
	N8	N8	M5b	M5b	B Clavatorella bermudezi	**15.73**				15.73						-
					** T Globorotalia miozea [Ind.]	**15.85**					15.85					I: C5Br.15
					B Praeorbulina circularis	**15.96**		15.96								NA: C5C.1n.95
15.97																
Burdigalian					B Globigerinoides diminutus	**16.06**				(16.06)						(SA: C5Cn.1n.7)
					B Globorotalia archeomenardii	**16.26**				16.26						-
			M5b/ M5a	**M5b/ M5a**	B Praeorbulina glomerosa sensu stricto	**16.27**				16.27						SA: C5Cn.1n.8
			M5a	M5a	B Praeorbulina curva	**16.28**		16.28								NA: base C5Cn.1n
	N8/N7	**N8/N7**	**M5a/ M4b**	**M5a/ M4b**	B Praeorbulina sicana	**16.38**				16.38						SA: C5Cn.2n.55
	N7	N7	M4b	M4b	T Globorotalia incognita	**16.39**					16.39					I: C5Cn.2n.5
			M4b/ M4a	**M4b/ M4a**	B Fohsella birnageae	**16.69**				16.69						SA: C5Cn.3n.1
					B Globorotalia miozea	**16.70**					16.70					I: C5Cn.3n.1
			M4a	M4a	B Globorotalia zealandica	**17.26**					17.26					I: C5Dn.9

CALIBRATION REFERENCES	COMMENTS
SA: Chaisson & Pearson (1997), Shackleton & Crowhurst (1997)	Astronomically tuned from South Atlantic ODP Sites 925 & 926.
SA: Shackleton & Crowhurst (1997), Shackleton et al. (1999), Turco et al. (2002)	Astronomically tuned from South Atlantic ODP Sites 925 & 926.
SA: Chaisson & Pearson (1997), Shackleton et al. (1999), Turco et al. (2002); NA: Berggren (1992), Wright & Miller (1992), Berggren et al. (1985a)	Astronomically tuned from South Atlantic ODP Sites 925 & 926. Berggren et al. (1995a) note that 'Globororotalia fohsi robusta' FAD is based on revised interpretation of magnetostratigraphy of North Atlantic Hole 563, in turn based on stable isotope studies in Wright & Miller (1992) = Chron C5An.2n (C5An.2n.5). Wade et al. (2011) on Fohsella lineage: "Taxonomic subdivision of this gradual chronocline into species and subspecies is inevitably subjective".
SA: Shackleton & Crowhurst (1997), Shackleton et al. (1999), Turco et al. (2002)	Astronomically tuned from South Atlantic ODP Sites 925 & 926.
NA: Berggren et al. (1985a), Wright & Miller (1992); SA: Chaisson & Pearson (1997), Shackleton & Crowhurst (1997), Shackleton et al. (1999)	Astronomically tuned from South Atlantic ODP Sites 925 & 926. Berggren et al. (1995a) assign Miocene foram 'Globorotalia fohsi s.str.' FAD is based on reinterpretation of Wright & Miller (1992) magnetostratigraphy at Hole 563, in turn based on stable isotope studies at Holes 747A and 608 = C5Ar.n1–2 (undiff.); 12.7 Ma = placed here as middle or Chron C5Ar.2r.8. Wade et al. (2011) on Fohsella lineage: "Taxonomic subdivision of this gradual chronocline into species and subspecies is inevitably subjective".
NA: Wright & Miller (1992); I: Berggren (1992)	Berggren et al. (1995a) -- 'Neogloboquadrina nympha' FAD is recorded in Holes 747A, 751, Indian Ocean Kerguelen Plateau assigned to polarity chron C5AB (base). In North Atlantic Hole 608, the FAD occurs in sample assigned to polarity chron C5AAr, just above C5ABn => averaged here as Chron C5ABn.5.
SA: Shackleton et al. (1999), Turco et al. (2002); I: Berggren (1992), Wright & Miller (1992), Berggren et al. (1985a)	Astronomically tuned from South Atlantic ODP Sites 925 & 926. Berggren et al. (1995a) listed LAD as much younger in C5An.1n or 11.9 Ma; therefore the cycle calibration needs comparison to other sites. Berggren et al's calibration was in Hole 747A (Kerguelen Plateau, south Indian Ocean); with magnetic stratigraphy reinterpreted in Wright & Miller (1992) = Chron C5An.1n. In the northeastern Atlantic (Site 982 and 918) "Top regular Globorotalia praescitula/zealandica group" has an age of ca. 13.8 Ma in Anthonissen, (2009b).
NA: Berggren et al. (1985a), Wright & Miller (1992); SA: Shackleton et al. (1999), Turco et al. (2002)	Astronomically tuned from Atlantic ODP Sites 925 & 926. Berggren et al (1995) assigned 'Globorotalia praefohsi' FAD to a much younger level, based on reinterpretation of magnetostratigraphy at North Atlantic Hole 563, in turn based on stable isotope studies at Holes 747A and 608 (Wright & Miller, 1992) = C5Ar.n1–2 (undiff.) = placed here at Chron C5Ar.2r.8. Note this North Atlantic FAD age-assignment was nearly identical to FAD of 'Gt. fohsi s.str.'
NA: Berggren et al. (1985a); SA: Shackleton et al. (1999), Pearson & Chaisson (1997)	Astronomically tuned from South Atlantic ODP Sites 925 & 926. Berggren et al. (1995a) used to define base of Mediterranean Zone Mt7, with age assignment of C5ADn (lower); 14.6 Ma = base C5ADn = nearly 1 myr older than South Atlantic assignment.
SA: Shackleton et al. (1999), Turco et al. (2002)	Astronomically tuned from South Atlantic ODP Sites 925 & 926, and considered LAD of "regular" by Lourens et al. (2004). Note that Wade et al. (2011) list this LAD without "regular" as "LAD C. bermudezi" following Pearson & Chaisson (1997) at 303.56 mcd in Hole 926A with an age of 13.8 Ma. However, detailed analysis of Hole 926A by Turco et al. (2002) resulted in a significantly higher "LAD C. bermudezi" at 268.74 mcd and an age of 12.0 Ma according to Lourens et al. (2004) astrochronology. Turco et al. (2002) did identify a "LRO" last regular occurrence of C. bermudezi at 299.34 mcd, approximating the Pearson & Chaisson observation. We therefore retain the "regular" for this event at 13.82 Ma, appreciating that the LAD at 12.0 Ma may not be a useful correlation level due to very rare abundances of this taxon.
SA: Shackleton et al. (1999), Pearson & Chaisson (1997)	Astronomically tuned from South Atlantic ODP Sites 925 & 926.
SA: Shackleton et al. (1999), Pearson & Chaisson (1997), Berggren et al. (1985a), Ryan et al. (1974)	Was used by Berggren et al (1995a) to define base of Serravallian stage. Astronomically tuned from South Atlantic ODP Sites 925 & 926. Berggren et al (1995a) assigned as C5B.1n; placed here at base of C5ADr = nearly 0.5 myr older than Atlantic-ODP cycle-strat assignment.
SA: Shackleton et al. (1999), Pearson & Chaisson (1997)	Astronomically tuned from South Atlantic ODP Sites 925 & 926.
SA: Shackleton et al. (1999), Pearson & Chaisson (1997), Berggren et al. (1983), Pujol (1983)	Astronomically tuned from Atlantic ODP Sites 925 & 926. Also in the South Atlantic (DSDP Leg 72) Berggren et al. (1995a) had assigned as Chron C5B.1n.
SA: Shackleton et al. (1999), Pearson & Chaisson (1997)	Astronomically tuned from South Atlantic ODP Sites 925 & 926.
SA: Berggren et al. (1983)	Berggren et al. (1995a) = C5B.1n; 14.8 Ma.
SA: Shackleton et al. (1999), Pearson & Chaisson (1997)	Astronomically tuned from South Atlantic ODP Sites 925 & 926. [In Lourens et al. (2004) GTS2004 Neogene foraminifer appendix it was accidentally marked as FAD, should be LAD].
SA: Poore et al. (1983), Berggren et al. (1985a)	Berggren et al. (1995a) used Orbulina suturalis (age of 15.09 Ma, based on mid-C5Bn.2n; which had 15.1 Ma in South Atlantic DSDP Leg 73) to define base of Zone M6. However, Lourens et al. (GTS2004) used FAD Orbulina universa FAD at 14.73 Ma based on astronomically tuned from Atlantic ODP Sites 925 & 926. Wade et al. (2011) argued for retaining the age estimate from Berggren et al. (1995a) due to "the rarity of Orbulina at the beginning of its range at Ceara Rise (Pearson and Chaisson, 1997)."
SA: Shackleton et al. (1999), Pearson & Chaisson (1997), Wade et al. (2011)	Astronomically tuned from South Atlantic ODP Sites 925 & 926. Was assigned as 14.89 Ma in Lourens et al. (GTS2004), but Wade et al. (2011) note the mistake in Shackleton et al. (1999) and propagated in that GTS2004 table where 14.8 Ma should read 15.8 Ma for FAD C. bermudezi. Calibrated to ATNTS2004 of Lourens et al. (2004) with an age of 15.73 Ma according to Wade et al. (2011).
I: Li et al. (1992)	Berggren et al. (1995a) -- 'Globorotalia miozea' LAD is calibrated in Hole 747A, Kerguelen Plateau, Indian Ocean; as within lower polarity chron C5B.1n or C5Br (lower); assigned as 15.9 Ma on CK'95 scale = C5Br.15.
NA: Berggren et al. (1985a), Wade et al. (2011)	Wade et al. (2011) calibrate FAD as 15.96 Ma. Berggren et al (1995) had a similar age using C5Cn.1n (upper); 16.0 Ma = C5C.1n.95. Berggren et al. (1985a) writes: "sequential appearance of praeorbulinid taxa occurs in "normal" (DSDP Hole 563) and expanded (DSDP Hole 558) sequence of anomaly 5C correlative in North Atlantic."
SA: Berggren et al. (1983), Berggren et al. (1985a)	Berggren et al. (1985a) writes: "FAD Globigerinoides diminutus occurs in Hole 516 [South Atlantic DSDP Leg 72] about 15 m below FAD Orbulina suturalis in an interval with no paleomagnetic data and about 5 m above FAD P. sicana in mid-part of Chron C5CN (Berggren et al., 1983)" = (C5Cn.2n); (16.1 Ma) = (C5Cn.1n.7). Further investigation at other sites is required to test this calibration.
SA: Shackleton et al. (1999), Pearson & Chaisson (1997)	Age estimated from South Atlantic ODP Sites 925 & 926 (Lourens et al., 2004)
SA: Shackleton et al. (1999), Pearson & Chaisson (1997), Berggren et al. (1983), Berggren et al. (1985a)	Age estimated from Atlantic ODP Sites 925 & 926 (Lourens et al., 2004). Berggren et al (1995a) uses to define base of Medit. Zone Mt5b, and assign to C5Cn.1n (upper) in South Atlantic Hole 516 DSDP Leg 72; 16.1 Ma = C5Cn.1n.8. Berggren et al. (1985a) writes: "FAD Praeorbulina glomerosa occurs in Hole 516 [South Atlantic DSDP Leg 72] about 15 m below FAD Orbulina suturalis in an interval with no paleomagnetic data and about 5 m above FAD P. sicana in mid-part of Chron C5CN (Berggren et al., 1983)"
NA: Berggren et al. (1985a), Wade et al. (2011)	Wade et al. (2011) astrochronology age projects as same chron-age calibration as Berggren et al. (1995) = mid-C5Cn; 16.3 Ma = base C5Cn.1n. Berggren et al. (1985a) writes: "sequential appearance of praeorbulinid taxa occurs in "normal" (DSDP Hole 563) and expanded (DSDP Hole 558) sequence of anomaly 5C correlative in North Atlantic."
SA: Berggren et al. (1983), Berggren et al. (1985a), Wade et al. (2011)	Age of 16.97 Ma was estimated from Atlantic ODP Sites 925 & 926; in Lourens et al. (2004) -- however this implied revised calibration to polarity chrons would seem to invert Medit. zones Mt4b and Mt5a of Berggren et al. (1995a). Berggren et al. (1995a) assign age as Chron C5Cn.2n (mid) and write: "LAD P. sicana occurs in interval of no paleomagnetic data about 20 m above FAD P. sicana which occurs in anom. 5C correlative in Hole 516 (Berggren et al., 1983)." Wade et al. (2011) write: 'Praeorbulina taxa are rare at Ceara Rise (Pearson and Chaisson, 1997) and therefore were not included in the revised calibration, and we have retained the ages reported in Berggren et al. (1995a) for FAD Orbulina suturalis (15.1 Ma), FAD Praeorbulina circularis, (16.0 Ma), FAD Praeorbulina curva (16.3 Ma) and Praeorbulina sicana (16.4 Ma)." We tentatively follow the Wade et al. (2011) assessment until further sites are investigated.
I: Berggren (1992)	Berggren et al. (1995) -- 'Globorotalia incognita' LAD is calibrated in Hole 747C, Kerguelen Plateau = Chron C5Cn.2n (mid).
SA: Berggren et al. (1983), Wade et al. (2011)	Wade et al. (2011) astrochronology age projects as same chron-age calibration as Berggren et al. (1995) -- 'Globorotalia birnageae' (former genus) FAD is calibrated in Hole 516, South Atlantic = ca. Chron C5Cn.3n.1. Subzone "M4b" is 'Globigerinoides bisphericus Partial Range Zone' (FAD of Globorotalia birnageae to FAD of Praeorbulina sicana) (Berggren et al., 1995). This short subzone spans the upper part of the 'N7' zone -- and now seems to be entirely WITHIN Zone M5 (after cycle-calibration). The FAD of G. birnageae is difficult to determine precisely because of its intergradation with Gl. trilobus s.str.
I: Berggren (1992)	Berggren et al. (1995) note 'Globorotalia miozea' FAD is recorded in Hole 751, Kerguelen Plateau, south Indian Ocean = C5Cn.3n (lower); 16.7 Ma = placed at C5Cn.3n.1, and used to define base of Medit. Zone Mt4.
I: Berggren (1992), Li et al. (1992)	Wade et al. (2011) astrochronology age projects as same chron-age calibration as Berggren et al. (1995) -- 'Globorotalia zealandica' is recorded in Indian Ocean Kerguelen Plateau Hole 747A (Li et al. 1992); and observed in lower part of polarity zone C5Dn in Hole 751 (Berggren 1992) = Chron C5Dn (upper) (C5Dn.9). Wade et al. (2011): Mistake in Berggren et al. (1995) where "*LAD G. zealandica*" should have read "FAD G. zealandica"

GTS2012 CHRONO-STRATIGRAPHY: Age in Ma / STAGE (AGE)	STANDARD TROPICAL-SUBTROPICAL BIOZONE (BIOCHRON): After Blow (1969, 1979), Kennett & Srinivasan (1983), Berggren et al. (1995a), Chaisson & Pearson (1997), Pearson & Chaisson (1997): Indo-Pacific	Atlantic	After Berggren et al. (1995a) and Wade et al. (2011) with emended M14: Indo-Pacific	Atlantic	BIOHORIZON (DATUM): T = Top/LAD; B = Base/FAD; X = Coiling change; ** = Regional age only	AGE (Ma) GTS2012 scaling: *This study*	Published Error	CALIBRATED AGES COMPARED GLOBALLY: North Atlantic (incl. Gulf Mex.) NA	Mediterranean M	South Atlantic SA	Indian Ocean I	South Pacific SP	Equatorial Pacific EP	Southern Ocean SO	North Pacific NP	CALIBRATION: Magnetochron/ Marine Isotope Stage (various authors, see ref's)
					T Globorotalia semivera	**17.26**				17.26						SA: C5Dn.9
	N7/N6	**N7/N6**	**M4a/ M3**	**M4a/ M3**	T Catapsydrax dissimilis	**17.54**		18.3		**17.54**						NA: C5Dn.9
					B Globigerinatella insueta sensu stricto	**17.59**				17.59						-
			M3	M3	** T Globoquadrina dehiscens forma spinosa [S. Atl.]	**17.61**				17.61						SA: C5Dr.6
					B Globorotalia praescitula	**18.26**				18.26						SA: C5En.5
					T Globoquadrina binaiensis	**19.09**				19.09						-
			M3/M2	**M3/M2**	B Globigerinatella sp.	**19.30**				19.30						-
	N6-N5	N6-N5			B Globoquadrina binaiensis	**19.30**				19.30						-
			M2	M2	B Globigerinoides altiaperturus	**20.03**			20.03							M: base C6r
(20.44)																
					T Tenuitella munda	**20.78**					20.78					I: C6Ar.8
					B Globorotalia incognita	**20.93**					20.93					I: C6Ar.4
					T Globoturborotalita angulisuturalis	**20.94**				20.94						SA: C6Ar.4
	N5/N4b	**N5/N4b**	**M2/ M1b**	**M2/ M1b**	T Paragloborotalia kugleri	**21.12**			21.5	**21.12**						M: C6Ar.6
					T Paragloborotalia pseudokugleri	**21.31**				21.31						SA: C6Ar.4
Aquitanian	N4b	N4b	M1b	M1b	B Globoquadrina dehiscens forma spinosa	**21.44**				21.44						SA: C6AAr.1n.5
					T Dentoglobigerina globularis	**21.98**				21.98						SA: C6Bn.1r.3
	N4b/ N4a	**N4b/ N4a**	**M1b/ M1a**	**M1b/ M1a**	B Globoquadrina dehiscens	**22.44**			22.44							M: C6Br.5
	N4a	N4a	M1a	M1a	T Globigerina ciperoensis	**22.90**				22.90						
					B Globigerinoides trilobus sensu lato	**22.96**				22.96						
	N4a/ P22	**N4a/ P22**	**M1a/ O7**	**M1a/ O7**	B Paragloborotalia kugleri	**22.96**		23.0		22.96						SA: base C6Cn.2n
23.03																
					T Globigerina euapertura	**23.03**					23.03					I: C6Cn.1n
			O7	O7	T Tenuitella gemma	**23.50**					23.50					I: C6Cr.7
Chattian	P22 (N3)	P22 (N3)			B common Globigerinoides primordius	**23.50**				23.50						SA: C6Cr.7
			O7/O6	**O7/O6**	B Paragloborotalia pseudokugleri	**25.21**							25.21			NA: C7n; EP: C8n.1n.3
			O6	O6	B Globigerinoides primordius	**26.12**		26.12								NA: C8r.7
	P22/ P21	**P22/ P21**	**O6/O5**	**O6/O5**	T Paragloborotalia opima sensu stricto	**26.93**				27.30						SA: C9n.9
(28.09)																
	P21 (N2)	P21 (N2)	**O5/O4**	**O5/O4**	T common Chiloguembelina cubensis	**28.09**			**28.09**		28.5					M: base C10n.1n; I: mid-C10n
	P21/ P20	**P21/ P20**	**O4/O3**	**O4/O3**	B Globigerina angulisuturalis	**29.18**		29.18		29.18			29.18			NA, SA, EP: top C11n
					B Tenuitellinata juvenilis	**29.50**					29.50					I: C11n.1r.5
Rupelian	P20 (N1)	P20 (N1)	O3	O3	T Subbotina angiporoides	**29.84**				29.84	29.84					SA & I: C11n.2n.3
	P20/ P19	**P20/ P19**	**O3/O2**	**O3/O2**	T Turborotalia ampliapertura	**30.28**		30.28		30.28						NA & SA: C11r.5
	P19	P19	O2	O2	B Paragloborotalia opima	**30.72**		30.72								NA: C12n.5

CALIBRATION REFERENCES	COMMENTS
SA: Berggren et al. (1983)	Calibrated in the South Atlantic DSDP Leg 72 with the same chron assignment as "*LAD of G. zealandica*" in Berggren et al' (1995a) based on Berggren et al. (1983). [Note that this "*LAD*" should read "FAD" G. zealandica according to Wade et al. (2011).]
NA: Berggren et al. (1985a); SA: Shackleton et al. (1999), Pearson & Chaisson (1997)	Age estimated from Atlantic ODP Sites 925 & 926; in Lourens et al. (GTS2004 Neogene foraminifer appendix). Berggren et al. (1995a) calibrate 'Catapsydrax dissimilis' LAD as recorded in Holes 558, 563, North Atlantic; and observed at top of polarity zone C5Dr in Hole 608 in interval of strong dissolution = Chron C5Dn (upper) = placed at C5Dn.9.
SA: Shackleton et al. (1999), Pearson & Chaisson (1997)	Age estimated from South Atlantic ODP Sites 925 & 926; in Lourens et al. (GTS2004 Neogene foraminifer appendix) -- Note that this is much higher FAD than estimated calibrations by Berggren et al. (1995a); which results in Zone M3 being nearly non-existent. Berggren et al. (1995a) -- 'Globigerinatella insueta' FAD has an inferred correlation to base of Chron C5En.
SA: Berggren et al. (1983), Pujol (1983)	Chron-age calibration by Berggren et al. (1995) -- 'Globoquadrina dehiscens forma spinosa' LAD is recorded in Hole 615, South Atlantic = placed at Chron C5Dr.6.
SA: Berggren et al. (1983)	Wade et al. (2011) astrochronology age projects as same chron-age calibration as Berggren et al. (1995a), who used this FAD to define base of Medit. Zone Mt3. 'Globorotalia praescitula' FAD is recorded in Indian Ocean Holes 747A, 748B, 751, North Atlantic Hole 608; and is recorded in lowest part of polarity zone C5En in South Atlantic Hole 516F = placed at Chron C5En.5.
SA: Shackleton et al. (1999), Pearson & Chaisson (1997)	Age estimated from South Atlantic ODP Sites 925 & 926.
SA: Shackleton et al. (1999), Pearson & Chaisson (1997)	Age estimated from South Atlantic ODP Sites 925 & 926.
SA: Shackleton et al. (1999), Pearson & Chaisson (1997)	Age estimated from South Atlantic ODP Sites 925 & 926.
M: Berggren et al. (1983), Pujol (1983), Montanari et al. (1991)	Used by Berggren et al. (1995a) as "working" base of Burdigalian stage -- but current (GTS2012) working definition for base of Burdigalian is FAD of nannofossil H. ampliaperta, as calibrated at Ceara Rise = about 0.4 myr older. Wade et al. (2011) astrochronology age projects as same chron-age calibration as Berggren et al. (1995a) -- 'Globigerinoides altiaperturus' FAD is recored in Hole 516F, South Atlantic; and is within in older part of polarity chron C6r in Contessa Highway section = set as base of Chron C6r. NOTE: Berggren's derived age implies that he assigned this FAD to the base of polarity chron C6A; thereby effectively defining the base of the Burdigalian as base of Chron C6A; a magnetochron event which would be more convenent and global than a biostratigraphic event.
I: Li et al. (1992), Wright & Miller (1992), Berggren et al. (1995a)	Wade et al. (2011) astrochronology age projects as same chron-age calibration as Berggren et al. (1995a) -- 'Tenuitella munda' LAD is recorded in Hole 747, Kergulen Plateau; with the magnetostratigraphy reinterpreted in Wright & Miller (1992) = ChronC6Ar.8.
I: Berggren (1992), Wright & Miller (1992), Berggren et al. (1995a)	Same as chron-age of Berggren et al. (1995a) -- 'Globorotalia incognita' FAD is recorded in Hole 747A; with magnetostratigraphy reinterpreted in Wright & Miller (1992) = placed at Chron C6Ar.4.
SA: Berggren et al. (1983), Wright & Miller (1992), Berggren et al. (1995a)	Chron-age calibration by Berggren et al. (1995a) -- 'Globoturborotalita angulisuturalis' LAD is recorded in Hole 516F; with magnetostratigraphy reinterpreted in Wright & Miller (1992) = placed at Chron C6Ar.4.
M: Montanari et al. (1991); SA: Shackleton et al. (1999), Pearson & Chaisson (1997), Berggren et al. (1993), Pujol (1993)	Age estimated from Atlantic ODP Sites 925 & 926; in Lourens et al. (GTS2004 Neogene foraminifer appendix). Berggren et al. (1995a) use this level as base of Medit. Zone Mt2, and note that Miocene foram 'Paragloborotalia kugleri' LAD is recorded in Hole 516F, South Atlantic, and at base of polarity zone C6An in Contessa Highway section = placed at Chron C6Ar.6.
SA: Shackleton et al. (1999), Pearson & Chaisson (1997), Berggren et al. (1983), Wright & Miller (1992), Berggren et al. (1995a)	Age estimated from Atlantic ODP Sites 925 & 926; in Lourens et al. (GTS2004 Neogene foraminifer appendix). Berggren et al. (1995) have similar age -- 'Paragloborotalia (formerly Globorotalia) pseudokugleri' LAD is recorded in Hole 516F; with magnetostratigraphy reinterpreted in Wright & Miller (1992) = placed at Chron C6Ar.4.
SA: Berggren et al. (1983), Pujol (1983)	Chron-age calibration by Berggren et al. (1995a) -- 'Globoquadrina dehiscens forma spinosa' FAD is recorded in Hole 516F, South Atlantic = placed at Chron C6AAr.1n.5.
SA: Berggren et al. (1983)	Chron-age calibration by Berggren et al. (1995a) -- 'Globoquadrina globularis' LAD is recorded in Hole 516F, South Atlantic = placed at Chron C6Bn.1r.3.
NA: Berggren et al. (1985a); M: Montanari et al. (1991); SA: Berggren et al. (1983), Pujol (1983), Poore et. al. (1983)	Wade et al. (2011) astrochronology age projects as essentially same chron-age calibration as Berggren et al. (1995a), who used to define Medit. Zone Mt1b -- Miocene foram 'Globoquadrina dehiscens' FAD is recorded in South Atlantic Hole 516F, North Atlantic Holes 558, 563; and Contessa Highway Section = placed at Chron C6Br.5.
SA: Shackleton et al. (1999), Pearson & Chaisson (1997)	Age estimated from South Atlantic ODP Sites 925 & 926; in Lourens et al. (2004).
SA: Shackleton et al. (1999), Pearson & Chaisson (1997)	Age estimated from South Atlantic ODP Sites 925 & 926; in Lourens et al. (2004).
NA: Berggren et al. (1985a); SA: Shackleton et al. (1999), Pearson & Chaisson (1997), Berggren et al. (1983)	FAD Paragloborotalia kugleri s.str. = M1 base; and approx. Oligocene/Miocene boundary. Age estimated from Atlantic ODP Sites 925 & 926; in Lourens et al. (GTS2004 Neogene foraminifer appendix). Berggren et al (1995a) notes that 'Globorotalia kugleri' FAD is recorded in Holes 558, 563 (North Atlantic) and 516F (South Atlantic) = base of Chron C6Cn.2n.
I: Li et al. (1992)	Chron-age calibration by Berggren et al. (1995a) -- 'Globigerina euapertura' LAD is recorded in Hole 747A within Chron C6Cn.1n.
I: Li et al. (1992)	Wade et al. (2011) astrochronology age projects same chron-age calibration as Berggren et al. (1995a) -- 'Tenuitella gemma' LAD is recorded in Hole 747A = Chron C6Cr.7.
SA: Berggren et al. (1983), Pujol (1983)	Wade et al. (2011) astrochronology age projects same chron-age calibration as Berggren et al. (1995a); recorded in South Atlantic Hole 516F = ca. Chron C6Cr.7.
NA & EP: Leckie et al. (1993)	Wade et al. (2011) astrochronology age projects SAME chron-age calibration as Berggren et al. (1995a); recorded in polarity zone C8n in Hole 803D (Ontong Java Plateau) and in C7n in Hole 628A (Little Bahama Bank) = placed at Chron C8n.1n.3.
NA: Leckie et al. (1993)	Wade et al. (2011) astrochronology age projects chron-age calibration as Berggren et al. (1995a) -- 'Globigerinoides primordius' FAD is recorded in Hole 628A (Little Bahama Bank); and their 26.7 Ma age projects as Chron C8r.7 on their polarity scale.
SA: Berggren et al. (1985b), Wade et al. (2011)	Wade et al. (2011) diagram projects LAD as middle of Chron C9n (used here), but unfortunately, do not specify the reference sites for this chron calibration. Berggren et al. (1995a) placed slightly higher (age estimate implied approx. C9n.9).
M: Coccione et al. (2008); I: Berggren et al. (1985b), Li et al. (1992)	Base of Chattian (working version) is base of Chron C10n.1n. Last common occurrence of C. cubensis has also been used to place base-Chattian (Coccioni et al., 2008, GSA Bull.; and in GTS2004). Coccioni et al. (2008; Italian sections) have this LCO at about 2/3rds up in Chron C10n (subchrons C10n.1n and .2n were not resolved). Berggren and Pearson (2005) indicate that the "LAD is recorded in mid-part of polarity zone C10n in Holes 747A (Berggren et al., 1985, Li et al., 1992), 749 (Berggren, 1992), in the mid-part of concatenated C9n and C10n in Hole 748B (Berggren, 1992), and associated approximately with C10n in Holes 689B and 690B (Stott & Kennett, 1990). "Whether this datum refers to extinction or strong reduction in numbers remains a moot point; records of discontinuouss presence in reduced numbers of this taxon into basal Miocene levels continue (Leckie et al., 1993)". Wade et al. (2011) conclude with assigning this LCO a numerical value within 0.05 myr of the base of Chron C10n.1n on their time scale; therefore, base of Chron C10n.1n is retained here.
NA, SA, EP: Berggren & Miller (1988), Leckie et al. (1993)	Berggren and Pearson (2005) estimate is similar to Berggren et al. (1995a) chron-age -- 'Globoturborotalita (formerly Globigerina) angulisuturalis' FAD, which defines base of P21a, is recorded by Berggren and Miller (1988) in South Atlantic Hole 516, North Atlantic Holes 558, 628A and west equatorial Pacific Site 803C. Their 29.4 Ma age was the top of Chron C11n on their polarity scale.
I: Li et al. (1992)	Chron-age calibration by Berggren et al. (1995a) -- 'Tenuitellinata juvenilis' FAD occurs within an interval of 'no polarity data' between two normal polarity intervals interpreted as parts of polarity chron C11 in Hole 747A . Their 29.7 Ma age is Chron C11n.1r.5 on their polarity scale.
SA: Berggren et al., (1985b), I: Berggren, (1992)	Chron-age calibration by Berggren et al. (1995a). This LAD is recorded in early part of polarity zone C11n in South Atlantic Hole 516F (Berggren et al., 1985b), Indian Ocean Hole 748B (Berggren, 1992) and Southern Ocean Holes 689B and 690B (Stott & Kennett, 1990). Their 30.0 Ma age is Chron C11n.2n.3 on their polarity scale.
NA: Berggren et al. (1985b), Stott & Kennett (1990); SA: Berggren et al. (1985b)	Oligocene foram 'Turborotalia ampliapertura' defines top of P19, and was formerly 'Globigerina ampliapertura' (e.g., Blow, 1969). Age-calibration from Berggren and Pearson (2005), which is similar as Berggren et al. (1995) -- LAD is recorded in polarity zone C11r in Holes 516F (South Atlantic) and 558 (North Atlantic) and St. Stephen's Quarry, Alabama borehole (North Atlantic). Their 30.3 Ma age is Chron C11r.5 on their polarity scale.
NA: Berggren et al. (1985b)	Berggren & Miller (1988, p.377) discuss 'opima' subspecies. Chron-age calibration by Berggren et al. (1995) -- FAD is recorded in Holes 558 and elsewhere. Their 30.6 Ma age is Chron C12n.5 on their polarity scale.

GTS2012 CHRONO-STRATIGRAPHY Age in Ma	STANDARD TROPICAL-SUBTROPICAL BIOZONE (BIOCHRON) After Blow (1969, 1979), Kennett & Srinivasan (1983), Berggren et al. (1995a), Chaisson & Pearson (1997), Pearson & Chaisson (1997)		After Berggren et al. (1995a) and Wade et al. (2011) with emended M14		BIOHORIZON (DATUM)	AGE (Ma) GTS2012 scaling		CALIBRATED AGES COMPARED GLOBALLY North Atlantic (incl. Gulf Mex.)	Mediterranean	South Atlantic	Indian Ocean	South Pacific	Equatorial Pacific	Southern Ocean	North Pacific	CALIBRATION Magnetochron/ Marine Isotope Stage
STAGE (AGE)	Indo-Pacific	Atlantic	Indo-Pacific	Atlantic	T = Top/LAD; B = Base/FAD; X = Coiling change; ** = Regional age only	*This study*	Published Error	NA	M	SA	I	SP	EP	SO	NP	(various authors, see ref's)
	P19/ P18	**P19/ P18**	**O2/O1**	**O2/O1**	T Pseudohastigerina naguewichiensis	**32.10**		32.10		32.10						NA & SA: C12r.5
33.89	P18	P18	O1	O1	B Cassigerinella chipolensis	**33.89**		33.89								NA: C13r.86
			O1/E16	**O1/E16**	T Hantkenina spp., Hantkenina alabamensis	**33.89**			33.89							M: C13r.85
			E16	E16	T common Pseudohastigerina micra	**33.89**		33.89			33.89					NA & I: C13r.85
	P18/ P17	**P18/ P17**			T Turborotalia cerroazulensis	**34.03**			34.03							M: C13r.75
Priabonian	**P17/ P16**	**P17/ P16**			T Cribrohantkenina inflata	**34.22**			34.23							M: C13r.6
	P16	P16	**E16/E15**	**E16/ E15**	T Globigerinatheka index	**34.61**			34.62		34.62					M & I: C13r.3
			E15	E15	T Turborotalia pomeroli	**35.66**			35.75							M: C15r.1
					B Turborotalia cunialensis	**35.71**			35.80							M: base C15r
	P16/ P15	**P16/ P15**			B Cribrohantkenina inflata	**35.87**			35.98							M: C16n.1n.1
(37.75)	**P15**	**P15**	**E15/ E14**	**E15/ E14**	T Globigerinatheka semiinvoluta	**36.18**			36.30							M: C16n.2n.8
			E14	E14	T Acarinina spp.	**37.75**			37.96							M: base C17n.1n
					T Subbotina linaperta	**37.96**					38.17					I: C17n.2n.6
					T Acarinina collactea	**37.96**					38.17					I: C17n.2n.6
			E14/ E13	**E14/ E13**	T Morozovelloides crassatus	**38.25**		38.46								NA: C17n.3n.5
Bartonian	**P15/ P14**	**P15/ P14**	E13	E13	B Globigerinatheka semiinvoluta	**38.62**			38.46							M: top C18n.1n
	P14	P14			T Acarinina mcgowrani	**38.62**		38.46								NA: C17n.3n.5
					T Planorotalites spp.	**38.62**			38.83							M: top C18n.1n
					T Acarinina primitiva	**39.12**								39.33		SO: C18n.1n.5
					T Turborotalia frontosa	**39.42**										Unknown region: C18n.1n.2
	P14/ P13	**P14/ P13**	**E13/ E12**	**E13/ E12**	T Orbulinoides beckmanni	**40.03**		40.21								NA: C18n.2n.25
	P13/ P12	**P13/ P12**	**E12/ E11**	**E12/ E11**	B Orbulinoides beckmanni	**40.49**										Unknown region: C18r.66
(41.15)	P12	P12	E11	E11	T Acarinina bullbrooki	**40.49**										Unknown region: C18r.66
			E11/ E10	**E11/ E10**	T Guembelitrioides nuttalli	**(42.07)**										Unknown region: (C19r.25)
			E10	E10	B Turborotalia pomeroli	**42.21**										Unknown region: C19r.1
					B Globigerinatheka index	**42.64**								42.68		SO: C20n.7
					B Morozovelloides lehneri	**43.15**										Unknown region: C20n.25
	P12/ P11	**P12/ P11**	**E10/E9**	**E10/E9**	T Morozovella aragonensis	**43.26**										Unknown region: C20n.15
Lutetian	**P11/ P10**	**P11/ P10**	**E9/E8**	**E9/E8**	B Globigerinatheka kugleri	**(43.88)**					(43.91)					I: (ca. C20r.75)
	P10/P9	**P10/P9**	E8	E8	B Hantkenina singanoae	**(44.49)**		ca. 44 Ma								(NA: ca. C20r.5)
	P9	P9			B Turborotalia possagnoensis	**45.49**										Unknown region: C20r.1
			E8/E7b	**E8/E7b**	B Guembelitrioides nuttalli	**(45.72)**										(NA: C20r.0 to C21n.9)

CALIBRATION REFERENCES	COMMENTS
NA: Berggren et al. (1985b), Miller et al. (1993); SA: Berggren et al. (1985b)	Berggren and Pearson (2005), which is similar to chron-age by Berggren et al. (1995a) -- Oligocene foram 'Pseudohastigerina spp.' LAD defines top of P18, and is recorded in South Atlantic Hole 516F, North Atlantic Holes 558, 563 and St. Stephen's Quarry, Alabama borehole. Earliest Oligocene foraminifer LAD of Psudohastigerina datum was noted as "C11r" in Berggren tables, but it seems to be a typographic error, because his age placement and chronologic order of this datum was at "32.0 Ma", hence in at Chron C12r..5 on their polarity scale (used here).
NA: Miller et al. (1993)	Marker for base of Rupelian; set as Chron C13r.86. Same chron-age calibration as Berggren et al. (1995a) -- Oligocene foram 'Cassigerinella chipolensis' FAD is recorded in St. Stephen's Quarry, Alabama borehole (Miller et al., 1993).
M: Coccione et al. (1988), Nocci et al. (1986)	Basal LAD = Marker for Eocene/Olig boundary = Chron C13.86. Berggren and Pearson (2005), which is similar as Berggren et al. (1995) -- the youngest of 3 normal "events" in polarity chron C13r in Contessa Highway section (Nocci et al., 1986) and above the normal "event" C13n.2n at Massignano (Coccione et al., 1988), where this LAD defines the Eocene/Oligocene boundary.
NA: Miller et al., (2008), Wade et al. (2011); I: Wade and Pearson (2008), Wade et al. (2011)	Wade et al. (2011): "We have added the HCO of Pseudohastigerina micra as a secondary marker for the Eocene/Oligocene boundary. Pseudohastigerina micra is common in upper Eocene sediments, and this form undergoes a significant size decrease coeval with the extinction of Hantkenina (Wade and Pearson, 2008; Wade and Olsson, 2009). This event appears to be coeval between the Indian Ocean (Wade and Pearson, 2008) and the Gulf of Mexico (Miller et al., 2008)." Same level as top of Hantkenina alabamensis.
M: Coccione et al. (1988), Nocci et al. (1986)	Same chron-age calibration as Berggren et al. (1995a) -- 'Turborotalia cerroazulensis" LAD defines top of P17, and is associated with youngest of 3 normal "events" in polarity chron C13r in Contessa Section (Nocci et al., 1986) and above normal "event" C13n.2n at Massignano (Coccione et al., 1988). It is not absolutely clear whether the LAD of 'Hantkenina spp.' and 'T. cerroazulensis' group are synchronous or sequential. Their 33.8 Ma age is Chron C13r.75 on their polarity scale.
M: Coccione et al. (1988), Nocci et al. (1986)	Same chron-age calibration as Berggren et al. (1995a) -- 'Cribrohantkenina inflata' LAD defines top of P16, and is located between LAD of 'Gl. index' and 'T. cunialensis' and 'Hantkenina' in mid-part of polarity chron C13r at Massignano (Coccione et al., 1988) and just below youngest of 3 normal "events" in C13r in Contessa Highway section (Nocci et al., 1986). Their 34.0 Ma age is Chron C13r.6 on their polarity scale. B. Wade (pers. comm.; Oct'11) -- "Zone P17 continues to be problematic as the zonal marker has been synonymized. A solution is to call this P16/P17."
M: Nocci et al., (1986), Premoli-Silva et al., (1988a); I: Berggren (1992)	One of the few taxa which has an LAD that is isochronous between low and high latitudes and provinces. Age calibration from Berggren and Pearson (2005, which as extended discussion) is similar to Berggren et al. (1995a) -- 'Globigerapsis index' LAD is essentially coincident with the LAD of 'Discoaster saipanensis' and 'Discoaster barbadoensis' and with lower of 3 normal "events" in polarity chron C13r in Contessa Highway section (Nocci et al., 1986, Premoli-Silva et al., 1988a). Their 34.3 Ma age is Chron C13r.3 on their polarity scale.
M: Coccione et al. (1988)	Chron-age calibration by Berggren et al. (1995a) -- 'Turborotalia pomeroli' LAD occurs with LAD of 'P. semiinvoluta' in polarity Chron C16n.2n in Massignano section, but their summary 35.3 Ma age is Chron C15r.1 on their polarity scale.
M: Nocci et al., (1986), Premoli-Silva et al., (1988a), Wade et al. (2011)	Wade et al. (2011) table assigns as 35.3 Ma, which is base of Chron C15r on their polarity scale. Similar to the chron-age calibration of Berggren et al. (1995a) -- 'Turborotalia cunialensis' FAD defines base of P16, and is just above (i.e. younger than) LAD of 'G. semiinvoluta'.
M: Nocci et al., (1986), Premoli-Silva et al., (1988a), Coccione et al. (1988)	Eocene foram 'Cribrohantkenina inflata' FAD was used previously as P15/P16 boundary (e.g., Blow, 1979; Berggren and Miller, 1988); but discontinued by Berggren et al. (1995a). Same chron-age calibration as Berggren et al. (1995a, which has extended discussion) -- their 35.5 Ma summary age corresponds to Chron C16n.1n.1 on their polarity scale.
M: Coccione et al. (1988), Berggren & Pearson (2005)	Located in Italian Massignano section within C15r; but Berggren & Pearson (2005) estimate as 35.8 Ma or at Chron C16n.2n.8 on their polarity scale.
M: Nocci et al. (1986), (1986), Premoli-Silva et al., (1988a), Berggren et al. (1985b)	Chron-age calibration by Berggren et al. (1995a) -- 'Acarinina spp.' LAD occurs in mid-C17n (Berggren et al., 1985b) or top of polarity chron C18n (Nocci et al., 1986, Premoli-Silva et al., 1988a) in the Apennines, which is about 1 myr older. Their 37.5 Ma age corresponds to base of Chron C17n.1n on their polarity scale.
I: Berggren (1992); Stott & Kennett (1990)	Chron-age calibration by Berggren et al. (1995a) -- 'Subbotina linaperta' is recorded in mid-C17n in Southern Ocean Hole 689B (Stott & Kennett, 1990) and in polarity zone C17n in Indian Ocean Hole 748B (Berggren, 1992). Their 37.7 Ma summary age corresponds to Chron C17n.2n.6 on their polarity scale.
I: Berggren (1992); Stott & Kennett (1990)	Chron-age calibration by Berggren et al. (1995a) -- 'Acarinina collactea' LAD is recorded in interval of uninterpretable paleomagnetic data just above level identified as polarity chron C17n in Indian Ocean Hole 748B (Berggren, 1992) and at level interpreted as C18n in Southern Ocean Hole 690B (Weddell Sea (Stott & Kennett, 1990). Their 37.7 Ma summary age corresponds to Chron C17n.2n.6 on their polairty scale.
NA: Wade et al. (2011)	Morozovelloides crassatus is the senior synonym of Morozovella spinulosa. Wade et al. (2011) assign as Chron C17n.3n.5; which is similar to chron-age by Berggren and Pearson (2005) and Berggren et al. (1995).
M: Nocci et al. (1986)	Eocene foram 'Porticulasphaera semiinvoluta' defines base of P15, and has a brief overlap with 'Truncorotaloides' (Blow, 1979). However, a brief, but distinct, separation of 'Acarinina' and 'G. semiinvoluta' is recorded in Nocchi et al. (1986, Fig. a). Wade et al. (2011) assign as 37.7 Ma, which is Chron C17n.3n.5 on that GTS04 polarity scale). This is slightly younger than in Berggren et al. (1995a), which was at the top of Chron C18n.1n on their polarity scale.
NA: Wade (2004)	Wade et al. (2011) (citing Wade, 2004) assign this LAD as same level as LAD of Morozovelloides crassatus based on investigations in the western North Atlantic (Blake Nose).
M: Nocci et al. (1986)	Chron-age calibration by Berggren et al. (1995a). Their 38.5 Ma is the top of Chron C18n.1n on their polarity scale.
SO: Stott & Kennett (1990); I: Berggren (1992)	Chron-age calibration as Berggren et al. (1995a) -- 'Acarinina primitiva' is recorded in mid-part of polarity chron C18n in Southern Ocean Hole 689 and in interval of uninterpretable paleomagnetic data just above probable C18n in Indian Ocean Hole 748B. Their 39.0 Ma age is Chron C18n.1n.5 on their polarity scale.
Berggren et al. (1985b)	Formerly "Subbotina frontosa". Wade et al. (2011) = Same chron-age calibration as Berggren et al. (1995a), who cited Berggren et al. (1985b). Their 39.3 Ma age is Chron C18n.1n.2 on their polarity scale.
NA: Wade (2004)	Defines base of P14 (or E13). Wade et al. (2011) has same age calibration as Berggren and Pearson (2005), which is at Chron C18n.2n.25 on their polarity scale.
Berggren et al. (1985b)	Defines P13 (= E12) base. Wade et al. (2011) has same chron-age calibration as Berggren et al. (1995a), which is Chron C18r.66 on their polarity scale.
Berggren et al. (1985b)	Berggren and Pearson (2005) use same chron-age calibration as Berggren et al. (1995a); and their 40.5 Ma is at Chron C18r.66 on their polarity scale.
Berggren & Pearson (2005)	DASHED: Wade et al. (2011): "The age of the HO of Guembelitrioides nuttalli in not well constrained and requires further study." Age assignment by Berggren and Pearson (2005) of 42.3 Ma is at Chron C19r.25 on their polarity scale.
Berggren et al. (1985b)	Chron-age calibration by Berggren et al. (1995a). Their 42.4 Ma age is at Chron C19r.1 on their polarity scale.
SO: Stott & Kennett (1990)	Chron-age calibration by Berggren et al. (1995a) -- 'Globigerinatheka index' FAD is recorded near polarity zone C20n/C20r boundary in Southern Ocean Hole 689B. Their 42.9 Ma age is at Chron C20n.7 on their polarity scale.
Berggren et al. (1985b)	Formerly "Morozovella lehneri". Wade et al. (2011) projects as same chron-age calibration as Berggren et al. (1995a) at Chron C20n.25.
Berggren et al. (1985b)	Defines top of P11 or E9. Assignment by Berggren and Pearson (2005) is same chron-age calibration as Berggren et al. (1995a). Their 43.6 Ma is Chron C20n.15 on their polarity scale.
I: Pearson et al. (2004), Wade et al. (2011)	DASHED -- Wade et al. (2011): "The age of the LO [FAD] of Globigerinatheka kugleri in not well constrained and there is a significant need for further study of this interval in continuous sections with magnetostratigraphy". They project as 43.4 Ma relative to GTS04 polarity => Chron C20r.75. Wade et al. (2011) reference the Tanzania study (Indian Ocean) by Pearson et al. (2004).
NA: Payros et al. (2009), Wade et al. (2011)	DASHED -- Wade et al. (2011) (citing Payros et al. (2009 = Lutetian GSSP in Spain) imply that the FAD of Hantkenina singanoae (which is the first Hantkenina spp., hence base of former zone P10) is at 43.5 Ma age "relative to GTS04 polarity", therefore at Chron C20r.7 (which would be 44.0 in GTS2012 scale). But dashed here at C20r.5 to avoid near-overlap with P11 that they imply begins at Chron C20r.75. [More study is needed of relative placement of these two FADs.]
Berggren et al. (1995a)	Chron-age calibration by Berggren et al. (1995a). Their 46.0 Ma age is at Chron C20r.1 on their polarity scale. No region or original calibration reference cited.
NA: Payros et al. (2009), Wade et al. (2011)	FAD of G. nuttalli is dashed at approximately the base of Chron C21r (or may be slightly lower at C21n.9) and is the base of revised zone E8 (Wade et al., 2011). This datum-chron correlation is nearly 3 million years higher than in Berggren & Pearson (2005). Wade et al. (2011) summarize: "In a detailed magnetobiostratigraphic study of an expanded lower–middle Eocene succession from the western Pyrenees, Payros et al. (2009) examined the divergence of planktonic foraminiferal Zones P9 and P10 and the standard zonation (BKSA95). Unlike BKSA95 which places the boundary between planktonic foraminiferal Zones P9 and P10 within calcareous nannofossil Zone NP14 (Subzone CP12a) and magnetic polarity Chron C22n/C21r boundary, they found the boundary to occur within Zone NP15 (=Zone CP13) and Chron C20r. ... The P9/P10 zonal boundary is therefore a surprising 3.1 myr younger than in BKSA95." In addition, the first occurrence of Hantkenina (alternate suggested marker for base of P10), which had been a working definition of base of Lutetian Stage, was found to be significantly higher (possibly near the top of Chron C20r; ca. another 2 myr higher). Calibration of these events needs further careful study.

GTS2012 CHRONO-STRATIGRAPHY	STANDARD TROPICAL-SUBTROPICAL BIOZONE (BIOCHRON)					AGE (Ma)		CALIBRATED AGES COMPARED GLOBALLY								CALIBRATION
Age in Ma	After Blow (1969, 1979), Kennett & Srinivasan (1983), Berggren et al. (1995a), Chaisson & Pearson (1997), Pearson & Chaisson (1997)		After Berggren et al. (1995a) and Wade et al. (2011) with emended M14		BIOHORIZON (DATUM)	GTS2012 scaling		North Atlantic (incl. Gulf Mex.)	Mediterranean	South Atlantic	Indian Ocean	South Pacific	Equatorial Pacific	Southern Ocean	North Pacific	Magnetochron/ Marine Isotope Stage
STAGE (AGE)	Indo-Pacific	Atlantic	Indo-Pacific	Atlantic	T = Top/LAD; B = Base/FAD; X = Coiling change; ** = Regional age only	*This study*	Published Error	NA	M	SA	I	SP	EP	SO	NP	(various authors, see ref's)
47.84			E7b	E7b												
Ypresian			E7b/ E7a	E7b/ E7a	B Turborotalia frontosa	48.31		48.30								NA: within C21r
			E7a/E6	E7a/E6	B Acarinina cuneicamerata	50.20					50.19					(I: C22r.33)
	P9/P8	P9/P8	E6	E6	B Planorotalites palmerae	50.20										(Unknown region: C22r.33)
	P8/P7	P8/P7	E6/E5	E6/E5	T Morozovella subbotinae	50.67										Unknown region: C23n.1n.8
	P7	P7	E5	E5	B Acarinina pentacamerata	50.67								50.65		SO: C23n.1n.8
	P7/P6b	P7/P6b	E5/E4	E5/E4	B Morozovella aragonensis	52.54				52.55						SA: C23r.1
	P6b	P6b	E4	E4	T Morozovella marginodentata	52.85										NA: C24n.1n.5
					T Morozovella lensiformis	53.14		53.08							53.08	NA & NP: C24n.1r.5
					T Morozovella aequa	54.20		54.15							54.15	NA & NP: C24r.93
	P6b/ P6a	P6b/ P6a	E4/E3	E4/E3	B Morozovella formosa	54.61									54.56	NP: C24r.8
	P6a	P6a	E3	E3	B Morozovella lensiformis	54.61		54.56							54.56	NA & NP: C24r.8
					T Subbotina velascoensis	55.07		55.04								NA: C24r.65
	P6a/P5	P6a/P5	E3/E2	E3/E2	T Morozovella velascoensis	(55.20)									(55.17)	(NP: C24r.61)
	P5	P5	E2	E2	T Morozovella acuta	55.39		55.36								NA: C24r.55
					B Morozovella gracilis	55.39		55.36								NA: C24r.55
					B Igorina broedermanni	55.39		55.36								NA: C24r.55
					B Morozovella marginodentata	55.54		55.52							55.52	NA & NP: C24r.5
			E2/E1	E2/E1	B Pseudohastigerina wilcoxensis	55.81										Unknown region: lower-mid C24r
			E1	E1	B Globanomalina australiformis	55.96								55.96		SO: lower-mid C24r
			E1/P5	E1/P5	B Acarinina sibaiyaensis	55.96										Unknown region: lower-mid C24r
55.96																
Thanetian	P5/P4	P5/P4	P5/P4c	P5/P4c	T Globanomalina pseudomenardii	57.10					57.10				57.10	I: base C24r
	P4	P4	P4c	P4c	B Morozovella subbotinae	57.10									57.10	NP: base C24r
					T Acarinina mckannai	57.66		57.66			(57.66)		(57.66)			NA: base C25n
					T Acarinina acarinata	57.66		57.66			(57.66)		(57.66)			NA: base C25n
			P4c/ P4b	P4c/ P4b	B Acarinina soldadoensis	57.79		57.79			(57.79)		(57.79)			NA: C25r.9
			P4b	P4b	B Acarinina coalingensis	57.79		57.79			(57.79)		(57.79)			NA: C25r.9
					B Morozovella aequa	57.79		57.79			(57.79)		(57.79)			NA: C25r.9
					T Acarinina subsphaerica	58.44		58.44			(58.44)		(58.44)			NA: C25r.4
59.24																
					B Acarinina mckannai	60.43		60.43			(60.43)		(60.43)			NA: C26r.6
			P4b/ P4a	P4b/ P4a	T Parasubbotina variospira	60.52		60.52					(60.52)			NA: C26r.57

CALIBRATION REFERENCES	COMMENTS
NA: Payros et al. (2009), Wade et al. (2011)	T. frontosa FAD (Base of Zone E7b) is "one of most signficant foraminifer events in Middle/Early Eocene boundary interval; and occurs 13 precession cycles (0.26 myr) above base of C21r at the Lutetian GSSP section (Lutetian GSSP; published in Episodes, 2011). [Wade et al. (2011) table has an age of 48.6 Ma; but this would project only the base of that Chron C21r on their polarity time scale.]
I: Hancock et al. (2002), Wade et al. (2011)	Acarinina cuneicamerata = base E7. Chron-age calibration by Berggren et al. (1995a): their 50.4 Ma is at Chron C22r.33 on their polarity scale. Wade et al. (2011) cites Hancock et al. (2002) Exmouth Plateau study.
Berggren et al. (1985b)	Planorotalites ("Astrorotalia" in some publications) palmerae = base P9. Chron-age calibration by Berggren et al. (1995a): their 50.4 Ma age is Chron C22r.33 on their polarity scale.
Berggren et al. (1995a), Berggren & Pearson (2005)	LAD of M. subbotinae defines top of E5 (base of E6), and LAD of M. formosa was top of P7 (base of P8). Chron-age from Berggren and Pearson (2005), which is same chron-age calibration as Berggren et al. (1995a) -- Eocene Foram Zone P8 base (LAD of 'Morozovella formosa') is drawn as uppermost C23n in chart in Aubry et al. (1988); and is shown with an age of 50.8 Ma in Berggren (1995, SEPM); but the source of these calibrations are not cited. Assignment = uppermost C23n = placed here as Chron C23n.1n.8
SO: Stott & Kennett (1990)	Chron-age calibration by Berggren et al. (1995a) -- 'Acarinina pentacamerata' FAD is recorded at top of polarity zone C23n in Southern Ocean Hole 689 . Their 50.8 Ma age is Chron C23n.1n.8 on their polarity scale.
NA: Snyder & Waters (1985), Berggren et al. (1995a); NP: Corfield (1987); SA: Corfield (1987)	Defines P7 (=E5) base. Berggren and Pearson (2005), which is same chron-age calibration as Berggren et al. (1995a) -- 'Morozovella argonensis' FAD is recorded in North Atlantic Hole 550, North Pacific Shatsky Rise Hole 577 and in Chron C23r in South Atlantic (Walvis Ridge) Hole 527. Their 52.3 Ma age is at Chron C23r.1 on their polarity scale. [*The Wade et al. (2011) table based on Berggren-Pearson (2005) lists the same 52.3 Ma, which suggests they didn't convert to the chron-ages of the GTS04 magnetic scale.*]
NA: Snyder & Waters (1985), Berggren et al. (1995a), Wade et al. (2011)	Wade et al. (2011) have a GTS2004-age of 52.8 Ma implying Chron C24n.1n.5 on their polarity scale. 'Morozovella marginodentata' LAD is recorded in North Atlantic Hole 550.
NA: Snyder & Waters (1985), Berggren et al. (1995a); NP: Corfield (1987)	Chron-age calibration by Berggren et al. (1995a) -- 'Morozovella lensiformis' LAD is recorded in North Atlantic Holes 550 and North Pacific Shatsky Rise Hole 577. Their 52.7 Ma age is at Chron C24n.1r.5 on their polarity scale.
NA: Snyder & Waters (1985), Berggren et al. (1995a); NP: Corfield (1987)	Chron-age calibration by Berggren et al. (1995a) -- 'Morozovella aequa' LAD is recorded in North Atlantic Hole 550; and within polarity zone C24n.n1 in North Pacific Shatsky Rise Hole 577. Their 53.6 Ma summary age is at Chron C24r.93 on their polarity scale.
NP: Corfield (1987), Wade et al. (2011)	Base of Zone P6b (= Zone E4 of Berggren-Pearson. 2005). Wade et al. (2011) projects as Chron C24r.8, which is similar to the chron-age calibration by Berggren et al. (1995a) -- 'Morozovella formosa formosa' FAD is calibrated in North Pacific Shatsky Rise Hole 577.
NA: Snyder & Waters (1985), Berggren et al. (1995a); NP: Corfield (1987), Wade et al. (2011)	P6b base of Berggren et al. (1995a) was either M. formosa or M. lensiformis. Chron-age calibration by Berggren et al. (1995a) -- 'Morozovella lensiformis' FAD is recorded in North Atlantic Hole 550 and North Pacific Shatsky Rise Hole 577. Wade et al. (2011) projects as Chron C24r.8.
NA: Snyder & Waters (1985), Berggren et al. (1995a)	Chron-age calibration by Berggren et al. (1995a) -- 'Subbotina velascoensis' LAD is recorded in North Atlantic Hole 550 at Chron C24r.65. Berggren and Pearson (2005) place LAD of S. velascoensis in Zone E3.
NP: Corfield (1987), Wade et al. (2011)	Wade et al. (2011) projects as 54.9 Ma relative to GTS04 polarity chrons, therefore at Chron C24r.61 (used here). Berggren and Pearson (2005), puts this LAD 0.2 myr higher than Berggren et al. (1995a) -- This LAD is recorded in younger part of Chron C24r in Hole 577 and in mid-C24r in Hole 577 (at 81.98 m). However, paleomagnetic data (Bleil, 1985, DSDP Leg 86: 449, Fig. 7b) suggests a hiatus occurs between ~82m & 83 m in Hole 577, which would have the effect of depressing (rendering older) the HO of 'M. velascoensis'. The HO of 'M. velascoensis' in Hole 577 is in Nannofossil zone CP8 (=NP9) (Monechi, 1985: 310, Table 5) which supports interpretation of hiatus in Hole 577, because the LO of 'M. valascoensis' is in Nannofossil Zone NP10. Furthermore, Monechi (1985: p.307) indicates absence of extremely condensed sedimentation across the Paleocene/Eocene boundary interval. Essentially, it is not yet possible to position this LAD proportionally within the long Chron C24r.
NA: Wade et al. (2011), Berggren et al. (1995a)	Wade et al. (2011) projects as 55.1 Ma => Chron C24r.55 on their polarity scale. Nearly same chron-age calibration as Berggren et al. (1995a) -- 'Morozovella acuta' LAD is recorded in North Atlantic Hole 550 , where it occurs midway between dated Ash Bed -17 (54.5 Ma) and +19 (54.0 Ma) [using older Ar-Ar monitor standards].
NA: Snyder & Waters (1985), Berggren et al. (1995a), Wade et al. (2011)	Wade et al. (2011) projects as ca. Chron C24r.55 on their polarity scale. Nearly same chron-age calibration as Berggren et al. (1995a), based on Morozovella formosa gracilis FAD recorded in North Atlantic Hole 550.
NA: Snyder & Waters (1985), Berggren et al. (1995a), Wade et al. (2011)	Wade et al. (2011) projects as ca. Chron C24r.55 on their polarity scale. Nearly same chron-age calibration as Berggren et al. (1995a), based on 'Muricella broedermanni' ('Igorina broedermanni') FAD recorded in North Atlantic Hole 550.
NA: Snyder & Waters (1985), Berggren et al. (1995a); NP: Corfield (1987), Wade et al. (2011)	Wade et al. (2011) projects as 55.1 Ma => Chron C24r.5 on their polarity scale = higher than chron-age calibration as Berggren et al. (1995a) -- 'Morozovella marginodentata' FAD is calibrated in North Atlantic Hole 550; but was interpreted in North Pacific Hole 577 at the top of polarity chron C25n.
Berggren & Pearson (2005)	Berggren and Pearson (2005) -- FAD is top (not base) of PETM carbon-isotope excursion. Wade et al. (2011) show PETM as having ca. 0.1 myr duration (rounded). Following offset used by Berggren and Pearson (2007) and Wade et al. (2011), this FAD is set as 0.15 myr above PE boundary (base of PETM).
SO: Stott & Kennett (1990)	Marker for Paleocene/Eocene boundary in high southern latitudes. Set here as equal to PE boundary. Discussion of calibration as Berggren et al. (1995a) -- Paleocene foram 'Globanomallina australiformis' FAD in high latitudes [= AP4/5 boundary of Stott and Kennett (1990)] is recorded in Holes 689B and 690B. This FAD is recorded also at Indian Ocean Sites 738, 747 and 748, but with no paleomagnetic control. Berggren et al (1995a) = lower-mid Chron C24r.
Berggren & Pearson (2005)	Berggren and Pearson (2005): marker for Paleocene/Eocene boundary. One of "Excursion taxa" at PE boundary, therefore age is set equal to that PE age = lower-mid Chron C24r.
I: Pospichal et al. (1991); NP: Corfield (1987)	LAD defines top of Zone P4. Berggren and Pearson (2005) have similar chron-age calibration (base of Chron C24r) as Berggren et al. (1995a) -- Paleocene foram 'Globanomalina pseudomenardii' was called "Planorotallites pseudomenardii' in Berggren & Miller (1988). The LAD (= base Zone P5) is recorded in North Pacific Shatsky Rise Hole 577 at either base of polarity chron C24r (Corfield, 1987) and or top C25n (Liu & Olsson pers comm., 1993) and in Indian Ocean Hole 752 at top C25n (Pospichal et al., 1991). Juxtaposition/overlap of Gl. pseudomenardii LAD and Gt. subbotinae FAD is observed in Holes 465 and 758 (Nederbragt & van Hinte, 1987).
NP: Corfield (1987)	M. subbotinae FAD is an alternate marker for base of zone P5 zone. This event was used to define the base of Foram Zone 'P6a' by Berggren and Miller (1988), but Berggren et al. (1995a) consider that the expression of this FAD is commonly delayed in cores due to a widespread dissolution event(s) during polarity chron C24r. Chron-age calibration by Berggren et al. (1995a) -- Paleocene foram 'Morozovella subbotinae' FAD is recorded in North Pacific Hole 577. Their 55.9 Ma age was base of Chron C24r on their polarity scale.
NA, I, EP: Berggren et al. (1995a)	Chron-age calibration (base of Chron C25n) by Berggren et al. (1995a) -- Paleocene foram 'Acarinina mckannai' LAD is calibrated in DSDP Hole 384 (N. Atl.); and at comparable levels but without paleomagnetic control in Holes 465 (Eq. Pac.) and 758A (Indian Ocean) (Berggren et al., 1995a).
NA, I, EP: Berggren et al. (1995a)	Chron-age calibration (base of Chron C25n) by Berggren et al. (1995a) -- 'Acarinina acarinata' LAD is calibrated in DSDP Hole 384 (N. Atl.); and at comparable levels but without paleomagnetic control in Holes 465 (Eq. Pac.) and 758A (Indian Ocean) (Berggren et al., 1995a).
NA, I, EP: Berggren et al. (1995a)	Paleocene foram 'Acarinina soldadoensis' FAD defines base of Zone P4c of Berggren et al. (1995a). Berggren and Pearson (2005) has same chron-age calibration (Chron C25r.9) as Berggren et al. (1995a) -- This datum is calibrated in DSDP Hole 384 (N. Atlantic); and at comparable levels but without paleomagnetic control in Holes 465 (Eq. Pacific) and 758A (Indian Ocean), in association with FAD of A. coalingensis-triplex group.
NA, I, EP: Berggren et al. (1995a)	Chron-age calibration by Berggren et al. (1995) -- FAD of Acar. coalingensis-triplex group (Wade et al'11 list this group as a simple A. coalingensis) is calibrated in DSDP Hole 384(N. Atlantic); and at comparable levels but without paleomagnetic control in Holes 465 (Eq. Pacific) and 758A (Indian Ocean), in association with FAD of Acar. soldadoensis. Berggren et al. (1995a) project as 56.5 Ma, which is Chron C25r.9 on their polarity scale.
NA, I, EP: Berggren et al. (1995a)	Chron-age calibration by Berggren et al. (1995) -- 'Morozovella aequa' FAD is calibrated in DSDP Hole 384 (N. Atlantic); and at comparable levels but without paleomagnetic control in Holes 465 (Eq. Pacific) and 758A (Indian Ocean), in association with FAD of A. coalingensis-triplex group. Berggren et al. (1995a) project as 56.5 Ma, which is Chron C25r.9 on their polarity scale.
NA, I, EP: Berggren et al. (1995a)	Chron-age calibration by Berggren et al. (1995a; who had as Acarinina subsphaerica (common)) -- This datum is calibrated in DSDP Hole 384 (N. Atlantic); and at comparable levels but without paleomagnetic control in Holes 465 (Eq. Pacific) and 758A (Indian Ocean), and occurs approximately midway in Zone P4. Berggren et al. (1995a) project as 57.1 Ma, which is Chron C25r.4 on their polarity scale. ['Acarinina subsphaerica' LAD had defined base of Zone P4b of Berggren et al. (1995a).]
NA, I, EP: Berggren et al. (1995a)	Wade et al. (2011) assigns the FAD with the same chron-age calibration (Chron C26r.6) as Berggren et al. (1995a), but only THIS species -- In Berggren'95: Paleocene forams 'Acarinina acarinata', 'Acarinina mckannai' and 'Acarinina subsphaerica' FADs are calibrated in DSDP Hole 384 (N. Atlantic); and at comparable levels but without paleomagnetic control in Holes 465 (Eq. Pacific) and 758A (Indian Ocean) according to Berggren et al. (1995a). The FAD's of three acarininids (acarinata, subsphaerica, mckannai) are observed to coincide closely with the FAD of Gl. pseudomenardii. Berggren et al. (1995a) project as 59.1 Ma or Chron C26r.6 on their polarity scale.
NA, EP: Berggren et al. (1995a)	LAD defines top of revised zone P4a (e.g., Berggren and Pearson, 2005). Berggren and Pearson (2005) assignment follows Berggren et al. (1995a) [as "P. variolaria" LAD and "P. varianta" LAD] who calibrated this event in North Atlantic Hole 384 and at comparable levels but without paleomagnetic control in Holes 465 (Eq. Pacific). Berggren et al. (1995a) project as 59.2 Ma or Chron C26r.57 on their polarity scale.

GTS2012 CHRONO-STRATIGRAPHY	STANDARD TROPICAL-SUBTROPICAL BIOZONE (BIOCHRON)					AGE (Ma)		CALIBRATED AGES COMPARED GLOBALLY								CALIBRATION
Age in Ma	After Blow (1969, 1979), Kennett & Srinivasan (1983), Berggren et al. (1995a), Chaisson & Pearson (1997), Pearson & Chaisson (1997)		After Berggren et al. (1995a) and Wade et al. (2011) with emended M14		BIOHORIZON (DATUM)	GTS2012 scaling		North Atlantic (incl. Gulf Mex.)	Mediterranean	South Atlantic	Indian Ocean	South Pacific	Equatorial Pacific	Southern Ocean	North Pacific	Magnetochron/ Marine Isotope Stage
STAGE (AGE)	Indo-Pacific	Atlantic	Indo-Pacific	Atlantic	T = Top/LAD; B = Base/FAD; X = Coiling change; ** = Regional age only	*This study*	Published Error	NA	M	SA	I	SP	EP	SO	NP	(various authors, see ref's)
Selandian			P4a	P4a	B Acarinina acarinata	**60.52**		60.52			(60.52)		(60.52)			NA: C26r.6
	P4/P3b	**P4/P3b**			B Acarinina subsphaerica	**60.52**		60.52			(60.52)		(60.52)			NA: C26r.6
	P3b	P3b	**P4a/ P3b**	**P4a/ P3b**	B Globanomalina pseudomenardii	**60.73**		**60.73**			(60.73)		58.4			NA: C26r.5; NP: upper C26r; EP: lower C25r
	P3b/ P3a	**P3b/ P3a**	**P3b/ P3a**	**P3b/ P3a**	B Igorina albeari	**61.33**		61.33			(61.33)		(61.33)			NA: C26r.3
	P3a	P3a	P3a	P3a	B Morozovella velascoensis	**61.33**		61.33			(61.33)		(61.33)			NA: C26r.3
61.61																
					B Acarinina strabocella	**61.77**								61.77		SO: C26r.15
					B Morozovella conicotruncata	**62.22**									62.22	NP: lower C26r
	P3/P2	**P3/P2**	**P3a/P2**	**P3a/P2**	B Morozovella angulata	**62.29**		**62.29**							62.2	NA: C27n.8; NP: lower C26r
	P2	P2	P2	P2	B Igorina pusilla	**62.29**	dashed								62.29	NP: base C26r
					B Morozovella praeangulata	**62.46**		62.60			62.60		62.60			NA, EP, I: C27r.92
Danian	**P2/P1c**	**P2/P1c**	**P2/P1c**	**P2/P1c**	B Praemurica uncinata	**62.60**		62.60			62.60		62.60			NA, EP, I: C27r.92
	P1c/ P1b	**P1c/ P1b**	**P1c/ P1b**	**P1c/ P1b**	B Globanomalina compressa	**63.90**		63.90						63.90		NA, SO: C28n.65
	P1b	P1b	P1b	P1b	B Praemurica inconstans	**63.90**		63.90						63.90		NA, SO: C28n.65
					B Parasubbotina varianta	**64.02**		64.02			(64.02)					NA: C28n.55
	P1b/ P1a	**P1b/ P1a**	**P1b/ P1a**	**P1b/ P1a**	B Subbotina triloculinoides	**65.25**		65.25								NA: C29n.6
	P1a/Pα	**P1a/Pα**	**P1a/Pα**	**P1a/ Pα**	T Parvularugoglobigerina eugubina	**65.72**										Unknown region: C29r.95
	Pα	Pα	Pα	Pα	B Parasubbotina pseudobulloides	**65.76**		65.72	65.72		65.72					NA, M, I: C29r.9
					B Parvularugoglobigerina extensa	**65.94**										-
	Pα/P0	**Pα/P0**	**Pα/P0**	**Pα/P0**	B Parvularugoglobigerina eugubina	**66.00**			66.00							-
66.04	P0	P0	P0	P0												

CALIBRATION REFERENCES	COMMENTS
NA, I, EP: Berggren et al. (1995a), Wade et al. (2011)	Wade et al'11 has this FAD as 0.1 myr older (hence projects as Chron C26r.57) than in chron-age placement in Berggren et al. (1995) (59.1 Ma, hence Chron C26r.6 on their scale) -- 'Acarinina acarinata', 'Acarinina mckannai' and 'Acarinina subsphaerica' FADs are calibrated in DSDP Hole 384; and at comparable levels but without paleomagnetic control in Holes 465 and 758A (Berggren et al., 1995, SEPM). The FAD's of three acarininids (acarinata, subsphaerica, mckannai) are observed to coincide closely with the FAD of Gl. pseudomenardii.
NA, I, EP: Berggren et al. (1995a), Wade et al. (2011)	Wade et al'11 has this FAD as 0.1 myr older (hence projects as Chron C26r.57) than in chron-age placement in Berggren et al. (1995) (59.1 Ma, hence Chron C26r.6 on their scale) -- 'Acarinina acarinata', 'Acarinina mckannai' and 'Acarinina subsphaerica' FADs are calibrated in DSDP Hole 384; and at comparable levels but without paleomagnetic control in Holes 465 and 758A (Berggren et al., 1995, SEPM). The FAD's of three acarininids (acarinata, subsphaerica, mckannai) are observed to coincide closely with the FAD of Gl. pseudomenardii.
NA: Berggren et al. (1995a), Berggren & Pearson (2005); I: Berggren et al. (1995a); EP: Saint Marc (1987); NP: Corfield (1987)	Summary calibration by Berggren and Pearson (2005) corresponds to Chron C26r.5 on their polarity scale.
NA: Berggren et al. (1995a)	Igorina albeari (=Muricella albeari = Globorotalia laevigata) FAD defines base of zone P3b. Berggren and Pearson (2005) has same chron-age calibration (Chron C26r.3) as Berggren et al. (1995a) -- This FAD occurs approximately mid-way in Zone P3 in Holes 463 (Eq. Pacific) and 758A (Indian Ocean) and in the mid-Chron C26r in Hole 384 (North Atlantic). Their 60.0 Ma age corresponds to Chron C26r.3 on their polarity scale.
NA, I, EP: Berggren et al. (1995a)	Chron-age calibration by Berggren et al. (1995a) -- 'Morozovella vellascoensis' FAD is calibrated in DSDP Hole 384 (N. Atlantic); and at comparable levels but without paleomagnetic control in Holes 465 (Eq. Pacific) and 758A (Indian Ocean) Their 60.0 Ma age corresponds to Chron C26r.3 on their polarity scale.
NA & SO: Berggren et al. (1985b); I: Berggren et al. (1985b), Huber (1991)	Chron-age calibration by Berggren et al. (1995a) -- 'Praemurica strabocella' FAD occurs in Holes 384 (N. Atlantic) and 758A (Indian Ocean) midway between FAD of M. angulata and FAD of I. albeari, and in Nannofossil Zone NP5. However, it is recorded in Hole 690 (Southern Ocean) in lower Zone CP5 (=NP6); in Hole 689B (Southern Ocean) in CP8 (=NP9) but equated with C26n; and in Hole 738 (Indian Ocean) at CP 4/5 (=NP 5/6) boundary. Their 60.5 Ma summary age corresponds to Chron C26r.15 on their polarity scale.
NA, I, EP: Berggren et al. (1995a); NP: Pospichal et al. (1991), Nederbragt & van Hinte (1987)	Chron-age calibration by Berggren et al. (1995a) -- 'Morozovella conicotruncata' FAD occurs at virtually the same level as FAD of M. angulata in Holes 384 (N. Atlantic), 465 (Eq. Pacific), 758A (Indian Ocean) and 577 (North Pacific), although the definite position in Hole 577 is difficult to determine because of absence of C27n (Bleil, 1985). Their 61.0 Ma age corresponds to base of Chron C26r on their polarity scale.
NA, I, EP, NP: Berggren et al. (1995a)	Former working definition for base of Selandian stage. Berggren and Pearson (2005) has same chron-age calibration (ca. Chron C27n.8) as Berggren et al. (1995a) -- Paleocene foram 'Morozovella angulata' FAD occurs in Hole 384 (North Atlantic) in upper C27n at <2m above FAD of 'P. uncinata') in Nannofossil Zone NP4. In Holes 465 (Eq. Pacific) and 758A (Indian Ocean) it is 2m above FAD of 'P. uncinata'. In Hole 577 (N. Pacific) within lower part of Chron C26r, but precise position difficult to determine because of absence of C27n (Bleil, 1985, DSDP Leg 86). Chron assignment = C27n (upper; placed here as C27n.8).
I, EP: Berggren et al. (1995a); NP: Pospichal et al. (1991), Nederbragt & van Hinte (1987)	Dashed. A former secondary marker for working definition for base of Selandian stage. But, "although this FAD happens to coincide with the FAD of M. angulata in DSDP 384 (used by Berggren et al., 1995a, for calibration), other records deviate from this." (Robert Speijer, pers. comm., Sept 2011). Berggren and Pearson (2005) has same chron-age calibration (ca. Chron C27n.8) as Berggren et al. (1995a), who had indicated that 'Muricella pusilla' (called 'Igorina pusilla' in Berggren & Miller, 1988) FAD is recorded in lower part of polarity chron C26r and top of Nannofossil Zone CP3 (= NP3) in Hole 577 (North Pacific), but precise position is difficult to determine because of absence of C27n and possible unconformity just above C28n at 102.95 m (see Bleil, 1985:449-450). This FAD occurs at same level as FAD of M. angulata and M. coniscotruncata in Holes 465 (Eq. Pacific) and 758A (Indian Ocean). Their 61.0 Ma age corresponds to base of Chron C26r on their polarity scale.
NA, I, EP: Berggren et al. (1995a); NP: Berggren et al. (1985b), Berggren & Pearson (2005)	Chron-age calibration by Berggren et al. (1995) -- 'Praemurica praeangulata' FAD is calibrated in DSDP Hole 384 (N. Atlantic); and at comparable levels but without paleomagnetic control in Holes 465 (Eq. Pacific) and 758A (Indian Ocean) [see comments for 'Moro. angulata'], and occurs at approximately the same level in Holes 738 and 750A. Their summary age assignment of 61.2 Ma corresponds to Chron C27n.2 on their scale.
NA, I, EP: Berggren et al. (1995a); NP: Berggren et al. (1985b), Berggren & Pearson (2005)	Praemurica uncinata' (formerly 'Morozovella uncinata') FAD defines base of Zone P2. Berggren and Pearson (2005) moved the FAD 0.4 myr older than in Berggren et al. (1995), who had an extended discussion of calibration problems in different sites. Berggren and Pearson (2005) summary age assignment of 61.37 Ma corresponds to Chron C27r.92 on their polarity scale.
NA: Berggren et al. (1995a), Wade et al. (2011); SO: Stott & Kennett (1990), Wade et al. (2011)	Wade et al. (2011) assigned as Chron C28n.65 on their polarity scale, following Berggren and Pearson (2005) and Berggren et al. (1995a): 'Praemurica inconstans' FAD is recorded in Southern Ocean Holes 689B and 690C and North Atlantic Hole 384 in mid-C28n. It also occurs at a comparable level (by correlation) in an interval with no polarity data just above incomplete polarity chron C28n in Indian Ocean Hole 750A.
NA: Berggren et al. (1995a), Wade et al. (2011); SO: Stott & Kennett (1990), Wade et al. (2011)	Coeval with FAD of G. compressa; therefore assigned as Chron C28n.65 (Wade et al., 2011). See comments for FAD of G. compressa.
NA & I: Berggren et al. (1995a)	Chron-age calibration by Berggren et al. (1995a) -- 'Parasubbotina varianta' is recorded in North Atlantic Hole 384 at mid-Nannofossil Zone NP2, and at comparable level by correlation in Indian Ocean Hole 750A . Their 63.0 Ma age corresponds to Chron C28n.55 on their polarity scale.
NA: Berggren et al. (1995a)	Defines base of zone P1b. Berggren and Pearson (2005) uses same chron-age calibration as Berggren et al. (1995a) -- 'Subbotina triloculinoides' FAD is recorded in Hole 384 (North Atlantic). Their 64.3 Ma age corresponds to Chron C29n.6 on their polarity scale.
Berggren et al. (1995a)	Defines base of zone P1a. Berggren and Pearson (2005) uses same chron-age calibration as Berggren et al. (1995a): The LAD, defining the top of Zone P-alpha, is near the FAD of P. pseudobulloides; and Berggren et al. (1995a) "admittedly somewhat arbitrarily" placed both events just below the polarity chron C29n/r boundary (p.146) = Chron C29r.95.
NA: Berggren et al. (1995a); M: Groot et al. (1989); I: Huber (1991)	Chron-age calibration by Berggren et al. (1995a) -- 'Parasubbotina pseudobulloides' is recorded in Hole 738 (Indian Ocean), Hole 384 (North Atlantic) and Agost, Spain (Mediterranean); first common occurrence (FCO) noted in older part of polarity chron C29N in Southern Ocean Holes 689, 690 (Stott & Kennett, 1990). Its FAD may slightly overlap LAD of P. eugubina (review in Berggren et al., 1995a). Chron assignment = just below C29r/n boundary = Chron C29r.9. [NOTE: In the Table 8 of Berggren et al. (1995a), P. pseudobulloides FAD was accidentally labled "P. eugubina FAD".]
Olsson et al. (1999), Wade et al. (2011)	Event according to Wade et al. (2011) citing Olsson et al.(1999). Their table places this FAD as 0.1 myr above base of Cenozoic.
M: Groot et al. (1989), Berggren et al. (1995a), Wade et al. (2011)	Defines base of Pa ('alpha' Greek letter). Berggren and Pearson (2005) uses same chron-age calibration as Berggren et al. (1995a) -- "Parvularugoglobigerina eugubina" FAD is estimated at about 0.03 myr after the end-Cretaceous event based on Spanish sections in Groot et al., (1990), discussed in Berggren, 1995a). Wade et al. (2011) assign a 0.04 myr duration for basal-Cenozoic zone P0 (used here).

Table A3.2. Cretaceous Planktonic Foraminiferal Biochronology.

GTS2012 CHRONO. STRATIGRAPHY	STANDARD TROPICAL-SUBTROPICAL BIOZONE (BIOCHRON)	BIOHORIZON (DATUM)	AGE (Ma)	CALIBRATION REFERENCES	COMMENTS
STAGE (AGE)	Composite from Robaszynski (1998); Huber et al. (2008); Huber & Leckie (2011); Petrizzo et al. (2011); and B. Huber and M.R. Petrizzo, pers. comm. (2011)	T = Top/LAD; B = Base/FAD	*This study*		
Maastrichtian	**P0/Plummerita hantkeninoides**	T Globotruncana spp., T Racemiguembelina fructicosa; Tops of other Cretaceous forams	**66.04**	Berggren et al. (1995b); Berggren and Pearson (2005)	End of Cretaceous. Ar-Ar age is recalibrated as 66.0 Ma in GTS2012. About Chron C29r.5 (Husson et al., 2011).
	P. hantkeninoides/P. hariaensis	T Abathomphalus mayaroensis	**66.35**	Huber et al. (2008); Robaszynski (1998, SEPM chart)	Brian Huber (written comm., Sept 2011): "Add Plummerita hantkeninoides zone with dashed base at topmost Maastrichtian using Li and Keller (1998) age (about 350 kyr below K/Pg)". If the base is marked by LAD of A. mayaroensis, then Robaszynski (1998, SEPM chart) had assigned as ~0.3 m.y. before K/T.
	Pseudoguembelina hariaensis	T Gansserina gansseri	**(66.49)**	Huber et al. (2008)	Chron C30n.95 estimated relative to GTS04 magnetics by Huber et al. (2008) at ODP sites on Blake Nose (central Atlantic). Dashed, due to 0.7 myr differences among holes (essentially at K/T in Hole 1049C with spherule layer overlying it).
		T Contusotruncana patelliformis	**66.72**	Huber et al. (2008)	Chron C30n.82 estimated relative to GTS04 magnetics by Huber et al. (2008) at ODP sites on Blake Nose (central Atlantic).
	P. hariaensis/A. mayaroensis	B Pseudoguembelina hariaensis	**67.3**	Huber et al. (2008)	Use of a Pseudoguembelina hariaensis Zone (base is FAD of species) of Caron'95 is recommended by Huber et al. (2008). Base is dashed by them at mid-Chron 30n.
	Abathomphalus mayaroensis	T Abathomphalus intermedia	**67.33**	P. Sikora (pers. comm., 2005)	Useful datum tabulated in EGI database (Paul Sikora, 2005), with age of middle of Chron C30n derived from composite standard.
		T Globotruncana linneiana	**68.37**	Huber et al. (2008)	Base of Chron C30r estimated relative to GTS04 magnetics by Huber et al. (2008) at ODP sites on Blake Nose (central Atlantic).
		T Globotruncana bulloides	**68.82**	Huber et al. (2008)	Chron C31n.5 estimated relative to GTS04 magnetics by Huber et al. (2008) at ODP sites on Blake Nose (central Atlantic).
		T Rugoglobigerina pennyi	**68.86**	P. Sikora (pers. comm., 2005)	Useful datum tabulated in EGI database (Paul Sikora, 2005), with age of 45% up in Chron C31n derived from composite standard.
		T Contusotruncana fornicata	**69.13**	Huber et al. (2008)	Chron C31n.15 estimated relative to GTS04 magnetics by Huber et al. (2008) at ODP sites on Blake Nose (central Atlantic).
	A. mayaroensis/R. fructicosa	B Abathomphalus mayaroensis	**69.18**	Huber et al. (2008)	A. mayaroensis zone begins just above base of Chron C31n (summarized by Huber et al. (2008), with implied placement as Chron C31n.1). A slightly higher Chron C31n.4 was in ODP 171B table (in turn from Erba et al, 1995) for the Tethyan FAD. The FAD of A. mayaroensis seems to be older in North Sea (Paul Sikora, EGI database, 2005). Robaszynski (1998, SEPM chart) correlation was approx. top of ammonite J. nebrascansis zone.
	Racemiguembelina fructicosa	B Pseudotextularia elegans	**69.55**	Huber et al. (2008)	Chron C31r.87 estimated relative to GTS04 magnetics by Huber et al. (2008) at ODP sites on Blake Nose (central Atlantic).
		B Planoglobulina acervulinoides	**(70.05)**	Huber et al. (2008)	DASH -- Chron C31r.65 estimated relative to GTS04 magnetics by Huber et al. (2008) at ODP sites on Blake Nose (central Atlantic). But this FAD is reported much earlier elsewhere (by ca. 2 myr) which Huber et al. (2008) suggests "result from differing taxonomic concepts of the species".
	(R. fructicosa/P. palpebra)	B Racemiguembelina fructicosa	**(70.14)**	Huber et al. (2008)	DASHED (diachronous): Upper third of Chron C31r. Huber et al. (2008); who interpolate FAD at Chron C31r.6 relative to GTS04 magnetics => C31r.6), which is same as in ODP 171B table (in turn from Erba et al, 1995). However, this FAD in Tethyan realm is calibrated in EGI database (Paul Sikora, 2005) as above the FAD of A. mayaroensis (2 myr young) => this Zone may not be applicable. "Latitudinal diachroneity" concluded by Huber et al. (2008) citing the later FAD in Chron C31n at DSDP 525 reported by Li and Keller (1998).
	Pseudoguembelina palpebra	T Globotruncana ventricosa	**(70.14)**	Robaszynski (1998, SEPM chart)	Dashed at Chron C31r.6.
		B Globotruncana lapparenti, Globotruncana linneiana	**70.90**	P. Sikora (pers. comm., 2005)	Useful datums tabulated in EGI database (Paul Sikora, 2005), with age at 25% up in Chron C31r derived from composite standard.
		B Contusotruncana contusa	**(71.01)**	Huber et al. (2008); P. Sikora (pers. comm., 2005)	DASH -- Chron C31r.2 relative to GTS04 magnetics by Huber et al. (2008). ODP 171B table (in turn from Erba et al, 1995) assigned higher at Chron C31r.6. This FAD is a poor marker (Paul Sikora, pers. commun., 2005).
		B Racemiguembelina powelli	**71.47**	Huber et al. (2008)	Chron C32n.1n.9 estimated relative to GTS04 magnetics by Huber et al. (2008)
	P. palpebra/G. gansseri	B Pseudoguembelina palpabra	**71.75**	Huber et al. (2008)	P. palpebra zone added by Huber et al. (2008) "because recognition of the G. gansseri and G. aegyptiaca zones is deemed unreliable (at least at ODP Leg 207 Blake Nose sites). Base is "upper Chron C32n.1r" (Brian Huber, written commun., Sept 11) -- assigned here as Chron C32n.1r.75 and coeval with FAD of nannofossil T. phacelosus (base of CC24, in their version).
	Gansserina gansseri	B Pseudoguembelina kempensis	**71.97**	Huber et al. (2008)	Chron C32.2n.98. estimated relative to GTS04 magnetics by Huber et al. (2008).
72.05					
	(G. gansseri/G. aegyptiaca)	B Gansserina gansseri	**(72.97)**	Huber et al. (2008); Robaszynski (1998, SEPM chart)	Huber et al. (2008) and Robaszynski (1998, SEPM chart) places at Chron C32n.2n.4; following Premoli Silva and Sliter (1994); but Brian Huber advised to dash it.
	Globotruncana aegyptiaca	B Planoglobulina acervulinoides	**(72.97)**	Huber et al. (2008); Robaszynski (1998, SEPM chart)	Huber et al. (2008) and Robaszynski (1998, SEPM chart) places at at same level as FAD of G. gansseri = Chron C32n.2n.4, citing Premoli Silva and Sliter (1994). BUT, iit s much higher at Leg 207 sites (by about 2 myr), which Huber et al. (2008) suggests "result from differing taxonomic concepts of the species".

GTS2012 CHRONO-STRATIGRAPHY	STANDARD TROPICAL-SUBTROPICAL BIOZONE (BIOCHRON)	BIOHORIZON (DATUM)	AGE (Ma)	CALIBRATION REFERENCES	COMMENTS
STAGE (AGE)	Composite from Robaszynski (1998); Huber et al. (2008); Huber & Leckie (2011); Petrizzo et al. (2011); and B. Huber and M.R. Petrizzo, pers. comm. (2011)	T = Top/LAD; B = Base/FAD	*This study*		
Campanian	**(G. aegyptiaca/G. havanensis)**	B Globotruncana aegyptiaca	**(74.00)**	Huber et al. (2008); Robaszynski (1998, SEPM chart)	DASH -- Huber et al. (2008) (also implied in Robaszynski (1998, SEPM chart)) places at mid-Chron C32r.1n according to their interpolation relative to GTS04 magnetics), although their text says "C32r.2r.5.
	Globotruncanella havanensis	B Pseudoguembelina excolata	**(74.00)**	Huber et al. (2008); Robaszynski (1998, SEPM chart)	DASH -- Robaszynski (1998, SEPM chart) has as same level as FAD of G. aegyptiaca (therefore, mid-Chron C32r.1n). Huber et al. (2008) "model age" table as same mid-Chron C32r.1n relative to GTS04 magnetics, although their text says "mid-C32r.2r" and they later indicate it might be as old as 74.4 Ma, based on Leg 207 sites (their Table 7).
	G. havanensis/R. calcarata	T Radotruncana calcarata	**75.71**	Huber et al. (2008)	Calcarata was previously assigned to genus Globotruncanita. Huber et al. (2008) project age corresponding to Chron C33n.75 (although their text implies C32n.92). Robaszynski (1998, SEPM chart) placed at Chron C33n.85.
	Radotruncana calcarata	B Pseudotextularia elegans	**(75.71)**	Robaszynski (1998, SEPM chart)	Robaszynski (1998, SEPM chart) places at same level (Chron C33n.85) as LAD of G. calcarata.
		B Globotruncanella havanensis	**(75.94)**	Robaszynski (1998, SEPM chart)	Dashed at about middle of G. calcarata zone.
	R. calcarata/C. plummerae	B Radotruncana calcarata	**76.18**	Huber et al. (2008)	R. calcarata was previously assigned to genus Globotruncanita. FAD at Chron C33n.67 (Huber et al., 2008, although their text had typo-error of "C33n.76").
	Contusotruncana plummerae	B Globotruncana ventricosa (consistent)	**77.66**	P. Sikora (pers. comm., 2005)	EGI calibration (Paul Sikora, 2005) is 40% up in Chron C33n. First consistent occurrence is a better marker for base of a G. ventricosa Zone; but Petrizzo et al. (2011) recommended its replacement (as shown here).
		T Archaeoglobigerina bosquensis	**77.79**	P. Sikora (pers. comm., 2005)	Useful datum tabulated in EGI database (Paul Sikora, 2005), with age of 2/3rds up in Middle Campanian derived from composite standard.
	C. plummerae/G. elevata	B Contusotruncana plummerae	**(79.20)**	Huber et al. (2008); Petrizzo et al. (2011)	Chron C33n.125 (1/8th up in chron) according to calibrations by Huber et al. (2008). Formerly called Globotruncana fornicata Plummer subsp. plummerae. Petrizzo et al. (2011) assigned C. plummerae zone to replace the previous G. ventricosa zone due to rarity of that basal marker.
	Globotruncanita elevata	B Globotruncana ventricosa (rare)	**(79.90)**	Robaszynski (1998, SEPM chart); Petrizzo et al. (2011)	Top of Chron C33r in Robaszynsk (1998, SEPM chart): In Huber et al. (2008), the "rare" FAD marked the base of the G. ventricosa Zone. However, the rare FAD is not a useful datum, because it remains rare until mid-Middle Campanian (Paul Sikora, 2005); therefore consistent occurrence is a better definition of the zone. Therefore, Petrizzo et al. (2011) removed Globotruncana ventricosa Zone, as shown here.
		B Pseudoguembelina costulata	**81.96**	Robaszynski (1998, SEPM chart)	Robaszynski (1998, SEPM chart) has at Chron C33r.45
		T Ventilabrella eggeri	**82.71**	Robaszynski (1998, SEPM chart)	Robaszynski (1998, SEPM chart) has at Chron C33r.25
		T Marginotruncana coronata	**82.89**	P. Sikora (pers. comm., 2005)	Useful datum tabulated in EGI database (Paul Sikora, 2005), with age of mid-Early Campanian (ca. 20% up in Chron C33r) derived from composite standard.
		T Whiteinella baltica	**83.08**	P. Sikora (pers. comm., 2005)	Useful datum tabulated in EGI database (Paul Sikora, 2005), with age just below LAD of M. coronata (ca. 15% up in Chron C33r) derived from composite standard.
		B Globotruncanita elevata	**83.64**	Robaszynski (1998, SEPM chart)	Base of Campanian in Robaszynski (1998, SEPM chart). Base of G. elevata Zone is marked both by LAD of D. asymetrica and FAD of G. elevata.
(83.64)	**G. elevata/D. asymetrica**				
Santonian	Dicarinella asymetrica	T Dicarinella asymetrica	**83.64**	Robaszynski (1998, SEPM chart)	Coeval with base of C33r. Top of Santonian in Robaszynski (1998, SEPM chart).
		T Dicarinella concavata, Sigalia deflaensis	**83.64**	Robaszynski (1998, SEPM chart)	Top of Santonian in Robaszynski (1998, SEPM chart)
		T Sigalia decoratissima decoratissima	**84.44**	Robaszynski (1998, SEPM chart)	Robaszynski (1998, SEPM chart) places at About 70% up between FAD of S. carpathica and LAD of S. deflaensis Forams
		T Sigalitruncana sigali	**84.95**	P. Sikora (pers. comm., 2005)	Useful datum of "mid-Santonian" tabulated in EGI database (Paul Sikora, 2005), with age derived from composite standard.
		T Sigalia carpathica	**85.44**	Robaszynski (1998, SEPM chart)	LAD of Sigalia carpathica (called Sigalia decoratissima carpathica, or spelled "carpatica" in some publications) -- Robaszynski (1998, SEPM chart) places at about one-third up between FAD of S. dec. carpathica and LAD of S. deflaensis.
(86.26)					
		B Sigalia carpathica	**86.32**	Lamoida et al., pending proposal for Santonian GSSP, 2011	The FAD of Sigalia carpathica (called Sigalia decoratissima carpathica, or spelled "carpatica" in some publications) was proposed as a secondary marker for the Coniacian/Santonian boundary during the Second International Symposium on Cretaceous Stage Boundaries (Lamolda and Hancock, 1996). Proposed GSSP in Olazagutia, Spain (Lamolda and Santonian Working Group, 2011) has FAD midway between FAD of L. cayeuxii nannofossil, and FAD of the "GSSP marker" inoceramid -- which is the scaling used here.
		T Whiteinella archaeocretacea	**(86.33)**	Gale et al. (2007)	Dashed -- Just above FAD of nannofossil L. cayeuxii at Ten Mile Creek candidate for base-Santonian GSSP (placed here as 0.05 myr above). However, according to EGI database (Paul Sikora, 2005), this LAD might be as high as the base of Middle Santonian; but this may be partly a base-Santonian definition problem.

GTS2012 CHRONO-STRATIGRAPHY	STANDARD TROPICAL-SUBTROPICAL BIOZONE (BIOCHRON)	BIOHORIZON (DATUM)	AGE (Ma)	CALIBRATION REFERENCES	COMMENTS
STAGE (AGE)	Composite from Robaszynski (1998); Huber et al. (2008); Huber & Leckie (2011); Petrizzo et al. (2011); and B. Huber and M.R. Petrizzo, pers. comm. (2011)	T = Top/LAD; B = Base/FAD	*This study*		
Coniacian		T Whiteinella paradubia	**(86.33)**	Gale et al. (2007)	Dashed -- Useful marker just below Inoceramus marker for base-Santonian (EGI database, Paul Sikora, 2005) and at Ten Mile Creek (Gale et al., 2007) (put here at LAD of Whiteinella at ca 0.05 myr above FAD of L. cayeuxii at Ten Mile Creek); but recorded above candidate GSSP level in Olazagutia, Spain (Lamolda and Santonian Working Group, 2011).
	(D. asymetrica/D. concavata)	B Dicarinella asymetrica	**(86.66)**	Gale et al. (2007)	Coniacian-Santonian boundary interval section of Ten Mile Creek (Gale et al., 2007) suggest ca. 75% up in ammonite S. depressus Zone or lower. Base-Santonian GSSP proposal for Olazagutia, Spain (Lamolda and Santonian Working Group, 2011) implies that this level would be in upper Coniacian. In contrast, Robaszynski (1998, SEPM chart) assigns as base of Middle Santonian (= base ammonite C. vermiformis zone). D. asymetrica Zone is range-zone of the index species.
	Dicarinella concavata	B Archaeoglobigerina bosquensis	**87.71**	P. Sikora (pers. comm., 2005)	Useful marker just above (ca. 0.15 myr) base of Middle Coniacian (EGI database, Paul Sikora, 2005)
		T. Whiteinella aprica	**87.86**	P. Sikora (pers. comm., 2005)	Useful marker at base of Middle Coniacian (EGI database, Paul Sikora, 2005)
		T Dicarinella hagni	**89.57**	P. Sikora (pers. comm., 2005)	Useful marker not too far above (ca. 0.2 myr used here) base of Coniacian (EGI database, relative placement according to Paul Sikora, 2005).
		B Pseudotextularia nuttalli	**89.77**	Robaszynski (1998, SEPM chart)	Potential marker for base of Coniacian according to Robaszynski (1998, SEPM chart).
(89.77)					
Turonian	**(D. concavata/M. schneegansi)**	B Dicarinella concavata	**(91.08)**	ODP Leg 171 explanatory notes	Dashed -- ODP Leg 171B placed near base of S. ferronensis ammonite zone (used here, on advice of Brian Huber, written comm., Sept 11). Robaszynski (1998, SEPM chart) placed at one-third up in ammonite P. hyatti zone or about 0.8 myr higher (hence about 2 myr below base-Coniacian).
	Marginotruncana schneegansi	B Marginotruncana coronata	**91.78**	P. Sikora (pers. comm., 2005)	Useful marker at mid-Middle Turonian (EGI database, Paul Sikora, 2005)
	M. schneegansi/H. helvetica	T Helvetoglobotruncana helvetica	**92.99**	UCL workshop on Late Cretaceous microfossil stratigraphy (June 2011)	Assigned coeval with FAD of nannofossil E. eximius at 80% up in ammonite M. nodosoides Zone. But that nanno FAD might be diachronous relative to LAD of H. helvetica (E. eximius is above that LAD at Demerara Rise Leg 207; but below the LAD at Tanzania and S. Atlantic) (M.R. Petrizzo at UCL workshop on Late Cretaceous microfossil stratigraphy, June 2011). Robaszynski (1998, SEPM chart) assigned FAD at about 1 myr higher.
	(H. helvetica/W. archaeocretacea)	B Helvetoglobotruncana helvetica	**(93.52)**	UCL workshop on Late Cretaceous microfossil stratigraphy (June 2011)	Dashed just above top of ammonite W. devonense zone (M.R. Petrizzo at UCL workshop on Late Cretaceous microfossil stratigraphy, June 2011); placed 1/3rd up in overlying P. flexuosum zone here. Robaszynski (1998, SEPM chart) had equivalent to the top of W. devonense zone.
93.9					
Cenomanian	W. archaeocretacea	B Dicarinella imbricata	**94.20**	P. Sikora (pers. comm., 2005)	Useful marker just below OAE2 event (EGI database, Paul Sikora, 2005). Assigned here as ca. 0.3 myr before base of Turonian
	(W. archaeocretacea/R. cushmani)	T Rotalipora cushmani	**(94.30)**	Robaszynski (1998, SEPM chart)	Not well calibrated, because apparent top at Gubbio is preservation artifact and placement in Morocco suggests diachronous (M.R. Petrizzo; at UCL Late Cretaceous microfossil workshop; June 2011). Robaszynski (1998, SEPM chart) places arbitrary (?) at 0.4 myr before top of Cenomanian (= top of Rotalipora genus?) -- used here. Caron (1985) assigns the base of this zone to the FAD of W. archaeocretacea, rather than LAD of R. cushmani.
	Rotalipora cushmani	B Whiteinella archaeocretacea, Heterohelix globulosa (=reussi)	**94.90**	Robaszynski (1998, SEPM chart)	Robaszynski (1998, SEPM chart) places these FADs at an arbitrary (?) 1.0 myr before top of Cenomanian.
		B Whiteinella aprica	**95.47**	P. Sikora (pers. comm., 2005)	Useful marker at base of Late Cenomanian (EGI database, Paul Sikora, 2005)
		T Thalmanninella reicheli, Parathalmanninella appenninica, Thalmanninella globotruncanoides	**95.47**	Robaszynski (1998, SEPM chart)	These were formerly called "Rotalipora reicheli, Rotalipora appenninica, Rotalipora globotruncanoides". Robaszynski (1998, SEPM chart) places at base of Upper Cenomanian
	R. cushmani/T. reicheli	B Rotalipora cushmani	**95.94**	Robaszynski (1998, SEPM chart)	Robaszynski (1998, SEPM chart) places at mid-point of ammonite Acanthoceras muldoonense zone.
	T. reicheli/T. globotruncanoides	B Thalmanninella reicheli	**96.08**	Robaszynski (1998, SEPM chart)	Formerly called "Rotalipora reicheli". Robaszynski (1998, SEPM chart) assigns as base of Middle Cenomanian.
	T. globotruncanoides/P. appenninica	B Thalmanninella globotruncanoides	**100.50**	Gale et al. (2011)	Formerly called "Rotalipora globotruncanoides". Primary marker of Cenomanian GSSP. Formerly called "R. greenhornensis"? Robaszynski (1998, SEPM chart) had assigned as latest Albian (mid-point of uppermost ammonite subzone, M. perinflatum), but this was prior to GSSP definition of boundary that truncated that zone.
100.5					
	Parathalmanninella appenninica	T Planomalina buxtorfi	**101.18**	Robaszynski (1998, SEPM chart)	Robaszynski (1998, SEPM chart) places at 1/5 up in ammonite M. periflatum zone.
		T Pseudothalmanninella ticinensis, Ticinella primula	**101.51**	Robaszynski (1998, SEPM chart)	Formerly called "Rotalipora ticinensis". Robaszynski (1998, SEPM chart) places LADs at 2/3rds up in ammonite M. rostratum zone.
		T Biticinella breggiensis	**101.59**	Robaszynski (1998, SEPM chart)	Robaszynski (1998, SEPM chart) places at 2/5 from base of ammonite M. rostratum zone.

GTS2012 CHRONO-STRATIGRAPHY	STANDARD TROPICAL-SUBTROPICAL BIOZONE (BIOCHRON)	BIOHORIZON (DATUM)	AGE (Ma)	CALIBRATION REFERENCES	COMMENTS
STAGE (AGE)	Composite from Robaszynski (1998); Huber et al. (2008); Huber & Leckie (2011); Petrizzo et al. (2011); and B. Huber and M.R. Petrizzo, pers. comm. (2011)	T = Top/LAD; B = Base/FAD	*This study*		
Albian	**P. appenninica/P. ticinensis**	B Parathalmanninella appenninica	**101.92**	Grippo et al. (2004); Gale et al. (2011)	Formerly called "Rotalipora appenninica". Base of planktonic foraminifer Rotalipora appenninica is 3.3 long-eccentricity cycles of 405-kyr (1.3 myr) below base of Cenomanian in Piobbico cycle stratigraphy (Grippo et al., 2004). Leckie et al. (2002) indicate that R. appenninica FAD is just slightly older/below the FAD of P. buxtorfi, but Robaszynski (1998, SEPM chart) and Bellier et al., (2000) show the events to be simultaneous. Robaszynski (1998, SEPM chart) had assigned as base of ammonite Stoliczkaia dispar Zone, which is M. rostratum Zone on current scale (Gale et al., 2011).
	Pseudothalmanninella ticinensis	B Planomalina buxtorfi	**101.92**	Grippo et al. (2004); Gale et al. (2011)	FAD is 3.5 cycles of 405 kyr in Piobbico and Col de Palluel cyclestratigraphy (Grippo et al, 2004; Gale et al., 2011). FAD of P. buxtorfi is a marker for base of R. appenninica Zone on some scales.
		T Pseudothalmanninella subticinensis	**102.42**	Robaszynski (1998, SEPM chart)	Formerly called"Rotalipora subticinensis". Robaszynski (1998, SEPM chart) assigned as middle of an ammonite C. auritus subzone, which is equated to the current M. fallax ammonite zone.
	P. ticinensis/P. subticinensis	B Pseudothalmanninella ticinensis	**103.94**	Grippo et al. (2004); Gale et al. (2011)	Formerly called "Rotalipora ticinensis". Base of zone is 8.5 long-eccentricity cycles (3.3 myr) below base of Cenomanian in Piobbico and Col de Palleul cyclostratigraphy (Grippo et al., 2004; Gale et al., 2011). Robaszynski (1998, SEPM chart) had assigned as 2/5 from base of H. varicosum s.z.
	(P. subticinensis/T. praeticinensis)	B Pseudothalmanninella subticinensis	**(106.98)**	Robaszynski (1998, SEPM chart)	Formerly called "Rotalipora subticinensis". Robaszynski (1998, SEPM chart) had assigned as FAD is at the base of the ammonite H. orbignyi subzone of the M. (M.) inflatum Zone (used here). Base of foraminifer R. subticinensis (base of that subzone) is not indicated in this Piobbico cyclostratigraphy; therefore, age estimate is dashed.
	T. praeticinensis/T. primula	B Ticinella praeticinensis	**107.39**	Grippo et al. (2004)	Base of T. praeticinensis is 17 long-eccentricity cycles below base of Cenomanian in Piobbico cycle stratigraphy (Grippo et al., 2004). The duration (of former zone usage to base of P. ticinensis) is 3.4 myr, nearly the same as the 3.3 myr used in Leckie et al. (2002). Note that Leckie et al. (2002) defined the T. praeticinensis foraminifer subzone as the base of B. breggiensis Zone, in turn defined by the base of that species; however, the actual FAD of T. praeticinensis is older than the base of B. breggiensis according to both Leckie et al. (2002) and Robaszynski (1998, SEPM chart). In contrast, the zonal schemes of Robaszynski (1998, SEPM chart) and Bellier et al. (2000) do not have a B. breggiensis zone, but define the base of T. praeticinensis zone by the first occurrence of the name species
	Ticinella primula	B Biticinella breggiensis	**107.59**	Robaszynski (1998, SEPM chart)	Base of ammonite M. inflatum Zone (Robaszynski, 1998, SEPM chart); which would imply a placement near FAD of T. praeticinensis. FAD of B. breggiensis is used as base of a B. breggiensis Zone in some scales
	T. primula/T. madecassiana	B Ticinella primula	**111.84**	Grippo et al. (2004); Huang et al. (2010d)	Base of foraminifer T. primula is 28 long-eccentricity 405-kyr cycles below base of Cenomanian (therefore 11.3 myr below) in Piobbico cycle stratigraphy (Grippo et al., 2004; Huang et al., 2010d). Robaszynski (1998, SEPM chart) had placed in middle of ammonite O. raulinianus subzone.
	(T. madecassiana/M. rischi)	B Ticinella madecassiana	**(112.40)**	Huber and Leckie (2011)	New zone replacing upper part of former H.planispira zone (Huber and Leckie, 2011). FAD is dashed at middle of that former zone.
(112.95)	Microhedbergella rischi				
Aptian	**(M. rischi/M. renilaevis)**	B Microhedbergella rischi	**(112.96)**	Huber and Leckie (2011)	Could be a definition for the base-Albian (easy to recognize). Huber and Leckie (2011) estimated as ca. 300 kyr above LAD of P. eubejaouaensis (used here); but Huber (pers. comm., Sept 2011) implied that this zone could be shorter. Zone is "accompanied, within a short stratigraphic thickness, by the extinction of several other Aptian hedbergellids, including Hd. infracretacea, Hd. occulta, Hd. gorbachikae, and Hd. excelsa, as well as Gl. aptiensis. Species diversity is very low," (Huber and Leckie, 2011)
	M. renilaevis/M. miniglobularis	B Microhedbergella renilaevis	**113.14**	Huang et al. (2010d); Petrizzo et al. (Newsletters on Stratigraphy, in press as of 2011)	Middle of Kilian level Petrizzo et al. (Newsletters on Stratigraphy, in press as of 2011), which is about 31.1 cycles below base-Cenomanian according to Huang et al. (2010d)). Mi renilaevis co-occurs with common Mi. miniglobularis at ODP-DSDP sites in S. Atlantic, North Atlantic and Indian Ocean; therefore is an important marker.
	M. miniglobularis/P. eubejaouaensis	B Microhedbergella miniglobularis	**113.26**	Huang et al. (2010d); Petrizzo et al. (Newsletters on Stratigraphy, in press as of 2011)	The FAD of this "new genera" Microhedbergella is when P. eubejaouaensis vanishes. The M. miniglobularis zone is a brief zone from "25 cm above LAD of P. eubejaouaensis" spanning 2 meters to the middle of the Kilian level in SE France (Petrizzo et al.; Newsletters on Stratigraphy, in press as of 2011)
	Paraticinella eubejaouaensis	T Paraticinella eubejaouaensis	**113.26**	Huang et al. (2010d); Petrizzo et al. (Newsletters on Stratigraphy, in press as of 2011)	This LAD was the base of a former H. planispira foraminifer zone; and occurs at 31.5 long-eccentricity cycles below base of Cenomanian in Piobbico cycle stratigraphy (Huang et al, 2010). Brian Huber (to J.Ogg; 15Aug11 "Note that there is no longer a planispira zone because true planispira did not evolve until the late Albian!") Calibrations of other authors are inconsistent -- the base of the H. planispira Zone has been placed above base of nannofossil P. columnata (Leckie et al., 2002), middle of ammonite H. jacobi zone of Late Aptian (Herle et al., 2004) and much higher as the middle of Lower Albian (Bellier et al., 2000) -- these and other assignments are briefly reviewed in Petrizzo et al. (Newsletters on Stratigraphy, in press as of 2011). However, Robaszynski (1998, SEPM chart) placed the FAD of H. planispira below the FAD of nannofossil P. columnata, as is indicated at Piobbico and in the reference sections in SE France; thereforehad assigned as 50% up in ammonite H. jacobi Zone.
	P. eubejaouaensis/H. infracretacea	B Paraticinella eubejaouaensis	**118.02**	Huang et al. (2010d)	FAD is 43 1/4 cycles (of 405 kyr) below base of Cenomanian in Piobbico cycle strat (Huang et al., 2010d). Robaszynski (1998, SEPM chart) had estimated as 3/5 up in ammonite D. nodosocostatum subzone. Formerly called Ticinella bejaouaensis.

GTS2012 CHRONO-STRATIGRAPHY	STANDARD TROPICAL-SUBTROPICAL BIOZONE (BIOCHRON)	BIOHORIZON (DATUM)	AGE (Ma)	CALIBRATION REFERENCES	COMMENTS
STAGE (AGE)	Composite from Robaszynski (1998); Huber et al. (2008); Huber & Leckie (2011); Petrizzo et al. (2011); and B. Huber and M.R. Petrizzo, pers. comm. (2011)	T = Top/LAD; B = Base/FAD	*This study*		
	H. infracretacea/G. algerianus	T Globigerinelloides algerianus	**118.93**	Huang et al. (2010d)	LAD of G. algerianus 45.5 cycles of 405-kyr below base-Cenomanian at Piobbico (Huang et al., 2011). This LAD is base of a "gap" zone that has been given different names (e.g., Hedbergella infracretacea, H. trocoidea, H. gorbachikae, or Planomalina cheniourensis Zone, depending on author). Brian Huber (pers. comm. to J.Ogg; 15Aug11 - "I now prefer using the Hedbergella infracretacea Zone since that species is so widespread and present within that stratigraphic interval." Robaszynski (1998, SEPM chart) had assigned this LAD much lower at just below top of ammonite P. nufieldiensis Zone.
	G. algerianus/G. ferreolensis	B Globigerinelloides algerianus	**122.17**	Huang et al. (2010d); Robaszynski (1998, SEPM chart)	Base Globigerinelloides algerianus = 53.5 cycles of 0.405 below base-Cenomaniann at Piobbico (Huang et al., 2010d). Robaszynski (1998, SEPM chart) had assigned as 3/5 up in ammonite E. martinoides Zone.
	G. ferreolensis/L. cabri	T Leupoldina cabri	**122.98**	Huang et al. (2010d); Robaszynski (1998, SEPM chart)	LAD of L. cabri (= base of G. ferreolensis zone) = 55.5 cycles of 405 kyr below base-Cenomanian in Piobbico core (Huang et al., 2010d). Robaszynski (1998, SEPM chart) had assigned as 1/5 up in ammoite E. martinoides Zone.
	"L. cabri (consistent)/L. cabri (rare)"	B Leupoldina cabri (consistent)	**123.89**	Huang et al. (2010d); Robaszynski (1998, SEPM chart)	Consistent L. cabri (and calibration of FAD in SE France, etc.) is 57.75 cycles of 405kyr below base-Cenomanian at Piobbico (Huang et al., 2010d). Robaszynski (1998, SEPM chart) had assigned as 2/3 from base of ammonite Deshayesites deshayesi Zone.
	L. cabri/G. blowi	B Leupoldina cabri (rare)	**125.71**	Premoli Silva et al. (1999); Huang et al. (2010d)	Base of L. cabri zone (using downward extension by Premoli Silva et al., 1999) = FAD of rare L.cabri = 62.25 cycles of 405 kyr below base of Cenomanian at Piobbico (just below base of Selli OAE1a) (Huang et al., 2010d). The apparent FAD of L. cabri in the middle of the ammonite D. deshayesi zone in the Bedoule section (Moullade et al., 1998) may be a delayed first common occurrence).
126.3	Globigerinelloides blowi				
Barremian	**G. blowi/H. similis**	B Globigerinelloides blowi	**128.73**	Robaszynski (1998, SEPM chart)	Estimated as about 1/4 below top of ammonite A. vandenheckii zone.
	H. similis/H. sigali	B Hedbergella similis	**130.37**	Robaszynski (1998, SEPM chart)	Base of N. nicklesi Zone.
130.77	Hedbergella sigali				
Hauterivian					
133.88					
Valanginian	**H. sigali/G. hoterivica**	B Hedbergella sigali/delrioensis	**134.22**	Robaszynski (1998, SEPM chart)	ca. 0.1 myr above middle of ammonite H. trinodosum subzone. Base of foraminifer H. sigali Zone.
139.4	Globuligerina hoterivica				
Berriasian	**G. hoterivica/C. gulekhensis**	B Globuligerina hoterivica	**141.93**	Robaszynski (1998, SEPM chart)	Base of ammonite M. paraminounum subzone. Called "Favusella" hoterivica in some studies.
(145.01)	Conoglobigerina gulekhensis				
Tithonian	**base C. gulekhensis**	B Conoglobigerina gulekhensis	**145.95**	Robaszynski (1998, SEPM chart)	Base of ammonite B. jacobi zone.

Table A3.3. Quaternary-Neogene Calcareous Nannofossil Biochronology.

GTS2012 CHRONO-STRATIGRAPHY	STANDARD TROPICAL-SUBTROPICAL BIOZONE (BIOCHRON)		BIOHORIZON (DATUM)	AGE (Ma)	CALIBRATED AGES COMPARED GLOBALLY					CALIBRATION
Age in Ma	Okada & Bukry (1980)	Martini (1971)		GTS2012 scaling	N. Atlantic (incl. Gulf Mex.)	Mediter-ranean	South Atlantic	Equatorial Pacific	North Pacific	Magnetochron/ Marine Isotope Stage
STAGE (AGE)	**CN Zones**	**NN Zones**	T = Top/LAD; B = Base/FAD; X = Crossover of dominance	***This study***	**NA**	**M**	**SA**	**EP**	**NP**	(various authors, see ref's)
Tarantian (Lt. Pleist.)	CN15	NN21	X Gephyrocapsa caribbeanica to E. huxleyi shift of dominance [in "transitional" waters]	**0.07**		0.07				O-18 stage 4; Chron C1n.91
			X Gephyrocapsa caribbeanica to E. huxleyi shift of dominance [in tropical and subtropical waters]	**0.09**		0.09				O-18 stage 5a-5b; Chron C1n.88
0.126										
Ionian (M. Pleist.)	**CN15/CN14b**	**NN21/NN20**	B Emiliania huxleyi	**0.29**		0.27	0.29	0.29		O-18 stage 8; Chron C1n.67
	CN14b/CN14a	**NN20/NN19**	T Pseudoemiliania lacunosa	**0.44**		0.47	0.44	0.44		O-18 stage 12; Chron C1n.4
	CN14a	NN19	T Gephyrocapsa sp.3	**0.61**		0.61				-
0.781										
Calabrian			T Reticulofenestra asanoi (common)	**0.91**		0.90	0.91			-
			T small Gephyrocapsa spp. dominance	**1.02**						-
			B Gephyrocapsa sp.3	**1.02**		0.97		1.02		O-18 stage 25; Chron C1r.1r (above Jaramillo)
	CN14a/CN13b		B medium (over 4 microns) Gephyrocapsa spp. reentrance (reemG event)	**1.04**		0.96	1.01	1.04		O-18 stage 29; Chron C1r.1n.5 (within Jaramillo)
	CN13b		B Reticulofenestra asanoi (common)	**1.14**		1.08	1.14			-
			T large (over 5.5 microns) Gephyrocapsa spp.	**1.24**		1.25	1.26	1.24		O-18 stage 37; Chron C1r.3r.97
			B small Gephyrocapsa spp. dominance	**1.24**				1.24		-
			T Helicosphaera sellii	**(1.26)**		1.26		1.34		Mid-Lat = O-18 stage 37; Low-Lat = upper O-18 stage 49, top Chron C1r.3r
			T Calcidiscus macintyrei	**1.60**		1.66	1.61	1.60		upper O-18 stage 55; = Chron C1r.3r.3.
			B large (over 5.5 microns) Gephyrocapsa spp.	**1.62**		1.62	1.56	1.46		upper O-18 stage 48; = Chron C1r.3r.55
	CN13b/CN13a		B medium (over 4 microns) Gephyrocapsa spp. (=bmG event)	**1.73**		1.73	1.69	1.67		transition O-18 stages59/60; = Chron C1r.3r.15 (above top of Olduvai).
1.806	CN13a									
Gelasian	**CN13a/CN12d**	**NN19/NN18**	T Discoaster brouweri	**1.93**		1.95	1.93	2.06		lowermost Olduvai; = base of Chron C2n.
	CN12d	NN18	T Discoaster triradiatus	**1.95**		1.95				base Chron C2n (lowermost Olduvai)
			B acme Discoaster triradiatus	**2.22**		2.22	2.14			~ C2r.2r.6 (early Matuyama)

CALIBRATION REFERENCES	COMMENTS
Berggren et al. (1995b) citing Rio et al. (1990), Thierstein et al. (1977)	Same chron-age scaling as Berggren et al. (1995b) = O-18 stage 4; 0.075 Ma; = C1n.91
Berggren et al. (1995b) citing Rio et al. (1990), Thierstein et al. (1977)	Same chron-age scaling as Berggren et al. (1995b) = O-18 stages 5a-5b; 0.09 Ma; = C1n.88
Lourens et al. (2004; Table A2.2)	Ubiquitous species, often forming blooms. Coccospheres often with multiple layers of coccoliths. Astronomically tuned in Pacific (ODP Legs 111 & 138) and Atlantic (Leg 154, Sites 925 & 926). Age is recorded slightly younger (0.27 myr) in Mediterranean. Berggren et al. (1995b) placed similarly= O-18 stage 8; Chron C1n.67.
Lourens et al. (2004; Table A2.2)	LAD defines top of NN19, therefore at base of NN20. Astronomically tuned in Pacific (ODP Legs 111 & 138) and Atlantic (Leg 154, Sites 925 & 926). Age is recorded slightly older (0.47 myr) in Mediterranean. Berggren et al. (1995b) placed similaly = O-18 stage 12; Chron C1n.4.
Lourens et al. (2004; Table A2.2)	Cycle-calibrated in Eastern Mediterranean.
Lourens et al. (2004; Table A2.2)	Astronomically tuned in Atlantic (Leg 154, Sites 925 & 926). LAD is at same time (0.90 Ma) in Mediterranean.
Lourens et al. (2004; Table A2.2)	Astronomically tuned in Pacific (ODP Legs 111 & 138).
Lourens et al. (2004; Table A2.2)	Astronomically tuned in Pacific (ODP Legs 111 & 138). FAD is recorded slightly later (0.97 Ma) in Mediterranean. Berggren et al. (1995b) placed similarly = O-18 stage 25; Chron C1r.1r.15 (above Jaramillo).
Lourens et al. (2004; Table A2.2)	Astronomically tuned in Pacific (ODP Legs 111 & 138). FAD recorded slightly later (1.01 Ma) at Atlantic Leg 154, Sites 925 & 926, and later (0.96 Ma) in Mediterranean. Berggren et al. (1995b) placed similarly = O-18 stage 29; Chron C1r.1n.5 (within Jaramillo).
Lourens et al. (2004; Table A2.2)	Astronomically tuned in Atlantic Leg 154, Sites 925 & 926. FAD is recorded slightly later (1.08 Ma).
Lourens et al. (2004; Table A2.2)	"tIG event" Astronomically tuned in Pacific (ODP Legs 111 & 138). LAD recorded slightly earlier (1.26 Ma) at Atlantic Leg 154, Sites 925 & 926, and (1.25 Ma) in Mediterranean. Berggren et al. (1995b) have same assignment = O-18 stage 37; Chron C1r.3r.97.
Lourens et al. (2004; Table A2.2)	Astronomically tuned in Pacific (ODP Legs 111 & 138).
Lourens et al. (2004; Table A2.2)	Astronomically tuned in eastern Mediterranean. This LAD is recorded slightly earlier (1.34 Ma) in ODP Legs 111 & 138 (Pacific). Berggren et al. (1995b) summarize as Mid-Lat = O-18 stage 37; Equat. Zone = upper O-18 stage 49, therefore at top of Chron C1r.3r. [NOTE: H. selli LAD is diachronous among regions]
Lourens et al. (2004; Table A2.2)	Astronomically tuned in Pacific ODP Legs 111 & 138 and Atlantic Leg 154, Sites 925 & 926. This LAD is recorded slightly earlier (1.66 Ma) in Mediterranean. Berggren et al. (1995) assigns as upper O-18 stage 55; therefore Chron C1r.3r.3.
Lourens et al. (2004; Table A2.2)	"bIG event" Astronomically tuned in eastern Mediterranean. This FAD is recorded slightly later (1.46 and 1.56 Ma) in Pacific ODP Legs 111 & 138 and in Atlantic Leg 154, Sites 925 & 926, respectively. Berggren et al. (1995b) assign FAD slightly younger, as upper O-18 stage 48; 1.46-1.48 Ma, therefore at Chron C1r.3r.55.
Lourens et al. (2004; Table A2.2)	Astronomically tuned in eastern Mediterranean. This FAD is recorded slightly later (1.67 and 1.69 Ma) in Pacific ODP Legs 111 & 138 and in Atlantic Leg 154, Sites 925 & 926, respectively. Pleistocene Nanno FAD of this Gephyrocapsa species was called "medium-o" by Berggren, Hilgen et al. (in earlier version of their 1995 GSA paper) and "oceanica s.l." by Leg 145 -- both lists give "Rio et al, in press" as source, and have same age. Berggren et al. (1995) assign as transition O-18 stages59/60; therefore at Chron C1r.3r.15 (above top of Olduvai).
Lourens et al. (2004; Table A2.2)	Astronomically tuned in Atlantic (Leg 154, Sites 925 & 926). This LAD of D. brouweri is recorded slightly earlier (1.95 Ma) in Mediterranean (refs. 11, 22-25), and significantly earlier (2.06 Ma) in Pacific ODP Legs 111 and 138. Berggren et al. (1995b) assign in lowermost Olduvai; 1.95 Ma; therefore at base of Chron C2n.
Berggren et al. (1995b) citing Rio et al. (1990), Backman and Pestiaux (1987)	Base of Chron C2n (lowermost Olduvai)
Lourens et al. (2004; Table A2.2)	Astronomically tuned in eastern Mediterranean. This Acme onset is recorded slightly later (2.14 Ma) in Atlantic Leg 154, Sites 925 & 926. Pliocene nannofossil 'Discoaster triradiatus' (onset acme) occurs at 2.25 Ma in Mediterranean in Berggren et al. (1995b) compilation; therefore, C2r.2r.6 (early Matuyama).

GTS2012 CHRONO-STRATIGRAPHY	STANDARD TROPICAL-SUBTROPICAL BIOZONE (BIOCHRON)		BIOHORIZON (DATUM)	AGE (Ma)	CALIBRATED AGES COMPARED GLOBALLY					CALIBRATION
Age in Ma	Okada & Bukry (1980)	Martini (1971)		GTS2012 scaling	N. Atlantic (incl. Gulf Mex.)	Mediter-ranean	South Atlantic	Equatorial Pacific	North Pacific	Magnetochron/ Marine Isotope Stage
STAGE (AGE)	**CN Zones**	**NN Zones**	T = Top/LAD; B = Base/FAD; X = Crossover of dominance	*This study*	**NA**	**M**	**SA**	**EP**	**NP**	(various authors, see ref's)
	CN12d/CN12c	**NN18/NN17**	T Discoaster pentaradiatus	**2.39**		2.51	2.39			~ Chron C2r.3r.25
	CN12c/CN12b	**NN17/NN16**	T Discoaster surculus	**2.49**		2.54	2.49	2.52		~ Chron C2r.3r.1
2.588	CN12b	NN16								
Piacenzian	**CN12b/CN12a**	NN16	T Discoaster tamalis (subtop)	**2.80**		2.80	2.80	2.87		~ Chron C2An.1n.6 (near top of Gauss)
Piacenzian	CN12a	NN16	T Sphenolithus spp. (subtop)	**3.54**		3.70	3.54	3.65		base Chron C2An.3n (base of Gauss)
3.600	CN12a	NN16								
Zanclean	**CN12a/CN11b**	**NN16/NN15**	T Reticulofenestra pseudoumbilicus	**3.70**		3.84	3.70	3.79		~ Chron C2Ar.7 (uppermost part of upper-reversed interval of Gilbert)
Zanclean	CN11b	**NN15/NN14**	T **Amaurolithus** tricorniculatus	**(3.92)**		(3.92)				-
Zanclean	CN11b	NN14	B Discoaster brouweri (subbottom)	**4.12**		4.12				-
Zanclean	**CN11b/CN11a**	**NN14/NN13**	B common Discoaster asymmetricus	**4.13**		4.12		4.13		~ Chron C2n.1n.8 (near top of Cochiti, or in upper Gilbert)
Zanclean	**CN11a/CN10c**	NN13	T Amaurolithus primus	**4.50**				4.50		M: Nunivak subchron; EP: ~ Chron C3n.3n.8) near top Sidufjall)
Zanclean	CN10c	NN13	T Reticulofenestra antarctica	**4.91**		4.91				-
Zanclean	CN10c	NN13	B Reticulofenestra pseudoumbilica, Discoaster ovata (subbottom)	**4.91**		4.91				-
Zanclean	CN10c	NN13	T Ceratolithus acutus	**5.04**			5.04	5.04		-
Zanclean	CN10c	NN13	X Ceratolithus acutus to C. rugosus (Atl.)	**5.05**			5.05			-
Zanclean	**CN10c/CN10b**	**NN13/NN12**	B Ceratolithus rugosus	**5.12**			5.05	5.12		M: near top Nunivak subchron; EP & SA: ~ Chron C3n.4n.5 (Thvera)
Zanclean	CN10b	NN12	T Triquetrorhabdulus rugosus	**5.28**			5.28			Chron C3r.85 (within uppermost part of Gilbert)
5.333	CN10b	NN12								
Messinian	CN10b	NN12	B Ceratolithus larrymayeri (sp.1)	**5.34**			5.34			-
Messinian	**CN10b/CN10a**	NN12	B Ceratolithus acutus	**5.35**			5.35	5.32		~ Chron C3r.85 (uppermost C3r in Gilbert)
Messinian	**CN10a/CN9d**	**NN12/NN11**	T Discoaster quinqueramus	**5.59**	5.54		5.58	5.59		middle of Chron C3r (Gilbert)
Messinian	**CN9d/CN9c**		T Nicklithus amplificus	**5.94**	6.00	5.94	5.98			-
Messinian	CN9c		X Nicklithus amplificus to Triquetrorhabdulus rugosus	**6.79**			6.79			-

CALIBRATION REFERENCES	COMMENTS
Lourens et al. (2004; Table A2.2)	LAD of D. pentaradiatus (quintatus) defines top of NN17, therefore at base of NN18. Astronomically tuned in Atlantic (Leg 154, Sites 925 & 926). This LAD is recorded earlier (2.51 Ma) in Mediterranean. Berggren et al. (1995b) places near Matuyama/Gauss boundary, or at about Chron C2r.3r.25.
Lourens et al. (2004; Table A2.2)	LAD of D. surculus defines top of NN16, therefore at base of NN17. Astronomically tuned in Atlantic (Leg 154, Sites 925 & 926). This LAD is recorded slightly earlier (2.52 Ma and 2.54 Ma) in Pacific ODP Legs 111 and 138 and in eastern Mediterranean, respectively. Berggren et al. (1995b) place near Matuyama/Gauss boundary at about Chron C2r.3r.1.
Lourens et al. (2004; Table A2.2)	Lourens et al (GTS2004 table) refers to this LAD as a "subtop". Astronomically tuned in Atlantic (Leg 154, Sites 925 & 926) and eastern Mediterranean. This LAD is recorded slightly earlier (2.87 Ma) in Pacific ODP Legs 111 and 138. Pliocene nannofossil 'Discoaster tamalis' (LAD) occurs at 2.73 Ma in Mediterranean in Berggren et al. (1995b) compilation, with main assignment at about Chron C2An.1n.6 (near top of Gauss).
Lourens et al. (2004; Table A2.2)	LAD is called a "subtop" in Lourens et al. (GTS2004 table). Astronomically tuned in Atlantic (Leg 154, Sites 925 & 926). This LAD is recorded slightly earlier (3.65 Ma and 3.70 Ma) in Pacific ODP Legs 111 and 138 and in eastern Mediterranean, respectively. According to Berggren, Hilgen, et al (1995), Pliocene nannofossil 'Sphenolithus' (LAD) occurs at in Mediterranean at base of Chron C2An.3n (base of Gauss).
Lourens et al. (2004; Table A2.2)	This LAD defines top of NN15, therefore at base of NN16. Astronomically tuned in Atlantic (Leg 154, Sites 925 & 926). This LAD is recorded slightly earlier (3.79 Ma and 3.84 Ma) in Pacific (ODP Legs 111 and 138) and in eastern Mediterranean, respectively. Acccording to Berggren et al. (1995b), Pliocene nannofossil Reticulofenestra pseudoumbilicus (LAD) occurs in Mediterranean at about Chron C2Ar.7 (uppermost part of upper-reversed interval of Gilbert).
Berggren et al. (1995b)	Mid-way between D. asymmetricus FAD & R. pseudoumbilicus LAD in the schematic chart of Berggren et al. (1995b). Not assigned an age in Lourens et al. (2004; Table A2.2).
Lourens et al. (2004; Table A2.2)	Lourens et al. (2004) referred to this FCO as a "subbottom". Astronomically tuned in eastern Mediterranean (as 4.12 Ma)
Lourens et al. (2004; Table A2.2)	Lourens et al. (2004) referred to this FCO as a "subbottom". Astronomically tuned in Pacific (ODP Legs 111 and 138) and in eastern Mediterranean (as 4.12 Ma). In Berggren et al. (1995b) compilation, Pliocene nannofossil 'Discoaster asymmetricus' (FCO) occurs at about Chron C2n.1n.8 (near top of Cochiti, or in upper Gilbert).
Lourens et al. (2004; Table A2.2)	Lourens et al (2004) referred to this FAD as a "subbottom". Astronomically tuned in Pacific (ODP Legs 111 and 138). In Berggren et al. (1995b), the Pliocene nannofossil Amaurolithus primus (LAD) occurs in Nunivak subchron (ca. 4.55 Ma) in Mediterranean; or near top of Sidufjall subchron (ca. 4.58 Ma) in oceanic areas; therefore at about Chron C3n.3n.8.
Lourens et al. (2004; Table A2.2)	Astronomically tuned in eastern Mediterranean.
Lourens et al. (2004; Table A2.2)	Astronomically tuned in eastern Mediterranean. Lourens et al. (2004) classify both FADs in Mediterranean as "subbottoms".
Lourens et al. (2004; Table A2.2)	Astronomically tuned in Pacific (ODP Legs 111 and 138) and at Atlantic Leg 154, Sites 925 & 926. Berggren et al. (1995b) assign as slightly older (~0.07 myr) than A. primus LAD.
Lourens et al. (2004; Table A2.2)	This transition is NOT the FAD of C. acutus, but apparently a change in Atlantic dominance. Astronomically tuned in Atlantic (Leg 154, Sites 925 & 926).
Lourens et al. (2004; Table A2.2)	Astronomically tuned in Pacific (ODP Legs 111 and 138). This FAD is reported later (5.05 Ma) in Atlantic Leg 154, Sites 925 & 926. In Berggren et al. (1995b), this datum is diachronous among regions -- Pliocene nannofossil Cerotolithus rugosus (FAD) occurs near top of Nunivak (4.5 Ma) in Mediterranean, but within Thvera (about Chron C3n.4n.5); 5.0-5.23 Ma in oceanic areas.
Lourens et al. (2004; Table A2.2)	Marker for base of Pliocene. Astronomically tuned in Atlantic (Leg 154, Sites 925 & 926). Berggren et al. (1995b) assign as about Chron C3r.85 (within uppermost part of Gilbert).
Lourens et al. (2004; Table A2.2)	Tabulated in Lourens et al. (2004), and would be a regional marker for base of Pliocene. Astronomically tuned in Atlantic (Leg 154, Sites 925 & 926).
Lourens et al. (2004; Table A2.2)	Base C. acutus is a marker for base of Pliocene. Astronomically tuned in Atlantic (Leg 154, Sites 925 & 926), but has essentially same age (5.32 Ma) in Pacific (ODP Legs 111 & 138). Berggren et al. (1995b) assigned to approximately Chron C3r.85 (uppermost part of C3r in Gilbert).
Lourens et al. (2004; Table A2.2)	Astronomically tuned in Pacific (ODP Legs 111 and 138); and reported at same time (5.58 Ma) in Atlantic Leg 154, Sites 925 & 926. Berggren. Kent, et al. (1995) assign as middle of Chron C3r (Gilbert).
Lourens et al. (2004; Table A2.2)	Genus for amplificus was "Amaurolithus" in Berggren et al. (1995a) table; and had been used as base of "Subzone NN11d". Astronomically tuned in eastern Mediterranean. This LAD (and LCO) is reported at same time (5.98 Ma) in Atlantic Leg 154, Sites 925 & 926. The LCO is at 6.12 Ma and 6.14 Ma in Pacific (ODP Legs 11 & 138) and in Mediterranean, respectively.
Lourens et al. (2004; Table A2.2)	Astronomically tuned in Atlantic (Leg 154, Sites 925 & 926).

GTS2012 CHRONO-STRATIGRAPHY	STANDARD TROPICAL-SUBTROPICAL BIOZONE (BIOCHRON)		BIOHORIZON (DATUM)	AGE (Ma)	CALIBRATED AGES COMPARED GLOBALLY					CALIBRATION
Age in Ma	Okada & Bukry (1980)	Martini (1971)		GTS2012 scaling	N. Atlantic (incl. Gulf Mex.)	Mediter-ranean	South Atlantic	Equatorial Pacific	North Pacific	Magnetochron/ Marine Isotope Stage
STAGE (AGE)	CN Zones	NN Zones	T = Top/LAD; B = Base/FAD; X = Crossover of dominance	*This study*	NA	M	SA	EP	NP	(various authors, see ref's)
	CN9c/CN9b	NN11	B Nicklithus amplificus	**6.91**		6.68	6.91			-
7.246	CN9b									
	CN9b/CN9a		B Amaurolithus primus, Amaurolithus spp.	**7.42**		7.42	7.36			-
	CN9a		T Discoaster loeblichii	**7.53**						Base of Chron C3Br
			B common Discoaster surculus	**7.79**			7.79			-
			B Discoaster quinqueramus	**(8.12)**					(8.12)	either in Chron C4 or the base of Chron C4n.2n
	CN9a/CN8	**NN11/NN10**	B Discoaster berggrenii	**8.29**		8.71	8.29			-
	CN8	NN10	T Minylitha convallis	**8.68**		8.68				(upper Chron C4n.2n)
			B Discoaster loeblichii	**8.77**						Base Chron C4r
			B paracme Reticulofenestra pseudoumbilicus	**8.79**		8.71	8.79			-
			T Discoaster bollii	**9.21**						Middle of Chron C4Ar.1r.
			B common Discoaster pentaradiatus	**9.37**		9.37				-
	CN8/CN7	**NN10/NN9**	T Discoaster hamatus	**9.53**		9.53	9.69			~ Chron C4Ar.2r.5
Tortonian	CN7	NN9	T Catinaster calyculus	**9.67**			9.67			~ base Chron C4An
			T Catinaster coalitus	**9.69**			9.69			-
			B Minylitha convallis	**9.75**		9.61	9.75			-
			X Discoaster hamatus to Discoaster neohamatus	**9.76**			9.76			-
			B Discoaster bellus	**10.40**		10.40				-
			X Catinaster calyculus to Catinaster coalitus	**10.41**			10.41			-
			B Discoaster neohamatus	**10.52**		9.87	10.52			-
	CN7/CN6	**NN9/NN8**	B Discoaster hamatus	**10.55**		10.18	10.55			-
	CN6	NN8	B common Helicosphaera stalis	**10.71**		10.71				-
			T common Helicosphaera walbersdorfensis	**10.74**		10.74				-
			B Discoaster brouweri	**10.76**		10.73	10.76			-
			B Catinaster calyculus	**10.79**			10.79			-
	CN6/CN5b	**NN8/NN7**	B Catinaster coalitus	**10.89**		10.73	10.89			-
	CN5b	NN7	T Coccolithus miopelagicus	**10.97**		10.97	11.02			-
			T Calcidiscus premacintyrei	**11.21**			11.21			-
			T common Discoaster kugleri	**11.58**		11.60	11.58			middle of Chron C5r.2n
11.608										
			T Cyclicargolithus floridanus	**11.85**					11.85	Chron C5r.3r.5

CALIBRATION REFERENCES	COMMENTS
Lourens et al. (2004; Table A2.2)	Genus for amplificus was "Amaurolithus" in Berggren et al. (1995a) table; and had been used for a Subzone NN11c (and "CN9c") of their zonal scheme. Astronomically tuned in Atlantic (Leg 154, Sites 925 & 926). This FAD occurs later (6.68 Ma) in eastern Mediterranean.
Lourens et al. (2004; Table A2.2)	LAD of Amaurolithus primus is also LAD of genus Amaurolithus. Astronomically tuned in eastern Mediterranean. This FAD is recorded slightly later (7.36 Ma) in Atlantic (Leg 154, Sites 925 & 926).
Berggren et al. (1995a) citing Raffi and Flores (1995) and Raffi et al. (1995)	Base of Chron C3Br.
Lourens et al. (2004; Table A2.2)	Astronomically tuned in Atlantic (Leg 154, Sites 925 & 926).
ODP Leg 145 Shipboard Scientific Party (1993)	Leg 145 compilation indicates either in Chron C4 or the base of Chron C4n.2n (for older FAD estimate; used here).
Lourens et al. (2004; Table A2.2)	Astronomically tuned in Atlantic (Leg 154, Sites 925 & 926).
Lourens et al. (2004; Table A2.2)	Astronomically tuned in eastern Mediterranean. However, Berggren et al. (1995a) calibrated this LAD as much higher; in upper Chron C4n.2n; at about 7.8 Ma.
Berggren et al. (1995a) citing Raffi and Flores (1995) and Raffi et al. (1995)	Base of Chron C4r.
Lourens et al. (2004; Table A2.2)	Lourens et al (2004) tabulate this "paracme" event with astronomical tuning in Atlantic (Leg 154, Sites 925 & 926). Nearly simultaneous (8.71 Ma) in eastern Mediterranean, where is defines base of zone MNN11.
Berggren et al. (1995a) citing Gartner (1992) and Miller et al. (1985)	Middle of Chron C4Ar.1r.
Lourens et al. (2004; Table A2.2)	Astronomically tuned in eastern Mediterranean. LAD may be 0.3 myr earlier in Indo-Pacific.
Lourens et al. (2004; Table A2.2)	Astronomically tuned in eastern Mediterranean. LAD occurs slightly earlier (9.69 Ma) at Atlantic Leg 154, Sites 925 & 926. Astronomical age implies approximately Chron C4Ar.2r.5.
Lourens et al. (2004; Table A2.2)	Astronomically tuned in Atlantic (Leg 154, Sites 925 & 926). Leg 145 compilation (as LAD of genus Catinaster) had assigned as approximately base of Chron C4An. Berggren, Kent, et al (1995) had this LAD as base of Subzones NN9b & CN7b.
Lourens et al. (2004; Table A2.2)	Astronomically tuned in Atlantic (Leg 154, Sites 925 & 926).
Lourens et al. (2004; Table A2.2)	Astronomically tuned in Atlantic (Leg 154, Sites 925 & 926). FAD occurs slightly younger (9.61 Ma) in eastern Mediterranean.
Lourens et al. (2004; Table A2.2)	Astronomically tuned in Atlantic (Leg 154, Sites 925 & 926).
Lourens et al. (2004; Table A2.2)	Astronomically tuned in eastern Mediterranean.
Lourens et al. (2004; Table A2.2)	Astronomically tuned in Atlantic (Leg 154, Sites 925 & 926).
Lourens et al. (2004; Table A2.2)	Astronomically tuned in Atlantic (Leg 154, Sites 925 & 926). FAD occurs much younger (9.87 Ma) in eastern Mediterranean.
Lourens et al. (2004; Table A2.2)	Astronomically tuned in Atlantic (Leg 154, Sites 925 & 926). FAD occurs much younger (10.18 Ma) in eastern Mediterranean. Berggren et al. (1995a) note that this datum is "controversial" and inconsistent, with different reported correlations to magnetostratigraphy. More checks on this calibration are needed.
Lourens et al. (2004; Table A2.2)	Astronomically tuned in eastern Mediterranean.
Lourens et al. (2004; Table A2.2)	Astronomically tuned in eastern Mediterranean.
Lourens et al. (2004; Table A2.2)	Astronomically tuned in Atlantic (Leg 154, Sites 925 & 926). FAD occurs simultaneously (10.73 Ma) in eastern Mediterranean.
Lourens et al. (2004; Table A2.2)	Astronomically tuned in Atlantic (Leg 154, Sites 925 & 926). Berggren, Kent, et al (1995) note that earlier magnetostratigraphic correlations were disputed.
Lourens et al. (2004; Table A2.2)	Astronomically tuned in Atlantic (Leg 154, Sites 925 & 926). FAD occurs slightly later (10.73 Ma) in eastern Mediterranean.
Lourens et al. (2004; Table A2.2)	Astronomically tuned in eastern Mediterranean. LAD is recorded slightly earlier (11.02 Ma) at Atlantic Leg 154, Sites 925 & 926.
Lourens et al. (2004; Table A2.2)	Astronomically tuned in Atlantic (Leg 154, Sites 925 & 926).
Lourens et al. (2004; Table A2.2)	Astronomically tuned in Atlantic (Leg 154, Sites 925 & 926); and simultaneous (11.60 Ma) in eastern Mediterranean. This "LCO" of Lourens et al. (2004) seems to correspond in calibration to the "LAD" of Berggren et al. (1995a) at middle of Chron C5r.2n.
ODP Leg 145 Shipboard Scientific Party (1993)	C. floridanus LCO is much earlier (13.28 Ma) in astronomical tuning from Atlantic and Mediterranean. Lourens et al. (2004) did not tabulate the LAD as a separate level. Therefore, LAD is placed using same magnetostratigraphic scaling as in the Leg 145 synthesis at Chron C5r.3r.5.

GTS2012 CHRONO-STRATIGRAPHY	STANDARD TROPICAL-SUBTROPICAL BIOZONE (BIOCHRON)		BIOHORIZON (DATUM)	AGE (Ma)	CALIBRATED AGES COMPARED GLOBALLY					CALIBRATION
Age in Ma	Okada & Bukry (1980)	Martini (1971)		GTS2012 scaling	N. Atlantic (incl. Gulf Mex.)	Mediter-ranean	South Atlantic	Equatorial Pacific	North Pacific	Magnetochron/ Marine Isotope Stage
STAGE (AGE)	CN Zones	NN Zones	T = Top/LAD; B = Base/FAD; X = Crossover of dominance	*This study*	NA	M	SA	EP	NP	(various authors, see ref's)
	CN5b/CN5a	**NN7/NN6**	B common Discoaster kugleri	**11.90**		11.90	11.86			~ lower Chron C5r.3r
			T Coronocyclus nitescens	**12.12**			12.12			-
			T regular Calcidiscus premacintyrei	**12.38**		12.38	12.45			-
			B common Calcidiscus macintyrei	**12.46**			12.46			-
Serravallian	CN5a	NN6	B Reticulofenestra pseudoumbilicus	**12.83**					12.83	~ base Chron C5Ar.2r
			B Triquetrorhabdulus rugosus	**13.27**	13.27					~ Chron C5AAr.5
			T common Cyclicargolithus floridanus	**13.28**		13.28	13.33			within Chron C5AAr
			B Calcidiscus macintyrei	**13.36**					13.36	base Chron C5An
	CN5a/CN4	**NN6/NN5**	T Sphenolithus heteromorphus	**13.53**		13.65	13.53			~ Chron C5ABr.6
13.82	CN4	NN5								
	CN4/CN3	**NN5/NN4**	T Helicosphaera ampliaperta	**14.91**			14.91			middle of Chron C5Br
Langhian			T abundant Discoaster deflandrei group	**15.80**			15.80			~ Chron C5Br
	CN3	NN4	B Discoaster signus	**15.85**					15.85	~ Chron C5Br.15
15.97										
			B Sphenolithus heteromorphus	**17.71**			17.71			Chron C5Dr.1
	CN3/CN2	**NN4/NN3**	T Sphenolithus belemnos	**17.95**			17.95			-
Burdigalian	**CN2/CN1c**	**NN3/NN2**	T Triquetrorhabdulus carinatus	**18.28**			18.28			-
			B Sphenolithus belemnos	**19.03**			19.03			Chron C6n.9
			B Helicosphaera ampliaperta	**20.43**			20.43			-
(20.44)										
	CN1c	NN2	X Helicosphaera euphratis to Helicosphaera carteri	**20.92**			20.92			-
			B common Helicosphaera carteri	**22.03**			22.03			-
Aquitanian			T Orthorhabdus serratus	**22.42**						Chron C6Br.5
			B Sphenolithus disbelemnos	**22.76**			22.76			-
	CN1c/CN1a-b	**NN2/NN1**	B Discoaster druggi (sensu stricto)	**22.82**		22.82				-
	CN1a-b	NN1	T Sphenolithus capricornutus	**22.97**					22.97	Chron C6Cn.2n.5
23.03										

CALIBRATION REFERENCES	COMMENTS
Lourens et al. (2004; Table A2.2)	Astronomically tuned in Atlantic (Leg 154, Sites 925 & 926); and simultaneous (11.90 Ma) in eastern Mediterranean. This "FCO" of Lourens et al. (2004) seems to correspond in calibration to the "FAD" of Berggren et al. (1995a), which they summarized as approximately lower Chron C5r.3r.
Lourens et al. (2004; Table A2.2)	Astronomically tuned in Atlantic (Leg 154, Sites 925 & 926).
Lourens et al. (2004; Table A2.2)	Last "regular" occurrence is astronomically tuned in eastern Mediterranean. This LRO is recorded slightly earlier (12.45 Ma) at Atlantic Leg 154, Sites 925 & 926.
Lourens et al. (2004; Table A2.2)	FCO is astronomically tuned in Atlantic (Leg 154, Sites 925 & 926).
ODP Leg 145 Shipboard Scientific Party (1993)	Same chron-age scaling as Berggren et al. (1995a) as approximately base of Chron C5Ar.2r
Berggren et al. (1995a) citing Gartner (1992)	Triquetrorhabdulus rugosus FAD is estimated as 12.5 Ma in Leg 145 table, but 13.2 Ma (0.7 m.y. older) in Berggren et al. (1995a), based on Hole 608 which is used here; therefore at Chron C5AAr.5 on their magnetic polarity scale.
Lourens et al. (2004; Table A2.2)	Last common occurrence is astronomically tuned in eastern Mediterranean. This LCO is recorded simultaneously as an "LAD" (13.33 Ma) at Atlantic Leg 154, Sites 925 & 926. Leg 145 compilation had chron-age assignment within Chron C5AAr.
ODP Leg 145 Shipboard Scientific Party (1993)	Base of Chron C5An.
Lourens et al. (2004; Table A2.2)	Astronomically tuned in Atlantic (Leg 154, Sites 925 & 926); but LAD is reported slightly earlier (13.654 Ma) in eastern Mediterranean. The offset, although only 100 kyr, causes a problem, because this nannofossil S. heteromorphus LAD had been a candidate for the "primary marker" for base-Serravallian GSSP as placed in the in the Mediterranean region. Berggren et al. (1995a) had a similar chron-age assignment of approximately Chron C5ABr.6.
Lourens et al. (2004; Table A2.2)	Astronomically tuned in Atlantic (Leg 154, Sites 925 & 926). Berggren et al. (1995a) had placed this LAD about 0.6 myr older as middle of Chron C5Br.
Lourens et al. (2004; Table A2.2)	Highest abundant occurrence is astronomically tuned in Atlantic (Leg 154, Sites 925 & 926). Leg 145 synthesis implied assignment in Chron C5Br.
ODP Leg 145 Shipboard Scientific Party (1993)	Leg 145 synthesis placed at about Chron C5Br.15, nearly same as the top of acme of D. deflandrei
Lourens et al. (2004; Table A2.2)	Lourens et al. (2004) assign this event as a first-common occurrence (FCO) that is astronomically tuned in Atlantic (Leg 154, Sites 925 & 926). This is essentially the FAD as compiled by Berggren et al. (1995a) which was projected as Chron C5Dr.1.
Lourens et al. (2004; Table A2.2)	Lourens et al (2004) assign this event as a LCO that is astronomically tuned in Atlantic (Leg 154, Sites 925 & 926). This is only slightly above the LAD, which was the original definition of the zone (a 0.1 myr difference according to J. Bergen, pers. commun., 2005).
Lourens et al. (2004; Table A2.2)	Dashed. Astronomically tuned in Atlantic (Leg 154, Sites 925 & 926); which is ~1myr younger than earlier estimates which ranged from 19.5 to 23.0 Ma. The very poorly constrained age on the LAD probably reflects diachroneity. Hodell & Woodruff (1994) have a similar conclusion, although their estimate for this event in Site 289 suggests that it may be reworked.
Lourens et al. (2004; Table A2.2)	Astronomically tuned in Atlantic (Leg 154, Sites 925 & 926). Berggren et al. (1995a) had similar assignment that projected as Chron C6n.9 in their magnetostratigraphic scale.
Lourens et al. (2004; Table A2.2)	Candidate as a primary marker for base of Burdigalian. Genus spelled as "Helicopontosphaera" in Lourens et al. (2004). Astronomically tuned in Atlantic (Leg 154, Sites 925 & 926).
Lourens et al. (2004; Table A2.2)	Astronomically tuned in Atlantic (Leg 154, Sites 925 & 926).
Lourens et al. (2004; Table A2.2)	Astronomically tuned in Atlantic (Leg 154, Sites 925 & 926).
ODP Leg 145 Shipboard Scientific Party (1993)	Also known as "Triquetrorhadulus serratus". Chron C6Br.5.
Lourens et al. (2004; Table A2.2)	Astronomically tuned in Atlantic (Leg 154, Sites 925 & 926). This species was not recognized in other schemes.
Lourens et al. (2004; Table A2.2)	Astronomically tuned in eastern Mediterranean (based on projecting astronomical tuning of long-period minima to base-Miocene), which is consistent with ODP Leg 154. However, smaller forms appear earlier, at end of Oligocene (J. Bergen and E. de Kaenel, pers. commun., 2005).
ODP Leg 145 Shipboard Scientific Party (1993)	Chron C6Cn.2n.5.

Table A3.4. Paleogene-Cretaceous Calcareous Nannofossil Biochronology

GTS2012 CHRONO-STRATIGRAPHY	STANDARD TROPICAL-SUBTROPICAL BIOZONE (BIOCHRON)		BIOHORIZON (DATUM)	AGE (Ma)	CALIBRATION REFERENCES	COMMENTS
STAGE (AGE)	CN-CP Zones: Okada & Bukry (1980); CC Zones:	NN-NP Zones: Martini (1971); NC Zones: Roth (1978,	T = Top/LAD; B = Base/FAD	This study		
Chattian	CN1a-b	NN1	T Sphenolithus delphix	23.11	Lourens et al. (2004; Table A2.2); Raffi et al. (2006)	Oligocene Nanno 'Spenolithus delphix LAD' may be similar to its 'top of acme'. Astronomically tuned in Atlantic (Leg 154, Sites 925 & 926).
	CN1a-b/CP19b	**NN1/NP25**	T Reticulofenestra bisecta (>10 micron)	23.13	ODP Leg 145 Shipboard Scientific Party (1993)	Chron C6Cn.2r.5. Leg 145 referred to this LAD as "top of acme"; but ODP Leg 171 scheme assigned as LAD. Global (not just high-latitude) marker for base of NN1. "Reticulofenestra bisecta" was called "Dictyococcites bisectus" in Perch-Nielsen (1985). The LAD of 'R. bisectus' is used to approximate the NP25/NN1 boundary in high latitudes; and was "substituted for H. recta, the marker of the NP15/NN1 boundary in Martini's (1971) zonal scheme" by Aubry (in Berggren, Kent, et al., 1995).
	CP19b	NP25	B Sphenolithus delphix	23.21	Lourens et al. (2004; Table A2.2); Raffi et al. (2006)	Oligocene Nanno 'Spenolithus delphix FAD' may be similar to its "onset of acme" (Leg 145 table). Astronomically tuned in Atlantic (Leg 154, Sites 925 & 926); this FAD placement is nearly 1 myr higher than tables in Leg 145 and Berggren et al. (1995a) where it was in upper Chron C6Cr.
			T Zygrhablithus bijugatus	23.76	Berggren et al. (1995a) citing Miller et al. (1985)	Approximately Chron C6Cr.3.
			T Sphenolithus ciperoensis	24.43	Blaj et al. (2009); P. Bown (pers. comm.. 2011)	DASH -- Cycle-calibrated as 24.43 Ma (1.40 myr prior to 23.03 Ma for base-Miocene) by Blaj et al. (2009) (Eq. Pac.); but could be higher, because this genera are not very common (Paul Bown to J.Ogg, June'11). Low-latitude marker for base of NN1. Used as base of CN1a in some publications (in which LAD of D. bisectus becomes base of CN1b, rather than the present combined subzones). Scaling by Berggren et al. (1995a) was base of Chron C6Cr.
			T Cyclicargolithus abisectus (common)	24.67	Lyle et al. (2002)	Cycle-calibrated as 24.67 Ma (1.64 myr prior to 23.03 Ma for base-Miocene) by Lyle et al. (2002). [From IODP Expedition 320/321 Scientists, 2010]
			B Tri. longus - > Triquetrorhabdulus carinatus (abundance shift)	24.67	IODP Expedition 320/321 Scientists (2010)	Crossover in dominance is same age as same as peak of T. carinatus acme in IODP Expedition 320/321 Scientists (2010) explanatory notes; and that event is Cycle-calibrated as 24.67 Ma (1.64 myr prior to 23.03 Ma for base-Miocene). Low-priority event.
			T Chiasmolithus altus	25.44	Berggren et al. (1995a) citing Wei and Thierstein (1991)	Same chron-age scaling as Berggren et al. (1995) = upper Chron C8n (C8n.2n.8)
			B Triquetrorhabdulus carinatus (common)	26.57	Blaj et al. (2009)	Cycle-calibrated as 26.57 Ma (3.54 myr prior to 23.03 Ma for base-Miocene) by Blaj et al.'09 (Eq. Pac.). Low-priority event.
	CP19b/CP19a	**NP25/NP24**	T Sphenolithus distentus	26.84	Blaj et al. (2009)	Oligocene Nanno 'Sphenolithus distentus' LAD is cycle-calibrated as 26.81 Ma (3.78 myr prior to 23.03 Ma for base-Miocene) by Blaj et al. (2009) (Eq. Pac.). This is younger than chron-age of C9n by Aubry (in Berggren, Kent, et al., 1995) at 27.5 Ma, but is consistent with mid-Chron C8n.2n correlation in Leg 145 table. The LCO was placed in Chron 9n in basal-Chattian compilation of Italian sections by Coccioni et al. (GSA Bull, 2008).
	CP19a	NP24	T Sphenolithus predistentus	26.93	Blaj et al. (2009)	Cycle-calibrated as 26.93 Ma (3.90 myr relative to 23.03 Ma for base-Miocene) by Blaj et al. (2009) (Eq. Pac.).
(28.09)						
Rupelian			T Sphenolithus pseudoradians	28.73	Berggren et al. (1995a) citing Poore et al. (1982)	DASH -- Not widely used. Indeed, calibrations of Sphenolithus are tricky in Oligocene (Paul Bown to J.Ogg, June 2011). IODP Expedition 320/321 Scientists (2010) table used same chron-age scaling as Berggren et al. (1995a) at middle of Chron C10r.
	(CP19a/CP18)	**(NP24/NP23)**	B Sphenolithus ciperoensis	29.62	Berggren et al. (1995a); Coccioni et al. (2008); P. Bown (pers. comm., 2011)	DASH -- Difficult to pin down age within the transition; therefore has a wide range fo calibrations. Cycle-calibrated as 27.14 Ma (4.11 myr relative to 23.03 Ma for base-Miocene) by Blaj et al.'09 (Eq. Pac.); but Paul Bown (to J.Ogg, June'11) advised to not use this very high placement, because this might be a very high level in the transition. Nanno 'Sphenolithus ciperoensis FAD' is assigned by Aubry (in Berggren, Kent, et al., 1995) within Subchron C11n.2n; but was placed at essentially base of Chron C11n.2n at proposed Chattian GSSP (Coccioni et al., GSA Bull., 2008). However, to avoid overlap with cycle-scaled "underlying" CP18 marker, the base of this NP24/CP19 is set slightly higher at Chron C11n.2n.8. This is about 1.3 myr older than estimate in Leg 145 table.
	(CP18/CP17)	NP23	B Sphenolithus distentus	30.00	Blaj et al. (2009)	DASH: Diachronous -- Oligocene Nanno 'Sphenolithus distentus FAD' is "a very inconsistent datum which may occur as low as Nanno Zone NP21 ... or as high as in Zone NP23" (Aubry, in Berggren et al., 1995). Cycle-scaled placement by Blaj et al. (2009; Eq. Pac.) is 30.00 Ma (6.97 myr relative to base-Miocene of 23.03 Ma; used here). Age estimates in Aubry's table range from 31.5 to 33.1 Ma (but the oldest level was "probably due to another species, newly called as "Sp. akropodus", being confused for it"; Paul Bown to J.Ogg, June'11). THe oldest FAD would imply that zone "CP18" begins before "CP17"!
	CP17		T Reticulofenestra umbilicus (south high lat.)	31.35	Berggren et al. (1995a) citing themselves and Miller et al. (1985)	Diachronous -- Oligocene Nanno 'Reticulofenestra umbilicus LAD' was specified as ">14 micron" in Leg 145 table. This LAD occurs earlier (32.3 Ma) in low-mid latitudes than in southern high-latitudes (31.3 Ma) = 1 myr diachroneity. For simplicity, the bases of the associated Zone NP23 and CP17 are drawn on the chart with the older mid-latitude level (32.3 Ma). Same chron-age scaling as Berggren,Kent, et al. (1995) as upper Chron C12r (C12r.85). Taxa "R. hillae" is considered to be a morphotype of R. umbilicus.
	(CP17/CP16c)	**(NP23/NP22)**	T Reticulofenestra umbilicus (low-mid lat.)	32.02	Blaj et al. (2009)	Diachronous -- Oligocene Nanno 'Reticulofenestra umbilicus LAD' was specified as ">14 micron" in Leg 145 and tables in IODP Expedition 320/321 Scientists (2010). Cycle-calibrated as 32.02 Ma (= 8.99 myr relative to 23.03 for base-Miocene) by Blaj et al.'09 (Eq. Pac.). This LAD occurs earlier in low-mid latitudes than in southern high-latitudes (31.3 Ma) = 1m.y. diachroneity. Same chron-age scaling as Berggren et al. (1995) = mid C12r; 32.3 Ma; [j,t] = C12r.35. Taxa "R.hillae" is considered to be a morphotype of R. umbilicus.
		NP22	T Isthmolithus recurvus (south high lat.)	32.49	Villa et al. (2008)	Diachronous -- Oligocene Nanno 'Isthmolithus recurvus LAD' is "one of the most inconsistent datums" (Aubry, in Berggren, Kent, et al., 1995), with estimated diachroneity of 1.3 myr. Cycle-calibrated as 32.49 Ma (9.46 myr prior to 23.03 Ma for base-Miocene) by Villa et al. (2008) (Kerguelen Plateau).
	CP16c/CP16b	**NP22/NP21**	T Coccolithus formosus	32.92	Blaj et al. (2009)	Also known as "Ericsonia formosa". Diachronous (Berggren et al. (1995a) indicate the LAD is 7 myr nearly earlier in south high lat!). Cycle-calibrated as 32.92 Ma (9.89 myr prior to 23.03 Ma for base-Miocene) by Blaj et al. (2009) (Eq. Pac.). This is same as chron-age scaling as Berggren et al. (1995) as revised by Leg 145 = lowermost Chron C12r.
	CP16b/CP16a	NP21	T Clausicoccus subdistichus (top of acme)	33.43	Berggren et al. (1995a) citing Premoli Silva et al. (1988)	Chron C13n.5. Oligocene Nanno 'Clausicoccus subdistichus' was called 'Ericsonia subdisticha' (Leg 145 table) and 'Ericsonia obrata' (Massignano section, Premoli Silva et al., 1988). [was labeled as "base of acme" in GTS04/08]
33.89	CP16a					
			T Reticulofenestra oamaruensis (south high lat.)	33.97	Villa et al. (2008)	Restricted to southern high latitudes. Cycle-calibrated as 33.97 Ma (10.84 myr prior to 23.03 Ma for base-Miocene) by Villa et al. (2008) (Kerguelen Plateau), used here. This is consistent with the chron-age scaling as Berggren et al. (1995) as revised by Leg 145 = uppermost Chron C13r.

GTS2012 CHRONO-STRATIGRAPHY	STANDARD TROPICAL-SUBTROPICAL BIOZONE (BIOCHRON)		BIOHORIZON (DATUM)	AGE (Ma)	CALIBRATION REFERENCES	COMMENTS
STAGE (AGE)	CN-CP Zones: Okada & Bukry (1980); CC Zones:	NN-NP Zones: Martini (1971); NC Zones: Roth (1978,	T = Top/LAD; B = Base/FAD	*This study*		
Priabonian	(CP16a/CP15)	(NP21/NP20-19)	T Discoaster saipanensis	34.44	Blaj et al. (2009)	Diachronous. Cycle-calibrated as 34.44 Ma (11.41 myr prior to 23.03 Ma for base-Miocene) by Blaj et al. (2009) (Eq. Pac.). This is consistent with the chron-age scaling as Berggren et al. (1995) as revised by Leg 145 = middle of Chron C13r.
	CP15	NP20-19	T Discoaster barbadiensis	34.76	Blaj et al. (2009)	Diachronous. Another marker for Zone CP16a base. Cycle-calibrated as 34.77 Ma (11.73 myr prior to 23.03 Ma for base-Miocene) by Blaj et al. (2009) (Eq. Pac.). This is consistent with the chron-age scaling as Berggren et al. (1995) as revised by Leg 145 = lower Chron C13r.
			T Reticulofenestra reticulata	35.40	Berggren et al. (1995a) explanatory notes (2011); IODP Expedition 320/321 Scientists (2010)	Diachronous. Berggren et al. (1995), as revised by Leg 145, assigns as upper C15r (C15r.75). IODP Expedition 320/321 Scientists (2010) has a similar 35.2 Ma, citing Bachman (1987), but didn't clarify the calibration to chrons.
			B Reticulofenestra oamaruensis (south high lat.)	35.54	Villa et al. (2008)	Restricted to high latitudes. Cycle-calibrated as 35.54 Ma (12.51 myr relative to 23.03 Ma for base-Miocene) by Villa et al. (2008) (Kerguelen Plateau). This is consistent with the chron-age scaling as Berggren et al. (1995) as revised by Leg 145 = middle of Chron C16n.1n.
			T Reticulofenestra reticulata (high lat.)	35.92	Villa et al. (2008)	Diachronous? Cycle-calibrated as 35.92 Ma (12.89 myr prior to 23.03 Ma for base-Miocene) by Villa et al. (2008) (Kerguelen Plateau). This is consistent with the chron-age scaling as Berggren et al. (1995) as revised by Leg 145 = Chron C16n.2n.4. However, IODP Expedition 320/321 Scientists (2010) used 35.2 Ma age citing Bachman (1987), but didn't give chron-scaling).
Chattian		NP20-19/NP18	B Isthmolithus recurvus	36.97	IODP Expedition 320/321 Scientists (2010); Berggren et al. (1995a)	DASH -- The base of NP19-20 is defined by the first occurrence of I. recurvus and falls in the magnetochron C16r according to Backman 1986 at Site 523 [as cited by IODP Expedition 320/321 Scientists (2010), FAD is 36.6 Ma, which would be base Chron C16r in Cande-Kent'95 scale (used here)]. Several papers report the FO of I. recurvus (higher) in the C16n.2n (i.e Sites 1090, Massignano etc.). [Simonetta Monechi, as relayed by N. Vandenberghe to J.Ogg, Jan 2011]. Need more checking on this datum!
		NP18/NP17	B Chiasmolithus oamaruensis (common)	37.32	IODP Expedition 320/321 Scientists (2010); Berggren et al. (1995a)	DASH -- IODP Expedition 320/321 Scientists (2010) assigns as 37.0 Ma, citing Berggren et al. (1995a) = middle of C17n.1n (used here). However, this FAD is placed at base of Chron C17n.2n in proposed Priabonian GSSP (Agnini et al., GSA Bull., 2011); but that is the lowest rare occurrence.
(37.75)						
Bartonian	CP15/CP14b	NP17	T Chiasmolithus grandis	37.98	Agnini et al. (2011)	DASH -- Middle of Chron C17n.2n at proposed Priabonian GSSP in Italy (Agnini et al., GSA Bull., 2011). This is one chron older than Berggren et al. (1995a) assignment as middle of Chron C17n.1n; but they indicate the LAD may occur approx. 2 myr earlier in Mediterranean. Normally, C. grandis does NOT overlap C. oarmaruensis. "Unfortunately, the LO of C. oamaruensis as well as the HO of C. grandis have a low degree of reproducibility in many areas because of their scarse abundances" (Agnini et al., GSA Bull., 2011).
	CP14b		B Chiasmolithus oamaruensis (rare)	38.09	Agnini et al. (2011)	This earliest (rare) FAD is placed at the base of Chron C17n.2n in proposed Priabonian GSSP (Agnini et al., GSA Bull., 2011); but normally, C. grandis does NOT overlap C. oarmaruensis. "Unfortunately, the LO of C. oamaruensis as well as the HO of C. grandis have a low degree of reproducibility in many areas because of their scarse abundances" (Agnini et al., GSA Bull., 2011)
			B Reticulofenestra bisecta (>10 micron)	38.25	IODP Expedition 320/321 Scientists (2010); Berggren et al. (1995a)	IODP Expedition 320/321 Scientists (2010) cite same chron-age scaling as Berggren et al. (1995) = middle of Chron C17n.3n. Also known as "Dictylococcites bisectus" (or bisecta).
	CP14b/CP14a	NP17/NP16	T Chiasmolithus solitus	40.40	IODP Expedition 320/321 Scientists (2010); Berggren et al. (1995a)	IODP Expedition 320/321 Scientists (2010) cite same chron-age scaling as Berggren et al. (1995) = upper Chron C18r (C18r.75). LAD reported higher (C18n) in Hole 748.
(41.15)						
Lutetian	CP14a	NP16	B Reticulofenestra reticulata	41.66	IODP Expedition 320/321 Scientists (2010); Berggren et al. (1995a)	IODP Expedition 320/321 Scientists (2010) cite same chron-age scaling as Berggren et al. (1995) = mid Chron C19r; but C19r.7 is used here to keep in same order to "underlying" datum.
			T Nannotetrina spp.	41.85	ODP Leg 145 Shipboard Scientific Party (1993)	IODP Expedition 320/321 Scientists (2010) cite 42.3 Ma based on Bachman (1987), but chron-age not given. Leg 145 places as upper Chron C19r. An assignment of Chron C19r.5 is used here to keep in relative order to adjacent datums.
	CP14a/CP13c		B Reticulofenestra umbilicus (>14 μm)	(41.94)	Berggren et al. (1995a); IODP Expedition 320/321 Scientists (2010)	DASH -- Berggren et al. (1995a) assign as middle of Chron C19r (C19r.4). Base of zone CP14 is FAD of R. umbilicus, but a further criteria of >10 μm is used for clarity rather than the pure FAD (Paul Bown to J.Ogg, June 2011). IODP Expedition 320/321 Scientists (2010) cite 42.5 Ma for >14 μm based on Bachman (1987), but chron-age is not given. Eocene Nanno 'R. umbilica' FAD is latitude dependent; and occurs as early as basal C20r in Contessa section (Italy).
	CP13c	NP16/NP15c	T Nannotetrina fulgens	(42.87)	Berggren et al. (1995a)	DASH -- Berggren et al. (1995a) assign as middle of Chron C20n.5. IODP Expedition 320/321 Scientists (2010) cite 43.4 Ma based on Bachman (1986), but chron-age not given. Base of NP16 was LAD of B. gladius, but rarely preserved (Paul Bown to J.Ogg, June2011); therefore IODP Leg 320-321 used LAD of N. fulgens as proxy => NP15 becomes the Range Zone of N. fulgens.
		NP15c	T Blackites gladius	(43.09)	Berggren et al. (1995a)	DASH -- Poor calibration. Berggren et al. (1995a) estimate as lower Chron C20n (C20n.3). Base of NP16 was LAD of B. gladius, but rarely preserved (Paul Bown to J.Ogg, June2011); therefore IODP Expedition 320/321 Scientists (2010) used LAD of N. fulgens as proxy.
			B Reticulofenestra umbilicus	43.32	Berggren et al. (1995a)	Basal Chron C20n (C20n.1).
	CP13c/CP13b	NP15c/NP15b	T Chiasmolithus gigas	44.12	Berggren et al. (1995a)	Poor calibration. Berggren et al. (1995a) estimate as upper Chron C20r (C20r.7). IODP Expedition 320/321 Scientists (2010) cite 44.0 Ma based on Bachman (1986), but chron-age not given.
	CP13b/CP13a	NP15b/NP15a	B Chiasmolithus gigas	45.49	Agnini et al. (2006)	Composite Fig.7 of Agnini et al. (2006) indicates placement at Chron C20r.1 (used here). IODP Expedition 320/321 Scientists (2010) assigned as 46.1 Ma (based on CK'95 scale) citing the same Agnini et al. (2006) but this is the base of Chron C20r on that CK'95 scale.
	CP13a	NP15a	T Discoaster sublodoensis (5-rayed)	(46.21)	Agnini et al. (2006); Berggren et al. (1995a)	DASH -- LAD of D. sublodoensis occurs slightly above FAD of N. fulgens (Aubry in Berggren et al. (1995a) had placed about 0.08 myr above FAD of N. fulgens). Therefore, re-calibration of N. fulgens by Agnini et al. (2006) at about Chron C21n.65 suggests this LAD would be at about Chron C21n.7 (used here pending clarification).
	CP13a/CP12b	NP15a/NP14b	B Nannotetrina fulgens	46.29	Agnini et al. (2006)	Agnini et al. (2006) Fig. 7 suggests placement at Chron C21n.65. This is 0.5 myr younger than chron-age of Berggren et al. (1995) =of lower-mid Chron C21n.
	CP12b	NP14b	T Discoaster lodoensis	(47.41)	ODP Leg 145 Shipboard Scientific Party (1993)	DOTTED -- ODP Leg 145 assigned as just below top of Chron C21r (C21r.95 used here). In contrast, IODP Expedition 320/321 Scientists (2010) assigned as 48.4 Ma (based on CK'95 scale) citing Agnini et al. (2006) which would be at C21r.6 on that scale. However, the nannofossils at the Lutetian GSSP (Molina et al., 2011) suggest that the LAD of D. lodoensis continues upward as rare occurrences that are above FAD of N. fulgens (about 1 myr higher!).
			T Blackites piriformis	47.73	Molina et al. (2011)	B. piriformis has a short range spanning base-Lutetian GSSP (Molina et al., 2011) -- 5 precessions cycles below, to 5 1/2 cycles above = +0.10 to -0.11 myr offsets.
			B Nannotetrina cristata, Nannotetrina spp.	47.73	Molina et al. (2011)	Nannotetrina cristata begins 5 1/2 precession cycles (0.11 myr) above Lutetian GSSP (base of B. inflata) at Lutetian GSSP (Molina et al., 2011). Coincides with LAD of B. piriformis. IODP Expedition 320/321 Scientists (2010) assigned as 48.0 Ma (based on CK'95 scale) citing Agnini et al. (2006) = base of Chron C21n on that scale. This is essentially same as the placement for a general "Nannotetrina" used by Leg 145 with FAD at base of Chron C21n.

GTS2012 CHRONO-STRATIGRAPHY	STANDARD TROPICAL-SUBTROPICAL BIOZONE (BIOCHRON)		BIOHORIZON (DATUM)	AGE (Ma)	CALIBRATION REFERENCES	COMMENTS
STAGE (AGE)	CN-CP Zones: Okada & Bukry (1980); CC Zones:	NN-NP Zones: Martini (1971); NC Zones: Roth (1978,	T = Top/LAD; B = Base/FAD	*This study*		
	CP12b/CP12a	**NP14b/NP14a**	B Blackites inflatus	47.84	Molina et al. (2011)	FAD of B. inflatus = GSSP marker (Feb'11 ratification) = set as 39 precession cycles (used 20kyr cycles here) from base of Chron C21r in GSSP section (Molina et al., 2011) = middle of Chron C21r (prob. C21r.55), which is similar chron-age scaling in Berggren et al. (1995a).
47.84	CP12a	NP14a				
Ypresian			B Blackites piriformis	47.94	Molina et al. (2011)	B. piriformis has a short range spanning base-Lutetian GSSP (Molina et al., 2011) -- 5 precessions cycles below, to 5 1/2 cycles above = +0.10 to -0.11 myr offsets.
	CP12a/CP11	**NP14a/NP13**	B Discoaster sublodoensis (5-rayed)	49.11	IODP Expedition 320/321 Scientists (2010); Agnini et al. (2006); Molina et al. (Episodes, in press as of 2011)	IODP Expedition 320/321 Scientists (2010) assigned as 49.5 Ma (based on CK'95 scale) citing Agnini et al. (2006) => Chron C21n.3 on that scale; used here. This is just slightly younger than the chron-age scaling of Berggren et al. (1995a) of base of C22n. The Lutetian GSSP nannofossils (Molina et al., Episodes, in press as of 2011) indicate the FAD may be as high as C21n.5.
	(CP11/CP10)	**NP13/NP12**	T Tribrachiatus orthostylus	50.50	IODP Expedition 320/321 Scientists (2010); Agnini et al. (2006); Berggren et al. (1995a)	IODP Expedition 320/321 Scientists (2010) assigned as 50.7 Ma (based on CK'95 scale) citing Agnini et al. (2006) => Chron C21r.1 on that CK'95 scale; used here; which is essentially same chron-age scaling as Berggren et al. (1995a). Eocene nanno "Tri. orthostylus" LAD (= base of Zone NP13) has a poorly defined age, and may be time-transgressive? (see discussion in Berggren, Kent, et al., 1985).
	CP10	NP12	B Dictyococcites, Reticulofenestra	50.50	IODP Expedition 320/321 Scientists (2010); Agnini et al. (2006)	IODP Expedition 320/321 Scientists (2010) assigned as 50.7 Ma (based on CK'95 scale), or coeval with LAD of Tri. orthostylus (base of NP13) citing Agnini et al. (2006) = base of Chron C21n on that CK'95 (used here).
	CP10/CP9b	**NP12/NP11**	B Discoaster lodoensis	53.70	IODP Expedition 320/321 Scientists (2010); Agnini et al. (2007)	IODP Expedition 320/321 Scientists (2010) assigned as 53.1 Ma (based on CK'95 scale) citing Agnini et al. (2007) who placed it at Chron C24n.3n.5 (used here). This is about 0.25 older than chron-age scaling of Berggren et al. (1995a) of the middle of Chron C24n.2r.
	(CP9b/CP9a)	**(NP11/NP10)**	T Tribrachiatus contortus	(54.17)	IODP Expedition 320/321 Scientists (2010); Agnini et al. (2006); Berggren et al. (1995a)	DASH -- LAD of T. contortus is not seen very often, therefore IODP Expedition 320/321 Scientists (2010) used FAD of Tri. orthostylus as alternate marker for base-NP11. Eocene Nanno evolution from Tribrachiatus contortus to Tribrachiatus orthostylus was formerly described as an abundance shift, but Aubry has subdivided it into different morphotypes. The event here was "Tribrachiatus contortus (Morphotype B)". Age assignment by Leg 320-321 of 53.5 Ma (on CK'95 scale) is based on Agnini et al. (2007) = Chron C24r.94 on that CK'95 scale (used here). This essentially the same chron-age scaling as Berggren et al. (1995a) of ca. C24r.9.
	CP9a	NP10	B Sphenolithus radians	54.17	IODP Expedition 320/321 Scientists (2010); Agnini et al. (2007)	Age assignment by IODP Expedition 320/321 Scientists (2010) of 53.5 Ma (on CK'95 scale) is based on Agnini et al. (2007) = Chron C24r.94 (used here) = same as NP11 base. This is about 0.3 myr older than assignment in ODP Leg 145 of the base of Chron C24n.
			B Tribrachiatus orthostylus	54.37	IODP Expedition 320/321 Scientists (2010); Agnini et al. (2006); Berggren et al. (1995a)	Alternate marker for base of NP11 (IODP Expedition 320/321 Scientists (2010); Paul Bown to J.Ogg, June 2011). Age assignment by Leg 320-321 of 53.7 Ma (on CK'95 scale) is based on Agnini et al. (2007) = C24r.875 on that CK'95 scale (used here), which is identical to the chron-age scaling of Berggren et al. (1995a).
			T Tribrachiatus bramlettei	54.42	Agnini et al. (2007)	Agnini et al. (2007) assigned as about Chron C24r.86 (used here). This is slightly younger than chron-age scaling of Berggren et al. (1995a) of ca. C24r.8.
			B Tribrachiatus contortus	54.76	Agnini et al. (2007); Berggren et al. (1995a)	Agnini et al. (2007) assigned as Chron C24r.75 (used here); which is similar to chron assignment by Berggren et al. (1995a). Eocene Nanno evolution from Tribrachiatus contortus to Tribrachiatus orthostylus was formerly described as an abundance shift, but M.P. Aubry has subdivided it into different morphotypes. The event of Agnini et al. (2007) was "Tribrachiatus contortus (Morphotype B)".
	CP9a/CP8b		B Discoaster diastypus	54.95	Agnini et al. (2007); Berggren et al. (1995a)	Agnini et al. (2007) assigned as Chron C24r.69 (used here); which is nearly same as relative chron-age scaling of Berggren et al. (1995) of ca. Chron C24r.75.
	CP8b		B Tribrachiatus bramlettei (common)	55.42	Agnini et al. (2007)	Agnini et al. (2007) assigned as Chron C24r.54 (used here); but indicate that lowest rare occurrence is as early as C24r.46 (their preference for the base of NP10).
			T Fasciculithus spp.	55.64	IODP Expedition 320/321 Scientists (2010); Agnini et al. (2007)	IODP Expedition 320/321 Scientists (2010) assigned as 53.1 Ma (based on CK'95 scale) citing Agnini et al. (2007) who placed it at Chron C24r.47 (used here). This is much lower than Leg 145 assignment of ca. Chron C24r.66.
			B Campylosphaera eodela (common)	55.81	IODP Expedition 320/321 Scientists (2010)	Agnini et al'07 who placed the lowest Common Occurrence (LCO) at 0.15 myr above PETM (used here). They used the Common C. eodela for their Base of CP8b; in contrast to the FAD used by IODP Expedition 320/321 Scientists (2010).
		NP10/NP9	B Tribrachiatus bramlettei	55.86	Aubry et al., 2007; P. Bown, pers. comm. 2011	Dashed at ca. 0.1 myr above PETM -- "base NP10 is just slightly above base Eocene." [Paul Bown based on Aubry et al. 2007, to J.Ogg, June'11,].
		NP9	B Rhomboaster spp.	55.96	Agnini et al. (2007); Berggren et al. (1995a)	Agnini et al. (2007) placed the LO of Rhomboaster spp. at the PETM (used here). This is similar to chron-age scaling of Berggren et al. (1995a) of ca. Chron C24r.3
55.96						
Thanetian	**CP8b/CP8a**		B Campylosphaera eodela	56.66	Agnini et al. (2007); Berggren et al. (1995a)	Also called "Cruciplacolithus" eodelus. FAD of C. eodelus closely precedes C-13 excursion in many DSDP sites. Agnini et al. (2007) assign it as 55.5 Ma on the CK'95 scale, implying at Chron C24r.14; which is same chron-age scaling as Berggren et al. (1995a).
	CP8a		T Ericsonia robusta	(56.78)	Raffi et al. (2005)	DASH -- Raffi et al. (2005) summary figure assigned as Chron C24r.1 (used here). Leg 320-321, citing Agnini et al. (2007), placed this LAD slightly lower at top of C25n. In contrast, Berggren et al. (1995a) had assigned as slightly higher (Chron C24r.3).
	CP8a/CP7	**NP9/NP8**	B Discoaster multiradiatus (common)	57.21	IODP Expedition 320/321 Scientists (2010); Agnini et al. (2007)	IODP Expedition 320/321 Scientists (2010) assigned as 56.0 Ma (based on CK'95 scale) citing Agnini et al. (2007) who placed it at Chron C25n.8 (chron placement used here). This is slightly higher than chron-age scaling of Berggren et al. (1995a) of about Chron C25n.4.
	CP7	NP8	B Discoaster multiradiatus (rare)	57.32	Agnini et al. (2007); Berggren et al. (1995a)	Agnini et al. (2007) assigned as Chron C25.6 (used here). This nearly same as the chron-age scaling of Berggren et al. (1995) of ca. Chron C25n.4.
			T Discoaster okadai	57.35	Agnini et al. (2007)	Agnini et al. (2007) assigned as Chron C25n.56.
			B Discoaster okadai	57.47	IODP Expedition 320/321 Scientists (2010); Agnini et al. (2007)	IODP Expedition 320/321 Scientists (2010) assigned as 56.2 Ma (based on CK'95 scale) citing Agnini et al. (2007) implying Chron C25n.34 on that CK'95 scale (chron placement used here). This is higher than chron-age scaling of Berggren et al. (1995) of upper mid-C25r.
			B Discoaster nobilis	(57.50)	IODP Expedition 320/321 Scientists (2010); Agnini et al. (2007)	DOTTED as alternate placement for base of NP8 also. IODP Expedition 320/321 Scientists (2010) assigned as 56.2 Ma (based on CK'95 scale) citing Agnini et al. (2007) implying placement at Chron C25n.3 on that CK'95 scale (used here). This is much higher than chron-age scaling of Berggren et al. (1995) of upper mid-Chron C25r.
		NP8/NP7	B Heliolithus riedelii	(58.70)	Berggren et al. (1995a)	DASHED -- FAD of H. riedelii is unreliable marker (not even tabulated in Agnini et al., 2007). Paleocene Nanno 'H. riedeli' FAD age estimated from it's placement 0.6 m.y. before FAD of D. nobilis in Berggren et al. (1995a) table. Chron-age scaling set as lower Chron C25r (C25r.2).

GTS2012 CHRONO-STRATIGRAPHY	STANDARD TROPICAL-SUBTROPICAL BIOZONE (BIOCHRON)		BIOHORIZON (DATUM)	AGE (Ma)	CALIBRATION REFERENCES	COMMENTS
STAGE (AGE)	CN-CP Zones: Okada & Bukry (1980); CC Zones:	NN-NP Zones: Martini (1971); NC Zones: Roth (1978,	T = Top/LAD; B = Base/FAD	*This study*		
		NP7	T Heliolithus kleinpellii	58.80	Agnini et al. (2007); Berggren et al. (1995a)	Agnini et al. (2007) placed top of Common H. kleinpelli at Chron C25r.12, which is the chron-age scaling as Berggren et al. (1995a).
	CN7/CP6	NP7/NP6	B Discoaster mohleri	58.97	IODP Expedition 320/321 Scientists (2010); Agnini et al. (2006); Berggren et al. (1995a)	IODP Expedition 320/321 Scientists (2010) assigned as 57.6 Ma (based on CK'95 scale) citing Agnini et al. (2007) who placed it at top of Chron C26n (C26n.95 used here) = same as observed at the Thanetian GSSP and same chron-age scaling as Berggren et al. (1995a).
59.24	CP6	NP6				
Selandian	CP6/CP5	NP6/NP5	B Heliolithus kleinpellii	59.54	Schmitz et al. (Episodes, in press as of 2011)	At Zumaya GSSP, this event occurs within uppermost Chron C26r (C26r.9 is used here) = same chron-age scaling as Berggren et al. (1995a) of ca. Chron C26r.85. However, IODP Expedition 320/321 Scientists (2010) assigned as 58.0 Ma (based on CK'95 scale) citing Agnini et al. (2007) who placed it at top of Chron C26r.
	CP5	NP5	B Heliolithus cantabriae	59.60	IODP Expedition 320/321 Scientists (2010); Agnini et al. (2007)	IODP Expedition 320/321 Scientists (2010) assigned as 58.3 Ma (based on CK'95 scale) citing Agnini et al. (2007) who placed it at Chron C29r.88 (used here). This is just slightly higher than chron-age scaling by Berggren et al. (1995a) of ca. Chron C26r.65.
			B Sphenolithus anarrhopus	59.68	IODP Expedition 320/321 Scientists (2010); Agnini et al. (2006); Berggren et al. (1995a)	IODP Expedition 320/321 Scientists (2010) assigned as 58.1 Ma (based on CK'95 scale) citing Agnini et al. (2007) who placed it at uppermost Chron C26r (C26n.95). BUT, this would put it above the NP6 boundary (approx. Chron C26r.9), therefore assigned here as Chron C26r.85 = same chron-age scaling as Berggren et al. (1995).
			T Fasciculithus pileatus	60.73	Berggren et al. (1995a)	Middle of Chron C26r. Cruciplacolithus tenuis LAD was coeval, but is now omitted (may go much higher than in table of Berggren, Kent, et al., 1995).
			B Chiamolithus consuetus	61.03	Berggren et al. (1995a)	Middle of Chron C26r (C26r.4).
	CP5/CP4	NP5/NP4	B Fasciculithus tympaniformis	61.51	Schmitz et al. (Episodes, in press as of 2011)	Cycle-strat of Selandian GSSP (summarized in Schmitz et al., in press as of 2011) indicates that this event is about 0.1 myr above base-Selandian (used here). This is about same chron-age scaling as Berggren et al. (1995) of ca. Chron C26r.4. In contrast, IODP Expedition 320/321 Scientists (2010) assigned as 59.9 Ma (based on CK'95 scale) citing Agnini et al. (2007) who placed it at Chron C29r.8. Coeval with FAD Neochiastozygus perfectus (North Sea).
	CP4	NP4	B Fasciculithus – 2nd radiation	61.59	Bernaola et al., 2009	FAD of F. ulii. Cycle strat indicates it is 0.02 myr above Selandian GSSP (used here) (Bernaola et al., 2009). The Selandian GSSP is defined at a sequence boundary 32 precession cycles above base of C26r. This is much older than chron-age scaling of Berggren et al. (1995) of lower Chron C26r.
61.61						
Danian			B Neochiastozygus perfectus	61.76	Bernaola et al., 2009	This event occurs 75% up between base of Chron C26r and the Selandian GSSP (Bernaola et al., 2009).
			B Sphenolithus primus	61.98	Bernaola et al., 2009; Berggren et al. (1995a)	This event occurs 40% up between base of Chron C26r and the Selandian GSSP (Bernaola et al., 2009), which projects as the same chron-age scaling as in Berggren et al. (1995) of lowermost C26r (C26r.1). Approx. FAD of Neochiastozygus saepes in high latitudes.
			B Chiasmolithus bidens / edentulus	62.07	Bernaola et al., 2009; Berggren et al. (1995a)	Just above base of Chron C26r (C26r.05) at Selandian GSSP at Zumaya; which is the same chron-age scaling as in Berggren et al. (1995a).
			B Fasciculithus – 1st radiation	62.13	Bernaola et al., 2009	Just above base of Chron C26r (C26r.03) at the Selandian GSSP at Zumaya.
	CP4/CP3	NP4/NP3	B Ellipsolithus macellus	63.25	Berggren et al. (1995a)	Paleocene Nanno 'E. macellus' FAD assigned to lower C27r by Berggren et al. (1985b; 1995a) (C27r.25 is used here). Leg 145 places nearly 1.6 myr higher (at of Chron C26r), citing Backman (1986). Berggren et al. (1995a) indicate that this is a solution-susceptible taxon; so the earlier FAD may be more appropriate. Defines NP4 base.
	CP2/CP1b	NP3/NP2	B Chiasmolithus danicus	64.81	Berggren et al. (1995a)	Middle of Chron C28r.
	CP1b/CP1a	NP2/NP1	B Cruciplacolithus tenuis	65.47	Berggren et al. (1995a)	This taxa might be C. intermedius (Paul Bown to J.Ogg, June2011). Defines NP2 base. Berggren et al. (1995a) assign as lower Chron C29n (C29n.3).
	CP1a	NP1	B Cruciplacolithus primus (3.5-5 μm)	65.76	Berggren et al. (1995a)	Medium-sized (3.5-5 μm). Can be used to subdivide Zone NP1. Berggren et al. (1995a) assign as uppermost C29r (C29r.9 used here).
			B Neobiscutum parvulum	(65.9)	Paul Bown (pers. comm., 2011)	DASHED – 1/4th up in zone CP1a.
66.04						
			T Micula murus, other Cret. nannos	66.04	Berggren et al. (1995a)	Base of Cenozoic (Chron C29r.5)
	CP1a/CC26b	NP1/NC23	B Biantholithus sparsus; B Calcisphere FLOOD	66.06	Berggren et al. (1995a)	Base of Cenozoic (Chron C29r.5). Zone NP1 begins with the acme of Thoracosphaera (calc. dinoflagellate), or is the interval from the top of the Cretaceous to the FAD of C. tenuis (base of NP2).
	CC26b	NC23	B Cribrosphaerella daniae (oldest possible)	66.76	Voigt et al. (EPSL, in press as of Dec, 2011); Burnett et al. (1998)	Chron C30n.8. Voigt et al. (submitted to EPSL in 2011) correlates this base of zone NC20d via carbon-13 curve to about Chron C30n.8 at Gubbio (used here). Burnett et al. (1998) schematically shows slightly younger relative to German belemnites as about middle of belemnite B. baltica / danica Zone (NW Europe). Von Salis (1998; SEPM) assigned FAD as synchronous with N. frequens in Tethys.
	CC26b/CC26a		B Micula prinsii	67.30	Huber et al. (2008)	Chron C30n.5. Zone CC26b (or NC20d) begins with FAD of M. prinsii in Chron C30n.5 (Huber et al., 2008; and B. Huber, pers. commun., June'11). Diachronous (p.16, ODP Leg 171B Expl. Notes); with Boreal FAD later than Tethyan (Von Salis, 1998, SEPM chart).
	CC26a/CC25c		B Ceratolithoides kamptneri	(67.84)	Huber et al. (2008)	Chron C30n.2. Assigned as coeval with FAD of N. frequens (dual markers for base of Zone CC26)
			B Nephrolithus frequens	67.84	Huber et al. (2008); Voigt et al. (EPSL, in press as of Dec, 2011)	Chron C30n.2. Zone CC26 begins with FAD of either N. frequens or C. kamptneri (in settngs proximal). Huber et al. (2008) compute as 67.44 Ma relative to GTS04 magnetics, implying Chron C30n.14; which is close to the Chron C30n.25 assignment via carbon-13 correlations by Voigt et al. (submitted to EPSL in 2011); therefore Chron C30n.2 is used here. Burnett et al. (1998) schematically shows as about 40% up in belemnite B. argenta / junior Zone (NW Europe), which is just slightly (0.2 myr) younger; but states that N. frequens has a "well-documented diachronous FO", which is youngest at northern high latitudes.
	(CC25c/CC25b)	(NC23/NC22)	B Micula murus	69.00	Huber et al. (2008)	Chron C31n.3. FAD of M. murus (base of Zone CC25c of Self-Trail, 2001) assigned by Huber et al. (2008) as 68.45 Ma (corrected) relative to GTS04 magnetics (upper Chron C31n) = assigned here as Chron C31n.3. Dashed – diachronous (p.16, ODP Leg 171B Expl. Notes).

GTS2012 CHRONO-STRATIGRAPHY	STANDARD TROPICAL-SUBTROPICAL BIOZONE (BIOCHRON)		BIOHORIZON (DATUM)	AGE (Ma)	CALIBRATION REFERENCES	COMMENTS
STAGE (AGE)	CN-CP Zones: Okada & Bukry (1980); CC Zones:	NN-NP Zones: Martini (1971); NC Zones: Roth (1978,	T = Top/LAD; B = Base/FAD	*This study*		
Maastrichtian	CC25b	NC22	B Cribrocorona gallica	69.00	Von Salis (1998, SEPM chart)	Chron C31n.3. Von Salis (1998, SEPM chart) shows as synchronous FAD with M. murus.
	CC25b/CC25a	**NC22/NC21**	B Lithraphidites quadratus	69.18	Huber et al. (2008); Voigt et al. (EPSL, in press as of Dec, 2011)	Chron C31n.1. Huber et al. (2008) interpolate as 68.70 Ma on GTS04 scale, implying Chron C31n.05. However, Voigt et al. (submitted to EPSL, June 2011) put it slightly higher (about C31n.15); therefore Chron C31n.1 is used here. Burnett et al. (1998) schematically shows as about 80% up in belemnite B. junior / tegulatus Zone (NW Europe), which is about 0.7 myr higher; and Voigt et al. (submitted, 2011) suggests that there is diachroneity.
	CC25a/CC24	NC21	T Reinhardtites levis	70.14	Voigt et al. (EPSL, in press as of Dec, 2011); Watkins (pers. commun., 2011)	Chron C31r.6. D. Watkins (pers. comm., 2011) had calibrated it in USA Western Interior as 1 precession cycle below a 70.06 Ma Ar-Ar age (hence 70.08, ignoring uncertainty on Ar-Ar age). Burnett et al. (1998) schematically shows as about 85% up in belemnite B. sumensis Zone (NW Europe); but Voigt et al (2011, submitted to EPSL) show it as midway in that belemnite Zone, synchronous with Gubbio, and correlate via carbon-13 curve to about Chron C31r.5 (70.3 Ma in GTS2012 magnetics). "Just above the FAD of planktonic foraminifer R. fructosa" (Huber et al., 2008). Bralower et al. (1995) suggest using this LAD to subdivide NC21 into a/b subzones.
			T Tranolithus orionatus	71.01	Huber et al. (2008); Voigt et al. (EPSL, in press as of Dec, 2011)	Chron C31r.2. Voigt et al. (2011, submitted to EPSL) shows this datum (base zone UC18) at about 15% up in belemnite B. sumensis Zone and projects via carbon-13 curve to Chron C31r.2 at Gubbio. Huber et al. (2008) project age as 70.6 Ma relative to GTS04 magnetics = C31r.16 (nearly same). Burnett et al.'98 schematicaly shows this LAD slightly higher at about 1/3rd up in belemnite B. sumensis Zone of NW Europe. Leg 207 uses this LAD for base of Zone CC24, rather than T. phacelosus ("most workers consider it difficult to distinquish T. orionatus from T. phacelosus, and consider the later species as the senior synonym -- J. Self-Trail", cited in Huber et al.,, 2008).
	CC24	**NC21/NC20**	T Uniplanarius trifidus	71.31	Burnett et al. (1998)	Uppermost belemnite Belem. obtusa zone (Burnett et al., 1998) (90% up used here), and correlated via Sr 87/86 to US Western Interior ages (lower ammonite B. grandis zone). Occurs ~0.75 myr above Maastrichtian GSSP at Tercis GSSP (see GTS2004). Called Quadrum trifidum (and T. trifidus) in some publications.
			T Broinsonia parca constricta	72.02	Burnett et al. (1998); Voigt et al. (EPSL, in press as of Dec, 2011)	Base of belemnite B. obtusa Zone (NW Europe) (Burnett et al., 1998). Voigt et al. (EPSL, submitted in 2011) have the same belemnite relationship, but suggest from carbon-13 curves that the datum can be significantly higher at Maastrichtian GSSP at Tercis (equivalent to TOP of obtusus instead) and in the magnetostratigraphic section at Gubbio.
72.05						
			T Monomarginatus quaternarius	(72.18)	Burnett et al. (1998)	About 40% up in belemnite B. pseudobtusa Belemnite Zone (NW Europe) (schematic placement in Burnett et al., 1998). However, James Bergen (pers. Commun.) inverts this sequence, with this M.quarternarius LAD below T. caistorensis LAD.
			T Tortolithus caistorensis	(72.28)	Burnett et al. (1998)	Base of belemnite B. pseudobtusa Belemnite Zone (NW Europe) (schematic placement in Burnett et al., 1998).
	(CC24/CC23)		T Tranolithus phacelosus	72.32	Huber et al. (2008)	Upper part of Chron C32n.2n. Huber et al. (2008) calibrate as 71.80 Ma relative to GTS04 magnetic scale => Chron C32n.2n.8. Commonly used for base of CC24; although Leg 207 used the higher LAD of T. orionatus.
			T Uniplanarius gothicus	(72.54)	Burnett et al. (1998)	Burnett et al.'98 schematically shows as about middle of belemnite B. lanceolata Zone (NW Europe) (schematic placement in Burnett et al., 1998), therefore, below Maastrichtian GSSP. However, this LAD was reported at that GSSP at Tercis at just below LAD of T. trifidus, implying about 0.5 myr above that GSSP.
		NC20	T Heteromarginatus bugensis	73.36	Burnett et al. (1998)	About 90% up in belemnite B. langei Zone of NW Europe (schematic placement in Burnett et al., 1998).
	CC23		T Reinhardtites anthophorus	(74.51)	Burnett et al. (1998)	About 70% up in belemnite B. langei Zone of NW Europe (schematic placement in Burnett et al., 1998).; which is similar to Boreal LAD of E. eximius. EGI database (2005) calibrates Tethyan LAD as 74.0 Ma on GTS04 timescale (therefore as Chron C33n.92; which would be 74.7 Ma in GTS12 magnetic scale). Also called "Rein. elegans".
			T Broinsonia parca	(74.59)	Huber et al. (2008)	Chron C33n.95. Huber et al. (2008) project the age as 73.90 Ma relative to GTS04 magnetic timescale, implying Chron C33n.95. Also called "Aspidolithus parcus" in some schemes. GTS04 Fig 19.1 shows LAD as much lower (mid-Chron C33n); but Von Salis (1998; SEPM) Tethyan/Boreal scales has much higher (4 myr above).
			B Eiffellithus parallelus	(74.74)	Burnett et al. (1998)	Burnett et al. (1998) schematically shows as about 75% up in ammonite E. jenneyi zone (Western Interior). Marker for base of Zone UC15e; but seems to locally occur above zone UC16-base.
			B Reinhardtites levis	(74.84)	Von Salis (1998, SEPM chart)	Middle of ammonite E. jennyi zone; but diachronous. FAD is shown equivalent to mid-jenneyi zone in Von Salis (1998; SEPM; column for Tethyan); but it is ~5 myr below FAD in the Boreal column. Burnett et al. (1998) also implied that its apparent FAD is much later in Boreal realm than in Tethyan/temperate; and schematically shows as about 50% up in belemnite B. minor zone of NW Europe.
	CC23/CC22		T Eiffellithus eximius	(75.93)	Huber et al. (2008); Voigt et al. (EPSL, in press as of Dec, 2011)	Chron C33n.7; but diachronous. Huber et al.(2008) project at 75.31 Ma relative to GTS04 magnetic timescale, hence at Chron C33n.7; which seems consistent with other Tethyan placements. However, in Boreal realm studies, Jackie Lees (Burnett et al.) assigns as upper Upper Campanian (ca. 74 Ma); and this offset is verified in cross-region C-13 studies by Voigt et al. (2011, submitted to EPSL).
	CC22/CC21	**NC20/NC19**	B Uniplanarius trifidum	76.82	Huber et al. (2008)	Chron C33n.55. Called Quadrum trifidum in some studies. Huber et al. (2008) review and select Chron C33n.55 as best calibration (76.29 Ma relative to GTS04 magnetics). Slighly below base of foraminifer G. calcarata zone in Tanzania (Petrizzo et al., 2011); but Bralower et al. (1995) suggested coeval with FAD of foraminifer G. calcarata, which may indicate uncertainty in Campanian inter-correlations.
Campanian	CC21		T Orastrum campanensis	77.08	Burnett et al. (1998)	About 25% up in belemnite B. langei zone of NW Europe (schematic placement in Burnett et al., 1998; used here), or perhaps at about middle of B. compressus Zone of N. America. However, Jim Bergen (pers. comm.) assigned O. campanensis LAD as above E. eximius LAD.
	CC21/CC20	NC19	B Uniplanarius sissinghii	77.61	Huber et al. (2008)	Chron C33n.4. Called Quadrum sissingh in some studies. Huber et al. (2008) cite Erba et al. (1995) as placing at Chron C33n.45; but their 77.10 Ma "relative to GTS04 magnetics" implies a slightly lower Chron C33n.41 (used here).
	CC20		B Prediscosphaera stoveri	78.75	Burnett et al. (1998)	About 50% up in belemnite B. minor zone of NW Europe (schematic placement in Burnett et al., 1998; used here), or about middle of B. stevensoni Zone of N. America. Von Salis (1998; SEPM) projected this FAD as nearly 5 myr older at 4/10 up in ammonite B. sp. (smooth) Zone.
	CC20/CC19	**NC19/NC18**	B Ceratolithoides aculeus	79.00	UCL workshop on Late Cretaceous microfossil stratigraphy (June 2011)	Just above base of foraminifer C. plummera Zone according to M.R. Petrizzo (June 2011 workshop notes); assigned here as 0.2 myr above. This FAD age-projection is very similar from estimate in ODP 171/207, which had coincided with middle of the ammonite B. perplexus (late) zone.
			B Heteromarginatus bugensis	79.01	Burnett et al. (1998)	Middle of ammonite "B. perplexus" Zone of N. America (schematic placement in Burnett et al., 1998).
	CC19		T Lithastrinus grillii	(79.73)	Burnett et al. (1998)	About 85% up in ammonite B. asperiformis Zone of N. America (schematic placement in Burnett et al., 1998; used here; but she notes that LAD is partially paleolatitude dependent). EGI database calibrates as base of Middle Campanian in Tethyan realm; whereas Jackie Lees (Burnett et al.) assigns as upper Lower Campanian.
		NC18	T Marthasterites furcatus	80.97	Von Salis (1998, SEPM chart)	Von Salis (1998; SEPM) had assigned LAD as base of Middle Campanian (base of ammonite B. obtusus zone).
	CC19/CC18		B Misceomarginatus pleniporus	80.97	Burnett et al. (1998)	Base of B. obtusus Zone of N. America (base of Middle Campanian) (schematic placement in Burnett et al., 1998).

GTS2012 CHRONO-STRATIGRAPHY	STANDARD TROPICAL-SUBTROPICAL BIOZONE (BIOCHRON)		BIOHORIZON (DATUM)	AGE (Ma)	CALIBRATION REFERENCES	COMMENTS
STAGE (AGE)	CN-CP Zones: Okada & Bukry (1980); CC Zones:	NN-NP Zones: Martini (1971); NC Zones: Roth (1978,	T = Top/LAD; B = Base/FAD	*This study*		
	CC18		B Ceratolithoides verbeekii	81.21	Burnett et al. (1998)	50% up in ammonite Baculites sp. (smooth) Zone of N. America (schematic placement in Burnett et al., 1998).
			B Bukryaster hayii	81.25	Burnett et al. (1998)	25% up in ammonite Baculites sp. (smooth) Zone of N. America (schematic placement in Burnett et al., 1998).
			B Broinsonia parca constricta	81.38	Burnett et al. (1998)	60% up in ammonite S. hippocrepis III Zone of N. America (schematic placement in Burnett et al., 1998). Called "Aspidolithus parcus constrictus" in some schemes.
	CC18/CC17	**NC18/NC17**	B Broinsonia parca parca	81.43	Burnett et al. (1998)	40% up in ammonite S. hippocrepis III Zone of N. America (schematic placement in Burnett et al., 1998). Called "Aspidolithus parcus parcus" in some schemes. Subzone UC14a of Burnett et al. (1998) is very brief.
	CC17	NC17	B Orastrum campanensis (FCO)	82.76	Burnett et al. (1998)	Base of ammonite S. hippocrepis I Zone of N. America (schematic placement in Burnett et al., 1998). Von Salis (1998; SEPM chart) assigned FAD as just above base of Campanian.
			B Arkhangelskiella cymbiformis	83.20	Burnett et al. (1998)	Arkhangelskiella cymbiformis (sensu Burnett et al., 1998) FAD is middle of ammonite S. leei III Zone of N. America (schematic placement in Burnett et al., 1998). [Rarely used by other workers]
(83.64)						
Santonian	**CC17/CC16**		B Calculites obscurus	84.08	ODP Leg 207 Shipboard Scientific Party (2004)	Near Late/Middle Santonian substage boundary (base of ammonite D. bassleri zone of N. America).
	CC16		T Zeugrhabdotus noeliae	85.28	Burnett et al. (1998)	About 85% up in Middle Santonian (U/M Santonian boundary is LAD of crinoid Marsupites testudinarius) (schematic placement in Burnett et al., 1998). A useful datum in both Boreal and Tethyan realms.
			T Lithastrinus septenarius	(85.56)	Burnett et al. (1998)	Base of crinoid Uintacrinus socialis Zone (in Chalk) (schematic placement in Burnett et al., 1998), set here as base of Middle Santonian. Von Salis (1998; SEPM chart) had assigned this LAD as just below (ca. 0.1 myr) the FAD of L. cayeuxii, but this LAD is significantly above in both GSSP candidates for the base-Santonian (Ten Mile Creek in Texas; and Olazagutia in Spain). Same as L. "moratus" of Varol, 1992.
(86.26)						
Coniacian	**CC16/CC15**	**NC17/NC16**	B Lucianorhabdus cayeuxii	86.38	Gale et al. (2007	Assigned here as ca. 85% up in Magadiceramus crenelatus inoceramid zone, based on Ten Mile Creek (Texas) candidate for base-Santonian GSSP in Gale et al. (2007) (although base of that inoceramid zone is not established). Schematic in Burnett et al. (1998) suggests about base of ammonite P. serratomarginatus ammonite Zone. In contrast, Boreal column in Von Salis (1998; SEPM chart) assigned FAD nearly 2 myr higher – in mid-Santonian !! Assignment as lowermost Santonian or uppermost Coniacian (depending on future GSSP definition, and correlation to W.Interior ammonite usage) is supported by Santonian GSSP studies.
	CC15	NC16	T Quadrum gartneri	(86.44)	Burnett et al. (1998)	Dashed schematically placed midway between FADs of L. grillii and L. cayeuxii (Burnett et al., 1998 who suggeststhis LAD is about 70% up in ammonite G. margae ammonite Zone).
			B Lithastrinus grillii	86.50	Gale et al. (2007	Assigned here as ca. 70% up in Magadiceramus crenelatus inoceramid zone, based on relative spacing of nannofossil events at Ten Mile Creek (Texas) candidate for base-Santonian GSSP in Gale et al. (2007) (although base of that zone is not established) and at Olazagutia, Spain (Lamolda and Santonian Working Group, 2011). Schematic diagram in Burnett et al. (1998) suggests about 40% up in ammonite G. margae Zone.
	(CC15/CC14)		B Reinhardtites anthophorus	(88.14)	Burnett et al. (1998)	Dashed as midway between FADs of M. stauropora and L. grillii (middle of nannofossil zone UC10) following schematic in Burnett et al. (1998). Burnett et al. (1998) notes that "FO of Reinhardtites anthophorus is an unreliable datum. Its inception, which is widely believed to be in the Coniacian, is somewhat obscure."
(89.77)						
Turonian	**CC14/CC13**	**NC16/NC15**	B Micula stauropora	89.77	Burnett et al. (1998)	Base of ammonite P. tridorsatum Zone (schematic placement in Burnett et al., 1998). Called as "Micula decussata" in some studies. FAD placed at Late/Middle Coniacian boundary in ODP Leg 171 and Tethyan column of Von Salis (1998, SEPM chart); but placed at Middle/Early Coniacian in that Boreal column.
	CC13	NC15	B Broinsonia parca expansa	89.95	Burnett et al. (1998)	Base of ammonite F. petrocoriensis Zone (schematic placement in Burnett et al., 1998). Same as "B. lacunosa" and "Aspidolithus parcus expansus" of some authors.
	CC13/CC12	**NC15/NC14**	B Marthasterites furcatus	90.24	Von Salis (1998, SEPM chart)	Von Salis (1998; SEPM chart) had assigned FAD as base of ammonite S. nigricollensis Zone of N. America.
	CC12	NC14	B Zeugrhabdotus biperforatus	90.71	Burnett et al. (1998)	Schematic diagram in Burnett et al. (1998) suggests about 1/6th up in "expanded" S. neptuni Zone (version used in GTS04 included "overlying" Prionocyclus germari Zone").
			B Lithastrinus septenarius (senso lato)	(91.78)	Burnett et al. (1998); UCL workshop on Late Cretaceous microfossil stratigraphy (June 2011)	About 20% up in R. ornatissimum subzone of ammonite C. woollgari Zone (schematic placement in Burnett et al., 1998). FAD is a long transition with different concepts of species (Dave Watkins; at UCL working group, June 2011) => "sensu lato" used here for the earlier FAD. A much higher placement at Middle/Early Coniacian boundary was shown by Von Salis (1998; SEPM chart) in Tethyan region; but this may be another taxonomic concept. Same as L. "moratus" of Varol, 1992.
			B Lucianorhabdus quadrifidus	92.26	Burnett et al. (1998)	About 25% up in R. kallesi subzone of ammonite C. woollgari Zone of N. America (schematic placement in Burnett et al., 1998).
	(CC12/CC11)	**(NC14/NC12-13)**	B Eiffellithus eximius	(92.99)	Burnett et al. (1998)	Assigned as 80% up in ammonite M. nodosoides Zone of N. America (schematic placement in Burnett et al., 1998). Generally considered as coeval with foram H. helvetica LAD; but relationship seems diachronous (above LAD at Demerara Rise Leg 207; within Helvetica zone at Tanzania and S. Atlantic. Dual-markers (Eiffellithus eximius, Lucianorhabdus maleformis) were used for base of zone CC12; but separated in Burnett et al. (1998). Von Salis (1998; SEPM) had assigned both FADs as middle of ammonite P. percarinatus Zone.
	CC11		B Kamptnerius magnificus	(92.99)	Burnett et al. (1998)	Burnett et al. (1998): "FO of consistently-occurring K. magnificus lies areound the FO of E. eximius." (used here). However, FAD ranges as low as lower-Lower Turonian and as high as lower-Middle Turonian (Kanungo, 2005; and pers. commun., Oct 2005 at EGI.)
	(CC11/CC10b)		B Quadrum gartneri	(93.55)	UCL workshop on Late Cretaceous microfossil stratigraphy (June 2011)	"Just above base-Turonian at top of ammonite W. devonense Zone" (Dave Watkins; UCL Late Cretaceous workshop; June 2011) – used here. This is lower (by about 0.5 myr) than schematic diagram in Burnett et al. (1998), which suggests about 1/3rd up in ammonite M. nodosoides Zone; but is similar to placement at base of M. nodosoides Zone [Sudeep Kanungo, UCL thesis, June 2005; and pers. commun., Oct 2005 at EGI]. In contrast, Boreal column in Von Salis (1998; SEPM chart) had placed Q. garneri FAD as much lower – in Cenomanian, projecting as 40% up in ammonite N. juddii Zone.
	CC10b		B Lucianorhabdus maleformis	93.55	Burnett et al. (1998)	Burnett et al. (1998) assigned at same level as Q. garneri FAD.
			B Marthasterites furcatus	(93.64)	Burnett et al. (1998)	Dashed midway between FADs of E. moratus and L. maleformis (middle of Subzone UC6b). Burnett et al. (1998) "Marthasterites furcatus first occurs here (in Zone UC6b) at higher paleolatitudes." but notes that "Personal observations have shown that this event is unreliable, the FO lying stratigraphically much lower at higher latitudes."

GTS2012 CHRONO-STRATIGRAPHY	STANDARD TROPICAL-SUBTROPICAL BIOZONE (BIOCHRON)		BIOHORIZON (DATUM)	AGE (Ma)	CALIBRATION REFERENCES	COMMENTS
STAGE (AGE)	CN-CP Zones: Okada & Bukry (1980); CC Zones:	NN-NP Zones: Martini (1971); NC Zones: Roth (1978,	T = Top/LAD; B = Base/FAD	This study		
		NC12-13	B Eprolithus moratus	(93.73)		Lowermost Turonian; middle of ammonite W. devonense Zone (or, about 1/3rd up in enlarged zone of W. coloradoense) of N. America (schematic placement in Burnett et al., 1998; used here). But also reported from below LAD of H. chiastia, or at base of foram W. archeocretea Foram zone in Pueblo (Colorado), implying uppermost Cenomanian (D. Dave Watkins; UCL Late Cretaceous workshop; June 2011). Same as "E. eptapetalus" of Varol (1992).
93.9						
Cenomanian	CC10b/CC10a		T Helenea chiastia	93.90	UCL workshop on Late Cretaceous microfossil stratigraphy (June 2011)	LAD occurs at base-Cenomanian GSSP level in Pueblo (Colorado) (Dave Watkins at UCL Late Cretaceous workshop; June 2011). Called "Microstaurus chiastius" in some studies. "Rare but cute" (D. Watkins).
	CC10a		B Quadrum intermedium	94.07	Burnett et al. (1998)	This is the 5-element Q. intermedium. Schematic diagram in Burnett et al. (1998) suggests about 50% up in ammonite N. juddii Zone of N. America.
			T Axopodorhabdus albianus	(94.20)	UCL workshop on Late Cretaceous microfossil stratigraphy (June 2011); Burnett et al. (1998)	LAD of A. albianus is above LAD of foraminifer Cushmani (M.R. Petrizzo, UCL workshop on Late Cretaceous microfossil stratigraphy, June 2011) => put arbitrary 0.1 myr above that Foram LAD. This is similar to placement in schematic diagram of Burnett et al. (1998) at about 85% up in ammonite M. geslinianum Zone; but reversed the "usual observed" sequence of LAD of A. albianus occurring after the LAD of R. asper (which is the basis of the nannofossil NK zones in this interval).
		(NC12 - 13/NC11)	T Rhagodiscus asper	(94.30)	Burnett et al. (1998)	Base of ammonite B. clydense Zone of N. America (schematic placement in Burnett et al., 1998). LAD of P. asper (R. asper) is used by Bralower et al (1995) to define base of zone NC13, and LAD of A. albianus for base of NC12, with ca. 1 myr separation. However, J. Bergen (pers. comm., 2007) assigned this LAD in lower Turonian (above LAD of M. chiastius).
		(NC11/NC10b)	T Lithraphidites acutus	(94.39)	Burnett et al. (1998)	About 60% up in ammonite M. geslinianum Zone of N. America (schematic placement in Burnett et al., 1998). Rare, and difficult to use (comments at UCL workshop on Late Cretaceous microfossil stratigraphy, June 2011).
		NC10b	T Cretarabdus striatus	94.44	Burnett et al. (1998)	About 45% up in ammonite M. geslinianum Zone of N. America (schematic placement in Burnett et al., 1998). Called "loriei" in some studies.
			T Cylindralithus biarcus	94.54	Burnett et al. (1998)	About 10% up in ammonite M. geslinianum Zone of N. America (schematic placement in Burnett et al., 1998). ODP Leg 207 placed just 0.1 myr below L. acutus LAD.
			T Corollithion kennedyi	94.64	Burnett et al. (1998)	90% up in ammonite C. guerangeri Zone of N. America (schematic placement in Burnett et al., 1998).
			T Gartnerago nanum	94.79	Burnett et al. (1998)	2/3rds up in ammonite C. guerangeri Zone of N. America (schematic placement in Burnett et al., 1998).
			T Staurolithites gausorhethium	95.02	Burnett et al. (1998)	1/3rd up in ammonite C. guerangeri Zone of N. America (schematic placement in Burnett et al., 1998).
			T Gartnerago theta	95.93	Burnett et al. (1998)	25% up in ammonite A. rhotomagense Zone of N. America (schematic placement in Burnett et al., 1998).
	CC10a/CC9b		B Lithraphidites acutus, Microrhabdulus decoratus	96.16	UCL workshop on Late Cretaceous microfossil stratigraphy (June 2011)	FAD of L. acutus occurs near Thatcher Limestone in Western Interior (Andy Gale at UCL workshop on Late Cretaceous microfossil stratigraphy, June 2011) => In Conlinoceras tarrantense (=Conlinoceras gilberti) ammonite zone. Assigned here as middle of that Zone. Base of CC10 has dual-markers (M. decoratus; L. acutus).
	CC9b		B Cylindralithus sculptus	97.31	Burnett et al. (1998)	50% up in ammonite M. dixoni Zone of N. America (schematic placement in Burnett et al., 1998).
			T Zeugrhabdotus xenotus	(97.73)	Burnett et al. (1998)	30% up in ammonite M. dixoni Zone of N. America (schematic placement in Burnett et al., 1998). "Shelf-preferring form, therefore subzone [IC2b; LAD of Z. xenotus to FAD of C. sculptus] might not be determinable in oceanic sequences".
			B Gartnerago segmentatum	98.26	Burnett et al. (1998)	Burnett et al. (1998) assigns as basal (but not base) of M. dixoni ammonite zone, so set as 5% up here. Called "G. obliquum" by some authors; but Burnett et al. (1998) indicates that true "obliquum" comes in within mid-UC2c, or about 1 myr higher. However, it is mainly a cold-realm taxa.
			T Gartnerago chiasta	99.94	Burnett et al. (1998)	70% up in ammonite N. carcitanense Zone of N. America (schematic placement in Burnett et al., 1998). Zone UC1c is very brief.
			T Watznaureria britannica	100.03	Burnett et al. (1998)	50% up in ammonite N. carcitanense Zone of N. America (schematic placement in Burnett et al., 1998).
		NC10b/NC10a	B Corollithion kennedyi	100.45	UCL workshop on Late Cretaceous microfossil stratigraphy (June 2011)	ODP Leg 171 recorded this FAD as 2 precession cycles (0.05 myr) above FAD of Rotalipora globotruncanoides (marker for the base of Cenomanian) (Dave Watkins at UCL workshop on Late Cretaceous microfossil stratigraphy, June 2011); used here. Burnett et al. (1998) assigned much higher -- 40% up in N. carcitanense subzone.
100.5						
Albian	CC9b/CC9a	NC10a	T Hayesites albiensis	100.84	Sudeep Kanungo, UCL thesis, June 2005	60% up in former ammonite S. dispar Zone (now interval from base of M. rostratum to top of M. perinflatum zones) [Sudeep Kanungo, UCL thesis, June 2005; and pers. commun., Oct 2005 at EGI]. Bown et al. (1998) schematically shows as 30% up in S. dispar Zone. Von Salis (1998; SEPM chart) assigned as middle of former M. perinflatum subzone.
	CC9a/CC8b	NC10a/NC9b	B Eiffellithus turriseiffelii	103.13	Gale et al. (2011)	2 FAD datums are used, depending on morphotype. FAD of large is base of M. fallax ammonite zone at Col de Palluel (Gale et al., 2011) (used here). FAD of small form is earlier. Bown et al. (1998) schematically shows as 90% up in M. inflatum Zone, which is same numerical-age level. Bralower et al. (1997) had put slightly lower -- coeval with FAD of foraminifer R. ticinensis. Von Salis (1998; SEPM) assigned Boreal FAD as 7/10 up in E. loricatus Z., E. meandrinus s.z. or Tropical as just below (0.15 myr) the base of H. orbignyi subzone.
	CC8b	NC9b/NC9a	B Eiffellithus monechiae	107.59	Bralower et al. (1997)	Coeval with FAD of foraminifer B. breggiensis by Bralower et al. (1997), but see note on calibration of that datum. Called E. cf. E. eximius in Bralower et al. (1997) chart.
		NC9a/NC8c	B Axopodorhabdus albianus	109.94	Grippo et al. (2004)	Base of nannofossil *A. albianus* (called "P." albianus in Grippo et al., 2004) is 23.3 long-eccentricity cycles below top of Albian (9.5 myr) => 109.1 Ma. This projects to base of ammonite P. steinmanni subzone.
	CC8b/CC8a	NC8c/NC8b	B Tranolithus orionatus (=T. phacelosus)	110.73	Bralower et al. (1997); Grippo et al. (2004)	25% up in foraminifer T. primula Zone. Bralower et al. (1997) charts schematically displayed this FAD in middle of T. primula Zone, but this would be inconsistent with cycle-scaling of overlying FAD of A. albianus (base of next higher NC zone) by Grippo et al. (2004), therefore placed at 25% up here. Taxa was "phacelosus" in older literature (e.g., von Salis, 1998).
	CC8a	NC8b	B Cribrosphaerella ehrenbergii	111.3	Von Salis (1998, SEPM chart)	70% up in Nanno Zone NC8b. Von Salis (1998; SEPM) had assigned FAD as middle of ammonite L. tardefurcata Zone.
		NC8b/NC8a	B Hayesites albiensis	112.65	Grippo et al. (2004); Huang et al. (2010d)	FAD is 30 cycles of 405 kyr below base of Cenomanian in Piobbico core (Huang et al., 2010d). Bralower et al. (1997) placed in middle of foraminifer H. planispira Zone (LAD of T. bejaouensis to FAD of T. primula).
(112.95)						
	(CC8a/CC7b)	(NC8a/NC7b)	B Prediscosphaera columnata (subcircular)	(112.95)	Mutterlose et al. (2003); Grippo et al. (2004); Huang et al. (2010d)	DASH -- transitional trend in morphology; and subcircular form used here. Base of this subcircular P. columnata is 30.75 long-eccentricity cycles below top of Albian (12.45 myr) (Grippo et al., 2004; Huang et al., 2010) and is used in GTS2012 as working definition for base of Albian. Coincides with base of ammonite L. schrammeni subzone at Vohrum, Germany (Mutterlose et al., 2003).

GTS2012 CHRONO-STRATIGRAPHY	STANDARD TROPICAL-SUBTROPICAL BIOZONE (BIOCHRON)		BIOHORIZON (DATUM)	AGE (Ma)	CALIBRATION REFERENCES	COMMENTS
STAGE (AGE)	CN-CP Zones: Okada & Bukry (1980); CC Zones:	NN-NP Zones: Martini (1971); NC Zones: Roth (1978,	T = Top/LAD; B = Base/FAD	*This study*		
			T Farhania varolii	113.45	J. Bergen (pers. comm., 2005)	About middle of ammonite H. jacobi Zone. 65% up between Niveau Pacquer and Niveau Jacobi (J. Bergen, pers. comm., 2005). Called Eprolithus varolii in some studies.
			B Acaenolithus viriosus	115.64	J. Bergen (pers. comm., 2005)	About base of ammonite H. jacobi Zone. 30% up between Niveau Pacquer and Niveau Jacobi (J. Bergen, pers. comm., 2005).
			B Nannoconus regularis	(116.23)	Von Salis (1998, SEPM chart)	Dashed – von Salis (1998; SEPM) had assigned FAD as top of ammonite N. nolani Zone, D. nodosocostatum subzone; but no details given.
		NC7b	T Nannoconus truitti (acme)	116.83	Von Salis (1998, SEPM chart)	Dashed – von Salis (1998; SEPM) had assigned FAD as base of ammonite N. nolani Zone, D. nodosocostatum subzone; but no details given. Uncertain calibration (J. Bergen, pers. comm., 2005).
			T Prediscosphaera spinosa	118.33	J. Bergen (pers. comm., 2005)	10% up in Nanno Zone NC7c (J. Bergen, pers. comm., 2005)
			B Rhagodiscus hamptonii	118.93	Bralower et al. (1997); J. Bergen (pers. comm., 2005)	Coeval with LAD of Foram G. algeriana (base of foraminifer H. trocoidea Foram Zone, which is also called H. gorbachikae Zone) (Bralower et al. (1997 – NOTE: J. Bergen (pers. comm., 2005) clarifies that Bralower et al. called this base-NC7c marker as Rhagodiscus achylostaurion. But true R. achyylostaurion has its FAD in Barremian.] Was a potential datum for base of a "NC7c" zone; but not that widespread (only SE France and India – Bown (in Kennedy et al., 2000a)).
			B Nannoconus truitti (acme)	121.34	Von Salis (1998, SEPM chart)	Von Salis (1998, SEPM chart) placed at ca. 1/3rd up in ammonite C. martinoides Zone (1/3rd up in E. subnodosocostatum Zone). Dashed as uncertain calibration (J. Bergen, pers. comm., 2005)
Aptian	CC7b	NC7b/NC7a	T Micrantholithus hoschulzii	122.25	Sikora and Bergen (2004); Bralower et al. (1997)	90% of foraminifer G. ferrolensis Zone (Sikora and Bergen, 2004), which is 90% between LAD of L. cabri and FAD of G. algeriana. Similar to placement by Bralower et al. (1997).
		NC7a	B Radiolithus planus	122.98	Erba (2004)	Base of foraminifer G. ferreolensis zone (Erba, 2004).
		NC7a/NC6b	B Eprolithus floralis	123.88	Erba (2004)	This marks the ""Nannoconid Return event"" (Erba, 2004) at about 2/3rds up in foraminifer L. cabri Zone (in broad sense). Important age control for scaling Early Aptian.
			T Nannoconus steinmanii	124.61	J. Bergen (pers. comm., 2005)	20% up in ammonite D. grandis subzone, which would be 60% up in D. deshayesi Zone (J. Bergen, pers. comm., 2005)
			B Braarudosphaera africana	124.98	Von Salis (1998, SEPM chart)	40% up in ammonite D. deshayesi zone (Von Salis, 1998; SEPM).
		NC6b	B Rhagodiscus angustus	(125.71)	Erba (2004)	Dashed at base of foraminifer L. cabri zone (Erba, 2004; although her diagram implies this L. cabri zone begins with rare occurences; used here). Also reported as at at base of anoxic interval near base of foraminifer "G. blowii Zone" (a different version of that zone) in SE France (Renard et al., 2005); at 56.5 cycles of 405kyr below base-Cenomanian at Piobbico core above top of Selli organic-rich OAE1a zone (Huang et al., 2010, but this seems poorly determined). In contrast, Von Salis (1998; SEPM) had assigned FAD as base of ammonite D. weissi Zone
			T Retecapsa angustiforata	125.83	J. Bergen (pers. comm., 2005)	60% up in ammonite D. weissi Zone (J. Bergen, pers. comm., 2005; based on La Bedoule-Cassis section)
		NC6b/NC6a	T Conusphaera rothii	125.95	J. Bergen (pers. comm., 2005)	20% up in ammonite D. weissi (J. Bergen, pers. comm., 2005; based on La Bedoule-Cassis section).
			B Stoverius achylosum	126.04	Bralower et al. (1997)	80% up in Nanno Zone NC6a (Bralower et al., 1997), or in upper-third of OAE1a anoxic event.
	CC7b/CC7a		B Rhagodiscus gallagheri	126.18	J. Bergen (pers. comm., 2005)	40% up in ammonite D. tuarkyricus (D. oglanlensis) Zone (J. Bergen, pers. comm., 2005; based on La Bedoule-Cassis section).
126.3		NC6a				
	CC7a		T NANNOCONID CRISIS	(126.28)		Also called Rucinolithus irregularis. 95% up in Chron M1n. SEPM'98 had assigned as just below (0.1 myr) the base of Aptian.
	CC7a/CC6	NC6a/NC5e	B Hayesites irregularis	126.40	Von Salis (1998, SEPM chart)	95% up in Chron M1n based on Von Salis (1998; SEPM) assignment as just below (ca. 0.1 myr) the base of Aptian. Called Rucinolithus irregularis in some studies.
Barremian	CC6	NC5e/NC5d	B Flabellites oblongus (consistent)	127.31	Bralower (1987); J. Bergen (pers. comm., 2005)	50% up in Chron M1n (Bralower, 1987). Von Salis (1998; SEPM) had assigned FAD as 1/10 up in ammonite A. vandenheckii Zone.
		NC5d	B Rhadodiscus achlyostaurion	127.66	Bergen (1994) and J. Bergen (pers. comm., 2005)	Middle of ammonite H. feraudianus Zone (Bergen (1994) and J. Bergen, 2005, based on SE France sections). Von Salis (1998; SEPM) had assigned as base of I. giraudi Z.
	CC6/CC5b	NC5d/NC5c	T Calcicalathina oblongata	130.08	Bralower (1987)	35% up in Chron M3r (Bralower, 1987; and Bergen (1994) and J. Bergen, pers. comm., 2005, based on projecting SE France sections).
130.77	CC5b	NC5c				
	(CC5b/CC5a)	(NC5c/NC5b)	T Lithraphidites bollii	(131.51)	Bralower (1987)	Dashed at ca. 75% of Chron M5r (Bralower, 1987; and Bergen (1994) and J. Bergen, pers. comm., 2005, based on projecting SE France sections). Von Salis (1998; SEPM) had assigned LAD as 6/10 up in P. angulicostata auct. Z.
	CC5a	NC5b/NC5a	B Rucinolithus terebrodentarius, R. windleyae	131.94	Bergen (1994); Von Salis (1998, SEPM chart); J. Bergen (pers. comm., 2005)	Dashed as upper (80% up) Chron M7n as average between Von Salis (1998; SEPM chart) and Bergen (1994, pers. comm., 2005), based on projecting the FAD near top of ammonite B. balearis Zone in SE France sections). The two taxa are combined by some specialists. Bralower subdivides NC5 with this datum.
	(CC5a/CC4b)	NC5a	T Speetonia colligata	(132.54)	Von Salis (1998, SEPM chart)	Dashed at ca. 25% up in ammonite P. ligatus Zone (slightly shifted down from Von Salis (1998; SEPM chart) to avoid overlap with subzone b.
		(NC5a/NC4)	T Cruciellipsis cuvillieri	(132.87)	Von Salis (1998, SEPM chart)	Dashed at ca. 55% up in ammonite S. sayni Zone (Von Salis, 1998; SEPM chart).
		(NC4/NC3)	T Tubodiscus verenae	(132.93)	Bergen (1994); Von Salis (1998, SEPM chart); J. Bergen (pers. comm., 2005)	Mid-Hauterivian (dashed at 40% of ammonite S. sayni Zone to avoid overlap with overlying zone NC5) (modified from Von Salis, 1998; SEPM chart). Defines base of NC4, but difficult to pinpoint LAD (Bergen (1994) and J. Bergen, pers. comm., 2005).
Hauterivian	CC4b		B Zeugrhabdotus scutula	132.99	Bergen (1994); Von Salis (1998, SEPM chart); J. Bergen (pers. comm., 2005)	25% up in ammonite S. sayni Zone (Von Salis, 1998, SEPM chart; and Bergen (1994) and J. Bergen, pers. comm., 2005, based on SE France sections).

GTS2012 CHRONO-STRATIGRAPHY	STANDARD TROPICAL-SUBTROPICAL BIOZONE (BIOCHRON)		BIOHORIZON (DATUM)	AGE (Ma)	CALIBRATION REFERENCES	COMMENTS
STAGE (AGE)	CN-CP Zones: Okada & Bukry (1980); CC Zones:	NN-NP Zones: Martini (1971); NC Zones: Roth (1978,	T = Top/LAD; B = Base/FAD	This study		
			T Eiffellithus striatus	133.09	Bergen (1994); Von Salis (1998, SEPM chart); J. Bergen (pers. comm., 2005)	Base of ammonite S. sayni Zone (Von Salis, 1998, SEPM chart; and Bergen (1994) and J. Bergen, pers. comm., 2005, based on SE France sections).
	CC4b/CC4a		B Lithraphidites bollii	133.53	Bergen (1994); Von Salis (1998, SEPM chart); J. Bergen (pers. comm., 2005)	25% up in ammonite C. loryi Zone (Von Salis, 1998, SEPM chart; and Bergen (1994) and J. Bergen, pers. comm., 2005, based on SE France sections). Bralower (1987) uses this to subdivide NC4, but it seems that base of NC4 (LAD of T. verenae) can be above this datum.
			T Eiffellithus windii	133.74	Bergen (1994); Von Salis (1998, SEPM chart); J. Bergen (pers. comm., 2005)	Middle of ammonite A. radiatus Zone (Von Salis, 1998, SEPM chart; and Bergen (1994) and J. Bergen, pers. comm., 2005, based on SE France sections).
133.88	CC4a					
			B Nannoconus bucheri	(133.98)	Bergen (1994); Von Salis (1998, SEPM chart); J. Bergen (pers. comm., 2005)	Just below (0.1 myr) the base of Hauterivian (Von Salis, 1998, SEPM chart; and Bergen (1994) and J. Bergen, pers. comm., 2005, based on SE France sections).
	CC4a/CC3b	NC3	B Eiffellithus striatus	134.98	Bergen (1994); Von Salis (1998, SEPM chart); J. Bergen (pers. comm., 2005)	Base of ammonite C. furcillata subzone in revised Tethyan ammonite zonation (was middle of ammonite H. trinodosum s.z. of N. pachydicranus Zone in older scheme) (Von Salis, 1998, SEPM chart; and Bergen (1994) and J. Bergen, pers. comm., 2005, based on SE France sections).
			T Rucinolithus wisei	(136.01)	Bergen (1994) and J. Bergen (pers. comm., 2005)	Dashed at 40% up in ammonite S. verrucosum Zone (older version, which included the N. peregrinus subzone) (Bergen (1994) and J. Bergen, pers. comm., 2005, based on SE France sections).
Valanginian	CC3b		B Zeugrhabdotus trivectis	136.92	Bergen (1994) and J. Bergen (pers. comm., 2005)	60% up in ammonite B. campylotoxus Zone (Bergen (1994) and J. Bergen, pers. comm., 2005, based on SE France sections).
			T Eiffellithus primus, C. deflanderi	137.05	Bergen (1994) and J. Bergen (pers. comm., 2005)	Middle of ammonite B. campylotoxus Zone (Bergen (1994) and J. Bergen, pers. comm., 2005, based on SE France sections).
	CC3b/CC3a		B Eiffellithus windii	137.55	Bergen (1994) and J. Bergen (pers. comm., 2005)	10% up in ammonite B. campylotoxus Zone (Bergen (1994) and J. Bergen, pers. comm., 2005, based on SE France sections). Von Salis (1998; SEPM) had assigned FAD as 8/10 up in T. pertransiens Zone.
	CC3a		B Micrantholithus speetonensis, R. dekaenelii	137.68	Bergen (1994) and J. Bergen (pers. comm., 2005)	Base of ammonite B. campylotoxus Zone (Bergen (1994) and J. Bergen, pers. comm., 2005, based on SE France sections).
	CC3a/CC2	NC3/NC2	B Calcicalathina oblongata	139.45	Bralower (1987); Bergen (1994) and J. Bergen (pers. comm., 2005)	20% up in Chron M14r (Bralower, 1987). Von Salis (1998; SEPM chart) had assigned FAD as 7/10 up in ammonite T. otopeta Zone, which is similar.
139.4						
			B Rucinolithus. wisei	140.12	Bergen (1994) and J. Bergen (pers. comm., 2005)	75% up in ammonite T. alpillensis Zone (J. Bergen, 1994). Von Salis (1998; SEPM chart) had assigned FAD as base of B. picteti subzone.
			B Tubodiscus verenae	(140.30)	Bergen (1994) and J. Bergen (pers. comm., 2005)	Dashed at middle of ammonite T. alpillensis Zone, poorly constrained (Bergen (1994) and J. Bergen, pers. comm., 2005, based on SE France sections).
	CC2	NC2	B Percivalia fenestrata	140.66	Bergen (1994); Von Salis (1998, SEPM chart); J. Bergen (pers. comm., 2005)	Base of ammonite T. alpillensis Zone (Bergen (1994); Von Salis (1998; SEPM chart) and J. Bergen, pers. comm., 2005, based on SE France sections). This FAD is used to subdivide NC2 (called NK2) to NC2a/2b in scheme of Bralower et al. (1989).
Berriasian			B Rhaqodiscus nebulosus, Diadorhombus rectus	143.14	Bralower et al. (1989); Bergen (1994)	Rh. nebulosus FAD is Middle of ammonite S. subalpina s.z. (Bergen, 1994; pers. comm., 2005). Diado. rectus FAD is 60% up in Chron M17r (Bralower, 1993), which is approximately synchronous.
			B Assipetra infracretacea	143.86	Bralower et al. (1989)	10% up in Chron M17r (Bralower et al., 1989). Von Salis (1998; SEPM) assigned FAD as 1/4 up in ammonite B. jacobi Zone, P. grandis s.z.
			B Markalius circumradiatus	144.00	Bralower et al. (1989)	Base of Chron M17r (Bralower et al., 1989). Von Salis (1998; SEPM) had assigned FAD just above (0.1 myr) the base of B. jacobi Z., P. grandis s.z.
	CC2/CC1	NC2/NC1	B Retecapsa angustiforata	144.93	Bergen (1994) and J. Bergen (pers. comm., 2005)	Base of ammonte P. grandis subzone (J. Bergen, 1994; and pers. comm., 2005). Called C. angustiforatus in some studies.
(145.01)						
Tithonian	CC1	NC1	B Nannoconus kamtneri & N. steinmannii	145.44	Bergen (1994); Von Salis (1998, SEPM chart); J. Bergen (pers. comm., 2005)	50% up in ammonite B. jacobi subzone (J. Bergen, 1994; Von Salis, 1998, SEPM chart; and J. Bergen, pers. comm., 2005). Bralower (1987) places at 30% up in Chron M17r, which would be much higher. However, Bergen's version might be the N. steinmanii minor of Bralower (1987) (J. Bergen, pers. comm., 2005). Von Salis (1998; SEPM) had assigned FAD as middle of B. jacobi Z., P. grandis s.z.
	base CC1	base NC1	T Haqius noelae	145.44	Bergen (1994) and J. Bergen (pers. comm., 2005)	coeval with FAD of N. steinmannii at 50% up in ammonite B. jacobi subzone (J. Bergen, 1994; and pers. comm., 2005). In contrast, Bralower (1987) placed much higher as 10% up in Chron M17r. Von Salis (1998; SEPM) had assigned FAD as 1/4 up in ammonite P. grandis subzone.

REFERENCES

Agnini, C., Fornaciari, E., Raffi, I., Rio, D., Röhl, U., Westerhold, T., 2007. High-resolution nannofossil biochronology of middle Paleocene to early Eocene at ODP Site 1262: implications for calcareous nannoplankton evolution. Marine Micropaleontology 64, 215–248. doi:10.1016/j.marmicro.2007.05.003.

Agnini, C., Muttoni, G., Kent, D.V., Rio, D., 2006. Eocene biostratigraphy and magnetic stratigraphy from Possagno, Italy: the calcareous nannofossil response to climate variability. Earth and Planet Science Letters 241, 815–830. doi:10.1016/j.epsl.2005.11.005.

Anthonissen, E.D., 2009b. Calibrating the Neogene of the Nordic Atlantic: an integrated microfossil biostratigraphy linking the geologic time scale with regional geology. Doctoral dissertation, University of Oslo, Norway, Nr. 829.

Anthonissen, E.D., 2008. Late Pliocene and Pleistocene biostratigraphy of the Nordic Atlantic region. Newsletters on Stratigraphy 43, 33–48.

Aubry, M.-P., 1993. Neogene allostratigraphy and depositional history of the DeSoto Canyon area, northern Gulf of Mexico. Micropaleontology 39, 327–366.

Backman, J., 1986. Late Paleocene to middle Eocene calcareous nannofossil biochronology from the Shatsky Rise, Walvis Ridge and Italy. Palaeogeography Palaeoclimatology Palaeoecology 57, 43–59. doi:10.1016/0031-0182(86)90005-2.

Backman, J., 1987. Quantitative calcareous nannofossil biochronology of middle eocene through early oligocene sediment from DSDP Sites 522 and 523. Abhandlungen der Geologischen Bundesanstalt (Austria) 39, 21–31.

Backman, J., Pestiaux, P., 1987. Pliocene Discoaster abundance variations. Deep Sea Drilling Project Site 606: biochronology and paleoenvironmental implications. In: Ruddiman, W.F., Kidd, R.B., et al. (Eds.), Initial Reports of the Deep Sea Drilling Project, 94, pp. 903–910.

Bellier, J.-P., Moullade, M., Huber, B.T., 2000. Mid-Cretaceous Planktonic Foraminifers from Blake Nose: revised biostratigraphic framework. In: Kroon, D., Norris, R.D., Klaus, A. (Eds.), Proceedings of the Ocean Drilling Program Scientific Results, 171B, Online; http://www-odp.tamu.edu/publications/171B-SR/chap-03/chap 2007/10/5. at 101.92.

Berggren, W.A., 1993. Neogene planktonic foraminiferal biostratigraphy of eastern Jamaica. Geological Society of America Memoir 182, 179–217.

Berggren, W.A., 1992. Neogene planktonic foraminifer magnetobiostratigraphy of the southern Kerguelen Plateau (Sites 747, 748, and 751). In: Wise Jr., S.W., Schlich, R., et al. (Eds.), Proceedings of the Ocean Drilling Program. Scientific Results, 120, pp. 631–647.

Berggren, W.A., Aubry, M.P., Hamilton, N., 1983. Neogene magnetobiostratigraphy of deep sea drilling project site 516 (Rio Grande Rise, South Atlantic). In: Barker, P., et al. (Eds.), Initial Reports of the Deep Sea Drilling Project, 72, pp. 675–713.

Berggren, W.A., Kent, D.V., Van Couvering, J.A., 1985a. The Neogene, Part 2. Neogene geochronology and chronostratigraphy. In: Snelling, N.J. (Ed.), The Chronology of the Geological Record. Geological Society of London Memoir, 10, pp. 211–260.

Berggren, W.A., Kent, D.V., Flynn, J.J., 1985b. Jurassic to Paleogene, Part 2. Paleogene geochronology and chronostratigraphy. In: Snelling, N.J. (Ed.), The Chronology of the Geological Record. Geological Society of London Memoir, 10, pp. 141–195.

Blaj, T., Backman, J., Raffi, I., 2009. Late eocene to oligocene preservation history and biochronology of calcareous nannofossils from paleo-equatorial Pacific Ocean sediments. Rivista Italiana Paleontologia e Stratigrafia 115, 67–84.

Bralower, T.J., 1987. Valanginian to Aptian calcareous nannofossil stratigraphy and correlation with the upper M-sequence magnetic anomalies. Marine Micropaleontology 11, 293–310.

Bylinskaya, M.E., 2005. Range and stratigraphic significance of the *Globorotalia crassaformis* plexus. Journal of Iberian Geology 31, 51–63.

Chaisson, W.P., Leckie, R.M., 1993. High-resolution Neogene planktonic foraminifer biostratigraphy of Site 806, Ontong Java Plateau (western Equatorial Pacific). In: Berger, W.H., Kroenke, L.W., Mayer, L.A., et al. (Eds.), Proceedings of the Ocean Drilling Program. Scientific Results, 130, pp. 137–178.

Chaisson, W.P., Pearson, P.N., 1997. Planktonic foraminifer biostratigraphy at Site 925: Middle Miocene–Pleistocene. In: Shackleton, N.J., Curry, W.B., Richter, C., Bralower, T.J. (Eds.), Proceedings of the Ocean Drilling Program. Scientific Results, 154, pp. 3–31.

Channell, J.E.T., Rio, D., Sprovieri, R., Glaçon, G., 1990. Biomagnetostratigraphic correlations from Leg 107 in the Tyrrhenian Sea. In: Kasten, K.A., Mascle, J., et al. (Eds.), Proceedings of the Ocean Drilling Program. Scientific Results, 107, pp. 669–682.

Chaproniere, G.C.H., Styzen, M.J., Sager, W.W., Nishi, H., Quinterno, P.J., Abrahamsen, N., 1994. Late neogene biostratigraphic and magnetostratigraphic synthesis. Proceedings of the Ocean Drilling Program. Scientific Results 135, 857–877.

Clement, B.M., Robinson, R., 1986. The magnetostratigraphy of Leg 94 sediments. In: Ruddimnan, W.R., Kidd, R.B., Thomas, E., et al. (Eds.), Initial Results of the Deep Sea Drilling Program, 94, pp. 635–650.

Erba, E., Premoli Silva, I., Watkins, D.K., 1995. Cretaceous calcareous plankton biostratigraphy of sites 872 through 879. In: Haggerty, J.A., Premoli Silva, I., Rack, F., McNutt, M.K. (Eds.), Proceedings of the Ocean Drilling Program. Scientific Results, 144, pp. 157–169.

Gartner, S., 1992. Miocene nannofossil chronology in the North Atlantic DSDP Site 608. Marine Micropaleontology 18, 307–331.

Glaçon, G., et al., 1990. Planktonic foraminiferal events and stable isotope records in the upper Miocene, Site 654. In: Kastens, K., Mascle, J., et al. (Eds.), Proceedings of the Ocean Drilling Program. Scientific Results, 107, pp. 415–427.

Groot, J.J., de Jonge, R.B.G., Langereis, C.G., ten Kate, G.H.Z., Smit, J., 1989. Magnetostratigraphy of the Cretaceous-Tertiary boundary at Agost (Spain). Earth and Planetary Science Letters 94, 385–397.

Hancock, H.J.L., Chaproniere, G.C., Dickens, G.R., Henderson, R.A., 2002. Early Paleocene planktic foraminiferal and carbon isotope stratigraphy, Hole 762C, Exmouth Plateau, northwest Australian margin. Journal of Micropalaeontology 21, 29–42.

Hays, J.D., Saito, T., Opdyke, N.D., Burckle, L.H., 1969. Pliocene-Pleistocene sediments of the equatorial Pacific: their paleomagnetic biostratigraphic and climatic record. Geological Society of America Bulletin 80, 1481–1514.

Hilgen, F.J., Krijgsman, W., 1999. Cyclostratigraphy and astrochronology of the Tripoli diatomite formation (pre-evaporite Messinian, Sicily, Italy). Terra Nova 11, 16–22.

Hodell, D.A., Curtis, J.H., Sierro, F.J., Raymo, M.E., 2001. Correlation of late Miocene to early pliocene sequences between the Mediterranean and North Atlantic. Paleoceanography 16, 164–178.

Hodell, D., Woodruff, F., 1994. Variations in the strontium isotopic ratio of seawater during the Miocene: Stratigraphic and geochemical implications. Paleoceanography 9, 405–426. doi: 10.1029/94PA00292.

Hodell, D.A., Kennett, J.P., 1986. Late Miocene–early Pliocene stratigraphy and paleoceanography of the South Atlantic and southwest Pacific Oceans: a synthesis. Paleoceanography 1, 285–311.

Huber, B.T., 1991. Maestrichtian planktonic foraminifer biostratigraphy and the Cretaceous/Tertiary boundary at hole 738C, Kerguelen Plateau (Southern Indian Ocean). In: Barron, J., Larsen, B., et al. (Eds.), Proceedings of the Ocean Drilling Program. Scientific Results, 119, pp. 451–465. doi:10.2973/odp.proc.sr.119.143.1991.

IODP Expedition 320/321 Scientists, 2010. Methods. In: Pälike, H., Lyle, M., Nishi, H., Raffi, I., Gamage, K., Klaus, A., the Expedition 320/321 Scientists (Eds.), Proceedings of the Integrated Ocean Drilling Program, 320/321. (Chapter 2) p. 80. doi:10.2204/iodp.proc.320321.102.2010.

Israelson, C., Spezzaferri, S., 1998. Strontium isotope stratigraphy from ocean drilling program sites 918 and 919. In: Larsen, H.C., Saunders, A.D., Clift, P.D. (Eds.), Proceedings of the Ocean Drilling Program. Scientific Results, 152, pp. 233–241.

Jenkins, D.G., Houghton, S.D., 1989. Neogene planktonic foraminiferal biostratigraphy of ODP Site 677, Panama Basin. Proceedings of the Ocean Drilling Program, Scientific Results 111, 289–293.

Joyce, J.E., Tjalsma, L.R.C., Prutzman, J.M., 1990. High-resolution planktic stable isotope record and spectral analysis for the last 5.35 m.y: Ocean Drilling Program Site 625 Northeast Gulf of Mexico. Paleoceanography 5, 507–529.

Kanungo, S., 2005. Biostratigraphy and paleoceanography of mid-Cretaceous calcareous nannofossils: studies from the Cauvery Basin, SE India; the gault clay formation, SE England; ODP Leg 171B, western North Atlantic and ODP Leg 198, northwest Pacific Ocean. Unpublished Ph.D. Thesis. University College, London, UK, pp. 260.

Keigwin Jr., L.D., 1982. Neogene planktonic foraminifera from deep sea drilling project sites 502 and 503. In: Prell, W.L., Gardner, J.V., et al. (Eds.), Initial Results of the Deep Sea Drilling Project, 68, pp. 269–288.

Krijgsman, W., Langereis, C.G., Daams, R., Van der Meulen, A., 1994. Magnetostratigraphic dating of the middle Miocene climate change in the continental deposits of the Aragonian type area in the Calatayud-Teruel basin (Central Spain). Earth and Planetary Science Letters 128, 513–526.

Langereis, C.G., Hilgen, F.J., 1991. The Rossello composite: A Mediterranean and global reference section for the early to early late Pliocene. Earth and Planetary Science Letters 104, 211–225.

Leckie, R.M., Farnham, C., Schmidt, M.G., 1993. Oligocene planktonic foraminifer biostratigraphy of hole 803D (Ontong Java Plateau) and hole 628A (Little Bahama Bank), and comparison with the southern high latitudes. In: Berger, W.H., Kroenke, L.W., Mayer, L.A., et al. (Eds.), Proceeding of the Ocean Drilling Program: Scientific Results, 130, pp. 113–136.

Li, L., Keller, G., 1998. Maastrichtian climate, productivity and faunal turnovers in planktic foraminifera in South Atlantic DSDP sites 525A and 21. Marine Micropaleontology 33, 55–86.

Li, Q., Radford, S.S., Banner, F.T., 1992. Distribution of microperforate tenuitellid planktonic foraminifers in holes 747A and 749B, Kerguelen Plateau. In: Wise Jr., S.W., Schlich, R., et al. (Eds.), Proceeding of the Ocean Drilling Program: Scientific Results, 120, pp. 569–594.

Lyle, M., Wilson, P.A., Janecek, T.R., et al., 2002. Proceedings of the Ocean Drilling Program. Initial Reports 199. doi:10.2973/odp.proc.ir.199.2002.

Miller, K.G., Feigenson, Wright, J.D., Clement, B.M., 1991. Miocene isotope reference section, deep sea drilling project site 608: an evaluation of isotope and biostratigraphic resolution. Paleoceanography 6, 33–52.

Miller, K.G., Thompson, P.R., Kent, D.V., 1993. Integrated late eocene-oligocene stratigraphy of the Alabama Coastal plain: correlation of hiatuses and stratal surfaces to glacioeustatic lowerings. Paleoceanography 8, 313–331.

Miller, K.G., Aubry, M.-P., Khan, J., Melillo, A.J., Kent, D.V., Berggren, W.A., 1985. Oligocene to miocene biostratigraphy, magnetostratigraphy, and isotopic stratigraphy of the western North Atlantic. Geology 13, 257–261.

Miller, K.G., Browning, J.V., Aubry, M.-P., Wade, B.S., Katz, M.E., Kulpecz, A.A., Wright, J.D., 2008. Eocene–Oligocene global climate and sea-level changes: St. Stephens Quarry, Alabama. Geological Society of America Bulletin 120, 34–53.

Miller, K.G., Wright, J.D., van Fossen, M.C., Kent, D.V., 1994. Miocene stable isotopic stratigraphy and magnetostratigraphy of Buff Bay, Jamaica. Geological Society of America Bulletin 106, 1605–1620.

Mix, A.C., Le, J., Shackleton, N.J., 1995. Benthic foraminiferal stable isotope stratigraphy of site 846: 0–1.8 Ma. In: Pisias, N.G., Mayer, L.A., Janecek, T.R., Palmer-Julson, A., van Andel, T.H. (Eds.), Proceeding of the Ocean Drilling Program. Scientific Results, 138, pp. 839–854.

Montanari, A., Deino, A., Coccioni, R., Langenheim, V.E., Capo, R., Monechi, S., 1991. Geochronology, Sr isotope analysis, magnetostratigraphy, and plankton stratigraphy across the Oligocene-Miocene boundary in the Contessa section (Gubbio, Italy). Newsletters on Stratigraphy 23, 151–180.

Nederbragt, A.J., Van Hinte, J.E., 1987. Biometric analysis of *Planorotalites pseudomenardii* (Upper Paleocene) at Deep Sea Drilling Project Site 605, Northwestern Atlantic. Initial Results of the Deep Sea Drilling Project 93, 577–591.

Nocchi, M., Parisi, G., Monaco, P., Monecki, S., Madile, M., Napoleoni, G., Ripepe, M., Orlando, M., Premoli Silva, I., Bice, D.M., 1986. The Eocene–Oligocene boundary in the Umbrian pelagic sequences, Italy. In: Pomerol, C., Premoli Silva, I. (Eds.), Terminal Eocene Events. Developments in Palaeontology and Stratigraphy, 9, pp. 24–40.

ODP Leg 145 Shipboard Scientific Party, 1993. Explanatory notes. In: Rea, D.K., Basov, I.A., Janecek, T.R., Palmer-Julson, A., et al. (Eds.), Proceedings of the Ocean Drilling Program. Initial Reports, 145, pp. 9–33.

ODP Leg 171B Shipboard Scientific Party, 1998. Explanatory notes. In: Norris, R.D., Kroon, D., Klaus, A., et al. (Eds.), Proceedings of the Ocean Drilling Program. Initial Reports, 171B, pp. 11–44.

ODP Leg 207 Shipboard Scientific Party, 2004. Explanatory notes. In: Erbacher, J., Mosher, D.C., Malone, M.J., et al. (Eds.), Proceedings of the Ocean Drilling Program. Initial Reports, 207, p. 94. doi:10.2973/odp.proc.ir.207.102.2004.

Payros, A., Orue-Etxebarria, X., Bernaola, G., Apellaniz, E., Dinarès-Turell, J., Tosquella, J., Caballero, F., 2009. Characterization and astronomically calibrated age of the first occurrence of *Turborotalia frontosa* in the Gorrondatxe section, a prospective Lutetian GSSP: implications for the Eocene time scale. Lethaia 42, 255–264.

Pearson, P.N., 1995. Planktonic foraminifer biostratigraphy and the development of pelagic caps on guyots in the Marshall Islands Group. In: Haggerty, J., Premoli Silva, I., Rack, F., McNutt, M.K. (Eds.), Proceeding of the Ocean Drilling Program. Scientific Results, 144, pp. 21–59.

Pearson, P.N., Chaisson, W.P., 1997. Late Paleocene to middle Miocene planktonic foraminifer biostratigraphy of the Ceara Rise. In: Shackleton, N.J., Curry, W.B., Richter, C., Bralower, T.J. (Eds.), Proceeding of the Ocean Drilling Program. Scientific Results, 154, pp. 33–68.

Pearson, P.N., Nicholas, C.J., Singano, J.M., Bown, P.R., Coxall, H.K., van Dongen, B.E., Huber, B.T., Karega, A., Lees, J.A., Msaky, E., Pancost, R.D., Pearson, M., Roberts, A.P., 2004. Paleogene and Cretaceous sediment cores from the Kilwa and Lindi areas of coastal Tanzania: Tanzania drilling project sites 1 to 5. Journal of African Earth Sciences 39, 25–62.

Poore, R.Z., Tauxe, L., Percival, S.F., Labrecque, J.L., Wright, R., Peterson, N.P., Smith, C.C., Tucker, P., Hsu, K.J., 1983. Late Cretaceous-Cenozoic magnetostratigraphy and biostratigraphy correlations of the South Atlantic Ocean: DSDP Leg 73. Palaeogeography Palaeoclimatology Palaeoecology 42, 127–149.

Poore, R.Z., Tauxe, L., Percival, S.F., LaBrecque, J.L., 1982. Late Eocene-Oligocene magnetostratigraphy and biostratigraphy at South Atlantic DSDP Site 522. Geology 10, 508–511.

Pospichal, J.J., et al., 1991. Cretaceous-Paleogene biomagnetostratigraphy of ODP Sites 752-755, Broken Ridge: a synthesis. In: Peirce, J., Weissel, J., et al. (Eds.), Proceedings of the Ocean Drilling Program. Scientific Results, 121, pp. 721–741.

Premoli Silva, I., Sliter, W.V., 1994. Cretaceous planktonic foraminiferal biostratigraphy and evolutionary trends from the Bottaccione Section, Gubbio, Italy. Palaeontographica Italiana 82, 2–90.

Premoli Silva, I., Orlando, M., Monechi, S., Madile, M., Napoleone, G., Ripepe, M., Oct. 1987. 1988. Calcareous plankton biostratigraphy and magnetostratigraphy at the Eocene-Oligocene transition in the Gubbio area. In: Premoli Silva, I., Coccioni, R., Montanari, A. (Eds.), Int. Subcomm. Paleog. Strat., Eocene/Oligocene Meeting, Ancona. Spec. Publ., II, 6, pp. 137–161.

Pujol, C., 1983. Cenozoic planktonic foraminiferal biostratigraphy of the southwestern Atlantic (Rio Grande Rise). In: Barker, P.F., Carlson, R.L., Johnson, D.A., et al. (Eds.), Initial Reports of the Deep Sea Drilling Project, 72, pp. 623–673.

Pujol, C., Duprat, J., 1983. Quaternary planktonic foraminifers of the southwestern Atlantic (Rio Grande Rise) Deep Sea Drilling Project Leg 72. In: Barker, P.F., Carlson, R.L., Johnson, D.A., et al. (Eds.), Initial Reports of the Deep Sea Drilling Project, 72, pp. 601–615.

Raffi, I., Flores, J.A., 1995. Pleistocene through Miocene calcareous nannofossils from Eastern Equatorial Pacific Ocean (Leg 138). In: Pisias, N.G., Mayer, L.A., Janecek, T.R., Palmer-Julson, A., van Andel, T.H. (Eds.), Proceedings of the Ocean Drilling Program. Scientific Results, 138, pp. 233–286.

Raffi, I., Backman, J., Pälike, H., 2005. Changes in calcareous nannofossil assemblages across the Paleocene/Eocene transition from the paleo-equatorial Pacific Ocean. Palaeogeography Palaeoclimatology Palaeoecology 226, 93–126.

Raffi, I., Rio, D., D'atri, A., Fornaciari, E., Rocchetti, S., 1995. Quantitative distribution patterns and biomagnetostratigraphy of middle and late miocene calcareous nannofossils from Equatorial Indian and Pacific Oceans (Legs 115, 130, and 138). In: Pisias, N.G., Mayer, L.A., Janecek, T.R., Palmer-Julson, A., van Andel, T.H. (Eds.), Proceedings of the Ocean Driling Program. Scientific Results, 138, pp. 479–502.

Rio, D., Raffi, I., Villa, G., 1990. Pliocene–Pleistocene calcareous nannofossil distribution patterns in the western Mediterranean. In: Kastens, K.A., Mascle, J., et al. (Eds.), Proceedings of the Ocean Drilling Program. Scientific Results, 107, pp. 513–533.

Ryan, W.B.F., Cita, M.B., Rawson, M.O., Burckle, L.H., Saito, T., 1974. A paleomagnetic assignment of Neogene stage boundaries and the development of isochronous datum planes between the Mediterranean, the Pacific and Indian Oceans in order to investigate the response of the world ocean to the Mediterranean 'Salinity Crisis'. Rivista Italiana di Paleontologia e Stratigrafia 80, 631–688.

Saint-Marc, P., 1987. Biostratigraphic and Paleocene benthic and planktonic foraminifers, Site 605, Deep Sea Drilling Project Leg 93. In: van Hinte, J.E., Wise Jr., S.W. (Eds.), Initial Reports of the Deep Sea Drilling Project, 93, pp. 539–547.

Saito, T., Burckle, L.H., Hays, J.D., 1975. Late Miocene to Pleistocene biostratigraphy of equatorial Pacific sediments. In: Saito, T., Burckle, L. (Eds.), Late Neogene Epoch Boundaries. Micropaleontology Special Paper, 1, pp. 226–244.

Sierro, F.J., Krijgsman, W., Hilgen, F.J., Flores, J.A., 2001. The abad composite (SE Spain): a Mediterranean reference section for the Messinian and the Astronomical Polarity Time Scale (APTS). Palaeogeography Palaeoclimatology Palaeoecology 168, 143–172.

Spezzaferri, S., 1998. Planktonic foraminifer biostratigraphy and paleoenvironmental implications of Leg 152 sites (East Greenland margin). In: Saunders, A.D., Larsen, H.C., Wise Jr., S.W. (Eds.), Proceedings of the Ocean Drilling Program. Scientific Results, 152, pp. 161–189.

Spiegler, D., Jansen, E., 1989. Planktonic foraminifer biostratigraphy of Norwegian Sea sediments; ODP Leg 104. Proceedings of the Ocean Drilling Program 104, 681–696.

Srinivasan, M.S., Sinha, D.K., 1992. Late Neogene planktonic foraminiferal events of the southwest Pacific and Indian Ocean: a comparison. In: Tsuchi, R., Ingle Jr., J.C. (Eds.), Pacific Neogene: Environment, Evolution and Events. Univ. Tokyo Press, Tokyo, pp. 203–220.

Stott, L.D., Kennett, J.P., 1990. Antarctic Paleogene planktonic foraminifer biostratigraphy: ODP Leg 113, Sites 689 and 690. In: Barker, P.F., Kennett, J.P., et al. (Eds.), Proceedings of the Ocean Drilling Program, 113, pp. 549–569.

Thierstein, H.R., Geitzenauer, K.R., Molfino, B., Shackleton, N.J., 1977. Global synchroneity of late Quaternary coccolith datum levels: validation by oxygen isotopes. Geology 5, 400–404.

Thompson, P.R., Bé, A.W.H., Duplessy, J.-C., Shackleton, N.J., 1979. Disappearance of pink-pigmented *Globigerinoides ruber* at 120,000 yr BP in the Indian and Pacific oceans. Nature 280, 554–558.

Turco, E., Bambini, A.M., Foresi, L.M., Iaccarino, S., Lirer, F., Mazzei, R., Salvatorini, G., 2002. Middle Miocene high resolution calcareous plankton biostratigraphy at Site 926 (Leg 154, equatorial Atlantic Ocean): paleoecological and paleobiogeographical implications. Geobios 35, 257–276.

Varol, O., 1992. Revision of the polycyclolithaceae and its contribution to cretaceous biostratigraphy. Newsletters on Stratigraphy 27, 93–127.

Villa, G., Fioroni, C., Pea, L., Bohaty, S., Persico, P., 2008. Middle Eocene–late Oligocene climate variability: calcareous nannofossil response at Kerguelen Plateau, Site 748. Marine Micropaleontology 69, 173–192. doi:10.1016/j.marmicro.2008.07.006.

Wade, B.S., 2004. Planktonic foraminiferal biostratigraphy and mechanisms in the extinction of *Morozovella* in the late Middle Eocene. Marine Micropaleontology 51, 23–38.

Wei, W., Thierstein, H.R., 1991. Upper Cretaceous and Cenozoic calcareous nannofossils of the Kerguelen Plateau (southern Indian Ocean) and Prydz Bay (East Antarctica). In: Barron, J., Larsen, B., et al. (Eds.), Proceedings of the Ocean Drilling Program. Scientific Results, 119, pp. 467–494.

Wright, J.D., Miller, K.G., 1992. Miocene stable isotope stratigraphy, Site 747, Kerguelen Plateau. In: Wise Jr., S.W., Schlich, R., et al. (Eds.), Proceedings of the Ocean Drilling Program. Scientific Results, 120, pp. 855–866.

Zhang, J., Miller, K.G., Berggren, W.A., 1993. Neogene planktonic foraminiferal biostratigraphy of the northeastern Gulf of Mexico. Micropaleontology 39, 299–326.

Index

Page numbers with "f" denote figures "t" tables.

C

F

G

H

I

J

K

L

M

N

Q

R

T